The Prokaryotes

A Handbook on Habitats, Isolation, and Identification of Bacteria

Edited by

MORTIMER P. STARR	University of California, Davis
HEINZ STOLP	University of Bayreuth
HANS G. TRÜPER	University of Bonn
ALBERT BALOWS	Centers for Disease Control, Atlanta
HANS G. SCHLEGEL	University of Göttingen

Volume II

Springer-Verlag

Berlin Heidelberg New York

Mortimer P. Starr, Department of Bacteriology, University of California, Davis, California 95616, U.S.A.

Heinz Stolp, Lehrstuhl für Mikrobiologie, Universität Bayreuth, 8580 Bayreuth, Federal Republic of Germany

Hans G. Trüper, Institut für Mikrobiologie, Rheinische Friedrich-Wilhelms-Universität, 5300 Bonn 1, Federal Republic of Germany

Albert Balows, Bacteriology Division, Centers for Disease Control, Atlanta, Georgia 30333, U.S.A.

Hans G. Schlegel, Institut für Mikrobiologie, Universität Göttingen, 3400 Göttingen, Federal Republic of Germany

Library of Congress Cataloging in Publication Data
Main entry under title:

The Prokaryotes.

  Bibliography: p.
  Includes index.
  1. Microbial ecology.   2. Bacteria—Identification.
3. Bacteriology—Cultures and culture media. I. Starr,
Mortimer P. [DNLM: 1. Bacteria. QW 4 P963]
QR100.P76   589.9   81-9192
                      AACR2

Printed in the United States of America.

9 8 7 6 5 4 3 2 1

ISBN 3-540-08871-7 Springer-Verlag Berlin Heidelberg New York
ISBN 0-387-08871-7 Springer-Verlag New York Heidelberg Berlin

Table of Contents

Volume II

List of Contributors

Jan R. Andreesen
Institut für Mikrobiologie
Rheinische Friedrich-Wilhelms-
    Universität
5300 Bonn 1
Federal Republic of Germany

Michel Aragno
Institut de Botanique
Université de Neuchâtel
2000 Neuchâtel 7
Switzerland

Albert Balows
Bacteriology Division
Centers for Disease Control
Atlanta, Georgia 30333
U.S.A.

Lane Barksdale
Department of Microbiology
New York University School of
    Medicine and Medical Center
New York, New York 10016
U.S.A.

John Bauld
Baas Becking Geobiological
    Laboratory
Canberra City, A.C.T.
Australia 2601

Linda Baumann
Department of Bacteriology
University of California
Davis, California 95616
U.S.A.

Paul Baumann
Department of Bacteriology
University of California
Davis, California 95616
U.S.A.

Jan-Hendrik Becking
Institute for Atomic Sciences in
    Agriculture
Wageningen
The Netherlands

Tom Bergan
Department of Microbiology
University of Oslo
P.O. Box 1108 Blindern
Oslo 3
Norway

Roger C. W. Berkeley
Department of Bacteriology
University of Bristol
Bristol BS8 1TD
United Kingdom

Hanno Biebl
Gesellschaft für
Biotechnologische Forschung mbH
3300 Braunschweig
Federal Republic of Germany

Charles E. Bland
Department of Biology
East Carolina University
Greenville, North Carolina 27834
U.S.A.

Edward J. Bottone
Department of Microbiology
Mount Sinai Hospital and the
    Mount Sinai School of Medicine
New York, New York 10029
U.S.A.

Kjell Bøvre
Kaptein W. Wilhelmsen og Frues
    Bakteriologiske Institutt
University of Oslo
Oslo 1
Norway

Don J. Brenner
Enteric Section
Centers for Disease Control
Atlanta, Georgia 30333
U.S.A.

Thomas D. Brock
Department of Bacteriology
University of Wisconsin
Madison, Wisconsin 53706
U.S.A.

George H. Brownell
Department of Cell and Molecular
    Biology
Medical College of Georgia
Augusta, Georgia 30901
U.S.A.

Marvin P. Bryant
Departments of Dairy Science and
    Microbiology
University of Illinois at Urbana-
    Champaign
Urbana, Illinois 61801
U.S.A.

Ercole Canale-Parola
Department of Microbiology
University of Massachusetts
Amherst, Massachusetts 01003
U.S.A.

J. Geoffrey Carr
Long Ashton Research Station
University of Bristol
Bristol BS18 9AF
United Kingdom

Gordon R. Carter
Department of Microbiology and
    Public Health
Michigan State University
East Lansing, Michigan 48824
U.S.A.

Richard W. Castenholz

Department of Biology
University of Oregon
Eugene, Oregon 97403
U.S.A.

David G. Chase

Cell Biology Laboratory
Veterans Administration Hospital
Sepulveda, California 91343
U.S.A.

Dieter Claus

Deutsche Sammlung von
   Mikroorganismen
3400 Göttingen
Federal Republic of Germany

John N. Couch

Department of Botany
University of North Carolina
Chapel Hill, North Carolina 27514
U.S.A.

Thomas Cross

Postgraduate School of Studies in
   Biological Sciences
University of Bradford
Bradford, West Yorkshire
   BD7 1DP
United Kingdom

Cecil S. Cummins

Department of Anaerobic
   Microbiology
Virginia Polytechnic Institute and
   State University
Blacksburg, Virginia 24061
U.S.A.

Howard Dalton

Department of Biological Sciences
University of Warwick
Coventry CV4 7AL
United Kingdom

Gregory A. Dasch

Medical Microbiology Branch
Naval Medical Research Institute
Bethesda, Maryland 20014
U.S.A.

Michael J. Davis

Department of Plant Pathology
Rutgers University
New Brunswick, New Jersey 08903
U.S.A.

Maria H. Deinema

Laboratory of Microbiology
Agricultural University
Wageningen
The Netherlands

Jozef De Ley

Laboratorium voor Microbiologie
   en microbiële Genetica
Rijksuniversiteit Gent
9000 Gent
Belgium

Robert B. Dienst

Department of Cell and Molecular
   Biology
Medical College of Georgia
Augusta, Georgia 30901
U.S.A.

Nicanor Domínguez

Instituto de Medicina Tropical
   "Daniel Carrión"
Universidad Nacional Mayor de
   San Marcos
Lima
Peru

Vulus R. Dowell, Jr.

Bacteriology Division
Centers for Disease Control
Atlanta, Georgia 30333
U.S.A.

Patrick R. Dugan

Department of Microbiology
Ohio State University
Columbus, Ohio 43210
U.S.A.

Martin Dworkin

Department of Microbiology
University of Minnesota—Medical
   School
Minneapolis, Minnesota 55455
U.S.A.

Henry T. Eigelsbach

Western Maryland College
Westminster, Maryland 21157
U.S.A.

Friedrich W. Ewald

Paul Ehrlich Institut
6000 Frankfurt am Main 70
Federal Republic of Germany

Richard Facklam

Bacteriology Division
Centers for Disease Control
Atlanta, Georgia 30333
U.S.A.

John J. Farmer III

Enteric Section
Centers for Disease Control
Atlanta, Georgia 30333
U.S.A.

Dilip Gadkari

Lehrstuhl für Mikrobiologie
Universität Bayreuth
8580 Bayreuth
Federal Republic of Germany

Friedrich Gehring

Lehrstuhl für Experimentelle
   Zahnheilkunde
Universitätsklinik und Poliklinik
   für Zahn-, Mund- und
   Kieferkrankheiten
Universität Würzburg
8700 Würzburg
Federal Republic of Germany

Arnold Geis

Institut für Mikrobiologie
Bundesanstalt für Milchforschung
2300 Kiel 1
Federal Republic of Germany

A. Graves Gillaspie, Jr.

Applied Plant Pathology
   Laboratory
Agricultural Research Service
U.S. Department of Agriculture
Beltsville, Maryland 20705
U.S.A.

Robert C. Good

Bacteriology Division
Centers for Disease Control
Atlanta, Georgia 30333
U.S.A.

Michael Goodfellow

Department of Microbiology
The Medical School, The
   University
Newcastle upon Tyne, NE1 7RU
United Kingdom

Joyce K. Gordon

Department of Microbiology
University of British Columbia
Vancouver, British Columbia
   V6T 1W5
Canada

Rainer Gothe

Institut für Parasitologie
Justus-Liebig-Universität Giessen
6300 Lahn-Giessen 1
Federal Republic of Germany

Gerhard Gottschalk

Institut für Mikrobiologie
Universität Göttingen
3400 Göttingen
Federal Republic of Germany

Francine Grimont

Service des Entérobactéries
Unité INSERM 199
Institut Pasteur
75724 Paris Cedex 15
France

Patrick A. D. Grimont

Service des Entérobactéries
Unité INSERM 199
Institut Pasteur
75724 Paris Cedex 15
France

Roger J. Gross

Division of Enteric Pathogens
Central Public Health Laboratory
London NW9 5HT
United Kingdom

David L. Gutnick

Department of Microbiology
Tel Aviv University
Ramat Aviv
Israel

Norman Hagen

Department of Microbiology
University Hospital
Tromsø
Norway

Hans H. Hanert

Botanisches Institut
Technische Universität
  Braunschweig
3300 Braunschweig
Federal Republic of Germany

J. Woodland Hastings

Biological Laboratories
Harvard University
Cambridge, Massachusetts 02138
U.S.A.

Leonard Hayflick

Children's Hospital Medical Center
Bruce Lyon Memorial Research
  Laboratory
Oakland, California 94609
U.S.A.

Robert B. Hespell

Department of Dairy Science and
  Microbiology
University of Illinois at Urbana-
  Champaign
Urbana, Illinois 61801
U.S.A.

Donald C. Hildebrand

Department of Plant Pathology
University of California
Berkeley, California 94720
U.S.A.

Edward O. Hill

Department of Pathology and
  Laboratory Medicine
Emory University School of
  Medicine and Emory University
  Hospital
Atlanta, Georgia 30322
U.S.A.

Gale B. Hill

Department of Microbiology
  and Immunology
Duke University Medical Center
Durham, North Carolina 27710
U.S.A.

Hans Hippe

Deutsche Sammlung von
  Mikroorganismen
3400 Göttingen
Federal Republic of Germany

Peter Hirsch

Institut für Allgemeine
  Mikrobiologie
Universität Kiel
2300 Kiel
Federal Republic of Germany

Tor Hofstad

Department of Microbiology
University of Bergen
5016 Haukeland Sykehus
Bergen
Norway

Johannes F. Imhoff

Institut für Mikrobiologie
Rheinische Friedrich-Wilhelms-
  Universität
D-5300 Bonn 1
Federal Republic of Germany

John L. Ingraham

Department of Bacteriology
University of California
Davis, California 95616
U.S.A.

Henry D. Isenberg

Department of Microbiology
Long Island Jewish-Hillside
  Medical Center
New Hyde Park, New York 11042
U.S.A.

Holger W. Jannasch

Woods Hole Oceanographic
  Institution
Woods Hole, Massachusetts 02543
U.S.A.

Randall M. Jeter

Department of Bacteriology
University of California
Davis, California 95616
U.S.A.

John L. Johnson

Department of Anaerobic
  Microbiology
Virginia Polytechnic Institute and
  State University
Blacksburg, Virginia 24061
U.S.A.

Russell C. Johnson

Department of Microbiology
University of Minnesota
Minneapolis, Minnesota 55455
U.S.A.

Dorothy Jones

Department of Microbiology
University of Leicester
Leicester LEl 7RH
United Kingdom

Ronald M. Keddie

Department of Microbiology
University of Reading
Reading, RGl 5AQ
United Kingdom

Donovan P. Kelly

Department of Environmental
  Sciences
University of Warwick
Coventry CV4 7AL
United Kingdom

Richard T. Kelly

Department of Pathology
Baptist Memorial Hospital
Memphis, Tennessee 38146
U.S.A.

Mogens Kilian

Department of Oral Biology
Royal Dental College
8000 Aarhus C
Denmark

Wesley E. Kloos

Department of Genetics
North Carolina State University
Raleigh, North Carolina 27650
U.S.A.

Miloslav Kocur

Czechoslovak Collection of
  Microorganisms
J. E. Purkyně University
662 43 Brno
Czechoslovakia

Johannes Krämer

Institut für Mikrobiologie
Rheinische Friedrich-Wilhelms-
  Universität
5300 Bonn 1
Federal Republic of Germany

Heinz E. Krampitz

Institut für Infektions- und
  Tropenmedizin
der Universität München
8000 München 40
Federal Republic of Germany

Julius P. Kreier
Department of Microbiology
The Ohio State University
Columbus, Ohio 43210
U.S.A.

Aloysius Krieg
Biologische Bundesanstalt für
    Land- und Forstwirtschaft
Institut für Biologische
    Schädlingsbekämpfung
6100 Darmstadt
Federal Republic of Germany

Noel R. Krieg
Department of Biology
Virginia Polytechnic Institute and
    State University
Blacksburg, Virginia 24061
U.S.A.

George P. Kubica
Laboratory Training and
    Consultation Division
Centers for Disease Control
Atlanta, Georgia 30333
U.S.A.

J. Gijs Kuenen
Laboratory of Microbiology
Delft University of Technology
Julianalaan 67A
2628 Delft
The Netherlands

Daisy A. Kuhn
Department of Biology
California State University
Northridge, California 91330
U.S.A.

Hans Jürgen Kutzner
Institut für Mikrobiologie
    der Technischen Hochschule
    Darmstadt
6100 Darmstadt
Federal Republic of Germany

Jan W. M. la Rivière
International Institute for Hydraulic
    and Environmental Engineering
2601 DA Delft
The Netherlands

Helge Larsen
Department of Biochemistry
University of Trondheim
7034 Trondheim-NTH
Norway

Hubert A. Lechevalier
Waksman Institute of Microbiology
Rutgers, The State University of
    New Jersey
Piscataway, New Jersey 08854
U.S.A.

Mary P. Lechevalier
Waksman Institute of Microbiology
Rutgers, The State University of
    New Jersey
Piscataway, New Jersey 08854
U.S.A.

Léon Le Minor
Institut Pasteur
Service des Entérobactéries
Unité INSERM 199
75724 Paris Cedex 15
France

Ralph A. Lewin
Scripps Institution of
    Oceanography
University of California, San
    Diego
La Jolla, California 92093
U.S.A.

Robert P. Lewis
Southern California Permanente
    Medical Group
Harbor City, California 90710
U.S.A.
Department of Medicine
University of California School of
    Medicine
Los Angeles, California 90024
U.S.A.

Barbara B. Lippincott
Department of Biological Sciences
Northwestern University
Evanston, Illinois 60201
U.S.A.

James A. Lippincott
Department of Biological Sciences
Northwestern University
Evanston, Illinois 60201
U.S.A.

Niall A. Logan
Department of Biological Sciences
Glasgow College of Technology
Glasgow G4 0BA
United Kingdom

George L. Lombard
Bacteriology Division
Centers for Disease Control
Atlanta, Georgia 30333
U.S.A.

Randolph E. McCoy
University of Florida Agricultural
    Research Center
Fort Lauderdale, Florida 33314
U.S.A.

Virginia G. McGann
U.S. Army Medical Research
    Institute of Infectious Diseases
Frederick, Maryland 21701
U.S.A.

Joan M. Macy
Department of Animal Science
University of California
Davis, California 95616
U.S.A.

Sarabelle Madoff
Department of Bacteriology
Massachusetts General Hospital
Boston, Massachusetts 02114
U.S.A.

Robert A. Mah
Division of Environmental and
    Nutritional Sciences
School of Public Health
University of California
Los Angeles, California 90024
U.S.A.

Lynn Margulis
Department of Biology
Boston University
Boston, Massachusetts 02215
U.S.A.

Gerald Masover
Children's Hospital Medical Center
Bruce Lyon Memorial Research
    Laboratory
Oakland, California 94609
U.S.A.

Margaret E. Meyer
Department of Epidemiology and
    Preventive Medicine
University of California
Davis, California 95616
U.S.A.

David E. Minnikin
Department of Organic Chemistry
University of Newcastle upon
    Tyne
Newcastle upon Tyne, NE1 7RU
United Kingdom

Richard L. Moore
Department of Microbiology and
    Infectious Disease
The University of Calgary
Calgary, Alberta T2N 4N1
Canada

Maurice O. Moss
Department of Microbiology
University of Surrey
Guildford, Surrey GU2 5XH
United Kingdom

Eppe Gerke Mulder

Laboratory of Microbiology
Agricultural University
Wageningen
The Netherlands

Kenneth H. Nealson

Scripps Institution of
  Oceanography
University of California, San Diego
La Jolla, California 92093
U.S.A.

John R. Norris

Cadbury Schweppes Limited
University of Reading
Whiteknights
Reading, Berkshire RG6 2AD
United Kingdom

Anthony G. O'Donnell

Department of Organic Chemistry
University of Newcastle upon Tyne
Newcastle upon Tyne NE1 7RU
United Kingdom

Frits Ørskov

Collaborative Centre for Reference
  and Research on Escherichia
  and Klebsiella (WHO)
State Serum Institute
2300 Copenhagen S
Denmark

Ida Ørskov

Collaborative Centre for Reference
  and Research on Escherichia
  and Klebsiella (WHO)
State Serum Institute
2300 Copenhagen S
Denmark

Leslie A. Page

Route 4, Box 326
Bay St. Louis, Mississippi 39520
U.S.A.
Formerly:
United States Department of
  Agriculture
Agricultural Research Service,
  North Central Region
National Animal Disease Center
Ames, Iowa 50010
U.S.A.

Norberto J. Palleroni

Chemical Research Division
Hoffmann-La Roche
Nutley, New Jersey 07110
U.S.A.

John L. Penner

Department of Medical
  Microbiology
University of Toronto
Toronto, Ontario M5G 1L5
Canada

Norbert Pfennig

Fakultät für Biologie
Universität Konstanz
7750 Konstanz
Federal Republic of Germany

James E. Phillips

Department of Veterinary Pathology
University of Edinburgh
Royal (Dick) School of Veterinary
  Studies
Edinburgh, EH9 IQH
Scotland, United Kingdom

Beverly K. Pierson

Department of Biology
University of Puget Sound
Tacoma, Washington 98416
U.S.A.

Margaret Pittman

Bureau of Biologics
Food and Drug Administration
Bethesda, Maryland 20205
U.S.A.

Louise B. Preer

Department of Biology
Indiana University
Bloomington, Indiana 47405
U.S.A.

Gerhard Pulverer

Hygiene-Institut der
  Universität Köln
5000 Köln 41
Federal Republic of Germany

Thomas J. Quan

Bureau of Laboratories
Centers for Disease Control
Fort Collins, Colorado 80522
U.S.A.

Harkisan D. Raj

Department of Microbiology
California State University
Long Beach, California 90840
U.S.A.

Hans Reichenbach

Gesellschaft für Biotechnologische
  Forschung
Abteilung Mikrobiologie
3300 Braunschweig-Stöckheim
Federal Republic of Germany

Rosmarie Rippka

Institut Pasteur
75724 Paris Cedex 15
France

Miodrag Ristic

Department of Veterinary
  Pathology and Hygiene
University of Illinois
Urbana, Illinois 61801
U.S.A.

David S. Roberts

Wellcome Animal Health Division
Kansas City, Kansas 66103
U.S.A.

Eugene Rosenberg

Department of Microbiology
Tel Aviv University
Ramat Aviv
Israel

Bernard Rowe

Division of Enteric Pathogens
Central Public Health Laboratory
London NW9 5HT
United Kingdom

Colin Ryall

Faculty of Medicine
University of Calabar
Calabar
Nigeria

Riichi Sakazaki

National Institute of Health
Kamiosaki, Shinagawa-ku,
  Tokyo 141
Japan

Vittorio Scardovi

Istituto di Microbiologia Agraria e
  Tecnica
Università degli Studi di Bologna
Bologna
Italy

Klaus P. Schaal

Hygiene-Institut der
  Universität Köln
5000 Köln 41
Federal Republic of Germany

Hans G. Schlegel

Institut für Mikrobiologie
Universität Göttingen
3400 Göttingen
Federal Republic of Germany

Karl-Heinz Schleifer

Lehrstuhl für Mikrobiologie
Technische Universität München
8000 München 2
Federal Republic of Germany

Heinz Schlesner

Institut für Allgemeine
  Mikrobiologie
Universität Kiel
2300 Kiel
Federal Republic of Germany

Jean M. Schmidt

Department of Botany and
  Microbiology
Arizona State University
Tempe, Arizona 85287
U.S.A.

Karin Schmidt

Institut für Mikrobiologie
Universität Göttingen
3400 Göttingen
Federal Republic of Germany

Milton N. Schroth

Department of Plant Pathology
University of California
Berkeley, California 94720
U.S.A.

Ramon J. Seidler

Department of Microbiology
Oregon State University
Corvallis, Oregon 97331
U.S.A.

M. Elisabeth Sharpe

National Institute for Research in
  Dairying
University of Reading
Shinfield, Reading RG2 9AT
United Kingdom

Charles C. Shepard

Leprosy and Rickettsia Branch,
  Virology Division
Centers for Disease Control
Atlanta, Georgia 30333
U.S.A.

Irving J. Slotnick

Cedars-Sinai Medical Center
Los Angeles, California 90048
U.S.A.

Robert M. Smibert

Department of Anaerobic
  Microbiology
Virginia Polytechnic Institute and
  State University
Blacksburg, Virginia 24061
U.S.A.

Michael R. Smith

U.S. Department of Agriculture
Agricultural Research Service
Western Regional Research
  Laboratory
Berkeley, California 94710
U.S.A.

James T. Staley

Department of Microbiology and
  Immunology
University of Washington
Seattle, Washington 98195
U.S.A.

Roger Y. Stanier

Institut Pasteur
75724 Paris Cedex 15
France

Mortimer P. Starr

Department of Bacteriology
University of California
Davis, California 95616
U.S.A.

Heinz Stolp

Lehrstuhl für Mikrobiologie
Universität Bayreuth
8580 Bayreuth
Federal Republic of Germany

Vera L. Sutter

Research Service
VA Wadsworth Medical Center
Los Angeles, California 90073
U.S.A.
Department of Medicine
University of California School of
  Medicine
Los Angeles, California 90024
U.S.A.

James R. Swafford

Department of Botany and
  Microbiology
Arizona State University
Tempe, Arizona 85287
U.S.A.

Jean Swings

Laboratorium voor Microbiologie
  en microbiële Genetica
Rijksuniversiteit Gent
9000 Gent
Belgium

Michael Teuber

Institut für Mikrobiologie
Bundesanstalt für Milchforschung
2300 Kiel 1
Federal Republic of Germany

Richard C. Tilton

Department of Laboratory
  Medicine
University of Connecticut Health
  Center
Farmington, Connecticut 06032
U.S.A.

Leleng P. To

Department of Biology
Boston University
Boston, Massachusetts 02215
U.S.A.

William C. Trentini

Biology Department
Mount Allison University
Sackville, N.B. EOA 3CO
Canada

Hans G. Trüper

Institut für Mikrobiologie
Rheinische Friedrich-Wilhelms-
  Universität
5300 Bonn 1
Federal Republic of Germany

Joseph G. Tully

Laboratory of Infectious Diseases
National Institute of Allergy and
  Infectious Diseases
Bethesda, Maryland 20205
U.S.A.

Olli H. Tuovinen

Department of Microbiology
Ohio State University
Columbus, Ohio 43210
U.S.A.

Frederica W. Valois

Woods Hole Oceanographic
  Institution
Woods Hole, Massachusetts 02543
U.S.A.

Neylan A. Vedros

School of Public Health
University of California
Berkeley, California 94720
U.S.A.

Anne K. Vidaver

Department of Plant Pathology
University of Nebraska
Lincoln, Nebraska 68583
U.S.A.

James M. Vincent

Department of Microbiology
The University of Sydney
Sydney, N.S.W. 2006
Australia

Anthony E. Walsby

Marine Science Laboratories
University College of North Wales
Gwynedd LL59 5EH
Wales, United Kingdom

Alastair C. Wardlaw
Department of Microbiology
Glasgow University
Glasgow G11 6NU
Scotland, United Kingdom

John B. Waterbury
Woods Hole Oceanographic
    Institution
Woods Hole, Massachusetts 02543
U.S.A.

Stanley W. Watson
Woods Hole Oceanographic
    Institution
Woods Hole, Massachusetts 02543
U.S.A.

Owen B. Weeks
Arts and Sciences Research Center
New Mexico State University
Las Cruces, New Mexico 88003
U.S.A.

Emilio Weiss
Chair of Science
Naval Medical Research Institute
Bethesda, Maryland 20014
U.S.A.

Elizabeth M. H. Wellington
Botany Department
Liverpool University
Liverpool L69 3BX
United Kingdom

Herbert J. Welshimer
Department of Microbiology
Virginia Commonwealth
    University
Richmond, Virginia 23298
U.S.A.

Robert F. Whitcomb
Insect Pathology Laboratory
Agricultural Research Service
U.S. Department of Agriculture
Beltsville, Maryland 20705
U.S.A.

Roger Whittenbury
Department of Biological Sciences
University of Warwick
Coventry CV4 7AL
United Kingdom

Friedrich Widdel
Fakultät für Biologie
Universität Konstanz
7750 Konstanz
Federal Republic of Germany

Wolfgang Wiessner
Abteilung für Experimentelle
    Phykologie
Universität Göttingen
3400 Göttingen
Federal Republic of Germany

Hazel W. Wilkinson
Bacteriology Division
Centers for Disease Control
Atlanta, Georgia 30333
U.S.A.

Stanley T. Williams
Botany Department
Liverpool University
Liverpool L69 3BX
United Kingdom

Georgi A. Zavarzin
Institute of Microbiology
Academy of Sciences of the
    U.S.S.R.
117312 Moscow
U.S.S.R.

SECTION M

The Families Enterobacteriaceae and Vibrionaceae

Some Marine Bacteria

Introduction to the Family Enterobacteriaceae

DON J. BRENNER

Sarles et al. (1956) used the following quotation to convey to their readers the astonishing adaptability of bacteria: "The blood of normal animals, tissues in the physiological interior of healthy animals and plants, deep layers of soil and rocks, and the pits of active volcanoes are about the only places where bacteria are not commonly found." As one of their readers, I was quite impressed.

Now, more than 20 years later, I remain impressed by the near ubiquity of members of the family Enterobacteriaceae, the enterobacteria or enteric bacteria. I have been charged with introducing the reader to these organisms and reviewing their taxonomy. I shall also attempt to discuss current and future trends and problems in the taxonomy and identification of these organisms.

The enterobacteria have been studied more thoroughly than any other group of organisms. The rapid generation time, ability to grow on minimal media, and ease with which *Escherichia coli* can be manipulated have made this organism the primary tool of the bacterial geneticist and molecular biologist. *E. coli* is, thus far, the major organism that can be used in deoxyribonucleic acid (DNA) recombinant experiments. Other species, including *Salmonella typhimurium, Shigella flexneri, Proteus mirabilis,* and *Serratia marcescens,* are also popular for genetic and molecular studies.

The distribution of enterobacteria is world wide and their host range includes animals from insects to man, fruits, vegetables, grains, flowering plants, and trees. They are found in water and soil, and can live as saprophytes, symbionts, epiphytes, and parasites.

Economic and Medical Importance

Many species of the family Enterobacteriaceae are of substantial economic importance. For example, various species of *Erwinia* and *Pectobacterium* cause blight, wilt, or soft rot disease in corn, potatoes, and many other crops, often destroying substantial amounts of the crops (Starr and Chatterjee, 1972). Both the commercial and tropical fish industries are severely affected by some diseases that enterobacteria cause. Salmon and trout hatcheries have problems with redmouth disease caused by the so-called RM organism (Ross, Rucker, and Ewing, 1966), which has recently been named *Yersinia ruckeri* (Ewing et al., 1978). *Edwardsiella tarda* is pathogenic for eels, catfish, and goldfish (Shotts and Snieszko, 1976). Shotts is of the opinion that several other members of the family Enterobacteriaceae will prove to be important fish pathogens (Shotts and Bullock, 1976).

The organism called *Arizona hinshawii* in the United States and *Salmonella arizonae* in Europe is pathogenic and highly fatal for poultry, especially turkeys (Edwards and Ewing, 1972; Williams, 1965). Arizonae cause a characteristic diarrheal syndrome in poultry that poultry microbiologists refer to as arizonosis. Arizonae are also pathogenic for snakes and other reptiles, often creating problems in zoos. Pullorum disease, caused by *Salmonella pullorum,* principally affects eggs and chicks with a high level of fatality (Von Roekel, 1965). (Three species of *Salmonella, S. typhi, S. cholerae-suis,* and *S. enteritidis* are recognized in the United States. All other types are reported as serotypes of *S. enteritidis. S. enteritidis* ser Pullorum is the complete designation. In this chapter, *S. pullorum, S. gallinarum,* etc., will be used with the understanding that they are serotypes of *S. enteritidis.)* Fowl typhoid is a septicemic disease of domesticated birds, especially chickens, caused by *Salmonella gallinarum* (Hall, 1965). By 1965, more than 115 serotypes of *Salmonella* had been isolated from fowl in the United States (Williams, 1965). By now that number has doubtless doubled and perhaps tripled. The distribution of salmonellosis in poultry is worldwide, although, as in human disease, certain serotypes are prevalent in some regions and absent in others. A mortality rate of 10–20% is normal in young birds, mostly during the first 2 weeks after hatching.

Sheep suffer from a variety of illnesses caused by members of the family Enterobacteriaceae (Jensen, 1974). Salmonellal abortion is usually caused by *S. abortus-ovis, S. typhimurium,* or *S. dublin;* these organisms also cause stillbirths and wool damage.

Colibacillosis, infantile diarrhea in lambs, is caused by enterotoxigenic strains of *Escherichia coli,* most of which contain the capsular antigen K99. K99 is thought to be a colonization factor that enables a strain to become established in the small intestine (see this Handbook, Chapter 89). A dysentery-like, hemorrhagic colitis in sheep is caused by *Salmonella typhimurium.*

More than 100 serotypes of *Salmonella* have been isolated from pigs; however only two serotypes, *S. cholerae-suis* and *S. typhi-suis,* have pigs as their primary host (Barnes and Sorensen, 1975). *S. cholerae-suis* has a wide host range, including man, but *S. typhi-suis* is rarely pathogenic to organisms other than pigs. Other serotypes frequently isolated from pigs with salmonellosis are *S. typhimurium* and *S. derby.* Young animals often have a septicemic salmonellosis. Older animals normally have a chronic diarrheal infection. *E. coli* infection in pigs may manifest as diarrhea in piglets or weanlings, or as an edema disease preceded by mild diarrhea (Bruner and Gillespie, 1973). Both forms are acute and highly fatal. The causative strains elaborate enterotoxin, and most of them contain a specific colonization factor, K88 (see this Handbook, Chapter 89).

Calf scours is an acute diarrheal disease that, if untreated, is usually fatal. As in pig and sheep diarrheal disease, specific serotypes of *E. coli* are associated with this disease (Bruner and Gillespie, 1973). The disease is due to an enterotoxin and K99. *E. coli* also causes septicemic disease in calves. Klebsiellae and *Citrobacter freundii* have been implicated as causative agents of bovine mastitis. Most salmonellosis in cattle is due to *S. dublin* and *S. typhimurium,* although 100 or more serotypes have been isolated (Ewing, 1969).

Salmonellae, especially *S. typhimurium, S. newport,* and *S. anatum,* cause outbreaks of enteritis with high mortality in horses (Ewing, 1969). Dogs and cats are often infected by salmonellae or are carriers of these organisms (Ewing 1969).

Some members of the family Enterobacteriaceae, such as shigellae and certain host-adapted serotypes of *Salmonella,* have a narrow host range. Other species, including salmonellae, *E. coli,* and yersiniae, infect or are carried by hosts ranging from insects to primates and man. A very incomplete list of animal hosts for salmonellae, yersiniae, and *Edwardsiella tarda* is presented in Table 1. Certain enterobacteria have been isolated from only a single animal host. These include *Proteus myxofaciens* from larvae of gypsy moths (Cosenza and Podgwaite. 1966) and *Escherichia blattae* from the hindgut of cockroaches (Burgess, McDermott, and Whiting, 1973). Other species such as *Yersinia pseudotuberculosis* seem to have a wide host range (Mair, 1968). Additional research will undoubtedly reveal a substantially broader host range for many species of the family

Enterobacteriaceae as well as identify new species with a limited host range.

Enterobacteria were divided into enteric pathogens and nonpathogens according to their ability to cause food- or waterborne diarrheal disease. Historically the pathogenic genera were *Salmonella* (and *Arizona*) and *Shigella.* Starting in the late 1940s, a group of *E. coli* serotypes was isolated from samples obtained during outbreaks of infantile diarrhea (Ewing, Tatum, and Davis, 1957; Kauffmann and DuPont, 1950; Taylor, Powell, and Wright, 1949). These are called enteropathogenic serotypes of *E. coli.* Enterotoxin activity in *E. coli* was first reported by De, Bhattacharya, and Sarkar (1956). Two types of enterotoxin, a heat-labile enterotoxin related to cholera enterotoxin (Gyles, 1974; Holmgren, Soderland, and Wadstrom, 1973; Smith and Sack, 1973) and a heat-stable enterotoxin (Dean et al., 1972; Smith and Gyles, 1970; Smith and Lingood, 1972) are well documented. A third type of enterotoxin, similar to that recently described in *Shigella flexneri* (O'Brien et al., 1977) has also been observed in *E. coli* (S. Formal, personal communication).

Strains of certain *E. coli* serotypes can also cause an invasive, *Shigella*-like, bloody diarrhea (Gemski and Formal, 1975; Sakazaki, Tamura, and Saito, 1967). Opinions vary as to the value of serotyping in the diagnosis of *E. coli* diarrhea, and very little is known about the relative importance of enteropathogenic, enterotoxigenic, and enteroinvasive strains of *E. coli* in most parts of the world (Farmer et al., 1977a; Gangarosa and Merson, 1977; Rowe, Gross, and Scotland, 1975; Rowe, Scotland, and Gross, 1977). These points are discussed in Chapter 89, this Handbook.

Yersinia enterocolitica and some rare strains of *Klebsiella pneumoniae* can also cause diarrhea. *Y. enterocolitica* strains cause acute enteritis, apparently because of both invasiveness and toxin production (J. Feeley, personal communication), and were recently implicated in a milk-borne outbreak of diarrhea and mesentery lymphadenitis (Center for Disease Control, 1977b). Toxigenic strains of *Klebsiella pneumoniae* frequently have been isolated from patients with tropical sprue (Klipstein et al., 1973). Toxigenic isolates of *Enterobacter cloacae* were also reported (Klipstein et al., 1973). Since the enterotoxins in *E. coli* are on transmissible plasmids (Gyles, So, and Falkow, 1974), it would not be surprising to find enterotoxin in other species.

Although far beyond the scope of this chapter, the etiology and epidemiology of foodborne and waterborne enteric disease outbreaks are extremely important and fascinating subjects that demand some comment. There were 1,064 foodborne disease outbreaks in the United States reported to the Center for Disease Control (CDC) between 1972 and 1974 (Center for Disease Control, 1976a). In a given

Table 1. Animal sources of *Edwardsiella tarda,* yersiniae, and salmonellae.

Edwardsiella tarda[a]	*Yersinia enterocolitica*[b]	*Yersinia pseudotuberculosis*[c]	*Salmonella*[d]
Cobra	Cat	Birds (30 species)	Bat
Cow	Chinchilla	Cat	Baboon
Eel	Cow	Coypu	Badger
Grass snake	Dog	Deer	Bird (37 species)
Lizard	Galago	Fox	Camel
Panther	Guinea pig	Guinea pig	Cat
Pig	Hare	Hedgehog	Chicken
Salmon	Horse	Marten	Cockroach
Skunk	Mink	Mink	Cow
Trout	Monkey	Mole	Deer
Turtle	Sheep	Mouse	Dog
Viper		Rabbit	Donkey
		Rat	Echidna
		Tick	Elephant
		Vole	Ferret
			Flea
			Fly
			Fox
			Frog
			Giraffe
			Goose
			Guinea pig
			Hedgehog
			Hippopotamus
			Horse
			Kangaroo
			Lizard (many species)
			Louse
			Mink
			Mole
			Monkey (many species)
			Mouse
			Oyster
			Pig
			Rabbit
			Rat
			Rhinoceros
			Sheep
			Shrew
			Sloth
			Snake (many species)
			Squirrel
			Tick
			Toad
			Turkey
			Turtle
			Vole

[a] Data from Brenner et al. 1969; D'Empaire, 1969; Shotts and Snieszko, 1976.
[b] Data from Bottone, 1977.
[c] Data from Mair, 1968.
[d] Data from Ewing, 1969; Taylor, 1968.

year, 30–60% of the agents responsible for these outbreaks are not identified. In 1974, salmonellae accounted for 17% of the etiologically confirmed outbreaks and for 5,500 (60%) of the cases. Shigellae accounted for 14 outbreaks with some 1,700 cases between 1972 and 1974 (Center for Disease Control, 1976a). These figures represent the tip of an iceberg because many outbreaks are not detected, most detected outbreaks are not reported to CDC, and, for *Salmonella,* it is estimated that each reported case represents 100 total cases.

Many foodborne outbreaks are caused by contaminated beef or beef products (Center for Disease Control, 1976c). Other common sources are pork, poultry, milk and milk products, eggs, and fish (Center for Disease Control, 1976b). The suddenness with which a rare *Salmonella* serotype can become a serious public health problem is exemplified by *S. agona* (Center for Disease Control, 1976c; Clark et al., 1973). *S. agona* was isolated from man only twice before 1970. By 1972, it was the eighth most frequently isolated serotype in the United States and, by 1975, the fifth most frequently isolated. This organism was the second most frequently isolated serotype in the United Kingdom in 1972 and was a major problem in The Netherlands, Israel, and other parts of the world. The pathway of infection was apparently from contaminated Peruvian fish meal to animals and then to man (Clark et al., 1973).

Species of enterobacteria not normally associated with outbreaks of diarrheal disease are often referred to as "nonpathogens". Nothing could be further from the truth. Most of these species are more properly termed opportunistic pathogens. Once established, they can cause a variety of infections, including urinary tract disease, pneumonia, septicemia, meningitis, and wound infection.

Enterobacteria were responsible for about 50% of nosocomial infections occurring in the United States during 1974 (Center for Disease Control, 1977a). Data presented in Table 2 indicate that *E. coli, Klebsiella, Enterobacter, Proteus, Providencia, Morganella,* and *Serratia marcescens* are the prime offenders.

The so-called compromised host is particularly susceptible to nosocomial infections. The catheterized patient, patients on immunosuppressants, burn patients, cancer patients, and elderly patients are vulnerable to opportunistic pathogens. To make matters worse, many of the organisms acquired in the hospital are multiply drug resistant.

Taxonomy and Nomenclature

The family Enterobacteriaceae may be defined as Gram-negative, oxidase-negative, asporogenous, non-acid-fast, rod-shaped bacteria. They are motile by peritrichous flagellae or are nonmotile; grow both aerobically and anaerobically; grow well on artificial media; produce acid and often gas fermentatively from glucose, other carbohydrates, and related compounds; are catalase positive except for *Shigella dysenteriae* 1; and reduce nitrates to nitrites except for some species of *Erwinia* (Cowan et al., 1974; Edwards and Ewing, 1972). There were many changes in both the taxonomy and nomenclature of the family Enterobacteriaceae from the seventh edition of *Bergey's Manual of Determinative Bacteriology* (Breed, Murray, and Smith, 1957) to the eighth edition (Cowan et al., 1974). These changes, presented in Table 3, were often a direct or indirect result of the monumental work of Edwards and Ewing (1972) and later of Ewing. These investigators carefully examined tens of thousands of cultures and (without the aid of a computer!) compiled percentage data for species and for biogroups within species. The availability of percentage data allowed classification based on quantitative rather than qualitative criteria.

Starting mainly in the mid-1960s, a number of novel approaches were used to help classify and identify enterobacteria. Among these were numerical or computer taxonomy, the use of additional biochemical tests, patterns of resistance to antibiotics (antibiograms), patterns of resistance to heavy metals, drugs, and other toxic substances (resistograms), sensitivity to specific bacteriophages, and

Table 2. Nosocomial infections caused by enterobacteria.[a]

Causal organism(s)	Percentage of total infections						
	Bacteremia	Surgical	Lower respiratory	Urinary tract	Cutaneous	Other	Total
Escherichia coli	16	17	12	32	10	18	23
Klebsiellae	13	6	16	10	9	7	9
Enterobacter sp.	5	5	10	4	3	4	5
Protei	4	8	7	13	10	7	10
Serratia spp.	4	1	3	2	1	1	2

[a] Data from Center for Disease Control, 1977a.

Table 3. Classification of members of the family Enterobacteriaceae in the seventh and eighth editions of *Bergey's Manual of Determinative Bacteriology* (Breed, Murray, and Smith, 1957; Cowan et al., 1974).

Seventh edition	Eighth edition	Seventh edition	Eighth edition
Escherichia coli	NC[a]	*Serratia marcescens*	NC
Escherichia aurescens	*E. coli*	*Serratia indica*	*S. marcescens*
Escherichia freundii	*C. freundii*	*Serratia plymuthica*	*S. marcescens*
Escherichia intermedia	*Citrobacter intermedius*	*Serratia kiliensis*	*S. marcescens*
Aerobacter aerogenes	*Enterobacter aerogenes*	*Serratia piscatorum*	*S. marcescens*
Aerobacter cloacae	*Enterobacter cloacae*	*Proteus vulgaris*	NC
NL[b]	*Hafnia alvei*	*Proteus mirabilis*	NC
		Proteus morganii	NC
Klebsiella pneumoniae	NC	*Proteus rettgeri*	NC
Klebsiella ozaenae	NC	*Proteus inconstans*	NC
Klebsiella rhinoscleromatis	NC	NL	*Edwardsiella tarda*
Paracolobactrum aerogenoides	*Klebsiella*	*Salmonella* (343 serotypes)	*Salmonella* (1540 serotypes)
Paracolobactrum intermedium	*Citrobacter*		
Paracolobactrum arizonae	*Salmonella*	*Shigella dysenteriae*	NC
Paracolobactrum coliforme	*E. coli*	*Shigella schmitzii*	*S. dysenteriae*
Alginobacter acidofaciens	NL	*Shigella arabinotarda*	*S. dysenteriae*
		Shigella boydii	NC
Erwinia amylovora	NC	*Shigella flexneri*	NC
Erwinia vitivora	*E. herbicola*	*Shigella alkalescens*	*E. coli*
Erwinia milletiae	*E. herbicola*	*Shigella sonnei*	NC
Erwinia cassavae	*E. herbicola*	*Shigella dispar*	*E. coli*
Erwinia salicis	NC	*Pasteurella pestis*	*Yersinia pestis*
Erwinia tracheiphila	NC	*Pasteurella pseudotuberculosis*	*Yersinia pseudotuberculosis*
NC	*E. nigrifluens*	NL	*Yersinia enterocolitica*
NC	*E. quercina*		
Erwinia chrysanthemi	NC		
Erwinia carnegieana	*E. carotovora*		
Erwinia dissolvens	not *Erwinia*		
Erwinia nimipressuralis	not *Erwinia*		
NL	*E. uredovora*		
NL	*E. stewartii*		
Erwinia atroseptica	*E. carotovora* var. *atroseptica*		
Erwinia ananas	*E. herbicola* var. *ananas*		
Erwinia aroideae	*E. carotovora*		
Erwinia citrimaculans	*E. herbicola*		
NL	*E. cypripedii*		
NL	*E. rhapontici*		

[a] No change.
[b] Not listed.

the determination of relatedness between bacteria by DNA hybridization. New biochemical tests, numerical techniques, and DNA hybridization have been of great help to taxonomists. Antibiograms have been used to great advantage in many clinical laboratories, while the potential of resistograms and specific bacteriophages remains largely untapped. DNA hybridization has become a powerful tool in the taxonomy of the family Enterobacteriaceae and in the classification of atypical isolates. Many of the taxonomic changes made or proposed since the eighth edition of *Bergey's Manual* (Cowan et al., 1974) was completed in 1970 were a direct result of DNA hybridization studies or were confirmed by these studies. Before these changes are discussed, it will be worthwhile to discuss both the biased taxonomic

views held by this author and the rationale behind DNA relatedness.

There seems to be significant resistance, particularly in medical circles, to taxonomic changes in the family Enterobacteriaceae and to the establishment of new species. Can we not assume that microbiologists, physicians, epidemiologists, and workers in related health fields agree that any branch of science is constantly changing as new data become available? If this assumption is valid, I can only conclude that these people do not consider bacterial taxonomy to be a science. Ironically, these same people definitely view the characterization of plasmids or viruses biochemically and or by DNA hybridization as science, but probably not as taxonomy. Obviously they cannot have it both ways. In fact, taxonomy is the science of classification—of bacteria, plasmids, or viruses.

Some comments are also in order with regard to the proliferation of species within the family Enterobacteriaceae. *Bergey's Manual* lists 36 species of enterobacteria, 12 of which are in the genus *Erwinia* (Cowan et al., 1974). *Bacillus* contains 48 species, *Mycobacterium* has 30, *Streptoverticillium* has 40, and, on the basis of antibiotic production, *Streptomyces* has 464. If one seeks divine guidance for the number of species "permitted" within a genus or a family, the family Enterobacteriaceae seem to be well within celestial limits. The problem in taxonomy is not how many species to define, but rather on what basis should species be defined. This is a scientific problem. A practical problem also exists. Different specialties require different bases for differentiating organisms. For instance, the medical microbiologist is concerned with pathogenicity, the phytopathologist with plant pathogenicity and host range, the industrial microbiologist with the purity and quantity of a specific metabolic product, and the geneticist with the presence or absence of a specific gene.

One solution is to have a "scientific taxonomy" for "scholars" and a separate, practical taxonomic scheme for each specialty group. I personally find this solution both distasteful and abhorrent; but, more importantly, it is confusing and impractical. Consider an example. Strains of *Vibrio cholerae* have more than 80 different O antigen groups. Strains of all O antigen groups are extremely similar both biochemically and genetically—to a greater extent than strains of *E. coli*. Strains from many of these O groups cause cholera, but only one O group, O1, contains the epidemic strains of cholera. Some clinical microbiologists and many epidemiologists recognize O1 and only O1 as *Vibrio cholerae* and refer to all other O groups as either nonagglutinating vibrios (NAG vibrios) or noncholera vibrios. Obviously confusion reigns when two groups attach different definitions to a species name.

Most taxonomists agree that a single characteristic, regardless of importance (pathogenicity, etc.), is not a valid basis for identification to the species level. There must be at least one taxonomic level that means the same thing to everyone. Logically that level should be species. A species should be defined scientifically on the basis of all available knowledge. A genus ideally should consist of a group of species that are related to one another, both genotypically and phenotypically, to a greater extent than they are to members of any other genus. Where this is not feasible, a genus should contain a group of phenotypically similar species.

Only one definition of a species should exist— one that connotes the same thing to everyone. *Escherichia coli* must mean the same thing regardless of serotype, biotype, toxigenicity, pathogenic-

ity, source of isolation, lysogenicity, plasmids, specific genetic markers, etc. These special markers should be defined at a subspecies level. The use of subspecific designations such as toxotype (ST or LT), pathotype, mutotype, and variety, in addition to the currently used biotype, serotype, and phage type, allows a complete description of strains of a given species without bastardizing the species concept.

But what is a species?! Species have always been and will continue to be designated arbitrarily. In the past, partly by necessity and partly because of the prejudice of the investigator, the criteria used in arbitrarily designating species varied greatly. Differences in pathogenicity, host range, or serology were deemed sufficient to create species. Most species are now designated on the basis of biochemical differences. How many biochemical differences are sufficient to create a new species? How many tests should be done? Should all biochemical differences be given equal weight or should they be weighted differently? There are no simple answers to these questions, which is probably why a uniform concept for designating species has not been proposed.

Some species of the family Enterobacteriaceae exhibit very little biochemical variability (*Edwardsiella tarda, Yersinia pseudotuberculosis*), whereas other species, including *E. coli*, contain hundreds of different biotypes. Determining the biochemical boundaries of variable species is difficult. The task is further complicated by the ability of enterobacteria to acquire plasmids that specify metabolic genes. For example, a chromosomal gene for hydrogen sulfide production has never been demonstrated in *E. coli*, nor have chromosomal genes for lactose fermentation been demonstrated in *Salmonella typhimurium*. We know that hydrogen sulfide–producing strains of *E. coli* and lactose-fermenting salmonellae exist, and that these functions are mediated by genes present on plasmids.

How then are we to identify biochemically aberrant strains, and how can we possibly arrive at a uniform definition of a species? The answer is found in a combined genotypic-phenotypic approach. To determine genotypic relatedness, double-stranded DNA from one organism is isotopically labeled and broken into fragments about the size of an average gene (2.5×10^5 daltons). The fragments are separated into single strands by heating and are reacted with similarly treated, unlabeled DNA preparations from the same strain and a series of other strains (Brenner, 1977). The reassociation of two strands of DNA is a very specific reaction that requires a high degree of nucleotide base sequence complementarity. Related DNA sequences will reassociate, regardless of their source, whereas unrelated sequences will not. DNA relatedness studies have been done on most groups of enterobacteria during the past 15 years (Brenner, 1973, 1974, 1977; Brenner and Cowie, 1968; Brenner and Falkow, 1971; Brenner, Fanning, and Steigerwalt, 1974a,b; Brenner, Martin, and Hoyer, 1967; Brenner, Steigerwalt, and Fanning, 1972; Brenner et al., 1969, 1972, 1973a,b, 1976, 1978; Crosa et al., 1973, 1974; Ewing et al., 1978; Ferragut and Leclerc, 1978; Gardner and Kado, 1972; Izard et al., 1979a,b, 1981; Jain, Radsak, and Mannheim, 1974; Leete, 1977; McCarthy and Bolton, 1963; Moore and Brubaker, 1975; Murata and Starr, 1974; Ritter and Gerloff, 1966; Seidler, Knittel, and Brown, 1975; Steigerwalt et al., 1975; Stoleru, Le Minor, and Lheritier, 1976).

Species were defined retrospectively after studies were done on most well-described species of the family Enterobacteriaceae. DNA relatedness among strains of a given species was usually 70% or higher, and this level of relatedness was *arbitrarily* chosen to define a species. The "genetic" definition of a species can now be enlarged to encompass five parameters: relatedness at conditions optimal for DNA reassociation; relatedness at conditions less than optimal for DNA reassociation, at which only highly complementary sequences can reassociate; stability of related sequences to heat; genome size; and guanine-plus-cytosine (G+C) content of DNA. Stability of reassociated DNA sequences to heat (thermal stability) is important because each degree decrease in thermal stability is caused by approximately 1% of unpaired nucleotide bases in related DNA sequences (Bonner et al., 1973). Thus the degree of similarity (or evolutionary divergence) in related DNA sequences can be approximated by comparing their thermal stability with that of sequences in which labeled and unlabeled DNA are from the same strain. For example, the definition of *E. coli* would be: a series of strains that are 70% or more related at conditions optimal for DNA reassociation; 60% or more related at conditions less than optimal for DNA reassociation; whose thermal stability of related sequences is within 4°C of homologous reassociated *E. coli* DNA; having a genome size of between 2.3×10^9 and 3.0×10^9 daltons; and a G+C content between 49 and 52% (Brenner et al., 1972a). After a decade of experience, we believe that a species is a group of strains with 70% or more relatedness at optimal conditions, with 55% or more relatedness at less than optimal conditions, and with a thermal stability of related sequences within 6°C of reassociated homologous DNA.

All strains tested from most species are in one DNA relatedness group. Medically important organisms were often assigned to species largely on the basis of pathogenicity. In these cases, several phenotypic species are in the same DNA relatedness group (see below). Alternatively, organisms of lesser medical relevance were often lumped into a single phenotypic species. Strains from these species often are contained in two or more DNA relatedness

groups. In some cases, the species defined on the basis of DNA relatedness studies differ from well-established species defined on the basis of phenotypic characters. Thus far, although recommendations for taxonomic changes have been made in several genera, they have not yet been proposed in the most important medical genera—*Salmonella*, *Shigella*, and *Escherichia*. The nomenclature in these genera is so deeply ingrained that acceptance of changes would be strongly resisted and would result in confusion. It is probable that certain indicated changes will be proposed in these genera in the future.

DNA hybridization determines relatedness in the entire genome—some 3,000 genes for enterobacteria. Relatedness values are not affected by mutations in one or in dozens of individual genes, by deletion mutations, or by the presence of plasmids, because the overall impact of these changes in DNA does not significantly affect the overall DNA relatedness between organisms.

A series of atypical strains is tested to determine the phenotypic boundaries of a DNA relatedness group. For example, a series of biochemically atypical strains of *E. coli* was tested for relatedness to a typical *E. coli* strain as shown in Table 4 (B. R. Davis, G. R. Fanning, S. D. Allen, and D. J. Bren-

Table 4. Biochemical limits of the *Escherichia coli* DNA relatedness group (B. R. Davis, R. G. Fanning, S. D. Allen, and D. J. Brenner, unpublished observations).

Atypical biochemical reactions[a]	Range of DNA relatedness to *E. coli* K-12 (%)
None	85–99
Lac⁻	92–100
Ado⁺	100
Ind⁺	90–93
Man⁻	93
Phe⁺	99–100
H₂S⁺	100
Cit⁺	95–100
Ure⁺	79–100
KCN⁺	97–100
Ino⁺, Ind⁻	100
H₂S⁺, Cit⁺	98
H₂S⁺, Lac⁻	96
H₂S⁺, Ind⁻	100
H₂S⁺, Ino⁺	94
H₂S⁺, Man⁻	99
MR⁻, Man⁻	91–100
Lys⁻, Arg⁻, Orn⁻	88–100
YP⁺	95
KCN⁺, YP⁺, Cel⁺	41–47

[a] Lac, lactose; Ado, adonitol; Ino, inositol; Man, mannitol; Phe, phenylalanine deaminase, Cit, Simmons' citrate; Ure, urease; Ind, indole, MR, methyl red; Lys, lysine decarboxylase; Arg, arginine dihydrolase; Orn, ornithine decarboxylase; YP, yellow pigment; Cel, cellobiose. Biochemical tests in this and all subsequent tables were done at 36 ± 1°C unless otherwise specified.

ner, unpublished observations). KCN-positive, yellow-pigmented, cellobiose-positive strains were only about 45% related to *E. coli*. It was subsequently shown that these strains formed a DNA-relatedness group distinct from all other enterobacteria. All of the other biochemically atypical strains were highly related to typical *E. coli*. We therefore know with certainty that all of the aberrant biochemical patterns except for the KCN-positive, yellow-pigmented, cellobiose-positive group belong to the species *E. coli,* and we have defined the phenotypic boundaries of *E. coli.* As strains with additional aberrant patterns are encountered, they will also be included or excluded from *E. coli* on the basis of genetic relatedness to typical *E. coli.* There is no standard for phenotypic diversity that can be applied to define a species. The phenotypic boundaries of some species *(E. coli, Y. enterocolitica)* are broad, whereas those of other species *(Edwardsiella tarda, Y. pseudotuberculosis)* are very narrow.

Unfortunately, most scientific books on bacterial taxonomy are partially out of date at the time of publication. This certainly was true of the eighth edition of *Bergey's Manual.* The eighth edition was published in 1974, but the chapter on the family Enterobacteriaceae was completed in 1970 (Cowan et al., 1974)! I have attempted to minimize this problem in the classification of enterobacteria as presented in Table 5. The current classification is given and compared with that in the eighth edition of *Bergey's Manual* (Cowan et al., 1974). In addition, I have presented the taxonomic changes that I anticipate will occur within the next 5 years. While these anticipated changes are based on data obtained in my laboratory and the laboratories of many of my colleagues, they are obviously somewhat biased to my thinking, which is not necessarily shared by all of my colleagues. For these reasons, all projected changes are clearly designated as such. They are intended to indicate trends and should not be taken as "gospel", especially by nonspecialists.

Escherichia and *Shigella*

There are no changes in the four established species of *Shigella* or in *Escherichia coli,* despite the fact that all of these organisms are one species genetically (Brenner et al., 1972, 1973a). A new species, *Escherichia ewing* (B. R. Davis, G. R. Fanning, S. G. Allen, and D. J. Brenner, unpublished observations), will be added to the genus *Escherichia.* This organism is approximately 45% related to *E. coli* (Table 4) (unless specified, all DNA relatedness values were obtained from reactions done at a 60°C incubation temperature, which is optimal for DNA reassociation). *E. ewing* differs from *E. coli* phenotypically because it is KCN positive, yellow-pigmented, and cellobiose positive. A third possible species of *Escherichia* is *E. fergusonii.* These strains differ from *E. coli* in reactions for lactose (−), adonitol (+), sorbitol (−), cellobiose (+), and mucate (−). They are about 56% related to *E. coli* by DNA hybridization (J. J. Farmer, G. R. Fanning, and D. J. Brenner, unpublished observations). An organism named *Escherichia blattae* was first iso-

Table 5. Present and anticipated classification of members of the family Enterobacteriaceae compared to classification in the eighth edition of *Bergey's Manual* (Cowan et al., 1974).

Present or anticipated	*Bergey's Manual,* eighth edition
Escherichia coli	NC[a]
(Escherichia ewing)[b]	NL[c]
(Escherichia fergusonii)[b]	NL
(Escherichia blattae)[b]	NL
Shigella dysenteriae	NC
Shigella flexneri	NC
Shigella boydii	NC
Shigella sonnei	NC
Edwardsiella tarda	NC
Salmonella—see Table 6	See Table 6
Arizona—see Table 6	See Table 6
Citrobacter freundii	NC
Citrobacter diversus	*Citrobacter intermedius* biotype b
(Citrobacter amalonaticus)[b]	*Citrobacter intermedius* biotype a
Klebsiella pneumoniae	NC
Klebsiella ozaenae	NC
Klebsiella rhinoscleromatis	NC
(Klebsiella oxytoca)[b]	Indole[+], gelatin[+] biotype of *K. pneumoniae*

Table 5. Present and anticipated classification of members of the family
Enterobacteriaceae compared to classification in the eighth edition of
Bergey's Manual (Cowan et al., 1974) *(Continued).*

Present or anticipated	*Bergey's Manual*, eighth edition
(unnamed indole⁻ *Klebsiella*)[b]	NL
(unnamed indole⁺ *Klebsiella*)[b]	NL
Enterobacter aerogenes	NC
(Enterobacter amnigena)[b]	NC
Enterobacter cloacae	NC
(Enterobacter sakazakii)[b]	Yellow-pigmented *E. cloacae*
Enterobacter agglomerans	*Ewinia herbicola, Erwinia stewartii* and *Erwinia uredovora*
(Enterobacter vulneris)[b]	NL
(Enterobacter gergoviae)[b]	NL
Hafnia alvei	NC
(Hafnia species 2; "*Obesumbacterium proteus*")[b]	NL
(Serratia ficaria)[b]	NL
(Serratia fonticola)[b]	NL
Serratia marcescens	NC
Serratia liquefaciens	NL
Serratia rubidaea (S. marinorubra)	NL
(Serratia plymuthica)[b]	NL
(Serratia odorifera)[b]	NL
Proteus mirabilis	NC
Proteus vulgaris	NC
Proteus myxofaciens	NL
Providencia rettgeri	*Proteus rettgeri*
Providencia alcalifaciens	*Proteus inconstans* subgroup a
Providencia stuartii	*Proteus inconstans* subgroup b
Morganella morganii	*Proteus morganii*
Erwinia amylovora	NC
Erwinia salicis	NC
Erwinia tracheiphila	NC
Erwinia nigrifluens	NC
Erwinia quercina	NC
Erwinia rubrifaciens	NC
(Erwinia mallotivora)[b]	NL
Pectobacterium carotovorum	*Erwinia carotovora*
Pectobacterium carnegieana	NL
Pectobacterium chrysanthemi	*Erwinia chrysanthemi*
Pectobacterium cypripedii	*Erwinia cypripedii*
Pectobacterium rhapontici	*Erwinia rhapontici*
Yersinia pestis	NC
Yersinia pseudotuberculosis	NC
Yersinia enterocolitica	NC
(Yersinia intermedia)[b]	NL
(Yersinia fredericksenii)[b]	NL
(Yersinia kristensenii)[b]	NL
(Yersinia ruckeri)[b]	NL
"*Hafnia*" species[c]	NL
Obesumbacterium proteus group[b]	NL
(*Kluyvera* species, as yet unnamed)[b]	NL
(*Kluyvera* species, as yet unnamed)[b]	NL
(*Rahnella aquatilus*)[b]	NL

[a] No change.
[b] Proposed or anticipated change. See text for details.
[c] Not listed.

lated by Burgess, McDermott, and Whiting (1973) from the hindgut of cockroaches found in hospitals and other environments in London. A similar organism was isolated independently in 1973 from cockroaches on Easter Island (G. Nogrady, personal communication). DNA relatedness studies (G. R. Fanning, F. W. Hickman, E. Cadet, G. Vaughn, and D. J. Brenner, in preparation) showed that the London and Easter Island isolates belonged to the same DNA relatedness group. *E. blattae* is also discussed in the section on *Hafnia* in this chapter.

Edwardsiella tarda

Edwardsiella tarda is the only species in the genus *Edwardsiella* for which strains are available. Strains of *E. tarda* exhibit very little biochemical or genetic variability. *E. tarda* is approximately 20% related to almost all other enterobacteria (Brenner, Fanning, and Steigerwalt, 1974a). *Edwardsiella anguillimortifera* (Hoshina, 1962; Sakazaki and Tamura, 1975) was proposed as a synonym with priority over *E. tarda*. The description of *E. anguillimortifera* differs from that of *E. tarda*. No strains are known that correspond to the description of *E. anguillimortifera* (see this Handbook, Chapter 90).

Salmonella and *Arizona*

There are more than 2,000 serotypes of *Salmonella* and *Arizona,* most of which are named and erroneously considered as species. There are three taxonomic schemes used for salmonellae and arizonae.

The three-species *Salmonella* recommendation of Ewing (Edwards and Ewing, 1972) is used in the United States. *S. cholerae-suis, S. typhi,* and *S. enteritidis* are the three recognized species. All salmonellae except *S. cholerae-suis* and *S. typhi* are considered serotypes of *S. enteritidis* (Table 6), for example, *S. enteritidis* serotype Typhimurium, *S. enteritidis* serotype Enteritidis, and *S. enteritidis* bioserotype Paratyphi-A (Edwards and Ewing, 1972). Arizonae are placed in a separate genus, *Arizona,* as *Arizona hinshawii.* Outside of the United States, a four-subgenera schema is used (Cowan et al., 1974; see this Handbook, Chapter 92). Subgenus I contains biochemically typical salmonellae, subgenus II and subgenus IV contain two different sets of biochemically atypical strains, and subgenus III contains all arizonae, designated *S. arizonae.* There is no designation of species. In the eighth edition of *Bergey's Manual,* 11 "types" of *Salmonella* are described without being officially designated as species (Cowan et al., 1974). These are shown in Table 6, in which the three schemes for *Salmonella* are compared.

DNA relatedness studies showed that all of the salmonellae and arizonae tested were 70% or more related, which is consistent with one genetic species (Crosa et al., 1973; Stoleru et al. 1976). It was possible to distinguish five relatedness subgroups corresponding to (i) subgenus I, (ii) subgenus II, (iii) subgenus IV, (iv) subgenus III strains with monophasic H antigens that are usually delayed fermenters of lactose, and (v) subgenus III strains that have diphasic flagellar antigens and ferment lactose rapidly.

Table 6. Classification of *Salmonella* and *Arizona.*

Four subgenera (Cowan et al., 1974)	*Arizona,* three-species *Salmonella* (Edwards and Ewing, 1972)
Subgenus I	
S. cholerae-suis	S. cholerae-suis
S. typhi	S. typhi
S. hirschfeldii (S. paratyphi C)	S. enteritidis ser. Paratyphi C
S. paratyphi A	S. enteritidis ser. Paratyphi A
S. schottmuelleri (S. paratyphi B)	S. enteritidis ser. Paratyphi B
S. typhimurium	S. enteritidis ser. Typhimurium
S. enteritidis	S. enteritidis ser. Enteritidis
S. gallinarum	S. enteritidis ser. Gallinarum
Subgenus II	
S. Salamae (also called S. dar-es-salaam)	S. enteritidis ser. Dar-es-salaam
Subgenus III	
S. arizonae	*Arizona hinshawii*
Subgenus IV	
S. houtenae (also called S. houten)	S. enteritidis ser. Houten

Neither the *Arizona* and three-species *Salmonella* scheme nor the four-subgenera scheme correlates well with the DNA data. A logical taxonomic solution would be a single species, with five subspecies representing the five DNA subgroups. Such a scheme is unlikely to gain acceptance in the forseeable future. It is important for bacteriologists to be familiar with both the four subgenera scheme used by the WHO International Reference Center for Salmonella and the *Arizona,* and the three-species *Salmonella* scheme because both schemes will be encountered. An up-to-date Kauffmann-White scheme of *Salmonella* serotypes is presented in the chapter on *Salmonella* (this Handbook, Chapter 92).

Citrobacter

The two species of *Citrobacter* described in *Bergey's Manual* (Cowan et al., 1974) are *C. freundii* and *C. intermedius. C. intermedius* biotype b (Table 5) has also been called *C. diversus (Ewing and Davis, 1972), C. koseri* (Fredericksen, 1970), and *Levinea malonatica* (Young et al., 1971. *Citrobacter intermedius* biotype a corresponds to *Levinea amalonatica.* Other names have also been suggested for the two species of *Levinea* (see this Handbook, Chapter 91).

Crosa et al. (1974) showed that *C. freundii, C. diversus (C. koseri, L. malonatica),* and *L. amalonatica* were about 60% related to one another. These three species should all be in the same genus. Therefore it will be proposed that *Levinea amalonatica* be transferred to the genus *Citrobacter* as *C. amalonaticus* (Table 5). Several biogroups in *C. freundii* are 60–70% related to typical strains of *C. freundii* (Crosa et al., 1974). Further study may reveal that some of these strains should be separate species. Additional groups of novel *Citrobacter*-like organisms described by Leclerc (Gavini, Lefebvre, and Leclerc, 1976; Leclerc and Buttiaux, 1965) are also candidates for new species.

Klebsiella

Initial studies indicated that strains of *Klebsiella pneumoniae, K. ozaenae,* and *K. rhinoscleromatis* were all in the same DNA relatedness group (Brenner, Steigerwalt, and Fanning, 1972b). These studies further showed that klebsiellae were easily separable from both motile and nonmotile strains of *Enterobacter aerogenes.* Data from two numerical taxonomic studies supported the assumed homogeneity of *Klebsiella pneumoniae* (Bascomb et al., 1971, Johnson et al., 1975). Neither the DNA relatedness study nor the numerical taxonomic studies included indole-positive, gelatin-positive strains. These were usually considered to be a biogroup of

K. pneumoniae (Edwards and Ewing, 1972), although a separate species for these organisms had been proposed (Lautrop, 1956). Jain, Radsak, and Mannheim (1974) showed that indole-positive strains of klebsiellae are 60% or less related to species of *Klebsiella* and *Enterobacter.* They tested four indole-positive strains, three of which were related at the species level. In addition to indole and gelatin, these strains differ from *K. pneumoniae* in their ability to degrade pectate (Von Riesen, 1976). The indole-positive, gelatin-positive (or delayed), pectate-positive strains represent a species distinct from *K. pneumoniae. Klebsiella oxytoca* was proposed for this species and is validly published (Lautrop, 1956; Table 5). Two additional DNA relatedness groups have been identified in klebsiellae isolated from environmental sources and occasionally from humans (Seidler, Knittel, and Brown, 1975; D. J. Brenner, unpublished). One of these groups is indole-positive, pectate-negative, and the other is indole-negative and differs from *K. pneumoniae* in its ability to grow at 10°C but not at 45°C (Table 5). These organisms are discussed in the chapter on environmental *Klebsiella* (this Handbook, Chapter 94). Recent data indicate that motile, acetoin-negative strains of *K. penumoniae* exist in water (Ferragut and Leclerc, 1978).

Enterobacter

The eighth edition of *Bergey's Manual* lists only two species of *Enterobacter, E. cloacae* and *E. aerogenes* (Cowan et al., 1974). There are currently six species known, one of which may actually represent as many as 10 separate species (Table 5). The biochemical differentiation of these new species will be discussed in the chapter on *Enterobacter* (this Handbook, Chapter 95).

Yellow-pigmented strains formerly included in *E. cloacae* were in a separate, highly related group (Steigerwalt et al., 1975). In addition to pigment differences, these strains differ from *E. cloacae* in their negative sorbitol and delayed positive DNase reactions, and they have been named *Enterobacter sakazakii* (J. J. Farmer, in preparation; Brenner, Fanning, and Steigerwalt, 1977). *Enterobacter agglomerans* is difficult to deal with taxonomically as well as diagnostically. The organisms now called *Enterobacter agglomerans* by most clinical bacteriologists were first isolated from plant sources almost 70 years ago. In the eighth edition of *Bergey's Manual* (Cowan et al., 1974), they constitute the Herbicola group in the genus *Erwinia.* There are three species, one of which contains two varieties: *Erwinia herbicola* var. *herbicola, Erwinia herbicola* var. *ananas, Erwinia uredovora,* and *Erwinia stewartii* (see this Handbook, Chapter 102). DNA hybridization studies reveal that strains of *Entero-*

bacter agglomerans belong to at least 10 DNA relatedness groups (D. J. Brenner, A. G. Steigerwalt, G. R. Fanning, unpublished observations). It is therefore likely that classification of this group will change substantially with the addition of several species. It is premature to predict whether these organisms will remain in *Erwinia* or *Enterobacter,* be placed in a single new genus, or be placed in more than one genus (see this Handbook, Chapter 95).

Enterobacter vulneris is another newly described species (J. K. Leete, A. M. Murlin, and D. J. Brenner, unpublished observation). *E. vulneris* are yellow-pigmented organisms that resemble the *E. agglomerans* group except for their ability to decarboxylate lysine and often arginine. Most *E. vulneris* strains were isolated from wound infections.

Enterobacter gergoviae is most similar to *E. aerogenes. E. gergoviae* is urease positive and is also separable from *E. aerogenes* by reactions for KCN (−) and sorbitol (−). *E. gergoviae* was first isolated in France and Africa (Richard et al., 1976) and subsequently in the United States (M. A. Asbury, unpublished observation). Most isolates are from urine, wounds, and blood (Richard et al., 1976).

Enterobacter amnigena, a newly described water isolate, was recently proposed by Izard et al. (1981). This species is phenotypically similar to *E. cloacae,* as is at least one other as yet unnamed water isolate (Gavini et al., 1976; Izard et al., 1979b).

Each of the *Enterobacter* species is discussed in some detail in the chapter on *Enterobacter* (this Handbook, Chapter 95), which should be consulted for biochemical characterization and differentiation of *Enterobacter* species.

Hafnia

Hafnia alvei has been shuttled back and forth between the genus *Hafnia* and the genus *Enterobacter* (*E. hafniae*). *H. alvei* is the only species described in the eighth edition of *Bergey's Manual* (Cowan et al., 1974). Barbe (1969) describes two types of *H. alvei* that differ in reactions for D-(−)-arabinose, arbutin, esculin, and salicin (Table 7). DNA hybridization experiments (Steigerwalt et al., 1975) revealed two DNA relatedness groups in *H. alvei.* Subsequent experiments confirmed these two relatedness groups (G. R. Fanning, F. W. Hickman, E. Cadet, G. Vaughn, and D. J. Brenner, in preparation). These hydridization groups did not correlate perfectly with Barbe's arbutin-esculin-salicin positive and negative groups. D-(−)-Arabinose fermentation was not tested. Although no single biochemical test was completely diagnostic in separating the first two hybridization groups, they could be separated by a series of reactions (Table 7). In all probability, further study will provide a firm basis for

Table 7. Biochemical differentiation and DNA relatedness of strains of *Hafnia alvei.*[a]

A. Barbe's biotypes (Barbe, 1969)

Test	*H. alvei* group 1	*H. alvei* group 2
D-(−)-Arabinose	+	−
Arbutin	−	+
Esculin	−	+
Salicin	−	+

B. DNA relatedness and differential biochemical reactions in *H. alvei*
(G. R. Fanning, F. W. Hickman, E. Cadet, G. Vaughn, and D. J. Brenner, in preparation).

Test	*H. alvei* DNA relatedness group		
	1	2	3
Hafnia-specific phage susceptibility	+	+	−
Voges-Proskauer	+	+	−
Jordan's tartrate	+	+	−
Catalase	+	+	−
D-Fructose	+	+	−
Motility at 24 h	−9	+100	+
Salicin	V45	−	+
Arbutin	V45	−	(+) or +
Esculin	V45	V(57)	+
Inulin	−(9)	V(71)	−
Sodium acetate	−(9)	V(71)	+
Urease	V45	−7	−

[a]All reactions carried out at 36 ± 1°C. See Table 8 for definition of symbols.

Table 8. Biochemical comparison of *Hafnia alvei*, *Escherichia blattae* and *Obesumbacterium proteus*.[a]

Test	H. alvei DNA group 1	H. alvei DNA group 2	E. blattae	O. proteus 1	O. proteus 2
Susceptibility to *Hafnia* phage	+	+	−	+	−
Motility at 24 h	−	+	−	−	−
Growth in KCN	+	+	−	V	−
Malonate	V82	V29	+	V50	−
Gas from D-glucose	+	+	+	−	−
D-Mannitol	+	+	−	+	−
Salicin	V45	−	−	+	−
L-(+)-Arabinose	+	+	+	V	−
Rhamnose	+	+	+	V	+
D-Xylose	+	+	+	−	V
Cellobiose	V(33)	V(67)	−	−	−
Esculin	V45	V(57)	−	+	−
Jordan's tartrate	+	+	−	V	V
ONPG	V67	+	−	V	−
Arbutin	V45	−	−	+	−
Voges-Proskauer	+	+	−	+	−
Inulin	−	V(71)	−	−	−

[a] All reactions carried out at 36 ± 1°C. +, 90% or more positive within 48 h; −, 0 to 9.9% positive within 48 h; V, variable; the number following V is the percentage positive; ( ), delayed positive, 3–7 days, the number in parentheses is the percentage delayed positive. Unless noted, acid production is the test used for carbohydrates and related compounds.

designating two species from the strains now included in *H. alvei* (Table 5).

A third hybridization group (Table 7) was formed by atypical "*H. alvei*" strains that were resistant to *Hafnia*-specific phage and differed from *H. alvei* groups 1 and 2 in reactions for Voges-Proskauer (−), Jordan's tartrate (−), catalase (−), and levulose (−). These organisms are more closely related to species of *Salmonella*, *Escherichia*, *Citrobacter*, and *Enterobacter* than to *Hafnia*. These organisms may eventually be placed in *Escherichia* or in a new genus. For now they are referred to as "*Hafnia*" species 3 (Table 5).

Priest et al. (1973) proposed the name *Hafnia protea* for two groups of strains previously called *Obesumbacterium proteus*. More recent work (Fanning, Hickman, Cadet, Vaughn, and Brenner, in preparation) indicates that *O. proteus* group 1 and group 2 are very different phenotypically (Table 8) and by DNA hybridization (Table 9). *O. proteus* group 1 is indistinguishable from *H. alvei* 1 on the basis of DNA relatedness. Biochemically it is sensitive to *Hafnia*-specific phage and appears to be an anaerogenic, xylose-negative biogroup of *H. alvei* group 1. *O. proteus* group 2 is biochemically very different from both *H. alvei* 1 and *O. proteus* 1 (Table 8). DNA relatedness between strains of *O. proteus* group 1 and group 2 is only 25–30%. In fact *O. proteus* group 2 is more than 50% related to *E. blattae* (Table 9).

Table 9. DNA relatedness among species of *Hafnia* and *Hafnia*-like organisms.[a]

Labeled DNA	Average percent relatedness						
	H. alvei 1	H. alvei 2	H. alvei 3	Escherichia blattae	Obesumbacterium proteus 1	O. proteus 2	E. coli
H. alvei 1	78	57	27	26	75	24	25
H. alvei 2	59	86	27	22	52	20	25
H. alvei 3	27	27	90	34	21	24	43
E. blattae	26	24	42	90	23	55	42
O. proteus 1	72	55	30	24	94	25	27
O. proteus 2	28	28	45	64	29	94	39

[a] These data are summarized from G. R. Fanning, F. W. Hickman, E. Cadet, G. Vaughn, and D. J. Brenner, in preparation. DNA hybridization reactions were done as described by Steigerwalt et al., 1975. The figures shown are average relatedness values obtained from 3 to 12 strains.

Table 10. DNA relatedness among species of *Serratia.*[a]

Labeled DNA	Average percent relatedness					
	S. marcescens	*S. liquefaciens*	*S. rubidaea*	*S. plymuthica*	*S. odorifera*	*S. fonticola*
S. marcescens	91	58	49	—	—	54
S. liquefaciens	56	78	45	—	—	47
S. rubidaea	50	42	73	—	—	43
S. plymuthica	45	55	29	88	36	35
S. odorifera	31	35	32	38	80	38
S. fonticola	50	—	—	—	—	94

[a] Data adapted from Steigerwalt et al., 1975 and Grimont et al., 1978. The figures shown are average relatedness values obtained from 3 to 12 strains.

To summarize, there are two DNA relatedness groups in *H. alvei,* one of which includes *O. proteus* group 1, that are susceptible to *Hafnia*-specific phage and phenotypically similar. These relatedness groups represent separate species, but presently are both considered as *H. alvei.* Three additional DNA relatedness groups, representing three biochemically separate species that are resistant to *Hafnia*-specific phage, are formed by strains of *E. blattae, O. proteus* group 2, and "*H. alvei*" group 3. These three species do not phylogenetically belong in the genus *Hafnia.* They may eventually be included in *Escherichia* or a new genus.

Serratia

If *Bergey's Manual* (Cowan et al., 1974) is used for comparison, the genus *Serratia* has at least four new species. *Serratia marcescens* is the only species listed in the eighth edition. *S. liquefaciens* was inexplicably omitted, and *S. rubidaea* (*S. marinorubra*) was described after the data for *Bergey's Manual* had been compiled. Two new species, *Serratia plymuthica* and *Serratia odorifera,* were recently described by Grimont and associates (Grimont et al., 1978; Grimont, Grimont, and Dulong de Rosnay, 1977; see this Handbook, Chapter 97). A fifth candidate for a species in *Serratia* is a lysine-positive, *Citrobacter*-like organism described by Leclerc (Gavini, Lefebvre, and Leclerc, 1976; Leclerc and Buttiaux, 1965), and named *Serratia fonticola* (Gavini, Ferragut, and Leclerc, 1979). Each of these species has been confirmed by DNA hybridization data (Grimont et al., 1978; Steigerwalt et al., 1975), a summary of which is shown in Table 10. There are two biogroups in *S. odorifera* that are about 70% interrelated. Additional species of *Serratia* may emerge when *S. liquefaciens* is studied more thoroughly. Table 11 shows a series of tests that are useful in separating *Serratia* species. It should be noted that *S. plymuthica* and the lysine-positive *Citrobacter*-like organism have been isolated almost exclusively from water. A detailed ecological and biochemical description of serratiae is given in the chapter on *Serratia* (this Handbook, Chapter 97). P. A. D. Grimont (personal communication) has isolated a new *Serratia* species from figs: *Serratia ficaria.*

Proteeae

The tribe Proteeae has usually been considered to include one genus with five species (Cowan et al., 1974) or two genera: *Proteus* with four species and

Table 11. Biochemical tests of value in separating species of *Serratia.*[a]

Test	*S. marcescens*	*S. liquefaciens*	*S. rubidaea*	*S. plymuthica*	*S. odorifera* BG1	*S. odorifera* BG2	*S. fonticola*
L-Arabinose	−	+	+	+	+	+	+
Raffinose	−	+	+	+	+	−	+
Lactose	−	V	+	(+)	(+)	(+)	+
D-Xylose	−	+	+	+	+	+	V
D-Sorbitol	+	+	−	V	+	+	+
Ornithine decarboxylase	+	+	−	−	+	−	+
Adonitol	+	−	+	−	(+)	(+)	+
Lysine decarboxylase	+	+	V	−	+	+	+
Malonate	−	−	V	−	−	−	+
Dulcitol	−	−	−	−	−	−	+

[a] See Table 8 for definition of symbols.

Providencia with two species (Edwards and Ewing, 1972; Tables 3 and 5). Phenotypic and genotypic differences among species of protei prompted several proposals to divide these organisms into three or more genera (Coetzee, 1972; Kauffmann, 1954). The most recent of these multigeneric proposals was based on comprehensive DNA relatedness studies, as well as biochemical and serological reactions (Brenner et al., 1978; Farmer et al., 1977b; Penner and Hennessy, 1977; Penner, Hinton, and Hennessy, 1975; Penner et al., 1976; Ursing, 1974).

According to these proposals (Brenner et al., 1978), the genus *Proteus* includes *P. mirabilis, P. vulgaris,* and *P. myxofaciens* (Table 5). *P. myxofaciens* was isolated from gypsy moth larvae (Cosenza and Podgwaite, 1966). *Proteus morganii* has been placed in a new genus: *Morganella,* as *Morganella morganii. M. morganii* strains are highly interrelated, have a G+C content that is markedly different from those of all other protei, and are not closely related to any other species of protei.

Six biogroups were described in the genus *Providencia* (Ewing, Davis, and Sikes, 1972). Biogroups 1–4 were *P. alcalifaciens* and biogroups 5 and 6 constituted *P. stuartii.* DNA hybridization reactions revealed three major relatedness groups among strains of *Providencia* (Brenner et al., 1978). Biogroups 1 and 2 were in one relatedness group. Biogroup 3 strains formed a second relatedness group; however biogroup 3 is presently retained within *P. alcalifaciens.* Biogroups 4–6 were inseparable from one another. Therefore biogroup 4 was transferred from *P. alcalifaciens* to *P. stuartii.*

Proteus rettgeri strains formed two major DNA relatedness groups. One relatedness group contained strains from *Proteus rettgeri* biogroups 1–4, and the second group was composed of biogroup 5 strains (Table 12). Biogroup 5 strains were inseparable by DNA relatedness from *Providencia stuartii.* Biogroup 5 of *P. rettgeri* also had *P. stuartii* O antigens and resembled *P. stuartii* biochemically except for being urease positive (Penner and Hennessy, 1977;

Penner, Hinton, and Hennessy, 1975; Table 13). It was shown that biogroup 5 strains of *Proteus rettgeri* gave rise to both urease-positive and urease-negative clones (Farmer et al., 1977b). These strains are, therefore, urease-positive *Providencia stuartii* in which urease production is most likely mediated by the presence of a plasmid. Biogroups 1–4 of *Proteus rettgeri* are much closer biochemically and genetically to *Providencia* species than to *Proteus* species. For these reasons, *Proteus rettgeri* was transferred to the genus *Providencia* as *Providencia rettgeri.* The reactions of value in separating urea-positive *Providencia stuartii* from *Providencia rettgeri* are seen in Table 13.

Erwiniae

The genus *Erwinia* was created as a catchall for plant pathogens (this Handbook, Chapters 4 and 102). Assignment to species was often based solely on the plant from which the organism was isolated. This practice created much confusion because host-range studies were rarely carried out, and many of these so-called species were in fact the same organism isolated from different sources. Much of this

Table 12. Biogroups of *Providencia* species.[a]

Species and biogroup	Gas from D-glucose	Acid from adonitol	Acid from i-inositol
Providencia alcalifaciens			
Biogroup 1	+	+	−
Biogroup 2	−	+	−
Biogroup 3	+	−	−
Providencia stuartii			
Biogroup 4	−	−	−
Biogroup 5	−	−	+
Biogroup 6	−	+	+

[a] Adapted from Ewing, Davis, and Sikes, 1972. See Table 8 for definition of symbols.

Table 13. Biochemical differentiation of *Providencia rettgeri* and *Providencia stuartii.*[a]

| Test | *P. rettgeri* biogroup | | | | *P. stuartii* | |
	1	2	3	4	Urease[+]	Urease[−]
Urease	+	+	+	+	+	−
Trehalose	−	−	−	V	+	+
D-Mannitol	+	+	+	+	V	V
Salicin	+	+	−	−	−	−
L-Rhamnose	−	+	+	−	−	−
Adonitol	+	+	+	+	−	−
D-Arabitol	+	+	+	+	−	−
i-Erythritol	V	V	V	V	−	−

[a] See Table 8 for definition of symbols.

confusion was cleared up as a result of systematic studies by Graham (1964) and Dye (1968, 1969a,b,c). More recently their results have been confirmed and extended somewhat by DNA relatedness studies (Brenner, Fanning, and Steigerwalt, 1974b; Brenner et al., 1973b; Gardner and Kado, 1972; Murata and Starr, 1974).

There is now general agreement about the identity of most species, but several nomenclaturally different schemes are used. Three groups of erwiniae may be distinguished by biochemical reactions and pathogenicity. The Amylovora group (true erwiniae, *Erwinia sensu stricto*) is most inactive metabolically, and requires organic nitrogen. These organisms cause necrotic or wilt diseases. The Carotovora group is more active metabolically, reduces nitrates to nitrites, and causes soft rot diseases. Some authors have placed these organisms in a separate genus, *Pectobacterium* (Brenner et al., 1973b; Waldee, 1945).

The Herbicola group contains both nitrate-positive and nitrate-negative species. They are present in soil and on plant surfaces as secondary and occasionally primary phytopathogens. Most isolates from plants are apparently yellow pigmented. These organisms also cause human disease. The human isolates are often nonpigmented. To further confuse the issue, human isolates have been classified in the genus *Enterobacter* as *Enterobacter agglomerans* (Ewing and Fife, 1972). DNA relatedness studies indicate at least 10 separate species within the *E. agglomerans*–Herbicola group complex (D. J. Brenner, A. G. Steigerwalt, G. R. Fanning, unpublished observations; Leete, 1977). It is important to realize that the *E. agglomerans*–Herbicola group complex will appear as *E. agglomerans* in much of

the medical literature and as species of *Erwinia* in plant literature (Tables 5 and 14). Further discussion of these organisms is found in the chapters of this Handbook on *Erwinia* (Chapter 102) and *Enterobacter* (Chapter 95).

The currently used taxonomic groupings for erwiniae are shown in Table 14. The only species that has not been tested for DNA relatedness is *Erwinia mallotivora* (Goto, 1976). There is no substantive difference in the treatment of the true erwiniae (Amylovora group) (Table 14). Dye (1968) has treated all of these organisms as varieties of *E. amylovora*. None of them show more than 50% relatedness to *E. amylovora* or to each other, and therefore separate species seem justified (Brenner et al., 1974b).

Dye (1969a) also treats all members of the Carotovora group as varieties of *Erwinia carotovora*. DNA relatedness data support separate species status for each of these organisms which are placed in the genus *Pectobacterium* (Brenner et al., 1973b). Most investigators do not recognize *Erwinia* (*Pectobacterium*) *carnegieana* as a valid species. It is true that many cultures labeled *E. carnegieana* are not *Erwinia* or are *E. carotovora;* however, there are strains that appear to be valid representatives of *E. carnegieana* (Brenner et al., 1973b). *Erwinia atroseptica* is now recognized as a variety of *E. carotovora* (Cowan et al., 1974; Dye, 1969a) and not a separate species. This fact is confirmed by DNA relatedness studies (Brenner et al., 1973b). There are indications of additional species within the *E. chrysanthemi* complex. Strains isolated from sugar cane, corn, and sweet potato can be separated by DNA hybridization (Brenner, Fanning, and Steigerwalt, 1977; Schaad and Brenner, 1977).

Table 14. Comparison of taxonomic groupings for erwiniae.

Erwinia Species	*Bergey's Manual* (Cowan et al., 1974)	Dye (1968, 1969a,b,c)	Brenner et al. (1973b, 1974b)
E. amylovora	*E. amylovora*	*E. amylovora*	*E. amylovora*
E. salicis	*E. salicis*	*E. amylovora* var. *salicis*	*E. salicis*
E. tracheiphila	*E. tracheiphila*	*E. amylovora* var. *tracheiphila*	*E. tracheiphila*
E. nigrifluens	*E. nigrifluens*	*E. amylovora* var. *nigrifluens*	*E. nigrifluens*
E. quercina	*E. quercina*	*E. amylovora* var. *quercina*	*E. quercina*
E. rubrifaciens	*E. rubrifaciens*	*E. amylovora* var. *rubrifaciens*	*E. rubrifaciens*
E. mallotivora	Not listed	Not tested	Not tested
E. herbicola var. *herbicola*	*E. herbicola* var. *herbicola*	*E. herbicola* var. *herbicola*	*Enterobacter agglomerans*
E. herbicola var. *ananas*	*E. herbicola* var. *ananas*	*E. herbicola* var. *ananas*	*Enterobacter agglomerans*
E. stewartii	*E. stewartii*	*E. stewartii*	*Enterobacter agglomerans*
E. uredovora	*E. uredovora*	*E. uredovora*	*Enterobacter agglomerans*
E. carotovora	*E. carotovora*	*E. carotovora* var. *carotovora*	*Pectobacterium carotovorum*
E. carnegieana	Not *Erwinia*	Not *Erwinia*	*Pectobacterium carnegieana*
E. chrysanthemi	*E. chrysanthemi*	*E. carotovora* var. *chrysanthemi*	*Pectobacterium chrysanthemi*
E. cypripedii	*E. cypripedii*	*E. carotovora* var. *cypripedii*	*Pectobacterium cypripedii*
E. rhapontici	*E. rhapontici*	*E. carotovora* var. *rhapontici*	*Pectobacterium rhapontici*
E. atroseptica	*E. carotovora* var. *atroseptica*	*E. carotovora* var. *atroseptica*	*Pectobacterium carotovorum*

Yersinia

As formally constituted, the genus *Pasteurella* included both oxidase-positive and oxidase-negative organisms. When *Pasteurella* was subdivided, the oxidase-positive species remained in *Pasteurella* or were placed in a new genus, *Francisella*. The oxidase-negative organisms were placed in a new genus, *Yersinia,* and were transferred to the family Enterobacteriaceae. The species affected were *Yersinia pestis* and *Yersinia pseudotuberculosis*. A third species, *Yersinia enterocolitica,* was added to this genus (Cowan et al., 1974; Fredericksen, 1964; Mollaret and Knapp, 1972).

Both *Y. pestis* and *Y. pseudotuberculosis* are biochemically homogeneous species. The two are separable by motility at 22°C, and by rhamnose and urease reactions (*Y. pestis* is negative and *Y. pseudotuberculosis* is positive for all three tests). Two dozen strains of *Y. pseudotuberculosis* were 80% or more related in DNA hybridization tests (Brenner et al., 1976). In this study, *Y. pseudotuberculosis* was 45–50% related to *Y. enterocolitica* and 10–20% related to other species of enterobacteria. *Y. pseudotuberculosis* and *Y. pestis* belong to the same DNA relatedness group with greater than 80% relatedness (Moore and Brubaker, 1975; Ritter and Gerloff, 1966).

Yersinia enterocolitica strains are quite heterologous in their biochemical reactions. Most investigators recognize at least five major biogroups within *Y. enterocolitica* (Bercovier et al., 1979; Bottone, 1977; Niléhn, 1969; Wauters, 1973). Y. *enterocolitica* strains can be divided by serotype, phage susceptibility pattern, source of isolation, geographical distribution, and pathogenicity (see this Handbook, Chapter 99). Several investigators have considered dividing *Y. enterocolitica* into two or more species. The need for additional genetic studies was expressed by the International Committee on Systematic Bacteriology Subcommittee on *Pasteurella,*

Yersinia, and *Francisella* (Mollaret and Knapp, 1972). The genetic relatedness studies and biochemical studies necessary to meaningfully classify the various groups of atypical *Y. enterocolitica* and *Y. enterocolitica*–like strains have now been done (Bercovier et al., 1979; Brenner et al., 1976; unpublished collaborative data from D. J. Brenner and A. G. Steigerwalt, Center for Disease Control, Atlanta, Georgia; H. Bercovier, J. M. Alonso, and H. H. Mollaret, Institut Pasteur, Paris, France; J. Ursing, Malmo General Hospital, Malmo, Sweden; G. R. Fanning and G. Vaughn, Walter Reed Army Institute of Research, Washington, D.C.). These workers have shown that *Y. enterocolitica* and the three newly proposed species (see below) are 50% or more interrelated, 45–60% related to *Y. pseudotuberculosis,* and 10–35% related to other species of enterobacteria.

There is unanimous agreement that biogroups 1A, 2, 3, and 4 (Table 15) are *Yersinia enterocolitica*. DNAs from members of these biogroups are 80% more related (Brenner et al., 1976; Table 17). Biogroup 5 is metabolically inactive. These strains are negative for trehalose, lipase, indole, xylose, and sucrose. The biogroup 5 strains are highly related to biogroups 1A, 2, 3, and 4 by DNA hybridization (Table 16) and are certainly members of *Yersinia enterocolitica*.

Biogroup 1B strains are sucrose negative. They are not known to be enteropathogenic and usually are isolated from water or other environmental sources (see this Handbook, Chapter 99). These strains are usually 60–70% and sometimes as much as 75% related to the other biogroups of *Y. enterocolitica* by DNA hybridization (Table 16). DNA reactions done at 75°C show that biogroup 1B strains are 40–50% related to other biogroups. Furthermore, the thermal stability of related DNA sequences formed between 1B and other biogroups is significantly less than related DNA sequences formed from different strains in biogroups 1A, 2, 3,

Table 15. Biogroups of *Yersinia enterocolitica* and proposed new species previously included in *Y. enterocolitica*.[a]

| Reaction | Y. enterocolitica biotype | | | | | | Y. intermedia | Y. fredericksenii |
	1A	2	3	4	5	1B		
Lipase	+	−	−	−	−	+	+	+
Indole	+	+	−	−	−	V	+	+
D-Xylose	+	+	+	−	−	+	+	+
Trehalose	+	+	+	+	−	+	+	+
DNase	−	+	+	+	+	−	−	−
Sucrose	+	+	+	+	−	−	+	+
L-Rhamnose	−	−	−	−	−	−	+ or (+)	+
Raffinose	−	−	−	−	−	−	+ or (+)	−
Melibiose	−	−	−	−	−	−	+	−

[a] Data adapted from Bercovier et al., 1979. See Table 8 for definition of symbols.

Table 16. DNA relatedness of *Yersinia enterocolitica* biogroups and proposed new species previously included in *Y. enterocolitica*.[a]

Labeled DNA	Percent relatedness				
	Y. enterocolitica biogroups 1A,2,3,4	*Y. enterocolitica* biogroup 5	*Y. kristensenii*	*Y. intermedia*	*Y. fredericksenii*
Y. enterocolitica biogroups 1A,2,3,4	80–100	80–95	60–70	50–60	50–60
Y. enterocolitica biogroup 5	80–95	75–100	60–70	50–60	50–60
Y. enterocolitica biogroup 1B	60–70	60–70	80–100	40–55	40–55
Y. intermedia	50–60	50–60	40–55	80–100	50–65
Y. fredericksenii	50–60	50–60	40–55	50–65	75–100

[a] These data are summarized from Brenner et al., 1976, and in preliminary form from the unpublished work of D. J. Brenner and A. G. Steigerwalt; H. Bercovier, J. M. Alonso, and H. H. Mollaret; J. Ursing; and G. R. Fanning.

and 4. It is clear that biogroup 1B is separable from *Y. enterocolitica,* although its relatedness to other biogroups approaches species level relatedness (70%). The name *Yersinia kristensenii* will be proposed for this group (H. Bercovier, J. Ursing, D. J. Brenner, A. G. Steigerwalt, G. R. Fanning, G. P. Carter, and H. H. Mollaret, unpublished data).

Two groups of rhamnose-positive strains have been included in *Y. enterocolitica.* One group ferments L-rhamnose, raffinose, and melibiose. It has been isolated from nosocomial infections, but not from the mesentery lymphadenitis, pseudoappendicitis, or gastroenteritis syndromes associated with *Y. enterocolitica* (Alonso et al., 1975; Bottone et al., 1974). The name *Yersinia intermedia* was suggested for these organisms (E. J. Bottone and B. Chester, personal communication) and proposed informally by Brenner (1979).

The name *Yersinia fredericksenii* was informally proposed for the second group of rhamnose-positive strains (Brenner, 1979). These organisms are found in water and soil and probably are rarely pathogenic (Alonso et al., 1975). They ferment rhamnose, but not raffinose or melibiose (see this Handbook, Chapter 99 for complete biochemical reactions for both rhamnose-positive groups). *Y. intermedia* and *Y. fredericksenii* are easily separable from one another, from *Y. enterocolitica* (Tables 15 and 16), and from *Y. pseudotuberculosis,* both biochemically and by DNA relatedness.

Yersinia ruckeri (Ewing et al., 1978), formerly called red mouth bacterium (Ross, Rucker, and Ewing, 1966), causes disease in salmon and trout. There is no known case of human infection caused by this organism. *Y. ruckeri* grows best at 22–25°C; some strains will not grow at 37°C. The biochemical reactions listed in Table 17 are from cultures incubated at 22–25°C. *Y. ruckeri* could have been included in either *Yersinia* or *Serratia* on the basis of biochemical and DNA hybridization data. It is

25–30% related to species from both genera. *Y. ruckeri* DNA contains about 48% G+C (Ewing et al., 1978). DNA from yersiniae contains 47–49% G+C, and DNA from *Serratia* species contains 56–59% G+C. Genetically, therefore, *Y. ruckeri* is closer to *Yersinia* than to *Serratia.*

The name *Yersinia philomiragia* was proposed for an organism that is pathogenic for muskrats (Jensen, Owen, and Jellison, 1969). A single strain of *Y. philomiragia* was tested for DNA relatedness to *Yersinia pestis, Pasteurella multocida, Francisella novicida,* and *F. tularensis* (Ritter and Gerloff, 1966). The relatedness values were about 3% to the *Francisella* species, 6% to *P. multocida,* and 24% to *Y. pestis.* Antigenic studies (Ohara, Sato, and Homma, 1974) show that *Y. philomiragia* is antigenically related to *F. tularensis* and *F. novicida,* but not to *Brucella abortus.* Species of *Pasteurella* and *Yersinia* were not included in this study. The biochemical reactions reported by Jensen, Owen, and Jellison (1969) for *Y. philomiragia* are shown in Table 18. It is not a *Yersinia* or a member of the family Enterobacteriaceae. It should be referred to as the Philomiragia bacterium until further studies allow it to be properly classified (Ursing, Steigerwalt, and Brenner, in preparation).

Additional, Newly Described Enterobacteria

There has been a seemingly logarithmic increase in the number of newly described species and groups within the family Enterobacteriaceae during the past 2 years. This is largely due to three factors: (i) increased use of polyphasic identification schemes including computer assisted identification and DNA hybridization; (ii) increased interest in nonhuman and environmental isolates; and (iii) the entry into the field of several excellent French research groups including those of Le Minor, Richard,

Table 17. Biochemical reactions at 22–25°C of 33 strains of *Yersinia ruckeri*.

Test or substrate	Percent positive at 48
Hydrogen sulfide (TSI)	0
Urease	0
Indole	0
Methyl red	97
Voges-Proskauer	3 (21)[a]
Simmon's citrate	3 (94)
Growth in KCN	27 (27)
Motility	85
Gelatin	52 (14)
Lysine decarboxylase	88 (12)
Arginine dihydrolase	3
Ornithine decarboxylase	100
Phenylalanine deaminase	0
D-Glucose acid	100
gas	12
Lactose	0 (12)[b]
Sucrose	0
D-Mannitol	100
Dulcitol	0
Salicin	0
Adonitol	0
i-Inositol	0
D-Sorbitol	0
L-Arabinose	0
Raffinose	0
L-Rhamnose	0
Malonate	0
Mucate	0
Christensen's citrate	76 (24)
Sodium acetate	0
Sodium alginate nutrient	0
utilization	0
Lipase, corn oil	55 (21)
Maltose	100
D-Xylose	0
Trehalose	100
Cellobiose	0
Glycerol	70 (15)
α-Methyl-D-glucoside	0
Erythritol	0
Esculin	0
Mannose	100
Melibiose	0
Amygdalin	0
β-Galactosidase	100
DNase	0
Nitrate to nitrite	85 (9)
Oxidation-fermentation	100
Oxidase	0
Pectate	0
Cetrimide	0 (24)
Pigment	0
Organic acids	
Citrate	0 (100)
D-Tartrate	0

[a] Figures in parentheses indicate percentages of delayed reactions (3 days or more).
[b] Tiny bubble to 5%.

Table 18. Biochemical reactions of the Philomiragia bacterium.[a]

Test	Reaction
Gram stain	−
Acid fast stain	−
Catalase	+
Motility (37°C)	−
Motility (room temperature)	−
Gelatin liquefaction	+
H$_2$S	−
Indole	−
Nitrate reduction	−
Urease	−
D-Glucose (acid)	+
D-Glucose (gas)	−
Maltose	+
Sucrose	+
Lactose	−
L-Rhamnose	−
Raffinose	−
L-Arabinose	−
D-Xylose	−
Mannose	−
Galactose	V
D-Fructose	V
D-Sorbitol	−
i-Inositol	−
Salicin	−
Glycerol	−
Dulcitol	−
Dextrin	−
Inulin	−
Starch	−

[a] Data adapted from Jensen, Owen, and Jellison (1969). See Table 8 for definition of symbols.

and the Grimonts; Mollaret, Bercovier, and Alonso; and Leclerc, Gavini, and Izard.

Some of the newer groups are:

1. *Rahnella aquatilus* (Izard et al., 1979a), isolated from water and nonpolluted soil. It is phenotypically most similar to *E. cloacae*.
2. *Kluyvera* (Fanning et al., 1979), which contains two species. They are very closely related phenotypically, and biochemically intermediate between *Citrobacter* and *Enterobacter*. Most isolated are from humans, with sputum being the most frequent source of isolation in the United States.
3. Two species of lipase-positive, DNase-negative organisms isolated from humans that will be proposed as a new genus (J. J. Farmer III and P. A. D. Grimont, personal communication).
4. *Yersinia* biogroups X1 and X2 that are separate from all described *Yersinia* species (H. Bercovier, personal communication).
5. More than a dozen new clinical groups are currently being described at the CDC (J. J. Farmer III, personal communication). These are usually phenotypically most like *Citrobacter*, *Enterobacter*, and *Serratia*.

Literature Cited

Alonso, J. M., Begot, J., Bercovier, H., Mollaret, H. H. 1975. Sur un group de souches de *Yersinia enterocolitica* fermentant le rhamnose. Médicine et Maladies Infectieuses **5**:490–492.

Barbe, J. 1969. Organisation methodique de l'étude des caracters enzymatiques des bactéries de la tribu des *Klebsielleae:* Application à la classification. Doctor of Pharmacy Thesis. Faculté Mixte de Médecine et de Pharmacie de Marseille.

Barnes, D. M., Sorensen, D. K. 1975. Salmonellosis, pp. 554–564. In: Dunne, H. W., Leman, A. D. (eds.), Diseases of swine, 4th ed. Ames: Iowa University Press.

Bascomb, S., Lapage, S. P., Curtis, M. A., Willcox, W. R. 1971. Numerical classification of the tribe *Klebsielleae*. Journal of General Microbiology **66**:279–295.

Bercovier, H., Alonso, J. M., Bentaiba, Z. N., Brault, J., Mollaret, H. H. 1979. Contribution to the definition and the taxonomy of *Yersinia enterocolitica*. Contributions to Microbiology and Immunology **5**:12–22.

Bonner, T. I., Brenner, D. J., Neufeld, B. R., Britten, R. J. 1973. Reduction in the rate of DNA reassociation by sequence divergence. Journal of Molecular Biology **81**:123–135.

Bottone, E. J. 1977. *Yersinia enterocolitica:* A panoramic view of a charismatic microorganism. CRC Critical Reviews in Microbiology **6**:211–241.

Bottone, E. J., Chester, B., Malowany, M., Allerhand, J. 1974. Unusual *Yersinia enterocolitica* isolates not associated with mesentery lymphadenitis. Journal of Applied Microbiology **27**:858–861.

Breed, R. S., Murray, E. G. D., Smith, N. R. 1957. Bergey's manual of determinative bacteriology, 7th ed, pp. 332–393; 395–400. Baltimore: Williams & Wilkins.

Brenner, D. J. 1973. Deoxyribonucleic acid reassociation in the taxonomy of enteric bacteria. International Journal of Systematic Bacteriology **23**:298–307.

Brenner, D. J. 1974. DNA reassociation for the clinical differentiation of enteric bacteria. Public Health Laboratory **32**:118–130.

Brenner, D. J. 1977. Characterization and clinical identification of *Enterobacteriaceae* by DNA hybridization. Progress in Clinical Pathology **7**:71–117.

Brenner, D. J. 1979. Speciation in *Yersinia*. Contributions to Microbiology and Immunology **5**:33–43.

Brenner, D. J., Cowie, D. B. 1968. Thermal stability of *Escherichia coli–Salmonella typhimurium* deoxyribonucleic acid duplexes. Journal of Bacteriology **95**:2258–2262.

Brenner, D. J., Falkow, S. 1971. Molecular relationships among members of the *Enterobacteriaceae*. Advances in Genetics **16**:81–118.

Brenner, D. J., Fanning, G. R., Steigerwalt, A. G. 1974a. Polynucleotide sequence relatedness in *Edwardsiella tarda*. International Journal of Systematic Bacteriology **24**:186–190.

Brenner, D. J., Fanning, G. R., Steigerwalt, A. G. 1974b. Deoxyribonucleic acid relatedness among erwiniae and other *Enterobacteriaceae:* ·The gall, wilt, and dry-necrosis organisms (genus *Erwinia* Winslow et al., *sensu stricto*). International Journal of Systematic Bacteriology **24**:197–204.

Brenner, D. J., Fanning, G. R., Steigerwalt, A. G. 1977. Deoxyribonucleic acid relatedness among erwiniae and other enterobacteria. II. Corn stalk rot bacterium and *Pectobacterium chrysanthemi*. International Journal of Systematic Bacteriology **27**:211–221.

Brenner, D. J., Martin, M. A., Hoyer, B. H. 1967. Deoxyribonucleic acid homologies among some bacteria. Journal of Bacteriology **94**:486–487.

Brenner, D. J., Steigerwalt, A. G., Fanning, G. R. 1972. Differentiation of *Enterobacter aerogenes* from klebsiellae by deoxyribonucleic acid reassociation. International Journal of Systematic Bacteriology **22**:193–200.

Brenner, D. J., Fanning, G. R., Johnson K. E., Citarella, R. V., Falkow, S. 1969. Polynucleotide sequence relationships among members of *Enterobacteriaceae*. Journal of Bacteriology **98**:637–650.

Brenner, D. J., Fanning, G. R., Skerman, F. J., Falkow, S. 1972. Polynucleotide sequence divergence among strains of *Escherichia coli* and closely related organisms. Journal of Bacteriology **109**:953–965.

Brenner, D. J., Fanning, G. R., Miklos, G. V., Steigerwalt, A. G. 1973a. Polynucleotide sequence relatedness among *Shigella* species. International Journal of Systematic Bacteriology **23**:1–7.

Brenner, D. J., Steigerwalt, A. G., Miklos, G. V., Fanning, G. R. 1973b. Deoxyribonucleic acid relatedness among erwiniae and other *Enterobacteriaceae:* The soft-rot organisms (Genus *Pectobacterium* Waldee). International Journal of Systematic Bacteriology **23**:205–216.

Brenner, D. J., Steigerwalt, A. G., Falcao, D. P., Weaver, H. E., Fanning, G. R. 1976. Characterization of *Yersinia enterocolitica* and *Yersinia pseudotuberculosis* by deoxyribonucleic acid hybridization and by biochemical reactions. International Journal of Systematic Bacteriology **26**:184–190.

Brenner, D. J., Farmer, J. J., III, Hickman, F. W., Asbury, M. A., Steigerwalt, A. G. 1977. Taxonomic and nomenclatural changes in *Enterobacteriaceae*. Atlanta: Center for, Disease Control.

Brenner, D. J., Farmer, J. J., III, Fanning, G. R., Steigerwalt, A. G., Klykken, P., Wathen, H. G., Hickman, F. W., Ewing, W. H. 1978. Deoxyribonucleic acid relatedness of *Proteus* and *Providencia* species. International Journal of Systematic Bacteriology **28**:269–282.

Bruner, D. W., Gillespie, J. H. 1973. Hagan's infectious diseases of domestic animals, 6th ed. Ithaca: Cornell University Press.

Burgess, N. R. H., McDermott, S. N., Whiting, J. 1973. Anaerobic bacteria occurring in the hind-gut of the cockroach, *Blatta orientalis*. Journal of Hygiene **71**:1–7.

Center for Disease Control. 1976a. Shigella surveillance report number 38, annual summary. Atlanta: Center for Disease Control.

Center for Disease Control. 1976b. Foodborne and waterborne disease outbreaks annual summary 1974. Atlanta: Center for Disease Control.

Center for Disease Control. 1976c. Salmonella surveillance annual summary 1975. Atlanta: Center for Disease Control.

Center for Disease Control. 1977a. National nosocomial infections study report, Annual Summary 1974. Atlanta: Center for Disease Control.

Center for Disease Control. 1977b. Morbidity and Mortality Weekly Report **26**:53–54.

Clark, G. C., Kaufmann, A. F., Gangarosa, E. J., Thompson, M. A. 1973. Epidemiology of an international outbreak of *Salmonella agona*. Lancet **ii**:490–493.

Coetzee, J. N. 1972. Genetics of the *Proteus* group. Annual Review of Microbiology **26**:23–54.

Cosenza, B. J., Podgwaite, J. D. 1966. A new species of *Proteus* isolated from larvae of the gypsy moth *Porthetria dispar* (L.). Antonie van Leeuwenhoek Journal of Microbiology and Serology **32**:187–191.

Cowan, S. T., Orskov, F., Sakazaki, R., Sedlak, J., Le Minor, L., Rohde, R., Carpenter, K. P., Orskov, I., Lautrop, H., Mollaret, H. H., Thal, E., Lelliot, R. A. 1974. *Enterobacteriaceae*, pp. 290–340. In: Buchanan, R. E., Gibbons, N. E. (eds.), Bergey's manual of determinative bacteriology, 8th ed. Baltimore: Williams & Wilkins.

Crosa, J. H., Brenner, D. J., Ewing, W. H., Falkow, S. 1973. Molecular relationships among the *Salmonelleae*. Journal of Bacteriology **115**:307–315.

Crosa, J. H., Steigerwalt, A. G., Fanning, G. R., Brenner, D. J. 1974. Polynucleotide sequence divergence in the genus *Citrobacter*. Journal of General Microbiology **83**:271–282.

D'Empaire, M. 1969. Les facteurs de croissance des *Edwardsiella tarda*. Annales de l'Institut Pasteur **116**:63–68.

De, S. N., Bhattacharya, K., Sarkar, J. K. 1956. A study of the pathogenicity of strains of *Bacterium coli* from acute and chronic enteritis. Journal of Pathology and Bacteriology **61**:201–209.

Dean, A. G., Ching, Y., Williams, R. G., Harden, L. B. 1972. Test for *Escherichia coli* enterotoxin using infant mice: Application in a study of diarrhea in children in Honolulu. Journal of Infectious Diseases **125**:407–411.

Dye, D. 1968. A taxonomic study of the genus *Erwinia*. I. The "Amylovora" group. New Zealand Journal of Science **11**:590–607.

Dye, D. 1969a. A taxonomic study of the genus *Erwinia*. II. The "Carotovora" group. New Zealand Journal of Science **12**:81–97.

Dye, D. 1969b. A taxonomic study of the genus *Erwinia*. III. The "Herbicola" group. New Zealand Journal of Science **12**:223–236.

Dye, D. 1969c. A taxonomic study of the genus *Erwinia*. IV. "Atypical" erwiniae. New Zealand Journal of Science **12**:833–838.

Edwards, P. R., Ewing, W. H. 1972. Identification of *Enterobacteriaceae*, 3rd ed. Minneapolis, Minnesota: Burgess.

Ewing, W. H. 1969. Excerpts from: An evaluation of the *Salmonella* problem. Center for Disease Control Publication, Atlanta, GA.

Ewing, W. H., Davis, B. R. 1972. Biochemical characterization of *Citrobacter diversus* (Burkey) Werkman and Gillen and designation of the neotype strain. International Journal of Systematic Bacteriology **22**:12–18.

Ewing, W. H., Davis, B. R., Sikes, J. V. 1972. Biochemical characterization of *Providencia*. Public Health Laboratory **30**:25–38.

Ewing, W. H., Fife, M. A. 1972. *Enterobacter agglomerans* (Beijerinck) comb. nov. (the Herbicola-Lathyri bacteria). International Journal of Systematic Bacteriology **22**:4–11.

Ewing, W. H., Ross, A. J., Brenner, D. J., Fanning, G. R. 1978. *Yersinia ruckeri sp. n.*, the RM bacterium. International Journal of Systematic Bacteriology **28**:37–44.

Ewing, W. H., Tatum, H. W., Davis, B. R. 1957. The occurrence of *Escherichia coli* serotypes associated with diarrheal disease in the United States. Public Health Laboratory **15**:118–138.

Fanning, G. R., Farmer, J. J., III, Parker, J. N., Huntley-Carter, G. P., Brenner, D. J. 1979. *Kluyvera:* A new genus in Enterobacteriaceae. Abstracts of the Annual Meeting of the American Society for Microbiology **1979**:100.

Farmer, J. J., III, Davis, B. R., Cherry, W. B., Brenner, D. J., Dowell, V. R., Jr., Balows, A. 1977a. "Enteropathogenic serotypes" of *Escherichia coli* which are really not. Journal of Pediatrics **90**:1047–1049.

Farmer, J. J., III, Hickman, F. W., Brenner, D. J., Schreiber, M., Rickenbach, D. G. 1977b. Unusual *Enterobacteriaceae:* "*Proteus rettgeri*" that "change" into *Providencia stuartii*. Journal of Clinical Microbiology **6**:373–378.

Fredericksen, W. 1964. A study of some *Yersinia pseudotuberculosis*-like bacteria ("*Bacterium enterocoliticum*" and "*Pasteurella* X"). Scandinavian Congress of Pathology and Microbiology Proceedings **14**:103–104.

Ferragut, C., Leclerc, H. 1978. Characterization of motile and acetoin-negative *Klebsiella pneumoniae* strains by DNA:DNA hybridization. Antonie van Leeuwenhoek Journal of Microbiology and Serology **44**:407–424.

Fredericksen, W. 1970. *Citrobacter koseri* (n.sp.). A new species within the genus *Citrobacter*, with a comment on the taxonomic position of *Citrobacter intermedium* (Werkman and Gillen). Spisy Prirodovedecke Faculty University J. E. Purkyne, Brne **47**:89–94.

Gangarosa, E. J., Merson, M. H. 1977. An epidemiologic assessment of the relevance of the so-called enteropathogenic serogroups of *Escherichia coli* in diarrhea. New England Journal of Medicine **296**:1210–1213.

Gardner, J. M., Kado, C. I. 1972. Comparative base sequence homologies of the deoxyribonucleic acid of *Erwinia* species and other *Enterobacteriaceae*. International Journal of Systematic Bacteriology **22**:201–209.

Gavini, F., Ferragut, C., Leclerc, H. 1976. Étude taxonomique d'entérobactéries appartenant ou apparentées au genre *Enterobacter*. Annales de Microbiologie **127B**:317–335.

Gavini, F., Lefebvre, B., Leclerc, H. 1976. Positions taxonomiques d'entérobactéries H_2S^- par rapport au genre *Citrobacter*. Annales de Microbiologie **127A**:275–295.

Gavini, F., Ferragut, C., Izard, D., Trinel, P. A., Leclerc, H., Lefebvre, B., Mossel, D. A. A. 1979. *Serratia fonticola,* a new species from water. International Journal of Systematic Bacteriology **29**:92–101.

Gemski, P., Formal, S. B. 1975. Shigellosis: An invasive infection of the gastrointestinal tract. Microbiology [English translation of Mikrobiologiya] **44**:165–169.

Goto, M. 1976. *Erwinia mallotivora* sp. nov., the causal organism of bacterial leaf spot of *Mallotus japonicus* Muell. Arg. International Journal of Systematic Bacteriology **26**:467–473.

Graham, D. C. 1964. Taxonomy of the soft rot coliform bacteria. Annual Review of Phytopathology **2**:13–43.

Gross, R. J., Scotland, S. M., Rowe, B. 1976. Enterotoxin testing of *Escherichia coli* causing epidemic infantile enteritis in the U. K. Lancet **i**:629–631.

Grimont, P. A. D., Grimont, F., Dulong de Rosnay, H. L. C. 1977. Taxonomy of the genus *Serratia*. Journal of General Microbiology **98**:39–66.

Grimont, P. A. D., Grimont, F., Richard, C., Davis, B. R., Steigerwalt, A. G., Brenner, D. J. 1978. Deoxyribonucleic acid relatedness between *Serratia plymuthica* and other *Serratia* species with a description of *Serratia odorifera* spec. nov. (type strain: ICPB 3995). International Journal of Systematic Bacteriology **28**:453–463.

Gyles, C. L. 1974. Relationships among heat-labile enterotoxins of *Escherichia coli* and *Vibrio cholerae*. Journal of Infectious Diseases **129**:277–283.

Gyles, C., So, M., Falkow, S. 1974. The enterotoxin plasmids of *Escherichia coli*. Journal of Infectious Diseases **130**:40–49.

Hall, W. J. 1965. Fowl typhoid, pp. 329–358. In: Biester, H. E., Schwarte, L. H. (eds.), Diseases of poultry, 5th ed. Ames, Iowa: Iowa State University Press.

Holmgren, J., Soderland, O., Wadstrom, T. 1973. Crossreactivity between heat-labile enterotoxins of *Vibrio cholerae* and *Escherichia coli* in neutralization tests in rabbit ileum and skin. Acta Pathologica et Microbiologica Scandinavica **81**:757–762.

Hoshina, T. 1962. On a new bacterium, *Paracolobactrum anguillimortiferum* n. sp. Bulletin of the Japanese Society of Science and Fisheries **28**:162–164.

Izard, D., Gavini, F., Trinel, P. A., Leclerc, H. 1979a. *Rahnella aquatilis,* nouveau membre de la famille des *Enterobacteriaceae*. Annales de Microbiologie **130A**:163–167.

Izard, D., Gavini, F., Trinel, P. A., Leclerc, H. 1979b. Étude d'un groupe d'Enterobacteriaceae (group H_1) apparenté à l'espèce *Enterobacter cloacae*. Canadian Journal of Microbiology **25**:713–718.

Izard, D., Gavini, F., Trinel, P. A., Leclerc, H. 1981. Deoxyribonucleic acid relatedness between *Enterobacter cloacae* and *Enterobacter amnigena* sp. nov. International Journal of Systematic Bacteriology **31**:35–42.

Jain, K., Radsak, K., Mannheim, W. 1974. Differentiation of the Oxytocum group from *Klebsiella* by deoxyribonucleic acid-deoxyribonucleic acid hybridization. International Journal of Systematic Bacteriology **24**:402–407.

Jensen, R. 1974. Diseases of sheep. Philadelphia: Lea & Febiger.

Jensen, W. I., Owen, C. R., Jellison, W. J. 1969. *Yersinia philomiragia* sp. n., a new member of the *Pasteurella* group of bacteria, naturally pathogenic for the muskrat *(Ondatra zibethica)*. Journal of Bacteriology **100**:1237–1241.

Johnson, R., Colwell, R. R., Sakazaki, R., Tamura, K. 1975.

Numerical taxonomic study of the *Enterobacteriaceae*. International Journal of Systematic Bacteriology **25**: 12–37.

Kauffmann, F. 1954. *Enterobacteriaceae,* 2nd ed. Copenhagen: Munksgaard.

Kauffmann, F., DuPont, A. J. 1950. *Escherichia* strains from infantile epidemic gastroenteritis. Acta Pathologica et Microbiologica Scandinavica **27**:552–564.

Klipstein, F. A., Holdeman, L. V., Corcino, J. J., Moore, W. E. C. 1973. Enterotoxigenic intestinal bacteria in tropical sprue. Annals of Internal Medicine **79**:632–636.

Lautrop, H. 1956. Gelatin-liquefying *Klebsiella* strains (*Bacterium oxytocum* Flugge). Acta Pathologica et Microbiologica Scandinavica **39**:375–384.

Leclerc, H., Buttiaux, R. 1965. Les *Citrobacter*. Annales de l'Institut Pasteur **16**:67–74.

Leete, J. K. 1977. Characterization of *Enterobacter agglomerans* by DNA hybridization and biochemical reactions. D. P. H. Thesis, University of North Carolina.

Mair, N. S. 1968. Pseudotuberculosis in free-living wild animals. Symposia of the Zoological Society of London **24**: 107–117.

McCarthy, B. J., Bolton, E. T. 1963. An approach to the measurement of genetic relatedness among organisms. Proceedings of the National Academy of Sciences of the United States of America **50**:156–164.

Mollaret, H. H., Knapp, W. 1972. International Committee on Systematic Bacteriology Subcommittee on the Taxonomy of *Pasteurella, Yersinia,* and *Francisella*. International Journal of Systematic Bacteriology **22**:401.

Moore, R. L., Brubaker, R. R. 1975. Hybridization of deoxyribonucleotide sequences of *Yersinia enterocolitica* and other selected members of *Enterobacteriaceae*. International Journal of Systematic Bacteriology **25**:336–339.

Murata, N., Starr, M. P. 1974. Intrageneric clustering and divergence of *Erwinia* strains from plants and man in the light of deoxyribonucleic acid segmental homology. Canadian Journal of Microbiology **20**:1545–1565.

Niléhn, B. 1969. Studies on *Yersinia enterocolitica* with special reference to bacterial diagnosis and occurrence in human acute enteric disease. Acta Pathologica et Microbiologica Scandinavica, Suppl. **206**:1–48.

Ohara, S., Sato, T., Homma, M. 1974. Serological studies on *Francisella tularensis, Francisella novicida, Yersinia philomiragia,* and *Brucella abortus*. International Journal of Systematic Bacteriology **24**:191–196.

O'Brien, A. D., Thompson, M. R., Gemski, P., Doctor, B. P., Formal, S. B. 1977. Biological properties of *Shigella flexneri* 2A toxin and its serological relationship to *Shigella dysenteriae* 1 toxin. Infection and Immunity **15**:796–798.

Penner, J. L., Hennessy, J. N. 1977. Reassignment of the intermediate strains of *Proteus rettgeri* biovar 5 to *Providencia stuartii* on the basis of the somatic (O) antigens. International Journal of Systematic Bacteriology **27**:71–74.

Penner, J. L., Hinton, N. A., Hennessy, J. N. 1975. Biochemical differentiation of *Proteus rettgeri*. Journal of Clinical Microbiology **1**:136–142.

Penner, J. L., Hinton, N. A., Whiteley, G. R., Hennessy, J. N. 1976. Variation in urease activity of endemic hospital strains of *Proteus rettgeri* and *Providencia stuartii*. Journal of Infectious Diseases **134**:370–376.

Priest, F. G., Somerville, H. J., Cole, J. A., Hough, J. S. 1973. The taxonomic position of *Obesumbacterium proteus,* a common brewery contaminant. Journal of General Microbiology **75**:295–307.

Richard, C., Joly, B., Sirot, J., Stoleru, G. H., Popoff, M. 1976. Étude de souches de *Enterobacter* appartenant à un groupe particulier proche de *E. aerogenes*. Annales de Microbiologie **127A**:545–548.

Ritter, D. B., Gerloff, R. K. 1966. Deoxyribonucleic acid hybridization among some species of the genus *Pasteurella*. Journal of Bacteriology **92**:1838–1839.

Ross, A. J., Rucker, R. R., Ewing, W. H. 1966. Description of a bacterium associated with redmouth disease of rainbow trout (*Salmo gairdneri*). Canadian Journal of Microbiology **12**:763–770.

Rowe, B., Gross, R. J., Scotland, S. M. 1975. Serotyping of enteropathogenic *Escherichia coli*. Lancet **ii**:925.

Rowe, B., Scotland, S. M., Gross, R. J. 1977. Enterotoxingenic *Escherichia coli* causing infantile enteritis in Britain. Lancet **i**:90–91.

Sakazaki, R., Tamura, K. 1975. Priority of the specific epithet *anguillimortiferum* over the specific epithet *tarda* in the name of the organism presently known as *Edwardsiella tarda*. International Journal of Systematic Bacteriology **25**:219–220.

Sakazaki, R., Tamura, K., Saito, M. 1967. Enteropathogenic *Escherichia coli* associated with diarrhea in children and adults. Japanese Journal of Medical Science and Biology **20**:387–399.

Sarles, W. B., Frazier, W. C., Wilson, J. B., Knight, S. G. 1956. Microbiology, general and applied, 2nd ed. p. 1. New York, Harper & Brothers.

Schaad, N. W., Brenner, D. 1977. A bacterial wilt and root rot of sweet potato caused by *Erwinia chrysanthemi*. Phytopathology **67**:302–308.

Seidler, R. J., Knittel, M. D., Brown, C. 1975. Potential pathogens in the environment: Cultural reactions and nucleic acid studies on *Klebsiella pneumoniae* from clinical and environmental sources. Applied Microbiology **29**:819–825.

Shotts, E. B., Bullock, G. L. 1976. Rapid diagnostic approaches in the identification of gram-negative bacterial diseases of fish. Fish Pathology **10**:187–190.

Shotts, E. B., Snieszko, S. F. 1976. Selected bacterial fish diseases, pp. 143–151. In: Page, L. A. (ed.), Wildlife diseases. New York: Plenum.

Smith, H. W., Gyles, C. L. 1970. The relationship between two apparently different enterotoxins produced by enteropathogenic strains of *Escherichia coli* of porcine origin. Journal of Medical Microbiology **3**:387–401.

Smith, H. W., Lingood, M. A. 1972. Further observation on *Escherichia coli* enterotoxins with particular regard to those produced by atypical piglet strains and by calf and lamb strains. The transmissible nature of these enterotoxins and of a K antigen possessed by calf and lamb strains. Journal of Medical Microbiology **5**:243–250.

Smith, H. W., Sack, R. B. 1973. Immunologic cross-reactions of enterotoxins from *Escherichia coli* and *Vibrio cholerae*. Journal of Infectious Diseases **127**:164–170.

Starr, M. P., Chatterjee, A. K. 1972. The genus *Erwinia*: Enterobacteria pathogenic to plants and animals. Annual Review of Microbiology **26**:389–426.

Steigerwalt, A. G., Fanning, G. R., Fife-Asbury, M. A., Brenner, D. J. 1975. DNA relatedness among species of *Enterobacter* and *Serratia*. Canadian Journal of Microbiology **22**:121–137.

Stoleru, G. H., Le Minor, L., Lheritier, A. M. 1976. Polynucleotide sequence divergence among strains of *Salmonella* subgenus IV and closely related organisms. Annales de Microbiologie **127A**:477–486.

Taylor, J. 1968. *Salmonella* in wild animals. Symposia of the Zoological Society of London **24**:53–73.

Taylor, J., Powell, B. W., Wright, J. 1949. Infantile diarrhea and vomiting: A clinical and bacteriological investigation. British Medical Journal **ii**:117–125.

Ursing, J. 1974. Biochemical study of *Proteus inconstans (Providencia)*. Occurrence of urease positive strains. Acta Pathologica et Microbiologica Scandinavica, Sect. B. **82**:527–532.

Von Riesen, V. L. 1976. Pectinolytic, indole-positive strains of *Klebsiella pneumoniae*. International Journal of Systematic Bacteriology **26**:143–145.

Von Roekel, H. V. 1965. Pullorum disease, pp. 220–259. In: Biester, H. E., Schwarte, L. H. (eds.), Diseases of poultry, 5th ed. Ames, Iowa: Iowa State University Press.

Waldee, E. L. 1945. Comparative studies of some peritrichous phytopathogenic bacteria. Iowa State Journal of Science 19:435–484.

Wauters, G. 1973. Correlation between ecology, biochemical behaviour and antigenic properties of Yersinia enterocolitica. Contributions to Microbiology and Immunology 2:38–41.

Williams, J. E. 1965. Paratyphoid and Arizona infections, pp. 260–328. In: Biester, H. E., Schwarte, L. H. (eds.), Diseases of poultry, 5th ed. Ames, Iowa: Iowa State University Press.

Young, V. M., Kenton, D. M., Hobbs, B. J., Moody, M. R. 1971. Levinea, a new genus of the family Enterobacteriaceae. International Journal of Systematic Bacteriology 21:58–63.

Escherichia coli

FRITS ØRSKOV

The species *Escherichia coli,* the only member of the genus *Escherichia* (Buchanan and Gibbons, 1974), constitutes probably the most extensively examined group of living organisms. It is named after the German bacteriologist Escherich, who described it as a normal inhabitant of the human intestine. For many years it was looked upon as just that and as an opportunistic pathogen. However, more recent findings have disclosed that special strains should be reckoned as bacteria pathogenic for animals and man. Because it is so easy to handle in the laboratory and because of the wealth of information accumulated about *Escherichia coli,* it has always been a favorite organism for all types of microbiological studies.

Habitats

Escherichia coli in the Normal Intestine

Escherichia coli's natural habitat is in the lower part of the intestine of most warm-blooded animals. From a few hours after birth and throughout life, a succession of different *E. coli* strains will inhabit the most distal part of the ileum and the whole of the colon. However, they make up only a small fraction—about 1% or less—of the bacteria found in the colon and thus in feces.

E. *coli* is an opportunistic pathogen and, because of its continuous presence in the gut, it has ample opportunity to cause extraintestinal diseases of any type when local or general immunity has been reduced in favor of the bacteria. Most common in man are urinary tract infections, of which a majority are caused by *E. coli* (Grüneberg, Leigh, and Brumfitt, 1968). By serological means it is possible to divide *E. coli* strains into an increasing number of serogroups and serotypes defined by O, K, and H antigens (see Identification). However, some serological groups and types are much more commonly found than others (Table 1). There are many indications that support the view that the serogroups and serotypes frequently found in extraintestinal diseases are the same as those found to be prevalent in the feces (prevalence theory). Other findings, how-ever, seem to point to the possibility that at least some serogroups and serotypes have a special association with extraintestinal disease (special pathogenicity theory). A final decision as to which of the two possibilities is right cannot be made on the present data and very likely the answer will not be an "either-or". The finding that 85% of *E. coli* strains isolated from neonatal meningitis have capsular antigen K1 (Sarff et al., 1975) points to a special role in virulence of this antigen, and the special prevalence of some O groups such as O6 in urinary tract infections (Grüneberg, Leigh, and Brumfitt, 1968) also seems to stress the special pathogenicity of such strains. Table 1 lists *E. coli* O groups from different extraintestinal diseases and from normal feces. Much information concerning the distribution of O groups among *E. coli* in the normal intestine has accumulated since Kauffmann and Perch (1943) and Perch (1944) made the first prospective examinations of the fecal flora of two persons followed over a period of many months. Sears and his co-workers (Sears, Brownlee, and Uchiyama, 1950; Sears and Brownlee, 1952) concluded from their examinations that the *E. coli* flora consisted of "resident strains", most often one or two that persist over long periods of time (weeks or months), and "transient strains", which stay for a few days or weeks only. Hartley, Clements, and Linton (1977), who in recent years have contributed much to the elucidation of these problems, propose to abandon the terms resident and transient. They suggest calling any strain found among 10 randomly isolated colonies from a fecal culture a *majority* strain, whether or not it is the predominant O group. Any additional O group found by examination of a greater number of colonies or by special selection procedures designed to detect strains present in low numbers may be called a *minority* strain. A majority or a minority resident is a serotype that is repeatedly isolated from an individual over a defined period of time.

The extent of the influence on the host organism of the normal fluctuating *E. coli* flora is not known. Germ-free animals or *E. coli*-free animals thrive well without *E. coli,* but it is known that the acquisition of the *Escherichia* flora, normally occurring right after birth, does influence the host in several

Table 1. *Escherichia coli* O groups from extraintestinal infections in man.[a]

Urinary tract infections	Septicemia	Other infections	Neonatal meningitis[b]	Feces (healthy adults and children)
O1, O2, O4, O6, O7, O8, O9, O11, O22, O25, O62, O75	O1, O2, O4, O6, O7, O9, O11, O18, O22, O25, O75	O1, O2, O4, O6, O8, O9, O11, O21, O62	O1, O6, O7, O16, O18, O83	O1, O2, O4, O6, O7, O8, O18, O25, O45, O75, O81

[a] Since the prevalence rates have been compiled from many different investigations, the O groups are listed numerically.
[b] The prevalent O groups listed here are characterized by having the same K antigen, K1, when found in this disease.

ways. Shedlofsky and Freter (1974) suggested and also produced evidence for the hypothesis that two main mechanisms control the population size that is achievable for a given species in the gut, namely (i) the competitive antagonistic interaction with other bacterial species in the normal intestinal flora and (ii) local immune mechanisms. Local immunity is probably not strongly active against those autochthonous bacteria that constitute the predominant flora of the intestine. The main function of local immunity should therefore be to protect against the potential invasive bacteria, e.g., *E. coli,* which are normally only present in low numbers due to the antagonistic action of the normal, predominant flora.

The highly toxic lipopolysaccharide is absorbed from the intestine in small amounts, but is normally to a great extent trapped and detoxified in the liver before entering the blood stream. However, when this mechanism is impaired, the lipopolysaccharide may spill over and exert its multiple biological effects, e.g., induction of O antibody production (Bjørneboe, Prytz, and Ørskov, 1972). Capsular polysaccharides of normal intestinal *E. coli* may also induce antibody formation, and it has been recently shown (Schneerson et al., 1972) that such anticapsular antibodies to the *E. coli* antigen K100 most likely are responsible for the anticapsular antibodies found in adult human serum against the cross-reacting *Haemophilus influenzae* type b capsule.

It is generally accepted that *E. coli* organisms are not normal inhabitants of soil and water, and there-

fore the finding of *E. coli* in water is considered an indication of fecal contamination (see Isolation).

Escherichia coli in Diarrheal Diseases

For many years it has been accepted that *Escherichia coli* plays a role in diarrheal diseases in man and domestic animals. Based on epidemiological findings, some special serogroups—the enteropathogenic types (EPEC)—were incriminated as the cause of serious outbreaks of diarrhea in infants (Table 2). These special serotypes, like O111:H2 and O55:H6, were rare in adults and in healthy infants who had no contact with children's institutions. They could be found in infants without causing disease, and the reason for their association with disease has not yet been disclosed. Recent findings show that some such strains produce a special type of toxin similar in some respects to that found in *Shigella dysenteriae* type 1 (O'Brien et al., 1977).

Other special serotypes that harbor plasmids determining particular enterotoxins are frequently found from diarrheal diseases in young pigs and calves, and these serotypes occur all over the world. Until recently, it was believed that some diarrheal serotypes could be labeled as pig strains and other as calf strains, but this animal species specificity is apparently not as strict as hitherto believed (Moon et al., 1977). The toxin-determining plasmids are transmissible and two toxins can be described: one toxin (LT), which is a protein, is heat-labile and

Table 2. *Escherichia coli* O groups, O:H types and O:K:H types from intestinal infections in man.

Infantile diarrhea	Diarrhea in adults and children	
Sporadic cases and outbreaks, mostly in institutions	Enterotoxigenic strains, sporadic cases and outbreaks	Dysentery-like disease, sporadic cases and outbreaks
EPEC[a]	ETEC	EIEC
O20, O26, O44, O55, O86, O111, O114, O119, O125, O126, O127, O128, O142, O158	O6:K15:H16, O8:K40:H9, O8:K25:H9, O11:H27, O15:H11, O20:H⁻, O25:K7:H42, O25:(K98)[b]:H⁻, O27:H7, O78:H11, O78:H12, O128:H7, O148:H28, O149:H10, O159:H20	O28ac, O112, O124, O136, O143, O144, O152, O164

[a] EPEC, ETEC, and EIEC are special *E. coli* serotypes.
[b] Not K98 but related to K98.

immunogenic; the other toxin (ST) is heat-stable, of low molecular weight, and nonimmunogenic. Similar plasmid-determined toxins have been isolated from human *E. coli* strains from cases of diarrhea in warmer climates—traveler's diarrhea—and from cholera-like disease in areas where cholera is endemic. Some of these strains belong to special serotypes (ETEC) (Ørskov et al., 1976; Ørskov and Ørskov, 1977; Scotland, Gross, and Rowe, 1977) (Table 2) that are different from those found in animal diarrhea. A third group of special serotypes (EIEC) are found to be the cause of human dysenterylike disease. While EPEC and ETEC strains are noninvasive and play their role in the small intestine, where they are found in large numbers during the acute phase of the disease, the EIEC strains invade, like *Shigella,* the epithelial cells of the colon where they cause local inflammation.

E. coli and its role in diarrheal diseases are reviewed in several recent papers: Moon (1974), Sack (1975), Sakazaki et al. (1974), Sojka (1965), and Voino-Yasenetsky and Bakacs (1977).

Isolation

Escherichia coli bacteria are chemoorganotrophic and facultatively anaerobic, i.e., under aerobic conditions they can use organic compounds as energy source and under anaerobic conditions energy is obtained from fermentation of carbohydrates. Most strains will grow on minimal media containing only ammonium salts and a carbohydrate source. No particular media, therefore, exist for the general isolation of *E. coli.* Traditionally, different media have been used by different groups of investigators. Medical microbiologists often use agar media selective for Enterobacteriaceae, i.e., simple media that, by addition of different compounds like bile salts, preferentially allow the growth of Enterobacteriaceae and some other Gram-negative rods. Traditionally, an indicator and a specific sugar are added. The sugar most frequently used is lactose, because the pathogenic Enterobacteriaceae, *Salmonella* and *Shigella,* are usually lactose negative and can therefore be easily and routinely distinguished from other assumed nonpathogenic enterobacterial species like *E. coli.* One such very useful medium, developed in the Media Department, Statens Seruminstitut, is bromothymol blue (BTB) agar:

Bromothymol Blue Agar Selective for Enterobacteriaceae

Peptone (Orthana Ltd., Copenhagen)	10 g
NaCl	5 g
Oxoid yeast extract L21	5 g
Distilled water	1,000 ml

The pH is adjusted to 8, 11 g Japan agar powder is added, and the preparation is autoclaved at 120°C for 20 min.

To the medium are added:

5% Maranil solution (Paste A75 [dodecylbenzolsulfonate], Henkel, West Germany)	1 ml
50% Sodium thiosulfate	2 ml
1% Bromothymol blue (Riedel de Häen, West Germany)	10 ml
33% Lactose	27 ml
33% Glucose	1.2 ml

pH is adjusted to 7.7–7.8. In order to obtain the optimal medium, the amount of glucose has to be adjusted for every new batch of yeast extract, peptone, and agar.

In our experience, BTB agar in particular has been very useful in the search for different clones among lactose-positive colonies because of the easily detectable differences in shades of the yellow-orange color in colonies corresponding to the different clones (i.e., if two colonies from the same specimen look alike, they very often are identical, while two colonies that look different often represent two clones).

Depending on the scope of the investigation, many different additional media are employed. Veterinary microbiologists, examining for enterotoxigenic *E. coli* from piglet diarrhea, use blood agar because it is known that enterotoxigenic *E. coli* strains from this disease usually carry a plasmid coded for hemolysin production (Hly). In hygienic laboratories, where one is looking for fecal contamination of water, primary isolation is often carried out on EMB agar, a lactose-peptone agar containing eosin and methylene blue. Descriptions of media and reagents used for *E. coli* and other Enterobacteriaceae can be found in several common laboratory manuals, e.g., Edwards and Ewing (1972), Sedlak and Rische (1968), and Lennette, Spaulding, and Truant (1974).

E. coli is easily kept alive in several common preservation media such as meat extract agar or egg yolk medium (Kauffmann, 1966). In the dark at 18–22°C, such cultures will be kept alive for 10 years or more. Often it is important to restrict the number of mutations, and for this purpose freeze-drying or preservation of the cultures in broth medium with 10% glycerol at −80°C is preferable.

Bacteriological analysis of water is regulated according to special standards somewhat different in different countries:

The "coliform" bacteria detected are subdivided by a limited number of tests, which include the methyl red and Voges-Proskauer tests, the ability to grow with citrate as a carbon source, the indole reaction, the ability to grow in MacConkey medium at 44°C, and the ability to liquefy gelatin. The so-called type 1 *E. coli* corresponds to the typical *E. coli* pattern as defined in the eighth edition of *Bergey's Manual of Determinative Bacteriology*, while most of the other reaction patterns belong to other genera and species like *Klebsiella, Enterobacter,* and *Citrobacter.* More details and references can be found in Chapter 92 of the manual of Topley and Wilson (Wilson and Miles, 1975) and in a recent review paper by Bonde (1977).

Identification

The identification of *Escherichia coli* is well established through many years and seldom gives rise to problems. This genus, which, according to *Bergey's Manual,* is also the type genus of the Enterobacteriaceae, conforms to the definitions of the family Enterobacteriaceae and the tribe *Escherichieae.* The primary subdivision of the Enterobacteriaceae into five primary groups is tabulated in Table 3, and the biochemical reactions given by *E. coli* with the test most frequently used can be found in Table 4.

The genera *Escherichia* and *Shigella* are closely related and are probably only kept as two separate genera for historical reasons. Several strains that show reaction patterns placing them between *Escherichia* and *Shigella* have given rise to much controversy, e.g., certain anaerogenic, nonmotile strains that in serological and other respects have been labeled as *Alkalescens-Dispar* strains. They should be considered as "loss-mutants" of certain *E. coli* strains and called *Escherichia.*

In recent years, a number of serologically well-defined strains isolated from patients with a dysenterylike disease conform to the *E. coli* definition. These strains cross-react serologically with certain *Shigella* serotypes and are, like *Shigella,* invasive into epithelial cells of the guinea pig cornea. They naturally belong in the *Escherichia* group, and an earlier reason to keep them separate, pathogenicity, has less force today than ever, because it is widely accepted that some bacterial diarrheal diseases are caused by or closely associated with certain *E. coli* strains.

Most of the classical biochemical identification tests are chromosomally determined and are quite stable. However, in recent years *E. coli* strains have been found to carry plasmids that determine single (or occasionally more) characters not fitting the accepted *E. coli* definition. Thus, H_2S-positive *E. coli* have been found in several laboratories (Lautrop, Ørskov, and Gaarslev, 1971; Layne et al., 1971). It is not known which selective forces account for the simultaneous isolation of strains with plasmid-determined H_2S characters in different parts of the world.

Other plasmid-determined traits characteristic of strains found under certain ecological conditions are seen in *E. coli* strains isolated from neonatal diarrhea in pigs. These strains usually have plasmid-determined hemolytic ability and also may be urease producers—another trait that does not fit the typical *E. coli* description. However, it is not yet known whether this capacity is plasmid determined.

In addition, it should be mentioned that *E. coli* strains from calves are frequently inositol positive, a fermentative property not found positive in *E. coli* from other sources. It has not yet been established that this aberrant fermentative character is plasmid determined.

DNA base ratio varies between 50 and 51 mol%. DNA reassociation examination of several typical *E. coli* strains shows some heterogeneity (Brenner et al., 1972a).

The rod-shaped morphology of *E. coli* is determined by the peptidoglycan of the cell wall. In electron microscopy, the typical Gram-negative cell wall

Table 3. Distinguishing characteristics of the five primary groups (tribes) of the Enterobacteriaceae.[a]

	Tribe I *Escherichieae*	Tribe II *Klebsielleae*	Tribe III *Proteeae*	Tribe IV *Yersinieae*	Tribe V *Erwinieae*
Fermentation pattern	Mixed acid	2, 3-Butanediol		Mixed acid	Mixed acid & 2, 3-butanediol
M.R.	+	D[b]	+	+	
V.P.	−	D	D	−	D
Phenylalanine deamination	−	−	+	−	D
Nitrate reduction	+	+	+	+	D
Urease	−	D	D	D	−
KCN, growth in	D	+	+	−	D
Optimal temp for growth	37C	37C	37C	30–37C	27–30C
G+C, %	50–53	52–59	39–42	45–47	50–58

[a] From Buchanan and Gibbons, 1974, with permission.

[b] D, different reactions given by different species of a genus.

Table 4. Main biochemical characters of primary groups I to IV.[a]

	Group I					Group II				Group III. Proteus	Group IV. Yersinia
	Escherichia	*Edwardsiella*	*Citrobacter*	*Salmonella*	*Shigella*	*Klebsiella*	*Enterobacter*	*Hafnia*	*Serratia*		
Catalase	+	+	+	+	D[b]	+	+	+	+	+	+
Oxidase	−	−	−	−	−	−	−	−	−	−	−
β-Galactosidase	+	−	+	D	d	+	+	+	+	−	+
Gas from glucose at 37°C	+	+	+	+	−	d	+	+	d	D	−
KCN (growth on)	−	−	+	D	−	+	+	+	+	+	−
Mucate (acid)	+	−	+	D	−	d	d	−	−		
Nitrate reduced	+	+	+	+	+	+	+	+	+	+	+
G+C, moles %	50–51			50–53		52–56	52–59	52–57	53–59	39–42 (one species = 50)	45–47
Carbohydrates (acid from)											
Adonitol	−	−	−	−	−	d	+	−	d	D	D
Arabinose	+	−	+	+		+	+	+	−	D	D
Dulcitol	d	−	d	D	d	d	−	−	−	−	+
Esculin	d	−	d	−			D	−	−	d	−
Inositol	−	−	−	d	−	+	D	−	d	D	D
Lactose	+ or ×	−	+ or ×	D	D	D	+	−	−	−	−
Maltose	+	+	+	+		+	+	+	+	D	+
Mannitol	+	−	+	+	D	+	+	+	+	D	+
Salicin	d	−	d	−	−	+	+	−	+	d	D
Sorbitol	+	−	+	+		+	+	−		−	D
Sucrose	d	−	d	−	D	+	+	− or ×	+	D	D
Trehalose	+	−	+	+		+	+	+	+	d	+
Xylose	d	−	+	+	D	+	+	+	d	D	D
Related C sources											
Citrate	−	−	+	+	−	d	+	+	+		−
Gluconate		−		−	−	+	+	+	+		
Malonate	−	−	d	D	−	D	+	−			−
d-Tartrate	d	−	+	D		d	−	−			D
M.R.	+	+	+	+	+	D	−	−	−	+	+
V.P.	−	−	−	−	−	D	+	+	D	d	−
Protein reactions											
Arginine	d	−	d	+	−	−	D	−	−	−	−
Gelatin hydrolysis	−	−	−	D	−	(d)	(+)	−	+	D	−
H₂S from TSI	−	+	D	+	−	−	−	−	−	D	−
Indole	+	+	D	−	D	d	−	−	−	D	D
Lysine decarboxylated	+	+	−	+		d	D	+	+	d	−
Ornithine	d	+	d	+	d	−	+	+	+	D	D
Urea hydrolyzed	−	−	(+)	−	−	d	(d)	−	−	D	D
Glutamic acid	−	−	−	−	−	−		−	−	D	D
Phenylalanine	−	−	−	−	−	−	−	−	−	+	−

[a] From Buchanan and Gibbons, 1974, with permission.

[b] D, different reactions given by different species of a genus; d, different reactions given by different strains of a species of serotype; ×, late and irregularly positive (mutative).

Table 5. Characteristics of *Escherichia coli* antigens.

	O	K	H	Fimbrial
Chemistry	Lipopolysaccharide	Polysaccharide (acidic)	Protein	Protein
Heat stability (100°C, 1 h)	Stable	Stable (hapten)[a]	Labile	Labile
Serological determination	Agglutination	Gel precipitation (agglutination)	Agglutination	Agglutination (gel precipitation)
Genetic determinant	Chromosome	Chromosome	Chromosome	Plasmid and chromosome
Number (1978)	164	103[b]	59	?[c]

[a] K polysaccharide antigens are nonimmunogenic after heat treatment (haptens).

[b] Of the 103 K antigens some have been deleted, 2 are proteins (K88 and K99) and 76 are acidic polysaccharides.

[c] Two fimbrial antigens have been given K numbers: K88 and K99. For further details and for a discussion of former K antigen definitions, see Ørskov et al. (1977).

can be seen. In the outer membrane proteins, lipoproteins and the lipopolysaccharide are found. External to this, many strains carry a polysaccharide capsule that is chemically and genetically independent of the lipopolysaccharide. Motile organisms have protein flagella. Many or all *E. coli* may possess proteinaceous fimbriae or pili.

Subdivision of *E. coli* is best carried out by serological methods. The techniques used are very similar to those used for other Enterobacteriaceae and were first developed by Kauffmann (1966). Three surface structures constitute the foundation for the serological methods: the lipopolysaccharide of the cell wall (O antigen), the polysaccharide capsule (K antigen), and the proteinaceous flagella (H antigen) (Table 5). Presently, more than 160 O antigens, 80 K antigens, and 50 H antigens are recognized and new ones are being added constantly (Ørskov et al., 1977). These antigens, which are chromosomally determined, can be found in many combinations, but some serotypes are more frequent in nature than others. The simultaneous finding of characteristic fermentation patterns in such well-characterized O:K:H serotypes isolated in different geographical areas seems to indicate that some clones of *E. coli* are widely disseminated (Ørskov et al., 1976; Ørskov and Ørskov, 1977).

In addition to the antigens described, there are some plasmid-determined, proteinaceous, antigenic pili or fimbriae, e.g., K88 and K99, that possess adhesion factors and play a role in the diarrhea caused by enterotoxigenic *E. coli* strains. The technical application of serology and biochemistry to *E. coli* and the background of this application have been described extensively by Edwards and Ewing (1972). The reader is also referred to other in-depth discussions of the *Escherichia* group such as Sedlak and Rische (1968) and Lennette, Spaulding, and Truant (1974). Extensive reviews of *E. coli* O and K antigens have recently been published by Ørskov et al. (1977) and Ørskov and Ørskov (1978). Bacte-

riophage and bacteriocin schemes for primary typing of *E. coli* have never been in general use, though they can be employed in special situations such as following nosocomial infections (Milch et al., 1977). It is possible to subdivide the O:K:H serotypes into bacteriophage subtypes, but such systems have not been accepted for general use. The fermentation patterns of the single strains are highly stable. Therefore, when a subdivision of serotypes is required for epidemiological reasons, the simplest and often the most efficient way may be to use the differences in fermentation patterns if the number of substrates tested is large enough to provide the different patterns sought. Most of the commercially available biochemical test systems do not appear to be useful in biotype determination.

Resistance to antibiotics is a common feature in *E. coli* strains, especially in situations where the use of such drugs has exerted a selective pressure. Most often resistance is determined by plasmid-carried resistant (R) determinants, several of which can be found in the same strain and often in the same plasmid. Multiresistant antibiotic patterns are transferable by conjugation.

Literature Cited

Bjørneboe, M., Prytz, H., Ørskov, F. 1972. Antibodies to intestinal microbes in serum of patients with cirrhosis of the liver. Lancet i: 58–60.

Bonde, G. J. 1977. Bacterial indication of water pollution, pp. 273–334. In: Droop, Jannash (eds.), Advances in aquatic microbiology, vol. 1. London, New York, San Francisco: Academic Press.

Brenner, D. J., Fanning, G. R., Skerman, F. J., Falkow, S. 1972a. Polynucleotide sequence divergenic among strains of *Escherichia coli* and closely related organisms. Journal of Bacteriology **109:**953–965.

Brenner, D. J., Fanning, G. R., Steigerwalt, A. G., Ørskov, I., Ørskov, F. 1972b. Polynucleotide sequence relatedness among three groups of pathogenic *Escherichia coli* strains. Infection and Immunity **6:**308–315.

Buchanan, R. E., Gibbons, N. E. (eds.). 1974. Bergey's manual of determinative bacteriology, 8th ed. Baltimore: Williams & Wilkins.

Edwards, P. R., Ewing, W. H. 1972. Identification of Enterobacteriaceae, 3rd ed. Minneapolis: Burgess Publishing Co.

Ewing, W. H., Martin, W. J. 1974. *Enterobacteriaceae*, pp. 189–221. In: Lennette, E. H., Spaulding, E. H., Truant, J. P. (eds.), Manual of clinical microbiology, 2nd ed. Washington, D. C.: American Society for Microbiology.

Grüneberg, R. N., Leigh, D. A., Brumfitt, W. 1968. *Escherichia coli* serotypes in urinary tract infection: Studies in domiciliary, antenatal and hospital practice, pp. 68–79. In: O'Grady, F., Brumfitt, W. (eds.), Urinary tract infection. London: Oxford University Press.

Hartley, C. L., Clements, H. M., Linton, K. B. 1977. *Escherichia coli* in the faecal flora of man. Journal of Applied Bacteriology **43**:261–269.

Kauffmann, F. 1966. The bacteriology of Enterobacteriaceae. Copenhagen: E. Munksgaard.

Kauffmann, F., Perch, B. 1943. Über die Koliflora des gesunden Menschen. Acta Pathologica et Microbiologica Scandinavica **20**:201–220.

Lautrop, H., Ørskov, I., Gaarslev, K. 1971. Hydrogen sulphide producing variants of *Escherichia coli*. Acta Pathologica et Microbiologica Scandinavica **79**:641–650.

Layne, P., Hu, A. S. L., Balows, A., Davis, B. R. 1970. Extra chromosomal nature of hydrogen sulfide production in *Escherichia coli*. Journal of Bacteriology **106**:1029–1030.

Lennette, E. H., Spaulding, E. H., Truant, J. P. 1974. Manual of clinical microbiology, 2nd ed., pp. 189–221. Washington, D.C.: American Society of Microbiology.

Milch, H., Czirok, E., Madar, J., Semjén, G. 1977. Characterization of *Escherichia coli* serogroups causing meningitis, sepsis and enteritis. II. Classification of *Escherichia coli* O78 strains by phage sensitivity, colicin type and antibiotic resistance. Acta Microbiologica Academiae Scientiarum Hungaricae **24**:127–137.

Moon, H. W. 1974. Pathogenesis of enteric diseases caused by *Escherichia coli*. Advances in Veterinary Science and Comparative Medicine **18**:179–212.

Moon, H. W., Nagy, B., Isaacson, R. E., Ørskov, I. 1977. Occurrence of K99 antigen on *Escherichia coli* isolated from pigs and colonization of pig ileum by K99$^+$ enterotoxigenic *E. coli* from calves and pigs. Infection and Immunity **15**:614–620.

O'Brien, A. D., Thompson, M. R., Cantley, J. R., Formal, S. B. 1977. Production of a *Shigella dysenteriae*-like toxin by pathogenic *Escherichia coli*. Abstracts of the Annual Meeting of American Society for Microbiology 1977:32.

Ørskov, F., Ørskov, I. 1978. Serotyping of Enterobacteriaceae with special emphasis on K antigen determination, pp. 1–77. In: Bergan, T., Norris, J. R. (eds.), Methods in microbiology, vol. 11. London: Academic Press.

Ørskov, F., Ørskov, I., Evans, D. J. Jr., Sack, R. B., Sack, D. A., Wadström, T. 1976. Special *Escherichia coli* serotypes among enterotoxigenic strains from diarrhea in adults and children. Medical Microbiology and Immunology **162**:73–80.

Ørskov I., Ørskov, F. 1977. Special O:K:H serotypes among enterotoxigenic *E. coli* strains from diarrhea in adults and children. Occurrence of the CF (colonization factor) antigen and of hemagglutinating abilities. Medical Microbiology and Immunology **163**:99–110.

Ørskov, I., Ørskov, F., Jann, B., Jann, K. 1977. Serology, chemistry and genetics of O and K antigens of *Escherichia coli*. Bacteriological Reviews **41**:667–710.

Perch, B. 1944. Weitere Untersuchungen über die Koliflora des gesunden Menschen. Acta Pathologica et Microbiologica Scandinavica **21**:239–247.

Sack, R. B. 1975. Human diarrheal disease caused by enterotoxigenic *Escherichia coli*. Annual Review of Microbiology **29**:333–353.

Sakazaki, R., Takamura, K., Nakamura, A., Kurata, T., Gohda, A., Takeuchi, S. 1974. Enteropathogenicity and enterotoxigenicity of human enteropathogenic *Escherichia coli*. Japanese Journal of Medical Science and Biology **27**:19–33.

Sarff, L., McCracken, G. H., Jr., Schiffer, M. S., Glode, M. P., Robbins, J. B., Ørskov, I., Ørskov, F. 1975. Epidemiology of *Escherichia coli* K1 in healthy and diseased newborns. Lancet **i**:1099–1104.

Schneerson, R., Bradshaw, M., Whisnant, J. K., Myerowitz, R. L., Parke, J. C., Robbins, J. B. 1972. An *Escherichia coli* antigen cross-reactive with the capsular polysaccharide of *Haemophilus influenzae* type b. Occurrence among known serotypes, and immunochemical and biologic properties of *E. coli* antisera toward *H. influenzae* type b. Journal of Immunology **108**:1551–1562.

Scotland, S. M., Gross, R. J., Rowe, B. 1977. Serotype-related enterotoxigenicity in *Escherichia coli* O6:H16 and O148:H28. Journal of Hygiene **79**:395–403.

Sears, H. J., Brownlee, I. 1952. Further observations on the persistence of individual strains of *Escherichia coli* in the intestinal tract of man. Journal of Bacteriology **63**:47–57.

Sears, H. J., Brownlee, I., Uchiyama, J. K. 1950. Persistence of individual strains of *Escherichia coli* in the intestinal tract of man. Journal of Bacteriology **59**:293–301.

Sedlak, J., Rische, H. (eds.). 1968. Enterobacteriaceae-Infektionen. Leipzig: Georg Thieme.

Shedlofsky, S., Freter, R. 1974. Synergism between ecological and immunologic control mechanisms of intestinal flora. Journal of Infectious Diseases **129**:296–303.

Sojka, W. J. 1965. *Escherichia coli* in domestic animals and poultry. Commonwealth Agricultural Bureaux, Franham Royal, Bucks, England.

Voino-Yasenetsky, M. V., Bakacs, T. (eds.). 1977. Pathogenesis of intestinal infections. Microbiological and pathological principles. Budapest: Akademiai Kiado.

Wilson, G. S., Miles, A. A. (eds.). 1975. Topley and Wilson: Principles of bacteriology, virology and immunity, 6th ed., pp. 2648–2660. London: Arnold & Co.

CHAPTER 90

The Genus *Edwardsiella*[1]

JOHN J. FARMER III

The name *Edwardsiella* was first introduced by Ewing et al. in 1965 to describe a new group of enteric bacteria that had previously been described under such vernacular names as "paracolon", Asakusa group (Sakazaki and Murata, 1962), or Bartholomew group (King and Adler, 1964). Undoubtedly, strains of *Edwardsiella* had been isolated long before this time, but published descriptions are so vague or incomplete that a retrospective identification cannot be made with certainty. Since 1965 several hundred reports in the literature have described various aspects of the genus. *Edwardsiella* causes opportunistic infections in humans and may also cause, or have a role in, diarrhea. They are found in animals, where they occasionally cause disease, and are frequently isolated from environmental waters.

Edwardsiella was coined to honor the American bacteriologist P. R. Edwards for his many contributions to enteric bacteriology (Edwards and Ewing, 1972). The best described species is *Edwardsiella tarda*, the name used in most reports in the literature. However, in 1975 Sakazaki and Tamura rediscovered a 1962 paper in which Hoshina described a new organism, *Paracolobactrum anguillimortiferum*, which is almost certainly an edwardsiella. They pointed out the similarities between Hoshina's organism (no type strain was designated by Hoshina, and apparently none of Hoshina's strains have survived) and *E. tarda*. However, since Table 1 shows that there are significant phenotypic differences between the two organisms, I will recognize two species of *Edwardsiella*, *E. tarda* and *E. anguillimortifera* (*anguillimortiferum* becomes *anguillimortifera* when it is transferred from *Paracolobactrum* to *Edwardsiella*). These names may represent the same organisms, but strains that fit Hoshina's original description must be isolated and compared with *E. tarda* before the issue can be resolved (Farmer, Brenner, and Clark, 1976; Sakazaki and Tamura, 1975). (The author would greatly appreciate receiving cultures that fit

Hoshina's description as given in Table 1, so that this comparison can be made.)

Edwardsiella tarda is a well-characterized species in the family Enterobacteriaceae. Brenner, Fanning, and Steigerwalt (1974) found that 20 *E. tarda* strains from five countries and a wide variety of human and animal sources were highly related by DNA-DNA hybridization. The mean relatedness (60°C) was 88.2% (standard deviation $=3.9$), with a range of 82–95%. The relatedness to other groups of Enterobacteriaceae was less than 30%. These data indicate that *Edwardsiella* deserves generic status in a phylogenetic classification of Enterobacteriaceae, which disagrees with the opinion of the late S. T. Cowan, who regarded *E. tarda* as "a biotype of *Escherichia coli*" or "less satisfactorily as a species, *Escherichia tarda*" (Cowan, 1975). It will be interesting to compare the DNA relatedness of *E. tarda* with that of *E. anguillimortifera* if authentic strains of the latter can be isolated.

Habitats

As a Human Pathogen—Cause of Diarrhea

Since *Edwardsiella tarda* was first described, investigators have speculated that it can cause diarrhea (Bhat, Myers, and Carpenter, 1967; Ewing et al., 1965). Some intriguing evidence comes from a study of the Orang Asli, a group of jungle-dwelling natives of West Malaysia (Gilman et al., 1971). There were 29 isolates of *E. tarda* among 208 patients hospitalized with bloody diarrhea, but only one isolate among 120 control individuals (hospital patients without diarrhea). An interesting relationship between *E. tarda* and the protozoan *Entamoeba histolytica* was also shown. Twenty-five of the patients with bloody diarrhea had both organisms, but four had *E. tarda* only. Twenty-four of the 25 patients with both organisms had significant antibody titers to a whole-cell antigen of *E. tarda* prepared by the method of Sakazaki (1967). A control group of 15 patients had no antibodies. All of the patients who were culture negative for *E. tarda* but positive for *Entamoeba histolytica* also had antibodies to

[1] This material was prepared in my official capacity as an employee of the United States Government. It is part of the public domain and may not be copyrighted.

Table 1. Differences in the phenotypic properties of *Edwardsiella tarda* and *Edwardsiella anguillimortifera*.

Property	E. tarda[a]	E. anguillimortifera[b]
Methyl red test	+	−
Phenylalanine required	−	+
Threonine required	−	+
Valine required	−	+
Cysteine required	+	−
Methionine required	+	−
Pathogenic for trout[c]	−	+

[a] Properties of *E. tarda* are based on data obtained in the Enteric Section, Center for Disease Control, and on numerous descriptions in the literature. The nutritional requirements are taken from the study by D'Empaire (1969).
[b] Properties of *E. anguillimortifera* are based on the original description of *Paracolobactrum anguillimortiferum* by Hoshina (1962).
[c] Meyer and Bullock (1973) found no pathogenicity for "fingerling brown trout" (*Salmo trutta*); Hoshina (1962) found pathogenicity for rainbow trout.

Edwardsiella. These data indicate that *E. tarda* may be involved in the pathogenesis of amoebic dysentery. Another explanation is that the presence of *E. tarda* is due to a change in the gut microenvironment and that the organism plays no role in diarrhea. Makulu, Gatti, and Vandepitte (1973) also found an association between *E. tarda* and *Entamoeba histolytica* in patients from Zaire with bloody diarrhea, but it was not as strong as in the previous study.

Edwardsiella tarda is rarely present in the feces of healthy people. Onogawa et al. (1976) in Japan found only one carrier among 97,704 food handlers and 25 carriers among 255,896 school children. Several studies indicate that the number of positive specimens detected depends upon the methods used in processing stool cultures, the geographic area of the study, and the season in which the survey is done (Iveson, 1973). These variables have not always been considered in studies on the relative incidence of *E. tarda* in patients with diarrhea and in controls. A higher isolation rate has invariably been found among the diarrhea patients (Bhat, Myers, and Carpenter, 1967; Ewing et al., 1965; Gilman et al., 1971; Makulu, Gatti, and Vandepitte, 1973; Nguyen-Van-Ai et al., 1975). It would appear that *E. tarda* can cause diarrhea, particularly in underdeveloped countries, but it appears not to be an "inherent pathogen", a status given to *Salmonella* and *Shigella*. The role of *Edwardsiella* in diarrhea needs further study. One promising technique is to test a patient's acute-phase and convalescent-phase sera against the strain of *E. tarda* isolated from the stool. Chatty and Gavan (1968) reported an instance in which *E. tarda* was isolated from a patient with diarrhea who had lived in Central and South America. A convalescent-phase serum from the patient had an antibody titer of 1:160 to both somatic and flagellar antigens, so *E. tarda* was incriminated as a

possible cause of the diarrhea. Similar studies are needed for all isolates of *E. tarda* from stools of people with diarrhea and from healthy controls.

As a Human Pathogen—Opportunistic Infections

Although *Edwardsiella tarda* has been well documented as an opportunistic pathogen, it is rarely found in most industrialized countries. It seldom causes meningitis, endocarditis, bacteremia, or urinary tract infections but is often isolated from wounds (Jordan and Hadley, 1969). A typical example was the report of Chatty and Gavan (1968) of a boy who struck a submerged log while swimming in a lake. A splinter entered his right thigh and eventually led to gas gangrene confirmed by the isolation of *Clostridium perfringens*. *E. tarda* was also isolated but probably only colonized the wound. Wound cultures often yield other bacteria in addition to *E. tarda*, so its role in these infections is difficult to define.

In Animals and the Environment

Edwardsiella tarda has been isolated from many animals including pets (Nguyen-Van-Ai et al., 1975), domestic animals (Owens, Nelson, and Addison, 1974), animals in zoos (Otis and Behler, 1973), rats (Nguyen-Van-Ai et al., 1975), aquatic animals and birds (White, Simpson, and Williams, 1973), fish (Nguyen-Van-Ai et al., 1975), frogs (Bartlett, Trast, and Lior, 1977), turtles (Otis and Behler, 1973), and marine animals (Nguyen-Van-Ai et al., 1975). It is also frequently found in the environment, particularly where these animals live (White, Simpson, and Williams, 1973). Most *E. tarda* isolates have come from stools or other specimens from healthy animals, but *E. tarda* can cause outbreaks of "red disease" in pond-cultured eels (Wakabayashi and Egusa, 1973) or of "emphysematous putrefactive disease" (gas-filled lesions in the muscles) of channel catfish (Meyer and Bullock, 1973). Isolated cases of septicemia have been reported in other animals (Chamoiseau, 1967).

Ecology and Epidemiology

The natural reservoir of *Edwardsiella tarda* appears to be the intestine of animals, whose feces disseminate the organism into the environment. Most human infections probably result from contact with *E. tarda* in the environment. Endogenous human infections, although probably rare, may occur if gut carriage has been established. Recently it has been found that much of the diarrhea in developing countries is caused by strains of *Escherichia coli* that have acquired plasmids which code for entero-toxin production and probably also for gut coloniza-tion. It is tempting to postulate a similar model to explain the association of *E. tarda* with diarrhea. In this model, most strains would not be able to cause diarrhea, but an occasional strain would acquire a plasmid that codes for enterotoxin production and gut colonization, thus explaining the higher inci-dence of *E. tarda* in patients with diarrhea than in controls. Strains of *E. tarda* from people without diarrhea would presumably lack one of the virulence factors. A second possible model explaining the role of *E. tarda* in dysentery also has an analogy in the species *Escherichia coli*. Only a few strains of the latter species are able to invade mucosal cells in the colon and cause the bloody diarrhea typical of dys-entery. The dysentery model for *E. tarda* postulates that only the strains that are endemic or occasionally epidemic in countries where sanitation is poor can cause dysentery. However, the association of *E. tarda* with diarrhea does not necessarily indicate a causal relationship, as Gilman et al. (1971) empha-sized.

Isolation

Most data on *Edwardsiella* isolation have come from culture surveys to detect *Salmonella* and *Shigella* rather than in efforts to isolate *E. tarda*. Similarly there has not been a systematic study to evaluate growth and survival of *E. tarda* in enrichments and on plating media commonly used in enteric bacteri-ology. *E. tarda* strains usually grow on these com-mon plating media (agars): blood (5% sheep), choc-olate, MacConkey, SS (*Salmonella-Shigella*), and desoxycholate citrate. However, they grow more slowly than other species of Enterobacteriaceae. Pure cultures grew on brilliant green and bismuth sulfite agar (Sakazaki, 1967), but Iveson (1973) found these two media useless in isolating *E. tarda* from feces.

Strains of *E. tarda* are often isolated from liquid enrichments such as tetrathionate and selenite F used to isolate *Salmonella,* and occasionally these enrich-ments result in a higher yield than does direct plating (Makulu, Gratti, and Vandepitte, 1973). Iveson (1973) described an efficient method for isolating *E. tarda* from stool cultures.

Isolation of *Edwardsiella tarda* (Iveson, 1973)

Specimens are first enriched (either at 37°C or 43°C) with strontium chloride B medium (0.5 g tryptone (Difco), 0.8 g NaCl, 0.1 g KH$_2$PO$_4$, 3.4 g strontium chloride, and 100 ml distilled water; sterilized by heating at 100°C for 30 min). After 24 h of incubation, plates of desoxycholate citrate agar are streaked.

This method was excellent for isolating *Salmo-nella, Arizona, Shigella,* and *Edwardsiella* from stool cultures, and could presumably be adapted to all types of specimens, including those from the environment.

Edwardsiella tarda has intrinsic resistance to the polypeptide antibiotic colistin, which can be used in isolation procedures. Muyembe, Vandepitte, and Desmyter (1973) showed that all *E. tarda* strains grew in the presence of 10 μg/ml colistin and that more than 80% grew in 100 μg/ml. Other metabolic properties of *E. tarda* could be considered in design-ing a differential and selective medium. Colistin could be added to peptone iron agar (Edwards and Ewing, 1972) so that organisms that grow and pro-duce black colonies (*E. tarda* is H$_2$S$^+$) would proba-bly be *E. tarda* or an H$_2$S$^+$ *Proteus* strain. A differ-ent approach would be to incorporate several carbohydrates (or related compounds) not fermented by *E. tarda* into a fermentation base such as MacConkey agar base without lactose (Difco). For example, the following could be added to the base: 10 μg/ml colistin, 1% D-mannitol, 1% D-xylose, 1% trehalose, and 1% adonitol. Most Enterobacteri-aceae would either be inhibited (only *Serratia, Pro-teus, Providencia,* and *Morganella* are colistin-resistant) or form red colonies because they ferment one or more of the four compounds. Colonies of *E. tarda* should be colorless. Many other approaches combining colistin with a differential biochemical reaction are feasible.

Identification

Biochemical Tests

Edwardsiella tarda is very easy to identify, in con-trast to many other species of Enterobacteriaceae. The biochemical reactions are very uniform (Table 2) and are similar to those of *E. anguillimortifera*, except for the methyl red test and nutritional re-quirements (Table 1) (D'Empaire, 1969). The bio-chemical reactions of edwardsiellae differ from those of any other species (Edwards and Ewing, 1972). Although strains of *E. tarda* are occasionally slightly atypical, they can be readily identified. *E. tarda* is superficially related (biochemically) to *Escherichia coli* and to *Salmonella-Arizona*, but is easily differentiated on the basis of a complete set of biochemical test results or on the basis of antibiotic susceptibility patterns.

Below the Species Level

Two independent serotyping schemes have been described for *Edwardsiella tarda*. Sakazaki (1967) recognized 17 O-antigen groups, 11 H-antigen groups, and 18 O–H combinations. Edwards and Ewing (1972) described a scheme (technical details

Table 2. Biochemical reactions of *Edwardsiella tarda.*[a]

Positive tests	
Indole (99.0)[b]	Gas production from D-glucose (99.2)
Methyl red	Fermentation of:
H_2S production (99.7)	D-glucose
Lysine decarboxylase	Maltose (99.1)
Ornithine decarboxylase (99.7)	$NO_3^- \rightarrow NO_2^-$
Motility (98.0)	Citrate (Christensen's) (99.0)

Negative tests		
Voges-Proskauer	Fermentation of:	
Citrate (Simmons')	Lactose	Raffinose
Urea	Sucrose (0.3)	L-rhamnose
Phenylalanine	D-mannitol	D-xylose
Arginine dihydrolase	Dulcitol	Trehalose (0.3)
Gelatin	Salicin	Cellobiose
KCN, growth in	Adonitol	α-CH$_3$-D-glucoside
Malonate	*i*-inositol	Erythritol
Tartrate (Jordan's)	D-sorbitol (0.3)	Esculin
Acetate	L-arabinose (9.4)	Mucate
Lipase (corn oil)		
DNase		
Oxidase		
Pigment production		
ONPG		

[a] All reactions are at 36°C except gelatin (22°C) and DNase (25°C) and occurred within 24 or 48 h (most occurred within 24 h). The data are taken from Edwards and Ewing (1972), whose methods are well documented. Data are based on 394 cultures.

[b] Numbers in parentheses indicate the percentage positive for a variable test if results are other than 100% or 0%.

have not been published) with 49 O antigens, 37 H antigens, and 148 O–H combinations among 394 cultures studied. Techniques such as bacteriocin production or susceptibility, bacteriophage typing, and biotyping have seldom been used for *Edwardsiella,* although Hamon et al. (1969) did demonstrate bacteriocin production and sensitivity.

Antibiotic Susceptibility

Strains of *Edwardsiella tarda* are intrinsically resistant to colistin. However, they are usually susceptible to other antimicrobials (Muyembe, Vandepitte, and Desmyter, 1973). Strains resistant to sulfonamides or other agents undoubtedly result from exposure to these or similar drugs. I could find no reports of R plasmid (R factor)–mediated antibiotic resistance among *E. tarda* strains, a fact that further suggests that man's contact with this organism is only occasional and that *E. tarda* infection might best be considered a zoonosis (a disease transmitted from animals to man).

Addendum

Recently, J. P. Hawke (*Journal of the Fisheries Research Board of Canada* **36:**1508–1512, 1979) de-scribed an "*Edwardsiella*-like" organism associated with generalized sepsis and death in pond-raised catfish. The organism resembled *Edwardsiella tarda* but was more active biochemically at 25°C than at 37°C, and was (at 37°C) indole negative, H_2S(TSI) negative, and nonmotile and did not produce gas during D-glucose fermentation. The organism was isolated in the United States from diseased catfish on farms in Alabama, Georgia, and Mississippi, and from a fish kill in a Maryland river. The new organism does not have a scientific name, and at present the vernacular names "bacterium 7752" or "*Edwardsiella* GA 7752" are being used.

The controversy on the names *Edwardsiella tarda* vs. *Edwardsiella anguillimortifera* continues. An update can be found in: Sakazaki, R., Tamura, K. 1978. *International Journal of Systematic Bacteriology* **28:**130–131.

Literature Cited

Bartlett, K. H., Trust, T. J., Lior, H. 1977. Small pet aquarium frogs as a source of *Salmonella*. Applied and Environmental Microbiology **33:**1026–1029.

Bhat, P., Myers, R. M., Carpenter, K. P. 1967. *Edwardsiella tarda* in a study of juvenile diarrhea. Journal of Hygiene **65:**293–298.

Brenner, D. J., Fanning, G. R., Steigerwalt, A. G. 1974. Poly-nucleotide sequence relationships in *Edwardsiella tarda*. International Journal of Systematic Bacteriology **24:**186–190.

Chamoiseau, G. 1967. Note sur le pouvoir pathogène d'*Edward-siella tarda*. Un cas de septicémie mortelle du pigeon. Revue d'Élevage et de Médecine Vétérinaire des Pays Tropicaux (Paris) **20:**493–495.

Chatty, H. B., Gavan, T. L. 1968. *Edwardsiella tarda*—identification and clinical significance. Report of two cases. Cleveland Clinic Quarterly **35:**223–228.

Cowan, S. T. 1974. Cowan and Steel's manual for the identifica-tion of medical bacteria, 2nd ed. London New York: Cam-bridge University Press.

D'Empaire, M. 1969. Les facteurs de croissance des *Edwardsiella tarda*. Annales de l'Institut Pasteur **116:**63–68.

Edwards, P. R., Ewing, W. H. 1972. Identification of Enterobac-teriaceae, 3rd ed. Minneapolis: Burgess Publishing.

Ewing, W. H., McWhorter, A. C., Escobar, M. R., Lubin, A. H. 1965. *Edwardsiella*, a new genus of Enterobacteriaceae based on a new species, *E. tarda*. International Bulletin of Bacteri-ological Nomenclature and Taxonomy **15:**33–38.

Farmer, J. J., III, Brenner, D. J., Clark, W. A. 1976. Proposal to conserve the specific epithet *tarda* over the specific epithet *anguillimortiferum* in the name of the organism known as *Edwardsiella tarda*. International Journal of Systematic Bac-teriology **26:**293–294.

Gilman, R. H., Madasamy, M., Gan, E., Mariappan, M., Davis, C. E., Kyser, K. A. 1971. *Edwardsiella tarda* in jungle diarrhoea and a possible association with *Entamoeba histol-ytica*. Southeast Asian Journal of Tropical Medicine and Pub-lic Health **2:**186–189.

Hamon, Y., Kayser, A., Le Minor, L., Maresz, J. 1969. Les bactériocines d'*Edwardsiella tarda*. Intérêt taxonomique de l'étude de ces antibiotiques. Comptes Rendus de l'Académie des Sciences, Series D **268:**2517–2520.

Hoshina, T. 1962. On a new bacterium, *Paracolobactrum anguillimortiferum* n. sp. Bulletin of the Society of Scientific Fisheries **28:**162–164.

Iveson, J. B. 1973. Enrichment procedures for the isolation of *Salmonella, Arizona, Edwardsiella* and *Shigella* from faeces. Journal of Hygiene **71:**349–361.

Jordan, G. W., Hadley, W. K. 1969. Human infections with *Edwardsiella tarda*. Annals of Internal Medicine **70:**283–288.

King, B. M., Adler, D. L. 1964. A previously undescribed group of Enterobacteriaceae. American Journal of Clinical Pathol-ogy **41:**230–232.

Makulu, A., Gatti, F., Vandepitte, J. 1973. *Edwardsiella tarda* infections in Zaire. Annales de la Société Belge de Médecine Tropicale (Bruxelles) **53:**165–172.

Meyer, F. P., Bullock, G. L. 1973. *Edwardsiella tarda,* a new pathogen of channel catfish (*Ictalurus punctatus*). Applied Microbiology **25:**155–156.

Muyembe, T., Vandepitte, J., Desmyter, J. 1973. Natural colistin resistance in *Edwardsiella tarda*. Antimicrobial Agents and Chemotherapy **4:**521–524.

Nguyen-Van-Ai, Nguyen-Duc-Hanh, Le-Tien-Van, Nguyen-Van-Le, Nguyen-Thi Lan-Huong 1975. Contribution a l'étude des *Edwardsiella tarda* isolés au Viet-Nam. Bulletin de la Société de Pathologie Exotique et de Ses Filiales (Paris) **68:**355–359.

Onogawa, T., Terayama, T., Zen-Yoji, H., Amano, Y., Suzuki, K. 1976. Distribution of *Edwardsiella tarda* and hydrogen sulfide-producing *Escherichia coli* in healthy per-sons. [In Japanese with English summary.] Journal of the Japanese Association for Infectious Diseases **50:**10–17.

Otis, V. S., Behler, J. L. 1973. The occurrence of *Salmonellae* and *Edwardsiella* in the turtles of the New York Zoological park. Journal of Wildlife Diseases **9:**4–6.

Owens, D.R., Nelson, S. L., Addison, J. B. 1974. Isolation of *Edwardsiella tarda* from swine. Applied Microbiology **27:**703–705.

Sakazaki, R. 1967. Studies on the Asakusa group of Enterobacte-riaceae (*Edwardsiella tarda*). Japanese Journal of Medical Science and Biology **20:**205–212.

Sakazaki, R., Murata, Y. 1962. The new group of the Enterobac-teriaceae, the Asakusa group. [In Japanese.] Japanese Jour-nal of Bacteriology **17:**616–617.

Sakazaki, R., Tamura, K. 1975. Priority of the specific epithet *anguillimortiferum* over the specific epithet *tarda* in the name of the organism presently known as *Edwardsiella tarda*. International Journal of Systematic Bacteriology **25:**219–220.

Wakabayashi, H., Egusa, S. 1973. *Edwardsiella tarda* (*Para-colobactrum anguillimortiferum*) associated with pond-cultured eel disease. Bulletin of the Japanese Society of Scientific Fisheries **39:**931–936.

White, F. H., Simpson, C. F., Williams, L. E., Jr. 1973. Isola-tion of *Edwardsiella tarda* from aquatic animal species and surface waters in Florida. Journal of Wildlife Diseases **9:**204–208.

CHAPTER 91

The Genus *Citrobacter*

JOHN J. FARMER III[1]

Bacteria of the genus *Citrobacter* are well known in many areas of microbiology. They are frequently isolated from clinical specimens and should be identified by the microbiologist. Citrobacters are also found in food, water, sewage, soil, and similar environments, and can be difficult to identify because they are similar to other species of Enterobacteriaceae. Although genetic analysis of *Citrobacter* has not been as thorough as that of *Escherichia coli,* techniques have recently become available to facilitate complete chromosome mapping (de Graaff, Krevning, and Stouthamer, 1974). Recently an L-asparaginase, used in cancer treatment, was purified from *Citrobacter* spp. This preparation can be used as a secondary treatment for cancer patients who make antibodies to the L-asparaginase from *E. coli,* which is sometimes used in treatment (Davidson et al., 1977). Sedlák's two excellent reviews (Sedlák, 1973; Sedlák et al., 1971) covering many aspects of *Citrobacter* are highly recommended to readers who require more information on this interesting genus.

The name *Citrobacter* was coined in 1932 by Werkman and Gillen (1932) for a heterogeneous group of "coli-aerogenes" intermediates that produced trimethylene glycol from glycerol. The name *Citrobacter* is derived from the Latin roots "citrus" and "bacter", which mean "lemon" and "rod", respectively. Thus a very literal definition for *Citrobacter* would be "a rod-shaped bacterium that utilizes citrate". In their original report, Werkman and Gillen listed seven species—*C. freundii, C. album, C. glycologenes, C. intermedium, C. decolorans, C. diversum,* and *C. anindolicum.* Only a few subcultures of the strains studied by Werkman and Gillen have survived, a fact that now complicates nomenclatural arguments about *Citrobacter.* The eighth edition of *Bergey's Manual of Determi-*

native *Bacteriology* (Buchanan and Gibbons, 1974) lists the following properties for *Citrobacter:* Gram-negative rods, motile (with peritrichous flagella), no capsules, citrate utilized as the sole source of carbon, glucose and other carbohydrates fermented with the production of acid and gas (carbon dioxide to hydrogen ratio is 1:1), lactose fermentation may be delayed or absent, trimethyleneglycol formed from glycerol, growth not inhibited by KCN, and ability to grow in the following: Muller's tetrathionate broth, Leifson's selenite broth, sodium deoxycholate–citrate agar, Wilson and Blair's bismuth sulfite medium, and Kristensen's brilliant green–phenol red agar.

Although cultures with the properties of *Citrobacter* have been known since the beginning of the bacteriologic era, only within the last few years have we had an adequate classification. Table 1 summarizes current nomenclature and lists some older synonyms for the three species. It is very difficult to interpret much of the *Citrobacter* literature before 1970 because of the difficulties with nomenclature and because of inadequate descriptions. We do know definitely that citrobacters have been isolated from man, a wide variety of animals, food, water, and soil. If these isolates could be tested today, many would belong to one of the three defined species, some would probably be new (yet undescribed) species of *Citrobacter,* and others would be species in other genera resembling *Citrobacter* (Leclerc and Buttiaux, 1965). *Citrobacter freundii* is the type species for the genus and has been widely accepted, but there is very little agreement among authorities about the nomenclature and classification of the other species in the "*Citrobacter* group". The name *Citrobacter intermedius* often appears in the literature, but the type strain in almost identical with *Citrobacter freundii* (Ewing and Davis, 1972). Thus, it is a junior synonym and should not be used. I have chosen to use the name *Citrobacter diversus* rather than *C. koseri* for the organism redescribed (original description was by Werkman and Gillen, 1932) by Ewing and Davis (1972) and by Frederiksen (1970). We are currently examining this nomenclatural question to determine whether *diversus* or *koseri* is more appropriate.

[1] In collaboration with Don J. Brenner for the nomenclatural proposal to transfer *Levinea amalonatica* to the genus *Citrobacter* where it thus becomes *Citrobacter amalonaticus* (Young, Kenton, Hobbs, and Moody 1971), Brenner and Farmer 1981, which is a new combination (*comb. nov.*).

This material was prepared in my official capacity as an employee of the United States Government. It is part of the public domain and may not be copyrighted.

Table 1. The nomenclature and classification of *Citrobacter* used by different authors.

Recommendation, this chapter	*Bergey's Manual* (Buchanan and Gibbons, 1974)	Cowan (1974)	Young et al. (1971)	Ewing and Davis (1972)
C. freundii[a]	*C. freundii*	*C. freundii*	*C. freundii*	*C. freundii*
C. diversus[b]	*C. intermedius* biotype b	*C. koseri*	*Levinea malonatica*	*C. diversus*
C. amalonaticus[c]	*C. intermedius* biotype a	*Levinea* spp.	*Levinea amalonatica*	*C. freundii*-H_2S^- indole$^+$ biogroup

[a] Synonyms for *C. freundii*—*Bacterium freundii, Escherichia freundii, Salmonella ballerup, Salmonella coli, Salmonella hormachei,* ''ballerup group'', ''Bethesda group'', ''Bethesda-ballerup group'', *Colobactrum freundii, Colobactrum intermedium, Paracolobactrum intermedium, Citrobacter intermedius, Citrobacter intermedium.*
[b] Other synonyms for *C. diversus*—*Citrobacter intermedius, Citrobacter intermedium, Escherichia intermedia, Paracolobactrum intermedium.*
[c] Other synonyms for *C. amalonaticus*—*Padlewskia* sp., *Citrobacter intermedius.*

In 1971 Young et al. proposed a new genus ''*Levinea*'' for organisms that were previously included in *Citrobacter*. Two new species were proposed, *Levinea malonatica* and *Levinea amalonatica* (Young et al., 1971). *Levinea amalonatica* is a synonym of the ''indole$^+$, H_2S^- biogroup of *Citrobacter freundii*'' described by Ewing and Davis (1972). Studies by Crosa et al. (1974) using DNA-DNA hybridization indicated that *Levinea amalonatica* is distinct from *C. freundii* and is a valid species. Table 2 shows that there are also several phenotypic differences between the three species of the genus. The classification used in Table 1 lists all three of these organisms as species of *Citrobacter*. An alternative is to place *C. diversus* and *C. amalonaticus* in a separate genus—*Levinea*. The reasons for doing this have been discussed by Young et al. (1971) and Richard (1972). However, the current classification in a single genus is based on the overall phenotypic relatedness and on DNA base sequence relatedness (as measured by DNA-DNA hybridization). By DNA-DNA hybridization, *C. diversus* and *C.*

amalonaticus are no more closely related to each other than they are to *C. freundii* (Crosa et al., 1974). These relationships are illustrated in Fig. 1 and form the bases for defining the genus *Citrobacter* with three species. In addition, the biochemical reactions of *C. amalonaticus* are much more similar to those of *C. freundii* than to those of *C. diversus* (Table 2). Brenner et al. (1977) proposed moving *Levinea amalonaticus* to *Citrobacter*, but the suggestion appeared in a Center for Disease Control (CDC) publication. In order to place this proposal before the scientific community and to establish its standing in nomenclature, Brenner and Farmer hereby propose transferring *L. amalonatica* to *Citrobacter*, where it becomes *Citrobacter amalonaticus* (Young, Kenton, Hobbs, and Moody, 1971), Brenner and Farmer, 1981. This is a new combination (comb. nov.). The description remains the same (Young et al., 1971) and the type strain remains ATCC 25405 (NCTC 10805). One reason that Richard argued in favor of the validity of the genus *Levinea* was the difference in the G+C ratios (unpublished values cited by

Table 2. Tests useful in differentiating the three species of *Citrobacter*.

Property	*C. freundii*	*C. diversus*	*C. amalonaticus*
Biochemical tests			
Indole production	$-$[a]	$+$	$+$
H_2S production	$+$	$-$	$-$
Malonate utilization	$-$	$+$	$-$
KCN (growth in)	$+$	$-$	$+$
Tyrosine-clearing	$-$	$+$	$-$
Adonitol (acid production)	$-$	$+$	$-$
Antibiogram			
Cephalothin (30 μg)[b]	10.9(2.9)[c]	23.5(1.2)	18.0(2.0)
Ampicillin (10 μg)	14.3(3.1)	7.1(1.2)	8.7(2.4)
Carbenicillin (100 μg)	24.1(0.9)	12.6(1.6)	16.5(2.4)

[a] Symbols: $-$, most (usually 90% or more) strains are negative at 24 or 48 h; $+$, most (usually 90% or more) strains are positive at 24 or 48 h.
[b] The number in parentheses is the strength of the antibiotic in the disk.
[c] The mean and standard deviation (in parentheses) of the zone of complete inhibition around the antibiotic disk.

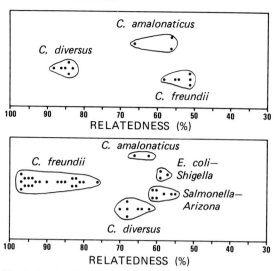

Fig. 1. DNA relatedness in the genus *Citrobacter*. Strain relatedness by DNA-DNA hybridization at 60°C. Top: The DNA of *Citrobacter diversus* ATCC 25408 was labeled and compared with DNA from strains of *C. freundii* and *C. amalonaticus*. Bottom: Labeled DNA from *C. freundii* strain CDC 460–61 was used. Points on the figure represent the percentage relatedness of the test strain to the reference strains. The figure was drawn according to the data of Crosa et al. (1974).

Richard, 1972) between *Levinea* (56–59 mol%) and *C. freundii* (50 mol%). Recent determinations (M. Mandel, personal communication) of the G+C base ratios are much closer than those just cited:

C. freundii (2 strains)	52–53 mol%
C. diversus (2 strains)	54 mol%
C. amalonaticus (2 strains)	54–57 mol%

The discrepancies between these two differing sets of values for the G+C ratios need resolution. Those who accept and use *Levinea* as a genus should use *Levinea diversa* (or *Levinea koseri*) rather than *Levinea malonatica* because the species name "*malonatica*" is clearly a junior synonym (Crosa et al., 1974) and has no standing in nomenclature. In the future, other new species will probably be added to *Citrobacter* as well as to *Enterobacter*, making the process of separating the two genera increasingly more difficult if they represent a continuum of evolution.

Habitats

As a Human Pathogen—Cause of Diarrhea

From antiquity to the end of World War II, most countries had outbreaks of "summer diarrhea" of varying severity and duration (Topley and Wilson, 1936). Many different bacterial species were stated to be probable etiological agents, and citrobacters probably had some role.

From 1943 to 1953, several reports suggested that citrobacters could cause diarrhea. Stuart et al. (1943) studied a heterogeneous group of enteric organisms (probably including several citrobacters) and found that these strains were present in almost pure cultures from patients with diarrhea and often could be traced to food handlers. In additon, four laboratory-acquired infections were caused by these strains. Barnes and Cherry (1946) and West and Edwards (1954) described several additional outbreaks of diarrhea in which citrobacters were implicated. The evidence was not conclusive and was complicated by the fact that many healthy people carry citrobacters as part of their normal flora. The monograph by West and Edwards (1954) seems to provide the most provocative early evidence for the etiological role of *Citrobacter* as a cause of diarrhea. They used serotyping to aid their analysis and found a good correlation (although it was not exact) between the presence of specific strains (serotypes) and epidemiological data. However, they concluded that the evidence was only suggestive and could not implicate *Citrobacter* as a definite cause of diarrhea.

Recently, several reports (Guerrant et al., 1976; Wadström et al., 1976) offered a possible explanation for the manner in which *C. freundii* and *C. diversus* may cause diarrhea. Culture filtrates of strains isolated from patients with diarrhea have caused cytopathological effects in standard tissue culture systems used to assay for enterotoxins. It is not known whether the *Citrobacter* strains contain "colonization factors" that are also probably required before the strain can colonize the small intestine and cause diarrhea. With *Escherichia coli*, enterotoxin production itself is probably not sufficient to cause disease without colonization. *Citrobacter* can rapidly lose the ability to cause this cytopathological effect in the tissue culture assay (Joy Wells, personal communication), so these data must be interpreted cautiously. Enterotoxin production would explain the earlier reports of outbreaks of diarrhea in man and animals (Swiderski and Jedrzejowski, 1976) and confirm the earlier work of Nestorescu et al. (1966), who showed the cytotoxic properties of culture filtrates from three "highly pathogenic" strains of *Citrobacter*. These separate lines of evidence indicate that citrobacters may occasionally have a role in diarrhea (Bărbulescu, 1972; Kalashinikova et al., 1974; Kalashinikova, Zhukovskaya, and Sheina, 1976; Nestorescu et al., 1964; Nestorescu et al., 1966; Popovici et al., 1964; Popovici et al., 1967). Diarrhea due to *Citrobacter* (both epidemics and sporadic cases) is probably rare, and most strains of *Citrobacter* are not enteric pathogens. Thus the situation with *Citrobacter* may be similar to that with *E. coli*, because most strains are normal inhabitants of the gut and do not cause diarrhea. However, strains may occasionally become pathogenic (perhaps by acquiring plasmids that code for enterotoxin production and gut colonization or by other pathogenic mechanisms) and cause diarrhea.

As a Human Pathogen—Opportunistic Infections

All three species of *Citrobacter* are opportunistic pathogens (Altmann et al., 1976; Booth and McDonald, 1971; Jones et al., 1973; Madrazo et al., 1975), with *C. freundii* being the most common. Citrobacters are frequently isolated from wound and respiratory tract specimens and are often present in mixed culture (Madrazo, Geiger, and Lauter, 1975; Slifkin and Engwall, 1969). Thus it is difficult to assess their clinical relevance, but most would agree that they are usually secondary rather than primary invaders and are often only colonizers. All three species are frequently isolated from feces (unpublished data from the Bacteriology Division, CDC), which provide a likely reservoir for endogenous infections. Citrobacters have been documented as the cause of urinary tract infection, but in the large series of Blazevic, Stemper, and Matsen (1972), they accounted for less than 0.5% of the infections. Citrobacters may occasionally cause bacteremia. The cumulative report from the United Kingdom in 1975 listed 19 cases (2 fatal) caused by *C. freundii*, 4 (none fatal) caused by *C. diversus* (*koseri*), and 6 (2 fatal) caused by *Citrobacter* sp. (Epidemiological Research Laboratory, 1976). For comparison, during the same period there were, 1,201 (231 fatal) bacteremias caused by *Escherichia coli*. Citrobacters rarely cause meningitis. When Tamborlane and Soto (1975) described a case in 1975 they commented that only 8 cases of neonatal meningitis caused by *C. diversus* and 3 cases caused by other *Citrobacter* had been reported in the literature. Most meningitis cases are sporadic, but occasionally outbreaks occur (Gross, Rowe, and Easton, 1973; Gwynn and George, 1973; Puentes et al., 1975; Ribeiro, Davis, and Jones, 1976). An outbreak of septicemia was caused by several different bacteria, including *C. freundii*, present in commercial intravenous fluids administered to patients (Center for Disease Control, 1973).

In Other Animals

Citrobacters are often isolated from animals. The list includes household pets, livestock, wild animals, birds, insects, marine animals, and many others (Edwards and Ewing, 1972; unpublished data from the Bacteriology Division, CDC). The organism involved in fish spoilage described by Carman and Levin (1977) as *Aeromonas phenologenes* is really a *Citrobacter*. Much of the data on the distribution of citrobacters has been obtained during the search for *Salmonella*. Systematic searches for citrobacters have seldom been attempted.

In the Environment

Because citrobacters require no vitamins or amino acids for growth (Keevil, Hough, and Cole, 1977; Machtiger and O'Leary, 1971), they are well adapted for growth and survival in the environment. They have been isolated from many water and soil samples (Leclerc and Buttiaux, 1965). Interestingly, many of these strains have different phenotypic characteristics from those of the three defined species of *Citrobacter*. A group that Leclerc and Buttiaux (1965) called a "lysine positive *Citrobacter*" was studied by Crosa et al. (1974) and found to be a new (unnamed) species of Enterobacteriaceae that was closer (by DNA hybridization) to *Serratia* than to *Citrobacter*. This example illustrates that much additional work is required on environmental isolates that do not resemble any of the currently defined species of Enterobacteriaceae. Environmental cultures labeled *Citrobacter* sp. or *Enterobacter* sp. are a heterogeneous group that will probably be divided into many new species as they are characterized (Brenner et al., 1977).

Isolation

Isolation of Citrobacters in General

Seldom have systematic efforts been made to isolate citrobacters selectively. In most instances, they are isolated on nonselective plating media and identified along with the other bacteria present in both clinical and environmental samples.

Citrobacters grow well on the plating media commonly used in enteric bacteriology. At 24 h, they form colonies 2–3 mm in diameter on media such as blood (5% sheep blood) agar or Trypticase soy agar. They also grow well on MacConkey agar, which can be used to inhibit Gram-positive bacteria and many other Gram-negative bacteria. On MacConkey agar, citrobacters can be red, pink, or colorless, depending on how rapidly they ferment lactose. Systematic studies (including plating efficiencies of many strains) have rarely been done, but citrobacters are often isolated from enrichments such as tetrathionate, tetrathionate-brilliant green, selenite, and sodium taurocholate, and on plating media such as bismuth sulfite and brilliant green (Sedlák, 1973), which were originally designed to isolate *Salmonella*. Many citrobacters were picked from these enrichments and plates as "suspect *Salmonella*". Citrobacters grow on citrate as the sole source of carbon, so Simmons' citrate agar (Edwards and Ewing, 1972) can be poured into Petri dishes and used as a partially selective medium. This approach can be particularly useful for specimens such as feces or sewage that predominately contain *Escherichia coli*. Because *E. coli* strains cannot utilize citrate as a carbon source (occasional strains grow but usually after several days), they are effectively eliminated by this procedure. Unfortunately, other citrate-utilizing bacteria will grow as well as *Citrobacter*, so each colony must be identified. This approach is more to eliminate *E. coli* and other citrate-negative organisms than to select for *Citrobacter* specifically.

Isolation of *Citrobacter freundii*

The two methods described above, use of Simmons' citrate agar and use of *Salmonella* plating media (Bismuth sulfite or brilliant green) and/or enrichments, have been used successfully in isolating *Citrobacter freundii* but neither medium is very specific.

Most strains of *C. freundii* produce hydrogen sulfide in peptone iron agar, but most other enteric bacteria are H_2S^-. Thus, pour-plates of peptone iron agar can be made from specimens, and black colonies (H_2S^+) are picked and screened as potential *C. freundii*. There will be both false-positive and false-negative results with all these methods. A good selective medium for *C. freundii* or *C. amalonaticus* has not been formulated.

Isolation of *Citrobacter diversus*

Citrobacter diversus can dissolve precipitated L-tyrosine suspended in Trypticase soy agar, which forms the basis for a useful isolation medium.

Isolation Medium for *Citrobacter diversus* (Sheth and Kurup, 1975, modified by the Enteric Section, CDC)

Component 1. In a 1,000-ml Erlenmeyer flask, mix the following ingredients:

Trypticase peptone	7.5 g
Phytone peptone	2.5 g
Agar-agar	7.5 g
Distilled water	500 ml

Autoclave at 121°C for 15 min; cool to 50°C and pour 18 ml into 100-×-13-mm Petri dishes. Allow the medium to harden completely.
Component 2. In a 1,000-ml Erlenmeyer flask, mix the following ingredients:

L-Tyrosine	1 g
Trypticase peptone	7.5 g
Phytone peptone	2.5 g
Agar-agar	7.5 g
Distilled water	250 ml

Add a magnetic stirring bar (1-inch long); autoclave at 121°C for 15 min; and cool in a 50°C water bath.
Component 3. In a 500-ml Erlenmeyer flask, add 1 g of L-tyrosine and 250 ml of distilled water. Suspend the tyrosine in water on a stirring hot plate. Tyrosine is not soluble at room temperature. Heat to boiling. The solution should become clear. Autoclave at 121°C for 15 min, and then immediately plunge it into ice water (0°C) for 20 min. Fine white crystals should precipitate.
Final medium. Put the flask of component 3 in a 50°C water bath for 20 min; then pour component 3 into component 2. Put this mixture on a stirring hot plate; stir it well and dispense (with a Cornwall syringe) 7 ml of this mixture onto the component 1 plates which were prepared previously and have hardened. Swirl plate so that overlay will cover plate. Important: Do this on a level surface. The plates are sealed in a plastic bag, and stored at 4°C until used.

On this medium, colonies of *C. diversus* are 2–3 mm in diameter at 24–30 h, and have clear zone around them where the precipitated tyrosine dissolved. Among the Enterobacteriaceae almost all strains of *C. diversus*, *Proteus*, *Providencia*, and *Morganella morganii* (formerly called *Proteus morganii*) are positive for tyrosine clearing. *C. diversus* can be differentiated from the other three cited genera, which are positive in the "phenylalanine test" (Edwards and Ewing, 1972). Colonies that clear tyrosine are tested for their oxidase reaction (*C. diversus* is negative) and picked to phenylalanine slants (and tested after 24 h at 36°C). Because *C. diversus* is usually resistant to ampicillin and carbenicillin, either antibiotic can be added to tyrosine agar to make it selective as well as differential. George (1973), in a similar approach, added 10 μg/ml of ampicillin to MacConkey agar to make it more selective for *C. diversus*.

Identification

DNA-DNA Hybridization

Before the DNA hybridization studies of Crosa et al. (1974), it was difficult to clearly establish species lines in *Citrobacter*. However, their genetic data clearly indicate that *Citrobacter diversus* and *Levinea malonatica* are different names for the same organism. Similarly *Levinea amalonatica* is not an indole+, H_2S^- biogroup of *C. freundii* as suspected by Ewing and Davis (1972), but is a separate species. Once these species lines became clear it was easy to devise a simple scheme to differentiate the three *Citrobacter* species (Table 2). The study by Crosa et al. also indicated that additional species of

Citrobacter probably exist or that the evolution of *Citrobacter* may be a "continuum". Some strains may represent evolutionary intermediates that have survived in a particular ecological niche. Similar difficulties in defining a species also occur with higher organisms.

Biochemical Reactions

Ewing and Davis (1971, 1972) tabulated the results from common biochemical tests on a large number of *Citrobacter freundii* and *C. diversus* strains. Richard, Brison, and Lioult (1972) did the same for *C. amalonaticus* cultures. The tests most useful in differentiating the three *Citrobacter* species are listed in Table 2. Some cultures appear to belong to the genus *Citrobacter* but do not closely fit in any of the three defined species. These cultures can be reported as "*Citrobacter* species". Other cultures are best reported simply as "unidentified", even though they resemble *Citrobacter* more than organisms in other genera. There is some merit in defining as "*Citrobacter* sp." strains that are methyl red (MR)$^+$, Voges-Proskauer (VP)$^-$, and citrate$^+$ but do not fit in other species of Enterobacteriaceae. Similarly "*Enterobacter* sp." could be used to designate strains that are MR$^-$ (occasionally MR$^+$), VP$^+$, and citrate$^+$, but do not fit other species. Most of the *Citrobacter* cultures from human clinical specimens fit in one of the three species, but many of the strains from soil, water, food, and the hospital environment do not. This is further evidence that additional species of *Citrobacter* need to be defined (Gavini, Lefebvre, and Leclerc, 1976).

Below the Species Level

A number of techniques have been used to differentiate strains of *Citrobacter*. Serological typing was investigated by West and Edwards (1954), who hoped that differentiation of *C. freundii* strains would help define their role in diarrhea. The original West-Edwards scheme had 32 O-antigen groups, 87 H-antigen groups, and 167 different O–H combinations among the 506 strains tested. The West-Edwards scheme was expanded by Sedlák and colleagues (1971) and now includes 48 O and 90 H antigens. More details of the scheme and its immunochemical basis can be found in Sedlák's (1973) review. Although antigenic analysis in case-control studies has implicated *C. freundii* as a cause of diarrhea in Eastern Europe, serotyping has not been used to the extent it has been with *Salmonella*.

Gross and Rowe (1974, 1975) recently proposed an antigenic scheme with seven *C. diversus* O antigens that should be useful in the epidemiological analysis of nosocomial infections caused by this emerging pathogen. (Strains of *C. amalonaticus* did not react in the *C. diversus* antisera.) Although outbreaks caused by *Citrobacter* are rare, serotyping should assist in defining the ecology and epidemiology of this group, just as it did for some of the more common enteric pathogens. Bacteriocin sensitivity and production (Hamon and Péron, 1966), bacteriophage susceptibility, and biotyping have been described for citrobacters, but no standardized schemes have been proposed, and the techniques have been used rarely (Sedlák, 1973).

Antibiotic Susceptibilities

Reports of antibiotic susceptibility for the *Citrobacter* groups are very difficult to analyze because of the different terms and methods used (Holmes et al., 1974). Most of the strains we have studied are susceptible to most antibiotics, which is not true for many other species of Enterobacteriaceae. Resistant strains have constituted less than 5% of the total. The most probable explanation is that *Citrobacter* strains are more often endemic than epidemic, because strains from outbreaks tend to be much more antibiotic resistant than those from sporadic cases. In our experience, based on the standardized single disk method of Bauer et al. (1966), most citrobacters are resistant to penicillin and susceptible to colistin, sulfadiazine, tetracycline, nalidixic acid, chloramphenicol, streptomycin, kanamycin, and gentamicin. (Some *C. amalonaticus* are "intermediate" to nalidixic acid, streptomycin, and tetracycline.) *C. freundii* is often susceptible or intermediate to ampicillin, but the other two species are resistant. Table 2 shows the differences in zone sizes for cephalothin, ampicillin, and carbenicillin, and illustrates how zone sizes can be useful in identification. Because there are many variables in antibiotic susceptibility testing, these data must be interpreted cautiously.

Literature Cited

Altmann, G., Sechter, I., Cahan, D., Gerichter, C. B. 1976. *Citrobacter diversus* isolated from clinical material. Journal of Clinical Microbiology **3**:390–392.

Barbulescu, E. 1972. Enterocolite acute asociate cu germeni de genul *Citrobacter*. Microbiologia Parazitologia Epidemiologia (Bucuresti) **17**:137–141.

Barnes, L. A., Cherry, W. B. 1946. A group of paracolon organisms having apparent pathogenicity. American Journal of Public Health **36**:481–483.

Bauer, A. W., Kirby, W. M. M., Sherris, J. C., Turck, M. 1966. Antibiotic susceptibility testing by a standardized single disk method. American Journal of Clinical Pathology **45**:493–496.

Blazevic, D. J., Stemper, J. E., Matsen, J. M. 1972. Organisms encountered in urine culture over a 10-year period. Applied Microbiology **23**:421–422.

Booth, E. V., McDonald, S. 1971. A new group of enterobacteria, possibly a new *Citrobacter* sp. Journal of Medical Microbiology **4**:329–336.

Brenner, D. J., Farmer, J. J., III, Hickman, F. W., Asbury, M. A., Steigerwalt, A. G. 1977. Taxonomic and nomenclatural changes in Enterobacteriaceae. Atlanta, Georgia: Center for Disease Control.

Buchanan, R. E., Gibbons N. E. (eds.) 1974. Bergey's manual of determinative bacteriology, 8th ed. Baltimore: Williams & Wilkins.

Carman, G. M., Levin, R. E. 1977. Partial purification and some properties of tyrosine phenol-lyase from *Aeromonas phenologenes* ATCC 29063. Applied Microbiology **33:**192–198.

Center for Disease Control 1973. Follow-up on septicemias associated with contaminated intravenous fluids—United States. Morbidity and Mortality Weekly Report **23:**115–116.

Cowan, S. T. 1974. Cowan and Steel's manual for the identification of medical bacteria, 2nd ed. London, New York: Cambridge University Press.

Crosa, J. H., Steigerwalt A. G., Fanning, G. R., Brenner, D. J. 1974. Polynucleotide sequence divergence in the genus *Citrobacter*. Journal of General Microbiology **83:**271–282.

Davidson, L., Burkom, M., Ahn, S., Chang, L.-C., Kitto, B. 1977. L-Asparaginases from *Citrobacter freundii*. Biochimica et Biophysica Acta **480:**282–294.

de Graaff, J., Kreuning, P. C., Stouthamer, A. H. 1974. Isolation and characterization of Hfr males in *Citrobacter freundii*. Antonie van Leeuwenhoek Journal of Microbiology and Serology **40:**161–170.

Edwards, P. R., Ewing, W. H. 1972. Identification of Enterobacteriaceae, 3rd ed. Minneapolis: Burgess Publishing Co.

Epidemiological Research Laboratory, Central Public Health Laboratory, Colindale. 1976. Quarterly Tabulations: Thirteen weeks 40–52, 1975. Communicable Disease Reports No. 40–52 (2nd. January 1976). London: Public Health Laboratory Service.

Ewing, W. H., Davis, B. R. 1971. Biochemical characterization of *Citrobacter freundii* and *Citrobacter diversus*. Atlanta, Georgia: Center for Disease Control.

Ewing, W. H., Davis, B. R. 1972. Biochemical characterization of *Citrobacter diversus* (Burkey) Werkman and Gillen and designation of the neotype strain. International Journal of Systematic Bacteriology **22:**12–18.

Frederiksen, W. 1970. *Citrobacter koseri* (n. sp.): A new species within the genus *Citrobacter*, with a comment on the taxonomic position of *Citrobacter intermedium* (Werkman and Gillen). Spisy Prirodovedecke Fakulty University J. E. Purkyne v Brne, Ser. K **47:**89–94.

Gavini, F., Lefebvre, B., Leclerc, H. 1976. Positions taxonomiques D'Entérobactéries H₂S⁻ par rapport au genre *Citrobacter*. Annales de Microbiologie (Paris) **127A:**275–295.

George, R. H. 1973. Neonatal meningitis caused by *Citrobacter koseri*. Journal of Clinical Pathology **26:**552.

Gross, R. J., Rowe, B. 1974. The serology of *Citrobacter koseri*, *Levinea malonatica*, and *Levinea amalonatica*. Journal of Medical Microbiology **7:**155–161.

Gross, R. J., Rowe, B. 1975. *Citroacter koseri*. I. An extended antigenic scheme for *Citrobacter koseri* (syn. *C. diversus*, *Levinea malonatica*). Journal of Hygiene **75:**121–127.

Gross, R. J., Rowe, B., Easton, J. A. 1973. Neonatal meningitis caused by *Citrobacter koseri*. Journal of Clinical Pathology **26:**138–139.

Guerrant, R. L., Dickens, M. D., Wenzel, R. P., Kapikian, A. Z. 1976. Toxigenic bacterial diarrhea: Nursery outbreak involving multiple bacterial strains. Journal of Pediatrics **89:**885–891.

Gwynn, C. M., George, R. H. 1973. Neonatal citrobacter meningitis. Archives of Diseases in Childhood **48:**455–458.

Hamon, Y., Péron, Y. 1966. Étude des bactériocines produites par les bactéries apparenant au groupe *Citrobacter*-Bethesda. Annales de l'Institut Pasteur (Paris) **111:**497–501.

Holmes, B., King, A., Phillips, I., Lapage, S. P. 1974. Sensitivity of *Citrobacter freundii* and *Citrobacter koseri* to cephalosporins and penicillins. Journal of Clinical Pathology **27:**729–733.

Jones, S. R., Ragsdale, A. R., Kutscher, E., Sanford, J. P. 1973. Clinical and bacteriologic observations on a recently recognized species of Enterobacteriaceae, *Citrobacter diversus*. Journal of Infectious Diseases **128:**563–565.

Kalashinikova, G. K., Zhukovskaya, N. A., Sheina, I. V. 1976. Bacteria of the *Citrobacter* genus as one of the possible indices of dysbacteriosis in intestinal diseases. Zhurnal Mikrobiologii, Epidemiologii i Immunobiologii **7:**50–60.

Kalashinikova, G. K., Lokosova, A. K., Sorokina, R. S., Brodova, A. I., Grivtsova, A. S., Batura, A. P., Raginskaya, V. P. 1974. Concerning the etiological role of bacteria belonging to *Citrobacter* and *Hafnia* genera in children suffering from disease accompanied by diarrhea and some of their epidemiological peculiarities. Zhurnal Mikrobiologii, Epidemiologii i Immunobiologii **51:**78–81.

Keevil, C. W., Hough, J. S., Cole, J. A. 1977. Prototrophic growth of *Citrobacter freundii* and the biochemical basis for its apparent growth requirements in aerated media. Journal of General Microbiology **98:**273–276.

Leclerc, H., Buttiaux, R. 1965. Les *Citrobacter*. Annales de l'Institut Pasteur de Lille **16:**67–74.

Machtiger, N. A., O'Leary, W. M. 1971. Nutritional requirements of *Arizona, Citrobacter,* and *Providencia*. Journal of Bacteriology **108:**948–950.

Madrazo, A., Geiger, J., Lauter, C. B. 1975. *Citrobacter diversus* at Grace Hospital, Detroit, Michigan. American Journal of Medical Science **270:**497–501.

Madrazo, A., Henderson, M. D., Baker, L., Vaitkevicius, V. K., Lauter, C. B. 1975. Massive empyema due to *Citrobacter diversus*. Chest **68:**104–106.

Nestorescu, N., Popovici, M., Szégli, L., Negut, A., Negut, M., Barbulescu, E. 1964. Untersuchung einiger hoch-pathogener atypischer Citrobacter-stämme. Zentralblatt für Bakteriologie, Parasitenkunde, Infektionskrankheiten und Hygiene, Abt. 1 Orig. **194:**443–450.

Nestorescu, N., Szégli, L., Popovici, M., Dima, V. 1966. Untersuchung einiger hoch-pathogener atypischer Citrobacter-Stämme. 2. Mitteilung: Cytotoxische Wirkung der Kulturfiltrate. Zentralblatt für Bakteriologie, Parasitenkunde, Infektionskrankheiten und Hygiene, Abt. Orig. **199:**46–52.

Popovici, M., Szégli, L., Soare, L., Negut, A., Dimitriu, N., Stanciu, V. 1964. Rôle des germes du groupe *Citrobacter* dans l'étiologie des toxiinfections alimentaires. Archives Roumaines de Pathologie Experimentale et de Microbiologie **23:**1005–1010.

Popovici, M., Szégli, L., Racovitză, C., Bădulescu, E., Florescu, D., Negut, M., Negut, A., Thomas, E., Masek, S. 1967. Über die aetiologische Bedeutung und Häufigkeit der Keime der Citrobactergruppe bei Enteritiden. Zentralblatt für Bakteriologie, Parasitenkunde, Infektionskrankheiten und Hygiene, Abt. 1, Orig. **204:**112–121.

Puentes, R., Cerda, M., Orellana, B., Lopez, T. M. 1975. Sepsis a *Citrobacter* en el lactante. Revista Chilena de Pediatriá **46:**211–217.

Ribeiro, C. D., Davis, P., Jones, D. M. 1976. *Citrobacter koseri* meningitis in a special care baby unit. Journal of Clinical Pathology **29:**1094–1096.

Richard, C. 1972. Le genre *Levinea* (famille des *Enterobacteriaceae*). Bulletin de Liaison de l'Association des Anciens Élèves et Diplomes de l'Institut Pasteur á Paris **54:**85–92.

Richard, C., Brison, B., Lioult, J. 1972. Étude taxonomique de "*Levinea*" nouveau genre de la famille des *Enterobactéries*. Annales de l'Institut Pasteur (Paris) **122:**1137–1146.

Sedlák, J. 1973. Present knowledge and aspects of *Citrobacter*. Current Topics in Microbiology and Immunology **62:**41–59.

Sedlák, J., Puchmayerová-Šlajsová, M., Keleti, J., Lüderitz, O. 1971. On the taxonomy, ecology and immunochemistry of genus Citrobacter. Journal of Hygiene, Epidemiology, Microbiology, and Immunology **15:**366–374.

Sheth, N. K., Kurup, V. P. 1975. Evaluation of tyrosine medium for the identification of *Enterobacteriaceae*. Journal of Clinical Microbiology **1**:483–485.

Slifkin, M., Engwall, C. 1969. The clinical significance of *Citrobacter intermedium*. American Journal of Clinical Pathology **52**:351–355.

Šourek, J., Aldová, E. 1976. Serotyping of strains belonging to the *Citrobacter-Levinea* group isolated from diagnostic material. Zentralblatt für Bakteriologie, Parasitenkunde, Infektionskrankhieiten und Hygiene, Abt. 1, Orig., Reihe A **234**:480–490.

Stuart, C. A., Wheeler, K. M., Rustigian, R., Zimmerman, A. 1943. Biochemical and antigenic relationships of the paracolon bacteria. Journal of Bacteriology **45**:101–119.

Swiderski, M., Jedrzejowski, A. 1976. Infekcyjna lekooporność i wirulencja pateczek *Citrobacter* z orgniska intoksykacji prosiat. [with English summary.] Medycyna Weterynaryjna **32**:665–668.

Tamborlane, W. V., Soto, E. V. 1975. *Citrobacter diversus* meningitis: A case report. Pediatrics **55**:739–741.

Topley, W. W. C., Wilson, G. S. 1936. Principles of Bacteriology and Immunology, 2nd ed, pp. 1245–1254. Baltimore: William & Wilkins.

Wadström, T., Aust-Kettis, A., Habte, D., Holmgren, J., Meeuwisse, G., Möllby, R., Söderlind, O. 1976. Enterotoxin-producing bacteria and parasites in stools of Ethiopian children with diarrhoeal disease. Archives of Disease in Childhood **51**:865–870.

Werkman, C. H., Gillen, G. F. 1932. Bacteria producing trimethylene glycol. Journal of Bacteriology **23**:167–182.

West, M. G., Edwards, P. R. 1954. The Bethesda-Ballerup group of paracolon bacteria. Public health monograph no. 22. relanta: U.S. Department of Health, Education, and Welfare. Center for Disease Control.

Young, V. M., Kenton, D. M., Hobbs, B. J., Moody, M. R. 1971. *Levinea,* a new genus of the family *Enterobacteriaceae*. International Journal of Systematic Bacteriology **21**:58–63.

The Genus *Salmonella*

LÉON LE MINOR

History

Long before the bacteriological era, typhoid—the first salmonellosis ever recognized—was defined with clinical signs and symptoms and anatomical changes.

Although pathologists (Petit and Serres, 1813) had noticed intestinal ulcerations predominating near the cecum, enteric fever was not clearly distinguished from enteritic tuberculosis until Bretonneau (1822) described what he named "dothienentérite" and pointed out its contagiousness. In a synthesis of earlier works, Louis (1829) grouped several described aspects of enteric fever under the name typhoid. Once the disease (typhoid) was clearly defined, its epidemic transmission and its pathogenesis could be studied, and prevention and treatment could be attempted. Budd (1856) postulated that each case of typhoid is epidemiologically linked to a former case, and that a "specific toxin" is disseminated with the patient's feces. Pettenkofer (1868) theorized about the role of underground water levels, and during the subsequent half-century prophylaxis was concerned with improvements in water supply systems and seafood regulations.

The typhoid bacillus was first observed by Eberth (1880) in spleen sections and mesenteric lymph nodes from a patient who died from typhoid. Koch (1880) confirmed the finding and Gaffky (1884) succeeded in cultivating the bacterium. At that time, due to the lack of differential characters, separation from other enteric bacteria was uncertain. Doubt concerning the validity of the aforementioned discoveries subsisted until Pfeiffer and Kolle (1896) and Gruber and Durham (1896) showed that the serum from an animal immunized with the typhoid bacillus agglutinated this same typhoid bacillus. Independently, Widal (1896) in Paris and Grunbaum (1896) in London showed that the serum of patients afflicted with typhoid agglutinates the typhoid bacillus. Serodiagnosis of typhoid was then possible.

In 1896, two isolates were recovered by Achard and Bensaude from two patients with clinical typhoid, but with a negative Widal serodiagnosis. They called the disease paratyphoid, and the bacteria "bacille paratyphique". A similar case was published by Gwyn (1898), who called the bacterium "paracoli bacillus". Schottmüller (1901) confirmed that "paracoli bacillus" and "bacille paratyphique", although responsible for clinical cases of typhoid, were distinct from the typhoid bacillus both culturally and serologically. He named the first organism "Paratyphus A" and the second "Paratyphus B". A new concept of a group of diseases caused by a group of bacteria had emerged.

Antigenic analysis began when Castellani (1902) described a method for absorbing antisera; somatic and flagellar antigens were differentiated (Smith and Reagh, 1903) and called O and H, respectively (Weil and Felix, 1918a,b, 1920). Andrewes (1922, 1925) showed that flagellar antigens could be diphasic, and Felix and Pitt (1934) found that a surface antigen (Vi) could prevent agglutination of the typhoid bacillus. The first antigenic scheme for *Salmonella* (White, 1926) was subsequently developed extensively by Kauffmann (1966b, 1978). The Kauffmann-White antigenic scheme contained 100 serotypes in 1941 (Kauffmann, 1941). In 1978, this scheme contained about 2,000 serotypes.

Other bacteria resembling "paratyphus" bacteria were isolated from diseased animals. Salmon and Smith (1885) isolated *Bacillus cholerae-suis* from pigs with hog cholera and thought they had isolated the infectious agent of hog cholera. Similar bacteria were isolated from food-borne intoxication (Gaffky and Paak, 1885; Gärtner, 1888) or animal disease (Loeffler, 1892). A bacterial genus was created to include these bacteria and was named *Salmonella* by Lignières (1900) (Salmonella Subcommittee, 1934).

Immunization against typhoid by using heat-killed bacteria was initiated by Chantemesse and Widal (1888) and Wright and Semple (1897).

Typhoid played the role of a pilot disease that attracted the interest of a great number of workers (bacteriologists, immunologists, hygienists).

Antigenic Structure

As with all Enterobacteriaceae, the genus *Salmonella* has three kinds of antigens with diagnostic application: somatic, flagellar, and surface.

CELL WALL (SOMATIC O) ANTIGENS. Somatic antigens are heat stable and alcohol resistant. Antigenic specificity is carried by the polysaccharide component of the protein-lipopolysaccharide endotoxin complex (Lüderitz, Staub, and Westphal, 1966; Lüderitz et al., 1971). Cross-absorption studies could individualize a large number of antigenic factors, 67 of which are or have been used for serological identification. O factors labeled with the same figure are closely related, although not always antigenically identical. According to their diagnostic relevance, O factors can be classified as:

1. Major O antigens, i.e., antigens that identify the O antigenic group (e.g., antigenic factor O:4 characterizes *Salmonella* O group B). Major O antigens are determined by locus rfb, located at 66 min on the *Salmonella typhimurium* LT2 chromosomal map.
2. Minor O antigens, i.e., antigens that have less or no discriminating value (e.g., all strains of *Salmonella* O groups A, B, and D have antigen O:12).

Other minor antigens result from a chemical modification of major antigens (and as a result may not have the same taxonomic weight as major antigens). For example, *Salmonella* O group B that possesses antigenic factors O:4 and O:12 may have the following minor antigens:

1. Antigenic factor O:5 results from acetylation of abequose present in the repeating units of the polysaccharide responsible for the specificity O:4,12.
2. Antigenic factor O:1 results from a modification due to phage conversion of the linkage (1-4 to 1-6) between galactose and glucose in the same repeating units.

Other minor O antigens are determined by phage conversion, e.g., O factor 14 in group C (6,7 changed into 6,7,14), and O factors 6,14 in group K (18 changed into 6,14,18). Cumulative effects of phage conversions may occur, as in *Salmonella* E group, in which the presence of phage ϵ_{15} turns antigen 3,10 (rhamnose-mannose-acetylgalactose-α-mannose) into 3,15, and phage ϵ_{34} turns antigen 3,15 (rhamnose-mannose-galactose-β-mannose) into 3,15,34 (addition of glucose to galactose).

Loss of the ability to build complete polysaccharide chains results in the R state. Several R states, resulting from different mutations, have been described (Mäkelä and Stocker, 1969). Change from S to R form also results in loss of both pathogenicity and autoagglutinability of strains. An exceptional T form, intermediate between S and R forms, has been described (Kauffmann, 1956, 1957).

SURFACE (ENVELOPE) ANTIGENS. Surface antigens, commonly observed in other genera (e.g., *Escherichia* and *Klebsiella*), may be found in some *Salmonella* serotypes. Surface antigens in *Salmonella* may mask O antigens, and the bacteria will not be agglutinated with O antisera. Only one specific surface antigen is known: Vi antigen. Heating at 100°C generally solubilizes Vi antigen, and heat-treated bacteria can be agglutinated with proper O antisera. Vi antigen occurs in only three *Salmonella* serotypes (out of about 2,000): *S. typhi*, *S. para-typhi-C*, and *S. dublin*. Strains of these three serotypes may or may not have the Vi antigen. The presence of Vi antigens is controlled by two genes, Vi A and Vi B, respectively located at 69 and 137 min on the *Salmonella typhimurium* chromosomal map (Sanderson, 1972).

FLAGELLAR (H) ANTIGENS. Flagellar antigens are heat-labile proteins. Tube agglutination with flagella-specific antisera is characteristic: agglutination is flaky (bacteria are loosely attached to each other by their flagella), quickly achieved (contact between flagella from different bacterial cells is quickly achieved), and can be dissociated by shaking (flagella break easily). Antiflagellar antibodies can immobilize bacteria with corresponding H antigens. The presence of flagella and motility are controlled by a number of genes located at 48 and 62 min on the *Salmonella typhimurium* chromosomal map.

A few *Salmonella* serotypes (e.g., *S. enteritidis*, *S. typhi*) can produce only one flagellar antigen (like any non-*Salmonella* enterobacterium); H antigen is then monophasic. Most *Salmonella* serotypes, however, can alternatively produce two H antigens; H antigen is then diphasic. For example, *S. typhimurium* bacterial cells can produce flagella with either antigen i or antigen 1,2. If a bacterial cell with H antigen i is picked up, the derived clone will consist of bacteria with i flagellar antigen. However, at a frequency of 10^{-3}–10^{-5}, bacterial cells with 1,2 flagellar antigen will appear. A culture of *S. typhimurium* on suitable semisolid agar medium will rapidly grow throughout the medium. If i-specific antiserum is added to a semisolid agar medium prior to inoculation, motility of bacterial cells with i flagellar antigen will be prevented. Bacterial cells with the other flagellar phase (1,2) may invade the medium. The reverse experiment (1,2-specific antiserum added) will result in immobilization of bacteria with antigen 1,2 and spread of bacteria with the other flagellar phase (i). This method (H antigen phase inversion) allows selection of bacteria with flagellar phase not corresponding to the antiserum. It is widely used for complete identification of *Salmonella* serotypes. Addition of i-specific and 1,2-specific antisera to semisolid agar media results in complete immobilization of *S. typhimurium*.

The H_1 gene, responsible for the production of

phase 1 flagellar antigen, is located at 62 min on the *Salmonella* chromosomal map. The H_2 gene—responsible for the production of phase 2 flagellar antigen—and vh_2—responsible for phase variation—are both located at 83 min on the chromosomal map. Certain *Salmonella* serotypes (*S. paratyphi-A, S. abortus-equi*) are monophasic because vh_2 gene is not functional (see Le Minor, 1971).

Phase 1 flagellar antigens were originally designated by lowercase letters, and phase 2 antigens were designated by numbers. Phase 2 was also called "nonspecific phase" (versus specific phase 1) because of coagglutinations between the first phase 2 antigens studied (e.g., 1,2; 1,5; 1,6; 1,7; all having factor 1 in common). When all letters of the alphabet were used, the letter z followed by a subscript was used to designate new phase 1 antigens (the latest described is z_{61}). Complexity of antigenic relationships led to the addition of letters to phase 2 numbers. Apparent awkwardness of antigen designation in the Kauffmann-White scheme is the result of historical development of knowledge.

Genetics of *Salmonella*

The genetic map of *Salmonella typhimurium* LT2 is not very different from that of *Escherichia coli* K-12 (Brachman, Low, and Taylor, 1976). F plasmid can be transferred to *S. typhimurium* and an Hfr strain of *S. typhimurium* may subsequently be selected. Conjugative chromosomal transfer may occur from *S. typhimurium* Hfr to *E. coli* or from *E. coli* Hfr to *S. typhimurium*. Chromosomal genes responsible for O, Vi, and H antigens could be transferred from one genus to the other (Iino and Lederberg, 1964).

As for most Enterobacteriaceae, *Salmonella* organisms may harbor "foreign" replicons: temperate phages (Le Minor, 1968) and plasmids. Plasmids in *Salmonella* may code for antibiotic resistance (resistance plasmids are frequent due to the selective pressure of antibiotic misuse), bacteriocins, metabolic characteristics such as lactose or sucrose fermentation (Le Minor, Coynault, and Pessoa, 1974; Le Minor et al., 1973), or nothing known. In taxonomy, it is important to distinguish between chromosomally or nonchromosomally controlled biochemical or antigenic characters. Transfer of a lactose or sucrose plasmid to a *Salmonella* should not change its taxonomic position (nor its name).

Taxonomy

The genus *Salmonella* is a member of the family Enterobacteriaceae. This genus is composed of bacteria related to each other both phenotypically and genotypically. *Salmonella* DNA base composition is 50–52 mol% G+C (Marmur, Falkow, and Mandel, 1963), similar to that of *Escherichia, Shigella,* and *Citrobacter*. The bacteria of the genus *Salmonella* are also at least 80% related to each other by polynucleotide sequence analysis. The genera closest to *Salmonella* with respect to DNA relatedness are *Escherichia, Shigella,* and *Citrobacter* (Cross et al., 1973). Similar relationships were found by numerical taxonomy (Johnson et al., 1975). However, in a classification based on carbon source utilization tests (Véron and Le Minor, 1975a,b), the genus *Shigella* was found to be distant from *Salmonella* because of a limited nutritional versatility of *Shigella*.

The current definition of the genus *Salmonella* rests on biochemical characteristics (this Handbook, Chapter 88). Kauffmann (1963, 1966a) subdivided the genus *Salmonella* into four subgenera: I, II, III, and IV. Most *Salmonella* strains recovered from man and warm-blooded vertebrates belong to subgenus I. Subgenera II and III (also called *Arizona*) are close to each other, and differentiation between these two subgenera is questionable. Subgenera II, III, and IV are more frequently associated with reptiles, in which they behave commensally. Among fifty thousand *Salmonella* strains received at the French National *Salmonella* Center, subgenera II, III, and IV were represented by only 0.16%, 0.53%, and 0.04% of strains, respectively.

Another taxonomic system was proposed (Edwards and Ewing, 1972) in which the genus *Salmonella* is limited to Kauffmann's subgenus I. Kauffmann's subgenus III is called genus *Arizona*, and strains that belong to subgenera II and IV are considered atypical strains of either *Salmonella* or *Arizona*.

Nomenclature

Salmonella nomenclature is a chronically controversial theme, as it does not follow the rules of the International Code of Nomenclature of Bacteria.

Due to their medical or veterinary role, the serotypes first isolated were given a specific name evoking that role: *S. typhi* (agent of typhoid), *S. paratyphi* ("paratyphoid" bacterium), *S. abortus-ovis* (agent of abortion in sheep), *S. typhimurium* (agent of disease of the mouse), etc. In fact, *Salmonella* serotypes first thought to be associated with an animal species were later found to be ubiquitous (*S. typhimurium, S. bovis-morbificans*). Names of new *Salmonella* serotypes were then chosen according to the geographic origin of the first isolated strains: *S. london, S. panama, S. stanleyville*. This tradition was continued with *Salmonella* subgenus I serotypes. New serotypes of subgenera II, III, and IV are only designated by their antigenic formulas.

Kauffmann (1966b) defined the species as a "group of related sero-fermentative phage-types". *Salmonella* serotypes are then elevated to the rank of species. In this system, a specific name should also be given to serotypes in other genera (e.g., *Escherichia coli* serotypes).

Edwards and Ewing (1972) recognized a genus *Arizona* distinct from the genus *Salmonella*. The genus *Arizona* was described with only one species: *A. hinshawii;* three species were retained in the genus *Salmonella: S. cholerae-suis, S. typhi,* and *S. enteritidis. Salmonella* serotypes commonly called *S. paratyphi-A, S. paratyphi-B,* and *S. typhimurium* should then be called *S. enteritidis* bioserotype paratyphi-A, *S. enteritidis* bioserotype paratyphi-B, and *S. enteritidis* bioserotype typhimurium.

Le Minor, Rohde, and Taylor (1970) proposed to subdivide the genus *Salmonella* into four species corresponding to Kauffmann's subgenera: *Salmonella kauffmannii* (subgenus I), *S. salamae* (subgenus II), *S. arizonae* (subgenus III), and *S. houtenae* (subgenus IV). Correct use of this nomenclature would give names such as: *S. kauffmannii* serotype typhi, *S. kauffmannii* serotype cholerae-suis, etc. The names *S. salamae* and *S. houtenae* derive from the name of the first serotype isolated in each of these species, i.e., *S. dar-es-salaam* and *S. houten,* respectively. The Kauffmann-White schema was built irrespective of the subdivision into four subgenera (or species). The same antigenic factor symbols are used for the entire *Salmonella* genus. It would not be wise to continue using particular special symbols for subgenus III (*Arizona*) antigens when these antigens are identical with or very closely related to subgenus I antigens. The eighth edition of *Bergey's Manual of Determinative Bacteriology* (Buchanan and Gibbons, 1974) kept a specific name for each *Salmonella* subgenus I serotype and recognized *S. salamae, S. arizonae,* and *S. houtenae* for subgenera II, III, and IV, respectively.

It is the author's opinion that, due to the lack of definition of a bacterial species, it is more important to have a universal agreement on how to recognize a *Salmonella* and a *Salmonella* serotype than to decide what should be a species, a bioserotype, or a serotype. Agreement on nomenclature, moreover, would ease scientific communication.

Habitats

The principal habitat of *Salmonella* is the intestinal tract of man and animals. *Salmonella* serotypes can be adapted to one particular host, be ubiquitous, or have a still unknown habitat.

Salmonella serotypes that are host adapted usually cause grave diseases with positive blood cultures. *S. typhi, S. paratyphi-A,* and *S. sendai* are strictly human serotypes. Salmonellosis in these cases is transmitted from man to man through fecal contamination of food. *S. gallinarum-pullorum, S. abortus-ovis,* and *S. typhi-suis* are, respectively, avian, ovine, and porcine *Salmonella* serotypes.

Such host-adapted serotypes cannot grow on minimal medium without growth factors (contrary to ubiquitous *Salmonella* serotypes).

Ubiquitous (non-host-adapted) *Salmonella* serotypes (e.g., *S. typhimurium*) cause very diverse clinical symptoms, from asymptomatic infection to grave typhoid-like syndrome in infants or highly susceptible animals (mice). In human adults, ubiquitous *Salmonella* organisms are mostly responsible for food-borne toxic infections.

The pathogenic role of a number of *Salmonella* serotypes is unknown. This is especially the case with serotypes from subgenera II, III, and IV; a number of these serotypes have been isolated only rarely (some only once) during a systematic search in cold-blooded animals.

Salmonella in Man

Pathogenic action of *Salmonella* depends upon the serotype, the strain, the dose, and the host status. Certain serotypes are highly pathogenic for man, while others are devoid of any pathogenic action. Strains of the same serotype may also differ in pathogenicity. An oral dose of at least 10^5 *Salmonella typhi* cells is needed to cause typhoid in 50% of human volunteers; and at least 10^9 *S. typhimurium* (oral dose) cells are needed to cause symptoms of a toxic infection. Infants, immunosupressed patients, and those affected with blood disease are more susceptible to *Salmonella* infection than healthy adults.

The pathogenesis of typhoid has been clarified by Reilly et al. (1935). Typhoid bacilli enter the human digestive tract, penetrate the intestinal mucosa (causing no lesion), and are stopped in the mesenteric lymph nodes. There, bacterial multiplication occurs and part of the bacterial population lyses. From the mesenteric lymph nodes, bacilli and endotoxin may be released into the blood stream. Release of bacilli is responsible for the septicemic phase of typhoid and for infectious metastasis (e.g., osteitis). Release of endotoxin is responsible for intestinal hemorrhages and perforations (through action on the abdominal sympathetic nervous system), and for cardiovascular "collapsus and tuphos" (a stuporous state—origin of the name typhoid) through action on the third ventriculus neurovegetative centers.

Salmonella excretion by human patients may continue long after clinical cure. Asymptomatic carriers are dangerous when unnoticed. About 5% of patients clinically cured from typhoid remain carriers for months or even years. Antibiotics are usually ineffective on *Salmonella* carriage (even if salmonellae are susceptible to them). The risk of *Salmonella* carriage is higher when patients have intestinal or urinary parasites such as *Bilharzia.*

Salmonellae survive sewage treatments if germicides are not added. A typical cycle of typhoid is as follows: Sewage from a community is directed to a sewage plant. Effluent from the sewage plant passes into a coastal river where edible shellfish (mussels, oysters) live. Shellfish concentrate bacteria as they filter several liters of water per hour. Ingestion by man of these seafoods (uncooked or superficially cooked) will cause typhoid. Evidence of such a cycle was given by the use of epidemiological markers, including phage typing (Rische, 1973). Salmonellae do not multiply in contaminated shellfish. If highly *Salmonella*-contaminated shellfish are placed in clean water for 2 weeks, the *Salmonella* organisms are not found anymore.

The incidence of typhoid—strictly a human disease—decreases when the level of development of a country increases (i.e., controlled water sewage systems, pasteurization of milk and dairy products). Where these hygienic conditions are missing, the probability of fecal contamination of water and food remains high, and so is the incidence of typhoid.

Food-borne *Salmonella* toxic infections are caused by ubiquitous *Salmonella* (e.g., *S. typhimurium*). About 12 h following ingestion of contaminated food (containing a sufficient number of *Salmonella*), symptoms appear (diarrhea, vomiting, fever) and last 2–5 days. Spontaneous cure usually occurs.

Salmonella may be associated with all kinds of food. Contamination of meat may originate from animal salmonellosis, but most often it results from contamination of muscles with the intestinal contents during evisceration of animals, washing, and transportation of carcasses. Surface contamination of meat is of little consequence, as cooking will sterilize it (although handling of contaminated meat may contaminate hands, tables, kitchenware, towels, etc.). However, when contaminated meat is ground, multiplication of *Salmonella* may occur inside, and if cooking is superficial, ingestion of this highly contaminated food will produce a *Salmonella* toxic infection. Toxic infection will follow ingestion of any food that allows multiplication of *Salmonella* (cream, mayonnaise, etc.), as a large number of salmonellae are needed to give symptoms. Prevention of *Salmonella* toxic infection relies on avoiding contamination (improvement of hygiene), preventing multiplication of *Salmonella* in food (constant storage of food at 4°C), and use of pasteurization or sterilization when possible (milk). Vegetables and fruits may carry *Salmonella* when contaminated with fertilizers of fecal origin, or when washed with polluted water.

Incidence of food-borne *Salmonella* intoxication/infection remains high in developed countries because of commercially prepared food or ingredients for food. Any contamination of commercially prepared food will result in a large-scale intoxication/infection. In underdeveloped countries, food-borne *Salmonella* intoxications are less spectacular

because of the smaller number of individuals simultaneously contaminated, and also because the bacteriological diagnosis of *Salmonella* toxic infection may not be available. However, incidence of *Salmonella* carriage in underdeveloped countries is high.

Salmonella epidemics occur among infants in pediatric wards. Frequency and gravity of these epidemics are affected by hygienic conditions, malnutrition, and excessive use of antibiotics that select for multiresistant strains.

Salmonella in Animals

Salmonelloses in farm animals are responsible for serious economic loss. In western Europe, abortion in sheep and cows, respectively caused by *S. abortus-ovis* and *S. dublin,* is far more frequent than abortion due to *Brucella*. Industrialization of animal production (e.g., chickens and turkeys), which keeps a large number of animals in a limited space, is prone to extensive enzootics. Contamination of meat for human consumption often occurs in slaughterhouses, water baths, or apparatuses for defeathering poultry. Rejection of contaminated meat by importing countries results in large economic loss.

Animal salmonelloses also imply contamination through contaminated food, and occasionally a respiratory penetration of *Salmonella* carried in the air (industrial breeding plants). Enzootic salmonellosis may be maintained in some areas by the survival of *Salmonella* in soil or by highly infected abortion products (e.g., with *S. abortus-ovis*).

Protein ingredients for animal feeding (e.g., meat or fish powders) are frequently contaminated. This contamination probably originates from defective hygiene in slaughterhouses and factories that produce these protein powders. When fishing is done in *Salmonella*-polluted water, *Salmonella* may be recovered from the fish (ears, scales, intestinal tract), but no *Salmonella* infection occurs in fish.

The incidence of salmonellosis in animals has been studied extensively but our knowledge is nonetheless incomplete. Domestic fowls and animals as well as wild birds, animals, and reptiles constitute an important reservoir of *Salmonella*. The presence of salmonellae in all types of animals may be a transient change in the bowl flora with no apparent illness or, at the other extreme, salmonellosis may be a persistent disease with high mortality (Edwards and Galton, 1967). Carriage or infection with salmonellae in animals constitutes an important economic factor in terms of high losses to farm and agricultural industries and as an important public health hazard. For example, poultry flocks numbering in the hundreds of thousands may be housed in conditions that preclude any attempt to reduce morbidity and mortality to the chickens and possibly to humans from those chickens that eventually come to the marketplace (Will, Diesch, and Pomeroy, 1973). In 1970–1971, it was estimated that 280,000 cases of pet turtle–associated salmonellosis were occurring annually in the United States (Center for Disease Control, 1972) due to lack of a certification program to insure that only salmonellae-free turtles could be sold as pets.

Salmonella in the Natural Environment

Salmonellae are disseminated in the natural environment (water, soil, sometimes plants used as food) through human or animal excretion. Man and animals (either wild or domesticated) can excrete *Salmonella* either when clinically diseased or after having had salmonellosis, if they remain carriers. *Salmonella* organisms do not seem to multiply significantly in the natural environment (out of digestive tracts), but they can survive several weeks in water and several years in soil if conditions of temperature, humidity, and pH are favorable (Delage, 1960). However, Thomason, Dodd, and Cherry (1977) advance the theory, based on some environmental studies, that *Salmonella bareilly* may multiply as a free-living organism on vegetation or in the soil and is dispersed to weather pools by rainfall. It appears that Fair and Morrison (1967) were correct when they stated that the existence of naturally occurring potable surface waters is a myth.

Worldwide Problem of Salmonellosis

Long distance transportation of people has contributed to the worldwide distribution of *Salmonella* strains. It is common in a number of countries to detect exotic *Salmonella* (unusual serotypes or phage types) following summer vacations.

World foodstuff export-import also contributes to the international spread of *Salmonella* strains. A vigilant control of these exchanges is necessary, as even such ingredients as butter colors, cacao, coconut powder, and spices have been found contaminated with *Salmonella*. A few serotypes are still geographically limited (Kelterborn, 1967; Van Oye, 1964), but such limits are not absolute and in no way definitive as international exchanges increase.

Isolation

Although most *Salmonella* strains are able to grow on minimal medium, they are usually isolated on complex media. The strategy of isolation depends on whether the sample originates from a normally sterile area (blood, cerebrospinal fluid [CSF]) or from an area with a rich bacterial flora (digestive tract, environmental samples, animal carcasses, etc.).

Nonselective Isolation

When the sample is from a normally sterile specimen (e.g., blood, CSF, bone marrow), bacterial cells (if any) are likely to be in low number in cases of salmonellosis. Furthermore, in cases of infection, the isolation method should allow recovery of any other infectious agents. Any commercially available nutrient broth may be used, including brain heart infusion broth and tryptic soy broth. One hundred milliliters of broth are then inoculated with 10 ml of the patient's blood. Castañeda flasks can be used in a multipurpose blood culturing method: A flat-sided 100-ml rectangular bottle contains an agar layer (e.g., tryptose agar) on one of the narrow side walls and 10 ml tryptose broth (containing 2% sodium citrate). The double medium is inoculated with 10 ml of the patient's blood. On the first day and every other day, the broth-blood mixture is allowed to flow over the agar layer. Otherwise, the bottle is incubated in an upright position. This method allows observation of colonies on the agar layer without opening the bottle. Growth is controlled microscopically and biochemical identification can proceed after examination of a Gram-stained preparation.

Enrichment Procedure and Selective Isolation

When the sample is from a specimen normally associated with a bacterial flora (fecal samples, autopsy samples, environmental samples), enrichment and selective isolation are needed.

Enrichment—i.e., an increased ratio of *Salmonella* cells to other bacterial cells—can be obtained using a number of liquid media (see Edwards and Ewing, 1972), of which modified tetrathionate broth and selenite broth are used worldwide.

A number of plating media have been devised for the isolation of *Salmonella*. Some media are differential and nonselective, i.e., they contain lactose with a pH indicator, but do not contain any inhibitor for non-*Salmonella* (e.g., bromocresol purple lactose agar). Other media are differential and slightly selective, i.e., in addition to lactose and a pH indicator, they contain an inhibitor for nonenterics (e.g., MacConkey agar [bile salts and violet crystal], Drigalski agar [violet crystal], eosin–methylene blue agar). The most commonly used media selective for *Salmonella* are SS agar (lactose, neutral red, thiosulfate, and ferric citrate, for differentiation; bile salts and brilliant green, as inhibitors), Wilson-Blair bismuth sulfite agar, Hektoen enteric medium (lactose, sucrose, salicin, bromothymol blue, Andrade's indicator, thiosulfate, and ferric ammonium citrate, for differentiation; bile salts and deoxycholate, as inhibitors), and Kristensen-Lester-Juergens brilliant green agar (lactose and phenol red, for differentiation; brilliant green, as inhibitor). All these media are commercially available. Selective media bearing the same name (e.g., SS agar) may have different selective powers according to their commercial origin (manufacturer) and the degree of quality control exercised by the manufacturer.

After overnight incubation at 37°C, selective and/or differential media need to be screened for "suspect" colonies. As most pathogenic *Salmonella* belong to subgenus I, suspect colonies will be medium or small colonies that are lactose negative and H_2S positive or negative. For example, H_2S-positive *Salmonella* will give uncolored colonies with a black center on SS agar, and H_2S-negative colonies will be colorless. Several "suspect" colonies need to be screened biochemically for *Salmonella*. A quick screening can be done by inoculating a whole colony into a few drops of urea broth and incubating at 37°C for 2 h. Several colonies are tested, and the biochemical characterization is continued only for urease-negative colonies (urease-positive colonies are likely to be *Proteus* spp.).

Identification

Antigenic cross-reactions between members of the family Enterobacteriaceae are such that no agglutination should be begun before any biochemical characterization. Media used for *Salmonella* identification are those used for identification of all Enterobacteriaceae. Formulas and instructions for use can be found elsewhere (Edwards and Ewing, 1972; Lennette, Spaulding, and Truant, 1974).

Most *Salmonella* strains are motile and peritrichous. *S. gallinarum-pullorum* is the constant nonmotile exception. However, nonmotile mutants (without flagella, or with nonfunctional flagella) may occur occasionally in any serotype. Most strains grow on nutrient agar as smooth colonies, 2–4 mm in diameter. Dwarf colonies (diameter 0.5–1 mm) are typically given by a few serotypes (e.g., *S. abortus-ovis, S. typhi-ovis*), or atypically by other serotypes. Most strains do not require any growth factor (prototrophic strains). However, auxotrophic strains occur, especially in host-adapted serotypes such as *S. typhi, S. paratyphi-A,* and *S. sendai* (all three strictly human), or *S. abortus-ovis* (strictly ovine).

Table 1 lists the characteristics shared by most *Salmonella* strains of subgenus I, Table 2 lists the particular characteristics of each *Salmonella* subgenus, and Table 3 lists characteristics shown by only a few serotypes. Atypical characters (e.g., lactose or sucrose fermentation) may be found in strains harboring metabolic plasmids (Le Minor, Coynault, and Pessoa, 1974; Le Minor et al., 1973).

Positive identification of *Salmonella* spp. should not be difficult to achieve. However, the organisms that are most often wrongly identified as *Salmonella* are *Hafnia, Citrobacter, Proteus mirabilis,* and *Pseudomonas putrefaciens (Alteromonas putrefaciens)*. *Hafnia alvei* (LDC-, ODC-, gas-positive) may resemble H_2S-negative *Salmonella* when grown at 37°C. At 20–30°C, ONPG and Voges-Proskauer tests are normally positive. Furthermore, a phage has been described that specifically lyses *Hafnia* strains (Guinée and Valkenburg, 1968). ONPG-negative strains of *Citrobacter* should be differentiated from *Salmonella* by their typical fecal odor, lack of lysine decarboxylase activity, and growth in KCN broth. *Proteus mirabilis* strains have no lysine decarboxylase, but have phenylalanine/tryptophan deaminase, urease, and often a positive Voges-Proskauer test. *Hafnia, Citrobacter,* and *Proteus mirabilis* are commonly found in fecal specimens. *Pseudomonas putrefaciens* occurs in spoiled food. It resembles *Salmonella* because of H_2S production and negative lactose, indole, and urease tests. However, this species is a strict aerobe, is oxidase positive, anaerogenic, lipase, and DNase positive, and has no lysine decarboxylase.

Use of a *Salmonella* polyvalent antiserum is of doubtful diagnostic usefulness because extensive cross-reaction occurs between most Enterobacteria-

Table 1. Characteristics shared by most *Salmonella* strains belonging to subgenus 1.

Characteristics present	Characteristics absent
Motile, Gram-negative bacteria	Acid from adonitol, sucrose,
Acid and gas from glucose,	salicin, lactose
mannitol, maltose, and sorbitol	ONPG test negative
Methyl red test positive	Indole test negative
Lysine decarboxylase	Urease
Ornithine decarboxylase	Voges-Proskauer test negative
H_2S from thiosulfate	Growth with KCN
Growth on Simmons' citrate	Phenylalanine and tryptophan
	deaminase
	Gelatin hydrolysis

Table 2. Characteristics of the four *Salmonella* subgenera.[a]

| | *Salmonella* subgenera | | | |
	I	II	III[b]	IV
Malonate	−	+	+	−
Gelatin	−	+	+	+
KCN	−	−	−	+
Dulcitol	+	+	−	−
ONPG test	−	− or ×[c]	+	−
D-Tartrate	+	− or ×	− or ×	− or ×

[a] From Kauffmann (1966a).
[b] Also called *Arizona*.
[c] Late and irregularly positive.

ceae. Use of phage Ol (Felix and Callow, 1943; Welkos, Schreiber, and Baer, 1974) may be more useful for food bacteriologists who need to screen a large number of colonies. Sensitivity and specificity of this test are satisfactory because, apart from *Salmonella* strains, Ol phage lyses only *Escherichia coli* strains (that are easily distinguished from *Salmonella*).

Serotyping can be initiated with the use of pooled sera and completed by the use of monospecific antisera. O-group antigens are identified first, and then minor O antigens. *Salmonella typhi* strains may be found nonagglutinable with O antisera; Vi-specific antiserum should then be used. Heating at 100°C for 10 min unmasks O antigen, and agglutination with O antisera may proceed. *S. typhi*, however, has typical biochemical characteristics that should allow prompt identification.

Completion of serotyping requires determination of both phase 1 and phase 2 flagellar (H) antigens. When the culture is a mixture of phase 1 and phase 2 bacterial cells, identification is achieved readily. Otherwise (if most flagella are in the same phase) phase inversion is required.

When the strain under study is not identified as a frequently occurring serotype, genus and subgenus biochemical identification should be rechecked.

Help from a national *Salmonella* center can be sought.

The Kauffmann-White scheme lists antigenic formulas of *Salmonella*. Antigens are given in this order: O factors, Vi (if present), H phase 1, and H phase 2. For example, *S. typhi* antigenic formula is 9,12,[Vi]:d:-, which means that *S. typhi* has O factors 9 (major) and 12 (minor), may or may not have Vi (indicated by brackets), has phase 1 flagellar antigen d, and has no phase 2 antigen. *S. paratyphi-B* antigenic formula is 1,4,[5],12:b:1,2. Underlining O factor 1 means it originates from phage conversion. O factor 5 is in brackets to indicate unconstant presence. H antigen is diphasic (b and 1,2). Factors either underlined or in brackets do not define a serotype.

About 2,000 *Salmonella* serotypes are listed in the Kauffmann-White scheme (there is a yearly increase of about 30 serotypes in the list). Serotypes are classified in groups according to major antigens. More antigenic factors have been observed that have no diagnostic value and are not listed in the Kauffmann-White scheme.

The most frequent O groups (received at the French National *Salmonella* Center) are B (62.9%), D (17.2%), C (11.2%), and E (5.4%). Twenty-six serotypes representing about 90% of *Salmonella* strains encountered (mostly in man) are listed with their antigenic formulas in Table 4. A more extended Kauffmann-White scheme can be found in *Bergey's Manual* (Buchanan and Gibbons, 1974).

Fluorescent-antibody (FA) techniques are as specific as agglutination. Use of *Salmonella* polyvalent antiserum has the same pitfalls in FA techniques as in agglutination. Screening for carriers of a given *Salmonella* serotype can be done with FA by using specific O antisera. Fecal smears can be used for this purpose. H antigens can also be detected using fluorescent antibodies, but smears must be prepared from cultures and not directly from fecal samples. In any case, FA techniques allow a screening for *Salmonella*, not a definite identification. Such identifi-

Table 3. Characteristics of a few *Salmonella* serotypes.

| | Serotypes | | | | |
	S. typhi	*S. paratyphi-A*	*S. cholerae-suis*	*S. gallinarum*[a]	*S. pullorum*[a]
Gas from glucose	−	+	+	−	+
H₂S	weak	−[b]	−[c]	−	−
Lysine decarboxylase	+	−	+	+	+
Ornithine decarboxylase	−	+	+	−	+
Simmons' citrate	−	−	(+)[d]	−	−
Motility	+	+	+	−	−

[a] *S. gallinarum* and *S. pullorum* are usually considered as the same serotype. Strains with intermediate biochemical characters occur.
[b] Strains from Asia are often H₂S positive.
[c] *S. cholerae-suis* var. *kunzendorf* is H₂S positive.
[d] Delayed positive.

Table 4. Abbreviated Kauffmann-White scheme of *Salmonella* serotypes.

Salmonella serotype	Somatic (O) antigen	Flagellar (H) antigen	
		Phase 1	Phase 2
Group A			
S. paratyphi-A	1,2,12	a	—
Group B			
S. paratyphi-B	1,4,[5],12	b	1,2
S. wien	4,12	b	1,w
S. saint-paul	1,4,[5],12	e,h	1,2
S. derby	1,4,12	f,g	—
S. agona	4,12	f,g,s	—
S. typhimurium	1,4,[5],12	i	1,2
S. bredeney	1,4,12,27	1,v	1,7
S. brandenburg	4,12	1,v	e,n,z_{15}
S. heidelberg	1,4,[5],12	r	1,2
S. coeln	4,[5],12	y	1,2
Group C$_1$			
S. cholerae-suis	6,7	c	1,5
S. montevideo	6,7	g,m,[p],s	—
S. thompson	6,7	k	1,5
S. infantis	6,7	r	1,5
Group C$_2$			
S. newport	6,8	e,h	1,2
S. bovis-morbificans	6,8	r	1,5
Group D$_1$			
S. typhi	9,12,[Vi]	d	—
S. enteritidis	1,9,12	g,m	—
S. dublin	1,9,12	g,p	—
S. panama	1,9,12	1,v	1,5
Group E$_1$			
S. anatum	3,10	e,h	1,6
S. zanzibar	3,10	k	1,5
S. london	3,10	1,v	1,6
Group E$_4$			
S. senftenberg	1,3,19	g,[s],t	—

cation would require isolation of the strain and biochemical characterization. However, the use of immunofluorescence for mass screening of foodstuffs, feces, and water offers an economical alternative to the conventional procedures (Cherry et al., 1975).

Subdivision of *Salmonella* Serotypes

The most clinically significant *Salmonella* serotypes can be subdivided for epidemiological purposes into biotypes, phage types, bacteriocin types, and antibiotypes.

Biotypes (biovars) are differentiated by biochemical characters such as D-xylose fermentation and tetrathionate reduction. For example, the most frequent biotype of *S. typhi* ferments xylose and reduces tetrathionate.

Phage types are determined by testing the susceptibility of a *Salmonella* strain to a collection of bacteriophages. Phages are used at a "routine test dilution" (RTD) calibrated on homologous bacteria. Phage typing of *Salmonella* strains that possess Vi antigen (*S. typhi, S. paratyphi-C,* and occasionally *S. dublin*) uses a collection of variants of Vi II phage (Craigie and Yen, 1938). Other phage-typing systems use different phages isolated from sewage or produced by lysogenic bacteria. Phage typing of *S. paratyphi-B* (Felix and Callow, 1943) and *S. typhimurium* (Anderson, 1964) is best performed by phage-typing centers. Phage typing is presently the most sensitive epidemiological marker for *Salmonella typhi* (96 phage types) and *S. typhimurium*

(125 phage types). Phage typing of other serotypes is occasionally devised to solve local epidemiological problems (Rische, 1973). Phage types may be changed by lysogenization or plasmid transfer (reviewed in Le Minor, 1968).

Bacteriocin types may be based on the production of, or susceptibility to, bacteriocins. *Salmonella* spp. may produce bacteriocins active on *Escherichia coli* (Fredericq, 1952).

Patterns of susceptibility to antibiotics (antibiogram) may also help to subdivide other infraspecific divisions.

Bacteriocin production and antibiotic resistance are most often determined by plasmids, and are then prone to variation (loss of plasmid). However, characterization of plasmids harbored by clinical strains may be of epidemiological interest when the study is limited in time and space.

Antibiotic Susceptibility

During the last decade, antibiotic resistance and multiresistance of *Salmonella* spp. have increased a great deal. The cause appears to be the increased and indiscriminate use of antibiotics in the treatment of man and animals and the addition of eutrophic antibiotics to the food of breeding animals. Plasmid-borne antibiotic resistance is very frequent among *Salmonella* strains involved in pediatric epidemics (*S. typhimurium, S. panama, S. wien, S. infantis*). Resistance to ampicillin, streptomycin, kanamycin, chloramphenicol, tetracycline, and sulfonamides is commonly observed. Colistin resistance has not yet been observed. Until 1972, *S. typhi* had remained susceptible to antibiotics, including chloramphenicol (the antibiotic most commonly used against typhoid). In 1972, a widespread epidemic in Mexico was caused by a chloramphenicol-resistant *S. typhi*. More chloramphenicol-resistant strains were then isolated in India, Thailand, and Vietnam (Anderson, 1975). Possible importation or appearance of chloramphenicol-resistant strains is a real threat. *Salmonella* strains should be systematically checked for antibiotic resistance to aid in the choice of an efficient drug when needed and to detect any change in antibiotic susceptibility of strains (either from animal or human source). Indiscriminate distribution and use of antibiotics should be discouraged.

Serological Diagnosis of *Salmonella* Infections

Agglutinins for *Salmonella* can be demonstrated in the serum of patients affected with serious salmonellosis (Widal's serodiagnosis). Anti-O and anti-H agglutinins are detected separately. O-antigenic suspensions are alcohol-killed *Salmonella* (H antigen is then destroyed), and H-antigenic suspensions are highly flagellated, formalin-killed *Salmonella*. Because of the wide diversity of *Salmonella* serotypes, serological diagnosis is normally limited to the detection of anti-O and anti-H agglutinins for *S. typhi, S. paratyphi-A, S. paratyphi-B,* and sometimes *S. paratyphi-C*. During typhoid, anti-O agglutinins can be detected after the 8th day of the disease and can attain a titer of about 1/400. Anti-O agglutinins quickly disappear after cure. Anti-H agglutinins can be detected after the 10th day and reach a titer of about 1/1,600. After recovery, H antibodies decrease to 1/200 and are still detectable several years after. The same observation applies to people who were immunized against typhoid and paratyphoid. Therefore it is wise to perform the Widal test on paired sera to detect a rise in titer that would indicate infection as opposed to past immunization.

Agglutinins are produced in the patient's body in response to bacterial antigens. When antigens from two different bacteria are related, cross-agglutination will occur (e.g., O antigen of *S. typhi* and *S. paratyphi-B,* with a common factor 12). The evolution of agglutinins in the course of the disease should be taken into account for a better interpretation of the serological test. Agglutinins may also be produced in response to any *Salmonella* infection including asymptomatic carriage, and the serological results will depend on the antigenic structure of the infecting *Salmonella* (e.g., infection with *S. enteritidis* will result in the production of anti-O agglutinins that react with *S. typhi* O antigens). Serological investigation is designed to help orient the diagnosis of a major occurrence of salmonellosis; this may be important in countries where proper isolation of bacteria is difficult to achieve, but serological testing requires the use of standardized antigens, tube agglutination procedures, and control sera. However, diagnosis may be certain only when the infectious agent has been isolated and identified.

National and International Surveillance of *Salmonella*

At the national level, surveillance of *Salmonella* is done by a *Salmonella* center usually designated by each country's government. A *Salmonella* phage-typing center and a *Shigella* center are often associated with the *Salmonella* center. National centers collect epidemiological information from clinical bacteriology, veterinary, food, and water control laboratories. These centers should provide assistance in the identification of infrequent serotypes or atypical strains, and in epidemiologically oriented studies (e.g., *S. typhi* phage typing). The national centers should be affiliated with those government offices (department or ministry) in charge of public health and agriculture, with foreign national centers (especially of countries that exchange travelers and food), and with the World Health Organization

(WHO). WHO also recognizes International and Regional Collaborating Centers for Reference and Research on *Salmonella* that collect information from national centers, keep a collection of reference strains (and dispatch them to national centers upon request), study unusual strains and those suspected of belonging to a new serotype, and publish a yearly supplement to the Kauffmann-White scheme and a regular, complete, up-to-date scheme.

Literature Cited

Achard, C., Bensaude, R. 1896. Infections paratyphoïdiques. Bulletin et Mémoires de la Société Médicale des Hôpitaux de Paris. **13:**820–833.

Anderson, E. S. 1964. The phage typing of salmonellae other than *S. typhi*, pp. 88–110. In: Van Oye, E. (ed.), The world problem of salmonellosis. The Hague: Junk.

Anderson, E. S. 1975. The problem and implications of chloramphenicol resistance in the typhoid bacillus. Journal of Hygiene **74:**289–299.

Andrewes, F. W. 1922. Studies on group-agglutination. I. The *Salmonella* group and its antigenic structure. Journal of Pathology and Bacteriology **25:**505–521.

Andrewes, F. W. 1925. Studies on group-agglutination. II. The absorption of agglutinins in the diphasic Salmonellas. Journal of Pathology and Bacteriology **28:**345–359.

Brachman, B. J., Low, K. B., Taylor, A. L. 1976. Recalibrated linkage map of *Escherichia coli* K-12. Bacteriological Reviews **40:**116–167.

Bretonneau, P. F. 1822. Traité de la dothénentérie et de la spécificité. Edited by L. Dubreuil-Chambardel from original (1822) manuscripts. 1922. Paris: Vigot.

Buchanan, R. E., Gibbons, N. E. (eds.). 1974. Bergey's manual of determinative bacteriology, 8th ed. Baltimore: Williams & Wilkins.

Budd, W. 1856. Cited in Kolle, W., Wassermann, A. 1903. Handbuch der pathogenen Mikroorganismen **2:**291. Jena: Gustav Fischer Verlag, 6 vols.

Castellani, A. 1902. Die Agglutination bei gemischter Infection und die Diagnose der letzteren. Zeitschrift für Hygiene und Infektionskrankheiten **40:**1–20.

Center for Disease Control. 1972. Morbidity and Mortality Weekly Report **21**(52).

Chantemesse, A., Widal, F. 1888. De l'immunité contre le virus de la fièvre typhoïde conférée par des substances solubles. Annales de l'Institut Pasteur **2:**55–59.

Cherry, W. B., Thomason, B. M., Gladden, J. B., Holsing, N, Muslen, A. M. 1975. Detection of salmonellae in foodstuffs, feces, and water by immunofluorescence. Annals of the New York Academy of Sciences **254:**350–368.

Craigie, J., Yen, C. H. 1938. The demonstration of types of *B. typhosus* by means of preparations of type II Vi phage. 2. The stability and epidemiological significance of V form types of *B. typhosus*. Canadian Public Health Journal **29:**484–496.

Crosa, J. F., Brenner, D. J., Ewing, W. H., Falkow, S. 1973. Molecular relationships among the *Salmonellae*. Journal of Bacteriology **115:**307–315.

Delage, B. 1960. Survie des salmonelles dans la terre. Bulletin de l'Académie Nationale de Médecine **144:**686–689.

Eberth, C. J. 1880. Die Organismen in den Organen bei Typhus abdominalis. Archiv für Pathologische Anatomie und Physiologie und für Klinische Medicin **81:**58–74.

Edwards, P. R., Ewing, W. H. 1972. Identification of Enterobacteriaceae, 3rd ed. Minneapolis: Burgess.

Edwards, P. R., Galton, M. M. 1967. Salmonellosis, pp. 1–64. In: Brandly, C. A., Cornelius, C. (eds.), Advances in veterinary medicine, vol. 11. New York, London: Academic Press.

Fair, J. F., Morrison, S. M. 1967. Recovery of bacterial pathogens from high quality surface water. Water Resources Research **3:**799–803.

Felix, A., Pitt, R. M. 1934. A new antigen of *B. typhosus*. Its relation to virulence and to active and passive immunisation. Lancet **ii:**186–191.

Felix, A., Callow, B. R. 1943. Typing of paratyphoid B bacilli by means of Vi bacteriophage. British Medical Journal **ii:**127–130.

Fair, J. F., Morrison, S. M. 1967. Recovery of bacterial pathogens from high quality surface water. Water Resources Research **3:**799–803.

Fredericq, P. 1952. Recherche des propriétés lysogènes et antibiotiques chez les *Salmonella*. Comptes Rendus de la Société de Biologie **146:**298–300.

Gaffky, C. 1884. Cited by Kauffmann, F. 1978. Das Fundament. Copenhagen: Munksgaard.

Gaffky, C., Paak, C. 1885. Cited by: Uhlenhuth, P., Hübner, E. 1913. In: Kolle, W., von Wassermann, A. (eds.), Handbuch der pathogenen Mikroorganismen, 2nd ed., vol. 3, p. 1037. Jena: Gustav Fischer Verlag.

Gärtner, A. 1888. Cited by: Kauffmann, F. 1978. Das Fundament. Copenhagen: Munksgaard.

Gruber, M., Durham, H. E. 1896. Eine neue Methode zur raschen Erkeunung des choleras vibrio und des Typhusbacillus. Münscher Medizinische Wochenschrift **43:**285–286.

Grunbaum, A. S. 1896. Preliminary note on the use of the agglutinative action of human action for the diagnosis of enteric fever. Lancet **16:**806–807.

Guinée, P. A. M., Valkenburg, J. J. 1968. Diagnostic value of a *Hafnia*-specific bacteriophage. Journal of Bacteriology **96:**564

Gwyn, N. B. 1898. On infection with a paracolon bacillus in a case with all the clinical features of typhoid fever. Johns Hopkins Hospital Bulletin **9:**54–56.

Iino, T., Lederberg, J. 1964. Genetics of *Salmonella*, pp. 110–142. In: Van Oye, E. (ed.), The world problem of salmonellosis. The Hague: Junk.

Johnson, R., Colwell, R. R., Sakazaki, R., Tamura, K. 1975. Numerical taxonomy study of the *Enterobacteriaceae*. International Journal of Systematic Bacteriology **25:**12–37.

Kauffmann, F. 1941. Die Bakteriologie der Salmonella-Gruppe. Copenhagen: Munksgaard.

Kauffmann, F. 1956. A new antigen of *S. paratyphi B* and *S. typhi-murium*. Acta Pathologica et Microbiologica Scandinavica **39:**299–304.

Kauffmann, F. 1957. On the T antigen of *Salmonella bareilly*. Acta Pathologica et Microbiologica Scandinavica **40:**343–344.

Kauffmann, F. 1963. Zur Differentialdiagnose der *Salmonella*-sub-genera I, II und III. Acta Pathologica et Microbiologica Scandinavica **58:**109–113.

Kauffmann, F. 1966a. Zur Klassifizierung und Nomenklatur der *Salmonella* sub-genera I-IV. Zentralblatt für Bakteriologie, Parasitenkunde, Infektionskrankheiten und Hygiene, Abt. 1:2, Referate **202:**482–483.

Kauffmann, F. 1966b. The bacteriology of *Enterobacteriaceae*. Baltimore: Williams & Wilkins.

Kauffmann, F. 1978. Das Fundament. Copenhagen: Munksgaard.

Kelterborn, E. 1967. *Salmonella* species. Leipzig: Hirzel.

Koch, R. 1881. Cited by: Kauffmann, F. 1978. Das Fundament. Copenhagen: Munksgaard.

Le Minor, L. 1968. Lysogénie et classification des *Salmonella*. International Journal of Systematic Bacteriology **18:**197–201.

Le Minor, L. 1971. Connaissances actuelles sur les déterminants génétiques des antigenes des *Salmonella*, pp. 341–348. In: Pérez-Miravete, A., Peláez, D. (eds.) Recent advances in microbiology. 10th International Congress of Microbiology. Mexico: Asociación Mexicana de Microbiología.

Le Minor, L., Coynault, C., Pessoa, G. 1974. Déterminisme plasmidique du caractère atypique "lactose positif" de souches de *S. typhimurium* et *S. oranienburg* isolées au Brésil lors d'épidémies de 1971 à 1973. Annales de Microbiologie **125A**:261–285.

Le Minor, L., Rohde, R., Taylor, J. 1970. Nomenclature des *Salmonella*. Annales de l'Institut Pasteur **119**:206–210.

Le Minor, L., Coynault, C., Rohde, R., Rowe, B., Aleksic, S. 1973. Localisation plasmidique du déterminant génétique du caractère atypique "saccharose+" des *Salmonella*. Annales de Microbiologie **124B**:295–306.

Lennette, E. H., Spaulding, E. H., Truant, J. P. (eds.). 1974. Manual of clinical microbiology, 2nd ed. Washington, D.C.: American Society for Microbiology.

Lignières, J. 1900. Maladies du porc. Bulletin de la Société Central de Médecine Vétérinaire **18**:389–431.

Loeffler, F. 1892. Quoted by: Kauffmann, F. 1978. Das Fundament. Copenhagen: Munksgaard.

Louis, P. C. A. 1829. Fièvre typhoïde. Paris:. Baillière.

Lüderitz, O., Staub, A. M., Westphal, O. 1966. Immunochemistry of O and R antigens of *Salmonella* and related *Enterobacteriaceae*. Bacteriological Reviews **30**:192–255.

Lüderitz, O., Westphal, O., Staub, A. M., Nikaido, H. 1971. Isolation and chemical and immunological characterization of bacterial lipopolysaccharides, pp. 145–223. In: Weinbaum, G., Kadis, S., Ajl, S. J. (eds.), Microbial toxins, vol. IV. Bacterial endotoxins. New York, London: Academic Press.

Mäkelä, P. H., Stocker, B. A. D. 1969. Genetics of polysaccharide biosynthesis. Annual Review of Genetics **3**:291–322.

Marmur, J., Falkow, S., Mandel, M. 1963. New approaches to bacterial taxonomy. Annual Review of Microbiology **17**:329–372.

Petit, A., Serres, E. R. A. 1813. Traité de la fièvre entéromésentérique observée, reconnue et signalée publiquement á l'Hôtel-Dieu de Paris dans les années 1811–1812–1813. Paris: Hacquart.

Pettenkofer, M. 1868. Cited in: Kolle, W., Wassermann, A. 1903. Handbuch der pathogenen Mikroorganismen **2**:291. Jena: Gustav Fischer Verlag, 6 vols.

Pfeiffer, R., Kolle, W. 1896. Zur Differentialdiagnose des Typhus-bacillus vermittels Serum der gegen Typhus immunisierten Thiere. Deutsche Medizinische Wochenschrift **22**:185–186.

Reilly, J., Rivalier, E., Compagnon, A., Laplane, R., du Buit, H. 1935. Sur la pathogénie de la dothienentérie. Le rôle du système nerveux végétatif dans la genèse des lésions intestinales. Annales de Médecine **37**:321–358.

Rische, H. 1973. Lysotypie. Infektionskrankheiten und ihre Erreger. Eine Sammlung von Monographien, vol. 14. Jena: Gustav Fischer Verlag.

Salmon, D. E. 1885. Report of the Chief of the Bureau, p. 228. In: First Annual Report of the Bureau of Animal Industry for the year 1884. Washington, D.C.: Government Printing Office.

Salmonella Subcommittee of the Nomenclature Committee of the International Society for Microbiology. 1934. The genus *Salmonella* Lignières 1900. Journal of Hygiene **34**:330–350.

Sanderson, K. E. 1972. Linkage map of *Salmonella typhimurium*. Bacteriological Reviews **36**:555–586.

Schottmüller, H. 1901. Weitere Mittheilungen über mehrere das Bild des Typhus bietende Krankheitsfälle, hervorgerufen durch typhusähnliche Bacillun. (Paratyphus). Zeitschrift für Hygiene und Infektionskrankheiten **36**:368–396.

Smith, T., Reagh, A. L. 1903. The non-identity of agglutinins acting upon the flagella and upon the body of bacteria. Journal of Medical Research **10**:83–119.

Thomason, B.M., Dodd, D. J., Cherry, W. B. 1977. Increased recovery of salmonellae from environmental samples enriched with buffered peptone water. Applied and Environmental Microbiology **34**:270–273.

Van Oye, E. (ed.) 1964. The world problem of salmonellosis. The Hague: Junk.

Véron, M., Le Minor, L. 1975a. Nutrition et taxonomie des *Enterobacteriaceae* et bactéries voisines. II. Résultats d'ensemble et classification. Annales de Microbiologie **126B**:111–124.

Véron, M., Le Minor, L. 1975b. Nutrition et taxonomie des *Enterobacteriaceae* et bactéries voisines. III. Caractères nutritionnels et différenciation des groupes taxonomiques. Annales de Microbiologie **126B**:125–147.

Weil, E., Felix, A. 1918a. Untersuchungen über die gewöhnlichen *Proteus* Stämme und ihre Beziehungen zu den X Stämme. Wiener Klinische Wochenschrift **11**:637–639.

Weil, E., Felix, A. 1918b. Über die Doppelnatur der Rezeptoren beim Paratyphus B. Wiener Klinische Wochenschrift **11**:986–988.

Weil, E., Felix, A. 1920. Über den Doppeltypus der Rezeptoren in der Typhus-Paratyphus-Gruppe. Zeitschrift für Immunitätsforschung und experimentelle Therapie **29**:24.

Welkos, S., Schreiber, M., Baer, H. 1974. Identification of *Salmonella* with the O-1 bacteriophage. Applied Microbiology **28**:618–622.

White, P. B. 1926. Further studies on the *Salmonella* group. Great Britain Medical Research Council Special Report No. 103. London: Her Majesty's Stationery Office.

Widal, F. 1896. Séro-diagnostic de la fièvre typhoïde. Bulletin et Mémoires de la Société Médicale des Hôpitaux de Paris **13**:561–566.

Widal, F., Sicard., A. 1897. Étude sur le sérodiagnostic et sur la réaction agglutinante chez les typhiques. Annales de l'Institut Pasteur **11**:353–432.

Will, L. A., Diesch, S. L., Pomeroy, B. S. 1973. Survival of *Salmonella typhimurium* in animal manure disposal in a model oxidation ditch. American Journal of Public Health **63**:322–326.

Wright, A. E., Semple, D. 1897. Remarks on vaccination against typhoid fever. British Medical Journal **1**:256–259.

CHAPTER 93

The Genus *Klebsiella* (Medical Aspects)

IDA ØRSKOV

Klebsiella is a genus in the family Enterobacteriaceae. In general, it is currently accepted in the medical literature that bacteria earlier known as *Aerobacter aerogenes (Bacterium lactis aerogenes)* are placed in the genus *Klebsiella* with *K. pneumoniae* (Friedländer's bacillus), as proposed many years ago by several authors, e.g., Edwards (1929) and Kauffmann (1949). However, there is no agreement with respect to subdivision of the genus. In *Bergey's Manual of Determinative Bacteriology*, eighth edition (Buchanan and Gibbons, 1974), Ørskov describes three species only: *K. pneumoniae, K. ozaenae,* and *K. rhinoscleromatis,* a classification that agrees with that of Edwards and Ewing (1972). This system is used by American and many western European workers, but only seldom in publications from the United Kingdom. As a consequence of Duguid's suggestion (1959) that saprophytic *Klebsiella aerogenes* strains were fimbriate, while pathogenic *K. pneumoniae, K. rhinoscleromatis,* and *K. ozaenae* strains were nonfimbriate, Cowan et al. (1960) reviewed the characters of these bacteria and concluded that six species could be recognized: *K. aerogenes, K. edwardsii, K. atlantae, K. pneumoniae (sensu stricto), K. ozaenae,* and *K. rhinoscleromatis.* Determination of fimbriation was not essential for species definition. Later Bascomb et al. (1971)—another group from the United Kingdom—defined six taxa in the genus *Klebsiella* based on numerical classification. One of these taxa was *K. pneumoniae (sensu stricto),* another one contained *K. aerogenes, K. edwardsii,* and *K. atlantae* and, in addition, indole-forming *Klebsiella* strains (*K. oxytocum*) that were excluded from the *Klebsiella* group by Cowan et al. (1960). In a recent publication on *Klebsiella* taxonomy by Barr (1977) it is stressed that Ørskov (1974) included *K. pneumoniae (sensu stricto)* in *K. pneumoniae (sensu lato).* The reason for this is that I consider it highly probable that the four capsulated strains—the fifth is noncapsulated—that make up the *K. pneumoniae (sensu stricto)* taxon by Bascomb et al. (1971) all belong to capsule type 3 like the strains examined by Cowan et al. (1960); in that case they might be descendents of the same bacterial cell—in other words belong to the same clone. If this is so, it appears more reasonable to consider such strains, which are VP- and KCN-negative, as belonging to a special serofermentative type or biotype. The same argument could be used in favor of giving up the species *K. rhinoscleromatis* (also capsule type 3) and, perhaps, the species *K. ozaenae,* although more than one clone is involved. In this connection it should be mentioned that Brenner, Steigerwalt, and Fanning (1973) found 80–90% DNA relatedness between *K. pneumoniae (sensu lato), K. ozaenae, K. rhinoscleromatis,* and *K. edwardsii.* No *K. pneumoniae (sensu stricto)* according to Cowan et al. (1960) was included in that study. Similarly, Jain, Radsak, and Mannheim (1974) reported that DNA reassociation values do not allow species differentiation within the true indole-negative *Klebsiella* strains.

It is pointed out above that different nomenclature of *Klebsiella* strains is used by different authors. Slopek and Durlakowa (1967) and Durlakowa, Lachowicz, and Slopek (1967) grouped *Klebsiella* strains in the same six taxa as Cowan et al. (1960), but the names and ranks were somewhat changed. However, in many places outside the United Kingdom (Richard, 1973) the system of Cowan et al. (1960) and Bascomb et al. (1971) is considered to give more disadvantages than advantages. According to Barr (1977), some British scientists, working in a clinical environment, have accepted the similarity of several *Klebsiella* species and refer to *K. aerogenes/oxytoca/edwardsii/atlantae* as an identifiable group that—as Barr says—on the basis of Bascomb's study (1971) would be adequately described as *K. aerogenes.* The result would be the definition of two species, *K. aerogenes* and *K. pneumoniae (sensu stricto)* in the United Kingdom, in conflict with a single species, *K. pneumoniae (sensu lato),* in other European countries and in the United States. Barr sees the resolution of this problem as a necessity and suggests that the United Kingdom adopt the usage of *K. pneumoniae (sensu lato)* to include *K. aerogenes/oxytoca/edwardsii/atlantae* and *pneumoniae* and retain the ability to refer independently to Friedländer's bacillus as *K. pneumoniae (sensu stricto).*

Another problem is caused by the indole-forming and gelatin-liquefying *Klebsiella* strains that have been differently named and placed throughout the years. The name *Bacterium oxytocum* Flügge was revived by Lautrop (1956) for such strains. In early editions of *Bergey's Manual* (Bergey, 1923, 1934) they were named *Aerobacter oxytocum,* but they were not recognized in the later editions. In the eighth edition, indole- and gelatin-positive strains were included in the species *K. pneumoniae* and thus only considered as biotypes. Also, Fife, Ewing, and Davis (1965) regarded these strains as biotypes. Cowan et al. (1960) included no such strains in their study, in contrast to Bascomb et al. (1971) who found them to belong in the *K. aerogenes/oxytoca/ edwardsii/atlantae* taxon. However, some authors (Kaluzewski, 1967; Stenzel, Bürger, and Mannheim, 1972) have found that these strains were sufficiently different from other *Klebsiella* strains to be regarded as a separate species. DNA relatedness studies indicated that indole- and gelatin-positive *Klebsiella* strains (the *oxytocum* group) represented a distinct DNA homology group that should be established as a new genus of Enterobacteriaceae (Jain et al., 1974). This finding has been confirmed by Brenner (personal communication).

As a consequence of the above discussion, it is logical to propose that *Klebsiella* should be considered as a genus with only one species, while another genus should comprise the indole- and gelatin-positive *Klebsiella*-like strains.

Habitats

Strains of *Klebsiella pneumoniae* occur in the respiratory tract and particularly in the normal intestine of man and animals, but usually only in low numbers. Furthermore they can be found in soil, dust, air, and water. Formerly, these bacteria were only rarely considered to be pathogenic, but during the last 25 years they have become increasingly recognized as causes of clinical infections, particularly in hospitalized patients. The main reason for this is most probably the large-scale usage of chemotherapeutics and antibiotics. The infecting organisms may be eradicated, but the antibiotics always exert a selective pressure in favor of those resistant bacteria that are present and that carry plasmids with resistance determinants to the drug used and to other drugs. Some organisms may transfer such R plasmids not only to their own kind but to other species as well, and later infections caused by such resistant bacteria may be difficult to treat.

In addition to being the cause of a small percentage of acute pneumonia cases in man, *K. pneumoniae* is an opportunistic pathogen that can give rise to all kinds of infections. A large majority of patients with such infections are generally elderly and already compromised by another disease or surgical procedure and most of them have received antibiotic ther-

apy (Finland, 1973). The most common *Klebsiella* infections are those that involve the urinary tract; a large proportion of them are of nosocomial origin caused by strains endemic within hospitals (Ørskov, 1952). Instances of bacteremia may be of particular interest, because it is not uncommon that such cases are fatal (Steinhauer et al., 1966). A high percentage of *Klebsiella* bacteremias may originate from nosocomial urinary tract infections; it should also be mentioned that cases of nosocomial bacteremia can be caused by klebsiellae present, as contaminants e.g., on the stoppers of infusion flasks (Le Bouar, 1972).

Most frequently capsule type 1 has been isolated from cases of acute pneumonia due to *K. pneumoniae;* types 2 and 3 are also commonly involved (Finland, 1967; Hyde and Hyde, 1943; Julianelle, 1941; Perlman and Bullowa, 1941; Solomon, 1937). Capsular types 1-6 are, on the whole, most frequently associated with various infections of the respiratory tract. Steinhauer et al. (1966) and Eickhoff, Steinhauer, and Finland (1966), who also found these types to be the most common respiratory *K. pneumoniae* strains, did not consider them to be acquired in the hospital, unlike strains isolated elsewhere, and found them to be relatively susceptible to antibiotics. Strains of capsular types 1-6 are rarely isolated from fecal or urinary specimens and may thus have a special affinity for epithelial cells of the respiratory tract or may be more easily destroyed in the gut. *Klebsiella* capsular types above type 6 (and of type 2) are often associated with urinary tract infections, but are capable of causing all kinds of infections in man. Some types more than others may be incriminated as hospital types. Some of the types that have been reported to be endemic within hospitals by authors from different countries are capsular type 2 (Ørskov, 1952; Richard, 1973; Steinhauer et al., 1966), types 8 and 9 (Comninos, 1977; Ørskov, 1952), and type 24 (Comninos, 1977; Ørskov, 1952; Richard, 1973; Steinhauer et al. 1966). Some *Klebsiella* strains of capsular types 1, 2, and 5 are particularly found in association with metritis in mares (Dimock and Edwards, 1926; Edwards, 1928; Platt, Atherton, and Ørskov, 1976).

K. rhinoscleromatis can be cultured from cases of scleroma (a chronic granulomatous disease of the upper respiratory tract occurring most commonly in the nose). Scleroma does not occur frequently but is endemic in certain areas in eastern and central Europe and in South and Central America. Recently it was reported (Rees and Gregory, 1977) that 10 strains isolated from patients with clinical signs of the nasal form of rhinoscleroma were classified according to their biochemical properties and capsular antigen types as either *K. ozaenae* type 4, *K. rhinoscleromatis* type 3, or *K. pneumoniae* type 3, and it was suggested that further studies might reveal additional types. It should be mentioned, however, that opinions differ as to whether *Klebsiella* is the primary cause of this disease.

Ozena, an atrophic rhinitis with an unpleasant smell, is considered to be associated with *K. ozaenae* of capsular types 4, 5, or 6.

Isolation

Like all members of the Enterobacteriaceae family, *Klebsiella* grows readily on ordinary media. Many such media are easily obtainable, e.g., nutrient agar or blood agar, or more differential plating media such as eosin–methylene blue (EMB) agar, MacConkey agar, or bromothymol blue (BTB) agar that contain an indicator and a sugar, most often lactose. At Statens Seruminstitut, Copenhagen, BTB is the preferred medium, the formulation of which is given in Chapter 35, this Handbook. For formulae and instructions for preparation of other media, see Sedlak and Rische (1968), Edwards and Ewing (1972), and Lennette, Spaulding, and Truant (1974). Capsulated *Klebsiella* colonies are easily recognized on the basis of their morphology because *Klebsiella*—in contrast to most other bacteria able to grow on these media—will form mucoid dome-shaped colonies of varying degrees of stickiness. In cases where dense growth of other Enterobacteriaceae occurs, the plate should be left for 3 to 5 days at room temperature, after which time the *Klebsiella*, if present, will easily be seen as small shiny elevations in the dense growth. Colonies that are indistinguishable from those of *Klebsiella* may be formed by *Enterobacter,* particularly *E. aerogenes,* and by *Escherichia coli* strains with capsular K antigens of the type called the A form (Kauffmann, 1944; Ørskov et al., 1977).

If mucoid colonies of different yellowish shades are present on a BTB plate, this may indicate the presence of different serological and/or biochemical types. However, individual colonies from the same strain may show similar variation in the shade of color.

Klebsiella strains can be conserved in meat extract agar or on an egg yolk medium (Kauffmann, 1966). Such cultures kept at room temperature in the dark will be alive for 10 years or more. Some workers prefer to keep the cultures as freeze-dried or at −80°C in a broth medium with 10% glycerol. The latter method is probably the best way to avoid mutations and loss of plasmids.

Identification

The main biochemical characters of the groups that make up the family Enterobacteriaceae, as well as the tests used in the differentiation of the three species within the genus *Klebsiella,* are given in *Bergey's Manual,* eighth edition (Tables 8.3 and 8.11).

Strains are isolated, of course, which do not conform to any recognized group and such "*Klebsiella*-like" strains may easily be classified differently by different workers. Many biochemical reactions can be used for the subdivision of *K. pneumoniae* into biotypes. Ørskov (1957) found 23 biotypes among indole- and/or gelatin-negative strains based on the dulcitol, adonitol, sorbose, urease, and organic acid (d-tartrate and sodium citrate) tests. Indole- and/or gelatin-positive strains belonged to 10 biotypes. The experience in our laboratory, where we receive presumptive *Klebsiella* strains for capsule type determination, has shown that the organisms most easily mistaken for *K. pneumoniae* are *Enterobacter aerogenes* and *Enterobacter cloacae.* Biochemical reactions of particular interest for differentiation in this respect are listed in Table 1. However, the following comments should be added. In our laboratory it is not unusual for the MR test to be positive or weakly positive; this may be explained by the fact that some strains produce acetoin and 2,3-butanediol very slowly and in such small amounts that the MR test remains positive or is weakly positive at the time when the VP test has become positive. A negative VP test can, in some cases, be explained by disappearance of the acetoin before the test is read. Some single biochemical reactions will frequently deviate from the normal *Klebsiella* pattern, e.g., a negative urease or a positive ornithine test. When a strain behaves like a *K. pneumoniae* according to the table, but liquefies gelatin (always slowly), the indole reaction is nearly always positive as well and the strain can be classified as biotype *K. oxytoca* (Hugh, 1959; Kaluzewski, 1967; Lautrop, 1956; Stenzel et al., 1972; Véron and Le Minor, 1975). *K. pneumoniae* is active with respect to fermentation of sugars, but strains of the *oxytocum* group are even more active (Lautrop, 1956) and are further characterized by low temperature ranges for growth and gas formation as compared to *K. pneumoniae* (Stenzel, Bürger, and Mannheim, 1972). According to Gavini et al. (1977), the strains that constitute the *K. oxytoca* class (one of four classes in the genus *Klebsiella*) do not grow at all at 44.5°C. The ability to digest polypectate by indole-positive *Klebsiella* has been reported by von Riesen (1976) who discusses the significance of this ability for identification of *oxytocum* group strains, particularly because the gelatin liquefaction may be undetected as it is a slow reaction. For these *oxytocum* group variants Jain et al. (1974) proposed the establishment of a new genus based on DNA homology studies, as mentioned in the introduction. Ørskov (1955, 1957) found many indole-positive *Klebsiella* strains that did not liquefy gelatin. Similar strains were examined by Gavini et al. (1977) who found them negative in the pectinolysis test, in contrast to the *K. oxytoca* strains; they emphasized that most indole-positive, gelatin-negative strains were not of human

Table 1. Main similar and distinguishing characters of *Klebsiella pneumoniae,
Enterobacter aerogenes, and Enterobacter cloacae*.

Character or test	K. pneumoniae	E. aerogenes	E. cloacae
Adonitol	+	+	−
Dulcitol	d[a]	−	−
Inositol	+	+	−
KCN	+	+	+
Gelatin	−	(+)[b]	(+)
Citrate	+	+	+
Methyl red	−	−	−
Voges-Proskauer	+	+	+
Urease	+	−	−
Lysine decarboxylase	+	+	−
Arginine dihydrolase	−	−	+
Ornithine	−	+	+
Motility	−	+	+
Capsule	+	+/−[c]	−/+

[a] d, different reactions by different strains.

[b] (+), a slow gelatin liquefaction.

[c] +/−, most strains of *E. aerogenes* have capsules, while most strains that we classify as
E. cloacae are noncapsulated (= −/+); +, positive reaction in standard test or presence
of characteristic; −, negative reaction in standard test or absence of characteristic.

origin. DNA relatedness studies of such strains are needed.

One other problem concerns the differentiation between *K. pneumoniae*, *K. ozaenae*, and *K. rhinoscleromatis*. *K. pneumoniae* is biochemically active, in contrast to *K. rhinoscleromatis*, which can be characterized biochemically as very inactive. It is thus simple to differentiate between these two groups. A positive reaction, e.g., the malonate test, found in a *K. rhinoscleromatis* strain will also be positive in most *K. pneumoniae* strains (with the exception of the MR test, which tells that *K. rhinoscleromatis* lacks the ability to produce acetoin and 2,3-butanediol). *K. ozaenae* strains will give different reactions in many tests, but they are always anaerogenic and VP-negative, similar to *K. rhinoscleromatis*. Therefore a very active strain may be classified as *K. pneumoniae* and a less active strain as *K. rhinoscleromatis*. Serologically *K. rhinoscleromatis* belongs to capsule type 3, while *K. ozaenae* strains are of capsule types 4, 5, and 6. This fact helps to classify these strains, which supposedly occur rarely outside the upper respiratory tract.

Also it should be mentioned that strains of *K. pneumoniae* have been reported to be able to fix nitrogen, i.e., to convert N_2 to NH_3 under anaerobic conditions. This ability has not been found in *K. ozaenae*, *K. rhinoscleromatis*, *E. aerogenes*, or *E. cloacae* strains (Mahl et al., 1965). The presence of N_2-fixing *Klebsiella* strains in the intestine of man and animals has also been demonstrated (Bergersen and Hipsley, 1970). The regulation and the genetics of N_2 fixation in *K. pneumoniae* have been reviewed by Brill (1975). The *nif* genes determining this character are located on the chromosome but can be mobilized by an R factor and transferred to other organisms.

Klebsiella strains possess O (lipopolysaccharide) antigens and K (capsular) antigens that are acidic polysaccharides (Nimmich, 1968). As the number of O antigen types is small compared with the K antigen types, the serologic type determination is based on examination of K antigens using several serologic techniques. The last K antigen type established was numbered K82 (Ørskov and Fife-Asbury, 1977). For details of *Klebsiella* antigens and the common procedures of capsule type determination, see Kauffmann (1949, 1966), Edwards and Ewing (1972), and Ørskov and Ørskov (1978).

A few of the K antigen type strains are indole- and/or gelatin-positive, as type 26. According to the above-mentioned considerations, such strains may in the future be moved to a new independent species (*K. oxytocum*); it should be stressed, however, that it is not only *K. oxytocum* strains that belong to these particular antigen types. All K types have been found in true *K. pneumoniae* strains and probably also in *K. oxytocum* strains. Usually capsulated *E. aerogenes* strains react with one or more *Klebsiella* antisera. It is our experience that particularly type 68 is frequently found among both *K. oxytocum* and *E. aerogenes* strains. Many of the latter group originate from horses (Platt et al., 1976).

Differentiation of *Klebsiella* strains by phages isolated from stool has been proposed by Slopek et al. (1967). Other phages that are specific for *Klebsiella* capsules have been isolated (Stirm et al. 1971; Stirm, personal communication), but a typing system with such phages is far from being worked out.

The base ratio (% G+C) for *Klebsiella* is in the range of 53–57% (De Ley, 1970). Brenner et al. (1973) found the relative reassociation DNA values between *K. pneumoniae* capsule type 2 and five *Klebsiella* strains to be 80–90%, but only 56% between the same strain and *E. aerogenes*.

Seidler, Knittel, and Brown (1975) studied 107 *Klebsiella* strains, some of which were *K. pneumoniae* reference cultures, but most of which were obtained from different habitats such as humans, trees, rivers, etc., and found a base ratio from 54% to 59% G+C and a range of DNA relative reassociation to a human *K. pneumoniae* reference strain from 5% to 100%, and they concluded that *K. pneumoniae* is more heterogeneous than previously thought. Seidler et al. state that in terms of cultural reactions, serotyping, and genetic analysis most environmental isolates are indistinguishable from strains of human origin. Biochemically, the most atypical group of strains comes from fir trees, but all strains of that group were confirmed—according to the authors—as *K. pneumoniae* serotype 68 by the Center for Disease Control, Atlanta, Georgia. All such strains should most probably not be classified as *K. pneumoniae*, because a reaction with *Klebsiella* type 68 antiserum does not warrant—as stated above—classifying a strain as *K. pneumoniae*. For a better understanding of the taxonomy of *Klebsiella*-like strains their origin is highly important, as emphasized by Gavini et al. (1977).

Most *Klebsiella* strains isolated from nosocomial infections contain R factors that carry resistance determinants, most often to several drugs. One of the most important factors associated with acquired institutional carriage or infection caused by *Klebsiella* is the initiation of proper antibiotic therapy. It was the finding of a striking uniformity in resistance pattern of *Klebsiella* strains isolated in 1950 from patients in the same ward that led to the demonstration of nosocomial infections with *Klebsiella* in urinary tract infections (Ørskov, 1952, 1954). This was at a time when the existence of R factors was unknown. Numerous reports have appeared throughout the years concerning these matters. Here I shall content myself with referring to the book *Multiple Drug Resistance* by Falkow (1975). This author points out that in general *Klebsiella* is a better recipient for R factors than, e.g., *Escherichia coli*, a fact that may have made *Klebsiella* a culprit in serious nosocomial epidemic disease.

Literature Cited

Barr, J. G. 1977. *Klebsiella:* Taxonomy, nomenclature and communication. Journal of Clinical Pathology **30**:943–944.

Bascomb, S., Lapage, S. P., Willcox, W. R., Curtis, M. A. 1971. Numerical classification of the tribe *Klebsielleae*. Journal of General Microbiology **66**:279–295.

Bergersen, F. J., Hipsley, E. H. 1970. The presence of N_2-fixing bacteria in the intestines of man and animal. Journal of General Microbiology **60**:61–65.

Bergey, D. H. (ed.). 1923. Bergey's manual of determinative bacteriology, pp. 206–207. Baltimore: Williams & Wilkins.

Bergey, D. H. (ed.). 1934. Bergey's manual of determinative bacteriology, 4th ed., pp. 357–358. London: Baillière, Tindall and Cox.

Brenner, D. J., Steigerwalt, A. G., Fanning, G. R. 1973. Differentiation of *Enterobacter aerogenes* from *Klebsiella* by deoxyribonucleic acid reassociation. International Journal of Systematic Bacteriology **22**:193–200.

Brill, W. J. 1975. Regulation and genetics of bacterial nitrogen fixation. Annual Review of Microbiology **29**:109–131.

Buchanan, R. E., Gibbons, N. E. (eds.). 1974. Bergey's manual of determinative bacteriology, 8th ed., pp. 290–340. Baltimore: Williams & Wilkins.

Comninos, G. 1977. Études biochimiques et sérologiques de 100 souches de *Klebsiella* isolées du Centre Hospitalier Universitaire Vaudois. Thèse. Zurich: Juris.

Cowan, S. T., Steel, K. J., Shaw, C., Duguid, J. P. 1960. A classification of the *Klebsiella* group. Journal of General Microbiology **23**:601–612.

De Ley, J. 1970. Reexamination of the association between melting point, buoyant density, and chemical base composition of deoxyribonucleic acid. Journal of Bacteriology **101**:738–754.

Dimock, W. W., Edwards, P. R. 1926–27. Genital infections in mares by an organism of the Encapsulatus group. Journal of the American Veterinary Medical Association **70**:469–480.

Duguid, J. P. 1959. Fimbriae and adhesive properties of *Klebsiella* strains. Journal of General Microbiology **21**:271–286.

Durlakowa, I., Lachowicz, Z., Slopek, S. 1967. Biochemical properties of *Klebsiella* bacilli. Archivum Immunologiae et Therapiae Experimentalis **15**:490–496.

Edwards, P. R. 1928. The relation of encapsulated bacilli found in metritis in mares to encapsulated bacilli from human sources. Journal of Bacteriology **15**:245–266.

Edwards, P. R. 1929. Relationships of the encapsulated bacilli with special reference to *Bacterium aerogenes*. Journal of Bacteriology **17**:339–353.

Edwards, P. R., Ewing, W. H. 1972. Identification of Enterobacteriaceae, 3rd ed. Minneapolis: Burgess.

Eickhoff, T. C., Steinhauer, B. W., Finland, M. 1966. The *Klebsiella-Enterobacter-Serratia* division. Biochemical and serologic characteristics and susceptibility to antibiotics. Annals of Internal Medicine **65**:1163–1179.

Falkow, S. 1975. Infectious multiple drug resistance. London: Pion.

Fife, M. A., Ewing, W. H., Davis, B. R. 1965. The biochemical reactions of the tribe *Klebsiellae*. Communicable Disease Center Monograph. Atlanta, Georgia: U.S. Department of Health, Education and Welfare, Public Center Service.

Finland, M. 1967. *Klebsiella* pneumonia (Friedländer's Bacillus pneumonia), pp. 159–161. In: Beeson, P. B., McDermott, W. (eds.), Cecil Loeb textbook of medicine. Philadelphia and London: W. B. Saunders.

Finland, M. 1973. Excursions into epidemiology: Selected studies during the past four decades at Boston City Hospital. Review. Journal of Infectious Diseases **128**:76–124.

Gavini, F., Leclerc, H., Lefebvre, B., Ferragut, C., Izard, D. 1977. Étude taxonomique d'enterobactéries appartenant ou apparentées au genre *Klebsiella*. Annales de Microbiologie (Annales de l'Institut Pasteur) **128B**:45–59.

Hugh, R. 1959. *Oxytoca* group organisms isolated from otopharyngeal region. Canadian Journal of Microbiology **5**:251–254.

Hyde, L., Hyde, B. 1943. Primary Friedländer pneumonia. American Journal of Medical Sciences **205**:660–675.

Jain, K., Radsak, K., Mannheim, W. 1974. Differentiation of the *Oxytocum* group from *Klebsiella* by deoxyribonucleic acid–deoxyribonucleic acid hybridization. International Journal of Systematic Bacteriology **24**:402–407.

Julianelle, L. A. 1941. The pneumonia of Friedländer's Bacillus. Annals of Internal Medicine **15**:190–206.

Kaluzewski, S. 1967. Taxonomic position of indole-positive strains of *Klebsiella*. Medycyna Dóswiadczalna i Mikrobiologia **19**:327–335.

Kauffmann, F. 1944. Zur Serologie der Coligruppe. Acta Pathologica et Microbiologica Scandinavica **21**:20–45.

Kauffmann, F. 1949. On the serology of the *Klebsiella* group. Acta Pathologica et Microbiologica Scandinavica **26**: 381–406.

Kauffmann, F. 1966. The bacteriology of Enterobacteriaceae. Copenhagen: Munksgaard.

Lautrop, H. 1956. Gelatin liquefying *Klebsiella* strains (*Bacterium oxytocum* Flügge). Acta Pathologica et Microbiologica Scandinavica **39**:375–384.

Le Bouar, Y. 1972. Enterobacteries, pp. 299–307. In: Daguet, G. (ed.), Techniques en bactériologie. 1. Aerobies. Paris: Flammerion.

Lennette, E. H., Spaulding, E. H., Truant, J. P. 1974. Manual of clinical microbiology, 2nd ed. Washington D.C.: American Society for Microbiology.

Mahl, M. C., Wilson, P. W., Fife, M. A., Ewing, W. H. 1965. Nitrogen fixation by members of the tribe *Klebsiellae*. Journal of Bacteriology **89**:1482–1487.

Nimmich, W. 1968. Zur Isolierung und qualitativen Bausteinanalyse der K-Antigene von Klebsiellen. Zeitschrift für Medizinische Mikrobiologie und Immunologie **154**:117–131.

Ørskov, F., Ørskov, I. 1978. Serotyping of Enterobacteriaceae with special emphasis on K antigen determination, pp. 1–77. In: Bergan, T., Norris, J. R. (eds.), Methods in microbiology, vol. 11. London: Academic Press.

Ørskov, I. 1952. Nosocomial infections with *Klebsiella* in lesions of the urinary tract. Acta Pathologica et Microbiologica Scandinavica, Suppl. **92**:259–271.

Ørskov, I. 1954. Nosocomial infections with *Klebsiella* in lesions of the urinary tract. II. Acta Pathologica et Microbiologica Scandinavica **35**:194–204.

Ørskov, I. 1955. The biochemical properties of *Klebsiella* (*Klebsiella-Aerogenes*) strains. Acta Pathologica et Microbiologica Scandinavica **37**:353–368.

Ørskov, I. 1957. Biochemical types in the *Klebsiella* group. Acta Pathologica et Microbiologica Scandinavica **40**:155–162.

Ørskov, I. 1974. *Klebsiella*, pp. 321–324. In: Buchanan, R. E., Gibbons, N. E. (eds.), Bergey's manual of determinative bacteriology, 8th ed. Baltimore: Williams & Wilkins.

Ørskov, I., Fife-Asbury, M. A. 1977. New *Klebsiella* capsular antigen K82 and the deletion of five of those previously assigned. International Journal of Systematic Bacteriology **27**:386–387.

Ørskov, I., Ørskov, F., Jann, B., Jann, K. 1977. Serology, chemistry and genetics of O and K antigens of *Escherichia coli*. Bacteriological Reviews **41**:667–710.

Perlman, E., Bullowa, J. G. M. 1941. Primary Bacillus Friedländer (*Klebsiella pneumoniae*) pneumonia; therapy of B. Friedländer B pneumonia. Archives of Internal Medicine **67**:907–920.

Platt, H., Atherton, J. G., Ørskov, I. 1976. *Klebsiella* and *Enterobacter* organisms isolated from horses. Journal of Hygiene **77**:401–408.

Rees, T. A., Gregory, M. M. 1977. Causative organisms in rhinoscleroma. Lancet **i**:650.

Richard, C. 1973. Étude antigénique et biochimique de 500 souches de *Klebsiella*. Annales de Biologie Clinique **31**:295–303.

Sedlak, J., Rische, H. 1968. Enterobacteriaceae Infektionen. Leipzig: Georg Thieme.

Seidler, R. J., Knittel, M. D., Brown, C. 1975. Potential pathogens in the environment: Cultural reactions and nucleic acid studies on *Klebsiella pneumoniae* from clinical and environmental sources. Journal of Applied Microbiology **29**:819–825.

Slopek, S., Durlakowa, I. 1967. Studies on the taxonomy of *Klebsiella* bacilli. Archivum Immunologiae et Therapiae Experimentalis **15**:481–487.

Slopek, S., Przondo-Hessek, A., Milch, A., Déak, S. 1967. A working scheme for bacteriophage typing of *Klebsiella* bacilli. Archivum Immunologiae et Therapiae Experimentalis **15**:589–599.

Solomon, S. 1937. Primary Friedländer pneumonia; report of 32 cases. Journal of the American Medical Association **108**:937–947.

Steinhauer, B. W., Eickhoff, T. C., Kislak, J. W., Finland, M. 1966. The *Klebsiella-Enterobacter-Serratia* division. Clinical and epidemiologic characteristics. Annals of Internal Medicine **65**:1180–1194.

Stenzel, W., Bürger, H., Mannheim, W. 1972. Zur Systematik und Differentialdiagnostik der *Klebsiella*-Gruppe mit besonderer Berücksichtigung der sogenannten *Oxytocum*-Typen. Zentralblatt für Bakteriologie, Parasitenkunde, Infektionskrankheiten und Hygiene, Abt. 1 Orig., Reihe A **219**:193–203.

Stirm, S., Bessler, W., Fehmel, F., Freund-Mölbert, E., Thurow, H. 1971. Isolation of spike-formed particles from bacteriophage lysates. Virology **45**:303–308.

Véron, M., Le Minor, L. 1975. Nutrition et taxonomie des *Enterobacteriaceae* et bactéries voisine. III. Caractères nutritionelles et différenciation des groupes taxonomiques. Annales de Microbiologie (Annales de l'Institut Pasteur) **126B**:125–147.

von Riesen, V. L. 1976. Pectinolytic, indole-positive strains of *Klebsiella pneumoniae*. International Journal of Systematic Bacteriology **26**:143–145.

CHAPTER 94

The Genus *Klebsiella* (Nonmedical Aspects)

RAMON J. SEIDLER

This chapter deals with the isolation, distribution, and properties of *Klebsiella* isolated from sources not directly related to infectious diseases or normal human flora. These klebsiellae have been identified as *K. pneumoniae* and the indole-positive *K. oxytoca*. *K. rhinoscleromatis*, *K. ozaenae*, and properties of other medically derived *Klebsiella* are discussed in Chapter 93, this Handbook.

Klebsiella are widely recognized as multiply antibiotic-resistant, opportunistic pathogens and as agents of bacteremias and genitourinary, respiratory, and other serious infections primarily in stressed patients (Eickhoff, 1971; Martin, Yu, and Washington, 1971; Selden et al., 1971). *Klebsiella* are also associated with outbreaks of mastitis in cattle and cause a variety of infections in other domestic and wild animals (Braman et al., 1973; Fox and Rohovsky, 1975; Merkt et al., 1975; Wyand and Hayden, 1973).

Through the years, *Klebsiella* have been isolated on numerous occasions from the environment as total coliforms, but little significance was attached to these because such isolates were originally classified as "*Aerobacter*" or "*Bacterium aerogenes*" (Beckwith, 1931; Edwards, 1929; Johnson and Levine, 1917; Nunez and Colmer, 1968). Generic differentiation was based on habitat of origin, which would designate a coliform of the indole, methyl red, Voges-Proskauer, citrate (IMViC) pattern $- - + +$ from sputum as *K. pneumoniae* and a biochemically similar isolate from soil as "*Aerobacter*" (Breed, 1957). Perhaps the first indication of an impending taxonomic and ecological enigma arose in 1929 (Edwards, 1929) when it was discovered that 5 of 29 "*Aerobacter*" obtained from soil could be agglutinated with type B capsular antiserum prepared against Friedländer's bacillus (*Klebsiella*).

In the more recent literature, a survey of the dinitrogen-fixing capabilities of "*Aerobacter*" led to the confirmation that many of these environmentally derived cultures were serotypable as *K. pneumoniae* (Mahl et al., 1965). Indeed, with the current phenotypic descriptions and revised nomenclature it is readily possible to distinguish *Klebsiella* from *Enterobacter* ("*Aerobacter*") based on motility, decarboxylases, and dihydrolase patterns (Edwards and Ewing, 1972; Hormaeche and Edwards, 1960). Using these tests, several research groups have subsequently confirmed the widespread occurrence of *Klebsiella* in a variety of natural habitats (Aho et al., 1974; Brown and Seidler, 1973; Dufour and Cabelli, 1976; Duncan and Razzell, 1972; Knittel, 1975; Matsen, Spindler, and Blosser, 1974; Newman and Kowalski, 1973).

The ubiquitous distribution of an animal and human opportunistic pathogen in high cell densities in environments apparently free of recent fecal contamination raises complex questions as to their health significance and, indeed, whether such isolates are *K. pneumoniae sensu stricto,* atypical *Enterobacter,* or perhaps new species of Enterobacteriaceae. As many as seven different IMViC patterns have been described for environmentally derived *Klebsiella,* while only three to four patterns are commonly observed among human and animal isolates (Brown and Seidler, 1973; Campbell and Roth, 1975; Dufour and Cabelli, 1976; Edwards and Ewing, 1972). These various patterns have added to taxonomic confusion, since IMViC results have been used (along with other minor phenotypic variations) to denote several species in the genus, including *K. pneumoniae, K. aerogenes, K. edwardsii,* nonmotile *Aerobacter aerogenes,* and *K. oxytoca* (Cowan, Steel, and Shaw, 1960). There now seems to be some general agreement that the first four species are synonyms of *K. pneumoniae* (Ørskov, 1974), but more recently a new genus (*Oxytocum*) has been suggested for the indole-producing *Klebsiella* (Jain, Radsak, and Mannheim, 1974).

Studies on the nucleic acids of bacteria that are phenotypically and serologically identifiable as *K. pneumoniae* have revealed great molecular heterogeneity (Seidler, Knittel, and Brown, 1975). The overall mean DNA base composition of some 40 isolates identified as *K. pneumoniae* of both medical and environmental origins ranged from 53.9 to 59.2 mol% G+C, whereas most described enteric species have only a 1–2 mol% G+C range (Hill, 1966). This heterogeneity was also reflected in differences in genome size and in levels of relative reassociation ranging from 5 to 100%. It was significant that some isolates from medical origins, plants, and industrial effluents exhibited low levels of relative reassociation with the human *K. pneumoniae* neotype strain (Knittel, 1975; Seidler, Knittel, and Brown, 1975). Thus, no correlation was found between culture origin and amount of relatedness to known medical isolates. Furthermore, no correlation was found between level of relative reassociation and IMViC pattern or any other phenotypic trait studied.

Several approaches have been used to attempt clarification of the ecological and taxonomic enigma presented by *Klebsiella* in the environment. For example, surveys of the response of *Klebsiella* in the fecal coliform test have illustrated that some 85% of the medical isolates and 16% or more of the environmentally derived isolates are positive (Bagley and Seidler, 1977; Dufour and Cabelli, 1976). The response of *Klebsiella* in the fecal coliform test was also shown to be stable, since identical reactions were obtained over 270 generations during growth at 35°C in sterile industrial effluent. Based on the high incidence of fecal coliform-positive *Klebsiella* among known medical isolates and their presence in the gastrointestinal tract of warm-blooded animals, it was concluded that environmentally derived *Klebsiella* should be considered as valid a fecal coliform as *Escherichia coli* (Bagley and Seidler, 1977).

Surveys on the virulence for mice of medical and environmentally derived *Klebsiella* have illustrated no significant differences between the isolates. Mean lethal dose values for cultures of human clinical, bovine mastitis, and environmental origins (drinking water, fresh vegetables, and sawdust) were 4.6×10^4, 1.5×10^4, and 4.2×10^4, respectively (Bagley and Seidler, 1978). A similar observation was made for *Klebsiella* of human and natural receiving water origins (Matsen, Spindler, and Blosser, 1974).

High klebsiellae densities in the botanical environment can be accounted for by the presence of specific growth-promoting substances that allow their rapid proliferation from small inocula (Talbot, Morrow, and Seidler, 1977). Isolates of both environmental and medical origins grow with equal vigor in industrial effluents, aqueous extracts of sawdust, and on the surfaces of fresh vegetables (Knittel et al., 1977; Talbot, Morrow, and Seidler, 1977). It was concluded that aquatic environments which become contaminated with botanical material serve as potential reservoirs to perpetuate the growth and spread of opportunistic *Klebsiella* pathogens (Knittel et al., 1977).

Habitats

Table 1 summarizes the nonhuman or animal environments from which klebsiellae have been isolated. Whenever possible, estimates of their cell densities have also been provided. It is immediately apparent that the highest *Klebsiella* cell densities are associated with botanical milieu.

Industrial effluents from wood products and from textile finishing plants have the highest *Klebsiella* cell densities. Rather extensive studies have been made on these wastes to ascertain the origin(s) and health significance of these high coliform and fecal coliform counts (Dufour and Cabelli, 1976; Huntley, Jones, and Cabelli, 1976; Knittel, 1975; Matsen, Spindler, and Blosser, 1974; Seidler, Knittel, and Brown, 1975). The greatest *Klebsiella* densities are found in lagoons and aeration basins designed to reduce the biological oxygen demand of these nutrient-rich effluents. Recent studies have confirmed the rapid and extensive growth by *Klebsiella* within the waste liquors from pulp mills (Fig. 1), where final cell densities can reach 10^5–10^7 *Klebsiella*/ml (Knittel et al., 1977).

It has also been illustrated that dinitrogen-fixing (acetylene reduction) activity is common in *Klebsiella* and other coliforms isolated from these botanical environments (Aho et al., 1974; Brown and Seidler, 1973; Mahl et al., 1965). For example, nitrogenase of *Klebsiella* and other coliform isolates from paper mill process waters was still active in the presence of 1.8% (vol/vol) oxygen and was generally not repressed by 500 mg/liter of various organic nitrogen compounds. It was therefore suggested that under conditions found in these industrial wastes, nitrogen fixation contributes significantly to the nitrogen economy of these cells (Neilson and Sparell, 1976).

Studies from several groups have shown that klebsiellae are actually associated with the raw materials processed by pulp, textile, and sugar cane mills. Coliforms including *Klebsiella* have been isolated from wood, sawdust, living trees, raw bailed

Table 1. Distribution of *Klebsiella* in the environment.[a]

Pulp/paper mill effluents	10^3–10^7/ml
Textile finishing plant effluents	up to 10^7/ml
Potato waste lagoons	up to 10^6/ml
Fresh vegetables and seeds	10^2–10^5/g
Salads	10^1–10^7/g
Lake water	10^{-1}–10^1/ml
Aquarium water	10^3–10^4/ml
Unprocessed bailed cotton	30–60/g
Finished drinking water	up to 1–2/ml

[a] Additional sources of *Klebsiella*: receiving waters, chlorinated water supplies, sugar cane and mill wastes, feedlot waste and runoff, seawater (estuary), soil, forest environment, living trees, sawdust, lumber.

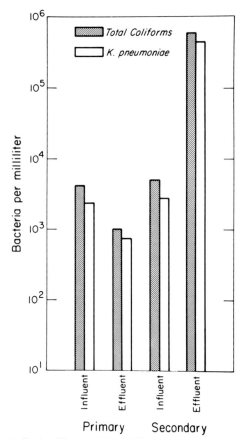

Fig. 1. Total coliforms and *Klebsiella pneumoniae* cell densities per ml observed in four stages of treatment of waste waters from a sulfite pulp mill. The raw wastes contain 10^3 or more *K. pneumoniae* per ml, which multiply to 10^6 per ml during the secondary treatment in the aeration basin. (From Knittel et al., 1977.)

can colonize the surfaces of cow teats (Rendos, Eberhart, and Kesler, 1975). It was suggested that such an environment is an important reservoir for these agents of coliform mastitis (Newman and Kowalski, 1973).

Other botanical material has recently been found to be the source of *Klebsiella* in certain drinking water supplies. This environment consists of reservoirs constructed of redwood that are used to store finished drinking water (Fig. 2). Redwood reservoirs are used throughout the western United States, Canada, and in other countries to store drinking water for rural communities, recreational areas, mobile home parks, etc. Coliform contamination was associated with improperly designed reservoirs with long water-retention periods and with inadequate chlorination. Under these conditions, an extensive microbial slime matrix may accumulate on the surfaces of some wooden staves, becoming obvious to the unaided eye (Fig. 3). The slime contains coliforms, including *Klebsiella,* and other bacteria embedded in mycelial stages of fungal growth. Occasionally, slime will slough off into the reservoir, enter the distribution system, and give rise to coliform counts that exceed U. S. Environmental Protection Agency standards for finished drinking water by as much as 10- to 40-fold (Seidler, Morrow, and Bagley, 1977).

Fecal coliform-positive and fecal coliform-negative *Klebsiella* have been recovered from the surfaces of fresh market produce and from salads served in schools, canteens, and hospitals (Brown and Seidler, 1973; Duncan and Razzell, 1972;

cotton, and from the surfaces of sugar cane plants (Bagley and Seidler, 1977; Campbell et al., 1976; Duncan and Razzell, 1972; Newman and Kowalski, 1973; Nunez and Colmer, 1968; Rendos, Eberhart, and Kesler, 1975). *Klebsiella* isolated from within living trees have been shown to exhibit high levels of dinitrogen-fixing activity, as described above for cultures obtained from the lagoons and aeration basins of pulp mills (Aho et al., 1974).

The source(s) of *Klebsiella* in these botanical materials has never been resolved, but studies have shown that some cultures are fecal coliforms. It is conceivable that *Klebsiella* could arise from the activities of insects, rodents, and other wild animals and birds. An occasional small inoculum in these materials could rapidly proliferate, since a variety of botanical materials contain the appropriate nutrients to support the growth of *Klebsiella,* including those isolates of fecal, disease, and environmental origins (Knittel et al., 1977). Sawdust bedding used for dairy herds has also been shown to contain substantial numbers of these opportunistic pathogens, which

Fig 2. A 50,000-gallon redwood reservoir used to store finished drinking water for a rural residential community.

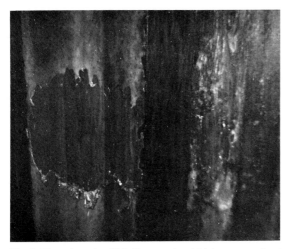

Fig. 3. Extensive microbial slime matrix which has accumulated on redwood staves below the normal level of water in a reservoir (photograph from Seidler, Morrow, and Bagley, 1977).

Shooter et al., 1971). It has been suggested that oral ingestion of *Klebsiella* from this botanical material may be the means by which the gastrointestinal tract of humans becomes colonized with these opportunistic pathogens (Montgomerie et al., 1970). Intestinal colonization by *Klebsiella* is a most likely reservoir of nosocomial infections (Selden et al., 1971).

Isolation

There are several selective/differential media available for the isolation and enumeration of *Klebsiella* from the environment. The mC medium (Dufour and Cabelli, 1975) was designed to enumerate total coliforms in seawater, and, in conjunction with in situ tests, can separate coliforms into component genera. This medium has also been used to isolate and quantitate total coliforms and *Klebsiella* in textile mill waste waters (Dufour and Cabelli, 1976). The advantage of this medium is its compatibility with the detection of in situ urease and tryptophanase activities accomplished by transferring the membrane filter to suitable substrates placed on absorbent pads. In this manner, *K. pneumoniae* (urease positive) can be enumerated and distinguished from *Escherichia coli* (indole positive, urease negative).

The mC medium does lack selectivity, since background microbial flora were only reduced about 100-fold over nonselective media. Thus, the authors recommended that seawater volumes of 10 ml or less be filtered. Specificity was high, since 95% of the presumptive coliform colonies were confirmed by using gas production in lactose broth at 35°C as a criterion. Only 6% false negatives were obtained on more than 2,500 isolates examined.

Isolation and Enumeration of *Klebsiella* (Dufour and Cabelli, 1975, 1976)

mC medium: "L-glutamate (Calbiochem), 0.4 g; L-cysteine hydrochloride (Calbiochem), 0.05 g; L-proline (Calbiochem), 0.04 g; L-tryptophan (Calbiochem), 0.06 g; yeast extract (Difco), 0.2 g; ammonium chloride (Fisher), 0.2 g; lactose (Fisher), 0.8 g; salt solution, 50 ml; and distilled water, 50 ml. . . . The salt solution was prepared as follows: NaCl, 0.2 g; KCl, 0.8 g; K_2HPO_4, 0.3 g; KH_2PO_4, 0.1 g; $MgSO_4 \cdot 7H_2O$, 0.02 g; and distilled water, 100 ml."

The ingredients were dissolved, the pH was adjusted to 7.5 with 1 N NaOH, 1.5 g of agar was added, and the mixture was autoclaved at 121°C for 15 min. Before cooling, 1 ml of a 1% aqueous solution (wt/vol) of aniline blue (Matheson, Coleman, and Bell) was added. The medium was allowed to stand for 5 min before being cooled to 55°C, after which the following unsterilized constituents were added per 100 ml of medium: sodium taurocholate, 50 mg; vancomycin, 0.1 g; and 2,4-dinitrophenol, 1 ml of a 0.1% (wt/vol) aqueous solution. The final pH of the medium was 7.1. The medium then was dispensed in 4-ml quantities into membrane filter plates (50 × 12 mm). Appropriate quantities of the aquatic samples were passed through 0.45-μm membrane filters, applied to the surface of the mC agar, and incubated at 35°C for 22 ± 2 h. All blue to blue-gray colonies were counted as coliforms.

In Situ Identification of *Klebsiella* Incubated at 22°C

After counting and marking of the coliform colonies, the membrane may be transferred from mC agar to an absorbent pad saturated with an aqueous solution consisting of 2% (wt/vol) urea and 0.01% (wt/vol) phenol red adjusted to a pH of 5.0 ± 0.2 with 6 N HCl. This substrate may be stored up to 1 week at 6–8°C. Coliform (blue) colonies that were urease positive took on a deep magenta color, whereas urease-negative coliform colonies remained blue. The filter may then be transferred from the in situ urea pad to a pad saturated with tryptophanase substrate for about a 20-min induction period. The filter is then placed on another pad saturated with indole reagent. Indole-positive colonies turn red or exhibit a red halo surrounding the blue colony within 10 min. The tryptophan substrate consists of 0.5 g of Trypticase and 0.2 g of tryptophan dissolved in 100 ml of distilled water; adjust to pH 7.2, filter-sterilize, and refrigerate until used. The indole reagent consists of 5.0 g of *para*-dimethylaminobenzaldehyde dissolved in 90 ml ethyl alcohol. Carefully add 10 ml concentrated hydrochloric acid.

Recently, Dufour and Lupo (1977) have developed a membrane filtration medium for enumeration of *Klebsiella* found in fresh and marine waters. This medium, designated mK, was highly selective and differential. It detected *Klebsiella* when they constituted less than 0.001% of the total coliform population. The medium was also specific for *Klebsiella*, since only 10% of the presumptive colonies were false positives and less than 5% were false negatives. The medium is based on fermentation of adonitol in the presence of the pH indicator phenol red and various inhibitors.

Membrane Filtration and Enumeration of *Klebsiella* (Dufour and Lupo, 1977)

Medium mK contains the following ingredients:

Adonitol	4 g
Agar	15 g
2 × Salt solution	500 ml
Distilled water	300 ml

The 2 × salt solution contains:

NaCl	2 g
KCl	8 g
K_2HPO_4	3 g
KH_2PO_4	1 g
$MgSO_4 \cdot 7H_2O$	0.2 g
Distilled water	1,000 ml

Mix and autoclave for 15 min at 121°C. Uric acid solution: uric acid, 0.3 g, and distilled water, 200 ml. Dissolve uric acid in a small volume of 1 N NaOH, add to distilled water, and readjust pH to 7.1 with 1 N HCl. Sterilize by filtration. The remaining solutions contain a pH indicator and inhibitors that are placed into sterile distilled water: phenol red, 0.1 g/10 ml; sodium taurocholate, 0.4 g/30 ml; carbenicillin, 0.005 g/ml. Add uric acid solution to cooled adonitol plus salts, then add phenol red and inhibitors; mix and dispense. Following incubation of the membrane filter (0.45-μm porosity) at 35°C for 48 h, count all nonpink colonies.

Brown and Seidler (1973) used a nitrogen-deficient agar medium and Difco m-Endo agar LES medium to isolate and enumerate *Klebsiella* from the surfaces of fresh market produce and vegetable seeds.

Isolation and Enumeration of *Klebsiella* (Brown and Seidler, 1973)

The outer surfaces of vegetables were aseptically peeled into sterile 0.01 M tris(hydroxymethyl) aminomethane (Tris) buffer (pH 7.5), covered, and shaken for 5 min at 22°C. Serial dilutions were plated onto the nitrogen deficient and m-Endo LES agar media using the spread plate technique. The modified Hino and Wilson nitrogen deficient medium was occasionally successful when plates were incubated at 37°C for 3–5 days under an atmosphere consisting of 80% nitrogen and 20% argon. Large mucoid colonies were picked and identified as *Klebsiella* using standard techniques (Edwards, and Ewing, 1972). Plates of m-Endo agar LES were incubated at 37°C for 24 h. Some one-third of the raised green sheen colonies appearing on the 10^{-2} dilution plates were subsequently identified as *K. pneumoniae*. Isolation of *Klebsiella* from commercial vegetable seeds was also demonstrated with the latter technique following a 24-h preincubation of seeds in sterile Tris buffer before plating onto m-Endo agar LES. Fresh vegetables and seed surfaces are good sources of dinitrogen-fixing *Klebsiella* since 20/40 isolates reduced acetylene under anaerobic conditions (Brown and Seidler, 1973).

Double violet agar was described by Campbell and Roth (1975) for the isolation and enumeration of *Klebsiella* from aquatic environments. As one becomes familiar with the various types of colony morphology the authors claim greater than 80% accuracy in the presumptive identification of *Klebsiella*.

Isolation of *Klebsiella* from Aquatic Environments (Campbell and Roth, 1975)

The composition of double violet agar is: Difco violet red bile agar, 41.5 g; methyl violet 2B (Allied Chemical Corporation), 2.0 g; Mix and heat to boiling; cool and pour plates. Water samples may be filtered through 0.45 μm membrane filters; incubation is for 24 h at 35°C. *E. coli*, *Pseudomonas*, *Aeromonas*, and *Proteus* do not grow or grow sparsely. *Klebsiella* colonies are glistening lavender and exhibit abundant growth. They can be distinguished, with experience, from *Enterobacter* species, which generally form smaller purple colonies.

The possibility of presumptive identification of *K. pneumoniae* on mFC medium was described by Stramer (1976). Using standard membrane filtration procedures (American Public Health Association, 1976), the author was able to distinctly recognize fecal coliform-positive (FC+) colonies formed by *Klebsiella*. Such colonies were described as atypical light blue, nucleated, and mucoid, and differed from the more typical dark blue colonies of *E. coli*. In a recent survey of the FC response of some 190 pure *Klebsiella* cultures, Bagley and Seidler (1977) described colony morphology of *Klebsiella* ranging from typical flat, dark blue, to raised, mucoid dark blue, to atypical cream-colored. Some 40% of the isolates exhibited flat blue colonies with some orange coloration. The use of mFC medium for the presumptive identification of *Klebsiella* is therefore of limited use and is not recommended for routine enumeration purposes since colony morphology may vary significantly. In addition, many environmentally derived klebsiellae are FC−, and mFC medium would not allow enumeration of all *Klebsiella* present in the sample.

Identification

As is the case with most genera of the Enterobacteriaceae, *Klebsiella* may be readily identified using conventional phenotypic traits (Edwards and Ewing, 1972). Typical useful reactions include the following: fermentation of lactose with production of typical coliform colonies on selective/differential media or by gas production in lactose at 35°C; lack of motility; negative reactions in ornithine decarboxylase and arginine dihydrolase media; positive reactions in lysine decarboxylase and urease media. Most other biochemical tests are either variable or of no key differential value.

Several other reactions were recently found to be useful in subdividing isolates that would otherwise identify as *K. pneumoniae*. These reactions included response in the fecal coliform test, growth in nutrient broth at 10°C, indole production, and pectin liquefaction (Naemura and Seidler, 1978). Table 2 summarizes the test results performed on *Klebsiella* of diverse environmental and human and animal origins. Over 90% of the isolates were placed into three groups. An inverse correlation was found between the growth response at 10°C and the operational definition of a fecal coliform. Low temperature growth also correlated well with the indole and pectin tests. Group I isolates, which were fecal coliform positive and negative in the other tests, corresponded to *K.*

pneumoniae sensu stricto; group II isolates phenotypically resemble group I but are distinguished by reciprocal responses in the fecal coliform test and growth at 10°C; these isolates grow most rapidly at 10°C and are genetically most divergent (DNA relative reassociation) from group I. The indole-producing, pectin-liquefying isolates of group III conform to the previously described *K. oxytoca* and exhibit intermediate levels of DNA relative reassociation to *K. pneumoniae* (Naemura, Bagley, and Seidler, 1977; Naemura and Seidler, 1978).

Literature Cited

Aho, P. E., Seidler, R. J., Evans, H. J., Raju, P. N. 1974. Distribution, enumeration, and identification of nitrogen-fixing bacteria associated with decay in living white fir trees. Phytopathology **64:**1413–1420.

American Public Health Association. 1976. Standard methods for the examination of water and wastewater, 14th ed. New York: American Public Health Association.

Bagley, S. T., Seidler, R. J. 1977. Significance of fecal coliform-positive *Klebsiella*. Applied and Environmental Microbiology **33:**1141–1148.

Bagley, S. T., Seidler, R. J. 1978. Comparative pathogenicity of environmental and clinical *Klebsiella*. Health Laboratory Science **15:**104–111.

Beckwith, T. D. 1931. The bacteriology of pulp slime. Journal of Bacteriology **22:**15–22.

Braman, S. K., Eberhart, R. J., Ashbury, M. A., Herman, G. J. 1973. Capsular types of *Klebsiella pneumoniae* associated with bovine mastitis. Journal of the American Veterinary Medical Association **162:**109–111.

Breed, R. S. 1957. *Klebsiella*, pp. 344–346. In: Breed, R. S., Murray, E. G. D., Smith, N. R. (eds.), Bergey's manual of determinative bacteriology, 7th ed. Baltimore: Williams & Wilkins.

Brown, C., Seidler, R. J. 1973. Potential pathogens in the environment: *Klebsiella pneumoniae*, a taxonomic and ecological enigma. Applied Microbiology **25:**900–904.

Campbell, L. M., Michaels, G., Klein, R. D., Roth, I. L. 1976. Isolation of *Klebsiella pneumoniae* from lake water. Canadian Journal of Microbiology **22:**1762–1767.

Campbell, L. M., Roth, I. V. 1975. Published inconsistencies in methyl red and Voges-Proskauer reactions of *Klebsiella pneumoniae*. International Journal of Systematic Bacteriology **25:**386–387.

Cowan, S. T., Steel, K. J., Shaw, C. 1960. A classification of the *Klebsiella* group. Journal of General Microbiology **23:**601–612.

Dufour, A. P., Cabelli, V. J. 1975. Membrane filter procedure for enumerating the component genera of the coliform group in seawater. Applied Microbiology **29:**826–833.

Dufour, A. P., Cabelli, V. J. 1976. Characteristics of *Klebsiella* from textile finishing plant effluents. Journal Water Pollution Control Federation **48:**872–879.

Dufour, A. P., Lupo, L. B. 1977. A membrane filter method for enumerating *Klebsiella* species. Abstracts of the Annual Meeting of the American Society for Microbiology **1977:**262.

Duncan, D. W., Razzell, W. E. 1972. *Klebsiella* biotypes among coliforms isolated from forest environments and farm produce. Applied Microbiology **24:**933–938.

Edwards, P. R. 1929. Relationships of the encapsulated bacilli with special reference to *Bacterium aerogenes*. Journal of Bacteriology **17:**339–353.

Table 2. Groups in the genus *Klebsiella* determined by indole (I), pectin (P), fecal coliform (FC), and 10°C growth responses.[a]

Group[b]	I	P	FC	10°C	No. isolates[c]	% of total
I	−	−	+	−	70	35
II	−	−	−	+	60	30
III	+	+	−	+	56	28
IV	+	−	−	+	10	5
V	−	+	−	+	6	3

[a] (From Naemura and Seidler, 1978).
[b] The *K. pneumoniae* neotype strain (ATCC 13883) is FC− but was placed in group I since it did not grow at 10°C and had high DNA relative reassociation with other group I isolates. ATCC cultures 13882 and 15574 are also members of group I; ATCC 13182 (*K. oxytoca*) is in group III.
[c] Five of the 202 isolates did not conform to the preceding patterns and were grouped as follows: 2 isolates I−, P−, FC−, 10°C−, group I; 2 isolates I−, P−, FC+, 10°C+, group II; 1 isolate I−, P+, FC+, 10°C+, group V.

Edwards, P. R., Ewing, W. H. 1972. Identification of *Enterobacteriaceae*, 3rd ed. Minneapolis: Burgess Publishing.

Eickhoff, T. C. 1971. Nosocomial infections due to *Klebsiella pneumoniae:* Mechanisms of intra-hospital spread, pp. 117–125. In: Brachman, P. S., Eickhoff, T. C., (eds.), Proceedings of the International Conference on Nosocomial Infection. Chicago: American Hospital Association.

Fox, J. G., Rohovsky, M. W. 1975. Meningitis caused by *Klebsiella* spp. in two Rhesus monkeys. Journal of the American Veterinary Medical Association **67:**634–636.

Hill, L. R. 1966. An index to deoxyribonucleic acid base composition of bacterial species. Journal of General Microbiology **44:**419–437.

Hormaeche, E., Edwards, P. R. 1960. A proposed genus *Enterobacter*. Internal Bulletin of Bacteriological Nomenclature and Taxonomy **10:**71–74.

Huntley, B. E., Jones, A. C., Cabelli, V. J. 1976. *Klebsiella* densities in waters receiving wood pulp effluents. Journal Water Pollution Control Federation **48:**1766–1771.

Jain, K., Radsak, K., Mannheim, W. 1974. Differentiation of the oxytocum group from *Klebsiella* by DNA/DNA hybridization. International Journal of Systematic Bacteriology **24:**402–407.

Johnson, B. R., Levine, M. 1917. Characteristics of coli-like microorganisms from the soil. Journal of Bacteriology **2:**379–401.

Knittel, M. D. 1975. Taxonomy of *Klebsiella pneumoniae* isolated from pulp/paper mill waste water. Environmental Protection Technology Series EPA-660/2-75-024, Corvallis, Oregon: U.S. Environmental Protection Agency.

Knittel, M. D., Seidler, R. J., Eby, C., Cabe, L. M. 1977. Colonization of the botanical environment by *Klebsiella* isolates of pathogenic origin. Applied and Environmental Microbiology **34:**557–563.

Mahl, M. C., Wilson, P. W., Fife, M. A., Ewing, W. H. 1965. Nitrogen fixation by members of the tribe *Klebsiella*. Journal of Bacteriology **89:**1482–1487.

Martin, W. J., Yu, P. K. W., Washington, J. A., II. 1971. Epidemiological significance of *Klebsiella pneumoniae*. A 3 month study. Mayo Clinic Proceedings **46:**785–793.

Matsen, J. M., Spindler, J. A., Blosser, R. O. 1974. Characterization of *Klebsiella* isolates from natural receiving waters and comparison with human isolates. Applied Microbiology **28:**672–678.

Merkt, H., Klug, E., Bohm, K. H., Weiss, R. 1975. Recent observations concerning *Klebsiella* infections in stallions. Journal of Reproduction and Fertility, Suppl. **23:**143–145.

Montgomerie, J. Z., Taylor, D. E. M., Doak, P. B., North, J. D. K., Martin, W. J. 1970. *Klebsiella* in faecal flora of renal-transplant patients. Lancet **ii:**787–792.

Naemura, L. G., Bagley, S. T., Seidler, R. J. 1977. Biochemical characterization of the *Klebsiella-Oxytoca* group. Abstracts of the Annual Meeting of the American Society for Microbiology **1977:**179.

Naemura, L. G., Seidler, R. J. 1978. Significance of low-temperature growth associated with the fecal coliform response, indole production, and pectin liquefaction in *Klebsiella*. Applied and Environmental Microbiology **35:**392–396.

Newman, L. E., Kowalski, J. J. 1973. Fresh sawdust bedding—a possible source of *Klebsiella* organisms. American Journal of Veterinary Research **34:**979–980.

Nielson, A. H., Sparell, L. 1976. Acetylene reduction (nitrogen fixation) by *Enterobacteriaceae* isolated from paper mill process waters. Applied and Environmental Microbiology **32:**197–205.

Nunez, W. J., Colmer, A. R. 1968. Differentiation of *Aerobacter-Klebsiella* isolated from sugar cane. Applied Microbiology **16:**1875–1890.

Ørskov, I. 1974. *Klebsiella,* pp. 321–324. In: Buchanan, R. E., Gibbons, N. E. (eds.), Bergey's manual of determinative bacteriology, 8th ed. Baltimore: Williams & Wilkins.

Rendos, J. J., Eberhart, R. J., Kesler, E. M. 1975. Microbial populations of teat ends of dairy cows and bedding materials. Journal of Dairy Science **58:**1492–1500.

Seidler, R. J., Knittel, M. D., Brown, C. 1975. Potential pathogens in the environment: Cultural reactions and nucleic acid studies on *Klebsiella pneumoniae* from clinical and environmental sources. Applied Microbiology **29:**819–825.

Seidler, R. J., Morrow, J. E., Bagley, S. T. 1977. *Klebsielleae* in drinking water emanating from redwood tanks. Applied and Environmental Microbiology **33:**893–900.

Selden, R., Lee, S., Wang, W. L. L., Bennett, J. V., Eickhoff, T. C. 1971. Nosocomial *Klebsiella* infections: Intestinal colonization as a reservoir. Annals of Internal Medicine **74:**657–664.

Shooter, R. A., Faiers, M. C., Cooke, E. M., Breaden, A. L., O'Farrell, S. M. 1971. Isolation of *Escherichia coli, Pseudomonas aeruginosa,* and *Klebsiella* from food in hospitals, canteens, and schools. Lancet **ii:**390–392.

Stenzel, W., Bürger, H., Mannheim, W. 1972. Zur Systematik und Differentialdiagnostik der Klebsiella-Gruppe mit besonderer Berücksichtigung der sogenannten Oxytocum-Typen. Zentralblatt für Bakteriologie, Parasitenkunde, Infektionskrankheiten und Hygiene, Abt. 1 Orig., Reihe A **219:**193–203.

Stramer, S. L. 1976. Presumptive identification of *Klebsiella pneumoniae* on m-FC medium. Canadian Journal of Microbiology **22:**1774–1776.

Talbot, H. W., Morrow, J. E., Seidler, R. J. 1977. Nutritional factors associated with the presence of *Klebsiella* in the environment. Abstracts of the Annual Meeting of the American Society for Microbiology **1977:**262.

Wyand, D. S., Hayden, D. W. 1973. *Klebsiella* infection in muskrats. Journal of the American Veterinary Medical Association **163:**589–591.

CHAPTER 95

The Genus *Enterobacter*

DON J. BRENNER

The genus *Enterobacter* did not exist when the seventh edition of *Bergey's Manual of Determinative Bacteriology* was published (Breed and Murray, 1957). The seventh edition contained the genus *Aerobacter* with *A. aerogenes* and *A. cloacae*. Yellow-pigmented strains of *Aerobacter aerogenes* were mentioned. The genus *Paracolobactrum* was a catchall for organisms resembling other species except that their lactose fermentation was delayed or occasionally did not occur. *Paracolobactrum aerogenoides* consisted of organisms now classified as *Enterobacter aerogenes, Enterobacter cloacae,* and *Hafnia alvei. Aerobacter aerogenes* strains were incorrectly described as usually nonmotile. In *Bergey's Manual,* seventh edition, there were six species of *Erwinia* that are part of the *Enterobacter agglomerans* complex (see this Handbook, Chapter 88). These are *E. vitivora, E. milletiae, E. cassavae, E. ananas, E. citrimaculans,* and *E. mangiferae*.

On the basis of descriptions given in the seventh edition, it was impossible to distinguish *Aerobacter aerogenes* from *Klebsiella pneumoniae*. In effect, if the isolates were from the respiratory tract, they were called *K. pneumoniae* and if they were from stools, urine, or environmental sources, they were called *A. aerogenes*.

Several proposals were made in an attempt to end this confusion. These are summarized by Hormaeche and Edwards (1960a,b) and Edwards and Ewing (1962, 1972). In the early 1960s several synonyms were used for species of *Enterobacter* (Table 1). Many taxonomists included *Serratia liquefaciens* and *Hafnia alvei* in the genus *Enterobacter*. Extensive biochemical characterization and deoxyribonucleic acid (DNA) relatedness studies showed that these organisms should not remain in *Enterobacter*. In 1963, *Cloacae* was rejected and *Enterobacter* officially accepted by the Judicial Commission, with *Enterobacter cloacae* designated as the type species (Hormaeche and Edwards, 1960a,b; Opinion 28, 1963).

Contributions to the section on Enterobacteriaceae in the eighth edition of *Bergey's Manual* were completed prior to 1970 (Cowan et al., 1974). This edition lists only *E. cloacae* and *E. aerogenes*.

Mention is made of *E. cloacae* strains that produce a nondiffusible yellow pigment. These have now been recognized as a new species, *Enterobacter sakazakii* (J. J. Farmer, in preparation; Brenner et al., 1977).

In the eighth edition of *Bergey's Manual* (Cowan et al., 1974) the Herbicola group within the genus *Erwinia* corresponds to the *Enterobacter agglomerans* complex (Ewing and Fife, 1972). The Herbicola group contains *Erwinia herbicola* var. *herbicola, Erwinia herbicola* var. *ananas, Erwinia uredovora,* and *Erwinia stewartii*. These organisms are discussed in Chapter 102 on *Erwinia* (this Handbook). They are also considered here because the plant pathologist and the medical microbiologist use a different battery of tests with which to describe these organisms. Furthermore, neither the Herbicola group nor the *Enterobacter agglomerans* concept of classification is adequate for this diverse group of organisms, which may contain 10 or more species. In the Center for Disease Control (CDC) Enteric Section, we report all these organisms as *E. agglomerans*. Eventually these species will be designated and placed in *Erwinia, Enterobacter,* both of these genera, or as species in a new genus.

Enterobacter gergoviae (Richard et al., 1976) and *Enterobacter vulneris* (J. K. Leete, A. M. Murlin, and D. J. Brenner, unpublished) are the names to be proposed for the two most recently described species.

Table 1. Nomenclatural designations for species of *Enterobacter* used in the 1960s.

Current usage	Synonyms
Enterobacter cloacae	*Aerobacter cloacae*
	Aerobacter A
	Cloaca A
Enterobacter aerogenes	*Aerobacter aerogenes*
	Aerobacter B
	Cloaca B
Serratia liquefaciens	*Enterobacter liquefaciens*
	Aerobacter C
Hafnia alvei	*Enterobacter hafniae*
	Enterobacter alvei

Habitats

The biochemical tests now used to detect lysine decarboxylase, arginine dihydrolase, and ornithine decarboxylase were introduced only about 20 years ago, and were not in general use until the mid-1960s. These tests are essential to separate species of *Enterobacter* from one another and from *Klebsiella pneumoniae*. A large number of clinical and water microbiologists do not subdivide the "*Klebsiella-Enterobacter*" group. In the absence of decarboxylase data, it is extremely difficult to critically evaluate the older data on incidence and habitat of *Enterobacter* species. This is particularly true for *E. aerogenes,* the early descriptions of which were misleading and whose nonmotile strains were impossible to distinguish from *K. pneumoniae* without the reaction for ornithine decarboxylase. *Enterobacter sakazakii, E. gergoviae,* and *E. vulneris* are newly described species for which relatively few data are available on prevalence and habitat. *Enterobacter amnigena* is a newly described species isolated from water and unpolluted soil (Izard et al., 1981). Gavini, Ferragut, and Leclerc (1976) have described at least one additional, as yet unnamed, species of *Enterobacter* whose habitat is water.

Enterobacter cloacae is found in the feces of man and animals, sewage, soil, and water (Breed and Murray, 1957). *E. aerogenes* has also been isolated from the intestine of man and animals, from milk and dairy products, and is widely distributed in nature (Breed and Murray, 1957). Specific sources of *E. cloacae* and *E. aerogenes* include skin from axilla, groin, and webs between digits (Aly and Maibach, 1977); animal hides and meat (Newton, Harrison, and Smith, 1977); healthy horses (Platt, Atherton, and Ørskov, 1976; Sakazaki and Namioka, 1960). *E. cloacae* has also been reported from cows, dogs, diseased pigs, horses, and chickens (Sakazaki and Namioka, 1960); from catfish in waters contaminated by sewage (Troast, 1975); from fatal septicemia in a chimpanzee (Schmidt and But-

ler, 1971); and from grasses, wheat, and corn (Nelson et al., 1976). *E. aerogenes* has also been isolated from sugarcane (Nunez and Colmer, 1968) and bananas (Starr and Mandel, 1969).

In addition to man, *Enterobacter agglomerans* has been isolated from a long list of plants, from wood-eating termites (Potrikus and Breznak, 1977), a beetle (Dye, 1969), deer (Graham and Hodgkiss, 1967; Muraschi, Friend, and Bolles, 1965), horses, pigs, calves, and dogs (Knud B. Pederson, personal communication). There are more than 35 synonyms for the organisms grouped as *Enterobacter agglomerans* (Dye, 1969; Ewing and Fife, 1971; Leete, 1977). Most of these names were given to strains isolated from various plant sources or soil. The names currently used are *Erwinia herbicola* var. *ananas, Erwinia herbicola* var. *herbicola, Erwinia uredovora,* and *Erwinia stewartii.*

The vast majority of *Enterobacter sakazakii, E. gergoviae,* and *E. vulneris* isolates are from humans. Any statement concerning the natural habitat of these species must therefore await the results of environmental searches.

Enterobacter species are usually considered opportunistic pathogens. They are indeed opportunistic and have become a serious cause of nosocomial infections. The incidence of *E. cloacae* in clinical specimens is assumed to be substantially greater than that of *E. aerogenes.* Many strains of *E. aerogenes* have probably been misidentified as *K. pneumoniae* and others mistakenly called *E. cloacae.* Recent data from several hospitals indicate that *E. aerogenes* is frequently isolated.

At Boston City Hospital, 200 strains of *Enterobacter,* 40% of which were *E. aerogenes,* were isolated in a 10-month period (Toala et al., 1970). Other studies indicate anywhere from a two to one ratio of *E. cloacae* to *E. aerogenes* (Dans et al., 1970; Edmondson and Sanford, 1967) to an almost two to one ratio of *E. aerogenes* to *E. cloacae* (Braunstein, Gibson, and Tucker, 1969). Most *Enterobacter* infections are hospital acquired (Buckwold, Hand, and Hansebout, 1977; Dans et al.,

Table 2. Source of human strains of *Enterobacter* species submitted to the Center for Disease Control between 1973 and 1977.

Species	Lung, throat and sputum	Urine	Wound and abscess	Blood	Stool	CSF[a]	Vagina	Ear, eye	Skin	Other	Total
Enterobacter cloacae	15	22	22	26	28	3	1			10	127
Enterobacter aerogenes	3	10	6	4	1					7	31
Enterobacter agglomerans[b]	29	36	44	36	13	3	4	9	8	11	193
Enterobacter agglomerans[c]	30	31	17	27	7	1				24	137
Enterobacter sakazakii	4	2	1	1	5	2				1	16
Enterobacter gergoviae	10	8	10	3			1		3	2	37
Enterobacter vulneris	4	17	2	1			1	1			26

[a] CSF = Cerebrospinal fluid.
[b] Center for Disease Control cultures from 1949 through 1970 (Ewing and Fife, 1971).
[c] Center for Disease Control cultures from 1971 through 1977.

1970; Edmondson and Sanford, 1967; Seidmon, Mosovich, and Neter, 1975). The possible risk associated with *E. cloacae* and *E. aerogenes* infections is exemplified by Buckwold, Hand, and Hansebout (1977), who found that they were the most frequent cause (8 of 23 cases) of hospital-acquired bacterial meningitis in neurosurgical patients. Five of these eight *Enterobacter* infections were fatal.

Hospital studies indicate that *E. cloacae* and *E. aerogenes* are most frequently isolated from sputum and next most frequently from urine (Edmondson and Sanford, 1967; Toala et al., 1970). The sources of human strains of *E. cloacae* and *E. aerogenes* submitted to the CDC between 1973 and 1977 are shown in Table 2. The numbers shown are probably not representative of the true incidence of these organisms and should, therefore, be used only as a guide. For instance, almost all stool and wound cultures are from single cases, but many of the blood isolates of *E. cloacae* were from one outbreak (Maki et al. 1976). These organisms are clearly not restricted to the respiratory and urinary tracts.

Enterobacter agglomerans was the principal organism (*E. cloacae* was also implicated) responsible for a nationwide outbreak of septicemia caused by contaminated closures on bottles of infusion fluids in the U.S. in 1970. There were 25 hospitals involved, with 378 cases and 40 deaths (Maki et al., 1976). It was this outbreak that stimulated clinical interest in this group of organisms and led to the designation of *Enterobacter agglomerans* for organisms that had happily resided in the genus *Erwinia* (and still do in *Bergey's Manual;* Cowan et al., 1974).

The first human isolates of *E. agglomerans* were reported in 1928 (see Ewing and Fife, 1971, for literature survey). In 1970, von Graevenitz reported results on 54 strains. The isolates were most frequently found in wounds, and next most frequently in sputum and urine. Gilardi and Bottone (1971) reported 61 isolates—mostly from wounds, blood, eyes, and the nasopharynx. Pien et al. (1972) reported 51 patient isolates in 18 months. Most were from wounds in the summer and autumn, and only seven were considered to be primary pathogens. Ewing and Fife (1971) summarized the sources of the cultures received at the CDC between 1949 and 1970 (Table 2). The most frequent human sources were wounds, blood, urine, and the respiratory tract. Similar results were obtained more recently by Richard (1974) in France, Ursing (1977) in Sweden, Geere (1977) in the United States, and at the CDC (Table 2).

Enterobacter sakazakii caused two fatal cases of meningitis in infants reported from England in 1961 (Urmenyi and Franklin) and a case of neonatal meningitis from Denmark in 1965 (Jøker, Nørholm, and Siboni). Gilardi and Bottone (1971) isolated four yellow-pigmented *Enterobacter* strains from wounds and urine. Two were called yellow *E. cloacae,* and two were called yellow *E. aerogenes.* They may have all been *E. sakazakii.* Only a few strains of *E. sakazakii* have been received at the CDC (Table 2). Of these, five isolates were from stools, four from sputum or the lungs, and two each from urine and spinal fluid.

Richard et al. (1976) reported the first isolates of *Enterobacter gergoviae* from France and Africa. Of 21 human isolates, 11 were from urine, 10 of which were multiply antibiotic-resistant strains from one hospital. Forty-four isolates of *E. gergoviae* have been studied at CDC. Seven of these were sent by C. Richard, and 28 strains were received from K. Tomfohrde of Analytab Products, Inc. The most frequent sources were wounds, throat and sputum, and urine (Table 2).

Enterobacter vulneris was first isolated in 1969 from the intestine of a cowbird (J. K. Leete, A. M. Murlin, and D. J. Brenner, unpublished data). One subsequent culture was from a rabbit and one from a hospital floor. The sources of 26 of 31 human isolates are given in Table 2. Seventeen of these are wound isolates.

Enterobacter species are not considered to be intestinal pathogens. These organisms can no longer be dismissed as potential diarrheal agents because both heat-labile (Wadström et al., 1976; Klipstein and Engert, 1977) and heat-stable enterotoxin activity (Klipstein and Engert, 1976; 1977) have been detected in *Enterobacter cloacae.*

Isolation

Media and tests used in the isolation of *Enterobacter* species are similar to those used for other Enterobacteriaceae. The choice of media is often governed by the type of specimen.

Specimens that cannot be immediately processed should be placed in a transport medium, such as buffered glycerol saline, to maintain a stable bacterial population and to prevent the overgrowth of any one organism. Specimens that may contain small numbers of bacteria should be incubated in a liquid medium before plating (Ewing and Martin, 1974). Specimens that contain mixed flora should be incubated in a liquid enrichment medium such as Gram-negative broth.

Plating media are categorized on the basis of their selectivity. Noninhibitory media support the growth of essentially all Enterobacteriaceae as well as a wide variety of other organisms. The most popular of these are blood agar and nutrient agar. Some laboratories prefer a differential, noninhibitory medium that supports the growth of all Enterobacteriaceae, but separates them based on their ability to utilize a given substrate. Examples are bromothymol blue–lactose agar and phenol red–lactose agar. These media must be used with caution since very few reactions are 100% positive or negative. Differential media that support almost all Enterobacteriaceae, but few other organisms, include MacConkey agar and eosin–methylene blue agar. There are no widely used moderately or highly selective differential media for *Enterobacter* species. One such medium, double violet lactose agar, was recently described (Campbell and Roth, 1975). It seems to favor the growth of *Klebsiella pneumoniae* and *E. cloacae* over that of other Enterobacteriaceae. Both *E. aerogenes* and *E. cloacae* are highly resistant to cephalosporins (Benner et al., 1965; Ramirez, 1968; Sherris, 1974; Toala et al., 1970). The incorporation of a cephalosporin into a differential medium would select for *E. aerogenes* and *E. cloacae*.

Many clinical laboratories find one or more complex solid media of value in the primary differentiation of isolated colonies. The most popular of these are triple sugar iron agar (TSI), Kligler's iron agar, lysine iron agar (LIA), and motility-indole-ornithine agar. For detailed discussions of transport, enrichment, and isolation media, the reader is referred to Ewing and Martin (1974), Edwards and Ewing (1972), and Vera and Dumoff (1974).

There are no special selective media used for the isolation of *Enterobacter agglomerans*. Most strains grow at 37°C, but many grow better at 30°C, and an occasional strain does not grow at 37°C. A combination of selective media for bacteria isolated from plants is probably the best choice for strains that will not grow on media designed for Enterobacteriaceae (Kado and Heskett, 1970).

Identification

Twenty-six tests, in addition to Gram stain, oxidase, nitrate reduction, and motility, are recommended for the identification of *Enterobacter* species in clinical laboratories (Ewing and Martin, 1974). Additional tests are often necessary for biotyping or in identifying atypical strains. Biochemical reactions given by *Enterobacter* species studied in the Enteric Section of the Bacteriology Division, CDC, are shown in Table 3. The percentages are presented as a guide, not as absolute values. The cultures seen at the CDC may not be representative of those seen in any single laboratory. For example, a disproportionately large number of atypical strains are sent to the CDC for identification.

It is extremely important to identify an organism on the basis of its entire biochemical profile, rather than to include it in or exclude it from consideration due to one or two atypical reactions. *Enterobacter* species are usually negative for H_2S, indole, phenylalanine deaminase, dulcitol, erythritol, and corn

oil. They are motile, reduce nitrates to nitrites, and ferment glucose, mannitol, arabinose, rhamnose, xylose, trehalose, cellobiose, maltose, melibiose, and ONPG. Decarboxylase test results vary between species and are of great help in differentiation. These and other tests useful in differentiating between species of *Enterobacter* and between *Enterobacter, Klebsiella pneumoniae,* and *Hafnia alvei* are shown in Table 4 (also see Brenner et al., 1977; Edwards and Ewing, 1972; Ewing and Fife, 1971; Ewing and Martin, 1974).

E. aerogenes and *K. pneumoniae* are easily separated by DNA hybridization (Brenner, Steigerwalt, and Fanning, 1972). *K. pneumoniae* has never been found to be motile, but nonmotile strains of *E. aerogenes* are occasionally seen. The practice of identifying only motile strains as *E. aerogenes* is a serious error (Matsen and Blazevic, 1969). These organisms should be separated by motility, ornithine decarboxylase, and urease reactions (Table 4), and also by antibiotic susceptibility. *E. cloacae* and *E. sakazakii* are separated from one another on the basis of pigment, sorbitol, and delayed DNase reactions. The organisms that are comprised in *E. agglomerans* share negative decarboxylase reactions and little else. It is often quite difficult to differentiate unpigmented strains of *E. agglomerans* from decarboxylase-negative *E. coli* strains. The cellobiose reaction is often helpful in these cases (*E. agglomerans* is usually positive).

Strains of three species of *Enterobacter* are usually yellow pigmented. All *E. sakazakii* strains studied so far are yellow pigmented. About two-thirds of *E. vulneris* strains and *E. agglomerans* strains are yellow pigmented. Yellow pigment production is often media dependent and temperature dependent, occurring well at 22°C to 25°C and often poorly or not at all at 37°C. This is also apparently true for members of the *E. agglomerans* group. A higher percentage of plant isolates than of clinical isolates are yellow pigmented. Some pigmented strains grown at 37°C produce unpigmented variants at a high rate. These unpigmented mutants are auxotrophic for thiamine and do not revert for either character. This phenomenon has not been observed when strains are grown at 28°C to 30°C (Chatterjee and Gibbins, 1971).

Sakazaki and Namioka (1960) developed a serotyping scheme for *E. cloacae*. They found 53 different O antigen groups, 56 different H antigen groups, and 79 different serotypes among 170 cultures. There were no cross-reactions between *E. cloacae* and *Klebsiella,* but many cross-reactions occurred between the *E. cloacae* O and H groups. Encapsulated strains of *E. aerogenes* and *E. cloacae* had positive quelling and agglutination reactions with *Klebsiella* capsular antisera (Edwards and Ewing, 1962). These reactions are not specific—each *Enterobacter* strain reacts with several *Klebsiella*

Table 3. Biochemical reactions of *Enterobacter* species.[a]

Test	*Enterobacter cloacae* Sign	%+	(%+)	*Enterobacter aerogenes* Sign	%+	(%+)	*Enterobacter agglomerans aerogenic* Sign	%+	(%+)	*Enterobacter agglomerans anaerogenic* Sign	%+	(%+)	*Enterobacter sakazakii* Sign	%+	(%+)	*Enterobacter gergoviae* Sign	%+	(%+)	*Enterobacter vulneris* Sign	%+	(%+)
Hydrogen sulfide (TSI)	−	1		−	1		−	1		−	1		−	1		−	1		−	1	
Urease	v	65		−	3		v	46		v	21	(10)	−	1		+	99		−	1	
Indole	−	1		−	1		v	32		v	17		v	16		−	1		−	1	
Methyl red	−	3		−	1		v	44		v	47		v	22		−	2		+	99	
Voges-Proskauer	+	99		+	99		v	59		v	66		+	97		+	99		−	1	
Citrate (Simmons')	+	99		+	94		v	59	(6)	v	62	(26)	+	99		+	99		−	1	
KCN	+	98		+	99		v	64		v	34		+	94		−	1		v[c]	47	(41)
Motility	+	95		+	97		v	87		v	88		+	94		+	90		+	99	
Gelatin (22°C)	(+)	1	(96)	−	1	(77)	−	6	(57)	−	4	(83)	−	1		−	1		−	1	
Lysine decarboxylase	−	1		+	99		−	1		−	1		−	1		+w	99		+d	91	(6)
Arginine dihydrolase	+	97		−	1		−	1		−	1		+	99		−	1		v	41	(56)
Ornithine decarboxylase	+	96		+	99		−	1		−	1		+	97		+	99		−	1	
Phenylalanine deaminase	−	1		−	1		v	17		v	30		−	1		−	1		−	1	
D-Glucose: acid	+	99		+	99		+	99		+	99		+	99		+	99		+	99	
D-Glucose: gas	+	99		+	99		+	99		−	1		+	97		+	99		+	99	
Acid from:																					
Lactose	+	94	(5)	+	92	(5)	+	94	(3)	v	33	(10)	+	99		+ or(+)	57	(41)	(+)	12	(87)
Sucrose	+	97		+	99		v	68	(4)	v	73	(2)	+	99		+	99		v	12	(74)
D-Mannitol	+	99		+	99		+	99		+	99		+	99		+	99		+	99	
Dulcitol	v	13		−	4		−	2	(1)	v	57		−	6		−	1		−	1	
Salicin	+ or(+)	76	(18)	+	99		+	99		v	60	(21)	+	99		+	99		+ or(+)	26	(73)
Adonitol	v	28		+	99		v	27		−	3		−	1		−	1		−	1	
i-Inositol	v	22	(12)	+	99		−	9	(3)	v	21	(6)	v	72		−	2	(2)	−	1	
D-Sorbitol	+	95		+	99		v	62		v	17		−	1		−	1		−	1	
L-Arabinose	+	99		+	99		+	97		+	97		+	99		+	98		+	99	
Raffinose	+	97		+	96		v	20	(5)	v	65	(2)	+	99		+	99		+	99	
L-Rhamnose	+	92	(2)	+	99		+	99		v	80	(6)	+	99		+	99		+	91	(3)
Malonate	v	81		v	75		v	59		v	67		v	16		+	99		v	88	
Mucate	v	76		+	95		v	75		v	37		−	1		−	1		+	99	
D-Xylose	+	98		+	99		+	96		+ or(+)	89	(5)	+	99		+	99		+	99	
Trehalose	+	99		+	99		+	99		+	99		+	99		+	99		+	99	
Cellobiose	+	99		+	99		+	94	(3)	v	43	(20)	+	99		+	99		+	99	
Maltose	+	99		+	99		+	99		+ or(+)	85	(5)	+	99		+	99		+	99	
α-CH₃-Glucoside	+ or(+)	84	(13)	+	96		v	21	(3)	−	4	(1)	+	94		−	2		v	26	
Erythritol	−	1		−	1		−	1		−	1		−	1		−	1		−	1	
Esculin	v	29		+	98		+	93		v	59	(24)	+	99		+	99		+ or(+)	12	(82)
Melibiose	+	99		+	99		N[b]			N[b]			+	99		+	98		+	99	
ONPG	+	99		+	99		+	90		+	90		+	99		+	98		+	99	
Glycerol	v	43	(45)	+	99		v	32	(37)	v	20	(30)	+	94		+	99		N[b]		
Tartrate (Jordan's)	−	28		v	89		v	10		−	1		−	1		+	99		−	1	
Acetate	N[b]			+	90		v	68	(6)	v	49	(29)	+	94		+	98	(2)	+ or(+)	9	(82)
Corn oil	−	1	(2)	−	1		−	1		−	1		−	1		−	1		−	1	
Deoxyribonuclease (25°C)	−	1		−	1		−	1		−	1		+ or (+)	10	(89)	−	1		−	1	
Nitrate to nitrite	+	99		+	99		+	97		v	82		+	99		+	99		+	99	
Yellow pigment	−	1		−	1		v	58		v	74		+	99		−	1		v	65	

[a] + = 90% or more positive within 48 h; (+) = 90% or more delayed positive reaction in 3–7 days; − = less than 10% positive; v = variable = 10.1 to 89.9% positive; w = weak reaction. All reactions done at 36°C ± 1°C except for gelatin (22°C) and deoxyribonuclease (25°C). No percentage is given lower than 1% or higher than 99% in order to stimulate workers to consider no single reaction to be inviolate.

[b] N = insufficient data.

[c] Positive strains are very weakly positive strains positive at 1–2 days are often negative at 3–4 days.

Table 4. Differentiation of *Enterobacter* species from each other and from *Klebsiella pneumoniae* and *Hafnia alvei*.[a]

Test	E. cloacae	E. aerogenes	E. agglomerans	E. sakazakii	E. gergoviae	E. vulneris	K. pneumoniae	H. alvei
Motility	+	+	V	+	+	+	−	+
Lysine decarboxylase	−	+	−	−	+W	+	+	+
Arginine dihydrolase	+	−	−	+	−	V	−	V
Ornithine decarboxylase	+	+	−	+	+	−	−	+
Urea	V	−	V	−	+	−	+	−
Methyl red	−	−	V	V	−	+	V	V
Voges-Proskauer	+	+	V	+	+	−	+	V
Citrate (Simmons')	+	+	V	+	+	−	+	V
KCN (Growth)	+	+	V	+	−	V	+	+
Adonitol	V	+	V	−	−	−	V	−
i-Inositol	V	+	V	V	−	−	+	−
D-Sorbitol	+	+	V	−	−	−	+	−
Mucate	V	+	V	−	−	+	+	−
Deoxyribonuclease	−	−	−	+ or (+)	−	−	−	−
Yellow pigment	−	−	V	+	−	V	−	−

[a] Reactions done at 36°C ± 1°C. + = 90% or more positive within 48 h; V = 10.1–89.9% positive within 48 h; (+) = positive in 3–7 days; − = 0–10% positive; W = weakly positive reaction.

capsule types. The biochemical structure of capsular antigens in *E. aerogenes* and *Klebsiella* is quite similar (Eriksen, 1965). Serological schemes are not available for other *Enterobacter* species.

Biotyping schemes are not routinely used to distinguish between strains of any *Enterobacter* species. Bacteriocin and phage typing schemes are not used.

Antimicrobial Susceptibility

Enterobacter cloacae and *E. aerogenes* have similar qualitative and quantitative responses to antibiotics (Benner et al., 1965; Finland et al., 1976a,b; Toala et al., 1970). Multiply resistant strains, especially to aminoglycosides and tetracyclines, are encountered in many hospitals. They are highly resistant to penicillins, except for carbenicillin (Finland et al., 1976b; Sherris, 1974; Toala et al., 1970). Resistance to penicillins is due to the presence of β-lactamases (Farrar and Newsome, 1976). *Enterobacter cloacae* and *E. aerogenes* are also highly resistant to most cephalosporins (Benner et al., 1965; Finland et al., 1976b; Ramirez, 1968; Sherris, 1974; Toala et al., 1970), again due to β-lactamases (Farrar and Newsome, 1976). Wild-type klebsiellae are susceptible to cephalosporins (Finland et al., 1976a,b; Sherris, 1974; Toala et al., 1970), but hospital strains are sometimes resistant (Benner et al., 1965; Ramirez, 1968). The most effective commonly used antibiotics against *E. cloacae* and *E. aerogenes* are gentamicin and kanamycin. Among the newer antibiotics, amikacin, sisomicin, and verdamicin are effective (Finland et al., 1976a,b). Acquisition of resistance is common and, therefore, average minimal inhibitory concentrations (Table 5) should be used only as

a guide in those cases where treatment must be initiated before a susceptibility profile is available.

Enterobacter agglomerans strains had minimal inhibitory concentrations indicating resistance (not inhibited by 25 µg/ml) to penicillin, erythromycin, and clindamycin; relative resistance (inhibited between 12.5 µg/ml and 25 µg/ml) to cephalothin, ampicillin, carbenicillin, and nitrofurantoin; and susceptibility (inhibited by less than 12.5 µg/ml) to other antibiotics (Pien et al., 1972). Generally, similar results were obtained in antibiotic disk sensitivity tests (Leete, 1977; Table 5). *Enterobacter vulneris* was also resistant to clindamycin, penicillin, carbenicillin, and erythromycin, and was susceptible to other antibiotics (Leete, 1977; Table 5).

Very little has been published on the antibiotic susceptibility of *E. sakazakii* and *E. gergoviae*. Gilardi and Bottone (1971) reported that four strains of pigmented *Enterobacter* (apparently *E. sakazakii*) were resistant to penicillin, novobiocin, erythromycin, lincomycin, and cephalothin, and susceptible to gentamicin, polymyxin B, kanamycin, neomycin, nalidixic acid, nitrofurazone, streptomycin, chloramphenicol, and tetracycline. Richard et al. (1976) reported that most strains of *E. gergoviae* were sensitive to antibiotics, but 10 strains isolated from a hospital outbreak were multiply resistant.

Literature Cited

Aly, R., Maibach, H. I. 1977. Aerobic microbial flora of intertrigenous skin. Applied and Environmental Microbiology **33**:97–100.

Benner, E. J., Micklewait, J. S., Brodie, J. L., Kirby, W. M. M. 1965. Natural and acquired resistance of *Klebsiella-Aerobacter* to cephalothin and cephaloridine. Proceedings of the Society for Experimental Biology and Medicine **119**:536–541.

Table 5. Antibiotic susceptibility patterns of *Enterobacter* species.[a]

Antibiotic	*Enterobacter* cloacae and aerogenes[b]	*Enterobacter* agglomerans[b]	*Enterobacter* vulneris[c]
Amikacin	S	S	S
Ampicillin	R	r	S
Carbenicillin	S	R	R
Cefamandole	NT	S	S
Cephalothin	R	r	S
Chloramphenicol	r	S	S
Clindamycin	R	R	R
Colistin	R	NT	NT
Erythromycin	R	R	R
Gentamicin	S	S	S
Kanamycin	S	S	S
Nafcillin	R	NT	NT
Nalidixic acid	S	S	S
Neomycin	S	NT	NT
Nitrofurantoin	R	R	S
Novobiocin	R	NT	NT
Penicillin	R	R	R
Polymyxin B	r	S	S
Rifampin	R	NT	NT
Sisomicin	S	NT	NT
Streptomycin	S	S	NT
Tetracycline	S	S	S
Tobramycin	NT	S	S
Trimethoprim	S	S	S
Vancomycin	NT	S	NT

[a] Data on *E. cloacae* and *E. aerogenes* from Finland et al., 1976a,b. Data on *E. agglomerans* from Pien et al., 1972. Data on *E. vulneris* from Leete, 1977.

[b] Minimal inhibitory concentration: resistant = R = not inhibited by 25 g/ml; moderately resistant = r = inhibited between 12.5 g/ml and 25 g/ml; susceptible = S = inhibited by less than 12.5 g/ml; NT = not tested.

[c] Zone sizes indicating resistance and sensitivity were interpreted according to the manufacturer's recommendation for each antibiotic tested.

Braunstein, H., Gibson, B. C., Tucker, E. B. 1969. Species distribution of the *Klebsiella-Enterobacter* group at the University of Kentucky Medical Center. Southern Medical Journal **62:**1491–1495.

Breed, R. S., Murray E. G. D. 1957. Family IV. Enterobacteriaceae Rahn, 1937, pp. 333, 341–345, 346–347, 350–351. In: Breed, R. S. Murray E. G. D., Smith, N. R. (eds.), Bergey's manual of determinative bacteriology, 7th ed. Baltimore: Williams & Wilkins.

Brenner, D. J., Steigerwalt, A. G., Fanning, G. R. 1972. Differentiation of *Enterobacter aerogenes* from klebsiellae by deoxyribonucleic acid reassociation. International Journal of Systematic Bacteriology **22:**193–200.

Brenner, D. J., Farmer, J. J., III, Hickman, F. W., Asbury, M. A., Steigerwalt, A. G. 1977. Taxonomic and nomenclatural changes in Enterobacteriaceae. Atlanta, Georgia: Center for Disease Control.

Buckwold, F. J., Hand, R., Hansebout, R. R. 1977. Hospital-acquired bacterial meningitis in neurosurgical patients. Journal of Neurosurgery **46:**494–500.

Campbell, L. M., Roth, I. L. 1975. Methyl violet: A selective agent for differentiation of *Klebsiella pneumoniae* from *Enterobacter aerogenes* and other Gram-negative organisms. Applied Microbiology **30:**258–261.

Chatterjee, A. K., Gibbins, L. N. 1971. Induction of nonpigmented variants of *Erwinia herbicola* by incubation at supraoptimal temperatures. Journal of Bacteriology **105:**107–112.

Cowan, S. T., Ørskov, F., Sakazaki, R., Sedlak, J., Le Minor, L., Rohde, R. Carpenter, K. P., Ørskov, I.,

Lautrop, H., Mollaret, H. H., Thal, E., Lelliott, R. A. 1974. Enterobacteriaceae, pp. 290–340. In: Buchanan, R. E., Gibbons, N. E., (eds.), Bergey's manual of determinative bacteriology, 8th ed. Baltimore: Williams & Wilkins.

Dans, P. E., Barrett, F. F., Casey, J. I., Finland, M. 1970. *Klebsiella-Enterobacter* at Boston City Hospital, 1967. Archives of Internal Medicine **125:**94–101.

Dye, D. 1969. A taxonomic study of the genus *Erwinia*. III. The "Herbicola" group. New Zealand Journal of Science **12:**223–236.

Edmondson, E. B., Sanford, J. P. 1967. The *Klebsiella-Enterobacter* (*Aerobacter*)-*Serratia* group: A clinical and bacteriological evaluation. Medicine **46:**323–340.

Edwards, P. R., Ewing, W. H. 1962. Identification of Enterobacteriaceae, 2nd ed. Minneapolis: Burgess Publishing Co.

Edwards, P. R., Ewing, W. H. 1972. Identification of Enterobacteriaceae, 3rd ed. Minneapolis: Burgess Publishing Co.

Eriksen, J. 1965. Immunochemical studies on some serological cross-reactions in the *Klebsiella* group. II. Structure of the capsular polysaccharides of *Klebsiella aerogenes* strain B. 1076/48 and "Enterobacter" strain 349. Acta Pathologica et Microbiologica Scandinavica **64:**511–521.

Ewing, W. H., Fife, M. A. 1971. *Enterobacter agglomerans*. The Herbicola-Lathyri bacteria. Atlanta, Georgia: Center for Disease Control.

Ewing, W. H., Fife, M. A.: 1972. *Enterobacter agglomerans* (Beijerinck) comb. nov. (the Herbicola-Lathyri bacteria). International Journal of Systematic Bacteriology **22:**4–11.

Ewing, W. H., Martin, W. J., 1974. Enterobacteriaceae, pp.

189–221. In: Lennette, E. H., Spaulding, E. H., Truant, J. P., (eds.), Manual of clinical microbiology, 2nd ed. Washington D. C.: American Society for Microbiology.

Farrar, W. E., Jr., Newsome, J. K. 1976. Diversity of β-lactamase activity among clinical isolates of Gram-negative bacilli. American Journal of Clinical Pathology **65:**570–574.

Finland, M., Garner, C., Wilcox, C., Sabath, L. D. 1976a. Susceptibility of "enterobacteria" to aminoglycoside antibiotics: Comparisons with tetracyclines, polymyxins, chloramphenicol, and spectinomycin. Journal of Infectious Diseases, Suppl. **134:**57–74.

Finland, M., Garner, C., Wilcox, C., Sabath, L. D. 1976b. Susceptibility of "enterobacteria" to penicillins, cephalosporins, lincomycins, erythromycin, and rifampin. Journal of Infectious Diseases **134:**S75–S96.

Gavini, F., Ferragut, C., Leclerc, H. 1976. Etude taxonomique d'entérobactéries appartenant ou apparentées au genre *Enterobacter*. Annales de Microbiologie (Institut Pasteur) **127B:**317–335.

Geere, I. W. 1977. *Enterobacter agglomerans:* The clinically important plant pathogen. Canadian Medical Association Journal **116:**517–519.

Gilardi, G. L., Bottone, E. 1971. *Erwinia* and yellow-pigmented *Enterobacter* isolates from human sources. Antonie van Leewenhoek Journal of Microbiology and Serology, **37:**529–535.

Graham, D. C., Hodgkiss, W. 1967. Identity of Gram negative, yellow pigmented, fermentative bacteria isolated from plants and animals. Journal of Applied Bacteriology **30:**175–189.

Hormaeche, E., Edwards, P. R. 1960a. A proposed genus *Enterobacter*. International Bulletin of Bacteriological Nomenclature and Taxonomy **10:**71–74.

Hormaeche, E., Edwards, P. R. 1960b. Proposal for the rejection of the generic name *Cloaca* Castellani and Chalmers, and proposal of *Enterobacter* as a generic name with designation of type species and of its type culture. International Bulletin of Bacteriological Nomenclature and Taxonomy **10:**75–76.

Izard, D., Gavini, F., Trinel, P. A., Leclerc, H. 1981. Deoxyribonucleic acid relatedness between *Enterobacter cloacae* and *Enterobacter amnigenus* sp. nov. International Journal of Systematic Bacteriology **31:**35–42.

Jøker, R. N., Nørholm, T., Siboni, K. E. 1965. A case of neonatal meningitis caused by a yellow *Enterobacter*. Danish Medical Bulletin **12:**128–130.

Kado, C. I., Heskett, M. G. 1970. Selective media for isolation of *Agrobacterium, Corynebacterium, Erwinia, Pseudomonas,* and *Xanthomonas*. Phytopathology **60:**969–976.

Klipstein, F. A., Engert, R. F. 1976. Partial purification and properties of *Enterobacter cloacae* heat-stable enterotoxin. Infection and Immunity **13:**1307–1314.

Klipstein, F. A., Engert, R. F. 1977. Immunological interrelationships between cholera toxin and the heat-labile and heat-stable enterotoxins of coliform bacteria. Infection and Immunity **18:**110–117.

Leete, J. K. 1977. Characterization of *Enterobacter agglomerans* by DNA hybridization and biochemical reactions. DPH thesis, University of North Carolina, Chapel Hill, North Carolina.

Maki, D. G., Rhame, F. S., Mackel, D. C., Bennett, J. V. 1976. Nationwide epidemic of septicemia caused by contaminated intravenous products. I. Epidemiologic and clinical features. American Journal of Medicine **60:**471–485.

Matsen, J. M., Blazevic, D. J. 1969. Characterization of ornithine decarboxylase-positive, nonmotile strains of the *Klebsiella-Enterobacter* group. Applied Microbiology **18:**566–569.

Muraschi, T. F., Friend, M., Bolles, D. 1965. *Erwinia*-like microorganisms isolated from animal and human hosts. Applied Microbiology **13:**128–131.

Nelson, A. D., Barber, L. E., Tjepkema, J., Russell, S. A., Powelson, R., Evans, H. J., Seidler, R. J. 1976. Nitrogen fixation associated with grasses in Oregon. Canadian Journal of Microbiology **22:**523–530.

Newton, K. G., Harrison, J. C. L., Smith, K. M. 1977. Coliforms from hides and meat. Applied and Environmental Microbiology **33:**199–200.

Nunez, W. J., Colmer, A. R. 1968. Differentiation of *Aerobacter-Klebsiella* isolated from sugarcane. Applied Microbiology **16:**1875–1878.

Opinion 28. 1963. Rejection of the bacterial generic name *Cloaca* Castellani and Chalmers and acceptance of *Enterobacter* Hormaeche and Edwards as a bacterial generic name with type species *Enterobacter cloacae* (Jordan) Hormaeche and Edwards. International Bulletin of Bacteriological Nomenclature and Taxonomy **13:**38.

Pien, F. D., Martin, W. J., Hermans, P. E., Washington, J. A., II. 1972. Clinical and bacteriologic observations on proposed species, *Enterobacter agglomerans* (Herbicola-Lathyri bacteria). Mayo Clinic Proceedings **47:**739–745.

Platt, H., Atherton, J. G., Ørskov, I. 1976. *Klebsiella* and *Enterobacter* organisms isolated from horses. Journal of Hygiene **77:**401–408.

Potrikus, C. J., Breznak, J. A. 1977. Nitrogen-fixing *Enterobacter agglomerans* isolated from guts of wood-eating termites. Applied and Environmental Microbiology **33:**392–399.

Ramirez, M. J. 1968. Differentiation of *Klebsiella-Enterobacter (Aerobacter)-Serratia* by biochemical tests and antibiotic susceptibility. Applied Microbiology **16:**1548–1550.

Richard, C. 1974. Une nouvelle espèce d'entérobactérie rencontrée in bactériologie médicale Enterobacter agglomerans (ex-Erwinia herbicola). Étude biochimique de 205 souches. Annales de Biologie Clinique **32:**341–346.

Richard, C., Joly, B., Sirot, J., Stoleru, G. H., Popoff, M.: 1976. Étude de souches de *Enterobacter* — appartenant a un groupe particulier proche de *E. aerogenes*. Annales de l'Institut Pasteur **127A:**545–548.

Sakazaki, R., Namioka, S. 1960. Serological studies on the *Cloaca (Aerobacter)* group of enteric bacteria. Japanese Journal of Medical Science and Biology **13:**1–12.

Schmidt, R. E., Butler, T. M. 1971. *Klebsiella-Enterobacter* infections in chimpanzees. Laboratory Animal Sciences **21:**946–949.

Seidmon, E. J., Mosovich, L. L., Neter, E. 1975. Colonization by *Enterobacteriaceae* of the respiratory tract of children with cystic fibrosis of the pancreas and their antibody response. Journal of Pediatrics **87:**528–533.

Sherris, J. C., 1974. Future needs, pp. 439–442. In: Lennette, E. H., Spaulding, E. H., Truant, J. P., (eds.), Manual of clinical microbiology, 2nd ed. Washington D. C.: American Society for Microbiology.

Starr, M. P., Mandel, M. 1969. DNA base composition and taxonomy of phytopathogenic and other enterobacteria. Journal of General Microbiology **56:**113–123.

Toala, P., Lee, Y. H., Wilcox, C., Finland, M.: 1970. Susceptibility of *Enterobacter aerogenes* and *Enterobacter cloacae* to 19 antimicrobial agents *in vitro*. American Journal of the Medical Sciences **260:**41–55.

Troast, J. L. 1975. Antibodies against enteric bacteria in brown bullhead catfish (*Ictalurus nebulosus,* LeSueur) inhabiting contaminated waters. Applied Microbiology **30:**189–192.

Urmenyi, A. M. C., Franklin, A. W. 1961. Neonatal death from pigmented coliform infection. Lancet **i:**313–315.

Ursing, J. 1977. Characterization of *Enterobacter agglomerans* (*Erwinia* spp.) from clinical specimens. Acta Pathologica et Microbiologica Scandinavica, Sect. B **85:**61–66.

Vera, H. D., Dumoff, M. 1974. Culture media, pp. 881–929. In: Lennette, E. H., Spaulding, E. H., Truant, J. P., (eds.), Manual of clinical microbiology, 2nd ed. Washington D.C.: American Society for Microbiology.

von Graevenitz, A. 1970. *Erwinia* species isolates. Annals of the New York Academy of Sciences **174:**436–443.

Wadström, T., Aust-Kettis, A., Habte, D., Holmgren, J., Meeuwisse, G., Möllby, R., Söderlind, O. 1976. Enterotoxin-producing bacteria and parasites in stools of Ethiopian children with diarrhoeal disease. Archives of Diseases in Childhood **51:**865–870.

CHAPTER 96

The Genus *Hafnia*

RIICHI SAKAZAKI

The bacteria of the genus *Hafnia* have been described under several names: *Bacillus asiaticus* and *Bacillus asiaticus* var. *mobilis* (Castellani, 1912 cited by Ewing and Fife, 1968), *Bacterium cadaveris* (Gale and Epps, 1943), biotype 32011 of the genus *Aerobacter* (Stuart et al., 1943), *Hafnia alvei* (Møller, 1954), *Enterobacter alvei* (Sakazaki, 1961), *Enterobacter* (*Aerobacter*) *aerogenes* subsp. *hafniae* (Ewing, 1963), and *Enterobacter hafniae* (Ewing and Fife, 1968). However, *Hafnia alvei* Møller 1954 is the only correct name for these bacteria. During his studies on amino acid decarboxylases among members of the family Enterobacteriaceae, Møller (1954) found a new group of bacteria in which a supposedly authentic strain of *Bacillus paratyphi-alvei* of Bahr (1919) was included. He proposed the name *Hafnia alvei* for this group of bacteria, since he considered that the Bahr strain ought to be regarded as the type species of this group. Ewing and Fife (1968) pointed out that the Bahr strain labeled *B. paratyphi-alvei* was not an authentic strain of the species, because biochemical reactions of the strain were not the same as those reported by Bahr (1919). They considered, therefore, that the specific epithet *alvei* was illegitimate. However, the generic name *Hafnia* and the specific epithet *alvei* were legitimate, since there was no doubt that the Bahr strain studied by Møller (1954) was a new bacterium at that time, whether characters of the strain were different from those described by Bahr (1919) or not. In addition to this, Sakazaki (1961) and Ewing and Fife (1968) suggested a combination *Enterobacter alvei* or *Enterobacter hafniae* for this bacterium. Since the generic name *Hafnia* Møller 1954 has priority over the name *Enterobacter* Hormaeche and Edwards 1960, these combinations were illegitimate. In numerical taxonomy studies, it was indicated that *Hafnia alvei* retained a position separate from *Enterobacter;* this indication appeared to justify the status of *Hafnia* as a separate genus rather than including in the genus *Enterobacter* (Gavini et al., 1976; Johnson et al., 1975). In DNA reassociation studies, Steigerwalt et al. (1976) indicated that *Hafnia alvei* showed only 11–26% association with either *Enterobacter* or *Klebsiella*.

Only a single species, *Hafnia alvei*, has been designated, although many biovars may be seen among strains in this genus.

Habitats

Earlier medical bacteriologists (Deacon, 1952; Eveland and Faber, 1953; Stuart and Rustigian, 1943; Stuart et al., 1943) incriminated biotype 32011 of paracolon bacteria, which is now classified within *Hafnia alvei* as a causative agent of intestinal disorders, since strains of this biotype were frequently isolated from patients with diarrhea. Harada, Shimizu, and Matsuyama (1957) reported the isolation of *H. alvei* from sporadic cases of gastroenteritis. Emslie-Smith (1961) reported a possible role of *H. alvei* in producing gastroenteritis. However, no conclusive evidence on enteropathogenicity of *H. alvei* has been obtained from any of those studies. On the other hand, Kauffmann (1954) described the organisms of this group to be probably nonpathogenic for humans. Matsumoto (1963) reported that *Hafnia* organisms were isolated from 13% of stool specimens from apparently healthy individuals using salmonella-shigella agar. Sakazaki (1966, unpublished data) found the hafniae in 42% of normal stool specimens.

Hafniae occur in various diseased conditions in man and animals and seem to be opportunistic pathogens. They are frequently found in clinical bacteriology not only in stool but also in sputum, urine, and specimens from wounds, abscesses, autopsies, the throat, and abdominal cavity (Sakazaki, unpublished data; Washington, Birk, and Ritts, 1971; Whitby and Muir, 1961). Englund (1969) reported a case of septicemia due to *Hafnia* spp. Kume (1962) described a case of equine abortion in which *Hafnia* spp. were isolated from the fetus and lochia in mostly pure culture, and a marked rise of agglutinin titer to the organisms was demonstrated in acute and convalescent sera collected from the dam. It is difficult, however, to assign any clear-cut clinical significance to this organism in most cases. Although hafniae are frequently encountered in clinical specimens, they are found in mixed cultures in most cases. Even in clinical specimens from which the organism is isolated in pure culture, some underlying illness or predisposing factor may be found. Washington, Birk, and Ritts (1971) reviewed the epidemiology of hafniae isolated in their laboratory. They concluded that the majority of the isolates originated in the respiratory tract and were considered to be commensals, while a few were secondary invaders. They also reported that previous administration of semisynthetic penicillins or cephalosporins was a common feature of patients who acquired the organism nosocomially.

The hafniae are distributed in animals and birds and sometimes in natural environments such as soil, sewage, and water. McClure, Eveland, and Kase (1957) reported the occurrence of *Paracolobactrum aerogenoides,* which is now called *Hafnia alvei,* in wild and caged birds, none of which appeared sick.

Isolation

No special media have been devised for the isolation of *Hafnia* organisms. No selective isolation media or less selective isolation media for enterobacteria, such as eosin-methylene blue, MacConkey, deoxycholate-lactose, xylose-lysine-deoxycholate (XLD), and Hektoen enteric agars are used for the isolation of the hafniae in routine laboratories. Highly selective isolation media such as salmonella-shigella, deoxycholate-citrate, and deoxycholate-citrate-lactose-sucrose (DCLS) agars are also available for this purpose, although about one-fourth of the *Hafnia* strains would be inhibited on these media (Sakazaki, unpublished). Hafniae may not grow on bismuth-sulfite agar. On isolation media mentioned above, the *Hafnia* organisms produce colorless colonies that closely resemble those of *Salmonella* because they are non-lactose fermenters. However, some strains of the organism may form red or pink colonies on the media containing sucrose, such as DCLS and XLD agars, because of acid production from this sugar. The colonies on agar plates may be translucent, circular, and low-convex, with a smooth surface and entire edge. The colonies on the less inhibitory agar plates are relatively large, with an average diameter 2–3 mm, but they vary considerably in size. Rarely does a strain produce mucoid colonies.

There are no selective enrichment broth media for the isolation of hafniae. Many strains of the organism may grow in selenite and tetrathionate broths, which are well-known enrichment media for the isolation of salmonellae, but some strains fail to grow in those broths. Although no special media are employed when dealing with clinical materials, Sakazaki (unpublished data) used a differential isolation medium, deoxycholate-lactose-sucrose-sorbitol agar, to isolate hafniae from stool specimens. In the early study of Sakazaki (1961), the *Hafnia* organisms were not included in his collection in which several thousand cultures of Enterobacteriaceae were isolated from healthy individuals. Later, however, Sakazaki (unpublished data) isolated hafniae from 42% of 598 normal stool specimens using this medium, while only 2% of the specimens showed growth of hafniae on MacConkey agar.

Deoxycholate-Lactose-Sucrose-Sorbitol Agar for Isolating *Hafnia*

The agar medium contains, per liter:

Yeast extract 5.0 g	Sucrose 5.0 g
Lactose 10.0 g	Sodium deoxycholate
D-Sorbitol 10.0 g	2.5 g
Sodium citrate 20.0 g	Ferric citrate 1.0 g
Neutral red 0.02 g	Agar 15.0 g
Trypticase 5.0 g	

The medium is adjusted to pH 7.4 and should not be overheated and autoclaved. Colorless colonies of the *Hafnia* cultures on this agar medium may be easily differentiated from other intestinal inhabitants such as *Escherichia, Citrobacter, Klebsiella, Enterobacter,* and *Serratia* which usually produce red or pink colonies because of fermentation of lactose, sucrose or/and sorbitol. Most of the *Hafnia* strains fail to ferment these sugars. The growth of coliforms is more or less inhibited on this medium.

Although some characteristics such as acetoin production are more active in culture at room temperature than at 35°C, the latter temperature is suitable for the incubation of the isolation plate culture. A good growth may be obtained after overnight incubation on each of the isolation agars.

Any commonly employed method is acceptable for the maintenance of *Hafnia* cultures. The hafniae will remain viable up to a year without subculture when the culture is stabbed into a tube with semisolid medium and the tube is sealed with a rubber or cork stopper that has been soaked in hot paraffin and then kept at room temperature in the dark. A semisolid nutrient agar may be used for this purpose. According to the author's experience, the following medium is an excellent choice.

Semisolid Nutrient Agar for *Hafnia* Maintenance

The medium contains, per liter: nutrient broth (Difco), 8.0 g; sodium chloride, 0.5 g; agar, 3.0 g; pH 7.2. The medium is tubed in 4–5 ml amount in a 13-×-100-mm tube and sterilized at 121°C for 15 min. The medium should not contain any carbohydrate.

Identification

The genus *Hafnia* is composed of Gram-negative, motile, peritrichously flagellated rods that conform to the definition of the family Enterobacteriaceae. Most cultures of the genus *Hafnia* are not encapsulated. There is some discrepancy on the report of DNA base composition of *Hafnia*. It was reported as 52–57 mol% in the eighth edition of *Bergey's Manual of Determinative Bacteriology* (Sakazaki, 1974), but as 48.0–48.7 mol% by Ritter and Gerloff (1966) and Starr and Mandel (1969). This difference may be due to strains and methods employed.

Only one species, *Hafnia alvei,* is recognized in the genus *Hafnia* at the present time, but the possibility of the presence of another species has been suggested from the results of numerical taxonomy studies by Johnson et al. (1975) and Gavini et al. (1976). The type strain of *H. alvei* is Stuart 32011; ATCC 13337; NCTC 8106.

Biochemical reactions given by 840 strains of the species of the author's collection are summarized in Table 1. On the results of some tests such as methyl red, Voges-Proskauer, and citrate utilization as a sole carbon source, the temperature of incubation has a remarkable effect. Methyl red and Voges-Proskauer tests may produce different reactions by different strains at 35°C. At 22–25°C, almost all strains give a negative reaction to methyl red test and a positive reaction to the Voges-Proskauer test. Citrate may be utilized by the majority of strains after 3–4 days at 22–25°C, whereas many strains may give a negative reaction to this test at 35°C. Motility is also increased when cultures are incubated at room temperature.

Møller (1954) and Kauffmann (1954) defined *Hafnia* organisms as hydrogen sulfide producers, since the majority of strains of *Hafnia* slightly darken ferric chloride-gelatin medium (Kauffmann, 1954) and SIM medium (Difco). However, they fail to blacken Kligler iron agar and triple sugar iron agar. Ewing (1960) pointed out, for the hydrogen sulfide test of the family Enterobacteriaceae, that sensitivity of the test should be poised at a certain level, and that either Kligler iron agar or triple sugar iron agar serves this purpose admirably, because each permits easy group differentiation within the family. Thus *H. alvei* organisms are now redefined to be hydrogen sulfide negative.

From the diagnostic laboratory standpoint, *H. alvei* should be differentiated from members of *Enterobacter* and *Serratia*. All strains of *H. alvei* may give + − + reactions in the decarboxylase tests of lysine, arginine, and ornithine, and can be differentiated from *Enterobacter cloacae* (typical patterns are − + +). Comparing *H. alvei* with *Enterobacter aerogenes* and *Serratia* species as well as *Enterobacter cloacae,* the failure of *H. alvei* to ferment

Table 1. Biochemical characteristics of 840 strains of *Hafnia alvei*.

Test or substrate	Sign[a]	% +	%(+)[b]
Indole	−	0	
Methyl red (22°C)	−	0.6	
Methyl red (37°C)	d	36.8	
Voges-Proskauer (22°C)	+	98.0	
Voges-Proskauer (37°C)	d	59.0	
Citrate utilization (Simmons) (22°C)	d	9.6	65.8
Citrate utilization (Simmons) (37°C)	d	0	62.5
Citrate utilization (Christensen)	+	2.8	68.4
Nitrate to nitrite	+	100	
H₂S (TSI)	−	0	
Urease (Christensen)	−	2	
Gelatinase	−	0	
Phenylalanine deaminase	−	0	
Lysine decarboxylase	+	100	
Arginine dihydrolase	−	0	
Ornithine decarboxylase	+	100	
KCN	+	98.2	
Motility	+	95.4	
Malonate utilization	d	83.4	
Sodium acetate utilization	d	62.0	
Organic acid (Kauffmann-Petersen):			
Citrate	+	99.4	
D-Tartrate	−	0	
Mucate	−	0	
ONPG	d	73.2	
Gas from glucose	+	100	
Fermentation:			
Arabinose	+	99.0	
Cellobiose	d	58.2	21.2
Glucose	+	100	
Lactose	−	0	8.0
Maltose	+	100	
Melibiose	−	4.2	1.0
Raffinose	−	0	
Rhamnose	+	82.0	17.2
Sucrose	d	8.4	68.0
Trehalose	+	100	
Xylose	+	98.4	
Adonitol	−	0	
Dulcitol	−	0	
Erythritol	−	2.0	
Mannitol	+	100	
Sorbitol	−	0	
Glycerol	+	100	
Salicin	d	15.8	4.2
Inositol	−	0	
Esculin hydrolysis	−	4.8	1.2
Lipase, corn oil	−	0	
Sodium alginate	−	0	
DNase	−	0	

[a] Symbols: +, 90% or more positive; −, 90% or more negative; d, different reactions (11–89% positive).
[b] Percentages of delayed positive reactions.

raffinose, sorbitol, adonitol, and inositol is a particularly valuable character, although some strains may be aberrant with respect to one or two of these sugar reactions. All strains of *H. alvei* possess no activity of lipase and DNase. Tests and substrates valuable in

Table 2. Biochemical differentiation of *Hafnia alvei, Enterobacter* spp., and *Serratia liquefaciens.*[a]

Test or substrate	H. alvei	E. cloacae	E. aerogenes	S. liquefaciens
Citrate (Simmons)	d	+	+	+
Gelatinase	−	(+)	(d)	+
Lysine decarboxylase	+	−	+	+
Arginine dihydrolase	−	+	−	−
Mucate	−	d	+	−
Fermentation:				
Lactose	−	+	+	(d)
Raffinose	−	+	+	+
Rhamnose	+	+	+	−
Inositol	−	d	+	+
Sorbitol	−	+	+	+
Salicin	d	d	+	+
Lipase, corn oil	−	−	−	+
DNase	−	−	−	d

[a] Symbols: +, 90% or more positive within 1 or 2 days; (+), 90% or more late positive; −, 90% or more negative; d, different reaction (11–89% positive); (d), 11–89% late positive.

the differentiation of *H. alvei, Enterobacter* spp., and *Serratia* spp. are given in Table 2.

Stuart et al. (1943) and Eveland and Faber (1953) emphasized a close similarity between biotype 32011, which is now classified within *H. alvei,* and members of the genus *Salmonella* in biochemical reactions. Although *H. alvei* fails to produce hydrogen sulfide, unlike *Salmonella,* most *H. alvei* cultures give positive reactions in lysine and ornithine decarboxylase tests and negative reactions in indole, urease, and lactose and sucrose fermentation tests, as well as the Voges-Proskauer test, which is performed with cultures incubated at 35°C. In addition to these, it has been reported that *Hafnia* cultures are frequently agglutinated by *Salmonella* O antisera (Eveland and Faber, 1953; Harada, Shimizu, and Matsuyama, 1957). *H. alvei* may also be misidentified with some non-lactose fermenter of *Escherichia coli.* Tests of value in the differentiation of *H. alvei, Salmonella* spp., and *E. coli* are listed in Table 3. For the biochemical test methods, refer to Edwards and Ewing (1974) and Cowan (1974).

Guinée and Valkenburg (1968) described a

Table 3. Biochemical differentiation of *Hafnia alvei,* H_2S-negative *Salmonella* spp., and *Escherichia coli.*[a]

Test or substrate	H. alvei	Salmonella spp.	E. coli
Voges-Proskauer, 22°C	+	−	−
Citrate (Simmons)	(d)	d	−
KCN	+	−	−
Fermentation:			
Lactose	−	−	+
Sucrose	−	−	d
Sorbitol	−	+	+

[a] Symbols: +, 90% or more positive within 1 or 2 days; −, 90% or more negative; d, different reaction (11–89% positive); (d), 11–89% late positive.

Hafnia-specific bacteriophage, phage 1672, that provides a reliable tool for the identification of *Hafnia* strains. They reported that the phage lysed all 100 strains of *H. alvei* tested, while it did not lyse the tested strains of *Enterobacter, Klebsiella, Citrobacter, Serratia,* and *Salmonella.*

Identification of Hafniae with Phage 1672 (Guinée and Valkenburg, 1968)

The specific phage 1672 was isolated from surface water with *Hafnia* strain 1672 as the propagating bacterium. A well-dried nutrient agar plate is surface-inoculated with a fresh broth culture of the strain to be tested. After decantation, the plate is allowed to dry at room temperature for 15 min. A drop of the undiluted phage 1672 is then spotted on the plate inoculated with a Pasteur pipette, and the plates are again allowed to dry. Readings are made after 16–20 h of incubation at 37°C. Clear plaques with a diameter of 1–2 mm are produced. The phage preparation is obtained after the usual purification and will contain around 10^9 plaque-forming units per ml. The phage is not inactivated by heating at 60°C for 30 min.

Serology of *H. alvei* was first studied by Stuart and Rustigian (1943), who divided 127 of 149 cultures of the biotype 32011 of bacteria into 8 serovars. Eveland and Faber (1953) studied serology of 58 cultures of "*Paracolobactrum aerogenoides*" belonging to the biotype 32011 and reported 21 somatic and 22 flagellar antigens within the cultures studied. Deacon (1952) carried out serological studies on 17 cultures of slow lactose-fermenting "*Aerobacter*" *cloacae,* including the biotype 32011, and distinguished 12 somatic and 6 flagellar antigens among them. However, cultures of the biotype 32011 studied by Stuart and Rustigian

(1943) and Eveland and Faber (1953) and those of "*Aerobacter*" *cloacae* studied by Deacon (1952) included not only *Hafnia* strains but also *Enterobacter cloacae* (Sakazaki and Namioka, 1957; Sakazaki, 1961). Therefore, the studies performed by those authors contributed little information to aid in the establishment of an antigenic schema of *H. alvei*. Serology of cultures that were biochemically well defined as members of the genus *Hafnia* was studied by Sakazaki (1961), and 29 O groups and 23 H antigens were established among 294 strains studied. Later, Matsumoto (1963, 1964) extended the antigens to 68 O-antigen groups and 34 H antigens. Some strains of *H. alvei* possess K antigens and alpha antigen (Deacon, 1952; Emslie-Smith, 1961; Sakazaki, 1961). Sakazaki (1961) suggested that the K antigen of *H. alvei* seemed to be a slime antigen; Deacon (1952) suggested the presence of the A (heat-stable) type of K antigens in strains he studied, but Sakazaki (1961) could not demonstrate such antigens within 294 cultures of *H. alvei*. Deacon (1952) also reported the diphasic variation in the H antigens of the strains he studied, but Sakazaki (1961) failed to observe such variation.

Stamp and Stone (1944), Sakazaki (1961), Matsumoto (1963), and Sedlák and Šlajsová (1966) demonstrated interrelationships between O antigens of *H. alvei* and those of certain *Enterobacter cloacae*, *Escherichia coli* and *Citrobacter freundii*. Eveland and Faber (1953) reported O-antigenic relationships between the 32011 group and *Salmonella*.

The majority of strains of *H. alvei* are susceptible to carbenicillin, streptomycin, gentamicin, kanamycin, chloramphenicol, tetracycline, polymyxin B, and nalidixic acid, but resistant to cephalothin and ampicillin. Washington, Birk, and Ritts (1971) noted a striking difference between *H. alvei* and *Serratia liquefaciens* with respect to susceptibility to ampicillin and polymyxin B. They reported that "most of the isolates of *Enterobacter* (*Serratia*) *liquefaciens* tested were susceptible to ampicillin at 20 mcg/ml, whereas none of the test strains of '*Enterobacter hafniae*' were; all of the tested strains of '*E. hafniae*' were susceptible to polymyxin B at 10 mcg/ml, whereas only 6% of the strains of *E. liquefaciens* were''.

Literature Cited

Bahr, L. 1919. Paratyfus hos Honningbien samt nogle undersøgelser verdrørende Forekomsten af Bakterierhenhorende til Coli-tyfus gruppen. i. Honningbiens tarm. Scandinavia Veterinar Tidskrift, 9:25–40, 45–60.

Castellani, A. 1912. Observations on some intestinal bacteria found in man. Zentralblatt für Bakteriologie, Parasitenkunde, Infektionskrankheiten, und Hygiene, Abt. 1, Orig. 65:262–269.

Cowan, S. T. 1974. Cowan and Steel's manual for the identification of medical bacteria, 2nd ed., pp. 137–180. London: Cambridge University Press.

Deacon, W. E. 1952. Antigenic study of certain slow lactose fermenting *Aerobacter cloacae* cultures. Proceedings of Society of Experimental Biology and Medicine 81:165–170.

Edwards, P. R., Ewing, W. H. 1974. Identification of *Enterobacteriaceae*, 3rd ed. Minneapolis: Burgess Publishing Co.

Emslie-Smith, A. H. 1961. *Hafnia alvei* strains possessing alpha antigen of Stamp and Stone. Journal of Pathology and Bacteriology 81:534–536.

Englund, G. W. 1969. Persistent septicemia due to *Hafnia alvei*: Report of a case. American Journal of Clinical Pathology 51:717–719.

Eveland, W. C., Faber, J. E. 1953. Antigenic studies of a group of paracolon bacteria (32011 group). Journal of Infectious Disease 93:226–236.

Ewing, W. H. 1960. Biochemical method for group differentiation. Atlanta: Communicable Disease Center Publication.

Ewing, W. H. 1963. An outline of nomenclature for the family *Enterobacteriaceae*. International Bulletin of Bacterial Nomenclature and Taxonomy, 13:95–110.

Ewing, W. H., Fife, M. A. 1968. *Enterobacter hafniae* (the Hafnia group). International Journal of Systematic Bacteriology 18:263–271.

Gale, E. F., Epps, H. M. R. 1943. Lysine decarboxylase: Preparation of enzyme and coenzyme. Nature 152:327–328.

Gavini, F., Ferragut, C., Lefebre, B., Leclerc, H. 1976. Étude taxonomique d'Entérobacteries appartenant ou apparentées au genre *Enterobacter*. Annual de Microbiologie 127B:317–335.

Guinée, P. A. M., Valkenburg, J. J. 1968. Diagnostic value of a *Hafnia*-specific bacteriophage. Journal of Bacteriology 96:564.

Harada, K., Shimizu, K., Matsuyama, T. 1957. *Hafnia* isolated from man. [In Japanese.] Gumma Journal of Medical Science 6:109–112.

Hormaeche, E., Edwards, P. R. 1960. A proposed genus of *Enterobacteriaceae*. International Bulletin of Bacterial Nomenclature and Taxonomy 10:71–77.

Johnson, R., Colwell, R. R., Sakazaki, R., Tamura, K. 1975. Numerical taxonomy study of the *Enterobacteriaceae*. International Journal of Systematic Bacteriology 25:12–37.

Kauffmann, F. 1954. *Enterobacteriaceae*, 2nd ed., pp. 257–259. Copenhagen: Ejnar Munksgaard.

Kume, T. 1962. A case of abortion possibly due to *Hafnia* organism. [In Japanese.] Journal of Hokkaido Veterinarian Association 6:1–4.

McClure, H. E., Eveland, W. C., Kase, A. 1957. The occurrence of certain *Enterobacteriaceae* in birds. American Journal of Veterinary Research 18:207–209.

Matsumoto, H. 1963. Studies on the *Hafnia* isolated from normal human. Japanese Journal of Microbiology 7:105–114.

Matsumoto, H. 1964. Additional new antigens of Hafnia group. Japanese Journal of Microbiology 8:139–141.

Møller, V. 1954. Distribution of amino acid decarboxylase in *Enterobacteriaceae*. Acta Pathologica et Microbiologica Scandinavica 35:259–277.

Ritter, D. B., Gerloff, R. K. 1966. Deoxyribonucleic acid hybridization among some species of the genus *Pasteurella*. Journal of Bacteriology 92:1838–1839.

Sakazaki, R. 1961. Studies on the Hafnia group of *Enterobacteriaceae*. Japanese Journal of Medical Science and Biology 14:223–241.

Sakazaki, R. 1974. *Hafnia*, pp. 325–326. In: Buchanan, R. E., Gibbons, N. E. (eds.), Bergey's manual of determinative bacteriology, 8th ed. Baltimore: Williams & Wilkins.

Sakazaki, R., Namioka, S. 1957. Biochemical studies on Voges-Proskauer positive enteric bacteria. Japanese Journal of Experimental Medicine 27:273–282.

Sedlák, J., Šlajsová, M. 1966. On the antigenic relationships of certain Citrobacter and Hafnia cultures. Journal of General Microbiology 43:151–158.

Stamp, L., Stone, D. M. 1944. An agglutinogen common to certain strains of lactose and non-lactose fermenting coliform bacilli. Journal of Hygiene **43:**266–272.

Starr, M. P., Mandel, M. 1969. DNA base composition and taxonomy of phytopathogenic and other enterobacteria. Journal of General Microbiology **56:**113–123.

Steigerwalt, A. G., Fanning, G. R., Fife, M. A., Brenner, D. J. 1976. DNA relatedness among species of *Enterobacter* and *Serratia*. Canadian Journal of Microbiology **22:**121–137.

Stuart, C. A., Rustigian, R. 1943. Further studies on one type of paracolon organisms. American Journal of Public Health **33:**1323–1325.

Stuart, C. A., Wheeler, K. M., Rustigian, R., Zimmermann, A. 1943. Biochemical and antigenic relationships of the paracolon bacteria. Journal of Bacteriology **45:**101–119.

Washington, J. A., II, Birk, R. J., Ritts, R. E., Jr. 1971. Bacteriologic and epidemiologic characteristics of *Enterobacter hafniae* and *Enterobacter liquefaciens*. Journal of Infectious Diseases **124:**379–386.

Whitby, J. L., Muir, G. G. 1961. Bacteriological studies of urinary tract infection. British Journal of Urology **33:**130–134.

CHAPTER 97

The Genus *Serratia*

PATRICK A. D. GRIMONT and FRANCINE GRIMONT

The genus *Serratia*, a member of the Enterobacteriaceae (this Handbook, Chapter 88), is a group of bacteria related to the type species *Serratia marcescens* both phenotypically and by polynucleotide sequence analysis. Some species and biotypes of *Serratia* produce a nondiffusible red pigment: prodigiosin, which is 2-methyl-3-amyl-6-methoxyprodigiosene (Williams and Hearn, 1967). The multiplication of red-pigmented *Serratia* was incriminated in the appearance of bloodlike spots (e.g., on bread, consecrated wafers [sacramental Hosts], and polenta) with rather disastrous sociological consequences. In this context, several scholars have traced the history of the genus *Serratia* back to antiquity (Gaughran, 1969; Harrison, 1924; Reid, 1936). However, several bacterial species outside the genus *Serratia* produce prodigiosin or prodigiosin-like pigments (D'Aoust and Gerber, 1974; Gandhi et al., 1973; Gauthier, 1976; Williams and Hearn, 1967) or many other kinds of red pigments, and the identity of microorganisms involved in these prodigious phenomena can only be surmised.

Bizio (1823) named *Serratia marcescens*, the red-pigmented microorganism he observed on polenta. Ehrenberg (1848) named *Monas prodigiosa*, a motile bacterium isolated from red spots on food. No cultures of these organisms were preserved, but the name *Serratia marcescens* was preferred over the *Erythrobacillus pyosepticus*—a culture of which was preserved as ATCC 275 (Fortineau, 1904)—by Breed and Breed (1924, 1927) and by the editors of *Bergey's Manual of Determinative Bacteriology* (Bergey et al., 1923). The name *S. marcescens* is now universally accepted and a neotype strain has been designated (Martinec and Kocur, 1961a).

At the start of this century, more than 76 nomenspecies had been described with red or pink pigmentation (Hefferan, 1903–1904), and 23 *Serratia* species were listed in the first edition of *Bergey's Manual* (Bergey et al., 1923). This number progressively decreased to 5 in the fifth edition of *Bergey's Manual* (Breed, Murray, and Smith, 1957), and later to one species: *S. marcescens* (Ewing, Davis, and Reavis, 1959; Martinec and Kocur, 1960, 1961a, b, c, d). The only *Serratia* species recognized in the eighth edition of *Bergey's Manual* was *S. marcescens* (Sakazaki, 1974). Several works, using numerical taxonomy or polynucleotide sequence relatedness, recently defined five species in the genus *Serratia*. These species (and synonyms) are:

1. *Serratia marcescens* Bizio 1823: "Typical" *S. marcescens* (Colwell and Mandel, 1965; Ewing, Davis, and Reavis, 1959; Martinec and Kocur, 1961a); *Serratia* pattern 1 (Fulton, Forney, and Leifson, 1959); *Serratia* biotype 1 (Bascomb et al., 1971); phenon A (Grimont and Dulong de Rosnay, 1972; Grimont et al., 1977); *S. marcescens* DNA hybridization group (Steigerwalt et al., 1976). The neotype strain is ATCC 13880 (Martinec and Kocur, 1961a). The type strains of *Bacillus indicus* (Eisenberg, 1886), *Erythrobacillus pyosepticus* (Fortineau, 1904), *Bacillus sphingidis* (White, 1923a), and *Serratia anolium* (Duran-Reynals and Clausen, 1937) are all referable to the taxonomic entity now known as *S. marcescens*. Strains labeled *S. marcescens* subsp. *kiliensis*, according to Ewing, Davis, and Johnson (1962), are Voges-Proskauernegative variants of *S. marcescens*.

2. *Serratia proteamaculans* (Paine and Stansfield 1919) Grimont, Grimont, and Starr 1978: *Erwinia proteamaculans* (Paine and Stansfield 1919) Dye 1966; *Enterobacter liquefaciens* (Grimes and Hennerty 1931) Ewing 1963; *Serratia liquefaciens* (Grimes and Hennerty 1931) Bascomb et al. 1971; phenon C1 (Grimont et al., 1977); *S. liquefaciens* hybridization group (Steigerwalt et al., 1976). The type strain is ICPB XP176 (ATCC 19323, NCPPB 245).

3. *Serratia plymuthica* (Lehmann and Neumann 1896) Breed, Murray, and Hitchens 1948: excluded from the genus *Serratia* (Ewing, Davis, and Reavis, 1959); "atypical" *S. marcescens* (Colwell and Mandel, 1965); *S. marcescens* var. *kiliensis* according to Martinec and Kocur (1961d) (*not* Ewing, Davis, and Johnson, 1962); *Serratia* III (Mandel and Rownd, 1964); *Serratia* pattern 2 (Fulton et al., 1959); atypical *S. rubidaea* (Ewing, Davis, and Fife, 1972; Ewing et al., 1973); and phenon C2 (Grimont et al., 1977). The type strain is CCM 640 (ATCC 183). The type strain of *Bacterium*

kiliense Lehmann and Neumann 1896 and *Serratia esseyana* Combe 1933 is *S. plymuthica*.

4. *Serratia marinorubra* ZoBell and Upham 1944: Prodigiosus VIII (Hefferan 1904); *Serratia* biotype 2 (Bascomb et al., 1971); *S. rubidaea* (Ewing, Davis, and Fife, 1972; Ewing et al., 1973); phenon B (Grimont and Dulong de Rosnay, 1972; Grimont et al., 1977); *S. rubidaea* DNA hybridization group (Steigerwalt et al., 1976). The reasons for preferring the epithet *marinorubra* over *rubidaea* are given elsewhere (Grimont et al., 1977). The type strain is NCTC 10912 (ATCC 27614).

5. *Serratia odorifera* Grimont et al. (1978): Strains similar to the unclustered *Serratia* strain 38 (Grimont and Dulong de Rosnay, 1972; Grimont et al., 1977). The type strain is ICPB 3995 (NCTC 11214, ATCC 33077).

The inclusion of the genus *Serratia* in the tribe Klebsielleae is no longer tenable. Studies on DNA relatedness, immunological cross-reaction between isofunctional enzymes, and the physical properties, regulation, and amino acid sequences of enzymes all showed the genus *Serratia* to be consistently different from the group composed of the genera *Escherichia, Shigella, Salmonella, Citrobacter, Klebsiella,* and *Enterobacter* (reviewed by Grimont and Grimont, 1978a).

Until the late 1950s, isolation of *Serratia* spp. from human patients was a rare curiosity. Later on, *S. marcescens* became more and more frequently involved in nosocomial infections. Nonpigmented *S. marcescens* strains are now a threat in surgical and intensive care units (Altemeier et al., 1969; Bodey, Rodriguez, and Smith, 1970; Cabrera, 1969; Clayton and von Graevenitz, 1966; Gale and Sonnenwirth, 1962).

Habitats

The long-confused status of *Serratia* taxonomy has prevented any precise knowledge of the habitat of *Serratia* species. After the aforementioned five *Serratia* species had been defined, it became apparent that they could have different habitats. Table 1 shows the distribution among species of *Serratia* strains isolated from water, plants, insects, and hospitalized human patients. Table 2 shows the distribution among biotypes of *S. marcescens* strains isolated from these same habitats.

Serratia in Water and Soil

Pigmented *Serratia* isolated from terrestrial waters most often are *S. marcescens* and *S. plymuthica* and less frequently *S. marinorubra*, whereas nonpigmented *Serratia* are essentially *S. proteamaculans (liquefaciens)* and *S. marcescens* biotypes A3 and

Table 1. Distribution among *Serratia* species of 1,273 strains isolated from four different habitats.

Serratia species	Habitats and number of isolates			
	Water[a]	Plants[b]	Insects[a]	Hospitalized humans[c]
S. marcescens	31	19	28	1,078
S. proteamaculans	6	30	19	27
S. plymuthica	19	5	0	0
S. marinorubra	5	3	1	2
S. odorifera	unk[d]	4	unk[d]	(1)[e]
Total for 5 species	61	60	48	1,107

[a] Isolated under nonselective conditions. Most strains from insects were isolated by O. Lysenko.
[b] Isolated using CT agar.
[c] Isolated at the Pellegrin Hospital (Bordeaux) from 1968 through 1975.
[d] The taxon *S. odorifera* was unknown at the time of the isolation.
[e] There was only partial awareness of the taxon *S. odorifera* at the time of the isolation.

A4 (Grimont, P. A. D., 1977). Water is probably the principal habitat of *S. plymuthica*. Nomenspecies resembling *S. plymuthica* (Breunig's Kiel bacillus, *Bacillus miniaceus, Serratia miquelii,* and *S. esseyana*) were also isolated from water. Seawater isolates belong to the same species as terrestrial water isolates. *Serratia marinorubra* is not a true marine species, as it shows no sodium requirement (P. Baumann, personal communication). Nonpigmented *S. marcescens* biotypes A5, A8abc, and TCT have not yet been found in water or soil (outside of hospitals).

In soil, *S. marcescens* might play a role in the biological cycle of metals by mineralizing organic iron and dissolving gold and copper (Parès, 1964). A mineralization role has also been attributed to cold-tolerant *Serratia* associated with low-moor peat (Janota-Bassalik, 1963).

Table 2. Distribution among *Serratia marcescens* biotypes of 1,139 strains isolated from four different habitats.

Biotype	Habitats and number of isolates			
	Water[a]	Plants[b]	Insects[a]	Hospitalized humans[c]
A1ab	8	8	3	2
A2ab/A6ab	14	2	15	78[d]
A3abcd	1	5	3	78
A4ab	6	3	4	277
A5/A8abc	0	0	2	497
TCT	0	0	1	134
Total for all biotypes	29	18	28	1,066

[a] Isolated under nonselective conditions. Most strains from insects were isolated by O. Lysenko.
[b] Isolated using CT agar.
[c] Isolated at the Pellegrin Hospital (Bordeaux) from 1968 through 1975.
[d] Seventy-two biotype A2b strains were isolated from infants in the Neonatology ward (mostly from feces).

Serratia on Plants

Serratia proteamaculans (liquefaciens) was once isolated from a leaf spot disease of the tropical plant *Protea cynaroides* (the King Protea) (Paine and Stansfield, 1919). However, the experimental lesions caused by *S. proteamaculans* on detached leaves of *Protea* suggest a hypersensitivity reaction (Paine and Berridge, 1921). Similar lesions were obtained with other species of *Serratia* (Grimont, Grimont, and Starr, 1978). Inoculation of *S. marcescens* ICPB 2875 and *S. marinorubra* ICPB 2881 on tobacco and bean leaves also produced a typical hypersensitivity reaction (Lakso and Starr, 1970). In the course of an ecological survey (P. A. D. Grimont, F. Grimont, and M. P. Starr, unpublished), *Serratia* spp. were associated with 30% of 209 plant specimens examined: tree leaves *(Eucalyptus, Pistacia, Acacia, Arbutus, Prunus, Pinus, Quercus)*, shrubs *(Ceanothus, Eriogonium)*, fruits (Calimyrna figs, coconuts), vegetables (tomatoes, leeks, green onions, lettuce, broccoli, artichokes, radishes, spinach, carrots, cauliflower), herbs *(Sorghum*, grass), mushrooms and toadstools, and moss. All five species were recovered with a predominance of *S. proteamaculans*. The only nonpigmented biotypes of *S. marcescens* recovered from plants were A3 and A4.

Figs of the Calimyrna variety (Smyrna variety adapted to California) deserve special mention. Calimyrna figs contain only pistillate flowers. In order to become edible, these figs need to be pollinated specifically by the fig wasp *Blastophaga psenes*. The fig wasp has a life cycle limited to caprifigs, inedible fruits produced by the caprifigtree. When a young female fig wasp makes her way out of a caprifig, she covers herself with pollen and a specific fungal and microbial flora (Phaff and Miller, 1961). The fig wasp will then enter an unpollenated caprifig of a new crop, pollenate it, oviposit in a pistillate flower, and die; or it may be carried by the wind to a Calimyrna fig—a cul-de-sac in the wasp life cycle—to which the fig wasp will bring pollen and the microbial/fungal flora in its desperate attempt to oviposit. Since 1927 (Caldis, 1927; Phaff and Miller, 1961; Smith and Hansen, 1931), a red-pigmented *Serratia* has been repeatedly isolated from caprifigs, pollenated Calimyrna figs, and fig wasps. We identified the pigmented isolates as *Serratia marcescens* biotype A1b. The fact that the Calimyrna fig is internally sterile until the fig wasp enters it provides an ecological niche for this *S. marcescens* biotype. The multiplication of *S. marcescens* in caprifigs or Calimyrna figs is limited and this organism does not cause fig spoilage.

Vegetables used in salads might bring *Serratia* strains to hospitals and contaminate the patient's digestive tract. *S. marcescens, S. liquefaciens,* and *S. marinorubra* were found in 29%, 28%, and 11% (respectively) of vegetable salads served in a hospital in Pittsburgh (Wright, Kominos, and Yee, 1976). Similar results were found in a Paris hospital (Loiseau-Marolleau and Laforest, 1976). However, that biotypes/serotypes found in patients are the same as those found in salads still needs to be proven.

Serratia in Insects

There is an extensive literature on *Serratia* associated with insects. This topic has been reviewed in detail (Bucher, 1963a; Grimont and Grimont, 1978a; Steinhaus, 1959). The insects involved belong to numerous species and genera of the orders Orthoptera (crickets and grasshoppers), Isoptera (termites), Coleoptera (beetles and weevils), Lepidoptera (moths), Hymenoptera (bees and wasps), and Diptera (flies). Taxonomic uncertainties make difficult a retrospective evaluation of the role of *Serratia* spp. in insect infections. Red-pigmented *Serratia* were easily recognized (although not determined to species with any certainty), whereas nonpigmented strains were often referred to genera other than *Serratia*. For example, *S. proteamaculans (liquefaciens)* strains were named *Bacillus noctuarum* (White, 1923b), *Bacillus melolonthae liquefaciens* (Paillot, 1916), *Paracolobactrum rhyncoli* (Pesson, Toumanoff, and Haradas, 1955), and *Cloaca* B type 71-12A (Bucher and Stephens, 1959); nonpigmented *S. marcescens* strains were named *Bacillus sphingidis* (White, 1923a) and *Bacillus apisepticus* (Burnside, 1928). The distribution among the various *Serratia* species, of 48 strains associated with insects (Grimont, Grimont, and Lysenko, 1979), is given in Table 1. The absence of *S. plymuthica* (a pigmented species) in this collection is unexplained. Steinhaus (1941) reported the presence of *S. plymuthica* in the gut of healthy crickets *(Neombius fasciatus),* but the taxonomic schemes in vogue at that time did not allow a definite identification of *S. plymuthica*. The red-pigmented *Serratia* associated with the fig wasp *Blastophaga psenes* and formerly identified as *S. plymuthica* (Phaff and Miller, 1961) were reidentified by us as *S. marcescens* biotype A1b (P. A. D. Grimont, F. Grimont, and M. P. Starr, unpublished). The rarity of *S. marinorubra* (also a pigmented species) in insects might be explained by the inability to produce chitinase—a virulence factor for insect-associated *Serratia* species (Lysenko, 1976).

Serratia marcescens and *S. proteamaculans (liquefaciens)* are considered potential insect pathogens (Bucher, 1960). They cause a lethal septicemia after penetration into the hemocoel. More than 70 species of insects were found susceptible to inoculation with *Serratia* (Bucher, 1963a). The lethal dose (LD_{50}) of inoculated *Serratia* was calculated for several insects (intrahemocoelic injection): 10–50 Serratia cells per grasshopper (Bucher, 1959), 5.1 cells per adult bollweevil (Slatten and Larson, 1967), 7.5 and 14.5 cells per third and fourth instar larva (respectively) of Lymantria dispar (Podgwaite and Cosenza, 1976), and 40 cells per Galleria mellonella larva (Stephens, 1959). The LD_{50} of ingested Serratia is much higher. The hemolymph of insects—normally bactericidal for nonpathogens—cannot prevent multiplication of potential pathogens (Stephens, 1963). Lecithinase, proteinase, and chitinase play a role in the virulence of Serratia for insects, and purified Serratia proteinase or chitinase is very toxic when injected into the hemocoel (Kaska, 1976; Lysenko, 1976). Serratia strains in the insect digestive tract probably originate from plants. The multiplication of Serratia strains in the insect digestive tract has not been quantitatively studied. Antibacterial substances in ingested leaves might interfere with bacterial multiplication, but Serratia strains were found resistant to these (Kushner and Harvey, 1962). How potential pathogens (e.g., Serratia) can enter the hemolymph from the gut is generally unknown. However, spontaneous gut rupture, which happens in about 10% of grasshoppers, may allow Serratia strains to invade the hemocoel (Bucher, 1959). Direct injection occurs when Itoplectis conquisitor contaminated with Serratia stings host pupae to oviposit into their body (Bucher, 1963b). No genuine epizootic of Serratia infection among insects has been observed in the field. However, Serratia epizootics are common among reared insects (Bucher, 1963a).

Serratia in Vertebrates

Serratia has been associated with chronic infections of cold-blooded vertebrates: nodular infection of Anolis equestris, the Cuban lizard (Duran-Reynals and Clausen, 1937); subcutaneous abscess of iguanid lizards (Boam et al., 1970); arthritis in the lizard Tupinambis tequixin (Ackerman, Kishimoto, and Emerson, 1971); ulcerative disease in the painted turtle Chrysemys picta (Jackson and Fulton, 1976). Serratia strains have also been recovered from healthy small green pet turtles Pseudemys scripta elegans (McCoy and Seidler, 1973) and from geckos and turtles in Vietnam (Capponi, Sureau, and Le Minor, 1956).

Poultry may be contaminated with Serratia. A deadly Serratia epizootic among chick embryos was observed in a Japanese hatchery (Izawa et al., 1971). The hens carried S. marcescens in their digestive tract, but were themselves unaffected. Contamination of chicken carcasses with S. liquefaciens (Lahellec et al., 1975) and spoilage of eggs by red-pigmented Serratia (Alford et al., 1950) have been reported. Serratia strains (mostly red-pigmented) are responsible for 0.2–1.5% of mastitis in cows (Barnum, Thackeray, and Fish, 1958; Roussel, Lucas, and Bouley, 1969; Wilson, 1963). Raw milk, therefore, may occasionally contain Serratia spp. S. proteamaculans (liquefaciens) is common in dairy products (Grimes and Hennerty, 1931). Serratia strains have been involved in septicemia in foals (Deom and Mortelmans, 1953), goats (Wijewanta and Fernando, 1970), and pigs (Brisou and Cadeillan, 1959); they have also been implicated in conjunctivitis of the horse (Carter, 1973) and abortion in cows (Smith and Reynolds, 1970).

In wild mammals, Serratia has been found in the anal sac of the red fox Vulpes vulpes (Gosden and Ware, 1976), and S. plymuthica in the gut of Microtus arvalis (C. Richard, personal communication).

Serratia in Man

The healthy human being is not likely to get infected by Serratia, whereas the hospitalized patient is frequently colonized or infected. S. marcescens is presently the only nosocomial species of Serratia (Table 1). S. proteamaculans and S. marinorubra are occasionally isolated from clinical specimens, but their pathogenic role is not established (Johnson and Ellner, 1974). No S. plymuthica has ever been involved in a significant infection, and the role of S. odorifera in clinical infections needs to be studied.

Among the different biotypes of S. marcescens, only the nonpigmented biotypes are a real threat in hospitals (Bodey, Rodriguez, and Smith, 1970; Clayton and von Graevenitz, 1966; Ewing, Johnson, and Davis, 1962; Farmer et al., 1976). Pigmented biotypes are rarely responsible for an outbreak unless the patients are grossly contaminated by means of nonsterile injection products or aerosols (Daschner and Senska-Euringer, 1975; Negut, Szegli, and Raiculescu, 1972; Sanders et al., 1970). Pigmented biotypes of S. marcescens may multiply in sputum and turn it red (pseudo-hemoptysis) (Gale, Lord, and Durham, 1957; Griffith, 1967; Robinson and Woolley, 1957; Woodward and Clark, 1913). They are also responsible for the red-diaper syndrome and asymptomatic colonization of the digestive tract in the newborn (Grimont, P. A. D., 1977; Hernandez Marchant, Rojas, and Arcaya, 1960; Waisman and Stone, 1958). Pigmented S. marcescens biotypes are frequently found in newborn feces but not in adult feces. Nonpigmented S. marcescens biotypes can be found in feces of hospitalized adults when highly selective media are used. S. marcescens was isolated from feces of 33–57% of Serratia-infected patients, and 2–5% of noninfected patients (Cate, 1972; Denis and Blanchard, 1975). We isolated nonpigmented biotypes in feces of 21% of patients in an infected ward (Urology), and in feces of 3% of the patients in other, less infected wards (unpublished). The relationship between the strains found in feces and those infecting the patient needs clarification.

Clinically, *Serratia* infections are not different from infections by other opportunistic pathogens (von Graevenitz, 1977): respiratory tract infection and colonization of intubated patients (e.g., Cabrera, 1969); urinary tract infection and colonization of patients with indwelling catheters (e.g., Maki et al., 1973); surgical wound infection or superinfection (e.g., Cabrera, 1969); and septicemia in patients with intravenous catheterization or complicating a local infection (e.g., Altemeier et al., 1969). A nonpigmented biotype is likely to enter a ward (or a hospital) through contaminated food (Wright, Kominos, and Yee, 1976), contaminated personnel (suggested by Schaberg et al., 1976), or transfer of a contaminated patient (personal observation). The *S. marcescens* strain will find in a ward a number of compromised patients, often exhibiting the same susceptibility to opportunistic infections: many patients in the ward may have a urinary catheter (especially in the urology ward), an endotracheal tube connected to a respirator (in an intensive care unit), a large burned area (in a burn unit), or a depression of immunological defenses (in a cancer ward); and they may receive "prophylactically" the same broad-spectrum antibiotics to which nonpigmented *S. marcescens* often are resistant. In a busy ward crowded with such patients, the *Serratia* strain would be disseminated among patients by the hands of personnel (Maki et al., 1973; Traub, 1972b). The patient's inanimate environment is frequently contaminated by the patient's strain, and might play a role in contaminating the hands of personnel. Hospital epidemiology of *S. marcescens* infections is not very different from that of other opportunistic Gram-negative bacteria. Recent reviews on this topic are available (Farmer et al., 1976; Schaberg et al., 1976; von Graevenitz, 1977).

Isolation

Underlying Principles

Pigmented species and biotypes of *Serratia* often give pink to red colonies on nutrient agar. A low-phosphate agar without glucose, such as peptone-glycerol agar (Difco peptone, 5 g; glycerol, 10 ml; Difco agar, 20 g; distilled water, 1 liter), is the best way to demonstrate pigmentation (Williams and Hearn, 1967). Although the genus *Serratia* includes all red-pigmented enterobacteria, identification of a

pink or red colony should be confirmed by biochemical tests because a few nonenterobacteria may produce prodigiosin (see Introduction). On nutrient agar, nonpigmented species or biotypes of *Serratia* give opaque-whitish, mucoid, or transparent smooth colonies. None of these traits is specific for the isolation of *Serratia* from a mixture of enteric bacteria. Colonies of *S. odorifera* give off a specific, musty odor. Detection of this odor may suggest the presence of at least one colony of that species on the plate.

The salt tolerance and relatively low minimal growth temperature of all *Serratia* species may help devise a means of enriching them. A nutrient broth containing 4% NaCl and incubated at 15–20°C will allow multiplication of *Serratia* but not of *Enterobacter cloacae* or *Aeromonas hydrophila* (P. A. D. Grimont, personal observation). This procedure has been used by us, but no systematic study of its effectiveness has been conducted.

Strains of the genus *Serratia* do not normally require addition of growth factors to a minimal medium. Thus, an enrichment medium for *Serratia* can be a minimal medium with a carbon source commonly used by *Serratia* spp. and uncommonly used by non-*Serratia* spp.

Production of extracellular gelatinase, lecithinase, and DNase are essential characteristics of the genus *Serratia* (Edwards and Ewing, 1972; Grimont et al., 1977). Agar media that show the presence of one or a combination of extracellular enzymes can be used to differentiate *Serratia* colonies.

Finally, resistance of *Serratia* spp. to several compounds, including colistimethate, cephalothin (Greenup and Blazevic, 1971), and thallium salts (Starr et al., 1976), can make enrichment or differential media selective for *Serratia*. However, some *S. plymuthica* or *S. marinorubra* strains may be more susceptible to colistimethate, cephalothin, or ampicillin than is *S. marcescens*.

No medium has yet been devised to selectively isolate one particular *Serratia* species (other than *S. marcescens*), although specific carbon sources or conditions may be found in taxonomic works (Grimont et al., 1977). Occasionally, a selective medium may be devised according to a specific pattern of antibiotic resistance displayed by a nosocomial strain of *S. marcescens* (Denis and Blanchard, 1975), but such media cannot be of general utility.

Selective Media Based on DNase Production and Antibiotic Resistance

Farmer, Silva, and Williams (1973) proposed the following deoxyribonuclease–toluidine blue–cephalothin (DTC) agar.

DTC Agar for Isolating *Serratia*
(Farmer, Silva, and Williams, 1973)

Deoxyribonuclease agar (BBL)	21 g
Agar (Difco)	2.5 g
Toluidine blue O	0.05 g
Distilled water	500 ml

Mix on a mechanical stirrer until the dye goes into solution, autoclave with a Teflon stirring bar at 121°C for 15 min, cool to 50°C, add 5 ml (500 mg) of cephalothin (Keflin injectible, Eli Lilly), stir, and dispense in sterile Petri dishes.

Typically, *Serratia* spp. grow on this blue DTC medium with a red halo extending several millimeters around the colonies. "False negatives" (no growth or no halo) are very rare among *S. marcescens* isolates, but several strains of *S. marinorubra* and *S. plymuthica* failed to give the proper reaction (Starr et al., 1976).

A medium with DNase test agar, toluidine blue, cephalothin (30 μg/ml), and colistimethate (30 μg/ml) has been proposed by Cate (1972); and a medium with DNase test agar, toluidine blue, egg yolk, and cephalothin (100 μg/ml) has been proposed by Goldin, Shaffer, and Brown (1969). The latter combines DNase and lecithinase detection.

Berkowitz and Lee (1973) devised the following medium for the isolation of *S. marcesens*.

DNase Medium for Isolating *Serratia marcescens*
(Berkowitz and Lee, 1973)

Deoxyribonuclease test agar with methyl green (Difco)	42 g
L-Arabinose	10 g
Phenol red	0.05 g
Methyl green 1%	4 ml
Distilled water	up to 1,000 ml

After autoclaving, add ampicillin (5 μg/ml), colistimethate (5 μg/ml), cephalothin (10 μg/ml), and amphotericin B (2.5 μg/ml).

Serratia colonies hydrolyze DNA, and the green component of the medium's dark color disappears around the colonies. *S. marcescens*, the only *Serratia* species unable to ferment L-arabinose, will give colonies surrounded by a red halo, whereas other *Serratia* species will give a yellow halo. The other *Serratia* species, however, may be inhibited by the antibiotic mixture. Wright, Kominos, and Yee (1976) used this medium to isolate and enumerate *S. marcescens* from vegetable salads.

Selective Media Based on Carbon-Source Utilization

A minimal medium with *meso*-erythritol as sole carbon source was devised by Slotnick and Dougherty (1972). A modification of this medium including an antiseptic, "Irgasan" (4',2'4'-trichloro-2-hydroxydiphenylether), has been proposed (Lynch and Kenealy, 1976). Unfortunately, *Serratia proteamaculans, S. plymuthica, S. odorifera* biotype 1, and the nosocomial biotypes A5, A8a, A8b, A8c, and TCT of *S. marcescens* cannot grow with erythritol as sole carbon source (Grimont et al., 1977; Grimont et al., 1978; Starr et al., 1976). Therefore, these media should not be used in epidemiological or ecological surveys unless the study is knowingly limited to erythritol-positive serratiae.

A minimal medium with *meso*-inositol as sole carbon source has been proposed for the selective isolation of *Klebsiella pneumoniae* and *Serratia* spp. (Legakis, Papavassiliou, and Xilinas, 1976). However, only a few genera and species have been studied by these authors, and it has been shown (Grimont et al., 1977; Véron and Le Minor, 1975) that *Enterobacter aerogenes, Enterobacter cloacae, Erwinia herbicola,* and pectinolytic *Erwinia* spp. can also grow on a minimal medium with inositol as sole carbon source.

A caprylate-thallous (CT) agar medium has been devised for the selective isolation of all *Serratia* species and biotypes. This CT agar, derived from M70 minimal medium (Véron, 1975), is made up as follows (Starr et al., 1976):

CT Agar for Selective Isolation of *Serratia*
(Starr et al., 1976)

First, a trace element solution (Véron, 1975) is prepared:

Distilled water	1,000 ml
H_3PO_4	1.96 g
$FeSO_4 \cdot 7H_2O$	0.0556 g
$ZnSO_4 \cdot 4H_2O$	0.0287 g
$MnSO_4 \cdot 4H_2O$	0.0223 g
$CuSO_4 \cdot 5H_2O$	0.025 g
$Co(NO_3)_2 \cdot 6H_2O$	0.003 g
H_3BO_3	0.0062 g

Store at 4°C (keeps well at least 1 year). Then the following solutions (solution A and solution B) are prepared.
Solution A:

$CaCl_2 \cdot 2H_2O$	0.0147 g
$MgSO_4 \cdot 7H_2O$	0.123 g
KH_2PO_4	0.680 g
K_2HPO_4	2.610 g
Trace element solution (see above)	10 ml
Caprylic acid	1.1 ml
Yeast extract Difco (5% wt/vol solution)	2 ml
Thallous sulfate	0.25 g
Distilled water	up to 500 ml

The pH is adjusted to 7.2 with NaOH. Autoclave at 110°C (or 120°C) for 20 min.

Solution B:

NaCl	7 g
$(NH_4)_2SO_4$	1 g
Agar (Difco)	15 g
Distilled water	500 g

The pH is adjusted to 7.2. Autoclave at the same time as solution A. After autoclaving, solutions A and B are mixed aseptically and the resulting medium is poured in thick layers into sterile, plastic Petri dishes (25–30 ml for Petri dishes 9–10 cm in diameter). The medium is no longer effective if it is remelted, but plates of CT agar keep well for several weeks at 4°C if contamination and desiccation are prevented.

CT agar is very useful for the isolation of *Serratia* spp. from feces, sputum, and any other polymicrobial clinical sample. *Serratia* colonies are apparent within 3 days, and further incubation allows colonies to get bigger (2–5 mm). Occasionally, *Providencia* spp., *Acinetobacter* spp., or *Pseudomonas* spp. may develop colonies on this medium. Other bacteria give only pinpoint colonies. When samples are from the natural environment or food, the results are less clear, especially if the sample brings nutrients. Several colonies must then be checked for DNase. Preenrichment in nutrient broth with 4% NaCl, incubated overnight at 20°C, can be used; about 0.1 ml of the enrichment culture is streaked onto CT agar. Broth that has supported overnight bacterial growth usually does not bring enough nutrients to affect adversely the selective isolation of *Serratia* spp. (P. A. D. Grimont, unpublished observation).

Identification

Usual methods for the identification of serratiae and other *Enterobacteriaceae* can be found in Edwards and Ewing (1972) and Lennette, Spaulding, and Truant (1974). However, following the pioneering example of Stanier, Palleroni, and Doudoroff (1966) with the pseudomonads, recent developments in enterobacterial taxonomy (Grimont et al., 1977; Véron and Le Minor, 1975) have demonstrated the usefulness of carbon source utilization tests. The methodology of carbon source utilization tests and other procedures that are not in general use, but are essential to *Serratia* identification, are detailed herein.

Carbon Source Utilization Tests

For carbon source utilization tests, we currently employ the same M70 medium (Véron, 1975) used in the caprylate thallous selective agar (see above), but omitting the yeast extract, thallous sulfate, and caprylate from solution A. Neutralized carbon sources (the purest brand available) are added to solution A to give a 0.1% final concentration (weight of the anion/vol) in the complete medium (0.2% in the case of carbohydrates). Sterilization of carbon source in aqueous solution is achieved by filtration through Millipore filters or by heating at 80°C for 20 min. In our experience, carbon source utilization tests (CSUT) gave reproducible and clear-cut results. Utilization of carbohydrates and polyalcohols often give more useful results than fermentation tests (FT). More than 95% of CSUT and FT using D-xylose, L-arabinose, trehalose, sorbitol, sucrose, and rhamnose gave the same results (Grimont, P. A. D. 1977). However, results with CSUT and FT are not parallel when adonitol, inositol, cellobiose, and lactose are studied. Whereas almost all *Serratia* strains can grow on inositol, acid from inositol was produced by only 76% of *S. marcescens*, 36% of *S. marinorubra*, 92% of *S. proteamaculans*, and 35% of *S. plymuthica* strains within 2 days. CSUT with adonitol is a very good test for separating *S. marcescens* from *S. proteamaculans* (99% vs. 0% positive strains, respectively), but FT with adonitol is a poor test for separating these two species (32% vs. 0% positive strains in 2 days, respectively).

Tetrathionate Reductase Test

The tetrathionate reductase test was found very useful for the identification of *Serratia* spp. The liquid medium of Le Minor et al. (1970) is composed of:

Tetrathionate Reductase Test for Identifying *Serratia* spp. (Le Minor et al., 1970)

$K_2S_4O_6$	5 g
Bromothymol blue (0.2% aqueous solution)	25 ml
Peptone water (peptone [Difco], 10 g; NaCl, 5 g; distilled water, 1 liter)	q.s. 1 liter

Adjust pH to 7.4. Filter-sterilize and dispense into 12-×-120-mm tubes (1 ml) or wells of sterile microculture plate (0.1 ml per well) for replica inoculation. After inoculation, we recommend adding a layer of sterile mineral oil, if microculture plates are used. Within 1 or 2 days, reduction of tetrathionate will acidify the medium, indicated by a yellow color. Negative tubes remain green or turn blue.

Voges-Proskauer Test

Results of the Voges-Proskauer test are variable with some *Serratia* spp., probably because end products of the butanediol pathway are metabolized (Grimont et al., 1977). Many strains of *S. plymuthica* and *S. odorifera* give negative results with O'Meara's method (O'Meara, 1931) and positive results with Richard's procedure (Richard, 1972). In this latter procedure, 0.5-ml amounts of Clark and Lubs medium (BBL) in 16-mm-wide test tubes are inoculated. After overnight incubation at 30°C, 0.5 ml α-naphthol (5 g α-naphthol in 100 ml absolute ethyl alcohol) and 0.5 ml NaOH 4 M are added. Tubes are examined for a red color.

Gas Production

Gas production differentiates *Serratia marcescens* from *S. proteamaculans* better when glucose agar is used rather than a liquid medium with a Durham inverted tube (Grimont et al., 1977). Glucose agar consists of nutrient agar (meat extract [Liebig], 3 g; yeast extract [Difco], 10 g; agar [BBL], 15 g; distilled water to 1 liter; pH 7.4) supplemented with 1% (wt/vol) glucose. Dispense into tubes (160 × 10 mm). Autoclave 20 min at 120°C, cool at 50°C, and inoculate before the agar sets. Tubes are examined for bubbles of gas for up to 3 days.

Gluconate Test

For the gluconate test, the medium of Haynes (1951) (tryptone [Difco], 1.5 g; yeast extract [Difco], 1 g; K_2HPO_4, 1 g; potassium gluconate, 40 g; distilled water, 1 liter; pH 7.0) is incubated for 2 days, and the presence of reducing derivatives is tested by adding a Clinitest tablet (Ames Co., Elkhart, Indiana). The reaction is considered positive when a light-green to rusty-yellow color develops. A deep blue color is given by a negative test.

Iodoacetate Test

For the iodacetate test, the medium of Lysenko (1961) (tryptone [Difco], 1 g; yeast extract [Difco], 1 g; NaCl, 5 g; K_2HPO_4, 0.3 g; glucose, 10 g; bromothymol blue 1.5% aqueous solution, 2 ml; distilled water, 1 liter; pH 7.2) is sterilized by boiling for 30 min on each of three successive days. Sodium iodoacetate (filter-sterilized) is added to a final concentration of 10^{-3} M (208 mg/liter). The inoculated medium is examined for acid production each day for 5 days. Controls without iodoacetate need to be included.

β-Xylosidase

The method of Brisou, Richard, and Lenriot (1972) for demonstrating β-xylosidase can conveniently be adapted for use with sterile microculture plates.

Aqueous (1% wt/vol) p-nitrophenyl-β-xyloside is dispensed into the wells in 0.05-ml amounts, followed by 0.05 ml of a fresh bacterial suspension in 0.25 M phosphate buffer, pH 7. Plates are examined for a yellow color after 24 h.

Acid and HCN Production from Amygdalin

Acid and HCN Production by *Serratia* from Amygdalin (Grimont, P. A. D., 1977)

Peptone (Oxoid L37)	10 g
Lab-Lemco	5 g
NaCl	5 g
Bromothymol blue, 0.05% solution	5 ml
Agar (Difco)	3 g
Distilled water	800 ml

After adjustment at pH 7.0, the medium is autoclaved 20 min at 120°C. Before the agar sets, add 200 ml of a 5% (wt/vol), filter-sterilized, aqueous solution of amygdalin (Sigma) and dispense in sterile test tubes with cotton plugs. Allow a short slant to form. Alkaline picrate reactive paper strips are prepared according to Sneath (1966). One yellow reactive paper is maintained by the cotton plug at the top of each inoculated tube. Development of a brick-red color witnesses the production of HCN. This detection is faster than acid production. Positive cultures should be taken out of the incubator to avoid diffusion of HCN to other tubes.

Susceptibility to 5-Fluorocytosine

For determining susceptibility to 5-fluorocytosine, an M70 minimal agar plate containing 0.2% (wt/vol) glucose is inoculated with a few drops of a culture in broth spread on the plate. After drying for 15 min at 37°C, a paper disk impregnated with 100 μg 5-fluorocytosine is placed on the plate. After overnight incubation, the plate is examined for an inhibition zone.

Other tests listed in Table 4 and Tables 6–10 can be performed as described by Grimont et al. (1977).

Conditions of Incubation

Best results are obtained when *Serratia* cultures are incubated at 30°C. At 37°C, pigmentation often fails to appear, and the Voges-Proskauer test is often negative. *S. plymuthica* may even fail to grow at 37°C. Otherwise, there is little difference between results of tests held at 30°C and those held at 37°C as far as *S. marcescens* is concerned.

Identification of *Serratia* at the Genus Level

Members of the genus *Serratia* share the characteristics defining the family Enterobacteriaceae. Only occasionally can a nitrate-negative strain be isolated. The properties that best define the genus *Serratia* are listed in Table 3. Although lipase activity on tributyrin or corn oil is listed, *S. odorifera* strains are only weakly lipolytic. A weak urease activity, lack of motility, or presence of a capsule are occasionally observed. Serratiae are clearly differentiated from *Klebsiella* spp., *Enterobacter aerogenes,* and *Enterobacter cloacae* by production of gelatinase, lipase, and DNase, and by growth on caprylate as sole carbon source. *Klebsiella oxytoca* may hydrolyze polygalacturonic acid (Starr et al., 1977; von Riesen, 1976). Some soft-rot *Erwinia* spp. produce extracellular proteinase and DNase, but these strains are pectinolytic, give a negative gluconate test, and cannot produce acid from glucose in the presence of iodoacetate (Grimont et al., 1977). *Erwinia herbicola* strains usually do not produce extracellular proteinases, lipases, and DNase; moreover, they give a negative gluconate test. More tests are available to differentiate each *Serratia* species from *Klebsiella* spp., *Enterobacter* spp., and *Erwinia* spp. "*Citrobacter*-like" organisms that show about 47% DNA relatedness to *Serratia* spp. (Crosa et al., 1974) have no proteinase or DNase, produce acid from dulcitol, and fail to grow on 4-aminobutyrate as sole carbon source (Gavini, Lefebvre, and Leclerc, 1976).

Identification of *Serratia* Species

The characteristics best allowing identification of each *Serratia* species are given in Table 4. Carbon source utilization tests are invaluable for unambiguous identification. Indole production by *S. odorifera* is not reproducibly observed when peptone water is used. A defined medium containing tryptophan (e.g., "urée-indole" medium, Institut Pasteur Production, Paris; or indole test in API 20E strips) gives consistently positive results with this species. With the use of classical tests (Edwards and Ewing, 1972), *S. plymuthica* may be confused either with *S. proteamaculans (liquefaciens)* or *S. marinorubra (rubidaea)*. Existence of *S. plymuthica* accounts for so-called atypical *S. liquefaciens* (lysine and ornithine decarboxylase negative), or atypical *S. rubidaea* (sorbitol positive). The new species *S. odorifera* can also be recognized in a number of "rhamnose-positive *S. liquefaciens*" strains. Carbon source utilization tests can help to sort out these "atypical" strains: after examining more than 3,000 *Serratia* strains, we have found no *S. proteamaculans* and no *S. plymuthica* strains that were able to grow on adonitol and erythritol, and no *S. marinorubra* strains able to grow on sorbitol as sole carbon source.

Use of commercial identification systems, such as API 20E (API Systems, La Balme-les-Grottes, France) is now widespread. The characteristics shown by *Serratia* species on API 20E are given in Table 5. With this system, some *S. marcescens*

Table 3. Characteristics defining the genus *Serratia*.[a]

Positive characteristics of the genus *Serratia*[b]	
Motile rods	ONPG hydrolyzed
Growth at 20°C in 1 day	DNase
Growth at pH 9	Lipase (tributyrin, corn oil)
Growth with 4% NaCl	Proteinase(s)
Acid from glucose in the presence of 10^{-3} M iodoacetate	Growth on minimal medium without addition of growth factors
Production of reducing compound(s) from gluconate	Carbon source utilization tests: D-alanine, L-alanine,
Voges-Proskauer test (Richard)	4-aminobutyrate, caprylate,
Acid from maltose	citrate, D-glucosamine,
Acid from mannitol	kynurenate, maltose,
Acid from salicin	L-proline, putrescine,
Acid from trehalose	tyrosine
	Growth on caprylate-thallous agar

Negative characteristics of the genus *Serratia*[b]	
Urease (Ferguson)	Acid from dulcitol
H₂S from thiosulfate	Carbon source utilization tests: butyrate,
Phenylalanine/tryptophan deaminase	dulcitol, 5-aminovalerate
Polygalacturonidase	Sodium ion requirement
Amylase	Anaerobic growth with KClO₃

[a] Grimont et al., 1977, 1978.
[b] More than 90% of each species giving the same reaction.

Table 4. Differential characteristics of the five *Serratia* species.[a]

Characteristic	*S. marcescens*	*S. proteamaculans*	*S. plymuthica*	*S. marinorubra*	*S. odorifera*
Red pigment	d	0	v	+	0
Musty, vegetable-like odor	0	0	0	v	+
Good growth at 5°C	0	+	+	0	+
Growth at 40°C	+	0	0	v	NT
Growth in 8.5% NaCl	v	v	0	+	NT
Growth with KCN	(+)	(+)	(v)	0	v
Indole	0	0	0	0	+
Tetrathionate reduction	d	+	0	0	0
Gas from glucose agar	0	+	v	0	0
β-Xylosidase	0	0	v	+	+
Acid from:					
Adonitol	v	0	0	+	(+)
L-Arabinose	0	+	+	+	+
Lactose	d	v	(+)	+	(+)
Melibiose	0	+	+	+	+
Raffinose	d	+	+	+	d
Rhamnose	0	0	0	0	+
Sorbitol	+	+	d	0	+
Sucrose	+	+	+	+	d
Xylose	0	+	+	+	+
HCN from amygdalin	0	(+)	+	d	NT
Lysine decarboxylase	+	+	0	d	+
Ornithine decarboxylase	+	+	0	0	d
Arginine decarboxylase	0	d	0	0	0
Chitin hydrolysis	+	d	+	0	0
Tween 80 hydrolysis	+	+	+	+	0
Growth on 5 μg/ml					
tetracycline	+	0	0	+	+
5-Fluorocytosine resistance	+	+	+	+	0
Carbon source utilization:					
Adonitol	+	0	0	+	+
L-Arabinose	0	+	+	+	+
D-Arabitol	0	0	0	+	0
Betaine	0	0	v	+	0
Benzoate	d	d	(v)	(+)	0
Cellobiose	0	(+)	+	+	+
meso-Erythritol	d	0	0	+	d
Lactose	d	(v)	m	+	m
Melibiose	0	+	+	+	+
α-Methylglucoside	0	(+)	(v)	+	0
Mucate	0	0	v	+	+
Nicotinate	(+)	+	(+)	0	+
Quinate	d	0	+	(+)	0
Raffinose	v	+	+	+	d
Rhamnose	0	0	0	0	+
Sorbitol	+	+	d	0	+
Sucrose	+	+	+	+	d
D-Tartrate	0	0	0	v	+
Trigonelline	d	0	0	+	+
D-Xylose	0	+	+	+	+

[a]Symbols: +, positive for 90% or more strains; 0, negative for 90% or more strains; v, positive for 10–89% strains; d, test used to differentiate biotypes; ( ), delayed reaction; m, growth of mutants.

strains may give a fermentative pattern identical with that of typical *S. proteamaculans* (melibiose- and arabinose-positive). Sorbitol-negative *S. plymuthica* strains may also resemble *S. marinorubra*. Rhamnose-negative *Serratia* strains are easily distinguished from *Klebsiella* and *Enterobacter* spp., which are rhamnose-positive. *S. odorifera* resembles *Enterobacter aerogenes* when API 20E or the equivalent standard tests are used. Detection of indole, evidence of a gelatinase, lack of gas production, and a typical musty, vegetable-like odor should correctly establish the identification.

Table 5. Characteristics of *Serratia* species on tests of the API 20E strips (28–35°C).[a]

Test	S. marcescens	S. proteamaculans	S. plymuthica	S. marinorubra	S. odorifera
ONPG	+	+	+	+	+
ADH	0	d	0	0	0
LDC	+	+	0	d	+
ODC	+	+	0	0	d
CIT	+	+	+	+	+
H_2S	0	0	0	0	0
URE	(v)[b]	0	0	0	0
TDA	0	0	0	0	0
IND	0	0	0	0	+
VP	+	+	v	+	+
GEL	+	+	+	+	+
GLU	+	+	+	+	+
MAN	+	+	+	+	+
INO	v	+	+	+	+
SOR	+	+	d	0	+
RHA	0	0[c]	0[c]	0	+
SAC	+	+	+	+	d
MEL	v[b]	+	+	+	+
AMY	+	+	+	+	+
ARA	v[b]	+	+	+	+

[a] Results at 37°C may be erratic with *S. plymuthica*. Symbols as in Table 4.
[b] Positive results are not confirmed by standard tests performed in tubes.
[c] Exceptional strains positive.

Identification of *Serratia* Biotypes

Biotypes in *Serratia marcescens, S. proteamaculans, S. plymuthica,* and *S. marinorubra* were first defined by numerical taxonomy (Grimont et al., 1977). We now recognize 19 biotypes in *S. marcescens,* 4 in *S. proteamaculans,* 3 in *S. plymuthica,* 3 in *S. marinorubra,* and 2 in *S. odorifera.* Characteristics of these biotypes are given in Tables 6–10. Carbon source utilization tests are major elements in the identification of these biotypes. Multipoint inoculation devices (e.g., Denley Multipoint Inoculator, Denley Instruments, Ltd., Bolney Cross, Bolney, Sussex, England), are most useful in biotyping 20 or more strains at a time. Differentiation between *S. proteamaculans* biotypes C1a and C1b may be superfluous; electrophoresis of proteinases (Grimont, Grimont, and Dulong de Rosnay, 1977) could not discriminate between these two biotypes. The significance of *S. plymuthica* biotype C2c is uncertain, since at present this biotype is constituted only by some historical strains.

Biotyping of *S. marcescens* is epidemiologically useful (Grimont and Grimont, 1978b). Pigmentation occurs only in 6 *S. marcescens* biotypes: A1a, A1b, A2a, A2b, A6a, and A6b. The 13 other *S. marcescens* biotypes correspond to nonpigmented strains: A3a, A3b, A3c, A3d, A4a, A4b, A5, A8a, A8b, A8c, TCT, TT, and TC. Biotypes A6b, A4b, TT, and TC are rarely found and their ecological-epidemiological significance is unknown. Nonpigmented biotypes A3abcd and A4a are ubiquitous, whereas nonpigmented biotypes A5, A8abc, and TCT seem restricted to hospitalized patients. Pigmented biotypes are ubiquitous.

Serotyping of *Serratia marcescens*

Hefferan (1906) did a pioneering work in studying antigenic relationships among the "prodigiosus" group of red-pigmented bacilli. However, the systematic inventory of somatic (O) antigens of *S. marcescens* began in 1957 (Davis and Woodward, 1957). Flagellar antigens (H) were described (Ewing, Davis, and Reavis, 1959) and an antigenic scheme was developed (Edwards and Ewing, 1972; Ewing, Davis, and Reavis, 1959; Le Minor and Pigache, 1977, 1978; Sedlák, Dlabač, and Motlikova, 1965; Traub and Kleber, 1977). The present system consists of 20 somatic antigens (O1 to O20) and 20 flagellar antigens (H1 to H20). Serotyping of *S. marcescens* is not easy. Cross-reactions occur between O antigens 2 and 3; 6 and 7; and 6, 12, and 14. Cross-reactions between O6 and O14 are so extensive that the epidemiological distinction between these two antigens might be questionable. A subdivision of O antigens into several factors has been proposed (Sedlák, Dlabač, and Motlikova, 1965).

To overcome the tediousness of flagellar agglutination, an immobilization test in semisolid agar has been described for determination of H antigens (Le Minor and Pigache, 1977). The technique is facilitated by the use of pools of nonabsorbed sera. No

Table 6. Identification of *Serratia marcescens* biotypes.[a]

	A1		A2		A3				A4		A5	A6		A8			TCT	TT	TC
	a	b	a	b	a	b	c	d	a	b		a	b	a	b	c			
Growth on:																			
m-Erythritol	+	+	+	+	+	+	+	+	+	+	0	+	+	0	0	0	0	0	0
Trigonelline	0	0	0	+	0	+	0	+	0	0	+	0	+	+	+	+	+	+	0
Quinate and/or																			
4-hydroxybenzoate	0	0	0	0	0	0	0	0	+	0	+	+	+	+	+	+	0	0	0
3-Hydroxybenzoate	0	0	0	0	+	+	0	0	0	0	0	0	0	0	+	0	0	0	0
Benzoate	+	+	0	0	0	0	0	0	0	0	0	0	0	0	0	0	0	0	0
DL-Carnitine	+	0	v	v	v	v	v	v	v	v	+	v	v	0	0	0	+	0	+
Lactose	0	0	0	0	0[b]	0	0[b]	0[b]	0[b]	0	0[b]	0	0	0	0[b]	+	0[b]	0[b]	0
Tetrathionate reduction	+	+	+	+	+	+	+	+	0	0	+	+	+	+	+	+	+	+	+
Red pigment	+[c]	+[c]	+	+	0	0	0	0	0	0	0	+[c]	+[c]	0	0	0	0	0	0

[a] Symbols as in Table 4.
[b] Lactose-positive strains occasionally isolated. Lactose differentiates between biotypes A8a and A8c.
[c] Nonpigmented strains occasionally isolated.
Data from Grimont and Grimont 1978b.

Table 7. Identification of *Serratia proteamaculans* (= *S. liquefaciens*) biotypes.[a]

	Biotype			
Trait	C1a	C1b	C1c[b]	C1d[c]
Growth on:				
Benzoate	0	0	0	+
Aconitate	0	0	+	0
L-Ornithine	+	+	0	0
Arginine decarboxylase	0	0	0	+
Chitin hydrolysis	+	0	0	0
Methyl red test	0	+	v	v
Triacetin, acid	+	+	(+)	0

[a] Symbols as in Table 4.
[b] The type strain of *S. proteamaculans* (ATCC 19323, ICPB XP176) corresponds to biotype C1c.
[c] The type strain of *S. liquefaciens* (ATCC 14460) corresponds to biotype C1d.

Table 8. Identification of *Serratia plymuthica* biotypes.[a]

	Biotype		
Trait	C2a	C2b	C2c
Growth on:			
Benzoate	+	v	0
Betaine	+	0[b]	0
Glycerol	+	+	0[b]
Aconitate	+	+	0
Acetate	+	+	0
Sorbitol[c]	+	v	+
β-Xylosidase	0	v	+

[a] Symbols as in Table 4.
[b] Exceptional strains positive.
[c] Same results with fermentation test.

Table 9. Identification of *Serratia marinorubra* biotypes.[a]

	Biotype		
Trait	B1	B2	B3
Growth on:			
D-Melezitose	v	+	+
Benzoate	+	+	v
Ethanol	0	+	0
D-Tartrate	+	0	v
Histamine	v	0	+
Voges-Proskauer (O'Meara)	+	0	v
Lysine decarboxylase	+	+	0
Malonate (Leifson)	+	+	0
HCN from amygdalin (4 days)	0	+	0

[a] Symbols as in Table 4.

Table 10. Identification of *Serratia odorifera* biotypes.[a]

	Biotype	
Trait	1[b]	2
Growth on:		
m-Erythritol	0	+
L-Ornithine	(+)	0
Sucrose	+	0
Ornithine decarboxylase	+	0
Acid from sucrose	+	0
Acid from raffinose	+	0[c]

[a] Symbols as in Table 4.
[b] The type strain ICPB 3995 corresponds to biotype 1.
[c] Some strains positive in 3–7 days.

anti-H:9 serum is needed in pools, as bacteria with H:9 antigen are immobilized by anti-H:8 and anti-H:10 sera. Once the unknown strain is immobilized by a serum pool, corresponding individual sera are used to achieve final identification of H antigens. The following absorbed sera need to be prepared: anti-H:8 absorbed by H:10; anti-H:10 absorbed by H:8; and anti-H:3 absorbed by H:10. Other cross-reactions are overcome by dilution of sera. The immobilization test works because H antigens are monophasic in *S. marcescens*.

Some striking correlations between antigenic composition and biotype have been shown (Grimont et al., 1979). Pigmented biotypes and nonpigmented biotypes are antigenically segregated. Serotyping subdivides the biotypes, but when two biotypes correspond to the same serotype, these biotypes usually differ by only one biochemical reaction. Biotypes can be united in "biogroups" to amplify biotype-serotype correlation (Table 11). These biogroups might be valid infraspecific taxons.

Few immunological comparisons between *S. marcescens* and other *Serratia* spp. or other genera have been conducted. Hefferan (1906) has shown that aerogenic species *Bacillus kiliensis, B. plymouthensis,* and *B. miniaceus* could be agglutinated by high dilutions of a same serum and these species constituted a group antigenically distinct from *B. prodigiosus.* Several *Enterobacter liquefaciens* strains could be agglutinated with *S. marcescens* O antisera (Ewing, Davis, and Reavis, 1959; Hamon, Le Minor, and Péron, 1970). Other enterobacteria may also react with these O antisera (Ewing, Davis, and Reavis, 1959). No antigenic relationship has been found between *Chromobacterium* spp. and *Serratia marcescens* (Sneath, 1960).

Bacteriocin Typing of *Serratia marcescens*

Two different schemes have been developed for typing *Serratia marcescens* strains by susceptibility to bacteriocins. Traub, Raymond, and Startsman (1971) found 37 strains producing bacteriocin (marcescin) out of 50 treated with mitomycin. Ninety-two percent of their strains were susceptible to one or more marcescins. Ten computer-selected marcescin-producing strains could classify 92% of the strains into 16 types. The number of types was later extended to 37, and 85% of hospital strains were found typable (Traub and Raymond, 1971). Little correlation was found between serotyping and typing by susceptibility to marcescins (Traub, 1972a). Farmer (1972b) independently developed another system of typing by susceptibility to marcescins. After mitomycin induction, 12 bacteriocin-producing strains were selected and the 12 marcescins could subdivide 93 strains into 79 types.

Table 11. Correlation between biotypes[a] and serotypes[b] in *Serratia marcescens*

Biogroups	Serotypes
A1a,b	O5:H2, O5:H3, O10:H6
A2a,b/A6a,b	O6:H2, O6:H3, O6:H10, O6:H13, O8:H3, O13:H5, O14:H2, O14:H3, O14:H8, O14:H10
A3a,b,c,d	O3:H5, O3:H11, O5:H15, O9:H11, O9:H17, O12:H5, O12:H9, O12:H11, O12:H17, O12:H20, O13:H17, O14:H5, O14:H20, O15:H3, O15:H5, O15:H8
A4a,b	O1:H1, O1:H4, O2:H1, O2:H8, O3:H1, O4:H1, O4:H4, O5:H1, O5:H6, O5:H8, O9:H1, O13:H1, O13:H13
A5/A8a,b,c	O3:H12, O4:H12, O6:H4, O6:H12, O8:H12, O14:H4, O14:H12, O15:H12
TCT/TT	O1:H7, O2:H7, O4:H7, O5:H7, O5:H19, O11:H4, O13:H7, O13:H12, O19:H14
TC	O10:H8, O20:H12

[a] Grimont et al. 1977, Grimont and Grimont, 1978b.
[b] Serotypes are listed only when at least two strains were studied and gave identical results. Data from Grimont et al. 1979.

Typing of *S. marcescens* by production of bacteriocin was also developed by Farmer (1972a). Twenty-three strains susceptible to bacteriocins were selected to serve as indicator strains. Each unknown *S. marcescens* strain is treated with mitomycin and the marcescin produced (if any) is identified according to the pattern of susceptibility demonstrated by indicator strains. This method could type 91% of the strains studied. Typing by production of bacteriocin is likely to be a more reliable epidemiological marker than typing by susceptibility to bacteriocins (Farmer, 1972a). No specific bacteriocin typing system has been developed for other species of *Serratia.*

Phage Typing of *Serratia*

The first phage typing scheme in *Serratia* was reported by Pillich, Hradečna, and Kocur (1964), who studied the susceptibility of 107 *Serratia* cultures to 10 phages, and described 7 groups. This system, however, was not used to type clinical strains. Hamilton and Brown (1972) selected 34 phages that could classify 90.6% of 204 strains into 23 phage groups and 71 phage types. They could show some correlation between phage typing and inositol fermentation and susceptibility to carbenicillin. Farmer (1975) has developed a phage typing system common to *Serratia marcescens, S. liquefaciens,* and *S. rubidaea (marinorubra).* With 74 phages, 95% of *S. marcescens* and 50% of *S. liquefaciens* and *S. rubidaea* could be typed. Phage typing was found convenient in subdividing prevalent serotypes (Negut, Davis, and Farmer, 1975).

A recent phage typing system has been developed for nosocomial strains isolated in Bordeaux (Grimont, F. 1977). Most of these *S. marcescens* strains gave phage susceptibility patterns that corresponded to phage types described by F. Grimont (1977); however, strains isolated in other cities of France were less often typable, and strains received from other countries were rarely typable in this system.

Bacteriocin or phage typing systems rarely survive unchanged beyond the original publications. The reason is that when more strains are studied (especially nonlocal strains), more strains are found with new patterns of susceptibility to phage or bacteriocin. The theoretical maximum number of types is 2^n when n bacteriocins or phages are used. For this reason, Farmer (1970) has proposed a simple code to transform typing patterns into a number.

Proteinase Zymograms of *Serratia marcescens*

By using agar gel electrophoresis, Grimont, Grimont, and Dulong de Rosnay (1977) and Grimont and Grimont (1978c) described seven proteinases produced by *S. marcescens* (labeled P_5, P_6, P_7, P_{9a}, P_{9b}, P_{11}, and P_{12}). Each *S. marcescens* strain can produce one to four proteinases. 651 *S. marcescens* strains, isolated in a period of 6 years in a French hospital, gave 33 different proteinase patterns (called zymotypes). However, 6 zymotypes accounted for 76% of the isolates. Proteinase zymograms could be a useful epidemiological marker (Grimont and Grimont, 1978c).

Acknowledgments

We are indebted to M. P. Starr for his constant help and advice.

Literature Cited

Ackerman, L. J., Kishimoto, R. A., Emerson, J. S. 1971. Nonpigmented *Serratia marcescens* arthritis in a Teju *(Tupinambis tequixin)*. American Journal of Veterinary Research **32**:823–826.

Alford, L. R., Holmes, N. E., Scott, W. T., Vickery, J. R. 1950. Studies on the preservation of shell eggs. I. The nature of wastage in Australian export eggs. Australian Journal of Applied Science **1**:208–214.

Altemeier, W. A., Culbertson, W. R., Fullen, W. D., McDonough, J. J. 1969. *Serratia marcescens* septicemia. A new threat in surgery. Archives of Surgery **99**:232–238.

Barnum, D. A., Thackeray, E. L., Fish, N. A. 1958. An outbreak of mastitis caused by *Serratia marcescens*. Canadian Journal of Comparative and Medical Veterinary Science **22**:392–395.

Bascomb, S., Lapage, S. P., Willcox, W. R., Curtis, M. A. 1971. Numerical classification of the tribe Klebsielleae. Journal of General Microbiology **66**:279–295.

Bergey, D. H., Harrison, F. C., Breed, R. S., Hammer, B. W., Huntoon, F. M. 1923. Bergey's manual of determinative bacteriology, 1st ed. Baltimore: Williams & Wilkins.

Berkowitz, D. M., Lee, W. S. 1973. A selective medium for isolation and identification of *Serratia marcescens*. Abstracts of the Annual Meeting of the American Society for Microbiology **1973**:105.

Bizio, B. 1823. Lettera di Bartolomeo Bizio al chiarissimo canonico Angelo Bellani sopra il fenomeno della polenta porporina. Biblioteca Italiana o sia Giornale di Letteratura Scienze a Arti **30**:275–295.

Boam, G. W., Sanger, V. L., Cowan, D. F., Vaughan, D. P. 1970. Subcutaneous abscesses in iguanid lizards. Journal of the American Veterinary Medical Association **157**:617–619.

Bodey, G. P., Rodriguez, U., Smith, J. P. 1970. *Serratia* sp. infections in cancer patients. Cancer **25**:199–205.

Breed, R. S., Breed, M. E. 1924. The type species of the genus *Serratia*, commonly known as *Bacillus prodigiosus*. Journal of Bacteriology **9**:545–557.

Breed, R. S., Breed, M. E. 1927. The genus *Serratia* Bizio. Centralblatt für Bakteriologie, Parasitenkunde und Infektionskrankheiten, Abt. 2 **71**:435–440.

Breed, R. S., Murray, E. G. D., Hitchens, A. P. 1948. Bergey's manual of determinative bacteriology, 6th ed. Baltimore: Williams & Wilkins.

Breed, R. S., Murray, E. G. D., Smith, N. R. 1957. Bergey's manual of determinative bacteriology, 7th ed. Baltimore: Williams & Wilkins.

Brisou, J., Cadeillan, J. 1959. Etude sur les *Serratia*. A propos de quatre souches isolées en médecine vétérinaire. Bulletin de l'Association des Diplomés de Microbiologie de la Faculté de Pharmacie de Nancy **75**:34–39.

Brisou, B., Richard, C., Lenriot, A. 1972. Intérêt taxonomique de la recherche de la β-xylosidase chez les *Enterobacteriaceae*. Annales de l'Institut Pasteur **123**:341–347.

Bucher, G. E. 1959. Bacteria of grasshoppers of western Canada. III. Frequency of occurence, pathogenicity. Journal of Insect Pathology **1**:391–405.

Bucher, G. E. 1960. Potential bacterial pathogens of insects and their characteristics. Journal of Insect Pathology **2**:172–195.

Bucher, G. E. 1963a. Nonsporulating bacterial pathogens, pp. 117–147. In: Steinhaus, E. A. (ed.), Insect pathology. An advanced treatise. vol. 2. New York, London: Academic Press.

Bucher, G. E. 1963b. Transmission of bacterial pathogens by the ovipositor of a hymenopterous parasite. Journal of Insect Pathology **5**:277–283.

Bucher, G. E., Stephens, J. M. 1959. Bacteria of grasshoppers of western Canada: I. The Enterobacteriaceae. Journal of Insect Pathology **1**:356–373.

Burnside, C. E. 1928. A septicemic condition of adult bees. Journal of Economic Entomology **21**:379–386.

Cabrera, H. A. 1969. An outbreak of *Serratia marcescens* and its control. Archives of Internal Medicine **123**:650–655.

Caldis, P. D. 1927. Etiology and transmission of endosepsis (internal rot) of the fruit of the fig. Hilgardia **2**:287–328.

Capponi, M., Sureau, P., Le Minor, L. 1956. Contribution à l'étude des salmonelles du Centre-Vietnam. Bulletin de la Société de Pathologie Exotique **49**:796–801.

Carter, G. R. 1973. Diagnostic procedures in veterinary microbiology, 2nd ed. Springfield, Illinois: Charles C. Thomas.

Cate, J. C. 1972. Isolation of *Serratia marcescens* from stools with an antibiotic plate, pp. 763–764. In: Hejzlar, M., Semonský, M., Masák, S. (eds.), Advances in antimicrobial and antineoplastic chemotherapy. Progress in research and clinical application. Proceedings of the 7th International Congress of Chemotherapy, vol. I/2. Munich, Berlin, Vienna: Urban & Schwarzenberg.

Clayton, E. D. W., von Graevenitz, A. 1966. Nonpigmented *Serratia marcescens*. Journal of the American Medical Association **197**:1059–1064.

Colwell, R. R., Mandel, M. 1965. Adansonian analysis and deoxyribonucleic acid base composition of *Serratia marcescens*. Journal of Bacteriology **89**:454–461.

Combe, E. 1933. Etude comparative de trois bactéries chromogènes à pigment rouge: *Serratia marcescens* Bizio (*Bacillus prodigiosus* Flügge), *Serratia kiliensis* Comité S.B.A. (Bacterium h Breunig), *Serratia esseyana* n. sp. Combe (Bacille rouge d'Essey Lasseur). Thèse de Pharmacie. University of Nancy, Nancy, France.

Crosa, J. H., Steigerwalt, A. G., Fanning, G. R., Brenner, D. J. 1974. Polynucleotide sequence divergence in the genus *Citrobacter.* Journal of General Microbiology **83:**271–282.

D'Aoust, J. Y., Gerber, N. N. 1974. Isolation and purification of prodigiosin from *Vibrio psychroerythreus.* Journal of Bacteriology **118:**756–757.

Daschner, F., Senska-Euringer, C. 1975. Kontaminierte infusionen als Ursache nosokomialer *Serratia marcescens* Sepsis bei Kindern. Deutsche Medizinische Wochenschrift **100:**2324–2328.

Davis, B. R., Woodward, J. M. 1957. Some relationships of the somatic antigens of a group of *Serratia marcescens* cultures. Canadian Journal of Microbiology **3:**591–597.

Denis, F. A., Blanchard, P. 1975. Enquête sur les porteurs intestinaux de *Serratia.* Possibilité d'infections d'origine endogène. Nouvelle Presse Médicale **4:**2114–2115.

Deom, J., Mortelmans, J. 1953. Etude d'une souche pathogène de *Serratia marcescens.* Revue d'Immunologie **17:**394–398.

Duran-Reynals, F., Clausen, H. J. 1937. A contagious tumor-like condition in the lizard *(Anolis equestris)* as induced by a new bacterial species, *Serratia anolium* (sp. n.) Journal of Bacteriology **33:**369–379.

Dye, D. W. 1966. A comparative study of some atypical "xanthomonads". New Zealand Journal of Science **9:**843–854.

Edwards, P. R., Ewing, W. H. 1972. Identification of Enterobacteriaceae, 3rd ed. Minneapolis: Burgess.

Ehrenberg, C. G. 1848. Communication presented at the plenary session of the Royal Prussian Academy of Sciences, October 26, pp. 346–362. Bericht über die zur Bekanntmachung geeigneten Verhandlungen der königliche preussische Akademie der Wissenschaften zu Berlin.

Eisenberg, J. 1886. Bakteriologische Diagnostik. Hülfs-Tabellen beim Praktischen Arbeiten. Hamburg, Leipzig: Leopold Voss.

Ewing, W. H. 1963. An outline of nomenclature for the family *Enterobacteriaceae.* International Bulletin of Bacteriological Nomenclature and Taxonomy **13:**95–110.

Ewing, W. H., Davis, B. R., Fife, M. A. 1972. Biochemical characterization of *Serratia liquefaciens* and *Serratia rubidaea.* Atlanta: Center for Disease Control.

Ewing, W. H., Davis, B. R., Johnson, J. G. 1962. The genus *Serratia:* Its taxonomy and nomenclature. International Bulletin of Bacteriological Nomenclature and Taxonomy **12:**47–52.

Ewing, W. H., Davis, B. R., Reavis, R. W. 1959. Studies on the *Serratia* group. Atlanta: Center for Disease Control.

Ewing, W. H., Johnson, J. G., Davis, B. R. 1962. The occurrence of *Serratia marcescens* in nosocomial infections. Atlanta: Center for Disease Control.

Ewing, W. H., Davis, B. R., Fife, M. A., Lessel, E. F. 1973. Biochemical characterization of *Serratia liquefaciens* (Grimes and Hennerty) Bascomb et al. (formerly *Enterobacter liquefaciens*) and *Serratia rubidea* (Stapp) comb. nov. and designation of type and neotype strains. International Journal of Systematic Bacteriology **23:**217–225.

Farmer, J. J., III. 1970. Mnemonic for reporting bacteriocin and bacteriophage types. Lancet **ii:**96.

Farmer, J. J., III. 1972a. Epidemiological differentiation of *Serratia marcescens:* Typing by bacteriocin production. Applied Microbiology **23:**218–225.

Farmer, J. J., III. 1972b. Epidemiological differentiation of *Serratia marcescens:* Typing by bacteriocin sensitivity. Applied Microbiology **23:**226–231.

Farmer, J. J., III. 1975. Lysotypie de *Serratia marcescens.* Archives Roumaines de Pathologie Experimentale et de Microbiologie **34:**189.

Farmer, J. J., III., Davis, B. R., Hickman, F. H., Presley, D. B., Bodey, G. P., Negut, M., Bobo, R. A. 1976. Detection of *Serratia* outbreaks in hospital. Lancet **ii:**455–459.

Farmer, J. J., III., Silva, F., Williams, D. R. 1973. Isolation of *Serratia marcescens* on deoxyribonuclease-toluidine blue-cephalothin agar. Applied Microbiology **25:**151–152.

Fortineau, L. 1904. *Erythrobacillus pyosepticus* et bactéries rouges. M.D. Thesis. Faculté de Médecine, Paris.

Fulton, M., Forney, C. E., Leifson, E. 1959. Identification of *Serratia* occurring in man and animals. Canadian Journal of Microbiology **5:**269–275.

Gale, D., Lord, J. D., Durham, N. C. 1957. Overgrowth of *Serratia marcescens* in respiratory tracts simulating hemoptysis. Report of a case. Journal of the American Medical Association **164:**1328–1330.

Gale, D., Sonnenwirth, A. C. 1962. Frequent human isolation of *Serratia marcescens.* Bacteriological and pathogenicity studies of twelve strains of *S. marcescens* recovered from nine patients during a six-month period. Archives of Internal Medicine **109:**414–421.

Gandhi, N. M., Nazareth, J., Divekar, P. V., Kohl, H., de Souza, N. J. 1973. Magnesidin, a novel magnesium-containing antibiotic. Journal of Antibiotics **26:**797–798.

Gaughran, E. R. L. 1969. From superstition to science: The history of a bacterium. Transactions of the New York Academy of Sciences **31:**3–24.

Gauthier, M. J. 1976. *Alteromonas rubra* sp. nov., a new marine antibiotic-producing bacterium. International Journal of Systematic Bacteriology **26:**459–466.

Gavini, F., Lefebvre, B., Leclerc, H. 1976. Positions taxonomiques d'entérobactéries H_2S^- par rapport au genre *Citrobacter.* Annales de Microbiologie **127A:**75–295.

Goldin, M., Shaffer, J. G., Brown, E. 1969. A new selective and differential medium for *Serratia.* Bacteriological Proceedings **1969:**96.

Gosden, P. E., Ware, G. C. 1976. The aerobic bacterial flora of the anal sac of the red fox. Journal of Applied Bacteriology **41:**271–275.

Greenup, P., Blazevic, D. J. 1971. Antibiotic susceptibilities of *Serratia marcescens* and *Enterobacter liquefaciens.* Applied Microbiology **22:**309–314.

Griffith, L. J. 1967. Significance of *Serratia marcescens* in medical bacteriology. Clinical Medicine **74:**24–29.

Grimes, M., Hennerty, A. J. 1931. A study of bacteria belonging to the sub-genus *Aerobacter.* Scientific Proceedings of the Royal Dublin Society, New Series **20:**89–97.

Grimont, F. 1977. Les bactériophages des *Serratia* et bactéries voisines. Taxonomie et lysotypie. Thèse de Pharmacie. University of Bordeaux II, Bordeaux, France.

Grimont, P. A. D. 1977. Le Genre *Serratia.* Taxonomie et approche écologique. Ph.D. Thesis. University of Bordeaux I, Bordeaux, France.

Grimont, P. A. D., Dulong de Rosnay, H. L. C. 1972. Numerical study of 60 strains of *Serratia.* Journal of General Microbiology **72:**259–268.

Grimont, P. A. D., Grimont, F. 1978a. The genus *Serratia.* Annual Review of Microbiology **32:**221–248.

Grimont, P. A. D., Grimont, F. 1978b. Biotyping of *Serratia marcescens* and its use in epidemiological studies. Journal of Clinical Microbiology **8:**73–83.

Grimont, P. A. D., Grimont, F. 1978c. Proteinase zymograms of *Serratia marcescens* as an epidemiological tool. Current Microbiology **1:**15–18.

Grimont, P. A. D., Grimont, F., Dulong de Rosnay, H. L. C. 1977. Characterization of *Serratia marcescens, S. liquefaciens, S. plymuthica* and *S. marinorubra* by electrophoresis of their proteinases. Journal of General Microbiology **99:**301–310.

Grimont, P. A. D., Grimont, F., Lysenko, O. 1979. Species and biotype identification of *Serratia* strains associated with insects. Current Microbiology **2:**139–142.

Grimont, P. A. D., Grimont, F., Starr, M. P. 1978. *Serratia proteamaculans* (Paine and Stansfield) comb. nov., a senior subjective synonym of *Serratia liquefaciens* (Grimes and Hennerty) Bascomb et al. International Journal of Systematic Bacteriology **28:**503–510.

Grimont, P. A. D., Grimont, F., Dulong de Rosnay, H. L. C., Sneath, P. H. A. 1977. Taxonomy of the genus *Serratia*. Journal of General Microbiology **98:**39–66.

Grimont, P. A. D., Grimont, F., Richard, C., Davis, B. R., Steigerwalt, A. G., Brenner, D. J. 1978. Deoxyribonucleic acid relatedness between *Serratia plymuthica* and other *Serratia* species with a description of *Serratia odorifera* sp. nov. (type strain: ICPB 3995). International Journal of Systematic Bacteriology **28:**453–463.

Grimont, P. A. D., Grimont, F., Le Minor, S., Davis, B., Pigache, F. 1979. Compatible results obtained from biotyping and serotyping in *Serratia marcescens*. Journal of Clinical Microbiology **10:**425–432.

Hamilton, R. L., Brown, W. J. 1972. Bacteriophage typing of clinically isolated *Serratia marcescens*. Applied Microbiology **24:**899–906.

Hamon, Y., Le Minor, L., Péron, Y. 1970. Les bactériocines d'*Enterobacter liquefaciens*. Intérêt taxonomique de leur étude. Comptes Rendus Hebdomadaires des Séances de l'Académie des Sciences, Série D **270:**886–889.

Harrison, F. C. 1924. The "miraculous" microorganism. Transactions of the Royal Society of Canada **18:**1–17.

Haynes, W. D. 1951. *Pseudomonas aeruginosa*—its characterization and identification. Journal of General Microbiology **5:**939–950.

Hefferan, M. 1904. A comparative and experimental study of bacilli producing red pigment. Centralblatt für Bakteriologie, Parasitenkunde und Infektionskrankheiten, Abt. 2 **11:**311–317, 397–404, 456–475, 520–540.

Hefferan, M. 1906. Agglutination and biological relationship in the prodigiosus group. Centralblatt für Bakteriologie, Parasitenkunde und Infektionskrankheiten Abt. 1, Orig. **41:**553–562.

Hernandez Marchant, R., Rojas, P. O., Arcaya, 1960. Sindrome del panal rojo. Revista Chilena de Pediátrica **31:**335–339.

Izawa, H., Nagabayashi, T., Kazuno, Y., Soekawa, M. 1971. Occurrence of death of chick embryos by *Serratia marcescens* infection. Japanese Journal of Bacteriology **26:**200–204.

Jackson, C. G., Jr., Fulton, M. 1976. A turtle colony epizootic apparently of microbial origin. Journal of Wildlife Disease **6:**466–468.

Janota-Bassalik, L. 1963. Psychrophiles in low-moor peat. Acta Microbiologica Polonica **12:**25–40.

Johnson, E., Ellner, P. D. 1974. Distribution of *Serratia* species in clinical specimens. Applied Microbiology **28:**513–514.

Kaska, M. 1976. The toxicity of extracellular proteases of the bacterium *Serratia marcescens* for larvae of greater wax moth, *Galleria mellonella*. Journal of Invertebrate Pathology **27:**271.

Kushner, D. J., Harvey, G. T. 1962. Antibacterial substances in leaves: Their possible role in insect resistance to disease. Journal of Insect Pathology **4:**155–184.

Lahellec, C., Meurier, C., Bennejean, G., Catsaras, M. 1975. A study of 5920 strains of psychrotrophic bacteria isolated from chickens. Journal of Applied Bacteriology **38:**89–97.

Lakso, J. U., Starr, M. P. 1970. Comparative injuriousness to plants of *Erwinia* spp. and other enterobacteria from plants and animals. Journal of Applied Bacteriology **33:**692–707.

Legakis, N. J., Papavassiliou, J. Th., Xilinas, M. E. 1976. Inositol as a selective substrate for the growth of klebsiellae and serratiae. Zentralblatt für Bakteriologie, Parasitenkunde, Infektionskrankheiten und Hygiene, Abt. 1 Orig., Reihe A **235:**453–458.

Lehmann, K. B., Neumann, R. 1896. Atlas und Grundriss der Bakteriologie und Lehrbuch der speciellen bakteriologischen Diagnostik, Teil II. Munich: Lehmann.

Le Minor, L., Chippaux, M., Pichinoty, F., Coynault, C., Piéchaud, M. 1970. Méthodes simples permettant de rechercher la tétrathionate-réductase en cultures liquides ou sur colonies isolées. Annales de l'Institut Pasteur **119:**733–737.

Le Minor, S., Pigache, F. 1977. Étude antigénique de souches de *Serratia marcescens* isolées en France. I. Antigènes H: Indi-

vidualisation de six nouveaux facteurs H. Annales de Microbiologie **128B:**207–214.

Le Minor, S., Pigache, F. 1978. Etude antigénique de souches de *Serratia marcescens* isolées en France. II. Caractérisation des antigènes O et individualisation de 5 nouveaux facteurs, fréquence des sérotypes et désignation des nouveaux facteurs H. Annales de Microbiologie **129B:**407–423.

Lennette, E. H., Spaulding, E. H., Truant, J. P. (eds.). 1974. Manual of clinical microbiology, 2nd ed. Washington, D. C.: American Society for Microbiology.

Loiseau-Marolleau, M. L., Laforest, H. 1976. Contribution à l'étude de la flore bactérienne des aliments en milieu hospitalier. Médecine et Maladies Infectieuses **6:**160–171.

Lynch, D. L., Kenealy, W. R. 1976. A selective medium for the isolation of *Serratia* sp. from raw sewage. Microbios Letters **1:**35–37.

Lysenko, O. 1961. *Pseudomonas*—an attempt at a general classification. Journal of General Microbiology **25:**379–408.

Lysenko, O. 1976. Chitinase of *Serratia marcescens* and its toxicity to insects. Journal of Invertebrate Pathology **27:**385–386.

McCoy, R. H., Seidler, R. J. 1973. Potential pathogens in the environment: Isolation, enumeration, and identification of seven genera of intestinal bacteria associated with small green pet turtles. Applied Microbiology **25:**534–538.

Maki, D. G., Hennekens, C. G., Phillips, C. W., Shaw, W. V., Bennet, J. V. 1973. Nosocomial urinary tract infection with *Serratia marcescens*: An epidemiologic study. Journal of Infectious Diseases **128:**579–587.

Mandel, M., Rownd, R. 1964. Deoxyribonucleic acid base composition in the *Enterobacteriaceae*: An evolutionary sequence, pp. 585–597. In: Leone, C. A. (ed.), Taxonomic biochemistry and serology. New York: Ronald Press.

Martinec, T. Kocur, M. 1960. The taxonomic status of *Serratia plymuthica* (Lehman and Neumann) Bergey et al. and of *Serratia indica* (Eisenberg) Bergey et al. International Bulletin of Bacteriological Nomenclature and Taxonomy **10:**247–254.

Martinec, T., Kocur, M. 1961a. The taxonomic status of *Serratia marcescens* Bizio. International Bulletin of Bacteriological Nomenclature and Taxonomy **11:**7–12.

Martinec, T., Kocur, M. 1961b. A taxonomic study of the members of the genus *Serratia*. International Bulletin of Bacteriological Nomenclature and Taxonomy **11:**73–78.

Martinec, T., Kocur, M. 1961c. Contribution to the taxonomic studies of *Serratia kiliensis* (Lehmann and Neumann) Bergey. International Bulletin of Bacteriological Nomenclature and Taxonomy **11:**87–90.

Martinec, T., Kocur, M. 1961d. Taxonomická studie rodu *Serratia*. Folia Facultatis Scientarum Naturalium Universitalis Purkynianae Brunensis **2:**1–77.

Negut, M., Davis, B. R., Farmer, J. J., III. 1975 Différenciation épidémiologique de *Serratia marcescens*: Comparaison entre lysotypie et sérotypie. Archives Roumaines de Pathologie Expérimentale et de Microbiologie **34:**189.

Negut, M., Szegli, L., Raiculescu, M. 1972. Germes du genre *Serratia* incriminés dans l'étiologie de certains cas de septicémie chez les enfants. Archives Roumaines de Pathologie Expérimentale et de Microbiologie **31:**225–234.

O'Meara, R. A. Q. 1931. A simple, delicate and rapid method of detecting the formation of acetylmethylcarbinol by bacteria fermenting carbohydrate. Journal of Pathology and Bacteriology **34:**401.

Paillot, A. 1916. Existence de plusieurs variétés et races de Coccobacilles dans les septicémies naturelles du Hanneton. Comptes Rendus Hebdomadaires des Séances de l'Académie des Sciences **163:**531–534.

Paine, S. G., Berridge, E. M. 1921. Studies in bacteriosis. V. Further investigation of a suggested bacteriolytic action in *Protea cynaroides* affected with the leaf-spot disease. Annals of Applied Biology **8:**20–26.

Paine, S. G., Stansfield, H. 1919. Studies in bacteriosis. III. A bacterial leaf spot disease of *Protea cynaroides*, exhibiting a host reaction of possibly bacteriolytic nature. Annals of Applied Biology **6:**27–39.

Parès, Y. 1964. Action de *Serratia marcescens* dans le cycle biologique des métaux. Annales de l'Institut Pasteur **107:**136–141.

Pesson, P., Toumanoff, C., Haradas, C. 1955. Etude des épizooties bactériennes observées dans les élevages d'insectes xylophages (*Rhyncolus porcatus* Germain, *Scolytus scolytus* Fabricius, *Scolytus (Scolytochelus) multistriatus* Marsham). Annales des Epiphyties **6:**315–328.

Phaff, H. J., Miller, M. W. 1961. A specific microflora associated with the fig wasp, *Blastophaga psenes* Linnaeus. Journal of Insect Pathology **3:**233–243.

Pillich, J., Hradečná, Z., Kocur, H. 1964. An attempt at phage typing in the genus *Serratia*. Journal of Applied Bacteriology **27:**65–68.

Podgwaite, J. D., Cosenza, B. J. 1976. A strain of *Serratia marcescens* pathogenic for larvae of *Lymantria dispar:* Infectivity and mechanisms of pathogenicity. Journal of Invertebrate Pathology **27:**199–208.

Reid, R. D. 1936. Studies on bacterial pigmentation. I. Historical considerations. Journal of Bacteriology **31:**205–210.

Richard, C. 1972. Méthode rapide pour l'étude des réactions de rouge de méthyle et Voges-Proskauer. Annales de l'Institut Pasteur **122:**979–986.

Robinson, W., Woolley, P. B. 1957. Pseudo-hemoptysis due to *Chromobacterium prodigiosum*. Lancet **i:**819.

Roussel, A., Lucas, A., Bouley, G. 1969. *Serratia marcescens (Bacillus prodigiosus)*, une bactérie de la pathologie comparée. Revue de Pathologie Comparée et de Médecine Expérimentale **6:**27–29.

Sakazaki, R. 1974. Genus IX. *Serratia* Bizio 1823, 288, p. 326. In: Buchanan, R. E., Gibbons, N. E. (eds.), In: Bergey's manual of determinative bacteriology, 8th ed. Baltimore: Williams & Wilkins.

Sanders, C. V., Jr., Luby, J. P., Johanson, W. G., Jr., Barnett, J. A., Sanford, J. P. 1970. *Serratia marcescens* infections from inhalation therapy medications: Nosocomial outbreak. Annals of Internal Medicine **73:**15–21.

Schaberg, D. R., Alford, R. H., Anderson, R., Farmer, J. J., III., Melly, M. A., Schaffner, W. 1976. An outbreak of nosocomial infection due to multiply resistant *Serratia marcescens:* Evidence of interhospital spread. Journal of Infectious Diseases **134:**181–188.

Sedlák, J., Dlabač, V., Motlikova, M. 1965. The taxonomy of the *Serratia* genus. Journal of Hygiene, Epidemiology, Microbiology, and Immunology **9:**45–53.

Slatten, B. H., Larson, A. D. 1967. Mechanism of pathogenicity of *Serratia marcescens* I. Virulence for the adult boll weevil. Journal of Invertebrate Pathology **9:**78–81.

Slotnick, I. J., Dougherty, M. 1972. Erythritol as a selective substrate for the growth of *Serratia marcescens*. Applied Microbiology **24:**292–293.

Smith, R. E., Hansen, H. N. 1931. Fruit spoilage of figs. University of California Agricultural Station, Bulletin **506:**1–84.

Smith, R. E., Reynolds, I. M. 1970. *Serratia marcescens* associated with bovine abortion. Journal of the American Veterinary Medicine Association **157:**1200–1203.

Sneatn, P. H. A. 1960. A study of the bacterial genus *Chromobacterium*. Iowa State Journal of Science **34:**243–500.

Sneath, P. H. A. 1966. Identification methods applied to *Chromobacterium*, pp. 15–20. In: Gibbs, B. M., Skinner, F. A. (eds.), Identification methods for microbiologists, vol. IA. London: Academic Press.

Stanier, R. Y., Palleroni, N. J., Doudoroff, M. 1966. The aerobic pseudomonads: A taxonomic study. Journal of General Microbiology **43:**159–271.

Starr, M. P., Grimont, P. A. D., Grimont, F., Starr, P. B. 1976. Caprylate-thallous agar medium for selectively isolating *Serratia* and its utility in the clinical laboratory. Journal of Clinical Microbiology **4:** 270–276.

Starr, M. P., Chatterjee, A. K., Starr, P. B., Buchanan, G. E. 1977. Enzymatic degradation of polygalacturonic acid by

Yersinia and *Klebsiella* species in relation to clinical laboratory procedures. Journal of Clinical Microbiology **6:**379–386.

Steigerwalt, A. G., Fanning, G. R., Fife-Asbury, M. A., Brenner, D. J. 1976. DNA relatedness among species of *Enterobacter* and *Serratia*. Canadian Journal of Microbiology **22:**121–137.

Steinhaus, E. A. 1941. A study of the bacteria associated with thirty species of insects. Journal of Bacteriology **42:** 757–789.

Steinhaus, E. A. 1959. *Serratia marcescens* Bizio as an insect pathogen. Hilgardia **28:**351–380.

Stephens, J. M. 1959. Immune responses of some insects to some bacterial antigens. Canadian Journal of Microbiology **5:**203–228.

Stephens, J. M. 1963. Bactericidal activity of hemolymph of some normal insects. Journal of Insect Pathology **5:**61–65.

Traub, W. H. 1972a. Bacteriocin typing of *Serratia marcescens* isolates of known serotype-group. Applied Microbiology **23:**979–981.

Traub, W. H. 1972b. Continued surveillance of *Serratia marcescens* infections by bacteriocin typing: Investigation of two outbreaks of cross-infection in an intensive care unit. Applied Microbiology **23:**982–985.

Traub, W. H., Kleber, I. 1977. Serotyping of *Serratia marcescens*. Evaluation of Le Minor's H-immobilization test and description of three new flagellar H antigens. Journal of Clinical Microbiology **5:**115–121.

Traub, W. H., Raymond, E. A. 1971. Epidemiological surveillance of *Serratia marcescens* infections by bacteriocin typing. Applied Microbiology **22:**1058–1063.

Traub, W. H., Raymond, E. A., Startsman, T. S. 1971. Bacteriocin (marcescin) typing of clinical isolates of *Serratia marcescens*. Applied Microbiology **21:**837–840.

Véron, M. 1975. Nutrition et taxonomie des *Enterobacteriaceae* et bactéries voisines. I. Méthode d'étude des auxanogrammes. Annales de Microbiologie **126A:**267–274.

Véron, M., Le Minor, L. 1975. Nutrition et taxonomie des *Enterobacteriaceae* et bactéries voisines. III. Caractéres nutritionnels et différenciation des groupes taxonomiques. Annales de Microbiologie **126B:**125–147.

von Graevenitz, A. 1977. The role of opportunistic bacteria in human disease. Annual Review of Microbiology **31:**447–471.

von Riesen, V. L. 1976. Pectinolytic, indole-positive strains of *Klebsiella pneumoniae*. International Journal of Systematic Bacteriology **26:**143–145.

Waisman, H. A., Stone, W. H. 1958. The presence of *Serratia marcescens* as the predominating organism in the intestinal tract of the newborn. The occurrence of the "Red Diaper Syndrome". Pediatrics **21:**8–12.

White, G. F. 1923a. Hornworm septicemia. Journal of Agricultural Research **26:**447–486.

White, G. F. 1923b. Cutworm septicemia. Journal of Agricultural Research **26:**487–496.

Wijewanta, E. A., Fernando, M. 1970. Infection in goats owing to *Serratia marcescens*. Veterinary Record **87:**282–284.

Williams, R. P., Hearn, W. R. 1967. Prodigiosin, pp. 410–432, 449–451. In: Gottlieb, D., Shaw, P. D. (eds.), Antibiotics. II. Biosynthesis. Berlin, Heidelberg, New York: Springer-Verlag.

Wilson, C. D. 1963. The microbiology of bovine mastitis in Great Britain. Bulletin de l'Office International des Epizooties **60:**533–551.

Woodward, H. M. M., Clark, K. B. 1913. A case of infection in man by "*Bacterium prodigiosum*". Lancet **i:**314.

Wright, C., Kominos, S. D., Yee, R. B. 1976. *Enterobacteriaceae* and *Pseudomonas aeruginosa* recovered from vegetable salads. Applied and Environmental Microbiology **31:**453–454.

ZoBell, C. E., Upham, H. C. 1944. A list of marine bacteria including descriptions of sixty new species. Bulletin of the Scripps Institution of Oceanography **5:**239–281.

The Tribe Proteeae

JOHN L. PENNER

The tribe Proteeae consists of a unique group of bacteria within the Enterobacteriaceae. These organisms have numerous features that clearly conform to the characteristics of this family but also show some striking differences. They are readily isolatable and occur frequently in pathological specimens submitted for bacteriological examination. The ability to swarm over the surface of agar media and the rapid hydrolysis of urea are well-known features, although swarming is limited essentially to the genus *Proteus* and not all Proteeae hydrolyze urea. They are distinguished from other Enterobacteriaceae by their remarkable ability to oxidatively deaminate a wide range of amino acids, affording them opportunities for special roles in the microbial ecology, not only of the parasitized host, but also in nature, where they are presumed to exist in the free-living state and thought to participate in the biogeochemical cycles.

Typical Proteeae, like other Enterobacteriaceae, are Gram-negative, peritrichously flagellated motile rods capable of fermenting D-glucose and growing in the presence of cyanide. Species within the tribe are separated by differences in biochemical reactions, in somatic (O) and flagellar (H) antigens, and in susceptibility to bacteriophage and bacteriocins.

The Proteeae are of concern to man primarily because they are often involved in serious infections that have been noted with higher frequency in recent years, accompanying the increase in use of antimicrobials and the increase in numbers of infection-susceptible patients in the hospital. Of particular concern is the resistance of some infections to antimicrobial therapy. Some of the most antibiotic-resistant strains known are found among the Proteeae.

The classification of the Proteeae has been the subject of much controversy. In the most recent proposals, seven species distributed among three genera are recognized (Brenner et al., 1978). The proposals were based on results obtained by deoxyribonucleic acid (DNA) relatedness tests and should resolve past taxonomic disagreements. The three genera are *Proteus, Providencia,* and *Morganella*. The genus *Proteus* includes the two familiar species, *Prot. vulgaris* and *Prot. mirabilis,* and *Prot. myxofaciens* described by Cosenza and Podgwaite (1966). The genus *Providencia* includes, in addition to *Prvd. alcalifaciens* and *Prvd. stuartii,* the *Prvd. rettgeri* formerly located in the genus *Proteus*. Bacteria previously known at *Prot. morganii* were assigned to the taxon *Morganella morganii*. This classification will be followed and the generic abbreviations recommended by Johnson, Rogosa, and Krichevsky (1976) for *Proteus* and *Providencia* (*Prot.* and *Prvd.*) will be used, but the first letter of *Morganella* will be the abbreviated form for that genus.

Isolation of the Proteeae

In the clinical laboratory, the growth of Proteeae from patient stool specimens is often considered a nuisance because they are not generally considered to be primary disease-producing pathogens and because they are often carried intestinally by apparently healthy people. *Proteus* strains can be a particular nuisance because many swarm over the surface of moist agar plates, making it difficult to isolate other bacteria in pure culture (Cowan, 1974). Media have been formulated to inhibit swarming by incorporating, in the media, either bile salts or an anionic detergent, Teepol (Jameson and Emberley, 1956); reducing the electrolytic content by not adding sodium chloride (Mackey and Sandys, 1966), or by increasing the concentration of the agar (Cowan, 1974). Proteeae grow on such media as discrete colonies.

Media that inhibit swarming and are used for the primary isolation of enteric bacteria are, in most cases, suitable for direct isolation of Proteeae. Some are more suitable for this purpose than others, but comparative studies to identify the most suitable medium for the primary isolation of Proteeae have not been reported. In studies on isolation media, interest has been mainly directed toward evaluating media for selection of *Salmonella* and *Shigella* (Hynes, 1942). Differences in the composition of the media could account for differences observed by Hynes (1942), who reported higher rates of isolation for *Proteus* on deoxycholate-citrate (DC) agar than on bismuth sulfite (BS) or MacConkey agar, and by Lanyi (1956), who noted higher rates of isolation on

BS than on DC or Endo agar. Rustigian and Stuart (1945) also noted high rates of isolation but they used eosin methylene blue (EMB) and Salmonella-Shigella (SS) agar.

In a few cases, primary isolation media have been developed specifically for Proteeae. Malinowski (1966) reported that *Proteus* could be differentiated from other non–lactose fermenters and from lactose fermenters on media containing β-alanine that is decarboxylated by Proteeae; Blake (1975) took advantage of the resistance of Proteeae to colistin in developing a medium consisting of cystine-lactose-electrolyte-deficient (CLED) agar containing 50 μg/ml of the antibiotic. A medium for the quantitative determination of urease-positive Proteeae in milk was described by Zarett and Doetsch (1949) and a medium with bile salts, lithium chloride, sodium thiosulfate, and sodium citrate was developed for primary isolation of *Proteus* by Xilinas, Papavassiliou, and Legakis (1975).

Deoxycholate-citrate agar was used for isolation of *Prvd. rettgeri* (Cook, 1948). Colonies of this species were differentiated from other enterics by the yellow-orange centers. Similar observations were made for colonies of *Providencia* strains by Buttiaux, Fresnoy, and Moriamez (1954) and Catsaras, Antoniewski, and Buttiaux (1965) but it is not clear whether their collections included only *Prvd. stuartii* or only *Prvd. alcalifaciens* or both species. Ridge and Thomas (1955) found that some *Providencia* did not grow on this medium, and Blake (1975) found it to be more inhibitory than colistin-containing CLED medium for *Prvd. rettgeri*.

The use of liquid enrichment media is recommended when feces are to be examined for Proteeae. Tetrathionate broth is commonly used for enrichment of *Salmonella* but *Proteus* is favored more by this medium than is *Salmonella* (Hynes, 1942). In a study on the intestinal carriage of *Proteus*, only 10% of the feces samples were positive when examined by direct plating on DC agar, but an increase to 25% was observed when plating was preceded by enrichment with tetrathionate broth (Hynes, 1942). Rustigian and Stuart (1945) used either tetrathionate or selenite broth for enrichment and EMB or SS media for plating and observed that enrichment increased the rate of isolation for *Prot. mirabilis* from 8.2% to 23.6%, for *Prot. vulgaris* from 0 to 2.7%, and for *M. morganii* from 1.8% to 10%. Since these differences are significant, it is necessary to refer to the method of isolation when quoting the incidence of Proteeae in human feces.

Identification and Differentiation of Species of the Proteeae

For preliminary differentiation of the Proteeae, colonies from isolation media such as MacConkey, DC, EMB, BS, SS, and CLED may be subcultured to multitest media. The ones commonly used are triple sugar iron (TSI), Kligler's iron, lysine iron agar (LIA), or motility-indole-ornithine media. Phenylalanine agar (or tryptophan agar) should also be inoculated. Other media for the rapid identification of Proteeae, particularly for the genus *Proteus*, have been reported and may be used in studies related to this genus (Huang, 1966; Phillips, 1966; Singer, 1950a,b; Vassiliadis and Politi, 1968).

The most distinguishing biochemical feature of the Proteeae is their ability to oxidatively deaminate a wide range of amino acids (Bernheim, Bernheim, and Webster, 1935; Drasar and Hill, 1974; Singer and Volcani, 1955; Stumpf and Green, 1944). The test that separates Proteeae from other Enterobacteriaceae is based on this remarkable enzymatic activity. The use of phenylalanine in the biochemical test series was advocated by Henriksen and Closs (1938) and Henriksen (1950), and the use of tryptophan was recommended by Richard (1966), Roland, Bourbon, and Szturm (1947), Singer and Volcani (1955), and Thibault and Le Minor (1957). Ewing, Davis, and Reavis (1957) showed that only 2% of Proteeae were negative when phenylalanine agar was used and that the agar medium was more satisfactory than the combined phenylalanine-malonate medium. The test is most useful in identifying the occasional urea-negative strains of *Proteus*, *Morganella*, and *Prvd. rettgeri*.

The hydrolysis of urea is a well-known reaction of *Proteus*, *Morganella*, and *Prvd. rettgeri*. A few strains of *Prvd. stuartii* and *Prvd. alcalifaciens* produce urease, as first observed by Stuart et al. (1943) in their study of a group of strains labeled 29911. These reactions were usually weak or delayed, unlike the small percentage of *Prvd. stuartii* that possess strong urease activity and that were included in biogroup 5 of *Prvd. rettgeri* (Penner, Hinton, and Hennessy, 1975), or included in the intermediate group by Ursing (1974). Furthermore, small percentages of *Prot. mirabilis*, *Prot. vulgaris*, *Prvd. rettgeri*, and *M. morganii* are urea negative, and other Enterobacteriaceae, mostly belonging to *Klebsiella* and *Citrobacter*, may be urea positive (Vuye and Pijck, 1973); therefore, the urea test is not absolutely reliable in differentiating Proteeae. Despite these limitations, the test is nevertheless most important because an overwhelming majority of the tribe are urease producers and give a clear positive result, usually within 6 h or less when Christensen's urea medium is used (Christensen, 1946). The test for urea hydrolysis should, therefore, be included in the list of reactions typical of the tribe.

The reactions of Proteeae in LIA are of particular interest because they are unlike those of other Enterobacteriaceae. The butt is acid but, because of the oxidative deamination of lysine, the slant is red with all Proteeae except *M. morganii,* thus differentiating them from other Enterobacteriaceae that produce either an alkaline or acid reaction (Edwards and Ewing, 1972). Positive reactions in LIA for hydrogen sulfide production indicate *Prot. mirabilis* or *Prot. vulgaris,* and a positive test for indole production, demonstrated by suspending from the stopper of the tube a strip of filter paper impregnated with Kovac's reagent, indicates a strain other than *Prot. mirabilis,* the only species of the tribe commonly unable to produce this compound from tryptophan.

Proteeae are typically lactose negative but the occurrence of lactose-positive strains has been reported. In early studies, only occasional isolates were noted (Proom, 1955; Shaw and Clarke, 1955; Singer and Bar-Chay, 1954). More recently, Suter et al. (1968) and Sutter and Foecking (1962) identified numerous lactose-positive strains among isolates of *Prot. mirabilis, Prot. vulgaris, Prvd. rettgeri,* and *M. morganii* and among *Providencia* strains not assigned to species. The precentage was highest among the latter group but this may be attributed, at least to some extent, to lactose-positive endemic strains such as those described by Richard, Popoff, and Pastor (1974) and Traub et al (1971). Falkow et al. (1964) demonstrated that lactose fermentation in *Prot. mirabilis* was attributable to a plasmid and that the occurrence of lactose-positive strains was due to in vivo spread of this factor.

The Voges-Proskauer test is negative for most Proteeae. However, about 16% of *Prot. mirabilis* are positive when the test is performed at 37°C and 52% of *Prot. mirabilis* and 11% of *Prot. vulgaris* may be positive at 22°C (Ewing and Martin, 1974). Proteeae are usually methyl red positive but it should be noted that clear positive results are more readily obtained for *Prvd. rettgeri* when incubated at room temperature for 4 days (Suassuna, Suassuna, and Ewing, 1961).

Proteeae are typically flagellated. Some strains of *Providencia* and *M. morganii,* however, may not form flagella at temperatures above 30°C (Coetzee and DeKlerk, 1964) and should therefore be cultured at lower temperatures to demonstrate motility. It should be noted that swarming at 37°C, with the production of concentric halos of spreading growth, is characteristic only of the genus *Proteus* (Sevin and Buttiaux, 1939). The nature of the swarming phenomenon in *Proteus* has been described in a comprehensive review (Williams, 1978). *M. morganii* and *Providencia* produce blunt pseudopodial outgrowths called "emanations" by Rauss (1936) and "arborizations" by Coetzee and de Klerk (1964) when cultured on 1% agar at 22°C. After 48 h, these may extend and spread to form a film or sheet, not unlike that of some *Proteus* strains (Coetzee and de Klerk, 1964). Observations of this phenomenon were also reported for *Prvd. rettgeri* by Rustigian and Stuart (1943b), for other *Providencia*

by Stuart, Wheeler, and McGann (1946), and for other Enterobacteriaceae by Stuart et al (1946).

The commonly observed reactions characteristic of typical Proteeae are shown in Table 1. For percentage values, the reader is referred to complilations reported by Brenner et al. (1978), Edwards and Ewing (1972), and Ewing and Martin (1974).

The biochemical reactions selected to show differentiation of the seven species are shown in Table 2. The genus *Proteus* is distinguished from *Providencia* and *Morganella* by production of hydrogen sulfide in TSI, the production of lipases, the ability to swarm, and the inability to attack either mannose or polyhydric alcohols. Acid production in polyhydric alcohols is a key characteristic of *Providencia.* The inability to grow on citrate (Simmons') and the ability to decarboxylate ornithine are important in differentiating *Morganella* from *Providencia* and *Proteus.*

The importance of the mannose fermentation test should not be overlooked. The genus *Proteus* is uniformly negative and *Providencia* and *Morganella* are almost uniformly positive in this substrate (Ewing and Martin, 1974; Ewing and Davis, 1972; Ewing, Davis, and Sikes, 1972; Rauss and Vörös, 1959).

Separation of *Proteus* into the three species, *Prot. mirabilis, Prot. vulgaris,* and *Prot. myxofaciens,* is accomplished with tests for indole production, maltose fermentation, and ornithine decarboxylation. The reactions for *Prot. myxofaciens* are according to Cosenza and Podgwaite (1966) and from the study of one strain by Brenner et al. (1978).

The genus *Providencia* is separated into its constituent species primarily on differences in their catabolism of polyhydric alcohols. *Prvd. alcalifaciens* strains produce acid in adonitol, *Prvd. stuartii* in inositol and, infrequently, in mannitol; but *Prvd. rettgeri* reacts positively in inositol, mannitol, adonitol, arabitol and, in most cases, in erythritol. *Prvd. stuartii* appears to be the only one of the three *Providencia* species that ferments trehalose. This substrate was used to show biochemical types among the strains of *Providencia* first by Stuart, Wheeler, and McGann (1946). Richard (1966) included reactions with this disaccharide in a scheme to differentiate biogroups 5 and 6 (*Prvd. stuartii*) from biogroups 1–4 (*Prvd. alcalifaciens*). Ursing (1974) and Penner et al. (1976b) found it a valuable reagent for separating urea-positive *Prvd. stuartii* from *Prvd. rettgeri.* Although not all *Prvd. rettgeri* strains ferment erythritol, it is nevertheless an important test because a positive reaction in this substrate is not produced by other strains of the tribe (Kauffmann, 1956a; Penner, Hinton, and Hennessy, 1975; Ursing, 1974). It is particularly useful in identifying urea-negative *Prvd. rettgeri.*

The tests in Table 2 were selected primarily for

Table 1. Biochemical reactions of Proteeae.

Test or substrate	Reaction[a]
Phenylalanine deaminase	+
Growth in KCN	+
NO$_3^-$ $\longrightarrow$ NO$_2^-$	+
Urea	+ or −
Motility	+
Oxidase	−
Malonate utilization	−
Voges-Proskauer	−
Methyl red	+
Lysine decarboxylase	−
Arginine dihydrolase	−
Ornithine decarboxylase	+ or −
Acid produced from:	
D-Glucose	+
Lactose	−
Dulcitol	−
D-Sorbitol	−
Melibiose	−

[a] Symbols: 90% or greater positive, +; 90% or greater negative, −; different reactions, + or −.

differentiation of Proteeae according to genus and species. Subdivision of the species into biochemical types is accomplished with additional tests. Biotypes have been described for *Prvd. alcalifaciens* and *Prvd. stuartii* by Ewing, Davis, and Sikes (1972), for *Prvd. rettgeri* by Penner, Hinton, and Hennessy (1975), and for *M. morganii* by Rauss and Vörös (1959). Extensive lists of tests and reactions of Proteeae have been compiled by Brenner et al. (1978), Ewing and Davis (1972), and Ewing, Davis, and Sikes (1972).

Strains of *Proteus* require nicotinic acid but not pentothenic acid for growth (Fildes, 1938). *Providencia* strains require neither of these vitamins (Proom, 1955; Proom and Woiwod, 1951) and *Morganella* strains require both (Pelczar and Porter, 1940). Decarboxylation of leucine and valine is apparently limited among the Enterobacteriaceae to *Prot. vulgaris*, *Prot. mirabilis*, and *M. morganii* (Ekladius, King, and Sutton, 1957; King, 1953; Proom and Woiwod, 1951). Lautrop (1974) indicated that separation of the genera within the Proteeae could be accomplished on the basis of these differences in nutritional requirements. Proteeae degrade tyrosine and a test based on this property has been suggested by Sheth and Kurup (1975) to separate them from most other Enterobacteriaceae. Another distinguishing feature of the Proteeae is production of a reddish-brown pigment when cultured on nutrient agar containing 5% tryptophan (Polster and Svobodova, 1964). This phenomenon was reported as unique among the Enterobacteriaceae but has neither received much attention nor been exploited as a character for identification and classification.

Requirements for essential nutrients, decarboxylation of leucine and valine, and pigment formation from tryptophan should be further investigated with the objective of developing simplified tests suitable for routine differentiation of Proteeae.

Media and Reagents for the Proteeae

The identification and classification of the Proteeae can, in all cases, be accomplished with media and

Table 2. Biochemical reactions to differentiate the species of the Proteeae.[a]

Substrate or test	Proteus mirabilis	Proteus vulgaris	Proteus myxofaciens	Providencia alcalifaciens	Providencia stuartii	Providencia rettgeri	Morganella morganii
Urea	+	+	+	−	+ or −	+	+
Citrate (Simmons')	+ or −	− or +	+	+	+	+	−
Indole	−	+	−	+	+	+	+
Swarming (37°C)	+	+	+	−	−	−	−
H$_2$S (on TSI)	+	+	−	−	−	−	−
Gelatin (22°C)	+	+	+	−	−	−	−
Lipase (corn oil)	+	+	+	−	−	−	−
L-Xylose	+	+	−	−	−	− or +	−
Maltose	−	+	−	−	−	−	−
α-Methyl-D-glucoside	−	+	+	−	−	−	+
D-Mannose	−	−	−	+	+	+	+
Ornithine decarboxylase	+	−	−	−	−	−	+
Trehalose	+	− or +	+	−	+	−	−
Inositol	−	−	−	−	+	+	−
D-Mannitol	−	−	−	−	− or +	+	−
Adonitol	−	−	−	+	− or +	+	−
D-Arabitol	−	−	−	−	−	+	−
Erythritol	−	−	−	−	−	+ or −	−

[a] Symbols: 90% or greater positive, +; 90% or greater negative, −; and different reactions, + or −.

reagents that are familiar to bacteriologists. Most of these are available from commercial suppliers.

Formulation of the media and methods of preparation are clearly presented in manuals used in diagnostic bacteriology (Cowan, 1974; Edwards and Ewing, 1972; Kauffmann, 1966; Vera and Dumoff, 1974).

Antimicrobial Susceptibility of the Proteeae

Some members of the tribe are among the most antibiotic-resistant bacteria known in the clinical laboratory. A feature common to all Proteeae species is intrinsic resistance to bacitracin and the polymyxins (Potee, Wright, and Finland, 1954; Shimizu, Iyobe, and Mitsuhashi, 1977; von Graevenitz and Nourbakhsh, 1972). Susceptible strains are rare and Li and Miller (1970) were prompted to point out that, if an organism was susceptible to colistin, it was almost certain not to be *Proteus*. Furthermore, strains of the tribe are resistant to erythromycin (Chiu and Hoeprich, 1961). In contrast, as many as 70–98% of Proteeae are susceptible to nalidixic acid (Huang and Chou, 1968; Li and Miller, 1970; von Graevenitz and Nourbakhsh, 1972).

Prot. mirabilis is the strain most frequently isolated from human infections and, fortunately, also the most susceptible to clinically useful antibiotics. Very resistant strains of this species, however, are occasionally isolated. Increasing resistance, particularly to the tetracyclines, has been noted in some hospitals (Chiu and Hoeprich, 1961; Potee, Wright, and Finland, 1954). Susceptibility to tetracyclines is mainly restricted to *Prot. vulgaris* and *M. morganii*. As many as 50% of *M. morganii* have been reported susceptible to doxycycline and as many as 25% of *Prot. vulgaris* may be susceptible to doxycycline and 41% to tetracycline (Li and Miller, 1970; von Graevenitz and Nourbakhsh, 1972). The majority of *Prot. mirabilis* are susceptible to chloramphenicol, the majority of the *Providencia* are resistant and *Prot. vulgaris* and *M. morganii* occupy an intermediate position with more than 50% susceptible. The percentage of Proteeae susceptible to nitrofurantoin has been reported as low as 5% and as high as 80%. Both susceptible and resistant species are found in each genus.

The antimicrobials most useful clinically against Proteeae are the penicillins and aminoglycosides. Because considerable variation in susceptibility occurs among the species, it is advantageous to consider each separately.

Prot. mirabilis is unique among the Proteeae in its susceptibility to penicillins but this was not realized in early studies in which the genus *Proteus* was not separated according to species. It was Potee, Wright, and Finland (1954) who demonstrated the moderate susceptibility of some strains of this species to penicillin. Their finding has been confirmed in numerous subsequent investigations and the percentage of susceptible strains has been shown to remain at high levels (Barber and Waterworth, 1964; Chiu and Hoeprich, 1961; Greenwood and O'Grady, 1969; Grossberg et al., 1962; Hook and Petersdorf, 1960; Huang and Chou, 1968; Li and Miller, 1970; von Graevenitz and Nourbakhsh, 1972). Although the percentage of strains resistant to ampicillin has been reported as low as 1.5% (von Graevenitz and Nourbakhsh, 1972), others have noted higher percentages of resistant strains, particularly among nosocomial strains (Grossberg et al., 1962; Li and Miller, 1970) and the necessity of performing susceptibility tests on each isolate has been stressed (Barry and Hoeprich, 1973; Shafi and Datta, 1975). Ampicillin is ineffective against strains that produce penicillinase but is very active against the majority that do not. Carbenicillin shows the same pattern of activity as ampicillin with strains of this species (Acred et al., 1967).

The majority of *Prot. mirabilis* strains are susceptible to the cephalosporins (Eykyn, 1971; Moellering and Swartz, 1976). Von Graevenitz and Nourbakhsh (1972) reported only 1% of 864 *Prot. mirabilis* strains resistant to cephalothin. It should be noted that the susceptibility of *Prot. mirabilis* is in sharp contrast to other Proteeae that are generally very resistant to cephalosporins.

Aminoglycosides are second to penicillins in antimicrobial activity against *Prot. mirabilis*. Kanamycin is more effective than streptomycin. The percentage of strains susceptible to kanamycin has been reported to be 82%, 91.7%, and 97.5% by Huang and Chou (1968), Li and Miller (1970), and von Graevenitz and Nourbakhsh (1972) in studies on 350, 326, and 864 strains, respectively. In the same studies, the percentages of strains susceptible to streptomycin were 68.2%, 71%, and 90%. Neomycin was found to be active on 93.7% by Huang and Chou (1968) but Chiu and Hoeprich found only 27% of their 52 strains susceptible and Grossberg et al. (1962) found all 27 of their strains from urinary tract infections to be resistant.

Gentamicin has become the preferred aminoglycoside in the treatment of Gram-negative infections resistant to therapy with penicillins and cephalosporins, and the vast majority of *Prot. mirabilis* strains are sensitive to this agent. von Graevenitz and Nourbakhsh(1972) found all 864 strains susceptible. It is important to note, however, that this species may acquire plasmids that cause high levels of resistant to numerous antibiotics. Shafi and Datta (1975) reported that gentamicin resistance in *Prot. mirabilis* was caused by two R factors, one coded for resistance to gentamicin and the other to both gentamicin and tobramycin. Amikacin is effective against many Gram-negative bacteria, including most strains of *Prot. mirabilis* and other Proteeae. Of particular importance is the high antimicrobial activity of this aminoglycoside against gentamicin-resistant strains. However, gentamicin-resistant *Prot. mirabilis* strains that are also resistant to amikacin have been reported (Briedis and Robson, 1976; Draser et al., 1976). Currently, *Prot. mirabilis* strains that are highly resistant to aminoglycosides are uncommon but it can be anticipated that resistant forms will be more frequently encountered, particularly in hsopital outbreaks, as increased resistance to aminoglycosides has usually accompanied the use of these agents.

Prot. vulgaris strains are distinguished from *Prot. mirabilis* because the former are highly resistant to penicillins and cephalosporins. The only penicillin that has shown significant activity against *Prot. vulgaris* is carbenicillin. Most strains are susceptible to aminoglycosides but small numbers (1.5–3%) are resistant to kanamycin or gentamicin and pose a problem in the therapeutic management of patients infected with them. Amikacin appears to be effective against some of these resistant strains.

Unfortunately, in numerous studies, *Prvd. alcalifaciens* and *Prvd. stuartii* were not considered separately and significant differences in antimicrobial susceptibilities remained undetected. Middleton (1958) separated strains into biochemical group 1 (*Prvd. alcalifaciens*) and biochemical group 2 (*Prvd. stuartii*). This allowed the observation that strains of the former group were susceptible to streptomycin and chloramphenicol and that strains of the latter were generally very resistant. Overturf, Wilkins, and Ressler (1974) also observed that *Prvd. stuartii* were markedly more resistant than *Prvd. alcalifaciens* but a particular group of 90 *Prvd. stuartii* from a burn unit were all resistant to gentamicin, kanamycin, tobramycin, and carbenicillin. Such uniformly high resistance is unusual and thus, the 90 isolates were more likely a reflection of a series of nosocomial infections attributable to one or a small number of resistant endemic strains in their hospital. Only amikacin was effective against all isolates of both species.

Prvd. stuartii is more closely related to *Prvd. rettgeri* than to *Prvd. alcalifaciens* in its resistance to antimicrobial agents. Highly resistant strains of both these closely-related species have been found to occur in nosocomial infections (see "The Genus *Providencia*: Infections in Man", this chapter). Amikacin appears to be particularly active on such resistant *Prvd. rettgeri* and *Prvd. stuartii* (Briedis and Robson, 1976; Dhawan et al., 1977; Drasar et al., 1976; Eickhoff and Ehret, 1977; Overturf, Wilkins, and Ressler, 1974).

M. morganii, like other Proteeae, is generally resistant to erythromycin, polymyxin B, and colistin, but susceptible to nalidixic acid. Many, but not all, of the strains are resistant to ampicillin and cephalo-

thin, but, in general, most are susceptible to kanamycin, gentamicin, and amikacin. Strains of *M. morganii* that are highly resistant to ampicillin and cloxacillin, however, may be very susceptible to mixtures of the two agents at low concentrations, but similarly resistant strains of *Prot. vulgaris* and *Prvd. rettgeri* are not susceptible to this synergistic combination (Bornside, 1968). Apart from this apparently species-specific feature, the patterns of resistance and susceptibility to other antimicrobials displayed by *M. morganii* resemble most closely those of *Prot. vulgaris*.

It is apparent that differences in intrinsic resistance exist among the species and thus it is important that the practice of grouping Proteeae under terms of convenience such as "indole-positive *Proteus*" or "*Providencia*" strains in antibiotic susceptibility studies be discontinued. Correct classification is necessary to draw attention to such species differences.

THE GENUS *PROTEUS*

The first account of bacteria known as *Proteus* was given by Hauser (1885). He gave this name to bacteria that reminded him of the Greek deity, Proteus, whose most distinguishing feature was his faculty of assuming different shapes. Hauser named one variety *vulgaris* because he thought it to be more common than the other, which he named *mirabilis*. Both varieties were noted to liquefy gelatin, the former more actively than the latter, and both underwent changes in colonial morphology from rounded colonies to irregularly shaped colonies, with amoeba-like processes at the peripheries, and, finally, to a thin film that spread over the surface of the medium. Wenner and Rettger (1919) found that the two varieties could be differentiated on the basis of maltose fermentation and selected the name *vulgaris* for maltose fermenters and *mirabilis* for those that could not attack this sugar. Moltke (1927) showed that the strains hydrolyzed urea and this reaction became accepted as a major taxonomic feature for *Proteus* and bacteria were included or excluded in the genus on the basis of this reaction (Elek, 1948; Rustigian and Stuart, 1941, 1943a, 1945; Stuart, Van Stratum, and Rustigian, 1945). As a result, the genus eventually consisted of four species, namely, *Prot. vulgaris*, *Prot. mirabilis*, *Prot. rettgeri*, and *Prot. morganii*.

It should be noted that Kauffmann (1951) and Kauffmann and Edwards (1952) regarded *Prot. mirabilis* and *Prot. vulgaris* as varieties of the same species that they named *Prot. hauseri*. Kauffmann recognized the two species (Kauffmann, 1953, 1956a,b) but later considered *Prot. vulgaris* and *Prot. mirabilis* as two subgenera of the genus *Proteus* (Kauffmann, 1966). In the latter classification, the species were not given names but were represented by their antigenic formulas. On the other hand, Borman, Stuart, and Wheeler (1944) were also of the view that only one species should be recognized for the genus but they selected the name *Prot. vulgaris*. Belyavin (1951) and Belyavin, Miles, and Miles (1951) also favored *Prot. vulgaris* as the only species and published studies under the title of *Prot. vulgaris* that were actually concerned with *Prot. mirabilis*. Finally, a third species, *Prot. myxofaciens,* was included in the genus. This species was proposed by Cosenza and Podgwaite (1966) but Lautrop (1974) reported it to be identical to *Erwinia herbicola*. However, on the basis of DNA relatedness tests it was concluded by Brenner et al. (1978) to be a valid species.

Habitats of *Proteus mirabilis* and *Proteus vulgaris*

Proteus mirabilis and *Prot. vulgaris* are widely distributed in nature and are thought to have an important role in the decomposition of organic matter (Cantu, 1911; Wilson and Miles, 1975). *Prot. mirabilis* is the more common of the two species in nature (Levine and Hoyt, 1945). Both species occur in polluted waters. In heavily polluted rivers, the *Proteus* populations are heterogeneous and consist of large numbers of different strains as has been demonstrated with the Dienes reaction (Sturdza, 1963). *Prot. mirabilis* can be frequently isolated from manure and garden soil and from the feces of mice, rats, monkeys, raccoons, dogs, cats, cattle, pigs, and birds (Cantu, 1911; Phillips, 1955a). *Prot. vulgaris* has been more frequently isolated than *Prot. mirabilis* only in pigs (Phillips, 1955a) and in snakes (Müller, 1972). *Prot. vulgaris* is the species found in packing-house wastes (Levine, 1942). Both species also occur in the intestines of some humans. Krikler (1953) reported an incidence of 27% (*Prot. mirabilis* 17%, *Prot. vulgaris* 5%, atypical *Proteus* 4%) in feces of healthy persons. Rustigian and Stuart (1945) found *Prot. mirabilis* in 24% and *Prot. vulgaris* in 3% of the fecal samples of apparently healthy individuals. In both studies, the results were obtained with the use of enrichment media. Sturdza (1965) found that among 45 carriers of *Proteus,* 33 carried one, seven carried two, four carried three and one carried four different strains as defined by the Dienes test. The role of *Proteus* in the ecology of the intestine is not well understood but it is known that it assists in the hydrolysis of urea although its contribution is minor in comparison to that of the large populations of urease-producing anaerobes (Brown, Hill, and Richards, 1971; Sabbaj, Sutter, and Finegold, 1971). It is, however, unique among the intestinal Enterobacteriaceae in production of

ammonia and keto acids through oxidative deamination of a wide spectrum of amino acids (Drasar and Hill, 1974). Their occurrence in the human intestine is of considerable medical importance because this habitat serves as a reservoir for bacteria that may become involved in autoinfections or cross-infections, particularly among infection-susceptible, hospitalized patients.

Proteus mirabilis and *Proteus vulgaris* Infections in Man

Early efforts to draw attention to the pathogenicity of *Proteus* were received with skepticism (Taylor, 1928). This attitude has largely dissipated over the course of recent years as these bacteria have become recognized as significant participants in Gram-negative infections, the subject of much concern since the introduction of antibiotic therapy (Yow, 1952). *Prot. mirabilis* is the most common infectious species of the Proteeae and accounts for 70–90% of all infections caused by the genus *Proteus* (Adler et al., 1971b; Grossberg et al., 1962; Lanyi, 1957; McMillan, 1972; Scott, 1960; von Graevenitz and Spector, 1969). It is the third most common species of Enterobacteriaceae isolated in the clinical laboratory (Hickman and Farmer, 1976). In a hospital population of geriatric patients, it was reported to be the leading pathogen (Li and Miller, 1970). Among urology and paraplegic patients, *Prot. mirabilis* was second only to *Escherichia coli* as a cause of Gram-negative infections (Scott, 1960; Søgaard, Zimmermann-Nielsen, Siboni, 1974). Infections with *Prot. mirabilis* and *Prot. vulgaris* are difficult to treat and generalized infections are often fatal (Noall, Sewards, and Waterworth, 1962).

Of the two species, *Prot. mirabilis* is more often found in nosocomial and community-acquired infections. Its prevalence in community-acquired infections has apparently remained fairly constant (Adler et al., 1971b), but a sharp increase in its prevalence in hospital-acquired infections has been observed (Adler et al. 1971b; Koch, 1956; Yow, 1952). Different views have been expressed on the manner in which nosocomial infections are contracted. Approximately one-quarter of the population have been reported to be intestinal carriers of *Proteus* (Krikler, 1953; Rustigian and Stuart, 1945). It is, therefore, conceivable that the patient himself could be the source of the infectious agent and evidence in support of the autoinfection route has been reported (de Louvois, 1969; Schwarz et al., 1969; Story, 1954). On the other hand, patients that are not carriers on admission to hospital may also contract *Proteus* infections, and case-to-case transmission or transmission from a common reservoir of the infecting strain has been postulated (Dutton and Ralston, 1957; Kippax, 1957). Typing systems capable of differentiating among strains with precision are required for such epidemiological studies. Several systems are in the process of being developed (see following sections).

The most common site of *Proteus* infections is the urinary tract. In the study of Adler et al. (1971b), 53% of the nosocomial and 63% of the community-acquired *Proteus* (predominantly *Prot. mirabilis*) were at this site. *Prot. mirabilis* urinary tract infections acquired outside the hospital are uncommon and their occurrence is often associated with a predisposing condition. Grossberg et al. (1962) found most isolates from community-acquired *Prot. mirabilis* urinary tract infections were from diabetics, and Wallace and Petersdorf (1971) suggested that the presence of *Prot. mirabilis* in the case of an initial urinary tract infection should alert suspicions of a structural abnormality. Catheterization, surgery, and other forms of urologic manipulation are factors predisposing patients to nosocomial urinary tract infections, particularly to opportunists like *Proteus* (Evans, 1974; Maki, 1972). The infected urinary tract is the most common source of Gram-negative bacteremia (Hodgin and Sanford, 1965; Maiztegui et al., 1965; McCabe and Jackson, 1962; Sullivan et al., 1973). *Proteus* bacteremias are difficult to treat and are often fatal (Abrams, 1948; Hewitt et al., 1965; Jepsen and Korner, 1975; Koch, 1956). The fatality rate of patients with *Proteus* bacteremias varies from 15% to 88% and is largely dependent upon the severity of the underlying conditions (Lewis and Fekety, 1969).

An often-mentioned factor contributing to the pathogenicity of *Proteus* in the urinary tract is the activity of urease in producing ammonia and elevating pH (Braude and Siemienski, 1960; MacLaren, 1968; Musher et al., 1975; Phillips, 1955b). Recent animal studies, using the urease inhibitor acetohydroxamic acid, confirmed the importance of the urease enzyme (Griffith, Musher, and Campbell, 1973; MacLaren, 1974; Musher et al., 1975). Other factors that remain to be defined are also necessary for virulence (Eudy, Burrous, and Sigler, 1971; Phillips, 1955b).

In addition to their involvement in infections of the urinary tract, *Proteus* species have been isolated from a variety of septic lesions (Lanyi, 1957; von Graevenitz and Spector, 1969). Under suitable conditions, these opportunists may be isolated from wounds, burns, respiratory tract, skin, and from infections of the eyes, nose, and throat. Evidence of a major role for *Prot. mirabilis* in infantile enteritis, particularly for strains of serotype O3, was provided by Lanyi (1956). He isolated *Proteus* species from 46.2% of infants with diarrhea. However, little significance has been attached to the isolation of *Proteus* from patients with diarrhea because approximately one-quarter of the population are intestinal carriers (Carpenter, 1964).

Although well known as opportunistic pathogens of the urinary tract, *Proteus* species are not commonly recognized for causing primary disease. However, in recent years an increase in rapidly fatal primary disease caused by *Proteus* species has been documented. Contamination of the neonatal umbilical stump with *Proteus* spp. may lead to highly fatal bacteremia and meningitis (Annotations, 1966; Becker et al., 1962; Burke et al., 1971; Levy and Ingall, 1967; Librach, 1968; Shortland-Webb, 1968). Moreover, the incidence of Gram-negative osteomyelitis is apparently increasing in parallel with other Gram-negative infections and *Prot. mirabilis* has been isolated from children and adults afflicted with this condition (Levy, O'Connor, and Ingall, 1967; Meyers et al., 1973).

Antigenic Structure and Serotyping

The somatic (O) and flagellar (H) antigens of *Proteus mirabilis* and *Prot. vulgaris* have been extensively examined. Winkle (1945) described a scheme of 13 somatic (O) and 8 flagellar (H) antigens. Kauffmann and Perch (1947) systematically analyzed the antigenic components of the three most common strains of the genus (X19, X2, XK) and established the basis for an antigenic scheme suitable for use in serotyping. This scheme was later enlarged to include 49 O groups that could be subdivided by means of 19 H antigens to produce 98 types (Perch, 1948). In an extended scheme, the first 30 O groups were analyzed in detail and 55 partial O antigens and 31 partial H antigens were defined among the 30 O and 16 H antigens. Numerous partial antigens were common to both *Prot. vulgaris* and *Prot. mirabilis* but in only a few cases did strains of both species share an O or H antigen composed of identical partial antigens. A strong association of a particular range of serotypes with each species was noted by Lanyi (1956). The strains of Winkle (1945) were examined by Perch (1948) and, except for one H antigen, were shown to possess antigens defined in the Perch scheme. Other investigators studied the serology of the genus *Proteus* but used sets of strains different from those of Perch (Belyavin, Miles, and Miles, 1951; Keating, 1956; Krikler, 1953). Some of the strains studied by Belyavin, Miles, and Miles (1951) were the same as some of the Perch scheme. In a serological study of colonial variation, three serological phases (A, B, and C) were described by Belyavin (1951). A rough variant was identified by Coetzee and Sachs (1960) in addition to Belyavin's three phases. The A phase consists of motile coccobacillary bacteria and is associated with typical swarming; the B phase is composed of nonmotile filaments, giant cells, and coccobacilli; and C phase consists of long, tangled filaments that often spread in a uniform film. Phases A and C, but neither B nor the rough form, possess O antigens and it is important, therefore, that the appropriate phases be selected in antigen preparations for serological studies.

In studies on *Prot. vulgaris* and *Prot. mirabilis* infections, Lanyi (1956, 1957) used the Perch scheme to serotype strains isolated from feces of infants with enteritis, from urinary tract infections, and from pathological specimens from other sites of infection. He showed that some serotypes, particularly O3 and O26, were more frequently involved in infections than were other types. De Louvois (1969) prepared antisera against Perch strains belonging to *Prot. mirabilis* and also found a higher frequency of the O3 serotype in human infections. Such studies indicate the epidemiological value of the Perch scheme and should encourage wider adoption of the scheme in differentiating strains in the reference laboratory.

Dienes Reaction

Dienes (1946) observed that swarms of unlike strains of *Proteus* failed to penetrate into each other, resulting in a sharp line of demarcation between the swarms of each strain when cultured on nutrient agar. The absence of this effect between two swarms was interpreted to signify that the swarming strains were the same. This observation has been exploited in the differentiation of *Proteus* strains for epidemiological purposes in the clinical laboratory and is referred to as the Dienes reaction (or test). When strains are tested against each other in this way, the observation of a line of demarcation is recorded as a Dienes-positive reaction and the strains are considered different. The absence of the line is recorded as a Dienes-negative reaction and the strains are considered the same with respect to this test.

The biological basis for the production of the line of demarcation is not well understood and much uncertainty exists regarding the specificity of the test. The suggestion of Krikler (1953, p. 72) that similar specificities of the flagellar (H) antigens were necessary for the bacteria to spread into one another without the line of inhibition was not supported by subsequent investigation. Skirrow (1969) and Šourek (1968) reported Dienes-negative reactions between strains differing in H antigens and de Louvois (1969) noted that serologically identical strains could be further subdivided by the Dienes test. Kippax (1957) found that strains of different biochemical types occasionally swarmed together without giving a line of demarcation. Skirrow (1969) suggested that the Dienes reaction could be due to bacteriocin-like substances but neither Šourek (1968) nor Tracy and Thomson (1972) could find support for this hypothesis. However, Senior (1977a) has provided convincing evidence that the Dienes phenomenon is mediated by bacteriocins and

that the Dienes type is dependent upon both the bacteriocin the strain produces and the bacteriocins to which it is sensitive.

Hickman and Farmer (1976) found that the Dienes test often failed to correlate with bacteriophage typing and, in some cases, gave equivocal results, making interpretations of epidemiological data more difficult. On occasion, false-positive results were obtained when a strain was tested against itself. On other occasions, repeated tests gave reactions opposite to those of the first test.

Further doubts concerning the reliability of the test have been expressed by France and Markham (1968) and Story (1954), who used the test in epidemiological studies. De Louvois (1969) suggested that, in the case of comparing randomly selected isolates, a Dienes-positive reaction could be taken as an indication that the strains were different but a Dienes-negative reaction should not be considered a reliable test of identity.

Despite these difficulties, the Dienes reaction can be a very useful test in the clinical laboratory when its limitations are recognized, and particularly when used in conjunction with one or more typing systems.

Bacteriophage Typing of *Proteus*

Proteus bacteriophage may be readily isolated from sewage and from many lysogenic strains of the genus (Coetzee, 1972; Vieu, 1963). Several descriptions of schemes for bacteriophage typing have been reported (France and Markham, 1968; Hickman and Farmer, 1976; Izdebska-Szymona, Monczak, and Lemczak, 1971; Pavlatou, Hassikou-Kaklamani, and Zantioti, 1965; Schmidt and Jeffries, 1974; Vieu, 1958; Vieu and Capponi, 1965). Most of the typing schemes were primarily concerned with *Prot. mirabilis,* although phage that lysed *Prot. vulgaris* were described. No typing scheme has gained wide acceptance. The most comprehensive study of bacteriophage typing for *Prot. mirabilis* was performed by Hickman and Farmer (1976). From a total of 82 bacteriophage, 23 were selected as a provisional set to provide epidemiological typing service for *Prot. mirabilis* at the Center for Disease Control, Atlanta, Georgia.

Bacteriocinogeny of *Proteus mirabilis* and *Proteus vulgaris*

The production of bacteriocins (proticins) by *Proteus mirabilis* and *Prot. vulgaris* has recently been investigated to determine if typing systems can be developed. Cradock-Watson (1965) showed that 139 of 229 (61%) *Prot. mirabilis* and 1 of 10 (10%) *Prot. vulgaris* strains produced bacteriocins in solid media. However, Dimitracopoulos, Tzannetis, and

Papavassiliou (1972) found bacteriocin production in their collection of strains to be rare. Tracy and Thomson (1972) were unable to type their *Proteus* strains when using the indicator strains of Cradock-Watson (1965) but found that some were typable when locally isolated strains were used as indicators of bacteriocin production. Senior (1977b) described a method of typing that was an extension of the one described by Cradock-Watson (1965). He recorded both the bacteriocin production and the bacteriocins to which strains under test were sensitive and referred to the scheme as a "production/sensitivity" typing scheme. He noted that there appeared to be no correlation between the bacteriocins to which a strain was susceptible and the type of bacteriocin produced by the strain. Using the technique of mitomycin C induction, Al-Jumaili (1975a) found 87% of *Prot. mirabilis* and 60% of *Prot. vulgaris* bacteriocinogenic and, with 12 strains selected as a standard set, he observed 48 sensitivity patterns among 1,000 isolates and, furthermore, demonstrated the applicability of bacteriocin typing in epidemiological investigations (Al-Jumaili, 1975b).

Proteus myxofaciens

Cosenza and Podgwaite (1966) isolated a slime-producing bacterium from living and dead larvae of the gypsy moth *Porthetria dispar* (L.) in New York and Connecticut. The authors found its biochemical reactions characteristic of the genus *Proteus* but not typical of either *Prot. vulgaris* or *Prot. mirabilis*. It swarmed on the surface of moist agar, deaminated phenylalanine, hydrolyzed urea, produced acid in glucose, sucrose, maltose, and trehalose, produced hydrogen sulfide and acetyl methyl carbinol, but did not ferment lactose or mannitol and did not produce indole or utilize citrate. The authors proposed a new species within the genus *Proteus* and, to denote its ability to produce slime, they selected the epithet myxofaciens. Lautrop (1974, p. 327) suggested that *Prot. myxofaciens* was identical to *Erwinia herbicola* (*Enterobacter agglomerans*), but Brenner et al. (1978) studied one strain of this new species and found their biochemical tests, in most cases, agreed with those the original authors. Furthermore, DNA relatedness studies showed that it was 50% related to *Prot. mirabilis* and *Prot. vulgaris* and they supported the recognition of *Prot. myxofaciens* as a valid species.

A striking feature of this new species is its ability to produce slime so extensively that it can cause Trypticase soy broth in which it is cultured to become so viscous that when the contents of the tube are poured out they emerge in the form of a plug. Although *Prot. myxofaciens* was isolated from the larvae of the gypsy moth, it has not been clearly established that the bacterium has a role in the production of disease. Further studies of this recent addition to the genus will surely be forthcoming to elucidate this point, and others, of interest.

THE GENUS *PROVIDENCIA*

The generic name proposed by Kauffmann and Edwards (1952) was derived from Providence, Rhode Island, where bacteria of this group were collected and studied under the label 29911 by Stuart et al. (1943) and Stuart, Wheeler, and McGann (1946). Cowan (1956) and Shaw and Clarke (1955) prefer the name *Proteus inconstans* because, like members of the genus *Proteus,* these bacteria deaminate phenylalanine and because a culture that appeared to belong to this group was described by Ornstein (1921) as *Bacillus inconstans*. *Bergey's Manual of Determinative Bacteriology* has continued to use this nomenclature (Lautrop, 1974) although the epithet *inconstans* has been cited as being invalid (Ewing, 1962). Strains of *Providencia* were divided into two biochemically different groups by Ewing, Tanner, and Dennard (1954) and later designated subgroups A and B. A culture of *Eberthella alcalifaciens* De Salles Gomes (ATCC 9886) was found by Suassuna and Suassuna (1960) to be a typical strain of *Providencia* and the epithet *alcalifaciens* was assigned to strains of subgroup A (Ewing, 1962). Members of subgroup B were assigned to a second species with the epithet *stuartii* first suggested by Buttiaux et al. (1954).

Prvd. rettgeri has only been recently included in this genus. It was first isolated in 1904 by Rettger, who sent it to be examined by Hadley, Elkins, and Caldwell (1918). They described the bacterium as a short, Gram-negative, nonmotile rod that produced acid from glucose, mannitol, adonitol, salicin, xylose, and mannose, but not from maltose, sucrose, lactose, arabinose, dextrin, inulin, dulcitol, or erythritol, and as incapable of indole production. St. John-Brooks and Rhodes (1923), however, reported a positive reaction for indole production. It was placed in the genus *Eberthella* (Bergey et al., 1923) until Weldin (1927) reassigned it to the genus *Shigella*. This classification was accepted (Bergey et al., 1939) until Neter (1942) proposed its removal from this genus because it had been reported to be motile by Edwards in a personal communication to Neter (1942). Rustigian and Stuart (1943a) studied bacteria described as atypical *Shigella* by Cope and Kilander (1942) and strains in their own collection labeled 33111 (Stuart et al., 1943) and found both groups biochemically similar and capable of hydrolyzing urea. On the basis of the latter reaction, they placed them in the genus *Proteus* and suggested the epithet *entericus* (Rustigian and Stuart, 1943a). This epithet was later withdrawn in favor of *rettgeri* on finding that these bacteria were the same as Rettger's isolate (Rustigian and Stuart, 1943b). This nomenclature was widely accepted but many supported a proposal by Kauffmann (1953) to establish a new genus (*Rettgerella*) for these bacteria.

Fulton and Curtis (1946) showed differences among the *rettgeri* strains in their reactions in rhamnose and salicin and four biogroups were described by Namioka and Sakazaki (1958) on the basis of reactions in these two substrates. Penner, Hinton, and Hennessy (1975) found that strains of these four biogroups almost invariably showed positive reactions in each of three polyhydric alcohols (mannitol, adonitol, and arabitol) and most also showed positive reactions in erythritol. Strains that did not produce acid in erythritol or in more than one of the other polyhydric alcohols were placed in the newly-defined biogroup 5. Strains in the latter biogroup resembled *Prvd. stuartii* except in their ability to hydrolyze urea and were later reassigned to *Prvd. stuartii* because they agglutinated specifically in antisera prepared against *Providencia* O-type strains (Penner and Hennessy, 1977) and were found inseparable from *Prvd. stuartii* in DNA relatedness tests (Brenner et al. 1978). The genus *Providencia* had been repeatedly defined as a urease-negative genus (Ewing, 1958, 1962; Rauss, 1962; Report, 1958). Thus, the exclusion from the genus of urease-positive strains had become the accepted practice in clinical laboratories. However, urease-positive *Providencia* had been noted in several studies (Carpenter, 1964; Ewing, Tanner, and Dennard, 1954; Shaw and Clarke, 1955; Stuart, Wheeler, and McGann, 1946; Stuart et al., 1943; Ursing, 1974). The broadening of the genus *Providencia* to include urease-positive isolates was, therefore, consistent with the findings reported in these studies.

It had become evident that the *rettgeri* strains were more closely related to the genus *Providencia* than to other Proteeae through detailed studies by several groups of investigators. Buttiaux et al. (1954) and Richard (1966) drew attention to the close similarities of their biochemical reactions. Proom and Woiwod (1951) and Proom (1955) showed that the *rettgeri* strains and more than half of their 55 Providence strains did not produce amines (isobutyl and isoamyl) as did *Prot. vulgaris*, *Prot. mirabilis*, and *Prot. morganii*. On the basis of this finding and other biochemical similarities, Proom (1955) proposed that a new genus (unnamed) be established for Providence and *rettgeri* strains to separate them from the genus *Proteus*. Smit and Coetzee (1967) examined the serological specificities of the phenylalanine deaminases and found cross-reactions between enzymes from *rettgeri* and Providence strains but absence of cross-reactions between them and enzymes extracted from other Proteeae and suggested that both groups should be placed in one genus, namely, *Rettgerella*. Cook (1948) described yellow-orange centered colonies on deoxycholate-citrate agar produced by *rettgeri* strains, and

Buttiaux, Frenoy, and Moriamez (1954) noted that this apparent pigmentation was produced only by *rettgeri* and Providence strains. The basis for the color was investigated by Catsaras, Antoniewski, and Buttiaux (1965) and reported to be the precipitation of ferric hydroxide as a result of the alkalinity produced by the growth of the bacteria on the medium. This was yet another feature illustrating similarity of the two groups and a feature by which they could be distinguished from other Enterobacteriaceae. Pichinoty et al. (1966) and Pichinoty and Piechaud (1968) described two forms (A and B) of nitrate reductase among Enterobacteriaceae. The A form of the enzyme was present in *rettgeri* strains and *Prvd. stuartii* but the rarer B form was present in *Prvd. alcalifaciens*. This indicated a relationship of *rettgeri* strains closer to *Prvd. stuartii* than to *Prvd. alcalifaciens*. Further convincing evidence for the close relationship between *rettgeri* strains and the genus *Providencia* was produced in DNA relatedness studies (Brenner, et al., 1978) and the authors proposed the transfer of the *rettgeri* strains to the genus *Providencia*.

Habitats of *Providencia*

Comprehensive studies on the habitats of these bacteria have not been reported. In studies on the habitats of *Proteus*, however, occasional isolates of *Providencia* were noted and from such scanty evidence it appears that there is at least some degree of overlapping in the habitats of *Proteus* and *Providencia* in nature.

The occurrence of *Providencia* in the normal human intestine is rare. Singer and Bar-Chay (1954) studied fecal strains of the Providence group. In this group, *Prvd. stuartii*, of course, was not differentiated from *Prvd. alcalifaciens*, and the incidence determined for these bacteria in human stool specimens cannot be related to one species or the other. However, it is significant that in 500 normal adults the incidence of Providence strains was only 0.6% and in 337 normal babies it was 2.3%. For *Prvd. rettgeri*, Blake (1975) observed an incidence of 1.0% in feces from 765 inpatients in a general hospital. It is evident from these limited studies that the healthy human intestine only rarely serves as a habitat for any of the three *Providencia* species.

Infections in Man by *Providencia*

Providencia has emerged from relative obscurity to prominence as a serious infectious agent of man only within the last two decades. Some strains of *Prvd. stuartii* and *Prvd. rettgeri* are resistant to many antimicrobial agents and have a strong propensity for causing nosocomial infections. These two features highlight their medical importance and outweigh the fact that they are only infrequently isolated in the clinical laboratory. Numerous studies concerned with their resistance and spread in the hospital have been described (Edwards et al., 1974; Iannini, Eickhoff, and LaForce, 1976; Lindsey et al., 1976; Milner, 1963; Omland, 1960; Schaberg, Weinstein, and Stamm, 1976; Traub et al., 1971; Washington et al., 1973). These studies were, of course, performed without the benefit of the current classification system, and urease-positive *Prvd. stuartii* strains were therefore included in the *Prvd. rettgeri* group. However, typical isolates of *Prvd. rettgeri* characterized according to reactions in Table 2 have been obtained from patients with urinary tract infections (Penner and Hennessy, 1979b). Furthermore, cross-infections due to a lactose-positive *Prvd. rettgeri*, typical in other reactions of the species according to Table 2, was described by Richard, Popoff, and Pastor (1974) and Traub et al. (1971).

Considerable evidence has accumulated on the occurrence of *Prvd. stuartii* in nosocomial urinary tract infections (Dobrey, 1971; Hamilton-Miller, Reynolds, and Brumfitt, 1974; Li and Miller, 1970; Overturf, Wilkins, and Ressler, 1974; Penner et al., 1979a; Report, 1977; Solberg and Matsen, 1971; Stickler and Thomas, 1976; Whiteley et al., 1977). Furthermore, *Prvd. stuartii* has caused serious infections in burn patients (Curreri et al., 1973; Overturf, Wilkins, and Ressler, 1974; Wenzel et al., 1976). Bacteremias due to this species have high fatality rates (Curreri et al., 1973; Janis, Evans, and Hoeprich, 1968; Klastersky et al., 1974; Milstoc and Steinberg, 1973; Solberg and Matsen, 1971). Keane and English (1975) noted increases in the number of *Prvd. stuartii* isolates, not only from burns and urines, but also from blood and sputum over a 3-year period.

Certain important differences exist between *Prvd. alcalifaciens* and the other *Providencia* species but they were not readily apparent from the early publications in which *Prvd. stuartii* and *Prvd. alcalifaciens* were classified in the paracolon or Providence groups. Galton et al. (1947) isolated strains from feces of patients suffering from diarrhea and sent them to Stuart, who studied them along with other fecal isolates and was persuaded that they were causative agents of enteritis and diarrhea (Stuart, Wheeler, and McGann, 1946). Most of their isolates gave negative reactions in trehalose and inositol but positive reactions in adonitol and, under current classification, would be *Prvd. alcalifaciens*. The fecal·isolates from children with gastroenteritis studied by Brown (1952) were inositol negative and, therefore, were most likely also *Prvd. alcalifaciens* and, of the 165 *Providencia* strains isolated by Bhat, Shanthakumari, and Meyers (1971) from preschool children with diarrhea, 155 were *Prvd. alcalifaciens*. This association of *Prvd. alcalifaciens* with

fecal isolates and diarrhea contrasts sharply with the association of *Prvd. stuartii* and *Prvd. rettgeri* with urological specimens and nosocomial urinary tract infections. The majority of *Prvd. stuartii* and *Prvd. rettgeri* strains in our collection were obtained from urine specimens and rarely from feces, while the *Prvd. alcalifaciens* strains were obtained predominantly from feces of patients in a pediatric hospital (Penner et al., 1979b). It should be noted that all but two or three of the 35 Providence strains of Brooke (1951) from urines, uterine secretions, etc., were trehalose and inositol positive and adonitol negative. They were, therefore, *Prvd. stuartii* and hence evidence for the association of this species with infections of the urinary tract was available much earlier than is generally acknowledged and at approximately the same period of time that *Prvd. alcalifaciens* was being recognized for its involvement in the gastrointestinal tract. On the other hand, Singer and Bar-Chay (1954) studied fecal isolates that were predominantly *Prvd. stuartii* (inositol and trehalose positive and adonitol negative) and found an incidence of 2.3% in 337 healthy children, but 10% in 116 infants with diarrhea. Therefore, it would appear that, generally, *Prvd. alcalifaciens* is associated with infections of the intestinal tract and *Prvd. stuartii* and *Prvd. rettgeri* with infections of the urinary tract, but rigidity should not be attached to this generalization as exceptions do occur. Further clarification of the individuality of the *Providencia* species will no doubt be realized when the delineation of species of this genus becomes routine, as advocated by Fields et al. (1967), Janis, Evans, and Hoeprich (1968), and Monto and Rantz (1965).

Antigenic Structure of The Genus *Providencia*

Reports on serotyping the species of *Providencia* are few, but increasing interest in the epidemiology of human *Providencia* infections has prompted studies on the antigenic structures of these bacteria to provide the basis for serotyping schemes.

Serology of *Providencia rettgeri*

Early reports on *Providencia rettgeri* noted that the species was serologically heterogeneous (Cope and Kasper, 1949; Rustigian and Stuart, 1943b; Stuart et al., 1943). References were made to complexities such as the occurrence of antigens incapable of being agglutinated but capable of binding antibody (Fulton and Curtis, 1946; Rustigian and Stuart, 1943b; Stuart et al., 1943), the interference of capsular antigens in agglutination reactions (Cope and Kasper, 1949; Omland, 1960), the apparent thermolability of O antigens (Fulton and Curtis, 1946), and the loss of antigenicity during laboratory maintenance of the cultures (Stuart et al., 1943). A comprehensive study on the antigenic structure of *Prvd. rettgeri* was reported by Namioka and Sakazaki (1958, 1959). Among 103 strains, they identified 45 types based on 34 somatic (O), 26 flagellar (H), and one capsular (K) antigens. Their study firmly established that the species was serologically heterogeneous and showed that serotyping could be applied in epidemiological studies but, as in previous studies, complexities in the serology of some of the strains were noted.

Using the passive hemagglutination technique, Penner and Hinton (1973) found that unilateral (nonreciprocal) reactions were caused by a thermolabile antigen common to numerous strains that were different in their O specificities. Absorption of O antisera with a selected strain abolished the unilateral reactions. To exclude involvement in agglutination reactions of the thermolabile component, Penner, Hinton, and Hennessy (1974) produced a serotyping system using only antigen preparations that had been autoclaved. In these investigations, the passive hemagglutination technique was employed in the analysis of the O specificities. The advantage of the technique is that it can be automated and is more sensitive than techniques that require bacterial cell suspensions for titrating antisera. From the data obtained in such O antigenic analyses, the absorptions necessary to produce specific typing antisera were clearly indicated. O antisera at low dilutions of 1:5 agglutinated only cell suspensions of homologous O-type strains, and rarely were isolates agglutinated in more than one typing antiserum. As more strains were collected, it became evident that the species was constituted of a large number of different O specificities and, therefore, the use of pooled antisera was initiated to facilitate serotyping (Penner, Hinton, and Hennessy, 1976). The number of O specificities reported in the latter study was 84 but the number has increased with further studies of new strains (author's unpublished observations). The applicability of the scheme in serotyping *Prvd. rettgeri* isolates has been demonstrated (Penner et al., 1976b; Washington et al., 1973).

Serology of *Providencia*
alcalifaciens and *Providencia stuartii*

In a study by Ewing et al. (1954) on the antigenic structure of the Providence group of bacteria, 56 somatic (O), 28 flagellar (H), and 2 capsular (K) antigens were defined. The scheme was not widely used, but its applicability in epidemiological studies was demonstrated in several reports (Bhat, Shantha-

kumari, and Meyers 1971; Solberg and Matsen, 1971; Stenzel, 1961). Penner et al. (1976a) used the O-type strains of this scheme to produce a system for serotyping on the basis of 62 O antigens, but additional O specificities were defined as new isolates were obtained from a wider range of sources.

Examinations of the O antigens of the three *Providencia* species showed, for the most part, that each of the species was characterized by its own set of O specificities and, although cross-reactions were observed between O types of the different species, they were relatively few in number and were caused mostly by cross-reacting "a" components between strains related through "a,b-a,c" relations. In this connection, it is interesting to note that all typable isolates of subgroup A (*Prvd. alcalifaciens*) in the study of Bhat, Shanthakumari, and Meyers (1971) agglutinated only in antisera against O-type strains of *Prvd. alcalifaciens* (O3, O8, O12, O16, O19, O21) and the isolates of subgroup B (*Prvd. stuartii*) agglutinated in antiserum against the O20 type strain that belongs to *Prvd. stuartii*. Facility in serotyping may therefore be accomplished by classifying *Providencia* according to species through biochemical tests before testing them in antisera prepared against O-type strains of the same species. As interest in the epidemiology of *Providencia* infections grows, it is expected that reference centers will extend their services to provide serotyping of this genus.

Bacteriophage of *Providencia*

Considerable information on the structure and genetics of *Providencia* bacteriophage has accumulated but studies directed toward the development of bacteriophage typing systems for the species of this genus have not been reported (Coetzee, 1972). However, there are good indications that such systems can be produced. Coetzee (1963) reported that bacteriophage could readily be isolated from sewage and from lysogenic strains and that such phage had lytic activity on *Prvd. rettgeri*, "Providence", and *Proteus*, but not on *M. morganii* strains.

Bacteriocinogeny in *Providencia*

Coetzee (1967) used 32 *Providencia rettgeri* and 37 Providence strains to test intraspecifically for bacteriocinogeny. Five of the Providence but none of the *rettgeri* strains produced bacteriocins spontaneously. One Providence bacteriocin was active on a *Prvd. rettgeri* strain. Bacteriocinogeny was shown in *Prvd. rettgeri* when bacterial cultures were induced with mitomycin C in Trypticase soy broth (Craddock and Traub, 1971). Despite this demonstration of bacteriocinogeny, no attempts to develop bacteriocin typing schemes for the three *Providencia* species have been reported.

MORGANELLA MORGANII

Morgan (1906) isolated a bacterium from the intestines of infants suffering from diarrhea and referred to it as Bacterium No. 1. The epithet *morganii* was assigned by Winslow, Kliger, and Rothberg (1919). The generic designation has undergone several changes. The bacterium has been placed in the *meta-coli* group (Bahr and Thomsen, 1912), in the genus *Salmonella* (Castellani and Chalmers, 1919), in the genus *Proteus* (Rauss, 1936; Yale, 1939), and in the genus *Morganella* (Fulton, 1943). Its location in the *meta-coli* group was supported by Thjøtta (1920) and Jordan, Crawford, and McBroom (1935), and its placement in *Salmonella* was advocated by Besson and de Lavergne (1921) and Magheru (1923). The results of the investigations conducted by Rauss (1936) gave the impetus to relocate the organism in the genus *Proteus*. He demonstrated that some strains could be induced to swarm at lower temperatures (20–28°C) on media with reduced concentration of agar and that a group of the strains shared flagellar (H) antigens with *Proteus vulgaris*. He suggested that the bacterium "belongs taxonomically to the genus *Proteus*" and Yale (1939) attributed the name *Proteus morganii* to him. However, Lessel (1971) pointed out that Rauss had not published the name and therefore Yale, not Rauss, should be cited as the author. This direction was followed in the eighth edition of *Bergey's Manual* (Lautrop, 1974).

Sevin and Buttiaux (1939) confirmed the findings of Rauss but noted that *morganii* strains never showed the regular and concentric halos characteristic of swarming *Proteus*.

Fulton (1943) recommended that the *morganii* strains, along with bacteria classified as *Bacterium columbensis,* be placed in a new genus and he proposed the name *Morganella*. Rustigian and Stuart (1941, 1943a, 1945) supported the inclusion of *morganii* strains in the genus *Proteus* primarily because of its urease production. Proom and Woiwod (1951) regarded the strains of *B. columbensis* as lactose-negative variants of *Escherichia coli* and the *morganii* strains as closely related to *Proteus* (particularly because of their amine production and nutritional requirements), and opposed the placement of *morganii* strains into a separate genus.

As a result of an extensive study of the biochemical and serological characteristics, Rauss (1962) concluded that *morganii* strains should be given generic status within the tribe Proteeae. As interest in the bacteria increased, evidence from several studies supported their separation from the genus *Proteus*. The base composition of the DNA of *morganii* strains was found to be 50% guanine plus cytosine, corresponding in composition to *Escherichia, Salmonella* and *Shigella* and not to *Prot. vulgaris* and *Prot. mirabilis* with 39% (Falkow, Ryman, and Washington, 1962). The urease (Guo and Liu, 1965) and phenylalanine deaminase enzymes (Smit and Coetzee, 1967) of *morganii* strains were found to have serological specificities distinct from similar enzymes of other Proteeae. In deoxyribonucleic acid relatedness studies, it was shown that *M. morganii* was related to enteric bacteria at a 20% level and at not more than 20% to other Proteeae (Brenner et al., 1978). This was strong phylogenetic evidence for reassignment to a separate genus and, therefore, they proposed the genus *Morganella* with *M. morganii* as the type species.

Habitats of *Morganella*

Systematic studies to determine the habitat of *Morganella morganii* have not been conducted, but it has been reported to occur in normal stools (Winslow, Kliger, and Rothberg, 1919) and in low frequency in the intestines of animals that have been examined for other bacteria. In this type of study, Phillips (1955a) isolated the organism from dogs, unspecified reptiles, and other mammals. Its occurrence was rare, however, and it was far outnumbered by *Proteus mirabilis* and *Prot. vulgaris*. Müller (1972) noted its rare occurrence also in snakes. In early studies it was reported to occur frequently in milk and in the intestines of cows, mice, and insects (Castellani and Douglas, 1932) but Jordan, Crawford, and McBroom (1935) felt that earlier methods of identification were inadequate and that confusion of *M. morganii* with slow lactose fermenters could have occurred. In their studies, *M. morganii* were obtained from patients with diarrhea but not from healthy children, adults, or captured wild animals.

Pathogenicity of *Morganella*

The first report on the pathogenicity of *Morganella morganii* was by Morgan (1906). He isolated the bacterium from infants suffering from diarrhea and from a nurse working with these patients. He noted that it was the only non-lactose-fermenting bacterium in her stool. In virulence tests, suspensions of bacteria were administered orally to rats and rabbits. Death, preceded by violent diarrhea, occurred within 24 h. Bacteria could be recovered before and after death from spleens of infected animals. Similar degrees of virulence were demonstrated in experiments by Tribondeau and Fichet (1916) and Magheru (1923). They used parenteral routes of administration and showed that the guinea pig was susceptible and could be used to demonstrate virulence of *M. morganii*. From seven of nine adults, severely ill with ulcerative colitis, only *M. morganii* was identified among the non-lactose-fermenting colonies by Thjøtta (1920). Recovery of some patients was associated with the disappearance of *M. morganii* in stool cultures. However, because agglutinins could not be detected in the patients' sera against homologous isolates, he did not see fit to conclude that *M. morganii* was the proved etiological agent. Further circumstantial evidence of a pathogenic role in human disease for these bacteria was provided by Rauss (1936). He isolated 48 strains from stools of patients suffering from dysentery-like illness and, in all cultures, he noted the absence of pathogenic agents such as the dysentery bacillus.

Sevin and Buttiaux (1939) isolated *Morganella morganii* in pure culture from a case of hematuria, thus directing attention to this species as an infectious agent also of the human urinary tract. With recent recognition of the increasing importance of Gram-negative infections, more attention has been given to the Proteeae as both primary and secondary or opportunistic agents of infection. As a result, greater efforts have been made toward more complete classification of the Proteeae in clinical laboratories and the occurrence of *M. morganii* in urine, blood, sputa, pus, and feces from patients with bacteremias, wound, urinary tract, and respiratory tract infections has been noted (Adler et al., 1971a,b; Chiu and Hoeprich, 1961; Lanyi, 1956; Potee, Wright, and Finland, 1954; von Graevenitz and Spector, 1969). In the clinical laboratory, the rate of isolation of *M. morganii* is generally low. From 363 pathological specimens (other than feces) Proteeae were isolated by Lanyi (1957) and 19 (5%) of these were *M. morganii*. A total of 1,207 Gram-negative isolates were cultured from urine specimens in a male urological ward over a 9-year period (Søgaard, Zimmermann-Nielsen, and Siboni, 1974) and of these, 26 (2.2%) were *M. morganii*. A similar value, 2.4%, can be calculated from the data of McMillan (1972) for the incidence in urine of elderly females. Von Graevenitz and Spector (1969) considered strains cultured from specimens other than feces as contributing to clinically evident infection if they occurred in pure culture or in the same numbers as other bacteria in mixed cultures at the site of infection. Under these criteria, *M. morganii* was considered a significant contributor when isolated from urine and a relatively frequent contributor when isolated from sputa and wounds. Most infections were hospital acquired by elderly patients often subjected to instrumentation, and therefore reflected the medical importance of *M. morganii* as an opportunistic secondary invader rather than as a primary pathogen, except in infections of the urinary tract.

Antigenic Structure of *Morganella*

The existing knowledge of the antigenic structure of *Morganella morganii* may be largely attributed to the extensive studies by Rauss and co-workers. In a series of investigations, an antigenic scheme based on somatic (O) and flagellar (H) antigens first described by Rauss and Vörös (1959) has been extended so that currently the number of defined serogroups is 42 and the number of serotypes is 75 (Rauss and Vörös, 1967; Rauss et al., 1975).

In another serological study (Sompolinsky, 1957), five O specificities were defined among 110 strains of *M. morganii*. Of particular interest was the finding of shared antigens between the most common O-type and *Shigella flexneri*.

The applicability of the scheme developed by Rauss and his colleagues in differentiating strains was demonstrated by them in their studies and by Penner and Hennessy (1979a), and the scheme will no doubt find more use when interest in the epidemiology of *M. morganii* infections will demand precise typing.

Bacteriophages of *Morganella morganii*

The isolation and lytic activity of *Morganella morganii* bacteriophage have been described in several reports (Schmidt and Jeffries, 1974; Taubeneck, 1962; Vieu, 1963). It was shown that these phage rarely lyse bacteria belonging to *Proteus* and *Providencia* and that bacteriophage isolated from the latter genera were not active on *M. morganii* (Coetzee, 1963).

Schmidt and Jeffries (1974) isolated seven *M. morganii* bacteriophage but they found 14 different lytic patterns among 26 bacterial strains and thus demonstrated the feasibility of typing this species. The plaques were small, hazy, or clouded, but, with experience, the results could be accurately interpreted. The success of their typing, although limited to small numbers of bacteriophage and bacterial strains, should, nevertheless, encourage more extensive studies toward the development of a standardized bacteriophage typing system for this species.

Bacteriocins of *Morganella morganii*

Coetzee (1967) found 12 bacteriocinogenic strains among 94 *Morganella morganii*. Producer strains were active on a number of strains of the same species and each morganocin could be distinguished by its own spectrum of activity. The author reported bacteriocinogenic activity on MacConkey but no activity on nutrient agar and suggested that previous failures to demonstrate bacteriocinogeny in this species were due to the media used. The importance of his study is that it indicated the possibility of bacteriocin typing and should encourage further studies to search for additional producer and indicator strains necessary to establish a standardized typing system.

Literature Cited

Abrams, H. L. 1948. Septicemia due to *Proteus vulgaris.* New England Journal of Medicine **238:**185–187.

Acred, P., Brown, D. M., Knudsen, E. T., Rolinson, G. N., Sutherland, R. 1967. New semi-synthetic penicillin active against *Pseudomonas pyocyanea,* Nature **215:**25–30.

Adler, J. L., Burke, J. P., Martin, D. F., Finland, M. 1971a. *Proteus* infections in a general hospital. I. Biochemical characteristics and antibiotic susceptibility of the organisms with special reference to proticine typing and the Dienes phenomenon. Annals of Internal Medicine **75:**517–530.

Adler, J. L., Burke, J. P., Martin, D. F., Finland, M. 1971b. *Proteus* infections in a general hospital. II. Some clinical and epidemiological characteristics with an analysis of 71 cases of *Proteus* bacteremia. Annals of Internal Medicine. **75:**531–536.

Annotations. 1966. Proteus: The quiet sort. Lancet **i:**1196.

Al-Jumaili, I. J. 1975a. Bacteriocine typing of *Proteus.* Journal of Clinical Pathology **28:**784–787.

Al-Jumaili, I. J. 1975b. An evaluation of two methods of bacteriocine typing of organisms of the genus *Proteus.* Journal of Clinical Pathology **28:**788–792.

Bahr, L., Thomsen, A. 1912. Fortgesetzte Untersuchungen über die Aetiologie der *Cholera infantum* (1910). Zentralblatt für Bakteriologie, Parasitenkunde und Infektionskrankheiten, Abt. 1 Orig. **66:**365–386.

Barber, M., Waterworth, P. M. 1964. Antibiotic sensitivity of *Proteus* species. Journal of Clinical Pathology **17:**69–74.

Barry, A. L., Hoeprich, P. D. 1973. In vitro activity of cephalothin and three penicillins against *Escherichia coli* and *Proteus* species. Antimicrobial Agents and Chemotherapy **4:**354–360.

Becker, A. H. 1962. Infections due to *Proteus mirabilis* in newborn nursery. American Journal of Diseases of Children **104:**355–359.

Belyavin, G. 1951. Cultural and serological phases of *Proteus vulgaris.* Journal of General Microbiology **5:**197–207.

Belyavin, G., Miles, E. M., Miles, A. A. 1951. The serology of fifty strains of *Proteus vulgaris.* Journal of General Microbiology **5:**178–196.

Bergey, D. H., Harrison, F. C., Breed, R. S., Hammer, B. W., Huntoon, F. M. 1923. Bergey's manual of determinative bacteriology, 1st ed. Baltimore: Williams & Wilkins.

Bergey, D. H., Breed, R. S., Murray, E. G. D., Hitchens, A. P. 1939. Bergey's manual of determinative bacteriology, 5th ed., Baltimore: William & Wilkins.

Bernheim, F., Bernheim, M. L. C., Webster, M. D. 1935. Oxidation of certain amino acids by "resting" *Bacillus proteus.* Journal of Biological Chemistry **110:**165–172.

Besson, A., de Lavergne, 1921. Sur le bacille de Morgan. Comptes Rendus Hebdomadaires des Séances et Mémoires de la Société de Biologie **84:**530–532.

Bhat, P., Shanthakumari, S., Meyers, R. M. 1971. The Providence group: Subgroups including biotypes and serotypes of fecal strains isolated in Vellore. Indian Journal of Medical Research **59:**1184–1189.

Blake, H. E. 1975. An epidemiological study of *Proteus rettgeri* infections based on serotyping, biotyping, and antibiotic sensitivity patterns. Ph. D. Thesis. University of Toronto, pp. 47–54.

Borman, E. K., Stuart, C. A., Wheeler, K. M. 1944. Taxonomy of the family *Enterobacteriaceae.* Journal of Bacteriology **48:**351–367.

Bornside, G. H. 1968. Synergistic antibacterial activity of ampicillin-cloxacillin mixtures against *Proteus morganii.* Applied Microbiology **16:**1507–1511.

Braude, I. A., Siemienski, J. 1960. Role of bacterial urease in experimental pyelonephritis. Journal of Bacteriology **80:**171–179.

Brenner, D. J., Farmer, J. J., Fanning, G. R., Steigerwalt, A. G., Klykken, P., Wathen, H. G., Hickman, F. W., Ewing, W. H. 1978. Deoxyribonucleic acid relatedness in species of *Proteus* and *Providencia.* International Journal of Systematic Bacteriology **28:**269–282.

Briedis, D. J., Robson, H. G. 1976. Comparative activity of netilmicin, gentamicin, amikacin and tobramycin against *Pseudomonas aeruginosa* and *Enterobacteriaceae.* Antimicrobial Agents and Chemotherapy **10:**592–597.

Brooke, M. S. 1951. Biochemical investigations on certain urinary strains of *Enterobacteriaceae: B. cloacae,* (2) "Providence". Acta Pathologica et Microbiologica Scandinavica **29:**1–8.

Brown, C. L., Hill, M. J., Richards, P. 1971. Bacterial ureases in uremic men. Lancet **ii:**406–408.

Brown, G. W. 1952. Anaerogenic paracolon bacilli associated with gastroenteritis in children. Medical Journal of Australia **2:**658–664.

Burke, J. P., Ingall, D., Klein, J. O., Gezon, H. M., Finland, M. 1971. *Proteus mirabilis* infections in a hospital nursery traced to a human carrier. New England Journal of Medicine **284:**115–121.

Buttiaux, R., Frenoy, R., Moriamez, J. 1954. Les caractères biochemiques du genre *Providencia.* Annales de l'Institut Pasteur de Lille **6:**62–79.

Buttiaux, R., Osteux, R., Fresnoy, R., Moriamez, J. 1954. Les propriétés biochemiques caractéristiques du genre *Proteus.* Inclusion souhaitable des *Providencia* dans celui-ci. Annales de l'Institut Pasteur **87:**375–386.

Cantu, C. 1911. Le Bacillus *Proteus* sa distribution dans la nature. Annales de l'Institut Pasteur **25:**852–864.

Carpenter, K. P. 1964. The Proteus-Providence group, pp. 13–24. In: Dyke, S. D. (ed.), Recent advances in clinical pathology, series IV. Boston: Little, Brown and Co.

Castellani, A., Chalmers, A. J. 1919. Manual of tropical medicine, 3rd ed., pp. 938–939. London: Bailliere, Tindall and Cox.

Castellani, A., Douglas, M. 1932. Serological studies on Bacillus Morgan No. 1. (*Salmonella Morgani,* Cast. and Chalm.). Journal of Tropical Medicine and Hygiene **35:**161–164.

Catsaras, M., Antoniewski, J., Buttiaux, R. 1965. Sur la production de colonies a centre orange par *Proteus rettgeri* et *Providencia* sur la gelose au desoxycholate-citrate-lactose. Annales de l'Institute Pasteur de Lille **16:**99–101.

Chiu, V. S. W., Hoeprich, P. D. 1961. Susceptibility of *Proteus* and Providence bacilli to 10 antimicrobial agents. American Journal of the Medical Sciences **241:**309–321.

Christensen, W. B. 1946. Urea decomposition as a means of differentiating *Proteus* and *Paracolon* cultures from each other and from *Salmonella* and *Shigella* types. Journal of Bacteriology **52:**461–466.

Coetzee, J. N. 1963. Lysogeny in *Proteus rettgeri* and the host-range of *P. rettgeri* and *P. hauseri* bacteriophages. Journal of General Microbiology **31:**219–229.

Coetzee, J. N. 1967. Bacteriocinogeny in strains of Providence and *Proteus morganii.* Nature **213:**614–616.

Coetzee, J. N. 1972. Genetics of the *Proteus* group. Annual Review of Microbiology **26:**23–54.

Coetzee, J. N., De Klerk, H. C. 1964. Effect of temperature on flagellation, motility and swarming of *Proteus*. Nature **202:**211–212.

Coetzee, J. N., Sachs, T. G. 1960. Morphological variants of *Proteus hauseri* Journal of General Microbiology **23:**209–216.

Cook, G. T. 1948. Urease and other biochemical reactions of the *Proteus* group. Journal of Pathology and Bacteriology **60:**171–181.

Cope, E. J., Kasper, J. A. 1949. A serological study of *Proteus rettgeri* and similar organisms. Journal of Bacteriology **57:**259–264.

Cope, E. J., Kilander, J. A. 1942. Study of atypical enteric organisms of the *Shigella* group. American Journal of Public Health **32:**352–354.

Cosenza, B. J., Podgwaite, J. D. 1966. A new species of *Proteus* isolated from larvae of the gypsy moth *Prothetria dispar* (L). Antonie van Leeuwenhoek Journal of Microbiology and Serology **32:**187–191.

Cowan, S. T. 1956. Taxonomic rank of *Enterobacteriaceae* 'Groups'. Journal of General Microbiology **15:**345–358.

Cowan, S. T. 1974. Cowan and Steel's manual for the identification of medical bacteria, 2nd ed., pp. 137–180. Cambridge: Cambridge University Press.

Craddock, M. E., Traub, W. H. 1971. Bacteriocinogeny in *Proteus rettgeri*. Experientia **27:**980.

Cradock-Watson, J. E. 1965. The production of bacteriocines by *Proteus* species. Zentralblatt für Bakteriologie, Parasitenkunde, Infectionskrankheiten und Hygiene, Abt. 1 Orig. **196:**385–388.

Curreri, W. P., Bruck, H. M., Lindberg, R. B., Mason, A. D., Pruitt, B. A. 1973. *Providencia stuartii* sepsis: A new challenge in the treatment of thermal injury. Annals of Surgery **177:**133–138.

de Louvois, J. 1969. Serotyping and the Dienes reaction on *Proteus mirabilis* from hospital infections. Journal of Clinical Pathology **22:**263–268.

Dienes, L. 1946. Reproductive processes in *Proteus* cultures. Proceedings of the Society for Experimental Biology and Medicine **63:**265–270.

Dhawan, V., Marso, E., Martin, W. J., Young, L. S. 1977. In vitro studies with netilmicin compared with amikacin, gentamicin and tobramycin. Antimicrobial Agents and Chemotherapy **11:**64–73.

Dimitracopoulos, G., Tzannetis, S., Papavassiliou, J. 1972. Bacteriocinogeny in *Proteus* Species. Zentralblatt für Bakteriologie, Parasitenkunde, Infektionskrankheiten und Hygiene. Abt. 1 Orig. **222:**227–231.

Dobrey, R. 1971. A hospital study of the Providence group with particular reference to subgrouping and the possible presence of R factors. Canadian Journal of Medical Technology. **33:**177–187.

Drasar, F. A., Farrell, W., Maskell, J., Williams, J. D. 1976. Tobramycin, amikacin, sissomicin and gentamicin resistant Gram-negative rods. British Medical Journal **2:**1284–1287.

Drasar, B. S., Hill, M. J. 1974. Human intestinal flora. New York: Academic Press.

Dutton, A. A. C., Ralston, M. 1957. Urinary tract infection in a male urological ward with special reference to the mode of infection. Lancet **I:**115–119.

Edwards, L. D., Cross, A., Levin, S., Landau, W. 1974. Outbreak of a nosocomial infection with a strain of *Proteus rettgeri* resistant to many antimicrobials. American Journal of Clinical Pathology **61:**41–46.

Edwards, P. R., Ewing, W. H. 1972. Identification of Enterobacteriaceae. 3rd ed, pp. 337–356. Minneapolis: Burgess.

Eickhoff, T. C., Ehret, J. M. 1977. In vitro activity of netilmicin compared with gentamicin, tobramycin, amikacin and kanamycin. Antimicrobial Agents and Chemotherapy **11:**791–796.

Ekladius, L., King, H. K., Sutton, G. R. 1957. Decarboxylation of neutral amino acids in *Proteus vulgaris*. Journal of General Microbiology **17:**602–619.

Elek, S. D. 1948. Rapid identification of *Proteus*. Journal of Pathology and Bacteriology **60:**183–192.

Eudy, W. W., Burrous, S. E., Sigler, F. W. 1971. Renal lysozyme levels in animals developing "sterile pyelonephritis". Infection and Immunity **4:**269–273.

Evans, A. T. 1974. Nosocomial infections and the urologist. Journal of Urology **111:**813–816.

Ewing, W. H. 1958. The nomenclature and taxonomy of the Proteus and Providence groups. International Bulletin of Bacteriological Nomenclature and Taxonomy **8:**17–22.

Ewing, W. H. 1962. The tribe *Proteeae:*Its nomenclature and taxonomy. International Bulletin of Bacteriological Nomenclature and Taxonomy **12:**93–102.

Ewing, W. H., Davis, B. R., Reavis, R. W. 1957. Phenylalanine and malonate media and their use in enteric bacteriology. Public Health Laboratory **15:**153–161.

Ewing, W. H., Davis, B. R., 1972. Biochemical characterization of the species of genus *Proteus*. Public Health Laboratory **30:**46–57.

Ewing, W. H., Davis, B. R., Sikes, J. V. 1972. Biochemical characterization of *Providencia*. Public Health Laboratory **30:**25–38.

Ewing, W. H., Martin, W. J. 1974. *Enterobacteriaceae*, pp. 189–221. In: Lennette, E. H., Spaulding, E. H., Truant, J. P. (eds.), Manual of clinical microbiology, 2nd. ed. Washington, D.C.: American Society for Microbiology.

Ewing, W. H., Tanner, K. E., Dennard, D. A. 1954. The Providence group: An intermediate group of enteric bacteria. Journal of Infectious Diseases **94:**134–140.

Eykyn, S. 1971. Use and control of cephalosporins. Journal of Clinical Pathology **24:**419–429.

Falkow, S., Ryman, I. R., Washington, O. 1962. Deoxyribonucleic acid base composition of *Proteus* and Providence organisms. Journal of Bacteriology **83:**1318–1321.

Falkow, S., Wohlieter, J. A., Citarella, R. V., Baron, L. S. 1964. Transfer of episomal elements of *Proteus*. II. Nature of lac+ *Proteus* strains isolated from clinical specimens. Journal of Bacteriology **88:**1598–1601.

Fields, B. N., Uwaydah, M. M., Kunz, L. J., Swartz, M. N. 1967. The so-called "Paracolon" bacteria. American Journal of Medicine **42:**89–106.

Fildes, P. 1938. The growth of *Proteus* on ammonium lactate plus nicotinic acid. British Journal of Experimental Pathology **19:**239–244.

France, D. R., Markham, N. P. 1968. Epidemiological aspects of *Proteus* infections with particular reference to phage typing. Journal of Clinical Pathology **21:**97–102.

Fulton, M. 1943. The identity of *Bacterium columbensis* Castellani. Journal of Bacteriology **46:**79–82.

Fulton, M., Curtis, S. F. 1946. A study of Rettger's *Bacillus* and related bacteria. Proceedings of the Society for Experimental Biology and Medicine **61:**334–338.

Galton, M. M., Hess, M. E., Collins, P. 1947. The isolation and distribution in Florida of an anaerogenic paracolon, Type 29911. Journal of Bacteriology **53:**649–651.

Greenwood, D., O'Grady, F. 1969. A comparison of the effects of ampicillin on *Escherichia coli* and *Proteus mirabilis*. Journal of Medical Microbiology **2:**435–441.

Griffith, D. P., Musher, D. M., Campbell, J. W. 1973. Inhibition of bacterial urease. Investigative Urology **11:**234–238.

Grossberg, S. E., Petersdorf, R. G., Curtin, J. A., Bennett, I. L. 1962. Factors influencing the species and antimicrobial resistance of urinary pathogens. American Journal of Medicine **32:**44–55.

Guo, M. M. S., Liu, P. V. 1965. Serological specificities of ureases of *Proteus* species. Journal of General Microbiology. **38:**417–422.

Hadley, P., Elkins, M. W., Caldwell, D. W. 1918. The colontyphoid intermediates as causative agents of disease in birds: I. The paratyphoid bacteria. Agricultural Experimental Station of the Rhode Island State College. Bulletin No. 174.

Hamilton-Miller, J. M. T., Reynolds, A. V., Brumfitt, W. 1974.

Apparent emergence of gentamicin-resistant *Providencia stuartii* during therapy with gentamicin. Lancet **ii**:527.

Hauser, G. 1885. Uber Faulnissbacterien. Leipzig: Verlag von F. C. W. Fogel.

Henriksen, S. D. 1950. A comparison of the phenylpyruvic acid reaction and the urease test in the differentiation of *Proteus* from other enteric organisms. Journal of Bacteriology **60**:225–231.

Henriksen, S. D., Closs, K. 1938. The production of phenylpyruvic acid by bacteria. Acta Pathologica et Microbiologica Scandinavica **15**:101–113.

Hewitt, C. B., Overholt, E. L., Finder, R. J., Patton, J. F. 1965. Gram-negative septicemica in urology. Journal of Urology **93**:299–302.

Hickman, F. W., Farmer III, J. J. 1976. Differentiation of *Proteus mirabilis* by bacteriophage typing and the Dienes reaction. Journal of Clinical Microbiology **3**:350–353.

Hodgin, U. G., Sanford, J. P. 1965. Gram-negative rod bacteremia. An analysis of 100 patients. American Journal of Medicine **39**:952–960.

Hook, E. W., Petersdorf, R. G. 1960. In vitro and in vivo susceptibility of *Proteus* species to the action of certain antimicrobial drugs. Johns Hopkins Medical Bulletin **107**: 337–348.

Huang, C. T. 1966. Multitest media for rapid identification of *Proteus* species with notes on biochemical reactions of strains isolated from urine and pus. Journal of Clinical Pathology **19**:438–442.

Huang, C. T., Chou, G. 1968. Drug sensitivity of *Proteus* species. Journal of Clinical Pathology **21**:103–106.

Hynes, M. 1942. The isolation of intestinal pathogens by selective media. Journal of Pathology and Bacteriology **54**:193–207.

Iannini, P. B., Eickhoff, T. C., La Force, F. M. 1976. Multidrug resistant *Proteus rettgeri:* An emerging problem. Annals of Internal Medicine **85**:161–164.

Izdebska-Szymona, K., Monczak, E., Lemczak, B. 1971. Preliminary scheme of phage typing *Proteus mirabilis* strains. Experimental Medicine and Microbiology **23**:18–22. [Originally published in Polish.]

Jameson, J. E., Emberley, N. W. 1956. A substitute for bile salts in culture media. Journal of General Microbiology **15**:198–204.

Janis, B., Evans, R. G., Hoeprich, P. D. 1968. Providence bacillus bacteremia and septicopyemia. American Journal of Medicine **45**:943–947.

Jepson, O. B., Korner, B. 1975. Bacteremia in a general hospital. A prospective study of 102 consecutive cases. Scandinavian Journal of Infectious Diseases **7**:179–184.

Johnson, R., Rogosa, M., Krichevsky, M. I. 1976. Abbreviations of names of genera suggested for coding microbiological data. International Journal of Systematic Bacteriology **26**:278–282.

Jordon, E. O., Crawford, R. R., McBroom, J. 1935. The Morgan bacillus. Journal of Bacteriology **22**:131–148.

Kauffmann, F. 1951. "*Enterobacteriaceae*" Copenhagen: Munksgaard.

Kauffmann, F. 1953. On the classification and nomenclature of *Enterobacteriaceae*. Rivista dell'Instituto Sieroterapico Italiano **28**:485–491.

Kauffmann, F. 1956a. On biochemical investigations of *Enterobacteriaceae*. Acta Pathologica et Microbiologica Scandinavica **39**:85–93.

Kauffmann, F. 1956b. A simplified biochemical table of *Enterobacteriaceae*. Acta Pathologica et Microbiologica Scandinavica **39**:103–106.

Kauffmann, F. 1966. The Bacteriology of *Enterobacteriaceae*, pp. 333–360. Baltimore: Williams & Wilkins.

Kauffmann, F., Edwards, P. R. 1952. Classification and nomenclature of *Enterobacteriaceae*. International Bulletin of Bacteriological Nomenclature and Taxonomy **2**:2–8.

Kauffmann, F., Perch, B. 1947. On the occurrence of *Proteus* X

strains in Denmark. Acta Pathologica et Microbiologica Scandinavica **24**:135–149.

Keane, C. T., English, L. F., Wise, R. 1975. *Providencia stuartii* infections Lancet **ii**:1045.

Keating, S. V. 1956. A biochemical and serological study of the genus *Proteus*. Medical Journal of Australia **2**:160–172.

King, H. K. 1953. The decarboxylation of valine and leucine by washed suspensions of *Proteus vulgaris*. Biochemical Journal **54**:xi.

Kippax, P. W. 1957. A study of *Proteus* infections in a male urological ward. Journal of Clinical Pathology **10**:211–213.

Klastersky, J., Bogaerts, A.-M., Noterman, J., Van Laer, E., Daneau, D., Mouawad, E. 1974. Infections caused by Providence bacilli. Scandinavian Journal of Infectious Diseases **6**:153–160.

Koch, M. L. 1956. Bacteremia due to bacterial species of the genera *Aerobacter, Escherichia, Paracolobactrum, Proteus,* and *Pseudomonas*. Antibiotic Medicine and Clinical Therapy **2**:113–121.

Krikler, M. S. 1953. The serology of *Proteus vulgaris*. Ph. D. Thesis. University of London, p. 37.

Lanyi, B. 1956. Serological typing of *Proteus* strains from infantile enteritis and other sources. Acta Microbiologica Academiae Scientiarum Hungaricae **3**:417–428.

Lanyi, B. 1957. Serological typing of *Proteus* strains. Sensitivity of serotypes to antibiotics. Acta Microbiologica Academiae Scientiarum Hungaricae **4**:447–457.

Lautrop, H. 1974. Genus X *Proteus*, pp. 327–330. In: Buchanan, R. E., Gibbons, N. E. (eds.), Bergey's manual of determinative bacteriology, 8th ed. Baltimore: Williams & Wilkins.

Lessel, E. F. 1971. Status of the name *Proteus morganii* and designation of the neotype strain. International Journal of Systematic Bacteriology **21**:55–57.

Levine, M. 1942. An ecological study of *Proteus*. Journal of Bacteriology **43**:33–34.

Levine, M. G., Hoyt, R. E. 1945. *Proteus* speciation. Journal of Bacteriology **49**:523.

Levy, H. L., Ingall, D. 1967. Meningitis in neonates due to *Proteus mirabilis*. American Journal of Diseases of Children **114**:320–324.

Levy, H. L., O'Connor, J. F., Ingall, D. 1967. Neonatal osteomyelitis due to *Proteus mirabilis*. Journal of the American Medical Association **202**:130–134.

Lewis, J., Fekety, F. R. 1969. *Proteus* bacteremia. Johns Hopkins Medical Journal **124**:151–156.

Li, K., Miller, C. 1970. Pathogenic bacteria and their sensitivity patterns in a hospital population of geriatric patients with chronic disease. Journal of the American Geriatrics Society **18**:286–294.

Librach, I. M. 1968. *Proteus* meningitis. Developmental Medicine and Child Neurology **10**:392–394.

Lindsey, J. O., Martin, W. T., Sonnenwirth, A. C., Bennet, J. V. 1976. An outbreak of nosocomial *Proteus rettgeri* urinary tract infection. American Journal of Epidemiology **103**:261–269.

McCabe, W. R., Jackson, G. G. 1962. Gram-negative bacteremia. 1. Etiology and Ecology. Archives of Internal Medicine **110**:847–855.

Mackey, J. P., Sandys, G. H. 1966. Diagnosis of urinary infections. British Medical Journal **i**:1173.

MacLaren, D. M. 1968. The significance of urease in *Proteus* pyelonephritis: A bacteriological study. Journal of Pathology and Bacteriology **96**:45–56.

MacLaren, D. M. 1974. The influence of acetohydroxamic acid on experimental *Proteus* pyelonephritis. Investigative Urology **12**:146–149.

McMillan, S. A. 1972. Bacteriuria of elderly women in hospitals. Occurrence and drug resistance. Lancet **ii**:452–455.

Magheru, A. 1923. Recherches expérimentales sur le Bacille de Morgan. Comptes Rendus Hepdomadaires des Séances et Memoires de la Société de Biologie **89**:643–645.

Maiztegui, J. I., Biegeleisen, J. Z. Cherry, W. B., Kass, E. H. 1965. Bacteremia due to gram-negative rods. A clinical, bacteriologic, serologic and immunofluorescent study. New England Journal of Medicine 272:222–229.

Maki, D. G., Hennekens, C. H., Bennet, J. V. 1972. Prevention of catheter-associated urinary tract infection. Journal of the American Medical Association 221:1270–1271.

Malinowski, F. 1966. A primary isolation medium for the differentiation of genus Proteus from other nonlactose and lactose fermenters. Canadian Journal of Medical Technology 28:118–121.

Meyers, B. R., Berson, B. L., Gilbert, M., Hirschman, S. Z. 1973. Clinical patterns of osteomyelitis due to gram-negative bacteria. Archives of Internal Medicine 131:228–233.

Middleton, J. E. 1958. The sensitivity in vitro of the Providence group of enteric bacteria to 14 antibiotics and nitrofurantoin. Journal of Clinical Pathology 11:270–272.

Milner, P. F. 1963. The differentiation of Enterobacteriaceae infecting the urinary tract. A study in male paraplegics. Journal of Clinical Pathology 16:39–45.

Milstoc, M., Steinberg, P. 1973. Fatal septicemia due to Providence group bacilli. Journal of the American Geriatrics Society 21:159–163.

Moellering, R. C., Swartz, M. N. 1976. The newer cephalosporins. New England Journal of Medicine 294:24–28.

Moltke, O. 1927. Contributions to the characterization and systematic classification of the Bac. proteus vulgaris (Hauser). Copenhagen: Levin and Munksgaard.

Monto, A. S., Rantz, L. A. 1965. Classification of so-called paracolon bacilli isolated from urinary tract infection. Journal of Laboratory and Clinical Medicine 65:64–70.

Morgan, H. de R. 1906. Upon the bacteriology of the summer diarrhoea of infants. British Medical Journal i:908–912.

Müller, H. E. 1972. The aerobic fecal flora of reptiles with special reference to the Enterobacteria of snakes. Zentralblatt für Bakteriologie, Parasitenkunde, Infektionskrankheiten und Hygiene, Abt. 1 Orig. 222:487–495.

Musher, D. M., Griffith, D. P., Yawn, D., Rossen, R. D. 1975. Role of urease in pyelonephritis resulting from urinary tract infection with Proteus. Journal of Infectious Diseases 131:177–181.

Namioka, S., Sakazaki, R. 1958. Étude sur les Rettgerella. Annales de l'Institute Pasteur 94:485–499.

Namioka, S., Sakazaki, R. 1959. New K antigen (C antigen) possessed by Proteus and Rettgerella cultures. Journal of Bacteriology 78:301–306.

Neter, E. 1942. The genus Shigella (dysentery bacilli and allied species). Bacteriological Reviews 6:1–36.

Noall, E. W. P., Sewards, H. F. G., Waterworth, P. M. 1962. Successful treatment of a case of Proteus septicemia. British Medical Journal 2:110–111.

Omland, T. 1960. Nosocomial urinary tract infections caused by Proteus rettgeri. Acta Pathologica et Microbiologica Scandinavica 48:221–230.

Ornstein, M. 1921. Zur Bakteriologie des Schmitzbazillus. Zeitschrift für Hygiene und Infektionskrankheiten 91:152–178.

Overturf, G. D., Wilkins, J., Ressler, R. 1974. Emergence of resistance of Providencia stuartii to multiple antibiotics: Speciation and biochemical characterization of Providencia. Journal of Infectious Diseases 129:353–357.

Pavlatou, M., Hassikou-Kaklamani, E., Žantioti, M. 1965. Lysotypie du genre Proteus. Annales de l'Institut Pasteur 108:402–407.

Pelczar, M. J., Porter, J. R. 1940. Pantothenic acid and nicotinic acid as essential growth substances for Morgan's bacillus. (Proteus morganii). Proceedings of the Society of Experimental Biology and Medicine 43:151–154.

Penner, J. L., Hennessy, J. N. 1977. Reassignment of the intermediate strains of Proteus rettgeri biovar 5 to Providencia stuartii on the basis of the somatic (O) antigens. International Journal of Systematic Bacteriology 27:71–74.

Penner, J. L., Hennessy, J. N. 1979a. O antigen grouping of Morganella morganii (Proteus morganii) by slide agglutination. Journal of Clinical Microbiology 10:8–13.

Penner, J. L., Hennessy, J. N. 1979b. Application of O-serotyping in a study of Providencia rettgeri (Proteus rettgeri) isolated from human and non-human sources. Journal of Clinical Microbiology 10:834–840.

Penner, J. L., Hinton, N. A. 1973. A study of the serotyping of Proteus rettgeri. Canadian Journal of Microbiology 19:271–279.

Penner, J. L., Hinton, N. A., Hennessy, J. 1974. Serotyping of Proteus rettgeri on the basis of O antigens. Canadian Journal of Microbiology 20:777–789.

Penner, J. L., Hinton, N. A., Hennessy, J. 1975. Biotypes of Proteus rettgeri. Journal of Clinical Microbiology 1:136–142.

Penner, J. L., Hinton, N. A., Hennessy, J. N. 1976. Evaluation of a Proteus rettgeri O-serotyping system for epidemiological investigation. Journal of Clinical Microbiology 3:385–389.

Penner, J. L., Hinton, N. A., Hennessy, J. N., Whiteley, G. R. 1976a. Reconstitution of the somatic (O-) antigenic scheme for Providencia and preparation of O-typing antisera. Journal of Infectious Diseases 133:283–292.

Penner, J. L., Hinton, N. A., Whiteley, G. R., Hennessy, J. N. 1976b. Variation in urease activity of endemic hospital strains of Proteus rettgeri and Providencia stuartii. Journal of Infectious Diseases 134:370–376.

Penner, J. L., Hinton, N. A., Duncan, I. B. R., Hennessy, J. N., Whiteley, G. R. 1979a. O-serotyping of Providencia stuartii isolates collected from twelve hospitals. Journal of Clinical Microbiology 9:11–14.

Penner, J. L., Fleming, P. C., Whiteley, G. R., Hennessy, J. N. 1979b. A study on O-serotyping Providencia alcalifaciens. Journal of Clinical Microbiology 10:761–765.

Perch, B. 1948. On the serology of the Proteus group. Acta Pathologica et Microbiologica Scandinavica 25:703–714.

Phillips, D. I. 1966. A stab medium for the rapid tentative identification of Proteus species. Journal of Medical Laboratory Technology 23:103–104.

Phillips, J. E. 1955a. In vitro studies of Proteus organisms of animal origin. Journal of Hygiene 53:26–31.

Phillips, J. E. 1955b. The experimental pathogenicity in mice of strains of Proteus of animal origin. Journal of Hygiene 53:212–216.

Pichinoty, F., Piéchaud, M. 1968. Recherche des nitrate-réductases bacteriennes A et B: Methodes. Annales de l'Institut Pasteur 114:77–98.

Pichinoty, F., Rigano, C., Bigliardi-Rouvier, J., Le Minor, L., Piéchaud, M. 1966. Recherche des nitrate-réductases A et B chez les Enterobacteriaceae. Annales de l'Institut Pasteur 110:126–130.

Polster, M., Svobodova, M. 1964. Production of reddish-brown pigment from DL-tryptophan by Enterobacteria of the Proteus-Providencia Group. Experientia 20:637–638.

Potee, K. G., Wright, S. S., Finland, M. 1954. In vitro susceptibility of recently isolated strains of Proteus to ten antibiotics. Journal of Laboratory and Clinical Medicine 44:463–477.

Proom, H. 1955. Amine production and nutrition in the Providence group. Journal of General Microbiology 13:170–175.

Proom, H., Woiwod, A. J. 1951. Amine production in the genus Proteus. Journal of General Microbiology 5:930–938.

Rauss, K. F. 1936. The systematic position of Morgan's bacillus. Journal of Pathology and Bacteriology 42:183–192.

Rauss, K. 1962. A proposal for the nomenclature and classification of the Proteus and Providence Groups. International Bulletin of Bacteriological Nomenclature and Taxonomy 12:53–64.

Rauss, K., Vörös, S. 1959. The biochemical and serological properties of Proteus morganii. Acta Microbiologica Academiae Scientiarum Hungaricae 6:233–248.

Rauss, K., Vörös, S. 1967. Five new serotypes of Morganella morganii. Acta Microbiologica Academiae Scientiarum Hungaricae 14:195–198.

Rauss, K., Puzova, H., Dubay, L., Velin, D., Doliak, M., Vörös, S. 1975. New serotypes of *Morganella morganii*. Acta Microbiologica Academiae Scientiarum Hungaricae **22:**315–321.

Report. 1958. Enterobacteriaceae Subcommittee of the Nomenclature Committee of the International Association of Microbiological Societies. *Enterobacteriaceae.* International Bulletin of Bacteriological Nomenclature and Taxonomy **8:** 25–70.

Report. 1977. Epidemiological Research Laboratory of the Public Health Laboratory Service: Unusual Infections in Intensive Care Unit. British Medical Journal **i:**111.

Richard, C. 1966. Caractères biochemiques des biotypes de *Providencia:* Leurs rapports avec le genre *Rettgerella.* Annales de l'Institut Pasteur **110:**105–114.

Richard, C., Popoff, M., Pastor, G. P. 1974. Étude bactériologique d'infections urinaires intrahospitalières a *Proteus rettgeri* fermentant le lactose. Annales de Biologie Clinique **32:**149–154.

Ridge, L. E. L., Thomas, M. E. M. 1955. Infection with the Providence type of paracolon bacillus in a residential nursery. Journal of Pathology and Bacteriology **69:**335–337.

Roland, F., Bourbon, D., Szturm, S. 1947. Différenciation rapide des entero-bactériacées sans action sur le lactose. Annales de l'Institut Pasteur **73:**914–916.

Rustigian, R., Stuart, C. A. 1941. Decomposition of urea by *Proteus.* Proceedings of the Society for Experimental Biology and Medicine **47:**108–112. (1941)

Rustigian, R., Stuart, C. A. 1943a. The biochemical and serological relationships of the organisms of the genus *Proteus.* Journal of Bacteriology **45:**198–199.

Rustigian, R., Stuart, C. A. 1943b. Taxonomic relationships in the genus *Proteus.* Proceedings of the Society for Experimental Biology and Medicine **53:**241–243.

Rustigian, R., Stuart, C. A. 1945. The biochemical and serological relationships of the organisms of the genus *Proteus.* Journal of Bacteriology **49:**419–436.

Sabbaj, J., Sutter, V. L., Finegold, S. M. 1970. Urease and deaminase activities of fecal bacteria in hepatic coma, pp. 181–185. In: Antimicrobial Agents and Chemotherapy—1970. Washington, D. C.: American Society for Microbiology.

St. John-Brooks, R., Rhodes, M. 1923. The organisms of the fowl typhoid group. Journal of Pathology and Bacteriology **26:**433–439.

Schaberg, D. R., Weinstein, R. A., Stamm, W. E. 1976. Epidemics of nosocomial urinary tract infection caused by multiply resistant gram-negative bacilli: Epidemiology and control. Journal of Infectious Diseases **133:**363–366.

Schmidt, W. C., Jeffries, C. D. 1974. Bacteriophage typing of *Proteus mirabilis, Proteus vulgaris* and *Proteus morganii.* Applied Microbiology **27:**47–53.

Schwarz, H. Schirmer, H. K. A., Ehlers, B., Post, B. 1969. Urinary tract infections: Correlation between organisms obtained simultaneously from the urine and feces of patients wtih bacteriuria and pyuria. Journal of Urology **101:**765–767.

Scott, T. G. 1960. The bacteriology of urinary infections in paraplegia. Journal of Clinical Pathology **13:**54–56.

Senior, B. W. 1977a. The Dienes phenomenon: Identification of the determinants of compatibility. Journal of General Microbiology **102:**235–244.

Senior, B. W. 1977b. Typing of *Proteus* strains by proticine production and sensitivity. Journal of Medical Microbiology **10:**7–17.

Sevin, A., Buttiaux, R. 1939. The characters and systematic position of Morgan's bacillus. Journal of Pathology and Bacteriology **49:**457–466.

Shafi, M. S., Datta, N. 1975. Infection caused by *Proteus mirabilis* strains with transferable gentamicin-resistance factors. Lancet **i:**1355–1357.

Shaw, C., Clarke, P. H. 1955. Biochemical classification of *Proteus* and Providence cultures. Journal of General Microbiology **13:**155–161.

Sheth, N. K., Kurup, V. P. 1975. Evaluation of tyrosine medium for the identification of *Enterobacteriaceae.* Journal of Clinical Microbiology **1:**483–485.

Shimizu, S., Iyobe, S., Mitsuhashi, S. 1977. Inducible high resistance to colistin in *Proteus* strains. Antimicrobial Agents and Chemotherapy **12:**1–3.

Shortland-Webb, W. R. 1968. *Proteus* and coliform meningoencephalitis in neonates. Journal of Clinical Pathology **21:**422–431.

Singer, J. 1950a. Culture of *Enterobacteriaceae.* I. A practical medium containing urea, tryptone, lactose and indicators. American Journal of Clinical Pathology **20:**880–883.

Singer, J. 1950b. Culture of *Enterobacteriaceae.* II. Use of urea triple-sugar agar. American Journal of Clinical Pathology **20:**884–885.

Singer, J., Bar-Chay, J. 1954. Biochemical investigation of Providence strains and their relationship to the *Proteus* Group. Journal of Hygiene **52:**1–8.

Singer, J., Volcani, B. E. 1955. An improved ferric chloride test for differentiating *Proteus*-Providence group from other *Enterobacteriaceae.* Journal of Bacteriology **69:**303–306.

Skirrow, M. B. 1969. The Dienes (mutual inhibition) test in the investigation of *Proteus* infections. Journal of Medical Microbiology **2:**471–477.

Smit, J. A., Coetzee, J. N. 1967. Serological specificities of phenylalanine deaminases of the Proteus-Providence group. Nature **214:**1238–1239.

Søgaard, H., Zimmermann-Nielsen, C., Siboni, K. 1974. Antibiotic-resistant gram-negative bacilli in a urological ward for male patients during a nine-year period: Relationship to antibiotic consumption. Journal of Infectious Diseases **130:**646–650.

Solberg, C. O., Matsen, J. M. 1971. Infections with Providence bacilli. A clinical and bacteriologic study. American Journal of Medicine **50:**241–246.

Sompolinsky, D. 1957. Relations antigéniques entra *Proteus morganii* et *Shigella flexneri.* Annales de l'Institut Pasteur **92:**343–349.

Šourek, J. 1968. On some findings concerning Diene's phenomenon in swarming *Proteus* strains. Zentralblatt für Bakteriologie, Parasitenkunde, Infektionskrankheiten und Hygiene, Abt. 1 Orig. **208:**419–427.

Stenzel, W. 1961. *Proteus inconstans* 013 H30 als Enteritiserreger bei Kleinkindern. Zentralblatt für Bakteriologie, Parasitenkunde, Infektionskrankheiten und Hygiene, Abt. 1 Orig. **182:**178–183.

Strickler, D. J., Thomas, B. 1976. Sensitivity of Providence to antiseptics and disinfectants. Journal of Clinical Pathology **29:**815–823.

Story, P. 1954. Proteus infections in hospital. Journal of Pathology and Bacteriology **68:**55–62.

Stuart, C. A., Van Stratum, E., Rustigian, R. 1945. Further studies on urease production by *Proteus* and related organisms. Journal of Bacteriology **49:**437–444.

Stuart, C. A., Wheeler, K. M., McGann, V. 1946. Further studies on one anaerogenic paracolon organism, type 29911. Journal of Bacteriology **52:**431–438.

Stuart, C. A., Wheeler, K. M., Rustigian, R., Zimmerman, A. 1943. Biochemical and antigenic relationships of the paracolon bacteria. Journal of Bacteriology **45:**101–109.

Stuart, C. A., Wheeler, K. M., McGann, V., Howard, I. 1946. Motility and swarming of some *Enterobacteriaceae.* Journal of Bacteriology **52:**519–525.

Stumpf, P. K., Green, D. E. 1944. L-Amino acid oxidase of *Proteus vulgaris.* Journal of Biological Chemistry **153:**387–399.

Sturdza, S. A. 1963. Beobachtungen über die mannigfaltigkeit der zur Proteusgruppe gehörenden Bakterienstamme. Zentralblatt für Bakteriologie, Parasitenkunde, Infektionskrankheiten und Hygiene, Abt. 1 Orig. **188:**530–536.

Sturdza, S. A. 1965. Über die heterogene beschaffenheit fäkaler *Proteus*-populationen. Zentralblatt für Bakteriologie, Parasitenkunde, Infektionskrankheiten und Hygiene, Abt. 1 Orig. **195**:489–500.

Suassuna, I., Suassuna, I. R. 1960. Sôbre a posiçãe sistemática de *Proteus americanus* Pacheco, *Proteus paraamericanus* Magalhães and Aragão E *Eberthella alcalifaciens* de Salles Gomes. Anals de Microbiologia **8**:161–168.

Suassuna, I., Suassuna, I. R., Ewing, W. H. 1961. The methyl red and Voges-Proskauer reactions of *Enterobacteriaceae*. Public Health Laboratory **19**:67–77.

Sullivan, N. M., Sutter, V. L., Mims, M. M., Marsh, V. H., Finegold, S. M. 1973. Clinical aspects of bacteremia after manipulation of the genitourinary tract. Journal of Infectious Diseases **127**:49–55.

Suter, L. S., Ulrich, E. W., Koelz, B. S., Street, V. W. 1968. Metabolic variations of *Proteus* in the Memphis area and other geographical areas. Applied Microbiology **16**:881–889.

Sutter, V. L., Foecking, F. J. 1962. Biochemical characteristics of lactose-fermenting *Proteus rettgeri* from clinical specimens. Journal of Bacteriology **83**:933–935.

Taubeneck, V. 1962. Beobachtungen an lysogenen *Proteus*-stammen. Zentralblatt für Bakteriologie, Parasitenkunde, Infektionskrankheiten und Hygiene, Abt. 1 Orig. **185**:416–418.

Taylor, J. F. 1928. *B. proteus* infections. Journal of Pathology and Bacteriology **31**:897–915.

Thibault, P., Le Minor, L. 1957. Méthods simples de recherche de la lysine-décarboxylase et de la tryptophane-desaminase a l'aide des milieux pour différenciation rapide des *Enterobacteriacees*. Annales de l'Institut Pasteur **92**:551–554.

Thjøtta, T. 1920. On the bacillus of Morgan No. 1—a metacolonbacillus. Journal of Bacteriology **5**:67–77.

Tracy, O., Thomson, E. J. 1972. An evaluation of three methods of typing organisms of the genus *Proteus*. Journal of Clinical Pathology **25**:69–72.

Traub, W. H., Craddock, M. E., Raymond, E. A., Fox, M., McCall, C. E. 1971. Characterization of an unusual strain of *Proteus rettgeri* associated with an outbreak of nosocomial urinary tract infection. Applied Microbiology **22**:278–283.

Tribondeau, L., Fichet, M. 1916. Note sur les dysenteries des Dardanelles. Annales de l'Institut Pasteur **30**:357–362.

Ursing, J. 1974. Biochemical study of *Proteus inconstans* (*Providencia*). Occurrence of urease positive strains. Acta Pathologica Microbiologica Scandinavica **82**:527–532.

Vassiliadis, P., Politi, G. 1968. Milieu combiné pour la recherche de l'urease et la transformation de la l-phénylalanine en acide phenylpyruvique. Annales de l'Institut Pasteur **114**:431–435.

Vera, H. D., Dumoff, M. 1974. Culture media, pp. 881–929. In: Lennette, E. H., Spaulding, E. H., Truant, J. P. (eds.), Manual of Clinical Microbiology, 2nd ed. Washington, D.C.: American Society for Microbiology.

Vieu, J. F. 1958. Note préliminaire sur la lysotypie de *Proteus hauseri* Zentralblatt für Bakteriologie, Parasitenkunde, Infektionskrankheiten und Hygiene, Abt. 1 Orig. **171**:612–615.

Vieu, J. F. 1963. Distribution de la lysogenie parmi les *Proteus* et les *Providencia*. Comptes Rendus Hebdomadaires des Séances de l'Academie des Sciences **256**:4317–4319.

Vieu, J. F., Capponi, M. 1965. Lysotypie des *Proteus* OX19, OXK, OX2 et OXL. Annales de l'Institut Pasteur **108**:103–106.

von Graevenitz, A., Spector, H. 1969. Observations on indolpositive *Proteus.* Yale Journal of Biology and Medicine **41**:434–445.

von Graevenitz, A., Nourbakhsh, M. 1972. Antimicrobial resistance of the genera *Proteus, Providencia* and *Serratia* with special reference to multiple resistance patterns. Medical Microbiology and Immunology **157**:142–148.

Vuye, A., Pijck, J. 1973. Urease activity of *Enterobacteriaceae:* Which medium to choose. Applied Microbiology **26**:850–854.

Wallace, J. J., Petersdorf, R. G. 1971. Urinary tract infections. Postgraduate Medicine **50**:138–144.

Washington, J. A., II, Senjem, D. H., Haldorson, A., Schutt, A. H., Martin, W. J. 1973. Nosocomially acquired bacteriuria due to *Proteus rettgeri* and *Providencia stuartii*. American Journal of Clinical Pathology **60**:836–838.

Weldin, J. C. 1927. The colon-typhoid group of bacteria and related forms. Relationships and classification. Iowa State College Journal of Science **I**:121–197.

Wenner, J. J., Rettger, L. F. 1919. A systematic study of the *Proteus* group of bacteria. Journal of Bacteriology **4**:331–353.

Wenzel, R. P., Hunting, K. J., Osterman, C. A., Sande, M. A. 1976. *Providencia stuartii*, a hospital pathogen: Potential factors for its emergence and transmission. American Journal of Epidemiology **104**:170–180.

Whiteley, G. R., Penner, J. L., Stewart, I. O., Stokan, P. C., Hinton, N. A. 1977. Nosocomial urinary tract infections caused by two O-serotypes of *Providencia stuartii* in one hospital. Journal of Clinical Microbiology **6**:551–554.

Williams, F. O. 1978. Nature of the swarming phenomenon in *Proteus*. Annual Reviews of Microbiology **32**:101–122.

Wilson, G. S., Miles, A. 1975. Topley and Wilson's principles of bacteriology, virology and immunity, 6th ed., pp. 887–900. London: Arnold.

Winkle, S. 1945. Zur Typendifferenzierung in der Gattung *Proteus* Hauser. Zentralblatt für Bakteriologie, Parasitenkunde und Infektionskrankheiten, Orig. **151**:494–501.

Winslow, C.-E. A., Kliger, I. J., Rothberg, W. 1919. Studies on the classification of the colon-typhoid group of bacteria with special reference to their reactions. Journal of Bacteriology **4**:429–503.

Xilinas, M. E., Papavassiliou, J. T., Legakis, N. J. 1975. Selective medium for growth of *Proteus*. Journal of Clinical Microbiology **2**:459–460.

Yale, M. W. 1939. Genus VI. *Proteus* Hauser. In: Bergey, D. H., Breed, R. S., Murray, E. G. D., Hitchens, A. P. (eds.), Bergey's manual of determinative bacteriology, 5th ed. Baltimore: Williams & Wilkins.

Yow, E. M. 1952. Development of *Proteus* and *Pseudomonas* infections during antibiotic therapy. Journal of the American Medical Association **149**:1184–1188.

Zarett, A. J., Doetsch, R. N. 1949. A new selective medium for the quantitative determination of members of the genus *Proteus* in milk. Journal of Bacteriology **57**:266.

CHAPTER 99

Yersinia enterocolitica and *Yersinia pseudotuberculosis*

EDWARD J. BOTTONE

Historically, the early destiny of mankind was bitterly fought and honed through the advent of scourges that decimated human lives. While the terror engendered by the premier member of such pestilence, the plague bacillus *Yersinia (Pasteurella) pestis,* has incipiently subsided on a worldwide basis, its downward slope has been intersected by the trajectory of interest in *Yersinia (Pasteurella) pseudotuberculosis* and *Y. ("Pasteurella* X") *enterocolitica,* two members of the genus *Yersinia* (Mollaret and Thal, 1974) that serve as the focal point for this chapter.

In 1883, Malassez and Vignal reported a disease in guinea pigs characterized by the production of a grayish-white ovoid nodules in the liver, spleen, and lungs, along with swollen mesenteric lymph nodes with caseous necrosis. The pathology was observed 6 days after the inoculation of an emulsified nodule with adjacent subcutaneous tissue derived from the forearm of a 4-year-old child who, 2 h earlier, presumably had died of tuberculous meningitis. Tubercle bacilli, however, were never demonstrated in the nodule. Continued serial transfer of pathological material from the initial guinea pig into another and so forth for six passages resulted in the repeated observation of nodular lesions at the point of inoculation and in viscera. Tubercle bacilli, however, were only recovered from a fifth-passage guinea pig that died 51 days after inoculation. Inoculum derived from the latter animal was introduced into two guinea pigs that were sacrificed after 8 and 20 days, respectively. Tubercle bacilli could only be demonstrated with the 20-day postinoculation animal. In the numerous instances in which tubercle bacilli could not be observed, Malassez and Vignal noted "des masses zoogloeique de microcoques" in the "forme ou espece de tuberculose sans bacilles". These findings in otherwise typical tuberculosis-like lesions puzzled the authors, who were critical of their own techniques and thought perhaps that the zoogloeic forms were another stage in the life cycle of the tubercle bacillus. Subsequently, Malassez and Vignal (1884) showed that the microorganisms causing "tuberculose zoogloeique" were distinct from Koch's bacillus.

Retrospectively, from that moment in 1883 to the present, the pathology described by Malassez and Vignal is consistent with later reports of epizootic tuberculosis-like disease in guinea pigs, to which Eberth (1885) ascribed the term "pseudotuberculosis"; Pfeiffer (1889) described one of the causative agents as *Bacillus pseudotuberculosis rodentium.* The question not resolved by Malassez and Vignal, and which still remains unanswered, is one of cause and effect. As *Y. pseudotuberculosis* is often present in guinea pigs, it is conceivable that the pathology noted by these investigators was a manifestation of a spontaneously arising, latent infection with *Y. pseudotuberculosis* activated by inoculation. Death after 6 days in the first-generation animals suggests an acute disease. Nevertheless, time has not diminished the enchantment of their words penned nearly a century ago and augmented by hand-crafted sketches.

The natural evolution of a microbial species is attended by a legacy of nomenclatural ascription applied to either connote a particular host range predilection or disease entity. In the case of *Y. pseudotuberculosis,* such a double entendre speckles the early microbiological-pathological history of this species. Malassez and Vignal's (1883) original description of the characteristic macroscopic pathology (pseudotuberculosis) in guinea pigs has become intimately associated both nomenclaturally and pathologically with this microbial species, although zoonotic pseudotuberculosis may be caused by morphologically, culturally, and biochemically distinct species (Wilson and Miles, 1964).

In a truly outstanding review of the first 57 years of the predominantly European experience with avian pseudotuberculosis, Beaudette (1940), beginning with Malassez and Vignal's paper, remarkably dissected and chronicled the indecision surrounding the actual causative agent of this entity. Although numerous investigators described at postmortem examination the typical macroscopic morphology of the tubercle-like nodules in liver and spleen, little uniformity existed regarding the nature of the causative agent. Variously the microorganism was described as a Gram-negative, gas-producing, cholera-like bacillus (Rieck, 1889), a Gram-positive anaerogenic bacillus (Freese, 1907), *Bacillus canariensis necrophorus* (Meissner and Schern, 1908), *B. pseudotuberculosis avium* (Lerche, 1927), and *B. parapestis* (Truche and Bauche, 1933). The latter designation to some extent reflected the confusion engendered by the biological similarity of the described pseudotuberculosis agent to the plague bacillus (Bessonova et al., 1963; Topping, Watts, and Lillie, 1938; Wayson, 1925). Even the basic disease entity underwent host-associated microbiological descriptions, i.e., canary cholera (Kern, 1896) and paracholera and parapest of turkeys (Beck and Huck, 1925; Truche and Bauche, 1933).

While the early history of *Yersinia pseudotuberculosis* evolved mainly through the manifest animal pathology, *Y. enterocolitica,* on the other hand, made its entrance as a bona fide human pathogen. In the United States in 1934, McIver and Pike recorded under the name *Flavobacterium pseudomallei* Whitmore a small Gram-negative coccobacillus isolated from a facial abscess of a 53-year-old farm dweller; the organism culturally and biochemically was ultimately recognized as *Y. enterocolitica*. The report of this manifestation was reinforced by Schleifstein and Coleman (1939), who called attention to the marked similarity between this isolate and four others and *Actinobacillus (Bacterium) lignieri* and especially *Y. (Pasteurella) pseudotuberculosis*. Because the microbiological properties were sufficiently different from the two latter species, Schleifstein and Coleman (1943) proposed the name *Bacterium enterocoliticum* for this "unidentified microorganism".

From this initial description, *Y. enterocolitica* underwent various nomenclature designations ranging from the practical *Pasteurella pseudotuberculosis* X (Hassig, Karrer, and Pusterla, 1949), "*Pasteurella pseudotuberculosis*" type b (Dickinson and Mocquot, 1961), *Pasteurella* X (Daniels and Goudzwaard, 1963; Knapp and Thal, 1963), "*Pasteurella pseudotuberculosis atypique*" (Wauters, 1970), to the provisional but highly romanticized "Les germes X" (Mollaret and Destombes, 1964).

Van Loghem (1944) was the first to suggest that the plague bacillus *(Y. pestis)* and its congener *Y. (Pasteurella) pseudotuberculosis* were distinct enough from the "hemorrhagic septicemia" group of *Pasteurella* to warrant separate generic status. He proposed the name *Yersinia* to commemorate the memory of A. J. Yersin, who first described the plague bacillus (Bibel and Chen, 1976; Mollaret, 1973). Frederiksen (1964) coupled Schleifstein and Coleman's "enterocoliticum" with *Yersinia* and introduced *Yersinia enterocolitica* into the family *Enterobacteriacae* (Mollaret and Thal, 1974; Thal, 1954). The wisdom of removing *Y. pestis, Y. pseudotuberculosis,* and *Y. enterocolitica* from the genus *Pasteurella* has been substantiated by numerical taxonomy (Smith and Thal, 1965) and DNA hybridization studies (Brenner et al., 1976; Domaradskij, Murchenkov, and Shimanjuk, 1973).

Yersinia pseudotuberculosis strains of serogroups 1 through 6 are phenotypically homogeneous and display upwards of an 80% deoxyribonucleic acid (DNA) relatedness (Brenner, 1978). *Y. enterocolitica,* however, shows great variability in phenotypic characteristics, among which rhamnose fermentation or lack of sucrose fermentation is unique. Such biochemical heterogeneity is also reflected in the genotypic diversity noted among strains. Brenner et al. (1976) have shown that *Y. enterocolitica* comprises at least four different DNA relatedness groups. The first corresponds to biochemically typical *Y. enterocolitica* (irrespective of indole production); the second encompasses strains differing from group 1 representatives only by fermenting rhamnose. Group 3 strains ferment, in addition to rhamnose, raffinose, melibiose, and α-methylglucoside, while group 4 strains are sucrose and acetylmethylcarbinol negative. DNA data indicate that while group 1 is truly representative of *Y. enterocolitica,* strains of groups 2, 3, and 4 are more closely related to yersiniae than to other members within the Enterobacteriaceae family (Brenner et al., 1976; 1977). The data of Brenner and colleagues further indicate that *Y. enterocolitica* is 40–60% related to *Y. pseudotuberculosis*.

In recent years, *Yersinia enterocolitica* – like organisms of DNA homology groups 2, 3, and 4 have been recovered more frequently from environmental and human sources (Alonso et al., 1975; Bottone and Robin, 1977; Bottone et al., 1974; Brenner et al., 1976; Hanna et al., 1976; Highsmith et al., 1977). For strains differing from typical *Y. enterocolitica* by fermenting rhamnose, melibiose, and raffinose, and utilizing sodium citrate, especially at 22°C, Brenner (1978) has proposed the term *Y. intermedia* sp. nov., a name originally suggested by B. Chester et al. (unpublished data, 1977). For isolates differing from typical *Y. enterocolitica* solely in their ability to ferment rhamnose (group 2), Brenner has proposed the name *Y. frederiksenii* in honor of W. Frederiksen. Awaiting taxonomic clarification are sucrose-negative varieties of *Y. enterocolitica*. Until such time, these latter strains are presently regarded as belonging to a sucrose-negative biogroup of *Y. enterocolitica* (Brenner et al., 1977).

It should be noted that speciation within the genus *Yersinia* has not closed with the above members. Indeed, Ewing et al. (1978) have proposed the name *Y. ruckeri* sp. nov. for the organism formerly designated the redmouth bacterium (Ross, Rucker, and Ewing, 1966), a term applied because of the inflammation produced by this species around the head and mouth of rainbow trout. Additionally, the exact taxonomic status of *Y. philomiragia* (Jensen, Owen, and Jellison, 1969), a natural and experimental pathogen of muskrats, is awaiting further biochemical and DNA relatedness studies.

Habitats

In Europe and in North America, *Yersinia pseudotuberculosis* is widespread throughout the animal kingdom, especially in mammalian and avian hosts (Topping, Watts, and Lillie, 1938). In a survey of 177 strains encountered in Great Britain, Mair (1965) showed that *Y. pseudotuberculosis* could be recovered from diverse animal sources ranging from farm animals, domestic pets, and experimental animals, particularly guinea pigs, to wild and captive animals. Among Mair's avian subjects, the turkey yielded the highest recovery rates. In North America, early investigators (Beaudette, 1940; Rosenwald and Dickinson, 1944; Wayson, 1925) and, more recently, Wetzler and Hubbert (1968b) have confirmed the prevalence of *Y. pseudotuberculosis* in rodents and fowl. Epizootics in flocks of turkeys have been accompanied by the same postmortem findings noted in guinea pigs (Blaxland, 1947). In turkeys, osteomyelitic lesions at the end of long bones have also been described (Wise and Uppal, 1972).

While early microbiological interest in *Y. pseudotuberculosis* focused upon the animal wastage induced by this species, attention to the closely related *Y. enterocolitica* was barely perceptible in scientific colloquia, even though Schleifstein and Coleman's (1939) report had appeared in the United States. Today we are aware that *Y. enterocolitica* also shares an animal habitat, and that the strains involved mainly in human infections, i.e., serotypes O:3, O:5,27, O:9, reside principally in swine vectors (Esseveld and Goudzwaard, 1973; Toma and Deidrick, 1975; Tsubokura, Otsuki, and Itagaki, 1973; Zen-Yoji et al., 1974). Other animal reservoirs of *Y. enterocolitica* literally transcend phylogenetic barriers and encompass such representatives as chinchillas (Becht, 1963; Vandepitte, Wauters, and Isebaert, 1973), rodents (Alonso and Bercovier, 1975; Kapperud, 1975), hares (Mollaret and Lucas, 1965), birds (Hubbert, 1972), cows (Inoue and Kurose, 1975), deer (Wetzler and Hubert, 1968a), frogs and snails (Botzler, Wetzler, and Cowan, 1968; Botzler et al., 1976), and brown trout (Kapperud and Johnsson, 1976). Environmental sources include water (Botzler et al., 1976; Keet, 1974; Lassen, 1972; Saari and Quan, 1975) and foods (Hanna et al., 1976; Lee, 1977; Schiemann and Toma, 1978). The majority of the nonporcine isolates have comprised a heterogeneous group of *Y. enterocolitica* and *Y. enterocolitica*–like strains displaying atypical biochemical and serological properties and infrequently associated with human

infections (Bottone, 1977). While interest in *Y. enterocolitica* has erupted vibrantly in the last decade, it must be remembered that this species was in our midst concomitant with *Y. pseudotuberculosis* (and *Y. pestis* for that matter).

Yersinia pseudotuberculosis and *Yersinia enterocolitica* are both Gram-negative, small, asporogenous coccobacilli that simultaneously share and are distinguished by facets of their geographic and host distribution, basic microbiology, and frequency of pathological manifestations in man and animals.

In its natural animal reservoir, *Y. pseudotuberculosis* is a common but underreported cause of sporadic zoonotic disease in both avian and mammalian species (Beaudette, 1940; Wayson, 1925; Wetzler and Hubbert, 1968) and is highly virulent for experimental animals. With either naturally acquired or experimentally induced infections, death ensues rapidly and visceral miliary "pseudotubercles" are apparent at necropsy.

Yersinia enterocolitica, by way of contrast, is seldom encountered among avian hosts and is not as spontaneous a zoonotic agent in nature. Instead, only isolated episodes of enzootic disease have occurred, principally in captive animals (Baggs et al., 1976; Quan et al., 1974). Experimental animal pathogenicity, while long recognized and well established for *Y. pseudotuberculosis*, has only recently served as a focal point of *Y. enterocolitica* research (Alonso et al., 1975; Baggs et al., 1976; Carter 1975; Maruyama, 1973). This latter aspect can be attributed mainly to the belief that an experimental animal model to study pathogenesis was not readily available (Mollaret and Guillon, 1965).

Retrospectively, Schleifstein and Coleman (1939) had described *Y. enterocolitica* virulence for white mice inoculated by the subcutaneous, peritoneal, and oral routes. Histologically, granulomatous lesions or nodules were consistently found in the liver and spleen of these animals and were composed of centrally situated aggregations of Gram-negative coccobacilli. Subsequent investigators have confirmed and extended these early observations (Alonso et al., 1975b; Bercovier et al., 1976; Carter, 1975; Quan et al., 1974).

Pathogenesis

While the earlier pioneering studies highlighted the similarity of macroscopic lesions in animals to those seen in human enteral infections, Une (1977a) has enriched our understanding of the microscopic events surrounding the pathogenesis of *Y. enterocolitica*. By experimental intraduodenal inoculation of rabbits, Une showed that pathogenic strains (i.e., serotypes O:3, O:9) penetrated epithelial linings of the intestinal mucosa, entered the reticuloendothelial tissue (lamina propria and lymph follicles), and underwent intracellular multiplication within mononuclear cells. The resultant enterocolitis was characterized by small focal ulcerations in the mucous membrane and granulomatous lesions (pseudotubercles) in the mesenteric lymph nodes, liver, and spleen. In severe infections, the granulomas in lymph follicles underwent necrosis and ulceration, and massive numbers or organisms (colonies) could be discerned in the necrotic centers. The exact route of distribution of organisms from the duodenal lumen to the mesenteric lymph nodes, liver, and spleen remained unclear to Une. However, he postulated their spread as either "by blood and or the lymph stream following penetration into the lamina propria."

Based on the observed penetration through epithelial linings and subsequent multiplication in reticuloendothelial tissues, Une (1977a) felt that *Y. enterocolitica* should be grouped with the invasive enteropathogenic species such as *Salmonella* and *Shigella* and some *Escherichia coli* (Savage, 1972). In contrast to *Shigella,* which multiply primarily within epithelial cells (Savage, 1972), *Y. enterocolitica* passes through epithelial cells to the lamina propria and lymph follicles and multiplies within mononuclear cells resident in these areas (Table 1).

Switching to the in vitro assay system of HeLa cells and rabbit peritoneal macrophages, Une (1977b) confirmed his in vivo observations that pathogenic strains of *Y. enterocolitica* have the ability to penetrate epithelial cells and to survive and multiply within macrophages. Lee et al. (1977) have essentially confirmed Une's observations.

Comparing *Y. pseudotuberculosis* with *Y. enterocolitica,* Une (1977c) showed that mortality rates among rabbits challenged by direct inoculation into the duodenal lumen was greater for the former organism and that the gross appearance and microscopic examination of the lesions were reflective of a more virulent infection with this organism than with *Y. enterocolitica*. Additionally, Une noted catarrhal inflammation and hemorrhage throughout the small intestine in the severest cases of *Y. pseudotuberculosis* infection, possibly due to the production of an exotoxin. While this feature was absent with *Y. enterocolitica* in this animal model, Okudaira, Fukuda, and Tamura (1972) have reported a case of acute regional ileocecitis that was histologically characterized by catarrhal inflammation of the intestine.

Une (1977a,b) and earlier Bovallius and Nilsson (1975) and Richardson and Harkness (1970) have shown that *Y. pseudotuberculosis* pathogenesis, like that of *Y. enterocolitica,* is intimately associated with penetration of epithelial linings, survival, and multiplication within host cells.

We have seen that experimental *Y. pseudotuber-*

culosis and *Y. enterocolitica* enteral infections are nearly coincident. Yet there exists a marked intrinsic ability of the former species to initiate disease, whereas for *Y. enterocolitica* this capability apparently resides with specific serotypes. In his extensive studies, Une (1977 a,b,c) showed that *Y. pseudotuberculosis* serotypes III and V, seldom encountered in human infections, were pathogenic for rabbits and invaded, survived, and multiplied within HeLa cells and macrophages in vitro. Bovallius and Nilsson (1975) showed that *Y. pseudotuberculosis* isolates possess the initial attribute for pathogenicity, namely, epithelial cell penetration.

Yersinia enterocolitica experimental animal pathogenicity and ability to penetrate, survive, and multiply within HeLa cells and monocytes is markedly interwoven with the source of the isolate, serotype, and to some extent biochemical behavior. Human isolates capable of producing classic syndromes (mesenteric lymphadenitis, terminal ileitis, septicemia), especially serotypes O:3, O:5,27, O:8, and O:9, produce experimental infections in conventional animals (Alonso et al., 1975b; Carter, 1975; Une, 1977a) and have the ability to infect epithelial cells (Lee et al., 1977; Une, 1977b; Une et al., 1977). *Y. enterocolitica* strains derived mainly from environmental sources such as animal, water, and food reservoirs lack animal pathogenicity and ability to penetrate epithelial cells (Lee et al., 1977; Quan et al., 1974; Une et al., 1977). Swine serotype O:3 isolates behave as human isolates with regard to pathogenicity and penetration (Une et al., 1977), whereas several biochemically and serologically diverse isolates, i.e., *Y. intermedia,* from human and environmental sources were devoid of invasiveness (Lee et al., 1977; Une, 1977a; Une et al., 1977).

Experimental *Y. enterocolitica* infection in animal models resembles naturally acquired human disease in severity and pathological findings. In cases of childhood *Y. enterocolitica* enteral infection, the major symptom observed is profuse watery diarrhea rather than the "pseudoappendicitis" syndrome. To account for these observations, Pai and Mors (1978) and Pai, Mors, and Toma (1978) have elucidated another facet in the yersiniosis enterocolitica enigma: enterotoxin production. These investigators showed that heat-stable enterotoxin production, as assessed in the infant mouse model and rabbit ileal loop assay, was more prevalent (90%) among human isolates of *Y. enterocolitica* as contrasted to (10%) for rhamnose-fermenting isolates. Characteristics of the enterotoxin were similar to those of that produced by enterotoxigenic strains of *E. coli* (Evans, Evans, and Pierce, 1973). Enterotoxin synthesis was enhanced by growth at 26°C and was absent in culture filtrates of cells grown at 37°C. Pai and Mors speculated that since entero-

Table 1. Comparison of enteroinvasive *Yersinia enterocolitica, Y. pseudotuberculosis, Salmonella, Shigella,* and *Escherichia coli.*

Characteristic	Species				
	Y. entero-colitica	*Y. pseudo-tuberculosis*	*Salmonella*	*Shigella*	*E. coli*
Epithelial cell penetration	Yes	Yes	Yes	Yes	Yes
Epithelial cell multiplication	No	No	?	Yes	Yes
Passage through intercellular junctions	No	No	Yes	No	No
Lamina propria invasion	Yes	Yes	Yes	Yes	Yes
Proliferation in macrophages	Yes	Yes	Yes	No	No

toxin production was absent at 37°C the role of this component in vivo, where body temperature approaches 37°C, remains to be more clearly elucidated through animal experimentation.

The possibility that enterotoxin production and epithelial cell penetration (Sereney Test) are plasmid mediated has been recently suggested by Zink et al. (1978). Working with a serotype O:8 strain isolated and maintained at 22°C (D. L. Zink, personal communication), these investigators showed the presence of two plasmids (41 Mdal, 35 Mdal) occurring singly or together. Only strains harboring the 41 Mdal unit are invasive. Loss of this plasmid either by growth at 35°C or by ethidium bromide curing results in loss of tissue invasiveness (Zink et al., 1980). As a clear-cut association between plasmid DNA content and heat-stable (ST) enterotoxin production could not be established, Zink and colleagues proposed that ST production in *Y. enterocolitica* is encoded by chromosomal genes. This finding is in contrast to the plasmid-mediated enterotoxin production reported for *E. coli* (Evans et al., 1975). In addition to temperature-dependent tissue invasiveness (Lee et al., 1977; Feeley et al., 1979) and enterotoxin production, *Y. enterocolitica* cells cultured at lower temperatures show increased resistance to the bactericidal activity of normal serum and display prolonged in vivo survival (Nilehn, 1973).

Clinically, the main overlapping manifestations of human infection with these microbial species are: (i) acute mesenteric lymphadenitis and terminal ileitis mimicking appendicitis (pseudoappendicitis); (ii) enterocolitis; (iii) acute typhoid-like septicemia or septicemia in compromised hosts; (iv) pseudotumoral matting of mesenteric lymph nodes; and (v) nonsuppurative sequelae such as erythema nodosum and arthritis (Mollaret, 1972). Postinfection myocarditis (Leino and Kalliomaki, 1974) and glomerulonephritis have also been described (Forsstrom et al., 1977; Friedberg, Larsen, and Dennenberg,

1978). Other human associations documented with the more versatile *Y. enterocolitica* include skin lesions (Lewis and Alexander, 1976; McIver and Pike, 1934), meningitis and panophthalmitis (Sonnenwirth, 1970), and subacute localized abscesses of liver and spleen (Rabson, Hallett, and Koornhoff, 1975).

Although the major clinical presentations of *Y. pseudotuberculosis* and *Y. enterocolitica* are superimposable, as shown in Tables 2, 3, and 4, there exist subtle distinctions in their frequency, age- and sex-related attack rate, pathology, diagnosis, and epidemiology. Postinfection nonsuppurative arthritis, for example, shows a predilection for adults and has a distinct geographic distribution associated with serobiophage type of the infecting *Y. enterocolitica* strain. Thus, the incidence is highest in Scandinavian countries, and these sequelae are related to infection due to serotype O:3, biotype 4, phage type 8, and serotype O:9 strains. In countries where serotype O:3, biotype 4, phage type 9A (South Africa), or phage type 9B (Canada) predominate, arthritis is not encountered as a common sequela (Bottone, 1977). The reader is referred to recent reviews (Ahvonen, 1972a,b; Bottone, 1977; Mollaret, 1972) for a more extensive treatment of other aspects of *Y. enterocolitica* infection, including their historical nuances. Of particular interest are the recent reports of Bradford, Noce, and Gutman (1974) and Vantrappen et al. (1977) regarding intestinal pathology.

Ecologically, *Y. enterocolitica*, in contrast to its more staid counterpart *Y. pseudotuberculosis*, is still evolving in its host adaptation. Indeed, there is emerging scientific and clinical evidence to support the contention advanced by Mollaret (1976) that *Y. enterocolitica* may be viewed as comprising two basic groups, the first composed of the so-called classic strains (ie., serotypes O:3, O:5,27, O:8, O:9), which are biologically stable, host adapted in

Table 2. Comparative clinical features of intestinal manifestations of *Yersinia pseudotuberculosis* and *Y. enterocolitica* infection.

Clinical presentation	*Y. pseudotuberculosis*	*Y. enterocolitica*
Acute adbominal syndrome Mesenteric lymphadenitis	Frequent; usually affects male children and young adults 2–24 years; lesser incidence among 5- to 12-year-olds. Enlarged lymph nodes singly or matted in ileocecal angle; microabscesses usually present. Histologically may show reticular cell proliferation with giant cells and eosinophils; appendix usually normal.	Seen mainly in adolescents and adults of both sexes; seldom complicates enteritis in children under 7 years. Lymph nodes enlarged, may be confluent in ileocecum; proliferation of large pyroninophilic cells in lymph nodes. Appendix usually normal; lymphoid hyperplasia, suppuration, and microabscesses may be found.
Terminal ileitis	Infrequent complication. Terminal ileum thickened and inflamed; usually only part of bowel involved. Focal mucosal ulcerations at site of Peyer's patches.	Frequent complication. Ileal lesions more severe, intensive hemorrhagic necrosis may occur.
Acute enteritis	Infrequent	Most frequent clinical syndrome, particularly in children. Usually self-limiting, although chronic course may ensue; mucosal ulcerations present during early disease. Located throughout gastrointestinal tract, especially at sites of lymphoid tissue within intestinal mucosa. Leukocytes in stool.

Table 3. Comparative and overlapping clinical features of *Yersinia* septicemia and postinfection sequelae.

Clinical presentation	*Y. pseudotuberculosis* and *Y. enterocolitica*
Septicemia	High mortality rate seen primarily in hosts compromised by underlying disease, i.e., cirrhosis, hemochromatosis of liver, or by immunosuppression. Fulminating typhoid-like syndrome seen in otherwise healthy hosts; metastatic abscesses in spleen; when in liver, resembles amebic hepatitis.
Arthritis	Acute and chronic, seen mainly in adults; affects predominantly weight-bearing joints. Syndrome more chronic in patients with HLA-B27 histocompatibility antigen. Commonly encountered with *Y. enterocolitica*. Acute arthritis seen with *Y. pseudotuberculosis* except in patients with HLA-B27.
Erythema nodosum	When due to *Y. enterocolitica*, higher incidence among women over 40 years; males under 20 years mainly affected by *Y. pseudotuberculosis*. Frequently concomitant with arthritis.

Table 4. Laboratory diagnosis of yersiniosis attributable to
Yersinia pseudotuberculosis and *Y. enterocolitica.*

Clinical presentation	*Y. pseudotuberculosis*	*Y. enterocolitica*
Acute abdominal syndrome		
Culture:		
Stool	Seldom positive.	Frequently positive.
Lymph nodes ⎫ Terminal ileum ⎬ Appendix ⎭	Frequently positive.	Frequently positive.
Peritoneal exudate	Insufficient data.	Insufficient data.
Blood	Seldom positive.	Seldom positive.
Serology	Antibodies present during acute phase. Short-lived, parallels course of disease. Antibody to types I, II, V specific, serotypes II, IV cross-react with groups B, D *Salmonella.*	Antibody titer absent or low during acute phase, rises rapidly as disease progresses. Titers may persist for years. Serotype O:9 cross-reacts with *Brucella* species and group H *Salmonella.*
Nonsuppurative sequelae		
Stool culture Serology	Seldom positive. High titers present that either diminish or persist with course of disease.	Frequently positive. High titers present, especially with serotype O:3 infections.

animals and man, and produce typical clinical syndromes associated with this species (Tables 2, 3, 4). These strains are also capable of epithelial cell penetration, intracellular survival, and multiplication.

The second group of *Yersinia enterocolitica* is exemplified by the biochemically and serologically atypical strains of which rhamnose-, raffinose-, and melibiose-fermenting varieties may serve as prototypes. These serologically unusual strains are mainly encountered from environmental sources and display biologically diverse temperature-related microbiological attributes (Bottone, 1977; Bottone et al., 1974; Chester et al., 1977). None of such strains tested to date in various laboratories is enteroinvasive. Apparently these strains are of restricted pathogenic potential, incapable of producing the more serious manifestations of yersiniosis (Table 5; Bottone, 1978). It may be that these yersiniae are still in the evolutionary stage of becoming host adapted in man and hence may ultimately contribute to the expanding spectrum of human yersiniosis. Certainly their continued "nonpathogenic" status will be tempered by their potential acquisition from other Enterobacteriaceae of extrachromosomal elements coding for colonization factors and; as noted with *Y. enterocolitica,* tissue invasiveness.

Culturally, the isolation of *Y. pseudotuberculosis* and *Y. enterocolitica* is embodied in the nature of the specimen undergoing bacteriological analysis and an awareness of the variability attending their growth patterns on commonly used isolation agar media for enteric bacteria. The recovery of either of these species from specimens obtained from their basic lesion in mesenteric lymph nodes, or from the blood, is nonproblematic. In contrast to these normally sterile sites, isolation from areas heavily colonized by other microbial species (ie., feces, appendix, terminal ileum) requires microbiological finesse; recognition of the pinpoint size and color of colonies on various media is paramount.

Yersinia pseudotuberculosis and *Yersinia enterocolitica* are distinct from other Enterobacteriaceae because of the paucity of growth on "enteric agar" after 24 h incubation at both 22°C and 37°C. This cultural feature may readily obscure their presence. Under both incubation conditions, colony sizes range from barely perceptible to pinpoint. Depending upon the growth medium, diameters of 0.1–2 mm may be achieved after 48 h incubation (Table 6; Bottone, 1977). Because of the smaller size of the yersiniae in comparison with other enteric pathogens such as *Salmonella* and *Shigella,* Wauters (1970) has recommended scanning agar surfaces (especially deoxycholate citrate lactose [DCL]) under stereoscopic microscopy with oblique lighting and the agar surface inclined 45°. Specifically regarding *Y. enterocolitica,* both Nilehn (1969) and Wauters (1970) have suggested incubating inoculated agar media between 25°C and 29°C, respectively, to enhance growth and subsequent viewing of colonies. This procedure will also aid isolation of *Y. pseudotuberculosis* (Wetzler, 1970).

Table 5. Comparisons of clinical, laboratory, and epidemiological properties of *Yersinia pseudotuberculosis*, *Y. enterocolitica*, and *Y. intermedia*.

Property	Y. pseudotuberculosis	Y. enterocolitica	Y. intermedia
Clinical syndromes	Mesenteric lymphadenitis, terminal ileitis, enteritis, septicemia, visceral abscesses, postinfection sequelae		Mild enteritis, superficial skin abscesses, conjunctivitis, cystitis
Enterovirulence	Epithelial cell invasion, multiplication within macrophages, enterotoxigenic (O:3)		Lack invasiveness or enterotoxigenicity[a]
Biochemically	Homogeneous	Homogeneous	Heterogeneous
Serotypes	I, II, IV, VI, III, V, rare in humans	O:3, O:5,27, O:8, O:9	Heterogeneous: O:14 O:16, O:17, NAG[b]
Phage types	Not available	8, 9A, 9B, 10	Nongroupable (phage group 10)
Host range	Animals, humans		Animals, humans, water, food

[a] Based upon a limited number of strains tested.
[b] NAG, nonagglutinable with 34 existing *Y. enterocolitica* antisera.

Table 6. Typical colonial characteristics of *Yersinia pseudotuberculosis*[a] and *Y. enterocolitica*[b] on blood and enteric agars.[c]

Medium	Y. pseudotuberculosis	Y. enterocolitica
Sheep blood agar	Grayish, low convex smooth, 1–2 mm, nonhemolytic	Gray-white, smooth, 1–2 mm, usually nonhemolytic
Deoxycholate	Pinpoint (48 h), yellowish-pink	Pinpoint, colorless
Endo	Pinpoint, colorless, smooth, pink-fuchsia (48 h)	0.5–1 mm, colorless
Eosin methylene blue	Pinpoint, colorless (pale violet), no metallic sheen	Pinpoint, lavender, metallic sheen (48 h)
Hektoen-Enteric (H-E)	Pinpoint, smooth, green (colorless) to yellow (48 h)	Pinpoint, frequently salmon (coliform) colored
MacConkey	Pinpoint, colorless, pink (48 h)	Pinpoint, slightly pink
Salmonella-shigella (SS)	Barely perceptible 24 h Pinpoint, pinkish (48 h)	Pinpoint, colorless
Xylose lysine deoxycholate	Barely perceptible 24 h smooth, colorless (grayish-white) to yellow (48 h)	Pinpoint, yellow (coliform), sculptured contours

[a] *Y. pseudotuberculosis* serotype I.
[b] *Y. enterocolitica* serotypes O:3, O:8, O:9. Rhamnose-fermenting strains do not grow well on H-E and SS agars.
[c] Twenty-four-hour incubation unless specifically stated otherwise.

Yersinia pseudotuberculosis, irrespective of serotype, retains a degree of uniformity regarding growth, or lack thereof, on agar substrates (bismuth sulfite, brilliant green). Indeed, because of such cultural constancy, a medium has been described (Morris, 1958) to enhance isolation from contaminated materials. This medium, modified by Paterson and Cook (1963), is especially useful for epidemiological surveys.

Medium for Isolating *Yersinia pseudotuberculosis* from Contaminated Materials (Morris, 1958; modified by Paterson and Cook, 1963)

Peptic digest of sheep blood	10.0 ml
Novobiocin (0.2%)	2.0 ml
Erythromycin (0.2%)	0.5 ml
Mycostatin (50,000 units/ml)	0.8 ml
Crystal violet (Gurr 548, 0.1%)	0.5 ml

These constituents were added to 200 ml of melted trypsinized meat agar at 50°C just before the plates were poured. The plates were inoculated on the surface with a 10% suspension of feces in phosphate-buffered solution (pH 7.6) and were incubated aerobically at 37°C for 24–48 h.

The challenge inherent in the recovery of *Y.*

enterocolitica from fecal samples resides in its cultural heterogeneity, which is a reflection of the medium used and the serological and biochemical profile of the isolate. On Endo, MacConkey, and, DCL agar plates, typical colorless colonies develop, indicative of a non–lactose fermenter. When inoculated on eosin methylene blue (EMB), xylose lysine deoxycholate (XLD), or Hektoen-Enteric (H-E) agars, colonies of strains belonging to the more commonly encountered serotypes in human infections (O:3, O:5,27, O:8, O:9) may be readily overlooked as ''coliform'' colonies because of their utilization of fermentable carbohydrates in these media (Table 6). Thus the production of a metallic sheen on EMB (sucrose fermented) and of yellow colonies on XLD (xylose and sucrose fermented) and H-E agars (sucrose fermented) precludes further processing under the assumption that one is not dealing with an ''enteric pathogen''. The other facet of isolation is the variability of *Y. intermedia* strains to initiate growth on ''enteric'' media. Thus, rhamnose-fermenting strains of unusual serotypes (O:14, O:16, O:17, nonagglutinable) develop sparsely or not at all on H-E, XLD, or salmonella-shigella (SS) agar (Bottone, 1977). Paradoxically, while biochemically typical *Y. enterocolitica* organisms grow readily on SS agar, one of three sucrose-negative isolates examined in our laboratory failed to grow on this medium. Because of such cultural intricacies, multiple enteric media must be inoculated to ensure recovery of *Y. enterocolitica*. As a consequence of such diversity, this species remains enshrouded in a veil of epidemiological darkness regarding carriage versus actual attack rates.

The recovery of *Y. pseudotuberculosis* and *Y. enterocolitica* from enteric sources may be markedly influenced by the number of organisms present and the diagnostic approach. In contrast to direct isolation from primary media, Paterson and Cook (1963) achieved a definitive increase in the recovery of *Y. pseudotuberculosis* from feces when specimens were kept in a phosphate-buffered solution at 3–4°C and subcultured at 7-day intervals for 28 days. Subsequently, cold enrichment in isotonic buffered saline (Greenwood et al., 1975) or in isotonic saline containing 25 g/ml potassium tellurite (Wetzler, 1970) has also been used successfully for *Y. enterocolitica* isolation.

Cold enrichment has revealed the presence of *Y. enterocolitica* in the stools of symptomatic patients in the absence of isolation by conventional methods (Eiss, 1975; Greenwood et al., 1975; Wilson, McCormick, and Feeley, 1976; Weissfeld and Sonnenwirth, 1980; E. J. Bottone, unpublished data). Growth in such a nutritionally deficient milieu under reduced temperature is apparently not uncommon, as Highsmith et al. (1977) have shown that various *Y. enterocolitica* strains derived mainly from well water are capable of growth in sterile distilled water devoid of added nutrients. Generation time for these isolates at 4°C was 8.4 h.

It should be noted, however, that cold enrichment of stools is of relative value for the recovery of ''classic'' *Y. enterocolitica* serotypes such as O:3, O:5,27, O:8, and O:9. It has been clearly established that during the acute gastroenteritis phase these yersiniae are present in large numbers and do not present isolation difficulties (Pai et al., 1979; Van Noyen, Vandepitte, and Wauters, 1980). Cold enrichment, however, does enhance recovery of *Y. enterocolitica* from the stools obtained from asymptomatic carriers or from those convalescing from *Y. enterocolitica* gastroenteritis (Pai et al., 1979). Interestingly, isolates recovered by cold enrichment from asymptomatic carriers belong mainly to biotype I. With the exception of serotype O:8 biotype I isolates, it appears that other biotype I isolates may not be human pathogens (Van Noyen, Vandepitte, and Wauters, 1980). These strains appear to lack invasiveness and do not elicit an antibody response.

Biochemically, *Yersinia pseudotuberculosis* and *Yersinia enterocolitica* shows convergent attributes necessary for ascription of a suspect isolate to either yersinial species. These consist of anaerogenic glucose fermentation, urease production, motility at 22°C but not at 37°C, and absence of phenylalanine deaminase, oxidase, lysine decarboxylase, and arginine dihydrolase activities (Table 7).

From this juncture, these microbial species disengage and course through two divergent biochemical profiles. On the one hand, *Y. pseudotuberculosis* remains biochemically stable, showing minimal variability among isolates irrespective of source, serotype, and incubation temperature. With *Y. enterocolitica*, however, there exists an echeloning of biochemical profiles, ranging from strains that are biochemically constant to those displaying an expanded biochemical profile. Either path is markedly influenced by the source of the isolate, its serotype, and incubation temperature. In the former category are the typical strains, especially those producing classic syndromes in which biochemical homogeneity resides within a given serological group (Wauters, 1973). These strains are well defined antigenically and in host specificity, being encountered mainly from the chinchilla (O:1), hare (O:2), and pig and man (O:3, O:5,27, O:8, O:9).

Outside of these homogeneous groups, there exist numerous *Y. enterocolitica* strains that originate from natural environmental sources, are serologically aberrant, and display marked temperature-dependent biochemical heterogeneity. Among the latter are the rhamnose-fermenting isolates and to a lesser extent, sucrose nonfermenters (Bottone, 1977; Chester et al., 1977; Hanna et al, 1976; Highsmith et al., 1977).

As a consequence of biochemical behavior, *Y.*

Table 7. Distinguishing characteristics of *Yersinia pseudotuberculosis, Y. enterocolitica,* rhamnose-fermenting, and sucrose-negative isolates.

Test	*Y. pseudotuberculosis*	*Y. enterocolitica*	Rhamnose positive[b]	Sucrose negative
TSI slant/butt	Alkaline/acid	Acid/acid	Acid/acid	Alkaline/acid
KIA slant/butt	Alkaline/acid	Alkaline/acid	Alkaline/acid	Alkaline/acid
Motility:				
22°C	+	+	+	+
37°C	0	0	0	0
Urease	+	+	+	+
Oxidase	0	0	0	0
Lysine decarboxylase	0	0	0	0
Arginine dihydrolase	0	0	0	0
Phenylalanine deaminase	0	0	0	0
Simmons' citrate	0	0	+(22°C) or 0	0
Indole	0	V	+	+(22°C)
β-Galactosidase	+(22°C)	+(22°C)	+(22°C)	+ or 0
Acetyl methyl carbinol:				
22°C	0(V + L)	+	+	0
37°C	0	0	0	0
Ornithine decarboxylase	0	+	+	+ or 0
Fermentation:				
Glucose	+	+	+	+
Lactose	0	0	0(+L)	0
Sucrose	0	+	+	0
Rhamnose	+	0	+(22°C)	0
Raffinose	0	0	+(22°C) or 0	0
Melibiose	+	0	+(22°C) or 0	0
α-Methylglucoside	0	0	+(22°C) or 0	0
Trehalose	+	+, 0	+	+ or 0

[a] 0, negative; +, positive; (+L), late positive; V, variable.

[b] Species designation *Y. intermedia* proposed by Brenner (1978) for isolates fermenting, in addition, raffinose, melibiose, and α-methylglucoside, and *Y. frederiksenii* for strains negative for these characters.

enterocolitica strains may be grouped (biotyped) according to their reactivity in various substrates. Such a classification schema was originally introduced by Nilehn (1969) and modified by Wauters (1973; Table 8).

It is emphasized that isolates displaying the salient constellation of features for consideration as *Y. pseudotuberculosis* or *Y. enterocolitica* be actively pursued biochemically and not be discarded merely because of discordant results vis-à-vis typical

Table 8. Biotyping schema for *Yersinia enterocolitica* (Wauters, 1970).[a]

Test	Biotypes				
	1	2	3	4	5
Lecithinase	+	0	0	0	0
Indole	+	(+)			
Lactose (acid OF medium)	+	+	+	0	0
Xylose, 48 h	+	+	+	0	0
Trehalose	+	+	+	+	0
Nitrate reduction	+	+	+	+	0
Ornithine decarboxylase	+	+	+	+	0
β-Galactosidase	+	+	+	+	0

[a] +, Positive; (+), positive at 29°C; 0, negative.

strains. Such isolates should be forwarded to reference laboratories for definitive identification. As urease activity is of cardinal importance in recognizing these two species, Christensen's urea agar is recommended for detection. Slants inoculated from primary cultures must be held a minimum of 48 h before being discarded as negative. Similarly, cognizance of the "coliform" type of reaction (acid slant/acid butt) that typical *Y. enterocolitica* renders on triple sugar iron agar (TSI) also aids isolation. To obviate this obscuring reaction, Kligler's iron agar (KIA) should be used for initial screening. In this agar-slant medium, *Y. enterocolitica,* as well as *Y. pseudotuberculosis,* yields an alkaline slant (lactose negative) and acid butt (glucose fermented). As this reaction could be indicative of a *Salmonella* or *Shigella* species, a more thorough biochemical characterization usually follows, which then could reveal a *Yersinia* species.

Serologically, based on the immunochemistry of somatic O antigens, *Y. pseudotuberculosis* may be divided into six serotypes, with types 1, 2, and 4 being further differentiated into subtypes 1A, 1B, 2A, 2B, 4A, and 4B. Smooth (S) and rough (R) somatic antigens are also present in the cell envelope (Davies, 1958). The latter antigen (serofactor 1) is an envelope or capsular protective antigen encountered in all serotypes and shared with *Y. pestis* (Davies, 1958). The S antigen (serofactor 13) is a lipopolysaccharide-like heat-stable surface component that is irregularly found in all serotypes (except 4B) and in some strains of *Y. enterocolitica* (Wetzler, 1970).

Virulent strains of *Y. pseudotuberculosis* freshly isolated from epizootics in guinea pigs possess both V (immunogenic protein) and W (nonprotective lipoprotein) antigens common to *Y. pestis* as well (Burrows and Bacon, 1960). V and W antigens act in concert to confer an antiphagocytic property on the microorganism. Strains lacking these antigens have considerably reduced virulence. The sharing of antigenic components between *Y. pseudotuberculosis* and *Y. pestis* has led to the speculation that the broad emergence of *Y. pseudotuberculosis* in European countries may have provided cross-reacting immunity to the plague bacillus. It is conceivable that the alternation of plague epidemic and quiescence may actually be related to the level of protective antibody acquired as a result of contact with *Y. pseudotuberculosis*.

Yersinia enterocolitica, in keeping with its other multifaceted properties, may be separated into 34 distinct serotypes based upon O-antigen content. These serotypes, however, do not represent the spectrum of antigenic types, as numerous strains that are nontypable have been forwarded to reference laboratories. Such isolates are mainly of environmental origin and only infrequently encountered from human sources.

Antibiotic susceptibility patterns of *Y. pseudotuberculosis* and *Y. enterocolitica* also show the constancy of the former species and the diversity intrinsic to *Y. enterocolitica*. In contrast to *Y. pseudotuberculosis*, which lacks any form of β-lactamase activity and is hence susceptible to penicillins and cephalosporins, *Y. enterocolitica* does possess β-lactamase, both constitutive and of extrachromosomal origin. This species is therefore resistant to ampicillin, cephalothin, carbenicillin, and penicillin (Cornelis, Wauters, and Vanderhaeghe, 1973). However, serotype O:8, unlike other more commonly occurring serotypes, is susceptible to ampicillin but displays a variable resistance to carbenicillin and cephalothin.

Isolates designated *Y. intermedia* by Brenner et al. (1979) appear to possess broad β-lactamase activity and are resistant to ampicillin, carbenicillin, cephalothin, and penicillin. These strains, in addition, show a temperature-dependent relationship regarding reactivity to ampicillin, cephalothin, colistin, penicillin, and streptomycin. At 22°C incubation, as compared with results obtained at 37°C, increased resistance is noted as evidenced by smaller zone diameters in the disk-agar diffusion method or by overt turbidity in the broth dilution assay

(Bottone and Robin, 1977; Chester and Stotzky, 1976). This phenomenon is observed to a lesser extent with chloramphenicol.

Yersinia pseudotuberculosis and *Yersinia enterocolitica* are usually inhibited in vitro by the tetracyclines, aminoglycosides, chloramphenicol, colistin, and, for *Y. enterocolitica*, by the combination of trimethoprim and sulfamethoxazole (Gutman et al., 1973).

Epidemiologically, *Y. pseudotuberculosis* and *Y. enterocolitica* show contrasting distribution patterns. While both agents are widespread globally, *Y. enterocolitica* serotypes, in contrast to *Y. pseudotuberculosis*, are geographically focalized. Although the incidence of cases of *Y. pseudotuberculosis* may show a clustering in a geographic locale (Bronson, May, and Ruebner, 1972), serotype 1, which accounts for over 90% of human infections, is not geographically demarcated but encountered worldwide (Wetzler, 1970). Human infections with *Y. enterocolitica* are caused mainly by serotypes O:3, O:8, and O:9. While serotypes O:3 and O:9 are typically European strains, serotype O:8 has been isolated exclusively in the United States. Indeed, within the serotype O:3 strains, a peculiar distribution exists relative to phage types. Serotype O:3 biotype 4 phage type 8 is the European strain, whereas phage type 9B is recovered solely in Canada and phage type 9A has been isolated extensively in South Africa (Mollaret, 1972).

Transmission of *Y. pseudotuberculosis* and *Y. enterocolitica* from their natural reservoirs to humans has been the subject of much debate. Spontaneously occurring outbreaks of *Y. pseudotuberculosis* disease in animals has been associated with the ingestion of contaminated foodstuff (Paterson and Cook, 1963) or possible contact with excreta from animals known to harbor this microorganism (Bronson, May, and Ruebner, 1972). The epidemiology of human *Y. pseudotuberculosis* infections, however, is still not resolved and only inferentially attributed to contact with infected wild or domesticated animals, or to foods contaminated with their excrements.

Speculation regarding the events leading to human acquisition of *Y. enterocolitica* is slowly being superseded by definitive epidemiological investigation linking transmission to animal, water, and food sources. While healthy carriers, either animal or human, have been suspected (Mollaret, 1972), the reports of Gutman et al. (1973) and Wilson, McCormick, and Feeley (1976) strongly suggest acquisition from sick household dogs, which harbored the same serotype of *Y. enterocolitica* recovered from symptomatic patients. Once the organism is introduced into a household setting, person-to-person transmission may ensue, as evidenced by the sequential onset of disease in other family members (Gutman et al, 1973).

Food-borne yersiniosis has long been suspected and only recently been proven to be a source of gastrointestinal illness. Black et al. (1978) reported an epidemic of *Y. enterocolitica* infection that occurred in upper New York State and affected 222 individuals, 36 of whom (all children) were hospitalized. Sixteen of these children underwent appendectomies for the "pseudoappendicitis" syndrome. Epidemiological investigation revealed the source of the causative serotype O:8 strain to be chocolate milk prepared by the hand mixing of chocolate syrup into previously pasteurized milk.

Although the above reports have elucidated the transmission aspects of *Y. enterocolitica,* unanswered still are factors relative to the increased incidence of yersiniosis during colder months (Mollaret, 1972), the age-related clinical symptomatology, and the nature of the infecting strain at the time of acquisition. The last parameter is of fundamental concern, as many of the microbiological and pathogenic factors (i.e., enteroinvasiveness, entertoxin production, intracellular survival) are temperature related. Thus, our knowledge of the bacterium's surface topography during its transition from vector (source) to a host in which disease ensues is crucial to a fuller understanding of yersinial pathogenesis. For instance, are colonization factor(s) acquired at lower growth temperatures that enable adherence to epithelial surfaces a prerequisite to penetration? Are such postulated factors lost after host establishment (increased temperature), as in vitro data seem to indicate (Lee et al., 1977; Une, 1977b)? Are biochemically and serologically diverse strains lacking these surface components and hence incapable of producing classic syndromes?

While such questions still remain unanswered, other tantalizing aspects of *Y. enterocolitica* infection continue to emerge. Among the latter are the recently described sequelae of acute glomerulonephritis (Forsstrom et al., 1977; Friedberg, Larsen, and Dennenberg, 1978) and the demonstration of antibodies to *Y. enterocolitica* in patients with various thyroid disorders (Bech et al., 1974; Shenkman and Bottone, 1976).

Such reports clearly indicate that the yersiniosis chapter is far from being concluded here but rather stands on a foundation of ever growing data seeking new interpretation.

Literature Cited

Ahvonen, P. 1972a. Human yersiniosis in Finland. I. Bacteriology and serology. Annals of Clinical Research **4:**30–38.

Ahvonen, P. 1972b. Human yersiniosis in Finland. II. Clinical features, Annals of Clinical Research **4:**39–48.

Alonso, J. M., Bercovier, H. 1975. Premiers isolements en France de *Yersinia enterocolitica* chez des micromammiferes sauvages. Medecine et Maladies Infectieuses **5:**180–181.

Alonso, J. M., Bejat, J., Bercovier, H., Mollaret, H. H. 1975a. Sur un groupe de souches de *Yersinia enterocolitica* fermentant le rhamnose. Medecine et Maladies Infectieuses **10:**490–492.

Alonso, J. M., Bercovier, H., Destombes, P., Mollaret, H. H. 1975b. Pouvoir pathogene experimental de *Yersinia enterocolitica* chez la souris athymique (NUDE). Annales de Microbiologie **126B:** 187–199.

Baggs, R. B., Hunt, R. D., Garcia, F. G., Hajema, E. M., Blake, B. J., Fraser, C. E. O. 1976. Pseudotuberculosis *(Yersinia enterocolitica)* in the owl monkey *(Aotus trivirgotus).* Laboratory Animal Science **26:**1079–1083.

Beaudette, F. R. 1940. A case of pseudotuberculosis in a blackbird. Journal of the American Veterinary Medical Association **97:**151–157.

Bech, K., Larsen, J. H., Hansen, J. H., Nerup, J. 1974. *Yersinia enterocolitica* infection and thyroid disorders. [Letter.] Lancet **ii:**951–952.

Becht, H. 1963. Untersuchungen über die Pseudotuberkulose beim chinchilla. Deutsche Tierärztliche Wochenschrift **69:**626.

Beck, A., Huck, W. 1925. Enzootische Erkrankungen von Truthühner und Kanarienvögeln durch Bakterien aus der Gruppe der hämorrhagischen Septikämie (Paracholera). Centralblatt für Bakteriologie, Parasitenkunde und Infektionskrankheiten Abt. 1 Orig. **95:**330–339.

Bercovier, H., Alonso, J. M., Destombes, P., Mollaret, H. H. 1976. Infection experimentale de souris axenique par *Yersinia enterocolitica.* Annales de Microbiologie **127A:**493–501.

Bessonova, A., Lenskaya, G., Molodtzova, P., Mossolova, O. 1936. Some cases of spontaneous transmutation of *B. pestis* into *B. pseudotuberculosis rodentium* Pfeiffer. [*In Russian.*] *Vestnik Microbiologii, Epidemiologii i Parazitologii* **15:**151–162.

Bibel, D. J., Chen, T. J. 1976. Diagnosis of plague: Analysis of the Yersin-Kitasato controversy. Bacteriological Reviews **40:**433–651.

Black, R. E., Jackson, R. J., Tsai, T., Medevesky, M., Shayegani, M., Feeley, J. C., MacLeod, K. I. E., Wakelee, A. M. 1978. Epidemic *Yersinia enterocolitica* infection due to contaminated chocolate milk. New England Journal of Medicine **298:**76–79.

Blaxland, J. D. 1947. *Pasteurella pseudotuberculosis* infection in turkeys. Veterinary Record **59:**317–318.

Bottone, E. J. 1977. *Yersinia enterocolitica:* A panoramic view of a charismatic microorganism. CRC Critical Reviews in Microbiology **5:**211–214.

Bottone, E. J. 1978. Atypical *Yersinia enterocolitica:* Clinical and epidemiological parameters. Journal of Clinical Microbiology **7:**562–567.

Bottone, E. J., Robin, T. 1977. *Yersinia enterocolitica:* Recovery and characterization of two unusual isolates from a case of acute enteritis. Journal of Clinical Microbiology **5:**341–345.

Bottone, E. J., Chester, B., Malowany, M. S., Allerhand, J. 1974. Unusual *Yersinia enterocolitica* isolates not associated with mesenteric lymphadenitis. Applied Microbiology **5:**858–861.

Botzler, R. G., Wetzler, T. F., Cowan, A. B. 1968. *Yersinia enterocolitica* and *Yersinia*-like organisms isolated from frogs and snails. Bulletin of Wildlife Diseases Association **4:**110–115.

Botzler, R. G., Wetzler, T. F., Cowan, A. B., Quan, T. J. Yersinae in pondwater and snails. Journal of Wildlife Diseases Association **12:**492–496.

Bovalius, Å., Nilsson, G. 1975. Ingestion and survival of *Yersinia pseudotuberculosis* in HeLa cells. Canadian Journal of Microbiology, **21:**1997–2007.

Bradford, W. D., Noce, P. S., Gutman, L. T. 1974. Pathologic features of enteric infection with *Yersinia enterocolitica.* Archives of Pathology **98:**17–22.

Brenner, D. J. 1979. Speciation in *Yersinia,* pp. 33–43. In: Carter, P. B., Lafleur, L., Toma, S. (eds.), Contributions to microbiology and immunology, vol. 5. Basel, New York: Karger.

Brenner, D. J., Farmer, J. J., III, Hickman, F. W., Asbury, M. A., Steigerwalt, A. G. 1977. Taxonomic and nomenclature changes in *Enterobacteriacae*. CDC Publication No. 78-8356. Atlanta: Center for Disease Control.

Brenner, D. J., Steigerwalt, A. G., Falcão, D. P., Weaver, R. E., Fanning, G. R. 1976. Characterization of *Yersinia enterocolitica* and *Yersinia pseudotuberculosis* by deoxyribonucleic acid hybridization and by biochemical reactions. International Journal of Systematic Bacteriology **26:**180–194.

Bronson, R. T., May, B. D., Ruebner, B. H. 1972. An outbreak of infection by *Yersinia pseudotuberculosis* in nonhuman primates. American Journal of Pathology **69:**289–308.

Burrows, T. W., Bacon, G. A. 1960. V and W antigens in strains of *Pasteurella pseudotuberculosis*. British Journal of Experimental Pathology **41:**38–44.

Carter, P. B. 1975. Pathogenicity of *Yersinia enterocolitica* for mice. Infection and Immunity **11:**164–170.

Chester, B., Stotzky, G. 1976. Temperature-dependent cultural and biochemical characteristics of rhamnose-positive *Yersinia enterocolitica*. Journal of Clinical Microbiology **3:**119–127.

Chester, B., Stotzky, G., Bottone, E. J., Malowany, M. S., Allerhand, J. 1977. *Yersinia enterocolitica*: Biochemical, serological, and gas-liquid chromatographic characterization of rhamnose, raffinose, melibiose, and citrate utilizing strains. Journal of Clinical Microbiology **6:**461–468.

Cornelis, G., Wauters, G., Vanderhaeghe, H. 1973. Presence de β-lactamase chez *Yersinia enterocolitica*. Annales de Microbiologie **124B:**139–152.

Daniels, J. J. H. M., Goudzwaard, C. 1963. Enkele stammern van een op *Pasteurella pseudotuberculosis* gelijkend niet geidentificiered species, geisoleerd bij knaagdieren. Tijdschrift voor Diergeneeskunde **88:**96–102.

Davies, D. A. L. 1958. The smooth and rough somatic antigens of *Pasteurella pseudotuberculosis*. Journal of General Microbiology **18:**118–128.

Dickinson, A. B., Mocquot, G. 1961. Studies on the bacterial flora of the alimentary tract of pigs. I. *Enterobacteriaceae* and other gram-negative bacteria. Journal of Applied Bacteriology **24:**252–284.

Domaradskij, I., Marchenkov, V., Shimanjuk, A. 1973. New data obtained in comparative studies of plague agent and related organisms, pp. 2–3. In: Winblad, S. (ed.), Contributions to microbiology and immunology, vol. 2. Basel, New York: Karger.

Eberth, C. J. 1885. Zwei mykosen des meerschweinchens. Archiv für Pathologische Anatomie und Physiologie und für Klinische Medizin **100:**15–27.

Eiss, J. 1975. Selective culturing of *Yersinia enterocolitica* at a low temperature. Scandinavian Journal of Infectious Diseases **7:**249–251.

Esseveld, H., Goudzwaard, C. 1973. On the epidemiology of *Y. enterocolitica* infections: Pigs as the source of infections in man, pp. 99–101. In: Winblad, S. (ed.), Contributions to microbiology and immunology, vol. 2. Basel, New York: Karger.

Evans, D. G., Evans, D. J., Jr., Pierce, N. F. 1973. Differences in the response of rabbit small intestine to heat-labile and heat-stable enterotoxins of *Escherichia coli*. Infection and Immunity **7:**873–880.

Evans, D. G., Silver, R. P., Evans, D. J., Jr., Chase, D. G., Gorbach, S. L. 1975. Plasmid-controlled colonization factor associated with virulence in *Escherichia coli* enterotoxigenic for humans. Infection and Immunity **12:**656–667.

Ewing, W. H., Ross, A. J., Brenner, D. J., Fanning, G. R. 1978. *Yersinia ruckeri* sp. nov., the redmouth (RM) bacterium. International Journal of Systematic Bacteriology **28:**37–44.

Feeley, J. C., Wells, J. G., Tsai, T. F., Puhr, N. D. 1979. Detection of enterotoxigenic and invasive strains of *Yersinia enterocolitica*, pp. 329–334. In: Carter, P. B., Lafleur, L., Toma, S. (eds.), Contributions to microbiology and immunology, vol. 5. Basel, New York: Karger.

Forsstrom, J. Viander, M., Lehtonen, A., Ekfors, T. 1977. *Yersinia enterocolitica* infection complicated by glomerulonephritis. Scandinavian Journal of Infectious Diseases **9:**253–256.

Frederiksen, W. 1964. A study of some *Yersinia pseudotuberculosis*-like bacteria ("*Bacterium enterocoliticum*" and "*Pasteurella X*"), pp. 103–104. In: Proceedings of the XIV Scandinavian Congress of Pathology and Microbiology, Oslo.

Freese 1907. Üeber seuchenhafte erkankungen mit septikamischen charakter bei kanarienvögeln. Deutsche Tierärztliche Wochenschrift **36:**501–505.

Friedberg, M., Larsen, S., Dennenberg, T. 1978. *Yersinia enterocolitica* and glomerulonephritis. [Letter.] Lancet **i:**498–499.

Greenwood, J. R., Flanigan, S. M., Pickett, M. J., Martin, W. J. 1975. Clinical isolation of *Yersinia enterocolitica*: Cold temperature enrichment. Journal of Clinical Microbiology **2:**559–560.

Gutman, L. T., Ottesen, E. A., Quan, T. J., Noce, P. S., Katz, S. L. 1973. An interfamilial outbreak of *Yersinia entrocolitica* enteritis. New England Journal of Medicine **288:**1372–1377.

Hanna, M. O., Zink, D. L., Carpenter, Z. L., Vanderzant, C. 1976. *Yersinia enterocolitica*-like organisms from vacuum-packaged beef and lamb. Journal of Food Science **41:**1254–1256.

Hassig, A., Karrer, J., Pusterla, F. 1949. Über pseudotuberkulose beim menschen. Schweizerische Medizinische Wochenschrift **79:**971–973.

Highsmith, A. K., Feeley, J. C., Skaliy, P., Wells, J. G., Wood, B. T. 1977. Isolation of *Yersinia enterocolitica* from well water and growth in distilled water. Applied and Environmental Microbiology **34:**745–750.

Hubbert, W. T. 1972. Yersiniosis in mammals and birds in the United States. Case reports and review. American Journal of tropical Medicine and Hygiene **21:**458–463.

Inoue, M., Kurose, M. 1975. Isolation of *Yersinia enterocolitica* from cow's intestinal contents and beef meat. Japanese Journal of Veterinary Science **37:**91–93.

Jensen, W. I., Owen, R., Jellison, W. J. 1969. *Yersinia philomiragia* sp. n., a new member of the *Pasteurella* group of bacteria, naturally pathogenic for the muskrat *(Ondatra zibethica)*. Journal of Bacteriology **100:**1237–1241.

Kapperud, G. 1975. *Yersinia enterocolitica* in small rodents from Norway, Sweden and Finland. Acta Pathologica et Microbiologica Scandinavica, Sect. B **83:**335–342.

Kapperud, G., Jonsson, B. 1976. *Yersinia enterocolitica* in brown trout *(Salmo Trutta L)* from Norway. Acta Pathologica et Microbiologica Scandinavica, Sect. B **84:**66–68.

Keet, E. E. 1974. *Yersinia enterocolitica* septicemia. Source of infection and incubation period identified. New York State Journal of Medicine **74:**2226–2230.

Kern. 1896. Eine neue infektiose krankheit der kanarienvögel (kanarienkcholera). Deutsche Zeitschrift für Tiermedizin und Vergleichende Pathologie **22:**171–180.

Knapp, W., Thal, E. 1963. Untersuchungen über die kulturell-biochemischen, serologischen, tierexperimentellen und immunologischen eigenschaften einer vorläufig "*Pasteurella X*" benannten bakterienart. Zentralblatt für Bakteriologie, Parasitenkunde, Infektionskrankheiten und Hygiene, Abt. 1 Orig. **190:**472–484.

Lassen, J. *Yersinia enterocolitica* in drinking water. Scandinavian Journal of Infectious Diseases **4:**125–127.

Lee, W. H. 1977. An assessment of *Yersinia enterocolitica* and its presence in foods. Journal of Food Protection **40:**486–489.

Lee, W. H., McGrath, P. P., Carter, P. H., Eide, E. L. 1977. The ability of some *Yersinia enterocolitica* strains to invade HeLa cells. Canadian Journal of Microbiology **23:**1714–1722.

Leino, R., Kalliomaki, J. K. 1974. Yersiniosis as an internal disease. Annals of Internal Medicine **81:**458–461.

Lerche, 1927. Die "Paracholera" der Puten und ihre Beziehung zun Pseudotuberkulose der Nagetiere. Centralblatt für Bakteriologie, Parasitenkunde und Infektionskrankheiten, Abt. 1, Orig. **104:**493–502.

Lewis, J. F., Alexander, J. 1976. Facial abscess due to *Yersinia enterocolitica*. American Journal of Clinical Pathology **66:**1016–1018.

McIver, M. A., Pike, R. M. 1934. Chronic glanders-like infection of face caused by an organism resembling *Flavobacterium pseudomallei* Whitmore, pp. 16–21. In: Clinical miscellany, Mary Imogene Bassett Hospital, vol. 1. Cooperstown, New York.

Mair, N. S. 1965. Sources and serological classification of 177 strains of *Pasteurella pseudotuberculosis* isolated in Great Britain. Journal of Pathology and Bacteriology **90:**275–278.

Malassez, L., Vignal, W. 1883. Tuberculose zoogloeique (forme ou espece de tuberculose sans bacilles). Archives de Physiologie Normale et Pathologique, Series 3 **2:**369–412.

Malassez, L., Vignal, W. 1884. Sur le microorganisme de la tuberculose zoogloeique. Archives de Physiologie Normale et Pathologique, Series 3 **4:**81–105.

Maruyama, T. 1973. Studies on biological characteristics and pathogenicity of *Yersinia enterocolitica*. 2. Experimental infections in monkeys. [In Japanese.] Japanese Journal of Bacteriology **28:**413–422.

Meissner, Schern. 1908. Die infektiöse nekrose bei den kanarienvögeln. Archiv für Wissenschaftliche und Praktische Tierheilkunde **34:**133–149.

Mollaret, H. H. 1972. *Yersinia enterocolitica* infection. A new problem in pathology. [Editorial translation]. Annales de Biologie Clinique **30:**1–6.

Mollaret, H. H. 1973. Alexandre Yersin tel qu'en lui-même enfin. Les révélations d'une correspondance inedite échelonnée de 1884 a 1926. La Nouvelle Presse Medicale **2:**2575–2580.

Mollaret, H. H. 1976. Contribution a l'etude epidemiologique des infections a *Yersinia enterocolitica*. III. Bilan provisoire des connaissances. Medecine et Maladies Infectieuses **6:**442–448.

Mollaret, H. H., Destombes, P. 1964. Les germes ''X'' en pathologie humaine. Presse Medicale **72:**2913–2915.

Mollaret, H. H., Guillon, J. C. 1965. Contribution a l'etude d'un nouveau groupe de germes (*Yersinia enterocolitica*) proches du bacille de Malassez et Vignal. II. Pouvoir pathogène experimental. Annales de l'Institut Pasteur **109:**603–613.

Mollaret, H. H., Lucas, A. 1965. Sur les particularités biochimiques des souches de *Yersinia enterocolitica* isolées chez les lievres. Annales de l'Institut Pasteur **108:**121–125.

Mollaret, H. H., Thal, E. 1974. *Yersinia*, pp. 330–332. In: Buchanan, R. E., Gibbons, N. E. (eds.), Bergey's manual of determinative bacteriology, 8th ed. Baltimore: Williams & Wilkins.

Morris, E. J. 1958. Selective media for some *Pasteurella* species. Journal of General Microbiology **19:**305–311.

Nilehn, B. 1969. Studies on *Yersinia enterocolitica*. Acta Patiologica et Microbiologica Scandinavica, suppl. **206:**1–48.

Nilehn, B. 1973. The relationship of incubation temperature to serum bactericidal effect, pathogenicity and *in vivo* survival of *Yersinia enterocolitica*, pp. 85–92. In: Winblad, S. (ed.), Contribution to microbiology and immunology, vol. 2. Basel, New York: Karger.

Okudaira, M., Fukuda, S., Tamura, S. 1972. Acute regional ileocecitis due to *Yersinia enterocolitica* in man. Report of a case. [In Japanese.] Transactiones Societatis Pathologicae Japanicae **61:**210–211.

Pai, C. H., Mors, V. 1978. Production of enterotoxin by *Yersinia enterocolitica* Infection and Immunity **19:**908–911.

Pai, C. H., Mors, V., Toma, S. 1978. Prevalence of enterotoxigenicity in human and nonhuman isolates of *Yersinia enterocolitica* from human stools. Journal of Clinical Microbiology **9:**712–715.

Pai, C. H., Sorger, S., Lafleur, L., Lackman, L., Marks, M. 1979. Efficacy of cold enrichment techniques for recovery of

Yersinia enterocolitica from human stools. Journal of Clinical Microbiology **9:**712–715.

Paterson, J. S., Cook, R. 1963. A method for the recovery of *Pasteurella pseudotuberculosis* from faeces. Journal of Pathology and Bacteriology **85:**241–242.

Pfeiffer, A. 1889. Ueber die bacilläre Pseudotuberculose bei Nagethieren. Leipzig: Verlag von Georg Thieme. [Reviewed by Dittrich. 1890. In: Centralblatt für Bakteriologie und Parasitenkunde **7:**219–221.]

Quan, T. J., Meek, J. L., Tsuchiya, K. R., Hudson, B. W., Barnes, A. M. 1974. Experimental pathogenicity of recent North American isolates of *Yersinia enterocolitica*. Journal of Infectious Diseases **129:**341–344.

Rabson, A. R., Hallett, A. F., Koornhof, H. J. 1975. Generalized *Yersinia enterocolitica* infection. Journal of Infectious Diseases **131:**447–451.

Richardson, M., Harkness, T. K. 1970. Intracellular *Pasteurella pseudotuberculosis:* Multiplication in cultured spleen and kidney cells. Infection and Immunity **2:**631–639.

Rieck, M. 1889. Eine infektiöse krankheit der kanarienvögel. Deutsche Zeitschrift für Tiermedizin und Vergleichende Pathologie **15:**Nos. 1 and 2, 68–80.

Rosenwald, A. S., Dickinson, E. M. 1944. A report on *Pasteurella pseudotuberculosis* infection in turkeys. American Journal of Veterinary Research **5:**246–249.

Ross, A. J., Rucker, R. R., Ewing, W. H. 1966. Description of a bacterium associated with redmouth disease of rainbow trout *(Salmo gairdneri)*. Canadian Journal of Microbiology **12:**763–770.

Saari, T. H., Quan, T. J. 1975. Waterborne *Yersinia enterocolitica* in Colorado. Abstracts of the Annual Meeting of the American Society for Microbiology **1975:**45.

Savage, D. C. 1972. Survival on mucosal epithelia, epithelial penetration and growth in tissues of pathogenic bacteria, pp. 25–29. In: Smith, H., Pearce, J. H. (eds.), Microbiol pathogenicity in man and animals. London: Cambridge University Press.

Schiemann, D. A., Toma, S. 1978. Isolation of *Yersinia enterocolitica* from raw milk. Applied and Environmental Microbiology **35:**54–58.

Schleifstein, J., Coleman, M. B. 1939. An unidentified microorganism resembling *A. lignieri* and *Past. pseudotuberculosis* and pathogenic for man. New York State Journal of Medicine **39:**1749–1753.

Schleifstein, J., Coleman, M. 1943. *Bacterium enterocoliticum,* p. 56. Annual Report of the Division of Laboratories and Research. New York State Department of Health.

Shenkman, L., Bottone, E. J. 1976. Antibodies to *Yersinia enterocolitica* in thyroid disease. Annals of Internal Medicine **85:**735–739.

Smith, J. E., Thal, E. 1965. A taxonomic study of the genus *Pasteurella* using a numerical technique. Acta Pathologica et Microbiologica Scandinavica **64:**213–223.

Sonnenwirth, A. C. 1970. Bacteremia with and without meningitis due to *Yersinia enterocolitica, Edwarsiella tarda, Comamonas terrigena* and *Pseudomonas maltophilia*. Annals of the New York Academy of Sciences **174:**488–502.

Thal, E. 1954. Untersuchungen über *Pasteurella pseudotuberkulosis*. Thesis. Berlingska Boktryckeriet, Lund.

Toma, S., Deidrick, V. R. 1975. Isolation of *Yersinia enterocolitica* from swine. Journal of Clinical Microbiology **2:**478–481.

Topping, N. H., Watts, C. E., Lillie, R. D. 1938. A case of human infection with *B. pseudotuberculosis rodentium*. Public Health Reports **53:**1340–1352.

Truche, C., Bauche, J. 1933. La pseudtuberculose chez la poule et le faisan. Bulletin de l'Academie Veterinaire de France **1:**43–46.

Tsubokura, M., Otsuki, K., Itagaki, K. 1973. Studies on *Yersinia enterocolitica* I. Isolation of *Y. enterocolitica* from swine. Japanese Journal of Veterinary Science **35:**419–424.

Une, T. 1977a. Studies on the pathogenicity of *Yersinia enter-*

ocolitica. I. Experimental infection in rabbits. Microbiology and Immunology **21**:349–363.

Une, T. 1977b. Studies on the pathogenicity of *Yersinia enterocolitica*. II. Interaction with cultured cells in vitro. Microbiology and Immunology **21**:365–377.

Une, T. 1977c. Studies on the pathogenicity of *Yersinia enterocolitica*. III. Comparative studies between *Y. enterocolitica* and *Y. pseudotuberculosis*. Microbiology and Immunology **21**:505–516.

Une, T., Zen-Yoji, H., Maruyama, T., Yanagawa, Y. 1977. Correlation between epithelial cell infectivity *in vitro* and O-antigen groups of *Yersinia enterocolitica*. Microbiology and Immunology **21**:727–729.

Vandepitte, J., Wauters, G., Isebaert, A. 1973. Epidemiology of *Yersinia enterocolitica* infections in Hungary, pp. 111–119. In: Winblad, S. (ed.), Contributions to microbiology and immunology, vol. 2. Basel, New York: Karger.

Van Loghem, J. J. 1944. The classification of the plague bacillus. Antonie van Leeuwenhoek Journal of Microbiology and Serology **10**:15–16.

Van Noyen, R., Vandepitte, J., Wauters, G. 1980. Nonvalue of cold enrichment of stools for isolation of *Yersinia enterocolitica* serotypes 3 and 9 from patients. Journal of Clinical Microbiology **11**:127–131.

Vantrappen, G., Agg, H. O., Ponette, E., Geboes, K., Bertrand, Ph. 1977. *Yersinia enteritis* and *enterocolitis*: Gastroenterological aspects. Gastroenterology **72**:220–227.

Wauters, G. 1970. Contribution a l'etude de *Yersinia enterocolitica*, Thèse d'agrégé. Universite Catholique de Louvain, Vander, Louvain.

Wauters, G. 1973. Correlation between ecology, biochemical behaviour, and antigenic properties of *Yersinia enterocolitica*, pp. 38–41. In: Winblad, S. (ed.), Contribution to microbiology and immunology, vol. 2. Basel, New York: Karger.

Wayson, N. E. 1925. A disease in wild rats with gross pathology resembling plague. Public Health Reports **40**:1975–1979.

Weissfeld, A. S., Sonnenwirth, A. C. 1980. *Yersinia enterocolitica* in adults with gastrointestinal disturbances: Need for cold enrichment. Journal of Clinical Microbiology **11**:196–197.

Wetzler, T. F. 1970. Pseudotuberculosis, pp. 449–468. In: Bodily, H. L., Updyke, E. L., Mason, J. O. (eds.), Diagnostic procedures for bacterial, mycotic and parasitic infections, 5th ed. New York: American Public Health Association.

Wetzler, T., Hubbert, W. T. 1968a. *Yersinia enterocolitica* in North America. Symposium Series in Immunobiological Standards **9**:343–356.

Wetzler, T. F., Hubbert, W. T. 1968b. *Pasteurella pseudotuberculosis* in North America. Symposium Series in Immunobiological Standardization **9**:33–44.

Wilson, G. S., Miles, A. A. 1964. In: Topley and Wilson's principles of bacteriology and immunity, vol. 1,2, 5th ed. London: Arnold.

Wilson, H. D., McCormick, J. B., Feeley, J. C. 1976. *Yersinia enterocolitica* infection in a 4 month old infant associated with infection in household dogs. Journal of Pediatrics **89**:767–769.

Wise, D. R., Uppal, P. K. 1972. Osteomyelitis in turkeys caused by *Yersinia pseudotuberculosis*. Journal of Medical Microbiology **5**:128–130.

Zen-Yoji, H., Sakai, S., Maruyama, T., Yanagawa, Y. 1974. Isolation of *Yersinia enterocolitica* and *Yersinia pseudotuberculosis* from swine, cattle and rats at an abattoir. Japanese Journal of Microbiology **18**:103–105.

Zink, D. K., Feeley, J. C., Wells, J. G., Vanderzant, C., Vickery, J. C., Roof, W. D., O'Donovan, G. A.: 1980. Plasmid mediated tissue invasiveness in *Yersinia enterocolitica*. Nature **283**:224–226.

Zink, D. L., Feeley, J. C., Wells, J. G., Vickery, J. C., O'Donovan, G. A., Vanderzant, C. 1978. Possible plasmid-mediated virulence in *Yersinia enterocolitica*. Tenth Annual Gulf Coast Molecular Biology Conference, Corpus Christi, Texas, January, 1978. Special Publication Number 5. Texas Journal of Science, O'Donovan, G. A., Womack, J. E. (eds.), Texas Academy of Science, Box 10979, San Angelo, Texas.

CHAPTER 100

Yersinia pestis

THOMAS J. QUAN

Yersinia pestis, the etiological agent of plague, is considered an infectious parasite, primarily of rodents. Human infections have resulted from the bite of an infective flea and from the handling of infected carcasses of rodents or other animals.

Although plague has not been a public health problem of great magnitude in the United States, it remains a significant threat for several reasons. Since its introduction in 1900, the agent has become established among various wild rodent hosts in large areas of the western United States. The human population of the western United States is expanding at a rapid rate, and the potential for direct or indirect contact between human and reservoir mammals is expanding proportionately. Because plague infection in humans has been relatively rare, physicians are not likely to consider plague in their differential diagnosis for acutely ill patients unless they practice in well-known plague enzootic areas or have had previous experience with the disease. Untreated or inappropriately treated bubonic plague usually disseminates throughout the body and may cause plague pneumonia; as a result, a small but definite possibility arises that the infection will become transmissible via airborne droplets. At this stage the intermediary services of an arthropod vector are no longer needed, and direct man-to-man transmission may occur. Primary pneumonic plague is highly infectious and rapidly fatal if untreated. Although epidemics of primary pneumonic plague have not occurred in recent years, at least 12 patients have developed pneumonia secondary to bubonic or septicemic plague in the United States during the period 1974–1977 (Center for Disease Control, 1976, 1977a, 1978). That no family contacts or attending medical personnel of these patients contracted primary pneumonias was probably due to early isolation of the patient or suitable antibiotic prophylaxis of exposed persons or both in most situations.

Fictional, but accurate, vivid descriptions of plague epidemics were written by such notable authors as Boccaccio, in *Decameron;* Defoe, in *Journal of the Plague Year;* and Camus, in *The Plague.* For more than a brief outline of the historical aspects of plague, the reader is referred to the excellent detailed accounts of plague that have been presented by Hirst (1953), Link (1955), Shrewsbury (1970), and Pollitzer (1954). The ancestral home of *Y. pestis* is believed to be the Central Plateau of the Yunnan Province of China, and plague has probably spread to other areas of the world from this ancient focus periodically throughout recorded history via established mercantile trade routes. Perhaps the first written record of human plague is that given in 1 Samuel v–vi, which reported occurrence of the disease in about 1320 B.C. Other historically important pandemics include the one that occurred during the reign of the Roman Emperor Justinian, 542–600 A.D.; the "Black Death" of the fourteenth century, which is said to have caused the death of fully one-fourth of the population of Europe; and pandemics of the fifteenth, sixteenth, and seventeenth centuries, among which were the "Great Plagues" of London and Marseilles.

The current pandemic was initiated in the ancestral Asian focus in the middle of the nineteenth century and slowly spread across the world. By 1900, the first recognized human plague in the United States was reported in San Francisco. Between 1900 and 1924, 493 human plague cases were recorded, most of which were thought to have been transmitted by the rat-to-flea cycle or via pneumonic dissemination from another patient (Kartman, 1970; Link, 1955). Since 1924, however, all naturally acquired human plague cases in this country have been associated directly or indirectly with wild rodent sources in the western United States. *Y. pestis* was introduced into several areas of the eastern United States via ship-borne rats. Several human plague cases were acquired in Galveston, Texas; New Orleans, Louisiana; and Pensacola, Florida; however, long-term foci have not been established in these areas. Evidence of animal plague has been found in 15 of the 17 contiguous western United States, but most of the recent human cases were acquired in New Mexico, Arizona, and California, although an occasional human case was acquired in other states (Colorado, Oregon, Idaho, or Utah) (Center for Disease Control, 1976, 1977a, 1978).

From 1950 through 1969, 38 human plague cases were reported in the United States—a yearly average of 1.9 cases. From 1970 through 1977, however, the incidence of plague in the human population has increased in this country (and in other areas as well). Plague was confirmed by bacteriological or serological tests in 13 individuals during 1970; 2 in 1971; 1 in 1972; 2 in 1973; 8 in 1974; 20 in 1975; 16 in 1976; and 19 in 1977; for a total of 81 cases and a yearly average of 10.1. Of the 81 patients with plague, 11 died of the infection, for a case fatality rate of 13.6%. This fatality rate may be considered quite high for a disease as easily cured as plague, but it may be attributed to several factors, not the least of which were (i) delay by the patient in seeking medical help and (ii) lack of recognition of the disease as plague.

As reflected by these figures, plague is not only of historical interest. It is a continuing threat to the public health in those areas where foci have been established and in areas where plague may be imported.

In addition to the United States, other locations where plague has been reported among human populations during recent years are the temperate regions of South America (Bolivia, Brazil, Ecuador, and Peru), Africa (Angola, Kenya, Lesotho, Libya, Madagascar, Namibia, Rhodesia, Tanzania, and Zaire), and Asia (Burma, Cambodia—also known as the Khmer Republic—and Vietnam) (Velimirovic, 1973, 1974; WHO, 1974–77). In some of these areas human infections with *Yersinia pestis* are acquired primarily from commensal rats or fleas, while in others infections result primarily from contact with native wild rodents or their fleas.

Plague has been one of the diseases of man definitely associated with unsanitary conditions. This association was made especially evident during the conflict in Vietnam in the late 1960s and early 1970s when thousands of human plague cases were reported yearly. War and the attendant destruction of urban areas, the displacement of the population and subsequent crowding into makeshift shelters, and the disruption of sanitary facilities created suitable, if not ideal, conditions for the man-to-man as well as the rat-to-rat spread of plague (Kartman, 1970).

War, however, is not required for the creation of ideal conditions for the maintenance and proliferation of *Y. pestis*. Part of the increased incidence of human plague in the United States may be attributed to human activities. For example, in the southwestern United States especially, there has been a strong movement among builders and residents to keep urban areas as "natural" as possible. Very little alteration of the environment is permitted, and natural vegetation is used in place of more typical lawn and garden landscaping. These conditions have permitted the native rodents to exist in close proximity to the human residents, and the possibilities for contact, direct or indirect, between the two are increased.

International Health Regulations (WHO, 1974) require that human plague cases and large-scale plague epizootics be reported. Signatory nations also are required to conduct plague surveillance and institute control measures to prevent possible international spread. It should be noted that modern transport facilities make it possible for a person in early, prepatent stages of plague to board an aircraft and literally travel around the world before symptoms appear. If he traveled to nonendemic areas, a new focus could be established. Although no patients are known to have acquired plague in the United States and then traveled beyond national borders, three patients are known to have traveled from enzootic areas of the United States to areas where human plague has never been seen (e.g., Colorado to eastern Texas; New Mexico to Massachusetts; and New Mexico to California; all three travelers died at their destination cities, with their illness undiagnosed as plague until after postmortem examination) (Center for Disease Control, 1976; Kartman, 1970).

Habitats

Yersinia pestis is best understood as a parasitic bacterium, and its habitat may be considered essentially dichotomous—partly within the mammalian host and partly within the arthropod vector. Additionally, some reports support the theory of a saprophytic existence in the soil of burrows occupied by plague-infected rodents. Whether or not this burrowing or telluric plague is a significant part of the cycle has yet to be resolved. *Y. pestis* has been recovered from soil samples collected in burrows up to 11 months after the occupants died of plague infection. The recovered organisms were fully virulent, because the method of isolation would not have yielded nonvirulent forms. It was hypothesized that this form of existence might account for the long-term conservation of the agent during interepizootic periods. Burrows of animals that have died of plague, if left unoccupied by new hosts, collapse about the plague-infected carcasses. Plague organisms seep into the soil as the carcass decomposes and, in a favorable environment, survive for prolonged periods. Eventually, another burrowing rodent ploughs through the contaminated soil and becomes infected by direct contact with plague organisms. The new host serves as a new reservoir and, as it becomes bacteremic, serves to infect its fleas. Should this new host die, its fleas will seek other hosts and the typical flea-borne cycle is reinstituted (Baltazard et al., 1963; Karimi, 1963; Mollaret, 1963; and Mollaret et al., 1963). While definitive proof has not been obtained for the telluric cycle, the reoccurrence of plague in areas where no activity has been seen for varying periods of times lends some support for the hypothesis.

The Mammalian Host

Pollitzer (1954) listed 199 species of 11 rodent and lagomorph families as hosts that are naturally infected with *Yersinia pestis* in various areas throughout the world. Among other mammals considered to be important natural, but occasional, hosts for *Y. pestis* were listed an additional 25 species including carnivores. While the classic plague pandemics and perhaps those of the current Asian and Southeast Asian foci have been associated primarily with commensal rodent-to-flea cycles, wild rodent plague has become fairly well established in many areas. In the United States, as mentioned in the Introduction of this chapter, all naturally acquired human plague cases that have occurred since 1924 have had as the ultimate sources of infection direct or indirect contact with plague-infected wild rodents. In the United States, the most commonly involved rodent species have been rock or ground squirrels (*Spermophilus variegatus, S. beecheyi, S. lateralis, S. beldingi*), chipmunks (*Eutamias* species), prairie dogs (*Cynomys* species), woodrats (*Neotoma* species), and rabbits (*Sylvilagus* species) (Center for Disease Control, 1976, 1977a, 1978). Humans have developed plague infections after exposure to various other mammalian or arthropod sources as well, but the species named above are the principal sources incriminated repeatedly during recent decades.

That human plague is only incidental to the rodent-to-flea cycle is demonstrated by large-scale epizootics among various rodent species without the occurrence of human cases, although ample exposure presumably occurred. Such epizootics happened in 1976 in the towns of Aspen and Vail, Colorado, among *Spermophilus richardsoni,* and in Denver and Fort Collins, Colorado, and Cheyenne, Wyoming, among tree squirrels, *Sciurus niger* (Center for Disease Control, 1977a).

Mammals have exhibited varying degrees of resistance to plague infection. Some species, e.g., kangaroo rats, are considered refractory; others, e.g., prairie dogs and house mice, are known to be very susceptible. In each mammalian population, however, different individuals have heterogeneous responses to plague infection, some surviving inoculation of very high numbers of *Y. pestis,* others succumbing after inoculation of relatively few organisms. This heterogeneous response is considered essential if the organism is to maintain a parasitic life style, for if every animal affected were to succumb, the organism would itself be destroyed.

The effects of *Yersinia pestis* infection in mammals range from asymptomatic to acutely and fulminately fatal. In most animals found dead of plague, there are no grossly visible pathological lesions. In murine animals, death is rapid (2–4 days after experimental inoculation) because of the action of the murine toxin of *Y. pestis.* In other rodents the dis-ease progresses more slowly, and in those surviving more than 4 or 5 days after inoculation, subserosal vascular engorgement and enlarged nodes with edema and hemorrhage may also be seen. The spleen is enlarged as much as four times normal size, is engorged with blood, and appears dark red. Sometimes microfoci (*Y. pestis* colonies) are apparent on the surface and on cut sections of the spleen and liver. Extensive lesions may be formed in the lungs if the animal has survived 5–7 days, and the involved lungs are edematous and hemorrhagic. If properly stained, smears of blood or other tissues will demonstrate myriad plague organisms.

In man, who might be classified as a moderately susceptible host, onset of symptoms is generally sudden and begins 2 to 7 days after exposure; the condition is marked by high fever, headache, myalgia, shaking chills, and pain in the inguinal or axillary areas. The pain is associated with the development of a "bubo," an inflamed lymph node, which may or may not be present early after onset. Buboes develop in the lymph node draining the area of the body where inoculation of the organism occurred either via flea bite or direct contact with infected carcasses.

At the point of entry, organisms are phagocytized by the patient's polymorphonuclear cells, in which *Y. pestis* cells are destroyed, and by mononuclear cells, in which *Y. pestis* not only survives but multiplies and produces a capsular envelope conferring resistance to further phagocytic action upon the organism (Cavanaugh and Randall, 1959). The infected mononuclear cells are carried to the regional lymph node(s), and as *Y. pestis* grows the lymph node(s) becomes inflamed and swells, forming the bubo. Usually buboes develop in the groin (femoral or inguinal nodes) after flea bite infection and in the axilla after direct (hand) contact. However, buboes have been reported in lymph nodes of many body regions, e.g., in cervical, popliteal, mesenteric, mediastinal, and epithrochlear areas.

If the infection is untreated, the bubo undergoes necrosis and abscess formation and disseminates infection via either the lymphatic channels or the blood stream to other organs throughout the body, e.g., spleen, liver, lungs. Once established in the lungs, the organisms may become transmissible to new hosts via the airborne droplet route. Close, fairly intimate contact apparently is necessary for direct man-to-man spread, because no secondary cases of primary plague pneumonia were associated with any of the 12 plague patients who developed pneumonia in 1974–1977 in the United States. Although preventing the development of infectious plague pneumonia is of prime importance from the public health standpoint, from the individual patient's standpoint, establishment of foci of infection in organs other than the lungs also creates problems. For example, central nervous system invasion results in plague meningitis, which is not as amenable to antibiotic treatment as uncomplicated plague. If infection is allowed to progress, disseminated intravascular coagulation problems and petechial hemorrhages develop, probably as a result of the action of one or more of the plague endotoxins. Tissue damage and necrosis ensue. When these events occur in the lungs, the patient's respiration is compromised and the patient eventually suffocates.

With adequate, appropriate antibiotic treatment, however, the organisms are destroyed and the patient survives. Appropriate drugs for treatment of plague include streptomycin and tetracycline for uncomplicated courses of infection; chloramphenicol is effective when there is meningeal involvement. In areas of the world where these antibiotics are not generally available, the sulfa drugs are still efficacious. Sulfamethoxazole-trimethroprim was used with success in Vietnam (Ai et al., 1973). Bacteriolytic drugs, e.g., streptomycin, must be administered carefully so that the patient is not overwhelmed by the release of large amounts of _Yersinia pestis_ endotoxin in a short period of time. Patients recovering from adequately treated plague have few, if any, residual difficulties. Some patients may have developed such large buboes that resorption of resolution by natural means is not possible and surgical incision and drainage is required. Recovery from plague probably confers long-term immunity to future plague infections. Should second infections occur, milder cases and fewer fatalities would be expected (Butler and Hudson, 1977).

Suspected plague patients should be treated with specific antibiotics as soon as specimens for laboratory processing have been collected. Treatment should not be delayed until laboratory confirmation of the diagnosis of plague is obtained.

Often a retrospective confirmation of the diagnosis is possible by serological tests when no bacterial culture has been isolated. Fourfold increases in titer to specific _Y. pestis_ antigens between sera drawn during acute and convalescent periods is considered definitive evidence of _Y. pestis_ infection. Currently, the best test for serodiagnosis of plague is the passive hemagglutination test performed with _Y. pestis_ fraction 1 antigen (Bahmanyar and Cavanaugh, 1976; Goldenberg, Hudson, and Kartman, 1970a). Other serological tests occasionally used include complement fixation or simple whole cell bacterial agglutination. Since _Y. pestis_ antigens are not commercially available, however, serum samples usually are sent to reference laboratories for testing.

The Flea Vector

Pollitzer (1954) summarized the vector efficiency studies of a number of researchers and stated that 13 species of commensal rodent fleas probably play important roles in maintaining and transmitting plague. Eighty-five species of wild rodent fleas found naturally infected or experimentally capable of being infected were listed as probable vectors. Undoubtedly many other species, if not all, would prove to have some measurable success in transmitting _Yersinia pestis_.

In the flea, _Y. pestis_ produces a truly enteric disease. The flea obtains the infection by feeding upon a bacteremic plague host. Early studies by Bacot and Martin (1914) and Douglas and Wheeler (1943) have shown that mechanical transmission can occur if the flea attempts to feed on a new host soon after contaminating its mouthparts, but biological transmission does not occur until a week or two after the original infective blood meal. Once the infective meal is obtained, _Y. pestis_ organisms begin to multiply in the flea's stomach. Growth continues until the stomach is effectively blocked. As the flea attempts to feed, blood from the new host is siphoned up to the block and regurgitated along with plague organisms back into the blood stream of the host. Because the block of plague organisms prevents the flea from obtaining necessary nutrition, the flea becomes increasingly hungry and attempts to feed many times, infecting each new host it encounters. If the block persists the flea dies of dehydration and starvation. Under proper climatic conditions of temperature, relative humidity, and water vapor pressure, the mass of organisms in the flea's stomach may break up and be passed by the flea, which then resumes normal feeding (Cavanaugh and Marshall, 1972).

Isolation

Yersinia pestis may be readily isolated from human patients by direct culture of (i) serial blood specimens, (ii) bubo aspirates, (iii) swabs taken from pustules or ulcers at the flea bite site, (iv) sputum from patients with plague pneumonia, or (v) tissue specimens (liver, spleen, lung, lymph node, bone marrow, etc.) obtained by biopsy, excision, or autopsy.

Y. pestis also is easily isolated from plague-infected animals if specimens of the same sort are obtained and processed either from the living but moribund animal or from animals that have been dead only a short time. When cultivated in vitro, _Y. pestis_ is not a good competitor among the myriad organisms contained in the gastrointestinal system; thus normal flora and the flora associated with decomposing flesh usually overgrow and mask the presence of _Y. pestis_.

Similarly, _Y. pestis_ may be recovered by direct culture of fleas if proper techniques are used, but because of overgrowth of the normal flora of the fleas direct culture is not usually done.

For any specimen, but especially for specimens

being examined for *Y. pestis* but considered heavily
contaminated with other organisms as well, an excellent means of recovering the plague organisms is
by animal inoculation. Suspect specimens are triturated with sterile physiological saline (sterile sand, if
necessary), and a small volume of the suspension is
inoculated into laboratory animals of known susceptibility to plague. Routes of inoculation may be intraperitoneal for specimens considered to contain
pure or almost pure cultures of *Y. pestis;* subcutaneous for specimens considered to be moderately
contaminated, e.g., flea suspensions; or percutaneous for heavily contaminated specimens, e.g., tissues excised from badly decomposed carcasses.
Regardless of the inoculation route chosen, if *Y.
pestis* is present in the suspension, this organism will
establish a progressive infection leading to the experimental animal's death, whereas the contaminating flora will be contained or destroyed or both at
the inoculation site. Tissues of the dead experimental animals are excised and processed bacteriologically to recover the agent.

Only the virulent forms of *Y. pestis* can be recovered by using animal inoculation techniques. If
atypical avirulent varieties are being sought, direct
culture of specimens on inhibitory or selective agars
should be used (Williams et al., 1978).

Y. pestis may be demonstrated in infected tissues
by microscopic examination of smears stained by
fluorescent-antibody (FA) techniques or more generally by polychromatic stains such as Wayson or
Giemsa. FA techniques have been developed that
require antisera prepared against either whole cell *Y.
pestis* antigens or a more specific antigen of the capsular envelope termed fraction 1 (Goldenberg, Hudson, and Kartman, 1970b). FA tests provide rapid,
presumptive diagnosis with fairly good reliability.
When smears containing *Y. pestis* are properly
stained by either the Wayson or the Giemsa technique, large bipolar-staining cells resembling closed
safety pins can be demonstrated.

Wayson stain is prepared by dissolving 0.20 g
basic fuchsin and 0.75 g methylene blue in 20 ml
95% ethanol (dye contents: 90%). The solution is
slowly poured into 200 ml 5% aqueous phenol.

Air-dried, methanol-fixed (3–5 min) smears are
stained for 10–20 s, rinsed with tap water, and
blotted dry.

Y. pestis is a Gram-negative rod or coccobacillus
that grows well, but slowly, on many commonly
used bacteriological media including those used for
the recovery of enteric pathogens. The organism is
not particularly fastidious, but optimal growth is
achieved with an enriched agar medium supplemented with 5% blood. Even on such enriched
media the colonies develop slowly, reaching pinpoint size (0.1–0.2 mm) within 24 h and 1.5–
2.0 mm within 48 h at the optimal incubation temperature of 28°C. Characteristically, colonies of vir

ulent *Y. pestis* have a "hammered-copper" appearance usually evident after incubation for 36–48 h at
28°C or 37°C and best seen with a hand lens or stereoscopic microscope. In stationary, pure broth cultures of *Y. pestis,* a stalactite type of growth is produced, with turbidity evident only at the bottom and
up one side of the tube. The broth is never fully
turbid unless the tube is shaken, and when the tube
is slightly agitated the visible growth falls to the
bottom. The cell yield after incubation for 24 hours
even in enriched fluid media seldom exceeds more
than $1-2 \times 10^8$ organisms/ml.

Gram-negative rods or coccobacilli that demonstrate bipolar staining with Wayson or Giemsa stain,
fluoresce with *Y. pestis*–specific fluorescent antibody conjugates, and have both the colonial morphology and pattern of growth in stationary pure
broth culture may be presumptively called *Y. pestis.*
Confirmatory identification tests may be performed
at the local laboratory, or the culture may be referred
to a regional or reference laboratory.

Identification

Confirmation of cultures as *Yersinia pestis* is based
on test results compatible with those listed in
Table 1.

In addition to the characteristics mentioned in the
preceding paragraphs for tentative identification of
Y. pestis, the following tests are used to confirm this
identification: sensitivity to *Y. pestis*–specific bacteriophage, pathogenicity for laboratory animals,
and agglutination with specific antiserum. Bacteriophage sensitivity tests may be performed quite simply by using paper strips impregnated with a bacteriophage suspension of high titer (samples available
from plague reference laboratories). Cultures of
known *Y. pestis,* suspect *Y. pestis,* known *Y. pseudotuberculosis* type IA, and perhaps other enteric
(non-*Yersinia*) organisms are streaked in parallel
lines across the surface of two blood agar plates. A
bacteriophage-impregnated paper strip is placed
across the lines of inoculation. One of the blood agar
plates is incubated at 37°C and the other at
20–22°C. Visible zones of lysis will be seen at the
junction of the paper strip and streaks of *Y. pestis*
growth on both plates. *Y. pseudotuberculosis* and an
occasional strain of other enteric bacteria will be
lysed only on the plate incubated at 37°C.

For agglutination tests, cultures of suspect *Y.
pestis* are suspended in saline and killed by adding
formalin to a final concentration of 1%. The formalinized suspension is allowed to stand at room temperature for at least an hour. This suspension, which
still must be treated as though viable, is centrifuged
to sediment the bacterial cells. The cells are then
washed twice with physiological saline, resuspended
in fresh physiological saline, and mixed with equal

Table 1. Characteristics and biochemical reactions of *Yersinia pestis*.[a]

Gram stain	Negative rods/coccobacilli	Size: $0.5-0.8 \times 1.0-2.0$ μm	
Motility	Negative	Capsule	Negative[b]
Oxidase	Negative	Acid, without gas, from:	
ONPG	Positive	Arabinose	Positive
Catalase	Positive	Cellobiose	Negative
Voges-Proskauer	Negative	Fructose	Positive
Indole	Negative	Fucose	Negative
Citrate utilization	Negative	Galactose	Positive
Urease	Negative[c]	Glucose	Positive
H$_2$S production	Negative	Lactose	Negative
Nitrate reduction	Positive[d]	Maltose	Positive
Tetrathionate reduction	Negative	Mannose	Positive
		Melezitose	Negative
Esculin hydrolysis	Positive	Melibiose	Variable
Gelatin hydrolysis	Negative	Raffinose	Negative
Growth in KCN	Negative	Sucrose	Negative
Growth on MacConkey agar	Positive	Trehalose	Positive
		Xylose	Variable
D-Tartrate	Positive	Adonitol	Negative
Litmus milk	No change	Dulcitol	Negative
Arginine dihydrolase	Negative	Erythritol	Negative
Lysine decarboxylase	Negative	Glycerol	Positive
Ornithine decarboxylase	Negative	Inositol	Negative
Pigmentation on CR or H Agar[e,f]	Positive	Mannitol	Positive
		Sorbitol	Positive
Production of:[f]			
Fraction 1	Positive	Amygdalin	Positive
VW antigens	Positive	Glycogen	Negative
Coagulase	Positive	Salicin	Variable
Fibrinolysin	Positive	Starch	Weakly positive
Production of pesticin 1[f]	Positive		
Sensitivity to Y.p. bacteriophage[f]	Positive		
Pathogenicity for lab mice[f]	Positive		

[a] Adapted from Mollaret and Thal (1974).
[b] True capsule lacking; envelope may be present.
[c] On initial isolation, some strains produce urease.
[d] Varieties of *Y. pestis* differ in reduction of nitrate and fermentation of glycerol.
[e] CR or H agar: Congo red agar or hemin agar. Virulent colonies are pigmented red or black respectively.
[f] Tests for special characteristics should be done by a plague reference laboratory.

volumes of dilutions of high-titered antiplague serum (Bahmanyar and Cavanaugh, 1976).

Pathogenicity tests in animals should be carried out only in facilities that permit biological containment. It is especially important that hematophagous arthropods be totally excluded from areas where plague-infected animals are held. Cultures of suspect *Y. pestis* are diluted in physiological saline to contain approximately 100, 1,000, or 10,000 organisms per 0.1 ml. Experimental animals (usually mice for economic reasons) of known susceptibility

to *Y. pestis* are inoculated subcutaneously with 0.1 ml of one of the various concentrations. If the cultures are virulent, *Y. pestis*-inoculated mice will become ill and die, usually within 2–4 days, rarely longer. Tissues from mice dying after inoculation are excised and cultured for recovery of the inoculated organism, or at least are examined by the FA test for presumptive evidence of plague as the cause of death.

Use of miniaturized systems or kits for the identification of human pathogens (for example, the

Minitek[1] system, BBL Microbiology Systems or the API 20E enteric identification kit) is increasing in clinical laboratories in the United States. When used properly, such systems can work well and can expedite the correct identification of *Y. pestis* cultures. It should be remembered, however, that *Y. pestis* grows slowly and may often require heavy inoculation of the test system or longer incubation than is recommended for the more common enteric bacteria.

In vitro tests that are used to characterize cultures of *Y. pestis* and that usually yield positive results only with virulent strains include pigmentation on Congo red or hemin agars; production of fraction 1 and VW antigens; production of coagulase and fibrinolysin enzymes; and production of pesticin 1 bacteriocin. Full details for preparing the various media and reagents and for performing these tests are given by Surgalla, Beesley, and Albizo (1970). None of these tests are necessary for the confirmation of a culture as *Y. pestis,* but they are valuable supplementary tools, especially in the absence of laboratory animals. Although some of these tests may be readily carried out at the clinical laboratory (e.g., coagulase, demonstration of F1 production by FA tests), because most require specialized media and/or reagents that might not be readily available, they are best performed at plague reference laboratories.

Cultures of *Y. pestis* are most likely to be incorrectly identified when a diagnosis of possible plague has not been considered or the microbiologist is not asked to test for *Y. pestis.*

Y. pestis is easily confused with *Y. pseudotuberculosis* (a very closely related species) and, on occasion, with indole-negative *Shigella* species, especially *S. sonnei.* Perhaps the easiest test for distinguishing *Y. pestis* from other species is sensitivity to *Y. pestis*–specific bacteriophage at 20–22°C. Biochemical tests that might be used to differentiate *Y. pestis* from *Y. pseudotuberculosis* include urease, motility at 25°C, and acid production from melibiose and rhamnose, all of which are usually positive for *Y. pseudotuberculosis* but negative for *Y. pestis.* Biochemical tests do not provide reliable means for distinguishing *Y. pestis* from some varieties of *Shigella,* so bacteriophage sensitivity tests and serological tests need to be used also.

Three varieties of *Y. pestis* were described by Devignat (1953) and were differentiated on the basis of glycerol fermentation and nitrate reduction test results. The three varieties are glycerol-positive, nitrate-positive *Y. pestis antiqua,* considered the ancestral variety; glycerol-positive, nitrate-negative *Y. pestis mediaevalis;* and glycerol-negative, nitrate-positive *Y. pestis orientalis.* All three varieties are isolated in different areas of the world. For example, *Y. pestis orientalis* has been the only variety isolated from plague-infected humans, animals, or fleas in the United States.

Other Studies

No definitive studies have been made yet on more than a few individual strains of *Yersinia pestis* with relation to DNA base ratios or DNA hybridization experiments to assess relatedness to other bacterial species. Mollaret and Thal (1974) stated that the G+C content for the genus *Yersinia* is in the range of 45.8–46.8 mol% (T_m).

DNA hybridization experiments were performed by Ritter and Gerloff (1966) on a single strain of *Y. pestis* (cited as 19SP, but probably 195/P), and Brenner et al. (1976) included a single unspecified *Y. pestis* strain in their work on the other *Yersinia* species. Results of the two studies indicated relationships of 87%, 43%, 24%, 22%, and 16%, respectively, to single strains of *Y. pseudotuberculosis, Y. enterocolitica, Y. philomiragia* (labeled as unidentified isolate "CRO 319-031" by Ritter and Gerloff), *Escherichia coli,* and *Pasteurella multocida.* Further work defining DNA base ratios and DNA homologies with a larger sample of *Y. pestis* isolates is certainly desirable.

Discontinuous acrylamide gel electrophoresis techniques to distinguish major protein variants were used by Hudson and Quan (1975) and Hudson, Quan, and Bailey (1976) for the analysis of possible interspecific and intraspecific differences among the Yersinieae. Qualitative electrophoretic characters that would allow differentiation of major biotypes or serotypes within species were not found, though identifying characters could be distinguished at the species level. Results of these studies, which included 161 strains of *Y. pestis* from nine distinct geographic areas, allowed the disposition of the strains into 11 related groups comprising 57 separate electrophoretypic patterns that corresponded very well with the geographic origins of the strains.

This type of study used in conjunction with DNA homology studies and/or the more common biotyping tests such as those by Devignat (1953), which described the three accepted varieties of *Y. pestis,* might allow for individualized characterization of *Y. pestis* cultures. If sufficient numbers of isolates were thoroughly studied, a pattern of definite epidemiological/epizootiological value might emerge.

Notes on Safety Aspects

Yersinia pestis has been classified by the Center for Disease Control (CDC) as an especially hazardous agent (Class III) requiring special facilities and pro-

---

[1] Use of trade names is for identification only and does not constitute endorsement by the Public Health Service or by the U.S. Department of Health, Education, and Welfare.

tective equipment for personnel (Center for Disease Control, 1977b). It certainly is recommended that such facilities and equipment be used if available, but the author feels that for routine procedures like those used in clinical laboratories to tentatively identify *Y. pestis,* these stringent precautions are unnecessary if the usual, accepted aseptic techniques are maintained. For research purposes, for growth of large volumes of bacteria for antigen production, or for animal experimentation, extra safety measures are mandatory.

A plague vaccine is commercially available but is not generally recommended except for persons who are at high risk, namely, laboratory personnel routinely working with *Y. pestis,* caretakers of infected animals, and possibly travelers to enzootic areas who would have a high degree of contact with reservoir species.

Literature Cited

Ai, N. V., Hanh, N. D., Diem, P. V., Le, N. V. 1973. Cotrimoxazole in bubonic plague. British Medical Journal **4:**108–109.

Bacot, A. W., Martin, C. J. 1914. Observations on the mechanism of the transmission of plague by fleas. Journal of Hygiene. Plague Supplement III:423–439.

Bahmanyar, M., Cavanaugh, D. C. 1976. Plague manual. Geneva, Switzerland: World Health Organization.

Baltazard, M., Karimi, Y., Eftekhari, M., Chamsa, M., Mollaret, H. H. 1963. La conservation de la peste en fayer invetere. Hypotheses de Travail. Bulletin de la Société de Pathologie Exotique **56:**1230–1241.

Brenner, D. J., Steigerwalt, A. G., Falcao, D. P., Weaver, R. E., Fanning, G. R. 1976. Characterization of *Yersinia enterocolitica* and *Yersinia pseudotuberculosis* by deoxyribonucleic acid hybridization and by biochemical studies. International Journal of Systematic Bacteriology 26:180–194.

Butler, T., Hudson, B. W. 1977. The serological response to *Yersinia pestis* infection. Bulletin of the World Health Organization **55:**39–42.

Cavanaugh, D. C., Marshall, J. D. 1972. The influence of climate on the seasonal prevalence of plague in the Republic of Viet Nam. Journal of Wildlife Diseases **8:**85–94.

Cavanaugh, D. C., Randall, R. 1959. The role of multiplication of *Pasteurella pestis* in mononuclear phagocytes in the pathogenesis of flea-borne plague. Journal of Immunology **83:**348–363.

Center for Disease Control. 1976. Vector-Borne Diseases Division 1975 Report. Chapter 20. Plague Activities. Ft. Collins, Colorado.

Center for Disease Control. 1977a. Vector-Borne Diseases Division 1976 Report. Part II. Plague Activities. Ft. Collins, Colorado.

Center for Disease Control. 1977b. Lab Safety at the Center for Disease Control. HEW Publication 77-8118. Atlanta, Georgia.

Center for Disease Control. 1978. Vector-Borne Diseases Division 1977 Report. Part II. Plague Activities. Ft. Collins, Colorado.

Devignat, R. 1953. La peste antique du Congo belge dans le

cadre de l'Aistoire et de la Geographie. Institut Royal Colonial Belge, Memoirs **23:**1–48.

Douglas, J. R., Wheeler, C. M. 1943. Sylvatic plague studies. II. The fate of *Pasteurella pestis* in the flea. Journal of Infectious Diseases **72:**18–30.

Goldenberg, M. I., Hudson, B. W., Kartman, L. E. 1970a. *Pasteurella* infections. 1. *Pasteurella pestis.* Hemagglutination, pp. 434–435. In: Bodily, H. L., Updyke, E. L., Mason, J. O. (eds.), Diagnostic procedures for bacterial, mycotic and parasitic infection, 5th ed. Washington, D.C.: American Public Health Association.

Goldenberg, M. I., Hudson, B. W., Kartman, L. E. 1970b. *Pasteurella* infections. 1. *Pasteurella pestis.* Fluorescent antibody examination, p. 436. In: Bodily, H. L., Updyke, E. L., Mason, J. O. (eds.), Diagnostic procedures for bacterial, mycotic and parasitic infection, 5th ed. Washington, D.C.: American Public Health Association.

Hirst, L. F. 1953. The conquest of plague, p. 436. London: Oxford University Press.

Hudson, B. W., Quan, T. J. 1975. Electrophoretic studies of the yersiniae. American Journal of Tropical Medicine and Hygiene **26:**968–973.

Hudson, B. W., Quan, T. J., Bailey, R. E. 1976. Electrophoretic studies of the geographic distribution of *Yersinia pestis* protein variants. International Journal of Systematic Bacteriology **26:**1–16.

Karimi, Y. 1963. Conservation naturelle de la peste dans le soil. Bulletin de la Societe de Pathologie Exotique **56:**1183–1186.

Kartman, L. 1970. Historical and oecological observations on plague in the United States. Tropical and Geographical Medicine **22:**257–275.

Link, V. 1955. A history of plague in the United States. Public Health Monograph No. 26. Washington, D.C.: United States Government Printing Office.

Mollaret, H. H. 1963. Conservation experimental de la peste dans le soil. Bulletin de la Société de Pathologie Exotique **56:**1168–1182.

Mollaret, H. H., Karimi, Y., Eftekhari, M., Baltazard, M. 1963. La peste de fouissement. Bulletin de la Société Pathologie Exotique **56:**1186–1193.

Mollaret, H. H., Thal, E. 1974. *Yersinia.* In: Buchanan, R. E., Gibbons, N. E. (eds.), Bergey's manual of determinative bacteriology, 8th ed. Baltimore: Williams & Wilkins.

Pollitzer, R. 1954. Plague. World Health Organization Monograph Series No. 22. Geneva: World Health Organization.

Ritter, D. B., Gerloff, R. K. 1966. Deoxyribonucleic acid hybridization among some of the species of the genus *Pasteurella.* Journal of Bacteriology **92:**1838–1839.

Shrewsbury, J. F. D. 1970. A history of bubonic plague in the British Isles. London: Cambridge University Press.

Surgalla, M. J., Beesley, E. D., Albizo, J. M. 1970. Practical applications of new laboratory methods for plague investigation. Bulletin of the World Health Organization **42:**993–997.

Velimirovic, B. 1973. Reappearance of plague in Khmer Republic (Formerly Cambodia). Zeitschrift fur Tropenmedizin und Parasitologie **24:**265–270.

Velimirovic, B. 1974. Review of the global epidemiological situation of plague since the last Congress of Tropical Medicine and Malaria in 1968. Zentralblatt fur Bakteriologie, Parasitenkunde, Infektionskrankheiten und Hygiene. Abt. 1 Orig., Reihe A **229:**127–133.

Williams, J. E., Harrison, D. N., Quan, T. J., Mullins, J. L., Barnes, A. M., Cavanaugh, D. C. 1978. Atypical plague bacilli-isolated from rodents, fleas, and man. American Journal of Public Health **68:**262–264.

World Health Organization. 1974. International Health Regulations (1969), 2nd Annotated ed. Geneva: World Health Organization.

World Health Organization. 1974–1977. Weekly Epidemiological Record. Plague in 1973; **49:**253–254, Plague in 1974; **50:**317–318, Plague in 1975; **51:**237–238, Plague in 1976; **53:**229–230.

CHAPTER 101

The Genus *Shigella*

BERNARD ROWE and ROGER J. GROSS

Historical Development

During the last 25 years of the nineteenth century, the differentiation was made between amoebic and bacillary dysentery. Lösch (1875) isolated amoebae from the stools of dysentery patients and, over the ensuing 15 years, the individuality of amoebic dysentery and bacillary dysentery was established (Councilman and Lafleur, 1891).

Shiga (1898) used bacterial culture to investigate an epidemic of acute dysentery in Japan; using the serum from one of the patients, he identified the epidemic agent in the feces of 34 out of 36 cases. He used a variety of laboratory animals to study the pathogenic effects of this organism, which he characterized by simple biochemical tests but erroneously described as slowly motile. Chantemesse and Widal (1888) identified an organism from epidemic dysentery, and although the organism was inadequately characterized, some years later their cultures were reexamined and found to be identical with *Bacillus dysenteriae* of Shiga.

Kruse (1900) investigated an outbreak in Germany and described an organism which he found in almost every case; the organism was nonmotile but in other respects resembled Shiga's bacillus. Subsequent work confirmed that the two organisms were identical and nonmotile; the organism was named the Shiga-Kruse bacillus. In another investigation of dysentery in an asylum, Kruse (1901) found other bacilli which resembled *B. dysenteriae* but which were serologically distinct. Because the cases were less severe than those caused by the Shiga-Kruse bacillus, the organism was called *Bacillus pseudo-dysentery*.

Flexner (1900) together with Strong and Musgrave (1900) investigated dysentery in American troops in the Philippines. The organisms isolated in the Philippines were studied in Germany by Martini and Lentz (1902). Antisera produced in goats showed that the strains of dysentery bacilli formed two groups. The first contained the bacillus of Shiga and Kruse; the second contained the Flexner-Manilla bacillus, the Strong and Musgrave bacillus, and the pseudo-dysentery strains of Kruse. Subsequently, biochemical testing for the ability to produce acid from mannitol showed that the first group was negative and the second positive.

Hiss and Russell (1903) isolated a bacillus from the stools of a fatal case of dysentery in a child. This organism fermented mannitol and thus differed from the Shiga-Kruse bacillus. Hiss and Russell called it the Y bacillus.

Kruse et al. (1907) made a detailed study of the pseudo-dysentery bacilli based on biochemical and serological investigations and divided them into races (A to H). One of these, type E, was lactose-fermenting and was eventually named *Bacillus dysenteriae* Sonne. Schmitz (1917) isolated a non-mannitol-fermenting, indole-positive organism from cases of dysentery. This organism differed serologically from the Shiga-Kruse bacillus.

During the 1914–1918 war, British workers made an intensive study of the bacteriology of dysentery. Murray (1918) proposed a classification which included four members: *Bacillus dysenteriae* Shiga, Schmitz, Flexner, and Kruse E (Sonne). Three types had relatively simple antigenic compositions, and were easily differentiated from each other (Shiga, Schmitz, and Sonne). A fourth group had complex antigenic structures and differentiation was difficult; this group was named the paradysentery or Flexner group. Andrewes and Inman (1919) improved the serological classification of the Flexner group by quantitative analysis of their antigens. They described four antigens, V, W, X, and Z; four types could be recognized, depending on which antigen was predominant. For example, Andrewes' type V contained large amounts of V antigen but small amounts of W, X and Z. In addition, the Y bacillus of Hiss and Russell was found to contain all four antigens in equal proportions.

Boyd (1932) made extensive studies of the mannitol-fermenting dysentery bacilli and attempted to develop a serotyping scheme which elaborated upon the V, W, X, and Z types of Andrewes and Inman. In addition to describing several new Flexner serotypes (103, P119, 88), he established the serological rationale whereby each type possessed a type-specific antigen together with varying amounts of group antigens. The types X and Y possessed group antigens but no specific antigens; they were loss variants having lost their type-specific antigens.

Boyd (1938) identified six specific types of mannitol-fermenting dysentery bacilli which did not possess any of the Flexner antigens; eventually these types became *Shigella boydii* 1 to 6. Large and Sankaran (1934) and Sachs (1943) studied large numbers of non-mannitol-fermenting organisms from dysentery patients and identified five serotypes which were distinct from the Shiga or Schmitz organisms; these became *Shigella dysenteriae* 3 to 7.

Ewing (1949) proposed a scheme for the nomenclature of *Shigella,* and the Enterobacteriaceae Subcommittee (1954) approved a classification based on his proposal. *Shigella* was adopted as the generic name and the genus was defined as Gram-negative, nonmotile, nonsporeforming rods corresponding to *Shigella dysenteriae* (Shiga's bacillus) in morphology and staining properties. The biochemical characteristics of the genus are described in the "Identification" section of this chapter.

In the modern (1954) classification, the genus *Shigella* is divided into four subgroups. Subgroup A, *Shigella dysenteriae:* All types do not ferment lactose or mannitol; seven types were described; Shiga and Schmitz are types 1 and 2, respectively. Subgroup B, *Shigella flexneri:* All types ferment mannitol but not lactose; each type possesses a type-specific antigen together with common or group antigens, which may be found in other types in the subgroup; six types were described with subtypes; loss variants with only group antigens exist—X and Y variants. Subgroup C, *Shigella boydii:* All types ferment mannitol but not lactose; each type possesses a type-specific antigen but none of the antigens of *S. flexneri;* 11 types were recognized. Subgroup D, *Shigella sonnei:* All types ferment lactose and mannitol. Note: The Enterobacteriaceae Committee (1958) accepted three additional types for inclusion into *S. dysenteriae* and four additional types for inclusion into *S. boydii.*

Habitats

Clinical Aspects

Shigella species are the causative agents of bacillary dysentery. Dysentery or "bloody flux" was recognized by Hippocrates as a clinically identifiable form of diarrhea, but the differentiation of the dysenteric diseases by etiological agents did not commence until Lösch (1875) isolated amoebae from a case of dysentery and reproduced the disease in dogs by injecting the amoebae into the rectum. About 20 years later, Shiga (1898) isolated the dysentery bacillus; before that time it was believed that all dysentery was amoebic. During the following 30 years, many different types of dysentery bacilli were identified,

and the clinical and epidemiological distinction between amoebic and bacillary dysentery was established. Amoebic dysentery tends to run a chronic clinical course, frequently with liver abscess complications. Although certain geographical areas, notably in the tropics, have a high endemicity, epidemics do not occur. In contradistinction, bacillary dysentery is usually of short duration, extraintestinal involvement is rare, and epidemics are common.

For humans, the infecting dose of *Shigella* varies from 10 to 100 organisms in contrast to the high infecting dose (10^5 cells) required for salmonella food-poisoning (Dupont and Hornick, 1973). Children are more susceptible to shigella infection than adults (Shaw, 1953); in an epidemic, children are more likely to suffer with diarrhea, whereas adults often become symptomless carriers. Age-sex correlations show that, except in the child-bearing ages, infection is less frequent among females (Thomas and Tillet, 1973).

The incubation period is usually between 2 and 3 days, but may be as brief as 12 h. The onset of symptoms is usually sudden, and frequently the initial symptom is abdominal colic. In all but the very mild cases there is pyrexia, which is accompanied by general prostration in severe cases. In a typical case of bacillary dysentery, there are numerous stools of small volume containing blood and mucus. In a mild case, the frequency of stools is about eight in 24 h, whereas in severe cases as many as 30 stools may be passed. In some cases, the typical dysentery stools do not occur; instead, watery stools may be produced and these may vary from mild diarrhea to a fulminating cholera-like disease with severe loss of fluids. In a typical case of dysentery the symptoms last about 4 days, but exceptional cases may continue for 10–14 days.

S. sonnei infection usually causes a mild disease with short-term diarrhea. The disease associated with *S. flexneri* tends to be more severe, while *S. boydii* and *S. dysenteriae* infections produce a wide range of severity. Shiga's bacillus has caused many epidemics of severe disease, but even so some infections are mild. Host factors play an important part in determining the clinical symptomatology. Case fatality is usually low except when infection occurs in babies less than 3 months old, in malnourished patients, or in psychiatric hospitals.

Extraintestinal complications are rare, but some cases develop a sterile polyarthritis and eye disorders such as conjunctivitis or iritis. These complications usually appear about 10–14 days after the onset of dysentery, and there is evidence to suggest that these are manifestations of autoimmune disease and are a form of Reiter's syndrome. A similar symptom complex is seen after infection with *Yersinia enterocolitica,* and patients with histocompatibility antigen HLA B27 seem particularly prone to these sequelae. Chronic complications following bacillary dysentery are rare but may include peripheral neuritis and intestinal stenosis.

During the acute stage of the disease, the shigella organisms are excreted in large numbers in the feces; during recovery the numbers fall, although the organisms may remain in the feces for several weeks after the symptoms have subsided (Thomas and Tillett, 1973).

TREATMENT.   Most cases of shigellosis, especially those due to *S. sonnei,* are mild and do not require antibiotic therapy. Symptomatic treatment with the maintenance of hydration and electrolyte balance is all that is required. Treatment with a suitable antibiotic is necessary in the very young, the aged, or the debilitated when suffering with severe dysentery. The evidence that antibiotic therapy reduces the period of excretion of the organisms seems equivocal.

In many countries, the incidence of antibiotic resistance among shigellae is high (Neu et al., 1975, Thomas and Tillett, 1973), and this resistance is frequently plasmid mediated and multiple. Therefore, it is necessary to determine the resistance pattern of the strain before giving antibiotics to treat a severe case of shigellosis. The routine treatment with antibiotics is to be avoided because it exerts a selective pressure that increases the incidence of plasmid-mediated resistance.

USE OF VACCINES.   Various vaccines have been evaluated in laboratory studies with monkeys and in challenge trials on humans. Parenteral immunization with killed vaccines has failed to produce protection in laboratory animals (Formal et al., 1967) or against natural disease (Higgins, Floyd, and Koder, 1955). A polyvalent oral vaccine of streptomycin-dependent strains of *S. flexneri* and *S. sonnei,* when given in large-scale trials to children and adults, gave significant protection but the application of the vaccine was limited because multiple high doses were required (Mel et al., 1971).

Pathogenesis

Shigellae are pathogens of man and other primates; although there have been occasional reports of infections in dogs, other animals are resistant to infection. Laboratory animals such as mice, rabbits, and guinea pigs may be infected orally but only after conditioning by starvation and by treatment with gastric antacids together with antiperistaltic agents at the time of challenge.

In humans the lesions of bacillary dysentery are usually limited to the rectum and large intestine, but in severe cases a part of the terminal ileum may be involved. The typical lesion is acute inflammation with ulceration limited to the epithelium; the bacteria rarely spread deeper than the lamina propria and bloodstream involvement is rare. Infections due to

S. sonnei rarely extend beyond the epithelial inflammation stage, but infections with Shiga's bacillus or *S. flexneri* strains frequently produce ulceration.

Investigations of the pathophysiology of dysentery, in which rhesus monkeys were challenged orally, showed that the essential prerequisites for the production of disease are penetration of the colonic epithelial cells followed by intraepithelial multiplication (Gemski et al., 1972). In monkeys with classical dysentery the only physiological defect was in the colon; in animals with simple diarrhea there was a net jejunal secretion, although ileal transport remained normal. Bacterial invasion was not seen in the ileum or jejunum, and the results suggested that a bacterial enterotoxin might be involved (Rout et al., 1975).

The invasive properties of *Shigella* have been investigated using the ability to produce keratoconjunctivitis in the eyes of guinea pigs (Sérèny test), and to invade HeLa cells in tissue culture (Ogawa et al., 1967). The rabbit ileal loop test has also been used as an experimental model. Recent work has demonstrated that strains of *S. dysenteriae* 1 and *S. flexneri* 2a produce toxins that are lethal to mice, enterotoxic for rabbit ileal loops, and cytotoxic for HeLa cells (O'Brien et al., 1977). This demonstration of related toxins from *S. dysenteriae* 1 and *S. flexneri* might imply that the enterotoxin has a role in the pathogenesis of bacillary dysentery. It was initially suggested that the enterotoxin of *S. dysenteriae* 1 did not stimulate adenyl cyclase as does the cholera enterotoxin and the heat-labile enterotoxin of *Escherichia coli,* but it has now been shown that under optimal assay conditions adenyl cyclase is stimulated by Shiga enterotoxin (Charney et al., 1976). Further work is needed and, in any case, there is no doubt that epithelial invasion and multiplication remain the main virulence factors.

Epidemiology

Bacillary dysentery has a global distribution but the highest incidence occurs in areas with poor hygienic conditions. Notably, these areas are in the developing countries of the tropics where bacillary dysentery frequently combines with malnutrition to cause high morbidity and mortality. In such areas, children have about a 50% chance of dying from diarrhea before reaching 7 years of age.

Humans form the reservoir of infection for *Shigella,* and reports of human infections originating in animals are epidemiological curiosities. Food-borne outbreaks occur, especially in the tropics and less frequently in developed countries (Coultrip, Beaumont, and Siletchnik, 1977). The food may be infected from the feces of food-handlers or, in some situations, flies may act as vectors. Water-borne disease is said to be particularly important during the rainy seasons in tropical countries, when sewage contamination of drinking water occurs. In developed countries, water-borne outbreaks of shigellosis are occasionally reported (Baine et al., 1975); a not uncommon situation is for a defect in the water-distribution system to allow cross-connection with the sewage pipes (Ross and Gillespie, 1952). Outbreaks have been reported from swimming in sewage-contaminated water, but such outbreaks are probably rare (Rosenberg et al., 1976).

It is unlikely that flies are vectors in developed countries with good sewage disposal. The most frequent means of spread of infection by *Shigella* is from person to person by feco-oral transmission. Cases of dysentery in the acute phase constitute the main risk because the liquid stools are teeming with *Shigella* and may easily contaminate environmental surfaces, especially in toilets. The solid stools of symptom-free excreters in the recovery phase are less likely to produce significant environmental contamination. The shigella organisms can survive for several weeks in cool, humid environments, such as toilets (Hutchinson, 1956), which may explain the main seasonal peak in the winter in the developed countries of the temperate zones. In contrast, the seasonal peak in the tropics is usually in the warm months and may be related to fly prevalence.

In northwest Europe and North America, common-source outbreaks are rare, although shigellosis remains endemic. Infant schools and day-care centers for children are the main foci; although the usual pattern is of low-grade endemicity, occasional brisk epidemics are not uncommon in such institutions (Weissman et al., 1974b). Children frequently transmit infection from school to home and thence to parents and siblings, who introduce the infection to other groups, thus maintaining the endemic cycle (Thomas and Tillett, 1973). Institutions for the mentally subnormal are also important foci of bacillary dysentery, and the control and eradication of the disease in these cases is even more difficult to achieve than in schools.

Despite the high incidence of infection in primary-school children and in the preschool age groups attending nursery classes, it is rare in the developed countries to find infants infected with *Shigella* (Haltalin, 1967). In contrast, in the developing countries shigella infections are common in infants at the time of weaning. It is likely that during this period the infant is more at risk to infection from gross environmental contamination, as well as from infected food.

S. sonnei and *S. flexneri* are jointly responsible for most cases of bacillary dysentery, but the different shigella types vary in distribution and incidence throughout the world. However, in many areas the paucity of laboratory facilities precludes reliable information. In tropical countries, many different types of *Shigella* are found and epidemics due to more than one type are common. As the level of hygiene in a country improves, *S. sonnei* becomes predominant. In England and Wales, it accounts for about 95% of infections; in the United States in 1976, 62% of infections were due to *S. sonnei* (Center for Disease Control, 1977). As *S. sonnei* begins to predominate, there is a change in the seasonal pattern of bacillary dysentery. *S. sonnei* tends to have its seasonal peak in the autumn-winter, whereas *S. flexneri* tends to peak in the warmer months.

Shiga's bacillus (*S. dysenteriae* 1) was the cause of severe and extensive epidemics in Japan during the last 10 years of the nineteenth century. Before World War I, Shiga's bacillus caused serious outbreaks in Europe and the Americas, but since the 1920s such outbreaks have been uncommon. In India, Asia, and the Middle East, dysentery caused by Shiga's bacillus has remained an endemic disease and occasionally epidemics still occur. During World War II, outbreaks affected British troops in the Middle East; Shiga infections also occurred in United States troops in Vietnam. In 1963, an outbreak of severe dysentery due to Shiga's bacillus occurred in Somalia (Cahill, Davies, and Johnson, 1966).

During 1969 and 1970, Shiga's bacillus caused a serious pandemic, which originated in Guatemala and spread to most Central American countries (Gangarosa et al., 1970; Mata et al., 1970). The epidemic strain possessed plasmid-mediated, multiple-drug resistance and the epidemic was characterized by high attack rates, morbidity, and mortality, especially in children. It has been reported that there were 120,000 cases and 13,000 deaths. Returning travelers imported infections into the United States; in 1970–1972 there were 140 cases, mostly in the border states with Mexico. In comparison, there were only 11 isolates between 1965 and 1968 and, in 1976, only 10 cases of Shiga dysentery occurred in the United States (Weissman et al., 1974a). A severe epidemic of Shiga dysentery, which showed many similarities to the Central American outbreak, commenced in Bangladesh in 1972 (Rahaman et al., 1974). The epidemic strain possessed multiple-drug resistance and many of the patients had serious clinical disease. Epidemic dysentery due to Shiga's bacillus had been absent from these two areas for a prolonged period. The outbreaks demonstrate that, when a virulent strain is introduced into populations with poor standards of hygiene, very serious pandemic disease may occur.

In Western Europe, sporadic cases of Shiga's dysentery are almost always in persons returning from abroad. In England and Wales fewer than 10 cases are reported each year, and most of these have a history of recent travel to Asia (especially the Indian subcontinent) or, less frequently, the Middle East or Africa (Rowe and Gross, 1976).

S. dysenteriae serotypes other than Shiga's bacillus and *S. boydii* are unusual (about 1–2%) in England and Wales (Rowe, Gross, and Allen, 1974) and the United States (Center for Disease Control, 1977), but are not uncommonly reported in Asia and the Middle East. For many geographical areas, reliable epidemiological data are not available.

Isolation

Food and Water

Because the minimum infecting dose of *Shigella* is small, its occurrence in food, milk, and water may be significant, even when only a small number of organisms are present. Reliable and effective enrichment methods are not available, however, and the true incidence of *Shigella* contamination of foodstuffs cannot be determined. The Gram-negative (GN) broth of Hajna (1955) has sometimes proved useful for the enrichment of *Shigella*, and it is recommended that the investigation of foodstuffs include an enrichment step using this medium. Subsequent steps in the isolation of *Shigella* from foodstuffs should follow the procedure recommended for the examination of fecal specimens.

Fecal Specimens

Whenever possible, freshly passed stools should be examined; if this is not possible, fecal swabs showing marked fecal staining may be used. The specimens should be collected during the acute stage of the disease and before any chemotherapy is started. Specimens should be examined as soon after collection as possible. Enrichment with GN broth may be advantageous, but isolation is normally by direct plating. If the specimen includes blood and mucus, these should be included in the portion to be examined.

Some strains grow poorly on inhibitory media and it is advisable to use both a relatively noninhibitory medium, such as MacConkey agar or eosin methylene blue (EMB) agar, and an inhibitory medium, such as deoxycholate citrate (DCA) agar or shigella-salmonella (SS) agar. Instructions for the preparation of these media are given by Edwards and Ewing (1972). The stool specimens are streaked onto the chosen media; after overnight incubation at 37°C, colonies that do not ferment lactose are selected for further examination. It should be noted that, even when stool specimens from acute dysentery are examined, there may be only a scanty growth of *Shigella*.

Identification

Biochemical Characteristics

The genus *Shigella* consists of nonmotile organisms that conform to the definition of the family Enterobacteriaceae (Edwards and Ewing, 1972). Their biochemical reactions are summarized in Table 1. Members of the genus do not produce hydrogen sulfide in triple sugar iron (TSI) agar; they do not produce urease and they do not utilize citrate in Simmons' medium or in Christensen's medium. They do not decarboxylate lysine or deaminate phenylalanine. Salicin, adonitol, and inositol are not fermented. Only *S. sonnei* strains commonly ferment lactose, usually after more than 24 h incubation, although lactose-fermenting strains of *S. flexneri* 2a (Trifonova, Bratoeva, and Tekelieva, 1974) and *S. boydii* 9 (Manolov, Trifonova, and Ghinchev, 1962) have been reported. In addition to these lactose-fermenting strains, strains of *S. dysenteriae* 1 give positive results in tests for β-galactosidase activity. Fermentation of sucrose, like that of lactose, is restricted to strains of *S. sonnei* and usually requires several days incubation. Only strains of *S. sonnei* and *S. boydii* 13 decarboxylate ornithine, and only certain biotypes of *S. flexneri* 4a utilize sodium acetate. The production of gas from glucose occurs only in certain biotypes of *S. flexneri* 6 (Table 2), although aerogenic strains of *S. boydii* 13 (Rowe, Gross, and van Oye, 1975) and *S. boydii* 14 (Carpenter, 1961) have been described. The ability to produce indole varies with serotype, but it is worth noting that strains of *S. dysenteriae* 1, *S. flexneri* 6, and *S. sonnei* are always negative.

The G+C content of *Shigella* DNA is 49–53 mol% (Normore, 1973). DNA-reassociation studies indicate that most *Shigella* strains share 80% or more of their nucleotide sequences, and a similar degree of relatedness exists between *Shigella* and *Escherichia coli* strains. Strains of *S. boydii* 13 average only about 65% relatedness to other *Shigella* and *E. coli* strains (Brenner et al., 1973).

The biochemical identification of *Shigella* is complicated by the similarity of some strains of other genera. In particular, strains of *Enterobacter hafniae*, *Providencia* sp., *Aeromonas* sp., and atypical *E. coli* frequently cause difficulties.

Strains of *E. coli* that do not ferment lactose or are anaerogenic are a common problem. Of particular interest are members of the Alkalescens–Dispar (A-D) group, which are defined as nonmotile, anaerogenic biotypes of *E. coli*. These are best differentiated from *Shigella* by means of the Christensen's citrate and lysine decarboxylase tests in which *Shigella* is always negative. Some highly atypical strains have caused difficulties, and Shmilovitz, Kretzer, and Levy (1974) have suggested the recog-

Table 1. Biochemical reactions of *Shigella*.

Test or substrate	Result[a]	Comments
β-Galactosidase	d	Strains of *S. dysenteriae* 1 and *S. sonnei* are positive; positive strains of *S. flexneri* 2a and *S. boydii* 9 have been described.
Simmons' citrate	−	
Christensen's citrate	−	
Sodium acetate	− or +	Some biotypes of *S. flexneri* 4a are positive, all other types are negative.
Arginine decarboxylase	d	
Lysine decarboxylase	−	
Ornithine decarboxylase	d	Strains of *S. boydii* 13 and *S. sonnei* are positive.
Gelatin	−	
Gluconate	−	
H₂S (TSI)	−	
Indole	d	Strains of *S. dysenteriae* 1, *S. flexneri* 6, and *S. sonnei* are always negative; *S. dysenteriae* 2 is always positive.
KCN	−	
Malonate	−	
MR	+	
VP	−	
P.P.A.	−	
Urease	−	
Motility	−	
Glucose:		
Acid	+	
Gas	− or +	Some biotypes of *S. flexneri* 6 are positive; positive strains of *S. boydii* 13 and 14 have been described.
Adonitol	−	
Cellobiose	−	
Dulcitol	d	
Inositol	−	
Lactose	d	Strains of *S. sonnei* are usually positive after several days' incubation; positive strains of *S. flexneri* 2a and *S. boydii* 9 have been described.
Mannitol	d	Strains of *S. dysenteriae* are negative, but one positive strain of *S. dysenteriae* 3 has been described; negative biotypes of *S. flexneri* 4a ("*S. rabaulensis,*" "*S. rio*") and *S. flexneri* 6 (Newcastle biotype) occur; negative biotypes of *S. sommei* occur rarely.
Salicin	−	
Sucrose	d	Strains of *S. sonnei* are usually positive after several days' incubation.
Xylose	d	

[a]− or +, Majority of strains negative; d, different reactions.

Table 2. Biotypes of *Shigella flexneri* 6.[a]

Biotype	Glucose	Mannitol	Dulcitol
Boyd 88	A	A	− or (A)
Manchester	AG	AG	− or (AG)
Newcastle	A or AG	−	− or (A) or (AG)

[a] Results in parentheses indicate delayed reactions; A, acid production; G, gas production; −, negative.

Table 3. Biochemical and antigenic characteristics of the four subgroups of the genus *Shigella*.

Characters	Species and serotypes	Main synonyms
Subgroup A	*S. dysenteriae* 1	*S. shigae*
Not mannitol fermenters;	2	*S. schmitzii*, *S. ambigua*
each serologically	3	*S. largei* Q771,
distinct		*S. arabinotarda*[a] A
	4	*S. largei* Q1167,
		S. arabinotarda B
	5	*S. largei* Q1030
	6	*S. largei* Q454
	7	*S. largei* Q902
	8	Serotype 599-52
	9	Serotype 58
	10	Serotype 2050-50
Subgroup B	*S. flexneri* 1a	V
Usually mannitol	1b	VZ
fermenters; members	2a	W
serologically related	2b	WX
to each other	3a	Z
	3b	
	3c	
	4a	103[b]
	4b	103Z
	5	P119
	6	Newcastle, Manchester, or
		Boyd 88 bacillus[c]
	X variant	X
	Y variant	Y
Subgroup C	*S. boydii* 1	170
Usually mannitol	2	P288
fermenters; each	3	D1
serologically distinct	4	P274
	5	P143
	6	D19
	7	Lavington 1, *S. etousae*
	8	Serotype 112
	9	Serotype 1296/7
	10	Serotype 430
	11	Serotype 34
	12	Serotype 123
	13	Serotype 425
	14	Serotype 2770-51
	15	Serotype 703
Subgroup D	*S. sonnei*	Duval's bacillus,
Mannitol fermenter, late		*B. ceylonensis* A
lactose and sucrose		
fermenter		

[a] See Table 1.
[b] Mannitol negative biotypes are sometimes known as *S. rabaulensis* or *S. rio*.
[c] See Table 2.

nition of an intermediate group to be known as Intermediate Shigella Coli Alkalescens Dispar (ISCAD). Stenzel (1978) proposed the inclusion of such strains in *Shigella* subgroup D and suggested that this subgroup should be renamed *S. metadysenteriae*. The situation is further complicated by the fact that some strains of *E. coli* share with *Shigella* the ability to cause bacillary dysentery and to cause keratoconjunctivitis of the guinea pig eye in the Sèrèny test (Sakazaki et al., 1974). Nevertheless, the Enterobacteriaceae Subcommittee of the International Committee on Bacteriological Nomenclature (Carpenter, 1963) has advised that pathogenicity should not be considered in the classification of Enterobacteriaceae and that strains with biochemical reactions which do not conform strictly to those of *Shigella* should be classified as atypical *E. coli*.

Serotyping Scheme

The genus *Shigella* is divided into four subgroups on the basis of biochemical and antigenic characteristics (Table 3). Each subgroup is divided into a number of different serotypes, which are distinguished by the presence of a specific somatic antigen. Antisera are used in slide-agglutination tests for the identification of these specific antigens, although the results should be confirmed using tube-agglutination techniques.

Within the subgroup *S. flexneri*, common or group antigens exist in addition to the specific or type antigens. Each group antigen occurs in more than one serotype, and subserotypes are distinguished by the presence of particular group antigens (Table 4). In practice, it is necessary in the identification of *S. flexneri* to be able to recognize group factors 7, 8 (X group factors), 3, 4 (Y group factors), and 6. When commercial antisera are used, this recognition is usually achieved by the provision of separate antisera for factors X and Y, while group factor 6 is included in antiserum for specific antigen III (Table 5). Variants of *S. flexneri* are occasionally found that lack specific antigen and therefore contain only group factor antigens. According to which

group factors are present, these may be identified either as X or Y variants.

The immunochemical and genetic basis of the complex antigenic structure of *S. flexneri* has been summarized by Petrovskaya and Bondarenko (1977). The lipopolysaccharide O antigen of all *S. flexneri* serotypes contains group antigens 3, 4 as a main primary structure. The type-specific antigens I, II, IV, and V and the group antigens 7, 8 are all the result of phage conversion of the 3, 4 antigens, resulting in the incorporation of α-glucosyl secondary side chains. Type-specific antigen III and group antigen 6 differ from the above antigens in that they contain acetyl groups. Nevertheless these antigens are also formed as a result of phage conversion of the 3, 4 antigens. The lipopolysaccharide O antigen of *S. flexneri* 6 differs considerably from that of other *S. flexneri* serotypes and it does not contain the immunochemical determinants of the 3, 4 antigens. Strains of *S. flexneri* 6 therefore resemble strains of *S. boydii* immunochemically, and Petrovskaya and Bondarenko recommend that they be reclassified as such.

In strains of *S. sonnei*, the somatic antigens may undergo a variation of form from *S. sonnei* I to *S. sonnei* II (Wheeler and Mickle, 1945). This variation resembles the smooth to rough (S to R) variation which is common among the Enterobacteriaceae. Separate antisera are available for the identification of the two forms, although it is of little importance to distinguish between them since cultures often consist of a mixture of S and R forms. The somatic antigen of *S. sonnei* I is identical to that

Table 4. Antigenic characteristics of *Shigella flexneri* (*Shigella* subgroup B).

Serotype	Subserotype[a]	Specific antigen	Group antigens
1	1a	I	1,2, 4,5, 9
	1b	IS	1,2, 4,5,6, 9
2	2a	II	1, 3,4,
	2b	II	1, 7,8,9
3	3a	III	1, 6,7,8,9
	3b	III	1, 3,4, 6
	3c	III	1, 6
4	4a	IV	1, 3,4 B
	4b	IV	1, 3,4, 6 B
5		V	1, 7,8,9
		V	1, 3,4
6		VI	1,2, 4
X variant			1, 7,8,9
Y variant			1, 3,4

[a] For explanation of undesignated subserotypes see footnote *c* to Table 5.

Table 5. Reactions of *Shigella flexneri* serotypes in diagnostic absorbed slide-agglutinating serums.

Serum		Serotype with simplified antigenic formula[a]													
Type	Agglutinins	1a I:2,4	1b I:S:6,2,4	2a II:3,4	2b II:7,8	3a[b] III:6:7,8	3b III:6:3,4	3c III:6...	4a IV:B:3,4	4b IV:B:6:3,4	5[c] V:7,8	5[c] V:(3,4)	6 VI:2,4	X −:7,8	Y −:3,4
1	I	++	++	−	−	−	−	−	−	−	−	−	−	−	−
2	II	−	−	++	++	−	−	−	−	−	−	−	−	−	−
3	III:6	−	+	−	−	++	++	++	−	+	−	−	−	−	−
4	IV:B	−	−	−	−	−	−	−	++	++	−	−	−	−	−
5	V	−	−	−	−	−	−	−	−	−	++	++	−	−	−
6	VI	−	−	−	−	++	−	−	−	−	−	−	++	−	−
X	7,8	−	−	−	++	−	−	−	−	−	+	−	−	++	−
Y	3,4	−/±	−/±	−/+	−	−	++	−	−/+	−/±	−	−/±	−	−	++

[a] Arabic numerals are used to designate serotypes but it is customary to use roman numerals to express *type-specific* antigens or agglutinins, and arabic numerals for group antigens or agglutinins.

[b] Occasional variants may also react in absorbed Y serum.

[c] Nomenclature of subserotypes is sub judice.

of the C27 type of *Aeromonas shigelloides,* and a C27 antiserum can be used for the identification of *S. sonnei* I strains (Bader, 1954).

In addition to the serotypes shown in Table 3, Ewing, Reavis, and Davis (1958) have described a number of provisional *Shigella* serotypes. These may be added to the serotyping scheme in the future, but in the meantime they remain subjudice and antisera for their identification is usually available only at reference laboratories.

Strains of *Shigella* occasionally occur which do not agglutinate with *Shigella* antisera in the unheated state. Such strains may become agglutinable after heating at 100°C for 30 min.

The serological identification of *Shigella* is complicated by the widespread sharing of antigens among the Enterobacteriaceae. In particular, most *Shigella* somatic antigens are identical to or related to one or another of the somatic antigens of *E. coli* (Edwards and Ewing, 1972; Rowe, Gross, and Guiney, 1976). It is therefore essential that serological and biochemical tests be interpreted together.

Routine Identification

Many screening procedures have been proposed for the routine identification of *Shigella,* and the method described here is that recommended by the authors.

Non-lactose-fermenting colonies are selected from the primary plating media and inoculated directly onto triple sugar iron (TSI) agar and Christensen's urea medium. Any organism that produces urease after 4–6 h incubation is probably *Proteus* sp. and can be discarded. Any organism that after overnight incubation is found to produce H_2S or is

urease positive or which acidifies the slant of the TSI agar may be excluded, although H_2S-producing strains may warrant investigation as possible *Salmonella* strains. In all other cases, the butt of the TSI agar should be examined and if there is acidification, indicating the fermentation of glucose, the TSI cultures should be plated for purity. At the same time the TSI culture may be examined serologically with *Shigella* antisera using slide-agglutination techniques (Edwards and Ewing, 1972).

Preliminary tests:

$$
\left.\begin{array}{rl}
\text{Glucose} & \text{A} \\
\text{H}_2\text{S} & - \\
\text{Sucrose and lactose} & -
\end{array}\right\} \text{TSI}
$$
$$
\text{Urea} \quad -
$$

It is important to remember that there is widespread sharing of antigens among the Enterobacteriaceae and the identification of organisms which are agglutinated by *Shigella* antisera should be confirmed using further biochemical tests, as follows:

Additional tests:

Simmons' citrate	−
Gluconate	−
P.P.A.	−
Lysine decarboxylase	−
Glucose	A
Lactose	−
Mannitol	(depends on subgroup)
Oxidase	−
Motility	−

N.B.: *S. sonnei* ferments lactose after extended incubation. *S. flexneri* 6 may produce gas from glucose.

Table 6. Typing *Shigella sonnei* by using colicine production as a marker.[a]

Indicator strain[b]	1a	1b	2	3	3a	4	5	6	7	8	9	10	11	12	13	14
2	+	+	−	+	+	+	+	−	−	+	+	+	−	+	−	+
56	+	+	+	+	+	+	+	+	−	−	+	+	−	+	+	+
17	+	+	+	+	+	+	+	−	+	+	+	+	−	−	+	+
2m	−	−	−	−	−	v	v	−	−	−	−	−	−	−	−	v
38	−	−	−	+	+	+	+	−	−	+	v	+	−	+	−	+
56/56	+	+	−	+	+	−	+	+	−	−	+	+	−	+	+	+
56/98	+	+	−	+	+	−	+	+	−	−	+	+	−	+	+	+
R1	−	−	−	+	+	+	+	−	−	+	+	+	−	+	−	+
R6	−	+	+	+	+	+	+	+	−	−	+	+	−	+	+	+
M19	+	+	−	+	−	−	+	−	−	−	−	+	−	−	−	−
2/7	−	−	−	+	+	+	+	−	−	−	−	−	−	−	−	+
2/64	−	−	−	+	+	+	+	−	−	−	−	−	−	−	−	+
2/15	−	−	−	+	+	+	+	−	−	+	+	+	−	+	−	+
R/5	−	−	−	+	+	−	−	−	−	−	−	−	−	−	−	−
Escherichia coli Pow	+	+	+	+	+	+	+	+	−	+	+	+	+	+	+	+

[a] +, Inhibition of indicator strains; v, variable reaction; −, no inhibition of indicator strains.
[b] All undesignated strains are *S. sonnei* except M19, which is *S. dysenteriae* 2.

Strains that are confirmed biochemically as *Shigella* but fail to agglutinate in the *Shigella* antisera even after boiling may belong to provisional serotypes not yet included in the serotyping scheme or to new serotypes not previously recognized. Such strains should be sent to a reference laboratory for further examination.

When commercial antisera are used it is important that the manufacturer's instructions be followed. The exact composition of the antisera depends on the manufacturer and the agglutination titers will depend on the temperature and length of incubation of tube-agglutination tests.

Colicine Typing of *Shigella sonnei*

Colicine typing is of value in epidemiological studies of *Shigella sonnei*. The current scheme is based on that of Abbott and Shannon (1958) and distinguishes 16 colicine types using 15 indicator strains. In this method, the organism under investigation is inoculated heavily in a broad streak across a blood agar plate and incubated at 37°C for 24 h. The growth is then removed from the agar by scraping with a glass slide, and the organisms remaining on the slide are killed with chloroform. The 15 indicator strains are then inoculated onto the plate in streaks at right angles to the original line of growth. After incubation for a further 8–12 h, the patterns of inhibition of growth of the indicator strains can be examined and compared with the chart (Table 6). It is important that colicine type strains be maintained for use as controls and colicine types 3 and 11 should be included in every batch of tests.

Literature Cited

Abbott, J. D., Shannon, R. 1958. A method for typing *Shigella sonnei*, using colicine production as a marker. Journal of Clinical Pathology **11:**71–77.

Andrewes, F. W., Inman, A. C. 1919. A study of the serological races of the Flexner groups of dysentery bacilli. Great Britain Medical Research Council Special Report Series No. 42.

Bader, R.-E. 1954. Über die Herstellung eines agglutinierenden Serums gegen die Rundform von *Shigella sonnei* mit einem Stamm der Gattung *Pseudomonas*. Zeitschrift für Hygiene und Infektionskrankheiten **140:**450–456.

Baine, W. B., Herron, C. A., Bridson, K., Barker, W. H., Lindell, S., Mallison, G. F., Wells, J. G., Martin, W. T., Kosuri, M. R., Carr, F., Volker, E. 1975. Waterborne shigellosis at a public school. American Journal of Epidemiology **101:**323–332.

Boyd, J. S. K. 1932. Further investigations into the character and classification of the mannite fermenting dysentery bacilli. Journal of the Royal Army Medical Corps **59:**331–342.

Boyd, J. S. K. 1938. The antigenic structure of the mannitol-fermenting group of dysentery bacilli. Journal of Hygiene Cambridge **38:**477–499.

Brenner, D. J., Fanning, G. R., Miklos, G. V., Steigerwalt, A. G. 1973. Polynucleotide sequence relatedness among *Shigella* species. International Journal of Systematic Bacteriology **23:**1–7.

Cahill, K. M., Davies, J. A., Johnson, R. 1966. Report on an epidemic due to *Shigella dysenteriae* type 1 in the Somalia interior. American Journal of Tropical Medicine and Hygiene **15:**52–56.

Carpenter, K. P. 1961. The relationship of the Enterobacterium A12 (Sachs) to *Shigella boydii* 14. Journal of General Microbiology **26:**535–542.

Carpenter, K. P. 1963. Report of the subcommittee on taxonomy of the *Enterobacteriaceae*. International Journal of Systematic Bacteriology **13:**69–93.

Center for Disease Control. 1977. Shigella surveillance. Report No. 39.

Chantemesse, A., Widel, F. 1888. Sur les microbes de la dysentery épidémique. Bulletin Academie Médicine **19:**522–529.

Charney, A. N., Gots, R. E., Formal, S. B., Giannella, R. A. 1976. Activation of intestinal mucosal adenylate cyclase by *Shigella dysenteriae 1* enterotoxin. Gastroenterology **70:**1085–1090.

Coultrip, R. L., Beaumont, W., Siletchnik, M. D. 1977. Outbreak of shigellosis—Fort Bliss, Texas. Morbidity and Mortality Weekly Report **26:**107–108.

Councilman, W. T., Lafleur, H. A. 1891. Amoebic dysentery. Johns Hopkins Hospital Reports **2:**393–548.

Dupont, H. L., Hornick, R. B. 1973. Clinical approach to infectious diarrhoeas. Medicine **52:**265–270.

Edwards, P. R., Ewing, W. H. 1972. Identification of *Enterobacteriaceae*. Minneapolis: Burgess.

Enterobacteriaceae Subcommittee Reports. 1954. International Bulletin of Bacteriological Nomenclature and Taxonomy **4:**1–94.

Ewing, W. H. 1949. Shigella nomenclature. Journal of Bacteriology **57:**633–638.

Ewing, W. H., Reavis, R. W., Davis, B. R. 1958. Provisional *Shigella* serotypes. Canadian Journal of Microbiology **4:**89–107.

Flexner, S. 1900. On the aetiology of tropical dysentery. Philadelphia Medical Journal **6:**414–424.

Formal, S. B., Maenza, R. M., Austin, S., LaBrec, E. H. 1967. Failure of parenteral vaccines to protect monkeys against experimental shigellosis. Society for Experimental Biology and Medicine Proceedings **125:**347–349.

Gangarosa, E. J., Perera, D. R., Mata, L. J., Mendizábal-Morris, C., Guzmán, G., Reller, L. B. 1970. Epidemic Shiga bacillus dysentery in Central America. II. Epidemiologic studies in 1969. Journal of Infectious Diseases **122:**181–190.

Gemski, G., Akio, T., Washington, O., Formal, S. B. 1972. Shigellosis due to *Shigella dysenteriae* 1: Relative importance of mucosal invasion versus toxin production in pathogenesis. Journal of Infectious Diseases **126:**523–530.

Hajna, A. A. 1955. A new enrichment broth medium for Gram-negative organisms of the intestinal group. Public Health Laboratory **13:**83–89.

Haltalin, K. C. 1967. Neonatal shigellosis. Report of 16 cases and review of the literature. American Journal of Diseases of Children **114:**603–611.

Higgins, A., Floyd, T., Koder, M. 1955. Studies in shigellosis III. A controlled evaluation of a monovalent shigella vaccine in a highly endemic environment. American Journal of Tropical Medicine and Hygiene **4:**281–288.

Hiss, P. H., Russell, F. F. 1903. A study of a bacillus resembling the bacillus of Shiga from a case of fatal diarrhoea in a child: With remarks on the recognition of dysentery, typhoid and allied bacilli. Medical News, New York **82:**289–295.

Hutchinson, R. I. 1956. Some observations on the method of spread of Sonne dysentery. Monthly Bulletin Ministry of Health and Public Health Laboratory Service **15:**110–118.

Kruse, W. 1900. Uber die Ruhr als Volkskrankheit und ihren Erreger. Deutsche Medizinische Wochenschrift **26:**637–679.

Kruse, W. 1901. Weitere Untersuchungen uber die Ruhr und Ruhrbazillen. Deutsche Medizinische Wochenschrift **27:**370–372.

Kruse, W., Rittershaus, Kemp, Metz. 1907. Dysenterie und

Pseudodysenterie. Zeitschrift für Hygiene und Infektionskrankheiten **57**:417–488.

Large, D. T., Sankaran, O. K. 1934. Dysentery among troops in Quetta. Part II. The non-mannite fermenting group of organisms. Journal of the Royal Army Medical Corps **63**:231–237.

Lösch, F. 1875. Massenhafte Entwicklung von Amoeben im Dickdarm. Virchows Archives **65**:196–211.

Manolov, D. G., Trifonova, A., Ghinchev, P. 1962. A new lactose-fermenting species of the *Shigella* genus. Journal of Hygiene, Epidemiology, Microbiology and Immunology, Praha **6**:422–427.

Martini, E., Lentz, O. 1902. Ueber die Differenzirung der Ruhrbacillen mittels der Agglutination. Zeitschrift für Hygiene und Infektionskrankheiten **41**:540–557.

Mata, L. J., Gangarosa, E. J., Caceres, A., Perera, D. R., Mejicanos, M. L. 1970. Epidemic Shiga bacillus dysentery in Central America. I. Etiologic investigations in Guatemala 1969. Journal of Infectious Diseases **122**:170–180.

Mel, D., Gangarosa, E. J., Radovanovic, M. L., Arsic, B. L., Litvinjenko, S. 1971. Studies on vaccination against bacillary dysentery. 6. Protection of children by oral immunization with streptomycin-dependent *Shigella* strains. Bulletin of the World Health Organization **45**:457–464.

Murray, E. G. D. 1918. An attempt at classification of Bacillus dysenteriae based on the examination of the agglutinating properties of fifty three strains. Journal of the Royal Army Medical Corps **31**:257–271, 353–398.

Neu, H. C., Cherubin, C. E., Longo, E. D., Winter, W. 1975. Antimicrobial resistance of *Shigella* isolated in New York City in 1973. Antimicrobial Agents and Chemotherapy **7**:833–835.

Normore, W. M. 1973. Guanine-plus-cytosine (GC) composition of the DNA of bacteria, fungi, algae and protozoa, pp. 585–740. In: Laskin, A. I., Lechevalier, H. A. (eds.), CRC handbook of microbiology. II. Microbial composition. Cleveland: CRC Press.

O'Brien, A. D., Thompson, M. R., Gemski, P., Doctor, B. P., Formal, S. B.: 1977. Biological properties of Shigella flexneri 2A toxin and its serological relationship to *Shigella dysenteriae* 1 toxin. Infection and Immunity **15**:796–798.

Ogawa, H., Nakamura, A., Nakaya, R., Mise, K., Honjo, S., Takasaka, M., Fujiwara, T., Imaizumi, K. 1967. Virulence and epithelial cell invasiveness of dysentery bacilli. Japanese Journal of Medical Science and Biology **20**:315–328.

Petrovskaya, V. G., Bondarenko, V. M. 1977. Recommended corrections to the classification of *Shigella flexneri* on a genetic basis. International Journal of Systematic Bacteriology **27**:171–175.

Rahaman, M. M., Huq, I., Dey, C. R., Kibriya, A. K. M. G., Curlin, G. 1974. Ampicillin-resistant Shiga bacillus in Bangladesh. Lancet **i**:406.

Rosenberg, M. L., Hazlet, K. K., Schaefer, I., Wells, J. G., Pruneda, R. C. 1976. Shigellosis from swimming. Journal of the American Medical Association **16**:1849–1852.

Ross, A. I., Gillespie, E. H. 1952. An outbreak of waterborne gastro–enteritis and Sonne dysentery. Monthly Bulletin of the Ministry of Health and the Public Health Laboratory Service **11**:34–36.

Rout, W. R., Formal, S. B., Gianella, R. A., Dammin, G. J. 1975. The pathophysiology of *Shigella* diarrhoea in the Rhesus monkey; intestinal transport, morphology and bacteriological studies. Gastroenterology **68**:270–278.

Rowe, B., Gross, R. J. 1976. Shiga dysentery in England and Wales. British Medical Journal **1**:532.

Rowe, B., Gross, R. J., Allen, H. A. 1974. *Shigella dysenteriae* and *Shigella boydii* in England and Wales during 1972 and 1973. British Medical Journal **iv**:641–642.

Rowe, B., Gross, R. J., Guiney, M. 1976. Antigenic relationships between Escherichia coli O antigens O149 to O163 and *Shigella* O antigens. International Journal of Systematic Bacteriology **26**:76–78.

Rowe, B., Gross, R. J., van Oye, E. 1975. An organism differing from *Shigella boydii* 13 only in its ability to produce gas from glucose. International Journal of Systematic Bacteriology **25**:301–303.

Sachs, A. 1943. A report on an investigation into the characteristics of new types of non-mannitol-fermenting bacilli isolated from cases of bacillary dysentery in India and Egypt. Journal of the Royal Army Medical Corps **80**:92–99.

Sakazaki, R., Tamura, K., Nakamura, A., Kurata, T., Gohda, A., Takeuchi, S. 1974. Enteropathogenicity and enterotoxigenicity of human enteropathogenic *Escherichia coli*. Japanese Journal of Medical Science and Biology **27**:19–33.

Schmitz, K. E. F. 1917. Ein neuer Typus aus der Gruppe der Ruhrbazillen als Erreger einer grosseren Epidemie. Zeitschrift für Hygiene und Infektionskrankheiten **84**:449–516.

Shaw, C. H. 1953. Sonne dysentery in Ipswich. A study of infection in the home. Monthly Bulletin of the Ministry of Health and the Public Health Laboratory Service **12**:44–53.

Shiga, K. 1898. Ueber den Dysenterie bacillus (Bacillus dysenteriae). Centralblatt für Bakteriologie, Parasitenkunde und Infektionskrankheiten **24**:913–918.

Shmilovitz, M., Kretzer, O., Levy, E. 1974. The anaerogenic serotype 147 as an etiologic agent of dysentery in Israel. Israeli Journal of Medical Science **10**:1425–1429.

Stenzel, W. 1978. Problems of *Escherichieae* systematics and the classification of atypical dysentery bacilli. International Journal of Systematic Bacteriology **28**:597–598.

Strong, R. P., Musgrave, W. E. 1900. Preliminary note regarding the aetiology of the dysenteries of Manila. Report Surgeon General of the Army, Washington.

Thomas, M. E. M., Tillett, H. E. 1973. Dysentery in general practice: A study of cases and their contacts in Enfield and an epidemiological comparison with salmonellosis. Journal of Hygiene Cambridge **71**:373–389.

Trifonova, A., Bratoeva, M., Tekelieva, R. 1974. Studies on biochemical variants of *Sh. flexneri*. I. Studies on the lactose positive variants of *Sh. flexneri* 2a. Zentralblatt für Bakteriologie, Parasitenkunde, Infektionskrankheiten und Hygiene. Abt. 1 Orig., Reihe A **226**:343–348.

Weissman, J. B., Marton, K. I., Lewis, J. N., Freidmann, C. T. H., Gangarosa, E. J. 1974a. Impact in the United States of the Shiga dysentery pandemic in Central America and Mexico. Review of surveillance data through 1972. Journal of Infectious Diseases **129**:218–223.

Weissman, J. B., Schmerler, A., Weiler, P., Filice, G., Godbey, N., Hansen, I. 1974b. The role of pre-school children and day-care centers in the spread of Shigellosis in urban communities. Journal of Paediatrics **84**:797–802.

Wheeler, K. M., Mickle, F. L. 1945. Antigens of *Shigella sonnei*. Journal of Immunology **51**:257–267.

The Genus *Erwinia*

MORTIMER P. STARR

The enterobacteria of the genus *Erwinia* have a great deal of practical and theoretical importance: They cause important diseases of plants; they form a significant component of the nonphytopathogenic epiphytic flora of plant surfaces; some *Erwinia* strains are opportunistic pathogens of man and other animals. This genus *Erwinia* was proposed originally as a repository for the peritrichous phytopathogenic bacteria by a committee (Winslow et al., 1917), which admitted candidly that there was then insufficient information for a more rational nomenclatural action. This taxonomic segregation of the phytopathogenic bacteria—that is, solely because they come from plant habitats—stemmed not only from the lack of scientific facts but also from the disciplinal insularity that existed because of the completely separate developments of plant and animal bacteriology (Starr, 1959, 1975, 1979; Starr and Chatterjee, 1972; this Handbook, Chapter 4).

Comparative studies of plant and animal enterobacteria in recent years (Brenner, Fanning, and Steigerwalt, 1972; Chatterjee and Starr, 1972a; Lakso and Starr, 1970; Murata and Starr, 1974; Sakazaki et al., 1976; Starr and Mandel, 1969; White and Starr, 1971) have in fact suggested that other taxonomic arrangements are perhaps indicated. However, these have not been effected because there is not yet a consensus concerning the most reasonable arrangements. The current "compromise" treatment of the plant-associated enterobacteria of the genus *Erwinia* in the eighth edition of *Bergey's Manual of Determinative Bacteriology* (Lelliott, 1974) perpetuates this atavistic segregation.

Erwinia is a genus in the family Enterobacteriaceae (Cowan, 1974; this Handbook, Chapter 88), a major bacterial group of considerable significance to human and animal biology as well as to experimental microbiology. However, it is clear—as might be expected from the aforementioned disciplinal insularity that gave birth to this genus—that *Erwinia* is heterogeneous in terms of pathogenic capacity as well as bacteriological phenotypic characteristics. The erwinias fall into several phenotypic clusters, which correlate loosely with groupings based on their molecular genetics (Brenner, Fanning, and

Steigerwalt, 1972, 1974; Brenner et al., 1973; Gardner and Kado, 1972; Murata and Starr, 1974). The four "natural" groups depicted by Dye (1968, 1969a,b,c) are generally accepted by the community of phytobacteriologists (Lelliott, 1974). One group (Dye, 1968; Graham, 1964) consists of *Erwinia* species with strong pectolytic capacities that enable them to cause soft-rots (that is, tissue-macerating diseases and storage rots) in a wide variety of plants; it is called the "carotovora-group" or the "soft-rot group" or the "pectolytic group"; or it is sometimes referred to the genus *Pectobacterium* (Waldee, 1945).

Another of Dye's (1969b) groups of the genus *Erwinia* consists of erwinias that (usually) produce a yellow, water-insoluble pigmentation; such organisms are commonly isolated from plants as pathogens or as nonphytopathogenic epiphytes. Most epiphytes belonging to this group are included in the species *Erwinia herbicola*, and those from diseased plants are named *Erwinia milletiae*, *E. ananas*, *E. lathyri*, *E. mangiferae*, *E. stewartii*, etc., in accordance with their claimed but sometimes dubious phytopathogenicity (Graham and Hodgkiss, 1967). Dye (1969b) contends that many of these yellow-pigmented "pathogens" are in fact nonphytopathogenic epiphytes, which have been confused with phytopathogens because they often outgrow the true pathogens on isolation plates. These organisms, when considered as a group, are called the "herbicola group" or the "herbicola-lathyri group" or the "yellow-pigmented *Erwinia* group". Yellow-pigmented and related nonpigmented *Erwinia* strains isolated from diseases in man and other animals (Cooper-Smith and von Graevenitz, 1978; Schneierson and Bottone, 1973; Starr and Chatterjee, 1972; von Graevenitz, 1977) are generally considered as belonging to the herbicola-lathyri group; we shall refer to them here as "human clinical erwinias" or "*Erwinia herbicola* from human sources"; they are also called *Enterobacter agglomerans* (Ewing and Fife, 1972) as well as by many other names (Graham and Hodgkiss, 1967).

The third of Dye's (1968) groups of *Erwinia*, the members of which form neither pectic enzymes nor yellow pigments but do cause dry-necrotic or wilt symptoms on plants, is often typified (perhaps unjustifiably) by *Erwinia amylovora*. It is called the "amylovora group" or the "white, nonpectolytic erwinias" or *"Erwinia sensu stricto"* or the "true erwinias". The fourth of Dye's (1969c) groups—which he terms "atypical erwinias"—is a miscellaneous assemblage. Various systematic arrangements, including placement of most *Erwinia* species other than *E. amylovora* (and its possible relatives) into other enterobacterial genera, have been recommended. Proposals of this sort have been considered by a number of writers (Brenner, Fanning, and Steigerwalt, 1972, 1974, 1977; Brenner et al., 1973; Grimont, Grimont, and Starr, 1978; Imbs et al., 1977; Murata and Starr, 1974; Starr and Chatterjee, 1972).

Habitats

As Plant Pathogens

Members of the genus *Erwinia* are best known from habitats in which they are causing frank diseases of plants; that is, situations in which these bacteria are the kinds of antagonistic symbionts (Starr, 1975; Starr and Chatterjee, 1972) that bring about infectious diseases of plants. The diseases they cause fall into three general types based on the predominant symptom-complex: (i) soft-rot diseases, labeled thus because of the extensive maceration of the plant tissues, presumably brought about by the marked ability of these bacteria to digest the pectic substances of the plant cell walls; (ii) necrotic diseases, in which the production by the bacteria of still ill-defined "toxins" (Buchanan and Starr, 1980; Goodman, Huang, and Huang, 1974; Sjulin and Beer, 1978; Strobel, 1977) seems to bring about the harm to the plant tissues; and (iii) wilt diseases, in which the bacteria (or their products) cause some blockage of the water-conducting system of the plants. Naturally, mixtures of these symptoms can occur, and this set of distinctions can best be viewed as a didactic device. Descriptions of such diseases can be found in Elliott (1951), Israilski (1955), Stapp (1956, 1958), Dowson (1957), Rangaswami (1962), Gorlenko (1965), and in many plant pathology textbooks.

As (Nonphytopathogenic) Epiphytes of Plants

Bacteria of the yellow-pigmented *Erwinia herbicola* group are very common on the surfaces of land plants, especially on the leaves and buds (Crosse, 1971; Dickinson and Preece, 1976; Gibbins, 1972; Last and Deighton, 1965; Leben, 1965; Preece and Dickinson, 1971). They seem to cause no detectable disease symptoms, although some members of this bacterial group (known by other names) have been reported to cause an assortment of often questionable disease conditions in plants (Graham and Hodgkiss, 1967). The nonphytopathogenic (epiphytic) erwinias do play some important role(s) in modifying the courses of diseases caused by other *Erwinia* species (Gibbins, 1972). For example, a reduction in the severity of the fire blight disease of pears and apples (caused by *Erwinia amylovora*) has been attributed to these epiphytic erwinias (Chatterjee, Gibbins, and Carpenter, 1969; Riggle and Klos, 1972), and the principle has been applied practically—but with different bacterial antagonists of *E. amylovora*—by Thomson et al. (1976) to reduce colonization and infection of pear flowers. The reported ability of some *Erwinia herbicola* strains to fix dinitrogen (Neilson, 1979; Papen and Werner, 1979) may have some relevance to the ability of *E. herbicola* to colonize a habitat so nitrogen-poor as the leaf surface.

Habitats Adjacent to Diseased Plants

Erwinia species have occasionally been sought and sometimes found in nonphytopathological habitats, usually those immediately adjacent to diseased plants. Such studies have involved asymptomatic plants (possibly already infected) which are near diseased ones (Miller and Schroth, 1972), insects (Stahl and Luepschen, 1977), infected weeds (Burr and Schroth, 1977), plant debris coming from infected plants (Burr and Schroth, 1977; Goto, 1972), packinghouse waters (Segall, 1971), and soil (Burr and Schroth, 1977; Meneley and Stanghellini, 1976; Schaad and Wilson, 1970). Very little systematic work has been done on this subject until quite recently, largely because of the lack in the past of suitable selective isolation media and the consequent uncertainty that the bacteria that had been isolated from such nonphytopathological habitats were indeed erwinias. One obtains the impression from the fragmentary literature that some erwinias persist in such habitats over extended periods and that others persist only for a relatively short time.

As Opportunistic Pathogens of Man and Other Animals

A considerable literature now exists on the occurrence of erwinias in various disease conditions in several kinds of animals including man (Schneierson and Bottone, 1973; Starr and Chatterjee, 1972; von Graevenitz, 1977). Essentially all such cases of animal infection have been attributed to members of the *Erwinia herbicola* group, which are called *Enterobacter agglomerans* by some medical bacteriologists (Ewing and Fife, 1972). It is certainly possible that other sorts of putative erwinias also cause disease in animals, but that—because of disciplinal insularity (Starr, 1979)—they are not recognized as *Erwinia* species. One possible candidate would be the non-phytopathogenic yet pectolytic enterobacteria (Bagley and Starr, 1979; Starr et al., 1977; von Riesen, 1976), called *Klebsiella oxytoca* or the "oxytocum" group of *Klebsiella pneumoniae* by medical and environmental bacteriologists (this Handbook, Chapters 93 and 94), some of which might well be related to bacteria referred to the genus *Erwinia* by plant bacteriologists.

In most cases of human infection by *Erwinia*, the patient has been compromised by one or another of an assortment of predisposing conditions (Schneierson and Bottone, 1973; von Graevenitz, 1977). Although plant strains of *Erwinia herbicola* are generally yellow pigmented, the erwinias isolated from such human clinical sources often are nonpigmented, possibly as a result of their exposure in the animal body to supraoptimal temperatures (Chatterjee and Gibbins, 1971). They are also generally resistant to many antimicrobial agents, probably as a result of the improper use of these drugs and the potential for the promiscuous conjugal transfer among *Erwinia* and other enterobacteria of genetic material specifying drug resistance (Chatterjee and Starr, 1972b).

In Industrial Fermentations

The antileukemic activity of bacterial L-asparaginase has directed attention to microbial production of this enzyme in industrial fermentations. Soft-rot erwinias seem to be well suited for this application (Cammack, Marlborough, and Miller, 1972; Grossowicz and Rasooly, 1972; Shifrin, Solis, and Chaiken, 1973). The *Erwinia* asparaginase is effective in treatment of acute lymphocytic leukemia. This therapeutic efficacy and an immunological specificity different from the L-asparaginase of *Escherichia coli* (an important factor if a patient develops hypersensitivity to one enzyme but not the other) have attracted much commercial interest to the *Erwinia* asparaginase by pharmaceutical companies in several countries. No published figures seem to be available on the extent of commercial production, nor are details about the industrial fermentation process seemingly provided to the general public.

Isolation

Enzymatic Basis for the Isolation of Soft-Rot Erwinias

Isolation of the pectolytic (soft-rot) erwinias is facilitated by their remarkable ability to digest the pectic substances of plants. Considered simplistically, these pectic substances—which constitute the primary cell wall material of higher plants—are linear biopolymers consisting mainly of methyl esters of pyranose galacturonic acid residues bonded through α-glycosidic linkages at the 1,4 positions. The various methyl esters as isolated from plants are called pectins; the nonesterified material is called polygalacturonic acid (polygalacturonate) or pectic acid (pectate); pectin, in its natural locus, is variously bonded to other plant components and this little known water-insoluble material is called protopectin (Joslyn, 1962).

All of these pectic substances can be attacked by an assortment of microbial pectic enzymes. The soft-rot erwinias invariably have a wide array of such pectic enzymes: pectin methyl esterase (Bonnet and Venard, 1975; Goto and Okabe, 1962), hydrolytic polygalacturonase (Nasuno and Starr, 1966), several sorts of polygalacturonic acid *trans*-eliminase or lyase (Bagley and Starr, 1979; Chatterjee et al., 1979; Garibaldi and Bateman, 1971; Hatanaka and Ozawa, 1973; Moran, Nasuno, and Starr, 1968a; Moran and Starr, 1969; Starr et al., 1977), oligogalacturonic acid *trans*-eliminase or lyase (Moran, Nasuno, and Starr, 1968b), as well as enzymes that further catabolize galacturonic acid itself (Kilgore and Starr, 1959). It is the tissue-macerating action of all or some of these enzymes that brings about the soft-rots of plants (Bateman and Basham, 1976; Bateman and Millar, 1966; Wood, 1955). While such pectic enzymes occur also in many other bacteria, the enterobacteria other than *Erwinia*—with the exception of *Yersinia enterocolitica, Y. pseudotuberculosis, Y. pestis,* and *Klebsiella oxytoca* (Bagley and Starr, 1979; Chatterjee et al., 1979; Starr et al., 1977; von Riesen, 1975, 1976)—are not pectolytic (Davis and Ewing, 1964; Lakso and Starr, 1970). The occurrence in soft-rot *Erwinia* cultures of such pectic enzymes is the basis of several methods used for their isolation.

Enrichment of Soft-Rot Erwinias

Although enrichment is usually not necessary when dealing with fresh, plant diseased material, it is often a profitable step when attempting to isolate soft-rot erwinias from soil or from heavily contaminated plant material. The procedure described by Meneley and Stanghellini (1976) is suitable for this purpose.

Enrichment of Soft-Rot Erwinias (Meneley and Stanghellini, 1976)

"To 25 g of soil in a 250-ml Erlenmeyer flask, the following sterilized ingredients were added aseptically in sequence":

Distilled water	225 ml
Sodium polypectate (Sunkist)	0.625 g

10% $(NH_4)_2SO_4$	2.5 ml
10% K_2HPO_4	2.5 ml
5% $MgSO_4 \cdot 7H_2O$	1.5 ml

During the addition of pectate, the soil-water mixture was stirred briskly to minimize coagulation. Remaining visible clumps of pectate were broken up with a spatula. After addition of salts, the entire mixture was stirred thoroughly to suspend the soil particles. Flasks were incubated anaerobically for 48 h at room temperature (about 24°C). The pH of the soil mixture ranged from 5.3 to 6.0, depending on the soil used.

"Prior to use, soluble sugars were removed from pectate by the following modification of Wood's (1955) technique: 500 ml of 60% ethanol (EtOH) were added to 100 g of sodium polypectate and autoclaved for 15–20 minutes. Subsequently, the pectate was washed three times with 250-ml aliquots of 60% EtOH (acidified to 5% HCl), rinsed three times with 150-ml aliquots of 95% EtOH, and dried at 60C for 12 hours."

Burr and Schroth (1977) report that enrichment of soft-rot erwinias can be effected by anaerobic incubation (BBL GasPak system; $H_2 + CO_2$) of crude cultures in a selective liquid pectate medium. Another enrichment procedure often practiced by plant pathologists is to transfer some of the usually contaminated, infected soft-rot material to another plant or plant part. Aseptically removed slices of potato tubers or carrot roots are commonly used for this purpose. As soon as the inoculated plant material starts to rot, material from the lesion is streaked onto one of the selective isolation media containing a pectic substance.

Selective Isolation Media for Soft-Rot Erwinias

Numerous solid media containing pectic substances are used for the selective isolation of the soft-rot erwinias. In some cases, the pectic substance itself—usually in the form of a calcium-stabilized gel—is the medium's solidifying agent (Starr, 1947). Current directions (Starr et al., 1977) for preparing this calcium-stabilized pectate (polygalacturonate) gel (PEC) medium follow. The previously used sodium ammonium pectate (No. 24 of Sunkist Growers, Ontario, California), a product that presently is not commercially available, was replaced by sodium polygalacturonate; either Sigma (St. Louis, Missouri) No. P-1879 or ICN (Cleveland, Ohio) No. 102921 is satisfactory.

Preparation of Calcium-Stabilized Pectate (PEC) Medium (Starr et al., 1977)

"Distilled water (100 ml) and a magnetic stirring bar were placed in a one-liter Erlenmeyer flask (situated on a heater fitted with a magnetic stirrer). The following ingredients were then added while stirring: 0.6 ml of 10.0% aqueous $CaCl_2 \cdot 2H_2O$; 1.0 ml of 0.1% aqueous bromothymol blue (BTB) in 6.4×10^{-4} N NaOH: 0.5 g of Difco yeast extract; and (very slowly, so that each particle is wetted) 3.0 g of sodium polygalacturonate (one of the brands mentioned above). After the polygalacturonate was uniformly wetted and suspended, the heater was turned on, and the temperature was brought almost to boiling with continuous stirring. The material has a strong tendency to foam (the reason that a rather large flask is specified); this foaming can be avoided by removing the flask from the heater before the solution actually boils. While the solution was hot and was stirred, its pH was adjusted (by monitoring the color of the BTB indicator) to 7.3 with 1 N NaOH (taking care not to overshoot, because polygalacturonate is quite labile at elevated pH). The medium was autoclaved (121°C, 15 min) and allowed to cool to about 95°C before opening the autoclave (again, to avoid foaming). It was then poured directly into sterilized petri dishes, which should be held at room temperature until the gel sets. PEC medium thus prepared will, after sterilization and gelling, usually be at pH 6.8 to 7.0 (light green in color). It is important to emphasize that the PEC medium cannot be remelted and, consequently, that the plates must be poured while the medium is still fluid, soon after it comes from the autoclave, since gelation may occur even at elevated temperatures. Surplus moisture on the covers of the petri dishes or the surfaces of the medium can be controlled by pouring the plates after the PEC medium has cooled for 5 to 15 min in a 50°C waterbath (the brand of polygalacturonic acid will dictate how long the medium can be held and cooled before irreversible gelation begins), by replacing the petri dish covers with dry ones, and/or by drying the plates for a day or two at 25°C. Plates of PEC medium can be stored for several weeks at room or refrigerator temperatures, sealed within the plastic sleeves in which plastic petri dishes are commonly packed. Because PEC medium cannot be remelted, if tubes of medium (slants or stabs) are required, the unsterilized PEC medium is dispensed into tubes, which are then autoclaved and allowed to gel in a slanting or upright position."

However, such pectate gel media must be observed frequently for discrete zones of liquefaction around colonies, otherwise the pectolytic action is usually so strong that the entire plate becomes liquefied in a short time and it is impossible to determine which colonies are the pectolytic ones. The better course is to supplement such pectate gels with a small amount of agar (usually, 0.3–0.5%); under these conditions, localized pectate liquefaction (sinking of the pectolytic colonies into the semisolid agar) occurs. Many such pectate semisolid agar media have been described; the task of making definitive recommendations is simplified by the study of O'Neill and Logan (1975), in which several such selective isolation media were compared in terms of their efficiency in the diagnosis and enumeration of soft-rot erwinias from soil and plant tissues. O'Neill and Logan (1975) recommend the recipe of Cuppels and Kelman (1974), as slightly modified by them.

Crystal Violet–Pectate Medium (Cuppels and Kelman, 1974, as modified by O'Neill and Logan, 1975)

''To 200 ml of boiling distilled water in a preheated blender, 3 ml of 10% $CaCl_2 \cdot 6H_2O$, 0.5 ml of crystal violet (0.75% aqueous) and 4.5 ml of N-NaOH were added together with 5 ml of 0.001% aqueous solution of Silicone D.C. Antifoam Emulsion (Hopkin and Williams). The ingredients were blended at high speed for 15 s then 1 g of $NaNO_3$, 0.05 g of yeast extract (Difco B-127), 0.4 g $MnSO_4 \cdot 4H_2O$, 2.0 g agar (Oxoid No. 3) and 15 g polygalacturonic acid (Sigma No. P-1879) were added together with 300 ml of boiling distilled water. The mixture was blended again at high speed for 30 s, poured into a 1 litre flask capped with aluminum foil and sterilized at 121° for 15 min. The medium was cooled to 60° and poured into sterile Petri dishes.''

Another such pectate semisolid agar medium is the PEC-SSA medium of Starr et al. (1977). To make PEC-SSA medium (Starr et al., 1977), 0.3% agar (Difco) is added to the PEC medium (described above) after the sodium polygalacturonate (either Sigma No. P-1879 or ICN No. 102921) is completely wetted, the agar and polygalacturonate are completely dissolved by further heating and stirring, the pH is adjusted to 7.3, and the PEC-SSA medium is autoclaved, dispensed, and stored—all as recommended above for PEC medium.

In using such pectate gel media, including the semisolid agar versions, properly diluted material is streaked or spread over the dried surfaces. When working with cultures, several are spotted on each plate. The ability to degrade polygalacturonic acid is scored by periodically observing the plates for liquefaction of the medium and/or sinking of the colonies.

Another procedure (Starr et al., 1977) that can be recommended, particularly for isolation of nonpectolytic mutant strains (Chatterjee and Starr, 1977), is based on the precipitability with 2 N HCl of undecomposed polygalacturonic acid in a polygalacturonic acid–yeast extract–agar (PEC-YA) medium. Pectolytic colonies in this procedure are surrounded by clear halos in a turbid background.

Preparation and Use of Pectate–Yeast Extract Agar (PEC-YA) Medium (Starr et al., 1977)

''The brand of polygalacturonic acid suitable for this procedure is very critical; neither Sigma no. P-1879 nor ICN no. 102921 was suitable; Sunkist Growers no. 24 and 3491 worked quite well, but neither product is now commercially available; the polygalacturonic acid no. P 21750 of Pfaltz-Bauer (Stamford, Conn.) is both suitable and available. The pectate-yeast extract-agar (PEC-YA) medium was prepared (following the same precautions given above for PEC medium) as follows: to 100 ml of cold distilled water (held in a one-liter Erlenmeyer flask, fitted with a magnetic stirring bar, that is situated on a magnetic stirrer-heater) were added slowly with continuous stirring 1.0 g of polygalacturonic acid (Sunkist Growers no. 3491 or Pfaltz-Bauer P 21750), 1.0 g of yeast extract (Difco), 1.0 ml of 0.1% aqueous BTB in 6.4×10^{-4} N NaOH, and 1.5 g of agar (Difco). After the components were completely wetted, the heater was turned on and the medium was brought almost to boiling with continuous stirring. When all of the components were dissolved, the pH was adjusted while stirring to approximately 7.3 with 1 N NaOH (taking care not to overshoot) by monitoring the color change of the BTB indicator dye and periodically checking with pH paper (pHydrion paper; Micro Essential Laboratory, Brooklyn, N.Y.). The medium was autoclaved at 121°C for 15 min. The PEC-YA medium was poured into plates after the medium had cooled to about 50°C. The pH of the medium after autoclaving was about 6.8 to 7.0 (light green in color). Because polygalacturonic acid is alkali-labile, especially at elevated temperatures, any desired adjustment of the PEC-YA medium to a higher pH should be effected, after autoclaving and cooling somewhat, with sterile 1 N NaOH solution while stirring continuously; in this case, it might be necessary to substitute a more appropriate pH indicator for the BTB. To use this PEC-YA medium, cultures were streaked or spotted on the surface and allowed to grow. The plates were then flooded with 2 N HCl, held at room temperature for a few minutes, and scored for the appearance of clear halos around and beneath the colonies in the otherwise turbid medium. Since the HCl-scoring method kills the organisms, it was necessary to prepare replicate plates if the colonies were to be recovered or if the time-course of pectolytic action was to be followed.''

A similar technique, in which the undecomposed polygalacturonic acid is precipitated with cetyltrimethylammonium bromide (Cetrimide), was devised by Jayasankar and Graham (1970) and used by Hankin, Zucker, and Sands (1971) and Burr and Schroth (1977). The procedures used by Burr and Schroth (1977) involved additional selective components and conditions, which appear to be advantageous for selective isolation of soft-rot erwinias.

PEC-YA medium at pH 8.0 has been used for isolating mutant strains of *Erwinia chrysanthemi* which no longer produce polygalacturonic acid *trans*-eliminase (Chatterjee and Starr, 1977, 1979b; Chatterjee et al., 1979) and the same medium at about pH 5.6 shows promise of utility in isolating mutant strains of this organism which no longer produce hydrolytic polygalacturonase (M. P. Starr and P. R. Dong, unpublished data). Hildebrand (1971) has used pectate media at both low pH and high pH for taxonomic purposes; the principle, however, is the same as our use for mutant selection: the differing pH optima for polygalacturonic acid *trans*-eliminase (about pH 8.5–9.0) and hydrolytic polygalacturonase (about pH 5.0–5.6).

Isolation of Yellow-Pigmented Erwinias from Plants and Animals

The plant erwinias that center about *Erwinia herbicola* usually form yellow colonies (Graham and Hodgkiss, 1967). Practically all epiphytic and phytopathogenic strains of this group are pigmented when isolated from plants; although most strains isolated from nosocomial infections in man are pigmented, a substantial minority of such strains are not pigmented (Ewing and Martin, 1974). Irreversible loss of pigmentation when *E. herbicola* and *E. stewartii* cultures are cultivated at superoptimal temperatures has been reported (Chatterjee and Gibbins, 1971; Garibaldi and Gibbins, 1975), and some such mechanism may be involved in the occurrence of nonpigmented human clinical *Erwinia* (*"Enterobacter agglomerans"*) isolates—coming as they do from the relatively elevated temperatures of the human body. The pigments of those yellow-pigmented *Erwinia* strains that have been studied to date are carotenoids, with β-carotene, β-cryptoxanthin, and zeaxanthin predominating (M. P. Starr, A. G. Andrewes, D. G. Gilliland, and others, unpublished observations).

Isolation of these yellow-pigmented *Erwinia* strains from plants is usually effected on nonselective media by looking for yellow colonies. This practice leads to a great deal of confusion with the other yellow-pigmented bacteria that occur in the same habitats (Dickinson and Preece, 1976; Gibbins, 1972; Goodfellow, Austin, and Dickinson, 1976; Preece and Dickinson, 1971). In medical diagnostic laboratories, the primary plating media ordinarily used for other enterobacteria (e.g., MacConkey agar or eosin–methylene blue agar) are

employed for isolating members of this human clinical *Erwinia* group (Ewing and Martin, 1974). No single one of the selective media described by Kado and Heskett (1970) can successfully isolate all members of this group, though combinations of these media might be used. To the best of my knowledge, there are no selective media now in use for isolating these organisms. However, dinitrogen-fixing strains of *E. herbicola* (Neilson, 1979; Papen and Werner, 1979) might well be selected on nitrogen-free media.

Isolation of *Erwinia amylovora* and Possible Relatives

Erwinia amylovora causes a disease, fire blight of pears and apples, of considerable economic significance (van der Zwet and Keil, 1979). Perhaps for this reason, some attention has been devoted to devising methods for its selective isolation from diseased plants and related habitats. Miller and Schroth (1972) reported a selective medium which served to isolate *E. amylovora* selectively from pear trees and on which *E. amylovora* grew as well as on nonselective potato-glucose-peptone agar.

Selective Medium for Isolating *Erwinia amylovora* (Miller and Schroth, 1972)

"The selective medium was prepared by adding the following compounds, in the order listed, per liter of distilled water:"

Mannitol	10.0 g
Nicotinic acid	0.5 g
L-Asparagine	3.0 g
K_2HPO_4	2.0 g
$MgSO_4 \cdot 7H_2O$	0.2 g
Sodium taurocholate (Difco)	2.5 g
Tergitol anionic 7 (sodium heptadecyl sulfate, Union Carbide)	0.1 ml
Nitrilotriacetic acid (NTA; 2% aqueous solution; NTA is first neutralized with ca. 0.73 g of KOH/g NTA)	10.0 ml
Bromothymol blue (H_2O-soluble; Matheson, Colman & Bell; 5% aqueous solution)	9.0 ml
Neutral red (H_2O-soluble; Matheson, Colman & Bell; 5% solution)	2.5 ml
Agar	20.0 g

"The medium was adjusted with 1 N NaOH (ca. 5 ml) to pH 7.2 to 7.3. The preparation was autoclaved at 121°C for 15 min. The pH of the medium after autoclaving should be ca. 7.4. Fifty mg Actidione (cycloheximide, Nutritional Biochemical Corp.), and 1.75 ml of a 1.0% solution of thallium nitrate (K & K Rare Chemicals) were added to the autoclaved medium."

Another selective medium for isolating *Erwinia amylovora* has been described by Crosse and Goodman (1973). Their medium contains (per liter of distilled water): 400 g sucrose, 30 g nutrient agar (BBL), 2 ml of a 0.1% solution of crystal violet in absolute alcohol, and 50 ml of a 0.1% solution of cycloheximide; the medium is sterilized by autoclaving, and poured into sterilized Petri dishes. Appropriate dilutions of the material to be examined for *E. amylovora* are streaked on the surfaces of the previously dried medium. This medium, which involves an unusually high (40%) sucrose content, is said to permit isolation of *E. amylovora* with high efficiency and selectivity from the mixed microflora that exists on apple leaf surfaces. Colonies of *E. amylovora* on this medium, when examined at 15 and 30 × magnification under oblique light, showed characteristic surface craters which are claimed to enable "positive identification of the pathogen".

Medium D3 of Kado and Heskett (1970) is reported to recover *E. amylovora* from laboratory mixtures of various plant bacteria with an efficiency of colony formation (relative to a nonselective medium) of 78%, and to allow recovery of *E. amylovora* added to unsterilized soil with an efficiency of 57%.

Medium D3 for Isolation of *Erwinia* Species (Kado and Heskett, 1970)

Medium D3 contains, per liter:

Sucrose	10 g
Arabinose	10 g
Casein hydrolysate	5 g
LiCl	7 g
Glycine	3 g
NaCl	5 g
$MgSO_4 \cdot 7H_2O$	0.3 g
Sodium dodecyl sulfate	50 mg
Bromothymol blue	60 mg
Acid fuchsin	100 mg
Agar	15 g

"The medium is adjusted to pH 8.2 with NaOH before autoclaving. The medium has a pH of 6.9–7.1 after autoclaving. . . . Medium D3 is selective for *Erwinia* species, which characteristically produce a red coloration of the medium; the intensity of the color depends on the species. Most of the soft-rotting group (*E. aroideae, E. atroseptica, E. carotovora,* etc.) characteristically produce a more intense color reaction than the amylovora group (e.g., *E. amylovora, E. quercina, E. tracheiphila, E. rubrifaciens,* etc.). This last group requires at least 48 hr before any reaction is noticeable. The color of the plate will eventually turn completely red-orange on prolonged growth of *Erwinia* species. The growth of other genera is usually suppressed, although some saprophytes (*Escherichia* species) may occasionally appear when soil is plated. Trials with unsterilized diseased tissues from naturally infected pear (fire blight) and walnut trees (bark canker) showed that this medium is selective for the respective causal organisms, *E. amylovora* and *E. rubrifaciens.* Some *Xanthomonas* species produce colonies after several days of incubation, but are easily distinguished by the dark-blue reaction surrounding their colonies and lack of strong acid production."

The aforementioned *Erwinia rubrifaciens,* which causes a canker disease of Persian walnuts, has come in for some separate attention in this connection. Schaad and Wilson (1970) report that the following selective medium was useful in isolating this organism from infected plants and from soil.

Selective Medium for Isolating *Erwinia rubrifaciens* (Schaad and Wilson, 1970)

The selective medium used for isolation of *E. rubrifaciens* had the following composition:

Glycerol	10 ml
Ammonium sulfate	5 g
Dipotassium phosphate	2 g
Eosin Y	0.4 g
Methylene blue	0.065 g
Cycloheximide (85–100% active ingredient)	250 mg
Novobiocin (pure)	40 mg
Neomycin sulfate (680 μg/mg)	40 mg
Agar	15 g
Distilled water	1 liter

"The many soil-inhabiting bacteria that require organic nitrogen to grow cannot utilize ammonium sulfate, on which *E. rubrifaciens* grows readily. Eosin–methylene blue helps in the diagnosis of *E. rubrifaciens,* and inhibits development of gram-positive bacteria. For enhanced selectivity, cycloheximide was added to suppress fungus growth, and novobiocin and neomycin were added to inhibit bacterial growth further. Each ingredient was added prior to autoclaving at 15-lb pressure for 15 min . . . Colonies of *E. rubrifaciens* became visible on this medium in 4 days at 27 C, and were 1–2 mm in diam in 6 days. They were circular, with entire margins and a convex surface. The dye has been absorbed to the extent that, by transmitted light, the colony was deep blue with translucent margins. When the colonies were numerous, the surrounding medium developed a greenish metallic sheen."

Procedures for isolating another member of the amylovora group, *Erwinia salicis,* the causal agent of watermark disease of cricket bat and other willows (*Salix* spp.), have been detailed by de Kam (1976).

Willow-Sucrose Agar (WSA) Medium for Isolating *Erwinia salicis* (de Kam, 1976)

> "100 g willow wood without bark was ground with water in a blender, shaken for 20 min, and filtered through cotton wool. The filtrate was made up to 1 litre and 50 g sucrose and 15 g agar were added; pH (about 5) was not adjusted. . . . Willow sucrose agar had the advantage that it inhibited certain contaminating organisms, notably *Bacillus cereus* var. *mycoides,* one of the most troublesome when isolations [of *Erwinia salicis*] are made from the soil. In addition, *E. salicis* was more easily recognised in the presence of sucrose since typical slimy colonies were produced. Recently it was found that the addition of 0.06% 'Lab Lemco' [Oxoid brand of beef extract] broth to this medium strongly stimulates the growth of *E. salicis,* whereas *B. cereus* var. *mycoides* is still inhibited."

Isolation of *Erwinia salicis* from Badly Contaminated or "Symptomless" Tissue (de Kam, 1976)

> "From newly diseased tissue the organism could be isolated readily by ordinary plating techniques. With decayed and dead tissue, or tissue still 'symptomless' but yet invaded with small numbers of bacteria, the following method was used. About 25 g of wood taken from several parts of a branch was completely ground in a blender with 150 ml demineralized water and shaken for 15 min. The liquid was then filtered through cotton wool and centrifuged, the pellet being plated out on WSA. After one week of incubation, typical colonies were subcultured."

Erwinia tracheiphila, the cause of a wilt disease of cucumbers and related plants, is somewhat difficult to isolate from plants. This organism occurs in a rather viscid and stringy slime in the vascular tissue of infected plants. Prend and John (1961) recommend injecting sterile water into the vascular tissue of a petiole in the direction of the infected leaf, withdrawing the diluted slime (containing the bacteria), and then plating this less viscid material on nutrient agar. A technique of this sort may have general utility in isolating other vascular wilt pathogens. It has been our experience that *E. tracheiphila* grows much better on yeast extract–glucose–calcium carbonate (YGC) agar (Lakso and Starr, 1970) than it does on nutrient (i.e., beef extract–peptone) agar, and its isolation from infected plants might be facilitated by using the YGC medium. The same might be said for *Erwinia salicis* (but see de Kam's [1976] selective procedures, given above). In fact, this YGC medium (sometimes called YDC agar, where "D" stands for dextrose, the old name for glucose) is a very useful general purpose one, suitable for isolating and maintaining practically all phytopathogenic bacteria.

Yeast Extract–Glucose–Calcium Carbonate (YGC) Agar (Lakso and Starr, 1970)

To one liter of distilled water, add:

Yeast extract (Difco)	10 g
Glucose	20 g
Calcium carbonate (finely divided, U.S.P. grade)	20 g
Agar	15 g

Dissolve by heating and stirring. Sterilize by autoclaving for 15 min at 121°C. If slants are required, distribute the medium (with frequent swirling; a magnetic stirring bar, sterilized with the medium, is a useful aid) into test tubes before autoclaving; after autoclaving, cool to approximately 50°C, swirl each tube vigorously to suspend the calcium carbonate uniformly, and cool the tubes in a slanted position very quickly (for example, in a cold room) so that the calcium carbonate does not settle out before the agar solidifies. For plates, vigorously swirl the flask containing the autoclaved and cooled medium before and during the pouring of the YGC agar into Petri dishes; the plates should be kept on a cool surface so that the agar solidifies before the calcium carbonate can settle out. It is very important that the $CaCO_3$ be finely divided; the cosmetic grade of precipitated chalk is best for this purpose.

Identification

Given the aforementioned circularity in the taxonomic practice, identification of a bacterium as a member of the genus *Erwinia* can be certain only when the culture has been isolated from a known plant disease situation or an immediately adjacent habitat. The axenic culture so isolated is checked for virulence on the same species of plant; methods suitable for this purpose are given in Chapter VI of Király et al. (1970). Somewhat equivalent plant systems are often used for determining virulence (see, for example, Buchanan and Starr, 1980; Lakso and Starr, 1970; Pugashetti and Starr, 1975; Starr and Dye, 1965). The reason for the caveat "somewhat" is that one cannot be certain that the harmfulness determined on a surrogate plant or on an excised plant part is the same as the full pathogenicity measured by inoculation into the entire plant that usually is infected by that bacterium. If the axenic culture (upon such experimental plant inoculation) causes the expected disease, and if it has the general bacteriological characteristics of a member of the Enterobacteriaceae (i.e., a Gram-negative, motile, peritrichously flagellated, aerobic or facultatively anaerobic, fermentative, rod-shaped bacterium), the assumption is made that the organism belongs to the *Erwinia* species that ordinarily causes the disease under consideration.

Assignment of an *Erwinia* culture (defined in the aforementioned ambiguous manner) to one or another of Dye's major groups of that genus can be made on the pectolytic capacity of the soft-rot erwinias, the yellow pigmentation of the plant members of the *Erwinia herbicola* group, and the lack of both of these traits in the members of the white, nonpectolytic *Erwinia amylovora* "group". In the present state of ignorance, further determination (that is, to species) is loaded with uncertainty, unless the determination is based on a known phytopathogenic capacity—whereupon this uncertainty is replaced with another (the taxonomic circularity already mentioned).

However, there are a few rays of hope. *Erwinia rhapontici* forms on a sucrose-peptone agar a pink diffusible pigment of unknown composition, and this pigmentation aids in the isolation and identification of this organism (Roberts, 1974). Formation of the unusual blue (bipyridyl) pigment, indigoidine, seems to distinguish *Erwinia chrysanthemi* from other soft-rot erwinias (Starr, Cosens, and Knackmuss, 1966).

Serology might provide a useful determinative and diagnostic tool in the genus *Erwinia*. However, there is no modern and comprehensive serological analysis of the entire genus or even of any one of Dye's major groups. Moreover, *Erwinia* antisera are seemingly not commercially available, rendering difficult the routine exploitation of what is known about *Erwinia* serology (Allan and Kelman, 1977; De Boer, Copeman, and Vruggink, 1979; de Kam, 1976; Elrod, 1941a,b, 1946; Goto, 1976; Lazar, 1972a,b; Schaad, 1979; Stanghellini et al., 1977; Vruggink and Maas Geesteranus, 1975; Wong and Preece, 1973; Yakrus and Schaad, 1979).

DNA-DNA homology (in vitro "hybridization") is, in theory at least, a possible determinative tool in *Erwinia*—as it already is actually in other bacterial groups—but the existing knowledge regarding *Erwinia* (Brenner, Fanning, and Steigerwalt, 1972, 1974, 1977; Brenner et al., 1973; Gardner and Kado, 1972; Murata and Starr, 1974) needs to be translated into simplified techniques for application to routine use in identification.

Taxonomically useful derivatives should eventually sprout, too, from the emerging knowledge about conjugational and plasmid genetics of *Erwinia* (Chatterjee, Behrens, and Starr, 1979; Chatterjee and Starr, 1972a,b, 1973a,b, 1977, 1979a,b, 1980; Coplin, 1978; Gibbins et al., 1976; Lacy and Leary, 1979; Lacy and Sparks, 1979; Pugashetti, Chatterjee, and Starr, 1978; Pugashetti and Starr, 1975). Similar predictions (or, at least hopes) might be ventured about metabolic pathways (Kilgore and Starr, 1959; Sutton and Starr, 1959, 1960; White and Starr, 1971), about bacteriophage and bacteriocin typing (Echandi and Moyer, 1979; Ritchie and Klos, 1979; Vidaver, 1976), and about many other

approaches. Any or all might become effective tools in *Erwinia* identification at some categorial level or another. Oh! that our knowledge about *Erwinia* were not so fragmentary!

When there is no knowledge about the plant habitat or the phytopathogenic capability, identification of a culture as a member of the genus *Erwinia* is now essentially impossible. In this context, examination of Table 8.1 of the eighth edition of *Bergey's Manual of Determinative Bacteriology* (Cowan, 1974) is quite instructive. There is not a single "distinguishing characteristic" of the erwinias that permits unambiguous placement into the genus *Erwinia* as opposed to placement into other enterobacterial genera. Tables 8.18, 8.19, and 8.20 of the same *Manual* (Lelliott, 1974) further illustrate the difficulty or practical impossibility of the situation.

The erwinias that occur in human disease—the so-called "*Enterobacter agglomerans*" (Ewing and Fife, 1972) of the medical bacteriologist—provoke similar difficulties in identification, but here the determination is somewhat simplified because one has "merely" to discriminate between the sorts of *Erwinia herbicola* that occur in human sources (not, at the moment, the entire genus *Erwinia*) and the other enterobacteria from man. Ewing and Martin (1974) summarize procedures that might be suitable for this purpose. However, this human clinical *Erwinia* group is by no means homogeneous and "very little justification is provided for *Enterobacter agglomerans* to be a species group as it is presently defined" (Sakazaki et al., 1976).

Literature Cited

Allan, E., Kelman, A. 1977. Immunofluorescent stain procedures for detection and identification of *Erwinia carotovora* var. *atroseptica*. Phytopathology **67:**1305–1312.

Bagley, S. T., Starr, M. P. 1979. Characterization of intracellular polygalacturonic acid *trans*-eliminase from *Klebsiella oxytoca*, *Yersinia enterocolitica*, and *Erwinia chrysanthemi*. Current Microbiology **2:**381–386.

Bateman, D. F., Basham, H. G. 1976. Degradation of plant cell walls and membranes, pp. 316–355. In: Heitefuss, R., Williams, P. H. (eds.), Physiological plant pathology. Berlin, Heidelberg, New York: Springer-Verlag.

Bateman, D. F., Millar, R. L. 1966. Pectic enzymes in tissue degradation. Annual Review of Phytopathology **4:**119–146.

Bonnet, P., Venard, P. 1975. Production de pectine-méthylestérase *in vitro* par différences souches d'*Erwinia* pectinolytiques. Annales de Phytopathologie **7:**51–59.

Brenner, D. J., Fanning, G. R., Steigerwalt, A. G. 1972. Deoxyribonucleic acid relatedness among species of *Erwinia* and between *Erwinia* species and other enterobacteria. Journal of Bacteriology **110:**12–17.

Brenner, D. J., Fanning, G. R., Steigerwalt, A. G. 1974. Deoxyribonucleic acid relatedness among erwiniae and other *Enterobacteriaceae:* The gall, wilt, and dry-necrosis organisms (genus *Erwinia* Winslow *et al., sensu stricto*). International Journal of Systematic Bacteriology **24:**197–204.

Brenner, D. J., Fanning, G. R., Steigerwalt, A. G. 1977. Deoxyribonucleic acid relatedness among erwiniae and other enterobacteria. II. Corn stalk rot bacterium and *Pectobacterium chrysanthemi*. International Journal of Systematic Bacteriology **27**:211–221.

Brenner, D. J., Steigerwalt, A. G., Miklos, G. V., Fanning, G. R. 1973. Deoxyribonucleic acid relatedness among erwiniae and other *Enterobacteriaceae*: The soft-rot organisms (genus *Pectobacterium* Waldee). International Journal of Systematic Bacteriology **23**:205–216.

Buchanan, G. E., Starr, M. P. 1980. Phytotoxic material from associations between *Erwinia amylovora* and pear tissue culture: Possible role in necrotic symptomatology of fireblight disease. Current Microbiology **4**:67–72.

Burr, T. J., Schroth, M. N. 1977. Occurrence of soft-rot *Erwinia* spp. in soil and plant material. Phytopathology **67**:1382–1387.

Cammack, K. A., Marlborough, D. I., Miller, D. S. 1972. Physical properties and subunit structure of L-asparaginase isolated from *Erwinia carotovora*. Biochemical Journal **126**:361–379.

Chatterjee, A. K., Behrens, M. K., Starr, M. P. 1979. Genetic and molecular properties of E-*lac*+, a transmissible plasmid of *Erwinia herbicola*, pp. 75–79. In: Station de Pathologie Végétale et Phytobactériologie (ed.), Proceedings of the IVth Conference on Plant Pathogenic Bacteria [Angers, 1978]. Beaucouzé: Institut National de la Recherche Agronomique.

Chatterjee, A. K., Gibbins, L. N. 1971. Induction of nonpigmented variants of *Erwinia herbicola* by incubation at supraoptimal temperatures. Journal of Bacteriology **105**:107–112.

Chatterjee, A. K., Gibbins, L. N., Carpenter, J. A. 1969. Some observations on the physiology of *Erwinia herbicola* and its possible implication as a factor antagonistic to *Erwinia amylovora* in the "fire-blight" syndrome. Canadian Journal of Microbiology **15**:640–642.

Chatterjee, A. K., Starr, M. P. 1972a. Genetic transfer of episomic elements among *Erwinia* species and other enterobacteria: F'*lac*+. Journal of Bacteriology **111**:169–176.

Chatterjee, A. K., Starr, M. P. 1972b. Transfer among *Erwinia* spp. and other enterobacteria of antibiotic resistance carried on R factors. Journal of Bacteriology **112**:576–584.

Chatterjee, A. K., Starr, M. P. 1973a. Transmission of *lac* by the sex factor E in *Erwinia* strains from human clinical sources. Infection and Immunity **8**:563–572.

Chatterjee, A. K., Starr, M. P. 1973b. Gene transmission among strains of *Erwinia amylovora*. Journal of Bacteriology **116**:1100–1106.

Chatterjee, A. K., Starr, M. P. 1977. Donor strains of the soft-rot bacterium *Erwinia chrysanthemi* and conjugational transfer of the pectolytic capacity. Journal of Bacteriology **132**:862–869.

Chatterjee, A. K., Starr, M. P. 1979a. Genetics and physiology of plant pathogenicity in *Erwinia amylovora*, pp. 81–88. In: Station de Pathologie Végétale et Phytobactériologie (ed.), Proceedings of the IVth Conference on Plant Pathogenic Bacteria [Angers, 1978]. Beaucouzé: Institut National de la Recherche Agronomique.

Chatterjee, A. K., Starr, M. P. 1979b. Genetics of pectolytic enzyme production in *Erwinia chrysanthemi*, pp. 89–94. In: Station de Pathologie Végétale et Phytobactériologie (ed.), Proceedings of the IVth Conference on Plant Pathogenic Bacteria [Angers, 1978]. Beaucouzé: Institut National de la Recherche Agronomique.

Chatterjee, A. K., Starr, M. P. 1980. Genetics of *Erwinia* species. Annual Review of Microbiology **34**:645–676.

Chatterjee, A. K., Buchanan, G. E., Behrens, M. K., Starr, M. P. 1979. Synthesis and excretion of polygalacturonic acid *trans*-eliminase in *Erwinia*, *Yersinia*, and *Klebsiella* species. Canadian Journal of Microbiology **25**:94–102.

Cooper-Smith, M. E., von Graevenitz, A. 1978. Nonepidemic *Erwinia herbicola (Enterobacter agglomerans)* in blood cultures: Bacteriological analysis of fifteen cases. Current Microbiology **1**:29–32.

Coplin, D. L. 1978. Properties of F and P group plasmids in *Erwinia stewartii*. Phytopathology **68**:1637–1643.

Cowan, S. T. 1974. Enterobacteriaceae, pp. 290–293. In: Buchanan, R. E., Gibbons, N. E. (eds.), Bergey's manual of determinative bacteriology, 8th ed. Baltimore: Williams & Wilkins.

Crosse, J. E. 1971. Interactions between saprophytic and pathogenic bacteria in plant disease, pp. 283–290. In: Preece, T. F., Dickinson, C. H. (eds.), Ecology of leaf surface micro-organisms. London, New York: Academic Press.

Crosse, J. E., Goodman, R. N. 1973. A selective medium for and a definitive colony characteristic of *Erwinia amylovora*. Phytopathology **63**:1425–1426.

Cuppels, D., Kelman, A. 1974. Evaluation of selective media for isolation of soft-rot bacteria from soil and plant tissue. Phytopathology **64**:468–475.

Davis, B. R., Ewing, W. H. 1964. Lipolytic, pectolytic, and alginolytic activities of *Enterobacteriaceae*. Journal of Bacteriology **88**:16–19.

De Boer, S. H., Copeman, R. J., Vruggink, H. 1979. Serogroups of *Erwinia carotovora* potato strains determined with diffusible somatic antigens. Phytopathology **69**:316–319.

de Kam, M. 1976. *Erwinia salicis*: Its metabolism and variability in vitro, and a method to demonstrate the pathogen in the host. Antonie van Leeuwenhoek Journal of Microbiology and Serology **42**:421–428.

Dickinson, C. H., Preece, T. F. (eds.). 1976. Microbiology of aerial plant surfaces. London: Academic Press.

Dowson, W. J. 1957. Plant diseases due to bacteria. Cambridge: Cambridge University Press.

Dye, D. W. 1968. A taxonomic study of the genus *Erwinia*. I. The "amylovora" group. New Zealand Journal of Science **11**:590–607.

Dye, D. W. 1969a. A taxonomic study of the genus *Erwinia*. II. The "carotovora" group. New Zealand Journal of Science **12**:81–97.

Dye, D. W. 1969b. A taxonomic study of the genus *Erwinia*. III. The "herbicola" group. New Zealand Journal of Science **12**:223–236.

Dye, D. W. 1969c. A taxonomic study of the genus *Erwinia*. IV. Atypical erwinias. New Zealand Journal of Science **12**:833–839.

Echandi, E., Moyer, J. W. 1979. Production, properties, and morphology of bacteriocins from *Erwinia chrysanthemi*. Phytopathology **69**:1293–1297.

Elliott, C. 1951. Manual of bacterial plant pathogens, 2nd ed. Waltham, Massachusetts: Chronica Botanica.

Elrod, R. P. 1941a. Serological studies of the Erwineae. I. *Erwinia amylovora*. Botanical Gazette **103**:123–131.

Elrod, R. P. 1941b. Serological studies of the Erwineae. II. Soft-rot group: With some biochemical considerations. Botanical Gazette **103**:266–279.

Elrod, R. P. 1946. The serological relationship between *Erwinia tracheiphila* and species of *Shigella*. Journal of Bacteriology **52**:405–410.

Ewing, W. H., Fife, M. A. 1972. *Enterobacter agglomerans* (Beijerinck) comb. nov. (the herbicola-lathyri bacteria). International Journal of Systematic Bacteriology **22**:4–11.

Ewing, W. H., Martin, W. J. 1974. Enterobacteriaceae, pp. 189–221. In: Lennette, E. H., Spaulding, E. H., Truant, J. P. (eds.), Manual of clinical microbiology, 2nd ed. Washington, D.C.: American Society for Microbiology.

Gardner, J. M., Kado, C. I. 1972. Comparative base sequence homologies of the deoxyribonucleic acids of *Erwinia* species and other *Enterobacteriaceae*. International Journal of Systematic Bacteriology **22**:201–209.

Garibaldi, A., Bateman, D. F. 1971. Pectic enzymes produced by *Erwinia chrysanthemi* and their effects on plant tissue. Physiological Plant Pathology **1**:25–40.

Garibaldi, A., Gibbins, L. N. 1975. Induction of avirulent variants in *Erwinia stewartii* by incubation at supraoptimal temperatures. Canadian Journal of Microbiology 21:1282–1287.

Gibbins, L. N. 1972. Relationships between pathogenic and non-pathogenic bacterial inhabitants of aerial plant surfaces, pp. 15–24. In: Maas Geesteranus, H. P. (ed.), Proceedings of the Third International Conference on Plant Pathogenic Bacteria. Wageningen: Centre for Agricultural Publishing and Documentation.

Gibbins, L. N., Bennett, P. M., Saunders, J. R., Grinsted, J., Connolly, J. C. 1976. Acceptance and transfer of R-factor RP1 by members of the "herbicola" group of the genus *Erwinia*. Journal of Bacteriology 128:309–316.

Goodfellow, M., Austin, B., Dickinson, C. H. 1976. Numerical taxonomy of some yellow-pigmented bacteria isolated from plants. Journal of General Microbiology 97:219–233.

Goodman, R. N., Huang, J.-S., Huang, P.-Y. 1974. Host-specific phytotoxic polysaccharide from apple tissue infected by *Erwinia amylovora*. Science 183:1081–1082.

Gorlenko, M. V. 1965. Bacterial diseases of plants, 2nd ed., revised and enlarged. [Translated from the 1961 Russian original.] Jerusalem: Israel Program for Scientific Translations.

Goto, M. 1972. The significance of the vegetation for the survival of plant pathogenic bacteria, pp. 39–52. In: Maas Geesteranus, H. P. (ed.), Proceedings of the Third International Conference on Plant Pathogenic Bacteria. Wageningen: Centre for Agricultural Publishing and Documentation.

Goto, M. 1976. *Erwinia mallotivora* sp. nov., the causal organism of bacterial leaf spot of *Mallotus japonicus* Muell. Arg. International Journal of Systematic Bacteriology 26:467–473.

Goto, M., Okabe, N. 1962. Studies on pectin-methyl-esterase secreted by *Erwinia carotovora* (Jones) Holland. Annals of the Phytopathological Society of Japan 27:1–9.

Graham, D. C. 1964. Taxonomy of the soft-rot coliform bacteria. Annual Review of Phytopathology 2:13–42.

Graham, D. C., Hodgkiss, W. 1967. Identity of Gram negative, yellow pigmented fermentative bacteria isolated from plants and animals. Journal of Applied Bacteriology 30:175–189.

Grimont, P. A. D., Grimont, F., Starr, M. P. 1978. *Serratia proteamaculans* (Paine and Stansfield) comb. nov., a senior subjective synonym of *Serratia liquefaciens* (Grimes and Hennerty) Bascomb et al. International Journal of Systematic Bacteriology 28:503–510.

Grossowicz, N., Rasooly, G. 1972. Production and purification of L-asparaginase from *Erwinia carotovora*, pp. 333–338. In: Terui, G. (ed.), Fermentation technology today. Proceedings of the IVth International Fermentation Symposium. Tokyo: Society of Fermentation Technology.

Hankin, L., Zucker, M., Sands, D.C. 1971. Improved solid media for the detection and enumeration of pectolytic bacteria. Applied Microbiology 22:205–209.

Hatanaka, C., Ozawa, J. 1973. Effect of metal ions on activity of exopectic acid transeliminase of *Erwinia* sp. Agricultural and Biological Chemistry 37:593–597.

Hildebrand, D. C. 1971. Pectate and pectin gels for differentiation of *Pseudomonas* sp. and other bacterial plant pathogens. Phytopathology 61:1430–1436.

Imbs, M. A., Bene, R., Girard, T., Goulon, C., Dixneuf, P. 1977. Taxonomie numérique des bactéries du genre *Erwinia*. Microbia 3:3–34.

Israilski, W. P. 1955. Bakterielle Pflanzenkrankheiten. [Translation into German of the 1952 Russian edition.] Berlin: Deutscher Bauernverlag.

Jayasankar, N. P., Graham, P. H. 1970. An agar plate method for screening and enumerating pectinolytic microorganisms. Canadian Journal of Microbiology 16:1023.

Joslyn, M. A. 1962. The chemistry of protopectin: A critical review of historical data and recent developments. Advances in Food Research 11:1–107.

Kado, C. I., Heskett, M. G. 1970. Selective media for isolation of *Agrobacterium, Corynebacterium, Erwinia, Pseudomonas,* and *Xanthomonas*. Phytopathology 60:969–976.

Kilgore, W. W., Starr, M. P. 1959. Catabolism of galacturonic

and glucuronic acids by *Erwinia carotovora*. Journal of Biological Chemistry 234:2227–2235.

Király, Z., Klement, Z., Solymosy, F., Vörös, J. 1970. Methods in plant pathology. Budapest: Akadémiai Kiadó.

Lacy, G. H., Leary, J. V. 1979. Genetic systems in phytopathogenic bacteria. Annual Review of Phytopathology 17:181–202.

Lacy, G. H., Sparks, R. B., Jr. 1979. Transformation of *Erwinia herbicola* with plasmid pBR322 deoxyribonucleic acid. Phytopathology 69:1293–1297.

Lakso, J. U., Starr, M. P. 1970. Comparative injuriousness to plants of *Erwinia* spp. and other enterobacteria from plants and animals. Journal of Applied Bacteriology 33:692–707.

Last, F. T., Deighton, F. C. 1965. The non-parasitic microflora on the surfaces of living leaves. The British Mycological Society Transactions 48:83–99.

Lazar, I. 1972a. Studies on the preparation of anti-*Erwinia* sera in rabbits, pp. 125–130. In: Maas Geesteranus, H. P. (ed.), Proceedings of the Third International Conference on Plant Pathogenic Bacteria. Wageningen: Centre for Agricultural Publishing and Documentation.

Lazar, I. 1972b. Serological relationships between the 'amylovora', 'carotovora' and 'herbicola' groups of the genus *Erwinia*, pp. 131–141. In: Maas Geesteranus, H. P. (ed.), Proceedings of the Third International Conference on Plant Pathogenic Bacteria. Wageningen: Centre for Agricultural Publishing and Documentation.

Leben, C. 1965. Epiphytic microorganisms in relation to plant disease. Annual Review of Phytopathology 3:209–230.

Lelliott, R. A. 1974. *Erwinia*, pp. 332–340. In: Buchanan, R. E., Gibbons, N. E. (eds.), Bergey's manual of determinative bacteriology, 8th ed. Baltimore: Williams & Wilkins.

Meneley, J. C., Stanghellini, M. E. 1976. Isolation of soft-rot *Erwinia* spp. from agricultural soils using an enrichment technique. Phytopathology 66:367–370.

Miller, T. D., Schroth, M. N. 1972. Monitoring the epiphytic population of *Erwinia amylovora* on pear with a selective medium. Phytopathology 62:1175–1182.

Moran, F., Nasuno, S., Starr, M. P. 1968a. Extracellular and intracellular polygalacturonic acid *trans*-eliminases of *Erwinia carotovora*. Archives of Biochemistry and Biophysics 123:298–306.

Moran, F., Nasuno, S., Starr, M. P. 1968b. Oligogalacturonide *trans*-eliminase of *Erwinia carotovora*. Archives of Biochemistry and Biophysics 125:734–741.

Moran, F., Starr, M. P. 1969. Metabolic regulation of polygalacturonic acid *trans*-eliminase in *Erwinia*. European Journal of Biochemistry 11:291–295.

Murata, N., Starr, M. P. 1974. Intrageneric clustering and divergence of *Erwinia* strains from plants and man in the light of deoxyribonucleic acid segmental homology. Canadian Journal of Microbiology 20:1545–1565.

Nasuno, S., Starr, M. P. 1966. Polygalacturonase of *Erwinia carotovora*. Journal of Biological Chemistry 241:5298–5306.

Neilson, A. H. 1979. Nitrogen fixation in a biotype of *Erwinia herbicola* resembling *Escherichia coli*. Journal of Applied Bacteriology 46:483–491.

O'Neill, R., Logan, C. 1975. A comparison of various selective isolation media for their efficiency in the diagnosis and enumeration of soft rot coliform bacteria. Journal of Applied Bacteriology 39:139–146.

Papen, H., Werner, D. 1979. N$_2$-fixation in *Erwinia herbicola*. Archives of Microbiology 120:25–30.

Preece, T. F., Dickinson, C. H. (eds.). 1971. Ecology of leaf surface microorganisms. London, New York: Academic Press.

Prend, J., John, C. A. 1961. Method of isolation of *Erwinia tracheiphila* and an improved inoculation technique. Phytopathology 51:255–258.

Pugashetti, B. K., Chatterjee, A. K., Starr, M. P. 1978. Isolation and characterization of Hfr strains of *Erwinia amylovora*. Canadian Journal of Microbiology 24:448–454.

Pugashetti, B. K., Starr, M. P. 1975. Conjugational transfer of genes determining plant virulence in *Erwinia amylovora*. Journal of Bacteriology 122:485–491.

Rangaswami, G. 1962. Bacterial plant diseases in India. Bombay: Asia Publishing House.

Riggle, J. H., Klos, E. J. 1972. Relationship of *Erwinia herbicola* to *E. amylovora*. Canadian Journal of Botany 50:1077–1083.

Ritchie, D. F., Klos, E. J. 1979. Some properties of *Erwinia amylovora* bacteriophages. Phytopathology 69:1078–1083.

Roberts, P. 1974. *Erwinia rhapontici* (Millard) Burkholder associated with pink grain of wheat. Journal of Applied Bacteriology 37:353–358.

Sakazaki, R., Tamura, K., Johnson, R., Colwell, R. R. 1976. Taxonomy of some recently described species in the family *Enterobacteriaceae*. International Journal of Systematic Bacteriology 26:158–179.

Schaad, N. W. 1979. Serological identification of plant pathogenic bacteria. Annual Review of Phytopathology 17:123–147.

Schaad, N. W., Wilson, E. E. 1970. Survival of *Erwinia rubrifaciens* in soil. Phytopathology 60:557–558.

Schneierson, S. S., Bottone, E. J. 1973. *Erwinia* infections in man. CRC Critical Reviews in Clinical Laboratory Sciences 4:341–355.

Segall, R. H. 1971. Selective medium for enumerating *Erwinia* species commonly found in vegetable packinghouse waters. Phytopathology 61:425–426.

Shifrin, S., Solis, B. G., Chaiken, I. M. 1973. L-Asparaginase from *Erwinia carotovora*. Journal of Biological Chemistry 248:3464–3469.

Sjulin, T. M., Beer, S. V. 1978. Mechanism of wilt induction by amylovorin in cotoneaster shoots and its relation to wilting of shoots infected by *Erwinia amylovora*. Phytopathology 68:89–94.

Stahl, F. J., Luepschen, N. S. 1977. Transmission of *Erwinia amylovora* to pear fruit by *Lygus* spp. Plant Disease Reporter 61:936–939.

Stanghellini, M. E., Sands, D. C., Kronland, W. C., Mendonca, M. M. 1977. Serological and physiological differentiation among isolates of *Erwinia carotovora* from potato and sugarbeet. Phytopathology 67:1178–1182.

Stapp, C. 1956. Bakterielle Krankheiten. Berlin, Hamburg: Verlag Paul Parey.

Stapp, C. 1958. Pflanzenpathogene Bakterien. Sorauer, P. (ed.), Handbuch der Pflanzenkrankheiten, vol. 2. Berlin, Hamburg: Verlag Paul Parey.

Starr, M. P. 1947. The causal agent of bacterial root and stem disease of guayule. Phytopathology 37:291–300.

Starr, M. P. 1959. Bacteria as plant pathogens. Annual Review of Microbiology 13:211–238.

Starr, M. P. 1975. A generalized scheme for classifying organismic associations, pp. 1–20. In: Jennings, D. H., Lee, D. L. (eds.), Symbiosis. Cambridge: Cambridge University Press. [Symposia of the Society for Experimental Biology 29:1–20.]

Starr, M. P. 1979. Plant-associated bacteria as human pathogens: Disciplinal insularity, ambilateral harmfulness, epistemological primacy. Annals of Internal Medicine 90:708–710.

Starr, M. P., Chatterjee, A. K. 1972. The genus *Erwinia*: Enterobacteria pathogenic to plants and animals. Annual Review of Microbiology 26:389–426.

Starr, M. P., Cosens, G., Knackmuss, H.-J. 1966. Formation of the blue pigment indigoidine by phytopathogenic *Erwinia*. Applied Microbiology 14:870–872.

Starr, M. P., Dye, D. W. 1965. Scoring virulence of phytopathogenic bacteria. New Zealand Journal of Science 8:93–105.

Starr, M. P., Mandel, M. 1969. DNA base composition and taxonomy of phytopathogenic and other enterobacteria. Journal of General Microbiology 56:113–123.

Starr, M. P., Chatterjee, A. K., Starr, P. B., Buchanan, G. E. 1977. Enzymatic degradation of polygalacturonic acid by *Yersinia* and *Klebsiella* species in relation to clinical laboratory procedures. Journal of Clinical Microbiology 6:379–386.

Strobel, G. A. 1977. Bacterial phytotoxins. Annual Review of Microbiology 31:205–224.

Sutton, D. D., Starr, M. P. 1959. Anaerobic dissimilation of glucose by *Erwinia amylovora*. Journal of Bacteriology 78:427–431.

Sutton, D. D., Starr, M. P. 1960. Intermediary metabolism of carbohydrate by *Erwinia amylovora*. Journal of Bacteriology 90:104–110.

Thomson, S. V., Schroth, M. N., Moller, W. J., Reil, W. O. 1976. Efficacy of bactericides and saprophytic bacteria in reducing colonization and infection of pear flowers by *Erwinia amylovora*. Phytopathology 66:1457–1459.

van der Zwet, T., Keil, H. I. 1979. Fire blight. A bacterial disease of rosaceous plants. Agriculture Handbook 510. Washington, D. C.: United States Department of Agriculture.

Vidaver, A. K. 1976. Prospects for control of phytopathogenic bacteria by bacteriophages and bacteriocins. Annual Review of Phytopathology 14:451–465.

von Graevenitz, A. 1977. The role of opportunistic bacteria in human disease. Annual Review of Microbiology 31:447–471.

von Riesen, V. L. 1975. Polypectate digestion of *Yersinia*. Journal of Clinical Microbiology 2:552–553.

von Riesen, V. L. 1976. Pectinolytic, indole-positive strains of *Klebsiella pneumoniae*. International Journal of Systematic Bacteriology 26:143–145.

Vruggink, H., Maas Geesteranus, H. P. 1975. Serological recognition of *Erwinia carotovora* var. *atroseptica*, the causal organism of potato blackleg. Potato Research 18:546–555.

Waldee, E. L. 1945. Comparative studies of some peritrichous phytopathogenic bacteria. Iowa State College Journal of Science 19:435–484.

White, J. N., Starr, M. P. 1971. Glucose fermentation endproducts of *Erwinia* spp. and other enterobacteria. Journal of Applied Bacteriology 34:459–475.

Winslow, C.-E. A., Broadhurst, J., Buchanan, R. E., Krumwiede, C., Jr., Rogers, L. A., Smith, G. H. 1917. The families and genera of the bacteria. Preliminary report of the Committee of the Society of American Bacteriologists on Characterization and Classification of Bacterial Types. Journal of Bacteriology 2:505–566.

Wong, W. C., Preece, T. F. 1973. Infection of cricket bat willow (*Salix alba* var. *caerulea*) Sm. by *Erwinia salicis* (Day) Chester detected in the field by the use of a specific antiserum. Plant Pathology 22:95–97.

Wood, R. K. S. 1955. Studies in the physiology of parasitism. XVIII. Pectic enzymes secreted by *Bacterium aroideae*. Annals of Botany 19:1–27.

Yakrus, M., Schaad, N. W. 1979. Serological relationships among strains of *Erwinia chrysanthemi*. Phytopathology 69:517–522.

The Genera *Vibrio, Plesiomonas,* and *Aeromonas*

RIICHI SAKAZAKI and ALBERT BALOWS

THE GENUS *VIBRIO*

Various kinds of Gram-negative, polarly flagellated rod-shaped bacteria were previously classified as the genus *Vibrio* in the family Spirillaceae. Thirty-four *Vibrio* species were described by Breed (1957) in the seventh edition of *Bergey's Manual of Determinative Bacteriology,* and 207 species names were listed in *Index Bergeyana* (Buchanan, Holt, and Lessel, 1966). Since the International Subcommittee on Taxonomy of Vibrios (1966) reported a provisional definition for the genus *Vibrio,* however, only a few species comforming to the definition have been retained in the genus *Vibrio.* In the eighth edition of *Bergey's Manual,* Shewan and Véron (1974) included five species (*V. cholerae, V. parahaemolyticus, V. anguillarum, V. fischeri,* and *V. costicola*) in the genus. As suggested by Eddy and Carpenter (1964), the genus *Vibrio* is currently classified in the family Vibrionaceae together with genera *Aeromonas, Plesiomonas, Photobacterium,* and *Lucibacterium* (Shewan and Véron, 1974). Of the five *Vibrio* species mentioned above, *V. cholerae* and *V. parahaemolyticus* are known to be human enteropathogens, and *V. anguillarum* is pathogenic for fish and shellfish.

Habitats

Vibrio cholerae

Vibrio cholerae is the only terrestrial species of the genus described at present, and is best known as the causative agent of cholera. Cholera has been epidemic in the delta of the Ganges and Brahmaputra rivers in eastern India and Bangladesh since the beginning of recorded history. Between 1817 and 1960, cholera extended from those areas in six pandemics to spread over most of the world. Another focus of cholera-like disease, which initially showed no endemic tendency, was discovered in 1937 on the island of Celebes in what is now Indonesia (de Moor, 1939). However, this disease began to spread from the Southwest Pacific in 1961 to the Middle East in 1965 and to North and West Africa

and Eastern Europe in 1970. This seventh pandemic of cholera is still continuing. It is caused by the El Tor biovar of *V. cholerae* originating from Indonesia, rather than by the classical cholera vibrio from the Indo-Bangladesh subcontinent. Even in India, the homeland of classical cholera, almost all cases are now caused by the El Tor biovar. The seventh pandemic of cholera has involved at least 45 areas in the world. In some countries, cholera has now become endemic, but it has not gained a foothold in countries with good sanitation even though imported cases and small outbreaks have been reported.

Typical cholera is characterized by a sudden onset of vomiting and painless diarrhea with characteristic rice-water stools, followed by suppressed renal function, thirst, cramp in the legs and abdomen, hoarse speech or even appoinia, and progressive weakness and collapse due to the marked dehydration and electrolyte imbalance. These symptoms are usually present within 5–12 h after the onset of diarrhea, and untreated patients often die within 24 h. It is recognized, however, that the cholera vibrio, especially the El Tor vibrio, actually produces many asymptomatic infections with mild rather than severe clinical symptoms. Mild infections cannot be distinguished from other diarrheal disease unless the vibrio is isolated and identified. Cholera caused by the El Tor biovar has a lower morbidity and mortality rate than that caused by classical *V. cholerae.* The ratio of severe to mild of asymptomatic infections is between 1:5 and 1:10 for classical cholera, but only from about 1:25 to 1:100 for El Tor cholera.

V. cholerae is transmitted by the oral ingestion of water or food contaminated with the vibrio. The vibrios multiply in the small intestine and produce an enterotoxin, which either stimulates the mucosal cells to secrete large quantities of isotonic fluid or increases the permeability of the vascular endothelium. Cholera enterotoxin activates the adenyl cyclase in the intestinal cell membrane, causing increased intestinal secretion, which effects electrolyte transport mediated by cyclic AMP. During the last decade, considerable progress has been made in understanding the production and action of cholera enterotoxin. The developments in this field have been reviewed by Finkelstein (1973, 1975).

There is no conclusive proof that cholera is transmitted by direct contact, although some epidemics have been attributed to this mode of spread. The most prevalent mode of transmission is contact with the environment, particularly through drinking contaminated water from rivers, wells, tanks, or canals. In such circumstances, only sporadic clinically apparent infections may occur over a long period of time, or explosive outbreaks sometimes occur in a brief time span. The latter epidemics are presumably caused by a heavily contaminated common vehicle, such as food. Felsenfeld (1965) implicated unpurified water, ice, eating utensils, sweet soft drinks, food contaminated after cooking or pasteurization, and fruits and vegetables washed with sewage-polluted water and eaten raw as the most important vehicles of infection. In general, poor environmental sanitation, particularly the lack of adequate supplies of clean potable water, seems to be of fundamental importance in the spread of cholera. Thus, cholera is usually associated with lower socioeconomic groups.

A group of vibrios that closely resemble *V. cholerae* but are not agglutinated with cholera antiserum have been recognized as "nonagglutinable" (NAG) or "noncholera" vibrios (NCV). These vibrios are now also classified as *V. cholerae* and should be considered as potential agents of cholera-like disease, i.e., the cholera syndrome. Outbreaks or epidemics of such disease caused by these vibrios have been reported by many investigators (Bäck, Ljunggren, and Smith, 1974; Chatterjee and Neogy, 1971; Gaines et al., 1954; Ko, Lutticken, and Pulverer, 1973; McIntyre, Feeley, and Greenough, 1965; Nacescu et al., 1974). The organisms have also been reported to be associated with a gastroenteritis-type disease (Aldová et al., 1968; Dakin et al., 1974; Zakhariev et al., 1976). Zinnaka and Carpenter (1972) and Ohashi, Shimada, and Fukumi (1972) reported that some strains of these vibrios produced an enterotoxin which could be neutralized by antiserum against cholera enterotoxin. An interesting assortment of infections caused by noncholera vibrios was reported by Hughes et al. (1978). Bhaskaran and Sinha (1971) demonstrated that O-antigen specificity was transferable from NAG vibrio to cholera vibrio by chromosomal hybridization.

The endemicity of cholera is maintained by a cycle that involves patients, excrement, and the environment. Infected individuals usually excrete vibrios for only a few days, although a few carriers were found to harbor *V. cholerae* for months or years (Azurin et al., 1967; Wallace et al., 1967). *V. cholerae* may survive in feces for a few weeks at low temperatures but is killed within a day or two at room temperature. Its viability outside the human body is of great epidemiological importance and depends upon many factors. *V. cholerae* appears to have only limited ability to survive in the environment (Felsenfeld, 1965; Pandit et al., 1967; Pesigan, Plantilla, and Rolda, 1967). Although water probably plays the most important role in the spread of cholera, *V. cholerae* survives for only 7–10 days in surface water and may survive longer in seawater. *V. cholerae* can survive for a long period of time in clothing or bedding, and may continue to contaminate water in which these items are washed. The El Tor vibrio tends to be more resistant to various environmental factors than is the classical cholera vibrio. It is highly probable that the El Tor vibrio has a still unknown survival mechanism that has contributed significantly to the duration and spread of the present pandemic.

Humans are the only known natural reservoir. However, Sanyal et al. (1974) reported that in epidemic areas domestic animals such as dogs, hens, and cows often carry *V. cholerae,* and they suggested that hydrobionts may take part in the maintenance of the endemicity. Bisgaard and Kristensen (1975) reported continuous isolation of *V. cholerae* other than the cholera vibrio from ducks and their surroundings.

Vibrio parahaemolyticus

Vibrio parahaemolyticus inhabits coastal and estuarine waters. It is now recognized as a cause of gastroenteritis in humans following consumption of seafoods, either directly or indirectly contaminated with the vibrio. *V. parahaemolyticus* was first found in Japan and has been reported to cause 50–70% of the cases of bacterial, food-borne enteritis in Japan (Ministry of Health and Welfare of Japan, 1974, 1975, 1976, 1977; Okabe, 1974). During the last 10 years, however, outbreaks of food-borne illness caused by this vibrio have been recognized not only in Southeast Asia and India (Atthasampunna, 1974; Bonang, Lintong, and Santoso, 1974; Chatterjee, Gorbach, and Neogy, 1970; Chatterjee, Neogy, and Gorbach, 1970; Chun, Seol, and Tak, 1974; Inaba, 1973; Joseph, 1974; Neumann et al., 1972a; Sakazaki et al., 1971; Sircar et al., 1976; Van and Tuan, 1974) but also in the United States (Barker, 1974), Australia (Battey et al., 1970), Africa (Bockemühl, Amédomé, and Triemer, 1972), and Europe (Ciufecu, Necescu, and Florescu, 1976; Hooper, Barrow, and McNab, 1974; Lázničková and Aldová, 1976; Zakhariev et al., 1976). An outbreak of *V. parahaemolyticus* infection among airline passengers and crew returning from the Orient was reported in England (Peffers et al., 1973). Enteritis caused by *V. parahaemolyticus* is almost always directly or indirectly associated with seafoods. Raw fish or shellfish are the most important sources of *V. parahaemolyticus* food-borne enteritis in

Japan, where the high incidence undoubtedly results from the national custom of eating raw fish. In other countries where seafoods are not usually eaten raw, cross-contamination of food by water or kitchen utensils is the most likely source of the infection (Barker, 1974; Barrow and Miller, 1976).

The outstanding features of *V. parahaemolyticus* food-borne illness are severe abdominal pain, diarrhea, nausea, vomiting, mild fever, and headache. This vibrio may cause very mild illness or a very severe, cholera-like illness that leads to dehydration or dysentery-like illness with stools containing blood and mucus; fulminating septicemia has been reported (Zide, Davis, and Ehrenkranz, 1974). The mortality rate is very low, with the most deaths occurring among elderly or otherwise debilitated persons.

Most food-borne illness caused by *V. parahaemolyticus* occurs during the warmer months. This seasonal prevalence of gastroenteritis is associated with the number of the vibrios present in the marine environment, which in turn depends upon temperature. *V. parahaemolyticus* is found in coastal seawater when the water is warmer than 15°C. The number of vibrios in the marine environment decreases markedly during the cold season (Colwell et al., 1974; Kaneko and Colwell, 1973; Nishio, Kida, and Shimouchi, 1967a,b). The vibrios may not be found in market fish in the winter, although they are present in most of the fish marketed in the summer (Barrow and Miller, 1972; Leistner and Hechelmann, 1974; Miyamoto et al., 1960, 1961b,c; Noguchi and Asakawa, 1967). It is probable, however, that *V. parahaemolyticus* survives in sediments in coastal and estuarine areas in the winter (Kaneko and Colwell, 1973; Nishio, Kida, and Shimouchi, 1967b). Kaneko and Colwell (1973) reported a close relationship between the distribution of *V. parahaemolyticus* and the zooplankton present in the seawater.

V. parahaemolyticus may also be found in fresh water, especially if it is brackish and contains much organic material (Yanagisawa and Takeuchi, 1975). Thus, certain industrial effluents and commercially sold marine products may contribute to the level of vibrio contamination (Barrow, 1973; Kristensen, 1974).

Kato et al. (1965) found that strains of *V. parahaemolyticus* isolated from patients with gastroenteritis were hemolytic under specific test conditions, whereas isolates of the vibrio from marine sources were nonhemolytic on an agar medium that contained human red blood cells. This property was named the Kanagawa phenomenon to denote the prefecture in Japan where it was first detected. Sakazaki et al. (1968) showed that 96% of the strains isolated from patients with gastroenteritis reacted positively in the Kanagawa test, whereas only 1% of isolates from sea fish did so. In feeding tests on human volunteers, only Kanagawa-positive strains caused gastroenteritis (Sakazaki et al., 1968; Sanyal and Sen, 1974). A thermostable extracellular substance is responsible for the Kanagawa reaction, although several hemolysins are found in *V. parahaemolyticus*. The purification, physicochemical nature, and biological activity of the Kanagawa hemolysin have been investigated extensively (Honda et al., 1976a,b; Obara, 1971; Takeda, Hori, and Miwatani, 1974; Takeda et al., 1975a,b Zen-Yoji et al., 1971, 1974, 1975). The Kanagawa hemolysin is a lethal toxin which exhibits cardiotoxicity. Although it is not yet known whether the Kanagawa hemolysin is associated directly with the enteropathogenicity of *V. parahaemolyticus*, it may only be present in pathogenic strains. However, the actual mechanism of enteropathogenicity, whether due to enterotoxin production, a single manifestation of the Kanagawa hemolysin, or both, is still not known.

In contrast to isolates from human patients, almost all strains isolated from marine sources, even from seafoods implicated in outbreaks of food-borne gastroenteritis, are Kanagawa negative, which may indicate that the production of Kanagawa hemolysin is plasmid mediated, but there is no evidence to support this suggestion. Another suggestion is that selective multiplication of Kanagawa-positive vibrios occurs in the human intestine even though they are initially greatly outnumbered by Kanagawa-negative vibrios (Sakazaki et al., 1974). Some outbreaks of food poisoning associated with Kanagawa-negative strains have been reported (Teramoto, Nakanishi, and Maejima, 1969; Zen-Yoji et al., 1970). Under certain conditions, Kanagawa-negative vibrios may multiply in the intestine and produce enteritis, although these isolates caused no illness in human volunteers (Sanyal and Sen, 1974).

The relation of *V. parahaemolyticus* to disease of fish is uncertain. However, in man, *V. parahaemolyticus* also has been isolated from infected skin or lesions on the hands, feet, eyes, and ears of fish handlers and swimmers (McSweeney and Forgan-Smith, 1977; Roland, 1970; von Graevenitz and Carrington, 1973; Twedt, Spaulding, and Hall, 1969).

Vibrio anguillarum

Vibrio anguillarum is a marine organism that causes a septicemic disease of marine fish. The disease is characterized by deep focal necrotizing myositis and subdermal hemorrhages. The vibrio has also been isolated from a number of marine invertebrates and fish. Tubiash, Colwell, and Sakazaki (1970) reported necrotic disease of larval and juvenile bivalve mollusks. *V. anguillarum* causes outbreaks of disease in fish farms which are responsible for considerable economic losses in many parts of the world

(Cisar and Fryer, 1969; Evelyn, 1971; Hastein, 1975; Holt, 1970; Hoshina, 1956, 1957; Ross, Martin, and Bressler, 1968; Smith, 1961). Outbreaks of infection with this vibrio occasionally occur at low temperatures, but the main predisposing factor of fish-farm outbreaks is usually a rise in water temperature to 15°C or above.

Other Marine Vibrios

Various kinds of named or unnamed, halophilic vibrios may be found in the marine environment and are occasionally associated with humans. *V. alginolyticus* (Sakazaki, 1968) is the most common vibrio isolated from coastal and estuarine water and seafoods and is sometimes associated with human wound and other infections (Pien, Lee, and Higa, 1977; Rubin and Tilton, 1975; Schmidt, Chmel and Cobbs, 1979). *V. metschnikovii* (Gameleia, 1888), which is a senior synonym of *V. proteus* (Buchner, 1885), was reported to be isolated from fecal specimens of man and animals and to cause gastroenteritis (cholera nostras) by previous investigators (Buchner, 1885; Finkler and Prior, 1884; Gameleia, 1888), but members of this species may often be found in seawater and seafoods. An unnamed vibrio, *Vibrio* sp. 1669, causes septicemic disease of marine fish like that reported to be caused by *V. anguillarum* by Harrell et al. (1976). Hollis et al. (1976) and Clark and Steigerwalt (1977) described a group of halophilic vibrios which were repeatedly isolated from blood cultures from human patients. A more recent report (Blake et al., 1979) suggests that this lactose-positive, halophilic vibrio is a potentially serious pathogen in systemic infections. *Vibrio* sp. 6330 comprises a group of halophilic vibrios which resemble the *V. parahaemolyticus* reported by Sakazaki et al. (1970b).

Isolation

Vibrio cholerae

Isolation of *Vibrio cholerae* is facilitated by its ability to grow at a high pH range, 9.0–9.6, and to ferment sucrose. Cholera is usually an emergency situation, since it is a disease of short incubation and duration which requires rapid diagnosis and treatment. The bacteriological diagnosis of cholera should be made on the basis of rapid detection of vibrios on the spot, if possible, and confirmation of the diagnosis, if necessary, by a reference laboratory. An important aspect to be managed by the local laboratory is the collection of specimens from patients with suspected cases and the inoculation of appropriate media. Of equal importance is that methodology, including media for the isolation and

identification of *V. cholerae,* be standardized, and, especially in nonepidemic areas, that laboratory workers be properly trained. The successful management of cholera often depends on the ability of the local laboratory to make a preliminary diagnosis.

Stool specimens are collected as soon as possible, preferably before any antibiotic treatment, and should be inoculated directly onto primary isolation media and enrichment broths as soon as possible after collection. If culturing must be delayed, the specimens should be collected and placed in a transport medium. If the specimens can be examined within 8 h, they should be collected in alkaline peptone water. Otherwise, a transport medium such as the one described by Cary and Blair (1964) can be used because *V. cholerae* remains viable in it for at least 5 days at ambient temperatures. Stool specimens collected for culturing *V. cholerae* must not be frozen or refrigerated for prolonged periods because low temperatures kill the vibrios.

Although numerous selective media have been devised for isolating *V. cholerae,* a combination of thiosulfate–citrate–bile salt–sucrose (TCBS) agar (Kobayashi et al., 1963) and Vibrio agar (Tamura, Shimada, and Prescott, 1971) for each specimen can be recommended. Alkaline bile salt agar, which is less selective and not a differential isolation medium, is commonly used in some endemic areas. However, much experience is required in using this agar medium. Although *V. cholerae* may be detected on alkaline bile salt agar with ease when the oblique light technique is applied, the technique is very time-consuming and therefore should not be used in observing numerous plates in the routine laboratory. Gelatin-taurocholate-tellurite agar is more selective than alkaline bile salt medium and is widely used, but caution is necessary in selecting the proper quality of gelatin and in standardizing the tellurite solution to prepare the medium. Tamura, Shimada, and Prescott (1971) carried out a comparative study of three media (TCBS agar, Vibrio agar, and alkaline bile salt agar) for isolating *V. cholerae* from 400 stool specimens obtained from persons with cholera in the field in Calcutta. They demonstrated that all positive cultures on alkaline bile salt agar were also positive on either TCBS or Vibrio agar, or on both media, while *V. cholerae* was more frequently detected with the latter two media than with the former. Their results suggest that alkaline bile salt agar is not an efficient medium for isolating *V. cholerae* unless the oblique light technique is used.

TCBS Agar for Isolating *Vibrio cholerae* (Kobayashi et al., 1963)

TCBS agar, which is now available commercially (Eiken, Nissui, BBL, Difco, Oxoid), was devised by investigators at Eiken Chemicals Company. The published formula from Eiken is as follows:

Yeast extract	5 g
Peptone	10 g
Sucrose	20 g
Sodium thiosulfate	10 g
Sodium citrate	10 g
Sodium cholate	3 g
Ox-gall	5 g
Sodium chloride	10 g
Ferric citrate	1 g
Thymol blue	0.04 g
Bromothymol blue	0.04 g
Agar	15 g
Distilled water	1 liter

The final pH of the medium is 8.6. The medium is heated to dissolve the ingredients, cooled to 45°C, and poured into Petri dishes. It must not be autoclaved.

TCBS agar is highly selective for *V. cholerae,* including the so-called NAG, and it strongly inhibits most enteric bacteria, pseudomonads, and Grampositive bacteria present in feces. It has been reported, however, that lot-to-lot and brand-to-brand variation adversely affects the ability of TCBS agar to inhibit normal fecal bacterial flora, although different lots or brands appear to support the growth of *V. cholerae* and *V. parahaemolyticus* equally well (McCormack et al., 1974).

V. cholerae forms small or medium-sized, faint yellow, sucrose-fermenting colonies on TCBS agar, but due to its high selectivity, some strains of *V. cholerae,* especially laboratory strains of the classical cholera vibrio, will not grow on TCBS agar. Some strains of *Proteus* and enterococci may grow, but their colonial morphology, including size, color, and opacity, is easily differentiated from that of *V. cholerae.*

Colonies of *V. cholerae* growing on TCBS agar are usually rather sticky and may not be agglutinated satisfactorily or are slow to agglutinate with group-specific O antiserum.

Vibrio Agar (Tamura, Shimada, and Prescott, 1972)

Dehydrated Vibrio agar is available from Nissui Seiyaku Company, Tokyo. According to Tamura, Shimada, and Prescott (1971), the medium contains per liter:

Yeast extract (Difco)	5 g
Trypticase (BBL)	4 g
Proteose peptone (Difco)	3 g
Sucrose	20 g
Sodium thiosulfate · 5H$_2$O	6.5 g
Sodium citrate · 2H$_2$O	10 g
Sodium deoxycholate	1 g
Sodium chloride	10 g
Ox-gall (depigmented solids)	5 g
Sodium lauryl sulfate	0.2 g
Water blue	0.2 g

| Cresol red | 0.02 g |
| Agar | 15 g |

The medium is adjusted to pH 8.5. The medium is heated to dissolve the ingredients, cooled to 45°C, and poured into Petri dishes. It must not be autoclaved.

Vibrio agar is less selective than TCBS agar but more selective than alkaline bile salt agar. Colonies of *V. cholerae* on *Vibrio* agar are bluish gray, translucent, and slightly sticky; they are larger than colonies of other organisms and are also larger than *V. cholerae* colonies on TCBS agar. Colonies of sucrose-fermenting contaminants, such as *Proteus* and *Aeromonas,* are bluer and smaller than those of *V. cholerae* on Vibrio agar and can be easily differentiated. The stickiness of *V. cholerae* colonies also helps to distinguish them from other organisms. Sometimes, *Klebsiella* spp. grow on Vibrio agar in large colonies, which are easily differentiated from those of *V. cholerae* because of the mucoid and opaque appearance of the former. Colonies of *V. cholerae* on Vibrio agar are easily agglutinated with *V. cholerae* O-1 antiserum.

In view of the importance of cholera, the isolation procedure for *V. cholerae* should always include the inoculation of an enrichment broth, even if the specimen is obtained during the acute phase of illness. Although numerous enrichment broths for *V. cholerae* are described, alkaline peptone water and Monsur's tellurite–bile salt broth (Monsur, 1963) are recommended.

Alkaline Peptone Water for Isolating *Vibrio cholerae*

Alkaline peptone water contains 1% peptone and 1% sodium chloride in distilled water and is adjusted to pH 8.6–9.0. Ten milliliters is dispensed into 30-ml, wide-mouthed, screw-capped bottles or tubes and autoclaved at 121°C for 15 min. About 0.5–1.0 ml or approximately 0.5 g of the stool specimen is inoculated into the medium, and after 6–8 h incubation at 35–37°C, a loopful of the broth culture is streaked onto both TCBS and Vibrio agars. The incubation period of alkaline peptone water culture is best limited to 8 h or less to prevent overgrowth by other fecal organisms. Secondary enrichment, in which a loopful of culture of primary alkaline peptone water is subcultured to another tube of the same broth medium and incubated again for 8 h, may also be recommended. This enrichment culture procedure is needed particularly for specimens from convalescent patients and contacts.

Monsur's tellurite–bile salt broth (Monsur, 1963) is a more selective enrichment medium than alkaline peptone water. Most fecal organisms other than *V. cholerae* and *V. parahaemolyticus* are inhibited in this medium. The procedure for this medium is the same as that for alkaline peptone water, but an overnight incubation instead of 8 h will give better results. This medium is particularly useful for secondary enrichment from the primary culture of alkaline peptone water for specimens from convalescent patients and contacts.

Monsur's Tellurite–Bile Salt Broth for Isolating *Vibrio cholerae* (Monsur, 1963)

> The broth is prepared by adding the following compounds, per liter of distilled water:
>
> | | |
> |---|---|
> | Peptone | 10 g |
> | Sodium chloride | 10 g |
> | Sodium taurocholate | 5 g |
> | Sodium carbonate | 1 g |
>
> The medium should be at pH 9.0–9.2. After autoclaving at 121°C for 15 min, potassium tellurite is added to give a final concentration of 1:100,000.

Vibrio parahaemolyticus

The diagnosis of gastroenteritis caused by *Vibrio parahaemolyticus* usually necessitates the isolation of the vibrios from stool specimens or vomitus of patients and from the suspected foods. Seawater and seafoods may also be cultured for the isolation of *V. parahaemolyticus*. Attempts to relate the culture isolated from patient specimens to the ingested food or other environmental sources are usually futile.

Detection of *V. parahaemolyticus* in stool specimens taken during the later stage of diarrhea may be difficult because the number of vibrios decreases rapidly as the patient recovers from gastroenteritis. The specimen to be cultured may be either a freshly evacuated stool or feces collected by means of rectal swabs; these should be cultured soon after collection. If culturing must be delayed, the specimens should be placed in a transport or holding medium. Cary-Blair transport medium (Cary and Blair, 1964) without additional salt is adequate to preserve stool specimens that contain *V. parahaemolyticus* (Neumann et al., 1972b; Sakazaki et al., 1971). Alkaline peptone water is also useful for preserving stool specimens which will be subcultured within 8 h after collection. The precautions to be followed when using Cary-Blair medium and alkaline peptone water for *V. parahaemolyticus* are the same as described for *V. cholerae*. When seafoods are to be used in attempts to isolate vibrios, specimens for culturing are collected from the body surface, bowel, and gills of fish. Shellfish should be homogenized in a nonaerosolizing blender.

V. parahaemolyticus grows well on a number of ordinary laboratory media with the addition of 2–5% sodium chloride. It may even grow on mannitol-salt agar, which is used primarily to isolate staphylococci (Carruthers and Kabat, 1976). However, TCBS agar (Kobayashi et al., 1963) makes the isolation of the vibrios much easier. Isolated colonies of *V. parahaemolyticus* on TCBS agar, after 18–24 h of incubation at 37°C, are round, moist, 2–3 mm in diameter, and have green or blue centers due to the uptake of alkaline bromothymol blue. *V. alginolyticus,* which is commonly found in seafoods, also grows very well on TCBS agar. Colonies of *V. alginolyticus* are larger than those of *V. parahaemolyticus* and are yellow due to the fermentation of sucrose. Some marine bacteria which are not usually found in human feces may also form yellow or sometimes blue colonies on this agar, although most of these are usually inhibited by TCBS agar.

When diarrheal stools can be cultured soon after collection, enrichment techniques are not necessary for *V. parahaemolyticus*. However, enrichment techniques may be needed to isolate *V. parahaemolyticus* from stools of convalescent patients. Alkaline peptone water and Monsur's tellurite-bile salt broth, which are used for the isolation of *V. cholerae*, are also satisfactory for selective enrichment of stool specimens for *V. parahaemolyticus*. Secondary enrichment may also facilitate the isolation of *V. parahaemolyticus* from patients who have received antimicrobial treatment.

A selective enrichment procedure is usually essential to isolate *V. parahaemolyticus* from seafoods and seawater in order to inhibit the growth of other marine organisms present. *V. alginolyticus* is one of the most commonly found vibrios that is present in numbers exceeding those of *V. parahaemolyticus* in seafoods and seawater. A selective enrichment broth that has been widely used is salt-polymyxin broth (Sakazaki, 1972).

Salt-Polymyxin Broth for Selective Enrichment of *Vibrio parahaemolyticus* (Sakazaki, 1972)

> Broth medium contains 0.3% yeast extract, 1% peptone, 2% sodium chloride, and 250 μg of polymyxin B per liter of distilled water. The medium is adjusted to pH 8.6–9.0. The broth is dispensed in about 10-ml amounts and autoclaved at 121°C for 15 min. Each tube in the broth is inoculated with approximately 1 g or 1 ml of the specimen to be cultured. After 8–12 h of incubation, a loopful of the enrichment broth culture should be plated onto the isolation agar. The medium is available commercially (Nissui Seiyaku, Eiken) in dehydrated form.

Salt-polymyxin broth inhibits not only enterobacteria, pseudomonads, and Gram-positive organisms, but also marine vibrios that are closely related to *V. parahaemolyticus*. When specimens from sea sources are cultured in this broth medium, an almost pure culture of *V. parahaemolyticus* is usually obtained, because *V. alginolyticus* usually cannot grow in this broth within the recommended 8 h of incubation at 37°C. Nakanishi, Murase, and Teramoto (1977) demonstrated that salt-polymyxin broth is superior to other enrichment broth media such as nutrient broth with 4% salt and glucose-salt-teepol broth (Akiyama, Takizawa, and Obara, 1964). Subculture onto TCBS agar from salt-polymyxin broth culture is recommended. If less-selective isolation media such as BTB-teepol agar (Akiyama et al., 1963) are used in isolating *V. parahaemolyticus* after enrichment, the resulting colonies may be difficult to identify. Nakanishi, Murase, and Teramoto (1977) concluded that *V. parahaemolyticus* could be more easily detected in seafoods with a combination of salt-polymyxin broth and TCBS agar, because they did not have to identify colonies biochemically with this combination. This approach is not recommended for workers with limited experience with *V. parahaemolyticus*.

Vibrio anguillarum

Vibrio anguillarum must have 1% or 2% sodium chloride in order to grow. Although there are no special media for isolating this vibrio, nutrient agar containing 2–3% salt added to the distilled water or 50% sterile seawater can be mixed with 50% distilled water. However, BTB-teepol agar (Akiyama et al., 1963) or a modified version of it is preferable for isolating *V. anguillarum* from pathological materials obtained from fish. The growth of most strains of *V. anguillarum* may be inhibited on TCBS agar.

BTB-Teepol Agar for Isolating *Vibrio anguillarum* (Akiyama et al., 1963)

This medium was originally developed to be used in isolating *V. parahaemolyticus* and contains the following:

Beef extract	5.0 g
Peptone	10.0 g
Sucrose	10.0 g
Sodium chloride	20.0 g
Teepol (Shell)	2.0 ml
Bromothymol blue	0.08 g
Agar	15 g
Distilled water	1 liter

The pH of the medium is ajdusted to 7.8. Instead of teepol, 0.1 ml of Tergitol 7 (Union Carbide) may be preferable because of its stability. The constituents are dissolved by gentle heating and autoclaving at 121°C for 15 min. Gram-positive bacteria are generally inhibited on this medium. *V. anguillarum* usually forms moderate-sized, yellow colonies as a result of sucrose fermentation. Because other marine organisms may grow

on this medium, further identification procedures are necessary to differentiate *V. anguillarum*.

Identification

The Subcommittee on Taxonomy of Vibrios (1966, 1972) recommended the amended provisional description of the genus *Vibrio*. This definition was expanded by Hugh and Sakazaki (1972). The minimal number of characters that Hugh and Sakazaki (1972) suggested for use in identifying *Vibrio* species, including *V. cholerae, V. parahaemolyticus, V. alginolyticus, V. albensis,* and *V. costicola,* includes: polar monotrichous; Gram-negative, asporogenous rod; positive indophenol-oxidase test; acid but no gas production from glucose under a petrolatum seal; acid from mannitol; positive in lysine- and ornithine-decarboxylase tests. The G+C ratio of these species is 40–50 mol%. However, the genus *Vibrio* comprises a poorly characterized group of organisms, some of which do not fit the description given above. Thus, sometimes it may be difficult to differentiate among *Vibrio* and other genera such as *Aeromonas*. It should also be noted that many halophilic marine vibrios with the same G+C ratio of DNA as that of the genus *Vibrio* appear as peritrichously flagellated rods in some cultures (Baumann, Baumann, and Mandel, 1971; Leifson, 1963). *V. metschnikovii* consists of a group of organisms that react negatively in the indophenol-oxidase test; and *V. anguillarum* reacts positively in the arginine-dihydrolase test but negatively in lysine- and ornithine-decarboxylase tests.

Vibrio cholerae

A group of vibrios have the same flagellar antigen and other characteristics as those of cholera vibrio. These organisms are distinguishable from the cholera vibrio and from each other by their characteristic somatic antigens. They are often isolated from patients with cholera syndrome, gastroenteritis, or other symptoms of a febrile illness. They are usually called NAG (nonagglutinable) vibrios because they are not agglutinated by O antiserum of cholera vibrio. They are indistinguishable from the cholera vibrio morphologically and biochemically.

Sakazaki, Gomez, and Sebald (1967) carried out a taxonomic study on these NAG vibrios and concluded that they should be classified as *V. cholerae*. This conclusion was supported by Citarella and Colwell (1970), who demonstrated a close polynucleotide relationship between the typical cholera and NAG vibrios. The Subcommittee on Taxonomy of Vibrios (1972) therefore recommended that the designation *V. cholerae* be expanded to include the NAG vibrios. Shewan and Véron (1974) regarded *V. albensis* and *V. proteus* (an illegitimate synonym of *V. metschnikovii*), as biovars of *V. cholerae*. However, *V. albensis* is a luminous bacterium which is biochemically different from *V. cholerae. V. metschnikovii* is an oxidase-negative, halophilic vibrio. Deoxyribonucleic acid hybridization studies by Citarella and Colwell (1970) indicated that *V. metschnikovii* was only 2% related to *V. cholerae*. Therefore, *V. albensis* and *V. metschnikovii* should be separate species from *V. cholerae*.

The rapid diagnosis of cholera would greatly aid in planning surveillance, control measures, and epidemiological investigations; therefore, suspicious colonies should be tested for agglutination directly with *V. cholerae* O-1 serum. At least 5–10 colonies plus a sweep from the confluent growth from an isolation agar plate should be tested before concluding that the culture does not contain cholera vibrio. If some agglutination occurs with the sweep from confluent growth, it should be subcultured in enrichment broth such as Monsur's broth, and from subsequent plating on solid media and suspicious colonies should be further examined.

Whether or not the colonies tested are agglutinated with O-1 serum, they should be tested with a battery of differential media for biochemical identification. The biochemical characteristics of *V. cholerae* are presented in Table 1, and the ones which constitute the minimum required to identify the organisms are noted. Reactions that differentiate *V. cholerae* from other fermentative organisms isolated from stool specimens are shown in Table 2.

V. cholerae possesses two antigenic components—O (somatic) and H (flagellar) antigens. The O antigen is thermostable and is not destroyed when treated at 100°C for 2 h. The H antigen of *V. cholerae* (including the NAGs) is reported to be the

Table 1. Biochemical characteristics of *Vibrio cholerae.*[a]

Test or substrate	Reaction	%+	%(+)
Oxidase*	+	100	
Indole	+	100	
Voges-Proskauer	d	85.2	
Citrate (Simmons')	(+)	0	86.3
Hydrogen sulfide (TSI)*	−	0	
Urease	−	0	
Gelatinase	+	100	
Phenylalanine deaminase	−	0	
Lysine decarboxylase*	+	99.9	
Arginine dihydrolase*	−	0	
Ornithine decarboxylase*	+	98.8	
Growth in peptone water:			
Without NaCl*	+	98.6	
With 8% NaCl	−	0	
Growth at 42°C	+	100	
ONPG	+	100	
Fermentation:			
Glucose, acid*	+	100	
Glucose, gas*	−	0	
Arabinose	−	0	
Lactose	(+)	1	98.2
Mannose	d	72.6	
Raffinose	−	0	
Rhamnose	−	0	
Sucrose	+	95.2	
Trehalose	+	100	
Xylose	−	0	
Adonitol	−	0	
Dulcitol	−	0	
Inositol*	−	0	
Mannitol	+	98.2	
Salicin	−	0	

[a] The data are based on the results from 838 strains. *, Minimal characteristics needed to identify *V. cholerae*; +, 90% or more positive; −, 90% or more negative; d, 11–89% positive reactions; (+), delayed reactions.

Table 2. Differentiation of *Vibrio cholerae* and related Gram-negative fermentative rods isolated from stool specimens.

Test or substrate	Cholera vibrio	"NAG" vibrio	Aero-monas	Plesio-monas	Entero-bacteriaceae
	Vibrio cholerae				
Oxidase	+	+	+	+	−
Lysine decarboxylase	+	+	−	+	d
Arginine dihydrolase	−	−	d	+	d
Ornithine decarboxylase	+	+	−	+	d
Hydrogen sulfide (TSI)	−	−	−	−	d
Gelatinase	+	+	+	−	d
Gas from glucose	−	−	d	−	d
Acid from:					
Mannitol	+	+	+	−	d
Mannose	+	d	+	−	d
Inositol	−	−	−	+	−
O/129 susceptibility	+	+	−	d	−
Agglutination with					
V. cholerae O-1 serum	+	−	−	−	−

same (Bhattacharyya, 1977; Gardner and Venkatra-man, 1935; Sakazaki et al., 1970a), but this view is not shared by Smith (1974). In any event, the O antigens are used to designate serovars in *V. cholerae* (Gallut, 1962; Gardner and Venkatraman, 1935). In 1965, Smith and Goodner, in a description of the classification of vibrios, gave numbers to 187 serovars of non-cholera-producing vibrios, including some aeromonads and marine-type vibrios. They recognized that their work was preliminary but pointed to the necessity of using O antigens to aid in the recognition of vibrios. Sakazaki et al. (1970a) and later Shimada and Sakazaki (1977) established 60 serovars of *V. cholerae*, with serovar 1 assigned to the true cholera vibrio. From these efforts, an international typing system based on O antigens should emerge.

Although the H antigen is not used in distinguishing serovars in *V. cholerae*, it presumably has some diagnostic significance because it is specific for *V. cholerae*. A method for preparing specific H antiserum was described by Bhattacharyya (1977). The rough (R) antigens of all the *V. cholerae* serovars are identical, and O antisera for most strains of *V. cholerae* (even for cholera vibrio) contain some R antibody, which may cause cross-reactions and thus lead to incorrect identifications (Shimada and Sakazaki, 1973). All of the diagnostic O antisera should be absorbed with an R strain to remove the R antibody.

A quantitative variation of specific O antigenic factors may occur in some serovars of *V. cholerae*. In serovar 1 of *V. cholerae*, O antigenic variants designated Ogawa, Inaba, and Hikojima are recognized. Although Nobechi (1923) represented the antigenic formulae of the Ogawa and Inaba strains as A, B, C and A, B, X, respectively, Heiberg (1935) and Kauffmann (1950) concluded that the antigenic differences among these variants were quantitative rather than qualitative. Their findings were recently confirmed and expanded by Sakazaki and Tamura (1971), who demonstrated that cholera vibrios possessed three somatic antigen factors, designated a, b, and c, and that Inaba strains were mutants that had lost factor b; thus, this b factor can be recognized as specific for the Ogawa form. Hikojima variants are regarded as intermediate between the Ogawa and Inaba variants.

Some incomplete Ogawa strains that contain a large amount of factor c may be wrongly identified as Inaba variants in agglutination tests. Such strains may be incompletely mutated Ogawa strains. In fact, all agglutinins in the Inaba antiserum could be removed by repeated absorption with an excess Ogawa culture, although the titer of the Ogawa agglutinins may not diminish when absorbed repeatedly with the Inaba cultures. It is difficult, therefore, to obtain pure Inaba-specific antiserum. In any case, the Inaba-specific antiserum could agglutinate

Ogawa strains weakly. Burrows et al. (1946) suggested an "A" variant, having only the factor a, but Sakazaki and Shimada (1972; unpublished) confirmed that their "A" variant was actually R. It has been frequently noted that the Inaba variant may be isolated usually late in the course of an outbreak originally caused by the Ogawa variant. However, the Ogawa strains may not be isolated during the course of an outbreak caused by the Inaba variant.

Antigenic conversion from the Ogawa to the Inaba was thought to be irreversible, but Nobechi, Nakano, and Nagao (1967) were able to demonstrate Ogawa-to-Inaba and Inaba-to-Ogawa serotypic conversion. Using gnotobiotic mice as an animal model in which to demonstrate the effect of immunological pressure, Sack and Miller (1969) and Miller et al. (1972) clearly showed gradual and progressive serotypic conversion in both directions. They used a freshly isolated rough variant that was passed three times through mice; smooth, Ogawa, Inaba, and Hikojima forms were isolated. This conversion may help to explain how the appearance of one serovar in an outbreak might be replaced with another, and also sheds light on how carriers may harbor a rough variant for months which then undergoes serovar conversion and initiates the chain of events leading to an outbreak of cholera.

Serovar 1 of *V. cholerae* is divided into two biovars according to their hemolytic activity on sheep red blood cells: *V. cholerae* biovar *cholerae* for the classical vibrio and *V. cholerae* biovar *eltor* for the El Tor vibrio. However, the conventional hemolysis method does not always provide reproducible results. More reliable hemolysis can be obtained by incubating blood-agar cultures at 37°C for 24 h anaerobically without CO_2 (Sakazaki, Tamura, and Murase, 1971). However, care must be exercised in distinguishing hemolysis from greening-lysis on these blood-agar plates. The tube hemolysis test described by Barua and Mukherji (1964) can also be used to obtain reproducible results (Sakazaki, Tamura, and Murase, 1971).

Tube Hemolysis Test (Barua and Mukherji, 1964)
Two milliliters of heart infusion broth with 1% glycerol are inoculated with a loopful of an overnight broth culture of the strain to be tested. After 24–48 h of incubation, 0.1 ml of a 10% suspension of fresh, washed sheep red blood cells in saline is added to the broth culture. The mixture is incubated in a water bath at 37°C for 2 h and left in a refrigerator at 4°C overnight. In readings taken the next morning, complete or partial hemolysis constitutes a positive test result. A heated control is necessary.

In addition to the hemolysis test, hemagglutination of chicken or sheep red blood cells, susceptibility to disks impregnated with polymyxin B (Finkelstein and Mukerjee, 1963), and lysis by group IV phage of Mukerjee (1963) are used to differentiate El Tor biovar from the classical cholera biovar (see Table 3 for a complete listing). Chicken or sheep red blood cells are checked for hemagglutination in slide tests in which a loopful of a heavy suspension of the growth from an agar slant culture is placed in saline for 18–24 h and then mixed with a loopful of a 20% suspension of washed red blood cells. The agglutination will take place within 10 min. The polymyxin B susceptibility test is carried out by using a 50-unit polymyxin B disk with the usual technique for other antimicrobial susceptibility tests. Although the test to demonstrate lysis by the group IV cholera phage is a useful test in differentiating the two biovars, it has its limitations because most routine bacteriology laboratories are not equipped to perform it; if facilities are available, the test can be easily done at the routine test dilution (RTD).

Because *V. cholerae* strains other than cholera vibrios have been implicated as the etiological agents of sporadic cases or outbreaks of cholera-like diarrheal disease, it is also important to be able to isolate and recognize these vibrios. The various serovars of those strains of *V. cholerae* can be determined at the National Institute of Health, Tokyo, the Center for Disease Control, Atlanta, and at the Public Health Laboratory Service, Maidstone, Kent, England.

Vibrio parahaemolyticus

Several synonyms, *Pasteurella parahemolytica* (*sic*) (Fujino, 1951), *Pseudomonas enteritis* (Takikawa, 1958), and *Oceanomonas enteritidis* and *Oceanomonas parahaemolytica* (Miyamoto, Nakamura, and Takizawa, 1961a), have been proposed, but the name *Vibrio parahaemolyticus* (Sakazaki, Iwanami, and Fukumi, 1963) has been most widely accepted.

In their early study, Sakazaki, Iwanami, and Fukumi (1963) divided *V. parahaemolyticus* into two subgroups. Later, however, subgroup 2 was reclassified as a new species for which the designation *V. alginolyticus* was proposed (Sakazaki, 1968). Although Shewan and Véron (1974) classified *V. alginolyticus* into biovar 2 of *V. parahaemolyticus* in the eighth edition of *Bergey's Manual,* the results of numerous taxonomic and DNA relatedness studies show that the two are actually separate species (Anderson and Ordal, 1972; Citarella and Colwell, 1970; Colwell, 1970; Hanaoka, Kato, and Amano, 1969; Staley and Colwell, 1973; Zen-Yoji et al., 1965). On the other hand, Baumann, Baumann, and Mandel (1971) suggested that *V. parahaemolyticus* and *V. alginolyticus* should be classified into the genus *Beneckea,* which was originally described by Campbell (1957) for certain marine

vibrios. However, the Subcommittee on Taxonomy of Vibrios (1975) stated that there are no known strains, including a type or neotype of the type species *B. labra* and that "the relationship of *B. labra* to *V. parahaemolyticus* cannot be objectively evaluated or established without a type or neotype strains of *B. labra.*" Thus, the name *V. parahaemolyticus* was recognized in the eighth edition of *Bergey's Manual,* but similar recognition of *V. alginolyticus* did not take place.

V. parahaemolyticus is a facultatively halophilic organism. In broth cultures, the vibrios have a sheathed, single polar flagellum, but young cultures on the surface of nutrient agar may have unsheathed, peritrichous flagella (Allen and Baumann, 1971). The identification of *V. parahaemolyticus* should be based on its physiological and biochemical properties.

Human isolates of *V. parahaemolyticus* can be easily identified when the organisms are grown on TCBS agar. Among the Gram-negative fermentative rods found in human stools, only *V. parahaemolyticus* produces large, green or blue colonies on this medium. Provisional identification of strains isolated from stool specimens can be made from recognizing this colony growth pattern on the TCBS plate and performing a few differential tests. Colonies suspected of being *V. parahaemolyticus* on the isolation plates are subcultured into triple sugar iron (TSI) agar and MR-VP broth, each of which contain 2% salt. The TSI slant is incubated overnight at 37°C, and the MR-VP broth is incubated at 25–30°C for 15–18 h. A culture of *V. parahaemolyticus* on TSI agar exhibits an acid butt with no gas and no blackening, and has an alkaline slant. *V. parahaemolyticus* reacts negatively in the Voges-Proskauer test. The oxidase test can be performed on growth from the red slant of TSI agar; *V. parahaemolyticus* reacts positively. Some additional tests, such as lysine decarboxylase and growth in two tubes of peptone water or nutrient broth (Difco), one with 8% salt and the other without salt, are necessary in identifying isolates grown on media other than TCBS agar.

Strains isolated from marine sources and foodstuffs must be further tested to differentiate them from related marine organisms. The characteristics of *V. parahaemolyticus* which must be used in identification (Hugh and Sakazaki, 1972) are shown in Table 4. Of the differential tests listed in this table, those for indole, the amino acid decarboxylases, and sugar fermentations are performed with the same basic media used for the enterobacteria, all of which should contain 2% salt.

Tests for halophilism, salt tolerance, and the ability to grow at 42°C are important in differentiating *V. parahaemolyticus* from similar marine organisms. These tests should be done with careful attention to avoid confusion.

Table 3. Differentiation of two biotypes of cholera vibrios.[a]

Test or substrate	Cholerae biovar		El Tor biovar	
	Reaction	%+	Reaction	%+
Hemolysis[b]	−	0	+	98.6
Chicken red cell agglutination	−	9.8	+	94.2
Susceptibility to:				
Phage IV	+	100	−	0
Polymyxin B (50 μg)	+	100	−	0
Voges-Proskauer	−	3.0	+	98.6

[a] The data are based on the results from 480 strains.
[b] Tested with sheep blood-agar plates grown anaerobically.

Table 4. Biochemical characteristics of *Vibrio parahaemolyticus*.[a]

Test or substrate	Reaction	%+	
		Human source	Marine source
Oxidase*	+	100	100
Indole	+	100	100
Voges-Proskauer*	−	0	0
Citrate (Simmons')	+	100	100
Hydrogen sulfide (TSI)	−	0	0
Urease	−	0	0
Gelatinase	+	100	100
Lysine decarboxylase*	+	97	96
Arginine dihydrolase*	−	0	0
Ornithine decarboxylase*	+	95	96
Growth in peptone water:			
Without NaCl*	−	0	0
With 8% NaCl*	+	98	100
With 10% NaCl*	−	0	0
Growth at 42°C	+	96	94
Fermentation:			
Glucose, acid*	+	100	100
Glucose, gas*	−	0	0
Arabinose	d	78	83
Cellobiose	(d)	63	44
Lactose	−	0	0
Mannose	+	100	100
Raffinose	−	0	0
Rhamnose	−	0.1	0
Sucrose*	−	0	2
Trehalose	+	100	100
Xylose	−	0	0
Adonitol	−	0	0
Dulcitol	−	0	0
Inositol	−	0	0
Mannitol	+	100	100
Salicin	−	0	0
Kanagawa reaction	d	98	0.5

[a] The data are based on the results from 2,354 strains. *, Minimal characteristics for identifying *V. parahaemolyticus*; +, 90% or more positive; −, 90% or more negative; d,
11–89% positive reactions; (d), Reaction varies with different strains, but if positive, reaction will occur after 48 h or more of incubation.

Tests for Halophilism and Salt Tolerance of *Vibrio parahaemolyticus*

A series of nutrient broth tubes, that contain 0.3% beef extract (Difco, BBL) and 0.5% peptone (tryptone [Difco] or Trypticase [BBL]) with no salt, 8% , and 10% salt are used in these tests. One percent tryptone (Difco) or Trypticase (BBL) broth solution can be substituted for nutrient broth. A loopful of an overnight broth culture to be tested is inoculated into 10-ml amounts of each medium and incubated for 12–18 h at 37°C. Only heavy or moderate growth is interpreted as positive.

Test for Ability of *Vibrio parahaemolyticus* to Grow at 42°C

One loopful of an overnight broth culture is inoculated into 5- to 10-ml amounts of nutrient broth containing 2% salt and then incubated in a water bath at 42°C for 24 h. Poor growth is considered negative. The incubator used for this test should be an agitated water bath that can be maintained at 42°C $\pm$ 0.02°C.

The criteria for differentiating *V. parahaemolyticus* from other marine vibrios are shown in Table 5. *V. alginolyticus* (Sakazaki, 1968) is found more frequently in coastal or temperate seawaters and sea fish than is *V. parahaemolyticus*. Sometimes the former is isolated from stool specimens of patients with gastroenteritis and from wound infections. *V. alginolyticus* grows well in peptone water containing 10% salt, ferments sucrose, and reacts positively in the Voges-Proskauer test, but it behaves similarly to *V. parahaemolyticus* in most other biochemical

tests. *Vibrio* sp. biovar 6330 (Sakazaki et al., 1970b) may often be mistakenly identified as *V. parahaemolyticus* in routine isolation procedures, because many strains of this unnamed vibrio do not ferment sucrose or produce acetoin; they react positively in lysine- and ornithine-decarboxylase tests. However, they do not grow in peptone water containing 8% salt. They are found only in marine sources.

Three antigenic components can be recognized in strains of *V. parahaemolyticus*. The O antigen is thermostable and not destroyed by heating at 121°C for 2 h. The K antigen is a capsular substance and its agglutinability is inactivated by heating. Newly isolated strains of the vibrio have well-developed K antigens and may not be agglutinated by the homologous O antiserum in the living state. The H antigens of the vibrio are all identical (Sakazaki, Iwanami, and Tamura, 1968; Terada, 1968), although the single polar flagellum is antigenically distinct from the peritrichous flagella (Shinoda et al., 1974).

An antigenic schema recognizing 11 O groups and 41 K antigens for *V. parahaemolyticus* was established by Sakazaki, Iwanami, and Tamura (1968). K antigens 2, 14, 16, 27, and 35 were subsequently excluded because they were found to be identical to others already recognized. Later, 61 K antigens were recognized (Committee on Serotyping of *Vibrio parahaemolyticus,* 1970; Terada et al., 1975; Tokoro, Toto, and Yamada, 1977; Yokota et al., 1977). The antigenic schema as of 1977 is shown in Table 6.

Serovars of *V. parahaemolyticus* can be determined in slide agglutination tests with diagnostic O

Table 5. Differentiation of *Vibrio parahaemolyticus* from other marine vibrios and from *Vibrio cholerae*.

Test or substrate	*V. parahaemolyticus*	*V. alginolyticus*	*V. anguillarum*	*V. metschnikovii*	*V. fischeri*	*Vibrio* sp. 6330[a]	*V. cholerae*
Growth in 0% NaCl	−	−	−	−	−	−	+
Growth in 6% NaCl	+	+	d	−	−	d	−
Growth in 8% NaCl	+	+	−	−	−	−	−
Growth in 10% NaCl	−	+	−	−	−	−	−
Growth at 42°C	+	+	−	−	−	d	+
Indole	+	+	d	+	+	+	+
Lysine decarboxylase	+	+	−	+	+	+	+
Arginine dihydrolase	−	−	+	+	−	−	−
Arabinose	d	−	d	−	−	−	−
Sucrose	−	+	d	+	+	d	+
Inositol	−	−	−	+	−	−	−
Oxidase	+	+	+	−	+	+	+

[a] Sakazaki et al. (1970b).

Table 6. Antigenic schema for *Vibrio parahaemolyticus* (1977).

O group	K antigen
1	1, 25, 26, 32, 38, 41, 56, 58
2	3, 28
3	4, 5, 6, 7, 29, 30, 31, 33, 37, 43, 45, 48, 54, 57, 58, 59
4	4, 8, 9, 10, 11, 12, 13, 34, 42, 49, 53, 55
5	15, 17, 30, 47, 60, 61
6	18, 46
7	19
8	20, 21, 22, 39
9	23, 44
10	19, 24, 52
11	36, 40, 50, 51

and K antisera commercially available from Toshiba Kagaku Kogyo Co. Ltd., Tokyo. For O agglutination tests, smooth, translucent colonies from overnight agar cultures are harvested in saline and heated at 100°C for 1 h. Washing the packed cells with 1% saline twice after they are centrifuged may increase O agglutinability of the culture. If there is no agglutination with any available O antisera, O-form (bright or translucent) colonies should be looked for on a nutrient agar plate using the oblique-light method. These colonies should be tested with O antisera. O agglutination may also occur after O-K cultures are autoclaved at 121°C for 2 h. For K antigen determination, smooth, opaque (but not mucoid) colonies should be subcultured onto nutrient agar. A dense suspension of live organisms in 1% saline is used for K agglutination tests. The antigenic schema of *V. parahaemolyticus* is based on strains isolated from human patients. Although the serovars of most strains isolated from diarrheal stools may be determined with existing, established O and K antisera, they cannot be used to identify serovars of many isolates from marine sources.

As mentioned above, the Kanagawa hemolysin may indicate an enteropathogenicity of a strain of *V. parahaemolyticus*. The accuracy and precision of Kanagawa test results depend on the medium used.

Kanagawa Test

The Kanagawa test is performed with Wagatsuma agar (Wagatsuma, 1968). The basal medium of Wagatsuma agar contains, per liter:

Yeast extract	5 g
Peptone (Difco)	10 g
Sodium chloride	70 g
D-Mannitol	5 g
Crystal violet	0.01 g
Agar	15 g

The pH is adjusted to 7.5. The medium is heated to boiling to dissolve the ingredients. The un-

autoclaved, melted basal medium is divided into two equal portions and kept at 50°C. One-tenth of the agar base volume of a 20% suspension of washed fresh human red blood cells is added to one portion of the medium, and one-tenth of the agar base volume of a 20% suspension of washed horse red blood cells is added to the other. Each blood-agar medium is thoroughly mixed and poured into Petri dishes. According to Wagatsuma (1968), other peptones should not be used instead of peptone (Difco). Autoclaving or otherwise overheating the basal medium may produce misleading results. No substitute should be used for human red blood cells. To perform the test, a loopful from an overnight broth culture is spot-inoculated onto a human red blood cell agar plate and onto a horse red blood cell agar plate, and the results are read after incubation at 37°C for 18–20 h. A positive Kanagawa reaction consists of a clear zone of beta-hemolysis around the growth of *V. parahaemolyticus* on the plate with human red blood cells but no hemolysis around the growth on the plate with the horse red blood cells. Alpha-hemolysis or discoloration of human red blood cells under or around the growth should be interpreted as negative, as is clear hemolysis on both the human red blood cell agar and the horse red blood cell agar. The test can only be used for *V. parahaemolyticus*. Other marine vibrios may show beta-hemolysis on Wagatsuma agar with human red blood cells, but such results are not related to their enteropathogenicity.

Vibrio anguillarum

Several names, *Vibrio piscium* (David, 1927), *V. piscium* var. *japonicus* (Hoshina, 1957), and *V. ichthyodermis* (Shewan, Hobbs, and Hodgkiss, 1960) were suggested for fish pathogens, but these organisms are now included in the species *V. anguillarum* (Kiehn and Pacha, 1969; Sebald and Véron, 1963; Simidu and Kaneko, 1973; Véron and Sebald, 1964).

Members of *V. anguillarum* are polarly monotrichated rods with facultative halophilism. Many of the biochemical properties of *V. anguillarum* vary from strain to strain, as is especially apparent in results of tests for indole production, citrate utilization as a sole carbon source, and fermentation of arabinose, lactose, cellobiose, sucrose, and trehalose (Evelyn, 1971; Hastein and Smith, 1977; Nybelin, 1975; Smith, 1961). In general, however, *V. anguillarum* reacts negatively in lysine- and ornithine-decarboxylase tests and positively in the arginine-dihydrolase test; grows in nutrient broth with 0.5–6% salt, but not in broth with 8% salt; and cannot grow at 42°C. The biochemical characteristics of *V. anguillarum* are shown in Table 7.

Nybelin (1975) attached importance to the ability to ferment sucrose and mannitol and to produce indole. He divided *V. anguillarum* into biovars A (sucrose +, mannitol +, indole +) and B (sucrose −, mannitol −, indole −). Smith (1961) described an additional biovar C (sucrose +, mannitol +, indole −), but Hastein and Smith (1977) recognized only two biovars of the vibrio after performing a computer analysis on 163 strains. The most consistent difference which they found between the two biogroups was that group I fermented arabinose and group II did not.

THE GENUS *PLESIOMONAS*

Ferguson and Henderson (1947) first described a motile coliform organism possessing a somatic antigen identical with that of *Shigella sonnei,* isolated from the feces of a patient whose clinical history was not available. They called the organism strain C27. Most early investigators were confident that the C27 organism was a member of the family Enterobacteriaceae. In fact, Cowan (1956) proposed a new species, *"Escherichia sonnei",* for this organism together with *Shigella sonnei.* Bader (1954) studied a culture of C27 organism possessing the *S. sonnei* O antigen, unaware that there were previous papers on this organism. He found that the organism showed lophotrichous flagellation, and classified it into the genus *Pseudomonas* with the name *Pseudomonas shigelloides.* Later, Ewing, Hugh, and Johnson (1961) transferred the organism to the genus *Aeromonas* because of its glucose-fermentative behavior. The generic name *Plesiomonas* was given by Habs and Schubert (1962), since the organism did not exhibit some important characters of either *Aeromonas* or *Vibrio.* This suggestion was supported by the DNA study of Sebald and Véron (1963), who demonstrated that its G+C content was 51 mol%, as opposed to that of *Aeromonas* DNA, which ranged from 57 to 63 mol%, and of *Vibrio,* which ranged from 40 to 50 mol%, although they suggested an illegitimate generic name, *Fergusonia,* for the C27 organisms. Establishing *Plesiomonas* as a separate genus was further justified by a numerical taxonomy study by Eddy and Carpenter (1964) who suggested the family Vibrionaceae for the genera *Vibrio, Aeromonas,* and *Plesiomonas.* Only a single species, *P. shigelloides,* was included in the genus *Plesiomonas.* The name *P. shigelloides* was adopted in the eighth edition of *Bergey's Manual of Determinative Bacteriology.* Hendrie, Shewan, and Véron (1971) proposed the transfer of *P. shigelloides* to the genus *Vibrio* because its phenotypic characters were closer to the genus *Vibrio* than to *Aeromonas.* However, this proposal has not been accepted and we prefer to regard this species as *P. shigelloides.*

Habitats of *Plesiomonas*

Plesiomonas shigelloides has been isolated from a variety of specimens of human and animal origin. Furthermore, it can be found in fresh water and sometimes in seawater.

It is suggested by Tsukamoto et al. (1978) that *P. shigelloides* is an inhabitant of fresh water. The presence of the organisms in water is more frequent in the warmer seasons than in winter. The organism is isolated more frequently from mud than from fresh water, which suggests that the optimum place for survival and growth of *P. shigelloides* is mud, which presumably contains sufficient nutrients. Although *P. shigelloides* has been isolated from seawater, especially in coastal areas, it appears that this is a transient rather than a permanent habitat (Zakhariev, 1971).

Table 7. Biochemical characteristics of *Vibrio anguillarum.*[a]

Test or substrate	Reaction	%+	%(+)
Oxidase	+	100	
Indole	d	71.7	
Voges-Proskauer	d	52.8	
Citrate (Simmons')	d	0	14.0
Hydrogen sulfide (TSI)	−	0	
Urease	−	0	
Gelatinase	+	100	
Phenylalanine deaminase	−	0	
Lysine decarboxylase	−	0	
Arginine dihydrolase	+	93.8	
Ornithine decarboxylase	−	0	
Growth in peptone water:			
Without NaCl	−	0	
With 6% NaCl	d	87.0	
With 8% NaCl	−	0	
Growth at 37°C	−	29.9	
Growth at 42°C	−	0	
ONPG	+	100	
Fermentation:			
Glucose, acid	+	100	
Glucose, gas	−	0	
Arabinose	d	64.0	
Lactose	(+)	3	96.8
Mannose	+	97.0	
Raffinose	−	0	
Rhamnose	−	0	
Sucrose	d	87.0	2.0
Trehalose	+	92.9	
Xylose	−	0	
Adonitol	−	0	
Dulcitol	−	0	
Inositol	−	0	
Mannitol	+	100	
Salicin	−	0	

[a] The data are based on the results from 322 strains. +, 90% or more positive within 1 or 2 days; −, 90% or more negative; d, 11–89% positive reactions; (+), delayed reactions, 3 or more days.

P. shigelloides has long attracted attention partly because it has often been isolated from human patients suffering from intestinal disorders (Aldová, Rakovský and Chovancová, 1966; Bhat, Shanthakumari, and Rajan, 1974; Cooper and Brown, 1968; Eddy and Carpenter, 1964; Ewing, Hugh, and Johnson, 1961; Fourquet et al., 1973; Geizer, Kopecký, and Aldová, 1966; Hori et al., 1966; Jandl and Linke, 1976; Osada and Shibata, 1956; Paučková and Fukalová, 1968; Sakazaki et al., 1959, 1971; Schmid, Velaudapillai, and Niles, 1954; Tsukamoto et al., 1978; Zajc-Satler, Dragas, and Kumelj, 1972). Most cases from which *P. shigelloides* is isolated are gastroenteritis, but dysentery-like symptoms were also reported (Osada and Shibata, 1956). Outbreaks or epidemics of acute gastroenteritis in which *P. shigelloides* was incriminated as the possible etiological agent were reported by Hori et al. (1966) and Tsukamoto et al. (1978). Sakazaki et al. (1959) described 15 strains that had come from three outbreaks of gastroenteritis. It has been found in association with other known enteropathogens such as *Shigella, Salmonella, Vibrio cholerae,* and *Vibrio parahaemolyticus* (Aldová, Rakovský, and Chovancová, 1966; Ueda, Yamazaki, and Hori, 1961; Vandepitte et al., 1957).

The organisms have been occasionally found in feces from healthy persons who had no clinical history. Bhat, Shanthakumari, and Rajan (1974) described that the prevalence of isolates from diarrheal patients was temporally unrelated to the presence of diarrhea, although *P. shigelloides* and *Aeromonas hydrophila* were isolated from 10% of stool specimens in an epidemic of diarrhea. Nevertheless, the isolation of *P. shigelloides* from feces of normal persons is rare compared with that from diarrheal patients (Catsaras and Buttiaux, 1965; Nakanishi, Leistner, and Hechelmann, 1969; Vandepitte, Makulu, and Gatti, 1974). Kosakai (1957) carried out feeding experiments with human volunteers; he used a strain of this species possessing an O antigen identical with that of *S. sonnei,* and which had been isolated from a patient with diarrhea. No clinical symptoms developed in any cases. A discrepancy is also found in enteropathogenicity tests with animal models. Ligated ileal loop tests in rabbits with some strains of *P. shigelloides* have so far given negative reactions; nor do these strains produce enterotoxins, as does *Escherichia coli* (Sanyal, Singh, and Sen, 1975; Tamura and Sakazaki, unpublished). Recently, however, Saraswathe, Sharma, and Sanyal (1980) reported that live cells and culture filtrates of six strains of *P. shigelloides* caused fluid accumulation in ileal loops in rabbits. Further studies are necessary to determine the role of this organism in the production of diarrheal disease.

P. shigelloides may be found as an opportunistic pathogen. Ewing, Hugh, and Johnson (1961) reported the isolation of two strains from blood and cerebrospinal fluid. Ellner and MacCarthy (1973) described the isolation of the organism from a patient with cellulitis who suffered from sickle cell anemia, and who had acquired the infection through a fish bone and developed fatal septicemia. Another isolation of the organism as an agent of cellulitis was reported by von Graevenitz and Mensch (1968).

The isolation of *P. shigelloides* from animals has also been reported. A culture studied by Bader (1954) was isolated from the stools of a dog affected with enteritis. One of four strains of *P. shigelloides* found in the Belgian Congo by Vandepitte et al. (1957) was from the spleen of a chimpanzee. Two of 21 strains collected by Sakazaki et al. (1959) were obtained from the mesenteric lymph nodes of dogs. *P. shigelloides* had also been found in monkeys, goats, sheep, cows, and cats (Ewing, Hugh, and Johnson, 1961; Habs and Schubert, 1962.)

Isolation of *Plesiomonas*

Plesiomonas shigelloides grows well on plating media that is highly selective for enteropathogens, such as Salmonella-Shigella (SS), deoxycholate-citrate-lactose (DCL), xylose-lysine-deoxycholate (XLD), Hektoen, and brilliant green agars, as well as on MacConkey and deoxycholate-lactose media, but poorly on bismuth sulfite agar. Strains that ferment lactose promptly produce red or pink colonies on these media. If the strain is a slow lactose fermenter, colonies may be colorless and resemble those of shigellae after 24 h incubation, and become pink after prolonged incubation. In the latter instance, they sometimes yield colorless colonies with red papillae.

Schubert (1977) reported that, among the selective media mentioned above, SS agar is the most favorable for growth of *P. shigelloides,* but that its indicator system does not permit differentiation of colonies of *P. shigelloides* from those of Enterobacteriaceae. He formulated an isolation agar, inositol–brilliant green–bile salts agar, to overcome these difficulties.

Inositol–Brilliant Green–Bile Salts (IBG) Agar
for Isolation of *Plesiomonas* (Schubert, 1977)
The medium is formulated as follows:

Proteose peptone (Difco)	10.0 g
Lab-Lemco beef extract (Oxoid)	5.0 g
Sodium chloride	5.0 g
Bile salts no. 3 (Difco)	8.5 g
Agar (Difco)	15.0 g
Brilliant green	0.00033 g
Neutral red	0.025 g
Distilled water	1 liter

The pH of the medium is adjusted to 7.2. The medium is dissolved by heating and plates are poured without autoclaving. After 48 h of incubation, most strains of plesiomonads produce whitish colonies that possess a more or less pinkish touch, depending on the distance to other colonies. Although inositol is fermented by *P. shigelloides*, acid formation is usually very weak. Colonies of *P. shigelloides* may be confirmed by a positive oxidase reaction.

Identification of *Plesiomonas*

Plesiomonas shigelloides is a Gram-negative, rod-shaped bacterium that averages $2 \times 0.7 \ \mu$m, but an occasional culture may show rather long rods or filaments. They are usually motile, but nonmotile strains are known. In motile cultures, polar tuft flagella (1 to 5) are produced. In addition to the polar flagella, occasional cells may yield lateral flagella in young cultures, especially agar cultures. No capsule is seen.

The organisms are facultatively anaerobic or aerobic bacteria that ferment glucose with no producton of gas. They grow well at 35°C. Physiological and biochemical characteristics of 280 strains of *P. shigelloides* in our collection are summarized in Table 8. The tests given in the table are those used for Enterobacteriaceae.

The positive oxidase test rapidly differentiates *P. shigelloides* from Enterobacteriaceae. As *Plesiomonas* is invariably anaerogenic, the organisms most likely to be confused with it are *Vibrio* and anaerogenic *Aeromonas* strains. Although lysine and ornithine decarboxylase and arginine dihydrolase tests may be of value in differentiating those anaerogenic strains, fermentation of inositol and mannitol are the most useful tests in rapidly distinguishing *P. shigelloides* from *Vibrio* and *Aeromonas*. These organisms are differentiated in Table 9.

P. shigelloides has no halophilism. Some authors reported hemolysis of few strains of this species grown on blood agar (Aldová, Rakovský, and Chovancová, 1966; Zajc-Satler, Dragas, and Kumelj, 1971), but such strains were not included in collections of Ewing, Hugh, and Johnson (1961) and ours.

In general, *P. shigelloides* is susceptible to cephalothin, chloramphenicol, mitomycin, rifampin, colistin, polymyxin B, nalidixic acid, and sulfamethoxazole-trimethoprim, and resistant to streptomycin and neomycin. Susceptibility to ampicillin, carbenicillin, tetracycline, kanamycin, amikacin, gentamicin, and tobramycin is variable.

P. shigelloides is sensitive to dryness. When the cultures are maintained on an agar slant in a tube with a cotton plug, they may be killed within a few days. However, if the cultures are maintained in semisolid medium in a screw-capped tube or a tube with a rubber or paraffin-cork stopper and are stored

at room temperature, they may survive for months or years.

The serology of *P. shigelloides* was studied by Sakazaki et al. (1959), who distinguished five somatic and four flagellar antigens in 29 strains without designating antigenic symbols. Quincke (1967) divided his cultures into 16 O-antigen groups without taking account of H antigens. Aldová and Geizer (1968) divided their 14 strains into 5 O groups and distinguished 5 H antigens among the strains. Recently, Shimada and Sakazaki (1978) established an

Table 8. Biochemical characteristics of *Plesiomonas shigelloides* (summary of 280 strains).

Test (substrate)	Reaction	%+	%(+)
Oxidase	+	100	
Catalase	+	100	
Indole	+	100	
Methyl red	+	100	
Voges-Proskauer (25°C)	−	0	
Citrate (Simmons')	−	0	
Nitrate to nitrite	+	100	
H$_2$S (Kligler)	−	0	
Urease (Christensen)	−	0	
Gelatinase (Kohn)	−	0	
Phenylalanine deaminase	−	0	
Lysine decarboxylase	+	99.3	0.7
Arginine dihydrolase	+	97.8	1.7
Ornithine decarboxylase	+	91.4	2.8
KCN	−	1.8	
Malonate	−	0	
D-Tartrate	−	0	
Mucate	−	0	
Lipase (Tween 80)	−	0	
DNase	−	0	
Gas from glucose	−	0	
Acid from carbohydrate:			
Arabinose	−	0	
Cellobiose	−	0	
Glucose	+	100	
Lactose	d	40.0	52.8
Maltose	d	62.8	
Melibiose	d	40.0	10.0
Raffinose	−	0	
Rhamnose	−	0	
Sucrose	−	0	2.8
Trehalose	+	100	
Xylose	−	0	
Adonitol	−	0	
Dulcitol	−	0	
Erythritol	−	0	
Mannitol	−	0	
Sorbitol	−	0	
Salicin	d	11.7	10.3
Inositol	+	98.2	1.8
Esculin hydrolysis	−	0	
ONPG	+	100	

+, 90% or more positive within 1 or 2 days; −, 90% or more negative; d, 11–89% positive reactions; (+), delayed reaction, 3 or more days.

Table 9. Differentiation of *Plesiomonas shigelloides* from
related organisms.[a]

Test (substrate)	Plesiomonas shigelloides		Aeromonas hydrophila		Vibrio cholerae	
Lysine decarboxylase	+	100	−	0	+	98
Arginine dihydrolase	+	98	d	87	−	0
Ornithine decarboxylase	+	92	−	0	+	100
Acid from:						
Mannitol	−	0	+	99	+	99
Sucrose	−	3	d	82	+	99
Inositol	+	99	−	0	−	0
Lipase (Tween 80)	−	0	+	99	+	100
Gelatin (Kohn)	−	0	d	81	+	99

[a] Number indicates percent positive; +, positive reaction within 1 or 2 days; −, negative reaction; d, variable reaction.

antigenic schema that included 40 serovars consisting of 30 O-antigen groups and 11 H antigens (Table 10).

For O-antigen determination, the cultures to be tested should be heated at 100°C for 2 h and washed twice, because the majority of strains may be O-inagglutinable in the live state. Actively motile cultures obtained by several passages through semisolid medium should be used for H-antigen determination.

P. shigelloides was noticed by earlier investigators because its O antigen was identical with that of *Shigella sonnei*. Hori et al. (1966) reported a close O-antigenic relationship between their isolates of this species and *Shigella dysenteriae* serovar 7. The O groups of *P. shigelloides* concerned are designated O17 and O22, respectively, in the antigenic schema of Shimada and Sakazaki (1978). Moreover, additional relationships are recognized between *P. shigelloides* O11 and *S. dysenteriae* serovar 8 and between *P. shigelloides* O23 and *S. boydii* serovar 13.

Whang, Heller, and Neter (1972) demonstrated the presence of the common antigen of Enterobacteriaceae in strains of *P. shigelloides*.

THE GENUS *AEROMONAS*

The genus *Aeromonas* Stanier 1943 includes two groups of organisms, a psychrophilic, nonmotile group and a mesophilic, usually motile group. The former group of bacteria is well defined as *A. salmonicida*. However, the classification and nomenclature of aeromonads of the latter group are still in a state of flux. Although many scientific names had been proposed for the motile aeromonads, Ewing, Hugh, and Johnson (1961) suggested that various specific names other than *A. hydrophila* should be discarded and that the bacteria previously described under these names should be considered as biovars

within a single species, *A. hydrophila* (Chester) Stanier 1943. In the eighth edition of *Bergey's Manual of Determinative Bacteriology*, Schubert (1974) differentiated two species for the motile aeromonads, *A. hydrophila* and *A. punctata*, with three subspecies *(hydrophila, anaerogenes,* and *proteolytica)* under *A. hydrophila* and two *(punctata* and *caviae)* under *A. punctata.* Recently, however, it

Table 10. Antigenic schema of *Plesiomonas shigelloides* (Shimada and Sakazaki, 1978).

O group	H antigen
1	1a,1b
2	1a,1c
3	2
4	3
5	4
6	3
7	2; nonmotile
8	3; 5
9	2
10	11
11	2; 5
12	2; 3; 9
13	2
14	4; 5
15	10
16	5
17	2; 6
18	2
19	2
20	2
21	7; 8
22	3; 5; 8
23	1a,1c
24	5; 8
25	3
26	1a,1c
27	3
28	3
29	2
30	1a,1c

has been indicated that *A. hydrophila* subsp. *proteolytica* should be removed from the genus *Aeromonas,* because of its halophilism and its DNA base composition, which differs from that of members of the genus *Aeromonas* (Boulanger, Lallier, and Cousineau, 1977; Gibson et al., 1977; Kleeberger, 1977; McCarthy, 1975; Popoff and Véron, 1976). Popoff and Véron (1976) suggested that *A. punctata* should be a later synonym of *A. hydrophila* and that a new species, *A. sobria,* be recognized on the basis of a numerical taxonomy study and DNA base composition. Although MacInnes, Trust, and Crosa (1979) considered that *A. hydrophila* and *A. sobria* are not genetically distinct groups, M. Popoff (personal communication) suggested from the results of DNA-DNA homology studies that these two species should be genetically heterogeneous. In addition, the latter investigator indicated that anaerogenic strains of *A. hydrophila* defined by Popoff and Véron (1976) constitute a species that is genetically separate from *A. hydrophila* and *A. sobria.*

In contrast, *A. salmonicida* could be demonstrated by MacInnes, Trust, and Crosa (1979) to be a genetically homogeneous group with very high DNA-DNA homology values, although Schubert (1974) had divided the species into three subspecies (*salmonicida, achromogenes,* and *masoucida*). The former authors suggested that these three subspecies did not warrant subspecies status.

In this section, therefore, the genus *Aeromonas* will for convenience be divided into four species, *A. hydrophila, A. sobria,* "*A. caviae*", and *A. salmonicida.* "*A. caviae*" is a temporary name for anaerogenic, motile aeromonads for present time, but *A. punctata* subsp. *caviae* (Schubert, 1967a), which corresponds to these aeromonads, is elevated to specific rank.

Habitats of *Aeromonas*

Members of the genus *Aeromonas* are microorganisms that occur widely in fresh waters. Leclerc and Buttiaux (1962) indicated that motile aeromonads constitute an important error in the enumeration of coliform in drinking water, since the organisms were present in 30% of 9,036 water samples yielding positive results with presumptive coliform examination. Neilson (1978) noted the isolation of motile aeromonads in high number from activated sludge samples. Schubert (1967b) who studied distribution of aeromonads in surface water, showed that a fecal origin could be excluded in searching for the origin of aeromonads in waste water, and that a large number of aeromonads were found in the mud from the siphons of sinks and in the first portion of the wastewater drainage system; this region may therefore be considered the main generation place of the organisms. Hazen et al. (1978) reported

a wide range of prevalence and distribution of *A. hydrophila* in natural waters in the United States and concluded that abundance of *A. hydrophila* in so many different systems would seem to indicate an important role for the organisms in natural aquatic processes.

A. salmonicida has a wide geographic distribution in freshwater fishes and produces frunculosis and bacteremia, particularly in salmon and trout, which is the most important bacterial disease in most salmon-farming countries. The organism is thought to survive between epizootics within the kidney and intestine of the carrier fish. It is released when the carrier is stressed by environmental factors, allowing dissemination of infection through the population (Mackie et al., 1930, 1933, 1935). Klonts, Yasutake, and Ross (1966) reported that virulent strains of *A. salmonicida* produce a leucocidin and have a lipopolysaccharide, which is also present in smaller amounts in avirulent strains. On the other hand, motile aeromonads produce a severe hemorrhagic syndrome (red-sore disease) in fish (Bullock, 1961; Shimizu, 1969; Shotts et al., 1972), although epizootics of the disease are not due to carrier fish (unlike those caused by *A. salmonicida*). Epizootics of the disease can be related to some other stress on fish, such as high temperature or low oxygen levels, in addition to infection with motile aeromonads from natural aquatic habitats. Hazen and Fliermans (1979) suggested that epizootics of red-sore disease may be enhanced by thermal effluent.

Aeromonads have for many years been of interest as the causative agent of "red leg" disease in frogs (Kulp and Borden, 1942). Rigney, Zilinsky, and Rouf (1978) studied the mechanism of pathogenicity of *A. hydrophila* in red leg disease, and they concluded that "red leg disease in frogs represents a complex interaction between endotoxin and hemolysin and that stress-producing factors other than the endotoxin might trigger disease production." Motile aeromonads have also been reported to be responsible for a number of diseases in reptiles such as snakes, turtles, and alligators (Esterabadi, Entessar, and Khan, 1973; Marcus, 1971; Nygaard, Bissett, and Wood, 1970; Shotts et al., 1972; Thal and Dinter, 1953; Vézina and Desrochers, 1971), guinea pigs (Scherago, 1937), and cattle (Wohlgemuth, Pierce, and Kilbride, 1971).

Aeromonads have been isolated on a number of occasions from human patients with metastatic myositis, cellulitis, gangrene, peritonitis, osteomyeritis, tonsilitis, pneumonia, bronchopneumonia, meningitis, endocarditis, and carcinoma, and they have been recovered from throat, sputum, tracheal aspirate, urine, blood, ascitic fluid, pleural fluid, pus, gall bladder, placenta, wounds, and abscesses (Beer, 1963; Bulger and Sherris, 1966; Caselitz and Günther, 1960; Caselitz and Maass, 1962; Caselitz, Hofmann, and Martinez-Silva, 1958; Conn, 1964;

Davis, Kane, and Garagusi, 1978; Dean and Post, 1967; Gilardi, Bottone, and Birnbaum, 1970; Hanson et al., 1977; Hill, Caselitz, and Moody, 1954; Joseph et al., 1979; Kjem, 1955; Ketover, Young, and Armstrong, 1973; Kok, 1967; Lopez, Quesada, and Saied, 1968; McCracken and Barkley, 1972; Pearson, Mitchell, and Hughes, 1972; Phillips, Bernhard, and Rosenthal, 1974; Qadri et al., 1976; Ramsay et al., 1978; Shackelford, Ratzan, and Shearer, 1973; Stephen et al., 1975; Tapper et al., 1975; von Graevenitz and Mensch, 1968; Washington, 1972; Zajc-Satler, 1972). In most cases, such infection with aeromonads is opportunistic. Septicemia is found in patients with leukemia, solid tumors under chemotherapy, and hepatobiliary disease, and in immunologically competent individuals (Abrams, Zierdt, and Brown, 1971; Davis, Kane, and Garagusi, 1978; DeFronzo, Murray, and Maddrey, 1973; von Graevenitz and Mensch, 1968). In those cases, the organism is presumably of intestinal origin. Recently, however, primary wound infections with motile aeromonads in polluted water have been reported (Davis, Kane, and Garagusi, 1978; Hanson et al., 1977; Joseph et al., 1979).

Aeromonas can be found in stools of 0.2–0.7% of apparently healthy individuals (Catsaras and Buttiaux, 1965; Lautrop, 1961; Paučková and Fukalová, 1968). However, much interest in their enteropathogenicity has been stimulated in recent years, since they have been repeatedly found to be associated with human diarrheal disease (Bhat, Shanthakumari, and Rajan, 1974; Chatterjee and Neogy, 1972; Chatterjee et al., 1976; Fritsche, Donn, and Hoffmann, 1975; Fukaya et al., 1962; Gilardi, 1967; Helm and Stille, 1970; Lautrop, 1961; Martinez-Silver, Guzmann-Urrego, and Caselitz, 1961; Paučková and Fukalová, 1968; Rosner, 1964; Sakazaki et al., 1971; Sanyal, Singh, and Sen, 1975; Simon and von Graevenitz, 1969; von Graevenitz and Mensch, 1968; Wadström, Ljungh, and Wretlind, 1976; Wassum, 1967). Although some patients present a short-lived, self-limited diarrhea, others show moderate or severe gastroenteritis with watery or bloody stool, abdominal pain, fever, and dehydration.

Using rabbit ileal loop models, Sanyal, Singh, and Sen (1975) first reported enteropathogenicity of *A. hydrophila.* Later, an enterotoxic activity was detected in culture supernatants of *A. hydrophila* by Wadström, Ljungh, and Wretlind (1976) and Ljungh, Popoff, and Wadström (1977) using tests of rabbit ileal loops, rabbit skin, and Y1 adrenal cells, and by Annapurna and Sanyal (1975, 1977) using rabbit ileal loop tests. *A. hydrophila* has been recognized as producing a variety of extracellular products, including enzymes and hemolysins. Bernheimer and Avigad (1974) studied an extracellular hemolysin termed "aerolysin" from *A. hydrophila,* and reported that the physical properties

of aerolysin showed considerable resemblance to those described for the exotoxin of *Pseudomonas aeruginosa.* Wadström, Ljungh, and Wretlind (1976) recognized that, although hemolysin and cytotoxin(s) in culture supernatants of *A. hydrophila* interfered in test systems for enterotoxin, enterotoxic activity to Y1 adrenal cells was demonstrated after inactivation of the hemolytic and cytotoxic factors by heating the test samples at 56°C for 10 min. Dubey and Sanyal (1978, 1979) characterized enterotoxin with crude preparation and reported that the crude enterotoxin preparations contained a non-dialyzable and heat-labile antigenic protein that caused the fluid accumulation in rabbit ileal loops, similar to that caused by *Vibrio cholerae* enterotoxin. On the other hand, Donta and Haddow (1978), who attempted to confirm the findings of enterotoxigenic strains of *A. hydrophila* and compare its toxic properties in tissue culture with those of *Vibrio cholerae* and *Escherichia coli,* reported that most strains of *A. hydrophila* studied demonstrated cytotoxic activity but that none of them was found to be enterotoxigenic. Cumberbach et al. (1979) tested 96 strains of *A. hydrophila* for enterotoxin such as that produced by *V. cholerae,* cytotoxin, and hemolysin. They found that 69% of the strains were both cytotoxic and hemolytic, but found no evidence of a separate enterotoxigenic activity in any of the strains. However, they found that cytotoxin production correlated with enteropathogenicity and concluded that the enteropathogenic potential is mediated by a cytotoxin. Cytotoxic activity appears to be a stable property and may not be associated with any plasmid (Cumberbach et al., 1979). Very similar findings were also presented by Asao et al. (1978).

Although there is a discrepancy in the presence of an enterotoxin categorized in *V. cholerae* and *E. coli,* it does appear that some strains of *A. hydrophila* have an ability to produce diarrhea by a cytotoxin. Most investigators described such aeromonads as *A. hydrophila,* but *A. sobria* and "*A. caviae*" may be included in this category. Fritsche, Dahn, and Hoffmann (1975) reported a case of acute gastroenteritis caused by *A. punctata* subsp. *caviae.* Boulanger, Lallier, and Cousineau (1977) reported that some strains of *A. hydrophila* and *A. sobria* isolated from fish produced positive rabbit ileal loop and/or suckling mouse reactions. They gave no details of relationship among the "enterotoxin", cytotoxin, and hemolysin of those strains, but suggested the presence of two different enterotoxigenic activities in those strains.

A significant correlation between the cytotoxic and hemolytic activities in some aeromonads has been reported (Asao et al., 1978; Cumberbach et al., 1979). This fact suggests that the two activities are different expressions of the same molecule. As mentioned before, however, Wretlind, Möllby, and

Wadström (1971) suggested that *A. hydrophila* produces two separate hemolysins that can be distinguished from "enterotoxin". Boulanger, Lallier, and Cousineau (1977) reported that *A. hydrophila* produced two types of hemolysin, whereas *A. sobria* produced only one type of hemolysin. On the other hand, Kobayashi et al. (1980, unpublished) recognized that the cytotoxic activity of *A. hydrophila* correlated with a hemolysin to ovine erythrocytes, whereas that of *A. sobria* was associated with a hemolysin to rabbit erythrocytes. In addition, they also found that the enterocytotoxic and hemolytic activities of *A. hydrophila* were neutralized by the antihemolysin produced with purified hemolysin preparation from a strain of *A. hydrophila,* but that those activities of *A. sobria* were not always neutralized with the antihemolysin of *A. hydrophila.*

Thus, further studies may be required to definitely establish the enteropathogenic mechanisms of aeromonads, although the enteropathogenicity of motile aeromonads has long been known.

Isolation of *Aeromonas*

Cultures of *Aeromonas salmonicida* grow well on/in ordinary media at 18–25°C but not at 35°C. Optimal temperature is about 22° C. Colonies of *A. salmonicida* on nutrient agar containing 0.1% tyrosine or phenylalanine produce a soluble, brown, melanin-like pigment. However, some nonpigmented strains may be present.

Motile aeromonads including *A. hydrophila, A. sobria,* and "*A. caviae*" do not require enriched media and they grow at 35°C. Many strains isolated from clinical specimens produce a wide hemolytic zone surrounding colonies on blood-agar plates. They can be isolated with plating media used for enterobacteria such as MacConkey and deoxycholate agars, on which most of them yield colorless colonies. A few strains may produce prompt lactose-fermenting colonies on these media. They may or may not grow on salmonella-shigella agar and a majority of strains are inhibited on thiosulfate–citrate–bile salt–sucrose (TCBS) agar.

Because any isolation medium for enterobacteria does not distinguish between colonies of *Aeromonas* and Enterobacteriaceae, several special media for the isolation of aeromonads from stool specimens have been proposed. Von Graevenitz and Zinterhofer (1970) described a supplemented DNase medium based on the production of DNase by *Aeromonas* strains.

Supplemented DNase Agar (von Graevenitz and Zinterhofer, 1970)

DNase test agar (BBL) is supplemented with 0.01% of toluidine blue and 30 µg/ml of ampicillin. The basal medium is dissolved by heating and toluidine blue and ampicillin are added after autoclaving. After 16–24 h of incubation, aeromonads produce colonies with a pink zone, more or less 1 mm from the edge. Ampicillin-susceptible *Escherichia coli* and *Proteus mirabilis* strains and most staphylococci are inhibited on this medium. Ampicillin-resistant enterobacteria can grow on this agar and some of them, such as *Serratia* strains, produce colonies showing a pink zone. However, colonies of *Aeromonas* may be easily differentiated from those of DNase-positive enterobacteria with the oxidase test.

Another isolation medium supplemented with ampicillin, PXA agar, was described by Rogol et al. (1978).

Pril-Xylose-Ampicillin (PXA) Agar (Rogol et al., 1978)

The medium consists of nutrient agar containing 1% xylose, 0.0025% phenol red, 30 µg/ml ampicillin, and 0.02% Pril (Böhme Fettchemie, GmbH, Duesseldorf), which is a quaternary ammonium detergent. Because aeromonads do not ferment xylose, they produce colorless colonies on this medium. Many enterobacteria may be inhibited by ampicillin, and even if they grow on this medium, they can be distinguished with yellow colonies caused by xylose fermentation. This medium prevents swarming of *Proteus* cultures.

Shotts and Rimler (1973) devised a selective differential isolation agar containing novobiocin for the isolation of *Aeromonas.*

Rimler-Shotts Agar (Shotts and Rimler, 1973)

L-Lysine hydrochloride	5.0 g
L-Ornithine hydrochloride	6.5 g
Maltose	3.5 g
Sodium thiosulfate	6.8 g
L-Cysteine hydrochloride	0.3 g
Bromothymol blue	0.03 g
Ferric ammonium citrate	0.8 g
Sodium deoxycholate	1.0 g
Novobiocin	0.005 g
Yeast extract	3.0 g
Sodium chloride	5.0 g
Agar	13.5 g
Distilled water	to 1 liter

The mixture is boiled for 1 min to dissolve components and, after pH is adjusted to 7.0, poured into plates. Four types of colonies may be obtained on this medium: a yellow colony, which indicates maltose fermentation; a yellow colony with a black center, which indicates maltose fermentation and H_2S production; a greenish-yellow or green colony, which indicates ly-sine and/or ornithine decarboxylation; and a green colony with a black center, indicating amino acid decarboxylation and H₂S production. *Aeromonas* produces only the first type of colonies. However, occasional strains of *Citrobacter freundii* and *Salmonella typhi* may also yield yellow colonies, so that all yellow colonies should be tested for oxidase activity. Gram-positive organisms and *Vibrio* may be inhibited by the presence of sodium deoxycholate and novobiocin. Since color changes occur due to pH reactions, colonial growth should be observed between the 20th and 24th h for maximal accuracy. Shotts and Rimler (1973) emphasized that this medium is effective in presumptive identification of *Aeromonas* strains with 94% accuracy. However, strains of aeromonads which are novobiocin sensitive will not be detected by this medium. Furthermore, so-called group F vibrios cannot be distinguished on this medium.

Identification of *Aeromonas*

Members of the bacterial genus *Aeromonas* are Gram-negative, rod-shaped bacteria, averaging $1.0-3.5 \times 0.4-1.0$ μm, with polar flagella when flagellated. In flagellated cultures, cells with two flagella at one pole are occasionally encountered in the predominantly monotrichous population. In addition, occasional cells in young cultures on agar medium yield lateral or peritrichous flagella which are generally shorter than the polar flagellum. No capsule is present.

Aeromonas is a facultatively anaerobic or aerobic organism that produces oxidase and catalase and ferments glucose with or without production of gas. Strains are differentiated from Enterobacteriaceae by means of flagellation and oxidase production and from *Pseudomonas* by fermentative metabolism of glucose. They are also distinguished from other members of the family Vibrionaceae by certain biochemical and physiological characteristics such as amino acid decarboxylase production, sodium dependence, some sugar fermentation, and sensitivity to vibriostatic agent O/129 (2:4-diamino 6:7-diisopropyl pteridine).

Aeromonas salmonicida

Aeromonas salmonicida is definitely a separate organism from other members of the genus *Aeromonas*. Smith (1963) suggested that this species be transferred to the genus *Necromonas*. *A. salmonicida* is nonmotile and psychrophilic, unlike other members of the genus *Aeromonas*. The majority of strains of this species produce diffusible brown pigment when tyrosine or phenylalanine is contained

in culture media, but some nonpigment variants are present; they gave positive reactions in tests of gelatinase, amylase, lecithinase, lipase, and arginine dihydrolase, and negative in tests of citrate utilization (Simmons'), urea decomposition, H₂S production (Kligler), and lysine and ornithine decarboxylations. They usually ferment arabinose, galactose, mannose, mannitol, and esculin, but lactose, raffinose, rhamnose, sucrose, trehalose, xylose, and salicin are not attacked. The Voges-Proskauer test is constantly negative. Some strains may produce indole. Most strains produce a small amount of gas from glucose. Although some aberrant strains may occur, *A. salmonicida* may be presumptively identified without detailed characterization when an organism isolated from fish reveals gram-negative, nonmotile rods which are oxidase positive and produce a soluble brown pigmentation on appropriate agar medium.

Aeromonas hydrophila, A. sobria, and "*A. caviae*"

Aeromonas species other than *A. salmonicida* are composed of mesophilic, motile organisms which grow well at 35°C. Physiological and biochemical characteristics of the three species are listed in Table 11. The tests given in the table are those used for Enterobacteriaceae.

Anaerogenic, Voges-Proskauer-negative aeromonads, which had been called *A. caviae* with the later synonym *A. formicans,* have been classified into *A. punctata* subsp. *caviae* in the eighth edition of *Bergey's Manual* (Schubert, 1974). Although recent taxonomic studies suggested, as mentioned before, that *A. punctata* is a later synonym of *A. hydrophila* (Popoff and Véron, 1976), DNA-DNA homology studies indicated that anaerogenic subspecies of the former fall into a group separate from *A. hydrophila* and *A. sobria* (M. Popoff, personal communication). For this reason, therefore, *A. punctata* subsp. *caviae* should be elevated to specific rank; the old name *A. caviae* may be most appropriate to describe this species for the present.

A. caviae is differentiated from *A. hydrophila* by its inability to produce gas from glucose and to produce acetylmethylcarbinol. Unlike the other two species, most strains of *A. caviae* reported in the table fermented citrate in Kauffmann-Petersen's organic acid medium, but it is not certain whether this reaction is constant with all strains of *A. caviae,* since the author has had little experience with these organisms and the 86 strains in the table were collected during a study on diarrheal disease in Calcutta (Sakazaki et al., 1971).

A. sobria may be differentiated from *A. hydrophila* and *A. caviae* by its inability to ferment glucosides such as salicin, arbutin, and esculin and to

Table 11. Biochemical and physiological characteristics of *Aeromonas hydrophila, A. sobria,* and *A. caviae.*

Test (substrate)	A. hydrophila (200)[a]		A. sobria (40)		A. caviae (86)	
	Sign[b]	% Positive	Sign	% Positive	Sign	% Positive
Oxidase	+	100	+	100	+	100
Catalase	+	100	+	100	+	100
Indole	+	90.5	+	100	+	90.6
Methyl red (35°C)	d	76.5	d	75.0	+	94.1
Voges-Proskauer (25°C)	d	85.5	d	37.5	−	0
Citrate (Simmons')	d	67.5	d	75.0	d	72.0
Citrate (K-P)	−	0.2	−	0	+	94.1
Nitrate to Nitrite	+	100	+	100	+	100
H$_2$S	−	0	−	0	−	0
Urease (Christensen)	−	0.2	−	0	−	0
Gelatinase	+	99.0	+	100	+	100
Phenylalanine deaminase	−	0	−	0	−	0
Lysine decarboxylase	−	0	−	0	−	0
Arginine dihydrolase	d	87.0	d	70.0	+	100
Ornithine decarboxylase	−	0	−	0	−	0
KCN	+	96.0	−	0	+	100
Malonate	−	0	−	0	−	0
D-Tartrate	−	0	−	0	−	0
Mucate	−	0	−	0	−	0
Lipase (Tween 80)	+	99.5	+	100	+	100
DNase	+	99.5	+	100	+	100
Gas from glucose	+	99.0	+	97.5	−	0
Acid from carbohydrate:						
Arabinose	d	66.0	d	17.5	d	87.2
Cellobiose	d	42.5	d	75.0	d	84.8
Glucose	+	100	+	100	+	100
Lactose	d	76.5	d	37.5	d	52.3
Maltose	+	100	+	100	+	100
Melibiose	−	1.5	−	2.0	−	0
Raffinose	−	2.5	−	0	−	0
Rhamnose	−	7.5	−	0	−	0
Sucrose	d	82.5	+	100	+	100
Trehalose	+	100	+	100	+	100
Xylose	−	0.2	−	0	−	0
Adonitol	−	0	−	0	−	0
Dulcitol	−	0	−	0	−	0
Erythritol	−	0	−	0	−	0
Mannitol	+	99.5	+	100	+	100
Sorbitol	d	11.0	−	2.0	−	1.1
Salicin	d	81.5	−	2.0	+	91.8
Arbutin	+	94.5	−	2.5	+	100
Inositol	−	0	−	0	−	0
Esculin hydrolysis	+	99.0	−	0	+	100
ONPG	+	100	+	100	+	100
Motility	+	99.5	+	100	+	100
G+C content (mol%)	61–63		57–60		61–62	

[a] Numeral in parentheses indicates the number of strains studied.
[b] Sign: +, 90% or more positive within 1 or 2 days; −, 90% or more negative; d, 11–89% positive.

grow in KCN medium. In addition to this, Popoff and Véron (1976) described utilizations of L-arabinose, salicin, L-arginine, and L-histidine as a sole carbon source to distinguish between *A. hydrophila* and *A. sobria,* as indicated in Table 12.

Kaper et al. (1979) devised a single tube medium for the presumptive identification of mesophilic aeromonads. This medium is useful for the routine diagnostic laboratory.

Aeromonas hydrophila (AH) Medium (Kaper et al., 1979)

The AH medium contains (per liter):

Proteose peptone (Difco)   5 g
Yeast extract (Difco)   3 g

Table 12. Differentiation between *Aeromonas hydrophila* and *A. sobria*.[a]

Test (substrate)	A. hydrophila	A. sobria
Esculin hydrolysis	+ (100)[b]	− (19)
Salicin fermentation	+ (86)	− (19)
Growth in KCN medium	+ (98)	− (19)
Utilization of a sole C source:		
L-Arabinose	+ (93)	− (19)
Salicin	+ (79)	− (23)
L-Arginine	+ (74)	− (8)
L-Histidine	+ (98)	− (8)

[a] After Popoff and Véron (1976).
[b] Numeral in parentheses indicates percent positive.

Tryptone (Difco)	10 g
L-Ornithine hydrochloride	5 g
Mannitol	1 g
Inositol	10 g
Sodium thiosulfate	0.4 g
Ferric ammonium citrate	0.5 g
Bromocresol purple	0.02 g
Agar	3 g

The pH of the medium is adjusted to 6.7. The medium is dissolved by heating, dispensed in 5-ml quantities in tubes (13 × 100 mm), and autoclaved at 121°C for 12 min. Cultures to be tested are inoculated into the medium by stabbing with a straight needle. The tubes are incubated at 35°C for 18–24 h; reactions are then recorded. For the detection of indole production, 3–4 drops of Kovács reagent are added. Mannitol and inositol fermentations, ornithine decarboxylation and deamination, indole producton, motility, and H_2S production can be recorded in a single tube of the medium. This medium enables rapid presumptive identification of *A. hydrophila* (and other two species) as well as good differentiation of *Klebsiella, Proteus,* and other enterobacteria. Most cultures of aeromonads in the AH medium give a yellow butt with a purple band at the top. They also give positive reactions, as a rule, in motility and indole tests but negative reactions in the H_2S test.

Some marine vibrios, such as *Vibrio anguillarum, V. metschnikovii,* and *Beneckea splendida,* may sometimes be confused with *Aeromonas,* because they are positive for the arginine dihydrolase test and negative in lysine and ornithine decarboxylase tests. In clinical bacteriology laboratories, so-called group F vibrios (Furniss et al., 1977; Lee, Donovan, and Furniss, 1978), which are one of the marine vibrios and occasionally associated with human diarrhea, can also interfere with the presumptive identification of *Aeromonas* because of similar reactions in amino acid decarboxylase tests. However, these marine organisms cannot grow in peptone water containing no salt and are sensitive to the vibriostatic agent O/129, as opposed to *Aeromonas,* which is salt independent and resistant to O/129.

The serology of *Aeromonas hydrophila (sensu lato)* was preliminarily studied by Ewing, Hugh, Johnson (1961), but the study was incomplete. No antigenic schema has so far been given for any species of the genus *Aeromonas.*

Literature Cited

Abrams, E., Zierdt, C. H., Brown, J. A. 1971. Observations on *Aeromonas hydrophila* septicaemia in a patient with leukaemia. Journal of Clinical Pathology **24**:491–492.

Akiyama, S., Takizawa, K., Obara, Y. 1964. Study on enrichment broth for *Vibrio parahaemolyticus.* [In Japanese.] Annual Report of Kanagawa Prefectural Institute of Public Health **13**:7–9.

Akiyama, S., Takizawa, K., Ichinoe, H., Enomoto, S., Kobayashi, T., Sakazaki, R. 1963. Application of teepol to isolation of *Vibrio parahaemolyticus.* [In Japanese.] Japanese Journal of Bacteriology **18**:255–256.

Aldová, E., Geizer, E. 1968. A contribution to the serology of *Aeromonas shigelloides.* Zentralblatt für Bakteriologie, Parasitenkunde, Infektionskrankheiten und Hygiene, Abt. 1 Orig. **207**:35–40.

Aldová, E., Rakovský, J., Chovancová, A. 1966. The microbiological diagnostics of *Aeromonas shigelloides* isolated in Cuba. Journal of Hygiene, Epidemiology, Microbiology and Immunology **10**:470–482.

Aldová, E., Lázničková, K., Štěpánková, E., Lietava, J. 1968. Isolation of nonagglutinable vibrios from an enteritis outbreak in Czechoslovakia. Journal of Infectious Diseases **118**:25–31.

Allen, R. D., Baumann, P. 1971. Structure and arrangement of flagella in species of the genus *Beneckea* and *Photobacterium fischeri.* Journal of Bacteriology **107**:295–302.

Anderson, R. S., Ordal, E. J. 1972. Deoxyribonucleic acid relationships among marine vibrios. Journal of Bacteriology **109**:696–706.

Annapurna, E., Sanyal, S. C. 1975. Studies on enteropathogenicity of *Aeromonas hydrophila* in an experimental model. Indian Journal of Preventive Society of Medicine **6**:234–237.

Annapurna, E., Sanyal, S. C. 1977. Enterotoxicity of *Aeromonas hydrophila.* Journal of Medical Microbiology **10**:317–323.

Asao, T., Kobayashi, K., Niihara, T., Shimada, T., Tamura, K., Sakazaki, R. 1978. Toxigenicity of *Aeromonas hydrophila.* [In Japanese.] Journal of Japanese Association for Infectious Diseases **52**:40–41. [Abstracts of the 52nd Annual Meeting, 1978.]

Atthasampunna, P. 1974. *Vibrio parahaemolyticus* food poisoning in Thailand, pp. 21–26. In: Fujino, T., Sakaguchi, G., Sakazaki, R., Takeda, Y. (eds.), International Symposium on *Vibrio parahaemolyticus.* Tokyo: Saikon.

Avtsyn, A. P., Shakhlamov, V. A., Trager, R. S., Kalinina, N. A., Balyn, I. R., Timashkevich, T. B., Polyakova, G. P. 1976. NAG infection in tadpoles of *Rana temporaria.* Bulletin of Experimental Biology and Medicine **82**:1045–1048.

Azurin, J. C., Kobari, K., Barua, D., Alvero, M., Gomez, C. Z., Dizzon, J. J., Nakano, E., Suplido, R., Ledesma, L. 1967. Long-term carrier of cholera: Cholera Dolores. Bulletin of the World Health Organization **37**:745–749.

Bäck, E., Ljunggren, A., Smith, H., Jr. 1974. Non-cholera vibrios in Sweden. Lancet **i**:723–724.

Bader, R.-E. 1954. Über die Herstellung eines agglutinierenden Serums gegen die Rundform von *Shigella sonnei* mit einem Stamm der Gattung *Pseudomonas.* Zeitschrift für Hygiene und Infektionskrankheiten **140:**450–456.

Barker, W. H., Jr. 1974. *Vibrio parahaemolyticus* outbreaks in the United States, pp. 47–52. In: Fujino, T., Sakaguchi, G., Sakazaki, R., Takeda, Y. (eds.), International symposium on *Vibrio parahaemolyticus.* Tokyo: Saikon.

Barrow, G. I. 1973. Marine microorganisms and food poisoning, pp. 181–196. In: Hobbs, B. C., Christian, J. H. B. (eds.), Microbiological safety and foods. London: Academic Press.

Barrow, G. I., Miller, D. C. 1972. *Vibrio parahaemolyticus:* A potential pathogen from marine sources in Britain. Lancet **i:**485–486.

Barrow, G. I., Miller, D. C. 1976. *Vibrio parahaemolyticus* and seafoods, pp. 181–195. In: Skinner, F. A., Carr, J. G. (eds.), Microbiology in agriculture, fisheries and food. London: Academic Press.

Barua, D., Mukherji, A. C. 1964. Observation on the El Tor vibrios isolated from cases of cholera in Calcutta. Bulletin of Calcutta School of Tropical Medicine **12:**147–148.

Battey, Y. M., Wallace, R. B., Allan, B. C., Keeffe, B. M. 1970. Gastroenteritis in Australia caused by a marine vibrio. Medical Journal of Australia **1:**430–433.

Baumann, P., Baumann, L., Mandel, M. 1971. Taxonomy of marine bacteria: The genus *Beneckea.* Journal of Bacteriology **107:**268–294.

Beer, H. 1963. Zur Diagnostik gramnegativer, aerober Stäbchen. Pathologica et Microbiologica **26:**607–634.

Bernheimer, A. W., Avigad, L. S. 1974. Partial characterization of aerolysin, a lytic exotoxin from *Aeromonas hydrophila.* Infection and Immunity **9:**1016–1021.

Bhaskaran, K., Sinha, V. B. 1971. Hybridization in vibrios. Indian Journal of Experimental Biology **9:**119–120.

Bhat, P., Shanthakumari, S., Rajan, D. 1974. The characterization and significance of *Plesiomonas shigelloides* and *Aeromonas hydrophila* isolated from an epidemic of diarrhoea. Indian Journal of Medical Research **62:**1051–1060.

Bhattacharyya, F. K. 1977. The agglutination reactions of cholera vibrios. Japanese Journal of Medical Science and Biology **30:**259–268.

Bisgaard, M., Kristensen, K. K. 1975. Isolation, characterization and public health aspects of *Vibrio cholerae NAG* isolated from a Danish duck farm. Avian Pathology **4:**271–276.

Blake, P. A., Merson, M. H., Weaver, R. E., Hollis, D. G., Heublein, P. C. 1979. Disease caused by a marine *Vibrio:* Clinical characteristics and epidemiology. New England Journal of Medicine **300:**1–5.

Bockemühl, J., Amédomé, A., Triemer, A. 1972. Gastroentérites cholériformes dues à *Vibrio parahaemolyticus* sur la côte du Togo (Afrique occidentale). Zeitschrift für Tropenmedizin und Parasitologie **23:**308–315.

Bonang, G., Lintong, M., Santoso, U. S. 1974. The isolation and susceptibility to various antimicrobial agents of *Vibrio parahaemolyticus* from acute gastroenteritis cases and from seafood in Jakarta, pp. 27–31. In: Fujino, T., Sakaguchi, G., Sakazaki, R., Takeda, Y. (eds.), International symposium on *Vibrio parahaemolyticus.* Tokyo: Saikon.

Boulanger, Y., Lallier, R., Cousineau, G. 1977. Isolation of enterotoxigenic *Aeromonas* from fish. Canadian Journal of Microbiology **23:**1161–1164.

Breed, R. S. 1957. Genus I. *Vibrio* Mueller 1773, pp. 229–248. In: Breed, R. S., Murray, E. G. D., Smith, N. R., (eds.), Bergey's manual of determinative bacteriology, 7th ed. Baltimore: Williams & Wilkins.

Buchanan, R. E., Holt, J. G., Lessel, E. F. 1966. Index Bergeyana. Baltimore: Williams & Wilkins.

Buchner, H. 1885. Uber die Koch schen und Finkler-Prior schen "Kommabaccillen". Sitzungsberichte Gesellschaft für Morphologie und Physiologie in München **1:**1–10.

Bulger, R. J., Sherris, J. C. 1966. The clinical significance of

Aeromonas hydrophila. Report of two cases. Archives of Internal Medicine **118:**562–564.

Bullock, G. L. 1961. The identification and separation of *Aeromonas liquefaciens* from *Pseudomonas fluorescens* and related organisms occurring in diseased fish. Applied Microbiology **9:**587–590.

Burrows, W., Mather, A. M., McGann, V. G., Wagner, S. M. 1946. Studies on immunity to Asiatic cholera. II. The O and H antigenic structure of the cholera and related vibrios. Journal of Infectious Diseases **79:**168–197.

Campbell, L. L. 1957. Genus V. *Beneckea* Campbell gen. nov., pp. 328–332. In: Breed, R. S., Murray, E. G. D., Smith, N. R. (eds.), Bergey's manual of determinative bacteriology, 7th ed. Baltimore: Williams & Wilkins.

Carruthers, M. M., Kabat, W. J. 1976. Isolation of *Vibrio parahaemolyticus* from fecal specimens on mannitol salt agar. Journal of Clinical Microbiology **4:**175–179.

Cary, S. G., Blair, E. B. 1964. New transport medium for shipment of clinical specimens. I. Fecal specimens. Journal of Bacteriology **88:**96–98.

Caselitz, F.-H., Günther, R. 1960. Weitere Beiträge zum Genus *Aeromonas.* Zentralblatt für Bakteriologie, Parasitenkunde, Infektionskrankheiten und Hygiene, Abt. 1 Orig. **178:**15–24.

Caselitz, F.-H., Hofmann, A., Martinez-Silva, R. 1958. Unbeschriebener Keim der Familie *Pseudomonadaceae* als Infektionserreger. Zentralblatt für Bakteriologie, Parasitenkunde, Infektionskrankheiten und Hygiene, Abt. 1 Orig. **170:**564–570.

Caselitz, F.-H., Maass, W. 1962. Aeromonasstämme als Krankheitserreger, ihre Empfindlichkeit gegenüber Antibiotika und Sulfonamiden. Deutsche Medizinische Wochenschrift **87:**198–200.

Catsaras, M., Buttiaux, R. 1965. Les *Aeromonas* dans les matières fécales humaines. Annales de l'Institut Pasteur de Lille **16:**85–88.

Chatterjee, B. D., Gorbach, S. L., Neogy, K. N. 1970. *Vibrio parahaemolyticus* and diarrhoea associated with non-cholera vibrios. Bulletin of the World Health Organization **42:**460–463.

Chatterjee, B. D., Neogy, K. N. 1971. Common biotypes of non-cholera vibrios from cases of diarrhoea. Indian Journal of Medical Research **57:**95.

Chatterjee, B. D., Neogy, K. N. 1972. Studies on *Aeromonas* and *Plesiomonas* species isolated from cases of choleraic diarrhoea. Indian Journal of Medical Research **60:**520–524.

Chatterjee, B. D., Neogy, K. N., Gorbach, S. L. 1970. Study of *Vibrio parahaemolyticus* from cases of diarrhoea in Calcutta. Indian Journal of Medical Research **58:**234–238.

Chatterjee, B. D., De, P. K., Sen, T., Misra, I. B. 1976. Cholera syndrome in Bengal. Lancet **i:**317.

Chun, D., Seol, S. Y., Tak, R. 1974. *Vibrio parahaemolyticus* in the Republic of Korea. American Journal of Tropical Medicine and Hygiene **23:**1125–1130.

Cisar, J. O., Fryer, J. L. 1969. An epizootic of vibriosis in Chinook salmon. Bulletin of the Wildlife Disease Association **5:**73–76.

Citarella, R. V., Colwell, R. R. 1970. Polyphasic taxonomy of the genus *Vibrio:* Polynucleotide sequence relationships among selected *Vibrio* species. Journal of Bacteriology **104:**434–442.

Ciufecu, C., Necescu, N., Florescu, D. 1976. Morphological, cultural and biochemical characteristics of *Vibrio parahaemolyticus,* isolated in Rumania from acute gastroenteritis. Zentralblatt für Bakteriologie, Parasitenkunde, Infektionskrankheiten und Hygiene, Abt. 1 Orig., Reihe A **234:**212–218.

Clark, W. A., Steigerwalt, A. G. 1977. Deoxyribonucleic acid reassociation experiments with a halophilic, lactose-fermenting vibrio isolated from blood cultures. International Journal of Systematic Bacteriology **27:**194–199.

Colwell, R. R. 1970. Polyphasic taxonomy of the genus *Vibrio:* Numerical taxonomy of *Vibrio cholerae, Vibrio parahaemolyticus,* and related *Vibrio* species. Journal of Bacteriology **104:**410–433.

Colwell, R. R., Kaneko, T., Staley, T., Sochard, M., Pickar, J., Wan, J. 1974. *Vibrio parahaemolyticus:* Taxonomy, ecology and pathogenicity, pp. 169–176. In: Fujino, T., Sakaguchi, G., Sakazaki, R., Takeda, Y. (eds.), International symposium on *Vibrio parahaemolyticus.* Tokyo: Saikon.

Committee on the Serotyping of *Vibrio parahaemolyticus.* 1970. New serotypes of *Vibrio parahaemolyticus.* Japanese Journal of Microbiology **14:**249–250.

Conn, H. O. 1964. Spontaneous peritonitis and bacteremia in Laennec's cirrhosis caused by enteric organisms. Annals of Internal Medicine **60:**568–580.

Cooper, R. G., Brown, G. W. 1968. *Plesiomonas shigelloides* in South Australia. Journal of Clinical Pathology **21:**715–718.

Cowan, S. T. 1956. Taxonomic rank of Enterobacteriaceae 'groups'. Journal of General Microbiology **15:**345–348.

Cumberbach, N., Gurwith, M. J., Langston, C., Sack, R. B., Brunton, J. L. 1979. Cytotoxic enterotoxin produced by *Aeromonas hydrophila:* Relationship of toxigenic isolates to diarrheal disease. Infection and Immunity **23:**829–837.

Dakin, W. P. H., Howell, D. J., Sutton, R. G. A., O'Keefe, M. F., Thomas, P. 1974. Gastroenteritis due to non-agglutinable (non-cholera) vibrios. Medical Journal of Australia **2:**487–490.

David, H. 1927. Ueber eine durch choleraähnliche Vibrionen hervorgerufene Fischseuche. Zentralblatt für Bakteriologie, Parasitenkunde und Infektionskrankheiten, Abt. 1 Orig. **102:**46–60.

Davis, W. A., II, Kane, J. G., Garagusi, V. F. 1978. Human *Aeromonas* infections: A review of the literature and a case report of endocarditis. Medicine (Baltimore) **57:**267–277.

Dean, H. M., Post, R. M. 1967. Fatal infection with *Aeromonas hydrophila* in a patient with acute myelogenous leukemia. Annals of Internal Medicine **66:**1177–1179.

DeFronzo, R. A., Murray, G. F., Maddrey, W. C. 1973. *Aeromonas* septicemia from hepatobiliary disease. American Journal of Digestive Disease **18:**323–331.

de Moor, C. E. 1939. Epidemic cholera in South Celebes caused by vibrio El Tor. Mededeelingen van den Dienst der Volksgezondheid in Nederlandsch-Indie **28:**320–355.

Donta, S. T., Haddow, A. D. 1978. Cytotoxic activity of *Aeromonas hydrophila.* Infection and Immunity **21:**989–993.

Dubey, R. S., Sanyal, S. C. 1978. Enterotoxicity of *Aeromonas hydrophila:* Skin responses and in vitro neutralization. Zentralblatt für Bakteriologie, Parasitenkunde, Infektionskrankheiten und Hygiene, Abt. 1 Orig., Reihe A **242:**487–499.

Dubey, R. S., Sanyal, S. C. 1979. Studies on the characterization and neutralization of *Aeromonas hydrophila* enterotoxin in the rabbit ileal loop model. Journal of Medical Microbiology **12:**345–352.

Eddy, B. P., Carpenter, K. P. 1964. Further studies on *Aeromonas.* II. Taxonomy of *Aeromonas* and C27 strains. Journal of Applied Bacteriology **27:**96–109.

Ellner, P. D., MacCarthy, L. R. 1973. *Aeromonas shigelloides* bacteremia: A case report. American Journal of Clinical Pathology **59:**216–218.

Esterabadi, A. H., Entessar, F., Khan, M. A. 1973. Isolation and identification of *Aeromonas hydrophila* from an outbreak of haemorrhagic septicemia in snakes. Canadian Journal of Comparative Medicine **37:**418–420.

Evelyn, T. P. T. 1971. First records of vibriosis in Pacific salmon cultured in Canada, and taxonomic status of the responsible bacterium, *Vibrio anguillarum.* Journal of Fisheries Research Board of Canada **28:**517–525.

Ewing, W. H., Hugh, R., Johnson, J. G. 1961. Studies on the *Aeromonas* group. Atlanta: U.S. Department of Health, Education and Welfare, Communicable Disease Center.

Felsenfeld, O. 1965. Notes on food, beverages and fomites contaminated with *Vibrio cholerae.* Bulletin of the World Health Organization **33:**725–734.

Ferguson, W. W., Henderson, N. D. 1947. Description of strain C27: A motile organism with the major antigen of *Shigella sonnei* phase I. Journal of Bacteriology **54:**179–181.

Finkelstein, R. A. 1973. Cholera. CRC Critical Reviews in Microbiology **2:**553–623.

Finkelstein, R. A. 1975. Immunology of cholera, pp. 137–196. In: Arbert, W. (ed.), Current topics in microbiology and immunology, vol. 69. Berlin: Springer-Verlag.

Finkelstein, R. A., Mukerjee, S. 1963. Hemagglutination: A rapid method for differentiating *Vibrio cholerae* and El Tor vibrios. Proceedings of the Society of Experimental Biology and Medicine **112:**355–359.

Finkler, D., Prior, J. 1884. Über den Bacillus der Cholera nostras und seine Cultur. Deutsche Medizinische Wochenschrift **10:**632–634.

Fourquet, R., Couturier, Y., Jamet, A., Griffet, P. 1973. Premières souches de *Plesiomonas shigelloides* isolées chez l'homme à Madagascar. Archive de l'Institut Pasteur Madagascar **42:**61–68.

Fritsche, D., Dahn, R., Hoffmann, G. 1975. *Aeromonas punctata* subsp. *caviae* als Erreger einer akuten Gastroenteritis. Zentralblatt für Bakteriologie, Parasitenkunde, Infektionskrankheiten und Hygiene, Abt. 1 Orig., Reihe A **233:**232–235.

Fujino, T. 1951. Bacterial food poisoning. [In Japanese.] Saishin Igaku **6:**263–271.

Fukaya, K., Nakamura, S., Takayama, H., Kitamoto, O., Sakazaki, R. 1962. Two cases of acute diarrhea due to *Aeromonas.* [In Japanese.] Journal of the Japanese Association for Infectious Diseases **36:**8–13.

Furniss, A. L., Lee, J. V., Donovan, T. J. 1977. Group F. A new vibrio? Lancet **ii:**565–566.

Gaines, S., Duangmani, C., Noyes, H. E., Occeno, T. 1964. Occurrence of non-agglutinable vibrios in diarrheal patients in Bangkok, Thailand. Journal of Microbiological Society of Thailand **8–10:**6–17.

Gallut, J. 1963. Contribution a l'étude du complexe antigénique "O" des vibrions. III. Recherches sur les agglutinogènes thermostables des vibrions NAG. Annales de l'Institut Pasteur **103:**1080–1097.

Gameleia, M. N. 1888. *Vibrio metschnikovii* (n. sp.) et rapports avec le microbe du cholera asiatica. Annales de l'Institut Pasteur (Paris) **2:**482–488.

Gardner, A. D., Venkatraman, K. V. 1935. The antigen of the cholera group of vibrios. Journal of Hygiene **35:**262–282.

Geizer, E., Kopecký, K., Aldová, E. 1966. Isolation of *Aeromonas shigelloides* in a child. Journal of Hygiene, Epidemiology, Microbiology and Immunology **10:**23–26.

Gibson, D. M., Hendrie, M. S., Houston, N. C., Hobbs, G. 1977. The identification of some Gram negative heterotrophic aquatic bacteria, pp. 135–159. In: Skinner, F. A., Shewan, J. M. (eds.), Aquatic microbiology. Society for Applied Bacteriology Symposium Series No. 6. London, New York: Academic Press.

Gilardi, G. L. 1967. Morphological and biochemical characteristics of *Aeromonas punctata (hydrophila, liquefaciens)* isolated from human sources. Applied Microbiology **15:**417–421.

Gilardi, G. L., Bottone, E., Birnbaum, M. 1970. Unusual fermentative, Gram-negative bacilli isolated from clinical specimens. II. Characterization of *Aeromonas* species. Applied Microbiology **20:**156–159.

Habs, H., Schubert, R. H. W. 1962. Über die biochemischen Merkmale und die taxonomische Stellung von *Pseudomonas shigelloides* (Bader). Zentralblatt für Bakteriologie, Parasitenkunde, Infektionskrankheiten und Hygiene, Abt. 1 Orig. **186:**316–327.

Hanaoka, M., Kato, Y., Amano, T. 1969. Complementary examination of DNA's among *Vibrio* species. Biken Journal **12:**181–185.

Hanson, P. G., Standridge, J., Jarrett, F., Maki, D. G. 1977. Freshwater wound infection due to *Aeromonas hydrophila.* Journal of the American Medical Association **238:**1053–1054.

Harrell, L. W., Novotny, A. J., Schiewe, M. H., Hodgins, H. O. 1976. Isolation and description of two vibrios pathogenic to Pacific salmon in Puget Sound, Washington. Fishery Bulletin **74:**447–449.

Hastein, T. 1975. Vibriosis in fish, a clinical, pathological and bacteriological study of the disease in a Norwegian fish farm. Ph.D. Thesis. University of Stirling, United Kingdom.

Hastein, T., Smith, J. E. 1977. A study of *Vibrio anguillarum* from farmed and wild fish using principal components analysis. Journal of Fish Biology **11:**69–75.

Hazen, T. C., Fliermans, C. B. 1979. Distribution of *Aeromonas hydrophila* in natural and man-made thermal effluents. Applied and Environmental Microbiology **38:**166–168.

Hazen, T. C., Fliermans, C. B., Hirsch, R. P., Esch, G. W. 1978. Prevalence and distribution of *Aeromonas hydrophila* in the United States. Applied and Environmental Microbiology **36:**731–738.

Heiberg, B. 1935. On the classification of *Vibrio cholerae* and the cholera-like vibrios. Copenhagen: Busk.

Helm, E. P., Stille, W. 1970. Akute Enteritis durch *Aeromonas hydrophila.* Deutsche Medizinische Wochenschrift **95:**18–24.

Hendrie, S. M., Shewan, J. M., Véron, M. 1971. *Aeromonas shigelloides* (Bader) Ewing et al.: A proposal that it be transferred to the genus *Vibrio.* International Journal of Systematic Bacteriology **21:**25–27.

Hill, K. R., Caselitz, F.-H., Moody, L. M. 1954. A case of acute metastatic myositis caused by a new organism of the family *Pseudomonadaceae.* West Indian Medical Journal **3:**9–11.

Hollis, D. G., Weaver, R. E., Baker, C. N., Thornsberry, C. 1976. A halophilic *Vibrio* species isolated from blood cultures. Journal of Clinical Microbiology **3:**425–431.

Holt, G. 1970. Vibriosis *(Vibrio anguillarum)* as an epizootic disease in rainbow trout *(Salmo gaidneri).* Acta Veterinaria Scandinavica **11:**600–603.

Honda, T., Goshima, K., Takeda, Y., Sugino, Y., Miwatani, T. 1976a. Demonstration of cardiotoxin activity of thermostable direct hemolysin (lethal toxin) by *Vibrio parahaemolyticus.* Infection and Immunity **13:**163–171.

Honda, T., Taga, S., Takeda, T., Hasibuan, M. A., Takeda, Y., Miwatani, T. 1976b. Identification of lethal toxin with thermostable direct hemolysin produced by *Vibrio parahaemolyticus* and some physicochemical properties of the purified toxin. Infection and Immunity **13:**133–139.

Hooper, W. L., Barrow, G. I., McNab, D. J. N. 1974. *Vibrio parahaemolyticus* food poisoning in Britain. Lancet **i:**1100–1102.

Hori, M., Hayashi, K., Maeshima, K., Kigawa, M., Miyasato, T., Yoneda, Y., Hagihara, Y. 1966. Food poisoning caused by *Aeromonas shigelloides* with an antigen common to *Shigella dysenteriae* 7. Journal of the Japanese Association for Infectious Disease **39:**441–448.

Hoshina, T. 1956. An epidemic disease affecting rainbow trout in Japan. Journal of the Tokyo University of Fisheries **42:**15–16.

Hoshina, T. 1957. Further observations of the causative bacteria of the epidemic disease like furunculosis of rainbow trout. Journal of the Tokyo University of Fisheries **43:**59–66.

Hugh, R., Sakazaki, R. 1972. Minimal number of characters for the identification of *Vibrio* species, *Vibrio cholerae,* and *Vibrio parahaemolyticus.* Public Health Laboratory **30:**133–137.

Hughes, J. M., Hollis, D. G., Gangarosa, E. J., Weaver, R. E. 1978. Non-cholera vibrio infections in the United States. Annals of Internal Medicine **88:**602–606.

Inaba, H. 1973. Investigation of gastroenteritis at San Lazaro Hospital in Manila. [In Japanese.] Japanese Journal of Public Health **20:**459–465.

Jandl, G., Linke, K. 1976. Report about 2 cases of gastroenteritis caused by *Plesiomonas shigelloides.* Zentralblatt für Bakteriologie, Parasitenkunde, Infektionskrankheiten und Hygiene, Abt. 1 Orig., Reihe A **236:**136–140.

Joseph, S. W. 1974. Observation on *Vibrio parahaemolyticus* in Indonesia, pp. 35–40. In: Fujino, T., Sakaguchi, G., Sakazaki, R., Takeda, Y. (eds.), International symposium on *Vibrio parahaemolyticus.* Tokyo: Saikon.

Joseph, S. W., Daily, O. P., Hunt, W. S., Seidler, R. J., Allen, D. A., Colwell, R. R. 1979. *Aeromonas* primary wound infection of a diver in polluted waters. Journal of Clinical Microbiology **10:**46–49.

Kaneko, T., Colwell, R. R. 1973. Ecology of *Vibrio parahaemolyticus* in Chesapeake Bay. Journal of Bacteriology **113:**24–32.

Kaper, J., Seidler, R. J., Lockman, H., Colwell, R. R. 1979. Medium for the presumptive identification of *Aeromonas hydrophila* and *Enterobacteriaceae.* Applied and Environmental Microbiology **38:**1023–1026.

Kato, T., Obara, Y., Ichinoe, H., Nagashima, K., Akiyama, S., Takizawa, K., Matsushima, A., Yamai, S., Miyamoto, Y. 1965. Grouping of *Vibrio parahaemolyticus* (biotype 1) by hemolytic reaction. [In Japanese.] Shokuhin Eisei Kenkyu **15:**83–86.

Kauffmann, F. 1950. On the serology of *Vibrio cholerae.* Acta Pathologica et Microbiologica Scandinavica **27:**283–299.

Ketover, B. P., Young, L. S., Armstrong, D. 1973. Septicemia due to *Aeromonas hydrophila:* Clinical and immunologic aspects. Journal of Infectious Diseases **127:**284–290.

Kiehn, E. D., Pacha, R. E. 1969. Characterization and relatedness of marine vibrios pathogenic to fish. Deoxyribonucleic acid homology and base composition. Journal of Bacteriology **100:**1248–1255.

Kjems, E. 1955. Studies on five bacterial strains of the genus *Pseudomonas.* Acta Pathologica et Microbiologica Scandinavica **36:**531–536.

Kleeberger, A. 1977. Taxonomische Untersuchungen an Aeromonaden aus Milch, Wasser und Hackelfleisch. Zeitschrift für Lebensmittel-Untersuchung und -Forschung **163:**44–47.

Klonts, G. W., Yasutake, W. T., Ross, A. J. 1966. Bacterial diseases of Salmonidae in the Western United States: Pathogenesis of furunculosis in rainbow trout. American Journal of Veterinary Research **27:**1455–1457.

Ko, H. L., Lutticken, R., Pulverer, G. 1973. Occurrence of non-cholera vibrios in West Germany. Deutsche Medizinische Wochenschrift **98:**1494–1499.

Kobayashi, T., Enomoto, S., Sakazaki, R., Kuwabara, S. 1963. A new selective isolation medium for the vibrio group (modified Nakanishi medium-TCBS agar). [In Japanese.] Japanese Journal of Bacteriology **18:**387–392.

Kok, N. 1967. *Aeromonas hydrophila s. liquefaciens* isolated from tonsillitis in man. Acta Pathologica et Microbiologica Scandinavica **71:**599–602.

Kristensen, K. K. 1974. Semiquantitative examinations on the contents of *Vibrio parahaemolyticus* in the sound between Sweden and Denmark, pp. 105–110. In: Fujino, T., Sakaguchi, G., Sakazaki, R., Takeda, Y. (eds.), International symposium on *Vibrio parahaemolyticus.* Tokyo: Saikon.

Kulp, W. L., Borden, D. G. 1942. Further studies of *Proteus hydrophila,* the etiological agent in "red leg" disease of frogs. Journal of Bacteriology **44:**673–685.

Lautrop, H. 1961. *Aeromonas hydrophila* isolated from human faeces and its possible pathological significance. Acta Pathologica et Microbiologica Scandinavica, Suppl. **144:**299–301.

Lázničková, K., Aldová, E. 1976. Human infection caused by *Vibrio parahaemolyticus* in Czechoslovakia. Journal of Hygiene, Epidemiology, Microbiology and Immunology **20:**374–376.

Leclerc, H., Buttiaux, R. 1962. Fréquence des *Aeromonas* dans les eaux d'alimentation. Annales de l'Institut Pasteur **103:**97–100.

Lee, J. V., Donovan, T. J., Furniss, A. L. 1978. The characterization, taxonomy and emended description of *Vibrio metschnikovii.* International Journal of Systematic Bacteriology **28:**99–111.

Leifson, E. 1963. Mixed polar and peritrichous flagellation of marine bacteria. Journal of Bacteriology **86:**166–167.

Leistner, L., Hechelmann, H. 1974. Occurrence and significance of *Vibrio parahaemolyticus* in Europe, pp. 83–90. In: Fujino, T., Sakaguchi, G., Sakazaki, R., Takeda, Y. (eds.), International symposium on *Vibrio parahaemolyticus*. Tokyo: Saikon.

Ljungh, A., Popoff, M., Wadström, T. 1977. *Aeromonas hydrophila* in acute diarrheal disease: Detection of enterotoxin and biotyping of strains. Journal of Clinical Microbiology **6:**96–100.

Lopez, J. F., Quesada, J., Saied, A. 1968. Bacteremia and osteomyelitis due to *Aeromonas hydrophila*. A complication during the treatment of acute leukemia. American Journal of Clinical Pathology **50:**587–591.

McCarthy, D. H. 1975. The bacteriology and taxonomy of *Aeromonas liquefaciens*. Technical Report Series No. 2. Weymouth, Dorset: Ministry of Agriculture, Fisheries and Food, Fish Disease Laboratories.

McCormack, W. M., Dewitt, W. E., Bailey, P. E., Morris, G. K., Sorharjono, P., Gangarosa, E. J. 1974. Evaluation of thiosulfate-citrate-bile salt-sucrose agar, a selective medium for the isolation of *Vibrio cholerae* and other pathogenic vibrios. Journal of Infectious Diseases **129:**497–500.

McCracken, A. W., Barkley, R. 1972. Isolation of *Aeromonas* species from clinical sources. Journal of Clinical Pathology **25:**970–975.

MacInnes, J. I., Trust, T. J., Crosa, J. H. 1979. Deoxyribonucleic acid relationships among members of the genus *Aeromonas*. Canadian Journal of Microbiology **25:**579–586.

McIntyre, O. R., Feeley, J. C., Greenough, W. B., III, Benenson, A. S., Hassan, S. I., Saad, A. 1965. Diarrhea caused by non-cholera vibrios. American Journal of Tropical Medicine and Hygiene **14:**412–418.

Mackie, T. S., Arkwright, J. A., Pryce-Tennant, T. E., Mottram, J. C., Johnston, W. D., Menzies, W. J. M. 1930. Furunculosis committee first interim report. Edinburgh: H.M.S.O.

Mackie, T. S., Arkwright, J. A., Pryce-Tennant, T. E., Mottram, J. C., Johnston, W. D., Menzies, W. J. M. 1933. Furunculosis committee second interim report. Edinburgh: H.M.S.O.

Mackie, T. S., Arkwright, J. A., Pryce-Tennant, T. E., Mottram, J. C., Johnston, W. D., Menzies, W. J. M. 1935. Furunculosis committee final report. Edinburgh: H.M.S.O.

McSweeney, R. J., Forgan-Smith, W. R. 1977. Wound infection in Australia from halophilic vibrios. Medical Journal of Australia **1:**896–897.

Marcus, L. C. 1971. Infectious diseases of reptiles. Journal of the American Veterinary Medical Association **159:**1626–1631.

Martinez-Silva, R., Guzmann-Urrego, M., Caselitz, F.-H. 1961. Zur Frage der Bedeutung von *Aeromonas* stämmen bei Säuglingsenteritis. Zeitschrift für Tropenmedizin und Parasitologie **12:**445–451.

Ministry of Health and Welfare of Japan. 1974. Statistics of food poisoning during 1973. [In Japanese.] Shokuhin Eisei Kenkyu **24:**708–721.

Ministry of Health and Welfare of Japan. 1975. Statistics of food poisoning during 1974. [In Japanese.] Shokuhin Eisei Kenkyu **25:**687–699.

Ministry of Health and Welfare of Japan. 1976. Statistics of food poisoning during 1975. [In Japanese.] Shokuhin Eisei Kenkyu **26:**850–863.

Ministry of Health and Welfare of Japan. 1977. Statistics of food poisoning during 1976. [In Japanese.] Shokuhin Eisei Kenkyu **27:**925–941.

Miyamoto, Y., Nakamura, K., Takizawa, K. 1961. Pathogenic halophiles. Proposals of a new genus "*Oceanomonas*" and the amended species names. Japanese Journal of Microbiology **5:**477–486.

Miyamoto, Y., Nakamura, K., Takizawa, K., Kodama, T. 1960. Carangina fish poisoning. I. [In Japanese.] Japanese Journal of Public Health **7:**587–592.

Miyamoto, Y., Nakamura, K., Takizawa, K., Kodama, T.
1961a. Carangina fish poisoning. II. [In Japanese.] Japanese Journal of Public Health **8:**673–678.

Miyamoto, Y., Nakamura, K., Takizawa, K., Kodama, T. 1961b. Carangina fish poisoning. III. [In Japanese.] Japanese Journal of Public Health **8:**703–707.

Monsur, K. A. 1963. Bacteriological diagnosis of cholera under field conditions. Bulletin of the World Health Organization **28:**387–389.

Mukerjee, S. 1963. Bacteriophage typing in cholera. Bulletin of the World Health Organization **28:**337–345.

Nacescu, N., Ciufecu, C., Nicoara, I., Florescu, D., Konrad, I. 1974. Morphological, cultural and biochemical characteristics of so-called NAG vibrios isolated from human feces and vomitus. Preliminary report. Zentralblatt für Bakteriologie, Parasitenkunde, Infektionskrankheiten und Hygiene, Abt. 1 Orig., Reihe A **229:**209–215.

Nakanishi, H., Leistner, L., Hechelmann, H. 1969. Das Vorkommen von enteropathogenen, gramnegativen Stäbchen in Patientenstühlen. Fleischwirtschaft **49:**1501.

Nakanishi, H., Murase, M., Teramoto, T. 1977. Salt-polymyxin broth, a new enrichment medium for *Vibrio parahaemolyticus*. [In Japanese.] Media Circle **22:**1–4.

Neilson, A. H. 1978. The occurrence of aeromonads in activated sludge: Isolation of *Aeromonas sobria* and its possible confusion with *Escherichia coli*. Journal of Applied Bacteriology **44:**259–264.

Neumann, D. A., Benenson, M. W., Hubster, E., Tuan, N. T. N., Van, L. T. 1972a. *Vibrio parahaemolyticus* in the Republic of Vietnam. American Journal of Tropical Medicine and Hygiene **21:**464–466.

Neumann, D. A., Benenson, M. W., Hubster, E., Tuan, N. T. N. 1972b. Cary-Blair, a transport medium for *Vibrio parahaemolyticus*. American Journal of Clinical Pathology **57:**33–34.

Nikkawa, T., Obara, Y., Yamai, S., Miyamoto, Y. 1972. Purification of a hemolysin from *Vibrio parahaemolyticus*. Japanese Journal of Medical Science and Biology **25:**197–200.

Nishio, T., Kida, M., Shimouchi, H. 1967a. Ecological studies on *Vibrio parahaemolyticus*. I. Distribution in sea water and sea mud. [In Japanese.] Medical Journal of Hiroshima University **15:**615–618.

Nishio, T., Kida, M., Shimouchi, H. 1967b. Ecological studies on *Vibrio parahaemolyticus*. II. Distribution in water and mud of estuary. [In Japanese.] Medical Journal of Hiroshima University **15:**619–622.

Nobechi, K. 1923. Contribution to the knowledge of *Vibrio cholerae*. III. Immunological studies upon the type of *Vibrio cholerae*. Scientific Report of the Government Institute for Infectious Diseases of Tokyo Imperial University **2:**43–88.

Nobechi, K., Nakano, E., Nagao, M. 1967. Studies on shifting of the serotypes of cholera vibrios. The first report: Studies in vitro, p. 119. In: Symposium on cholera. Sponsored by the U.S.-Japan Cooperative Medical Science Program, Office of International Research, National Institutes of Health, Palo Alto, California, 26–28 July 1967. Bethesda, Maryland: Office of International Research, National Institutes of Health.

Noguchi, M., Asakawa, Y. 1967. Distribution of *Vibrio parahaemolyticus*, pp. 313–323. In: Fujino, T., Fukumi, H. (eds.), *Vibrio parahaemolyticus*. [In Japanese.] Tokyo: Naya Shoten.

Noguchi, M., Asakawa, Y. 1967. Distribution of *Vibrio parahaemolyticus*, pp. 313–323. In: Fujino, T., Fukumi, H. (eds.), *Vibrio parahaemolyticus*. [In Japanese.] Tokyo: Naya Shoten.

Nybelin, O. 1975. Untersuchungen über den bei Fischenkrankheitserregenden Spaltpilz *Vibrio anguillarum*. Meddn Statens Undersögerser Försogsanstalt Sottwattenfisket **8:**5–62.

Nygaard, G. S., Bissett, M. L., Wood, R. M. 1970. Laboratory infection of aeromonads from man and other animals. Applied Microbiology **19:**618–620.

Obara, Y. 1971. Studies on hemolytic factors of *Vibrio para*-

haemolyticus. II. The extraction of hemolysin and its properties. [In Japanese.] Journal of the Japanese Association for Infectious Disease **45**:392–398.

Ohashi, M., Shimada, T., Fukumi, H. 1972. In vitro production of enterotoxin and hemorrhagic principle by *Vibrio cholerae*, NAG. Japanese Journal of Medical Science and Biology **25**:179–194.

Okabe, S. 1974. Statistical review of food poisoning in Japan—especially that by *Vibrio parahaemolyticus*, pp. 5–8. In: Fujino, T., Sakaguchi, G., Sakazaki, R., Takeda, Y. (eds.), International symposium on *Vibrio parahaemolyticus*. Tokyo: Saikon.

Osada, A., Shibata, I. 1956. On the fraternal diarrhoea supposedly caused by the paracolon organisms with the major somatic antigens of *S. sonnei*. Acta Paediatrica Japonica **60**:739–742.

Pandit, C. G., Pal, S. C., Murti, G. V. S., Misra, B. S., Murty, D. K., Shrivastava, J. B. 1967. Survival of *Vibrio cholerae* biotype El Tor in well water. Bulletin of the World Health Organization **37**:681–685.

Paučková, V., Fukalová, A. 1968. Occurrence of *Aeromonas hydrophila* and *Aeromonas shigelloides* in feces. Zentralblatt für Bakteriologie, Parasitenkunde, Infektionskrankheiten und Hygiene, Abt. 1 Orig. **216**:212–216.

Pearson, T. A., Mitchell, C. A., Hughes, W. T. 1972. *Aeromonas hydrophila* septicemia. American Journal of Diseases of Children **123**:579–582.

Peffers, A. S. R., Bailey, J., Barrow, G. I., Hobbs, B. C. 1973. *Vibrio parahaemolyticus* and international air travel. Lancet **i**:143–145.

Pesigan, T. P., Plantilla, J., Rolda, M. 1967. Applied studies on the viability of El Tor vibrios. Bulletin of the World Health Organization **37**:779–786.

Phillips, J. A., Bernhard, H. E., Rosenthal, S. G. 1974. *Aeromonas hydrophila* infections. Pediatrics **53**:110–112.

Pien, F., Lee, K., Higa, H. 1977. *Vibrio alginolyticus* infections in Hawaii. Journal of Clinical Microbiology **5**:670–672.

Popoff, M., Véron, M. 1976. A taxonomic study of the *Aeromonas hydrophila–Aeromonas punctata* group. Journal of General Microbiology **94**:11–22.

Qadri, S. M. H., Gordon, P., Wende, R. D., Williams, R. P. 1976. Meningitis due to *Aeromonas hydrophila*. Journal of Clinical Microbiology **3**:102–104.

Quincke, G. 1967. Untersuchungen über die O-Antigene der Plesiomonaden. Archiv für Hygiene **151**:525–529.

Ramsay, A. M. B., Rosenbaum, B. J., Yarbrough, C. L., Hotz, J. A. 1978. *Aeromonas hydrophila* sepsis in a patient undergoing hemodialysis therapy. Journal of the American Medical Association **239**:128–129.

Rigney, M. M., Zilinsky, J. W., Rouf, M. A. 1978. Pathogenicity of *Aeromonas hydrophila* in red leg disease in frogs. Current Microbiology **1**:175–179.

Rogol, M., Sechter, I., Grinberg, L., Gerichter, Ch. B. 1979. Pril-xylose-ampicillin agar, a new selective medium for the isolation of *Aeromonas hydrophila*. Journal of Medical Microbiology **12**:229–232.

Roland, F. P. 1970. Leg gangrene and endotoxin shock due to *Vibrio parahaemolyticus*—an infection acquired in New England coastal waters. New England Journal of Medicine **282**:1306.

Rosner, R. 1964. *Aeromonas hydrophila* as the etiologic agent in a case of severe gastroenteritis. American Journal of Clinical Pathology **42**:402–404.

Ross, A. J., Martin, J. E., Bressler, V. 1968. *Vibrio anguillarum* from an epizootic in rainbow trout (*Salmo gairdneri*) in the U.S.A. Bulletin of Office of International Epizootics **69**:1139–1148.

Rubin, S. J., Tilton, R. C. 1975. Isolation of *Vibrio alginolyticus* from wound infections. Journal of Clinical Microbiology **2**:556–558.

Sakazaki, R. 1968. Proposal of *Vibrio alginolyticus* for the biotype 2 of *Vibrio parahaemolyticus*. Japanese Journal of Medical Science and Biology **21**:359–362.

Sakazaki, R. 1972. Control of contamination with *Vibrio parahaemolyticus* in seafoods and isolation and identification of the vibrio, pp. 375–385. In: Hobbs, B. C., Christian, J. H. B. (eds.), The microbiological safety of food. London: Academic Press.

Sakazaki, R., Gomez, C. Z., Sebald, M. 1967. Taxonomical studies of the so-called NAG vibrios. Japanese Journal of Medical Science and Biology **20**:265–280.

Sakazaki, R., Iwanami, S., Fukumi, H. 1963. Studies on the enteropathogenic, facultatively halophilic bacteria, *Vibrio parahaemolyticus*. I. Morphological, cultural and biochemical properties and its taxonomical position. Japanese Journal of Medical Science and Biology **16**:161–188.

Sakazaki, R., Iwanami, S., Tamura, K. 1968. Studies on the enteropathogenic, facultatively halophilic bacteria, *Vibrio parahaemolyticus*. II. Serological characteristics. Japanese Journal of Medical Science and Biology **21**:313–324.

Sakazaki, R., Shimada, T. 1972. Serovars of *Vibrio cholerae* identified during 1970–1975. Japanese Journal of Medical Science and Biology **30**:279–282.

Sakazaki, R., Tamura, K. 1971. Somatic antigen variation in *Vibrio cholerae*. Japanese Journal of Medical Science and Biology **24**:93–100.

Sakazaki, R., Tamura, K., Murase, M. 1971. Determination of the hemolytic activity of *Vibrio cholerae*. Japanese Journal of Medical Science and Biology **24**:83–91.

Sakazaki, R., Namioka, S., Nakaya, R., Fukumi, H. 1959. Studies on the so-called paracolon C27 (Ferguson). Japanese Journal of Medical Science and Biology **12**:355–363.

Sakazaki, R., Tamura, K., Kato, T., Obara, Y., Yamai, S., Hobo, K. 1968. Studies on the enteropathogenic, facultatively halophilic bacteria, *Vibrio parahaemolyticus*. III. Enteropathogenicity. Japanese Journal of Medical Science and Biology **21**:325–331.

Sakazaki, R., Tamura, K., Gomez, C. Z., Sen, R. 1970a. Serological studies on the cholera group of vibrios. Japanese Journal of Medical Science and Biology **23**:13–20.

Sakazaki, R., Tamura, K., Ikuta, K., Sebald, M. 1970b. Taxonomic studies on marine vibrios, pp. 583–593. In: Iizuka, H., Hasegawa, T. (eds.), Culture collections of microorganisms. Proceedings of the International Conference on Culture Collections, Tokyo, Oct. 7–11, 1968. Baltimore: University Park Press.

Sakazaki, R., Tamura, K., Prescott, L. M., Bencic, Z., Sanyal, S. C., Sinka, R. 1971. Bacteriological examination of diarrheal stools in Calcutta. Indian Journal of Medical Research **59**:1025–1034.

Sakazaki, R., Tamura, K., Nakamura, A., Kurata, T., Gohda, A., Kazuno, Y. 1974. Studies on enteropathogenic activity of *Vibrio parahaemolyticus* using ligated gut loop model in rabbits. Japanese Journal of Medical Science and Biology **27**:35–43.

Sakurai, J., Matsuzaki, A., Miwatani, T. 1973. Purification and characterization of thermostable direct hemolysin of *Vibrio parahaemolyticus*. Infection and Immunity **8**:775–780.

Sakurai, J., Bahavar, M. A., Jinguji, Y., Miwatani, T. 1975. Interaction of thermostable direct hemolysin of *Vibrio parahaemolyticus* with human erythrocytes. Biken Journal **8**:187–192.

Sakurai, J., Honda, T., Jinguji, Y., Arita, M., Miwatani, T. 1976. Cytotoxic effects of thermostable direct hemolysin produced by *Vibrio parahaemolyticus* on FL cells. Infection and Immunity **13**:876–883.

Sanyal, S. C., Sen, P. C. 1974. Human volunteer study on the pathogenicity of *Vibrio parahaemolyticus*, pp. 227–235. In: Fujino, T., Sakaguchi, G., Sakazaki, R., Takeda, Y. (eds.), International symposium on *Vibrio parahaemolyticus*. Tokyo: Saikon.

Sanyal, S. C., Singh, S. J., Sen, P. C. 1975. Enteropathogenicity of *Aeromonas hydrophila* and *Plesiomonas shigelloides*. Journal of Medical Microbiology **8**:195–198.

Sanyal, S. C., Singh, S. J., Tiwari, I. C., Sen, P. C., Marwah,

S. M., Hazarika, U. R., Singh, H., Shimada, T., Sakazaki, R. 1974. Role of household animals in maintenance of cholera infection in a community. Journal of Infectious Diseases **130:**575–579.

Saraswathe, B., Sharma, P., Sanyal, S. C. 1980. Enteropathogenicity of *Plesiomonas shigelloides*—a neoenteropathogen. Asian Journal of Infectious Disease, in press.

Scherago, M. 1937. An epizootic septicemia of young guinea pigs caused by *Pseudomonas caviae* n. sp. Journal of Infectious Diseases **60:**245–250.

Schmid, E. E., Velaudapillai, T., Niles, G. R. 1954. Study of paracolon organisms with the major antigen of *Shigella sonnei*, form I. Journal of Bacteriology **68:**50–52.

Schmidt, U., Chmel, H., Cobbs, C. 1979. *Vibrio alginolyticus* infections in humans. Journal of Clinical Microbiology **10:**666–668.

Schubert, R. H. W. 1967a. The taxonomy and nomenclature of the genus *Aeromonas* Kluyver and van Niel 1936. Part II. Suggestions on the taxonomy and nomenclature of the anaerogenic aeromonads. International Journal of Systematic Bacteriology **17:**273–279.

Schubert, R. H. W. 1967b. Das Vorkommen der Aeromonaden in oberirdischen Gewässern. Archiv für Hygiene und Bakteriologie **150:**688–708.

Schubert, R. H. W. 1974. Genus *Aeromonas* Kluyver and van Niel, pp. 345–349. In: Buchanan, R. E., Gibbons, N. E. (eds.), Bergey's manual of determinative bacteriology, 8th ed. Baltimore: Williams & Wilkins.

Schubert, R. H. W. 1977. Über den Nachweis von *Plesiomonas shigelloides* Habs und Schubert 1962 und ein Elektivmedium, den Inositol-Brilliantgrün-Gallesalz-Agar. Ernst-Rodewalt-Archiv **4:**97–103.

Sebald, M., Véron, M. 1963. Teneur en bases de l'ADN et classification des vibrios. Annales de l'Institut Pasteur de Lille **105:**897–910.

Shackelford, P. G., Ratzan, S. A., Shearer, W. T. 1973. Ecthyma gangrenosum produced by *Aeromonas hydrophila*. Journal of Pediatrics **83:**100–101.

Shewan, J. M., Hobbs, G., Hodgkiss, W. 1960. A determinative schema for the identification of certain genera of gram-negative bacteria, with special reference to the Pseudomonadaceae. Journal of Applied Bacteriology **23:**379–390.

Shewan, J. M., Véron, M. 1974. Genus I. *Vibrio* Pacini 1854, pp. 340–345. In: Buchanan, R. E., Gibbons, N. E. (eds.), Bergey's manual of determinative bacteriology, 8th ed. Baltimore: Williams & Wilkins.

Shimada, T., Sakazaki, R. 1973. R antigen of *Vibrio cholerae*. Japanese Journal of Medical Science and Biology **26:**155–160.

Shimada, T., Sakazaki, R. 1977. Additional serovars and inter-O antigenic relationships of *Vibrio cholerae*. Japanese Journal of Medical Science and Biology **30:**275–277.

Shimada, T., Sakazaki, R. 1978. On the serology of *Plesiomonas shigelloides*. Japanese Journal of Medical Science and Biology **31:**135–142.

Shimizu, R. 1969. Studies on pathogenic properties of *Aeromonas liquefaciens*. I. Production of toxic substance to eel. Bulletin of the Japanese Society of Scientific Fisheries **35:**55–63.

Shinoda, S., Honda, T., Takeda, Y., Miwatani, T. 1974. Antigenic difference between polar monotrichous and peritrichous flagella of *Vibrio parahaemolyticus*. Journal of Bacteriology **120:**923–928.

Shotts, E. B., Rimler, R. 1973. Medium for the isolation of *Aeromonas hydrophila*. Applied Microbiology **26:**550–553.

Shotts, E. B., Gaines, J. L., Martin, L., Prestwood, A. K. 1972. *Aeromonas*-induced death among fish and reptiles in a eutrophic inland lake. Journal of American Veterinary Medical Association **161:**603–607.

Simidu, E., Kaneko, E. 1973. A numerical taxonomy of *Vibrio* and *Aeromonas* from normal and diseased fish. Bulletin of the Japanese Society of Scientific Fisheries **39:**689–703.

Simon, G., von Graevenitz, A. 1969. Intestinal and water-borne infections due to *Aeromonas hydrophila*. Public Health Laboratory **27:**159–162.

Sircar, B. K., Deb, B. C., De, S. P., Ghosh, A., Pal, S. C. 1976. Clinical and epidemiological studies on *Vibrio parahaemolyticus* infection in Calcutta (1975). Indian Journal of Medical Research **64:**1576–1580.

Smith, H. L., Jr. 1974. Antibody responses in rabbits to infections of whole cell, flagella, and flagellin preparations of cholera and noncholera vibrios. Applied Microbiology **27:**375–378.

Smith, H. L., Jr., Goodner, K. 1965. On the classification of vibrios, pp. 4–8. In: Bushnell, O. A., Brookhyser, C. S. (eds.), Proceedings of the Cholera Research Symposium, Honolulu, Hawaii. Washington, D.C.: Government Printing Office.

Smith, I. W. 1961. A disease of finnock due to *Vibrio anguillarum*. Journal of General Microbiology **24:**247–252.

Smith, I. W. 1963. The classification of '*Bacterium salmonicida*'. Journal of General Microbiology **33:**263–274.

Staley, T. E., Colwell, R. R. 1973. Deoxyribonucleic acid reassociation among members of the genus *Vibrio*. International Journal of Systematic Bacteriology **23:**316–332.

Stephen, S., Rao, K. N. A., Kumar, M. S., Indrani, R. 1975. Human infection with *Aeromonas* species: Varied clinical manifestation. Annals of Internal Medicine **83:**368–369.

Subcommittee on Taxonomy of Vibrios. 1966. Minutes of IAMS Subcommittee on the Taxonomy of Vibrios. International Journal of Systematic Bacteriology **16:**135–142.

Subcommittee on Taxonomy of Vibrios. 1972. Report (1966–1970) of the Subcommittee on Taxonomy of Vibrios to the International Committee on Nomenclature of Bacteria. International Journal of Systematic Bacteriology **22:**123.

Subcommittee on Taxonomy of Vibrios. 1975. Minutes of the closed meeting, 3 September 1974. International Journal of Systematic Bacteriology **25:**389–391.

Takeda, Y., Hori, Y., Miwatani, T. 1974. Demonstration of a temperature-dependent inactivating factor of the thermostable direct hemolysin in *Vibrio parahaemolyticus*. Infection and Immunity **10:**6–10.

Takeda, Y., Hori, Y., Taga, S., Sakurai, J., Miwatani, T. 1975a. Characterization of the temperature-dependent inactivating factor of the thermostable direct hemolysin in *Vibrio parahaemolyticus*. Infection and Immunity **12:**449–454.

Takeda, Y., Takeda, T., Honda, T., Sakurai, J., Ohtomo, N., Miwatani, T. 1975b. Inhibition of hemolytic activity of thermostable direct hemolysin of *Vibrio parahaemolyticus* by ganglioside. Infection and Immunity **12:**931–933.

Takikawa, I. 1958. Studies on pathogenic halophilic bacteria. Yokohama Medical Bulletin **2:**313–322.

Tamura, K., Shimada, S., Prescott, L. M. 1971. Vibrio agar: A new plating medium for isolation of *Vibrio cholerae*. Japanese Journal of Medical Science and Biology **24:**125–127.

Tapper, M. L., McCarthy, L. R., Mayo, J. B., Armstrong, D. 1975. Recurrent *Aeromonas* sepsis in a patient with leukemia. American Journal of Clinical Pathology **64:**525–530.

Terada, Y. 1968. Serological studies of *Vibrio parahaemolyticus*. II. Flagellar antigens. [In Japanese.] Japanese Journal of Bacteriology **23:**767–771.

Terada, Y., Yokoo, Y., Nakanishi, H., Teramoto, T. 1975. A new K antigen of *Vibrio parahaemolyticus* isolated from patients with gastroenteritis in Kobe City. [In Japanese.] Japanese Journal of Bacteriology **30:**515–516.

Teramoto, T., Nakanishi, H., Maejima, K. 1969. Kanagawa reaction of *Vibrio parahaemolyticus* isolated from food poisoning. [In Japanese.] Modern Media **15:**215–216.

Thal, E., Dinter, Z. 1953. Zur Pathogenität der Stammgruppe "455" (*Enterobacteriaceae*) für die Maus. Nordisk Veterinärmedicin **5:**855–858.

Tokoro, M., Toto, K., Yamada, F. 1977. A new K antigen of *Vibrio parahaemolyticus*. [In Japanese.] Japanese Journal of Bacteriology **32:**393–394.

Tsukamoto, T., Kinoshita, Y., Shimada, T., Sakazaki, R. 1978. Two epidemics of diarrhoeal disease possibly caused by *Plesiomonas shigelloides.* Journal of Hygiene **80:**275–280.

Tubiash, H. S., Colwell, R. R., Sakazaki, R. 1970. Marine vibrios associated with bacillary necrosis, a disease of larval and juvenile bivalve mollusks. Journal of Bacteriology **103:**272–273.

Twedt, R. M., Spaulding, P. L., and Hall, H. E. 1969. Morphological, cultural, biochemical, and serological comparison of Japanese strains of *Vibrio parahaemolyticus* with related cultures isolated in the United States. Journal of Bacteriology **98:**511–518.

Ueda, S., Yamazaki, S., Hori, M. 1963. The isolation of paracolon C27 and halophilic organisms from an outbreak of food poisoning. Japanese Journal of Public Health **10:**67–70.

Van, L. T., Tuan, N. T. N. 1974. *Vibrio parahaemolyticus* isolated among diarrhea patients, pp. 15–20. In: Fujino, T., Sakaguchi, G., Sakazaki, R., Takeda, Y. (eds.), International symposium on *Vibrio parahaemolyticus.* Tokyo: Saikon.

Vandepitte, J., Makulu, A., Gatti, F. 1974. *Plesiomonas shigelloides.* Survey and possible association with diarrhoea in Zaire. Annales de la Société Belge de Médecine Tropicale **54:**503–513.

Vandepitte, J., Ghysels, G., Goethem, V. H., Marrecau, N. 1957. Sur les coli bactéries abérrantes ayant l'antigène somatique de *Shigella sonnei* en phase I. Annales de la Société Belge de Médecine Tropicale **37:**737–742.

Véron, M., Sebald, M. 1964. Sur la teneur en bases de l'ADN et al position taxonomique de *Vibrio ichthyodermis* Shewan, Hobbs et Hodgkiss 1960. Annales de l'Institut Pasteur **107:**422–423.

Vézina, R., Desrochers, R. 1971. Incidence d'*Aeromonas hydrophila* chez la perche, *Perca flavescens* Mitchill. Canadian Journal of Microbiology **17:**1101–1103.

von Graevenitz, A., Carrington, G. O. 1973. Halophilic vibrios from extraintestinal lesions in man. Infection **1:**54–58.

von Graevenitz, A., Mensch, A. H. 1968. The genus *Aeromonas* in human bacteriology. New England Journal of Medicine **278:**245–249.

von Graevenitz, A., Zinterhofer, L. 1970. The detection of *Aeromonas hydrophila* in stool specimens. Health Laboratory Science **7:**124–127.

Wadström, T., Ljungh, A., Wretlind, B. 1976. Enterotoxin, haemolysin and cytotoxic protein in *Aeromonas hydrophila* from human infections. Acta Pathologica et Microbiologica Scandinavica, Sect. B **84:**112–114.

Wagatsuma, S. 1968. On a medium for hemolytic reaction. [In Japanese.] Media Circle **13:**156–162.

Wallace, C. K., Pierce, N. F., Anderson, P. N., Brown, T. C., Lewes, G. W., Sanyal, S. N., Segre, G. V., Waldman, R. H. 1967. Probable gallbladder infection in convalescent cholera patients. Lancet **i:**865–868.

Whang, H. Y., Heller, M. E., Neter, E. 1972. Production by *Aeromonas* of common enterobacterial antigen and its possible taxonomic significance. Journal of Bacteriology **110:**161–164.

Washington, J. A., II. 1967. *Aeromonas hydrophila* in clinical bacteriologic specimens. Annals of Internal Medicine **76:**611–614.

Wassum, I. 1967. Ein Fall von *Aeromonas*-Enteritis. Zentralblatt für Arbeitsmedizin **17:**313–314.

Wohlgemuth, K. R., Pierce, R. L., Kilbride, C. A. 1972. Bovine abortion associated with *Aeromonas hydrophila.* Journal of the American Veterinary Medical Association **160:**1001–1002.

World Health Organization. 1969. Outbreak of gastroenteritis by nonagglutinable (NAG) vibrios. Weekly Epidemiological Record **44:**10.

Wretlind, B., Möllby, R., Wadström, T. 1971. Separation of two hemolysins from *Aeromonas hydrophila* by isoelectric focusing. Infection and Immunity **4:**503–505.

Yanagisawa, F., Takeuchi, T. 1975. Halophilic bacteria. [In Japanese.] Shokuhin Eisei Kenkyu **7:**11–18.

Yokota, Y., Tokoro, M., Nishiyama, I., Terada, Y. 1977. A new serovar of *Vibrio parahaemolyticus.* [In Japanese.] Japanese Journal of Bacteriology **32:**509–510.

Zajc-Satler, J. 1972. Morphological and biochemical studies of 27 strains belonging to the genus *Aeromonas* isolated from clinical sources. Journal of Medical Microbiology **5:**263–265.

Zajc-Satler, J., Dragas, A. Z., Kumelj, M. 1972. Morphological and biochemical studies of 6 strains of *Plesiomonas shigelloides* from clinical sources. Zentralblatt für Bakteriologie, Parasitenkunde, Infektionskrankheiten und Hygiene, Abt. 1 Orig., Reihe A **219:**514–521.

Zakhariev, Z. A. 1971. *Plesiomonas shigelloides* isolated from sea water. Journal of Hygiene, Epidemiology, Microbiology and Immunology **15:**402–404.

Zakhariev, Z., Tyujekchiev, T., Valkov, V., Todeva, M. 1976. Food poisoning caused by parahaemolytic and NAG vibrios after eating meat products. Journal of Hygiene, Epidemiology, Microbiology and Immunology **20:**150–156.

Zen-Yoji, H., Sakai, S., Terayama, T., Kudoh, Y., Itoh, T., Benoki, M., Nagasaki, M. 1965. Epidemiology, enteropathogenicity and classification of *Vibrio parahaemolyticus.* Journal of Infectious Diseases **115:**436–444.

Zen-Yoji, H., Sakai, S., Kudoh, Y., Itoh, T., Terayama, T. 1970. Antigenic schema and epidemiology of *Vibrio parahaemolyticus.* Health Laboratory Science **7:**100–108.

Zen-Yoji, H., Hitokoto, H., Morozumi, S., Le Clair, R. A. 1971. Purification and characterization of a hemolysin produced by *Vibrio parahaemolyticus.* Journal of Infectious Diseases **123:**665–667.

Zen-Yoji, H., Kudoh, Y., Igarashi, H., Ohta, K., Fukai, K., 1974. Purification and identification of enteropathogenic toxin "a" produced by *Vibrio parahaemolyticus* and their biological and pathological activities, pp. 237–243. In: Fujino, T., Sakaguchi, G., Sakazaki, R., Takeda, Y. (eds.), International symposium on *Vibrio parahaemolyticus.* Tokyo: Saikon.

Zen-Yoji, H., Kudoh, Y., Igarashi, H., Ohta, K., Hoshino, T. 1975. An enteropathogenic toxin of *Vibrio parahaemolyticus,* pp. 263–272. In: Hasegawa, T. (ed.), Proceedings of the First Intersectional Congress of IAMS, vol. 4. Tokyo: Science Council of Japan.

Zide, N., Davis, J., Ehrenkranz, N. J. 1974. Fulminating *Vibrio parahaemolyticus* septicemia. A syndrome of erythema multiforme, hemolytic anemia, and hypotension. Archives of Internal Medicine **133:**479–481.

Zinnaka, Y., Carpenter, C. C. J., Jr. 1972. An enterotoxin produced by noncholera vibrios. Johns Hopkins Medical Journal **131:**403–411.

CHAPTER 104

The Marine Gram-Negative Eubacteria: Genera *Photobacterium, Beneckea, Alteromonas, Pseudomonas,* and *Alcaligenes*

PAUL BAUMANN and LINDA BAUMANN

The present volume is an eloquent testimony to the wealth of morphological, physiological, and ecological information concerning prokaryotes found in terrestrial and freshwater habitats. The rapid advances in genetic, biochemical, and immunological methods for the study of similarity have given us new insight into the evolutionary relationships of several bacterial groups (Baumann and Baumann, 1978; Cocks and Wilson, 1972; London, 1977; Mandel, 1969; Palleroni, 1975; Stanier, 1971). In contrast, we have little information of comparable depth and magnitude concerning marine prokaryotes. The reasons for this are multiple. Perhaps the major obstacle has been the fact that most microbiologists with an interest in the structural, physiological, and genetic attributes of bacteria have not had ready access to the expensive facilities which would allow the collection of samples from the open ocean. Conversely, microbiologists with access to such facilities have been primarily concerned not with individual organisms but with the activities of microbial populations and their effect on the gross transformation of matter. An additional problem has been the lack of recognition that there are bacterial species unique to the oceans which are distinguished from most terrestrial isolates by their specific ionic requirements. The recognition and acceptance of these attributes of marine bacteria has been hindered by (i) the fact that many offshore environments are heavily contaminated with terrestrial organisms which may predominate in some enrichment cultures, (ii) the confusion between salt requirement (a stable attribute of marine species) and salt tolerance (a property that is often present in or that may be acquired by both marine and terrestrial bacteria), and (iii) the widespread misconception that a terrestrial medium can be made suitable for the cultivation of all marine bacteria by the addition of 3% NaCl, the simplicity of the modification suggesting that salt requirement is itself a trivial attribute. An additional factor has been the comforting but deceptive ease with which many marine isolates can be assigned to existing genera, primarily on the basis of gross morphology and relation to oxygen and not necessarily on the basis of natural relationship.

The first comprehensive study of bacteria indigenous to the oceans was performed by Bernhard Fischer (1894) of the University of Kiel. His major conclusions have been confirmed and extended by numerous investigators who have shown that the majority of the heterotrophic bacterial flora of the open oceans consists of Gram-negative, straight or curved rods or spirals which are usually motile by means of flagella (Baumann, Baumann, and Mandel, 1971; Baumann et al., 1971, 1972; MacLeod, 1965, 1968; Pfister and Burkholder, 1965; Sieburth, 1979). Most of these organisms are eubacteria in that they have rigid cell walls. Fischer (1894) made the important observation that the highest plate counts were obtained when sea water or 3% NaCl was included in the nutrient medium. This finding was subsequently interpreted to be primarily an osmotic phenomenon since many marine bacteria were found to lyse in dilute media (Harvey, 1915; Pratt, 1974). Richter (1928), however, clearly demonstrated a specific Na$^+$ requirement for the growth of a marine luminous bacterium and, in addition, showed that complex media were not suitable for establishing this requirement since they are contaminated with inorganic ions (for reviews dealing with the early literature see Larsen, 1962; MacLeod, 1965, 1968). Since these results were overlooked and complex media continued to be used, considerable controversy persisted concerning the presence and stability of this attribute. MacLeod independently reestablished the early findings of Richter and developed synthetic media for the purpose of testing the presence of a Na$^+$ requirement (MacLeod, 1968). The extensive application of these methods has since established that all or most Gram-negative marine bacteria have a specific requirement for Na$^+$ (Baumann, Baumann, and Mandel, 1971; Baumann, Baumann, and Reichelt, 1973; Baumann et al., 1971, 1972; Hidaka and Sakai, 1968; MacLeod, 1965, 1968). Furthermore, the work of MacLeod (1968) and others (reviewed by Pratt, 1974) has indicated the stability of this requirement. Using 31 different marine isolates, including 7 species of facultative anaerobes (G+C contents in their DNAs of 39–48 mol%), 12 species of nonfermentative marine organisms (G+C range of 30–68 mol%), and 1 strain of a marine host-independent bdellovibrio (Taylor et al., 1974), Reichelt and Baumann (1974) studied the effect of NaCl concentration on growth rate and cell yield in media containing 50 mM Mg^{2+} and 10 mM Ca^{2+} (marine medium) and 2 mM Mg^{2+} and 0.55 mM Ca^{2+} (terrestrial medium). The optimum growth rates and cell yields in the marine medium ranged from 70 to 300 mM NaCl, while the optima in the terrestrial medium ranged from 100 to 460 mM. In many strains, the higher concentrations of Mg^{2+} and Ca^{2+} present in the marine medium reduced the amount of NaCl required for optimal growth rate and yield and decreased the generation time. Some strains failed to grow unless the medium contained the higher Mg^{2+} and Ca^{2+} concentrations, indicating that the addition of 3% NaCl to a terrestrial medium will not make it suitable for the cultivation of many common marine bacteria. The extent to which the Na^+ requirement may be partially reduced by other ions differs considerably among marine isolates (MacLeod, 1965, 1968; Pratt, 1974). In the case of *Alteromonas haloplanktis* (the organism used in the extensive studies of MacLeod and his collaborators), Li^+ has little or no sparing effect, while in the case of *Beneckea parahaemolytica* this ion is able to considerably reduce the concentration of Na^+ required (Morishita and Takada, 1976).

Detailed studies of the physiological basis of the Na^+ requirement have so far been restricted to a strain of the species *Alteromonas haloplanktis*. In this organism, Na^+ is essential for: (i) the function of all the examined permease systems which include those involved in the uptake of amino acids, tricarboxylic acid cycle intermediates, galactose, orthophosphate, and K^+ (Fein and MacLeod, 1975; Thompson and MacLeod, 1973, 1974; Thompson, Costerton, and MacLeod, 1970; Wong, 1969) and (ii) the maintenance of the integrity of the cell wall (Forsberg, Costerton, and MacLeod, 1970). Conclusion (i) has been extended to the marine species *Photobacterium fischeri* (Drapeau, Matula, and MacLeod, 1966; Wong, 1969) and to *Pseudomonas doudoroffii*, which requires 75 mM Na^+ for the optimal rate of uptake of D-fructose (L. Baumann, unpublished observations). Conclusion (ii) has also been extended to *Alteromonas espejiana*, the host of the lipid-containing, marine phage PM-2 (Diedrich and Cota-Robles, 1974; Espejo and Canelo, 1968). In a recent survey of the effect of Na^+ on the integrity of the cell wall, Laddaga and MacLeod (1977) found that in the absence of this ion a weakening of the outer membrane occurred in 14 out of 20 marine strains examined. The totality of these observations suggest that the requirement for Na^+ by marine bacteria is a complex, multigenic trait which would not be readily lost by mutation. In contrast to the marine bacteria and the extreme halophiles which require over 3 M Na^+ (Larsen, 1962), the growth of most Gram-negative terrestrial organisms does not appear to be Na^+ dependent. Where a requirement has been demonstrated, it has generally been found to be considerably lower than that observed in marine bacteria and may only be present under certain conditions of cultivation (Kodama and Tanaguchi, 1976; Reichelt and Baumann, 1974). An interesting exception is the rumen bacteria, which live in an environment having a relatively high concentration of this ion (Caldwell and Hudson, 1974; Reichelt and Baumann, 1974). The level of the Na^+ requirement and its stability imply that marine bacteria would not be able to colonize most terrestrial habitats. Conversely, there is considerable evidence that Gram-negative terrestrial bacteria do not survive in the marine environment (Jannasch, 1968; Mitchell and Morris, 1969; Moebus, 1972). These observations suggest an ecological separation of Gram-negative marine and terrestrial organisms as a consequence of specific adaptations to their respective habitats. In the case of marine bacteria, this attribute was probably acquired as a result of physiological adaptations to life in an environment having a relatively constant ionic composition. In this context, it is curious that many marine bacteria appear to grow better at 50–75% seawater concentration (Gundersen, 1976) and that the optimal concentration of Na^+ (70–300 mM) for the growth of a number of marine isolates is considerably lower than the Na^+ concentration in sea water (450–480 mM) (Reichelt and Baumann, 1974). No obvious correlation has been observed between the source of isolation of the strains and the amount of Na^+ necessary for optimal growth.

Another fundamental question has concerned the existence of bacterial species unique to the marine environment. Stanier (1941) succinctly formulated this problem and proposed an experimental solution, the essence of which was a comparison of the bacterial flora performing similar biological functions (e.g., mineralization of simple organic compounds) in both marine and terrestrial habitats. The application of this approach has been facilitated by the existence of relatively specific enrichment methods for the isolation of terrestrial pseudomonads as well as by the extensive phenotypic characterization of these organisms, which allows the ready identification of species (Palleroni, 1975; Palleroni and Doudoroff, 1972; Stanier, Palleroni, and Doudoroff, 1966). In utilizing enrichment methods for the isolation of marine bacteria, it is essential to use seawater samples obtained aseptically at locations where contamination by terrestrial organisms is minimal or absent. Using these precautions, it could be readily demonstrated that when a marine inoculum was used in enrichment cultures selective for certain species of terrestrial pseudomonads, the resulting flora consisted of facultative anaerobes and nonfermentative organisms that were different from terrestrial species (Baumann, Baumann, and Mandel, 1971; Baumann, Baumann, and Reichelt, 1973; Baumann et al.,

1971, 1972; Reichelt and Baumann, 1973b). These results established that the mineralization of simple organic compounds in the ocean is performed by a bacterial flora that is different from that performing these functions in terrestrial habitats. This conclusion as well as our previous speculations concerning the ecological separation of marine and terrestrial bacteria is based in part on an extensive phenotypic characterization of approximately 800 Gram-negative marine eubacteria, most of which were obtained from coastal and open ocean waters as well as from clinical samples and the surfaces and intestinal contents of marine fish and squid. We have deliberately avoided estuarine habitats since these environments are complex with respect to the diversity of habitats, salinity, availability of nutrients, and contamination by terrestrial organisms. It should be noted that the frequent designation of a species as "estuarine" has no real conceptual meaning since it is not yet known whether there are any species indigenous to estuaries or other coastal habitats. Since the open oceans contain relatively low concentrations of bacteria relative to those found directly offshore, the inability to detect a particular species in the open ocean may simply be due to the limitation of sample size.

About 50% of the oceans (by area) reach depths where the pressures range from 380 to 1,100 atm (ZoBell, 1963). Consequently, an important question has been the possible existence of bacteria that

Table 1. Some major properties and subdivisions of the Gram-negative, rod-shaped, marine eubacteria.

Genera	Fermentation	Mol% G+C	Polar flagella	Sheathed polar flagella	Peritrichous flagella	Aspartokinase group[a]	Presence of 1-phosphofructokinase in D-fructose-grown cells[b]	Constituent species and/or groups
Photobacterium	+	39–44	+	+ / −	−	3 isofunctional enzymes	+	P. fischeri, P. logei
								P. phosphoreum, P. leiognathi, P. angustum
Beneckea	+	45–54	+	+	+[c],−			B. harveyi, B. campbellii, B. parahaemolytica, B. alginolytica, B. natriegens, B. vulnifica B. splendida, B. pelagia, B. nigrapulchrituda, B. anguillarum, B. proteolytica, B. gazogenes, group E-3
Alteromonas	−	40–50	+	−	−	ND[d]	−	A. macleodii, A. haloplanktis, A. luteoviolaceus, A. rubra, A. citrea, A. espejiana, A. undina
						IV		A. communis, A. vaga
Pseudomonas	−	54–64	+	−	−	II		Groups B-1, B-2, I-2
						V		Groups H-1, I-1
						III	+	P. doudoroffii
							−	P. nautica[e], group G-1
								P. marina
Alcaligenes	−	52–68	−	−	+	I	+	A. pacificus, A. cupidus, A. venustus, A. aestus
Unassigned	−	28–33	+	−	−	ND	ND	Group H-2

[a] Data from Baumann and Baumann (1973a, 1974).
[b] Indicates absence of 1-phosphofructokinase or inability to grow on D-fructose. Data from Baumann, P., and Baumann (1975), Gee, Baumann, and Baumann (1975), and Sawyer, Baumann, and Baumann (1977).
[c] Some strains on solid media make unsheathed peritrichous flagella in addition to the sheathed polar flagellum.
[d] ND, not determined.
[e] Some strains of this species have 1–4 lateral flagella with a wavelength different from that of the polar flagellum.

are specifically adapted to life at high hydrostatic pressures and would be inhibited or killed by exposure to 1 atm. The resolution of this problem has recently been undertaken by Jannasch and his collaborators, who have constructed special samplers that allow the cultivation of marine microorganisms at the hydrostatic pressures at which they were sampled without introducing any steps necessitating decompression (Jannasch and Wirsen, 1977; Jannasch, Wirsen, and Taylor, 1976). Although detailed investigations of this problem have only recently begun, the results so far published strongly suggest that if microorganisms specifically adapted to high pressures do exist, their numbers constitute a relatively small proportion of the total metabolically active, heterotrophic, bacterial flora of the deep oceans.

Over 90% of the marine environment (by volume) has a temperature below 5°C (ZoBell, 1963). Consequently, it is also of considerable interest to know whether the psychrophilic isolates from the ocean differ from previously characterized mesophiles only in their relation to temperature or whether they actually constitute different species. An answer to this question should be readily obtained by application of the various methods used for the characterization of marine bacteria.

The marine, Gram-negative, heterotrophic eubacteria considered in this chapter are a large and diverse group consisting of straight or curved rods that are motile by means of flagella. These organisms can be subdivided into (i) the facultative anaerobes (genera *Beneckea* and *Photobacterium*) and (ii) a large and heterogeneous group of nonfermentative organisms (genera *Alcaligenes, Alteromonas,* and *Pseudomonas*). The facultative anaerobes have been given the informal designation "the marine enterobacteria" since they share a number of distinctive properties with the terrestrial enterobacteria, which include such ecologically diverse genera as *Escherichia, Enterobacter, Serratia, Erwinia,* and *Aeromonas* (Baumann and Baumann, 1977; Stanier, Adelberg, and Ingraham, 1976; Starr and Chatterjee, 1972). The nonfermentative organisms share a number of gross structural and physiological properties with the terrestrial species of *Pseudomonas* and *Alcaligenes* (Davis, Stanier, and Doudoroff, 1970; Stanier, Palleroni, and Doudoroff, 1966). Some of the properties of these genera are presented in Table 1; morphological representatives are given in Figs. 1–40. The present discussion will not include the marine agar decomposers, some of which are Gram-negative eubacteria (Humm, 1946; Stanier, 1941), or the marine spirilla which are considered in Chapter 52, this Handbook. For additional information on marine microorganisms and the marine environment, the reader is referred to Sieburth (1979).

Habitats

Since detailed taxonomic studies allowing identification of species of Gram-negative marine eubacteria are just beginning, generalizations concerning habitats and distribution of most species cannot be made. One notable exception is *Beneckea parahaemolytica,* the causative agent of a human gastroenteritis contracted from the consumption of contaminated seafoods. Epidemiological data have indicated that this organism has a worldwide distribution (Fujino et al., 1974; Miwatani and Takeda, 1976). Another exception is the marine luminous bacteria, some of which are able to enter into symbiotic association with marine animals. The latter topic is considered in Chapter 105, this Handbook. In Table 2 we have compiled the sources and geographical locations from which some of the strains have been isolated. Rather rigid criteria have been used; the isolates included have been identified by means of an extensive phenotypic characterization and/or by means of in vitro DNA/DNA homology studies. Application of these criteria is necessary since the identification of strains from marine sources has, in the past, been problematic (see "Identification" section, this chapter). It is of considerable interest that strains from geographically diverse locations, which have been assigned to the same species on the basis of phenotypic similarities, were found to have in vitro DNA/DNA homologies ≥81% (Table 2). Most of the remaining species of marine enterobacteria (Table 1, Fig. 44) and the nonfermentative marine eubacteria (Table 1) were isolated from the open ocean 10–35 miles off the coast of Oahu, Hawaii, at depths ranging from surface waters to 1,300 m, or off the coast of Oahu at sites that had little or no obvious terrestrial contamination. An exception to this is *Alteromonas haloplanktis,* 10 strains of which were isolated from coastal North American waters and 1 strain from a bottom sample from the Indian Ocean obtained at 3,000 m. In addition, 1 strain of *Pseudomonas marina* was obtained off Massachusetts and strains of *Pseudomonas nautica* and group I-2 were obtained from coastal waters of California.

Recently, a number of studies have appeared dealing with the ecology of luminous bacteria in sea water off the coast of San Diego, California (Ruby and Nealson, 1978), the eastern Mediterranean, and the Gulf of Elat (Shilo and Yetinson, 1979; Yetinson and Shilo, 1979), as well as from two stations in the open ocean in the North Atlantic and over the Puerto Rico Trench (Ruby, Greenberg, and Hastings, 1980). The results of these studies show a species-specific pattern influenced by season, depth, geographical locale, and salinity. A major importance of these investigations has been the fact that they represent ecological studies in

Table 2. Sources and geographical distribution of *Photobacterium* and some species of *Beneckea*

Species	Unusual properties	Sources of isolation	Geographical origin of samples	Geographical origin of strains studied by in vitro DNA/DNA homology[a]	References
Beneckea parahaemolytica	Causative agent of a gastroenteritis contacted by the consumption of contaminated seafoods; may cause localized tissue infections	Stool samples of patients suffering from gastroenteritis, localized tissue infections; clams, shrimp, crab, shellfish; coastal sea water	Coastal regions having tropical and temperate climates	Japan, east and west coasts of North America, Gulf of Mexico	Anderson and Ordal (1972); Baumann, Baumann, and Mandel (1971); Baumann, Baumann, and Reichelt (1973); Casellas, Caria, and Gerghi (1977); Fujino et al. (1974); Reichelt, Baumann, and Baumann (1976); Staley and Colwell (1973)
Beneckea alginolytica	Swarms on solid complex medium; may cause localized tissue infections	Sea water; surfaces of fish; wounds	Coastal regions having tropical and temperate climates	Japan, east and west coasts of North America, Gulf of Mexico, Hawaii	Anderson and Ordal (1972); Baumann, Baumann, and Mandel (1971); Baumann, Baumann, and Reichelt (1973); Casellas, Caria, and Gerghi (1977); Golten and Scheffers (1975); Pien, Lee, and Higa, (1977); Reichelt, Baumann, and Baumann (1976); Rubin and Tilton (1975); Ryan (1976); Staley and Colwell (1973); Stephen et al. (1978)
Beneckea vulnifica	May cause localized infections as well as a fatal septicemia	Human blood, localized infections	East and west coasts of North America, Gulf of Mexico, Japan, Hawaii, Belgium	East and west coasts of North America, Gulf of Mexico	Baumann, Baumann, and Reichelt (1973); Blake et al. (1979); Clark and Steigerwalt (1977); Hollis et al. (1976); Matsuo et al. (1978); Mertens et al. (1979); Reichelt, Baumann, and Baumann (1976)
Beneckea anguillarum	Causative agent of "red disease" of eels; diseases of salmon, sea trout, and cod	Diseased fish and eels	Japan, west coast of North America, Scotland, Denmark	Japan, west coast of North America, Scotland, Denmark	Anderson and Ordal (1972); Baumann, Bang, and Baumann (1978); Harrell et al. (1976); Schiewe, Crosa, and Ordal (1977); Sindermann (1966, 1970); Smith (1961)
Beneckea natriegens	Nutritional versatility	Coastal sea water	Hawaii, Gulf of Mexico	Hawaii, Gulf of Mexico	Baumann, Baumann, and Mandel (1971); Payne, Eagon, and Williams (1961); Reichelt, Baumann, and Baumann (1976)

Organism	Characteristics	Habitat	Distribution	Distribution	References
Beneckea harveyi	Most strains are able to luminesce	Surfaces of fish and squid; coastal sea water, open ocean	East and west coasts of North America, Puerto Rico, Portugal, Israel, Salton Sea (California), Hawaii, New Guinea	East and west coasts of North America, Puerto Rico, Israel	Reichelt and Baumann (1973b); Reichelt, Baumann, and Baumann (1976); Ruby and Nealson (1978); Ruby, Greenberg, and Hastings (1980); Yetinson and Shilo (1979). See Chapter 105, this Handbook
Beneckea splendida biotype I	Luminescence	Coastal sea water	East coast of North America, Denmark	East coast of North America, Denmark	Reichelt, Baumann, and Baumann (1976). See Chapter 105, this Handbook
Photobacterium phosphoreum	Luminescence; capacity to enter into a symbiotic association	Luminous organs; surfaces of fish, squid, octopus; intestinal contents of fish; coastal sea water; open ocean	Japan, New Guinea. Hawaii, west coast of North America, Spain, Denmark, New Zealand	Japan, Hawaii	Herring (1975); Reichelt and Baumann (1973b); Reichelt, Baumann, and Baumann (1976); Ruby and Morin (1978); Ruby and Nealson (1978); Ruby, Greenberg, and Hastings (1980); Singleton and Skerman (1973). See Chapter 105, this Handbook
Photobacterium leiognathi	Luminescence; capacity to enter into a symbiotic association	Luminous organs; surfaces of fish, squid, octopus; intestinal contents of fish; coastal sea water; open ocean	Japan, New Guinea, Siam, Indonesia, Philippines, Australia, Hawaii, Gulf of Mexico, Israel, Indian Ocean	Siam, Indonesia, Hawaii, Gulf of Mexico, Indian Ocean	Bassot (1975); Boisvert, Chatelain, and Bassot (1967); Reichelt and Baumann (1973b, 1975); Reichelt, Baumann, and Baumann (1976); Reichelt, Nealson, and Hastings (1977); Ruby, Greenberg, and Hastings (1980); Yetinson and Shilo (1979). See Chapter 105, this Handbook
Photobacterium fischeri	Luminescence; capacity to enter into a symbiotic association	Luminous organs; surfaces of fish and squid; coastal sea water, open ocean	East and west coasts of North America, Denmark, Salton Sea (California), Hawaii, Israel, Australia	East and west coasts of North America, Hawaii	Fitzgerald (1977); Reichelt and Baumann (1973b); Reichelt, Baumann, and Baumann (1976); Ruby and Nealson (1978); Ruby, Greenberg, and Hastings (1980); Yetinson and Shilo (1979). See Chapter 105, this Handbook
Photobacterium logei	Luminescence	Exoskeleton lesions of tanner crabs; intestinal contents of fish; scallops; marine sediments	West coast of North America, Arctic sediments, New Zealand	DNA homology not determined	Bang, Baumann, and Nealson (1978); Baross, Tester, and Morita (1978); Singleton and Skerman (1973)

[a] The DNA/DNA homology of these strains was 81%.

which marine bacteria could be identified to the species level. Shilo and Yetinson (1979) were able to observe a correlation between the ecology of the luminous organisms and their physiological attributes.

Methods of Isolation

Most enrichment cultures are incubated at room temperature (18–22°C), while cultures on Petri plates are incubated at 25°C. These temperatures are primarily a matter of convenience and should be modified (usually reduced, as in the case of the luminous bacteria) to suit the particular needs of the investigator.

Media

Artificial Sea Water (ASW) (MacLeod, 1968)

NaCl	400 mM
$MgSO_4 \cdot 7H_2O$	100 mM
KCl	20 mM
$CaCl_2 \cdot 2H_2O$	20 mM

Dissolve the salts separately and combine.

Basal Medium (BM)

Tris-HCl (pH 7.5)	50 or 100 mM
NH_4Cl	19 mM
$K_2HPO_4 \cdot 3H_2O$	0.33 mM
$FeSO_4 \cdot 7H_2O$	0.1 mM
Half-strength ASW (1/2 ASW)	

The sole or principal carbon and energy source is usually provided at a concentration of 0.1–0.2% (wt/vol for solids and vol/vol for liquids). Suggested methods for the sterilization of a variety of organic compounds (autoclaving or filtration) have been given by Palleroni and Doudoroff (1972). For the cultivation of amino acid–requiring organisms, BM is supplemented with 1 mg/liter each of L-alanine, L-arginine, L-asparagine, L-aspartate, L-cysteine, L-glutamate, L-glutamine, glycine, L-histidine, L-isoleucine, L-leucine, L-lysine, L-methionine, L-phenylalanine, L-proline, L-serine, L-threonine, L-tryptophan, L-tyrosine, and L-valine. The amino acids are filter-sterilized and added to the autoclaved medium.

Basal Medium Agar (BMA)

Prepare equal volumes of double-strength BM and double-strength agar in distilled water (40 g/liter refined agar), sterilize by autoclaving, and combine before pouring plates. The carbon and energy source (unless labile or volatile) is added to the double-strength BM prior to autoclaving. Filter-sterilized labile or volatile compounds are added to BMA which has been cooled to about 41°C prior to the pouring of plates. Some volatile substrates (e.g., geraniol, n-hexadecane, naphthalene, phenol) are not added to the medium but are placed on sterile filter paper in the lids of the inverted Petri plates, then incubated in air-tight containers.

Yeast Extract Broth (YEB)

BM containing 5 g/liter yeast extract instead of the single carbon and energy source.

Yeast Extract Agar (YEA)

YEB containing 20 g/liter agar.

Luminous Medium (LM)

BM containing 50 mM Tris-HCl (pH 7.5), 0.3% glycerol, 5 g/liter yeast extract, 5 g/liter tryptone, 1 g/liter $CaCO_3$, 20 g/liter agar.

Marine Agar (MA)

Formula 2216 of ZoBell (1941) available commercially from Difco.

It should be stressed that the ASW base of MacLeod (1968) may not be suitable for the cultivation of some marine organisms that may require additional mineral components. A number of different artificial seawater formulations have been compiled by Kinne (1976). In addition, although many marine bacteria are not adversely affected by Tris buffer, the compound may prove toxic for some strains.

Isolation from Sea Water

It is essential that sea water be collected aseptically using sterile samplers. In general, this presents little difficulty when samples of surface waters are collected. A convenient way of obtaining sea water from different depths is by use of the Niskin butterfly sampler (General Oceanics, Miami, Florida, USA). The collected samples can be used either for enrichment cultures or for direct isolation.

Enrichment Cultures

In the case of aerobic enrichments, 500 ml of sea water are added to a sterile 2-liter Erlenmeyer flask containing 25 ml of 1 M Tris-HCl (pH 7.5), 0.5 g NH_4Cl, 38 mg $K_2HPO_4 \cdot 3H_2O$, 14 mg $FeSO_4 \cdot 7H_2O$, and 0.5–1.0 g or ml of the carbon and energy source. (Acidic or basic carbon sources may necessitate readjustment of the pH to 7.5.) Depending on the source of the sample and the organic compound used to support growth, both the sample volume and the volume of the concentrated solution may be increased or decreased. The flasks are observed for signs of growth for up to 10 days, at which time they are streaked on BMA containing 0.1% of the same carbon and energy source as was used in the enrichment.

Enrichments for denitrifiers are performed in sterile 500-ml reagent bottles containing the concentrated solution described above as well as 1.5 g NaNO$_3$ and 1 g or 1 ml of a nonfermentable carbon and energy source. The bottles are filled to the top with the sea water sample, stoppered to exclude air bubbles, and incubated. When signs of growth and gas production are evident (within 10 days of incubation), the culture is streaked on BMA containing 0.1% of the carbon and energy source used in the enrichment. In some cases it is advisable to use 1-liter reagent bottles in order to increase the volume of the sea water sample.

Relatively specific enrichments have been found for a number of marine species. Perhaps the best one is an enrichment for *Pseudomonas nautica* which frequently predominates in enrichment cultures for denitrifiers that contain 0.2% Na-butyrate as the carbon and energy source. Aerobic enrichments with 0.2% Na-butyrate select for an unidentified spirillum. The use of aerobic enrichments containing 0.2% *m*-hydroxybenzoate usually results in the predominance of *Alteromonas communis* or *A. vaga*. Enrichments containing 0.2% chitin, incubated aerobically for 5–10 days, often contain a blue-black sediment associated with the chitin particles. Such enrichments, when streaked on BMA containing 0.2% lactose, usually yield *Beneckea nigrapulchrituda*, a chitin-decomposer which forms colonies containing crystals of blue-black pigment (Figs. 41 and 42) (Baumann et al., 1971). *Beneckea alginolytica* can be readily obtained from aerobic enrichments containing 0.5% yeast extract and 0.5% tryptone (instead of the single organic carbon and energy source). When streaked on YEA or MA, this organism swarms in the manner of *Proteus*. Single colonies can be obtained by streaking on either a complex medium containing 4% agar or on minimal medium such as BMA with 0.2% glycerol.

Direct Isolation

Samples of sea water (5–300 ml) are filtered through 0.22- or 0.45-μm, 47-mm-diameter nitrocellulose filters which are placed on Petri plates containing either a complex medium (YEA or MA) or BMA with 0.1% of the carbon and energy source. The size of the filtered sample will depend on the source of the sea water and the composition of the medium used for direct isolation. After an incubation of 2–10 days at 25°C, colonies are picked and restreaked on homologous media. *Beneckea nigrapulchrituda* (Figs. 41 and 42) and *Alteromonas macleodii* can be frequently obtained by direct isolation on plates containing BMA with 0.2% lactose.

Many agar decomposers produce a broad but barely perceptible indentation in the agar surrounding the colony. Consequently, it is advisable to streak a pure culture on BMA without an added carbon and energy source and containing appropriate supplements for growth factor–requiring organisms. Growth on this medium indicates that the isolate is able to utilize agar. We have as yet failed to obtain an organism from the marine environment that is able to utilize Tris as a carbon and energy source, although some organisms appear to utilize this compound as a poor nitrogen source (I. P. Crawford, K. H. Nealson, personal communication).

Isolation from Surfaces and Intestinal Contents of Fish

Sterile cotton-tipped applicator sticks are used to swab the gills, mouth, and rectal region as well as other surfaces of the fish. A plate of complex medium (such as YEA or MA) is inoculated by making a line of 4- to 5-cm length at one edge of the Petri dish. The inoculum is spread by means of streaking with a loop. A 2- to 3-cm portion of the fish intestine is dissected out and placed onto a sterile Petri plate. Gentle pressure with sterile forceps or an applicator stick generally forces out some of the intestinal contents, which are then streaked on complex medium. In both cases, the plates are incubated 1–4 days and observed for colonies which are purified by streaking on homologous medium. Some marine strains are capable of extensive swarming on complex media which makes it impossible to isolate single colonies. Swarming can be prevented by raising the concentration of agar to 4%.

Isolation of Luminous Bacteria

Bacteria able to luminesce are common in the marine environment. The detection of luminescence by a simple visual examination poses a considerable number of problems since the intensity of the emitted light varies greatly with different isolates and is furthermore affected by the medium used for cultivation and the age of the cells. A satisfactory medium used for the observation of luminescence is LM (a medium based on an unpublished formulation of M. Doudoroff). Strains should be streaked on this medium and incubated at 15° and 25°C with periodic examinations at 12–36 h after streaking. In many strains luminescence is relatively dim and short-lived, so frequent observation (every 3 h) is suggested. It is important to have plates with isolated colonies, since in the case of many dim strains only the isolated colonies luminesce. The examination of the LM plates for luminous isolates should be performed in complete darkness and the eyes should be dark-adapted for at least 10 min. The picking of luminous bacteria from marine animals or from plates having both luminous and nonluminous strains is greatly facilitated by the use of sterile toothpicks and a low intensity light bulb (5–10 W) connected to a rheostat. When an area of luminescence is observed, a sterile toothpick should be positioned roughly over the site. The current is switched on and the intensity of the light is gradually increased until the specimen is just barely visible. By fixing one's eyes on the luminous spot and gradually increasing the intensity of the light it becomes apparent which area contains the luminous organisms. The toothpick is quickly touched to this site, an LM plate is inoculated by making a line 4–5 cm long at the edge of the Petri plate, and the light is quickly turned off. By using this procedure, the investigator's eyes do not have to be repeatedly dark-adapted and, more importantly, it is possible to actually see the area from which the inoculum is picked. A large number of plates can be inoculated with toothpicks and the inocula subsequently spread by streaking with a loop.

In the following section we will deal with the direct isolation of luminous strains from sea water and the surfaces and intestinal contents of marine animals, as well as isolation from enrichments. Methods for the isolation of luminous strains from the light organs of marine animals are described in Chapter 105, this Handbook.

Isolation from Sea Water

In some coastal waters, the concentration of luminous bacteria may be sufficient to allow detection in a 0.1- to 0.3-ml sample spread onto an LM plate. With larger volumes of sea water, aliquots of up to 300 ml may be filtered through 0.22- or 0.45-μm, 47-mm-diameter nitrocellulose filters which are subsequently placed onto Petri plates containing LM. Since crowded conditions tend to inhibit luminescence and since nonluminous bacteria greatly outnumber the luminous isolates, it is important that the filter contain a relatively sparse bacterial population. It is not recommended that the soft agar overlay method be used for enumeration of luminous bacteria, since brief exposure to 41°C (the temperature of the molten agar used for the overlay) may kill some strains of *Photobacterium phosphoreum* and *P. logei*.

Isolation from Surfaces and Intestinal Contents

Fresh squid and octopus, which have been kept on ice in fish markets, often have luminous spots when examined in the dark; luminous spots are rare on fresh fish. The isolation of luminous organisms from such specimens, from the surfaces of marine animals, as well as from the intestinal contents of fish, is performed as described in the preceding sections of this chapter, "Isolation from Surfaces and Intestinal Contents of Fish" and "Isolation of Luminous Bacteria", with the substitution of LM for MA or YEB. In general, either the intestinal contents of fish have a relatively large population of luminous bacteria or these organisms appear to be absent.

Isolation by Enrichment (based on suggestions of M. Doudoroff)

In some cases, visible regions of luminescence can be obtained on fresh fish, squid, or octopus by half-submerging the specimen in a shallow layer of ASW and incubating 10–18 h at 12–15°C. The luminous sites are touched with sterile toothpicks and inoculated onto LM plates. This method is relatively specific for *P. phosphoreum*.

Preservation

Most of the species considered in this chapter can be maintained on MA slants at 18°C and transferred monthly. After each transfer, the cultures are allowed to grow at 25°C for 1–2 days and again placed at 18°C. Most of the strains do not survive well on MA slants kept at 4°C. Luminous species, with one exception, are maintained on LM slants as described above. *P. phosphoreum* is best preserved on LM kept at 4°C; after each monthly transfer, the strains are incubated at 18°C for 1 day prior to storage at 4°C. Strains of groups H-2 and I-2 grow better on BMA containing 0.2% Na-lactate than on MA or YEB and are maintained on the former medium. Some isolates of *A. macleodii, A. haloplanktis,* and *B. campbellii* tend to acquire growth factor requirements after prolonged cultivation on MA. Consequently, it may be advisable to maintain these strains on BMA containing 0.2% D-glucose or, in the case of *B. campbellii,* 0.2% glycerol.

Most of the strains studied have been lyophilized and kept at 4°C. With a few exceptions, viable cells could be recovered after 3–5 years of storage. For the preparation of lyophils, the growth from a fresh slant is suspended in about 0.5 ml of a sterile solution consisting of one-quarter-strength ASW, 5 g/liter yeast extract, and 5 g/liter peptone (adjusted to pH 7.5) and transferred into a lyophil tube which is subsequently dipped into a mixture of dry ice and acetone and placed under vacuum for 10–12 h. The lyophils are reconstituted by suspending the powder in about 0.5 ml YEB and streaking a portion of the liquid onto YEA while the remainder is inoculated into a tube containing 4 ml of the same medium. Growth is generally observed after 1–2 days incubation at 25°C; for *P. phosphoreum* a lower temperature (15–18°C) should be used.

Identification

Comprehensive studies on the taxonomy of the marine eubacteria are relatively recent and few in number. In the past there has been little practical reason for the characterization of marine species, most investigators being content with the assignment of strains to preexisting genera on the basis of gross morphology and relation to oxygen. Traditionally, the major taxonomic interest has centered on the luminous bacteria (Hendrie, Hodgkiss, and Shewan, 1970) as well as on strains pathogenic for marine animals (Sindermann, 1966, 1970). Perhaps the major impetus for the characterization of common marine bacteria has come from the realization that a facultatively anaerobic marine species *(Beneckea parahaemolytica)* is the causative agent of a gastroenteritis contracted from the consumption of contaminated seafood (Fujino et al., 1974; Miwatani and Takeda, 1976). This organism is of particular importance in the Orient, where raw fish is a common part of the diet. Once it was realized that *B. parahaemolytica* was a marine organism and consequently required NaCl for growth, its identification from stool samples of patients suffering from gastroenteritis was a relatively straightforward task, since a few phenotypic properties were adequate for the distinction of this pathogen from common fecal flora as well as from other organisms that were capable of causing gastroenteritis. A major simplification of the identification was rendered by the diseased host who selected for the pathogen. The identification of *B. parahaemolytica* directly from marine samples is considerably more complex since there has been no preselection of this species and the investigator must differentiate the isolate from other common marine bacteria (Baumann, Baumann, and Reichelt, 1973). This obvious point has been overlooked in the application of a widely used scheme for the identification of *B. parahaemolytica,* which was primarily based on experience with clinical samples (Sakazaki, Iwanami, and Fukumi, 1963). The implication of this scheme is that facultative anaerobes of marine origin can be subdivided into three species, *B. parahaemolytica, B. alginolytica,* and *B. anguillarum.* The latter became a catch-all for those strains that could not be assigned to the first two species. The obvious inadequacy of this diagnostic scheme was clearly indicated by the *in vitro* DNA/DNA homology studies of Anderson and Ordal (1972), which showed that (i) only 3 of the 12 strains isolated from marine sources and identified as *B. parahaemolytica* had significant homology

with authentic strains of this species and (ii) on the basis of DNA homology, a legitimate species which included the type strain of *B. anguillarum* could be readily recognized. The identification of *B. parahaemolytica* from samples of the marine environment has been placed on a considerably firmer footing by the extensive phenotypic and genotypic characterization of facultative anaerobes of marine origin (Baumann and Baumann, 1977; Reichelt, Baumann, and Baumann, 1976). The results of these studies have shown which traits are useful in distinguishing *B. parahaemolytica* from other common marine bacteria and have indicated that a number of studies dealing with the distribution of this pathogen were deficient in that they failed to differentiate *B. parahaemolytica* from other common marine isolates (Baumann, Baumann, and Reichelt, 1973). An additional impetus for the taxonomic studies of marine bacteria has been the extensive biochemical investigations dealing with luminescence of marine isolates (reviewed by Hastings and Nealson, 1977). Although the mechanisms of luminescence appear similar, the differences noted in different strains has made it of considerable interest to perform a phenotypic characterization of luminous isolates and establish their relationship to one another as well as to nonluminous species. Furthermore, ecological studies dealing with luminous bacteria as well as studies of their symbiotic associations (Table 2) have required the delineation of species and a relatively simple means for their identification (see Chapter 105, this Handbook).

Methodology

Most of the methods used for the identification of marine species (Baumann, Baumann, and Mandel, 1971; Baumann et al., 1972; Reichelt and Baumann, 1973a,b) have been derived from those of Stanier, Palleroni, and Doudoroff (1966) as well as Palleroni and Doudoroff (1972). The majority of the tests involving growth of the isolates have been performed at an incubation temperature of 25°C. It should be stressed that for the characterization of some marine bacteria (e.g., psychrophiles that may include luminous isolates), the temperature should be considerably reduced. Such a reduction may affect the nutritional spectrum in that fewer or more organic compounds may be utilized at the lower temperature (Ingraham, 1962).

GROWTH ON DEFINED MEDIA AND Na⁺ REQUIREMENT. Before attempting the identification of a marine isolate, it is advisable to test for the presence of a Na⁺ requirement as well as for the ability to grow without added organic growth factors. Both properties can be tested by comparing the growth of the organism in two different media, one consisting of BM with 0.1% glycerol, 0.1% K-acetate, and

0.1% K$_2$-succinate (10 ml in a 50-ml Erlenmeyer flask) and a second, similar medium in which all of the Na$^+$ has been replaced by equimolar amounts of K$^+$ (MacLeod, 1968). The cultures are incubated on a shaker for 2–3 days, at which time the optical density is measured at 540 nm. If the strain is a marine isolate having no organic growth factor requirements, the optical density in the Na$^+$-containing medium is generally over 0.2, while the optical density in the K$^+$-containing medium is usually below 0.01. Acetate, succinate (K-salts), and glycerol are used as sources of carbon and energy since all the organisms so far tested utilize one or more of these compounds. If growth is not detected in either medium, alternate carbon sources (e.g., pyruvate, D-glucose) should be tested or both media should be supplemented with a mixture of amino acids and vitamins (see "Methods of Isolation", this chapter; MacLeod, 1968) to determine whether the organism has a growth factor requirement. A more detailed determination of the level of Na$^+$ required for optimum growth rate and cell yield can be performed as described by Reichelt and Baumann (1974).

Three important tests used in the placement of a strain into one of several major categories will be described in some detail. These three tests are the ability to ferment D-glucose, the ability to accumulate intracellular poly-β-hydroxybutyrate (PHB), and flagellation.

FERMENTATION. The ability to ferment D-glucose is tested by the use of two media. The first (F-1) contains YEB, 100 mM Tris-HCl (pH 7.5), 1% D-glucose, and 2 g/liter agar. After boiling to melt the agar, 10-ml aliquots of the medium are dispensed into test tubes, autoclaved, and inoculated by means of a stab. About 5 ml of 2% agar (autoclaved and cooled to 41°C) are carefully layered over the medium to make an agar plug. The test tube is observed for turbidity and gas production for a period of 4 days. Facultative anaerobes, which may or may not produce gas, will exhibit good turbidity; strict aerobes will not grow under these conditions.

An additional check is performed in the second medium (F-2), which is similar to F-1 but differs in the inclusion of 1 g/liter Na-thioglycolate and the omission of the agar plug. The medium is inoculated by means of a stab and examined at 24 and 48 h. Facultative anaerobes grow throughout the medium while strict aerobes grow only at the surface. At 48 h, the pH of the medium is determined; the facultative anaerobes usually reduce the pH to a value below 5.8, whereas with the strict aerobes the pH remains at about 7.5. The combined use of both F-1 and F-2 media gives results which are considerably less ambiguous than those obtained with various media containing pH indicators. An additional refinement, which may be of use in some cases, is the inclusion of controls consisting of both F-1 and F-2

media lacking D-glucose. Occasional isolates may be found which cannot utilize D-glucose but grow aerobically at the expense of other carbohydrates. Such strains should be tested for their ability to ferment these sugars using the methods described.

ACCUMULATION OF PHB. The ability to accumulate PHB as an intracellular reserve product is present in many species of marine eubacteria. This trait is of considerable taxonomic importance, especially in the present scheme for the differentiation of nonfermentative marine eubacteria. In general, the simplest way to observe the accumulation of PHB is to grow the cells in BM which is limited for nitrogen [0.02% (NH$_4$)$_2$SO$_4$] and contains excess DL-β-hydroxybutyrate (0.4%). The culture is examined daily by means of a phase-contrast microscope, for a period of 4 days, for the presence of bright intracellular granules characteristic of PHB (Fig. 10, 13). These granules can also be stained with Sudan Black as described by Burdon (1946). In some cases, difficulties are encountered since the granules may be small. In addition, some marine bacteria, especially in old cultures, have involution forms containing inclusions that might be confused with PHB. A relatively easy chemical identification of PHB may be performed by alkaline hypochlorite digestion of the cells followed by the solubilization of PHB in chloroform and precipitation with acetone (Williamson and Wilkinson, 1958). A quantitative estimation of the extracted PHB can subsequently be made by the spectrophotometric method described by Slepecky and Law (1960).

Although most strains which accumulate PHB can utilize the monomer (β-hydroxybutyrate), notable exceptions are the species *Photobacterium phosphoreum*, *P. leiognathi*, and *P. angustum* (Reichelt and Baumann, 1973b; Reichelt, Baumann, and Baumann, 1976). These organisms, when grown in BM containing 0.2% D-glucose, accumulate PHB in early stationary phase of growth to such an extent that, often, the entire cell appears bright under the phase-contrast microscope. In the case of the three species of *Photobacterium*, the accumulation of PHB is not an important diagnostic trait. It is clear from these results, however, that in the characterization of any new species, the ability to accumulate PHB must be tested using various carbon sources and the observations should be conducted at different stages of the growth cycle.

FLAGELLATION. The mode of flagellar insertion is an important diagnostic trait for the differentiation of the nonfermentative eubacteria; the peritrichously flagellated strains are assigned to the genus *Alcaligenes*, while the polarly flagellated strains are assigned to *Pseudomonas* or *Alteromonas* (Table 1). Flagellation is also important in the generic assignments of marine enterobacteria. A distinctive property of some species of *Beneckea* is their shift in flagellation with different conditions of cultivation—cells which are polarly flagellated in liquid medium become peritrichously flagellated when transferred to a solid medium (Figs. 1–3, 8, 9). A relatively simple and generally reliable method for the staining of flagella is that of Leifson (1960). Only the slight modifications and precautions necessary for the application of this method to marine bacteria will be considered.

For the staining of flagella of strains grown in liquid medium, a loopful of cells from a fresh slant is inoculated into 5 ml of YEB and incubated overnight on a shaker. About 0.2 ml of the fully grown culture is transferred into 5 ml of YEB and when a light turbidity is detected the culture is poured into a centrifuge tube containing 0.08 ml of neutralized (pH 7.5) 37% formaldehyde (formalin). About 15 ml of distilled water are slowly added to the centrifuge tube. After a light centrifugation (about 1 min at 3,000 × g), the supernatant is decanted and the liquid drained; the walls of the inverted centrifuge tube are rinsed briefly with distilled water. About 0.3 ml of distilled water is added to the pellet, which is allowed to resuspend without agitation. After washing the cells two additional times, the final suspension is carefully poured into a clean test tube, and distilled water is slowly added to give a lightly turbid suspension (optical density of about 0.1 at 540 nm). The subsequent staining operations are performed as described by Leifson (1960). For the staining of flagella of strains grown on solid medium, a loopful of cells from a fresh slant is spread onto a YEA plate making a patch of about 3–4 cm². After an overnight incubation, 1 ml of a solution containing 50 mM Tris-HCl (pH 7.5), 1/2 ASW, and 0.57% neutralized formaldehyde is pipetted onto the cells, which are gently suspended by rotating the plate very slowly, in one direction, while holding a bent glass rod over the patch of cells. The resulting suspension is decanted against a glass rod into a centrifuge tube (do not pipette), and washed as described for cells grown in liquid medium. These precautions are especially necessary in the case of species of *Beneckea* with peritrichous flagella which are readily removed from the cells by shearing forces (Baumann, Baumann, and Mandel, 1971). In some cases, a reduction or an increase in formaldehyde concentration (0.25- to 10-fold variation of the amounts described above) may give better results for strains harvested from both solid and liquid media. It is best to use the lowest possible concentration, since high concentrations of formaldehyde may result in straight flagella lacking their characteristic wave form.

Although the Leifson method gives unambiguous results in most cases, some problems have been encountered with *Pseudomonas marina*, groups B-1, B-2, and I-2 (Baumann et al., 1972). When stained by the Leifson procedure, these organisms appeared to have single polar flagella, polar tufts, single lateral flagella, lateral tufts, and, rarely, degenerately peritrichous or peritrichous flagellation (Figs. 29, 39, and 40). An examination by means of the electron microscope indicated that whenever the flagellar insertions were visible, they were polar (Figs. 36–38). It was also observed that the flagella that occurred in polar tufts had a tendency to curl and bend backwards toward the cell, giving rise to arrangements that could explain some of the varied results observed with the Leifson method (Figs. 38 and 39). In many cases the polar flagella were intertwined, forming a single bundle, which could give the appearance of a single polar or a single lateral flagellum in a Leifson preparation (Figs. 29 and 36). Unfortunately, in the case of these organisms the insertion of the flagella could only be established after a reasonably extensive electron microscopic examination (for additional examples of Leifson preparations and electron micrographs illustrating this problem see Baumann et al., 1972). Therefore, considerable caution should be exercised in interpreting the flagellation of cells that appear to give considerable variability as to the distribution of the flagella in Leifson preparations. An additional complication may arise out of the fact that some strains have tubular projections (Fig. 2) which are evaginations of the outer membrane of the cell wall (Allen and Baumann, 1971; Baumann et al., 1972). However, these evaginations are usually readily distinguishable from flagella since they are straight and lack the wave form characteristic of flagella.

MORPHOLOGY. When examined during exponential phase of growth in YEB or other relatively simple media, the cells of Gram-negative marine eubacteria generally appear to be regular, straight or curved rods (Figs. 1–35). Many species give rise to involution forms in early stationary phase, a tendency particularly noticeable in some strains of *Photobacterium*. For this reason, it is important to examine the morphology of marine bacteria in exponential phase of growth in liquid media. Many of these involution forms are spherical and have been called "spheroplasts" or "round bodies" (Felter, Colwell, and Chapman, 1969; Levin and Vaughn, 1968). These designations are unfortunate since they imply a regularity in the frequently bizarre shapes observed in old cultures (Felter, Colwell, and Chapman, 1969; Kennedy, Colwell, and Chapman, 1970). In *Desulfovibrio aestuarii* and *Nitrospina gracilis*, the formation of "spheroplasts" and "round bodies" was accompanied by a loss of viability (Levin and Vaughn, 1968; Watson and Waterbury, 1971). Baker and Park (1975) have shown that the formation of these structures in stationary phase cultures of a *Vibrio* sp. correlates with a decrease in the amount of peptidoglycan and loss of viability. The

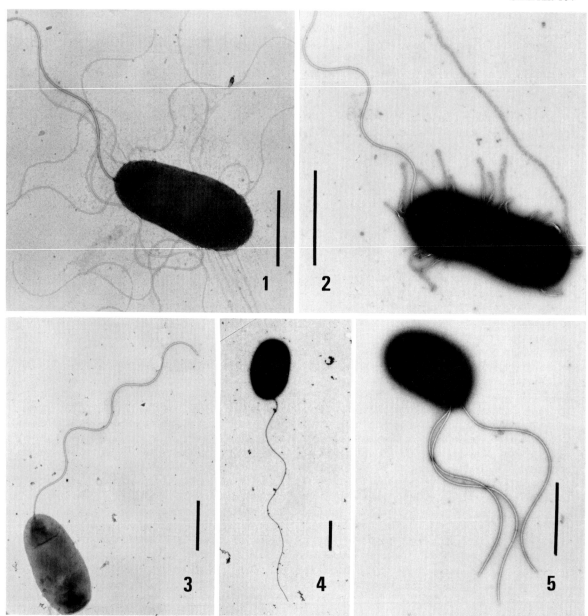

Figs. 1–5. Electron micrographs of representative morphological types of the marine enterobacteria. Marker indicates 1 μm. (1) *Beneckea parahaemolytica* grown on solid medium. (2) *B. campbellii* grown in liquid medium. (3) *B. parahaemolytica* grown in liquid medium. (4) *Photobacterium phosphoreum*. (5) *P. fischeri*. [Figs. 1–3 from Allen and Baumann (1971), with permission. Figs. 4 and 5 courtesy of R. D. Allen.]

early stages in the formation of involution forms in marine enterobacteria observed by phase contrast microscopy and electron microscopy (R. D. Allen, P. Baumann, unpublished observations) resemble the K$^+$-depleted, plasmolyzed cells of *Alteromonas haloplanktis* (Thompson, Costerton, and MacLeod, 1970). Since the intracellular K$^+$ content of this organism is involved in maintenance of turgor and since energy is required for K$^+$ accumulation and maintenance in the cell (Thompson and MacLeod, 1973, 1974), it is possible that one of the early manifestations of cell death is a loss of the ability to maintain a high intracellular K$^+$ concentration leading to plasmolysis and resulting in structures which subsequently become "spheroplasts", "round bodies", or other involution forms. Colwell (1973) has stated that the formation of "round bodies" is a characteristic of the genus *Vibrio* (defined to include species of *Beneckea*), a view which is untenable since these involution forms are observed in a wide variety of organisms, including strict aerobes of marine origin (Baumann et al., 1972), species of *Photobacterium*, and the nitrifying bacterium, *Nitrospina gracilis*.

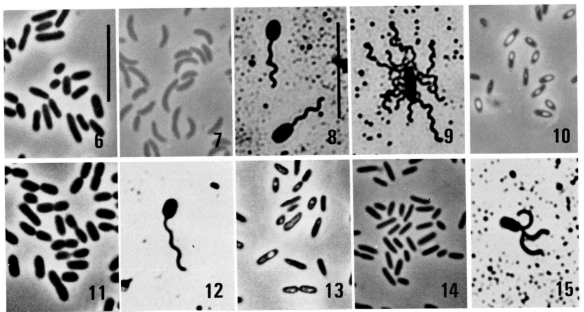

Figs. 6–15. Phase-contrast micrographs and Leifson flagella stains of representative morphological types of marine enterobacteria. Marker in Fig. 6 indicates 10 μm and is applicable to all the phase-contrast micrographs; marker in Fig. 8 indicates 10 μm and is applicable to all the Leifson flagella stains. (6) *Beneckea parahaemolytica*. (7) *B. splendida* biotype II. (8) *B. parahaemolytica* grown in liquid medium. (9) *B. parahaemolytica* grown on solid medium. (10) *B. nereida* containing granules of PHB. (11) *Photobacterium phosphoreum*. (12) *P. angustum*. (13) *P. leiognathi* containing granules of PHB. (14 and 15) *P. fischeri*. [Figs. 6, 7, 10, and 15 from Baumann et al., 1971, with permission. Figs. 8 and 9 from Baumann, P., and Baumann, 1973, with permission.]

NUTRITIONAL SCREENING. The assignment of strains to species is, to a considerable extent, based on a nutritional analysis which tests the ability of the isolates to utilize different organic compounds as sole or principal sources of carbon and energy. Organisms with organic growth factor requirements can often be screened on media supplemented with low levels of amino acids (e.g., *Alteromonas espejiana, A. undina,* and some strains of *Photobacterium phosphoreum*), 0.01% yeast extract (*Beneckea anguillarum* biotype II), or 0.005% yeast extract and 0.005% tryptone (some strains of *P. logei*). In order to expedite the nutritional screening, use is made of the replica plating technique of Lederberg and Lederberg (1952). A master plate is made by inoculating approximately 5-mm² patches of cells onto YEA. After an overnight incubation, this master plate may be used to make submaster plates (Stanier, Palleroni, and Doudoroff, 1966) which are subsequently replicated onto 10 different media. The first medium is BMA lacking an added carbon and energy source, followed by 8 plates of BMA containing different carbon and energy sources, and a final YEA plate. The first plate tests for any growth due to carry-over of nutrients from the submaster plate or due to impurities in the agar, while the last plate tests for the presence of an adequate inoculum. The plates are examined every other day for a total of 6 days and scored for growth. Using a purified grade of agar, the first plate has always lacked visible growth, even when low levels of amino acids were included. We have used a maximum of 16 patches per plate and suggest that in the testing of facultative anaerobes on fermentable substrates the master plate be restricted to 4 well-separated patches. Major physiological types (strict aerobes, facultative anaerobes) should not be intermixed on the same master but segregated onto different masters. Some strains of *Beneckea* swarm on complex solid media (such as YEA) but not on a minimal medium. In the case of these strains, the master plate is prepared with BMA containing 0.2% glycerol.

EXTRACELLULAR ENZYMES. The production of an extracellular chitinase, alginase, amylase, gelatinase, and lipase is determined by inoculating the appropriate solid medium with a patch of cells and observing for a zone of hydrolysis beyond the limits of growth; as many as four different strains can be tested on one plate. Chitinase production is tested on YEA plates overlayed with 10–15 ml of modified YEA, containing about 5 g/liter colloidal chitin (prepared as described by Berger and Reynolds, 1958) and 2.5 g/liter yeast extract. To test for alginase activity, colloidal chitin is replaced by 20 g/liter Na-alginate. An overlay method is not needed to detect the presence of an amylase, gelatinase, and lipase; to test for these extracellular enzymes, YEA is supplemented with 2 g/liter starch,

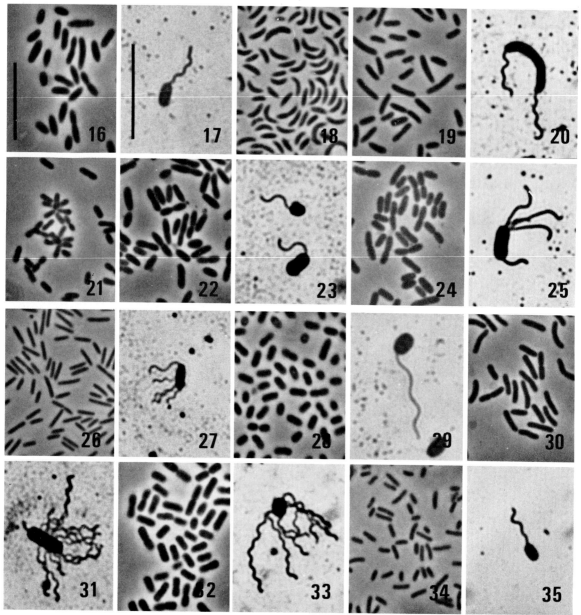

Figs. 16–35. Phase-contrast micrographs and Leifson flagella stains of representative species and groups of nonfermentative marine eubacteria. Marker in Fig. 16 indicates 10 μm and is applicable to all the phase-contrast micrographs; marker in Fig. 17 represents 10 μm and is applicable to all the Leifson flagella stains. (16 and 17) *Alteromonas macleodii*. (18) *A. communis*. (19) *A. vaga*. (20) *A. communis*. (21) Group B-1. (22 and 23) Group I-2. (24 and 25) *Pseudomonas doudoroffii*. (26 and 27) *P. nautica*. (28 and 29) *P. marina*. (30 and 31) *Alcaligenes cupidus*. (32 and 33) *A. venustus*. (34 and 35) Group H-2. [Fig. 35 from Baumann et al., 1972, with permission.]

50 g/liter gelatin, and 10 ml/liter polyethylene sorbitan monooleate (Tween-80), respectively. The hydrolysis of chitin or alginate results in a zone of clearing, whereas lipase activity on Tween-80 plates is detected by the appearance of a precipitate of calcium oleate; all three are observed for a period of 6 days. After an incubation of 48 h, the starch plates are flooded with Lugol's iodine solution (Stanier, Palleroni, and Doudoroff, 1966) and the gelatin plates with acidic mercuric chloride (Skerman,

1967) in order to visualize the unhydrolyzed polymer.

DENITRIFICATION. The ability to denitrify is tested in a medium similar to F-1 (see section on "Fermentation", this chapter) but differing in the substitution of 0.1% succinate, 0.1% acetate, and 0.1% lactate for 1% D-glucose and the addition of 3 g/liter $NaNO_3$. After a 48-h incubation in 5 ml of the medium, the cells are inoculated into 10 ml of the same

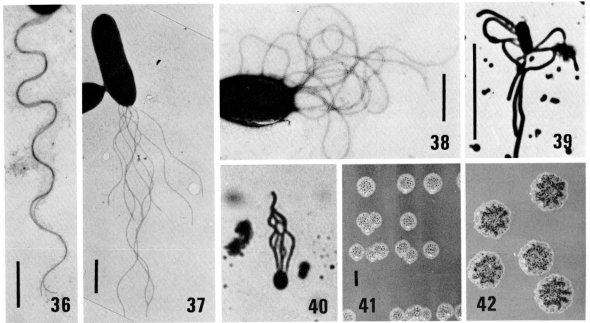

Figs. 36–42. (36–40) Electron micrographs and Leifson flagella stains of *Pseudomonas marina* illustrating some of the arrangements which may lead to difficulties in interpreting Leifson stains. Marker in Figs. 36, 37, and 38 represents 1 µm; marker in Fig. 39 represents 10 µm and is applicable to all the Leifson flagella stains. (36–38) *P. marina* strain 143. (39 and 40) *P. marina* strain 140. Figs. 41 and 42 illustrate the granules of blue-black pigment embedded in colonies of *Beneckea nigrapulchrituda*, strains 153 and 156, respectively. Marker in Fig. 41, which represents 1 mm, is also applicable to Fig. 42. [Figs. 39 and 40 and Figs. 41 and 42 from Baumann et al., 1972, and Baumann et al., 1971, respectively, with permission.]

medium by means of a stab and overlayed with an agar plug. The culture is observed for turbidity and gas production for a period of 6 days. In the past, only gas producers have been scored positive for denitrification, a practice that may result in an underestimation of the number of strains positive for this trait (see Chapter 73, this Handbook).

ARGININE DIHYDROLASE. The presence of a constitutive arginine dihydrolase system is determined by the procedure of Stanier, Palleroni, and Doudoroff (1966) modified by testing for L-ornithine production (Ratner 1962) rather than L-arginine consumption. Additional modifications necessary for the application of this test to marine bacteria are the use of BM containing a suitable carbon and energy source (usually 0.2% glycerol or succinate) for the growth of the organism and the use of 1/2 ASW containing 50 mM Tris-HCl (pH 7.5) in any steps involving dilution or suspension of the cells. Although this method has the advantage of specificity since it tests for the anaerobic conversion of L-arginine to L-ornithine, it is unsuitable for large-scale routine use. For the latter purpose we have modified the method of Thornley (1960) which tests for an arginine-dependent increase in alkalinity under anaerobic growth conditions. This procedure suffers from the disadvantage of not being a specific test for the arginine dihydrolase system since an increase in pH can also be due to a decarboxylation or

deamination of L-arginine, as is the case with *Photobacterium phosphoreum* and *P. leiognathi*, which do not have a constitutive arginine dihydrolase. The value of the method is that it can be used to easily detect potentially positive strains which should then be checked for the anaerobic production of L-ornithine from L-arginine.

Arginine Dihydrolase Medium (ADH)

BM lacking Tris-HCl and containing 10 g/liter L-arginine, 1 g/liter peptone, 10 mg/liter phenol red, and 2 g/liter agar. After boiling to melt the agar and adjusting the pH to 7.0, 3-ml aliquots are dispensed into test tubes, autoclaved for 10 min, and inoculated by means of a stab. About 2 ml of 2% agar (autoclaved and cooled to 41 °C) are carefully layered over the medium to make an agar plug. A parallel tube, identical except for the omission of the L-arginine, is used as a control. Both cultures are incubated for 4 days and observed, periodically, for a difference in color due to alkali production in the L-arginine-containing medium.

MISCELLANEOUS TESTS. The following tests are described in detail by Stanier, Palleroni, and Doudoroff (1966), and only the slight modifications necessary for their application to the characterization of marine bacteria will be mentioned. A method of

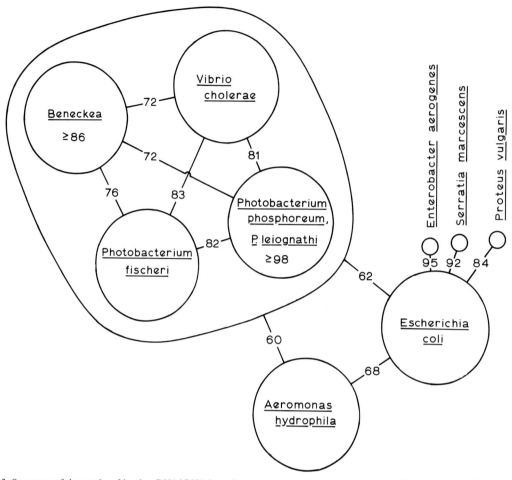

Fig. 43 Summary of the results of in vitro DNA/rRNA homology studies among marine and terrestrial enterobacteria (Baumann and Baumann, 1976). Values represent average percent homologies. Unpublished observations of C. R. Woese, G. E. Fox, and L. B. Zablen, involving comparative cataloging of 16S rRNA (Fox, Pechman, and Woese, 1977) from 8 species closely related or identical to those used in the rRNA homology studies of Baumann and Baumann (1976), are with one exception consistent with the results of the latter study indicating a high degree of agreement between the two approaches. The sole difference was in the value between *Photobacterium leiognathi* and *P. fischeri* which, by extrapolation from the 16S rRNA data, should have been related by about 82% rRNA homology, a number approximately 10% lower than that obtained by Baumann and Baumann (1976). This lower value has been used in Fig. 43; all the remaining data are from Baumann and Baumann (1976).

some importance is the oxidase reaction, which appears to test for the presence of cytochrome *c* (Stanier, Palleroni, and Doudoroff, 1966). Strains of some species of marine bacteria contain low levels of cytochrome *c* and may give a slow or negative oxidase reaction. Addition of a drop of toluene, prior to the oxidase reagent, often results in an accelerated reaction and sometimes reverses a negative result (Baumann et al., 1972; Reichelt and Baumann, 1973b; Reichelt, Baumann, and Baumann, 1976). Consequently, this modification should be tried on all strains which initially give a negative reaction. In the determination of the mechanism of aromatic ring cleavage (*m* or *o*), the cells are grown on BMA containing 0.05–0.1% of the appropriate aromatic compound. Temperature relationships are established using LM for the luminous

bacteria and YEB for the remaining species. The strains are inoculated into test tubes containing 5 ml of the medium and incubated in a water bath shaker set at the appropriate temperature.

The Marine Enterobacteria

The genera *Beneckea* and *Photobacterium* resemble the terrestrial enterobacteria in a number of properties, which include: (i) the distinctive pattern of regulation of aspartokinase activity by means of three isofunctional enzymes (Baumann and Baumann, 1973a; Cohen, Stanier, and Le Bras, 1969), (ii) the utilization of D-glucose via a constitutive Embden-Meyerhof pathway and D-gluconate via an inducible Entner-Doudoroff pathway (Baumann, Baumann, and Reichelt, 1973; Eagon and Wang, 1962; Kubota et al., 1979; Reichelt and Baumann, 1973b), (iii) the utilization of D-fructose by means of an inducible 1-phosphofructokinase and the Embden-Meyerhof pathway (Gee, Baumann, and Baumann, 1975), and (iv) the fermentation of D-glucose to end products characteristic of the mixed acid fermentation, with or without the production of small amounts of acetoin and 2,3-butylene glycol (Doudoroff, 1942b). The studies of Crawford and his collaborators have further substantiated the relationship between *Beneckea* and the terrestrial enterobacteria by showing that the organization and regulation of genes for tryptophan biosynthesis in *B. harveyi* are similar to those of the terrestrial enterobacteria (Bieger and Crawford, 1978; Crawford, 1975). (For a review concerning the marine enterobacteria see Baumann and Baumann, 1977.)

In order to assign species of marine enterobacteria to genera on the basis of evolutionary relationships, we have performed studies of rRNA homology by means of the in vitro rRNA/DNA hybridization technique (Baumann and Baumann, 1976). The results obtained by this method (unlike those obtained from in vitro DNA/DNA hybridization) provide a measure of distant relationship since the genes coding for rRNA are conserved to a greater extent than most of the remaining genome. A summary of these results, which is presented in Fig. 43, indicates that species of *Beneckea* are as distantly related to *Photobacterium phosphoreum* and *P. leiognathi* as they are to *P. fischeri* and the structurally and physiologically similar pathogenic species *Vibrio cholerae* (the type species of the genus *Vibrio*). Since a number of physiological, morphological, and genetic properties indicates a generic separation between *Beneckea* and the species *P. phosphoreum* and *P. leiognathi*, internal consistency dictates the same with respect to *Beneckea* and *V. cholerae* (Fig. 43). The latter generic separation is also supported by studies of electrophoretic mobilities of superoxide dismutases of species of *Beneckea* and *V. cholerae* (Bang, Woolkalis, and Baumann, 1978). Similarly, the results of the rRNA homology studies suggest a separate generic status for *P. fischeri*, distinct from *P. phosphoreum* and *P. leiognathi* (Fig. 43), a course of action we are not willing to undertake pending the completion of studies of enzyme evolution in this group of organisms.

The genus *Beneckea* is composed of straight or curved rods (Figs. 1–3, 6, and 7) which are oxidase-positive, facultative anaerobes, fermenting D-glucose with the production of acid and, with the exception of *B. gazogenes*, no gas[1] (Baumann, L., and Baumann, 1973a; Baumann, Baumann, and Mandel, 1971; Baumann, Baumann, and Reichelt, 1973; Baumann et al., 1971; Harwood, 1978; Reichelt and Baumann, 1973b). Some species pro-

duce acetoin as well as 2,3-butylene glycol; none is capable of denitrification. Except for *B. anguillarum* biotype II, essentially all of the strains of *Beneckea* have no growth factor requirements. All species grow at 30°C; none grows at 45°C, while some grow at 4°C or 40°C. Only two species are pigmented—*B. nigrapulchrituda* which produces a water-insoluble, blue-black pigment (Figs. 41 and 42) and *B. gazogenes* which produces prodigiosin. When grown in liquid medium, all species of *Beneckea* have a single sheathed polar flagellum (Figs. 2, 3, and 8). When grown on solid medium, some species have additional unsheathed peritrichous flagella (Figs. 1 and 9) (Allen and Baumann, 1971; Baumann and Baumann, 1973; Baumann, Baumann, and Mandel, 1971; Baumann, Baumann, and Reichelt, 1973; de Boer, Golten, and Scheffers 1975a, b; McCarthy, 1975; Reichelt and Baumann, 1973b; Ulitzur, 1974, 1975; Yabuuchi et al., 1974). The flagellins of the polar and peritrichous flagella differ in their amino acid compositions as well as in their immunological properties (Shinoda et al., 1974a, b, 1976). Mutant studies have shown that the peritrichous flagella are essential for swarming on solid media, while the polar flagella are necessary for motility in liquid media (Shinoda and Okamoto, 1977). Some species of *Beneckea* accumulate PHB as an intracellular reserve product (Fig. 10) and produce a constitutive arginine dihydrolase. Most species synthesize an extracellular amylase, gelatinase, lipase, and chitinase; some strains make an extracellular alginase. Different species are able to utilize a total of 14–67 organic compounds as sole sources of carbon and energy; these include pentoses, hexoses, disaccharides, sugar acids, sugar alcohols, C_2-C_{10} monocarboxylic fatty acids, tricarboxylic acid cycle intermediates, and amino acids. Species utilizing aromatic compounds degrade the intermediate protocatechuate by means of a *m* cleavage. None of the species utilizes cellulose, formate, C_6-C_{10} dicarboxylic acids, L-isoleucine, L-valine, L-lysine, L-tryptophan, purines, pyrimidines, or *n*-hexadecane. Some of the species luminesce. The G+C content in the DNA of most *Beneckea* species is 45–48 mol%; *B. proteolytica* and strains of group E-3 have G+C contents of 51 and 54 mol%, respectively. The type species of *Beneckea* is *B. campbellii* (Ad Hoc Committee, 1980).

---

[1] A group of marine bacteria recently described by Lee, Donovan, and Furniss (1978) and Lee, Shread, and Furniss (1978) and designated group F contains some strains which produce gas during the fermentation of glucose. Studies of superoxide dismutase evolution by means of the microcomplement fixation technique have indicated that with antiserum prepared to the enzyme from *Beneckea alginolytica* strains of group F are more related to this species than to *Vibrio cholerae* (S. S. Bang, personal communication).

Species of the genus *Photobacterium* are facultative anaerobes which ferment D-glucose with the production of acid; some species produce gas as well as acetoin and 2,3-butylene glycol (Hendrie, Hodgkiss, and Shewan, 1970; Reichelt and Baumann, 1973b, 1975). None of the species denitrifies or grows at 40°C; some grow at 4°C, and all are capable of growth at 20°C. Two species contain organic growth factor–requiring strains: strains of *P. phosphoreum* may require L-methionine, either alone or in combination with a few other amino acids (Doudoroff, 1942a; Reichelt and Baumann, 1973b), whereas some strains of *P. logei* appear to have rather extensive organic growth factor requirements. Most species synthesize an extracellular chitinase, while some produce an amylase (Fitzgerald, 1977), lipase, and gelatinase; none produces an alginase. The nutritional versatility of species of *Photobacterium* is considerably less than that of *Beneckea;* only 7–22 organic compounds can be utilized as sole or principal sources of carbon and energy, including pentoses, hexoses, disaccharides, and a few sugar acids, sugar alcohols, tricarboxylic acid cycle intermediates, and amino acids. Most species luminesce. The G+C contents in the DNAs of the species of *Photobacterium* span the range of 39–44 mol%. On the basis of structural properties, *in vitro* DNA/DNA homology and rRNA homology (Table 1, Figs. 43 and 44), species of this genus can be subdivided into two groups. The first consists of *P. phosphoreum, P. leiognathi,* and *P. angustum* (Fig. 11), which are motile by 1–3 unsheathed polar flagella (Figs. 4 and 12) and accumulate PHB as an intracellular reserve product (Fig. 13) (Eberhard and Rouser, 1971; Reichelt and Baumann, 1973b). These three species contain oxidase-positive and -negative strains; some of the latter become oxidase positive by pretreatment with toluene (see section on "Identification", this chapter). Both the oxidase-positive and -negative strains were found to have low levels of cytochrome *c* (Reichelt and Baumann, 1973b; Reichelt, Baumann, and Baumann, 1976). The second group consists of the species *P. fischeri* (Fig. 14) and *P. logei,* which are oxidase positive, motile by 2–11 sheathed polar flagella (Figs. 5 and 15), and unable to accumulate PHB. These two species, unlike the remaining species of *Photobacterium,* produce a yellow cell-associated pigment. The type species of *Photobacterium* is *P. phosphoreum* (Ad Hoc Committee, 1980).

A summary of the results of a study of genetic relationships among strains of *Beneckea* and *Photobacterium,* by means of in vitro DNA/DNA hybridization, is presented in Fig. 44. Some of the properties of use in identifying species of these two genera are presented in Table 3; a table separating *B. gazogenes* from species of *Beneckea* has been published by Harwood (1978). Unfortunately, there is no set of nutritional characters which clearly distinguishes species of *Beneckea* from *Photobacterium.* A trait of some use for this purpose is the utilization of D-alanine, a property absent in *Photobacterium* and *B. anguillarum* biotype II but present in all but a few of the remaining strains of *Beneckea* (Table 3). Table 3 is primarily intended as a guide from which the investigator may wish to select those properties that are distinctive for a particular species or make a more suitable table for the differentiation of species having a common property (e.g., luminescence). For example, an inspection of Table 3 will indicate that the ability to make peritrichous flagella on solid medium grow at 40°C, produce amylase, utilize D-galactose, butyrate, ethanol, L-serine, L-leucine, L-arginine, and putrescine, and the inability to accumulate PHB, produce a constitutive arginine dihydrolase, utilize sucrose, sorbitol, *p*-hydroxybenzoate, or δ-aminovalerate is diagnostic for *B. parahaemolytica.* This combination of traits distinguishes *B. parahaemolytica* from *B. vulnifica* as well as *B. harveyi,* all of which share the ability to grow at 40°C and the inability to utilize sucrose, two traits which were once considered to be diagnostic for *B. parahaemolytica* (Sakazaki, 1968; Sakazaki, Iwanami, and Fukumi, 1963). Similarly, another set of diagnostic traits can be extracted from Table 3 for the identification of the human pathogen *B. vulnifica* or the fish and eel pathogen *B. anguillarum* biotypes I and II.

In our initial phenotypic characterization of *Beneckea vulnifica* (Table 3), all the strains tested were found to be unable to utilize lactose as a sole source of carbon and energy. Hollis et al. (1976) have, however, subsequently described this species as fermenting this sugar. Our recent unpublished observations have, in part, clarified the discrepancy between the two studies. Wild-type strains of *B. vulnifica* do not utilize lactose; however, some strains may acquire the ability to utilize this sugar by mutation. (Despite extensive attempts to obtain spontaneous lactose-utilizing mutants, only 4 out of 6 strains tested yielded isolates capable of growth on lactose.) Since the fermentation tests of Hollis et al. (1976) are performed in complex medium, sufficient substrate is available to support the growth of a population of cells large enough to potentially contain spontaneous lactose-utilizing mutants. It should be noted that two strains of *B. vulnifica* common to our study and that of Twedt, Spaulding, and Hall (1969) were scored lactose-negative by the latter authors using a method similar to that of Hollis et al. (1976).

The ability of *B. alginolytica* to swarm on solid complex medium (containing 2% agar) has been used as a trait distinguishing this species from *B. parahaemolytica* (Sakazaki, 1968; Sakazaki, Iwanami, and Fukumi, 1963). This property should be regarded with considerable caution, since (i) some strains of *B. parahaemolytica* are capable of limited swarming (Baumann, Baumann, and

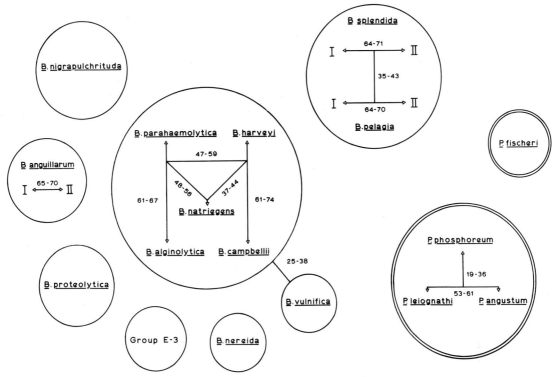

Fig. 44. Summary of the results of in vitro DNA/DNA homology studies among marine enterobacteria. Roman numerals indicate biotypes. Strains within each species or biotype are related by DNA homologies of over 80%. The internal homology of the two strains of Group E-3 has not been determined. Numbers indicate the range of DNA homologies between the various species or biotypes. Species or groups of species which are not interconnected are related by DNA homologies of less than 30%. Double circles have been drawn around species of *Photobacterium* to distinguish them from species of *Beneckea*. Figure redrawn from Baumann and Baumann (1977). Data for *B. anguillarum* are normalized averages derived from Schiewe, Crosa, and Ordal (1977).

Reichelt, 1973; Twedt, Spaulding, and Hall, 1969), (ii) a nonswarming strain with all the phenotypic properties of *B. alginolytica* has been shown to be identical to swarming strains of this species by in vitro DNA/DNA hybridization studies (Reichelt, Baumann, and Baumann, 1976), and (iii) swarming is affected by the moistness of the medium. Some strains of *B. harveyi* and the single isolate of *B. proteolytica* are also capable of swarming on complex solid media.

An important test for the identification of *B. alginolytica*, *B. anguillarum* biotype I, and *B. proteolytica* is the Voges-Proskauer reaction, which detects the presence of acetoin (and/or diacetyl), an intermediate in the pathway leading to the formation of 2,3-butylene glycol. Some strains produce low amounts of this intermediate so that it is advisable to perform the test (Skerman, 1967) on cultures grown in 10 ml YEB supplemented with 2% D-glucose and incubated for 3 and 5 days (Baumann, Baumann, and Mandel, 1971). If only a slight color is detected, a quantitative assay for acetoin and/or diacetyl should be performed as described by Eggleton, Eldsen, and Gough (1943) and Neish (1952). The application of this procedure has eliminated misinterpretation of positive reactions, since some strains,

which do not appear to produce acetoin and/or diacetyl, may give a slight color that could be interpreted as a positive Voges-Proskauer reaction (Baumann, Baumann, and Mandel, 1971; Reichelt and Baumann, 1973b). Neither the formation of acetoin and/or diacetyl nor the formation of 2,3-butylene glycol is useful for the differentiation of *P. phosphoreum* and *P. leiognathi* (Table 3) (Reichelt and Baumann, 1973b).

The ability to luminesce is a property which at the present time appears to be restricted to the species *P. fischeri, P. logei, P. phosphoreum, P. leiognathi, B. spendida* biotype I, and *B. harveyi*. With the exception of *B. harveyi*, which contains many nonluminous strains phenotypically and genotypically indistinguishable from luminous isolates, and *P. logei*, which contains one nonluminous isolate, all the remaining species consist of only luminescent strains. The latter situation may simply be a consequence of the fact that these strains have been isolated on the basis of their ability to emit light. Therefore, the inability to luminesce should not, a priori, exclude strains from these species. The number of luminous isolates studied to date as well as the sources sampled are not sufficient to establish that luminous strains are restricted to four species of

Table 3. Some distinguishing properties of species of *Photobacterium* and *Beneckea*.[a]

Trait	*P. fischeri* (12)[b]	*P. logei* (11)	*P. leiognathi* (30)	*P. angustum* (5)	*P. phosphoreum* (79)	*B. harveyi* (85)	*B. campbellii* (44)	*B. parahaemolytica* (132)	*B. alginolytica* (31)	*B. natriegens* (5)	*B. vulnifica* (14)	*B. splendida* I[c] (4)	*B. splendida* II (15)	*B. pelagia* I (7)	*B. pelagia* II (4)	*B. nereida* (6)	*B. anguillarum* I (20)	*B. anguillarum* II (5)	*B. nigrapulchrituda* (14)	*B. proteolytica* (1)	Group E-3 (2)
Mol% G+C	40	40	43	41	42	45–48	45–48	45–48	45–48	45–48	45–48	45–48	45–48	45–48	45–48	45–48	45–48	45–48	45–48	51	54
Peritrichous[d]	+	+	−	−	−	+	+	+	+	+	+	+	+	+	+	+	+	+	+	+	+
Sheathed polar flagella[e]	+	+	+	+	+	+	+	+	+	+	+	+	+	+	+	+	+	+	+	+	+
Number of polar flagella[e]	2–8	2–5	1–3	1–3	1–3	1	1	1	1	1	1	1	1	1	1	1	1	1	1	1	1
Curved rods[f]	1	1	1	2	−	−	−	2	−	−	13	+	9	−	+	−	−	d	d	−	−
PHB accumulation	−	−	+	+	+	−	−	−	−	+	−	+	−	−	+	+	+	+	+	−	+
Arginine dihydrolase	+	10	+	−	+	−	−	−	−	−	−	+	−	−	−	+	+	+	7	−	+
Luminescence	+	−	+	−	+	61	−	−	−	−	−	+	−	−	−	−	−	−	−	−	+
Gas from D-glucose	−	−	2	−	71	−	−	−	−	−	−	−	−	−	−	−	−	−	−	−	−
Production of acetoin and/or diacetyl	−	−	15	−	71	−	−	−	+	−	−	−	−	−	−	−	+	−	−	+	−
Growth at 4°C	−	+	−	−	75	−	−	−	−	−	−	3	−	1	1	3	−	−	−	−	−
Growth at 35°C	8	−	28	4	−	+	43	+	+	+	+	−	−	−	−	−	+	−	−	−	+
Growth at 40°C	−[h]	−	−	−	−	34	43	+	+	2	+	−	−	−	−	−	−	−	−	−	+
Amylase	11	−	24	2	−	+	−	+	+	2	+	+	13	+	+	5	19	+	13	+	+
Lipase	11	9	2	2	−	+	+	+	+	+	+	+	+	+	+	−	18	+	+	+	+
Gelatinase	1	−	−[i]	4	−	84	+	+	+	4	+	+	−	+	+	1	+	−	−	+	−
Alginase	−	−	−	−	−	26	−	−	+	−	−	2	−	−	+	−	−	−	−	−	−
D-Xylose	−	−	−	+	−	−	−	−	−	−	−	−	−	−	−	−	−	−	−	−	−
L-Arabinose	−	−	−	−	−	4	−	91	−	+	−	−	−	−	+	−	−	−	−	−	+
D-Galactose	+	+	+	+	78	71	−	128	5	+	+	+	+	+	+	5	16	9	+	+	+
Sucrose	1	1	−	3	−	52	43	−	+	+	−	3	−	+	+	−	+	−	−	−	−
Maltose	11	+	−	3	78	+	43	131	+	3	+	+	+	+	+	+	+	+	+	+	+
Cellobiose	11	+	−	−	−	+	12	8	1	3	+	1	−	+	+	−	16	+	5	+	+
D-Gluconate	−	10	+	+	68	84	3	+	+	3	+	+	−	+	+	+	+	+	5	−	+
D-Glucuronate	−	−	−	−	41	+	1	82	+	3	+	−	−	−	−	−	−	−	+	+	−
Acetate	−	−	25	+	−	79	17	128	+	+	+	3	13	+	+	5	+	+	+	+	+
Propionate	−	−	−	−	−	+	40	+	+	+	+	+	−	+	+	+	17	+	+	+	+
Butyrate	−	−	−	−	−	−	−	106	+	+	−	−	−	−	−	+	17	−	−	−	+

Heptanoate	−	−	−	−	+	−	−	+	−	−	+	+
Pelargonate	−	−	−	−	+	−	−	+	−	1	+	+
DL-Lactate	+	+	13	+	+	+	+	+	13	+	+	+
α-Ketoglutarate	1	−	−	+	130	+	+	−	+	+	+	+
Sorbitol	−	−	−	−	−	−	−	−	−	18	+	−
Inositol	−	−	−	2	−	−	−	−	−	12	−	−
Ethanol	−	−	5	+	117	+	5	−	−	−	−	1
p-Hydroxybenzoate	−	−	−	+	88	+	+	−	11	−	−	−
Glycine	−	45	−	+	88	+	+	+	11	−	+	+
D-α-Alanine	+	84	−	+	+	+	+	+	14	+	+	+
L-Serine	3	72	29	+	117	+	+	2	10	18	+	+
L-Leucine	−	−	−	4	129	+	+	+	−	−	−	1
L-Glutamate	20	31	9	+	130	+	+	+	9	+	+	+
L-Arginine	−	72	−	+	119	+	30	+	−	−	+	1
L-Citrulline	−	14	−	+	2	+	−	+	−	−	−	1
γ-Aminobutyrate	−	−	−	+	−	−	−	+	−	−	−	+
δ-Aminovalerate	−	−	−	+	−	+	−	+	−	−	−	+
L-Proline	+	29	43	+	131	+	26	+	14	19	+	+
Putrescine	−	−	−	−	112	+	26	+	−	−	+	+
Sarcosine	−	−	−	−	−	−	−	−	−	−	−	+
Other traits	j	j	−	k	−	−	−	−	−	i	l	

[a] +, All strains positive; −, all strains negative; numbers indicate number of positive strains; boldface numbers indicate that the number represents 80% or more of the strains. Data from Bang, Baumann, and Nealson (1978); Baumann, Baumann, Baumann, and Mandel (1971); Reichelt, Baumann, and Baumann (1976).

[b] Number in parentheses refers to number of strains studied.

[c] Roman numerals refer to biotypes.

[d] Strains having unsheathed peritrichous flagella in addition to a sheathed polar flagellum when grown on solid medium.

[e] Determined in liquid medium.

[f] +, All curved rods; −, all straight rods; numbers indicate number of curved rods.

[g] Straight rods during exponential phase of growth becoming curved in stationary phase.

[h] Fitzgerald (1977) and Ruby and Nealson (1976) have found a few strains of P. fischeri to be positive for this trait. The inclusion of these data in numerical form is not possible since it is not clear how many of the strains were independent isolates.

[i] Reichelt, Nealson, and Hastings (1977) have found some strains of P. leiognathi to be positive for this trait. The inclusion of these data in numerical form is not possible since it is not clear how many of the strains were independent isolates.

[j] Produces a yellow, cell-associated pigment.

[k] Unlike other species, B. natriegens is able to utilize L-rhamnose, benzoate, malonate, spermine, and hippurate as sole sources of carbon and energy.

[l] Produces an extracellular blue-black pigment.

Photobacterium and two species of *Beneckea*. With this reservation in mind, it is possible to make a simplified diagnostic table from the data presented in Table 3 for the identification of luminous bacteria (see Chapter 105, this Handbook). Some problems may be encountered in the differentiation of *B. harveyi* from *B. spendida* biotype I. These two species, although genotypically distinct (Fig. 44), are only distinguishable by a small number of phenotypic traits (flagellation on solid medium, cell curvature, and arginine dihydrolase; Table 3). A similar problem exists with the species *P. fischeri* and *P. logei,* which are only distinguished by growth at 4°C and D-gluconate utilization (Table 3). A property not listed in Table 3 which is useful for the differentiation of these two species is growth at 30°C; none of the strains of *P. logei* grew at this temperature, while 11 out of 12 strains of *P. fischeri* were able to grow at 30°C.

The taxonomic treatment of the marine enterobacteria in the eighth edition of *Bergey's Manual of Determinative Bacteriology* (Buchanan and Gibbons, 1974) was formalized prior to the availability of the extensive data from DNA/DNA and DNA/rRNA homology studies and before the very distinctive flagellation of some species of *Beneckea* was recognized. Consequently, there are considerable differences between the classification presented in this chapter and that found in *Bergey's Manual,* which assigns *B. parahaemolytica, B. anguillarum,* and *P. fischeri* to the genus *Vibrio* and *B. harveyi* to *Lucibacterium* (for a detailed discussion see Baumann and Baumann, 1977). The results of the DNA/rRNA homology studies are in disagreement with these assignments since they indicate a close relationship between *B. parahaemolytica, B. harveyi,* and *B. anguillarum* and support a generic separation between species assigned to *Beneckea, Photobacterium,* and *V. cholerae* (Fig. 43) (Baumann and Baumann, 1976). In addition, the results of the in vitro DNA/DNA homology studies indicate a close relationship between *B. parahaemolytica* and *B. harveyi* (Fig. 44). The assignment of the latter species to *Beneckea* and not *Lucibacterium* is predicated by the rules of priority. Finally, the assignment of *B. proteolytica* to *Aeromonas hydrophila* subspecies *proteolytica* in the eighth edition of *Bergey's Manual* is again untenable due to the results of the in vitro DNA/rRNA homology studies which indicate that this species is closely related to other species of *Beneckea* and is distinct from *Aeromonas hydrophila.*

The Nonfermentative Marine Eubacteria

The nonfermentative marine eubacteria considered in this section comprise a large and diverse group of organisms consisting of polarly or peritrichously flagellated, straight or curved rods, with G+C

contents in their DNAs of 28–68 mol% (Baumann et al., 1972; Chan et al., 1978; Reichelt and Baumann, 1973a). Representative morphological types are presented in Figs. 16–40. Examination by means of the electron microscope has indicated that none of the strains has sheathed flagella. In addition, unlike some species of *Beneckea,* their flagellation does not appear to be affected by conditions of cultivation. An extensive phenotypic characterization of these organisms has indicated that on the basis of overall similarity the strains could be assigned to a number of species and groups that are distinguishable from one another by multiple, independent, phenotypic traits. (The assignment of a group designation to small clusters of phenotypically related strains was meant as a provisional measure pending further study of additional strains.) Most of the isolates have no organic growth factor requirements and exhibit considerable diversity with respect to their nutritional spectrum, being able to utilize between 11 and 85 organic compounds as sole sources of carbon and energy. These compounds include carbohydrates, monocarboxylic and dicarboxylic acids, amino acids, amines, aromatic compounds, purines, and pyrimidines, as well as *n*-hexadecane. Aromatic compounds, if utilized, are degraded by means of either an *o* or *m* cleavage. Species able to utilize D-glucose appear to degrade this hexose via the Entner-Doudoroff pathway (Baumann and Baumann, 1973b; Sawyer, Baumann, and Baumann, 1977). None of the strains utilizes cellulose, formate, oxalate, or methanol. One species *(P. nautica)* is a vigorous denitrifier. With the exception of a number of phenotypically related species in the genus *Alteromonas,* most of the nonfermentative eubacteria do not produce an extracellular amylase, gelatinase, lipase, chitinase, or alginase. None of the strains has a constitutive arginine dihydrolase. All grow at 30°C; some grow at 4 or 40°C. Most strains, when grown on solid medium, have no obvious pigmentation; some strains of *P. doudoroffii* produce a brown, water-soluble pigment on complex medium, and some strains of group I-2 make a blue-black, water-insoluble pigment, the production of which is enhanced on minimal medium.

The results of extensive studies on the marine enterobacteria permit generic assignments which are based, in part, on genetic relationships. Studies of the nonfermentative marine eubacteria are not as advanced and, consequently, it was necessary to assign these strains to existing genera on the basis of flagellation and the G+C contents of their DNAs and not on the basis of known natural relationships. The inclusion of marine and terrestrial species in one genus was a provisional measure which may have to be modified after studies of the relationship between these organisms. A summary of the generic assignments of the nonfermentative marine species and groups is presented in Table 1.

Table 4. Some distinguishing properties of species of *Alteromonas, Pseudomonas nautica,* and group H-2 (organisms which do not accumulate PHB and have single polar flagella).[a]

Trait	A. communis (33)[b]	A. vaga (17)	A. macleodii (21)	A. haloplanktis[c] (25)	A. espejiana[d] (18)	A. undina[d] (8)	A. rubra[e] (3)	A. luteoviolaceus[f] (16)	A. citrea[g] (3)	P. nautica[h] (34)	Group H-2 (4)
Mol% G+C	45–48	46–50	44–47	41–44	43–44	43–44	46–48	40–43	41–45	57–62	28–33
Straight rods[i]	−	+	+	+	+	−	+	+	+	+	3
Ring cleavage[j]	m	m	−	−	−	−	−	ND[k]	−	o	−
Oxidase	+	−	+	+	+	+	+	+	+	+	+
Denitrification	−	−	−	−	−	−	ND	ND	ND	29	−
Growth at 35°C	+	+	+	21	8	−	+	+	1	+	+
Growth at 40°C	+	−	15	−	−	−	−	−	1	29	−
Amino acid(s) required	−	−	−	6	+	+	+	ND	+	−	−
Amylase	−	−	19	4	+	5	+	15	+	3	−
Gelatinase	−	−	20	+	+	+	+	+	+	1	−
Lipase	−	−	+	+	+	+	+	+	+	+	
Alginase	−	−	3	−	+	−	ND	ND	ND	−	−
Chitinase	−	−	−	16	−	+	−	−	1	−	−
D-Glucose	+	+	+	+	+	+	+	+	+	+	
D-Mannose	29	+	−	21	12	−	+	−	+		
D-Galactose	11	15	+	5	+	−	−	ND	−	−	−
Sucrose	−	1	+	23	+	+	−	−	−	−	−
Cellobiose	−	14	+	−	9	−	−	−	−	−	−
Melibiose	−	−	20	−	+	−	ND	ND	ND	−	−
Lactose	−	−	+	1	+	−	−	−	−	−	−
Salicin	−	−	+	−	−	−	−	ND	−	−	−
D-Gluconate	+	+	18	1	−	−	−	ND	−	−	−
N-Acetylglucosamine	−	15	7	23	−	+	ND	ND	ND	−	−
Succinate	+	+	−	+	−	+	−	ND	−	31	+
Fumarate	+	+	−	+	−	+	−	ND	−	31	+
DL-Malate	+	+	−	−	−	−	−	ND	−	29	−
DL-β-Hydroxybutyrate	+	7	12	13	−	−	−	ND	−	30	−
DL-Glycerate	4	−	+	−	−	−	ND	ND	ND	−	−
Citrate	+	+	−	21	17	−	−	ND	−	12	−
Aconitate	+	16	−	21	17	−	ND	ND	ND	1	−
Erythritol	−	+	−	−	−	−	−	ND	−	−	−
Mannitol	+	+	9	11	+	−	−	−	−	−	−
Glycerol	+	+	+	−	−	−	−	−	−	−	−
Ethanol	+	−	17	20	−	+	ND	ND	ND	22	−
γ-Aminobutyrate	+	16	−	−	−	−	ND	ND	ND	−	−
L-Tyrosine	−	−	18	23	+	+	ND	ND	ND	−	−
Sorbitol, α-Ketoglutarate, m-Hydroxybenzoate	+	+	−	−	−	−	−	ND	−	−	−
n-Hexadecane	−	−	−	−	−	−	ND	ND	ND	28	−
Other traits[l]	+	+	−	−	−	−	ND	ND	ND	−	−
Pigmentation	−	−	−	−	−	−	+[m]	14[n]	+[o]	−	−

[a] +, All strains positive; −, all strains negative; numbers indicate number of positive strains; numbers in boldface indicate that the number represents 80% or more of the strains. Unless otherwise indicated data are from Baumann et al. (1972).   [b] Number in parentheses refers to number of strains studied.   [c] Data from Baumann et al. (1972), Chan et al. (1978), and Reichelt and Baumann (1973a).   [d] Data from Chan et al. (1978).   [e] Data from Gauthier (1976b) and Gerber and Gauthier (1979).   [f] Data from Gauthier (1976a,b).   [g] Data from Gauthier (1977).   [h] Some strains of *P. nautica* have 1–3 lateral flagella.   [i] +, Straight rods; −, curved rods.   [j] Mechanism of aromatic ring cleavage by species capable of growth on aromatic compounds.   [k] ND, not determined. [l] Utilization of saccharate, *meso*-inositol, p-hydroxybenzoate, quinate, and sarcosine.   [m] Prodigiosin.   [n] Violacein.   [o] Lemon-yellow noncarotenoid pigment.

Table 5. Some distinguishing properties of species and groups of *Pseudomonas* which accumulate PHB.[a]

Trait	*P. doudoroffii* (11)[b]	*P. marina* (7)	Groups B-1, B-2 (7)	Group I-2 (5)	Group H-1 (4)	Group I-1 (4)	Group G-1 (11)
Mol% G+C	55–60	62–64	58–64	61–64	54–57	56–57	55–57
Number of polar flagella	1–3	2–5	4–6	3–6	1	1–2	1
Ring cleavage[c]	o	−	o	o	m	−	m
Oxidase	+	−	+	+	+	+	+
Growth at 4°C	−	**6**	−	−	−	−	−
Growth at 35°C	+	+	**6**	2	+	−	+
D-Ribose	4	+	−	−	−	−	−
D-Glucose	−	+	+	1	+	−	−
D-Galactose	−	+	+	−	−	−	−
D-Fructose	**10**	+	+	−	+	−	−
D-Gluconate	−	+	2	−	−	−	−
N-Acetylglucosamine	−	−	+	−	−	−	−
Valerate	−	**6**	+	3	−	+	**8**
Pelargonate	−	+	−	−	−	−	2
Glycolate	**10**	−	3	−	−	−	−
DL-Glycerate	**10**	+	4	−	−	−	−
Aconitate	+	+	**6**	−	3	−	**10**
Mannitol	−	+	+	−	+	−	−
Glycerol	−	+	+	+	−	−	−
Ethanol	**10**	5	−	−	+	+	+
Phenylacetate	−	−	2	−	−	−	**10**
Quinate	−	−	1	−	+	−	5
Glycine	+	−	+	−	−	3	−
L-Threonine	8	−	+	+	−	2	−
L-Ornithine	+	5	**6**	4	−	+	10
δ-Aminovalerate	+	−	−	−	−	+	3
L-Histidine	+	−	+	4	−	−	−
L-Tyrosine	−	+	+	4	−	−	+
Ethanolamine	−	−	+	4	1	2	−
Sarcosine	+	**6**	+	+	−	+	−
Allantoin	+	−	3	−	−	−	−

[a] +, All strains positive; −, all strains negative; numbers indicate number of positive strains; boldface numbers indicate that the number represents 80% or more of the strains. Data from Baumann et al. (1972).

[b] Number in parentheses refers to number of strains studied.

[c] Ring cleavage for strains capable of growth on aromatic compounds.

Many polarly flagellated strict aerobes of marine origin have G+C contents in their DNAs of 40–50 mol%. Since strict aerobes with a G+C content in this range appear to be rare or absent in terrestrial habitats, and since it seemed unreasonable to extend the G+C range characteristic of *Pseudomonas* to encompass these organisms, they were given a new generic designation, *Alteromonas*. Although the disadvantages of creating a genus solely on the basis of G+C content were apparent, it was thought that further work might reveal additional properties that would allow a better generic definition. These ex-

pectations have been partially realized by the work of Gauthier (1976a, 1976b, 1977) and Chan et al. (1978). The genus *Alteromonas* appears to be composed of two phenotypically distinct groups consisting of closely related species: (i) *A. communis* and *A. vaga* and (ii) *A. macleodii* (the type species of *Alteromonas*), *A. haloplanktis*, *A. luteoviolaceus*, *A. rubra*, *A. citrea*, *A. espejiana*, and *A. undina*. The latter species share a number of distinctive properties such as the presence of extracellular hydrolases and a peculiar pattern of utilization of tricarboxylic acid cycle intermediates and glycerol

(Table 4). Further study of these organisms and the possible exclusion of *A. communis* and *A. vaga* may allow a better definition of *Alteromonas*. The remaining species and groups of the polarly flagellated nonfermentative marine eubacteria had, with one exception, a G+C content of 54–64 mol% and were assigned to the genus *Pseudomonas*. The exception was group H-2, which had a G+C content of 28–33 mol% and was left unassigned. Although most strains of *Pseudomonas nautica* had single polar flagella, some isolates also had 1–4 lateral flagella with a wavelength different from that of the polar flagellum (Fig. 27). A similar situation has been noted in some strains of the terrestrial species *Pseudomonas stutzeri* and *P. mendocina* (Palleroni and Doudoroff, 1972). All the peritrichously flagellated species had G+C contents of 52–68 mol% and were placed in the genus *Alcaligenes*.

Studies of the pathways utilized in the catabolism of D-fructose (Baumann, L., and Baumann, 1975; Baumann, P., and Baumann, 1975; Sawyer, Baumann, and Baumann, 1977), as well as the pattern of regulation of aspartokinase activity (Baumann and Baumann, 1974), have provided some insight into the relationships among the species of nonfermentative marine eubacteria. One grouping cuts across the present generic assignments (Table 1), since all four species of *Alcaligenes* and *P. marina* were found to share a number of distinctive properties, including the pattern of regulation of aspartokinase activity and the presence of inducible phosphoenolpyruvate: D-fructose phosphotransferase and 1-phosphofructokinase activities in cell-free extracts of D-fructose-grown cells (Sawyer, Baumann, and Baumann, 1977). These common properties strongly argue against the generic separation of the four species of *Alcaligenes* and *P. marina,* a separation that was based on their differences in flagellation. With the exception of *P. doudoroffii,* which had a distinctive pathway of D-fructose and D-ribose catabolism (Baumann, L., and Baumann, 1975; Baumann, P., and Baumann, 1975) as well as a unique pattern of aspartokinase activity, none of the remaining species and groups of *Pseudomonas* and *Alteromonas* that were able to grow on D-fructose had a 1-phosphofructokinase. The mode of regulation of aspartokinase activity in species and groups of marine *Pseudomonas* suggested additional relationships and indicated the heterogeneity of this genus (Table 1). The similarity of groups B-1, B-2, and I-2 was previously suspected on the basis of their flagellation as well as their tendency to form rosettes (Fig. 21). The heterogeneity of the marine pseudomonads suggests that, as in the case of the terrestrial *Pseudomonas,* the marine members of this genus consist of groups of species which are quite diverse and are only united by two structural properties (rod shape and polar flagellation) and their obligately respiratory metabolism (Palleroni and

Table 6. Some distinguishing properties of species of *Alcaligenes* (peritrichously flagellated organisms).[a]

Trait	*A. pacificus* (6)[b]	*A. cupidus* (5)	*A. venustus* (14)	*A. aestus* (6)
Mol% G+C	67–68	60–63	52–55	57–58
Ring cleavage[c]	*o*	*o*	*o*	−
Oxidase	+	−	+	+
Growth at 4°C	−	−	+	−
L-Arabinose	−	+	−	−
D-Mannose	−	+	−	−
Saccharate	−	+	−	−
Suberate	−	−	−	+
Sebacate	−	−	1	+
Glycolate	−	+	−	−
Aconitate	**5**	+	**13**	−
Mannitol	−	+	**12**	+
δ-Aminovalerate	+	**4**	+	−
L-Histidine	+	**4**	**12**	−
L-Tyrosine	+	+	+	−
DL-Kynurenine	+	−	−	−
Ethanolamine	−	−	**13**	−
Benzylamine	**5**	−	−	−
Putrescine	+	+	+	−
Sarcosine	+	+	+	−
Allantoin	**5**	+	+	−

[a] +, All strains positive; −, all strains negative; numbers indicate number of positive strains; boldface numbers indicate that the number represents 80% or more of the strains. Data from Baumann et al. (1972).
[b] Number in parentheses refers to number of strains studied.
[c] Ring cleavage for strains capable of growth on aromatic compounds.

Doudoroff, 1972; Palleroni et al., 1973). As can be seen from this brief overview of the nonfermentative marine eubacteria, considerable work will be necessary before generic assignments of these species can be made on the basis of some degree of natural relationship.

For purposes of identification, we have compiled, in Table 4, those traits that distinguish species and groups of polarly flagellated, rod-shaped organisms not accumulating PHB. This category includes all the species of *Alteromonas, P. nautica,* and the as yet unassigned group H-2. Distinguishing traits for species and groups of rod-shaped, polarly flagellated organisms which accumulate PHB are presented in Table 5. All these species and groups are members of the genus *Pseudomonas;* Table 6 lists the diagnostic traits for the differentiation of species of *Alcaligenes* that are assigned to this genus on the basis of their peritrichous flagellation.

Literature Cited

Ad Hoc Committee of the Judicial Commission of the ICSB 1980. Approved lists of bacterial names. International Journal of Systematic Bacteriology 30:225–420.

Allen, R. D., Baumann, P. 1971. Structure and arrangement of flagella in species of the genus Beneckea and Photobacterium fischeri. Journal of Bacteriology 107:295–302.

Anderson, R. S., Ordal, E. J. 1972. Deoxyribonucleic acid relationships among marine vibrios. Journal of Bacteriology 109:696–706.

Baker, D. A., Park, R. W. A. 1975. Changes in morphology and cell wall structure that occur during growth of Vibrio sp. NCTC 4716 in batch culture. Journal of General Microbiology 86:12–28.

Bang, S. S., Baumann, P., Nealson, K. H. 1978. Phenotypic characterization of Photobacterium logei (sp. nov.), a species related to P. fischeri. Current Microbiology 1:285–288.

Bang, S. S., Woolkalis, M. J., Baumann, P. 1978. Electrophoretic mobilities of superoxide dismutases from species of Photobacterium, Beneckea, Vibrio, and selected terrestrial enterobacteria. Current Microbiology 1:371–376.

Baross, J. A., Tester, P. A., Morita, R. Y. 1978. Incidence, microscopy, and etiology of exoskeleton lesions in the tanner crab, Chionoecetes tanneri. Journal of the Fisheries Research Board of Canada 35:1141–1149.

Bassot, J. M. 1975. Les organes lumineux à bactéries symbiotiques de quelques téléostéens Léiognathides. Archives de Zoologie Expérimentale et Générale 116:359–373.

Baumann, L., Baumann, P. 1973a. Regulation of aspartokinase activity in the genus Beneckea and marine, luminous bacteria. Archiv für Mikrobiologie 90:171–188.

Baumann, L., Baumann, P. 1973b. Enzymes of glucose catabolism in cell-free extracts of non-fermentative marine eubacteria. Canadian Journal of Microbiology 19:302–304.

Baumann, L., Baumann, P. 1974. Regulation of aspartokinase activity in non-fermentative, marine eubacteria. Archives of Microbiology 95:1–18.

Baumann, L., Baumann, P. 1975. Catabolism of D-fructose and D-ribose by Pseudomonas doudoroffii. II. Properties of 1-phosphofructokinase and 6-phosphofructokinase. Archives of Microbiology 105:241–248.

Baumann, L., Baumann, P. 1976. Study of relationship among marine and terrestrial enterobacteria by means of in vitro DNA/ribosomal RNA hybridization. Microbios Letters 3:11–20.

Baumann, L., Baumann, P. 1978. Studies of relationship among terrestrial Pseudomonas, Alcaligenes, and enterobacteria by an immunological comparison of glutamine synthetase. Archives of Microbiology 119:25–30.

Baumann, L., Baumann, P., Mandel, M., Allen, R. D. 1972. Taxonomy of aerobic marine eubacteria. Journal of Bacteriology 110:402–429.

Baumann, P., Bang, S. S., Baumann, L. 1978. Phenotypic characterization of Beneckea anguillara biotypes I and II. Current Microbiology 1:85–88.

Baumann, P., Baumann, L. 1973. Phenotypic characterization of Beneckea parahaemolytica: A preliminary report. Journal of Milk and Food Technology 36:214–219.

Baumann, P., Baumann, L. 1975. Catabolism of D-fructose and D-ribose by Pseudomonas doudoroffii. I. Physiological studies and mutant analysis. Archives of Microbiology 105:225–240.

Baumann, P., Baumann, L. 1977. Biology of the marine enterobacteria: Genera Beneckea and Photobacterium. Annual Review of Microbiology 31:39–61.

Baumann, P., Baumann, L., Mandel, M. 1971. Taxonomy of marine bacteria: The genus Beneckea. Journal of Bacteriology 107:268–294.

Baumann, P., Baumann, L., Reichelt, J. L. 1973. Taxonomy of marine bacteria: Beneckea parahaemolytica and Beneckea alginolytica. Journal of Bacteriology 113:1144–1155.

Baumann, P., Baumann, L., Mandel, M., Allen, R. D. 1971. Taxonomy of marine bacteria: Beneckea nigrapulchrituda sp. n. Journal of Bacteriology 108:1380–1383.

Berger, L. R., Reynolds, D. M. 1958. The chitinase system of a strain of Streptomyces griseus. Biochimica et Biophysica Acta 29:522–534.

Bieger, C. D., Crawford, I. P. 1978. Genes of tryptophan biosynthesis in the marine luminous bacterium Beneckea harveyi. Abstracts of the Annual Meeting of the American Society for Microbiology 1978:160.

Blake, P. A., Merson, M. H., Weaver, R. E., Hollis, D. G., Heublein, P. C. 1979. Disease caused by a marine Vibrio. Clinical characteristics and epidemiology. New England Journal of Medicine 300:1–5.

Boisvert, H., Chatelain, R., Bassot, J. M. 1967. Étude d'un Photobacterium isolé de l'organe lumineux de poissons Leiognathidae. Annales de l'Institut Pasteur 112:520–524.

Buchanan, R. E., Gibbons, N. E. (eds.). 1974. Bergey's manual of determinative bacteriology, 8th ed. Baltimore: Williams & Wilkins.

Burdon, K. L. 1946. Fatty material in bacteria and fungi revealed by staining dried, fixed slide preparations. Journal of Bacteriology 52:665–678.

Caldwell, D. R., Hudson, R. F. 1974. Sodium, an obligate growth requirement for predominant rumen bacteria. Applied Microbiology 27:549–552.

Casellas, J. M., Caria, M. A., Gerghi, M. E. 1977. Aislamiento de Vibrio parahaemolyticus, a partír de cholgas y mejillons en Argentina. Revista de la Asociacion Argentina de Microbiologia 58:41–53.

Chan, K. Y., Baumann, L., Garza, M. M., Baumann, P. 1978. Two new species of Alteromonas—A. espejiana and A. undina International Journal of Systematic Bacteriology 28:217–222.

Clark, W. A., Steigerwalt, A. G. 1977. Deoxyribonucleic acid reassociation experiments with a halophilic, lactose-fermenting vibrio isolated from blood cultures. International Journal of Systematic Bacteriology 27:194–199.

Cocks, G. T., Wilson, A. C. 1972. Enzyme evolution in the Enterobacteriaceae. Journal of Bacteriology 110:793–802.

Cohen, G. N., Stanier, R. Y., Le Bras, G. 1969. Regulation of the biosynthesis of amino acids of the aspartate family in coliform bacteria and the pseudomonads. Journal of Bacteriology 99:791–801.

Colwell, R. R. 1973. Vibrio and Spirilla, pp. 97–104. In: Laskin, A. I., Lechevalier, H. A. (eds.), CRC handbook of microbiology, vol. I. Cleveland, Ohio: CRC Press.

Crawford, I. P. 1975. Gene rearrangements in the evolution of the tryptophan synthetic pathway. Bacteriological Reviews 39:87–120.

Davis, D. H., Stanier, R. Y., Doudoroff, M. 1970. Taxonomic studies on some gram negative polarly flagellated "hydrogen bacteria" and related species. Archiv für Mikrobiologie 70:1–13.

de Boer, W. E., Golten, C., Scheffers, W. A. 1975a. Effects of some physical factors on flagellation and swarming of Vibrio alginolyticus. Netherlands Journal of Sea Research 9:197–213.

de Boer, W. E., Golten, C., Scheffers, W. A. 1975b. Effects of some chemical factors on flagellation and swarming of Vibrio alginolyticus. Antonie van Leeuwenhoek Journal of Microbiology and Serology 41:385–403.

Diedrich, D. L., Cota-Robles, E. H. 1974. Heterogeneity in lipid composition of the outer membrane and cytoplasmic membrane of Pseudomonas BAL-31. Journal of Bacteriology 119:1006–1018.

Doudoroff, M. 1942a. Studies on the luminous bacteria. I. Nutritional requirements of some species with special reference to methionine. Journal of Bacteriology 44:451–459.

Doudoroff, M. 1942b. Studies on luminous bacteria. II. Some

observations on the anaerobic metabolism of facultatively anaerobic species. Journal of Bacteriology **44**:461–467.

Drapeau, G. R., Matula, T. I., MacLeod, R. A. 1966. Nutrition and metabolism of marine bacteria. XV. Relation of Na⁺-activated transport to the Na⁺ requirement of a marine pseudomonad for growth. Journal of Bacteriology **92**:63–71.

Eagon, R. G., Wang, C. H. 1962. Dissimilation of glucose and gluconic acid by *Pseudomonas natriegens*. Journal of Bacteriology **83**:879–886.

Eberhard, A., Rouser, G. 1971. Quantitative analysis of the phospholipids of some marine bioluminescent bacteria. Lipids **6**:410–414.

Eggleton, P., Eldsen, S. R., Gough, N. 1943. The estimation of creatine and diacetyl. Biochemical Journal **37**:526–529.

Espejo, R. T., Canelo, E. S. 1968. Properties and characterization of the host bacterium of bacteriophage PM2. Journal of Bacteriology **95**:1887–1891.

Fein, J. E., MacLeod, R. A. 1975. Characterization of neutral amino acid transport in a marine pseudomonad. Journal of Bacteriology **124**:1177–1190.

Felter, R. A., Colwell, R. R., Chapman, G. B. 1969. Morphology and round body formation in *Vibrio marinus*. Journal of Bacteriology **99**:326–335.

Fischer, B. 1894. Die Bakterien des Meeres nach den Untersuchungen der Plankton-Expedition unter gleichzeitiger Berücksichtigung einiger älterer und neuerer Untersuchungen. Ergebnisse der Plankton-Expedition der Humbolt-Stiftung **4**:1–83.

Fitzgerald, J. M. 1977. Classification of luminous bacteria from the light organ of the Australian Pinecone fish, *Cleidopus gloriamaris*. Archives of Microbiology **112**:153–156.

Forsberg, C. W., Costerton, J. W., MacLeod, R. A. 1970. Separation and localization of cell wall layers of a Gram-negative bacterium. Journal of Bacteriology **104**:1338–1353.

Fox, G. E., Pechman, K. R., Woese, C. R. 1977. Comparative cataloging of 16S ribosomal ribonucleic acid: Molecular approach to procaryotic systematics. International Journal of Systematic Bacteriology **27**:44–57.

Fujino, T., Sakaguchi, G., Sakazaki, R., Takeda, Y. (eds.). 1974. International symposium on *Vibrio parahaemolyticus*. Tokyo: Saikon.

Gauthier, M. J. 1976a. Morphological, physiological and biochemical characteristics of some violet pigmented bacteria isolated from seawater. Canadian Journal of Microbiology **22**:138–149.

Gauthier, M. J. 1976b. *Alteromonas rubra* sp. nov., a new marine antibiotic-producing bacterium. International Journal of Systematic Bacteriology **26**:459–466.

Gauthier, M. J. 1977. *Alteromonas citrea* sp. nov., a new Gram-negative yellow pigmented bacterium isolated from seawater. International Journal of Systematic Bacteriology **27**:349–354.

Gee, D. L., Baumann, P., Baumann, L. 1975. Enzymes of D-fructose catabolism in species of *Beneckea* and *Photobacterium*. Archives of Microbiology **103**:205–207.

Gerber, N. N., Gauthier, M. J. 1979. New prodigiosin-like pigment from *Alteromonas rubra*. Applied and Environmental Microbiology **37**:1176–1179.

Golten, C., Scheffers, W. A. 1975. Marine vibrios isolated from water along the Dutch coast. Netherlands Journal of Sea Research **9**:351–364.

Gundersen, K. 1976. Cultivation of micro-organisms. 3.1. Bacteria, pp. 301–356. In: Kinne, O. (ed.), Marine ecology, vol. 3, part 1. London: John Wiley & Sons.

Harrell, L. W., Novotny, A. J., Schiewe, M. H., Hodgins, H. O. 1976. Isolation and description of two vibrios pathogenic to Pacific salmon in Puget Sound, Washington. Fishery Bulletin **74**:447–449.

Harvey, E. N. 1915. The effect of certain organic and inorganic substances upon light production by luminous bacteria. Biological Bulletin of the Marine Biology Laboratory, Woods Hole **29**:308–311.

Harwood, C. S. 1978. *Beneckea gazogenes* sp. nov., a red, facultatively anaerobic, marine bacterium. Current Microbiology **1**:233–238.

Hastings, J. W., Nealson, K. H. 1977. Bacterial bioluminescence. Annual Review of Microbiology **31**:549–595.

Hendrie, M. S., Hodgkiss, W., Shewan, J. M. 1970. The identification, taxonomy and classification of luminous bacteria. Journal of General Microbiology **64**:151–169.

Herring, P. J. 1975. Bacterial bioluminescence in some argentinoid fishes, pp. 563–572. In: Barnes, H. (ed.). Proceedings of the 9th European Marine Biology Symposium. Aberdeen: Aberdeen University Press.

Hidaka, T., Sakai, M. 1968. Comparative observation of the inorganic salt requirements of the marine and terrestrial bacteria. Bulletin of the Misaki Marine Biology Institute, Kyoto University **12**:125–149.

Hollis, D. G., Weaver, R. E., Baker, C. N., Thornsberry, C. 1976. Halophilic *Vibrio* species isolated from blood cultures. Journal of Clinical Microbiology **3**:425–431.

Humm, H. J. 1946. Marine agar-digesting bacteria of the south Atlantic coast. Duke University Marine Laboratory Bulletin **3**:45–75.

Ingraham, J. L. 1962. Temperature relationships, pp. 285–288. In: Gunsalus, I. C., Stanier, R. Y. (eds.). The bacteria, vol. IV. New York: Academic Press.

Jannasch, H. W. 1968. Competitive elimination of *Enterobacteriaceae* from sea water. Applied Microbiology **16**:1616–1618.

Jannasch, H. W., Wirsen, C. O. 1977. Retrieval of concentrated and undecompressed microbial populations from the deep sea. Applied and Environmental Microbiology **33**:642–646.

Jannasch, H. W., Wirsen, C. O., Taylor, C. D. 1976. Undecompressed microbial populations from the deep sea. Applied and Environmental Microbiology **32**:360–367.

Kennedy, S. F., Colwell, R. R., Chapman, G. B. 1970. Ultrastructure of a marine psychrophilic *Vibrio*. Canadian Journal of Microbiology **16**:1027–1031.

Kinne, O. 1976. Cultivation of marine organisms: Water-quality management and technology, pp. 19–36. In: Kinne, O., (ed.). Marine ecology, vol. 3, part 1. London: John Wiley & Sons.

Kodama, T., Tanaguchi, S. 1976. Sodium-dependent growth and respiration of a nonhalophilic bacterium, *Pseudomonas stutzeri*. Journal of General Microbiology **96**:17–24.

Kubota, Y., Iuchi, S., Fujisawa, A., Tanaka, S. 1979. Separation of four components of the phosphoenolpyruvate: Glucose phosphotransferase system in *Vibrio parahaemolyticus*. Microbiology and Immunology **23**:131–146.

Laddaga, R. A., MacLeod, R. A. 1977. Relation of salts to the integrity of the outer membrane of some Gram-negative marine bacteria. Abstracts of the Annual Meeting of the American Society for Microbiology **1977**:186.

Larsen, H. 1962. Halophilism, pp. 297–342. In: Gunsalus, I. C., Stanier, R. Y. (eds.). The bacteria, vol. IV. New York: Academic Press.

Lederberg, J., Lederberg, E. M. 1952. Replica plating and indirect selection of bacterial mutants. Journal of Bacteriology **63**:399–406.

Lee, J. V., Donovan, T. J., Furniss, A. L. 1978. Characterization, taxonomy, and emended description of *Vibrio metschnikovii*. International Journal of Systematic Bacteriology **28**:99–111.

Lee, J. V., Shread, P., Furniss, A. L. 1978. The taxonomy of group F organisms: Relationships to *Vibrio* and *Aeromonas*. Journal of Applied Bacteriology **45**:ix.

Leifson, E. 1960. Atlas of bacterial flagellation. New York: Academic Press.

Levin, R. E., Vaughn, R. H. 1968. Spontaneous spheroplast formation by *Desulfovibrio aestuarii*. Canadian Journal of Microbiology **14**:1271–1276.

London, J. 1977. A demonstration of evolutionary relationships among the lactic acid bacteria by an immunochemical study of malic enzyme and fructose diphosphate aldolase, pp. 58–88. In: Salton, M. R. J. (ed.). Immunochemistry of enzymes. London: John Wiley & Sons.

McCarthy, D. H. 1975. *Aeromonas proteolytica*—a halophilic aeromonad? Canadian Journal of Microbiology 21:902–904.

MacLeod, R. A. 1965. The question of the existence of specific marine bacteria. Bacteriological Reviews 29:9–23.

MacLeod, R. A. 1968. On the role of inorganic ions in the physiology of marine bacteria. Advances in Microbiology of the Sea 1:95–126.

Mandel, M. 1969. New approaches to bacterial taxonomy: Perspective and prospects. Annual Review of Microbiology 23:239–274.

Matsuo, T., Kohno, S., Ikeda, T., Saruwatari, K., Ninomiya, H. 1978. Fulminating lactose-positive *Vibrio* septicemia. Acta Pathologica Japonica 28:937–948.

Mertens, A., Nagler, J., Hansen, W., Gepts-Friedenreich, E. 1979. Halophilic, lactose-positive *Vibrio* in a case of fatal septicemia. Journal of Clinical Microbiology 9:233–235.

Mitchell, R., Morris, J. C. 1969. The fate of intestinal bacteria in the sea, pp. 811–817. In: Jenkins, S. H. (ed.). Advances in water pollution research. Oxford: Pergamon Press.

Miwatani, T., Takeda, Y. 1976. *Vibrio parahaemolyticus*. A causative bacterium of food poisoning. Tokyo: Saikon.

Moebus, K. 1972. Bactericidal properties of natural and synthetic sea water as influenced by addition of low amounts of organic matter. Marine Biology 15:81–88.

Morishita, H., Takada, H. 1976. Sparing effects of lithium ion on the specific requirement for sodium ion for growth of *Vibrio parahaemolyticus*. Canadian Journal of Microbiology 22:1263–1268.

Neish, A. C. 1952. Analytical methods for bacterial fermentations. Saskatoon, Canada: National Research Council of Canada.

Palleroni, N. J. 1975. General properties and taxonomy of the genus *Pseudomonas*, pp. 1–36. In: Clarke, P. H., Richmond, M. H. (eds.). Genetics and biochemistry of *Pseudomonas*. London: John Wiley & Sons.

Palleroni, N. J., Doudoroff, M. 1972. Some properties and taxonomic subdivisions of the genus *Pseudomonas*. Annual Review of Phytopathology 10:73–100.

Palleroni, N. J., Kunisawa, R., Contopoulou, B., Doudoroff, M. 1973. Nucleic acid homologies in the genus *Pseudomonas*. International Journal of Systematic Bacteriology 23:333–339.

Payne, W. J., Eagon, R. G., Williams, A. K. 1961. Some observations on the physiology of *Pseudomonas natriegens, nov. spec.* Antonie van Leeuwenhoek Journal of Microbiology and Serology 27:121–128.

Pfister, R. M., Burkholder, P. R. 1965. Numerical taxonomy of some bacteria isolated from antarctic and tropical sea waters. Journal of Bacteriology 90:863–872.

Pien, F., Lee, K., Higa, H. 1977. *Vibrio alginolyticus* infections in Hawaii. Journal of Clinical Microbiology 5:670–672.

Pratt, D. 1974. Salt requirements for growth and function of marine bacteria, pp. 3–15. In: Colwell, R. R., Morita, R. Y., (eds.). Effect of the ocean environment on microbial activities. Baltimore: University Park Press.

Ratner, S. 1962. Transaminidase, pp. 843–848. In: Colowick, S. P., Kaplan, N. O. (eds.). Methods in enzymology, vol. 5. New York: Academic Press.

Reichelt, J. L., Baumann, P. 1973a. Change of the name *Alteromonas marinopraesens* to *Alteromonas haloplanktis* comb. nov. and assignment of strain ATCC 23821 and strain c-A1 of De Voe and Oginsky to this species. International Journal of Systematic Bacteriology 23:438–441.

Reichelt, J. L., Baumann, P. 1973b. Taxonomy of the marine, luminous bacteria. Archiv für Mikrobiologie 94:283–330.

Reichelt, J. L., Baumann, P. 1974. Effect of sodium chloride on growth of heterotrophic marine bacteria. Archives of Microbiology 97:329–345.

Reichelt, J. L., Baumann, P. 1975. *Photobacterium mandapamensis* Hendrie et al., a later subjective synonym of *Photobacterium leiognathi* Boisvert et al. International Jour-

nal of Systematic Bacteriology 25:208–209.

Reichelt, J. L., Baumann, P., Baumann, L. 1976. Study of genetic relationships among marine species of the genera *Beneckea* and *Photobacterium* by means of in vitro DNA/DNA hybridization. Archives of Microbiology 110:101–120.

Reichelt, J. L., Nealson, K., Hastings, J. W. 1977. The specificity of symbiosis: Pony fish and luminescent bacteria. Archives of Microbiology 112:157–161.

Richter, O. 1928. Natrium: Ein notwendiges Nährelement für eine marine mikroaerophile Leuchtbakterie. Anzeiger der Akademie der Wissenschaften, Wien, Mathematische-Naturwissenschaftliche Klasse 101:261–292.

Rubin, S. J., Tilton, R. C. 1975. Isolation of *Vibrio alginolyticus* from wound infections. Journal of Clinical Microbiology 2:556–558.

Ruby, E. G., Greenberg, E. P., Hastings, J. W. 1980. Planktonic marine luminous bacteria: Species distribution in the water column. Applied and Environmental Microbiology, 39:302–306.

Ruby, E. G., Morin, J. G. 1978. Specificity of symbiosis between deep-sea fishes and psychrotrophic luminous bacteria. Deep-Sea Research 25:161–167.

Ruby, E. G., Nealson, K. H. 1976. Symbiotic association of *Photobacterium fischeri* with the marine luminous fish *Monocentris japonica*: A model of symbiosis based on bacterial studies. Biological Bulletin of the Marine Biology Laboratory, Woods Hole 151:574–586.

Ruby, E. G., Nealson, K. H. 1978. Seasonal changes in the species composition of luminous bacteria in nearshore sea water. Limnology and Oceanography 23:530–533.

Ryan, W. J. 1976. Marine vibrios associated with superficial septic lesions. Journal of Clinical Pathology 29:1014–1015.

Sakazaki, R. 1968. Proposal of *Vibrio alginolyticus* for the biotype 2 of *Vibrio parahaemolyticus*. Japanese Journal of Medical Science and Biology 21:359–362.

Sakazaki, R., Iwanami, S., Fukumi, H. 1963. Studies on the enteropathogenic, facultatively halophilic bacteria, *Vibrio parahaemolyticus*. I. Morphological, cultural and biochemical properties and its taxonomic position. Japanese Journal of Medical Science and Biology 16:161–188.

Sawyer, M. H., Baumann, P., Baumann, L. 1977. Pathways of D-fructose and D-glucose catabolism in marine species of *Alcaligenes, Pseudomonas marina*, and *Alteromonas communis*. Archives of Microbiology 112:169–172.

Schiewe, M. H., Crosa, J. H., Ordal, E. J. 1977. Deoxyribonucleic acid relationships among marine vibrios pathogenic to fish. Canadian Journal of Microbiology 23:954–958.

Shilo, M., Yetinson, T. 1979. Physiological characteristics underlying the distribution patterns of luminous bacteria in the Mediterranean Sea and the Gulf of Elat. Applied and Environmental Microbiology 38:577–584.

Shinoda, S., Okamoto, K. 1977. Formation and function of *Vibrio parahaemolyticus* lateral flagella. Journal of Bacteriology 129:1266–1271.

Shinoda, S., Honda, T., Takeda, Y., Miwatani, T. 1974a. Antigenic difference between polar monotrichous and peritrichous flagella of *Vibrio parahaemolyticus*. Journal of Bacteriology 120:923–928.

Shinoda, S., Miwatani, T., Honda, T., Takeda, Y. 1974b. Antigenicity of flagella of *Vibrio parahaemolyticus*, pp. 193–197. In: Fujino, T., Sakaguchi, G., Sakazaki, R., Takeda, Y. (eds.). International Symposium on *Vibrio parahaemolyticus*. Tokyo: Saikon.

Shinoda, S., Kariyama, R., Ogawa, M., Takeda, Y., Miwatani, T. 1976. Flagellar antigens of various species of the genus *Vibrio* and related genera. International Journal of Systematic Bacteriology 26:97–101.

Sieburth, J. M. 1979. Sea microbes. New York: Oxford University Press.

Sinderman, C. J., 1966. Diseases of marine fish. London: Academic Press.

Sinderman, C. J. 1970. Principal diseases of marine fish and shellfish. New York: Academic Press.

Singleton, R. J., Skerman, T. M. 1973. A taxonomic study by computer analysis of marine bacteria from New Zealand waters. Journal of the Royal Society of New Zealand 3:129–140.

Skerman, V. B. D. 1967. A guide to the identification of the genera of bacteria, 2nd ed. Baltimore: Williams & Wilkins.

Slepecky, R. A., Law, J. H. 1960. A rapid spectrophotometric assay of alpha, beta unsaturated acids and beta-hydroxy acids. Analytical Chemistry 32:1697–1699.

Smith, I. W. 1961. A disease of finnock due to *Vibrio anguillarum*. Journal of General Microbiology 24:247–252.

Staley, T. E., Colwell, R. R. 1973. Deoxyribonucleic acid reassociation among members of the genus *Vibrio*. International Journal of Systematic Bacteriology 23:316–332.

Stanier, R. Y. 1941. Studies on marine agar digesting bacteria. Journal of Bacteriology 42:527–559.

Stanier, R. Y. 1971. Toward an evolutionary taxonomy of the bacteria, pp. 595–604. In: Pérez-Miravete, A., Peláez, D. (eds.). Recent advances in microbiologia. Mexico, D. F.: Associacion Mexicana de Microbiologia.

Stanier, R. Y., Adelberg, E. A., Ingraham, J. 1976. The microbial world, 4th ed. pp. 612–629. Englewood Cliffs, New Jersey: Prentice-Hall.

Stanier, R. Y., Palleroni, N. J., Doudoroff, M. 1966. The aerobic pseudomonads: A taxonomic study. Journal of General Microbiology 43:159–271.

Starr, M. P., Chatterjee, A. K. 1972. The genus *Erwinia*: Enterobacteria pathogenic to plants and animals. Annual Review of Microbiology 26:389–426.

Stephen, S., Vaz, A. L., Chandrashekara, I., Rao, K. N. A. 1978. Characterization of *Vibrio alginolyticus (Beneckea alginolytica)* isolated from the fauna of Arabian sea. Indian Journal of Medical Research 68:7–11.

Taylor, V. I., Baumann, P., Reichelt, J. L., Allen, R. D. 1974. Isolation, enumeration, and host range of marine bdellovibrios. Archives of Microbiology 98:101–114.

Thompson, J., Costerton, J. W., MacLeod, R. A. 1970. K⁺-dependent deplasmolysis of a marine pseudomonad plasmolyzed in a hypotonic solution. Journal of Bacteriology 102:843–854.

Thompson, J., MacLeod, R. A. 1973. Na⁺ and K⁺ gradients and α-aminoisobutyric acid transport in a marine pseudomonad. Journal of Biological Chemistry 248:7106–7111.

Thompson, J., MacLeod, R. A. 1974. Potassium transport and the relationship between intracellular potassium concentration and amino acid uptake by cells of a marine pseudomonad. Journal of Bacteriology 120:598–603.

Thornley, M. J. 1960. The differentiation of *Pseudomonas* from other Gram-negative bacteria on the basis of arginine metabolism. Journal of Applied Bacteriology 23:37–52.

Twedt, R. M., Spaulding, L., Hall, H. E. 1969. Morphological, cultural, biochemical, and serological comparisons of Japanese strains of *Vibrio parahaemolyticus* with related cultures isolated in the United States. Journal of Bacteriology 98:511–518.

Ulitzur, S. 1974. Induction of swarming in *Vibrio parahaemolyticus*. Archives of Microbiology 101:357–363.

Ulitzur, S. 1975. The mechanism of swarming of *Vibrio alginolyticus*. Archives of Microbiology 104:67–71.

Watson, S. W., Waterbury, J. B. 1971. Characteristics of two marine nitrite oxidizing bacteria, *Nitrospina gracilis nov. gen. nov. sp.* and *Nitrococcus mobilis nov. gen. nov. sp.* Archiv für Mikrobiologie 77:203–230.

Williamson, D. H., Wilkinson, J. K. 1958. The isolation and estimation of poly-β-hydroxybutyrate inclusions in *Bacillus* species. Journal of General Microbiology 19:198–209.

Wong, P. T. S. 1969. Studies on the mechanism of Na⁺ dependent transport in marine bacteria. Ph. D. Thesis. McGill University, Montreal, Canada.

Yabuuchi, E., Miwatani, T., Takeda, Y., Arita, M. 1974. Flagellar morphology of *Vibrio parahaemolyticus* (Fujino et al.) Sakazaki, Iwanami and Fukumi 1963. Japanese Journal of Microbiology 18:295–305.

Yetinson, T., Shilo, M. 1979. Seasonal and geographic distribution of luminous bacteria in the Eastern Mediterranean Sea and the Gulf of Elat. Applied and Environmental Microbiology 37:1230–1238.

ZoBell, C. E. 1941. Studies on marine bacteria. I. The cultural requirements of heterotrophic aerobes. Journal of Marine Research 4:42–75.

ZoBell, C. E. 1963. Domain of the marine microbiologist, pp. 3–24. In: Oppenheimer, C. H. (ed.). Marine microbiology. Springfield, Illinois: Charles C Thomas.

Addendum

The results of recent studies of the evolution of glutamine synthetase and superoxide dismutase have necessitated a revision of the generic assignments for the marine enterobacteria (Baumann, Bang, and Baumann, 1980; Baumann et al., 1980). Species assigned to *Beneckea* as well as the species *Photobacterium fischeri* and *P. logei* are now placed into *Vibrio* while the genus *Photobacterium* is restricted to *P. phosphoreum, P. leiognathi,* and *P. angustum*. The following is a list of species of marine vibrios with the designations used in this chapter given in parentheses: *Vibrio fischeri (P. fischeri), V. logei (P. logei), V. harveyi (Beneckea harveyi), V. campbellii (B. campbellii), V. parahaemolyticus (B. parahaemolytica), V. alginolyticus (B. alginolytica), V. natriegens (B. natriegens), V. vulnificus (B. vulnifica), V. splendidus (B. splendida), V. pelagius (B. pelagia), V. nereis (B. nereida), V. anguillarum (B. anguillarum), V. nigripulcritudo*

(B. nigrapulchrituda), and *V. proteolyticus (B. proteolytica)*. Two additional species of *Alteromonas* have been recently described: *A. aurantia* (Gauthier and Breittmayer, 1979) and a luminous species, *A. hanedai* (Jensen et al., 1980).

Baumann, L., Bang, S. S., Baumann, P. 1980. Study of relationship among species of *Vibrio, Photobacterium,* and terrestrial enterobacteria by an immunological comparison of glutamine synthetase and superoxide dismutase. Current Microbiology 4:133–138.

Baumann, P., Baumann, L., Bang, S. S., Woolkalis, M. J. 1980. Reevaluation of the taxonomy of *Vibrio, Beneckea,* and *Photobacterium*: Abolition of the genus *Beneckea*. Current Microbiology 4:127–132.

Gauthier, M. J., Breittmayer, V. A. 1979. A new antibiotic-producing bacterium from seawater: *Alteromonas aurantia* sp. nov. International Journal of Systematic Bacteriology 29:366–372.

Jensen, M. J., Tebo, B. M., Baumann, P., Mandel, M., Nealson, K. H. 1980. Characterization of *Alteromonas hanedai* (sp. nov.), a nonfermentative luminous species of marine origin. Current Microbiology 3:311–315.

The Symbiotic Luminous Bacteria

J. WOODLAND HASTINGS and KENNETH H. NEALSON

The bioluminescent bacteria have long intrigued microbiologists, in particular because it is not completely evident why they should emit light. In fact, what is the function of bacterial luminescence? One explanation is that the light emission is of no direct value to the bacteria themselves, but is of importance as a light source to a partner in a mutualistic, symbiotic relationship. In return, the luminous bacteria are provided with a protected niche and nutrients. There are many clear examples of such specific symbiotic relationships, most of these involve a vertebrate (fish) host that carries an apparently pure culture of a given luminous bacterial species in a specially adapted tissue compartment called a light organ (Harvey, 1957b; Herring and Morin, 1978; Nealson and Hastings, 1979). To our knowledge, these relationships are biologically unique; no other such mutualisms are known to exist that involve vertebrates that maintain pure cultures of prokaryotes in specialized organs. However, those species of luminous bacteria that inhabit light organs are not limited to this existence; they appear to possess an impressive ecological versatility, and the luminescence may have other functions. Thus, this chapter will focus not only on the biology of specific light organ symbionts, but also on the spectrum of symbiotic and nonsymbiotic modes of existence of the luminous bacteria.

Life Styles and Modes of Existence

Are there species of luminous bacteria that are not involved in symbiotic relationships? Some biologists have assumed that free-living isolates might simply be escapees from the symbiotic state, perhaps growing opportunistically as saprophytes. On the other hand, other workers, dating back to Beijerinck (1889, 1916), have dealt primarily with organisms isolated directly from the ocean or as saprophytes, and have given little consideration to the question of symbiosis. From recent taxonomic studies, it is now clear that, among marine luminous bacteria, there are at least two major generic groups, *Photobacterium* and *Beneckea (Vibrio)* (Baumann and Baumann,

1977; Reichelt and Baumann, 1973; Baumann and Baumann, this Handbook, Chapter 104). In our discussions here, we will use the generic name *Beneckea*, but it should be noted that there is some controversy regarding the validity of this genus, and some workers feel it should be referred to as *Vibrio* (see discussion in Hastings and Nealson, 1977). All bacteria that have been isolated from light organs as symbionts are species of *Photobacterium*, while no *Beneckea* species has been identified in this niche. All species from both genera occur free in the ocean and may be isolated directly from seawater samples. Thus, if the light organ is considered to be the only symbiotic niche, one could, at our present state of knowledge, designate the *Photobacterium* species as symbiotic, and the *Beneckea* species as free-living (Hastings and Mitchell, 1971).

However, if symbiosis is used in its original broad sense as defined by de Bary (1879) and by many contemporary microbiologists (Stanier, Adelberg, and Ingraham, 1976; Starr, 1975), then probably all species of luminous bacteria may be considered to be involved in symbiotic relationships, be they mutualistic, parasitic, or commensal. Only in the mutualistic light organ symbioses is the relationship species specific and of obvious value to the host. In the other cases, indeed in the saprophytic mode as well, the function of the light emission has not been well established, and some authors have considered that it is of no use to the bacteria (Harvey, 1952; McElroy and Seliger, 1962). However, the luminescence may well have a positive selection value in these cases, that relates to the propagation of the bacteria. The light emission of the bacteria growing on a substrate, be it a parasitized crustacean, the surface of a dead fish or squid, or a fecal pellet, could serve to attract organisms to feed on the material, to the ultimate end of propagating the bacteria (Hastings and Nealson, 1977; Nealson and Hastings, 1979; Robison and Morin, 1977). Ingested luminous bacteria survive in the gut and are expelled into the ocean (Ruby, 1977). Irrespective of the function or even the activity of luminescence in associations other than light organs, it is clear that parasitic, commensal, and saprophytic relationships may be taxonomically nonspecific, and that all luminous species can flourish opportunistically in these modes.

The marine luminous genera, *Photobacterium* and *Beneckea (Vibrio)*, are described in detail in this Handbook, Chapter 104, along with other marine Gram-negative eubacteria. In this chapter, we will focus specifically on the bioluminescent species of these genera as well as the newly described terrestrial genus, *Xenorhabdus* (Thomas and Poinar, 1979), with special reference to their physiological properties, and the function of light emission, emphasizing the major differences between the species.

Distinctions Between Species

Although all species of luminous bacteria utilize similar pathways and enzyme systems to achieve light emission (Fig. 1), the *Photobacterium* and *Beneckea* species can be distinguished by a simple and quick assay of the luciferase in crude cell extracts (Hastings and Mitchell, 1971; Hastings et al., 1969; Nealson, 1978). When dodecanal (12 carbon aldehyde) is used in the reaction in vitro, the decay of luminescence is relatively fast with luciferases from any of the *Photobacterium* species, and about five to ten times slower with *Beneckea* luciferases (Fig. 4). For isolates of a given species, the decay rates are quite similar, within 10% of a mean value (Fitzgerald, 1979; K. H. Nealson, unpublished).

It is also useful to consider the control of luciferase synthesis in relation to the different species. Luciferase is not a constitutive enzyme; in certain species (but probably not all strains; Katznelson and Ulitzur, 1977) of both genera it is inducible, and subject to several other complex controls that oper-

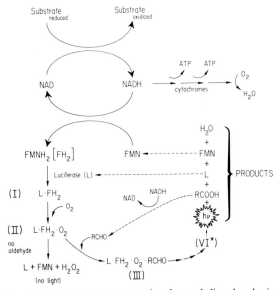

Fig. 1. Substrates, enzymes, and pathways believed to be involved in the bacterial bioluminescent reaction.

ate at the level of transcription (Hastings and Nealson, 1977; Nealson, Platt, and Hastings, 1970). The induction mechanism, which may be significant in terms of the ecology of some of the luminous species, is actually an autoinduction. The luminous bacteria produce a small molecule (autoinducer) that accumulates in the medium and induces the synthesis of the components of the luminous system. For *Photobacterium fischeri*, autoinducer has been partially purified and shown to be a highly specific small molecule (Eberhard, 1972; Magner, Eberhard, and Nealson, 1972; Nealson, 1977). It induces synthesis of the luminescent system for *P. fischeri* strains, but not for strains of other species. Similar studies with crude inducer preparations indicate that inducer activity for *Beneckea harveyi* luminescence is produced by several different *Beneckea* species, both luminous and nonluminous (Greenberg, Hastings, and Ulitzur, 1978).

Ecologically, the requirement for an inducer could be significant. For bacteria free in sea water, external autoinducer concentrations remain low, and luciferase synthesis will be repressed, thus conserving energy. Conversely, for associated bacteria (saprophytes, parasites, or symbionts), autoinducer could accumulate to levels sufficient for derepression of the synthesis of the luminous system to occur. Such a mechanism may be viewed as an adaptation to alternate environments, where luminescence has advantage in one, and no advantage, or even disadvantage, in the other (Hastings and Nealson, 1977). The environment that favors the development of the luminescent system is relatively nonspecific. It requires only that the bacteria be "enclosed", so that inducer will accumulate. In an open system, such as with bacteria that are free in sea water (Booth and Nealson, 1975), the bioluminescent system is not synthesized (Rosson and Nealson, 1979) so there is no need to discuss the function of light emission.

Inducible enzymes are ones that are synthesized and have a function under some conditions and not others (Magasanik, 1970), and the synthesis of such enzymes is often subject to catabolite repression. In the *Beneckea* species, luciferase synthesis is subject to catabolite repression by glucose, and this repression is reversed by cyclic AMP (Nealson, Eberhard, and Hastings, 1972). On the other hand, although there are inhibitory effects of glucose in some of the *Photobacterium* species, no permanent catabolite repression has been found (Makemson, 1973; Ruby and Nealson, 1976). The ecological significance of these distinctions is not immediately clear, although it may relate to the fact that the light organ symbionts receive nutrients from the host via the blood stream which may be high in glucose. Repression of luminescence could be important in the guts of fishes, where the product of chitin digestion (*N*-acetylglucosamine) could repress and thus block the development of luminescence. However, because all luminous species can presumably occupy the gut, it is not possible at this time to postulate an ecological significance for differences in catabolite repression between the species as related to this niche. It is worth noting that in many species of fish, the gut tracts possess pigmentation that effectively absorbs any light emission from gut contents (Herring and Morin, 1978).

Another feature that provides some distinction between luminous species is the effect of oxygen on growth and luciferase synthesis (Nealson and Hastings, 1977). Although oxygen is an absolute requirement for luminescence, only low levels are required. Bacteria continue to emit light at a maximum rate even at oxygen concentrations in the medium of about 0.25% of air saturation (Harvey, 1952). In some luminous species, growth at low oxygen tensions actually increases the amount of luciferase per cell. In *P. fischeri* and *P. phosphoreum*, the synthesis of the bioluminescent system still occurs at oxygen concentrations at which aerobic growth is greatly inhibited. On the other hand, in *B. harveyi* and in *P. leiognathi,* growth and luciferase synthesis are both blocked, and to the same extent, at low oxygen tensions (Nealson and Hastings, 1977). As with catabolite repression, the ecological significance of the oxygen effect is not altogether clear. The differences between *P. leiognathi* and other *Photobacterium* species suggest that the physiological conditions in the light organs might be distinctly different and appropriately correlated. For the nonspecific niches, such a hypothesis is clearly not fruitful.

Another unusual feature of the metabolism of luminous bacteria that may be related to their symbiotic association in light organs is the excretion of pyruvate during aerobic growth on a variety of sugars (Ruby and Nealson, 1977). Among the *Photobacterium* species, *P. fischeri* is the most active in this regard, returning as much as 35% of the glucose carbon to the growth medium as pyruvate. *P. phosphoreum* is less active, and no *P. leiognathi* strains have been found to excrete pyruvate. *Beneckea* species, both luminous and nonluminous, are quite variable in this property, but many do excrete pyruvate (Ruby and Nealson, 1977); *B. harveyi* does so on a wide variety of sugars, including N-acetylglucosamine (Ruby, 1977).

Finally, there are significant nutritional differences between the *Photobacterium* and *Beneckea* species. *Beneckea* species are more cosmopolitan and nutritionally versatile, capable of utilizing 30–45 (depending on the strain) of 147 different organic compounds tested as carbon and energy sources, whereas *Photobacterium* species utilize only between 7 and 22, depending on species and strain (Reichelt and Baumann, 1973). In general, freshly isolated luminous bacteria of both genera,

including those isolated directly from light organs, will grow prototrophically in a minimal medium. The exception to this may be *P. phosphoreum*, where a considerable fraction (about 45%) of the strains tested by Reichelt and Baumann (1973) exhibited some growth requirement, most commonly methionine. Only a few (about 5%) of the other three species required growth factors (the type strain of *P. leiognathi* required methionine; Reichelt and Baumann, 1975) and all of these had been maintained in culture collections for several years. Ruby and Nealson (1978) found no auxotrophic strains among some 2,300 freshly isolated strains from coastal water near San Diego, California (*B. harveyi*, 651; *P. fischeri*, 1,601; *P. phosphoreum*, 16).

Habitats

In the marine environment, both *Photobacterium* and *Beneckea* are widely distributed. In addition to their occurrence in sea water, they exist in a variety of habitats associated with living and dead animal material, including specific exosymbioses (Table 1). There are also suggestions that the luminous bacteria occur as endosymbionts in squids and tunicates (Buchner, 1965; Harvey, 1952; Leisman, Cohn, and Nealson, 1980; Mackie and Bone, 1978). The existence of nonmarine luminous bacteria is known from both the older literature and very recent studies (Khan and Brooks, 1976; Poinar and Thomas, 1977), but knowledge concerning habitats and physiology is very limited. For instance, *Xenorhabdus luminescens* (Thomas and Poinar, 1979) has thus far been found only in the alimentary tract of *Heterorhabditis bacteriophora* (a parasitic nematode) and inside insects attacked by the nematode.

Sea Water

Although the distributional patterns are not yet understood, it is clear that all marine luminous species live free in sea water. Coastal water near San Diego, California, was found to contain between 10^3 and 6×10^3 luminous bacteria per liter measured in monthly samples over a 2-year period (Ruby and Nealson, 1978). Some were evidently associated with particulate matter, but many were unattached; filtration of sea water through an 8-μm membrane filter decreased the colony count only between 10 and 50% (Ruby, 1977). The most prevalent luminous species isolated in the summer were in the genus *Beneckea*, while *Photobacterium fischeri* was found to predominate in the winter. Differences in the species composition of summer and winter luminous populations have also been found in Woods Hole, Massachusetts, coastal seawater samples (E. G. Ruby and J. W. Hastings, unpublished).

Table 1. Habitats of bioluminescent bacteria.

Mode	Habitat or host	Bacterial species[a]
Free-living	Sea water	All *Photobacterium* and *Beneckea* species
Saprophytic	Meat, fish, salt meats, wounds	All *Photobacterium* and *Beneckea* species
Parasitic	Many marine crustaceans, some terrestrial and freshwater forms	*P. fischeri, P. phosphoreum* and *B. harveyi.* Others cultured, not identified
Commensal	Digestive tracts of marine fish and invertebrates. Outer surfaces of marine animals.	All *Photobacterium* and *Beneckea* species
Light organ exosymbionts	Teleost fishes (11 families)	All *Photobacterium* species. No *Beneckea* species. Some isolates have not yet been successfully cultured
	Squids	*P. fischeri*[b]; possibly some endosymbionts
Light organ endosymbionts[c]	Tunicates	Not cultured, not identified
Symbiotic/parasitic	Nematode/caterpillar	*Xenorhabdus luminescens*, a new nonmarine genus[d]

[a] Only the luminous species of these genera are considered. Probably all luminous genera have nonluminous species.
[b] G. Leisman and K. H. Nealson (unpublished). Many different isolates have been cultured, assigned names, and found to have characteristics that distinguish them from ordinary marine luminous bacteria (Buchner, 1965; Harvey, 1952). Among the other names used were *Vibrio pierantonii, Coccobacillus pierantonii,* and *Bacillus sullasepia.*
[c] The occurrence of luminous endosymbionts has not been firmly established, although recent reports by Mackie and Bone (1978) support it.
[d] Thomas and Poinar, 1979.

Again, *Beneckea* appears to be most abundant in the summer months, while in the winter, only *P. fischeri,* and to a lesser extent, *P. phosphoreum* occur. It is not clear what factor or factors are responsible for these seasonal variations in species composition of the luminous populations, but such variations had been noted by Beijerinck (1889, 1916) in water samples from the North Sea. In warmer waters, *P. leiognathi* are often found (Hastings and Mitchell, 1971; Katznelson and Ulitzur, 1977; Shilo and Yetinson, 1979), and complex distributional patterns involving *B. harveyi, P. leiognathi,* and *P. fischeri* have been described in both the Mediterranean Sea and the Gulf of Elat (Yetinson and Shilo, 1979). It is postulated that several factors control the distribution, including bacterial responses to photooxidation, temperature, salinity, and nutrient levels.

Differences in species abundance have also been found in vertical sampling of open ocean waters to depths of 6,000 m. *B. harveyi* occurs near the surface and *P. phosphoreum* in the deeper waters of the mid-Atlantic; maximum numbers of the latter occur at 600 m, while total numbers of heterotrophic bacteria decrease with depth (Ruby, Greenberg, and Hastings, 1980). Similar observations have been made by Orndorff and Colwell (1980) in the Sargasso Sea and by Nealson (unpublished) in the northern tropical Pacific.

Saprophytes, Enrichments

If a swab from the surface of a marine animal (but not seaweed, or other surfaces) is streaked on seawater nutrient agar plates, luminous colonies will almost always be present among those colonies that

appear. In fact, the time-honored method for isolating luminous bacteria involves allowing bacterial growth on the surface of a fresh (dead) fish or squid, followed by inspection and visual selection in the dark for luminous colonies. Both *Beneckea* and *Photobacterium* species may be so obtained; the temperature of incubation of the fish may result in the predominance of certain luminous species. Meat, in general, is susceptible to overgrowth by luminous bacteria; meats in storage have commonly been observed to emit light due to a growth of luminous bacteria (Harvey, 1952). Such observations have been less frequent since the advent of refrigeration. The bacteria are reputed to be harmless, and, in fact, to be an indication that no putrefaction has occurred. There are also old accounts of the occurrence of luminous bacteria in open human wounds, reported especially from battlefield hospitals during the 19th century (Harvey, 1957a). Again, the presence of luminous bacteria was taken as a good sign of wound healing. Luminous bacteria are also characteristically found on the surfaces of crabs, lobsters, and other crustaceans, often associated with necrotic lesions (Baross, Tester, and Morita, 1977).

Parasites

Perhaps the area receiving least study in recent times has been that of the luminous bacterial parasites. Nevertheless, this may be a very important ecological niche. The older literature is extensive, and is reviewed by Harvey (1952). From these reports, it is clear that luminous bacteria infect a variety of marine crustaceans. In general, it has been possible to isolate and grow the infecting organism in pure culture. Inman (1926) presented evidence that luminous bacteria were common inhabitants of the intestinal tracts of sand fleas. He proposed that the parasitic forms were merely opportunists, taking advantage of a weakened host to establish a cellular infection. However, little is known of the actual mechanism of infection, or even the location (extra- or intracellular) of the parasites.

From the published data it is not possible to identify the bacterial species responsible for the infections. However, one of us (K.H.N.) has recently made identifications of luminous bacterial parasites infecting marine crustaceans in the San Diego area. These experiments have shown that the same host may be parasitized by different bacterial species of both *Photobacterium* and *Beneckea,* apparently correlated with the free-living isolates by season. To further emphasize the lack of specificity of this niche, it can be shown that the parasitic bacteria have a wide host range and are capable of infecting many different crustaceans (Giardi, 1889; Inman, 1926). Infected animals invariably have a dense culture of luminous bacteria in the hemolymph.

There are also many reports of parasitic infections of terrestrial organisms. Recent studies link the long recognized parasitism of caterpillars by luminous bacteria (Harvey, 1952) with a mutualistic symbiosis between a nonmarine species of bacteria (see below), and a specific genus of nematode (Khan and Brooks, 1976; Poinar, Thomas, and Hess, 1977). The parasitic habitat may thus be a more common and significant one than now recognized.

Commensal Luminous Bacteria

Quantitatively, the most important habitat of luminous bacteria may be that of the gut tracts of marine animals. It is not uncommon to find between 5×10^6 and 5×10^7 luminous bacteria per milliliter of gut material (Ruby, 1977; Ruby and Morin, 1979), and it may be that many marine fishes carry these bacteria as major enteric forms (Spencer, 1961). All marine species of luminous bacteria produce extracellular chitinase (Reichelt and Baumann, 1973; Spencer, 1961) and are thus conceivably important as gut symbionts in the digestion of chitin (Robison and Morin, 1977; ZoBell and Rittenberg, 1938). J. Baross (personal communication) has observed that the gut contents of flatfish *(Parophyrs vetulus* and *Citarichthys sordidus)* contained about 1×10^8 bacteria per gram, all of which were luminous and chitinoclastic. Ruby (1977) studied *Oxyjulis californica, Chromis punctipinnis,* and *Argyropelecus hemigymnus* from the waters off southern California and found *B. harveyi, P. fischeri,* and *P. phosphoreum* in different samples, often with just one of the species predominating, and appar-

ently correlated with season and/or water temperature. Fecal pellets were luminescent and also contained large numbers of viable luminous bacteria and extractable luciferase. Luminescent fecal pellets have also been reported from the Antarctic cod (Raymond and DeVries, 1976) and *Sergestes similis* (Warner, Latz, and Case, 1979), although no luminous bacteria were cultured in these cases. Other reports indicate that the involvement of luminous bacteria as digestive tract symbionts is not limited to fish; invertebrates such as mussels, scallops, and crabs also carry such bacteria (Inman, 1926; J. Baross, personal communication; P. O. Dunlap and J. G. Morin, personal communication).

Specific Symbionts with Marine Fishes

A great number of marine fishes and some squid emit light by harboring luminous bacteria in structured and sometimes highly adapted light organs. Unlike the more nonspecific associations, where the attraction of feeders may be a general function of the light emission, each of the specific symbioses appears to involve one or more particular functions serving the physiology and behavior of the host. Moreover, the association is species specific, with each group of fish being associated with only one bacterial species (Fitzgerald, 1977; Reichelt, Nealson, and Hastings, 1977; Ruby, 1977; Ruby and Morin, 1978; Ruby and Nealson, 1976). Furthermore, mutual benefits accrue; the bacteria are supplied with nutrients and a protected environment, while the fish is supplied with light, which it may use for one or more purposes, to attract prey, assist in escaping or diverting predation, and for communication (Morin et al., 1975).

An example of the first function, attracting prey, is provided by the angler fishes *(Lophiformes, Ceratiodia).* These animals possess a structure (esca) with a luminous bait to attract would-be predators who are then converted into prey (O'Day, 1974; Pietsch, 1974). The luminescence is due to symbiotic bacteria cultured in the light organ at the tip of the projecting structure. As yet, there has been no success in culturing these symbionts (O'Day, 1974).

Light emission is used by fish to startle, frighten or divert predators. One effective way to avoid predators is to render oneself invisible. In reflected light this can be achieved by matching the background color and reflectivity. An organism such as a fish that can be silhouetted against the downwelling light cannot use such a technique, but emitting light of a color and intensity that matches that which it shadows from above would indeed achieve the end. Recent experiments with fish (Warner, Latz, and Case, 1979) and squid (Young and Roper, 1976) have provided convincing evidence to support this

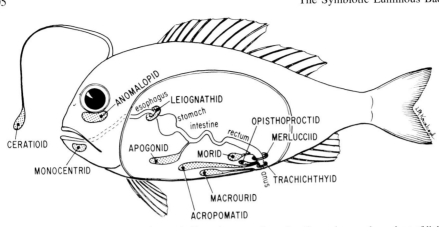

Fig. 2. "Ichthylicht", a diagrammatic fish used here to indicate the approximate locations, sizes, and openings of light organs of the various groups of luminescent fishes (see Table 2 for details).

hypothesis. The preponderant ventral location of the light organs (Fig. 2) is suggestive. In the leiognathids, the physical structure and optical mechanisms are remarkably complex. Light is produced by bacteria maintained in a small organ that surrounds and communicates with the esophagus. Some of the light is directed into the internally reflective swim bladder, whence it passes to part or all of the posterior ventral surface by means of light-conducting fibers and tissues (biological fiber optics) and reflective devices. There are other light-conducting pathways to the anterior ventral surface. The result is a diffuse and very even illumination over the entire ventral surface. Subject to appropriate control and vertical positioning of this fish, the intensity of the ventral surface luminescence can match that of the downwelling light.

Fish may also utilize luminescence for intraspecies communication and for illumination of their surroundings in order to see. The anomalopid (flashlight) and monocentrid (pinecone) fish are possible examples. The light organs, located on the head, are subject to rapid on-off control either by means of rotation or shutters (anomalopids) or by slow changes via chromatophores (monocentrids).

A list of teleost fish families and genera known to harbor luminous bacteria is given in Table 2. The bacteria themselves emit light continuously; control of the luminescence by the fish involves chromatophores or shutters. The bacterial symbionts are extracellular, being maintained in canals or tubules. The organs are extensively vascularized, presumably for the supply of nutrients and oxygen to the bacteria. In the tubules, which are in close proximity to the vascular system, the bacteria characteristically grow to dense cultures, as illustrated in Fig. 3. Viable counts of bacteria from such organs typically yield from 10^9 to 10^{10} luminous bacteria per milliliter of organ fluid (Hastings and Mitchell, 1971; Ruby and Morin, 1978; Ruby and Nealson, 1976). The uniform appearance of the bacteria in the light

organs and colonial isolates therefrom had long suggested that the bacteria in a given organ are a single, homogeneous culture (Bassot, 1975; Haneda, 1966), an assumption recently confirmed on the basis of taxonomic analyses of light organ populations of a variety of luminous fishes (see above). The shallow water and tropical fishes that experience warmer waters have bacterial symbionts that will tolerate higher temperatures (e.g., *P. fischeri* or *P. leiognathi*). The deep-sea and midwater fishes, on the other hand, harbor *P. phosphoreum* organisms as their symbionts; this species is well suited for growth at low temperatures. Included in Table 2 are the bacterial species associated with each group of fish.

The luminous organs differ with respect to location on the host, openings to the exterior, and individual structural components (Table 2). Fig. 2 is a composite drawing showing the variety of locations of light organs that harbor luminous bacteria. Fig. 3 illustrates different magnifications of the light organ of *Monocentris japonicus*.

A feature revealed by recent ultrastructural studies of some light organs is the occurrence of large, rather unusual mitochondria-containing cells. These have been observed in anomalopids (Kessel, 1978) and in monocentrids (Fig. 3; Tebo, Linthicum, and Nealson, 1979), but not in leiognathids (Bassot, 1975). Their function may involve pyruvate removal, possibly related to the maintenance of a low or proper oxygen tension as proposed by Kessel (1978), or perhaps to energy generation or to some other process.

Squid Symbioses

Herring (1977) lists 69 genera of luminous squids in 19 families (and three orders); only two of these families have genera that use luminous bacteria as their source of light. All the others have their own

Table 2. Luminous bacteria and their teleost symbiotic hosts.

Bacterial symbionts	Host fish families[a]	Fish habitat[b]	Light presentation and control[c]	References
Photobacterium fischeri	Monocentridae	Shallow–100 m, temperate	Direct, chromatophores or mechanical covers	Graham, Paxton and Cho, 1972; Haneda, 1966; Okada, 1926; Ruby and Nealson, 1976
Photobacterium leiognathi	Leiognathidae[d]	Warm, shallow, tropical	Indirect, ventral glow, chromatophores, shutters and reflectors	Bassot, 1975; Boisvert, Chatelain, and Bassot, 1967; Haneda, 1940; Haneda and Tsuji, 1976; Reichelt, Nealson, and Hastings, 1977
	Apogonidae	Warm, shallow	Indirect, ventral glow, chromatophores	Fitzgerald, 1979; Haneda, 1950; Iwai, 1958; Reichelt and Baumann, 1973
Photobacterium phosphoreum	Macrouridae	Deep, cold	Direct, ventral, chromatophores	Okamura, 1970; Ruby and Morin, 1978; Singleton and Skerman, 1973; Yasaki and Haneda, 1935
	Merluccidae	Deep, cold	Indirect, ventral, chromatophores	Haneda and Yoshiba, 1970; Reichelt and Baumann, 1973
	Opisthoproctidae	Deep, cold	Indirect, ventral, chromatophores	Bertelsen and Munk, 1964; Herring, 1975; Ruby and Morin, 1978
	Trachichthydae	Deep, cold	Indirect, ventral, chromatophores	Haneda, 1957; Reichelt and Baumann, 1973; Woods and Sonoda, 1973
	Moridae	Deep, cold	Indirect, ventral, chromatophores	Marshall and Cohen, 1973; J.G. Morin and K.H. Nealson (unpublished)
Unidentified symbiont	Acropomatidae	Deep, cold	Indirect, ventral	Haneda, 1950; Yasaki and Haneda, 1936
Noncultivable symbionts	Anomalopidae	Shallow-deep	Direct, mechanical shutter, or organ rotation	Bassot, 1968; Haneda and Tsuji, 1971b; Kessel, 1978; McCosker, 1977
	Ceratiodiae	Deep	Direct, chromatophores	Hansen and Herring, 1977; O'Day, 1974; Pietsch, 1974
Beneckea harveyi splendida *Vibrio cholerae* (biotype *albensis*)	No *Beneckea* or *Vibrio* species have yet been isolated as symbionts from light organs.			

[a] The known luminous genera of these families are the following: Monocentridae (*Monocentris, Cleidopus*); Leiognathidae (*Leiognathus, Gazza, Secutor*); Apogonidae (*Apogon, Siphamia*); Macrouridae (*Cetonurus, Coelorhynchus, Hymenocephalus, Lepidorhynchus, Malacocephalus, Nezumia, Odontomaerus, Sphagemacrurus, Trachonorus, Ventrifossa*); Moridae (*Brosmiculus, Gadella, Physiculus, Tripterophycis*); Merluccidae (*Steindachneria*); Trachichthydae (*Paratrachichthys, Holostethus*); Acropomatidae (*Acropoma*); Amomalopidae (*Anomalops, Kryptophaneron, Photoblepharon*); Ceratioideae (includes several families of angler fish).
[b] "Deep" is used to designate any depth below the thermocline and may be different for different fishes. The important habitat parameter with regard to symbiosis may be temperature.
[c] Light-emitting systems of fish have been classed by Haneda (1950) as direct or indirect, depending on whether the light shines directly to the exterior or passes via optical elements.
[d] Bacteria from at least one species have been reported not to grow on artificial media (Haneda and Tsuji, 1976).

chemical mechanisms, although in many cases these have not been characterized. In the family *Loliginidae*, the genera *Loligo* and *Uroteuthis* have luminous species with bacterial symbionts, as do the genera *Sepiola, Rossia, Euprymna*, and *Rondoletiola* in the family Sepiolidae. Several other genera in these families have been suspected of using bacterial symbionts as a source of light, but unequivocal evidence has not been presented. These genera include *Doryteuthis, Heteroteuthis, Sepiolina*, and *Spirula*. On a recent cruise, G. Leisman (unpublished) demonstrated that the bacterial symbiont of *Euprymna scalopes* is *P. fischeri*. Also, although luminous bacteria could not be cultured from the light organ of *Heteroteuthis hawaiiensis*, it was nevertheless possible to detect bacterial luciferase (Leisman, Cohn, and Nealson, 1980).

In this latter case, it is possible, as suggested by Buchner (1965), that some squids harbor bacterial endosymbionts. This has been neither substantiated nor disproved. In the well-described cases of squid symbiosis, the anatomical adaptations are similar among the squids. The bacteria are maintained in paired organs in the mantle cavity lying against the ink sac near the anus. Although the light is commonly displayed as a luminous secretion into the surrounding medium, it can be seen, in at least some species, through the walls of the organ in the animal

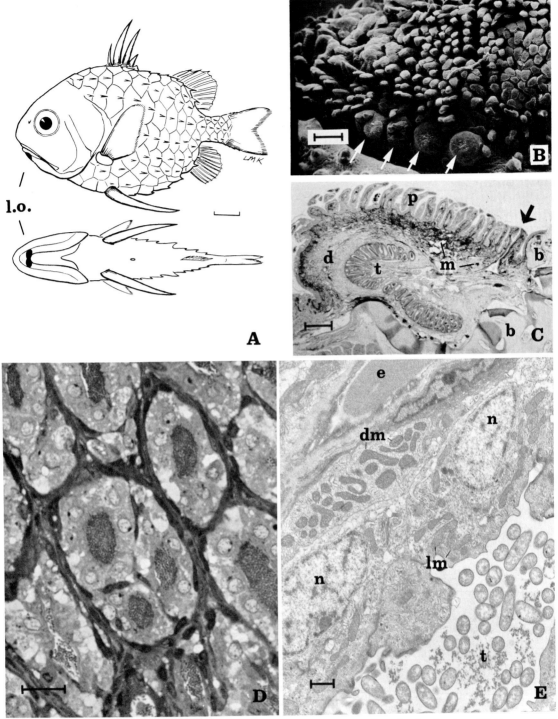

Fig. 3. *Monocentris japonicus*. (A) Line drawing of fish and ventral view of lower jaw, showing location of light organs (lo). Bar = 1.0 cm. (B) Scanning electron micrograph of the dorsal surface of the light organ. Numerous dermal papillae can be seen. The emissary ducts from the light organ emerge at the tips of the four large dermal papillae (arrows). Bar = 0.2 mm. (C) Light micrograph of a sagittal section of the lower jaw. m, melanocytes; t, tubules with bacteria; b, mandibular bone; d, dermal layer; p, dermal papillae. Arrow points to emissary duct. Bar = 50 μm. (D) Light micrograph showing the light organ tubules filled with bacteria. Tubules are lined with a single layer of cuboidal epithelial cells that display loose nuclear chromatin and prominent nucleoli supported by connective tissue cells. Blood capillaries are sparse and not readily visible. Bar = 15 μm. (E) Electron micrograph showing the major features of tubule epithelium. Epithelial cells that make up the lining of the tubules have light-staining mitochondria (lm) with fine cristae. Epithelial cells that are further away from the tubule lumen next to the blood capillaries have dark-staining mitochondria (dm) with thick cristae. t, tubule containing luminous bacteria; e, erythrocyte visible in capillary; n, nucleus of tubule epithelium cells. Bar = 1 μm.

itself (Herring, 1977). Luminescence discharged into the water is presumably used to frighten and confuse other organisms in the dark, where the usual black ink would of course be ineffective. This and a number of other cases were studied extensively during the early part of the century (Buchner, 1965; Herring, 1977, 1978; Kishitani, 1932), but there have been only a few reports since then (Haneda, 1956; Haneda and Tsuji, 1971a). Bacterial isolations have been made by several different laboratories from many different hosts, but with the exception of the one unpublished report discussed above, the symbionts have not been identified.

Other Marine Symbioses

In his 1952 book, Harvey discusses, but largely discounts, the possibility that luminous tunicates harbor endosymbiotic bacteria as their source of light. Both pyrosomes and salps contain intracellular "bacteria-like bodies", but these have not been cultured. Buchner (1965) states that symbiotic luminous bacteria are definitely involved, and that they are large ($2–3~\mu m \times 10–30~\mu m$) sporeformers. The spores are reported to be found intracellularly and are hypothesized to be involved with infection of eggs and transmission of the symbionts. Both their size and their capacity to form spores indicate that these "bacteria" would be a different taxonomic group from the other marine luminous bacteria, and suggest that if such a symbiotic association occurs in tunicates, it is fundamentally different from other systems. Galt (1978) and Herring (1978) both report that a bacterial system is not involved in tunicate luminescence, while Mackie and Bone (1978) believe, on the basis of ultrastructural analyses, that bacteria are involved. Recent biochemical studies (Leisman, Cohn, and Nealson, 1980) indicate that bacterial luciferase is present in extracts of pyrosomes, supporting the contention that the light is of symbiotic origin.

Nonmarine Luminous Bacteria

Although luminous bacteria are essentially ubiquitous in the marine environment and are not at all so in fresh water, some apparently nonmarine forms have been isolated (Khan and Brooks, 1976; Poinar, Thomas, and Hess, 1977). There are numerous reports of luminous bacteria having been isolated as parasites or saprophytes from freshwater hosts, but many of these relationships could presumably involve marine bacteria growing on the high ionic strength of the body fluids or tissues of the host (Harvey, 1952; Y. Haneda, personal communication). In fact, there is little doubt that marine luminous bacteria can infect or be caused to infect many

different nonmarine hosts including crustaceans, insects, and even frogs (Harvey, 1952). There appear to be at least two distinct types of nonmarine luminous bacteria, although one, *Vibrio cholerae*, biotype *albensis*, isolated from the Elbe River (Harvey, 1952) is represented by only one known strain. It requires very low sodium ion concentrations for growth and luminescence, and was therefore classed as nonmarine by Reichelt and Baumann (1973). The other example is the case of the parasitic bacteria of terrestrial insects, most commonly caterpillars.

The older literature describes many examples of land animals becoming infected with luminous bacteria; the well-documented hosts include mole crickets, mayflies, ants, wood lice, and millipedes (Harvey, 1952). Caterpillars were also known to be prone to infection, which has now been shown to involve a specific symbiosis and parasitism (Khan and Brooks, 1976; Poinar, Thomas, and Hess, 1977). Luminous bacteria are carried symbiotically in the gut of a nematode that in its juvenile stage attacks caterpillars and other insects. After boring through the gut wall, the nematode releases the symbiotic bacteria as an inoculum of the hemolymph; the bacteria grow there and are pathogenic for the caterpillar. The life cycle of the nematode is completed in the hemocoel and luminous bacteria are incorporated into the progeny. The caterpillars become luminous during the early stages of the infection, with a constant glow that decreases as the nematodes mature (Poinar et al., 1980). The function of the luminescence is not known, although it may serve to disperse the nematode progeny either by attracting other susceptible insects, or perhaps feeding organisms. Free bacteria ingested by the insects are not capable of killing the hosts, and without the bacteria the nematode cannot complete its life cycle. However, the bacteria can be cultured outside the host and are highly virulent when injected directly into the hemocoel; as few as 40 bacteria per caterpillar are required to achieve 50% kill (LD_{50}).

The bacteria have properties that differ strikingly from the *Photobacterium* and *Beneckea* species, and have been placed in a new genus called *Xenorhabdus* (Thomas and Poinar, 1979). They are unusually large ($5–10~\mu m$ in length), pigmented, chitinase and oxidase negative, and prefer low salt (growth is inhibited by 3% NaCl), suggesting that they are truly nonmarine. However, the light-emitting system of these bacteria involves a mechanism similar to that shown in Fig. 2; their luciferases exhibit slow decay kinetics with dodecanal (Poinar et al., 1980). It is not possible to state whether the many other examples of land animals being infected by luminous bacteria are due to similar nonmarine forms, but without inquiry they should not be assumed to be marine forms.

CHAPTER 105

Isolation

The media and basic methodology for the isolation and growth of luminous bacteria are described elsewhere in this Handbook, Chapter 104. It is easy to obtain the bacteria, which occur in many marine and some nonmarine habitats. Culturing them is usually routine; a simple sea water complete (SWC) medium supports growth of all known marine species that can be cultured. SWC medium contains 5 g peptone, 3 g yeast extract, and 3 ml glycerol per liter of 75% sea water. Glucose is not recommended because (i) it represses luminescence in *Beneckea* and some *Photobacterium* species and (ii) it leads to the excretion of pyruvic acid in many isolates, resulting in a lowered pH and early cessation of both luminescence and growth.

Light organ symbionts have generally been isolated on similar media, but some hosts (anomalopids [Haneda and Tsuji, 1971b], ceratioids [O'Day, 1974], and some leiognathids [Haneda and Tsuji, 1976]) have symbionts that have not yet been grown in host-free culture; these may require special media. Usually, however, the light organ symbionts are cultivable, and many different symbionts have been studied. Because the luminous organs communicate with the exterior, where contaminating bacteria (luminous and nonluminous) occur, isolations should be made from living or fresh specimens whenever possible.

The organ should be excised, and the surface should be sterilized with 70% ethanol, rinsed in sterile sea water, and then opened to remove the bacteria. The method of choice is to remove a known volume of organ fluid with a sterile micropipette (1–5 μl), plate a series of dilutions (sterile 3% NaCl), and count colonies to determine the density of viable luminous bacteria in the organ. Alternatively, the organ fluid can be streaked directly onto agar plates. All symbionts that have been cultivated grow well on SWC agar (1.2% agar). More defined media are described by Nealson (1978) and in this Handbook, Chapter 104.

Once the organ is opened, the luminescence of the symbiotic bacteria should be clearly visible in a dim room. In cases where the emission is very dim, the use of rheostat-controlled red lamp will allow dark adaptation of the worker's eyes and can be adjusted so that the luminous spots can be seen and then picked with a sterile toothpick. Such isolations will invariably yield pure cultures of luminous bacteria. The incubation temperature may be an important consideration, especially with isolates from deep-sea fishes, since some of these bacteria may be psychrophiles favoring low temperatures (4–10°C; Makemson, 1973).

As a final point in the isolation of symbionts, it should be mentioned that cultures often lose their ability to luminesce (Hastings and Nealson, 1977; Keynan and Hastings, 1961). Thus, it is wise to preserve the isolates as soon as possible after isolation. Standard preservation techniques, including maintenance media and lyophilization, are discussed by Nealson (1978) and in this Handbook, Chapter 104. One effective method not mentioned in the latter reference is that of quick-freezing and cold storage. The cultures are grown to midlog phase in nutrient seawater broth, diluted with an equal volume of sterile glycerol, quick-frozen in a dry ice-acetone bath, and stored in the frozen state (−20°C or below). This method is acceptable for all species.

Identification

The property of bioluminescence in marine bacteria is confined to only two genera, *Beneckea* and *Photobacterium*, and all known cultivable light organ symbionts are in the latter genus. The taxonomic properties of these genera and the identification of the bacteria are discussed in Chapter 104 of this Handbook.

It is possible to distinguish between these genera on the basis of the kinetic properties of their light-emitting enzymes (luciferases). Cells are scraped from plates after overnight growth and lysed osmotically by suspension in a low-ionic-strength lysis buffer (10^{-2} M ethylenediaminetetraacetate, 10^{-3} M dithiothreitol, adjusted to pH 7.0 with NaOH). These extracts are assayed for luciferase activity in vitro. With dodecanal as the aldehyde in the assay, results similar to those shown in Fig. 4 are seen; *Beneckea* luciferases have slow decay kinetics, while all *Photobacterium* luciferases display fast kinetics. The decay rates of *Beneckea* and *Photobacterium* enzymes differ by a factor of 5–10, and so can be used for a rapid screening. Within a species, the range of variation is small, of the order of 10–15% around a mean value (Fitzgerald, 1979; K. H. Nealson, unpublished). Several hundred extracts can be assayed in a few hours, because only a few seconds of decay need to be observed to distinguish between fast and slow kinetics.

Once the genus has been determined, using the "luciferase kinetics" method, species identifications can be made. With *Beneckea* species, extensive biochemical and nutritional comparisons are required (this Handbook, Chapter 104). With *Photobacterium* species, the task is easier; the information can be gathered by making only ten different solid media and replica plating the strains to all ten (Table 3).

The "luciferase kinetics" method can be used even with bacterial symbionts that cannot be cultured outside the host. For example, the luciferase extracted from the light organs of anomalopids

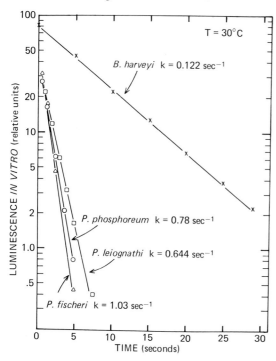

Fig. 4. Decay kinetics of luciferases from different species of luminous bacteria. Bright bacteria are lysed by suspension in cold lysis buffer (10 mM ethylenediaminetetraacetate; 1 mM dithiothreitol; pH 7.0). Twenty microliters of lysate is added to a scintillation vial containing 10 μl of aldehyde suspension (0.1 ml of dodecanal in 10 ml of water) in 1 ml of 10 mM phosphate buffer (pH 7.1) in a light-tight chamber, and 1 ml of FMNH$_2$ is injected. The resulting flash of light is recorded on a strip chart recorder and plotted on semilog paper to determine rate constants. When dodecanal is used, luciferase from *Beneckea harveyi* displays slow-decay kinetics, while *Photobacterium* species show a fast decay. Because the decay is very temperature dependent, controls of known species must be done. Typically, a 6- to 10-fold difference in decay rates is seen, so that the genera can be easily distinguished even without replotting the data.

Table 3. Diagnostic taxonomic characters of the luminous bacteria.[a]

	Beneckea harveyi	Photobacterium fischeri	P. leiognathi	P. phosphoreum	Amount of carbon source (%)	Time of scoring (days)
Growth at 4°C[b]	−	(−)[c]	−	+	−	
Growth at 35°C[b]	+	(+)	(+)	−	−	1
Amylase	+	(−)	−	−	−	4
Lipase	+	−	−	−	−	2,4,6
Gelatinase	+	−	(−)	−	−	4
Growth on:						
Maltose[b]	+	+	−	+	0.2	3
Cellobiose[b]	+	+	−	−	0.2	3
Gluconate[b]	+	(−)	+	+	0.1	3
Glucuronate[b]	+	−	−	(−)	0.1	3
Mannitol[b]	+	+	−	−	0.1	5
Proline	+	+	+	(−)	0.1	3
Lactate[b]	+	−	+	(−)	0.2	5
Pyruvate[b]	+	(+)	+	−	0.1	
Acetate[b]	+	(+)	+	−	0.05	10
Propionate	+	−	−	−	0.05	10
Heptanoate	+	−	−	−	0.05	10
L-Tyrosine	+	−	−	−	0.4	10
Mol % (G+C)	46.5 ± 1.3	39.8 ± 1.1	41.5 ± 0.7	42.9 ± 0.5		
Flagellation[d]	SP Pr	SP	P	P		
Polar flagella	1	2–8	1–3	1–3		
PHB accumulated[e]	−	−	+	+		
Gas from glucose	−	−	(−)	+		
Luciferase kinetics	Slow	Fast	Fast	Fast		

[a] Adapted from the methods of Reichelt and Baumann (1973).
[b] For rapid screening method, tests performed on strains having fast enzyme.
[c] Parentheses indicate that the trait exhibits a small degree of variability.
[d] S, sheathed flagella; P, polar flagella; Pr, peritrichous flagella; SP Pr, a conversion from one mode of flagellation to the other.
[e] PHB, Poly-β-hydroxybutyric acid.

(*Photoblepharon, Anomalops,* and *Kryptophaneron*) yielded slow (*Beneckea*-type) kinetics (Leisman, Cohn, and Nealson, 1980). On the basis of morphological studies, Kessel (1978) noted that the symbionts of *Photoblepharon* do not fit into the current taxonomic scheme of Reichelt, Baumann, and Baumann (1976), and thus might be a new species.

Perspectives and Problems

We have described here a biological phenomenon that has its focus in specialized organs and occurs mostly in marine fish carrying as symbionts bioluminescent bacteria that provide a light source for their hosts. This appears to be the only example of such a highly specific relationship between a prokaryote and a vertebrate. Although a high degree of species specificity between the two partners has been recently demonstrated, there is no information concerning the way(s) in which such relationships are established (possibly during development) and maintained in the adult, free of contaminants.

Symbiotic relationships involving luminous bacteria have previously been known only with marine bacteria, but a case of nonmarine luminous bacteria that are symbiotic with a terrestrial host has been recently described. In this case no specialized light organ is involved, and it may well be that such nonmarine luminous bacteria and their symbioses are more common than previously thought.

Recent results with salps (J. W. Hastings, unpublished) and pyromes and squids (Leisman, Cohn, and Nealson, 1980) suggest that these invertebrates contain symbiotic bacteria that are highly integrated, possibly intracellular. If so, the luminous bacterial symbioses represent a wide spectrum of associations, and may offer good systems for the study of the evolution of symbiosis.

A new perspective that we have discussed deals with the way in which luminous bacteria occupy other habitats not associated with light organ symbiosis, and apparently compete there with some success.

We have not discussed in this chapter the new findings that the gene(s) for luciferase apparently occur in several bacteria presently classed as nonluminous species. From five species of nonluminous *Beneckea* active bacterial luciferase has been extracted in small amounts (Nealson and Walton, 1978). Nonluminous species also produce a substance active in inducing the synthesis of luciferase in *B. harveyi* (Greenberg, Hastings, and Ulitzur, 1978). The luciferase and other related genes may thus be far more ubiquitous than previously envisioned, and one can obviously speculate on possible alternate enzymatic functions for luciferase as well as on the potential for its becoming expressed as active luciferase in the nonluminous species.

Addendum

It has now been proposed that the genus *Beneckea* be abolished and that its species be moved into the genus *Vibrio* (Baumann et al., 1980), and that two species formerly assigned to the genus *Photobacterium* (*P. fischeri* and *P. logei*) be moved into the genus *Vibrio*. The species assignments in all cases remain the same.

The autoinducer for *Photobacterium* (now *Vibrio*) *fischeri* has been purified, identified, and synthesized. The active material is *N*-(3-oxohexanoyl)-3-aminodihydro-2-(3*H*)-furanone (Eberhard et al., 1981).

Literature Cited

Baross, J., Tester, P. A., Morita, R. Y. 1977. Microbiology and pathology of necrotic lesions in the Tanner crab, *Chionoecetes tanneri* Rathbun. Abstracts of the Annual Meeting of the American Society for Microbiology **1977**:230.

Bassot, J.-M. 1968. Les organes lumineux à bactéries symbiotiques de téléostéen *Anomalops*. Données histologiques. Bulletin Société Zoologie **93**:569–581.

Bassot, J.-M. 1975. Les organes lumineux à bactéries symbiotiques de quelques téléostéens leiognathides. Archive Zoologie Expérimentale Générale **116**:359–373.

Baumann, P., Baumann, L. 1977. Biology of the marine enterobacteria: Genera *Beneckea* and *Photobacterium*. Annual Review of Microbiology **31**:39–61.

Baumann, P., Baumann, L., Bang, S. S., Woolkalis, M. J. 1980. Reevaluation of the taxonomy of *Vibrio, Beneckea,* and *Photobacterium*: Abolition of the genus *Beneckea*. Current Microbiology **4**:127–133.

Beijerinck, M. W. 1889. Le *Photobacterium luminosum*, bactérie luminesce de la Mer du Nord. Archives Neerlandaises des Sciences Exactes et Naturelles **23**:401–405.

Beijerinck, M. W. 1916. Die Leuchtbakterien der Nordsee im August und September. Folia Microbiologia **4**:15–40.

Bertelsen, E., Munk, O. 1964. Rectal light organs in the argentinoid fishes *Opisthoproctus* and *Winteria*. Dana Report, No. 62. Carlsberg Foundation.

Boisvert, H., Chatelain, R., Bassot, J.-M. 1967. Étude d'un *Photobacterium* isolé de l'organe lumineux de poissons Leiognathidae. Annales de l'Institut Pasteur **112**:520–524.

Booth, C. R., Nealson, K. H. 1975. Light emission by luminous bacteria in the open ocean. Biophysical Journal **15**:56.

Buchner, P. 1965. Endosymbiosis of animals with plant microorganisms. New York: John Wiley & Sons.

de Bary, A. 1879. Die Erscheinung der Symbiose. Strassburg: Trübner.

Eberhard, A. 1972. Initiation and activation of bacterial luciferase synthesis. Journal of Bacteriology **190**:1101–1105.

Eberhard, A., Burlingame, A. L., Eberhard, C., Kenyon, G. L., Nealson, K. H., Oppenheimer, N. J. 1981. Structural identification of autoinducer of *Photobacterium fischeri* luciferase. Biochemistry **20**:2444–2449.

Fitzgerald, J. M. 1977. Classification of luminous bacteria from the light organ of the Australian Pinecone fish, *Cleidopus gloriamaris*. Archives of Microbiology **112**:153–156.

Fitzgerald, J. M. 1979. Studies on the taxonomy and bioluminescence of some luminous marine bacteria. Ph.D. Thesis. Monash University, Victoria, Australia.

Galt, C. P. 1978. Bioluminescence: Dual mechanism in a planktonic tunicate produces brilliant surface display. Science **200**:70–72.

Giardi, M. A. 1889. Sur l'infection phosphorescente des Talitres et autres Crustacés. Comptes Rendus Hebdomadaires des Séances de l'Académie des Sciences **109**:503–506.

Graham, P. H., Paxton, J. R., Cho, R. Y. 1972. Characterization of luminous bacteria from the light organs of the Australian Pinecone fish *(Cleidopus gloriamaris)*. Archiv für Mikrobiologie **81**:305–308.

Greenberg, E. P., Hastings, J. W., Ulitzur, S. 1978. Induction of luciferase synthesis in *Beneckea harveyi* by other marine bacteria. Abstracts of the Annual Meeting of the American Society for Microbiology **1978**:102.

Haneda, Y. 1940. On the luminescence of the fishes belonging to the family Leiognathidae of the tropical Pacific. Palao Tropical Biological Station Studies **2**:29–30.

Haneda, Y. 1950. Luminous organs of fish which emit light indirectly. Pacific Science **4**:214–227.

Haneda, Y., 1956. Squid producing and abundant luminous secretion found in Suruga Bay, Japan. Scientific Reports of the Yokosuka City Museum **1**:27–30.

Haneda, Y. 1957. Observations on luminescence in the deep-sea fish, *Paratrachichthys prosthemius*. Scientific Reports of the Yokosuka City Museum **2**:15–22.

Haneda, Y. 1966. On a luminous organ of the Australian Pinecone fish, *Cleidopus gloriamaris*, pp. 547–555. In: Johnson, F., Haneda, Y. (eds.), Bioluminescence in progress. Princeton, New Jersey: Princeton University Press.

Haneda, Y., Tsuji, F. I. 1971a. Descriptions of some luminous squids from the water of northern New Guinea collected by the R/V Tagula. Scientific Reports of the Yokosuka City Museum **18**:29–33.

Haneda, Y., Tsuji, F. I. 1971b. Light production in the luminous fishes *Photoblepharon* and *Anomalops* from the Banda Islands. Science **173**:143–145.

Haneda, Y., Tsuji, F. I. 1976. The luminescent systems of pony-fishes. Journal of Morphology **150**:539–552.

Haneda, Y., Yoshiba, S. 1970. On a luminous substance of acanthine fish, *Steindachnaria argentea*, from the Gulf of Mexico. Scientific Reports of the Yokosuka City Museum **16**:1–4.

Hansen, K., Herring, P. J. 1977. Dual bioluminescent systems in the anglerfish genus *Linophryne* (Pisces: Ceratioidea). Journal of Zoology **182**:103–124.

Harvey, E. N. 1952. Bioluminescence. New York: Academic Press.

Harvey, E. N. 1957a. A history of luminescence. Memoirs of the American Philosophical Society, vol. 44. Philadelphia: American Philosophical Society.

Harvey, E. N. 1957b. The luminous organs of fishes, pp. 345–355. In: Brown, M. E. (ed.), The physiology of fishes. New York: Academic Press.

Hastings, J. W., Mitchell, G. W. 1971. Endosymbiotic bioluminescent bacteria from the light organ of pony fish. Biological Bulletin **141**:261–268.

Hastings, J. W., Nealson, K. H. 1977. Bacterial bioluminescence. Annual Review of Microbiology **31**:549–595.

Hastings, J. W., Weber, R., Friedland, J., Eberhard, A., Mitchell, G. W., Gunsalus, A. 1969. Structurally distinct bacterial luciferases. Biochemistry **8**:4681–4689.

Herring, P. J. 1975. Bacterial bioluminescence in some argentinoid fishes, pp. 563–572. In: Barnes, H. (ed.), Proceedings of the 9th Environmental Marine Biology Symposium. Aberdeen: Aberdeen University Press.

Herring, P. J. 1977. Luminescence in cephalopods and fish. symposium of the Zoological Society of London **38**:127–159.

Herring, P. J. 1978. Bioluminescence of invertebrates other than insects, pp. 190–240. In: Herring, P. J. (ed.), Bioluminescence in action. London, New York: Academic Press.

Herring, P. J., Morin, J. G. 1978. Bioluminescence in fishes, pp. 272–329. In: Herring, P. J. (ed.), Bioluminescence in action. London: Academic Press.

Inman, O. L. 1926. A pathogenic luminous bacterium. Biological Bulletin **53**:197–200.

Iwai, T. 1958. A study of the luminous organ of the apogonid fish *Siphamia versicolor* (Smith and Radcliffe). Journal of the Washington Academy of Science **48**:267–270.

Katznelson, R., Ulitzur, S. 1977. Control of luciferase synthesis in a newly isolated strain of *Photobacterium leiognathi*. Archives of Microbiology **115**:347–351.

Kessel, M. 1978. The ultrastructure of the relationship between the luminous organ of the teleost fish *Photoblepharon palpebratus* and its symbiotic bacteria. Cytobiologie **15**:145–158.

Keynan, A., Hastings, J. W. 1961. The isolation and characterization of dark mutants of luminescent bacteria. Biological Bulletin **121**:375.

Khan, A., Brooks, W. M. 1976. A chromogenic bioluminescent bacterium associated with the entomophilic nematode *Chromonema heliothidis*. Journal of Invertebrate Pathology **29**:253–261.

Kishitani, T. 1932. Studien über Leuchtsymbiose von japanischen Sepien. Okajimas. Folia Anatomica Japonica **10**:315–418 und Stud. Tokugawa Institute **2**:315–418.

Leisman, G., Cohn, D. H., Nealson, K. H. 1980. Bacterial origin of luminescence in marine animals. Science **208**:1271–1273.

McCosker, J. E. 1977. Flashlight fishes. Scientific American **236**:106–114.

McElroy, W. D., Seliger, H. 1962. Origin and evolution of bioluminescence, pp. 91–101. In: Kasha, M., Pullman, B. (eds.), Horizons in biochemistry. New York: Academic Press.

Mackie, G. O., Bone, Q. 1978. Luminescence and associated effector activity in *Pyrosoma* (Tunicata: Pyrosomida). Proceedings of the Royal Society of London **202**:483–495.

Magasanik, B. 1970. Glucose effects: Inducer exclusion and repression, pp. 189–219. In: Beckwith, J. R., Zipser, E. (eds.), The lactose operon. New York: Cold Spring Harbor.

Magner, J., Eberhard, A., Nealson, K. 1972. Characterization of bioluminescent bacteria by studies of the inducers of luciferase synthesis. Biological Bulletin **143**:469.

Makemson, J. C. 1973. Control of *in vivo* luminescence in a psychrophilic marine *Photobacterium*. Archiv für Mikrobiologie **93**:347–358.

Marshall, B. B., Cohen, E. M. 1973. Anacathini (Gadiformes), characters and synopsis of families, pp. 476–495. In: Cohen, E. M. (ed.), Fishes of the western North Atlantic. New Haven, Connecticut: Sears Foundation of Marine Research.

Morin, J. G., Harrington, A., Nealson, K. H., Krieger, N., Baldwin, T. O., Hastings, J. W. 1975. Light for all reasons: Versatility in the behavioral repertoire of the flashlight fish. Science **190**:74–76.

Nealson, K. H. 1977. Autoinduction of bacterial luciferase. Occurrence, mechanism and significance. Archives of Microbiology **112**:73–79.

Nealson, K. H. 1978. Isolation, identification and manipulation of luminous bacteria, pp. 153–166. In: DeLuca, M. (ed.), Methods in enzymology, vol. LII. New York: Academic Press.

Nealson, K. H., Eberhard, A., Hastings, J. W. 1972. Catabolite repression of bacterial bioluminescence: Functional implications. Proceedings of the National Academy of Sciences of the United States of America **69**:1037–1076.

Nealson, K. H., Hastings, J. W. 1977. Low oxygen is optimal for luciferase synthesis in some bacteria. Ecological implications. Archives of Microbiology **112**:9–16.

Nealson, K. H., Hastings, J. W. 1979. Bacterial bioluminescence: Its control and ecological significance. Microbiological Reviews **43**:496–518.

Nealson, K. H., Platt, T., Hastings, J. W. 1970. The cellular control of the synthesis and activity of the bacterial luminescent system. Journal of Bacteriology **104**:313–322.

Nealson, K. H., Walton, D. S. 1978. Luciferase in non-luminous species of *Beneckea*. Abstracts of the Annual Meeting of the American Society for Microbiology **1978**:102.

O'Day, W. T. 1974. Bacterial luminescence in the deep-sea

anglerfish *Oneirodes acanthias* (Gilbert, 1915). Los Angeles County Museum of Natural History, Contributions in Science **255**:1–12.

Okada, Y. K. 1926. On the photogenic organ of the knight-fish (*Monocentris japonicus* [Houttuyn]). Biological Bulletin **50**:356–373.

Okamura, O. 1970. Studies on the macrourid fishes of Japan, morphology, ecology and phylogeny. Report of the Usa Marine Biological Station **17**:1–179.

Orndorff, S. A., Colwell, R. R. 1980. Distribution and identification of luminous bacteria from the Sargasso Sea. Applied and Environmental Microbiology **39**:983–987.

Pietsch, T. W. 1974. Osteology and relationships of ceratioid anglerfishes of the family Oneirodidae, with a review of the genus *Oneirodes lutken*. Los Angeles County Museum of Natural History, Science Bulletin **18**:1–113.

Poinar, G. O. Jr., Thomas, G. M., Hess, R. 1977. Characteristics of the specific bacterium associated with *Heterorhabditis bacteriophora*. Nematologica **23**:97–102.

Poinar, G. O. Jr., Thomas, G., Haygood, M., Nealson, K. H. 1980. Growth and luminescence of the symbiotic bacteria associated with the terrestrial nematode, *Heterorhabditis bacteriophora*. Soil Biology and Biochemistry **12**:5–10.

Raymond, J. A., DeVries, A. L. 1976. Bioluminescence in McMurdo Sound, Antarctica. Limnology and Oceanography **21**:599–602.

Reichelt, J. L., Baumann, P. 1973. Taxonomy of the marine, luminous bacteria. Archiv für Mikrobiologie **94**:283–330.

Reichelt, J. L., Baumann, P. 1975. *Photobacterium mandapanensis* Hendrie et al., a later subjective synonym of *Photobacterium leiognathi* Boisvert et al. International Journal of Systematic Bacteriology **25**:208–209.

Reichelt, J. L., Baumann, P., Baumann, L. 1976. Study of genetic relationships among marine species of the genera *Beneckea* and *Photobacterium* by means of in vitro DNA/DNA hybridization. Archives of Microbiology **110**:101–120.

Reichelt, J. L., Nealson, K., Hastings, J. W. 1977. The specificity of symbiosis: Pony fish and luminescent bacteria. Archives of Microbiology **112**:157–161.

Robison, B. H., Morin, J. G. 1977. Luminous bacteria in the alimentary tracts of midwater fishes. Abstracts of the Annual Meeting of the Western Society of Naturalists **1977**:71a.

Rosson, R. A., Nealson, K. H. 1979. Control of bacterial bioluminescence. Abstracts of the American Society of Photobiology **1979**:150–151.

Ruby, E. G. 1977. Ecological associations of marine luminous bacteria. Ph.D. Thesis. University of California, San Diego, California.

Ruby, E. G., Greenberg, E. P., Hastings, J. W. 1980. Planktonic marine luminous bacteria: Species distribution in the water column. Applied and Environmental Microbiology **39**:302–306.

Ruby, E. G., Morin, J. G. 1978. Specificity of symbiosis between deep-sea fishes and psychrotrophic luminous bacteria. Deep-Sea Research **25**:161–167.

Ruby, E. G., Morin, J. G. 1979. Luminous enteric bacteria of marine fishes: A study of their distribution, densities, and dispersion. Applied and Environmental Microbiology **38**:406–411.

Ruby, E. G., Nealson, K. H. 1976. Symbiotic association of *Photobacterium fischeri* with the marine luminous fish *Monocentris japonicus*: A model of symbiosis based on bacterial studies. Biological Bulletin **151**:574–586.

Ruby, E. G., Nealson, K. H. 1977. Pyruvate production and excretion by the luminous marine bacteria. Applied and Environmental Microbiology **34**:164–169.

Ruby, E. G., Nealson, K. H. 1978. Seasonal changes in the species composition of luminous bacteria in nearshore seawater. Limnology and Oceanography **23**:530–533.

Shilo, M., Yetinson, T. 1979. Physiological characteristics underlying the distribution patterns of luminous bacteria in the Mediterranean Sea and the Gulf of Elat. Applied and Environmental Microbiology **38**:577–584.

Singleton, R. J., Skerman, T. M. 1973. A taxonomic study of computer analysis of marine bacteria from New Zealand waters. Proceedings of the Academy of Natural Sciences, Philadelphia **78**:245–247.

Spencer, R. 1961. Chitinoclastic activity in the luminous bacteria. Nature **190**:938.

Stanier, R. Y., Adelberg, E. A., Ingraham, J. L. 1976. The microbial world, 4th ed. Englewood Cliffs, New Jersey: Prentice-Hall.

Starr, M. P. 1975. A generalized scheme for classifying organismic associations. Symposium of the Society for Experimental Biology **29**:1–20.

Tebo, B. M., Linthicum, D. S., Nealson, K. H. 1979. Luminous bacteria and light emitting fish: Ultrastructure of the symbiosis. BioSystems **11**:269–280.

Thomas, G. M., Poinar, G. O. Jr. 1979. *Xenorhabdus* gen. nov., a genus of entomopathogenic, nematophilic bacteria of the family Enterobacteriaceae. International Journal of Systematic Bacteriology **29**:352–360.

Warner, J. A., Latz, M. I., Case, J. F. 1979. Cryptic bioluminescence in a midwater shrimp. Science **203**:1109–1110.

Woods, L. P., Sonoda, P. M. 1973. Order Berycomorphi (Beryciformes), pp. 263–396. In: Cohen, D. M. (ed.), Fishes of the western North Atlantic. New Haven, Connecticut: Sears Foundation for Marine Research.

Yasaki, Y., Haneda, Y. 1935. On the luminescence of the sea fishes, family Macrouridae. [In Japanese.] Journal of Applied Zoology **7**:165–177.

Yasaki, Y., Haneda, Y. 1936. Über einen neuen Typus von Leuchtorgan im Fische. Proceedings of the Imperial Academy (Tokyo) **12**:55–57.

Yetinson, T., Shilo, M. 1979. Seasonal and geographic distribution of luminous bacteria in the eastern Mediterranean Sea and the Gulf of Elat. Applied and Environmental Microbiology **37**:1230–1238.

Young, R. E., Roper, C. F. E. 1976. Bioluminescent countershading in midwater animals: Evidence from living squid. Science **191**:1046–1048.

ZoBell, C. E., Rittenberg, S. C. 1938. The occurrence and characteristics of chitinoclastic bacteria in the sea. Journal of Bacteriology **35**:275–287.

SECTION N

Miscellaneous, Facultatively Anaerobic, Gram-Negative, Rod-Shaped Bacteria

CHAPTER 106

The Genus *Zymomonas*

J. GEOFFREY CARR

These bacteria are little known outside the world of alcoholic fermentations. They resemble the yeasts in being able to metabolize large quantities of simple sugars quantitatively to ethanol and carbon dioxide. Very few bacteria produce ethanol as a major metabolic end product, a property that marks zymomonads as unusual. The organisms' other characteristics are not unusual. They are fairly large Gram-negative rods, usually in pairs but also in very characteristic clusters; in the early stages of growth they are intensely motile by means of polar flagella. These bacteria are microaerophilic, growing vigorously in liquid medium without any special precautions but growing slowly on agar unless oxygen is restricted. The resemblance to yeast-like metabolism extends only as far as the substrates used and end products produced, because the mechanism of yeast fermentation is by way of the Embden-Meyerhof-Parnas pathway, whereas that of the zymomonads is by the Entner-Doudoroff route (Entner and Doudoroff, 1952).

The bacteria were first mentioned by Barker and Hillier (1912) as the causal organism of a disorder of low-acid ciders made in southwestern England. Although they never named the organism, it is quite clear from the description that the cause of "cider sickness" was one of a group of bacteria now ascribed to the genus *Zymomonas*. Organisms of this kind have received, over the years, many names. Lindner (1928a,b) isolated an organism from the Mexican drink pulque and named it *Termobacterium mobile.* Later, Kluyver and Hoppenbrouwers (1931) renamed it *Pseudomonas lindneri,* but this was later changed by Kluyver and van Neil (1936) to *Zymomonas mobile,* because of its marked motility. Shimwell (1937) discovered the organism in beer and assigned to it the temporary name of *Achromobacter anaerobium.* By 1950, Shimwell (1950) suggested a new genus *Saccharomonas* to be included in the family Pseudomonadaceae. The genus was to contain two species, *Saccharomonas lindneri* and *Saccharomonas anaerobia,* to include the beer and cider organisms. Later, Millis (1956) compared several zymomonads that included two strains of the cider sickness bacillus, a strain labeled *Achromobacter anaerobium,* and three strains des-ignated *Termobacterium mobile.* She showed that the cider strains and the one called *Achromobacter anaerobium* fermented fructose and glucose, whereas the three strains of *Termobacterium mobile* fermented sucrose in addition. Millis (1956) stated that the name *Zymomonas* (Kluyver and van Niel, 1936) should take precedence over Shimwell's (1950) proposal to call the organisms *Saccharomonas.* She also proposed that the "cider sickness" bacillus be called *Zymomonas anaerobia* var. *pomaceae.* Thus, after Millis's (1956) publication the situation was that there were two species (i) *Zymomonas mobile* (Kluyver and van Neil, 1936) as the type species and (ii) the beer organism, *Saccharomonas anaerobia* (Shimwell, 1950), which became *Zymomonas anaerobia.* In addition, she named the strain from cider as a subspecies *Zymomonas anaerobia* var. *pomaceae.* The major difference between the two species was the inability of *Zymomonas anaerobia* to ferment sucrose.

Dadds, Martin, and Carr (1973) published a paper throwing considerable doubt on the division of the two species *Zymomonas mobilis* and *Zymomonas anaerobia* because they had discovered that the ability to ferment sucrose was adaptive. They stated that a difference of one sugar fermentation was hardly sufficient to justify the existence of two species. They did not, however, put this forward as a formal proposal because too few strains had been examined to justify so radical a change at that time. The discovery of the adaptive nature of sucrose fermentation was later confirmed by Richards and Corbey (1974). In the eighth edition of *Bergey's Manual of Determinative Bacteriology* (Buchanan and Gibbons, 1974), no mention was made in the section on *Zymomonas* (Carr, 1974) of the earlier paper by Dadds, Martin, and Carr (1973). The reason for this reflects the difference in time it takes to publish a comprehensive volume such as *Bergey's Manual* and the less complicated task of publishing a scientific paper. The information in *Bergey's Manual* about the genus *Zymomonas* had been prepared a considerable time before the facts stated by Dadds, Martin, and Carr (1973) were known. Unfortunately, the *Manual* had gone too far along the route to publication to include the new information.

The present situation concerned with the classification and nomenclature of these organisms has been reviewed in some considerable detail by De Ley and Swings (1976) and Swings and De Ley (1977). These workers have examined a comprehensive selection of *Zymomonas* strains and carried out detailed experiments upon them. As a result, they have twice put forward proposals for the nomenclature of these bacteria (De Ley and Swings, 1976; Swings and De Ley, 1977). The second of these is quite simple and states that the genus *Zymomonas* should have as its type species *Zymomonas mobilis* and that it should contain two subspecies, namely *mobilis* and *pomaceae*. Swings and De Ley (1977) emphasize throughout their paper the marked differences between the "cider sickness" strains and those from other sources. In this author's opinion there are sufficient differences between subspecies *mobilis* and *pomaceae* for them to be raised to the rank of species. This is further reinforced by the work of Swings and Van Pee (1977) who show ten differences between *Z. mobilis* subsp. *mobilis* and *Z. mobilis* subsp. *pomaceae* ranging from their ability to grow in certain media to their infrared spectra. In their first proposal on nomenclature, De Ley and Swings (1976) very properly referred to the original discoverers of the zymomonads (Barker and Hillier, 1912). It is regrettable that they have chosen to drop the names of these pioneers in their second proposal (Swings and De Ley, 1977).

The relationships of the genus *Zymomonas* with other groups of bacteria are not clear. Swings and De Ley (1977) regard them as being related to the acetic acid bacteria and suggest that some time in the past they were aerobic.

Habitats

Ciders and Perries

One of the first written descriptions of "cider sickness" was by Lloyd (1903) in which he noted the presence of "sulphuretted hydrogen" in sick ciders. There was, however, no mention of its being microbiological in origin until Barker (1908) made the following statements:

"It usually makes its appearance during the middle or latter half of the summer, hot weather probably favouring its development. In some cases the liquid becomes very turbid while in others it remains clear but throws down a heavy deposit. A large evolution of gas occurs hence the trouble is usually referred to as "second fermentation", which is distinguishable from normal alcoholic fermentation whether 'primary' or 'secondary' by the development of a peculiar disagreeable odour and flavour, due possibly, in part at least, to the formation of

acetaldehyde. The cider is rendered almost unsaleable. . .''

Barker (1908) then goes on to say that the disorder is due to a specific organism, that it is transmissible from "sick" to sound cider. He also notes that the organism is most often found in ciders with residual sugars either undergoing slow fermentation or where fermentation has ceased completely. Further points raised were that the organism's growth was encouraged by low acidity and elevated temperature. This is an accurate description of the disorder that requires no additions even after 70 years. Barker continued to publish work on "cider sickness" until the 1950s. For a comprehensive survey of this subject, readers are referred to Swan (1953).

One of the most useful pieces of work on the organisms of "cider sickness" was by Millis (1951, 1956), who not only reexamined the technology of the disorder but also defined its bacteriology. It was Millis who saw the connection between the "cider sickness" organism and the genus *Zymomonas*. This work will be mentioned later.

As a result of the work carried out in the early days the technology of "cider sickness" became well understood and established the pattern of English cider making. The fact that this organism can only utilize fructose and glucose was used to good effect in the prevention of "cider sickness". Vigorous fermentations were maintained by the additon of yeast nutrients that stimulated fermentation and quickly reduced the sugar content to nil. It is still the practice in England to store ciders "dry" and only to sweeten them just before sale. These methods, coupled with an overall increase in juice acidity, have virtually eradicated "cider sickness". It therefore came as something of a surprise when this organism was discovered by Carr and Passmore (1971) in apple pulp prepared for juice extraction. This was the first and only occasion that this organism has been isolated from any source in the cider industry other than "sick cider".

In France, the method of cider making sometimes involves the storage of low-acid, partly fermented juices that contain residual sugars. As Millis (1951) showed, these are the ideal conditions for the development of "cider sickness". It is, therefore, surprising that only passing reference has been made to the occurrence of this disorder in French ciders. Warcollier (1928) refers to *la pousse,* which describes a disorder involving the release of gas (probably "cider sickness"). This term has been discarded in favor of *le framboisé,* a term that applies to a disorder characterized by the occurrence of excess acetaldehyde. There is no doubt that when Auclair (1955) used the term *framboisé* he described a classical outbreak of "cider sickness". Pollard (1959) suggested that the term *le framboisé* described two phenomena, one caused by the "sickness" organism and the other by acetic acid bacte-

ria. More recently Drilleau (1977) has made a similar statement, except that the second type of
framboisé is said to be due to the presence of lactic
and acetic acid bacteria. Such a phenomenon has
never been experienced in English ciders, although
excess aldehyde has occasionally occurred in perries
without the intervention of a zymomonad.

Beer

Shimwell (1937) first isolated a *Zymomonas* strain
from the bristles of brushes in cask-washing ma-
chines and brewery yards, suggesting that some
development of the organisms took place outside the
brewery and that it came back by way of contaminated empty casks. The work of Dadds (1971) confirmed this and his work will be discussed later.
Shimwell (1948) had by this time given the beer
organism the temporary name of *Achromobacter
anaerobium*. He described it as being anaerobic,
Gram-negative, indifferent to hop antiseptic, and
capable of growing in a wide range of beers. The
spoilage it caused was described as being characterized by a dense silky turbidity accompanied by an
unpleasant odor and flavor suggesting rotten apples.
This spoilage was accompanied by a copious evolution of CO_2. He did not mention the production of
H_2S as did Lloyd (1903) for cider and later Rainbow
(1971) for beer. Shimwell (1948) said of this organism: "There is, however, one species of Gram
negative bacterium which can not only develop in
beers of normal pH and hop rate, but which is by far
the most dangerous of all beer spoilage bacteria".
He based this statement on the difficulty of controlling the organism because of its tolerance of low pH
and very rapid growth, rendering the beer undrink-
able.

Ault (1965) emphasized the presence of
Zymomonas strains in keg beers. This is one of the
more recent methods of dispensing beers in stainless
steel vessels in a carbonated condition and expelling
them from the keg by a top pressure of CO_2. It is
suggested by Ault (1965) that beer prepared in this
way is particularly vulnerable to the growth of
Zymomonas strains. He mentions the damage done
to the flavor and reiterates the earlier descriptions of
the production of excess acetaldehyde and H_2S. Ault
(1965) cites an example of this organism being
found in returned kegs, but gives the primary source
of infection in this outbreak as an improperly sterilized keg-filling apparatus.

Although there is not a very extensive literature
on the technology of *Zymomonas* in brewing, British
beers have been affected by the growth of this organism for quite a considerable time. In contrast, beers
brewed in mainland Europe appear to be unaffected
by *Zymomonas*, which may be explained in one of
two ways: (i) a failure to recognize the organism

when present or (ii), as Swings and De Ley (1977)
surmise, the typical lager fermentation carried out at
temperatures ranging from 8 to 12°C inhibits the
growth of the organism. One reference pertaining to
mainland Europe is the use of a *Zymomonas* strain in
the manufacture of "near" beer in Austria (Lüers,
1932).

In response to the need for rapid identification of
this organism in breweries, Dadds (1971, 1972)
devised a selective cultural method for growing the
bacteria and at the same time developed an immunofluorescence technique. Although neither method
was quite as specific as had been hoped for, a combination of the two produced a more rapid means of
detection than had previously existed for
Zymomonas strains in beer.

Other Sources of *Zymomonas*

While beer and cider are now consumed universally,
the other fermented beverages in which these organisms are found have a more restricted appeal. Its
occurrence in various fermented beverages does,
however, illustrate the wide geographical distribution of this organism. One drink that is popular in
many parts of the world is palm wine, which is made
in tropical countries. It is made from the fermenting
palm sap. The species of palm used are as diverse as
the countries in which they grow, and the method
and site of sap extraction from the plant are also
very varied. *Zymomonas* strains have been discovered in Java (Roelofsen, 1941), Nigeria (Faparusi,
1974; Okafor, 1975), and Zaire (Van Pee and
Swings, 1971). In the various accounts given it is not
clear whether the *Zymomonas* sp. present is the fermenting organism or whether it plays a subsidiary
role to the yeasts.

Sources other than palm wine from which this
organism has been isolated are as follows: fermenting Mexican agave sap and agave flowers (Lindner,
1928a), fermenting Brazilian sugar cane (Gonçalves
de Lima et al., 1970), fermenting Trinidadian cocoa
beans (Ostovar and Keeney, 1973), and Spanish
honey and honeybees (Ruiz-Augueso and Rodriguez-Navarro, 1975). Thus *Zymomonas* strains are
widely distributed and are mainly found in fermenting or potential fermenting situations.

Survival of *Zymomonas* Strains

When discussing *Zymomonas* strains of palm wine,
Swings and De Ley (1977) state that zymomonads
"are extremely well adapted to this ecological
niche" and continue by explaining that the right
nutrients are present and the organisms are resistant
to alcohol. This explains how and why the organisms grow but says nothing about their survival.

Fermentation is a transitory condition and when it is finished virtually no sugar remains. How then does this organism survive?

Some light has been thrown on survival by Millis (1951). She found that apple juice yeast extract medium infected with strains of *Zymomonas* and stored at room temperature yielded 100% viable cultures at 9 weeks, 69% at 13 weeks, and 0% (all were dead) at 17 weeks. Dadds (1972) showed that in soil near the brewery inoculated with *Zymomonas* the organisms survived about 14 days. When steel plates were inoculated with *Zymomonas* it was noted that the factor controlling death was the relative humidity. The higher the moisture content, the more rapidly death occurs. It is thus still not known how zymomonads survive to reinfect breweries or cider manufacturies.

Aspects of Biochemistry and Nutrition

It has been mentioned earlier that these organisms metabolize glucose, thereby gaining energy by way of the Entner-Doudoroff glycolytic pathway. These organisms were first shown to have such a pathway by Gibbs and De Moss (1951, 1954) who by labeling glucose in various positions showed that the end products of its metabolism were labeled differently from those of the Embden-Meyerhof-Parnas pathway. They concluded that *Pseudomonas lindneri*, as they called this particular strain of *Zymomonas*, had a different glycolytic pathway. Further light was shed on this by Dawes, Ribbons, and Large (1966) who showed the presence of enzymes able to produce the key compound 2-keto-3-deoxy-6-phosphogluconate and those able to split it to form pyruvate and glyceraldehyde-3-phosphate. This confirmed the presence of the Entner-Doudoroff pathway which has been illustrated by Carr (1968).

A further interesting point has been noted by Belaich and Senez (1965) about a strain of *Zymomonas mobilis*. They showed that it had an absolute requirement for pantothenate. If a synthetic medium plus pantothenate was used to culture this organism, growth was reduced to 50% of that obtained in a complex medium. They also showed that under these conditions the organisms were energetically uncoupled, which means that they produced energy from glucose faster than it was utilized for the synthesis of other materials. Belaich and Senez (1965) concluded that complex media must contain a factor that keeps energy production and utilization in balance. Swings and De Ley (1977) have also reported a requirement for biotin in some strains of these organisms. During the course of this work Belaich and Senez (1965) showed that the strain of

Z. mobilis they investigated produced from 1 mol of glucose 1.58 mol of ethyl alcohol, 1.7 mol of CO_2, and 0.2 mol of lactate. Similar results were obtained by McGill, Dawes, and Ribbons (1965) using a strain designated *Z. anaerobia*. They showed that their organisms produced 1.8 mol of ethyl alcohol, 1.9 mol of CO_2, and some acetaldehyde from 1 mol of glucose. Dawes, Ribbons, and Rees (1966), working with a strain of *Z. mobilis*, showed that the metabolism of sucrose was slow and that it resulted in the production of a levan.

A point of practical importance with these organisms is their insensitivity to sulphur dioxide. Millis (1956) showed that one strain of *Z. anaerobia* subsp. *pomaceae* was able to grow in the presence of 500 ppm SO_2, a concentration which far exceeds those permitted as a preservative in ciders and beers. One other characteristic that is also of practical importance in the beverage industry is these organisms' ability to form H_2S which, added to their production of acetaldehyde, makes zymomonads producers of unpleasant off-flavors and aromas in drinks such as beer and cider.

Isolation

These are not difficult organisms to isolate, providing a suitable medium is available. To isolate these organisms certain procedures must be followed. The best growth temperature is 25–30°C, although slower growth can be achieved at temperatures above and below this optimum. Liquid culture may be grown in air, but if visible growth is required rapidly on agar, oxygen must be restricted. Extremely slow growth will occur on plates incubated aerobically, showing that the organisms are microaerophilic rather than truly anaerobic. A highly suitable medium for the zymomonads of ciders and perries is apple juice yeast extract medium.

Apple Juice–Yeast Extract Medium (AJYE) (Carr, 1953)

Add 10 g per liter of Difco yeast extract to an apple juice prepared from a culinary or dessert variety and adjust the pH to 4.8 with NaOH. To ensure a firm gel, after autoclaving at 114°C for 10 min, add agar at a concentration of 30 g/liter.

Beer zymomonads are often isolated on malt agar.

Malt Extract–Yeast Extract–Glucose–Peptone Medium (MYGP) (Wickerham, 1951; modified by Dadds, 1972)

Add 3 g yeast extract, 3 g malt extract, 5 g peptone, and 10 g glucose to 1 liter water. Autoclave at 120°C for 15 min. Dadds' (1972) modifications are to double the amount of glucose and to adjust the pH to 4.0 with lactic acid. For solid medium, 20 g/liter of agar is added.

Many substrates in which zymomonads grow naturally make suitable isolation media, particularly if fortified with yeast extract and glucose.

Since the natural substrates are not universally available, it is essential to have alternative media, common to most laboratories, on which these organisms will grow.

Yeast Water Agar (Dadds, 1972)
Add the following:

Yeast extract	4 g
Glucose	20 g
Casein hydrolysate	5 g
Agar	15 g
KH_2PO_4	0.55 g
KCl	0.40 g
$CaCl_2$	0.125 g
$MgCl_2 \cdot 7H_2O$	0.125 g
$FeCl_3 \cdot 6H_2O$	0.0025 g
$MnSO_4 \cdot 4H_2O$	0.0025 g
Bromocresol green	1 ml of a 2.2% alcoholic solution
Distilled water	to 1 liter

Other isolation media are known, but for determination of the organism's nutritional requirements the reader's attention is directed to Swings and De Ley (1977).

Identification

Once *Zymomonas* has been isolated on the appropriate medium its identification is a relatively simple matter. In the eighth edition of *Bergey's Manual,* the genus *Zymomonas* is listed as being of uncertain affiliation, which means that its differences from other groups are so marked as to place these organisms on their own.

If a series of sugar fermentations with Durham tubes is prepared at an acid pH, *Zymomonas* spp. may be readily identified. A convenient pH is 4.8, which is within the bromocresol green indicator range. A suitable buffered basal medium is that prepared by Green and Gray (1950), but Dadds's (1972) yeast water medium minus glucose would probably be as suitable.

Inoculation of *Zymomonas* spp. into media containing one of a range of sugars will lead to growth and gas production only when glucose, fructose, or possibly sucrose is present. All other sugars would be unaffected. The indicator will not register any acidity. As far as is known, no other bacteria behave in this manner. Thus, other tests that might be ap-

plied, such as Gram stain, motility, production of H_2S, production of ethanol, acetaldehyde, etc., are merely confirmatory that the organisms examined are *Zymomonas* spp.

For most investigators, the tests described above are probably as far as it is necessary to go to identify the organisms as *Zymomonas* spp. If, however, a finer distinction is to be made, tests listed by Carr (1974) in *Bergey's Manual* and by Swings and Van Pee (1977) may be applied. These include the type of cytochromes present, G+C ratios, DNA homology, infrared spectra, growth in the presence of various toxic materials and at different temperatures, the most stimulatory amino acids, and tolerance to oxygen.

Literature Cited

Auclair, B. 1955. Le framboisé des cidres. Industries Alimentaires et Agricoles **72:**185–191.

Ault, R. G. 1965. Spoilage bacteria in brewing—a review. Journal of the Institute of Brewing **71:**376–391.

Barker, B. T. P., Ettel, J. 1908. Cider sickness, p. 30. In: National Fruit and Cider Institute, Long Ashton, Bristol. Report for the years 1903–1912. Bath: William Lewis & Son.

Barker, B. T. P., Hillier, V. F. 1912. Cider sickness. Journal of Agricultural Science **5:**67–85.

Belaich, J. P., Senez, J. C. 1965. Influence of aeration and of pantothenate on growth yields of *Zymomonas mobilis.* Journal of Bacteriology **89:**1195–1200.

Buchanan, R. E., Gibbons, N. E. (eds.). 1974. Bergey's manual of determinative bacteriology, 8th ed. Baltimore: Williams & Wilkins.

Carr, J. G. 1953. The lactic acid bacteria of cider: I. Some organisms responsible for the malo-lactic fermentation, pp. 144–150. The annual report of the Agricultural and Horticultural Research Station (The National Fruit and Cider Institute, Long Ashton, Bristol). Bristol: Mendip Press.

Carr, J. G. 1968. Biological principles in fermentation, p. 32. London: Heinemann.

Carr, J. G. 1974. Genus *Zymomonas,* pp. 352–353. In: Buchanan, R. E., Gibbons, N. E. (eds.), Bergey's manual of determinative bacteriology, 8th ed. Baltimore: Williams & Wilkins.

Carr, J. G., Passmore, S. M. 1971. Discovery of the 'cider sickness' bacterium *Zymomonas anaerobia* in apple pulp. Journal of the Institute of Brewing **77:**462–466.

Dadds, M. J. S. 1971. The detection of *Zymomonas anaerobia,* pp. 219–222. In: Shapton, D. A., Board, R. G. (eds.), Isolation of anaerobes. The Society for Applied Bacteriology Technical Series No. 5. London, New York: Academic Press.

Dadds, M. J. S. 1972. Detection and survival of *Zymomonas* in breweries. Ph.D. thesis. University of Bath, Bath, England.

Dadds, M. J. S., Martin, P. A., Carr, J. G. 1973. The doubtful status of the species *Zymomonas anaerobia* and *Z. mobilis.* Journal of Applied Bacteriology **36:**531–539.

Dawes, E. A., Ribbons, D. W., Large, P. J. 1966. The route of ethanol formation in *Zymomonas mobilis.* Biochemical Journal **98:**795–803.

Dawes, E. A., Ribbons, D. W., Rees, D. A. 1966. Sucrose utilization by *Zymomonas mobilis:* Formation of a levan. Biochemical Journal **98:**804–812.

De Ley, J., Swings, J. 1976. Phenotypic description, numerical analysis, and proposal for an improved taxonomy and nomenclature of the genus *Zymomonas* Kluyver and van Niel 1936. International Journal of Systematic Bacteriology **26:**146–157.

Drilleau, J. F., 1977. Le framboisé dans les cidres. Bios **7:**37–44.

Entner, N., Doudoroff, M. 1952. Glucose and gluconic acid oxidation of *Pseudomonas saccharophila*. Journal of Biological Chemistry **196:**853–862.

Faparusi, S. I. 1974. Microorganisms from oil palm tree (*Elaeis guineensis*) tap holes. Journal of Food Science **39:**755–757.

Gibbs, M., De Moss, R. D. 1951. Ethanol formation in *Pseudomonas lindneri*. Archives of Biochemistry and Biophysics **34:**478–479.

Gibbs, M., De Moss, R. D. 1954. Anaerobic dissimilation of C^{14}-labelled glucose and fructose by *Pseudomonas lindneri*. Journal of Biological Chemistry **207:**689–694.

Gonçalves de Lima, O., De Araújo, J. M., Schumacher, I. E., Cavalcanti Da Silva, E. 1970. Estudos de microorganismos antagonistas presentes nas bedidas fermentadas usadas pelo povo do Recife I Sobre uma variedade de *Zymomonas mobilis* (Lindner) (1928) Kluyver e van Niel (1936): *Zymomonas mobilis* var. *recifencis* (Gonçalves de Lima, Araújo, Schumacher and Cavalcanti) (1970), isolada de bebida popular denominada "caldo-de-cana picado". Revista do Instituto de Antibioticos Universidade do Recife **10:**3–15.

Green, S. R., Gray, P. P. 1950. A differential procedure applicable to bacteriological investigation in brewing. Wallerstein Laboratories Communications **13:**357–368.

Kluyver, A. J., Hoppenbrouwers, W. J. 1931. Ein merkwürdiges Gärungsbakterium: Lindner's *Termobacterium mobile*. Archiv für Mikrobiologie **2:**245–260.

Kluyver, A. J., van Niel, C. B. 1936. Prospects for a rational system of classification of bacteria. Zentralblatt für Bakteriologie, Parasitenkunde und Infektionskrankheiten, Abt. 2 **94:**369–403.

Lindner, P. 1928a. Gärungstudien über Pulque in Mexiko. Bericht des Westpreussischen Botanisch-Zoologischen Vereins **50:**253–255.

Lindner, P. 1928b. Atlas der mikroskopischen Grundlagen der Gärungskunde mit besonderer Berücksichtigung der biologischen Betriebskontrolle, 3rd ed. Berlin: Verlagsbuchhandlung Paul Parey.

Lloyd, F. J. 1903. Report on the results of investigations into cider-making, p. 107. London: HMSO for the Board of Agriculture and Fisheries.

Lüers, P. 1932. Ueber Möglichkeiten der Herstellung neuerer Getränke im Brauereibetrieb. Wochenschrift für Brauerei **49:**73–79.

McGill, D. J., Dawes, E. A., Ribbons, D. W. 1965. Carbohydrate metabolism and growth yield coefficients of *Zymomonas anaerobia*. Biochemical Journal **94:**44–45.

Millis, N. F. 1951. Some bacterial fermentations of cider. Ph.D. Thesis. University of Bristol, Bristol, England.

Millis, N. F. 1956. A study of the cider-sickness bacillus—a new variety of *Zymomonas anaerobia*. Journal of General Microbiology **15:**521–528.

Okafor, N. 1975. Microbiology of Nigerian palm wine with particular reference to bacteria. Journal of Applied Bacteriology **38:**81–88.

Ostovar, K., Keeney, P. G. 1973. Isolation and characterization of microorganisms involved in the fermentation of Trinidad's cacao beans. Journal of Food Science **38:**611–617.

Pollard, A. 1959. Le framboisé du cidre et la "cider sickness". Industries Alimentaires et Agricoles **76:**537.

Rainbow, C. 1971. Spoilage organisms in breweries. Process Biochemistry **6:**15–17, 26.

Richards, M., Corbey, D. A. 1974. Isolation of *Zymomonas* from primed beer. Journal of the Institute of Brewing **80:**241–244.

Roelofsen, P. A. 1941. De alkoholbakterie in arensap. Natuurwetenschappelijk Tijdschrift voor Nederlandsch-Indie **101:**274.

Ruiz-Argueso, T., Rodriguez-Navarro, A. 1975. Microbiology of ripening honey. Applied Microbiology **30:**893–896.

Shimwell, J. L. 1937. Study of a new type of beer disease bacterium (*Achromobacter anaerobium* sp. nov.) producing alcoholic fermentation of glucose. Journal of the Institute of Brewing **43:**501–509.

Shimwell, J. L. 1948. Brewing bacteriology. V. Gram-negative wort, yeast, and beer bacteria. Wallerstein Laboratories Communications **11:**135–145.

Shimwell, J. L. 1950. *Saccharomonas,* a proposed new genus for bacteria producing a quantitative alcoholic fermentation of glucose. Journal of the Institute of Brewing **56:**179–182.

Swan, H. S. D. 1952. Author index to papers in the Long Ashton annual reports from 1907–1952, pp. 229–302. In: The Annual Report of the Agricultural and Horticultural Research Station (The National Fruit and Cider Institute), Long Ashton, Bristol. Bath: Mendip Press.

Swings, J., De Ley, J. 1977. The biology of *Zymomonas*. Bacteriological Reviews **41:**1–46.

Swings, J., Van Pee, W. 1977. Infra-red spectroscopy of *Zymomonas* cells. Journal of General and Applied Microbiology **23:**297–301.

Van Pee, W., Swings, J. 1971. Chemical and microbiological studies on Congolese palm wine (*Elaeis guineensis*). East African Agricultural and Forestry Journal **36:**311–314.

Warcollier, G. 1928. La cidrerie, 3rd ed. p. 380. Paris: Bailliere.

Wickerham, L. J. 1951. Taxonomy of yeasts. U.S. Department of Agriculture Technical Bulletin No. 1029. Washington, D.C.: United States Department of Agriculture.

CHAPTER 107

The Genus *Chromobacterium*

MAURICE O. MOSS and COLIN RYALL

Color has always held a special fascination for man, and bacteriologists are no exception; so, despite the medical insignificance of many of them, bacteria producing colored colonies have been widely studied since the earliest days of bacteriology. At first the name *Chromobacterium* was used for organisms producing brightly colored colonies—red and yellow as well as purple. Subsequently the name was restricted to those producing purple and violet colonies, although it was soon appreciated that these colors may be due to a number of chemically unrelated pigments. The genus is now restricted further to those organisms in which the purple coloration is due to the pigment violacein (Fig. 1). It is now widely appreciated that the property of color may reflect an ability to synthesize a single pigment and, although the property may be useful in the recognition of a species or group of species, it should not be considered as necessarily indicating a close relationship between organisms producing the same pigment.

Purple-pigmented bacteria have been described since the end of the nineteenth century as causing discolorations of a wide range of natural substrates, such as potato (Schroeter, 1872), sheep wool (Fraser and Mulcock, 1956; Seddon, 1937), egg white (Bergonzini, 1879), pig bladder (Zopf, 1883), milk (Hueppe, 1884) and, most recently, chilled poultry (Cox, 1975). Since the report of a violet bacterium causing a septicemic infection in buffalo (Woolley, 1904, 1905), there have been a number of accounts of disease associated with chromobacteria in mammals, including man.

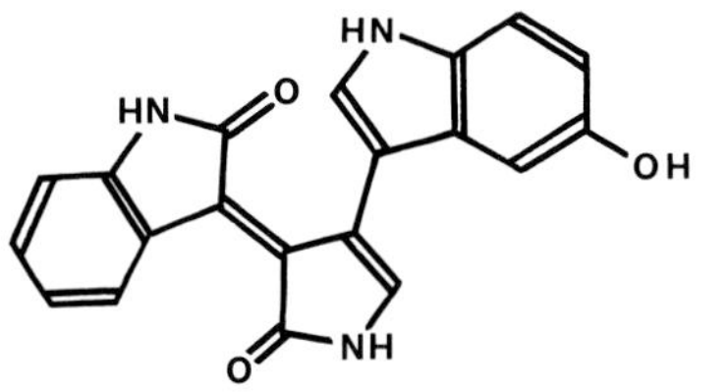

Fig. 1. Structure of the purple pigment violacein, produced by *Chromobacterium*.

The genus *Chromobacterium* was first given a restricted definition by Buchanan (1918), but a great deal of confusion existed until studies such as those of Cruess-Callaghan and Gorman (1933), Conn (1938), Gilman (1953), Leifson (1956), Sneath (1956), and Eltinge (1957). Sneath proposed that the accepted isolates of *Chromobacterium* be referred to two species only: *Chromobacterium violaceum*, based on the mesophilic fermentative group, and *C. lividum*, based on the psychrophilic nonfermentative group. Both the genus and the two species are fully described by Sneath (1974) and are the subject of an exhaustive study including many aspects, such as pigmentation, flagellation, biochemical properties, morphology, serology, and pathogenicity (Sneath, 1960). At the time of the earlier studies Sneath (1956) commented that: "The systematic position of the genus is uncertain; it possesses many characteristics of the Pseudomonadaceae and some characteristics of the Enterobacteriaceae. It may be closely allied to *Vibrio*. In practice it is the production of violacein which is the main distinguishing feature in *Chromobacterium*. This is unsatisfactory since unpigmented strains, which may be common in nature, are probably never assigned to the genus; there may be nonchromogenic species which should be placed in the genus." Certainly the results of an Adansonian study of organisms included in the family Rhizobiaceae, as it was earlier conceived, provided strong evidence that *Chromobacterium* should be separated from the other genera of the family (Moffett and Colwell, 1968).

It is almost incredible that 20 years later the position is still essentially the same, with the genus being placed with other genera of uncertain affiliation among the Gram-negative, facultatively anaerobic rods in the eighth edition of *Bergey's Manual of Determinative Bacteriology* (Sneath, 1974), despite one of the two fully described species being obligately aerobic—unless grown in the presence of nitrate.

On several occasions authors have suggested that the two species are not closely related and Sneath

(1957) commented, from the results of a numerical taxonomic study, that the two groups of *Chromobacterium* are as distantly related as *Pseudomonas* is to *Serratia*. Indeed it has been proposed that each of the two species of *Chromobacterium* is more closely related to members of other genera than to the other *Chromobacterium* species and that each should be considered as a separate genus among the Pseudomonadaceae. The predominant obstacles to any changes have been the importance of violacein production in the assignment of strains to the genus and the difficulty of breaking up a genus that contains only two accepted species.

The production of violacein has become less important since the recognition of some marine violacein-producing isolates that could not be satisfactorily classified in the genus *Chromobacterium* and have been referred to as *Alteromonas* (Gauthier, 1976). Furthermore, Sivendra, Lo, and Lim (1975) have demonstrated that it is possible to identify nonpigmented strains of *Chromobacterium violaceum* isolated from pond water samples, despite the fact that it was not possible to induce pigment production. Colorless strains of *Chromobacterium lividum* have also been described, isolated from abattoir effluent and able to grow on meat surfaces at low temperatures (Etherington et al., 1976). It is of particular interest that these strains are strongly proteolytic, a property usually associated with the mesophilic species and one of the characters used to differentiate the two *Chromobacterium* species. Recently a group of freshwater fermentative psychrophilic strains were described and assigned to a new species, *C. fluviatile* (Moss, Ryall, and Logan, 1978). These strains are characterized by the production of very thin, spreading colonies, and, although they share many biochemical characteristics with *C. violaceum*, they have a much lower mol% G+C content. With all these reservations in mind it is still necessary, for the present, to accept the convenience of the genus *Chromobacterium* as a group of Gram-negative, oxidase-positive, catalase-positive, chemoorganotrophic, motile, round-ended rods that always have a polar flagellum, may have one or more peritrichous flagella, and usually produce the purple pigment violacein. They are nutritionally undemanding and are capable of growing on ordinary peptone-based media without any requirement for complex growth factors. Within this genus, the fermentative mesophilic organisms, which produce hydrogen cyanide, are referred to as *Chromobacterium violaceum*, including those strains that do not produce acid from carbohydrate anaerobically and had been called *C. laurentium* by Leifson (1956). The oxidative psychrophiles have all been referred to as *C. lividum,* including a number of strains with distinctive membranous or gelatinous colonies previously referred to as *C. amethystinum* or *C. membranaceum.*

Habitats

Soil and Water

Chromobacteria have been isolated from the soil and water of many parts of the world from polar regions to the tropics. Although *Chromobacterium lividum* is more commonly isolated from temperate and cooler regions and *C. violaceum* from tropical regions, Morris (1954) reported isolating both mesophiles and psychrophiles from the soil and water of Trinidad, whereas Christensen and Cook (1970) reported the isolation of both mesophiles and psychrophiles from the humic layers of certain Alberta soils without, unfortunately, giving any details of the isolates that might allow their identification.

Members of the British Antarctic Survey have recently shown that purple-pigmented isolates, presumed to be chromobacteria, may form a significant proportion of the flora of soils and lakes of Antarctica (D. D. Wynn-Williams, personal communication, 1977).

From a semiquantitative study of chromobacteria in soil and freshwater sources such as rivers, creeks, and springs, Corpe (1951) concluded that the soil was the natural habitat of at least some of his isolates. During an extensive study of chromobacteria in a British lowland river, and the associated soils and bottom sediments, Ryall (1976) confirmed that some strains were indeed more frequently isolated from the soil. He also showed that many of the isolates from the river water samples were of soil origin, but that some strains, especially those now described as *Chromobacterium fluviatile,* may have a more direct association with river water. It has been implied several times that the appearance of chromobacteria in rivers is due to contamination by soil, usually following rainfall (Calderini, 1925; Rice, 1938). Certainly this could account for the observation that chromobacteria are most frequently isolated from rivers during the winter, when they are usually carrying a high level of silt following heavy rainfall.

Although they have been isolated from soils so frequently, it has not been possible to suggest what ecological niche chromobacteria occupy. Whether or not they have any specific role in the rhizosphere is still uncertain. Although Holding (1960) found no evidence of such an association, chromobacteria have been reported from rhizosphere soil (Hussain and Vančura, 1970), and Dudchenko, Ulyashavo, and Ivankevych (1973) recorded higher numbers in the rhizosphere of some crop plants than in that of perennial grass.

Despite a long series of studies of the soils of New Zealand and the Pacific Islands, Stout (1960, 1961a,b, 1962, 1971a,b) found the distribution of chromobacteria uneven and perplexing, and only a single isolate was in fact recorded from the Pacific

Islands. It has not been possible, from these studies, to associate the organism with any particular ecological niche.

There is a strong possibility that chromobacteria are more closely associated with moist soils containing a relatively high content of organic material than with dryer soils with low organic content (Christensen and Cook, 1970; Morris, 1954). Many isolates of chromobacteria, especially those from the soil, can reduce nitrate actively (Eltinge, 1956), and they may play an important role in denitrification processes. Denitrification, or nitrate respiration, usually requires anaerobic conditions and such conditions are frequently associated with increased levels of organic materials. Chromobacteria have been isolated from muddy water (Nunnally and Dunlop, 1968), sewage effluent (Jordan, 1890), activated sludge plant (Zharova, Kordyan, and Lazurkevich, 1974), and farm water supplies (Thomas and Thomas, 1947).

The function of the purple pigment violacein has been studied from several points of view, and it has been variously suggested that the pigment serves as a protectant against visible and ultraviolet radiation, as an intermediate in respiratory processes, and as a mechanism for removing excess tryptophan or indole compounds. The pigment does have demonstrable antibiotic activity and it has provided some protection against predation by protozoa. Both *Chromobacterium violaceum* and violacein itself were rapidly lethal to certain ciliate protozoa (Burbanck, 1942; Kidder and Stuart, 1939). Singh (1942, 1945) showed that amoebae and flagellate protozoa would not consume *C. violaceum* but were killed by the bacteria. Groscop and Brent (1964) have shown that violacein itself is toxic to soil amoebae, and Curds and Vandyke (1966) found *C. violaceum* to be toxic to *Vorticella microstoma*.

The bacteriophagous nematode *Pelodera chitwoodi* has rejected *Chromobacterium* species (Joshi et al., 1974). The antagonistic activity of violacein against a wide range of protozoa and microscopic metazoa that feed on bacteria will almost certainly have an influence on the survival of chromobacteria in soil and water, although this has not been demonstrated directly in these environments. Pigment production does not appear to contribute to the pathogenicity of *Chromobacterium violaceum* in mammals (Sivendra and Tan, 1977).

One of the most striking characteristics of colonies of some strains of *Chromobacterium* growing on solid media is the production of tough, membranous growth with gelatinous texture; samples of these colonies are difficult to obtain without removing the entire colony on the needle or loop. This property is particularly characteristic of soil isolates and is due, at least in part, to the production of an extracellular polysaccharide (Corpe, 1960). It has been suggested by Martin and Richards (1963) that the polysaccharide of *Chromobacterium* may be important in the formation of soil aggregates, because it is resistant to microbial decomposition and persists after the bacteria producing it have died and been degraded.

Plants

Although the role of *Chromobacterium* in the rhizosphere is still ambiguous, some strains are symbiotic with certain plants. Bettelheim, Gordon, and Taylor (1968) isolated strains of *Chromobacterium lividum* from germinating seeds of *Psychotria nairobiensis* (Rubiaceae) and *Ardisia crispa* (Myrsinaceae). Using immunofluorescence techniques, they were able to demonstrate the presence of this organism in the tissues of the leaf nodules formed by these plants.

In a critical review of bacteria reported to be associated with the leaf nodules of the families Myrsinaceae and Rubiaceae, Horner and Lersten (1972) carefully defined leaf nodules and considered that bacteria proved to be inhabiting such nodules are part of a population maintained by the host plant in the vegetative and floral buds and are passed to successive generations through the seeds. The reviewers concluded that *Chromobacterium* seemed to be a good candidate as a true leaf-nodule symbiont but were not able to confirm its role (Lersten and Horner, 1976). The strains isolated by Bettelheim, Gordon, and Taylor (1968) produce cytokinins (Rodriguez-Pereira et al., 1972) and provide the host plant (*Ardisia* sp.) with those growth factors that it is unable to produce itself.

It has been reported that maize seedlings inoculated with *C. violaceum* produced a significantly increased yield of dry plant material (Hussain and Vančura, 1970).

Although there have not been many reports of chromobacteria interacting with fungi, they occasionally cause discoloration of the gills and flesh of agarics, and strains of psychrophilic chromobacteria have been implicated in the stimulation of sporangium formation by *Phytophthora cinnamomi* (Zentmyer, 1965). However, it must be added that Broadbent, Baker, and Aust (1974) found no evidence that *Chromobacterium* spp. were involved in the induction of the formation of sporangia during their study of the association of bacteria with the reproduction of *Phytophthora*.

Animals

Chromobacteria have been isolated, presumably in a nonpathogenic role, from the intestine of the larvae of St. Mark's fly *(Bibio marci);* they were considered to have originated from the soil on which the larvae feed (Szabo, Marton, and Buti, 1969). Chromobacteria have also been isolated from the feces of the oriental cockroach *(Blatta orientalis)* (Longfellow, 1913), the digestive tract of the phytophagous aquatic beetle *Tropisternus lateralis nimbatus* (Wooldridge and Wooldridge, 1972), and the gills and crystalline style of the freshwater Venus mussel *(Meretrix meretrix)* (Kisitani and Sumiyosi, 1939). In a study of the "oosphere" flora of stream-incubating eggs of *Oncorrhynchus keta* and *O. gorbuscha* (Pacific salmon) Bell, Hoskins, and Hodgkiss (1972) obtained only a single isolate of *Chromobacterium* from live eggs that seemed to have a fairly specific flora of *Cytophaga. Pseudomonas* and *Chromobacterium* were more readily isolated from simulated eggs made of polyethylene spheres that were left in the same water, and from the river water itself.

Although all attempts to infect laboratory animals with *Chromobacterium lividum* have been unsuccessful, the mesophilic species *C. violaceum* has frequently been recorded from infections in mammals, including man—often with fatal consequences. Such infections have only been reported from tropical and subtropical regions (Johnson, DiSalvo, and Steuer, 1971; Joseph et al., 1971), a finding that reflects the natural distribution of the organism in the soil and water of warm countries. It is probable that these infections arise from contamination by or ingestion of water or soil containing the organism.

Sneath (1960) has reviewed in detail the earlier literature describing cases of chromobacteriosis. *C. violaceum* is perhaps best considered as an opportunistic pathogen causing an uncommon, usually septicemic disease that has much in common with melioidosis, caused by *Pseudomonas pseudomallei*, another soil-borne saprophyte. Infections of swine have been particularly well studied (Laws and Hall, 1964; Sippel, Medina, and Atwood, 1954; Wijewanta and Wettimuny, 1969). Groves et al. (1969) describe the deaths of nine Malayan gibbons *(Hylobates)* and a Malayan sun bear *(Helarctos malayanus)* at the National Zoo, Kuala Lumpur, attributed to *C. violaceum*. In at least four of the cases infection was via cuts, abrasions, and possibly oral wounds. Chromobacteriosis has recently been diagnosed in an Assam macaque *(Macaca assamensis),* the animal dying within 4 days of the diagnosis. Subsequent necropsy revealed extensive hepatic necrosis, and *Chromobacterium violaceum* was isolated from blood, liver, lungs, spleen, and kidneys (McClure and Chang, 1976).

Ognibene and Thomas (1970) describe 2 cases of American soldiers suffering fatal infections of *C. violaceum* in Vietnam and note that, of the 16 fatal cases recorded in the literature up to that time, 9 had occurred in Vietnam. The organism was readily isolated from a major water source in that country.

Clinically the disease is variable. Sneath et al. (1953) recorded cases of mild diarrhea and subsequent recovery, although most cases in man and animals have been fatal. Death may occur after a few days or as long as a year following infection (Johnson, DiSalvo, and Steuer, 1971). Abscesses at several sites have been recorded (Ognibene and Thomas, 1970; Soule, 1939) and are often followed by septicemia. It is the septicemia that usually leads to rapid death (Darrasse et al., 1955; Nunnally and Dunlop, 1968).

A strain of *Chromobacterium violaceum* has recently been isolated from a fatally infected patient and shown to produce a β-lactamase (Farrar and O'Dell, 1976). Like the enzyme produced by *Pseudomonas aeruginosa*, the *C. violaceum* enzyme was found to be primarily active against cephalosporins and sensitive to inhibition by cloxacillin. It was also shown to be inducible by benzylpenicillin.

The disease responds to tetracyclines if given at an early enough stage. Once general septicemia has become established treatment is usually too late, although Victorica, Baer, and Ayoub (1974) have reported the successful treatment of systemic chromobacteriosis in a 17-year-old girl.

Isolation

Chromobacteria are relatively undemanding in their nutritional requirements and will grow on ordinary laboratory media such as nutrient agar. *Chromobacterium lividum* is obligately aerobic, but even the facultative *C. violaceum* is best grown on surface-spread plates because the production of violacein, by which they are most readily recognized, requires the presence of oxygen. Sneath (1960) considered that blood-based agar was more suitable for isolates from medical sources because small inocula of *C. violaceum* are often inhibited by the presence of peroxides that may occur in freshly prepared nutrient agar.

In the presence of large numbers of other bacteria, chromobacteria do not grow well or produce their characteristic pigment so readily on ordinary media. Some form of enrichment procedure or selective medium is required for the isolation of chromobacteria from most soil and water samples.

The enrichment technique of Corpe (1960) has been widely used: "Five gram samples of soil placed in sterile Petri dishes and soaked with sterile distilled water. Sterile polished or precooked rice grains were sprinkled over the surface and the plates incubated at 23–25°C for five days."

The majority of strains of *Chromobacterium* can utilize citrate as the sole source of carbon, although *C. violaceum* and *C. fluviatile* do so slowly. Citrate ammonium salts agar containing (g/liter) citric acid, 2; NaCl, 1; MgSO$_4$ · 7H$_2$O, 0.2; (NH$_4$) H$_2$PO$_4$, 1; K$_2$HPO$_4$, 1; Ionagar no. 2 (Oxoid) is fairly selective for *Chromobacterium lividum* and possibly other chromobacteria, but pigmentation is extremely poor and it is difficult to distinguish chromobacteria from other organisms able to grow on this medium.

After an extensive survey of the effect of antibiotics on the growth of chromobacteria, Ryall and Moss (1975) described a selective medium for the counting of chromobacteria in water samples based on quarter-strength nutrient agar. The medium contained (g/liter): Lab-Lemco beef extract (Oxoid), 0.25; yeast extract, 0.5; peptone, 1.25; NaCl, 1.25; agar, 20. To the molten agar (45°C) were added filter-sterilized solutions of colistin and sodium deoxycholate to give final concentrations of 15 μg/ml and 0.3 mg/ml, respectively.

When it was required to use the medium for counting chromobacteria in samples of soil, cycloheximide was also added to give a final concentration of 30 μg/ml to reduce the level of fungal contamination.

Keeble and Cross (1977) have since described an improved selective medium for counting chromobacteria in soil and water. It is based on Bennett's agar and contains (g/liter): yeast extract (Difco), 1; Lab-Lemco beef extract (Oxoid), 1; Casitone (Difco), 2; glucose, 10; agar (Lab. M no. 2), 18. To the molten medium were added filter-sterilized solutions of neomycin hydrochloride, cycloheximide, and nystatin to give a final concentration of 50 μg/ml of each. This medium has given not only increased counts, but also much improved pigmentation. It is interesting to note that Darrasse et al. (1955) reported the stimulation of pigmentation by low concentrations of neomycin, and Ryall (1976) recorded that zones of inhibition produced by antibiotic disks containing neomycin were surrounded by a zone of intense pigmentation on plates of chromobacteria.

Identification

Members of the genus *Chromobacterium* have been traditionally distinguished from members of other genera by the production of the nondiffusible purple pigment violacein (Fig. 1). The presence of the pigment is readily confirmed by the following properties:

1. Insolubility in water and chloroform but solubility in ethanol.
2. An ethanolic solution gives a green color on the addition of 10% aqueous sulfuric acid.
3. An ethanolic solution gives a green and then reddish brown color on the addition of 10% aqueous NaOH solution.
4. A comparison of the absorption spectra in ethanol of crude pigments from *Chromobacterium* with that of pure violacein (Fig. 2) (Johnson and Beer, 1971).

Violacein-producing isolates from soil and fresh water can usually be identified as either *C. lividum* or *C. violaceum* with the group of tests described by Sneath (1966, 1974), and this group of tests can also be used to distinguish those organisms referred to as *C. fluviatile* and producing thinly spreading, purple-colored colonies on solid media (Table 1) (Moss, Ryall, and Logan, 1978).

The production of violacein has made it more difficult to determine the oxidase reaction of chromobacteria. They are now recognized as oxidase positive (Bascomb et al., 1973; Sneath, 1974), and methods have been described for easily determining the oxidase reaction of purple-pigmented bacteria (Dhar and Johnson, 1973; Sivendra, Lo, and Lim, 1975).

As discussed earlier, the production of violacein may no longer be considered as specific to this

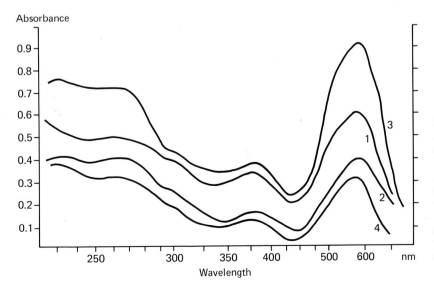

Fig. 2. Absorption spectra in ethanol of purified violacein and of crude pigments extracted from strains of *Chromobacterium*. Abscissa represents wavelength; ordinate represents absorbance. Curve 1, violacein (Johnson and Beer, 1971). Curve 2, *C. violaceum* NCTC 9757. Curve 3, *C. lividum* NCTC 9796. Curve 4, *C. fluviatile* mom 163, University of Surrey.

Table 1. Major differential characters of species of *Chromobacterium*.[a]

Character	C. violaceum	C. lividum	C. fluviatile
Growth at 4°C	−	+	+
Growth at 37°C	+	−	−
Cyanide produced	+	−	−
Turbidity from egg yolk	+	−	(+)
Acid from trehalose	+	−	+
Acid from arabinose	−	+	−
Acid from xylose	−	+	−
Casein hydrolysis	+	− or (+)	+
Esculin hydrolysis	−	+	−
Glucose fermented (O-F test)[b]	F	O	F
Arginine decarboxylase	+	−	−

[a] (+), weakly positive.
[b] F, fermented; O, oxidized.

genus and, in addition to the common experience of the facile isolation of colorless variants from laboratory-stocked strains of *Chromobacterium,* wild-type isolates of nonpigmented strains of both *C. violaceum* and *C. lividum* have been successfully identified. Routinely it would be expected that nonpigmented strains of *C. lividum* would be identified as *Pseudomonas;* indeed, with the key of Hendrie and Shewan (1966) such an organism might be identified as *P. fragi.* Nonpigmented strains of *C. violaceum* might be associated with *Aeromonas,* or even *Vibrio,* although in the authors' limited experience, chromobacteria are resistant to the vibriostat 2,4-diamino-6,7-diisopropyl pteridine (0/129). Sivendra and Tan (1977) have shown that colorless strains of *C. violaceum* are just as virulent for mice as pigmented strains.

Another character that has been associated with the genus *Chromobacterium* is the production of two distinct types of flagella, a weakly staining single polar flagellum and one or more normally staining lateral flagella (Fig. 3). The flagella are not only visibly distinct (Leifson, 1956), but they have been shown to be antigenically different (Sneath, 1974). The lateral flagella are most common in young bacteria from cultures grown on solid media. Leifson (1956) noted that strains having lateral flagella often clumped and adhered to sides of tubes. Mixed flagellation is a recognized characteristic of strains of oxidase-positive, Gram-negative rods in genera such as *Pseudomonas* and *Vibrio.* Recent studies on the flagellation of *Vibrio alginolyticus* have given rise to

the suggestion that the two types of flagella have different roles to play in the natural history of the bacteria, the polar flagellum being an organelle of motility, the lateral flagella organelles of adhesion (DeBoer, Golten, and Scheffers, 1975).

A number of reactions that may distinguish members of the genus *Chromobacterium* from each other and from other genera of oxidase-positive, Gram-negative, polarly flagellate rods are shown in Table 2, and, although further work is required to establish the true affinities of the purple-colored bacteria, it is certain that the different species fit into different parts of the spectrum of variation among bacteria.

The question of further identification within the genus has not been discussed in any detail. Although

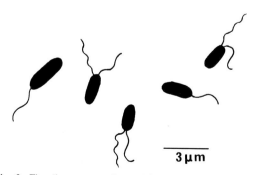

Fig. 3. Flagellar patterns observed in *Chromobacterium fluviatile.* Drawings made from a smear stained by the method of Leifson.

Table 2. A comparison of nonpigmented chromobacteria with other oxidase-positive, Gram-negative, motile rods.[a]

Character	*C. lividum*	*C. violaceum*	*C. fluviatile*	*Pseudomonas*	*Alteromonas*	*Aeromonas*	*Vibrio*	*Plesiomonas*
G+C (mol%)	65–72	63–68	50–52	58–70	43–55	57–63	40–50	51
Glucose fermented (O-F test)	O	F	F	O	NR	F	F	F
$NO_3^- \rightarrow NO_2^-$	+	+	+	+	−	+	+	+
O/129 sensitivity	−	−	−	−	−	−	+	+
Gelatin hydrolysis	(±)	+	+	(+−)	(+−)	+	+	−
Esculin hydrolysis	+	−	−	−	(+−)	(+−)	−	−
Citrate utilization	+	+	+	(+−)	−	(+−)	(+−)	−
Arginine dihydrolase	−	(±)	−	(+−)	−	+	(+−)	+
Arginine decarboxylase	−	+	−	+	(+−)	+	−	+
Lysine decarboxylase	−	−	−	NR	(+−)	−	+	+
Ornithine decarboxylase	−	−	−	NR	(+−)	−	+	+
Polar flagella	1	1	1	1 or >1	1	1	1	>1
Lateral flagella	+	+	+	−(+)	−(+)	−(+)	−(+)	−

[a] Abbreviations: O, oxidative; F, fermentative; NR, no reaction in Hugh and Leifson medium. Symbols: +, positive for most strains or species; −, negative for most strains or species; (±), variable within species; (+−), variable within genus; −(+), usually absent, detected in some species.

Sneath (1974) gave good reasons for including psychrophiles producing membranous colonies, which may have been called *C. amethystinum* or *C. membranaceum* in the past, in the single species *C. lividum*, Moss, Ryall, and Logan (1978) gave some evidence for reseparating these strains, but went no further than to refer to them as "atypical *C. lividum*". Our experience has been that such strains are more frequently isolated from the soil, are generally less reactive in tests for acid from sugars, and

are particularly inactive in the production of acid from polyols. Table 3 gives some reactions that may be useful in separating these strains from "typical" strains of *C. lividum*.

An important paper on the taxonomy of chromobacteria has recently appeared from the laboratories of J. De Ley. Using the relatively new techniques of hybridizing ribosomal RNA and DNA, De Ley, Segers, and Gillis (1978) have obtained two parameters for describing the relatedness of isolates of

Table 3. Differentiation of "atypical *Chromobacterium lividum*" strains from "typical" *Chromobacterium* strains.[a]

Test	*C. lividum*	"Atypical *C. lividum*"	*C. violaceum*	*C. fluviatile*
Colony texture	Butyrous	Membranous	Butyrous	Thin, butyrous spreading
Gas from nitrate	(±)	+	−	−
Hydrolysis of starch	−	+	−	(±)
Acid from:				
Mannitol	+	−	−	−
Sorbitol	+	−	(±)	−
Inositol	+	−	(±)	−
Glycerol	+	−	(±)	(±)

[a] Symbols: +, positive for most strains; −, negative for most strains; (±), variable within species.

bacteria; the temperature at which 50% of the hybrid material was denatured and the percentage of ribosomal RNA which would bind to DNA.

Using these two parameters, the authors confirmed that strains of *Chromobacterium lividum* and *C. violaceum* formed two tight separate clusters which seemed less related to each other than do many other genera one to another. They considered that this new evidence, with that derived from studies of the phenotypes, justifies separating out two genera.

Their suggestion is to retain the genus *Chromobacterium* Bergonzini, suitably amended, for *C. violaceum*, i.e., the mesophilic fermentative organisms, and to create a new genus, *Janthinobacterium*, for those isolates which previously had been identified correctly as *C. lividium*, i.e., the psychrophilic oxidative organisms. They also present evidence that the isolate described as *C. amethystinum* ATCC 6915 should also be referred to *Janthinobacterium lividum*. The type and only species of *Chromobacterium* is *C. violaceum* (neotype strain NCTC 9757 = NCIB 9131 = ATCC 12472), whereas the type and only species of *Janthinobacterium* is *J. lividum* (neotype strain NCTC 9796 = NCIB 9130 = ATCC 12473).

These suggestions leave the organism described as *C. fluviatile* in an isolated position, thus emphasizing the need for further work.

Acknowledgments

The authors would like to acknowledge the encouragement given by P. H. A. Sneath and the inspiration of his own monumental studies on this genus. We thank N. A. Logan and B. Sveshtarova de Gomez for their enthusiastic help and the Natural Environmental Research Council for the award of a studentship to C. Ryall.

Literature Cited

It should be noted that nomenclature before 1960 is confused and, frequently, references to *Chromobacterium violaceum* would now be interpreted as *Chromobacterium lividum*.

Bascomb, S., Lapage, S. P., Curtis, M. A., Wilcox, W. R. 1973. Identification of reference strains. Journal of General Microbiology **77**:291–315.

Bell, G. R., Hoskins, G. E., Hodgkiss, W. 1972. Aspects of the characterisation, identification, and ecology of the bacterial flora associated with the surface of stream incubating Pacific salmon (*Oncorrhynchus*) eggs. Journal of the Fisheries Research Board of Canada **28**:1511–1525.

Bergonzini, C. 1879. I Bacteri. Annuario della Società dei Naturalisti in Modena, Series 2, Anno **13**:19–100.

Bettelheim, K. A., Gordon, J. F., Taylor, J. 1968. The detection of a strain of *Chromobacterium lividum* in the tissues of cer-

tain leaf nodulated plants by the immunofluorescence technique. Journal of General Microbiology **54**:177–184.

Broadbent, P., Baker, K. F., Aust, J. 1974. Association of bacteria with sporangium formation and breakdown of sporangia in *Phytophthora* sp. Australian Journal of Agricultural Research **25**:139–145.

Buchanan, R. E. 1918. Studies in the nomenclature and classification of the bacteria. V. Subgroups and genera of the Bacteriaceae. Journal of Bacteriology **3**:27–61.

Burbanck, W. D. 1942. Physiology of the ciliate *Colpidium colpoda*. I. The effect of various bacteria as food on the division rate of *Colpidium colpoda*. Physiological Zoology **15**:342–362.

Calderini, M. 1925. Richerche sui batteri violacei delle acque. Annali d'Lgiene Sperimentale **35**:765–784.

Christensen, P. J., Cook, F. D. 1970. The microbiology of Alberta muskeg. Canadian Journal of Soil Science **50**:171–178.

Conn, H. J. 1938. Taxonomic relationships of certain non-spore forming rods in soil. Journal of Bacteriology **36**:320–321.

Corpe, W. A. 1951. A study of the wide spread distribution of *Chromobacterium* species in soil by a simple technique. Journal of Bacteriology **62**:515–517.

Corpe, W. A. 1960. The extracellular polysaccharide of gelatinous strains of *Chromobacterium violaceum*. Canadian Journal of Microbiology **6**:153–163.

Cox, N. A. 1975. Isolation and identification of a genus, *Chromobacterium*, not previously found on processed poultry. Applied Microbiology **29**:864.

Cruess-Callaghan, G., Gorman, M. J. 1933. On the characteristics of *Bacterium violaceum* (Schroter) and some allied species of violet bacteria. Scientific Proceedings of the Royal Dublin Society, New Series **21**:213–221.

Curds, C. R., Vandyke, J. M. 1966. The feeding habits and growth rates of some fresh-water ciliates found in activated sludge plants. Journal of Applied Ecology **3**:127–137.

Darrasse, H., Mazaud, R., Guidicelli, P., Camain, R. 1955. Un cas humain mortel d'infection à *Chromobacterium violaceum* (Septicémie et multiples abcès hépatique). Bulletin de la Societé de Pathologie Exotique **48**:704–713.

DeBoer, W. E., Golten, C., Scheffers, W. A. 1975. Effects of some physical factors on flagellation and swarming of *Vibrio alginolyticus*. Netherlands Journal of Sea Research **9**:197–213.

De Ley, J., Segers, P., Gillis, M. 1978. Intra and intergeneric similarities of *Chromobacterium* and *Janthinobacterium* ribosomal ribonucleic acid cistrons. International Journal of Systematic Bacteriology **28**:154–168.

Dhar, S. K., Johnson, R. 1973. The oxidase activity of *Chromobacterium*. Technical Methods **26**:304–306.

Dudchenko, V. H., Ulyashavo, R. M., Ivankevych, N. P. 1973. Species composition of microflora in rotation and continuous seedlings of main crops. Mikrobiolohichnyi Zhurnal (Kiev) **35**:560–564.

Eltinge, E. T. 1956. Nitrate reduction in the genus *Chromobacterium*. Antonie van Leeuwenhoek Journal of Microbiology and Serology **22**:139–144.

Eltinge, E. T. 1957. Status of the genus *Chromobacterium*. International Journal of Bacteriological Nomenclature and Taxonomy **7**:37–44.

Etherington, D. J., Newman, P. B., Dainty, R. H., Partridge, S. M. 1976. Purification and properties of the extracellular metallo-proteinases of *Chromobacterium lividum* (NC1B 10926). Biochimica et Biophysica Acta **445**:739–752.

Farrar, W. E., O'Dell, N. M. 1976. β-Lactamase activity in *Chromobacterium violaceum*. Journal of Infectious Diseases **134**:290–293.

Fraser, I. E. B., Mulcock, A. P. 1956. Staining of wool by bacterial pigments. Nature **177**:628–629.

Gauthier, M. J. 1976. Morphological, physiological and biochemical characteristics of some violet-pigmented bacteria isolated from seawater. Canadian Journal of Microbiology **22**:138–149.

Gilman, J. P. 1953. Studies on certain species of bacteria assigned to the genus *Chromobacterium*. Journal of Bacteriology **65**:48–52.

Groscop, J. A., Brent, M. M. 1964. The effects of selected strains of pigmented microorganisms on small free-living amoebae. Canadian Journal of Microbiology **10**:579–584.

Groves, M. G., Strauss, J. M., Abbas, J., Davis, C. E. 1969. Natural infections of gibbons with a bacterium producing a violet pigment *(Chromobacterium violaceum)*. Journal of Infectious Diseases **120**:605–610.

Hendrie, M. S., Shewan, J. M. 1966. The identification of certain *Pseudomonas* species, pp. 1–6. In: Gibbs, B. M., Skinner, F. A. (eds.), Identification methods for microbiologists, Part A. New York, London: Academic Press.

Holding, A. J. 1960. The properties and classification of the predominant Gram-negative bacteria occurring in soils. Journal of Applied Bacteriology **23**:515–525.

Horner, H. T., Lersten, N. R. 1972. Nomenclature of bacteria in leaf nodules of the families *Myrsinaceae* and *Rubiaceae*. International Journal of Systematic Bacteriology **22**: 117–122.

Hueppe, F. 1884. Untersuchungen über die Zersetzungen der Milch durch Mikroorganismen. Mittheilungen aus dem Kaiserlichen Gesundheitsamte zu Berlin **2**:309–371.

Hussain, A., Vančura, V. 1970. Formation of biologically active substances by rhizosphere bacteria and their effects on plant growth. Folia Microbiologica **15**:468–478.

Johnson, E. A., Beer, R. J. S. 1971. Violacein, spectrum No. J8/2. In: Perkampus, H. M., Sandeman, I., Timmons, C. J. (eds.), UV atlas of organic compounds, vol. 5. London: Butterworth; Weinheim/Bergstr.: Verlag Chemie.

Johnson, W. M., DiSalvo, A.F., Steuer, R. R. 1971. Fatal *Chromobacterium violaceum* septicaemia. American Journal of Clinical Pathology **56**:400–406.

Jordan, E. O. 1890. A report on certain species of bacteria observed in sewage. Report of the Massachusetts Board of Public Health; Report on Water Supply and Sewerage, part **2**:821–844.

Joseph, P. G., Sivendra, R., Anwar, M., Ong, S. F. 1971. *Chromobacterium violaceum* infection in animals. Kajian Veterinarie, Malaysia—Singapore **3**:55–66.

Joshi, M. M., Wilt, G. R., Cody, R. M., Chopra, B. K. 1974. Detection of a *Vibrio* sp. by the bacteriophagous nematode *Pelodera chitwoodi*. Journal of Applied Bacteriology **37**:419–426.

Keeble, J. R., Cross, T. 1977. An improved medium for the enumeration of *Chromobacterium* in soil and water. Journal of Applied Bacteriology **43**:325–327.

Kidder, G. W., Stuart, C. A. 1939. Studies in Ciliates. I. The role of bacteria in the growth and reproduction of *Colpoda*. Physiological Zoology **12**:329–340.

Kisitani, T., Sumiyosi, T. 1939. Über den Entwicklung zyklus einer Violacein bildenden Bakterie. Journal of Science of the Hiroshima University, Series B, Division 2, Botany **3**:153–164.

Laws, L., Hall, W. T. R. 1964. *Chromobacterium violaceum* infections in a pig. Queensland Journal of Agricultural Science **20**:499–513.

Leifson, E. 1956. Morphological and physiological characteristics of the genus *Chromobacterium*. Journal of Bacteriology **71**:393–400.

Lersten, N. R., Horner, H. T., Jr. 1976. Bacterial leaf nodule symbiosis in angiosperms with emphasis on Rubiaceae and Myrsinaceae. Botanical Reviews **42**:145–214.

Longfellow, R. C. 1913. The common house roach as a carrier of disease. American Journal of Public Health **3**:58–61.

McClure, H. M., Chang, J. 1976. *Chromobacterium violaceum* infection in a nonhuman primate *(Macaca assamensis)*. Laboratory Animal Science **26**:807–810.

Martin, J. P., Richards, S. J. 1963. Decomposition and binding action of polysaccharide from *Chromobacterium violaceum* in soil. Journal of Bacteriology **85**:1288–1294.

Moffett, M. L., Colwell, R. R. 1968. Adansonian analysis of the Rhizobiaceae. Journal of General Microbiology **51**:245–266.

Morris, M. B. 1954. Some notes on the affinity between strains of *Chromobacterium violaceum* and *Chromobacter ianthinum* isolated from Trinidad soils. Transactions of the Fifth International Congress of Soil Science, Leopoldville **3**:107–112.

Moss, M. O., Ryall, C., Logan, N. A. 1978. The classification and characterization of chromobacteria from a lowland river. Journal of General Microbiology **105**:11–21.

Nunnally, R. M., Dunlop, W. H. 1968. Fatal septicaemia due to *Chromobacterium janthinum*. Journal of the Louisiana Medical Society **120**:278–280.

Ognibene, A. J., Thomas, E. 1970. Fatal infection due to *Chromobacterium violaceum* in Vietnam. American Journal of Clinical Pathology **54**:607–610.

Rice, J. W. 1938. A distribution study of *Chromobacterium lividum*. Journal of Bacteriology **36**:667–668.

Rodriguez-Pereira, A. S., Houwen, P. J. W., Deurenberg-Vos, H. W. J., Pey, E. B. F. 1972. Cytokinins and the bacterial symbiosis of *Ardisia* species. Zeitschrift für Pflanzen Physiologie **68**:170–177.

Ryall, C. 1976. The ecology of *Chromobacterium* in freshwater. Ph.D. Thesis. University of Surrey, Surrey, England.

Ryall, C., Moss, M. O. 1975. Selective media for the enumeration of *Chromobacterium* spp. in soil and water. Journal of Applied Bacteriology **38**:53–59.

Schroeter, J. 1872. Ueber einige durch Bacterien gebildete Pigmente. Beiträge zur Biologie der Pflanzen **1**:109–126.

Seddon, H. R. 1937. Bacterial colouration of wool, in studies on the sheep blowfly problem. Science Bulletin of the Department of Agriculture, New South Wales, No. **54**:96–110.

Singh, B. N. 1942. Toxic effects of certain bacterial metabolic products on soil protozoa. Nature **149**:168.

Singh, B. N. 1945. The selection of bacterial food by soil amoebae, and the toxic effects of bacterial pigments and other products on soil protozoa. British Journal of Experimental Pathology **26**:316–325.

Sippel, W. L., Medina, G., Atwood, M. B. 1954. Outbreaks of disease in animals, associated with *Chromobacterium violaceum*. I. The disease of swine. Journal of the American Veterinary Medical Association **124**:467–471.

Sivendra, R., Lo, H. S., Lim, K. T. 1975. Identification of *Chromobacterium violaceum:* pigmented and non-pigmented strains. Journal of General Microbiology **90**:21–31.

Sivendra, R., Tan, S. H. 1977. Pathogenicity of non-pigmented cultures of *Chromobacterium violaceum*. Journal of Clinical Microbiology **5**:514–516.

Sneath, P. H. A. 1956. Cultural and biochemical characteristics of the genus *Chromobacterium*. Journal of General Microbiology **15**:70–98.

Sneath, P. H. A. 1957. The application of computers to taxonomy. Journal of General Microbiology **17**:201–226.

Sneath, P. H. A. 1960. A study of the bacterial genus *Chromobacterium*. Iowa State Journal of Science **34**:243–500.

Sneath, P. H. A. 1966. Identification methods applied to *Chromobacterium*, pp. 15–20. In: Identification methods for microbiologists, part A, Gibbs, B. B., Skinner, F. A. (eds.). New York, London: Academic Press.

Sneath, P. H. A. 1974. *Chromobacterium* Bergonzini 1881. In: Bergey's manual of determinative bacteriology, 8th ed. pp. 354–357. Buchanan, R. E., Gibbons, N. E. (eds.). Baltimore: The Williams & Wilkins Co.

Sneath, P. H. A., Whelan, J. P. F., Singh, R. B., Edwards, D. 1953. Fatal infection by *Chromobacterium violaceum*. Lancet **ii**:276–277.

Soule, M. H. 1939. A study of two strains of *B. violaceus* isolated from human beings. American Journal of Pathology **15**:592–595.

Stout, J. D. 1960. Bacteria of soil and pasture leaves at Claudelands showgrounds. New Zealand Journal of Agricultural Research **3**:413–430.

Stout, J. D. 1961a. A bacterial survey of some New Zealand forest lands, grasslands and peats. New Zealand Journal of Agricultural Research **4:**1–30.

Stout, J. D. 1961b. Biological and chemical changes following scrub burning on a New Zealand hill soil. 4. Microbiological changes. New Zealand Journal of Science **4:**740–752.

Stout, J. D. 1962. The antibiotic relationships of some free living bacteria. Journal of General Microbiology **27:**209–219.

Stout, J. D. 1971a. The distribution of soil bacteria in relation to biochemical activity and pedogenesis. I. General introduction and factors affecting populations at Taita Experimental Station, New Zealand. New Zealand Journal of Science **14:**816–833.

Stout, J. D. 1971b. The distribution of soil bacteria in relation to biochemical activity and pedogenesis. II. Soils of some Pacific islands. New Zealand Journal of Science **14:**834–850.

Szabo, I., Marton, M., Buti, I. 1969. Intestinal microflora of the larvae of St. Mark's fly. IV. Studies on the intestinal bacterial flora of a larva-population. Acta Microbiologica Academiae Scientiarum Hungaricae. **16:**381–397.

Thomas, S. B., Thomas, B. F. 1947. Some observations on the bacterial flora of farm water supplies. Proceedings of the Society of Applied Bacteriology **2:**65–69.

Victorica, B., Baer, H., Ayoub, E. M. 1974. Successful treatment of systemic *Chromobacterium violaceum* infection. Journal of the American Medical Association **230:**578–580.

Wijewanta, E. A., Wettimuny, S. G. de S. 1969. *Chromobacterium violaceum* infection in pigs. Research in Veterinary Science **10:**389–390.

Wooldridge, D. P., Wooldridge, C. R. 1972. Bacteria from the digestive tract of the phytophagous aquatic beetle *Tropisternus lateralis nimbatus* (Coleoptera, Hydrophilidae). Environmental Entomology **1:**522–534.

Woolley, P. G. 1904. A report on *Bacillus violaceum* var. *manilae,* a pathogenic microorganism. U.S. Department Interior Bureau Government Laboratories Bulletin 15.

Woolley, P. G. 1905. *Bacillus violaceus manilae* (A pathogenic microorganism). Johns Hopkins Hospital Bulletin **16:**89–93.

Zentmyer, G. A. 1965. Bacterial stimulation of sporangium production in *Phytophthora cinnamomi.* Science **150:**1178–1179.

Zharova, L. G., Kordyum, V. A., Lazurkevich, E. V. 1974. Biological system for antipollution treatment of liquid human excretions. In: Zatula, D. G., Rotmistrov, M. N. (eds.), Biochemical purification of waste waters. Proceedings of the 1st Republic Conference (Kiev, 20–21 March 1972). [In Russian.] Kiev: Naukova Dumka.

Zopf, W. 1883. Die Spaltpiltze, 1st ed. In: Schenk, A. (ed.), Handbuch der Botanik, p. 68. Breslau: Trewendt.

CHAPTER 108

The Genus *Flavobacterium*

OWEN B. WEEKS

Flavobacterium is a genus of pigmented, rod-shaped, nonfermentative bacteria (Weeks, 1974) that are commonly encountered in microbiological laboratories. Primary emphasis was placed upon the property of pigmentation when the genus was first formed (Bergey et al., 1923), and this practice has continued even though pigmentation is not restricted to *Flavobacterium* but is shared by genetically diverse bacteria (Weeks, 1969). Emphasis upon pigmentation for taxonomic assignment to *Flavobacterium* has given the genus a dubious reputation (McMeekin, Patterson, and Murray, 1972; Weeks, 1969) and one investigator has said it is a "collecting heap" (Christensen, 1977a, p. 1601). The genus has served too frequently as a repository for pigmented bacteria possessing the general attributes of *Flavobacterium* but which had not been subjected to detailed taxonomic characterization. A specific example of an uncritical taxonomic assignment to *Flavobacterium* is *F. heparinum*, in which a specialized physiological ability, degradation of heparin, was studied and identification with the genus was based only upon general phenetic qualities (Payza and Korn, 1956). In this instance, the taxonomic convenience of a generally defined genus was taken advantage of without regard for taxonomic health. Pigmentation is a useful phenetic property, especially if the chemical nature of the coloring matter is known, but it should have no more prominence than other phenetic qualities. Taxonomic heterogeneity and general uncertainty have characterized *Flavobacterium* from its conception and the history of the genus is a record of proposals to achieve credibility for the genus.

Flavobacterium was formed in 1923 as a genus of the family Bacteriaceae that encompassed the rod-shaped, non-endospore-forming, chemoorganotrophic bacteria (Bergey et al., 1923). Most of the pigmented bacteria of the family were segregated in a tribe, Chromobactereae, which contained four genera of aerobic bacteria separated on the basis of color. These were *Serratia, Flavobacterium, Chromobacterium,* and *Pseudomonas,* respectively the red, yellow, purple, and green fluorescent strains. The concept of *Flavobacterium* was scarcely altered in the successive editions of *Bergey's Manual of Determinative Bacteriology* until the sixth (Bergey et al., 1939), which eliminated from the genus the least well-described species, the polarly flagellated strains, and those known to be Gram-positive. The designated type species, *F. aquatile,* was redescribed as nonmotile to replace its earlier description as peritrichously flagellated. According to Breed (Bergey et al., 1939), this redescription was necessary to correct an error which had been made in the first edition. The generic description was enlarged in 1955 and *F. aquatile* strain Taylor (ATCC 11947 = NCIB 8649) was proposed to represent the type species (Weeks, 1955). The selection was officially proposed as the neotype later (Sneath and Skerman, 1966).

Stanier (1947) recognized that the cytophagas have more than casual phenetic resemblance to pigmented, Gram-negative eubacteria such as *Flavobacterium*. This is an unresolved problem which has dominated taxonomic consideration of *Flavobacterium*. Differentiation of flavobacteria from cytophagas has depended primarily upon demonstration of the cytophagal movement and colonial translocation characteristic of the latter bacteria, but absence of these features has not deterred assignment of flavobacterial species to *Cytophaga* (Mitchell, Hendrie, and Shewan, 1969). The investigations of Hayes (1963) and of Floodgate and Hayes (1963) illustrate the problem posed by Stanier and the consequences of indiscriminant assignment of pigmented bacteria to *Flavobacterium*. Of the 61 strains assigned to *Flavobacterium* by earlier investigations, critical review indicated that 32 could be considered nonmotile flavobacterial strains, 21 showed spreading growth on agar medium and could be designated cytophagas, and the remainder (8) could be considered strains of *Pseudomonas, Vibrio, Corynebacterium,* or bacteria of unknown taxonomic affiliation.

The question of *Flavobacterium* versus *Cytophaga* (and therefore *Flexibacter*) has been discussed extensively (Christensen, 1977a; McMeekin, Patterson, and Murray, 1972; Mitchell, Hendrie, and Shewan, 1969; Weeks, 1969). An answer would be more nearly possible if two issues were resolved. These are the heterogeneity of the genus and the differentiation of nonmotile flavobacteria from cytophagas. A primary requirement for the resolution of both issues is an acceptable definition of *Flavobacterium*. The definition in *Bergey's Manual,* eighth edition (Weeks, 1974) originated in that proposed by Weeks (1955). Neither of the above issues was apparent in 1955, but both were considered for the later publication. The heterogeneity of the cultures that were included in the genus (Weeks, 1974) is shown by the two disparate ranges of DNA base ratios reported; 30–42 and 63–70 mol% (G+C). The lower range of values is characteristic of the nonmotile cultures of species of *Flavobacterium;* unfortunately, this range is also typical of both *Cytophaga* and *Flexibacter* (Leadbetter, 1974).

Differentiation of flavobacteria and the cytophagas has been considered in detail recently by Christensen (1977a). It depends greatly upon the presence or absence of gliding, cellular movements that result in colonial spreading, a veil-like growth (Henrichsen, 1972). Other phenetic differences have received less attention. Gliding motility may be difficult to demonstrate since it is dependent upon experimental conditions (Hayes, 1963; Henrichsen, 1972). Spreading colonial growth is not an exclusive property of the cytophaga group and can be produced by cellular movements other than gliding.

Henrichsen (1972) has offered precise definitions of six types of cellular movements that result in colonial spreading, i.e., surface translocation. Three of these, including gliding, could explain the spreading type of growth observed in some flavobacterial colonies. Spreading due to cytophagal gliding entails the oriented, cooperative effort of many cells and produces a "micromorphological pattern" (Henrichsen, 1972, p. 491). Henrichsen differentiated the cellular movements by microscopic observation of margins of the spreading bacterial colony. Movement in hanging-drop preparations was considered not a reliable method to demonstrate gliding motility.

In some studies of flavobacteria, microscopic examination of spreading colonies was used to decide whether gliding motion occurred (McMeekin, Patterson, and Murray, 1971; McMeekin, Stewart, and Murray, 1972), but not in others (Hayes, 1963). Presumptive evidence should not be used to decide that a culture demonstrates cytophagal gliding, especially when gliding is the basis for assignment of a flavobacterial strain to the cytopohaga group. Other phenetic properties could be useful in deciding whether a nonmotile, flavobacterial culture should be placed with the cytophagas or remain in limbo in an arbitrary *Flavobacterium–Cytophaga* complex (Callies and Mannheim, 1978; McMeekin, Patterson, and Murray, 1971).

Freshly isolated cytophagas have unusual ability to use a great variety of complex natural polymers, i.e., proteins, DNA, RNA, cell walls, lipids, cellulose, agar, chitin, starch, alginates, keratins, porphyrans, pectin, for nutrients. This ability is not a general property of flavobacterial species, although individual strains may hydrolyze starch, casein, chitin, and gelatin.

Two chemotaxonomic properties may eventually prove useful in distinguishing cytophagas and nonmotile flavobacteria. These are the chemical nature of the respiratory quinones, combined with ability to use fumarate as an electron acceptor under anaerobic conditions (Callies and Mannheim, 1978) and the chemical nature of the pigments (Achenbach and Kohl, 1977; Achenbach, Kohl, and Reichenbach, 1977; Achenbach et al., 1974). Comprehensive studies of 49 cultures representing *Flavobacterium, Cytophaga, Flexibacter,* and the *Flavobacterium–Cytophaga* complex showed there were two chemotaxonomic groups: one contained ubiquinones and could not use fumarate as an electron acceptor; the other contained menaquinones and could use fumarate as an electron acceptor. Structure studies of the pigments have shown *Cytophaga johnsonae* and *Flexibacter elegans* contain highly substituted, nonisoprenoid polyene pigments, whereas the two flavobacteria that have been studied contain carotenoids. *Flavobacterium aquatile* strain Taylor (ATCC 11947) contains zeaxanthin as its principal carotenoid (O. B. Weeks, unpublished study) and *Flavobacterium* sp. 0147 (R1519) also contains the β-carotene-zeaxanthin carotenoid system (McDermott, Britton, and Goodwin, 1973). Additional studies will be necessary to establish the utility of pigment structures in differentiating flavobacteria and cytophaga.

There have been few definitions of *Flavobacterium* offered as alternatives to that appearing in *Bergey's Manual,* eighth edition (Weeks, 1974), and the taxonomic treatments that have been proposed are essentially similar except for that of Callies and Mannheim (1978). Mitchell, Hendrie, and Shewan (1969) recognized two subgroups of *Flavobacterium,* one for the nonmotile strains not identifiable as *Cytophaga* and the other for the rarely encountered, peritrichously motile cultures. Hayes (1963) from numerical taxonomic studies of marine bacteria recommended that *Flavobacterium* be reserved for nonmotile strains and a separate genus be used to accommodate the peritrichously flagellated cultures. Later, DNA base analyses were made of representative cultures of the collection (Hayes, Wilcox, and Parish, 1977). The data could be arranged into two different ranges: 30.8–39.5 and 51.4–63 mol% G+C. About 83% of the selected cultures had their values in the lower range and of these about 65% were nonmotile, nongliding in character. Both flagellated and nonmotile strains occurred in the high G+C range. McMeekin, Patterson, and Murray (1971) and McMeeken, Stewart, and Murray (1972), from the studies of pigmented cultures that had been isolated from food and water, recommended that the cultures which could not be placed in either *Flavobacterium* or *Cytophaga* (*sensu stricto*) be arranged into a *Flavobacterium-Cytophaga* complex. Brisou (1958) proposed a genus *Empedobacter* for nonmotile flavobacteria and would reserve *Flavobacterium* for motile strains. Perhaps the most recent taxonomic proposal is that of Callies and Mannheim (1978). Species of *Flavobacterium* would be those containing ubiquinones and not using fumarate as an electron acceptor in anaerobic conditions, whereas the cytophagas would contain menaquinones and utilize the fumarate. A taxonomic arrangement based upon this proposal would be very different from the present one.

Flavobacterium may have been used indiscriminately as a taxon of convenience and therefore have acquired an uncertainty but this will not be solved by transferring problem strains to the Cytophagaceae. It seems more reasonable to accept *Flavobacterium* as it is now presented (Weeks, 1974). Of the 12 species currently listed, doubts have been expressed about the suitability of the reference cultures for the nomenclatural type species, *F. aquatile* (ATCC 11947), as well as for *F. devorans* (ATCC 10829) and *F. rigense* strain F18. The culture of *F. aquatile* was said by Perry (1973) to show gliding movements. He based his judgement upon examination of hanging drop preparations; the strain did not produce spreading growth on agar plates, although such growth was previously reported by Mitchell, Hendrie, and Shewan (1969) when cultures were incubated at 15°C. The cultures representing *F. devorans* and *F. rigense* are said to be polar flagellates (E. Yabuuchi and M. Hendrie, personal communications) and therefore inappropriate to represent these species. The present recommendation is to eliminate from the genus all species not represented by reference cultures (Weeks, 1974) and, on this basis, both *F. devorans* and *F. rigense* have precarious status.

Holmes, Snell, and Lapage (1977) have identified 9 recent clinical isolates as strains of *F. odoratum* and recommended one of Stutzer's original strains as the neotype, i.e., *F. odoratum* (ATCC 4651 = NCTC 11036). In a separate study, Holmes, Snell, and Lapage (1978) identified another taxonometric group as *F. breve* and recommended a centrotypic strain as the neotype, i.e., NCTC 11099, to replace the present reference culture, ATCC 14234, which was not removed from the taxon.

F. oceanosedimentum is the name proposed to represent some flavobacterial cultures isolated from marine sediments (Carty and Litchfield, 1978). The culture *Flavobacterium* sp. Hoffmann-La Roche and Co., 0147 (R1519) has been used extensively in studies of carotenoid biosynthesis (McDermott, Britton, and Goodwin, 1973) and is a noteworthy strain. Pichinoty et al. (1976) have isolated an un-

usual denitrifying bacterium from soil by an enrichment technique. It is chemoorganotrophic, obligately aerobic, nonmotile, pigmented, and photochromogenic; it requires sugars as sources of carbon and energy and depends upon either NO_2^-, N_2O, or O_2 as an electron acceptor. It has a low DNA base ratio (40.8 mol% G+C), but did not exhibit cytophagal gliding or colonial spreading. The bacterium was designated *Flavobacterium* sp. (No. CIP 12-75, Institute Pasteur). *F. pectinovorum* (ATCC 19366 = NCIB 9059) is now considered a strain of *Cytophaga johnsonae* (Christensen, 1977b). Finally, *F. thermophilum,* which was proposed for an extremely thermophilic bacterium, has been renamed *Thermus thermophilus* (Oshima and Imahori, 1974).

The current literature contains numerous references to *Flavobacterium* species which are Gram positive and no longer placed in the genus proper but are included in an appendix to the genus. Most of these are coryneform in nature and many have been proposed for industrial processes for which patents have been granted.

Habitats

Yellow-pigmented, nonfermentative (or a delayed reaction), Gram-negative, motile or nonmotile, rod-shaped bacteria, which have been placed in *Flavobacterium* or termed flavobacteria, have been isolated from fresh and marine waters, soil and ocean sediments, foods and food-processing plants, and clinical materials. Probably these bacteria are widely distributed in nature and especially common in water, which would explain their seeming omnipresence.

Nonclinical Sources

Flavobacterial strains have been isolated most commonly from freshwater and marine environments, and almost any general bacteriological survey of such habitats has reported their presence (ZoBell, 1944). The Adansonian studies of Hayes (1963) and of Floodgate and Hayes (1963) used marine strains isolated from fish surfaces, sea water, and marine mud; the samples were taken from the north Atlantic region, western North American coast, and Florida. A comprehensive investigation of dairy and meat-processing industries also showed numerous flavobacteria (McMeekin, Patterson, and Murray, 1971; McMeekin, Stewart, and Murray, 1972). The bacteria have also been found in chlorinated cooling water of vegetable canning plants and were the cause of characteristic spoilage following post-sterilization contamination (Bean and Everton, 1969).

Clinical Sources

Flavobacteria have consistently been found among the nonfermentative, or late fermentative, Gram-negative bacteria isolated from clinical specimens such as blood, urine, infected wounds, and feces (Pickett and Manclark, 1970; Picket and Pedersen, 1970a,b; Tatum, Ewing, and Weaver, 1974). Their frequency of occurrence is usually low, 1% or less, and their pathogenicity, low-grade or doubtful. Three species and an unnamed group represent these bacteria: *F. meningosepticum* (King, 1959); *F. odoratum* (Holmes, Snell, and Lapage, 1977); *F. breve* (Holmes, Snell, and Lapage, 1978); and Group IIb of King (Tatum, Ewing, and Weaver, 1974).

F. meningosepticum is associated with a sometimes-fatal meningitis of infants and has been isolated from their throats, spinal fluid, and blood as well as from throats of normal adults (King, 1959). *F. odoratum* and *F. breve* represent present-day isolations of long-established species and were among the nonfermentative bacilli identified in a computer-assisted program of the Central Public Health Laboratory, London (Holmes, Snell, and Lapage, 1977, 1978). *F. breve* was found six times during a 10-year period in a collection of 1,700 clinical strains; *F. odoratum* was found nine times among 1,500 strains. Both species were obtained from urine, blood, and infected wounds and were considered nonpathogenic. Group IIb flavobacteria also are found infrequently but consistently from the same types of clinical samples and are not regarded as pathogenic. Group IIb bacteria can be differentiated from the named clinical flavobacteria and have been the subject of a recent comprehensive study (Price, 1977).

Isolation

Nonclinical *Flavobacterium*

Flavobacteria are chemoorganotrophic and essentially not difficult to isolate, although maintenance of the cultures sometimes presents a problem. Species with specialized metabolic abilities, such as *F. heparinum* (Payza and Korn, 1956) and *Flavobacterium* sp. Pichinoty (Pichinoty et al., 1976), require selective enrichment procedures for primary isolation.

General studies do not require enrichment procedures and usually nutrient agar–type media are used. Studies of marine, pigmented bacteria, for example, have used seawater-agar media, such as the following (Hayes, 1963; wt/vol):

Beef extract (Lab Lemco)	1.0%
Peptone (Evans)	1.0%

Agar (Difco)	1.5%
Aged sea water + distilled water (vol/vol)	3% + 1%
pH 7.2–7.3	

Isolation of yellow-pigmented bacteria from food and food-processing equipment used a similar medium (McMeekin, Patterson, and Murray, 1971; wt/vol):

Beef extract (Oxoid L20)	1.0%
Peptone (Oxoid L37)	1.0%
NaCl	0.5%
Agar (Oxoid No. 3)	1.2%

Incubation temperatures were similar to the environments, i.e., 20–25°C, and incubation times were about 4 days.

Weeks (1955) used medium M1, which contains lesser amounts of nutrients, for both isolation and maintenance of flavobacterial culture. Medium M1 contains (Difco ingredients):

Proteose peptone	0.5%
Yeast extract	0.1%
Beef extract	0.2%
NaCl	0.3%
Agar	1.2%
pH 7.2–7.4	

It is not unusual to find that media containing relatively large amounts of the individual nutrients, such as those used by Hayes (1963) and McMeekin, Patterson, and Murray (1971), are not as well suited to maintenance of flavobacterial cultures as are media with small concentrations. An excellent study relating to maintenance was done by Christensen and Cook (1972) and dealt primarily with isolation of cytophagas but flavobacteria were included. In general, media containing small amounts of nutrients were superior. For example, the medium (P.M.Y.A. II) contained (wt/vol):

Peptonized milk	0.1%
Yeast extract	0.02%
Sodium acetate	0.002%
Agar	1.5%

This medium was excellent for detection of cytophagal gliding and colonial swarming, which would be useful in differentiating flavobacteria and cytophaga.

Clinical *Flavobacterium*

Tatum, Ewing, and Weaver (1974) have described in detail the procedures for isolation of clinical flavobacteria, such as those of Group IIb and *Flavobacterium meningosepticum*. Their work was part of a large program to isolate and characterize miscellaneous Gram-negative bacteria associated with clinical samples. The methods now constitute a standardized procedure and were developed over a 24-year period, during which almost 4,500 of the bacterial specimens were identified. The tests primarily are of cultural properties and permit the isolation of the clinical flavobacteria and assignment to one of the categories. Primary plating of a clinical specimen usually is done upon blood, chocolate, MacConkey, or eosin–methylene blue agar but other media may be used. Incubation temperature is 35–37°C, but the bacteria will grow at room temperatures (20–25°C). Holmes, Snell, and Lapage (1977, 1978) used nutrient agar in their studies of *F. breve* and *F. odoratum*. Pickett and his colleagues have used cystine-Trypticase agar (BBL) for maintenance of cultures of the clinical flavobacteria. Long-term preservation was done in brucella broth (BBL) containing 10% glycerol and the cultures were stored at −50°C (Pickett and Pederson, 1970a).

Identification

Assignment of a culture to *Flavobacterium* now rests primarily upon the attributes ascribed to the genus (Weeks, 1974). Since these are quite general, the personal judgment of the investigator is a major contribution, especially for the nonclinical strains which have not been studied as thoroughly as the clinical strains.

Clinical isolates probably can usually be easily identified by following the general procedure described by Tatum, Ewing, and Weaver (1974) and using their differential tables as well as those arranged by Holmes, Snell, and Lapage (1977, 1978). The unknown could be placed in Group IIb or, possibly, in *F. meningosepticum, F. breve,* or *F. odoratum. F. meningosepticum* is closely related to Group IIb (Owen and Lapage, 1974; Price, 1977), as *F. breve* and *F. odoratum* probably are. Group IIb appears to be a phenetic conglomerate, and the three species components of it. The three species are each represented by a substantial number of strains, each of which is apparently homogeneous genetically as well as phenetically. Studies of DNA reassociation substantiate this. Owen and Snell (1976) have shown that the six serovars of *F. meningosepticum* have a high degree of genetic interrelatedness; apparently serovars of *F. odoratum* shows genetic interrelationship also (B. Holmes, personal communication). *F. meningosepticum,* however, did not show genetic interrelationship with *F. aquatile, F. breve, F. odoratum,* and strains of Group IIb (Owen and Snell, 1976).

Identification of the nonclinical flavobacteria seems to be essentially a matter of deciding whether nonmotile flavobacteria are cytophagas. In this connection, it seems necessary to know whether or not the isolate demonstrates gliding movement and colonial translocation in the sense of Henrichsen (1972), the DNA base ratio, and whether the culture can utilize a variety of complex macromolecules in its nutrition. Cultures that display properties of typical cytophagas obviously present no problem since these would not be flavobacteria; cultures not having such properties could be considered flavobacteria. Isolates that do not display cytophagal gliding but do demonstrate colonial translocation would not be cytophagas according to Henrichsen (1972). Cultures that are believed to display gliding motion but which do not show colonial translocation probably should not be assigned to Cytophagaceae. Christensen (1977b) believes that colonial spreading would occur in such cultures if the correct experimental conditions were used. Therefore, such cultures should be placed in the appendix of *Flavobacterium* among the species incertae sedis and further studies should be made. The studies of Christensen and Cook (1972) might be useful in these cases.

The type of pigment contained in a flavobacterial culture cannot be determined from electronic absorption spectra. The two most likely molecular classes would be the carotenoids and aryl polyenes. Since carotenoids also have polyene chromophores, spectra of the two cannot be distinguished. This inability is clearly illustrated by the work with the *xanthomonas* carotenoid, the structure of which was proven to be an aryl polyene ester (Andrewes et al., 1973).

Literature Cited

Achenbach, H., Kohl, W. 1977. Die Hauptpigmente aus *Cytophaga johnsonae*-Zwei neue Pigmente vom Flexirubin-Typ. Tetrahedron Letters, No. **12:**1061–1062.

Achenbach, H., Kohl, W., Reichenbach, H. 1977. 5-Chlorflexirubin, ein Nebenpigment aus *Flexibacter elegans.* Justus Liebigs Annalen der Chemie **1:**1–7.

Achenbach, H., Kohl, W., Reichenbach, H., Kleinig, H. 1974. Zur Struktur des Flexirubins. Tetrahedron Letters, No. **30:**2555–2556.

Andrewes, A. G., Hertzberg, S., Liaaen-Jensen, S., Starr, M. P. 1973. Xanthomonas pigments. 2. The *Xanothomonas* ''carotenoids''—non-carotenoid brominated aryl-polyene esters. Acta Chemica Scandinavica **27:**2383–2395.

Bean, P. G., Everton, J. R. 1969. Observations on the taxonomy of chromogenic bacteria isolated from cannery environments. Journal of Applied Bacteriology **32:**51–59.

Bergey, D. H., Breed, R. S., Murray, E. G. D., Hitchens, A. P. 1939. Bergey's manual of determinative bacteriology, 6th ed. Baltimore: Williams & Wilkins.

Bergey, D. H., Harrison, F. C., Breed, R. S., Hammer, B. W., Huntoon, F. M. 1923. Bergey's manual of determinative bacteriology. Baltimore: Williams & Wilkins.

Brisou, J. 1958. Étude de quelques *Pseudomonadaceae.* Bordeaux, France: Baillet.

Callies, E., Mannheim, W. 1978. Classification of the *Flavobacterium-Cytophaga* complex on the basis of respiratory quinones and fumarate respiration. International Journal of Systematic Bacteriology **28:**14–19.

Carty, C. E., Litchfield, C. D. 1978. Characterization of a new marine sedimentary bacterium as *Flavobacterium oceanosedimentum* sp. nov. International Journal of Systematic Bacteriology **28:**561–566.

Christensen, P. 1977a. The history, biology, and taxonomy of the *Cytophaga* group. Canadian Journal of Microbiology **23:**1599–1653.

Christensen, P. 1977b. Synonymy of *Flavobacterium pectinovorum* Dorey with *Cytophaga johnsonae* Stanier. International Journal of Systematic Bacteriology **27:**122–132.

Christensen, P. J., Cook, F. D. 1972. The isolation and enumeration of cytophagas. Canadian Journal of Microbiology **18:**1933–1940.

Floodgate, G. D., Hayes, P. R. 1963. The adansonian taxonomy of some yellow pigmented marine bacteria. Journal of General Microbiology **30:**237–244.

Hayes, P. R. 1963. Studies on marine flavobacteria. Journal of General Microbiology **30:**1–19.

Hayes, P. R., Wilcock, A. P. D., Parish, J. H. 1977. Deoxyribonucleic acid base composition of flavobacteria and related Gram negative yellow pigmented rods. Journal of Applied Bacteriology **43:**111–115.

Henrichsen, J. 1972. Bacterial surface translocation: A survey and a classification. Bacteriological Reviews **36:**478–503.

Holmes, B., Snell, J. J. S., Lapage, S. P. 1977. Revised description, from clinical isolates, of *Flavobacterium odoratum* Stutzer and Kwaschnina 1929, and a designation of the neotype strain. International Journal of Systematic Bacteriology **27:**330–336.

Holmes, B., Snell, J. J. S., Lapage, S. P. 1978. Revised description, from clinical strains, of *Flavobacterium breve* (Lustig) Bergey et al. 1923 and proposal of the neotype strain. International Journal of Systematic Bacteriology **28:**201–208.

King, E. O. 1959. Studies on a group of previously unclassified bacteria associated with meningitis in infants. American Journal of Clinical Pathology **31:**241–247.

Leadbetter, E. R. 1974. *Cytophagales*, pp. 99–112. In: Buchanan, R. E., Gibbons, N. E. (eds.), Bergey's manual of determinative bacteriology, 8th ed. Baltimore: Williams & Wilkins.

McDermott, J. C. B., Britton, G., Goodwin, T. W. 1973. Effect of inhibitors on zeaxanthin synthesis in a *Flavobacterium*. Journal of General Microbiology **77:**161–171.

McMeekin, T. A., Patterson, J. T., Murray, J. G. 1971. An initial approach to the taxonomy of some Gram negative yellow pigmented rods. Journal of Applied Bacteriology **34:**699–716.

McMeekin, T. A., Stewart, D. B., Murray, J. G. 1972. The adansonian taxonomy and the deoxyribonucleic acid base ratio composition of some gram-negative, yellow pigmented rods. Journal of Applied Bacteriology **35:**129–137.

Mitchell, T. G., Hendrie, M. S., Shewan, J. M. 1969. The taxonomy, differentiation and identification of *Cytophaga* species. Journal of Applied Bacteriology **32:**40–50.

Oshima, T., Imahori, K. 1974. Description of *Thermus thermophilus* (Yoshida and Oshima) comb. nov., a nonsporulating thermophilic bacterium from a Japanese thermal spa. International Journal of Systematic Bacteriology **24:**102–112.

Owen, R. J., Lapage, S. P. 1974. A comparison of King's group IIb of *Flavobacterium* with *Flavobacterium meningosepticum*. Antonie van Leeuwenhoek Journal of Microbiology and Serology **40:**255–264.

Owen, R. J., Snell, J. J. S. 1976. Deoxyribonucleic acid reassociation in the classification of flavobacteria. Journal of General Microbiology **93:**89–102.

Payza, N., Korn, E. D. 1956. The degradation of heparin by bacterial enzymes. I. Adaptation and lyophilized cells. Journal of Biological Chemistry **223:**853–858.

Perry, L. B. 1973. Gliding motility in some non-spreading flexibacteria. Journal of Applied Bacteriology **36:**227–232.

Pichinoty, F., Bigliardi-Rouvier, J., Mandel, M., Greenway, B., Méténier, G., Garcia, J. L. 1976. The isolation and properties of a denitrifying bacterium of the genus *Flavobacterium*. Antonie van Leeuwenhoek Journal of Microbiology and Serology **42:**349–354.

Pickett, M. J., Manclark, C. R. 1970. Nonfermentative bacilli associated with man: I. Nomenclature. American Journal of Clinical Pathology **54:**155–163.

Pickett, M. J., Pedersen, M. M. 1970a. Nonfermentative bacilli associated with man: II. Detection and identification. American Journal of Clinical Pathology **54:**164–177.

Pickett, M. J., Pedersen, M. M. 1970b. Characterization of saccharolytic nonfermentative bacteria associated with man. Canadian Journal of Microbiology **16:**351–362.

Price, K. A. 1977. A study of the taxonomy of flavobacteria isolated in clinical laboratories. Ph.D. Thesis. University of California, Los Angeles, California.

Sneath, P. H. A., Skerman, V. B. D. 1966. A list of type and reference strains of bacteria. International Journal of Systematic Bacteriology **16:**1–133.

Stanier, R. Y. 1947. Studies on non-fruiting myxobacteria. I. *Cytophaga johnsonae* n. sp. a chitin-decomposing myxobacterium. Journal of Bacteriology **53:**297–315.

Tatum, H. W., Ewing, W. H., Weaver, R. E. 1974. Miscellaneous Gram-negative bacteria, pp. 270–294. In: Lennette, E. H., Spaulding, E. H., Truant, J. P. (eds.). Manual of clinical microbiology, 2nd ed. Washington, D.C.: American Society for Microbiology.

Weeks, O. B. 1955. *Flavobacterium aquatile* (Frankland and Frankland) Bergey et al., type species of the genus *Flavobacterium*. Journal of Bacteriology **69:**649–658.

Weeks, O. B. 1969. Problems concerning the relationships of cytophagas and flavobacteria. Journal of Applied Bacteriology **32:**13–18.

Weeks, O. B. 1974. *Flavobacterium*, pp. 357–364. In: Buchanan, R. E., Gibbons, N. E. (eds.), Bergey's manual of determinative bacteriology, Baltimore: Williams & Wilkins.

ZoBell, C. E., Upham, H. C. 1944. A list of marine bacteria including descriptions of sixty new species. Bulletin of the Scripps Institute of Oceanography **5:**239–292.

The Genus *Haemophilus*

MOGENS KILIAN

The genus *Haemophilus* is a group of Gram-negative, rod-shaped bacteria that have a tendency to show varying degrees of pleomorphism depending upon the growth conditions. The bacteria of this genus are aerobic or facultatively anaerobic and are heterofermentative. The name of the genus comes from the early discovery that this group of bacteria has a distinct requirement for some accessory growth factors present in blood.

The genus was founded on the "influenza-bacillus" isolated by Richard Pfeiffer in 1892, which originally was regarded to be the cause of influenza. Although totally without responsibility for that disease, Pfeiffer's organism still carries the name *Haemophilus influenzae*. Today *H. influenzae*, as well as other *Haemophilus* species, is recognized as an important pathogen in its own right (Turk and May, 1965; Zinnemann, 1960).

Since its foundation by the American Committee on Classification and Nomenclature (Winslow et al., 1920), the genus *Haemophilus* has for some time encompassed organisms now included in the genera or species *Alcaligenes*, *Bordetella*, *Moraxella*, *Pasteurella*, *Bacteroides melaninogenicus*, *Corynebacterium pyogenes*, *Corynebacterium vaginale*, and *Eikenella corrodens*. However, at present there is general agreement that the genus is restricted to bacteria with a distinct requirement for hemin or certain other porphyrins (X factor) and/or nicotinamide adenine dinucleotide (NAD) (V factor), or certain definable coenzyme-like substances (Zinnemann, 1967).

The genus *Haemophilus* has been included in the family Brucellaceae, or its predecessor Parvobacteriaceae, but is now described in the eighth edition of *Bergey's Manual of Determinative Bacteriology*, like other genera of those families, under "Genera of Uncertain Affiliation". Johnson and Sneath (1973), as a result of a numerical taxonomic study, found that *Haemophilus* is phenotically very similar to the genera *Pasteurella* and *Actinobacillus*, but is well separated from the nonfermentative genera *Brucella*, *Bordetella*, and *Moraxella*.

The genus includes 13 well-defined species, 2 of which require both X and V factors (*H. influenzae*, *H. haemolyticus*), 4 require X factor only (*H.*

haemoglobinophilus, *H. influenzae-murium*, *H. aphrophilus*, *H. ducreyi*), and 7 require V factor only (*H. parainfluenzae*, *H. paraphrophilus*, *H. segnis*, *H. parasuis*, *H. pleuropneumoniae*, *H. paragallinarum*, *H. avium*). Two of these species, *H. segnis* and *H. avium*, have been described recently (Hinz and Kunjara, 1977; Kilian, 1976a; Kilian and Theilade, 1978). There have been some doubts concerning the true nature of the species *H. ducreyi*. Apparently, both Gram-negative and Gram-positive organisms have been given that name (Kilian and Theilade, 1975). However, certain Gram-negative bacteria, which can be isolated from the disease soft chancre, do possess the key characteristics of the genus *Haemophilus* (Kilian, 1976a).

Several other *Haemophilus* species are described in the literature. In addition to most of those mentioned above, Zinnemann and Biberstein (1974) in *Bergey's Manual* list *H. suis*, *H. gallinarum*, *H. parahaemolyticus*, and *H. paraphrohaemolyticus*, and, as species incertae sedis, *H. ovis*, *H. putoriorum*, *H. citreus*, *H. piscium*, and *H. aegyptius*.

H. suis and *H. gallinarum* were the original names given to organisms causing disease in pigs and fowl, respectively. Both of these groups of bacteria were found to require both the X and V factors by the original authors (Delaplane, Erwin, and Stuart, 1934; Lewis and Shope, 1931). However, since their descriptions, every *Haemophilus* organism isolated from the same diseases has required only the V factor; thus the organisms have more recently been named *H. parasuis* and *H. paragallinarum*, according to the proposal of Biberstein and White (1969). At present, no well-described isolates with the original characteristics of *H. gallinarum* seem to exist, whereas a recent publication (Biberstein, Gunnarsson, and Hurvell, 1977) indicates that true *H. suis* strains may in fact be isolated from swine.

Strains of the species *H. parahaemolyticus* and *H. paraphrohaemolyticus* have recently been shown to be virtually identical to strains of *H. parainfluenzae* except for their hemolytic activity (Kilian, 1976a; Sneath and Johnson, 1973). This hemolytic activity is variable (Kilian, 1976b) and is, therefore, a poor taxonomic marker in *Haemophilus*.

However, the invalidation of this character as a criterion for establishing the two species is not based upon its instability, but upon the fact that no other independent character supports their separation from *H. parainfluenzae*. An exception are hemolytic strains isolated from pleuropneumonia in pigs. Such strains were included in the species *H. parahaemolyticus* in *Bergey's Manual* (Zinnemann and Biberstein, 1974), but a recent taxonomic study (Kilian, Nicolet, and Biberstein, 1978) clearly indicates that they deserve specific status and should be named *H. pleuropneumoniae* according to the proposal of Shope (1964).

Bacteria with the proposed names *H. ovis, H. putoriorum,* and *H. citreus* were originally isolated from sheep, ferrets, and deer, respectively. However, none of these species has been recovered recently, and information on their characteristics is very limited (Zinnemann and Biberstein, 1974). The species *H. piscium* was proposed by Snieszko, Griffin, and Friddle (1950) for organisms causing ulcerations in trout. Recently, however, *H. piscium* was shown not to belong to the genus *Haemophilus* (Kilian, 1976a). *H. aegyptius* (Koch-Weeks bacillus) seems to be separated from biotype III of *H. influenzae* only by its ability to hemagglutinate, and should probably be regarded as a variety of that biotype (Kilian, 1976a; Kilian et al., 1976).

H. aphrophilus represents a special taxonomic problem. Although usually X-requiring on primary isolation, the organism does possess the enzymatic capacities to synthesize the X factor (Kilian, 1976a; White and Granick, 1963), and upon subcultivation the requirement is often absent (Boyce, Frazer, and Zinnemann, 1968; Sutter and Finegold, 1970). As the species is closely related to *H. paraphrophilus,* it is undoubtedly best maintained in the genus *Haemophilus*. However, this will probably require a revision of the definition of the genus suggested by the Subcommittee on the Taxonomy of *Haemophilus*

(Zinnemann, 1967). Such a revision would imply the possible affiliation with the genus *Haemophilus* of such seemingly closely related organisms as "*Haemophilus somnus*" (Bailie, 1969), "*Haemophilus agni*" (Kennedy et al., 1958), and *Actinobacillus actinomycetemcomitans,* which do not require the growth factors X or V.

Habitats

Members of the genus *Haemophilus* are strict parasites. They form part of the indigenous flora of the mucous membranes of the human upper respiratory tract, mouth, and sometimes the vagina (Fleming and Maclean, 1930; Kilian, 1976a; Kilian, Heine-Jensen, and Bülow, 1972; Kilian and Schiøtt, 1975; Sims, 1970; Turk and May, 1965). The same applies to pigs (Harris, Ross, and Switzer, 1969; Little, 1970), various species of fowl (Grebe and Hinz, 1975), and monkeys (Kilian and Rölla, 1976). *Haemophilus* organisms have also been isolated from healthy dogs (Kristensen, 1922; Balish et al., 1977), cattle (Little and Pritchard, 1977), cats (Rivers and Bayne-Jones, 1923), rats (Harr, Tinsley, and Weswig, 1969; Kilian 1976a), mice (Csukás, 1976), deer (Diernhofer, 1949), sheep (Mitchell, 1925), guinea pigs (Kristensen, 1922), and ferrets (Kairies, 1935). However, detailed information on the carrier rates in many of these animals is lacking, and most of the isolated organisms are poorly described.

The *Haemophilus* species found in man have very specific ecological preferences. Whereas *H. parainfluenzae* is ubiquitously present in the mouth and upper respiratory tract, *H. influenzae* is confined to the mucous membranes behind the palatinal arches (Table 1). In contrast, the species *H. aphrophilus, H. paraphrophilus,* and *H. segnis* seem to prefer the surfaces of teeth, where they may form part of the microbial deposits known as dental plaque (Kilian,

Table 1. Percentage distribution into species of *Haemophilus* strains isolated from infectious diseases and normal floras of man.[a]

Origin	No. of strains examined	H. influenzae biotypes				H. haemolyticus	H. parainfluenzae	H. segnis	H. paraphrophilus	H. aphrophilus
		I	II	III	IV					
Meningitis and epiglottitis	157	94%	5%	0	1%	0	0	0	0	0
Healthy respiratory tract	107	8%	26%	13%	7%	8%	37%	1%	0	0
Oral cavity	649	0	0	0	0	0	74%	19%	2%	5%
Conjunctivitis	104	9%	41%	46%	0	0	4%	0	0	0

[a] Figures of the table are compiled from Kilian and Schiøtt (1975), Kilian (1976a), Kilian et al. (1976), and Kilian, Sørensen, and Frederiksen (1979).

1976a; Kilian and Schiøtt, 1975; Kraut, Finegold, and Sutter, 1972). In saliva, the total number of haemophili amounts to 3.6×10^7 (Kilian and Schiøtt, 1975; Sims, 1970). The factors determining the ability of haemophili to colonize different surfaces are unknown, but some recent findings may provide possible explanations. *H. aphrophilus* and *H. paraphrophilus* both have an innate capacity to adhere to smooth surfaces in vitro (Kilian and Schiøtt, 1975), a trait that might account for their predilection for surfaces of teeth. A neuraminidase activity has been demonstrated in several *Haemophilus* species by Tuyau and Sims (1974) and Müller and Hinz (1977). This enzyme might substantiate an attachment to sialic acid–containing receptors in the glycoproteins covering mucosal surfaces.

Some of the *Haemophilus* species are important pathogens in humans as well as in various animal species. *H. influenzae* is among the three leading causes of bacterial meningitis and is also implicated in epiglottitis, chronic bronchitis, conjunctivitis, and occasionally pneumonia. The vast majority of strains causing meningitis and epiglottitis possess a capsule of serotype b. Furthermore, such strains belong to biotype I, and may be separated from the majority of *H. influenzae* strains found indigenously in the upper respiratory tract (Table 1). The clinical importance of *H. influenzae* has been comprehensively treated in the book by Turk and May (1965), which also provides an extensive bibliography on the subject. The four species that belong to the normal flora of the human mouth (Table 1) occasionally cause bacterial endocarditis and brain abscesses.

Serious and often fatal infections in pigs and fowl caused by *H. pleuropneumoniae, H. parasuis,* and *H. paragallinarum* occur frequently and have considerable economic implications (Biberstein and Cameron, 1961; Hinz, 1973; Nicolet, 1968; Page, 1962). Natural *Haemophilus* infections have, furthermore, been observed in mice (Kairies and Schwartzer, 1936), rats (Harr, Tinsley, and Weswig, 1969), sheep (Mitchell, 1925), cattle (Firehammer, 1959), deer (Biberstein, Gunnarsson, and Hurvell, 1977), and cats (J. Nicolet, personal communication).

Besides their importance in human and veterinary medicine, haemophili have been valuable tools to biochemists. Much of the present knowledge on the structure and function of the respiratory chain has been gained through studies of hemin-dependent and -independent strains of *Haemophilus* (Gilder and Granick, 1947; Lwoff and Lwoff, 1937a; Smith and White, 1962; White, 1963; White and Granick, 1963; White and Smith, 1962).

The more recent discovery of the so-called restriction endonucleases (class II), first isolated by Smith and his colleagues (Kelly and Smith, 1970; Smith and Wilcox, 1970) from a strain of *H. influenzae* serotype d, has pointed to a revolution in molecular biology. These site-specific endodeoxyribonucleases allow molecular biologists to isolate specific fragments of chromosomes. Since 1970, more than 80 different restriction endonucleases have been discovered in various microorganisms. However, the genus *Haemophilus* has been the most productive of all genera so far examined; 22 different enzymes have been isolated from a total of 29 strains. Roberts (1976), in his extensive review on restriction endonucleases, provides detailed information on strains producing these enzymes and the enzymes' specific cleavage sites on the DNA molecule.

Isolation

Growth Requirements

Haemophili are fastidious but grow luxuriantly on sufficiently rich media. Only *Haemophilus ducreyi* is notably difficult to cultivate in vitro. As mentioned in the Introduction, the key characteristic of haemophili is their requirement for either one or both of the two growth factors known as X and V. The X factor is hemin, which is used by the bacteria as prosthetic group in the iron-containing respiratory cytochromes and in heme enzymes such as catalase and peroxidase (Lwoff and Lwoff, 1937a). Whereas *H. parainfluenzae* and the other hemin-independent *Haemophilus* species form hemin by the classical biosynthetic pathway (Fig. 1) (Granick and Mauzerall, 1961), the hemin-requiring organisms have lost their enzymatic capacities to convert δ-aminolevulinic acid to protoporphyrin. However, the enzyme that catalyzes the final insertion of iron into the protoporphyrin ring to convert it to hemin is still present in most X-requiring strains. This accounts for the fact that protoporphyrin IX and certain other iron-free porphyrins usually may be substituted for hemin as supply of X factor (Brumfitt, 1959; Gilder and Granick, 1947). Occasional strains do, however, require the ultimate protoporphyrin-iron complex, hemin (Kilian et al., 1976; White and Granick, 1963).

When incubated anaerobically, *H. influenzae* does not form cytochromes (White, 1963); the hemin requirement, therefore, is greatly reduced, if not totally absent (Gilder and Granick, 1947). *H. influenzae* is thus capable of shifting into a pure anaerobic metabolism. In contrast, *H. parainflu-*

δ-AMINOLEVULINIC ACID ⟶ PORPHOBILINOGEN ⟶
PORPHYRINS ⟶ PROTOPORPHYRIN IX ⟶ HEMIN
Fe^{++}

Fig. 1. The main steps in the biosynthesis of hemin.

enzae is much less flexible and continues to form the cytochrome system under anaerobic conditions. The growth of this species in an oxygen-free atmosphere is, therefore, dependent on nitrate (Sinclair and White, 1970; White and Smith, 1962), which takes over the role of oxygen as the final electron acceptor.

The V factor has been identified as nicotinamide adenine dinucleotide (NAD or NADP) (Lwoff and Lwoff, 1937b). However, as with the X factor, various precursors may also satisfy the V requirement (Shifrine and Biberstein, 1960). NAD serves as coenzyme for one of the three major classes of oxidation-reduction enzymes, the pyridine-linked dehydrogenases. This growth factor is, therefore, found in all types of cells, including blood and yeast cells. Some bacteria, such as staphylococci and pseudomonads, produce NAD in excess of their own requirement and excrete it (or, in fact, a yet unidentified riboside precursor [Shifrine and Biberstein, 1960]). When such bacteria grow on an agar plate, the growth factor diffuses into the medium and thereby allows growth of V-dependent haemophili in its vicinity. This is known as the "satellite phenomenon" (Fig. 2) and is widely used for the presumptive identification of *Haemophilus* strains. However, it should be noted that satellite growth is not necessarily indicative of a V-factor requirement. In fact, bacteria used as feeder strains excrete several potential growth factors, including the X factor. Symbiotic growth of an organism is thus solely indicative of a growth requirement not being satisfied by the medium. For example, "*H. somnus*" does produce satellite growth on media devoid of blood or yeast

hydrolysate; nevertheless, this organism does not respond to the growth factors X and V (Kennedy et al., 1960). Strains of *H. parasuis* (Kilian, 1976a) and *H. suis* (Biberstein, Gunnarsson, and Hurvell, 1977) likewise show satellite growth on chocolate agar around a feeder organism, despite the fact that this medium contains both the X and V factors.

The minimal amounts of the two growth factors required in cultivation media have been estimated in several studies. Strains of *H. influenzae* grow well in the presence of $0.2-1.0$ μg NAD per ml of medium, whereas strains of *H. parainfluenzae* require $1.0-5.0$ μg (Evans, Smith, and Wicken, 1974). The minimal hemin requirement of *H. influenzae* has been estimated to range between 0.1 and 10 μg per ml medium, depending on the size of the inoculum and other unidentified factors (Biberstein and Spencer, 1962; Brumfitt, 1959; Evans, Smith, and Wicken, 1974; Gilder and Granick, 1947). However, considerably higher amounts of hemin have recently been found to be required by two other *Haemophilus* species. Strains of *H. influenzae-murium* grow well with $80-100$ μg hemin per ml (Csukás, 1976), and 200 μg hemin is required as additive per ml gonococcal agar base in order to obtain growth of *H. ducreyi* equivalent to that on chocolate agar (Hammond et al., 1978a).

Addition of serum (1%) to fluid culture media promotes growth of the species *H. paragallinarum* and *H. parasuis* (Biberstein, Gunnarsson, and Hurvell, 1977; Hinz and Kunjara, 1977). Page (1962) has suggested that serum acts as an absorbent of growth inhibitors. In fact, autoclaved proteose peptone has been found to be toxic to strains of *H. influenzae* (Evans and Smith, 1974). This toxicity can be neutralized by adding sodium dithionite (0.1 g/liter). A similar effect may be obtained with albumin (0.19 g/liter) or sodium oleate (4.8 mg/liter) (Butler, 1962).

Optimal growth of haemophili takes place at $35-37°C$ and under aerobic conditions. An exception is *H. ducreyi*, which grows significantly better at 33°C. Particularly on primary isolation, some species seem to require incubation in air supplemented with 5–10% carbon dioxide (Hinz and Kunjara, 1977; Khairat, 1940; Zinnemann et al., 1968). However, it is not clear whether this growth promotion could instead be an effect of the increased humidity obtained in a jar (Frazer, Zinnemann and Boyce, 1969; Sutter and Finegold, 1970).

Growth Media

In theory, all fresh media based on meat extracts or meat digests, or media that include yeast extract, contain both X and V factors, though usually in insufficient quantities. The X factor is heat stable, whereas the V factor is destroyed by the sterilizing

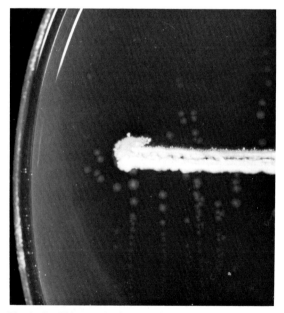

Fig. 2. Satellitic growth of *Haemophilus* colonies in the vicinity of *Staphylococcus aureus*.

procedure applied to most culture media. Blood of human and animal origin contains both growth factors in considerable quantities. But in ordinary blood agar, only the X factor is present in sufficient amounts to allow growth of *H. influenzae*. The V factor is imprisoned in the blood cells and is not directly available to the bacteria. Furthermore, NAD liberated by lysis of some of the blood cells will usually be destroyed by certain heat-labile substances present in most sera (Holt, 1962; Krumwiede and Kuttner, 1938; Waterworth, 1955). Krumwiede and Kuttner (1938) demonstrated such substances in sheep, goat, bovine, and human blood, but not in the blood of the rat, rabbit, or guinea pig. Their nature and heat lability vary with the species of animal from which the blood is derived, and this can have important practical consequences. For example, if human blood is used instead of horse blood, the medium may fail to give satisfactory growth of haemophili unless the blood has been kept at 75°C for about 30 min or heated to 90°C for at least 15 min (Turk and May, 1965).

Blood agar, the principal medium of clinical bacteriologists, allows satisfactory growth of *H. influenzae* and other V-dependent species only when the agar plate is cross-inoculated with a *Staphylococcus* species or another V-factor-excreting microorganism. Blood agar is, therefore, valuable for the detection of bacteria growing as satellite colonies in cultures of clinical samples. Another advantage of this medium is that it allows the detection of hemolytic activity. With potentially hemolytic haemophili, the clearest zones are obtained on media containing 5% bovine or sheep blood, whereas horse blood is less satisfactory (Kilian, 1976b). If the medium is to be used to demonstrate the special hemolytic phenomenon known as the CAMP reaction (see below), either bovine or sheep blood is required.

Without a feeder organism, blood agar is by no means a satisfactory medium for haemophili. Far better results are obtained in media in which the blood cells have been lysed by gentle heating (chocolate agar and Levinthal's agar) or by peptic digestion (Fildes' medium). In such media hemin and NAD are present in approximate quantities of 500 mg and 150 mg per liter, respectively.

Chocolate agar is the most generally applicable medium for isolation and cultivation of haemophili. The main principle in its preparation is that the heating of the blood must be sufficient to allow liberation of the V factor and to inactivate V-factor-splitting enzymes without destroying the V factor itself. The following procedure has been found useful.

Chocolate Agar for Isolating Haemophili

Prepare Difco blood agar base or brain heart infusion agar according to the manufacturer's instructions and autoclave at 120°C for 15 min. Allow the medium to cool to about 90°C, add 10% defibrinated horse or bovine blood, and shake gently until the medium takes on a chocolate color. Pour in slopes or plates.

For the isolation of fastidious haemophili from conjunctivae or chancroid lesions (*H. ducreyi*), higher isolation rates are obtained when the chocolate agar is enriched with 1% IsoVitaleX (BBL) (Hammond et al., 1978b; Vastine et al., 1974).

Levinthal's agar has the advantage of being colorless and transparent, which makes it an excellent medium for differentiating between encapsulated and nonencapsulated strains. There are several modifications of the original procedure described by Levinthal (1918). In our hands, the following slight variation of the procedure described by Turk and May (1965) has given good results.

Levinthal's Agar for Isolating Haemophili (Turk and May, 1967)

1. Prepare Difco brain heart infusion according to the manufacturer's instructions and autoclave at 120°C for 15 min.
2. Slowly add defibrinated or oxalated horse blood to a concentration of 10%, shaking continually. Bring to a boil in a waterbath and keep gently boiling until the mixture consists of brown flakes and particles in clear fluid. Allow the mixture to cool and the larger particles to settle, then filter the supernatant through sterile filter paper or glass wool. To ensure sterility, the medium may then be brought to the boiling point by a short heating. This product is known as Levinthal stock.
3. Prepare Difco proteose no. 3 with 30 g/liter agar, autoclave, and mix with an equal amount of Levinthal stock. Pour into plates.

An agar medium without blood but with the same characteristics as those of Levinthal's agar may be prepared in the following way: Prepare Difco brain heart infusion agar containing 2% proteose peptone no. 3 (Difco), autoclave, and allow to cool to about 55°C. Add 4 ml (~10 mg/liter) of hemin stock solution and 10 ml (~25 mg/liter) of NAD stock solution. For the cultivation of some species (see above), it may be necessary to increase supplementation of hemin to 40–80 ml of stock solution (~100–200 mg hemin per liter of medium).

Hemin stock solution is added to blood-free media when used for the cultivation of hemin-requiring species. A stock solution prepared in the following way may be stored for several months: Add 100 mg of hemin HCl to 40 ml of 0.2 M KOH in 50% ethanol. Store in a dark bottle in a refrigerator. For use, add required amount before or after sterilization of the medium.

NAD stock solution is added aseptically to media after sterilization and cooling to about 55°C. The solution is prepared in the following way: Dissolve 100 mg of nicotinamide adenine dinucleotide (= diphosphopyridine nucleotide) in 40 ml of water. Sterilize by filtration and store in a refrigerator for up to 3 weeks or at −20°C.

Broth media may be prepared in the same way as the agar media described above, except that the agar is omitted. We find Levinthal stock to be a highly satisfactory medium. In the experience of Alexander (1958), the best fluid medium for culture of *H. influenzae* is that made by mixing Levinthal stock with 3 volumes of neopeptone broth.

Selective Isolation of *Haemophilus*

For the purpose of surveying the incidence of *Haemophilus* species in material from the respiratory tract or other mucous membranes, a selective medium is required. Haemophili are relatively slow growing and may, in mixed cultures, easily be overgrown. Thus, respiratory samples often give rise to heavy growth of haemophili on selective media in cases where no apparent *Haemophilus* colonies are detectable by routine cultivation methods (Hovig and Aandahl, 1969; Kilian, Heine-Jensen, and Bülow, 1972).

When designing a selective medium, one may take advantage of the resistance of *Haemophilus* species to bacitracin (Love and Finland, 1954; Sutter and Finegold, 1970). Due to its effectiveness in inhibiting a wide range of other pharyngeal and oral organisms, bacitracin has proved to be highly valuable as a selective agent for haemophili (Baber, 1963; Crawford, Barden, and Kirkman, 1969; Hovig and Aandahl, 1969; Klein and Blazevic, 1970). Some pharyngeal and oral *Neisseria* species will grow on media containing bacitracin, but the colonies produced by these organisms are easily distinguished from colonies of haemophili. More important, however, *Actinobacillus actinomycetem-comitans* and *Eikenella corrodens,* which have the same ecological preferences as haemophili, grow on bacitracin-containing media. With the exception of the corroding type of *E. corrodens,* both of these two species have colonial morphologies very similar to those of some *Haemophilus* species (Kilian and Schiøtt, 1975).

Bacitracin may be added to all types of agar media. The concentrations used by various authors differ from 5 to 18.9 units per ml of medium. In our experience, chocolate agar with 18.9 units (= 300 μg) per ml, as proposed by Hovig and Aandahl (1969), is a highly satisfactory medium for the selective isolation of haemophili from the respiratory tract (Kilian, Heine-Jensen, and Bülow,

1972) and oral cavity (Kilian and Schiøtt, 1975). The counts of haemophili in saliva obtained on this medium are identical with those found by Sims (1970), who used chocolate agar with a combination of bacitracin (10 units/ml) and cloxacillin (5 μg/ml).

Defined Media

Media that may be used in metabolic and genetic studies of haemophili have been described by Herbst and Snell (1949), Butler (1962), Wolin (1963), Talmadge and Herriott (1960), and Herriott et al. (1970).

Conservation of Cultures

Plate or tube cultures of haemophili will usually survive for only 1 week without subcultivation. Survival is better at room temperature than at 4°C. The most satisfactory way of conserving *Haemophilus* strains is by lyophilization. Overnight cultures in Levinthal's broth or heavy suspensions in skim milk of chocolate agar cultures can be lyophilized, stored at room temperature, and recovered when required by adding Levinthal's broth. By this method, *Haemophilus* strains can be preserved for at least 25 years.

Levinthal's broth cultures, incubated for 24–48 h at 37°C, remain viable at −70°C for at least 2 years when stored in sealed glass ampules. For short-term storage, Turk (1964) has described a simple procedure by which most strains of *H. influenzae* remain viable for 4 weeks at room temperature. Chocolate agar slopes in screw-capped bottles are heavily inoculated, incubated overnight, and stored unopened at room temperature. For details on medium, see Turk (1964).

Identification

As the genus *Haemophilus* is defined at present (Zinnemann, 1967), the main criterion for placing an unknown aerobic or facultatively anaerobic, Gram-negative coccobacillus or rod in that genus is a requirement for either one or both of the two growth factors X and V. In addition to this, the universal ability of haemophili to reduce nitrate will distinguish them from such bacteria as bordetellae and *Cardiobacterium hominis.*

For a long time, differentiation of the *Haemophilus* species has largely been based on the demonstration of accessory growth factor requirements, ability to lyse blood cells, and enhancement of growth under increased CO_2 tension (Zinnemann and Biberstein, 1974). Few conventional bacteriological tests were applied with success in studies of haemophili. Besides the fact that media used in such tests usually do not support growth of haemophili, there was a seeming inconsistency of traditional biochemical reactivity within the *Haemophilus* species. Thus, Zinnemann and Biberstein (1974), in *Bergey's Manual,* state that "the usefulness of sugar reactions in characterizing *Haemophilus* species is disputed." Therefore, the consequent paucity of distinguishing characters has long rendered an exact identification of isolated *Haemophilus* organisms very difficult. The main reason for this difficulty has been the overriding weight placed on the accessory growth factor requirements as a means of identification combined with the fact that the methods used to determine these growth requirements have not been reliable (Evans and Smith, 1972; Kilian, Heine-Jensen, and Bülow, 1972).

As previously mentioned in this chapter, the satellite phenomenon (Fig. 2) is widely used as a means of detecting the V-factor requirement. This phenomenon may be demonstrated on an agar medium lacking the V factor by cross-inoculating the plate with an appropriate feeder strain (i.e., *Staphylococcus* or *Pseudomonas* species). More unequivocal results may, however, be obtained by the application of a V-factor-impregnated paper disk. Such disks are commercially available or may be prepared by adding 10 μl of NAD stock solution (see "Growth Media", this chapter) to small sterile filter paper strips or disks. Media in which all ingredients have been autoclaved are consistently free of V factor, whereas ordinary blood agar contains varying amounts of V factor. A convincing satellite phenomenon is, therefore, sometimes difficult to achieve on ordinary blood agar plates. Far better results are obtained on a blood agar medium in which the blood (5–10%) has been added before autoclaving. This medium is completely devoid of V factor but otherwise satisfies all growth requirements of the *Haemophilus* species, including the X factor. In contrast, blood-free media often do not support growth of *H. parainfluenzae* strains unless supplemented with sodium oleate and thiamine (Evans and Smith, 1972).

Demonstration of the X-factor requirement is often accomplished by comparative testing for growth on a second medium without blood, or by demonstrating growth around a paper disk impregnated with hemin. An alternative to this is the method of Cooper and Attenborough (1968), which detects growth and glucose fermentation as a function of hemin supplementation. However, all these methods have the same inherent pitfalls. Probably all complex media that will otherwise support growth of haemophili contain traces of the X factor. Furthermore, even when particular care is being exercised, X factor is often carried over with the inoculate. The X-factor requirement may be determined more accurately by carrying out serial subcultivations on defined agar media with and without hemin (Kilian et al., 1972), but this method is both expensive and time-consuming.

A much more accurate and direct way of determining the X-factor requirement was first suggested by Biberstein, Mini, and Gills (1963), who showed that hemin-independent *Haemophilus* strains excrete porphobilinogen (PBG) and porphyrins—intermediates in the hemin biosynthetic pathway (Fig. 1)—when supplied with δ-aminolevulinic acid (ALA). In contrast, X-requiring strains do not excrete these compounds because they lack the enzymes responsible for their synthesis (White and Granick, 1963). Both PBG and porphyrins may be visualized by simple procedures (Schwartz et al., 1960). These findings have been utilized in a rapid and simple test for X-factor requirement:

Porphyrin Test for *Haemophilus*
X-Factor Requirement (Kilian, 1974)

Substrate: δ-Aminolevulinic acid hydrochloride (Sigma), 2 mM, and MgSO$_4$, 0.8 mM, in 0.1 M phosphate buffer (Sørensen) at pH 6.9. The substrate is distributed in 0.5-ml quantities in small glass tubes.

Inoculation: Suspend a heavy loopful of bacteria (about 1 mg) from an agar plate culture in the substrate.

Reading: After inoculation for 4 h at 37°C, expose to Wood's light (360 nm), preferably in a dark room. A red fluorescence from the bacterial cells and/or from the fluid is indicative of porphyrins, i.e., the strain is independent of the X factor.

Alternative method of reading: Add 0.5 ml of Kovacs' reagent (p-dimethylaminobenzaldehyde, 5 g, amyl alcohol, 75 ml, and concentrated HCl, 25 ml), shake vigorously, and allow phases to separate. A red color in the lower water phase is indicative of PBG, which means that the strain is independent of the X factor. Using this method of reading, an inoculated "substrate" without ALA must be included as a negative control. Kovacs' reagent also gives a red color with indole. Although a color reaction due to indole will be present in the upper alcohol phase, indole-positive strains of *H. influenzae* may be mistakenly identified as being X-factor independent (Lund and Blazevic, 1977).

It should be emphasized that the excretion of PBG and porphyrins under the conditions of the porphyrin test is not a universal character of bacteria belonging to other groups. Therefore, a negative result obtained with a "non-*Haemophilus*" organism does not necessarily imply that it is X dependent.

The advantage of the porphyrin test is stressed by the results of a taxonomic study of 426 *Haemophilus* strains in which the method was applied for the determination of the X-factor requirement (Kilian, 1976a). When the strain collection was subdivided on the basis of the results obtained with the prophyrin test, a perfect correlation with fermentation reactions, DNA base compositions, and other characters was revealed. Furthermore, by correlating the identification of 257 *Haemophilus* strains obtained by the use of the disk method and the porphyrin test, it was found that 45 strains (18%) were incorrectly identified by the former method (M. Kilian and K. R. Eriksen, unpublished results). The most frequent mistake (12%) was that strains of *H. influenzae* were incorrectly identified as *H. parainfluenzae*.

The above-mentioned taxonomic study by Kilian (1976a) showed that fermentation reactions are highly valuable for the separation of the *Haemophilus* species (Table 2). Whereas X-factor-independent species ferment sucrose, strains of *H. influenzae* never do so, contrary to the information given in *Bergey's Manual* (Zinnemann and Biberstein, 1974). Other carbohydrates that are of particular value as a means of identification are glucose, lactose, xylose, mannitol, and ribose (not deoxyribose as incorrectly stated by Kilian [1976a]). Fermentation reactions may be studied in phenol red broth base (Difco) with 1% of the carbohydrates and supplemented with the two accessory growth factors as described above. In this medium, most strains of *H. parainfluenzae*, *H. aphrophilus*, *H. paraphrophilus*, and *H. haemolyticus* release hydrogen and carbon dioxide during fermentation of glucose (Kilian, 1976a). This may be demonstrated by inserting an inverted Durham tube in the glucose broth. Fermentation reactions are read after incubation for 2–5 days. Biberstein, Gunnarsson, and Hurvell (1977) have used a micro-method for the determination of fermentation reactions that is simple and more rapid than the traditional one. However, as preliminary studies indicate that the results obtained by the use of the two methods do not always agree (Biberstein, Gunnarsson, and Hurvell, 1977; Kilian, Nicolet, and Biberstein, 1978), a more extensive study is required in order to establish an identification scheme based on the use of the micro-method.

Rapid micro-methods may be used with advantage for detection of tryptophanase (indole), urease, amino acid decarboxylases, and glycosidases. The reactions are valuable for the identification of the species and for their subdivision into biotypes (Table 2). The methods used by Kilian (1976a) are described below. All substrates are used in 0.5-ml quantities and are inoculated by suspending a heavy loopful of bacteria from an agar plate culture. The

results are read after incubation for 4 h at 37°C. The incubation may be prolonged for up to 24 h.

Micromethods for Identifying *Haemophilus* spp.
Indole test (Clarke and Cowan, 1952):
Substrate: 0.1% L-tryptophan in 1/15 M phosphate buffer (Sørensen) at pH 6.8.

Reading: Add 1 volume of Kovacs' reagent and shake. Red color in the upper alcohol phase indicates the presence of indole.

Urease test (Lautrop, 1960):

Substrate:		
Distilled water	100 ml	
KH_2PO_4	0.1 g	
K_2HPO_4	0.1 g	
NaCl	0.5 g	
Phenol red 1:500	0.5 ml	

Adjust the pH to 7.0 with 5 N NaOH, autoclave, and add, as a filter-sterilized solution, 10.4 ml urea (20% aqueous solution) (phenol red 1:500: phenol red, 0.2 g, in distilled water, 92 ml, and NaOH, 8 ml).

Reading: Red color indicates urease activity.

Amino acid decarboxylase activities are tested in Møller's medium (1955), which is commercially available (Difco). A purple color indicates decarboxylase activity.

Beta-galactosidase (ONPG) test (Bülow, 1964; modified):

Substrate: 0.1%   2-nitrophenyl-β-D-galactopyranoside in 1/15 M phosphate buffer (Sørensen) at pH 8.0.

Reading: Yellow color indicates β-galactosidase activity.

Other glycosidases, including α-fucosidase, may be detected by the methods described by Kilian and Bülow (1976).

Alternative methods: The tests for indole, urease, ornithine and lysine decarboxylases, and β-galactosidase may also be performed by using the strips of the Pathotec system (General Diagnostics, Warner-Lambert Co.).

Testing of strains for enhancement of growth by carbon dioxide may be performed by comparing growth on lightly inoculated duplicate sets of chocolate agar incubated with and without 10% CO_2 added to the incubation atmosphere. The hemagglutinating activity as found in some isolates from conjunctivitis (Davis, Pittman, and Griffiths, 1950; Kilian et al., 1976), and some isolates from fowl (Iritani, Hidaka, and Katagiri, 1977) and swine (Bakos, 1955) may be determined as described by Davis, Pittmann, and Griffitts (1950). The synergistic effect of the hemolysin of *H. pleuropneumoniae* and the *Staphylococcus aureus* β-toxin (CAMP reaction) may be demonstrated as described by Kilian

Table 2. Biochemical and physiological characteristics of the *Haemophilus* species and related bacteria.[a]

	X factor required[b]	V factor required	Indole	Urease	Ornithine decarboxylase	Lysine decarboxylase	Hemolysis	CAMP reaction	Glucose, acid	Glucose, gas	Sucrose, acid	Lactose, acid	Xylose, acid	Ribose, acid	Mannitol, acid	Galactose, acid	Sorbitol, acid	β-Galactosidase (ONPG)	α-Glucosidase (PNPG)	α-Fucosidase	Catalase	CO₂ enhances growth	Alkaline phosphatase	Nitrate reduction	G+C mol%
H. influenzae																									
biotype I	+	+	+	+	+	d	–	–	+	–	–	–	+	+	–	+	–	–	–	–	+	–	+	+	39
biotype II	+	+	+	+	–	–	–	–	+	–	–	–	d	+	–	+	–	–	–	–	+	–	+	+	
biotype III	+	+	–	+	–	–	–	–	+	–	–	–	d	+	–	+	–	–	–	–	+	–	+	+	
biotype IV	+	+	–	+	+	–	–	–	+	–	–	–	+	+	–	+	–	–	–	–	d	–	+	+	
biotype V	+	+	+	–	+	–	–	–	+	–	–	–	+	+	–	+	–	–	–	–	+	–	+	+	
H. haemolyticus	+	+	d	+	–	–	+	–	+	d	–	–	d	+	–	+	–	–	–	–	+	–	+	+	38
H. ducreyi	+	–	–	–	–	–	–	–	–	–	–	–	–	–	–	–	–	–	–	–	–	–	+	+	38
H. haemoglobinophilus	+	–	+	–	–	–	–	–	+	–	+	–	+	d	+	+	·	d	–	–	+	–	–	+	38
H. influenzae-murium	+	–	–	–	–	–	–	+	·	+	–	–	–	·	–	–	d	–	+	·	·	+			
H. parainfluenzae																									
biotype I	–	+	–	–	+	–	–	–	+	d	+	–	–	–	–	+	–	+	–	–	d	–	+	+	40
biotype II	–	+	–	+	+	–	d	–	+	+	+	–	–	–	–	+	–	d	–	–	d	–	+	+	
biotype III	–	+	–	+	–	–	d	–	+	–	+	–	–	–	–	+	–	d	–	–	+	d	+	+	
H. segnis	–	+	–	–	–	–	–	–	w	–	w	–	–	–	–	w	–	d	–	–	d	–	+	+	44
H. aphrophilus	–[c]	–	–	–	–	–	–	–	+	+	+	+	–	+	–	+	–	+	d	–	–	+	+	+	42
H. paraphrophilus	–	+	–	–	–	–	–	–	+	+	+	+	–	+	–	+	–	+	d	–	–	+	+	+	42
H. parasuis	–	+	–	–	–	–	–	–	+	–	+	d	–	+	d	·	·	d	d	+	+	d	+	+	42
H. paragallinarum	–	+	–	–	–	–	–	–	+	–	+	d	+	+	+	–	+	+	–	d	–	+	+	+	42
H. avium	–	+	–	–	–	–	–	–	+	–	+	d	d	·	d	+	d	d	+	–	+	–	+	+	42
H. pleuropneumoniae	–	+	–	+	–	–	+	+	+	–	+	d	+	–	+	+	–	+	–	–	d	–	+	+	42
Haemophilus sp. (Firehammer, 1959)	–	+	–	–	·	·	+	·	+	·	+	–	+	·	+	–	+	·	·	·	·	·	·	+	
"H. somnus"	–	–[d]	+	–	·	·	·	–	+	–	d	–	+	·	–	–	+	·	·	·	–	+	·	+	37
"H. piscium"	–	–	–	–	–	–	–	–	+	–	+	–	–	–	–	–	·	·	w	–	–	–	–	–	55
Actinobacillus actino-																									
mycetem-comitans	–	–	–	–	–	–	–	+	d	–	–	d	·	+	+	–	–	–	–	–	+	+	+	+	43
Eikenella corrodens	–	–	–	–	+	+	–	–	–	–	–	–	–	–	–	–	–	–	–	–	–	+	–	+	56
Cardiobacterium hominis	–	–	+	–	–	–	–	–	+	–	+	–	–	·	d	–	+	–	–	·	–	d	·	–	62
Pasteurella multocida	–	–	+	–	+	d	–	–	+	–	+	–	d	·	+	+	+	–	+	·	+	–	+	+	37–40

[a] Data compiled from Kilian (1976a), Firehammer (1959), Frederiksen (1973), Kennedy et al. (1960), Csukás (1976), Hinz and Kunjara (1977), Kilian, Nicolet, and Biberstein (1978), and Slotnick and Dougherty (1964). Biochemical properties of nonclassified *Haemophilus* strains from humans are described by Ryan (1968) and Kilian (1976a), from a variety of avian species by Grebe and Hinz (1975), from rats by Kilian (1976a), and from swine by Biberstein, Gunnarsson, and Hurvell (1977) and Kilian, Nicolet, and Biberstein (1978). +, more than 90% of strains positive; –, more than 90% of strains negative; d, 11–89% of strains positive; w, weak reaction; ·, data not available.

[b] X-factor requirement as determined by the porphyrin test (Kilian, 1974).

[c] X factor is usually required on primary isolation, but porphyrin test is always positive.

[d] Isolates of "H. somnus" show satellitic growth on blood-free media, but are unaffected by growth factors X and V.

(1976b). Capsule production may be detected on Levinthal's agar by observing the distinctive iridescent appearance of colonies when examined with the aid of strong lighting obliquely transmitted through the plate. Typing of encapsulated strains of *H. influenzae* into the Pittman serotypes a through f may be performed by slide agglutination or capsule swelling tests using commercially available typing sera (Hyland Laboratories, Difco, Oxoid).

A diagnostic scheme based on the above characters is presented in Table 2. The table includes all *Haemophilus* species, some species that probably do not belong to the genus but have been tentatively placed there, and selected organisms related to the genus *Haemophilus*. A list of proposed reference strains of the *Haemophilus* species is provided in Table 3.

Table 3. Reference strains of the *Haemophilus* species.

Species		Strain number[a]	Status
H. influenzae biotype I	(serotype a)	NCTC 8466	
H. influenzae biotype I	(serotype b)	NCTC 7279	
H. influenzae biotype I	(serotype f)	NCTC 8473	
H. influenzae biotype II	(serotype c)	NCTC 8469	
H. influenzae biotype II		NCTC 8143	Type strain (neotype)
H. influenzae biotype III		NCTC 4560 = ATCC 19418	
H. influenzae biotype IV	(serotype d)	NCTC 8470	
H. influenzae biotype IV	(serotype e)	NCTC 10479	
H. aegyptius		NCTC 8502 = ATCC 11116	Type strain (neotype)
H. haemolyticus		NCTC 10659	Type strain (neotype)
H. haemoglobinophilus		NCTC 1659	Type strain (neotype)
H. ducreyi		CIP 542	Type strain (neotype)
H. parainfluenzae biotype I		NCTC 7857	Type strain (neotype)
H. parainfluenzae biotype II		NCTC 10665	
H. parainfluenzae biotype III		HK 303	
H. parahaemolyticus		NCTC 8479 = ATCC 10014	
H. paraphrohaemolyticus		NCTC 10670 = ATCC 29240	Type strain (holotype)
H. paraphrophilus		NCTC 10557 = ATCC 29241	Type strain (holotype)
H. aphrophilus		NCTC 5906	Type strain (holotype)
H. segnis		NCTC 10977 = CCM 6052	Type strain (holotype)
H. parasuis		NCTC 4557 = ATCC 19417	
H. pleuropneumoniae		ATCC 27088 = CCM 5869	Type strain (neotype)
H. paragallinarum		ATCC 29545	Type strain (neotype)
H. avium		ATCC 29546	Type strain (neotype)

[a] NCTC, National Collection Type Cultures, Colindale, London; CIP, Collection de l'Institut Pasteur, Paris; HK, collection studied by Kilian (1976a); ATCC, American Type Culture Collection, Rockville, Maryland; CCM, Czechoslovak Collection of Microorganisms, Brno.

Literature Cited

Alexander, H. E. 1958. The *Hemophilus* group, pp. 470–485. In: Dubos, R. J. (ed.), Bacterial and mycotic infections in man, 3rd ed. Philadelphia: J. P. Lippincott.

Baber, K. G. 1963. A selective medium for the isolation of *Haemophilus* from sputum. Journal of Medical Laboratory Technology **26:**391–396.

Bailie, W. E. 1969. Characteristics of *Hemophilus somnus* (new species). Ph.D. dissertation. Kansas State University, Manhattan, Kansas.

Bakos, K. 1955. Studien über *Haemophilus suis* mit besonderer Berücksichtigung der serologischen Differenzierung seiner Stämme. Thesis. Uppsala University, Uppsala, Sweden.

Balish, E., Cleven, D., Brown, J., Yale, C. E. 1977. Nose, throat, and fecal flora of beagle dogs housed in "locked" or "open" environments. Applied and Enviromental Microbiology **34:**207–221.

Biberstein, E. L., Cameron, H. S. 1961. The family *Brucellaceae* in veterinary research. Annual Review of Microbiology **15:**93–118.

Biberstein, E. L., Gunnarsson, A., Hurvell, B. 1977. Cultural and biochemical criteria for the identification of *Haemophilus* spp. from swine. American Journal of Veterinary Research **30:**7–12.

Biberstein, E. L., Mini, P. D., Gills, M. G. Action of *Haemophilus* cultures on δ-aminolevulinic acid. Journal of Bacteriology **86:**814–819.

Biberstein, E. L., Spencer, P. D. 1962. Oxidative metabolism of *Haemophilus* species grown at different levels of hemin supplementation. Journal of Bacteriology **84:**916–920.

Biberstein, E. L., White, D. C. 1969. A proposal for the establishment of two new *Haemophilus* species. Journal of Medical Microbiology **2:**75–78.

Boyce, J. M. H., Frazer, J., Zinnemann, K. 1968. The growth requirements of *Haemophilus aphrophilus*. Journal of Medical Microbiology **2:**55–62.

Brumfitt, W. 1959. Some growth requirements of *Haemophilus influenzae* and *Haemophilus pertussis*. Journal of Pathology and Bacteriology **77:**95–100.

Bülow, P. 1964. The ONPG test in diagnostic bacteriology. I. Methodological investigations. Acta Pathologica et Microbiologica Scandinavica **60:**376–386.

Butler, L. O. 1962. A defined medium for *Haemophilus influenzae* and *Haemophilus parainfluenzae*. Journal of General Microbiology **27:**51–60.

Clarke, P. H., Cowan, S. T. 1952. Biochemical methods for bacteriology. Journal of General Microbiology **6:**187–197.

Cooper, R. G., Attenborough, I. D. 1968. An indicator method for the detection of bacterial X and V factor dependence. Australian Journal of Experimental Biology and Medical Sciences **45:**803–806.

Crawford, J. J., Barden, L., Kirkman, J. B. 1969. Selective culture medium to survey the incidence of *Haemophilus* species. Applied Microbiology **18:**646–649.

Csukás, Z. 1976. Reisolation and characterization of *Haemophilus influenzaemurium*. Acta Microbiologica Academiae Scientiarum Hungaricae **23:**89–96.

Davis, D. J., Pitmann, M., Griffitts, J. J. 1950. Hemagglutination by the Koch-Weeks bacillus (*Hemophilus aegyptius*). Journal of Bacteriology **59:**427–431.

Delaplane, J. P., Erwin, L. E., Stuart, H. O. 1934. A hemophilic bacillus as the cause of an infectious rhinitis (coryza) of fowls. Agricultural Experimental Station. Rhode Island State College Bulletin No. 244, 1–12.

Diernhofer, K. 1949. Haemophile Bakterien im Geschlechtstrakt des Rindes. Wiener tierärztliche Monatsschrift **36:**582–588.

Evans, N. M., Smith, D. D. 1972. The effect of the medium and source of growth factors on the satellitism test for *Haemophilus* species. Journal of Medical Microbiology 5: 509–514.

Evans, N. M., Smith, D. D. 1974. The inhibition of *Haemophilus influenzae* by certain agar and peptone preparations. Journal of Medical Microbiology 7:305–310.

Evans, N. M., Smith, D. D., Wicken, A. J. 1974. Haemin and nicotinamide adenine dinucleotide requirements of *Haemophilus influenzae* and *Haemophilus parainfluenzae*. Journal of Medical Microbiology 7:359–365.

Firehammer, B. D. 1959. Bovine abortion due to *Haemophilus* species. Journal of the American Veterinary Medical Association 135:421–422.

Fleming, A., Maclean, I. H. 1930. On the occurrence of influenza bacilli in the mouths of normal people. British Journal of Experimental Pathology 11:127–134.

Frazer, J., Zinnemann, K., Boyce, J. M. H. 1969. The effect of different environmental conditions on some characters of *Haemophilus paraphrophilus*. Journal of Medical Microbiology 2:563–566.

Frederiksen, W. 1973. *Pasteurella* taxonomy and nomenclature. Contributions to Microbiology and Immunology 2:170–176.

Gilder, H., Granick, S. 1947. Studies on the Hemophilus group of organisms. Quantitative aspects of growth on various porphin compounds. Journal of General Physiology 31:103–117.

Granick, S., Mauzerall, D. 1967. The metabolism of heme and chlorophyll, pp. 525–625. In: Greenberg, D. M. (ed.), Metabolic pathways, vol. 3. New York: Academic Press.

Grebe, H. H., Hinz, K. H. 1975. Vorkommen von Bakterien der Gattung Haemophilus bei verschiedenen Vogelarten. Zentralblatt für Veterinärmedizin Reihe B 22:749–757.

Hammond, G. W., Lian, C.-J., Wilt, J. C., Albritton, W. L., Ronald, A. R. 1978a. Determination of the hemin requirement of *Haemophilus ducreyi*: Evaluation of the porphyrin test and media used in the satellite growth test. Journal of Clinical Microbiology 7:243–246.

Hammond, G. W., Lian, C.-J., Wilt, J. C., Ronald, A. R. 1978b. Comparison of specimen collection and laboratory techniques for the isolation of *Haemophilus ducreyi*. Journal of Clinical Microbiology 7:39–43.

Harr, J. R., Tinsley, I. J., Weswig, P. H. 1969. Haemophilus isolated from a rat respiratory epizootic. Journal of the American Veterinary Medical Association 155:1126–1130.

Harris, D. L., Ross, R. F. and Switzer, W. P. 1969. Incidence of certain microorganisms in nasal cavities of swine in Iowa. American Journal of Veterinary Research 30:1621–1624.

Herbst, E. J., Snell, E. E. 1949. The nutritional requirements of *Haemophilus parainfluenzae* 7901. Journal of Bacteriology 58:379–386.

Herriott, R. M., Meyer, E. Y., Vogt, M., Modan, M. 1970. Defined medium for growth of *Haemophilus influenzae*. Journal of Bacteriology 101:513–516.

Hinz, K. H. 1973. Beitrag zur Differenzierung von *Haemophilus*-stämmen aus Hühnern. 1. Mitteilung: Kulturelle und biochemische Untersuchungen. Avian Pathology 2:211–229.

Hinz, K. H., Kunjara, C. 1977. *Haemophilus avium*, a new species from chickens. International Journal of Systematic Bacteriology 27:324–329.

Holt, L. B. 1962. The growth factor requirements of *Haemophilus influenzae*. Journal of General Microbiology 27:317–322.

Hovig, B., Aandahl, E. H. 1969. A selective method for the isolation of *Haemophilus* in material from the respiratory tract. Acta Pathologica et Microbiologica Scandinavica 77:676–684.

Iritani, Y., Hidaka, S., Katagiri, K. 1977. Production and properties of hemagglutinin of *Haemophilus gallinarum*. Avian Diseases 21:39–49.

Johnson, R., Sneath, P. H. A. 1973. Taxonomy of *Bordetella* and related organisms of the families *Achromobacteraceae*, *Brucellaceae* and *Neisseriaceae*. International Journal of Systematic Bacteriology 23:381–404.

Kairies, A. 1935. Influenzaerkrankungen bei Frettchen und Beschreibung eines *Bacterium influenzae putoriorum multiforme*. Zeitschrift für Hygiene und Infektionskrankheiten 117:12–17.

Kairies, A., Schwartzer, K. 1936. Studien zu einer bakteriellen Influenza der Mäuse and Beschreibung eines "Bacterium influenzae murium". Zentralblatt für Backteriologie, Parasitenkunde, und Infektionskrankheiten, Abt. 1 Orig. 137: 351–359.

Kelly, T. J., Smith, H. O. 1970. A restriction enzyme from *Haemophilus influenzae*. II. Base sequence of the recognition site. Journal of Molecular Biology 51:393–409.

Kennedy, P. C., Biberstein, E. L., Howarth, J. A., Frazier, L. M., Dungworth, D. L. 1960. Infectious meningoencephalitis in cattle, caused by a *Haemophilus*-like organism. American Journal of Veterinary Research 21:403–409.

Kennedy, P. C., Frazier, L. M., Theilen, G. H., Biberstein, E. L. 1958. A septicemic disease of lambs caused by *Haemophilus agni* (new species). American Journal of Veterinary Research 19:645–654.

Khairat, O. 1940. Endocarditis due to a new species of *Haemophilus*. Journal of Pathology and Bacteriology 50:497–505.

Kilian, M. 1974. A rapid method for the differentiation of *Haemophilus* strains. The porphyrin test. Acta Pathologica et Microbiologica Scandinavica, Sect. B 82:835–842.

Kilian, M. 1976a. A taxonomic study of the genus *Haemophilus*. With the proposal of a new species. Journal of General Microbiology 93:9–62.

Kilian, M. 1976b. Haemolytic activity of *Haemophilus* species. Acta Pathologica et Microbiologica Scandinavica, Sect. B 84:339–341.

Kilian, M., Bülow, P. 1976. Rapid diagnosis of *Enterobacteriaceae*. I. Detection of bacterial glycosidases. Acta Pathologica et Microbiologica Scandinavica, Sect. B 84:245–251.

Kilian, M., Heine-Jensen, J., Bülow, P. 1972. *Haemophilus* in the upper respiratory tract of children. A bacteriological, serological and clinical investigation. Acta Pathologica et Microbiologica Scandinavica, Sect. B 80:571–578.

Kilian, M., Nicolet, J., Biberstein, E. L. 1978. Biochemical and serological characterization of *Haemophilus pleuropneumoniae* (Matthews and Pattison 1961) Shope 1964 and proposal of a neotype strain. International Journal of Systematic Bacteriology 28:20–26.

Kilian, M., Rölla, G. 1976. Initial colonization of teeth in monkeys as related to diet. Infection and Immunity 14:1022–1027.

Kilian, M., Schiøtt, C. R. 1975. Haemophili and related bacteria in the human oral cavity. Archives of Oral Biology 20:791–796.

Kilian, M., Sørensen, I., Frederiksen, W. 1979. Biochemical characterics of 130 recent isolates from *H. influenzae* meningitis. Journal of Clinical Microbiology 9:409–412.

Kilian, M., Theilade, J. 1975. Cell wall ultrastructure of strains of *Haemophilus ducreyi* and *Haemophilus piscium*. International Journal of Systematic Bacteriology 25:351–356.

Kilian, M., Theilade, J. 1978. Amended description of *Haemophilus segnis* Kilian 1977. International Journal of Systematic Bacteriology 28:411–415.

Kilian, M., Mordhorst, C. H., Dawson, C. R., Lautrop, H. 1976. The taxonomy of haemophili isolated from conjunctivae. Acta Pathologica et Microbiologica Scandinavica, Sect. B 84:132–138.

Klein, M., Blazevic, D. J. 1970. Evaluation of a selective medium for isolation of *Haemophilus* from respiratory cultures. American Journal of Medical Technology 36:97–106.

Kraut, M. S., Finegold, S. M., Sutter, V. L. 1972. Detection of *Haemophilus aphrophilus* in the human oral flora with a selective medium. Journal of Infectious Diseases 126: 189–192.

Kristensen, M. 1922. Investigations into the occurrence and classification of the haemoglobinophilic bacteria. Thesis. Copenhagen: Levin & Munksgaard Publishers.

Krumwiede, E., Kuttner, A. G. 1938. A growth inhibitory substance for the influenza group of organisms in the blood of various animal species. Journal of Experimental Medicine 67:429–441.

Lautrop, H. 1960. Laboratory diagnosis of whooping-cough or Bordetella infections. Bulletin of the World Health Organization 23:15–31.

Levinthal, W. 1918. Bacteriologische und serologische Influenzastudien. Zeitschrift für Hygiene und Infektionskrankheiten 86:1–24.

Lewis, P. A., Shope, R. E. 1931. Swine influenza. II. A hemophilic bacillus from the respiratory tract of infected swine Journal of Experimental Medicine 54:361–371.

Little, T. W. A. 1970. Haemophilus infection in pigs. Veterinary Record 87:399–402.

Little, T. W. A., Pritchard, D. G. Cited by: Pritchard, D. G., Macleod, N. S. M. 1977. Veterinary Record 100:126–127.

Love, B. D., Finland, M. 1954. Susceptibility of recently isolated strains of Haemophilus influenzae to eleven antibiotics in vitro. Journal of Pediatrics 45:531–537.

Lund, M. E., Blazevic, D. J. 1977. Rapid speciation of Haemophilus with the porphyrin production test versus the satellite test for X. Journal of Clinical Microbiology 5:142–144.

Lwoff, A., Lwoff, M. 1937a. Role physiologique de l'hemine pour Haemophilus influenzae Pfeiffer. Annales de l'Institut Pasteur 59:129–136.

Lwoff, A., Lwoff, M. 1937b. Studies on codehydrogenases. II. Physiological function of growth factor "V." Proceedings of the Royal Society, B 122:360–373.

Mitchell, C. A. 1925. Haemophilus ovis (nov. spec.) as the cause of a specific disease of sheep. Journal of the American Veterinary Medical Association 68:8–18.

Møller, V. 1955. Simplified test for some amino acid decarboxylases and for the arginine dihydrolase system. Acta Pathologica et Microbiologica Scandinavica 36:158–172.

Müller, H. E., Hinz, K.-H. 1977. Über das Vorkommen von Neuraminidase und N-Acetyl-neuraminat-Pyruvat-Lyase bei human-pathogenen Haemophilus-Arten. Zentralblatt für Bakteriologie, Parasitenkunde, Infektionskrankheiten und Hygiene, Abt. I. Orig., Reihe A 239:231–239.

Nicolet, J. 1968. Sur l'hemophilose du porc. I. Identification d'un agent frequent: Haemophilus parahaemolyticus. Pathology and Microbiology 31:215–225.

Page, L. A. 1962. Haemophilus infections in chickens. American Journal of Veterinary Research 23:85–95.

Pfeiffer, R. 1892. Vorläufige Mitteilungen über die Erreger der Influenza. Deutsche Medizinische Wochenschrift 18:28.

Rivers, T. M., Bayne-Jones, S. 1923. Influenza-like bacilli isolated from cats. Journal of Experimental Medicine 37:131–138.

Roberts, R. J. 1976. Restriction endonucleases. Critical Reviews in Biochemistry 4:123–164.

Ryan, W. J. 1968. An X-factor requiring Haemophilus species. Journal of General Microbiology 52:275–286.

Schwartz, S., Berg, M. H., Bossenmaier, I., Dinsmore, H. 1960. Determination of porphyrins in biological materials, pp. 221–293. In: Glick, D. (ed.), Methods of biochemical analysis, vol. 8. New York: Interscience Publishers.

Shifrine, M., Biberstein, E. L. 1960. A growth factor for Haemophilus species secreted by a pseudomonad. Nature 187:623.

Shope, R. E. 1964. Porcine contagious pleuropneumonia. I. Experimental transmission, etiology, and pathology. Journal of Experimental Medicine 119:357–368.

Sims, W. 1970. Oral haemophili. Journal of Medical Microbiology 3:615–625.

Sinclair, P. R., White, D. C. 1970. Effect of nitrate, fumarate, and oxygen on the formation of the membrane-bound electron transport system of Haemophilus parainfluenzae. Journal of Bacteriology 101:365–372.

Slotnick, I. J., Dougherty, M. 1964. Further characterization of an unclassified group of bacteria causing endocarditis in man: Cardiobacterium hominis gen. et sp. n. Antonie van Leeuwenhoek Journal of Microbiology and Serology 30:261–272.

Smith, H. O., Wilcox, K. W. 1970. A restriction enzyme from Haemophilus influenzae: I. Purification and general properties. Journal of Molecular Biology 51:379–391.

Smith, L., White, D. C. 1962. Structure of the respiratory chain system as indicated by studies with Haemophilus parainfluenzae. Journal of Biological Chemistry 237:1337–1341.

Sneath, P. H. A., Johnson, R. 1973. Numerical taxonomy of Haemophilus and related bacteria. International Journal of Systematic Bacteriology 23:405–418.

Snieszko, S. F., Griffin, P. J., Friddle, S. B. 1950. A new bacterium (Haemophilus piscium n. sp.) from ulcer disease of trout. Journal of Bacteriology 59:699–710.

Sutter, V. L., Finegold, S. M. 1970. Haemophilus aphrophilus infections: Clinical and bacteriologic studies. Annals of the New York Academy of Sciences 174:468–487.

Talmadge, M. B., Herriott, R. M. 1960. A chemically defined medium for growth, transformation, and isolation of nutritional mutants of Haemophilus influenzae. Biochemical and Biophysical Research Communications 2:203–206.

Turk, D. C. 1964. Short term storage of Haemophilus influenzae. Journal of Clinical Pathology 17:297–300.

Turk, D. C., May, J. R. 1967. Haemophilus influenzae: Its clinical importance. London: English Universities Press.

Tuyau, J. E., Sims, W. 1974. Neuraminidase activity in human oral strains of haemophili. Archives of Oral Biology 19:817–820.

Vastine, D. W., Dawson, C. R., Hoshiwara, I., Yoneda, C., Daghfous, T., Messadi, M. 1974. Comparison of media for the isolation of Haemophilus species from cases of seasonal conjunctivitis associated with severe endemic trachoma. Applied Microbiology 28:688–691.

Waterworth, P. M. 1955. The stimulation and inhibition of the growth of Haemophilus influenzae on media containing blood. British Journal of Experimental Pathology 36:186–194.

White, D. C. 1963. Respiratory systems in the hemin-requiring Haemophilus species. Journal of Bacteriology 85:84–96.

White, D. C., Granick, S. 1963. Hemin biosynthesis in Haemophilus. Journal of Bacteriology 85:842–850.

White, D. C., Smith, L. 1962. Hematin enzymes of Haemophilus parainfluenzae. Journal of Biological Chemistry 237:1332–1336.

Winslow, C. E. A., Broadhurst, J., Buchanan, R. E., Krumwiede, C., Rogers, L. A., Smith, G. H. 1920. The families and genera of the bacteria. Final report of the society of American bacteriologists on characterization and classification of bacterial types. Journal of Bacteriology 5:191–229.

Wolin, H. E. 1963. Defined medium for Haemophilus influenzae type b. Journal of Bacteriology 85:253–254.

Zinnemann, K. 1960. Haemophilus influenzae and its pathogenicity. Ergebnisse der Mikrobiologie, Immunitätsforschung und Experimentellen Therapie 33:307–368.

Zinnemann, K. 1967. Report of the subcommittee on the taxonomy of Haemophilus (1962–66). International Journal of Systematic Bacteriology 17:165–166.

Zinnemann, K., Biberstein, E. L. 1974. Genus Haemophilus Winslow, Broadhurst, Buchanan, Krumwiede, Rogers and Smith 1917, 561, pp. 364–368. In: Buchanan, R. E., Gibbons, N. E., (eds.), Bergey's manual of determinative bacteriology, 8th ed. Baltimore: Williams & Wilkins.

Zinnemann, K., Rogers, K. B., Frazer, J., Boyce, J. M. H. 1968. A new V-dependent Haemophilus species preferring increased CO_2 tension for growth and named Haemophilus paraphrophilus, nov. sp. Journal of Pathology and Bacteriology 96:413–419.

The Genus *Pasteurella*

GORDON R. CARTER

Bacteria included in the genus *Pasteurella* are commensals and occasional pathogens of many species of domestic and wild animals. Until very recently, the genus included the species formerly known as *Pasteurella pestis*, *P. pseudotuberculosis*, *P. enterocolitica*, *P. tularensis*, and *P. novicida*. The first three now constitute the genus *Yersinia* and have been placed in the Enterobacteriaceae. The two remaining species, the tularemia agents, which clearly did not belong in the *Pasteurella* genus, have been given the generic name *Francisella*.

The following species are recognized in the eighth edition of *Bergey's Manual of Determinative Bacteriology* (Buchanan and Gibbons, 1974): *Pasteurella multocida*, *Pasteurella haemolytica*, *Pasteurella pneumotropica*, and *Pasteurella ureae*.

The nonfermentative organism called *Pasteurella anatipestifer* does not properly belong in the genus *Pasteurella* and is correctly placed for the present in the category *species incertae sedis*. The unofficial species *P. gallinarum* and *P. aerogenes* are of minor significance in veterinary microbiology.

The principal characteristics of members of the genus are: small Gram-negative rods or coccobacilli, nonmotile, facultatively anaerobic, catalase and oxidase positive, fermentative (except *P. anatipestifer*) producing acid but no gas from a number of carbohydrates.

The genus *Pasteurella* was established by Trevisan in 1887. In *Bergey's Manual* (Buchanan and Gibbons, 1974) it is listed in Part 8 under the heading "Gram-negative Facultatively Anaerobic Rods: Genera of Uncertain Affiliation". The type species of the genus has had many names over the years, beginning with *P. cholerae-gallinarum* (Trevisan, 1887). They are listed and referenced in *Index Bergeyana* (Buchanan, Holt, and Lessel, 1966). The names in current use are *P. septica* (Topley and Wilson, 1929), which is used mainly in Britain and Ireland, and *P. multocida*. The latter name, which had been introduced earlier, was proposed by Rosenbusch and Merchant (1939). Because *P. multocida* has had wide acceptance, it has prevailed over the original name *P. cholerae-gallinarum*.

Bacteria identical with what Newson and Cross (1932) named *P. haemolytica* were first referred to by Jones (1921) as *Bacillus bovisepticus* Group 1. It is now agreed that the strains recovered from ewes with mastitis by Marsh (1932) and called *P. mastitidis* were actually *P. haemolytica*. The importance and wide occurrence of this species in animals became evident from the studies that followed these early reports.

Smith (1959, 1961) reported on the occurrence of two different types of *P. haemolytica*, which he called types A and T. These varieties possessed different biochemical, cultural, and serological characteristics, which will be detailed under *Identification*. The differences are such that Frederiksen (1973) thought they might be regarded as two separate species. This contention was supported by the DNA-RNA hybridization studies of Biberstein and Francis (1968).

Pasteurella pneumotropica was first described by Jawetz (1950). Frederiksen (1973) has delineated three biotypes within the species on the basis of differences in several biochemical reactions.

The organism generally referred to as *P. anatipestifer* is included in *Bergey's Manual* (Buchanan and Gibbons, 1974), under the genus *Pasteurella*, as a *species incertae sedis*. It was formerly called *Pfeifferella anatipestifer* (Hendrickson and Hillbert, 1932).

Frederiksen (1973) has pointed out that *P. gallinarum* differs more from *P. haemolytica* than from *P. multocida* and *P. pneumotropica*. Its apparent stability and predilection for chickens and turkeys would seem to support a claim for species status.

McAllister and Carter (1974) gave the name *P. aerogenes* to an organism they isolated from the feces of swine. Subsequently it was pointed out by W. Frederiksen (personal communication, 1975) that what appeared to be the identical organism had been described earlier by Dickinson and Mocquot (1961) and called group X type b—presumably belonging to the genus *Actinobacillus*. Determination of the correct genus for this organism will require further study.

Habitats

As Commensals

The species *Pasteurella multocida*, *P. haemolytica*, and *P. pneumotropica* are found as commensals on the mucous membranes of the upper respiratory and digestive tracts of a considerable proportion of healthy animals. Smith (1955) cultured the noses and tonsils of 111 dogs and recovered *P. multocida* from 54% of the tonsils and 10% of the noses. In the same report, Smith (1955) reviewed the literature to 1955 on carrier rates in cattle, water buffaloes, swine, cats, and rats. The rates ranged from 3.5 to 7% for water buffaloes and cattle to 90% for cats. *P. multocida* can be recovered from the gingiva of a large percentage of cats. Saphir and Carter (1976) obtained cultures of *P. multocida* and *Pasteurella*-like organisms from the gingiva of 11 of 50 normal dogs. As mentioned previously, *P. multocida* has a remarkably broad host range and it would seem likely that its commensal state is similarly widespread.

Ordinarily, *P. multocida* does not occur as a commensal in humans. Occasional carriers have been detected among individuals such as veterinary students and laboratory workers who are exposed to animals and cultures. Although respiratory and "internal" infections occur most frequently in people associated with farm animals, a study of the carrier rate does not appear to have been carried out.

Pasteurella haemolytica has not been recovered from nearly as many species of animals as *P. multocida*. Smith (1955) reviewed the several reports on the occurrence of *P. haemolytica* in healthy cattle and sheep. The carrier rate ranged from 6% for the tonsils of cattle to 40% for the nasal passages of sheep. Twenty-eight of 41 normal lambs and 23 of 36 normal ewes yielded *Pasteurella* organisms from throat cultures (Hamdy, Pounden, and Ferguson, 1959). *P. haemolytica* accounted for the larger portion of these cultures. This species is also recovered from poultry, but there appears to be no information available on its occurrence as a commensal. Just as *P. multocida* is more common in the nasopharynx of cattle, *P. haemolytica* is more prevalent as a commensal in sheep.

In the report in which *P. pneumotropica* was first described, mention was made of the occurrence of the organism in a latent form in the respiratory tract of laboratory mice, rats, and guinea pigs (Jawetz, 1950). The carrier rate ranged from a low of 8.3% to a high of 100% in these species. Brennan, Fritz, and Flynn (1965) examined large numbers of laboratory mice and found that the carrier rate was approximately 95%. Other workers have commented on the frequent presence of *P. pneumotropica* as a commensal in the upper respiratory tract of laboratory animals. Van Dorssen, de Smidt, and Stam (1964)

recovered *P. pneumotropica* from 13 of 100 healthy cats examined.

The questionable pathogen *P. ureae* has thus far only been reported from humans. That it occurs as a rather uncommon commensal of the human upper respiratory tract is indicated by several reports. In one study it was found in 1% of the sputum specimens (Jones and O'Connor, 1962). Henriksen and Jyssum (1960, 1961) recovered it from the nasal passages of three individuals and later from the upper respiratory tracts of seven.

The carrier rate or commensalism of *P. anatipestifer* does not appear to have been formally investigated. Heddleston (1975) states in a discussion of the disease that the organism can be isolated from the upper respiratory passages of birds with no apparent infection.

As Pathogens of Animals

The species *Pasteurella multocida* includes strains of considerable antigenic, cultural, biochemical, and host diversity. On the basis of these differences, Carter (1976) has proposed the division of the species into five biotypes: the mucoid, the hemorrhagic septicemia, the porcine, the canine, and the feline biotypes. Many attempts have been made to classify *P. multocida* serologically, and the accumulated results have revealed considerable serological complexity with a number of different serological or antigenic varieties.

It is now clear that four different serologically specific capsular substances occur. These were originally identified by an indirect hemagglutination procedure (Carter, 1955). On the basis of serological differences in these substances, four types were designated, viz., types A, B, D, and E (Carter, 1955, 1961). The proposed mucoid biotype includes the type A cultures, the hemorrhagic septicemia biotype, the type B and E cultures, and the porcine biotype, the type D cultures.

The occurrence of different somatic or O group antigens was first clearly recognized by Namioka and Murata (1961a,b), using an agglutinin absorption procedure, and later by Heddleston et al. (1972) with an agar gel precipitin test. Namioka and Murata (1961b) designated their somatic varieties by numbers which they combined with the capsular type A, B, D, or E to fully designate a serotype. This scheme appears to have solved the problem of identifying serotypes although the procedures are somewhat laborious. Carter (1972) was able to simplify the identification of O group antigens by decapsulating the mucoid type A strains with hyaluronidase prior to using a simple agglutination procedure. The principal serological varieties and the diseases and hosts with which they are associated are summarized in Table 1.

Table 1. Serological varieties of *Pasteurella multocida* and the animal diseases with which they are commonly associated.

Capsule	O Groups[a]	Diseases and most common hosts
Type A	Many	Fowl cholera. Respiratory and other infections in farm and other animals.
Type B	Two	Hemorrhagic septicemia in cattle, water buffaloes (principally Middle and Far East), and bison.
Type D	Many	Most common in respiratory diseases of swine, but also in infections in many animal species.
Type E	One	Hemorrhagic septicemia in cattle in Central Africa.

[a] Identified thus far.

The diseases caused by *P. multocida* can be divided into two broad categories: those in which this organism is the primary cause, e.g., fowl cholera and "epizootic" hemorrhagic septicemia; and the many infectious processes in which *P. multocida* is a secondary invader, i.e., secondary to a primary "breaching" agent such as a virus or a mycoplasma. The latter category includes many sporadic infections, such as pneumonia, sinusitis, mastitis, infections of the central nervous system, abortion, and also important pneumonic infections of cattle, sheep, and swine, which may affect small to large numbers of individuals. The division of the diseases into these broad categories is probably an oversimplification in that some of the sporadic infections and some of the pneumonic diseases may on occasion be primary.

The strains of *P. multocida* that have been called the canine and feline biotypes are the predominant strains recovered from dogs and cats (Carter, 1976). They have not been defined serologically and are weak pathogenically as compared with the varieties causing disease in farm animals. They are almost always involved as secondary invaders in various primary processes or in bite infections.

Although there has been little study of the occurrence of serotypes in various human infections, what evidence there is indicates that the mucoid (type A) and the porcine biotype (type D) are the most common causes of sporadic respiratory or internal infections such as sinusitis, pneumonia, endocarditis, bronchiectasis, appendicitis, various abscesses, central nervous system infections, and terminal septicemias (Carter, 1967). Type B and E cultures have not yet been recovered from humans. Many of the human infections would appear to be of the secondary type.

The most common source of human infections due to *P. multocida* are from dog, cat, and wild animal bites. Cat scratches may also result in infection. Presumably the pasteurellae associated with the teeth, gingiva, and oral cavity are planted by the teeth or otherwise introduced into the damaged tissue. These infections, although frequently severe, rarely lead to septicemia.

Pasteurella haemolytica cultures have been classified serologically in essentially the same manner as *P. multocida*. There are capsular types, identified by an indirect hemagglutination procedure, and somatic types, identified by an agglutination procedure (Biberstein et al., 1960). The capsular types have been designated with numbers and the somatic types with capital letters. The occurrence of disease and serotypes vis-à-vis the two principal varieties of *P. haemolytica*, called types A and D and referred to earlier, are as follows (Biberstein and Francis, 1968):

Type A	Type D
Capsular types 1,2,5,6,7,8,9,11, and 12	Capsular types 3,4, and 10
Major somatic types A and B	
Predominant cause of pneumonia of cattle and sheep and septicemia of newborn lambs	Predominant in septicemia of lambs 3 months of age or older

Pasteurella haemolytica, like *P. multocida*, has the capacity to produce primary disease, e.g., septicemia in lambs, as well as secondary infections. The former species is very important as one of the causes of pneumonic pasteurellosis of cattle and sheep. Although infections occur in other species, e.g., swine, they are not common. Infections have not been reported in human beings.

Cultures described as *P. haemolytica* are recovered from chickens and turkeys with respiratory disease and salpingitis. These strains are notable for their particulary wide zones of beta hemolysis, and differ also in other respects (Heddleston, 1975) from conventional cultures of *P. haemolytica* from cattle and sheep.

No attempt appears to have been made to determine if different serotypes of *P. pneumotropica* occur. The organism has a low potential for causing disease and when it does, infections most commonly involve the respiratory tract and particularly the lungs of mice and rats. It would seem that when associated with pneumonia it is most likely a secondary invader—as it is in chronic murine pneumonia in the rat. Other pathological conditions in mice from which it has been recovered are conjunctivitis, metritis, urocystitis, abscesses, and dermatitis (Brennan, Fritz, and Flynn, 1965). The same authors recovered cultures from a variety of disease processes in dogs, a rat, a hamster, and a kangaroo rat. Human infections with *P. pneumotropica* have been referred to by Olson and Meadows (1969). They included two dog- and one cat-bite infections, and an association of the organism with rhinitis and tonsilitis in a young boy. Rogers et al. (1973) have described a fatal septicemia in man.

Pasteurella ureae is the only member of the genus whose apparent natural habitat is man. The organism is only weakly pathogenic and usually occurs secondary to trauma or another infectious or pathological process. It is recovered most frequently from humans with chronic bronchitis or bronchiectasis (Jones and O'Connor, 1962). Other diseases with which it has been associated are a fatal case of meningitis (Wang and Haiby, 1966), sinusitis and ozena (Henriksen and Jyssum, 1960; Omland and Henriksen, 1961), fatal endocarditis (Doty, Loomus, and Wolf, 1963), and pneumonia (Starkebaum and Plorde, 1977).

Pasteurella anatipestifer was recently shown to be serologically heterogeneous (Harry, 1964) by an agglutination procedure. It causes an economically important acute or chronic septicemic disease of 1- to 8-week-old domestic ducklings. Infections have also been reported in turkeys, quail, waterfowl, and pheasants (Heddleston, 1975).

The pathogenicity of *P. gallinarum* is low. It is considered a secondary invader when isolated from disease processes in chickens and turkeys associated with the respiratory system (Heddleston, 1975).

Pasteurella aerogenes appears to be a normal inhabitant of the pig's intestinal tract (Dickinson and Mocquot, 1961; McAllister and Carter, 1974). Only two cultures, one from aborted swine fetuses and one from a bite infection in a human attributed to a boar, may have had pathogenic properties (McAllister and Carter, 1974).

Isolation

The *Pasteurella* species are usually recovered in clinical microbiology laboratories where it is customary to plate many clinical specimens directly onto blood agar. Media such as Trypticase soy blood agar base (BBL) and tryptose blood agar base (Difco) containing 5% sheep or bovine blood are widely used in the United States. Comparable blood agar media are available and used worldwide. All of the *Pasteurella* species are facultatively anaerobic, growing well usually within 24–48 h at 35–37°C in air or in air containing 5% carbon dioxide.

The recovery of *Pasteurella* species from clinical specimens does not present any difficulty unless the material has become contaminated with other organisms such as *Proteus* species. This happens occasionally when specimens are taken from animals that have been dead for several hours, particularly during warm weather. It may also occur when the clinical materials are not adequately refrigerated while in transit to the laboratory. When plates are overgrown with *Proteus,* the specimen should be replated on blood agar containing 4% agar. This medium controls spreading and allows the isolation of *Pasteurella.*

If specimens are grossly contaminated, *Pasteurella*

multocida (except some canine strains) can frequently be recovered by mouse or rabbit inoculation. Mice or rabbits are inoculated subcutaneously with graded doses of a 5–10% suspension of tissues in nutrient broth. Swabs can be agitated in nutrient broth to release bacteria and the broth then inoculated subcutaneously into mice or rabbits. The doses of broth can range from 0.1 to 0.5 ml for mice and from 0.5 to 2.0 ml for rabbits. Blood is taken from animals prior to or shortly after death for culture on blood agar. After death and necropsy, organs are plated on blood agar. The other pasteurellae are not sufficiently pathogenic for laboratory animals to be recovered in this manner.

Media such as Trypticase soy agar or tryptose agar without blood or serum may be used to propagate all of the pasteurellae, but blood agar is preferred for initial isolation. Growth in these base media is increased if 0.3% yeast extract (Difco) is added. Media other than blood agar that have been used for primary isolation are:

> *Pasteurella multocida*—dextrose starch agar (Difco)
>
> *P. haemolytica*—dextrose starch agar, MacConkey agar
>
> *P. pneumotropica*—dextrose starch agar
>
> *P. ureae*—dextrose starch agar
>
> *P. anatipestifer*—Trypticase soy agar, chocolate agar; 5–10% CO_2 aids growth
>
> *P. gallinarum*—dextrose starch agar
>
> *P. aerogenes*–MacConkey agar, tryptose agar

Dextrose starch agar was first introduced by Heddleston, Watko, and Rebers (1964) to study the colonial variants of *P. multocida.*

Selective Media for *Pasteurella* Species

Selective media for the isolation of pasteurellae have received little attention, no doubt because, generally speaking, in routine diagnostic work they do not present a problem in recovery. However, under certain circumstances, such as surveys involving specimens containing mixed populations of bacteria, it could be advantageous to employ selective media. The pasteurellae are infrequently recovered from the feces and intestinal mucosa and this may be due to the markedly preponderant growth of enterobacteria obtained on blood agar and other media. Selective media could possibly increase the number of isolations of pasteurellae from these mixed populations.

Morris (1958) has described two selective media for the isolation of *P. multocida* and *P. haemolytica.*

Selective Medium for *Pasteurella multocida* (Morris, 1958)

The base medium was tryptic digest of beef agar (TMA) with 5% (vol/vol) peptic sheep's blood added. The preparation of a digest medium and

Filde's peptic digest of blood are described in detail by Cruickshank (1965). To the base medium the following are added: neomycin, 2.5 μg/ml; potassium tellurite, 2.5 μg/ml; tyrothricin, 10 μg/ml; and actidione, 100 μg/ml.

Selective Medium for *Pasteurella haemolytica* (Morris, 1958)

The base medium was tryptose agar (Difco) (TA) with 5% (vol/vol) peptic sheep's blood added (Cruickshank, 1965). To this base medium were added neomycin, 1.5 μg/ml; novobiocin, 2 μg/ml; and actidione, 100 μg/ml.

It would seem likely that a conventional blood agar medium could be substituted for the TMA and TA used by Morris (1958).

Das (1958) designed a medium for the isolation of *P. multocida* from contaminated specimens that consisted of the following: stock nutrient agar, 1,000 ml; crystal violet (0.1%), 2 ml; cobalt chloride (10%), 6 ml; and esculin, 1.09 ml. The inclusion of an antifungal agent was not deemed necessary. Namioka and Murata (1961a) devised a medium called YPC agar for their serological and antigenic studies of *P. multocida*.

Nutrition and Cultivation

PASTEURELLA MULTOCIDA. Webster and Baudisch (1925) observed that growth of a rabbit culture from small inocula was promoted by blood and some other iron-containing compounds. In support of this observation, the author has noted that it is sometimes difficult to initiate growth from small inocula and reconstituted freeze-dried organisms on media such as dextrose starch agar and tryptose agar. In contrast, growth from these sources can be readily obtained on blood agar.

Berkman, Saunders, and Koser (1940) and Berkman (1942) studied the requirements of cultures for several accessory growth factors. Pantothenic acid and nicotinamide were required for growth, and the latter could not be replaced with nicotinic acid. Several cultures required the ''butyl factor'' which was thought to be biotin. Jordan (1952a) observed that in addition to nicotinamide and pantothenic acid, thiamine was essential for her murine culture.

Jordan (1952a) also found that blood could be replaced by catalase, sodium sulfite, hematin, and other compounds capable of catalyzing the decomposition of hydrogen peroxide. She also noted that blood and the compounds that would replace it were not required for anaerobic growth. Jordan (1952b) concluded that the mechanism of hydrogen peroxide breakdown did not necessarily explain the effect of hematin on growth because the latter did not affect the elaboration of the enzyme in stationary cultures.

Jordan (1952b) observed that in a chemically defined medium with glucose, galactose, and maltose as the only sources of carbon, aerobic growth was inhibited to a variable degree; however, lactic acid and sucrose were adequate sources of carbon for aerobic growth.

A chemically defined medium for *Pasteurella multocida* has been described by Watko (1966). It is a modification of the chemically defined medium designed for Gram-negative organisms by McKenzie et al. (1948). The modified medium supported the growth of nine cultures through 10 transfers. There were no changes in the biochemical characteristics of the organisms and growth could be initiated with two or three organisms. Wessman and Wessman (1970) have described chemically defined media for *P. multocida* and *P. ureae,* and have compared the thiamine requirements of these species with *P. haemolytica.* In summary, their observations were as follows: *P. multocida* was grown in a medium consisting of 17 amino acids, plus inorganic salts, citrate, nicotinamide, pantothenate, thiamine or thiamine monophosphate, adenine, guanine, uracil, and energy source; the medium for *P. ureae* was the same except that it contained an additional amino acid. Six strains of *P. multocida,* 5 of *P. ureae,* and 2 of *P. haemolytica* grew well with normal amounts of thiamine; however, 9 of 11 strains of *P. haemolytica* could satisfy their thiamine requirement with free thiamine only if large amounts of the vitamin were provided. All strains used thiamine monophosphate or thiamine pyrophosphate efficiently. *Pasteurella multocide* grew well with a combination of the thiamine moieties, 4-methyl-5-(β-hydroxyethyl)thiazole and 2-methyl-4-amino-5-hydroxymethyl pyrimidine, or with their phosphate esters; *P. haemolytica* grew only if the phosphates were provided.

Sterne and Hutchinson (1958), employing a continuous culture apparatus, found that the addition of autodigest of pancreas to their basal broth medium provided the greatest yield of a type 1 (Roberts, 1947) hemorrhagic septicemic strain. The propagation of type 1 strains for vaccine production has been described in detail by Bain (1963) in his monograph on hemorrhagic septicemia. He, too, refers to the greatest yields being obtained with the addition of autodigest of pancreas.

PASTEURELLA HAEMOLYTICA. Wessman (1965) described a casein hydrolysate medium for *Pasteurella haemolytica.* The optimum concentration of casein hydrolysate necessary for optimal growth was 1.5–2.0%, depending upon the carbon source supplied. Essential vitamins were calcium pantothenate, nicotinamide, and thiamine. The same investigator (Wessman, 1966) subsequently described a chemically defined medium for *P. haemolytica* that yielded a density of 2 × 10^{10} cells per ml after 16 h of incubation. Some of the differences in growth requirements between *P. multocida* and *P. haemolytica* have been referred to above under the former species

PASTEURELLA UREAE. Chemically defined media for *Pasteurella ureae* and *P. multocida* were referred to above (Wessman and Wessman, 1970). It is of interest that all strains of *P. ureae* required uracil, guanine, and adenine, while only certain cultures of *P. multocida* required them. Wessman and Wessman (1972), in their development of a chemically defined medium for P. ureae, observed that the species grew best in a medium containing 16 amino acids, some of which were essential and others stimulatory. There was a requirement for uracil plus two purines, and the vitamins nicotinamide and pantothenate.

Identification

Cultural and Morphological Characteristics

PASTEURELLA MULTOCIDA. Round, grayish colonies ranging from 1 to 3 mm in diameter are seen after 24 h of incubation. The type A mucoid cultures yield large, moist, mucoid colonies with flowing margins. The type B, D, and E colonies are relatively nonmucoid and smooth. Canine strains produce the smallest colonies. Subculture of fresh cultures often results in the appearance of several different colonial variants (Carter, 1957). All typical cultures of *P. multocida* possess a characteristic musty odor.

PASTEURELLA HAEMOLYTICA. This is the only species to produce distinctive zones of beta hemolysis. Bovine blood is preferred for optimum hemolysis. Good growth is obtained after 24 h of incubation. Colonies are round, grayish, and in size approximate the colonies of the nonmucoid cultures of *P. multocida*. Lactose-fermenting strains of *P. haemolytica* yield small, round, pink to red colonies on MacConkey agar.

PASTEURELLA PNEUMOTROPICA. Colonies are indistinguishable from those of the nonmucoid cultures of *P. multocida*.

PASTEURELLA UREAE. Round, moderately mucoid colonies, 1 to 2 mm in diameter, appear in 24 h. A greenish discoloration and partial hemolysis is observed around colonies on blood agar in 48 h.

PASTEURELLA ANATIPESTIFER. In 24 h, colonies are round, smooth, and translucent, resembling those of nonmucoid cultures of *P. multocida*.

PASTEURELLA AEROGENES. On blood agar after 24 h colonies are circular, smooth, entire, convex, and translucent, with a diameter of approximately 1 mm. After 24 h on MacConkey agar, colonies are indistinguishable from those of *Salmonella*. On further incubation, colonies develop a faint pinkish color.

None of the species has distinctive cellular morphology. Cultures are rather pleomorphic, but freshly isolated strains consist mainly of small coccobacillary forms. Generally speaking, the coccoid rods range in size from 0.25 μm to 0.4 μm by 0.6 μm to 2.5 μm. Some bacillary forms of varying length are seen in early cultures and these forms increase as strains are subcultured.

Differential Characteristics

All of the species grow well in commonly used broths and in conventional media employed for differential biochemical tests. They grow best at 35–37°C, are nonmotile, and produce oxidase. Most cultures of the various species produce catalase but only *P. anatipestifer* produces gelatinase. The latter is also distinctive in not reducing nitrate. All produce some hydrogen sulfide when lead acetate strips are employed. With these characteristics and those listed in Table 2, the various species can be differentiated.

As mentioned previously, *P. multocida* is a heterogeneous species, and several different varieties or

Table 2. Differential characteristics of *Pasteurella* species.[a]

	MacConkey agar	Indole	Urease	Fermentation			
				Glucose	Lactose	Sucrose	Mannitol
P. multocida	−*	+	−	A	(−)	A	(A)
P. haemolytica	+	−	−*	A	(A)	A	A
P. pneumotropica	−*	+	+	A	(A)	A	−
P. gallinarum	−	−	−	A	−	A	−
P. anatipestifer	−	−	+	−	−	−	−
P. ureae	−	−	+	A	−	A	A
P. aerogenes	+	−	+	AG	−	AG	−

[a] Asterisk indicates some exceptions; parentheses indicate most strains.

Table 3. Differential features of Types A and T of *Pasteurella haemolytica*.[a]

Type A	Type T
1. Acid produced from arabinose (10 days)	1. No acid produced from arabinose (10 days)
2. No acid produced from trehalose (10 days)	2. Acid produced from trehalose (10 days)
3. Highly susceptible to penicillin *in vitro*	3. Relatively resistant to penicillin *in vitro*
4. Cultures die off rapidly	4. Cultures survive longer
5. Small grey colonies	5. Larger colonies with tan-colored centers
6. Capsular serotypes 1,2,5,6,7,8,9,11, and 12	6. Capsular types 3, 4 and 10
7. Major somatic serotypes A and B	7. Major somatic serotypes C and D
8. Predominant in pneumonia of cattle and sheep, and septicemia of newborn lambs	8. Predominant in septicemia of lambs 3 months of age or older
9. Common in nasopharynx of normal cattle and sheep	9. Uncommon in nasopharynx of normal cattle and sheep

[a] Biberstein and Francis, 1968.

biotypes can be identified (Carter, 1976). Mammalian cultures of *P. haemolytica* were shown by Smith (1959, 1961) to belong to one of two categories referred to as types A and T. Biberstein and Francis (1968) have tabulated the characteristics that distinguish these types (see Table 3). Frederiksen (1973) divided cultures of *P. pneumotropica* into three biotypes on the basis of different reactions in several biochemical tests.

Results of decarboxylation tests in Moeller's medium are not available for all the *Pasteurella* species; however, differential data of Weaver, Tatum, and Hollis (1972) provide the following information:

	P. multocida	P. haemolytica	P. pneumotropica
Lysine decarboxylase	−	−	+ or −
Arginine dihydrolase	−	−	−
Ornithine decarboxylase	+[a]	−	+

[a] Some exceptions.

A recurring problem in the clinical microbiology laboratory is the occurrence of *Pasteurella*-like organisms that do not fit precise species descriptions because of one or more aberrant reactions. These cultures are recovered from various animal species. Saphir and Carter (1976) recovered several of them from the gingiva of dogs. There have also been reports of *Pasteurella* organisms (other than *P. aerogenes*) producing gas from carbohydrates (Rogers and Elder, 1967). Weaver, Tatum, and Hollis (1972) have listed the differential characteristics of 44 cultures they called *Pasteurella* n. sp. I or "gas" (multocida?). These strains, many of which were

recovered from dog-bite infections in man, are possibly a biotype of *P. pneumotropica*.

Literature Cited

Bain, R. V. S. 1963. Hemorrhagic septicemia. Food and agricultural studies no. 64. Rome: Food Agricultural Organization of the United Nations.

Berkman, S. 1942. Accessory growth factor requirements of the members of the Genus *Pasteurella*. Journal of Infectious Diseases **71**:201–211.

Berkman, S., Saunders, F., Koser, S. A. 1940. Accessory growth factor requirements of some members of the *Pasteurella* group. Proceedings of the Society for Experimental Biology and Medicine **44**:68–70.

Biberstein, E. L., Francis, C. K. 1968. Nucleic acid homologies between the A and T types of *Pasteurella haemolytica*. Journal of Medical Microbiology **1**:105–108.

Biberstein, E. L., Gills, M., Knight, H. 1960. Serological types of *Pasteurella hemolytica*. Cornell Veterinarian **50**:283–300.

Brennan, P. C., Fritz, T. E., Flynn, R. J. 1965. *Pasteurella pneumotropica*: Cultural and biochemical characteristics and its association with disease in laboratory animals. Laboratory Animal Care **15**:307–312.

Buchanan, R. E., Gibbons, N. E. (eds.). 1974. Bergey's manual of determinative bacteriology, 8th ed. Baltimore: Williams & Wilkins.

Buchanan, R. E., Holt, J. G., Lessel, E. F., Jr. (eds.). 1966. Index Bergeyana. Baltimore: Williams & Wilkins.

Carter, G. R. 1955. Studies on *Pasteurella multocida*. I. A hemagglutination test for the identification of serological types. American Journal of Veterinary Research **16**:481–484.

Carter, G. R. 1957. Studies on *Pasteurella multocida*. II. Identification of antigenic and colonial characteristics. American Journal of Veterinary Research **18**:210–213.

Carter, G. R. 1961. A new serological type of *Pasteurella multocida* from Central Africa. Veterinary Record **73**:1052.

Carter, G. R. 1967. Pasteurellosis: *Pasteurella multocida* and *Pasteurella hemolytica*. Advances in Veterinary Science **11**:321–379.

Carter, G. R. 1972. Simplified identification of somatic varieties of *Pasteurella multocida* causing fowl cholera. Avian Diseases **16**:1109–1114.

Carter, G. R. 1976. A proposal for five biotypes of *Pasteurella multocida*. 19th Annual Proceedings of the American Association of Veterinary Diagnosticians: 189–196.

Cruickshank, R. (ed.). 1965. Medical microbiology, 11th ed. Baltimore:Williams & Wilkins.

Das, M. S. 1958. Studies on *Pasteurella septica* (*Pasteurella multocida*). Observations on some biophysical characters. Journal of Comparative Pathology and Therapeutics **68**:288–294.

Dickinson, A. B., Mocquot, G. 1961. Studies on the bacterial flora of the alimentary tract of pigs. I. Enterobacteriaceae and other Gram-negative bacteria. Journal of Applied Bacteriology **24**:252–284.

Doty, G. L., Loomus, G. N., Wolf, P. L. 1963. *Pasteurella* endocarditis. New England Journal of Medicine **268**:830–832.

Frederiksen, W. 1973. *Yersinia, Pasteurella* and *Francisella*, pp. 170–176. In: Winblad, S. (ed.), Contributions to microbiology and immunology. Basel: Karger.

Hamdy, A. H., Pounden, W. D., Ferguson, L. C. 1959. Microbial agents associated with pneumonia in slaughtered lambs. American Journal of Veterinary Research **20**:87–90.

Harry, E. G. 1964. *Pasteurella* (*Pfeifferella*) *anatipestifer* serotypes isolated from cases of anatipestifer septicemia in ducks. Veterinary Record **84**:649–657.

Heddleston, K. L. 1975. Pasteurellosis, pp. 38–51. In: Hitchner, S. B., Domermuth, C. H., Purchase, H. G., Williams, J. E. (eds.), Isolation of avian pathogens. Ithaca, New York: Arnold Printing Co.

Heddleston, K. L., Goodson, T., Leibovitz, J., Angstrom, C. I. 1972. Serological and biochemical characteristics of *Pasteurella multocida* from free-flying birds and poultry. Avian Diseases **16**:729–734.

Heddleston, K. L., Watko, L. P., Rebers, P. A. 1964. Dissociation of a fowl cholera strain of *Pasteurella multocida*. Avian Diseases **8**:649–657.

Hendrickson, J. M., Hillbert, K. F. 1932. A new and serious septicemic disease of young ducks with a description of causative organism, *Pfeifferella anatipestifer*. Cornell Veterinarian **22**:239–252.

Henriksen, S. D., Jyssum, K. 1960. A new variety of *Pasteurella hemolytica* from the human respiratory tract. Acta Pathologica et Microbiologica Scandinavica **50**:443.

Henriksen, S. D., Jyssum, K. 1961. A study of some pasteurella strains from the human respiratory tract. Acta Pathologica et Microbiologica Scandinavica **51**:354–368.

Hommez, J., Devriese, L. A. 1976. *Pasteurella aerogenes* isolations from swine. Zentralblatt für Veterinär medizin, Reihe B **23**:265–268.

Jawetz, E. 1950. A pneumotropic pasteurella of laboratory animals. 1. Bacteriological and serological characteristics of the organism. Journal of Infectious Diseases **86**:172–183.

Jones, D. M., O'Connor, P. M. 1962. *Pasteurella haemolytica* var. *ureae* from human sputum. Journal of Clinical Pathology **15**:247–248.

Jones, F. S. 1921. A study of *Bacillus bovisepticus*. Journal of Experimental Medicine **34**:561–577.

Jordan, R. M. M. 1952a. The nutrition of *Pasteurella septica*. I. The action of haematin. British Journal of Experimental Pathology **33**:27–35.

Jordan, R. M. M. 1952b. The nutrition of *Pasteurella septica*. II. The formation of hydrogen peroxide in a chemically-defined medium. British Journal of Experimental Pathology **33**:36–45.

McAllister, H. A., Carter, G. R. 1974. An aerogenic pasteurella-like organism recovered from swine. American Journal of Veterinary Research **35**:917–922.

McKenzie, D., Stradler, M., Booth, J., Oleson, J. J., Subbarow, Y. 1948. The use of synthetic medium as an in vitro test of possible chemotherapeutic agents against Gram-negative bacteria. Journal of Immunology **60**:283–294.

Marsh, H. 1932. Mastitis in ewes, caused by an infection with a pasteurella. Journal of the American Veterinary Medical Association **81**:376–382.

Morris, E. J. 1958. Selective media for some *Pasteurella* species. Journal of General Microbiology **19**:305–311.

Namioka, S., Murata, M. 1961a. Serological studies on *Pasteurella multocida*. I. A simplified method for capsule typing of the organism. Cornell Veterinarian **51**:498–507.

Namioka, S., Murata, M. 1961b. Serological studies of *Pasteurella multocida*. II. Characteristics of somatic (O) antigen of the organism. Cornell Veterinarian **51**:507–521.

Newsom, I. E., Cross, F. 1932. Some bipolar organisms found in pneumonia in sheep. Journal of the American Veterinary Medical Association **80**:711–719.

Olson, J. R., Meadows, T. R. 1969. *Pasteurella pneumotropica* infection resulting from a cat bite. Journal of Clinical Pathology **51**:709–710.

Omland, T., Henriksen, S. D. 1961. Two new strains of *Pasteurella hemolytica* var. *ureae* isolated from the respiratory tract. Acta Pathologica et Microbiologica Scandinavica **53**:117–120.

Roberts, R. S. 1947. An immunological study of *Pasteurella septica*. Journal of Comparative Pathology and Therapeutics **57**:261–278.

Rogers, B. T., Anderson, J. C., Palmer, C. A., Henderson, W. G. 1973. Septicaemia due to *Pasteurella pneumotropica*. Journal of Clinical Pathology **26**:396–398.

Rogers, R. J., Elder, J. K. 1967. Purulent leptomeningititis in a dog associated with an aerogenic *Pasteurella multocida*. Australian Veterinary Journal **43**:81–82.

Rosenbusch, C. T., Merchant, I. A. 1939. A study of the hemorrhagic septicemia *Pasteurellae*. Journal of Bacteriology **37**:69–89.

Saphir, D. A., Carter, G. R. 1976. Gingival flora of the dog with special reference to bacteria associated with bites. Journal of Clinical Microbiology **3**:344–349.

Smith, G. R. 1959. Isolation of two types of *Pasteurella haemolytica* from sheep. Nature **183**:1132–1133.

Smith, G. R. 1961. The characteristics of two types of *Pasteurella haemolytica* associated with different pathological conditions in sheep. Journal of Comparative Pathology and Therapeutics **81**:431–440.

Smith, J. E. 1955. Studies on *Pasteurella septica*. I. The occurrence in the nose and tonsils of dogs. Journal of Comparative Pathology and Therapeutics **65**:239–245.

Starkebaum, G. A., Plorde, J. J. 1977. Pasteurella pneumonia: Report of a case and review of the literature. Journal of Clinical Microbiology **5**:332–335.

Sterne, M., Hutchinson, I. 1958. The production of bovine haemorrhagic septicaemia vaccine by continuous culture. British Veterinary Journal **114**:176–179.

Topley, W. W. C., Wilson, G. S. 1929. The principles of bacteriology and immunity, 1st ed. London: Edward Arnold and Co.

Trevisan, V. 1887. Sul micrococco della rabia e sulla possibilità di riconoscere durante il periodo d'incubazione, dall'esame del sangue della persona morsicata, si ha contratta l'infezione rabbica. Rend 1st Lombardo Ser II **20**:88–105.

Van Dorssen, C. A., de Smidt, A. C., Stam, J. W. E. 1964. Over het voorkomen van *Pasteurella pneumotropica* bij katten. Tijdschrift voor Diergeneeskunde **89**:674–682.

Wang, W. L. L., Haiby, G. 1966. Meningitis caused by *Pasteurella ureae*. Journal of Clinical Pathology **45**:562–565.

Watko, L. P. 1966. A chemically defined medium for growth of *Pasteurella multocida*. Canadian Journal of Microbiology **12**:933–937.

Weaver, R. E., Tatum, H. W., Hollis, D. G. 1972. The identification of unusual Gram-negative bacteria, table 3, preliminary revision. King, E. O. (ed.). Atlanta:Center for Disease Control.

Webster, L. T., Baudisch, O. 1925. Biology of *Bacterium Lepisepticum*. II. The structure of some iron compounds which influence the growth of certain bacteria of the hemophilic, anaerobic, and hemorrhagic septicemia groups. Journal of Experimental Medicine **42:**473– 482.

Wessman, G. E. 1965. Cultivation of *Pasteurella haemolytica* in a casein hydrolysate medium. Applied Microbiology **13:**426– 431.

Wessman, G. E. 1966. Cultivation of *Pasteurella haemolytica* in a chemically defined medium. Applied Microbiology **14:**597– 602.

Wessman, G. E., Wessman, G. W. 1970. Chemically defined media for *Pasteurella multocida* and *Pasteurella ureae,* and a comparison of their thiamine requirements with those of *Pasteurella ureae* in a chemically defined medium. Canadian **16:**751– 757.

Wessman, G. E., Wessman, G. 1972. Requirements for growth of *Pasteurella ureae* in a chemically defined medium. Canadian Journal of Microbiology **18:**107– 109.

The Genus *Actinobacillus*

JAMES E. PHILLIPS

Organisms of the genus *Actinobacillus* were first described by Lignières and Spitz (1902) in actinomycotic lesions in cattle in Argentina. The authors referred to this organism as ''l'actinobacille'' from its morphology and the association with the pathological lesion. In 1910, Brumpt proposed the name *Actinobacillus lignieresii*.

This genus has been used as a repository for ''species for which no obvious home can be found'' (Cowan, 1974), but it is intended to deal in detail here only with those two species included in the eighth edition of *Bergey's Manual of Determinative Bacteriology* (Phillips, 1974), viz., *Actinobacillus lignieresii* and *Actinobacillus equuli,* together with *Actinobacillus suis* (van Dorssen and Jaartsveld, 1962) which has now been more fully characterized and can be regarded as a distinct species (Frederiksen, 1973; Kim, 1976; Mair et al., 1974). Various organisms have been described as actinobacilli, but here they have not been included in (nor excluded from) the genus. *A. actinomycetemcomitans* (Klinger, 1912) is found in association with actinomycetes in lesions of actinomycosis in man. *A. actinoides* (Smith, 1918), a cause of pneumonia in calves, bears great similarities to *Streptobacillus moniliformis* and should probably be placed in the same group. Arseculeratne (1961, 1962) described *A. capsulatus* from a natural infection in laboratory rabbits. Ross et al. (1972) isolated, from sows, actinobacilli that were related to, but not identical with, the generally accepted species of the genus. Hacking and Sileo (1977) have recovered a hemolytic actinobacillus from waterfowl and have shown it to have close similarities with *A. lignieresii* and *Pasteurella ureae.*

The close relationship of organisms of the genus *Actinobacillus* and those of the *Pasteurella* group may give rise to difficulties in identification and has certainly presented taxonomic problems. Mráz (1969), in a comparative study of *A. lignieresii* and *P. haemolytica,* has drawn attention to the close similarity between these two species (similarity index = 95% on 60 characters) and has proposed that *P. haemolytica* should be renamed *Actinobacillus haemolyticus.* Mráz, Vladík, and Boháček

(1976) have also examined strains of an organism, originally described by Kohlert (1968) under the name *Pasteurella salpingitidis,* and have proposed that this be renamed *Actinobacillus salpingitidis.*

There are some species that have previously been included in the genus but which should now be excluded. *A. mallei* has been removed to the genus *Pseudomonas.* The organism described by Baynes and Simmons (1960) from cases of ovine epididymitis, *A. seminis,* should not be considered in the genus *Actinobacillus.*

Habitats

Members of the genus *Actinobacillus* are encountered most often as pathogens causing a variety of conditions in animals, especially domesticated stock. They also occur as commensal organisms in the alimentary, respiratory, and genital tracts of normal animals. They are to be regarded as opportunistic pathogens, there usually being some factor present that assists entry of the organism into the susceptible tissues. The diseases caused by the actinobacilli are usually sporadic in nature, but where the trigger factor is common to a group of animals, several individuals in that group may be affected.

Actinobacillus lignieresii

The classical disease in cattle caused by *Actinobacillus lignieresii* is wooden tongue, a chronic granulomatous lesion affecting the tongue and other soft tissues of the head and upper alimentary tract together with the associated lymph nodes (Bosworth, 1923; Davies and Torrance, 1930; Thompson, 1933; Till and Palmer, 1960). Lesions have been described in the lungs and liver (Davies and Torrance, 1930), in the pleura (Misdorp, 1963), and in the heart (Thornton, 1976). Subcutaneous lesions affecting the skin of various areas of the body have been recorded (Hebeler, Linton, and Osborne, 1961; Mawditt and Greenham, 1962).

In sheep, the lesions are more usually suppurative in character, often with involvement of the skin or lungs (pyobacillosis). The infecting organism was first described by Christiansen (1917) as *Bacterium purifaciens,* but the identity of this organism with *A. lignieresii* was established by Tunnicliff (1941) and Taylor (1944). The organism has been reported causing epididymo-orchitis (Laws and Elder, 1969a) and mastitis (Laws and Elder, 1969b) in sheep.

Infection of animals other than cattle and sheep with *A. lignieresii* is not common, but lingual lesions have been reported in the dog by Fletcher, Linton, and Osborne (1956) and Kemenes and Markói (1959), epidural abscess in a horse by Chladeck and Ruth (1976), and in ducks it has been recovered from cases of salpingitis (Bisgaard, 1975). The presence of *A. lignieresii* associated with disease in man has been reported (Pathak and Ristic, 1962; Thompson and Willius, 1932).

A. lignieresii has been demonstrated as a commensal organism in the mouth of healthy cattle (Phillips, 1964), in normal bovine rumens (Phillips, 1961), and in rumens of healthy sheep (Phillips, 1966). It is likely that such organisms constitute the source of infection, with entry occurring through minor wounds produced in the epithelial surfaces of the upper alimentary tract. The presence of agents that may cause mechanical damage of the epithelium has been found associated with multiple cases of actinobacillosis in groups of cattle (Campbell et al., 1975; Gerring, 1947; Hebeler, Linton, and Osborne, 1961; Nakazawa et al., 1977) and in sheep (Davis and Stiles, 1939; Hayston, 1948; Thomas, 1931).

A. lignieresii is worldwide in distribution and clinical infections have been reported from all continents.

Actinobacillus equuli

Actinobacillus equuli is a pathogen of horses found in association with various clinical conditions, especially in young animals. The most usual syndrome in foals within the first few days of life is an acute septicemic infection ("sleepy foal disease") that may become chronic ("joint ill") with lesions of purulent nephritis and purulent arthritis (Dimock, Edwards, and Bruner, 1947). In adult horses it may be found associated with septicemia (Magnusson, 1919; Mráz, Zakopal, and Matovšek, 1968; Zakopal and Nesvadba, 1968), purulent nephritis (Meyer, 1910), endocarditis (Innes, Berger, and Francis, 1950; Svenkerud and Iversen, 1949; Vallée et al., 1974), meningitis (Weidlich, 1955), chronic alveolar emphysema (Larsen, 1974), and abortion (Webb, Cockram, and Pryde, 1976).

The organism is also pathogenic for swine. Piglets are most often affected, but older animals may show lesions. The lesions that may be found include abortion (Werdin et al., 1976), arthritis (Pedersen, 1977), endocarditis (Ashford and Shirlaw, 1962; Jones and Simmons, 1971), meningoencephalitis (Terpstra and Akkermans, 1955), metritis (Edwards and Taylor, 1941), and septicemia (Magnusson, 1931; Windsor, 1973).

A. equuli has been recognized in horses as part of the normal bacterial flora of the intestinal tract (Cottew and Francis, 1954; Laudien, 1923) and has been isolated from the tonsillar region (Dimock, Edwards, and Bruner, 1947; Jarmai, 1929), and from the tracheal mucus (Kim, Phillips, and Atherton, 1976) of healthy horses. The occurrence of *A. equuli* in normal swine, however, has not been documented.

A. equuli has not been widely recognized as a pathogen in other animal species, but Moon, Barnes, and Higbee (1969) isolated it from monkeys, and du Plessis, Cameron, and Langen (1967) and Osbaldiston and Walker (1972) described outbreaks of enteritis of calves in which *A. equuli* predominated in the intestinal flora. Vallée (1959) reported *A. equuli* causing disease in rabbits and Vallée et al. (1960) recorded its association with skin lesions in a dog.

The opportunistic character of *A. equuli* as a pathogen is recognized, especially in the case of young animals, where it is often the weak individual that is liable to succumb to infection.

Actinobacillus suis

The porcine actinobacillus, *Actinobacillus suis,* is found as a pathogen of all ages of swine with an acute septicemic form (Mair et al., 1974; Zimmermann, 1964), sometimes with pneumonia (van Dorssen and Jaartsveld, 1962) or nephritis (Bouley, 1966), or a more chronic form with arthritis (van Dorssen and Jaartsveld, 1962). It has also been recognized as a pathogen of horses (Veterinary Investigation Service, 1975). In a number of cases of disease in horses reported to be due to *A. equuli,* the infecting organism has undoubtedly been *A. suis* (Bell, 1973; Carter, Marshall, and Jolly, 1971; Cottew and Francis, 1954) and it is likely that other unnamed actinobacilli from horses (Larsen, 1974) are also this species.

The presence of *A. suis* in normal swine has not been reported, but the hemolytic actinobacilli recovered from irradiated swine by Wetmore et al. (1963) may have been commensal strains of *A. suis* that had assumed the pathogenic role in the stressed hosts. Cutlip, Amtower, and Zinober (1972) isolated actinobacilli from the tonsils of normal pigs in a herd in which a case of septic embolic actinobacillosis had occurred.

The presence of *A. suis* in normal horses has been reported (Kim, Phillips, and Atherton, 1976).

Isolation

Representatives of the genus *Actinobacillus* grow readily on the enriched media usually employed for the isolation of pathogens from animal tissues. In most fresh tissues, they will usually be found as the sole or predominating organisms.

Isolation of *Actinobacillus lignieresii* from Lesions in the Tongue and Associated Lymph Nodes of Cattle
Medium—Hartley's digest broth:

Ox heart (minced)	3,000 g
Water	5,000 ml
Anhydrous sodium carbonate (0.8% solution)	5,000 ml
Pancreatin	50 g
Concentrated hydrochloric acid	80 ml

Mix the minced meat and water and heat to 80°C. Add the sodium carbonate and cool to 45°C. Add pancreatin and incubate at 45°C for 4 h, stirring frequently. When digestion is complete, add the hydrochloric acid and steam at 100°C for 30 min. Cool to room temperature and add N caustic soda to bring the pH to 8.0. Boil for 25 min to precipitate phosphates and filter while hot. Allow to cool and adjust to pH 7.5. Sterilize by autoclaving at 121°C for 15 min.

Prepare horse blood agar by adding 1% agar (Oxoid no. 1) to the digest broth, autoclave to sterilize, cool to 50°C and add 5% oxalated horse blood. Distribute into sterile Petri plates.
Preparation of tissue:
Incise to the center of the lesion using aseptic technique and withdraw samples of the scanty volumes of pus with an inoculating loop or capillary pipette.
Inoculation of medium:
Spread the inoculum so as to give well-isolated colonies at one side of the plate.
Incubation:
Incubate at 37°C for 18–24 h. The addition of 5–10% carbon dioxide to the atmosphere usually improves the growth for primary isolation, but most strains will grow adequately without this addition.

The ability of most strains of actinobacilli to grow on MacConkey medium (Mráz, 1975) might be of use for isolation, especially if cultures on this medium are set up in parallel with those on blood or serum agar. When working with *A. suis,* the use of sheep blood agar has the advantage that colonies of this species, being hemolytic, are more easily selected.
The isolation of actinobacilli from mixed popula-

tions (e.g., from the surface of the tongue and from the contents and epithelial and mucosal surfaces of the alimentary tract) may prove difficult because of the overgrowth of the actinobacillus colonies by other more strongly growing bacteria. In these situations, selective media may be of advantage. Two selective media have been described, that of Till and Palmer (1960) for the isolation of *A. lignieresii* from the mouths of normal cattle and that of Phillips (1961, 1964) used to recover *A. lignieresii* from the rumen and mouths of normal cattle.

Selective Medium for *Actinobacillus lignieresii* (Till and Palmer, 1960)
Prepare Filde's peptic digest of blood according to the method of Cruickshank (1965) as follows:

"Sodium chloride, 0.85% aqueous solution	150 ml
Hydrochloric acid	6 ml
Defibrinated sheep blood	50 ml
Pepsin	1 g
Sodium hydroxide, 20% aqueous solution	about 12 ml
Chloroform	0.5 ml

Mix the saline, acid, blood and pepsin in a stoppered bottle and heat at 55°C for 2–24 hours. Add sodium hydroxide until a sample of the mixture diluted with water gives a permanganate red colour with cresol red indicator. Add pure hydrochloric acid drop by drop until a sample of the mixture shows almost no change of colour with cresol red but a definite red tint with phenol red. It is important to avoid excess of acid. Add chloroform and shake the mixture vigorously."

The complete medium consists of:

Hartley's digest broth	900 ml
Agar	10 g
*Filde's peptic digest	100 ml
*Oleandomycin phosphate	20 mg
*Neomycin sulfate	1.5 mg

The medium is dispensed in Petri plates.
Swabs taken from the surface of the bovine tongue are inoculated onto the medium and the inoculum is spread. Incubate the plates overnight at 37°C and then select colonies of the correct morphology for further examination.

Although Till and Palmer (1960) did not succeed in isolating *A. lignieresii* from the surface of the tongues of normal cattle, they showed that their medium would support the growth of known strains

---

*These components are added after the basic nutrient agar has been sterilized and cooled to 50°C.

of *A. lignieresii* while giving a high degree of inhibition of contaminating organisms.

Selective Medium for *Actinobacillus lignieresii* (Phillips, 1961, 1964)

Phillips' medium incorporates an antifungal agent and is prepared as follows:

Prepare a stock solution of oleandomycin phosphate (5 mg/ml) in sterile distilled water, distribute in 0.2-ml amounts and store frozen at −20°C. After thawing, prepare a working solution by adding 9.8 ml sterile distilled water. Prepare a stock suspension of nystatin (200 units/ml) in sterile distilled water, distribute in 1-ml amounts, and store frozen at −20°C. Both these stocks will store satisfactorily for at least 2 months. The final medium is prepared by adding the antibiotics to horse blood agar, to give final concentrations of oleandomycin (1 μg/ml) and nystatin (200 units/ml) as follows:

Hartley's digest agar	93 ml
Horse blood (oxalated)	5 ml
Oleandomycin phosphate working solution	1 ml
Nystatin stock suspension	1 ml

Melt the agar base and cool to 50°C. Add the blood and antibiotics and pour into sterile Petri plates.

Heavy suspensions of actinobacilli may be prepared by growing the organisms on Hartley's digest agar in Roux flasks at 37°C for 24 h and washing off the growth into sterile saline or other suitable medium using sterile glass beads to dislodge the growth. An alternative method is to grow the organisms in a flask of Hartley's digest broth in a shallow layer and incubate at 37°C for 24 h in a shaking bath (120 strokes/min).

Actinobacilli rapidly lose their viability when stored on solid media (e.g., blood agar) for more than 4 days. Cultures suspended in sterile rabbit serum or sterile 20% peptone solution and dried from the frozen state will remain viable in sealed ampules for many years. An alternative method of keeping stock cultures is as heavy suspensions in rabbit serum or 20% peptone solution stored at −70°C in a low-temperature cabinet.

Identification

All members of the genus *Actinobacillus* are markedly pleomorphic, showing short bacillary or coccobacillary forms interspersed with coccal elements lying in close association with the rods and giving the "Morse code" form described in *A. lignieresii* (Phillips, 1960). On media containing glucose or

Table 1. Characteristics of the genus *Actinobacillus*.

Positive:	Growth on MacConkey medium
	Reduction of nitrates
	Production of β-galactosidase and urease
	Fermentation (acid only) of glucose, levulose, mannose, galactose, xylose, maltose, sucrose, and dextrin
Negative:	Production of indole
	Fermentation of inulin, inositol, dulcitol, and adonitol

maltose, longer almost filamentous forms may be seen, and often chains of short bacillary elements are present. In the case of *A. equuli* and *A. suis,* extracellular material staining faintly pink with Gram's stain is often seen.

Colonies of all three species are sticky, especially on primary isolation, but those of *A. lignieresii* lose this characteristic on subculture. *A. suis* and *A. equuli* are very sticky (*suis* less so than *equuli*) and colonies are firmly adherent to the underlying medium. In broth cultures, the sticky nature of these two species is also apparent.

All members of the genus have the common characteristics shown in Table 1. These characteristics would not exclude some members of the *Pasteurella* group and, indeed, differentiation between the two genera may be difficult, with isolates not falling clearly into the recognized species. Close similarities of the two genera have been pointed out (Hacking and Sileo, 1977; Mráz, 1969, 1975; Smith, 1974).

The differential characteristics of the three species of *Actinobacillus* are set out in Table 2. There is evidence, however, that a degree of overlapping between species occurs, especially between *A. equuli* and *A. suis*. The occurrence of strains isolated from

Table 2. Differential characteristics of the three *Actinobacillus* species.

	A. lignieresii	*A. equuli*	*A. suis*
Fermentation of:			
Lactose	+ slow	+	+
Mannitol	+	+	−
Salicin	−	−	+
Cellobiose	−	−	+
Melibiose	−	+	+
Trehalose	−	+	+
Hydrolysis of:			
Hippurate	−	+	+
Esculin	−	−	+
Hemolysis on sheep blood agar	−	−	+
Pigment production	−	−	+

horses having characters similar to those of *A. suis* has been reported (Kim, Phillips, and Atherton, 1976), differing from *A. equuli* mainly in their hydrolysis of esculin, fermentation of cellobiose and salicin and nonfermentation of mannitol, and in being hemolytic on sheep blood agar. However, Mráz, Zakopal, and Matovšek (1968) drew attention to the existence of strains that, while being strongly hemolytic on sheep blood agar, are similar in all other respects to *A. equuli*, and this possibility should be borne in mind in identification. The equine strains examined by Kim, Phillips, and Atherton (1976) have, on further investigation (Kim, 1976), shown one difference from those of porcine origin: unlike the porcine strains, they do not produce pigment. The pigment, creamy-yellow in color, can best be demonstrated by washing centrifuged broth cultures and observing the color of the deposit.

Gelatinase activity was considered by Phillips (1974) to be a characteristic feature of *A. equuli*, based upon the examination of a small number of strains from the National Collection of Type Cultures, using the gelatin agar method of Frazier (1926). Examination of a wider selection of strains (Kim, 1976) has shown that many are gelatinase negative, although positive strains are encountered from time to time (Frederiksen, 1973; Meyer, 1910; Vallée et al., 1974).

The three species can be separated into a number of antigenic types on the basis of their heat-stable somatic antigens. In *A. lignieresii*, at least 6 such types have been described (Phillips, 1967), and *A. equuli* presents at least 28 types (Kim, 1976). *A. suis* has not been investigated so thoroughly, but there is evidence of cross relationships occurring between this species and *A. equuli* (Kim, 1976; Vallée et al., 1974). Antigenic cross relationships have also been demonstrated between *A. lignieresii* and *A. suis* (Ross et al., 1972; Vallée et al., 1974) and between *A. lignieresii* and *A. equuli* (Wetmore et al., 1963).

Some minor antigenic cross relationships between members of the genus *Actinobacillus* and those of the *Pasteurella* group, especially *P. haemolytica*, have been noted, but in general there are no strong antigenic cross-reactions between these two genera (Mráz, 1977; Ross et al., 1972).

Determination of DNA base ratios does not give useful differentiation between species, the G+C values lying within the range 40.0–42.6 mol% for *A. lignieresii* and *A. equuli* (Boháček and Mráz, 1967). There is also satisfactory generic differentiation with *Pasteurella* in which the G+C values lie in the range 36.5–43 mol% (Smith, 1974).

The use of electrophoresis in polyacrylamide to demonstrate distinct differences in the cell proteins of the three species of *Actinobacillus* has been reported (Ross et al., 1972; Vallée et al., 1974) and may be of value in differentiation.

Literature Cited

Arseculeratne, S. N. 1961. A preliminary report on actinobacillosis as a natural infection in laboratory rabbits. Ceylon Veterinary Journal **9**:5–8.

Arseculeratne, S. N. 1962. Actinobacillosis in joints of rabbits. Journal of Comparative Pathology **72**:33–39.

Ashford, W. A., Shirlaw, J. F. 1962. A case of verrucose endocarditis in a piglet caused by an organism of the *Actinobacillus* genus. Veterinary Record **74**:1417–1418.

Baynes, I. D., Simmons, G. C. 1960. Ovine epididymitis caused by *Actinobacillus seminis* n. sp. Australian Veterinary Journal **36**:454–459.

Bell, J. C. 1973. *Actinobacillus equuli* infection in pigs. Veterinary Record **92**:543–544.

Bisgaard, M. 1975. Characterization of atypical *Actinobacillus lignieresii* isolated from ducks with salpingitis and peritonitis. Nordisk Veterinaermedicin **27**:378–383.

Boháček, J., Mráz, O. 1967. Basengehalt der Desoxyribonukleinsäure bei den Arten Pasteurella haemolytica, Actinobacillus lignieresii und Actinobacillus equuli. Zentralblatt für Bakteriologie, Parasitenkunde, Infektionskrankheiten und Hygiene, Abt. 1 Orig. **202**:468–478.

Bosworth, T. J. 1923. The causal organisms of bovine actinomycosis. Journal of Comparative Pathology **36**:1–22.

Bouley, G. 1966. Étude d'une souche d'*Actinobacillus suis* [Van Dorssen et Jaartsveld] isolée en Normandie. Recueil de Médecine Vétérinaire **142**:25–29.

Brumpt, E. 1910. Precis de parasitologie, 1st ed. Paris: Masson & Co.

Campbell, S. G., Whitlock, R. H., Timoney, J. F., Underwood, A. M. 1975. An unusual epizootic of actinobacillosis in dairy heifers. Journal of the American Veterinary Medical Association **166**:604–606.

Carter, P. L., Marshall, R. B., Jolly, R. D. 1971. A haemolytic variant of *Actinobacillus equuli* causing an acute septicaemia in a foal. New Zealand Veterinary Journal **19**:264–265.

Chladeck, D. W., Ruth, G. R. 1976. Isolation of *Actinobacillus lignieresi* from an epidural abscess in a horse with progressive paralysis. Journal of the American Veterinary Medical Association **168**:64–66.

Christiansen, M. 1917. En ejendommelig pyaemisk Lidelse hos Faar. Maanedsskrift for Dyrlaeger **29**:449–458.

Cottew, G. S., Francis, J. 1954. The isolation of *Shigella equuli* and *Salmonella newport* from normal horses. Australian Veterinary Journal **30**:301–304.

Cowan, S. T. 1974. Cowan and Steel's Manual for the identification of medical bacteria, 2nd ed. London: Cambridge University Press.

Cruickshank, R. 1965. Medical microbiology, 11th ed. Edinburgh: E. & S. Livingstone.

Cutlip, R. C., Amtower, W. C., Zinober, M. R. 1972. Septic embolic actinobacillosis of swine: A case report and laboratory reproduction of the disease. American Journal of Veterinary Research **33**:1621–1626.

Davies, G. O., Torrance, H. L. 1930. Observations regarding the etiology of actinomycosis in cattle and swine. Journal of Comparative Pathology **43**:216–233.

Davis, C. L., Stiles, G. W. 1939. Actinobacillosis in rams. Journal of the American Veterinary Medical Association **95**:754–756.

Dimock, W. W., Edwards, P. R., Bruner, D. W. 1947. Infections of fetuses and foals, pp. 1–39. Bulletin of the Kentucky Agricultural Experiment Station, no. 509.

du Plessis, J. L., Cameron, C. M., Langen E. 1967. Focal necrotic pneumonia and rumeno-enteritis in Afrikaner calves. Journal of the South African Veterinary Medical Association **38:**121–128.

Edwards, P. R., Taylor, E. L. 1941. *Shigella equirulis* infection in a sow. Cornell Veterinarian **31:**392–393.

Fletcher, R. B., Linton, H., Osborne, A. D. 1956. Actinobacillosis of the tongue of a dog. Veterinary Record **68:**645–646.

Frazier, W. C. 1926. A method for the detection of changes in gelatin due to bacteria. Journal of Infectious Diseases **39:**302–309.

Frederiksen, W. 1973. *Pasteurella* taxonomy and nomenclature, pp. 170–176. In: Winblad, S. (ed), Contributions to microbiology and immunology, vol 2, *Yersinia, Pasteurella* and *Francisella*. Basel: Karger.

Gerring, J. C. 1947. An unusually high incidence of actinobacillosis in cattle following the burning-off of peat country. Australian Veterinary Journal **23:**122–124.

Hacking, M. A., Sileo, L. 1977. Isolation of a hemolytic *Actinobacillus* from waterfowl. Journal of Wildlife Diseases **13:**69–73.

Hayston, J. T. 1948. Actinobacillosis in sheep. Australian Veterinary Journal **24:**64–66.

Hebeler, H. F., Linton, A. H., Osborne, A. D. 1961. Atypical actinobacillosis in a dairy herd. Veterinary Record **73:**517–521.

Innes, J. R. M., Berger, J., Francis, J. 1950. Subacute bacterial endocarditis with pulmonary embolism in a horse associated with *Shigella equirulis*. British Veterinary Journal **106:**245–250.

Jarmai, K. 1929. Viskosseptikämien bei älteren Fohlen und erwachsenen Pferden. Deutsche Tierärztliche Wochenschrift **37:**517–519.

Jones, J. E. T., Simmons, J. R. 1971. Endocarditis in the pig caused by *Actinobacillus equuli:* A field and an experimental case. British Veterinary Journal **127:**25–29.

Kemenes, F., Markói, B. 1959. Aktinobacillus lignieresi okozta tályog kutya szájüregében. Magyar Állatorvosok Lapja **14:**31–32.

Kim, B. H. 1976. Studies on *Actinobacillus equuli*. Ph.D. thesis. University of Edinburgh.

Kim, B. H., Phillips, J. E., Atherton, J. G. 1976. *Actinobacillus suis* in the horse. Veterinary Record **98:**239.

Klinger, R. 1912. Untersuchungen über menschliche Aktinomykose. Centralblatt für Bakteriologie, Parasitenkunde und Infektionskrankheiten, Abt. 1 Orig. **62:**191–200.

Kohlert, R. 1968. Untersuchungen zur Ätiologie der Eileiterentzündung beim Huhn. Monatshefte für Veterinärmedizin **23:**392–395.

Larsen, J. L. 1974. Isolation of haemolytic actinobacilli from horses. Acta Pathologica et Microbiologica Scandinavica, Sect. B. **82:**453–454.

Laudien, L. 1923. Kotuntersuchungen bei Pferden auf die Anwesenheit des Bacterium pyosepticum equi und von Paratyphusbazillen. Inaugural Dissertation, Hannover.

Laws, L., Elder, J. K. 1969a. Ovine epididymo-orchitis caused by *Actinobacillus lignieresi*. Australian Veterinary Journal **45:**384.

Laws, L., Elder, J. K. 1969b. Mastitis in sheep caused by *Actinobacillus lignieresi*. Australian Veterinary Journal **45:**401–403.

Lignières, J., Spitz, G. 1902. L'Actinobacillose. Bulletin et mémoires. Société Centrale de Médecine Vétérinaire **20:**487–535 and 546–565.

Magnusson, H. 1919. Joint-ill in foals: Etiology. Journal of Comparative Pathology **32:**143–182.

Magnusson, H. 1931. *Bacterium viscosum equi* [Adsersen] in suckling pigs and its relation to *Bacillus polymorphus suis* [Degen] in focal interstitial nephritis in swine. Eleventh International Veterinary Congress, London **3:**488–507.

Mair, N. S., Randall, C. J., Thomas, G. W., Harbourne, J. F., McCrea, C. T., Cowl, K. P. 1974. *Actinobacillus suis* infection in pigs. A report of four outbreaks and two sporadic cases. Journal of Comparative Pathology **84:**113–119.

Mawditt, A. L., Greenham, L. W. 1962. A case of subcutaneous bovine actinobacillosis. Veterinary Record **74:**290–292.

Meyer, K. F. 1910. Experimental studies on a specific purulent nephritis of Equidae, pp. 122–158. Transvaal Department of Agriculture. Report of the Government Veterinary Bacteriologist 1908/9.

Misdorp, W. 1963. Lesions resembling tuberculosis in Dutch slaughter cattle. Tijdschrift voor Diergeneeskunde **88:**575–587.

Moon, H. W., Barnes, D. M., Higbee, J. M. 1969. Septic embolic actinobacillosis. A report of two cases in New World monkeys. Veterinary Pathology **6:**481–486.

Mráz. O. 1969. Vergleichende Studie der Arten Actinobacillus lignieresii und Pasteurella haemolytica. III. Actinobacillus haemolyticus (Newsom und Cross, 1932) comb. nov. Zentralblatt für Bakteriologie, Parasitenkunde, Infektionskrankheiten und Hygiene, Abt. 1 Orig. **209:**349–364.

Mráz, O. 1975. Differentiation possibilities between pasteurellae and actinobacilli. Acta Veterinaria Brno **44:**105–113.

Mráz, O. 1977. Antigenní vztahy mezi Pasteurelami a Aktinobacily. Veterinární Medicína **22:**121–132.

Mráz, O., Vladík, P., Boháček, J. 1976. Actinobacilli in domestic fowl. Zentralblatt für Bakteriologie, Parasitenkunde, Infektionskrankheiten und Hygiene, Abt. 1 Orig., Reihe A **236:**294–307.

Mráz, O., Zakopal, J., Matovšek, Z. 1968. The finding of a haemolytic variant of *Actinobacillus equuli* as the cause of septicaemia in a herd of breeding mares. Sborník Vysoké školy zemědělské v Brně [řada B] **37:**263–275.

Nakazawa, M., Azuma, R., Yamashita, T., Iwao, T., Uchimura, M. 1977. Collective outbreaks of bovine actinobacillosis. Japanese Journal of Veterinary Science **39:**549–557.

Osbaldiston, G. W., Walker, R. D. 1972. Enteric actinobacillosis in calves. Cornell Veterinarian **62:**364–371.

Pathak, R. C., Ristic, M. 1962. Detection of an antibody to *Actinobacillus lignieresi* in infected human beings and the antigenic characterization of isolates of human and bovine origin. American Journal of Veterinary Research **23:**310–314.

Pedersen, K. B. 1977. *Actinobacillus* infektioner hos svin. Nordisk Veterinaermedicin **29:**137–140.

Phillips, J. E. 1960. The characterisation of *Actinobacillus lignieresi*. Journal of Pathology and Bacteriology **79:**331–336.

Phillips, J. E. 1961. The commensal role of *Actinobacillus lignieresi*. Journal of Pathology and Bacteriology **82:**205–208.

Phillips, J. E. 1964. Commensal actinobacilli from the bovine tongue. Journal of Pathology and Bacteriology **87:**442–444.

Phillips, J. E. 1966. *Actinobacillus lignieresi:* A study of the organism and its association with its hosts. D. V. M. & S. thesis. University of Edinburgh.

Phillips, J. E. 1967. Antigenic structure and serological typing of *Actinobacillus lignieresi*. Journal of Pathology and Bacteriology **93:**463–475.

Phillips, J. E. 1974. *Actinobacillus*, pp. 373–377. In: Buchanan, R. E., Gibbons, N. E. (eds.), Bergey's manual of determinative bacteriology, 8th ed. Baltimore: Williams & Wilkins.

Ross, R. F., Hall, J. E., Orning, A. P., Dale, S. E. 1972. Characterization of an *Actinobacillus* isolated from the sow vagina. International Journal of Systematic Bacteriology **22:**39–46.

Smith, J. E. 1974 *Pasteurella*, pp. 370–373. In: Buchanan, R. E., Gibbons, N. E. (eds.), Bergey's manual of determinative bacteriology, 8th ed. Baltimore: Williams & Wilkins.

Smith, T. 1918. A pleomorphic bacillus from pneumonic lungs of calves simulating Actinomyces. Journal of Experimental Medicine 28:333–343.

Svenkerud, R. R., Iversen, L. 1949. Shigella equirulis [B. viscosum equi] som årsak til klappeendocarditis hos hest. Nordisk Veterinärmedicin 1:227–232.

Taylor, A. W. 1944. Actinobacillosis in sheep. Journal of Comparative Pathology 54:228–237.

Terpstra, J. I., Akkermans, J. P. W. M. 1955. Opmerkingen bij het Jaarverslag van 1954. Tijdschrift voor Diergeneeskunde 80:741–751.

Thomas, A. D. 1931. Actinobacillosis and other complications in sheep which may arise from the feeding of prickly pear [Opuntia spp.]. Report on Veterinary Research. Department of Agriculture, Union of South Africa 17:215–229.

Thompson, L. 1933. Actinobacillosis of cattle in the United States. Journal of Infectious Diseases 52:223–229.

Thompson, L., Willius, F. A. 1932. Actinobacillus bacteremia. Journal of the American Medical Association 99:298–301.

Thornton, H. 1976. Unusual pathological condition. Acute actinobacillotic myocarditis in an ox. Rhodesian Veterinary Journal 7:38.

Till, D. H., Palmer, F. P. 1960. Review of actinobacillosis with a study of the causal organism. Veterinary Record 72:527–534.

Tunnicliff, E. A. 1941. A study of Actinobacillus lignieresi from sheep affected with actinobacillosis. Journal of Infectious Diseases 69:52–58.

Vallée, A. 1959. Isolement de Bacterium viscosum equi chez deux lapins domestiques. Recueil de Médecine Vétérinaire 135:821–822.

Vallée, A., Durieux, J., Durieux, M., Virat, B. 1960. Étude d'une pyodermite particulièrement rebelle chez le chien isolement d'Actinobacillus equuli associé à un staphylocoque. Bulletin de l'Académie Vétérinaire de France 33:153–156.

Vallée, A., Tinelli, R., Guillon, J. C., Priol, A. le, Cuong, T. 1974. Étude d'un actinobacillus isolé chez un cheval. Recueil de Médecine Vétérinaire 150:695–700.

van Dorssen, C. A., Jaartsveld, F. H. J. 1962. Actinobacillus suis [novo species], een bij het varken voorkomende bacterie. Tijdschrift voor Diergeneeskunde 87:450–458.

Veterinary Investigation Service. 1975. Few specimens received at V I centres. Veterinary Record 97:319 and 336.

Webb, R. F., Cockram, F. A., Pryde, L. 1976. The isolation of Actinobacillus equuli from equine abortion. Australian Veterinary Journal 52:100–101.

Weidlich, N. 1955. Seltener Verlauf einer Pyoseptikum-Infektion bei einem dreijährigen Hengst [Rhinitis mit anschliessender Meningitis]. Berliner und Münchener Tierärztliche Wochenschrift 68:303–304.

Werdin, R. E., Hurtgen, J. P., Bates, F. Y., Borgwardt, F. C. 1976. Porcine abortion caused by Actinobacillus equuli. Journal of the American Veterinary Medical Association 169:704–706.

Wetmore, P. W., Thiel, J. F., Herman, Y. F., Harr, J. R. 1963. Comparison of selected Actinobacillus species with a hemolytic variety of Actinobacillus from irradiated swine. Journal of Infectious Diseases 113:186–194.

Windsor, R. S. 1973. Actinobacillus equuli infection in a litter of pigs and a review of previous reports on similar infections. Veterinary Record 92:178–180.

Zakopal, J., Nesvadba, J. 1968. Aktinobacilläre Septikämie als Massenerkrankung in einer Zuchtstutenherde. Zentralblatt für Veterinärmedizin, Reihe A 15:41–59.

Zimmermann, T. 1964. Untersuchungen über die Actinobazillose des Schweines. 1. Mitteilung: Isolierung und Charakterisierung des Erregers. Deutsche Tierärztliche Wochenschrift 71:457–461.

CHAPTER 112

The Genus *Cardiobacterium*

IRVING J. SLOTNICK

The original strains of the genus *Cardiobacterium* were isolated from the blood of persons ill with subacute bacterial endocarditis within a 10-month period, over a broad geographical area of the United States (Tucker et al., 1962). Unable to place them in any known taxon, King (1964) arbitrarily assigned these to "group II D", one of several unclassified or ill-defined categories on hand at that time at the Center for Disease Control. This group housed those fastidious, Gram-negative, fermentative rods that were oxidase positive but were unable to grow on MacConkey agar (King, 1964).

Their generic uniqueness was then established (Slotnick and Dougherty, 1964), and the name *Cardiobacterium hominis* was proposed. The criticism (some justified) of the particular placement of the genus *Cardiobacterium* in the current (eighth) edition of *Bergey's Manual of Determinative Bacteriology* (Buchanan and Gibbons, 1974) should not detract from the considerable usefulness of the description presented there. In view of the relatively close temporal relation of the original four cases and the relative ease of their culture, it is surprising that *Cardiobacterium hominis* was not encountered previously. It is more than likely that the close resemblance of the *Cardiobacterium* organisms to *Streptobacillus, Pasteurella, Actinobacillus,* and *Hemophilus* organisms had resulted in their misidentification in the past. Since 1962, infections with *C. hominis* are being reported regularly (Jobanputra and Moysey, 1977; Savage et al., 1977). Although primarily associated with endocarditis in patients with preexisting cardiovascular defects or disease, *C. hominis* organisms are being encountered in an ever increasing number of diseased sites, including the genital tract, cheek, mandible, empyema fluid, and spinal fluid (R. Weaver, personal communication). The exact extent of morbidity and mortality is not known, but human infection is more common and potentially more serious than previously observed.

Habitats

Cardiobacterium hominis is part of the indigenous commensal human respiratory flora (Slotnick and Dougherty, 1964; Slotnick, Mertz, and Dougherty, 1964). None of the persons examined in the group harboring organisms in the nose or throat presented any evidence of past or current history of rheumatic heart disease, endocarditis, or related cardiovascular disorders. Interestingly, in this particular set of random persons examined for respiratory carriage of *C. hominis,* the organisms were repeatedly detectable over a period of several months in the majority of the individuals in about the same proportions with respect to other respiratory flora. In one of the adult female subjects, *C. hominis* was actually the predominant throat organism (I. J. Slotnick, unpublished observations). Transient carriage, either acquisition or disappearance of *C. hominis,* was also encountered. It was also clear that *C. hominis* is distributed in the respiratory tract among persons of all ages and without predilection for sex.

Fluorescent antibody smear analysis has implicated *C. hominis* as a possible resident of the human intestinal tract, but cultural isolation from stool specimens was extremely difficult and unsuccessful because of lack of any selective medium to hold back overgrowth by the generally large numbers of enteric bacteria present.

C. hominis does not appear in midstream or catheterized urine, but is occasionally found in cervical-vaginal cultures. These data support the opinion that *C. hominis,* under unknown conditions, is a transient resident of the genitourinary tract. The infrequent incidence of *C. hominis* there may be explained in that the genital tract offers a very unfavorable anatomic site for the survival of this organism, which is known to be very sensitive to chemical inhibition. Other factors may certainly prevail. *C. hominis* is not found in the respiratory secretions, urine, or stool of laboratory rats, rabbits, and guinea pigs examined by culture and fluorescent antibody staining. Furthermore, short-term feeding experiments fail to establish these organisms in the oropharyngeal or gastrointestinal flora of these same animals (Slotnick, unpublished observations).

Role As a Pathogen

Human disease caused by *Cardiobacterium hominis* principally involves cardiac pathology (Jobanputra and Mosey, 1977; Midgley et al., 1970; Perdue, Dorney, and Ferrier, 1968; Snyder and Ellner, 1969; Tucker et al., 1962; Weiner and Werthamer, 1975;). The low intrinsic virulence of these organisms is in keeping with the usual subacute course of the infective endocarditis they produce. Most frequently the patient has a history of preexisting cardiac lesions, but the organisms can cause aortic and mitral valve damage in some patients who present no evidence of previous cardiovascular defects or disease. Infection with *C. hominis* can result in cerebral emboli, with the development of a fatal mycotic aneurysm (Laguna, Derby, and Chase, 1975) or the subsequent appearance of a peripheral arterial embolus to the femoral artery and a mycotic aneurysm at this site (Perdue, Dorney, and Ferrier, 1968). Not unexpectedly, as has been the case with so many emerging pathogens, a greater number of other infected sites are being encountered. It is assumed that illness produced by *C. hominis* arises via endogenous infestation, probably from a respiratory focus. In the case of *C. hominis* endocarditis described by Jobanputra and Mosey (1977), dental sepsis is the most likely origin of the organisms. The lack of any known animal host and failure to detect *C. hominis* after extensive environmental sampling (soil, water, and several hospital sites) speak against an exogenous route of invasion (Slotnick, unpublished observations). Exotoxins are not produced and other virulence factors remain unknown. *C. hominis* is not pathogenic for any of the common laboratory animals. The immune mechanisms operative in hosts infected with *C. hominis* remain to be elucidated. Second attacks or chronic cases have not been observed. Humoral antibodies are not regularly formed in normal adult throat carriers of *C. hominis*.

Serum of endocarditis patients may give high titers in agglutination and complement fixation tests. Midgley et al. (1970) report a case with admission titers of 320 and 640, respectively, in agglutination and complement fixation tests. Serum obtained 9 months later gave an agglutination titer of 160, showing little change, whereas the complement fixation titer had fallen considerably, to 20. Neither of these sera produced agglutination with a variety of other organisms isolated in their laboratory; conversely, numerous sera from other patients failed to agglutinate *C. hominis* in slide agglutination tests.

Isolation

Principles

Cardiobacterium hominis is an aerobic, facultatively anaerobic, weakly fermentative, chemoorganotroph. Good growth is most easily obtained when cultures are incubated at 35°C in humidors containing filter paper pads saturated with water. The requirement for high humidity is growth limiting. In addition, atmospheres of 3–5% CO_2 are preferable. Unquestionably, blood is the customary specimen of investigative choice, but it is important to recognize that body fluids, exudates, wounds, and tissues may bear examination in specific instances. A useful enrichment medium is not available.

Procedure for Processing Specimens

Blood:
Inoculate into conventional blood culture broths gassed to contain an atmosphere of 10% CO_2 and 0.25% sodium polyanethol sulfonate ("liquoid") added as anticoagulant. Although citrate is not inhibitory, other organisms to be selected may be sensitive to citrate; therefore this anticoagulant should be avoided. Incubate at 35°C for at least 7 days.

Body fluids, exudates, tissue:
Plate and streak out an aliquot of fluid, minced tissue, or swab specimen directly onto each of 3 blood agar plates, incubating one in air, one in a CO_2 incubator (candle jar), and one in the anaerobe jar. Incubate at 35°C for a minimum of 4 days.

Media

In addition to usual laboratory infusion broths, *Cardiobacterium hominis* grows well on Trypticase soy agar (BBL) or tryptose blood agar (Difco) with or without 5% blood enrichment, chocolate agar, PPLO Agar (Difco), cystine heart agar, Casman agar base (BBL), and heart infusion agar. Moderate growth occurs on nutrient agar, and Lowenstein-Jensen medium. The organism does not grow on MacConkey agar, Simmons citrate agar, SS agar, tellurite agar base, potato dextrose agar, Sabouraud dextrose agar, phenylethyl alcohol blood agar, EMB, or Endo agar. The organisms grow poorly, if at all, on triple sugar iron agar or Kligler iron agar.

Growth Support Medium for *Cardiobacterium hominis* (Slotnick and Dougherty, 1965).
The nutrient solution has the following composition:

Distilled water   890 ml
Glucose   5 g

Arginine	160 mg
Glutamic acid	200 mg
Glycine	176 mg
Histidine	126 mg
Leucine	426 mg
Proline	100 mg
Threonine	276 mg
Valine	185 mg
Tyrosine	35 mg
Vitamin solution	10 ml
Buffered salts solution	100 ml

Buffered salts solution (g/liter):

$MgSO_4 \cdot 7H_2O$	4
$MnSO_4 \cdot H_2O$	0.3
NaCl	5
Na_2PHO_4	284
KH_2PO_4	272
$ZnSO_4 \cdot 7H_2O$	0.4
$FeSO_4 \cdot 7H_2O$	4
$CuSO_4 \cdot 5H_2O$	0.05

Vitamin solution (mg/liter):

Ca pantothenate	1
Nicotinamide	1
Pyridoxine HCl	2
Thiamine HCl	1
Biotin	0.1

The pH is adjusted to 7.0. Sterilization is done by filtration. After inoculation, stationary cultures are incubated at 35°C in the CO_2 incubator, 5–10% CO_2.

Antibiotic Susceptibility

Cardiobacterium hominis is susceptible to a wide range of antibiotics at pharmacological dose levels. Penicillin or streptomycin, either singly or in combination, should be considered as the drugs of choice. In case of allergy or other contraindications for these agents, any of the broad-spectrum antibiotics can be used. Drug resistance in treatment has not been a problem.

Conservation of Cultures

Stock cultures for permanent maintenance are grown for 48 h at 35°C in tryptose blood agar (Difco) enriched with 5% human blood. Sterile glycerol is then added to a final concentration of 10%. Aliquots (1–2 ml) of the culture are distributed into screw-cap vials and stored in deep freeze at −70°C. Cultures stored in this manner remain viable for a minimum of 2 years. If the thawing of stock vial is minimal, it may be refrozen and used several times. Skim milk suspensions (10%) of *C. hominis* survive lyophilization, which is a recommended alternative method for long-term storage.

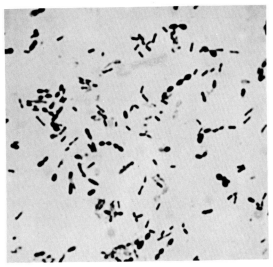

Fig. 1. *Cardiobacterium hominis* tear-drop morphology and chaining tendency. ×1,800.

Identification

Definitive identification of *Cardiobacterium hominis* is materially simplified using specific fluorescent antibody staining (Slotnick, Mertz, and Dougherty, 1964). Since anti–*C. hominis* conjugate is not commercially available and it is not likely that the smaller laboratory will wish to prepare its own, reliable identification can be made on the basis of the tinctorial reactions, morphology (Figs. 1, 2, and 3), cultural growth patterns, and the biochemical profile of *C. hominis* cited in Table 1 (Buchanan and Gibbons, 1974; Slotnick and Dougherty, 1964). Carbohydrate utilization test media may require enrichment with serum or ascitic fluid to obtain sufficient growth. Detection of indole may require xylol extraction.

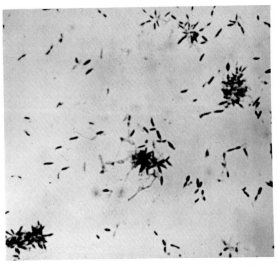

Fig. 2. *Cardiobacterium hominis* rosette clusters in a Gram-stained preparation of a blood agar culture. ×1,800.

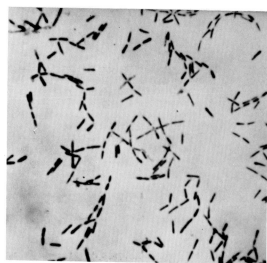

Fig. 3. *Cardiobacterium hominis.* Culture with singular lack of pleomorphic cell types. Rod-shaped cells predominate. ×1,800.

Staining Reactions

These organisms appear as Gram-negative rods, 0.5 μm wide and 1.0–2.2 μm long, arranged singly, in pairs, short chains, and clusters. They may be uniform or present as pleomorphic rods with one or both ends enlarged, tear-drop forms, or rosette

Table 1. Identifying characteristics of *Cardiobacterium hominis.*

Oxidase test	+
MacConkey	–
Voges-Proskauer test	–
Catalase	–
Indole	+/weak
Citrate	–
Nitrate reduction	–
H₂S TSI	–
Pbac	+/weak
ONPG	–
Urease	–
ADH/LDC/ODC	–
TSI, acid butt	+ or NG[a]
Gelatinase	–
Hippurate hyd.	–
Carbohydrates:	
Arabinose	–
Lactose	–
Maltose	+
Mannitol	d[b]
Raffinose	.
Salicin	–
Sorbitol	+
Sucrose	+
Trehalose	–
Xylose	–
Galactose	–
Glucose	+

[a] NG, No growth.
[b] d, Some strains positive—may be delayed.

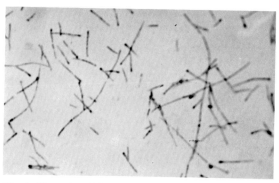

Fig. 4. Illustration of *Cardiobacterium hominis* "lollipop" variant. ×1,800.

clusters. A singular morphological variant occurs with a circular and enlarged terminal portion, giving the cells a "lollipop" appearance (Fig. 4). Savage (1977) recently related pleomorphism to yeast extract content of the medium. Organisms appear homogeneously stained and of uniform dimensions on blood agar containing yeast extract. Non-acid-fast sudanophilic and metachromatic inclusions are demonstrably nonsporeforming. Thin sections of *C. hominis* show a cell wall of Gram-negative type with a cytoplasm containing numerous intrusive membranes disposed around the periphery of the cells, especially at the poles (Reyn, Birch-Anderson, and Murray, 1971). The cell wall consists of a unit membrane sandwiched between dense outer and inner layers. Polar caps are seen on the ends of cells. The substructure of the surface layer is unusual for a Gram-negative bacterium.

Growth and Culture

Cardiobacterium hominis colonies on blood agar media are punctiform after 24 h and attain a maximum size of 1–2 mm in 48–72 h. Colonies are circular, convex, smooth, entire, glistening, opaque, and butyrous. In older cultures, the center of the colony appears more opaque and grayish white in contrast to the edges. A slight greening of blood agar develops around dense areas of growth in 2–3 days, becoming brownish with further incubation. Clear hemolytic zones are not observed on agar plates that contain 5% rabbit or sheep blood.

Literature Cited

Buchanan, R. E., Gibbons, N. E. 1974. Bergey's manual of determinative bacteriology, 8th ed. Baltimore: Williams & Wilkins.

Jobanputra, R. S., Moysey, J. 1977. Endocarditis due to *Cardiobacterium hominis.* Journal of Clinical Pathology **30:**1033–1036.

King, E. O. 1964. The identification of unusual pathogenic

Gram-negative bacteria. Atlanta, Georgia: Center for Disease Control.

Laguna, J., Derby, B. M., Chase, R. 1975. *Cardiobacterium hominis* endocarditis with cerebral mycotic aneurysm. Archives of Neurology **32:**638–639.

Midgley, J., LaPage, S. P., Jenkins, B. A. G., Barrow, G. I., Roberts, M. E., Buck, A. G. 1970. *Cardiobaterium hominis* endocarditis. Journal of Medical Microbiology **3:**91–98.

Perdue, C. D., Dorney, E. R., Ferrier, F. 1968. Embolomycotic aneurysm associated with bacterial endocarditis due to *Cardiobacterium hominis*. American Surgeon **34:**901–904.

Reyn, A., Birch-Anderson, A., Murray, R. G. E. 1971. The fine structure of *Cardiobacterium hominis*. Acta Pathologica et Microbiologica Scandinavica, Sect. B **79:**51–60.

Savage, D. D., Kagan, R. L., Young, A. N., Horvath, A. E. 1977. *Cardiobacterium hominis* endocarditis: Description of two patients and characterization of the organism. Journal of Clinical Microbiology **5:**75–80.

Slotnick, I. J., Dougherty, M. 1964. Further characterization of an unclassified group of bacteria causing endocarditis in man: *Cardiobacterium hominis* gen. et sp. n. Antonie van Leeuwenhoek Journal of Microbiology and Serology **30:**261–272.

Slotnick, I. J., Dougherty, M. 1965. Unusual toxicity of riboflavin and flavin mononucleotide for *Cardiobacterium hominis*. Antonie van Leeuwenhoek, Journal of Microbiology and Serology **31:**355–360.

Slotnick, I. J., Mertz, J. A., Doughterty, M. 1964. Fluorescent antibody detection of human occurrence of an unclassified bacterial group causing endocarditis. Journal of Infectious Diseases **114:**503–505.

Snyder, A. I., Ellner, P. D. 1969. *Cardiobacterium hominis* endocarditis. New York State Journal of Medicine **69:**704–705.

Tucker, D. N., Slotnick, I. J., King, E. O., Tynes, B., Nicholson, J., Crevasse, L. 1962. Endocarditis caused by a *Pasteurella*-like organism: Report of four cases. New England Journal of Medicine **267:**913–916.

Weaver, R., C. D. C., Atlanta, Georgia. Personal Communication, (1977).

Weiner, M., Werthamer, S. 1975. *Cardiobacterium hominis* endocarditis: Characterization of the unusual organisms and review of the literature. American Journal of Clinical Pathology **63:**131–134.

CHAPTER 113

The Genus *Streptobacillus*

IRVING J. SLOTNICK

The genus *Streptobacillus* houses a single member, its type species, *Streptobacillus moniliformis* (Buchanan and Gibbons, 1974). Other than speculation, sufficient data are lacking to determine whether or not this monotype is merely a representative of a hitherto unrecognized or misclassified group of organisms. Pleomorphism and complex nutrition have been important factors contributing to the profusion of distinguishing cognomens affixed to *S. moniliformis* isolates by early investigators. The recognized illegitimate designation, *Streptobacillus moniliformis*, quite descriptive of the distinctive, beaded, "necklace-shaped" forms appearing in culture, has been retained because of common usage (Buchanan, Holt, and Lessel, 1966). Synonyms are *Actinomyces muris, Actinobacillus multiformis, Haverhillia multiformis, Harverhillia moniliformis*, and *Streptothrix muris ratti*. It is generally agreed that all of these organisms are identical and that the generic name of *Streptobacillus* be used for all of these bacilli.

Primary interest in the streptobacilli originated because these organisms, albeit rarely, produce serious human and animal morbidity and significant mortality in untreated human hosts (Lambe et al. 1973; McCormack, Kaye, and Hook, 1967). However, the discovery by Klieneberger (1935) of L-forms in cultures of *S. moniliformis* focused added interest, which continues to be of enormous biomedical impact. It is well known that all strains of *S. moniliformis* can undergo a reversible conversion to transitional phase variants, in vivo as well as in vitro, i.e., organisms undergoing change to a wall-defective phase or reverting back to bacterial form. These microbial variants are postulated to play a role in disease and, indeed, variant forms of bacteria, cultured by special techniques, have been implicated as etiological agents (see this Handbook, Chapter 166).

The streptobacilli are still relatively unexplored. Major dilemmas still exist regarding their taxonomic status, defined nutritional requirements, pleomorphism, and biochemical mechanisms of pathogenicity.

Habitats

Streptobacillus moniliformis is a normal resident of the oropharynx of wild or laboratory rats, mice, and other rodents. At least one-half of otherwise healthy laboratory rats harbor streptobacilli in their nasopharyngeal or conjunctival secretions as normal flora (Strangeways, 1933). Although the rat and secondarily the mouse are the chief reservoirs, the weasel, squirrel, dog, cat, and pig have been occasionally associated with human streptobacillary disease, but the role of these animals as natural hosts for these organisms is not really known (Altemeier, Snyder, and Howe, 1945). Healthy animal carriers do not play any apparent role in transmission of the disease among themselves: Streptobacilli were not found in the mouth washings of healthy mice living in close contact with rats, while they could be demonstrated in the mouths of mice fed heavy suspensions of the organisms at least 2 days after feeding (Freundt, 1956). Arthropods are not known hosts or vectors for streptobacilli. Distribution is worldwide (Altemeier, Snyder, and Howe, 1945).

Streptobacillary infection in the human is equated with clinical illness. Human colonization with or without a carrier state has never been established. Although possible, it is of little epidemiologic consequence based on the infrequency of human isolates and lack of person-to-person transmission. Nor have isolations of streptobacilli from nonliving environments been realized. The contamination of food, food products (milk), and water via rodent detritus provides a source for spread to susceptible hosts (Parker and Hudson, 1926). Excretion of streptobacilli through the kidney was demonstrated by Levaditi (1932), who cultivated the bacilli from urine bladder aspirates. Other evidence (Freundt, 1956) suggests, however, that this method is not very reliable and that attempts to recover the streptobacilli from voided urine do not lend support to the theory that urine is an important source of infection. Excretion of streptobacilli with feces is not likely to occur, as the organism could not be demonstrated in the stomach or intestines of mice after feeding of cultures and only exceptionally in the bile of septicemic mice.

Streptobacilli as Human Pathogens

Two diseases are included under the general term of rat-bite fever; one is caused by *Streptobacillus moniliformis,* the other, sodoku, is caused by *Spirillum minor* (formerly *minus*) (see this Handbook, Chapter 52). Similarities in their clinical presentation and epidemiology can lead to diagnostic confusion but there are sufficient differences in their manifestations and cultural characteristics to make them quite distinguishable (Altemeier, Snyder, and Howe, 1945). Streptobacilli harbored in the nasopharyngeal secretions of an infected animal are usually transmitted to the human via biting. *S. moniliformis* infection is known to occur without contact with rats, as in the Haverhill, Massachusetts, 1926 milk-borne epidemic. The disease under these circumstances has been termed Haverhill fever. The term "Haverhill fever" connotes nonrodent bite but is used synonymously with streptobacillary rat-bite fever (Lambe et al., 1973; Parker and Hudson, 1926). Blood from an experimental laboratory animal has infected a human subject. Infection has also occurred in persons working or living in rat-infested buildings without reference to direct animal contact.

Streptobacillary rat-bite fever is an acute, febrile, prostrating illness with significant mortality, when untreated, ranging up to 10%. Persons with cardiovascular deficits and immunologically compromised patients are particularly vulnerable to sequelae of pneumonia or endocarditis (Lambe et al., 1973; McCormack, Kaye, and Hook, 1967; Watkins, 1946). An incubation period, usually less than 10 days but ranging from a few hours to three weeks, is followed by sudden illness with chills, fever, vomiting, malaise, and severe headache. Normal healing of bite wounds usually takes place but in some cases it may break down later to form an ulcer that runs a chronic course, often painful and prolonged and occasionally ending in a subcutaneous abscess. Alternate remissions and febrile episodes of irregular duration may persist for periods of a few weeks to months. Shortly after onset of symptoms, there is usually an accompanying nonpruritic, morbilliform, maculopapular, or petechial rash which is most marked on the extremities but can become generalized. Arthritic and arthralgic symptoms to varying degrees are frequent, with excess joint fluid and painful swellings of several joints. Approximately one-half of the patients develop a nonsuppurative polyarthritis, which is a hallmark of this disease. Clinical signs and symptoms usually regress after 1–2 weeks but, without appropriate antimicrobial therapy, this may be prolonged due to numerous complications.

Penicillin is the agent of choice in the treatment of bacterial-phase streptobacillary fever. Lack of inhibition of L-phase streptobacilli renders penicillin of no use for their elimination. Tetracycline, erythromycin, and streptomycin are useful alternative drugs in cases of penicillin-resistant streptobacilli and penicillin allergy. In addition to these antibiotics, Lambe et al. (1973) reported the organism to be susceptible to chloramphenicol, kanamycin, lincomycin, and gentamicin. McCormack, Kaye, and Hook (1967) advise that serum bactericidal levels be at least 1:4 to assure adequacy of therapy.

Streptobacilli as Animal Pathogens

Epizootic streptobacillosis has arisen among laboratory mice, turkey flocks, and guinea pig stocks. *Streptobacillus moniliformis* infection in mice develops as an acute generalized infection that results either in the rapid death of the animal or in a chronic disease condition characterized by polyarthritis (Freundt, 1956). Organisms are readily isolated from joint fluid and blood in spontaneous cases of streptobacillary arthritis in mice. A study of host-streptobacillus relationships (Savage, 1972) indicates that multiple factors operate in experimental streptobacillosis in mice. These include the host vascular system anatomy, the manner of growth of the infectious agent, some predilection of the organism for the joints, weak host immune response, and resistance of *S. moniliformis* to phagocyte destruction.

Streptobacillary tendon-sheath infection in the turkey was first reported by Boyer, Bruner, and Brown (1958), in which about 10% of the breeder toms in a flock and about 100 showed lameness. The animals were unable to stand and remained recumbent. They continued to eat and drink sparingly but lost weight and eventually died. An incompletely controlled rat problem supported the presumption that the infestation was secondary to rat bites. This presumption was strengthened by Yamamoto and Clark (1966) who further detailed the epizootiology of the natural disease in turkeys. They showed that an isolate of *S. moniliformis* from a rat trapped in the vicinity of the turkey compound was similar to the turkey isolate in all respects, including shared common antigens.

A high incidence of cervical adenitis was recently observed among stocks of guinea pigs at several research laboratories (Fleming, 1976). The guinea pigs, all obtained from a single supplier, developed abscesses that yielded pure cultures of *S. moniliformis.* Previous reports had already shown *S. moniliformis* to be a cause of cervical adenitis in the guinea pig. Interestingly, these isolates, unlike those from rats and mice, only grew under strict anaerobic conditions.

Isolation

Principles

Streptobacillus moniliformis is characteristically a nutritionally complex aerobe and facultative anaerobe, requiring blood, serum, or ascitic fluid media supplementation. Growth does not occur or cannot be sustained in infusion, nutrient, and other broths unless 10–20% of one of these body fluids is added. Although CO_2 generally does not stimulate growth, the CO_2 incubator or candle jar are the choice environments for cultures because CO_2 is a requirement for fresh isolates and some laboratory strains. Particular strains are obligate anaerobes. It is also important that incubation of cultures on the surface of agar media take place in an atmosphere saturated with moisture. Broth cultures should be transferred every 24–48 h to avoid any marked fall in pH, which will render cultures nonviable (Buchanan and Gibbons, 1974; Fleming, 1976; Rogosa, 1974). Fastidious growth, unknown minimal nutritional requirements, and lack of selective agents combine to preclude practical enrichment technique.

Procedure for Processing Specimens (Rogosa, 1974)

Blood.

Mix equal volumes (up to 10 ml) of blood with sterile 2.5% sodium citrate. Pack the cells by centrifugation for 25–30 min. Inoculate broth tubes with 0.1 ml of sediment. Prepare the plate inoculum by taking a small volume of sedimented cells and mix with an equal volume of broth. Place 0.1 ml of this mixture on the surface of each agar plate and distribute the inoculum by gently tilting the plate in several directions. (If screw-cap tubes are used, these should fit loosely.) Incubate in the CO_2 incubator or candle jar at 35°C.

Joint and body fluids.

Mix equal volumes of fluid and 2.5% sodium citrate to avoid clotting. Inoculate broth with 1 ml or more fluid. Prepare plates as described above.

Pus, exudate, or cutaneous eruption.

Inoculate by swabbing if the sample is limited.

Media

Thioglycolate, cooked meat medium, and any of the blood culture broths routinely used in the clinical laboratory (Bartlett, Ellner and Washington, 1974) serve as excellent basal broths for the growth of *Streptobacillus moniliformis*. However, the medium recommended by Rogosa (1974) is preferable because it favors the growth of a wide variety of L-phase and bacterial phase, nutritionally fastidious organisms.

Growth Medium for *Streptobacillus moniliformis* (Rogosa, 1974):

"Dissolve 40 g of dehydrated Heart Infusion Agar (Difco) in 850 ml of deionized, distilled water; adjust the pH to 7.6 with 5 N NaOH; dispense in 85-ml volumes into screw-cap bottles; sterilize by autoclaving for 15 min at 121 C; just before pouring plates add 10 ml of sterile horse serum (previously heated for 30 min at 56 to 60 C) and also add 5 ml of sterile 10% (wt/vol) solution of yeast extract (Oxoid, Difco, or BBL) previously adjusted to pH 7.0 and sterilized by filtration through a Seitz-type pad ($0.01-\mu$m pore size) or through a Millipore or similar filter ($0.45-\mu$m pore size); pour plates in sterile disposable 60×15 mm plastic petri dishes (Falcon Plastics, Los Angeles, Calif.).

"A similar broth medium containing 25 g of dehydrated Heart Infusion Broth (Difco) instead of 40 g of dehydrated Heart Infusion Agar (Difco), but with all other ingredients and medium preparation identical with the agar medium, is recommended."

Conservation of Cultures

Freshly grown broth cultures in screw-cap tubes may be kept frozen at 20–70°C and are viable for several years (Rogosa, 1974). *Streptobacillus moniliformis* survives lyophilization and this may be a preferred method for long-term storage. At 3–4°C, serum broth cultures are viable 7–10 days; sealed serum agar cultures are viable 14–15 days (Buchanan and Gibbons, 1974).

Serology

Agglutinins to *Streptobacillus moniliformis* appear in patients' sera and are of considerable diagnostic value. They are titered with formalin-treated cell suspensions of strain ATCC 14647 (Rogosa, 1974), using the usual tube agglutination technique. Titers of 1:80 or greater are considered diagnostic, but a rise in titer using paired sera taken at 5 to 10-day intervals is necessary to confirm recent infection, as titers of 1:80 may persist at least 2 years. Titers ranging from 1:1,280 to 1:5,120 have been recorded (Lambe et al. 1973).

Identification

Specific fluorescent staining of an isolate with anti–*Streptobacillus moniliformis* conjugate (Lambe et al., 1973) offers a means for rapid, definitive identification. In the absence of such a conjugate, which is not readily available, morphology and

a

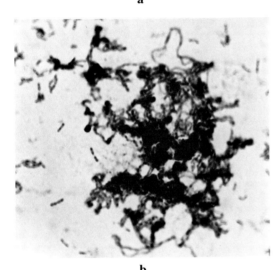

b

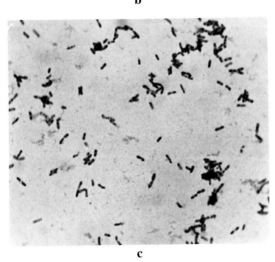

c

Fig. 1. (a) Illustration of pleomorphic morphology of *Streptobacillus moniliformis*. Brain heart infusion broth granule. Gram stain, × 1,200. (b) Illustration of pleomorphic morphology of *S. moniliformis*. Thioglycolate broth granule. Gram stain, × 1,200. (c) *S. moniliformis* showing tendency to more regular, rodlike morphology. Blood agar plate growth. Gram stain, × 1,200.

Table 1. Distinguishing features of *Streptobacillus moniliformis.*[a]

Microscopic appearance
  Gram-negative, highly pleomorphic rods, frequently forms chains or long, curved, looped filaments (100–150 μm) with oval to elongated bulbous, yeastlike swellings and beaded forms.[b]
Growth appearance
  In broth: Discrete colonies resembling "fluff balls" or "bread crumbs". Colonies may be seen resting on surface of sedimented red cells; clumps may adhere to sides or deposit on bottom of tubes.[c]
  On solid media: Blood agar: small, round, gray, translucent; appear on 2nd or 3rd day; in depths of blood agar, colonies appear in 5–7 days. Alpha or nonhemolytic. Serum agar: small, round, colorless, granular; low-power colony examination shows filamentous forms at periphery.
L-phase colonies
  "Fried egg" appearance; transfer only by agar block cutout.
Biochemical reactions
  Carbohydrate fermentation: glucose, fructose, maltose, salicin, starch.
  Negative reactions: catalase, oxidase, indole, nitrate, urease, gelatinase.
G+C ratio
  24–26 mol%.

[a] Data from Buchanan and Gibbons, 1974; Lambe et al., 1973; Rogosa, 1974.
[b] See Fig. 1.
[c] See Fig. 2.

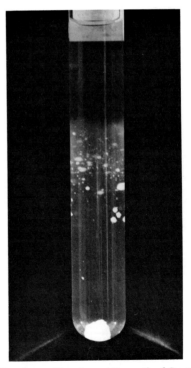

Fig. 2. Granular or "bread crumb" growth of *Streptobacillus moniliformis* in broth.

staining reactions, growth requirements and patterns, together with a biochemical profile, serve to identify the organism (Table 1).

Literature Cited

Altemeier, W. A., Snyder, H., Howe, G. 1945. Penicillin therapy in ratbite fever. Journal of the American Medical Association 127:270–273.

Bartlett, R. C., Ellner, P. D., Washington, J. A., II. 1974. Blood cultures, Cumitech 1. Washington, D. C.: American Society for Microbiology.

Boyer, C. I., Jr., Bruner, D. W., Brown, J. A. 1958. A *Streptobacillus,* the cause of tendon-sheath infection in turkeys. Avian Diseases 2:418–427.

Buchanan, R. E., Gibbons, N. E., (eds.). 1974. Bergey's manual of determinative bacteriology, 8th ed. Baltimore: Williams & Wilkins.

Buchanan, R. E., Holt, J. G., Lessel, E. F., Jr. 1966. Index Bergeyana. Baltimore: Williams & Wilkins.

Fleming, M. P. 1976. *Streptobacillus moniliformis* isolations from cervical abscesses of guinea pigs. Veterinary Record 99:256.

Freundt, E. A. 1956. *Streptobacillus moniliformis* infection in mice. Acta Pathologica et Microbiologica Scandinavica 38:231–245.

Klieneberger, E. 1935. The natural occurrence of pleuropneumonia-like organisms in apparent symbiosis with *Streptobacillus moniliformis* and other bacteria. Journal of Pathology and Bacteriology 40:93–105.

Lambe, D. W., McPhedran, A. M., Mertz, J., Stewart, P. 1973. *Streptobacillus moniliformis* isolated from a case of Haverhill fever: Biochemical characterization. American Journal of Clinical Pathology 60:854–860.

Levaditi, C., Selbie, R. F., Schoen, R. 1932. Le rhumatisme infectieux spontane de la souris provogue par le *Streptobacillus moniliformis.* Annales de l'Institut Pasteur de Lille 48:308–343.

McCormack, R. C., Kaye, D., Hook, E. W. 1967. Endocarditis due to *Streptobacillus moniliformis.* Journal of the American Medical Association 200:77.

Parker, F., Jr., Hudson, N. P. 1926. The etiology of Haverhill fever erythema arthiticum epidemicum. American Journal of Pathology 2:357–359.

Rogosa, M. 1974. *Streptobacillus moniliformis* and *Spirillum minor,* pp. 326–332. In: Lennette, E. H., Spaulding, E. H., Truant, J. P., (eds.), Manual of clinical microbiology, 2nd ed. Washington, D.C.: American Society for Mcrobiology.

Savage, N. L. 1972. Host parasite relationships in experimental *Streptobacillus moniliformis* arthritis in mice. Infection and Immunity 5:183–190.

Strangeways, W. L. 1933. Rats as carriers of *Streptobacillus moniliformis.* Journal of Pathology and Bacteriology 37:45–51.

Watkins, C. G. 1946. Ratbite fever. Journal of Pediatrics 28:429–448.

Yamamoto, R., Clark, G. T. 1966. *Streptobacillus moniliformis* infection in turkeys. Veterinary Record 79:95–100.

CHAPTER 114

The Genus *Calymmatobacterium*

ROBERT B. DIENST and GEORGE H. BROWNELL

Calymmatobacterium is a genus presently classified in the family *Brucellaceae*. Only one species, *Calymmatobacterium granulomatis,* has been described (Buchanan and Gibbons, 1974). The bacteria are pleomorphic rods, 1–2 μm in length, with rounded ends, occurring singly and in clusters. They exhibit single or bipolar condensation of chromatin, the latter giving rise to the characteristic closed "safety-pin" forms. The bacteria are usually encapsulated and readily demonstrated by Wright's stain as blue bacillary bodies surrounded by well-defined, dense, pinkish capsules. They are Gram-negative and non-motile.

History

The coccobacillary microbes first observed by Donovan in 1905 were frequently referred to as Donovan bodies when seen in tissue smears from patients with granulomatous lesions in the inguinal region. The Donovan bodies were later called *Calymmatobacterium granulomatis* by Aragao and Vianna in 1913; when Anderson, DeMonbreun, and Goodpasture first isolated the organisms in 1944 in the yolk sacs of chick embryos, they described the etiologic agent for granuloma inguinale as *Donovania granulomatis* (1945).

Calymmatobacterium granulomatis has been proved to be the causal agent of granuloma inguinale (Dienst, Greenblatt, and Sanderson, 1938). Presently the disease is referred to as donovanosis because initial lesions have been diagnosed in skin areas other than around the genital organs. The organism is pathogenic only for man, and infection cannot be produced in laboratory animals.

Donovanosis is seen throughout the world and is endemic in many areas including the United States. This infection is encountered primarily in the dark-skinned races and occurs mostly in regions where the climate is warm and humid for several months of the year. Most researchers support the contention that donovanosis is not primarily a venereal disease as such, but is an infection resulting from intimate contamination and poor hygiene.

Isolation

At present the only source for *Calymmatobacterium granulomatis* is isolation from the lesions of donovanosis. An organism reportedly isolated from human feces (Goldberg, 1962) has antigenic similarities to *C. granulomatis*.

The organism was originally cultivated by inoculating the yolk sacs of chick embryos with exudate from patients with donovanosis (Anderson, 1943). Later a pure culture of *C. granulomatis* was isolated by inoculating fresh egg yolk medium with exudate aspirated from a pseudo-bubo of a patient (Dienst, Greenblatt, and Chen, 1948). The isolate could be subcultured and maintained in the same culture medium. Examination of the subcultures revealed large numbers of encapsulated organisms consistent with the morphology of the Donovan bodies as seen in donovanosis.

Several other media have been reported for cultivation of *C. granulomatis*. Successful cultivation has been reported on coagulated egg yolk slants known as "Dulaney slants" (Dulaney, Guo, and Packer, 1948), as well as on a semisynthetic medium used for cultivation of laboratory strains (Goldberg, 1959).

There are several factors of importance for cultivation of *C. granulomatis:* first, the maintenance of a low oxidation-reduction potential; second, the requirement of a growth factor found in egg yolk; and third, a semisolid medium containing 0.12% agar (Dienst, Greenblatt, and Chen, 1948). The optimum temperature for cultivation is 37°C.

Identification

At present, three procedures are available to substantiate the clinical diagnosis of donovanosis: (i) demonstrating the presence of characteristic Donovan bodies in the diseased tissue either by stained tissue smears or by biopsy; (ii) demonstrating the presence of complement-fixing antibodies in the serum of the patient; (iii) demonstrating a positive skin reaction of the patient to an intradermal inocu-

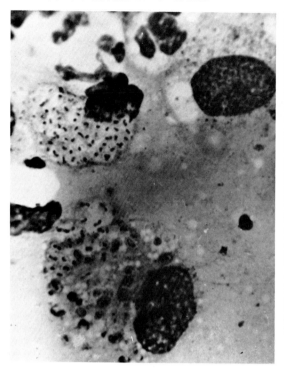
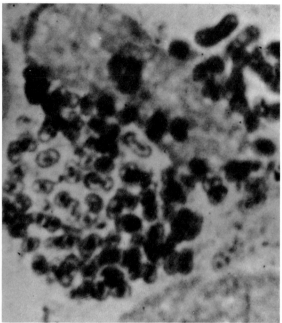

Figs. 1 and 2. Macrophages filled with *Calymmatobacterium granulomatis,* the so-called Donovan bodies.

lation of *Calymmatobacterium granulomatis* antigen. Of these tests, the examination of tissue smears stained by Wright's or Giemsa stain is the most accurate and certainly the simplest. Exudate from infected tissue, when collected, smeared, and stained properly, demonstrates to an experienced microscopist characteristic intracellular microorganisms in the large mononuclear monocytes that are specific for donovanosis (Dienst, Greenblatt, and Chen, 1948). Figs. 1 and 2 show the morphological and intracytoplasmic characteristics of *C. granulomatis.*

Proper treatment of donovanosis may be expected to effect complete cure. Most of the broad-spectrum antibiotics, in doses of 2 g daily for 10–15 days, or streptomycin, in doses of 4 g daily for 5 days, have proved adequate in most instances. A few cases, however, may require a second course of therapy with the same or a different antibiotic (Greenblatt *et al.,* 1958).

There are no protective antibodies produced by *C. granulomatis* in the patient. Once the infection occurs the disease persists chronically and may spread through the lymphatics to all tissue unless treated with antibiotics. The patient does produce specific sensitizing antibodies, as shown by skin testing (Chen, Dienst, and Greenblatt, 1949). Antibodies can be detected in patient serum by complement fixation procedures (Dulaney and Packer, 1947). The test antigens used included pus from granulomatous lesions, whole or ruptured *C. granulomatis* organisms, and boiled or extracted egg yolk medium after growth of *C. granulomatis.*

Literature Cited

Anderson, K. 1943. The cultivation from granuloma inguinale of microorganisms having the characteristics of Donovan bodies in yolk sac of chick embryos. Science **97:**560–561.

Anderson, K. A., DeMonbreun, W. A., Goodpasture, E. W. 1945. An etiologic consideration of *Donovania granulomatis* cultivated from granuloma inguinale (three cases) in embryonic yolk. Journal of Experimental Medicine **81:**25–40.

Aragao, H., Vianna G. 1913. Resquizas sober o *Granuloma venereo* (Untersuchungen ueber das *Granuloma venereum*) Mem. Inst. Oswaldo Cruz Rio de Janeiro **5:**211–238.

Buchanan, R. E., Gibbons, N. E. (eds.). 1974. Bergey's Manual of Determinative bacteriology, 8th ed. Baltimore: Williams & Wilkins.

Chen, C. H., Dienst, R. B., Greenblatt, R. B. 1949. Skin reaction of patients to *Donovania granulomatis.* American Journal of Syphilis, Gonorrhea, and Venereal Diseases **33:**60–64.

Dienst, R. B., Greenblatt, R. B., Sanderson, E. S. 1938. Cultural studies on the "Donovan Bodies" of granuloma inguinale. Journal of Infectious Diseases **62:**112–114.

Dienst, R. B., in collaboration with Greenblatt, R. B., Chen, C. H. 1948. Laboratory diagnosis of Granuloma Inguinale and studies on the cultivation of the Donovan Body. American Journal of Syphilis, Gonorrhea, and Venereal Diseases **32:**301–306.

Dulaney, A. D., Packer, H. 1947. Complement fixation studies with pus antigen in granuloma inguinale. Proceedings of the Society for Experimental Biology and Medicine **65:**254–256.

Dulaney, A. D., Guo, K., Packer, H. 1948. *Donovania granulomatis:* cultivation, antigenic preparation, and immunological tests. Journal of Immunology **59:**335–340.

Greenblatt, R. B., Baldwin, K. R., Dienst, R. B. 1958. The minor venereal diseases: Their diagnosis and treatment. Clinical Obstetrics and Gynecology, vol. 2, Quarterly Report 2, 519–563.

Goldberg, J. 1959. Studies on granuloma inguinale. IV. Growth requirements of *Donovania granulomatis* and its relationship to the natural habitat of the organism. British Journal of Venereal Diseases **35**:266–268.

Goldberg, J. 1962. Studies on granuloma inguinale. V. Isolation of a bacterium resembling *Donovania granulomatis* from the faeces of a patient with granuloma inguinale. British Journal of Venereal Diseases **38**:99–102.

Anaerobic, Gram-Negative, Rod-Shaped Bacteria

CHAPTER 115

The Anaerobic Way of Life of Prokaryotes

GERHARD GOTTSCHALK

Molecular oxygen, in appreciable amounts, is found only in those areas on earth that are in direct contact with air or are inhabited by organisms carrying out an oxygenic photosynthesis. The solubility of oxygen in water is low. In equilibrium with air of 1 atm and at 20°C, water will contain approximately 9 mg/liter of dissolved oxygen; this oxygen is rapidly consumed by aerobic organisms, so that deeper layers of many waters and soils—especially if they are rich in organic compounds—and of mud and sludge are practically anaerobic. Nevertheless, these areas are inhabited by numerous organisms that fulfill the important ecological role of converting insoluble organic material to soluble compounds and gases that can circulate back into aerobic regions. Other important anaerobic habitats are the rumen and the intestinal tract.

Prokaryotes that can live in the above-mentioned environments are either phototrophs, which, of course, can only flourish if light is available, or anaerobic chemotrophs. Frequently, the latter group of organisms is regarded as very primitive in that they can only carry out simple fermentations. However, here it must be considered that the anaerobes are a very heterogeneous assembly of organisms. With respect to their relationship to aerobic metabolism three groups of organisms capable of growth in an anaerobic environment can be envisaged:

1. Organisms that can carry out a nitrate- or nitrite-dependent respiration. These organisms are true aerobes, but when exposed to an anaerobic environment they can use nitrate or nitrite as alternate electron acceptors, and the electron transport from NADH to these acceptors is coupled to the phosphorylation of adenosine 5-diphosphate, as is the electron transport to oxygen.
2. Organisms that are facultative aerobes. The enterobacteria are the most prominent representatives of this group. These organisms grow as typical aerobes in the presence of oxygen; in its absence they carry out fermentations.
3. Obligate anaerobic bacteria that are characterized by the inability to synthesize a respiratory chain

with oxygen as terminal electron acceptor and to oxidize organic substrates to carbon dioxide and water. They are restricted to life without oxygen.

Clearly, members of groups 1 and 2 are not that typical for the anaerobic way of life because they are also equipped for aerobic metabolism. What about the organisms of group 3—are they so much different from a facultative aerobe such as *Escherichia coli* or an obligate aerobe such as *Azotobacter vinelandii?* Are they really more primitive than these bacteria? What are their achievements as the consequence of their permanent existence under anaerobic conditions? In the following, the properties of the obligate anaerobic bacteria will be discussed under these aspects.

Morphology and Macromolecules of Obligate Anaerobes

Among the obligate anaerobes, a diversity of cell form and ultrastructure similar to the one of aerobic organisms is found. Rods, cocci, vibrios, spirilla, coryneforms, and sarcinas are common; the cells are motile or nonmotile. A number of obligate anaerobes contain intracytoplasmic membrane systems (Cho and Doy, 1973; Zeikus, 1977). It is also noteworthy that the most complex structural change observed in prokaryotes—the formation of endospores—is a characteristic feature of the genera *Clostridium, Desulfotomaculum,* and *Sporolactobacillus.* The metamorphosis of a fragile vegetative cell into a highly resistant spore structure is, therefore, not a privilege of aerobes. Bacteriophages specific for obligate anaerobes are also known (McClung, 1956).

With the exception of the methane bacteria, the general composition of the cell walls of obligate anaerobes is comparable with that of other groups of bacteria. The walls contain peptidoglycan. Grampositive, nonsporeforming anaerobes contain a multilayered peptidoglycan with interpeptide bridges that is considered as the more primitive type (Schleifer and Kandler, 1972). The peptidoglycan of the clostridia, however, is of the directly cross-linked type that also occurs in the most highly evolved prokaryotes, the myxobacteria and cyanobacteria (Schleifer and Kandler, 1972). Teichoic acids occur in Gram-positive aerobes and obligate anaerobes (Archibald, 1974). The composition of the outer membrane and of the external layers of Gram-negative bacteria seems to be more closely associated with other parameters of the habitat of the organisms (e.g., salinity, pH, temperature) than with aerobiosis or anaerobiosis (Costerton, Ingram, and Cheng, 1974; Sleytr and Glauert, 1976). The methane bacteria, however, form a distinct group among the anaerobes because their cells walls are devoid of muramic acid, glucosamine, and D-glutamate, the main structural component of their walls being a heteropolysaccharide (Kandler and Hippe, 1977). Another remarkable difference between the methane bacteria and all other bacteria has been revealed in studies of the sequence of the 16 S ribosomal RNAs of methane bacteria. These RNAs are only very distantly related to the corresponding macromolecules of other bacteria (Fox et al., 1977; Woese and Fox, 1977).

The G+C content of DNA has become an important tool in the classification of prokaryotes; it may vary between 23 and 73 mol%. On the average, the obligate anaerobes are found preferentially in the lower range of this scale. For the clostridia, the G+C content ranges from 23 to 43 mol%; the corresponding values for fusobacteria and streptococci are 26–34 and 33–42 mol%, respectively. The *Desulfovibrio* species and the propionibacteria, however, are in the upper region (46–61 and 64–67 mol%, respectively), so that a low G+C content cannot be considered as a characteristic feature of all obligate anaerobes.

From the two storage materials frequently accumulated by microorganisms, glycogen (or glycogen-like polysaccharides) and poly-β-hydroxybutyrate, the latter does not seem to be common among anaerobes. To the author's knowledge, its presence has only been demonstrated in *Clostridium botulinum* (Emeruwa and Hawirko, 1973). Glycogen, however, is readily stored by many obligate anaerobes (Gavard and Milhaud, 1952; Hungate, 1963; Laishly, Brown, and Otto, 1974; Strasdine, 1968). In *Clostridium pasteurianum,* one of the two enzymes involved in polysaccharide synthesis, ADP-glucose pyrophosphorylase, is not subject to allosteric control (Robson, Robson, and Morris, 1974), whereas glycogen synthesis in aerobic and phototrophic organisms is regulated at the level of this enzyme (Preiss, 1977). This has been interpreted as evidence for the primitive nature of the enzyme system involved in reserve material synthesis in *Clostridium* (Robson, Robson, and Morris, 1974).

With respect to the morphology and the composition of macromolecules, the only known major differences between obligate anaerobes and other prokaryotes concern the cell wall and the 16 S RNA of the methane bacteria. Also, it should be mentioned that the obligate anaerobes are found mainly at the lower end of the range of the G+C content observed in prokaryotes, and that glycogen or glycan-like polysaccharides are more typical storage materials of obligate anaerobes than poly-β-hydroxybutyrate.

Biosynthesis of Small Molecules in Obligate Anaerobes

The highest biosynthetic capacity has undoubtedly to be ascribed to organisms that are able to synthesize all their cell material from carbon dioxide, an inorganic electron donor, and minerals. Such potent organisms are found among the phototrophic bacteria (e.g., Chromatiaceae) and the aerobic chemolithotrophs (e.g., thiobacilli and Nitrobacteraceae). These bacteria use the Calvin cycle for CO_2 fixation; they contain the key enzymes of this cycle: ribulose-1,5-bisphosphate carboxylase and phosphoribulokinase (McFadden, 1973). C-autotrophic organisms sensu stricto are also found among the obligate anaerobes. *Methanobacterium thermoautrophicum* grows in a mineral medium under an atmosphere containing molecular hydrogen and carbon dioxide (Zeikus and Wolfe, 1972). Other methane bacteria require vitamins and additional organic compounds for growth, but these organisms still derive a large percentage of the cell carbon from CO_2 (Zeikus, 1977). This is also true for the acetogenic bacteria *Clostridium aceticum* and *Acetobacterium woodii,* which ferment H_2 and CO_2 to acetate (Balch et al., 1977; Wieringa, 1940). Approximately 30% of the cell carbon originates from CO_2 in organisms growing on C_2 compounds—in *Clostridium kluyveri* growing on ethanol and acetate (Tomlinson and Barker, 1954) and in sulfate-reducing bacteria growing on ethanol plus sulfate (Sorokin, 1966).

In all the anaerobes mentioned, the Calvin cycle is not involved in CO_2 fixation. However, some sort of a CO_2-fixation cycle must, of course, operate in organisms such as *Methanobacterium thermoautrophicum.* Although its complete nature remains to be elucidated, studies in various laboratories have shown that one type of CO_2-fixation reaction is particularly important in anaerobes, the reductive carboxylation of CoA esters to the corresponding α-oxo acids:

$$\text{acetyl-CoA} + CO_2 \xrightarrow[\text{ferredoxin}]{\text{reduced}} \text{pyruvate} + \text{CoA}$$

$$\text{succinyl-CoA} + CO_2 \xrightarrow{\text{reduced ferredoxin}} \alpha\text{-oxoglutarate} + CoA$$

Pyruvate is synthesized this way by *C. kluyveri* (Andrew and Morris, 1965), by methane bacteria (Zeikus et al., 1977), by mixed rumen bacteria (Emmanuel and Milligan, 1973), and by sulfate-reducing bacteria (Buchanan, 1973; Suh and Akagi, 1966). α-Oxoglutarate synthesis from succinyl-CoA has been demonstrated in *Bacteroides ruminicola* (Allison and Robinson, 1970) and also in methane bacteria (Zeikus et al., 1977). Branched-chain α-oxo acids as precursors of the amino acids valine, leucine, and isoleucine can also be formed from the corresponding CoA esters by reductive carboxylations (Allison, 1969).

With regard to the importance of the carboxylation reactions discussed in the preceding paragraph, an interesting parallel exists between the chemotrophic obligate anaerobes and the phototrophic green sulfur bacteria (Chlorobiaceae). These organisms also lack the key enzymes of the Calvin cycle (Buchanan and Sirevåg, 1976; Quandt et al., 1977); pyruvate and α-oxoglutarate synthases are being used for CO_2 fixation (Evans and Buchanan, 1965; Quandt, Pfennig, and Gottschalk, 1978; Sirevåg and Ormerod, 1970). A reductive tricarboxylic acid cycle has been shown not to operate in the form proposed (Beuscher and Gottschalk, 1972; Evans, Buchanan, and Arnon, 1966), and a cyclic pathway of CO_2 fixation is not known for these organisms.

Another characteristic reaction of obligate anaerobes is the reduction of CO_2 to formate. The electron donor used may be reduced ferredoxin (*Clostridium pasteurianum*), NADPH (*C. thermoaceticum*), or of unknown structure (*C. formicoaceticum*) (Andreesen and Ljungdahl, 1974; Leonhardt and Andreesen, 1977; Thauer, 1972; Thauer, Fuchs, and Jungermann, 1977). Some formate dehydrogenases of anaerobes contain tungsten instead of molybdenum (Ljungdahl and Andreesen, 1975).

Like aerobic organisms, many obligate anaerobic bacteria employ reactions of the tricarboxylic acid cycle for glutamate synthesis via citrate and α-oxoglutarate. It is somewhat surprising that organisms such as clostridia and sulfate-reducing bacteria contain citrate synthase—an enzyme that seems to be so characteristic for aerobic metabolism. Some of them, e.g., *Clostridium kluyveri* and *Desulfovibrio vulgaris*, contain a special type (re-citrate synthase) that differs from the usual si-type in its stereospecificity (Gottschalk, 1968; Gottschalk and Barker, 1967). Anaerobes, however, do not contain a complete tricarboxylic acid cycle. Sulfate-reducing bacteria lack α-oxoglutarate dehydrogenase and succinyl-CoA synthetase (Lewis and Miller, 1977).

It has been mentioned that *Methanobacterium thermoautrophicum* does not require any growth factors. Likewise, a number of organotrophic anaer-

obes grow on simple media containing a carbon and energy source and minerals. To give just one example, *Bacteroides amylophilus* grows on maltose and minerals in a CO_2 atmosphere (Hamlin and Hungate, 1956). Other obligate anaerobes require a small number of growth factors such as biotin, p-aminobenzoic acid, and pantothenic acid. However, the majority of these organisms exhibit rather complex growth factor requirements. This, of course, has something to do with their natural environment, which usually is very rich in vitamins, amino acids, etc. The lactic acid bacteria have such a limited biosynthetic capacity that they can only grow if all the monomers needed for macromolecule biosynthesis are provided by the environment.

It then can be concluded that it is characteristic of obligate anaerobes to employ reductive carboxylation reactions for biosyntheses. Presumably, the Calvin cycle does not occur among chemotrophic obligate anaerobes. Many obligate anaerobes exhibit rather complex growth factor requirements.

Energy Metabolism of Obligate Anaerobes and the Anaerobic Food Chain

The discussion of the macromolecular architecture and the biosynthetic pathways of the obligate anaerobes has revealed some interesting but not really dramatic differences in comparison with other prokaryotes. What about the anaerobic energy metabolism? Clearly, the ability to carry out fermentations is not limited to obligate anaerobes. These processes are also conducted by facultative anaerobes and, under the appropriate conditions, even by eukaryotes such as yeasts and protozoa. It has often been maintained in the past that the inability to perform electron transport phosphorylation and to synthesize cytochromes may be a useful criterion for characterizing obligate anaerobes. This criterion is not applicable. The occurrence of cytochromes was demonstrated in propionibacteria in 1942 (Chaix and Fromageot), in sulfate-reducing bacteria in 1956 (Postgate), in *Bacteroides ruminicola* in 1962 (White, Bryant, and Caldwell), and recently even in clostridia (Gottwald et al., 1975). *Bacteroides fragilis* will only grow well if hemin is added to the medium, because hemin is required for the biosynthesis of cytochromes (Macy, Probst, and Gottschalk, 1975; Varel and Bryant, 1974).

It is a fact that several obligate anaerobes gain ATP only by substrate-level phosphorylation (e.g., most clostridia and lactic acid bacteria). However, others also employ electron transport phosphorylation as a means of ATP production (Thauer, Jungermann, and Decker, 1977). This is true for sulfate- and sulfur-reducing bacteria. The reduction of sulfate to sulfite is associated with the conversion of ATP to AMP. Since the reduction of sulfate to sulfide by no means requires ATP, the reduction of sulfite to sulfide must be coupled to the phosphorylation of at least two ADPs. Presumably, even more than two ATPs are produced in this electron transport process, because recently sulfate-reducing bacteria have been described that grow with sulfate, H_2, CO_2, and acetate (Badziong, Thauer, and Zeikus, 1978). *Desulfotomaculum acetoxidans* can oxidize acetate with the concomitant reduction of sulfate to sulfide (Widdel and Pfennig, 1977). Accordingly, this fermentation can only be understood if the reduction of sulfate to sulfide brings about a net production of ATP. Another organism that is rich in cytochromes and must gain ATP by electron transport phosphorylation is *Desulfuromonas acetoxidans*. It couples the reduction of elemental sulfur to the oxidation of acetate (Pfennig and Biebl, 1976). *Spirillum* 5175 grows at the expense of sulfur or sulfite reduction with molecular hydrogen (Wolfe and Pfennig, 1977).

Methanogenic bacteria and acetogenic bacteria (*Clostridium aceticum*, *Acetobacterium woodii*) convert molecular hydrogen and carbon dioxide to methane and acetic acid, respectively:

$$4 H_2 + CO_2 \rightarrow CH_4 + 2 H_2O$$
$$4 H_2 + 2 CO_2 \rightarrow CH_3\text{-}COOH + 2 H_2O$$

It is unlikely that these organisms can couple the reduction of CO_2 with substrate-level phosphorylation, so it has to be assumed that they also produce ATP by electron transport phosphorylation (Barker, 1956; Mah et al., 1977; Schoberth, 1977; Wolfe, 1971; Zeikus, 1977).

A number of bacteria that form succinate as an end product or produce propionate via succinate synthesize ATP by electron transport phosphorylation. The reaction that is taken advantage of for this purpose is the fumarate reductase reaction (Kröger, 1977). This enzyme system is located in the membrane and is associated with redox carriers (in most cases menaquinones and cytochromes) that take part in the electron transfer from H_2, formate, lactate, glycerol phosphate, or NADH to fumarate (Fig. 1). That in several anaerobes this electron transfer is coupled to the phosphorylation of ADP has become evident from a number of observations.

Vibrio succinogenes grows on fumarate and H_2,

the only product being succinate (Wolin, Wolin, and Jacobs, 1961):

$$\text{fumarate} + H_2 \rightarrow \text{succinate}$$

Clearly, this organism must gain ATP in the fumarate reductase system. Several organisms that ferment carbohydrates to succinate, propionate, acetate, and CO_2 exhibit unusually high growth yields as compared with saccharolytic clostridia or lactic acid bacteria. *Propionibacterium freudenreichii* forms 65 g of cells (dry weight) per mole of glucose (De Vries, van Wyck-Kapteyn, and Stouthamer, 1973), *Selenomonas ruminantium* forms 62 g (Hobson and Summers, 1972), and *Bacteroides fragilis* 50 g (Macy, Probst, and Gottschalk, 1975), whereas the yields of the lactic acid bacteria are on the order of 15–30 g per mole of glucose (Decker, Jungermann, and Thauer, 1970). Finally, ATP synthesis coupled to the reduction of fumarate with H_2 has been demonstrated using particulate fractions of *Desulfovibrio gigas* (Barton, LeGall, and Peck, 1970) and *Vibrio succinogenes* (Kröger, 1977).

From the above discussion, it is apparent that the mechanism of ATP synthesis by electron transport phosphorylation is known in obligate anaerobes. An important difference from the aerobes is, of course, that the obligate anaerobes *per definitionem* cannot use oxygen as the terminal electron acceptor of their electron transport chain and that they do not contain the elaborate and ''long'' chains that are required for an energy-conserving electron transport from donors such as NADH to the highly electropositive oxygen. The electron acceptors used by anaerobes (CO_2, sulfate, fumarate) are much more electronegative than oxygen (Fig. 2), and, as a consequence, the ATP yield per two electrons is much lower than with

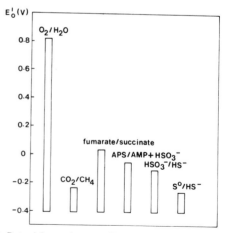

Fig. 2. Potential span between H_2 and the electron acceptors of electron transport chains. APS = adenosine 5′-phosphosulfate, which is reduced to HSO_3^- + AMP. This is the first redox reaction in sulfate reduction. The data were taken from Thauer, Jungermann, and Decker (1977).

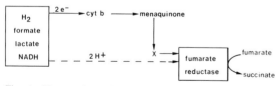

Fig. 1. The membrane-bound fumarate reductase system. X, unknown carrier.

oxygen. Instead of the water and carbon dioxide produced by aerobes, the energy metabolism of anaerobes is associated with the production of large amounts of reduced carbon or sulfur compounds (and carbon dioxide). Despite these important differences it is obvious that many of the existing obligate anaerobes are equipped with a well-advanced energy metabolism and that only some groups of anaerobes carry out fermentation processes resembling those which might have been predominant in the early stages of life on earth (e.g., lactic acid fermentation).

It must be mentioned in this connection that evolution has also created new anaerobic habitats such as the rumen and the gut (Clarke and Bauchop, 1977; Hungate, 1966, 1975). The challenge of these habitats has probably contributed much to the diversity as we find it among the various obligate anaerobes today. This diversity of catabolic and fermentation pathways is quite impressive. It reflects the major achievement of the anaerobes, which is represented in a complicated food chain able to bring about the conversion of practically every biological compound to methane and CO_2 or, with sulfate as electron acceptor, to sulfide and CO_2.

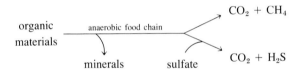

It is important that many bacterial species participate in this conversion. This gives flexibility to the system and ensures that the overall process can proceed under a great variety of conditions: at low or high temperatures, at different pH values, in marine habitats or in fresh water, and in the presence of widely varying nutrient concentrations.

The end products of the breakdown of organic material by the anaerobic food chain are, as we have seen, gaseous or soluble compounds. This is of great importance because these compounds can be freed from the anaerobic environment by diffusion, and thus the carbon cycle and the cycles of the other bioelements can be completed.

The anaerobic food chain starts with the fermentation of biological macromolecules, which usually are converted to low-molecular-weight compounds by exoenzymes. Many anaerobes are able to utilize cellulose (e.g., Clostridium cellobioparum, Clostridium thermocellum, Ruminococcus albus), starch (e.g., Bacteroides amylophilus), and other polysaccharides. The corresponding disaccharides and monosaccharides are taken up by the organisms and, depending on the enzymes available to the bacterial species involved, are fermented to organic acids, alcohols, carbon dioxide, and molecular hydrogen. For the breakdown of hexoses, the Embden-Meyerhof

pathway is usually employed by obligate anaerobes. A number of organisms, however, use other pathways. The heterofermentative lactic acid bacteria such as Leuconostoc mesenteroides and Lactobacillus brevis oxidize glucose first to xylulose 5-phosphate, which then undergoes a phosphoketolase reaction to yield acetyl phosphate and 3-phosphoglyceraldehyde. These compounds are converted to the products ethanol and lactate. Bifidobacterium bifidum uses a pathway for sugar degradation different from the two mentioned above. It also involves phosphoketolase reactions and yields 2 mol lactate and 3 mol acetate from 2 mol glucose (for literature on the fermentation of carbohydrates see Doelle, 1975; Wood, 1961).

Many Clostridium, Peptococcus, and Bacteroides species are proteolytic. The breakdown of amino acids by anaerobes usually proceeds along pathways different from those used by aerobes. The latter use oxidative pathways that lead to the formation of acetyl-CoA or intermediates of the tricarboxylic acid cycle. In anaerobes, two types of amino acid fermentation are known:

1. A number of single amino acids are degraded by rather complicated and elaborate pathways, e.g., glycine by Peptococcus anaerobius, glutamate by Clostridium tetanomorphum, and lysine by Clostridium sticklandii.
2. Several amino acids are fermented in pairs, a fermentation type discovered by Stickland and called the Stickland reaction. Here the oxidation of one particular amino acid is coupled to the reduction of a second one:

alanine + 2 H_2O → acetate + NH_3 + CO_2 + 4 H
2 glycine + 4 H → 2 acetate + NH_3

The reduction of glycine to acetate is a rather complex reaction. The corresponding enzyme system is associated with the membrane and consists of several proteins including a selenoprotein. Amino acids such as alanine, leucine, valine, and histidine are preferably used as H donors, whereas glycine, proline, ornithine, and arginine function as H acceptors. The Stickland reaction is carried out by several proteolytic clostridia such as C. sporogenes, C. sticklandii, C. histolyticum, and C. botulinum (for literature on the fermentation of amino acids see Barker, 1961; Elsden, Hilton, and Waller, 1976). Purines and pyrimidines, which are supplied by the anaerobic decomposition of organic material, also serve as substrates for several obligate anaerobes, which ferment these compounds to acetate, glycine, ammonia, formate, and CO_2. Clostridium acidiurici and C. cylindrosporum are very specialized, in that they grow only with guanine, urate, xanthine, and hypoxanthine; C. oroticum utilizes orotic acid as well as several carbohydrates (for literature on the

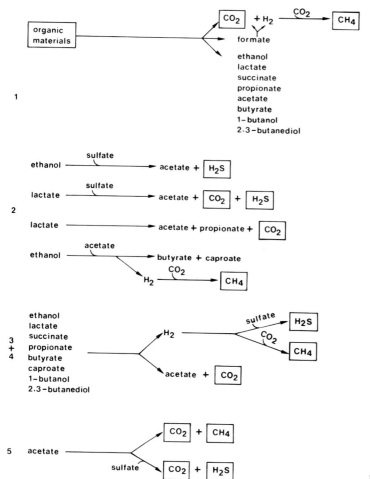

Fig. 3. The anaerobic food chain.

fermentation of purines and pyrimidines see Barker, 1961; Vogels and Van der Drift, 1976).

The fermentations discussed above constitute the first step of the anaerobic food chain and bring about the conversion of complex organic material to acids, alcohols, CO_2, and H_2. As is indicated in Fig. 3, these products serve as substrates for other reactions in a second step of the food chain. Lactate is fermented to CO_2, acetate, and propionate by the propionibacteria, *Veillonella* species, and *Megasphaera elsdenii*. *Clostridium kluyveri* carries out an ethanol-acetate fermentation that yields butyrate, caproate, and H_2. Sulfate-reducing bacteria utilize lactate, L-malate, and ethanol and form acetate, CO_2, and H_2S (for literature see Thauer, Jungermann, and Decker, 1977).

The next two steps of the anaerobic food chain can only be understood if they are considered together. The basis for them is that methanogenic and sulfate-reducing bacteria are able to consume H_2 so effectively that its partial pressure is kept below 10^{-3} atm in anaerobic environments. Under these conditions, the oxidation of organic compounds to acetate and molecular hydrogen becomes thermodynamically feasible (Hungate, 1967; Wolin, 1976). Clearly,

organisms that carry out these oxidations are difficult to study in pure culture. They will only grow if H_2 is continuously removed, and the organisms can more conveniently be studied in mixed cultures with methanogenic bacteria or sulfate-reducing bacteria (Bryant et al., 1977; Mah et al., 1977; Zeikus, 1977). A mixed culture of the so-called S organism with a methane bacterium will, for instance, oxidize ethanol to acetate plus H_2 (Bryant et al., 1967):

$$2 \text{ ethanol} + 2 \text{ H}_2\text{O} \rightarrow 2 \text{ acetate} + 4 \text{ H}_2$$
$$CO_2 + 4 \text{ H}_2 \rightarrow CH_4 + 2 \text{ H}_2\text{O}$$

Hydrogen gas does not accumulate, and the term used to describe this type of syntrophic metabolism is "interspecies hydrogen transfer". Another example illustrates the benefit of interspecies hydrogen transfer. *Ruminococcus albus,* in pure culture, will ferment glucose to ethanol, acetate, CO_2, and H_2:

$$\text{glucose} \rightarrow \text{acetate} + \text{ethanol} + 2 \text{ CO}_2 + 2 \text{ H}_2$$

In the presence of a methanogenic organism (or *Vibrio succinogenes,* which reduces fumarate with H_2 to succinate), only acetate and CO_2 are formed as carbon compounds (Wolin, 1976):

$$glucose \rightarrow 2\ acetate + 2\ CO_2 + 4\ H_2$$

In mixed culture, *Ruminococcus albus* gains 4 ATPs per glucose as compared with 3 ATPs per glucose in pure culture.

It should be emphasized that processes of this type are of great importance in nature. They are not well known yet because they are so difficult to study in the laboratory. Recently an organism has been described that oxidizes fatty acids such as butyrate, caproate, and octanoate to acetate and H_2, provided that the H_2 produced is consumed by a methanogenic or sulfate-reducing bacterium (Mc Inerney, Bryant, and Pfennig, 1978). In many of these processes, more than two organisms are probably involved. It has, for instance, been observed that benzoate is converted to methane and CO_2 by a consortium consisting of more than three different species (Ferry and Wolfe, 1976).

Although not all the processes involved are understood yet, it can be assumed that steps 3 and 4 of the food chain bring about the conversion of the products of steps 1 and 2 to acetate and CO_2. The hydrogen liberated is used to reduce CO_2 and sulfate, respectively. Apparently in a side process, H_2 and CO_2 yield acetate instead of methane. This fermentation is carried out by *Clostridium aceticum* and *Acetobacterium woodii* (Balch et al., 1977).

In the final step of the anaerobic food chain, acetate is oxidized to CO_2. Three organisms are known so far that carry out fermentations involving the oxidation of acetate. *Desulfotomaculum acetoxidans* (Widdel and Pfennig, 1977) and *Desulfuromonas acetoxidans* (Pfennig and Biebl, 1976) employ sulfate or elemental sulfur as electron acceptor. *Methanosarcina barkeri* ferments acetate to CO_2 and methane (Mah et al., 1977; Stadtman, 1967).

The conclusion of this chapter is that mechanisms involved in ATP synthesis by obligate anaerobes are principally the same as those of aerobes. The ATP yields per mole of substrate degraded are much smaller as compared with those of aerobes. What really constitutes the anaerobic way of life is that the anaerobes together, in a very complex series of fermentative processes, bring about the complete oxidation of organic material. Instead of oxygen or nitrate, which are rapidly consumed by aerobes, the obligate anaerobes use sulfate (and other sulfur compounds) or carbon dioxide as terminal electron acceptors.

Oxygen Sensitivity of Anaerobes

The reduction of oxygen may lead to the formation of three different products: water, hydrogen peroxide, or the superoxide anion:

$$O_2 + 4\ H^+ + 4\ e^- \rightarrow 2\ H_2O$$

$$O_2 + 2\ H^+ + 2\ e^- \rightarrow H_2O_2$$
$$O_2 + 1\ e^- \rightarrow O_2^-$$

In aerobic respiration, water, of course, is the major reduction product, but small amounts of hydrogen peroxide and the superoxide anion are also formed when reduced flavoproteins or iron sulfur proteins come into contact with oxygen. Both compounds are toxic to organisms, and aerobes contain an appropriate protective mechanism to overcome the deleterious effects of hydrogen peroxide and the superoxide anion. This mechanism consists of the two enzymes superoxide dismutase and catalase.

$$O_2^- + O_2^- + 2\ H^+ \xrightarrow{\text{superoxide dismutase}} H_2O_2 + O_2$$
$$H_2O_2 \xrightarrow{\text{catalase}} H_2O + \tfrac{1}{2}\ O_2$$

Aerobes usually contain both enzymes (McCord, Keele, and Fridovich, 1971; Morris, 1976). Some exceptions have been reported. The aerobe *Bacillus popilliae,* for instance, lacks catalase (Costilow and Keele, 1972). It may, however, be replaced in its function by peroxidases that oxidize NADH or other reduced compounds at the expense of H_2O_2:

$$XH_2 + H_2O_2 \xrightarrow{\text{peroxidase}} X + 2\ H_2O$$

Superoxide dismutase seems to be indispensable to aerobes.

Among the obligate anaerobic bacteria, organisms are found that are more or less aerotolerant. Many of these organisms (e.g., a number of lactic acid bacteria) have been shown to contain superoxide dismutase and to lack catalase (McCord, Keele, and Fridovich, 1971; Morris, 1976). A number of anaerobes, however, are decidedly intolerant of oxygen. Most noteworthy in this connection are the methane bacteria. Their cultivation in the laboratory requires special precautions (Hungate, 1969); also, many clostridial species and sulfate-reducing bacteria will not grow even at low oxygen tensions. Obligate anaerobes are very rich in flavoproteins, which readily react with oxygen to yield hydrogen peroxide and the superoxide anion. Because these organisms contain very low superoxide dismutase activities, if any, they are subject to the deleterious effects of the reduction products of oxygen.

It must be pointed out in this connection that the simple exclusion of molecular oxygen is not sufficient to provide good conditions of growth for strict anaerobes. In addition, they require a low redox potential in their environment and growth media supplemented with compounds such as ascorbate, hydrogen sulfide, sodium thioglycolate, or cysteine.

It could be argued that the oxygen intolerance of many anaerobes is a disadvantage for these organisms. In fact, this property does make the cultivation and investigation of these microorganisms in the laboratory difficult. However, they flourish in nature and are abundant in many habitats, which signifies the limits of oxygen-dependent metabolism.

Concluding Remarks

This article has attempted to show that the anaerobes have successfully used the time from their first appearance on earth some 3.5×10^9 years ago until now to take advantage of the potentialities of their habitat. Their major achievement is that they have developed a system of fermentations that takes advantage of two electron acceptors that are abundant under anaerobic conditions: carbon dioxide and sulfate. The utilization of these acceptors required the acquisition of electron transport phosphorylation as a means of ATP synthesis.

Literature Cited

Allison, M. J. 1969. Biosynthesis of amino acids by ruminal microorganisms. Journal of Animal Sciences **29:**797–807.

Allison, M. J., Robinson, I. M. 1970. Biosynthesis of α-ketoglutarate by the reductive carboxylation of succinate in *Bacteroides ruminicola.* Journal of Bacteriology **104:**50–56.

Andreesen, J. R., Ljungdahl, L. G. 1974. Nicotinamide adenine dinucleotide phosphate-dependent formate dehydrogenase from *Clostridium thermoaceticum:* Purification and properties. Journal of Bacteriology **120:**6–14.

Andrew, J. G., Morris, J. G. 1965. The biosynthesis of alanine by *Clostridium kluyveri.* Biochimica et Biophysica Acta **97:**176–179.

Archibald, A. R. 1974. The structure, biosynthesis and function of teichoic acid, pp. 53–90. In: Rose, A. H., Tempest, D. W. (eds.), Advances in microbial physiology, vol. 11. London, New York: Academic Press.

Badziong, W., Thauer, R. K., Zeikus, J. G. 1978. Isolation and characterization of *Desulfovibrio* growing on hydrogen plus sulfate as the sole energy source. Archives of Microbiology **116:**41–49.

Balch, W. E., Schoberth, S., Tanner, R. S., Wolfe, R. S. 1977. *Acetobacterium,* a new genus of hydrogen-oxidizing carbon dioxide-reducing anaerobes, and characterization of *Acetobacterium woodii* sp. nov. International Journal of Systematic Bacteriology **27:**355–361.

Barker, H. A. 1956. Bacterial fermentations. New York: John Wiley & Sons.

Barker, H. A. 1961. Fermentations of nitrogenous organic compounds, pp. 151–207. In: Gunsalus, I. C., Stanier, R. Y. (eds.), The bacteria, vol. 2. New York, London: Academic Press.

Barton, L. L., LeGall, J., Peck, H. D., Jr. 1970. Phosphorylation coupled to oxidation of hydrogen with fumarate in extracts of the sulfate reducing bacterium, *Desulfovibrio gigas.* Biochemical and Biophysical Research Communications **41:**1036–1042.

Beuscher, N., Gottschalk, G. 1972. Lack of citrate lyase—the key enzyme of the reductive carboxylic acid cycle in *Chlorobium thiosulfatophilum* and *Rhodospirillum rubrum.* Zeitschrift für Naturforschung **27b:**967–973.

Bryant, M. P., Wolin, E. A., Wolin, M. J., Wolfe, R. S. 1967. *Methanobacillus omelianskii,* a symbioassociation of two species of bacteria. Archiv für Mikrobiologie **59:**20–31.

Bryant, M. P., Campbell, L. L., Reddy, C. A., Crabill, M. R. 1977. Growth of *Desulfovibrio* in lactate or ethanol media low in sulfate in association with H_2-utilizing methanogenic bacteria. Applied and Environmental Microbiology **33:** 1162–1169.

Buchanan, B. B. 1973. Ferredoxin and carbon assimilation, pp. 129–150. In. Lovenberg, W. (ed.), Iron-sulfur proteins, vol. 1. New York, London: Academic Press.

Buchanan, B. B., Sirevåg, R. 1976. Ribulose 1,5-diphosphate carboxylase and *Chlorobium thiosulfatophilum.* Archives of Microbiology **109:**15–19.

Chaix, P., Fromageot, C. 1942. Les cytochromes de *Propionibacterium pentosaceum.* Bulletin de la Société de Chemie Biologique **24:**1125–1127.

Cho, K. Y., Doy, C. H. 1973. Ultrastructure of the obligately anaerobic bacteria *Clostridium kluyveri* and *C. acetobutylicum.* Australian Journal of Biological Sciences **26:**547–558.

Clarke, R. T. J., Bauchop, T. 1977. Microbial ecology of the gut. London: Academic Press.

Costerton, J. W., Ingram, J. M., Cheng, K.-J. 1974. Structure and function of the cell envelope of Gram-negative bacteria. Bacteriological Reviews **38:**87–110.

Costilow, R. N., Keele, B. B., Jr. 1972. Superoxide dismutase in *Bacillus popilliae.* Journal of Bacteriology **111:**628.

Decker, K., Jungermann, K., Thauer, R. K. 1970. Energy production in anaerobic organisms. Angewandte Chemie International Edition in English **9:**138–158.

de Vries, W., van Wyck-Kapteyn, W. M. C., Stouthamer, A. H. 1973. Generation of ATP during cytochrome-linked anaerobic electron transport in propionic acid bacteria. Journal of General Microbiology **76:**31–41.

Doelle, H. W. 1975. Bacterial metabolism. New York, San Francisco, London: Academic Press.

Elsden, S. R., Hilton, M. G., Waller, J. M. 1976. The end products of the metabolism of aromatic amino acids by clostridia. Archives of Microbiology **107:**283–288.

Emeruwa, A. C., Hawirko, R. Z. 1973. Poly-β-hydroxybutyrate metabolism during growth and sporulation of *Clostridium botulinum.* Journal of Bacteriology **116:**989–993.

Emmanuel, B., Milligan, L. P. 1973. Reductive carboxylation of acetyl phosphate by cell-free extracts of mixed rumen microorganisms. Journal of General Microbiology **77:**537–539.

Evans, M. C. W., Buchanan, B. B. 1965. Photoreduction of ferredoxin and its use in carbon dioxide fixation by a subcellular system from a photosynthetic bacterium. Proceedings of the National Academy of Sciences of the United States of America **63:**1420–1425.

Evans, M. C. W., Buchanan, B. B., Arnon, D. I. 1966. A new ferredoxin-dependent carbon reduction cycle in a photosynthetic bacterium. Proceedings of the National Academy of Sciences of the United States of America **55:**928–934.

Ferry, J. G., Wolfe, R. S. 1976. Anaerobic degradation of benzoate to methane by a microbial consortium. Archives of Microbiology **107:**33–40.

Fox, G. E., Magrum, L. J., Balch, W. E., Wolfe, R. S. 1977. Classification of methanogenic bacteria by 16 S ribosomal RNA characterization. Proceedings of the National Academy of Sciences of the United States of America **74:**4537–4541.

Gavard, R., Milhaud, G. 1952. Sur un polysaccharide isole de *Clostridium butyricum.* Annales de l'Institut Pasteur **82:**471–483.

Gottschalk, G. 1968. The stereospecificity of the citrate synthase in sulfate-reducing and photosynthetic bacteria. European Journal of Biochemistry **5:**346–351.

Gottschalk, G., Barker, H. A. 1967. Presence and stereospecificity of citrate synthase in anaerobic bacteria. Biochemistry **6:**1027–1034.

Gottwald, M., Andreesen, J. R., LeGall, J., Ljungdahl, L. G. 1975. Presence of cytochrome and menaquinone in *Clostridium formicoaceticum* and *Clostridium thermoaceticum.* Journal of Bacteriology **122:**325–328.

Hamlin, L. J., Hungate, R. E. 1956. Culture and physiology of a starch digesting bacterium (*Bacteroides amylophilus* n. sp.) from the bovine rumen. Journal of Bacteriology, **27**:548–554.

Hobson, P. N., Summers, R. 1972. ATP-pool and growth yield in *Selenomonas ruminantium*. Journal of General Microbiology **70**:351–360.

Hungate, R. E. 1963. Polysaccharide storage and growth efficiency in *Ruminococcus albus*. Journal of Bacteriology **86**:848–854.

Hungate, R. E. 1966. The rumen and its microbes. New York, London: Academic Press.

Hungate, R. E. 1967. Hydrogen as an intermediate in the rumen fermentation. Archiv für Mikrobiologie **59**:158–164.

Hungate, R. E. 1969. A roll tube method for cultivation of strict anaerobes, pp. 117–132. In: Norris, J. R., Ribbons, D. W. (eds.), Methods in microbiology, vol 3 B. New York, London: Academic Press.

Hungate, R. E. 1975. The rumen microbial ecosystem. Annual Review of Ecology and Systematics **6**:39–66.

Kandler, O., Hippe, H. 1977. Lack of peptidoglycan in the cell walls of *Methanosarcina barkeri*. Archives of Microbiology **113**:57–60.

Kröger, A. 1977. Phosphorylative electron transport with fumarate and nitrate as terminal hydrogen acceptors, pp. 61–93. In: Haddock, B. A., Hamilton, W. A. (eds.), Microbial energetics. Cambridge: Cambridge University Press.

Laishley, E. J., Brown, R. G., Otto, M. C. 1974. Characteristics of a reserve α-glucan from *Clostridium pasteurianum*. Canadian Journal of Microbiology **20**:559–562.

Leonhardt, U., Andreesen, J. R. 1977. Some properties of formate dehydrogenase, accumulation and incorporation of [185]W-tungsten into proteins of *Clostridium formicoaceticum*. Archives of Microbiology **115**:277–284.

Lewis, A. J., Miller, J. D. A. 1977. The tricarboxylic acid pathway in *Desulfovibrio*. Canadian Journal of Microbiology **23**:916–921.

Ljungdahl, L. G., Andreesen, J. R. 1975. Tungsten, a component of active formate dehydrogenase of *Clostridium thermoaceticum*. FEBS Letters **54**:279–282.

McClung, L. S. 1956. The anaerobic bacteria with special reference to the genus *Clostridium*. Annual Reviews of Microbiology **10**:173–192.

McCord, J. M., Keele, B. B., Jr., Fridovich, I. 1971. An enzyme-based theory of obligate anaerobiosis: the physiological function of superoxide dismutase. Proceedings of the National Academy of Sciences of the United States of America **68**:1024–1027.

McFadden, B. A. 1973. Autotrophic CO_2 assimilation and the evolution of ribulose-diphosphate carboxylase. Bacteriological Reviews **37**:289–319.

McInerney, M. J., Bryant, M. P., Pfennig, N. 1978. An anaerobic bacterium that oxidizes fatty acids in syntrophic association with H_2-utilizing bacteria. Abstracts of the Annual Meeting of the American Society for Microbiology **1978**:94.

Macy, J., Probst, I., Gottschalk, G. 1975. Evidence for cytochrome involvement in fumarate reduction and adenosine 5'-triphosphate synthesis by *Bacteroides fragilis* grown in the presence of hemin. Journal of Bacteriology **123**:436–442.

Mah, R. A., Ward, D. M., Baresi, L., Glass, T. L. 1977. Biogenesis of methane. Annual Reviews of Microbiology **31**:309–341.

Morris, J. G. 1976. Oxygen and the obligate anaerobe. Journal of Applied Bacteriology **40**:229–244.

Pfennig, N., Biebl, H. 1976. *Desulfuromonas acetoxidans* gen. nov. and sp. nov., a new anaerobic sulfur-reducing, acetate-oxidizing bacterium. Archives of Microbiology **110**:3–12.

Postgate, J. R. 1956. Cytochrome c_3 and desulphoviridin; pigments of the anaerobe *Desulphovibrio desulphuricans*. Journal of General Microbiology **14**:545–572.

Preiss, J. 1977. Regulation of adenosine diphosphate glucose pyrophosphorylase. Advances in Enzymology **46**:317–381.

Quandt, L., Pfennig, N., Gottschalk, G. 1978. Evidence for the key position of pyruvate synthase in the assimilation of CO_2 by *Chlorobium*. FEMS Microbiology Letters **3**:227–230.

Quandt, L., Gottschalk, G., Ziegler, H., Stichler, W. 1977. Isotope discrimination by phototrophic bacteria. FEMS Microbiology Letters **1**:125–128.

Robson, R. L., Robson, R. M., Morris, J. G. 1974. The biosynthesis of granulose by *Clostridium pasteurianum*. Biochemical Journal **144**:503–511.

Schleifer, K. H., Kandler, O. 1972. Peptidoglycan types of bacterial cell walls and their taxonomic implications. Bacteriological Reviews **36**:407–477.

Schoberth, S. 1977. Acetic acid from H_2 and CO_2. Archives of Microbiology **114**:143–148.

Sirevåg, R., Ormerod, J. G. 1970. Carbon dioxide fixation in green sulphur bacteria. Biochemical Journal **120**:399–408.

Sleytr, U. B., Glauert, A. M. 1976. Ultrastructure of the cell walls of two closely related clostridia that posses different regular arrays of surface subunits. Journal of Bacteriology **126**:869–882.

Sorokin, Y. I. 1966. Role of carbon dioxide and acetate in the biosynthesis by sulfate-reducing bacteria. Nature **210**:551–552.

Stadtman, T. C. 1967. Methane fermentation. Annual Reviews of Microbiology **21**:121–142.

Strasdine, G. A. 1968. Amylopectin accumulation in *Clostridium botulinum* type E. Canadian Journal of Microbiology **14**:1059–1062.

Suh, B., Akagi, J. M. 1966. Pyruvate-carbon dioxide exchange reaction of *Desulfovibrio desulfuricans*. Journal of Bacteriology **91**:2281–2285.

Thauer, R. K. 1972. CO_2-reduction to formate by NADPH. The initial step in the total synthesis of acetate from CO_2 in *Clostridium thermoaceticum*. FEBS Letters **27**:111–115.

Thauer, R. K., Fuchs, G., Jungermann, K. 1977. The role of iron-sulfur proteins in formate metabolism, pp. 121–156. In: Lovenberg, W. (ed.), Iron-sulfur proteins, vol. 3. New York, London: Academic Press.

Thauer, R. K., Jungermann, K., Decker, K. 1977. Energy conservation in chemotrophic anaerobic bacteria. Bacteriological Reviews **41**:100–180.

Tomlinson, N., Barker, H. A. 1954. Carbon dioxide and acetate utilization by *Clostridium kluyveri*. I. Influence of nutritional conditions on utilization patterns. Journal of Biological Chemistry **209**:585–595.

Varel, V. H., Bryant, M. P. 1974. Nutritional features of *Bacteroides fragilis* subsp. *fragilis*. Applied Microbiology **18**:251–257.

Vogels, G. D., Van der Drift, C. 1976. Degradation of purines and pyrimidines by microorganisms. Bacteriological Reviews **40**:403–468.

White, D. C., Bryant, M. P., Caldwell, D. R. 1962. Cytochrome-linked fermentation in *Bacteroides ruminicola*. Journal of Bacteriology **84**:822–828.

Widdel, F., Pfennig, N. 1977. A new anaerobic, sporing, acetate-oxidizing, sulfate-reducing bacterium, *Desulfotomaculum* (emend.) *acetoxidans*. Archives of Microbiology **112**:119–122.

Wieringa, K. T. 1940. The formation of acetic acid from carbon dioxide and hydrogen by anaerobic spore-forming bacteria. Antonie van Leeuwenhoek Journal of Microbiology and Serology **6**:251–262.

Woese, C. R., Fox, G. E. 1977. Phylogenetic structure of the prokaryotic domain: The primary kingdoms. Proceedings of the National Academy of Sciences of the United States of America **74**:5088–5090.

Wolfe, R. S. 1971. Microbial formation of methane, pp. 107–146. In: Rose, A. H., Wilkinson, J. F. (eds.), Advances in microbial physiology, vol. 6. London, New York: Academic Press.

Wolfe, R. S., Pfennig, N. 1977. Reduction of sulfur by Spirillum 5175 and Syntrophism with *Chlorobium*. Applied and Environmental Microbiology **33**:427–433.

Wolin, M. J. 1976. Interactions between H$_2$-producing and methane-producing species, pp. 141–150. In: Schlegel, H. G., Gottschalk, G., Pfennig, N. (eds.), Microbial production and utilization of gases. Göttingen: E. Goltze.

Wolin, M. J., Wolin, E. A., Jacobs, N. J. 1961. Cytochrome-producing anaerobic vibrio, *Vibrio succinogenes,* sp. n. Journal of Bacteriology **81:**911–917.

Wood, W. A. 1961. Fermentation of carbohydrates and related compounds, pp. 59–149. In: Gunsalus, I. C., Stanier, R. Y. (eds.), The bacteria, vol. 2. New York, London: Academic Press.

Zeikus, J. G. 1977. The biology of methanogenic bacteria. Bacteriological Reviews **41:**514–541.

Zeikus, J. G., Wolfe, R. S. 1972. *Methanobacterium thermoautotrophicum* sp. n., an anaerobic, autotrophic, extreme thermophile. Journal of Bacteriology **109:**707–713.

Zeikus, J. G., Fuchs, G., Kenealy, W., Thauer, R. K. 1977. Oxidoreductases involved in cell carbon synthesis of *Methanobacterium thermoautotrophicum.* Journal of Bacteriology **132:**604–613.

CHAPTER 116

Pathogenic Members of the Genus *Bacteroides*

VULUS R. DOWELL, JR., and GEORGE L. LOMBARD

The genus *Bacteroides,* as classified in the eighth edition of *Bergey's Manual of Determinative Bacteriology* (Buchanan and Gibbons, 1974), contains 22 species and a number of subspecies of obligately anaerobic, nonsporeforming, Gram-negative bacilli. Although little is known about the pathogenicity of some, it appears that only a small number of species contain microorganisms that commonly are associated with disease in man or lower animals (Balows, 1974; Gorbach and Bartlett, 1974a; Smith, 1975). This chapter is concerned primarily with the *Bacteroides* species commonly isolated from properly collected clinical materials associated with disease in humans and some associated with disease in animals. Other *Bacteroides* species are discussed in this Handbook, Chapter 117.

Bacteroides species commonly isolated from human clinical specimens include members of the *Bacteroides fragilis* group *(B. distasonis, B. fragilis, B. thetaiotaomicron, B. vulgatus),* members of the *Bacteroides melaninogenicus* group *(B. asaccharolyticus, B. melaninogenicus* subsp. *intermedius),* *Bacteroides ureolyticus,* and members of the Center for Disease Control (CDC) groups F-1 and F-2, which are similar in many respects to *B. ureolyticus* (Table 1). In addition to these, we will discuss *Bacteroides oralis* and *Bacteroides pneumosintes,*

which are infrequently associated with human disease; the recently proposed species *Bacteroides disiens* and *Bacteroides bivius* (Holdeman and Johnson, 1977); and *Bacteroides nodosus,* one of the principal agents of foot rot in sheep (Smith, 1975).

Recent studies (Bartlett et al., 1976; Chow, Cunningham, and Guze, 1976; Tally et al., 1975) have revealed that the obligate anaerobes can conveniently be divided into the moderate and strict anaerobe groups described by Loesche (1969). Loesche found that strict anaerobes, such as *Succinivibrio dextrinosolvens, Butyrivibrio fibrisolvens, Treponema macrodentium, Treponema denticola, Treponema oralis, Clostridium haemolyticum, Selenomonas ruminantium,* and *Lachnospira multiparus,* were sensitive to oxygen and would not form surface colonies on an agar medium if the atmosphere contained more than 0.5% oxygen. On the other hand, representatives of the moderate anaerobe group *(B. fragilis, B. melaninogenicus, B. oralis, Fusobacterium nucleatum, Clostridium novyi* type A, and *Peptostreptococcus elsdenii)* formed colonies in an atmosphere that contained oxygen concentrations as high as 3%. In the same study, Loesche observed a significant difference in oxygen tolerance of the strict and moderate anaerobes. Strict anaerobes showed a significant decrease in numbers after exposure to air

Table 1. Bacteroides commonly isolated from clinical materials and associated with disease in humans.

Present designation	Former designation[a]
Bacteroides ureolyticus	*Bacteroides corrodens*
CDC Group F-1[b]	
CDC Group F-2[b]	
Bacteroides distasonis[c]	*B. fragilis* subsp. *distasonis*
Bacteroides fragilis[c]	*B. fragilis* subsp. *fragilis*
Bacteroides thetaiotaomicron[c]	*B. fragilis* subsp. *thetaiotaomicron*
Bacteroides vulgatus[c]	*B. fragilis* subsp. *vulgatus*
Bacteroides asaccharolyticus[d]	*B. melaninogenicus* subsp. *asaccharolyticus*
Bacteroides melaninogenicus subsp. *intermedius*[d]	

[a]Buchanan and Gibbons, 1974.
[b]Dowell and Hawkins, 1974.
[c]Cato and Johnson, 1976.
[d]Finegold and Barnes, 1977.

for 20 min but could still be recovered in small numbers after exposure for 1 h. The moderate anaerobes showed little decrease in numbers after 100 min of exposure to air, and some (e.g., *B. fragilis*) could be exposed to air for as long as 6 h without a significant decrease in numbers.

It now appears that most bacteroides isolated from properly selected and properly collected clinical materials from humans with disease are moderate anaerobes as defined by Loesche (Bartlett et al., 1976; Gorbach and Bartlett, 1974a,b,c; Tally et al., 1975). Although present in the normal microbiota of humans, especially in the large intestine, strict anaerobes are seldom, if ever, associated with disease in humans. Some, for example, *C. haemolyticum* and *Clostridium novyi* type B, are occasionally isolated from pathologic specimens associated with disease in domestic and wild animals (Smith, 1975).

Habitats

Normal Microbiota of Humans and Animals

Members of the family Bacteroidaceae, including the genera *Bacteroides, Fusobacterium,* and *Leptotrichia* (Buchanan and Gibbons, 1974), are widely distributed in nature. Their principal habitats are the mucous membranes of the oral cavity, gastrointestinal tract, and genitourinary orifices of humans and a variety of lower animals. Bacteroides have been isolated from a variety of animals including: buffaloes, cats, chickens, cattle, deer, dogs, gazelles, goats, guinea pigs, hamsters, horses, mice, monkeys, pigs, rattlesnakes, rabbits, raccoons, rats, sheep, and western toads (Bruner and Gillespie, 1973; Buchanan and Gibbons, 1974; Finegold, 1977; Hungate, 1966; Langworth, 1977; Prevot, 1966; Rosebury, 1962; Skinner and Carr, 1974; Smith, 1975).

Various workers have emphasized the numerical importance of the bacteroides in human feces (Eggerth and Gagnon, 1933; Finegold, 1977; Holdeman, Cato, and Moore, 1977; Lewis and Rettger, 1940; Misra, 1938; Moore, Cato, and Holdeman, 1969; Riddell, Morton, and Murray, 1953; Rosebury, 1962; Rubner, 1957; Smith and Crabb, 1961; Subrzycki and Spaulding, 1962; van Houte and Gibbons, 1966; Weiss and Rettger, 1937). Members of the *B. fragilis* group (*B. distasonis, B. fragilis, B. ovatus, B. thetaiotaomicron, B. vulgatus*) (Cato and Johnson, 1976) are especially prevalent in the large intestine of humans, where they may outnumber the aerobic and facultatively anaerobic, Gram-negative bacilli by a factor of 1,000 to 1 (Finegold, 1977). Members of the *B. fragilis* group also reside in the mucous membranes of the normal male and female genitalia, but not of the oral cavity (Finegold, 1977; Gibbons, 1974; Skinner and Carr,

1974; Smith, 1975). On the other hand, *B. oralis* (Buchanan and Gibbons, 1974; Loesche, Socransky, and Gibbons, 1964) is more common in the oral cavity and genitourinary orifices than in the colon (Finegold, 1977; Loesche, 1974).

Although *B. fragilis* (*B. fragilis* subsp. *fragilis;* Buchanan and Gibbons, 1974) is less numerous than other members of the *B. fragilis* group in human feces (Finegold, 1977; Moore and Holdeman, 1974; Werner, 1973), it is much more commonly associated with disease than are the other species (*B. distasonis, B. ovatus, B. thetaiotaomicron, B. vulgatus*). *B. ovatus* is usually found in large numbers in the colon contents or feces of humans, but this species is seldom, if ever, associated with disease in man (Balows, 1974; Dowell and Hawkins, 1974; Holdeman, Cato, and Moore, 1977).

Members of the *B. melaninogenicus* group are commonly found on the mucous membranes of the oral cavity, gastrointestinal tract, and genitourinary orifices of humans (Burdon, 1928; Finegold, 1977; van Houte and Gibbons, 1966). These organisms are ubiquitously present in the oral cavity of adults, where they comprise approximately 5% of the total quantity of bacteria cultivable from the gingival crevice area. They comprise a much smaller proportion of the bacteria on the surfaces of the tongue and cheeks, where streptococci predominate (Gibbons, 1974; Loesche, 1974). These microorganisms are also a part of the normal microbiota of the large intestine of humans (Finegold and Miller, 1968; Smith, 1975) and are commonly isolated from the normal vagina and urethra of females and the urethra of males (Burdon, 1928; Finegold, 1977). Members of the *B. melaninogenicus* group are normal inhabitants of the alimentary tract of various domestic and wild animals (Lev, 1958; Smith, 1975; Smith and Crabb, 1961) and appear to be the most commonly isolated anaerobic, nonsporeforming, Gram-negative bacilli from animal clinical materials (Berg and Loan, 1975; Biberstein, Knight, and England, 1968; Bruner and Gillespie, 1973). Many strains of organisms in the *B. melaninogenicus* group require vitamin K or related components and hemin for growth (Gibbons and MacDonald, 1960; Lev, 1958). A number of bacteria synthesize vitamin K compounds in sufficient quantities to support the growth of *B. melaninogenicus* in vitro and in vivo (Biberstein, Knight, and England, 1968; Bishop and King, 1962; Gibbons, 1974; Gibbons and Engle, 1964; Gibbons et al., 1963). The requirement for vitamin K is probably the reason that *B. melaninogenicus* is seldom found alone in pathologic materials (Smith, 1975).

Little is known about the ecology of *B. ureolyticus* (*B. corrodens*). It appears to be a normal inhabitant of the oral cavity, gastrointestinal tract, and genitourinary orifices of humans (Holdeman, Cato, and Moore, 1977; Jackson et al., 1971). The species *B.*

corrodens, as originally described by Eiken (1958), included facultatively anaerobic as well as obligately anaerobic, Gram-negative, "corroding" bacilli. Since then, facultative anaerobes with these characteristics have been placed in the genus *Eikenella* (Jackson and Goodman, 1972). Riley, Tatum, and Weaver (1973) compared microaerophilic bacteria designated HB-1 by King and Tatum (1962) with the type strain of *B. corrodens* (ATCC 23834, NCTC 10696), a facultative anaerobe, and found that the microorganisms were biochemically and serologically identical. Organisms of CDC group F-1 and CDC group F-2 have a number of characteristics in common with *B. ureolyticus,* but they do not pit the surface of blood agar (Dowell and Hawkins, 1974). Members of the F-1 and F-2 groups have been isolated from a variety of clinical materials (V. R. Dowell and G. L. Lombard, unpublished data), but we have little information on the habitats of the microorganisms.

The organism *B. pneumosintes* was initially described by Olitsky and Gates during studies of nasopharyngeal secretions from patients with influenza (Olitsky and Gates, 1921, 1922). They found it to be an extremely small, obligately anaerobic, Gram-negative rod, $0.15-0.3$ μm in length, which passed through Berkefeld V and N filters and formed small colonies on the surface of 5% rabbit blood agar after incubation in a Brown (1921) anaerobic jar for $7-10$ days. After completing a later study, Mills, Haobley, and Dochez (1928) concluded that these and other filter-passing, anaerobic organisms described by Olitsky and Gates and by Gates and McCartney in their studies of influenza and the common cold were normal flora of the upper respiratory tract and had no etiological relationship to the common cold. Although *B. pneumosintes* is occasionally isolated from clinical materials (Dowell and Hawkins, 1974), little information is available on the role of the organism in health or disease. Bartlett et al. (1974b) isolated *B. pneumosintes* from only one patient in their carefully conducted study of 100 cases of anaerobic

pleuropulmonary disease, and the organism was mixed with other bacteria in that case.

Although *B. disiens* and *B. bivius* have been isolated from a variety of human clinical materials (Holdeman, Cato, and Moore, 1977; Holdeman and Johnson, 1977), it is difficult to draw any conclusions as to their habitat(s) or their pathogenicity on the basis of available information. Both species appear to have a predilection for female and male genitalia (Table 2). Holdeman and Johnson (1977), in their report, did not mention other microorganisms accompanying *B. disiens* and *B. bivius* in the clinical materials.

The following information related to the isolation of *B. bivius, B. disiens,* and other obligately anaerobic, nonsporeforming, Gram-negative bacilli from the vaginal flora of premenopausal women at Duke University Medical Center was supplied by Gale B. Hill (pers. comm., April 7, 1978; (Table 3). In a recent study, the bacterial flora of 65 premenopausal women with no signs of clinical infection who were entering the hospital for vaginal hysterectomy was analyzed. A swab of the posterior fornix was immediately transported to the laboratory in an anaerobic specimen collector and cultured for aerobic and anaerobic bacteria. Of these 65 specimens, 63% (41/65) had one or more species of anaerobic, Gram-negative bacilli. *Bacteroides bivius* was the most frequent isolate and was found in 34% (22/65) of these women. Next in frequency were *B. disiens* (18%) and *B. melaninogenicus* (19%, including all former subspecies). Members of the *B. fragilis* group (14%, 9/65) included *B. fragilis* and *B. vulgatus* (three isolates each), *B. ovatus* (two isolates), and *B. thetaiotaomicron* (one isolate).

The only natural habitat of *B. nodosus* (Buchanan and Gibbons, 1974), the chief agent of ovine foot rot, appears to be the feet of sheep and cattle. It can exist in the soil for only a few weeks, and does not appear to be a part of the animals' intestinal microbiota (Smith, 1975). Ovine foot rot is characterized by inflammation of the interdigital skin and hoof

Table 2. Sources of the *Bacteroides disiens* and *Bacteroides bivius* strains studied by Holdeman and Johnson (1977).

Species	No. of strains	Female	Male
Bacteroides disiens	6 5	Bartholin abscess, cervical swab, pelvic abscess, vaginal swab, fluids (posthysterectomy), groin (postadrenalectomy)	Stomach, perineum, penile ulcer, scrotal cyst, penis wound
Bacteroides bivius	11 3	Endometrium, chest fluid, cervicovaginal swab from normal individual, breast abscess, transabdominal hysterectomy, blood (septic abortion), blood, thigh wound, peritoneal fluid (pelvic inflammatory disease), vaginal discharge, postcesarian section	Penis drainage, urethral discharge, cyst drainage

Table 3. Anaerobic Gram-negative bacilli isolated from vaginas of 65 premenopausal women.[a]

Species	No. of women	% Incidence
Bacteroides bivius	22	34
Bacteroides disiens	12	18
Bacteroides melaninogenicus subsp. *intermedius*	7	11
Bacteroides asaccharolyticus	4	6
Bacteroides melaninogenicus subsp. *melaninogenicus*	1	2
Bacteroides ruminicola subsp. *brevis*	4	6
Bacteroides coagulans	3	5
Bacteroides capillosus	3	5
Bacteroides vulgatus	3	5
Bacteroides fragilis	3	5
Bacteroides oralis	2	3
Bacteroides ovatus	2	3
Bacteroides thetaiotaomicron	1	2
Bacteroides sp.	24	37
Fusobacterium naviforme	5	8
Fusobacterium nucleatum	4	6
Fusobacterium gonidiaformans	2	3
Fusobacterium russii	2	3
Fusobacterium sp.	1	2

[a]Personal communication, Gale B. Hill, Duke University, 1978.

matrix, which eventually leads to separation of the hoof from the soft tissues (Thorley, 1976). The disease is caused by a mixed bacterial infection of the uncornified epidermis with *B. nodosus (Fusiformis nodosus;* Wilson and Miles, 1964) and *Fusobacterium necrophorum (Fusiformis necrophorus;* Egerton, Roberts, and Parsonson, 1969; Roberts and Egerton, 1969). Other bacteria, such as *Corynebacterium pyogenes* and *Treponema penortha,* may also participate in the disease (Smith, 1975; Wilson and Miles, 1964). Predisposing factors to the development of foot rot in sheep include traumatic injury of interdigital tissue, prolonged wet conditions, penetration of the interdigital skin by larvae of *Strongyloides papillosus* (Katitch et al., 1967), and exposure of sheep to lower than usual temperatures (Smith, 1975).

Diseases of Humans and Animals

Members of the genus *Bacteroides* have been found in association with a plethora of diseases in humans, ranging from relatively benign to severe with fatal outcome (Table 4). Essentially, all are endogenous infections involving microorganisms from the oral cavity, gastrointestinal tract, or genitourinary tract of the patient. Polymicrobic infections involving bacteroides are more common than infections involving one *Bacteroides* species alone, and some of the microbial mixtures appear to act synergistically (Gorbach and Bartlett, 1974a,b,c). Examples of synergism between bacteroides and other bacteria have been demonstrated experimentally (Altemeier, 1938, 1940; Roberts, 1969; Roberts and Egerton, 1969; Socransky and Gibbons, 1965), and synergis-

tic infections involving bacteroides are probably quite common in humans.

Patients with bacteroides infection have invariably been compromised by one or more predisposing factors (Finegold, Massh, and Bartlett, 1971). Factors which may result in lowering oxygen levels and the oxidation-reduction potential of tissue so that bacteroides can multiply include trauma, vascular constriction, necrosis, and concomitant infection with other bacteria. Common predisposing factors include surgery; an underlying illness such as leukemia, carcinoma, diabetes mellitus, arteriosclerosis, or alcoholism; and the use of antibiotics, X- or gamma-ray irradiation, immunosuppressants, or corticosteroids in therapy (Felner and Dowell, 1971).

Complications of a bacteroides infection may include bacteremia, thromboembolism, metastatic abscess formation, necrosis, and various other disorders which can contribute to the severity of the disease (Finegold, 1977; Gorbach and Bartlett, 1974a,b,c). Elaboration of heparinase (Gesner and Jenkins, 1964) and stimulation of coagulation by bacteroides (Bjornson and Hill, 1973) may contribute to thromboembolic complications in patients with bacteroides infection.

Isolation

Each step in the procedure for laboratory diagnosis of bacteroides infections must be properly performed and carefully controlled to avoid generating misleading information for the physician or veterinarian. These steps include:

Table 4. Diseases of humans involving *Bacteroides*.

Disease	Selected references
Central nervous system:	
Brain abscess	Beerens and Tahon-Castel, 1965
Meningitis, extradural or subdural empyema	Heineman and Braude, 1963; McFarlan, 1943; Swartz and Karchmer, 1974
Eye, nose, throat, dental:	
Otitis media	Swartz and Karchmer, 1974
Chronic sinusitis	Frederick and Braude, 1974
Dental infections	Dormer and Babett, 1972; Sabristan and Grigsby, 1972; Socransky and Gibbons, 1965; Loesche, 1974
Pleuropulmonary:	
Aspiration pneumonia	Bartlett et al., 1974b
Empyema	Bartlett et al., 1974a; Beerens and Tahon-Castel, 1965; Sullivan et al., 1973
Lung abscess	Bartlett et al., 1974b; Beerens and Tahon-Castel, 1965
Intraabdominal:	
Intraabdominal infections in general	Dack, 1940; Nobles, 1973; Finegold and Rosenblatt, 1976; Gorbach and Bartlett, 1974a,b,c
Appendicitis with peritonitis	Altemeier, 1938; Lanz and Tavel, 1904
Liver abscess	Altemeier, 1974; Sabbaj, Sutter, and Finegold, 1972
Obstetric-gynecological:	
Bartholin gland abscess	Parker and Jones, 1966
Endometritis	Swenson et al., 1973
Pelvic abscess	Swenson et al., 1973; Thadepalli, Gorbach, and Keith, 1973
Post abortal sepsis	Pearson and Anderson, 1970; Rotheram and Shick, 1969; Thadepalli, Gorbach, and Keith, 1973
Tuboovarian abscess	Altemeier, 1940; Ledger, Campbell, and Willson, 1968; Pearson and Anderson, 1970; Swenson et al., 1973; Thadepalli, Gorbach, and Keith, 1973
Urinary tract:	
Various infections in compromised patients	Finegold, 1974; Martin and Segura, 1974
Cardiovascular:	
Bacteremia	Chow and Guze, 1974; Felner and Dowell, 1971; Sonnenwirth, 1974; Washington, 1973
Endocarditis	Felner, 1974; Felner and Dowell, 1970; Finegold, 1977; Nastro and Finegold, 1973; Weinstein and Rubin, 1973
Thromboembolic disease	Altemeier, Hill, and Fuller, 1969; Gorbach and Bartlett, 1974a
Bone and joint:	
Osteomyelitis	Finegold, 1977; Leake, 1972; Ziment, Miller, and Finegold, 1968
Purulent arthritis	Finegold, 1977; Ziment, Davis, and Finegold, 1969
Miscellaneous:	
Breast abscess	Finegold, 1977
Decubitus ulcers	Felner and Dowell, 1970, 1971; Nastro and Finegold, 1973
Perirectal abscess	Finegold, 1977; Gorbach and Bartlett, 1974a
Pilonidal sinus infection	Pearson and Smiley, 1968

1. Selection, collection, and transport of clinical specimens.
2. Microscopic examination of the clinical materials.
3. Selection of appropriate primary isolation media.
4. Inoculation of media.
5. Use of anaerobic systems.
6. Incubation of cultures (period of incubation, temperature of incubation, gaseous atmosphere, humidity).
7. Subculture of colonies to obtain pure cultures.
8. Identification of pure-culture isolates.

Table 5. Examples of appropriate and inappropriate clinical materials for isolating *Bacteroides*.

Appropriate

1. Aspirated pus.
2. Tissue (by biopsy, surgical removal, or at autopsy).
3. Fluids (cerebrospinal, pleural, ascitic, pericardial, synovial).
4. Transtracheal aspirates.
5. Direct lung aspirates.
6. Urine collected by suprapubic aspiration.

Inappropriate

1. Surface swab samples from encrusted surfaces of abscesses or burn eschars.
2. Samples of lesions or abscesses adjacent to skin or mucous membranes which have not been properly decontaminated before sampling to exclude normal flora and surface contaminants.
3. Dry swab samples.
4. Drainage from an intestinal fistula obviously contaminated with bowel contents or feces.

Various authors recently have emphasized the need for proper selection, collection, and transport of clinical specimens and the value of microscopic examination in the laboratory diagnosis of anaerobic bacterial infections. The following references are recommended for specific details related to these subjects: Bartlett et al., 1978; Dowell, 1974, 1975, 1977; Dowell and Hawkins, 1974; Finegold, 1977; Finegold et al., 1974, 1976; Hill, 1978; Sutter, Vargo, and Finegold, 1975; Watt and Collee, 1974. Appropriate and inappropriate materials for the isolation of bacteroides are listed in Table 5.

Prompt examination and culture of the clinical materials after they reach the laboratory are essential to prevent gross changes in the microbial populations of the specimens. Low temperatures have been reported as being detrimental to some anaerobes, e.g., *Bacteroides fragilis* and *Clostridium perfringens* (Hagen, Wood, and Hashimoto, 1977). We hold, however, that clinical specimens in anaerobic transport systems not processed within 2–3 h after collection should be refrigerated to prevent undue multiplication of microorganisms and the possibility of overgrowth by organisms such as members of the family Enterobacteriaceae, if present (Mena et al., 1978). Others (Finegold, 1977; Finegold et al., 1974) have already expressed this view.

A plethora of selective and nonselective media have been used to isolate bacteroides from clinical materials. These include a variety of agar media in plates, tubes, and bottles and liquid media in tubes and various other types of containers.

Nonselective Solid Media

Blood agar prepared in various ways is the most commonly used nonselective plating medium. The diversity of ingredients used for preparing blood agar medium is illustrated in Table 6, which gives the ingredients of 12 different blood agar media used by various investigators in recent years for the cultivation of bacteroides. This is by no means an exhaustive list of the various blood agar formulations that have been used for this purpose. It is well known that varying the amino acid, essential vitamin, vitamin K, hemin, carbohydrate, or blood (animal species it is obtained from and composition) content of blood agar can markedly affect the rapidity of growth, colony characteristics, microscopic morphology, pigment production, and hemolysis of anaerobic bacteria. For this reason, we feel that standardization of blood agar and other media for isolation and identification of anaerobes is definitely needed to allow better comparison of results between laboratories (Dowell et al., 1977).

CDC Anaerobe Blood Agar (BA) (Dowell et al., 1977)

The CDC anaerobe blood agar (BA), medium 11 in Table 6, is prepared by adding the following ingredients to 1,000 ml of demineralized water and heating them until they are dissolved:

Trypticase soy agar (BBL)[1]	40.0 g
Agar (additional)	5.0 g
Yeast extract (Difco)	5.0 g
Hemin	5.0 mg
Vitamin K_1 (3 phytylmenadione)	10.0 mg
L-Cystine	400.0 mg

The hemin and L-cystine are dissolved in 5 ml of 1 N sodium hydroxide before they are added to the other ingredients. The vitamin K_1 is added from a stock solution containing 1 g of 3 phytylmenadione (ICN Pharmaceuticals, Inc., Cleveland, Ohio) plus 99 ml of absolute ethanol.

After the ingredients are dissolved, the pH is adjusted to 7.5 and the medium is autoclaved at 121°C for 15 min. When it cools to 48°C, 50 ml of sterile, defibrinated sheep blood is added, and the mixture is stirred and dispensed into 15-×-100-mm plastic Petri dishes. After they cool to room temperature, plates of the solidified medium are placed in cellophane bags (Dixie Packaging Co., Greenville, South Carolina) and are stored in a refrigerator at 4°C for up to 6 weeks. Before they are used, the refrigerated plates are held in an anaerobic glove box or an anaerobic jar for 6 to 24 h and then are placed in a nitrogen holding jar as described previously (Dowell, 1977).

This medium supports adequate growth of essentially all obligately anaerobic bacteria encountered

---

[1] Use of trade names is for identification only and does not constitute endorsement by the Public Health Service or by the U. S. Department of Health, Education, and Welfare.

Table 6. Ingredients of various nonselective blood agar media used for cultivation of *Bacteroides*.

Medium	Base	Supplements	Blood	Reference
1	Supplemented blood agar base (BBL)	Menadione (0.5 μg/ml)	2 to 5% Sheep or rabbit	Holland, Hill, and Altemeier, 1977
2	Supplemented brain heart infusion agar	Yeast extract (0.5%); vitamin K$_1$ (10 μg/ml); hemin (5.0 μg/ml)	5% Sheep	Finegold, Shepherd, and Spaulding, 1977
3	Supplemented brain heart infusion agar	Yeast extract (0.5%); vitamin K$_1$ (1.0 μg/ml); hemin (5 μg/ml); cysteine-HCl · H$_2$O (0.5 g/liter); agar (1.0 g/liter)	Not specified (0.4 ml blood per 10 ml of base medium)	Holdeman, Cato, and Moore, 1977
4	Supplemented brain heart infusion agar	Vitamin K$_1$ (10 μg/ml)	5% Sheep	Rosenblatt, 1976
5	Supplemented brain heart infusion agar	Yeast extract, Difco (0.5%); vitamin K (aqua-methyton, Merck, Sharp and Dohme) (0.5 μg/ml); hemin (5.0 μg/ml); cysteine-HCl · H$_2$O, (0.3 mg/ml); agar (0.1%)	5% Sheep	Hanson and Martin, 1976
6	Supplemented Brucella agar (Pfizer)	Vitamin K$_1$ (10 μg/ml)	5% Sheep	Bartlett et al., 1976; Rosenblatt, 1976; Sutter, Vargo, and Finegold, 1975
7	Columbia agar	None	10% Human or horse	Watt and Collee, 1974
8	Supplemented Columbia agar (modified)	Cysteine-HCl · H$_2$O (0.5 g/liter); palladium chloride (0.33 g/liter); dithiothreitol (0.1 g/liter); menadione (0.5 μg/ml); hemin (5.0 μg/ml)	5% Sheep	Ellner, Granato, and May, 1973
9	Supplemented Trypticase soy agar	Yeast extract (Difco) (0.5%); hemin (5 μg/ml); menadione (0.5 μg/ml)	5% Rabbit	Dowell and Hawkins, 1974; Starr, Killgore, and Dowell, 1971
10	Trypticase soy agar	None	5% Sheep	Spaulding et al., 1974
11	Supplemented Trypticase soy agar	Yeast extract (Difco) (0.5%); agar (0.5%); hemin (5 μg/ml); L-cystine (0.4 g/liter); vitamin K$_1$ (10 μg/ml)	5% Sheep or 5% rabbit	Dowell et al., 1977
12	Supplemented Schaedler agar	Menadione (0.5 μg/ml)	5% Rabbit	Starr, Killgore, and Dowell, 1971

in clinical materials, including the bacteroides discussed in this chapter. We have found that it supports better growth of various anaerobes (especially *Clostridium novyi* type B, *C. haemolyticum, Fusobacterium necrophorum* and certain anaerobic cocci) than does the Trypticase soy–yeast extract–blood agar (medium 9 in Table 6) used routinely in the CDC Anaerobe Section until recently (Dowell and Hawkins, 1974). Our experience with Schaedler blood agar (Starr, Killgore, and Dowell, 1971) and the report by Moore (1968) prompted us to supplement the Trypticase soy–yeast extract–blood agar (medium 9, Table 6) with 400 mg of L-cystine per liter (medium 11, Table 6), which greatly improved growth of *C. novyi* type B, *C. haemolyticum, F. necrophorum,* and certain anaerobic cocci.

Various *Bacteroides* species are known to require hemin for growth (Caldwell et al., 1965; Gibbons and MacDonald, 1960; Holdeman and Johnson, 1977; Jackson et al., 1971; Khairat, 1967; Lev, 1958; Lev, Kuedell, and Milford, 1971; Macy, Probst, and Gottschalk, 1975; Quinto, 1962, 1966;

Quinto and Sebald, 1964; Quinto, Sebald, and Prevot, 1963; Rizza et al., 1968; Smibert and Holdeman, 1976; Sperry, Appleman, and Wilkins, 1977; Varel and Bryant, 1974). Wilkins et al. (1976) reported that inhibition of *Bacteroides fragilis* on blood agar prepared with certain base media was reversed by including hemin in the medium. Catalase production by members of the *Bacteroides fragilis* group is also dependent upon the presence of hemin in the medium (Dowell and Lombard, 1977; Gregory, Kowalski, and Holdeman, 1977; Quinto and Sebald, 1964; Stargel et al., 1976; Stargel, Lombard, and Dowell, 1978). Quinto and Sebald (1964) were the first to report the effect of hemin on catalase production by *B. fragilis*. They found that hemin stimulated the growth of three strains of *Ristella pseudoinsolita* (now classified as *B. fragilis;* Buchanan and Gibbons, 1974) and that all three of the strains produced catalase if cultivated in a medium containing hemin. The production of catalase by *B. fragilis* is repressed in a medium containing fermentable carbohydrate, regardless of the presence of hemin (Gregory, Kowalski, and Holdeman, 1977; Stargel et al., 1976).

In addition to blood agar, various other nonselective solid media are commonly used for cultivating bacteroides as well as other obligate anaerobes (Aranki et al., 1969; Beerens and Tahon-Castel, 1965; Finegold, 1977; Holdeman, Cato and Moore, 1977; Phillips and Sussman, 1974; Prévot and Fredette, 1966; Quinto and Sebald, 1964; Schaedler, Dubos, and Costello, 1965; Smith, 1975; Sutter, Vargo, and Finegold, 1975; Suzuki, Ushijima, and Ichinose, 1966; Ueno et al., 1974; Werner, 1973). Use of prereduced anaerobically sterilized (PRAS) brain heart infusion agar-supplemented (BHIA) in roll-streak tubes is presently recommended by Holdeman et al (1977) for isolating various obligate anaerobes, including bacteroides, from clinical materials.

Brain Heart Infusion Agar (BHIA)

This medium is prepared by dispensing 10 ml of prereduced brain heart infusion broth-supplemented (BHI) into tubes that contain 0.25 g of agar and autoclaving as described for PRAS media (Holdeman, Cato, and Moore, 1977). The BHI contains the following ingredients:

Brain heart infusion broth (dehydrated)	37.0 g
Yeast extract	5.0 g
Cysteine-HCl·H_2O	0.5 g
Hemin solution	10.0 ml
Vitamin K_1 solution	0.2 ml
Resazurin solution	4.0 ml
Distilled water	1,000.0 ml

Hemin solution is prepared by dissolving 50 mg of hemin in 1 ml of 1 N sodium hydroxide, diluting to 100 ml with distilled water, and autoclaving at 121°C for 15 min.

Vitamin K_1 solution consists of 0.15 ml vitamin K_1 dissolved in 30 ml of 95% ethanol. Store in a brown bottle in the refrigerator.

Resazurin solution is prepared by dissolving 25 mg resazurin in 100 ml of distilled water (Holdeman, Cato, and Moore, 1977).

Nonselective Liquid Media

Nonselective liquid media are commonly recommended as back-ups for the solid media used for isolation of anaerobes and to allow isolation of slow-growing species that may not form visible colonies on agar media without prolonged incubation. Liquid media are also commonly used for isolation of bacteroides and other anaerobes from the blood of patients with bacteremia (Dowell, 1974; Dowell and Hawkins, 1974; Finegold, 1977; Holdeman, Cato, and Moore, 1977; Smith, 1975; Sutter, Vargo, and

Finegold, 1975; Watt and Collee, 1974). Two of the most commonly used liquid media for isolating anaerobes are enriched thioglycolate broth (THIO) and chopped meat glucose medium (CMG). Various formulations for preparing thioglycolate broth and chopped meat media have been published. Some are unsatisfactory for cultivation of bacteroides and other fastidious anaerobes. For this reason, specific instructions are given for preparation of the THIO and CMG we use in the CDC Anaerobe Section (Dowell et al., 1977).

Enriched Thioglycolate (THIO) Medium
(Dowell et al., 1977)

Weigh out 30 g of thioglycolate medium without indicator (BBL 0135C) and suspend in 1,000 ml of distilled water. Add 0.5 ml of 1% hemin solution (1 g of hemin dissolved in 5 ml of 1 N sodium hydrozide and diluted to 100 ml with distilled water) and 0.1 ml of 1% vitamin K_1 solution (1 g of vitamin K_1; 3-phytylmenadione, ICN; in 99 ml of absolute ethanol).

Heat to dissolve ingredients completely (allow medium to boil for at least 1–2 min). Adjust pH to 7.2 and dispense in 7-ml quantities into 15-×-90-mm tubes with one-piece plastic caps (Catalog number 949-1040, Rochester Scientific Co., Rochester, New York). Autoclave at 121°C for 15 min.

After the tubes cool and with caps loose, pass them into an anaerobic glove box so that the atmosphere of approximately 85% N_2, 10% H_2, 5% CO_2 replaces air in tubes. Fasten caps securely so that they are air-tight and remove the tubes from the glove box. Store the THIO in a refrigerator at 4°C or at ambient temperature.

If used as described by Dowell and Hawkins (1974) and incubated anaerobically, this medium should give good growth from a small inoculum of essentially all commonly isolated bacteroides from clinical samples (Dowell et al., 1977). It is not necessary to boil the medium if it is prepared as described above.

Chopped Meat Glucose (CMG) Medium
(Dowell et al., 1977)

Medium contains:

Lean ground beef	500.0 g
Distilled water	1,000.0 ml
Sodium hydroxide (1 N solution)	25.0 ml
Trypticase (BBL)	30.0 g
Yeast extract (Difco)	5.0 g
K_2HPO_4	5.0 g
D-Glucose	3.0 g
Hemin (1% solution)	0.5 ml
Vitamin K_1 (1% alcoholic solution)	0.1 ml

L-Cysteine   0.5 g

The 1% hemin solution is prepared by dissolving 1 g of hemin in 5 ml of 1 N NaOH and diluting to 100 ml with distilled water.

The 1% alcoholic solution of vitamin K_1 is prepared by dissolving 1 g of vitamin K_1 (3-phytylmenadione, ICN) in 99 ml of absolute ethanol.

Preparation: Obtain fresh lean beef; remove excess fat and connective tissue and grind in a meat grinder (fine grind). Mix 500 g of the ground beef with 1,000 ml of distilled water and 25 ml of 1 N NaOH. Heat to boiling while stirring. After mixture is cool, refrigerate overnight at 4°C. Skim remaining fat off surface of mixture. Filter the mixture through two layers of gauze. Retain the meat particles and the liquid filtrate. Add enough distilled water to the filtrate to give a final volume of 1,000 ml. Add all of the ingredients except the L-cysteine to the filtrate. Heat until the ingredients dissolve completely. Cool to less than 50°C and add the L-cysteine. Mix to dissolve it completely. Adjust the pH of the broth to 7.4. Wash the meat particles several times with distilled water to remove excess NaOH and spread particles thinly on a clean towel to partially dry. With a small scoop, dispense about 0.5 g of the meat particles into 15-×-90-mm screw-capped tubes. Add 7 ml of the enriched broth filtrate to each tube. Autoclave the tubes at 121°C for 15 min. After the tubes cool and with the caps loose, pass them into an anaerobic glove box so that an atmosphere of approximately 85% N_2, 10% H_2, 5% CO_2 replaces the air in the tubes. After the caps are tightened securely, remove the tubes from the glove box. Store the CMG tubes in a refrigerator at 4°C or at ambient temperature.

CMG should support good growth from a small inoculum of essentially all commonly isolated bacteroides from clinical samples if used as described by Dowell and Hawkins (1974) and incubated anaerobically. There is no need to boil the medium before it is used.

Recommended Media

SELECTIVE MEDIA. Examples of various selective media, their ingredients, and references to papers which describe the use of the media for isolation of bacteroides are given in Table 7. It is obvious from the diversity of ingredients in the 10 media listed that there is little agreement on what selective medium(s) should be used for isolating the microorganisms from clinical specimens. We have found that the CDC anaerobe blood agar supplemented with

kanamycin and vancomycin (KVA) and CDC anaerobe blood agar supplemented with paromomycin and vancomycin (PVA) are especially useful for isolating *Bacteroides* from mixed bacterial populations. The CDC anaerobe blood agar with phenethylalcohol (PEA) is also useful in isolating bacteroides because it inhibits growth of facultatively anaerobic, Gram-negative bacteria but allows the growth of Gram-negative as well as Gram-positive obligately anaerobic bacteria (Dowell et al., 1977; Dowell, Hill, and Altemeier, 1964).

NONSELECTIVE MEDIA. We recommend that at least the following media or their equivalent be used for all types of clinical specimens (except blood) regardless of the anatomical source. These media, prepared as described by Dowell et al., 1977, are now available commercially in the United States from various companies: (i) one plate of BA incubated anaerobically; (ii) one plate of regular blood agar incubated in a candle extinction jar or a carbon dioxide incubator; (iii) one tube of THIO and/or one tube of CMG incubated anaerobically.

Inoculation of Media

Liquid sample. If a liquid sample, such as pus aspirated from an abscess, is being tested, use a capillary pipette and inoculate the liquid medium(s) near the bottom with one or two drops of the material and place one drop on each plating medium. Streak with a platinum or stainless steel inoculating loop to obtain isolated colonies. Also prepare smears for Gram stain and other microscopic examinations.

Solid tissue specimens. Mince specimen with sterile scissors or scalpel, add sufficient broth such as THIO to emulsify the specimen, add sterile sand if necessary, and grind with a sterile mortar and pestle or a tissue grinder. Treat like a liquid specimen as described above.

Swabs. If more than one swab sample is obtained, inoculate the liquid media directly with one swab and use separate swabs for inoculating plating media and preparing slides for microscopic examination. If only one swab sample is received, prepare a suspension by scrubbing the material from the swab in a small amount (0.5–1 ml) of THIO and treat as a liquid specimen.

In addition to the nonselective media described above, it is advantageous to use selective media when the clinical specimen is likely to contain a mixture of microorganisms. The choice of the selective media to be used can be based on the anatomical source of the specimen and the microscopic appearance of the Gram-stained direct smear of the clinical material as illustrated in Table 8. This table lists the

Table 7. Examples of selective media used for isolating *Bacteroides*.

Medium	Base	Supplements	Reference
1. Kanamycin–vancomycin laked blood agar (LKV)	Brucella agar (Pfizer)	Vitamin K_1 (10 μg/ml); vancomycin (7.5 μg/ml); kanamycin (10 μg/ml); laked sheep blood (5%)	Rosenblatt, 1976; Sutter, Vargo, and Finegold, 1975; Finegold, Shepherd, and Spaulding, 1977
2. Kanamycin–vancomycin blood agar	Brucella agar (Pfizer)	Vitamin K_1 (10 μg/ml); vancomycin (7.5 μg/ml); kanamycin (100 μg/ml)	Sutter, Vargo, and Finegold, 1975
3. Kanamycin blood agar	Blood agar base (BBL)	Menadione (0.5 μg/ml); kanamycin (100 μg/ml); sheep blood (2–5%)	Holland, Hill, and Altemeier, 1977
4. Neomycin blood agar	Columbia agar	Neomycin (70 μg/ml); human or horse blood (10%)	Watt and Collee, 1974
5. Neomycin laked blood agar	Brucella agar	Menadione (10 μg/ml); neomycin (100 μg/ml); laked sheep blood (5%)	Bartlett et al., 1976
6. Kanamycin–vancomycin laked blood agar	Brucella agar	Menadione (10 μg/ml); kanamycin (75 μg/ml); vancomycin (7.5 μg/ml); laked sheep blood (5%)	Bartlett et al., 1976
7. Kanamycin–vancomycin (on KV)	Modified Columbia agar	Cysteine-HCl · H_2O (0.5 g/liter); palladium chloride (0.33 g/liter); dithiothreitol (0.1 g/liter); menadione (0.5 μg/liter); hemin (5 μg/ml); kanamycin (75 mg/liter); vancomycin (7.5 mg/liter); 5% laked sheep blood	Ellner, Granato, and May, 1973
8. Phenethylalcohol blood agar	Phenethyl alcohol agar (BBL)	Menadione (0.5 μg/ml); hemin (5.0 μg/ml); rabbit blood (5%)	Dowell, Hill, and Altemeier, 1964; Dowell et al., 1977
9. CDC anaerobe blood agar with kanamycin and vancomycin (KV)	Supplemented Trypticase soy agar	Yeast extract ([Difco] 0.5%); agar (0.5%); hemin (5 μg/ml); vitamin K_1 (10 μg/ml); L-cystine (0.4 g/liter); kanamycin (100 μg/ml); vancomycin (7.5 μg/ml); sheep blood (5%)	Dowell et al., 1977
10. CDC anaerobe blood agar with paromomycin and vancomycin (PV)	Supplemented Trypticase soy agar	Agar (0.5%); yeast extract ([Difco] 0.5%); hemin (5 μg/ml); vitamin K_1 (10 μg/ml); L-cystine (0.4 g/liter); laked sheep blood (5%); paromomycin (100 μg/ml); vancomycin (7.5 μg/ml)	Dowell et al., 1977

media appropriate for isolating aerobes, facultative anaerobes, and microaerophiles as well as obligate anaerobes.

Anaerobic Systems

Numerous devices have been described for the cultivation of anaerobic bacteria from the early days of bacteriology (Sonnenwirth, 1972). The systems most widely used now are (i) jars (evacuation-replacement and GasPak [BBL] gas generator types), (ii) roll-streak tube systems in which PRAS media are used, and (iii) anaerobic glove boxes (Dowell and Hawkins, 1974). A number of investigators (Killgore et al., 1973; Phillips and Sussman, 1974; Rosenblatt, Fallon, and Finegold, 1973) have found that jar techniques are comparable to the more

sophisticated glove box and roll-streak tube methods for recovery of commonly encountered anaerobes (such as the *Bacteroides* species described in this chapter) if the clinical specimens are selected, collected, and transported to the laboratory properly. The choice of the anaerobic system to be used in a particular clinical laboratory should be based on the size of the laboratory, availability of laboratory space, specimen workload, cost of the equipment and media, and the technical capabilities of the laboratory personnel. Regardless of the anaerobic system, the following conditions are required for optimal recovery of bacteroides and other obligate anaerobes:

1. Proper selection, collection, and transport of clinical materials.

Table 8. Guide to selection of media for primary culture of clinical materials from various sources.[a]

Source of specimen	Air	Candle extinction jar or CO$_2$ incubator	Anaerobic system
		Media and incubation conditions, 35°C	
Central nervous system		Blood agar, chocolate agar	BA, THIO
Eye, ear, oropharyngeal	MAC	Blood agar, chocolate agar, PEA	BA, PEA, THIO
Pulmonary	MAC	Blood agar, chocolate agar	BA, PEA, THIO
Intraabdominal	MAC	Blood agar, PEA, (TM)	BA, PEA, (KV), THIO
Genitourinary	MAC	Blood agar, PEA, TM	BA, PEA, (KV), THIO
Muscle tissue	MAC	Blood agar, PEA	BA, PEA, NEY, CMG
Bone marrow		Blood agar, chocolate agar	BA, THIO
Miscellaneous body fluids		Blood agar, chocolate agar	BA, THIO

[a] Adapted from Dowell (1975).

BA, CDC Anaerobe blood agar; MAC, MacConkey agar; PEA, CDC anaerobe blood agar with phenethyl alcohol; TM, modified Thayer-Martin agar; KV, CDC anaerobe blood agar with kanamycin and vancomycin; NEY, neomycin egg yolk agar; THIO, enriched thioglycolate broth; CMG, chopped meat glucose medium; (   ), optional.

2. Processing of specimens with minimal exposure to atmospheric oxygen.
3. Use of freshly prepared or properly reduced stored media.
4. Proper use of the anaerobic system, including use of an active catalyst to allow effective removal of oxygen.

Use of nitrogen holding jars (Dowell, 1977) in conjunction with anaerobic jars and glove boxes is highly recommended. These jars will minimize exposure of uninoculated media and the anaerobes to atmosphere oxygen during inspection and inoculation procedures.

Incubation of Cultures

It is generally recommended that primary plating media for isolation of bacteroides should be incubated for at least 48 h and preferably for 3–5 days to obtain maximum recovery of the microorganisms (Dowell and Hawkins, 1974; Spaulding et al., 1974). If anaerobic jars are opened too soon, some of the slow-growing anaerobes (such as the more fastidious bacteroides) may fail to form colonies because of exposure to air. Liquid cultures (THIO and CMG) as well as those grown on solid media should be incubated anaerobically to minimize exposure to oxygen and to allow maximal recovery of anaerobes, especially from specimens containing small numbers of the microorganisms. Unless growth is apparent (at which time Gram-stained preparations are inspected and subcultures are made,

if necessary) the THIO and CMG cultures should be held for at least 2 weeks before they are discarded as negative.

Observation of Colonies and Determination of the Oxygen Tolerance of Isolates

After incubation under anaerobic conditions, all of the colony types that appear on primary plating media should be carefully observed under a dissecting microscope, and a Gram-stained smear of each should be examined. In addition, their relationship to oxygen must be determined by comparing the ability of the isolate to form colonies on BA when incubated in air, in a candle extinction jar, and in an anaerobic system (Dowell and Hawkins, 1974). Hemolysis of the blood in the medium and fluorescence of colonies under a Wood's Lamp are also recorded (Dowell and Lombard, 1977). If colonies are sufficiently separated, a tube of THIO or CMG should be inoculated with each colony type to serve as a source of inoculum for differential media after anaerobic incubation at 35°C for 18–24 h.

Identification

The family Bacteroidaceae as presently classified in the eighth edition of *Bergey's Manual* (Buchanan and Gibbons, 1974) contains Gram-negative, obligately anaerobic, nonsporeforming, bacilli which are either nonmotile or motile by means of peritrichous flagella. These can be differentiated to the genus level as follows:

Key to Genera of Bacteroidaceae

1. Butyric acid is a major metabolic product, succinic acid is not produced . . . . *Fusobacterium*
2. Lactic acid is usually only major product . . . . *Leptotrichia*
3. No butyric acid produced in absence of iso acids, succinic acid is produced . . . . *Bacteroides*

Presumptive Identification

Several reports have shown that approximately half of the anaerobic bacteria isolated from proper clinical specimens of humans were Gram-negative, nonsporeforming bacilli (Martin, 1971; Stokes, 1958; Zabransky, 1970). Because of the frequency with which these microorganisms are encountered in clinical specimens, a number of investigators have advocated the use of presumptive grouping techniques as a practical approach to the identification of *Bacteroides* and *Fusobacterium* species.

USE OF ANTIBIOTIC DISKS. Sutter and Finegold (1971) demonstrated that differences in susceptibility to antibiotics on paper disks (colistin, 10 μg; erythromycin, 60 μg; kanamycin, 1,000 μg; neomycin, 1,000 μg; penicillin, 2 units; and rifampin, 15 μg) in combination with other characteristics such as colonial morphology, pigment production, growth in bile, esculin hydrolysis, and reactions on egg yolk agar were useful in identifying most of the commonly encountered *Bacteroides* and *Fusobacterium* species.

USE OF BILE AGAR AND KANAMYCIN DISKS. A simple method for identifying members of the *Bacteroides fragilis* group was described by Vargo, Korzeniowski, and Spaulding (1974). Because the growth of the *B. fragilis* group is not inhibited by bile and because these anaerobes are resistant to a high concentration of kanamycin, these authors recommended the use of tryptic soy agar with 2% (wt/vol) oxgall in plates, inoculated with several colonies from a 48- to 72-h blood agar culture. At the same time, another blood agar plate is inoculated with several colonies and the inoculum is evenly distributed over the surface of the blood agar medium with a sterile swab previously moistened in brain heart infusion broth. A filter paper disk containing 1,000 μg of kanamycin is placed on the surface of the inoculated medium. Both plates are incubated in an anaerobic atmosphere and examined after 24 h for growth on the bile agar plate and for inhibition of growth around the kanamycin disk. Of

the 190 strains of anaerobes they tested (150 isolates of *Bacteroides* and *Fusobacterium* species; 40 isolates of various Gram-positive anaerobes), only members of the *B. fragilis* group were resistant to the concentration of kanamycin used and were able to grow on the bile agar.

USE OF OXGALL DISKS WITH KANAMYCIN. Draper and Barry (1977) reported a modification of the Vargo, Korzeniowski, and Spaulding (1974) procedure in which they used paper disks impregnated with oxgall (25 mg per disk) and 1,000-μg disks of kanamycin on Brucella blood agar. The surface of the blood agar was inoculated with a swab which had been dipped in a thioglycolate broth culture adjusted to the turbidity of a MacFarland No. 0.5 standard, and the two disks were placed on the inoculated medium with sterile forceps. The plates were incubated in an anaerobic atmosphere and then examined after 24 h and 48 h incubation. An isolate was considered resistant to kanamycin if the zone of growth inhibition around the disk was less than 12 mm in diameter, and resistant to bile if there was any growth of the microorganism immediately surrounding the disk containing bile. Of 158 isolates of anaerobes from clinical specimens tested, 57% were anaerobic, gram-negative, nonsporeforming bacilli, and 78% of these were identified as members of the *B. fragilis* group.

USE OF KANAMYCIN-ESCULIN-BILE AGAR (KEB). Chan and Porschen (1977) found another selective medium, kanamycin-esculin-bile agar, to be useful for presumptive identification of the *Bacteroides fragilis* group. The KEB contains 1,000 μg of kanamycin/ml, 0.5% esculin, 0.05% ferric ammonium citrate, and 20% bile (2% oxgall) in Trypticase soy agar (BBL). This medium was found to inhibit the growth of other obligately anaerobic bacteria, as well as most facultative anaerobes, but allowed growth of the *B. fragilis* group. These investigators were able to correctly identify 175 of 176 isolates from clinical materials as members of the *B. fragilis* group, and most of the isolates (165/175 or 94%) were identified within 48 h.

We routinely use a large battery of differential tests to characterize and identify anaerobe isolates in the CDC Anaerobe Reference Laboratory (Dowell and Hawkins, 1974). On the basis of data accumulated from our tests we have found certain characteristics to be especially useful for identifying anaerobes (Table 9). Recognizing that it is impractical for clinical and public health laboratories to use a large number of tests in the identification of anaerobe isolates, we developed a procedure for presumptive identification of the commonly encountered *Bacteroides* and *Fusobacterium* species (Dowell and Lombard, 1977), which is based on the characteristics listed in Table 9.

Table 9. Characteristics especially useful in identification of anaerobic bacteria.

1. Relation to oxygen
2. Colony characteristics
3. Hemolysis of rabbit and sheep erythrocytes
4. Pitting of medium
5. Fluorescence of colonies with ultraviolet light (Wood's lamp)
6. Gram reaction
7. Microscopic features (morphology, spores, flagella, etc.)
8. Motility
9. Catalase
10. Reactions on egg yolk agar (lecithinase, lipase, proteolysis)
11. Esculin hydrolysis
12. Casein hydrolysis
13. Gelatin hydrolysis
14. Fermentation of certain carbohydrates
15. Urease
16. Nitrate reduction
17. Metabolic products (GLC)
18. Growth in THIO
19. Growth in presence of bile
20. Growth in presence of antibiotics (penicillin, rifampin, kanamycin)
21. Toxin neutralization tests

DOWELL-LOMBARD PRESUMPTIVE IDENTIFICATION PROCEDURE. After an obligately anaerobic, non-sporeforming, Gram-negative rod is isolated in pure culture (THIO or CMG), the presumptive identification procedure requires the use of one anaerobe blood agar plate and one presumpto quadrant plate, which contains four different media (LD agar, LD egg yolk agar, LD bile agar, and LD esculin agar). In addition one blank paper disk, one kanamycin disk (1 mg), one rifampin disk (15 μg) and one penicillin disk (2 units) are required (Table 10). Detailed instructions for preparing and storing the media required for the presumptive identification procedure are given by Dowell and Lombard (1977) and Dowell et al. (1977).

As shown in Table 9, with this minimal number of media it is possible to determine more than 20 characteristics that are useful for identification of *Bacteroides* and *Fusobacterium* species commonly associated with human disease.

Presumptive Identification of *Bacteroides* Species (Dowell and Lombard, 1977)

Inoculation of Media. As soon as the anaerobe has been isolated on a solid medium, either a turbid cell suspension (equal to a MacFarland No. 1 nephelometer standard or greater) in LD broth prepared from isolated colonies or a young active culture in THIO or CMG can be used for inoculating the various media.

1. Place one or two drops of cell suspension or broth culture on each quadrant of the pre-

sumpto plate and streak three-fourths of the medium with the capillary pipette.
2. Place a sterile, blank, paper disk 1/4 inch in diameter on the LD agar near the outer periphery of the quadrant. This disk is used in the test for indole after 48 h incubation.
3. Evenly inoculate the surface of an anaerobe-blood-agar plate with a sterile swab that has been dipped in the cell suspension or culture.
4. Place the antibiotic disks (penicillin 2 units, rifampin 15 mcg, kanamycin 1 mg) on the blood-agar plate with sterile forceps. Evenly space the disks so that zones of inhibition will not overlap.

Incubation. Incubate the presumpto plate and the anaerobe blood agar plate with antibiotic disks in an anaerobic system such as an anaerobic glove box or an anaerobe jar (e.g., GasPak jar) at 35°C for 48 h. With fast-growing anaerobes such as the *B. fragilis* group, preliminary observation of the plates may be made after 24 h of incubation, and tentative identification of the isolate may be possible in many instances.

Observation and Interpretation of Results with the Presumpto Plate.
1. LD Agar.
   a. Note and record the degree of growth on LD Agar (light, moderate, heavy).
   b. Test for indole by adding 2 drops of para-

Table 10. Media and characteristics of cultures which can be identified in procedure for presumptive identification of *Bacteroides* and *Fusobacterium* species.

Media	Characteristics
Blood agar	Colony characteristics; hemolysis, pigment; fluorescence with ultraviolet light (Wood's lamp); pitting of agar; cellular morphology; Gram reaction; spores; motility (wet mount); inhibition of growth by penicillin, rifampin, or kanamycin
Enriched thioglycolate medium	Rapidity of growth, appearance of growth, gas production, odor, cellular morphology
Presumpto plate:	
LD agar	Indole, growth on LD medium, catalase[a]
LD esculin agar	Esculin hydrolysis, H_2S, catalase
LD egg yolk agar	Lipase, lecithinase, proteolysis
LD bile agar	Growth in presence of 20% bile (2% oxgall), formation of an insoluble precipitate under and immediately surrounding growth

[a] The catalase test can be performed by adding 3% hydrogen peroxide to the growth on LD agar, but reactions after addition of H_2O_2 to catalase-positive cultures are more vigorous on LD esculin agar.

dimethylaminocinammaldehyde (PACA) in 10% (vol/vol) aqueous HCl to the paper disk on the LD medium. Observe for the development of a blue or bluish green color in the disk within 30 s, which indicates a positive reaction for indole. Development of another color (pink, red, violet) or no color is negative for indole.

2. LD Egg Yolk Agar.

   a. Formation of a zone of insoluble precipitate in the medium surrounding the bacterial colonies is indicative of lecithinase production. This zone is best seen with transmitted light.

   b. The presence of an iridescent sheen "pearly layer" on the surface of colonies and on the medium immediately surrounding the bacterial growth (best demonstrated with reflected light) is indicative of lipase production.

   c. Clearing of the medium in the vicinity of the bacterial growth indicates proteolysis.

3. LD Esculin Agar.

   a. Esculin hydrolysis is indicated by the development of a reddish brown to dark brown color in the esculin agar surrounding the bacterial growth after exposure of the presumpto plate to air for at least 5 min. Further evidence of esculin hydrolysis is obtained by examining the esculin agar quadrant under a Wood's lamp. Esculin agar exhibits a bright blue fluorescence under the ultraviolet light which is not present after the esculin is hydrolyzed.

   b. Blackening of the bacterial colonies on the esculin agar indicates H_2S production. The blackening dissipates rapidly after colonies are exposed to air. Therefore, the bacterial growth should be observed for blackening under anaerobic conditions (anaerobic glove box) or immediately after anaerobic jars are opened in air. If no apparent blackening of growth is noted, individual colonies should be observed with a stereomicroscope with oblique light for darkening of growth (light tan to brown).

   c. To test for hydrogen peroxide degradation as an indication of catalase, expose the plates to air for at least 30 min and then flood the esculin agar quadrant with a few drops of fresh, 3% hydrogen peroxide. Sustained bubbling after addition of the H_2O_2 is interpreted as a positive reaction for catalase. In some cases rapid bubbling may not be evident until after 30–60 s.

4. LD Bile Agar.

   a. Compare the degree of bacterial growth on the LD bile agar with that on the plain LD agar and record as I (growth less than the LD agar control) or E (growth equal to or greater than on LD agar control).

   b. Using transmitted light, observe for the presence or absence of an insoluble white precipitate underneath or immediately surrounding the bacterial growth. If in doubt, inspect under a stereomicroscope using transmitted light.

5. Inhibition by antibiotics on anaerobe blood agar. Observe the growth for zones of inhibition around the antibiotic disks and record as follows:

   a. Penicillin, 2-unit disk: record as S (zone of inhibition 12 mm or greater in diameter) or R (inhibition zone less than 12 mm).

   b. Rifampin, 15-μg disk: record as S (zone of inhibition 15 mm or larger) or R (inhibition zone less than 15 mm).

   c. Kanamycin, 1-mg disk: record as S (zone of inhibition 12 mm or greater) or R (inhibition zone less than 12 mm in diameter).

The characteristics exhibited by commonly encountered *Bacteroides* species tested by this procedure are shown in Table 11.

USE OF MICROTECHNIQUES. If additional information on the biochemical characteristics of an isolate is desired, several techniques have been described in which small amounts of differential media are placed in tubes or microtiter trays and are inoculated heavily with the test culture (Holland et al., 1977; Morgan, Liv, and Smith, 1976; Schreckenberger and Blazevic, 1974, 1976). Commercially prepared microsystems have received wide acceptance in recent years for identification of anaerobes (Hansen and Stewart, 1976; Moore, Sutter, and Finegold, 1975; Nord, Dahlback, and Wadstrom, 1975; Stargel et al., 1976; Stargel, Lombard, and Dowell, 1978; Starr et al., 1973). However, it is generally agreed that these systems may require supplementary tests such as catalase, lecithinase, lipase, motility, and identification of metabolic products by gas-liquid chromatography for definitive identification of certain anaerobes.

At the annual meeting of the American Society for Microbiology in New Orleans, Louisiana, May 1977, G. L. Lombard and M. D. Stargel reported on a study in which they modified the spot indole test described by Sutter, Vargo, and Finegold (1975). This modified test procedure was simple and specific and was more sensitive in detecting indole with both the API-20A and the Minitek microsystem than were the methods recommended by the manufacturers of these microsystems. The procedure used in this study was to remove a drop of culture suspension, previously incubated in an anaerobic atmos-

Table 11. Characteristics of commonly isolated *Bacteroides* species.[a]

Characteristic	*B. distasonis*	*B. fragilis*	*B. thetaiotaomicron*	*B. vulgatus*	*B. ureolyticus*	CDC F-1	CDC F-2	*B. asaccharolyticus*	*B. melaninogenicus* subsp. *intermedius*
Anaerobe blood agar:									
Relation to O₂	OA	OA	OA	OA	OA	OA	OA	OA	OA
Colonies	Convex, semi-opaque, entire edge	Convex, mottled surface, entire edge	Convex, opaque, entire edge	Convex, semi-opaque, entire edge	Pin point, convex, irregular edge	Pin point, convex, entire edge	Pin point, convex, entire edge	Small to medium, convex, entire edge	Small to medium, convex, entire edge
Hemolysis, sheep blood	–	–	–	–	–	–	–	+	+
Hemolysis, rabbit blood	–	–	–	–	–	–	–	+	+
Black pigment	–	–	–	–	–	–	–	+	+
Red fluorescence	–	–	–	–	+	–	–	+	+
Pitting of agar	–	–	–	–	–	–	–	–	–
Cellular morphology on blood agar	Small rods, variable in length	Small rods, variable in length	Small rods, variable in length	Small rods, variable in length	Small, slim rods, variable in length	Small, slim rods, variable in length	Small, slim rods, variable in length	Tiny coccoid rods	Tiny coccoid rods
Gram reaction	–	–	–	–	–	–	–	–	–
Spores	–	–	–	–	–	–	–	–	–
Motility	–	–	–	–	–	–	–	–	–
Penicillin (2 U disc)	R	R	R	R	S	S	S	S	S
Rifampin (15 mcg disc)	S	S	S	S	S	S	S	S	S
Kanamycin (1 mg disc)	R	R	R	R	S	S	S	R	R
Enriched thioglycolate:									
Rapidity of growth	Moderate to rapid	Moderate to rapid	Moderate to rapid	Moderate to rapid	Slow	Slow	Slow	Moderate	Moderate
Appearance	Flocculent	Flocculent	Flocculent	Flocculent	Cloudy	Cloudy	Cloudy	Cloudy	Cloudy
Gas	–⁺	–⁺	–⁺	–⁺	–	–	–	–	–
Odor	Butyrous	Butyrous	Butyrous	Butyrous				Acrid	Acrid
Cellular morphology	Medium rods, vacuolated	Medium rods, vacuolated	Medium rods, vacuolated	Medium rods, vacuolated	Small slim rods	Small slim rods	Small slim rods	Coccoid rods with some pleomorphic forms	Coccoid rods with some pleomorphic forms
Presumpto plate:									
Growth on LD agar	Moderate	Moderate	Moderate	Moderate	Light	Light	Light	Moderate	Moderate
Indole	–	–	+	–	–	–	–	+	+
Lecithinase	–	–	–	–	–	–	–	–	–
Lipase	–	–	–	–⁺	–	–	–	–	–⁺
Proteolysis	–	–	+	–	–	–	–	–	–
Esculin hydrolysis	+	+	+	–	–	–	–	–	–
H₂S	–	–	V	–⁺	–	–	–	–	–
Catalase	+	+	–	–	–	–	–	–	–
20% bile agar, growth	E	E	E	E	I	I	I	E or I	I
Precipitate in 20% bile agar	+	+	–	–	–	–	–	–	–

[a] OA, obligate anaerobe; +, positive reaction in > 90% of strains tested; –, negative reaction in > 90% of strains tested; –⁺, usually negative but may exhibit a positive reaction; R, resistant; S, sensitive; E, equal or greater than growth on LD agar control; I, less than growth on LD agar control; V, variable reaction.

phere at 35°C for 48 h, from the cupule used for the detection of esculin hydrolysis and to place the drop on a filter paper saturated with 1% PACA in 10% (vol/vol) concentrated HCl. A deep blue to aqua blue color, which develops usually within 15 s, indicates the presence of indole.

The modified PACA test was positive for 250 of 253 indole-positive strains of anaerobic bacteria and had a 98.8% agreement with the conventional tube test for indole performed with Ehrlich's reagent after 5–7 days of incubation. Nine of 18 strains of *Bacteroides thetaiotaomicron* and 1 of 2 strains of *B. ovatus* were negative for indole in both of the microsystems when the test was performed as recommended by the manufacturers. However, 17 of the 18 strains of *B. thetaiotaomicron* and both strains of *B. ovatus* were positive by the modified PACA method.

Definitive Identification of *Bacteroides*

The two most commonly used standardized procedures for definitive identification of anaerobes in the United States are those outlined by Holdeman, Cato, and Moore (1977) and Dowell and Hawkins (1974). The prereduced, anaerobically sterilized (PRAS) media used in the VPI Anaerobe Laboratory (Holdeman, Cato, and Moore, 1977) are prepared, sterilized, and stored under oxygen-free gas in rubber-stoppered tubes. All bacteriological manipulations, such as inoculation and sampling of cultures, are performed under a stream of oxygen-free carbon dioxide by inserting a sterile cannula and passing a gentle stream of gas into the neck of the tube until the rubber stopper is replaced.

The media used by the Center for Disease Control for anaerobes are prepared in the conventional manner, but are gassed after they are autoclaved. All tubed media are then allowed to cool with the caps loose and are passed into an anaerobic glove box so that the atmosphere of approximately 85% N_2, 5% CO_2, and 10% H_2 replaces the air in the tubes. After the caps are tightened securely in the glove box, the tubed media are removed and stored either at ambient temperature or in a refrigerator, at 4°C. The media are carefully inoculated with a Pasteur pipette in the conventional manner without introducing excess air during inoculation. The inoculated culture media are then placed in an anaerobic glove box or anaerobic jar with caps loose for incubation (Dowell et al., 1977). To save space in the anaerobic glove box's incubator, one can tighten the caps of the tubes, remove them from the glove box, and incubate in a conventional incubator at 35°C. Detailed differential characteristics of the *Bacteroides* species commonly associated with human disease compiled from data obtained with the CDC techniques are shown in Tables 12-17.

THE *BACTEROIDES FRAGILIS* GROUP. The *Bacteroides fragilis* group (*B. distasonis, B. fragilis, B. thetaiotaomicron, B. vulgatus*). Colonies of the four species of this group are very similar on the surface of anaerobe blood agar. The colonies are usually 1–3 mm in diameter after 48 h incubation, circular, convex, entire, and semi-opaque. When viewed with a stereoscopic microscope under oblique lighting, they usually show concentric rings in the internal structure of the colony. This mottled effect is almost always present in the colonies of *B. fragilis* and *B. distasonis* isolates and is less commonly observed in *B. vulgatus* and *B. thetaiotaomicron* colonies. The cells from colonies of the *B. fragilis* group on anaerobe blood agar are usually uniformly stained Gram-negative small rods with rounded ends, somewhat variable in length. However, in a liquid medium (THIO especially) the cells are quite variable in length, stain irregularly and may appear to have clear vacuoles in some cells. In some instances, the vacuoles cause enlarged swollen areas, which can be mistaken for endospores.

Table 12 shows the composite reactions of 34 *B. distasonis*, 832 *B. fragilis*, 142 *B. thetaiotaomicron*, and 150 *B. vulgatus* strains when tested according to CDC procedures. The metabolic products detected in peptone yeast extract glucose (PYG) broth cultures of these are shown in Table 13.

Other characteristics of the *B. fragilis* group which are useful in differentiating them from other *Bacteroides* species are:

1. Most isolates of the *B. fragilis* group will produce catalase provided that they are grown in a medium which contains hemin and does not contain glucose or another fermentable carbohydrate. The test for catalase should be performed within 48–72 h after inoculation of the medium because the catalase activity (degradation of hydrogen peroxide) of a culture diminishes after prolonged incubation. The ability to degrade hydrogen peroxide is exhibited by essentially all strains of *B. fragilis* and *B. distasonis*, and the *B. vulgatus* strains that hydrolyze esculin, but the reaction is variable with *B. thetaiotaomicron* and *B. ovatus* strains and negative with *B. vulgatus* strains that do not hydrolyze esculin (Stargel, Lombard, and Dowell, 1978; Wilkins et al., 1978).

2. All members of the *B. fragilis* group, including *B. distasonis, B. fragilis, B. thetaiotaomicron, B. vulgatus*, and *B. ovatus*, are able to grow in the presence of bile, and the growth of some in liquid media is stimulated by the addition of bile. As noted by Dowell and Lombard (1977), many but not all strains of *B. fragilis* form a precipitate in broth or solid media containing bile. On the other hand, the growth of *B. oralis, B. rumino-*

Table 12. Detailed characteristics of commonly encountered *Bacteroides* species of the *B. fragilis* group.

Species	B. distasonis		B. fragilis		B. thetaiotaomicron		B. vulgatus	
No. of strains	34		832		142		50	
Reaction	Sign	% +	Sign	% +	Sign	% +	Sign	% +
Aerobic growth	−	0	−	3	−	4	−	2
Motility	−	0	−	0	−	0	−	0
Hemolysis	−	6	−	9	−	8	−⁺	16
Glucose	A	100	A	100	A	100	A	100
Mannitol	−	0	−	0	−	0	−	0
Lactose	A	100	A	100	A	100	A	100
Sucrose	A	100	A	100	A	100	A	100
Maltose	A	100	A	100	A	100	A	100
Salicin	V	42	−	1	−ᴬ	25	−	6
Glycerol	−	3	−	0	−	3	−	0
D-Xylose	A	100	A	93	A	98	A	98
L-Arabinose	V	53	−	0	A	96	A	90
Starch	V	65	A⁻	86	A	94	A⁻	88
Starch hydrolysis	V	65	V⁻	86	+	94	+⁻	88
Mannose	A	100	A	98	A	100	A	100
Rhamnose	V	65	−	0	A	92	A	100
Trehalose	A	100	−	0	V	70	−	0
Esculin hydrolysis	+	100	+	99	+	98	−⁺	22
Milk-clot	+⁻	88	+	95	+	90	+⁻	87
Gelatin	−	0	−	1	−	1	−	0
Indole	−	0	−	0	+	100	−	0
H₂S	−	0	−	1	−	1	−	0
Catalaseᵃ	−⁺	21	−	10	−⁺	13	−	4
Urease	−	0	−	0	−	0	−	0
Nitrate reduction	−	0	−	0	−	0	−	0

ᵃ A, acid; +, positive; −, negative; V, variable; −⁺, usually negative, some positive; +⁻, usually positive, some negative; A⁻, usually acid, some negative; −ᴬ, usually negative, some acid.

ᵇ Brain heart infusion agar, not supplemented with hemin.

cola, and a number of other *Bacteroides* species is inhibited on media containing bile.

3. As shown in Table 12, 57 of 843 (7%) of the *B. fragilis* strains we examined did not ferment D-xylose or L-arabinose, but exhibited other phenotypic characteristics which were identical to the more common biotype of *B. fragilis* that ferments D-xylose but not L-arabinose. None of the 57 strains was inhibited in media containing 20% bile, and all of them were resistant to a 2-unit penicillin disk. Without testing for susceptibility to penicillin and growth in the presence of 20% bile, these strains could have been mistakenly identified as *B. oralis*.

4. Succinic acid production by members of the *B. fragilis* group requires the presence of hemin in the growth medium. All members of the *B. fragilis* group produce detectable amounts of

Table 13. Metabolic acids detected in PYG broth by GLC of various members of the *Bacteroides fragilis* group.ᵃ

Species	B. distasonis		B. fragilis		B. thetaiotaomicron		B. vulgatus	
Organic acid	Sign	% +	Sign	% +	Sign	% +	Sign	% +
Acetic	+	100	+	100	+	100	+	100
Propionic	+⁻	76	V	58	V	71	+⁻	88
Isobutyric	−	0	−⁺	11	−⁺	11	−	8
Butyric	−	0	−	0	−	0	−	0
Isovaleric	V	53	V	52	V	28	−⁺	22
Valeric	−	0	−	0	−	0	−	0
Isocaproic	−	0	−	0	−	0	−	0
Caproic	−	0	−	0	−	0	−	0
Lactic	V	55	+⁻	89	V	74	V	81
Succinic	+⁻	77	+⁻	75	+⁻	77	V	81

ᵃ +, Positive; −, negative; V, variable; +⁻, usually positive, some negative; −⁺, usually negative, some positive.

succinic acid in PYG medium supplemented with 5 μg of hemin per ml (G. L. Lombard and V. R. Dowell, Jr., unpublished data).

BACTEROIDES UREOLYTICUS, BACTEROIDES CDC GROUP F-1 AND CDC GROUP F-2. Colonies of *Bacteroides ureolyticus* on blood agar are pinpoint to 1 mm in size, circular, entire, low convex to slightly umbonate, and translucent. When the growth on blood agar is viewed with a dissecting microscope equipped with an oblique light source, the colonies are situated in a shallow pit or crater in the agar. This type of growth is commonly referred to as corroding or pitting of the agar. An occasional strain of *B. ureolyticus* may be observed in which the growth swarms or spreads from the center of the colony. Cells of *B. ureolyticus* from blood agar or broth media appear as small, thin rods slightly variable in length. Growth may be inhibited if the medium is not supplemented with hemin (5–25 μg/ml).

Colonies of *Bacteroides* CDC group F-1 and F-2 on blood agar are pinpoint to 1 mm in size, circular, entire, low convex, and translucent. Pitting of the surface of blood agar has not been observed with either of these two groups. Cells from blood agar or from broth media of the *Bacteroides* CDC groups F-1 and F-2 appear as small, thin rods slightly varia-

ble in length. No vacuoles are found in cells from THIO cultures as exhibited by members of the *B. fragilis* group.

Detailed biochemical characteristics of this group of microorganisms are shown in Tables 14 and 15.

Additional characteristics of *B. ureolyticus* to be considered in differentiating them from other microorganisms include:

1. *B. ureolyticus* should not be confused with the obligately anaerobic strains of Gram-negative bacilli that we have arbitrarily designated as *Bacteroides* CDC Group F-1 and F-2 or with the microaerophilic species *Eikenella corrodens*. The latter species commonly pits the surface of blood agar but exhibits optimal growth in a CO_2 incubator with 5–10% CO_2. *E. corrodens* may also be inhibited if the medium is not supplemented with hemin (5–25 μg/ml). Typical strains of *E. corrodens* are asaccharolytic, oxidase positive by the Kovac technique, and negative for catalase, urease, and gelatin liquefaction. All strains reduce nitrate to nitrite (Jackson and Goodman, 1972).

2. Only when tested by the special procedures described by Jackson and Goodman (1978) will *B. ureolyticus* consistently hydrolyze urea, reduce

Table 14. Detailed characteristics of commonly encountered *Bacteroides* species or groups.[a]

Species	*Bacteroides* CDC Gp F-1		*Bacteroides* CDC Gp F-2		*B. ureolyticus*	
No. of strains	70		50		21	
Reaction	Sign	% +	Sign	% +	Sign	% +
Aerobic growth	−	0	−	0	−	0
Motility	−	0	−	0	−	0
Hemolysis	−	0	−	0	−	0
Glucose	−	0	−	0	−	0
Mannitol	−	0	−	0	−	0
Lactose	−	0	−	0	−	0
Sucrose	−	0	−	0	−	0
Maltose	−	0	−	0	−	0
Salicin	−	0	−	0	−	0
Glycerol	−	0	−	0	−	0
D-Xylose	−	0	−	0	−	0
L-Arabinose	−	0	−	0	−	0
Starch	−	0	−	0	−	0
Starch hydrolysis	−	0	−	0	−	0
Mannose	−	0	−	0	−	0
Rhamnose	−	0	−	0	−	0
Trehalose	−	0	−	0	−	0
Esculin hydrolysis	−	0	−	0	−	0
Milk-clot	−	0	−	0	−	0
Gelatin	−	0	−	0	−	0
Indole	−	0	−	0	−	0
H₂S	−	3	−⁺	12	−	7
Catalase	−	0	−⁺	23	−	0
Urease	−	0	−	0	+	100
Nitrate reduction	−	0	+	100	+	100

[a] −, Negative; +, positive; − ⁺, usually negative, some positive.

Table 15. Metabolic acids detected in PYG broth by gas-liquid chromatography of various asaccharolytic *Bacteroides* species or groups.[a]

Species/Group	*Bacteroides* CDC Gp F-1		*Bacteroides* CDC Gp F-2		*B. ureolyticus*	
Organic acid	Sign	% +	Sign	% +	Sign	% +
Acetic	+	100	+	100	+	100
Propionic	−	0	−	0	−	0
Isobutyric	−	0	−	0	−	0
Butyric	−	0	−	0	−	0
Isovaleric	−	0	−	0	−	0
Valeric	−	0	−	0	−	0
Isocaproic	−	0	−	0	−	0
Caproic	−	0	−	0	−	0
Lactic	−[+]	21	−	0	−	0
Succinic	V	63	V	54	+[−]	84

[a] +, Positive; −, negative; V, variable; −[+], usually negative, some positive; +[−], usually positive, some negative.

nitrate to nitrite, liquify gelatin weakly, and produce oxidase in detectable amounts.

3. Growth of *B. ureolyticus* is greatly stimulated when formate and fumarate are added to broth medium as described by Smibert and Holdeman (1976).

BACTEROIDES MELANINOGENICUS GROUP—*B. ASACCHAROLYTICUS* AND *B. MELANINOGENICUS* SUBSP. *INTERMEDIUS*. Colonies of the two species are quite similar on blood agar. The colonies are 1–3 mm in size, circular, entire, convex, and opaque. The colonies usually first appear to be slightly gray; upon continued incubation, the pigmentation intensifies to a shiny, jet black color. Before the colonies develop the black pigment, they will show a bright brick red fluorescence when exposed to long-wave ultraviolet light. This fluorescence will diminish as the pigmentation increases, and in fully pigmented cultures fluorescence is often not seen. The black pigment is a hemin derivative which is formed only when the medium contains blood (hemoglobin) (Schwabacher, Lucas, and Rimington, 1947). The pigment develops more rapidly on lysed blood agar than on solid media containing whole blood. Cells from blood agar medium usually appear as small coccoid rods, but in broth cultures they are usually longer and pleomorphic in shape and size, and may exhibit vacuoles. Detailed points of these two species are shown in Tables 16 and 17.

Some key characteristics that are useful in identifying these species are:

1. Certain other microorganisms such as *Peptococcus niger*, *Peptococcus micros*, *Peptostreptococcus anaerobius*, and some *Streptococcus* species may produce dark brown colonies on a blood agar medium. This pigmentation is not dependent on the presence of hemoglobin in the medium and may be observed on media other than blood agar.

2. *B. asaccharolyticus* and *B. melaninogenicus*

require hemin and vitamin K compounds, or both, in the medium for good growth (Gibbons and MacDonald, 1960). Sodium succinate, an additional growth factor, can be used to replace hemin in the presence of vitamin K and can also reduce the amount of vitamin K required in the presence of hemin (Lev, Kuedall, and Milford, 1971).

3. In peptone–yeast extract–glucose (PYG) broth (Moore, Cato, and Holdeman, 1969) supplemented with hemin and vitamin K, *B. asaccharolyticus* strains produce detectable amounts of acetic, propionic, isobutyric, butyric, and isovaleric acids and no succinic acid. *B. melaninogenicus* subsp. *intermedius* strains produce detectable amounts of acetic and succinic acids, and some produce isobutyric and isovaleric acids but do not produce detectable amounts of propionic or butyric acids (Table 17).

4. *B. melaninogenicus* subsp. *intermedius* strains usually produce indole and do not hydrolyze esculin, which helps to differentiate them from strains of *B. melaninogenicus* subsp. *melaninogenicus* that hydrolyze esculin and do not produce indole.

5. After subculture a few times, many strains of the *B. melaninogenicus* group tend to lose their ability to hemolyze blood and produce black colonies. This loss appears to be a common trait of *B. melaninogenicus* subsp. *melaninogenicus* strains we have examined but has not been observed with strains of *B. asaccharolyticus* and *B. melaninogenicus* subsp. *intermedius*. An occasional strain of *B. melaninogenicus* subsp. *melaninogenicus* has been observed that will form lightly pigmented and nonpigmented colonies. Both types of colonies, however, show typical vivid brick-red fluorescence when viewed with long-wave ultraviolet light.

As mentioned previously, certain other *Bacteroides* species may be isolated from human clinical

Table 16. Detailed characteristics of commonly encountered members of the *Bacteroides melaninogenicus* group.

Species	*B. asaccharolyticus*		*B. melaninogenicus* subsp. *intermedius*	
No. of strains	16		14	
Reaction	Sign	% +	Sign	% +
Aerobic growth	0	0	−	0
Motility	−	0	−	0
Hemolysis	+	100	+	100
Glucose	−	0	+	100
Mannitol	−	0	−	0
Lactose	−	0	V	64
Sucrose	−	0	V	71
Maltose	−	0	V	64
Salicin	−	0	−	0
Glycerol	−	0	−	0
D-Xylose	−	0	−	9
L-Arabinose	−	0	−	0
Starch	−	0	A	100
Starch hydrolysis	−	0	+	100
Mannose	−	0	A⁻	79
Rhamnose	−	0	−	0
Trehalose	−	0	−	0
Esculin hydrolysis	−	0	−	0
Milk digested	+⁻	88	+⁻	86
Gelatin	+	94	+	92
Indole	+⁻	82	+⁻	86
H₂S	−	0	−	0
Catalase	−	0	−	0
Urease	−	0	−	0
Nitrate reduction	−	0	−	0

[a] +, Positive; −, negative; +⁻, usually positive, some negative; V, variable.

specimens. These isolates are rare and constitute only a small percentage of the *Bacteroides* species isolated from properly collected specimens.

BACTEROIDES BIVIUS AND *B. DISIENS*.   These two recently described species (Holdeman and Johnson, 1977) have many similar characteristics but vary in their ability to ferment certain carbohydrates. Colonies of *B. bivius* and *B. disiens* on blood agar are 1–2 mm in size, circular, entire, convex, translucent, and nonhemolytic. The colonies do not develop black or tan pigmentation, and only an occasional strain may form colonies that will fluoresce light orange to pink when viewed under long-wave ultraviolet light. The cells of *B. bivius* and *B. disiens* vary in length, ranging from short, almost coccoidal

Table 17. Metabolic acids detected in PYG broth by gas-liquid chromatography of commonly encountered members of the *Bacteroides melaninogenicus* group.[a]

Species	*B. asaccharolyticus*		*B. melaninogenicus* subsp. *intermedius*	
Organic acid	Sign	% +	Sign	% +
Acetic	+	100	+	100
Propionic	+	100	−	0
Isobutyric	+	100	V	37
Butyric	+	100	−	0
Isovaleric	+	100	V	73
Valeric	−	0	−	0
Isocaproic	−	0	−	0
Caproic	−	0	−	0
Lactic	−	0	−	0
Succinic	−	0	+	100

[a] +, Positive; −, negative; V, variable.

rods to long rods that may be slightly curved. Both species are saccharolytic and proteolytic; gelatin and milk are digested. Indole, urease, lecithinase, lipase, and catalase are not produced, and esculin is not hydrolyzed. Both species ferment glucose, maltose, and starch but do not ferment mannitol, sucrose, salicin, xylose, arabinose, rhamnose, or trehalose. *B. divius* strains ferment lactose and mannose but strains of *B. disiens* do not. Metabolic products produced in PYG broth by both species are primarily acetic and succinic acids, with some strains showing trace amounts of isovaleric acid. Both *B. bivius* and *B. disiens* require hemin for growth (Holdeman and Johnson, 1977).

BACTEROIDES CAPILLOSUS. These microorganisms form colonies on blood agar that are usually pinpoint to 1 mm in size, circular, convex, entire, translucent, and nonhemolytic. The cells may vary markedly in length and width, and pleomorphism of cells, with curved filaments and irregular staining, is common in liquid media. They are nonmotile. On ordinary differential media, strains of *B. capillosus* hydrolyze esculin and are asaccharolytic. However, glucose is usually fermented and certain other carbohydrates such as lactose, maltose, mannose, sucrose, starch, and xylose may be weakly fermented if the medium is supplemented with Tween 80. Gelatin is not hydrolyzed, and nitrate is not reduced; indole, urease, lecithinase, lipase, and catalase are not produced. Milk may be clotted but is not digested. Metabolic products produced in PYG broth are acetic and succinic acids (Buchanan and Gibbons, 1974; Holdeman, Cato, and Moore, 1977; Smith, 1975).

BACTEROIDES PUTREDINIS. Colonies on the surface of blood agar are usually less than 1 mm in size, circular, entire, convex, translucent, and nonhemolytic. The colonies do not develop black or tan pigmentation and do not fluoresce when viewed with long-wave ultraviolet light. The cells from solid and liquid media are of moderate size and unremarkable in appearance. Strains of *B. putredinis* are consistently asaccharolytic and hydrolyze casein and gelatin. Indole is formed. A weak catalase reaction may be exhibited. Nitrate is not reduced; lecithinase, lipase, and urease are not produced; and esculin is not hydrolyzed. Metabolic products produced in PYG medium are acetic, propionic, isobutyric, butyric, and isovaleric acids. (Buchanan and Gibbons, 1974; Holdeman, Cato, and Moore, 1977; Smith, 1975).

BACTEROIDES PNEUMOSINTES. Even after extended incubation, colonies on blood agar are usually less than 0.5 mm in size; they are circular, entire, convex, transparent, and nonhemolytic. Cells of this species are extremely small coccoid rods that are barely discernible with a light microscope and are capable of passing through Berkefeld V and Chamberland filters. The cells often stain poorly by the conventional Gram procedure. Most strains of *B. pneumosintes* are quite fastidious and require the medium to be supplemented with serum, blood, or other body fluid for best growth. They are rather inert in differential media and do not ferment carbohydrates, liquify gelatin, or digest milk. Nitrate is not reduced to nitrite; esculin is not hydrolyzed; and lecithinase, lipase, indole, urease, and catalase are not produced. Only trace amounts of acetic acid are produced in PYG broth (Buchanan and Gibbons, 1974; Dowell and Hawkins, 1974; Smith, 1975).

BACTEROIDES NODOSUS. This microorganism is primarily of importance in veterinary medicine and is known to be involved with foot rot disease in sheep and cattle, as discussed above in this chapter. Colonies on blood agar are 0.5–2.0 mm in size after 48 h of incubation, circular, entire, convex, translucent, and nonhemolytic. Some may pit the surface of blood agar. The cells vary in length and may show a terminal swelling. The swollen forms are usually more common in cells observed in direct smears of clinical materials than in culture media (Buchanan and Gibbons, 1974). *B. nodosus* is a proteolytic organism which digests milk and hydrolyzes gelatin. Nitrate is not reduced; carbohydrates are not fermented; and indole, urease, catalase, lecithinase, and lipase are not produced. Metabolic products in PYG medium include acetic, propionic, and succinic acids. A practical FA technique for identifying *B. nodosus* was reported by Roberts and Walker (1971). Using this technique, they were able to identify cases of foot rot disease in cattle and sheep of seven different countries.

Literature Cited

Altemeier, W. A. 1938. The bacterial flora of acute perforated appendicitis with peritonitis: A bacteriologic study based upon one hundred cases. Annuals of Surgery **107:**517–528.

Altemeier, W. A. 1940. The anaerobic streptococci in tuboovarian abscess. American Journal of Obstetrics and Gynecology **39:**1038–1042.

Altemeier, W. A. 1974. Liver abscess: The etiologic role of anaerobic bacteria, pp. 387–398. In: Balows, A., DeHaan, R. M., Dowell, V. R., Jr., Guze, L. B. (eds), Anaerobic bacteria: Role in disease, Springfield, Illinois: Charles C. Thomas.

Altemeier, W. A., Hill, E. O., Fuller, W. D. 1969. Acute and recurrent thromboembolic disease: New concept of etiology. Annuals of Surgery **170:**547–558.

Aranki, A., Syed, S. A., Kenney, E. B., Freter, R. 1969. Isolation of anaerobic bacteria from human gingiva and mouse cecum by means of a simplified glove box procedure. Applied Microbiology **17:**568–576.

Balows, A. 1974. Anaerobic bacteria perspectives, pp. 3–6. In: Balows, A., DeHaan, R. M., Dowell, V. R., Jr., Guze, L. B. (eds.), Anaerobic bacteria: Role in disease. Springfield, Illinois: Charles C Thomas.

Bartlett, J. G., Gorbach, S. L., Thadepalli, H., Finegold, S. M. 1974a. The bacteriology of empyema. Lancet **i:**338–340.

Bartlett, J. G., Gorbach, S. L., Tally, F. P., Finegold, S. M. 1974b. Bacteriology and treatment of primary pulmonary lung abscess. American Review of Respiratory Diseases **109:**510–518.

Bartlett, J. G., Sullivan-Sigler, N., Louie, T. J., Gorbach, S. L. 1976. Anaerobes survive in clinical specimens despite delayed processing. Journal of Clinical Microbiology **3:**133–136.

Bartlett, R. C., Allen, V. D., Blazevic, D. J., Dolan, C. T., Dowell, V. R., Jr., Gavan, T. L., Inhorn, S. L., Lombard, G. L., Matsen, J. M., Melvin, D. M., Sommers, H. M., Suggs, M. T., West, B. S. 1978. Clinical microbiology, pp. 871–1005. In: Inhorn, S. L. (ed.), Quality assurance practices for health laboratories. Washington, D.C.: American Public Health Association.

Beerens, H., Tahon-Castel, M. 1965. Infections humaines a bactéries anaérobies non-toxigenes. Brussels Presses Academiques Europeenes.

Berg, J. N., Loan, R. W. 1975. *Fusobacterium necrophorum* and *Bacteroides melaninogenicus* as etiologic agents of foot rot in cattle. American Journal of Veterinary Research **36:**1115–1122.

Biberstein, E. L., Knight, H. D., England, K. 1968. *Bacteroides melaninogenicus* in diseases of domestic animals. Journal of the American Veterinary Medical Association **153:**1045–1049.

Bishop, D. H. L., King, H. K. 1962. Ubiquinone and vitamin K in bacteria. 2 Intracellular distribution in *Escherichia coli* and *Micrococcus lysodeikticus*. Biochemical Journal **85:**550–554.

Bjornson, H. S., Hill, E. O. 1973. *Bacteroidaceae* in thromboembolic disease: Effects of cell wall components on blood coagulation in vivo and in vitro. Infection and Immunity **8:**911–918.

Brown, J. H. 1921. An improved anaerobe jar. Journal of Experimental Medicine **33:**677–681.

Bruner, D. W., Gillespie, J. H. 1973. Hagan's infectious diseases of domestic animals, 6th edition. Ithaca: Comstock.

Buchanan, R. E., Gibbons, N. E. (eds). 1974. Bergey's manual of determinative bacteriology, 8th ed. Baltimore: Williams & Wilkins.

Burdon, K. L. 1928. *Bacterium melaninogenicum* from normal and pathologic tissues. Journal of Infectious Diseases **42:**161–171.

Caldwell, D. R., White, D. C., Bryant, M. P., Doetsch, R. N. 1965. Specificity of the heme requirement for growth of *Bacteroides ruminicola*. Journal of Bacteriology **90:**1645–1654.

Cato, E. P., Johnson, J. L. 1976. Reinstatement of species rank for *Bacteroides fragilis*, *B. ovatus*, *B. distasonis*, *B. thetaiotaomicron*, and *B. vulgatus*: Designation of neotype strains for *Bacteroides fragilis* (Veillon and Zuber) Castellani and Chalmers and *Bacteroides thetaiotaomicron* (Distaso) Castellani and Chalmers. International Journal of Systematic Bacteriology **26:**230–237.

Chan, P. C. K., Porschen, R. K. 1977. Evaluation of kanamycin-esculin bile agar for isolation and presumptive identification of *Bacteroides fragilis* group. Journal of Clinical Microbiology **6:**528–529.

Chow, A. W., Guze, L. B. 1974. Bacteroidaceae bacteremia: Clinical experience with 112 patients. Medicine **53:**93–126.

Chow, A. W., Cunningham, P. J., Guze, L. B. 1976. Survival of anaerobic and aerobic bacteria in a nonsupportive gassed transport system. Journal of Clinical Microbiology **3:**128–132.

Dack, G. M. 1940. Non-sporeforming anaerobic bacteria of medical importance. Bacteriological Reviews **4:**227–259.

Dormer, B. J. J., Babett, J. A. 1972. Orofacial infection due to bacteroides a neglected pathogen. Journal of Oral Surgery **30:**658–660.

Dowell, V. R., Jr. 1974. Collection of clinical specimens and primary isolation of anaerobic bacteria, pp. 9–20. In: Balows, A., DeHaan, R. M., Dowell, V. R., Jr., Guze, L. B. (eds.), Anaerobic bacteria: Role in disease. Springfield, Illinois: Charles C. Thomas.

Dowell, V. R., Jr. 1975. Wound and abscess specimens, pp. 70–81. In: Balows, A. (ed.), Clinical microbiology. How to start and when to stop. Springfield, Illinois: Charles C. Thomas.

Dowell, V. R., Jr. 1977. Clinical veterinary anaerobic bacteriology, pp. 1–12. Atlanta: Center for Disease Control.

Dowell, V. R., Jr., Hawkins, T. M. 1974. Laboratory methods in anaerobic bacteriology, CDC laboratory manual. U. S. Department of Health, Education and Welfare Publication No. (CDC) 74-8272. Atlanta: Center for Disease Control.

Dowell, V. R., Jr., Hill, E. O., Altemeier, W. A. 1964. Use of phenethyl alcohol in media for isolation of anaerobic bacteria. Journal of Bacteriology **88:**1811–1813.

Dowell, V. R., Jr., Lombard, G. L. 1977. Presumptive identification of anaerobic nonsporeforming Gram-negative bacilli, pp. 1–13. Atlanta: Center for Disease Control.

Dowell, V. R., Jr., Lombard, G. L., Thompson, F. S., Armfield, A. Y. 1977. Media for isolation, characterization and identification of obligately anaerobic bacteria, pp. 1–46. Atlanta: Center for Disease Control.

Draper, D. L., Barry, A. L. 1977. Rapid identification of *Bacteroides fragilis* with bile and antibiotic disks. Journal of Clinical Microbiology **5:**439–443.

Egerton, J. R., Roberts, D. S., Parsonson, J. M. 1969. The aetiology and pathogenesis of ovine foot-rot. I. A histological study of the bacterial invasion. Journal of Comparative Pathology **79:**207–219.

Eggerth, A. H., Gagnon, B. H. 1933. The *Bacteroides* of human feces. Journal of Bacteriology **25:**389–413.

Eiken, M. 1958. Studies on an anaerobic, rod-shaped, Gram-negative microorganism: *Bacteroides corrodens*, n. sp. Acta Pathologica et Microbiologica Scandinavica **43:**404–416.

Ellner, P. D., Granato, P. A., May, C. B. 1973. Recovery and identification of anaerobes: A system suitable for the routine clinical laboratory. Applied Microbiology **26:**904–913.

Felner, J. M. 1974. Infective endocarditis caused by anaerobic bacteria, pp. 345–352. In: Balows, A., DeHaan, R. M., Dowell, V. R., Jr., Guze, L. B. (eds.), Anaerobic bacteria: Role in disease. Springfield, Illinois: Charles C. Thomas.

Felner, J. M., Dowell, V. R., Jr. 1970. Anaerobic bacterial endocarditis. New England Journal of Medicine **283:**1188–1192.

Felner, J. M., Dowell, V. R., Jr. 1971. "Bacteroides bacteremia." American Journal of Medicine **50:**787–796.

Finegold, S. M. 1974. Intra-abdominal, genito-urinary, skin and soft tissue infections due to non-sporing anaerobic bacteria, pp. 160–188. In: Phillips, I., Sussman, M. (eds.), Infection with non-sporing anaerobic bacteria. London: Churchill Livingstone.

Finegold, S. M. 1977. Anaerobic bacteria in human disease. New York, San Francisco, London: Academic Press.

Finegold, S. M., Barnes, E. M. 1977. Report of the ICSB taxonomic subcommittee on Gram-negative anaerobic rods. Proposal that the saccharolytic and asaccharolytic strains at present classified in the species *Bacteroides melaninogenicus* (Oliver and Wherry) be reclassified in two species as *Bacteroides melaninogenicus* and *Bacteroides asaccharolyticus*. International Journal of Systematic Bacteriology **27:**388–391.

Finegold, S. M., Massh, V. H., Bartlett, J. G. 1971. Anaerobic infections in the compromised host, pp. 123–134. In: Brachman, P., Eickhoff, T. C., (eds.), Proceedings of an International Conference on Hospital Infections, Atlanta, 1970. Chicago: American Hospital Association.

Finegold, S. M., Miller, L. G. 1968. Normal fecal flora of adult humans. Bacteriological Proceedings **1968:**93.

Finegold, S. M., Rosenblatt, J. E. 1973. Practical aspects of anaerobic sepsis. Medicine **52:**311–322.

Finegold, S. M., Shepherd, W. E., Spaulding, E. H. 1977.

Cumitech 5. Practical anaerobic bacteriology: Cumulative techniques and procedures in clinical microbiology, pp. 1–14. Washington, D.C.: American Society for Microbiology.

Finegold, S. M., Sutter, V. L., Attebery, H. R., Rosenblatt, J. E. 1974. Isolation of anaerobic bacteria, pp. 365–375. In: Lennette, E. H., Spaulding, E. H., Truant, J. P. (eds.), Manual of clinical microbiology, 2nd ed. Washington, D.C.: American Society for Microbiology.

Finegold, S. M., Rosenblatt, J. E., Sutter, V. L., Attebery, H. R. 1976. Scope monograph on anaerobic infections, 3rd ed., pp. 1–69. Thomas, B. A. (ed.), Kalamazoo, Michigan: Upjohn.

Frederick, J., Braude, A. J. 1974. Anaerobic infection of the paranasal sinuses. New England Journal of Medicine **290:**135–137.

Gesner, B. M., Jenkins, C. R. 1964. Production of heparinase by *Bacteroides*. Journal of Bacteriology **81:**595–604.

Gibbons, R. J. 1974. Aspects of the pathogenicity and ecology of the indigenous oral flora of man, pp. 267–285. In: Balows, A., DeHaan, R. M., Dowell, V. R., Jr., Guze, L. B. (eds.), Anaerobic bacteria: Role in disease. Springfield, Illinois: Charles C. Thomas.

Gibbons, R. J., Engle, L. P. 1964. Vitamin K compounds in bacteria that are obligate anaerobes. Science **146:**1308–1309.

Gibbons, R. J., MacDonald, J. B. 1960. Hemin and vitamin K compounds as required factors for the cultivation of certain strains of *Bacteroides melaninogenicus*. Journal of Bacteriology **80:**164–170.

Gibbons, R. J., Socransky, S. S., Sawyer, S., Kapsimalis, B., MacDonald, J. B. 1963. The microbiota of the gingival crevice of man. II. The predominant cultivable organisms. Archives of Oral Biology **8:**281–289.

Gorbach, S. L., Bartlett, J. G. 1974a. Anaerobic infections. New England Journal of Medicine **290:**1177–1184.

Gorbach, S. L., Bartlett, J. G. 1974b. Anaerobic infections. New England Journal of Medicine **290:**1237–1245.

Gorbach, S. L., Bartlett, J. G. 1974c. Anaerobic infections. New England Journal of Medicine **290:**1289–1294.

Gregory, E. M., Kowalski, J. B., Holdeman, L. V. 1977. Production and some properties of catalase and superoxide dismutase from the anaerobe *Bacteroides distasonis*. Journal of Bacteriology **129:**1298–1302.

Hagen, J. C., Wood, W. S., Hashimoto, T. 1977. Effect of temperature on survival of *Bacteroides fragilis* subsp. *fragilis* and *Escherichia coli* in pus. Journal of Clinical Microbiology **6:**567–570.

Hansen, S. L., Stewart, B. J. 1976. Comparison of API and Minitek to Center for Disease Control methods for the biochemical characterization of anaerobes. Journal of Clinical Microbiology **4:**227–231.

Hanson, C. W., Martin, W. J. 1976. Evaluation of enrichment, storage, and age of blood agar medium in relation to its ability to support growth of anaerobic bacteria. Journal of Clinical Microbiology **4:**394–399.

Heineman, H. S., Braude, A. J. 1963. Anaerobic infection of the brain. American Journal of Medicine **35:**682–697.

Hill, G. B. 1978. Effects of storage in an anaerobic transport system on bacteria in known polymicrobic mixtures and in clinical specimens. Journal of Clinical Microbiology **8:**680–688.

Holdeman, L. V., Cato, E. P., Moore, W. E. C. (eds.). 1977. Anaerobe laboratory manual, 4th ed. Blacksburg, Virginia: Virginia Polytechnic Institute and State University Anaerobe Laboratory.

Holdeman, L. V., Johnson, J. L. 1977. *Bacteroides disiens* sp. nov. and *Bacteroides bivius* sp. nov. from human clinical infections. International Journal of Systematic Bacteriology **27:**337–345.

Holland, J. W., Hill, E. O., Altemeier, W. A. 1977. Numbers and types of anaerobic bacteria isolated from clinical specimens since 1960. Journal of Clinical Microbiology **5:**20–25.

Holland, J. W., Gagnet, S. M., Lewis, S. A., Stauffer, L. R. 1977. Clinical evaluation of a simple, rapid procedure for the presumptive identification of anaerobic bacteria. Journal of Clinical Microbiology **5:**416–426.

Hungate, R. E. 1966. The rumen and its microbes. New York, London: Academic Press.

Jackson, F. L., Goodman, Y. E. 1972. Transfer of the facultatively anaerobic organism *Bacteroides corrodens* Eiken to a new genus, *Eikenella*. International Journal of Systematic Bacteriology **22:**73–77.

Jackson, F. L., Goodman, Y. E., Bel, F. R., Wong, P. C., Whitehouse, R. L. S. 1971. Taxonomic status of facultative and strictly anaerobic "corroding bacilli" that have been classified as *Bacteroides corrodens*. Journal of Medical Microbiology **4:**171–184.

Jackson, F. L., Goodman, Y. E. 1978. *Bacteroides ureolyticus*, a new species to accommodate strains previously identified as "*Bacteroides corrodens*, anaerobic." Journal of Clinical Microbiology **8:**197–200.

Katitch, R. V., Csetkovitch, L. J., Voukitchevitch, Z., Parjevitch, D. J. 1967. Contribution à l étude de l etiologie du pietin du mouton. Bulletin Office International Des Epizooties **67:**1–10.

Khairat, O. 1967. *Bacteroides corrodens* isolated from bacteriaemias. Journal of Pathology and Bacteriology **94:**29–40.

Killgore, G. E., Starr, S. E., DelBene, V. E., Whaley, D. N., Dowell, V. R., Jr. 1973. Comparison of three anaerobic systems for the isolation of anaerobic bacteria from clinical specimens. American Journal of Clinical Pathology **59:**552–559.

King, E. O., Tatum, H. W. 1962. *Actinobacillus actinomycetemcomitans* and *Haemophilus aphrophilus*. Journal of Infectious Diseases **111:**85–94.

Langworth, B. F. 1977. *Fusobacterium necrophorum:* Its characteristics and role as an animal pathogen. Bacteriological Reviews **41:**373–390.

Lanz, O., Tavel, E. 1904. Bactériologie de l'appendicité. Revue de Chouirgie (Paris) **30:**43–58.

Leake, D. L. 1972. *Bacteroides* osteomyelitis of the mandible: A report of two cases. Oral Surgery **34:**585–588.

Ledger, W. J., Campbell, C., Willson, J. R. 1968. Postoperative adnexal infections. Obstetrics and Gynecology **31:**83–89.

Lev, M. 1958. Apparent requirement for vitamin K of rumen strains of *Fusiformis nigrescens*. Nature **181:**203–204.

Lev, M., Kuedell, K. C., Milford, A. F. 1971. Succinate as a growth factor for *Bacteroides melaninogenicus*. Journal of Bacteriology **108:**175–178.

Lewis, K. H., Rettger, L. F. 1940. Non-sporulating anaerobic bacteria of the intestinal tract. I. Occurrence and taxonomic relationships. Journal of Bacteriology **40:**287–307.

Loesche, W. J. 1969. Oxygen sensitivity of various anaerobic bacteria. Applied Microbiology **18:**723–727.

Loesche, W. J. 1974. Dental infections, pp. 409–434. In: Balows, A., DeHaan, R. M., Dowell, V. R., Jr., Guze, L. B. (eds.), Anaerobic bacteria: Role in disease. Springfield, Illinois: Charles C. Thomas.

Loesche, W. J., Socransky, S. S., Gibbons, R. J. 1964. *Bacteroides oralis* proposed new species isolated from the oral cavity of man. Journal of Bacteriology **88:**1329–1337.

McFarlan, A. M. 1943. The bacteriology of brain abscess. British Medical Journal **ii:**643–644.

Macy, J., Probst, J., Gottschalk, G. 1975. Evidence for cytochrome involvement in fumarate reduction and adenosine 5′-triphosphate synthesis by *Bacteroides fragilis* grown in the presence of hemin. Journal of Bacteriology **123:**436–442.

Martin, W. J. 1971. Practical method for isolation of anaerobic bacteria in the clinical laboratory. Applied Microbiology **22:**1168–1171.

Martin, W. J., Segura, J. W. 1974. Urinary tract infections due to anaerobic bacteria, pp. 359–367. In: Balows, A., DeHaan, R. M., Dowell, V. R., Jr., Guze, L. B. (eds.), Anaerobic bacteria: Role in disease. Springfield, Illinois: Charles C. Thomas.

Mena, E., Thompson, F. S., Armfield, A. Y., Dowell, V. R., Jr., Reinhardt, D. J. 1978. Evaluation of Port-A-Cul™ transport system for the protection of anaerobic bacteria. Journal of Clinical Microbiology **8**:28–35.

Mills, K. C., Haobley, G. S., Dochez, A. R. 1928. Studies in the common cold. II. A study of certain Gram-negative, filter-passing anaerobes of the upper respiratory tract. Journal of Experimental Medicine **47**:193–206.

Misra, S. S. 1938. A note on the predominance of the genus *Bacteroides* in human feces. Journal of Pathology and Bacteriology **46**:204–206.

Moore, H. B., Sutter, V. L., Finegold, S. M. 1975. Comparison of three procedures for biochemical testing of anaerobic bacteria. Journal of Clinical Microbiology **1**:15–24.

Moore, W. B. 1968. Solidified media suitable for the cultivation of *Clostridium novyi* type B. Journal of General Microbiology **53**:415–423.

Moore, W. E. C., Cato, E. P., Holdeman, L. V. 1969. Anaerobic bacteria of the gastrointestinal flora and their occurrence in clinical infections. Journal of Infectious Diseases **119**: 641–649.

Moore, W. E. C., Holdeman, L. V. 1974. The human fecal flora: The normal flora of 20 Japanese-Hawaiians. Applied Microbiology **27**:961–979.

Morgan, J. R., Liu, P. Y. K., Smith, J. A. 1976. Semi-microtechnique for the biochemical characterization of anaerobic bacteria. Journal of Clinical Microbiology **4**:315–318.

Nastro, L. J., Finegold, S. M. 1973. Endocarditis due to anaerobic Gram-negative bacilli. American Journal of Medicine **54**:482–496.

Nobles, E. R. 1973. *Bacteroides* infections. Annuals of Surgery **177**:601–606.

Nord, C. E., Dahlback, A., Wadstrom, T. 1975. Evaluation of a test kit for identification of anaerobic bacteria. Medical Microbiology and Immunology **161**:239–242.

Olitsky, P. K., Gates, F. L. 1921. Experimental studies of the nasopharyngeal secretions from influenza patients. IV. Anaerobic cultivation. Journal of Experimental Medicine **33**:713–729.

Olitsky, P. K., Gates, F. L. 1922. Experimental studies of the nasopharyngeal secretions from influenza patients. VIII. Further observations on the culture and morphological characters of *Bacterium pneumosintes*. Journal of Experimental Medicine **35**:813–821.

Parker, R. T., Jones, C. P. 1966. Anaerobic pelvic infections and developments in hyperbaric oxygen therapy. American Journal of Obstetrics and Gynecology **96**:645–658.

Pearson, H. E., Anderson, G. V. 1970. *Bacteroides* infections and pregnancy. Obstetrics and Gynecology **35**:31–36.

Pearson, H. E., Smiley, D. F. 1968. *Bacteroides* in pilonidal sinuses. American Journal of Surgery **115**:336–338.

Phillips, I., Sussman, M. (eds.). 1974. Infection with nonsporing anaerobic bacteria, pp. 1–234. Edinburgh: Churchill Livingstone.

Prévot, A. R., Fredette, V. 1966. Manual for the classification and determination of the anaerobic bacteria. Philadelphia: Lea and Febiger.

Quinto, G. 1962. Nutrition of five *Bacteroides* strains. Journal of Bacteriology **84**:559–562.

Quinto, G. 1966. Amino acid and vitamin requirements of several *Bacteroides* strains. Applied Microbiology **14**:1022–1026.

Quinto, G., Sebald, M. 1964. Identification of three hemin-requiring *Bacteroides* strains. American Journal of Medical Technology **30**:318–384.

Quinto, G., Sebald, M., Prévot, A. R. 1963. Études sur le pouvoir pathogène de *Ristella pseudoinsolita*. Role de l'hémine dans sa croissance. Annales de l'institut Pasteur **105**:455–459.

Riddell, M. I., Morton, H. S., Murray, E. G. D. 1953. The value of dihydrostreptomycin in preoperative preparation of the gut. American Journal of Medical Science **225**:535–546.

Roberts, D. S., Walker, P. D. 1971. Fluorescein-labelled antibody for the diagnosis of foot rot. Veterinary Record **92**:70–71.

Riley, P. S., Tatum, H. W., Weaver, R. E. 1973. Identify of HB-1 of King and *Eikenella corrodens* (Eiken) Jackson and Goodman. International Journal of Systematic Bacteriology **23**:75–76.

Rizza, V., Sinclair, P. R., White, D. C., Cuorant, P. R. 1968. Electron transport system of the protoheme-requiring anaerobe *Bacteroides melaninogenicus*. Journal of Bacteriology **96**:665–671.

Roberts, D. S. 1969. Synergic mechanisms in certain mixed infections. Journal of Infectious Diseases **120**:720–724.

Roberts, D. S., Egerton, J. R. 1969. The aetiology and pathogenesis of ovine foot rot. II. The pathogenic association of *Fusiformis nodosus* and *F. necrophorus*. Journal of Comparable Pathology **79**:217–227.

Roberts, D. S., Walker, P. D. 1971. Fluorescein-labelled antibody for the diagnosis of foot rot. Veterinary Research **92**:70–71.

Rosebury, T. 1962. Microorganisms indigenous to man. New York, Toronto, London: McGraw-Hill.

Rosenblatt, J. E. 1976. Isolation and identification of anaerobic bacteria. Human Pathology **7**:178–186.

Rosenblatt, J. E., Fallon, A., Finegold, S. M. 1973. Comparison of methods for isolation of anaerobic bacteria from clinical specimens. Applied Microbiology **25**:77–85.

Rotheram, E. B., Jr., Schick, S. F. 1969. Nonclostridial anaerobic bacteria in septic abortion. American Journal of Medicine **46**:80–89.

Rubner, B. 1957. The effect of chlortetracycline on the fecal flora of patients with and without cirrhosis of the liver. Journal of Pathology and Bacteriology **73**:429–437.

Sabbaj, J., Sutter, V. L., Finegold, S. M. 1972. Anaerobic pyogenic liver abscess. Annuals of Internal Medicine **77**:629–638.

Sabriston, C. B., Grigsby, W. R. 1972. Anaerobic bacteria from the advanced periodontal lesion. Journal of Periodontology **43**:199–201.

Schaedler, R. W., Dubos, R., Costello, R. 1965. The development of the bacterial flora in the gastrointestinal tract of mice. Journal of Experimental Medicine **122**:59–66.

Schwabacher, H., Lucas, D. R., Rimington, C. 1947. *Bacterium melaninogenicum* —a misnomer. Journal of General Microbiology **1**:109–120.

Schreckenberger, P. C., Blazevic, D. J. 1974. Rapid methods for biochemical testing of anaerobic bacteria. Applied Microbiology **28**:759–762.

Schreckenberger, P. C., Blazevic, D. J. 1976. Rapid fermentation testing of anaerobic bacteria. Journal of Clinical Microbiology **3**:313–317.

Skinner, F. A., Carr, J. G. (eds.). 1974. The normal microbial flora of man. London, New York: Academic Press.

Smibert, R. M., Holdeman, L. V. 1976. Clinical isolates of anaerobic Gram-negative rods with formate-fumarate energy metabolism: *Bacteroides corrodens*, *Vibrio succinogenes* and unidentified strains. Journal of Clinical Microbiology **3**:432–437.

Smith, L. DS. 1975. The pathogenic anaerobic bacteria, 2nd ed. Springfield, Illinois: Charles C. Thomas.

Smith, H. W., Crabb, W. E. 1961. Faecal bacterial flora of animals and man: Its development in young. Journal of Pathology and Bacteriology **82**:53–66.

Socransky, S. S., Gibbons, R. J. 1965. Required role of *Bacteroides melaninogenicus* in mixed anaerobic infections. Journal of Infectious Diseases **115**:247–253.

Sonnenwirth, A. C. 1972. Evolution of anaerobic methodology. American Journal of Clinical Nutrition **25**:1295–1298.

Sonnenwirth, A. C. 1974. Incidence of intestinal anaerobes in blood cultures, pp. 157–171. In: Balows, A., DeHaan, R. M., Dowell, V. R., Jr., Guze, L. B. (eds.), Anaerobic bacteria: Role in disease. Springfield, Illinois: Charles C. Thomas.

Spaulding, E. H., Vargo, V., Michaelson, T. C., Swenson, R. M. 1974. A comparison of two procedures for isolating anaerobic bacteria from clinical specimens, pp. 37–46. In: Balows, A., DeHaan, R. M., Dowell, V. R., Jr., Guze, L. B. (eds.), Anaerobic bacteria: Role in disease. Springfield, Illinois: Charles C. Thomas.

Sperry, J. F., Appleman, M. D., Wilkins, T. D. 1977. Requirement of heme for growth of *Bacteroides fragilis*. Applied and Environmental Microbiology **34**:386–390.

Stargel, M. D., Lombard, G. L., Dowell, V. R., Jr. 1978. Alternative procedures for identification of anaerobic bacteria. American Journal of Medical Technology **44**:709–722.

Stargel, M. D., Thompson, F. S., Phillips, S. E., Lombard, G. L., Dowell, V. R., Jr. 1976. Modification of the Minitek miniaturized differentiation system for characterization of anaerobic bacteria. Journal of Clinical Microbiology **3**:291–301.

Starr, S. E., Killgore, G. E., Dowell, V. R., Jr. 1971. Comparison of Schaedler agar and trypticase soy-yeast extract agar for the cultivation of anaerobic bacteria. Applied Microbiology **22**:655–658.

Starr, S. E., Thompson, F. S., Dowell, V. R., Jr., Balows, A. 1973. Micromethod system for identification of anaerobic bacteria. Applied Microbiology **25**:713–717.

Stokes, E. J. 1958. Anaerobes in routine diagnostic cultures. Lancet **i**:668–670.

Sullivan, K. M., O'Toole, R. D., Fisher, R. H., Sullivan, K. N. 1973. Anaerobic empyema thoracis. The role of anaerobes in 226 cases of culture-proven empyemas. Archives of Internal Medicine **131**:521–527.

Sutter, V. L., Finegold, S. M. 1971. Antibiotic disk susceptibility tests for rapid presumptive identification of Gram-negative anaerobic bacilli. Applied Microbiology **21**:13–20.

Sutter, V. L., Vargo, V. L., Finegold, S. M. 1975. Wadsworth anaerobic bacteriology manual, 2nd ed., pp. 1–106. Los Angeles: Wadsworth Hospital Center, Veterans Administration and Department of Medicine UCLA.

Suzuki, S., Ushijima, T., Ichinose, H. 1966. Differentiation of *Bacteroides* from *Sphaerophorus* and *Fusobacterium*. Japanese Journal of Microbiology **10**:193–200.

Swartz, M. N., Karchmer, A. W. 1974. Infections of the central nervous system, pp. 309–325. In: Balows, A., DeHaan, R. M., Dowell, V. R., Jr., Guze, L. B. (eds.), Anaerobic bacteria: Role in disease. Springfield, Illinois: Charles C. Thomas.

Swenson, R. M., Michaelson, T. C., Daly, M. J., Spaulding, E. H. 1973. Anaerobic bacterial infections of the female genital tract. Obstetrics and Gynecology **42**:538–541.

Tally, F. P., Stewart, P. R., Sutter, V. L., Rosenblatt, J. E. 1975. Oxygen tolerance of fresh clinical anaerobic bacteria. Journal of Clinical Microbiology **1**:161–164.

Thadepalli, H., Gorbach, S. L., Keith, L. 1973. Anaerobic infections of the female genital tract: Bacteriologic and therapeutic aspects. American Journal of Obstetrics and Gynecology **117**:1034–1040.

Thorley, C. M. 1976. A simplified method for the isolation of *Bacteroides nodosus* from ovine foot-rot and studies on its colony morphology and serology. Journal of Applied Bacteriology **40**:301–309.

Ueno, K., Sugihara, P. T., Brichnell, K. S., Atteberry, H. R., Sutter, V. L., Finegold, S. M. 1974. Comparison of characteristics of Gram-negative anaerobic bacilli isolated from feces of individuals in Japan and the United States, pp. 135–148. In: Balows, A., DeHaan, R. M., Dowell, V. R., Jr., Guze, L. B. (eds.), Anaerobic bacteria: Role in disease. Springfield, Illinois: Charles C. Thomas.

van Houte, J., Gibbons, R. J. 1966. Studies of the cultivable flora of normal human feces. Antonie van Leeuwenhoek Journal of Microbiology and Serology **32**:212–222.

Varel, V. H., Bryant, M. P. 1974. Nutritional features of *Bacteroides fragilis* subsp. *fragilis*. Applied Microbiology **28**:251–257.

Vargo, V., Korzeniowski, M., Spaulding, E. H. 1974. Tryptic soy bile-kanamycin test for the identification of *Bacteroides fragilis* Applied Microbiology **27**:480–483.

Washington, J. A., II. 1973. Bacteremia due to anaerobic unusual and fastidious bacteria, pp. 47–60. In: Sonnenwirth, A. C. (ed.), Bacteremia: Laboratory and clinical aspects. Springfield, Illinois: Charles C. Thomas.

Watt, B., Collee, J. G. 1974. Practical approaches to the isolation and identification of clinically important non-sporing anaerobes, pp. 7–19. In: Phillips J., Sussman, M., (eds.), Infection with non-sporing anaerobic bacteria. Edinburgh: Churchill Livingstone.

Weinstein, L., Rubin, R. H. 1973. Infective endocarditis. Progress in Cardiovascular Diseases **16**:239–274.

Weiss, J. E., Rettger, L. F. 1937. The Gram-negative *Bacteroides* of the intestine. Journal of Bacteriology **33**:423–434.

Werner, H. 1973. Experimentelle Infektionen durch Bakteroidazeen, pp. 185–229. In: Eichler, O. (ed.), Handbuch der experimentellen Pharmakologie, vol. XVI/11B. Berlin, Heidelberg, New York: Springer-Verlag.

Wilkins, T. D., Chalgren, S. L., Jemenez-Ulate, F., Drake, C. R., Jr., Johnson, J. L. 1976. Inhibition of *Bacteroides fragilis* on blood agar plates and reversal of inhibition by added hemin. Journal of Clinical Microbiology **3**:359–363.

Wilkins, T. D., Wagner, D. L., Veltri, B. J., Jr., Gregory, E. M. 1978. Factors affecting production of catalase by *Bacteroides*. Journal of Clinical Microbiology **8**:553–557.

Wilson, G. S., Miles, A. A. (eds.). 1964. Topley and Wilson's principles of bacteriology and immunity, 5th ed. Baltimore: Williams & Wilkins. page 2145

Zabransky, R. J. 1970. Isolation of anaerobic bacteria from clinical specimens. Mayo Clinic Proceedings **45**:256–264.

Ziment, I., Davis, A., Finegold, S. M. 1969. Joint infection by anaerobic bacteria: A case report and review of the literature. Arthritis and Rheumatism **12**:627–634.

Ziment, I., Miller, L. G., Finegold, S. M. 1968. Nonsporulating anaerobic bacteria in osteomyletis. Antimicrobial Agents and Chemotherapy—1967, pp. 77–85. Ann Arbor, Michigan: American Society for Microbiology.

Zubrzycki, L, Spaulding, E. H. 1962. Studies on the stability of the normal human fecal flora. Journal of Bacteriology **83**:968–974.

CHAPTER 117

Nonpathogenic Members of the Genus *Bacteroides*

JOAN M. MACY

Habitats

Nonpathogenic strains of *Bacteroides* are the common inhabitants of the alimentary tract of warm-blooded animals and account for a major portion of the microflora in certain parts of the gut. They inhabit those areas of the gastrointestinal tract where there is little or no host digestive capability (e.g., mouth, cecum, colon, rumen) (Gorbach and Levitan, 1970; Hungate, 1966; Hungate, 1977). Except for the mouth, these areas are essentially fermentation chambers where ingested food that is either not digested or not digestible by the host is metabolized by these and other organisms along with protein and carbohydrate of host origin. In most instances, certain products of the fermentation are then utilized by the host (e.g., microbial protein and volatile fatty acids [VFAs]).

Rumen-Reticulum

One such example of a fermentation chamber is the rumen-reticulum. Because the animal itself lacks digestive enzymes capable of degrading plant material (e.g., cellulose, hemicellulose, pectin) the microorganisms in the rumen-reticulum ferment this material to VFAs, with concomitant formation of more microbes. The VFAs are absorbed from the rumen and used by the animal, while a continuous portion of the protein-rich microorganisms is degraded and absorbed in the small intestine after being killed by the acid in the abomasum.

Among the organisms of the rumen, three *Bacteroides* species are present that fulfill important functions. The first of these, *Bacteroides succinogenes,* is one of the more important cellulolytic organisms in the rumen. Not only is it among the predominant bacteria cultured from the rumen of cows fed various diets (accounting for 4.6–20% of the total isolated) (Bryant and Burkey, 1953b; Bryant and Doetsch, 1954), but it is an active cellulose-fermenting organism and is the only rumen cellulolytic bacterium able to actively break down undegraded cotton fiber (Halliwell and Bryant, 1963).

While *B. succinogenes* does not utilize hemicellulose as a carbon and energy source, it does, nonetheless, degrade it (Coen and Dehority, 1970; Collings and Yokoyama, 1980; Dehority, 1973). It has also been reported that this organism degrades and utilizes pectin (Bryant and Doetsch, 1954; Dehority, 1973; Gradel and Dehority, 1972).

A second important rumen bacteroide is *Bacteroides amylophilus,* an organism able to grow only on starch or the di- and oligosaccharides derived from starch (Hamlin and Hungate, 1956). Hamlin and Hungate (1956) found that this organism occurred occasionally in large numbers when a high grain ration was provided; similarly, Blackburn and Hobson (1962) found it was prevalent in sheep fed a ration high in starch (10^6–10^8/ml). *B. amylophilus* is also actively proteolytic (Blackburn, 1968; Blackburn and Hobson, 1962).

The last major rumen bacteroide, *Bacteroides ruminicola,* is one of the most numerous and versatile fermenters of carbohydrate in the rumen, representing 6–19% (1.5–3.8×10^8/ml) of the total carbohydrate-fermenting bacteria present in cattle fed rations as different as alfalfa hay, alfalfa silage, and a wheat straw and grain mixture (Bryant and Burkey, 1953b; Bryant et al., 1958). This organism also ferments starch and xylan; some strains are able to degrade and utilize pectin (Dehority, 1969; Dehority, 1973; Gradel and Dehority, 1972) as well as some types of isolated hemicellulose (Coen and Dehority, 1970; Dehority, 1973).

Human Alimentary Tract; Colon, Mouth

Unlike ruminants in man and other carnivores the relationship between the alimentary tract and intestinal microbes is one of competition for ingested food (Hungate, 1977). Thus, it is not surprising to learn that the *Bacteroides* species are found primarily in the lower ileum and colon (Gorbach and Levitan, 1970), as stomach acidity and gastrointestinal motility assure low numbers of microbes in the upper portions of the small intestine (Gorbach and Levitan, 1970; Macy et al., 1978) where most of an ingested meal is digested and absorbed by the host. The carbohydrate and protein substrates available to microorganisms in the lower ileum and colon are primarily undigestible dietary plant materials (i.e., "dietary fiber" such as cellulose, hemicellulose, pectin, etc.), mucins, and epithelial cells. Thus the colon is also a fermentation chamber, with the nature of the fermentation determined by the types of substrates available.

Due to the difficulty in obtaining samples, much of the work concerned with the microbes of the lower intestine has been done with fecal samples, and it is assumed that this material is representative of the colonic contents from which it is derived. Some of the most extensive studies concerned with the isolation and characterization of anaerobes, using very good anaerobic conditions, have been done by Holdeman, Moore, and co-workers (Holdeman, Good, and Moore, 1976; Moore and Holdeman, 1974). In two different studies, they found that *Bacteroides* species accounted for 20–30% of the total cultivable flora, with *B. fragilis* subsp. *vulgatus* being the most frequently encountered of all organisms isolated (3–6 × 10^{10}/g feces). *B. fragilis* subsp. *thetaiotaomicron, distasonis,* and *a* were also detected in high numbers; while *B. fragilis* subsp. *fragilis, b,* and *ovatus* were occasionally isolated. It should be pointed out that species rank has recently been given to the organisms of subsp. *fragilis, ovatus, distasonis, thetaiotaomicron,* and *vulgatus* (Cato and Johnson, 1976).

With regard to the metabolic activities of colon *Bacteroides* species and their role in the gut, these organisms appear to be involved in polysaccharide (derived from plant material of the gut) degradation in the colon; many strains have been shown to utilize xylans, noncellulosic glucans, pectins, galactomannans, mucopolysaccharides, mucin, and glycoprotein (Salyers, 1979). In addition, a cellulose-degrading *Bacteroides* species has been isolated from human feces (Betian et al., 1977). Thus, it seems possible that colon bacteroides might be either directly or indirectly involved in bringing about the positive effects that result from an increase of fiber in the diet. It is also interesting to note that many of the colon *Bacteroides* species are able to transform bile salts (Aries and Hill, 1970; Drasar and Hill, 1974; Hylemon and Sherrod, 1975; MacDonald et al., 1975; Midtvedt and Norman, 1967; Sherrod and Hylemon, 1977; Stellewag and Hylemon, 1976) and steroids (Drasar and Hill, 1974) into various metabolites, some of which have been implicated in the etiology of colon cancer (Alcantara and Speckmann, 1976; Drasar and Hill, 1974; Hill et al., 1971; Mastromarino, Reddy, and Wynder, 1976; Renwick and Drasar, 1976).

Mention should also be made of the *Bacteroides* species found in the mouth; these include *B. oralis, B. ochraceus, B. corrodens, B. melaninogenicus* subsp. *intermedius* and *melaninogenicus,* and *B. asaccharolyticus* (Gibbons and van Houte, 1975; Holdeman, Cato, and Moore, 1977; Holdeman and Moore, 1974).

Alimentary Tracts Having a Cecum (or Ceca)

Like the rumen-reticulum, a cecum is a specialized fermentation chamber found in animals that consume large amounts of plant material. However, since much of the food of animals having a cecum can be digested by gut enzymes, the fermentation chamber is found posterior to the stomach at the junction of the small and large intestines; thus, the actual fermentation of fiber occurs in the combined cecum and large intestine (Hungate, 1977).

Among mammals, the flora of the rat and mouse cecum is probably the most well studied. Morotomi et al. (1975) reported that in conventional rats bacteroides were only rarely detected in the lower small intestine, while the cecum contained 10^8 bacteroides per gram of contents. Tannock (1977) was able to culture 10^8–10^9 bacteroides per gram of mouse cecal contents and found that *B. thetaiotaomicron* and organisms resembling *B. fragilis* appeared to be the dominant bacteroides in the mouse cecum. Other *Bacteroides* spp. included organisms resembling *B. ruminicola* subsp. *ruminicola* and *brevis, B. distasonis,* and organisms resembling *B. ochraceus.* Finally, Harris, Reddy, and Carter (1976), studying the anaerobic flora of the mouse large intestine, found 10^{10}–10^{11} bacteroides per gram of contents; these organisms were composed of six groups similar to the following previously described species: *B. clostridiiformis* subsp. *clostridiiformis, B. coagulans, B. fragilis, B. furcosus, B. pneumosintes,* and also a group not similar to any previously described *Bacteroides* spp.

In comparison with the rumen, little work has been done concerning either the fermentation that occurs in the cecum of rats and mice or the involvement of the *Bacteroides* species therein. Based upon the level of plant material in the diet, it would be expected that cellulose and other plant materials are fermented in the cecum-colon. Only the degradation of cellulose has been examined, and while most studies indicate it is not fermented (Lang and Briggs, 1976), Yang, Manoharan, and Young (1969) have shown that increased concentrations of cellulose in the diet resulted in the production of more volatile fatty acids in the cecum-colon of rats. In addition, Juhr and Haas (1976) detected cellulase activity in both the rat and mouse cecum. Although little understood, the cecal-colon fermentation results, nonetheless, in products qualitatively similar to those formed during the rumen fermentation; like the ruminant, the animals with ceca make use of the fermentation products (i.e., VFAs are absorbed

[Elsden et al., 1946; McBee, 1977] and the protein of microbial bodies is gained through coprophagy [Hungate, 1977]).

Chickens and turkeys have two ceca, that contain a flora (in 5-week-old chicks) consisting primarily of anaerobes, 40% of which are *Bacteroides* species (primarily *B. fragilis, B. hypermegas* and *B. clostridiiformis*) (Barnes and Impey, 1970; Barnes and Impey, 1972). In the ceca of 5-week-old chicks, Salanitro, Blake, and Muirhead (1974) found that 18.6% of the flora consisted of the *Bacteroides* species *B. clostridiiformis, B. hypermegas,* and *B. fragilis;* while 12.8% of the cecal flora in 14-day-old chicks were *Bacteroides* species (Salanitro et al., 1978).

Isolation

Nonselective Isolation—Roll Tubes

For nonselective isolation of strictly anaerobic nonpathogenic species of *Bacteroides,* it is necessary to use a medium that simulates in vitro the physical and chemical environment in which the organisms live. Such a "habitat-simulating" medium was first used by Hungate to isolate rumen microbes. His medium was prepared and maintained in an anaerobic atmosphere, a low E_h was established by addition of reducing compounds, a CO_2-HCO_3^- buffer (similar in concentration to that found in the rumen) was included, and rumen fluid accounted for a third of the medium (Hungate, 1950; Hungate, 1966). The method used to prepare this medium, as well as its use in isolation and culturing of strict anaerobes, was described by Hungate in 1950; more recently it has been presented again but in more detail (Hungate, 1969). Since its original description, Hungate's technique, or the "roll-tube technique", has been modified in many ways to meet the needs of various investigators (Balch and Wolfe, 1976; Bryant, 1972; Bryant and Burkey, 1953a; Bryant and Robinson, 1961; Caldwell and Bryant, 1966; Eller, Crabill, and Bryant, 1971; Latham and Sharpe, 1971; Macy, Snellen, and Hungate, 1972; Miller and Wolin, 1974). For example, Holdeman, Cato, and Moore (1977) have used it as the basis for developing very efficient methods for large-scale rapid isolation and identification of anaerobes.

Since all anaerobic roll-tube methods have been described in detail elsewhere (Balch and Wolfe, 1976; Barnes, 1979; Bryant, 1972; Clarke, 1977; Finegold et al., 1975; Holdeman, Cato, and Moore, 1977; Hungate, 1969), the following discussion concerning media preparation and isolation of nonpathogenic bacteroides will include only a general description of this "technique".

In its most simple form, the "roll-tube technique" requires certain basic equipment. Of utmost importance is a source of O_2-free gas; for the purpose of habitat simulation, the gas normally used for the isolation of nonpathogenic bacteroides is CO_2 (the rumen bacteroides require CO_2 for growth [Caldwell, Keeney, and van Soest, 1969; Dehority, 1971; Hamlin and Hungate, 1956]). This is accomplished by passing CO_2 through a vertical glass column packed with copper turnings, heated to 300–350°C, and reduced with hydrogen gas; such columns can either be purchased (Holdeman, Cato, and Moore, 1977) or constructed (Bryant, 1972; Bryant and Robinson, 1961; Hungate, 1969; Latham and Sharpe, 1971). If preferred, a solution of chromous acid or a Deoxy Gas Purifier (Matheson Gas Products; gas must contain 5% H_2) may be used instead of a copper column (Holdeman, Cato, and Moore, 1977; Finegold et al., 1975).

The remainder of the equipment consists simply of two sterile gassing needles and a sterile cotton-plugged Pasteur pipette from which the tip has been removed. The gassing needles are made from 1- or 2-ml glass Luer-lock syringes to which 6-inch, 21-gauge needles are attached; the plunger and flange of the syringes are removed and the barrels filled with cotton. After autoclaving, the "gassing needles" are connected to rubber tubing (preferably butyl rubber) through which the O_2-free gas is flowing.

MEDIA FOR RUMEN BACTEROIDES

Hungate's "Habitat-Simulating" Medium (Hungate, 1969)

The following stock solutions are prepared:
Mineral solution A (containing in g/liter):

NaCl	6.0
KH_2PO_4	3.0
$(NH_4)_2SO_4$	3.0
$MgSO_4$ (anhydrous)	0.6
$CaCl_2$ (anhydrous)	0.6

Mineral solution B (containing in g/liter):

K_2HPO_4	3.0

Resazurin solution: 0.1% solution in distilled water.
Cysteine-HCl solution: 3.0 g of cysteine-HCl are dissolved in 100 ml deoxygenated distilled water (boiled 2 min and cooled under a stream of CO_2 or N_2) in a round-bottom flask. The flask can then be autoclaved, after closing with a stopper that is wired into place, or portions of the solution can be transferred anaerobically to roll tubes, followed by autoclaving.
Substrate solutions: Normally, 10% solutions of soluble sugars are prepared in deoxygenated distilled water (N_2).
Rumen fluid: Rumen contents can be obtained from animals via a stomach tube or a fistula. Unless fistula sampling is done using the type of

apparatus described by Hungate (Hungate, 1950), the contents are strained through several layers of cheesecloth to remove large pieces of plant material. The rumen fluid is then centrifuged and the supernatant used for medium preparation. "Clarified rumen fluid" can also be used (see below for description of Medium 10 preparation).

In preparing Hungate's medium, the following are combined in a flask, preferably a round-bottom flask (for 1 liter of medium):

Mineral solution A	167 ml
Mineral solution B	167 ml
Resazurin	1 ml
Distilled water	398 ml
(65 ml "extra" H_2O	
added, as approximately 10% of the	
medium volume is lost during boiling)	

Agar is also added if required (2%, or less if Ionagar is used), and the medium is boiled approximately 2 min to remove oxygen (or until the agar is dissolved). A stream of O_2-free CO_2 is then passed into the flask via a gassing needle, and the medium is allowed to cool to room temperature (or if the medium contains agar it is placed in a water-bath and cooled to 45–47°C). Once cooled, 333 ml of rumen fluid is added, and appropriate volumes of the medium can then be dispensed into thick-walled culture tubes. This is accomplished as follows:

1. A stream of O_2-free CO_2 is passed into a tube, via a gassing needle, to displace the air.
2. O_2-free CO_2 is drawn into a 5- or 10-ml serological pipette, to which a rubber mouth tube is attached; this is best done using the gas in the medium flask.
3. After the required amount of medium is pulled into the pipette, the pipette is closed by squeezing the rubber tube, then the medium is released into the gassing roll-tube by opening the pipette again.
4. The tube is closed by inserting a butyl rubber stopper in the top, followed by removal of the gassing needle. The stopper is then pushed into place using a twisting motion.
5. The filled tubes are autoclaved in a "press" (Holdeman, Cato, and Moore, 1977; Hungate, 1969), in order to hold the stoppers in place.

It should be noted that when filling the tubes with a pipette, a rubber pipette-filler can be used instead of a mouth tube (Bryant, 1972); also, instead of pipette, a 5- or 10-ml glass syringe with a 6-inch, 13-gauge needle can be used.

Before inoculation of the roll-tube medium, sterile cysteine (0.1 ml/10 ml), $NaHCO_3$ (0.5 ml/10 ml), and substrate(s) are added using a sterile 1-ml syringe fitted with a 21-gauge nee-

dle. The syringe and needle are first rinsed with sterile O_2-free CO_2 that is passed through a sterile cotton-plugged Pasteur pipette from which the tip has been removed.

When growing cellulolytic rumen bacteroides, 0.1–1% cellulose is included in the medium. The cellulose is prepared as a suspension by pebble-milling Whatman No. 1 filter paper (12 h to 3 days, depending on the thickness of the suspension and size of the pebble mill); a portion of the suspension is added to the original medium mixture of salts, resazurin, and water (the amount of water added must be appropriately reduced).

In addition to the medium of Hungate, a number of media have been developed over the years by Bryant and co-workers for the cultivation of rumen microbes (Bryant and Burkey, 1953a; Bryant and Robinson, 1961; Bryant and Robinson, 1962; Caldwell and Bryant, 1966). Perhaps the most widely used of these media is Medium 10 of Caldwell and Bryant (1966).

Medium 10 for Cultivation of Rumen Bacteroides (Caldwell and Bryant, 1966)
The following stock solutions are prepared:
Mineral solution 1 (containing in g/liter):

K_2HPO_4	6.0

Mineral solution 2 (containing in g/liter):

KH_2PO_4	6.0
$(NH_4)_2SO_4$	12.0
NaCl	12.0
$MgSO_4 \cdot 7H_2O$	2.5
$CaCl_2 \cdot 2H_2O$	1.6

Volatile fatty acid solution:

Acetic acid	17 ml
Propionic acid	6 ml
Butyric acid	4 ml
n-Valeric acid	1 ml
Isovaleric acid	1 ml
Isobutyric acid	1 ml
DL-α-Methylbutyric acid	1 ml

A number of bacteroides require or are stimulated by VFAs (Holdeman, Cato, and Moore, 1977; Macy and Probst, 1979).
Resazurin: 0.1% solution.
Hemin solution: 0.01 g is dissolved in 100 ml of 5×10^{-3} M (0.002%) NaOH or in a mixture of ethyl alcohol and 0.2 M KOH [1:1, (vol/vol)]. Hemin is required for growth, or stimulates growth, of many bacteroides (Holdeman, Cato, and Moore, 1977; Macy and Probst, 1979).
Cysteine · HCl-Na_2S solution: 2.5 g of cysteine · HCl are dissolved in water and the pH is brought to 11 with NaOH. Then, 2.5 g of

$Na_2S \cdot 9H_2O$ is dissolved, and the volume brought up to 100 ml. This solution is then transferred to a round-bottom flask and boiled, under N_2. After cooling, the flask is stoppered and autoclaved (Holdeman, Cato, and Moore, 1977). Na_2CO_3 solution: 8 g of Na_2CO_3 are dissolved in 100 ml of water (preferably deoxygenated water under CO_2) in a round-bottom flask under an atmosphere of 100 CO_2. The flask is stoppered and autoclaved.

In preparing the medium, the following are combined in a round-bottom flask (for 1 liter of medium):

Mineral solution 1	37.5 ml
Mineral solution 2	37.5 ml
VFA solution	3.1 ml
Resazurin	1.0 ml
Hemin solution	10.0 ml
Glucose	0.5 g
Cellobiose	0.5 g
Soluble starch	0.5 g
Yeast extract	0.5 g
Trypticase	2.0 g
Distilled water	950.0 ml
(100 ml "extra" because of the approximate 10% lost due to boiling)	

Because of the presence of the VFAs, the pH of this medium must be adjusted to 6.5. Agar is then added, if required (2%, or less if Ionagar is used), and the medium is boiled approximately 2 min (or until the agar is dissolved). The flask is then closed immediately with a rubber stopper that is wired into place. After autoclaving, the medium is cooled in a water bath to 45–47°C if agar is included, or to a lower temperature (normally ambient temperature) when agar is not present. The rubber stopper is aseptically removed and gassing of the medium with sterile CO_2 is begun. The Na_2CO_3 (50 ml) is then added and appropriate volumes of medium are transferred aseptically to sterile culture tubes (that are being gassed with CO_2), which are then stoppered with butyl rubber stoppers; the transfer is accomplished with a sterile, cotton-plugged, glass, serological pipette with an attached mouth tube. Immediately before inoculation, the sterile cysteine-Na_2S solution is aseptically and anaerobically added to the medium (0.1 ml/10 ml) using a sterile 1-ml syringe and 21-gauge needle that has been rinsed with CO_2.

The medium can also be prepared as a "rumen fluid medium" (RFM) by replacing yeast extract, Trypticase, hemin, and the VFA mixture with 40% clarified rumen fluid; the amount of water added when preparing the medium is reduced to 550 ml. Clarified rumen fluid is prepared by boiling rumen fluid for a short time under CO_2 and autoclaving under CO_2 for 5 min, with subsequent centrifugation at $25,000 \times g$ for 20 min (Bryant and Robinson, 1961).

When growing the cellulolytic rumen bacteroides, glucose, cellobiose, and soluble starch are replaced by 0.1–1% cellulose. The cellulose is prepared in the manner already described (see Hungate's "Habitat-Simulating" Medium, above). Because of the addition of cellulose, the amount of distilled water added when preparing the medium must be appropriately reduced.

It should be pointed out that Bryant's method of preparing medium differs from Hungate's in a number of ways. Instead of using Na_2CO_3, Hungate adds filter-sterilized $NaHCO_3$ to his medium. In addition, rather than dispensing the medium after autoclaving, he transfers medium to culture tubes prior to autoclaving; this is of advantage in that it eliminates any possibility of contamination. The disadvantage of transfer prior to sterilization is that special thick-walled tubes must be used to prevent breakage during autoclaving, and these stoppered tubes must be held in a special "press" during autoclaving so that the stoppers remain in place. Hungate's method also involves more manipulations, since $NaHCO_3$ as well as cysteine-HCl and sugars must be added to the roll-tubes before inoculation. Of course, this problem can be eliminated simply by adding 5 g of $NaHCO_3$ to the medium after it has been boiled and cooled, prior to adding the medium to tubes; because Hungate's roll-tubes are thick-walled, they will not explode during autoclaving. Finally, Hungate prepares and sterilizes sugar substrates separately, and these are added to the medium prior to inoculation; such a procedure prevents "caramelization" of sugars that tends to take place when they are autoclaved as part of the medium, as in the procedure followed by Bryant and co-workers. Of course, it should be noted that, except for his use of $NaHCO_3$, Hungate's medium can certainly be made using Bryant's method, and, conversely, Medium 10 can be made according to Hungate's method.

Bryant and Robinson have also developed a defined medium for growth of rumen bacteria, a medium in which rumen bacteroides grow very well (Bryant, 1974; Bryant and Robinson, 1962).

Bryant-Robinson's Medium for Cultivation of Rumen Bacteroides (Bryant and Robinson, 1962)

The following stock solutions are prepared:
Mineral solution (containing in g/liter):

KH_2PO_4	18
NaCl	18
$CaCl_2 \cdot 6H_2O$	0.53
$MgCl_2 \cdot 6H_2O$	0.4
$MnCl_2 \cdot 4H_2O$	0.2
$CoCl_2 \cdot 6H_2O$	0.2

$FeSO_4 \cdot 7H_2O$ 0.08
$(NH_4)_2SO_4$ 8.0

Vitamin solution (containing in mg/100 ml):

Thiamine-HCl	20.0
Ca-D-pantothenate	20.0
Nicotinamide	20.0
Riboflavin	20.0
Pyridoxine-HCl	20.0
p-Aminobenzoic acid	1.0
Biotin	0.25
Folic acid	0.25
Vitamin B_{12}	0.1

This solution is filter-sterilized.
Volatile fatty acid solution:

Acetic acid	36.0 ml
Isobutyric acid	1.8 ml
n-Valeric acid	2.0 ml
DL-α-Methylbutyric acid	2.0 ml
Isovaleric acid	2.0 ml

Hemin solution, resazurin solution, cysteine · HCl-Na$_2$S solution, Na$_2$CO$_3$ solution: as for Medium 10, above.

In preparing the medium, the following are combined (for 1 liter of medium):

Mineral solution	50.0 ml
VFA solution	4.5 ml
Resazurin	1.0 ml
Hemin solution	10.0 ml
Glucose, cellobiose, or maltose	5.0 g
L-Methionine	0.08 g
Distilled water	975.0 ml

Further medium preparation is as for Medium 10, except that 5 ml of the filter-sterilized vitamin solution is added to the medium before it is dispensed. Addition of glucose and cellobiose or maltose at this time is also advisable (50 ml of a 10% solution prepared with deoxygenated water [N$_2$]).

MEDIA FOR CECAL, COLON BACTEROIDES. To isolate and grow the anaerobic organisms from chicken ceca, Barnes and Impey (Barnes, 1979; Barnes and Impey, 1970) found that Medium 10 could be used with good results if supplemented with chicken fecal and liver extracts. Salanitro, Fairchilds and Zgornicki (1974) also found that this supplemented Medium 10 gave the best results when compared with the rumen fluid medium (RFM) of Bryant and Robinson (1961) or with unsupplemented Medium 10 (Caldwell and Bryant, 1966). Presumably, use of Medium 10, the medium of Varel and Bryant (1974) (described below), or the medium of Caldwell and Arcand (1974) (described below), supplemented

with an extract of cecal material from the animal under study, would also prove to be good media for the isolation and growth of cecal or colon bacteroides; if such material is not available, rumen fluid might also serve as a good supplement.

Varel and Bryant (1974) have developed a minimal medium for growth of *Bacteroides fragilis*.

Varel-Bryant Medium for Growth of *Bacteroides fragilis* (Varel and Bryant, 1974)

The following stock solutions are prepared:
Mineral solution 3 (containing in g/liter):

KH_2PO_4	18
NaCl	18
$CaCl_2 \cdot 2H_2O$	0.53
$MgCl_2 \cdot 6H_2O$	0.4
$MnCl_2 \cdot 4H_2O$	0.2
$CoCl_2 \cdot 6H_2O$	0.02
$(NH_4)_2SO_4$	0.8
$FeSO_4 \cdot 7H_2O$	0.08

Hemin solution, Na$_2$CO$_3$ solution, resazurin solution: as for Medium 10.
B_{12} solution: 0.01 mg per 100 ml distilled water; filter-sterilize.
Cysteine · HCl solution: 2.5 g of cysteine · HCl · H$_2$O dissolved in 100 ml deoxygenated distilled water (under N$_2$).

In preparing the medium the following are combined (for 1 liter of medium):

Solution 3	50 ml
Resazurin	1 ml
Hemin solution	10 ml
Glucose	5 g
Distilled water	960 ml

Further medium preparations are as for Medium 10, except that 20 ml of the cysteine · HCl solution and 5 ml of the B_{12} solution are added to the medium with the Na$_2$CO$_3$ before tubes are filled; addition of glucose (50 ml of a 10% solution, prepared with deoxygenated water under N$_2$) to the medium at this time is preferable. Methionine can be substituted for B_{12} (7.5 μg/ml).

Varel and Bryant (1974) observed that *B. vulgatus, B. thetaiotaomicron,* and *B. distasonis* strains could also grow in this medium, except that one strain tested of *B. distasonis* had an absolute requirement for methionine and no requirement for B_{12}; a strain of *B. ovatus* required the presence of Casitone (0.2%) in the medium for growth. A medium similar to that of Varel and Bryant (1974) was developed by Caldwell and Arcand (1974), but it also included yeast extract, volatile fatty acids, menadione, and a vitamin mixture similar to that used in the Bryant-Robinson medium for growth of rumen bacteroides. This medium permitted substantial growth of many *Bacteroides* species including

B. clostridiiformis subsp. *clostridiiformis* and a number of other pathogenic bacteroides. Therefore, addition of Casitone, yeast extracts, VFAs, and a vitamin mixture to Varel-Bryant medium would probably allow the use of this medium for growth of most strains of *B. thetaiotaomicron, B. vulgatus, B. ovatus, B. distasonis,* and a number of the other species found in the cecum or colon.

B. corrodens has been grown by Smibert and Holdeman (1976) in the PYG (peptone–yeast extract–glucose) medium of Holdeman, Cato, and Moore (1977).

PYG Medium (Holdeman, Cato, and Moore, 1977)

The following stock solutions are prepared:
Salts solution (containing in g/liter):

CaCl$_2$ (anhydrous)	0.2
MgSO$_4$ (anhydrous)	0.2
K$_2$HPO$_4$	1.0
KH$_2$PO$_4$	1.0
NaHCO$_3$	10.0
NaCl	2.0

Resazurin solution: as for Medium 10.
Vitamin K solution: Dissolve 0.15 ml vitamin K$_1$ in 30 ml 95% ethyl alcohol.
Hemin Solution: Dissolve 50 mg hemin in 100 ml of 10^{-2} N NaOH; sterilize by autoclaving. In preparing the medium the following are combined (for 1 liter of medium):

Salts solution	40 ml
Resazurin	1 ml
Peptone	5 g
Trypticase	5 g
Yeast extract	10 g
Glucose	10 g
Distilled water	970 ml

The medium is boiled and then cooled under CO$_2$. After adjusting the pH to 7.0 with NaOH, 0.2 ml of vitamin K solution, 10 ml of hemin solution, and 0.5 g of cysteine · HCl · H$_2$O are added, and appropriate volumes of the medium are transferred anaerobically (CO$_2$) to thick-walled roll-tubes; these are then sterilized, while being held in a press. Before inoculation, 0.5 ml of a 10% fumarate solution and 0.3 ml of a 10% formate solution are added to the medium. Aspartate (0.3%), asparagine (0.6%), or malate (0.3%) can be substituted for fumarate; molecular hydrogen can replace formate in the medium. Presence of the CO$_2$-HCO$_3^-$ buffer does not appear to be necessary for growth, as the organisms grew under 100% N$_2$ in brain heart infusion medium (Smibert and Holdeman, 1976). Because *B. corrodens* grows on fumarate, malate, aspartate, or asparagine plus either formic acid or molecular hydrogen, glucose need not be included in the

medium (i.e., PY medium instead of PYG can be used).

MEDIA FOR ORAL BACTEROIDES.   These organisms are normally grown in a nondefined medium; studies concerning a defined medium have not been published. With regard to a nondefined medium, all of the oral bacteroides will grow in the PYG medium of Holdeman, Cato, and Moore (above) (as will all the nonpathogenic bacteroides). Additionally, Rizza et al. (1968) have described a medium for growth of *B. asaccharolyticus* and *B. melaninogenicus.*

Medium for Growth of *Bacteroides asaccharolyticus* and *B. melaninogenicus* (Rizza et al., 1968)

The medium contains (gas phase 5% CO$_2$):

Trypticase	27.0 g
Yeast extract	3.0 g
NaCl	2.0 g
K$_2$HPO$_4$	2.5 g
K$_2$CO$_3$	2.0 g
Vitamin K solution (Holdeman et al., 1977)	0.2 ml
Hemin solution (Holdeman, Cato, and Moore, 1977)	10.0 ml

Hemin, menadione, or both are required for growth, or enhance growth, of most oral bacteroides (Gibbons and MacDonald, 1960; Holdeman and Moore, 1974).

Based upon the above recipe, it would seem that *B. melaninogenicus* and *B. asaccharolyticus* would also grow in Medium 10 to which vitamin K has been added; however, the effect of VFAs and 100% CO$_2$ on growth of this organism is not known.

OBTAINING A SAMPLE FOR INOCULATION.   As has been mentioned, rumen fluid samples are usually obtained from a fistulated animal or via a stomach tube. When the fistulated animal is the source, an apparatus similar to that described by Hungate (1950) can be used; with larger animals it is possible to simply remove rumen contents with one's hand. Once the sample is removed it should be kept anaerobic and should be cultured as soon as possible after collection.

Holdeman, Cato, and Moore (1977) have described a simple and effective method for the collection of human fecal material. Samples are placed in a plastic bag as it is being flushed with CO$_2$. A more interesting method has been introduced by P. J. van Soest (personal communication), who has designed an "anaerobic toilet" for the collection of fecal specimens.

Cecal material from various animals should be obtained as quickly as possible from a sacrificed animal, and should be kept under anaerobic conditions; the contents are best removed from the cecum while in an anaerobic chamber, if such is available.

INOCULATION. For quantitative enumeration and isolation of bacteroides, a determined amount of the sample is mixed well, under anaerobic conditions, with a known amount of salts solution; usually this is accomplished so that a 1:10 dilution of the original material is obtained. For example, Bryant and co-workers used a Waring blender to mix a 1:10 dilution of rumen fluid samples in salts solution (carried out under a stream of CO_2). For fecal samples, Holdeman, Cato, and Moore (1977) thoroughly kneaded the material in their closed plastic sample bag and then removed approximately 1 g of this material, while gassing with CO_2, to a preweighed tube containing 9.0 ml of anaerobic salts solution plus a few glass beads. After weighing the tube, the contents were mixed well on a Vortex mixer.

Once mixed, the first 1:10 dilution is further diluted using a 1-ml syringe plus 21-gauge needle rinsed with the appropriate gas, until 10 or 11 serial 1:10 dilutions have been made. It is also possible to make 1:100 dilutions, thus requiring fewer tubes of dilution salts solution; the salts solution used for the dilution should be identical to the growth medium but without substrates, rumen fluid, or other such supplements (Bryant and Burkey, 1953a). Finally, 0.5 ml of the higher dilutions are inoculated into tubes of melted agar medium (at 45–47°C) containing 4.5 ml of agar medium per tube. Each inoculated tube is then placed on its side and "rolled" (preferably in ice or cold water), either manually or using a mechanical "tube-roller", until the agar has solidified on the inner surface of the tube.

ISOLATION AND PURIFICATION. Colonies are best transferred using either a platinum-iridium inoculating needle flattened at the tip and bent at a right angle near the end (Bryant, 1972; Hungate, 1969) or a Pasteur pipette drawn out to a capillary with a diameter approximately equal to that of the colony to be picked and with the glass bent at a right angle near the tip (see Hungate, 1969, for a detailed description of how to prepare such pipettes). If the latter method is used, the air in the Pasteur pipette is replaced with the appropriate gas by means of a mouth tube to which the pipette is attached; while passing a stream of O_2-free gas into the tube via a sterile gassing needle, the colony is then drawn carefully into the pipette (Hungate, 1969) and is transferred to a tube of fresh medium that is also being gassed. The colony is either placed on the surface of the agar in a prerolled tube or, after a small amount of liquid is drawn into the tip of the Pasteur pipette, is expelled into fresh medium (molten agar). The Pasteur pipette method assures protection of the inoculum from air during transfer, but it is a considerably more difficult procedure and such "protection" is usually not required when transferring *Bacteroides* species.

The picked colony can then either be streaked on the surface of agar (Bryant, 1972; Holdeman, Cato, and Moore, 1977) or be serially diluted in molten agar (45–47°C) (Hungate, 1969) by transferring only the "dead space" in the end of the 1-ml syringe plus 21-gauge needle to subsequent tubes. Subculturing of colonies should be repeated if the first subculture does not prove to be pure.

Nonselective Isolation—Other Methods

Although minor changes may be necessary, basically all the media (+ agar) described for roll-tubes can also be poured into Petri dishes for use in either anaerobic glove boxes or anaerobic jars. In this connection, it should be remembered that the gaseous atmosphere in either of these "containers" is rarely 100% CO_2; thus the concentration of bicarbonate or carbonate in the medium must be reduced accordingly in order to maintain the correct pH.

ANAEROBIC GLOVE BOX. In addition to the "roll-tube" method, anaerobic chambers (glove boxes) are also used for the cultivation of strictly anaerobic bacteria, and have the advantage that standard bacteriological techniques can be employed. It appears that the best type of glove box is that made of flexible vinyl, as described by Aranki et al. (Aranki and Freter, 1972; Aranki et al., 1969); if proper precautions are taken, even methanogens are not killed when handled in this chamber (Balch and Wolfe, 1976). To prepare media outside the anaerobic chamber, either palladium chloride or palladium black can be incorporated into the media (Aranki and Freter, 1972). Additionally, prereduced media can be prepared outside the anaerobic glove box and then can be poured into Petri dishes under anaerobic conditions after having been brought inside the chamber (Clarke, 1977).

ANAEROBIC JARS. In contrast to roll tubes and anaerobic glove boxes, use of conventional anaerobic methods (anaerobic jars, etc.) allows access of air to the specimens and media. Since such exposure will of course kill oxygen-sensitive anaerobes, only the least-sensitive *Bacteroides* species can be grown in this way. In fact, Holdeman, Cato, and Moore (1977) do not recommend the use of anaerobic jars when isolating organisms of the normal flora. For this reason, techniques concerned with the use of anaerobic jars will not be further discussed here. For a thorough description of these methods the reader should consult Holdeman, Cato, and Moore (1977) and Finegold et al. (1975).

Selective Isolation

There are no selective media designed for specific isolation of the rumen bacteroides. Starch or cellulose can be added in an attempt to select for *Bacteroides amylophilus* and *B. succinogenes*, respectively; however, other starch- or cellulose-degrading rumen bacteria will also grow. Since many cecal-colon bacteroides are known to be resistant to certain antibiotics, media containing these antibiotics have been used in a selective manner (Finegold, 1977; Finegold, Miller, and Posnick, 1965; Finegold, Sugihara, and Sutter; 1971). Provided that hemin and menadione are present in the medium, the combination kanamycin (75 μg/ml)-vancomycin (7.5 μg/ml) can be used to select for what Finegold et al. (1965) called *B. fragilis* and *B. melaninogenicus*. To select for the whole group of bacteroides of the cecum-colon, the paromomycin (100 μg/ml)-vancomycin (7.5 μg/ml) pair is appropriate (Finegold, Miller, and Posnick, 1965).

Enrichment

Very few attempts have been made to enrich for *Bacteroides* species. *B. amylophilus* might possibly be enriched for by including starch in Hungate's rumen-fluid medium or in Medium 10; although, unless the rumen fluid inoculum is significantly diluted before inoculation into enrichment media, the fast-growing but less numerous starch-degrading organisms (e.g., *Streptococcus bovis*) would predominate. Additionally, *B. succinogenes* would undoubtedly be enriched for (along with other cellulose-degrading organisms) if cellulose were included in the same media. Similar enrichments might also be made for starch- and cellulose-degrading bacteroides of the cecum and colon. Certainly, *B. corrodens* could easily be enriched for by including fumarate (or either malate, aspartate, or asparagine) plus formate (or H$_2$) in the appropriate medium (Smibert and Holdeman, 1976). Isolation of organisms from an enrichment would then be accomplished using the techniques already described, with inclusion in the medium of the enrichment substrate.

Maintenance of Cultures

Certain cultures of *Bacteroides* species can be lyophilized using normal procedures (Finegold et al., 1975; Latham and Sharpe, 1971); lyophilization is especially successful if cultures are prepared in an anaerobic chamber so as not to expose them to oxygen. Cultures can also be maintained as agar stab cultures (Bryant and Burkey, 1953a; Bryant and Robinson, 1962) or may be stored at −65°C if suspended in "skimmed milk, 10% bovine albumin, fetal calf serum or defibrinated blood" (Finegold et al., 1975).

Identification

As has already been discussed, the nonpathogenic bacteroides are found in the alimentary tracts of man and other animals, where they can account for a major portion of the normal flora. As described in the eighth edition of *Bergey's Manual of Determinative Bacteriology,* these organisms are of the genus *Bacteroides* and the family Bacteroidaceae. Organisms of this family are non-spore-forming, Gram-negative, strictly anaerobic rods that are peritrichously flagellated or nonmotile (Holdeman and Moore, 1974). In addition to the genus *Bacteroides,* the genera *Fusobacterium* and *Leptotrichia* are also included in the family. Since the publishing of the eighth edition of *Bergey's Manual,* Moore, Johnson, and Holdeman (1976) have recommended that the family Bacteroidaceae be emended so that additional genera exhibiting monotrichous or lophotrichous flagellation are included.

Determination of whether an organism of the family Bacteroidaceae belongs in the genus *Bacteroides* or in one of the other genera described in the eighth edition of *Bergey's Manual* is done primarily on the basis of fermentation products formed. *Bacteroides* species produce a number of different acids, including succinic, acetic, formic, lactic, propionic, and butyric, whereas lactic acid is the only major product of *Leptotrichia* species, and organisms of the genus *Fusobacterium* produce butyric acid as a major product (with no isobutyric or isovaleric acids).

Because of the exhaustive work concerning the taxonomy of the genus *Bacteroides* (both pathogenic and nonpathogenic) that has been done by Holdeman, Moore, and co-workers, their consideration of the genus in the eighth edition of *Bergey's Manual* (Holdeman and Moore, 1974) and in their *Anaerobe Laboratory Manual* (Holdeman, Cato, and Moore, 1977) should be consulted when attempting to phenotypically characterize and identify any organisms belonging to the genus *Bacteroides.* A recently published chapter by Barnes (1979) will also be helpful.

In addition to phenotypic characterization, other approaches have been employed to aid in classifying organisms of the genus *Bacteroides.* These include determination of mol% G+C (Table 1), measurement of DNA-DNA homologies (Johnson, 1973; Johnson, 1978; Johnson and Ault, 1978), and use of serology (recently reviewed by Hofstad, 1979). Based upon studies using these different approaches, several changes have been made in the classification of the genus *Bacteroides* since the printing of the eighth edition of *Bergey's Manual.*

Table 1. Mol% G+C of the DNA from nonpathogenic *Bacteroides* species.

Bacteroides sp.	Mol% G+C	Reference
B. fragilis (71)[a]	42[b]	Johnson, 1978
B. fragilis H	42.3	Reddy and Bryant, 1977
B. fragilis D	40.7	Reddy and Bryant, 1977
B. fragilis 2044	42.7	Reddy and Bryant, 1977
B. fragilis ATCC 25285	42	Cato and Johnson, 1976
B. vulgatus (43)[a]	41[b]	Johnson, 1978
B. vulgatus ATCC 8482[c]	41	Cato and Johnson, 1976
B. ovatus (36)[a]	41[b]	Johnson, 1978
B. ovatus ATCC 8483[c]	40	Cato and Johnson, 1976
B. thetaiotaomicron (29)[a]	42[b]	Johnson, 1978
B. thetaiotaomicron ATCC 29184	42	Cato and Johnson, 1976
B. distasonis (14)	44[b]	Johnson, 1978
B. distasonis ATCC 8503[c]	44	Cato and Johnson, 1976
Homology group 3452-A (19)[a]	41[b]	Johnson, 1978
B. uniformis (21)[a]	46[b]	Johnson, 1978
Homology group T4-1 (9)[a]	44–45[b]	Johnson, 1978
Homology group subsp. *a* (15)[a]	45[b]	Johnson, 1978
B. eggerthii (6)[a]	44.5[b]	Johnson, 1978
B. microfusus (3)[a]	59.5–60.3	Kaneuchi and Mitsuoka, 1978
B. microfusus Q-1[d] (ATCC 29728)	60.3	Kaneuchi and Mitsuoka, 1978
B. amylophilus 70	40.3	Reddy and Bryant, 1977
B. amylophilus ATCC 29744	42.1	Cato, Moore, and Bryant, 1978
B. succinogenes A3C	49.1	Reddy and Bryant, 1977
B. succinogenes B21A	47.6	Reddy and Bryant, 1977
B. succinogenes ATCC 19169[c]	42.8	Cato, Moore, and Bryant, 1978
B. disiens ATCC 29426[d]	40	Holdeman and Johnson, 1977
B. bivius ATCC 29303[d]	40	Holdeman and Johnson, 1977
B. multiacidus ATCC 27723[d]	57.3	Mitsuoka et al., 1974
B. corrodens (4)[a]	28–29.7	Smibert and Holdeman, 1976
B. ruminicola		
subsp. *brevis* GA 33, 118B	50.3	Reddy and Bryant, 1977
subsp. *ruminicola* 23	49.1	Reddy and Bryant, 1977
B. oralis 7CM	42.1	Reddy and Bryant, 1977
B. oralis J1	50.9	Reddy and Bryant, 1977
B. hypermegas ATCC 25560[c]	35	Holdeman, Cato, and Moore, 1977
B. bivius ATCC 29303	40	Holdeman, Cato, and Moore, 1977
B. asaccharolyticus		
CR₂a	49.5	Reddy and Bryant, 1977
NCTC 9337	54.2/50.0; 53.0	Finegold and Barnes, 1977; van Steenbergen, de Soet, and de Graaff, 1979
ATCC 25260	53/53.8; 53.2; 50.5	Finegold and Barnes, 1977; van Steenbergen, de Soet, and de Graaff, 1979; Shah et al., 1976
VPI 4199	53/50.7; 54.1	Finegold and Barnes, 1977; van Steenbergen, de Soet, and de Graaff, 1979
W83	48.2; 45.6	van Steenbergen, de Soet, and de Graaff, 1979; Shah et al., 1976
W50	48.6; 45.2	van Steenbergen, de Soet, and de Graaff, 1979; Shah et al., 1976
2848	48.0; 46.3	van Steenbergen, de Soet, and de Graaff, 1979; Shah et al., 1976
381	47.8	van Steenbergen, de Soet, and de Graaff, 1979
B536	54.2; 50.9	van Steenbergen, de Soet, and de Graaff, 1979; Shah et al., 1976

Table 1. Mol% G+C of the DNA from nonpathogenic *Bacteroides* species. (*Continued*)

Bacteroides sp.	Mol% G+C	Reference
B. melaninogenicus subsp.		
melaninogenicus		
ATCC 25845	40/42.2; 41.3; 38.5	Finegold and Barnes, 1977; van Steenbergen, de Soet, and de Graaff, 1979; Shah et al., 1976
WAL 2728	40.8; 36.4	van Steenbergen, de Soet, and de Graaff, 1979; Shah et al., 1976
VPI 9343	40.3; 38.0	van Steenbergen, de Soet, and de Graaff, 1979; Shah et al., 1976
intermedius		
ATCC 25846	45	Finegold and Barnes, 1977
ATCC 25611	44	Finegold and Barnes, 1977
NCTC 9336	42/42.7; 42.7	Finegold and Barnes, 1977; van Steenbergen, de Soet, and de Graaff, 1979
NCTC 9338	41.2	Finegold and Barnes, 1977
T588	41.0; 43.6	van Steenbergen, de Soet, and de Graaff, 1979; Shah et al., 1976
LH 100	41.6; 42.0	van Steenbergen, de Soet, and de Graaff, 1979; Shah et al., 1976
levii		
JP 2	45.3	van Steenbergen, de Soet, and de Graaff, 1979
Lev	46.7; 48.4	van Steenbergen, de Soet, and de Graaff, 1979; Reddy and Bryant, 1977

[a] Number in parentheses is number of strains examined.
[b] Average of strains.
[c] Neotype strain.
[d] Type strain.

1. In the eighth edition of *Bergey's Manual,* a group of similar organisms was placed together in the species *Bacteroides fragilis,* which was divided into five subspecies. Since then, DNA-DNA homology studies (Johnson, 1973) have shown that these subspecies should be given species rank (Cato and Johnson, 1976); three additional homology groups were also detected among the "*B. fragilis*" group (Johnson, 1978).

2. In the eighth edition of *Bergey's Manual,* both asaccharolytic and saccharolytic strains were classified together in the species *B. melaninogenicus,* with the subspecies *asaccharolyticus, intermedius,* and *melaninogenicus.* In 1977, Holdeman, Cato, and Moore proposed a fourth subspecies, *levii* (Holdeman, Cato, and Moore, 1977). Also in 1977, the asaccharolytic strains were assigned to a separate species because they differ significantly from the other groups of *B. melaninogenicus,* both in mol% G+C (see Table 1) and other properties (Finegold and Barnes, 1977).

3. Many organisms of the *B. clostridiiformis* subsp. *clostridiiformis* group have been shown to have spores (Kaneuchi et al., 1976), and thus have been placed in the genus *Clostridium* (Cato and Salmon, 1976; Holdeman, Cato, and Moore, 1977; Kaneuchi et al., 1976).

4. A new genus of the family Bacteroidaceae,

Pectinatus, has been described (Lee, Mabee, and Jangaard, 1978).

The habitats of the nonpathogenic bacteroides and the roles these organisms play in their various environments have been discussed briefly (see "Habitats"). Also interesting are some of the nutritional and physiological characteristics of these organisms. Because this has recently been reviewed, it will not be discussed further here (Macy and Probst, 1979). However, it is interesting to note that both the nutritional and physiological characteristics of these organisms reflect how they have evolved to greatly influence, and become dependent upon, the environment in which they live.

Acknowledgments

I would like to express my appreciation to I. Probst for her criticisms of the manuscript and her many helpful suggestions.

Literature Cited

Alcantara, E. N., Speckmann, E. W. 1976. Diet, nutrition, and cancer. American Journal of Clinical Nutrition **29:** 1035–1047.

Aranki, A., Freter, R. 1972. Use of anaerobic glove boxes for the cultivation of strictly anaerobic bacteria. American Journal of Clinical Nutrition **25**:1329–1334.

Aranki, A., Syed, S. A., Kenney, E. B., Freter, R. 1969. Isolation of anaerobic bacteria from human gingiva and mouse cecum by means of a simplified glove box procedure. Applied Microbiology **17**:568–576.

Aries, V., Hill, M. J. 1970. Degradation of steroids by intestinal bacteria. II. Enzymes catalyzing the oxidoreduction of the 3-α, 7-α and 12-α-hydroxyl groups in cholic acid and the dehydroxylation of the 7-α-hydroxyl group. Biochimica et Biophysica Acta **202**:535–543.

Balch, W. E., Wolfe, R. S. 1976. New approach to the cultivation of methanogenic bacteria: 2-Mercaptoethanesulfonic acid (HS-CoM)-dependent growth of *Methanobacterium ruminantium* in a pressurized atmosphere. Applied and Environmental Microbiology **32**:781–791.

Barnes, E. M. 1979. Methods for the characterization of the *Bacteroidaceae*, pp. 189–200. In: Skinner, F. A., Lovelock, D. W. (eds.), Identification methods for microbiologists. New York: Academic Press.

Barnes, E. M., Impey, C. S. 1970. The isolation and properties of the predominant anaerobic bacteria in the ceca of chickens and turkeys. British Poultry Science **11**:467–481.

Barnes, E. M., Impey, C. S. 1972. Some properties of the nonsporing anaerobes from poultry caeca. Journal of Applied Bacteriology **35**:241–251.

Betian, H. G., Linehan, B. A., Bryant, M. P., Holdeman, L. V. 1977. Isolation of cellulolytic *Bacteroides* sp. from human feces. Journal of Applied and Environmental Microbiology **33**:1009–1010.

Blackburn, T. H. 1968. Protease production by *Bacteroides amylophilus* strain H18. Journal of General Microbiology **53**:27–36.

Blackburn, T. H., Hobson, P. N. 1962. Further studies on the isolation of proteolytic bacteria from the sheep rumen. Journal of General Microbiology **29**:69–81.

Bryant, M. P. 1972. Commentary on the Hungate technique for culture of anaerobic bacteria. American Journal of Clinical Nutrition **25**:1324–1328.

Bryant, M. P. 1974. Nutritional features and ecology of predominant anaerobic bacteria of the intestinal tract. American Journal of Clinical Nutrition **27**:1313–1319.

Bryant, M. P., Burkey, L. A. 1953a. Cultural methods and some characteristics of some of the more numerous groups of bacteria in the bovine rumen. Journal of Dairy Science **36**:205–217.

Bryant, M. P., Burkey, L. A. 1953b. Numbers and some predominant groups of bacteria in the rumen of cows fed different rations. Journal of Dairy Science **36**:218–224.

Bryant, M. P., Doetsch, R. N. 1954. A study of actively cellulolytic rod-shaped bacteria of the bovine rumen. Journal of Dairy Science **37**:1176–1183.

Bryant, M. P., Robinson, I. M. 1961. An improved non-selective culture medium for ruminal bacteria and its use in determining diurnal variation in numbers of bacteria in the rumen. Journal of Dairy Science **44**:1446–1456.

Bryant, M. P., Robinson, I. M. 1962. Some nutritional characteristics of predominant culturable ruminal bacteria. Journal of Bacteriology **84**:605–614.

Bryant, M. P., Small, N., Bouma, C., Chu, H. 1958. *Bacteroides ruminicola* n. sp. and *Succinimonas amylolytica* the new genus and species. Journal of Bacteriology **76**:15–23.

Caldwell, D. R., Arcand, C. 1974. Inorganic and metal-organic growth requirements of the genus *Bacteroides*. Journal of Bacteriology **120**:322–333.

Caldwell, D. R., Bryant, M. P. 1966. Medium without rumen fluid for non-selective enumeration and isolation of rumen bacteria. Applied Microbiology **14**:794–801.

Caldwell, D. R., Keeney, M., van Soest, P. J. 1969. Effects of carbon dioxide on growth and maltose fermentation by *Bacteroides amylophilus*. Journal of Bacteriology **98**:668–676.

Cato, E. P., Johnson, J. L. 1976. Reinstatement of species rank for *Bacteroides fragilis*, *B. ovatus*, *B. distasonis*, *B. thetaiotaomicron*, and *B. vulgatus*. Designation of the neotype strains of *Bacteroides fragilis* (Veillon and Zuber) Castellani and Chalmers and *Bacteroides thetaiotaomicron* (Distaso) Castellani and Chalmers. International Journal of Systematic Bacteriology **26**:230–237.

Cato, E. P., Moore, W. E. C., Bryant, M. P. 1978. Designation of neotype strains for *Bacteroides amylophilus* Hamlin and Hungate 1956 and *Bacteroides succinogenes* Hungate 1950. International Journal of Systematic Bacteriology **28**:491–495.

Cato, E. P., Salmon, C. W. 1976. Transfer of *Bacteroides clostridiiformis* subsp. *clostridiiformis* (Burri and Ankersmit) Holdeman and Moore and *Bacteroides clostridiiformis* subsp. *girans* (Prévot) Holdeman and Moore to the genus *Clostridium* as *Clostridium clostridiiforme* (Burri and Ankersmit) comb. nov.: Emendation of description and designation of neotype strain. International Journal of Systematic Bacteriology **26**:205–211.

Clarke, R. T. J. 1977. Methods for studying gut microbes, pp. 1–33. In: Clarke, R. T. J., Bauchop, T. (eds.), Microbial ecology of the gut. New York: Academic Press.

Coen, J. A., Dehority, B. A. 1970. Degradation and utilization of hemicellulose from intact forages by pure cultures of rumen bacteria. Applied Microbiology **20**:362–368.

Collings, G. F., Yokoyama, M. T. 1980. Gas-liquid chromatography for evaluating polysaccharide degradation by *Ruminococcus flavefaciens* C94 and *Bacteroides succinogenes* S86. Applied and Environmental Microbiology **39**:566–571.

Dehority, B. A. 1969. Pectin-fermenting bacteria isolated from the bovine rumen. Journal of Bacteriology **99**:189–196.

Dehority, B. A. 1971. Carbon dioxide requirement of various species of rumen bacteria. Journal of Bacteriology **105**:70–76.

Dehority, B. A. 1973. Hemicellulose degradation by rumen bacteria. Federation Proceedings **32**:1819–1825.

Drasar, B. S., Hill, M. J. 1974. Human intestinal flora. New York, London: Academic Press.

Eller, C., Crabill, M. R., Bryant, M. P. 1971. Anaerobic roll tube media for non-selective enumeration and isolation of bacteria in human feces. Applied Microbiology **22**:522–529.

Elsden, S. R., Hitchcock, M. W. S., Marshall, R. A., Phillipson, A. T. 1946. Volatile acid in the digesta of ruminants and other animals. Journal of Experimental Biology **22**:191–202.

Finegold, S. M. 1977. Anaerobic bacteria in human disease. London and New York: Academic Press.

Finegold, S. M., Barnes, E. M. 1977. Report of the ICSB taxonomic subcommittee on gram-negative anaerobic rods. Proposal that the saccharolytic and asaccharolytic strains at present classified in the species *Bacteroides melaninogenicus* (Oliver and Wherry) be reclassified in two species as *Bacteroides melaninogenicus* and *Bacteroides asaccharolyticus*. International Journal of Systematic Bacteriology **27**:388–391.

Finegold, S. M., Miller, A. B., Posnick, D. J. 1965. Further studies on selective media for *Bacteroides* and other anaerobes. Ernährungsforschung **10**:517–528.

Finegold, S. M., Sugihara, P. T., Sutter, V. L. 1971. Use of selective media for isolation of anaerobes from humans, pp. 99–108. In: Shapton, D. A., Board, R. G. (eds.), Isolation of anaerobes. London, New York: Academic Press.

Finegold, S. M., Sutter, V. L., Attebery, H. R., Rosenblatt, J. E. 1975. Isolation of anaerobic bacteria, pp. 365–375. In: Lennette, E. H., Spaulding, E. H., Truant, J. P. (eds.), Manual of clinical microbiology, 2nd ed. Washington, D.C.: American Society of Microbiology.

Gibbons, R. J., MacDonald, J. B. 1960. Hemin and vitamin K compounds as required factors for the cultivation of certain strains of *Bacteroides melaninogenicus*. Journal of Bacteriology **80**:164–170.

Gibbons, R. J., van Houte, J. 1975. Bacterial adherence in oral microbial ecology. Annual Review of Microbiology **29:**19–44.

Gorbach, S. L., Levitan, R. 1970. Intestinal flora in health and in gastrointestinal diseases, pp. 252–275. In: Glass, G. B. J. (ed.), Progress in gastroenterology, vol. 2. New York: Grune & Stratton.

Gradel, C. M., Dehority, B. A. 1972. Fermentation of isolated pectin and pectin from intact forages by pure cultures of rumen bacteria. Applied Microbiology **23:**332–340.

Halliwell, G., Bryant, M. P. 1963. The cellulolytic activity of pure strains of bacteria from the rumen of cattle. Journal of General Microbiology **32:**441–448.

Hamlin, L. J., Hungate, R. E. 1956. Culture and physiology of a starch-digesting bacterium *Bacteroides amylophilus* n. sp. from the bovine rumen. Journal of Bacteriology **72:**548–554.

Harris, M. A., Reddy, C. A., Carter, G. C. 1976. Anaerobic bacteria from the large intestine of mice. Applied and Environmental Microbiology **31:**907–912.

Hill, M. J., Crowther, J. S., Drasar, B. S., Hawksworth, G., Aries, V. C., Williams, R. E. O. 1971. Bacteria and aetiology of cancer of large bowel. Lancet **i:**95–100.

Hofstad, T. 1979. Serological responses to antigens of *Bacteroidaceae*. Microbiological Reviews **43:**103–115.

Holdeman, L. V., Cato, E. P., Moore, W. E. C. (eds.). 1977. Anaerobe laboratory manual, 4th ed. Blacksburg, Virginia: Anaerobe Laboratory, Virginia Polytechnic Institute and State University.

Holdeman, L. V., Good, I. J., Moore, W. E. C. 1976. Human fecal flora: Variation in bacterial composition within individuals and a possible effect of emotional stress. Applied and Environmental Microbiology **31:**359–375.

Holdeman, L. V., Johnson, J. L. 1977. *Bacteroides disiens* sp. nov. and *Bacteroides bivius* sp. nov. from human clinical infections. International Journal of Systematic Bacteriology **27:**337–345.

Holdeman, L. V., Moore, W. E. C. 1974. Gram-negative anaerobic bacteria. Family I. Bacteroidaceae, Genus I. *Bacteroides*. pp. 384–404. In: Buchanan, R. E., Gibbons, N. E. (eds.) Bergey's manual of determinative bacteriology, 8th ed. Baltimore: Williams & Wilkins.

Hungate, R. E. 1950. The anaerobic mesophilic cellulolytic bacteria. Bacteriological Reviews **14:**1–49.

Hungate, R. E. 1966. The rumen and its microbes. New York: Academic Press.

Hungate, R. E. 1969. A roll tube method for cultivation of strict anaerobes, pp. 117–132. In: Norris, J. R., Ribbons, D. W. (eds.), Methods in microbiology, vol. 3B. London, New York: Academic Press.

Hungate, R. E. 1977. Microbial activities related to mammalian digestion and absorption of food, pp. 131–149. In: Spiller, G. A., Amen, R. J. (eds.), Fiber and human nutrition. New York, London: Plenum.

Hylemon, P. B., Sherrod, J. A. 1975. Multiple formes of 7-α-hydroxysteroid dehydrogenase in selected strains of *Bacteroides fragilis*. Journal of Bacteriology **122:**418–424.

Johnson, J. L. 1973. Use of nucleic-acid homologies in the taxonomy of anaerobic bacteria. International Journal of Systematic Bacteriology **23:**308–315.

Johnson, J. L. 1978. Taxonomy of the bacteroides. I. Deoxyribonucleic acid homologies among *Bacteroides fragilis* and other saccharolytic *Bacteroides* species. International Journal of Systematic Bacteriology **28:**245–256.

Johnson, J. L., Ault, D. A. 1978. Taxonomy of the bacteroides. II. Correlation of phenotypic characteristics with deoxyribonucleic acid homology groupings for *Bacteroides fragilis* and other saccharolytic *Bacteroides* species. International Journal of Systematic bacteriology **28:**257–268.

Juhr, N. C., Haas, A. 1976. Cellulolytische Activität im Cacuminhalt von Maus, Ratte, Goldhamster, Meerschweinchen und Kaninchen. Zeitschrift für Versuchstierkunde **18:**129–140.

Kaneuchi, C., Mitsuoka, T. 1978. *Bacteroides microfusus,* a new species from the intestines of calves, chickens, and Japanese quails. International Journal of Systematic Bacteriology **28:**478–481.

Kaneuchi, C., Watanabe, K., Terada, A., Benno, Y., Mitsuoka, T. 1976. Taxonomic study of *Bacteroides clostridiiformis* subsp. *clostridiiformis* (Burri and Ankersmit) Holdeman and Moore and of related organisms: Proposal of *Clostridium clostridiiformis* (Burri and Ankersmit) comb. nov. and *Clostridium symbiosum* (Stevens) comb. nov. International Journal of Systematic Bacteriology **26:**195–204.

Lang, J. H., Briggs, G. M. 1976. The use and function of fiber in diets of monogastric animals, pp. 151–169. In: Spiller, G. A., Amen, R. J. (eds.), Fiber in human nutrition. New York, London: Plenum.

Latham, M. J., Sharpe, M. E. 1971. The isolation of anaerobic organisms from the bovine rumen, pp. 133–147. In: Shapton, D. A., Board, R. G. (eds.), Isolation of anaerobes. London, New York: Academic Press.

Lee, S. Y., Mabee, M. S., Jangaard, N. O. 1978. *Pectinatus,* a new genus of the family *Bacteroidacea*. International Journal of Systematic Bacteriology **28:**582–594.

McBee, R. H. 1977. Fermentation in the hindgut, pp. 185–222. In: Clarke, R. T. J., Bauchop, T. (eds.), Microbial ecology of the gut. New York: Academic Press.

MacDonald, I. A., Williams, C. N., Mahony, D. E., Christie, W. M. 1975. NAD- and NADP-dependent 7-α-hydroxysteroid dehydrogenases from *Bacteroides fragilis*. Biochimica et Biophysica Acta **384:**12–24.

Macy, J. M., Probst, I. 1979. The biology of gastrointestinal bacteroides. Annual Review of Microbiology **33:**561–594.

Macy, J. M., Snellen, J. E., Hungate, R. E. 1972. Use of syringe methods for anaerobiosis. The American Journal of Clinical Nutrition **25:**1318–1323.

Macy, J. M., Yu, I., Caldwell, C., Hungate, R. E. 1978. Reliable sampling method for analysis of the ecology of the human alimentary tract. Applied and Environmental Microbiology **35:**113–120.

Mastromarino, A., Reddy, B. S., Wynder, E. L. 1976. Metabolic epidemiology of colon cancer: Enzymatic activity of fecal flora. American Journal of Clinical Nutrition **29:**1455–1460.

Midtvedt, T., Norman, A. 1967. Bile acid transformations by microbial strains belonging to genera found in intestinal contents. Acta Pathologica et Microbiologica Scandinavica **71:**629–638.

Miller, T. L., Wolin, M. J. 1974. A serum bottle modification of the Hungate technique for cultivating obligate anaerobes. Applied Microbiology **27:**985–987.

Mitsuoka, T., Terada, A., Watanabe, K., Uchida, K. 1974. *Bacteroides multiacidus,* a new species from the feces of humans and pigs. International Journal of Systematic Bacteriology **24:**35–41.

Moore, W. E. C., Holdeman, L. V. 1974. Human fecal flora: The normal flora of 20 Japanese-Hawaiians. Applied Microbiology **27:**961–979.

Moore, W. E. C., Johnson, J. L., Holdeman, L. V. 1976. Emendation of *Bacteroidaceae* and *Butyrivibrio* and descriptions of *Desulfomonas,* gen. nov. and ten new species in the *Desulfomonas, Butyrivibrio, Eubacterium, Clostridium,* and *Ruminococcus*. International Journal of Systematic Bacteriology **26:**238–252.

Morotomi, M., Watanabe, T., Suegara, N., Kawai, Y., Mutai, M. 1975. Distribution of indigenous bacteria in the digestive tract of conventional and gnotobiotic rats. Infection and Immunity **11:**962–968.

Reddy, C. A., Bryant, M. P. 1977. Deoxyribonucleic acid base composition of certain species of the genus *Bacteroides*. Canadian Journal of Microbiology **23:**1252–1256.

Renwick, A. G., Drasar, B. S. 1976. Environmental carcinogens and large bowel cancer. Nature **263:**234–235.

Rizza, V., Sinclair, P. R., White, D. C., Courant, P. R. 1968. Electron transport system of the protoheme-requiring anaerobe *Bacteroides melaninogenicus*. Journal of Bacteriology **96:**665–671.

Salanitro, J. P., Blake, I. G., Muirhead, P. A. 1974. Studies on the cecal microflora of commercial broiler chickens. Applied Microbiology **28**:439–447.

Salanitro, J. P., Fairchilds, I. G., Zgornicki, Y. D. 1974. Isolation, culture characteristics, and identification of anaerobic bacteria from the chicken cecum. Applied Microbiology **27**:678–687.

Salanitro, J. P., Blake, I. G., Muirhead, P. A., Maglio, M., Goodman, J. R. 1978. Bacteria isolated from the duodenum, ileum, and cecum of young chicks. Applied and Environmental Microbiology **35**:782–790.

Salyers, A. A. 1979. Energy sources of major intestinal fermentative anaerobes. American Journal of Clinical Nutrition **32**:158–163.

Shah, H. N., Williams, R. A. D., Bowden, G. H., Hardie, J. M. 1976. Comparison of the biochemical properties of *Bacteroides melaninogenicus* from human dental plaque and other sites. Journal of Applied Bacteriology **41**:473–492.

Sherrod, J. A., Hylemon, P. B. 1977. Partial purification and characterization of NAD-dependent 7-α-hydroxysteroid dehydrogenase from *Bacteroides thetaiotaomicron*. Biochimica et Biophysica Acta **486**:351–358.

Smibert, R. M., Holdeman, L. V. 1976. Clinical isolates of anaerobic gram-negative rods with a formate-fumarate energy metabolism: *Bacteroides corrodens*, *Vibrio succinogenes*, and unidentified strains. Journal of Clinical Microbiology **3**:432–437.

Stellewag, E. J., Hylemon, P. B. 1976. Purification and characterization of bile salt hydrolase from *Bacteroides fragilis* subsp. *fragilis*. Biochimica et Biophysica Acta **452**:165–176.

Tannock, G. W. 1977. Characteristics of *Bacteroides* isolated from cecum of conventional mice. Applied and Environmental Microbiology **33**:745–750

van Steenbergen, T. J. M., de Soet, J. J., de Graaff, J. 1979. DNA base composition of various strains of *Bacteroides melaninogenicus*. FEMS Microbiology Letters **5**:127–130.

Varel, V. H., Bryant, M. P. 1974. Nutritional features of *Bacteroides fragilis* subsp. *fragilis*. Applied Microbiology **18**:251–257.

Yang, M. G., Manoharan, K., Young, A. K. 1969. Influence and degradation of dietary cellulose in cecum of rats. Journal of Nutrition **97**:260–264.

The Genus *Fusobacterium*

TOR HOFSTAD

The genus *Fusobacterium* includes several species of obligately anaerobic, nonsporeforming, motile or nonmotile, Gram-negative rods. Some are slender, spindle-shaped bacilli, others are pleomorphic rods with parallel sides and rounded ends. Their habitat is the mucous membranes of man and animals.

During the last decade of the nineteenth century, several authors, among them Miller (1889), Plaut (1894), and Vincent (1896, 1899, 1904), observed spindle-shaped, or fusiform, bacilli in material from the diseased or healthy human mouth. Veillon and Zuber (1898), Lewkowicz (1901), and Ellermann (1904) were the first to cultivate fusiform bacilli. The organism cultured by Veillon and Zuber grew at room temperature, whereas those described by Lewkowicz and Ellermann apparently were more fastidious. Loeffler, in 1884, observed pleomorphic rods in diphtheric lesions of calves and doves. The same organism, identifiable with *F. necrophorum* of today, was cultured by Bang (1890–1891) from necrotic lesions of a number of domestic animals, and by Schmorl (1891) from an epizootic in rabbits. There is hardly a more confused chapter in bacterial taxonomy than that which deals with the organisms now included in the genus *Fusobacterium*. Fusiform bacilli that are not fusobacteria have been described as such.

The more pleomorphic fusobacteria without tapering ends have been described under different generic names. Examples are *Bacteroides, Sphaerophorus, Bacterium, Necrobacterium, Pseudobacterium, Bacillus, Actinomyces, Corynebacterium, Ristella,* and *Zuberella*.

The family name Bacteroidaceae was used by Pribram (1929) for strictly anaerobic rods. Ten years earlier, Castellani and Chalmers (1919) had proposed that the genus *Bacteroides* should contain obligately anaerobic bacilli that did not form spores. Eggerth and Gagnon (1933) and Weiss and Rettger (1937) excluded the Gram-positive rods from the genus. The generic name *Fusobacterium* was proposed by Knorr (1923) for obligately Gram-negative bacilli that were fusiform. Prévot (1938), who argued that the generic name *Fusobacterium* (and also the names Bacteroidaceae and *Bacteroides*) was invalid, used the term *Sphaerophorus* for the non-

motile, pleomorphic fusobacteria, and the term *Fusiformis* for the fusobacteria that had tapered ends. The seventh edition of *Bergey's Manual of Determinative Bacteriology* (Breed, Murray, and Smith, 1957) divided the family Bacteroidaceae into three genera: *Bacteroides* defined as rods with rounded ends, *Fusobacterium* defined as rods with tapering ends, and *Sphaerophorus* defined as rods with rounded ends that showed a marked pleomorphism and where filaments were common.

Cell morphology had so far been the main criterion for the classification of the nonsporeforming anaerobic rods. In more recent years, their biochemical properties have been studied more thoroughly, particularly by Beerens, Castel, and Fievez (1962), Sebald (1962), Werner (1972a, 1972b), Werner, Neuhaus, and Hussels (1971) and Holdeman and Moore (1972). Their work led to the inclusion of *Sphaerophorus* in genus *Fusobacterium*. This genus then comprises all anaerobic nonsporeforming, Gram-negative rods whose major metabolic product is butyric acid. Further, *Leptotrichia* (this Handbook, Chapter 119) was reestablished as a genus for the saccharolytic fusiform bacilli producing lactic acid as the only major fermentation product.

Up to now, 13 species of *Fusobacterium* are adequately described (Table 1). The species isolated most frequently from man and animals are *F. nucleatum* and *F. necrophorum*, respectively.

Habitats

All *Fusobacterium* species are parasites of man and animals. A doubtful exception may be the very rare *F. aquatile,* the first isolation of which was from filtered and chlorinated river water (Spray and Laux, quoted by Prévot, Turpin, and Kaiser, 1967).

The main human habitats of the different *Fusobacterium* species are listed in Table 1.

F. nucleatum is a constant member of the oral microflora of adults, but has also been isolated from the oral cavity of predentate children (Hurst, 1957; McCarthy, Snyder, and Parker, 1965). Occasionally, *F. naviforme* has been found in the mouth or the upper respiratory tract (Holdeman and Moore,

Table 1. The main human sources of *Fusobacterium* isolates.[a]

Species	Normal flora		Clinical specimens
	Mouth	Gastrointestinal tract	
F. nucleatum	+		+
F. necrophorum		+	+
F. aquatile		+	
F. glutinosum			+
F. gonidiaformans			+
F. mortiferum		+	+
F. naviforme		+	+
F. necrogenes		+	
F. plauti		+	
F. perfoetens[b]			
F. prausnitzii		+	
F. russii		+	+
F. varium		+	+

[a] Data mainly from Holdeman and Moore (1972).
[b] Isolated only twice from man (Tissier, 1905).

1972). The isolation of *F. necrophorum* from pleuropulmonary infections suggests that also this organism is able to live as a parasite on the mucous membranes of the oral cavity and upper respiratory tract of man. The number of fusobacteria per milliliter of saliva has been estimated to be 5.6×10^4 (Richardson and Jones, 1958). In different surveys, fusobacteria have been found to make up from 0.4 to 7% of the cultivable dental plaque flora (Hardie and Bowden, 1974). There are, however, great individual variations. In four patients with gingivitis, 9% of the microbial isolates from subgingival plaque samples were fusobacteria (Williams, Pantalone, and Sherris, 1976). Hadi and Russel (1969) reported a mean viable count of *F. nucleatum* per gram wet weight of gingival plaque material from patients with advanced chronic periodontal disease and acute ulcerative gingivitis of 3.3×10^7 and 9.3×10^7, respectively. In subjects with healthy gingivae, the corresponding figure was 5.7×10^6. A significant increase of *F. nucleatum* in gingival plaque with an increasing degree of gingival inflammation has also been reported by Van Palenstein Helderman (1975).

Fusobacterium makes up a small part of the fecal microflora of man, with individual variation ranging from about 7% to less than 1% (Finegold, Attebery, and Sutter, 1974; Finegold et al., 1975b; Holdeman, Good, and Moore, 1976; Moore and Holdeman, 1974; van Houte and Gibbons, 1966). The most prevalent species seem to be *F. prausnitzii, F. russi,* and *F. mortiferum.* Both the number of fusobacteria in feces and the relative frequency of the different species are influenced by the diet (Finegold, Attebery, and Sutter, 1974; Maier et al., 1974; Peach et al., 1974). Thus, Japanese on a traditional diet rich in carbohydrate have a relatively high number of *F. necrophorum* in feces (Ohtani, 1970a; Ueno et al., 1974).

The occurrence of fusobacteria on the mucous membranes of the genitourinary tract is virtually unknown. Fusobacteria were not present in the normal microflora of the cervix of 30 healthy females examined by Gorbach et al. (1973), and Hite, Hesseltine, and Goldstein (1947) found no fusobacteria in the healthy vaginas of pregnant women. Fusobacteria, particularly *F. necrophorum*, were, however, present in the vagina of pregnant women with trichomoniasis and in the postpartum uterus. Spaulding and Rettger (1937) found *F. nucleatum* in the normal vagina but not in the vagina of pregnant women. Davis and Pilot (1922) and Brams, Pilot, and Davis (1923) isolated fusiform bacilli (and spirochetes) from the clitoris region in females, and from preputial secretions of 50 of 100 men. *F. gonidiaformans* has been isolated from urine aspirated from the female bladder after urethral milking (Bran, Levisan, and Kaye, 1972). The normal habitat of this organism is possibly the large bowel. The human habitat of the extremely rare *F. glutinosum* is not known.

F. necrophorum is a normal inhabitant of the alimentary tract of cattle, horses, sheep, and pigs. Fuller and Lev (1964), in a study of the Gram-negative bacteria of the pig alimentary tract, found *F. necrophorum* to be present from the age of 43 days. Aalbæk (1972) isolated the organism from the colon of porkers in numbers up to 10^3 per gram of wet material, but in a considerably higher number in the ileum, cecum, and colon of pigs with experimental enteritis. *F. necrophorum* has also been found in infections or in feces of other animals, e.g., mules (Nolechek, 1918), goats (Jensen, 1913), reindeer (Horne, 1898–1899), antelope (Mettam and Carmichael, 1933), kangaroos (Bang, 1890–1891), dogs (Jensen, 1913), rabbits (Cameron and Williams, 1926; Schmorl, 1891), rats (Lewis and

Rettger, 1940), chickens (Jensen, 1913), and apes (Dack, Dragstedt, and Heinz, 1937; Dack, Heinz, and Dragstedt, 1935). It has been mentioned as occurring also in buffaloes, cats, guinea pigs, mice, snakes, tortoises, and fowl (Simon and Stovell, 1969; Weinberg, Nativelle, and Prévot, 1937).

Less is known about the presence in animals of the other *Fusobacterium* species. Fusiform bacilli, probably *F. nucleatum,* have been isolated from the alimentary tract of pigs (Aalbæk, 1972) and mice (Syed, 1972), and from the oral cavity and the throat of monkeys (Krygier et al., 1973; Pratt, 1927; Slanetz and Rettger, 1933), dogs (Slanetz and Rettger, 1933), cats (Prévot et al., 1951), rabbits (Pratt, 1927; Slanetz and Rettger, 1933), and guinea pigs (Pratt, 1927; Spaulding and Rettger, 1937). Terada, Uchida, and Mitsuoka (1976) isolated *F. necrogenes, F. aquatile,* and *F. mortiferum* from pig feces, and *F. perfoetens* has been found in feces of a piglet (Van Assche and Wilssens, 1977). *F. necrogenes* is a member of the cecal flora of poultry (Holdeman and Moore, 1972), and *F. plauti* has been isolated from the chicken cecum (Barnes and Impey, 1974). *F. russii* has been mentioned as being part of the normal microflora of mice and pigs, as well as of the rumen flora of cattle (Smith, 1975).

F. necrophorum is able to deconjugate bile salts (Shimada, Bricknell, and Finegold, 1969). Otherwise the physiological role of the fusobacteria in the gastrointestinal tract of man and animals is unknown.

Next to *Bacteroides fragilis* and *B. melaninogenicus, F. nucleatum* is the Gram-negative anaerobic organism encountered most often in human infections. Also, *F. necrophorum* is clearly pathogenic in man. Before the advent of antibiotics and other antimicrobial drugs, this organism was frequently isolated from suppurative infections of the oral cavity and the upper respiratory tract, and from pleuropulmonary infections. Reviewing the literature concerning anaerobic pleuropulmonary infections, Finegold (1977) found that *F. necrophorum* accounted for 24% of all anaerobic bacteria isolated from 358 cases. Today the organism is less commonly isolated from human infections. *F. nucleatum* is usually isolated in mixture with other anaerobic and/or facultative organisms. However, on a percentage basis, *F. nucleatum* and *F. necrophorum* have been isolated in pure culture from pyogenic infections more frequently than have other anaerobic bacteria (Beerens and Tahon-Castel, 1965; Werner and Pulverer, 1971; Bartlett, Sutter, and Finegold, 1974). Other species of *Fusobacterium* are occasionally isolated from clinical specimens, and nearly always in mixed culture.

Pathogenic fusobacteria are in particular isolated from inflammatory processes accompanied by necrosis and ulceration. They are most frequently found in pleuropulmonary infections and in abscesses of the brain and the liver. In a large series of patients observed in the time period 1958–1974, *F. nucleatum* was isolated in 28 of 72 cases with anaerobic pneumonitis or necrotizing pneumonia, in 19 of 45 cases of anaerobic lung abscess, in 19 of 70 cases of aspiration pneumonia, and in 16 of 83 cases of empyema (Finegold, 1977). Fifty of 248 isolates from 142 patients with brain abscesses due to anaerobic bacteria were fusobacteria (literature survey made by Finegold, 1977). In 31 cases of liver abscesses amenable to surgery, *F. nucleatum* and *F. necrophorum* accounted for 6 and 13%, respectively, of all bacterial isolates (Altemeier, 1974). In 67 cases of intraabdominal sepsis *Fusobacterium* species accounted for 7% of the anaerobic bacterial isolates and 4% of the total bacteria isolated (Finegold et al., 1975a). The *Fusobacterium* species isolated from genital tract infections in 200 females were 6% (Chow, Marshall, and Guze, 1975). There were 4 isolated *Fusobacterium* species from 56 patients with septic abortion (Finegold et al., 1975a). *F. nucleatum* and *F. necrophorum* may also be encountered in blood cultures, particularly when the oropulmonary tract is the portal of entry (Felner and Dowell, 1971).

F. nucleatum is invariably present in Plaut-Vincent's angina (fusospirochetal angina), and in acute necrotizing ulcerative gingivitis. *F. necrophorum* is almost always found in lesions of ulcerative colitis, but not as a causative agent.

F. necrophorum is an animal pathogen that is frequently isolated from necrotic and gangrenous lesions in cattle, sheep, and pigs, and less frequently from other animals. Carnivorous animals appear to be resistant. The most common manifestations of diseases associated with *F. necrophorum* are liver abscess (hepatic necrobacillosis) and foot rot.

Liver abscesses are especially encountered in heavily fattened cattle. Ninety percent of such abscesses contain *F. necrophorum* as the only organism or in combination with other organisms (Hussein and Shigidi, 1974; Kanoe et al., 1976; Newsom, 1938; Simon and Stovell, 1971). The disease is associated with inflammation of the forestomachs, presumably caused by irritating substances produced by fermentation of the high caloric feed and by foreign bodies (Jensen and Mackay, 1965). *F. necrophorum* present in the stomach content is thought to gain entry to the vascular system through the injured mucosa. Liver abscesses have thus been produced experimentally in cattle and sheep by intraportal injections of viable cells of a bovine isolate of *F. necrophorum* (Jensen, Flint, and Griner, 1954).

Foot rot is a bacterial dermatitis of ungulates characterized by separation of the hoof due to an inflammatory disruption of its epidermal matrix.

The disease is especially encountered in sheep and cattle. Foot rot in sheep is caused by *Bacteroides nodosus* in combination with *Fusobacterium necrophorum*. *B. nodosus* is the means by which the basic disease is transferred from one animal to another (Beveridge, 1941), whereas *F. necrophorum* is essential for the inflammatory destruction of tissue (Egerton, Roberts, and Parsonson, 1969; Roberts and Egerton, 1969). Injury to the foot and damp soil are predisposing factors (Graham and Egerton, 1968). The primary cause of epizootic foot rot in cattle has not been found. *F. necrophorum* is present in the lesions as a concurrent pathogen or a secondary invader, and is responsible for the major part of the tissue destruction. Berg and Loan (1975) induced typical lesions in cattle by the intradigital or intradermal inoculation of *F. necrophorum* alone or in combination with *Bacteroides melaninogenicus*. When pure cultures of *F. necrophorum* were inoculated, both organisms were recovered from the lesions.

Another disease associated with *F. necrophorum* is calf "diphtheria", which is a necrotic laryngitis in calves up to 2 years of age. *F. necrophorum* is also involved in several other suppurative or gangrenous processes in domestic animals, such as interdigital dermatitis and heel abscess in sheep (Parsonson, Egerton, and Roberts, 1967; Roberts et al., 1968), neonatal bacteremias in calves and lambs, necrotic enteritis of pigs, necrotic rhinitis of growing pigs, and oral infections in several animals.

F. necrophorum infection in domestic animals may present an economic problem in meat-producing countries (Langworth, 1977; Panel report, 1973). In the United States, hepatic necrobacillosis is the major reason for the condemnation of livers in cattle (Jensen and Mackay, 1965).

Foot rot is a common cause of lameness in beef and dairy cattle, and the incidence of the disease in a flock of sheep may approach 100%. Fortunately the disease can be satisfactorily controlled in sheep by carefully examining the feet at the beginning of the dry season, when the disease does not spread, and then eliminating all animals that show evidence of the disease (Bruner and Gillespie, 1973). In other countries, *F. necrophorum* infection is of less economic importance. Necrobacillosis was thus observed in Norway during 1974–1975 in only four herds of pigs and one herd of sheep (Central Bureau of Veterinary Statistics of Norway, 1976).

The natural infections have verified the infectivity and invasiveness in animals of *F. necrophorum* and in man of *F. nucleatum* and *F. necrophorum*. Experimental investigations in animals have shown that synergistic mechanisms may be of importance in the pathogenesis of mixed infections involving pathogenic fusobacteria (Hamp and Mergenhagen, 1963; Hill, Osterhout, and Pratt, 1974; Kaufman

et al., 1972; Onderdonk et al., 1976; Roberts, 1967a,b). *F. nucleatum* and *F. necrophorum*, as well as other *Fusobacterium* species, possess a cell wall lipopolysaccharide with the characteristics of an endotoxin (Garcia, Charlton, and McKay, 1975b; Hofstad and Kristoffersen, 1971; Sveen, Hofstad, and Milner, 1977; Warner et al., 1975). The presence in *F. necrophorum* of exotoxins with leukocidal and hemolytic activities has been demonstrated (Garcia, Alexander, and McKay, 1975a; Roberts, 1967a,b).

Isolation

The fusobacteria are heterotrophic organisms that grow readily on ordinary solid media, such as Brucella blood agar and brain heart infusion agar, and in fluid media with a base of meat extract and proteose peptone, Casitone, or tryptone. Most species are not particularly demanding with regard to a low oxidation-reduction potential. Thus, strains of *Fusobacterium nucleatum* were able to grow in an oxygen tension of up to 6% (Loesche, 1969). They are, however, fairly readily killed by exposure to air. This is possibly due to their susceptibility to hydrogen peroxide and is especially noticeable when thioglycolate or cysteine HCl is incorporated into media that are exposed to air before inoculation and incubation.

The fermentative ability of the fusobacteria is restricted and they are not proteolytic enough to digest coagulated proteins. *F. necrophorum* has an absolute need for proline-containing polypeptides (Wahren and Holme, 1973). Free amino acids other than proline are readily utilized.

The fusobacteria are susceptible to many of the commonly used antibiotics. They are, however, relatively resistant to vancomycin and neomycin. *F. varium* and *F. mortiferum* are resistant to rifampicin (rifampin). The growth of these two organisms is not inhibited by bile, to which other species of *Fusobacterium* are susceptible. Similar to several other Gram-negative bacteria, the fusobacteria will grow in the presence of low concentrations of various dyes.

Sampling

Fusobacterium nucleatum is best isolated from saliva or centrifuged salivary deposits, and from the crevice or pocket that exists between the gingiva and the tooth surface. Sampling from the crevice area is performed by the use of sterile filter paper points (absorbent dental points), which are gently inserted into the crevice. Saliva or salivary deposits may be inoculated to the medium either directly or after being resuspended in a reducing diluent such as the serum-containing diluent of Bowden and Hardie (1971) or the WAL diluent (Sutter et al., 1975). The infected tapering end of the paper point is streaked on a small area of the surface of the solid medium, and further spreading of the deposited material is carried out by a wire loop. Because of their presence in small numbers, isolation of other *Fusobacterium* species from their natural habitats in man can be difficult. Detailed directions for collection, transport, and processing of fecal specimens have been given by Sutter et al. (1975). It is essential that the specimens are thoroughly homogenized and adequately diluted in a reducing diluent before inoculation. This applies also to the isolation of *F. necrophorum* from the intestinal tract of animals.

Isolation Under Nonselective Conditions

When present in clinical specimens, *Fusobacterium nucleatum* and *F. necrophorum* and the less commonly isolated *Fusobacterium* species are usually recovered on solid nonselective media. If the colonies are carefully inspected together with liberal use of the Gram stain, and subculturing is promptly performed, isolation does not present much of a problem.

Nonselective isolation of fusobacteria from their natural habitats on the mucous membranes in man and animals is laborious and time-consuming. Such isolation attempts should be avoided in those instances where isolation on selective media is possible. However, isolation under nonselective conditions seems to be the most reliable method for examination of viable cells in normal flora specimens. For this purpose the roll tube method (Holdeman and Moore, 1972; Moore, 1966) or the use of a glove box (Aranki et al., 1969) is to be recommended. By inoculating roll tubes with 1 ml each of 10^8, 10^9, and 10^{10} dilutions of homogenized feces, bacterial species, among them fusobacteria, present in numbers as low as 3×10^{10} per gram of fecal dry matter (0.06% of the fecal bacterial population), were counted (Moore and Holdeman, 1974).

Selective Isolation

Media formulations have been developed and evaluated for selective isolation of fusobacteria from human (Baird-Parker, 1957; Finegold, Sugihara, and Sutter, 1971; Ohtani, 1970b; Omata and Disraely, 1956; Sutter, Sugihara, and Finegold, 1971) and animal (Fales and Teresa, 1972a) sources.

The media of Omata and Disraely (1956) and Baird-Parker (1957) are recommended for the isolation of oral fusobacteria. In both media the selective agents are dyes and antibiotics.

FM Selective Medium for Isolation of Fusobacteria (Omata and Disraely, 1956)

The basal medium contains the following ingredients, expressed as percentages, wt/vol, in distilled water:

Casitone	1.5
Yeast extract	0.5
Glucose	0.5
Sodium chloride	0.5
L-Cystine	0.075
Crystal violet	0.001
Streptomycin	0.001
Agar	1.5
pH	7.2

The autoclaved medium, without streptomycin, is enriched with either 5% sterile horse serum, 5% sterile human ascitic fluid, or corn starch. The starch, when used, is incorporated into the basal medium upon preparation. The horse serum and the ascitic fluid are added to the sterile Petri dishes prior to pouring. Streptomycin, diluted 1:1,000, is sterilized by filtration and added to the plates at the time of pouring.

Omata and Disraely (1956) found horse serum or ascitic fluid to be superior to starch as the growth-promoting supplement.

Selective Isolation Medium for Oral Fusobacteria (Baird-Parker, 1957)

The basal medium contains the following ingredients, expressed as percentages, wt/vol, in distilled water:

Proteose peptone	1
Yeast extract	0.1
Lab Lemco powder	0.3
Glucose	0.5
Soluble starch	0.2
Sodium nitrate	0.1
Cysteine HCl	0.05
Disodium hydrogen phosphate	0.5
Agar	1.5
pH	7.6

One part in 15,000 of ethyl violet and 100 μg/ml of bacitracin are added aseptically to the autoclaved and molten medium.

Baird-Parker's medium is selective for *F. nucleatum*, whereas the medium of Omata and Disraely also permits the growth of *Leptotrichia buccalis*. Baird-Parker's medium may be made selective for both organisms by the replacement of bacitracin by 15 μg/ml neomycin sulfate or 20 μg/ml streptomycin sulfate. It is essential that individual batches of the dyes be checked for their suitability.

Ohtani (1970a) modified the FM medium of Omata and Disraely. He replaced the streptomycin with 100 μg/ml neomycin and used pepsin-digested cow's blood as the enrichment instead of horse serum. The modified medium, which inhibits *Bacteroides* and Gram-positive bacteria, is well suited for the isolation of fusobacteria from feces.

Media selective for different groups of anaerobic organisms have been developed by Finegold and collaborators (Finegold, Sugihara, and Sutter, 1971; Sutter, Sugihara, and Finegold, 1971; Sutter et al., 1975). The prime use of these media is the isolation of anaerobic bacteria from clinical specimens, but they may also be employed for studies of the normal flora. The selectivity of these media depends on the addition of combinations of antibiotics.

Neomycin-Vancomycin Blood Agar
(Finegold, Sugihara, and Sutter, 1971; Sutter et al., 1975)

A total of 100 μg/ml of neomycin base is added to Brucella blood agar before autoclaving. After autoclaving, 7.5 μg/ml of vancomycin is added aseptically to the medium in addition to 10 μg/ml of vitamin K_1 and 5% sheep blood. Neomycin stock solution: 1 g of neomycin base activity is dissolved in 10 ml of sterile phosphate buffer, pH 8.0. This makes a final concentration of 100,000 μg/ml. The solution can be stored at 4°C for up to 1 year. Vancomycin stock solution: 0.075 g of vancomycin base activity is dissolved in 5 ml of 0.05 N HCl and 5 ml of sterile distilled water is added. This gives a final concentration of 500 μg/ml. The solution can be stored at 4°C for up to a month and at −20°C for up to 1 year. The preparation of the stock solutions of neomycin, vancomycin, and rifampin is described by Sutter et al. (1975).

This medium yields a reasonably good growth of *Fusobacterium* species, but *Bacteroides* will grow on the medium in varying degrees. Also *Veillonella* will grow on the medium.

The medium of Sutter, Sugihara, and Finegold (1971) is selective for *F. varium* and *F. mortiferum,* and can be used for isolation of these species from feces or from other sources. The selectivity of the medium depends on the addition of rifampin to standard blood agar.

Rifampin Blood Agar for Isolation of Fusobacteria
(Sutter, Sugihara, and Finegold, 1971; Sutter et al., 1975)

A total of 50 μg/ml of rifampicin (rifampin) is added to Brucella blood agar at the time of pouring plates. Rifampin stock solution: 0.1 g of rifampin is dissolved in 20 ml of absolute ethyl alcohol, and 80 ml of sterile distilled water is added. This gives a final concentration of 1,000 μg/ml. The solution can be stored at 4°C for up to 2 months.

F. varium and *F. mortiferum* grow freely on the medium, while the growth of *Bacteroides* and most other organisms present in human feces in high numbers is inhibited.

A medium selective for the isolation of *F. necrophorum* from bovine liver abscesses was reported by Fales and Teresa (1972a). The medium is based on the Trypticase and egg yolk medium of McClung and Toabe (1947), and contains crystal violet and phenethyl alcohol as selective agents.

Isolation Medium for *Fusobacterium necrophorum*
(Fales and Teresa, 1972a)

The medium has the following composition, g/415.0 ml of distilled water:

Trypticase	16.0
Biosate	4.0
Thiotone	2.0
Glucose	0.5
$MgSO_4$ (5% solution)	0.1 ml
Na_2HPO_4	2.5
Agar	8.3
pH	7.3

After autoclaving, the basal medium is cooled to 50°C, and 1.35 ml (0.27% vol/vol) of phenethyl alcohol is added. One egg yolk mixed with an equal volume of a 0.9% saline solution (total volume, approximately 45.0 ml) is blended with the basal medium, and then 11.5 mg of crystal violet dissolved in 25.0 ml of sterile distilled water is added. Finally, the volume is adjusted to 500 ml with sterile distilled water.

Small colonies of *Proteus* species appearing on the medium are easily distinguished from the larger colonies (1.5–1.7 mm in diameter after 48 h of incubation) of *F. necrophorum.*

As previously mentioned, the different selective media designed for isolation of *Fusobacterium* species allow other organisms to grow to a varying extent. In order to gain experience with these media it is important, therefore, to check the different colony types by Gram-staining.

Axenic Cultivation and Maintenance

Fusobacterium strains can be maintained by weekly serial subcultures on blood agar or by lyophilization. Batch cultivation is best performed in a nutrient broth with a tryptone base and supplemented with yeast extract (0.3%), glucose (0.25%), and cysteine HCl (0.1%), or in the selective media bases. If nar-

Table 2. Antibiotic disk identification of anaerobic Gram-negative rods.[a]

Bacterial group	Colistin 10 μg	Erythromycin 60 μg	Kanamycin 1,000 μg	Penicillin 2 units	Vancomycin 5 μg	Rifampin 5 μg
Bacteroides fragilis	R	S	R	R	R	S
B. melaninogenicus *B. oralis* *B. assacharolyticus*	V	S	R	S	V	S
B. corrodens	S	S	S	S	R	S
Fusobacterium mortiferum *F. varium*	S	R	S	S	R	R
Most other *Fusobacterium* spp.	S	S	S	S	R	S

[a] Modified from Sutter et al. (1975). S, sensitive, zones ≥10 mm; R, resistant, zones <10 mm; V, variable.

row-necked, well-filled containers are used, prereduced anaerobically sterilized (PRAS) media are usually not necessary.

F. necrophorum has been grown in continuous culture with glucose as the growth-limiting factor (Wahren, Bernholm, and Holme, 1971). Maximal cell yields (3.5 mg/ml, dry weight) were achieved at dilution rates between 0.19 h^{-1} and 0.40 h^{-1}, at a pH of 6.8, and at temperatures of 33–36°C.

Identification

Fusobacterium nucleatum has a characteristic cell morphology that makes a presumptive identification easy. The cells are Gram-negative, slender, spindle-shaped bacilli with sharply pointed ends, often appearing in pairs and end-to-end. Most cells are 5–10 μm long, but shorter and longer rod forms may be seen. The fusiform cells of *Leptotrichia buccalis* are thicker and usually larger (distinguishing characters are given in Table 1 of Chapter 119, this Handbook). Colonies of *F. nucleatum* on blood agar are low convex, glistening, and slightly irregular in form.

A preliminary grouping of anaerobic Gram-negative bacilli can be made from their susceptibility to specified antibiotics (Table 2). *Bacteroides fragilis,*

the most common anaerobic human isolate, is saccharolytic, and its growth is either stimulated or unaffected by 20% bile. The colonies of *B. melaninogenicus* and *B. asaccharolyticus* are characteristically pigmented, and those of *B. corrodens* are corroding. Definite identification of *Fusobacterium* species is based mainly on biochemical tests and gas-liquid chromatography (GLC) of acid end products. For this purpose, PRAS media based on peptone and yeast extract (Holdeman and Moore, 1972) are recommended. The distinguishing characters of *F. mortiferum* and *F. varium* are shown in Table 3, and those of the other *Fusobacterium* species in Table 4. Other characteristics of *F. necrophorum*: (i) they are Gram-negative pleomorphic cells that may be curved and often have spherical enlargements; (ii) free coccoid bodies and especially filaments are common; (iii) colonies are circular, rough, and often β-hemolytic; (iv) large amounts of gas are commonly produced in fluid media. Additional tests have to be set up for isolates of fusobacteria that do not fit the species listed in Tables 3 and 4 (see Holdeman and Moore, 1972).

If employed with GLC and a few supplementary tests, micromethod, multitest systems can be used for the identification of fusobacteria, as well as other anaerobes, in clinical specimens (Hansen and Stewart, 1976; Nord, Dahlbäck, and Wadstrøm, 1975;

Table 3. Characteristics of *Fusobacterium mortiferum* and *F. varium*.[a]

Characteristic	F. mortiferum	F. varium
Production of indole	−	+
Hydrolysis of esculin	+	−
Effect of bile on growth	None, or stimulation	None, or stimulation
Propionate from threonine	+	+
Fatty acids from PYG (GLC)	Acetic, propionic, butyric; sometimes formic, lactic, isovaleric, succinic	Acetic, butyric, lactic; sometimes succinic, propionic

[a] From Sutter et al., 1975, with permission. PYG, peptone-yeast extract-glucose broth; GLC, gas-liquid chromatography, +, positive reaction for majority of strains; −, negative reaction for majority of strains.

Table 4. Characteristics of *Fusobacterium nucleatum, F. necrophorum, F. gonidiaformans,* and *F. naviforme.*[a]

	F. nucleatum	F. necrophorum	F. gonidiaformans	F. naviforme
Production of indole	+	+	+	+
Reduction of nitrate	−	−	−	−
Fermentation of:				
glucose	−	−	+	−
levulose	+	−	−	−
mannose	−	−	−	
Hydrolysis of:				
esculin	−	−	+	−
starch	−	−	+	−
Effect of bile on growth	Inhibition	Inhibition	Inhibition	Inhibition
Production of lipase	−	+/−	−	−
Propionate from lactate	−	+	−	−
Propionate from threonine	+	+	+	−
Fatty acids from PYG (GLC)	Acetic, propionic, butyric, succinic; sometimes formic, lactic	Acetic, propionic, butyric; sometimes formic, succinic, lactic	Acetic, propionic, butyric; sometimes formic, succinic, lactic	Acetic, propionic, butyric, lactic; sometimes formic, succinic

[a] Reproduced from Sutter et al., 1975, with permission. See Table 3 legend.

Starr et al., 1973). The fluorescent-antibody technique has been used for the identification of fusobacteria in clinical specimens from man (Griffin, 1970; Stauffer et al., 1975) and *F. necrophorum* in bovine liver abscesses (Fales and Teresa, 1972b). Simon (1975) has described a hemagglutination inhibition test for rapid identification of *F. necrophorum*.

Addendum

Fusobacteria (*Fusobacterium mortiferum, F. nucleatum, F. varium,* and *F. necrophorum*) have been isolated in significant numbers from the gastrointestinal tract of grass carps maintained under defined culture conditions on pelleted diets and on aquatic weeds (Trust et al., 1979). A new selective medium for isolation of *F. nucleatum* from human periodontal pockets has recently been described (Walker et al., 1979).

Literature Cited

Aalbæk, B. 1972. Gram-negative anaerobes in the intestinal flora of pigs. Acta Veterinaria Scandinavica **13:**228–237.

Altemeier, W. A. 1974. Liver abscess: The etiologic role of anaerobic bacteria, pp. 387–398. In: Balows, A., DeHaan, R. M., Dowell, V. R., Jr., Guze, L. B. (eds.), Anaerobic bacteria: Role in disease. Springfield, Illinois: Charles C Thomas.

Aranki, A., Syed, S. A., Kenney, E. B., Freter, R. 1969. Isolation of anaerobic bacteria from human gingiva and mouse cecum by means of a simplified glove box procedure. Applied Microbiology **17:**568–576.

Baird-Parker, A. C. 1957. Isolation of *Leptotrichia buccalis* and *Fusobacterium* species from oral material. Nature **180:** 1056–1057.

Bang, B. 1890–1891. Om aarsagen til lokal nekrose. Maanedskrift for Dyrlæger **2:**235–259.

Barnes, E. M., Impey, C. S. 1974. The occurrence and properties of uric acid decomposing anaerobic bacteria in the avian caecum. Journal of Applied Bacteriology **37:**393–409.

Bartlett, J. G., Sutter, V. L., Finegold, S. M. 1974. Anaerobic pleuropulmonary disease: Clinical observations and bacteriology in 100 cases, pp. 327–344. In: Balows, A., DeHaan, R. M., Dowell, V. R., Jr., Guze, L. B. (eds.), Anaerobic bacteria: Role in disease. Springfield, Illinois: Charles C Thomas.

Beerens, H., Castel, M. M., Fievez L. 1962. Classification des *Bacteroidaceae*, p. 120. Abstracts of the VIII International Congress for Microbiology, Montreal.

Beerens, H., Tahon-Castel, M. M. 1965. Infectiones Humaines à Bactéries Anaérobies non Toxigènes. Brussels: Presses Académiques Européennes.

Berg, J. N., Loan, R. W. 1975. *Fusobacterium necrophorum* and *Bacteroides melaninogenicus* as etiologic agents of footrot in cattle. American Journal of Veterinary Research **36:**1115–1122.

Beveridge, W. I. B. 1941. Foot rot in sheep: A transmissible disease due to infection with *Fusiformis nodosus* (n. sp.). Council for Scientific and Industrial Research, Commonwealth of Australia, Bulletin 140.

Bowden, G. H., Hardie, J. M. 1971. Anaerobic organisms from the human mouth, pp. 177–205. In: Shapton, D. A., Board, R. G., (eds.), Isolation of anaerobes. Society for Applied Bacteriology Technical Series No. 5. London, New York: Academic Press.

Brams, J., Pilot, I., Davis, D. J. 1923. Studies of fusiform bacilli and spirochetes. II. Their occurrence in normal preputial secretions and in erosive and gangrenous balanitis. Journal of Infectious Diseases **32:**159–166.

Bran, J. L., Levison, M. E., Kaye, D. 1972. Entrance of bacteria into the female urinary bladder. New England Journal of Medicine **286:**626–629.

Breed, R. S., Murray, E. G. D., Smith, N. R. (eds.). 1957. Bergey's manual of determinative bacteriology, 7th ed. Baltimore: Williams & Wilkins.

Bruner, D. W., Gillespie, J. H. 1973. Hagan's infectious diseases of domestic animals: With special reference to etiology, diagnosis, and biologic therapy, 6th ed. Ithaca, N.Y., London: Cornell University Press.

Cameron, G. R., Williams, F. E. 1926. An epidemic affecting stock rabbits. Journal of Pathology and Bacteriology **29:**185–188.

Castellani, A., Chalmers, A. J. 1919. Manual of tropical medicine. Baltimore: William Wood & Company.

Central Bureau of Statistics of Norway. 1976. Veterinary statistics 1975, Oslo.

Chow, A. W., Marshall, J. R., Guze, L. B. 1975. Anaerobic infections of the female genital tract: Prospects and perspectives. Obstetrical and Gynecological Survey **30:**477–494.

Dack, G. M., Dragstedt, L. R., Heinz, T. E. 1937. Further studies on *Bacterium necrophorum* isolated from cases of chronic ulcerative colitis. Journal of Infectious Diseases **60:**335–355.

Dack, G. M., Heinz, T. E., Dragstedt, L. R. 1935. Ulcerative colitis. Study of bacteria in the isolated colons of three patients by cultures and by inoculation of monkeys. Archives of Surgery **31:**225–240.

Davis, D. J., Pilot, I. 1922. Studies of *Bacillus fusiformis* and Vincent's spirochete. I. Habitat and distribution of these organisms in relation to putrid and gangrenous processes. Journal of the Americal Medical Association **79:**944–951.

Egerton, J. R., Roberts, D. S., Parsonson, I. M. 1969. The aetiology and pathogenesis of ovine foot-rot. I. A histological study of the bacterial invasion. Journal of Comparative Pathology and Therapeutics **79:**207–216.

Eggerth, A. H., Gagnon, B. H. 1933. The bacteroides of human feces. Journal of Bacteriology **25:**389–413.

Ellermann, V. 1904. Über die Kultur der fusiformen Bacillen. Centralblatt für Bakteriologie, Parasitenkunde und Infektionskrankheiten, Abt. 1 Orig. **37:**729–730.

Fales, W. H., Teresa, G. W. 1972a. A selective medium for the isolation of *Sphaerophorus necrophorus*. American Journal of Veterinary Research **33:**2317–2321.

Fales, W. H., Teresa, G. W. 1972b. Fluorescent antibody technique for identifying isolates of *Sphaerophorus necrophorus* of bovine hepatic abscess origin. American Journal of Veterinary Research **33:**2323–2329.

Felner, J. M., Dowell, V. R., Jr. 1971. "Bacteroides" bacteremia. American Journal of Medicine **50:**787–796.

Finegold, S. 1977. Anaerobic bacteria in human disease. New York, San Francisco, London: Academic Press.

Finegold, S. M., Attebery, H. R., Sutter, V. L. 1974. Effect of diet on human fecal flora: Comparison of Japanese and American diets. American Journal of Clinical Nutrition **27:**1456–1469.

Finegold, S. M., Sugihara, P. T., Sutter, V. L. 1971. Use of selective media for isolation of anaerobes from humans, pp. 99–108. In: Shapton, D. A., Board, R. G. (eds.), Isolation of anaerobes. Society for Applied Bacteriology Technical Series No. 5. London, New York: Academic Press.

Finegold, S. M., Bartlett, J. G., Chow, A. W., Flora, D. J., Gorbach, S. L., Harder, E. J., Tally, F. P. 1975a. Management of anaerobic infections. Annals of Internal Medicine **83:**375–389.

Finegold, S. M., Flora, D. J., Attebery, H. R., Sutter, V. L. 1975b. Fecal bacteriology of colonic polyp patients and control patients. Cancer Research **35:**3407–3417.

Fuller, R., Lev, M. 1964. Quantitative studies on some of the Gram-negative anaerobic bacteria in the pig alimentary tract. Journal of Applied Bacteriology **27:**434–438.

Garcia, M. M., Alexander, D. C., McKay, K. A. 1975. Biological characterization of *Fusobacterium necrophorum* cell fractions in preparation for toxin and immunization studies. Infection and Immunity **11:**609–616.

Garcia, M. M., Charlton, K. M., McKay, K. A. 1975. Characterization of endotoxin from *Fusobacterium necrophorum*. Infection and Immunity **11:**371–379.

Gorbach, S. L., Menda, K. B., Thadepalli, H., Keith, L. 1973.

Anaerobic microflora of the cervix in healthy women. American Journal of Obstetrics and Gynecology **117:**1053–1055.

Graham, N. P. H., Egerton, J. R. 1968. Pathogenesis of ovine foot-rot: The role of some environmental factors. Australian Veterinary Journal **44:**235–240.

Griffin, M. H. 1970. Fluorescent antibody techniques in the identification of the Gram-negative nonsporeforming anaerobes. Health Laboratory Science **7:**78–83.

Hadi, A. W., Russel, C. 1969. Fusiforms in gingival material. Quantitative estimations from normal individuals and cases of periodontal disease. British Dental Journal **126:**83–84.

Hamp, E. G., Mergenhagen, S. E. 1963. Experimental intracutaneous fusobacterial and fusospirochetal infections. Journal of Infectious Diseases **112:**84–99.

Hansen, S. L., Stewart, B. J. 1976. Comparison of API and Minitek to Center for Disease Control methods for the biochemical characterization of anaerobes. Journal of Clinical Microbiology **4:**227–231.

Hardie, J. M., Bowden, G. H. 1974. The normal flora of the mouth, pp. 47–83. In: Skinner, F. A., Carr, J. G. (eds.), The normal microbial flora of man. Society for Applied Bacteriology Symposium Series No. 3. London, New York: Academic Press.

Hill, G. B., Osterhout, S., Pratt, P. C. 1974. Liver abscess production by non-spore-forming anaerobic bacteria in a mouse model. Infection and Immunity **9:**599–603.

Hite, K. E., Hesseltine, H. C., Goldstein, L. 1947. A study of the bacterial flora of the normal and pathological vagina and uterus. American Journal of Obstetrics and Gynecology **53:**233–240.

Hofstad, T., Kristoffersen, T. 1971. Preparation and chemical characteristics of endotoxic lipopolysaccharide from three strains of *Sphaerophorus necrophorus*. Acta Pathologica et Microbiologica Scandinavica, Sect. B **79:**385–390.

Holdeman, L. V., Good, I. J., Moore, W. E. C. 1976. Human fecal flora: Variation in bacterial composition within individuals and a possible effect of emotional stress. Applied and Environmental Microbiology **31:**359–376.

Holdeman, L. V., Moore, W. E. C. 1972. Anaerobe laboratory manual. Blacksburg, Virginia: VPI Anaerobe Laboratory, Virginia Polytechnic Institute and State University.

Horne, H. 1898–1899. Renens klovsyge. Norsk Veterinaer-Tidsskrift **10–11:**97–110.

Hurst, V. 1957. *Fusiformis* in the infant mouth. Journal of Dental Research **36:**513–515.

Hussein, H. E., Shigidi, M. T. A. 1974. Isolation of *Sphaerophorus necrophorus* from bovine liver abscesses in the Sudan. Tropical Animal Health and Production **6:**253–254.

Jensen, C. O. 1913. Die vom Nekrosebacillus (Bacillus necroseos) hervorgerufenen Krankheiten, pp. 234–250. In: Kolle, W., von Wassermann, A. (eds.), Handbuch der pathogenen Mikroorganismen, vol. 6. Jena: Gustav Fischer Verlag.

Jensen, R., Flint, J. C., Griner, L. A. 1954. Experimental hepatic necrobacillosis in beef cattle. American Journal of Veterinary Research **15:**5–14.

Jensen, R., Mackay, D. R. 1965. Diseases of feedlot cattle. Philadelphia: Lea & Febiger.

Kanoe, M., Imagawa, H., Toda, M., Sato, A., Inoue, M., Yoshimoto, Y. 1976. Bacteriology of bovine hepatic abscesses. Japanese Journal of Veterinary Science **38:**263–268.

Kaufman, E. J., Mashimo, P. A., Hausmann, E., Hanks, C. T., Ellison, S. A. 1972. Fusobacterial infection: Enhancement by cell free extracts of *Bacteroides melaninogenicus* possessing collagenolytic activity. Archives of Oral Biology **17:**577–580.

Knorr, M. 1923. Über die fusospirilläre Symbiose, die Gattung *Fusobacterium* (K. B. Lehmann) und *Spirillum sputigenum*. II. Mitteilung. Die Gattung *Fusobacterium*. Zentralblatt für Bakteriologie, Parasitenkunde und Infektionskrankheiten, Abt. 1 Orig. **89:**4–22.

Krygier, G., Genco, R. J., Mashimo, P. A., Hausmann, E. 1973. Experimental gingivitis in *Macaca speciosa* monkeys: Clinical, bacteriological and histological similarities to human gingivitis. Journal of Periodontology **44**:454–463.

Langworth, B. F. 1977. *Fusobacterium necrophorum:* Its characteristics and role as animal pathogen. Bacteriological Reviews **44**:373–390.

Lewis, K. H., Rettger, L. F. 1940. Non-sporulating anaerobic bacteria of the intestinal tract. Journal of Bacteriology **40**:287–307.

Lewkowicz, X. 1901. Recherches sur la flore microbienne de la bouche des nourrissons. Archives de Médicine Experimentale et d'Anatomie Pathologique **13**:633–660.

Loeffler, F. 1884. Bacillus der Kälberdiphterie. Mittheilungen aus dem Kaiserlichen Gesundheitsamte **2**:493—499.

Loesche, W. J. 1969. Oxygen sensitivity of various anaerobic bacteria. Applied Microbiology **18**:723–727.

McCarthy, C., Snyder, M. L., Parker, R. B. 1965. The indigenous oral flora of man. I. The new-born to the 1-year-old infant. Archives of Oral Biology **10**:61–70.

McClung, L. J., Toabe, R. 1947. The egg yolk plate reaction for the presumptive diagnosis of *Clostridium sporogenes* and certain species of the gangrene and botulinum groups. Journal of Bacteriology **53**:139–147.

Maier, B. R., Flynn, M. A., Burton, G. C., Tsukakawa, R. K., Hentges, D. J. 1974. Effects of a high-beef diet on bowel flora: A preliminary report. American Journal of Clinical Nutrition **27**:1470–1474.

Mettam, R. W. M., Carmichael, J. 1933. Necrobacillosis in recently captured antelope in Uganda. Journal of Comparative Pathology and Therapeutics **46**:16–24.

Miller, W. D. 1889. Die Mikroorganismen der Mundhöhle. Die örtlichen und allgemeinen Erkrankungen welche durch dieselben hervorgerufen werden. Leipzig: Georg Thieme Verlag.

Moore, W. E. C. 1966. Techniques for routine culture of fastidious anaerobes. International Journal of Systematic Bacteriology **16**:173–190.

Moore, W. E. C., Holdeman, L. V. 1974. Human fecal flora: The normal flora of 20 Japanese-Hawaiians. Applied and Environmental Microbiology **27**:961–979.

Newsom, I. E. 1938. A bacteriologic study of liver abscesses in cattle. Journal of Infectious Diseases **63**:232–233.

Nolechek, W. F. 1918. Necrobacillosis in horses and mules. Journal of the American Veterinary Medical Association **54**:150–155.

Nord, C.-E., Dahlbäck, A., Wadstrøm, T. 1975. Evaluation of a test kit for identification of anaerobic bacteria. Medical Microbiology and Immunology **161**:239–242.

Ohtani, F. 1970a. Selective isolation media for strictly anaerobic, non-sporulating Gram-negative rods. Japanese Journal of Bacteriology **25**:222–232.

Ohtani, F. 1970b. Selective media for the isolation of Gram-negative anaerobic rods. Part II. Distribution of Gram-negative anaerobic rods in feces of normal human beings. Japanese Journal of Bacteriology **25**:292–299.

Omata, R. R., Disraely, M. N. 1956. A selective medium for oral fusobacteria. Journal of Bacteriology **72**:677–680.

Onderdonk, A. B., Bartlett, J. G., Louie, T., Sullivan-Seigler, N., Gorbach, S. L. 1976. Microbial synergy in experimental intra-abdominal abscess. Infection and Immunity **13**:22–26.

Panel Report. 1973. Foot rot among cattle. Modern Veterinary Practice **54**:63–65.

Parsonson, I. M., Egerton, J. R., Roberts, D. S. 1967. Ovine interdigital dermatitis. Journal of Comparative Pathology and Therapeutics **77**:309–313.

Peach, S., Fernandez, F., Johnson, K., Drasar, B. S. 1974. The non-sporing anaerobic bacteria in human faeces. Journal of Medical Microbiology **7**:213–221.

Plaut, H. C. 1894. Studien zur Bakteriellen Diagnostik der Diphterie und der Anginen. Deutsche Medizinische Wochenschrift **20**:920–923.

Pratt, J. S. 1927. On the biology of *B. fusiformis.* Journal of Infectious Diseases **41**:461–466.

Prévot, A. R. 1938. Études de systématique bactérienne. III.

Invalidité du genre *Bacteroides* Castellani et Chalmers. Demembrement et reclassification. Annales de l'Institut Pasteur **60**:285–307.

Prévot, A. R., Goret, P., Joubert, L., Tardieux, P., Aladame, N. 1951. Recherches bactériologiques sur une infection purulente d'allure actinomycosique chez le chat. Annales de l'Institut Pasteur **81**:85–88.

Prévot, A. R., Turpin, A., Kaiser, P. 1967. Les bactéries anaérobies. Paris: Dunod.

Pribram, E. 1929. A contribution to the classification of microorganisms. Journal of Bacteriology **18**:361–394.

Richardson, R. L., Jones, M. 1958. A bacteriolytic census of human saliva. Journal of Dental Research **37**:697–709.

Roberts, D. S. 1967a. The pathogenic synergy of *Fusiformis necrophorus* and *Corynebacterium pyogenes.* I. Influence of the leucocidal exotoxin of *F. necrophorus.* British Journal of Experimental Pathology **48**:665–673.

Roberts, D. S. 1967b. The pathogenic synergy of *Fusiformis necrophorus* and *Corynebacterium pyogenes.* II. The response of *F. necrophorus* to a filterable product of *C. pyogenes.* British Journal of Experimental Pathology **48**:674–679.

Roberts, D. S., Egerton, J. R. 1969. The aetiology and pathogenesis of ovine foot-rot. II. The pathogenic association of *Fusiformis nodosus* and *F. necrophorus.* Journal of Comparative Pathology and Therapeutics **79**:217–227.

Roberts, D. S., Graham, N. P. H., Egerton, J. R., Parsonson, I. M. 1968. Infective bulbous necrosis (heel abscess) of sheep, a mixed infection with *Fusiformis necrophorus* and *Corynebacterium pyogenes.* Journal of Comparative Pathology and Therapeutics **78**:9–17.

Schmorl, G. 1891. Über ein pyogenes Fadenbacterium (*Streptothrix cuniculi*). Deutsche Zeitschrift für Tiermedizin und Vergleichende Pathologie **17**:375–408.

Sebald, M. 1962. Étude sur les bactéries anaérobies gram-négatives asporulées. Thèse de l'Université Paris.

Shimada, K., Bricknell, K. S., Finegold, S. M. 1969. Deconjugation of bile acids by intestinal bacteria: Review of literature and additional studies. Journal of Infectious Diseases **119**:273–281.

Simon, P. C. 1975. A simple method for rapid identification of *Sphaerophorus necrophorus* isolates. Canadian Journal of Comparative Medicine and Veterinary Science **39**:349–353.

Simon, P. C., Stovell, P. L. 1969. Diseases of animals associated with *Sphaerophorus necrophorus:* Characteristics of the organism. Veterinary Bulletin **39**:311–315.

Simon, P. C., Stovell, P. L. 1971. Isolation of *Sphaerophorus necrophorus* from bovine hepatic abscesses in British Columbia. Canadian Journal of Comparative Medicine and Veterinary Science **35**:103–106.

Slanetz, L. W., Rettger, L. F. 1933. A systematic study of the fusiform bacteria. Journal of Bacteriology **26**:599–617.

Smith, L. D. 1975. The pathogenic anaerobic bacteria, 2nd ed. Springfield, Illinois: Charles C Thomas.

Spaulding, E. H., Rettger, L. F. 1937. The *Fusobacterium* genus. I. Biochemical and serological classification. Journal of Bacteriology **34**:535–548.

Starr, S. E., Thompson, F. S., Dowell, V. R., Jr., Balows, A. 1973. Micromethod system for identification of anaerobic bacteria. Applied Microbiology **25**:713–717.

Stauffer, L. R., Hill, E. O., Holland, J. W., Altemeier, W. A. 1975. Indirect fluorescent antibody procedure for the rapid detection and identification of *Bacteroides* and *Fusobacterium* in clinical specimens. Journal of Clinical Microbiology **2**:337–344.

Sutter, V. L., Sugihara, P. T., Finegold, S. M. 1971. Rifampin-blood-agar as a selective medium for the isolation of certain anaerobic bacteria. Applied Microbiology **22**:777–780.

Sutter, V. L., Vargo, V. L., Finegold, S. M., Bricknell, K. S. 1975. Wadsworth anaerobic bacteriology manual. Los Angeles: Department of Continuing Education in Health Sciences University Extension, and the School of Medicine, UCLA.

Sveen, K., Hofstad, T., Milner, K. C. 1977. Lethality for mice and chick embryos, pyrogenicity in rabbits and ability to gelate lysates from amoebocytes of *Limulus polyphemus* by lipopolysaccharides from *Bacteroides, Fusobacterium* and *Veillonella*. Acta Pathologica et Microbiologica Scandinavica, Sect. B **85**:388–396.

Syed, S. A. 1972. Biochemical characteristics of *Fusobacterium* and *Bacteroides* species from mouse cecum. Canadian Journal of Microbiology **18**:169–174.

Terada, A., Uchida, K., Mitsuoka, T. 1976. Die Bacteroidaceenflora in den Faeces von Schweinen. Zentralblatt für Bakteriologie, Parasitenkunde, Infektionskrankheiten und Hygiene, Abt. 1 Orig., Reihe A **234**:362–370.

Tissier, H. 1905. Répartition des microbes dans l'intestin du nourrison. Annales de l'Institut Pasteur **19**:109–123.

Trust, T. J., Bull, L. M., Currie, B. R., Buckley, J. T. 1979. Obligate anaerobic bacteria in the gastrointestinal microflora of the grass carp (*Ctenopharyngodon idella*), goldfish (*Carassius auratus*), and rainbow trout (*Salmo gairdneri*). Journal of the Fisheries Research Board of Canada **36**:1174–1179.

Ueno, K., Sugihara, P. T., Brichnell, K. S., Attebery, H. R., Sutter, V. L., Finegold, S. M. 1974. Comparison of characteristics of Gram-negative anaerobic bacilli isolated from feces of individuals in Japan and the United States, pp. 135–148. In: Balows, A., DeHaan, R. M., Dowell, V. R., Jr., Guze, L. B. (eds.), Anaerobic bacteria: Role in disease. Springfield, Illinois: Charles C Thomas.

Van Assche, P. F., Wilssens, A. T. 1977. *Fusobacterium perfoetens* (Tissier) Moore and Holdeman 1973: Description and proposed neotype strains. International Journal of Systematic Bacteriology **27**:1–5.

van Houte, J., Gibbons, R. J. 1966. Studies of the cultivable flora of normal human feces. Antonie van Leeuwenhoek Journal of Microbiology and Serology **32**:212–222.

Van Palenstein Helderman, W. H. 1975. Total viable count and differential count of *Vibrio (Campylobacter) sputorum, Fusobacterium nucleatum, Bacteroides ochraceus* and *Veillonella* in the inflamed and non-inflamed human gingival crevice. Journal of Periodontal Research **10**:294–305.

Veillon, A., Zuber, A. 1898. Recherches sur quelques microbes strictement anaérobies et leur rôle en pathologie. Archives de Médicine Expérimentale et d'Anatomie Pathologique **10**:517–545.

Vincent, H. 1896. Sur l'etiologie et sur les lesions anatomo-pathologique de la pourriture d'Hôpital. Annales de l'Institut Pasteur **10**:488–510.

Vincent, H. 1899. Recherches bactériologiques sur l'angine a *Bacillus fusiformis*. Annales de l'Institut Pasteur **13**:609–620.

Vincent, H. 1904. Étiologie de la stomatite ulcéro-membraneuse primitive. Comptes Rendus des Séances de la Societé de Biologie et de ses Filiales **56**:311–313.

Wahren, A., Bernholm, K., Holme, T. 1971. Formation of proteolytic activity in continuous culture of *Sphaerophorus necrophorus*. Acta Pathologica et Microbiologica Scandinavica, Sect. B **79**:391–398.

Wahren, A., Holme, T. 1973. Amino acid and peptide requirement of *Fusiformis necrophorus*. Journal of Bacteriology **116**:279–284.

Walker, C. B., Ratliff, D., Muller, D., Mandell, R., Socransky, S. S. 1979. Medium for selective isolation of *Fusobacterium nucleatum* from human periodontal pockets. Journal of Clinical Microbiology **10**:844–849.

Warner, F., Fales, W. H., Sutherland, M. C., Teresa, G. W. 1975. Endotoxin from *Fusobacterium necrophorum* of bovine hepatic abscess origin. American Journal of Veterinary Research **36**:1015–1019.

Weinberg, M., Nativelle, R., Prévot, A. R. 1937. Les microbes anaérobies. Paris: Masson et Cie.

Weiss, J. E., Rettger, L. F. 1937. The Gram-negative bacteroides of the intestine. Journal of Bacteriology **33**:423–434.

Werner, H. 1972a. A comparative study of 55 *Sphaerophorus* strains. Differentiation of 3 species: *Sphaerophorus necrophorus, Sph. varius* and *Sph. freundii*. Medical Microbiology and Immunology **157**:299–314.

Werner, H. 1972b. Anaerobierdifferenzierung durch gaschromatographische Stoffwechselanalysen. Zentralblatt für Bakteriologie, Parasitenkunde, Infektionskrankheiten und Hygiene, Abt. 1 Orig., Reihe A **220**:446–451.

Werner, H., Neuhaus, F., Hussels, H. 1971. A biochemical study of fusiform anaerobes. Medical Microbiology and Immunology **157**:10–16.

Werner, H., Pulverer, G. 1971. Haufigkeit und medizinische Bedeutung der eiterregenden *Bacteroides* und *Sphaerophorus*-arten. Deutsche Medizinische Wochenschrift **96**:1325–1329.

Williams, B. L., Pantalone, R. M., Sherris, J. C. 1976. Subgingival microflora and periodontitis. Journal of Periodontal Research **11**:1–18.

CHAPTER 119

The Genus *Leptotrichia*

TOR HOFSTAD

The oral cavity of man is the natural habitat of several different kinds of filamentous organisms, including the Gram-negative anaerobic bacterium *Leptotrichia buccalis*. This organism, which is the only species of the genus *Leptotrichia*, was among the first microorganisms to be recognizably described and drawn in the letters of Antonie van Leeuwenhoek.

The generic name *Leptotrichia* was used by Trevisan (1879) for filamentous organisms found in the human mouth. The species designation *buccalis* had been used several years earlier by Robin (1853), who used the name *Leptothrix buccalis* for filamentous forms that he had observed in wet mounts of tooth scrapings. Wherry and Oliver (1916) were able to cultivate the organism, which they called *Leptothrix innominata* in accordance with Miller (1889), but the first adequate description of *Leptotrichia buccalis* was given by Thjøtta, Hartmann, and Bøe in 1939. These authors, as well as Bøe (1941) and Bøe and Thjøtta (1944), found that the organism had much in common with the fusobacteria and, consequently, should be classified as a Gram-negative anaerobic bacterial species. Later investigators (Davis and Baird-Parker, 1959; Hamilton and Zahler, 1957; Kasai, 1961, 1965) confirmed and extended the cultural and biochemical findings of Thjøtta, Hartmann, and Bøe (1939), but were of the opinion that *L. buccalis* was a Gram-positive organism. However, the ultrastructure of *L. buccalis* is that of a typical Gram-negative bacterium (Hofstad and Selvig, 1969). *Leptotrichia* is now classified as the third genus in the family Bacteroidaceae. Its capacity to produce lactic acid as the only major acid from glucose clearly distinguishes it from *Bacteroides* and *Fusobacterium*. *L. buccalis* is identical with *L. innominata* of Prévot, Turpin, and Kaiser (1967).

In the past, *L. buccalis* has frequently been confused with *Fusobacterium* species (Hine and Berry, 1937; Spaulding and Rettger, 1937a,b). The name *Leptotrichia* has also been used (Bibby and Berry, 1939; Kligler, 1915; Morris, 1954) for the facultative Gram-positive organism termed *Leptotrichia*

dentium by Davis and Baird-Parker (1959) and now classified as *Bacterionema matruchotii* (Gilmour, Howell, and Bibby, 1961). A filamentous organism described by Theilade and Gilmour (1961) and provisionally named *Leptotrichia aerogenes* by Hofstad (1967) is identical with the Gram-positive organism *Eubacterium saburreum*.

Habitats

The oral cavity of man is the only habitat of *Leptotrichia buccalis* reported in the literature. Recently, the organism has been isolated from the female genitourinary tract (C.-E. Nord, personal communication). The principal source of the organism in the oral cavity is the dental plaque, i.e., the bacterial deposit that forms on the tooth surface and at the gingival margin. The concentration of *L. buccalis* in plaque material is uncertain. Slack and Bowden (1965) found the number of *L. buccalis* organisms in 1-day-old experimental plaque to be less than 0.01% of the total viable count; the number increased to 2.3% in 14-day-old plaque. Hillman, Van Houte, and Gibbons (1970) have shown that *L. buccalis* adheres to untreated enamel powder and to enamel powder coated with human saliva. However, the occurrence of the organism in the mouth is not solely dependent on tooth eruption, since *L. buccalis* can be isolated from the dorsum of the tongue, and, occasionally, from predentate infants (McCarthy, Snyder, and Parker, 1965).

Leptotrichia buccalis has no known pathogenicity for man, but being a highly saccharolytic organism it may participate in the development of tooth decay. The organism has cell wall lipopolysaccharide with the characteristics of an endotoxin (Gustafson et al., 1966). Antibodies reacting with this lipopolysaccharide are present in normal human sera (Falkler and Hawley, 1976; Mergenhagen, de Araujo, and Varah, 1965). These are IgM antibodies (Hawley and Falkler, 1976) and may be included among the so-called natural antibodies.

Isolation

Leptotrichia buccalis is a heterotrophic organism with unknown and presumably complex nutritional requirements. It grows well at 37°C on solid media like brucella blood agar and brain heart infusion agar. Incubation in an anaerobic atmosphere containing 5–10% CO_2 is needed for successful isolation and optimal growth. Upon repeated transfers, some strains grow under microaerophilic conditions. The addition of cysteine hydrochloride, a fermentable carbohydrate, and serum to ordinary standard media is essential for optimal growth in fluid cultures.

Sampling

Leptotrichia buccalis can be isolated from saliva or centrifuged salivary deposits, dental plaque, and the soft tissues of the mouth. It is best isolated from plaque between adjacent teeth (interstitial plaque) and the crevice or pocket that exists between the gingiva and the surface of the tooth. Plaque samples are taken by a sterile metal instrument that can be used to scrape the surface of the tooth. Sampling from the gingival crevice is best performed by the use of sterile filter paper points that are gently inserted into the crevice.

Centrifuged salivary deposits or plaque material can be inoculated onto the medium either directly or after being suspended and diluted in a reducing diluent such as the WAL diluent (Sutter et al., 1975) or the serum-containing diluent of Bowden and Hardie (1971). The inoculum on the tapering end of the paper point is streaked over a small area of the surface of the solid medium, and further spreading of the deposited material carried out by a sterile wire loop.

Isolation Under Nonselective Conditions

Because of its characteristic colonial morphology (see below), *Leptotrichia buccalis* can be isolated from mixed cultures on blood agar or brain heart infusion agar plates. Especially useful are the basal media described by Baird-Parker (1957; see this Handbook, Chapter 118) and Kasai (1961). The medium of Kasai has the following composition:

Tryptone	20.0 g/liter
Yeast extract	2.0 g/liter
Soluble starch	20.0 g/liter
K_2HPO_4	5.0 g/liter
Sodium chloride	5.0 g/liter
Cysteine hydrochloride	5.0 g/liter

The isolation of *L. buccalis* is indirectly favored by starch, which is an unavailable energy source for many of the other organisms present in the inoculated material. It is advisable to prepare 1:10 dilutions of the saliva or plaque samples and spread 0.01–0.05 ml on each freshly prepared plate. Incubation should be carried out at 37°C for 2 or more days in an anaerobic atmosphere containing 5–10% CO_2. The GasPak system can be used.

Selective Isolation

Addition to the dye-containing medium of Baird-Parker (1957; see this Handbook, Chapter 118) of 0.05 g per liter of sulfathiazole makes it selective for *Leptotrichia buccalis*. It is essential to check individual lot numbers of the ethyl violet dye for its suitability. The medium of Omata and Disraely (1956; see this Handbook, Chapter 118) is also useful. This medium, which is selective for the isolation of oral fusobacteria, also supports growth of *L. buccalis*. The growth is not quantitative on the se-

Table 1. Differentiation of *Leptotrichia buccalis* from *Fusobacterium nucleatum* and *Eubacterium saburreum*.

Characteristic	L. buccalis	F. nucleatum	E. saburreum
Microscopic morphology	Gram-negative,[a] 5–15 μm long, thick fusiform bacilli; filaments common	Gram-negative, 3 to 10–15 μm long, slender fusiform bacilli; filaments uncommon	Gram-positive,[b] 5–20 μm long bacilli, rounded or blunt ends; bulbous swellings and filaments common
Colonial morphology (blood agar)	Smooth, shiny, rhizoid, or convoluted	Smooth, convex, "flecked" appearance	Rough, rhizoid, adherent; or smooth, rhizoid or convoluted, nonadherent
Terminal pH from glucose	<5.4	>5.8	<5.4
Production of indole	–	+	+
Production of gas	–	–	+
Foul odor	–	+	–
Predominant fatty acids from glucose (GLC)	Lactic	Butyric	Acetic and butyric

[a] Often Gram-positive in young cultures.
[b] Often Gram-negative in old cultures.

lective media; therefore, the material used for inoculation may best be used dispersed and diluted either 1:10 or undiluted.

Axenic Cultivation and Maintenance

Leptotrichia buccalis can be maintained by weekly serial subculture on blood agar. Viable cells are stored at −70°C or, less safely, in the lyophilized state. Mass cultivation is performed in nutrient broth supplemented with yeast extract (0.3%), glucose (0.5%), cysteine hydrochloride (0.1%), and serum (5%). The use of well-filled, narrow-necked bottles or similar containers makes the use of prereduced anaerobically sterilized (PRAS) media unnecessary.

Identification

A reliable identification of *Leptotrichia buccalis* can be made from cellular and colonial morphology and, in addition, from a few biochemical tests (Table 1).

Young colonies of *L. buccalis* are colorless, smooth and often shiny, raised and with a filamentous edge; the colonies are described as "medusa-head" colonies. Following prolonged incubation, the filamentous edge may disappear and the surface becomes convoluted, resembling that of a human brain. The colonies are nonadherent to the medium and after 2 days of incubation on blood agar measure 2–3 mm in diameter. On the basal media of Kasai (1961) and Baird-Parker (1957) the colonies are smaller, but more distinct. Two- to three-day-old colonies on blood agar of the Gram-positive oral organism *Eubacterium saburreum* may be mistaken for *L. buccalis*. When grown on the selective medium of Omata and Disraely (1956), it is possible to confuse *L. buccalis* colonies with those of *F. nucleatum*. In Gram-stained specimens, *L. buccalis* is 0.8- to 1.5-μm-wide and 5- to 15-μm-long fusiform rods, commonly occurring in pairs with the adjacent ends flattened (Fig. 1). Filaments are seen in old cultures. The organism is Gram-negative, often with Gram-positive granules, but may be wholly Gram-positive in very young cultures. In young cultures, therefore, the cells of *L. buccalis* are not unlike those of *Eubacterium saburreum*. However *E. saburreum* never has tapered ends.

Leptotrichia buccalis has been identified in plaque material using the fluorescent-antibody technique (Baboolal, 1968).

Literature Cited

Baboolal, R. 1968. Identification of filamentous micro-organisms of the human dental plaque by immuno-fluorescence. Caries Research **2:**273–280.

Baird-Parker, A. C. 1957. Isolation of *Leptotrichia buccalis* and *Fusobacterium* species from oral material. Nature **180:**1056–1057.

Bibby, B. G., Berry, G. P. 1939. A cultural study of filamentous bacteria obtained from the human mouth. Journal of Bacteriology **38:**263–274.

Bøe, J. 1941. Fusobacterium: Studies on its bacteriology, serology and pathogenicity. Skr. Norske Videnskaps-Akademi. I. Matematisk-naturvitenskapelig klasse No. 9, Oslo, 1941.

Bøe, J., Thjøtta, T. 1944. The position of *Fusobacterium* and *Leptotrichia* in the bacteriological system. Acta Pathologica et Microbiologica Scandinavica **21:**441–450.

Bowden, G. H., Hardie, J. M. 1971. Anaerobic organisms from the human mouth, pp. 177–205. In: Shapton, D. A., Board, R. C., (eds.), Isolation of anaerobes, The Society for Applied Bacteriology Technical Series no. 5. London, New York: Academic Press.

Davis, G. H. G., Baird-Parker, A. C. 1959. *Leptotrichia buccalis*. British Dental Journal **106:**70–73.

Falkler, W. A., Jr., Hawley, C. E. 1976. Antigens of *Leptotrichia buccalis*: I. Their serologic reaction with human sera. Journal of Periodontal Research **10:**211–215.

Gilmour, M. N., Howell, A., Jr., Bibby, B. G. 1961. The classification of organisms termed *Leptotrichia* (*Leptothrix*) *buccalis*. I. Review of the literature and proposed separation into *Leptotrichia buccalis* Trevisan, 1879 and *Bacterionema* gen. nov., *B. matruchotii* (Mendel, 1919) comb. nov. Bacteriological Reviews **25:**131–141.

Gustafson, R. L., Kroeger, A. V., Gustafson, J. L., Vaichulis, E. M. K. 1966. The biological activity of *Leptotrichia buccalis* endotoxin. Archives of Oral Biology **11:**1149–1162.

Hamilton, R. D., Zahler, S. A. 1957. A study of *Leptotrichia buccalis*. Journal of Bacteriology **73:**386–393.

Hawley, C. E., Falkler, W. A., Jr. 1976. Antigens of *Leptotrichia buccalis*. II. Their reaction with complement fixing IgM in human sera. Journal of Periodontal Research **10:**216–223.

Hillman, J. D., Van Houte, J., Gibbons, R. J. 1970. Sorption of bacteria to human enamel powder. Archives of Oral Biology **15:**899–903.

Hine, M. K., Berry, G. P. 1937. Morphological and cultural studies of the genus *Fusiformis*. Journal of Bacteriology **34:**517–533.

Hofstad, T. 1967. An anaerobic oral filamentous organism possibly related to *Leptotrichia buccalis*. I. Morphology, some physiological and serological properties. Acta Pathologica et Microbiologica Scandinavica **69:**543–548.

Hofstad, T., Selvig, K. A. 1969. Ultrastructure of *Leptotrichia buccalis*. Journal of General Microbiology **56:**23–26.

Kasai, G. J. 1961. A study of *Leptotrichia buccalis*. I. Morphol-

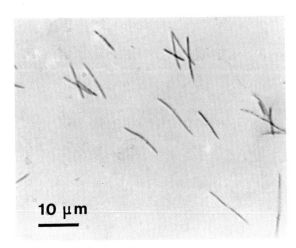

10 μm

Fig. 1. *Leptotrichia buccalis*. Blood agar plate.

ogy and preliminary observations. Journal of Dental Research **40**:800–811.

Kasai, G. J. 1965. A study of *Leptotrichia buccalis*. II. Biochemical and physiological observations. Journal of Dental Research **44**:1015–1022.

Kligler, I. J. 1915. A biochemical study and differentiation of oral bacteria with special reference to dental caries. Journal of the Allied Dental Societies **10**:282–330.

McCarthy, C., Snyder, M. L., Parker, R. B. 1965. The indigenous oral flora of man—I. The newborn to the 1-year-old infant. Archives of Oral Biology **10**:61–70.

Mergenhagen, S. E., de Araujo, W. C., Varah, E. 1965. Antibody to *Leptotrichia buccalis* in human sera. Archives of Oral Biology **10**:29–33.

Miller, W. D. 1889. Die Mikroorganismen der Mundhöhle. Die örtlichen und allgemeinen Erkrankungen welche durch dieselben hervorgerufen werden. Leipzig: Georg Thieme Verlag.

Morris, E. O. 1954. The bacteriology of the oral cavity. V. *Corynebacterium* and Gram-positive filamentous organisms. British Dental Journal **97**:29–36.

Omata, K. R., Disraely, M. N. 1956. A selective medium for oral fusobacteria. Journal of Bacteriology **72**:677–680.

Prévot, A. R., Turpin, A., Kaiser, P. 1967. Les bactéries anaérobies. Paris: Dunod.

Robin, C. 1853. Histoire naturelle des végétaux parasites qui croissent sur l'homme et sur les animaux vivants. Paris: J.-B. Baillière.

Slack, G. L., Bowden, G. H. 1965. Preliminary studies of experimental dental plaque *in vivo*, pp. 193–215. In: Hardwick, J. L., Held, H. R., König, K. G. (eds.), Advances in Fluorine Research and Dental Caries Prevention, vol. 3, Proceedings of the 11th Congress of the European Organization for Research on Fluorine and Dental Caries Prevention, Sandefjord, Norway 7th–9th July, 1964. Oxford-London-Edinburg-New York-Paris-Frankfurt: Pergamon Press.

Spaulding, E. H., Rettger, L. F. 1937a. The *Fusobacterium* genus. I. Biochemical and serological classification. Journal of Bacteriology **34**:535–548.

Spaulding, E. H., Rettger, L. F. 1937b. The *Fusobacterium* genus. II. Some observations on growth requirements and variation. Journal of Bacteriology **34**:549–563.

Sutter, V. K., Vargo, V. L., Finegold, S. M., Bricknell, K. S. 1975. Wadsworth anaerobic bacteriology manual. Los Angeles, California: Department of Continuing Education in Health Sciences University Extension, and the School of Medicine, UCLA.

Theilade, E., Gilmour, M. N. 1961. An anaerobic oral filamentous microorganism. Journal of Bacteriology **81**:661–666.

Thjøtta, T., Hartmann, O., Bøe, J. 1939. A study of *Leptotrichia trevisan*. History, morphology, biological and serological characteristics. Skr. Norske Videnskaps-Akademi. I. Matematisk-naturvitenskapelig klasse No. 5, Oslo, 1939.

Trevisan, V. 1879. Prime linee d'introduzione allo studio dei Batterj italiani. Rendiconti dell'istituto lombardo di scienze, Ser. 2 **12**:133–151.

Wherry, W. B., Oliver, W. W. 1916. *Leptothrix innominata* (Miller). Journal of Infectious Diseases **19**:299–303.

The Genera *Butyrivibrio, Succinivibrio, Succinimonas, Lachnospira,* and *Selenomonas*

ROBERT B. HESPELL and MARVIN P. BRYANT

Habitats

Butyrivibrio

Members of the genus *Butyrivibrio* are anaerobic, motile rods with tapered ends (Fig. 2A). They characteristically ferment glucose with butyric acid production. Butyrivibrios are common inhabitants of not only the bovine rumen, but probably are present in the rumens of all ruminants and have been repeatedly isolated from various species in many geographical areas. Butyrivibrios have also been shown to be present in high numbers (10^6–10^7 per g) in fecal material obtained from rabbits, horses, and humans (Brown and Moore, 1960). Members of this genus are quite versatile, as reflected by the rather large variations between strains in the energy sources used, fermentation products made, and nutritional growth requirements (Brown and Moore, 1960; Bryant and Robinson, 1962; Bryant and Small, 1956a), as well as in serological properties (Margherita and Hungate, 1963; Margherita, Hungate, and Storz, 1964). Most strains are able to degrade and ferment xylan, other hemicelluloses, and pectin. Some strains also digest cellulose. In addition, strains capable of anaerobically degrading complex heterocyclic compounds of the bioflavonoid type have been selectively isolated from the rumen (Cheng et al., 1969, 1970). This wide biochemical diversity of *Butyrivibrio* strains suggests these bacteria might be considered as the anaerobic equivalents of the metabolically diverse aerobic bacteria of the *Pseudomonas* genus. Along with contributing to the degradation of proteins, a major function of butyrivibrios appears to be the breakdown of starch and fibrous plant materials. *Butyrivibrio* fermentative activites provide relatively large amounts of butyric acid, which can be readily metabolized by both monogastric and polygastric animals.

Lachnospira, Succinivibrio, and Succinimonas

Bacteria of these genera are normally present as a small proportion of the total bacterial population in the bovine rumen of adult sheep and cattle fed a variety of diets. It is quite reasonable to suspect they also are present in the rumen of other ruminants and perhaps also in the cecum of herbivores. Under certain conditions, some members of these genera can be a major constituent of the bacterial flora of the rumen. *Lachnospira* are the main pectin fermenters in animals on diets high in pectin, e.g., legume pasture or bloat-provoking ladino clover. *Lachnospira* strains accounted for 16–31% of the total bacterial isolates obtained from rumen contents (Bryant et al., 1960). *Succinivibrio* species characteristically ferment dextrin and partially hydrolyze starch. These organisms are commonly found in bovine ruminants fed high grain diets and may represent between 13 and 25% of the total isolates from rumen contents of animals being fed high grain rations (Bryant, Robinson, and Lindahl, 1961; Wozny et al., 1977). The functional role of *Succinimonas* in the bovine rumen is in the fermentation of starch and/or its related hydrolytic products; but other starch-using bacteria (e.g. *Bacteroides amylophilus, Bacteroides ruminicola, Butyrivibrio fibrisolvens*) invariably outnumber *Succinimonas*. Succinimonads are usually not found in animals on high or wholly roughage rations (alfalfa silage; wheat straw), but are found in significant numbers when some grain is included in these diets (Bryant et al., 1958).

Selenomonas

Selenomonads are morphologically distinct bacteria that often can be easily differentiated from other bacteria by light microscopic observations with wet mount preparations made from natural habitats. The organisms appear as kidney-crescent- or helically shaped cells (Fig. 2E) which have an active, tumbling type of motility. Selenomonads have been observed and/or isolated primarily from three habitats: the rumen, the human mouth, and the ceca of rodents. Traditionally, speciation within *Selenomonas* has been based upon the habitat from which the strain was isolated, but this criterion, although helpful, has been largely supplanted by much more adequate cytological and biochemical criteria (see Identification section, Kingsley and Hoeniger, 1973).

Selenomonads from the rumen are usually considered and shown to be strains of *S. ruminantium*. Rumen selenomonads are routinely observed and isolated from rumen contents of cows and sheep that have been fed one of a wide variety of rations (Bryant, 1956; Hobson and Mann, 1961; Prins, 1971). Generally, they are more numerous in animals being fed rations containing more rapidly fermentable carbohydrates such as high grain as opposed to silage or straw rations (Caldwell and Bryant, 1966). The general functional role of rumen selenomonads is the fermentation of soluble carbohydrates and starch to mainly acetic, propionic, and lactic acids plus CO_2 and traces of H_2. However, rumen selenomonads also have other, and perhaps equally important, specialized roles. The studies indicate rumen selenomonads are among the most important members of the glycerol-fermenting flora of the rumen of sheep (Hobson and Mann, 1961) and cattle (Bryant, 1956). Many rumen selenomonad strains can ferment lactate, a compound not readily fermented by many rumen bacteria. These lactate-fermenting strains are usually designated as *S. ruminantium* subspecies *lactilytica*. Selenomonads also grow well at lower pH values than many other rumen bacteria can; thus, under conditions such as bloat and/or high grain feeding, which can lead to a lower ruminal pH and increase of lactic acid formation, a substantial increase in the number of rumen selenomonads is a common occurrence. Finally, although little is known about the bacterial species involved in the hydrolysis of urea in the rumen (Wozny et al., 1977), urease-producing strains of *S. ruminantium* have been shown to be in high numbers in rumen contents when enumerated on media containing urea as the major added nitrogen source (John et al., 1974). Many selenomonad strains are also capable of catabolizing certain amino acids (John, Isaacson, and Bryant, 1974).

Selenomonads were probably first observed by Leeuwenhoek in gingival scrapings from the human mouth (Dobell, 1932). Originally designated as *Spirillum sputigenum* by Miller (1887) and by subsequent workers, the present classification system places these organisms in the genus *Selenomonas* as *S. sputigena* (Buchanan and Gibbons, 1974). *Selenomonas sputigena* is part of the normal indige-nous microflora of the human gingival crevices and often is more abundant in those persons having clinically detectable gingivitis or periodontal disease. However, no specific functional roles can presently be assigned to *S. sputigena* in these diseases or in the general biochemical ecology associated with the dental microflora other than fermentation of carbohydrates to propionic and acetic acids.

Gram-negative organisms having the cell morphology and flagellar arrangement typically found with selenomonads have been observed in polluted river water (Leifson, 1960) and in the ceca of several kinds of animals. They have been enumerated and isolated from the cecum contents of the thirteen-lined ground squirrel *(Citellus tridecemlineatus)* and were shown to be present in numbers as high as 10^9–10^{10} per g of contents (Barnes and Burton, 1970). The selenomonads constituted a major portion (18% or more) of the total viable cell counts obtained from the cecum contents of both active and hibernating ground squirrels. All strains fermented glucose and some strains could, in addition, ferment starch, but further biochemical characteristics of the strains were not examined. Obviously then, selenomonads are of major importance to the overall microbial ecology and the fermentation of organic matter in the ground squirrel cecum, but their specific contributions have yet to be established.

The cecum of rats can also harbor large numbers of selenomonads. The study by Ogimoto (1972) indicated that about 5% of the bacterial colonies from 10^8 to 10^9 dilutions of rat cecum contents were selenomonads. Fourteen strains were isolated on a medium similar to NS-SD medium (see Isolation section). All of the strains fermented any one of several carbohydrates, including cellobiose, and propionate was a major fermentation product. The volatile fatty acids found in the ceca of eighteen individual rats were also analyzed and the data indicated that propionate along with acetate and butyrate were the major fermentation acids present, along with smaller amounts of *n*-valerate and isovalerate (Ogimoto, 1972). In addition, by using freeze-fracture electron microscopic techniques, selenomonads were seen to colonize the rat cecum submucosa. Overall, it is apparent that selenomonads must contribute significantly to the microbial fermentative activity in the rat cecum. However, their precise contributions with respect to a number of parameters, such as variation in numbers with dietary conditions or with rat strains, have yet to be elucidated.

In addition to squirrels and rats, selenomonads have been observed by many investigators to be in the cecal contents of guinea pigs. It is clear from the published photomicrographs (e.g., Kingsley and Hoeniger, 1973; Robinow, 1954) that the observed organisms are in fact *Selenomonas* species. These organisms were designated as *Selenomonas palpitans* (Simons, 1922), but since the organisms have never been isolated or grown in pure culture, this classification is doubtful (Buchanan and Gibbons, 1974). For the same reasons, determination of the fermentative and other roles of selenomonads in the guinea pig cecum as well as extensive comparisons to *S. ruminantium* or *S. sputigena* are not yet possible. Nevertheless, electron micrographs of *"S. palpitans"* (Kingsley and Hoeniger, 1973) suggest that this organism does differ in some respects in overall cellular morphology and flagellar arrangement from the former two species.

Isolation

Anaerobic Techniques

Most investigators will agree that the isolation, identification, and manipulation of strictly anaerobic bacteria require methodology which has two basic tenets: (i) the removal and elimination of oxygen from all environments to which the anaerobic bacteria are exposed, and (ii) the maintenance in liquid or solid culture media of low oxidation-reduction potentials, usually by use of chemical agents such as glutathione, L-cysteine, dithiothreitol, thioglycolate, or sodium sulfide. It has generally been thought that the exclusion of oxygen is the more important aspect and this has been confirmed by the results of some recent studies on some anaerobic intestinal bacteria including *Bacteroides* species (Onderdonk et al., 1976; Walden and Hentges, 1975). A similar or greater sensitivity to oxygen probably exists with many, if not all, strictly anaerobic bacteria present in ecosystems such as the rumen and the gastrointestinal tracts of mammals. Thus, it cannot be overemphasized that a major technical prerequisite to isolating strictly anaerobic bacteria is the exclusion of oxygen from all environments to which these bacteria may be exposed.

The majority of the various methodologies used for the isolation and cultivation of anaerobes incorporate some aspects of the anaerobic techniques described almost 30 years ago by Hungate (1950). The information contained in the more recent excellent article by Hungate (1969) on the general theory and methodology of these anaerobic techniques, along with the improvements made on the original method indicated by Bryant (1972), will give the interested investigator a technical background for cultivating not only the bacteria described in this chapter, but also most other anaerobes. From several standpoints, we recommend the Hungate anaerobic techniques with the Bryant modifications (see Fig. 1A–F) as still being the best ones available: (i) the equipment needs are simple and relatively inexpensive, (ii) the methods employed are extremely effective in removing oxygen rapidly and are routinely used for the growth of the strictest anaerobes

known—the methanogenic bacteria (e.g., Edwards and McBride, 1975; Paynter and Hungate, 1968; Smith, 1966), and (iii) they can be easily modified to adapt them to the specific requirements of many bacteria or for clinical studies (see Holdeman and Moore, 1977).

With respect to this last point, the Hungate techniques have been modified in numerous ways. For work with methanogenic bacteria, needles and syringes have been used (Paynter and Hungate, 1968; Smith, 1966). Miller and Wolin (1974) have described a methodology based on the use of serum bottles and hypodermic syringes. The methanogens can be grown using glove box techniques (Edwards and McBride, 1975), and Balch and Wolfe (1976) have developed a system for growing these organisms under a pressure of 2–3 atm. In recent years, there has been an increasingly wider use of anaerobic glove boxes for handling anaerobic bacteria. Glove boxes of the type made of flexible vinyl plastic as described by Aranki et al. (1969) and Aranki and Freter (1972) are available commercially (Coy Manufacturing Company, Ann Arbor, Michigan, U.S.A.). Clearly, the glove box system is vastly superior to the use of Brewer jars or their modern equivalents, since the bacteria in these latter systems are exposed to relatively high oxygen concentrations before complete anaerobiosis is attained. Although the glove box system is considerably more expensive than the equipment needed for the Hungate technique, it does have certain advantages: (i) the ability to carry out with ease and rapidity many of the standard "bench top" bacterial manipulations; (ii) it enables the researcher to pour and utilize standard agar plates for cloning of bacteria as well as for plating of bacteria that cannot survive the molten agar used with shake or roll tubes; (iii) provision of a large-scale anaerobic chamber for operating microbiological or biochemical equipment such as a chemostat or fraction collector. Experience in our laboratories and those of others has indicated that the maximum benefit can be obtained by combining various modifications of both the Hungate techniques and the glove box system. For instance, this combination of techniques has been found to be excellent for the cultivation of methanogenic bacteria (Balch and Wolfe, 1976; Edwards and McBride, 1975). In our laboratories, using standard Hungate techniques (Bryant, 1972) we routinely prepare media in round-bottom flasks and make the appropriate additions to the flask after sterilization by autoclaving (Fig. 1A–F). After the wires are removed and the rubber stopper taped into place, the flask is brought into the glove box through the entry lock. Subsequently, within the glove box, agar plates are poured and allowed to solidify. The moisture introduced into the glove box atmosphere (90:10 vol/vol Ar:H_2 or N_2:H_2) is removed by trays of silica gel which can be reused after drying in an external oven

(150°C, 3–6 h). The plates are then appropriately inoculated and transferred to an incubation chamber composed of a modified food pressure cooker which, in turn, is provided with the correct atmosphere by use of a gassing manifold as described by Balch and Wolfe (1976). The pressure cooker chamber may be incubated in an incubator housed within the glove box or be removed to an external incubator. Most operations in the glove box requiring the use of pipettes can easily be done through the use of automatic pipettes having disposable tips (e.g., Pipetman or Eppendorf types). The tips are autoclaved in standard test tubes with metal closures and are brought into the glove box at least 24 h prior to use to remove the trace amounts of oxygen always present in or on the surface of all plasticware. Many other combinations of glove box/Hungate techniques are possible, depending upon the particular needs of the laboratory work and the inventiveness of the investigator.

General Isolation Principles

The two growth media and appropriate solutions described below are prepared using the previously described anaerobic techniques. Both nonselective media are of a general nature and can be used directly or with slight modifications to isolate and grow a wide variety of anaerobic bacteria (besides those discussed in this chapter) present in highly anaerobic habitats such as the rumen, the gastrointestinal tract of animals, sewage sludge digesters, or aquatic sediments. These bacterial isolates can include species of *Bacteroides* (Betian et al., 1977; Dehority and Grubb, 1977), *Ruminococcus* (Bryant and Burkey, 1953; Herbeck and Bryant, 1974), anaerobic spirochetes (Bryant, 1952), phototrophs (Uffen, Sybesma, and Wolfe, 1971), sulfate-reducing bacteria (Bryant et al., 1977), and all types of bacteria in the gastrointestinal tract (Holdeman and Moore, 1977). The rumen fluid–containing medium is preferable for initial isolations, since the rumen fluid provides many defined and undefined nutrients which often may be needed by some strains that may be unusually fastidious upon initial isolation. However, rumen fluid may not always be available to the investigator, and in most instances it can be replaced by the appropriate supplements as indicated in the formulation of the non-rumen-fluid medium (NS-SD) given below. This type of medium is quite suitable for the isolation and enumeration of many predominant bacteria in the rumen (Caldwell and Bryant, 1966) and in human feces (Eller, Crabill, and Bryant, 1971).

At the present time, no really effective procedures are available, except for the isolation of *Selenomonas ruminantium,* to selectively isolate the various bacterial species discussed in this chapter.

Isolation of a particular species depends mainly upon the picking of a colony having the appropriate colony and bacterial cell characteristics (see Identification section) when grown on a nonselective medium (e.g., NS-CRF or NS-SD) in anaerobic roll tubes or agar plates. Slight modifications, such as the type of carbohydrates added to the two nonselective media below can be of some aid in the isolation. These modifications will be discussed in the Identification section.

Selective Isolation of *Selenomonas*

Enrichment and selective procedures for the isolation of selenomonads have been developed only for the rumen organisms that are considered a single species, *S. ruminantium* (Buchanan and Gibbons, 1974). As described by Tiwari, Bryant, and Wolfe (1969), rumen selenomonads can be selectively isolated using a medium (SS medium) with mannitol as the only added carbohydrate. The selective factors of the SS medium are: (i) the use of mannitol as the main energy source, since few rumen bacteria can ferment this sugar; (ii) the lower pH value of 6.0 does not greatly affect selenomonad growth, but is inhibitory to growth of many rumen bacteria; and (iii) the medium contains no added heme or branched-chain volatile acids, and either or both of these compounds are essential for growth of some of the more numerous rumen bacteria. Substitution of glycerol or lactate for mannitol in the SS medium should allow enrichment or even selective isolation of glycerol-fermenting selenomonads that are regarded as *S. ruminantium* variety *lactilytica* (Hobson and Mann, 1961). Since most selenomonad cells are relatively large, some physical procedures such as differential centrifugation (Prins, 1971) can be employed to increase the numbers of selenomonads relative to other bacteria in an inoculum obtained from the rumen or other sources. Alternatively to or in conjunction with differential centrifugation procedures, smaller bacteria can be removed by filtering the suspension through cellulose ester filter disk (pore diameter of ca. 0.6 μm). However, both of these physical selective methods are of limited value and are not applicable for quantitative recoveries.

Although no strictly selective media are known for the isolation of oral selenomonads, enrichment of these organisms is possible. MacDonald and Madlener (1957) examined several methods for isolating the oral selenomonad *S. sputigena* (*Spirillum sputigenum*). These authors found that the inclusion of 0.01% (wt/vol) sodium lauryl sulfate or 0.15% (wt/vol) sodium oleate in a complex medium (SLS medium) that also contains 10% (vol/vol) sheep serum did not affect the growth of *S. sputigena*, but markedly inhibited the growth of other microorganisms present in gingival scrapings. Using media with 1% (wt/vol) agar, *S. sputigena* growth appears as a film of spreading surface growth in which one or more contaminating organisms may also be present. Cloned and pure selenomonad strains can be obtained by picking from the outermost edge of this film and subculturing by streaking onto agar plates or by dilution through roll tubes. Both the serum and fatty acid components of the medium are of value for the initial isolation of oral selenomonad strains, but these components are not required for *S. sputigena* growth; Kingsley and Hoeniger (1973) have shown that several oral selenomonad strains grow very well in MPB broth medium which is similar to the NS-SD medium given below.

Selenomonads present in the guinea pig cecum have been called *Selenomonas palpitans* (Simons, 1922), but this species has never been grown in pure culture. The selenomonads present in the cecum of ground squirrels have been isolated in high numbers and grown axenically (Barnes and Burton, 1970) by using medium 10 (Caldwell and Bryant, 1966), which is similar to the NS-SD medium indicated below. Selenomonads in the rat cecum have been studied and enumerated by Ogimoto (1972), who used nonselective isolation methods and a medium similar to NS-SD, but without added volatile acids.

Preparing Growth Media for Cultivating *Butyrivibrio, Succinivibrio, Succinimonas, Lachnospira,* and *Selenomonas*

1. Basic minerals solution:

KH$_2$PO$_4$	18 g
NaCl	18 g
CaCl$_2$	0.4 g
MgCl$_2$ · 6H$_2$O	0.4 g
MnCl$_2$ · 6H$_2$O	0.2 g
CoCl$_2$ · 6H$_2$O	0.2 g

Dissolve above ingredients in distilled water and bring to final volume of 1 liter. Completely stable when stored at 5°C.

2. Trace minerals:

ZnSO$_4$ · 7H$_2$O	20 mg
CuSO$_4$ · 5H$_2$O	5 mg
AlK(SO$_4$) · 12H$_2$O	2 mg
H$_3$BO$_3$	10 mg
NaMoO$_4$ · 2H$_2$O	10 mg
Na$_2$SeO$_3$	10 mg
NiCl$_2$ · 6H$_2$O	5 mg
Na$_2$EDTA	50 mg

Dissolve each of the ingredients separately in distilled water, combine in the indicated order, and bring to a final volume of 100 ml with distilled water. Stable at room temperature.

3. Volatile fatty acid (VFA) solution:

Acetic acid	17 ml
Propionic acid	6 ml
n-Butyric acid	4 ml
iso-Butyric acid	1 ml
n-Valeric acid	1 ml
iso-Valeric acid	1 ml
DL-α-Methylbutyric acid	1 ml

The VFA mixture is prepared using reagent-grade acids and after combining, the mixture is taken to pH 7.0 with aqueous NaOH and brought to a final volume of 100 ml with distilled water. The mixture is stable at 5°C.

4. Cysteine-sulfide reducing mixture:

L-Cysteine · HCl	2.5 g
Na$_2$S · 9H$_2$O	2.5 g

The L-cysteine · HCl is dissolved in distilled water and pH is adjusted to 10.0 with NaOH. The Na$_2$S · 9H$_2$O is added and, after dissolution, the mixture is brought to a final volume of 100 ml with distilled water. The mixture is equilibrated with under O$_2$-free N$_2$ prior to sterilization by autoclaving. This mixture is stable for at least 4–5 weeks at room temperature when kept O$_2$-free, but should be discarded when a white precipitate (L-cysteine) forms.

5. Resazurin solution:

Resazurin	100 mg
Distilled water	100 ml

This solution is quite stable at room temperature. Resazurin serves as an oxidation-reduction indicator for the culture media, being colorless below and pink/red above about a −47 mV potential at pH 7.0 (Hungate, 1966).

6. Na$_2$CO$_3$ solution:

Na$_2$CO$_3$	
Distilled water	100 ml

This solution is equilibrated with and dispensed under O$_2$-free CO$_2$ after sterilization by autoclaving. Stable.

7. Hemin solution:

Hemin	100 mg
50% (vol/vol) ethanol	50 ml
0.05 N NaOH	50 ml

This solution is stable for several months when kept at 5°C.

8. Clarified rumen fluid:
Rumen contents are collected via a stomach tube or through a fistula from the ruminant animal. The animal should be fed on a diet of high quality hay with limited grain and the rumen contents collected about 6 h or more after feeding. The contents are filtered through two layers of cheesecloth and the filtrate centrifuged (16,000 × *g*, 30 min, 15–22°C) to remove the smaller plant particles and microorganisms. The filtration and centrifugation should be done shortly after col-

lection. The resultant supernatant fluid is the clarified rumen fluid and is autoclaved after dispensing in screw-cap bottles. This clarified rumen fluid is stored refrigerated and prior to its addition to media, any precipitated material is removed by centrifugation.

Some investigators prefer to use in media rumen fluid that still contains many of the indigenous bacteria, but which has been freed of the larger protozoa and forage particles by centrifugation (1,000 × g, 10 min). Due to the turbidity of the resultant medium with this type of rumen fluid, roll tubes should be made (see below) with about 4–5 ml of medium per tube; otherwise the small, pinpoint colonies cannot be readily detected. Although some studies have shown higher colony counts with media having bacteria-containing rumen fluid (Grubb and Dehority, 1976), other studies have indicated that deletion of the bacteria in the rumen fluid does not decrease the colony counts (Bryant and Robinson, 1961; Thorley, Sharpe, and Bryant, 1968).

9. Mineral solution S:

KH_2PO_4	12 g
$(NH_4)_2SO_4$	6 g
NaCl	12 g
$MgSO_4 \cdot 7H_2O$	2.5 g
$CaCl_2 \cdot 2H_2O$	1.6 g

Dissolve the above ingredients in distilled water and bring to a final volume of one liter. Completely stable when stored at 5°C.

Preparing Nonselective Isolation Media with (NS-CRP) or without (NS-SD) Clarified Rumen Fluid

The component solutions are mixed in the following order:

	Medium	
Component	With clarified rumen fluid (NS-CRF)	Without clarified rumen fluid (NS-SD)
Clarified rumen fluid	40 ml	—
Glucose	0.05 g	0.05 g
Cellobiose	0.05 g	0.05 g
Soluble starch	0.05 g	0.05 g
Xylose	0.05 g	0.05 g
Basic minerals solution	5 ml	5 ml
Trace minerals solution	—	0.5 ml
Resazurin solution	0.1 ml	0.1 ml
VFA solution	—	1.0 ml
Hemin solution	—	0.1 ml
Yeast extract	—	0.05 g
Trypticase	—	0.2 g
$(NH_4)_2SO_4$	0.05 g	0.05 g

Agar	2.0 g	2.0 g
Distilled water	69 ml	88 ml
Na_2CO_3 solution	5.0 ml	5.0 ml
Cysteine-sulfide solution	1.0 ml	1.0 ml

The preparation of the anaerobic media is done using the Hungate techniques (Hungate, 1969) as modified by Bryant (1972); these articles should be consulted before attempting the following routines. All of the medium ingredients, except the reducing agents, Na_2CO_3, and heat-labile components, are dissolved in distilled water and the mixture is adjusted to pH 6.5 with NaOH or KOH. The medium is made to volume (minus the total volume of the ingredients to be added after autoclaving) and placed into a round-bottom flask having a volume one-third to one-half greater than the medium. A stream of the appropriate O_2-free gas (usually CO_2) is passed (ca. 500 cc/min) through the flask via a bent gassing needle hung on the lip of the flask's neck while the medium is brought to boiling for 2–3 min over a flame. The flask is removed from the flame and immediately stoppered with a black rubber stopper while the gassing needle is concomitantly removed so that no air enters the flask (Fig. 1A,B). For agar-containing media, the medium can be first heated to near boiling in the flask in a steamer prior to gassing of the flask and boiling. The stopper is wired into place, the medium is autoclaved, and then the flask is allowed to cool to about 55°C or less (or to 47°C if agar is present). The flask is aseptically opened with concomitant insertion of the gassing needle into the neck, so that the small amount of air which enters upon opening is quickly flushed out and air is kept out of the flask during subsequent manipulations by continual flushing with the O_2-free gas. The sterile O_2-free solutions of cysteine-sulfide, Na_2CO_3, or other ingredients are then added to the medium using a sterile cotton-plugged glass pipet having an attached mouth tube. The medium is then anaerobically dispensed using this pipet into sterile rubber-stoppered glass tubes (Fig. 1C–F). For roll tubes or a liquid medium requiring a large gas phase (e.g., for methanogenic bacteria), we use Bellco 18-×-150-mm disposable culture tubes with black No. 1 stoppers, containing 5 to 9 ml of medium. For slants and routine cultures, 13-×-100-mm disposable tubes with No. 00 stoppers and containing about 3–4 ml of medium are used.

Quantitative Enumeration and Isolation of Rumen and Intestinal Bacteria

For quantitative enumeration and isolation of bacteria, the particular sample (e.g., rumen contents or feces) can be first mixed in a Waring blender under an O_2-free gas with an anaerobic

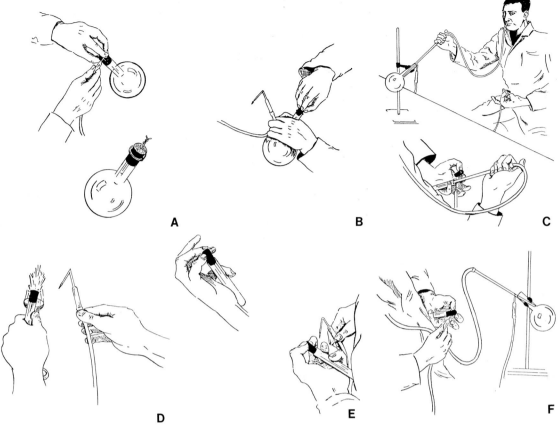

Fig. 1. Manipulation used in anaerobic techniques. (A) Position of hands during closing of flask containing medium and appearance of flask prepared for autoclaving; (B) aseptic removal of stopper from autoclaved flask; (C) transfer of liquid from flask to test tube; (D) holding and flaming of test tube and gassing pipette prior to opening of test tube; (E) removal of stopper (left) prior to insertion of gassing pipette (right) into test tube; and (F) Removal of gassing pipette and insertion of stopper after transfer of liquid into test tube. [From Caldwell, 1965, with permission.]

dilution solution (i.e., NS-CRF or NS-SD medium minus the rumen fluid and carbohydrates) and then serially diluted in this solution. Anaerobic plates or roll tubes are inoculated from these dilutions such that 15–100 colonies are obtained per plate or tube. For roll tubes, the tubes are rolled under cold water immediately after inoculation to minimize any possible heat stress to the bacteria caused by the temperature of the molten agar. For accurate enumeration, the colony counts should be made after 3–7 days of incubation (Bryant and Robinson, 1961). The colonies can be picked and transferred using a bent inoculating needle made of platinum-iridium alloy wire. In the primary isolation, the colony transfers should be made as stabs to the base of a slant of the appropriate medium since the growth of extreme anaerobes often can fail with primary transfers to broth. Cloning and culture purity can then be established by diluting into or streaking onto further roll tubes (Holdeman and Moore, 1977) or streaking anaerobic plates. For colonies that are very difficult to transfer, bent Pasteur pipettes with a mouth tube can be used (see Hungate, 1950).

Preparing Selective Medium for Isolation of *Selenomonas* (Selenomonad Selective Medium-SS Medium)

Mannitol	0.2 g
Trypticase	0.5 g
Yeast extract	0.1 g
Sodium acetate	0.1 g
$FeSO_4 \cdot 7H_2O$	0.1 g
L-Cysteine · HCl	0.08 g
Mineral solution S	4 ml
n-Valeric acid	0.05 ml
Distilled water	93.5 ml
8% wt/vol Sodium carbonate	2.5 ml

This SS medium has a final pH of 5.9–6.1 after the sodium carbonate solution (sterilized separately and added to the medium after autoclaving) is added. The medium is prepared, dispensed, and used under O_2-free carbon dioxide as described above.

Maintenance of Cultures

Stock cultures may be kept as stabs into the base of agar slants of NS-CRF or NS-SD media containing 1% wt/vol agar and 0.1% wt/vol of each carbohydrate. After inoculation, the slants are incubated at 37°C until visible growth is evident. The slants can be subsequently stored at room temperature and transferred to new slants within 6 days. Long-term storage of cultures may be accomplished by lyophilization or by storage of the slants in an ultra-low-temperature freezer (-60°C or more) and transferred at 12-month intervals to fresh slants. Alternatively, viable cell suspensions can be maintained for at least several years without transfer by storage in ampules kept in liquid nitrogen (Hespell and Canole-Parola, 1970). For liquid nitrogen storage, cells are harvested by centrifugation from broth cultures in the middle or late exponential growth phase. The cell pellets are resuspended in 5–10 times their pellet volume of uninoculated media, which has been supplemented (5% vol/vol) with filter-sterilized dimethylsulfoxide. Aliquots (0.5–1 ml) are dispensed anaerobically via syringes into ampules which are then sealed with a torch or with serum caps. The ampules are placed onto metal holders or "canes" and immersed in 95% ethanol contained in a graduated cylinder. Ampules showing the presence of a leak, as indicated by an efflux of gas and/or filling with ethanol, are discarded. The ampules in the ethanol bath are cooled for 8–16 h in a freezer (-30°C to -90°C) and immediately transferred in their holders to a liquid nitrogen refrigerator (-196°C). For recovery, the ampule is quickly removed from the liquid nitrogen refrigerator and immediately immersed in 500 ml of 32–35°C water to insure rapid thawing. The thawed ampule is immediately opened, an aliquot removed via a syringe, and a serial dilution in 10-fold steps to ca. 10^{-4}–10^{-5} in fresh broth medium is made. Generally, actively growing cultures are obtained from overnight incubation of the dilution tubes. If contamination of the stock culture is suspected, an aliquot from one of the dilution tubes made be used to streak an agar plate or to inoculate a tube of agar medium which in turn can be used to make a roll tube or pour plate.

Identification

Butyrivibrio

At the present time, no clear-cut selective growth media or selective procedures exist for the isolation of Butyrivibrio species. However, several possibilities exist for enhancement of the number of Butyrivibrio colonies in respect to total colonies observed on certain types of semiselective media. The observation that the majority of rumen bacterial strains capable of degrading the bioflavonoid rutin were Butyrivibrio strains (Cheng et al., 1969) suggests that the use of rutin in enrichment cultures or as the sole added carbohydrate in the nonselective media described above may result in a preferential isolation of Butyrivibrio. Another substrate which could be used in place of rutin would be plant saponins. Gutierrez, Davis, and Lindahl (1959) have shown that the majority of rumen bacterial strains isolated with media containing alfalfa saponin were mainly Butyrivibrio and Bacteroides strains. Alternatively, the use of complex polysaccharides such as cellulose, dextrin, pectin, or xylan as the sole added carbohydrate in nonselective media for plating or for roll tubes can be of some selective value. For instance, Shane, Gouws, and Kistner (1969) reported that Butyrivibrio constituted some 70% of the cellulolytic isolates obtained from rumen contents of sheep on a selected diet (low-protein teff hay). These researchers used finely ground, acid-treated cotton wool cellulose as the sole added energy source. In most instances when cellulose is used as the growth substrate, butyrivibrio colonies are those that are surrounded by only slight or weak zones of cellulose degradation as compared with the zones around the cellulolytic cocci (usually Ruminococcus species) or nonmotile, pleomorphic rods (usually Bacteroides species), when rumen contents are used as the source of inoculum. Strains designated as B385-like bacteria have been isolated from ruminants fed diets of rapidly fermentable carbohydrates and these strains are physiologically similar to butyrivibrios that were in low numbers in these animals (Bryant et al., 1961). However, the B385-like bacteria are somewhat larger than butyrivibrios and have tufts of flagella, indicating these organisms do not belong to the Butyrivibrio genus (Bryant, 1956).

Butyrivibrio strains display a wide range of variability in both morphological and physiological characteristics. Because of this variability and the lack of definitive information on genetic/molecular characteristics (e.g., G+C content of DNA; nucleic acid hybridization) of various strains, it is difficult to determine what factors can be used to delineate natural species patterns. All Butyrivibrio strains are now classified as a single species, B. fibrisolvens, (Bryant and Small, 1956a; Buchanan and Gibbons, 1974). Recently, Moore, Johnson, and Holdeman (1976) have described a new species, B. crossatus, and have recommended the family Bacteroidaceae be amended to include the genus Butyrivibrio. B. crossatus, isolated from human rectal or fecal material, differs from B. fibrisolvens in that it has lophotrichous flagella, does not produce gas, and can ferment only a limited number of carbohydrates (Moore, Johnson, and Holdeman, 1976). Eubacterium rectale is part of the fecal flora of humans and many strains are very similar to butyrivibrios as pointed out by Moore and Holdeman (1974). It is clear that ultrastructure studies are needed on the B385-like bacteria, *B. crossatus,* and *E. rectale* strains to determine whether these organisms also possess the thin Gram-positive cell wall of *B. fibrisolvens* (see below). If so, a reclassification of all these organisms into the genus *Eubacterium* (?) may be necessary. In the following discussion on the identifying characteristics for *Butyrivibrio,* we shall also consider that only one species exists, *B. fibrisolvens,* even though we recognize that further speciation and reclassification may eventually be needed.

In rumen fluid–carbohydrate media (e.g., NS-CRF), surface colonies are 2–4 mm in diameter, entire, slightly convex, translucent, and light tan in color. Some strains may form rough colonies that have filamentous edges and are lighter tan in color. Subsurface colonies are lenticular to ''Y'' shaped, and double lens–shaped colonies are not uncommon. In cellulose-containing media, the colonies of cellulose-digesting strains are surrounded by a zone of cellulose degradation, variable both in zone size and extent of digestion (usually weak). These latter colonies typically are lenticular to triangular in shape. Most strains grow rapidly in broth cultures and often tend to produce a flocculent sediment, whereas some strains produce a granular sediment that readily adheres to glass surfaces. Since growth is quite rapid (doubling times of 2 h or less) and much acid production occurs, cell characteristics and motility should be monitored with young cultures and in media with a low amount of added energy source.

The classic appearance of *Butyrivibrio* is a small, motile, slightly curved rod (0.4–0.6 × 2–5 μm long) with tapered ends. Some strains are almost spindle-shaped and have bluntly pointed ends (Fig. 2A). The cells show a translational motility characterized by a rapid or intense vibrating movement. Motility is by means of monotrichous flagellation, with the flagellum attached terminally or, less frequently, subterminally. Although often only a few cells in a preparation will show motility, truly nonmotile strains have not been well documented. The lack of motility in an isolated strain can be due to cultured conditions. Bryant and Small (1956a) reported that 15 strains were nonmotile when grown in a rumen fluid–glucose-cellobiose medium, but all strains were motile in the same medium if cellobiose was deleted and the glucose concentration decreased. This would suggest that a low pH due to acid production may inhibit motility, which is also consistent with the inhibition of growth by pH values of 6.3 or less (Gill and King, 1958). Optimal growth of most strains occurs with media of pH 6.5–7.2 in the temperature range of 30 to about 45°C, but no growth occurs at 20°C or at 50°C.

The initial early study of *Butyrivibrio* (Bryant and

Small, 1956a) as well as later studies (Brown and Moore, 1960; Cheng et al., 1969; Gill and King, 1958; Shane, Gouws, and Kistner, 1969) have reported that *Butyrivibrio* strains invariably appear as Gram-negative cells when stained by conventional procedures and viewed by light microscopy. Lipoteichoic acids are present only in Gram-positive bacteria, but Hewett, Wicken, and Knox (1976) isolated lipoteichoic acid from *B. fibrisolvens.* Subsequently, Cheng and Costerton (1977b) clearly established by electron microscope observations that the *B. fibrisolvens* cell wall is of a Gram-positive morphological type, but the peptidoglycan layer was rather thin (12–18 nm) as compared with that of most Gram-positive bacteria (30 to 50 nm). Thus, *Butyrivibrio* are structurally Gram-positive bacteria, but due to the thin cell wall, they stain in a Gram-negative pattern. As noted earlier, this unique cell wall characteristic may well prove to be of significant value in identifying *Butyrivibrio* strains and assigning newly isolated bacterial strains to this genus.

The production of large amounts of *n*-butyric acid from the fermentation of carbohydrates is the major fermentative characteristic that can be used in placing a newly isolated strain into the genus *Butyrivibrio.* Butyric acid is a major product made from the fermentation of both complex (dextrins, pectin, starch, xylan, cellulose) and simple (mainly glucose, fructose, cellobiose, maltose, sucrose, lactose) carbohydrates. Based mainly on the work of Shane, Gouws, and Kistner (1969), many butyrivibrios can be placed into two groups based on their fermentative patterns. Most strains are of the group I type which use exogenous acetic acid and produce formate, lactate, butyrate, and H_2 as the major fermentative products. The exogenous acetic acid is used in the formation of butyric acid (Latham and Legakis, 1976; van Gylswyk, 1976). On the other hand, group 2 strains produce formate, acetate, butyrate, and H_2 as major products. These two groups can also be differentiated by their growth responses to added volatile fatty acids (Roche et al., 1973). Propionate inhibits the growth of group I strains, but stimulates growth of group II strains. Addition of acetate and/or branched-chain volatile fatty acids can overcome the propionate inhibition with most group I strains. Many *Butyrivibrio* strains produce small amounts of ethanol, but no strain produces succinate. Although some strains require CO_2 to initiate growth, all strains studied do produce CO_2.

With respect to nitrogen requirements, ammonia is the nitrogen source utilized by most *Butyrivibrio* strains, but many strains also use mixtures of amino acids or complex nitrogen sources (casein hydrolysate, peptone). In the presence of growth-limiting levels of ammonia, one or more amino acids may be used or required (Gill and King, 1958). However, ammonia supports considerably more growth than

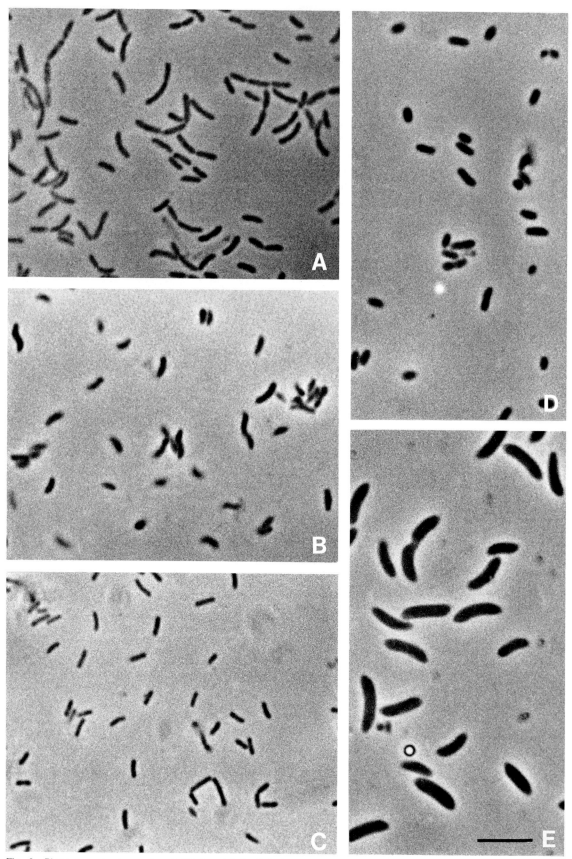

Fig. 2. Phase-contrast micrographs of (A) *Butyrivibrio fibrisolvens* strain A38, (B) *Succinivibrio dextrinosolvens* strain 22B, (C) *Lachnospira multiparus* strain 40, (D) *Succinimonas amylolytica* strain $B_2 4$, and (E) *Selenomonas ruminantium* strain GA192. All micrographs are at the same magnification and marker bar equals 5 μm.

an equivalent amount of peptide or amino acid nitrogen when these compounds are growth-limiting (Bryant, 1973). Generally, it appears that amino acid nitrogen is not used effectively by *B. fibrisolvens* unless supplied as a complex mixture. Only a few strains are proteolytic (as measured by gelatin liquefaction).

Few strains produce acetoin or hydrogen sulfide, but no strains reduce nitrate or produce catalase or indole. The growth of most strains is stimulated or requires the presence of one or more B-vitamins and acetate. The growth of a few strains may be stimulated by one or more volatile fatty acids (isovaleric, *n*-valeric, isobutyric, 2-methylbutyric). The *B. fibrisolvens* type strain (D-1; ATCC 19171) and some other strains can grow on chemically defined media containing these compounds (see Bryant, 1973; Roche et al., 1973).

Lachnospira

All isolated strains are considered to constitute a single species: *Lachnospira multiparus* (Buchanan and Gibbons, 1974). Some rather biochemically atypical isolates have been described (Akkado and Blackburn, 1963; Blackburn and Hobson, 1962), but the exact relationship(s) of these strains to *Lachnospira* has not been firmly established. *Lachnospira* are motile, weakly Gram-positive, curved rods (Fig. 2C) which can be easily distinguished from *Butyrivibrio* strains by their colony morphology and fermentation products. With rumen fluid–carbohydrate media (e.g., NS-CRF), the surface colonies of *Lachnospira* are large (up to 5 mm in diameter), flat, white, and filamentous. Colonies within agar media are very distinctive, appearing as woolly balls, and the cells are curved to helical rods. Hence, the name *Lachnospira*, derived from the Greek noun *lachnos* (woolly hair) and the Latin noun *spira* (coil), was given to these organisms (Bryant and Small, 1956b). The organisms commonly appear singly or in pairs of curved rods (0.4–0.6 μm × 2–4 μm long) with somewhat bluntly pointed ends (Fig. 2C). Usually, a few long chains or filaments are observed and the cells within these structures are only slightly curved and have distinctly rounded ends. Although *Lachnospira* was originally described as having polar, monotrichous flagellation (Bryant and Small, 1956b), it was found that the flagellum is subterminally or laterally attached (Buchanan and Gibbons, 1974; Leifson, 1960). When viewed by light microscopy as wet mount preparations, most cells show some translational motility that is characterized by frequent tumbling and movement in circular patterns. Cells taken from cultures in the early to middle exponential growth phase almost invariably stain Gram positively, but this property is rapidly lost and many

cells appear Gram negative when taken from heavily turbid broth cultures or older colonies. It is important that both motility and staining properties of isolated strains be made shortly after growth becomes visible in broths or on slants.

Most strains ferment glucose, fructose, sucrose, cellobiose, esculin, pectin, and salicin. D-Xylose is weakly or variably fermented. L-Arabinose, galactose, glycerol, lactose, maltose, mannitol, trehalose, inositol, cellulose, dextrin, gum arabic, and starch are not fermented. The main fermentation products produced from glucose are formic, acetic and lactic acids, ethanol, CO_2, and some H_2. The amount of acetoin varies. Biochemical tests for nitrate reduction, catalase, and indole or hydrogen sulfide formation are negative for all strains studied thus far.

In liquid media, the cell growth is visible as a flocculent sediment. The organism does grow readily over the range of 30–45°C with no growth at 22°C or at 50°C. All strains usually grow well in semidefined media (e.g., NS-SD; Caldwell and Bryant, 1966) or wholly defined media as described by Bryant and Robinson (1962). This latter medium was composed of only glucose, B-vitamins, ammonia, L-cysteine, bicarbonate, and minerals. Added CO_2 (or bicarbonate), acetate, amino acid, or peptide mixtures can stimulate growth (Bryant and Robinson, 1962). Ammonia, amino acid mixtures, or complex nitrogen sources (Trypticase, Casitone) are used as nitrogen sources. All strains examined require B-vitamins and with one strain studied in detail (Emery, Smith, and Fai To, 1957), biotin and *para*-aminobenzoic acid were required.

Succinivibrio

These organisms characteristically form light tan, entire, translucent surface colonies only 1–2 mm in diameter. Some colonies may be irregular, raised, and yellowish. Subsurface colonies are equally small, lenticular, and often difficult to detect. In media of low agar concentration (0.2 to 0.6% wt/vol), a white, diffuse, furry-like growth may occur, as expected for a motile organism. The cells invariably stain Gram negatively and are motile. The motility is translational with the cells showing a rapid vibrating movement. The flagellation arrangement is polar and monotrichous. The cells of most strains appear singly or in pairs of small, curved rods, usually 0.3–0.5 μm wide by 2–4 μm long and have pointed ends (Fig. 2B). Helical or twisted filaments of 2–4 coils and composed of 2 or more cells are commonly seen with many newly isolated strains. These helical filaments often show semirotational motility. This helical characteristic varies with strains and can be lost with successive subcultures in the laboratory, resulting in cells which appear as single or pairs of very slightly curved to straight rods (Bryant and Small, 1956b). Strains can form pleomorphic shapes—usually large, swollen, unevenly staining rods, 3–5 μm wide by 12–50 μm long (Wilson, 1953; S. M. Wilson and S. R. Elsden, personal communication, 1956) to spherical bodies (Scardovi, 1963). This unusual morphology generally occurs with adverse cultural conditions, particularly in media devoid or low in bicarbonate and/or yeast extract (Bryant, 1959). In liquid media, the cells of most strains tend to form helical filaments and clumps, often resulting in a flocculent sediment.

A large production of acetic and succinic acids with an uptake of large amounts of carbon dioxide is the fermentation pattern typically observed with *Succinivibrio* strains fermenting glucose (Bryant and Small, 1956b; Scardovi, 1963). Smaller amounts of formic acid are made along small, variable amounts of lactic acid. Propionic acid, butyric acid, and hydrogen gas are not made. Most strains actively ferment dextrin, galactose, D-xylose, and maltose. Some strains can also ferment esculin, salicin, cellobiose, fructose, L-arabinose, or mannitol. No strains that have been isolated can ferment cellulose, xylan, inulin, trehalose, lactose, inositol, or glycerol, and starch is not hydrolyzed (however, see Hungate, 1966). The enzymatic pathway for carbohydrate fermentation is the Embden-Meyerhof pathway (Scardovi, 1963). Carbon dioxide is required for the growth of all strains and it is primarily used for the formation of succinate via carboxylation of phosphoenolpyruvate to oxaloacetate (Scardovi, 1963).

Succinivibrio strains do not ferment amino acids or peptides, are not proteolytic, are negative for gelatin liquefaction, and have been shown (Bladen, Bryant, and Doetsch, 1961) not to produce ammonia from Trypticase. Except one strain (see Scardovi, 1963), most strains grow readily in media containing yeast extract and Trypticase but lacking rumen fluid (e.g., NS-SD). The growth of one strain when grown in a semidefined medium required both casein hydrolysate and B-vitamins but growth was increased greatly by the addition of ammonia (Bryant and Robinson, 1962). Since this strain can effectively utilize free amino acid mixtures as a nitrogen and carbon source (Bryant and Robinson, 1963), it seems probable that amino acids are major nitrogen sources used by succinivibrios. Good growth occurs at incubation temperatures from 30 to 39°C, but no growth occurs at 22°C or 45°C. All strains are negative for standard biochemical tests for nitrate reduction, catalase, and production of indole, hydrogen sulfide, or acetoin. It should be noted that succinivibrio-like organisms have been isolated from human blood specimens (Porschen and Chan, 1976; Southern, 1975). These isolates possessed lophotrichous flagellation and did not ferment xylose, arabinose, mannitol, or cellobiose. It is

questionable then whether these isolates are strains of *Succinivibrio dextrinosolvens*. Presently, all rumen strains are considered to constitute a single species: *Succinivibrio dextrinosolvens* (Bryant and Small, 1956b; Buchanan and Gibbons, 1974). Based on the characteristics of strains studied thus far, this classification appears quite reasonable until new isolates and/or further definitive studies (e.g., determination of G+C content of DNA) are carried out.

Succinimonas

Succinimonads produce colonies that are somewhat similar to those made by succinivibrios, but the latter organisms are distinguishable by their vibrioid and helical cell morphology. Succinimonad surface colonies on rumen fluid–carbohydrate media are light tan in color, translucent, smooth, convex, and are only 0.6–1.5 mm in diameter even after several days of incubation. Colonies produced within agar media are lenticular in shape. The typical cell morphology is that of a short, oval rod or coccobacillus, 1–1.5 μm by 1.2–3 μm long (Fig. 2D). The cells are never curved and the cell ends are well-rounded, never pointed. The cells always stain Gram negatively and can appear singly, in pairs, or as clumps (especially in older colonies). With wet mounts, most cells show varying degrees of translational motility, which stops rapidly upon exposure to air. Cells which have become attached to the glass surface can show a rotational movement about one cell end. The flagellar arrangement is polar and monotrichous. Some cells can have one or more intracellular granules. These granules usually lie near the cell periphery and are somewhat more distinct when viewed by phase microscopy. On occasion, staining will show a bipolar arrangement of these granules. The nature of these granules is unknown, but they may be glycogen granules that are found with many other rumen bacteria (e.g., see Cheng and Costerton, 1977a). However, it should be noted that no extracellular capsular material is evident when cells of most strains were viewed by light microscopy. Growth in liquid media is even and disperse. For reasons that are not entirely clear, growth in liquid media is relatively light and cultures with dense turbidity are not obtained.

Succinimonads ferment a limited range of carbohydrates, namely starch, dextrin, glucose, and maltose. The major fermentation products are acetic and succinic acids along with a large uptake of carbon dioxide (Bryant et al., 1958). No production of hydrogen gas, formate, lactate, butyrate, or ethanol occurs with the fermentation of glucose. Trace amounts of acetoin and/or propionate are sometimes produced.

Nutritional studies of the type strain and one other showed that both could be grown in a relatively simple, chemically defined medium containing glucose, bicarbonate, B-vitamins, minerals, ammonia as nitrogen source and sulfide as sulfur source. Amino acid mixtures or Casitone were stimulatory but could not replace ammonia, which was essential as the main nitrogen source. Acetate was highly stimulatory in concentrations to 30 mM whether or not an amino acid mixture was added. Cysteine could replace sulfide as sulfur source in media containing amino acids but was inhibitory in media in which ammonia was the sole nitrogen source. A B-vitamin mixture is necessary but the specific essential ones have not been studied.

All known strains do not hydrolyze gelatin, and neither peptides nor amino acids are fermented, even with strains isolated from rumen contents of animals on high protein diets (Bladen, Bryant, and Doetsch, 1961). No production of hydrogen sulfide, indole, or catalase has been observed. Growth occurs at 30–39°C, but not at 22°C or 45°C. Since the strain differences are rather minimal, all studied strains are considered to be a single species, *Succinimonas amylolytica* (Bryant and Small, 1956b; Buchanan and Gibbons, 1974).

Selenomonas

Strains are easily identified to the genus level because of their active tumbling motility and typical morphology consisting of Gram-negative, crescentic, vibrioid or helical rods (Fig. 2E), with a tuft of flagella that arises on the concave side of the cell. Longer cells with several tufts of flagella may be seen. The earlier controversy over the arrangement of flagella and generic designation (Bryant, 1956; Lessel and Breed, 1954; MacDonald and Madlener,

1957) was dispelled by the work of Kingsley and Hoeniger (1973), who proved that in both the oval and rumen species the flagella arise from a specialized structure located on the concave side of the cell through which binary fission occurs. They emphasized the need to study morphology in media with low energy sources and without certain nutrient deficiencies that may inhibit cell division or increase spheroplast formation as well as cause abnormalities in apparent flagellar arrangement. In laboratory culture, cells are usually 0.9–1.1 by 3.0–6.0 nm, although larger and smaller forms exist. They occur singly, in pairs, and in short chains. Oral strains sometimes form clumps and may be more helical than rumen strains.

The organisms are fermentative and glucose is fermented with production of acetate, propionate, and CO_2 and/or lactate. The amount of lactate produced depends on the strain and on culture conditions (Scheifinger, Linehan, and Wolin 1975). H_2 is produced by most if not all strains but is often difficult to detect (Holdeman and Moore, 1977; Scheifinger, Latham, and Wolin, 1975). Lactate-fermenting strains produce mainly propionate, acetate, and CO_2 (Bryant, 1956). They are strict anaerobes, and catalase is not produced. The temperature optimum is 35–40°C with a maximum near 45°C and minimum at 20–30°C. They grow at a lower pH (4.5–5.0) than many anaerobic, Gram-negative, motile rods.

Some features of use in differentiating *S. ruminantium, S. sputigena,* and the rat cecal strains of Ogimoto (1972) are shown in Table 1. The rat cecal strains have not been studied in detail but seem to be very similar to *S. ruminantium* except that none of them produced sulfide from cysteine.

Table 1. Some features of *Selenomonas ruminantium, Selenomonas sputigena,* and *Selenomonas* sp. from the rat cecum.

Group	Cellobiose[a]	Dulcitol	Salicin	H_2S[b]	G+C[c]
S. sputigena[d]	−	−	−	−	60.6
S. ruminantium[e]	+	±	+	±	54.0
Rat cecal selenomonad[f]	+	ND	+	−	ND

[a] Production of acid from carbohydrates; none of the strains, −; all of the strains +; most strains, ±; not determined, ND.

[b] H_2S production from cysteine.

[c] Mol% guanine plus cytosine in the DNA.

[d] *S. sputigena* also ferments dextrin, glycerol, esculin, galactose, arabinose, xylose, glucose lactose, sucrose, maltose, inulin, mannitol, and starch. Strains vary in fermentation of trehalose and raffinose and they do not ferment sorbitol, inositol, xylan, or cellulose. Indole and acetoin are not produced, gelatin is not liquefied, and nitrite is produced from nitrate.

[e] *S. ruminantium* also ferments fructose, esculin, glucose, and maltose, and often ferments trehalose, raffinose, dextrin, glycerol, lactate, sorbitol, inositol, galactose, arabinose, xylose, lactose, sucrose, inulin, mannitol, and starch. Strains do not ferment gum arabic, xylan, or cellulose. Indole and acetoin are not produced and gelatin is not liquefied. Some strains reduce nitrate to nitrite and some reduce nitrite on to ammonia.

[f] The rat cecal strains also fermented arabinose, xylose, glucose, sucrose, and mannitol, and varied on starch and in nitrate reduction to nitrite.

Some additional features of *S. ruminantium* follow. After 3–4 days of incubation on nonselective rumen fluid media (e.g., NS-CRF medium), colonies are large (3- to 6-mm diameter), smooth, entire, slightly convex, and light tan in color. With the mannitol selective medium (SS medium), similar colonies are observed, but the color is generally white and the formation of a gray to black area due to FeS production starting in the center of the colony is quite common. Many *S. ruminantium* strains have carbohydrate granules in the cytoplasm and cells may be strongly iodophilic (Prins, 1971), but no capsular material is present.

The nutrient requirements of *S. ruminantium* are quite simple and most, if not all, strains can be grown in a chemically defined medium containing glucose, minerals, B vitamins, ammonia, sulfide, a volatile acid such as *n*-valerate, and carbon dioxide (Bryant and Robinson, 1962; Kanegasaki and Takahashi, 1967; John, Isaacson, and Bryant, 1974; Tiwari, Bryant, and Wolfe, 1969). Sulfide or cysteine serves as sole sulfur source and ammonia or certain single amino acids such as cysteine, serine, threonine, aspartate, histidine, glutamate, and valine, or the purines adenine and uric acid may serve as sole nitrogen sources. Some strains utilize urea or nitrate as sole nitrogen source but most do not (John, Isaacson, and Bryant, 1974). Most nitrate-reducing strains do not reduce nitrite to ammonia and, thus, cannot use nitrate as sole nitrogen source. Biotin and *p*-aminobenzoic acid satisfy the vitamin requirements and when lactate is the energy source, aspartate, malate, or fumarate are necessary, and aromatic amino acids may be stimulatory to growth (Linehan, Scheifinger, and Wolin, 1978). *n*-Valerate or similar fatty acids are required for growth of some strains when glucose is the energy source, but this requirement may be lost and does not occur when glycerol or lactate is the energy source (Kanegasaki and Takahashi, 1967; Linehan, Scheifinger, and Wolin, 1978).

The rumen strains have been divided into three subspecies. Strains which ferment glycerol and lactate are designated *Selenomonas ruminantium* subsp. *lactilytica* (Bryant, 1956; Hobson, Mann, and Smith, 1962). Large strains measuring 2–3 by 5–10 μm that retain this size in pure culture were designated *S. ruminantium* subsp. *bryanti* (Prins, 1971). These strains usually do not produce H_2S or ferment arabinose, xylose, galactose, lactose, or dulcitol.

The features of the human oral organism, *S. sputigena* have not been studied in as much detail as *S. ruminantium*. Although *S. sputigena* has mostly been isolated using blood agar media, these media support rather poor growth of the organism. Best growth has been observed with thioglycolate media (Difco) and with media similar to NS-SD medium (Kingsley and Hoeniger, 1973). The colonies of newly isolated *S. sputigena* strains on blood agar (2% wt/vol) media are generally small (0.5–1.2 mm in diameter), smooth, convex, and gray to gray-yellow in color. The larger colonies tend to have an irregular edge with a translucent appearance. A partially mottled area within a colony is not unusual. Growth of the organisms in broth media may be turbid, but a flocculent to granular sediment often occurs.

Extensive, detailed examinations of the fermentative, nutritional, or other physiological properties of *S. sputigena* strains have not been done as of yet.

Other than identification by microscopic observations, the identification of species of selenomonads isolated from cecal contents of rodents poses several problems. The selenomonad in the guinea pig cecum, *S. palpitans*, has not been isolated and grown in pure culture, despite attempts by numerous workers. It is not clear why this is the case, particularly since this has been accomplished using media similar to NS-CRF and NS-SD media with selenomonad strains found in the ceca of rats and squirrels. *Selenomonas palpitans* may have a larger number of flagella (up to 22 have been seen), and they are bunched together in a single area rather than forming a line of insertion as is the case with *S. ruminantium* and *S. sputigena* (Kingsley and Hoeniger, 1973). Selenomonads have been enumerated by colony counts in the cecal contents of squirrels (Barnes and Burton, 1970), but studies to gain information on features of the isolates were not carried out. Whether the guinea pig or squirrel isolates are similar to the rat cecal isolates of Ogimoto (1972) is not known. The latter organism seems very similar to *S. ruminantium*. It is obvious that further isolations and comparative study of the features of selenomonads from the various ecosystems are needed.

Literature Cited

Akkada, A. R. A., Blackburn, T. H. 1963. Some observations on the nitrogen metabolism of rumen proteolytic bacteria. Journal of General Microbiology **31:**461–469.

Aranki, A., Freter, R. 1972. Use of anaerobic glove boxes for the cultivation of strictly anaerobic bacteria. American Journal of Clinical Nutrition **25:**1329–1334.

Aranki, A., Syed, S. A., Kenney, E. B., Freter, R. 1969. Isolation of anaerobic bacteria from human gingiva and mouse cecum by means of a simplified glove box procedure. Applied Microbiology **17:**568–576.

Balch, W. E., Wolfe, R. S. 1976. New approach to the cultivation of methanogenic bacteria: 2-Mercaptoethanesulfonic acid (HS-CoM)-dependent growth of *Methanobacterium ruminantium* in a pressurized atmosphere. Applied and Environmental Microbiology **32:**781–791.

Barnes, E. M., Burton, G. C. 1970. The effect of hibernation on the caecal flora of the thirteen-lined ground squirrel (*Citellus tridecemlineatus*). Journal of Applied Bacteriology **33:**505–514.

Betian, H. G., Linehan, B. A., Bryant, M. P., Holdeman, L. V. 1977. Isolation of a cellulolytic *Bacteroides* sp. from human feces. Applied and Environmental Microbiology **33:**1009–1010.

Blackburn, T. H., Hobson, P. N. 1962. Further studies on the isolation of proteolytic bacteria from the sheep rumen. Journal of General Microbiology **29:**69–81.

Bladen, H. A., Bryant, M. P., Doetsch, R. N. 1961. A study of bacterial species from the rumen which produce ammonia from protein hydrolysate. Journal of Applied Microbiology **9:**175–180.

Brown, D. W., Moore, W. E. C. 1960. Distribution of *Butyrivibrio fibrisolvens* in nature. Journal of Dairy Science **43:**1570–1574.

Bryant, M. P. 1952. The isolation and characteristics of a spirochete from the bovine rumen. Journal of Bacteriology **64:**325–335.

Bryant, M. P. 1956. The characteristics of strains of *Selenomonas* isolated from bovine rumen contents. Journal of Bacteriology **72:**162–167.

Bryant, M. P. 1959. Bacterial species of the rumen. Bacteriological Reviews **23:**125–153.

Bryant, M. P. 1972. Commentary on the Hungate technique for culture of anaerobic bacteria. American Journal of Clinical Nutrition **25:**1324–1328.

Bryant, M. P. 1973. Nutritional requirements of the predominant rumen cellulolytic bacteria. Federation of American Societies for Experimental Biology Proceedings **32:**1809–1813.

Bryant, M. P., Burkey, L. A. 1953. Cultural methods and some characteristics of some of the more numerous groups of bacteria in the bovine rumen. Journal of Dairy Science **36:**205–217.

Bryant, M. P., Robinson, I. M. 1961. An improved nonselective culture medium for ruminal bacteria and its use in determining diurnal variations in numbers of bacteria in the rumen. Journal of Dairy Science **44:**1446–1456.

Bryant, M. P., Robinson, I. M. 1962. Some nutritional characteristics of predominant culturable ruminal bacteria. Journal of Bacteriology **84:**605–614.

Bryant, M. P., Robinson, I. M. 1963. Apparent incorporation of ammonia and amino acid carbon during growth of selected species of ruminal bacteria. Journal of Dairy Science **46:**150–154.

Bryant, M. P., Robinson, I. M., Lindahl, I. L. 1961. A note on the flora and fauna in the rumen of steers fed a bloat-provoking ration and the effect of penicillin. Applied Microbiology **9:**511–515.

Bryant, M. P., Small, N. 1965a. The anaerobic monotrichous butyric acid-producing curved rod-shaped bacteria of the rumen. Journal of Bacteriology **72:**16–21.

Bryant, M. P., Small, N. 1965b. Characteristics of two new genera of anaerobic curved rods isolated from the rumen of cattle. Journal of Bacteriology **72:**22–26.

Bryant, M. P., Barrentine, B. F., Sykes, J. F., Robinson, I. M., Shawver, C. V., Williams, L. W. 1960. Predominant bacteria in the rumen of cattle on bloat-provoking Ladino clover pasture. Journal of Dairy Science **43:**1435–1444.

Bryant, M. P., Campbell, L. L., Reddy, C. A., Crabill, M. R. 1977. Growth of *Desulfovibrio* sp. in lactate or ethanol media low in sulfate in association with H_2-utilizing methanogenic bacteria. Applied and Environmental Microbiology **33:**1162–1169.

Bryant, M. P., Small, N., Bouma, C., Chu, H. 1958. *Bacteroides ruminicola* n. sp. and *Succinimonas amylolytica.* The new genus and species. Journal of Bacteriology **76:**15–23.

Buchanan, R. E., Gibbons, N. E. (eds.). 1974. Bergey's manual of determinative bacteriology, 8th ed. Baltimore: Williams & Wilkins.

Caldwell, D. R. 1965. The specificity of the hemin growth requirement of *Bacteroides ruminicola* subspecies *ruminicola.* M. S. Thesis. University of Maryland.

Caldwell, D. R., Bryant, M. P. 1966. Medium without rumen fluid for nonselective enumeration and isolation of rumen bacteria. Applied Microbiology **14:**794–801.

Cheng, K.-J., Costerton, J. W. 1977a. Ultrastructure of cell envelopes of bacteria of the bovine rumen. Applied Microbiology **29:**841–849.

Cheng, K.-J., Costerton, J. W. 1977b. Ultrastructure of *Butyrivibrio fibrisolvens:* A Gram-positive bacterium? Journal of Bacteriology **129:**1506–1512.

Cheng, K.-J., Jones, G. A., Simpson, F. J., Bryant, M. P. 1969. Isolation and identification of rumen bacteria capable of anaerobic rutin degradation. Canadian Journal of Microbiology **15:**1365–1371.

Cheng, K.-J., Krishnamurty, H. G., Jones, G. A., Simpson, F. J. 1970. Identification of products produced by the anaerobic degradation of naringin by *Butyrivibrio* sp. C_3. Canadian Journal of Microbiology **16:**129–131.

Dehority, B. A., Grubb, J. A. 1977. Glucose-1-phosphate as a selective substrate for enumeration of *Bacteroides* species in the rumen. Applied and Environmental Microbiology **33:**998–1001.

Dobell, C. 1960. Antonie van Leeuwenhoek and his little animals. New York: Dover Publications.

Edwards, T., McBride, B. C. 1975. New method for the isolation and identification of methanogenic bacteria. Applied Microbiology **29:**540–545.

Eller, C., Crabill, M. R., Bryant, M. P. 1971. Anaerobic roll tube media for nonselective enumeration and isolation of bacteria in human feces. Applied Microbiology **22:**522–529.

Emery, R. S., Smith, C. K., Fai To, L. 1957. Utilization of inorganic sulfate by rumen microorganisms. II. The ability of single strains of rumen bacteria to utilize inorganic sulfate. Applied Microbiology **5:**363–366.

Gill, J. W., King, K. W. 1958. Nutritional characteristics of a butyrivibrio. Journal of Bacteriology **75:**666–673.

Grubb, J. A., Dehority, B. A. 1976. Variation in colony counts of total viable anaerobic rumen bacteria as influenced by media and cultural methods. Applied and Environmental Microbiology **31:**262–267.

Gutierrez, J., Davis, R. E., Lindahl, I. L. 1959. Characteristics of saponin-utilizing bacteria from the rumen of cattle. Applied Microbiology **7:**304–308.

Herbeck, J. L., Bryant, M. P. 1974. Nutritional features of the intestinal anaerobe *Ruminococcus bromii.* Applied Microbiology **28:**1018–1022.

Hespell, R. B., Canale-Parola, E. 1970. *Spirochaeta litoralis* sp. n., a strictly anaerobic marine spirochete. Archiv für Mikrobiologie **74:**1–18.

Hewett, M. J., Wicken, A. J., Knox, K. W., Sharpe, M. E. 1976. Isolation of lipoteichoic acids from *Butyrivibrio fibrisolvens.* Journal of General Microbiology **94:**126–130.

Hobson, P. N., Mann, S. O. 1961. The isolation of glycerol-fermenting and lipolytic bacteria from the rumen of the sheep. Journal of General Microbiology **25:**227–240.

Hobson, P. N., Mann, S. O., Smith, W. 1962. Serological tests of a relationship between rumen selenomonads *in vitro* and *in vivo.* Journal of General Microbiology **29:**265–270.

Holdeman, L. V., Moore, W. E. C. Anaerobe laboratory manual. Blacksburg, Virginia. Virginia Polytechnic Institute and State University Anaerobe Laboratory.

Hungate, R. E. 1950. The anaerobic mesophilic cellulolytic bacteria. Bacteriological Reviews **14:**1–49.

Hungate, R. E. 1966. The rumen and its microbes. New York: Academic Press.

Hungate, R. E. 1969. A roll tube methods for cultivation of strict anaerobes, pp. 117–132. In: Norris, J. R., Gibbons, D. W., (eds.), Methods in microbiology, vol. 3B. New York: Academic Press.

John, A., Isaacson, H. R., Bryant, M. P. 1974. Isolation and characteristics of a ureolytic strain of *Selenomonas ruminantium.* Journal of Dairy Science **57:**1003–1014.

Kanegasaki, S., Takahashi, H. 1967. Function of growth factors for rumen microorganisms. I. Nutritional characteristics of

Selenomonas ruminantium. Journal of Bacteriology **93**: 456–463.

Kingsley, V. V., Hoeniger, J. F. M. 1973. Growth, structure, and classification of *Selenomonas.* Bacteriological Reviews **37**:479–521.

Latham, M. J., Legakis, N. J. 1976. Cultural factors influencing the utilization or production of acetate by *Butyrivibrio fibrisolvens.* Journal of General Microbiology **94**:380–388.

Leifson, E. 1960. Atlas of bacterial flagellation. New York: Academic Press.

Lessel, E. F., Breed, R. S. 1954. Selenomonas Boskamp, 1922—a genus that includes species showing an unusual type of flagellation. Bacteriological Reviews **18**:165–168.

Linehan, B., Scheifinger, C. C., Wolin, M. J. 1978. Nutritional requirements of *Selenomonas ruminantium* for growth on lactate, glycerol, or glucose. Applied and Environmental Microbiology **35**:317–322.

MacDonald, J. B., Madlener, E. M. 1957. Studies on the isolation of Spirillum sputigenum. Canadian Journal of Microbiology **3**:679–686.

Margherita, S. S., Hungate, R. E. 1963. Serological analysis of *Butyrivibrio* from the bovine rumen. Journal of Bacteriology **86**:855–860.

Margherita, S. S., Hungate, R. E., Storz, H. 1964. Variation in rumen *Butyrivibrio* strains. Journal of Bacteriology **87**: 1304–1308.

Miller, T. L., Wolin, M. J. 1974. A serum bottle modification of the Hungate technique for cultivating obligate anaerobes. Applied Microbiology **27**:985–987.

Miller, W. D. 1887. Über pathogene Mundpilze. Inaugural Dissertation, Berlin.

Moore, W. E. C., Holdeman, L. V. 1974. Human fecal flora: The normal flora of 20 Japanese-Hawaiians. Applied Microbiology **27**:961–979.

Moore, W. E. C., Johnson, J. L., Holdeman, L. V. 1976. Emendation of *Bacteroidaceae* and *Butyrivibrio* and descriptions of *Desulfomonas* gen. nov. and ten new species in the genera *Desulfomonas, Butyrivibrio, Eubacterium, Clostridium,* and *Ruminococcus.* International Journal of Systematic Bacteriology **26**:238–252.

Ogimoto, K. 1972. Über *Selenomonas* aus dem Caecum von Ratten. Zentralblatt fur Bakteriologie, Parasitenkunde, Infektionskrankheiten und Hygiene, Abt. 1 Orig., Reihe A **221**:467–473.

Onderdonk, A. B., Johnston, J., Mayhew, J. W., Gorbach, S. L. 1976. Effect of dissolved oxygen and Eh on *Bacteroides fragilis* during continuous culture. Applied and Environmental Microbiology **31**:168–172.

Paynter, M. J. B., Hungate, R. E. 1968. Characterization of *Methanobacterium mobilis* sp. n., isolated from the bovine rumen. Journal of Bacteriology **95**:1943–1951.

Porschen, R. K., Chan, P. 1977. Anaerobic vibrio-like organisms cultured from blood: *Desulfovibrio desulfuricans* and *Succinivibrio* species. Journal of Clinical Microbiology **5**:444–447.

Prins, R. A. 1971. Isolation, culture, and fermentation characteristics of *Selenomonas ruminantium* var. *bryanti* var. n. from the rumen of sheep. Journal of Bacteriology **105**:820–825.

Robinow, C. F. 1954. Addendum to: *Selenomonas* Boskamp 1922—a genus that includes species showing an unusual type of flagellation, by E. F. Lessel, Jr. and R. S. Breed. Bacteriological Reviews **18**:168.

Roché, C., Albertyn, H., van Gylswyk, N. O., Kistner, A. 1973. The growth response of cellulolytic acetate-utilizing and acetate-producing butyrivibrios to volatile fatty acids and other nutrients. Journal of General Microbiology **78**:253–260.

Scardovi, V. 1963. Studies in rumen bacteriology. I. A succinic acid producing vibrio: Main physiological characters and enzymology of its succinic acid forming system. Annali di Microbiologia **13**:171–187.

Scheifinger, C. C., Latham, M. J., Wolin, M. J. 1975. Relationship of lactate dehydrogenase specifity and growth rate to lactate metabolism by *Selenomonas ruminantium.* Applied Microbiology **30**:916–921.

Scheifinger, C. C., Linehan, B., Wolin, M. J. 1975. H_2 production by *Selenomonas ruminantium* in the absence and presence of methanogenic bacteria. Applied Microbiology **29**: 480–483.

Shane, B. S., Gouws, L., Kistner, A. 1969. Cellulolytic bacteria occurring in the rumen of sheep conditioned to low-protein teff hay. Journal of General Microbiology **55**:445–457.

Simons, H. 1922. Ueber *Selenomonas palpitans* n. sp. Centralblatt für Bakteriologie, Parasitenkunde und Inflektionskrankheiten, Abt. 1 Orig. **87**:50.

Smith, P. H. 1966. The microbial ecology of sludge methanogenesis. Developments in Industrial Microbiology **7**: 156–174.

Southern, P. M. 1975. Bacteremia due to *Succinivibrio dextrinosolvens.* American Journal of Clinical Pathology **64**: 540–543.

Thorley, C. M., Sharpe, M. E., Bryant, M. P. 1968. Modification of the rumen bacterial flora by feeding cattle ground and pelleted roughage as determined with culture media with and without rumen fluid. Journal of Dairy Science **51**:1811–1816.

Tiwari, A. D., Bryant, M. P., Wolfe, R. S. 1969. Simple method for isolation of *Selenomonas ruminantium* and some nutritional characteristics of the species. Journal of Dairy Science **52**:2054–2056.

Uffen, R. L., Sybesma, C., Wolfe, R. S. 1971. Mutants of *Rhodosprillum rubrum* obtained after long-term anaerobic, dark growth. Journal of Bacteriology **108**:1348–1356.

van Gylswyk, N. O. 1976. Some aspects of the metabolism of *Butyrivibrio fibrisolvens.* Journal of General Microbiology **97**:105–111.

Walden, W. C., Hentges, D. J. 1975. Differential effects of oxygen and oxidation-reduction potential on the multiplication of three species of anaerobic intestinal bacteria. Applied Microbiology **30**:781–785.

Wilson, S. M. 1953. Some carbohydrate-fermenting organisms isolated from the rumen of sheep. Journal of General Microbiology **9**:i–ii.

Wozny, M. A., Bryant, M. P., Holdeman, L. V., Moore, W. E. C. 1977. Urease assay and urease-producing species of anaerobes in the bovine rumen and human feces. Applied and Environmental Microbiology **33**:1097–1104.

Gram-Negative Cocci and Related Rod-Shaped Bacteria

The Genus *Neisseria*

NEYLAN A. VEDROS

The family Neisseriaceae comprises four genera: *Neisseria, Branhamella, Moraxella,* and *Acinetobacter* (Reyn, 1974). These genera share common characteristics, e.g., plump coccal or coccobacillary shape, cells arranged in pairs, and a Gram-negative staining reaction. These similarities were noted early (Audureau, 1940; Henriksen, 1952) and have been the motivation for intensive effort by many investigators to define the taxonomic relationships within the family (see review by Henriksen, 1976).

The genus *Neisseria* (Trevisan 1885, 105. Nom. cons. opin. 13, Jud. Comm. 1954, 153) is composed of six species, *N. gonorrhoeae, N. meningitidis, N. sicca, N. subflava* (including *perflava* and *flava*), *N. flavescens,* and *N. mucosa,* and 10 *species incertae sedis, N. animalis, N. canis, N. caviae, N. cinerea, N. cuniculi, N. denitrificans, N. lactamicus, N. ovis, N. suis,* and *N. elongata.* The inclusion of *N. elongata,* a rod-shaped organism, in the genus *Neisseria* is still under consideration, although the G+C content and genetic compatibility data strongly support its inclusion (Bøvre and Holten, 1970). Similarly, the removal of *N. ovis* and *N. caviae* from the genus and their placement with *Branhamella catarrhalis* (the "false" *Neisseria*) are also strongly supported by DNA base composition and genetic transformation data (Bøvre, 1967; Bøvre, Fiandt, and Szybalski, 1969; Henriksen and Bøvre, 1968) and enzymatic differences (Holten, 1973). Although many other *Neisseria* species appear in the literature (e.g., *N. crassa, N. winogradskyi, N. gibboni*), these organisms are not included in the genus because they do not exist any more, are incompletely described, or have been included in other genera.

The genus *Neisseria* was named for Albert Neisser, who discovered *N. gonorrhoeae* in 1879. This discovery was soon followed by the isolation of *N. meningitidis* by Weichselbaum in 1887; *N. sicca, N. subflava,* and *N. mucosa* by von Lingelsheim in 1906; and *N. flavescens* by Branham in 1930. The other *Neisseria* species, mainly of animal origin, have been isolated more recently. The characteristics describing the six species isolated from humans are described by Reyn (1974). The other *Neisseria* species are described in Table 1.

Habitats

In Humans

The principal habitat of those *Neisseria* species isolated from humans is the mucous membrane surfaces. *N. meningitidis, N. sicca, N. subflava,* and *N. mucosa* are found predominantly at the junction of the hard and soft palates in the posterior nasopharynx of healthy individuals. The majority of studies have been conducted with *N. meningitidis;* it was observed that 5–7% of healthy individuals ("carriers") harbor the organisms in interepidemic periods, 17–20% during epidemics (Faucon et al., 1970), and up to 100% in military camps during periods of high disease incidence. It appears that the organism is part of the transient normal flora in some individuals, while others are chronic carriers who continually shed the organisms into the environment. The intracellular nature of the meningococci in carriers was demonstrated by the fluorescent antibody technique (Sanborn and Vedros, 1966), but the immunogenic stimulation was variable (Vedros, Hunter, and Rust, 1966). Whether the carrier represents the antigenic challenge in nature that results in man's natural, high immunity to meningococcal disease is not known, but the role of bactericidal activity in the serum of those age groups susceptible to disease has been demonstrated (Goldschneider, Gotschlich, and Artenstein, 1969). Only rarely does the meningococcus cross the mucous membrane and result in disease. This event is usually preceded by rhinopharyngitis or a concomitant upper respiratory viral infection (Edwards et al., 1977).

Neisseria flavescens appears to be a rare saprophyte. It was first isolated from the spinal fluid of patients during a meningitis outbreak (Branham, 1930). Other reports of isolations from the urogenital tract (Wax, 1950) and the cerebrospinal fluid (Prentice, 1957; Radke and Cunningham, 1949) lacked complete bacteriological data to correctly define the isolates as *N. flavescens.* The most extensive study was conducted by Berger and Husmann (1972). They isolated *N. flavescens* from the throat of one soldier out of 506 tested. The isolate agglutinated weakly with rabbit antiserum to *N. flavescens* strain NCTC 8263 produced 12 years earlier. In the author's experience, dozens of extensive carrier surveys for meningococci using nonselective media have not resulted in a single isolate of *N. flavescens*. Any discussion, therefore, of the taxonomic classification of this species must be considered within the context of the limited strains available for study. *N. mucosa, N. mucosa* var. *heidelbergensis,* and *N. sicca* are normal inhabitants of the nasopharynx (Reyn, 1974). *N. mucosa* can move down into the lungs and result in pneumonia in children (Véron, Thibault, and Second, 1959), whereas *N. mucosa* var. *heidelbergensis* can be found as part of the normal flora of young adults (Berger, 1971). Recently, *N. mucosa* var. *heidelbergensis* has been isolated as part of the normal flora on mucous membrane tissues in the airways of the dolphin (Vedros, Johnston, and Warren, 1973). The development of a selective medium (Thayer and Martin, 1966) permitted extensive carrier surveys for meningococcal carriers in the late 1960s. It was frequently noted that a Gram-negative, cytochrome oxidase–positive diplococcus would grow on this selective medium but contained a yellow pigment compared with the typical transparent meningococcus. Laboratories did not routinely include lactose in their battery of sugars, but when it was included, this pigmented colony was observed to produce acid from glucose, maltose, and lactose, and to be serologically indistinguishable from Group B *N. meningitidis*. Similar strains had been described in 1934 (Jessen, 1934) and in greater detail later (Mitchell, Rhoden, and King, 1965). A large number of strains were studied by biochemical and serological analysis and designated as a new species, *Neisseria lactamicus* (Hollis, Wiggins, and Weaver, 1969). The species epithet was subsequently changed to *lactamica* for taxonomic purposes (Catlin, 1971).

Apparently, *N. lactamica* is part of the flora of the human nasopharynx. It was initially thought to be restricted to young children (Hollis et al., 1970), but can be isolated with lesser frequency in adults (Pykett, 1973). Occasionally the organism will penetrate the mucous membranes and beyond, causing a frank meningitis (Lauer and Fisher, 1976) or a septicemia (Wilson and Overman, 1976). The clinical significance of this lactose-utilizing *Neisseria* species remains to be proven, but its virulence appears to be much lower for humans than that of *N. meningitidis*. It may, however, be involved in several disease processes in children (Wilson and Overman, 1976).

The human nasopharynx serves as an excellent habitat for many *Neisseria*. During periods of high disease incidence, pure cultures of $>10^6$ meningococci can be obtained on nonselective media from many persons who have had close contact with the infected individual. The posterior nasopharynx is the primary site for isolation, but organisms can also be obtained from the surface of the tongue and from the teeth and gums. Other *Neisseria* species, such as *N. perflava,* can also be routinely isolated during these periods (Vedros, Robinson, and Gutekunst, 1966). These organisms are continually shed into the environment, probably in droplet-size particles (>10 μm). Considerable efforts have been made to study the environmental factors that influence the spread of the meningococci. The meningococci have been found to be stable as aerosols at very low ($<10\%$) or very high ($>90\%$) relative humidities (N. A. Vedros, unpublished observations), and there is a negative correlation between absolute humidity and disease incidence (Molineaux, 1970): the lower the absolute humidity, the higher the disease incidence.

The primary habitat for *N. gonorrhoeae* is the mucous membrane surfaces of the genital tissues, followed by (with less frequency) the rectum and pharynx. As with meningococci in the nasopharynx, the gonococci have an intimate host-cell relationship, with varying degrees of antibody response. Although the host is considered to be "infected" and "contagious", overt infections or symptomatic disease are not common, particularly in females. These target tissue habitats may be due simply to the manner in which the gonococci are transmitted among individuals. Once the organism penetrates the epithelial barrier, it can spread to surrounding glands, fallopian tubes, epididymis, the blood, synovial membranes in joints, and the heart. Under certain circumstances, the gonococci can invade the conjunctiva, resulting in gonococcal ophthalmia (Rein, 1977). This disease occurs most frequently in the newborn delivered of infected mothers, unless appropriate preventive measures are taken.

In the normal female cervix and uterus, both *Branhamella catarrhalis* and *N. sicca* are present, while *N. meningitidis* is found in the abnormal or venereally infected female (N. L. Fiumarra, personal communication). Why the gonococci are unique among the *Neisseria* in their attachment to and penetration of genital mucous membrane tissues in under intense investigation (see review by Watt and Ward, 1977). It is generally accepted that *Neisseria* species (especially the gonococci) resident on stratified squamous epithelium cannot penetrate this cell barrier. The organisms, however, may be held and trapped on these cells, as recently demonstrated with gonococci on vaginal squamous epithelium (Evans, 1977).

In Animals

The nasopharynx of domestic and experimental animals is the natural habitat for many *Neisseria* species. *N. caviae* and *N. animalis* have been isolated from the throats of guinea pigs (Berger, 1960; Pelczar, 1953), *N. canis* from the throat of dogs (Berger,

1962), and *N. cuniculi* from the throat of rabbits (Berger, 1962). Closely related taxonomically to *N. caviae* is *N. ovis*, which has been isolated routinely from keratoconjunctivitis lesions in sheep (Lindquist, 1960) and cattle. Experience has shown that if one uses intense effort, *Neisseria* species can be found resident in the nasopharynx of many animals, including primates. The relationship of these bacteria to the oral epithelial cells appears to be a loose and transient association, with no detectable antibodies to the homologous isolate.

Isolation

Human Pathogens

NEISSERIA MENINGITIDIS. Isolates of *N. meningitidis* may be obtained from spinal fluid, blood, expressed fluid from petechial lesions, or the oral cavity. On occasion, meningococci have been isolated from the cervix, urethra, and rectum of individuals thought to be venereally infected. Clinical materials such as blood or spinal fluid present no problem with mixed flora and can be cultured directly on the nonselective chocolate agar medium (CA) or Mueller-Hinton agar (M-H) containing 3% defibrinated sheep blood. Specimens from areas with a bacterial flora must be cultured on a selective medium such as Thayer-Martin or modified Thayer-Martin medium. The surface of the agar medium should be dry but not wrinkled. Prepared media stored in the refrigerator, tightly wrapped in plastic bags, usually maintain the proper surface moisture but must be prewarmed above 25°C before use. Since the viability of meningococci may be destroyed by excessive drying, chilling, autolytic enzymes, or unfavorable pH, it is important that all culture procedures be carried out as quickly as possible after obtaining the clinical specimen. The inoculated agar plates should be incubated at 36–37°C in air with increased CO_2 content (2–10%) and moisture. Commercial CO_2 incubators or candle jars with moistened paper in the bottom are satisfactory. For primary isolation, moisture (70–80%) is more important than CO_2. Growth should be evident within 18–20 h and negative plates should be reincubated another 24 h. The possibility of mixed bacterial meningitis involving *N. meningitidis* with *Haemophilus influenzae*, *Streptococcus pneumoniae*, or *Escherichia coli* must be considered (Herwig, Middelkamp, and Hartmann, 1963). It is critical, therefore, that well-isolated, distinctive colonies be obtained. Specimens of blood or spinal fluid should be obtained before chemotherapy is initiated. It is preferable to inoculate 5–10 ml blood drawn at the patient's bedside into 50–100 ml prewarmed Trypticase soy broth (TSB). A blood culture bottle containing TSB under vacuum with added CO_2 has proven very satisfac-

tory. The broth is incubated 35–37°C and inspected daily (up to 7 days) for evidence of turbidity, hemolysis, or discrete colonies on the sedimented red blood cells. Subcultures can then be made on agar media for identification. It is not practical to handle spinal fluid the way blood is handled; therefore, the spinal fluid sediment should be cultured directly on CA and a suitable blood agar plate. For a more complete discussion on the management of clinical specimens, see Isenberg et al. (1974).

Because of mixed flora, isolation from the nasopharynx requires a selective medium. Thayer-Martin medium (T-M) is satisfactory.

Selective Medium for *Neisseria meningitidis* (Thayer and Martin, 1966)

Beef, infusion from	300.0 g
Acid hydrolysate of casein	17.5 g
Starch	1.5 g
Agar	17.0 g
Distilled water	1.0 liter

Final pH, 7.4. Dispense and autoclave at 116–121°C for 15 min. Cool to 45–50°C and add: Vancomycin, 3 μg/ml; colistin, 7.5 μg/ml; nystatin, 12.5 units/ml.

For mass screening for meningococcal "carriers" or familial contacts of patients with overt disease, cotton swabs of the posterior nasopharynx are placed in 2–3 ml of TSB and cultured on duplicate plates of M-H agar medium. The M-H agar of one of the pair of plates contains 0.05 mg/ml sulfadiazine; the other, plain M-H agar. This technique, properly controlled, provides simultaneous results regarding the number of carriers versus those with strains resistant to the sulfadiazine, one of the preferred chemotherapeutic agents for treating carriers.

Several semidefined and defined media have been developed to study nutritional requirements and genetic competence of the meningococci (Frantz, 1942; Grossowicz, 1945; Jyssum, 1965; Mueller and Hinton, 1941). A highly satisfactory defined medium was developed by Catlin and Schloer (1962).

Defined Medium for Meningococci (Catlin and Schloer, 1962)

Solution A:	
NaCl	5.85 g
KCl	0.186 g
NH_4Cl	0.401 g
Na_2HPO_4	1.065 g
KH_2PO_4	0.170 g
$Na_3C_6H_5O_7 \cdot 2H_2O$	0.647 g
Distilled Water	350 ml
Solution B:	
$MgSO_4 \cdot 7H_2O$	0.616 g

MnSO$_4$·H$_2$O (0.15 M)	0.05 ml
Distilled water	50 ml

Solution C:

L-Arginine·HCl	0.300 g
L-Cysteine·HCl·H$_2$O	0.010 g
Glycine	0.100 g
Monosodium glutamate	1.100 g
Distilled water	25.0 ml

Solution D:

Sodium lactate	25.0 ml
Distilled water	50.0 ml

Solution E:

Purified agar	10.0 g
Distilled water	500 ml

Solution F:

CaCl$_2$·2H$_2$O (1.0 M)	0.5 ml

Sodium lactate is prepared from reagent lactic acid (85%): to 23.0 g chilled in an ice bath, 5.0 N NaOH slowly added with stirring to pH 7.4 (approximately 40.5 ml), distilled H$_2$O added to final volume of 100 ml.

Autoclave solutions B, E, and F (121–123°C for 15–18 min); filter-sterilize solutions A, C, and D. Mix solutions A, B, C, and D; adjust pH to 7.4 ± 0.5 with 1 N NaOH; at 50°C, add solution E then solution F while rotating container; pour solution into sterile Petri dishes.

NEISSERIA GONORRHOEAE. Specimens for isolation of the gonococci from females are obtained with cotton swabs principally from the cervix and secondarily from the rectum, urethra, and oropharynx. Urethral exudate is the main source of specimens in heterosexual males, and the rectum and oropharynx in homosexual males.

All specimens are cultured directly modified on T-M medium (Martin and Lester, 1971), incubated at 35–36°C in air with increased CO$_2$ content (2–10%) and moisture. Growth is evident by 20–24 h, but in up to 40% of isolations, it may not be apparent for 48 h (Kellogg, 1974).

In 5–10% of females with proctitis, the rectum is the primary site for isolation (Schmale, Martin, and Domescik, 1969). In males this is limited exclusively to homosexuals. Cotton swab specimens are obtained from the crypts just inside the rectal ring, avoiding fecal material, and cultured directly on modified T-M medium supplemented with trimethoprim to inhibit *Pseudomonas* and *Proteus* species (Martin and Lester, 1971). The bacteriological diagnosis of gonococcal systemic infections requires isolation and identification of the gonococci from the tissues involved. In suspected gonococcal arthritis, synovial fluid is aspirated, mixed with an equal volume of TSB, and centrifuged at high speed (e.g., 10,000 × *g* for 15 min), and the sediment is cul-

tured on enriched chocolate agar and modified T-M. Blood from gonococcemia is inoculated (approximately 5 ml) into biphasic medium consisting of 20 ml TSB containing 0.5% added glucose over a layer of agar base (GC medium base) and incubated at 35–36°C with increased CO$_2$ content (2–10%) for up to 10 days. The culture should be examined daily for turbidity and/or discrete colonies and subcultured on CA for identification.

Several defined media have been developed for nutritional and genetic studies of the gonococci (Catlin, 1973; Hunter and McVeigh, 1970). The medium developed by LaScolea and Young (1974) is described below:

Defined Medium for Gonococci (LaScolea and Young, 1974)

Part I

L-Arginine·HCl	250 μg/ml
L-Asparagine·H$_2$O	125 μg/ml
L-Cysteine·HCl·H$_2$O	1,200 μg/ml
L-Cystine	50 μg/ml
L-Glutamic acid	188 μg/ml
L-Isoleucine	125 μg/ml
L-Methionine	190 μg/ml
L-Proline	250 μg/ml
L-Serine	125 μg/ml
Hypoxanthine	2.5 μg/ml
Uracil	5.0 μg/ml
Glucose	7,500 μg/ml
Glycerol	0.125% (vol/vol)
Oxaloacetate	250 μg/ml
Pyruvic acid	0.125% (vol/vol)
Sodium acetate	1,500 μg/ml
Sodium citrate	1,125 μg/ml
Biotin	4.0 μg/ml
Calcium pantothenate	20.0 μg/ml
Nicotinamide adenine dinucleotide	10.0 μg/ml
Thiamine HCl	20.0 μg/ml
Thiamine pyrophosphate chloride	20.0 μg/ml
CaCl$_2$·2H$_2$O (1 mg/ml)[1]	5 μl/ml
Fe(NO$_3$)$_3$·9H$_2$O (1 mg/ml)[1]	5 μl/ml
KCl	300 μg/ml
K$_2$HPO$_4$	10,500 μg/ml
KH$_2$PO$_4$	4,500 μg/ml
K$_2$SO$_4$	900 μg/ml
MgCl$_2$·6H$_2$O	450 μg/ml
NH$_4$Cl	300 μg/ml
NaCl	5,250 μg/ml
NaHCO$_3$	1,000 μg/ml
Na$_2$SO$_3$	750 μg/ml
Sodium thioglycolate	25 μg/ml
Distilled water	123 ml

Part II

Soluble starch	1,000 μg/ml

Agar	10,000 μg/ml
Water	500 ml
Final pH, 7.4	

Autoclave Part II (15 psi 15 min), cool to 50°C. Part I adjusted to pH 7.20–7.25 with 6 N NaOH; equilibrate for 45 min at 50°C, filter (0.20-μm filter), and add to Part II. Pour in sterile Petri dishes. Glass distilled mineral water should be used for preparation of all solutions and the glassware used in experiments should be cleaned with nitric acid.[1]

OTHER *NEISSERIA* SPECIES. *Neisseria* species other than *N. meningitidis* and *N. gonorrhoeae* can be obtained on primary isolation on nutrient agar (NA). Experience has indicated that specimens from the nasopharynx, when cultured on CA, had a higher recovery rate than those cultured on NA. The CA plates are incubated at 36–37°C in air for 24–48 h. Dissimilar colonies that are oxidase positive (to be described later) are subcultured on NA. The process is tedious and demanding, but as yet there is no selective medium available for isolation of only the *Neisseria* species from a mixed flora in the nasopharynx. Modified T-M medium is satisfactory for the meningococci and gonococci but inhibitory for all other *Neisseria* species. Generally, only a few colonies of *Neisseria* are isolated from the nasopharynx of domestic and experimental animals, but they may be abundant, as seen with *N. perflava* in humans (Vedros, Hunter, and Rust, 1966).

Identification

There are at present very few biological parameters available to reproducibly and accurately identify *Neisseria* species. The Subcommittee on Neisseriaceae of the International Committee on Taxonomy is attempting to develop a set of minimum criteria for taxa of the genus *Neisseria*. The report of this Committee should be forthcoming soon. The criteria for identification described below are those in general use in clinical laboratories plus other characteristics that may become incorporated in the final identification scheme.

Morphology and Colonial Characteristics

With the exception of *Neisseria elongata*, members of the genus *Neisseria* are Gram-negative diplococci arranged in pairs with flattened, adjacent sides (coffee bean shaped). The cells range in diameter from 0.6 to 1.0 μm, depending on growth conditions, and divide in two planes. One plane is at right angles to the other, often resulting in tetrad formation in young cultures (3–5 h) when observed in wet mounts under phase-contrast microscopy. This characteristic can be used to differentiate the cocci from short rods that divide in only one plane and form short chains or pairs. Some Gram-staining variability has been noted among isolates from animals and gonococcal strains.

Fimbriae or pili have been observed in strains of *N. meningitidis*, *N. perflava*, *N. subflava*, and *N. gonorrhoeae* (Wistreich and Baker, 1971). The presence of pili on the gonococci and their role in initial attachment of the organism to mucosal cells have been the subject of intense study (Watt and Ward, 1977). Pili also play a prominent role in the inter- and intraspecific transfer of plasmids coded for β-lactamase production.

Neisseria colonies on agar surfaces vary from species to species with regard to pigment and consistency. The majority are entire with either smooth, granular, or butyrous consistency. *N. subflava*, *N. flavescens*, and *N. lactamica* have a yellow pigment, whereas *N. sicca* is noted for its dry, wrinkled appearance. The meningococci typically have a transparent appearance. The variation in colonial morphology of the gonococci has been used to differentiate and classify this species in much more detail than other *Neisseria* species. Small dew drops, glistening colonies designated T_1 and T_2, were found to be virulent for man (Kellogg et al., 1963) and to possess pili (Jephcott, Reyn, and Birch-Anderson, 1971). Colonies that were large, flat, and dull in appearance were designated T_3 and T_4. These colonies were avirulent for man and did not contain pili. The association of virulence with colonial morphology, in addition to variations in colonial color, has been the subject of intense study (see review by Swanson, 1977). All of the *Neisseria* species, with the exception of *N. meningitidis*, produce a pitting of agar surfaces.

Biochemical Reactions

The production of acid from various mono- and disaccharides has been the major biochemical reaction for identifying and classifying *Neisseria* (Reyn, 1974) (Table 1). Heavy inocula of bacteria are added to cystine Trypticase agar (CTA) containing a suitable pH indicator and 1% concentration of a given carbohydrate. After 24–48 h, the production of a yellow (or acid) reaction is considered positive. Experience has shown that many *Neisseria* isolates grow poorly in CTA and may give erroneous results. A modification of the rapid sugar fermentation confirmation (RFC) of Kellogg and Turner (1973) has been developed for routine identification of all *Neisseria* isolates (Vedros, 1978). The RFC uses preformed enzymes rather than growth for acid production, gives results in 2–4 h, and has been partic-

---

[1] Anhydrous compound used.

Table 1. Characteristics of certain *Neisseria* species.[a]

Organism	Pigment	Glucose	Maltose	Sucrose	Fructose	Mannose	Lactose	Polysaccharide from sucrose	NO$_4$ to NO$_3$	NO$_2$ to N	DNase	G+C (mol%)[b]	Comments
N. animalis	−	+	−	+	(+)	−	−	+	−	+	−	51.4	
N. cuniculi	(+)	−	−	−	−	−	−	−	−	−	+	44.2	
N. canis	Y	−	−	−	−	−	−	−	+	−	(+)	49.8	
N. caviae	−	−	−	−	−	−	−	−	+	+	(+)	46.0	Hemolytic
N. ovis	−	−	−	−	−	−	−	−	+	−	+	46.3	Hemolytic
N. elongata	−	−	−	−	−	−	−	−	−	+	(+)	54.3	Hemolytic
N. cinerea	−	−	−	−	−	−	−	−	−	+	+	ND	
N. denitrificans	−	+	−	+	+	+	+	+	−	+	+	ND	
N. lactamica	Y	+	+	−	−	−	+	−	−	+	(+)	ND	ONPG positive

[a] Symbols: −, no reaction; +, positive reaction; (+), variable; Y, yellow; ND, not tested.
[b] Determined by CsCl gradient technique.

ularly useful in identifying atypical clinical isolates and nonviable cultures received in the laboratory. The presence of cytochrome oxidase is typical of all *Neisseria* species. One or two drops of a fresh, 1% solution of tetramethyl-*p*-phenylenediamine is added onto a colony, and a change in color from violet to purple indicates a positive reaction. The color change usually occurs within 10–15 s, and any delay indicates that the isolate is probably not *Neisseria*. *N. sicca* can be differentiated from *N. perflava* by the detection of an iodine-positive metabolic product in the former when grown in Trypticase soy agar without sucrose (Berger and Catlin, 1975). Analysis of the cellular and free lipopolysaccharides indicates that *N. perflava*, *N. subflava*, and *N. flava* are distinct species and that *N. canis* and *N. subflava* are identical (Johnson, Perry, and McDonald, 1976).

In addition to Gram stain, cell morphology, oxidase reaction, and production or failure to produce acid from sugars, a test that is consistent with regard to classification of *Neisseria* is the reduction of nitrate to nitrite and nitrite to nitrogen gas. Both saccharolytic and nonsaccharolytic species can be subdivided on the basis of their nitrate- and nitrite-reducing capabilities. Such a classification scheme has been proposed by Berger (1967), but has yet to be confirmed by others.

More recently, analysis by gas-liquid chromatography (GLC) of hyroxy acids produced by *Neisseria* species shows promise in providing another taxonomic criterion (Brooks et al., 1972). Profiles were obtained for *N. meningitidis*, *N. gonorrhoeae*, *N. haemolysans*, *N. lactamica*, *N. perflava*, *N. flava*, *N. mucosa*, *B. catarrhalis*, *N. ovis*, and *N. caviae*. With GLC, analysis for fatty alcohols released from waxes indicates that waxes are not present in the "true" *Neisseria* but are present in all "false" *Neisseria* (Bryn, Jantzen, and Bøvre, 1977).

Serology

The use of serology in the identification and classification of any bacterium requires definitive knowledge of the antigenic determinants involved. Within the genus *Neisseria*, attention has focused on the meningococci and gonococci, for purposes of epidemiology and control of these important human diseases. Other *Neisseria* species are considered to be homogeneous with regard to antigenic types, but very little research has been conducted with these species.

In 1950, the meningococci were classified based on agglutination into serogroups A, B, C, and D (Branham, 1953). Since then, serogroups designated X, Y(E), Z, 29E(Z'), and W-135 have been identified (Hollis, Wiggins, and Schubert, 1968; Slaterus, Ruys, and Sieberg, 1963; Vedros, Ng, and Culver, 1968). Specific polysaccharides have been isolated and characterized from the surface of these various serogroups to explain their serological specificity (Vedros, 1978). These important studies have resulted in the successful use of polysaccharide vaccines against groups A and C meningococcal infections (Gotschlich, 1975). Within group B, several serotypes have been identified (Frasch and Chapman, 1973), as well as within Group C (Apicella, 1974). Prototype strains for each serogroup except 29E and standard methods for producing antisera and conducting the agglutination procedure have been established by the World Health Organization (Vedros, 1978).

The existence of distinct antigenic populations among the gonococci has been postulated because of the lack of immunity in man to repeated infections. Hence, no standardized or universally accepted serological classification schemes exist, although considerable efforts have been made in this area. Among the most promising are the use of the outer membrane protein complex that has been used to classify the gonococci into 16 serotypes (Johnston, Holmes, and Gotschlich, 1976), and the establishment of both serotypes and immunotypes (Arko et al., 1976; Wong et al., 1976; Wong et al., 1979). Other serotypes may be possible with either or both approaches. In addition, important advances have been made in understanding the immunochemistry of the gonococcus and the chemical nature of gonococcal lipopolysaccharides (Perry, Diena, and Ashton, 1977). The combination of these two approaches may result in a satisfactory serological classification of the gonococci.

Bacteriocins have been isolated from meningococci by induction with mitomycin C (1 μg/ml) or ultraviolet light (Kingsbury, 1966). *N. perflava, N. subflava, N. flavescens,* and two isolates of group A *N. meningitidis* were susceptible to the same bacteriocin isolated from *N. meningitidis*. Further epidemiological studies indicated that group C *N. meningitidis* could be divided into 11 patterns and group B into 32 patterns based on their sensitivity to meningocins from 14 indicator strains (Counts, Seeley, and Beaty, 1971). Bacteriocins have also been isolated from the gonococci (Flynn and McEntegart, 1972), but have been of limited use in epidemiological studies and for establishing a typing pattern. Further, there is a basic question as to whether the gonococci produce true bacteriocins or whether the inhibition is due to phospholipase A (Senff et al., 1976). Although bacteriocinogeny has been a useful tool in identifying certain serologically homogeneous bacteria during epidemics, its use in classification or identification of *Neisseria* species requires further study.

Bacteriophages have been isolated from throat washings and lysogenic organisms for *N. perflava* and *N. flavescens* (Stone, Culbertson, and Powell, 1956; Phelps, 1967) and in one instance phages have been isolated from *N. meningitidis* (Cary and Hunter, 1967). These phages have a limited host range and further study is needed to indicate their usefulness in identifying and taxonomically classifying *Neisseria* species.

The identification and subsequent classification of the species in the genus *Neisseria* are in a fluid state and will require further study before a satisfactory scheme is developed. For the present, Gramstaining reaction, cell morphology, cytochrome oxidase reaction, production of acid from various sugars, and nitrate and nitrite reduction are being used routinely in most laboratories. Added to these criteria are DNA base ratios, genetic compatibility by transformation, fatty acid analysis, and DNA hybridization. Selection of suitable prototype strains and collaborative efforts by several laboratories should provide enough data to define the species within the genus *Neisseria* and the genera within the family Neisseriaceae.

Literature Cited

Apicella, M. A. 1974. Identification of a subgroup antigen on the *N. meningitidis* group C capsular polysaccharide. Journal of Infectious Diseases **129**:147–153.

Arko, R. J., Wong, K. H., Bullard, J. C., Logan, L. C. 1976. Immunological and serological diversity of *Neisseria gonorrhoeae:* Immunotyping of gonococci by cross-protection in guinea pig subcutaneous chambers. Infection and Immunity **14**:1293–1296.

Audureau, A. 1940. Etude du genre *Moraxella*. Annales de l'Institut Pasteur **64**:126–166.

Berger, U. 1960. *Neisseria animalis* n. sp. Zeitschrift für Hygiene und Infektionskrankheiten **147**:158–161.

Berger, U. 1962. Uber das Vorkommen von Neisserien bei einigen Tieren. Zeitschrift für Hygiene **148**:445–457.

Berger, U. 1967. Zur Systematik der Neisseriaceae. Zentralblatt für Bakteriologie, Parasitenkunde, Infektionskrankheiten und Hygiene, Abt. I Orig., Reihe A **265**:241–248.

Berger, U. 1971. *Neisseria mucosa* var. *heidelbergensis*. Zeitschrift für Medizinische Mikrobiologie und Immunologie **156**:154–158.

Berger, U., Catlin, B. W. 1975. Biochemical differentiation between *N. sicca* and *N. perflava*. Zentralblatt für Bakteriologie, Parasitenkunde, Infektionskrankheiten und Hygiene, Abt. 1 Orig., Reihe A **232**:129–130.

Berger, U., Husmann, D. 1972. Untersuchengen zum normalen Vorkommen von *Neisseria flavescens* (Branham, 1930). Medical Microbiology and Immunology **158**:121–127.

Bøvre, K. 1967. Transformation and DNA base composition in taxonomy with special reference to recent studies in *Moraxella* and *Neisseria*. Acta Pathologica et Microbiologica Scandinavica **69**:123–144.

Bøvre, K., Fiandt, M., Szybalski, W. 1969. DNA base composition of *Neisseria, Moraxella,* and *Acinetobacter* as determined by measurement of buoyant density in CsCl gradients. Canadian Journal of Microbiology **15**:335–338.

Bøvre, K., Holten, E. 1970. *Neisseria elongata* sp. nov., a rod-shaped member of the genus *Neisseria*. Re-evaluation of cell shape as a criterion in classification. Journal of General Microbiology **60**:67–75.

Branham, S. E. 1930. A new meningococcus-like organism (*Neisseria flavescens* n. sp.) from epidemic meningitis. Public Health Reports **45**:I, 845–849.

Branham, S. 1953. Serological relationships among meningococci. Bacteriological Reviews **17**:175–188.

Brooks, J. B., Kellogg, D. S., Thacker, L., Turner, E. M. 1972. Analysis by gas chromatography of hydroxy acids produced by several species of *Neisseria*. Canadian Journal of Microbiology **18**:157–168.

Bryn, K., Jantzen, E., Bøvre, K. 1977. Occurrence and patterns of waxes in Neisseriaceae. Journal of General Microbiology **102**:33–43.

Cary, S. G., Hunter, D. H. 1967. Isolation of bacteriophages active against *Neisseria meningitidis*. Journal of Virology **1**:538–542.

Catlin, B. W. 1971. Report (1966–1970) of the Subcommittee on the Taxonomy of the Neisseriaceae to the International Committee on Nomenclature of Bacteria. International Journal of Systematic Bacteriology **21**:154–155.

Catlin, B. W. 1973. Nutritional profiles of *Neisseria gonorrhoeae, Neisseria meningitidis* and *Neisseria lactamica* in chemically defined media and the use of growth requirements for gonococcal typing. Journal of Infectious Diseases **128**:178–194.

Catlin, B. W., Schloer, G. M. 1962. A defined agar medium for genetic transformation of *Neisseria meningitidis*. Journal of Bacteriology **83**:470–474.

Counts, G. W., Seeley, L., Beaty, H. N. 1971. Identification of an epidemic strain of *Neisseria meningitidis* by bacteriocin typing. Journal of Infectious Diseases **124**:26–32.

Edwards, E. A., Devine, L. F., Sengbusch, C. H., Ward, H. W. 1977. Immunological investigations of meningococcal disease. III. Brevity of group C acquisition prior to disease occurrence. Scandinavian Journal of Infectious Diseases 9:105–110.

Evans, B. A. 1977. Ultrastructural study of cervical gonorrhoea. Journal of Infectious Diseases 136:248–255.

Faucon, R., Le Fevre, M., Menard, M., Millan, J. 1970. La meningite cerebrospinale a Fes en 1966–1967. II. L'épidémie de Fes E.–Les porteurs de germes. Medicine Tropicale 30:470–476.

Flynn, J., McEntegart, M. J. 1972. Bacteriocins from Neisseria gonorrhoeae and their possible role in epidemiological studies. Journal of Clinical Pathology 25:60–61.

Frantz, I. D., Jr. 1942. Growth requirements of the meningococcus. Journal of Bacteriology 46:757–761.

Frasch, C. E., Chapman, S. S. 1973. Classification of Neisseria meningitidis group B into distinct serotypes. III. Application of a new bactericidal inhibition technique to distribution of serotypes among cases and carriers. Journal of Infectious Diseases 127:149–154.

Goldschneider, I., Gotschlich, E. C., Artenstein, M. S. 1969. Human immunity to the meningococcus. I. The role of humoral antibodies. Journal of Experimental Medicine 129:1307–1326.

Gotschlich, E. C. 1975. Development of polysaccharide vaccines for the prevention of meningococcal infections. Monographs in Allergy 9:245–258.

Grossowicz, N. 1945. Growth requirements and metabolism of Neisseria intracellularis. Journal of Bacteriology 50:109–115.

Henriksen, S. D. 1952. Moraxella: Classification and taxonomy. Journal of General Microbiology 6:318–328.

Henriksen, S. D. 1976. Moraxella, Neisseria, Branhamella, and Acinetobacter. Annual Review of Microbiology 30:63–83.

Henriksen, S. D., Bøvre, K. 1968. The taxonomy of the genera Moraxella and Neisseria. Journal of General Microbiology 51:387–392.

Herwig, J. C., Middelkamp, J. N., Hartmann, A. F. 1963. Simultaneous mixed bacterial meningitis in children. Journal of Pediatrics 63:76–83.

Hollis, D. G., Wiggins, G. L., Schubert, J. H. 1968. Serological studies of ungroupable Neisseria meningitidis. Journal of Bacteriology 95:1–4.

Hollis, D. G., Wiggins, G. L., Weaver, R. E. 1969. Neisseria lactamicus sp. n., a lactose fermenting species resembling Neisseria meningitidis. Applied Microbiology 17:71–77.

Hollis, D. G., Wiggins, G. L., Weaver, R. E., Schubert, J. H. 1970. Current status of lactose fermenting Neisseria. Annals of the New York Academy of Sciences 174:444–449.

Holten, E. 1973. Glutamic dehydrogenases in genus Neisseria. Acta Pathologica et Microbiologica Scandinavica, Sect. B 81:49–58.

Hunter, K. M., McVeigh, I. 1970. Development of a chemically defined medium for growth of Neisseria gonorrhoeae. Antonie van Leeuwenhoek Journal of Microbiology and Serology 35:305–316.

Isenberg, H. D., Washington, J. A., II, Balows, A., Sonnenwirth, A. C. 1974. Collection, handling, and processing of specimens, pp. 59–88. In: Lennette, E. H., Spaulding, E. H., Truant, J. P. (eds.), Manual of clinical microbiology, 2nd ed. Washington, D. C.: American Society for Microbiology.

Jephcott, A. E., Reyn, A., Birch-Anderson, A. 1971. Neisseria gonorrhoeae. III. Demonstration of presumed appendates to cells from different colony types. Acta Pathologica et Microbiologica Scandinavica, Sect. B 79:437–439.

Jessen, J. 1934. Studien über grammnegative Kokken. Zentralblatt für Bakteriologie, Parasitenkunde, Infektionskrankheiten und Hygiene, Abt. 1 Orig. 113:75–88.

Johnson, K. G., Perry, M. B, McDonald, I. J. 1976. Studies of the cellular and free lipopolysaccharides from Neisseria canis and N. subflava. Canadian Journal of Microbiology 22:189–196.

Johnston, K. H., Holmes, K. K., Gotschlich, E. C. 1976. The serological classification of the Neisseria gonorrhoeae. I. Isolation of the outer membrane complex responsible for serotype specificity. Journal of Experimental Medicine 143:741–758.

Jyssum, K. 1965. Isolation of auxotrophs of Neisseria meningitidis. Acta Pathologica et Microbiologica Scandinavica 63:435–444.

Kellogg, D. S., Jr. 1974. Neisseria gonorrhoeae (gonococcus), p. 127. In: Lennette, E. H., Spaulding, E. H., Truant, J. P. (eds.), Manual of clinical microbiology, 2nd ed. Washington, D.C.: American Society for Microbiology.

Kellogg, D. S., Jr., Peacock, W. L., Deacon, W. E., Brown, L., Pirkle, C. I. 1963. Neisseria gonorrhoeae. I. Virulence genetically linked to clonal variation. Journal of Bacteriology 85:1274–1279.

Kellogg, D. S. Jr., Turner, E. M. 1973. Rapid fermentation confirmation of Neisseria gonorrhoeae. Applied Microbiology 25:550–552.

Kingsbury, D. T. 1966. Bacteriocin production by strains of Neisseria meningitidis. Journal of Bacteriology 91:1696–1699.

LaScolea, L. J., Young, F. E. 1974. Development of a defined minimal medium for the growth of Neisseria gonorrhoeae. Applied Microbiology 28:70–76.

Lauer, B. A., Fisher, C. E. 1976. Neisseria lactamica meningitis. American Journal of Diseases of Children 130:198–199.

Lindquist, K. A. 1960. Neisseria species associated with infectious keratoconjunctivitis of sheep. Neisseria ovis nov. spec. Journal of Infectious Diseases 106:162–165.

Martin, J. E., Jr., Lester, A. 1971. Transgrow, a medium for transport and growth of Neisseria gonorrhoeae and Neisseria meningitidis. Health Reports 86:30–33.

Mitchell, M. S., Rhoden, D. L., King, E. O. 1965. Lactose-fermenting organisms resembling Neisseria meningitidis. Journal of Bacteriology 90:560.

Molineaux, L. E. R. 1970. Climate and the epidemiology of meningococcal meningitis. Ph.D. thesis. University of California, Berkeley, California.

Mueller, J. H., Hinton, H. 1941. A protein free medium for primary isolation of the meningococcus and gonococcus. Proceedings of the Society for Experimental Biology and Medicine 48:330–333.

Pelczar, M. J., Jr. 1953. Neisseria caviae nov. spec. Journal of Bacteriology 65:744.

Perry, M. B., Diena, B. B., Ashton, F. E. 1977. Lipopolysaccharides of Neisseria gonorrhoeae pp. 286–301. In: Roberts, R. B. (ed.), The gonococcus. New York: John Wiley.

Phelps, L. N. 1967. Isolation and characterization of bacteriophages for Neisseria. Journal of General Virology 1:529–532.

Prentice, A. W. 1957. Neisseria flavescens as a cause of meningitis. Lancet I:613–614.

Pykett, A. H. 1973. Isolation of Neisseria lactamicus from the nasopharynx. Journal of Clinical Pathology 26:399–400.

Radke, R. A., Cunningham, G. C. 1949. A case of meningitis due to Pseudomonas aeruginosa (Bacillus pyocyaneus) and Neisseria flavescens with recovery. Journal of Pediatrics 35:99–101.

Rein, M. F. 1977. Epidemiology of gonococcal infections pp. 1–31. In: Roberts, R. B. (ed.), The gonococcus. New York: John Wiley.

Reyn, A. 1974. Gram-negative cocci and coccobacilli, pp. 427–444. In: Buchanan, R. E., Gibbons, N. E. (eds.), Bergey's manual of determinative bacteriology, 8th ed. Baltimore: Williams & Wilkins.

Sanborn, W. R., Vedros, N. A. 1966. Possibilities of application of complement fixation, indirect hemagglutination and fluorescent antibody tests to the epidemiology of meningococcal infections. Health Laboratory Science **3**:111–117.

Schmale, J. D., Martin, J. E., Jr., Domescik, G. 1969. Observation on the culture diagnosis of gonorrhoea in women. Journal of the American Medical Association **210**:312–314.

Senff, L. M., Wegener, W. S., Brooke, G. F., Finnerty, W. R., Makula, R. A. 1976. Phospholipid composition and phospholipase A activity of *Neisseria gonorrhoeae*. Journal of Bacteriology **127**:874–880.

Slaterus, K. W., Ruys, A. C., Sieberg, I. G. 1963. Types of meningococci isolated from carriers and patients in a nonepidemic period in the Netherlands. Antonie van Leeuwenhoek Journal of Microbiology and Serology **29**:265–271.

Stone, R. L., Culbertson, C. G., Powell, H. M. 1956. Studies on a bacteriophage active against a chromogenic *Neisseria*. Journal of Bacteriology **71**:516–520.

Swanson, J. 1977. Surface components associated with gonococcal-cell interaction, pp. 369–401. In: Roberts, R. B. (ed.), The gonococcus. New York: John Wiley.

Thayer, J. D., Martin, J. E. 1966. Improved medium selective for cultivation of *N. gonorrhoeae* and *N. meningitidis*. Public Health Reports **81**:559–562.

Vedros, N.A. 1978. Serology of meningococcus, pp. 293–314. In: Norris, J. R., Bergen, T. (eds.), Methods in microbiology, vol. 10. New York: Academic Press.

Vedros, N. A., Hunter, D. H., Rust, J. H., Jr. 1966. Studies on immunity in meningococcal meningitis. Military Medicine **131**:1413–1417.

Vedros, N. A., Johnston, D. G., Warren, P. I. 1973. *Neisseria* species isolated from dolphins. Journal of Wildlife Diseases **9**:241–244.

Vedros, N. A., Ng, J., Culver, G. 1968. A new serological group (E) of *Neisseria meningitidis*. Journal of Bacteriology **95**:1300–1304.

Vedros, N. A., Robinson, P. J., Gutekunst, R. A. 1966. Isolation and characterization of a bacterial inhibitor from human throat washings. Proceedings of the Society for Experimental Biology **122**:249–253.

Véron, M., Thibault, P., Second, L. 1959. *Neisseria mucosa* (Diplococcus mucosus Lingelsheim). Annales de l'Institut Pasteur **97**:497–510.

Watt, P. J., Ward, M. E. 1977. The interaction of gonococci with human epithelial cells, pp. 355–368. In: Roberts, R. B. (ed.), The gonococci. New York: John Wiley.

Wax. L. 1950. The identity of Neisseria other than the gonococcus isolated from the genito-urinary tract. Journal of Venereal Disease Information **31**:208–213.

Wilson, H. D., Overman, T. L. 1976. Septicemia due to *Neisseria lactamica*. Journal of Clinical Microbiology **4**:214–215.

Wistreich, G. A., Baker, F. F. 1971. The presence of fimbriae (pili) in three species of *Neisseria*. Journal of General Microbiology **65**:167–173.

Wong, K. H., Arko, R. J., Logan, L. C., Bullard, J. C. 1976. Immunological and serological diversity of *Neisseria gonorrhoeae*: Gonococcal serotypes and their relationship with immunotypes. Infection and Immunity **14**:1297–1301.

Wong, K. H., Arko, R. J., Schalla, W. O., Steurer, F. J. 1979. Immunological and serological diversity of *Neisseria gonorrhoeae*: Identification of new immunotypes and highly protective strains. Infection and Immunity **23**:717–722.

The Family Neisseriaceae: Rod-shaped Species of the Genera *Moraxella, Acinetobacter, Kingella,* and *Neisseria,* and the *Branhamella* Group of Cocci

KJELL BØVRE and NORMAN HAGEN

Increased use of genetic-molecular taxonomic methods during the last two decades has made clear the unsatisfactory status of the previous classification and nomenclature of the Neisseriaceae. The most easily cultivable and the most biochemically active bacteria had been subjected to excessive naming and splitting into a confusing series of genera and species, the majority of which may not deserve separate status. The less easily cultivable and the biochemically inactive organisms tended to be lumped into species or subspecies definitions that covered several of the presently recognized species, and often crossed the generic boundaries that are now becoming generally accepted. Consequently, the possibilities of exact identification have been strongly limited in the past, and information is incomplete or uncertain on occurrence, distribution in habitats, and natural roles, including pathogenicity, for many of the species as now defined. This presentation will, to some extent, concentrate on more recent investigations, including work in progress. For comprehensive surveys of older literature, the reader is referred to Berger (1963) and Henriksen (1973).

Supplementary discussion and references on taxonomy may be found in a recent review by Henriksen (1976).

Present Classification and Nomenclature

The outlines of the present classification within Neisseriaceae are shown in Table 1. All these organisms are nonflagellated, Gram-negative bacteria, strictly or preferentially aerobic. The genus *Acinetobacter* is the only completely oxidase-negative group of the family and consists of one rod-shaped species. The genus *Moraxella* primarily contains rod-shaped species, but in this presentation includes the coccal species *Moraxella catarrhalis, M. caviae, M. cuniculi,* and *M. ovis* (previously assigned to genus *Neisseria*). These cocci may be considered as members of the subgenus *Branhamella* of genus *Moraxella*. If so, the subgenus name may be included in the species name in parentheses, i.e., *Moraxella (Branhamella) catarrhalis* and correspondingly for the other coccal species.

Table 1. Classification within Neisseriaceae, with cellular shape and group ranges of G+C (mol%) of DNA.

Genus	Subgenus	Species	Cellular shape	G+C[a] (mol%)
Acinetobacter		*calcoaceticus*	rods	38–47
Moraxella	*Moraxella*	*lacunata, bovis, nonliquefaciens, osloensis, phenyl-pyruvica, atlantae*	rods	40–47.5
	Branhamella	*catarrhalis*[b] *caviae, cuniculi,*[c] *ovis*	cocci	40–47.5
Kingella		*kingae, indologenes, denitrificans*	rods	44.5–55
Neisseria		Coccal species		46.5–52
		elongata	rods	53–53.5
Unknown		*"M." urethralis*	rods	46–47

[a] Results with different methods compiled from literature (see text).
[b] For this species, *Branhamella* has been proposed also as the genus (Catlin, 1970).
[c] Not yet formally transferred to *Moraxella* (or *Branhamella*).

Several authors feel that *Branhamella* should constitute a separate genus, as formally proposed only for the *catarrhalis* species. The name *Branhamella catarrhalis* is therefore also a correct designation, the choice depending on taxonomic opinion. In *Bergey's Manual of Determinative Bacteriology,* eighth edition, *Branhamella* appears as a genus with one species, *B. catarrhalis,* whereas the three other coccal species are listed among species incertae sedis (Reyn, 1974).

The genus *Kingella,* consisting of rod-shaped species, has its representation in *Bergey's Manual,* eighth edition, as *M. kingii* (i.e., *Kingella kingae*) among the species incertae sedis (Lautrop, 1974). The two other species of *Kingella* are not definitely proved relatives of *K. kingae.*

The genus *Neisseria* consists mainly of coccal species, which are treated in Chapter 121 of this Handbook. The rod-shaped *Neisseria elongata* is listed among the species incertae sedis in *Bergey's Manual,* eighth edition (Reyn, 1974).

Included in Table 1 is "*Moraxella*" *urethralis,* another rod-shaped organism which is listed among species incertae sedis in *Bergey's Manual,* eighth edition (Lautrop, 1974). This species appears to belong in the family Neisseriaceae, but not to any of the presently defined genera.

Excluded from Table 1 are several new species of genus *Moraxella,* both rod-shaped and coccal ones, and an additional rod-shaped entity, which will certainly end up as a new species of the genus *Neisseria.* The circumscriptions of these organisms have not yet been published. Also excluded is a group of psychrophilic, oxidase-positive rods, which is presently under study and which is considered to belong to the family.

Progress of Genetic Classification

In 1961, Catlin and Cunningham detected two main genetic groups within the old genus *Neisseria.* One species, "*N.*" *catarrhalis,* was distinguished from the true *Neisseria* species by having no genetic compatibility with them in genetic transformation with streptomycin resistance (Str^r) as marker, in addition to a clearly distinct overall base composition of DNA. Bøvre (1963) first reported Str^r transformation results showing that the rod-shaped *Moraxella* was related to the coccal *catarrhalis* organism. This report was independently confirmed by Catlin (1964). The transfer of genus *Moraxella* to the family *Neisseriaceae* was formally proposed by Henriksen and Bøvre (1968b), and this proposal has been generally accepted.

Table 2 shows the affinities between various rod-shaped species of genus *Moraxella,* as based on experiments with Str^r transformation (Bøvre, 1964a, b, 1965a, c, d; Bøvre and Henriksen, 1967a, b; Bøvre et al., 1976). *M. lacunata* unifies the previous *M. lacunata* and *M. liquefaciens,* in response to observations of particularly close genetic affinities (Bøvre, 1965c; Henriksen and Bøvre, 1968b). *M. bovis* is now considered to comprise *M. equi* (Hughes and Pugh, 1970), because of close relationship in genetic transformation (unpublished). *M. lacunata, M. bovis,* and *M. nonliquefaciens* may be referred to as "classical moraxellae" or as the "*M. lacunata* group". The homogeneity of this group was confirmed by pulse-RNA (mRNA)-DNA hybridization, which showed interspecies affinities ranging from 13.1 to 34.2% of intrastrain autologous reactions. The affinities between *M. nonliquefaciens* and *M. osloensis*/*M. phenylpyruvica* were distinctly lower (1.2–2%) in such hybridization (Bøvre, 1970a). *M. nonliquefaciens* was previously, to a large extent, lumped with *M. osloensis* and "*M.*" *urethralis* as *Mima polymorpha* var. *oxidans* (Bøvre, 1965d; Lautrop, Bøvre, and Frederiksen, 1970), a concept which has been eliminated from modern taxonomy (Henriksen, 1963; Judicial Commission, 1971; Pickett and Manclark, 1965).

As shown in Table 3, the rod-shaped organisms of genus *Moraxella,* particularly the *M. lacunata* group, have close affinities in Str^r transformation to the coccal species that have been named "false

Table 2. Affinities in Str^r transformation between established rod-shaped species of genus *Moraxella.*[a]

Donor organisms	Ratios of interstrain to autologous transformation, with recipient organisms				
	M. lacunata	*M. bovis*	*M. nonliquefaciens*	*M. osloensis*	*M. atlantae*
M. lacunata	0.1–1	1×10^{-2}	2×10^{-3}	4×10^{-5}	$-<10^{-6}$
M. bovis	1×10^{-3}	0.1–1	3×10^{-3}	1×10^{-5}	$-<10^{-6}$
M. nonliquefaciens	5×10^{-3}	2×10^{-3}	0.3–1	6×10^{-6}	$-<10^{-6}$
M. osloensis	4×10^{-6}		$+<10^{-5}$	0.3–1	$-<10^{-6}$
M. atlantae			$+<10^{-6}$	9×10^{-6}	0.5–0.8
M. phenylpyruvica			$+<10^{-6}$	6×10^{-6}	3×10^{-6}[b] $(-<10^{-6})$

[a] After Bøvre (1964b, 1965a, c, d, 1967a), Bøvre and Henriksen (1967b), Bøvre et al. (1976), and unpublished results; average values.
$+<$, Transformants detected; $-<$, transformants not detected, followed by estimated upper limit of ratio.
[b] Result with only one donor strain; results with other strains as shown in parentheses.

Table 3. Affinities in Str^r transformation between the subgenera *Moraxella* and *Branhamella* of the genus *Moraxella,* compared with mutual affinities of species within *Branhamella.*[a]

| Donor organisms | Ratios of interstrain to autologous transformation, with recipient organisms | | | | |
| | | *Moraxella (Branhamella)* | | | |
	catarrhalis	*cuniculi*	*ovis*	(new, pigs)[b]	(new, sheep)[b]
M.(B.) catarrhalis		2×10^{-3}	4×10^{-5}	8×10^{-3}	5×10^{-4}
M.(B.) cuniculi	1×10^{-3}		2×10^{-4}	2×10^{-2}	2×10^{-4}
M.(B.) ovis	1×10^{-5}	2×10^{-4}		1×10^{-3}	5×10^{-3}
M.(B.) (new, pigs)	6×10^{-4}	5×10^{-3}	9×10^{-4}		5×10^{-4}
M.(B.) (new, sheep)	8×10^{-4}	2×10^{-4}	6×10^{-3}	1×10^{-3}	
M.(B.) caviae	2×10^{-5}	5×10^{-5}	5×10^{-4}	5×10^{-4}	3×10^{-4}
M. (new rod, cattle)[c]	5×10^{-4}	4×10^{-4}	2×10^{-3}	5×10^{-4}	1×10^{-3}
M. lacunata	5×10^{-5}	1×10^{-4}	2×10^{-4}	2×10^{-4}	1×10^{-3}
M. bovis	4×10^{-5}	3×10^{-5}	8×10^{-4}	1×10^{-4}	1×10^{-3}
M. nonliquefaciens	5×10^{-5}	7×10^{-5}	9×10^{-5}	2×10^{-5}	3×10^{-4}
M. (new rod, pigs)[d]	5×10^{-5}	2×10^{-5}	8×10^{-6}	6×10^{-6}	5×10^{-4}
M. osloensis	2×10^{-5}	1×10^{-5}	3×10^{-5}	1×10^{-5}	1×10^{-5}
M. phenylpyruvica	8×10^{-6}	2×10^{-5}	1×10^{-5}	3×10^{-5}	6×10^{-5}
M. atlantae	8×10^{-6}	2×10^{-5}	1×10^{-5}	3×10^{-5}	4×10^{-5}

[a] Partly preliminary results, with figures based on average of observations or only single experiments (Bøvre, 1965b, 1967a; Bøvre and Henriksen, 1967b; Bøvre et al., 1976; unpublished studies).
[b] Nonhemolytic.
[c] Nonhemolytic, not serum-liquefying.
[d] Hemolytic, not serum-liquefying.

neisseriae'' in the past. This group includes ''*Neisseria*'' *catarrhalis,* ''*N.*'' *caviae,* ''*N.*'' *ovis* and, as has been shown recently, ''*N.*'' *cuniculi* (Bøvre, 1965b, 1967a; Bøvre and Henriksen, 1967b; Bøvre et al., 1976; unpublished). The affinities are partly as high as or higher than between the coccal species mutually, and also distinctly higher than between some of the rod-shaped *Moraxella* species (see Table 2). It should be noted that the property of hemolysis in *Moraxella bovis* has been transferred with great efficiency to a nonhemolytic variant of *Moraxella (Branhamella) ovis* in genetic transformation, which strengthens the other findings of close relationship (Bøvre, 1967a). It has therefore been proposed to include these coccal species in genus *Moraxella* (Bøvre and Henriksen, 1967b).

The overlapping compatibilities were later consolidated by mRNA-DNA hybridization. Whereas the mutual affinities between the *catarrhalis,* *caviae,* and *ovis* species could be expressed as 2.5–5.5% of autologous reactions, the affinities between the cocci and the *M. lacunata* group ranged from 2.4 to 9.7% (Bøvre, 1970a). Catlin's proposal in 1970 to create a new genus, *Branhamella,* for ''*N.*''*catarrhalis* does not correspond well with this closeness of rod-shaped *Moraxella* species and the whole group of ''false neisseriae''. As illustrated by the Str^r transformation data in Tables 2 and 3 and by the hybridization results, the case appears stronger genetically for including these cocci in genus *Moraxella* than for keeping *M. osloensis, M. phenyl-pyruvica,* and *M. atlantae* in the same genus as the *M. lacunata* group. However, it has recently been proposed to divide the genus *Moraxella* into two subgenera: subgenus *Moraxella* for the rod-shaped organisms, and subgenus *Branhamella* for the coccal species belonging genetically to the genus (Bøvre, 1979). This step has been taken mainly to facilitate communication among those adhering to either taxonomy, without suggesting a genus distinction based mainly on one or two characters, the cellular shape and mode of division.

The distribution and genetic relations in Str^r transformation of oxidase-positive rods and cocci found as parasites in animals are now being extensively studied, particularly organisms with affinities to *Moraxella* but also rods related to *Neisseria* (see below). Several new homogeneous entities have been detected (unpublished). The preliminary transformation results with four of them are included in Table 3. There is no particularly close relationship between either of the two new rod-shaped entities and the other rod-shaped species (not tabulated).

Transformation studies of the genus *Acinetobacter* by Juni (1972, 1978), mainly by application of nutritional DNA markers, have shown that this genus is rather homogeneous, although significant heterogeneity was indicated by the use of certain markers. There was no indication of relatedness to other genera in this type of transformation. Our preliminary results with Str^r transformation indicate, even more strongly, evolutionary deviations within *Acinetobacter* (Bøvre et al., 1976; unpublished). Donors of this genus have distinct, but generally very low, Str^r transformation activities on *Bran-hamella* and *M. osloensis* recipients. The *Acineto-bacter* donors are usually somewhat less active on an *M. (B.) catarrhalis* recipient than are *M. osloensis* donors, and at least 10 times less active on an *M. (B.) ovis* recipient than is *M. osloensis* (Bøvre, 1967b).

In competitive DNA-DNA hybridization, evidence was provided for DNA homology between a large number of strains of *Acinetobacter* (Johnson, Anderson, and Ordal, 1970). However, the genus could be divided in several subgroups with intergroup relations of some strains down to 8% homology. Low DNA homologies were also found between some strains of *Acinetobacter* and *Moraxella* species (up to 12% for *M. osloensis*). The significance of this low-degree intergeneric affinity was emphasized by competitive ribosomal RNA-DNA hybridization (Johnson, Anderson, and Ordal, 1970). In this type of experiment, which reflects the degree of base sequence similarity in the conserved ''ribosomal'' part of the genomes (Doi and Igarshi, 1965; Dubnau et al., 1965), 66–69% homologies were found between *Acinetobacter* on the one hand and *M. osloensis*/*M. (B.) catarrhalis* on the other.

The hybridization and transformation results (Johnson, Anderson, and Ordal, 1970; Juni, 1972) constitute the main basis of the present unification into one genus, *Acinetobacter*, of the bacteria previously named *Bacterium anitratum*, *Cytophaga anitrata*, *C. lwoffi*, *Diplococcus mucosus*, *Herellea caseolytica*, *H. saponiphilum*, *H. vaginicola*, *Micrococcus calcoaceticus*, *M. cerificans*, *Mima polymorpha*, *Moraxella glucidolytica*, *M. lwoffi*, *Neisseria winogradskyi*, and by various epithets under the genera *Achromobacter* and *Alcaligenes*. It is not settled, however, whether the genus *Acinetobacter* should have species in addition to the one presently recognized, *Acinetobacter calcoaceticus* (Lautrop, 1974). The probability is high that more than one species will be established in the future, when more data are collected on genetic groups in relation to other properties. A candidate for a separate species is *A. lwoffii*, comprising one group of asaccharolytic strains (Lautrop, 1974; see this chapter, ''The Species'').

The results reviewed indicate that *Acinetobacter* and oxidase-positive organisms (genus *Moraxella*) may be considered correctly as genera of the same family.

Kingella kingae (previously *Moraxella kingii* and *M. kingae*) was distinguished as a new species belonging to a new genus of Neisseriaceae partly on the basis of results with Str^r transformation, which indicated no genetic affinity to other oxidase positives (Henriksen and Bøvre, 1968a, 1976). Recent results with the same technique may indicate a very low affinity of *K. kingae* to genus *Neisseria*, consistent with their allocation in the same family (Bøvre et al., 1977a), but no affinity to *Acinetobacter* (unpublished). The other two species of the new genus, *K. indologenes* and *K. denitrificans* (Snell and Lapage, 1976), were not included on the basis of genetic studies and have only incompletely been studied in this way (see this chapter, ''The Species'').

From the first studies by Catlin and Cunningham (1961), it is known that most species of genus *Neisseria* are closely interrelated genetically (after removal of *Branhamella*). Like *Kingella*, genus *Neisseria* shows no compatibility with any of the subgenera of genus *Moraxella* or *Acinetobacter* in Str^r transformation (Bøvre, 1956b, 1967a; Bøvre and Holten, 1970; Bøvre et al., 1977a; Catlin and Cunningham, 1961; unpublished), except for a few possible transformant colonies with *M. (B.) catarrhalis* as donor (Siddiqui and Goldberg, 1975).

The finding of close and overlapping genetic affinities between rod-shaped and coccal entities of genus *Moraxella* indicates that cellular shape and mode of division are not necessarily such fundamental taxonomic criteria as formerly believed. That, genetically, a genus may contain both rod-shaped and coccal species was shown also by the detection of *N. elongata*. Although rod-shaped, this species has very high genetic affinities to all coccal entities of genus *Neisseria* that have been tested in Str^r transformation (Bøvre and Holten, 1970; Bøvre et al., 1977a). As shown in Table 4, these affinities may be on the same order of magnitude as between the coccal species of the genus.

Table 4. Mutual affinities in Str^r transformation of *Neisseria elongata*, *N. meningitidis*, and *N. subflava*.[a]

Donor organisms	Ratios of interstrain to autologous transformation, with recipient organisms		
	N. elongata	*N. meningitidis*	*N. subflava*[b]
N. elongata		3×10^{-2}	1×10^{-1}
N. meningitidis	1×10^{-2}		3×10^{-1}
N. subflava[b]	2×10^{-2}	5×10^{-2}	

[a] After Bøvre and Holten (1970), Bøvre et al. (1977a), and Catlin and Cunningham (1961).
[b] Designated *N. flava* in original publications.

Other studies have revealed that the *N. elongata* concept is rather heterogeneous in terms of Sm^r transformation affinities (Bøvre, Fuglesang, and Henriksen, 1972), and also that there is one more, separate, rod-shaped entity within genus *Neisseria* (see this chapter, "The Species, *Neisseria elongata*"). No affinity in Str^r transformation has been detected between *N. elongata* and other genera of Neisseriaceae, except for a possible, very low affinity to *Kingella kingae* (Bøvre et al., 1977a). In mRNA-DNA hybridization, the affinity of *N. elongata* to *N. subflava* (*N. flava*) was found to be 11% of an autologous reaction, whereas the affinity to *M. nonliquefaciens* was only 0.4% (Bøvre, 1970a).

The species "*Moraxella*" *urethralis* (invalidly named) has no distinct Str^r transformation affinity to other Neisseriaceae species, including *Acinetobacter calcoaceticus*, *Neisseria elongata*, *Kingella kingae*, and several *Moraxella* species tested (Bøvre et al., 1976, 1977a; Lautrop, Bøvre, and Frederiksen, 1970; unpublished). Juni (1977) reported on some activity of this organism on *M. (B.) catarrhalis*, *M. osloensis* and *Acinetobacter* recipients in Str^r transformation. A marginal activity on an *M. (B.) catarrhalis* recipient has been confirmed by one of us (unpublished). Competitive DNA-DNA hybridization showed no homology between a strain of "*M.*" *urethralis* (ATCC 17960) and *Acinetobacter* or *M. osloensis* (Johnson, Anderson, and Ordal, 1970).

Str^r transformation has recently been performed with donors from a group of psychrophilic, oxidase-positive rods isolated from poultry, fish, and human clinical material. The group appears to have affinity to *Branhamella* similar to that of *Acinetobacter*. It has little affinity to the latter, but more distinct although weak compatibility with *M. osloensis*. This study, which is part of a collaborative investigation under the Subcommittee on *Moraxella* and Allied Bacteria, has not yet been published, but the bacterial group has been described phenotypically (Bøvre et al., 1974; Thornley, 1967 (phenon 3); see also this chapter, "The Species, *Moraxella phenylpyruvica*").

Table 1 shows the present ranges of G+C (mol%) in DNA of groups within Neisseriaceae, compiled from determinations with different methods (Baumann, Doudoroff, and Stanier, 1968a; Bøvre, Fiandt, and Szybalski, 1969; Bøvre and Holten, 1970; Bøvre et al., 1976, 1977a; Catlin and Cunningham, 1964a,b; De Ley, 1968; Hill, 1966; Johnson, Anderson, and Ordal, 1970; Lautrop, Bøvre, and Frederiksen, 1970; Reyn, 1974; Snell and Lapage, 1976). In general, the ranges of values for the various groups are wide, permitting no sharp distinction between them in most instances. However, when considered together with the results of other studies on relationship, the data have considerable supporting taxonomic value. The subgenera

Moraxella and *Branhamella* of genus *Moraxella* have identical overall ranges of mol% G+C, consistent with their closeness and overlapping compatibilities in genetic transformation and nucleic acid hybridization, and their separation from genus *Neisseria* is relatively clear also in terms of G+C content. Further, the high G+C content of *N. elongata* favors its allocation in the genus *Neisseria*. It should be noted that within each mol% range of G+C, several species may have similar base ratios (see this chapter, "The Species"), and that each species may show a scatter of values up to 3 mol% even when studied with the same method in one laboratory.

Correspondence of Genetic and Other Criteria at the Genus Level

It has been shown by gas-liquid chromatography that the qualitative and quantitative distribution of whole-cell fatty acids in Neisseriaceae generally reflects the spectrum of genetic affinities (Bøvre et al., 1976; Jantzen et al., 1974a,b, 1975, 1978; Lambert et al., 1971; Lewis, Weaver, and Hollis, 1968). By numerical treatment of such data (Jantzen et al., 1974b, 1975), the genera *Acinetobacter*, *Neisseria* and *Moraxella* form three distinct clusters, corresponding to the genetic separation. The subgenera *Moraxella* and *Bramhamella* overlap in the genus *Moraxella* cluster, and the rod-shaped *N. elongata* fits well within the genus *Neisseria*, all as expected from the genetic data. *Kingella kingae* appears separate, but has some association with the *Neisseria* strains in these terms (*K. indologenes* and *K. denitrificans* not examined). "*M.*" *urethralis* is clearly distinguished from the clusters mentioned, although the fatty acid composition bears resemblance to that of Neisseriaceae in general.

Further, the study of cellular waxes in Neisseriaceae has given results consistent with the genetic data. Thus waxes are regularly absent in *Neisseria* and present in species of the genus *Moraxella* (including *Branhamella*), except *M. phenylpyruvica* (Bøvre et al., 1976; Bryn, Jantzen, and Bøvre, 1977; Jantzen et al., 1978).

Gas chromatographic examination of cellular monosaccharides has also revealed corresponding information. Heptose is regularly present in *Neisseria* and absent in *Branhamella* and in rod-shaped moraxellae (Jantzen, Bryn, and Bøvre, 1976).

Studies by Baumann, Doudoroff, and Stanier (1968a,b) on nutritional and other physiological properties of *Acinetobacter* and *Moraxella* (including subgenus *Branhamella*) have given results largely corresponding with the genetic investigations. The taxonomic relevance of distinguishing sharply between *Acinetobacter* and the other genera of Neisseriaceae on the basis of the negative oxidase reactions of the former has been substantiated (Juni, 1972, 1978). The occurrence of carbonic anhydrase as a distinguishing trait of "true" *Neisseria* species (Berger and Issi, 1971) and the studies of glycolytic enzymes by Holten (1973, 1974a,b) have confirmed the basic differences between *Neisseria* and *Branhamella* expected from their genetic incompatibility. Also, the genetic distinction of *Kingella* from *Moraxella* is clearly reflected in physiological traits. Thus, the former is consistently catalase negative, has some growth ability anaerobically, is saccharolytic, and reduces nitrite (Table 5). The distinction of *Kingella* (*K. kingae* and *K. indologenes*) from other Neisseriaceae, including *Neisseria*, was demonstrated in studies of enzymes of nucleic acid metabolism (Jyssum and Bøvre, 1974). Further, biochemical investigations by Berger and Piotrowski (1974) have confirmed *N. elongata* as a "true" *Neisseria*. Thus, it possesses a carbonic anhydrase. Its affinity to the genus *Neisseria* was further reflected in the studies of glycolytic enzymes referred to above.

Habitats and Pathogenicity

For readily understandable reasons, several of the genetic entities recently circumscribed have not yet been systematically searched for with nondisputable identification methods in the various possible habitats. An impending task is the clarification of host distributions. The genetic identity of what is apparently the same bacteria from different sources is often uncertain. Thus, there are demonstrated examples of nonspecies identity with reference strains where a species has been reported from a new host. Such challenged identifications will probably increase as more isolates are scrutinized (see this chapter, "The Species"). Genetically confirmed cases of different hosts harboring the same species have been established as for *Moraxella* (*Branhamella*) *ovis,* that occur in sheep, cattle, and horses. An intraspecies relationship also has been found between *M. bovis* (cattle) and *M. equi* (horse) (unpublished). However, in some such cases, genetic deviation has been observed between strains from different hosts. One example is a slightly reduced affinity between human and guinea pig isolates of *M. lacunata* (Bøvre, 1965c). Given the many new species entities now becoming recognized in animal hosts for which we have not yet easily observable phenotypic criteria, it appears necessary to be very careful when considering parasitism in different hosts by the same well-defined Neisseriaceae species. Such demands for advanced identification may be less for studies of distribution of an organism in various ecological niches of the same host, but the need should not be underestimated.

All of the commonly recognized members of Neisseriaceae treated here, except *Acinetobacter,* are mainly or exclusively parasites of humans and animals; *Acinetobacter* is generally regarded as a primarily free-living group. It has not been ruled out, however, if some entities of acinetobacters are primary parasites of particular animal hosts. Although none of the organisms has a pathogenic potential comparable to *Neisseria gonorrhoeae* and *N. meningitidis,* some of the species have been relatively firmly associated with disease, e.g., *M. lacunata* and *M. bovis* as causes of keratoconjunctivitis (or conjunctivitis) in man and cattle, respectively. Some species, e.g., *M. osloensis* and *M. phenylpyruvica,* may have a considerable potential for septicemic, meningeal, or pyogenic infections. Both *M. osloensis* and *M. phenylpyruvica* have been isolated from blood and CSF cultures. Usually, these organisms are considered saprophytes or commensals; rarely do they provoke or contribute to disease and usually do so when introduced artificially or secondarily as a consequence of other ailments into the circulation or deeper tissues (preferentially in the compromised host), at otherwise damaged surfaces, or, more speculatively, by the mere spread to normally sterile mucosae (see this chapter, "The Species").

A relatively recent finding is the frequent occurrence of fimbriation in strains newly isolated from the parasitic state. Fimbriation may be associated with distinct colony types. Fimbria-associated adherence may determine parasitism specifically and at the same time be a prerequisite for virulence, if any, to be expressed by natural infection. This association is indicated in experimental inoculation of the bovine eye with fimbriated and nonfimbriated variants of *M. bovis.* (Bøvre, Bergan, and Frøholm, 1970; Bøvre and Frøholm, 1972a; Bøvre, Fuglesang, and Henriksen, 1972; Bøvre et al., 1977a; Frøholm, 1978; Frøholm and Bøvre, 1972; Henriksen, 1969a; Henriksen and Bøvre, 1969; Henriksen and Holten, 1976; Pedersen, 1970, 1973; Pedersen, Frøholm, and Bøvre, 1972; Figs. 1 and 2).

Isolation

Routinely employed blood agar medium will permit growth of all species. Some strains, such as *Moraxella lacunata* of the small-colony type and *Kingella denitrificans,* grow better on media with heated blood.

Blood Agar Medium

Tryptose blood agar base (Oxoid)	30 g (agar 12 g)
Glucose	1 g
Citrated human blood	50 ml
Distilled water to	1,000 ml
pH adjusted to 7.6	

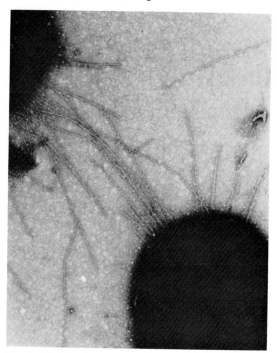

Fig. 1. Fimbriated variant of *Moraxella bovis*. Negative stain electron microscopy. × 25,000. [From Bøvre and Frøholm, 1972a, with permission.]

When not otherwise stated, colony descriptions in the section on "The Species" are based on growth on this or similar media. The same medium without blood (and glucose) will support growth of most strains (hitherto only except *M. lacunata*, ATCC 17967). Most strains (hitherto except small-colony type *M. lacunata*, and occasionally *Kingella kingae*)

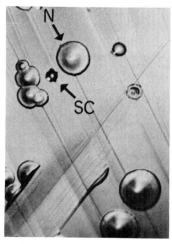

Fig. 2. *Moraxella bovis*, 48-h blood agar culture. Fimbriated (SC) and nonfimbriated (N) colony types, the former corroding the agar, as seen after scraping off the growth. × 5.3. [From Bøvre and Frøholm, 1972a, with permission.]

will grow on heart infusion broth (Difco). Mueller-Hinton broth (Difco) supplemented with 0.5% yeast extract (Difco), with or without agar, appears to support growth of all strains, although weakly for some. The addition of serum may improve growth in such cases. A mineral salt medium with ammonium ions and acetate (or other simple carbon source) will support growth of most *Acinetobacter, M. osloensis,* and "*M.*" *urethralis* strains (Baumann, 1968; Baumann, Doudoroff, and Stanier, 1968a, b; Bøvre and Henriksen, 1976; Bøvre et al., 1977b; Juni, 1972; Lautrop, 1974; Lautrop, Bøvre, and Frederiksen, 1970; Warskow and Juni, 1972) but is rarely used in isolation from clinical specimens. For this purpose, blood agar media are generally preferred, because of their superior differential properties and broad coverage of the bacteria in this family.

Selective procedures and media have been used to clarify the distribution of some species, but are generally not well developed. From soil and water, *Acinetobacter* is selectively grown by the use of relatively acid mineral medium with nitrate as nitrogen source and simple carbon sources, under vigorous aeration (Baumann, 1968). Some species of oxidase-positive rods may be selected for on Thayer-Martin type of media, such as *K. denitrificans* (Hollis, Wiggins, and Weaver, 1972; Snell and Lapage, 1976), *M. osloensis* and possibly *K. kingae* (Bøvre et al., 1977b). In addition to this, specially designed antibiotic-containing media have been used for the selection of *M. nonliquefaciens* and *M. lacunata* (van Bijsterveld and Winkler, 1970), and for *M. osloensis* and *Neisseria elongata* (Berger and Falsen, 1976). For *M. phenylpyruvica*, a selective medium containing tellurite and taurocholate has been devised (Snell, Hill, and Lapage, 1972).

Incubation may generally be performed at 33(−35)°C in a humid atmosphere, using a tight box with water covering the bottom. Some strains of *M. nonliquefaciens* do not grow at 37°C in an ordinary dry atmosphere, and it is advisable to lower the temperature even at high humidity. *Acinetobacter* often requires incubation below 37°C (Baumann, 1968; see below). For the group of oxidase-positive, psychrophilic rods mentioned previously, incubation below 33°C is almost always necessary. A parallel incubation at room temperature will cover their needs. Incubation in air with CO_2 supplementation is not required for any species, although low concentrations (2–4%) of CO_2 may improve growth of *M. lacunata* strains (Tatum, Ewing, and Weaver, 1974). One species is known for which 10% CO_2 reduces growth ("*M.*" *urethralis*, unpublished).

Identification

Genetic Transformation

The species *Kingella indologenes, K. kingae, Moraxella atlantae, M. bovis, M. lacunata, M. nonliquefaciens, M. osloensis, "M." urethralis, M. (Branhamella) catarrhalis, M. (B.) cuniculi, M. (B.) ovis, Neisseria elongata,* and several new species of *Moraxella (*and *Neisseria),* as well as the genus *Acinetobacter,* can be unequivocally identified by genetic transformation (Bøvre, 1964b, 1965a,b,c,d, 1970b; Bøvre, Fuglesang, and Henriksen, 1972; Bøvre et al., 1976, 1977b, unpublished results; Brooks and Sodeman, 1974; Henriksen and Bøvre, 1968a; Juni, 1972, 1974, 1977). This identification depends on the existence of transformable or competent strains, with the ability of efficient DNA uptake. Since competence can be lost on subcultivation, it is important in some cases of diagnostic transformation that competence may be associated with distinct colony types, consisting of fimbriated cells, e.g., in *M. bovis, M. nonliquefaciens, K. kingae,* and *N. elongata* (Bøvre and Frøholm, 1970, 1971, 1972b; Bøvre et al., 1977a; see Figs. 1 and 2). The identification of hitherto incompetent species can be aided by their activities as donors on recipients of related species, as for *M. phenylpyruvica* (Bøvre and Henriksen, 1967b; Bøvre et al., 1977b).

In one technical modification of such identification, reference Str^r-transforming DNAs are tested on a competent isolate as recipient. The final identification is based on comparison of autologous transformation activity (with DNA from a mutant of the isolate) and the various reference donor activities. One-third or more of the autologous activity indicates species identity, but lower values may also be considered in heterogeneous species.

Another modification of genetic identification by transformation is independent of the competence of the isolate. In this procedure, reference recipients maintained in the laboratory are used. Transforming DNA is extracted from the isolate (in nutritional-marker transformation) or a mutant thereof (in Str^r transformation). In the former case, auxotrophic recipients are exposed to DNA of the identified strain, which may be simply a crude lysate produced within 30 min from as little material as one bacterial colony or slightly more. Prototrophic transformant yield gives the species of the isolate within 1 or 2 days. This method has been successfully used in organisms with ability to grow on defined synthetic media: *Acinetobacter* (Brooks and Sodeman, 1974; Juni, 1972, 1978), *M. osloensis* (Juni, 1974), and *"M."urethralis* (Juni, 1977). The use of such crude lysate DNA has also been adapted to identification by Str^r transformation (unpublished). Usually, however, DNA has been extracted and partly purified from a mutant of the isolate (Bøvre, 1964a) before application in Str^r transformation.

Even if transfer of a "ribosomal" marker like Str^r occurs, with decreased frequency, between related species and sometimes between genera, very simple, "nonquantitative", Str^r transformation procedures may be applied for the initial allocation of an isolate to a group, e.g., to *Moraxella* or *Neisseria* (no interaction) (see references above). Quantitative or semiquantitative assays of transformants in relation to an autologous parallel are necessary for exact identification at the level of species, however. In nutritional marker transformation, the proper choice of marker employed is critical, since some such markers tend to be transferred with considerable frequency between closely related species (Janik, Juni, and Heym, 1976; Siddiqui and Goldberg, 1975). It is felt that quantitation would be ideal also in this type of genetic identification as a safety measure against the lumping of closely related entities that may not have been defined. An Str^r transformation system may be easier to quantitate than nutritional marker transformation and has a wider applicability to fastidious bacterial species.

Determination of Fatty Acid Pattern

Gas-liquid chromatography of whole-cell methanolysates may easily provide a tentative allocation of an isolate to group or species based on the cellular fatty acid pattern (Bøvre et al., 1976, 1977b; Jantzen et al., 1974a, b; 1975; 1978). These and the authors' unpublished studies of large collections of strains of each of several species identified genetically have shown a remarkable reproducibility of results in most instances, and the approach has become a valuable aid in exact identification. The method is particularly useful for distinction of species that have relatively separate fatty acid patterns, e.g., *Moraxella osloensis* and *"M." urethralis.* Pitfalls exist, however, since an isolate may belong to a still-undescribed species with similar fatty acid pattern (see this chapter, "The Species"). In difficult cases of identification, the method may be used primarily and appropriate transformation tests performed secondarily.

Several groups of oxidase-positive, nonmotile, Gram-negative rods have been excluded from the family Neisseriaceae on the basis of large amounts of branched-chain fatty acids, which may indicate their relations to section I of genus *Flavobacterium* (McMeekin and Shewan, 1978; Moss and Dees, 1978; Weeks, 1974). The species *M. saccharolytica* (Flamm, 1956; Owen and Snell, 1976) and *M. anatipestifer* (Bruner and Fabricant, 1954; Henriksen, 1973) share this trait (unpublished studies in cooperation with M. Bisgaard, K. Bryn, E. Jantzen, and O. Spanne).

Use of Cultural-Biochemical Criteria

Testing for the oxidase reaction is essential in the initial stage of identification in these groups of bacteria. Main emphasis has to be laid upon the kind of oxidase tests performed (Bøvre and Henriksen, 1976). There are two principal methods, one employing the reagent dimethyl-*p*-phenylenediamine and the other, tetramethyl-*p*-phenylenediamine. The dimethyl reagent in more distinguishing and less sensitive (Ellingworth, M'Leod, and Gordon, 1929). A positive reaction with the dimethyl reagent, dropped on 18- to 20-h-old agar culture, consists of a purplish color turning black within 1–2 min. The black reaction penetrates the whole colony within a few minutes and is permanent. This kind of reaction is seen with *Moraxella* (including *Branhamella*), *Neisseria*, *Kingella kingae*, and "*M.*" *urethralis*, as well as in *Flavobacterium* (and *Pseudomonas*) and a few other genera. The fading reaction seen with the tetramethyl reagent is more widely distributed, being seen in the organisms mentioned as well as in *K. indologenes*, *K. denitrificans*, *Eikenella corrodens*, *Cardiobacterium hominis*, *Pasteurella*, *Haemophilus*, *Brucella*, and other organisms, which are most often negative or only weakly reacting with the dimethyl reagent (Cowan, 1974; Ellingworth, M'Leod, and Gordon, 1929; Lapage, 1974; Snell and Lapage, 1976; van Bijsterveld, 1970; authors' experience).

Because of its negative reactions in both tests for oxidase, *Acinetobacter calcoaceticus* is relatively easy to identify within Neisseriaceae, as long as there are still no other tetramethyloxidase-negative species or problems with oxidase-negative variants of oxidase positives in the family. The nonfastidiousness and wide biochemical activities of *A. calcoaceticus* also makes it easily distinguishable from other bacteria by methods applied for nonfastidious, nonfermentative organisms in general, including commercially available diagnostic devices (Dowda, 1977; Gilardi, 1973; Oberhofer, 1979; Oberhofer, Rowen, and Cunningham, 1977; Otto and Pickett, 1976; Pickett and Pedersen, 1970; von Graevenitz and Grehn, 1976). It is considered preferable, however, to add a procedure of genetic identification in scientific work when new habitats or hosts are being explored.

For the oxidase-positive species treated here, the conventional diagnostic criteria and tests are insufficiently developed and less uniformly applicable (Berger, 1963; Berger and Piotrowski, 1974; Bøvre and Henriksen, 1976; Bøvre et al., 1976; Cowan, 1974; Henriksen, 1973; Lautrop, 1974; Reyn, 1974; Riley, Hollis, and Weaver, 1974; Snell, Hill, and Lapage, 1972; Snell and Lapage, 1976; Tatum, Ewing, and Weaver, 1974). Due to the relative biochemical inactivity of many of these bacteria, a few phenotypic criteria have gained particular importance. There is a danger, however, that some of these key reactions reflect the genetic concepts less reliably than hitherto believed. This possibility may have its theoretical origin in a partial preselection, by means of the same key reactions, of those groups of bacteria first reported as representative. Thus, with the increasing number of random isolates now being examined genetically, some key characteristics appear less firm than before, with the consequent need for adjustment of conventional criteria for the various species. For example, activities of phenylalanine and tryptophan deaminase may be absent in *M. phenylpyruvica* (Bøvre et al., 1977b; unpublished). In this case the finding of other criteria, apparently more stably associated with genotype, has been of importance (see this chapter, "The Species, *Moraxella phenylpyruvica*"). The need for such a development, guided by the genetic identity, is also becoming evident for *M. osloensis*, where some strains lack the key characteristic of being able to grow on minimal medium with the production of poly-β-hydroxybutyrate inclusions (Baumann, Doudoroff, and Stanier, 1968a; Bøvre et al., 1977b; Juni, 1974). Several other species belonging to these groups, under the standard test conditions for Enterobacteraceae and nonfermentative, Gram-negative bacteria in general, are unable to express activities for which they might be equipped. Thus, in the case of *K. kingae* and *K. indologenes*, a base medium consisting of Mueller-Hinton broth (Difco) with 0.5% yeast extract (Difco) is satisfactory to render a nitrite reduction test positive (unpublished, authors' experiments), whereas both these species are reported negative in tests with a more simple basal medium. For universal use, the system for cultural-biochemical identification must therefore be carefully standardized to cover all the needs of the most "difficult" species of the family.

Suboptimal conditions of the tests, which may be said to occur frequently at present, are probably one important reason for discrepancies between laboratories when studying these bacteria (Bøvre et al., 1974). The *Minimal Standards for Description of New Taxa Within the Genera Moraxella and Acinetobacter* (Bøvre and Henriksen, 1976) was a first attempt to meet the methodological requirements for identification. The standards, presented as a proposal by the Subcommitte on *Moraxella* and Allied Bacteria of the International Committee on Systematic Bacteriology, have already been improved in some respects (Bøvre et al., 1976), but are still far from optimal. The need for standards covering *Kingella* and the differentiation of *Moraxella* from *Neisseria* is obvious.

Table 5 presents some species characteristics in common tests of these oxidase-positive organisms. The table is primarily based on the authors' published and unpublished results but contains also information from Berger (1963), Berger and

Table 5. Characteristics of *Moraxella* (including *Branhamella*) and of *Kingella* species, *Neisseria elongata*, and "*Moraxella*" *urethralis*.[a]

	M. lacunata	*M. bovis*	*M. nonliquefaciens*	*M. osloensis*	*M. phenylpyruvica*	*M. atlantae*	*M.(B.) catarrhalis*	*M.(B.) caviae*	*M.(B.) cuniculi*	*M.(B.) ovis*	*K. kingae*	*K. indologenes*	*K. denitrificans*	*N. elongata*	"*M.*" *urethralis*
Oxidase (tetramethyl reagent)	+	+	+	+	+	+	+	+	+	+	+	w(−)	w(−)	+	+
Oxidase (dimethyl reagent)	+	+	+	+	+	+	+	+	+	+	+	w(−)	w(−)	+	+
Catalase	+	−+	+	+	+	+	+	+	+	+	+	+	+	−(w,+)	+
Cocci(C)/rods(R)[b]	R	R	R	R	R	R	C	C	C	C	R	R	R	R	R
Anaerobic growth[b]	−	−	−	−	−	−	−	−	−	−	w(−)	w	w	w(−)	−
Acid from glucose	−	−	−	−	−	−	−	−	−	−	+	+	+	−(w)	−
Hemolysis	+	+−	−	−	−	−	−	w	−	+−	+	−	−	−	−
Serum liquefaction	+	+(−)	−	+(−)	+(−)	−	−	−	−	−	−	−	−	−	+(−)
Growth on minimal medium[c]	−	−(+)	−	+(−)	−(+)	−	−	−	−	−	−	−	−	−	−
Growth at 5°C	−	−(+)	−	−	+(−)	−	−	−	−	−	−	−	−	−	−
Growth at 6% NaCl	−	−	−	−(w)	+(−)	−	−	−	−	−	−	−	−	−	−
Urease	−	−	−	−	+−	−	−	−	−	−	−	−(w)	w	−(w)	−
Phenylalanine deaminase	−	−	−	−	+(−)	−	−	−	−	−	−	+	−	−	−
Indole	−	−	+	−	−	−	−	−	−	−	−	−	−	−	−
Nitrate reduction	+	−(+)	+	−+	+(−)	−	+(−)	+	−	+(−)	−	+	+	−(+)	−
Nitrite reduction	+	+	+	+[d]	−	−	−+	+[d]	+[d]	−+	+[d]	+[d]	+[d]	+[d]	+[d]

[a] Mainly authors' own results (methods in Bøvre and Henriksen [1976] and Bøvre et al. [1976]; see text), with additional information from other studies (see text). +, Distinctly positive; −, negative; w, weak reaction; double symbols, most frequent reaction followed by alternative reaction; symbol(s) in parentheses, infrequent alternative reaction(s). The table may have to be revised in the future.

[b] Blood agar, GasPak Anaerobic System (BBL).

[c] Mineral salts + NH4+ + acetate.

[d] Base medium: Mueller-Hinton broth (Difco) with 0.5% yeast extract (Difco).

Piotrowski (1974), Tatum, Ewing, and Weaver (1974), Berger and Falsen (1976), and Snell and Lapage (1976), in addition to some other publications referred to in this chapter. Discrepancies can be found, e.g., on nitrite reduction by *K. kingae* and *K. indologenes*. Our results, which are tabulated, were based on the procedure mentioned above. The results of growth under anaerobic conditions tabulated were obtained with blood agar cultures incubated at 33°C for 2 days in GasPak Anaerobic System (BBL).

The table will have to be adjusted when more genetically identified strains of each species have been tested in the same set of optimally developed procedures. The complications arising when the many new, undescribed entities are included must also be considered. The genetic approach may ultimately prove to be the simplest and cheapest, and it will certainly be the most reliable way to exact identification of several of these species in the future. Anyway, a genetic proof of the identity of most of these organisms, when reported from new habitats or hosts, is preferable and within reach in view of the unique potency of genetic transformation in this field.

The Species

Acinetobacter calcoaceticus (Beijerinck 1911)
Baumann, Doudoroff, and Stanier 1968
Type strain ATCC 23055

SOURCES, HABITATS, AND PATHOGENICITY. Frequently isolated from soil, water, sewage, contaminated food, and mucosal and outer surfaces of animals (including fish) and man. Has been encountered as a presumed causal or contributory agent in almost any kind of infectious disease process, including frequent reports of meningitis and septicemia in man and septicemia and abortion in animals; possibility of contamination of the specimens complicates source evaluation, however. Main natural habitats are soil and water, where it seems to be important in the degradation of organic matter. Pathogenicity is generally low, but significant clinical roles must be considered in hosts and organs with reduced natural resistance and through iatrogenic infection (Baumann, 1968; Baumann, Doudoroff, and Stanier, 1968b; Bøvre et al., 1974; Carter, Isoun, and Heahey, 1970; Henriksen, 1973; Juni, 1972, 1978; Lautrop, 1974; Samuels et al., 1972). Possible occurrence of more than one ecological entity of *Acinetobacter* or subspecific entities of *A. calcoaceticus* having special parasitic or pathogenic potentials in animals and man has not been ruled out and deserves attention.

MAIN PHENOTYPIC CHARACTERISTICS. Usually medium thick (0.7–0.8 μm), sometimes plump (more than 1 μm), often coccoid rods. Colonies relatively large (2–3 mm in 24 h) and opaque, but strains of the *A. lwoffii* type (see below) may be somewhat smaller and less opaque, resembling *M. osloensis* colonies. Often unable to grow at 37°C. Although most strains may grow at 33°C, incubation temperatures from 20 to 30°C are preferred in some studies. Strictly aerobic, oxidase negative (see this chapter "Identification"), and catalase positive. Pronounced physiological and nutritional activity and diversity, either saccharolytic or nonsaccharolytic. Generally able to grow on mineral salt media with ammonium ions and acetate, lactate, or β-hydroxybutyrate (some exceptional strains), without production of poly-β-hydroxybutyrate inclusions. *A. lwoffii,* a term often used for nonsaccharolytic strains of *A. calcoaceticus,* is phenotypically heterogeneous but contains a subunit that may be close to deserving a separate subspecies or species status. This unit has relatively restricted metabolic capacity, including inability to utilize citrate (exceptions) and hydrolyze gelatin, is nonhemolytic, and shows a neutral (not alkaline) or very weak acid reaction on glucose-peptone media (Baumann, 1968; Baumann, Doudoroff, and Stanier, 1968b; Bøvre, 1967b; Henriksen, 1973; Juni, 1972, 1978; Lautrop, 1974; Thornley, 1967). Usually resistant to penicillin, based on β-lactamase production (Gilardi, 1973; Lautrop, 1974; authors' studies of strains presented by Samuels et al., 1972). Cellular lipid composition distinct from other Neisseriaceae, with a heterogeneity which does not appear to correspond to subgroups based on other criteria (Bryn, Jantzen, and Bøvre, 1977; Jantzen et al., 1975, 1978).

GENETIC IDENTIFICATION. *A. calcoaceticus* strains are rarely transformable, but are easily identified by genetic transformation (nutritional marker or Str^r), using the strain Bd4 isolated by Juni (1972) from soil as recipient. A rather pronounced genetic heterogeneity of this species revealed by nucleic acid hybridization and quantitative transformation against the soil isolate warrants further search for competent strains among animal and human isolated, to provide a basis for a critical analysis of genetic units within *A. calcoaceticus* as now conceived (see this chapter, "Introduction").

Moraxella lacunata (Eyre 1900) Lwoff 1939
Comprises the previous *M. liquefaciens* (Henriksen, 1969b; Henriksen and Bøvre, 1968b)
Neotype strain ATCC 17967 (= NCTC 11011)

SOURCES, HABITATS, AND PATHOGENICITY. Isolated from inflamed eyes (conjunctivitis with or without keratitis), as well as from healthy eyes, nose, and throat, and from maxillary sinus aspirate in chronic sinusitis and blood in endocarditis, all from man. Also isolated (genetically verified) from the conjunctivae of healthy guinea pigs. Isolates named *M. liquefaciens* from other animals, e.g., from cattle (Wilcox, 1970a,b), have not yet been examined genetically. Previously frequently encountered from human sources, now only rarely isolated; the reason for this shift is unknown. Main natural habitat is probably the upper respiratory tract and conjunctivae of man, but the host distribution should be further examined. The organism appears to have been a significant agent in conjunctivitis and keratitis of man in the past. However, despite reproduction of the disease by experimental inoculation of healthy individuals in some cases, the organism is generally considered of low primary pathogenicity, depending on predisposing factors (Axenfeld, 1897; Bøvre, 1965c, 1970b; Henriksen, 1973; Kaffka, 1964; Lautrop, 1974; Morax, 1896; Ryan, 1964; Silberfarb and Lawe, 1968; Steen and Berdal, 1950; van Bijsterveld, 1972, 1973).

MAIN PHENOTYPIC CHARACTERISTICS. Medium thick to plump rods (0.8–1.2 μm), coccoid to distinctly bacillary with partly very long cells. Colonies on blood agar translucent, either small (0.1–0.2 mm) as seen in the type strain, or larger (up to 3 mm in 48 h) as seen in strains originally named *M. liquefaciens* (e.g., ATCC 17952 = NCTC 7911). The guinea pig isolates have intermediately sized colonies. Strains of the small-colony type show distinctly improved growth on chocolate (heated blood) agar and may not grow on rich media without serum, other body fluids, or oleic acid. Grows with dark zones around the colonies on chocolate agar, is nonhemolytic, and liquefies inspissated bovine serum (Loeffler serum slants) and gelatin, provided the basal medium supports growth. Considered as highly susceptible to penicillin; some genetically unidentified strains have been resistant (Bøvre, 1965c; Henriksen, 1973; Lautrop, 1974). Cellular lipid composition similar to *Moraxella* in general, but the few strains examined are distinguishable from other species of the genus (Bryn, Jantzen, and Bøvre, 1977; Jantzen et al., 1947b, 1978).

GENETIC IDENTIFICATION. Transformable strains known and identification by Str^r transformation possible. The species is genetically slightly heterogeneous, apparently reflecting three phenotypic clusters of strains (small- and large-colony type human isolates, and guinea pig strains). The mutual affinities of these entities are still within 10% of autologous reactions (Bøvre, 1965c). *M. lacunata* is genetically closely related to *M. bovis* and *M. non-liquefaciens,* which makes quantitative or semi-quantitative procedures necessary for this type of identification (Table 2). A new group of serum-liquefying, nonhemolytic rods from the respiratory tract of pigs (not listed in Table 3) is not closely related to *M. lacunata* genetically, but has high affinity to the

Moraxella (new rod, pigs) of Table 3, which is nonliquefying and hemolytic (unpublished transformation studies on material supplied by K. B. Pedersen). A possibility of nonliquefying variants of *M. lacunata,* analogous to the situation in *M. bovis,* may be explored by quantitative genetic transformation. G+C content of DNA, 40–44.5 mol%.

Moraxella bovis (Hauduroy et al. 1937) Murray 1948
Neotype strain ATCC 10900 (Henriksen, 1971)

On a genetic and phenotypic basis, *Moraxella equi* (Hughes and Pugh 1970) is considered to belong to the same species (see below). A similar organism *(M. caprae)* isolated from keratoconjunctivitis in goats (Pande and Sekariah, 1960) has no standing in approved nomenclature because of the lack of a representative strain.

SOURCES, HABITATS, AND PATHOGENICITY. Isolated most frequently from bovine eyes in cases of infectious keratoconjunctivitis, but also from unaffected eyes and nasal cavity of cattle, and horses' eyes in cases of conjunctivitis *(M. equi).* Animal sources other than the bovine eye insufficiently examined. Has not been recorded from humans except once from blood. Main natural habitats are probably the upper respiratory tracts and conjunctivae of cattle and possibly of some other animal species. Pathogenicity has been debated. The organism has been clearly associated with some outbreaks of infectious bovine keratoconjunctivitis but, in numerous other outbreaks studied, *M. bovis* has not, or has only infrequently, been encountered. A significant potential pathogenicity is indicated by reproduction of the disease by conjunctival instillation of *M. bovis* cultures in cattle, although the experiments have been irregularly positive and often dependent on accessory factors, such as damage to the epithelium by ultraviolet irradiation and simultaneous infections by other agents. Attempts to establish conjunctival colonization have been successful in sheep and mice, but without development of keratoconjunctivitis in the sheep. With isolates *(M. equi)* from horses' eyes with conjunctivitis, the disease has been reproduced in horses, but not in cattle, whereas bovine isolates have had no effect on horses. Nonhemolytic variants of bovine *M. bovis* appear to lack pathogenic capacity, whereas fimbriation of the cells (see below) appears to be a prerequisite for conjunctival colonization and, indirectly, for elicitation of disease (Barner, 1952; Henriksen, 1973; Hughes and Pugh, 1970; Hughes, Pugh, and McDonald, 1965; Jones and Little, 1923; Pedersen, 1970, 1973; Pedersen, Frøholm, and Bøvre, 1972; Pugh and Hughes, 1968; Pugh, Hughes, and McDonald, 1966, 1968; Pugh, Hughes, and Packer, 1970; Wilcox, 1968; R. E. Weaver, personal communication).

MAIN PHENOTYPIC CHARACTERISTICS. Cells like those of *M. lacunata* or more slender rods. Colony-type variation associated with fimbriation of the cells (Figs. 1 and 2). The fimbriated spreading-corroding (SC) type of colony corrodes the agar, may show distinct spreading, and expresses twitching motility. This type of colony is spontaneously agglutinating and usually impossible to emulsify evenly. *M. bovis* is typically hemolytic with a wide zone of β-hemolysis around the colonies. Lack of hemolysis has been observed in primary cultures and among later progeny of bovine isolates. Equine isolates *(M. equi)* have hitherto been nonhemolytic. In general, *M. bovis* is distinguished from other rod-shaped *Moraxella* species by the hemolysis of most bovine isolates and the liquefaction of Loeffler serum slants and gelatin. However, the latter reactions may fail in some strains or be too slow for easy detection. Frequently catalase negative. Highly susceptible to penicillin (Bøvre, 1965a, 1967a; Bøvre and Frøholm, 1972a; Henrichsen, Frøholm, and Bøvre, 1972, Hughes and Pugh, 1970; Lautrop, 1974; Pedersen, 1973; Pugh and Hughes, 1968; authors' unpublished experiments). The cellular lipid composition is similar to the *Moraxella* pattern in general and not clearly distinguishable from that of some human parasites, such as *M. nonliquefaciens* (Bryn, Jantzen, and Bøvre, 1977; Jantzen et al., 1974b, 1978).

GENETIC IDENTIFICATION. Easily identified by Str^r transformation, making use of the association between genetic competence and the fimbriated (SC) colony type (Bøvre and Frøholm, 1971, 1972b). Closely related genetically to other species of the *M. lacunata* group. *M. bovis* appears to be slightly heterogeneous as expressed in genetic transformation, with interstrain affinities down to 10% of an autologous reaction in some experiments, However, there is as yet no indication that the *M. equi* isolate has lower affinity to bovine isolates than is observed between some of the latter mutually (unpublished observations). G+C content of DNA, 41–44.5 mol%.

Moraxella nonliquefaciens (Scarlett 1916) Lwoff 1939
Definition revised by Bøvre and Henriksen (1967a) Neotype strain 4663/62 (ATCC 19975 = NCTC 10464)

SOURCES, HABITATS, AND PATHOGENICITY. Isolated from specimens of the respiratory tract of man, particularly the nasal cavity, but also eyes, throat, bronchi, and lungs (Bøvre, 1964b, 1970b; Henriksen, 1958, 1973; Kaffka, 1964; Kaffka and Blödorn, 1959; Tatum, Ewing, and Weaver, 1974; van Bijsterveld, 1972). Reported from the genitourethral tract of man (Kittnar et al., 1975), but no isolate

from this area has yet been included in genetic identification attempts. Isolates named *M. nonliquefaciens* from animals, e.g., from cattle (Wilcox, 1970a,b) or dogs (Bailie, Stowe, and Schmitt, 1978), should also be examined genetically for confirmation of these habitats. Main natural habitat is most probably the human nasal cavity, where it was identified by genetic transformation in 18% of unselected swab specimens from outpatients in an otorhinolaryngeal department (Bøvre, 1970b). This incidence corresponds to that found by van Bijsterveld (1972). Kaffka and Blödorn (1959) found it much more frequently in the noses of children (44% of healthy individuals and 60% of children with respiratory ailments). It was found distinctly less frequently in the pharynx in this and other investigations. *M. nonliquefaciens* may be considered as a well-established parasite with good adaptation to its human host; it rarely if ever causes disease (Henriksen, 1973; Kaffka, 1964). A contribution to disease by secondary invasion and multiplication on damaged epithelium outside the natural ecological niche has not been ruled out, however (see van Bijsterveld, 1973). Interestingly, Berenesi and Mészáros (1960) found presumed *M. nonliquefaciens* in bronchoscopic material from 23% of bronchiectasic patients. It has relatively often been recovered in a strongly mucoid form (genetically verified, [Bøvre, 1964b]) from cases of ozaena and chronic bronchitis, but such strains have also been isolated from healthy individuals (Bottone and Allerhand, 1968; Henriksen, 1951, 1973; Kaffka, 1955).

MAIN PHENOTYPIC CHARACTERISTICS. Plump rods, typically $1–1.5 \times 1.5–2.5$ μm, with nearly square ends, predominantly in pairs or short chains. Variation of cell shape may be pronounced, often with giant forms and filaments. Colonies 1–1.5 mm after 24 h and 2–3 mm after 48 h of incubation, typically low convex or nearly flat, often with a domed, opaque center and a flat, translucent periphery. Some strains are strongly mucoid. Apparently more fastidious than other *Moraxella* species, except the small-colony type of *M. lacunata*. Thus, the species does not grow on Hugh and Leifson's OF medium. May be distinguished from most strains of *M. osloensis* and *M. phenylpyruvica* by growth tests in mineral medium with ammonium ions and acetate, and by the urease and phenylalanine/tryptophan deaminase reactions, which are all negative with *M. nonliquefaciens*. Recently isolated strains often grow as an SC colony type, associated with fimbriation and twitching motility. May rapidly dissociate into the noncorroding (N) type of colony of nonfrimbriated cells. Considered to be uniformly highly susceptible to penicillin (Bøvre, 1964b; Bovre, Bergan, and Frøholm, 1970; Bøvre and Frøholm, 1972a; Bøvre and Henriksen, 1967a; Henrichsen,

Frøholm, and Bøvre, 1972; Henriksen, 1973; Henriksen and Bøvre, 1969; Lautrop, 1974). Cellular lipid composition generally similar to that of other moraxellae, but distinguishable from other species of human origin presently known (Bryn, Jantzen, and Bøvre, 1977; Jantzen et al., 1974b, 1978).

GENETIC IDENTIFICATION. Easily identified by Str^r transformation. Competence associated with the SC colony type of fimbriated cells (Bøvre and Frøholm, 1970, 1971, 1972b). Closely related to *M. lacunata* and *M. bovis,* which makes quantitative or semi-quantitative procedures necessary for their genetic distinction. Only distantly related to *M. osloensis, M. phenylpyruvica,* and *M. atlantae* (Table 2). G+C content of DNA, 40–44 mol%.

Moraxella osloensis Bovre and Henriksen 1967
Type strain A1920 (ATCC 19976 = NCTC 10465)

SOURCES, HABITATS, AND PATHOGENICITY. Isolated from blood, cerebrospinal fluid, urine, pyogenic processes (joints, bursae, cutaneous manifestations), upper respiratory tract, and genitourethral specimens of man (Berger and Falsen, 1976; Bøvre, 1965d, 1970b; Bøvre and Henriksen, 1967a; Bøvre et al., 1977b; Butzler et al., 1974; Feigin, San Joaquin, and Middelkamp, 1969; Hansen et al., 1974; Henriksen, 1973; Juni, 1974; Lautrop, 1974; Tatum, Ewing, and Weaver, 1974; unpublished). Not yet isolated with certainty from nonhuman sources. Main natural habitat is possibly the human pharynx. Isolated from epipharynx of 4.2% of healthy adults by the use of selective medium that contains lincomycin (Berger and Falsen, 1976), isolated only occasionally in the nose (Bøvre, 1970b), and in the genitourethral tract of 0.6% of patients with gonorrhoea-like ailments (Bøvre et al., 1977b). In most situations, probably a harmless parasite, but it may have a considerable potential pathogenicity (e.g., as a cause of pyogenic manifestations and septicemia). Its pathogenic roles should be critically studied.

MAIN PHENOTYPIC CHARACTERISTICS. Medium thick to plump rods (as of *M. lacunata*), often coccoid or diplococcoid, sometimes with fusiform shape. May be indistinguishable from coccal neisseriae in direct smears of pus, and by microscopy of primary cultures. Very often development of fusiform, thick, partially Gram-positive cells after growth in sublethal concentrations of penicillin. The same procedure most often results in formation of longer rods in other rod-shaped *Moraxella* species and in *N. elongata,* and has no elongation effect on coccal *Neisseria* and *Branhamella* (Bøvre and Holten, 1970; Bøvre et al., 1976, 1977b; Catlin, 1975; Fig. 3). Usually grows with medium-sized colonies, somewhat smaller than of *M. nonlique-*

Fig. 3. *Moraxella osloensis,* Gram-stained preparation after growth close to a depot of penicillin. Note fusiform cells and retention of the stain (black cells or parts of cells). × 970. [From Bøvre et al., 1977b, with permission.]

faciens and without the distinct colony-type variation of the latter (see above). Usually (genetically verified exceptions) grows in mineral salt media with ammonium ions and acetate or β-hydroxybutyrate, resulting in intracellular inclusions of poly-β-hydroxybutyrate stainable with Sudan Black B (Baumann, Doudoroff, and Stanier, 1968a; Bøvre and Henriksen, 1976; Bøvre et al., 1977b; Juni, 1974). Usually slightly less susceptible to penicillin than other oxidase-positive Neisseriaceae, often with minimal inhibitory concentrations (MIC) around 0.1 U/ml (Baumann, Doudoroff, and Stanier, 1968a; Bøvre et al., 1977b). Penicillin-resistant isolates have been described (Hansen et al., 1974), which produce β-lactamase (authors' unpublished observations). Although the cellular lipid composition is of the *Moraxella* type generally, the fatty acid pattern is distinct and reproducible. Gas chromatography is therefore a powerful aid in the identification of this species (Bøvre et al., 1977b; Jantzen et al., 1974b, 1975, 1978).

GENETIC IDENTIFICATION. At present, *M. osloensis* is considered genetically homogeneous and well separated from other species of Neisseriaceae. Also, transformable strains occur frequently. Therefore, genetic identification by transformation is simple (Table 2; Bøvre et al., 1977b; Juni, 1974; see general section of "Identification"). Recently, however, a phenotypically almost identical strain from cattle showed a genetic affinity to the human isolates similar to the mutual relations between *M. bovis* and *M. nonliquefaciens* (unpublished; see

Table 2). It is possible, therefore, that *M. osloensis* as now recognized may have closely related species or subspecies in its vicinity, which strengthens the need for a quantitative procedure when transformation is used for identification. G+C content of DNA, 43–46 mol%.

Moraxella phenylpyruvica Bøvre and Henriksen 1967
Neotype strain 2863 (ATCC 23333 = NCTC 10526)
Includes the previous *M. polymorpha* (Flamm, 1957; Bøvre and Henriksen, 1967b)

SOURCES, HABITATS, AND PATHOGENICITY. Isolated from blood, cerebrospinal fluid, urine, genitourethral tract, wound exudate, abscess, pleural fluid and, occasionally, the nose of humans; from respiratory tract, genital tract, and brain of sheep and cattle; from intestine of a goat; and from genital tract of pigs. Main habitat is unknown, but the relative frequency of isolation indicates that animals are important reservoirs. The host range is still insufficiently examined, however. The pathogenicity of *M. phenylpyruvica* is unknown, but relatively numerous isolates from human blood and cerebrospinal fluid, and an apparent association with genital tract ailments in animals, warrant further studies of this species as an important potential pathogen (Bøvre and Henriksen, 1967b; Snell, Hill, and Lapage, 1972; Tatum, Ewing, and Weaver, 1974; authors' unpublished studies).

MAIN PHENOTYPIC CHARACTERISTICS. Medium to plump rods (generally like *M. lacunata*). Usually grows with relatively small colonies (0.2–0.5 mm in 24 h and 1 mm in 48 h). The colonies often appear somewhat opaque, with a very slight pink hue, probably due to blood-pigment accumulation. Distinct but slow growth at 4–10°C. Growth at high salt and bile concentrations (MIC for NaCl most often 9 g/100 ml and for bile salts, 5 g/100 ml), and growth stimulation by bile salts up to 4 g/100 ml. Urease production most often positive and, with some exceptions, phenylalanine/tryptophan deaminase positive. A strain that lacks both urease and the deaminase activities has been observed (supplied by U. Berger). Although most strains are highly susceptible to penicillin, several strains of human and animal origin have been found penicillin resistant and to produce β-lactamase (Bøvre and Henriksen, 1967b; Bøvre et al., 1976, 1977b; Snell, Hill, and Lapage, 1972; Tatum, Ewing, and Weaver, 1974; authors' unpublished studies). *M. phenylpyruvica* has a characteristic, although slightly variable, whole-cell fatty acid profile, but which is close to that of *M. atlantae* (Jantzen et al., 1974b; Bøvre et al., 1976, 1977b). The species is distinguished from all other presently recognized *Moraxella* entities by its lack of cellular waxes (Bryn, Jantzen, and Bøvre, 1977).

Another unclassified group of oxidase-positive rods within Neisseriaceae resembles *M. phenylpyruvica* in its ability to grow both at ordinary incubation and refrigerator temperatures, as well as on high salt and bile concentrations, and in showing strongly positive phenylalanine/tryptophan deaminase reactions. This group grows with much larger, opaque colonies and contains saccharolytic strains (supplied by E. Falsen). Some of these properties are shared by the previously mentioned group of psychrophilic, oxidase-positive rods (Bøvre et al., 1974; see this chapter, "Introduction"). Their mutual relationship has not been examined, but they appear similar also in fatty acid composition, distinct from *M. phenylpyruvica* (unpublished).

GENETIC IDENTIFICATION. Transformable strains not yet detected. The species can therefore only be used as donor in genetic transformation, where it shows very low affinities to other rods (Table 2). A genetic exclusion of *M. atlantae* by transformation in combination with gas chromatographic fatty acid analysis is useful for exact identification. The possibility of genetic subunits of *M. phenylpyruvica* as presently conceived has to be studied. G + C content of DNA, 42.5–43.5 mol%.

Moraxella atlantae Bøvre et al. 1976
Type strain 5118 (ATCC 29525 = NCTC 11091)
Most strains described have previously been allocated to King's group 3 (Bovre et al., 1976; Tatum, Ewing, and Weaver, 1974)

SOURCES, HABITATS, AND PATHONGENICITY. Isolated from blood, cerebrospinal fluid, and spleen of humans. Not reported from animals, but occurrence insufficiently examined. Natural habitats and pathogenicity not defined (see references above).

MAIN PHENOTYPIC CHARACTERISTICS. Variably sized, often plump, diplococcobacillary cells, of dimensions like *M. lacunata's*. Colonies usually very small, half of *M. phenylpyruvica* size (0.1–0.3 mm in 24 h, 0.2–0.5 mm in 48 h). Two main colony variants, one hemispherical with an even outline, the other more flat with a tendency to form a spreading zone. Pitting of the agar often seen, most pronounced beneath the latter type of colonies. As in *M. phenylpyruvica*, the colonies may appear slightly pink. Hitherto always nitrate negative, distinct from *M. nonliquefaciens* and most strains of *M. phenylpyruvica*. Distinguished from most strains of *M. phenylpyruvica* by being phenylalanine/tryptophan deaminase negative, without ability to grow at refrigerator temperature and high salt/bile concentrations. Stimulated by lower concentrations of bile, however. Susceptible to penicillin, but with somewhat high MIC values (0.05–0.1 U/ml), as *M. osloensis* (Bøvre et al., 1976). Cellular lipid composition may simulate that of *M. phenyl-pyruvica* but includes waxes (see *M. phenylpyruvica;* Bøvre et al., 1976; Bryn, Jantzen, and Bøvre, 1977). Cells of *M. atlantae* contain mannose, distinct from the monosaccharide contents of other rod-shaped *Moraxella* species (Jantzen, Bryn, and Bøvre, 1976).

GENETIC AFFINITIES. Usually transformable and easily identified by genetic transformation (Table 2). G + C content of DNA, 46.5–47.5 mol%.

Moraxella (Branhamella) catarrhalis (Frosch and Kolle 1896) Henriksen and Bøvre 1968; (Catlin 1970) Bøvre 1979
Neotype strain Ne 11 (ATCC 25238 = NCTC 11020)

SOURCES, HABITATS, AND PATHOGENICITY. Most frequently isolated from normal and inflamed nasal mucosa, inflammatory secretions of middle ear and maxillary sinus, and bronchial aspirate and sputum in bronchitis and pneumonia, all in man. Less frequently found in the normal pharynx, conjunctivae, and genitourethral tract (Arkwright, 1907; Axelson and Brorson, 1973; Berger, 1963; Bøvre, 1970b; Coffey, Martin, and Booth, 1967; Ghon, Pfeiffer, and Sederl, 1902; McNeely, Kitchens, and Kluge, 1976; May, 1953; Ninane, Joly, and Kraytman, 1978). Also reported from cerebrospinal fluid and blood cultures in meningitis and endocarditis (Clarke and Haining, 1936; original observations and reviews by Zinke [1936] and Cocchi and Ulivelli [1968]). However, the identification of some of the isolates are not convincingly reported (occasionally described as rods and with atypical colony texture [see below]) and may have been too dependent on negative carbohydrate acidification tests, which are seen sometime also with *N. meningitidis* (Arkwright, 1909; Bøvre, 1969). The organism has also been reported from animal sources, but it is an open question whether any of these isolates belong to the same species as defined genetically (see below). Main natural habitat is the nasal cavity of man, where it occurs with about the same frequency as *M. nonliquefaciens* (Bøvre, 1970b). It is particularly frequent in the nose of infants, where an incidence of about 60% has been recorded (Arkwright, 1907). It has been found in pure culture in 7% of acute otitis media exudate of young children (Coffey, Martin, and Booth, 1967) and has been encountered in transtracheal aspirates from 7% of patients suffering from acute exacerbation of chronic bronchitis (Ninane, Joly, and Kraytman, 1978). On the basis of findings like this, several authors express firm belief in the pathogenicity of *M. (B.) catarrhalis*. Others, like Henriksen (1976), consider it as a well-adapted parasite with very low or doubtful potential pathogenicity. Its tendency to grow, often in pure culture,

from inflammatory secretions in areas of the respiratory tract that are normally sterile, does not necessarily mean much more than an intraluminal spread of a harmless parasite. Still, a pathogenic role as an extranasal invader, particularly in otitis media of infants and respiratory disease of the compromised host, may be possible and should be studied further. A comparative study of *M. (B.) catarrhalis* and *M. nonliquefaciens* would be of interest, since these two organisms appear to behave much in the same way in relation to the human host.

MAIN PHENOTYPIC CHARACTERISTICS. Coccal cells of variable size, often as diplococci with adjacent sides flattened. Grows with rather opaque, raised (hemispherical) colonies that are smaller than *M. nonliquefaciens* colonies after 24 h of incubation (about 1 mm). May become several mm in diameter and more flat, convex, on prolonged incubation. Most often a characteristically friable colony texture; the intact colony is movable on the agar surface by pressure from the side, without the coherent, sticky texture of most "true" neisseriae. At present, easy to distinguish from other oxidase positives in the human host (Berger, 1963; Bøvre, 1965b; Catlin, 1970; Reyn, 1974), but the differential diagnosis of present and future animal species of subgenus *Branhamella* may become extremely difficult (see Table 5 and below). Usually highly susceptible to penicillin (Bøvre, 1965b), but strains resistant on the basis of beta-lactamase production have been reported (Malmvall, Brorson, and Johnsson, 1977; Ninane, Joly, and Kraytman, 1978). The cellular lipid composition shows the general traits of the genus *Moraxella*, but the fatty acid pattern is reproducibly distinguishable from other species of subgenus *Branhamella* now recognized (Bryn, Jantzen, and Bøvre, 1977; Jantzen et al., 1974b; author's unpublished studies of 58 strains).

GENETIC IDENTIFICATION. Frequently transformable and easily identified by quantitative or semi-quantitative Str^r transformation (Table 3). Nonhemolytic branhamellae isolated from cattle, sheep, horse, and pigs that have been hitherto tested in genetic transformation and are partly included in the table do not belong to the *M. (B.) catarrhalis* entity (unpublished). The genetic affinities of *catarrhalis* isolates reported from dogs (Bailie, Stowe, and Schmitt, 1978) and the nonhemolytic and hemolytic variants of this organism reported from bovine eyes by Wilcox (1970a, b) have not yet been determined. Strain NCTC 4103, which is not included in the results of Table 3, is genetically separate from the neotype strain (Bøvre, 1965b; Catlin and Cunningham, 1964a) and may represent a separate, undescribed, species. G+C content of DNA, 40–43 mol%.

Moraxella (Branhamella) caviae (Pelczar 1953)
Henriksen and Bøvre 1968, Bøvre 1979
Type strain ATCC 14659 = NCTC 10293

SOURCES, HABITATS, AND PATHOGENICITY. Insufficiently examined. Isolated from pharynx and mouth of healthy guinea pigs (Berger, 1962; Pelczar, 1953; Pelczar, Hajek, and Faber, 1949).

MAIN PHENOTYPIC CHARACTERISTICS. Similar to *M. (B.) ovis* in several respects. Originally considered as a pigmented species, but pigmentation has not been confirmed. Weakly hemolytic. Penicillin susceptible. Cellular lipid composition close to that of other branhamellae, but apparently distinguishable (Baumann, Doudoroff, and Stanier, 1968a; Berger, 1963; Bøvre, 1965b; Bryn, Jantzen, and Bøvre, 1977; Jantzen et al., 1974b; Pelczar, 1953; Pelczar, Hajek, and Faber, 1949).

GENETIC IDENTIFICATION. The genetic separation indicated in Table 3 shows that identification by Str^r transformation would be possible, provided a competent strain is found. The type strain, which is the only one examined in transformation, is incompetent (Bøvre, 1965b). G+C content of DNA, 44.5–47.5 mol%.

Moraxella (Branhamella) cuniculi
Identical with *Neisseria cuniculi* Berger 1962. The transfer to genus *Moraxella* or to *Branhamella* has not yet been formally proposed, but the designation *Branhamella* has been used by Snell and Lapage (1976)
Type strain ATCC 14688 = NCTC 10297

SOURCES, HABITATS, AND PATHOGENICITY. Insufficiently examined. Has been isolated from the mouth of healthy rabbits (Berger, 1962). Similar isolates from hares (supplied by J. J. S. Snell) have not yet been examined for genetic compatibility. Strains named *N. cuniculi* var. *gigantea,* from the human vagina and from a dog's mouth (Berger, 1962, 1963), no longer exist (U. Berger, personal communication).

MAIN PHENOTYPIC CHARACTERISTICS. Biochemically inactive, including a negative nitrate-reduction test (Berger, 1962, 1963). Cellular fatty acid composition similar to the pattern of *Moraxella* in general (unpublished experiments with the type strain). This similarity is in contrast to the results of Lambert et al. (1971), who found the same strain similar to "true" *Neisseria* in these terms.

GENETIC IDENTIFICATION. The type strain is transformable and identification is possible by quantitative Str^r transformation (unpublished; Table 3). G+C content of DNA, 44.5 mol%.

Moraxella (Branhamella) ovis (Lindqvist 1960)
Henriksen and Bøvre 1968, Bøvre 1979
Proposed neotype strain 199/55 (ATCC 33078 = NCTC 11227)

SOURCES, HABITATS, AND PATHOGENICITY. Isolated from eyes of sheep and cattle, nasal cavity of sheep and of a horse, and from the mouth of a sheep. Main habitats are not fully known but evidently comprise the upper respiratory tract, including the conjunctival sac, of ruminants. *M. (B.) ovis* has been a frequent finding in cultures from eyes in infectious keratoconjunctivitis of sheep and also, together with *M. bovis,* from cases of keratoconjunctivitis in cattle. However, the occurrence also on healthy conjunctivae indicates that the pathogenicity, if any, is dependent on accessory factors. The low pathogenicity has been confirmed by mostly unsuccessful attempts to reproduce the disease by experimental infection of ovine and bovine eyes (Fairlie, 1966; Lindqvist, 1960; Pedersen, 1972; Spradbrow, 1968, 1971; Spradbrow and Smith, 1967; unpublished genetic identification of bovine and equine isolates).

MAIN PHENOTYPIC CHARACTERISTICS. Cells like those of *M. (B.) catarrhalis,* sometimes with distinct tetrade formation. Somewhat larger colonies than of *M. (B.) catarrhalis,* grayish white, more opaque than *M. bovis* colonies. Colonies usually surrounded by a narrow zone of β-hemolysis. Nonhemolytic variants may arise spontaneously in cultures, and genetically confirmed, nonhemolytic isolates of *M. (B.) ovis* have been observed in primary cultures from sheep and cattle. The only isolate yet studied from a horse was also nonhemolytic. Such isolates may be difficult to distinguish by present phenotypic methods from other, not yet described nonhemolytic branhamellae, e.g., from sheep and pigs (Table 3). Highly susceptible to penicillin (Bøvre, 1965b; unpublished; references cited above). The cellular lipid composition is similar to that of *Moraxella* in general, but appears distinguishable when a standardized procedure of fatty acid analysis is used (Bryn, Jantzen, and Bøvre 1977; Jantzen et al., 1974b).

GENETIC IDENTIFICATION. Frequently transformable and easily identified by Str^r transformation, which must follow a quantitative or semiquantitative procedure because of the relatively close genetic affinities to some other branhamellae (Table 3). The genetic relations of *M. (B.) ovis* to the nonhemolytic and hemolytic *catarrhalis* varieties described from eyes of cattle (Wilcox, 1970a,b) have not yet been critically examined. G+C content of DNA, 44.5–46.5 mol%.

Kingella kingae (Henriksen and Bøvre 1968)
Henriksen and Bøvre 1976
Type strain 4177/66 (ATCC 23330 = NCTC 10529)

SOURCES, HABITATS, AND PATHOGENICITY. Most frequently isolated from throat specimens and blood cultures, but also from the genitourethral tract, nose, abscess, bone lesion, and joint. Only human isolates reported as yet. Main natural habitat may be the human pharynx, but other possible sources should be examined for determination of habitat spectrum. The pathogenicity is unknown, but probably low (Bøvre et al., 1977b; Henriksen, 1969a, 1973; Henriksen and Bøvre, 1968a; Tatum, Ewing, and Weaver, 1974).

MAIN PHENOTYPIC CHARACTERISTICS. Relatively slender to medium thick rods (0.5–0.8/μm) with square ends. Occur in pairs and have a tendency to form chains. Colonies small to medium-sized (0.1–0.6 mm in 24 h, 1–2 mm in 48 h), translucent to semiopaque. Frequently grows in primary cultures as pits or small corroded areas in the agar medium. This form represents an SC colony type (see above), which consists of fimbriated cells that exhibit twitching motility, and which may show variation into a N colony type of nonfimbriated cells. A narrow zone of β-hemolysis surrounds the colonies. Acid is produced from glucose and maltose, but not sucrose. Highly susceptible to penicillin (Frøholm and Bøvre, 1972; Henrichsen, Frøholm, and Bøvre, 1972; Henriksen, 1969a, 1973; Henriksen and Bøvre, 1968a). Cellular lipid composition distinct from other Neisseriaceae examined, with a certain resemblance to genus *Neisseria,* including a lack of waxes (Jantzen et al., 1974b, 1978; Bryn, Jantzen, and Bøvre, 1977).

GENETIC IDENTIFICATION. Frequently transformable; competence is associated with the SC colony type (Bøvre and Frøholm, 1971, 1972b). Hitherto no genetically closely related species among Neisseriaceae observed (*N. denitrificans* not examined), and genetic identification by Str^r transformation is therefore simple (Bøvre et al., 1977b; Henriksen and Bøvre, 1968a; see this chapter, "Introduction"). G+C content of DNA, 44.5–47.5 mol%.

Kingella indologenes Snell and Lapage 1976
Type strain ATCC 25869 = NCTC 10717

SOURCES, HABITATS, AND PATHOGENICITY. Isolated from human angular conjunctivitis (van Bijsterveld, 1970) and corneal abscess (Sutton et al., 1972). Natural habitats and pathogenic potential largely unknown.

MAIN PHENOTYPIC CHARACTERISTICS. Medium thick to slender rods, more slender than *K. kingae* in some studies. Colonies comparable to those of *K. kingae,* but may be smaller and more opaque (nonhemolytic). Corroding colonies observed in one strain (see below). Only weakly positive or negative oxidase reaction with the dimethyl reagent (see general section on "Identification"). Produces indole. Acid produced from glucose, maltose (irregularly), and sucrose. Penicillin susceptible (Bøvre et al., 1974; Henriksen, 1973; Snell and Lapage, 1976; Sutton et al., 1972; van Bijsterveld, 1970; unpublished observations).

GENETIC IDENTIFICATION. Corroding clones of NCTC 10883 (strain of Sutton et al. [1972]) are transformable. This strain shows intraspecies relationship with the type strain in Str^r transformation. Thus far, *K. indologenes* has shown no transformation affinity to *K. kingae (K. denitrificans* not examined) or other genera of the family (Bøvre et al., 1977a; unpublished studies in collaboration with J. E. Fuglesang). The species therefore appears to be easily identified by genetic transformation.

Kingella denitrificans Snell and Lapage 1976
Identical with group TM-1 of CDC (Hollis, Wiggins, and Weaver, 1972; Tatum, Ewing, and Weaver, 1974)
Type strain NCTC 10995

SOURCES, HABITATS, AND PATHOGENICITY. Using a selective medium such as Thayer-Martin medium *K. denitrificans* is frequently encountered in pharyngeal specimens from healthy adults and children (Hollis, Wiggins, and Weaver, 1972). The upper respiratory tract may therefore be its main natural habitat. Pathogenicity probably low or none for humans. Should be examined further with respect to habitats and hosts and clinical role.

MAIN PHENOTYPIC CHARACTERISTICS. Cells similar to, or distinctly more slender, than *K. kingae.* Colonies small, may be minute (similar to *M. atlantae*), translucent. Only weakly positive or negative oxidase reaction with the dimethyl reagent (see general section on "Identification"). Acid produced from glucose, maltose (irregularly), and not from sucrose (Hollis, Wiggins, and Weaver, 1972; Snell and Lapage, 1976; Tatum, Ewing, and Weaver, 1974).

GENETIC IDENTIFICATION. Not examined for transformable strains or genetic affinities to other species. Studies on relations between *K. denitrificans, K. indologenes, K. kingae, Neisseria elongata, Eikenella corrodens,* and *Cardiobacterium hominis* particularly warranted (see Snell and Lapage, 1976). G+C content of DNA, 54–55 mol%.

Neisseria elongata Bøvre and Holten 1970
Type strain M2 (ATCC 25295 = NCTC 10660)

SOURCES, HABITATS, AND PATHOGENICITY. *N. elongata* has been isolated from pharynx in healthy individuals and in pharyngitis and from bronchial aspirates, pus from perimandibular abscesses, genitourethral tract, and urine. Hitherto only recorded from humans. Main habitat may be the human pharynx, since it has been found in epipharyngeal specimens of 8.5% of healthy individuals, when selected for on a medium that contains lincomycin. Should be investigated further for habitat spectrum. At present, the species appears to be a largely harmless parasite (Berger and Piotrowski, 1974; Bøvre, Fuglesang, and Henriksen, 1972; Bøvre and Holten, 1970; Bøvre et al., 1977b; Hansen, Schoutens, and Yourassowsky, 1975; Henriksen and Holten, 1976; author's unpublished investigations).

MAIN PHENOTYPIC CHARACTERISTICS. Short and relatively slender rods (ca. 0.5 μm in diameter), with marked elongation effect of sublethal concentrations of penicillin. Grows with colonies largely similar to those of saprophytic coccal neisseriae, without distinct pigmentation (grayish white to slightly yellowish), often with clay-like, coherent consistency. Variation in colony form observed, with an SC type of colony consisting of fimbriated cells, and another N type with sparsely or non-fimbriated cells (Bøvre, Fuglesang, and Henriksen, 1972; Bøvre et al., 1977a; Henriksen and Holten, 1976). Distinguished from *Moraxella* by its less plump cells, nitrite reduction, and frequently negative catalase reaction, and from *Kingella* by its most frequently absent saccharolytic properties and its generally larger and more opaque colonies. Highly susceptible to penicillin (references above). Two subspecies other than that represented by the type strain proposed: subsp. *glycolytica* (Henriksen and Holten, 1976), which is strongly catalase positive and has a smooth colony consistency and which initially showed a distinct acid production from glucose; and subsp. *intermedia* (Berger and Falsen, 1976) which is distinctly catalase positive and shows a certain immunological difference from other strains. To what extent such differences may reflect more than a natural variation among strains of one and the same genetic unit is not known; however, *N. elongata* has a cellular fatty acid pattern of the genus *Neisseria* type, which is distinct from all other rod-shaped species hitherto examined (Jantzen et al., 1974b, 1978). The subspecies *glycolytica* is indistinguishable from the type strain of *N. elongata* by this means (Bøvre et al., 1977a). The species lacks cellular waxes (Bryn, Jantzen, and Bøvre, 1977) and contains heptose (Jantzen, Bryn, and Bøvre, 1976), as do coccal *Neisseria* species.

GENETIC IDENTIFICATION. Frequently transformable; the competence is associated with the SC colony type where such variation has been studied. The subspecies *glycolytica* shows identity reactions with the type strain of *N. elongata* in quantitative Str^r

transformation (Bøvre et al., 1977a). The subspecies *intermedia* has not been examined genetically. In genetic identification of *N. elongata*, the following facts are important: (i) the species, as now conceived, is rather heterogeneous in terms of Str^r transformation (affinities down to 2% of autologous reactions (Bøvre, Fuglesang, and Henriksen, 1972); and (ii) there exists another group of oxidase-positive rods (from dogs and dog bites), which has genetic affinity to the "true" genus *Neisseria* and which is clearly separate from *N. elongata* (unpublished studies in cooperation with W. E. Bailie, E. Falsen, and R. E. Weaver). Thus, although the genetic separation of the genera *Neisseria* and *Moraxella* makes it easy to apply transformation in the allocation of oxidase-positive rods to either group, the more complicated exact quantitation of the affinities is necessary also to identify an isolate as *N. elongata*, or the "relative" mentioned (see this chapter, "Introduction").

"Moraxella" urethralis Lautrop, Bøvre, and Frederiksen 1970

Largely identical with Group M-4 of Tatum, Ewing, and Weaver (1974). Representative strains ATCC 17960 and C1098 (Juni, 1977; Lautrop, Bøvre, and Frederiksen, 1970)

SOURCES, HABITATS, AND PATHOGENICITY. Most strains of *"M." urethralis* have been isolated from urine and the female genital tract. No animal isolate recovered. The main natural habitat may be the human genitourethral tract. Its potential as a pathogen is unknown (Hansen, Schoutens, and Yourassowsky, 1975; Juni, 1977; Lautrop, 1974; Lautrop, Bøvre, and Frederiksen, 1970; Riley, Hollis, and Weaver, 1974; Tatum, Ewing, and Weaver, 1974).

MAIN PHENOTYPIC CHARACTERISTICS. Small rods (ca. 0.5×1–1.5 μm), clearly distinct from the larger cells of rod-shaped moraxellae. Grows rather slowly (colonies ca. 0.5 mm or less in 24 h, and 1.5–2 mm in 48 h), with a tendency to grow best in areas of the blood agar with heaviest inoculum. The colonies are more overtly white than in other groups treated here, including *Acinetobacter*. Usually able to grow on simple mineral medium with ammonium ions and acetate (or β-hydroxybutyrate) as carbon source (occasional exceptions), with production of poly-β-hydroxybutyrate inclusions. May be distinguished from *M. osloensis* by its morphology and its reduction of nitrite. Susceptible to penicillin (see references above). Its cellular fatty acid composition is unique among hitherto known species of the family Neisseriaceae (Jantzen et al., 1974b, 1975, 1978), which makes gas chromatographic identification reliable (24 strains examined; unpublished). There are no waxes in the cells (Bryn, Jantzen, and Bøvre, 1977).

GENETIC IDENTIFICATION. Competence in genetic transformation infrequently observed and of low degree. Apparently has no distinct genetic affinity to other species of the family Neisseriaceae and may belong to another, not defined, genus of this family. May be identified by nutritional-marker transformation of the strain C1098 or, less easily, by Str^r transformation (Juni, 1977; Lautrop, Bøvre, and Frederiksen, 1970; unpublished results in cooperation with E. Juni; see this chapter, "Introduction", and general section of "Identification").

Acknowledgments

Our more recent investigations referred to have been supported in part by the Norwegian Research Council for Science and the Humanities (Grant C.17.14–20).

Literature Cited

Arkwright, J. A. 1907. On the occurrence of the *Micrococcus catarrhalis* in normal and catarrhal noses and its differentiation from other Gram-negative cocci. Journal of Hygiene **7**:145–154.

Arkwright, J. A. 1909. Varieties of the meningococcus with special reference to a comparison of strains from epidemic and sporadic sources. Journal of Hygiene **9**:104–121.

Axelson, A., Brorson, J. E. 1973. The correlation between bacteriological findings in the nose and maxillary sinus in acute maxillary sinusitis. Laryngoscope **83**:2003–2011.

Axenfeld, T. 1897. Ueber die chronische Diplobacillenconjunctivitis. Zentralblatt für Bakteriologie, Parasitenkunde, Infektionskrankheiten und Hygiene, Abt. 1 Orig. **21**:1–9.

Bailie, W. E., Stowe, E. C., Schmitt, A. M. 1978. Aerobic bacterial flora of oral and nasal fluids of canines with reference to bacteria associated with bites. Journal of Clinical Microbiology **7**:223–231.

Barner, R. D. 1952. A study of *Moraxella bovis* and its relation to bovine keratitis. American Journal of Veterinary Research **13**:132–144.

Baumann, P. 1968. Isolation of *Acinetobacter* from soil and water. Journal of Bacteriology **96**:39–42.

Baumann, P., Doudoroff, M., Stanier, R. Y. 1968a. Study of the *Moraxella* group. I. Genus *Moraxella* and the *Neisseria catarrhalis* group. Journal of Bacteriology **95**:58–73.

Baumann, P., Doudoroff, M., Stanier, R. Y. 1968b. A study of the *Moraxella* group. II. Oxidase-negative species (genus *Acinetobacter*). Journal of Bacteriology **95**:1520–1541.

Beijerinck, M. W. 1911. Pigments as products of oxidation by bacterial action. Proceedings of the Royal Academy of Sciences (Amsterdam) **13**:1066–1077.

Berenesi, G., Mészáros, G. 1960 *Moraxella duplex nonliquefaciens,* ein im Brochialsekret häufig nachweisbarer Keim. Zentralblatt für Bakteriologie, Parasitenkunde, Infektionskrankheiten und Hygiene, Abt. 1 Orig. **178**:406–408.

Berger, U. 1962. Über das Vorkommen von Neisserien bei einigen Tieren. Zeitschrift für Hygiene und Infektionskrankheiten, Medizinische Mikrobiologie, Immunologie und Virologie **148**:445–457.

Berger, U. 1963. Die anspruchslosen Neisserien. Ergebnisse der Mikrobiologie, Immunitätsforschung und Experimentelle Therapie **36**:97–167.

Berger, U., Falsen, E. 1976. Über die Artenverteilung von *Moraxella* und *Moraxella*-ähnlichen Keimen im Nasopharynx gesunder Erwachsener. Medical Microbiology and Immunology **162**:239–249.

Berger, U., Issi, R. 1971. Resistenz gegen Acetazolamid als taxonomisches Kriterium bei *Neisseria.* Archiv für Hygiene und Bakteriologie **154**:540–544.

Berger, U., Piotrowski, H. D. 1974. Die biochemische Diagnose von *Neisseria elongata* (Bøvre und Holten, 1970). Medical Microbiology and Immunology **159**:309–316.

Bottone, E., Allerhand, J. 1968. Association of mucoid encapsulated *Moraxella duplex* var. *nonliquefaciens* with chronic bronchitis. Applied Microbiology **16**:315–319.

Bøvre, K. 1963. Affinities between *Moraxella* spp. and a strain of *Neisseria catarrhalis* as expressed by transformation. Acta Pathologica et Microbiologica Scandinavica **58**:528.

Bøvre, K. 1964a. Studies on transformation in *Moraxella* and organisms assumed to be related to *Moraxella.* 1. A method for quantitative transformation in *Moraxella* and *Neisseria,* with streptomycin resistance as the genetic marker. Acta Pathologica et Microbiologica Scandinavica **61**:457–473.

Bøvre, K. 1964b. Studies on transformation in *Moraxella* and organisms assumed to be related to *Moraxella.* 2. Quantitative transformation reactions between *Moraxella nonliquefaciens* strains, with streptomycin resistance marked DNA. Acta Pathologica et Microbiologica Scandinavica **62**:239–248.

Bøvre, K. 1965a. Studies on transformation in *Moraxella* and organisms assumed to be related to *Moraxella.* 3. Quantitative streptomycin resistance transformation between *Moraxella bovis* and *Moraxella nonliquefaciens* strains. Acta Pathologica et Microbiologica Scandinavica **63**:42–50.

Bøvre, K. 1965b. Studies on transformation in *Moraxella* and organisms assumed to be related to *Moraxella.* 4. Streptomycin resistance transformation between asaccharolytic *Neisseria* strains. Acta Pathologica et Microbiologica Scandinavica **64**:229–242.

Bøvre, K. 1965c. Studies on transformation in *Moraxella* and organisms assumed to be related to *Moraxella.* 5. Streptomycin resistance transformation between serum-liquefying, nonhemolytic moraxellae, *Moraxella bovis* and *Moraxella nonliquefaciens.* Acta Pathologica et Microbiologica Scandinavica **65**:435–449.

Bøvre, K. 1965d. Studies on transformation in *Moraxella* and organisms assumed to be related to *Moraxella.* 6. A distinct group of *Moraxella nonliquefaciens*-like organisms (the 19116/51 group). Acta Pathologica et Microbiologica Scandinavica **65**:641–652.

Bøvre, K. 1967a. Studies on transformation in *Moraxella* and organisms assumed to be related to *Moraxella.* 7. Affinities between oxidase positive rods and neisseriae, as compared with group interactions on both sides. Acta Pathologica et Microbiologica Scandinavica **69**:92–108.

Bøvre, K. 1967b. Studies on transformation in *Moraxella* and organisms assumed to be related to *Moraxella.* 8. The relative position of some oxidase negative, immotile diplobacilli (*Achromobacter*) in the transformation system. Acta Pathologica et Microbiologica Scandinavica **69**:109–122.

Bøvre, K. 1969. Identification of an asaccharolytic *Neisseria* strain causing meningitis. Acta Pathologica et Microbiologica Scandinavica **76**:148–149.

Bøvre, K. 1970a. Pulse-RNA-DNA hybridization between rod-shaped and coccal species of the *Moraxella-Neisseria* groups. Acta Pathologica et Microbiologica Scandinavica, Sect. B **78**:565–574.

Bøvre, K. 1970b. Oxidase positive bacteria in the human nose. Incidence and species distribution, as diagnosed by genetic transformation. Acta Pathologica et Microbiologica Scandinavica, Sect. B **78**:780–784.

Bøvre, K. 1979. Proposal to divide the genus *Moraxella* Lwoff 1939 emend. Henriksen and Bøvre 1968 into two subgenera, subgenus *Moraxella* (Lwoff 1939) Bøvre 1979 and subgenus

Branhamella (Catlin 1970) Bøvre 1979. International Journal of Systematic Bacteriology **29**:403–406.

Bøvre, K., Bergan, T., Frøholm, L. O. 1970. Electron microscopical and serological characteristics associated with colony type in *Moraxella nonliquefaciens*. Acta Pathologica et Microbiologica Scandinavica, Sect. B **78**:765–779.

Bøvre, K., Fiandt, M. Szybalski, W. 1969. DNA base composition of *Neisseria*, *Moraxella*, and *Acinetobacter*, as determined by measurement of buoyant density in CsCl gradients. Canadian Journal of Microbiology **15**:335–338.

Bøvre, K., Frøholm, L. O. 1970. Correlation between the fimbriated state and competence of genetic transformation in *Moraxella nonliquefaciens* strains. Acta Pathologica et Microbiologica Scandinavica. Sect. B **78**:526–528.

Bøvre, K., Frøholm, L. O. 1971. Competence of genetic transformation correlated with the occurrence of fimbriae in three bacterial species. Nature New Biology **234**:151–152.

Bøvre, K., Frøholm, L. O. 1972a. Variation of colony morphology reflecting fimbriation in *Moraxella bovis* and two reference strains of *M. nonliquefaciens*. Acta Pathologica et Microbiologica Scandinavica, Sect. B **80**:629–640.

Bøvre, K., Frøholm, L. O. 1972b. Competence in genetic transformation related to colony type and fimbriation in three species of *Moraxella*. Acta Pathologica et Microbiologica Scandinavica, Sect. B **80**:649–659.

Bøvre, K., Frøholm, L. O., Henriksen, S. D., Holten, E. 1977a. Relationship of *Neisseria elongata* subsp. *glycolytica* to other members of the family *Neisseriaceae*. Acta Pathologica et Microbiologica Scandinavica, Sect. B **85**:18–26.

Bøvre, K., Fuglesang, J. E., Hagen, N., Jantzen, E., Frøholm, L. O. 1976. *Moraxella atlantae* sp. nov. and its distinction from *Moraxella phenylpyruvica*. International Journal of Systematic Bacteriology **26**:511–521.

Bøvre, K., Fuglesang, J. E., Henriksen, S. D. 1972. *Neisseria elongata*. Presentation of new isolates. Acta Pathologica et Microbiologica Scandinavica, Sect. B **80**:919–922.

Bøvre, K., Fuglesang, J. E., Henriksen, S. D., Lapage, S. P., Lautrop, H., Snell, J. J. S. 1974. Studies on a collection of Gram-negative bacterial strains showing resemblance to moraxellae: examination by conventional bacteriological methods. International Journal of Systematic Bacteriology **24**:438–446.

Bøvre, K., Hagen, N., Berdal, B. P., Jantzen, E. 1977b. Oxidase positive rods from cases of suspected gonorrhoea. A comparison of conventional, gas chromatographic and genetic methods of identification. Acta Pathologica et Microbiologica Scandinavica, Sect. B **85**:27–37.

Bøvre, K., Henriksen, S. D. 1967a. A new *Moraxella* species, *Moraxella osloensis*, and a revised description of *Moraxella nonliquefaciens*. International Journal of Systematic Bacteriology **17**:127–135.

Bøvre, K., Henriksen, S. D. 1967b. A revised description of *Moraxella polymorpha* Flamm 1957, with a proposal of a new name, *Moraxella phenylpyrouvica* for this species. International Journal of Systematic Bacteriology **17**:343–360.

Bøvre, K., Henriksen, S. D. 1976. Minimal standards for description of new taxa within the genera *Moraxella* and *Acinetobacter*: proposal by the Subcommittee on *Moraxella* and Allied Bacteria. International Journal of Systematic Bacteriology **26**:92–96.

Bøvre, K., Holten, E. 1970. *Neisseria elongata* sp. nov., a rod-shaped member of the genus *Neisseria*. Re-evaluation of cell shape as a criterion in classification. Journal of General Microbiology **60**:67–75.

Brooks, K., Sodeman, T. 1974. Clinical studies on a transformation test for identification of *Acinetobacter* (*Mima* and *Herellea*). Applied Microbiology **27**:1023–1026.

Bruner, D. W., Fabricant, J. 1954. A strain of *Moraxella anatipestifer* (*Pfeifferella anatipestifer*) isolated from ducks. Cornell Veterinarian **44**:461–464.

Bryn, K., Jantzen, E., Bøvre, K. 1977. Occurrence and patterns of waxes in *Neisseriaceae*. Journal of General Microbiology **102**:33–43.

Butzler, J. P., Hansen, W., Cadranel, S., Henriksen, S. D. 1974. Stomatitis with septicemia due to *Moraxella osloensis*. Journal of Pediatrics **84**:721–722.

Carter, G. R., Isoun, T. T., Heahey, K. K. 1970. Occurrence of *Mima* and *Herellea* species in clinical specimens from various specimens. Journal of the American Veterinary Medical Association **156**:1313–1318.

Catlin, B. W. 1964. Reciprocal genetic transformation between *Neisseria catarrhalis* and *Moraxella nonliquefaciens*. Journal of General Microbiology **37**:369–379.

Catlin, B. W. 1970. Transfer of the organism named *Neisseria catarrhalis* to *Branhamella* gen. nov. International Journal of Systematic Bacteriology **20**:155–159.

Catlin, B. W. 1975. Cellular elongation under the influence of antibacterial agents: Way to differentiate coccobacilli from cocci. Journal of Clinical Microbiology **1**:102–105.

Catlin, B. W., Cunningham, L. S. 1961. Transforming activities and base contents of deoxyribonucleate preparations from various neisseriae. Journal of General Microbiology **26**:303–312.

Catlin, B. W., Cunningham, L. S. 1964a. Genetic transformation of *Neisseria catarrhalis* by deoxyribonucleate preparations having different average base compositions. Journal of General Microbiology **37**:341–352.

Catlin, B. W., Cunningham, L. S. 1964b. Transforming activities and base composition of deoxyribonucleates from strains of *Moraxella* and *Mima*. Journal of General Microbiology **37**:353–367.

Clarke, R. M., Haining, R. B. 1936. *Neisseria catarrhalis* endocarditis. Annals of Internal Medicine **10**:117–121.

Cocchi, P., Ulivelli, A. 1968; Meningitis caused by *Neisseria catarrhalis*. Acta Paediatrica Scandinavica **57**:451–453.

Coffey, J. D., Martin, A. D., Booth, H. N. 1967. *Neisseria catarrhalis* in exudate otitis media. Archives of Otolaryngology **86**:403–406.

Cowan, S. T. 1974. Cowan and Steel's manual for the identification of medical bacteria, 2nd ed. Cambridge: Cambridge University Press.

De Ley, J. 1968. DNA base composition and taxonomy of some *Acinetobacter* strains. Antonie van Leeuwenhoek Journal of Microbiology and Serology **34**:109–114.

Doi, R. H., Igarashi, R. T. 1965. Conservation of ribosomal and messenger ribonucleic acid cistrons in *Bacillus* species. Journal of Bacteriology **90**:384–390.

Dowda, H. 1977. Evaluation of two rapid methods for identification of commonly encountered nonfermenting or oxidase-positive, Gram-negative rods. Journal of Clinical Microbiology **6**:605–609.

Dubnau, D., Smith, I., Morell, P., Marmur, J. 1965. Gene conservation in *Bacillus* species. I. Conserved genetic and nucleic acid base sequence homologies. Genetics **54**:491–498.

Ellingworth, S., M'Leod, J. W., Gordon, J. 1929. Further observations on the oxidation by bacteria of compounds of the para-phenylene diamine series. Journal of Pathology and Bacteriology **32**:173–183.

Eyre, J. W. 1900. A clinical and bacteriological study of diplobacillary conjunctivitis. Journal of Pathology and Bacteriology **6**:1–13.

Fairlie, G. 1966. The isolation of a haemolytic *Neisseria* from cattle and sheep in the north of Scotland. Veterinary Record **78**:649–650.

Feigin, R. D., San Joaquin, V., Middelkamp, J. N. 1969. Septic arthritis due to *Moraxella osloensis*. Journal of Pediatrics **75**:116–117.

Flamm, H. 1956. *Moraxella saccharolytica* (sp. n.) aus dem Liquor eines Kindes mit Meningitis. Zentralblatt für Bakteriologie, Parasitenkunde, Infektionskrankheiten und Hygiene, Abt. 1 Orig. **166**:498–502.

Flamm, H. 1957. Eine weitere neue Species des Genus *Moraxella*

M. polymorpha sp. n. Zentralblatt für Bakteriologie, Parasitenkunde, Infektionskrankheiten und Hygiene, Abt. 1 Orig. **168**:261–267.

Frøholm, L. O. 1978. Bacterial fimbriae (pili) and associated features. A review. National Institute of Public Health Annals (Norway) **1**:35–47.

Frøholm, L. O., Bøvre, K. 1972. Fimbriation associated with the spreading-corroding colony type in *Moraxella kingii.* Acta Pathologica et Microbiologica Scandinavica, Sect. B **80**:641–648.

Frosch, P., Kolle, W. 1896. Die Mikrokokken, pp. 154–155. In: Flügge, C. (ed.), Die Mikroorganismen, 3rd ed., Sect. 2. Leipzig: Vogel.

Ghon, A., Pfeiffer, H., Sederl, H. 1902. Der *Micrococcus catarrhalis* (R. Pfeiffer) als Krankheitserreger. Zeitschrift für Klinische Medizin **44**:262–295.

Gilardi, G. L. 1973. Nonfermentative gram-negative bacteria encountered in clinical specimens. Antonie van Leeuwenhoek Journal of Microbiology and Serology **39**:229–242.

Hansen, W., Schoutens, E., Yourassowsky, E. 1975. Isolement de souches microbiennes peu connues et récemment décrites: *Neisseria elongata* et *Moraxella urethralis.* Annales de Microbiologie **126A**:401–404.

Hansen, W., Butzler, J. P., Fuglesang, J. E., Henriksen, S. D. 1974. Isolation of penicillin and streptomycin resistant strains of *Moraxella oslensis.* Acta Pathologica et Microbiologica Scandinavica, Sect. B **82**:318–322.

Hauduroy, P., Ehringer, G., Urbain, A., Guillot, G., Magrou, J. 1937. Dictionnaire des bactéries pathogénes, p. 247. Paris: Masson et Cie.

Henrichsen, J., Frøholm, L. O., Bøvre, K. 1972. Studies on bacterial surface translocation. 2. Correlation of twitching motility and fimbriation in colony variants of *Moraxella nonliquefaciens, M. bovis* and *M. kingii.* Acta Pathologica et Microbiologica Scandinavica, Sect. B **80**:445–452.

Henriksen, S. D. 1951. *Moraxella duplex* var. *nonliquefaciens* as a cause of bronchial infection. Acta Pathologica et Microbiologica Scandinavica **29**:258–262.

Henriksen, S. D. 1958. *Moraxella duplex* var. *nonliquefaciens,* habitat and antibiotic sensitivity. Acta Pathologica et Microbiologica Scandinavica **43**:157–161.

Henriksen, S. D. 1963. *Mimeae.* The standing in nomenclature of the names of this tribus and of its genera and species. International Bulletin of Bacteriological Nomenclature and Taxonomy **13**:51–57.

Henriksen, S. D. 1969a. Corroding bacteria from the respiratory tract. 1. *Moraxella kingii.* Acta Pathologica et Microbiologica Scandinavica **75**:85–90.

Henriksen, S. D. 1969b. Proposal of a neotype strain for *Moraxella lacunata.* International Journal of Systematic Bacteriology **19**:263–265.

Henriksen, S. D. 1971. Designation of a neotype strain for *Moraxella bovis* (Hauduroy et al.) Murray. International Journal of Systematic Bacteriology **21**:28.

Henriksen, S. D. 1973. *Moraxella, Acinetobacter,* and the *Mimeae.* Bacteriological Reviews **37**:522–561.

Henriksen, S. D. 1976. *Moraxella, Neisseria, Branhamella,* and *Acinetobacter,* Annual Review of Microbiology **30**:63–83.

Henriksen, S. D., Bøvre, K. 1968a. *Moraxella kingii* sp. nov., a haemolytic, saccharolytic species of the genus *Moraxella.* Journal of General Microbiology **51**:377–385.

Henriksen, S. D., Bøvre, K. 1968b. The taxonomy of the genera *Moraxella* and *Neisseria.* Journal of General Microbiology **51**:387–392.

Henriksen, S. D., Bøvre, K. 1969b. Corroding and spreading colonies in *Moraxella nonliquefaciens.* Acta Pathologica et Microbiologica Scandinavica **76**:459–463.

Henriksen, S. D., Bøvre, K. 1976. Transfer of *Moraxella kingae* Henriksen and Bøvre to the genus *Kingella* gen. nov. in the family *Neisseriaceae.* International Journal of Systematic Bacteriology **26**:447–450.

Henriksen, S. D., Holten, E. 1976. *Neisseria elongata* subsp. *glycolytica* subsp. nov. International Journal of Systematic Bacteriology **26**:478–481.

Hill, L. R. 1966. An index to deoxyribonucleic acid base compositions of bacterial species. Journal of General Microbiology **44**:419–437.

Hollis, D. G., Wiggins, G. L., Weaver, R. E. 1972. An unclassified Gram-negative rod isolated from the pharynx on Thayer-Martin medium (selective agar). Applied Microbiology **24**:772–777.

Holten, E. 1973. Glutamate dehydrogenases in genus *Neisseria.* Acta Pathologica et Microbiologica Scandinavica, Sect. B **81**:49–58.

Holten, E. 1974a. Glucokinase and glucose 6-phosphate dehydrogenase in genus *Neisseria.* Acta Pathologica et Microbiologica Scandinavica, Sect. B **82**:201–206.

Holten, E. 1974b. Immunological comparison of NADP-dependent glutamate dehydrogenase and malate dehydrogenase in genus *Neisseria.* Acta Pathologica et Microbiologica Scandinavica, Sect. B **82**:849–859.

Hughes, D. E., Pugh, G. W. 1970. Isolation and description of a *Moraxella* from horses with conjunctivitis. American Journal of Veterinary Research **31**:457–462.

Hughes, D. E., Pugh, G. W., McDonald, T. J. 1965. Ultraviolet radiation and *Moraxella bovis* in the etiology of bovine infectious keratoconjunctivitis. American Journal of Veterinary Research **26**:1331–1338.

Janik, A., Juni, E., Heym, G. A. 1976. Genetic transformation as a tool for detection of *Neisseria gonorrhoeae.* Journal of Clinical Microbiology **4**:71–81.

Jantzen, E. Bryn, K., Bøvre, K. 1974a. Gas chromatography of bacterial whole cell methanolysates. IV. A procedure for fractionation and identification of fatty acids and monosaccharides of cellular structures. Acta Pathologica et Microbiologica Scandinavica, Sect. B **82**:753–766.

Jantzen, E., Bryn, K., Bøvre, K. 1976. Cellular monosaccharide patterns of *Neisseriaceae.* Acta Pathologica et Microbiologica Scandinavica, Sect. B **84**:177–188.

Jantzen, E., Bryn, K., Bergan, T., Bøvre, K. 1974b. Gas chromatography of bacterial whole cell methanolysates. V. Fatty acid composition of neisseriae and moraxellae. Acta Pathologica et Microbiologica Scandinavica, Sect. B **82**:767–779.

Jantzen E., Bryn, K., Bergan, T., Bøvre, K. 1975. Gas chromatography of bacterial whole cell methanolysates. VII. Fatty acid composition of *Acinetobacter* in relation to the taxonomy of *Neisseriaceae.* Acta Pathologica et Microbiologica Scandinavica, Sect. B **83**:569–580.

Jantzen, E., Bryn, K., Hagen, N., Bergen, T., Bøvre, K. 1978. Fatty acids and monosaccharides of *Neisseriaceae* in relation to established taxonomy. National Institute of Public Health Annals (Norway) **1**:59–71.

Johnson, J. L., Anderson, R. S., Ordal, E. J. 1970. Nucleic acid homologies among oxidase-negative *Moraxella* species. Journal of Bacteriology **101**:568–573.

Jones, F. S., Little, R. B. 1923. An infectious ophtalmia of cattle. Journal of Experimental Medicine **38**:139–148.

Judicial Commission, 1971. Opinion 40. Rejection of the names *Mima* DeBord and *Herellea* DeBord and of the specific epithets *polymorpha* and *vaginicola* in *Mima ploymorpha* DeBord and *Herellea vaginicola* DeBord, respectively, International Journal of Systematic Bacteriology **21**:105–106.

Juni, E. 1972. Interspecies transformation of *Acinetobacter:* Genetic evidence for a ubiquitous genus. Journal of Bacteriology **112**:917–931.

Juni, E. 1974. Simple genetic transformation assay for rapid diagnosis of *Moraxella osloensis.* Applied Microbiology **27**:16–24.

Juni, E. 1977. Genetic transformation assays for identification of strains of *Moraxella urethralis.* Journal of Clinical Microbiology **5**:227–235.

Juni, E. 1978. Genetics and physiology of *Acinetobacter*. Annual Review of Microbiology **32**:349–371.

Jyssum, S., Bovre, K. 1974. Search for thymidine phosphorylase, nucleoside deoxyribosyltransferase and thymidine kinase in *Moraxella, Acinetobacter,* and allied bacteria. Acta Pathologica et Microbiologica Scandinavica, Sect. B **82**:57–66.

Kaffka, A. 1955. Zum Vorkommen und zur Pathogenität von *Moraxella duplex* var. *nonliquefaciens*. Zentralblatt für Bakteriologie Parasitenkunde, Infektionskrankheiten and Hygiene, Abt. 1 Orig. **164**:451–457.

Kaffka A. 1964. Zur Taxonomie und Pathogenität der Moraxellen. Archiv für Hygiene und Bakteriologie **148**:379–387.

Kaffka, A., Blödorn, R. 1959. *Moraxella duplex* var. *nonliquefaciens*, ein im Respirationstrakt häufig vorkommender Keim. Zentralblatt für Bakteriologie, Parasitenkunde, Infektionskrankheiten und Hygiene, Abt. 1 Orig. **174**:594–600.

Kittnar, E., Petrášová, S., Hejzlar, Kaňka, J. 1975. The problem of pathogenicity of moraxellae in the urogenital tract of women. Journal of Hygiene, Epidemiology, Microbiology and Immunology **19**:286–292.

Lambert, M. A., Hollis, D. G., Moss, C. W., Weaver, R. E., Thomas, M. L. 1971. Cellular fatty acids of nonpathogenic *Neisseria*. Canadian Journal of Microbiology **17**:1491–1502.

Lapage, S. P. 1974. Genus *Cardiobacterium* Slotnick and Dougherty 1964, 271, pp. 377–378. In: Buchanan, R. E., Gibbons, N. E. (eds.), Bergey's manual of determinative bacteriology, 8th ed. Baltimore: Williams & Wilkins.

Lautrop, H. 1974. Genus III. *Moraxella* Lwoff 1939, 173. Genus IV. *Acinetobacter* Brisou and Prévot 1954, 727, pp. 433–438. In: Buchanan, R. E., Gibbons, N. E., (eds.), Bergey's manual of determinative bacteriology, 8th ed. Baltimore: Williams & Wilkins.

Lautrop, H., Bovre, K., Frederiksen, W. 1970. A *Moraxella*-like microorganism isolated from the genito-urinary tract of man. Acta Pathologica et Microbiologica Scandinavica, Sect. B **78**:255–256.

Lewis, V. J., Weaver, R. E., Hollis, D. G. 1968. Fatty acid composition of *Neisseria* species as determined by gas chromatography. Journal of Bacteriology **96**:1–5.

Lindqvist, K. 1960. A *Neisseria* species associated with infectious keratoconjunctivitis of sheep—*Neisseria ovis* nov. spec. Journal of Infectious Diseases **106**:162–165.

Lwoff, A. 1939. Revision et démembrement des *Hemophilae* le genre *Moraxella* nov. gen. Annales de l'Institut Pasteur **62**:168–176.

McMeekin, T. A., Shewan, J. M. 1978. A review. Taxonomic strategies for *Flavobacterium* and related genera. Journal of Applied Bacteriology **45**:321–332.

McNeely, D. J., Kitchens, C. S., Kluge, R. M. 1976. Fatal *Neisseria (Branhamella) catarrhalis* pneumonia in an immunodeficient host. American Review of Respiratory Disease **114**:399–402.

Malmvall, B.-E., Brorson, J.-E., Johnsson, J. 1977. In vitro sensitivity to penicillin V and β-lactamase production of *Branhamella catarrhalis*. Journal of Antimicrobial Chemotherapy **3**:374.

May, J. R. 1953. The bacteriology of chronic bronchitis. Lancet **ii**:534–537.

Morax, V. 1896. Note sur un diplobacille pathogene pour la conjonctivite humaine. Annales de l'Institut Pasteur **10**:337–345.

Moss, C. W., Dees, S. B. 1978. Cellular fatty acids of *Flavobacterium meningosepticum* and *Flavobacterium* species group IIb. Journal of Clinical Microbiology **8**:772–774.

Murray, E. G. D. 1948. Genus II. *Moraxella* Lwoff, pp. 590–592. In: Breed, R. S., Murray, E. G. D., Hitchens, A. P. (eds.), Bergey's manual of determinative bacteriology, 6th ed. Baltimore: Williams & Wilkins.

Ninane, G., Joly, J., Kraytman, M. 1978. Bronchopulmonary infection due to *Branhamella catarrhalis:* 11 cases assessed by transtracheal puncture. British Medical Journal **1**:276–278.

Oberhofer, T. R. 1979. Comparison of the API 20E and Oxi/Ferm systems in identification of nonfermentative and oxidase-positive fermentative bacteria. Journal of Clinical Microbiology **9**:220–226.

Oberhofer, T. R., Rowen, J. W., Cunningham, G. F. 1977. Characterization and identification of Gram-negative, nonfermentative bacteria. Jounal of Clinical Microbiology **5**:208–220.

Otto, L. A., Pickett, M. J. 1976. Rapid method for identification of Gram-negative, nonfermentative bacilli. Journal of Clinical Microbiology **3**:566–575.

Owen, R. J., Snell, J. J. S. 1976. Deoxyribonucleic acid reassociation in the classification of flavobacteria. Journal of General Microbiology **93**:89–102.

Pande, P. G., Sekariah, P. C. 1960. A preliminary note on the isolation of *Moraxella caprae* nov. sp. from an outbreak of infectious keratoconjunctivitis in goats. Current Science **29**:276–277.

Pedersen, K. B. 1970. *Moraxella bovis* isolated from cattle with infectious keratoconjunctivitis. Acta Pathologica et Microbiologica Scandinavica, Sect. B **78**:429–434.

Pedersen, K. B. 1972. Isolation and description of a haemolytic species of *Neisseria* (*N. ovis*) from cattle with infectious keratoconjunctivitis. Acta Pathologica et Microbiologica Scandinavica, Sect. B **80**:135–139.

Pedersen, K. B. 1973. Infectious keratoconjunctivitis in cattle. Ph. D. thesis. Royal Veterinary and Agricultural University, Copenhagen.

Pedersen, K. B., Frøholm, L. O., Bøvre, K. 1972. Fimbriation and colony type of *Moraxella bovis* in relation to conjunctival colonization and development of keratoconjunctivitis in cattle. Acta Pathologica et Microbiologica Scandinavica, Sect. B **80**:911–918.

Pelczar, M. J. 1953. *Neisseria caviae* nov. spec. Journal of Bacteriology **65**:744.

Pelczar, M. J., Hajek, J. P., Faber, J. E. 1949. Characterization of *Neisseria* isolated from the pharyngeal region of guinea pigs. Journal of Infectious Disease **85**:239–242.

Pickett, M. J., Manclark, C. R. 1965. Tribe *Mimeae*. An illegitimate epithet. American Journal of Clinical Pathology **43**:161–165.

Pickett, M. J., Pedersen, M. M. 1970. Characterization of saccharolytic nonfermentative bacteria associated with man. Canadian Journal of Microbiology **16**:351–362.

Pugh, G. W., Hughes, D. E. 1968. Experimental bovine infectious keratoconjunctivitis caused by sunlamp irradiation and *Moraxella bovis* infection: Correlation of hemolytic ability and pathogenicity. American Journal of Veterinary Research **29**:835–839.

Pugh, G. W., Hughes, D. E., McDonald, T. J. 1966. The isolation and characterization of *Moraxella bovis*. American Journal of Veterinary Research **27**:957–962.

Pugh, G. W., Hughes, D. E., McDonald, T. J. 1968. Keratoconjunctivitis produced by *Moraxella bovis* in laboratory animals. Journal of Veterinary Research **29**:2057–2061.

Pugh, G. W., Hughes, D. E., Packer, R. A. 1970. Bovine infectious keratoconjunctivitis: interactions of *Moraxella bovis* and infectious bovine rhinotracheitis virus. American Journal of Veterinary Research **31**:653–662.

Reyn, A. 1974. Genus I. *Neisseria* Trevisan 1885, 105. Genus II. *Branhamella* Catlin 1970, 157, pp. 428–432. In: Buchanan, R. E., Gibbons, N. E. (eds.), Bergey's manual of determinative bacteriology, 8th ed. Baltimore: Williams & Wilkins.

Riley, P. S., Hollis, D. G., Weaver, R. E. 1974. Characterization and differentiation of 59 strains of *Moraxella urethralis* from clinical specimens. Applied Microbiology **28**:355–358.

Ryan, W. J. 1964. *Moraxella* commonly present on the conjunctiva of guinea pigs. Journal of General Microbiology **35**:361–372.

Samuels, S. B., Pittman, B., Tatum, H. W., Cherry, W. B. 1972. Report on a study set of moraxellae and allied bacteria. International Journal of Systematic Bacteriology **22**:19–38.

Scarlett, M. 1916. Infections cornéennes a diplobacilles. Notes sur deux diplobacilles non encore décrits *(Bacillus duplex non liquefaciens* et *Bacillus duplex josefi).* Annales d'Oculistique **153:**100–111.

Siddiqui, A., Goldberg, I. D. 1975. Intrageneric transformation of *Neisseria gonorrhoeae* and *Neisseria perflava* to streptomycin resistance and nutritional independence. Journal of Bacteriology **124:**1359–1365.

Silberfarb, P. M., Lawe, J. E. 1968. Endocarditis due to *Moraxella liquefaciens.* Archives of Internal Medicine **86:**512–513.

Snell, J. J. S., Hill, L. R., Lapage, S. P. 1972. Identification and characterization of *Moraxella phenylpyruvica.* Journal of Clinical Pathology **25:**959–965.

Snell, J. J. S., Lapage, S. P. 1976. Transfer of some saccharolytic *Moraxella* species to *Kingella* Henriksen and Bovre 1976, with descriptions of *Kingella indologenes* sp. nov. and *Kingella denitrificans* sp. nov. International Journal of Systematic Bacteriology **26:**451–458.

Spradbrow, P. B. 1968. The bacterial flora of the ovine conjunctival sac. Australian Veterinary Journal **44:**117–118.

Spradbrow, P. 1971. Experimental infection of the ovine cornea with *Neisseria ovis.* Veterinary Record **88:**615–616.

Spradbrow, P. B., Smith, I. D. 1967. The isolation of organisms resembling *Neisseria ovis* from ovine keratoconjunctivitis. Australian Veterinary Journal **43:**40.

Steel, K. J. 1961. The oxidase reaction as a taxonomic tool. Journal of General Microbiology **25:**297–306.

Steen, E., Berdal, P. 1950. Chronic rhinosinusitis caused by *Moraxella liquefaciens* Petit. Acta Oto-Laryngologica (Stockholm) **38:**31–34.

Sutton, R. G. A., O'Keefe, M. F., Bundock, M. A., Jeboult, J., Tester, M. P. 1972. Isolation of a new *Moraxella* from a corneal abscess. Journal of Medical Microbiology **5:**148–150.

Tatum, H. W., Ewing, W. H., Weaver, R. E. 1974. Miscellaneous Gram-negative bacteria, pp. 270–294. In: Lennette, E. H., Spaulding, E. H., Truant, J. P. (eds.), Manual of clinical microbiology, 2nd ed. Washington D.C.: American Society for Microbiology.

Thornley, M. J. 1967. A taxonomic study of *Acinetobacter* and related genera. Journal of General Microbiology **49:**211–257.

van Bijsterveld, O. P. 1970. New *Moraxella* strain isolated from angular conjunctivitis. Applied Microbiology **20:**405–408.

van Bijsterveld, O. P. 1972. The incidence of *Moraxella* on mucous membranes and the skin. American Journal of Ophthalmology **74:**72–76.

van Bijsterveld, O. P. 1973. Host-parasite relationship and taxonomic position of *Moraxella* and morphologically related organisms. American Journal of Ophthalmology **76:**545–554.

van Bijsterveld, O. P. 1973. Host-parasite relationship and taxonomic position of *Moraxella* and morphologically related organisms. American Journal of Ophthalmology **76:**545–554.

von Graevenitz, A., Grehn, M. 1976. Züchtung und Differenzierung obligat aerober gramnegativer Stäbchen aus menschlichem Untersuchungsmaterial. Schema zur Anwendung in der Routine-Diagnostik. Zentralblatt für Bakteriologie, Parasitenkunde, Infektionskrankheiten und Hygiene, Abt. 1 Orig., Reihe A **236:**513–530.

Warskow, A. L., Juni, E. 1972. Nutritional requirements of *Acinetobacter* strains isolated from soil, water, and sewage. Journal of Bacteriology **112:**1014–1016.

Weeks, O. B. 1974. Genus *Flavobacterium* Bergey et al. 1923, 97, pp. 357–364. In: Buchanan, R. E., Gibbons, N. E. (eds.), Bergey's manual of determinative bacteriology, 8th ed. Baltimore: Williams & Wilkins.

Wilcox, G. E. 1968. Infectious bovine kerato-conjunctivitis: a review. Veterinary Bulletin **38:**349–360.

Wilcox, G. E. 1970a. Bacterial flora of the bovine eye with special reference to the *Moraxella* and *Neisseria.* Australian Veterinary Journal **46:**253–257.

Wilcox, G. E. 1970b. An examination of *Moraxella* and related genera commonly isolated from the bovine eye. Journal of Comparative Pathology **80:**65–73.

Zinke, W. 1936. Über *Micrococcus-catarrhalis*-Meningitis. Zeitschrift für Kinderheilkunde **58:**236–246.

CHAPTER 123

The Genus *Lampropedia*

MORTIMER P. STARR

Clones of *Lampropedia hyalina* are quite distinctive in appearance—so distinctive, in fact, that this prokaryote is one of very few that can instantly be recognized by a glimpse under the microscope, on the basis of the following features: (i) the clones consist of relatively large cells, seemingly spherical but actually somewhat elongate in shape; (ii) the cells bear prominent, glistening inclusions (usually one per cell, sometimes more); (iii) the cells are ordered into rectangular or square tablets or sheets one cell thick (Fig. 1). Although the sheeting and

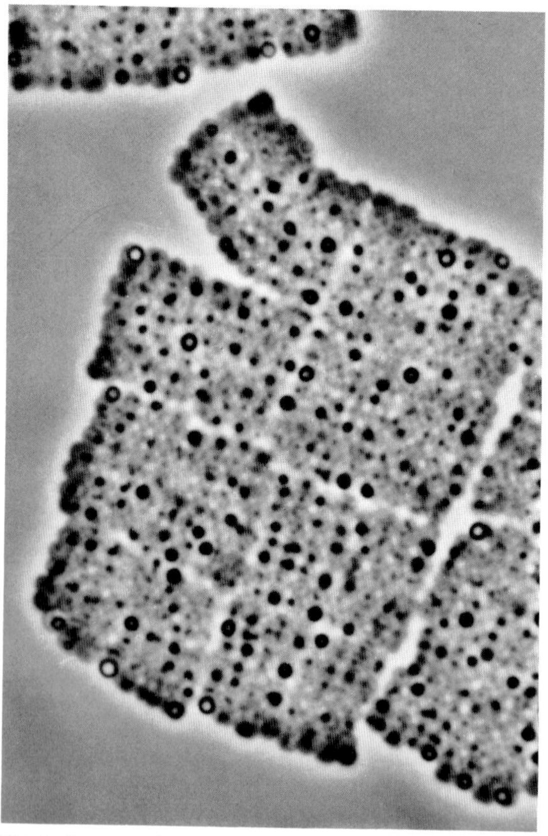

Fig. 1. *Lampropedia hyalina* grown on 1.0% Difco yeast extract at 30°C for 3 days. Note the typical subdivided ("window pane"), rectangular, sheeting format and the prominent poly-β-hydroxybutyrate inclusions. Phase-contrast, × 2,200. [From Kuhn and Starr, 1965, with permission.]

shining format—commemorated in the etymologies of the generic and specific names—attracted recorded attention to this organism as early as 1886 (Schroeter, 1886), some seventy years elapsed until the first isolation and axenic cultivation of *L. hyalina* was reported (Pringsheim, 1955). At least three other species have been assigned to the genus *Lampropedia* (De Toni and Trevisan, 1889); however, *L. hyalina* is the only species concerning which there is any currently reliable information (including axenic cultures). Hence, this treatment (with the exception of a few words in the final section) deals only with *Lampropedia hyalina*.

Habitats

Any consideration of the habitats of *Lampropedia* must take into account the significant fact that essentially all reports regarding occurrence of *Lampropedia* have been based solely on sightings—incident to microscopic examination of various natural samples—of the characteristic, monolayered, rectangular or squarish sheets (tablets) of seemingly spheroidal cells. Such morphological findings from natural material do not, of course, establish either (i) that the morphologically delineated creatures in these several sightings are taxonomically the same organism as *Lampropedia* (i.e., identical in all respects, including morphology), or (ii) that the sighting of the putative *Lampropedia* constitutes any firm demonstration that such natural material actually constitutes the normal habitat of *Lampropedia* in any strictly ecological sense.

With these caveats out of the way—but very much in mind—it can be reported that *Lampropedia hyalina* was found by Schroeter (1886) in swamp water and in highly eutrophic waters polluted with wastes from sugar refineries. De Toni and Trevisan (1889) offer stagnant water and partly submerged aquatic plants as habitats for *L. hyalina* and other named species of the genus *Lampropedia* listed by them. Kolkwitz (1909) reported extensive pellicles of *L. hyalina* developing on muddy water samples (polluted from neighboring habitations) held in the laboratory. Pringsheim (1955), who was the first to report isolation of an axenic clone of *L. hyalina,* found it in brown-colored liquid manure from a dairy barnyard. Similar reports, attributed to still other early workers, are listed by Breed, Murray, and Hitchens (1948) and Seeley (1974).

All in all, one could summarize the habitat picture emerging from the foregoing recital by saying that *Lampropedia* seems to be an inhabitant of aerobic and highly eutrophic waters, mud, and manure— were it not for the frequent reports of its occurrence also in such anaerobic habitats as the rumen and intestines of various animals. *Lampropedia* has often been seen in rumen fluid (Clarke, 1979; Eadie, 1962; Hungate, 1966; Smiles and Dobson, 1956); it is sometimes called the "window-pane sarcina" in papers on rumen biology. *Lampropedia* has been reported to occur in the intestinal contents and feces of certain herbivorous reptiles and in their intestinal nematodes (Schad, Knowles, and Meerovitch, 1964). The aforementioned caveat has particular merit in this context. Hungate (1966) has neatly summarized the situation regarding the sheeting bacteria observed microscopically in such anaerobic situations: "The regularly arranged plates often seen in the rumen . . . are *Lampropedia*. . . . It has not been definitely established that *Lampropedia* grows in the rumen. It usually occurs only in small numbers and could conceivably enter with the feed. If it does grow in the rumen it may be at the expense of the limited quantities of swallowed oxygen in the gas overlying the rumen contents."

Isolation

Lampropedia is by no means an uncommon organism, in the sense that reports about its occurrence are certainly not rare. Perhaps it is "not rare" because *Lampropedia* is morphologically conspicuous (Starr and Skerman, 1965) and thus attracts attention out of proportion to its actual incidence! Nevertheless, there is only one major report (Pringsheim, 1955) regarding isolation of *L. hyalina* in axenic culture. One other isolation is noted—almost en passant—in connection with otherwise unpublished observations of J. Kirschner (Hungate, 1966).

In both these cases, it would appear that enrichment of a sort was practiced. The natural samples— liquid manure in Pringsheim's (1955) case and rumen fluid in Kirschner's (Hungate, 1966)—were allowed to stand under aerobic conditions for some time. Under these conditions, surface films rich in *Lampropedia* were formed; pure cultures could be isolated from these surface films. The experiences of both investigators are related in what follows.

Enrichment and Isolation of *Lampropedia hyalina* (Pringsheim, 1955)

"Plating of the original material failed to produce colonies of Lampropedia owing to the overwhelming number of other bacteria, but single tablets transferred with a capillary pipette to soil-water cultures with starch or a wheat grain on the bottom gave rise to mixed cultures. Lampropedia tablets appeared after a few days in great numbers in the surface film and could be transferred again with a capillary to agar plates where they were streaked out with a loop. After one day at room temperature a few isolated colonies were observed on agar with 0.1% sodium acetate, 0.2% Difco yeast extract and 0.1% Difco tryptone. It was found later that acetate was not necessary to obtain growth of this bacterium, but served only to neutralize the otherwise slightly acid medium. Lampropedia multiplies only in neutral or weak alkaline media. By repeating the plating procedure, pure cultures were obtained. They multiplied still better at higher concentrations of nutrient substances, for instance 0.5% Difco Bacto peptone or Difco tryptone + 0.5% Difco yeast extract, pH 7, or similar mixtures as used for other bacteria. The most luxuriant growth was observed on the medium devised for *Caryophanon latum* (Pringsheim and Robinow, 1947) and composed as follows: sodium acetate 0.1%, beef extract 0.1%, yeast extract 0.2%, agar 1.5%."

Enrichment and Isolation of *Lampropedia* from Rumen Fluid (J. Kirschner, unpublished; cited by Hungate, 1966)

"Julius Kirschner, working at Davis, isolated this organism from rumen contents of a fistulated heifer. Preliminary enrichment occurred when rumen fluid was left standing for several days in an open Erlenmeyer flask at room temperature. A thick layer of *Lampropedia* accumulated at the surface. Streaking on aerobic plates of rumen fluid agar gave isolated colonies. The *Lampropedia* did not grow under anaerobic conditions and may be obligately aerobic."

Identification

From the earliest times, membership in the genus *Lampropedia* has been decided on purely morphological grounds: the rectangular, sheeting format; the seemingly spherical cells (actually, the cells are somewhat elongated in the plane perpendicular to the major dimension of the sheet, and somewhat flattened at the points of contact with adjacent cells); the prominent, glistening inclusion bodies. After Pringsheim (1955) isolated axenic cultures of *Lampropedia hyalina* (and distributed them generously), further definition was provided by his publication and cultures. Only one or two strains of *L. hyalina* (other than those isolated by Pringsheim) are publicly available; essentially all modern publications on *Lampropedia* have used Pringsheim's strains of *L. hyalina* or derivatives of them. In the present state of knowledge, there is no way to distinguish *L. hyalina* from the other named species of the genus *Lampropedia* listed by De Toni and Trevisan (1889) and (as species incertae sedis) by Seeley (1974); no information on these species is available beyond the vague nineteenth century descriptions. Hungate (1966) has suggested that the formal name of the bacterium referred to herein as *Lampropedia hyalina* Schroeter 1886 should more correctly—because of the principle of priority mandated in scientific nomenclature—be designated as *Lampropedia merismopedioides* (List 1885) Hungate 1966. However, this quite plausible idea was not even noted in the most recent taxonomic treatment of the genus *Lampropedia* (Seeley, 1974).

The sheeting format is a relatively rare property of prokaryotes. It can be seen in a somewhat re-stricted form in the tetrads of *Gaffkya* and *Pediococcus;* more extensive sheets—akin in general arrangement to those of *Lampropedia*—are found in *Pedioplana* (Wolff, 1907), *Thiopedia* (Hirsch, 1977), and *Merismopedia* (Pringsheim, 1966). Both *Thiopedia* and *Merismopedia* are photosynthetic; *Lampropedia* is not photosynthetic, although it was at one time mistakenly placed with the photosynthetic purple sulfur bacteria (Breed, Murray, and Hitchens, 1948). Pringsheim (1966) considered that *Lampropedia* might be an apochlorotic form of *Merismopedia*—a somewhat far-fetched notion, especially since the cells of most varieties of the cyanobacterium ("blue-green alga") *Merismopedia* are reported to be substantially larger (Echlin, 1965; Maruyama, 1967) than those of *Lampropedia*. The fixity of the sheeting format of *Merismopedia* (but not of *Lampropedia*) was called into question by Stanier et al. (1971) on the basis of their demonstration that *Aphanocapsa* strains can produce *Merismopedia*-like (i.e., sheeting) growth forms when cultivated in a mechanically undisturbed fashion under a diurnal light-dark cycle.

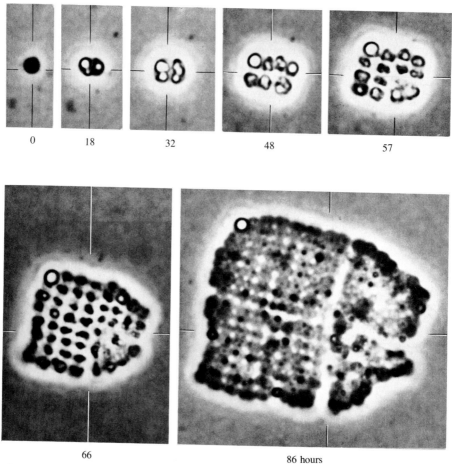

Fig. 2. Clone formation of *Lampropedia hyalina* growing in a sealed growth chamber (Starr and Kuhn, 1962) on 1.0% Difco yeast extract agar at room temperature. Cell division and formation of the poly-β-hydroxybutyrate along two axes alternating synchronously by 90 degrees. Phase-contrast, × 24,000. [From Kuhn and Starr, 1965, with permission.]

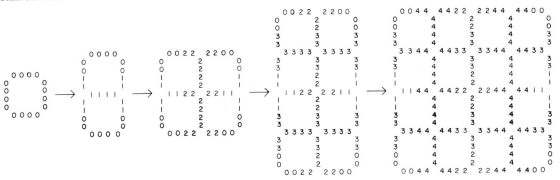

Fig. 3. Model of wall formation during growth of *Lampropedia hyalina* stemming from continuously observed preparations such as shown here in Fig. 2. Numbers indicate successive divisions. [From Kuhn and Starr, 1965, with permission.]

Pedioplana was described by Wolff (1907) as flagellated and actively motile; *Lampropedia* is definitely aflagellated and usually considered to be nonmotile. However, Pringsheim states that some of the bacterial aggregates have a capacity for "swimming in a sort of unbalanced swaying movement" and Seeley (1974) speaks of "a flickering movement of groups of cells in rapidly growing cultures." Neither of these movements is the vigorous translocational motility shown by typical flagellated bacteria; they are of the sort described in other bacteria as "twitching" (this Handbook, Chapters 1 and 122) and do deserve further attention. Nevertheless, apart from this discrepancy about flagellation and motility, *Pedioplana* and *Lampropedia* are superficially alike; in fact, Seeley (1974) has expressed the view that they are "synonymous"—a judgment which needs further scrutiny, including cultivation of Wolff's flagellated organism.

The geneological and ultrastructural bases of the sheeting format of *Lampropedia* have come in for some attention. In one study (Kuhn and Starr, 1965), the development of *Lampropedia hyalina* clones from single cells was followed in microslide cultures (Fig. 2). The typical rectangular or squarish sheets are formed by growth and division along two sets of growth axes, synchronized alternately at 90 degrees. The third division plane seems totally forbidden in *Lampropedia* (unlike, for example, the situation in sarcinae) and, at various times, one or the other of the two permitted division planes seems to be "shut off" by immediately preceding events (Fig. 3). The prominent intracytoplasmic inclusions, chemically identified as poly-β-hydroxybutyrate (Kuhn and Starr, 1965), follow the same division patterns as the cells; the formation of new polymer granules precedes cell division. Aging and a number of environmental factors (e.g., pH, nutrient concentration, temperature) were shown to upset these regular morphogenesis events in *Lampropedia*. The resulting disarray included increase in the number and mass of the poly-β-hydroxybutyrate inclusions, distension and lysis of cells, and disorderly cellular

arrangements (Fig. 4) which bore little resemblance to the typical sheets (Kuhn and Starr, 1965).

The ultrastructural basis for the cellular juxtapositions in *Lampropedia* that are visible as sheets (or tablets) has been reported independently by two groups (Chapman, Murray, and Salton, 1963; Murray, 1963; Pangborn and Starr, 1966) with quite similar results. What follows (including the illustrations) is drawn mainly from the paper of Pangborn

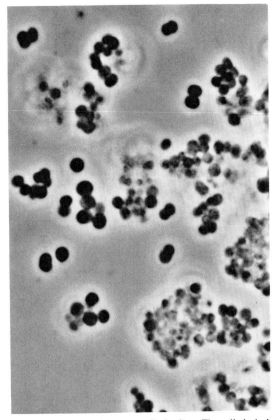

Fig. 4. Old culture of *Lampropedia hyalina*. The cells lack the poly-β-hydroxybutyrate inclusions and the sheeting cell arrangement has disintegrated. Phase-contrast, × 2,200. [From Kuhn and Starr, 1965, with permission.]

and Starr (1966). Ultrathin sections, precisely oriented either parallel to or perpendicular to the plane of growth, were cut and examined by transmission electron microscopy. In addition to the cytological features typical of Gram-negative prokaryotes, these preparations reveal the prominent poly-β-hydroxybutyrate inclusions, the three uniquely structured layers found external to the outer cell membrane ("the cell wall"), and the true cellular shape (somewhat cylindrical rather than strictly spherical). These features, including the three extra-wall layers, are depicted here in Figs. 5, 6, and 7. Murray's (1963) use of a nonsheeting strain of *L. hyalina,* in addition to a sheeting strain, has provided additional information, since two of the extra-wall layers seem to be lacking in the nonsheeting strain. Additional details about these and other ultrastructural features, based upon examination of wall fragments, are reported by Chapman, Murray, and Salton (1963).

The usual phenotypic traits of *Lampropedia hyalina*—based mainly on Pringsheim's (1955) paper or examination by Puttlitz and Seeley (1968) of Pringsheim's cultures—are listed by Seeley (1974). In addition to its distinctive morphological features (treated in the foregoing section), *L. hyalina* is seen as an aerobic, cytochrome-containing, chemoorganotroph; auxotrophic for two or three vitamins; limited in acceptable energy sources to a few Krebs cycle intermediates or close derivatives; and able to use any of several amino acids or ammonium chloride as sole nitrogen source. M. Mandel (personal communication) reported the G+C content of *L. hyalina* DNA to be 61 mol% (buoyant density).

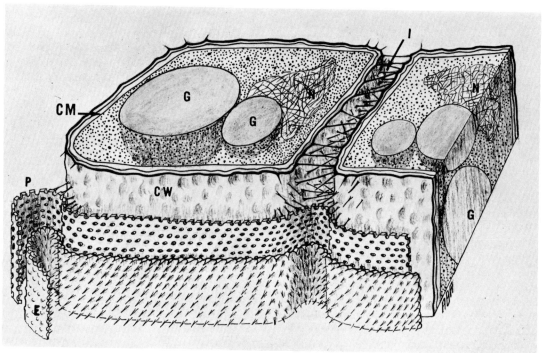

Fig. 5. Artist's reconstruction of portions of two cells of *Lampropedia hyalina,* based on serial stereomicrographs, showing the cytological features and unique surface layers. Abbreviations: CW, cell wall; CM, cytoplasmic membrane; E, echinulate layer; G, PHB granule; I, intercalated zone; N, nucleus; P, perforate layer. [From Pangborn and Starr, 1966, with permission.]

Fig. 6. Transverse section of *Lampropedia hyalina* (i.e., cut parallel to the growth plane of the sheet). Fixed by the method of Ryter and Kellenberger (1958), preceded by exposure to osmic acid vapors and followed by poststaining with uranyl acetate in methanol. The cells have the cytology typical of Gram-negative bacteria, and a squared shape in this orientation (the rounded, juxtaposed cell surfaces being flattened). The intercalated, perforate, and echinulate layers can be seen between the cells. See Fig. 5 for abbreviations. Bar = 1.0 μm. [From Pangborn and Starr, 1966, with permission.]

Fig. 7. Longitudinal section of *Lampropedia hyalina* (i.e., cut perpendicular to the growth plane of the sheet), prepared as stated for Fig. 6. Note the elongate shape of the cells in this orientation (i.e., the *Lampropedia* cell is not really spherical, but rather is cylindrical, in shape). The growth medium had been at the bottom, and the air at the top, of the sheet shown in the micrograph. The perforate and echinulate layers are seen on both surfaces surrounding the cell packet. Bar = 1.0 μm. [From Pangborn and Starr, 1966, with permission.]

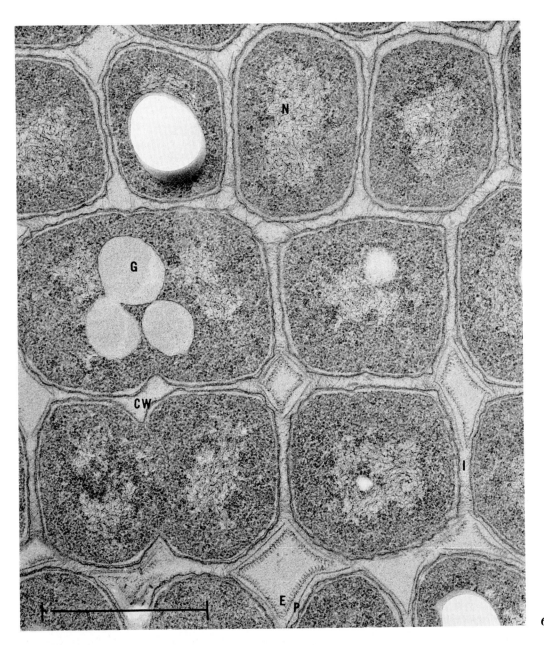

6

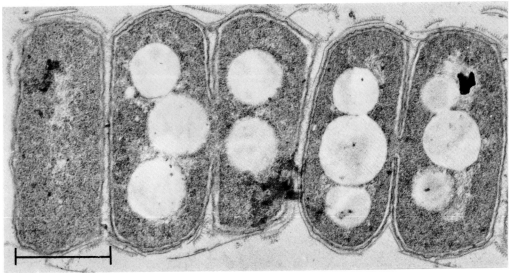

7

Literature Cited

Breed, R. S., Murray, E. G. D., Hitchens, A. P. 1948. Bergey's manual of determinative bacteriology, 6th ed., pp. 23 and 844. Baltimore: Williams & Wilkins.

Chapman, J. A., Murray, R. G. E., Salton, M. R. J. 1963. The surface anatomy of *Lampropedia hyalina*. Proceedings of the Royal Society B **158**:498–513.

Clarke, R. T. J. 1979. Niche in pasture-fed ruminants for the large rumen bacteria *Oscillospira, Lampropedia,* and Quin's and Eadie's ovals. Applied and Environmental Microbiology **37**:654–657.

De Toni, J. B., Trevisan, V. 1889. Schizomycetaceae Naeg., pp. 923–1087. In: Saccardo, P. A. (ed.), Sylloge fungorum omnium hucusque cognitorum, vol. 8. [Reprinted in 1944 by Edwards Brothers, Inc., Ann Arbor, Michigan.]

Eadie, J. M. 1962. The development of rumen microbial populations in lambs and calves under various conditions of management. Journal of General Microbiology **29**:563–578.

Echlin, P. 1965. The fine structure of unicellular blue green algae from the genus *Merismopedia*. Reports of the Proceedings of the Society for General Microbiology **1965**:i–ii. [Follows last page (429) of Volume 39 (1965) of Journal of General Microbiology.]

Hirsch, P. 1977. Ecology and morphogenesis of *Thiopedia* spp. in ponds, lakes, and laboratory cultures, pp. 13–15. In: Codd, G. A., Stewart, W. D. P. (eds.), Proceedings of the Second International Symposium on Photosynthetic Prokaryotes, Dundee, Scotland, 1976.

Hungate, R. E. 1966. The rumen and its microbes. New York, London: Academic Press.

Kolkwitz, R. 1909. Schizomycetes, pp. 1–186. In: Kryptogamenflora der Mark Brandenburg und angrenzender Gebiete. Band V, Pilze I. Leipzig: Verlag von Gebrüder Borntraeger. [The title page of the bound volume is dated 1915, but a note in the front matter states that the relevant signature appeared in 1909.]

Kuhn, D. A., Starr, M. P. 1965. Clonal morphogenesis of *Lampropedia hyalina*. Archives of Microbiology **52**:360–375.

Maruyama, K. 1967. Blue-green algae in the Alaskan Arctic. Bulletin of the National Science Museum (Tokyo) **10**:221–239 + 6 plates.

Murray, R. G. E. 1963. Role of superficial structures in the characteristic morphology of *Lampropedia hyalina*. Canadian Journal of Microbiology **9**:593–600.

Pangborn, J., Starr, M. P. 1966. Ultrastructure of *Lampropedia hyalina*. Journal of Bacteriology **91**:2025–2030.

Pringsheim, E. G. 1955. *Lampropedia hyalina* Schroeter 1886 and *Vannielia aggregata* n.g., n.sp., with remarks on natural and on organized colonies in bacteria. Journal of General Microbiology **13**:285–291.

Pringsheim, E. G. 1966. *Lampropedia hyalina* Schroeter, eine apochlorotische *Merismopedia* (Cyanophyceae). Kleine Mitteilungen über Flagellaten und Algen. XII. Archiv für Mikrobiologie **55**:200–208.

Pringsheim, E. G., Robinow, C. F. 1947. Observations on two very large bacteria, *Caryophanon latum* Peshkoff and *Lineola longa* (nomen provisorium). Journal of General Microbiology **1**:267–278.

Puttlitz, D. H., Seeley, H. W., Jr. 1968. Physiology and nutrition of *Lampropedia hyalina*. Journal of Bacteriology **96**:931–938.

Ryter, A., Kellenberger, E. 1958. Etude au microscope électronique de plasmas contenant de l'acide déoxyribonucléique. Zeitschrift für Naturforschung **13b**:597–605.

Schad, G. A., Knowles, R., Meerovitch, E. 1964. The occurrence of *Lampropedia* in the intestines of some reptiles and nematodes. Canadian Journal of Microbiology **10**:801–804.

Schroeter, J. 1886. Die Pilze Schlesiens, pp. 129–256. In: Cohn, F. (ed.), Kryptogamen-Flora von Schlesien, dritter Band, Erste Hälfte. Breslau: J. U. Kern's Verlag. [The title page of the bound volume is dated 1889, but a note in the front matter states that the relevant signatures appeared in 1886.]

Seeley, H. W., Jr. 1974. Genus *Lampropedia* Schroeter 1886, 151, pp. 440–441. In: Buchanan, R. E., Gibbons, N. E. (eds.), Bergey's manual of determinative bacteriology, 8th ed. Baltimore: Williams & Wilkins.

Smiles, J., Dobson, M. J. 1956. Direct ultra-violet and ultra-violet negative phase-contrast micrography of bacteria from the stomachs of the sheep. Journal of the Royal Microscopical Society, Series III **75**:244–253.

Stanier, R. Y., Kunisawa, R., Mandel, M., Cohen-Bazire, G. 1971. Purification and properties of unicellular blue-green algae (order Chroococcales). Bacteriological Reviews **35**:171–205.

Starr, M. P., Kuhn, D. A. 1962. On the origin of V-forms in *Arthrobacter atrocyaneus*. Archiv für Mikrobiologie **42**:289–298.

Starr, M. P., Skerman, V. B. D. 1965. Bacterial diversity: The natural history of selected morphologically unusual bacteria. Annual Review of Microbiology **19**:407–454.

Wolff, M. 1907. *Pedioplana haeckel* n.g., n.sp., und *Planosarcina schaudinni* n.sp., zwei neue bewegliche Coccaceen. Zentralblatt für Bakteriologie, Parasitenkunde, Infektionskrankheiten und Hygiene, Abt. 2 **18**:9–26.

Gram-Positive Cocci

CHAPTER 124

The Genus *Micrococcus*

KARL-HEINZ SCHLEIFER, WESLEY E. KLOOS, and MILOSLAV KOCUR

Many classifications of micrococci have been suggested during the last 30 years (Abd-el-Malek and Gibson, 1948; Hucker, 1948; Hucker and Breed, 1957; Kocur and Martinec, 1962; Pohja, 1960; Shaw, Stitt, and Cowan, 1951). However, none of them have been widely accepted by microbiologists, as they did not solve the problem of the clear-cut separation of micrococci from staphylococci (Cowan, 1962). The question of the differentiation of staphylococci from micrococci has been one of the key problems for many years. As criteria for their successful differentiation were not known, all these organisms were classified into one genus, i.e., either *Micrococcus* (Hucker, 1948) or *Staphylococcus* (Shaw, Stitt, and Cowan, 1951).

Evans, Bradford, and Niven (1955) were the first to show that staphylococci could grow and produce acid from glucose anaerobically, while micrococci could not. This experience was applied by Baird-Parker in his classification scheme for the differentiation of staphylococci (Baird-Parker, 1963, 1965) and in the suggestion of a standard glucose test for the separation of micrococci from staphylococci (International Subcommittee on Taxonomy of Staphylococci and Micrococci, 1965). Needless to say, further studies have demonstrated that this test by itself is not sufficient to provide a clear-cut separation. The application of new methods, in particular chemical ones, threw new light on the classification of these Gram-positive, catalase-positive cocci.

Habitats

As Suspected Transient Populations and Contaminants from Various Sources

Micrococci have been isolated sporadically from a variety of sources such as soil, estuarine mud, beach sand, marine and fresh water, plants, meat and dairy products, fomites, dust, and air, to name a few (Aaronson, 1955; Abd-el-Malek and Gibson, 1948; Baird-Parker, 1962; Doeringer and Dugan, 1973; Kitchell, 1962; Pearse, Humm, and Wharton, 1942; Pohja, 1960; Streby-Andrews and Kloos, 1971;

ZoBell and Upham, 1944; W. E. Kloos and others, unpublished studies). The predominant species isolated were *Micrococcus varians* and *Micrococcus luteus*. In early reports, these species were also classified as other species of micrococci, sarcinae, or staphylococci. Furthermore, several species of staphylococci (e.g., *Staphylococcus saprophyticus*, *Staphylococcus cohnii*, and *Staphylococcus xylosus*) were often misclassified as micrococci.

With the exception of animal and dairy products, most of the sources just listed contained small isolated populations of micrococci. We would, therefore, seriously question them as primary habitats. By contrast, micrococci are found more consistently and usually in larger populations on mammalian skin (Carr and Kloos, 1977; Glass, 1973; Kloos, Tornabene, and Schleifer, 1974; Kloos, Zimmerman, and Smith, 1976; Marples, 1965; Noble and Somerville, 1974), which we suspect is a primary natural habitat. It is certainly a rich source of micrococci. If our assumption is correct, many strains isolated from various other sources may have been, initially, contaminants disseminated by host animal carriers. For example, in support of this contention, W. E. Kloos and L. F. Kloos (unpublished studies) have found that sand taken from public swimming areas at several North Carolina ocean beaches often contained small populations of *M. luteus,* whereas sand taken from remote ocean island beaches, frequented by small numbers of indigenous animals and a few fishermen, only occasionally contained populations of *M. varians* and almost never *M. luteus,* a species common on human skin, but rare on the skin of other animals.

As Transient and Resident Populations on Mammalian Skin

Most humans support populations of micrococci on their skin. In a relatively recent study by Kloos, Tornabene, and Schleifer (1974), it was shown that 96% of 115 persons, living in 18 different states in the USA, carried cutaneous populations of micrococci. This percentage is somewhat higher than those reported earlier by others, which is perhaps due to the larger number of sites (10–16) sampled per individual and better information available on the recognition of micrococci on the basis of colony morphology. The percent of individuals carrying various *Micrococcus* species was as follows: *M. luteus*, 90%; *M. varians*, 75%; *M. lylae*, 33%; *M. nishinomiyaensis*, 28%; *M. kristinae*, 25%; *M. roseus*, 15%; *M. sedentarius*, 13%; *M. agilis*, 4%. *M. luteus* populations were usually relatively large and occupied, on the average, 51% of the skin sites sampled on individuals carrying this species.

In a temporal study by Kloos and Musselwhite (1975), it was found that micrococci usually constituted from 1 to 20% of the total aerobic bacteria isolated from the head, legs, and arms, but less than 1% of those isolated from the high bacterial density areas of the nares and axillae. Eighty percent of the individuals in the temporal study maintained micrococci on the head, legs, and arms over a period of 1 year. Where strain (or clonal) populations could be resolved, it was determined that certain strains of *M. luteus*, *M. varians*, *M. kristinae*, and *M. sedentarius* persisted on specific individuals for up to 1 year—a condition that strongly suggested a resident status. Several strains of *M. luteus* that were monitored for longer periods persisted for up to 2 1/2 years. Most strains appeared to be more transient and were isolated once or for periods up to 6 months. In a recent temporal study of the micrococci of infant skin, Carr and Kloos (1977) reported an increase in the occurrence of micrococci with increasing age up to 10 weeks. Micrococci were rarely isolated from infants less than 1 week of age.

Micrococci have been isolated from the skin of a variety of other mammals, including squirrels of the genus *Sciurus*, rats, raccoons, opossums, horses, swine, cattle, dogs, and various primates (Kloos, Zimmerman, and Smith, 1976, unpublished studies). The predominant species found on nonhuman mammals studied to date was *M. varians*. *M. luteus* was rarely isolated from nonhuman mammals. Other *Micrococcus* species found on the skin of humans appear to have a narrow host range, as their populations have not yet been discovered on other mammals. Most strains of these specialized species required one or more amino acids and vitamins for growth (Farrior and Kloos, 1975, 1976). We are currently examining what appear to be two different subspecies of *M. varians* that have been recently discovered on the skin of Old World monkeys and on the skin of New World monkeys and great apes, respectively. They are phenotypically distinct from *M. varians* strains isolated from lower mammals and humans, and from each other, although they exhibit the same L-Lys-L-Ala$_{3-4}$ peptidoglycan type and galactosamine in their cell wall polysaccharides as other *M. varians* strains (K.-H. Schleifer, unpublished studies). Future nucleic acid hybridization studies with these strains should clarify their taxo-

nomic status. The distribution of micrococci on other animals (e.g., birds, reptiles, amphibians, and fish) has not yet been adequately determined.

As Opportunistic Pathogens of Man

Micrococci have been reported to be associated with various infections, especially those of the urinary tract in man (Meers, Whyte, and Sandys, 1975; Mitchell, 1968; Roberts, 1967; Sellin et al., 1975; Telander and Wallmark, 1975); however, as a result of updated classification in the eighth edition of *Bergey's Manual of Determinative Bacteriology* (Baird-Parker, 1974), recommendations made by the International Committee on Systematic Bacteriology (ICSB), Subcommittee on the Taxonomy of Staphylococci and Micrococci (1976), and further examination of many of these organisms by various members of this Subcommittee, all organisms from such infections tested to date have been shown to be staphylococci.

Recently, two preliminary reports in the USA have indicated an association of *Micrococcus kristinae* with human infections. In one of the cases, large numbers of micrococci were isolated in pure culture from the patient's spinal fluid and in mixed culture with staphylococci from blood (J. C. McLaughlin, personal communication). It is interesting, and perhaps of some value, to note that strains isolated from both cases grew more rapidly aerobically, as well as anaerobically, than did *M. kristinae* populations isolated from normal skin in our previous studies. Although *M. kristinae* was not conclusively established as the etiological agent, association of this species with infections warrants further investigation, particularly with regard to situations arising in compromised patients (e.g., those receiving immunosuppressors).

Economic Significance

Micrococci have been reported to be used in the processing of fermented meat products to improve their color, aroma, flavor, and keeping quality (Kitchell, 1962; Niinivaara and Pohja, 1957; Pohja, 1960). However, based on an updated classification of these organisms, most strains used in processing have proven to be staphylococci (e.g., *Staphylococcus xylosus* and other species) (W. E. Kloos, unpublished studies). The true micrococci that have been useful in meat processing have proved to be *Micrococcus varians*. Species such as *Micrococcus luteus* and *M. varians* have been widely used in industries as assay organisms to test various antibiotics such as penicillin G and its derivatives, chloramphenicol, carbomycin, clindamycin, erythromycin, lincomycin, oleandomycin, novobiocin, chlortetracycline, bacitracin, rifamycin, vancomycin, and polymyxin in body fluids, feeds, milk, and pharmaceutical preparations (Bowman, 1957; Coates and Argoudelis, 1971; Grove and Randall, 1955; Kirshbaum and Arret, 1959; Simon and Yin, 1970); the vitamin biotin (Aaronson, 1955); and the enzyme lysozyme (Dickman and Proctor, 1952).

The relatively large amounts of long-chain aliphatic hydrocarbons present in most micrococci (Kloos, Tornabene, and Schleifer, 1974; Morrison, Tornabene, and Kloos, 1971; Tornabene, Morrison, and Kloos, 1970) may have some economic value in the future, especially if they can be economically extracted and processed into useful oils.

Isolation

Direct Isolation from Skin Under Nonselective Conditions

Various semiquantitative procedures have been described for isolating aerobic bacteria from human skin (Kloos and Musselwhite, 1975; Pachtman, Vicher, and Brunner, 1954; Smith, 1970; Williamson, 1965; Williamson and Kligman, 1965). The procedure described by Kloos and Musselwhite (1975) has been used for the isolation of micrococci and is suitable for use with human as well as other mammalian skin.

Isolation of Micrococci from Human and Other Mammalian Skin (Kloos and Musselwhite, 1975)
   "Sterile cotton swabs were moistened with a detergent containing 0.1% Triton X-100 (Packard) in 0.075 M phosphate buffer, pH 7.9 (Williamson, 1965), and rubbed vigorously, with rotation, over approximately 8-cm² sites. Swabbing was performed for 5 s on sites of the forehead, cheek, chin, nares, and axillae that usually contained large populations of bacteria and for 15 s on sites of the arms and legs that usually contained relatively small populations. Swabs taken from the forehead, cheek, chin, external naris, arms, and legs were immediately applied directly on agar media [standard agar plate, 100-mm diameter] by rubbing, with rotation, over the entire surface for two consecutive times. Swabs taken from the anterior naris and axillae were immediately rinsed once in 5 ml of detergent, and the rinse was applied to the surface of agar media. Later, during the course of the study, we observed that adults often contain populations of bacteria on the forehead, cheek, chin, and external naris that were too large to be analyzed by inoculating swabs directly onto media. In these instances, samples taken from a single swab rinse proved to be more satisfactory and produced well isolated colonies.

"The isolating medium (P agar) of Naylor and Burgi (1956) was nonselective and had the following composition: peptone (Difco), 10 g; yeast extract (Difco), 5 g; sodium chloride, 5 g; glucose, 1 g; agar (Difco), 15 g; and distilled water, 1,000 ml. . . . Inoculated agar media were incubated under aerobic conditions at 34 C for 4 days, at which time colonies were counted and recorded according to morphology and pigment. Subcultures [of selected colonies showing *Micrococcus* morphology] were stored at 4 C. For convenience, the original isolation plates could be stored at 4 C for 2 to 3 weeks prior to the isolation of cultures."

This procedure can be modified for use with other mammalian skin as follows.

Isolation of Micrococci from Nonhuman Mammalian Skin (Kloos, Zimmerman, and Smith, 1976)
   "Sites on the body were exposed for swabbing by parting the pelage away from the site area. Swabbing was performed for 5 s on sites in the anterior naris and ventral pouch (opossums only) that usually contained large populations of bacteria and for 15 s on sites of the forehead, forelimbs, hindlimbs, abdomen, and back that usually contained relatively small populations. Swabs taken from the forehead, forelimbs, hindlimbs, abdomen, and back were immediately applied directly on agar media by rubbing, with rotation, over the entire surface for two consecutive times. Swabs taken from the anterior naris and ventral pouch were immediately rinsed once in 5 ml of detergent and then applied to the surface of agar media." We have recently found that rinsing in detergent is also necessary for swabs taken from the face, back, abdomen, and perineum of many Old World monkeys and great apes.

Direct Isolation from Skin Under Selective Conditions

For shipping agar plates between collecting points or for use with nonhuman mammals, P agar was supplemented with the mold inhibitor cycloheximide (50 μg/ml) (Kloos, Tornabene, and Schleifer, 1974; Kloos, Zimmerman, and Smith, 1976). Many mammals, because of their close contact with soil and foliage, carry large populations of fungi on their skin. If allowed to grow, molds may rapidly cover the surface of unsupplemented P agar plates, making the isolation and characterization of bacteria very difficult. If *Bacillus* species are numerous on skin, it may be necessary to supplement P agar with 7% sodium chloride to inhibit their large or spreading colonies. However, such a high amount of sodium chloride will interfere with distinctive *Micrococcus* species colony morphology and *Micrococcus* colonies will not be easily distinguished from *Staphylococcus* colonies.

A selective medium for isolating micrococci (and corynebacteria) in the presence of populations of staphylococci has been described by Curry and Borovian (1976). This nitrofuran-containing medium, known as FTO agar, permits the growth of micrococci and prevents the growth of staphylococci. It is particularly useful for sampling areas of the skin such as the nares, axillae, and perineum where *Staphylococcus* populations are usually very large.

FTO Agar Medium for Direct, Selective Isolation of Micrococci from Skin (Curry and Borovian, 1976)

> "FTO agar is conveniently prepared in 700-ml amounts in 1-liter Erlenmeyer flasks. Tryptic soy agar (Difco) or Trypticase soy agar (BBL), fortified with 0.1% yeast extract and 0.5% Tween 80 (sorbitan monooleate), is the basal medium of Smith that fosters the growth of corynebacteria. . . . After autoclaving and cooling to 48°C, 0.1% of a 0.5% acetone stock of Oil Red O and 10% of a 0.05% acetone stock of Furoxone were added. To prevent flocculation, the latter was added slowly from a 100-ml graduate into swirling agar. Flasks were then left open or loosely covered in the water bath, to allow acetone volatilization prior to pouring and hardening plates on a level surface."

The plates can be incubated at 34–37°C for 3–4 days for adequate development of *Micrococcus* colonies. To reduce competition and crowding by lipophilic corynebacteria, Tween 80 should be omitted from the formula.

Identification

Separation of Staphylococci from Micrococci

The characters that separate staphylococci from micrococci are listed in Table 1. Marked differences in the DNA base and cell wall compositions and menaquinone patterns permit an accurate separation of the members of the two genera. However, these characters cannot be determined in the routine laboratory; therefore, other characters that can be easily analyzed are also listed in Table 1. A selective medium, FTO agar (see above), can be used to distinguish micrococci from staphylococci.

A simple test for the separation of staphylococci from micrococci was described by Schleifer and Kloos (1975b). It is based on the ability of staphylococci to produce acid aerobically from glycerol in the presence of 0.4 μg/ml of erythromycin and on their sensitivity to lysostaphin. Recent studies on the isolation of micrococci and staphylococci from dry sausage have indicated that the addition of erythromycin to the glycerol medium is not absolutely necessary (U. Fischer and K.-H. Schleifer, unpublished results). Only strains of *Micrococcus kristinae* and a few strains of *M. roseus* produce small amounts of acid aerobically from glycerol, but these organisms can be easily distinguished from staphylococci by their convex colony profile and characteristic colony pigment (Kloos, Tornabene, and Schleifer, 1974).

Lysostaphin Test for Separating Micrococci from Staphylococci (Schleifer and Kloos, 1975a)

> An agar overlay was prepared by adding 0.1 ml of a saline cell suspension (approximately 10⁷ colony-forming units per ml) to a tube containing 3 ml of fluid:

Peptone (Difco)   10 g

Table 1. Separation of staphylococci from micrococci.[a]

Character	*Staphylococcus*	*Micrococcus*	Reference
Anaerobic fermentation of glucose	+ (±,−)	− (±,+)	ICSB Subcommittee, 1976; Evans and Kloos, 1972
DNA base composition (mol% G+C)	30–38	66–73	Kocur, Bergan, and Mortensen, 1971
Cell wall composition:			
Peptidoglycan	> 2 moles Gly/ Mole Lys	> 2 moles Gly/ Mole Lys	Schleifer and Kandler, 1972; Schleifer and Kocur, 1973
Teichoic acid	+	−	Schleifer, 1973; Schleifer and Kloos, 1975a; Kloos and Schleifer, 1975
Menaquinones	Normal	Hydrogenated	Jeffries et al., 1969
FTO agar	No growth	Growth	Curry and Borovian, 1976
Resistance to lysostaphin	− (+)	+	Klesius and Schuhardt, 1968; Schleifer and Kloos, 1975b
Acid from glycerol-erythromycin medium	+ (−)	− (±)	Schleifer and Kloos, 1975b

[a] +, Positive; ±, weak; −, negative. Symbols in parentheses denote a character frequency of less than 30%.

Yeast extract (Difco)	5 g
Sodium chloride	5 g
Glucose	1 g
Agar (Difco)	7.5 g
Distilled water	1,000 ml

The contents were mixed thoroughly and then poured on the surface of a dry peptone–yeast extract–glucose agar plate (1.5% agar instead of 0.75%). A drop of sterile lysostaphin solution (200 μg/ml) was placed on the inoculated agar and incubated for 24 h (staphylococci can usually be recognized within 6–12 h). Lysostaphin susceptibility was interpreted according to the following spot-inhibition scheme: +, sensitive (complete growth inhibition); ±, slightly resistant (partial growth inhibition); and −, resistant (no visible growth inhibition).

The composition of the growth medium, in particular the source of peptone, is important for the lysostaphin test (Schleifer and Kloos, 1975b). To render all staphylococci lysostaphin sensitive, one has to cultivate them in a medium containing sufficient glycine. Peptone prepared from meat (e.g., peptone from Difco) exhibits a high glycine content, whereas peptone prepared from casein is much lower in glycine. Thus, media that contain peptone from casein have to be supplemented with 0.2–0.3% of glycine for the lysostaphin test.

Anaerobic Growth in a Glucose-Containing Medium

The use of shake cultures in a semisolid thioglycolate medium, as described by Evans and Kloos (1972), is recommended.

Anaerobic Growth in a Glucose-Containing Medium for Separating Micrococci from Staphylococci (Evans and Kloos, 1972)

"Medium. The medium selected for routine use was Brewer's fluid thioglycolate medium (Difco) with the addition of 0.3% agar (total of 0.35% agar in final medium). The medium contains glucose as an energy source, sodium thioglycolate to help maintain a low redox potential, and methylene blue as a redox indicator.

"Inoculation. For routine screening of cultures, one loopful of a 24-hr culture in Trypticase soy broth (BBL) was transferred into a tube of the sterile test medium that had been steamed and cooled to 50 C. (Approximately 8 ml of medium was used per 16-mm tube.) After the inoculum had been gently mixed with the loop, the medium was allowed to solidify at room temperature.

"Incubation. An incubation temperature of 35 C was selected to provide a rapid response from the staphylococci without being too high for most of the micrococci. Overnight incubation was sufficient for the staphylococci and most of the micrococci, but up to 72 hr was required for some of the slow-growing cultures.

"Observations. The tubes were examined by looking through them toward, but not directly at, a light while slowly rotating the tube. The location and intensity of the zones of growth were noted."

Symbols of anaerobic growth. ++, dense uniform; +, gradient of growth, from dense to weak down the tube; ±, individual colonies; −, only aerobic growth.

This method provides more reliable results for the separation of micrococci from staphylococci than the standard method for anaerobic glucose utilization proposed by the International Subcommittee on Taxonomy of Staphylococci and Micrococci (1965). As with any single test system, however, it is not completely satisfactory, e.g., strains of *Micrococcus kristinae* and *M. varians* can be misclassified as staphylococci and strains of *Staphylococcus hominis* and *S. cohnii* can be misclassified as micrococci. To obtain an accurate separation, it is recommended that the lysostaphin test be combined with the tests on glycerol degradation and anaerobic growth in a glucose-containing medium.

Two very recent publications describe simple test methods for the separation of staphylococci from micrococci. The first system uses polyvalent phages isolated from coagulase-negative staphylococci, which are quantitatively adsorbed by all staphylococci but not by micrococci (Schumacher-Perdreau, Pulverer, and Schleifer, 1978).

The other test, introduced by Seidl and Schleifer (1978), permits a serological separation of staphylococci from micrococci based on the different peptidoglycan types. Antisera were prepared against synthetic immunogens consisting of pentaglycine residues, which are covalently linked with their carboxyl termini to human serum albumin. The antisera reacted specifically with the N-terminal domain of oligoglycine peptides. Therefore, these antisera agglutinated strongly all peptidoglycans of staphylococci, because they all contain oligoglycine interpeptide bridges with N-terminal glycine residues. Peptidoglycan preparations of micrococci did not react with these antisera.

Serological Separation of Micrococci from Staphylococci (Seidl and Schleifer, 1978)

The serological detection of intact staphylococci is performed in the following way: Five to 10 ml of an overnight-grown culture is centrifuged in an Eppendorf microcentrifuge at $10,000 \times g$ for 2 min. Sedimented cells are extracted with 1.5 ml of 10% trichloroacetic acid at about 100°C for 30 min and spun off again. The sediment is washed twice with 1.5 ml distilled water, suspended in 1.5 ml 0.1 M phosphate buffer (pH 7.8) that contained trypsin (20 mg/100 ml), incubated at 37°C for 1 h, and centrifuged. The sediment is washed and then suspended in 10 ml borate-NaCl buffer (pH 8.2; 850 mg NaCl, 50 ml 0.1 M boric acid, 5.9 ml 0.1 M NaOH, distilled water to 100 ml). The suspension is diluted 1:100 with borate-NaCl buffer and 1 ml of the diluted cell suspension is added to a previously prepared mixture of 0.1 ml latex suspension (Difco), 0.5 ml purified antiserum (γ globulin fraction), and 9.4 ml borate-NaCl buffer. The agglutination can be read after incubating the reaction mixture at 4°C for 2 h.

If antisera are commercially available, this serological separation would be the simplest way to detect staphylococci and separate them, not only from micrococci, but also from other bacteria.

Differentiation of *Micrococcus* Species

Baird-Parker's latest classification of the family Micrococcaceae (1974) included three species in the genus *Micrococcus*. The genus *Micrococcus* includes Gram-positive spheres occurring in tetrads and in irregular clusters that are usually nonmotile and nonsporing. They are catalase-positive and aerobic with a strictly respiratory metabolism. The G+C content in the DNA ranges from 65 to 75%. Currently, nine species of micrococci are recognized in the genus *Micrococcus* (Table 2). The descriptions of six of them were revised and amended (Kloos, Tornabene, and Schleifer, 1974; Kocur and Martinec, 1972; Kocur and Páčová, 1970; Kocur, Páčová, and Martinec, 1972; Kocur and Schleifer, 1975; Kocur, Schleifer, and Kloos, 1975; Schleifer, Kloos, and Moore, 1972). Three new species of micrococci have been described quite recently. A moderately halophilic species, *M. halobius,* isolated from unrefined solar salt, was described by Onishi and Kamekura (1972). *Micrococcus lylae* and *M. kristinae* isolated from human skin were described by Kloos, Tornabene, and Schleifer (1974).

The application of new tests and chemical analysis of cells considerably contributed to the classification of micrococci. Micrococci may be differentiated into nine species by means of tests mentioned in Table 2. Their pigment production and colony morphology may be used as a simple test for their presumptive identification (Kloos, Tornabene, and Schleifer, 1974). Certain difficulties may occur in the differentiation of *M. luteus* and *M. lylae,* as both species have several features in common. However, *M. lylae* can be distinguished from *M. luteus* by cream-white or unpigmented colonies, lack of growth on inorganic nitrogen agar, lysozyme resist-

ance, and cell wall peptidoglycan. The most common yellow pigmented species, *M. luteus* and *M. varians,* differ in acid production from glucose, nitrate reduction, lysozyme susceptibility, growth on inorganic nitrogen agar and on Simmons' citrate agar, and oxidase reaction. *M. roseus* differs from other species in having pink colonies, nitrate reduction, and an inability to hydrolyze gelatin.

Micrococcus agilis differs significantly from other micrococci in several features. It possesses flagella, is psychrophilic, and exhibits β-galactosidase activity. *M. kristinae* is a clearly separated species, too. It produces acid from glucose and mannose aerobically and has a positive reaction to acetoin and esculin hydrolysis. *M. kristinae* produces unique wrinkled growth on purple agar (Difco) containing 1% maltose. The orange-pigmented *M. nishinomiyaensis* may be further distinguished from *M. kristinae,* which forms very pale orange colonies, by growth on 7.5% NaCl agar, acetoin production, and esculin hydrolysis. *M. sedentarius* differs from other *Micrococcus* species by being resistant to penicillin and methicillin, producing often water-soluble exopigment, growing very slowly, and reacting positively to the arginine dihydrolase test. *M. halobius* can be separated from other species, as it requires at least 5% NaCl for growth. Information on only one strain of this species has been reported to date.

There is little information about the actual relationships within the genus *Micrococcus*. According to recent DNA-DNA hybridization studies by Ogasawara-Fujita and Sakaguchi (1976), there is no close genetic relationship among members of the species *M. luteus, M. roseus,* and *M. varians*. Studies by Kloos and co-workers (Kloos, 1969; Kloos, Tornabene, and Schleifer, 1974) on the genetic exchange among micrococci demonstrated a genetic relationship between *M. luteus* and *M. lylae*. Transformation of an auxotrophic *M. luteus* strain occurred at about 5% of homologous values with DNA isolated from *M. lylae*. This low transformation rate suggests a considerable divergence of these two species, but nevertheless indicates a significant genetic relationship. DNA homology studies between the type strains of *M. luteus* and *M. lylae* support these findings. The DNA homology values were in the range of 40–50% at the optimal temperature, whereas, under the same hybridization conditions, values of only 10–18% were obtained between *M. luteus/M. lylae* and *M. kristinae* or *M. varians* (Schleifer, Heise, and Meyer, 1979). The interrelationship between *M. luteus* and *M. lylae* has also been demonstrated by comparative immunological studies (Rupprecht and Schleifer, 1977). Double-immunodiffusion tests and quantitative microcomplement fixation assay were performed with crude extracts or catalase-enriched preparations from various *Micrococcus* species and antisera

Table 2. Abbreviated scheme for the differentiation of species of the genus *Micrococcus*.[a]

Species	Major pigment[b]	Water-soluble exopigment	Growth on Simmons citrate agar	Growth on inorganic nitrogen agar	Acetoin	Nitrate reduction	Oxidase	Aerobic acid from		Lysozyme susceptibility[c]	Arginine dihydrolase	β-Galactosidase	Esculin hydrolysis	Peptidoglycan type	Amino sugar in cell wall polysaccharide
								Glucose	Mannose						
M. luteus	Y>CW	−	−	+>±,−	−	−>+	±,+	−	−	S	−	−	−	L-Lys-peptide subunit	Mannosamin-uronic acid
M. lylae	CW,—	−	−		−>±	−>±	±,+	−	−	SR	−	−	−	L-Lys-Asp	Galactosamine
M. varians	Y	−	+>±,−	−>±	±,−	+>±	±,−	+	−	R	−	−	−	L-Lys-L-Ala$_{3-4}$	Galactosamine
M. roseus	PR>OR	−	−>±	−	±>−	+>±	±,−	+,±	−	SR–R	−	−	+	L-Lys-L-Ala$_{3-4}$	Galactosamine
M. agilis	R	−	ND	ND	+	−>±	+	++	++	R	−	−	−	L-Lys-Thr-L-Ala$_3$	Glucosamine
M. kristinae	PO	−	−	−	+	−>±	±,+	++	−	R	−	−+	+	L-Lys-L-Ala$_3$	Glucosamine
M. nishino-miyaensis	O	−>+	−	±,−	−>±	+,±,−	±,+	−>±	−	SR–R	−	−	−	L-Lys-L-Ser$_2$-D-Glu	Galactosamine
M. sedentarius	CW>BY	+,±	ND	ND	−	−	−	−>±	ND	S–SR	+,−	−	−	Uncertain	—
M. halobius	−	−	ND	ND	−	−	+	−>±		R	+	+	−	ND	ND

[a] A single listed symbol denotes a character frequency of about 70–100%; the notation > denotes ''a frequency greater than''; a comma between symbols denotes nearly equal frequency. Symbols: ++, strong positive; +, positive; ±, weak; −, negative; ND, not determined.

[b] BY, buttercup yellow; CW, cream-white; O, orange; OR, orange red; PO, pale orange; PR, pastel red; R, red; —, unpigmented.

[c] S, susceptible (minimal inhibitory concentration [MIC]: below 5 μg/ml); SR, slightly resistant (MIC: 5–50 μg/ml); R, resistant (MIC: above 100 μg/ml).

against purified catalase of *M. luteus.* A rather good cross-reaction was found in the immunodiffusion test between *M. luteus* and *M. lylae,* whereas preparations of other micrococci did not react *(M. kristinae, M. nishinomiyaensis, M. roseus, M. varians)* with antisera against *M. luteus* catalase. The immunological distance values obtained by microcomplement fixation assay coincide with the immunodiffusion studies. These findings indicate that an interrelationship exists not only between *M. luteus* and *M. lylae* but also (a rather weak relationship) between *M. luteus* and *M. varians.* Strains of other *Micrococcus* species reacted with antiserum against *M. luteus* catalase as weakly as nonrelated strains. In contrast to the genus *Staphylococcus,* the genus *Micrococcus* comprises a rather heterogeneous group of organisms which, with the exception of *M. luteus, M. lylae,* and to some extent *M. varians,* are not detectably connected through epigenetic or genetic relationships.

Literature Cited

Aaronson, S. 1955. Biotin assay with a coccus, *Micrococcus sodonensis,* nov. sp. Journal of Bacteriology **69:**67–69.

Abd-el-Malek, Y., Gibson, T. 1948. Studies on the bacteriology of milk. II. The staphylococci and micrococci of milk. Journal of Dairy Research **15:**249–260.

Baird-Parker, A. C. 1962. The occurrence and enumeration, according to a new classification, of micrococci and staphylococci in bacon and on human and pig skin. Journal of Applied Bacteriology **25:**352–361.

Baird-Parker, A. C. 1963. A classification of micrococci and staphylococci based on physiological and biochemical tests. Journal of General Microbiology **30:**409–427.

Baird-Parker, A. C. 1965. The classification of staphylococci and micrococci from world-wide sources. Journal of General Microbiology **38:**363–387.

Baird-Parker, A. C. 1974. Family *Micrococcaceae,* pp. 478–489. In: Buchanan, R. E., Gibbons, N. E. (eds.), Bergey's manual of determinative bacteriology, 8th ed. Baltimore: Williams & Wilkins.

Bowman, F. W. 1957. Test organisms for antibiotic microbial assays. Antibiotics and Chemotherapy **7:**639–640.

Carr, D. L., Kloos, W. E. 1977. Temporal study of the staphylococci and micrococci of normal infant skin. Applied and Environmental Microbiology **34:**673–680.

Coates, J. H., Argoudelis, A. D. 1971. Microbial transformation of antibiotics: Phosphorylation of clindamycin by *Streptomyces coelicolor* Muller. Journal of Bacteriology **108:**459–464.

Cowan, S. T. 1962. An introduction to chaos, or the classification of micrococci and staphylococci. Journal of Applied Bacteriology **25:**324–340.

Curry, J. C., Borovian, G. E. 1976. Selective medium for distinguishing micrococci from staphylococci in the clinical laboratory. Journal of Clinical Microbiology **4:**455–457.

Dickman, S. R., Proctor, C. M. 1952. Factors affecting the activity of egg white lysozyme. Archives of Biochemistry and Biophysics **40:**364–372.

Doeringer, R. H., Dugan, P. R. 1973. Growth relationship between the blue-green alga *Anacystis nidulans* and *Sarcina flava* in mixed culture. Abstracts of the Annual Meeting of the American Society for Microbiology **1973:**45.

Evans, J. B., Bradford, W. L., Niven, C. F. 1955. Comments concerning the taxonomy of the genera *Micrococcus* and *Staphylococcus.* International Bulletin of Bacteriological Nomenclature and Taxonomy **5:**61–66.

Evans, J. B., Kloos, W. E. 1972. Use of shake cultures in a semisolid thioglycolate medium for differentiating staphylococci from micrococci. Applied Microbiology **23:**326–331.

Farrior, J. W., Kloos, W. E. 1975. Amino acid and vitamin requirements of *Micrococcus* species isolated from human skin. International Journal of Systematic Bacteriology **25:**80–82.

Farrior, J. W., Kloos, W. E. 1976. Sulfur amino acid auxotrophy in *Micrococcus* species isolated from human skin. Canadian Journal of Microbiology **22:**1680–1690.

Glass, M. 1973. Sarcina species on the skin of the human forearm. Transactions of St. John's Hospital Dermatological Society **59:**56–60.

Grove, D. C., Randall, W. A. 1955. Assay Methods of Antibiotics. Antibiotics Monographs No. 2. New York: Medical Encyclopedia.

Hucker, G. J. 1948. Genus I. *Micrococcus,* pp. 235–246. In: Breed, R. S., Murray, E. G. D., Hitchens, P. (eds.), Bergey's manual of determinative bacteriology, 6th ed. Baltimore: Williams & Wilkins.

Hucker, G. J., Breed, R. S. 1957. Genus *Micrococcus,* pp. 455–464. In: Breed, R. S., Murray, E. G. D., Smith, N. R. (eds.), Bergey's manual of determinative bacteriology, 7th ed. Baltimore: Williams & Wilkins.

International Subcommittee on Taxonomy of Staphylococci and Micrococci. 1965. Recommendation. International Bulletin of Bacteriological Nomenclature and Taxonomy **15:**109–110.

International Committee on Systematic Bacteriology, Subcommittee on the Taxonomy of Staphylococci and Micrococci. 1976. Minutes of the Meeting, Appendix 1: Identification of Staphylococci. International Journal of Systematic Bacteriology **26:**333–334.

Jeffries, L., Cawthorne, M. A., Harris, M., Cook, B., Diplock, A. T. 1969. Menaquinone determination in the taxonomy of Micrococcaceae. Journal of General Microbiology **54:**365–380.

Kirshbaum, A., Arret, B. 1959. Outline of details for assaying the commonly used antibiotics. Antibiotics and Chemotherapy **9:**613–617.

Kitchell, A. G. 1962. Micrococci and coagulase negative staphylococci in cured meats and meat products. Journal of Applied Bacteriology **25:**416–431.

Klesius, P. H., Schuhardt, V. T. 1968. Use of lysostaphin in the isolation of highly polymerized deoxyribonucleic acid and in the taxonomy of aerobic *Micrococcaceae.* Journal of Bacteriology **95:**739–743.

Kloos, W. E. 1969. Transformation of *Micrococcus lysodeikticus* by various members of the family *Micrococcaceae.* Journal of General Microbiology **59:**247–255.

Kloos, W. E., Musselwhite, M. S. 1975. Distribution and persistence of *Staphylococcus* and *Micrococcus* species and other aerobic bacteria on human skin. Applied Microbiology **30:**381–395.

Kloos, W. E., Schleifer, K. H. 1975. Isolation and characterization of staphylococci from human skin. II. Description of four new species: *Staphylococcus warneri, Staphylococcus capitis, Staphylococcus hominis* and *Staphylococcus simulans.* International Journal of Systematic Bacteriology **25:**62–79.

Kloos, W. E., Tornabene, T. G., Schleifer, K. H. 1974. Isolation and characterization of micrococci from human skin, including two new species: *Micrococcus lylae* and *Micrococcus kristinae.* International Journal of Systematic Bacteriology **24:**79–101.

Kloos, W. E., Zimmerman, R. J., Smith, R. F. 1976. Preliminary studies on the characterization and distribution of *Staphylococcus* and *Micrococcus* species on animal skin. Applied and Environmental Microbiology **31:**53–59.

Kocur, M., Bergan, T., Mortensen, N. 1971. DNA base composition of Gram-positive cocci. Journal of General Microbiology **69:**167–183.

Kocur, M., Martinec, T. 1962. A taxonomic study of the genus *Micrococcus* [in Czechoslovakian.] Folia Facultais Scientiarum Naturalium Universitatis Purkynianae Brunensis **3:**3–121.

Kocur, M., Martinec, T. 1972. Taxonomic status of *Micrococcus varians* Migula 1900 and designation of the neotype strain. International Journal of Systematic Bacteriology **22:**228–232.

Kocur, M., Páčová, Z. 1970. The taxonomic status of *Micrococcus roseus* Flügge, 1886. International Journal of Systematic Bacteriology **20:**233–240.

Kocur, M., Páčová, Z., Martinec, T. 1972. Taxonomic status of *Micrococcus luteus* (Schroeter 1872) Cohn 1872, and designation of the neotype strain. International Journal of Systematic Bacteriology **22:**218–223.

Kocur, M., Schleifer, K. H. 1975. Taxonomic status of *Micrococcus agilis* Ali-Cohen 1889. International Journal of Systematic Bacteriology **25:**294–297.

Kocur, M., Schleifer, K. H., Kloos, W. E. 1975. Taxonomic status of *Micrococcus nishinomiyaensis* Oda 1935. International Journal of Systematic Bacteriology **25:**290–293.

Marples, M. J. 1965. The ecology of the human skin. Springfield, Illinois: Charles C Thomas.

Meers, P. D., Whyte, W., Sandys, G. 1975. Coagulase-negative staphylococci and micrococci in urinary tract infections. Journal of Clinical Pathology **28:**270–273.

Mitchell, R. G. 1968. Classification of *Staphylococcus albus* strains isolated from the urinary tract. Journal of Clinical Pathology **21:**93–96.

Morrison, S. J., Tornabene, T. G., Kloos, W. E. 1971. Neutral lipids in the study of the relationships of members of the family *Micrococcaceae*. Journal of Bacteriology **108:**353–358.

Naylor, H. B., Burgi, E. 1956. Observations on abortive infections of *Micrococcus lysodeikticus* with bacteriophage. Virology **2:**577–593.

Niinivaara, F. P., Pohja, M. S. 1957. Erfahrungen über die Herstellung von Rohwurst mittels einer Bakterienreinkultur. Fleischwirtschaft **9:**789–790.

Noble, W. C., Somerville, D. A. 1974. Microbiology of Human Skin. London, Philadelphia, Toronto: W. B. Saunders.

Ogasawara-Fujita, N., Sakaguchi, K. 1976. Classification of micrococci on the basis of deoxyribonucleic acid homology. Journal of General Microbiology **94:**97–106.

Onishi, H., Kamekura, M. 1972. *Micrococcus halobius* sp. n. International Journal of Systematic Bacteriology **22:**233–236.

Pachman, E. A., Vicher, E. E., Brunner, M. J. 1954. The bacteriologic flora in seborrhoeic dermatitis. Journal of Investigative Dermatology **22:**389–397.

Pearse, A. S., Humm, H. J., Wharton, G. W. 1942. Ecology of sand beaches at Beaufort, N.C. Ecological Monographs **12:**137–178.

Pohja, M. S. 1960. Micrococci in fermented meat products. Classification and description of 171 different strains. Acta Agralia Fennica **96:**1–80.

Roberts, A. P. 1967. *Micrococcaceae* from the urinary tract in pregnancy. Journal of Clinical Pathology **20:**631–632.

Rupprecht, M., Schleifer, K. H. 1977. Comparative immunological study of catalases in the genus *Micrococcus*. Archives of Microbiology **114:**61–66.

Schleifer, K. H. 1973. Chemical composition of staphylococcal cell walls, pp. 13–32. In: (Jeljaszewicz, J., Hryniewicz, W.

(eds.), Contributions to microbiology and immunology, vol. 1. Staphylococci and staphylococcal infections. Basel: S. Karger.

Schleifer, K. H., Heise, W., Meyer, S. A. 1979. Deoxyribonucleic acid hybridization studies among micrococci. FEMS Microbiology Letters **6:**33–36.

Schleifer, K. H., Kandler, O. 1972. Peptidoglycan types of bacterial cell walls and their taxonomic implications. Bacteriological Reviews **36:**407–477.

Schleifer, K. H., Kloos, W. E. 1975a. Isolation and characterization of staphylococci from human skin. I. Amended descriptions of *Staphylococcus epidermidis* and *Staphylococcus saprophyticus* and descriptions of three new species: *Staphylococcus cohnii, Staphylococcus haemolyticus* and *Staphylococcus xylosus*. International Journal of Systematic Bacteriology **25:**50–61.

Schleifer, K. H., Kloos, W. E. 1975b. A simple test system for the separation of staphylococci from micrococci. Journal of Clinical Microbiology **1:**337.

Schleifer, K. H., Kloos, W. E., Moore, A. 1972. Taxonomic status of *Micrococcus luteus* (Schroeter 1872) Cohn 1872: Correlation between peptidoglycan type and genetic compatibility. International Journal of Systematic Bacteriology **22:**224–227.

Schleifer, K. H., Kocur, M. 1973. Classification of staphylococci based on chemical and biochemical properties. Archiv für Mikrobiologie **93:**65–85.

Schumacher-Perdreau, F., Pulverer, G., Schleifer, K. H. 1979. The phage adsorption test: A simple method for the differentiation between staphylococci and micrococci. Journal of Infectious Diseases **138:**392–395.

Seidl, P. H., Schleifer, K. H. 1978. Rapid test for the serological separation of staphylococci from micrococci. Applied and Environmental Microbiology **35:**479–482.

Sellin, M. A., Cooke, D. I., Gillespie, W. A., Sylvester, D. G. H., Anderson, J. D. 1975. Micrococcal urinary tract infections in young women. Lancet **ii:**570–572.

Shaw, C., Stitt, J. M., Cowan, S. T. 1951. Staphylococci and their classification. Journal of General Microbiology **5:**1010–1023.

Simon, H. J., Yin, E. J. 1970. Microbioassay of antimicrobial agents. Applied Microbiology **19:**573–579.

Smith, R. F. 1970. Comparative enumeration of lipophilic and nonlipophilic cutaneous diphtheroids and cocci. Applied Microbiology **19:**254–258.

Streby-Andrews, M. E., Kloos, W. E. 1971. Amino acid auxotrophy in natural strains of *Micrococcus luteus*. Bacteriological Proceedings **1971:**27.

Telander, B., Wallmark, G. 1975. *Micrococcus* subgroup 3—a common cause of acute urinary tract infection in women. Lakartidningen **72:**1967.

Tornabene, T. G., Morrison, S. J., Kloos, W. E. 1970. Aliphatic hydrocarbon contents of various members of the family *Micrococcaceae*. Lipids **5:**929–937.

Williamson, P. 1965. Quantitative estimation of cutaneous bacteria, pp. 3–11. In: Maibach, H. I., Hildick-Smith, G. (eds.), Skin bacteria and their role in infection. New York, Sydney, Toronto, London: McGraw-Hill.

Williamson, P., Kligman, A. M. 1965. A new method for the quantitative investigation of cutaneous bacteria. Journal of Investigative Dermatology **45:**498–503.

ZoBell, C. E., Upham, C. 1944. A list of marine bacteria including descriptions of sixty new species. Bulletin of Scripps Institute of Oceanography, University of California **5:**239–292.

CHAPTER 125

The Genus *Staphylococcus*

WESLEY E. KLOOS and KARL-HEINZ SCHLEIFER

Bacteria of the genus *Staphylococcus* have received a great deal of attention over the past century, and for good reason. The most notorious species, *Staphylococcus aureus,* causes a wide variety of common and annoying, as well as serious, infections in man and other animals. Their enterotoxins are responsible for the most common food poisoning. Their versatility in producing antibiotic-resistant populations has stimulated bacteriologists, geneticists, and biochemists alike to develop strategies to outmaneuver their attack. Members of the genus *Staphylococcus,* in addition to posing a threat to our health, happen to form a relatively large component of our cutaneous flora. The relationship between staphylococci and their hosts is most intimate and one we cannot ignore, especially when natural barriers are compromised by injury, surgical procedures, and placement of internal prostheses, and defense mechanisms are reduced by immunosuppressive therapy. On the other side of the ledger, these organisms, together with other resident flora, may form a defensive barrier to invasion by more serious pathogens. In keeping with the theme of this handbook, we will focus our attention on the habitats occupied by staphylococci, isolation procedures, and current developments in taxonomy. This will necessitate our overlooking some of the other active areas of *Staphylococcus* research. Our treatment and references by no means exhaust the subject, and we hope that the reader will further explore the many texts and papers cited. In this chapter, we will not belabor the long and complex history of the taxonomy of staphylococci, but rather will focus on key developments made in systematics from the sixties to the present. Excellent reviews of the older historical accounts have been written by Elek (1959), Cowan (1962), and Baird-Parker (1965b).

Elek (1959) clearly demonstrated an understanding of a major problem facing the taxonomy and future systematics of staphylococci when he made the following statement: "The medical bacteriologist is not very concerned as a rule with classification or nomenclature. To him the broad division into pathogens and nonpathogens overrides other considerations, and further subdivision matters only if it aids the recognition of a given pathogen or enables him to trace the source of infection. The biologist aims to fit the pathogen into the general order of living creatures, caring little about disease and the microbial features that enter into its causation. . . . The classification of staphylococci is much bedevilled by this fundamental difference in aim."

As a result of the long influence of medical and food bacteriologists on the taxonomy of Gram-positive, catalase-positive cocci, two very important classifications were made. At the genus level, staphylococci were separated from micrococci, and at the species level, *S. aureus* was separated from the coagulase-negative staphylococci. In the early sixties, Baird-Parker (1962, 1963, 1965a,b) conducted several large taxonomic studies of staphylococci and micrococci isolated from various sources. He classified these organisms on the basis of several simple and expedient characters. Two of his groups, though redefined today by new systematics, essentially became the species *S. epidermidis sensu stricto* and *S. saprophyticus,* which are perhaps the most medically significant of the coagulase-negative staphylococci. Biochemical and biological approaches to staphylococcal taxonomy followed shortly. Systematic bacteriologists began to look deeper into the molecular structure, immunology, genetics, and ecology of staphylococci; these investigations have led to the introduction of new species and subspecies and a more nearly complete view of natural populations. Much of the work is still preliminary, though it is gaining momentum and today encompasses the activities of many laboratories.

Habitats

As Suspected Transient Populations and Contaminants from Various Sources

Staphylococci have been isolated sporadically from a wide variety of sources, many of which are shared by micrococci and other bacteria. These sources include soil, beach sand, marine and fresh water, plant surfaces and products, feeds, meat, poultry, and dairy products; various surfaces of cooking ware, utensils, furniture, blankets, carpets, clothing, etc. in homes and hospitals; dust and air in homes, hospitals, animal facilities, or other inhabited areas; swimming pools; and paper currency, to name a few (Abd-el-Malek and Gibson, 1948; Baird-Parker, 1962; Buhles, 1969; reviewed by Elek, 1959; Hambraeus and Ransjo, 1976; Hare, 1962; Kitchell, 1962; Kloos, Schleifer, and Noble, 1976c; reviewed by Nahmias and Shulman, 1972; Noble, Lidwell, and Kingston, 1963; Patterson, 1966; Perry, 1969; Sharpe, Neave, and Reiter, 1962; Wilkoff, West-brook, and Dixon, 1969; Williams, 1967). With the exception of animal products, most of the sources just mentioned contained relatively small, isolated populations of staphylococci. Animal products and dust, air, and fomites in inhabited areas are most probably contaminated by human- or other animal-host-supported populations of staphylococci. The sporadic occurrence of staphylococci in soil, beach sand, natural waters, and on plants is more difficult to explain. We currently question these sources as primary habitats and favor the notion that many strains isolated from such sources were, initially, contaminants disseminated by animal host carriers. Staphylococci are found more consistently and in larger populations on mammalian skin (including the anterior nares), which we suspect is the primary natural habitat, and are also found, to some extent, in other areas such as the conjunctiva, throat, mouth, mammary glands, and intestinal tract (Carr and Kloos, 1977; Devriese and Oeding, 1976; reviewed by Elek, 1959; Hajek and Marsalek, 1976; Kligman, 1965; Kloos and Musselwhite, 1975; Kloos, Musselwhite, and Zimmerman, 1976; Kloos and Schleifer, 1975a; Kloos, Schleifer, and Noble, 1976; Kloos, Schleifer, and Smith, 1976; Kloos, Zimmerman, and Smith, 1976; Lacey, 1975; Live, 1972; Marples, 1965; Noble and Somerville, 1974). Staphylococci have been isolated from birds, though information on avian *Staphylococcus* populations is very scanty (see references just mentioned).

In preliminary studies, W. E. Kloos and L. F. Kloos (unpublished studies) isolated small numbers of *S. xylosus* and *S. sciuri* subsp. *sciuri* from beach sand and marsh grass *(Spartina alternaflora)* located along the bayside shore of Ocracoke and Hatteras islands, off the coast of North Carolina. Both of these species are found on a wide variety of different mammals, including those indigenous to these islands. They were not isolated from oceanside sand samples. One attractive explanation for the presence of staphylococci in sand and on marsh grass would be that they were deposited there by local mammals washing or feeding along the peaceful, bayside shores. However, for at least these two species, we must not rule out the possibility that sand and the surfaces of water plants might serve as suitable pri-mary habitats. For example, Emmett and Kloos (1975, 1979) have determined that the nutritional requirements for *S. xylosus* and *S. sciuri* subsp. *sciuri* are very simple. For most strains of *S. xylosus* and all strains of *S. sciuri* subsp. *sciuri* studied to date, excellent growth was obtained in media containing an inorganic nitrogen source, e.g., $(NH_4)_2SO_4$, as the sole source of nitrogen. On the other hand, host-specific *Staphylococcus* species, or those with a narrow host range, required vitamins and various amino acids as sources of nitrogen. *S. xylosus* and *S. sciuri* subsp. *sciuri* are the only species of staphylococci that exhibit a form of motility on the surface of soft agar media, though flagella have not yet been observed (W. E. Kloos and others, unpublished studies). Furthermore, these species grow luxuriantly, produce weak acid from a wide range of carbohydrates, and are quite halotolerant (Kloos, Schleifer, and Smith, 1976). Based on the above characters, it would appear that *S. xylosus* and *S. sciuri* can adapt to a wide variety of habitats and perhaps are more "primitive" and free-living than other staphylococci. At this point, we cannot be sure that one or both of these species did not evolve in such habitats prior to contact with animals and today may represent living "fossils" of the genus *Staphylococcus*.

As Transient and Resident Populations on Human Skin

Numerous reports have been made on the occurrence of *Staphylococcus* on human skin and together suggest that nearly all people support populations of this genus. Most accounts have dealt with the distribution and carrier rate of the most potentially pathogenic, coagulase-positive species, *S. aureus*. Noble and Somerville in their text *Microbiology of Human Skin* (1974) have made a comprehensive review of the literature pertaining to the carriage of *S. aureus*. In the anterior nares, generally regarded as the headquarters region or site of reproduction for this species, carrier rates were reported to range from less than 10% to more than 40% in normal adults outside of hospitals. Hospital patients, as a group, have carrier rates greater than those of nonhospitalized persons and the rate may increase considerably with prolonged hospitalization (Noble et al., 1964). Nasal carrier rates are highest in infants, plateau throughout most of the adult life, and reach their lowest levels in geriatrics (Noble and Somerville, 1974). *S. aureus* can colonize the nares of some individuals for long periods of time, from one to several years, whereas in other individuals carriage is for much shorter periods (Cameron, 1970; Kay, 1963; Kloos and Musselwhite, 1975; Roodyn, 1960). *S. aureus* can be isolated from other regions of the body such as the throat, perineum, axillae, intestinal tract, toeweb, and normal glabrous skin, though usually not as consistently or in as large a number as from the nares (Noble, 1969; Noble, Valkenburg, and Wolters, 1967; Solberg, 1965; Williams, 1963).

Until very recently, the other *Staphylococcus* species found on humans have not received as much attention as *S. aureus*. Most early studies of the skin carriage of coagulase-negative staphylococci used the classification schemes of Baird-Parker (1963, 1965a,b). Investigators using these schemes reported true members of the genus *Staphylococcus* as either staphylococci or micrococci, depending upon their ability or inability to ferment glucose in the absence of oxygen. Various coagulase-negative staphylococci were often lumped into the species *S. epidermidis* or *S. albus*. In these early studies, coagulase-negative staphylococci were isolated from essentially all areas of human skin and usually in large numbers (reviewed by Marples, 1965, and Noble and Somerville, 1974).

Based on new systematic studies of the genus *Staphylococcus,* up to nine different coagulase-negative species could be identified on human skin (Kloos and Schleifer, 1975a,b; Schleifer and Kloos, 1975). In these studies, it was reported that 100% of 40 persons, living in North Carolina and New Jersey, carried cutaneous populations of staphylococci. The percentage of individuals carrying various *Staphylococcus* species was as follows: *S. epidermidis*, 100%; *S. hominis*, 100%; *S. haemolyticus*, 78%; *S. saprophyticus*, 70%; *S. capitis*, 65%; *S. aureus*, 52%; *S. warneri*, 52%; *S. xylosus*, 42%; *S. cohnii*, 35%; *S. simulans*, 12%.

Using the updated taxonomy, Kloos and Musselwhite (1975) studied the distribution and temporal persistence of *Staphylococcus* species on the skin of persons ranging from 2 to 61 years of age and found that they often composed greater than 50% of the aerobic bacteria isolated from the head, nares, and axillae, and usually from 10 to 70% of those isolated from the legs and arms. Obligate anaerobic, micro-aerophilic, or slightly aerotolerant bacteria present on skin were not included in the calculation of these percentages. The most prevalent and persistent *Staphylococcus* species on human skin were *S. epidermidis sensu stricto* and *S. hominis,* followed by *S. haemolyticus, S. capitis,* and *S. aureus.*

Staphylococcus epidermidis and *S. aureus* usually account for greater than 90% of the staphylococci living in the nares. *S. hominis* and *S. epidermidis* usually account for greater than 95% of the staphylococci living in the axillae. These two species produce very large resident populations in most warm, moist, nutritional regions of the cutaneous habitat, with the notable exception that *S. hominis* is only occasionally found in the nares, and then in small to moderate-sized populations.

Both species usually compose greater than 65% of the staphylococci found on normal glabrous skin—most of the skin area. Certain strain (clonal) populations of these species persisted on individuals for at least 1 year and, in extended studies, it could be demonstrated that several strains persisted for at least 3 years. In the study by Kloos and Musselwhite (1975), *S. haemolyticus* usually composed less than 2% of the staphylococci isolated from the nares and axillae and from 1 to 30% of those isolated from the head, legs, and arms. More recent studies by Carr and Kloos (1977) have suggested that *S. haemolyticus* is more prevalent on infants than on children or adults, and usually composes a slightly higher percentage of the total staphylococci than *S. hominis*. Fifty percent of children and adults in the earlier temporal study maintained relatively persistent populations of *S. haemolyticus* on normal glabrous skin. *S. capitis* often produces moderate-sized populations on the head, especially in adults. This species is also found in smaller populations on the arms and legs. *S. warneri* is only occasionally found on human skin and then only as small populations. *S. simulans, S. saprophyticus, S. cohnii,* and *S. xylosus* are also only occasionally found on human skin. The population size of these species may vary from very small to relatively large, depending on the individual carrying them. Some persons carry persistent populations of these species on normal glabrous skin for periods of at least 1 year, though in the case of *S. xylosus,* such populations may be transient and continually replenished by contact with pets or farm animals known to maintain large populations of this species. On very rare occasions one or several colony-forming units of *S. intermedius* and *S. sciuri* subsp. *sciuri* have been isolated from the skin of individuals with dogs as pets. The species *S. epidermidis sensu stricto, S. hominis, S. capitis,* and *S. simulans* appear to be host-specific for humans. *S. epidermidis* has on rare occasions been isolated from farm animals having close contact with man.

Some very recent studies by Evans, Mattern, and Hallem (1978) and Kilpper and Schleifer (1978) have suggested the presence of an anaerobic *Staphylococcus* species living on human skin, particularly as large populations on the forehead. Although this organism has been classified as *Peptococcus saccharolyticus* by several investigators, DNA-DNA hybridization, cell wall peptidoglycan, and protein homology studies by Schleifer and co-workers indicate a close relationship to members of the genus *Staphylococcus.*

Growing populations of staphylococci on human skin appear to be largely localized around openings of hair follicles and in superficial levels of follicular canals, though some microcolonies may be scattered over the skin surface (Lovell, 1945; Marples, 1965; Montes and Wilborn, 1969).

As Populations on Other Mammalian Skin

Numerous studies have been concerned with characterizing staphylococci isolated from domestic animals, though few have estimated carrier rates, residency status, or population size in the cutaneous habitat. Like the situation described above with humans, attention was focused on the most potentially pathogenic, coagulase-positive species.

BOVINE STAPHYLOCOCCI. Cattle *(Bos taurus)* can carry populations of *Staphylococcus aureus* on udders and in quarter milk, especially if they are suffering from mastitis or are in herds with afflicted animals (Hajek and Marsalek, 1969a; Wallace, Quisenberry, and Tanimoto, 1962; reviewed by Live, 1972). Parisi and Baldwin (1963) have estimated the carriage rate of *S. aureus* in various herds in Ohio to be in the range of 4.6–62.5%. *S. aureus* has been infrequently isolated from the nares of cattle; some studies have even reported the complete absence of this species in the nares of all cows or bulls tested (Elek, 1959; Kloos, Musselwhite, and Zimmerman, 1976; Kloos, Zimmerman, and Smith, 1976; Moeller et al., 1963). The occurrence of this species on other cutaneous areas has not yet been thoroughly assessed. Some strains of *S. aureus* isolated from quarter milk and/or the nares have been shown to be characteristic of the human biotype A, which may have originated from dairy attendants or farm personnel coming in contact with the animals (Courter and Galton, 1962; Hajek and Marsalek, 1969a, Moeller et al., 1963; Oeding, 1973; Skaggs and Nicol, 1961; Wallace et al., 1962). The bovine biotype C is the more usual one isolated from cattle and can be distinquished from the human and other biotypes on the bases of immunological, biochemical, and phage-typing properties (Hajek and Marsalek, 1971b, 1976; Oeding, 1973; Oeding et al., 1971). Recently, Devriese and Oeding (1976) reported the presence of small numbers of *S. aureus* biotype B strains in the quarter milk of cattle suffering from mastitis. This biotype is prevalent on pigs and poultry, suggesting possible interchange between the bacteria of these animals.

Brown et al. (1967) and Devriese and Oeding (1975) have reported on the characterization of coagulase-positive staphylococci that have been isolated from the udders of cattle and resemble the species *S. hyicus* (Devriese et al., 1978). The distribution and carriage of this species on cattle have not yet been adequately assessed. Coagulase-negative species on cattle have received much less attention, though their presence has been confirmed (reviewed by Holmberg, 1973). Preliminary studies (Kloos, Musselwhite, and Zimmerman, 1976; Kloos, Zimmerman, and Smith, 1976) have indicated that *S. xylosus* can be found in large populations on the hairy cutaneous areas of cattle and that

this species often composes a large percentage of the total staphylococci present.

PORCINE STAPHYLOCOCCI. Pigs *(Sus scrofa)* can on occasion carry populations of *Staphylococcus aureus* in their nares and on hairy skin surfaces (Baird-Parker, 1962; Devriese and Oeding, 1976; Hajek and Marsalek, 1970; Kloos, Musselwhite, and Zimmerman, 1976; Smith and Bettge, 1972; Kloos, Zimmerman, and Smith, 1976). Most strains isolated were characteristic of biotype B, predominant on swine and poultry (Devriese and Oeding, 1976; Hajek and Marsalek, 1971b, 1976; Oeding et al., 1972). *S. aureus* biotypes A and C, characteristic of humans and cattle, respectively, have been less frequently isolated on pigs, and may reflect interchange of staphylococci between man and animals coming in close contact. It is not uncommon for farms raising cattle to also raise pigs, and vice versa. A major group of porcine staphylococci that has been associated with exudative epidermitis in pigs has been given the various names *Micrococcus hyicus* (Sompolinsky, 1953), *S. epidermidis* (Baird-Parker, 1965a,b; Devriese and Oeding, 1975), and *S. hyicus* (Devriese, 1977; Devriese et al., 1978; Jones, Deibel, and Niven, 1963; L'Ecuyer, 1967). *S. hyicus,* which may be regarded as a partially coagulase-positive species, appears to be quite common in the nares and on the hairy cutaneous areas of pigs, though comprehensive studies on the ecology of this species have not yet been reported. Porcine coagulase-negative species have not yet been studied in much detail. They are, however, distinctly different from human species (Devriese et al., 1978; Kloos, Musselwhite, and Zimmerman, 1976; Kloos, Zimmerman, and Smith, 1976; Smith and Bettge, 1972). Occasionally, small populations of *S. epidermidis sensu stricto, S. haemolyticus,* and *S. saprophyticus* characteristic of human strains have been isolated from pig skin, a finding that may reflect some limited interchange between staphylococci of different hosts.

OVINE STAPHYLOCOCCI. Sheep *(Ovis aries)* and their relatives, like the other artiodactyls just discussed, have been known to carry populations of *Staphylococcus aureus,* though few studies have been reported. Watson (1965) and Hajek and Marsalek (1971b, 1976) have isolated this species from the nares and udders of sheep. The predominant strains isolated were characteristic of biotype C, the major biotype also found in cattle (Hajek and Marsalek, 1971b; Oeding, Hajek, and Marsalek, 1976). On rare occasions, human biotype A and canine biotype E have been isolated from sheep, again suggesting possible transfer of staphylococci between various hosts coming in contact with these animals. Both sheep and goats *(Capra)* have relatively large cutaneous populations of *S. xylosus* and

S. sciuri subsp. *lentus,* a newly proposed subspecies that appears to be relatively host specific (Kloos, Musselwhite, and Zimmerman, 1976; Kloos, Schleifer, and Smith, 1976; Kloos, Zimmerman, and Smith, 1976; unpublished studies).

EQUINE STAPHYLOCOCCI. Horses can on occasion carry nasal populations of *Staphylococcus aureus* that presumably have diverse origins. Most strains studied to date have been characteristic of the human biotype A, porcine biotype B, or bovine and sheep biotype C (Hajek, Marsalek, and Herna, 1974; Oeding, Hajek, and Marsalek, 1974). It remains a mystery whether or not a specific biotype or subspecies of *S. aureus* exists that is predominantly adapted to horses. There is, however, a second co-agulase-positive species, *S. intermedius* (previously designated *S. aureus* biotype E), that is more commonly found in the nares than *S. aureus* (Hajek, 1976; Hajek and Marsalek, 1971b, 1976; Oeding et al., 1974; Meyer and Schleifer, 1978; Schleifer et al., 1976). Horses may carry large populations of *S. xylosus* on hairy cutaneous areas (Kloos, Musselwhite, and Zimmerman, 1976; Kloos, Zimmerman, and Smith, 1976). Other coagulase-negative species occurring on horses have not yet been adequately characterized.

CARNIVORE STAPHYLOCOCCI. Many studies have been concerned with staphylococci of the domestic dog *(Canis familiaris).* Dogs only occasionally carry populations of *Staphylococcus aureus.* Those that are present are usually characteristic of the human biotype A (Devriese and Oeding, 1976; Hajek and Marsalek, 1969b, 1971b, 1976; reviewed by Live, 1972; Oeding, 1973; Rountree, Freeman, and Johnston, 1956) and most likely originate from human handlers. The primary coagulase-positive species found in the nares and on the hairy cutaneous areas of dogs is *S. intermedius* (Hajek, 1976; Schleifer et al., 1976). Canine *S. intermedius* strains were previously designated *S. aureus* biotype E (Hajek and Marsalek, 1971b; Oeding, 1973; Oeding et al., 1970) or *Staphylococcus* sp. 1 (Kloos, Musselwhite, and Zimmerman, 1976; Kloos, Zimmerman, and Smith, 1976). This species can be found in relatively large populations on canine skin and can on occasion be transferred to the skin of human handlers living with dogs in their home (Kloos, Musselwhite, and Zimmerman, 1976; unpublished studies). It is very interesting that other carnivores such as mink *(Mustela lutreola)* (Hajek and Marsalek, 1976; Hajek, Marsalek, and Hubacek, 1972; Oeding, 1973; Oeding, Hajek, and Marsalek, 1973), foxes *(Vulpes* sp.) (Hajek and Marsalek, 1976), and raccoons *(Procyon lotor)* (Kloos, Musselwhite, and Zimmerman, 1976; Kloos, Zimmerman, and Smith, 1976) also commonly carry *S. intermedius* populations on their skin. Carnivore coagulase-negative

Staphylococcus species have not yet been studied in much detail. Preliminary studies have indicated that dogs and raccoons occasionally carry small populations of *S. sciuri* subsp. *sciuri* and usually carry medium to large populations of *S. xylosus* (Kloos, Musselwhite, and Zimmerman, 1976, Kloos, Zimmerman, and Smith, 1976; unpublished studies). Raccoons have also been shown to carry a *Staphylococcus* species that appears to be phenotypically intermediate between *S. xylosus* and *S. cohnii,* we have tentatively designated this species *Staphylococcus* sp. 3. Future nucleic acid hybridization studies should help to clarify the taxonomic position of this group of staphylococci.

RODENT STAPHYLOCOCCI. Natural populations of staphylococci on common laboratory rodents such as mice *(Mus musculus),* rats *(Rattus norvegicus),* guinea pigs *(Cavia porcellus),* and hamsters *(Cricetus mesocricetus)* have not been studied in much detail. A few reports have indicated carriage of coagulase-positive and -negative staphylococci in the nares and on body and tail skin (Buhles, 1969; Flynn, 1959; Hard, 1966; Rountree et al., 1956; Totten, 1958). By contrast, numerous reports have been made on experimental staphylococcal infections in rodents, especially mice (reviewed by Anderson, 1976; Noble and Somerville, 1974; and Smith, Warren, and Johnson, 1976). In preliminary studies (Kloos, Musselwhite, and Zimmerman, 1976; Kloos, Schleifer, and Smith, 1976; Kloos, Zimmerman, and Smith, 1976), very small numbers of *Staphylococcus aureus* and *S. intermedius* were on rare occasions isolated from wild eastern gray squirrels *(Sciurus carolinensis).* The *Staphylococcus* species found most frequently on the skin of wild eastern gray squirrels, southern flying squirrels *(Glaucomys volans),* the Central American squirrel *(Sciurus granatensis),* and spiny rats *(Proechimys semispinosus)* were *S. sciuri* subsp. *sciuri, S. xylosus,* and the unnamed *Staphylococcus* sp. 3.

RABBIT AND HARE STAPHYLOCOCCI. *Staphylococcus aureus* has been isolated from natural outbreaks of staphylococcal disease in both wild and domestic rabbits *(Sylvilagus* and *Oryctolagus* species) (Adlam et al., 1976; Cheatum, 1941; McCoy and Steenbergen, 1969). Coagulase-negative *Staphylococcus* species of hares and rabbits have not been studied in detail, though as a group they have been shown to occur on fur and skin (Buhles, 1969). Hajek and Marsalek (1971a) isolated *S. aureus* from the nares of approximately 5% of 462 European hares *(Lepus europaeus)* killed in hunts. Most of the strains were classified as biotype D, which appears to be unique to hares (Hajek and Marsalek, 1971a,b, 1976; Oeding, 1973; Oeding et al., 1973). One-fifth of the total *S. aureus* strains isolated were characteristic of biotype E, normally found on dogs. These may have been transferred to the killed hares by dogs used in the hunt.

NONHUMAN PRIMATE STAPHYLOCOCCI. Our preliminary studies on the staphylococci of captive and some wild nonhuman primates have generated some interesting ecological observations. The most notable find was that *Staphylococcus warneri* and *S. haemolyticus*, normally found on humans but not on other lower mammals studied to date, are species that are commonly found on the skin of nonhuman primates. A simian subspecies of *S. warneri* has proved to be the predominant species in the nares of most nonhuman primates when *S. aureus* is not present in large populations. The structure of *Staphylococcus* species populations on these primates appears to be approaching that found on man.

Prosimians of the genera *Lemur, Microcebus,* and *Galago* carried relatively large populations of *S. cohnii* and small to medium populations of *S. warneri, S. haemolyticus, S. xylosus,* and *S. sciuri* on hairy areas of the skin.

New World monkeys of the genera *Alouatta, Saimiri, Lagothrix,* and *Ateles* often carried relatively large populations of *S. aureus* in their nares and on hairy cutaneous areas. On some presumably healthy wild animals, *S. aureus* populations composed greater than 90% of the total staphylococci isolated. On occasion, captive *Saimiri* carried small populations of *S. intermedius*, though we must question the origin of this species, as dogs were maintained in the same facilities. *S. warneri, S. haemolyticus, S. xylosus,* and *S. sciuri* were carried in small to medium-sized populations on hairy areas of the skin. *S. cohnii* and *S. saprophyticus* were isolated only occasionally from skin.

Old World monkeys of the genera *Macaca, Cercocebus, Cercopithecus, Erythrocebus, Papio,* and *Presbytix* often carried relatively large populations of *S. aureus* in the nares. These monkeys also carried relatively large populations of *S. warneri* and small to medium-sized populations of *S. haemolyticus* on hairy areas of the skin. *S. xylosus, S. cohnii, S. saprophyticus,* and *S. sciuri* were isolated only occasionally from skin. *Macaca,* in addition, carried a possible new species, which we have tentatively designated *Staphylococcus* sp. 7.

Great apes of the genera *Pongo, Gorilla,* and *Pan* often carried small to medium-sized populations of *S. aureus* and relatively large populations of *S. warneri* and *S. haemolyticus* in the nares. *S. warneri* and *S. haemolyticus* were also carried as relatively large populations in the hairy cutaneous areas. *S. warneri* produced very large populations in the axillae of *Pan.* It is evident that *S. warneri* has been largely displaced by *S. epidermidis* in humans. Most of the great apes carried small populations of *S. cohnii* and *S. saprophyticus* and, only occasionally, small numbers of *S. xylosus.*

As Populations on Avian Skin

Comprehensive systematic studies of the staphylococci of birds have not been reported. However, several reports have been made on the coagulase-positive staphylococci occurring in the nares, body skin, and various infections of chickens (Devriese and Oeding, 1976; Devriese et al., 1972; Hajek and Marsalek, 1971b; Smith and Crab, 1960). *S. aureus* strains of biotypes A and B have been isolated occasionally from chickens (Devriese and Oeding, 1976; Hajek and Marsalek, 1971b, 1976), although their origin is not always clear. Biotype A strains probably originated from humans and some of the biotype B strains may have originated from other domestic mammals, notably pigs. *S. hyicus* strains have also been isolated from chickens (Devriese and Oeding, 1975; Devriese et al., 1978). Pigeons have been known to carry populations of *S. intermedius* (designated previously as *S. aureus* biotype F) in their nares (Devriese and Oeding, 1976; Hajek, 1976; Hajek and Marsalek, 1969c, 1971b, 1976; Oeding, 1973; Oeding et al., 1970; Schleifer et al., 1976). Avian *Staphylococcus* populations remain largely unexplored and should provide a considerable challenge to even the most ambitious systematic bacteriologists.

As Opportunistic Pathogens of Man and Other Animals

Staphylococcus aureus, since its early discovery by Ogston (1881) as a causative agent of acute suppurative diseases in man, continues to be a major cause of mortality and a variety of infections. Among the major human infections caused by this species are furuncles (boils), carbuncles, impetigo, toxic epidermal necrolysis (scalded skin syndrome), pneumonia, osteomyelitis, acute endocarditis, myocarditis, pericarditis, enterocolitis, mastitis, cystitis, prostatitis, cervicitis, cerebritis, bacteremia, and abscesses of the muscle, skin, urogenital tract, central nervous system, and various intraabdominal organs. Food poisoning is often attributed to staphylococcal enterotoxin. Comprehensive reviews on the nature of human infections caused by *S. aureus* can be found in the following texts: *Staphylococcus Pyogenes and Its Relation to Disease* (Elek, 1959), *The Staphylococci* (Cohen, 1972), *Staphylococci and Staphylococcal Infections: Recent Progress* (Jeljaszewicz, 1976), *Microbiology of Human Skin* (Noble and Somerville, 1974), and *Staphylococci and Staphylococcal Diseases* (Jeljaszewicz, 1976).

Staphylococcus aureus is also capable of producing a variety of infections in animals other than man. The more common natural ones reported include mastitis, synovitis, furuncles, suppurative dermatitis, abscesses in various organs, pyemia, and septicemia. Reference may be made to the above-mentioned texts and, in addition, Stableforth and Galloway's *Infectious Diseases of Animals: Diseases Due to Bacteria* (1959), Heidrich and Renk's *Diseases of the Mammary Glands of Domestic Animals* (1967), and the veterinary literature for information on the nature of *S. aureus* infections in animals. *S. hyicus* has been implicated in exudative epidermitis in pigs (Devriese, 1977; Devriese and Oeding, 1975; L'Ecuyer, 1966, 1967; Underhahl, Grace, and Twiehaus, 1965), and *S. intermedius* appears to be the causative agent of canine otitis externa (Devriese and Oeding, 1976).

Historically, the coagulase-positive species have been regarded as opportunistic pathogens, whereas the coagulase-negative species (some of which were called micrococci) have been generally regarded as nonpathogens. This view is changing, particularly with respect to human infections. Coagulase-negative species, particularly *S. epidermidis sensu stricto*, may be responsible for as many as 1–10% of all cases of human bacterial endocarditis (Geraci, Hanson, and Giuliani, 1968; Pulverer and Halswick, 1967; Pulverer and Pillich, 1971; Quinn, Cox, and Drake, 1966; Speller and Mitchell, 1973; Watanakunakorn and Hamburger, 1970). Bacteremia produced by coagulase-negative staphylococci occurs frequently after cardiac surgery and neurosurgery and after implantation of artificial internal prostheses (Andriole and Lyons, 1970; Black, Challacombe, and Ockenden, 1965; Callaghan, Cohen, and Stewart, 1961; Cluff et al., 1968; Dobrin et al., 1976; Fokes, 1970; Holt, 1971; Lam, McNeish, and Gibson, 1969; Moncrieff et al., 1973; Parisi, 1973; Quinn, Cox, and Fisher, 1965; Rames et al., 1970). Bacteremia produced by these organisms may persist indefinitely, cause tissue and organ injury, and present therapeutic problems. Holt (1972) concluded that ventriculoatrial shunts are most often infected or colonized with *S. epidermidis* of B-P subgroup II (Baird-Parker, 1965a) or biotype 1 (Baird-Parker, 1974). Baird-Parker's *S. epidermidis* biotype 1 would encompass a large percentage (ca. 80%) of *S. epidermidis sensu stricto* strains. Urinary tract infections such as acute pyelonephritis or cystitis may also be produced by coagulase-negative staphylococci (Bailey, 1973; Jakubicz and Borowski, 1976; Mabeck, 1969; Meers, Whyte, and Sandys, 1975; Mitchell, 1968; Mortensen, 1969; Oeding and Diagranes, 1976; Pulverer and Pillich, 1971; Roberts, 1967; Sellin et al., 1975; Telander and Wallmark, 1975; Torres-Pereira, 1962). The majority of these infections were once believed to be produced by micrococci of Baird-Parker's *Micrococcus* subgroup 3; however, these organisms have now been reclassified as staphylococci in the eighth edition of *Bergey's Manual of Determinative Bacteriology* (Buchanan and Gibbons, 1974) and most strains studied belong to the species *Staphylococcus*

saprophyticus. Using updated taxonomy, Nord et al. (1976) found that *S. epidermidis sensu stricto* was the predominant coagulase-negative species isolated from wound and several other pyogenic infections. *S. haemolyticus* and *S. hominis* were associated with a moderate percentage of wound and other pyogenic infections. *S. saprophyticus* was the predominant coagulase-negative species isolated from urinary tract infections, followed by *S. epidermidis* and *S. haemolyticus* in frequency of occurrence. Fleurette and co-workers (Brun, Fleurette, and Forey, 1977; Fleurette and Brun, 1977) have found that *S. epidermidis sensu stricto* was the predominant coagulase-negative species isolated from various infections, with the species *S. saprophyticus, S. capitis, S. haemolyticus,* and *S. hominis* occurring in infections at a much lower frequency.

According to the veterinary literature, coagulase-negative species may also be associated with bovine mastitis (Holmberg, 1973; Lee and Frost, 1970; Rose and McDonald, 1973; Stabenfeldt and Spencer, 1966).

Isolation

Isolation of *Staphylococcus aureus* from Foods

Staphylococcus aureus is an organism that is highly susceptible to heat treatment and most sanitizing agents. Hence, when it or its enterotoxins are found in processed foods, poor sanitation is usually indicated. This species has been confirmed to be the causative agent of many cases of severe food poisoning (reviewed by Bergdoll, 1972; Bergdoll and Bennett, 1976; Minor and Marth, 1972). Its identity in foods, therefore, is of major concern. The presence of other *Staphylococcus* species, some of which may be used inadvertently in food processing, is generally of less concern today, though Breckenridge and Bergdoll (1971) have reported an outbreak of food poisoning by a coagulase-negative staphylococcus. Detailed procedures for preparing food samples for analysis, isolating and enumerating *S. aureus,* and detecting staphylococcal enterotoxins in foods can be found in the recent texts *Compendium of Methods for the Microbiological Examination of Foods* (Speck, 1976) and *FDA Bacteriological Analytical Manual for Foods* (Food and Drug Administration, 1976). We will include here several of the recommended procedures for isolating *S. aureus* from foods.

NONSELECTIVE ENRICHMENT PROCEDURES.   For processed foods containing a small number of cells that may have been injured, e.g., as a result of heating, freezing, desiccation, or storage, the following procedure is suitable.

Nonselective Enrichment of *Staphylococcus aureus* (Heidelbaugh et al., 1973)

"A 50-ml sample of FS [food slurry] was transferred into 50 ml of double-strength Trypticase soy broth (TSB) and incubated at 35C for 2 hr. Then, 100 ml of single-strength TSB containing 20% NaCl was added to yield a final salt concentration of 10%. After incubation at 35C for 24 hrs ± 2 hr, 0.1 ml of the TSB culture was spread on each of two plates of Vogel and Johnson agar and incubated at 35C. Plates were examined after 24 and 48 hr for the presence of black colonies with yellow zones. Two or more typical representative colonies were transferred to brain heart infusion (BHI) tubes and incubated at 35C for 24 hr. The remainder of each colony was removed with a loop and emulsified in 0.2 ml of BHI, and 0.5 ml of coagulase plasma was added, mixed, and incubated in a 35C water bath for 4 hr. At the end of 4 hr [tubes were examined for the presence of firm clots, a positive reaction], negative tubes were noted, and [for these] the coagulase test was repeated with the 24-hr culture."

It should be noted that rabbit coagulase plasma is preferable over other plasmas, e.g., human or bovine plasma. Coagulase test tubes should be examined again at 6 and 24 h, if clots are not produced by 4 h. With this procedure, false-positive tests may occur with mixed cultures, but this will most likely be avoided if only well-isolated colonies typical of *S. aureus* are chosen. On rare occasions, coagulase-negative *S. aureus* strains may be isolated, but these can be confirmed on the basis of other physiological and biochemical tests (Kloos and Schleifer, 1975a,b). Report results as *S. aureus* present or absent in 5 g of food.

The presence of other coagulase-positive staphylococci such as *S. intermedius* and certain strains of *S. hyicus* in foods may be regarded as a possible threat to human health, though their ability to produce enterotoxins has not yet been adequately assessed. These species can be distinguished from *S. aureus* on the basis of several key characters (see Identification).

Vogel and Johnson Agar for Nonselective Enrichment of *Staphylococcus aureus* (Leininger, 1976)

"Basal medium: Trypticase or Tryptose

(pancreatic digest of casein)	10.0 g
Yeast extract	5.0 g
D-Mannitol	10.0 g
Dipotassium phosphate	5.0 g
Lithium chloride · 6H$_2$O	6.0 g
Glycine	10.0 g

Agar	16.0 g
Phenol red	0.025 g
Distilled water	1.0 liter

Suspend ingredients in distilled water, heat to boiling with frequent agitation to dissolve ingredients. Dispense into bottles or flasks, 98 ml per container, and autoclave 15 min. at 121C. Final pH 7.2 ± 0.2. Complete medium: Cool molten basal medium to 45 to 50C. Add 2 ml sterile 1% potassium tellurite solution to each 98 ml container and mix thoroughly just prior to pouring. Pour 15 to 18 ml into sterile Petri dishes."

Baird-Parker agar has gained wide acceptance and may be used as an alternative to Vogel and Johnson agar.

Baird-Parker Agar for Nonselective Enrichment of *Staphylococcus aureus* (Food and Drug Administration, 1976)

"Basal medium:

Tryptone	10 g
Beef extract	5 g
Yeast extract	1 g
Sodium pyruvate	10 g
Glycine	12 g
Lithium chloride · 6H$_2$O	5 g
Agar	20 g
Distilled water	950 ml

"Suspend the ingredients in distilled water, heat to boiling to dissolve completely. Dispense 95 ml portions in screw-cap bottles. Autoclave for 15 min. at 121C. Final pH, 7.0 ± 0.2 at 25C. Store for not more than one month at 4 ± 1C.

"For egg yolk (EY) enrichment soak eggs in aqueous mercuric chloride (1:1000) for not less than 1 minute. Rinse in sterile water and dry with a sterile cloth. Aseptically crack eggs, and separate whites and yolks. Blend yolk and sterile physiological saline solution (3 + 7 v/v) in high speed blender for 5 sec. and mix 50 ml blended egg yolk to 10 ml of filter-sterilized 1% potassium tellurite. Mix and store at 2 to 8C until used.

"Complete medium: Add 5 ml of prewarmed (45-50C) EY enrichment to 95 ml of molten basal medium that has been adjusted to 45-50C. Mix well and pour 15–18 ml into sterile 15 × 100 mm petri dishes. Store plates of complete medium at 4 ± 1C for no longer than 48 hr. before use. The medium should be densely opaque. On this medium, colonies of *S. aureus* are typically circular, smooth, convex, moist, 2 to 3 mm in diameter on uncrowded plates, gray to jet-black, frequently with a light colored margin surrounded by an opaque zone (ppt) and frequently with an outer clear zone; colonies have buttery to gummy consistency when touched with an inoculating needle.''

The complete Baird-Parker medium can be stabilized to store for up to 1 month, at 4°C, by omitting sodium pyruvate from the basal medium and adding 0.5 ml of 20% (wt/vol) sodium pyruvate solution to stored, undried plates; the plates are dried at 50°C and then inoculated (Holbrook, Anderson, and Baird-Parker, 1969).

SELECTIVE ENRICHMENT PROCEDURES. For raw food ingredients and unprocessed foods expected to contain <100 *S. aureus* cells/g and a large population of competing species, Baer, Gray, and Orth (1976) recommended using the procedures outlined by the Association of Official Analytical Chemists (1970, 1975).

Selective Enrichment of Most-Probable-Number *Staphylococcus aureus* (Baer, Gray, and Orth, 1976)

"Inoculate 3 (or 5) tubes of trypticase soy or tryptic soy broth containing 10% NaCl with 1 ml aliquots of decimal dilutions of each sample. The highest dilution must give a negative end-point. Incubate tubes 48 ± 2 hr. at 35–37C. Using a 3 mm loop, transfer one loopful from each tube showing positive growth (turbidity) to a plate of Baird-Parker agar medium with a properly dried surface. Streak the inoculum to obtain isolated colonies. Incubate plates 45–48 hr. at 34–37C. Pick at least one of each colony type typical of *S. aureus* and transfer into a small tube containing 0.2–0.3 ml of BHI broth and emulsify thoroughly. Inoculate an agar slant of suitable maintenance medium, e.g., tryptic soy agar, with a loopful of suspension. Incubate BHI culture suspension and slants 18–24 hr. at 35–37C.''

Use the brain heart infusion culture suspension for the coagulase test and slant culture for ancillary tests, including catalase activity, anaerobic utilization of glucose and mannitol, lysostaphin susceptibility, and thermostable nuclease production. Report the *S. aureus* per gram as the most probable number (MPN) per gram according to MPN values. Tables used to calculate MPN values can be found in several reference manuals, e.g., *FDA Bacteriological Analytical Manual for Foods* (Food and Drug Administration, 1976) and *Compendium of Methods for the Microbial Examination of Foods* (Speck, 1976). Recently, Brewer, Martin, and Ordal (1977) have reported on the addition of filter-sterilized catalase or pyruvate to Trypticase soy broth (TSB) containing

10% NaCl in an MPN technique for good recovery of heat-stressed *S. aureus* cells from foods.

DIRECT SURFACE PLATING PROCEDURES. For the detection of *Staphylococcus aureus* in raw or processed foods, Baer et al. (1976) recommend the direct surface plating procedure of the AOAC (1975). Direct plate counting of *S. aureus* is sometimes preferred by laboratories, especially when there is little chance of interference by competing species. In addition, plate counting is generally regarded as somewhat more accurate for the enumeration of staphylococci than MPN procedures.

Surface Plating of *Staphylococcus aureus* (Baer, Gray, and Orth, 1976)

"Plating of two or more decimal dilutions may be required to obtain plates with the desired number of colonies. For each dilution [of food homogenate] to be plated, aseptically transfer 1 ml of sample suspension to triplicate plates of Baird-Parker agar and distribute the 1 ml inoculum equitably over the triplicate plates (e.g., 0.4 ml–0.3 ml–0.3 ml). The sensitivity of this procedure may be increased by using larger inoculum volumes (>1 ml) distributed over >3 replicate plates. Spread the inoculum over the surface of the agar using sterile, bentglass streaking rods. Avoid the extreme edges of the plate. Retain the plates in an upright position until the inoculum is absorbed by the medium. Invert plates and incubate 45 to 48 hrs. at 35 to 37C. Select plates containing 20 to 300 colonies, unless plates at only lower dilutions (>200 colonies) have colonies with the typical appearance of *S. aureus*. When plates at the lowest dilution plated contain <20 colonies, they may be used. If several types of colonies are observed which appear to be *S. aureus,* count the number of colonies of each type and record counts separately. Select one or more colonies of each type counted and test for coagulase production. Add the number of *S. aureus* colonies on triplicate plates and multiply the total by the sample dilution factor. Report as *S. aureus*/g of product tested.''

Other laboratories have reported the successful use of tellurite polymyxin egg yolk agar (Crisley, Angelotti, and Foter, 1964) and KRANEP agar (Sinell and Baumgart, 1966) in place of Baird-Parker agar.

Isolation of *Staphylococcus aureus* from Water

The presence of *Staphylococcus aureus* in recreational waters, swimming pools, water that might be added to foods, and hydrotherapy pools poses a potentially serious threat to human health (reviewed by Evans, 1977). Staphylococci are somewhat resistant to halogen disinfectants. For this reason, significant numbers of these organisms would remain viable for extended periods of time in inadequately treated bathing places. According to Evans the enumeration of either total staphylococci or of *S. aureus* would seem to provide a good index of the level of contamination of recreational water by swimmers and the effectiveness of filtration and chlorination processes. The American Public Health Association (APHA), American Water Works Association (AWWA), and Water Pollution Control Federation (WPCF) have jointly published a manual, *Standard Methods for the Examination of Water and Wastewater* (Tares et al., 1971) in which is described a (tentative) membrane filter technique for the recovery and enumeration of *S. aureus*. This technique employs a standard coliform membrane filter apparatus and procedure which is described in detail in the manual.

Tentative Membrane Filter Technic for Recovery and Enumeration of *Staphylococcus aureus* (Tares et al., 1971)

"Filter samples of water through sterile membrane to attain 40–100 colonies on the membrane surface. Amounts varying from 100 to 10, 1, 0.1 or 0.01 ml may be necessary, depending on the number of organisms present. Transfer the filter directly to the broth-saturated [2 ml of M-Staphylococcus broth] pad in the petri dish, avoiding air bubbles. Invert culture plates and incubate at 35 ± 0.5 C for 48 hr. If no pigmented colonies develop, hold plates an additional 48 hr at room temperature before discarding as negative. Colonies produced by *Staphylococcus aureus* are generally yellow-gold in color. Record densities as yellow-gold colonies per 100 ml. Typical yellow-gold colonies should be fished for [Gram] staining and coagulase-testing."

Other staphylococci may also produce yellowish pigment on the M-Staphylococcus medium and some strains of *S. aureus* are unpigmented. Hence, the medium used here is not totally satisfactory for the presumptive identification of *S. aureus*.

M-Staphylococcus Broth (Tares et al., 1971)

Tryptone or polypeptone	10.0 g
Yeast extract	2.5 g
Lactose	2.0 g
Mannitol	10.0 g
K_2HPO_4	5.0 g
NaCl	75.0 g
Distilled water	1.0 liter

The medium was autoclaved at 121°C for 15 min. The pH should be 7.0 after sterilization.

Isolation of Staphylococci from Clinical Specimens

PREPARATION OF BLOOD CULTURES. Blood cultures are usually collected when there is a sudden increase in the patient's pulse rate and temperature, onset of chills, a change in sensorium, prostration, and hypotension. Another indication is intermittent fever associated with heart murmur. Timing in collection of certain cultures may be quite critical. Blood cultures are useful for establishing the etiological agent in endocarditis, but here the bacteremia is usually continuous and timing for collection is not critical. The procedure for collecting blood cultures described by Isenberg et al. (1974) in the American Society for Microbiology (ASM) *Manual of Clinical Microbiology* is adequate for the isolation of staphylococci from blood.

Collecting Blood Cultures for Isolation of Staphylococci (Isenberg et al., 1974)

"In patients with suspected bacterial endocarditis, three blood cultures have been shown to be sufficient to isolate the etiologic agent in nearly all instances. These should be collected separately at no less than hourly intervals within a 24-h period of time. In intermittent bacteremias, three separate blood cultures within 24 to 48 h are usually sufficient to isolate the etiologic agent. . . . In patients who have received antimicrobics prior to blood collection, a total of four to six separate blood cultures may be necessary . . . collection [should] be performed aseptically, first by cleansing the skin with 70 to 95% alcohol and, secondly, by applying 2% iodine [or iodophor] in concentric fashion to the venipuncture site . . . [and waiting] at least one minute. . . . [It is recommended that 1 to 5 ml of blood in infants and children and 10 to 20 ml in adults be collected for each culture.] In most instances the blood is inoculated directly into culture media at the patient's bedside, either with a syringe and needle or a transfer set."

Ivler (1974) recommends adding 3 to 5 ml venous blood to 50 ml of a medium such as tryptose-phosphate broth, TSB, or BHI broth, which is then incubated at 37°C. For persons receiving penicillin, penicillinase should be added to the culture bottle at the time the inoculum is taken. Cultures should be examined later on the same day and daily thereafter up to 7 days for evidence of turbidity, hemolysis, or discrete colonies. Gram-stained smears and aerobic and anaerobic subcultures of suspected positive cultures should be prepared immediately.

Isolation from Pus, Purulent Fluids, Sputum, Urine, and Positive Blood Cultures

The recommendations of Ivler (1974) are satisfactory for the isolation of staphylococci from various clinical specimens.

Isolation of Staphylococci from Clinical Specimens (Ivler, 1974)

"Streak specimens directly on a blood-agar plate and inoculate a tube of thioglycolate broth. Incubate at 37 C. Specimens from patients receiving penicillin may be treated with penicillinase to inactivate the drug prior to inoculation of the medium. . . . On blood-agar, abundant growth occurs in 18 to 24 h. Colonies will generally be 1 to 3 mm in diameter; they are usually opaque, circular, smooth, and raised, with a butyrous consistency. Colonies conforming to this description should be Gram-stained and observed for typical morphology."

Most staphylococci of medical interest will produce growth in the upper, as well as the lower, anaerobic portions of commercially available thioglycolate broth.

The composition of blood agar, as outlined by Vera and Dumoff (1974) in the ASM *Manual of Clinical Microbiology,* follows.

Preparation of Blood Agar (Vera and Dumoff, 1974)

Beef heart muscle infusion	375 g
Tryptose or thiotone peptic digest of animal tissue	10 g
Sodium chloride	5 g
Agar	15 g
Distilled or demineralized water	1 liter
pH 7.3 or 6.8	

"Autoclave at 121 C for 15 min. Cool to 50 C and [for complete medium] add 5% sterile defibrinated rabbit blood." Sheep or bovine blood may be used in place of rabbit blood; human blood should not be used for the preparation of blood agar. Blood agar base is readily available commercially.

Isolation from Feces

When large populations of competing genera are present in clinical specimens, e.g., feces, mannitol salt agar or phenylethyl alcohol agar should be inoculated, in addition to blood agar (Ivler, 1974). These selective media will inhibit the growth of bacteria of many other genera.

Mannitol Salt Agar (Chapman, 1945)

Beef extract	1.000 g
Peptone or polypeptone	10.000 g
Sodium chloride	75.000 g
Mannitol	10.000 g
Agar	15.000 g
Phenol red	0.025 g
Distilled water	1.000 liter

Final pH 7.4. Heat to dissolve and autoclave at 121°C for 15 min.

Phenylethyl Alcohol Agar, Brewer and Lilly's Medium (Vera and Dumoff, 1974)

Pancreatic digest of casein USP	15 g
Papaic digest of soya meal USP	5 g
Sodium chloride	5 g
Phenylethyl alcohol	2.5 g
Agar	15.0 g
Distilled water	1.0 liter
Final pH 7.3.	

Dissolve with heat, autoclave at 118 C for 15 min, cool, and pour into petri dishes. Add 5% defibrinated sheep blood if desired."

Both mannitol salt agar and phenylethyl alcohol agar are readily available commercially.

Isolation of Staphylococci from Skin

Several different basic methods are available for isolating staphylococci and other aerobic bacteria from skin (reviewed by Noble and Somerville, 1974). Washing or swabbing methods disperse cutaneous bacteria to provide samples of uniform composition. They break up large aggregates or microcolonies on skin into smaller colony-forming units (CFU) and in some cases single cells. Impression methods estimate the number of microcolonies or aggregates of bacteria on the skin surface. Biopsy methods can determine the location of bacteria in microniches on skin. Most sampling of aerobic bacteria on skin has been performed using swabbing methods. The swab technique described by Kloos and Musselwhite (1975) has been used for the isolation of staphylococci and micrococci and is suitable for use with human as well as other mammalian skin. The procedures and agar medium used for this technique are described in this Handbook, Chapter 124.

Identification

Criteria and procedures used to distinguish members of the genus *Staphylococcus* from members of the genus *Micrococcus* are described and discussed in this Handbook, Chapter 124. Staphylococci are Gram-positive, catalase- and benzidine-test-positive cocci. Their cell walls contain teichoic acid and their cell wall peptidoglycans contain more than 2 moles of glycine per mole of lysine. The DNA base composition is in the range 30–39 mol% G+C. Menaquinones are normal. Most strains demonstrate some susceptibility to lysostaphin, produce acid aerobically from glycerol and D-(+)-glucose, and are facultative with respect to oxygen requirements.

The Differentiation of Coagulase-Positive Species

The coagulase reaction and the production of heat-stable deoxyribonuclease are regarded as the main criteria for the separation of *Staphylococcus aureus* from other staphylococci. Studies by Hajek and Marsalek (1971b), however, have indicated that the coagulase-positive staphylococci are a heterogeneous group and can be divided into six different biotypes. The majority of the isolates from man belong to biotype A, from poultry and swine to biotype B, from bovines and sheep to biotype C, from hares to biotype D, from dogs, horses, and minks to biotype E, and from foxes and pigeons to biotype F. Strains of biotypes E and F are quite distinct from *S. aureus* strains belonging to biotypes A–D (Schleifer et al., 1976) and they are reclassified into a new species designated *S. intermedius* (Hajek, 1976). There are also a few strains of coagulase-positive staphylococci that are related to *S. hyicus* (Devriese et al., 1978). The main characteristics for differentiating among these coagulase-positive staphylococci are compiled in Table 1. An important criterion for their differentiation is the cell wall composition. Strains of *S. aureus* and *S. hyicus* do not contain or contain only traces of L-serine in their peptidoglycan, whereas most strains of *S. intermedius* exhibit serine. Ribitol teichoic acid, which is a typical constituent of cell walls of *S. aureus,* is replaced by glycerol teichoic acid in *S. intermedius* and *S. hyicus. S. aureus* strains usually ferment glucose to DL-lactic acid, whereas *S. intermedius* and *S. hyicus* strains produce only L-lactate. There are also differences as to the allosteric specificity of L-lactate dehydrogenase, the esterase pattern, and the catabolism of lactose or galactose. DNA-DNA hybridization and protein homology studies also demonstrated the heterogeneity of the coagulase-positive staphylococci. Less than 15% DNA homology (relative binding at supra-optimal temperatures) was detected between strains of *S. aureus* and strains of *S. intermedius* or *S. hyicus*. Antisera against catalase of *S. aureus* ATCC 12600 exhibited identical reaction in the double immunodiffusion test with crude extracts of other strains of *S. aureus* biotypes A–D, but revealed only partial identity with crude extracts of *S. intermedius* or *S. hyicus* strains. This was also confirmed by microcomplement fixation tests. Members of the species *S. aureus* showed low immunological distance values, whereas members of the species *S. intermedius* and *S. hyicus* produced high

immunological distance values. *S. intermedius* and the coagulase-positive strains of *S. hyicus* can also be separated from *S. aureus* by their inability to anaerobically ferment mannitol and to produce acid from maltose and acetylmethylcarbinol. Many strains of *S. intermedius* and *S. hyicus* also lack clumping factor activity in the slide coagulase test, though canine strains of *S. intermedius* are usually clumping factor positive and protein A negative (Blobel and co-workers, unpublished studies).

The Differentiation of Coagulase-Negative Species

There are at least 10 coagulase-negative *Staphylococcus* species currently recognized. They may be distinguished from one another on the basis of various morphological, physiological, and biochemical characteristics as well as on DNA nucleotide sequence relationships determined by DNA-DNA hybridization. Coagulase-negative staphylococci can be resolved at several different taxonomic levels. For example, with rare exception, *Staphylococcus xylosus, S. cohnii, S. saprophyticus,* and *S. sciuri* can be separated from *S. epidermidis, S. hominis, S. haemolyticus, S. warneri, S. capitis,* and *S. simulans* on the basis of novobiocin resistance (Kloos and Schleifer, 1975a,b; Schleifer and Kloos, 1975; Kloos, Schleifer, and Noble, 1976; Kloos, Schleifer, and Smith, 1976). Furthermore, the novobiocin-resistant species can often be distinguished from novobiocin-susceptible species by the absence or small numbers of amino acid requirements for growth (Emmett and Kloos, 1975, 1979).

SPECIES GROUPS.   Further subdivision can be made into four species groups based on major phenotypic differences as well as some new evidence on DNA relationships (Table 2).

The *Staphylococcus sciuri* species group is presently composed of one species and two subspecies, *S. sciuri* subsp. *sciuri* and *S. sciuri* subsp. *lentus* (Kloos, Schleifer, and Smith, 1976). Too few strains of the latter subspecies have been studied in detail to be certain that the subspecies rank is most appropriate. DNA reassociation experiments between the two subspecies have not yet been reported. Nevertheless, this species group can be distinguished from other staphylococci on the basis of a unique cell wall peptidoglycan type, L-Lys-L-Ala-Gly$_4$, production of acid from a wide variety of carbohydrates, colony morphology, and a combination of several other properties. The relative binding of DNA of representative strains of various species to *S. sciuri* strain SC 116 DNA at an optimal criterion of 60°C was very low and in the range of 15–21% (Schleifer, Meyer, and Rupprecht, 1979).

The *Staphylococcus saprophyticus* species group

Table 1. Differentiation of coagulase-positive staphylococci.[a]

Characters	S. aureus	S. intermedius	Coagulase-positive strains related to S. hyicus[b]	Reference
Biotypes	A, B, C, D	E, F	—	Hajek and Marsalek, 1971b
Peptidoglycan type	L-Lys-Gly$_{5-6}$	L-Lys-Gly$_{4-5}$, L-Ser	L-Lys-Gly$_{5-6}$	Schleifer et al., 1976; Devriese et al., 1978
Teichoic acid	Ribitol	Glycerol	Glycerol	
Lactic acid	D, L	L	L	
Lactate dehydrogenase	D- and L-LDH	FDP-dependent L-LDH	ND	Schleifer et al., 1976
Esterase pattern	Simple	Complex	ND	Schleifer et al., 1978
Lactose degradation	Via tagatose-6-P	Normal	ND	
DNA-DNA homology values with S. aureus	82–100%	10–15%	10–15%	Schleifer, 1978; Devriese et al., 1978
Reaction with antiserum against catalase of S. aureus				
Double immunodiffusion	Identity	Partial identity	Partial identity	M. Rupprecht and K. H. Schleifer, unpublished results
Immunological distance units	5–8	60–70	40–50	
Coagulase (rabbit plasma)	+	+	+	Hajek, 1976; Devriese et al., 1978
Heat-stable DNase	+	+	+	
Mannitol aerobically	+	+	–	
Mannitol anaerobically	+	–	–	
Maltose aerobically	+	–	–	
Clumping factor	+	– (+)	–	
Acetylmethylcarbinol production	+	–	–	

[a] ND, not determined; +, species possesses the character; (+), some strains, particularly those isolated from dogs, possess the character; –, species does not show evidence of the character.

[b] The species S. hyicus is composed of at least two recognizable subspecies. S. hyicus subsp. hyicus contains both coagulase-positive and coagulase-negative strains, whereas the other subspecies to be proposed by Devriese et al. (1978) contains only coagulase-negative strains.

Table 2. Differentiation of coagulase-negative *Staphylococcus* species groups.[a]

Characters	S. sciuri	S. saprophyticus	S. epidermidis	S. simulans	Reference
Peptidoglycan type	L-Lys-L-Ala-Gly$_4$	L-Lys-Gly$_{5-6}$ L-Lys-Gly$_{4-5}$, L-Ser	L-Lys-Gly$_{3-5}$, L-Ser	L-Lys-Gly$_{5-6}$	Schleifer and Kocur, 1973 Kloos and Schleifer, 1975a Schleifer and Kloos, 1975 Kloos, Schleifer, and Smith, 1976
Teichoic acid	Glycerol	Glycerol, Glycerol + Ribitol	Glycerol	Glycerol	Kloos and Schleifer, 1975a Schleifer and Kloos, 1975 Kloos, Schleifer, and Noble, 1976 Kloos, Zimmerman, and Smith, 1976
Novobiocin (MIC $\geq$ 1.6 μg/ml) resistance	+	+	−	−	
Lysostaphin (MIC $\geq$ 100 μg/ml) resistance	+>−	−>+	+	−	
Acid aerobically from:					
D-(+)-Cellobiose	+	−	−	−	
L-(+)-Arabinose	+,−	+,−	−	−	
D-(+)-Melezitose	−,+	−	−,+	−	
Xylitol	−	+,±,−	−	−	
Acetylmethylcarbinol production	−,±	+,±,−	+,±,−	−>±	
Growth at 15°C	+>±,−	+	−,+	+>±	
Growth on 15% NaCl agar	±>+,−	±,+>−	−,±>+	+>−	
Amino acid requirements	0–1	0–3	5–12	4–7	Emmett and Kloos, 1975, 1979
Surface "motility" on soft (0.4–0.75%) agar	±	+,−	−	−	W. E. Kloos and others, unpublished results

[a] Symbols for characters (unless noted otherwise): +, positive; ±, weak; −, negative.

is presently composed of *S. saprophyticus, S. xylosus,* and *S. cohnii* (Schleifer and Kloos, 1975; Kloos, Schleifer, and Noble, 1976). This species group can be distinguished from other staphylococci based on the combined characters of novobiocin resistance, colony morphology, cell wall composition, carbohydrate reactions, and growth properties at different salt concentrations and at different temperatures. Preliminary DNA reassociation reactions performed at 60°C have indicated a lower relative binding between species of the *S. saprophyticus* species group and other staphylococci (25–30%) than between species of this species group (42–55%) (Schleifer, Meyer, and Rupprecht, 1979; unpublished studies).

The *Staphylococcus epidermidis* species group is presently composed of *S. epidermidis, S. hominis, S. haemolyticus, S. warneri,* and *S. capitis* (Kloos and Schleifer, 1975a; Kloos, Schleifer, and Noble, 1976). This species group can be distinguished from other staphylococci on the basis of the combined characters of novobiocin susceptibility (only rare novobiocin-resistant strains have been encountered), higher L-serine content in cell wall peptidoglycan, colony morphology, higher resistance to lysostaphin, carbohydrate reactions, amino acid requirements, and growth properties at different salt concentrations and at different temperatures. Preliminary DNA reassociation reactions performed at 55 and 60°C have generally indicated a lower relative binding between species of the *S. epidermidis* species group and most other staphylococci (20–30%) than between species of this species group (35–61%) (W. E. Kloos, unpublished studies; Schleifer, Meyer, and Rupprecht, 1979).

The *Staphylococcus simulans* species group presently contains the single species *S. simulans.* It can be distinguished from other staphylococci primarily on the basis of a combination of characters including susceptibility to novobiocin, carbohydrate reactions, cell wall composition, colony morphology, and absence of acetylmethylcarbinol (acetoin) production (Kloos and Schleifer, 1975a; Kloos, Schleifer, and Noble, 1976). Preliminary DNA reassociation reactions performed at 60°C between *S. simulans* strain ATCC 27848 and other species showed low relative binding in a range of 23–30% (Schleifer, Meyer, and Rupprecht, 1979).

SPECIES AND SUBSPECIES.   Coagulase-negative staphylococci can also be classified at the species and, in certain instances, at the subspecies levels (Table 3). *S. sciuri* subsp. *sciuri* can be readily identified by its ability to produce acid, aerobically, from D-(+)-fucose and D-(+)-cellobiose within 24–48 h, novobiocin resistance, and distinctive colony morphology on P agar (see this Handbook, Chapter 124, for medium composition) and purple agar base (Difco) supplemented with 1% (wt/vol)

maltose (Kloos, Schleifer, and Smith, 1976). On the maltose agar medium, colonies are very large (>8 mm in diameter by 4 days' incubation at 34–36°C), slightly raised, and translucent to nearly transparent; acid is usually only weakly produced on this medium.

The tentative subspecies *S. sciuri* subsp. *lentus* can be readily identified by its ability to produce acid, aerobically, from D-(+)-cellobiose and usually from D-(+)-fucose and raffinose, slow growth, and distinctive colony morphology on P agar (Kloos, Schleifer, and Smith, 1976; Kloos, Zimmerman, and Smith, 1976). Colonies are small and often glistening and slimy; cells are often arranged in tetrads.

S. xylosus can be readily identified by its ability to produce acid, aerobically, from D-(+)-xylose and usually L-(+)-arabinose, novobiocin resistance, distinctive colony morphology (in common rough strains) on P agar, and usually positive nitrate reduction and phosphatase activity (Schleifer and Kloos, 1975). Based on preliminary DNA-DNA hybridization studies, this species appears to be more closely related to *S. saprophyticus* than other species (Schleifer, Meyer, and Rupprecht, 1979).

S. saprophyticus can be readily identified by its novobiocin resistance; ability to produce acid, aerobically, from sucrose and D-mannitol and usually xylitol, but not from D-(+)-xylose, L-(+)-arabinose, and D-(+)-mannose; and usually negative nitrate reduction and phosphatase activity (Schleifer and Kloos, 1975). *S. saprophyticus* contains a significantly larger percentage (12.1 ± 2.8%) of C_{16} fatty acids in the total cellular fatty acid fraction compared with other species (Durham and Kloos, 1978). *Staphylococcus cohnii* can be readily identified by its novobiocin resistance; inability to produce acid from sucrose, D-(+)-xylose, L-(+)-arabinose; usual ability to produce acid, aerobically, from D-(+)-mannose and D-mannitol; and usually negative nitrate reduction and phosphatase activity (Schleifer and Kloos, 1975).

S. epidermidis can be readily identified by its distinctive colony morphology on P agar; inability to produce acid from D-(+)-xylose, L-(+)-arabinose, D-mannitol, and D-(+)-trehalose; ability to produce acid, aerobically, from sucrose and usually D-(+)-mannose; novobiocin susceptibility; strong anaerobic growth in thioglycolate; and usually positive nitrate reduction and phosphatase activity (Kloos and Schleifer, 1975a; Schleifer and Kloos, 1975). Preliminary DNA-DNA hybridization studies suggest that *S. epidermidis* has diverged a considerable amount from other species or vice versa. Relative binding values at optimal criteria of 55 and 60°C were in a range of 29–54% between *S. epidermidis* and other species within the *S. epidermidis* species group. At the stringent criterion of 70°C, relative binding values dropped to 5–19%.

S. capitis can be readily identified by its distinc-

Table 3. Differentiation of coagulase-negative *Staphylococcus* species.[a]

Characters	S. sciuri subsp. sciuri	S. xylosus	S. sapro-phyticus	S. cohnii	S. epidermidis	S. capitis	S. hominis	S. haemo-lyticus	S. warneri	S. simulans
Peptidoglycan type	L-Lys-L-Ala-Gly₄	L-Lys-Gly₅₋₆	L-Lys-Gly₄₋₅, L-Ser	L-Lys-Gly₅₋₆	L-Lys-Gly₄₋₅, L-Ser	L-Lys-Gly₃₋₄, L-Ser	L-Lys-Gly₃₋₄, L-Ser	L-Lys-Gly₃₋₄, L-Ser	L-Lys-Gly₃₋₄, L-Ser	L-Lys-Gly₅₋₆
Teichoic acid components[b]	Gly GluN	Rib + Gly GluN	Rib + Gly GluN	Gly (Rib) (Glu) GluN	Gly Glu	Gly	Gly GluN	Gly GluN	Gly Glu	Gly GalN
Lactic acid (isomer)	L	L>D	DL	L,DL	L	L>D	DL,D	D	DL	DL,L
Esterases (elect, mobility)	Fast	Variable	Slow + mid	Variable	2 Mid	Slow, mid, fast	Slow + mid	Fast	Fast	Slow + 2 mid
Colony diameter (mm; 3 days 34°C + 2 days room temp)	7–11	5–10.5	5–8	5–9	2.5–4.5	1–3	3–4.5	4.5–8	3–5	5–7
Anaerobic growth (thioglycolate)[c]	±>+	Variable	+>±,±C	±>+,±C	+	+>±,±C	c>±,−	±C>+,±	+	+
Novobiocin (MIC ≥ 1.6 µg/ml) resistance	+[d]	+	+	+	−	−	−	−	−	−
Acid, aerobically from:										
D-(+)-Cellobiose or fucose	+	−	−	−	−	−	−	−	−	−
D-(+)-Xylose/L-(+)-arabinose	ara +, −	+	−	−	−	−	−	−	−	−
Sucrose	+	+	+	+	+	+	+	+	+	+
Maltose	±>−	+>±	+>±	+>±	+	−	+	+>−	+>±	−
α-Lactose	−>±	+>±,−	+>−	−,+	+>−	−	+,−	+,−	+>−	+>−
D-(+)-Trehalose	+>±	+	+	+	−	−	+>±	−>±,+	+>±	+>−
D-(+)-Melezitose	+,−	−	−	+	−>±,+	−	+,−	−>±,+	−>±,+	+>−
D-Mannitol	+	+>±,−	+>−	+>±,−	−	+	+,−	+,−	±>−	+>−
Xylitol	−	−>+,±	+,±>−	−>±,+	−	−	−	−	−	+>−
Hemolysis (bovine blood)	−	−>±	−	−>±	−>±	+>±,−	−>±	+>±	−>±	+>±,−
Nitrate reduction	+>±	+>−	−	−>±	+>±,−	+>±,−	+>±,−	+>±	−>+,±	+,−
Phosphatase activity	+>±,−	+>±,−	−>±,+	−>±	+	−>±	−>±	−>±	−>+,±,+	±>−
Growth on 15% NaCl agar	±,+>−	±,+>−	±>+	±>+,−	−	−,±	−,−>+	±,−>+	±>+,−	±>−

[a] Reference for the data shown in the table include: Kloos and Schleifer, 1975a, 1975b; Schleifer and Kloos, 1975; Kloos, Schleifer, and Noble, 1976; Kloos, Schleifer, and Smith, 1976; Zimmerman and Kloos, 1976.

[b] Gly, glycerol; Rib, ribitol; GluN, glucosamine; Glu, glucose; GalN, galactosamine.

[c] Symbols for anaerobic growth: +, dense uniform; ±, gradient of growth from dense to light down tube; ±C, large individual colonies to; c, small individual colonies to absence of visible growth.

[d] Symbols for characters (unless noted otherwise): +, positive; ±, weak; −, negative.

tive colony morphology on P agar and maltose agar media; inability to produce acid from D-(+)-xylose, L-(+)-arabinose, D-(+)-trehalose, α-lactose, maltose, and D-(+)-melezitose; usual ability to produce acid, aerobically, from D-(+)-mannose and D-mannitol; and novobiocin susceptibility (Kloos and Schleifer, 1975a). Preliminary DNA-DNA hybridization studies suggest that S. capitis has also diverged considerably from other species. Relative binding values at 55 and 60°C were in the range of 33–54% between S. capitis and other species of the S. epidermidis species group. At 70°C, relative binding values dropped to 4–15%.

The species S. hominis, S. haemolyticus, and S. warneri are not as easily distinguished from one another by simple, expedient tests as other species are. S. hominis can be identified by its colony morphology and pigment patterns on P agar and maltose agar media (though this requires some experience); ability to produce acid, aerobically, from β-D-(−)-fructose, maltose, and sucrose, and usually D-(+)-trehalose, D-(+)-turanose, and D-(+)-melezitose, but usually not from D-(+)-mannose and D-mannitol and not from D-(+)-xylose and L-(+)-arabinose; novobiocin susceptibility; usually poor or no detectable anaerobic growth in thioglycolate; weak or no hemolysis of bovine blood; weak or negative phosphatase activity; weak or positive nitrate reduction; and usually no growth on 15% NaCl agar medium (Kloos and Schleifer, 1975a). The cells of this species are often arranged in tetrads. The esterases and catalases of S. hominis are electrophoretically distinct from those of other species (Zimmerman, 1976; Zimmerman and Kloos, 1976). Preliminary DNA-DNA hybridization studies suggest that S. hominis has diverged to a considerable extent from other species, although there is some indication that this species may be more closely related to S. haemolyticus than to other species. Relative binding values at 55 and 60°C between S. hominis and S. haemolyticus DNA were in the range of 42–60%; whereas, values with DNA from other species of the S. epidermidis species group were in the range of 31–48%. However, at 70°C, relative binding values were in the range of 6–19% and, at this criterion, no distinction could be made between the above two subgroups.

S. haemolyticus can be identified by its usually large colony size (compared to S. hominis and S. warneri), moderate hemolytic activity on bovine blood, and many discrete colonies produced in anaerobic portions of a semisolid thioglycolate medium; ability to produce acid, aerobically, from maltose, sucrose, and D-(+)-trehalose, but usually not from D-(+)-mannose or D-(+)-melezitose and not from D-(+)-xylose and L-(+)-arabinose; novobiocin susceptibility; and usually positive nitrate reduction and weak or negative phosphatase activity (Schleifer and Kloos, 1975). A significant percent-

age of strains (30–40%) do not produce acid from β-D-(−)-fructose. Preliminary DNA-DNA hybridization studies suggest that S. haemolyticus has diverged to a considerable extent from other species and in addition, some divergence can be detected between strains of S. haemolyticus isolated from humans and other primate hosts. Relative binding values at 55°C were in the range of 36–52% between S. haemolyticus and other species of the S. epidermidis species group. At 70°C, relative binding values dropped to 10–16%. Relative binding values at 55°C between human S. haemolyticus DNAs were in the range of 79–100%, whereas between human and nonhuman primate S. haemolyticus DNA, the values were lower, in the range of 65–77% (Kloos and Wolfshohl, 1979). Relative binding at 70°C was even more revealing, where the human-nonhuman primate values dropped to 28–51%. The thermal stability of DNA duplexes formed between human and nonhuman primate S. haemolyticus strains was also notably less than that found in DNA duplexes between human strains.

Human S. warneri strains can be identified by their colony morphology and pigment pattern on P agar and maltose agar media; usual inability to produce acid from D-(+)-mannose, D-(+)-galactose, α-lactose, and D-(+)-turanose; inability to produce acid from D-(+)-xylose and L-(+)-arabinose, but ability to produce acid, aerobically, from β-D-(−)-fructose, sucrose, and D-(+)-trehalose; novobiocin susceptibility; good anaerobic growth in thioglycolate; and usually weak or negative nitrate reduction and phosphatase activity (Kloos and Schleifer, 1975a). Human S. warneri strains can also be distinguished from other staphylococci by their high levels of C_{20} fatty acids (48.3 ± 2.1% of total cellular fatty acids) and unique C_{22} fatty acid component (Durham and Kloos, 1978). S. warneri and S. epidermidis can be resolved from other species on the basis of anaerobic growth stimulation by either pyruvate or uracil, individually (Evans, 1976). Simian S. warneri strains can be distinguished from human strains on the basis of colony morphology, delayed acid production from α-lactose and D-(+)-galactose, the failure to produce acid from D-(+)-ribose, and an increased percentage of strains producing acid from D-(+)-melezitose, but failing to produce acid from D-(+)-trehalose. Preliminary DNA-DNA hybridization studies suggest that S. warneri has diverged considerably from other species and, in addition, moderate divergence can be detected between the human and nonhuman primate subspecies. Relative binding values at 55 and 60°C were in the range of 35–60% between S. warneri and other species of the S. epidermidis species group. At 70°C, relative binding values dropped to 10–13%. Relative binding values at 55°C between DNA from different human S. warneri strains or between DNA from different simian S. warneri strains were in the range of 87–100%, whereas between human and simian *S. warneri* DNA, the values were significantly lower, in the range of 50–69%. Relative binding at 70°C between human and other primate *S. warneri* DNA dropped to 19–35%. Such relative binding values would suggest a borderline case between subspecies and separate species status of these two groups. Here we favor the former classification.

S. simulans can be identified by its colony morphology on P agar and maltose agar media; inability to produce acid or production of only weak acid, aerobically, from maltose and D-(+)-galactose; inability to produce acid from D-(+)-turanose, D-(+)-melezitose, D-(+)-xylose, and L-(+)-arabinose, but ability to produce acid, aerobically, from β-(−)-fructose, α-lactose, sucrose, and usually D-mannitol; novobiocin susceptibility; good anaerobic growth in thioglycolate; and usually weak or negative acetoin production and positive nitrate reduction (Kloos and Schleifer, 1975a). *S. simulans* showed to all other staphylococci tested rather low relative DNA binding values of 23–30% at 60°C.

Literature Cited

Abd-el-Malek, Y., Gibson, T. 1948. Studies on the bacteriology of milk. II. The staphylococci and micrococci of milk. Journal of Dairy Research **15**:249–260.

Adlam, C., Thorley, C. M., Ward, P. D., Collins, M. 1976. Staphylococcal mastitis in rabbits: Description of a natural outbreak of the disease and attempts to reproduce the disease in the laboratory, pp. 761–771. In: Jeljaszewicz, J. (ed.), Staphylococci and staphylococcal diseases. Stuttgart, New York: Gustav Fischer Verlag.

Anderson, J. C. 1976. The contribution of the mouse mastitis model to our understanding of staphylococcal infection, pp. 783–790. In: Jeljaszewicz, J. (ed.), Staphylococci and staphylococcal diseases. Stuttgart, New York: Gustav Fischer Verlag.

Andriole, V. T., Lyons, R. W. 1970. Coagulase-negative *Staphylococcus*. Annals of the New York Academy of Sciences **174**:533–544.

Association of Official Analytical Chemists 1970. Official methods of analysis, 11th ed, pp. 843–844. Washington, D.C.: Association of Official Analytical Chemists.

Association of Official Analytical Chemists 1974. Changes in official methods of analysis made at the eighty-eighth annual meeting, October 14–17, 1974. Journal of the Association of Official Analytical Chemists **58**:416–417.

Baer, E. F., Gray, R. J. H., Orth, D. S. 1976. Methods for the isolation and enumeration of *Staphylococcus aureus*, pp. 374–386. In: Speck, M. L. (ed.), Compendium of methods for the microbiological examination of foods. Washington, D.C.: American Public Health Association.

Bailey, R. R. 1973. Significance of coagulase negative staphylococci in urine. Journal of Infectious Diseases **127**:179–184.

Baird-Parker, A. C. 1962. The occurrence and enumeration, according to a new classification, of micrococci and staphylococci in bacon and on human and pig skin. Journal of Applied Bacteriology **25**:352–361.

Baird-Parker, A. C. 1963. A classification of micrococci and staphylococci based on physiological and biochemical tests. Journal of General Microbiology **30**:409–427.

Baird-Parker, A. C. 1965a. The classification of staphylococci and micrococci from world-wide sources. Journal of General Microbiology **38**:363–387.

Baird-Parker, A. C. 1965b. Staphylococci and their classification. Annals of the New York Academy of Sciences **128**:4–25.

Baird-Parker, A. C. 1974. The basis for the present classification of staphylococci and micrococci. Annals of the New York Academy of Sciences **236**:7–14.

Bergdoll, M. S. 1972. The enterotoxins, pp. 301–331. In: Cohen, J. O. (ed.), The staphylococci. New York, London, Sydney, Toronto: John Wiley & Sons.

Bergdoll, M. S., Bennett, R. W. 1976. Staphylococcal enterotoxins, pp. 387–416. In: Speck, M. L. (ed.), Compendium of methods for the microbiological examination of foods. Washington, D.C.: American Public Health Association.

Black, J. A., Challacombe, D. N., Ockenden, B. G. 1965. Nephrotic syndrome associated with bacteremia after shunt operations for hydrocephalus. Lancet **ii**:921–924.

Breckinridge, J. C., Bergdoll, M. S. 1971. Outbreak of food-borne gastroenteritis due to a coagulase-negative enterotoxin-producing staphylococcus. New England Journal of Medicine **284**:541–543.

Brewer, D. G., Martin, S. E., Ordal, Z. J. 1977. Beneficial effects of catalase or pyruvate in a most-probable-number technique for the detecting of *Staphylococcus aureus*. Applied and Environmental Microbiology **34**:797–800.

Brown, R. W., Sandvik, O., Scherer, R. K., Rose, D. L. 1967. Differentiation of strains of *Staphylococcus epidermidis*. Journal of General Microbiology **47**:273–287.

Brun, Y., Fleurette, J., Forey, F. 1977. Identification biochimique des staphylocoques coagulase negatifs par micromethode. In: Staphylocoques, pp. 21–22. Abstracts of the Meeting of the Société Française de Microbiologie, September 17–18, 1977, Lyon.

Buchanan, R. E., Gibbons, N. E. (eds.) 1974. Bergey's manual of determinative bacteriology, 8th ed. Baltimore: Williams & Wilkins.

Buhles, W. C. 1969. Airborne staphylococcic contamination in experimental procedures on laboratory animals. Laboratory Animal Care **19**:465–469.

Callaghan, R. P., Cohen, S. J., Stewart, G. T. 1961. Septicemia due to colonization of Spitz-Holter valves by staphylococci: Five cases treated with methicillin. British Medical Journal **i**:860–962.

Cameron, A. S. 1970. Staphylococcal epidemiology in Antarctica. Journal of Hygiene **68**:43–52.

Carr, D. L., Kloos, W. E. 1977. Temporal study of the staphylococci and micrococci of normal infant skin. Applied and Environmental Microbiology **34**:673–680.

Chapman, G. H. 1945. The significance of sodium chloride in studies of staphylococci. Journal of Bacteriology **50**:201–203.

Cheatum, E. L. 1941. Lymphadenitis in New York cottontails. Journal of Wildlife Management **5**:304–308.

Cluff, L. E., Reynolds, R. C., Page, D. L., Breckenridge, J. L. 1968. Staphylococcal bacteremia and altered host resistance. Annals of Internal Medicine **69**:859–873.

Cohen, J. O. (ed.) 1972. The staphylococci. New York, London, Sydney, Toronto: John Wiley & Sons.

Courter, R. D., Galton, M. M. 1962. Animal staphylococcal infections and their public health significance. American Journal of Public Health **52**:1818–1827.

Cowan, S. T. 1962. An introduction to chaos, or the classification of staphylococci and micrococci. Journal of Applied Bacteriology **25**:324–340.

Crisley, F. D., Angelotti, R., Foter, M. J. 1964. Multiplication of *Staphylococcus aureus* in synthetic cream fillings and pies. Public Health Report **79**:369–376.

Devriese, L. A. 1977. Isolation and identification of *Staphylococcus hyicus*. American Journal of Veterinary Research **38**:787–792.

Devriese, L. A., Oeding, P. 1975. Coagulase and heat-resistant nuclease producing *Staphylococcus epidermidis* strains from animals. Journal of Applied Bacteriology **39**:197–207.

Devriese, L. A., Oeding, P. 1976. Characteristics of *Staphylococcus aureus* strains isolated from different animal species. Research in Veterinary Science **21**:284–291.

Devriese, L. A., Devos, A. H., Beumer, J., Maes, R. 1972. Characterization of staphylococci isolated from poultry. Poultry Science **51**:389–397.

Devriese, L. A., Hajek, V., Oeding, P., Meyer, S. A., Schleifer, K. H. 1978. *Staphylococcus hyicus* (Sompolinsky 1953) comb. nov. and *Staphylococcus hyicus* subsp. *chromogenes* subsp. nov. International Journal of Systematic Bacteriology **28**:482–490.

Dobrin, R. S., Day, N., Michael, A., Vernier, R. L., Fish, A. J., Quie, P. G. 1976. Studies of the immune response and renal injury associated with chronic coagulase-negative staphylococcal bacteremia, pp. 137–140. In: Jeljaszewicz, J. (ed.), Staphylococci and staphylococcal infections. Stuttgart, New York: Gustav Fischer Verlag.

Durham, D. R., Kloos, W. E. 1978. A comparative study of the total cellular fatty acids of *Staphylococcus* species of human origin. International Journal of Systematic Bacteriology **28**:223–228.

Elek, S. D. 1959. Staphylococcus pyogenes and its relation to disease. Edinburgh, London: E. & S. Livingstone Ltd.

Emmett, M., Kloos, W. E. 1975. Amino acid requirements of staphylococci isolated from human skin. Canadian Journal of Microbiology **21**:729–733.

Emmett, M., Kloos, W. E. 1979. The nature of arginine auxotrophy in cutaneous populations of staphylococci. Journal of General Microbiology **110**:305–314.

Evans, C. A., Mattern, K. L., Hallam, S. L. 1978. Isolation and identification of *Peptococcus saccharolyticus* from human skin. Journal of Clinical Microbiology **7**:261–264.

Evans, J. B. 1976. Anaerobic growth of *Staphylococcus* species from human skin: Effects of uracil and pyruvate. International Journal of Systematic Bacteriology **26**:17–21.

Evans, J. B. 1977. Coagulase positive staphylococci as indicators of potential health hazards from water, pp. 126–130. In: Hoadley, A. W., Dutka, B. J. (eds.), Bacterial indicators and health hazards associated with water. Philadelphia: American Society for Testing and Materials.

Fleurette, J., Brun, Y. 1977. Pouvoir pathogene des staphylocoques coagulase negatifs. In: Staphylocoques, pp. 13–14. Abstracts of the Meeting of the Société Française de Microbiologie, September 17–18, 1977, Lyon.

Flynn, R. J. 1959. Studies on the aetiology of ringtail of rats. Proceedings of the Animal Care Panel **9**:155–160.

Fokes, E. C. 1970. Occult infections of ventriculoatrial shunts. Journal of Neurology **33**:517–523.

Food and Drug Administration. 1976. FDA bacteriological analytical manual for foods. Washington, D.C.: Association of Official Analytical Chemists.

Geraci, J. E., Hanson, K. C., Giuliani, E. R. 1968. Endocarditis caused by coagulase-negative staphylococci. Mayo Clinic Proceedings **43**:420–434.

Hajek, V. 1976. *Staphylococcus intermedius*, a new species isolated from animals. International Journal of Systematic Bacteriology **26**:401–408.

Hajek, V., Marsalek, E. 1969a. A study of staphylococci of bovine origin. *Staphylococcus aureus* var. *bovis*. Zentralblatt für Bakteriologie, Parasitenkunde, Infektionskrankheiten und Hygiene, Abt. 1 Orig. **209**:154–160.

Hajek, V., Marsalek, E. 1969b. A study of staphylococci isolated from the upper respiratory tract of different animal species. I. Biological properties of *Staphylococcus aureus* strains of canine origin. Zentralblatt für Bakteriologie, Parasitenkunde, Infektionskrankheiten und Hygiene, Abt. 1 Orig. **212**:60–67.

Hajek, V., Marsalek, E. 1969c. A study of staphylococci isolated from the upper respiratory tract of different animal species. II.

Biochemical properties of *Staphylococcus aureus* strains of pigeon origin. Zentralblatt für Bakteriologie, Parasitenkunde, Infektionskrankheiten und Hygiene, Abt. 1 Orig. **212**:67–73.

Hajek, V., Marsalek, E. 1970. A study of staphylococci isolated from the upper respiratory tract of different animal species. III. Physiological properties of *Staphylococcus aureus* strains of porcine origin. Zentralblatt für Bakteriologie, Parasitenkunde, Infektionskrankheiten und Hygiene, Abt. 1 Orig. **214**:68–74.

Hajek, V., Marsalek, E. 1971a. A study of staphylococci isolated from the upper respiratory tract of different animal species. IV. Physiological properties of *Staphylococcus aureus* strains of hare origin. Zentralblatt für Bakteriologie, Parasitenkunde, Infektionskrankheiten und Hygiene, Abt. 1 Orig. **216**:168–174.

Hajek, V., Marsalek, E. 1971b. The differentiation of pathogenic staphylococci and a suggestion for their taxonomic classification. Zentralblatt für Bakteriologie, Parasitenkunde, Infektionskrankheiten und Hygiene, Abt. 1 Orig., Reihe A **217**:176–182.

Hajek, V., Marsalek, E. 1976. Evaluation of classificatory criteria for staphylococci, pp. 11–21. In: Jeljaszewicz, J. (ed.), Staphylococci and staphylococcal diseases. Stuttgart, New York: Gustav Fischer Verlag.

Hajek, V., Marsalek, E., Harna, V. 1974. A study of staphylococci isolated from the upper respiratory tract of different animal species. VI. Physiological properties of *Staphylococcus aureus* strains from horses. Zentralblatt für Bakteriologie, Parasitenkunde, Infektionskrankheiten und Hygiene, Abt. 1 Orig., Reihe A **229**:429–435.

Hajek, V., Marsalek, E., Hubacek, J. 1972. A study of staphylococci isolated from the upper respiratory tract of different animal species. V. Physiological properties of *Staphylococcus aureus* strains from mink. Zentralblatt für Bakteriologie, Parasitenkunde, Infektionskrankheiten und Hygiene, Abt. 1 Orig., Reihe A **222**:194–199.

Hambraeus, A., Ransjo, V. 1976. Clothes-born transmission of staphylococci—attempts to control *Staphylococcus aureus* infections in a burns unit, pp. 981–987. In: Jeljaszewicz, J. (ed.), Staphylococci and staphylococcal diseases. Stuttgart, New York: Gustav Fischer Verlag.

Hard, G. C. 1966. Staphylococcal infection of the tail of the laboratory rat. Laboratory Animal Care **16**:421–429.

Hare, R. 1962. Dispersal of staphylococci, pp. 75–86. In: Williams, R. E. O., Shooter, R. A. (eds.), Infection in hospitals. Philadelphia: F. A. Davis.

Heidelbaugh, N. D., Rowley, D. B., Powers, E. M., Bourland, C. T., McQueen, J. L. 1973. Microbiological testing of skylab foods. Applied Microbiology **25**:55–61.

Heidrich, H. J., Renk, W. 1967. Diseases of the mammary glands of domestic animals (translated by L. W. Van Den Heever). Philadelphia: W. B. Saunders.

Holbrook, R., Anderson, J. M., Baird-Parker, A. C. 1969. The performance of a stable version of Baird-Parker's medium for isolating *Staphylococcus aureus*. Journal of Applied Bacteriology **32**:187–192.

Holmberg, O. 1973. *Staphylococcus epidermidis* isolated from bovine milk: Biochemical properties, phage sensitivity, and pathogenicity for the udder. Acta Veterinaria Scandinavica, Suppl. **45**:1–144.

Holt, R. J. 1971. The colonization of ventriculo-atrial shunts by coagulase-negative staphylococci, pp. 81–87. In: Finland, M., Marget, W., Bartmann, K. (eds.), Bayer-Symposium. III. Bacterial infections: Changes in their causative agents. Trends and possible basis. New York, Heidelberg, Berlin: Springer-Verlag.

Holt, R. J. 1972. The pathogenic role of coagulase negative staphylococci. British Journal of Dermatology, Suppl. 8 **86**:42–49.

Isenberg, H. D., Washington, J. A., Balows, A., Sonnenwirth, A. C. 1974. Collection, handling, and processing of specimens, pp. 59–88. In: Lennette, E. H., Spaulding, E. H., Truant, J. P. (eds.), Manual of clinical microbiology, 2nd ed. Washington, D.C.: American Society for Microbiology.

Ivler, D. 1974. *Staphylococcus*, pp. 91–95. In: Lennette, E. H., Spaulding, E. H., Truant, J. P. (eds.), Manual of clinical microbiology, 2nd ed. Washington, D.C.: American Society for Microbiology.

Jakubicz, P., Borowski, J. 1976. *Staphylococcus epidermidis* and micrococci as an etiological agent in urinary tract infection, pp. 141–143. In: Jeljaszewicz, J. (ed.), Staphylococci and staphylococcal diseases. Stuttgart, New York: Gustav Fischer Verlag.

Jeljaszewicz, J. (ed.). 1976. Staphylococci and staphylococcal diseases. Stuttgart, New York: Gustav Fischer Verlag.

Jones, D., Deibel, R. H., Niven, C. F. 1963. Identity of *Staphylococcus epidermidis*. Journal of Bacteriology **85:**62–67.

Kay, C. R. 1963. Staphylococcal nasal carriage in the family. Journal of the College of General Practitioners **6:**47–50.

Kilpper, R., Schleifer, K. H. 1978. *Peptococcus saccharolyticus:* An anaerobic staphylococcus? Abstracts of the XII International Congress of Microbiology **12:**86.

Kitchell, A. G. 1962. Micrococci and coagulase negative staphylococci in cured meats and meat products. Journal of Applied Bacteriology **25:**416–431.

Kligman, A. M. 1975. The bacteriology of normal skin, pp. 13–31. In: Maibach, H. I., Hildick-Smith, G. (eds.), Skin bacteria and their role in infection. New York, Sydney, Toronto, London: McGraw-Hill.

Kloos, W. E., Musselwhite, M. S. 1975. Distribution and persistence of *Staphylococcus* and *Micrococcus* species and other aerobic bacteria on human skin. Applied Microbiology **30:**381–395.

Kloos, W. E., Musselwhite, M. S., Zimmerman, R. J. 1976. A comparison of the distribution of *Staphylococcus* species on human and animal skin, pp. 967–973. In: Jeljaszewicz, J. (ed.), Staphylococci and staphylococcal diseases.Stuttgart, New York: Gustav Fisher Verlag.

Kloos, W. E., Schleifer, K. H. 1975a. Isolation and characterization of staphylococci from human skin. II. Descriptions of four new species: *Staphylococcus warneri, Staphylococcus capitis, Staphylococcus hominis,* and *Staphylococcus simulans.* International Journal of Systematic Bacteriology **25:**62–79.

Kloos, W. E., Schleifer, K. H. 1975b. Simplified scheme for routine identification of human *Staphylococcus* species. Journal of Clinical Microbiology **1:**82–88.

Kloos, W. E., Schleifer, K. H., Noble, W. C. 1976. Estimation of character parameters in coagulase-negative *Staphylococcus* species, pp. 23–41. In: Jeljaszewicz, J. (ed.), Staphylococci and staphylococcal diseases. Stuttgart, New York: Gustav Fischer Verlag.

Kloos, W. E., Schleifer, K. H., Smith, R. F. 1976. Characterization of *Staphylococcus sciuri* sp. nov. and its subspecies. International Journal of Systematic Bacteriology **26:**22–37.

Kloos, W. E., Wolfshohl, J. F. 1979. Evidence of deoxyribonucleotide sequence divergence between staphylococci living on human and other primate skin. Current Microbiology **3:**167–173.

Kloos, W. E., Zimmerman, R. J., Smith, R. F. 1976. Preliminary studies on the characterization and distribution of *Staphylococcus* and *Micrococcus* species on animal skin. Applied and Environmental Microbiology **31:**53–59.

Lacey, R. W. 1975. Antibiotic resistance plasmids of *Staphylococcus aureus* and their clinical importance. Bacteriological Reviews **39:**1–32.

Lam, C. N., McNeish, A. S., Gibson, A. A. M. 1969. Nephrotic syndrome associated with complement deficiency and *Staphylococcus albus* bacteremia. Scotland Medical Journal **14:**86–88.

L'Ecuyer, C. 1966. Exudative epidermitis in pigs. Clinical stud-ies and preliminary transmission trials. Canadian Journal of Comparative Medicine and Veterinary Science **30:**9–16.

L'Ecuyer, C. 1967. Exudative epidermitis in pigs. Bacteriological studies on the causative agent. *Staphylococcus hyicus.* Canadian Journal of Comparative Medicine and Veterinary Science **31:**243–247.

Lee, C. S., Frost, A. J. 1970. Mastitis in slaughtered cows. I. Udder infection. Australian Veterinary Journal **46:**201–203.

Leininger, H. V. 1976. Equipment, media, reagents, routine tests and strains, pp. 10–94. In: Speck, M. L. (ed.), Compendium of methods for the microbiological examination of foods. Washington, D.C.: American Public Health Association.

Live, I. 1972. Staphylococci in animals: Differentiation and relationship to human staphylococcosis, pp. 443–456. In: Cohen, J. O. (ed.), The staphylococci. New York, London, Sydney, Toronto: John Wiley & Sons.

Lovell, D. L. 1945. Skin bacteria. Their location with reference to skin sterilization. Surgery, Gynecology, and Obstetrics **80:**174–177.

Mabeck, C. E. 1969. Significance of coagulase-negative staphylococcal bacteriuria. Lancet **ii:**1150–1152.

McCoy, R. H., Steenbergen F. 1969. Staphylococcus epizootic in Western Oregon cottontails. Bulletin of the Wildlife Disease Association **5:**11.

Marples, M. J. 1965. The ecology of the human skin. Springfield: Charles C Thomas.

Meers, P. D., Whyte, W., Sandys, G. 1975. Coagulase-negative staphylococci and micrococci in urinary tract infections. Journal of Clinical Pathology **28:**270–273.

Meyer, S. A., Schleifer, K. H. 1978. Deoxyribonucleic acid reassociation in the classification of coagulase-positive staphylococci. Archives of Microbiology **117:**183–188.

Minor, T. E., Marth, E. H. 1972. *Staphylococcus aureus* and staphylococcal food intoxications. A review. IV. Staphylococci in meat, bakery products, and other foods. Journal of Milk and Food Technology **35:**228–241.

Mitchell, R. G. 1968. Classification of *Staphylococcus albus* strains isolated from the urinary tract. Journal of Clinical Pathology **21:**93–96.

Moeller, R. W., Smith, I. M., Shoemaker, A. C., Tjalma, R. A. 1963. Transfer of hospital staphylococci from man to farm animals. Journal of the American Veterinary Medical Association **142:**613–617.

Moncrieff, M. W., Glasgow, E. F., Arthur, L. J. H., Hargreaves, H. M. 1973. Glomerulonephritis associated with *Staphylococcus albus* in a Spitz-Holter valve. Archives of Disease in Childhood **48:**69–71.

Montes, L. F., Wilborn, W. H. 1969. Location of bacterial skin flora. British Journal of Dermatology, Suppl. 1 **81:**23–26.

Mortensen, N. 1969. Studies in urinary infections. III. Biochemical characteristics of coagulase-negative staphylococci associated with urinary tract infections. Acta Medica Scandinavica **186:**47–51.

Nahmias, A. J., Shulman, J. A. 1972. Epidemiologic aspects and control methods, pp. 483–502. In: Cohen, J. O. (ed.), The staphylococci. New York, London, Sydney, Toronto: John Wiley & Sons.

Noble, W. C. 1969. Skin carriage of the *Micrococcaceae.* Journal of Clinical Pathology **22:**249–253.

Noble, W. C., Lidwell, O. M., Kingston, D. 1963. The size distribution of airborne particles carrying micro-organisms. Journal of Hygiene **61:**385–391.

Noble, W. C., Somerville, D. A. 1974. Microbiology of human skin. London, Philadelphia, Toronto: W. B. Saunders.

Noble, W. C., Valkenburg, H. A., Wolters, C. H. L. 1967. Carriage of *Staphylococcus aureus* in random samples of a normal population. Journal of Hygiene **65:**567–573.

Noble, W. C., Williams, R. E. O., Jevons, M. P., Shooter, R. A. 1964. Some aspects of nasal carriage of staphylococci. Journal of Clinical Pathology **17:**79–83.

Nord, C.-E., Holta-Oie, S., Ljungh, A., Wadstrom, T. 1976.

Characterization of coagulase-negative staphylococcal species from human infections. In: Jeljaszewicz, J. (ed.), Staphylococci and staphylococcal diseases. Stuttgart, New York: Gustav Fischer Verlag.

Oeding, P. 1973. Wall teichoic acids in animal *Staphylococcus aureus* strains determined by precipitation. Acta Pathologica et Microbiologica Scandinavica, Sect. B **81**:327–336.

Oeding, P., Digranes, A. 1976. *Staphylococcus saprophyticus:* Classification and infections, pp. 113–117. In: Jeljaszewicz, J. (ed.), Staphylococci and staphylococcal diseases. Stuttgart, New York: Gustav Fischer Verlag.

Oeding, P., Hajek, V., Marsalek, E. 1973. A comparison of antigenic structure and phage pattern with biochemical properties of *Staphylococcus aureus* strains isolated from hares and mink. Acta Pathologica et Microbiologica Scandinavica, Sec. B **81**:567–570.

Oeding, P., Hajek, V., Marsalek, E. 1974. A comparison of antigenic structure and phage pattern with biochemical properties of *Staphylococcus aureus* strains isolated from horses. Acta Pathologica et Microbiologica Scandinavica, Sect. B **82**:899–903.

Oeding, P., Hajek, V., Marsalek, E. 1976. A comparison of antigenic structure and phage pattern with biochemical properties of *Staphylococcus aureus* strains isolated from sheep. Acta Pathologica et Microbiologica Scandinavica, Sect. B **84**:61–65.

Oeding, P., Marandon, J. L., Hajek, V., Marsalek, E. 1970. Comparison of antigenic structure and phage pattern with biochemical properties of *Staphylococcus aureus* strains isolated from dogs and pigeons. Acta Pathologica et Microbiologica Scandinavica, Sect. B **78**:414–420.

Oeding, P., Marandon, J. L., Hajek, V., Marsalek, E. 1971. A comparison of phage pattern and antigenic structure with biochemical properties of *Staphylococcus aureus* strains isolated from cattle. Acta Pathologica et Microbiologica Scandinavica, Sect. B **79**:357–364.

Oeding, P., Marandon, J. L., Meyer, W., Hajek, V. Marsalek, E. 1972. A comparison of phage pattern and antigenic structure with biochemical properties of *Staphylococcus aureus* strains isolated from swine. Acta Pathologica et Microbiologica Scandinavica, Sect. B **80**:525–533.

Ogston, A. 1881. Report on micro-organisms in surgical diseases. British Medical Journal **i**:369–375.

Parisi, J. T. 1973. *Staphylococcus epidermidis,* an emerging pathogen. Missouri Medicine **70**:243–244.

Parisi, J. T., Baldwin, J. N. 1963. The incidence and persistence of certain strains of *Staphylococcus aureus* in dairy herds. American Journal of Veterinary Research **24**:551–556.

Patterson, J. T. 1966. Characteristics of staphylococci and micrococci isolated in a bacon curing factory. Journal of Applied Bacteriology **29**:461–469.

Perry, J. J. 1969. Isolation of *Staphylococcus epidermidis* from tobacco. Applied Microbiology **17**:647.

Pulverer, G., Halswick, R. 1967. Coagulase-negative staphylokokken *(Staphylococcus albus)* als Krankheitserreger. Deutsche Medizinische Wochenschrift **92**:1141–1145.

Pulverer, G., Pillich, J. 1971. Pathogenic significance of coagulase-negative staphylococci, pp. 91–96. In: Finland, M., Marget, W., Bartmann, K. (eds.), Bayer-Symposium. III. Bacterial infections: Changes in their causative agents. Trends and possible basis. New York, Heidelberg, Berlin: Springer-Verlag.

Quinn, E. L., Cox, F., Drake, E. H. 1966. Staphylococcal endocarditis: Disease of increasing importance. Journal of the American Medical Association **196**:815–818.

Quinn, E. L., Cox, F., Fisher, M. 1965. The problem of associating coagulase-negative staphylococci with disease. Annals of the New York Academy of Sciences **128**:428–442.

Rames, L., Wise, B., Goodman, J. R., Piel, C. F. 1970. Renal disease with *Staphylococcus albus* bacteremia: A complication of ventriculoatrial shunts. Journal of the American Medical Association **212**:1671–1677.

Roberts, A. P. 1967. Micrococcaceae from the urinary tract in pregnancy. Journal of Clinical Pathology **20**:631–632.

Roodyn, L. 1960. Epidemiology of staphylococcal infections. Journal of Hygiene **58**:1–10.

Rose, D. L., McDonald, J. S. 1973. Isolation and host range studies of *Staphylococcus epidermidis* and *Micrococcus* spp. bacteriophage. American Journal of Veterinary Research **34**:125–128.

Rountree, P. M., Freeman, B. H., Johnston, K. G. 1956. Nasal carriage of *Staphylococcus aureus* by various domestic and laboratory animals. Journal of Pathology and Bacteriology **72**:319–321.

Schleifer, K. H., Hartinger, A., Gotz, F. 1978. Occurrence of D-tagatose-6-phosphate pathway of D-galactose metabolism among staphylococci. FEMS Microbiology Letters **3**: 9–11.

Schleifer, K. H., Kloos, W. E. 1975. Isolation and characterization of staphylococci from human skin. I. Amended descriptions of *Staphylococcus epidermidis* and *Staphylococcus saprophyticus* and descriptions of these new species: *Staphylococcus cohnii, Staphylococcus haemolyticus,* and *Staphylococcus xylosus.* International Journal of Systematic Bacteriology **25**:50–61.

Schleifer, K. H., Kocur, M. 1973. Classification of staphylococci based on chemical and biochemical properties. Archiv für Mikrobiologie **93**:65–85.

Schleifer, K. H., Meyer, S. A., Rupprecht, M. 1979. Relatedness among coagulase-negative staphylococci: Deoxyribonucleic acid reassociation and comparative immunological studies. Archives of Microbiology **122**:93–101.

Schleifer, K. H., Schumacher-Perdreau, F., Gotz, F., Popp, B. 1976. Chemical and biochemical studies for the differentiation of coagulase-positive staphylococci. Archives of Microbiology **110**:263–270.

Sellin, M. A., Cooke, D. I., Gillespie, W. A., Sylvester, D. G. H., Anderson, J. D. 1975. Micrococcal urinary tract infections in young women. Lancet **ii**:570–572.

Sharpe, M. E., Neave, F. K., Reiter, B. 1962. Staphylococci and micrococci associated with dairying. Journal of Applied Bacteriology **25**:403–415.

Sinell, H. J., Baumgart, J. 1966. Selektionahrboden zur isolierung von staphylokokken aus lebensmitteln. Zentralblatt für Bakteriologie, Parasitenkunde, Infektionskrankheiten und Hygiene, Abt. 1 Orig. **197**:447–461.

Skaggs, J. W., Nicol, P. K. 1961. Role of domestic animals in staphylococcal disease. Biennial Conference of State and Territorial Epidemiologists, Communicable Disease Center, Atlanta, Georgia.

Smith, H. W., Crab, W. E. 1960. The effect of diets containing tetracyclines and penicillin on the *Staphylococcus aureus* flora of the nose and skin of pigs and chickens and their human attendants. Journal of Pathology and Bacteriology **79**:243–249.

Smith, I. M., Warren, G. H., Johnson, V. L. 1976. The use of the mouse model in testing antibiotics for the treatment of *Staphylococcus aureus* infections in man, pp. 791–808. In: Jeljaszewicz, J. (ed.), Staphylococci and staphylococcal diseases. Stuttgart, New York: Gustav Fischer Verlag.

Smith, R. F., Bettge, C. L. 1972. Comparative characteristics of human and porcine staphylococci and their differentiation in burn xenografting procedures. Applied Microbiology **24**: 929–932.

Solberg, C. O. 1965. A study of carriers of *Staphylococcus aureus* with special regard to quantitative bacterial estimations. Acta Medica Scandinavica, Suppl. I **178**:1–96.

Sompolinsky, D. 1953. De l'impétigo contagiosa suis et du *Micrococcus hyicus* n. sp. Schweizer Archiv für Tierheilkunde **95**:302–309.

Speck, M. L. (ed.) 1976. Compendium of methods for the microbiological examination of foods. Washington, D.C.: American Public Health Association.

Speller, D. C. E., Mitchell, R. G. 1973. Coagulase-negative

staphylococci causing endocarditis after cardiac surgery. Journal of Clinical Pathology **26**:517–524.

Stabenfeldt, G. H., Spencer, G. R. 1966. The lesions in bovine udders shedding nonhemolytic coagulase-negative staphylococci. Veterinary Pathology **3**:27–39.

Stableforth, A. W., Galloway, I. A. 1959. Infectious diseases of animals. Vol. 2: Diseases due to bacteria. London, New York: Academic Press.

Tares, M. J., Greenberg, A. E., Hoak, R. D., Rand, M. C. (eds.) 1971. Standard methods for the examination of water and wastewater, 13th ed. New York: American Public Health Association.

Telander, B., Wallmark, G. 1975. Micrococcus subgroup 3—a common cause of acute urinary tract infection in women. Lakartidningen **72**:1967.

Torres-Pereira, A. 1962. Coagulase-negative strains of *Staphylococcus* possessing antigen 51 as agents of urinary tract infections. Journal of Clinical Pathology **15**:252–253.

Totten, M. 1958. Ringtail in newborn Norway rats—a study of the environmental temperature and humidity on incidence. Journal of Hygiene **56**:190–196.

Underhahl, N. R., Grace, O. D., Twiehaus, M. J. 1965. Porcine exudative epidermitis: Characterization of bacterial agent. American Journal of Veterinary Research **26**:617–624.

Vera, H. D., Dumoff, M. 1974. Culture media, pp. 881–929. In: Lennette, E. H., Spaulding, E. H., Truant, J. P. (eds.), Manual of clinical microbiology, 2nd ed. Washington, D.C.: American Society for Microbiology.

Wallace, G. D., Quisenberry, W. B., Tanimoto, R. H. 1962. Bacteriophage type 80/81 staphylococcal infection in human beings associated with mastitis in dairy cattle. American Journal of Public Health **52**:1309–1317.

Watanakunakorn, C., Hamburger, M. 1970. *Staphylococcus epidermidis* endocarditis complicating a Starr-Edwards prosthesis: A therapeutic dilemma. Archives of Internal Medicine **126**:1014–1018.

Watson, W. A. 1965. The carriage of pathogenic staphylococci by sheep. Veterinary Record **77**:477–480.

Wilkoff, L. J., Westbrook, L., Dixon, G. J. 1969. Factors affecting the persistence of *Staphylococcus aureus* on fabrics. Applied Microbiology **17**:268–274.

Williams, R. E. O. 1963. Healthy carriage of *Staphylococcus aureus:* Its prevalence and importance. Bacteriological Reviews **27**:56–71.

Williams, R. E. O. 1967. Airborne staphylococci in the surgical ward. Journal of Hygiene **65**:207–217.

Zimmerman, R. J. 1976. Comparative zone electrophoresis of catalase of *Staphylococcus* species isolated from mammalian skin. Canadian Journal of Microbiology **22**:1691–1698.

Zimmerman, R. J., Kloos, W. E. 1976. Comparative zone electrophoresis of esterases of *Staphylococcus* species isolated from mammalian skin. Canadian Journal of Microbiology **22**:771–779.

CHAPTER 126

The Genus *Planococcus*

MILOSLAV KOCUR and KARL-HEINZ SCHLEIFER

Although the genus *Planococcus* was already described in 1894 (Migula, 1894), it still belongs to the less known genera of bacteria. Migula (1900) revised and amended the description of this genus and included four species of flagellated, nonspore-forming, aerobic cocci. Unfortunately, most of the other authors abandoned the genus *Planococcus*. Krasil'nikov (1949) was the only one to include this genus in his system of bacteria. The original cultures of planococci have been lost, and those authors who isolated new strains of Gram-positive, flagellated cocci placed them in the genus *Micrococcus*, as they were very similar in cultural and biochemical characteristics to true micrococci. *M. aquivivus* and *M. eucinetus*, originally described by ZoBell and Upham (1944) and Leifson (1964), are examples of such species.

Kocur et al. (1970) revised and amended the genus *Planococcus* Migula 1894 and recommended transferring the species *M. aquivivus* and *M. eucinetus* to this genus. The names of these species were thus objective synonyms of *Planococcus citreus,* the type species of the genus. Recently another species, *Planococcus halophilus,* has been described (Novitsky and Kushner, 1976).

Several authors contributed to a better knowledge of planococci. The determination of the DNA base composition revealed two groups among *Planococcus* species: one with 39–42 mol% G+C and the other with 47–51 mol% G+C (Boháček, Kocur, and Martinec, 1967, 1968). Their cell ultrastructure is similar to that of other Gram-positive cocci (Novitsky and Kushner, 1976; M. Kocur, unpublished data). The chemical cell wall analysis of *P. citreus* strains showed that their cell wall peptidoglycans were of the L-Lys-D-Glu type (Schleifer and Kandler, 1970), while in the peptidoglycan of *P. halophilus,* m-Dpm prevailed (Novitsky and Kushner, 1976). A serological examination of *P. citreus* strains proved no antigenic relationship to staphylococci and micrococci (Oeding, 1971). No teichoic acid was found in their cell walls (Endresen and Oeding, 1973). Planococci contain normal menaquinones (Jeffries et al., 1969; Yamada et al., 1976). The phospholipid pattern of planococci is similar to

that of *Sporosarcina ureae* (Komura, Yamada, and Komagata, 1975).

Habitats

So far, planococci have been isolated from marine environments, sea water (ZoBell and Upham, 1944), marine clams (Leifson, 1964), fish-brining tanks (Georgala, 1957), salted mackerel (Novitsky and Kushner, 1976), and frozen shrimps and prawns (K. Komagata, personal communication).

Isolation

No selective medium for the isolation of planococci has been devised until now. For their isolation, seawater agar or nutrient agar with 10% NaCl may be used.

Seawater agar

Beef extract	10 g
Agar	20 g
Tap water	250 ml
Peptone	10 g
Sea water	750 ml
pH 7.2	

Planococci can be cultivated on seawater agar or on nutrient agar with 1–10% NaCl. Cultures of planococci may be maintained on slant agar at 4°C (survival for 3 months) or in a freeze-dried state.

Identification

The genus *Planococcus* includes Gram-positive spheres occurring in pairs or tetrads. Each cell possesses 1–3 flagella. They are nonsporing, catalase positive, and aerobic; they produce yellow-brown pigment and do not attack carbohydrates. Their G+C content in the DNA ranges from 39 to 51 mol%.

Planococci may be clearly separated from other

genera of the family Micrococcaceae by several characters (see this Handbook, Chapters 124 and 125). Although they are a clearly separated group, certain difficulties may occur when differentiating them from micrococci, as they have some cultural and biochemical features in common. Their ability to grow on nutrient agar with 12% NaCl and the production of yellow-brown pigment are the only simple ways to differentiate them from micrococci. Sometimes it is necessary to examine the cell wall or the G+C content in the DNA. The present data show that planococci are related to the genus *Sporosarcina*, as they have several features in common, such as motility, G+C content in the DNA, cell wall peptidoglycan, and phospholipid pattern (Kocur, 1974; Kocur et al., 1970; Komura, Yamada, and Komagata, 1975; Schleifer and Kandler, 1970). *Sporosarcina ureae* differs from planococci in spore formation and urease production.

Literature Cited

Boháček, J., Kocur, M., Martinec, T. 1967. DNA base composition and taxonomy of some micrococci. Journal of General Microbiology **46**:369–376.

Boháček, J., Kocur, M., Martinec, T. 1968. Deoxyribonucleic acid base composition of some marine and halophilic micrococci. Journal of Applied Bacteriology **31**:215–219.

Endresen, C., Oeding, P. 1973. Purification and characterization of serologically active cell wall substances from *Planococcus* strains. Acta Pathologica et Microbiologica Scandinavica, Sect. B **81**:571–575.

Georgala, D. L. 1957. Quantitative and qualitative aspects of the skin flora of North Sea cod and the effect thereon of handling on ship and on shore, p. 118. Ph.D. Thesis. University of Aberdeen.

Jeffries, L., Cawthorne, M. A., Harris, M., Cook, B., Diplock, A. T. 1969. Menaquinone determination in the taxonomy of Micrococcaceae. Journal of General Microbiology **54**:365–380.

Kocur, M. 1974. Genus *Planococcus*, pp. 489–490. In: Buchanan, R. E., Gibbons, N. E. (eds.), Bergey's manual of determinative bacteriology, 8th ed. Baltimore: Williams & Wilkins.

Kocur, M., Páčová, Z., Hodgkiss, W., Martinec, T. 1970. The taxonomic status of the genus *Planococcus* Migula 1894. International Journal of Systematic Bacteriology **20**:241–248.

Komura, I., Yamada, K., Komagata, K. 1975. Taxonomic significance of phospholipid composition in aerobic Gram-positive cocci. Journal of General and Applied Microbiology **21**:97–107.

Krasil'nikov, N. A. 1949. Opredělitel baktěrij i aktinomycetov. Moscow: Akademii Nauk SSSR.

Leifson, E. 1964. *Micrococcus eucinetus* n. sp. International Journal of Systematic Bacteriology **14**:41–44.

Migula, W. 1894. Über ein neues System der Bakterien. Arbeiten aus dem Bakteriologischen Institut der Technischen Hochschule zu Karlsruhe **1**:235–238.

Migula, W. 1900. System der Bakterien. Jena: Gustav Fischer.

Novitsky, T. J., Kushner, D. J. 1976. *Planococcus halophilus* sp.nov., a facultatively halophilic coccus. International Journal of Systematic Bacteriology **26**:53–57.

Oeding, P. 1971. Serological investigations of *Planococcus* strains. International Journal of Systematic Bacteriology **21**:323–325.

Schleifer, K. H., Kandler, O. 1970. Amino acid sequence of the murein of *Planococcus* and other *Micrococcaceae*. Journal of Bacteriology **103**:387–392.

Yamada, Y., Inouye, G., Tahara, Y., Kondo, K. 1976. The menaquinone system in the classification of aerobic Gram-positive cocci in the genera *Micrococcus*, *Staphylococcus*, *Planococcus*, and *Sporosarcina*. Journal of General and Applied Microbiology **22**:227–236.

ZoBell, C. E., Upham, H. C. 1944. A list of marine bacteria including description of sixty new species. Bulletin of the Scripps Institution of Oceanography Technical Series **5**:239–292.

The Family Streptococcaceae (Medical Aspects)

RICHARD FACKLAM and HAZEL W. WILKINSON

The Streptococcaceae are Gram-positive, cyto-chrome-negative, coccoid bacteria that usually grow in chains of various lengths (hence, "strepto"), but sometimes form tetrads. Under certain growth conditions, they are elongated and therefore appear rodlike when Gram stained. Although morphological characteristics are useful, they cannot always be used to differentiate Streptococcaceae and Micrococcaceae. The most definitive difference between the two is that catalase is present in cultures of Micrococcaceae and absent in those of Streptococcaceae. Unfortunately, simply adding hydrogen peroxide to an agar slant culture is not specific for catalase, and therefore the occasional peroxidase-producing Streptococcus or Aerococcus and the catalase-negative Micrococcus and Staphylococcus are not differentiated with the relatively simple catalase test. However, they can be differentiated with the benzidine test (Deibel and Evans, 1960). A positive benzidine reaction indicates that the strain contains cytochromes and thus is a member of the Micrococcaceae. A negative benzidine reaction indicates that a Gram-positive coccus is a member of the Streptococcaceae.

Whereas Peptococcaceae are true anaerobes, Streptococcaceae are not, even though they usually grow better in a reduced oxygen atmosphere and thus are sometimes called "microaerophilic" or "aerotolerant anaerobic." Because the medically important Streptococcaceae are homofermentative, clinical laboratories with gas-liquid chromatography (GLC) equipment, when trying to determine whether an organism is a member of this family, can use analysis of spent media to determine whether the end product of glucose fermentation is primarily lactic acid (if so, homofermentation, hence, Streptococcus or Aerococcus; if not, heterofermentation, hence, Peptococcus) (Holdeman and Moore, 1974). Unfortunately, agreement has not yet been reached on the proper way to classify such exceptional strains as homofermentative anaerobic Gram-positive cocci.

The Streptococcaceae are divided into five genera in the eighth edition of Bergey's Manual of Determinative Bacteriology (Buchanan and Gibbons, 1974): Aerococcus, Gemella, Leuconostoc, Pediococcus, and Streptococcus. However, the taxonomic status of Gemella and Pediococcus is uncertain. Gemella species may actually be viridans streptococci (Facklam, 1977), and Pediococcus species are probably either aerococci or streptococci (Whittenbury, 1965). Leuconostoc strains have not been found in humans. Therefore, the Streptococcaceae isolated from humans need only be identified as streptococci or aerococci (Facklam and Smith, 1976).

The G+C content of the DNA of the Streptococcaceae, which is 32–44 mol%, is of little value in differentiating members of this family. The limited DNA hybridization studies done so far have shown both homology (Weissman et al., 1966) and heterology (Coykendall, 1977) among strains of the same species and heterology between species of Streptococcus (Weissman et al., 1966).

Habitats

The Streptococcaceae most commonly isolated from human infections are listed in Table 1. Lancefield's group A (Streptococcus pyogenes) was once considered the only Streptococcus of sufficient medical importance to identify definitively because of the nonsuppurative sequelae of primary pharyngeal or impetiginous group A infections. Poststreptococcal glomerulonephritis may follow either impetigo or pharyngitis, whereas acute rheumatic fever is known to sometimes follow pharyngitis but not impetigo. Postoperative wound infections with S. pyogenes can also lead to life-threatening complications such as cellulitis, erysipelas, fasciitis, gangrene, or septicemia. Anal carriage of S. pyogenes by hospital personnel is not uncommon and has been the apparent source of several nosocomial outbreaks (Schaffner et al., 1969). Asymptomatic colonization occurs more often in the pharynx, and S. pyogenes causes approximately 25% of all upper respiratory tract infections. Similarly, these organisms may be carried asymptomatically on the skin or may be the primary pathogen in impetiginous lesions.

In contrast, group B Streptococcus (S. agalactiae) organisms are found more often in the genital and intestinal tracts of apparently healthy human adults and infants (Wilkinson, 1978). It is not unusual to find group B streptococci in one-third of the vaginal cultures of third-trimester pregnant women, a reservoir that is apparently the inoculum by which many infants are colonized at birth on superficial sites (skin, umbilicus) and in body cavities (pharynx, external ear canal, nares, rectum). Nosocomial transmission may increase the colonization rate of infants in a hospital nursery to as much as 65% (Paredes et al., 1977). Most colonized infants (and adults) remain asymptomatic, but an estimated 1% of infants colonized with group B streptococci succumb to an invasive infection. When onset of disease occurs within the first few days after birth, sepsis and respiratory distress are predominant findings, with a 60–75% fatality rate (Yow, 1975). When onset occurs later, the infection frequently localizes to the meninges, with a 14–18% fatality rate. Adults may also be stricken by streptococcal group B meningitis and septicemia, although less often than neonates.

Group C includes three streptococcal species, all of which cause human infections. Of the three, *S. equisimilis* is found most often, frequently in cultures from the pharynx and occasionally in cultures from impetiginous lesions. Whether *S. equisimilis* causes pharyngitis is debatable. Group C streptococci have never been proved to cause acute glomerulonephritis or rheumatic fever. The natural habitat of *S. equi* and *S. zooepidemicus*, which are found primarily in animals other than humans, is unknown. Several isolates of each of these strains have been obtained from the sputum, pus, and blood of patients with abscesses and pneumonia.

All five group D streptococcal species have been isolated from the human intestinal tract. Approximately 20% of the cases of subacute bacterial endocarditis and 10% of urinary tract infections are caused by group D streptococci. The resistance of enterococcal species (*S. faecalis, S. faecium,* and *S. durans*) to penicillin and the resulting necessity of treating patients with systemic infections with combined antibiotic therapy rather than penicillin alone makes the differentiation of enterococcal isolates mandatory. *S. faecalis* is isolated frequently, *S. faecium* occasionally, and *S. durans* rarely from patients with endocarditis. Whether *S. avium,* which has been isolated from human urine and feces, causes human infection is unknown. Among the nonenterococcal species of group D streptococci, *S. bovis* has been isolated from a substantial number of patients with endocarditis. Because the organisms are susceptible to penicillin, patients with systemic group D infections are usually treated with penicillin alone. Although group D streptococci have been isolated from wound infections, they are not usually believed to be the etiological agent (Horovitz and von Graevenitz, 1977).

The microaerophilic group F streptococci (*S. anginosus*) have been isolated from the pharynx and intestinal tract of healthy individuals and from deep wounds, internal abscesses, and body fluids of infected patients. It is likely that strains were identified as anaerobic streptococci before gas-liquid chromatography (GLC) techniques were used to separate them from the streptococci. Strains lacking the group F antigen but physiologically identical to group F strains are isolated occasionally from human infections.

Group G streptococci have no universally accepted species designation although they are occasionally called "*S. canis*". They colonize the pharynx but, as is true with group C streptococci, their ability to cause pharyngitis is debatable. An M-protein antigen (generally associated with virulence among group A streptococci) was found in one group G strain (Maxted and Potter, 1967), and future studies may indicate the necessity of identifying group G strains in addition to group A isolates from the throat and from skin lesions. Group G streptococci have been isolated from patients with impetigo and from those with wound infections.

Streptococci belonging to serological groups other than A through G exist but rarely cause human infections. The remaining streptococci are known collectively as the "viridans" or "alpha" streptococci (sometimes incorrectly designated *S. viridans*) because they are not beta-hemolytic on blood agar. Although they do not have well-defined group antigens, some species contain antigens reactive with several group antisera. The viridans streptococci differ from pneumococci by being resistant to solubilization by bile and from enterococci by not growing in 6.5% NaCl broth or at 10°C and by being, for the most part, susceptible to penicillin. This susceptibility has probably been responsible for an overall lack of interest among clinical microbiologists in the taxonomy of viridans streptococci. Ten species have been described, but little is known about the pathogenicity of some of them. They occur normally in the pharynx, and some species (*S. mutans* and *S. sanguis*) are thought to be the primary cause of dental caries. Viridans streptococci are isolated more often than any other bacteria from bacterial endocarditis, causing between 50% and 70% of all cases.

Once considered a separate genus (*Diplococcus*), *Streptococcus pneumoniae* can be isolated from the pharynx of 30–70% of apparently normal humans. It is the bacterium most frequently isolated from patients with bacterial pneumonia and strikes debilitated patients with special severity. Recently, interest in pneumococcal vaccines has increased because of the worldwide isolation of pneumococcal strains with increased resistance to penicillin, and a polyvalent vaccine containing capsular polysaccharides of the 14 most prevalent capsular types is now available.

Table 1. The medically important Streptococcaceae (human): nomenclature, habitat, and pathogenicity.

Serological group	Species	Habitat	Diseases	Selected references
A	Streptococcus pyogenes	Pharynx Rectum Skin	Acute endocarditis Cellulitis Conjunctivitis Erysipelas Fasciitis Glomerulonephritis Impetigo Otitis media Pharyngitis Pneumonia Rheumatic fever Septicemia Wound infections	Burech, Koranyi, and Haynes, 1976 Dillon and Dudding, 1970 Hable et al., 1973 Markowitz and Taranta, 1972 Nelson, 1972 Portnoy et al., 1974 Rammelkamp, 1955 Richman, Breton, and Goldman, 1977 Schaffner et al., 1969 Wannamaker, 1970 Wannamaker, 1972
B	S. agalactiae	Pharynx Rectum Skin Genital tract	Abscesses Acute endocarditis Cellulitis Conjunctivitis Empyema Impetigo Meningitis Neonatal sepsis Osteomyelitis Otitis media Pericarditis Peritonitis Pneumonia Puerperal sepsis Respiratory distress Septic arthritis Urinary tract infections Wound infections	Aber et al., 1976 Anthony and Okada, 1977 Anthony and Concepcion, 1975 Belgaumker, 1975 Eickhoff et al., 1964 Howard and McCracken, 1974 McCracken, 1973 Mhalu, 1976 Paredes et al., 1977 Patterson and Hafeez, 1976 Wilkinson, 1978 Yow, 1975
C	S. equisimilis	Pharynx Skin	Acute endocarditis Pharyngitis Pneumonia Septicemia Wound infections	Armstrong et al., 1970 Benjamin and Parriello, 1976 Feingold, Stagg, and Kunz, 1966 Koshi, 1971 Lawrence and Cobb, 1972 Sanders, 1963
C	S. equi	Unknown	Abscesses	
C	S. zooepidemicus	Unknown	Pneumonia	
D	(Enterococci) S. faecalis S. faecium S. durans	Intestinal tract Vagina	Biliary infections Meningitis Peritonitis Pyelonephritis Subacute endocarditis Urinary tract infections	Bayer et al., 1977 Facklam, 1972 Gross et al., 1976 Gross et al., 1976 Horovitz and von Graevenitiz, 1977 Moellering, Watson, and Kunz, 1974 Sabbaj, Sutter, and Finegold, 1977 Thornsberry, Baker, and Facklam, 1974 Toala et al., 1969 Watanakunakorn and Glotzbecker, 1977
D/Q	S. avium	Intestinal tract	Unknown	
D	(Nonenterococcal) S. bovis	Intestinal tract	Meningitis Subacute endocarditis Urinary tract infections	

Table 1. The medically important Streptococcaceae (human): nomenclature, habitat, and pathogenicity. (*Continued*)

Serological group	Species	Habitat	Diseases	Selected references
F	*S. anginosus*	Intestinal tract Pharynx Vagina	Abscesses Empyema Meningitis Peritonitis Pneumonia Wound infections	Banatyne and Randall, 1977 Koepke, 1965 Poole and Wilson, 1976 Wort, 1975
G		Intestinal tract Pharynx Skin Vagina	Impetigo Pharyngitis Septicemia Wound infections	Baker, 1974 Belcher et al., 1975 Blevins et al., 1968 Duma et al., 1969 Hill et al., 1969 Maxted and Potter, 1967 Reinarz and Sanford, 1965

None	"Viridans"
	S. acidominimus
	S. anginosus-
	  constellatus
	S. MG-*intermedius*
	S. mitis
	S. morbillorium
	S. mutans
	S. salvarius
	S. sanguis I
	S. sanguis II
	S. uberis
		Intestinal tract
		Oral cavity
			Abscesses
			Caries
			Empyema
			Subacute
			  endocarditis
			Urinary tract
			  infections
			Wound infections
				Bateman, Eykyn, and Phillips, 1975
				Facklam, 1977
				Finland and Barnes, 1970
				Guthof, 1960
				Horstmeier and Washington, 1973
				Kast, 1971
				Lerner, 1975
				Mejare and Edwardsson, 1975
				Melmed, Katz, and Bank, 1976
				Parker and Ball, 1976

None	*S. pneumoniae*	Pharynx
		Vagina
			Conjunctivitis
			Otitis
			Pericarditis
			Peritonitis
			Pneumonia
			Septic arthritis
			Septicemia
			Subacute
			  endocarditis
				Burech et al., 1975
				Davies, 1977
				Epstein, Calia, and Gabuzda, 1968
				Finland and Barnes, 1977
				Kauffman, Watanakunakorn, and
				  Phair, 1976
				Kilbourn, 1973
				Klein, 1975
				Mausbach and Cho, 1976
				Mufson, 1974
				Newlands, 1976
				Paredes et al., 1976
				Schlossberg, Zacarias, and
				  Shulman, 1975

None	*Aerococcus*
	  viridans
		Unknown	Osteomyelitis
			Subacute
			  endocarditis
			Urinary tract
			  infections
				Colman, 1967
				Evans and Kerbaugh, 1970
				Parker and Ball, 1976
				Untereker and Hanna, 1976

Very little is known about *Aerococcus viridans* at this time. In their recent study of streptococcal systemic infections, Parker and Ball (1976) reported that aerococci were found in 1% of the cases. These organisms have also been isolated from air samples, but not from asymptomatic humans. Additional taxonomic and epidemiological studies are needed to understand the medical importance of these organisms.

Isolation

Transport

A throat swab is usually examined only for the presence of beta-hemolytic streptococci or *Corynebacterium diphtheriae* and, occasionally, for *Haemophilus influenzae* (for young children). Specimens taken from patients at risk may need additional investigation. If streptococci are the only pathogens suspected and if no more than 2 h elapse between the time the swab is taken and the time it is examined in the laboratory, no special transport media are needed. Streptococci survive in a dry environment for this length of time and can be transported to the laboratory in a paper envelope or a sterile test tube. If the swab is not to be processed until the next day, or if other pathogens such as those from wound infections must be considered, a holding medium (for example, Stuart's or Amies') should be used. If the swab is to be in transit for more than 1 day, the silica gel or the dry filter paper transport system should be used. These systems can be used for both throat and skin swabs. The necessary materials are available commercially. A modified silica gel transport system can be made, however, by placing enough silica gel crystals in a 15-×-125-mm screw-cap tube to cover the cotton tip of the swab. The tube is autoclaved and then dried in a hot air oven.

Growth, Isolation, and Determination of Hemolysis

Because most Streptococcaceae are fastidious, they must be grown in a rich organic medium. If beta-hemolytic streptococci are to be isolated, the medium must contain blood but not reducing sugars (the latter inhibit hemolysis). Most commercial blood agar bases are free of reducing sugars. Those commonly used include Trypticase soy, proteose peptone, brain heart infusion, heart infusion, neo-peptone infusion, and Todd-Hewitt agars. Hemolysis is affected by differences in animal blood only for some strains of *Streptococcus faecalis* (group D), which are beta-hemolytic on equine, human, and rabbit blood agar but are alpha-hemolytic on sheep blood agar. Microscopic examination is necessary regardless of the blood source. If human blood is used, each lot must be tested with control group A strains to make sure that the inhibitors present in some samples (such as type-specific antibodies, anti-streptolysin O antibodies, antibiotics, and citrates) inhibit neither the organisms' growth nor their hemolysins. Because colonies of *Haemophilus hemolyticus* are indistinguishable from those of beta-hemolytic streptococci, sheep blood is recommended for throat cultures. This blood lacks sufficient amounts of pyridine nucleotides (V factor) to support the growth of *H. hemolyticus*.

Concentrations of blood affect the size of the area of red blood cell lysis (zone size) only and not a decision on the type of hemolysis. If streak plates are used, lower concentrations of blood may make it difficult to distinguish alpha from beta hemolysis. Higher concentrations of blood may cause beta-hemolytic strains to appear nonhemolytic, unless the

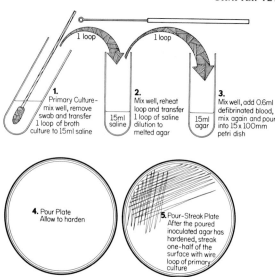

Fig. 1. Preparation of blood agar pour plates for isolation of beta-hemolytic streptococci.

agar is cut or stabbed with growth from the surface. The best blood agar plates for primary isolation contain 5% defibrinated blood in agar approximately 4 mm deep. One milliliter of defibrinated blood is added to 20 ml of melted agar, mixed well, and poured into a 15-×-100-mm Petri dish. Reference or control cultures having known hemolytic activity should be included in tests so that media and procedures can be controlled.

Preparation of Blood Agar Pour and Streak-Stab Plates for Isolation of Beta-Hemolytic Streptococci

The technique for preparing pour plates and pour-streak plates is outlined in Fig. 1. The swab is incubated in broth for 2–24 h at 35°C. Fifteen milliliters of agar are melted, cooled to, and then held at 50–55°C. If the culture has been incubated for 24 h, one loopful of it is transferred to 15 ml of saline to dilute the bacteria (step 1). After the ingredients have been mixed, one loopful of the saline-diluted suspension is transferred to 15 ml of melted agar (step 2). Six-tenths of a milliliter of defibrinated blood is then added to the agar (step 3). The contents of the tube are mixed well by swirling, poured into a 15-×-100-mm Petri dish, and allowed to harden (step 4). Step 5 is optional; one-half of the surface of the hardened blood agar plate is streaked with a drained loop of the primary culture to allow surface as well as subsurface growth. If the culture is from 2 to 6 h old, the saline dilution step (step 1) is omitted. Instead, one loopful of the culture is transferred directly to the melted agar, the blood is added, and then the rest of the procedure is followed.

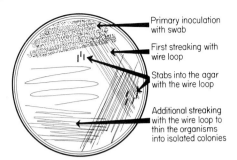

Fig. 2. Preparation of a blood agar streak-stab plate for isolation of beta-hemolytic streptococci.

The procedure for inoculating a blood agar streak-stab plate is shown in Fig. 2. The swab is rolled over the surface of approximately one-sixth of the agar so that the entire swab makes contact with the agar. The plate is then streaked to thin the organisms into isolated colonies. Afterward, the wire loop is stabbed into the agar to the bottom of the plate at an angle perpendicular to the surface to make a clean cut without ragged edges. One set of stabs should be made in an area of the agar plate that has not been streaked so that the streptococcal hemolysis is easier to read with a microscope. Ideally, pour plates are used to determine hemolysis, but by reading hemolysis next to a stab microscopically, the same hemolysis criteria used for pour plates can be applied to the streak-stab plates.

Pour plates can be incubated in any atmosphere because at least some subsurface colonies are initially anaerobic. Oxygen is driven out of the agar during melting and stays out until after the streptococci lyse the erythrocytes. Anaerobiosis is necessary for the detection of some group A strains whose hemolysins (streptolysin O) are inhibited by oxygen.

Streak plates, in contrast, should be incubated anaerobically. (A GasPak system or a mixture of 5–10% CO_2 with 90–95% N_2 in an anaerobic chamber can be used.) If it is not possible to incubate streak plates anaerobically, they should be incubated in a normal atmosphere rather than in a candle jar or a CO_2 incubator. Increased concentrations of CO_2 in the presence of oxygen increase peroxide production by some streptococci. Peroxide, in turn, inhibits red blood cell lysis, and therefore beta-hemolytic streptococci appear to be alpha-hemolytic. Streak plates that have been stabbed can be incubated in any atmosphere, the least favorable one being an atmosphere of increased CO_2.

Hemolysis is relatively unimportant in isolating group D streptococci. Instead, esculin can be added to the base medium as an indicator because these organisms can hydrolyze esculin.

Selection, enrichment, or selective-enrichment techniques are being used with increasing frequency for primary isolation of streptococci when quantitative information is not sought. There are contradictory opinions as to whether patients should be treated regardless of the numbers of infecting organisms (Horn, Goodpasture, and Jackson, 1974; Stollerman, 1962). If colony counts on primary isolation plates are considered necessary, enrichment is unnecessary. If, on the other hand, the objective is to detect even low numbers of streptococci, enrichment (incubation of the inoculum in broth), selection (incubation in an environment more conducive to the growth of streptococci than to unwanted organisms), or selective-enrichment may be advantageous.

Enrichment

Todd-Hewitt (THB), Trypticase soy, brain heart infusion, and heart infusion broths are commonly used for enrichment. Streptococcal groups A and B can be identified within 4 h, even in the presence of large numbers of contaminating organisms, by immunofluorescence (Moody, Ellis, and Updyke, 1958; Romero and Wilkinson, 1974). Otherwise, swabs can be incubated in THB overnight. Blood agar pour plates are then prepared from the broth culture; if contaminants overgrow the culture, selective blood agars (i.e., Columbia colistin–nalidixic acid, phenethyl alcohol) can be used in isolating the streptococci (Aber et al., 1976).

Selective Media Technique

Various selective agars are listed in Table 2. Because high concentrations of some inhibitors may also inhibit streptococcal growth (e.g., gentamicin), a nonselective plate should be used simultaneously with the selective media. The source of the specimen may dictate which selective agent to use. For example, inhibitors of Gram-negative rods (GNR) are not usually necessary for throat swabs but are very useful for rectal, vaginal, or wound specimens. Conversely, an inhibitor (such as crystal violet) of staphylococci, which are often found in the throat, would be useful in streptococcal throat swab cultures. Sulfamethoxazole and trimethoprim (SXT) have also been used in blood agar to inhibit staphylococci, viridans streptococci, and GNR.

Investigators attempting to recover beta-hemolytic streptococci have usually reported higher isolation rates with gentamicin in blood agar than with nonselective media (Black and Van Buskirk, 1973; Murray et al., 1976b). The latter authors reported that the improved recovery rate was due to better growth of non-group A rather than of group

Table 2. Selective agar media for isolation of Streptococcaceae.

Organisms selected	Base medium	Inhibitors (concn)	Organisms inhibited[a]	References
Beta-hemolytic strep-tococci	Columbia blood agar	Gentamicin (5.5 μg/ml)	GNR Staph	Black and Van Buskirk, 1973
	Trypticase soy blood agar	Gentamicin (5.0 μg/ml)	GNR Staph	Murray et al., 1976b
	Trypticase soy blood agar	Neomycin (30 μg/ml)	GNR Staph	Blanchette and Lawrence, 1967
	Tryptose blood agar	Neomycin (30 μg/ml) Nalidixic acid (15 μg/ml)	GNR Staph Neiss	Vincent, Gibbons, and Gaafar, 1971 Randolph et al., 1976
	Hartley's digest blood agar	Neomycin (4.25 μg/ml) Fusidic acid (0.5 μg/ml) Polymyxin B (17 units/ml)	GNR Staph	Lowbury, Kidson, and Lilly, 1964
	Protein digest-soy blood agar	Neomycin (30 μg/ml) Nalidixic acid (15 μg/ml) Amphotericin B (3 μg/ml)	GNR Staph Fungi	Freeburg and Buckingham, 1976
	Trypticase soy blood agar	Crystal violet (1 μg/ml)	Staph	Taplin and Lansdell, 1973
Group A and B beta-hemolytic streptococci	Trypticase soy sheep blood agar	Sulfamethoxazole (23.75 μg/ml) Trimethoprim (1.25 μg/ml)	GNR Staph Viridans	Gunn, 1977
Group B streptococci	Todd-Hewitt sheep blood agar	Nalidixic acid (15 μg/ml) Polymyxin B (10 μg/ml) Crystal violet (1 μg/ml)	GNR Staph	Gray, Pass, and Dillon, 1979
Group D streptococci	Bile-esculin azide agar	NaN$_3$ (0.25 g/l)	GNR	Sabbaj, Sutter, and Finegold, 1971
Pneumococci	Trypticase soy blood agar	Gentamicin (5.0 μg/ml)	GNR Staph	Dilworth et al., 1975
	Commercial sheep blood agar	Gentamicin (5.0 μg/ml)	GNR Staph	Sondag et al., 1977 Converse and Dillon, 1977
Gram-positive cocci	Columbia blood agar	Colistin (10 μg/ml) Nalidixic acid (15 μg/ml)	GNR	Ellner et al., 1966
	Trypticase-peptone blood agar	Phenylethyl alcohol (0.25%)	GNR	Dayton et al., 1974

[a] GNR, Gram-negative rods; Staph, staphylococci; Neiss, *Neisseria*.

A streptococci. The use of different agar bases may be responsible for some of the discrepant results. Selective agar containing gentamicin has been reported to substantially improve the recovery rates of pneumococci from the oropharynx (Dilworth et al., 1975; Sondag et al., 1977) and group B streptococci from vaginal swabs (Baker et al., 1973). Data based on growth rates, however, suggest that there is an initial inhibition of group B growth by this agent, which was shown to be relatively unstable in the same study (Gray, Pass, and Dillon, 1977).

Bile esculin–azide agar is an excellent isolation medium for group D streptococci. Azide inhibits GNR, and esculin makes group D streptococcal colonies appear black. Columbia colistin–nalidixic acid agar (CNA) and phenethyl alcohol agar (PEA) have been used successfully to isolate Gram-positive cocci from heavily contaminated swabs (anal, vaginal, etc). The advantage of CNA over PEA is that

hemolysis is visible on blood agar plates containing the former.

Selective-Enrichment Techniques

Table 3 lists several selective-enrichment broths. Using them in cultures provides the advantages of enrichment and selection by providing optimal conditions for streptococcal growth while inhibiting the growth of competitors. The most frequently used selective agents have been sodium azide (to inhibit GNR) and crystal violet (to inhibit staphylococci). A common mistake in preparing selective-enrichment broths is using the same concentration of inhibitors for broth as for selective agar. The fact that some inhibitors diffuse more widely in broth may dictate that their concentrations be reduced (Gray, Pass, and Dillon, 1977).

Table 3. Selective-enrichment media for isolation of Streptococcaceae.

Organisms selected	Base medium	Inhibitors (concn)	Organisms inhibited[a]	References
Beta-hemolytic streptococci	Infusion broth (Pike's medium)	NaN$_3$ (1:16,000) Crystal violet (1:500,000)	GNR Staph	Pike, 1945
	Modified heart infusion broth	NaN$_3$ (0.01%) Crystal violet (0.002%) NaCl (2.5%)	GNR Staph	Nakamizo and Sata, 1972
	Modified heart infusion broth	NaN$_3$ (0.01%) Camphor (0.002%) NaCl (2.75%)	GNR Staph	Nakamizo and Sata, 1972
	Modified heart infusion broth	NaN$_3$ (0.05%) Crystal violet (0.0022%) 8-Hydroxyquinoline-5-sulfonic acid (0.017%) NaCl (2.8%)	GNR Staph Viridans	Sato, 1972
Group B streptococci	Todd-Hewitt broth	Nalidixic acid (15 μg/ml) Gentamicin (8 μg/ml)	GNR Staph	Baker, Clark, and Barrett, 1973
	Todd-Hewitt broth	Nalidixic acid (15 μg/ml) Polymyxin (1 μg/ml) Crystal violet (0.1 μg/ml)	GNR Staph	Gray, Pass, and Dillon, 1979
Enterococci	*Streptococcus faecalis* broth	NaN$_3$ (0.5 g/l)	GNR	Hajna and Perry, 1943
	Buffered azide-glucose-glycerol broth	NaN$_3$ (0.5 g/l)	GNR	Hajna, 1951

[a] GNR, Gram-negative rods; Staph, staphylococci.

Identification

The major differential characteristics of the Strepto-coccaceae are listed in Table 4. Once isolated in pure culture, a catalase-negative, Gram-positive coccus is tested for hemolysis in pour plates (prefer-able, Fig. 1) or streak-stab plates (Fig. 2). If the strain is beta-hemolytic (see Determination of He-molysis and Fig. 3 for description), it is extracted and tested for serological group with groups A, B, C, D, F, and G antisera. If the strain is not beta-hemolytic, it is extracted and tested only with groups B and D antisera. Groups A, C, F, and G antisera often cross-react with alpha-hemolytic or nonhemo-lytic streptococci (Facklam, 1977). Streptococci that are not beta-hemolytic and are not members of groups B or D are identified with physiological tests. Group D streptococci can also be identified pre-sumptively with the bile esculin test. The 6.5% NaCl tolerance test differentiates enterococci from group D streptococci that are not enterococci (Fack-lam, 1973). Viridans streptococci and pneumococci are not beta-hemolytic, do not react with groups B or D antisera, and do not grow in 6.5% NaCl broth, but the latter are soluble in bile salts, whereas the former are not.

The aerococci are not beta-hemolytic, do not react with streptococcal grouping antisera, are insol-uble in bile, and grow in 6.5% NaCl broth. The fact that some aerococci blacken bile esculin means that they must be differentiated from enterococci on the basis of cellular arrangement.

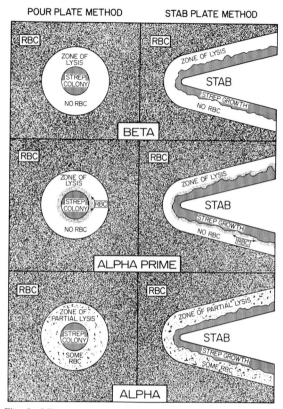

Fig. 3. Microscopic determination of streptococcal hemolysis.

Strains in each category in Table 4 are tested fur-ther for species identification.

Table 4. Differentiation of Streptococcaceae.[a]

Category	Most common cellular arrangement	Hemolysis	Streptococcal group antigen	Hydrolysis of bile esculin	Growth in 6.5% NaCl	Bile solubility
Beta-hemolytic "groupable" strepto-cocci	Chains	Beta	A, B, C, F, and G	−	(−)	−
Group D streptococci Enterococci	Short chains, diplo	Alpha Beta None	D	+	+	
Nonenterococci	Short chains, diplo	Alpha None	D	+	−	−
Viridans streptococci	Chains	Alpha None	(None)	(−)	−	−
Pneumococci	Diplo, short chains	Alpha	None	−	−	+
Aerococci	Tetrads, single cells	Alpha None	None	V	+	−

[a] +, Positive; −, negative; ( ), occasional exception; V, variable.

Determination of Hemolysis

In 1919, J. H. Brown described the hemolysis surrounding streptococcal subsurface colonies in blood agar pour plates. Brown's criteria also apply to streak-stab plates and to streak plates (only if the latter are incubated anaerobically) viewed microscopically. Peroxide-producing beta-hemolytic streptococci can appear alpha-hemolytic on the surface of blood agar plates incubated aerobically (Noble and Vosti, 1971). Peroxide-producing alpha-hemolytic streptococci may inhibit the expression of beta hemolysis produced by group A streptococci on the surface of blood agar plates (Holmberg and Hallander, 1973; LeBien and Bromel, 1975). These facts are especially important when hemolysis is determined on primary isolation plates from throat swabs because alpha-hemolytic streptococci are part of the normal throat flora, but beta-hemolytic throat cultures are potentially pathogenic.

Beta, alpha, and alpha-prime hemolysis can be differentiated at a magnification of approximately 60× around the subsurface growth in blood agar pour plates (Fig. 3, left side) and in blood agar stab plates (Fig. 3, right side). If a zone of complete lysis is formed from the edge of the streptococcal growth outward, the organism is beta-hemolytic. However, if intact erythrocytes are present next to the bacterial growth within a zone of complete lysis, the organism is a viridans *Streptococcus* and the hemolysis is called "alpha-prime" or "wide-zone alpha". This type of hemolysis can be seen only with the microscope. If a zone of partial lysis forms next to the streptococcal growth, the strain is alpha-hemolytic.

Extraction and Grouping

Beta-hemolytic streptococci are identified by extracting the carbohydrate antigen from the streptococcal cells and allowing the extract to react with group-specific precipitating antisera. These techniques were first described by Lancefield in 1933, and attempts since that time to extend the serological grouping system to the alpha-hemolytic streptococci have failed except with groups D and N. Results from examining more than 20,000 streptococcal strains isolated from humans and submitted to the Streptococcus Reference Laboratory, Center for Disease Control (CDC), Atlanta, Georgia, for over 10 years show that 1% of the beta-hemolytic strains are nonreactive with groups A, B, C, F, and G antisera. Only one group A variant, one group A intermediate, and two group L streptococci were identified during this period. No beta-hemolytic streptococci had positive precipitin reactions with groups E, H, K, M, and O antisera; extracts of several non-beta-hemolytic strains did. In addition, during the same 10-year period, 98% of 665 group D streptococcal strains (including several that were beta-hemolytic) reacted with group D antiserum. Therefore, over 99% of the beta-hemolytic streptococci isolated from humans can be serologically identified with groups A, B, C, D, F, and G antisera. The other beta-hemolytic groups (E, L, M, P, U, and V) are rarely if ever associated with human infections.

The alpha-hemolytic or nonhemolytic streptococci should be extracted and tested only with groups B and D antisera despite the fact that groups H, K, N, O, Q, R, S, and T are also alpha-hemolytic. The latter groups, which are usually identified as viridans streptococci, may also react with groups A, C, F, and G antisera (Facklam, 1977). Except for groups B and D, beta-hemolytic strains cannot be identified with physiological tests, and alpha-hemolytic or nonhemolytic strains cannot be identified serologically. For example, beta-hemolytic strains of groups A, C, F, and G sometimes resemble the viridans streptococci physiologically, and strains of eight different viridans species sometimes react with one antiserum (group F). Table 5 shows the beta-hemolytic streptococcal groups commonly isolated from human infections. Several extraction procedures that have been used for serological grouping are listed and compared in the following paragraphs.

Hot Hydrochloric Acid Extraction of Streptococci (Lancefield, 1933)

1. Grow pure cultures in 30 ml Todd-Hewitt or other suitable broth overnight 35°–37°C.
2. Pack the cells by centrifugation.
3. Discard the supernatant fluid and add 1 drop of 0.04% metacresol purple indicator and

Table 5. Serological groups of beta-hemolytic streptococci commonly isolated from human infections.

Group	*Streptococcus* species	Fermentation of:	
		Trehalose	Sorbitol
A	S. pyogenes		
B	S. agalactiae		
C	S. equisimilis	+	−
C	S. zooepidemicus	−	+
C	S. equi	−	−
F	S. anginosus		
G	—		

about 0.3 ml 0.2 N HCl (in 0.85% NaCl) to the sedimented cells. Mix well and transfer the suspension to a Kahn tube. If the suspension is not definitely pink (pH 2.0–2.4), add another drop or so of 0.2 N HCl.

4. Place the tube in a boiling water bath for 10 min. Shake the tube several times.
5. Remove the tube from the water bath and centrifuge.
6. Decant the supernatant fluid into a clean Kahn tube.
7. Neutralize by adding 0.2 N NaOH (in distilled water) drop by drop until the extract is slightly purple (pH 7.4–7.8). A deep purple indicates that the pH is too high, which may cause nonspecific cross-reactions. Although it is better not to add more salts or increase the volume, a back titration with 0.2 N HCl may be necessary.
8. Centrifuge the extract and decant the supernatant fluid into a small screw-capped vial. Add 1 drop of a 1:500 dilution of Merthiolate (1% in 1.4% Na borate) and store at 4°C or −20°C.

Hot Formamide Extraction of Streptococci (Fuller, 1938)

1. Grow strains in 5 ml Todd-Hewitt or other suitable broth overnight at 35–37°C.
2. Pack the cells by centrifugation.
3. Discard the supernatant fluid, add 0.1 ml formamide to the cells, and mix.
4. Place the tube in an oil bath heated to 150°C for 15 min.
5. Cool the tube in running tap water and add to it 0.25 ml acid-alcohol (95 parts anhydrous alcohol and 5 parts 2N HCl). Shake the tube to mix.
6. Centrifuge the tube and decant the supernatant fluid into a small tube.
7. Add 0.25 ml acetone; shake the tube to mix.
8. Centrifuge the tube and discard the supernatant fluid.
9. Add 1 ml saline and 1 drop of phenol red indicator to the precipitate. Shake and neutralize with a trace of sodium carbonate powder. Remove any insoluble precipitate by centrifugation.

Autoclave Extraction of Streptococci (Rantz and Randall, 1955)

1. Grow cells in 30 ml Todd-Hewitt or other suitable broth overnight at 35–37°C.
2. Pack the cells by centrifugation.
3. Discard the supernatant fluid, add 0.5 ml 0.85% NaCl solution to the cells, and shake to suspend the cells.

4. Autoclave the tube for 15 min at 121°C.
5. Centrifuge the tube.
6. Decant the supernatant fluid into a clean, sterile container.

Nitrous Acid Extraction of Streptococci (El Kholy, Wannamaker, and Krause, 1974)

1. Grow cells on the surface of a blood agar plate or in 5 ml Todd-Hewitt or other suitable broth overnight at 35°–37°C.
2. If 5 ml of broth is used, recover the cells by centrifugation and add 1 drop of saline to the packed cells. If an agar plate is used, recover the cells by adding 1 or 2 drops of saline to the plate, scraping the growth free, and transferring the suspension to a small tube.
3. Add 2 drops of 4 M $NaNO_2$ solution (276 g $NaNO_2$ per liter of distilled water) to the cell suspension.
4. Add 1 drop of glacial acetic acid and mix well.
5. Allow to react for 15 min at room temperature.
6. Add 1 drop of metacresol purple indicator and adjust the pH to 7.4 with 1 N NaOH.
7. Centrifuge to clarify.

Streptomyces albus Enzyme Extraction of Streptococci (Maxted, 1948)

1. Grow strains on a blood agar plate overnight at 37°C.
2. Pipette 0.25 ml *Streptomyces albus* enzyme solution (available commercially) into a small test tube (12 × 75 mm or smaller).
3. Scrape a large loopful of growth from the blood agar plate and suspend it in the enzyme solution.
4. Place the tube in a 45°C water bath until the solution is clear (about 90 min).
5. Cool tube to room temperature and centrifuge for 10 min.
6. Decant supernatant fluid into a clean container.

S. albus enzyme should be resuspended in the volume specified by the manufacturer and stored at −20° to −70°C.

Pronase B Enzyme Extraction of Streptococci (Ederer et al., 1972)

1. Grow strains on a blood agar plate overnight at 35–37°C.
2. Prepare borate buffer by adding 525 ml borate solution (12.404 g boric acid dissolved in 100 ml N NaOH and diluted to 1 liter with distilled water) to 475 ml 0.1 N HCl and 10 ml 1 M $CaCl_2$.

3. Prepare buffered enzyme solution (20 mg Pronase B/ml of borate buffer). Dispense in 0.5-ml portions in 13-×-75-mm tubes; cork and store at −20 or −70°C.
4. Remove all growth from the plate with a swab. Place the swab in a tube containing buffered enzyme solution and squeeze the swab as dry as possible by rotating it against the side of the tube. The suspension should be cloudy.
5. Place at 35–45°C for 2 h.
6. Centrifuge tube for 15–30 min.
7. Decant supernatant into clean, sterile container.

Streptomyces albus–Lysozyme Enzyme Extraction of Streptococci (Watson, Moellering, and Kunz, 1975)

1. Grow strains on a blood agar plate overnight at 35–37°C.
2. Prepare enzyme mixture by adding the *Streptomyces albus* enzyme to 5 ml of a solution of lysozyme (5 mg/ml distilled water). Centrifuge the solution to clarify and store in 0.5-ml quantities in 10-×-75-mm cork-stoppered tubes at −20°C.
3. Transfer the growth from the blood agar plate to the enzyme solution (0.5 ml) with a sterile swab. Mix the swab with the solution and rotate it against the side of the tube to remove as much of the moisture as possible. Discard the swab.
4. Incubate the enzyme-cell mixture in a water bath at 45–50°C for 90 min.
5. Centrifuge to clarify and decant into a clean container.

There are certain advantages and disadvantages to each extraction method. For example, the Lancefield hot acid technique is standard for grouping and must be used for typing groups A (Swift, Wilson, and Lancefield, 1943) and B streptococci. It is the only technique with which to extract the protein type-specific antigens in addition to the carbohydrate (groups A, B, C, F, and G) and teichoic acid (groups D and N) antigens. However, it is somewhat more complex and time-consuming than are other methods.

The Fuller hot formamide technique is also relatively complex and time-consuming and, like the Lancefield technique, can be used to extract all group antigens. However, the Fuller technique cannot be used when groups A and B streptococci are to be typed because it destroys their protein type-specific antigens.

The Rantz-Randall autoclave technique is relatively simple and can be used for grouping. In fact, the Lancefield, Fuller, and Rantz-Randall techniques were equally effective with CDC grouping antisera in the identification of 130 strains of groups A, B, C, D, F, and G streptococci.

El Kholy's nitrous acid technique is simple and reportedly effective in the identification of groups A, B, C, and G streptococci. However, because few stock strains and no clinical isolates of groups D and F streptococci were tested, these results must be further evaluated.

Maxted's *Streptomyces albus* enzyme extraction, although easy to perform, can be used to group only A, B, C, F, and G streptococci. The group D antigen is not extracted satisfactorily by the *S. albus* enzyme.

Ederer's Pronase B enzyme extraction, like the *S. albus* enzyme technique, is easy to perform but does not extract either the group D or the group F antigen as satisfactorily as do the Lancefield, Fuller, Rantz-Randall, and Watson techniques. Only for group A streptococci were satisfactory extractions obtained 100% of the time in CDC tests.

In contrast, Watson's *Streptomyces albus*–lysozyme enzyme technique extracts the group antigen of groups A, B, C, D, F, and G streptococci. Unfortunately, the reagents for this technique are more expensive than those for any of the other techniques.

The effectiveness of all extraction techniques depends largely on the quality of the antisera used in the precipitin test. With potent, specific antisera, all techniques work well within the limits of the tests just described. Control streptococcal strains should be used to test each new lot of commercial antiserum. Some antisera are of notoriously poor quality. The Lancefield, Fuller, Rantz-Randall, and Watson techniques can be used to extract and identify groups A, B, C, D, F, and G streptococci if good antisera are used. The El Kholy and Maxted techniques can be used to identify groups A, B, C, F, and G streptococci, but an alternative procedure must be used to identify group D streptococci. Ederer's technique does not work well unless antisera of exceptionally high quality are used. Cost, complexity, time requirements, and efficiency of extraction are all factors that influence a choice of extraction technique.

The Lancefield precipitin technique originally involved layering the extract under the antiserum, a method that was satisfactory for hyperimmune antisera. Because CDC and commercial antisera are usually not as potent as those specified for the Lancefield procedure, however, we layer the extract over the antiserum (Harrell and George, 1972).

Identification of Streptococci by the Capillary Precipitin Test

1. Dip the capillary tube (vaccine capillary tube with 1.2- to 1.5-mm outside diameter, Kimble borosilicate glass, both ends open; lightly fire-polished) into serum (in screw-capped

vial) until a column about 1 cm long has been drawn in by capillary action. (To maintain sterility of the serum specimens, sterilize the capillary tubes and keep them sterile at the lower end until after the serum is taken up.)

2. Holding the tube carefully so that air does not enter it, wipe it with facial tissue.

3. Dip tube into extract until an amount equal to that in the serum column is drawn up. If an air bubble separates serum and extract, discard tube and repeat.

4. Wipe tube carefully. Fingerprints, serum, or extract on the outside of the tube may simulate or obscure a positive reaction.

5. Plunge the lower end of the tube into plasticine until a small plug fills the opening. Do not let the reactants mix. The plasticine plug (at the same end of the tube as the reactants) will hold the reactants in place while the tube is inverted. Alternatively, hold a finger over the end of the tube until step 6 is completed.

6. Invert tube and insert it gently into the plasticine-filled groove of the rack.

7. Examine in bright light against a dark background. If a white precipitate appears within 5 min the reaction is strongly positive; weaker reactions develop more slowly. Precipitates that do not appear before 30 min should be disregarded.

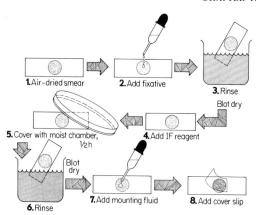

Fig. 4. Preparation of a slide for the fluorescent-antibody test (via direct immunofluorescence staining) to identify groups A and B streptococci.

Identification of Groups A and B by Immunofluorescence Staining

Distinct advantages of immunofluorescence (IF) over conventional identification techniques include rapidity (several hours rather than several days), sensitivity (detection of small numbers of streptococci in mixed cultures), and detection of nonhemolytic group B streptococci (often missed with conventional techniques). The procedures established by Moody, Ellis, and Updyke (1958) and Moody et al. (1963; see also Cherry and Moody, 1965, for pertinent field evaluations and modifications) for identifying group A streptococci by direct IF have become routine in many clinical laboratories that process numerous throat swabs. The IF reagent, or "conjugate", is composed of appropriate dilutions of anti-group-carbohydrate immunoglobulin attached to fluorescein isothiocyanate (FITC). Nonspecific staining of *Staphylococcus aureus* or members of other streptococcal groups, which cannot be distinguished from group A streptococci morphologically, can be avoided by adding either unlabeled streptococcal group C antiserum or unlabeled nonimmune serum to the conjugate. The "blocking" of nonspecific staining is explained theoretically by the demonstrated nonimmune binding of unlabeled immunoglobulin (by the Fc portion) to staphylococcal protein A or certain streptococcal receptors (Kronvall, 1973), which can prevent the nonspecific binding of the IF reagent to these receptors (i.e., by the Fc of the conjugated antiserum).

The group B streptococcal IF procedure, described by Romero and Wilkinson (1974), is less widely used because specific reagent is not commercially available. The group B conjugate must contain, in addition to unlabeled nonimmune serum (for the same reasons as for the group A IF procedure), both group B and type-specific FITC-labeled antibodies so that all serotypes are stained.

To be specific, the group A and group B conjugates should not contain antibodies reactive with the several R protein antigens found among many streptococcal groups (Wilkinson, 1972). The general procedure outlined in Fig. 4 for direct IF staining may be used to identify both groups A and B streptococci. Streptococcal cells can be grown on blood agar, suspended in buffer, and placed directly on glass slides, or can be grown in THB and then placed on glass slides. The latter enrichment is often preferred for pharyngeal or vaginal swabs because of the short incubation time required (2–6 h) for detection. Group B cells are usually washed only briefly to avoid removing surface antigens needed for correct identification. Specimens can be fixed by heat, ethanol, or formalin. Although several modifications have been described (Ederer and Chapman, 1972; Freeburg, 1970; Martin and Bigwood, 1969), only the most commonly used group A and group B IF staining procedures are presented.

Direct Immunofluorescence Staining to Identify Group A Streptococci (Moody, Ellis, and Updyke, 1958)

Prepare 0.01 M phosphate-buffered saline (PBS), pH 7.6, by diluting stock solution (12.36 g Na_2HPO_4, 1.80 g $NaH_2PO_4 \cdot H_2O$, and 85.00 g NaCl per liter of distilled water) 1:10 in distilled water. Prepare buffered glycerol mounting fluid by mixing 9 parts of glycerol with 1 part of carbonate buffer, pH 9.0. The latter is prepared by adding 100 ml of a solution containing 4.2 g NaHCO$_3$ to a solution of 5.3 g Na$_2$CO$_3$ in 100 ml of water until the pH is adjusted to 9.0. Inoculate the test organisms into 1 ml THB and incubate at 35–37°C until the broth is turbid (2–6 h usually required for positive throat swabs). Centrifuge the turbid broth, resuspend the cell sediment in approximately 1 ml PBS, recentrifuge the suspension, resuspend the cells in several drops of PBS, and prepare a smear of the cell suspension on an immunofluorescence slide. Allow the smear to air-dry, add several drops of 95% ethanol as a fixative, allow the smear to air-dry, and add several drops of IF reagent to it. Incubate the slide for 1/2 h at room temperature in a moist chamber (easily prepared by fitting half of a 15-cm Petri dish with moist filter paper). Tap off excess conjugate, then rinse the slide first in PBS (in a container for 10 min) and then briefly in distilled water. Allow the slide to air-dry or blot gently to dry and add a drop of mounting fluid and a cover slip. Examine smears with 95–100× oil immersion lens on a fluorescence microscope equipped for transmitted light with an HBO-200 mercury arc lamp, a 3-mm Schotts BG-12 primary filter, and a 1-mm Schotts OG-1 barrier filter (or comparable fluorescence assembly). Group A streptococcal cells should stain with a 3+ or 4+ intensity and should appear as fluorescent, yellow-green "doughnuts". Other streptococcal groups and *Staphylococcus aureus* cells should be invisible or stain with no more than a 2+ intensity.

Direct Immunofluorescent Staining to Identify Group B Streptococci (Romero and Wilkinson, 1974)

Prepare PBS, pH 7.6, and mounting fluid, pH 9.0, as described in the procedure for identifying group A streptococci. Inoculate the test organisms into 1 ml THB and incubate at 35°C until the broth is turbid (often in 2–6 h). Prepare a smear of the cell suspension on an immunofluorescence slide and allow it to air-dry at room temperature. Add several drops of 95% ethanol as fixative and allow it to evaporate, or add several drops of undiluted Formalin and after 5 min rinse the slide with PBS and then with distilled water (see group A procedure). Air-dry or blot dry, apply IF reagent to the smear, and incubate the slide in a moist chamber for 1/2 h at room temperature. Tap off excess conjugate, rinse and blot the slide (as described under group A procedure), add a drop of mounting fluid and coverslip to the smear, and look for IF-positive streptococcal cells that fluoresce in ultraviolet light (see description under group A procedure). All group B serotypes should stain equally well.

Differentiation of Group D Streptococci

Once an isolate is identified serologically as a group D *Streptococcus,* at least 10 physiological tests are needed for its species identification (Table 6). Because some group D strains react atypically in several tests, the "best fit" is determined or, as described by Deibel (1964), a "spectrum analysis" is done. For example, variants of *S. faecalis* that do not hydrolyze arginine and those that do not form acid in carbohydrate broths are still identified as *S. faecalis* and not as a new species. Additional differential tests are given by Deibel (1964) and Facklam (1972).

It is absolutely necessary to demonstrate that *Streptococcus bovis* variant and *S. equinus* have the group D antigen (preferably by extraction and serogrouping). These two strains cannot be distinguished physiologically from some of the viridans streptococci (e.g., *S.* MG-*intermedius*). Although serogrouping is preferable for the other group D species, some strains of *S. faecium, S. durans, S. avium,* and *S. bovis* have too little group D antigen to be detected, in which case a presumptive test (e.g., bile esculin) can be used. Then the strain can be identified as *S. faecium, S. durans, S. avium,* or *S. bovis* on the basis of physiological characteristics.

In our studies with group D streptococci isolated from human infections, we found that nearly 73% of the strains had typical physiological characteristics and 20% of the strains were atypical in one test. The remaining 7% were difficult to identify to species (Facklam, 1972).

There seem to be demographic differences in the relative distribution of group D species in human infections. *S. faecalis* predominates among the isolates from the United States (Facklam, 1972), whereas *S. bovis* is less common; the reverse is true in Great Britain (Parker and Ball, 1976). The other group D species—*S. faecium, S. durans, S. avium,* and *S. bovis* variant—occur much less frequently, and *S. equinus* is found rarely, if at all, in humans.

Differentiation of Viridans Streptococci

If the bile esculin test in used as a replacement rather than as an adjunct to serology, some viridans streptococci will be misidentified as group D. It is absolutely necessary to rule out group D before considering viridans species. Again, species are best differentiated by considering the results of the physiological tests in toto rather than attempting to use a key or flow chart. The major characteristics of the group are: (i) they are not beta-hemolytic on blood agar (except for *S. mutans,* occasionally); (ii) they

Table 6. Differentiation of group D streptococci.[a]

Streptococcus species	Hydrolysis of:		Utilization of pyruvate	Growth at 10°C	Growth in 6.5% NaCl	Acid from:				
	Arginine	Starch				Mannitol	Sorbitol	Sorbose	Arabinose	Lactose
S. faecalis	+	−	+	+	+	+	+	−	−	+
S. faecium	+	−	−	+	+	+	−	−	+	+
S. durans	+	−	−	+	+	−	−	−	−	+
S. avium	−	−	+	−	+	+	+	+	+	+
S. bovis	−	+	−	−	−	+	−	−	−	+
S. bovis variant	−	−	−	−	−	−	−	−	−	+
S. equinus	−	−	−	−	−	−	−	−	−	−

[a] +, Positive reaction; −, negative reaction.

do not have group B or D antigens; (iii) they are not soluble in bile; and (iv) most strains do not grow in 6.5% NaCl broth. Subsequent differential tests are listed in Table 7. Some reactions vary among some species and are stable among others. Therefore, reactions in Table 7 should be used only as a guide for differentiating the viridans species in a way similar to that shown in Table 6 for differentiating group D streptococci. Almost 97% of the viridans streptococci isolated from human infections in the United States can be identified with the tests listed in Table

7 (Facklam, 1977). A different battery of tests and different species nomenclature have been used by others. From a recent comparison of the two systems (Facklam, 1977), it is clear that more data are needed to resolve differences. Among the viridans species isolated from patients with endocarditis, S. sanguis biotype I was the most common species identified in two recent studies (Facklam, 1977; Parker and Ball, 1976). S. mutans was the second most common in the latter and third most common in the former study. S. sanguis biotype II was the

Table 7. Differentiation of viridans streptococci.[a]

Streptococcus species	Hemolysis on blood agar	Hydrolysis of:			Acid in:				Extracellular polysaccharide in 5% sucrose:	
		Hippurate	Esculin	Arginine	Mannitol	Inulin	Lactose	Raffinose	Broth	Agar
S. mutans	α, β, γ	−	(+)	−	+	+	+	(+)	D	D
S. uberis	α, γ	+	+	(−)	+	+	+	(+)	N	N
S. salivarius	(γ)	−	(+)	−	−	(+)	+	(+)	N	(L)
S. sanguis I	(α)	−	+	(+)	−	+	+	V	(D)	(D)
S. MG-intermedius	α, γ	−	+	(+)	−	−	+	V	(N)	(N)
S. sanguis II	(α)	−	−	−	−	−	+	+	V	V
S. mitis	(α)	−	−	−	−	−	+	−	N	(N)
S. acidominimus	α, γ	+	−	V	−	−	−	−	N	N
S. anginosus-constellatus	α, γ	−	(+)	(+)	−	−	−	−	N	N
S. morbillorium	α, γ	−	−	−	−	−	−	−	N	N

[a] +, Positive reaction; −, negative reaction; V, variable reaction; ( ), occasional exception; D, dextran; L, levan; N, none.

second most common in Facklam's study of endo-carditis patients. In the same two studies, *S. milleri* (British nomenclature, Parker and Ball, 1976) or *S. MG-intermedius* and *S. anginosus-constellatus* strains (American nomenclature, Facklam, 1977) were found most often in brain abscesses. *S. morbillorium,* formerly called *Peptococcus morbillorium* or *Gemella hemolysans,* constituted slightly more than 3% of the viridans streptococci isolated from all human specimens studied and less than 1% of the viridans isolates from endocarditis patients (Facklam, 1977).

Identification of Pneumococci

Streptococcus pneumoniae is a part of the normal flora of the human oral cavity and must be differentiated from the other viridans streptococci when pneumococcal disease is suspected. Gram stains of body fluids (e.g., sputa, aspirates) are of little value when used alone (Merrill et al., 1973).

Streptococcus pneumoniae grows in infusion base media supplemented with 5% defibrinated or citrated blood. Approximately 8% of the strains require increased carbon dioxide for growth during primary isolation (Austrian and Collins, 1966). Colonial morphology on blood agar plates is characteristic and sufficiently stable to permit tentative identification. Surface colonies are small, mucoidal, opalescent, and flattened with entire edges surrounded by a zone of greenish discoloration of the blood agar medium. In contrast, the viridans streptococcal surface colonies are smaller, gray to whitish gray, and opaque with entire edges; they may be surrounded by a small zone of greenish discoloration of the blood agar medium. Typical pneumococcal colonies are larger and appear more watery than viridans colonies. A dissecting or broad-field microscope (25 to 50× magnification) should be used to examine the colonies.

Tentatively identified pneumococcal colonies can be serologically identified by the Quellung reaction. Cells from a single colony, suspended in a drop of physiological saline or various body fluids (including cerebral spinal fluid, peritoneal fluid, transtracheal aspirates, and sputa) can be examined directly with the Quellung test, or a broth culture can be examined. Suitable infusion broths include tryptose phosphate, Todd-Hewitt, and thioglycolate.

Quellung Test for Identification of Pneumococci (Austrian, 1976)

1. Place a small drop of culture, cell suspension, or body fluid on a glass slide.
2. Add a loopful (1 mm) of antiserum and mix well.
3. Add a small loopful of saturated aqueous methylene blue dye and mix.
4. Place a cover slip over the mixture and after 10 min examine microscopically with the oil immersion lens.
5. To avoid antigen excess, which may cause negative reactions, prepare slides so that each microscopic field contains 50–100 cells.
6. To obtain the oblique illumination needed to examine the slide, manipulate the concave mirror so that only about one-third of the light passes through the condenser at low power (10×).

Minor modifications are given in detail by Austrian (1976) and Lund (1960).

A positive Quellung reaction is the result of the binding of the pneumococcal capsular polysaccharide with type-specific antiserum. The corresponding change in refractive index causes the capsule to appear swollen. Actually, it just becomes more visible. A typical Quellung reaction is shown in Fig. 5. The pneumococcal cell, stained dark blue, is surrounded by a sharply demarcated halo that represents the outer edge of the capsule. The light transmitted through the capsule appears brighter than either the pneumococcal cell or the background of the slide. Single cells, pairs, chains, and even clumps of organisms may have positive Quellung reactions.

Pneumococci are identified with the Quellung test and a battery of antisera. The Statens Serum Institute, Copenhagen, Denmark, produces polyvalent ("Omni") antiserum with 83 different type-specific antibodies and polyvalent pooled antisera (nine pools, A through I) with each containing 7 to 11 type-specific antisera. In addition, the Institute produces type- and group-specific antisera, the latter containing from one to four types (Lund, 1963, 1966). Other commercial sources of type, group,

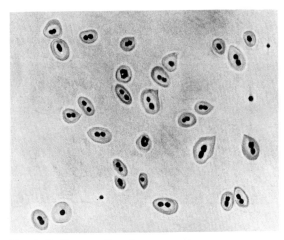

Fig. 5. Quellung reaction of pneumococci.

and pooled antisera are sporadically available, with the antibody composition of the pools varying among the different sources. Therefore, instructions supplied with each antiserum must be followed. Unless typing is required for epidemiological purposes, Omni and pooled sera are used to identify pneumococci. The presence of cross-reactive antibodies in Omni sera (Austrian, 1976) means that colonial and Gram stain morphology must also be relied upon.

Identification of Aerococci

No single characteristic is typical of the aerococci, which makes identifying them difficult. The two most typical characteristics are the arrangement of cells in tetrads and the fact that they grow in 6.5% NaCl broth—but some streptococci react comparably. The tests listed in Table 8 can be used to identify the aerococci. They are usually alpha-hemolytic, grow in 6.5% NaCl broth but not in broth incubated at 10°C or 45°C, and hydrolyze hippurate but not arginine. Atypical strains do not hydrolyze hippurate. Two species of aerococci may be found in humans (Facklam and Smith, 1976)—one that hydrolyzes hippurate and one that does not. On blood agar, the aerococci resemble the viridans streptococci, whereas in bile esculin and salt tolerance tests they resemble the group D streptococci. Most aerococci are susceptible to bacitracin, whereas most group D and viridans streptococci are not. This characteristic is a useful adjunct to those listed in Table 8.

Colman (1967) described aerococcus-like organisms from the blood and urine of patients with subacute bacterial endocarditis (SBE) and urinary tract infections (UTI). Parker and Ball (1976) reported aerococci in 1.3% of their SBE patients.

Presumptive Identification of Streptococci

Although streptococci are most accurately and completely identified by grouping beta-hemolytic strains (and alpha-hemolytic or nonhemolytic groups B and D, and pneumococci) and physiologically identifying to species other alpha-hemolytic and nonhemolytic strains, they can also be identified approximately 95% of the time with presumptive tests. Small laboratories may not have sufficient time or funds for serological grouping and complete physiological testing and may find the alternative methods listed in Table 9 especially useful.

The source of the infecting strain, its colonial morphology, and its hemolytic activity on primary blood agar media are the characteristics used in determining which tests to perform. For example, hemolysis and bacitracin susceptibility results are usually sufficient to differentiate the beta-hemolytic group A streptococci from the beta-hemolytic non-group A streptococci from throat and skin lesions. Similarly, determining optochin susceptibility or bile solubility on suspected pneumococcal colonies is sufficient for the presumptive identification of pneumococci. Regardless of hemolysis, group B streptococci are presumptively identified by either hippurate hydrolysis or by the CAMP test, and group D are presumptively identified by the bile esculin and salt tolerance tests. Streptococci not identified as groups A, B, D, or pneumococci are simply called "beta-hemolytic not group A, B, or D" or "viridans streptococci".

Table 8. Differentiation of aerococci from viridans streptococci and group D enterococci.[a]

Organism	Characteristic									
	Cellular arrangement	Hemolysis	Group D serological reaction	Bile-esculin reaction	Growth in 6.5% NaCl broth	Growth at 10°C	Growth at 45°C	Hippurate hydrolysis	Arginine hydrolysis	Peroxidase
Aerococci	Tetrads	α	−	V	+	−	−	(+)	−	V
Viridans streptococci	Chains	α or γ	−	(−)	−	−	V	(−)	V	−
Group D enterococci	Short chains, diplos	α β γ	+	+	+	+	+	(−)	+	(−)

[a] +, Positive reaction; −, negative reaction; V, variable reaction; ( ), occasional exception.

Table 9. Presumptive identification of streptococci.[a]

Category	Characteristic							
	Hemolysis	Bacitracin susceptibility	Susceptibility to sulfatrimethoprim sulfoxazole	CAMP reaction	Hippurate hydrolysis	Bile-esculin reaction	Growth in 6.5% NaCl broth	Optochin susceptibility and bile solubility
Group A	Beta	+	−	−	−	−	−	−
Group B	(Beta)	(−)	−	+	+	−	V	−
Beta-hemolytic streptococci, not groups A, B, or D	Beta	(−)	+	−	−	−	−	−
Group D, enterococcus	Alpha Beta None	−	−	−	(−)	+	+	−
Group D, not enterococcus	Alpha None	−	V	−	−	+	−	−
Viridans	Alpha None	(−)	+	−	(−)	(−)	−	−
Pneumococcus	Alpha	V	?	−	−	−	−	+

[a] +, Positive reaction or susceptible; −, negative reaction or resistant; V, variable reaction; ( ), exceptions occasionally occur.

HEMOLYSIS. It is important to emphasize the fact that hemolysis is the most important characteristic to determine in the presumptive identification of streptococci (see Determination of Hemolysis).

BACITRACIN. The bacitracin susceptibility test can be used for presumptive differentiation of beta-hemolytic group A and nongroup A streptococci. Several important factors affect bacitracin test results. Commercially available disks can be used to differentiate beta-hemolytic group A streptococci and other beta-hemolytic streptococci but only if they are differential disks (0.04 units) rather than sensitivity disks (which are too concentrated). A heavy inoculum is advisable, because if the inoculum is too light, non-group A streptococci will appear to be susceptible to bacitracin. A pure culture must be used. Bacitracin differential disks placed on primary plates inoculated with throat swabs identified only 50–65% of the group A streptococci (Murray et al., 1976a; Sprunt, Vail, and Asnes, 1974). Only beta-hemolytic streptococci should be tested, because many alpha-hemolytic streptococci, including pneumococci, are inhibited by the bacitracin differential disk. Each new lot of disks should be tested with known strains of group A and non-group A streptococci so that lot-to-lot variation and stability can be determined. Any size zone of inhibition must be interpreted as positive, even though several authors and a technical bulletin recommend measuring zone size. However, the originator of the test (Maxted,

1953) did not specify that zones had to be of certain size, and no experimental data are available to show that diameters must be measured in order to differentiate group A from non-group A streptococci. Furthermore, false-positive results are potentially less harmful than false-negatives. Approximately 5% of the group A streptococci in the Ederer and Chapman (1972) study had zones less than 10 mm in diameter. Facklam et al. (1974) identified 99.5% of the group A streptococci without considering zone size.

A pure beta-hemolytic culture (three or four colonies or a loopful of an overnight broth culture) is streaked onto a 15-×-100-mm blood agar plate. The bacitracin disk is placed in the center of the streaked area and the plate is incubated overnight at 35–37°C. A zone of inhibition of streptococcal growth around the bacitracin disk (Fig. 6) indicates that the strain can be reported as ''beta-hemolytic, presumptive group A by bacitracin'' or ''beta-hemolytic Streptococcus, presumptively not group A by bacitracin''. Up to 6% of the beta-hemolytic group B streptococci of human origin are also susceptible to bacitracin. Confusing members of the two groups can be avoided by using a physiological test to identify the group B streptococci presumptively. Three such tests are described below.

SXT. Disks containing 1.25 mg trimethoprim and 23.75 mg sulfamethoxazole (SXT) can be used in presumptively separating groups A and B streptococci from the other beta-hemolytic groups. Groups A and B must then be differentiated from each other by additional testing. SXT disks are available commercially. Sheep red blood cells and Trypticase soy agar base must be used to make the plates for the SXT test, and it is helpful to use the same criteria for inoculum size and interpretation of zones of inhibition for SXT as are used for bacitracin. Groups A and B streptococci are uniformly resistant to SXT (as depicted in Fig. 6) when incubated in normal atmosphere. Beta-hemolytic, SXT-resistant, bacitracin-susceptible streptococci are presumptively identified as group A, and those that are bacitracin-resistant are presumptively identified as group B. Beta-hemolytic, SXT-susceptible, bacitracin-resistant strains are labeled non-group A or B beta-hemolytic streptococci. An occasional group A streptococcal strain is susceptible to both the bacitracin and SXT disks (Gunn, 1976).

CAMP. A more specific test for the presumptive identification of group B streptococci is the CAMP test. This test is performed by making a single streak of the Streptococcus specimen perpendicular to but not touching (1 cm apart) a streak of beta-lysin-producing Staphylococcus aureus strain on a sheep or bovine blood agar plate. Washed sheep red blood cells that have been resuspended in sterile physiological saline and Trypticase soy agar are recom-

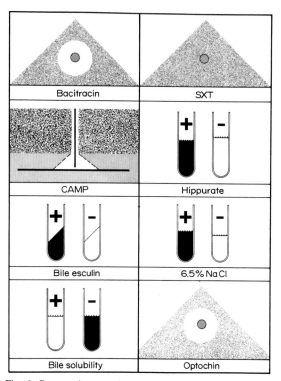

Fig. 6. Presumptive tests for identifying streptococci.

mended. Control strains of groups A, B, C, and G streptococci should be used, especially when plates are obtained commercially. The inoculated plates should be incubated in a candle jar or in normal atmosphere but not anaerobically. Some group A streptococci are positive when incubated in the candle jar, and even more strains are positive when incubated anaerobically. Group B streptococci produce a substance ("CAMP factor") that enlarges the zone of lysis produced by the *Staphylococcus* culture to form a typical arrowhead or flame-shaped clearing at the juncture of the two organisms (Fig. 6) (Darling, 1975).

Bacitracin-negative, CAMP-positive, beta-hemolytic streptococci can be reported as presumptive group B streptococci. Bacitracin-positive, CAMP-positive, beta-hemolytic streptococci are either group A or group B. The two groups can then be differentiated by hemolysis. Group B streptococci have smaller hemolysis zones on the surface and in the stab than do group A streptococci. Bacitracin-positive, CAMP-positive group A and group B streptococci are found frequently enough to make this distinction necessary.

Bacitracin-positive, CAMP-negative, beta-hemolytic streptococci are presumptive group A streptococci. Bacitracin-negative, CAMP-negative, beta-hemolytic streptococci are beta-hemolytic streptococci, not group A or B.

Nonhemolytic group B streptococci are usually CAMP positive; therefore, nonhemolytic streptococci that are bile-esculin negative and CAMP positive can be presumptively identified as nonhemolytic group B streptococci.

A modified CAMP test can be done with beta-lysin impregnated filter paper disks ("CAMP-disks", Wilkinson, 1977). The disk is placed on a sheep blood agar plate approximately 1 cm from the end of the streptococcal streak. After overnight aerobic incubation, a crescent-shaped clearing of the dark beta-lysin zone surrounding the disk provides presumptive evidence that the *Streptococcus* is group B. Perhaps in the future commercial disks will be available, thus eliminating the need for an actively growing beta-lysin-producing *Staphylococcus aureus* strain.

SODIUM HIPPURATE. A third presumptive test that can be used to identify group B streptococci is the hippurate hydrolysis test (Facklam et al., 1974). The formulas for the medium and reagents are described later in this chapter.

Inoculate the broth with two or three colonies of beta-hemolytic streptococci. Incubate overnight at 35°C and centrifuge the tube. To 0.2 ml of supernatant fluid add 0.8 ml ferric chloride reagent and mix well. If a heavy precipitate remains longer than 10 min, the test is positive and the *Streptococcus* is presumptively group B (Fig. 6). If the test is nega-

tive or weakly positive, the growth tube should be incubated for another 24 h and the test repeated.

All group B streptococci, whether they are beta-hemolytic or nonhemolytic, hydrolyze hippurate and therefore react positively. Group D streptococci may also react positively, but are identified by the bile-esculin (BE) test. Beta-hemolytic streptococci other than group B (groups A, C, G, and F) do not hydrolyze hippurate. The alpha-hemolytic streptococci, *S. uberis* and *S. acidominimus*, which also hydrolyze hippurate, are rarely found in human infections. However, *S. uberis* is found in bovine infections and therefore must be identified by additional tests. The BE-negative, hippurate-positive, beta-hemolytic streptococci can be reported as "presumptive group B streptococci by hippurate hydrolysis".

Several other presumptive tests for group B streptococci have been described, but none are as reliable as the CAMP or hippurate hydrolysis tests.

BILE ESCULIN. Many tests have been described to identify the group D streptococci presumptively, but the most accurate is probably the bile-esculin (BE) test (Facklam, 1973; Facklam and Moody, 1970; Gross, Houghton, and Senterfit, 1975). The formula for the medium is described later in this chapter.

Inoculate an agar slant of BE medium with two or three colonies and incubate the slant at 35°C. If more than half the slant is blackened within 24–48 h (Fig. 6), the test is positive and the *Streptococcus* is presumptively identified as group D. If less than half the slant is blackened or no blackening occurs within 24–48 h, the test is negative.

Although all group D streptococci are BE positive, this test cannot be used to differentiate enterococci, which are penicillin resistant, from susceptible nonenterococcal strains. It is necessary to differentiate the strains in order to treat enterococcal infections with a combination of antibiotics.

SALT TOLERANCE (6.5% NaCl). The enterococcal species (*S. faecalis*, *S. faecium*, *S. durans*, and *S. avium*) are easily differentiated from the nonentero-coccal species (*S. bovis* and *S. equinus*) by the 6.5% NaCl tolerance test (Facklam, 1973). The formula for the medium is given later in this chapter.

Inoculate two or three streptococcal colonies into 6.5% NaCl broth and incubate at 35°C. Examine for growth (indicated by turbidity and *sometimes* change in indicator color) in 24, 48, and 72 h (Fig. 6). Growth within 72 h indicates that the strain is salt tolerant (i.e., positive) and it can be identified as an enterococcus if it is also BE-positive or if it is serologically group D.

The enterococci usually grow heavily and cause an indicator change within 24 h, although some strains grow only after 48 h and some do not cause an indicator change even after 72 h. About 80% of group B streptococci also grow in this medium;

some change the indicator. Beta-hemolytic groups A, C, F, and G usually do not grow in this medium, although salt-tolerant group A streptococci do occasionally. The alpha-hemolytic, nongroupable streptococci (viridans) do not grow in 6.5% NaCl broth, nor do group D species *S. bovis* and *S. equinus*. Cultures that give positive salt tolerance tests and that are not presumptive group B or group D streptococci should be tested for purity by streaking the growth from the salt-tolerance test broth onto a blood agar plate and comparing the morphology to that of the original strain. If the morphology differs, a Gram stain and a catalase test should be performed.

An alpha-hemolytic or nonhemolytic strain that grows in 6.5% NaCl broth and reacts negatively in the bile-esculin test may be an *Aerococcus*. It is confirmed as such by Gram stain (tetrad cellular arrangement) and by the fact that it is negative in the arginine hydrolysis and in 10 and 45°C tests (see Table 8).

BILE SOLUBILITY.   The bile solubility test can be used for the presumptive differentiation of pneumococci from the other viridans streptococci (Lund, 1959).

Inoculate a broth (Todd-Hewitt, Trypticase soy, etc.) with two or three suspect colonies and incubate it overnight at 35°C in an atmosphere of increased CO_2 (CO_2 incubator or candle jar). Place 0.5 ml of the broth culture in each of two 13-×-100-mm test tubes. Add one or two drops of phenol red indicator and neutralize with 1 N NaOH. Add 0.5 ml 2% Na deoxycholate (bile) to one tube and 0.5 ml saline to the other. Incubate tubes at 35°C and examine periodically for up to 2 h. A clearing of turbidity in the bile tube and not in the saline control tube indicates a positive test—that is, the pneumococcal cells have been lysed (''solubilized'', Fig. 6).

Occasionally, the turbid broth only partially clears. For example, 83% of 263 pneumococcal strains were completely lysed by 2% Na deoxycholate and the remainder partially lysed, whereas when the concentration of deoxycholate was increased to 10%, 86% of the strains were completely lysed (unpublished). Therefore, either 2% or 10% deoxycholate can be used, but another test (Optochin) is needed to allow a more accurate presumptive identification of pneumococci. Partially soluble, alpha-hemolytic, Optochin-susceptible streptococci are presumptively identified as pneumococci. Resistant strains are viridans streptococci.

OPTOCHIN.   Optochin susceptibility is used for presumptive differentiation of alpha-hemolytic viridans streptococci and pneumococci (Bowers and Jeffries, 1955).

Two or three suspect colonies are streaked onto a quarter plate, and the Optochin disk is placed in the upper third of the streaked area. The plate is incubated overnight in a candle jar or CO_2 incubator (35°C). Cultures do not grow as well in normal atmosphere, and larger zones of inhibition occur. If a 6-mm disk is used, a zone of inhibition at least 14 mm in diameter is considered positive for pneumococci (Fig. 6). A diameter between 6 and 14 mm is questionable, and the strain is presumptively identified as a pneumococcus only if it is bile soluble. For 10-mm Optochin disks, a zone of inhibition at least 16 mm in diameter is positive, and strains with inhibition zones between 10 and 16 mm should be tested for bile solubility.

Results of testing 362 pneumococcal and 100 viridans strains indicate that slightly more pneumococci were correctly identified with the 10-mm disk (98.4% with zones at least 16 mm) than with the 6-mm disk (97.4% with zones at least 14 mm or 93.4% with zones at least 16 mm), but that 2% of the viridans streptococci were partially inhibited (11- to 15-mm zones) by the larger disks (unpublished data).

Formulas for, and Interpretation of, Physiological Tests

Acid in Carbohydrate Broths (Arabinose, Inulin, Lactose, Mannitol, Raffinose, Sorbitol, Sorbose, Trehalose)

1. Heart infusion broth, 22.5 g, in 900 ml distilled water.
2. Carbohydrate, 10 g, in 100 ml distilled water.
3. Indicator, 1 ml (1.6 g bromocresol purple in 100 ml 95% ethanol).

Mix 1, 2, and 3 together and dispense in 3-ml amounts in 13-×-100-mm screw-capped tubes. Sterilize in an autoclave for 10 min at 121°C. A positive reaction is recorded when the indicator changes from purple to yellow.

Arginine Hydrolysis (Moeller's Decarboxylase Medium)

1. Peptone (Orthana special), 5 g.
2. Beef extract, 5 g.
3. Indicator, 0.625 ml (1.6 g bromocresol purple, in 100 ml 95% ethanol).
4. Cresol red, 2.5 ml (0.2%, prepared by grinding 0.5 g cresol red powder to a fine powder, adding 26.2 ml 0.01 N NaOH, and diluting to 250 ml with distilled water).
5. Pyridoxal, 5 mg.
6. L-Arginine, 10 g (if D,L-arginine is used, add 20 g).
7. Distilled water, 1,000 ml.

Adjust pH to 6.0–6.5 and dispense in 3-ml amounts in 13-×-100-mm screw-capped tubes. Sterilize in an autoclave for 10 min at 121°C.

Immediately after inoculation, add a layer (about 10 mm) of sterile mineral oil. A positive reaction is recorded when the indicators turn violet to reddish violet. Yellow does not indicate a positive reaction—it indicates an acid reaction rather than deamination.

Bile Esculin

1. Add 23 g of nutrient agar to 400 ml H_2O and mix well; heat until colloidal.
2. Add 40 g Oxgall to 400 ml H_2O and mix well; heat into solution.
3. Add 0.5 g ferric citrate to 100 ml H_2O and mix well; heat into solution.
4. Combine solutions 1, 2, and 3 and mix well; heat to 100°C for 10 min.
5. Sterilize in autoclave at 121°C for 15 min. (This is the base medium.)
6. Cool to 50°C.
7. Prepare 1 g esculin solution (add 1 g esculin to 100 ml H_2O, heat gently to dissolve, and sterilize by filtration, Seitz or Millipore).
8. Aseptically, add sterile esculin solution to base medium and mix well; dispense into 16-×-125-mm screw-capped tubes and slant tubes.

The original medium also contained horse serum (50 ml) added to the base medium. However, we found in a controlled study that this addition was not necessary (Facklam, 1973). Also, at least one commercial source (Difco) adds the esculin to the base medium, as described in step 8. The dehydrated medium sold by Difco can be resuspended, tubed and autoclaved, slanted, and used with excellent results.

BE medium can be used to identify group D streptococci. All group D streptococci (including all enterococci) will blacken the BE slant, usually within 48 h. Most non-group D streptococci do not blacken the medium.

Esculin Hydrolysis

Heart infusion agar	40 g
Esculin	1 g
Ferric chloride	0.5 g
Distilled H_2O	1,000 ml

Heat to dissolve ingredients, dispense in 13-×-100-mm screw-capped tubes, and sterilize in an autoclave for 15 min at 121°C. Slant tubes for cooling period. Esculin hydrolysis is indicated when the medium turns black.

Glucan Production in Sucrose Agar

Heart infusion agar	40 g
Sucrose	50 g
Distilled H_2O	1,000 ml

Sterilize in an autoclave for 15 min at 121°C. Cool to 55°C and pour into 13-×-100-mm Petri dishes (approximately 10 ml in each).

Glucan production typical of *Streptococcus sanguis* and *S. mutans* results in adherent growth on the agar which can be either highly refractile or white and dry. Levan production, typical of *S. salivarius,* results in opaque, gummy, non-adherent growth. Typical *S. bovis* growth is similar to that of *S. salivarius* but is somewhat less gummy and rarely adheres to the medium. Large or small colonies that are mucoidal and non-adherent are considered to be negative or have no extracellular polysaccharide production.

Glucan Production in Sucrose Broth

Solution A:

NIH thio broth	28.5 g
K_2HPO_4	10.0 g
Sodium acetate	12.0 g
Distilled H_2O	500 ml

Solution B:

Sucrose	50.0 g
Distilled H_2O	500 ml

Sterilize solutions separately in an autoclave for 15 min at 121°C. Cool to 55°C, mix solutions, and dispense in 16-×-125-mm screw-capped tubes in 5-ml amounts. Glucan production is indicated when the broth is partially or completely gelled—a typical *S. sanguis* reaction. Glucan production is also indicated when gelatinous, adherent deposits form on the bottom and walls of the growth tube—a typical *S. mutans* reaction. An increase in viscosity of the broth indicates the production of slime (unknown polysaccharide), typical of *S. bovis*. Negative reactions are recorded when no gelling, deposit, or increase in viscosity occurs.

Growth at 10°C and 45°C

Heart infusion broth	25 g
Dextrose	1 g
Indicator (1.6 g bromo-cresol purple in 100 ml 95% ethanol)	1 ml
Distilled H_2O	1,000 ml

Dispense in 5-ml amounts in 16-×-125-mm screw-capped tubes. Sterilize in an autoclave for 15 min at 121°C. A positive reaction is recorded when growth is indicated by a color change from purple to yellow or by frank growth in the tube.

Pyruvate Utilization

Tryptone	10 g
Yeast extract	5 g
K_2HPO_4	5 g

NaCl	5 g
Sodium pyruvate	10 g
Bromothymol blue	0.104 g
Distilled water	1,000 ml

Check the pH and adjust to 7.1–7.4 if necessary. Dispense in 13-×-100-mm screw-capped tubes. Sterilize in an autoclave 15 min at 121°C. A positive reaction is recorded when the indicator changes from green to definite yellow. Yellow-green indicates a weak reaction and should be regarded as negative utilization of pyruvate.

Sodium Chloride (6.5%) Tolerance

Heart infusion broth	25 g
NaCl	60 g
Indicator (1.6 g bromo-cresol purple in 100 ml 95% ethanol)	1 ml
Dextrose	1 g
Distilled water	1,000 ml

Add all reagents together up to 1,000 ml (final volume). Dispense in 15-×-125-mm screw-capped tubes and sterilize in an autoclave 15 min at 121°C. A positive reaction is recorded when the indicator changes from purple to yellow or when growth is obvious even though the indicator does not change.

Sodium Hippurate Hydrolysis

Heart infusion broth	25 g
Na hippurate	10 g
Distilled H_2O	1,000 ml

Sterilize in an autoclave for 15 min at 121°C after dispensing in 15-×-125-mm screw-capped tubes. Tighten caps to prevent evaporation.

Ferric Chloride Reagent

| $FeCl_3 \cdot 6H_2O$ | 12 g |
| 2% Aqueous HCl | 100 ml |

(2% aq HCl is made by adding 5.4 ml of concentrated HCl (37%) to 94.6 ml H_2O.) Inoculate with two or three colonies of beta-hemolytic streptococci, incubate at 35°C for 20 h or longer, centrifuge the medium to pack the cells, and pipette 0.8 ml of the clear supernate into a Kahn tube. Add 0.2 ml of the ferric chloride reagent to the Kahn tube and mix well. If a heavy precipitate remains longer than 10 min, the test is positive.

Starch Hydrolysis

Heart infusion agar	40 g
Soluble starch	20 g
Distilled H_2O	1,000 ml

Warm to dissolve, and sterilize in an autoclave for 15 min at 121°C. Cool to 55°C and pour into sterile Petri dishes. Hydrolysis of starch is determined by flooding the surface of the plate with Gram's iodine 48 h after inoculation and incubation at 35°C. A zone of hydrolysis appears colorless and a dark blue to purple indicates that the starch has not been hydrolyzed.

Literature Cited

Aber, R. C., Allen, N., Howell, J. T., Wilkinson, H. W., Facklam, R. R. 1976. Nosocomial transmission of group B streptococci. Pediatrics 58:346–353.

Anthony, B. F., Concepcion, N. F. 1975. Group B Streptococcus in a general hospital. Journal of Infectious Diseases 132:561–567.

Anthony, B. F., Okada, D. M. 1977. The emergence of group B streptococci in infections of the newborn infant. Annual Review of Medicine 28:355–369.

Armstrong, D., Blevins, A., Louria, D. B., Henkel, J. S., Moody, M. D., Sukany, M. 1970. Groups B, C, and G streptococcal infections in a cancer hospital. Annals of the New York Academy of Sciences 174:511–522.

Austrian, R. 1976. The Quellung reaction: A neglected microbiologic technique. Mount Sinai Journal of Medicine 43:699–709.

Austrian, R., Collins, P. 1966. Importance of carbon dioxide in the isolation of pneumococci. Journal of Bacteriology 92:1281–1284.

Baker, C. J. 1974. Unusual occurrence of neonatal septicemia due to group G streptococcus. Pediatrics 53:568–569.

Baker, C. J., Clark, D. J., Barrett, F. F. 1973. Selective broth medium for isolation of group B streptococci. Applied Microbiology 26:884–885.

Bannatyne, R. M., Randall, C. 1977. Ecology of 350 isolates of group F streptococcus. American Journal of Clinical Pathology 67:184–186.

Bateman, N. T., Eykyn, S. J., Phillips, I. 1975. Pyogenic liver abscess caused by Streptococcus milleri. Lancet i:657.

Bayer, A. S., Seidel, M. S., Yoshikawa, T. T., Anthony, B. F., Guze, L. B. 1976. Group D enterococcal meningitis. Clinical and therapeutic considerations with report of three cases and review of the literature. Archives of Internal Medicine 136:883–886.

Belcher, D. W., Afoakwa, S. N., Osei-Tutu, E., Wurapa, F. K., Osei, L. 1975. Non-group-A streptococci in Ghanaian patients with pyoderma. Lancet ii:1032.

Belgaumker, T. K. 1975. Impetigo neonatorum congenita due to group B beta-hemolytic streptococcus infection. Journal of Pediatrics 86:982–983.

Benjamin, J. T., Perriello, V. A., Jr. 1976. Pharyngitis due to group C hemolytic streptococci in children. Journal of Pediatrics 89:254–256.

Black, W. A., Van Buskirk, F. 1973. Gentamicin as a selective agent for the isolation of beta haemolytic streptococci. Journal of Clinical Pathology 26:154–156.

Blanchette, L. P., Lawrence, C. 1967. Group A streptococcus screening with neomycin blood agar. American Journal of Clinical Pathology 48:441–443.

Blevins, A., Armstrong, D., Louria, B., Moody, M., Sukany, M. 1968. Incidence of streptococci of the Lancefield group G in patients at a cancer hospital. Bacteriological Proceedings 1968:101–102.

Bowers, E. F., Jeffries, L. R. 1955. Optochin in the identification of Str. pneumoniae. Journal of Clinical Pathology 8:58–60.

Brown, J. H. 1919. The use of blood agar for the study of streptococci. Rockefeller Institute for Medical Research Monograph No. 9. New York: The Rockefeller Institute for Medical Research.

Buchanan, R. E., Gibbons, N. E. (eds.). 1974. Bergey's manual of determinative bacteriology, 8th ed. Baltimore: Williams & Wilkins.

Burech, D. L., Koranyi, K. I., Haynes, R. E. 1976. Serious group A streptococcal diseases in children. Journal of Pediatrics **88:**972–974.

Burech, D. L., Koranyi, K., Haynes, R. E., Kramer, R. N. 1975. Pneumococcal bacteremia associated with gingival lesions in infants. American Journal of Diseases in Children **129:**1283–1284.

Cherry, W. B., Moody, M. D. 1965. Fluorescent-antibody techniques in diagnostic bacteriology. Bacteriological Reviews **29:**222–250.

Colman, G. 1967. Aerococcus-like organisms isolated from human infections. Journal of Clinical Pathology **20:**294–297.

Converse, G. M., III, Dillon, H. G., Jr. 1977. Epidemiological studies of *Streptococcus pneumoniae* in infants: Methods of isolating pneumococci. Journal of Clinical Microbiology **5:**293–296.

Coykendall, A. L. 1977. Proposal to elevate the subspecies of *Streptococcus mutans* to species status based on their molecular composition. International Journal of Systematic Bacteriology **27:**26–30.

Darling, C. L. 1975. Standardization and evaluation of the CAMP reaction for the prompt, presumptive identification of *Streptococcus agalactiae* (Lancefield group B) in clinical material. Journal of Clinical Microbiology **1:**171–174.

Davies, J. N. P. 1977. Mortality from pneumococcal meningitis. Lancet i:255.

Dayton, S. L., Chipps, D. D., Blasi, D., Smith, R. F. 1974. Evaluation of three media for selective isolation of Gram-positive bacteria from burn wounds. Applied Microbiology **27:**420–422.

Deibel, R. H. 1964. The group D streptococci. Bacteriological Reviews **28:**330–366.

Deibel, R. H., Evans, J. B. 1960. Modified benzidine test for the detection of cytochrome-containing respiratory systems in microorganisms. Journal of Bacteriology **79:**356–360.

Dillon, H. C., Dudding, B. A. 1970. Streptococcal infections, chap. 6. In: Kelley, V. C. (ed.), Brennemann's practice of pediatrics, vol. II. Hagerstown, Maryland: Harper & Row.

Dilworth J. A., Stewart, P., Gwaltney, J. M., Jr., Hendley, J. O., Sande, M. A. 1975. Methods to improve detection of pneumococci in respiratory secretions. Journal of Clinical Microbiology **2:**453–455.

Duma, R. J., Weinberg, A. N., Medrek, T. F., Kunz, L. J. 1969. Streptococcal infections. A bacteriologic and clinical study of streptococcal bacteremia. Medicine **48:**87–127.

Ederer, G. M., Chapman, S. S. 1972. Simplified flourescent-antibody staining method for primary plate isolates of group A streptococci. Applied Microbiology **24:**160–161.

Ederer, G. M., Herrmann, M. M., Bruce, R., Matsen, J. M., Chapman, S. S. 1972. Rapid extraction method with pronase B for grouping beta-hemolytic streptococci. Applied Microbiology **23:**285–288.

Eickhoff, T. C., Klein, J. O., Daly, A. K., Ingall, D., Finland, M. 1964. Neonatal sepsis and other infections due to group B beta-hemolytic streptococci. New England Journal of Medicine **271:**1221–1228.

El Kholy, A., Wannamaker, L. W., Krause, R. M. 1974. Simplified extraction procedure for serological grouping of beta-hemolytic streptococci. Applied Microbiology **28:**836–839.

Ellner, P. D., Stoessel, C. J., Drakeford, E., Vasi, F. 1966. A new culture medium for medical bacteriology. American Journal of Clinical Pathology **45:**502–504.

Epstein, M., Calia, F. M., Gabuzda, G. J. 1968. Pneumococcal peritonitis in patients with postnecrotic cirrhosis. New England Journal of Medicine **278:**69–73.

Evans, J. B., Kerbaugh, M. A. 1970. Recognition of *Aerococcus viridans* by the clinical microbiologist. Health Laboratory Science **7:**76–77.

Facklam, R. R. 1972. Recognition of group D streptococcal species of human origin by biochemical and physiological tests. Applied Microbiology **23:**1131–1139.

Facklam, R. R. 1973. Comparison of several laboratory media for presumptive identification of enterococci and group D streptococci. Applied Microbiology **26:**138–145.

Facklam, R. R. 1977. Physiological differentiation of viridans streptococci. Journal of Clinical Microbiology **5:**184–201.

Facklam, R. R., Moody, M. D. 1970. Presumptive identification of group D streptococci: The bile-esculin test. Applied Microbiology **20:**245–250.

Facklam, R. R., Smith, P. B. 1976. The Gram positive cocci. Human Pathology **7:**187–194.

Facklam, R. R., Padula, J. F., Thacker, L. G., Wortham, E. C., Sconyers, B. J. 1974. Presumptive identification of group A, B, and D streptococci. Applied Microbiology **27:**107–113.

Feingold, D. S., Stagg, N. L., Kunz, L. J. 1966. Extrarespiratory streptococcal infections. Importance of the various serologic groups. New England Journal of Medicine **275:**356–361.

Finland, M., Barnes, M. W. 1970. Changing etiology of bacterial endocarditis in the antibacterial era. Experiences at Boston City Hospital 1933–1965. Annals of Internal Medicine **72:**341–348.

Finland, M., Barnes, M. W. 1977. Changes in occurrence of capsular serotypes of *Streptococcus pneumoniae* at Boston City Hospital during selected years between 1935 and 1974. Journal of Clinical Microbiology **5:**154–166.

Freeburg, P. W. 1970. Rapid fluorescent-antibody stain technique with group A streptococci. Applied Microbiology **19:**940–942.

Freeburg, P. W., Buckingham, J. M. 1976. Evaluation of the Bacti-lab streptococci culture systems for selective recovery and identification of streptococci. Journal of Clinical Microbiology **3:**443–448.

Fuller, A. T. 1938. The formamide method for the extraction of polysaccharides from haemolytic streptococci. British Journal of Experimental Pathology **19:**130–139.

Gray, B. M., Pass, M. A., Dillon, H. C. Jr. 1979. Laboratory and field evaluation of selective media for isolation of group B streptococci. Journal of Clinical Microbiology **9:**466–470.

Gross, K. C., Houghton, M. P., Senterfit, L. B. 1975. Presumptive speciation of *Streptococcus bovis* from human sources by using arginine and pyruvate tests. Journal of Clinical Microbiology **1:**54–60.

Gross, P. A., Harkavy, L. M., Barden, G. E., Flower, M. F. 1976. The epidemiology of nosocomial enterococcal urinary tract infection. American Journal of the Medical Sciences **272:**75–81.

Gunn, B. A. 1976. SXT and Taxo A disks for presumptive identification of group A and B streptococci in throat cultures. Journal of Clinical Microbiology **4:**192–193.

Gunn, B. A., Ohashi, D. A., Gaydos, C. A., Holt, E. S. 1977. Selective and enhanced recovery of group A and B streptococci from throat cultures with sheep blood agar containing sulfamethoxazole and trimethoprim. Journal of Clinical Microbiology **5:**650–655.

Guthof, O. 1960. Zur Frage der Indentität pathogener vergrünender Streptokokken mit den "normalen Mundstreptokokken". Zeitschrift für Hygiene und Infektionskrankheiten **146:**425–432.

Hable, K. A., Horstmeier, C., Wold, A. D., Washington, J. A., II. 1973. Group A β-hemolytic streptococcemia. Mayo Clinic Proceedings **48:**336–339.

Hajna, A. A. 1951. A buffered azide glucose-glycerol broth for presumptive and confirmative tests for fecal streptococci. Public Health Laboratory **9:**80–81.

Hajna, A. A., Perry, C. A. 1943. Comparative study of presumptive and confirmative media for bacteria of the coliform group and for fecal streptococci. American Journal of Public Health and the Nation's Health **33:**550–556.

Harrell, W. K., George, J. R. 1972. Quantitative measurement of precipitating antibodies in streptococcal grouping antisera by the single radial immunodiffusion technique. Applied Microbiology 23:1047–1052.

Hill, H. R., Caldwell, G. G., Wilson, E., Hager, D., Zimmerman, R. A. 1969. Epidemic of pharyngitis due to streptococci of Lancefield group G. Lancet: ii:371–374.

Holdeman, L. V., Moore, W. E. C. 1974. New genus, Coprococcus, twelve new species, and emended descriptions of four previously described species of bacteria from human feces. International Journal of Systematic Bacteriology 24:260–277.

Holmberg, K., Hallander, H. O. 1973. Production of bactericidal concentrations of hydrogen peroxide by Streptococcus sanguis. Archives of Oral Biology 18:423–434.

Horvitz, R. A., von Graevenitz, A. 1977. A clinical study of the role of enterococci as sole agents of wound and tissue infection. Yale Journal of Biology and Medicine 50:391–395.

Horn, K. A., Goodpasture, H. C., Jackson, H. 1974. Sore throats: cultures and treatment. Journal of the American Medical Association 227:799–800.

Horstmeier, C., Washington, J. A., II. 1973. Microbiological study of streptococcal bacteremia. Applied Microbiology 26:589–591.

Howard, J. B., McCracken, G. H., Jr. 1974. The spectrum of group B streptococcal infections in infancy. American Journal of Diseases in Children 128:815–818.

Kast, A. 1971. Comparative statistical investigations regarding incidence, etiology and topography of subacute bacterial endocarditis. Japanese Circulation Journal 35:1203–1212.

Kauffman, C. A., Watanakunakorn, C., Phair, J. P. 1976. Pneumococcal arthritis. Journal of Rheumatology 3:4:409–419.

Kilbourn, J. P. 1973. Streptococcus pneumoniae in chronic bronchitis. Lancet i:1009.

Klein, J. O. 1975. Pneumococcal bacteremia in the young child. American Journal of Diseases in Children 129:1266–1267.

Koepke, J. P. 1965. Meningitis due to Streptococcus anginosus (Lancefield group F). Journal of the American Medical Association 193:739–740.

Koshi, G. 1971. Serologic groups of hemolytic streptococci in human infections. Indian Journal of Medical Research 59:394–400.

Kronvall, G. 1973. A surface component in group A, C, and G streptococci with non-immune reactivity for immunoglobulin G. Journal of Immunology 111:1401–1406.

Lancefield, R. C. 1933. A serological differentiation of human and other groups of hemolytic streptococci. Journal of Experimental Medicine 57:571–595.

Lawrence, M. S., Cobbs, C. G. 1972. Endocarditis due to group C streptococci. Southern Medical Journal 65:487–489.

LeBien, T. W., Bromel, M. C. 1975. Antibacterial properties of a peroxidogenic strain of Streptococcus mitior (mitis). Canadian Journal of Microbiology 21:101–103.

Lerner, P. I. 1975. Meningitis caused by Streptococcus in adults. Journal of Infectious Diseases, Suppl. 131:9–16.

Lowbury, E. J. L., Kidson, A., Lilly, H. A. 1964. A new selective blood agar medium for Streptococcus pyogenes and other haemolytic streptococci. Journal of Clinical Pathology 17:231–235.

Lund, E. 1959. Diagnosis of pneumococci by the optochin and bile tests. Acta Pathologica et Microbiologica Scandinavica 47:308–315.

Lund, E. 1960. Laboratory diagnosis of Pneumococcus infections. Bulletin of the World Health Organization 23:5–13.

Lund, E. 1963. Polyvalent, diagnostic pneumococcus sera. Acta Pathologica et Microbiologica Scandinavica 59:533–536.

Lund, E. 1966. Omni-serum. A diagnostic Pneumococcus serum, reacting with the 82 known types of pneumococcus. Acta Pathologica et Microbiologica Scandinavica 68:458–460.

Markowitz, M., Taranta, A. 1972. Problems and pitfalls in the management of streptococcal infections. Excerpta Medica, Special Issue, 3–11.

Martin, A. J., Bigwood, R. F., Jr. 1969. Rapid fluorescent-antibody staining technique. Applied Microbiology 17:14–16.

Mausbach, T. W., Cho, C. T. 1976. Pneumonia and pleural effusion. Association with influenza A virus and Staphylococcus aureus. American Journal of Diseases in Children 130:1005–1006.

Maxted, W. R. 1948. Preparation of streptococcal extracts for Lancefield grouping. Lancet ii:255–256.

Maxted, W. R. 1953. The use of bacitracin for identifying group A hemolytic streptococci. Journal of Clinical Pathology 6:224–226.

Maxted, W. R., Potter, E.: 1967. The presence of type 12 M-protein antigen in group G streptococci. Journal of General Microbiology 49:119–125.

McCracken, G. H., Jr. 1973. Group B streptococci: The new challenge in neonatal infections. Journal of Pediatrics 82:703–706.

Mejàre, B., Edwardsson, S. 1975. Streptococcus milleri (Guthof); an indigenous organism of the human oral cavity. Archives of Oral Biology 20:757–762.

Melmed, S., Katz, J., Bank, H. 1976. Streptococcal pancarditis. Chest 69:108–110.

Merrill, C. W., Gwaltney, J. M., Jr., Hendley, J. O., Sande, M. A. 1973. Rapid identification of pneumococci. Gram stain vs. the Quellung reaction. New England Journal of Medicine 288:510–512.

Mhalu, F. S. 1976. Infection with Streptococcus agalactiae in a London hospital. Journal of Clinical Pathology 29:309–312.

Moellering, R. C., Jr., Watson, B. K., Kunz, L. J. 1974. Endocarditis due to group D streptococci. Comparison of disease caused by Streptococcus bovis with that produced by the enterococci. American Journal of Medicine 57:239–250.

Moody, M. D., Ellis, E. C., Updyke, E. L. 1958. Staining bacterial smears with fluorescent antibody. IV. Grouping streptococci with fluorescent antibody. Journal of Bacteriology 75:553–560.

Moody, M. D., Siegel, A. C., Pittman, B., Winter, C. C. 1963. Fluorescent-antibody identification of group A streptococci from throat swabs. American Journal of Public Health and the Nation's Health 53:1083–1092.

Mufson, M.A., Kruss, D. M. Wasil, R. E. Metzger, W. I. 1974. Capsular types and outcome of bacteremic pneumococcal disease in the antibiotic era. Archives of Internal Medicine 134:505–510.

Murray, P. R., Wold, A. D., Hall, M. M., Washington, J. A. II. 1976a. Bacitracin differentiation for presumptive identification of group A β-hemolytic streptococci: Comparison of primary and purified plate testing. Journal of Pediatrics 89:576–579.

Murray, P. R., Wold, A. D., Schreck, C. A., Washington, J. A., II. 1976b. Effects of selective media and atmosphere of incubation on the isolation of group A streptococci. Journal of Clinical Microbiology 4:54–56.

Nakamizo, Y., Sato, M. 1972. New selective media for the isolation of Streptococcus hemolyticus. American Journal of Clinical Pathology 57:228–235.

Nelson, J. D. 1972. The bacterial etiology and antibiotic management of septic arthritis in infants and children. Pediatrics 50:437–440.

Newlands, W. J. 1976. Poststapedectomy otitis media and meningitis. Archives of Otolaryngology 102:51–54.

Noble, R. C., Vosti, K. L. 1971. Production of double zones of hemolysis by certain strains of hemolytic streptococci of groups A, B, C, and G on heart infusion agar. Applied Microbiology 22:171–176.

Paredes, A., Taber, L. H., Yow, M. D., Clark, D., Nathan, W. 1976. Prolonged pneumococcal meningitis due to an organism with increased resistance to penicillin. Pediatrics 58:378–381.

Paredes, A., Wong, P., Mason, E. O., Taber, L. G., Barrett, F. F.1977. Nosocomial transmission of group B streptococci in a newborn nursery. Pediatrics 59:679–682.

Parker, M. T., Ball, L. C. 1976. Streptococci and aerococci associated with systemic infections in man. Journal of Medical Microbiology 9:275–302.

Patterson, M. J., Hafeez, A. E. B. 1976. Group B streptococci in human disease. Bacteriological Reviews 40:774–792.

Pike, R. M. 1945. The isolation of hemolytic streptococci from throat swabs. Experiments with sodium azide and crystal violet in enrichment broth. American Journal of Hygiene 41:211–220.

Poole, P. M., Wilson, G. 1976. Infection with minute-colony-forming β-hemolytic streptococci. Journal of Clinical Pathology 29:740–745.

Portnoy, J., Mendelson, J., Dechene, J. P., Shragovitch, I. 1974. An outbreak of streptococcal wound sepsis: contamination of the wound during operation. Canadian Anaesthetists' Society Journal 21:498–502.

Rammelkamp, C. H., Jr. 1955. The natural history of streptococcal infections. Bulletin of the New York Academy of Medicine 31:103–112.

Randolph, M. F., Redys, J. J., Cope, J., Morris, K. E. 1976. Streptococcal pharyngitis. Evaluation of a new diagnostic kit for clinic and office use. American Journal of Diseases in Children 130:171–172.

Rantz, L. A., Randall, E. 1955. Use of autoclaved extracts for haemolytic streptococci for serological grouping. Stanford Medical Bulletin 13:290–291.

Reinarz, J. A., Sanford, J. P. 1965. Human infections caused by non-group A or D streptococci. Medicine 44:81–96.

Richman, D. D., Breton, S. J., Goldmann, D. A. 1977. Scarlet fever and group A streptococcal surgical wound infection traced to an anal carrier. Journal of Pediatrics 90:387–390.

Romero, R., Wilkinson, H. W. 1974. Identification of group B streptococci by immunofluorescence staining. Applied Microbiology 28:199–204.

Sabbaj, J., Sutter, V. L., Finegold, S. M. 1971. Comparison of selective media for isolation of presumptive group D streptococci from human feces. Applied Microbiology 22:1008–1011.

Sanders, V. 1963. Bacterial endocarditis due to a group C beta hemolytic streptococcus. Annals of Internal Medicine 58:858–861.

Sato, M. 1972. A new selective enrichment broth for detecting beta-hemolytic streptococci in throat cultures: Quinoline derivate and three percent salt as an additional agent to Pike's inhibitors. Japanese Journal of Microbiology 16:538–540.

Schaffner, W., Lefkowitz, L. B., Jr., Goodman, J. S., Koenig, M. G. 1969. Hospital outbreak of infections with group A streptococci traced to an asymptomatic anal carrier. New England Journal of Medicine 280:1224–1225.

Schlossberg, D., Zacarias, F., Shulman, J. A. 1975. Primary pneumococcal pericarditis. Journal of the American Medical Association 234:853.

Sondag, J. E., Morgens, R. K., Hoppe, J. E., Marr, J. J. 1977. Detection of pneumococci in respiratory secretions: Clinical evaluation of gentamicin blood agar. Journal of Clinical Microbiology 5:397–400.

Sprunt, K., Vail, D., Asnes, R. S. 1974. Identification of Streptococcus pyogenes in a pediatric outpatient department: A practical system designed for rapid results and resident teaching. Pediatrics 54:718–723.

Stollerman, G. H. 1962. The role of the selective throat culture for beta hemolytic streptococci in the diagnosis of acute pharyngitis. American Journal of Clinical Pathology 37:36–40.

Swift, H. F., Wilson, A. T., Lancefield, R. C. 1943. Typing group A hemolytic streptococci by M precipitin reactions in capillary pipettes. Journal of Experimental Medicine 78:127–133.

Taplin, D., Lansdell, L. 1973. Value of desiccated swabs for streptococcal epidemiology in the field. Applied Microbiology 25:135–138.

Thornsberry, C., Baker, C. N., Facklam, R. R. 1974. Antibiotic susceptibility of Streptococcus bovis and other group D streptococci causing endocarditis. Antimicrobial Agents and Chemotherapy 5:228–233.

Toala, P., McDonald, A., Wilcox, C., Finland, M. 1969. Susceptibility of group D streptococcus (enterococcus) to 21 antibiotics in vitro, with special reference to species differences. American Journal of the Medical Sciences 258:416–430.

Untereker, W. J., Hanna, B. A. 1976. Endocarditis and osteomyelitis caused by Aerococcus viridans. Mount Sinai Journal of Medicine 43:248–252.

Vincent, W. F., Gibbons, W. E., Gaafar, H. A. 1971. Selective medium for the isolation of streptococci from clinical specimens. Applied Microbiology 22:942–943.

Wannamaker, L. W. 1970. Differences between streptococcal infections of the throat and of the skin. New England Journal of Medicine 283:78–85.

Wannamaker, L. W. 1972. Perplexity and precision in the diagnosis of streptococcal pharyngitis. American Journal of Diseases in Children 124:352–358.

Watanakunakorn, C., Glotzbecker, C. 1977. Synergism with aminoglycosides of penicillin, ampicillin and vancomycin against non-enterococcal group-D streptococci and viridans streptococci. Journal of Medical Microbiology 10:133–138.

Watson, B. K., Moellering, R. C., Jr., Kunz, L. J. 1975. Identification of streptococci: Use of lysozyme and Streptomyces albus filtrate in the preparation of extracts for Lancefield grouping. Journal of Clinical Microbiology 1:274–278.

Weissman, S. M., Reich, P. R., Somerson, N. L., Cole, R. M. 1966. Genetic differentiation by nucleic acid homology. IV. Relationships among Lancefield groups and serotypes of streptococci. Journal of Bacteriology 92:1372–1377.

Whittenbury, R. 1965. A study of some pediococci and their relationship to Aerococcus viridans and the enterococci. Journal of General Microbiology 40:97–106.

Wilkinson, H. W. 1972. Comparison of streptococcal R antigens. Applied Microbiology 24:669–670.

Wilkinson, H. W. 1977. CAMP-disk test for presumptive identification of Group B streptococci. Journal of Clinical Microbiology 6:42–45.

Wilkinson, H. W. 1978. Group B streptococcal infection in humans. Annual Review of Microbiology 32:41–57.

Wort, A. J. 1975. Observations on group-F streptococci from human sources. Journal of Medical Microbiology 8:455–457.

Yow, M. 1975. Epidemiology of group B streptococcal infections, 1975. Progress in Clinical Biological Research 3:159–166.

CHAPTER 128

The Genus *Streptococcus* and Dental Diseases

FRIEDRICH J. GEHRING

An interesting aspect of oral microbiology is the fact that the very first bacteria were found in dense masses on the teeth of humans, almost 300 years ago, by Antonie van Leeuwenhoek (Dobell, 1960). This specific field of research (Berger and Hummel, 1964; Burnett, Scherp, and Schuster, 1976; Nolte, 1973) can thus be traced back directly to the beginnings of microbiology. The microorganisms of the oral cavity of humans and their relation to dental diseases were described extensively for the first time by Miller (1890). He is regarded as the true father of oral microbiology. His microbiological work on dental caries and the resulting arguments are, for the most part, valid to this day, a rare event in the natural sciences.

Enormous quantities of different types of microorganisms colonize the oral cavity of humans, and establish themselves on the mucosal surfaces of the cheek, the gingiva, the tongue, in the gingival crevices, and on the teeth. In the saliva, one can find about 10^8 organisms per ml, which corresponds to a well-grown bacterial broth culture. These microorganisms partly represent the mixed flora from different regions of the oral cavity. However, most of these microorganisms are present on the surfaces of the teeth, and these bacterial accumulations, usually termed bacterial or dental plaque, contain about 10^{11} organisms per g wet weight. This number matches the density of a bacterial colony on a solid culture medium.

The high humidity in the oral cavity of humans and temperatures of 35–37°C, as well as the nutrients remaining on the teeth after eating, provide an excellent environment for the proliferation of most oral microorganisms. Many of these bacteria encompassing several genera and species within a given genus constitute the indigenous oral flora. A substantial part of the total oral flora, in different areas of the oral cavity, is represented by the streptococci and the Gram-positive filamentous organisms (*Actinomyces* species). Microorganisms of the genera *Bacterionema, Corynebacterium, Leptotrichia, Nocardia,* or *Rothia* are usually also present. Other well-known oral organisms are lactobacilli and Gram-negative cocci, such as *Neisseria* and *Veillonella* species. The most frequent Gram-negative rods are obligatory anaerobic types, namely, members of the genera *Bacteroides, Fusobacterium,* and *Spirillum.* Many bacterial strains are merely transients of the oral cavity; they are not able to establish themselves because of unsuitable growth conditions. In general it can be said that the large majority of bacteria that have been identified has also been isolated at one time or another from the oral cavity of man. Only limited general statements on the qualitative and quantitative composition of the oral flora are possible, because it varies not only from individual to individual, but also within the individual over time. The bacteriologic analysis of an oral test sample is valid only for the moment of sampling and the prevailing conditions at the time.

Numerous oral microorganisms still cannot be cultured, as their growth requirements are not known. Thus it should not be expected that "the specific pathogen" of a pathological process in the oral cavity of humans, in which the participation of bacteria is suspected, can be readily found. The composition of the total flora is too complex. Also it always seems to be the total flora that takes part in pathological process of most important dental diseases, although individual organisms or groups can play a prominent role by the influence made by certain metabolic processes on the progress of the disease.

Dental diseases are usually chronic infections, which have a slow course and are generally not very serious. Table 1 presents the most important dental diseases, together with those microorganisms which can be looked upon primarily as causal agents. The microorganisms of the indigenous oral flora live in a more or less stable biological balance with one another and with the host. As a result of pathological processes, this balance is disturbed.

The oral bacteria possess many toxic factors, which make them pathogenic under certain conditions. For example, enzymic activities (Schultz-Haudt, 1964), such as hyaluronidase, proteases, collagenase, and neuraminidase, may exert activity on the host. Another group of harmful substances are the acidic metabolic end products, particularly organic acids, that originate from carbohydrate hydrolysis. The members of the genus *Streptococcus*

Table 1. Important dental diseases and the microorganisms with potential involvement.

Dental disease	Mircroorganisms potentially involved	Literature
Dental caries	Members of the total plaque micro-biota:	Gibbons and van Houte (1975a,b)
Pulpitis	*Streptococcus*, especially *Streptococcus mutans; Lactobacillus; Actinomyces;* and others	Stiles et al. (1976)
Periodontal disease Gingivitis Periodontitis Acute, necrotizing ulcerative gingivitis (Plaut-Vincent's infection)	Members of the total plaque micro-biota: *Streptococcus; Peptostreptococcus; Staphylococcus; Actinomyces; Corynebacterium; Bacteroides*, especially *Bacteroides melaninogenicus; Fusobacterium; Nocardia; Leptotrichia; Lactobacillus; Veillonella; Spriochaeta;* and others	Socransky (1970) Carlsson (1971)

have several metabolic activities that may be harmful to the host. They have an ecological advantage during colonization of the mouth over obligatory aerobic or anaerobic strains, primarily because of their facultative anaerobic nature. A detailed treatment of this genus, in relation to dental diseases, follows. Other aspects regarding the genus *Streptococcus* are treated in Chapters 127 and 129 of this Handbook.

In the eighth edition of *Bergey's Manual of Determinative Bacteriology* (Buchanan and Gibbons, 1974), 21 species of the well-established genus *Streptococcus,* one of the five genera of the family Streptococcaceae, are described. Most streptococci possess group-specific antigens, which, according to Lancefield (1933), provide the basis for their classification into several serological groups: A,B,C,D,E,F,G,H,K,L,M,N,O,P,Q,R,S,T,U (Buchanan and Gibbons, 1974; Hahn, Heeschen, and Tolle, 1970; Seelemann, 1954). Sherman (1937) classified the streptococci into four general groups (pyogenes, viridans, lactic, enterococcus) according to certain physiological characteristics. Although this scheme does not occur in the recent key to the genera of the family Streptococcaceae by Buchanan and Gibbons (1974), it still has a limited usefulness for the classification of streptococci.

The most important and most numerous streptococci in the oral cavity of humans belong to the viridans group, and are subdivided by Carlsson (1967a), Guggenheim (1968), and Gibbons (1972) into the species *S. salivarius, S. sanguis, S. mitis,* and *S. mutans*. Hardie and Bowden (1976a) and Colman and Williams (1972) list *S. salivarius, S. sanguis, S. mitior* (*mitis, viridans*), *S. milleri*, and *S. mutans* as species of streptococci. These last two species do not occur in *Bergey's Manual* (Buchanan and Gibbons, 1974). Each of these species is serologically heterogeneous and has characteristic physiological properties, which make their identification and classification possible.

In recent years, the interest of dental microbiologists in the viridans streptococci has increased, so that further taxonomic progress can be expected (Colman and Williams, 1972). For a long time the classification and nomenclature of these streptococci were considered be extremely confusing (Morris, 1954). Their taxonomic position is now somewhat better described (Hardie and Bowden, 1976a), but not totally satisfactory.

Streptococci establish themselves in the oral cavity of humans within the first few months of life, and later colonize different regions of the oral cavity in great numbers (Andrewes and Horder, 1906; Carlsson, 1967a; Gibbons, 1972; Gibbons et al., 1964a; Gordon, 1905; Gordon and Gibbons, 1966; Guggenheim, 1968; Morris, 1954; Socransky and Manganiello, 1971; van Houte, Gibbons, and Banghart, 1970).

Streptococci constitute between 30 and 60% of the total oral bacterial flora. After the introduction of modern germfree animal technique into biological research, germfree animals were used in ever increasing numbers in the etiological investigations of dental caries (Green, Blackmore, and Drucker, 1973). Using germfree rats, Orland et al. (1954) were the first to show that without bacteria caries does not occur. The part played by streptococci in caries etiology was studied in the course of numerous gnotobiotic (Nolte, 1973) and conventional animal experiments (König, 1966; Navia, 1977). It could be demonstrated that *S. faecalis, S. liquefaciens, S. mitis, S. salivarius, S. sanguis,* and *S. mutans* are potential cariogenic representatives of this genus. Among them, *S. mutans* always proved to be the most cariogenic. *S. mutans* was originally isolated from dentin samples which were taken from carious human teeth (Clarke, 1924), and was recognized again in the 1960s due to the rapid development of the gnotobiotic technics.

Fitzgerald and Keyes (1960) were not able to classify these streptococci at first, and assumed that they occupied an intermediary position between the enteric and lactic groups of the genus. Carlsson (1967a) then reintroduced the epithet *"mutans"* for this species. Since then, this name has been used with greater frequency, although a number of synonyms—such as "streptococci similar or resembling *S. mutans*"; "caries-inducing", "-inducive", or "-active streptococci"; and "plaque- or glucan-forming streptococci"—suggest that the taxonomic position is still uncertain. In *Bergey's Manual* (Deibel and Seeley, 1974), *S. mutans* has not yet been described. It has been mentioned only in connection with the description of *S. salivarius:* "Carlsson (1967a) has suggested a relation between dental caries and *S. mutans* (Clarke, 1924), an organism quite similar to *S. salivarius*. *S. mutans* has not yet been extensively studied and compared with *S. salivarius*." Despite this taxonomic deficiency, no species in the field of oral microbiology has been studied to such an extent over the past 20 years (Stiles, Loesche, and O'Brien, 1976).

Habitats of *Streptococcus* in Relation to Dental Caries

In the oral cavity of every human, those streptococci which occur most frequently have a "primary ecological niche" (Socransky and Manganiello, 1971) and colonize distinct areas. For example, *S. salivarius* is normally found on the tongue (Krasse, 1954).

Generally, bacteria found in the salivary flora are those washed off from various oral surfaces by the saliva (Gibbons et al., 1964b). As a result, large amounts of *S. mitis* are present in saliva—as this bacterium is also washed off the surface of the tongue, the buccal mucous membrane, and the plaque on teeth, where it forms an essential part of the local flora (Gibbons, 1972; Mejàre and Edwardsson, 1975). In contrast, *S. sanguis* (Carlsson, 1965a, 1967a; Rosan, Lai, and Listgarten 1976), *S. mutans* (Gibbons and van Houte, 1975a; Gibbons et al., 1974), and *S. milleri* (Mejàre and Edwardsson, 1975) are typical inhabitants of the plaque (McHugh, 1970) dental surfaces, and the gingival crevices, and are seldom found in other areas of the oral cavity. For example, *S. mutans* and *S. sanguis* do not establish themselves in the mouths of infants before teeth have emerged (Carlsson et al., 1970).

To be more specific, *S. mutans* is not equally distributed on dental surfaces but locally is limited to areas which are recognized to be highly suscepti-

ble to caries (Gibbons et al., 1974; Shklair, Keene, and Cullen, 1974; Street, Goldner, and LeRiche, 1976). Accordingly, this species can be isolated primarily from human enamel caries lesions (Littleton, Kakehashi, and Fitzgerald, 1970; Loesche and Syed, 1973; Loesche et al., 1975; Sumney and Jordan, 1974).

The close association between the local occurrence of *S. mutans* in the human oral cavity and caries can be ascribed primarily to specific metabolic processes, which this species possesses in contrast to other oral streptococci. These metabolic processes play an essential role in the pathogenesis of dental caries. Numerous investigations showed that *S. mutans* reacted on sucrose with the formation of extracellular polysaccharides (primarily polyglucans) of a sticky nature (Dahlqvist et al., 1967; Gehring, 1972; Gibbons, 1968; Gibbons and Nygaard, 1968; Guggenheim and Schroeder, 1967; Inoue and Smith, 1976; Newbrun, 1972; Nisizawa et al., 1976; van Houte, 1976; Wood and Critchley, 1966). The extracellular polysaccharides promote plaque formation and possibly can offer a selective advantage during colonization of smooth dental surfaces (Gibbons and van Houte, 1973, 1975b). Like dextrans, these high-molecular glucose polymers contain α-1,6 linkages, as well as α-1,3 linkages, which are responsible for their insolubility in water.

S. mutans is a highly acidogenic microorganism (Onose and Sandham, 1976; Ranke, B., and Ranke, E., 1971). Acidic metabolic end products of carbohydrate hydrolysis, in connection with the formation of glutinous polyglucans, promote the continuous local action of acids upon dental enamel and dentin. This can eventually lead to initial caries. Thus sugars represent "key substances" in the formation of dental plaque and the etiology of the most important dental diseases (summary and review by Mäkinen, 1974).

Furthermore, *S. mutans,* as well as *S. sanguis* and *S. salivarius* strains, form glycogen-like intracellular polysaccharides. A lack of carbohydrates leads to the degradation of these intracellular polysaccharides with the formation of acids, and thereby intensifies the cariogenic potential (Bramstedt and Lusty, 1968; Hamilton, 1976; Tanzer et al., 1976; van Houte, Winkler, and Jansen, 1969).

S. mutans is found worldwide among the different races and societies of man. Obviously *S. mutans* can be detected wherever one searches for it in the oral cavity of humans (Shklair, 1973). In different parts of the world, individual biotypes (Shklair, 1973) or serotypes such as a–e (E) according to Bratthall (1970) dominate. Bio- or serotypes c, d, and e (E) occur most often, whereas types a and b are less frequent (Bratthall, 1972a; Keene et al., 1977a; Shklair, 1973).

S. mutans inhabits plaques of humans without requiring large amounts of sucrose. Also unrefined sucrose is present in numerous plants, which have always been part of man's diet. Thus *S. mutans* must have existed in the oral cavity of humans for thousands of years, i.e., long before refined sucrose could be produced (Coykendall, 1976). Living conditions of *S. mutans* must have improved tremendously with the increasing consumption of sweets, which are produced out of refined sucrose, principally in modern industrial societies. The increasing multiplication and distribution of these organisms, as a result of this development, could possibly be responsible for the present devastating dimensions of caries distribution. Almost 100% of the population of central and northern Europe and North America suffer from this disease. The health services of the countries concerned spend enormous sums of money annually for the treatment of caries and its consequences.

However, the significance of *S. mutans'* cariogenicity should not be overestimated, as the dominating species of dental plaque are mostly *S. sanguis* and *S. mitis* (Carlsson, 1967a; Mejàre and Edwardsson, 1975), which, like *S. salivarius,* are much less cariogenic than *S. mutans* in animal experiments (Fitzgerald, 1968; Guggenheim, 1968). Like *S. mutans,* most *S. sanguis* and *S. salivarius* strains also produce extracellular polysaccharides from sucrose. But these again are less important for caries etiology than the glutinous polysaccharides produced by *S. mutans.* The total plaque microbiota is additionally composed of a number of other species, which may also contribute to the demineralizing acid production (Theilade, E. and Theilade, J., 1975). The correlations between the composition of bacterial plaque flora and the type, amount, and frequency of food intake are of a very complex nature. The associations of these relationships, apart from the increasing numbers of aciduric bacteria as a result of high carbohydrate diets, especially sucrose, are still unknown (Bibby, 1976).

Streptococci as Potential Microorganisms of Periodontal and Other Dental Diseases

As in the case of dental caries, the pathological process of all forms of periodontal diseases is presumably determined by a number of distinct metabolic processes of *Streptococcus mutans, S. sanguis, S. mitis, S. salivarius,* and *S. milleri* (Carlsson, 1971; Eastoe, Picton, and Alexander, 1971; Ellen, 1976; Gold, 1969; Mejàre and Edwardsson, 1975; Newman et al., 1976; Socransky, 1970), which primarily originate from the dental plaque microbiota. However, the Gram-negative anaerobic rods seem to play a greater role in these pathological processes than the streptococci (Keene and Shklair, 1973; Simon et al., 1972).

In clinically healthy as well as inflamed gingival tissues and on the surface of the sulcular epithelium, *Streptococcus pyogenes* could also be identified (Takeuchi et al., 1974). Clear-cut relationships between well-defined microorganisms and the initiation of periodontal lesions in humans are unknown (Ellen, 1976). The most numerous organisms found in periodontal pockets, during the advanced stage of periodontal diseases, were strains of *S. mitis,* composing 1/3 of the total pocket flora (Dwyer and Socransky, 1968). *Streptococcus mitis* as well as *S. mutans, S. sanguis,* and *S. salivarius,* can also be isolated from infected dental root canals (Mejàre, 1974, 1975b). *S. milleri* (Guthof, 1956), which is found in larger numbers on the teeth than in any other part of the oral cavity (Mejàre and Edwardsson, 1975), can also participate in infections of the root canal (Mejàre, 1975b).

Enterococci have been isolated principally from the gingival crevice area (Gibbons et al., 1963), but they rarely comprise at least 10% of the streptococci in this site. Gold, Jordan, and van Houte (1975) found that *S. faecalis* was the most frequently isolated species from the human mouth. *S. faecalis, S. faecium,* and other group D streptococci occur frequently in infected dental root canals (Mejàre, 1975a) or other oral lesions (Nord and Wadström, 1973). Other members of the Lancefield (1933) groups of streptococci are also found in plaques (Ranke et al., 1967) or in connection with infections in the oral cavity (Hahn, Heeschen, and Tolle, 1970).

It should be noted that *S. mutans, S. sanguis, S. salivarius, S. mitis,* and enterococci (mostly of the Lancefield group D) can also be considered as possible etiological agents for subacute bacterial endocarditis (S.B.E.) (Harder et al., 1974; Rogers, 1976a; Waddy, 1976).

In general the above-mentioned dental diseases, the most important of which are dental plaque infections, are not only multifactorial processes in relation to the microorganisms, but also with regard to the participation of other pathogenic factors.

A basic requirement for prophylaxis against dental diseases is an effective oral hygiene program, which would deprive the oral bacteria of much of their nutritional requirements for growth and multiplication.

Animals as Habitat of *Streptococcus*

Streptococci constitute a significant percentage of the indigenous oral flora of certain laboratory animals, e.g., rats, hamsters, and monkeys (Nolte, 1973). These experimental animals presumably acquired their infection from man. Rats are favored as experimental animals in caries research. Strains of *S. salivarius, S. sanguis,* and *S. mutans* could, for example, be isolated from their molar fissures

(Huxley, 1972). The *S. mutans* strains found in the oral cavity of rats and hamsters represent the same biotypes occurring in plaque of human (Gehring et al., 1976).

The occurrence of *S. mutans* other than in man and certain laboratory animals has also been reported. Two types of *S. mutans* also found in man and laboratory hamsters could be isolated from the mouths of wild rats feeding on sugar cane in the areas of Florida where this plant is extensively cultivated (Coykendall, Specht, and Samol, 1974). Other *S. mutans* strains were isolated from wild rats found in a landfill dump (Coykendall et al., 1976).

Isolation of Oral Streptococci

Nutritional Requirements

For the most part, the culture media ordinarily used for the isolation of facultative anaerobic streptococci are complex and contain: a carbohydrate source, amino acids, proteins or peptides, vitamins, pyrimidines, purines, inorganic ions, and fatty acids; an elevated CO_2-tension is also required during incubation.

The growth of oral streptococci such as *S. mutans, S. sanguis, S. salivarius, S. mitis,* and *S. milleri* in chemically defined media has been investigated by several researchers (Carlsson, 1970, 1972; Griffith and Melville, 1974; Inward, Upstone, and van Houte, 1970; Lawson, 1971; Osborne et al., 1976; Terleckyj, Willett, and Shockman, 1975) and is the basis for the study of nutrition, physiology, macromolecular synthesis, etc. These investigations show that the nutritional requirements vary considerably among oral streptococci and depend on the culture conditions of the inoculation, i.e., the inoculum used, aerobic or anaerobic growth, and possibly other factors. For example, given ideal growth conditions in vitro, *S. mutans* has a mean generation time of 40 min, whereas in the human mouth their growth is much slower (Ellwood, 1976).

Collection of Bacterial Samples from the Oral Cavity of Humans

Representative samples of the oral microbiota can be obtained from all areas of the mouth (Hardie and Bowden, 1974a, 1976b): from dental plaque on tooth surfaces (buccal, lingual, occlusal); from interstitial spaces between adjacent teeth; from the gingival crevice region; from soft areas such as the gingiva, tongue, and cheek; and from stimulated (by chewing inert material such as paraffin or a rubber band for some minutes) or unstimulated saliva collected directly from the mouth by movements of the tongue, lips, and cheeks.

Under controlled conditions the development of dental plaque can be achieved by the following method: the natural teeth must be cleaned until no visible surface deposits or integuments remain, and are then allowed to develop plaque for varying periods of one to several days, without brushing the teeth. After the desired period the tooth surfaces are air-dried, and plaque samples are obtained from some or all teeth with a sterile instrument (e.g., dental scalers, sterile Gracey curettes, or dental floss held in a plastic holder) by scraping the tooth surfaces. A detailed description of collection and identification of different streptococci from the deep areas of carious dentin is given by Edwardsson (1975). Cotton swabs can be used to sample the soft tissue areas.

For bacterial analysis of the various samples from the oral cavity, the specimens are deposited in small screw-cap vials, which hold 1–5 ml of a reduced transport fluid medium (RTF). This preserve the viability of the microorganisms without allowing the growth of any species (Hoover and Newbrun, 1977; Loesche, Hockett, and Syed, 1972; Rundell et al., 1973; Syed and Loesche, 1972).

Reduced Transport Fluid (RTF) Medium (Syed and Loesche, 1972)

RTF medium contains, per 100 ml: 0.5 ml of stock solution 8% Na_2CO_3, filter-sterilized with a 0.22-μm filter; 1.0 ml of 0.1 M ethylenediaminetetraacetate stock solution; 2.0 ml of 1% DTT stock solution (DL-Dithiothreitol, Cleland's Reagent, Sigma Chemical Co.), filter-sterilized with 0.22-μm filter; 7.5 ml of 0.6% K_2HPO_4 stock solution; and 7.5 ml of stock mineral solution containing 0.6% KH_2PO_4, 1.2% NaCl, 1.2% $(NH_4)_2SO_4$, and 0.24% $MgSO_4$.

This RTF medium is recommended for holding a population of streptococci stable when maintained at room temperature. When RTF is not available storage at 10°C is in any case better than storage at room temperature. If the plaque samples are to be analyzed in the laboratory immediately, a suspension in sterile phosphate buffer (pH 7.2), physiological saline, or thioglycolate broth (Difco) is suitable.

Enzymatic Basis for the Isolation of Extracellular Polysaccharide-Forming Streptococci

While extra- and intracellular enzymatic activities are responsible for sucrose hydrolysis, two other enzymatic mechanisms of *Streptococcus mutans, S. sanguis,* and *S. salivarius* contribute to the formation of dental caries; the synthesis of important extracellular polysaccharides from sucrose. A glucosyl transferase catalyzes polyglucans of the dextran type, with the release of free fructose, and a fructosyl transferase forms polyfructans of the levan type, with the release of glucose. *S. mutans* and *S. sanguis* mainly produce glucans, whereas *S. salivarius* produces more fructans (Chassy et al., 1976; Schachtele, Loken, and Schmitt, 1972). These extracellular polysaccharides have been shown to be both morphologically and chemically heterogeneous (Guggenheim and Schroeder, 1967; Inoue and Smith, 1976; Tanzer, Chassy, and Krichevsky, 1972). The extracellular glucans produced by *S. mutans* and *S. sanguis* can be separated into water-soluble and -insoluble fractions. Most of the glucans are highly branched at the C-3 and C-6 positions of the D-glucopyranose residues, except from some which contain a small quantity of α-1,4 glucopyranosyl linkages. The solubility of these substances presumably depends on structural features, such as the number of α-1,3 glucopyranosyl linkages. For instance, the $(1,3)$-α-D-glucanases of fungal and bacterial origin raise considerably the solubility of glucans produced by oral streptococci (Walker and Hare, 1977). With *S. mutans* the insoluble compounds prevail, while *S. salivarius* is completely without insoluble polysaccharides (Nisizawa, Imai, and Araya, 1977; Nisizawa et al., 1976). The polysaccharides of the above-mentioned three species have a tendency to form clusters of numerous molecules that tend to clump together. These clusters have a width of 24.8–28.5 nm and a height of 4.1–5.2 nm (Newbrun, Lacy, and Christie, 1971). It should be mentioned that these extracellular substances, the polysaccharides, are macromolecules with a molecular weight of 16–23 million (Newbrun and Baker, 1968).

Isolation of Extracellular Polysaccharide-Forming Streptococci

The relative proportions of insoluble and soluble polyglucans and of soluble polyfructans synthesized by these streptococci from sucrose determine the characteristic morphology of the colonies on culture media. As a rule, an agar medium with 5% sucrose as the carbohydrate source is used for isolation. For this purpose, Mitis Salivarius agar, dehydrated (Difco, B 298-MS-agar), is usually used. This medium inhibits the growth of most bacteria, except streptococci, because it contains trypan blue, crystal violet, and tellurite. The incorporation of polymyxin B sulfate in MS-agar, at a concentration of 100 U/ml, does not affect the growth of oral streptococci, but effectively suppresses the growth of Gram-negative organisms, which occasionally are present in test samples (Fitzgerald and Adams, 1975).

The collected samples of mixed flora of the human mouth are homogenized with a glass mortar and pestle for a few minutes, or dispersed by sonic oscillation for 5–15 s (with a microprobe tip if available), in either RTF medium, phosphate buffer, saline, or Ringers solution. Subsequently, the samples are diluted so that dilutions of 1:10, 1:100, and up to 1:1,000,000 are reached. Over the surface of MS-agar plates, 0.1 ml from each dilution is spread with the aid of bent glass rods. The plates are then incubated aerobically or in an atmosphere of 95% nitrogen or hydrogen with 5% carbon dioxide at 37°C for 48 h (the BBL GasPak anaerobic system is well suited for this purpose). At the end of the incubation, the appearance of the colonies on the MS-agar plates can be studied under a stereomicroscope at a magnification of 8–60×, in transmitted and reflected light.

On the aerobiocally incubated MS-agar, *Streptococcus sanguis* colonies form zooglea less than 2 mm in diameter, with deformation of the surrounding agar surface (Carlsson, 1965a). The consistency of the zooglea is rubbery and its attachment to the medium is so firm that, at best, only part of it can be removed from the medium with a platinum loop.

The colonies of *S. salivarius* on aerobically incubated MS-agar are highly convex, smooth ("gumdrop"-like) or rough, and more than 2–5 mm in diameter (Carlsson, 1965b; Chapman, 1944).

S. mutans, incubated anaerobically for two days on MS-agar, varies between rough, smooth, or mucoid colonies (Edwardsson, 1968, 1970). Jordan, Krasse, and Möller (1968) described the colonies of *S. mutans* isolated from hamsters and grown on MS-agar: "highly convex to pulvinate, the edge is undulated, the colony is opaque and its surface is finely granular, reminiscent of frosted glass. A varying amount of watery exudate is sometimes formed and this can be seen either as a drop on the colony or as discrete pool surrounding the colony".

On phenol red agar base (Difco) containing 5% sucrose, *S. mutans* variants also show characteristic colony forms (Gehring, 1972, 1976; Gehring et al., 1975). This culture medium is transparent and clear so that the various colony forms can be photographed well (Fig. 1).

The initial presumptive identification based on colony morphology, on culture medium with 5% sucrose, may be relatively simple after considerable experience in culturing streptococci which produce extracellular polysaccharides. But an incorrect identification can be the result of inexperience, since *S. mutans, S. sanguis,* and *S. salivarius* produce several colony types. As a general rule, identification based on colony morphology should always be supported by supplementary tests. For this purpose, typical colonies can be picked from the solid culture media with a sterile wire loop, inoculated in a broth medium, e.g., NIH-thioglycolate broth (Difco) or Todd Hewitt broth (Difco). Incubate overnight or longer and check the resulting growth by testing

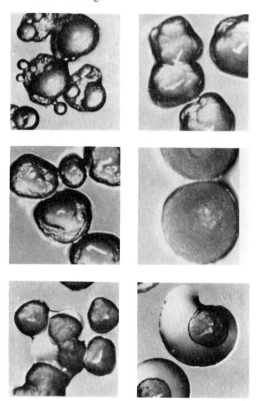

Fig. 1. Typical colonies of *Streptococcus* mutans on phenol red agar base (Difco), anaerobically cultivated at 37°C for 48 h.

some of the relevant physiological characteristics important for each species' identification. For instance, a simple test for the detection of glucose or fructose polymers in the liquid culture medium can be of use for a preliminary differentiation between polyglucan- (*S. sanguis, S. mutans*) and polyfructan-forming (*S. salivarius*) organisms. This test was first described by White and Niven (1946), and has been introduced to the routine identification of streptococci (Hahn, Heeschen, and Tolle, 1970). The improved procedure recommended by Guthof (1970) is as follows:

Detection of Dextrans and Levans (Guthof, 1970)
   Liquid culture medium: 1.5% tryptone solution, 0.5% NaCl, 0.1% yeast extract, 3% sucrose. Inoculate the test cultures and incubate for 1–2 days to yield distinctly visible turbid growth. Each culture tube is centrifuged and 0.1 ml of the supernatant of each culture is added to each of three tubes and mixed with 0.3 ml of 10% sodium acetate. Add to tube 1 a 0.8-fold volume (= 0.32 ml) of acetone, to tube 2 a 1.2-fold volume (= 0.48 ml) of ethanol, and to tube 3 a 1.5-fold volume (= 0.6 ml) of methanol. Shake each tube well for 1–3 min and observe. Flocculation in all three tubes or only in acetone and turbidity in the alcohols indicate dextran. If none of the tubes shows flocculation, an additional

1.3-fold volume of ethanol is added to tube 2, and an additional 1.5-fold volume of methanol to tube 3. After repeated shaking, take a second reading. Flocculation in both tubes indicates the presence of levan. To demonstrate the simultaneous production of dextran and levan, the tubes with methanol are centrifuged after dextran precipitation; the supernatant fluid is pipetted off, a 1.5-fold volume of methanol is added and after shaking again, flocculation indicates levan.

Solid Selective Isolation Media for *S. mutans*

A selective culture medium for *S. mutans* was described by Carlsson (1967b). It resembles MS-agar, but contains, in addition, a sulfonamide.

Solid Selective Medium (MC-agar) for Isolation of *Streptococcus mutans* (Carlsson, 1967b)
   This medium contains, per 1 liter distilled H_2O: 40 g diagnostic sensitivity test (DST) agar base (Oxoid), 50 g sucrose, 0.075 g trypan blue, and 0.0008 g crystal violet. The rehydrated medium is autoclaved, according to the directions for DST agar base. Cool to 45–50°C and add 1.0 ml of Chapman tellurite solution (Difco) and 1.0 g of sulfadimetine (5-ml ampule of Elkosin, Ciba), mix thoroughly and pour in Petri dishes.

Samples from the human oral cavity were anaerobically cultured on this medium and strains of *S. mutans* formed zoogleic colonies. The number of colonies of other organisms was reduced in most cases to less than 5% of the number growing on ordinary MS-agar. Other streptococci strains (*S. sanguis, S. salivarius, S. mitis*, etc.) than those of *S. mutans* are sensitive to the sulfonamide.

MS-agar is a logical starting point for the development of further selective culture media. Gold, Jordan, and van Houte (1973) modified MS-agar by adding 0.2 U/ml bacitracin and increasing the sucrose concentration to 20%. These selective agents allowed the undiminished recovery of *S. mutans* by completely inhibiting the streptococcal flora normally encountered on this medium. Small numbers of enterococci can also be detected on this medium, but they are easily differentiated on the basis of their colonial morphology.

In addition, *S. mutans* can tolerate high concentrations of sucrose (Gehring 1968, 1977; Ikeda and Sandham, 1972). Numerous strains show the ability to grow in 40% sucrose in a liquid culture medium and are able to lower the pH to about 4.0.

Rapid Differentiation of Colonies of *Streptococcus mutans* from Other Streptococci
   A differentiation between *S. mutans* colonies and colonies of other streptococci on MS-agar can also be achieved by spraying a solution of 10%

mannitol and 4% 2,3,5-triphenyltetrazolium chloride (TTC) on the plates after incubation (Gold, Jordan, and van Houte, 1974). If the TTC is reduced, it will stain the *S. mutans* colonies to dark pink, due to the mannitol-1-phosphate dehydrogenase–mediated hydrolysis of mannitol (Brown and Wittenberger, 1973) to acids by *S. mutans* (Carlsson, 1968).

Liquid Selective Isolation Media— Enrichment of *S. mutans*

Since in various oral samples and different subjects *S. mutans* is present only in very low numbers, its detection on solid culture media is difficult, if not impossible. Therefore selective enrichment procedures in liquid culture media are applied in entirely qualitative tests.

Selective Color-Test Medium for Isolation of *Streptococcus mutans* (Shklair and Walter, 1976) This medium contains, per 100 ml distilled H_2O:

Thioglycolate medium (without carbohydrate or indicator— Difco)	2.4 g
Lactoalbumin	0.25 g
Mannitol	0.5 g
Thallium acetate	0.025 g
Crystal violet	0.0001 g
Bromocresol purple as indicator	

Oral samples inoculated in screw-cap test tubes filled with 5 ml of this medium are incubated up to 7 days at 37°C. A color change in the medium from purple to yellow is presumptive evidence for the presence of *S. mutans*. The rate of the color change usually depends on the number of *S. mutans* present in the sample. To confirm the presence or absence of *S. mutans* the samples are plated on MS-agar when a color change is noted, or after 7 days if there is no color change. This color test can be useful in large-scale epidemiological studies for the presence or absence of *S. mutans*.

The selective effectiveness of several other diagnostic broths was compared by Loesche and Bhat (1976). They found that low levels of *S. mutans* in oral samples can be easily detected with the liquid media they tested.

Isolation of Other Streptococci Species

Streptococci such as *S. mitis* and *S. milleri,* which cannot synthesize extracellular polysaccharides from sucrose, grow in less characteristic colony forms on MS-agar. For further identification they have to be subcultured several times. Generally, the isolation

of streptococci species is carried out on blood plates, e.g., brain-liver-heart (Difco) supplemented with 1.5% agar (wt/vol) and 5% (vol/vol) human bank, sheep, or horse blood. *S. milleri* and related strains that grow on MS-agar form smooth or rough colonies (Mejàre and Edwardsson, 1975).

The conservation of all streptococci strains presents no problem, as they are extremely resistant to freezing and drying and can be stored in a lyophilized state for years.

Identification of Oral Streptococci Related to Dental Diseases

The general bacteriological characteristics of the oral streptococci are: Gram-positive cocci; appear in pairs or chains; nonmotile, catalase negative, oxidase negative; ferment carbohydrates, produce no gas, and are facultative anaerobes. All strains readily ferment glucose, mannose, galactose, fructose, sucrose, lactose, and maltose, but do not ferment inositol, ribitol, dulcitol, arabinose, xylose, or glucosamine. Sodium hippurate is not hydrolyzed. Although it is still difficult to identify and classify different species of the streptococci, several recent studies could improve their taxonomic position (Carlsson, 1968; Colman and Williams, 1972, 1973; Drucker and Melville, 1971; Guggenheim, 1968; Hahn, Heeschen, and Tolle, 1970; Hardie and Bowden, 1974a, 1976b; Rogers, 1969).

Streptococcus mutans Clarke, 1924

The principal identification criteria are listed in Table 2. A preliminary diagnosis of *Streptococcus mutans* usually is made from the characteristic morphology of its colonies on 5% sucrose containing culture medium. A serological identification can be achieved in a manner consistent with Bratthall (1970), who described five serological groups which he classified as a, b, c, d, (not to be mistaken for the serological groups of streptococci by Lancefield [1933] with capital letters) and e (= Lancefield's E). Perch, Kjems, and Ravn (1974) established two new serotypes, f and g, and another distinct type, SL-1.

The identification and quantitation of *S. mutans* by fluorescent antibody techniques are described by several authors (Bratthall, 1972b; Jablon, Ferrer, and Zinner, 1976; Loesche and Grenier, 1976; Thomson, Little, and Hageage, 1976). But generally the serological classification of *S. mutans* strains is still a delicate method, and the outcome of serotyping depends to a great extent on the selection of appropriate (i.e., specific) and sensitive antisera (Perch, Kjems, and Ravn, 1974).

Table 2. Properties of oral *Streptococcus* species.[a]

Biochemical test	S. sanguis (Types I and II)	S. salivarius (Types I and II)	S. mutans[b] (several sero-, geno-, and biotypes)	S. mitis (S. mitior, S. viridans)	S. milleri[b] (Streptococcus sp. MG)
Acid from:					
Mannitol	1,2,4,6,7 −	1,2,4,6,7 −	1,2,4,6,7,8,9 +	1,2,6,7,8 −	2,6,8,9 −
Sorbitol	1,2,3,4,6,7 −	1,2,3,4,5,6,7 −	1,2,4,5,6,7,8 +	1,2,3,5,6,7,8 −	2,6,7,8 −
Raffinose	2,3,6 +	1,2,3,5,6,9 +	1,4,5,6 +	2,3,5,6,8 −	2,6,8,9 +
Melibiose	1,5 +	1,4,5 −	1,4,5 +	9 −	9 −
Trehalose	1,2,3,4,6,9 +	1,2,3,4,9 +	1,2,4,5,6,9 +	1,2,3,5,8 −	2,8,9 +
Salicin	1,2,3,5,9 +	1,2,3,5,9 +	1,2,4,5,6 +	1,2,8,9 +	2,6,8,9 +
Inulin	1,3,5,9 +	1,2,3,5 +	1,2,5,6 +	1,2,3,5,9 −	2,9 −
Hydrolysis of:					
Esculin	2,4,5,7 +	1,2,4,5,6,7 +	1,2,4,5,6,7 +	1,2,5,6,7,8,9 −	2,6,7,8,9 +
Arginine	1,2,4,6,7,9 +	1,2,4,6,7 −	1,2,4,6,7 −	1,2,7,9 −	2,7,8,9 +
Extracellular polysaccharide from sucrose:					
Mainly glucan (dextran)	2,3,6 +	3,6 −	6,7 +		2,6,8 −
Mainly fructan (levan)	6 −	6,3 +	6,7 −		2,6,8 −
Hemolysis forms	1,3,4,5,6 α	1,3,4,5,6 α,γ	1,4,6 γ	1,3,5,6,8 α	6,8 α,β,γ
Hydrogen peroxide	7,9 +	7 −	7,8 −	7,8,9 +	7,8 +
Acetoin from glucose	2,7 −	2,5,7 d	2,5,7,8 +	2,7,8 +	2,6,7,8 +
Growth on: 10% bile-agar	2,4,5,6 +	2,4,5 +	2,5,6 +	2,6 −	2,6,7,8 +
40% bile-agar	2,4,6 d	2 d	2,4,6 d	2,6,8 −	2,6 +
Growth in 4% NaCl broth	2,6 −	2,6 −	1,2,6,8,9 +	2,6,8 −	2,6,8,9 −

[a] Literature: (1) Carlsson (1968); (2) Colman and Williams (1972); (3) Deibel and Seeley (1974); (4) Edwardsson (1968); (5) Guggenheim (1968); (6) Hahn et al. (1970); (7) Hardie and Bowden (1976a); (8) Mejàre and Edwardsson (1975); (9) Mejàre (1975b). Symbols: +, usually positive; −, usually negative; d, variable reactions; α, green zone around colonies on blood agar; β, clear, colorless zone around colonies on blood agar; γ, without reaction on blood agar.

[b] Not listed in *Bergey's Manual*, eighth edition (Buchanan and Gibbons, 1974).

Each type is characterized by more than one specific antigen (Coykendall, 1977; Facklam, 1974). The exact chemical nature of the antigens of *S. mutans* types is not yet known and cross-reactions between some types further complicate the picture.

By qualitative analysis of cell wall carbohydrate components (Hardie and Bowden, 1974b; for methods see also Colman and Williams, 1965), three patterns of reducing sugars were detected: (i) glucose, galactose and rhamnose; (ii) galactose and rhamnose; and (iii) glucose and rhamnose. The quantitative and qualitative chemical composition of purified cell walls, derived from several strains representing the four major serotypes (a, b, c, d) of *S. mutans,* was investigated by Bleiweis et al. (1976) and Cooper, Chorpenning, and Rosen (1975). They found that these cell walls contain 6.7% protein, 8.9% glycerol teichoic acid, 33.6% non-peptidoglycan polysaccharide, and 49.9% peptidoglycan. Drucker, Shuttleworth, and Melville (1968) determined that alanine, glutamine, and lysine were the main amino acids. The four serological groups of Bratthall (1970), a–d, correspond with the four genetic groups established by Coykendall (1971), which differ in their DNA base ratios (mol% G+C).

A differentiation of *S. mutans* into four new species, with detailed descriptions, was suggested (Coykendall, 1974, 1977): *Streptococcus rattus, Streptococcus cricetus, Streptococcus sobrinus,* and *Streptococcus ferus.* However, the difficulties encountered in the preparation of monospecific antisera and their not being available commercially, have led many authors to identify *S. mutans* strains on the basis of simple biochemical tests.

Facklam (1974) demonstrated that the taxon *S. mutans* is physiologically well defined and that a division into additional species is at present not necessary. As major criteria for the identification, he lists the inability of Lancefield (1933) extracts of *S. mutans* to react with streptococcal group D antisera (obtainable from Difco or Deutsche Wellcome GmbH, 3006 Grossburgwedel), the formation of gelatinous glucan deposits in liquid culture media containing 5% sucrose, and acid production from mannitol. Facklam (1974) further classifies numerous *S. mutans* reference strains into eight biotypes and subdivides the strains principally on the basis of positive or negative acid production from raffinose, sorbitol, lactose, and esculin.

Perch, Kjems, and Ravn (1974) distinguished 3 biotypes of *S. mutans* which are characterized by arginine and salicin degradation and the hemolysis forms. Like Facklam, these authors do not recommend a formal subdivision of the species. Shklair and Keene (1974) developed a useful biochemical scheme for the classification of *S. mutans* into 5

biotypes, "a"–"e", which somehow correspond with Bratthall's serotypes a–e (E) (1970). The differentiation of biotypes is based on the fermentation of mannitol (with and without 2 U/ml bacitracin), sorbitol, raffinose, melibiose, and the production of ammonia from arginine. A comparative review of physiological reactions of different serotypes of *S. mutans* was published by Hardie and Bowden (1976a).

Various types of *S. mutans* can also be characterized by bacteriocin production and susceptibility profiles to the bacteriocins. It has been suggested that this could be useful, apart from identification tests, in epidemiological and ecological studies (Hamada and Ooshima, 1975; Kelstrup and Gibbons, 1969; Paul and Slade, 1975; Rogers, 1975, 1976b, 1976c). There are indications to believe that the bacteriocins ("mutacins") from *S. mutans* contain at least two kinds of inhibitory substances, ranging from molecules with low molecular weights to higher-molecular-weight, protein-lipid complexes. The production of, and susceptibility to, the bacteriocin-like substances, which was shown to be influenced by the medium (Rogers, 1972), might be closely related to the colonial variations of *S. mutans* (Yamamoto et al., 1975).

Little is known about bacteriophages of *S. mutans.* Greer et al. (1971) found that some strains independently released morphologically identical phage after ultraviolet light and mitomycin induction. It must be emphasized again that *S. mutans* is not described as an independent species in the eighth edition of *Bergey's Manual* (Buchanan and Gibbons, 1974). But the International Subcommittee on Nomenclature for Streptococci reported their consensus: "*Streptococcus mutans* should be conserved as a valid species and NCTC strain 10449 should be designated as the neotype strain" (1971).

Due to intensive work in the dental sciences it has, in the meantime, become quite clear that exceptions can be found even for the most significant differentiating characteristics of *S. mutans.* Examples are the absence of fermentation of mannitol or sorbitol (Carlsson, 1968; Coykendall, 1976; Perch, Kjems, and Ravn, 1974), and the lack of extracellular glucan formation from sucrose (DeStoppelaar, 1971a; Facklam, 1974; Rogers, 1969). The summary by Hardie and Bowden (1976a) compares seven different published accounts of selected physiological tests (fermentation of mannitol, sorbitol, raffinose, esculin, dextran, etc.) of several *S. mutans* strains.

Additional subdivision of *S. mutans* strains into sero- or biotypes could further complicate the taxonomic outlines of the hitherto best-defined species among oral streptococci, and should be held in abeyance as long as a well-established description has not been published in a new edition of *Bergey's Manual of Determinative Bacteriology*.

Streptococcus sanguis White and Niven, 1946

Streptococcus sanguis is a well-known *Streptococcus* species (Buchanan and Gibbons, 1974; Colman and Williams, 1972) which can be identified according to the criteria reviewed in Table 2. It forms characteristic colonies on semisolid culture media containing 5% sucrose (Carlsson, 1965a; Hehre and Neill, 1946; Niven, Kiziuta, and White, 1946). In 5% sucrose broth, the extracellular polysaccharide formation leads to increasing viscosity of the medium and the reaction of the culture fluid with *Pneumococcus* type 2 antiserum is strongly positive, suggesting a relatively high percentage of α-1, 6 linkages in the synthesized polysaccharide (DeStoppelaar, 1971b). *S. sanguis* strains growing on blood agar, are greening streptococci and their identification can be achieved by using numerous well-defined characteristics (Carlsson, 1965a; Colman and Williams, 1972; DeStoppelaar, 1971b; Guggenheim, 1968; Hahn, Heeschen, and Tolle, 1970). The physiological nature of this species is somewhat more heterogeneous than that of *S. mutans,* the main differences being the formation of nonadhesive polysaccharides from sucrose, α-hemolysis with greening, and the lack of mannitol, sorbitol and Na-hippurate catabolism. A comparison of some of the more important taxonomic properties of *S. sanguis,* tested by several investigators, is found in the summary by Hardie and Bowden (1976a).

Washburn, White, and Niven (1946) described two types of antigens (I and II), for *S. sanguis*. Most strains possess type I and only a few strains produce both, type I/II, antigens. Types I and I/II bear the Lancefield group H antigen; however the isolated type II does not. The uncertain serological identification of *S. sanguis* strains with group H antisera has been known for a long time (Porterfield, 1950) and has not yet been elucidated. The formation of extracellular polysaccharide from sucrose also seems uncertain with some strains. Colman and Williams (1972) suggested that certain strains that do not synthesize polysaccharides should remain taxonomically in the *S. sanguis* species, whereas other strains with positive polysaccharide formation (*S. sanguis,* type II) should be associated with *S. mitior*. The differentiation between *S. sanguis,* type I, and *S. mitior* can be achieved by a number of tests (Hardie and Bowden, 1976a), e.g., the hydrolysis of arginine and esculin, which is positive for *S. sanguis,* type I, and negative for *S. mitior*. *S. sanguis* I, *S. sanguis* II, and *S. mitior* (that does not form dextran) have been compared in detail by Colman and Williams (1972).

Streptococcus salivarius Andrewes and Horder, 1906

The identification criteria for *Streptococcus salivarius* can be found in Table 2. Its description is usually based on Sherman, Niven, and Smiley (1943) and is found in *Bergey's Manual,* eighth edition (Buchanan and Gibbons, 1974). From sucrose, *S. salivarius* chiefly produces the fructose polymer levan (Sherman, Niven, and Smiley, 1943). Due to this remarkable property, which also determines the typical form of its colonies on MS-agar, it can easily be distinguished from *S. mutans* and *S. sanguis,* both of which mainly synthesize glucose polymers.

Serologically, *S. salivarius* has no group-specific antigens, and it, too, shows two types (I and II) of antigen reactions. Both types form levan, but only type I reacts positively to Lancefield group K antiserum. In each case, the component responsible for type specificity is a cell wall polysaccharide of galactose, glucose, rhamnose, and a trace of glucosamine (Montague and Knox, 1968). The serological relationships to streptococci group K and others are as yet unclear (Willers, Ottens, and Michel, 1964). According to a comparison of some of the physiological properties that were tested by different investigators and are relevant to the identification of *S. salivarius* by Hardie and Bowden (1976a), next to levan formation, further significant attributions of this species are the absence of hemolysis and hydrolysis of arginine as well as the positive fermentation of salicin and esculin.

Streptococcus mitis Andrewes and Horder, 1906

This bacterium, as listed in *Bergey's Manual,* eighth edition (Buchanan and Gibbons, 1974), is named and identified according to different aspects which are summarized in Table 2. A critical review of these aspects was published by Seeleman and Obiger (1958). The description of *S. mitis* is based on Sherman, Niven, and Smiley (1943) and comprises a heterogeneous group that has no classifying characteristics. Other species' names that are associated with *S. mitis* are *S. mitior* (Colman and Williams, 1972), which has recently been used more frequently (Hardie and Bowden, 1974a, 1976a) and *Streptococcus viridans* (Seeleman and Obiger, 1958), a name still used by German authors (Hahn, Heeschen, and Tolle, 1970). Colman and Williams (1965, 1972) prefer the name *S. mitior* because the cell walls of a number of strains contain ribitol and rhamnose, in contrast to *S. sanguis, S. salivarius, S. mutans* and *S. milleri*. In their opinion this term should be applied to strains which are α-hemolytic, do not metabolize arginine, esculin, mannitol, and sorbitol, and which form hydrogen peroxide. The uncertainties in connection with the classification of *S. mitis* (*mitior/viridans*) are demonstrated by the positive serological reactions to group K and group O antisera. Usually *S. mitis* (*mitior/viridans*) does not synthesize polyglucans from sucrose, except for a few strains that form dextran and are similar to *S. sanguis* type II (Colman and Williams, 1972).

Streptococcus milleri Guthof, 1956

Streptococcus milleri is not listed in *Bergey's Manual*, eighth edition (Buchanan and Gibbons, 1974). But recently interest has focused again on this member of the oral flora (Colman and Williams, 1972; Hardie and Bowden, 1974a, 1976a; Mejàre, 1975b; Mejàre and Edwardsson, 1975). *S milleri* seems to be related to *Streptococcus anginosus* (Andrewes and Horder, 1906) and *Streptococcus* sp. MG (Mirick et al., 1944). Serologically it is heterogeneous, and individual strains react to antisera of the Lancefield groups A, C, F, and G. In contrast to *S. mutans*, *S. sanguis*, and *S. salivarius*, the species *S. milleri* does not produce extracellular polysaccharides from sucrose. By growth on sulfonamide-containing MC-agar and 7.5% bile blood agar, *S. milleri* can easily be differentiated from *S. mitis* (*mitior*). The latter does not grow on these media and produces hydrogen peroxide.

The susceptibility of oral streptococci to antibiotics was described by Carlsson (1968), Mejàre (1975a), and Sukchotiratana, Linton, and Fletcher (1975). Chloramphenicol is an effective antibiotic, while polymyxin B and streptomycin are without effect (Carlsson, 1968). Penicillin and clindamycin also have marked effects on the normal oral streptococcal flora, whereas cephalexin is ineffective (Sukchotiratana, Linton, and Fletcher, 1975). Furthermore, growth inhibition caused by bacitracin, sulfaisodimidine, optochin, and nitrofurazone can help to differentiate species of biotypes (Colman, 1968; Colman and Williams, 1972; Mejàre, 1975b; Mejàre and Edwardsson, 1975; Shklair and Keene, 1974).

Many strains of the mentioned streptococci species related to dental diseases are available from the ATCC in the USA, the NCTC in London, and the Bundesanstalt für Milchforschung—Streptokokkenzentrale—D-23 Kiel, BRD. This last institution also identifies and classifies as far as possible those streptococci isolates of different origins that are sent to it.

Literature Cited

Andrewes, F. W., Horder, T. J. 1906. A study of the streptococci pathogenic for man. Lancet **ii:**708–713.

Berger, U., Hummel, K. 1964. Einführung in die Mikrobiologie und Immunologie unter besonderer Berücksichtigung der Mundhöhle. 2nd ed., p. 412. Munich, Berlin: Urban & Schwarzenberg.

Bibby, B. G. 1976. Influence of diet on the bacterial composition of plaques, pp. 477–490. In: Stiles, H. M., Loesche, W. J., O'Brien, T. C. (eds.), Proceedings ''Microbial Aspects of Dental Caries'', vol. 2. Washington, D.C., London: Information Retrieval.

Bleiweis, A. S., Taylor, M. C., Deepak, J., Brown, T. A., Wetherell, J. R. 1976. Comparative chemical compositions of cell walls of *Streptococcus mutans*. Journal of Dental Research **55** (Special Issue A):103–108.

Bramstedt, F., Lusty, C. J. 1968. The nature of the intracellular polysaccharides synthesised by streptococci in the dental plaque. Caries Research **2:**201–213.

Bratthall, D. 1970. Demonstration of five serological groups of streptococcal strains resembling *Streptococcus mutans*. Odontologisk Revy **21:**143–152.

Bratthall, D. 1972a. Demonstration of *Streptococcus mutans* strains in some selected areas of the world. Odontologisk Revy **23:**401–410.

Bratthall, D. 1972b. Immunofluorescent identification of *Streptococcus mutans*. Odontologisk Revy **23:**1–20.

Brown, A. T., Wittenberger, C. L. 1973. Mannitol and sorbitol catabolism in *Streptococcus mutans*. Archives of Oral Biology **18:**117–126.

Buchanan, R. E., Gibbons, N. E. (eds.). 1974. Bergey's manual of determinative bacteriology, 8th ed. Baltimore: Williams & Wilkins.

Burnett, G. W., Scherp, H. W., Schuster, G. S. 1976. Oral microbiology and infectious disease, 4th ed. Baltimore: Williams & Wilkins.

Carlsson, J. 1965a. Zooglea-forming streptococci, resembling *Streptococcus sanguis*, isolated from dental plaque in man. Odontologisk Revy **16:**348–358.

Carlsson, J. 1965b. Effect of diet on presence of *Streptococcus salivarius* in dental plaque and saliva. Odontologisk Revy **16:**336–347.

Carlsson, J. 1967a. Presence of various types of non-haemolytic streptococci in dental plaque, and in other sites of the oral cavity. Odontologisk Revy **18:**55–74.

Carlsson, J. 1967b. A medium for isolation of *Streptococcus mutans*. Archives of Oral Biology **12:**1657–1658.

Carlsson, J. 1968. A numerical taxonomic study of human oral streptococci. Odontologisk Revy **19:**137–160.

Carlsson, J. 1970. Chemically defined medium for growth of *Streptococcus sanguis*. Caries Research **4:**297–304.

Carlsson, J. 1971. Bacterial populations associated with the periodontium, p. 244. In: Eastoe, J. E., Picton, D. C. A., Alexander, A. G. (eds.), The prevention of periodontal disease. London: Henry Kimpton.

Carlsson, J. 1972. Nutritional requirements of *Streptococcus sanguis*. Archives of Oral Biology **17:**1327–1332.

Carlsson, J., Grahnen, H., Johnsson, G., Wikner, S. 1970. Establishment of *Streptococcus sanguis* in the mouths of infants. Archives of Oral Biology **15:**1143–1148.

Chapman, G. H. 1944. The isolation of streptococci from mixed cultures. Journal of Bacteriology **48:**113–114.

Chassy, B. M., Beall, J. R., Bielawski, R. M., Porter, E. V., Donkersloot, J. A. 1976. Occurrence and distribution of sucrose-metabolizing enzymes in oral streptococci. Infection and Immunity **14:**408–415.

Clarke, J. K. 1924. On the bacterial factor in the etiology of dental caries. British Journal of Experimental Pathology **5:**141–147.

Colman, G. 1968. The application of computers to the classification of streptococci. Journal of General Microbiology **50:**149–158.

Colman, G., Williams, R. E. O. 1965. The cell walls of streptococci. Journal of General Microbiology **41:**375–387.

Colman, G., Williams, R. E. O. 1972. Taxonomy of some human viridans streptococci, pp. 281–299. In: Wannamaker, L. W.,

Matsen, J. M. (eds.), Streptococci and streptococcal diseases. New York, London: Academic Press.

Colman, G., Williams, R. E. O. 1973. Identification of human streptococci, pp. 293–303. In: Dyke, S. C. (ed.), Recent advances in clinical pathology. Edinburgh, London: Churchill Livingstone.

Cooper, H. R., Chorpenning, F. W., Rosen, S. 1975. Preparation and chemical composition of the cell walls of *Streptococcus mutans*. Infection and Immunity **11**:823–828.

Coykendall, A. L. 1971. Genetic heterogeneity in *Streptococcus mutans*. Journal of Bacteriology **106**:192–196.

Coykendall, A. L. 1974. Four types of *Streptococcus mutans* based on their genetic, antigenic and biochemical characteristics. Journal of General Microbiology **83**:327–338.

Coykendall, A. L. 1976. On the evolution of *Streptococcus mutans* and dental caries, pp. 703–712. In: Stiles, H. M., Loesche, W. J., O'Brien, T. C. (eds.), Proceedings "Microbial aspects of dental caries", vol. 3. Washington, D.C., London: Information Retrieval.

Coykendall, A. L. 1977. Proposal to elevate the subspecies of *Streptococcus mutans* to species status, based on their molecular composition. International Journal of Systematic Bacteriology **27**:26–30.

Coykendall, A. L., Specht, P. A., Samol, H. H. 1974. *Streptococcus mutans* in a wild, sucrose-eating rat population. Infection and Immunity **10**:216–219.

Coykendall, A. L., Bratthall, D., O'Connor, K., Dvarskas, R. A. 1976. Serological and genetic examination of some nontypical *Streptococcus mutans* strains. Infection and Immunity **14**:667–670.

Dahlqvist, A., Krasse, B., Olsson, I., Gardell, S. 1967. Extracellular polysaccharides formed by caries-inducing streptococci. Helvetica Odontologica Acta **11**:15–21.

Deibel, R. H., Seeley, H. W., 1974. Streptococcaceae, pp. 490–509. In: Buchanan, R. E., Gibbons, N. E. (eds.), Bergey's manual of determinative bacteriology, 8th ed. Baltimore: Williams & Wilkins.

DeStoppelaar, J. D. 1971a. Decreased cariogenicity of a mutant of *Streptococcus mutans*. Archives of Oral Biology **16**:971–975.

DeStoppelaar, J. D. 1971b. *Streptococcus mutans, Streptococcus sanguis* and dental caries. Dissertation. Utrecht, Drukkerij Elinkwijk, the Netherlands.

Dobell, C. 1960. Antony van Leeuwenhoek and his "little animals", p.435. New York: Dover.

Drucker, D. B., Melville, T. H. 1971. The classification of some oral streptococci of human or rat origin. Archives of Oral Biology **16**:845–853.

Drucker, D. B., Shuttleworth, C. A., Melville, T. H. 1968. A quantitative analysis of the cell wall amino acids of cariogenic and non-cariogenic streptococci. Archives of Oral Biology **13**:937–940.

Dwyer, D. M., Socransky, S. S. 1968. Predominant cultivable microorganisms inhabiting periodontal pockets. British Dental Journal **124**:560–564.

Eastoe, J. E., Picton, D. C. A., Alexander, A. G. 1971. The prevention of periodontal disease. London: Henry Kimpton.

Edwardsson, S. 1968. Characteristics of caries-inducing human streptococci resembling *Streptococcus mutans*. Archives of Oral Biology **13**:637–646.

Edwardsson, S. 1970. The caries-inducing property of variants of *Streptococcus mutans*. Odontologisk Revy **21**:153–157.

Edwardsson, S. 1975. Bacteriological studies on deep areas of carious dentine. Odontologisk Revy, Suppl. 32 **25**:143.

Ellen, R. P. 1976. Microbiological assays for dental caries and periodontal disease susceptibility, pp. 3–23. In: Melcher, A. H., Zarb, G. A. (eds.), Oral sciences reviews, vol. 8. Munksgaard, Copenhagen: Villadsen & Christensen.

Ellwood, D. C. 1976. Chemostat studies of oral bacteria, pp. 785–798. In: Stiles, H. M., Loesche, W. J., O'Brien, T. C. (eds.), Proceedings "Microbial aspects of dental caries", vol. 3. Washington, D.C., London: Information Retrieval.

Facklam, R. R. 1974. Characteristics of *Streptococcus mutans* isolated from human dental plaque and blood. International Journal of Systematic Bacteriology **24**:313–319.

Fitzgerald, R. J., Keyes, P. H. 1960. Demonstration of the etiologic role of streptococci in experimental caries in the hamster. Journal American Dental Association **61**:9–19.

Fitzgerald, R. J. 1968. Dental caries research in gnotobiotic animals. Caries Research **2**:139–146.

Fitzgerald, R. J., Adams, B. O. 1975. Increased selectivity of mitis-salivarius agar containing polymyxin. Journal of Clinical Microbiology **1**:239–240.

Gehring, F. 1968. Über das Wachstum einiger Zahnplaquestreptokokken auf festen und in flüssigen Nährmedien mit verschiedenen Zuckerkonzentrationen. Deutsche Zahnärztliche Zeitschrift **23**:914–923.

Gehring, F. 1972. Extrazelluläre Polysaccharide bildende Streptokokken aus Zahnplaques und ihre Beziehung zur Zahnkaries. Munich: Carl Hanser Verlag.

Gehring, F. 1976. Über den Einsatz des Leitz-Auflichtilluminators Ultropak zum Leitz-Orthoplan-Mikroskop bei der Diagnose bestimmter Streptokokken in der mikrobiologischen Kariesforschung. Leitz Mitteilungen für Wissenschaft und Technik **6**:268–273.

Gehring, F. 1977. Mikrobiologische Untersuchungen im Rahmen der "Turku sugar studies". Deutsche Zahnärztliche Zeitschrift, Suppl. 1, **32**:84–88.

Gehring, F., Mäkinen, K. K., Larmas, M., Scheinin, A. 1975. Turku sugar studies. X. Occurrence of polysaccharide-forming streptococci and ability of the mixed plaque microbiota to ferment various carbohydrates. Acta Odontologica Scandinavica, Suppl. 70 **33**:223–237.

Gehring, F., Karle, E. J., Patz, J., Felfe, W., Bradatsch, U. 1976. Vorkommen von *Streptococcus mutans*-Varianten bei Mensch und Versuchstier. Deutsche Zahnärztliche Zeitschrift **31**:18–21.

Gibbons, R. J. 1968. Formation and significance of bacterial polysaccharides in caries etiology. Caries Research **2**:164–171.

Gibbons, R. J. 1972. Ecology and cariogenic potential of oral streptococci, pp. 371–385. In: Wannamaker, L. W., Matsen, J. M. (eds.), Streptococci and streptococcal diseases. New York, London: Academic Press.

Gibbons, R. J., Nygaard, M. 1968. Synthesis of insoluble dextran and its significance in the formation of gelatinous deposits by plaque-forming streptococci. Archives of Oral Biology **13**:1249–1262.

Gibbons, R. J., Kapsimalis, B., Socransky, S. S. 1964. The source of salivary bacteria. Archives of Oral Biology **9**:101–103.

Gibbons, R. J., van Houte, J. 1973. On the formation of dental plaques. Journal of Periodontology **44**:347–360.

Gibbons, R. J., van Houte, J. 1975a. Dental caries. Annual Review of Medicine **26**:121–136.

Gibbons, R. J., van Houte, J. 1975b. Bacterial adherence in oral microbial ecology. Annual Review of Microbiology **29**:19–44.

Gibbons, R. J., Socransky, S. S., Sawyer, S., Kapsimalis, B., MacDonald, J. B. 1963. The microbiota of the gingival crevice area of man. II. The predominant cultivable organisms. Archives of Oral Biology **8**:281–289.

Gibbons, R. J., Socransky, S. S., de Araujo, W. C., van Houte, J. 1964. Studies of the predominant cultivable microbiota of dental plaque. Archives of Oral Biology **9**:365–370.

Gibbons, R. J., Depaola, P. F., Spinell, D. M., Skobe, Z. 1974. Interdental localization of *Streptococcus mutans* as related to dental caries experience. Infection and Immunity **9**:481–488.

Gold, W. 1969. Dental caries and periodontal disease considered as infectious diseases. Advances in Applied Microbiology **11**:135–157.

Gold, O. G., Jordan, H. V., van Houte, J. 1973. A selective medium for *Streptococcus mutans*. Archives of Oral Biology **18**:1357–1364.

Gold, O. G., Jordan, H. V., van Houte, J. 1974. Identification of

Streptococcus mutans colonies by mannitol-dependent tetrazolium reduction. Archives of Oral Biology **19**:271–272.

Gold, O. G., Jordan, H. V., van Houte, J. 1975. The prevalence of enterococci in the human mouth and their pathogenicity in animal models. Archives of Oral Biology **20**:473–477.

Gordon, M. H. 1905. A ready method of differentiating streptococci and some results already obtained by its application. Lancet **ii**:1400–1403.

Gordon, D. F., Gibbons, R. J. 1966. Studies of the predominant cultivable flora of the tongue. Archives of Oral Biology **11**:627–633.

Greer, S. B., Hsiang, W., Musil, G., Zinner, D. D. 1971. Viruses of cariogenic streptococci. Journal of Dental Research **50**:1594–1604.

Green, R. M., Blackmore, D. K., Drucker, D. B. 1973. The role of gnotobiotic animals in the study of dental caries. British Dental Journal **134**:537–540.

Griffith, C. J., Melville, T. H. 1974. Growth of oral streptococci in a chemostat. Archives of Oral Biology **19**:87–90.

Guggenheim, B. 1968. Streptococci of dental plaques. Caries Research **2**:147–163.

Guggenheim, B., Schroeder, H. E. 1967. Biochemical and morphological aspects of extracellular polysaccharides produced by cariogenic streptococci. Helvetica Odontologica Acta **11**:131–152.

Guthof, O. 1956. Über pathogene "vergrünende Streptokokken". Zentralblatt für Bakteriologie, Parasitenkunde, Infektionskrankheiten und Hygiene, Abt. 1 Orig. **166**:553–564.

Guthof, O. 1970. Vorkommen und Nachweis von Ektopolysacchariden bei Streptokokken. Zentralblatt für Bakteriologie, Parasitenkunde, Infektionskrankheiten und Hygiene, Abt. 1 Orig. **215**:435–440.

Hahn, G., Heeschen, W., Tolle, A. 1970. *Streptococcus*. Eine Studie zur Struktur, Biochemie, Kultur und Klassifizierung. Kieler Milchwirtschaftliche Forschungsberichte **22**:333–546.

Hamada, S., Ooshima, T. 1975. Production and properties of bacteroicins (mutacins) from *Streptococcus mutans*. Archives of Oral Biology **20**:641–648.

Hamilton, I. R. 1976. Intracellular polysaccharide synthesis by cariogenic microorganisms, pp. 683–701. In: Stiles, H. M., Loesche, W. J., O'Brien, T. C. (eds.), Proceedings "Microbial aspects of dental caries", vol. 3. Washington, D.C., London: Information Retrieval.

Harder, E. J., Wilkowske, C. J., Washington, J. A., Geraci, J. E. 1974. *Streptococcus mutans* endocarditis. Annals of International Medicine **80**:364–368.

Hardie, J. M., Bowden, G. H. 1974a. The normal microbial flora of the mouth, pp. 47–83. In: Skinner, F. A., Carr, J. G. (eds.), The normal microbial flora of man. London, New York: Academic Press.

Hardie, J. M., Bowden, G. H. 1974b. Cell wall and serological studies on *Streptococcus mutans*. Caries Research **8**:301–316.

Hardie, J. M., Bowden, G. H. 1976a. Physiological classification of oral viridans streptococci. Journal of Dental Research **55**(Special Issue A):166–176.

Hardie, J. M., Bowden, G. H. 1976b. The microbial flora of dental plaque: Bacterial succession and isolation considerations, pp. 63–87. In: Stiles, H. M., Loesche, W. J., O'Brien, T. C. (eds.), Proceedings "Microbial aspects of dental caries", vol. 1. Washington, D.C., London: Information Retrieval.

Hehre, E. J., Neill, J. M. 1946. Formation of serologically reactive dextrans by streptococci from subacute bacterial endocarditis. Journal of Experimental Medicine **83**:147–162.

Hoover, C. I., Newbrun, E. 1977. Survival of bacteria from human dental plaque under various transport conditions. Journal of Clinical Microbiology **6**:212–218.

Huxley, H. G. 1972. The recovery of microorganisms from the fissures of rat molar teeth. Archives of Oral Biology **17**:1481–1485.

Ikeda, T., Sandham, H. J. 1972. A high-sucrose medium for the

identification of *Streptococcus mutans*. Archives of Oral Biology **17**:781–783.

Inoue, M., Smith, E. E. 1976. Extracellular polysaccharides formed by dextransucrase isozymes of cariogenic microorganisms, pp. 665–682. In: Stiles, H. M., Loesche, W. J., O'Brien, T. C. (eds.), Proceedings "Microbial aspects of dental caries", vol. 3. Washington, D.C., London: Information Retrieval.

International Committee on Nomenclature of Bacteria Subcommittee on Streptococci and Pneumonocci. 1971. Minutes of Meeting, 7 August 1970. International Journal of Systematic Bacteriology **21**:172–173.

Inward, P. W., Upstone, D., van Houte, J. 1970. Nutritional requirements of oral streptococci, pp. 217–224. In: McHugh, W. D. (ed.), Dental plaque. Edinburgh, London: E. & S. Livingstone.

Jablon, J. M., Ferrer, T., Zinner, D. D. 1976. Identification and quantitation of *Streptococcus mutans* by the fluorescent antibody technique. Journal of Dental Research **55** (Special Issue A):76–79.

Jordan, H. V., Krasse, B., Möller, A. 1968. A method of sampling human dental plaque for certain "caries-inducing" streptococci. Archives of Oral Biology **13**:919–927.

Keene, H. J., Shklair, I. L. 1973. Absence of an association between *Streptococcus mutans* and gingivitis in caries-free naval recruits. Journal of Periodontology **44**:705–708.

Keene, H. J., Shklair, I. L., Mickel, G. J., Wirthlin, M. R. 1977a. Distribution of *Streptococcus mutans* biotypes in five human populations. Journal of Dental Research **56**:5–10.

Keene, H. J., Shklair, I. L., Anderson, D. M., Mickel, G. J. 1977b. Relationship of *Streptococcus mutans* biotypes to dental caries prevalence in Saudi Arabian naval men. Journal of Dental Research **56**:356–361.

Kelstrup, J., Gibbons, R. J. 1969. Bacteriocins from human and rodent streptococci. Archives of Oral Biology **14**:251–258.

König, K. G., 1966. Möglichkeiten der Kariesprophylaxe beim Menschen und ihre Untersuchung im kurzfristigen Rattenexperiment. Bern, Stuttgart: Verlag Hans Huber.

Krasse, B. 1954. The proportional distribution of *Streptococcus salivarius* and other streptococci in various parts of the mouth. Odontologisk Revy **5**:203–211.

Lancefield, R. C. 1933. A serological differentiation of human and other groups of hemolytic streptococci. Journal of Experimental Medicine **57**:571–595.

Lawson, J. W. 1971. Growth of cariogenic streptococci in chemically defined medium. Archives of Oral Biology **16**:339–342.

Littleton, N. W., Kakehashi, S., Fitzgerald, R. J. 1970. Recovery of specific "caries-inducing" streptococci from carious lesions in the teeth of children. Archives of Oral Biology **15**:461–463.

Loesche, W. J., Bhat, M. 1976. Evaluation of diagnostic broths for *Streptococcus mutans*, pp. 291–301. In: Stiles, H. M., Loesche, W. J., O'Brien, T. C. (eds.), Proceedings "Microbial aspects of dental caries", Sp. Suppl. Microbiology Abstracts, vol. 1. Washington, D.C., London: Information Retrieval.

Loesche, W. J., Grenier, E. 1976. Detection of *Streptococcus mutans* in plaque samples by direct fluorescent antibody test. Journal of Dental Research **55** (Special Issue A):87–93.

Loesche, W. J., Hockett, R. N., Syed, S. A. 1972. The predominant cultivable flora of tooth surface plaque removed from institutionalized subjects. Archives of Oral Biology **17**:1311–1325.

Loesche, W. J., Syed, S. A. 1973. The predominant cultivable flora of carious plaque and carious dentine. Caries Research **7**:201–216.

Loesche, W. J., Rowan, J., Straffon, L. H., Loos, P. J. 1975. Association of *Streptococcus mutans* with human dental decay. Infection and Immunity **11**:1252–1260.

McHugh, W. D. 1970. Dental plaque, p. 298. Edinburgh, London: E. & S. Livingstone.

Mäkinen, K. K. 1974. Sugars and the formation of dental plaque, pp. 645–687. In: Sipple, H. L., McNutt, K. W. (eds.), Sugars in nutrition. New York: Academic Press.

Mejàre, B. 1974. The incidence and significance of *Streptococcus sanguis, Streptococcus mutans* and *Streptococcus salivarius* in root canal cultures from human teeth. Odontologisk Revy **25**:359–378.

Mejàre, B. 1975a. *Streptococcus faecalis* and *Streptococcus faecium* in infected dental root canals at filling and their susceptibility to azidocillin and some comparable antibiotics. Odontologisk Revy **26**:193–204.

Mejàre, B. 1975b. Characteristics of *Streptococcus milleri* and *Streptococcus mitior* from infected dental root canals. Odontologisk Revy **26**:291–308.

Mejàre, B., Edwardsson, S. 1975. *Streptococcus milleri* (Guthof); an indigenous organism of the human oral cavity. Archives of Oral Biology **20**:757–762.

Miller, W. D. 1973. The micro-organisms of the human mouth. The local and general diseases which are caused by them. Unaltered reprint of the original work by W. D. Miller published in 1890 in Philadelphia. Basel, Munich, Paris, London, New York, Sydney: Karger.

Mirick, G. S., Thomas, L., Curnen, E. C., Horsfall, F. L. 1944. Studies on a non-hemolytic streptococcus isolated from the respiratory tract of human beings. I. Biological characteristics of *Streptococcus* MG. Journal of Experimental Medicine **80**:391–406.

Montague, E. A., Knox, K. W. 1968. Antigenic components of the cell wall of *Streptococcus salivarius*. Journal of General Microbiology **54**:237–246.

Morris, E. O. 1954. The bacteriology of the oral cavity. III. *Streptococcus*. British Dental Journal **96**:95–108.

Navia, J. M. 1977. Animal models in dental research. Alabama: University of Alabama Press.

Newbrun, E. 1972. Extracellular polysaccharides synthesized by glucosyltransferases of oral streptococci. Caries Research **6**:132–147.

Newbrun, E., Baker, S. 1968. Physico-chemical characteristics of the levan produced by *Streptococcus salivarius*. Carbohydrate Research **6**:165–170.

Newbrun, E., Lacy, R., Christie, T. M. 1971. The morphology and size of the extracellular polysaccharides from oral streptococci. Archives of Oral Biology **16**:863–872.

Newman, M. G., Socransky, S. S., Savitt, E. D., Propas, D. A., Crawford, A. 1976. Studies of the microbiology of periodontosis. Journal of Periodontology **47**:373–379.

Nisizawa, T., Imai, S., Araya, S. 1977. Methylation analysis of extracellular glucans produced by *Streptococcus mutans* strain JC 2. Archives of Oral Biology **22**:281–285.

Nisizawa, T., Imai, S., Akada, H., Hinoide, M., Araya, S. 1976. Extracellular glucans produced by oral streptococci. Archives of Oral Biology **21**:207–213.

Niven, C. F., Jr., Kiziuta, Z., White, J. C. 1946. Synthesis of polysaccharide from sucrose by *Streptococcus S.B.E.* Journal of Bacteriology **51**:711–716.

Nolte, W. A. 1973. Oral microbiology, 2nd ed. Saint Louis, Missouri: Mosby.

Nord, C.-E., Wadström, T. 1973. Characterization of haemolytic enterococci isolated from oral infections. Acta Odontologica Scandinavica **31**:387–393.

Onose, H., Sandham, H. J. 1976. pH Changes during culture of human dental plaque streptococci on mitis-salivarius agar. Archives of Oral Biology **21**:291–296.

Orland, F. J., Blayney, J. R., Harrison, R. W., Reyniers, J. A., Trexler, P. C., Wagner, M., Gordon, H. A., Luckey, T. D. 1954. Use of the germfree animal technic in the study of experimental dental caries. I. Basic observations on rats reared free of all microorganisms. Journal of Dental Research **33**:147–174.

Osborne, R. M., Lamberts, B. L., Meyer, T. S., Roush, A. H. 1976. Acrylamide gel electrophoretic studies of extracellular

sucrose-metabolizing enzymes of *Streptococcus mutans*. Journal of Dental Research **55**:77–84.

Paul, D., Slade, H. D. 1975. Production and properties of an extracellular bacteriocin from *Streptococcus mutans* bacteriocidal for group A and other streptococci. Infection and Immunity **12**:1375–1385.

Perch, B., Kjems, E., Ravn, T. 1974. Biochemical and serological properties of *Streptococcus mutans* from various human and animal sources. Acta Pathologica et Microbiologica Scandinavica, Sect. B **82**:357–370.

Porterfield, J. S. 1950. Classification of the streptococci of subacute bacterial endocarditis. Journal of General Microbiology **4**:92–101.

Ranke, B., Ranke, E. 1971. Untersuchungen zur Kariogenität extrazelluläre Polysaccharide bildender Streptokokken aus menschlicher Zahnplaque. Deutsche Zahnärztliche Zeitschrift **26**:29–36.

Ranke, E., Ranke, B., Ahrens, G., Heeschen, W. 1967. Plaqueflora und Zahnkaries. 1. Mitteilung: Vorkommen α-hämolysierender und vergrünender Streptokokken. Deutsche Zahnärztliche Zeitschrift **22**:883–890.

Rogers, A. H. 1969. The proportional distribution and characteristics of streptococci in human dental plaque. Caries Research **3**:238–248.

Rogers, A. H. 1972. Effect of the medium on bacteriocin production among strains of *Streptococcus mutans*. Applied Microbiology **24**:294–295.

Rogers, A. H. 1975. Bacteriocin types of *Streptococcus mutans* in human mouths. Archives of Oral Biology **20**:853–858.

Rogers, A. H. 1976a. The oral cavity as a source of potential pathogens in focal infection. Oral Surgery, Oral Medicine, and Oral Pathology **42**:245–248.

Rogers, A. H. 1976b. Bacteriocinogeny and the properties of some bacteriocins of *Streptococcus mutans*. Archives of Oral Biology **21**:99–104.

Rogers, A. H. 1976c. Bacteriocin patterns of strains belonging to various serotypes of *Streptococcus mutans*. Archives of Oral Biology **21**:243–249.

Rosan, B., Lai, C. H., Listgarten, M. A. 1976. *Streptococcus sanguis:* A model in the application in immunochemical analysis for the in situ localization of bacteria in dental plaque. Journal of Dental Research **55** (Special Issue A), 124–141.

Rundell, B. B., Thomson, L. A., Loesche, W. J., Stiles, H. M. 1973. Evaluation of a new transport medium for the preservation of oral streptococci. Archives of Oral Biology **18**:871–878.

Schachtele, C. F., Loken, A. E., Schmitt, M. K. 1972. Use of specifically labeled sucrose for comparison of extracellular glucan and fructan metabolism by oral streptococci. Infection and Immunity **5**:263–266.

Schultz-Haudt, S. D. 1964. Biochemical aspects of periodontal disease. International Dental Journal **14**:398–406.

Seelemann, M. 1954. Biologie der Streptokokken, 2 ed. Nürnberg: Verlag Hans Carl.

Seelemann, M., Obiger, G. 1958. Biologie, Klassifizierung und Nomenklatur der sog. vergrünenden Streptokokken. Nürnberg: Verlag Hans Carl.

Sherman, J. M. 1937. The streptococci. Bacteriological Reviews **1**:1–97.

Sherman, J. M., Niven, C. F., Jr., Smiley, K. L. 1943. *Streptococcus salivarius* and other nonhemolytic streptococci of the human throat. Journal of Bacteriology **45**:249–263.

Shklair, I. L. 1973. *Streptococcus mutans* and the epidemiology of dental caries. Department of Health, Education, and Welfare Publication No. (NIH) 74-286: 7-13, Washington, D.C. U. S. Government Printing Office.

Shklair, I. L., Keene, H. J. 1974. A biochemical scheme for the separation of the five varieties of *Streptococcus mutans*. Archives of Oral Biology **19**:1079–1081.

Shklair, I. L., Keene, H. J., Cullen, P. 1974. The distribution of

Streptococcus mutans on the teeth of two groups of naval recruits. Archives of Oral Biology **19**:199–202.

Shklair, I. L., Walter, R. 1976. Evaluation of a selective medium-color test for *Streptococcus mutans*. Journal of Dental Research **55** (Special Issue B): B122.

Simon, B. I., Goldman, H. M., Ruben, M. P., Broitman, S., Baker, E. 1972. The role of endotoxin in periodontal disease. IV. Bacteriologic analyses of human gingival exudate as related to the quantity of endotoxin and clinical degree of inflammation. Journal of Periodontology **43**:468–475.

Socransky, S. S. 1970. Relationship of bacteria to the etiology of periodontal disease. Journal of Dental Research **49**:203–222.

Socransky, S. S., Manganiello, S. D. 1971. The oral microbiota of man from birth to senility. Journal of Periodontology **42**:485–494.

Stiles, H. M., Loesche, W. J., O'Brien, T. C. (eds.). 1976. Proceedings "Microbial aspects of dental caries". Sp. Suppl. Microbiology Abstracts, vol. 1–3. Washington D.C., London: Information Retrieval.

Street, C. M., Goldner, M., LeRiche, W. H. 1976. Epidemiology of dental caries in relation to *Streptococcus mutans* on tooth surfaces in 5-year-old children. Archives of Oral Biology **21**:273–275.

Sukchotiratana, M., Linton, A. H., Fletcher, J. P. 1975. Antibiotics and the oral streptococci of man. Journal of Applied Bacteriology **38**:277–294.

Sumney, D. L., Jordon, H. V. 1974. Characterization of bacteria isolated from human root surface carious lesions. Journal of Dental Research **53**:343–351.

Syed, S. A., Loesche, W. J. 1972. Survival of human dental plaque flora in various transport media. Applied Microbiology **24**:638–644.

Takeuchi, H., Sumitani, M., Tsubakimoto, K., Tsutsui, M. 1974. Oral microorganisms in the gingiva of individuals with periodontal disease. Journal of Dental Research **53**:132–136.

Tanzer, J. M., Chassy, B. M., Krichevsky, M. I. 1972. Sucrose metabolism by *Streptococcus mutans*, SL-I. Biochimica et Biophysica Acta **261**:379–387.

Tanzer, J. M., Freedman, M. L., Woodiel, F. N., Eifert, R. L., Rinehimer, L. A. 1976. Association of *Streptococcus mutans* virulence with synthesis of intracellular polysaccharide, pp. 597–616. In: Stiles, H. M., Loesche, W. J., O'Brien, T. C. (eds.), Proceedings "Microbial aspects of dental caries", vol. 3. Washington, D.C., London: Information Retrieval.

Terleckyj, B., Willett, N. P., Shockman, G. D. 1975. Growth of several cariogenic strains of oral streptococci in a chemically defined medium. Infection and Immunity **11**:649–655.

Theilade, E., Theilade, J. 1975. Role of plaque in the etiology of periodontal disease and caries, pp. 23–49. In: Melcher, A. H., Zarb, G. A. (eds.), Oral sciences reviews, vol. 9. Munksgaard, Copenhagen: Villadsen & Christensen.

Thomson, L. A., Little, W., Hageage, G. J. 1976. Application of fluorescent antibody methods in the analysis of plaque samples. Journal of Dental Research **55** (Special Issue A):80–86.

van Houte, J. 1976. Oral bacterial colonization: Mechanisms and implications, pp. 3–32. In: Stiles, H. M., Loesche, W. J., O'Brien, T. C. (eds.), Proceedings "Microbial aspects of dental caries", vol. 1. Washington, D.C., London: Information Retrieval.

van Houte, J., Gibbons, R. J., Banghart, S. 1970. Adherence as a determinant of the presence of *Streptococcus salivarius* and *Streptococcus sanguis* on the tooth surface. Archives of Oral Biology **15**:1025–1035.

van Houte, J., Winkler, K. C., Jansen, H. M. 1969. Iodophilic polysaccharide synthesis, acid production and growth in oral streptococci. Archives of Oral Biology **14**:45–61.

Waddy, J. 1976. Bacterial endocarditis: A cardiologist's view of dental involvement. Oral Surgery, Oral Medicine, and Oral Pathology **42**:240–244.

Walker, G. J., Hare, M. D. 1977. Metabolism of the polysaccharides of human dental plaque. Part II. Purification and properties of *Cladosporium resinae* (1→3)-α-D-glucanase, and the enzymic hydrolysis of glucans synthesised by extracellular D-glucosyltransferases of oral streptococci. Carbohydrate Research **58**:415–432.

Washburn, M. R., White, J. C., Niven, C. F., Jr. 1946. *Streptococcus S.B.E.*: Immunological characteristics. Journal of Bacteriology **51**:717–722.

White, J. C., Niven, C. F., Jr. 1946. *Streptococcus S.B.E.*: A streptococcus associated with subacute bacterial endocarditis. Journal of Bacteriology **51**:717–722.

Willers, J. M. N., Ottens, H., Michel, M. F. 1964. Immunochemical relationship between *Streptococcus* MG, F III and *Streptococcus salivarius*. Journal of General Microbiology **37**:425–431.

Wood, J. M., Critchley, P. 1966. The extracellular polysaccharide produced from sucrose by a cariogenic streptococcus. Archives of Oral Biology **11**:1039–1042.

Yamamoto, T., Imai, S., Nisizawa, T., Araya, S. 1975. Production of, and susceptibility to, bacteriocin-like substances in oral streptococci. Archives of Oral Biology **20**:389–391.

The Family Streptococcaceae (Nonmedical Aspects)

MICHAEL TEUBER and ARNOLD GEIS

In their nonmedical aspects, the Streptococcaceae are distinguished by important contributions to basic microbiology, genetics and molecular biology, and general and microbial biochemistry as well as to food science and biotechnology. The present day significance is in the large-scale use of *Streptococcus, Pediococcus,* and *Leuconostoc* species for industrial fermentations including dairy products, sausages, and wine, respectively.

This development was initiated in 1873 by Joseph Lister who was attempting to prove Pasteur's germ theory of fermentative changes. In his experiments with boiled milk as a nutrient medium he obtained by chance the first bacterial pure culture. It is worthwhile to recall in the context of this handbook his original discussion of this discovery marking the dawn of bacterial taxonomy:

Admitting then that we had here to deal with only one bacterium, it presents such peculiarities both morphologically and physiologically as to justify us, I think, in regarding it a definite and recognizable species for which I venture to suggest the name Bacterium lactis. This I do with diffidence, believing that up to this time no bacterium has been defined by reliable characters. Whether this is the only bacterium that can occasion the lactic acid fermentation, I am not prepared to say.

The accepted identity of this bacterium is now *Streptococcus lactis* (Löhnis, 1909). Pasteur's and Lister's concepts and experiments led Hermann Weigmann in Kiel and W. Storch in Copenhagen independently to introduce pure starter cultures of lactic acid bacteria including lactic streptococci into the dairy field for the fermentation and ripening of milk, cream, and cheese (Weigmann, 1905–1908).

In the area of genetics and molecular biology, Griffith's work on the virulence of smooth forms and the avirulence of rough variants of *Streptococcus (Diplococcus) pneumoniae* (1928) resulted in the discovery of the genetic transformability of living bacteria by extracts from heat-killed ones. The identification of deoxyribonucleic acid (DNA) as the physical and chemical carrier of the genetic information in these extracts by Avery, MacLeod, and McCarthy (1944) sparked the explosive progress of molecular biology in the last four decades.

A novel development constitutes the conjugal transfer of genetic information in oral streptococci (LeBlanc et al., 1978) and *Streptococcus lactis* subsp. *diacetylactis* (Kempler and McKay, 1979). The presence of plasmids in all investigated lactic streptococci (Pechmann and Teuber, 1980), which bear the information for technologically important functions like metabolism of lactose, fermentation of citrate, and proteolytic activity, gives an explanation for the long known genetic instability of these properties (Efstathiou and McKay, 1976). Although transduction and transformation experiments have been successfully performed with lactic streptococci (McKay and Baldwin, 1978), the knowledge of the genetics of these economically interesting microorganisms is still too rudimentary to allow an adequate and necessary genetic analysis. The situation is further complicated by the fact that many strains of lactic streptococci are lysogenic (Reiter and Kirikova, 1976) and produce bacteriocins (Kozak, Bardowski, and Dobrzanski, 1977) and antibiotics like the polypeptide nisin (Tagg, Dajani, and Wannamaker, 1976).

The industrial application of streptococci—mostly under nonsterile conditions—has accentuated the problem of severe incidences of phage attacks during the manufacture of fermented milk and cheese (Lawrence, 1978; Lembke et al., 1980). The practice of employing only mixed or multiple-strain starter cultures has only brought limited success.

Under these circumstances, the main problem in the isolation, characterization, and identification of a particular bacterium is not so much a matter of distinguishing the species level—though difficult enough in some cases—but of identifying a particular strain that has the wanted technological properties.

We confine ourselves in this chapter to:

1. The lactic streptococci used in the dairy industry.
2. The leuconostocs in use in the dairy and wine industry and those commonly found in fermented vegetables.
3. The pediococci encountered in the spoilage of beer and those used as starter cultures by the meat industry.

Although all these members of the Streptococcaceae

are commonly found on plant material, their role in the ecology of the plant microflora is not well investigated.

A thorough and competent treatment of all aspects of the streptococci has recently been attained in a Symposium of the Society for Applied Bacteriology (Skinner and Quesnel, 1978). The classic paper of Orla-Jensen (1919) still provides a basic description of the lactic streptococci.

Habitats

Lactic Streptococci

The lactic streptococci comprise the species *Streptococcus lactis, S. lactis* subsp. *diacetylactis, S. cremoris,* and *S. thermophilus. S. lactis* and *S. lactis* subsp. *diacetylactis* have commonly been detected directly or following enrichment in plant material, including fresh and frozen corn, corn silks, navy beans, cabbage, lettuce, peas, wheat middlings, grass, clover, potatoes, cucumbers, and cantaloupe (Sandine, Radich, and Elliker, 1972). Lactic streptococci are usually not found in fecal material or soil. Only small numbers occur on the surface of the cow and in its saliva. Since raw cow's milk consistently contains *S. lactis,* and to a much lesser extent *S. lactis* subsp. *diacetylactis* and *S. cremoris,* it is tempting to suggest that lactic streptococci enter the milk from the exterior of the udder during milking and from the fodder, which may be the primary source of infection. *S. cremoris* and *S. thermophilus* have hitherto not been isolated with certainty from habitats other than milk, fermented milk, cheese, and starter cultures.

Out of 34,935 streptococcal specimens from man and animals which were sent to the Streptokokkenzentrale Kiel for identification during the period from 1965 to 1977, 1,433 strains were classified as *S. lactis,* 40 as *S. cremoris,* 22 as *S. lactis* subsp. *diacetylactis,* and 104 as *S. thermophilus* (Hahn and Tolle, 1979). This low count of lactic streptococci within a large body of pathogenic and fecal streptococci reflects the accepted view that lactic streptococci are apathogenic to man and animal and that their incidence may be due to contamination from food and fodder, respectively.

In the horse stomach, where an active lactic fermentation takes place, 3×10^8 lactic streptococci have been counted per gram of content (Giesecke and Henderickx, 1973).

The most important habitat for lactic streptococci, however, concerns the dairy industry. As shown in Table 1, lactic streptococci are employed in single and mixed cultures for the production of all different kinds of cheeses, fermented milks, cultured butter, and casein.

These cultures fulfill the following functions:

1. Fermentation of lactose to lactate. The resulting lowered pH values (4.0–5.6) as compared to milk (6.6–6.7) prevent or retard growth of spoilage bacteria, especially *Clostridium, Staphylococcus,* Enterobacteriaceae, and psychrophilic Gram-negative bacteria like *Pseudomonas.* If the isoelectric point of casein at pH 4.6–4.8 is approached, it is precipitated. This effect is used to curdle milk in the production of cottage cheese, quarg, sour milk, yogurt, and casein. Starter bacteria for this purpose are *S. lactis, S. cremoris,* and *S. thermophilus* (besides lactobacilli).

2. Formation of aroma components. The most prominent compound is diacetyl, the characteristic flavor component of cultured butter. It is principally produced by *Leuconostoc cremoris (citrovorum)* in association with *S. lactis* subsp. *diacetylactis* and/or *S. cremoris* (Sharpe, 1979). Evolution of gas (CO_2) from citrate in milk is performed by *L. cremoris (citrovorum)* and *S. lactis* subsp. *diacetylactis,* which induce eye formation in cheese but also unwanted floating of the curd in the manufacture of cottage cheese or quarg if an unbalanced mixed starter is used (see Table 1).

3. Ripening of cheese. The proteases built by lactic streptococci influence mainly the texture and softness of the cheese body. Strains exhibiting too active proteolytic enzymes may induce a bitter taste due to bitter peptides.

The lactic streptococci and lactobacilli needed by the dairy and meat industries are supplied by firms and institutes that specialize in the production of starter cultures. In the Federal Republic of Germany, for example, at least 12 local, national, and international commercial enterprises produced and distributed the cultures for the manufacture in 1978 of 721,050 tons of cheese, about 300,000 tons of fermented milk products, and 564,000 tons of butter.

The cultures are typically provided in either of three forms:

1. Fresh, fluid cultures containing about 0.5×10^9 viable streptococci per milliliter.
2. Freeze-dried (lyophilized) cultures containing about 2×10^9 viable streptococci per gram.
3. Deep-frozen, concentrated cultures containing about 7×10^9 viable streptococci per milliliter.

Fresh cultures deteriorate during improper shipment and storage. The lyophilized material is convenient to ship and may be kept at 5–10°C for at least 6 months. Deep-frozen, concentrated cultures can be shipped and stored indefinitely in liquid nitrogen. After shipment in dry ice and storage at −45°C, keeping-time is at least 1 month (Jespersen, 1977).

Fluid and freeze-dried cultures are the seed mate-

Table 1. Mesophilic and thermophilic starters in the dairy industry.[a]

Composition of culture		Products
1. Mesophilic bacteria		
1.1. *Streptococcus cremoris*	95–98%	Cheese type without eye forma-
Streptococcus lactis	2–5%	tion, hard pressed cheese:
		Cheddar, Gouda, Edam; soft
		ripened: Camembert
1.2. *Streptococcus cremoris*	95%	Cottage cheese, quarg, fer-
Leuconostoc cremoris	5%	mented milks, cheese types
		with few or small eyes
1.3. *Streptococcus cremoris*	85–90%	Like 1.2.
S. lactis subsp. *diacetylactis*	3%	
S. lactis	3%	
L. cremoris	5%	
1.4. *S. cremoris*	70–75%	Cultured butter, fermented milk:
S. lactis subsp. *diacetylactis*	15–20%	buttermilk; cheese types with
S. lactis	1–5%	round eyes
L. cremoris	2–5%	
1.5. *S. lactis* (ropy strain)		Taette milk
1.6. *S. cremoris*		Casein
1.7. *S. lactis*		Kefir
Lactobacillus brevis		
Leuconostoc spp.		
2. Thermophilic starters		
2.1. *Streptococcus thermophilus*	50%	Yogurt
Lactobacillus bulgaricus	50%	
2.2. *Streptococcus thermophilus*		Swiss-type cheeses: Emmental,
Lactobacillus helveticus		Grana, Soviet cheese
Lactobacillus lactis		
(or *L. bulgaricus*)		
Propionibacterium spp.		
(optional)		

[a] The quantitative composition has been taken from the culture catalog of one leading supplier.

rial for the bulk starters to be prepared in the factory, whereas deep-frozen, concentrated cultures are preferred for direct vat inoculation but are more expensive than the conventional cultures.

Traditional cheese production proceeds in open vats. The milk needed for cheese-making—e.g., raw milk for Emmental or Grana, pasteurized for most others—cannot be a sterile environment. It may contain a variable number (up to more than 10^6/ml) of a variety of Gram-positive, acid-producing bacteria and increasingly Gram-negative, psychrophilic microorganisms due to an increase of refrigeration during production, transport, and workup of milk. Usually, starter cultures overgrow the endogenous microflora of milk which may, however, considerably contribute to the final quality, taste, and aroma of cheese.

The main microbiological problem is the common infection with bacteriophages which leads to a slowdown or in severe cases to a breakdown of the lactic fermentation and the ripening process. The sources of infection are the used milk (some phages survive pasteurization and disinfection), infected whey from the environment of the cheese vat and starter cultures containing lysogenic strains (Lawrence, 1978). To overcome the buildup of a bacteriophage pool in a particular cheese plant, rotation of the used starter cultures has been employed. Since starter cultures differing in their phage spectrum usually also differ in their fermentation properties, the industry prefers multiple-strain starter cultures (Limsowtin, Heap, and Lawrence, 1977). A recent technological advance is the development of the already mentioned, deep-frozen, concentrated starter cultures for direct vat inoculation. In this case, no mother cultures and bulk starters have to be grown in the factory itself, thus preventing a possible contamination of the starters a priori with homemade phages. In addition, closed vats have been developed to minimize contamination and reinfection with phages.

The use of deep-frozen starters would also eliminate the possibility that important technological functions coded for by plasmid DNA (lactose metabolism, protease activity) are being lost during the preparation of mother cultures and bulk starters by microbiologically untrained personnel. Although a large body of knowledge exists on cheese starters,

Table 2. Geographical distribution of the production of cheese and butter in 1978 (in metric tons).[a]

	Cheese	Butter and ghee
Developed market economies		
North America	1,942,300	563,154
Western Europe	4,108,622	2,207,290
Oceania	196,490	345,322
Others	138,295	83,890
Developing market economies		
Africa	36,296	56,731
Latin America	627,072	205,137
Near East	615,531	292,803
Far East	15,641	848,171
Others	—	1,100
Centrally planned economies		
Asia	212,958	99,450
Eastern Europe and USSR	2,590,508	2,268,802

[a] Source: FAO Production Yearbook, vol. 32, 1979.

the microbial ecology of cheese production and ripening is still not fully understood (Lawrence, Thomas, and Terzaghi, 1976; Stadhouders, 1975).

The economical significance of lactic streptococci is fundamental for the food industry. The world production of cheese in 1978 was estimated by the FAO at 10,483,713 tons, that of butter and ghee at 6,971,850 tons (Food and Agriculture Organization of the United Nations, 1979). Table 2 demonstrates that production is highest in the technologically developed parts of the world. Together with the amounts of fermented milks for which no accurate estimates were available, we may assume for 1978 a total of 2×10^{10} kg of dairy products manufactured with the aid of lactic streptococci and lactobacilli. Since the number of lactic streptococci easily reaches 10^9 viable units per gram of cheese or yogurt (Stadhouders, 1975), the streptococcal biomass handled annually by the dairy industry can be calculated to be on the order of 10^5 tons if we include the biomass present in typical by-products like cheese whey and buttermilk and assume a weight of 10^{-12} g per viable bacterium. Based on the wholesale prices for butter ($240–$360 per 100 kg) and cheese ($235–$255 per 100 kg cheddar and gouda, respectively) as recorded by the FAO for 1978 (Food and Agriculture Organization of the United Nations, 1979, Table 113, items 104–108), a value of the world output of fermented milk products of about $50 billion ($5 \times 10^{10}$) can be implicated. This amount compares to a value of about $67 billion of the 1978 wine production, also estimated from the FAO statistics.

Leuconostoc

Bacteria of the genus Leuconostoc have been found in a great number of natural and man-made habitats. Numerous strains have been isolated from grass, herbages, and silage (Garvie, 1960; Whittenbury, 1966) and from wine and grape leaves (Peynaud, 1968; Weiler and Radler, 1970). Leuconostoc species, especially L. oenos, are responsible for the malolactic fermentation of wine (Kunkee, 1967) which is decisive for the quality and the taste of wine (Dittrich, 1977). Leuconostocs play an essential role in the fermentation of vegetables like sauerkraut and cucumbers. The sequence of predominant bacterial species in the fermentation of sauerkraut is initiated by Leuconostoc mesenteroides carrying out the first part of the lactic fermentation (Pederson, 1960).

The ability of two species, L. dextranicum and L. cremoris (citrovorum), to ferment the citric acid of milk and to produce the flavor compound diacetyl has led to their incorporation into starter cultures for buttermilk, butter, quarg, and cheese (Sharpe, 1979).

Starter cultures containing Leuconostoc or other lactic acid bacteria have also been developed for the fermentation of eggs. Removal of glucose before the drying of the eggs is the apparent purpose of this fermentation (Galluzzo, Cotterill, and Marshall, 1974).

Leuconostoc species have been isolated from a great variety of spoiled food. Their tolerance to high sugar concentrations (up to 60% in L. mesenteroides) permits the organisms to grow in syrup and ice cream mixes. Leuconostoc is well known as a spoilage bacterium in sugar refineries. It is well adapted to growth in particular stages of sugar processing, causing harmful economic effects in the industry. The growth of these microorganisms can result in significant losses of sucrose, corrosion due to acid production, and formation of dextran gums which generate physical troubles in the production process (Tilbury, 1975). The ability of L. mesenteroides to synthesize dextrans from sucrose by an extracellular dextransucrase has been exploited for the production of dextrans on an industrial scale (Lawson and Sutherland, 1978).

Pediococcus

Organisms described as pediococci ("beer sarcina") have been studied for a long time mainly in relation to problems in the brewing industry (Balcke, 1884; Shimwell and Kirkpatrick, 1939). These microorganisms are potentially dangerous beer spoilers in that they are microaerophilic and are relatively tolerant of alcohol and the antibiotic hop resins. Spoilage of beer by Pediococcus cerevisiae is prominent by the secretion of voluminous capsular material which may turn beer viscous and ropy. Pediococcal contamination as a cause of diacetyl formation in beer has also been reported (Rainbow, 1975). In modern brewing, the pediococcus problem has vanished due to advanced food technology, especially cleaning in place (CIP) techniques for the fermentation equipment, postfermentation pasteurization, and aseptic filling.

Slime-forming strains of *P. cerevisiae* have been isolated from outbreaks of ropiness in cider plants (Beech and Carr, 1977).

Pediococci are commonly found in a great variety of fermenting plant materials like silage (Günther and White, 1961), sauerkraut (Pederson, 1960), cucumbers (Pederson and Albury, 1950), and olives (Vaughn, 1975). Weiler and Radler (1970) reported the isolation of 29 strains of *Pediococcus* from 102 wines of different origin and from grape leaves. These microorganisms have also been detected in proteinaceous foods such as fresh and cured meat, raw sausages (Reuter, 1970a,b), fresh and marinated fish (Blood, 1975), and poultry as well as in cheese (Dacre, 1958).

The isolation of pediococci from the rumen of cows (Baumann and Foster, 1956) and from feces of turkey (Harrison and Hansen, 1950) has been reported.

In the production of various kinds of raw sausages and bacon, starter cultures of *P. cerevisiae* are used in some European countries and the USA (Schiefer, 1978a,b; Deibel, Wilson, and Niven, 1961). As a result of the growth of the starter culture, the pH value of the raw material decreases rapidly. Due to this low pH value, contamination by potentially pathogenic bacteria like *Staphylococcus aureus* and members of the family Enterobacteriaceae is reduced. In addition, the use of starter cultures in the production of meat products shortens the production time and improves the quality of the product and the bacteriological safety of the production process (Schiefer, 1978a).

Enrichment and Isolation

Lactic acid bacteria are nutritionally fastidious. They all require complex media for optimal growth. In synthetic media, all strains of lactic acid streptococci require amino acids like isoleucine, valine, leucine, histidine, methionine, arginine, and proline and vitamins (niacin, Ca-pantothenate, and biotin) (Anderson and Elliker, 1953).

Isolation from Plant Material

Plant material like grass and herbages is the natural source of lactic acid bacteria. Ensilage allows the enrichment of lactic streptococci, leuconostocs, and pediococci. The isolation procedure takes advantage of the sequential growth of the above mentioned genera.

Ensilage Enrichment of Lactic Streptococci, Leuconostocs, and Pediococci (Whittenbury, 1965a)

Grasses and other plant material are collected and cut in pieces as aseptically as possible. The prepared material is then placed into sterile glass tubes and compressed. Fifty grams of material are enough to fill a 3-×-20-cm tube. The tubes are sealed in a way that permits gas under pressure to escape but prevents the entry of oxygen. A number of silages are prepared and incubated at 30°C. Tubes are opened beginning on the second day. The silage is removed and placed in flasks of sterile water, which are vigorously shaken. These suspensions can then either be streaked directly onto agar plates or diluted and pour-plated. The 2- to 3-day-old silages are the best sources for streptococci and leuconostocs; 4- to 7-day-old silages are best for pediococci.

A modification of the above method has been reported by Weiler and Radler (1970). Grape leaves are homogenized with the same amount of acetate buffer (pH 5.4; 0.2 M). The homogenate is placed into sterile tubes. The tubes are sealed and incubated at 30°C.

Isolation from Dairy Products

The main problem in isolating lactic acid bacteria from dairy products is a proper dissolution or dispersion of the solid or semisolid fat-containing material. Suitable methods are described by Olson, Anderson, and Sellars (1978).

Isolation of Lactic Acid Bacteria from Cheese Other Than Cottage Cheese (Olson, Anderson, and Sellars, 1978)

"Using aseptic technique, thoroughly comminute or mix each sample until representative portions can be removed. Heat 99-ml dilution blanks of sterile, freshly prepared (less than 7 days old), aqueous 2% sodium citrate to 40°C. Aseptically transfer 11 g of cheese to a sterile blender container previously warmed at 40°C and add the warmed sodium citrate blank. Mix for 2 min at a speed sufficient to emulsify the sample properly, invert the container to rinse particles from the interior walls, and remix for approximately 10 seconds. Inadvertent heating to temperatures in excess of 40°C from friction in agitation may occur with some mechanical blenders. This should be determined before use of the blender so corrective action can be taken, if needed. If heating is unavoidable with equipment available, mixing periods of less than 2 min may be used provided that complete emulsification is obtained. The 1:10 dilution should be plated or further diluted immediately, great care being taken to avoid air bubbles or foam.

"As an alternative method, 1 g ± 10 mg is rapidly weighed into a pre-sterilized 177-ml (6 oz) Whirl-Pak bag or its equivalent. Close the bag, transfer it to a flat surface and macerate the contents into a fine paste by rolling a 15 × 125-mm test tube or similar cylindrical object over the bag. The sample should not be forced into the corners or the ties seal area of the bag. Open the bag and add 9 ml of 2% sodium citrate at 40°C. Reclose the bag and roll the contents, as described above, to form a fine emulsion and proceed with plating immediately. Enumeration of bacterial species, such as lactic streptococci, that form chains may not be feasible with this method. However, this problem was not evident in a collaborative study involving analysis of Cheddar and Romano cheeses in nine laboratories in the United States."

Isolation of Lactic Acid Bacteria from Cottage Cheese (Olson, Anderson, and Sellars, 1978)

"Place the sterile blender container on a proper balance and tare. With a sterile spatula, mix contents of the cottage cheese container; or if in a tightly closed plastic sample pouch, gently knead and mix the enclosed curd. Aseptically remove the cover of the blender and place it upside down on the balance beside the container. Weigh 11 g of cottage cheese into the sterile container. Add 99 ml of warmed (40°C) sterile 2% sodium citrate solution as described in (1) [Isolation from Cheese Other Than Cottage Cheese] to disperse and dilute the cottage cheese curd. Proceed as in (1) and plate appropriate dilutions immediately. The alternative bag method described in (1) may also be used."

Isolation of Lactic Acid Bacteria from Cultured Milk, Cultured Cream, Yogurt, Acidophilus Milk, Bulgarian Buttermilk, and Similar Cultured or Acidified Semifluid Products (Olson, Anderson, and Sellars, 1978)

"After thoroughly mixing the sample, weigh 11 g of product into a sterile wide-mouth container, add 99 ml of sterile buffered distilled water (40°C), shake until a homogeneous dispersion is obtained, and withdraw appropriate amounts of this 1:10 dilution for plating or further dilution. The sample may be dispersed in 2% sodium citrate with a mechanical blender as described in (2) [Isolation from Cottage Cheese]."

Lactic Streptococci

Unfortunately, no satisfactory selective medium is available for the isolation of lactic streptococci. Two media, both commercially available, are generally accepted to give reliable growth of these organisms. The medium proposed by Elliker, Anderson, and Hannesson (1956) is widely used for the isolation and enumeration of lactic streptococci. M 17 medium (Terzaghi and Sandine, 1975), a complex medium supplemented by 1.9% β-disodium glycerophosphate, resulted in improved growth of lactic streptococci.

Elliker Agar Medium for Isolation of Lactic Streptococci (Elliker, Anderson, and Hannesson, 1956)

Medium contains, per liter:

Tryptone	20.0 g
Yeast extract	5.0 g
Gelatin	2.5 g
Dextrose	5.0 g
Lactose	5.0 g
Sucrose	5.0 g
Sodium chloride	4.0 g
Sodium acetate	1.5 g
Ascorbic acid	0.5 g
Agar	15.0 g

The medium has a pH of 6.8 before autoclaving.

This medium is probably the most cited for the isolation and growth of lactic streptococci, although it is unbuffered. This disadvantage can be overcome by the addition of suitable buffer substances. Recently, it was demonstrated that addition of 0.4% (wt/vol) of diammonium phosphate improves the enumeration of lactic streptococci on Elliker agar. Colony counts were up to about eight times greater due to improved buffering capacity (Barach, 1979).

M 17 Medium for Isolation of Lactic Streptococci (Terzaghi and Sandine, 1975)

Phytone peptone	5.0 g
Polypeptone	5.0 g
Yeast extract	5.0 g
Beef extract	2.5 g
Lactose	5.0 g
Ascorbic acid	0.5 g
β-Disodium glycerophosphate	19 g
1.0 M $MgSO_4 \cdot 7H_2O$	1.0 ml
Glass-distilled water	1 liter

The medium is sterilized at 121°C for 15 min. The pH of the broth is 7.1. Solid medium contains 10 g agar per liter of medium.

This medium is useful for the isolation of all strains of *S. cremoris, S. lactis, S. lactis* subsp. *diacetylactis,* and *S. thermophilus* and mutants of those strains lacking the ability to ferment lactose.

Lactose-fermenting streptococci form colonies 3–4 mm in diameter, whereas the mutant strains form minute colonies (less than 1 mm in diameter).

In contrast to Terzaghi and Sandine (1975), Shankar and Davies (1977) demonstrated the suppression of *Lactobacillus bulgaricus* in M 17 medium. The majority of these *Lactobacillus* strains failed to grow in this medium adjusted to pH 6.8. Since M 17 medium supported good growth of *S. thermophilus,* it can be used for the selective isolation of this microorganism from yogurt.

Lactic streptococcal bacteriophages can be efficiently demonstrated and distinguished on M 17 agar. Plaques larger than 6 mm in diameter could be observed as well as turbid plaques, indicating lysogeny (Terzaghi and Sandine, 1975).

Leuconostoc and *Pediococcus*

Just as for the lactic streptococci, no selective media are available for the isolation of leuconostocs or pediococci. Numerous different media have been proposed for this purpose. *Leuconostoc* and *Pediococcus* species were isolated from ensilaged plant material on glucose–yeast extract agar and on acetate agar.

Glucose–Yeast Extract Agar for Isolation of *Leuconostoc* and *Pediococcus* (Whittenbury, 1965a)

Medium contains, per liter:

Glucose	5.0 g
Yeast extract	5.0 g
Peptone	5.0 g
Meat extract	5.0 g
Agar	15 g
pH 6.5	

Acetate Agar for Isolation of *Leuconostoc* and *Pediococcus* (Whittenbury, 1965b)

This agar is a modification of the medium proposed by Keddie (1951) as being selective for lactobacilli.

Meat extract	50 g
Peptone	5.0 g
Yeast extract	5.0 g
Glucose	10 g
Tween 80	0.5 ml
Tap water	900 ml

The pH is adjusted to 5.4. Medium is autoclaved at 121°C for 15 min. Before plating, 100 ml of sterile 2 M acetic acid–sodium acetate buffer (pH 5.4) is added.

Yeast Extract–Glucose–Citrate Broth for Isolation of *Leuconostoc* (Garvie, 1967)

YGC has been suggested for the isolation and cultivation of leuconostocs from different sources. The medium is prepared by adding the following compounds to 1 liter distilled water:

Peptone	10.0 g
Lemco	10.0 g
Yeastrel	5.0 g
Glucose	10.0 g
Triammonium citrate	5.0 g
Sodium acetate	2.0 g
$MgSO_4 \cdot 7H_2O$	0.2 g
$MnSO_4 \cdot 4H_2O$	0.05 g
Tween 80	0.05 ml

The pH is adjusted to 6.7 and the medium is autoclaved at 121°C for 15 min.

Due to their ability to start growing at relatively low pH values, several acidic media have been employed for the isolation of leuconostocs (*L. oenos*) and of pediococci from wine (Garvie, 1967; Weiler and Radler, 1970).

Acidic Tomato Broth (ATB) for Isolation of *Leuconostoc* and *Pediococcus* (Garvie, 1967)

Medium contains, per liter:

Peptone	10.0 g
Yeastrel	5.0 g
Glucose	10.0 g
$MgSO_4 \cdot 7H_2O$	0.2 g
$MnSO_4 \cdot 4H_2O$	0.05 g
Tomato juice	25% (vol/vol)

The pH is 4.8 and the medium is autoclaved at 121°C for 15 min. Before use, a solution of cysteine hydrochloride sterilized by filtration is added to a final concentration of 0.05% (wt/vol).

Isolation of Lactic Acid Bacteria from Wine (Weiler and Radler, 1970)

Medium contains per liter:

Peptone	5.0 g
Yeast extract	5.0 g
Glucose	10.0 g
Diammonium hydrogencitrate	2.0 g
Sodium acetate · 3H$_2$O	5.0 g
Tween 80	1.0 g
KH_2PO_4	5.0 g
$MgSO_4 \cdot 7H_2O$	0.5 g
$MnSO_4 \cdot 4H_2O$	0.2 g
$FeSO_4 \cdot 7H_2O$	0.05 g
Agar	15.0 g

The pH is 5.3–5.4. Medium is sterilized 15 min at 121°C. To inhibit growth of yeasts, the medium is supplemented with sorbic acid at a final concentration of 0.05%.

Medium for Isolation of Pediococci from Beer (F. Eschenbecher, Weihenstephan, personal communication)

Medium contains 100 ml unhopped beer wort, fermented by *Saccharomyces cerevisiae*, and 200 ml of MRS broth (de Man, Rogosa, and Sharpe, 1960).

The thallium-acetate agar (Barnes, 1956) has been used by Reuter (1970a) for the isolation of pediococci and leuconostocs from meat and meat products. This medium is especially suitable for the isolation of both groups of bacteria from a mixed flora which is dominated by lactobacilli (Reuter, 1970a).

Enumeration of Citrate-Fermenting Bacteria in Lactic Starter Cultures and Dairy Products

To control gas and aroma (diacetyl) production in the fermentation of various dairy products, it is important to know the quantitative composition of the used starter cultures. *Leuconostoc* species and *Streptococcus lactis* subsp. *diacetylactis* are components of many mesophilic starter cultures (Table 1). These organisms are able to ferment citrate with concomitant production of CO_2 and diacetyl.

For the collective enumeration of leuconostocs and *S. lactis* subsp. *diacetylactis* in starters and fermented dairy products, a whey agar containing calcium lactate and Casamino Acids (WACCA) has been introduced by Galesloot, Hassing, and Stadhouders (1961).

WACCA 0.5% Medium for Enumeration of *Leuconostoc* and *Streptococcus lactis* subsp. *diacetylactis* (Galesloot, Hassing, and Stadhouders, 1961)

Composition of the medium:
Dissolve in 1 liter of whey 5 g Ca-lactate · 5H₂O, 7 g Casitone, and 0.5% yeast extract. Adjust to pH 7.3 with Ca(OH)₂-suspension. Steam for 30 min. Filtrate. Adjust to pH 7.1 with a NaOH solution. Add 1 ml MnSO₄ solution (40 mg MnSO₄ · 4H₂O/100 ml). Dissolve 15 g agar. Clarify with 5 g albumin. Sterilize for 15 min 110°C (15 ml/tube).

Preparation of whey:
Add to 1 liter of high-temperature-short-time (HTST)-pasteurized fresh skim milk at 30°C, 0.3 ml 35% CaCl₂ solution and 0.3 ml commercial rennet (strength 1:10,800). Cut coagulum after 30 min at 30°C. Filtrate after 2 h at 45°C.

Preparation of Ca-citrate suspension:
Suspend 28 g Ca-citrate (Merck) in a 100-ml 1.5% carboxymethylcellulose solution, prepared at 45°C. Allow to precipitate for 2 h at 45°C. The supernatant is steamed for 30 min.

Application:
Add 0.3–0.7 ml Ca-citrate suspension per 15 ml WACCA 0.5% (48°C). The amount to be added has to be adapted to the type of starter under investigation. Incubation 5 days at 25°C. Count after 2, 3, and 5 days.

A different medium developed by Nickels and Leesment (1964) for the same purpose gives comparable results.

For the selective isolation and enumeration of *Leuconostoc* strains from mixed strain starter cultures, the following medium has been proposed.

HP Medium for Enumeration of *Leuconostoc* (Pearce and Halligan, 1978)

Phytone	20.0 g
Yeast extract	6.0 g
Beef extract	10.0 g
Tween 80	0.5 g
Ammonium citrate	5.0 g
FeSO₄ · 7H₂O	0.04 g
MgSO₄ · 7H₂O	0.2 g
MnSO₄ · 4H₂O	0.05 g
Glucose (sterilized separately)	10.0 g

The addition of tetracycline (0.13 µg/ml) to the medium selectively inhibits the growth of streptococci and therefore allows the direct enumeration of *Leuconostoc* species in mixed-strain cultures.

Production and Preservation of Starter Cultures

As already mentioned, liquid, dried, and frozen starter cultures are in use (Jespersen, 1977; Pallasdies, 1976; Wigley, 1977). Starter cultures must have a high survival rate of microorganisms coupled with optimum activity for the desired technological performance, e.g., the fermentation of lactose to lactate, controlled proteolysis of casein, and production of aroma compounds like diacetyl. Since the genes for lactose and citrate fermentation as well as for certain proteases are located on plasmids (Kempler and McKay, 1979), continuous culture has not been successful since fermentation defective variants easily develop (Lawrence, Thomas, and Terzaghi, 1976). In most instances, pasteurized or sterilized skim milk is the basic nutrient medium for the large-scale production of starter cultures because it assures that only lactic streptococci fully adapted to the complex medium milk develop. For liquid starter cultures, the basic milk medium may be supplemented with yeast extract, glucose, lactose, and chalk. To obtain optimum activity and survival, it may be necessary to neutralize the formed lactic acid

Table 3. Typical procedure for the preservation and utilization of freeze-dried yogurt starters (Tamime and Robinson, 1976).

Stage 1: Starter propagated in autoclaved reconstituted skim milk powder (16% solids); inoculum rate 2% and incubation for 4 h at 42°C
Stage 2: Starter held at 5°C for 18 h, and 2–3 ml dispensed in sterilized glass vials; vials frozen at −40°C for 2–3 h
Stage 3: Cultures freeze-dried for up to 36 h
Stage 4: Vials closed under vacuum and sealed
Stage 5: Freeze-dried cultures stored at 5°C until required
Stage 6: Activation procedure
a. Content of vial emptied into 250 ml autoclaved reconstituted skim milk powder (9% solids), and incubated at 42°C for 12–15 h
b. For full activity, subcultured once or twice in the same propagating medium, using 2% inoculum and incubation at 42°C for 4 h; ratio of *Streptococcus thermophilus* to *Lactobacillus bulgaricus* maintained at 1 to 1

by addition of sodium or ammonium hydroxide. Since many strains of lactic streptococci produce hydrogen peroxide during growth under microaerophilic conditions, it has been beneficial to add catalase to the growth medium, thus leading to cell densities of more than 10^{10} viable units per milliliter of culture (Stanley, 1977).

Of course, many important details of the art of producing starter cultures are not being published for the protection of natural commercial interests. One published example for the manufacture of freeze-dried yogurt starters is described in Table 3. In contrast to this procedure, which uses reconstituted skim milk, the media for the production of concentrated starters are clarified by proteolytic digest of skim milk with papain or bacterial enzymes to avoid precipitation of casein in the separators used to collect the streptococcal biomass (Stanley, 1977).

Batch Fermentation for Concentrated Starters (Stanley, 1977)

"The (pilot) plant consists of a 40 gal and 100 gal batch pasteurization tank for use in medium preparation. Medium is pumped via a single, centrally placed peristaltic pump through a pasteurizer and into the 40 gal fermenter. This fermenter is fitted with pH control equipment, a stirrer for efficient mixing and hot and cold water jackets for temperature control. The pasteurized medium is aseptically inoculated with appropriate starter and incubated for 16–18 h at 22°C. A positive air pressure is maintained in the fermenter throughout this time. When grown, the culture is rapidly cooled in the fermenter, and harvested in a self-cleaning clarifier. The concentrated cells are automatically reconstituted by the desludging action of the separator and are pumped into the

CCV (cell collection vessel) where gentle agitation ensures homogeneity of the product and a cooling device maintains a refrigeration temperature. From the CCV the concentrate passes to the microflow cabinet where it is packed semi-automatically in syringes via a syringe filler unit. These syringes are then frozen under vapour phase liquid nitrogen and stored until ready for use.''

Further valuable details have been disclosed in a recent symposium "The manufacture and use of starters for the dairy industry" (Cox, 1977; Gordon and Shapton, 1977; Jespersen, 1977; Lewis, 1977; Osborne, 1977; Wigley, 1977).

Identification

Lactic streptococci, pediococci, and leuconostocs are Gram-positive, microaerophilic cocci which lack the cytochromes of the respiratory chain (Deibel and Seeley, 1974; Garvie, 1974; Kitahara, 1974). They can be simply differentiated from each other by the main fermentation products from glucose (see Table 4).

Lactic Streptococci

The common morphology consists of spherical or ovoid cells, 0.5–1 μm in diameter, in pairs or more-or-less long chains (Figs. 1–3). The lactic streptococci are in practice differentiated by their growth behavior at different temperatures (Table 5) into the mesophilic species *Streptococcus lactis* (Löhnis), *Streptococcus lactis* subsp. *diacetylactis* (Deibel and Seeley), and *Streptococcus cremoris* (Orla-Jensen) and the thermophilic species *Streptococcus thermophilus* (Orla-Jensen). *S. thermophilus* is differentiated at a first glance from the fecal streptococci by its inability to grow in the presence of 6.5% NaCl. Problems may arise in the identification of *S. lactis* subsp. *diacetylactis* if the plasmid-coded fermentation of citrate is lost. The data for the G+C content of DNA have only been reported in one Ph.D. thesis (Knittel, 1965).

Since the mesophilic species differ only in a few properties (Table 5), the G+C data are meaningless

Table 4. Differentiation scheme for streptococci, pediococci, and leuconostocs.

Fermentation products of glucose	
1. L-(+)-Lactic acid	*Streptococcus*
2. D-(−)-Lactic acid, CO$_2$, acetic acid, ethanol	*Leuconostoc*
3. DL-Lactic acid	*Pediococcus*

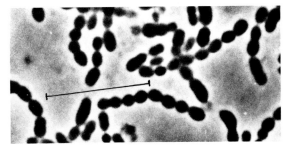

Fig. 3. Phase-contrast micrograph of *Streptococcus lactis* (strain forming chains). Bar = 10 μm.

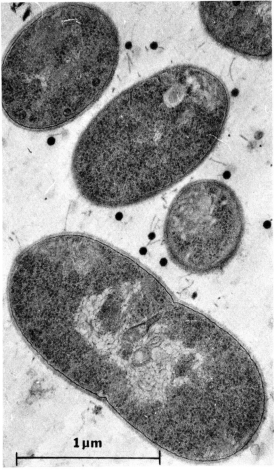

Fig. 1. Electron micrograph of thin sections of *Streptococcus lactis* at an early stage after infection with a specific bacteriophage. The large cell shows the typical features of a Gram-positive bacterium. It clearly demonstrates the almost rodlike appearance of a cell just starting to divide (compare Fig. 2 and 3). [Courtesy of Jürgen Lembke, Kiel.]

for this differentiation. DNA-DNA hybridization has been performed with one strain of *S. lactis* and *S. thermophilus* showing 5–14% homology (Ottogalli, Galli, and Dellaglio, 1979). However, that the three mesophilic species are closely related is also implicated by the observation that many bacteriophages cross the "species" line and attack strains of all three species (Lembke, et al., 1980). A phage typing system is presently being evaluated. Also, the plasmid pattern so far investigated does not allow a species differentiation (Pechmann and Teuber, 1980). An interesting, recent approach is differentiation on the basis of protein patterns after gel electrophoresis of soluble cell extracts (Jarvis and Wolff, 1979). By this method, classification of closely related strains seems possible. Collins and Jones (1979) have shown that the mesophilic lactic streptococci contain menaquinones with nine isoprene units as major component, in contrast to demethylmenachinones with nine isoprene units and menachinones with eight isoprene units in group D streptococci. At the moment, it is not possible to assess the number of different mesophilic lactic streptococci strains existing in dairies and starter cultures all over the world.

A differentiation which is very helpful for the everyday use in the dairy industry has been achieved by the development of the following agar medium.

Differential Enumeration of Mesophilic Lactic Streptococci (Reddy et al., 1972)

"An agar medium containing arginine and calcium citrate as specific substrates, diffusible (K_2HPO_4) and undiffusible ($CaCO_3$) buffer systems, and bromocresol purple as the pH indicator was developed to differentiate among lactic streptococci in pure and mixed cultures. Milk was added as the sole source of carbohydrate (lactose) and to provide growth-stimulating factors. Production of acid from lactose caused developing bacterial colonies to seem yellow. Subsequent arginine utilization by *Streptococcus lactis* and *S. diacetilactis* liberated ammonia, resulting in a localized pH shift back toward neutrality and a return of the original purple indicator

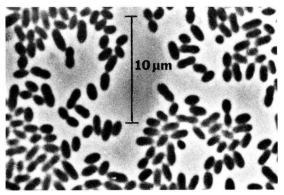

Fig. 2. Phase-contrast micrograph of *Streptococcus lactis* (strain forming pairs of ovoid cells).

Table 5. Physiological and other properties of lactic *Streptococcus* species used for their identification and differentiation.

	S. lactis	S. lactis subsp. *diacetylactis*	S. cremoris	S. thermophilus
Growth at 10°C	+	+	+	−
Growth at 40°C	+	+	−	+
Growth at 45°C	−	−	−	+
Growth in 4% NaCl	+	+	−	− (2% NaCl)
Growth in 6.5% NaCl	−	−	−	−
Growth at pH 9.2	+	+	−	ND
Growth with methylene blue (0.1% in milk)	+	+	−	−
Growth in the presence of bile (40%)	+	+	+	−
NH_3 from arginine	+	+	−	−
CO_2 from citrate	−	+	−	−
Diacetyl and acetoin	−	+	−	−
Fermentation of maltose	+	+	Rarely	−
Hydrolysis of starch	−	−	−	+
Heat resistance (30 min 60°C)	Variable	Variable	Variable	+
Serological group[a]	N	N	N	Ungrouped
Peptidoglycan type[b]	L-Lys-D-Asp	ND	L-Lys-D-Asp	L-Lys-L-Ala$_{2-3}$
G+C content of DNA[c]	33.8–36.9	33.6–34.8	35.0–36.2	40

[a] Lancefield, 1933.

[b] Schleifer and Kandler, 1972.

[c] Knittel, 1965.

hue. The effects of production of acid from lactose and ammonia were fixed around individual colonies by the buffering capacity of $CaCO_3$. After 36 hr at 32°C in a candle oats jar, colonies of *S. cremoris* were yellow, whereas colonies of *S. lactis* and *S. diacetilactis* were white. *S. diacetilactis,* on further incubation, utlized suspended calcium citrate, and, after 6 days, the citrate-degrading colonies exhibited clear zoning against a turbid background, making them easily distinguishable from the colonies of the other two species. The medium proved suitable for quantitative differential enumeration when compared with another widely used general agar medium for lactic streptococci.

Agar composition

"The differential medium contained 0.5% tryptone, 0.5% yeast extract, 0.25% Casamino Acids, 0.5% L-arginine-hydrochloride, 0.125% K_2HPO_4, 1% calcium citrate, 1.5% carboxy methyl cellulose, and 1.5% agar. Just before the pouring of plates, 5.0 ml of sterile, reconstituted, nonfat milk (11% solids), 10 ml of sterile 3% (w/v) $CaCO_3$ in distilled water, and 2.0 ml of sterile 0.1% bromocresol purple in distilled water were added to every 100 ml of sterile agar (melted and tempered to 55°C) and mixed to obtain homogeneity. After these additions, the medium pH was 5.9 ± 0.1."

Agar preparation

"The amount of agar required for 1 liter of the medium was suspended in 500 ml of distilled water and steamed until dissolution. In another glass beaker containing 500 ml of distilled water, 10 g of calcium citrate and 15 g of CMC were suspended and heated while being stirred until a homogeneous, white, turbid suspension was formed. The two portions were mixed together in a separate stainless-steel vessel containing the required quantities of tryptone, yeast extract, Casamino Acids, K_2HPO_4, and arginine. The mixture was covered and steamed for 15 min. The pH of the medium after steaming was adjusted to 5.6 with 6 N HCl. The agar was then dispensed into bottles in 100-ml quantities and sterilized at 121°C for 15 min."

Minor component strains (<5%), however, are difficult to detect. An advanced modification of this medium has been reported (Mullan and Walker, 1979).

Leuconostoc

In the eighth edition of *Bergey's Manual,* the genus *Leuconostoc* is divided into six generally accepted species (Garvie, 1974b). Hybridization studies performed by Garvie (1976), which are summarized in Table 6, show four genetic groups. The first three groups are formed by *L. lactis, L. paramesenteroides,* and *L. oenos,* respectively. The latter is acidophilic; it grows at pH values to acidic for other leuconostocs to start growing. *L. oenos* cannot form dextran from sucrose (Garvie, 1967). The hybridization data suggest a very close relationship between *L. mesen-*

Table 6. Homology of the DNAs of some leuconostoc species.[a]

Bacteria and NCDO No.	Hybridization (%) of labeled DNA from:		
	523	546	803
L. oenos 1674	15	11	—
L. mesenteroides 523	100	36	9
L. dextranicum 529	110	22	—
L. cremoris 543	66	29	—
L. lactis 546	49	100	46
L. paramesenteroides 803	6	13	100

[a] Data are from Garvie (1976). NCDO, National Collection of Dairy Organisms.

teroides, L. dextranicum, and L. cremoris. These species therefore belong to one genetic group. L. dextranicum and L. cremoris may also only be subspecies of L. mesenteroides (Garvie, 1979).

The physiological and some other properties of the members of this group are given in Table 7. More detailed information on physiological and morphological properties of the leuconostocs is given by Garvie (1960, 1967, 1974b,) and Whittenbury (1966). Phase-contrast micrographs of L. mesenteroides and L. cremoris are presented in Figs. 4 and 5, respectively.

Pediococcus

For the classification of the pediococci, several differing systems have been described. These systems are based on morphological and biochemical (Garvie, 1974a; Nakagawa and Kitahara, 1959) as well as serological criteria (Coster and White, 1964). The reported results are contradictory. Garvie claimed that the name Pediococcus cerevisiae had been applied to two different species of pediococci. It was requested to designate P. damnosus Clausen as the type species of Pediococcus (Garvie, 1974a). During the preparation of this manuscript, this sugges-

tion was accepted and published by the International Committee on Systematic Bacteriology:

"Suggestions by E. J. Garvie regarding the status of the genus Pediococcus were accepted and approved. The name P. cerevisiae was considered invalid, and the old name P. damnosus was reinstated. P. acidilactici, P. parvulus, and P. halophilus were retained. The species P. urinae-equi was rejected, but P. pentosaceus was accepted; however, the proposed neotype strain for the latter was rejected" (Subcommittee on Lactobacilli and Related Microorganisms, 1979).

By this decision, the description of the species of the genus Pediococcus as printed in the eighth edition of Bergey's Manual (Kitahara, 1974) is no longer valid in all points. The best information available at the moment on the genus Pediococcus is given in Tables 8 and 9. The peptidoglycan type is Lys-D-Asp, an important criterion in separating Pediococcus from Leuconostoc and Aerococcus (Schleifer and Kandler, 1972). The use of the commercially available API system for the identification of lactobacilli has recently been shown to be appli-

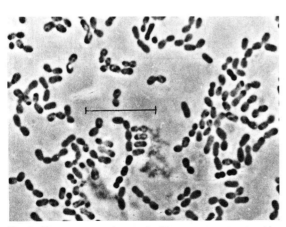

Fig. 4. Phase-contrast micrograph of Leuconostoc mesenteroides (DSM 20343). Bar = 10 μm.

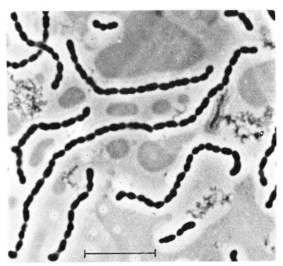

Fig. 5. Phase-contrast micrograph of Leuconostoc cremoris (DSM 20346). Bar = 10 μm.

Table 7. Physiological and other properties of *Leuconostoc mesenteriodes* NCDO 523, *L. dextranicum* NCDO 529, and *L. cremoris* NCDO 543.[a]

Property	*L. mesenteroides* NCDO 523	*L. dextranicum* NCDO 529	*L. cremoris* NCDO 543
Morphology	Gram-positive coccus or coccobacillus, usually in pairs	Gram-positive coccus in pairs and chains	Gram-positive coccus or coccobacillus in long chains of pairs
Reaction in litmus milk	Slightly acid	—	Slightly acid
Reaction in yeast, glucose, litmus, milk	Acid, clot, gas, and reduction	Acid, clot, and reduction	Acid, clot, and slight reduction
Growth at:			
10°C	+	+	+
37°C	+	+	−
39°C	+	+	−
45°C	−	−	−
Growth in:			
3% NaCl	+	−	−
6.5% NaCl	+	−	−
Dextran from sucrose	+	+	−
Citrate used	−	+	+
Acetoin formed in citrate medium	−	+	+
Survival at 55°C for:			
15 min	+	−	−
30 min	−	−	−
Growth at pH 4.8	−	−	−
Acid formed from:			
Arabinose	+	−	−
Xylose	+	−	−
Fructose	+	+	−
Glucose	+	+	+
Galactose	+	−	+
Mannose	+	+	−
Cellobiose	+	−	−
Lactose	+	−	+
Maltose	+	−	−
Sucrose	+	+	−
Trehalose	+	+	−
Melibiose	+	−	−
Raffinose	+	−	−
Esculin	+	−	−
Salicin	+	−	−
Mannitol	+	−	−
Requirement for:			
Uracil	−	−	+
Riboflavin	−	+	+
Pyridoxal	+	+	+
Nicotinic acid	+	+	+
Thiamine	+	+	+
B_{12}	−		−
Folic acid	−	+	+
Pantothenate	+	+	+
Guanine + adenine + uracil + xanthine	−		+
Cell wall peptide	Lys-ser-(ala)γ	Lys-ser-(ala)γ	
LDH group	D	D	D
G-6-PDH group	d	d	d
G+C content (mol%) of DNA	38.5	39.4	38.6
Buoyant density	38.7		38.5

[a] Garvie, 1979. NCDO, National Collection of Dairy Organisms.

Table 8. Characters of established species of the genus *Pediococcus* showing the principal differentiating factors.[a]

Property	P. damnosus	P. acidi-lactici	P. pentosaceus	P. parvulus
Morphology	Cocci occuring singly, in pairs, and in tetrads			
Gram stain	+	+	+	+
Optimum pH for growth initiation	5.0–5.5	6.0	7.0	Not known
Maximum pH for growth initiation	6.5	8.0	8.2	Not known
Growth at pH 9.0	−	−	−	−
Final pH in glucose broth	3.5–3.8	3.5–3.8	3.5–3.8	3.9–5.5
Growth at pH 8.0	−	+	+	Not known
Optimum temperature	25–30	40	35	30
Maximum temperature	35	55	42–45	40–42
Growth at 45°C	−	+	+	−
Growth at 50°C	−	+	−	−
Growth at 8.0% NaCl	−	+	+	−
Tolerance of hop antiseptic	+	−	−	Not known
CO_2 requirement	+	−	−	−
O_2 requirement	Facultative	Facultative	Facultative	Facultative
Type of lactic acid produced	DL	DL	DL	DL
Hydrolysis of:				
Arginine	−	+	+	−
Esculin	+	+	+	+
Acid produced from:				
Arabinose	−	±	+	−
Xylose	−	±	±	−
Fructose	+	+	+	±
Galactose	±	+	+	
Glucose	+	+	+	+
Mannose	+	+	+	
Rhamnose	−	±	±	
Cellobiose	+	±	+	
Lactose	±	± (slow)	± (slow)	±
Maltose	+ (slow)	±	+	±
Melibiose	−	±	±	
Melezitose	−	±	−	
Sucrose	±	±	±	−
Trehalose	+	±	+	±
Raffinose	−	−	−	−
Glycerol	−	slow	slow	−
Mannitol	−	−	−	
Sorbitol	−	−	−	−
Salicin	±	+	+	±

[a] Garvie, 1974a.

Table 9. DNA/DNA homology of some members of the genus *Pediococcus*.[a]

	20281	20283	20284	20332	20289	20235	20339
P. pentosaceus DSM 20281	100	97	16	7	8	6	4
P. pentosaceus subsp. intermedius DSM 20283	98	100	19	6	7	5	3
P. acidilactici DSM 20284	19	20	100	7	6	6	3
P. parvulus DSM 20332	7	7	7	100	34	8	4
P. damnosus (cerevisiae) DSM 20289	6	8	7	31	100	7	4
P. dextrinicus DSM 20335	4	4	5	5	4	100	4
P. halophilus DSM 20339	2	2	2	6	2	6	100

[a] Data from Back and Stackebrandt, 1978. DSM, Deutsche Sammlung für Mikroorganismen.

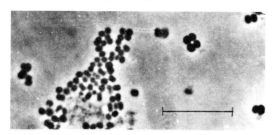

Fig. 6. Phase-contrast micrograph of *Pediococcus damnosus* (*cerevisiae*) (DSM 20331). Bar = 10 μm.

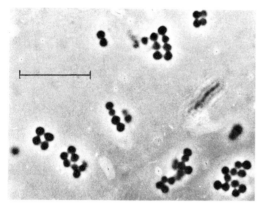

Fig. 7. Phase-contrast micrograph of *Pediococcus dextrinicus* (DSM 20335). Bar = 10 μm.

cable for the characterization of pediococci (Dolezil and Kirsop, 1977). The typical tetrad morphology of *P. damnosus* and *P. dextrinicus* is shown in Figs. 6 and 7.

Literature Cited

Avery, O. T., MacLeod, C. M., McCarthy, M. 1944. Studies on the chemical nature of the substance inducing transformation of pneumococcal types. I. Induction of transformation by a desoxyribonucleic acid fraction isolated from pneumococcus type III. Journal of Experimental Medicine **79:**137–158.

Anderson, A. W., Elliker, P. R. 1953. The nutritional requirements of lactic streptococci isolated from starter cultures. I. Growth in a synthetic medium. Journal of Dairy Science **36:**161–167.

Back, W., Stackebrandt, E. 1978. DNS/DNS-Homologiestudien innerhalb der Gattung *Pediococcus*. Archives of Microbiology **118:**79–85.

Balcke, J. 1884. Über häufig vorkommende Fehler in der Bierbereitung. Wochenschrift Brauerei **1:**181–195.

Barach, J. T. 1979. Improved enumeration of lactic acid streptococci on Elliker agar containing phosphate. Applied and Environmental Microbiology **38:**173–174.

Barnes, E. M. 1956. Methods for the isolation of fecal streptococci (Lancefield Group D) from bacon factories. Journal of Applied Bacteriology **19:**193–203.

Baumann, H. E., Foster, E. M. 1956. Characteristics of organisms isolated from the rumen of cows fed high and low roughage rations. Journal of Bacteriology **71:**333–338.

Beech, F. W., Carr, J. G. 1977. Cider and Perry, pp. 139–313.

In: Rose, A. H. (ed.), Economic microbiology, vol. 1. Alcoholic beverages. London, New York: Academic Press.

Blood, R. M. 1975. Lactic acid bacteria in marinated herring, pp. 195–208. In: Carr, J. G., Cutting, C. V., Whiting, G. C. (eds.), Lactic acid bacteria in beverages and food. London, New York: Academic Press.

Collins, M. D., Jones, D. 1979. The distribution of isoprenoid quinones in streptococci of serological groups D and N. Journal of General Microbiology **114:**27–33.

Coster, E., White, H. R. 1964. Further studies on the genus Pediococcus. Journal of General Microbiology **37:**15–31.

Cox, W. A. 1977. Characteristics and use of starter cultures in the manufacture of hard pressed cheese. Journal of the Society of Dairy Technology **30:**5–15.

Dacre, J. C. 1958. A note on the pediococci in New Zealand cheddar cheese. Journal of Dairy Research **25:**414–417.

Deibel, R. H., Wilson, G. D., Niven, C. F., Jr. 1961. Microbiology of meat curing. IV. A lyophilized *Pediococcus cerevisiae* starter culture for fermented sausage. Applied Microbiology **9:**239–243

Deibel, R. H., Seeley, H. W., Jr. 1974. Genus I. *Streptococcus* Rosenbach, pp. 490–509. In: Buchanan, R. E., Gibbons, N. E. (eds.), Bergey's manual of determinative bacteriology, 8th ed. Baltimore: Williams & Wilkins.

de Man, J. C., Rogosa, M., Sharpe, M. E. 1960. A medium for the cultivation of lactobacilli. Journal of Applied Bacteriology **23:**130–135.

Dittrich, H. H. 1977. Mikrobiologie des Weines. Stuttgart: Verlag Eugen Ulmer.

Dolezil, L., Kirsop, B. H. 1977. The use of A. P. I. Lactobacillus system for the characterization of pediococci. Journal of Applied Bacteriology **42:**213–217.

Efstathiou, J. D., McKay, L. L. 1976. Plasmids in *Streptococcus lactis:* Evidence that lactose metabolism and proteinase activity are plasmid linked. Applied and Environmental Microbiology **32:**38–44.

Elliker, P. R., Andersen, A. W., Hannesson, G. 1956. An agar medium for lactic acid streptococci and lactobacilli. Journal of Dairy Science **39:**1611–1612.

Food and Agriculture Organization of the United Nations. 1979. 1978 FAO production yearbook, vol. 32. FAO Statistics Series No. 22. Rome: Food and Agriculture Organization.

Galesloot, Th. E., Hassing, F., Stadhouders, J. 1961. Agar media voor het isoleren en tellen van aromabacterien in zuursels. Netherlands Milk and Dairy Journal **15:**127–150.

Galluzzo, S. J., Cotterill, O. J., Marshall, R. T. 1974. Fermentation of whole egg by heterofermentative streptococci. Poultry Science **53:**1575–1584.

Garvie, E. I. 1960. The genus *Leuconostoc* and its nomenclature. Journal of Dairy Science **27:**283–292.

Garvie, E. I. 1967. *Leuconostoc oenos* sp. nov. Journal of General Microbiology **48:**431–438.

Garvie, E. I. 1974a. Nomenclature problems of the pediococci. Request for an opinion. International Journal of Systematic Bacteriology **24:**301–306.

Garvie, E. I. 1974b. Genus II. *Leuconostoc* van Tieghem, pp. 510–513. In: Buchanan, R. E., Gibbons, N. E. (eds.), Bergey's manual of determinative bacteriology, 8th ed. Baltimore: Williams & Wilkins.

Garvie, E. I. 1976. Hybridization between the deoxyribonucleic acids of some strains of heterofermentative lactic acid bacteria. International Journal of Systematic Bacteriology **26:**116–122.

Garvie, E. I. 1979. Proposal of neotype strains for *Leuconostoc mesenteroides* (Tsenkovskii) van Tieghem, *Leuconostoc dextranicum* (Beijerinck) Hucker and Pederson, and *Leuconostoc cremoris* (Knudsen and Sørensen) Garvie. International Journal of Systematic Bacteriology **29:**149–151.

Giesecke, D., Henderickx. 1973. Biologie und Biochemie der mikrobiellen Verdauung, p. 294. Munich, Bern, Vienna: BLV Verlagsgesellschaft.

Günther, H. L., White, H. R. 1961. The cultural and physiological characters of the pediococci. Journal of General Microbiology **26**:185–197.

Gordon, J. F., Shapton, N. 1977. Characteristics and use of starters for the manufacture of yoghurt, cottage cheese, cultured buttermilk and other fermented products. Journal of the Society of Dairy Technology **30**:15–22.

Griffith, F. 1928. Significance of pneumococcal types. Journal of Hygiene **27**:113–159.

Hahn, G., Tolle, A. 1979. Ergebnisse aus der Streptokokken-Zentrale in Kiel von 1965 bis 1977—Ein Überblick. Zentralblatt für Bakteriologie, Parasitenkunde, Infektionskrankheiten und Hygiene, Abt. 1 Orig., Reihe A **244**:427–438.

Harrison, A. P., Hansen, P. A. 1950. The bacterial flora of the cecal feces of healthy turkeys. Journal of Bacteriology **54**:197–210.

Jarvis, A. W., Wolff, J. M. 1979. Grouping of lactic streptococci by gel electrophoresis of soluble cell extracts. Applied and Environmental Microbiology **37**:391–398.

Jespersen, N. J. T. 1977. The use of commercially available concentrated starters. Journal of the Society of Dairy Technology **30**:47–51.

Keddie, R. M. 1951. The enumeration of lactobacilli on grass and silage. Proceedings of the Society of Applied Bacteriology **14**:157–160.

Kempler, G. M., McKay, L. L. 1979. Genetic evidence for plasmid-linked lactose metabolism in Streptococcus lactis subsp. diacetylactis. Applied and Environmental Microbiology **37**:1041–1043.

Kitahara, K. 1974. Genus III. Pediococcus Balcke, pp. 513–515. In: Buchanan, R. E., Gibbons, N. E. (eds.), Bergey's manual of determinative bacteriology, 8th ed. Baltimore: Williams & Wilkins.

Knittel, M. D. 1965. Genetic homology and exchange in lactic acid streptococci. Ph.D. Thesis. Oregon State University.

Kozak, W., Bardowski, J., Dobrzański, W. T. 1978. Lactostrepcins—acid bacteriocins produced by lactic streptococci. Journal of Dairy Research **45**:247–257.

Kunkee, R. E. 1967. Malo-lactic fermentation, pp. 235–279. In: Perlman, D. (ed.), Advances in applied microbiology, vol. 9. London, New York: Academic Press.

Lancefield, R. C. 1933. A serological differentiation of human and other groups of hemolytic streptococci. Journal of Experimental Medicine **57**:571–595.

Lawrence, R. C. 1978. Action of bacteriophages on lactic acid bacteria: Consequences and protection. New Zealand Journal of Dairy Science and Technology **13**:129–136.

Lawrence, R. C., Thomas, T. D., Terzaghi, B. E. 1976. Reviews on the progress of dairy science: Cheese starters. Journal of Dairy Research **43**:141–193.

Lawson, C. J., Sutherland, I. W. 1978. Polysaccharides, pp. 327–392. In: Rose, A. H. (ed.), Economic microbiology, vol. 2. Primary products of metabolism. London, New York: Academic Press.

LeBlanc, D. J., Hawley, R., Lee, L. N., St. Martin, E. J. 1978. "Conjugal" transfer of plasmid DNA among oral streptococci. Proceedings of the National Academy of Sciences of the United States of America **75**:3484–3487.

Lembke, J., Krusch, U., Lompe, A., Teuber, M. 1980. Isolation and ultrastructure of bacteriophages of group N (lactic) streptococci. Zentralblatt für Bakteriologie—International Journal of Microbiology and Hygiene, Abt. 1 Orig., Reihe C **1**:79–91.

Lewis, J. E. 1977. Starter manufacture at individual cheese factories. Journal of the Society of Dairy Technology **30**:32–35.

Limsowtin, G. K. Y., Heap, H. A., Lawrence, R. C. 1977. A multiple starter concept for cheesemaking. New Zealand Journal of Dairy Science and Technology **12**:101–106.

Lister, J. 1873. A further contribution to the natural history of bacteria and the germ theory of fermentative changes. Quarterly Journal of Microbiological Science **13**:380–408.

Löhnis, F. 1909. Die Benennung der Milchsäurebakterien. Zentralblatt für Bakteriologie, Parasitenkunde, Infektionskrankheiten und Hygiene, Abt. 2 **22**:553–555.

McKay, L. L., Baldwin, K. A. 1978. Stabilization of lactose metabolism in Streptococcus lactis C2. Applied and Environmental Microbiology **36**:360–367.

Mullan, M. A., Walker, A. L. 1979. An agar medium and a simple streaking technique for the differentiation of the lactic streptococci. Dairy Industries International **44**:13–17.

Nakagawa, A., Kitahara, K. 1959. Taxonomic studies on the genus Pediococcus. Journal of General and Applied Microbiology **5**:95–126.

Nickels, C., Leesment, H. 1964. Methode zur Differenzierung und quantitativen Bestimmung von Säureweckerbakterien. Milchwissenschaft **19**:374–378.

Olson, N. F., Anderson, R. F., Sellars, R. 1978. Microbiological methods for cheese and other cultured products. In: Marth, E. H. (ed.), Standard methods for the examination of dairy products, 14th ed. Washington, D.C.: American Public Health Association.

Orla-Jensen, S. 1919. The lactic acid bacteria. Copenhagen: Host & Son.

Osborne, R. J. W. 1977. Production of frozen concentrated cheese starters by diffusion culture. Journal of the Society of Dairy Technology **30**:40–44.

Ottogalli, G., Galli, A., Dellaglio, F. 1979. Taxonomic relationships between Streptococcus thermophilus and some other streptococci. Journal of Dairy Research **46**:127–131.

Pallasdies, K. 1976. Der Einsatz von tiefgefrorenen Bakterienkonzentraten für die Herstellung von unterschiedlichen mildsauren Sauermilcherzeugnissen. Deutsche Milchwirtschaft **27**:563–576.

Pearce, L. E., Halligan, A. C. 1978. Cultural characteristics of Leuconostoc strains from cheese starters, pp. 520–521. 20th International Dairy Congress France 1978. Paris: Congrilait.

Pechmann, H., Teuber, M. 1980. Plasmid pattern of group N (lactic) streptococci. Zentralblatt für Bakteriologie—International Journal of Microbiology and Hygiene, Abt. 1 Orig., Reihe C **1**:133–136.

Pederson, C. S. 1960. Sauerkraut. In: Chichester, C. O., Mrak, E. M., Steward, G. F. (eds.), Advances of food research, vol. 10. London, New York: Academic Press.

Pederson, C. S., Albury, M. N. 1950. Effect of temperature upon bacteriological and chemical changes in fermenting cucumbers. New York State Agricultural Experimentation Station Bulletin **744**:1.

Peynaud, E. 1968. Études récent sur les bactéries lactique du vin, pp. **1**:219–256. In: Fermentation et vinification. [2e Symposium International d'Oenologie, Bordeaux-Cognac, 1967.] Paris: INRA.

Rainbow, C. 1975. Beer spoilage lactic acid bacteria, pp. 149–158. In: Carr, J. G., Cutting, C. V., Whiting, G. C. (eds.) Lactic acid bacteria in beverages and food London, New York: Academic Press.

Reddy, M. S., Vedamuthu, E. R., Washam, C. J., Reinbold, G. W. 1972. Agar medium for differential enumeration of lactic streptococci. Applied Microbiology **24**:947–952.

Reiter, B., Kirikova, M. 1976. The isolation of a lysogenic strain from a multiple strain starter culture. Journal of the Society of Dairy Technology **29**:221–225.

Reuter, G. 1970a. Laktobazillen und eng verwandte Mikroorganismen in Fleisch und Fleischerzeugnissen. 3. Mitteilung: Abgrenzung der einzelnen Keimgruppen. Fleischwirtschaft **50**:1081–1084.

Reuter, G. 1970b. Laktobazillen und eng verwandte Mikroorganismen in Fleisch und Fleischerzeugnissen. 4. Mitteilung: Die Ökologie von Laktobazillen, Leuconostoc-Spezies und Pediococcen. Fleischwirtschaft **50**:1397–1399.

Sandine, W. E., Radich, P. C., Elliker, P. R. 1972. Ecology of the lactic streptococci. A review. Journal of Milk and Food Technology **35**:176–184.

Schleifer, K. H., Kandler, O. 1972. Peptidoglycan types of bacterial cell walls and their taxonomic implications. Bacteriological Reviews **36**:407–477.

Schiefer, G., Schöne, R. 1978a. Rohwurstherstellung mittels Starterkulturen. Die Nahrung **22**:419–424.

Schiefer, G., Schöne, R. 1978b. Herstellung von Pökelware unter Verwendung von Starterkulturen. Fleisch **32**:215–216.

Shankar, P. A., Davies, F. L. 1977. A note on the suppression of *Lactobacillus bulgaricus* in media containing β-glycerophosphate and application of such media to selective isolation of *Streptococcus thermophilus* from yoghurt. Journal of the Society of Dairy Technology **30**:28–30.

Sharpe, M. E. 1979. Lactic acid bacteria in the dairy industry. Journal of the Society of Dairy Technology **32**:9–18.

Shimwell, J. L., Kirkpatrick, W. F. 1939. New light on the ''Sarcina'' question. Journal of the Institute of Brewing **45**:137–141.

Skinner, F. A., Quesnel, L. B. (eds.). 1978. Streptococci. The Society for Applied Bacteriology Symposium Series 7. London, New York, San Francisco: Academic Press.

Stadhouders, J. 1975. Microbes in milk and dairy products. An ecological approach. Netherlands Milk and Dairy Journal **29**:104–126.

Stanley, G. 1977. The manufacture of starters by batch fermentation and centrifugation to produce concentrates. Journal of the Society of Dairy Technology **30**:36–39.

Subcommittee on Lactobacilli and Related Microorganisms of the International Committee on Systematic Bacteriology. 1979. International Journal of Systematic Bacteriology **29**:435.

Tagg, J. R., Dajani, A. S., Wannamaker, L. W. 1976. Bacteriocins of Gram-positive bacteria. Bacteriological Reviews **40**:722–756.

Tamime, A. Y., Robinson, R. K. 1976. Recent developments in the production and preservation of starter cultures for yogurt. Dairy Industries International **41**:408–411.

Terzaghi, B. E., Sandine, W. E. 1975. Improved medium for lactic streptococci and their bacteriophages. Applied Microbiology **29**:807–813.

Tilbury, R. H. 1975. Occurrence and effects of lactic acid bacteria in the sugar industry, pp. 177–191. In: Carr, J. G., Cutting, C. V., Whiting, G. C. (eds.), Lactic acid bacteria in beverages and food. London New York: Academic Press.

Vaughn, R. H. 1975. Lactic acid fermentation of olives with special reference to California conditions, pp. 307–323. In: Carr, J. G., Cutting, C. V., Whiting, G. C. (eds.), Lactic acid Bacteria in beverages and food. London, New York: Academic Press.

Whittenbury, R. 1965a. The enrichment and isolation of lactic acid bacteria from plant material. Zentralblatt für Bakteriologie, Parasitenkunde, Infektionskrankheiten und Hygiene, Abt. 1 Suppl. **1**:395–398.

Whittenbury, R. 1965b. A study of some pediococci and their relationship to aerococcus viridans and the enterococci. Journal of General Microbiology **40**:97–106.

Whittenbury, R. 1966. A study of the genus *Leuconostoc*. Archiv für Mikrobiologie **53**:317–327.

Weigmann, H. 1905–1908. Das Reinzuchtsystem in der Butterbereitung und in der Käserei, pp. 293–309. In: Lafar, F. (ed.), Handbuch der Technischen Mykologie, vol. 2. Mykologie der Nahrungsmittelgewerbe. Jena: Gustav Fischer Verlag.

Weiler, H. G., Radler, F. 1970. Milchsäurebakterien aus Wein und von Rebenblättern. Zentralblatt für Bakteriologie, Parasitenkunde, Infektionskrankheiten und Hygiene, Abt. 2 Orig. **124**:707–732.

Wigley, R. C. 1977. The use of commercially available concentrated starters. Journal of the Society of Dairy Technology **30**:45–47.

The Anaerobic Cocci

GALE B. HILL

The anaerobic cocci are widely distributed as normal endogenous flora in man and animals. Their primary locations are the skin and the upper respiratory, the gastrointestinal, and the urogenital tracts. Certain of the anaerobic cocci are opportunistic pathogens and can cause infection in man and animals, but the potential pathogenicity, if any, of many species is unknown. Generally speaking, anaerobic bacteria are unable to survive and multiply in the presence of atmospheric oxygen; but like other anaerobes, the anaerobic cocci exhibit a wide range of sensitivity to oxygen that depends on the physical and chemical circumstances of exposure.

The study of anaerobic bacteria has been limited in the past by lack of adequate and convenient anaerobic culture methods. Procedures for growing anaerobic bacteria on solid media in an anaerobic environment were technically difficult and inefficient or just not available. The isolation of pure cultures was difficult because of the use of broth cultures and the propensity of anaerobic bacteria to grow in association with other anaerobic or facultative species. As a result, many descriptions and characterizations of anaerobic bacteria were actually based on mixed cultures, or the media and cultural environments employed were so different that the same species were reported on numerous occasions and given different names.

This lack of simple and efficient cultural techniques and identification schemes discouraged widespread study of the nonsporeforming anaerobes. Nevertheless, some clinicians and microbiologists published on the occurrence of the anaerobic cocci in human disease in the late 1800s and early 1900s, but their findings did not gain general acceptance. In the 1970s, renewed interest was brought about by improvements in anaerobic cultivation techniques and procedures for identification and by the availability of antibiotics with a broad anaerobic spectrum. These improvements have resulted in a more universal acceptance of the role of anaerobic bacteria in human infection and have stimulated interest in understanding these organisms as members of the endogenous flora of man and other animals.

Although our knowledge of the anaerobic cocci has expanded greatly in recent years, much remains to be done. Presently, there are different schemes of classification and identification, which retard the collection and sorting of data gathered on anaerobic cocci isolated from infection or from the normal flora. A universally accepted standard system of nomenclature with uncomplicated, effective identification procedures is needed.

Discussion of the anaerobic cocci in this chapter will be based on the taxonomy and nomenclature as described by Holdeman and Moore (1974), Gossling and Moore (1975), and Holdeman, Cato, and Moore (1977). Genera to be discussed are the anaerobic Gram-positive cocci—*Peptococcus, Peptostreptococcus, Streptococcus, "Gaffkya", Ruminococcus, Sarcina, Coprococcus*—and the Gram-negative cocci—*Veillonella, Acidaminococcus, Megasphaera,* and *Gemmiger.* The name *"Gaffkya"* is used with quotation marks because it has been rejected as a legitimate name without provision of an accepted name for the organism. Species of *Streptococcus* may be obligately anaerobic or microaerophilic, yet certain strains may occasionally grow under aerobic conditions. This group is included in the anaerobic cocci since these organisms grow best under an anaerobic environment and may not be isolated unless anaerobic conditions are used for primary cultivation.

Habitats

Animal Endogenous Flora and Environmental Species

The anaerobic cocci are present on the mucocutaneous surfaces of man and other animals, including the skin and the upper respiratory, gastrointestinal, and urogenital tracts. Anaerobic cocci are also present on cereal grains and in soil, sea sediment, and mud. Our knowledge is incomplete regarding the location of particular species in man and animals, but it is known that certain species have a predilection for particular body sites. Species of anaerobic, Gram-positive cocci are found on the skin of humans in relatively high concentrations (Holdeman, Cato, and

Moore, 1977; Rosebury, 1962; Sutter, Vargo, and Finegold, 1975; Thomas and Hare, 1954.)

The anaerobic cocci are located in numerous sites within the upper respiratory tract and mouth. Reports of their relative incidence compared to other flora depend on the sites sampled and the efficiency of the anaerobic plating methods used. Anaerobic, Gram-positive cocci have been found to constitute up to 13% (Hardie and Bowden, 1974) and *Veillonella* from 5% (Liljemark and Gibbons, 1971) to approximately 16% (Hardie and Bowden, 1974) of the cultivable flora of human saliva. Isolations of *Veillonella* from the predominant cultivable flora of dental plaque have varied widely but range up to 28% (Hardie and Bowden, 1974). In the human gingival region, recovery of anaerobic Gram-positive cocci ranged from approximately 7% to 15%, whereas anaerobic Gram-negative cocci constituted approximately 10% of the total cultivable flora (Hardie and Bowden, 1974). *Veillonella* has been reported to make up 8–15% of the cultivable flora of the tongue surface, which was correlated with its greater ability to adhere to tongue surfaces than to other sites in the mouth (Gibbons, 1974; Liljemark and Gibbons, 1971). Oral specimens from rats, hamsters, guinea pigs, and rabbits also contained *Veillonella* (Rogosa, 1964).

Many of the anaerobic cocci have been isolated from the gastrointestinal tract or rumen of man and other animals. *Peptostreptococcus productus,* in particular, and *Coprococcus eutactus* and *Ruminococcus bromii* were generally among the most common and predominant anaerobic coccal isolates in the fecal flora of North American male subjects and 20 Japanese-Hawaiians studied, but significant variation in fecal flora among individuals in each of the two study populations was apparent (Holdeman, Good, and Moore, 1976). Frequent isolations have also been reported for various species of *Ruminococcus, Coprococcus, Peptostreptococcus, Peptococcus,* and strains of *Acidaminococcus fermentans, Gemmiger formicilis,* and *Megasphaera elsdenii* (less frequent). Some of these isolations were in relatively high concentrations; the range was from 10^8 to 10^{10} per gram dry fecal weight (Attebery, Sutter, and Finegold, 1972; Holdeman, Good, and Moore, 1976; Sugihara et al., 1974). Although certain studies have reported that changes in human diet do not dramatically alter the anaerobic flora in individuals (Attebery, Sutter, and Finegold, 1972; Holdeman, Good, and Moore, 1976), the isolation of *Sarcina ventriculi* was quite common in different populations all on a vegetarian diet, but uncommon in persons living on mixed diets (Crowther, 1971). Although some of the species of anaerobic bacteria, such as *Ruminococcus albus, R. flavefaciens, R. bromii, P. productus, M. elsdenii, A. fermentans, Veillonella parvula,* and *Streptococcus intermedius,* are found in the large intestine of man and the intes-

tinal tract or rumen of animals, other species in man and animals are dissimilar and have not been characterized and named as yet (Attebery, Sutter, and Finegold, 1972, 1974; Barnes et al., 1977; Dehority, 1977; Giesecke, Wiesmayr, and Ledinek, 1970; Harris, Reddy, and Carter, 1976; Holdeman, Cato, and Moore, 1977; Holdeman, Good, and Moore, 1976; Sutter, Vargo, and Finegold, 1975).

Anaerobic cocci make up a significant proportion of the flora of the vagina and cervix of women. Anaerobic, Gram-positive cocci occur in approximately 80% of premenopausal women; the incidence appears to be higher in pregnancy and, particularly, immediately postpartum (Hill et al., unpublished; Goplerud, Ohm, and Galask, 1976). Generally species of *Peptococcus* are most common in the vagina, occurring in 36–64% of women, whereas *Peptostreptococcus* species have been reported in 23–31% (Bartlett et al., 1977; Levison et al., 1977; G. B. Hill et al., unpublished). Isolation of *Veillonella* from the vagina has been reported from 8% or 9% of women (Bartlett, et al., 1977; Levison et al., 1977) ranging to 20% (Hill et al., unpublished data). *Acidaminococcus* is found very infrequently. Mean viable counts of *Peptococcus* species have been reported as $10^{8.7}$ colony-forming units/g of vaginal secretions (Bartlett et al., 1977) and 10^7 colony-forming units/ml (Levison et al., 1977). Aggregate data did not reflect significant differences in the particular species of cocci or in their concentrations among the vaginal and endocervical flora, but paired samples from the vagina and cervix of individuals demonstrated that different species were found in the two sites (Bartlett et al., 1978).

The rate of isolation and the types of anaerobic cocci in the cervix have varied somewhat in different reports (Ohm and Galask, 1975; Gorbach et al., 1973; Bartlett et al., 1978). The major species of anaerobic cocci described in studies of vaginal and cervical flora are *Peptococcus asaccharolyticus, Peptostreptococcus anaerobius, Peptococcus prevotii, Peptococcus magnus,* and *"Gaffkya" anaerobia* (Bartlett et al., 1977; Goplerud et al., 1976; Ohm and Galask, 1975; Hill et al., unpublished data). *Streptococcus* species occur less frequently. Numerous isolates of anaerobic, Gram-positive cocci from the genital flora cannot be identified using current taxonomic schemes.

Human Infection

Early studies on the anaerobic cocci were made through the interest of investigators in obstetrics who noted the high incidence of anaerobic, Gram-positive cocci in puerperal infections. Until recently, all anaerobic Gram-positive cocci were referred to primarily as anaerobic "streptococci"; in the following discussion of older literature, the term anaerobic "streptococci" does not refer specifically to anaerobic organisms in the genus *Streptococcus*.

Anaerobic cocci have been described from human disease since 1893 when Veillon reported the isolation under anaerobic conditions of Gram-positive cocci from a Bartholin's abscess. He noted that the foul smell of the infected pus was also obtained in cultures of these organisms (Veillon, 1893). Additional investigations near the turn of the century by Menge, Krönig, Wegelius, and Schottmüller emphasized the role of anaerobic, Gram-positive cocci in puerperal infections. Schwarz and Dieckmann (1926) cultured the blood and uterus of 165 patients with suspected puerperal infections, and isolated organisms identified as *Streptococcus putridus* (present designation, *Peptostreptococcus anaerobius*) from 46 cultures and other anaerobic "streptococci" from 21 cultures. From their own experience and the previous work of Schottmüller, these authors emphasized that, although the anaerobic cocci were frequently involved in localized puerperal infections associated with a foul-smelling lochia, in certain instances these organisms could infect the uterine veins and cause thrombophlebitis, septicemia, and, occasionally, generalized peritonitis.

In another study of the streptococci associated with puerperal infection, Harris and Brown (1929) isolated facultative or anaerobic "streptococci" in pure culture or in mixture with other organisms from 113 (67%) of 168 uterine cultures. They pointed out that, although aerobic (facultative), β-hemolytic streptococci were responsible for the most serious type of puerperal infection, infections due to anaerobic "streptococci" were more frequently observed. The source of the aerobic, β-hemolytic streptococci appeared to be exogenous, whereas the anaerobic, nonhemolytic "streptococci" could be isolated from the vagina and cervix of normal nonpregnant, pregnant, parturient, and puerperal women, indicating that these endogenous organisms had the potential for invasion of the upper genital tract. Harris and Brown also tested for fermentation of numerous sugars, gas formation in cooked meat broth, coagulation of milk, and liquefaction of gelatin in an attempt to improve classification of the anaerobic "streptococci" and to compare their isolates with descriptions published by Prévot in 1924. Their work emphasized the diversity among the anaerobic cocci involved in infection.

Colebrook (1930) additionally provided strong evidence for the pathogenicity of the anaerobic, Gram-positive cocci in puerperal fever by demonstrating that two-thirds of maternal septicemias during the study period were due to a heterogeneous group of anaerobic "streptococci" (in pure culture or mixed with other organisms), whereas only one-third were due to the facultative, β-hemolytic *Streptococcus pyogenes*. Mortality associated with these anaerobic "streptococcal" septicemias was 39%, with septic thrombophlebitis demonstrated in a portion of those cases. Brown (1940) confirmed the higher incidence (more than twice) of anaerobic "streptococci" in contrast to hemolytic facultative streptococci in 31 fatal cases of puerperal infection. Although the foregoing authors all emphasized the necessity for performing anaerobic as well as aerobic cultures and did commendable anaerobic bacteriology with the rather cumbersome methods then available, anaerobic culture methods still did not become widely established.

Most of the earlier publications had emphasized the prevalence of anaerobic cocci in obstetric disease, but their occurrence in other infected sites began to receive attention. McDonald, Henthorne, and Thompson (1937) reported on 23 cases in which anaerobic "streptococci" were recovered at necropsy from the blood or other sites in infections involving the intestinal tract, lung, meninges, and other tissues. The organisms were recovered in pure culture from 11 of the 23 cases. Isolates were examined by sugar fermentations, production of hemolysis on blood agar, and other tests; agglutination studies demonstrated that certain strains belonged to the same serological group. In an excellent review, Sandusky et al. (1942) discussed the wide variety of human infections in which the anaerobic cocci were involved and presented data from 170 surgical lesions in which anaerobic Gram-positive cocci were isolated (17% in pure culture). These authors delineated the types of anaerobic, facultative, and aerobic bacteria with which the anaerobic "streptococci" were associated in mixed cultures, and pointed out that the anaerobic "streptococci" were part of the normal endogenous flora of the mouth and the gastrointestinal and urogenital tracts. From these sites, they could become invasive due to changes in the bacteria themselves, or in host defense mechanisms, or in conjunction with other microorganisms in a symbiotic infection.

From 1930 to 1950, numerous authors, for example Meleney and Altemeier, published on the importance of all anaerobic bacteria, including the anaerobic cocci, in a variety of infections. Thomas and Hare (1954) demonstrated anaerobic cocci from different types of infections and attempted to devise an approach to classification and identification of strains by convenient laboratory methods, recognizing the heterogeneity of the anaerobic cocci that occur in infections. Stokes (1958) provided additional impetus for the use of anaerobic culture methods. In her studies, anaerobic, Gram-positive cocci were present in 208 of 357 infections in which anaerobes were found in mixed culture and in 100 of 139 infections in which anaerobic species were found in pure culture. Bornstein et al. (1964) reported that 75% of the anaerobic "streptococci" isolated from specimens submitted to the hospital bacteriology laboratory were etiologically associated with infection. Finegold et al. (1968) reported on anaerobic cocci from 81 infections, including bacteremia, brain abscess, necrotizing pneumonia, empyema, intraabdominal abscess, liver abscess, peritonitis, wound infections, pelvic abscess, endometritis, perirectal abscess, soft-tissue infection, myositis, osteomyelitis, and subacute bacterial endocarditis. Anaerobic cocci were isolated in pure culture in one-third of the 81 cases, in mixture with other anaerobic bacteria in 16 cases, and in mixture with a variety of facultative or aerobic bacteria in the remainder. At the same time, Hill, Lewis, and Altemeier (1968) reported on 585 clinical specimens from a surgical service, of which 45% contained one or more anaerobic bacteria; *Peptostreptococcus* species were found in 23%, and *Peptococcus* species in 10% of these infections. Moore, Cato, and Holdeman (1969) discussed the anaerobic species, including the anaerobic cocci found in human feces, and the isolation of certain of these species from human infection. Martin (1974) reported that, of 14,839 unselected, culturally positive specimens, anaerobic bacteria were isolated from 49% and anaerobic cocci made up 27% of the anaerobic isolates. A more detailed study of the anaerobic, Gram-positive cocci isolated in the same hospital (Pien, Thompson, and Martin, 1972) demonstrated a high frequency of *Peptostreptococcus,* and particularly *Peptococcus,* in surgical wound infections primarily related to abdominal and pelvic procedures. The frequency of isolation of the anaerobic cocci has steadily increased with improved anaerobic culture procedures and will be discussed further in relation to specific sites of infection.

The common sites of infection for anaerobic cocci largely derive from their existence as endogenous flora, since they may spread contiguously to adjacent tissue or through the circulatory system when normal host defenses are sufficiently compromised to allow invasion. Reduction of the oxygen content and oxidation-reduction potential of tissue may result from impairment of the vascular supply and/or necrosis of tissue in many ways, e.g., surgery, malignancy, trauma, shock, edema, vasoactive drugs, irradiation, cytotoxic chemotherapy, foreign bodies, vascular disease, and others. Facultative and aerobic bacteria can utilize available oxygen and also provide reducing conditions for anaerobic growth. As with other anaerobic bacteria, only particular species of anaerobic cocci appear to have the potential to invade tissue under most circumstances and are isolated repeatedly from infections, whereas other species are found infrequently. The species of anaerobic cocci that may be found in human infection are listed in Table 1. Other organisms, which might rarely be isolated from infection but are believed to primarily be restricted to the normal flora (i.e., species of *Coprococcus, Rumino-*

Table 1. Species of anaerobic cocci commonly isolated from human infections.

Gram-positive:	*Peptococcus asaccharolyticus*[a]
	Peptococcus magnus[a]
	Peptococcus prevotii[a]
	Peptostreptococcus anaerobius[a]
	Peptostreptococcus micros
	Streptococcus intermedius[a]
	Streptococcus constellatus
	Streptococcus morbillorum
	"*Gaffkya*" *anaerobia*[a]
Gram-negative:	*Veillonella parvula*[a]
	Acidaminococcus fermentans

[a]Most frequent isolates.

coccus, Sarcina, Megasphaera, and *Gemmiger*), are not included.

Most infections involving anaerobic bacteria are polymicrobic or mixed infections composed of different species of anaerobic bacteria or a mixture of aerobic and/or facultative and anaerobic organisms. In the following discussion of major sites of infection for anaerobic cocci, it should be understood that the cocci are often present with other bacteria and usually cannot be designated as the sole etiological agent. However, anaerobic cocci are occasionally found in pure culture in blood, abscesses, and other sites of infection. Further information on the clinical aspects of anaerobic infections and the anaerobic cocci is available in the following references: Balows et al., 1974; Finegold, 1977; Gorbach and Bartlett, 1974; Lambe, Vroon, and Reitz, 1974; Smith, 1975.

INFECTIONS OF THE HEAD AND NECK. The anaerobic cocci are involved in a wide variety of infections about the head and neck, including suppurative thrombophlebitis of the internal jugular vein, chronic otitis media and mastoiditis, chronic sinusitis, brain abscess, a wide variety of oral infections (e.g., peridontal disease, Ludwig's angina, putrid infections of the mouth, and fusospirochetal disease), and various space infections (abscesses) of the lower head and neck (Bartlett and Gorbach, 1976; Finegold, 1977; Frederick and Braude, 1974; Heineman and Braude, 1963).

Anaerobic "streptococci" were the predominant anaerobic organisms in brain abscesses (Heineman and Braude, 1963), whereas veillonella were much less frequently isolated. The source of anaerobic cocci in brain abscess undoubtedly derives from the association of these bacteria with chronic otitis media, mastoiditis, and sinusitis and their contiguous spread into the central nervous system from these sites. Hematogenous spread following dental extractions or from anaerobic pleuropulmonary infections also provide a source of the anaerobic cocci in brain abscess. In chronic sinusitis, specimens obtained surgically yielded anaerobic "streptococci" in 28 and veillonella in 14 of 62 specimens that were positive for bacterial growth. A heavy growth of anaerobic "streptococci" was often obtained (Frederick and Braude, 1974). In 15 space infections, mainly along the mandible, anaerobic cocci were frequently obtained, particularly *Streptococcus* (previously designated *Peptostreptococcus*) *intermedius* and *Peptococcus prevotii* (Bartlett and Gorbach, 1976). Anaerobic organisms that often coexist with the anaerobic cocci in the infections just discussed are *Bacteroides melaninogenicus* and *Fusobacterium* species, particularly *F. nucleatum*.

PLEUROPULMONARY INFECTIONS. The importance of anaerobic bacteria in pulmonary disease was recognized very early and numerous studies on the etiology were performed in the early 1900s. Mixed anaerobic species were often isolated and the roles of the anaerobic "streptococci", fusobacteria, and spirochetes observed in "fusospirochetal" infections were controversial. Experimental simulation of the aspiration of anaerobic mouth flora demonstrated the importance of this mechanism in the production of pulmonary disease (Smith, 1927, 1932).

It is now recognized that anaerobic bacteria are important causes of pneumonitis, lung abscess, necrotizing pneumonia, and empyema, although mixed anaerobic and aerobic species are often present. However, exclusively anaerobic species are more commonly seen in these infections than exclusively aerobic and facultative bacteria; Gram-positive anaerobic cocci accounted for 33% of the total anaerobic isolates identified from 70 patients (Bartlett and Finegold, 1972). Of 45 cases of lung abscess involving anaerobic bacteria species of *Peptostreptococcus* and *Peptococcus* were isolated from 14, microaerophilic *Streptococcus* from 6, and *Veillonella* from 4 cases. Pure cultures of anaerobic, Gram-positive cocci were present in 8 cases. Other common anaerobic isolates were *F. nucleatum* and *B. melaninogenicus*. Bacteriological data from other types of pulmonary infection were similar except that anaerobic, Gram-positive cocci were isolated from a very high proportion of pneumonitis, in which there were also three cases with pure cultures of *Veillonella*. In empyema, microaerophilic streptococci were most numerous among the isolates of Gram-positive cocci; 3 were pure-culture infections (Bartlett and Finegold, 1974). In a prospective study of 54 cases of pulmonary infection following aspiration, the majority of pure-culture infections involved anaerobic, Gram-positive cocci, particularly peptostreptococci; *Veillonella* were seen less frequently and in mixed culture (Bartlett et al., 1974). In these studies, the anaerobic cocci were reported by genus and no further species identification was given.

Aspiration of mouth flora is generally responsible for these types of pleuropulmonary infections. Anaerobic organisms should be suspected in the presence of a putrid odor of sputum or other discharge, tissue necrosis, chronic or subacute presentation, and various settings that predispose to aspiration, such as alcoholism or general anesthesia. Additional underlying conditions observed are dental infections, other extrapulmonary anaerobic disease, pulmonary embolus with infarction, bronchogenic carcinoma, and bronchiectasis. Prospective studies using properly collected specimens, that avoid contamination with normal flora e.g., percutaneous transtracheal aspirates, thoracotomy specimens, pleural fluid, and blood cultures, reveal anaerobes in 71–94% of these types of infections (Bartlett and Finegold, 1974). Pathogenic facultative bacteria most commonly isolated with anaerobes are *Staphylococcus aureus* and members of the Enterobacteriaceae, particularly *Escherichia coli*.

ABDOMINAL INFECTIONS. The important role of anaerobic bacteria in intraabdominal infections, such as appendicitis and peritonitis, was demonstrated in the early 1900s (see review and investigations by Altemeier, 1938). The currently accepted rate of isolation of anaerobes in these infections is 90–96% (Gorbach and Bartlett, 1974; Finegold, 1977). The infectious complications generally result from spillage of fecal material into the peritoneal cavity from appendicitis, diverticulitis, inflammatory bowel disease, surgery, penetrating abdominal trauma, or cancer. Often the first clinical manifestation is peritonitis, which is followed by abscess formation and localization of the infection in survivors. Sandusky et al. (1942) provided a strong argument for the importance of the anaerobic "streptococci" in surgical infections by demonstrating 29 cases from which anaerobic, Gram-positive cocci were isolated in pure culture. More recently, 50 strains of anaerobic, Gram-positive cocci were isolated from 95 postoperative wound infections, a rate of isolation which was similar to the bacteroides group (Hoffmann and Gierhake, 1969). Peptostreptococci, along with *Bacteroides fragilis* and clostridial species, were frequently isolated in infections subsequent to penetrating abdominal wounds (Thadepalli, Gorbach, and Keith et al., 1973a). Among the anaerobic bacteria isolated from intraabdominal abscesses and other infections, anaerobic, Gram-positive cocci often were second only to *Bacteroides* species, particularly *B. fragilis*, in rate of isolation and accounted for a number of positive blood cultures. Other common coisolates were clostridia and aerobic and facultative organisms, primarily *E. coli*, the enterobacter-klebsiella group, enterococci, *Staphylococcus epidermidis*, *Proteus*, and *Pseudo-*

monas (Altemeier et al., 1973; Lorber and Swenson, 1975; Gorbach, Thadepalli, and Norsen, 1974).

LIVER ABSCESS.   The occurrence of microaerophilic and anaerobic, Gram-positive cocci in liver abscesses had been only sporadically reported (Patterson et al., 1967) until Sabbaj, Sutter, and Finegold (1972) reported on an incidence of anaerobic bacteria in liver abscesses of 45% and emphasized that the true incidence was probably much higher. In their series, among 50 strains of bacteria recovered from 25 cases of liver abscesses involving anaerobic bacteria, 20 strains of anaerobic or microaerophilic cocci were isolated along with 17 strains of anaerobic, Gram-negative rods. Their review of past literature demonstrated a high isolation rate of anaerobic or microaerophilic, Gram-positive cocci, but less than the rate of anaerobic, Gram-negative rods. The significant increase in the isolation rates of anaerobes in general and of anaerobic, Gram-positive cocci in particular in pyogenic liver abscesses is illustrated by an isolation rate for anaerobic, gram-positive cocci of 26% during earlier years of a 17-year study, compared to 50% using improved anaerobic culture procedures in the latter 2 years of the study (Altemeier, 1974). Pylephlebitis or cholangitis often precipitate development of liver abscesses; intestinal malignancy or inflammation, intestinal perforation, direct extension of perinephritic, subdiaphragmatic, pancreatic, or lung abscesses, and spread of distant infection by lymphatics or blood stream may also cause these infections. Common coisolates in these infections are fusobacteria, facultative streptococci, *E. coli,* and *Bacteroides,* particularly *B. fragilis.*

OBSTETRIC AND GYNECOLOGICAL INFECTIONS.   The obstetric and gynecological literature prior to 1960 had clearly delineated an important role of anaerobic, Gram-positive cocci in diseases of the female genital tract (see above). Use of more rigorous transport methods and of improved media and anaerobic cultural environments has dramatically increased the isolation of anaerobic, Gram-positive cocci and other anaerobes from these infections. The older literature indicated that anaerobic, Gram-positive cocci were the predominant pathogens in obstetric and gynecological disease, but the less efficient cultural procedures failed to isolate concomitant anaerobic, Gram-negative bacilli and other anaerobic organisms that were probably also present. The overall recovery rates of anaerobic, Gram-positive cocci and Gram-negative cocci from several investigations with a total of 200 patients with female genital tract infections were 37% and 4%, respectively, of the total 356 anaerobic isolates. Major isolates were *Peptostreptococcus anaerobius, Streptococcus intermedius, Peptococcus prevotii,*

and *Peptococcus magnus* (Chow, Marshall, and Guze, 1975; Finegold et al., 1975).

Anaerobic, Gram-positive cocci are isolated from approximately 50% to 80% of gynecological infections, a rate similar to that of anaerobic, Gram-negative bacilli. *Veillonella* are found much less frequently. Included among gynecological infections are vulvovaginal pyogenic infections, tubo-ovarian abscess, acute salpingitis, chronic pelvic inflammatory disease, pelvic thrombophlebitis, and postoperative infections, particularly after vaginal hysterectomy (Chow et al., 1975; Eschenbach et al., 1975; Finegold et al., 1975; Gorbach and Bartlett, 1974; Hall et al., 1967; Ledger et al., 1977; Parker and Jones, 1966; Sweet, 1975; Swenson and Lorber, 1977; Swenson et al., 1973; Thadepalli, Gorbach, and Keith, 1973). Anaerobic, Gram-positive cocci are also frequent pathogens in obstetric infections, including septic abortion, amnionitis, and endometritis, and in postoperative infections following cesarean section (Chow, Marshall, and Guze, 1977; Chow et al., 1975; Finegold et al., 1975; Gibbs et al., 1975; Gorbach and Bartlett, 1974; Ledger et al., 1977; Pearson and Anderson, 1970; Rotheram, 1974; Sweet and Ledger, 1973; Swenson and Lorber, 1977; Thadepalli et al., 1973). The following conditions predispose to obstetric and gynecological infections with microaerophilic and anaerobic cocci that are normally present in the genital tract (endocervix, vagina, external genitalia): prolonged labor, premature rupture of membranes, hemorrhage during labor, extensive manipulations, spontaneous or induced abortion, gonococcal salpingitis, vaginal or endocervical stenosis, cauterization, malignancy, surgery, and intrauterine contraceptive devices. Common co-isolates with anaerobic cocci are *B. fragilis* and other *Bacteroides* species and facultative bacteria such as *E. coli,* streptococci (not group A), and occasionally diphtheroids and coagulase-negative staphylococci. Anaerobic, Gram-positive cocci are occasionally isolated in pure culture.

SOFT-TISSUE INFECTIONS.   Anaerobic, Gram-positive cocci are either commonly found or are essential etiological agents for a large variety of soft-tissue infections. Many are mixed infections that may include other anaerobes, particularly the anaerobic, Gram-negative bacilli, *Bacteroides fragilis* and *B. melaninogenicus,* and/or facultative bacteria such as the Enterobacteriaceae, especially *Escherichia coli,* streptococcal species, and *Staphylococcus aureus.* The specific bacteria found largely depends on the area of the body infected and, therefore, the source of the organisms. Clinical syndromes may appear similar and yet may derive from completely different bacteria as etiological agents. These infections of skin and soft tissue commonly occur after surgery, traumatic injury, or ischemia associated with diabetes or other vascular disease; the infections frequently include extensive tissue necrosis, extension along subcutaneous and fascial planes, gas production, and a foul odor. Certain of these infections are recognizable syndromes: progressive bacterial synergistic gangrene—a synergistic infection of a microaerophilic streptococcus and *S. aureus* that produces progressive ulceration of the skin and subcutaneous tissue; fusospirochetal infection of the skin—a progressive, chronic, deep infection with multiple sinuses and a foul-smelling exudate that usually occurs after human bite wounds and commonly involves anaerobic "streptococci" in addition to fusiform bacilli and spirochetes; synergistic necrotizing cellulitis—a rapidly spreading necrotic infection of the skin and connective tissue with pain, fever, and systemic toxicity that may be caused by clostridia or a mixed bacterial flora; necrotizing fascitis—a severe infection with systemic toxicity, deep extension into the fascia, and a high mortality rate that may be caused by β-hemolytic, facultative streptococci or a mixed flora; streptococcal gangrene—a serious infection with pain, systemic toxicity, development of metastatic foci, and extensive necrosis of subcutaneous tissue and skin that may simulate the appearance of clostridial myonecrosis but is usually caused by β-hemolytic, facultative streptococci (commonly group A) but may also be produced by the anaerobic, Gram-positive cocci (Altemeier and Culbertson, 1948; Anderson, Marr, and Jaffe, 1972; Chambers, Bond, and Morris, 1974; Finegold, 1977; Giuliano et al., 1977; Gorbach and Bartlett, 1974; MacLennan, 1943, 1962; Meleney, 1931, 1933; Puhvel and Reisner, 1974; Smith, 1932; Stone and Martin, 1972).

BACTEREMIA. Anaerobic bacteremias, mostly from abdominal and pelvic infections, make up approximately 10% of all bacteremias in a general hospital. Bacteremias due to anaerobic, Gram-positive cocci occur more frequently in obstetric populations; but the mortality rate is lower than in bacteremias from abdominal infections, particularly when there is underlying disease. These organisms may also cause subacute, anaerobic bacterial endocarditis. *Veillonella* is infrequently isolated from blood cultures; when observed, it is usually in polymicrobic mixtures. Most reports have listed only *Peptostreptococcus* species. Since the methods of identification of each genus *(Peptostreptococcus, Peptococcus, Streptococcus)* have differed widely among investigators, the relative frequency of isolation of each genus cannot be accurately estimated (Chow and Guze, 1974; Chow, Marshall, and Guze, 1977; Felner and Dowell, 1970; Finegold, 1977; Ledger et al., 1975; Rotheram, 1974; Sonnenwirth, 1974; Wilson et al., 1972).

Experimental Animal Infections

In the first published description of infection by anaerobic, Gram-positive cocci (Veillon, 1893), the organism produced abscesses when injected subcutaneously into mice and guinea pigs. However, subsequent studies of the pathogenicity of anaerobic cocci for experimental animals were usually negative (Harris and Brown, 1929; McDonald et al., 1937; Weiss and Mercado, 1938). The lack of success in demonstrating pathogenicity was probably due to the technical problems of anaerobic cultivation at that time and the injection of insufficient concentrations of viable organisms.

Nevertheless, a number of early studies successfully demonstrated pathogenicity of anaerobic, Gram-positive cocci for animals and pioneered work on the synergistic nature of many polymicrobic infections involving the anaerobic cocci. Production of aspiration pneumonia and lung abscess in animals by intratracheal inoculation of pyorrhea exudate from humans demonstrated the infectivity of a "fusospirochetal mixture" that included anaerobic "streptococci"; the pathology in animals was similar to that observed in humans following aspiration of oral material (Smith, 1927, 1932). Using a variety of animal species, the potential pathogenicity of anaerobic, Gram-positive cocci and other anaerobes in the etiology of other diseases, such as progressive synergistic gangrene (Meleney, 1931), appendicitis peritonitis (Altemeier, 1942), genital tract infections (Hite, Locke, and Hesseltine, 1949; Steinhorn, 1945), and skin necrosis (Mergenhagen, Thonard, and Scherp, 1958), was described. These synergistic mixtures often involved other anaerobic bacteria, particularly species of *Bacteroides* or *Fusobacterium*, facultative organisms, such as *Escherichia coli*, *Staphylococcus aureus*, or *S. epidermidis*, or products of those organisms.

More recently, synergism has been demonstrated with human clinical isolates of *Peptostreptococcus anaerobius*, *S. intermedius*, and *Peptococcus asaccharolyticus*, each in mixture with another anaerobic species, in the production of intrahepatic and intraabdominal infections in mice (Hill, Osterhout, and Pratt, 1974; Hill, 1975). Synergism in the production of infection between *Veillonella alcalescens* (present designation *parvula*) and *S. intermedius* isolated from the gingival crevice area has also been demonstrated (Walker, West and Nitzan, 1977). There are a few examples of pathogenicity of anaerobic, Gram-positive cocci in pure culture (Mergenhagen, 1958; Veillon, 1893; Hill, 1975, unpublished data). As with human infections, different anaerobic, Gram-positive cocci mixed with a variety of other organisms (aerobic and/or anaerobic) can cause lesions that appear alike pathologically in experimental animals.

Isolation

Specimen Collection and Transport

The anaerobic cocci can be isolated from normal flora and from infected sites in man and other animals; certain species may also be found in soils, muds, sea water, marine sediment, and on cereal grains. This discussion of isolation techniques will emphasize the medically related aspects.

A major consideration in specimen collection from man and other animals concerns the avoidance of the normal flora of the upper respiratory, gastrointestinal, and urogenital tracts in obtaining specimens for clinical diagnostic purposes (see "Animal Endogenous Flora", preceding section). A diagnostic interpretation cannot be made based on isolation of anaerobic organisms from these sites since anaerobes, including anaerobic cocci, are normally present. This restriction eliminates culture of expectorated sputum, throat swabs, vaginal secretions, or feces. Exceptions to this guideline are appropriate if an association between organisms present in a normal flora site and production of a disease syndrome is suspected. Acceptable specimens for meaningful clinical data on the presence of anaerobic cocci include any normally sterile body fluid (pleural, sinusitis, joint, peritoneal), surgical specimens from normally sterile sites, blood cultures, abscess contents, and transtracheal and cul-de-sac aspirates. For the study of normal flora in a particular site (e.g., dental plaque, feces, or rumen) special sampling procedures may be required to obtain a composite bacteriological picture (Balows et al., 1974; Bryant, 1959; Dowell and Hawkins, 1974; Finegold, 1977; Holdeman, Cato, and Moore, 1977; Hungate, 1950, 1966; Shapton and Board, 1971; Skinner and Carr, 1974; Smith, 1975; Sutter, Vargo, and Finegold, 1975).

The anaerobic cocci are inactivated by oxygen, and specimens must be protected from oxygen exposure (and drying) from the time of collection, optimally, through processing in the laboratory. Certain anaerobic cocci, e.g., *Peptostreptococcus anaerobius,* are quite susceptible to inactivation by oxygen during transport to the laboratory; others, e.g., *Peptococcus magnus* and the streptococcal species such as *S. intermedius,* may be more tolerant to oxygen (Hill, 1978; Wilkins and Jimenez-Ulate, 1975).

Immediate transport to the laboratory (within approximately 10 min) and plating of the specimen is most effective for recovery of anaerobes. The greater the time interval between collection and culture, the more important that the specimen be protected from oxygen exposure. Transport containers, which provide an anaerobic (oxygen-free) environment, can be used for specimens during the transport interval. Various anaerobic transport systems

are available to accommodate liquid specimens (Port-A-Cul Vial, BBL Microbiology Systems; Anaport, Scott Laboratories), swab specimens (Port-A-Cul Tube, BBL; Anaerobic Culturette, Marion Scientific Corporation) or swabs, liquid, or tissue specimens (Anaerobic Specimen Collector, Becton-Dickinson; Anaswab, Scott Laboratories). A mini anaerobic jar has also been described for tissue specimens (Sutter, Vargo, and Finegold, 1975). Transport of liquid specimens in a syringe that is capped after expelling all air is also suitable.

Transport media have been utilized in the past, and certain reduced transport media (Syed and Loesche, 1972) appear to be efficacious for maintenance of overall numbers of anaerobes from dental plaque. There is concern, however, that transport media may selectively support certain organisms in mixed specimens so that the proportion among the species changes with time. For the same reason, it is not advisable to transport specimens in nutritive media, particularly if there is any delay anticipated in plating the specimens. Many anaerobic transport systems are designed simply to provide an anaerobic atmosphere without the use of transport media.

Certain transport systems have been demonstrated to be effective in maintenance of viability of anaerobes in pure or mixed cultures (Chow, Cunningham, and Guze, 1976; Hill, 1978; Mena et al., 1978). However, since most clinical specimens contain nutritive materials, storage for more than 8–12 h of polymicrobic specimens, particularly those that contain facultatively anaerobic bacteria in addition to anaerobes, may result in overgrowth of certain species and significant population changes (Hill, 1978). It is often suggested that specimens containing anaerobes should be transported or stored at room temperature because chilling or refrigeration may be detrimental to certain species (Hagen, Wood, and Hashimoto, 1976; Holdeman, Cato, and Moore, 1977). However, refrigeration of specimens contained in an anaerobic transport system might prevent overgrowth and population changes in a bacterial mixture and be harmless to most species (Mena et al., 1978). Further evaluation is needed.

Anaerobic Culture

The anaerobic cocci require special incubation methods to provide a low oxygen tension which will allow their growth. Sensitivity to oxygen among the anaerobic cocci varies, however. A strain of *Megasphaera elsdenii* was determined to be a moderate anaerobe (defined as maximum growth of $pO_2 < 3\%$) which grew best in an atmosphere of $\leqslant 1\%$ oxygen (Loesche, 1969). Among *Peptococcus* and *Peptostreptococcus* species, five of eight strains and three of six strains tested, repectively, were found to be strict anaerobes that required an oxygen concentration of < 0.4% for growth (Talley et al., 1975). The remaining species were moderate anaerobes, except for one strain of *Peptostreptococcus* which was aerotolerant (growth in ≥ 5% oxygen). *Ruminococcus*, *Gemmiger*, and *Sarcina* generally are strict anaerobes, requiring a very low oxygen concentration and media that has received minimal exposure to oxygen.

Members of the genus *Streptococcus* may be obligately anaerobic or aerotolerant (sometimes termed microaerophilic). *Streptococcus intermedius*, *S. constellatus*, and *S. morbillorum*, previously included in *Peptococcus* or *Peptostreptococcus*, are often aerotolerant; *S. hansenii* (Holdeman and Moore, 1974) and *S. pleomorphus* (Barnes et al., 1977) are obligately anaerobic. The microaerophilic streptococci often grow initially only under anaerobic conditions, but become more oxygen tolerant with continued subculture in the laboratory and then will grow in a CO_2 incubator or candle jar or, occasionally, aerobically. Certain strains may grow initially in a CO_2 incubator and, less frequently, in air. However, growth of these aerotolerant strains is more luxuriant under anaerobic conditions.

The anaerobic cocci also vary in the length of time they can survive exposure to air (Tally et al., 1975). The survival time also depends on the conditions of exposure, e.g., the presence of blood in media, growth phase of the bacteria, and so on (see also "Specimen Collection and Transport").

There are three major methods for anaerobic incubation: (i) the anaerobic jar, (ii) the roll tube, and (iii) the anaerobic glove box. Extremely oxygen-sensitive anaerobes that may be isolated from normal flora sites can be grown in the latter two environments. The anaerobic jar method is not optimal for extremely oxygen-sensitive and strict anaerobes since some oxygen exposure occurs during manipulation of cultures at the bench; also, when the GasPak (BBL) gas-generator method is used, there is a delay before anaerobic conditions are achieved. It appears, however, that the anaerobic cocci and other anaerobes that are common in human disease are sufficiently tolerant to oxygen that the anaerobic jar method (including the GasPak gas generator) is adequate for use (Dowell, 1972; Rosenblatt, Fallon, and Finegold 1973). Each of the three anaerobic incubation systems has advantages and disadvantages, and the choice of system must be based on the particular research or clinical application and other considerations of cost, convenience, space, and so on. A brief description of these anaerobic systems and references for further technical details are given below.

ANAEROBIC JAR. In this system, glass, metal, or plastic jars (e.g., Torbal, GasPak [BBL]) with a leak-proof closure, which may be vented or unvented, are constructed to hold Petri dishes and tubed media. Each jar is filled with an anaerobic gas mixture by using a GasPak Disposable Hydrogen + Carbon Dioxide Generator Envelope (BBL) or by evacuating the air from the jar with a vacuum source and replacing the volume with an anaerobic gas mixture (10% hydrogen, 5–10% carbon dioxide, and remainder nitrogen) from gas cylinders. The residual oxygen in the jar is reduced to water by a palladium catalyst in the presence of hydrogen. The older Brewer jars required an electrical connection for reduction of oxygen. Jars containing inoculated media are then placed in a standard incubator.

The holding jar technique described by Martin (1971) can be used to prestore plates under anaerobic conditions before inoculation and to prevent prolonged exposure of anaerobic bacteria to air before and after examination and transfer of cultures. The holding jar technique significantly enhances the effectiveness of the anaerobic jar method. Detailed descriptions of anaerobic jar techniques may be found in Dowell and Hawkins (1974) and Sutter, Vargo, and Finegold (1975).

ROLL TUBE. This system was developed by Hungate (1950) for use with rumen bacteria. The technique has been modified by Moore, Holdeman, and associates (Holdeman, Cato, and Moore, 1977) for application to clinical specimens as well as normal flora studies. This method includes preparation, sterilization, and storage of media under oxygen-free gas to maintain a low oxidation-reduction potential and to prevent any oxidation of media ingredients. Broth and solid media are placed in tubes or vials with oxygen-impermeable stoppers. Whenever the stopper is removed during inoculation or transfer of cultures, a gentle stream of oxygen-free gas is directed into the tube to exclude air. Alternatively, careful syringe injections can be made through specially designed stoppers without introducing air (Bryant, 1972; Hungate, 1950; Macy, Snellen, and Hungate, 1972). The media can be placed in a standard incubator and examined at any time since each tube holds its own anaerobic atmosphere.

ANAEROBIC GLOVE BOX. This system consists of a closed chamber made of flexible or rigid plastic, provided with gloves to allow manipulation within the chamber (Aranki and Freter, 1972; Jones, Whaley, and Dever, 1977; Sutter, Vargo, and Finegold, 1975). The chamber is filled with an oxygen-free gas mixture (10% hydrogen, 5–10% CO_2, and remainder nitrogen), and a palladium catalyst is used to reduce any oxygen to water. The air within an entry lock is evacuated and replaced with oxygen-free gas when material is passed in and out of the chamber. Depending on its size, the chamber may be used only for inoculation and transfer of cultures and with the inoculated media incubated in anaerobic jars in an external incubator, or the chamber itself may be heated to 37°C to additionally serve as an incubator, or the chamber may be sufficiently large to contain separate incubators. Different models of anaerobic glove boxes are commercially available. This system provides for a continuous oxygen-free environment during processing of specimens and culture of anaerobic bacteria.

Primary Isolation

GENERAL CONSIDERATIONS. Since the anaerobic cocci are most often found in polymicrobic mixtures, primary inoculation of specimens onto solid media is essential to obtain information on the frequency of isolation, the concentration, and the relative proportions of these organisms in material from infected and normal flora sites in man and animals or from other environmental sources. Primary inoculation only into broth often results in selective enrichment of the least fastidious, most rapidly growing organisms in the bacterial mixture, resulting in loss of information on the original composition of the specimen. Studies on specific organisms, however, may require enrichment cultures or selective media to facilitate isolation and increase the rate of recovery from mixed specimens. Often, more exacting control of low oxygen tensions and reduced and chemically enriched media are required for primary isolation, whereas organisms may be more aerotolerant and less fastidious after subsequent subcultures in the laboratory.

Another general consideration for isolation of anaerobic cocci is the different growth rates of various species. Whereas *Peptostreptococcus anaerobius* and *Streptococcus intermedius,* for example, grow quite rapidly in broth, other cocci such as *Peptostreptococcus micros* grow more slowly and never achieve high turbidity. Likewise, visible colonies of more rapidly growing species may appear on solid media within 24 h, while the more slowly growing organisms may be obvious only after 3–4 days. It is important to reexamine anaerobic media to obtain the more slowly growing organisms.

A dissecting microscope is extremely helpful for discerning differences in colonial morphology and for isolating the numerous species of anaerobic cocci that may be obtained from a single source. The colonial morphology among different species of *Peptostreptococcus, Streptococcus,* and *Peptococcus* may demonstrate only subtle differences and is not usually very distinctive. Colonies often appear white to grey, round, convex, opaque (periphery sometimes may appear translucent), shiny, with an entire edge. Most strains are nonhemolytic.

Identification procedures must be performed on pure cultures. Certain anaerobic cocci tend to grow in mixtures, and it may be difficult to isolate them or to obtain adequate growth in pure culture. To determine that a culture is pure, cellular morphology of organisms grown in broth and solid media must be carefully observed, and colony morphology must be observed on purity plates, preferably with a dissecting microscope. When colony variation occurs, evidence that the culture is pure can be obtained by reisolating each colonial type in broth and subsequently streaking each isolate on plates to demonstrate that the colonial variants reappear and all have the same identification pattern. A definitive test for aerotolerance is necessary to be sure the isolate is truly anaerobic and not facultatively anaerobic. Lack of growth on blood agar media, incubated under aerobic conditions or in a CO_2 incubator (or candle jar), and growth under anaerobic conditions indicate that the organism is truly anaerobic. Exceptions occur with certain streptococcal species, which may initially be obligately anaerobic but which become microaerophilic or aerotolerant after subsequent subcultures. Some may even grow initially in a CO_2-enriched environment or, infrequently, in air.

GROWTH REQUIREMENTS. The anaerobic cocci are chemoorganotrophic and their nutritional requirements are complex (Bryant, 1959; Gossling and Moore, 1975; Holdeman, Cato, and Moore, 1977; Holdeman and Moore, 1974; Hungate, 1966; Rogosa, 1974a,b; Smith, 1975). Growth factors may include certain vitamins and other cofactors, amino acids, heme, and reducing compounds such as cysteine. Carbon dioxide is often stimulatory or required. Growth of many anaerobic cocci is enhanced by the presence of Tween 80 (final concentration in medium, 0.02%) or oleic acid, which may enhance carbohydrate fermentation or proteolytic action. A fermentable carbohydrate may also enhance the growth of strains of *Peptostreptococcus, Streptococcus, Acidaminococcus,* and *Peptococcus,* although these organisms do not require fermentable carbohydrates for growth and can use amino acids and peptones as the main energy source. Organisms included in the genera *Ruminococcus, Sarcina, Gemmiger,* and *Coprococcus* (carbohydrate either required or highly stimulatory) require a fermentable carbohydrate for growth and cannot use peptones or amino acids as the sole source of energy. A mixture of carbohydrates, including starch and cellobiose, is normally used rather than just glucose, since glucose may not be fermented by all strains. For example, ruminococci may ferment cellobiose and not glucose; they often can digest cellulose. Addition of rumen fluid or fecal extracts may be necessary for growth of cocci obtained from the rumen or intestine (e.g., *Ruminococcus, Gemmiger*). The growth requirements of strains of *Ruminococcus* from the intestine or rumen have been identified in chemically defined media (Herbeck and Bryant, 1974; Slyter and Weaver, 1977). These studies also point up the nutritional interdependence of anaerobic organisms in these ecosystems. Organisms in the genus *Veillonella* do not ferment carbohydrates but can metabolize lactate, pyruvate, oxaloacetate, malate, and fumarate. Media that contains peptones and amino acids with the necessary vitamins and cofactors do not support growth without the presence of pyruvate or other substrates mentioned above. *Megasphaera elsdenii* can utilize lactate, glucose, maltose, and fructose for growth, whereas growth on other substrates is variable or negative; growth is enhanced in the presence of starch and CO_2. The optimal pH initially in media for growth of most anaerobic cocci is near pH 7 (range 6–8), but *Sarcina* is capable of growth over a very broad pH range (1–9.8). Most anaerobic cocci grow best at a temperature of 35–37°C, but many can grow over a wider range (25–40°C). *Gemmiger* and some *Megasphaera* and *Ruminococcus* strains can grow at temperatures up to 45°C.

NONSELECTIVE MEDIA. Highly supplemented solid media are used for nonselective isolation of anaerobic cocci from clinical sources or from normal flora. Enriched media such as brucella agar, Columbia agar, or brain heart infusion agar are often used with whole defibrinated sheep or rabbit blood (5%) and other supplements added. Media should contain yeast extract, which provides necessary vitamins and cofactors, and hemin (5 μg/ml final concentration) in addition to other supplements mentioned earlier that may be required by specific genera. Supplemented brain heart infusion agar is normally used for the roll-tube method without addition of blood, although 4% laked blood can be added. Rumen fluid–glucose–cellobiose agar is used for primary isolation of intestinal or rumen bacteria; Medium 10 may be used as a synthetic substitute for rumen fluid (Holdeman, Cato, and Moore, 1977). The specific formulations for these and other media as used for isolation from clinical or normal flora sources and additional references are described by Sutter, Vargo, and Finegold (1975), Dowell et al. (1977), Holdeman, Cato, and Moore (1977), Shapton and Board (1971), Bryant (1959), and Hungate (1966).

When clinical specimens are cultured, broth and solid media are both inoculated for primary isolation to grow organisms present in very low numbers in pure culture or to recognize species which otherwise might be missed on solid media alone because they are present in lower concentrations or grow more slowly than other bacteria in mixed populations. Supplemented chopped meat medium, containing glucose or a mixture of carbohydrates (Holdeman, Cato, and Moore, 1977; Dowell et al., 1977), and supplemented thioglycolate medium (Dowell et al., 1977; Sutter, Vargo, and Finegold, 1975) are suitable for this purpose. A number of blood-culture broth media bottled under CO_2 gas appear to be satisfactory for growth of anaerobic cocci, e.g., supplemented, prereduced brain heart infusion–yeast extract broth (Scott), supplemented peptone broth (Becton-Dickinson & Co.), thioglycolate medium 135C (BBL), Thiol broth (Difco), and Columbia broth. Certain of these blood culture media contain sodium polyanethol sulfonate (SPS, Liquoid), which is useful to enhance isolation of many anaerobes and other organisms from blood by inhibiting antibacterial substances. SPS may also inhibit the growth of *Peptostreptococcus anaerobius* (Graves, Morello, and Kocka 1974). Gelatin present in supplemented peptone broth protects *P. anaerobius* from the inhibiting effects of SPS (Wilkins and West, 1976). Sodium amylosulfate or sodium anethol sulfonate may be available and do not appear to inhibit anaerobic cocci (Holdeman, Cato, and Moore, 1977; Sutter. Vargo, and Finegold, 1975).

SELECTIVE MEDIA. Selective media may aid in isolation of anaerobic cocci from heavily mixed populations, particularly when these cocci are present in lower concentrations or grow more slowly than the other organisms in the mixture. Neomycin (final concentration of 100 μg/ml in media) or another aminoglycoside antibiotic, or phenylethyl alcohol may be added to agar medium to obtain the anaerobic cocci. These additives inhibit the facultative, Gram-negative bacilli that often occur with anaerobic cocci but allow growth of a variety of other facultative and anaerobic bacteria. The addition of vancomycin (final concentration 7.5 μg/ml) to media containing neomycin is selective for *Veillonella* (and anaerobic, Gram-negative bacilli) since vancomycin inhibits Gram-positive species (Finegold, Sugihara, and Sutter 1971; Sutter, Vargo, and Finegold 1975). More specific selective media for the anaerobic cocci would be desirable.

Maintenance of Pure Cultures

It is preferable to freeze or lyophilize pure cultures of anaerobic cocci for long-term storage rather than to perform subcultures at intervals. A variety of methods appear to work satisfactorily for either type of storage if a rather heavy suspension of young, actively growing bacteria is used. Organisms can be used from either broth or solid media although heavier suspensions are normally obtained by removing organisms from the surface of plates or slants. Unless the isolate requires a fermentable carbohydrate for growth, it is preferable to use a carbohydrate-free broth medium, such as supplemented chopped meat without glucose. To freeze cultures grown on solid media, growth is harvested from a supplemented blood-agar slant or plate and suspended in sterile, defibrinated rabbit blood

(Dowell and Hawkins, 1974), skim milk (Sutter, Vargo, and Finegold 1975), or inactivated horse serum. Other sources of sera would probably be satisfactory.

Storage of Pure Cultures of Anaerobic Cocci by Freezing

In our laboratory, the organisms are harvested from the surface of supplemented blood agar plates using a dry swab, which is then mixed in 1 ml of inactivated horse serum on a Vortex mixer to remove the bacteria from the swab. Approximately 0.3 ml of this heavy suspension is then placed into a small tube (6 × 50 mm), and the tube is sealed by inserting a cotton swab into the tube, cutting the excess portion of the swab handle off, and pouring molten Vaspar (combination of petrolatum and paraffin) into the top of the tube to provide a tight closure when the Vaspar solidifies. These steps are performed within an anaerobic glove box, which provides an anaerobic atmosphere in the storage tube. Tubes are then frozen in an ultra–deep freeze at −70°C. These cultures normally remain viable for prolonged periods (at least several years).

Another freezing method is to quick-freeze cultures in a bath of 95% alcohol and dry ice for storage at −20° to −42°C. These cultures normally remain viable for 3–6 months, and many organisms may survive as long as 1 year. Tubes containing an anaerobic gas mixture (prepared in an anaerobic glove box or gassed out) and fitted with oxygen-impermeable butyl rubber stoppers (black 000 stoppers fit 10- or 12-×-75-mm tubes) maintain viability of frozen cultures during storage better than tubes prepared in air and sealed with cotton stoppers.

Lyophilization is probably most effective for long-term storage of the majority of anaerobic species. Lyophilization of cultures grown in plain chopped meat medium without carbohydrate has been successful (L. Holdeman, personal communication). Alternatively, organisms harvested from slants or plates can be suspended in inactivated horse serum, a combination of horse serum and skim milk, or in skim milk only. A more complete description of lyophilization procedures may be found in Dowell and Hawkins (1974). Stock cultures prepared for storage by either freezing or lyophilization should be examined for purity before and after performing the procedure.

Identification

Taxonomy

The first classification scheme for the anaerobic cocci was offered by Prévot in 1925. Most of the early attempts at classification were heavily dependent on cellular or colonial morphology in deep agar cultures or on plates. Certain investigators also used fermentation and other biochemical reactions to increase the sensitivity of schemes of identification. Immunological techniques were used less frequently. In 1952, a more extensive classification of the anaerobic cocci was published (Hare et al., 1952), which deemphasized reliance on morphology and expanded the use of biochemical reactions; this work was later extended (Thomas and Hare, 1954). At present, a number of classification schemes exist, including those of Prévot, translated by Fredette (1966), Rogosa (1974a,b) in *Bergey's Manual of Determinative Bacteriology,* eighth edition, Dowell and Hawkins (1974), Holdeman, Cato, and Moore, (1977), and Sutter, Vargo, and Finegold (1975).

Taxonomy of the anaerobic cocci has been changed continuously in recent years. In 1971, the family Peptococcaceae was suggested to contain the Gram-positive genera *Peptococcus, Peptostreptococcus, Ruminococcus,* and *Sarcina* (Rogosa, 1971a,c). During the same year, the Gram-negative cocci, including *Veillonella, Acidaminococcus,* and *Megasphaera,* were transferred to the family Veillonellaceae (Rogosa, 1971b,d). More recently, two new genera of anaerobic cocci (*Gemmiger* and *Coprococcus*) have been described, and species previously classified in *Peptococcus* and *Peptostreptococcus* were changed to *Streptococcus* with emended species descriptions (Gossling and Moore, 1975; Holdeman and Moore, 1974).

Further change is required to arrive at agreement on the genus and species of strains presently available, and newly isolated types will demand continual revision in the future. Among the various existing taxonomic schemes, organisms may be referred to a single species in one instance and split into two or more species in another classification; organisms with similar reactions may be given different names. For instance, some species described in *Bergey's Manual* by Rogosa (1974a,b) are not included by Holdeman, Cato, and Moore (1977) or are listed under a different name and vice versa. Certain organisms are recognizable in all of the classification schemes, but the nomenclature may still be different (Smith, 1975; Wells and Field, 1976). Cocci are often isolated, particularly from the normal flora, that do not fit any published scheme of identification. More complete data from base composition studies (mol% G+C), DNA homology studies, analysis of cellular material (e.g., cell wall or fatty acid composition), and serological relationships are needed in addition to the more commonly used techniques of biochemical reactions and analysis of products to delineate taxonomic relationships and to define more clearly both genera and species. Until the taxonomic relationships are more clearly defined and suitable identification techniques are developed, correlation of particular species or strains of anaerobic cocci with disease states or other ecological impacts will be slow. The nomenclature and scheme of classification utilized in this chapter is that of Holdeman, Cato, and Moore (1977), whose manual primarily emphasizes the cocci that may occur in clinical specimens.

Identification to Genus

Assignment of anaerobic cocci to specific genera can normally be accomplished by observing the Gram reaction, determining whether a fermentable carbohydrate is required for growth, and by analyzing metabolic end products—volatile and nonvolatile acids and alcohols (Holdeman, Cato, and Moore, 1977). Characteristics that define the genera of anaerobic cocci are presented in Table 2. *Gemmiger* often stains Gram negative, but it is variable in Gram stain and may appear Gram positive. It can be differentiated from other genera by its requirement for a fermentable carbohydrate, by the

production of butyric acid, and by the morphological appearance of budding (Gossling and Moore, 1975). Members of the genera *Veillonella* and *Megasphaera* may also demonstrate a variable Gram stain and at times appear to be Gram positive. Gram-positive cocci may lose Gram positivity with age as is commonly exemplified by the facultative streptococci. Interpretation of the Gram stain for identification of anaerobic cocci may also be a problem with certain cocci (e.g., *Peptostreptococcus productus, P. anaerobius*), which appear sufficiently elongated to resemble coccobacillary, Gram-positive bacilli. Gram stains from a variety of media, broth and solid, and examination of cultures at various ages will often be sufficient to decide on the correct cellular morphology and Gram-stain reaction, but additional biochemical tests or analysis of end products may occasionally be required. Susceptibility or resistance to certain antibiotics may also be helpful in differentiating Gram-positive and Gram-negative organisms (Sutter, Vargo, and Finegold, 1975). None of the anaerobic cocci produces spores (*Sarcina* is a possible exception; Rogosa, 1974b) or flagella, and the organisms are nonmotile.

Table 2. Differential characteristics of the genera of anaerobic cocci.[a]

Genus	Cell wall type/ Gram-stain reaction	Cellular arrangement, morphology	Carbohydrates fermented	Lactate fermented	Peptones or amino acids, primary source energy and N	Produce volatile acids with >3 carbons	G+C (mol%)
Peptococcus	g+/g+	Diplococci, clumps, clusters, short chains, tetrads	−(+)	−(+)	+	v	36–37
Peptostreptococcus	g+/g+	Diplococci, short to long chains	+(−)	−	+	v	34
Streptococcus[b]	g+/g+	Diplococci, short to long chains	+	−	−	−	37–39
"*Gaffkya*"	g+/g+	Diplococci, clumps clusters, tetrads	+	−(+)	−	+	
Ruminococcus[c]	g+/g+	Diplococci, short to long chains	+ r	−	−	−	40–45
Sarcina	g+/g+	Cubic packets large cells	+ r	−	−	v	29–31
Coprococcus	g+/g+	Diplococci, short to long chains	+ sr	−(+)	−	+	39–42
Veillonella	g−/g−(g+)	Diplococci, clumps, short chains	−	+	−	−	40–44
Acidaminococcus	g−/g−	Diplococci, clumps	−	−	+	+	57
Megasphaera	g−/g−(g+)	Diplococci, clumps, chains, large cells	+	+	−	+	53–54
Gemmiger	g−/g−(g+)	Diplococci, chains, appearance of budding	+ r(s)	−	−	+	59

[a]References: Gossling and Moore, 1975; Holdeman and Moore, 1974; Holdeman et al., 1977; Rogosa, 1974a,b; Smith, 1975. Symbols: +, positive; −, negative; v, variable; g, Gram; s, stimulatory; r, required; sr, stimulatory or required. Where two symbols noted, e.g., g−(g+), the first symbol denotes the usual reaction and the second symbol denotes the less frequent reaction.
[b]*Streptococcus* strains may be obligately anaerobic, microaerophilic, or aerotolerant. Produce lactic acid as sole major acid product. Production of traces of butyric acid reported for *Streptococcus pleomorphus* (Barnes et al., 1977).
[c]*Ruminococcus* strains usually digest cellulose.

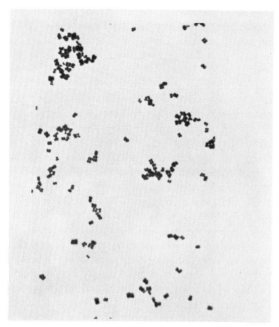

Fig. 1 *Peptococcus magnus* stained from supplemented blood agar. × 1,000.

Cellular arrangement has been suggested as a method to separate *Peptococcus* strains that occur singly, in pairs, and in clumps or clusters of cells (Fig. 1) from *Peptostreptococcus* strains and, particularly, *Streptococcus* strains that occur singly, in pairs, and in long chains (Fig. 2). Although cellular arrangement may be helpful, both *Peptococcus* and *Peptostreptococcus* strains may occur in short chains. Cellular arrangement may be helpful in rec-

ognition of *Gemmiger*, as previously mentioned, *Sarcina ventriculi* (cubic packets), *Megasphaera elsdenii* (large cells), *Peptostreptococcus micros* (tiny cells), and "*Gaffkya*" *anaerobia* (often in tetrads) although some *Peptococcus* strains may also appear in tetrads similar to "*Gaffkya*" (Fig. 1).

The use of catalase to separate *Peptococcus* from *Peptostreptococcus*, as can be demonstrated for *Staphylococcus* (catalase-positive) and facultative *Streptococcus* sp. (catalase-negative), has been suggested (Dowell and Hawkins, 1974). But Holdeman, Cato, and Moore (1977) and Rogosa (1974b) believe the catalase reaction is too variable to make this distinction. Preliminary results indicate that sensitivity to novobiocin of *Peptostreptococcus* species may be useful for differentiation from *Peptococcus* species, which are resistant (Wren, Eldon, and Dakin, 1977). *Veillonella* (Fig. 3) and *Acidaminococcus* species possess lipopolysaccharides with characteristic endotoxic activity in their cell walls similar to other Gram-negative bacteria (Hewett, Knox, and Bishop, 1971; Hofstad and Kristoffersen, 1970; Mergenhagen, Hampp, and Scherp, 1961). Ultraviolet light can be used to differentiate *Veillonella*, which displays a red fluorescence, from *Acidaminococcus* and *Megasphaera*, which lack fluorescence (Chow, Patton, and Guze, 1975). To decide whether an isolate requires a fermentable carbohydrate, and thus belongs to the genera *Ruminococcus*, *Coprococcus* (may only be stimulatory), *Gemmiger*, or *Sarcina*, the pure culture must be transferred several times in media without carbohydrate and the resultant growth then compared with growth of the organism in media containing a mixture of carbohydrates.

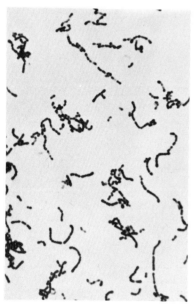

Fig. 2. *Peptostreptococcus anaerobius* stained from supplemented blood agar. × 1,000.

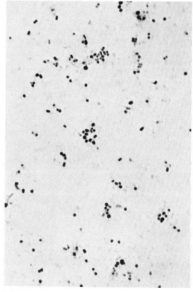

Fig. 3. *Veillonella parvula* stained from supplemented blood agar. × 1,000.

Identification to Species

Species identification among the anaerobic cocci is presently accomplished using carbohydrate fermentations and other tests, such as indole, nitrate reduction, liquefaction of gelatin, catalase reaction, and utilization of lactate or pyruvate. The addition of Tween 80 may be required to enhance growth of certain of the cocci and to obtain characteristic biochemical reactions (Holdeman, Cato, and Moore, 1977). Many of the clinically important species of cocci in the genera *Peptococcus* and *Peptostreptococcus*, however, are asaccharolytic and practical identification in the laboratory may often depend on a single positive reaction. Analysis of metabolic products by gas-liquid chromatography may aid in the differentiation of species. The chromatographic pattern of *Peptostreptococcus anaerobius*, for example, is sufficiently unique to enable its presumptive identification taken with the microscopic appearance. Some of the cocci produce gas in culture, usually consisting of CO_2 and H_2 and less frequently H_2S, but this characteristic may be variable. Recent data indicate that differentiation among the cocci is improved by use of deoxyribonuclease, tetrathionate reductase, Voges-Proskauer, and arginine dihydrolase tests (Vargo et al., 1977). Sodium polyanethol sulfonate disks are inhibitory for *P. anaerobius* and can be used as an aid in identification (Wideman et al., 1976). As previously mentioned, colonial morphology is not particularly helpful in differentiation among the clinically important species of Gram-positive cocci, although cellular arrangement may sometimes be helpful.

In general, the different taxonomic schemes also include descriptions of specific methods and media that are used for determination of sugar fermentations and biochemical tests. The most accurate identification of strains with a particular taxonomic scheme is thus obtained by using the media and procedures on which the scheme is based, although testing the same strains by other identification systems generally yields 90% or greater agreement among the biochemical reactions taken as a group. Commercial miniaturized systems for anaerobic identification appear promising (Stargel, Lombard, and Dowell, 1978). These miniaturized tests are more easily stored than conventional differential media prepared in larger volumes, and quality control is simplified (e.g., API 20A, Analytab, and Minitek, BBL). New identification systems, miniaturized or other, should preferably be standardized by determining the actual biochemical reactions of known anaerobic species (from a particular taxonomic scheme) using the new system.

Additional information on the status of certain species in the genera *Peptococcus* and *Peptostreptococcus* has been provided by analysis of their long-chain fatty acids (Wells and Field, 1976) and of the primary structure of the peptidoglycan types in the cell walls of *Peptococcus* species (Schleifer and Nimmermann, 1973). Data from these studies suggest changes in the taxonomic status of some species and support for the taxonomic treatment of other species as variously described in *Bergey's Manual* (Rogosa, 1974b) and by Holdeman, Cato, and Moore, (1977). One area of agreement in the analyses of cell-wall material and long-chain fatty acids was support for removal of *Peptococcus saccharolyticus* from its present position in *Peptococcus* (Holdeman, Cato, and Moore, 1977) as suggested by Rogosa (1974b) who did not, however, indicate an alternate taxonomic location. Further studies of this nature would be valuable. Serological studies suggest that species-specific antisera can be produced for species such as *Peptostreptococcus anaerobius* and *Peptococcus magnus* among others (Graham and Falkler, 1978; Porschen and Spaulding, 1974; Wong, Catena, and Hadley, 1978). These serological studies could offer simplified methods of identification in the clinical laboratory, and may also aid in clarifying taxonomy.

Antibiotic Susceptibility

LABORATORY SUSCEPTIBILITY TESTING. Recognition of the importance of anaerobic bacteria in infections has fostered the development of methods to test the antibiotic susceptibility of these organisms. Standardized methods for susceptibility testing of aerobic and facultative bacteria cannot be transposed for use with anaerobes because the results may be altered by differences in test conditions, such as the supplemented media, the anaerobic gaseous environment, and the generally slower growth rates of anaerobic bacteria. Since the anaerobic cocci are generally susceptible to clinically achievable levels of the penicillins, cephalosporins, chloramphenicol, clindamycin, and metronidazole (except for streptococcal species which are more resistant to this drug), clinical laboratories and physicians have for the most part relied on published susceptibility patterns provided by large, clinical or research, anaerobic bacteriology laboratories, and routine susceptibility testing has not been performed. Standardized methods for susceptibility testing of anaerobes are being developed, and it is anticipated that techniques can be further simplified.

A new reference technique for determination of the minimum inhibitory concentration (MIC) by an agar dilution test has been developed and evaluated under the auspices of the National Committee for Clinical Laboratory Standards (USA). The complete procedure has not been published, but preparation of the medium has been described (Wilkins and

Chalgren, 1976; Zabransky and Hauser, 1977). The medium is designed to support the growth of a wide variety of anaerobic bacteria, including the anaerobic cocci. Approximately 90% of clinical isolates will grow on this medium, but certain isolates, particularly from normal flora, may not. The anaerobic cocci that should be used for quality control to test for growth on the Wilkins–Chalgren media are deposited in the American Type Culture Collection as *Peptococcus asaccharolyticus* ATCC 29743, *P. magnus* ATCC 29328, and *P. variabilis* ATCC 14956 (this organism is considered to be *P. magnus* by Holdeman, Cato, and Moore, 1977). The agar dilution method is not intended for routine use in clinical laboratories but for a reference for development of more simplified tests or for periodic monitoring of antibiotic susceptibility patterns of anaerobic bacteria. Microdilution techniques in broth appear promising. The modified broth-disk method (Wilkins and Thiel, 1973) and a broth dilution method (Stalons and Thornsberry, 1975) are presently accepted methods for testing antibiotic susceptibility in clinical laboratories.

SUSCEPTIBILITY DATA AND CLINICAL USE. The anaerobic cocci are, in general, susceptible to the antibiotics used to treat anaerobic infections (Finegold et al., 1975). In a recent evaluation of in vitro susceptibilities of anaerobic bacteria to antibiotics (Sutter and Finegold, 1976), strains of *Peptococcus, Peptostreptococcus, Streptococcus* (microaerophilic and anaerobic), *Veillonella,* and *Acidaminococcus* were generally susceptible to achievable blood levels of the penicillins. Infrequently, high dose levels of penicillin G (16–32 μg/ml) were required for strains of *Peptostreptococcus* and *Streptococcus*. Occasionally, strains of *Peptostreptococcus* also required high levels of ampicillin, and rare Gram-negative cocci were resistant at 16 μg/ml. Carbenicillin was also quite active against the cocci, although rarely strains of *Peptostreptococcus, Streptococcus,* and Gram-negative cocci required 128 μg/ml. Cephalothin was effective against all strains tested at clinically achievable levels, as was the newer cephamycin, cefoxitin. Chloramphenicol was also active against all strains of cocci tested at 8 μg/ml or less, a level readily achievable clinically. *Peptostreptococcus* strains and the Gram-negative cocci were quite sensitive to clindamycin, but 9 of 58 strains of *Peptococcus* were resistant and required 64 μg/ml and higher for inhibition. All strains of *Streptococcus* were inhibited at clinically achievable levels (8 μg/ml) of clindamycin. The anaerobic cocci, like other anaerobic bacteria, demonstrate increasing resistance to tetracycline, and most of the anaerobic cocci tested were only moderately sensitive. Minocycline and doxycycline were more active than the parent compound. Metronidazole, which is active only against anaerobic bacteria, is particularly effective against the Gram-negative, anaerobic bacilli. The Gram-negative cocci tested were uniformly susceptible as were strains of *Peptococcus* and *Peptostreptococcus,* with rare exception. Streptococci, however, were often resistant, and 50% required greater than 128 μg/ml for inhibition. These streptococci included some strains that became aerotolerant after laboratory cultivation but were nonetheless grouped with the obligate anaerobes as is customary.

Although the susceptibility patterns among the anaerobic cocci are generally predictable, organisms that are resistant to the antibiotics used for treatment of anaerobic infections have been isolated. Thus, antibiotic susceptibility should preferably be verified in serious infections, particularly if anaerobic cocci are present in pure culture or in instances of endocarditis or osteomyelitis, for example, where MICs may also be required. Increased use of metronidazole may create a need for an increase in routine antibiotic screening for resistance among the anaerobic streptococci. However, in a mixed anaerobic infection a clinical response sometimes may be achieved without using a drug specifically effective against a *Peptococcus,* a *Peptostreptococcus* or, particularly, a *Veillonella* isolate. Often, drugs effective against only some of the bacteria in a mixture will clear an infection, probably by disrupting the synergistic interrelationships among the bacterial members.

Literature Cited

Altemeier, W. A. 1938. The bacterial flora of acute perforated appendicitis with peritonitis. Annals of Surgery **107:** 517–528.

Altemeier, W. A. 1942. The pathogenicity of the bacteria of appendicitis peritonitis. An experimental study. Surgery **11:**374–384.

Altemeier, W. A. 1974. Liver abscess: The etiologic role of anaerobic bacteria, pp. 387–398. In: Balows, A., DeHaan, R. M., Guze, L. B., Dowell, V. R., Jr. (eds.), Anaerobic bacteria—role in disease. Springfield, Illinois: Charles C Thomas.

Altemeier, W. A., Culbertson, W. R. 1948. Acute non-clostridial crepitant cellulitis. Surgery, Gynecology and Obstetrics **87:**206–212.

Altemeier, W. A., Culbertson, W. R., Fullen, W. D., Shook, C. D. 1973. Intra-abdominal abscesses. American Journal of Surgery **125:**70–79.

Anderson, C. B., Marr, J. J., Jaffe, B. M. 1972. Anaerobic streptococcal infections simulating gas gangrene. Archives of Surgery **104:**186–189.

Aranki, A., Freter, R. 1972. Use of anaerobic glove boxes for the cultivation of strictly anaerobic bacteria. In: Luckey, T. D., Floch, M. H. (eds.), Intestinal microecology. Bethesda, Maryland: The American Society for Clinical Nutrition. Reprinted from the American Journal of Clinical Nutrition **25:**1329–1334.

Attebery, H. R., Sutter, V. L., Finegold, S. M. 1972. Effect of a partially chemically defined diet on normal human fecal flora. In: Luckey, T. D., Floch, M. H. (eds.), Intestinal microecology. Bethesda, Maryland: The American Society for Clinical Nutrition. Reprinted from The American Journal of Clinical Nutrition 25:1391–1398.

Attebery, H. R., Sutter, V. L., Finegold, S. M. 1974. Normal human intestinal flora, pp. 81–97. In: Balows, A., DeHaan, R. M., Dowell, V. R., Jr., Guze, L. B. (eds.), Anaerobic bacteria—role in disease. Springfield, Illinois: Charles C Thomas.

Balows, A., DeHaan, R. M., Guze, L. B., Dowell, V. R., Jr. (eds.). 1974. Anaerobic bacteria—role in disease. Springfield, Illinois: Charles C Thomas.

Barnes, E. M., Impey, C. S., Stevens, B. J. H., Peel, J. L. 1977. *Streptococcus pleomorphus* sp. nov.: An anaerobic streptococcus isolated mainly from the caeca of birds. Journal of General Microbiology 102:45–53.

Bartlett, J. G., Finegold, S. M. 1972. Anaerobic pleuropulmonary infections. Medicine 51:413–450.

Bartlett, J. G., Finegold, S. M. 1974. Anaerobic infections of the lung and pleural space. American Review of Respiratory Disease 110:56–77.

Bartlett, J. G., Gorbach, S. L. 1976. Anaerobic infections of the head and neck. Otolaryngologic Clinics of North America 9:655–678.

Bartlett, J. G., Gorbach, S. L., Finegold, S. M. 1974. The bacteriology of aspiration pneumonia. American Journal of Medicine 56:202–207.

Bartlett, J. G., Onderdonk, A. B., Drude, E., Goldstein, C., Anderka, M., Alpert, S., McCormack, W. M. 1977. Quantitative bacteriology of the vaginal flora. Journal of Infectious Diseases 136:271–277.

Bartlett, J. G., Moon, N. E., Goldstein, P. R., Goren, B., Onderdonk, A. B., Polk, B. F. 1978. Cervical and vaginal bacterial flora: Ecologic niches in the female lower genital tract. American Journal of Obstetrics and Gynecology 130:658–661.

Bornstein, D. L., Weinberg, A. N., Swartz, M. N., Kunz, L. J. 1964. Anaerobic infections—review of current experience. Medicine 43:207–232.

Brown, T. K. 1940. Puerperal infection. American Journal of Surgery 48:164–168.

Bryant, M. P. 1959. Bacterial species of the rumen. Bacteriological Reviews 23:125–153.

Bryant, M. P. 1972. Commentary on the Hungate technique for cultivation of anaerobic bacteria. In: Luckey, T. D., Floch, M. H. (eds.), Intestinal microecology. Bethesda, Maryland: The American Society for Clinical Nutrition. Reprinted from The American Journal of Clinical Nutrition 25:1324–1328.

Chambers, C. H., Bond, G. F., Morris, J. H. 1974. Synergistic necrotizing myositis complicating vascular injury. Journal of Trauma 14:980–984.

Chow, A. W., Cummingham, P. J., Guze, L. B. 1976. Survival of anaerobic and aerobic bacteria in a nonsupportive gassed transport system. Journal of Clinical Microbiology 3:128–132.

Chow, A. W., Guze, L. B. 1974. *Bacteroidaceae* bacteremia: Clinical experience with 112 patients. Medicine 53:93–126.

Chow, A. W., Marshall, J. R., Guze, L. B. 1975. Anaerobic infections of the female genital tract: Prospects and perspectives. Obstetrical and Gynecological Survey 30:477–494.

Chow, A. W., Marshall, J. R., Guze, L. B. 1977. A double-blind comparison of clindamycin with penicillin plus chloramphenicol in treatment of septic abortion. The Journal of Infectious Diseases 135, Supplement: S35–S39.

Chow, A. W., Patten, V., Guze, L. B. 1975. Rapid screening of *Veillonella* by ultraviolet fluorescence. Journal of Clinical Microbiology 2:546–548.

Chow, A. W., Malkasian, K. L., Marshall, J. R., Guze, L. B. 1975. The bacteriology of acute pelvic inflammatory disease. American Journal of Obstetrics and Gynecology 122:876–879.

Colebrook, L. 1930. Infection by anaerobic streptococci in puerperal fever. British Medical Journal 2:134–137.

Crowther, J. S. 1971. *Sarcina ventriculi* in human faeces. Journal of Medical Microbiology 4:343–350.

Dehority, B. A. 1977. Cellulolytic cocci isolated from the cecum of guinea pigs *(Cavia porcellus)*. Applied and Environmental Microbiology 33:1278–1283.

Dowell, V. R., Jr. 1972. Comparison of techniques for isolation and identificaion of anaerobic bacteria. American Journal of Clinical Nutrition 25:1335–1343.

Dowell, V. R., Jr., Hawkins, T. M. 1974. Laboratory methods in anaerobic bacteriology—CDC laboratory manual. Atlanta: U.S. Department of Health, Education, and Welfare.

Dowell, V. R., Jr., Lombard, G. L., Thompson, F. S., Armfield, A. Y. 1977. Media for isolation, characterization, and identification of obligately anaerobic bacteria. Atlanta: U.S. Department of Health, Education, and Welfare.

Eschenbach, D. A., Buchanan, T. M., Pollock, H. M., Forsyth, P. S., Alexander, E. R., Lin, J., Wang, S., Wentworth, B. B., McCormack, W. M., Holmes, K. K. 1975. Polymicrobial etiology of acute pelvic inflammatory disease. New England Journal of Medicine 293:166–171.

Felner, J. M., Dowell, V. R., Jr. 1970. Anaerobic bacterial endocarditis. New England Journal of Medicine 283:1188–1192.

Finegold, S. M. 1977. Anaerobic bacteria in human disease. New York: Academic Press.

Finegold, S. M., Miller, A. B., Sutter, V. L. 1968. Anaerobic cocci in human infection. Bacteriological Proceedings 1968:94.

Finegold, S. M., Sugihara, P. T., Sutter, V. L. 1971. Use of selective media for isolation of anaerobes from humans, pp. 99–108. In: Shapton, D. A., Board, R. G. (eds.), Isolation of anaerobes. New York: Academic Press.

Finegold, S. M., Bartlett, J. G., Chow, A. W., Flora, D. J., Gorbach, S. L., Harder, E. J., Tally, F. P. 1975. Management of anaerobic infections. Annals of Internal Medicine 83:375–389.

Frederick, J., Braude, A. I. 1974. Anaerobic infection of the paranasal sinuses. New England Journal of Medicine 290:135–137.

Gibbons, R. J. 1974. Aspects of the pathogenicity and ecology of the indigenous oral flora of man, pp. 267–285. In: Balows, A., DeHaan, R. M., Dowell, V. R., Jr., Guze, L. B. (eds.), Anaerobic bacteria—role in disease. Springfield, Illinois: Charles C Thomas.

Gibbs, R. S., O'Dell, T. N., MacGregor, R. R., Schwarz, R. H., Morton, H. 1975. Puerperal endometritis: A prospective microbiologic study. American Journal of Obstetrics and Gynecology 121:919–925.

Giesecke, D., Wiesmayr, S., Ledinek, M. 1970. *Peptostreptococcus elsdenii* from the caecum of pigs. Journal of General Microbiology 64:123–126.

Giuliano, A., Lewis, F., Hadley, K., Blaisdell, F. W. 1977. Bacteriology of necrotizing fascitis. American Journal of Surgery 134:52–57.

Goplerud, C. P., Ohm, M. J., Galask, R. P. 1976. Aerobic and anaerobic flora of the cervix during pregnancy and the puerperium. American Journal of Obstetrics and Gynecology 126:858–868.

Gorbach, S. L., Bartlett, J. G. 1974. Anaerobic infections. New England Journal of Medicine 290:1177–1184, 1237–1245, 1289–1294.

Gorbach, S. L., Thadepalli, H. Norsen, J. 1974. Anaerobic microorganisms in intraabdominal infections, pp. 399–407. In: Balows, A., DeHaan, R. M., Guze, L. B., Dowell, V. R., Jr. (eds.), Anaerobic baceria—role in disease. Springfield, Illinois: Charles C Thomas.

Gorbach, S. L., Menda, K. B., Thadepalli, H., Keith, L. 1973. Anaerobic microflora of the cervix in healthy women. American Journal of Obstetrics and Gynecology 117:1053–1055.

Gossling, J., Moore, W. E. C. 1975. *Gemmiger formicilis*, n. gen., n. sp., an anaerobic budding bacterium from intestines.

International Journal of Systematic Bacteriology **25**:202–207.

Graham, M. B., Falkler, W. A., Jr. 1978. An extractable carbohydrate antigen shared by strains of *Peptostreptococcus anaerobius*. Abstracts of the Annual Meeting of the American Society for Microbiology **1978**:288.

Graves, M. H., Morello, J. A., Kocka, F. E. 1974. Sodium polyanethol sulfonate sensitivity of anaerobic cocci. Applied Microbiology **27**:1131–1133.

Hagen, J. C., Wood, W. S., Hashimoto, T. 1976. Effect of chilling on the survival of *Bacteroides fragilis*. Journal of Clinical Microbiology **4**:432–436.

Hall, W. L., Sobel, A. I., Jones, C. P., Parker, R. T. 1967. Anaerobic postoperative pelvic infections. Obstetrics and Gynecology **30**:1–7.

Hardie, J. M., Bowden, G. H. 1974. The normal microbial flora of the mouth, pp. 47–83. In: Skinner, F. A., Carr, J. G., (eds.), The normal microbial flora of man. New York: Academic Press.

Hare, R., Wildy, P., Billett, F. S., Twort, D. N. 1952. The anaerobic cocci: Gas formation, fermentation reactions, sensitivity to antibiotics and sulphonamides. Classification. Journal of Hygiene **50**:295–319.

Harris, J. W., Brown, J. H. 1929. Clinical and bacteriological study of 113 cases of streptococcic puerperal infection. Bulletin of the Johns Hopkins Hospital **44**:1–31.

Harris, M. A., Reddy, C. A., Carter, G. R. 1976. Anaerobic bacteria from the large intestine of mice. Applied and Environmental Microbiology **31**:907–912.

Heineman, H. S., Braude, A. I. 1963. Anaerobic infection of the brain. American Journal of Medicine **35**:682–697.

Herbeck, J. L., Bryant, M. P. 1974. Nutritional features of the intestinal anaerobe *Ruminococcus bromii*. Applied Microbiology **28**:1018–1022.

Hewett, M. J., Knox, K. W., Bishop, D. G. 1971. Biochemical studies on lipopolysaccharides of *Veillonella*. European Journal of Biochemistry **19**:169–175.

Hill, E. O., Lewis, S., Altemeier, W. A. 1968. Incidence and significance of nonsporulating anaerobes in surgical infections. Bacteriological Proceedings **1968**:94.

Hill, G. B. 1978. Effects of storage in an anaerobic transport system on bacteria in known polymicrobial mixtures and in clinical specimens. Journal of Clinical Microbiology **8**:680–688.

Hill, G. B. 1975. Further experimental evaluation of pathogenicity and synergism among selected anaerobic bacteria, p. 60. In: 15th Interscience Conference on Antimicrobial Agents and Chemotherapy. Washington, D. C.: American Society for Microbiology. [Abstract.]

Hill, G. B., Osterhout, S., Pratt, P. C. 1974. Liver abscess production by non-spore-forming anaerobic bacteria in a mouse model. Infection and Immunity **9**:599–603.

Hite, K. E., Locke, M., Hesseltine, H. C. 1949. Synergism in experimental infections with nonsporulating anaerobic bacteria. Journal of Infectious Diseases **84**:1–9.

Hoffmann, K., Gierhake, F. W. 1969. Postoperative infection of wounds by anaerobes. German Medical Monthly **14**:31–33.

Hofstad, T., Kristoffersen, T. 1970. Chemical composition of endotoxin from oral *Veillonella*. Acta Pathologica et Microbiologica Scandinavica, Sect. B **78**:760–764.

Holdeman, L. V., Cato, E. P., Moore, W. E. C. (eds.). 1977. Anaerobe laboratory manual, 4th ed. Blacksburg, Virginia: Anaerobe Laboratory, Virginia Polytechnic Institute and State University.

Holdeman, L. V., Good, I. J., Moore, W. E. C. 1976. Human fecal flora: Variation in bacterial composition within individuals and a possible effect of emotional stress. Applied and Environmental Microbiology **31**:359–375.

Holdeman, L. V., Moore, W. E. C. 1974. New genus, *Coprococcus*, twelve new species, and emended descriptions of four previously described species of bacteria from human feces. International Journal of Systematic Bacteriology **24**:260–277.

Hungate, R. E. 1950. The anaerobic mesophilic cellulolytic bacteria. Bacteriological Reviews **14**:1–63.

Hungate, R. E. 1966. The rumen and its microbes. New York: Academic Press, Inc.

Jones, G. L., Whaley, D. N., Dever, S. M. 1977. Use of the flexible anaerobic glove box. Atlanta: U.S. Department of Health, Education, and Welfare, Public Health Service.

Lambe, D. W., Vroon, D. H., Rietz, C. W. 1974. Infections due to anaerobic cocci, pp. 585–599. In: Balows, A., DeHaan, R. M., Dowell, V. R., Jr., Guze, L. B. (eds.), Anaerobic bacteria—role in disease. Springfield, Illinois: Charles C. Thomas.

Ledger, W. J., Norman, M., Gee, C., Lewis, W. 1975. Bacteremia on an obstetric-gynecologic service. American Journal of Obstetrics and Gynecology **121**:205–212.

Ledger, W. J., Gee, C. L., Lewis, W. P., Bobitt, J. R. 1977. Comparison of clindamycin and chloramphenicol in treatment of serious infections of the female genital tract. Journal of Infectious Diseases **135**:S30–S34.

Levison, M. E., Corman, L. C., Carrington, E. R., Kaye, D. 1977. Quantitative microflora of the vagina. American Journal of Obstetrics and Gynecology **127**:80–85.

Liljemark, W. F., Gibbons, R. J. 1971. Ability of *Veillonella* and *Neisseria* species to attach to oral surfaces and their proportions present indigenously. Infection and Immunity **4**:264–268.

Loesche, W. J. 1969. Oxygen sensitivity of various anaerobic bacteria. Applied Microbiology **18**:723–727.

Lorber, B., Swenson, R. M. 1975. The bacteriology of intra-abdominal infections. Surgical Clinics of North America **55**:1349–1354.

McDonald, J. R., Henthorne, J. C., Thompson, L. 1937. Role of anaerobic streptococci in human infections. Archives of Pathology **23**:230–240.

MacLennan, J. D. 1943. Streptococcal infection of muscle. Lancet May **8**:582–584.

MacLennan, J. D. 1962. The histotoxic clostridial infections of man. Bacteriological Reviews **26**:177–276.

Macy, J. M., Snellen, J. E., Hungate, R. E. 1972. Use of syringe methods for anaerobiosis. In: Luckey, T. D., Floch, M. H. (eds.), Intestinal microecology. Bethesda, Maryland: The American Society for Clinical Nutrition. Reprinted from The American Journal of Clinical Nutrition **25**:1318–1323.

Martin, W. J. 1971. Practical method for isolation of anaerobic bacteria in the clinical laboratory. Applied Microbiology **22**:1168–1171.

Martin, W. J. 1974. Isolation and identification of anaerobic bacteria in the clinical laboratory. Mayo Clinic Proceedings **49**:300–308.

Meleney, F. L. 1931. Bacterial synergism in disease processes with a confirmation of the synergistic bacterial etiology of a certain type of progressive gangrene of the abdominal wall. Annals of Surgery **94**:961–981.

Meleney, F. L. 1933. A differential diagnosis between certain types of infectious gangrene of the skin. Surgery, Gynecology and Obstetrics **56**:847–867.

Mena, E., Thompson, F. S., Armfield, A. Y., Dowell, V. R., Jr., Reinhardt, D. J. 1978. Evaluation of Port-A-Cul transport system for protection of anaerobic bacteria. Journal of Clinical Microbiology **8**:28–35.

Mergenhagen, S. E., Hampp, E. G., Scherp, H. W. 1961. Preparation and biological activities of endotoxins from oral bacteria. Journal of Infectious Diseases **108**:304–310.

Mergenhagen, S. E., Thonard, J. C., Scherp, H. W. 1958. Studies on synergistic infections I. Experimental infections with anaerobic streptococci. Journal of Infectious Diseases **103**:33–44.

Moore, W. E. C., Cato, E. P., Holdeman, L. V. 1969. Anaerobic bacteria of the gastrointestinal flora and their occurrence in clinical infections. Journal of Infectious Diseases **119**:641–649.

Ohm, M. J., Galask, R. P. 1975. Bacterial flora of the cervix from 100 prehysterectomy patients. American Journal of Obstetrics and Gynecology **122**:683–687.

Parker, R. T., Jones, C. P. 1966. Anaerobic pelvic infections and developments in hyperbaric oxygen therapy. American Journal of Obstetrics and Gynecology **96**:645–659.

Patterson, D. K., Ozeran, R. S., Glantz, G. J., Miller, A. B., Finegold, S. M. 1967. Pyogenic liver abscess due to microaerophilic streptococci. Annals of Surgery **165**:362–376.

Pearson, H. E., Anderson, G. V. 1970. Bacteroides infections and pregnancy. Obstetrics and Gynecology **35**:31–36.

Pien, F. D., Thompson, R. L., Martin, W. J. 1972. Clinical and bacteriologic studies of anaerobic gram-positive cocci. Mayo Clinic Proceedings **47**:251–257.

Porschen, R. K., Spaulding, E. H. 1974. Fluorescent antibody study of the gram-positive anaerobic cocci. Applied Microbiology **28**:851–855.

Prévot, A. R. 1924. Les streptocoques anaérobies. Thesis. Paris: Amidie Legrand.

Prévot, A. R. 1966. Manual for the classification and determination of the anaerobic bacteria. Philadelphia: Lea and Febiger.

Puhvel, S. M., Reisner, R. M. 1974. Dermatologic anaerobic infections (including acne), pp. 435–450. In: Balows, A., DeHaan, R. M., Dowell, V. R., Jr., Guze, L. B. (eds.), Anaerobic bacteria—role in disease. Springfield, Illinois: Charles C Thomas.

Rogosa, M. 1964. The genus *Veillonella*. Journal of Bacteriology **87**:162–170.

Rogosa, M. 1971a. *Peptococcaceae*, a new family to include the Gram-positive, anaerobic cocci of the genera *Peptococcus*, *Peptostreptococcus*, and *Ruminococcus*. International Journal of Systematic Bacteriology **21**:234–237.

Rogosa, M. 1971b. Transfer of *Peptostreptococcus elsdenii* Gutierrez et al. to a new genus, (*Megasphaera* (*M. elsdenii* (Gutierrez et al.) comb. nov.). International Journal of Systematic Bacteriology **21**:187–189.

Rogosa, M. 1971c. Transfer of *Sarcina* Goodsir from the family *Micrococcaceae* Pribram to the family *Peptococcaceae* Rogosa. International Journal of Systematic Bacteriology **21**:311–313.

Rogosa, M. 1971d. Transfer of *Veillonella* Prevot and *Acidaminococcus* Rogosa from *Neisseriaceae* to *Veillonellaceae* fam. nov., and the inclusion of *Megasphaera* Rogosa in *Veillonellaceae*. International Journal of Systematic Bacteriology **21**:231–233.

Rogosa, M. 1974a. Gram-negative anaerobic cocci, pp. 445–449. In: Buchanan, R. E., Gibbons, N. E. (eds.), Bergey's manual of determinative bacteriology, 8th ed. Baltimore: Williams & Wilkins.

Rogosa, M. 1974b. *Peptococcaceae*, pp. 517–528. In: Buchanan, R. E., Gibbons, N. E. (eds.), Bergey's manual of determinative bacteriology, 8th ed. Baltimore: Williams & Wilkins.

Rosebury, T. 1962. Microorganisms indigenous to man. New York: McGraw-Hill.

Rosenblatt, J. E., Fallon, A., Finegold, S. M. 1973. Comparison of methods for isolation of anaerobic bacteria from clinical specimens. Applied Microbiology **25**:75–85.

Rotheram, E. B. 1974. Septic abortion and related infections of pregnancy, pp. 369–378. In: Balows, A., DeHaan, R. M., Guze, L. B., Dowell, V. R., Jr. (eds.), Anaerobic bacteria—role in disease. Springfield, Illinois: Charles C Thomas.

Sabbaj, J., Sutter, V. L., Finegold, S. M. 1972. Anaerobic pyogenic liver abscess. Annals of Internal Medicine **77**:629–638.

Sandusky, W. R., Pulaski, E. J., Johnson, B. A., Meleney, F. L. 1942. The anaerobic nonhemolytic streptococci in surgical infections on a general surgical service. Surgery, Gynecology and Obstetrics **75**:145–156.

Schleifer, K. H., Nimmermann, E. 1973. Peptidoglycan types of strains of the genus *Peptococcus*. Archiv für Mikrobiologie **93**:245–258.

Schwarz, O., Dieckmann, W. J. 1926. Anaerobic streptococci:

Their role in puerperal infection. Southern Medical Journal **19**:470–479.

Shapton, D. A., Board, R. G. (eds.). 1971. Isolation of anaerobes. New York: Academic Press.

Skinner, F. A., Carr, J. G. (eds.). 1974. The normal microbial flora of man. New York: Academic Press.

Slyter, L. L., Weaver, J. M. 1977. Tetrahydrofolate and other growth requirements of certain strains of *Ruminococcus flavefaciens*. Applied and Environmental Microbiology **33**:363–369.

Smith, D. T. 1927. Experimental aspiratory abscess. Archives of Surgery **14**:232–239.

Smith, D. T. 1932. Oral spirochetes and related organisms in fusospirochetal disease. Baltimore: Williams & Wilkins.

Smith, L. DS. 1975. The pathogenic anaerobic bacteria, 2nd ed. Springfield, Illinois: Charles C Thomas.

Sonnenwirth, A. C. 1974. Incidence of intestinal anaerobes in blood cultures, pp. 157–171. In: Balows, A., Dehaan, R. M., Guze, L. B., Dowell, V. R., Jr. (eds.), Anaerobic bacteria-role in disease. Springfield, Illinois: Charles C Thomas.

Stalons, D. R., Thornsberry, C. 1975. Broth-dilution method for determining the antibiotic susceptibility of anaerobic bacteria. Antimicrobial Agents and Chemotherapy **7**:15–21.

Stargel, M. D., Lombard, G. L., Dowell, V. R., Jr. 1978. Alternative procedures for identification of anaerobic bacteria. American Journal of Medical Technology **44**:709–722.

Steinhorn, S. R. 1945. The possible role of bacterial synergism in puerperal infections due to anaerobic streptococci. American Journal of Obstetrics and Gynecology **50**:63–68.

Stokes, E. J. 1958. Anaerobes in routine diagnostic cultures. Lancet **i**:668–670.

Stone, H. H., Martin, J. D., Jr. 1972. Synergistic necrotizing cellulitis. Annals of Surgery **175**:702–711.

Sugihara, P. T., Sutter, V. L., Attebery, H. R., Bricknell, K. S., Finegold, S. M. 1974. Isolation of *Acidaminococcus fermentans* and *Megasphaera elsdenii* from normal human feces. Applied Microbiology **27**:274–275.

Sutter, V. L., Finegold, S. M. 1976. Susceptibility of anaerobic bacteria to 23 antimicrobial agents. Antimicrobial Agents and Chemotherapy **10**:736–752.

Sutter, V. L., Vargo, V. L., Finegold, S. M. 1975. Wadsworth anaerobic bacteriology manual, 2nd ed. Los Angeles: Anaerobic Bacteriology Laboratory, Wadsworth Hospital Center and the School of Medicine at the University of California at Los Angeles.

Sweet, R. L. 1975. Anaerobic infections of the female genital tract. American Journal of Obstetrics and Gynecology **122**:891–901.

Sweet, R. L., Ledger, W. J. 1973. Puerperal infectious morbidity. American Journal of Obstetrics and Gynecology **117**:1093–1100.

Swenson, R. M., Lorber, B. 1977. Clindamycin and carbenicillin in treatment of patients with intraabdominal and female genital tract infections. Journal of Infectious Diseases **135**:S40–S45.

Swenson, R. M., Michaelson, T. C., Daly, M. J., Spaulding, E. H. 1973. Anaerobic bacterial infections of the female genital tract. Obstetrics and Gynecology **42**:538–541.

Syed, S. A., Loesche, W. J. 1972. Survival of human dental plaque flora in various transport media. Applied Microbiology **24**:638–644.

Tally, F. P., Stewart, P. R., Sutter, V. L., Rosenblatt, J. E. 1975. Oxygen tolerance of fresh clinical anaerobic bacteria. Journal of Clinical Microbiology **1**:161–174.

Thadepalli, H., Gorbach, S. L., Keith, L. 1973. Anaerobic infections of the female genital tract; Bacteriologic and therapeutic aspects. American Journal of Obstetrics and Gynecology **117**:1034–1040.

Thadepalli, H., Gorbach, S. L., Broido, P. W., Norsen, J., Nyhus, L. 1973. Abdominal trauma, anaerobes, and antibiotics. Surgery, Gynecology and Obstetrics **137**:270–276.

Thomas, C. G. A., Hare, R. 1954. The classification of anaerobic cocci and their isolation in normal human beings and pathological processes. Journal of Clinical Pathology **7:**300–304.

Vargo, V. L., Bollard, P. M., Hodinka, N. E., D'Amato, R. F. 1977. Developmental study for the biochemical identification of anaerobic gram-positive cocci. Abstracts of the Annual Meeting of the American Society for Microbiology **1977:**60.

Veillon, M. A. 1893. Sur un microcoque anaerobie trouvé dans des suppurations fétides. Comptes Rendus de la Société de Biologie **45:**807–809.

Walker, C., West, S. E. H., Nitzan, D. 1977. The pathogenic components of the microbiota of the gingival crevice area of man. Abstracts of the Annual Meeting of the American Society for Microbiology **1977:**17. Washington, D.C.: American Society for Microbiology.

Weiss, C., Mercado, D. G. 1938. Studies of anaerobic streptococci from pulmonary abscesses. Journal of Infectious Diseases **62:**181–185.

Wells, C. L., Field, C. R. 1976. Long-chain fatty acids of peptococci and peptostreptococci. Journal of Clinical Microbiology **4:**512–521.

Wideman, P. A., Vargo, V. L., Citronbaum, D., Finegold, S. M. 1976. Evaluation of the sodium polyanethol sulfonate disk test for the identification of *Peptostreptococcus anaerobius*. Journal of Clinical Microbiology **4:**330–333.

Wilkins, T. D., Chalgren, S. 1976. Medium for use in antibiotic susceptibility testing of anaerobic bacteria. Antimicrobial Agents and Chemotherapy **10:**926–928.

Wilkins, T. D., Jimenez-Ulate, F. 1975. Anaerobic specimen transport device. Journal of Clinical Microbiology **2:**441–447.

Wilkins, T. D., Thiel, T. 1973. Modified broth-disk method for testing the antibiotic susceptibility of anaerobic bacteria. Antimicrobial Agents and Chemotherapy **3:**350–356.

Wilkins, T. D., West, S. E. H. 1976. Medium-dependent inhibition of *Peptostreptococcus anaerobius* by sodium polyanetholsulfonate in blood culture media. Journal of Clinical Microbiology **3:**393–396.

Wilson, W. R., Martin, W. J., Wilkowske, C. J., Washington, J. A. 1972. Anaerobic bacteremia. Mayo Clinic Proceedings **47:**639–646.

Wong, M., Catena, A., Hadley, W. K. 1978. Antigenic relationships among *Peptostreptococcus* species. Abstracts of the Annual Meeting of the American Society for Microbiology **1978:**288.

Wren, M. W. D., Eldon, C. P., Dakin, G. H. 1977. Novobiocin and the differentiation of peptococci and peptostreptococci. Journal of Clinical Pathology **30:**620–622.

Zabransky, R. J., Hauser, K. J. 1977. Stability of antibiotics in Wilkins-Chalgren anaerobic susceptibility testing medium after prolonged storage. Antimicrobial Agents and Chemotherapy **12:**440–441.

SECTION R

Gram-Positive, Asporogenous Rod-Shaped Bacteria

CHAPTER 131

The Genus *Lactobacillus*

M. ELISABETH SHARPE

The genus *Lactobacillus* is a member of the family Lactobacillaceae, the type species being *Lactobacillus delbrueckii* (Opinion 38; Judicial Commission, 1971). The genus has been extensively studied from aspects of taxonomy and identification, biochemistry and metabolism, and nutritional requirements. Lactobacilli are indispensable in the manufacture and preservation of many fermented food products and are involved in the protection of young animals against intestinal infections.

Habitats

Lactobacilli have complex nutritional requirements, needing to be supplied with carbohydrates, amino acids, peptides, fatty acids or fatty acid esters, salts, nucleic acid derivatives, and vitamins. Their ATP-generating metabolism is fermentative and they produce large amounts of lactic acid and small amounts of other end products. This mode of ATP-generating metabolism is not changed in the presence of O_2. Lactobacilli grow in a variety of habitats, wherever high levels of soluble carbohydrate, protein breakdown products, vitamins, and a low oxygen tension occur. They are aciduric or acidophilic, different species have adapted themselves to grow under widely different environmental conditions, and their production of high levels of lactic acid lowers the pH of the substrate and suppresses many other bacteria; these factors account for the wide distribution of lactobacilli and their successful establishment in many markedly different habitats.

Man and Animals

Soon after birth the mouth and intestinal tract of man and animals become the habitat of a profuse and diverse microflora. The balance of this microflora, including the lactobacilli, is influenced by a variety of factors including diet, animal physiology, and immunological factors. In the normal intestine a symbiotic relationship develops between the host and the lactobacillus microflora.

ORAL CAVITY. In the human oral cavity, a mainly anaerobic environment in which Gram-negative anaerobes and streptococci predominate, lactobacilli constitute only about 1% of the flora in the healthy mouth. Studies of tooth surfaces of infants and young children (Carlsson, Grahnen, and Jonsson, 1975) have shown lactobacilli to be present only in very small numbers or as transients in the mouth. The lactobacillus flora developing in 2- to 5-year-old children was at a much lower level than the streptococcal one and consisted mostly of *L. casei* and *L. casei* subsp. *rhamnosus,* with an occasional isolate of *L. acidophilus* and *L. fermentum. L. casei* was found more often in children with carious lesions than in those without them. Rogosa et al. (1953), in their classical paper on the identification of oral lactobacilli, identified 500 strains isolated from saliva specimens of 130 school children. *L. casei* and *L. fermentum* were the predominant species present in 59% and 45% of the samples, respectively, *L. acidophilus* in 22%, *L. brevis* in 17%. *L. buchneri, L. salivarius, L. plantarum,* and *L. cellobiosus* occurred less frequently. Those findings have been confirmed by other studies on children and adults (Camilleri, and Bowen, 1963; Takei, et al., 1971 [cited in London, 1976]) in which similar species were isolated and *L. casei* and *L. fermentum* were found to predominate. All strains of lactobacilli isolated from deep dental plaque by Shovell and Gillis (1972) were *L. casei* and other work has confirmed that this species is the prevalent lactobacillus in plaque. It is now recognized that the salivary flora is not representative of the microbial composition of different areas of the mouth; dental plaque contributes little to this flora (Hardie and Bowden, 1974), so levels of lactobacilli in the saliva may be misleading. Detailed studies of dental plaque show that it is initiated largely by organisms capable of adhering to tooth surfaces, including *Streptococcus mutans,* and consists partly of extracellular glucose polymers produced by the oral streptococci. Only at this stage may lactobacilli multiply within this matrix. Surveys of noncarious dental plaque show that lactobacilli are present only in small numbers of 5–12% or may not be isolated at all (Hardie and Bowden, 1974). The presence of unrestricted carious cavities considerably increases the lactobacillus count, and *L. casei* in particular appears to be associated with such lesions. However, as the low pH (initiated by the streptococci) found in carious cavities favors lactobacilli, the high count may well be the result of caries and not the cause (Hardie and Bowden, 1974).

INTESTINAL TRACT. Conditions influencing and controlling the intestinal microflora of homeothermic and poikilothermic animals are discussed by Clarke and Bauchop (1977); of man in particular, by Drasar and Hill (1974); and of the gastrointestinal ecosystem, by Savage (1977). In homeothermic animals, including man, stomach acidity influences the lactobacillus flora. In some studies it is difficult to assess the proportion of lactobacilli in the total intestinal flora, because where conventional anaerobic techniques have been used for enumeration and isolation, often no distinction has been made between anaerobic lactobacilli and bifidobacteria. The two have been regarded as synonymous, although lactobacilli and bifidobacteria in fact belong to different genera (*Bergey's Manual,* 1974).

In the neonatal mammal—such as the human infant, piglet, and calf—the intestinal tract is sterile at birth. It is rapidly colonized from the mouth and rectum, usually initially with the mother's vaginal and perianal flora. In detailed studies of a wide variety of young animals, Smith (1965a, 1971) showed that at birth the gastrointestinal (GI) tract becomes flooded with multiplying bacteria, including coliforms, clostridia, and other anaerobes, followed rapidly by lactobacilli. The balance of the flora is then controlled by acid secretion in the stomach (although this is buffered to some extent by the milk imbibed) and ingestion of immunoglobulins and other protective factors in the mother's milk (reviewed by Reiter, 1978), so that lactobacilli become the dominant organisms and other groups rapidly decline. Lactobacilli predominate in the stomach and small intestine, but in the jejunum and large intestine strict anaerobes such as bacteroides are the majority flora and lactobacilli constitute only 0.07–1% of the total flora. This desirable predominance of lactobacilli in the upper intestine, which is established on suckling, helps to prevent the potentially lethal diarrhea or scouring that occurs in young animals when enteropathogenic coliforms proliferate in the upper GI tract. The mechanisms by which lactobacilli suppress the rest of the bacterial flora is not fully understood but is probably partly due to lactic acid production and partly to inhibitory systems present in the raw milk, including the lactoperoxidase system, which can be activated by H_2O_2-producing lactobacilli (Reiter, 1978). In the young animal, lactobacilli usually become the principal component of the stomach and small intestine, developing from about 10^4 to 10^8/g intestinal contents, highest numbers being obtained in the lower part of the small intestine. In early weaned, usually domestic animals, the protective systems present in the mother's milk are absent and enteropathogenic coliforms may multiply, causing disease and death (Smith, 1971).

In the young calf Contrepois and Gouet (1973) confirmed the findings of Smith (1965a) of the dominance of lactobacilli in the stomach and upper GI tract, although they also observed varying numbers of strict anaerobes to be present from the stomach onwards. From the abdominal fluid of the milk-fed calf, isolates have been identified as *Lactobacillus lactis* and *L. fermentum* (M. E. Sharpe, unpublished data). It is of interest that the author's collection of lactobacilli includes strains of *L. lactis, L. bulgaricus,* and *L. fermentum* previously isolated from the dried stomach of the young calf (vell), indicating that the milk-fed calf stomach is their natural habitat.

Lactobacilli are indigenous to the stomach of rodents and pigs and to the crop of chickens, thickly colonizing areas of the epithelium in neonatal animals (reviewed by Savage, 1977). This adhering flora, which is continuously shedding organisms, is instrumental in controlling colonization and composition of the indigenous lactobacillus flora of the GI tract. Colonization is host-specific, i.e., only strains of lactobacilli isolated from birds will adhere to the chicken crop (Fuller, 1973) and only strains isolated from the rat will adhere to the rat stomach epithelium (Kawai and Shegara, 1977). In the neonatal piglet Fuller, Barrow, and Brooker (1978) found *L. fermentum, L. acidophilus,* and *L. salivarius,* sometimes associated with streptococci, to be colonizing the stomach epithelial cells; these species were also present in the duodenal contents, at levels of 10^7 to 10^8/g (Barrow, Fuller, and Newport, 1977). In the chicken crop, biotypes of *L. salivarius* and *L. fermentum* were the species adhering to the epithelium (Fuller, 1973).

In the human infant, sampling problems prevent microbial studies of the GI tract and only the fecal flora can be analyzed. As with other young animals, the mother's milk protects the infant against neonatal coliform infections and results in the development of a fecal flora containing high numbers of bifidobacteria and low numbers of lactobacilli. In bottle-fed infants, while numbers of bifidobacteria are maintained, higher levels of lactobacilli are present (Mitsuoka and Kaneuchi, 1977). This fecal flora is unlikely to reflect the lactic acid bacterial flora of the upper GI tract. Various aspects of the influence of feeding on the microflora are discussed by Reiter (1978). While there are extensive data on enumeration and identification of bifidobacteria in infants' feces, little is known of the lactobacilli present. In the feces of 66 infants aged 3–220 days, including breast- and bottle-fed babies, Mitsuoka,

Hayakawa, and Kimura (1975) found *L. acidophilus, L. fermentum,* and *L. salivarius* to be present at levels varying from 10^3 to 10^{10}/g feces. These they regarded as the indigenous species. Sharpe (unpublished data) isolated *L. fermentum* from the feces of 6 of 10 breast-fed infants, all less than 7 days old.

An extensive study of the flora of the alimentary tract (Smith, 1965b) included viable counts of lactobacilli—not further identified—in a wide range of homeothermic and some poikilothermic animals. In most monogastric animals, especially those on a cereal diet, lactobacilli are the majority flora of the stomach and small intestine. In herbivorous animals, however, they occur in large numbers only in the cecum and large intestine. Large numbers of lactobacilli are usually found in the anterior part of the stomach where the pH range is 4.3–6.4 in different animal species, but in the posterior part the pH range of 2.2–4.2 is low enough to reduce the numbers considerably. Lactobacilli increase progressively from duodenum to ileum from ca. 10^5 to 10^6/g, high numbers of 10^8 to 10^9/g being found in the large intestine, but, as with the young animal, bacteroides predominate here. In poikilothermic animals, such as frogs, fish, and tortoises, no lactobacilli were isolated. Many other investigators (cited by London, 1976) give levels of lactobacilli, not further identified, in the GI tract of dogs, pigs, ponies, cattle, mice, and turkeys. More detailed studies show that in the adult pig *L. fermentum, L. acidophilus, L. cellobiosus,* and *L. salivarius* predominate in the feces (Fuller et al., 1960; Mitsuoka, 1969). Detailed studies of the ileum, cecum, and colon epithelial tissues of beagle dogs (Davis et al., 1977) showed that they were harboring a wide variety of lactobacilli: *L. acidophilus* in the highest numbers, then *L. leichmannii, L. plantarum,* and *L. fermentum.* The much smaller numbers of many other species isolated may have been transients. Mitsuoka, Kimura, and Kobayashi (1976) isolated *L. acidophilus, L. salivarius,* and *L. fermentum,* including a number of different biotypes, from the feces of dogs. In the rat, *L. acidophilus* and *L. salivarius* predominated (Raibaud et al., 1973), some thermobacteria being able to ferment pentoses and one isolate being ureolytic (Moreau, Ducluzeau, and Raibaud, 1976). Kawai and Shegara (1977) isolated a strain of *L. fermentum* that could colonize rat stomach epithelial cells, and in the mouse, Roach, Savage, and Tannock (1977) isolated several different groups of unclassified thermobacteria, some also able to ferment ribose, which were the indigenous lactobacilli colonizing the stomach.

In the ruminant animal, the rumen begins to function when a particulate diet is fed. Anaerobic lactobacilli have been isolated as part of the majority flora from ruminating calves (Bryant et al., 1958), and similar organisms have been isolated from older animals and adult cattle (discussed by Sharpe, Hill,

and Lapage, 1973). The level of lactobacilli and bifidobacteria, usually not further differentiated, is dependent on the diet. On high-fiber diets lactobacilli may constitute 1–2% of the total flora, whereas on high-carbohydrate diets they may constitute 15–20% of the total (Latham, Sharpe, and Sutton, 1971). Little is known of the lactobacillus flora of the GI tract.

In man, with a pH of 3.0 in the stomach, the lactobacilli are likely to be transients from the saliva and food (Drasar and Hill, 1974; Reuter, 1965). However, Bernhardt (1974) isolated a wide variety of strains from gastric juices, mainly *L. acidophilus* and *L. fermentum.* Only strains of *L. acidophilus* were found in gastric juices with a pH less than 3.0. In the upper duodenum, small numbers of lactobacilli are present, which increase during passage down the small intestine, being the majority flora present. Reuter (1965), using an automatic capsule, showed that they consisted mainly of *L. acidophilus, L. fermentum,* and *L. salivarius* together with some anaerobic lactobacilli. In the large intestine, which is reflected in the fecal flora, lactobacilli are in the minority (Drasar and Hill, 1974). Lactobacilli isolated in the feces, usually at levels of between 10^4 and 10^9/g, consist of *L. acidophilus, L. fermentum, L. salivarius,* and, more irregularly, *L. lactis, L. casei, L. plantarum, L. brevis,* and *L. buchneri* (Reuter, 1965). By feeding people diets high in lactobacilli, tracing these last five species to the feces, and then excluding lactobacilli from the diet, Reuter (1965) was able to show that these five species were all transients. These results, together with those of other workers (Finegold et al., 1977; Mitsuoka, Hayakawa, and Kimura, 1975; Moore and Holdeman, 1974) indicate that *L. acidophilus, L. fermentum,* and *L. salivarius* are the predominant lactobacilli in the lower human intestine and feces, although *L. plantarum* and *L. leichmannii* may sometimes be prevalent. Different biotypes may be present, which are not the same as those found in animals (Mitsuoka, Hayakawa, and Kimura, 1975). Some isolates of the above species behaved as strict anaerobes (Moore and Holdeman, 1974), indicating the need for good anaerobic cultural techniques for isolation. Anaerobic lactobacilli have been isolated from the human GI tract, one strain being identified as *L. ruminis* (Sharpe, Hill, and Lapage, 1973) others being *L. rogosa, L. minutus,* and *L. crispatus* (Moore and Holdeman, 1974). Anaerobic lactobacilli may at times be part of the majority anaerobic flora of the feces, present at levels of 10^{10}/g feces (Mitsuoka and Ohno, 1977). In the large intestine, however, simple sugars are not available and organisms must rely on sugars derived from GI tract mucins and plant material (Salyers et al., 1977), and it seems likely that lactobacilli must rely on other organisms' degrading these complex carbohydrates to obtain their energy sources. Some strains of *L.*

acidophilus produce antibiotic substances (Vincent, Veomell, and Riley, 1959) inhibitory to other microorganisms and it has been suggested that implantation of the GI tract with large quantities of such a strain might repress undesirable intestinal Gram-negative organisms (discussed by Sandine et al., 1972). Drasar and Hill (1974), however, consider such implantation most unlikely to occur, as bacteria constituting the indigenous flora are resistant to change and adapted to their situation. The ability of a metabolite from a strain of *L. bulgaricus* to metabolize the effect of enterotoxin from *Escherichia coli* pathogenic to pigs might be of use in protecting early weaned piglets (Mitchell and Kenworthy, 1976).

The lactobacillus microflora of insects, which could be interesting plant vectors, is little known. Kvasnikov, Kotljar, and Vasileva (cited by London, 1976) isolated strains of *L. casei* and *L. cellobiosus* from the honeybee, silkworm moth, and from fruit flies; Ruiz-Arguesco and Rodriquez-Navarro (1975) found *L. viridescens* in the stomach of the honeybee.

PATHOGENIC LACTOBACILLI.   Some lactobacilli, particularly *L. casei* subsp. *rhamnosus,* have the potential to cause disease conditions in humans, such as subacute bacterial endocarditis, systemic septicemia, and abscesses. Other species sometimes associated with such conditions are *L. acidophilus, L. plantarum,* and occasionally *L. salivarius.* Sharpe, Hill, and Lapage (1973) and Berger (1974) describe a number of cases and discuss pathogenic aspects. It is considered that the site of invasion may often be the oral cavity, and that in some instances the lactobacilli may be secondary invaders.

HUMAN VAGINA.   In the healthy adult woman, the pH of 4 to 5 in the vagina protects against invasion with *Trichomonas vaginalis, Candida,* or other infections. This low pH is maintained by a microflora that includes large numbers of lactic acid bacteria such as lactobacilli and bifidobacteria and is considered to be largely due to fermentation of the glycogen present in the vaginal mucosa. Rogosa and Sharpe (1960), who cite the early literature, identified isolates from normal nonpregnant women as *L. acidophilus* (67%), *L. fermentum* (19%), *L. casei* subsp. *rhamnosus* (10%), and *L. cellobiosus* (4%). Other studies have also shown *L. acidophilus* and *L. fermentum* to be the dominant lactobacillus species present; *L. casei, L. plantarum, L. brevis, L. delbrueckii, L. lactis, L. bulgaricus, L. leichmannii,* and *L. salivarius* have also been isolated on occasions (Lenzner, 1966; Wylie and Henderson, 1969). Studying the transmission of lactobacilli from mother to child at the time of delivery, Carlsson and Gothefors (1975) isolated lactobacilli from the vaginas of 8 of 13 mothers examined: 44% of strains were *L. acidophilus* and 56% were *L. jensenii,* the

species isolated by Gasser, Mandel, and Rogosa (1970) from vaginal discharges. J. Narvus (personal communication) found that 23% of normal women harbor *L. jensenii.* Although many different species have been isolated from the vagina, *L. acidophilus* appears to predominate. The work of Carlsson and Gothefors (1975) and Levison et al. (1977) indicates that many women have more than one lactobacillus species present. There are few data on the numbers present in the vagina although Levison et al. (1977) found that they occurred at more than 10^5/ml of vaginal secretion.

Classically, lactobacilli have been regarded as able to ferment the glycogen of the vaginal epithelium. However, Rogosa and Sharpe (1960) found that only some of their isolates fermented glycogen (weakly and as a variable and adaptable characteristic), and Wylie and Henderson (1969) found only 3 of their 42 isolates to be positive. Stewart Tull's finding (1964) that vaginal strains of *L. acidophilus* could ferment glycogen only in the presence of normal human serum (containing a glycogenase) suggested that vaginal lactobacilli could utilize only the products of glycogen breakdown. It appears that some strains of lactobacillii can ferment glycogen but they are more likely to obtain available carbohydrate from enzymic conversion of this polysaccharide by tissues or possibly by other microorganisms.

Milk and Dairy Products

In milk and some dairy products, lactobacilli occur by chance and may have little effect, or they may bring about desirable or undesirable reactions. In other instances, often together with streptococci, they are deliberately inoculated as starters for large-scale commercial milk fermentation processes where they produce the required acidity and flavor components.

MILK.   Aseptically drawn raw milk contains no lactobacilli when it leaves the udder, but contamination with these organisms rapidly occurs, from the dairy utensils, dust, grass, silage, and other feeding stuffs. Milk is an ideal substrate for bacterial growth, but conditions that allow contamination and multiplication favor other organisms, and lactobacilli are usually outgrown. In the U.K., single-herd milks produced under good hygienic conditions contain small numbers of lactobacilli, >1 to 50/ml, whereas bulked herd market milks usually contain about 10^3/ml. Species present include *L. casei, L. plantarum, L. brevis, L. coryniformis, L. curvatus,* and occasionally *L. buchneri, L. lactis,* and *L. fermentum* (reviewed by Abo-Elnaga and Kandler, 1965a; Sharpe, 1962a). Raw ewe's milk contains the same species (Chomakov and Kirov, 1975). Heat treatment of the milk destroys most lactobacilli and full pasteurization (161°C/17 s) usually destroys all lactobacilli present, although heat-resistant strains may develop in a dairy. Such heat-treated milks, when used for processing, rapidly become recontaminated from the creamery environment with the same species of lactobacilli (reviewed by Sharpe, 1962a).

CHEESE. Lactobacilli are found in all types of cheese. With hard cheese such as Cheddar they are initially present in small numbers and gradually increase during ripening to become the dominant flora, rising to levels of about $10^6 - 10^8$/g cheese. Species present include the mesophilic *L. casei*, *L. plantarum*, *L. brevis*, *L. buchneri*, and many different biotypes of unclassified streptobacteria (reviewed by Sharpe, 1962a). These organisms are able to multiply at the low pH of 5.2 and in the salt concentration of 4.5% found in the cheese curd, while other organisms gradually die out. This rise in numbers of lactobacilli during ripening suggested that they were concerned with typical Cheddar cheese flavor production (reviewed by Fryer, 1969). Extensive work has shown that this is not the case (Law, Castanon, and Sharpe, 1976), although different strains may contribute to distinctive cheese flavor overtones. In the manufacture of Swiss cheese, lactobacilli are deliberately inoculated as starters and usually consist, together with a thermophilic streptococcus, of thermophilic lactobacilli such as *L. helveticus* or *L. lactis*, which withstand the high scald temperature required for making this cheese (Biede, Reinbold, and Hammond, 1976). These lactobacilli soon die out during ripening, and are replaced by the same mesophilic species as in Cheddar cheese (Veaux et al., 1974). *L. helveticus* is also used as a starter in Grana, Gorgonzola, and Parmesan cheese production (Bottazzi, Vescova, and Dellaglio, 1973). The rind of Stilton cheese also contains large numbers of lactobacilli, mainly *L. plantarum*, *L. casei*, and *L. brevis*.

FERMENTED MILK PRODUCTS. Yogurt is now produced on a large scale commercially (over one million tons in E.E.C. countries in 1976) by the incubation of milk with a mixed starter of *Lactobacillus bulgaricus* and *Streptococcus thermophilus* (Davis, 1975; Tamine, 1977). These two organisms multiply in associative growth, the former hydrolyzing protein to amino acids and dipeptides to stimulate the growth of the streptococcus (Shankar, 1978), and the latter producing favorable reducing and low pH conditions. Higher levels of lactic acid and the required flavorful acetaldehyde are produced when the two organisms are grown together (reviewed by Moon and Reinbold, 1976; Sharpe, 1972). Slime-producing strains of *L. bulgaricus* are valued as stabilizers in the yogurt (Davis, 1975). Acidophilus milk, fermented by the intestinal species *L. acidophilus*, is thought to have therapeutic properties,

probably due to production of an antibiotic (see below). In some countries, this organism is also incorporated in yogurt. Other fermented milks include kefir, in which a streptococcus, a yeast, and a heterofermentative lactobacillus produce CO_2, ethanol, and lactic acid, and the Japanese yakult, which is fermented by *L. casei* (Watanabe et al., 1970). Strains of *L. bulgaricus* and *L. lactis* are also used for large-scale production of lactic acid from pasteurized cheese whey (Poznanski et al., 1974).

SPOILAGE. Lactobacilli are so universally present in milk and dairy products that generally only strains having unusual characteristics cause spoilage. In liquid milk, slime-producing strains of *Lactobacillus casei*, *L. brevis*, *L. bulgaricus*, and *L. acidophilus* occasionally produce ropiness and *L. maltaromicus* may produce a malty flavor (Miller, Morgan, and Libbey, 1974). The citrate-utilizing species *L. casei* and the heterofermentative *L. brevis* may produce excessive CO_2, giving rise to unwanted gas pockets in cheese and blowing of packaged cheeses (Fryer, Sharpe, and Reiter, 1970; Keller and Jaarsman, 1975). Slime-forming strains of *L. plantarum* can multiply in cheese-pickling brines, causing ropiness, and a salt-tolerant streptobacterium multiplying in rennet has caused serious texture and flavor defects in Dutch cheese (Stadhouders and Veringa, 1967). Orange-pigmented strains (*L. plantarum* subsp. *rudensis* or *L. brevis* subsp. *rudensis*) may multiply in hard cheese (Sharpe, 1962a) and in brined white cheese (Chomakov, 1962).

Interactions, either stimulating or inhibiting, have been observed between lactobacilli and other organisms in milk and dairy products (Branen and Keenan, 1969; Gudkov and Sharpe, 1965; Khandak and Gudkov, 1977; Nieuwenhof, Stadhouders, and Hup, 1969). Some strains of *L. acidophilus* grown in milk produce an antibiotic (not lactic acid or H_2O_2) that is inhibitory to a wide range of Gram-positive and Gram-negative bacteria (Shahani, Vakil, and Kilara, 1977).

Meat and Meat Products

Lactobacilli proliferating in meats and meat products are able to grow at low temperatures, sometimes at $1-2°C$, are stimulated by increased CO_2 levels, and are salt tolerant, often multiplying in the presence of 10% NaCl. In fermented sausages, which contain added sucrose, lactobacilli or other lactic acid bacteria such as pediococci are essential for the curing process and are often added to the meat mix as starters (Deibel, Niven, and Wilson, 1961). They contribute to flavor, consistency, and keeping quality by producing organic acids (mainly lactic) and traces of other carbohydrate breakdown products, including acetoin. In the absence of an added starter, unclassified streptobacteria, *Lactobacillus plantarum, L. farciminis,* and *L. brevis* predominate (Reuter, 1970). Growth of lactobacilli on other types of meat generally induces spoilage. Increased use of vacuum packaging for wholesale and retail distribution of meat has emphasized the significance of lactobacilli as spoilage agents. Vaccum packaging creates a favorable environment for them: the increased CO_2 tension encourages their growth and suppresses that of aerobic organisms, particularly at refrigeration temperature (Kitchell and Shaw, 1975). With vacuum-packed fresh meats, large numbers of lactobacilli may develop, varying in numbers with the type of meat (Kitchell and Shaw, 1975). The level of carbohydrate available to the organisms may be a limiting factor (Gill, 1976). The work of Newton and Gill (1978) indicates that the predominance of lactobacilli on chilled meats may be due to production of an inhibitor which is neither lactic acid nor H_2O_2. Lactobacilli isolated are mainly unclassified streptobacteria. In vacuum-packed bacon, Kitchell and Shaw (1975) discuss the effects of increased CO_2 concentration, low temperature, salt, nitrite, and smoking on the ratios of lactobacilli and micrococci; all these factors except high salt levels favor lactobacilli. Strains isolated are mainly unclassified streptobacteria, *L. viridescus, L. brevis,* and *L. buchneri.* Sliced continental vacuum-packed meats such as frankfurters and sausages have similar contamination problems, unclassified streptobacteria being the most common isolates, while *L. brevis, L. viridescens,* and *L. plantarum* also occur. Cured sausages, frankfurters, bologna, and hams are particularly susceptible to spoilage (Reuter, 1975). Curing brines appear to be the source of lactobacilli contaminating hams (reviewed by Sharpe, 1962b). Types of spoilage of meats include souring due to lactic acid, greening that is caused by H_2O_2 reacting with meat pigments, slime production, gas formation, and off flavors. Reuter (1975) details characteristics of different spoilage species.

Large numbers of unclassified organisms, now regarded as unclassified streptobacteria, develop on chicken meat stored at $1-5°C$ (Barnes, 1976; Thornley and Sharpe, 1959). These may cause spoilage due to souring when shelf life is extended by controlling growth of pseudomonads by the use of O_2-impermeable film. In turkeys wrapped in such film and stored at $1°C$, these lactobacilli form a significant part of the spoilage flora.

Plant Material

The scanty data available on the occurrence of lactobacilli on plants suggest that only small numbers are present on intact plant material (reviewed by Keddie, 1959). Stirling and Whittenbury (1963) found that lactic acid bacteria were rarely present on living tissue of plants, and that leuconostocs consti-

tuted 80% of the isolates, and lactobacilli only 10%. Species isolated included atypical streptobacteria and betabacteria, *Lactobacillus plantarum, L. fermentum,* and also, from a wide variety of plants in a subtropical area (Mundt and Hammer, 1968), small numbers of *L. brevis,* and occasionally *L. casei, L. viridescens, L. cellobiosus,* and *L. salivarius.* The antibacterial effect of extracts of some higher plants, often due to 1,4-naphthaquinone derivatives, might contribute to this sparsity, although saponin-containing plants (which include grasses) were not inhibitory (Shcherbanovsky, Luko, and Kapelev, 1975). Mundt and Hammer (1968) consider that plants are not a natural reservoir for lactobacilli. On cut or bruised tissue, lactobacilli become more prevalent (Stirling and Whittenbury, 1964).

SILAGE. Lactobacilli are of considerable economic importance in the acidification of silage made from many different crops (for references see Keddie, 1959). The lactic acid formed during their growth prevents undesirable clostridial multiplication. Only after streptococci and leuconostocs have multiplied do lactobacilli, together with pediococci, become the majority silage microflora (Langston, Bouma, and Conner, 1962). Species chiefly isolated have been *Lactobacillus plantarum, L. brevis, L. buchneri,* unclassified streptobacteria, *L. coryniformis, L. curvatus, L. casei, L. fermentum, L. acidophilus,* and *L. salivarius* (Abo-Elnaga and Kandler, 1965a; Azeezullah et al., 1973; Keddie, 1959; Langston and Bouma, 1960). The presence of *L. fermentum, L. acidophilus,* and *L. salivarius* in some silages suggests intestinal waste as their source (Mundt and Hammer, 1968). The ability of strains of *L. casei* but not *L. plantarum* and *L. brevis* to attack grass fructosans (Kleeberger and Kuhbauch, 1976) is not of any apparent advantage to this species, as *L. plantarum* and *L. brevis* are much more prevalent in ensiled grasses. Inoculation of silage with strains of *L. plantarum* or other lactobacilli has not consistently improved its quality (Ohyama, Masaki, and Morichi, 1973).

SUGAR PROCESSING. In the sugar industry, lactobacilli are spoilage organisms where their growth is at the expense of sucrose production and their copious slime formation has deleterious physical effects on processing. With cane sugar, most spoilage is caused by leuconostocs. However, sugar-tolerant (able to multiply in 15% sucrose) acidophilic strains of lactobacilli, consisting mainly of *L. confusus* (Sharpe, Garvie, and Tilbury, 1972) and occasionally *L. plantarum* and *L. casei,* multiply in cane juice, causing souring and deterioration of canes. Most of these strains, including *L. plantarum* and *L. casei,* produce large amounts of dextran from the sucrose (Tilbury, 1975). They come from the cane itself and from contaminated equipment. Similar spoilage occurs with beet sugar production (Tilbury, 1975), where strains isolated include *L. casei, L. plantarum, L. cellobiosus,* and *L. fermentum* (Kvasnikov, Kotljar, and Vasileva, 1976).

FRUIT JUICES. Extensive growth of lactobacilli in fruit juices and other citrus products causes off flavor spoilage due to production of high levels of diacetyl (Christensen and Pederson, 1958), or blowing of canned grapefruit (Juven, 1976). Species responsible are *L. brevis* and *L. plantarum,* which in this milieu can multiply at a pH of less than 3.5 and at a temperature of 10°C (Juven, 1976; Murdock and Hatcher, 1975).

FERMENTED VEGETABLES. Lactobacilli participate in the preservation of vegetables in pickling brines where salt-tolerant strains multiply, producing a low pH and imparting characteristic flavor to the pickle. Such fermentations usually consist of a sequential fermentation by lactic acid bacteria. Leuconostocs initiate growth and reduce the pH to about 4.0 and are then succeeded by the more acid-tolerant pediococci and *Lactobacillus plantarum; L. brevis* may multiply more slowly. Such processes include cucumber pickling (Etchells, Fleming, and Bell, 1975), sauerkraut making (Stamer, 1975), and fermentation of olives (Vaughn, 1975). The species present and their exact sequence vary from one process to another, being controlled in industrial production by salt concentration, temperature, and sometimes by inoculation with a suitable lactobacillus starter, usually *L. plantarum.* In sauerkraut, spoilage may be caused by strains of *L. brevis* that impart a red color to the white cabbage (Stamer, 1975).

Fermented Beverages

Such products as wine, cider, and beer are poor media for the growth of lactobacilli, being impoverished in many of the nutrients present in the original fruit extract or grain by growth of the yeast used to bring about the required fermentation. Yet acid- and ethanol-tolerant strains have adapted themselves to grow, albeit slowly, in these habitats, sometimes evolving unusual enzyme processes to meet the requirements of their environment. Such lactobacilli are usually slow growing, sensitive to pH's over 6.5 with an optimum pH for growth of 4.0–5.0, and able to ferment few sugars.

WINE. Lactobacilli commonly occur in many types of wine, despite the high level of ethanol, low pH of 3.2–3.8, and the added SO_2 present. They may contribute favorably to the wine-making process or may cause spoilage. Their ability to metabolize organic acids such as malic, citric, and tartaric may affect the final product in a desirable or undesirable way. Lactobacilli are concerned particularly with the malo-lactic fermentation in which L-malic acid is converted to L-lactic acid and CO_2. This fermentation, which causes a rise in pH of the wine, is beneficial in an acid wine because it decreases the acidity and produces traces of flavorful compounds such as diethyl succinate. In low-acid wines, this process is detrimental and must be controlled. The malo-lactic fermentation is of great economic importance to the wine-making industry and the kinetics of the reaction have been intensively studied (reviewed by Radler, 1975). *Leuconostoc* spp., particularly *Leuc. oenos* occur more commonly than lactobacilli in wines; *Leuc. oenos* is particularly concerned with the malo-lactic fermentation. Most lactobacilli from wines decompose malate, and a variety of malo-lactic fermenting species have been isolated. They include mainly *Lactobacillus plantarum* and unclassified streptobacteria and the heterofermentative species *L. brevis, L. buchneri, L. hilgardii, L. trichodes, L. fructivorans, L. desidiosus,* and *L. yamanashiensis* (Barre, 1969; Chalfan, Goldberg, and Mateles, 1977; Nonomura and Oara, 1967; Peynaud and Domercq, 1967, 1970; Pilone, Kunkee, and Webb, 1966). The same species of lactobacilli are found in French, Spanish, German, Australian, Californian, and Japanese wines. From high-temperature (40–43°C) fermenting grape musts, however, Barre (1978) has isolated thermobacteria, some of which closely resemble *L. acidophilus.*

Other effects of lactobacilli in wine are due to production of diacetyl from citric acid, flavorful in traces, but causing spoilage if present in excess; spoilage such as bitterness caused by excess formation of mannitol from fermentation of fructose; and occasional flocculent growth of *L. trichodes* (Amerine and Kunkee, 1968). The source of malo-lactic bacteria in wine is uncertain (Kunkee, 1967). They have been detected only sporadically and in small numbers on grapes and grape leaves and may well be part of the established flora of the winery itself.

CIDER. The indigenous microflora of a cider factory is composed partly of lactobacilli. As with wine lactobacilli, only strains that have adapted themselves to survive a low pH, low levels of C and N compounds, and the presence of increasing ethanol will form part of the permanent flora. Many of these selected strains, particularly the heterofermentative species, can metabolize quinic acid, present at relatively high levels in cider-apple juice, and malic and citric acids (Carr, 1959). Lactobacilli may multiply in stored ciders, where they bring about several changes. They may take part in the malo-lactic fermentation—discussed with the wine lactobacilli—which will often be beneficial to the flavor of the cider; they may also metabolize citrate and pyruvate, yielding acetate, lactate, and acetoin. The metabolizing of fructose plus quinic acid to acetate, CO_2, and dihydroxyshikimate is of special interest (Whiting, 1975), since the acetate formed is detrimental to flavor.

Heterofermentative cider isolates are usually *Lactobacillus brevis.* They have an optimum pH of 4.0–5.0 and metabolize actively only at these low pHs. Fructose is preferentially utilized, glucose often only weakly (Carr, 1959). Slime-forming strains cause ropiness by production of polysaccharide from glucose, fructose, or maltose but not from sucrose (Millis, 1951). Homofermentative species are mainly strains of *L. plantarum* and also *L. yamanashiensis* (*L. mali*) (Carr et al., 1977) which does not, however, utilize quinic acid.

BEER. In the brewery, lactobacilli are always spoilage organisms, growing in the beer and producing unwanted off flavors and turbidity. Heterofermentative strains are prevalent and preferentially ferment maltose. They grow poorly on glucose unless an arginine supplement is present when the arginine is utilized as an energy source (Rainbow, 1975). They ferment a narrow range of carbohydrates, have complex nutritional requirements, and are able to tolerate the pHs of 3.8–4.3 obtained in their environment. Strains are usually *L. brevis, L. buchneri, L. fermentum,* or the slime-forming *L. vermiforme.* Occasionally, strains of *L. plantarum* and *L. casei* subsp. *alactosus* have been found (Kirsop and Dolezil, 1975). Their growth causes a silky turbidity in the beer and off flavors such as diacetyl (Scherrer, 1972), while slime-forming strains cause ropiness (Williamson, 1959). Their tolerance to the hop resin humolene, unusual for Gram-positive organisms, and the rapidity with which such resistance is acquired (Richards and Macrae, 1964) suggest that they have become adapted to the brewery as their natural habitat. Only with certain continental and kaffir beers are lactobacilli utilized in the fermentation process, a culture of *L. delbrueckii* being added to the mash to increase acidity (Whiting, 1975).

GRAIN MASHES. During the manufacture of malt whisky, lactobacilli may multiply to reach high numbers during the fermentation process itself. In contrast to brewing, the malt is not boiled and lactobacilli are often present in the malted barley. Thermophilic strains may multiply during fermentation when a rich supply of nutrients is readily available. This rapidly depresses the pH to such an extent that the activity of the debranching enzymes and residual amylases of the yeast are inhibited, fermentation is not completed, and a much lower yield of alcohol results (MacKenzie and Kenny, 1965; Simpson, 1968). Species of lactobacilli isolated

from distillery fermentations include *L. fermentum, L. brevis, L. casei, L. leichmannii, L. delbrueckii,* and *L. plantarum* (Bryan-Jones, 1975).

Lactobacillus–Yeast Food Fermentations

The heterofermentative lactobacillus, *Lactobacillus sanfrancisco,* which contributes to the manufacture of sourdough bread is another example of adaptation (Kline and Sugihara, 1971). Added as a mixed starter with a yeast, it ferments only maltose and requires an unidentified nutritional factor present in fresh yeast but not in the dried product (Sugihara and Kline, 1975). In Panettone cake, another sourdough product fermented by a naturally acquired flora of yeast and lactobacilli, the lactobacilli consist of *L. cellobiosus* and *L. plantarum* (Galli and Ottogalli, 1973). Both species ferment the usual range of sugars, including glucose and arabinose. Here, however, the substrate includes sugar and eggs and it has been unnecessary for the organisms to adapt to a nutritionally restricted environment.

In soya sauce manufacture (Wood et al., 1975), the lactobacillus has had to adapt itself quite differently. Here the homofermentative *L. delbrueckii* multiplies in an 18% brine solution containing soy bean and wheat flour on which, before brining, *Aspergillus* was previously grown. The enzymatic activities of the mold have released fermentable sugars, peptides, and amino acids into the brine so that the lactobacillus has a rich nutrient medium. It becomes the dominant organism in the brine until the pH reaches 5.0, when acidophilic halophilic yeasts take over.

LACTOBACILLUS-FERMENTED NOVEL FOODS. As soya bean products are increasingly used as food analogs, the feasibility of fermented products is now considered. Lactobacilli selected for the manufacture of fermented soya products must be able to utilize sucrose, the main carbohydrate present. Such strains include *Lactobacillus acidophilus, L. cellobiosus,* and *L. plantarum* (Mital, Steinkraus, and Naylor, 1974). *Lactobacillus acidophilus* has given promising results in producing fermented soy milks, but *L. bulgaricus* strains initially required addition of glucose or lactose (Wang, Kradej, and Hesseltine, 1974). The use of selected lysine-excreting mutants of lactobacilli to ferment lysine-deficient products such as soya, and thus increase the lysine content, has been suggested. Wild-type, lysine-excreting mutants of *L. acidophilus* and *L. bulgaricus* have been used to ferment soya bean milk to soy yogurt with increased lysine content, while mutants of *L. plantarum* have been used for corn silage (Sands and Hankin, 1976).

GENERAL COMMENT. Lactobacilli appear to come from two very different types of natural habitats: (i) man and animals, and (ii) plant material. In man and animals, they may dominate the flora, as in the upper gastrointestinal tract, or, as in the oral cavity and lower intestine, they may form only a minority of the total, strictly anaerobic flora. No particular adaptation appears to have occurred in lactobacilli in these environments. On plant material, lactobacilli occur in very small numbers, being dominated by the more successful leuconostocs. However, the species discussed in this section, which have adapted themselves to grow in various substrates containing 10–18% NaCl, 15% sucrose, or 15% ethanol, at pH 3.0 or at 1°C, are likely to have originated from such plant material. In the man-manipulated environments in which they occur, they have adapted themselves to such environmental extremes, this adaptation sometimes involving acquisition of special metabolic capacities that have endowed them with unique abilities to occupy their particular niches.

Isolation

Isolation media for lactobacilli must take into account the aciduric or acidophilic nature of these organisms and their complex nutritional requirements. In some cases species have adapted to extreme environmental conditions and can only grow on media that simulate their natural habitat. All media must contain adequate growth factors, which usually include yeast extract, peptone, vitamins, manganese, acetate, and often the stimulatory Tween 80 (discussed by Rogosa and Sharpe, 1959; Rogosa, Mitchell, and Wiseman, 1951). A low pH, ranging between 4.5 and 6.2, favors growth. In some habitats, particularly spoilage situations, lactobacilli may constitute the only organism present; usually they occur together with other organisms, which may include other lactic acid bacteria such as streptococci, pediococci, and leuconostocs. In other situations lactobacilli may be only a minority flora.

Media

When lactobacilli are the majority flora, the general purpose nonselective medium MRS agar (de Man, Rogosa, and Sharpe, 1960) can often be used for isolation. This medium is discussed in detail by Sharpe and Fryer (1965) and compared with the somewhat similar medium APT (Evans and Niven, 1951), which is commonly used for isolating *Lactobacillus viridescens* and other lactobacilli from meat products.

Nonselective MRS Agar for Isolating Lactobacilli (de Man, Rogosa, and Sharpe, 1960)

> For 1 liter of medium, dissolve 15 g agar in distilled water by steaming. Add:

Oxoid peptone	10 g
Meat extract	10 g
Yeast extract	5 g
K_2HPO_4	2 g
Diammonium citrate	2 g
Glucose	20 g
Tween 80	1 g
Na acetate	5 g
$MgSO_4 \cdot 7H_2O$	0.58 g
$MnSO_4 \cdot 4H_2O$	0.28 g

> Adjust pH to 6.2–6.4. Distribute the medium in convenient amounts and sterilize at 121°C for 15 min.

Nonselective APT Medium for Isolating Lactobacilli (Evans and Niven, 1951)

> For 1 liter of medium, dissolve 15 g agar in distilled water by steaming. Add:

Tryptone	10 g
Yeast extract	5 g
K_2HPO_4	5 g
Na citrate	5 g
NaCl	5 g
Glucose	10 g
Tween 80	1 g
$MgSO_4 \cdot 7H_2O$	0.8 g
$MnCl_2 \cdot 4H_2O$	0.14 g
$FeSO_4 \cdot 7H_2O$	0.04 g

> Adjust pH to 6.7 to 7.0. Distribute the medium in covenient amounts and sterilize at 121°C for 15 min.

When lactobacilli occur in conjunction with other organisms, selective media are necessary. The most widely used of these for the past 25 years is the acetate medium (SL) of Rogosa, Mitchell, and Wiseman (1951). In this and other similar media, the selective action is based on a low pH of 5.4, a high concentration of acetate ions (inhibitory to many other organisms), and the presence of the growth stimulatory Tween 80. These media are further described and discussed by Sharpe (1960). Care is needed in preparing SL medium, particularly with regard to the final pH. If this is higher than 5.4, the growth of streptococci is not inhibited. Dehydrated preparations can be purchased commercially, but the final pH should be checked. The acetate–acetic acid buffer mixture described originally (Rogosa, Mitchell, and Wiseman, 1951) tends to give a higher pH than claimed. Accordingly the present procedure is advised (D. J. Jayne-Williams, personal communication).

Selective SL medium for Isolating Lactobacilli
(Rogosa, Mitchell, and Wiseman, 1951)
For 1 liter of medium:

BBL Trypticase	10 g
Yeast extract	5 g
KH_2PO_4	6 g
Diammonium citrate	2 g
$MgSO_4 \cdot 7H_2O$	0.58 g
$MnSO_4 \cdot 4H_2O$	0.28 g
Glucose	10 g
Arabinose	5 g
Sucrose	5 g
Tween 80	1 g
Na acetate $\cdot$ 3H$_2$O	2.5 g
Glacial acetic acid (to titrate medium to pH 5.4)	99.5%
Agar	15 g

Dissolve the agar separately by steaming in 500 ml distilled water. Dissolve all the other ingredients except acetate and acetic acid in 300 ml distilled water without heating, then add this to the melted agar and steam for a further 5 min; excessive heating at this stage must be avoided. Dissolve the Na acetate in about 15 ml distilled water, without heating, and add acetic acid as a 10% vol/vol aqueous solution to pH 5.4. Make up volume to 20 ml. Add this buffer mixture to the hot basal medium and mix well. Cool a small portion and check with a glass electrode that the pH is 5.4. If too high, adjust with further acetic acid. While hot, distribute the medium in convenient amounts in sterile screw-capped bottles; no further sterilization is given. The medium should be a clear, light straw color, giving a firm gel. For use, dissolve in free-flowing steam. Avoid repeated melting and cooling. The addition of arabinose and sucrose as well as glucose allows the growth of strains that preferentially ferment these sugars.

This SL medium is recommended for isolation of a wide range of lactobacilli. However, the common meat spoilage species *L. viridescens* will not grow on SL agar (Rogosa and Sharpe, 1959); neither will the large numbers of unclassified Gram-positive isolates, now regarded as lactobacilli, from chilled chicken meat (Thornley and Sharpe, 1959), nor isolates from very acidic environments. Most streptococci and other organisms are inhibited, but the growth of some pediococci and leuconostocs (dairy and fermented vegetable sources), bifidobacteria (intestinal sources), and yeasts may occur. As these pediococci and leuconostocs have metabolic characteristics in common with lactobacilli and many cause similar changes in a product, their detection may be useful (Sharpe, 1962a).

ORAL CAVITY, INTESTINE, AND VAGINA.   SL medium was designed initially for selective isolation of lactobacilli from oral and intestinal sources (Rogosa, Mitchell, and Wiseman, 1951) and has remained the medium of choice. Bifidobacteria and an occasional streptococcus may also grow, and colonies may have to be further identified.

MILK AND DAIRY PRODUCTS.   SL medium is used for isolation of lactobacilli from milk, cheese, and many fermented milks. Cheese starter streptococci are completely suppressed when enumerating cheese samples. Some leuconostocs and pediococci, often found in milk and cheese, are not inhibited and colonies may have to be further identified. SL may not be optimum for some thermobacteria from dairy sources (Naylor, 1956; Rogosa, Mitchell, and Wiseman, 1951); for selective isolation from yogurt, M16 agar (Terzaghi and Sandine, 1975), with pH adjusted to 5.6 with 1.0 M acetic acid, has been used successfully (Davies, Shankar, and Underwood, 1977).

MEAT AND MEAT PRODUCTS.   The use of APT rather than MRS for the isolation of *L. viridescens* and other lactobacilli from meat is probably traditional rather than necessary, as these organisms also grow profusely on MRSA. Kitchell and Shaw (1975), discussing media for isolation from meats, suggest MRS in addition to APT, incorporating 0.1% thallous acetate with the pH adjusted to 5.5, or SL adjusted to pH 5.8.

FERMENTED VEGETABLES AND SILAGE.   For silage isolates SL medium is used. Keddie (1951) designed a medium similar to SL for this purpose. For the low-pH, fermented vegetable processes (cucumber, sauerkraut, olives, etc.) SL is suggested.

FRUIT JUICE.   For spoilage organisms from orange juice Juven (1976) recommends APT medium.

FERMENTED BEVERAGES.   Isolations from wine, cider, beers, and fermented grain mashes, where lactobacilli have adapted to extremely specialized environments, require quite different types of media. It may be necessary to include some of the natural substrate to provide the unknown growth factors necessary for strains which have become particularly adapted to their environment. Often tomato juice can replace these specific growth factors. It may also be necessary to suppress such aciduric organisms as yeasts, molds, and acetobacters (early work cited by Sharpe, 1960) with inhibitory agents.

WINE.   For isolation of the slow-growing lactobacilli from wines—both those taking part in the malo-lactic fermentation and spoilage strains—tomato juice and yeast extract are highly stimulatory and should be included in the medium. The pH should not exceed 5.0. Yoshizumi (1975) suggests the following medium:

Tomato Juice Medium for Isolating Lactobacilli from Wine (Yoshizumi, 1975)
For 1 liter of medium, dissolve in distilled water:

Glucose	10 g
Yeast extract	5 g
Polypeptone	5 g
KH_2PO_4	0.5 g
KCl	0.125 g
$CaCl_2 \cdot 2H_2O$	0.125 g
NaCl	0.125 g
$MgSO_4 \cdot 7H_2O$	0.125 g
$MnSO_4 \cdot 4H_2O$	0.003 g
Bromocresol green	0.03 g
Agar	15 g
Canned tomato juice	150 ml

Steam the agar to dissolve first. Adjust the final pH to 5.0. Sterilize at 121°C for 15 min. A fungiostat should be added to inhibit the growth of yeasts. The authors mention Eurocidin or Kabicidin (100 mg/liter). If these cannot be obtained, cycloheximide (100 mg/liter) or sorbic acid (1.2 g/liter) have been used (Chalfan, Goldberg, and Mateles, 1977). It is essential to cultivate under anaerobic conditions for isolation from the later stage of fermentation (Yoshizumi, 1975).

For isolation of lactobacilli from some wines it is necessary to use a grape-based medium with added yeast extract and a pH of 3.2–4.5, depending on the wine being examined (Castino, Usseglio-Tomasset, and Gandini, 1975; Chalfan, Goldberg, and Mateles, 1977).

CIDER. An apple juice–based medium is recommended (Carr and Davies, 1970) consisting of apple juice + 1% yeast extract, specific gravity adjusted to 1,040, and the pH value to 4.8 with NaOH. The addition of 3% agar is necessary to ensure a firm gel.

BEER. For isolation of spoilage-causing brewery lactobacilli, many selective media have been used (Hsu and Taparowsky, 1977). Boatwright and Kirsop (1976) described a sucrose agar medium and confirmed the accepted usefulness of cycloheximide, polymixin B, and phenyl ethanol in suppressing yeasts and Gram-negative bacteria. This sucrose agar compared favorably with other brewery media and could be used to cultivate a wide range of lactobacilli, but pediococci and leuconostocs also grew.

Sucrose Agar for Brewery Isolates (Boatwright and Kirsop, 1976)
For 1 liter of medium, dissolve in distilled water:

Sucrose	50 g
Oxoid peptone	10 g
Yeast extract	5 g
NaCl	5 g
$MnSO_4 \cdot 4H_2O$	0.5 g
$MgSO_4 \cdot 7H_2O$	0.5 g
Tween 80	0.1 g
$CaCO_3$	3 g
Bromocresol green	20 mg
Agar	20 g

Steam the agar first to dissolve it. Final pH is adjusted to 6.2. Sterilize at 121°C for 15 min after distributing in convenient amounts. Microbial inhibitors are added to the molten agar just before pouring plates, cycloheximide as a Seitz-filtered sterile solution to give a final concentration of 10 μg/ml, and 2-phenyl ethanol without dilution or sterilization, final concentration 0.3%.

Savel (1975), using a fermented wort medium and phenyl ethanol to suppress the Gram-negative bacteria, added 10 mg/liter $CdCl_2$, which selectively suppressed the pediococci and allowed lactobacilli to grow. This effect of $CdCl_2$ is an interesting observation which requires further attention. It is recognized that not all beer spoilage lactobacilli will grow on any one medium and it may be necessary to use a basal medium to which beer from the actual brewery concerned is added, to provide the correct growth factor.

GRAIN MASHES. MRS agar was found to be unsatisfactory for isolations and a medium based on a mixture of filter-sterilized malt extract and yeast autolysate has been developed (Bryan-Jones, 1975).

SAKE. For isolation of spoilage lactobacilli from rice wines it is necessary to use media containing D-mevalonic acid, a required growth factor (*Bergey's Manual,* 1974).

SOURDOUGH. *Lactobacillus sanfrancisco* in sourdough bread will not grow on APT, SL, or tomato juice–based media (Kline and Sugihari, 1971) and requires freshly made yeast extract, a high level (3–4%) of maltose, and a pH of 5.6 (Sugihari and Kline, 1975).

GASEOUS ENVIRONMENT. Most lactobacilli grow better either anaerobically or in the presence of increased CO_2 tension, particularly on first isolation. Agar plates should be incubated in an atmosphere of 90% H_2 + 10% CO_2. Surface plating is recommended so that different colonial types can be observed if present, often indicating the presence of more than one species or biotype. For isolation of anaerobic intestinal lactobacilli, poured, dried plates must be prereduced by overnight incubation in 90% H_2 + 10% CO_2. When selective media are used, particularly the SL medium, care must be taken not to dry the plates too long or the concentration of acetate at the agar surface may inhibit the lactobacilli. Isolates from animal sources and some dairy products are incubated at 37°C and from other habitats at 30°C, or at 22°C from low-temperature sources.

Axenic Culture and Maintenance

AXENIC CULTIVATION. Once isolated, unless there are special growth requirements, many species of lactobacilli can be cultured in MRS broth (de Man, Rogosa, and Sharpe, 1960) or maintained for short periods on MRS agar slopes or stabs. For anaerobic lactobacilli 0.05% cysteine should be added, the broth steamed just before use, and organisms cultured under 90% H_2 + 10% Co_2. For some strains or species of lactobacilli, particularly heterofermentative ones, a carbohydrate other than glucose, such as maltose or fructose, gives better growth. It might be useful to use the triple sugars as in SL medium. Wine isolates may still require 5% tomato juice or grape juice in the medium, with the pH adjusted to 5.0; cider isolates are cultivated on the apple juice based medium previously described; *Lactobacillus trichodes* requires 10% ethanol.

CONSERVATION OF CULTURES. For preservation for 3–6 months, strains can be stored in yeast glucose litmus milk (YGLM) + calcium carbonate (Bryan-Jones, 1975; Sharpe and Fryer, 1965). To reconstituted skim milk powder or fresh skim milk add litmus, final concentration 0.01%; yeast extract, 0.2%; glucose, 1%; liver extract, 0.25%; and calcium carbonate, 5%. Tube in 10-ml amounts, sterilize at 121°C for 10 min. Before use tubes should be incubated for 1 week to check for sterility. This medium is not suitable for the acidophilic heterofermentative isolates from wines and cider. Stab cultures in tomato juice agar at pH 5.0 are preferred.

FREEZE DRYING. Methods of lyophilization of lactic acid bacteria have been discussed by Radulovic (1971) and Morichi (1974). Although the addition of various cryoprotective agents to the cell suspension fluid cells has been suggested (reviewed by Bousfield and MacKenzie, 1976), horse serum glucose is quite satisfactory. Using the method described by Phillips, Latham, and Sharpe (1975), centrifuged packed cells from vigorously growing broth cultures are resuspended in 1 ml sterile horse serum containing 7.5% glucose and freeze-dried using the standard techniques of Lapage et al. (1970). Ampules are sealed under vacuum and stored at 6–10°C. Provided that the broth cultures are not overgrown, this simple technique has resulted in excellent preservation of strains. Many are still viable after 10–20 years, although others require more frequent relyophilization (E. I. Garvie, National Collection of Dairy Organisms, personal communication). Differences in viability are strain- rather than species-specific.

Identification

Lactobacilli are identified as Gram-positive, nonsporeforming rods, catalase negative, usually nonmotile, that do not usually reduce nitrate and that utilize glucose fermentatively (*Bergey's Manual*, 1974). They may be either homofermentative, producing more than 85% lactic acid from glucose, or heterofermentative, producing lactic acid, CO_2, ethanol, and/or acetic acid. Other organisms found in some of the same habitats and often growing on the same selective media as lactobacilli are leuconostocs, pediococci, bifidobacteria, or occasionally streptococci. In theory these should all be morphologically distinguishable from lactobacilli, but in practice this is not always so. Some heterofermentative lactobacilli grow as coccobacilli and are usually differentiated from morphologically similar leuconstocs by their production of ammonia from arginine (except *L. viridescens* and *L. confusus*) by not fermenting trehalose and by forming DL- and not DL-(−)-lactic acid from glucose. Some homofermentative anaerobic lactobacilli from intestinal sources may be club shaped and thus morphologically similar to bifidobacteria; to distinguish them it must be confirmed that they do not produce CO_2 and that their major fermentation end product from glucose is lactic acid, whereas bifidobacteria, which also do not produce CO_2, produce mainly acetic acid (2 acetate: 1 lactate). Pediococci, also isolated on several selective media for lactobacilli, can be differentiated morphologically, as can streptococci (see Sharpe, 1978).

Subgenera

The genus *Lactobacillus* was divided by Orla-Jensen (1919, 1943) into three main subgroups, *Thermobacterium*, *Streptobacterium*, and *Betabacterium*, on the basis of optimal growth temperatures and fermentation end products. The genus is still divided in this way, the groups having been confirmed by other tests (Rogosa, 1970). While Rogosa (*Bergey's Manual*, 1974) does not mention these subgroups by name, the three main subdivisions in his chapter on lactobacilli are, as he previously stated (Rogosa, 1970), based on this concept. Other recent reviews (London, 1976; Sharpe, 1974) recognize the subgroups and numerical taxonomic studies; e.g., Wilkinson and Jones (1977) group the lactobacilli into three phena corresponding largely to the subgenera. It has been shown (Buyze, van den Hamer, and de Haan, 1957; van den Hamer, 1960; Williams, 1971) that *Thermobacterium* and *Streptobacterium* contain fructose-1,6-diphosphate aldolase while *Betabacterium* species do not. Streptobacteria and betabacteria also contain glucose-6-phosphate dehydrogenase (G6PDH) and 6-phosphogluconate dehydrogenase (6PGDH); in the thermobacteria some species contain both G6PDH and 6PGDH, while in others (*L. delbrueckii, L. lactis,* and *L. leichmannii*) these two enzymes are either present in very small amounts or not detected (discussed by Williams, 1975).

The three subgenera are well established, and Table 1 shows simple biochemical tests which can be used to divide them. In practice, betabacteria are distinguished from the homofermentative lactobacilli by the production of CO_2 from glucose. Further confirmation is obtained by a thiamine requirement for growth (Rogosa, Franklin, and Perry, 1961), the absence of FDP aldolase, and the production of mannitol as an end product of the fermentation of fructose (Eltz and Vandemark, 1960).

A simple and useful method for detection of mannitol is described by Chalfan, Levy, and Mateles (1975) although in this paper there is some confusion as, unfortunately, the positive and negative results have been transposed. Also many heterofermentative, but few homofermentative lactobacilli produce ammonia from arginine in a standard test with a high level of glucose (2%) present.

Homofermentative strains are divided into thermobacteria and streptobacteria by growth temperatures and fermentation of ribose and gluconate (Rogosa, 1970), the groups being listed as IA and IB, respectively, by Rogosa (*Bergey's Manual,* 1974). *Betabacterium* is further divided into IIA and IIB, the latter group being less active biochemically, acidophilic, and ethanol tolerant. Rogosa's division of these into groups II and III (*Bergey's Manual,* 1974) seems to imply a much wider division than can be assessed on present data, so they are listed here as IIA and IIB. Each subgroup contains a number of well-defined species (Table 1), most of those listed being described in *Bergey's Manual.*

Species can generally be differentiated by simple biochemical and physiological tests (Tables 2–5). Methods of identification have been described and

Table 1. Subdivision of genus *Lactobacillus*.

Test[a]	I. Homofermentative		II. Heterofermentative	
	Glucose fermented almost entirely (over 85%) to lactic acid (LA)		Glucose fermented to LA (50%) + CO_2 + acetic acid + ethanol	
1[b]	−		+	
2	−		+	
3	+		−	
4	−		+	
	Thermobacterium 1A	*Streptobacterium* 1B	*Betabacterium* II	
5[b]		d[c]	} dependent on species	
6[b]	−	+	+ usually	
7[b]	−	+	+	
8[b]	−	+	All form DL-LA from glucose	
			IIA	IIB acidophilic, ethanol tolerant, inactive to most carbohydrates
	L. acidophilus	*L. casei*	*L. fermentum*	*L. hilgardii*
	L. helveticus	*L. plantarum*	*L. cellobiosus*	*L. trichodes*
	L. bulgaricus	*L. xylosus*	*L. brevis*	*L. fructivorans*
	L. lactis	*L. curvatus*	*L. buchneri*	*L. desidiosus*
	L. delbrueckii	*L. coryniformis*	*L. viridescens*	*L. heterohiochi*
	L. leichmannii	*L. homohiochi*	*L. confusus*	
	L. salivarius	*L. yamanashiensis*		
	L. jensenii	*L. farciminis*		
		L. alimentarius		
	L. ruminis } anaerobic *L. vitulinus*			

[a]1, CO_2 from glucose; 2, thiamine required for growth; 3, fructose diphosphate aldolase present; 4, fructose fermented to produce mannitol; 5, growth at 45°C; 6, growth at 15°C; 7, ribose fermented; 8, CO_2 from gluconate.
[b]Simple test used for initial identification of isolates.
[c]d, Variable reaction.

discussed by Rogosa et al. (1953), Rogosa and Sharpe (1959), and Sharpe (1962c, 1978). Determination of lactic acid isomers formed from glucose is often extremely useful for identification although more time-consuming than other simple tests. Total lactic acid can be determined chemically and L-(+)-lactic acid enzymatically (Garvie, 1967) and D-(−)-acid taken as the difference. The presence of *meso*-diaminopimelic acid (*meso*-DAP) in the peptidoglycan, discussed later, can now be simply detected and distinguishes some species (Schleifer and Kandler, 1972). Strains unidentifiable at species level can usually be assigned at least to the appropriate subgroup; in some cases further identification may be unnecessary and functional grouping of isolates may be more important, i.e., detection of all strains producing slime, H_2O_2 or showing extreme acid or ethanol tolerance.

THERMOBACTERIUM.   Ten species are presently recognized (Table 2), two of them being anaerobic. *Lactobacillus jugurt* is now recognized as a subspecies of *L. helveticus,* and it is also suggested (*Bergey's Manual,* 1974) that *L. lactis* and *L. bulgaricus* may be variants of a single species. The presence of purple granules in methylene blue stained cells of *L. lactis* and *L. bulgaricus* rapidly differentiate these species from *L. helveticus. Lactobacillus jensenii* is phenotypically almost identical to *L. leichmannii,* and can only be distinguished by complex tests described later (Gasser, Mandel, and Rogosa, 1970).

STREPTOBACTERIUM.   In addition to the classical species *Lactobacillus plantarum* and *L. casei* with its subspecies *alactosus* (does not ferment lactose) and *rhamnosus,* two other subspecies of *L. casei* are recognized (Abo-Elnaga and Kandler, 1965a): *L. casei* subsp. *tolerans,* a heat-resistant organism isolated from pasteurized milk, and subsp. *pseudoplantarum,* which, unlike *L. casei,* produces DL-lactic acid. Five other species are tabulated (Table 3), *L. xylosus, L. curvatus,* and *L. coryniformis* designated by Abo-Elnaga and Kandler (1965a) and *L. homohiochi* by Kitahara, Kaneko, and Goto (1957). *L. yamanashiensis* is now recognized to be the same as *L. mali* and has the prior name (Carr et al., 1977). As we have seen, many isolates from dairy, meat, and other food and plant sources are frequently identified only as strains of unclassified streptobacteria. These strains often have less active fermentation patterns than *L. casei* and *L. plantarum* and it may be possible to identify them as one of these later named species. *L. curvatus, L. coryniformis,* and the two newer subspecies of *L. casei* may be the same as atypical strains isolated by Keddie (1959) from silage and herbage (Rogosa, *Bergey's Manual,* 1974). There is a great need for broader taxonomic studies of unclassified strepto-

bacteria from all the above-mentioned habitats. Few comparative investigations have been made.

BETABACTERIUM.   Group IIA contains well-recognized, fermentatively active species (Table 4). It is now considered that *Lactobacillus fermentum* and *L. cellobiosus* are closely related to each other (Gasser, 1970; Sharpe, 1974; Williams, 1975), and *L. buchneri* may be a subspecies of *L. brevis* as it is distinguished mainly by the fermentation of melezitose. Further work is required to decide whether the slime-forming group *L. vermiforme* merits the species status which Sharpe, Garvie, and Tilbury (1972) consider appropriate. There are many common properties between *L. viridescens* and *L. confusus* and *Leuconostoc dextranicum* and *Leuconostoc mesenteroides,* respectively, including slime production and sugar fermentation patterns, and they are considered to be closely related (Garvie, 1976; Sharpe, Garvie, and Tilbury, 1972), although the leuconostocs form only D-(−)-lactic acid and have a single lactate dehydrogenase (LDH) while these lactobacilli form DL-lactic acid and have both a D-(−)- and L-(+)-LDH. *L. confusus* was originally named *L. coprophilus* subsp. *confusus* (Holzapfel and Kandler, 1969), but the original strains of *L. coprophilus* are no longer extant and no new strains have been isolated; therefore *L. confusus*—a widely distributed and well-recognized organism—has been raised to species level (Sharpe, Garvie, and Tilbury, 1972). Group IIB (Table 5) comprises species which are inert to most carbohydrates, able to grow at pH's as low as 3.2, and are organic acid and ethanol tolerant. They have been less widely studied than some of the others and appear to occur in more restricted habitats, although as Carr (1967) suggests, this may be due to workers' confining their observations to isolates from a single habitat. It is suggested by Dakin and Radwell (1971) that *L. fructivorans,* from acetic acid preserves, and *L. trichodes,* from wine, are the same species. Peynaud and Domercq (1970) in a study of 250 fresh isolates from grapes, musts, and wines suggest that these isolates should be identified as *L. fructivorans, L. desidiosus, L. hilgardii,* and *L. brevis* mainly on fermentation of arabinose and xylose, fermentation patterns for these two sugars being − −, − +, + −, and + + for the four species, respectively.

The designated species of the three subgenera are further confirmed by other more fundamental phenotypic studies and by genotypic data. Numerous papers and reviews have appeared on all these aspects (for general reviews on the taxonomy of the lactobacilli see London, 1976; Rogosa, 1970; Rogosa and Sharpe, 1959; Sharpe, 1962c, 1974; Williams, 1975). Most of these further taxonomic studies confirm the identity of *Lactobacillus* species as shown in Tables 1–5, i.e., in general all strains of a species react in the same way and show the same

Table 2. Differentiating characteristics in species of *Thermobacterium* (group IA *Bergey's Manual*, 1974).[a]

Lactobacillus species	Presence of granules	NH₃ from arginine	Lactic acid isomer	Esculin hydrolysis	Amygdalin	Cellobioses	Galactose	Lactose	Maltose	Mannitol	Melibiose	Salicin	Sorbitol	Sucrose	Trehalose	Vitamin requirement[b]				*meso*-DAP[c] in cell wall peptidoglycan
																Riboflavin	Pyridoxal	Folic acid	Thymidine	
L. delbrueckii	-	d	D(-)	-	-	-	w	-	d	-	-	-	-	+	-	+	-	-	+	-
L. leichmannii	+	d	D(-)	+	+	+	-	+	+	-	-	+	-	+	+	-	-	+	-	-
L. lactis	+	-	D(-)	d	-	-	+	+	+	-	-	+	-	+	+	+	-	-	-	-
L. bulgaricus	+	-	D(-)	-	-	-	+	+	-	-	-	-	-	-	-	+	-	-	-	-
L. helveticus	-	-	DL	+	+	+	+	+	+	-	-	+	-	+	d	+	+	-	-	-
L. acidophilus	-	-	DL	+	+	+	+	+	+	-	d	d	+	+	+	+	+	+	-	-
L. salivarius	-	-	L(+) and DL	d	-	-	-	-	+	+	+	+	+	+	+	+	-	+	-	-
L. jensenii	-	+	D(-)	+	+	+	+	-	+	-	+	+	-	+	+	-	-	+	-	-
L. ruminis	-	-	DL[d]	+	+	+	+	±	+	-	+	+	d	+	-	.	+	+	.	+
L. vitulinus	-	-	D(-)	+	+	+	+	+	+	-	+	+	d	+	d	.	.	+	.	+

[a] None ferments ribose, arabinose, xylose, melezitose; all ferment glucose. +, Positive reaction; -, negative reaction; d, variable reaction; w, weak reaction; ., no data available.

[b] +, Requirements; -, no requirement.

[c] *meso*-Diaminopimelic acid.

[d] L(+), 95%; D(-), 5%.

Table 3. Differentiating characteristics of species of *Streptobacterium* (group IIB in *Bergey's Manual*, 1974).[a]

Lactobacillus species	Growth at 45°C	Lactic acid isomer	Esculin hydrolysis	Amygdalin	Arabinose	Cellobiose	Galactose	Lactose	Maltose	Mannitol	Melezitose	Melibiose	Raffinose	Rhamnose	Ribose	Xylose	Vitamin requirements Pyridoxal	Vitamin requirements Folic acid	*meso*-DAP in cell wall peptidoglycan
L. casei subsp. *casei*	−	L(+)[b]	+	+	−	+	+	+	+	+	+	−	−	−	+	−	+	+	−
L. casei subsp. *rhamnosus*	+	L(+)[b]	+	+	−	+	+	+	+	+	+	−	−	+	+	−	+	+	−
L. plantarum	∓	DL	+	+	d	+	+	+	+	+	+	+	+	−	+	∓	−	−	+
L. xylosus	−	L(+)	−	+	−	+	+	−	+	+	−	−	−	−	+	+	·	·	−
L. curvatus	−	DL	+	−	−	+	+	w	+	−	−	−	−	d	+	−	·	·	−
L. coryniformis	−	DL or D(−)	−	−	−	−	+	−	+	+	−	d	d	d	−	−	·	·	−
L. homohiochi	−	D(−)	·	−	−	−	±	−	+	+	−	−	d	−	−	−	·	·	−
L. yamanashiensis	−	·	+	−	−	d	±	−	+	+	−	−	−	+	−	−	·	·	+
L. farciminis[c]	−	DL or L(+)	d	·	−	d	+	d	w	−	−	−	·	·	·	−	·	·	−
L. alimentarius	−	L(+)	+	·	−	+	+	+	+	−	−	−	−	·	·	−	·	·	−

[a] All ferment fructose, glucose, and mannose. +, Positive reaction; −, negative reaction; d, variable reaction; w, weak reaction; ·, no data available.
[b] May form a very small amount (1–5%) of D(−)-lactic acid.
[c] None, except some strains of *L. farciminis*, produces NH_3 from arginine under the specific test conditions (2% glucose present; Sharpe, Fryer, and Smith, 1966).

Table 4. Differentiating characteristics of species of *Betabacterium* IIA (group II of *Bergey's Manual*, 1974).[a]

Lactobacillus species	Growth at 15°C	Growth at 45°C	NH₃ from arginine	Slime from sucrose	Esculin hydrolysis	Amygdalin	Arabinose	Cellobiose	Mannitol	Melezitose	Melibiose	Raffinose	Xylose	Riboflavin	Folic acid
						\<-- Fermentation of --\>								\<-- Vitamin requirements --\>	
L. fermentum	−	+	+	−	−	−	d	−	−	−	+	+	d	−	−
L. cellobiosus	±	∓	+	−	+	+	+	+	−	−	+	+	d	−	−
L. brevis	+	−	+	−	d	−	+	−	w	−	+	w	d	−	+
L. buchneri	+	−	+	−	d	−	+	−	w	+	+	w	d	d	d
L. viridescens	+	−	−	d	−	−	−	−	−	−	−	−	−	+	+
L. confusus	+	+	d	+	+	+	±	+	−	−	−	−	+	d	+

[a] All ferment ribose, maltose, and fructose. None ferments rhamnose or sorbitol. Legend as for Table 2.

Table 5. Differentiating characteristics of species of *Betabacterium* IIB (group III of *Bergey's Manual*, 1974).[a]

Lactobacillus species	Growth at 15°C	Growth at 25°C	Growth in 15% EtOH	Growth below pH 4.0	Arabinose	Fructose	Glucose	Maltose	Ribose	Xylose	Malate	Citrate
					\<-- Fermentation of --\>						\<-- Dissimulated --\>	
L. hilgardii[b]	−	+	+	+	+	+	+	+	+	+	+[c]	+
L. trichodes[d]	w	+	+[d]	+	−	+	+	d	.	−	+[c]	−
L. fructivorans[b]	−	+	+	+	−	+	w	d	−	−	+[c]	−
L. desidiosus	+	+	+	.	+	w	w	−	.	−	.	d
L. heterohiochi	.	+	+	.	−	+	+	−	+	−	.	.

[a] None grows at 45°C; none ferments amygdalin, cellobiose, lactose, mannitol, mannose, melezitose, raffinose, rhamnose, salicin, sorbitol, or trehalose; esculin not hydrolyzed. Legends as for Table 2.
[b] Growth greatly stimulated by CO₂ for initial isolation.
[c] Dissimilate malate particularly at pH's below 3.8.
[d] Ethanol highly stimulatory for growth, many strains produce only trace growth without it.

characteristics. In some cases, however, investigations have shown that an apparently phenotypically homogeneous species contains more than one genotype.

Nutritional Requirements

Rogosa, Franklin, and Perry (1961) showed that vitamin requirements were consistent within species (Tables 2, 3, 4). Whereas further work has in general confirmed their findings, Abo-Elnaga and Kandler (1965b), extending the range of species examined, observed that in some species strains might have variable reactions although most results were in agreement with those of Rogosa, Franklin, and Perry (1961). Ledesma et al. (1977) however found some differences, particularly with riboflavin, and the three heterofermentative strains tested did not require thiamine. Such differences were probably due to medium composition. Ledesma et al. used a completely synthetic medium instead of one based on acid-hydrolyzed casein (Rogosa, Franklin, and

Perry, 1961), and it is important to use the medium and technique specified by Rogosa, Franklin, and Perry for identification tests.

Cell Wall and Cell Membrane Composition

PEPTIDOGLYCAN TYPES OF CELL WALLS. Investigations of the structure of the cell wall have shown that different types of peptidoglycans may be present and that these may be characteristic of certain subgroups or species. Kandler (1970) and Schleifer and Kandler (1972) have shown that the amino acid sequences of the interbridge peptide, which is covalently linked to two adjacent peptidoglycans, are useful characteristics for classification of lactobacilli. Using partially hydrolyzed cell walls, chromatography, and amino acid analysis, the sequence of amino acids can be studied quantitatively. Table 6 shows that this method is useful for differentiating the two anaerobic species of lactobacilli from the other thermobacteria, by the presence of *meso*-DAP in the former

Table 6. Peptidoglycan types of cell walls of *Lactobacillus*.[a]

Subgroup or species	Peptidoglycan type
Thermobacterium	
All except anaerobic species	L-Lys-D-Asp
L. ruminis, L. vitulinus	*meso*-DAP
Streptobacterium	
L. casei and subspecies	
L. coryniformis	L-Lys-D-Asp
L. curvatus, L. xylosus	
L. plantarum, L. yamanashiensis	*meso*-DAP
Betabacterium	
L. fermentum, L. cellobiosus	L-Orn-D-Asp
L. brevis, L. buchneri	
L. fructivorans, L. hilgardii	L-Lys-D-Asp
L. trichodes, L. desidiosus	
L. confusus	L-Lys-L-Ala
L. viridescens	L-Lys-L-Ala-L-Ser

[a] From data of Schleifer and Kandler (1972), Williams (1975).

and L-Lys-D-Asp type in the latter, and, similarly, *Lactobacillus plantarum* and *L. yamanashiensis* from other streptobacteria. In the betabacteria, *L. fermentum* and *L. cellobiosus* contain L-Orn-D-Asp (discussed by Williams, 1975; Williams and Sadler, 1971). All the other betabacteria contain L-Lys-D-Asp type except *L. viridescens* and *L. confusus* which contain L-Lys-L-Ala-L-Ser and L-Lys-L-Ala, respectively, types commonly found in leuconostocs. Although not species-specific, these characteristic structures are useful differentiations and can now be used qualitatively. For the detection of *meso*-DAP, whole cells are hydrolyzed and a single-dimension chromatographic method is used for identification (Schleifer and Kandler, 1972).

ANTIGENIC DETERMINANTS.   Serological studies of lactobacilli, using acid extracts of whole cells and precipitin methods, have shown that many strains can be assigned to seven groups based on specific antigenic determinants (summarized by Rogosa and Sharpe, 1959). Groups A, D, F, and G are species-specific, *Lactobacillus casei* strains fall into groups

B and C, but group E includes both *L. lactis* and *L. bulgaricus* and the heterofermentative *L. brevis* and *L. buchnerii*—also closely related to each other, but differing widely in their characteristics from the first two species. It is now known that *L. helveticus* does not cross-react with *L. casei* group B (reported by Knox, 1963, tabulated by Botazzi, 1977, and by London, 1976). The strain of *L. helveticus* used, NCTC 6375, was misclassified and has long since been redesignated as *L. casei* subsp. *rhamnosus*. As shown in Table 7, the carbohydrate polymers released by the acid method of extraction and responsible for specificity of groups A–F are of three types: cell wall polysaccharides, cell wall teichoic acids, and membrane teichoic acids. Much detailed work has been done on these types of antigens (reviewed by Knox and Wicken, 1976; Sharpe, 1970). Studies on teichoic acid in *L. fermentum* led to the significant discovery of lipoteichoic acid (reviewed by Knox and Wicken, 1973). Separation of lactobacilli into different serological groups has been a useful confirmation of physiologically characterized species. However, cross-reactions may occur, due to similar glycosyl substituents (Knox and Wicken, 1976) or to stimulation of antibodies against the glycerol teichoic acid backbone common to all membrane lipoteichoic acids (Sharpe et al., 1973a; Shimohashi, Kamiyama, and Arai, 1973). *Lactobacillus acidophilus* has proved difficult to classify serologically, probably being a heterogeneous species representing at least three different biotypes and serological groups (Gasser, 1970; Sharpe, 1970). Shimohashi and Mutai (1977) have divided the species into four serological groups that are consistent with the four biotypes determined by Mitsuoka (1969).

Electrophoretic Mobilities of Enzymes

The electrophoretic mobilities of individual enzymes of lactobacilli have been investigated and the implications of these findings discussed by London (1976). Morichi, Sharpe, and Reiter (1968) found

Table 7. Group antigens of *Lactobacillus*.[a]

Species	Group	Antigen	Location	Determinant
L. helveticus	A	GTA	Wall-membrane	α-Glc
L. casei	B	Polysaccharide	Wall	α-Rha
L. casei	C	Polysaccharide	Wall	β-Glc
L. plantarum	D	RTA	Wall	α-Glc
L. lactis				
L. bulgaricus	E	GTA	Wall	
L. brevis				
L. buchneri				
L. fermentum	F	GTA	Membrane	α-Gal
L. salivarius	G	?	?	?

[a] GTA, glycerol teichoic acid; RTA, ribitol teichoic acid; Glc, D-glucosyl; Rha, L-rhamnosyl; Gal, D-galactosyl.

that esterase patterns were variable. Most strains of *Lactobacillus casei* had identical single bands of enzyme, *L. lactis* and *L. leichmannii* each had species-specific patterns but variation occurred among strains of other species. However, Gasser (1970) found that the electrophoretic mobilities of NAD-dependent (NADd) and NAD-independent (NADi) lactic dehydrogenases (LDH) with both D-(−)- and L-(+)-lactic acids as substrates gave some useful indications of species relationships. *L. fermentum* and *L. cellobiosus* had identical single bands of NADd D-(−)-LDH, while *L. brevis* and *L. buchneri* showed different band patterns. *L. lactis*, *L. delbrueckii*, and *L. leichamanni* LDHs all had the same mobilities, but *L. bulgaricus* showed a different species-specific single band. *L. salivarius* and *L. jensenii* each had species-specific patterns (Gasser, Mandel, and Rogosa, 1970); *L. acidophilus* was divided into three different groups. *L. plantarum* strains all had the same pattern. This work was extended and included techniques to detect LDHs of *L. casei* and *L. curvatus* (Stetter and Kandler, 1973). The L-(+)- and D-(−)-LDHs of *L. confusus* and *L. viridescens* have been shown to be electrophoretically the same, differing from those of other heterofermentative lactobacilli (Garvie, 1975; Sharpe, Garvie, and Tilbury, 1972). Williams and Sadler (1971) examined the electrophoretic mobility of G6PD and found that in heterofermentative lactobacilli *L. fermentum* and *L. cellobiosus* had an identical single band while *L. viridescens*, *L. brevis*, and *L. buchneri* had more complex patterns. Species-specific patterns were found only with *L. lactis*, *L. leichmannii*, *L. salivarius*, and *L. plantarum* (Williams, 1971). This technique is useful with a limited number of species, and has also indicated relationships later confirmed by other techniques, such as that between *L. fermentum* and *L. cellobiosus*, and the differentiation of *L. leichmannii* and *L. jensenii*.

Enzyme Homology Studies

Immunological analysis of homologous enzymes of lactobacilli has been used to obtain taxonomic information in relation to their evolution. Gasser and Gasser (1971) prepared specific antisera against the D-LDHs from *Lactobacillus leichmanii*, *L. jensenii*, and *L. fermentum* and L-LDH from *L. acidophilus*, and tested them against LDHs from many homo- and heterofermentative lactobacilli. These antisera reacted with most of the LDHs, indicating that the bacteria in this large group are derived from a common ancestor. As with other studies, *L. leichmannii*, *L. delbrueckii*, *L. lactis*, and *L. bulgaricus* were closely related to each other, four subgroups of *L. acidophilus* were differentiated and found to cluster with *L. jensenii*, while *L. helveticus* formed a different group. Other such work has been concerned mainly with relationships between lactobacilli and other lactic acid bacteria (London, 1976).

Genotypic Relationships

DETERMINATION OF DNA BASE COMPOSITION. With the lactobacilli much useful data has emerged for definition of species and confirmation of phenotypic groupings. Although it is widely recognized that differences in G+C content indicate that organisms are not closely related, similarity or identity of G+C content does not necessarily mean that the strains are closely related. However, with lactobacilli the results obtained from determination of DNA base composition have in most cases confirmed the phenotypic species. Table 8 shows a compilation of analyses as reported in *Bergey's Manual* (1974). Results have been reported by Gasser and Mandel (1968), Miller, Sandine, and Elliker (1970), Dellaglio, Bottazzi, and Trovatelli (1973), Dellaglio, Botazzi, and Vescovo (1975), Garvie, Zezula, and Hill (1974), Sharpe et al. (1973b), and Carr et al.

Table 8. Mean base composition of DNA of *Lactobacillus*.[a]

Homofermenters		Heterofermenters		Heterofermenters	
Thermobacterium	mol% G+C	*Betabacterium*	mol% G+C	*Streptobacterium*	mol% G+C
L. salivarius	34.7	*L. fructivorans*	39.4	*L. yamanashiensis*	32.0
L. jensenii	36.1	*L. hilgardii*	40.3	*L. xylosus*	39.4
L. acidophilus	36.7	*L. confusus*	41	*L. curvatus*	43.9
L. helveticus	39.3	*L. trichodes*	42.7	*L. coryniformis*	45.0
L. bulgaricus	50.3	*L. viridescens*	37–42	*L. plantarum*	45.0
L. lactis	50.3	*L. brevis*	42–46	*L. homohiochi*	46.0
L. delbrueckii	50.0	*L. buchneri*	44.8	*L. casei*	46.4
L. leichmannii	50.8				
L. ruminis	43.7	*L. fermentum*	53.4		
L. vitulinus	34–37	*L. cellobiosus*	53.1		

[a] Results cited by Rogosa (*Bergey's Manual*, 1974), determined by buoyant density.

(1977). Slight variations in values have been found between different laboratories and different techniques, and an occasional aberrant strain has been reported (probably due to incorrectly named strains). Williams (1975) tabulates the results of different workers. In general, results are remarkably consistent. The wide span of G+C content in the genus *Lactobacillus* (32–50 mol%) indicates that this group has undergone considerable evolutionary diversification.

Thermobacteria species fall into two groups, one with G+C content of 34–38 mol%, the other with a much higher one of 48–50 mol%. The obligate homofermenter group of Williams (1971) comprised three of these latter species, *L. leichmannii*, *L. lactis*, and *L. delbrueckii*, but not *L. bulgaricus*. Most streptobacteria have values over a range of 39–46 mol%. It is of interest that *L. yamanashiensis*—allocated by numerical taxonomy studies (Wilkinson and Jones, 1977) to a separate phenon outside the three main subgenera—although phenotypically a member of the streptobacteria, has a much lower G+C content than other members of this subgenus. With the betabacteria, *L. fermentum* and *L. cellobiosus,* species commonly found in animal habitats, have higher values than the other heterofermentative species. This technique has also been useful in detecting heterogeneity among phenotypically similar organisms: wide differences in electrophoretic mobilities of LDHs of *L. leichmannii*, which divided the species into two groups were confirmed by the finding of significantly different G+C values of 36 and 50 mol% between the two groups. This second group (G+C of 36 mol%) is now designated as a separate species *L. jensenii* (Gasser, Mandel, and Rogosa, 1970), see Table 2.

DNA HYBRIDIZATION STUDIES. Genotypic relationships have been investigated among species of bacteria, using DNA-DNA or DNA-RNA hybridization studies, and thus evaluating the extent of structural homology between genomes. Using such techniques, natural relationships between lactobacilli have been determined and genotypic species defined. Simonds, Hansen, and Lakshmanan (1971) showed an 86–100% degree of hybridization between DNA of *Lactobacillus bulgaricus* and *L. lactis,* and little or none between *L. bulgaricus* and *L. helveticus*. Dellaglio, Bottazzi, and Trovatelli (1973) confirmed these findings showing that *L. helveticus* and its biotype *L. jugurt* shared between 80 and 100% DNA homology and *L. lactis* and *L. bulgaricus* shared between 72 and 100%, but there was no relationship between *L. helveticus* and the last two species. These authors also showed the correlation between DNA homology, G+C content, and serological group in 13 fresh isolates of thermobacteria of varying sugar fermentation patterns. The examination of fresh isolates, only rarely undertaken

by taxonomists who tend to use laboratory or collection strains, enhances the value of this work. Miller, Sandine, and Elliker (1971), using a wider variety of species and DNA-RNA homology, obtained less clear-cut results, although there was 98–100% homology between *L. leichmannii* and *L. lactis*. In his review, Bottazzi (1977) mentions a close DNA-DNA relationship between *L. leichmannii, L. lactis, L. bulgaricus,* and *L. delbrueckii*. With the streptobacteria, the detailed studies of Dellaglio, Bottazzi, and Vescovo (1975) on DNA-DNA hybridization found little relationship between *L. casei* and *L. plantarum*. *L. plantarum* strains formed one large homogeneous group and one smaller one with only a 40–50% DNA homology with the first group. *L. casei* contained three distinct genotypes with only a low level of homology between the DNA of *L. casei* subsp. *L. rhamnosus* and the other subspecies of this group. From these complex relationships it is suggested (London, 1976) that *L. casei* has already undergone extensive genetic diversion. The general DNA-DNA hybridization relationships among *L. corniformis, L. curvatus,* and *L. xylosus* justify their status as separate species (Dellaglio et al., 1975). Among the betabacteria, Miller, Sandine, and Elliker (1971) found a close relationship by DNA-RNA homology between *L. fermentum* and *L. cellobiosus*. A similar close DNA-DNA homology was found for these two species by Bottazzi (1977) who also found DNA relationships between *L. fructivorans* and *L. trichodes* (previously suggested on phenotypic data by Dakin and Radwell [1971]), and between *L. brevis, L. vermiforme,* and *L. hilgardii*. With the anerobic lactobacilli (Sharpe and Dellaglio, 1977), strains falling into *L. ruminis* showed 95–100% DNA homology; however, *L. vitulinus*, a phenotypically homogeneous species was found to be a genotypically heterologous group containing three different genotypes. Thus, in many instances DNA hybridization studies have confirmed the phenotypic grouping, but in others such investigations have revealed the presence of more than one genotype within a phenotypic group.

Other Species of Lactobacilli

Species so far not described in this chapter, because there are fewer data available on them, include the following.

LACTOBACILLUS CATENAFORME, L. CRISPATUS, L. MINUTUS, AND *L. ROGOSAE.* These four species are all described by Holdeman, Cato, and Moore (1977). They are obligate anaerobes, isolated from clinical material such as abscesses, or from normal human feces. The first three of these species have been transferred to the genus *Lactobacillus* from *Catanabacterium* and *Eubacterium*. *L. catenaforme* and *L. crispatus* grow at 45°C; ferment amygdalin, cellobiose, fructose, glycogen, glucose, lactose, and sucrose, and sometimes maltose but not pentoses, mannitol or melezitose; and hydrolyze starch. *L. catenaforme* ferments salicin, forms D-(−)-lactic acid, and has a G+C content of 31 mol%. *L. crispatus* may ferment starch, but not salicin, and forms DL-lactic acid. These two are the only species of lactobacilli known to hydrolyze starch. *L. minutus* ferments only glucose and sometimes produces NH₃ from arginine, grows at 45°C, and forms DL-lactic acid. *L. rogosae* is usually motile and homofermentative and has an optimum growth temperature of 37°C, with no growth at 15°C and little growth at 45°C. It ferments fructose with little or no fermentation of other carbohydrates, and has a G+C content of 59 mol%.

LACTOBACILLUS REUTERI. Early results (Kandler, 1970) of amino acid composition of peptidoglycans indicated that *Lactobacillus fermentum* contained L-Lys-D-Asp. However, Williams and Sadler (1971) found that the peptidoglycan of many strains of *L. fermentum,* like that of *L. cellobiosus,* contained Orn. Since then Kandler and Stetter (1973) have shown that the species once recognized as *L. fermentum* contained two different species, *L. fermentum,* with L-Orn-D-Asp in the cell wall and a G+C content of 53 mol%, and a suggested new species, *L. reuteri,* with L-Lys-D-Asp in the cell wall and, a G+C content of 40–41 mol%. Both grow at 45°C, and the sugar fermentation patterns of the two organisms are the same. Further work is necessary, however, to confirm that this is not a thermophilic subspecies of *L. brevis.*

LACTOBACILLUS SANFRANCISCO. This heterofermentative lactobacillus ferments maltose, producing lactic and acetic acids and CO₂. It does not ferment galactose, lactose, sucrose, raffinose, rhamnose, or ethanol. It grows at 15°C but not at 45°C (Kline and Sugihari, 1971). It has a G+C content of 38–39 mol% (Sriranganathan et al., 1973).

PENTOSE-UTILIZING THERMOPHILIC HOMOFERMENTATIVE LACTOBACILLI. Strains isolated from the mouse stomach (Roach, Savage, and Tannock, 1977), from rat intestine (Raibaud et al., 1973), and from high-temperature-fermenting grape musts (Barre, 1978) are homofermentative and able to grow at 45°C, but unlike thermobacteria, ferment ribose and arabinose. Barre (1978) considers that since his strains produce almost entirely lactate from pentoses and do not ferment gluconate they utilize a different pathway from streptobacteria. Further taxonomic studies of these organisms are necessary.

GENERAL COMMENT. Simple biochemical and physiological tests for identification of lactobacilli

into species are in general confirmed by other phenotypic and by genotypic studies. In some instances, where species can only be differentiated by complex techniques not always readily available, the worker must be guided by the purpose of his or her investigations. In taxonomic and classification studies, such distinctions or relationships are of the greatest interest. For more applied aspects where identification of many strains is often required, this may not be practicable, and differentiation of functional groups, mentioned earlier in this chapter, may be more useful. Comparisons with culture collection or neotype strains is helpful and the comparative studies and naming of neotype strains done by the *Lactobacillus* Subcommittee (Hansen, 1968) has been the basis of much of the present scheme of identification of lactobacilli.

Holdeman, Cato, and Moore (1977) have adapted a different method for identification of lactobacilli using a dichotomous key and giving, in alphabetical order of species, a table of biochemical characteristics. The key does not differentiate homofermentative from heterofermentative species, nor are the subgenera recognized. However, the detailed table of carbohydrate fermentation patterns is useful.

Bacteriophage

Lytic streptococcus bacteriophages are of common occurrence in cheese manufacture, killing the starter streptococci essential for production of lactic acid. This is much less of a problem with lactobacilli used in milk fermentations. However, virulent (lytic) phages specific for strains of *Lactobacillus lactis, L. bulgaricus, L. helveticus,* and *L. helveticus* subsp. *jugurt* used for Swiss-type cheese making have been reported (Kiuru and Tybeck, 1955; E. Tybeck, personal communication) and Sozzi and Maret (1975) have characterized a phage specific for a strain of *L. helveticus* isolated from Emmenthal cheese whey. Phage problems have occasionally been encountered in other milk fermentations, causing slow acid production. Phages for *L. lactis* or *L. bulgaricus* have been isolated from commercial yogurts by Pette and Chevalier (cited by Mocquot and Hurel, 1970); Peake and Stanley, 1978; Sozzi, 1977, and Watanabe et al. (1970) describe a virulent phage active against *L. casei,* isolated from yakult. Other isolations of phages or lactobacilli have been mainly from saliva or sewage. Coetzee, de Klerk, and Sacks (1960) attempted to initiate a phage typing scheme, but understandably in view of the many different species involved, they did not obtain clear-cut results. There have been few reports of lysogeny in lactobacilli until about 1970. A survey of 345 strains by Coetzee and de Klerk (1962) showed that only two strains of *L. fermentum* exhibited lysogenicity when induced by ultraviolet radiation. Treatment with mitomycin C (MC) has induced phage liberation in *L. acidophilus* (de Klerk and Hugo, 1970) and *L. salivarius* (Tohyama et al., 1972). Yokokura et al. (1974), covering a wider range of organisms, found that 40 strains of 7 species out of 148 strains belonging to 15 species were lysed when treated with MC: phage-like particles were seen by electron microscopy in 31/40 lysates. These included lysates of strains of *L. acidophilus*, *L. helveticus*, *L. casei*, *L. plantarum*, *L. brevis*, *L. buchnerii*, and *L. fermentum*. Phage-like particles from ten of these, mostly from *L. casei*, produced plaques and were considered to be active phages, while others were defective phages, unable to produce plaques. Stetter (1977) considered that the narrow host range strain specific of a phage for *L. casei* SL-1 might be due to immunity caused by lysogeny, as 17 of 21 strains of streptobacteria were lysogenized when induced with MC, some of these phages being defective. Lysogeny appeared to be a common phenomenon within the subgenus *Streptobacterium*. Morphology of lactobacillus phages has been studied by several workers and some differences relating to species noted (de Klerk and Hugo, 1970; de Klerk, Coetzee, and Fourie, 1965; Peake and Stanley, 1978; Sakurai, Tahahashi, and Arai, 1970; Sozzi, 1977; Stetter, 1977; Tohyama et al., 1972; Watanabe et al., 1970). Most of the phages have polyhedral heads and long tails which may be sheathless or have contractile or noncontractile sheaths. This morphology is similar to that of streptococcal phages. However, phages of the two phenotypically and genotypically related species *L. lactis* and *L. bulgaricus* have a different tail structure not previously reported in lactobacilli (Peake and Stanley, 1978; Sozzi, 1977) with a long tail having transverse rods along the main axis. Such unusual structures have only been observed with a *Bacillus cereus* phage and a temperate phage of *Actinomyces erythreus* (Tikhonenko, 1970). Studies of receptor sites have shown that D-galactose and L-rhamnose comprise a phage receptor for a strain of *L. casei* serological group B. Addition of these sugars inactivated the phage, and a phage-resistant mutant lacked galactosamine in its surface component (Yokokura, 1977). This effect is interesting, as this species of *Lactobacillus* has no teichoic acid in its cell wall, so wall teichoic acid is not involved here in absorption as in some other Gram-positive bacteria, e.g., staphylococci. Using isolated cell wall fragments of host strain *L. casei* PL-1, Watanabe, Takesue, and Ishihashi (1977) have shown that cell walls of *L. casei* may have incomplete receptors. The mode of action of a defective phage of *L. salivarius*, able to kill both lysogenic and nonlysogenic strains of *L. salivarius*, was similar in action to that of colicin (Tohyama, 1973) and these defective phages may be similar to bacteriocins. Studies on phages have therefore been concerned mainly with morphology, mode of action, and receptor sites, not with phage typing for identification.

Bacteriocins

Bacteriogenic strains of lactobacilli have been found among a number of different species, both homo- and heterofermentative (reviewed by Tagg, Dajani, and Wannamaker, 1976). Bacteriocin typing of large numbers of isolates belonging to the thermobacteria, streptobacteria, and betabacteria (Filippov, 1976a,b; Filippov and Rubanenko, 1977) showed that many strains were sensitive to the range of bacteriocins used and that species could often be subdivided into numerous types, not differentiated by biotyping. This might be useful for specialized identification of strains from specific sources, but is unlikely to contribute to general identification.

Literature Cited

Abo-Elnaga, I. G., Kandler, O. 1965a. Zur Taxonomie der Gattung *Lactobacillus* Beijerinck. I. Das Subgenus *Streptobacterium* Orla Jensen. Zentralblatt für Bakteriologie, Parasitenkunde, Infektionskrankheiten und Hygiene, Abt. 2 **119:**1–36.

Abo-Elnaga, I. G., Kandler, O. 1965b. Zur Taxonomie der Gattung *Lactobacillus* Beijerinck. III. Das Vitaminbedürfnis. Zentralblatt für Bakteriologie, Parasitenkunde. Infektionskrankheiten und Hygiene, Abt. 2 **119:**661–672.

Amerine, M. A., Kunkee, R. E. 1968. Microbiology of winemaking. Annual Review of Microbiology **22:**323–358.

Azeezullah, M., Sharma, A. K., Landon, K. C., Srinirasan, R. A. 1973. Microflora of oats and berseem silages. Indian Journal of Dairy Science **26:**232–236.

Barnes, E. M. 1976. Microbiological problems of poultry at refrigerator temperatures—a review. Journal of the Science of Food and Agriculture **27:**777–782.

Barre, P. 1969. Taxonomie numérique de lactobacilles isolates du vin. Archiv für Mikrobiologie **68:**74–86.

Barre, P. 1978. Identification of thermobacteria and homofermentative thermophilic pentose-utilizing lactobacilli from high temperature fermenting grape musts. Journal of Applied Bacteriology **44:**125–129.

Barrow, P. A., Fuller, R., Newport, M. J. 1977. Changes in the microflora and physiology of the anterior intestinal tract of pigs weaned at 2 days with special reference to the pathogenesis of diarrhea. Infection and Immunity **18:**586–595.

Berger, U. 1974. Pathogenicity of lactobacilli. Deutsche Medizinische Wochenschrift **99:**1200–1203.

Bernhardt, H. 1974. Presence of the genus *Lactobacillus* (Beijerinck) in the human stomach. Zentralblatt für Bakteriologie, Parasitenkunde, Infektionskrankheiten und Hygiene, Abt. 1 Orig., Reihe A **226:**479–490.

Bergey's Manual of Determinative Bacteriology, 8th ed. 1974. Buchanan, R. E., Gibbons, N. E. (eds.). Baltimore: Williams & Wilkins.

Biede, S. L., Reinbold, G. W., Hammond, E. G. 1976. Influence of *Lactobacillus bulgaricus* on microbiology and chemistry of Swiss cheese. Journal of Dairy Science **59:**854–858.

Boatwright, J., Kirsop, B. H. 1976. Sucrose agar—a growth medium for spoilage organisms. Journal of the Institute of Brewing **82:**343–346.

Bottazzi, V. 1977. Microbiologia dei fermenti lattici. 3.

Tassonomia: Recenti acquisizioni. Journal of the Italian Dairy Science Association **28**:213–232.

Bottazzi, V., Vescova, M., Dellaglio, F. 1973. Microbiology of Grana cheese. IX. Characteristics and distribution of *Lactobacillus helveticus* biotypes in natural whey cheese starter. Scienza e Technica Lattierocasearia **24**:23–39.

Bousfield, I. J., MacKenzie, A. R. 1976. Inactivation of bacteria by freeze drying, pp. 329–344. In: Skinner, F. A., Hugo, W. B. (eds.), Inhibition and inactivation of vegetative microbes. London, New York, San Francisco: Academic Press.

Branen, A. L., Keenan, T. W. 1969. Growth stimulation of *Lactobacillus* species by lactic streptococci. Applied Microbiology **17**:280–285.

Bryan-Jones, G. 1975. Lactic acid bacteria in distillery fermentations, pp. 165–175. In: Carr, J. G., Cutting, C. V., Whiting, G. C. (eds.), Lactic acid bacteria in beverages and foods. London, New York, San Francisco: Academic Press.

Bryant, M. P., Small, N., Bouma, C., Robinson, I. 1958. Studies on the composition of the ruminal flora and fauna of young calves. Journal of Dairy Science **41**:1747–1767.

Buyze, G., van den Hamer, C. J. A., de Haan, P. G. 1957. Correlation between hexose-monophosphate shunt, glycolytic system and fermentation-type in lactobacilli. Antonie van Leeuwenhoek Journal of Microbiology and Serology **23**:345–350.

Carlsson, J., Gothefors, L. 1975. Transmission of *Lactobacillus jensenii* and *Lactobacillus acidophilus* from mother to child at the time of delivery. Journal of Clinical Microbiology **1**:124–128.

Carlsson, J., Grahnen, H., Jonsson, G. 1975. Lactobacilli and streptococci in the mouth of children. Caries Research **9**:333–339.

Carr, J. G. 1959. Some special characteristics of the cider lactobacilli. Journal of Applied Bacteriology **22**:377–383.

Carr, J. G. 1967. Bacteries lactiques des boissons fermentées, leur écologie et leur classification, pp. 175–189. "Fermentations et Vinifications" 2ᵉ Symposium International d'Oenologie Bordeaux—Cognac.

Carr, J. G., Davies, P. A. 1970. Homofermentative lactobacilli of ciders including *Lactobacillus mali* nov. spec. Journal of Applied Bacteriology **33**:768–774.

Carr, J. G., Davies, P. A., Dellaglio, F., Vescovo, M., Williams, R. A. D. 1977. The relationship between *Lactobacillus mali* from cider and *Lactobacillus yamanashiensis* from wine. Journal of Applied Bacteriology **42**:219–228.

Castino, M., Usseglio-Tomasset, L., Gandini, A. 1975. Factors which affect the spontaneous inclination of the malo-lactic fermentation in wines. The possibility of transmission by inoculation and its effect on organoleptic properties, pp. 139–148. In: Carr, J. G., Cutting, C. V., Whiting, G. C. (eds.), Lactic acid bacteria in beverages and food. London, New York, San Francisco: Academic Press.

Chalfan, Y., Goldberg, I., Mateles, R. I. 1977. Isolation and characterization of malo-lactic bacteria from Israeli red wines. Journal of Food Science **42**:939–944.

Chalfan, Y., Levy, R., Mateles, R. I. 1975. Detection of mannitol formation by bacteria. Applied Microbiology **30**:476.

Chomakov. H. V. 1962. Rusty spots in brined white cheese. [In Russian, with English summary.] Mikrobiologiya **31**:726–730.

Chomakov, Kh., Kirov, N. 1975. Lactic acid bacteria in raw milk and White pickled cheese. Dairy Science Abstracts **37**:134.

Christensen, M. D., Pederson, C. S. 1958. Factors affecting diacetyl production by lactic acid bacteria. Applied Microbiology **6**:319–322.

Clarke, R. T. J., Bauchop, T. (eds.). 1977. Microbial ecology of the gut. London, New York, San Francisco: Academic Press.

Coetzee, J. N., de Klerk, H. C. 1962. Lysogeny in the genus *Lactobacillus*. Nature **187**:348–349.

Coetzee, J. N., de Klerk, H. C., Sacks, T. G. 1960. Host-range of *Lactobacillus* bacteriophages. Nature **187**:348–349.

Contrepois, M., Gouet, Ph. 1973. La microflore du tube digestif du jeune veau préruminant: Dénombrement de quelques groupes bacteriens a différents niveaux du tube digestif. Annales de Recherche Vétérinaire **4**:161–170.

Dakin, J. C., Radwell, J. Y. 1971. Lactobacilli causing spoilage of acetic acid preserves. Journal of Applied Bacteriology **34**:541–545.

Davies, F. L., Shankar, P. A., Underwood, H. M. 1977. The use of milk concentrated by reverse osmosis for the manufacture of yoghurt. Journal of the Society of Dairy Technology **30**:23–27.

Davis, C. P., Cleven, D., Balish, E., Yale, C. E. 1977. Bacterial association in the gastrointestinal tract of Beagle dogs. Applied and Environmental Microbiology **34**:194–206.

Davis, J. G. 1975. The microbiology of yoghurt, pp. 245–263. In: Carr, J. G., Cutting, C. V., Whiting, G. C. (eds.), Lactic acid bacteria in beverages and food. London, New York, San Francisco: Academic Press.

Deibel, R. H., Niven, C. F., Jr., Wilson, G. D. 1961. Microbiology of meat curing. III. Some microbiological and related technological aspects in the manufacture of fermented sausages. Applied Microbiology **9**:156–161.

de Klerk, H. C., Coetzee, J. N., Fourie, J. T. 1965. The fine structure of lactobacillus bacteriophages. Journal of General Microbiology **38**:35–38.

de Klerk, H. C., Hugo, N. 1970. Phage-like structures from *Lactobacillus acidophilus*. Journal of General Virology **8**:231–234.

Dellaglio, F., Bottazzi, V., Trovatelli, L. D. 1973. Deoxyribonucleic acid homology and base composition in some thermophilic lactobacilli. Journal of General Microbiology **74**:289–297.

Dellaglio, F., Bottazzi, V., Vescovo, M. 1975. Deoxyribonucleic acid homology among *Lactobacillus* species of the subgenus *Streptobacterium* Orla-Jensen. International Journal of Systematic Bacteriology **25**:160–172.

de Man, J. C., Rogosa, M., Sharpe, M. E. 1960. A medium for the cultivation of lactobacilli. Journal of Applied Bacteriology **23**:130–135.

Drasar, B. S., Hill, M. J. 1974. Human intestinal flora. London, New York, San Francisco: Academic Press.

Eltz, R. W., Vandemark, P. J. 1960. Fructose dissimilation by *Lactobacillus brevis*. Journal of Bacteriology **79**:763–776.

Etchells, J. L., Fleming, H. P., Bell, T. A. 1975. Factors influencing the growth of lactic acid bacteria during the fermentation of brined cucumbers, pp. 281–305. In: Carr, J. G., Cutting, C. V., Whiting, G. C. (eds.), Lactic acid bacteria in beverages and food. London, New York, San Francisco: Academic Press.

Evans, J. B., Niven, C. F. 1951. Nutrition of the heterofermentative lactobacilli that cause greening of cured meat products. Journal of Bacteriology **62**:599–603.

Filippov, V. A. 1976a. Sensitivity of some species of lactobacilli of subgenus *Betabacterium* to bacteriocins of lactobacilli of various species. Antibiotiki **21**:1075–1078.

Filippov, V. A. 1976b. Bacteriocin typing of lactobacilli of the *Streptobacterium* subgenus. Zhurnal Mikrobiologii Epidemiologii, i Immunobiologii **53**:86–88.

Filippov, V. A., Rubanenko, E. B. 1977. Differentiation of some species of lactobacilli of the *Thermobacterium* subgenus by the spectra of bacteriocin sensitivity. Zhurnal Mikrobiologii Epidemiologii i Immunobiologii **54**:46–50.

Finegold, S. M., Sutter, V. L., Sugihara, P. T., Elder, H. A., Lehmann, S. M., Phillips, R. L. 1977. Fecal microbial flora in Seventh Day Adventist population and control subjects. American Journal of Clinical Nutrition **30**:1781–1792.

Fryer, T. F. 1969. Microflora of Cheddar cheese and its influence on cheese flavour. Dairy Science Abstracts **31**:471–490.

Fryer, T. F., Sharpe, M. E., Reiter, B. 1970. Utilization of milk citrate by lactic acid bacteria and "blowing" of film wrapped cheese. Journal of Dairy Research **37**:17–28.

Fuller, R. 1973. Ecological studies on the lactobacillus flora associated with the crop epithelium of the fowl. Journal of Applied Bacteriology **36**:131–139.

Fuller, R., Barrow, P. A., Brooker, B. E. 1978. Bacteria associated with the gastric epithelium of neonatal pigs. Applied and Environmental Microbiology **35**:582–591.

Fuller, R., Newland, L. G. M., Briggs, C. A. E., Braude, R., Mitchell, K. G. 1960. The normal intestinal flora of the pig. IV. The effect of dietary supplements of penicillin, chlortetracycline or copper sulphate on the faecal flora. Journal of Applied Bacteriology **23**:195–205.

Galli, A., Ottogalli, G. 1973. Aspetti della microflora degli impasti per panettone. Annali di Microbiologia ed Enzimologia **23**:39–49.

Garvie, E. I. 1967. The production of L(+) and D(−) lactic acid in culture of some lactic acid bacteria with a special study of *Lactobacillus acidophilus* NCDO 2. Journal of Dairy Research **34**:31–38.

Garvie, E. I. 1975. Some properties of gas-forming lactic acid bacteria and their significance in classification, pp. 339–349. In: Carr, J. G., Cutting, C. V., Whiting, G. C. (eds.), Lactic acid bacteria in beverages and food. London, New York, San Francisco: Academic Press.

Garvie, E. I. 1976. Hybridization between the deoxyribonucleic acids of some strains of heterofermentative lactic bacteria. International Journal of Systematic Bacteriology **26**:116–122.

Garvie, E. I., Zezula, V., Hill, V. A. 1974. Guanine plus cytosine content of the deoxyribonucleic acid of the leuconostocs and some heterofermentative lactobacilli. International Journal of Systematic Bacteriology **24**:248–251.

Gasser, F. 1970. Electrophoretic characterization of lactic dehydrogenases in the genus *Lactobacillus*. Journal of General Microbiology **62**:223–239.

Gasser, F., Gasser, C. 1971. Immunological relationships among lactic dehydrogenases in the genera *Lactobacillus* and *Leuconostoc*. Journal of Bacteriology **106**:113–125.

Gasser, F., Mandel, M. 1968. Deoxyribonucleic acid base composition of the genus *Lactobacillus*. Journal of Bacteriology **96**:580–588.

Gasser, F., Mandel, M., Rogosa, M. 1970. *Lactobacillus jensenii* sp. nov., a new representative of the subgenus *Thermobacterium*. Journal of General Microbiology **62**:219–222.

Gill, C. O. 1976. Substrate limitation of bacterial growth at meat surfaces. Journal of Applied Bacteriology **41**:401–410.

Goudkov, A. V., Sharpe, M. E. 1965. Clostridia in dairying. Journal of Applied Bacteriology **28**:63–73.

Hansen, P. A. 1968. Type strains of *Lactobacillus* species. A report by the Taxonomic Subcommittee on lactobacilli and closely related organisms. International Committee on Nomenclature of Bacteria of the International Association of Microbiological Societies. Rockville, Maryland: American Type Culture Collection.

Hardie, J. M., Bowden, G. H. 1974. The normal microflora of the mouth, pp. 47–83. In: Skinner, F. A., Carr, J. G. (eds.), The normal microflora of man. London: Academic Press.

Holdeman, L. V., Cato, E. P., Moore, W. E. C. (eds.). 1977. Anaerobe laboratory manual, 4th ed. Blacksburg, Virginia: Virginia Polytechnic Institute Anaerobe Laboratory.

Holzapfel, W., Kandler, O. 1969. Zur Taxonomie der Gattung *Lactobacillus* Beijerinck. VI. *Lactobacillus coprophilus* subsp. *confusus* nov. subsp., eine neue Unterart der Untergattung *Betabacterium*. Zentralblatt für Bakteriologie, Parasitenkunde, Infektionskrankheiten und Hygiene, Abt. 2 **123**:658–666.

Hsu, W. P., Taparowsky, J. A. 1977. Growth response of common brewery bacteria to different media. Brewers Digest **52**:48–53.

Juven, B. J. 1976. Bacterial spoilage of citrus products at pH lower than 3.5. Journal of Food Protection **39**:819–822.

Kandler, O. 1970. Amino acid sequence of the murein and taxonomy of the genera *Lactobacillus*, *Bifidobacterium*, *Leuconostoc* and *Pediococcus*. International Journal of Systematic Bacteriology **20**:491–507.

Kandler, O., Stetter, K. 1973. Der Beitrag Neuerer Biochemischer Merkmale für die Systematic der Laktobazillen. Symposium Technische Mikobiologie 501–506.

Kawai, Y., Shegara, N. 1977. Specific adhesion of lactobacilli to keratinized epithelial cells of the rat stomach. American Journal of Clinical Nutrition **30**:1777–1780.

Keddie, R. M. 1951. The enumeration of lactobacilli on grass and in silage. Proceedings of the Society for Applied Bacteriology **14**:157–160.

Keddie, R. M. 1959. The properties and classification of lactobacilli isolated from grass and silage. Journal of Applied Bacteriology **22**:403–416.

Keller, J. J., Jaarsman, J. 1975. Lactobacilli and gas formation of film wrapped Cheddar cheese. South African Journal of Dairy Technology **7**:183–185.

Khandak, R. N., Gudkov, A. V. 1979. Isolation and identification of lactobacilli antagonistic to butyric acid bacteria and coliforms. Dairy Science Abstracts **39**:668.

Kirsop, B. H., Dolezil, L. 1975. Detection of lactobacilli in brewing, pp. 159–164. In: Carr, J. G., Cutting, C. V., Whiting, G. C. (eds.), Lactic acid bacteria in beverages and food. London, New York, San Francisco: Academic Press.

Kitahara, K., Kaneko, T., Goto, O. 1957. Taxonomic studies on the hiochi-bacteria, specific saprophytes of sake. I. Isolation and grouping of bacterial strains. Journal of General and Applied Microbiology **3**:102–110.

Kitchell, A. G., Shaw, B. G. 1975. Lactic acid bacteria in fresh and cured meat, pp. 209–220. In: Carr, J. G., Cutting, C. V., Whiting, G. C. (eds.), Lactic acid bacteria in beverages and food. London, New York, San Francisco: Academic Press.

Kiuru, U. J. T., Tybeck, E. 1955. Characteristics of bacteriophages active against lactic acid bacteria in Swiss cheese. Suomen Kemistilehti **28**:57–62.

Kleeberger, A., Kuhbauch, W. 1976. Decomposition of grass-fructosans by lactobacilli from silage. Zentralblatt für Bakteriologie, Parasitenkunde, Infektionskrankheiten und Hygiene, Abt. 2 **131**:398–404.

Kline, L., Sugihara, T. F. 1971. Microorganisms of the San Francisco sour dough bread process. II. Isolation and characterization of undescribed bacterial species responsible for souring activity. Applied Microbiology **21**:459–465.

Knox, K. W. 1963. Isolation of group specific products from *Lactobacillus casei* and *L. casei* var. *rhamnosus*. Journal of General Microbiology **31**:59–72.

Knox, K. W., Wicken, A. J. 1973. Immunological properties of teichoic acids. Bacteriological Reviews **37**:215–257.

Knox, K. W., Wicken, A. J. 1976. Grouping and cross-reacting antigens of oral lactic acid bacteria. Journal of Dental Research **55**:A116–A122.

Kunkee, R. E. 1967. Malo-lactic fermentation. Advances in Applied Microbiology **9**:235–279.

Kvasinov, E. I., Kotljar, A. N., Vasileva, Z. A. 1976. Species composition and certain peculiarities in physiology of bacteria isolated in sugar production. Mikrobiologichnii Zhurnal **38**:434–438.

Langston, C. W., Bouma, C. 1960. A study of the microorganisms from grass silage. II. The lactobacilli. Applied Microbiology **8**:223–234.

Langston, C. W., Bouma, C., Conner, R. M. 1962. Chemical and bacteriological changes in grass silage during the early stages of fermentation. II. Bacteriological changes. Journal of Dairy Science **45**:618–624.

Lapage, S. P., Shelton, J. E., Mitchell, T. G., Mackenzie, A. R. 1970. Culture collections and the preservation of bacteria. In: Norris, J. R., Ribbons, D. M. (eds.), Methods in microbiology, 3A. London and New York: Academic Press.

Latham, M. J., Sharpe, M. E., Sutton, J. D. 1971. The microbial flora of the rumen of cows fed hay and high cereal rations and its relationship to the rumen fermentation. Journal of Applied Bacteriology **34**:425–434.

Law, B. A., Castanon, M., Sharpe, M. E. 1976. The effect of non-starter bacteria on the chemical composition and the flavour of Cheddar cheese. Journal of Dairy Research **43**:117–125.

Ledesma, O. V., de Ruiz Holgado, A. P., Oliver, G., de Giori, G. S., Raibaud, P., Galpin, J. V. 1977. A synthetic medium for comparative nutritional studies of lactobacilli. Journal of Applied Bacteriology **42**:123–133.

Lenzner, A. A. 1966. Some results of the investigation of lactobacilli of human microflora. Survey of Research in Medicine, Tartu State University, 1940–1965 **191**:69–75.

Levison, M. E., Corman, L. C., Carrington, E. R., Kaye, D. 1977. Quantitative microflora of the vagina. Americal Journal of Obstetrics and Gynaecology **127**:80–85.

London, J. 1976. The ecology and taxonomic status of the lactobacilli. Annual Review of Microbiology **30**:279–301.

MacKenzie, K. G., Kenny, M. C. 1965. Non-volatile organic acid and pH changes during the fermentations of distiller's wort. Journal of the Institute of Brewing **71**:160–165.

Miller, A., III, Morgan, M. E., Libbey, L. M. 1974. *Lactobacillus maltaromicus,* a new species producing a malty aroma. International Journal of Systematic Bacteriology **24**:346–354.

Miller, A., III, Sandine, W. E., Elliker, P. R. 1970. Deoxyribonucleic acid base composition of lactobacilli determined by thermal denaturation. Journal of Bacteriology **102**:278–280.

Miller, A., III, Sandine, W. E., Elliker, P. R. 1971. Deoxyribonucleic acid homology in the genus *Lactobacillus.* Canadian Journal of Microbiology **17**:625–634.

Millis, N. F. 1951. Some bacterial fermentations in cider. Thesis. University of Bristol, Bristol, England.

Mital, B. K., Steinkraus, K. W., Naylor, W. B. 1974. Growth of lactic acid bacteria in soy milk. Journal of Food Science **39**:1018–1022.

Mitchell, I. de G., Kenworthy, R. 1976. Investigations on a metabolite from *Lactobacillus bulgaricus* which neutralizes the effect of enterotoxin from *Escherichia coli* pathogenic for pigs. Journal of Applied Bacteriology **41**:163–174.

Mitsuoka, T. 1969. Vergleichende Untersuchungen über die Lactobazillen aus den Faeces von Menschen, Schweinen und Hühnern. Zentralblatt für Bakteriologie, Parasitenkunde, Infektionskrankheiten und Hygiene, Abt. 1 Orig. **210**:32–51.

Mitsuoka, T., Hayakawa, K., Kimura, N. 1975. The fecal flora of man. III. Communication: The composition of lactobacillus flora of different age groups. Zentralblatt für Bakteriologie, Parasitenkunde, Infektionskrankheiten und Hygiene, Abt. 1 Orig., Reihe A **232**:499–511.

Mitsuoka, T., Kaneuchi, C. 1977. Ecology of the bifidobacteria. American Journal of Clinical Nutrition **30**:1799–1810.

Mitsuoka, T., Kimura, N., Kobayashi, A. 1976. Studies on the composition of the fecal flora of healthy dogs with the special references of lactobacillus flora and bifidobacterium flora. Zentralblatt für Bakteriologie, Parasitenkunde, Infektionskrankheiten und Hygiene, Abt. 1 Orig., Reihe A **235**:485–493.

Mitsuoka, T., Ohno, K. 1977. Fecal flora of man. V. Communication: The fluctuations of the fecal flora of the healthy adult. Zentralblatt für Bakteriologie, Parasitenkunde, Infektionskrankheiten und Hygiene, Abt. 1 Orig., Reihe A **238**:228–236.

Mocquot, G., Hurel, C. 1970. The selection and use of microorganisms for the manufacture of fermented and acidified milk products. Journal of the Society of Dairy Technology **23**:130–142.

Moon, N. J., Reinbold, G. W. 1978. Commensalism and competition in mixed cultures of *Lactobacillus bulgaricus* and *Streptococcus thermophilus.* Journal of Milk and Food Technology **39**:337–341.

Moore, W. E. C., Holdeman, L. V. 1974. Human fecal flora: The normal flora of 20 Japanese-Hawaiians. Applied Microbiology **27**:961–979.

Moreau, M.-C., Ducluzeau, R., Raibaud, P. 1976. Hydrolysis of urea in the gastrointestinal tract of 'monaxenic'' rats: Effect of immunization with strains of ureolytic bacteria. Infection and Immunity **13**:9–15.

Morichi, T. 1974. Preservation of lactic acid bacteria by freeze drying. Japan Agricultural Research Quarterly **8**:171–176.

Morichi, T., Sharpe, M. E., Reiter, B. 1968. Esterases and other soluble proteins of some lactic acid bacteria. Journal of General Microbiology **53**:405–414.

Mundt, J. O., Hammer, J. L. 1968. Lactobacilli on plants. Applied Microbiology **16**:1326–1330.

Murdock, D. I., Hatcher, W. S. 1975. Growth of microorganisms in chilled orange juice. Journal of Milk and Food Technology **38**:393–396.

Naylor, J. 1956. A note on the use of Rogosa and modified Rogosa media for the enumeration and culture of lactobacilli. Journal of Applied Bacteriology **19**:102–104.

Newton, K. G., Gill, C. O. 1978. The development of the anaerobic spoilage flora of meat stored at chill temperatures. Journal of Applied Bacteriology **44**:91–95.

Nieuwenhof, F. F. J., Stadhouders, J., Hup, G. 1969. Stimulating effects of lactobacilli on the growth of propionibacteria in cheese. Netherlands Milk and Dairy Journal **23**:287.

Nonomura, H., Ohara, Y. 1967. Die Klassifikation der Apfelsäure-Milchsäure-Bakterien. Mitteilungen-Höheren Bundeslehrund Versuchsanstalt für Weinund Obstbau, Klosterneuburg **17**:449–465.

Ohyama, Y., Masaki, S., Morichi, T. 1973. Effects of inoculation of *Lactobacillus plantarum* and addition of glucose at ensiling on the silage quality. Japanese Journal of Zootechnical Science **44**:404–410.

Orla-Jensen, S. 1919. The lactic acid bacteria. Copenhagen: Andr. Fred Host and Son.

Orla-Jensen, S. 1943. The lactic acid bacteria. Copenhagen: Munksgaard.

Peake, S. E., Stanley, G. 1978. Partial characterization of a bacteriophage of *Lactobacillus bulgaricus* isolated from yoghurt. Journal of Applied Bacteriology **44**:321–323.

Peynaud, E., Domercq, S. 1967. Etude de quelques bacilles homolactiques isolés de vins. Archiv für Mikrobiologie **57**:255–270.

Peynaud, E., Sapis-Domercq, S. 1970. Étude de deux cent cinquante souches de bacilles hétérolactiques isolés de vins. Archiv für Mikrobiologie **70**:348–360.

Phillips, B. A., Latham, M. J., Sharpe, M. E. 1975. A method for freeze drying rumen bacteria and other strict anaerobes. Journal of Applied Bacteriology **38**:319–322.

Pilone, G. J., Kunkee, R. E., Webb, A. D. 1966. Chemical characterization of wines fermented with various malo-lactic bacteria. Applied Microbiology **14**:608–615.

Poznanski, S. E., Kornacki, K., Smietana, Z., Rymaszenski, J., Surazynski, A., Chojnowski, W. 1974. New method for lactic acid manufacture using ion exchangers. Dairy Science Abstracts **36**:222.

Radler, F. 1975. The metabolism of organic acids by lactic acid bacteria, pp. 17–27. In: Carr, J. G., Cutting, C. V., Whiting, G. C. (eds.), Lactic acid bacteria in beverages and food. London, New York, San Francisco: Academic Press.

Radulovic, D. 1971. The effect of some factors on survival of *Lactobacillus acidophilus* during and after freeze drying. Mikrobiologija **8**:189–195.

Raibaud, P., Galpin, J. V., Ducluzeau, R., Mocquot, G., Oliver, G. 1973. Le genre *Lactobacillus* dans le tube digestif du rat. I. Caractères des souches homofermentaires isolées de rats holo- et gnotoxéniques. Annales de Microbiologie **124A**:83–109.

Rainbow, C. 1975. Beer spoilage lactic acid bacteria, pp. 149–158. In: Carr, J. G., Cutting, C. V., Whiting, G. C. (eds.), Lactic acid bacteria in beverages and food. London,

New York, San Francisco: Academic Press.

Reiter, B. 1978. Antimicrobial systems in milk. Journal of Dairy Research **45**:131–147.

Reuter, G. 1965. The incidence of lactobacilli in food products and their behavior in the human intestinal tract. Zentralblatt für Bakteriologie, Parasitenkunde, Infektionskrankheiten und Hygiene, Abt. 1 Orig. **197**:468–487.

Reuter, G. 1970. Lactobacilli and closely related microorganisms in meat and meat products. 2. Characterization of isolated lactobacilli strains. Die Fleischwirtschaft **50**:954–962.

Reuter, G. 1975. Classification problems, ecology and some biochemical activities of lactobacilli of meat products, pp. 221–229. In: Carr, J. G., Cutting, C. V., Whiting, G. C. (eds.), Lactic acid bacteria in beverages and food. London, New York, San Francisco: Academic Press.

Richards, M., Macrae, R. M. 1964. The significance of the use of hops in regard to the biological stability of beer. II. The development of resistance to hop resins by strains of lactobacilli. Journal of the Institute of Brewing **70**:484–488.

Roach, S., Savage, D. C., Tannock, G. W. 1977. Lactobacilli isolated from the stomach of conventional mice. Applied and Environmental Microbiology **33**:1197–1203.

Rogosa, M. 1970. Characters used in the classification of lactobacilli. International Journal of Systematic Bacteriology **20**:519–533.

Rogosa, M., Franklin, J. G., Perry, K. D. 1961. Correlation of the vitamin requirements with cultural and biochemical characters of *Lactobacillus* spp. Journal of General Microbiology **25**:473–482.

Rogosa, M., Mitchell, J. A., Wiseman, R. F. 1951. A selective medium for the isolation and enumeration of oral and faecal lactobacilli. Journal of Bacteriology **62**:132–133.

Rogosa, M., Sharpe, M. E. 1959. An approach to the classification of the lactobacilli. Journal of Applied Bacteriology **22**:329–340.

Rogosa, M., Sharpe, M. E. 1960. Species differentiation of human vaginal lactobacilli. Journal of General Microbiology **23**:197–201.

Rogosa, M., Wiseman, R. F., Mitchell, J. A., Disraely, M. 1953. Species differentiation of oral lactobacilli from man including descriptions of *Lactobacillus salivarius* nov. spec. and *Lactobacillus cellobiosus* nov. spec. Journal of Bacteriology **65**:681–699.

Ruiz-Arguesco, T., Rodriguez-Navarro, A. 1975. Microbiology of ripening honey. Applied Microbiology **30**:893–896.

Sakurai, T., Takahashi, T., Arai, H. 1970. The temperate phages of *Lactobacillus salivarius* and *Lactobacillus casei*. Japanese Journal of Microbiology **14**:333–336.

Salyers, A. A., West, S. E. H., Vercellotti, J. R., Wilkins, T. D. 1977. Fermentation of mucins and plant polysaccharides by anaerobic bacteria from the human colon. Applied and Environmental Microbiology **34**:529–533.

Sandine, W. E., Muralidhara, K. S., Elliker, P. R., England, D. C. 1972. Lactic acid bacteria in food and health: A review with special reference to enteropathogenic *Escherichia coli* as well as certain enteric diseases and their treatment with antibiotics and lactobacilli. Journal of Milk and Food Technology **35**:691–702.

Sands, D. C., Hankin, L. 1976. Fortification of foods by fermentation with lysine-excreting mutants of lactobacilli. Journal of Agriculture and Food Chemistry **24**:1104–1106.

Savage, D. C. 1977. Microbial ecology of the gastrointestinal tract. Annual Review of Microbiology **31**:107–133.

Savel, J. 1975. Correct medium for determining lactobacilli. Kvasny Prumysl **21**:169–171.

Scherrer, A. 1972. Formation and analysis of diacetyl, 2,3-pentanedione, acetoin and 2,3-butanediol in wort and beer. Wallerstein Laboratories Communications **35**:5–33.

Schleifer, K. H., Kandler, O. 1972. Peptidoglycan types of bacterial cell walls and their taxonomic implications. Bacteriological Reviews **36**:407–477.

Shahani, K. M., Vakil, J. R., Kilara, A. 1977. Natural antibiotic activity of *Lactobacillus acidophilus* and *bulgaricus*. II. Isolation of Acidophilin from *L. acidophilus*. Cultured Dairy Products Journal **12**:8–11.

Shankar, P. A., Davies, F. L. 1978. Amino acid and peptide utilization by *Streptococcus thermophilus* in relation to yogurt manufacture. Streptococci. Society for Applied Bacteriology Symposium Series **7**:402–403.

Sharpe, M. E. 1960. Selective media for the isolation and enumeration of lactobacilli. Laboratory Practice **9**:223–227.

Sharpe, M. E. 1962a. Enumeration and studies of lactobacilli in food products. Dairy Science Abstracts **24**:165–171.

Sharpe, M. E. 1962b. Lactobacilli in meat products. Food Manufacturer **37**:582–589.

Sharpe, M. E. 1962c. Taxonomy of the lactobacilli. Dairy Science Abstracts **24**:109–118.

Sharpe, M. E. 1970. Cell wall and cell membrane antigens used in the classification of lactobacilli. International Journal of Systematic Bacteriology **20**:509–518.

Sharpe, M. E. 1972. The relation of the microflora to the flavour of some dairy products, 64–79. Proceedings of the 3rd Nordic Aroma Symposium.

Sharpe, M. E. 1974. Recent aspects of taxonomy of the lactobacilli. Il ruolo terapeutico e nutrizionale dei lattobacilli: Seminario Internazionale, Roma 10, 15 Fondazione Giovanni Lorenzini.

Sharpe, M. E. 1979. Identification of lactic acid bacteria. In: Skinner, F. A. (ed.), Identification methods for microbiologists. Technical Series 14. London, New York, San Francisco: Academic Press.

Sharpe, M. E., Dellaglio, F. 1977. Deoxyribonucleic acid homology in anaerobic lactobacilli and in possibly related species. International Journal of Systematic Bacteriology **27**:19–21.

Sharpe, M. E., Fryer, T. F. 1965. Media for lactic acid bacteria. Laboratory Practice **14**:697–701.

Sharpe, M. E., Fryer, T. F., Smith, D. G. 1966. Identification of lactic acid bacteria. In: Gibbs, B. M., Skinner, F. A. (eds.), Identification methods for microbiologists, part 1. London, New York: Academic Press.

Sharpe, M. E., Garvie, E. I., Tilbury, R. H. 1972. Some slime-forming heterofermentative species of the genus *Lactobacillus*. Applied Microbiology **23**:389–397.

Sharpe, M. E., Hill, L. R., Lapage, S. P. 1973. Pathogenic lactobacilli. Journal of Medical Microbiology **6**:281–286.

Sharpe, M. E., Brock, J. H., Knox, K. W., Wicken, A. J. 1973a. Glycerol teichoic acid as a common antigenic factor in lactobacilli and some other Gram-positive organisms. Journal of General Microbiology **74**:119–126.

Sharpe, M. E., Latham, M. J., Garvie, E. I., Zirngibl, J., Kandler, O. 1973b. Two new species of *Lactobacillus* isolated from the bovine rumen, *Lactobacillus ruminis* sp. nov. and *Lactobacillus vitulinus* sp. nov. Journal of General Microbiology **77**:37–49.

Shcherbanovsky, L. R., Luko, Yu. A., Kapelev, I. G. 1975. Study of the antimicrobial effect of higher plants on lactic acid bacteria. Mikrobiologichnii Zhurnal **37**:629–634.

Shimohashi, H., Kamiyama, K., Arai, H. 1973. Studies on group antigens of *Lactobacillus casei*. Japanese Journal of Microbiology **17**:205–210.

Shimohashi, H., Mutai, M. 1977. Specific antigens of *Lactobacillus acidophilus*. Journal of General Microbiology **103**:337–344.

Shovell, F. E., Gillis, R. E. 1972. Biochemical and antigenic studies of lactobacilli isolated from deep dentinal caries. II. Antigenic aspects. Journal of Dental Research **51**:583–587.

Simonds, J., Hansen, P. A., Lakshmanan, S. 1971. Deoxyribonucleic acid hybridization among strains of lactobacilli. Journal of Bacteriology **107**:382–384.

Simpson, A. C. 1968. Manufacture of Scotch malt whisky. Process Biochemistry **3**:9–12.

Smith, H. W. 1965a. The development of the flora of the alimentary tract in young animals. Journal of Pathology and Bacteriology 90:495–513.

Smith, H. W. 1965b. Observations on the flora of the alimentary tract of animals and factors affecting its composition. Journal of Pathology and Bacteriology 89:95–122.

Smith, H. W. 1971. The bacteriology of the alimentary tract of domestic animals suffering from *Escherichia coli* infection. Annals of the New York Academy of Sciences 176:110–125.

Sozzi, T. 1977. L'infezione da batteriofago. II Latte, Anno 2, NI 31–36.

Sozzi, T., Maret, R. 1975. Isolation and characterization of *Streptococcus thermophilus* and *Lactobacillus helveticus* phages from Emmenthal starters. Lait 55:269–288.

Sriranganathan, N., Seidler, R. J., Sandine, W. E., Elliker, P. R. 1973. Cytological and deoxyribonucleic acid-deoxyribonucleic acid hybridization studies on *Lactobacillus* isolates from San Francisco sourdough. Applied Microbiology 25:461–470.

Stadhouders, J., Veringa, H. A. 1967. Texture and flavour defects in cheese caused by bacteria from contaminated rennet. Netherlands Milk and Dairy Journal 21:192–207.

Stamer, J. R. 1975. Recent developments in the fermentation of sauerkraut, pp. 267–280. In: Carr, J. G., Cutting, C. V., Whiting, G. C. (eds.), Lactic acid bacteria in beverages and food. London, New York, San Francisco: Academic Press.

Stetter, K. O. 1977. Evidence for frequent lysogeny in lactobacilli: Temperate bacteriophages within the subgenus *Streptobacterium*. Journal of Virology 24:685–689.

Stetter, K. O., Kandler, O. 1973. Formation of DL-lactic acid by lactobacilli and characterization of a lactic acid racemase from several streptobacteria. Archiv für Mikrobiologie 94:221–247.

Stewart-Tull, D. E. S. 1964. Evidence that vaginal lactobacilli do not ferment glycogen. American Journal of Obstetrics and Gynecology 88:676–679.

Stirling, A. C., Whittenbury, R. 1963. Sources of the lactic acid bacteria occurring in silage. Journal of Applied Bacteriology 26:86–90.

Sugihara, T. F., Kline, L.: 1975. Further studies on a growth medium for *Lactobacillus sanfrancisco*. Journal of Milk and Food Technology 38:667–672.

Tagg, J. R., Dajani, A. S., Wannamaker, L. W. 1976. Bacteriocins of Gram-positive bacteria. Bacteriological Reviews 40:722–756.

Tamine, A. Y. 1977. The behaviour of different starter cultures during the manufacture of yoghurt from hydrolysed milk. Dairy Industries International 42:7–11.

Terzaghi, B. E., Sandine, W. E. 1975. Improved medium for lactic streptococci and their bacteriophages. Applied Microbiology 29:807–813.

Thornley, M. J., Sharpe, M. E. 1959. Microorganisms from chicken meat related to both lactobacilli and aerobic spore formers. Journal of Applied Bacteriology 22:368–376.

Tikhonenko, A. Š. 1970. Ultrastructure of bacterial viruses, pp. 95–99. New York, London: Plenum Press.

Tilbury, R. H. 1975. Occurrence and effects of lactic acid bacteria in the sugar industry, pp. 177–191. In: Carr, J. G., Cutting, C. V., Whiting, G. C. (eds.), Lactic acid bacteria in beverages and food. London, New York, San Francisco: Academic Press.

Tohyama, K. 1973. Studies on temperate phages of *Lactobacillus salivarius*. II. Mode of action of defective phage 208 on nonlysogenic and homologous lysogenic *Lactobacillus salivarius* strains. Japanese Journal of Microbiology 17:173–180.

Tohyama, K., Sakurai, T., Arai, H., Oda, A. 1972. Studies on temperate phages of *Lactobacillus salivarius*. I. Morphological, biological, serological properties of newly isolated temperate phages of *Lactobacillus salivarius*. Japanese Journal of Microbiology 16:385–395.

van den Hamer, C. J. A. 1960. The carbohydrate metabolism of the lactic acid bacteria. Thesis. University of Utrecht, Netherlands.

Vaughn, R. H. 1975. Lactic acid fermentation of olives with special reference to California conditions, pp. 307–323. In: Carr, J. G., Cutting, C. V., Whiting, G. C. (eds.), Lactic acid bacteria in beverages and food. London, New York, San Francisco: Academic Press.

Veaux, M., Accolas, J.-P., Vassal, L., Auclair, J. 1974. Development of lactic acid bacteria and propionic acid bacteria during manufacture and ripening of Gruyère cheese. XIX International Dairy Congress 1E, 420.

Vincent, J. G., Veomett, R. C., Riley, R. F. 1959. Antibacterial activity associated with *Lactobacillus acidophilus*. Journal of Bacteriology 78:477–484.

Wang, H. L., Kradej, L., Hesseltine, C. W. 1974. Lactic fermentation of soy bean milk. Journal of Milk and Food Technology 37:71–73.

Watanabe, K., Takesue, S., Ishibashi, K. 1977. Reversibility of the adsorption of bacteriophage PL-1 to the cell walls isolated from *Lactobacillus casei*. Journal of General Virology 34:189–194.

Watanabe, K., Takesue, S., Jin-Nai, K., Yoshikawa, T. 1970. Bacteriophage active against the lactic acid beverage-producing bacterium *Lactobacillus casei*. Applied Microbiology 20:409–415.

Whiting, G. C. 1975. Some biochemical and flavour aspects of lactic acid bacteria in ciders and other alcoholic beverages, pp. 69–85. In: Carr, J. G., Cutting, C. V., Whiting, G. C. (eds.), Lactic acid bacteria in beverages and food. London, New York, San Francisco: Academic Press.

Wilkinson, B. J., Jones, D. 1977. A numerical taxonomic survey of *Listeria* and related bacteria. Journal of General Microbiology 98:399–421.

Williams, R. A. D. 1971. Cell wall composition and enzymology of lactobacilli. Journal of Dental Research 50:1104–1117.

Williams, R. A. D. 1975. A review of biochemical techniques in the classification of lactobacilli, pp. 351–367. In: Carr, J. G., Cutting, C. V., Whiting, G. C. (eds.), Lactic acid bacteria in beverages and food. London, New York, San Francisco: Academic Press.

Williams, R. A. D., Sadler, S. 1971. Electrophoresis of glucose-6-phosphate dehydrogenase, cell wall composition and the taxonomy of heterofermentative lactobacilli. Journal of General Microbiology 65:351–358.

Williamson, D. H. 1959. Studies on lactobacilli causing ropiness in beer. Journal of Applied Bacteriology 22:392–402.

Wood, B. J., Cardenas, O. S., Yong, F. M., McNulty, D. W. 1975. Lactobacilli in production of soy sauce, sour-dough bread and Parisian barm, pp. 325–335. In: Carr, J. G., Cutting, C. V., Whiting, G. C. (eds.), Lactic acid bacteria in beverages and food. London, New York, San Francisco: Academic Press.

Wylie, J. G., Henderson, A. 1969. Identity and glycogen-fermenting ability of lactobacilli isolated from the vagina of pregnant women. Journal of Medical Microbiology 2:363–366.

Yokokura, T. 1977. Phage receptor material in *Lactobacillus casei*. Journal of General Microbiology 100:139–145.

Yokokura, T., Kodaira, S., Ishiwa, H., Sakurai, T. 1974. Lysogeny in lactobacilli. Journal of General Microbiology 84:277–284.

Yoshizumi, H. 1975. A malo-lactic bacterium and its growth factor, pp. 87–102. In: Carr, J. G., Cutting, C. V., Whiting, G. C. (eds.), Lactic acid bacteria in beverages and food. London, New York, San Francisco: Academic Press.

CHAPTER 132

The Genus *Listeria* and Related Organisms

HERBERT J. WELSHIMER

The generic name *Listeria* has evolved after a number of changes since Murray, Webb, and Swann (1926) described a small, Gram-positive rod as the causative agent of a 1924 epizootic among rabbits and guinea pigs in the Cambridge University animal colonies. The production of a pronounced mononucleosis in experimentally infected rabbits led to the name *Bacterium monocytogenes*. One year later, Pirie (1927) unknowingly isolated and described the same organism from the gerbil *Tatera iobengulae* in South Africa. Pirie, observing marked liver involvement in experimentally infected animals, named the organism *Listerella hepatolytica*. After recognition of the dual name for the same organism, the species was named *Listerella monocytogenes;* however, with the discovery that the name *Listerella* had been applied to a mycetoon and a foraminifer, Pirie suggested changing the name to *Listeria monocytogenes* (1940).

The sixth edition of *Bergey's Manual of Determinative Bacteriology* (Breed, Murray, and Hitchins, 1948), as well as the seventh edition (Breed, Murray, and Smith, 1957), ranked the genus *Listeria* with a single species *Listeria monocytogenes* in the family Corynebacteriaceae. Studies on intergeneric relationship of *L. monocytogenes* (Stuart and Welshimer, 1974) to members of the Corynebacteriaceae using molecular methods and examining cardinal phenotypic characters did not justify the inclusion of *Listeria* in the family. A new family, Listeriaceae, was proposed (Stuart and Welshimer, 1974) and is still sub judice. The eighth edition of *Bergey's Manual* (Buchanan and Gibbons, 1974) does not recognize the family Corynebacteriaceae but presents *Listeria* as a genus of uncertain affiliation in a section headed "Gram-Positive, Asporogenous Rod-Shaped Bacteria". Four species of *Listeria* are described in the eighth edition of *Bergey's Manual* (Seeliger and Welshimer, 1974): *L. monocytogenes, L. denitrificans* (Sohiier, Benazet, and Piéchaud, 1948), *L. grayi* (Errebo Larsen and Seeliger, 1966), and *L. murrayi* (Welshimer and Meredith, 1971). Of these four species, only *L. monocytogenes* is associated with diseases of man and animals. Based on differences of growth characteristics, biochemical reactions, DNA base composition, and low molecular

relatedness to other species of *Listeria* (Stuart and Welshimer, 1973), *L. denitrificans* clearly is misplaced. The subcommittee of the International Committee on Systematic Bacteriology (ICSB) studying *Listeria* (LSC) and related organisms is suggesting that *L. denitrificans* be designated as *nomen generum perplexum*. The nonpathogenic species *L. grayi* and *L. murrayi* are so similar that the classification of *L. murrayi* as a subspecies of *L. grayi* seems appropriate. The designation of a new genus, *Murraya,* has been proposed to encompass these nonpathogenic organisms (Stuart and Welshimer, 1974). Pending the recommendation of the LSC and ICSB, *L. grayi* and *L. murrayi* subsequently may be designated *Murraya grayi* subsp. *grayi* and *Murraya grayi* subsp. *murrayi*, respectively. For purposes of this article the term "listeria" will include both generic designations.

In contrast to the *Listeria* strains isolated from active clinical infections, many of the *Listeria* strains isolated from healthy individuals and inanimate sources are nonhemolytic, nonpathogenic for laboratory animals, and incapable of evoking a monocytosis in rabbits. These organisms have been reported as *L. monocytogenes*. As the number of isolations of the nonpathogenic *L. monocytogenes* increases there is concern that the significance of the clinical and epidemiological role of *L. monocytogenes* may be minimized; consequently, efforts have been directed toward disclosing additional characteristics useful for differentiating the nonpathogenic *L. monocytogenes* from the classical pathogen. Significant differential properties have been described and are included in this chapter; the new species name *Listeria innocua* (still sub judice) was proposed by H. P. R. Seeliger before the LSC and participants at the International Symposium on Listeriosis held in Bulgaria in 1977.

Ivanov (Iwanow, 1962) described a different organism isolated from infected sheep in Bulgaria. The organism has some cardinal properties in common with *L. monocytogenes* and some distinctly different characteristics. It is biochemically less active. It has the same "H" and "O" antigens as found in the various *L. monocytogenes* serovars, but the antigenic formula sets it apart to be designated as serovar 5. A striking feature is the very pronounced beta type of hemolysis produced on blood agar plates, in contrast to the weak hemolysis of *L. monocytogenes*. The organism will be referred to as *Listeria* serovar 5 (L-s5).

Habitats

Listeriosis, a term encompassing the various clinical manifestations of listeric infections, provided the stimulus to study *L. monocytogenes* and reveal the existence of related organisms. For the first two decades after the discovery of *L. monocytogenes*, listeriosis was considered primarily a disease of animals and within the purview of veterinary medicine.

Listeria monocytogenes has been isolated from more than 37 species of mammals and 17 species of birds (Gray and Killinger, 1966), involving a geographical distribution from the Arctic to South Africa. Sporadic and epizootic outbreaks in herds of cattle, sheep, goats, and swine attain significant economic proportions since the mortality rate is high.

As late as 1951, listeriosis in man rarely was recognized; at that time, less than 100 cases had been reported (Seeliger, 1961). The occurrence of infection in humans is usually sporadic; however, rare epidemics have been confined to two locations, both in Germany, viz., the areas of Bremen and Halle. The Halle epidemic in 1966 consisted of 279 cases of listeriosis involving 203 newborn and premature infants and 76 mothers and resulted in 130 deaths (Ortel, 1968). Meningeal involvement, frequently accompanied by septicemia, is the most commonly encountered clinical manifestation. Listeriosis in humans usually occurs in one of three predisposing states, viz., in the fetus or neonate, in the female as a complication of pregnancy, and in the adult, 40 years or older, who is physiologically compromised by some underlying illness, alcoholism, malignancy, or immunosuppressive procedures. Intrauterine infection of the fetus results in fetal death, premature delivery of a stillborn fetus, or delivery of an acutely ill septicemic infant with disseminated listeriosis and meningitis (Erdmann, 1963).

Although originally conceived as an animal-borne disease, only rarely has direct transmission of listeriosis from animal to man been validated. As viewed by Seeliger (1961, 1972, 1976) and others (Kampelmacher, Mass, and van Noorle Jansen, 1975; Maupas et al., 1975) listeriosis is neither a zoonosis nor an anthroponosis. A common source of infection rather than an adaptation of *L. monocytogenes* to man or to animals is suggested by the similarity in the clinical syndromes, histological changes, involvement of the central nervous system, intrauterine infection, and predilection for individuals of particular age group and physical state. Neonatal listeriosis is now recognized as an important manifestation of human disease with *L. monocytogenes* (Albritton, Wiggins, and Feeley, 1976).

Until the middle 1960s, *L. monocytogenes* was exclusively associated with active clinical infections of man or animals. Since that time, listeriae have been isolated from sources other than clinical cases. By employing appropriate enrichment and selective procedures, listeriae have been isolated from the feces of healthy humans and animals, and from the tonsils of healthy, slaughtered pigs (Höhne, Loose, and Seeliger, 1975; Ralovich et al., 1971). In Wales, where there were no reported cases of listeriosis in records dating to the previous 15 years, *L. monocytogenes* was isolated from the feces of 0.6% of healthy humans examined and from the carcasses of 53% of fresh and frozen commercial chickens (Kwantes and Isaac, 1975). *L. monocytogenes* has been recovered from extracorporal sources such as silage, soil, decaying vegetation, and effluents (Gray, 1960; Kampelmacher and van Noorle Jansen, 1975; Weiss and Seeliger, 1975; Welshimer, 1968; Welshimer and Donker-Voet, 1971). Since listeriae have been isolated from these nonclinical sources, the concept of *L. monocytogenes* as a "saprophytic pathogen with an opportunistic mode of spread" (Meyer, 1961) now becomes increasingly attractive, along with the hypothesis (Seeliger, 1961) that the natural habitat of *Listeria monocytogenes* might be that of a soil or plant organism. A number of isolants of listeriae from feces of healthy humans and animals and from nature have been designated as *L. monocytogenes;* however, unlike *L. monocytogenes* isolated from active clinical infections, many of these are nonhemolytic and nonpathogenic. These nonpathogenic strains constitute the newly proposed species, *Listeria innocua* (Seeliger and Schoofs, 1977).

Murraya grayi subsp. *grayi* and *M. grayi* subsp. *murrayi* are nonpathogenic for laboratory animals. *M. grayi* subsp. *grayi* has been isolated from animal feces and from nature (Errebo Larsen and Seeliger, 1966; Welshimer and Meredith, 1971), and *M. grayi* subsp. *murrayi* has been isolated from soil and decayed vegetation (Welshimer and Merredith, 1971). Suggestive of a common and possible saprophytic heritage is the observation that the pathogenic *L. monocytogenes* shares with the nonpathogenic listeriae a similar DNA base composition, the ability to grow at low temperatures (4–10°C), the loss of motility at 37°C, and many similar biochemical properties.

Isolation

The listeriae multiply over a wide temperature range, with good growth at 25–37°C and slow growth at 4°C. *Listeria* colonies on 5% sheep blood agar prepared with an infusion base or tryptose blood agar base (Difco) resemble *Streptococcus* colonies. With the exception of the slower growing L-s5, which requires 72 h incubation, colonies 0.5–1.5 mm in diameter develop after 24 h incubation at 37°C. *L. monocytogenes* and L-s5 produce a beta type of hemolysis on sheep blood agar, whereas *L. innocua* and the other listeriae are nonhemolytic. In contrast to *L. monocytogenes,* which produces a very narrow zone of hemolysis (sometimes necessitating scraping the colony off the surface of the medium for detection), the L-s5 produces a broad zone of clear hemolysis.

Tryptose agar (Difco) is extensively used for cultivation of listeriae. Tryptose blood agar base (Difco) to which 1% glucose is added produces more luxurious and rapid growth. At 37°C for 24 h, the colonies are 0.3–1.5 mm in diameter (longer incubation period and smaller colonies with L-s5), translucent, round and smooth with entire margins, and appear blue-green by reflected light. The clear colorless tryptose media are excellent for the application of Gray's oblique lighting technique (1957) for the examination and selection of *Listeria* colonies from mixed flora. For best viewing, the plates should not be poured too thick. The lid is removed and the plate is examined at 10× to 20× magnification with a binocular scanning microscope. Colonies are illuminated from below by obliquely transmitted light. A microscope lamp, with filter removed, is placed in front of the microscope and tilted at a 45° angle to direct the light to a flat mirror placed on the bench in front of the microscope. The light is reflected from the mirror at an angle of 45° to provide oblique illumination of the specimen. The attached microscope mirror must be removed or tilted in such a way that it does not interfere with the passage of light from the flat mirror. *Listeria* colonies appear reticulated, with a distinctive blue-green cast that facilitates their selection in the midst of numerous other colonies.

Enrichment Procedures

Subcultures of *Listeria monocytogenes* grow well on most routinely employed media; however, primary isolation from infected tissues and from natural sources may be unsuccessful, as noted in the early studies of Murray, Webb, and Swann (1926), Gill (1937), and later by others. Gray found that clinical specimens from suspected listeric infections that failed to yield *L. monocytogenes* on primary plating would do so after holding at 4°C. Gray et al. (1948) macerated brain tissue from naturally infected cows, maintained it in tryptose broth at 4°C for 3 months, and cultured it at intervals. They demonstrated increased numbers of *L. monocytogenes* as the material was maintained at 4°C for longer intervals.

The combination of the cold enrichment tech-

nique and use of oblique lighting for the examination of colonies is basic for the isolation of *L. monocytogenes* and other listeriae from specimens likely to have mixed flora and from specimens in which the listeriae are undetectable on primary culture. Many workers have adopted the procedure of the Dutch investigators Kampelmacher and van Noorle Jansen (1969), who found that a higher percentage of early isolations could be obtained by the use of Stuart's medium (Stuart, 1959) in conjunction with the cold enrichment of specimens in tryptose broth.

Isolation of Listeriae Using Stuart's Medium (Stuart, 1959) and Cold Enrichment in Tryptose Broth (Kampelmacher, Huysinga, and van Noorle Jansen, 1972; Kampelmacher and van Noorle Jansen, 1969)

A 3-g sample is suspended in 30 ml of tryptose phosphate (Difco) broth and held at 4°C. Approximately 0.1 g of the specimen is collected on an alginate swab, introduced into a tube of Stuart transport medium (Difco) and held at 22°C for 1 week. The alginate swab is removed and suspended in 10 ml of sterile Ringer-Calgon solution (Wilson) to dissolve the alginate. Plates of blood agar, tryptose agar, and selective media are each inoculated with 1 loopful of the suspension and incubated at 37°C. At intervals, samples are taken from the refrigerated tryptose phosphate broth with sterile alginate swabs and treated as described above. The enrichment broth should be held for a minimum of 1 month and examined at weekly intervals, but it is preferable to hold the specimen for 6 months with monthly subcultures. Unfortunately, the length of time for cold enrichment is arbitrary; Kampelmacher and van Noorle Jansen (1969) isolated listeriae from some specimens only after cold enrichment for 30 months. Plates are observed after 24, 48, and 72 h of incubation. Blood agar plates are examined for hemolytic colonies and the other plates are microscopically examined by oblique lighting for typical listeriae colonies after incubating. This procedure is adaptable to other specimens.

Bockemühl et al. (1974) have developed a modified Stuart medium, *Listeria* Transport Enrichment Medium (LTE), which has both selective and enrichment properties.

Listeria Transport Enrichment (LTE) Medium (Bockemühl et al., 1974)

Sodium glycerophosphate	10.0 g
Sodium thioglycolate	1.0 g
Ca Cl₂	0.1 g
Nalidixic acid	0.04 g
Agar	2.0 g
Distilled water	1,000 ml

After sterilization (15 min at 120°C) the medium is cooled to about 70°C. Acridine dye ("Acriflavin neutral" or "Proflavinhemisulfat", Cassella-Farbwerk, Frankfurt/M., Federal Republic of Germany), is dissolved in distilled water, autoclaved for 15 min at 120°C and added to give a final concentration of 10 μg/ml (2 ml of a 0.5% solution). Final pH ± 7.4. (See note below regarding titration of dyes.) The medium is dispensed in sterile screw-capped tubes or bottles (10-ml content) to at least 4/5 of their volume. Inoculate with about 1 g of specimen. The inoculated LTE medium is held for 6–8 h at 22°C and subcultured. The LTE medium is also inoculated from the specimen undergoing cold enrichment at 4°C in tryptose phosphate broth with samples being removed at intervals up to 6 months. Plates of blood agar, a selective medium and tryptose agar are inoculated both from the tryptose broth and LTE medium and the colonies on the clear medium are examined microscopically by oblique illumination after 24, 48, and 72 h incubation.

Selective Media

There have been many different selective media proposed for the isolation of *Listeria monocytogenes* (Ralovich, 1975), with varying degrees of efficacy. The most successful selective media tested by field trials employ a combination of two inhibitors, viz., nalidixic acid, which inhibits Gram-negative bacteria, and an acridine dye at a concentration that inhibits many Gram-positive bacteria without suppressing the listeriae.

The Trypaflavin–Nalidixic Acid–Serum–Agar (TNSA) developed by Ralovich et al. (1971) has been used extensively for the isolation of listeriae from specimens that have been preenriched.

Trypaflavin–Nalidixic Acid–Serum–Agar (TNSA) Medium for Isolating Listeriae from Preenriched Specimens (Ralovich et al., 1971; B. Ralovich, personal communication)

Peptone (Difco)	10.0 g
Beef extract (Difco)	3.0 g
Ionagar no. 2	12.0 g
H$_2$O	926 ml

Adjust to pH 7.2–7.4, autoclave 121°C for 20 min, and cool to 50°C before aseptically adding the following:

1. 50 ml of bovine serum which has been inactivated by heating at 57°C for 30 min.
2. 20 ml of 0.2% solution of nalidixic acid to give a final concentration of 40 μg/ml (prepare by dissolving the nalidixic acid in boiling water at pH 8 and filter-sterilize).

3. Trypaflavin (Bayer, Germany; Pflatz and Bauer, Flushing, New York) to give a final concentration of 35 μg/ml (see note below) prepared by adding 3.5 ml of a 1% sterile solution.
4. The medium is poured sufficiently thin (13 ml per 100-mm-diameter Petri dish) to permit good viewing by oblique illumination. Viewed at this depth after incubation at 37°C for 1–2 days, the listeria colonies appear yellowish-green and they are the largest colonies on the plates among colonies of possible survivors of other species.

Ortel (1972), in a modification of the trypaflavin-nalidixic acid medium, eliminated the bovine serum.

Modified TNSA Medium for Isolating Listeriae (Ortel, personal communication)

The base medium consists of 20 g of tryptose (Difco), 1 g of dextrose, 5 g of NaCl, 15 g of agar (Difco); 975 ml of distilled water, nalidixic acid, 20 ml of 0.2% to give final concentration of 40 μg/ml (dissolve nalidixic acid in boiling water at pH 8). Adust pH 7.2–7.4. Autoclave at 121°C for 15 min. Cool to 55°C and add 4.5 ml of 1% filter-sterilized trypaflavin (Hoechst) to give 45 μg/ml. Pour 13 ml of medium per 100-mm-diameter Petri plate. Inoculate plates with cold enriched culture and observe colonies by obliquely transmitted light after 24, 48, and 72 h incubation at 37°C.

Bockemühl, Seeliger, and Kathke (1971) employ an acridine-nalidixic acid-agar medium that is rather simply prepared.

Acridine—Nalidixic Acid—Agar Medium for Isolating Listeriae (Bockemühl, Seeliger, and Kathke, 1971)

The base medium per liter of distilled water contains: 41 g of tryptose agar (Difco), 0.04 g of nalidixic acid, and distilled water to make 1,000 ml. After sterilization for 15 min at 120°C, the medium is cooled to 70°C. The acridine dye dissolved in distilled water and autoclaved for 15 min at 120°C is added to the base medium. Acriflavin neutral, 25 μg/ml (Cassella-Farbwerk), gave the most consistent results (H. P. R. Seeliger, personal communication). The medium is thoroughly mixed and about 13 ml are poured into dishes. After inoculation the medium is incubated for 24 h at 37°C and subsequently for 48 h at room temperature. The plates are microscopically examined by oblique lighting for the presence of light green to blue-green colonies.

Note: The concentrations of the dyes given in the formulas of the various enrichment and selective media are those designated by the investigators; however, the bacteriostatic action of different lots of the same dye from the same manufacturer may vary considerably and may or may not inhibit the listeriae as well as other Gram-positive organisms. At one extreme, Ralovich et al. (1971) found that 50 μg/ml of medium was a satisfactory concentration of trypaflavin, whereas Ortel (1972) found that 50 μg/ml of his trypaflavin inhibited the listeriae and 45 μg/ml was favorable. At the other extreme, Kampelmacher, Huysinga, and van Noorle Jansen (1972) used only 8/μg of trypaflavin per ml of medium, a concentration that Berger and Pietsch (1975) found inhibitory to the listeriae; Berger and Pietsch found that 6 μg of trypaflavin per ml of medium was optimal for survival of the listeriae and inhibitory to other organisms. It is essential that each batch of dye be titrated in the otherwise complete medium in order to establish the maximum concentration of dye permitting the development of *Listeria* colonies.

Maintenance

Tryptose blood agar base (Difco) with 1% added glucose (TBG) is excellent where rapid and luxurious growth is desired, and it is satisfactory for viewing colonies by oblique lighting. Stock cultures may be carried on tryptose agar (Difco) slants in screw-capped tubes that are incubated at 34–37°C and removed for refrigeration as soon as growth is evident. Transfer at 4- to 6-week intervals.

Conservation of *Listeria* Cultures

Inoculate a slant of TBG and incubate at 34–37°C for 18 h. Emulsify the growth of the entire slant in 0.2 ml of brain heart infusion (BHI) broth (Difco). With a Pasteur pipette apply just enough of the thick suspension to wet a sterile glass bead (with hole) placed in a sterile, cotton-plugged (10 × 75 mm) glass tube. The plug is trimmed and pushed into the tube 1.5 cm from the bottom. Four or five granules of a desiccant (Drierite or silica gel) are added. The tube is drawn with a glass torch and constricted to form an ampule. This is connected to a vacuum pump for 30 min and then sealed at the constriction with a gas flame while under vacuum. To reconstitute, break the vacuum at the tip of the ampule and ring midway at the cotton plug with a glass cutter. Remove the bead with a sterile inoculating wire and place in a tube of BHI broth to incubate. Strains of *Listeria* have been stored in a 4°C refrigerator for 7–10 years in satisfactory condition.

Identification

The listeriae comprising the species *Listeria monocytogenes*, *Listeria innocua*, *Murraya grayi* (*L. grayi*, *L. murrayi*), and the Ivanov serovar (L-s5) show the following characteristics: Gram-positive small rods; nonsporeforming; non-acid-fast; catalase-positive; oxidase-negative; aerobic and facultatively anaerobic; fermentative in sugars, producing acid without gas. Most strains are motile when cultivated at 20–25°C, but nonmotile at 37°C; rare strains of *L. monocytogenes* have been isolated without demonstrable motility at either temperature. Seeliger (personal communication) on examination of all such strains submitted to him found that, although nonmotile, they did possess flagella.

The catalase and motility tests are useful for screening purposes. Coccobacillary forms of the listeriae observed on primary isolation and on microscopic examination of stained smears of clinical preparations have been confused with streptococci as well as the colonies of both hemolytic and nonhemolytic *Streptococcus* species. A positive catalase test readily excludes the streptococci. Remove colonies with an inoculating loop and rub into a drop of H_2O_2 (3% concentration or greater) placed on a slide and observe for vigorous bubbling. Motility excludes most species of small, Gram-positive rods. Two tubes of BHI broth (Difco) are inoculated directly from isolated colonies and incubated at 37°C and 22–25°C. Hanging drop preparations may be observed microscopically (440× or under oil immersion) at the first trace of birefringence, which may be as soon as 4–6 h at 37°C and 18 h at 22–25°C. Few if any organisms will be motile at 37°C, whereas there will be many motile organisms at the lower temperature, with a number of the rods exhibiting a tumbling movement.

A combination of two procedures, hemolysis on sheep blood agar plates and the acidification of mannitol broth, provides a simple means for differentiating the four species of *Listeria*. A beta type of hemolysis is produced by *L. monocytogenes* and the L-s5 organism, whereas *L. innocua* and *M. grayi* are nonhemolytic. The faint hemolysis of *L. monocytogenes* and the pronounced hemolysis of L-s5 are distinctive. A weak or questionable hemolysis of *L. monocytogenes* can be resolved by using a modified CAMP test (Brzin and Seeliger, 1975; Darling, 1975; Groves and Welshimer, 1977). Streak the *Listeria* and a beta toxin-producing *Staphylococcus aureus* at right angles to each other (1–2 mm apart) on a sheep blood agar plate (tryptose blood agar base [Difco] or Trypticase soy agar [BBL] with 5% sheep blood). Incubate at 37°C in a candle extinction jar, and examine at 24 and 48 h for enhanced lysis along the *Listeria* streak within the weakly lytic zone of the beta toxin surrounding the implanted *Staphylococcus*. The nonhemolytic *L. innocua* and *M. grayi* can be differentiated by inoculating mannitol broth (1% mannitol in purple sugar broth base [Difco]). Both subspecies of *M. grayi* (synonyms *L. grayi* and *L. murrayi*) acidify the mannitol in 24–48 h at 37°C, while *L. innocua* is nonreactive.

Detailed descriptions of the biochemical activities are presented in the eighth edition of *Bergey's Manual of Determinative Bacteriology* (Buchanan and Gibbons, 1974); however, *L. innocua* is not distinguished from *L. monocytogenes* but is included as a nonhemolytic, avirulent strain of *L. monocytogenes*. Also *Murraya grayi* subsp. *grayi* and *Murraya grayi* subsp. *murrayi* appear as *L. grayi* and *L. murrayi*, respectively. It will be noted that these organisms are all active on a variety of carbohydrates in contrast to the Ivanov L-s5 organism, which is much less reactive metabolically and fails to acidify trehalose, salicin, maltose, and dextrin (Donker-Voet, 1972), all of which are acidified by the other species. Animal pathogenicity and antigen composition of the listeriae are helpful but not essential for the routine identification of the organisms, although these features are very important in the classification of the organisms and in consideration of the epidemiology and medical aspects. Animal pathogenicity tests may be performed by intraperitoneally inoculating 16- to 18-gram white mice with an overnight suspension of 2×10^9 organisms. The mice are observed for 2 weeks, although most animals will die within a period of 5 days or less and the organisms may be observed and recovered from the liver, spleen, and blood. Also, pathogenicity may be determined by the Anton eye test, which is conveniently performed on rabbits or guinea pigs. A few drops of overnight BHI broth culture are instilled into the conjunctiva of the animal or rubbed on the everted eyelid with a swab. The pathogenic organisms initiate a purulent discharge within 24–48 h.

Four basic serotypes of *Listeria monocytogenes* were originally described by Paterson (1940). The Seeliger–Donker-Voet antigenic scheme (Table 1.) (Donker-Voet 1972; Seeliger, 1975, 1976) has expanded to 17 serovars that encompass *L. monocytogenes* and other species as well. The presence of the unique "O" antigen factors X and XI associated with the lack of hemolytic activity and lack of pathogenicity has strengthened the justification for the classification of *L. innocua* as a new species (Seeliger and Schoofs, 1977). The "H" antigen factor E is unique to the genus *Murraya*. The Ivanov serovar 5 strain shares antigens in common with *L. monocytogenes* serovars, but its lack of biochemical homogeneity with the other serovars necessitates further study for appropriate taxonomic assignment.

Literature Cited

Albritton, W. L., Wiggins, G. L., Feeley, J. C. 1976. Neonatal listeriosis: Distribution of serotypes in relation to age at onset of disease. Journal of Pediatrics **88:**481–483.

Table 1. The Seeliger–Donker-Voet scheme for serovars of the listeria group of organisms.[a]

Species	Serovar designation	O antigens	H antigens
Listeria monocytogenes	1/2a	I II (III)[b]	AB
	1/2b	I II (III)	ABC
	1/2c	I II (III)	BD
	3a	II (III) IV	AB
	3b	II (III) IV	ABC
	3c	II (III) IV	BD
	4a	(III) (V) VII IX	ABC
	4ab	(III) V VI VII IX	ABC
	4b	(III) V VI	ABC
	4c	(III) V VII	ABC
	4d	(III) (V) VI VIII	ABC
	4e	(III) V VI (VIII) (IX)	ABC
(Ivanov serovar)	5	(III) (V) VI VIII X	ABC
	7?	(III) XII XIII	ABC
Listeria innocua[c]	6a[d]	(III) V VI VII IX XV	ABC
	6b[e]	(III) V VI VII IX X XI	ABC
Murraya grayi subsp. *grayi*		(III) XII XIV	E
Murraya grayi subsp. *murrayi*		(III) XII XIV	E

[a] Known antigenic formulae as of December, 1977. Courtesy of H. P. R. Seeliger.
[b] Parentheses indicate irregularly occurring O antigen factors.
[c] Additional antigen combinations are known but not listed.
[d,e] Formerly designated as 4f and 4g, respectively.

Berger, U., Pietsch, U. 1975. Untersuchungen zur Häufigkeit der Listerien-Keimträger. Medical Microbiology and Immunology **161**:63–71.

Bockenmühl, J., Seeliger, H. P. R., Kathke, R. 1971. Acridinfarbstoffe in Selektivnährböden zur Isolierung von *Listeria monocytogenes*. Medical Microbiology and Immunology **157**:84–95.

Bockenmühl, J., Feindt, E., Höhne, K., Seeliger, H. P. R. 1974. Acridinfarbstoffe in Selektivnährböden zur Isolierung von *Listeria monocytogenes*. II. *Modifiziertes Stuart-Medium:* Ein neues *Listeria*-Transport-Anreicherungsmedium. Medical Microbiology and Immunology **159**:289–299.

Breed, R. S., Murray, E. G. D., Hitchens, A. P. (eds.). 1948. Bergey's manual of determinative bacteriology, 6th ed. Baltimore: Williams & Wilkins.

Breed, R. S., Murray, E. G. D., Smith, N. R. (eds.). 1957. Bergey's manual of determinative bacteriology, 7th ed. Baltimore: Williams & Wilkins.

Brzin, B., Seeliger, H. P. R. 1975. A brief note on the CAMP phenomenon in *Listeria*, pp. 34–37. In: Woodbine, M. (ed.), Problems of listeriosis. Leicester, England: Leicester University Press.

Buchanan, R. E., Gibbons, N. E. (eds.). 1974. Bergey's manual of determinative bacteriology, 8th ed. Baltimore: Williams & Wilkins.

Darling, C. L. 1975. Standardization and evaluation of the CAMP reaction for the prompt, presumptive identification of *Streptococcus agalactiae* (Lancefield group B) in clinical material. Journal of Clinical Microbiology **1**:171–174.

Donker-Voet, J. 1972. *Listeria monocytogenes:* Some biochemical and serological aspects. Acta Microbiologica Academiae Scientiarum Hungaricae **19**:287–291.

Erdmann, G. 1963. Pediatric problems in listeriosis research, pp. 266–276. In: Gray, M. L. (ed.), Second symposium on listeric infection. Bozeman, Montana: Montana State College.

Errebo Larsen, H., Seeliger, H. P. R. 1966. A mannitol fermenting *Listeria: Listeria grayi* sp.n., pp. 35–39. In: Proceedings of the third international symposium on listeriosis, Bilthoven, The Netherlands.

Gill, D. A. 1937. Ovine bacterial encephalitis (Circling Disease) and the bacterial genus *Listerella*. Australian Veterinary Journal **13**:46–56.

Gray, M. L. 1957. A rapid method for the detection of colonies of *Listeria monocytogenes*. Zentralblatt für Bakteriologie, Parasitenkunde, Infektionskrankheiten und Hygiene, Abt. 1 Orig. **169**:373–377.

Gray, M. L. 1960. Isolation of *Listeria monocytogenes* from oat silage. Science **132**:1767–1768.

Gray, M. L., Killinger, A. H. 1966. *Listeria monocytogenes* and listeric infections. Bacteriological Reviews **30**:309–382.

Gray, M. L., Stafseth, H. J., Thorp, F., Jr., Sholl, L. B., Riley, W. F., Jr. 1948. A new technique for isolating listerellae from the bovine brain. Journal of Bacteriology **55**:471–476.

Groves, R. D., Welshimer, H. J. 1977. Separation of pathogenic from apathogenic *Listeria monocytogenes* by three in vitro reactions. Journal of Clinical Microbiology **5**:559–563.

Höhne, K., Loose, B., Seeliger, H. P. R. 1975. Recent findings of *Listeria monocytogenes* in slaughter animals of Togo (West Africa), pp. 127–233. In: Woodbine, M. (ed.), Problems of listeriosis. Leicester, England: Leicester University Press.

Iwanow, I. 1962. Untersuchungen über die Listeriose der Schafe in Bulgarien. Monatshefte für Veterinär Medizin **17**:729–736.

Kampelmacher, E. H., Huysinga, W. Th., van Noorle Jansen, L. M. 1972. The presence of *Listeria monocytogenes* in feces of pregnant women and neonates. Zentralblatt für Bakteriologie, Parasitenkunde, Infektionskrankheiten und Hygiene Ab. 1 Orig. Reihe A **222**:258–262.

Kampelmacher, E. H., Maas, D. E., van Noorle Jansen, L. M. 1975. Occurrence of *Listeria monocytogenes* in faeces of pregnant women with and without direct animal contact, pp.

214–216. In: Woodbine, M. (ed.), Problems of listeriosis. Leicester, England: Leicester University Press.

Kampelmacher, E. H., van Noorle Jansen, L. M. 1969. Isolation of *Listeria monocytogenes* from faeces of clinically healthy humans and animals. Zentralblatt für Bakteriologie, Parasitenkunde, Infektionskrankheiten und Hygiene Abt. 1 Orig. **211**:353–359.

Kampelmacher, E. H., van Noorle Jansen, L. M. 1975. Occurrence of *L. monocytogenes* in effluents, pp. 66–70. In: Woodbine, M. (ed.), Problems of listeriosis. Leicester, England: Leicester University Press.

Kwantes, W., Isaac, M. 1975. Listeria infection in West Glamorgan, pp. 112–114. In: Woodbine, M. (ed.), Problems of listeriosis. Leicester, England: Leicester University Press.

Maupas, P., Bind, J. L., Chiron, J. P., Darchis, J. P. 1975. Epidemiologic and pathogenic conception of animal and human listeriosis, pp. 221–228. In: Woodbine, M. (ed.), Problems of listeriosis. Leicester, England: Leicester University Press.

Meyer, K. F. 1961. Forward. In: Seeliger, H. P. R. Listeriosis. New York: Hafner Publishing Co.

Murray, E. G. D., Webb, R. A., Swann, M. B. R. 1926. A disease of rabbits characterized by a large mononuclear leucocytosis, caused by a hitherto undescribed bacillus *Bacterium monocytogenes* (n.sp.). Journal of Pathology and Bacteriology **29**:407–439.

Ortel, S. 1968. Bakteriologische, serologische und epidemiologische Untersuchungen während einer Listeriose-Epidemie. Das Deutsche Gesundheitswesen **16**:753–759.

Ortel, S. 1972. Experience with nalidixic acid-trypaflavin agar. Acta Microbiologica Acadamiae Scientiarum Hungaricae **19**:363–365.

Paterson, J. S. 1940. The antigenic structure of organisms of the genus *Listerella*. Journal of Pathology and Bacteriology **51**:427–440.

Pirie, J. H. H. 1927. A new disease of veld rodents "Tiger River Disease". Publications of the South Africa Institute for Medical Research **3**:163–186.

Pirie, J. H. H. 1940. *Listeria:* Change of name for a genus of bacteria. Nature **145**:264.

Ralovich, B. 1975. Selective and enrichment media to isolated *Listeria*, pp. 286–294. In: Woodbine, M. (ed.), Problems of listeriosis. Leicester, England: Leicester University Press.

Ralovich, B., Forray, A., Mérö, E., Málovics, H., Százados, I. 1971. New selective medium for isolation of *L. monocytogenes*. Zentralblatt für Bakteriologie, Parasitenkunde, Infektionskrankheiten und Hygiene, Abt. 1 Orig. **216**:88–91.

Seeliger, H. P. R. 1961. Listeriosis. New York: Hafner Publishing Co.

Seeliger, H. P. R. 1962. New outlook on the epidemiology and epizoology of listeriosis. Acta Microbiologica Academiae Scientiarum Hungaricae **19**:273–286.

Seeliger, H. P. R. 1975. *Serovars of Listeria monocytogenes* and other Listeria species, pp. 27–29. In: Woodbine, M. (ed.), Problems of listeriosis. Leicester, England: Leicester University Press.

Seeliger, H. P. R. 1976. Notion acteuelle sur l'épidémiologie de la listériose. Médecine et Maladies Infectieuses **6**:6–14.

Seeliger, H. P. R., Schoofs, M. 1977. Serological analysis of nonhemolyzing strains of *Listeria* sp. Abstracts VIIth International Symposium on the Problems of Listeriosis. Varna, Bulgaria, Sept. 23–27, 1977.

Seeliger, H. P. R., Welshimer, H. J. 1974. *Listeria*, pp. 593–596. In: Buchanan, R. E., Gibbons, N. E. (eds.), Bergey's manual of determinative bacteriology, 8th ed. Baltimore: Williams & Wilkins.

Sohier, R., Benazet, F., Piéchaud, M. 1948. Sur un germe du genre *Listeria* apparemment non pathogène. Annales de l'Institut Pasteur **74**:54–57.

Stuart, R. D. 1959. Transport medium for specimens in Public Health bacteriology. Public Health Reports **74**:431–438.

Stuart, S. E., Welshimer, H. J. 1973. Intrageneric relatedness of

Listeria pirie. International Journal of Systematic Bacteriology **23:**8–14.

Stuart, S. E., Welshimer, H. J. 1974. Taxonomic reexamination of *Listeria pirie* and transfer of *Listeria grayi* and *Listeria murrayi* to a new genus *Murraya*. International Journal of Systematic Bacteriology **24:**177–185.

Weis, J., Seeliger, H. P. R. 1975. Incidence of *Listeria monocytogenes* in nature. Applied Microbiology **30:**29–32.

Welshimer, H. J. 1968. Isolation of *Listeria monocytogenes* from vegetation. Journal of Bacteriology **95:**300–303.

Welshimer, H. J., Donker-Voet, J. 1971. *Listeria monocytogenes* in nature. Applied Microbiology **21:**516–519.

Welshimer, H. J., Meredith, A. L. 1971. *Listeria murrayi* sp.n.: A nitrate-reducing mannitol-fermenting *Listeria*. International Journal of Systemic Bacteriology **21:**3–7.

CHAPTER 133

The Genus *Erysipelothrix*

FRIEDRICH W. EWALD

The first member of this group was the bacillus of mouse septicemia, *Erysipelothrix murisepticus;* it was found by Koch in 1876 in the blood of mice that had been infected with putrefied meat (Koch, 1878). In 1882 Pasteur and Oumas described a slender, curved rod that they isolated from swine showing symptoms of swine erysipelas. Probably this was the same organism as that found by Koch. In 1882 Loeffler observed similar organisms in a pig that had died of swine erysipelas. He gave us the first accurate description of the swine erysipelas microorganism, and the published results of his work demonstrated that the disease in swine is caused by a bacillus identical to, or closely resembling, the mouse septicemia bacillus of Koch (Loeffler, 1886). Another organism closely related to *Erysipelothrix* was found by Rosenbach in cases of human erysipeloid. He was the first to use the term "erysipeloid" for the disease in man. Three species were distinguished by him, *E. murisepticus, E. porci,* and *E. erysipeloides* (Rosenbach, 1909). While at first strains of murine, porcine, and human origin were considered separately, it is now generally believed that these are almost identical variants of a single species. Subsequent studies made on the relationships of the microorganisms isolated from the above sources show that the strains are identical. The species name *Erysipelothrix porci* (Rosenbach, 1909) was antedated by *Bacterium rhusiopathiae* (Migula, 1900), and the new combination *Erysipelothrix rhusiopathiae* (Buchanan, 1918) was proposed.

Despite the fact that the etiological agent of the diseases swine erysipelas, erysipeloid, and mouse septicemia is a single species, the complexity of the situation is illustrated by the fact that at least 36 species names have been given to this species in the literature.

It is a rule of bacteriologic nomenclature that the name of an organism that was the first used must have priority. The first name was *Erysipelothrix insidiosa,* given by Trevisan in 1885. Therefore, the 1957 edition of *Bergey's Manual of Determinative Bacteriology* has included the three species *Erysipelothrix rhusiopathiae, Erysipelothrix murisepticus,* and *Erysipelothrix erysipeloides* in a single species, which has been given the name *Erysipelothrix insidiosa* (Langford and Hansen, 1953, 1954). But the epithet *insidiosus* or *insidiosa* did not appear in the literature for more than 60 years, until the 1953 proposal of Langford and Hansen. Shuman and Wellmann (1966) therefore proposed to the Judicial Commission of the International Committee on Nomenclature of Bacteria that the species name *E. insidiosa* (Trevisan) (Langford and Hansen) be regarded as "forgotten" epithet and rejected. The Editorial Secretary for the Judicial Commission therefore declared: The specific epithet *rhusiopathiae* in the scientific name of the organism known as *Erysipelothrix rhusiopathiae* (Migula) (Buchanan) is conserved against the specific epithet *insidiosa* (basionym: *Bacillus insidiosus,* Trevisan) and against all other specific epithets applied to this organism.

Habitats

Occurrence and Distribution

Erysipelothrix is widely distributed in nature. It may be found in sewage effluent from abattoirs and in feces from different animals suffering from erysipelas, as well as from animals not suspected of *Erysipelothrix* diseases (Teschner, Behrens, and Weiss, 1976). Most frequently, *Erysipelothrix* may be found in pigs because it is the cause of a widely distributed and common disease known as swine erysipelas; less frequently it has been found in polyarthritis in sheep, joint ill in lambs (Tontis et al., 1977), and infections in dogs (Goudswaard et al., 1973), horses, cattle, and other domestic animals. It is of some economic importance also in turkeys, peacocks, geese (Polner et al., 1972), pinnipeds (Sweeny, 1974), egg-laying chickens (Bisgaard and Olsen, 1975), partridges (Pettit, Gough, and Truscott, 1976), and zoo birds (Decker and Lindauer, 1977). There are also infections in whales (Provost, 1976) and in neonatal calves (Rebhuhn, 1976) and mixed infections in swine with *Salmonella* (Harrington and Ellis, 1975) or especially *Streptococcus* (Ewald, 1960). Ehrsam (1958) reports about many other animals as carriers of *E. rhusiopathiae.*

The repeated occurrence of swine erysipelas, sometimes with several years elapsing between epizootics, led early investigators to believe that the causative organism persisted in the soil. But this is a question of temperature and milieu.

Temperature exerts the most marked effect on viability. The organism prefers alkaline pH, especially with a high content of organic matter. Soil, food, and water are readily contaminated by sick animals because feces and urine contain the bacteria. Soil that may retain viable and virulent organisms (Topping, 1937) serves as an excellent medium for conveying the infection to the animal body by way of the digestive tract. Contaminated surface waters, rodents, wild birds, and insects (Wellmann, 1950, 1954) may also be responsible for transmitting the disease from one farm to another. Wood and Packer (1972) found the organisms in 50 of 133 samples of soil and manure collected from pig farms, even though the disease had not been observed for at least 5 years.

In a dried stage, the organisms remain viable for many years. *E. rhusiopathiae* has been isolated from fish meal used for the preparation of animal foods, which are therefore potential sources of infection. Although it is sometimes found in the slime surrounding the body of various fish, there is no evidence that the fish are naturally infected during life. The slime, however, is exposed to contamination later from trawler holds, fish boxes, and rats and mice on the docks (Murase et al., 1959; Schoop and Stoll, 1966).

The presence of *E. rhusiopathiae* in the tonsils of swine is very common. One can find the bacterium there as a saprophyte and, under certain conditions, as a pathogen. The role in the transmission of the disease played by healthy animals that harbor the organism may be questionable, but most of these organisms belong to a nonpathogenic group of *E. rhusiopathiae,* or are only facultative pathogens. Such strains may lead to subclinical infection of swine. One has to distinguish them from virulent strains, especially from those belonging to serotype A, which occur only sporadically (Wellmann, 1955).

Erysipelothrix rhusiopathiae is distributed throughout the world. One can find it in Africa (van Damme and Devriese, 1976), in Australia (Bond, 1976), and also in sylvatic mammals in northwestern Canada (Langford and Dorward, 1977), Mexico (Olivares et al., 1971), Brazil (Castro et al., 1972), and Japan (Hashimoto, Yoshida, and Sugawara, 1974). In Europe, the infection in swine appears in epizootic proportions in certain countries during the hot months from spring until summertime; sporadic outbreaks are observed during the colder months of the year. The disease has been reported in China and Japan, as well as in the United States. In North America, the disease in swine is most commonly

seen in the late summer months; chronic forms of swine erysipelas have been recognized throughout the entire year. The history of swine erysipelas in the United States is of some interest because, as Karlson (1967) reports, prior to 1930 it had been stated by many writers that the disease does not exist in the United States. In all likelihood the disease was not uncommon but was not clinically or bacteriologically recognized. Additional historical data on *E. rhusiopathiae* infections and its occurrence in America may be found in the paper by Shuman (1964).

Animal Diseases

Natural *Erysipelothrix* infections of swine occur by the oral route. The infection is commonly seen in three different forms: a mild form known as "red fever", "diamond skin disease", or swine erysipelas", in which skin involvement predominates. The name "diamond skin disease" refers to the appearance of reddish to purplish rhomboid spots known as "diamonds" in the skin. There is also an acute, severe form of septicemia. In such cases the sick animals may die within a few days. Less frequently a chronic form of infection develops, with local arthritis or endocarditis. *Erysipelothrix* infections in swine are often serious and fatal and may be, therefore, of considerable economic importance. The infection of commercial turkey or chicken flocks is characterized primarily by involvement of the long bones, from which positive cultures may be obtained. (A review is given by Levine, 1965.) There exist many publications dealing with the techniques of serological investigations of pigs, sanguination, isolation, hemagglutination-inhibition test, and agglutination. A good discussion on these was given by Heuner (1957) and Wellmann and Heuner (1959). For the treatment of swine that have already contracted the disease, the injection of antiserum from horse or swine and antibiotics is recommended. Active immunization with bacterins is commonly used to prevent the disease. Some strains, most of them belonging to serotype B, produce a soluble protective antigen in broth culture. This soluble antigen can be adsorbed to alum and combined with a formalized bacterial suspension. Such vaccines give very satisfactory results in the field. The vaccines and sera should be compared with standard vaccines and standard sera. Live vaccines may also be applied using attenuated or avirulent strains. The vaccines may be lyophilized. The potency of a bacterin depends not only on the immunogenic properties of the antigen used but also on the number of organisms/ml in the vaccine. Therefore, it is important for the producers to develop vaccines with a high content of microorganisms. Feist, Flossmann, and Erler, (1976) give the following medium, which is recommended for growing the bacteria with a high bacterial yield without using serum and without loss of antigenicity.

High Yield Medium for Growing *Erysipelothrix* (Feist, Flossmann, and Erler, 1976)

The medium contains per liter:

$Na_2HPO_4 \cdot 12H_2O$	18 g
Glucose	6.0 g
Peptone S	5.0 g
Yeast extract	5.0 g
L-Arginine-HCl	0.5 g
Tween 80	0.5 ml

The medium has a pH of 7.8–8.0 after autoclaving.

In recent years it has been economically advantageous to vaccinate large numbers of pigs with bi- or trivalent vaccines against erysipelas and Aujeszky's disease (Vasilev et al., 1976) or against erysipelas, Aujeszky's disease, and plague (Tsymbal et al., 1975). A newer method of vaccination in large farms is the use of aerosol immunization (Khasanov and Selivanov, 1976; Petzold et al., 1976; Wasiński, 1976). A review on erysipelas immunization is given by Eissner and Ewald (1973) and Bairey and Vogel (1973).

Many new aspects about the pathogenic significance of *E. rhusiopathiae* and the etiology of *Erysipelothrix* diseases in their acute and chronic phases have been provided in recent years, by the cooperation of a group of scientists of the "Tierärztliche Hochschule", Hannover, Federal Republic of Germany. These comparative experimental investigations served also as a model for studying the etiology and pathology of inflammatory rheumatic diseases (Böhm et al., 1975a,b; Schulz et al., 1977; Trautwein, Seidler, and Schulz, 1976). For information about the induction of chronic erysipelas see also White, Puls, and Hargrave (1975) and White, Mirikitani, and Hargrave (1976); experimental studies were also done by Aoki (1976), Hadler (1976), and, with emphasis on physiological aspects, Steinmann and Hsu (1976), Timoney (1976a,b), and Georgiev (1976).

Rats are the predominant experimental animal used for inducing chronic forms of erysipelas (Sakuma et al., 1975; Toshkov et al., 1976). The rapid development of arthritis may serve as a model for testing the therapeutic effect of various antirheumatoid preparations and antibiotics.

Mice and pigeons are very susceptible to *Erysipelothrix* infection; therefore, they are the preferred laboratory animals for testing the virulence of a strain. It is possible to infect swine by the oral route with virulent strains; but there are strains that are highly virulent for mice but not for pigs. Such strains often belong to the group N. The best methods for proving virulence for swine are cutaneous inoculation (Fortner and Dinter, 1944) or conjunctival infection (Möhlmann, Maas, and Meese, 1961). Such methods are also useful for testing the resistance, susceptibility, or the state of immunity of an animal. Using the cutaneous method of infection, the relationship between local skin reactions and generalized infection can easily be recognized (Wellmann, 1966).

Allergic skin tests in the pig have been studied by Böhm and Trautwein (1972).

Erysipelas in Man

Erysipelothrix rhusiopathiae is the etiological agent of erysipeloid, a common skin infection in man. Occasionally it may be confused with human erysipelas of streptococcal origin (Ewald, 1964, 1965). Erysipelas infections are largely limited to veterinarians, fish handlers, packing plant workers, and others who handle animal products. Man acquires the infection by contact with infected animals or contaminated fish or animal products such as meat, hides, bones, and manure. Generally the disease is limited to the fingers, hands, or arms, where the organisms have gained entrance via small cuts or abrasions. Such erysipeloid lesions are characterized by a sharply defined, slightly elevated purplish-red zone that extends peripherally. Often the regional lymph nodes are enlarged, and sometimes an associated arthritis occurs. Spontaneous healing requires about 4 weeks; second attacks may occur. Sometimes the disease may continue for months. Healing can be hastened by the administration of antibiotics and antiserum. Usually sulfonamides are of no value. Only rarely does the infection lead to systemic dissemination of the agent into the body, causing arthritis, meningitis, or endocarditis. *Erysipelothrix* endocarditis is a severe disease leading to widespread destruction of the involved cardiac valves. In a report of Volmer and Hasler (1976) 16 out of 28 patients with *Erysipelothrix* endocarditis died; 11 of them died in spite of treatment with antibiotics. Anamnestical studies showed that in most fatal cases the patients were excessive users of alcohol. For additional information on endocarditis, see Baird and Benn (1975), and Freland, (1977); for information about bacterial infections of the skin, erysipelas drug therapy, and erysipeloid diagnosis, see also Vandersteen (1974).

Isolation

Visualization and Morphology

Erysipelothrix rhusiopathiae stains well with ordinary dyes. It is Gram positive, but easily decolorized by acetone alcohol. The organism has no metachromatic granules; it is not acid fast. Sometimes one can see beaded rods when *E. rhusiopathiae* is stained with Gram's stain. Colonies occur in smooth and rough forms. Despite the great variability in morphological appearance, there are some typical characteristics. In the smooth colony form (S-form), "forma typica", microscopically the organism appears as a small, slightly curved, slender rod with rounded ends about 0.8–2.0 µm long and 0.2–0.4 µm wide (Fig. 1a). Young cultures consist of very short rods, the so-called cell retraction forms (Wiidik, 1952). These typical forms vary with the medium on which the organisms are growing and with many other factors such as pH, atmospheric conditions, chemical agents, and so on. The organisms may be isolated or in groups or small packets but without any specific position. In the rough colony form (R-form), microscopic long filaments up to 60 µm or more in chains or in entangled masses predominate. In smears from blood and tissue, especially in acute forms of erysipelas diseases such as septicemia, one can see organisms of the smooth colony form; in chronic cases of erysipelas with arthritis or endocarditis the rough form appears, sometimes with single organisms (Fig. 2a). *E. rhusiopathiae* is nonmotile, noncapsulated, and nonsporeforming.

Fully developed smooth colonies after 24 h incubation at 37°C are very small (1.0–1.5 mm in diameter), convex, and circular, with a smooth glistening surface and entire edge; they are water-clear, transparent, and amorphous. On further incubation they show some increase in size and the transparency disappears especially in the center of the colonies (Fig. 1b). Rough colonies are larger and flatter without any transparence; they are opaque and have uneven surfaces. The elongated filaments are the cause of the curled structure and fimbriated edge of such colonies (Fig. 2b). In contrast to smooth colonies, rough colonies tend to become confluent. In broth, the short rods (S-form) produce a slight uniform turbidity without pellicles and very little sediment. The rough organisms grow as floccules of various size and tend to settle in a very short time. These masses on the bottom of the tubes are difficult to disintegrate by shaking. Colonies on blood agar on occasion show a greening of the blood, then a slight clearing zone around the colonies. In gelatin stab cultures, *E. rhusiopathiae* produces growth confined to the line of inoculation. In less than 48 h, all strains in the S-form develop a typical growth just like a "pipe cleaner" (Ewald, 1964) (Fig. 3). During the days following further incubation, one can see the so-called test-tube brush type of growth. There is only a small hole of liquefaction of the gelatin on the top along the stab after further incubation.

Erysipelothrix rhusiopathiae has a strong tendency to dissociate from the S-form to the R-form. During such processes, changes or loss of typical

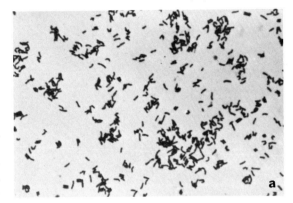

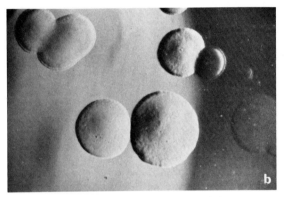

Fig. 1a,b. *Erysipelothrix rhusiopathiae*, smooth form.

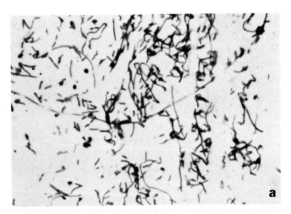

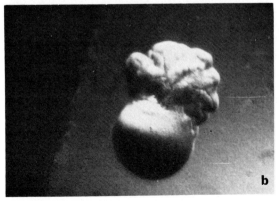

Fig. 2a,b. *Erysipelothrix rhusiopathiae*, smooth and rough forms.

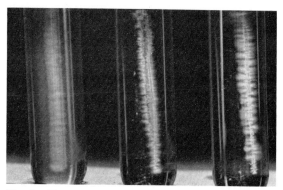

Fig. 3. Three *Erysipelothrix* strains on ferrochloride-gelatin medium after incubation of 48 h at 20°C. On the left, smooth strain; in the middle and on the right, dissociated strains.

properties often occur. The organisms change not only their morphological structure and cultural behavior but also often virulence and antigenic properties. Such dissociations have been described by Ewald (1962) (O- and o-variants); Wawrzkiewicz (1964) (S-, R-, and X-forms); Marica and Sîrmon (1969); and Marica and Vasiu (1970) (m-variants). Most of these and other studies revealed that despite the loss of several properties, the presence of type-specific polysaccharide antigens remains unaffected. During the last few years the so-called L-forms have been described, especially by Bulgarian authors (Cherepova et al., 1975; Nikolov, 1975; Todorov, 1976a,b). Electron microscopic studies on the structure of isolated L-forms have been carried out by Cherepova, Mihailova, and Gulubov (1975). Some years before, Stuart (1972) also demonstrated an electron microscopic examination of L forms and Pachas and Currid (1974) reported about the process of induction, subcultivation, adaptation to different osmolarity media, and morphological studies.

Growth Conditions and Biological Properties

Erysipelothrix rhusiopathiae grows scantily on ordinary culture media. The growth is improved by the addition of glucose (not more than 0.5%) or serum, especially horse serum, to the nutrient broth or agar. The presence of tryptophan enhances the development of bacterial growth. Riboflavin and small amounts of oleic acid are also recommended. Some workers prefer the use of Tween 80 (Feist, Flossmann, and Erler, 1976). The microorganism is microaerophilic, a characteristic that is obvious from the tendency of fresh cultures to grow beneath the surface of an agar shake culture; older cultures grow well under normal atmospheric conditions. Growth occurs at room temperature but is best at 37°C (limits are 15–42°C); the optimal pH is 7.4 (limits are 6.8 to 8.2). For a nonsporing microorganism, *E. rhusiopathiae* is remarkably resistant and in a dried stage

this resistance persists sometimes for many years. Extremely hot temperature or exposure to sunlight diminishes its viability; it remains alive in the soil at low temperatures. It is readily destroyed by boiling. *E. rhusiopathiae* is resistant to salting, pickling, and smoking for several months; it may remain viable in cadavers for up to 4 months due to their alkaline milieu. Certain strains have been found to survive and even multiply in concentrations of phenol that are destructive to many other microorganisms; 2% formalin solution, 1% NaOH, 3.5% kreolin, and 5% phenol will destroy the organisms after some minutes. They are resistant to sulfonamide preparations and polymyxin B; they are very susceptible to several antibiotics, such as penicillin, terramycin, or streptomycin (Sneath, Abott, and Cunliffe, 1951). The relatively high resistance to neomycin (Füzi, 1963a,b), the different susceptibilities to penicillin and streptomycin (Woodbine, 1950) and to tylosine (McDonald and Biberstein, 1974) and the high tolerance to sodium azide (Packer, 1943) are remarkable.

Erysipelothrix has only weak fermentative activity. It does not form indole and does not reduce nitrates. There is little or no change in litmus milk. The Voges-Proskauer, methyl red, neutral red, and methylene blue reductions are negative; the catalase and urease tests are also negative. There is no hydrolization of natrium hippurate and esculin, no growth in Koser's citrate, and no (or after prolonged incubation, only a slight) liquefaction of gelatin. *Erysipelothrix* produces acid without gas (with gas in glucose, lactose, and maltose; Sikes, 1965) in dextrose, lactose, fructose, and galactose; there is no or only a delayed or not constant reaction ($\pm$) in mannose($\pm$), maltose($\pm$), saccharose($\pm$), mannitol, dextrin, glycerin, salicin, dulcitol, inositol, arabinose($\pm$), xylose ($\pm$), rhamnose($\pm$), trehalose, raffinose, and starch. There is no reaction in inulin, adonitol, glycogen, amygdalin, cellobiose, melibiose, melezitose, sorbitol, sorbose, and erythritol (Sneath, Abott, and Cunliffe, 1951). Using a suitable medium like the ferric chloride–gelatin medium and healthy cultures, all *Erysipeothrix* strains produce hydrogen sulfide (Ewald, 1964).

There is no hemolysis of horse, swine, rabbit, guinea pig or human red blood cells. Cultures of serotype B will agglutinate chicken red blood cells. When fresh guinea pig serum is added to this complex, the chicken red blood cells (RBC) lyse, which indicates that the *Erysipelothrix*-RBC complex is capable of inducing an alternative pathway of complement activation (Dinter, Diderholm, and Rockborn, 1976). A narrow green zone of hemolysis may occur around deep colonies in blood agar with some strains (slight α-hemolysis), but there is never typical β-hemolysis.

Most strains produce hyaluronidase. There seems to be a correlation between antigen structure, virulence, and hyaluronidase production: virulent strains (most of them belong to serogroup A) are good hyaluronidase producers (Ewald, 1957). There is no correlation between virulence and fermentative activities. Neuraminidase seems to be another factor contributing to the pathogenicity of some *Erysipelothrix* strains (Nikolev and Abrashev, 1976). But it must be produced in high quantities. There appears to be a good correlation between virulence and neuraminidase production, but there are also virulent strains producing no or only small quantities of neuraminidase (Krasemann and Müller, 1975). The biosynthesis of the enzyme depends on the growth phase of the bacteria. The production of neuraminidase reaches its maximum at the end of the logarithmic phase of growth (Abrashev and Zamfirova, 1976). A significant correlation was found between the average neuraminidase production in different media and virulence. Active immunization with purified neuraminidase demonstrates only a poor protective effect in experimentally infected mice (Müller and Krasemann, 1976). Neuraminidase-neutralizing antibodies have been revealed in pigs with chronic forms of erysipelas (Müller and Seidler, 1975). For information on the Michaelis constants of neuraminidases of pathogenic and nonpathogenic microorganisms see Müller, von Nicolei, and Zilliken (1975); for material on the inhibition of bacterial neuraminidase by different anions see Rau and Müller (1975).

An endotoxin (phenol–water extract) in bacteria causing swine erysipelas was detected by Leimbeck et al., (1975). Its toxicity was demonstrated in chicken embryos. In some strains there was a correlation between the virulence for pigs and the toxicity of their extracts for chicken embryos; other strains did not show this relationship (Leimbeck and Böhm, 1975).

Dissemination and Isolation of *Erysipelothrix*

There is no problem in isolating *Erysipelothrix* from the blood of an infected animal with septicemia or other acute forms of erysipelas. Due to generalization, the bacteria are disseminated in the whole body and can be isolated from the organs, especially from the lymph nodes, kidneys, lungs, and spleen. The slender rods may also be found in smears and secretions. In cases of endocarditis one can observe a widespread destruction of the involved cardiac valves with masses of fibrinogen and clumps of bacteria. In chronic cases of arthritis, the same situation may be seen. It may be difficult to find the organisms in cases of erysipeloid; therefore, it is very important to take samples of tissue and secretions from the subcutaneous parts of the skin, because the organisms are located only in the deeper parts of the skin.

Seidler, Trautwein, and Böhm (1971) could demonstrate by direct and indirect fluorescent antibody tests the localization and identity of *Erysipelothrix*. They described the preparation of the antiserum, fractionation of the serum, purification of the conjugate by gel-filtration and ion-exchange chromatography, and the use of conjugates in the direct and indirect tests. The indirect immunofluorescence test may also be used for diagnosis of porcine erysipelas (Avilag, Unzucta, and Olgiun, 1972) or human erysipelas (Heggers, Buddington, and McAllister, 1974). There is a good conformity between the fluorescent-antibody technique and the cultural method for the detection of *Erysipelothrix* in primary broth cultures (Harrington, Wood, and Hulse, 1974). Sakuma et al. (1973) could demonstrate the dissemination with a whole body autobacteriography of infected mice. Whole body sagittal sections were placed on selective agar plates containing antibiotics and incubated. After 48 h, the plates were treated histochemically by the nitroblue tetrazolium method for vital staining to observe the colonies. The authors used a medium with the additives: sodium azide (NaN_3) 200 mg/liter, crystal violet (0.01%) 2 ml/liter, kanamycin 400 μg/ml, and neomycin 50 μg/ml. Under such conditions, even very small colonies of *E. rhusiopathiae* were distinguishable and without any contamination.

Sometimes a small number of bacteria with low viability may occur in chronic cases of erysipelas. Then enrichment cultures are needed. Longer periods of incubation and the use of serum from horse, calf, or swine as an enrichment in the broth increase the chance to isolate the bacteria. Often one has to isolate the microorganisms from heavily contaminated material, such as arthritic joints of swine or from feces. Then the small colonies that may not be seen during the first days of incubation may be transferred to suitable differential media for isolation and identification (Ewald, 1960; Wood, 1965; Wood and Packer, 1972; Zimmermann, 1963).

In vivo methods may help to isolate *Erysipelothrix*. For this, suspect clinical material is inoculated by scarification or subcutaneously in mice. These animals are very susceptible to *Erysipelothrix* and usually die a few days after inoculation. One can then isolate the organisms from the kidneys or spleen. When the possibility of contamination with other bacteria highly pathogenic for mice exists, such as *Pasteurella,* a selective medium must be used before passage into animals.

Selective Media for Detection and Enrichment

Packer (1943) could easily isolate *Erysipelothrix* from contaminated tissues by using a medium with sodium azide and crystal violet. But with such media it is impossible to differentiate very small streptococcal colonies from those of *Erysipelothrix*. Füzi (1963a) recommended blood agar with 100 μg/ml of neomycin or 400 μg/ml of kanamycin. These antibiotics prevent the growth of most other bacteria but do not inhibit *Erysipelothrix*. Böhm (1971) described two selective media using kanamycin, crystal violet, sodium azide, and liquefied phenol in concentrations in which other commonly encountered bacteria are inhibited but are tolerated by *Erysipelothrix*. The solid medium contains sucrose and waterblue. Streptococci that occasionally multiply in spite of the inhibitory substances may be differentiated from *Erysipelothrix*. Most of the *Streptococcus* strains form acid from sucrose, and one can see blue colonies; *Erysipelothrix* does not ferment sucrose and its colonies are colorless.

Selective Liquid Enrichment Medium for *Erysipelothrix* (Böhm, 1971)

Bouillon (meat from swine)	
twice concentrated	500.0 ml
Peptone (Witte)	30.0 g
Tap water	500.0 ml

Dissolve by autoclaving for 30 min and filtrate through paper, then adjust to pH 7.6 with NaOH.

Sodium phosphate (primary)	0.2 g
Sodium chloride	0.3 g
Glucose	10.0 g
Crystal violet, stock solution 1:1,000	10.0 g
Sodium azide, stock solution 1:25	10.0 g

Mix well; sterilization should be done by autoclaving for 15 min at 121°C. After cooling to 60°C, add:

Phenol liquefactum (it does not dissolve in cold broth)	1.0 g
Kanamycin sulfate, stock solution	2.0 ml
Serum (from swine or calf) filtrated sterile	50.0 ml

(Add the serum after the phenol).

To isolate the microorganisms from the above selective bouillon, the following solid medium is recommended.

Selective Solid Enrichment Medium for *Erysipelothrix* (Böhm, 1971)

Bouillon (meat of swine)	
twice concentrated	500.0 ml
Peptone (Witte)	10.0 g
Tap water	500.0 ml
Sodium chloride	5.0 ml

Adjust to pH 7.4 with N NaOH; dissolve by autoclaving for 20 min; control the pH and adjust for a second time if necessary; heat for a short time and filter through paper.

| Agar agar pulverized | 11.0 g |
| Sodium azide, stock solution 1:25 | 10.0 ml |

Sterilize by autoclaving 15 min at 121°C; after cooling to 60°C, add:

Kanamycin sulfate, stock solution	2.0 ml
Sucrose	50.0 g
Waterblue, 1% sterile solution	10.0 ml
Phenol liquefactum	1.0 ml

Add the phenol before the medium is cooled and pour into sterile Petri dishes. The medium must have only a blue shine; if not it should not be used.

Identification

Antigenic Structure, Serology, and Clinical Composition

Cross-agglutination with absorbed and nonabsorbed antisera prepared in rabbits by injection of normal or heat-treated bacteria and precipitation tests (ring precipitation or agar gel precipitation) using sera and cell extracts demonstrate species-specific thermolabile and type-specific thermostable antigens. The type-specific antigens are polysaccharide-polypeptide-simplexes with hapten character. For more details see the review of Eissner and Ewald (1973). There are two main serological groups, A and B, with different predominant type-specific antigens, also labeled A and B. The most virulent strains causing acute erysipelas belong to group A; strains of group B have been isolated primarily from chronic cases. Many other type-specific strains have been found and grouped using serological methods with type-specific, acid-soluble antigens or autoclaved bacteria and type-specific rabbit-erysipelas antisera. Fish are the hosts of such organisms, which may also live on the tonsils of domestic animals. Following the first designation, such strains are labeled C,D,E, and so on. According to Kucsera (1972) many such labeled types are recognizable among strains from different sources in different parts of the world. After his serological investigation of several strains of different origin, he proposed a new system for designating the serotypes in order to put an end to the resulting confusion in designations used for all these serotypes. He proposed that arabic numerals instead of letters be applied to serotypes according to the chronological order in which the serotypes are described (Kucsera, 1973). Strains without any remarkable type-specific antigen belong to the so-called group N and are designated with the letter N. Such strains are often old laboratory strains or nonpathogenic saprophytes or only virulent for mice but not for pigs. There is no antigenic difference between strains of human and animal origin.

The hemagglutinin of *Erysipelothrix* seems to be firmly bound to the bacterial cell, presumably as a constituent of the cell wall (Dinter, Diderholm, and Rockborn, 1976). The occurrence of *Erysipelothrix* bacteriophages is of some scientific interest. They may help in the diagnosis of *Erysipelothrix* type-specificity, for some of them have type-specific characters (Brill and Polityńska, 1961; Polityńska-Banaś, 1969; Revenko, 1968; Valerianov, Toshkoff, and Cholakova, 1976).

Investigations with cell wall preparations revealed the same type-specific hapten as from the intact bacteria. The amino acid composition of the cell wall included murein (Feist, 1972), composed of lysine, serine, and alanine. The content of reducing carbohydrates and amino sugars was similar to that of the cell wall of other Gram-positive bacteria. The following neutral sugars and acid monosaccharides were identified: galactose, glucose, arabinose, xylose, ribose, glucose-6-phosphate, and galactose-6-phosphate. From the monosaccharide patterns, three chemotypes were established. Although no relationship was found between chemotype and serotype, it is possible to differentiate *Erysipelothrix* by means of quantitative comparison of the content of reduced carbohydrates, glucose, and galactosamine. In 80% of the strains examined, quantitative analysis of the chemical composition of the cell wall agreed with the serological differentiation (Erler, 1971, 1972a,b). *Erysipelothrix* contained 3–4% DNA and 7–15% RNA, the amount of each varying from strain to strain at different stages of growth. There was no difference between serological types in the DNA base composition (38–40 mol% G+C). It was similar to that of *Listeria monocytogenes,* but very different from that of *Corynebacterium* (Flossmann and Erler, 1972).

For information on the protein composition as determined by electrophoresis and electrofocus patterns, the G+C content of their respective deoxyribonucleic acids, and carbohydrate fermentation, see White and Mirikitani (1976). For investigations of decarboxylases of amino acids, see Nicolov and Mihailova (1975).

Differential Diagnosis of *Erysipelothrix*

Occasionally it may be desirable to differentiate *Listeria monocytogenes* strains from *Erysipelothrix rhusiopathiae* strains. *Listeria* strains are very susceptible to neomycin (< 2 μg neomycin/ml), *Erysipelothrix* strains are highly resistant (> 100 μg neomycin/ml) when examined in surface cultures on blood agar with a paper disk method (Füzi, 1963b).

Petrov (1971) reports another simple method for differentiation of *Listeria* from *Erysipelothrix:* The addition of 8.5% NaCl to meat peptone broth inhibits the growth of *Erysipelothrix* and most saprophytes without affecting the growth of *Listeria* after incubation at 37°C. Both organisms may readily be distinguished by their different resistance to sodium azide, crystal violet, and "Mavekal" (disodium hexadecyl disulfonate) (Füzi and Pillis, 1962). There are also media for differentiation that contain TTC (2,3,5-triphenyltetrazolium chloride) or potassium tellurite, methylene blue, litmus milk, or prontosil. They show the higher reducing powers of *Listeria* in contrast to *Erysipelothrix* (Table 1).

A study of the morphology, growth requirements, biochemical properties, serological characteristics, and animal pathogenicity of several *Listeria, Pasteurella,* and *Erysipelothrix* strains isolated from pathological samples during bacteriological meat inspection led Butko (1972) to the opinion that there are many possibilities for error in the differentiation of these organisms. He recommends the following differential laboratory tests: motility, Gram staining, growth on agar with potassium tellurite, H_2S and indole production, catalase activity, type of hemolysis on blood agar, immunofluorescence with specific antiserum conjugates, production of conjunctivitis in rabbits, and pathogenicity by the intraperitoneal route for guinea pigs. He concluded that owing to the unreliability of any of these tests alone, reliable results can only be obtained by using all the above tests.

Taxonomy of *Erysipelothrix*

Jones (1975) studied 233 strains of coryneform bacteria using 173 morphological, physiological, and biochemical tests. The bacteria were grown on a soil extract medium that allowed growth of all strains, and all were incubated at 30°C. The results were subjected to computer analysis. The majority of the strains grouped into eight main clusters. In the first cluster, named A, *Lactobacillus, Listeria, Microbacterium thermosphactum,* and *Streptococcus faecalis* were assembled. In the second cluster, designated B, *Erysipelothrix* and *Streptococcus pyogenes* were collected. Based on those clusters she proposed it would be better to transfer the members of clusters A and B to the family Lactobacillaceae. The grouping of the species *Microbacterium thermosphactum* and the genera *Listeria* and *Erysipelothrix* with representatives of the Lactobacillaceae reaffirms the evidence from previous numerical taxonomic studies by Davis et al. (1969). Stuart and Pease (1972) who concluded on physiological grounds that *Listeria* bore the closest relationship to fecal streptococci; Flossmann and Erler (1972), on data derived from enzyme and DNA base ratio stud-

Table 1. Some of the described differences between *Erysipelothrix* and *Listeria*.

	E. rhusiopathiae	L. monocytogenes
Morphology	Slender, curved rods; nonmotile	Thicker; motile
Agar growth	Small glistening colonies without any blue shine	Very soft colonies with opal shining
Gelatin stab	"Pipe cleaner" growth along the stab	Irregular cloudy or filiform growth along the stab
β-Hemolysis	Negative	Positive
Hemolysin	Negative	Positive
Hemagglutination	Positive	Negative
Biochemical activities	Ferments fewer sugars, has fewer active reducing powers	Ferments more sugars, has more active reducing powers
Prontosil reduction	Negative	Positive
TTC^a reduction	Negative	Positive
MR and VP reactions	Both negative	Usually both positive
Hydrogen sulfide (H$_2$S)	Positive	Negative
Esculin fermentation	Negative	Positive
Catalase production	Negative	Positive
Growth at 4°C	Negative	Positive
Susceptibility to NaCl	High	Low
Susceptibility to Neomycin	Low	High
Pathogenicity	Kills pigeons but not guinea pigs; causes only a mild conjunctivitis in rabbits	Kills guinea pigs but not pigeons; causes a severe Keratoconjunctivitis in rabbits

^{a} 2,3,5-Triphenyltetrazolium chloride.

ies, concluded that *Erysipelothrix* bacteria showed a closer relationship to the Lactobacillaceae than to the Corynebacteriaceae. Comparison of the fatty acids of *Listeria* with those of *Erysipelothrix* and *Corynebacteria* led to fundamental differences in their fatty acid patterns, useful for differentiation and identification of these three genera (Tadayon and Carroll, 1971).

No common antigens were found by Pleszczyńska (1972) between strains of *Listeria monocytogenes* and *Erysipelothrix insidiosa* by using immunodiffusion and passive hemagglutination tests. Chromatography and infrared spectrophotometry of polysaccharides revealed differences between the two organisms. Additionally, no common precipitins were found between four polysaccharide and nucleoprotein fractions of the two organisms.

Paper and thin-layer chromatography of acid hydrolysates of the purified cell wall material showed that *Erysipelothrix* bacteria are distinguishable from *Listeria*, as the former contain lysine and glycine in the cell wall (Mann, 1969).

Literature Cited

Abrashev, I., Zamfirova, K. 1976. Dynamics in the accumulation of the enzyme neuraminidase depending on the growth phases of *Erysipelothrix insidiosa*. Acta Microbiologica, Virologica et Immunologica (Sofia) **4**:27–32.

Aoki, S. 1976. Experimental studies of chronic rheumatoid arthritis. Experimental models of arthritis. Japanese Journal of Clinical Medicine **34**:644–650.

Avilag, C., Unzucta, B. B. De., Olguin, R. F. 1972. Indirect immunofluorescence test for the diagnosis of porcine erysipelas. Veterinaria (Mexico) **3**:33–39.

Baird, P. J., Benn, R. 1975. *Erysipelothrix* endocarditis. Medical Journal of Australia **ii**:743–745.

Bairey, M. H., Vogel, J. H. 1973. Erysipelas immunizing product review, pp. 340–344. Proceedings of the Annual Meeting of the United States Animal Health Association.

Bisgaard, M., Olsen, P. 1975. Erysipelas in egg-laying chickens: Clinical, pathological and bacteriological investigations. Avian Pathology **4**:59–71.

Böhm, K. H. 1971. Neue Selektivnährböden für Rotlaufbakterien. Zentralblatt für Bakteriologie, Parasitenkunde, Infektionskrankheiten und Hygiene. Abt. 1 Orig., Reihe A **218**:330–334.

Böhm, K. H., Trautwein, G. 1972. Allergische Hautteste beim experimentellen Rotlauf des Schweines. Zentralblatt für Veterinärmedizin, Reihe B **19**:540–554.

Böhm, K. H., Franke, F., Messow, C., Schulz, L.-Cl., Trautwein, G., Weiland, F. 1975a. Versuche zur Erzeugung von chronischem Rotlauf beim Schwein. I. Totantigenversuche. Zentralblatt für Veterinärmedizin, Reihe B **22**:35–46.

Böhm, K. H., Franke, F., Baehr, K. H., Hazem, A. S., Schulze, W. 1975b. Versuche zur Erzeugung von chronischem Rotlauf beim Schwein. II. Lebendantigenversuch zur Standardisierung einer ModellKrankheit. Zentralblatt für Veterinärmedizin, Reihe B **22**:556–595.

Bond, M. P. 1976. Polyarthritis of pigs in Western Australia: The role of *Erysipelothrix rhusiopathiae*. Australian Veterinary Journal **52**:462–467.

Brill, J., Polityńska, E. 1961. Die Differenzierung von *Erysipelothrix rhusiopathiae*-Stämmen mit Bakteriophagen.

Zentralblatt für Bakteriologie, Parasitenkunde, Infektionskrankheiten und Hygiene, Abt. 1 Orig. **181**:473–477.

Buchanan, R. E. 1918. Studies in the nomenclature and classification of the bacteria. Journal Bacteriology **3**:27–61.

Butko, M. P. 1972. Differentiation of *Listeria*, *Erysipelothrix* and *Pasteurella* during bacteriological meat inspection. Problemy Veterinarnoi Sanitarii **41**:86–96.

Castro, A. F., de Pestana, Trabulsi, L. R., Campedelli, O., Troise, C. 1972. Characteristics of strains of *Erysipelothrix rhusiopathiae* isolated in Brasil. Revista de Microbiologica (Sao Paulo) **3**:11–24.

Cherepova, N., Mihailova, L., Gulubov, S. 1975. Electron microscopic studies on altered forms of *Erysipelothrix rhusiopathiae* obtained under the effect of novobiocin. Zentralblatt für Bakteriologie, Parasitenkunde, Infektionskrankheiten und Hygiene, Abt. 1 Orig., Reihe A **233**:245–252.

Cherepova, N., Nikolov, P., Gulubov, S., Mihailova, L., Abrachev, I. Illieva, K. 1975. Studies on the R-forms of *Erysipelothrix insidiosa*. Acta Microbiologica, Virologica et Immunologica (Sofia) **2**:13–21.

Davis, G. H. G., Fomin, L., Wilson, E., Newton, K. G. 1969. Numerical taxonomy of *Listeria*, *Streptococcus* and possibly related bacteria. Journal of General Microbiology **57**:333–348.

Decker, R. A., Lindauer, R. 1977. *Erysipelothrix* infection in two east African crowned cranes *(balearica regulorum gibbericeps)* and wood duck *(aix sponsa)*. Avian Diseases **21**:326–327.

Dinter, Z., Diderholm, H., Rockborn, G. 1976. Complement-dependent haemolysis following haemagglutination by *Erysipelothrix rhusiopathiae*. Zentralblatt für Bakteriologie, Parasitenkunde, Infektionskrankheiten und Hygiene, Abt . 1 Orig., Reihe A **236**:533–535.

Ehrsam, H. R. 1958. Epidemiologie und Bekämpfung des Schweinerotlaufs. Schweizer Archiv für Tierheilkunde **100**:202–208.

Eissner, G., Ewald, F. W. 1973. Rotlauf. In: Bieling, R., Kathe, J., Köhler, W., Mayr, A. (eds.), Infektionskrankheiten und ihre Erreger. Eine Sammlung von Monographien, vol. 13. Jena: Gustav Fischer Verlag.

Erler, W. 1971. Serologische, chemische und immunchemische Untersuchungen an Rotlaufbakterien. VIII. Die Neutralzucker der Zellwände. Archiv für Experimentelle Veterinärmedizin **25**:503–512.

Erler, W. 1972a. Serologische, chemische und immunchemische Untersuchungen an Rotlaufbakterien. IX. Die Aminozucker der Zellwände. Archiv für Experimentelle Veterinärmedizin **26**:797–807.

Erler, W. 1972b. Serologische, chemische und immunchemische Untersuchungen an Rotlaufbakterien. X. Die Differenzierung der Rotlaufbakterien nach chemischen Merkmalen. Archiv für Experimentelle Veterinärmedizin **26**:809–816.

Ewald, F. W. 1957. Das Hyaluronidase-Bildungsvermögen von Rotlaufbakterien. Monatshefte für Tierheilkunde **9**:333–341.

Ewald, F. W. 1960. Differentialdiagnostische Gesichtspunkte bei der Diagnose des Schweinerotlaufs unter besonderer Berücksichtigung des Vorkommens von *Streptokokken*. Archiv für Lebensmittelhygiene **11**:97–102.

Ewald, F. W. 1962. Über die Dissoziation von *Erysipelothrix rhusiopathiae*. III. Mitteilung und Schluss: Die Virulenz dissoziierter Rotlaufbakterien. Monatsschrift für Tierheilkunde **14**:260–267.

Ewald, F. W. 1964. Bakteriologie und Serologie humaner Rotlauferkrankungen unter besonderer Berücksichtigung des Ferrochlorid-Gelatine-Mediums nach Kaufmann. Arbeiten aus dem Paul-Ehrlich-Institut, Frankfurt am Main, Heft 61, 29–46.

Ewald, F. W. 1965. Erysipeloid und septische Rotlaufkomplikationen. Münchener Medizinische Wochenschrift **107**:365–369.

Feist, H. 1972. Serologische, chemische und immunchemische Untersuchungen an Rotlaufbakterien. XII. Das Murein der

Rotlaufbakterien. Archiv für Experimentelle Veterinärmedizin **26**:825–834.

Feist, H., Flossmann, K.-D., Erler, W. 1976. Einige Untersuchungen zum Nährstoffbedarf der Rotlaufbakterien. Archiv für Experimentelle Veterinärmedizin **30**:49–57.

Flossmann, K.-D., Erler, W. 1972. Serologische, chemische und immunchemische Untersuchungen an Rotlaufbakterien. XI. Isolierung und Charakterisierung von Desoxyribonukleinsäuren aus Rotlaufbakterien. Archiv für Experimentelle Veterinärmedizin **26**:817–824.

Fortner, J., Dinter, Z. 1944. Ist das Rotlaufbakterium der alleinige Erreger des Schweinerotlaufs? Zeitschrift für Infektionskrankheiten der Haustiere **60**:157–179.

Freland, C. 1977. Les infections a *Erysipelothrix rhusiopathiae*. Revue générale a propos de 31 cas de septicémies avec endocarditis relevés dans la littérature. Pathologie et Biologie **25**:345–352.

Füzi, M. 1963a. Über die selektive Züchtung der Schweinerotlaufbakterien. Zentralblatt für Bakteriologie, Parasitenkunde, Infektionskrankheiten und Hygiene, Abt. 1 Orig. **188**:387–392.

Füzi, M. 1963b. A neomycin sensitivity test for the rapid differentiation of *Listeria monocytogenes* and *Erysipelothrix rhusiopathiae*. Journal of Pathology and Bacteriology **85**:524–525.

Füzi, M., Pillis, I. 1962. Die Differenzierung der *Listeria* monocytogenes und *Erysipelothrix rhusiopathiae*. Zentralblatt für Bakteriologie, Parasitenkunde, Infektionskrankheiten und Hygiene, Abt. 1 Orig. **186**:556–561.

Georgiev, D. 1976. Histochemical studies of the content of acid and alkaline phosphatases in the synovial membranes and periarticular tissue of rats infected with *Erysipelothrix insidiosa*. Acta Microbiologica, Virologica and Immunologica (Sofia) **4**:11–15.

Goudswaard, J., Hartmann, E. G., Janmaat, A., Huismann, G. H. 1973. *Erysipflothrix rhusiopathiae* strain 7, a causative agent of endocarditis and arthritis in the dog. Tijdschrift voor Diergeneeskunde **98**:416–423.

Hadler, N. M. 1976. A pathogenetic model for erosive synovitis: Lessons from animal arthritides. Arthritis and Rheumatism **19**:256–266.

Harrington, R., Jr., Ellis, E. M. 1975. *Salmonella* and *Erysipelothrix* infection in swine. American Journal of Veterinary Research **36**:1379–1380.

Harrington, R., Jr., Wood, R. L., Hulse, D. C. 1974. Comparison of a fluorescent antibody technique and cultural method for the detection of *Erysipelothrix rhusiopathiae* in primary broth cultures. American Journal of Veterinary Research **35**:461–462.

Hashimoto, K., Yoshida, Y., Sugawara, H. 1974. Serotypes of *Ersipelothrix insidiosa* from swine, fish, and birds in Japan. National Institute of Animal Health Quarterly (Japan) **14**:113–120.

Heggers, I. P., Buddington, R. S., Mc.Allister, H. A. 1974. *Erysipelothrix* endocarditis diagnosis by fluorescence microscopy. Report of a case. American Journal of Clinical Pathology **62**:803–806.

Heuner, F. 1957. Zur Technik serologischer Rotlauf-Untersuchungen, Berliner und Münchener Tierärztliche Wochenschrift **70**:461–462.

Jones, D. 1975. A numerical taxonomic study of coryneform and related bacteria. Journal of General Microbiology **87**:52–96.

Karlson, A. G. 1967. The genus *Erysipelothrix*, pp. 466–474. In: Merchant, I. A., Packer, R. A. (eds.), Veterinary bacteriology and virology, 7th ed. Ames, Iowa: Iowa State University Press.

Khasanov, Ch. G., Selivanov, A. V. 1976. Aerogenic simultaneous vaccination of swine against plague, erysipelas and Aujeszky's disease. Veterinariia **43**:42–44.

Koch, R. 1878. Untersuchungen über die Ätiologie der Wundinfektionskrankheiten. Leipzig: Vogel.

Krasemann, C., Müller, H. E. 1975. Die Virulenz von *Erysipelo-*

thrix rhusiopathiae-Stämmen und ihre Neuraminidase-Produktion. Zentralblatt für Bakteriologie, Parasitenkunde, Infektionskrankheiten und Hygiene, Abt. 1 Orig., Reihe A **231**: 206–213.

Kucsera, G. 1972. Comparative study on special serotypes of *Erysipelothrix rhusiopathiae* strains isolated in Hungary and abroad. Acta Veterinaria Academiae Scientiarum Hungaricae **22**:251–261.

Kucsera, G. 1973. Proposal for standardization of the designations used for serotypes of *Erysipelothrix rhusiopathiae* (Migula) Buchanan. International Journal of Systematic Bacteriology **23**:184–188.

Langford, E. V., Dorward, W. J. 1977. *Erysipelothrix insidiosa* recovered from sylvatic mammals in Northwestern Canada during examinations for rabies and anthrax. Canadian Veterinary Journal **18**:101–104.

Langford, G. C., Hansen, P. A. 1953. *Erysipelothrix insidiosa*. Riass. Commun. VI. Congresso Internazionale di Microbiologia **1**:1–300, Roma.

Langford, G. C., Hansen, P. A. 1954. The species of *Erysipelothrix*. Antonie van Leeuwenhoek Journal of Microbiology and Serology **20**:87–92.

Leimbeck, R., Böhm, K. H. 1975. Untersuchungen über toxische Bestandteile von Rotlaufbakterien (*Erysipelothrix rhusiopathiae*). 1. Mitteilung: Toxizitätsprüfung eines Phenol-Wasser-Extraktes an Hühnerembryonen. Zentralblatt für Bakteriologie, Parasitenkunde, Infektionskrankheiten und Hygiene, Abt. 1 Orig., Reihe A **230**:367–378.

Leimbeck, R., Böhm, K.-H., Ehard, H., Schulz, L. Cl. 1975. Untersuchungen über toxische Bestandteile von Rotlaufbakterien (*Erysipelothrix rhusiopathiae*). 2. Mitteilung: Nähere Charakterisierung eines extrahierten Endotoxins. Zentralblatt für Bakteriologie, Parasitenkunde, Infektionskrankheiten und Hygiene, Abt. 1 Orig., Reihe A **232**: 266–286.

Levine, N. D. 1965. Erysipelas (Geflügelrotlauf), pp. 461–469, 1271–1276. In: Biester, H. E., Schwarte, L. H. (eds.), Disease of poultry, 5th ed. Ames, Iowa: Iowa State Univ. Press.

Loeffler, F.: 1886. Experimentelle Untersuchungen über Schweinerotlauf. Arbeiten aus dem Kaiserlichen Gesundheitsamt **1**:46–55.

McDonald, M. C., Biberstein, E. L. 1974. Determination of bacterial susceptibility to tylosin by single-disk agar diffusion tests. American Journal of Veterinary Research **35**:1563–1565.

Mann, S. 1969. Über die Zellwandbausteine von *Listeria monocytogenes* und *Erysipelothrix rhusiopathiae*. Zentralblatt für Bakteriologie, Parasitenkunde, Infektionskrankheiten und Hygiene, Abt. 1 Orig., Reihe A **209**:510–518.

Marcia, D., Sîrmon, E. 1969. Mutagene Wirkung einiger interferierender Bakterien auf *Erysipelothrix insidiosa*. Isolierung einer neuen *Erysipelothrix insidiosa*-Variante. Berliner und Münchener Tierärztliche Wochenschrift **82**:170–173.

Marica, D., Vasiu, A. 1970. Die Eigenschaften von *Erysipelothrix insidiosa* m-Varianten. Zentralblatt für Veterinärmedizin **17**:721–729.

Migula, W. 1900. System der Bakterien. Handbuch der Morphologie, Entwicklungsgeschichte und Systematik der Bacterium. Jena: G. Fischer Verlag.

Möhlmann, H., Maas, A., Meese, M. 1961. Untersuchungen und Vorschläge zur Prüfung der Rotlauf-adsorbat-Vakzine. Archiv für Experimentelle Veterinärmedizin **15**:150–182.

Müller, H. E., Krasemann, Ch. 1976. Immunität gegen *Erysipelothrix rhusiopathiae*-Infektion durch aktive Immunisierung mit homologer Neuraminidase. Zeitschrift für Immunitätsforschung **151**:237–241.

Müller, H. E., Seidler, D. 1975. Über das Vorkommen Neuraminidase-Neutralisierender Antikörper bei chronisch rotlaufkranken Schweinen. Zentralblatt für Bakteriologie, Parasitenkunde. Infektionskrankheiten und Hygiene, Abt. 1 Orig., Reihe A **230**:51–58.

Müller, H. E., von Nicolei, H., Zilliken, F. 1975. Michaelis-Konstanten von Neuraminidasen bei pathogenen und apathogenen Mikroorganismen. Zeitschrift für Naturforschung (C) **30**:417–419.

Murase, N., Suzuki, K., Isayama, Y., Murate, M. 1959. Studies on the typing of *Erysipelothrix rhusiopathiae*. III. Serological behaviours of the strains isolated from the body surface of marine fishes and their epizootiological significance in swine erysipelas. Japanese Journal of Veterinary Science **21**:215–218.

Nikolov, P. 1975. On the virulence and the antigenic structure of chlornitromycin-resistant variants of *Erysipelothrix rhusiopathiae*. Acta Microbiologica, Virologica et Immunologica (Sofia) **1**:57–61.

Nikolov, P., Abrashev, I. 1976. Comparative studies of the neuraminidase activity of *Erysipelothrix insidiosa*. Activity of virulent strains and avirulent variants of *Erysipelothrix insidiosa*. Acta Microbiologica, Virologica et Immunologica (Sofia) **3**:28–31.

Nikolov, P., Mihailova, L. 1975. Investigations of decarboxylases of amino acids in *Erysipelothrix insidiosa*. Acta Microbiologica, Virologica et Immunologica (Sofia) **1**:78–82.

Olivares, O. J., Martell, D. M. A., Torres, B. J., Flores, A. H. 1971. General characteristics of twenty strains of *Erysipelothrix insidiosa* isolated in Mexico. Técnica Pecuaria en Mexico **18**:32–39.

Pachas, W. N., Currid, V. R. 1974. L-form induction, morphology, and development in two related strains of *Erysipelothrix rhusiopathiae*. Journal of Bacteriology **119**:576–582.

Packer, R. A. 1943. The use of sodium acide and crystal violet in a selective medium for *Erysipelothrix rhusiopathiae* and *Streptococci*. Journal of Bacteriology **46**:343–349.

Pasteur, M., Oumas, M. 1882. Sur le rouget, ou mal rouge des porcs. Extrait d'une Lettre, Comptes Rendus Hebdomadaires des Séances de l'Académie des Sciences Paris, **95**:1120–1121.

Petrov, O. V. 1972. New ways of distinguishing *Listeria* from *Erysipelothrix* and other related micro-organisms. Veterinariia **39**:198–202.

Pettit, J. R., Gough, A. W., Truscott, R. B. 1976. *Erysipelothrix rhusiopathiae* infection in chukar partridge (alectoris graeca). Journal of Wildlife Diseases **12**:254–255.

Petzold, K., Floer, W., von Benten, C., Stuehmer, A. 1976. Experiments with a model of aerosol immunization of mice and swine against *Erysipelothrix insidiosa*. Developments in Biological Standardization **33**:57–62.

Pleszczyńska, E. 1972. Comparative studies on *Listeria* and *Erysipelothrix*. I. Analysis of whole antigens. II. Analysis of antigen fractions. Polskie Archiwum Weterynaryjne 15, Fasc. **3**:463–471, 473–481.

Polityńska-Banaś, E. 1969. Investigations on the bacteriocynogeny phaenomenon among *Erysipelothrix insidiosa* strains. Bulletin Instytut weterynarii W Pulawach (Pulawy, Poland) **13**:11–13.

Polner, T., Gajdács, G., Kemenes, F., Kucsera, G. 1972. Mortality among geese caused by *Erysipelothrix rhusiopathiae*. Ref.: The Veterinary Bulletin 42, No. 6778.

Provost, A. 1976. Infection des cétacés par le bacille du rouget du porc. La Nouvelle Presse Médicale **5**:276–277.

Rau, W., Müller, H. E. 1975. Über die Hemmung bakterieller Neuraminidasen durch verschiedene Anionen. Experientia **31**:515–516.

Rebhuhn, W. C. 1976. *Erysipelothrix insidiosa* septicaemia in neonatal calves. Veterinary Medicine and Small Animal Clinician **71**:684–686.

Revenko, I. P. 1968. Bacteriophage des Schweinerotlauferregers. Veterinariia (Moskva) **45**:25–27.

Rosenbach, F. J. 1909. Experimentelle, morphologische und klinische Studien über krankheitserregende Mikroorganismen des Schweinerotlaufs, des Erysipeloids und der Mäusesepticämie. Zeitschrift für Hygiene und Infektionskrankheiten **63**:343–371.

Sakuma, S., Sakuma, M., Okaniwa, A., Sato, Y. 1973. Detection of *Erysipelothrix insidiosa* in mice by whole body autobacteriographie. National Institute of Animal Health Quarterly (Tokyo) **13:**54–58.

Sakuma, S., Doi, K., Okawa, H., Okaniwa, A. 1975. Articular lesions in experimental *Erysipelothrix insidiosa* infection in rats. National Institute of Animal Health Quarterly (Tokyo) **15:**86–93.

Schoop, G., Stoll, L. 1966. Die auf Fischen vorkommenden *Erysipelothrix*-Typen. Zeitschrift für Medizinische Mikrobiologie und Immunologie **152:**188–197.

Schulz, L.-Cl., Drommer, W., Ehard, H., Hertrampf, B., Leibold, W., Messow, C., Mumme, J., Trautwein, G., Ueberschaer, S., Weiss, R., Winkelmann, J. 1977. Pathogenetische Bedeutung von *Erysipelothrix rhusiopathiae* in der akuten und chronischen Verlaufsform der Rotlaufarthritis. Deutsche Tierärztliche Wochenschrift **84:**107–111.

Seidler, D., Trautwein, G., Böhm, K.-H. 1971. Nachweis von *Erysipelothrix insidiosa* mit fluoreszierenden Antikörpern. Zentralblatt für Veterinärmedizin B, **18:**280–292.

Shuman, R. D. 1964. Swine erysipelas, chap. 24. In: Dunn, H. W., (ed.), Diseases of swine, 2nd ed. Ames, Iowa: Iowa State University Press.

Shuman, R. D., Wellmann, G. 1966. Status of the species name *Erysipelothrix rhusiopathiae* with request for an opinion. International Journal of Systematic Bacteriology **16:**195–196.

Sikes, D. 1965. Some biochemic properties of a smooth colony of *Erysipelothrix insidiosa* used for antigen production in the tube-test. American Journal of Veterinary Research **26:**636–640.

Sneath, P. H. A., Abott, J. D., Cunliffe, A. C. 1951. The bacteriology of erysipeloid. British Medical Journal **iii:**1063–1066.

Steinmann, C. R., Hsu, K. 1976. Specific detection and semi-quantitation of microorganisms in tissue by nucleic acid hybridization. II. Investigation of synovia from pigs with chronic *Erysipelothrix* arthritis. Arthritis and Rheumatism **19:**38–42.

Stuart, M. R. 1972. A note on the occurrence of core-like structures in association with *Erysipelothrix rhusiopathiae*. Journal of General Microbiology **73:**571–572.

Stuart, M. R., Pease, P. E. 1972. A numerical study on the relationships of *Listeria* and *Erysipelothrix*. Journal of General Microbiology **73:**551–565.

Sweeny, J. C. 1974. Common diseases of pinnipeds. Journal of the American Veterinary Medical Association **165:**805–810.

Tadayon, R. A., Carroll, K. K. 1971. Effect of growth conditions on the fatty acid composition of *Listeria monocytogenes* and comparison with the fatty acids of *Erysipelothrix* and *Corynebacterium*. Lipids **6:**820–825.

Teschner, U., Behrens, H., Weiss, R. 1976. Untersuchungen über das Vorkommen von *Erysipelothrix rhusiopathiae* im Schafkot. Berliner und Münchener Tierärztliche Wochenschrift **89:**441–443.

Timoney, J. F., Jr. 1976a. Erysipelas arthritis in swine: Concentrations of complement and third component of complement in synovia. American Journal of Veterinary Research **37:**5–8.

Timoney, J. F., Jr. 1976b. Erysipelas arthritis in swine: Lysosomal enzyme levels in synovial fluids. American Journal of Veterinary Research **37:**295–298.

Todorov, T. 1976a. Induction of L-forms of *Erysipelothrix insidiosa* using antibiotics and lysozyme. Acta Microbiologica, Virologica et Immunologica (Sofia) **4:**39–45.

Todorov, T. 1976b. Persistence of L-forms of *Erysipelothrix insidiosa* in the organisms of experimentally inoculated albino mice. Acta Microbiologica, Virologica et Immunologica (Sofia) **4:**46–51.

Tontis, A., Koenig, H., Luginbuehl, H., Nicolet, J., Glättli, H. R. 1977. Zur chronischen Rotlauf-Polyarthritis beim Lamm. Deutsche Tierärztliche Wochenschrift **84:**113–116.

Topping, L. E. 1937. The predominant microorganisms in soils.

I. Description and classification of the organisms. Zentralblatt für Bakteriologie, Parasitenkunde, Infektionskrankheiten und Hygiene, Abt. 2 **97:**289–304.

Toshkov, A. S., Noeva, K., Georgiev, D., Slavcheva, E., Shirova, L., Mihailova, L., Kyurkchiev, S. 1976. Chronic persistent *Erysipelothrix insidiosa* infections in rats. Acta Microbiologica, Virologica et Immunologica (Sofia) **3:**3–15.

Trautwein, G., Seidler, D., Schulz, L.-Cl., Drommer, W., Weiss, R., Böhm, K. H. 1976. Immunpathologie und Pathogenese der chronischen Rotlauf-Polyarthritis des Schweines. Zeitschrift für Rheumatologie **35:**217–239.

Trevisan, V.: Caratteri di alcuni nuovi generi di Batteriacee. Atti della Accademia Fisio-Medico-Statistica in Milano, Ser. 4, 3, 92–107 (1885). International Bulletin of Bacteriological Nomenclature and Taxonomy **2:**11–29.

Tsymbal, A. M., Lysenko, I. P., Kovalenko, V. T., Serbinenko, T. N., Kurilev, B. S. 1975. Simultaneous immunization of suckling pigs against plague, Aujeszky's disease and erysipelas. Veterinariia **42:**57–59.

Valerianov, T. S., Toschkoff, Al., Cholakova, S. 1976. Biological properties of *Erysipelothrix* phages isolated from lysogenic cultures. Acta Microbiologica, Virologica et Immunologica (Sofia) **3:**32–38.

van Damme, L. R. Devriese, L. A. 1976. The presence of *Erysipelothrix rhusiopathiae* in the tonsils of swine and in the larynx of chickens in Ruanda (Central Africa). Zentralblatt für Veterinärmedizin, (B) **23:**74–78.

Vandersteen, P. R. 1974. Bacterial infections of the skin. Minnesota Medicine **57:**838–843.

Vasilev, V., Stoev, J., Simeonov, S., Jotov, M., Kotsev, T. S. 1976. Production and testing of a live lyophilized bivalent vaccine against Aujeszky's disease and erysipelas in swine. Veterinarno-Meditsinski Nauki (Sofia) **13:**74–78.

Volmer, J., Hasler, G. 1976. *Erysipelothrix* Endocarditis. Deutsche Medizinische Wochenschrift **101:**1672–1674.

Wasiński, K. 1976. Aerosol immunization of pigs against erysipelas. Medycyna Weterynaryjna **32:**719–721.

Wawrzkiewicz, K. 1964. Dissociation forms of *Erysipelothrix insidiosa*. (Poln.) Acta Microbiologica Polonica **13:**45–54.

Wellmann, G. 1950. Rotlaufübertragung durch verschiedene blutsaugende Insektenarten auf Tauben. Zentralblatt für Bakteriologie, Parasitenkunde, Infektionskrankheiten und Hygiene, Abt. 1 Orig. **155:**109–115.

Wellmann, G. 1954. Rotlaufinfektionsversuche an wilden Mäusen, Sperlingen, Hühnern und Puten. Tierärztliche Umschau **9:**269–273.

Wellmann, G. 1955. Die subklinische Rotlaufinfektion und ihre Bedeutung für die Epidemiologie des Schweinerotlaufs. Zentralblatt für Bakteriologie, Parasitenkunde, Infektionskrankheiten une Hygiene, Abt. 1 Orig. **162:**265–274.

Wellmann, G. 1966. Beobachtungen bei der Rotlauf-Immunisierung von Schweinen. II. Der Wert verschiedener Infektionsmethoden für den Belastungsversuch. Berliner und Münchener Tierärztliche Wochenschrift **79:**474–477.

Wellmann, G., Heuner, F. 1959. Beziehungen zwischen serologisch nachweisbaren Antikörpern und der Immunität beim Schweinerotlauf. Zentralblatt für Bakteriologie, Parasitenkunde, Infektionskrankheiten und Hygiene, Abt. 1 Orig. **175:**373–387.

White, T. G., Mirikitani, F. K. 1976. Some biological and physicalchemical properties of *Erysipelothrix rhusiopathiae*. Cornell Veterinarian **66:**152–163.

White, T. G., Mirikitani, F. K., Hargrave, P. 1976. The effect of a bacterial extract on synovial cells in tissue culture. In vitro **12:**702–707.

White, T. G., Puls, J. L., Hargrave, P. 1975. Production of synovitis in rabbits by fractions of a cell-free extract of *Erysipelothrix rhusiopathiae*. Clinical Immunology and Immunopathology **3:**531–540.

Wiidik, R. W. 1952. Die wissenschaftlichen Grundlagen der aktiven Immunisierung gegen den Schweinerotlauf mit dem schwedischen avirulenten Impfstoff. Monatsschrift für Tierheilkunde **4**:145–210.

Wood, R. L. 1965. A selective liquid medium utilizing antibiotics for isolation of *Erysipelothrix insidiosa*. American Journal of Veterinary Research **26**:1303–1308.

Wood, R. L., Packer, R. 1972. Isolation of *Erysipelothrix rhusiopathiae* from soil and manure of swine-raising prem-

ises. American Journal of Veterinary Research **33**:1611–1620.

Woodbine, M. 1950. *Erysipelothrix rhusiopathiae*. Bacteriology and chemotherapy. Bacteriological Reviews **14**:161–178.

Zimmermann, G. 1963. Zur Untersuchung und Beurteilung rotlaufkranker und verdächtiger Schlachtschweine. Tierärztliche Umschau **18**:114–119.

The Genus *Caryophanon*

WILLIAM C. TRENTINI

Trichome-forming bacteria (Starr and Skerman, 1965) are a diverse group of genera that have as a common property the arrangement of individual cells in chains bounded by a common cell wall. The cells usually are closely appressed and wider than they are long. Often each completed cell unit within the actively growing trichome will show the growth of one or more developing septa, the completion of which results in new cells and extension of the trichome length. The multicellular trichomes often can divide or "break" at intervals to form shorter reproductive units, sometimes called "hormogonia", a rather ill-defined term (Stanier and Cohen-Bazire, 1977). Most trichome genera are Gram-negative and motility is by gliding. Many are phototrophic (cyanobacteria).

Caryophanon trichomes (Figs. 1 and 2) are Gram positive, motile by peritrichous flagella (Fig. 3), and chemoorganotrophic. The trichomes are uniseriate, show no branching or false branching, and appear uniform in width, except for slightly tapered terminal cells. Endospores are not formed and there is no sheath or capsule. In old cultures, in poor media, or with specific growth conditions leading to lysis, rounded bodies (spheroids) are formed but show no definite pattern of formation among trichomes (W. C. Trentini, unpublished results). The significance of spheroids is unknown, but it was hypothesized that they are part of a *Caryophanon* "life cycle" (Peshkov, 1939) and/or are resistant bodies (Weeks and Kelley, 1958). There are two recognized species, *Caryophanon latum* and *Caryophanon tenue*. The former is approximately 3 μm in diameter and produces more than one new septum per cell unit. *C. tenue* is about 1.5 μm in diameter and forms only one new septum per cell unit. The taxonomic position of *C. tenue* still remains questionable and will be discussed. Almost all investigations have been confined to isolates of *C. latum,* the morphology of which is strikingly sensitive to "growth conditions" and other equally unexplained factors. Both organisms are isolated from cattle manure; *C. tenue* has been isolated mainly in the USSR.

Historically, *C. latum* played a significant role in the development of bacterial cytology in the 1940s and 1950s. Because of the large size of the trichome and numerous stainable "nuclei", *Caryophanon* provided a visible model in the "great mitosis" debate (Tuffery, 1955). For the uninitiated, the spirit of this debate can be instructive as a pre-molecular biology phenomenon (Robinow, 1956). The genus *Caryophanon* has been the subject of a recent review (Trentini, 1978).

Habitat

Presentation of Evidence

The natural habitat of *Caryophanon* still is not understood with full clarity. *C. latum* was isolated first in 1937 and *C. tenue* in 1938 from "fresh cow manure" (Peshkov, 1939). Subsequent investigators have used cattle manure as the primary source of these organisms (Gershenfeld and Lam, 1953; Kele, 1970; Moran and Witter, 1976; Peshkov and Marek, 1973; Pringsheim and Robinow, 1947; Provost and Doetsch, 1962; Smith and Trentini, 1972; Trentini and Machen, 1973; Tuffery, 1955; Weeks and Kelley, 1958).

However, there are reports of finding *C. latum* in or on sources other than cattle manure: sewage, both crude and at various stages of purification (see discussion by R. A. Fox in Tuffery, 1953); oral cavity of dogs (Saphir and Carter, 1976); and "from decaying *Pleurotus* on the stump of a tree" (see p. 598, Buchanan and Gibbons, 1974). This author has found no reference to *Caryophanon* in the sewage microbiology literature. The organism found in the oral cavity of dogs was certainly *Simonsiella* (Nyby et al., 1977). The observation by Buchanan has not been verified.

Caryophanon latum also has been found associated with manure from goats (Kele, 1970), sheep (Kele, 1970; Trentini and Machen, 1973), and pigs (Trentini and Machen, 1973). These findings may be misleading, for the positive samples of goat and sheep manure were taken from barns or pastures that also had cattle present. The one positive sample of pig manure was incubated in the laboratory where

a

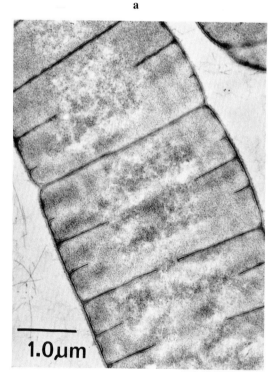

b

Fig. 1. *Caryophanon latum.* (a) Isolate D3. Freeze-etched trichomes illustrating closely appressed septal furrows. Notice that the distance between septal furrows does not appear constant. (b) Isolate J1. Longitudinal thin section showing that many septa are at various stages of completion. *C. latum* characteristically shows more than one dividing septum per cell unit. (Preparation of freeze-etched material by H. E. Gilleland, Jr.)

Caryophanon studies were in progress. Subsequent samples were negative when incubated in a room free of *Caryophanon* experiments.

Because *C. latum* was found in barn gutters, dust and air samples, bedding straw, barnyard soil, and in the air of occupied cattle pastures, it can be assumed that cross-contamination of animal manure with *Caryophanon* from cattle sources was highly probable (Trentini and Machen, 1973; see also Dean, 1963; Kele, 1970, for reference to some of the above-mentioned sources).

Assuming that *Caryophanon* is associated specifically with cattle manure, the major question then becomes whether these microbes are natural or transient residents of the bovine digestive tract, or whether they are inoculated onto voided manure from other sources.

Caryophanon have not been isolated from rectal or intestinal samples from slaughtered cattle (Pringsheim and Robinow, 1947; Trentini and Machen, 1973) or from cattle saliva (Kele, 1970; Trentini and Machen, 1973), teeth scrapings, or anal swabs (Kele, 1970). Kele (1970) found no *Caryophanon* in rumen samples. However, Dean (1963) reported ''abundant'' numbers in one sample of rumen fluid. Trentini and Machen (1973) found six out of seven rumen samples negative when enriched under various conditions. The one positive sample was taken from a stanchioned cow without regard to asepsis and transported over a period of 2 days before enrichment. The negative samples were taken aseptically at the site and enriched for several days. Several eminent rumen microbiologists, based upon routine observation over the years of rumen fluid used in their own investigations, have not verified the presence of *Caryophanon* in rumen fluid (R. E. Hungate, personal communications).

Caryophanon was not seen in very fresh cattle manure collected aseptically (Kele, 1970) and was more easily isolated from pasture samples of 1 to 2 days (Kele, 1970; Pringsheim and Robinow, 1947; Trentini and Machen, 1973; Weeks and Kelley, 1958). Only 15% of ''free catch'' samples from stanchioned cattle were positive for *Caryophanon* following enrichment (Trentini and Machen, 1973). Old, field-dried cattle manure was usually negative for these organisms (Pringsheim and Robinow, 1947; Trentini and Machen, 1973; Weeks and Kelley, 1958). However, Kele (1970) successfully isolated *C. latum* from old cattle manure.

Spring soil samples from a pasture that contained cattle the previous year were negative following enrichment (Trentini and Machen, 1973).

Conclusion

From the evidence presented, it can be seen that the natural habitat of *Caryophanon* is not known definitively. Kele (1970) concluded that *C. latum* is a secondary contaminant of cattle manure and can survive on small fragments of dried cattle manure in nature. However, he also stated that the true habitat of the organism is ruminant manure. Trentini and Machen (1973) differed in their conclusions. They considered *C. latum* to be a specific, natural, and transient resident of cattle manure, dispersed to new droppings by contaminated air (small, dried manure particles?) and probably also by cattle movement or various insects. Both Kele and Trentini and Machen agreed that *Caryophanon* was not a natural resident of the animal. Similar habitat studies have not been undertaken with *C. tenue*.

Isolation

Principles

From the preceding discussion, it is obvious that *Caryophanon* has been isolated routinely from cattle manure. Enrichment is not complicated or difficult because these organisms can be identified morphologically with the phase-contrast microscope. However, success will depend upon sampling manure of proper age and frequent examination of the enrichment samples. Fresh droppings regularly yielded positive results if gathered from the barn gutters commonly found with stanchioned cattle. Pasture samples were enriched most successfully following 1 to 2 days of field aging, provided rainfall was minimal and temperature was above freezing (for discussion pertaining to *C. latum,* see Trentini and Machen, 1973). Following enrichment, selective isolation was achieved by plating directly from peak enrichments onto cow dung agar containing streptomycin sulfate (Sm); a more laborious procedure involving differential centrifugation and filtration yielded up to 60–80% *Caryophanon* colonies when the last suspension was plated onto the cow dung–streptomycin agar medium (Smith and Trentini, 1972).

Enrichment Procedures

The following enrichment procedure was based upon the observation (Pringsheim and Robinow, 1947) that the numbers of *Caryophanon* "usually increased greatly when the material was kept in its own moisture in closed jars at room temperature for a day or two".

Enrichment of *Caryophanon* (Smith and Trentini, 1972)

"Samples of about 2-day-old cow dung were collected from local pastures in 600- to 800-ml beakers and covered with aluminum foil. All samples were returned to the laboratory within 30 min. Single-distilled water was added, and the beakers were gently swirled to effect a dung 'slurry' at the surface. After an average of 16- to 24-h incubation at room temperature, some beakers contained large numbers of typical, actively motile *C. latum* trichomes and trichome-chains in material taken from the liquid surface and viewed with the phase microscope. Attempts at isolation were made only on samples in which the quantities of *C. latum* had reached a level designated as a 'peak'; i.e., a wet mount made from the surface slurry material and viewed under phase contrast at 400× contained an average of at least three to four trichomes and (or) trichome-chains per field (units/field). More concentrated peaks with 20–30 units/field were observed on various occasions.''

The same procedure can be used for fresh droppings collected from barn gutters where cattle are stanchioned. To check for the presence of *Caryophanon* in materials other than cattle manure simply add the test material to sterilized cattle manure having a slurry surface (Trentini and Machen, 1973).

Selective Isolation of *Caryophanon* (Smith and Trentini, 1972)

"Surface slurry material from peak samples was carefully skimmed off and filtered once through 16 layers of cheesecloth to remove large particles. The solid matter trapped by the cheesecloth was rinsed with about 10 ml single-distilled water and filtered again by squeezing the pad gently. The combined filtrate was centrifuged at 4C in a Sorvall RC2-B at 1500 rpm for 5 min using an SS-34 centrifuge head. The pellet was suspended in 20 ml of nutrient wash solution (0.01% $MgSO_4\cdot 7H_2O$, 0.1% yeast extract, 0.01% anhydrous sodium acetate; pH 7.5). Further centrifugation and wash were repeated as above three times. The final resuspended pellet was washed through a 60-ml Buchner funnel (coarse grade) with an additional 10 ml of nutrient solution to further eliminate remaining large particles. The filtrate thus obtained was washed by vacuum through a membrane filter, 8 μ pore diameter, 47 mm diameter (Millipore Filter Corp., Bedford, Mass.) with about 400 ml nutrient solution. Care was taken to keep the filter moist throughout this procedure to prevent drying injury to the trichomes. The membrane filter, along with 2 ml of nutrient solution, was agitated at full speed on a vortex mixer (Deluxe Mixer, Scientific Products, Evanston, Ill.) until the paper appeared clean. Dilutions of 10^{-2} to 10^{-5} of the final suspension were spread on 25% cow dung agar plates containing 80 μg/ml Sm. After incubation at room temperature for 48 h, typical *C. latum* colonies were easily distinguished. Pure cultures of each isolate were obtained by restreaking on 25% cow dung agar without Sm.''

Preparation of Cow Dung Agar
(Smith and Trentini, 1972)

"Cow dung agar was prepared by mixing fresh (less than about 2- to 3-h old) cow dung and double-distilled water in a 1:3 v/v ratio (25%) for 2–3 min in a Waring Blendor (model 700B, Waring Products Co., Winsted, Conn.) to break up the large dung particles and ensure thorough dispersion. Bacto-agar was then added to a concentration of 1.5%." The agar was melted while being continuously stirred and autoclaved for standard time and temperature.

Axenic Cultivation and Maintenance

Caryophanon grows well on cattle manure agar. Morphology and motility are preserved to a larger extent when 0.5–1% lactalbumin hydrolysate is used as a supplement. It should be obvious that "growth", when defined for a morphologically unusual organism, should include the criterion of mass increase, particularly, mass increase that preserves the morphological integrity of the organism as close as possible to that seen in the original enrichment condition. It is indeed unfortunate that most investigators of *Caryophanon* chose to ignore the obvious (most recently, Moran and Witter, 1976).

Several semisynthetic media for *C. latum* have appeared in the literature over the years (Kele and McCoy, 1971; Pringsheim and Robinow, 1947; Provost and Doetsch, 1962; Smith and Trentini, 1973). All employ acetate as the major carbon source. Indeed, acetate may well be the only major carbon source utilizable by *C. latum* (Kele and McCoy, 1971; Provost and Doetsch, 1962; Trentini, unpublished results). Biotin and thiamine are required (Provost and Doetsch, 1962).

Based upon this author's criteria for defining growth and following numerous attempts to supplement growth with various commercial peptones, many carbon sources, amino acids, purines, pyrimidines, minerals, and pH manipulation, the following semisynthetic medium (LAVMm2) is being used in the author's laboratory at the time of writing. It is a modification of the media of Kele and McCoy (1971) and Smith and Trentini (1973).

Semisynthetic Medium LAVMm2 for *Caryophanon latum* (W. C. Trentini, unpublished)

The medium contains:

Biotin	0.02 μg/ml
Thiamine hydrochloride	0.05 μg/ml
Cupric acetate	0.04 μg/ml
$FeSO_4$	0.152 μg/ml
$CaCl_2$	11.1 μg/ml
$MgCl_2 \cdot 6H_2O$	20.3 μg/ml
Nitrilotriacetic acid	19.1 μg/ml
Sodium acetate	5.0 mg/ml
Lactalbumin hydrolysate, tissue culture grade	10.0 mg/ml

The medium was brought to pH 7.5 in double distilled water with anhydrous Na_2CO_3 prior to autoclaving, giving a final pH of 8.0–8.1 after sterilization at standard time and temperature. The medium can be buffered with 0.05 M Tris, but avoid excessive use of phosphate buffer due to detrimental morphological effects induced by phosphate. A solid medium was prepared with the addition of 1.5% of agar.

However, morphology and motility of *Caryophanon* are, indeed, very difficult to maintain (Fig. 2). These properties are best seen—in decreasing order—in enrichment culture, on cattle manure–lactalbumin hydrolysate agar, on cattle manure agar, on LAVMm2 agar. Growth in liquid LAVMm2 was satisfactory for some isolates, but at best it can be considered as simply the most adequate liquid medium to date. Average doubling times in this medium are approximately 70 min. Although there was a report of a defined, synthetic medium for *C. latum* (Kele and McCoy, 1971), it has not proved adequate to permit growth—as defined merely by mass increase—of various *C. latum* isolates in our laboratory. Liquid cultures need aeration, and we have used both shaken and sparged conditions. As a general rule, shaken cultures seemed to produce slightly more stable conditions than did sparged cultures. It must be stressed, however, that morphology, motility, and mass increase of *Caryophanon* are dependent on the particular isolate being used by the investigator, as well as the aforementioned nutritional and environmental observations. Recently, I have concluded that the value of lactalbumin hydrolysate itself as a major supplement is variable depending on the commercial source of the peptone and the particular batch or control number from a given company.

Our cultures are grown either at room temperature or quantitatively at 25°C. Although Moran and Whitter (1976) have stated that 35°C is the optimal temperature for growth of *C. latum,* these results were based only on the rate of change in colony diameter on a single medium, a criterion not acceptable by itself for organisms of unusual morphology. The medium used probably did not support natural trichome morphology and was chosen simply on the basis of supporting rapid colonial growth. Based on our own studies (Smith and Trentini, 1973) and subsequent observations, this author must conclude that the optimal growth temperature for *Caryophanon* is

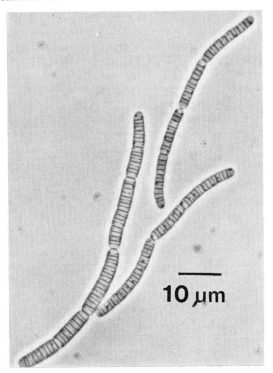

a

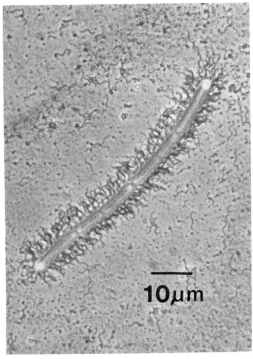

not 35°C, and that finding optimal growth conditions for *C. latum* must await further research that will account for natural trichome morphology as well as mass or colony diameter increase.

Caryophanon organisms are readily conserved by standard techniques such as refrigeration, sterile mineral oil overlay, and lyophilization.

Identification

General

Caryophanon organisms initially are identified according to their morphology and habitat as obligately aerobic, Gram-positive, peritrichously flagellated (Fig. 3), asporogenous, multicellular bacilli (trichomes) (Figs. 1 and 2) that can be enriched from cattle manure. Peshkov (1939, 1940) originally described two species, *C. latum* and *C. tenue*, based upon trichome width and number of dark and light bands per trichome. The dark bands were shown to be cell wall; the light bands were cytoplasm that contained nuclear material (Pringsheim and Robinow, 1947). Almost all investigations have been confined to *C. latum*. In the eighth edition of *Bergey's Manual* (Buchanan and Gibbons, 1974), the genus *Caryophanon* is listed under Part 16: Gram-Positive, Asporogenous Rod-Shaped Bacteria: Genus of Uncertain Affiliation. *C. tenue* was

Fig. 2. Comparison of morphology of two isolates of *Caryophanon latum*, J1(a) and D3(b). Both isolates grown on an agar surface (Trentini and Gilleland, 1974). J1 tends to form trichome chains. D3 trichomes are singular or paired. J1 trichomes are wider than those of D3. J1 was motile by peritrichous flagella while D3 was nonmotile and lacked flagella.

b

Fig. 3. *Caryophanon latum*, isolate J1, stained by the procedure of Mayfield and Inniss (1977). Note the peritrichous flagella.

mentioned but, taking into account the discussion of Weeks and Kelley (1958), this species was thought to be possibly a cultivation artifact of *C. latum*. Whether delineation of more than one species is justified in the genus is still quite unclear.

Caryophanon tenue

Caryophanon tenue had a narrow trichome diameter (1.5 μm) and "fewer" cells per trichome. Each cell of *C. tenue* produced only one cross septum (Peshkov and Marek, 1973). Phage have been isolated that lyse *C. tenue* but not *C. latum* (Peshkov, Marek, and Shadrina, 1973; Peshkov, Tikhonenko, and Marek, 1966). The description of phage development differed within the two species (Peshkov, Marek, and Shadrina, 1973). Peshkov and Marek (1973) claim that ultrastructurally in thin sections a difference between species can be seen in: (i) the profile of the external layer contour, (ii) the density of ribosome packing, and (iii) the interrelationship between mesosomes, cytoplasmic membrane, and the nuclear material. However, most of the above observations could be explained by presence of lysogenic phage, fixation and embedding artifacts, and the growth rate or stage of growth of the cultures used. Adcock, Seidler, and Trentini (1976) have shown that *C. tenue* has a significantly lower mol% G+C (41.2–41.6) than *C. latum*, a smaller calculated genome size (900–1,000 × 10⁶ daltons), and only 13–30% relative DNA-DNA reassociation with *C. latum*. If the three isolates of *C. tenue* used were, indeed, the designated organism, then it becomes quite clear that the two species are justified. However, the isolates of *C. tenue* used, although similar in physiology and antibiotic resistance to *C. latum*, morphologically did *not* resemble *C. tenue* as described by Peshkov, even when grown in sterile cattle manure slurry. The long narrow filaments had only one or two cross septa. However, these studies did show that *C. tenue, as we have access to it, is not* simply a narrow form of *C. latum* (Buchanan and Gibbons, 1974). It was very unfortunate that *C. tenue* cultures could not be obtained directly from the primary investigators in the USSR.

Caryophanon latum

The trichomes of effectively grown *Caryophanon latum* were 2.5–3.2 μm in diameter and 10–20 μm in length and were obviously multicellular (Figs. 1 and 2). Each cell in a trichome formed multiple cross septa (Peshkov and Marek, 1973; Trentini and Gilleland, 1974). Trichome chains often contained 2–6 trichomes. Thirty-six isolates from various geographic regions showed a narrow range (44.0–45.6 mol% G+C), a DNA-DNA relative reassociation of 78–92%, and a calculated genome

size of 1,100–1,200 × 10⁶ daltons (Adcock, Seidler, and Trentini, 1976). All isolates were uniform in their limited metabolic abilities and their antibiotic resistance pattern (see also Smith and Trentini, 1972). All isolates appeared to be inactive to many standard biochemical tests.

The peptidoglycan compositions of *C. latum* and *C. tenue* were very similar, containing a molar ratio of glutamic acid:alanine:lysine:muramic acid of about 2:2:1:1. One of the glutamic acid residues was bound to the ε-amino group of lysine and was responsible for the cross linking of the peptide subunit. The configuration of the glutamic acid is not known presently (K. H. Schleifer, personal communication; see also Becker, Wortzel, and Nelson, 1967). Septal and wall peptidoglycans were differentially attacked by both egg white lysozyme and lysozyme from *Chalaropsis* (Trentini and Murray, 1975).

To date, no lipid biochemistry (fatty acids, glycerides, phospholipids, glycolipids, fatty alcohols, sulfolipids, peptidolipids, waxes, and hydrocarbons) has been published for *Caryophanon* (Lechevalier, 1977). There are no enzyme or genetic studies in the literature. The characteristic yellow pigment of *C. latum* colonies (Provost and Doetsch, 1962) has not been identified, but is presumably a carotenoid.

In conclusion, *Caryophanon* (especially *C. latum*) is enriched and isolated easily from cattle manure. Growth and maintenance of these organisms is without difficulty, but very special care and attention must be devoted to growth with natural morphology. A fully reliable liquid growth condition must be developed to obtain quantitative physiological data for *Caryophanon*. Study of the general biology of *C. tenue* must be initiated. Conclusive habitat studies, phage isolation for typing purposes, potential of gene transfer, and enzymatic analysis may at a future date allow a more definitive paper to be written on the habitat, ecology, and taxonomy of this genus.

Literature Cited

Adcock, K. A., Seidler, R. J., Trentini, W. C. 1976. Deoxyribonucleic acid studies in the genus *Caryophanon*. Canadian Journal of Microbiology **22**:1320–1327.

Becker, B., Wortzel, E. M., Nelson, J. H., III. 1967. Chemical composition of the cell wall of *Caryophanon latum*. Nature **213**:300.

Buchanan, R. E., Gibbons, N. E. (eds.). 1974. Bergey's manual of determinative bacteriology, 8th ed. Baltimore: Williams & Wilkins.

Dean, D. S. 1963. Response of *Caryophanon latum* to oxygen. Journal of Bacteriology **85**:249–250.

Gershenfeld, L., Lam, G. T. 1953. The effect of certain antiseptics on *Caryophanon latum* Peshkoff. American Journal of Pharmacy **125**:5–34.

Kele, R. A. 1970. Investigations on the nutrition, morphogenesis and habitat of *Caryophanon latum*. Ph.D. Thesis. University of Wisconsin, Madison, Wisconsin.

Kele, R. A., McCoy, E. 1971. Defined liquid minimal medium for *Caryophanon latum*. Applied Microbiology **22**:728–729.

Lechevalier, M. P. 1977. Lipids in bacterial taxonomy—a taxonomist's view. CRC Critical Reviews in Microbiology **5**:109–210.

Mayfield, C. I., Inniss, W. E. 1977. A rapid, simple method for staining bacterial flagella. Canadian Journal of Microbiology **23**:1311–1313.

Moran, J. W., Witter, L. D. 1976. Effect of temperature and pH on the growth of *Caryophanon latum* colonies. Canadian Journal of Microbiology **22**:1401–1403.

Nyby, M. D., Gregory, D. A., Kuhn, D. A., Pangborn, J. 1977. Incidence of *Simonsiella* in the oral cavity of dogs. Journal of Clinical Microbiology **6**:87–88.

Peshkov, M. A. 1939. Cytology, karyology and cycle of development of new microbes—*Caryophanon latum* and *Caryophanon tenue*. Comptes Rendus (Doklady) de l'Académie des Sciences de l'URSS **25**:244–247.

Peshkov, M. A. 1940. Phylogenesis of new microbes *Caryophanon latum* and *Caryophanon tenue*–organisms which are intermediate between blue-green algae and the bacteria. Zhurnal Obshchei Biologii (Moscow) **1**:613–618.

Peshkov, M. A., Marek, B. I. 1973. Fine structure of *Caryophanon latum* and *Caryophanon tenue* Peshkoff. Microbiology [English translation of Mikrobiologiya] **41**:941–945.

Peshkov, M. A., Marek, B. I., Shadrina, I. A. 1973. Intracellular development of phage in trichomes of *Caryophanon latum* and *Caryophanon tenue*. Microbiology [English translation of Mikrobiologiya] **42**:89–94.

Peshkov, M. A., Tikhonenko, A. S., Marek, B. I. 1966. A bacteriophage against the multicellular microorganism *Caryophanon tenue* Peshkoff. Microbiology [English translation of Mikrobiologiya] **35**:577–581.

Pringsheim, E. G., Robinow, C. F. 1947. Observations on two very large bacteria, *Caryophanon latum* Peshkoff and *Lineola longa* (nomen provisorium). Journal of General Microbiology **1**:267–278.

Provost, P. J., Doetsch, R. N. 1962. An appraisal of *Caryophanon latum*. Journal of General Microbiology **28**:547–557.

Robinow, C. F. 1956. The chromatin bodies of bacteria. Bacteriological Reviews **20**:207–242.

Saphir, D. A., Carter, G. R. 1976. Gingival flora of the dog with special reference to bacteria associated with bites. Journal of Clinical Microbiology **3**:344–349.

Smith, D. L., Trentini, W. C. 1972. Enrichment and selective isolation of *Caryophanon latum*. Canadian Journal of Microbiology **18**:1197–1200.

Smith, D. L., Trentini, W. C. 1973. On the gram reaction of *Caryophanon latum*. Canadian Journal of Microbiology **19**:757–760.

Stanier, R. Y., Cohen-Bazire, G. 1977. Phototrophic prokaryotes: The cyanobacteria. Annual Review of Microbiology **31**:225–274.

Starr, M. P., Skerman, V. B. D. 1965. Bacterial diversity: The natural history of selected morphologically unusual bacteria. Annual Review of Microbiology **19**:407–454.

Trentini, W. C. 1978. Biology of the genus *Caryophanon*. Annual Review of Microbiology **32**:123–141.

Trentini, W. C., Gilleland, H. E., Jr. 1974. Ultrastructure of the cell envelope and septation process in *Caryophanon latum* as revealed by thin section and freeze-etching techniques. Canadian Journal of Microbiology **20**:1435–1442.

Trentini, W. C., Machen, C. 1973. Natural habitat of *Caryophanon latum*. Canadian Journal of Microbiology **19**:689–694.

Trentini, W. C., Murray, R. G. E. 1975. Ultrastructural effects of lysozymes on the cell wall of *Caryophanon latum*. Canadian Journal of Microbiology **21**:164–172.

Tuffery, A. A. 1953. The morphology and systematic position of *Caryophanon*. Atti del VI Congresso Internazionale di Microbiologia **1**:104.

Tuffery, A. A. 1955. Nuclear changes in the growth cycle of *Caryophanon latum*. Experimental Cell Research **9**:182–185.

Weeks, O. B., Kelley, L. M. 1958. Observations on the growth of the bacterium *Caryophanon latum*. Journal of Bacteriology **75**:326–330.

SECTION S

Endospore-Forming Bacteria

CHAPTER 135

The Genera *Bacillus* and *Sporolactobacillus*

JOHN R. NORRIS, ROGER C. W. BERKELEY, NIELL A. LOGAN, and ANTHONY G. O'DONNELL

THE GENERA OF ENDOSPOREFORMING BACTERIA

The ability to form endospores is widely distributed among bacteria, but the largest group of endospore-forming bacteria is the family Bacillaceae, in which five genera are currently recognized (Buchanan and Gibbons, 1974). Traditionally, any bacterium forming endospores was placed in one of the family's two main genera, *Bacillus* and *Clostridium,* unless it was markedly different from them. As a result, the taxonomy of these two genera is currently unsatisfactory. A wide diversity of properties is exhibited by the organisms that each contains, and this is emphasized by their wide ranges of G+C content: 32–68 mol% in the case of *Bacillus* and 23–43 mol% in *Clostridium.* Proposals to improve the taxonomy of both genera, by subdivision, have been made but in both cases there is inadequate information available at present. Increased knowledge of these two genera, however, and studies of atypical, endosporeforming bacteria have resulted in the recognition of the three other genera with this ability.

It is now realized that the only known endospore-forming coccus, *Sarcina ureae,* produces spores similar to those found in the genus *Bacillus* (MacDonald and MacDonald, 1962; Mazanec, Kocur, and Martinec, 1965; Thompson and Leadbetter, 1963) and it was proposed (MacDonald and MacDonald, 1962) that it should be renamed *Sporosarcina ureae* and placed in the family Bacillaceae.

Kitahara and Suzuki (1963) isolated endospore-formers which resembled *Lactobacillus* species in many respects. The production of spores similar to those of *Bacillus* (Kitahara and Lai, 1967), and the presence of diaminopimelic acid in the cell wall peptidoglycan (Kandler, 1967) showed a relation to Bacillaceae and the new genus *Sporolactobacillus* was placed in this family in the eighth edition of *Bergey's Manual of Determinative Bacteriology* (Buchanan and Gibbons, 1974).

Campbell and Postgate (1965) proposed the genus *Desulfotomaculum* to contain thermophilic, sulfate-reducing endosporeformers, and this genus was also included in the Bacillaceae in the eighth edition of *Bergey's Manual* (Buchanan and Gibbons, 1974).

A number of other endosporeformers have been isolated from the alimentary tracts of animals and from other environments and placed by their authors in several genera: *Anisomitus, Arthromitus, Bacillospira, Coleomitus, Entomitus, Fusosporus, Metabacterium, Oscillospira,* and *Sporospirillum.* Further information on these genera of uncertain taxonomic position is available in the eighth edition of *Bergey's Manual* (Buchanan and Gibbons, 1974).

The spores produced by several members of the Micromonosporaceae also show strong resemblance to those of *Bacillus.* It was proposed by Cross and Goodfellow (1973) that the genus *Thermoactinomyces* should be restricted to actinomycete species with a cell wall containing diaminopimelic acid (type III) and with the ability to form endospores.

Key to Genera of Endosporeforming Bacteria
  I. Cells rod-shaped
    A. Aerobic or facultative, usually producing catalase. *Bacillus* (Buchanan and Gibbons, 1974; Gordon, Haynes, and Pang, 1973; this chapter; this Handbook, Chapter 136)
    B. Microaerophilic, not producing catalase. *Sporolactobacillus* (Buchanan and Gibbons, 1974; Kitahara and Suzuki, 1963; this chapter)
    C. Anaerobic
      1. Not reducing sulfate to sulfide. *Clostridium* (Wilson and Miles, 1975; this Handbook, Chapter 137; this Handbook, Chapter 138)
      2. Reducing sulfate to sulfide, *Desulfotomaculum* (Buchanan and Gibbons, 1974; Campbell and Postgate, 1965)
 II. Cells spherical, in packets
    Sporosarcina (Norris, 1981; this Handbook, Chapter 139)
III. Cells occurring as hyphae
    Thermoactinomyces (Cross and Unsworth, 1981; this Handbook, Chapter 157)

The Bacterial Endospore

Endospores are resistant stages in the life cycle of several Gram-positive bacteria. They are formed inside the vegetative cell (sporangium) in response to nutritional deprivation and enable the organism to survive without metabolizing (cryptobiosis). These specialized structures are more resistant than the parental, vegetative, cells to the lethal effects of heat, drying, freezing, radiation or toxic chemicals; this probably accounts for the ubiquitous distribution of the endosporeforming bacteria. Of these specialized properties, heat resistance has received most attention because of its importance in sterilization procedures. Nevertheless, the major ecological role of spores is probably to enable survival in dry conditions.

Spores are formed by the invagination of the protoplasmic membrane of the vegetative cell and enclosure of the genetic material. A thin spore membrane and a thicker cortex are laid down between the two layers formed by the invaginating membrane (Murrell, 1967; Szulmajster, 1979). The rigid spore coat is synthesized outside the cortex and its "keratin-like" properties are believed to be responsible for the resistance to staining and harmful chemicals. Circumstantial evidence supports the hypothesis that heat and radiation resistances of bacterial endospores result partly from the relative dehydration of the spore protoplast or core (Gould and Dring, 1975). However, the means by which the dehydration occurs is not as yet fully understood. Lewis, Snell, and Burr (1960) and Alderton and Snell (1963) discussed the hypothesis that a relatively dry core might be achieved through compressive contraction of the surrounding cortex. It was pointed out that the cortex could either contract or expand to compress the core. The former hypothesis, that of a contracted cortex, has become popular despite its inability to account for certain anomalies which can be explained by the expanded cortex theory (Gould and Dring, 1975). For detailed discussions of the structure of the bacterial endospore see Tipper and Gauthier (1972) and Slepecky (1978).

Spore formation is the result of a series of genetically controlled regulatory changes and is looked on

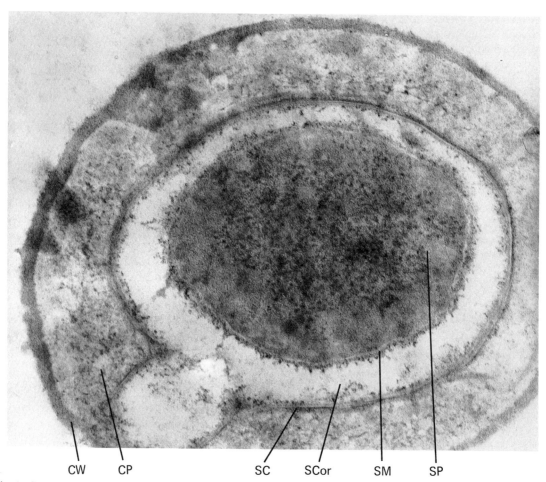

CW   CP   SC   SCor   SM   SP

Fig. 1. Cross-section of *Bacillus megaterium* containing a spore and showing the sporangium (cell), cell protoplast (CP) and wall (CW), spore coat (SC), spore cortex (SCor), spore membrane (SM), and spore protoplast (SP). × 120,000.

as a primitive form of cellular differentiation, since it has several features in common with cellular development in higher organisms (Piggott and Coote, 1976; Szulmajster, 1979).

THE GENUS *BACILLUS*

The aerobic, sporeforming bacilli that constitute the genus *Bacillus* were among the earliest of the bacteria to be described, and the genus has played a major role in the development of microbiology. The genus was created in 1872 by Cohn, who renamed Ehrenberg's inadequately described *Vibrio subtilis* as *Bacillus subtilis* as the first species. There is some doubt about the first observation of the spores themselves. Pasteur described endospores in the bacteria causing flacherie in silkworms but apparently did not realize that they were heat resistant or that they were responsible for the longevity of his bacterium. Heat resistance of spores was recognized in *B. subtilis* by Cohn in 1877, the year which saw Koch's classical demonstration of the development cycle of *B. anthracis* from vegetative cell to resistant spore and back to vegetative cell (Lechevalier and Solotorovsky, 1965).

Koch's description of the anthrax bacillus focused attention on the genus and on the nature of the endospore. The resistance of the spores to drying, their ubiquitous distribution in soil and dust, and the ease with which most species would grow on the media used in microbiological laboratories soon led to the isolation of many new species. The resulting nomenclatural confusion and the gradual introduction of order into the genus have been described in a captivating review by Gordon (1981), which is essential reading for anyone seriously interested in the genus *Bacillus*.

Most members of the genus are not difficult to isolate or to cultivate; the heat- and drought-resistant spores assist both isolation and maintenance of cultures in the laboratory.

Only the anthrax bacillus causes serious invasive disease in man, although several species are occasionally implicated as secondary or minor pathogens. There are, however, indications that such secondary pathogenesis is becoming more frequently recognized and perhaps actually more frequent, possibly in response to the increased use of immunosuppressant therapy regimes. *B. cereus* is a common cause of food poisoning. Several species are pathogenic for insects.

The majority of *Bacillus* species, however, are harmless saprophytes that play, by virtue of their considerable powers of proteolysis and ability to break down complex polysaccharides, an important role in the cycling of nutrients in nature. The ability to break down complex organic molecules and the heat resistance of the spores combine to render the genus perhaps the most important group of spoilage and contaminating organisms affecting man's industrial activities. Much effort in the food industry and elsewhere is spent in attempting to eliminate spores from products in which they may germinate and grow with disastrous consequences.

The physiological processes leading to the formation of the endospore are complex (Slepecky, 1978), one of the early stages being the production of an antibiotic. Several antibiotics, such as bacitracin, gramicidin, and polymyxin, for example, are produced commercially from members of the genus and to these products must be added valuable enzymes such as subtilysin, amylase, and several proteinases.

In addition to exhibiting a wide range of physiological activities, the genus *Bacillus* includes some of the largest bacteria known, several species exceeding 1 μm in diameter and 10–12 μm in length. They have, therefore, played an important part in the development of our understanding of the structure of bacterial cells. The process of spore formation itself has provided a model for the study of differentiation, which the extensive genetic and biochemical studies of recent years have shown to be unexpectedly complex (Szulmajster, 1979).

The genus today is in a state of transition and presents a fascinating series of problems for the taxonomist. As microbiologists have explored more extreme environments (acid, alkaline, high and low temperatures, saline), they have isolated new aerobic sporeformers whose relationships to existing species and, in some cases, status in the genus are far from clear. With the advent of techniques for the direct investigation of the genotype (DNA hybridization and base-composition determination) has come the realization that the genus, as defined at present, includes bacteria of widely differing genetic composition. The DNA base ratios of the various species, for instance, cover a range from just above 30 mol% to just below 70 mol% G+C, far wider than is normally considered reasonable within a single genus. Such problems are beyond the scope of this chapter, but basic information and references are provided to enable the serious researcher to gain entry to this fascinating subject.

No treatment of the aerobic, sporeforming bacteria can fail to acknowledge the debt owed by microbiology to the pioneering work of Ruth Gordon and her colleagues in bringing order into the genus *Bacillus*. In three classical monographs (Gordon, Haynes, and Pang, 1973; Smith, Gordon, and Clark, 1946, 1952) and in numerous other publications they have laid the foundation of tests and interpretations for our present and future understanding of the genus. This chapter rests heavily on that foundation, a fact which the authors are pleased to acknowledge.

Habitats of *Bacillus*

Soil and Water Saprophytes

Aerobic, sporeforming bacteria have been isolated from many different environments. They are most commonly found as inhabitants of the soil, *Bacillus cereus* being the most frequently isolated species. It has been suggested, however, that the predominance of *B. cereus* reflects its ability to "crowd out" other species during enrichment procedures rather than its actual numerical superiority in the original soil sample (Laubach, Rice, and Ford, 1916).

Soil is a heterogeneous environment in which conditions and nutrient supply may vary from micrometer to micrometer, allowing the coexistence of a great diversity of microorganisms (Veldkamp, 1970). In the region of the rhizosphere, there is an increase in the amounts of carbohydrates, amino acids, and vitamins that nurture the more fastidious microorganisms. Nakayama and Yanoshi (1967a) reported the successful isolation from the rhizosphere of *Bacillus* species capable of producing lactic acid; they argued that this location was probably the primary habitat of these bacteria. They also proposed that in this habitat motile organisms have an advantage in that they can more efficiently scavenge for the meager amounts of nutrient substances excreted from root hairs (Nakayama and Yanoshi, 1967a).

Microbial interactions in the rhizosphere are believed to be responsible for the occurrence of reticular chlorosis (frenching). This affliction of field-grown tobacco plants occurs erratically, and soils that support plants with such symptoms can be decontaminated by partial sterilization (Steinberg, 1950). In an earlier paper, Steinberg (1947) suggested that the frenching of tobacco was caused by the action of diffusates of presumably nonpathogenic bacteria. He examined some 60 species and strains, of which two seemed most effective in causing an approximation of the symptoms of frenching in tobacco seedlings—*B. cereus* Frankland and Frankland and *Erwinia carotovora* Jones. Oriental Zanthi tobacco seedlings growing aseptically on nutrient agar developed reticular chlorosis of their leaves prior to leaf narrowing when *B. cereus* was stab-inoculated into the agar.

The presence of *Bacillus* species in the environment of the tobacco plant need not always be detrimental to the tobacco industry, as was shown by English, Bell, and Berger (1967). Of 22 thermophilic strains of bacteria isolated by swabbing broad-leaf tobacco and incubating at 55°C, 20 were aerobic, endosporeforming rods of the genus *Bacillus*. These organisms belonged to the following species: *B. subtilis,* eight strains; *B. coagulans,* five strains; *B. megaterium,* four strains; *B. circulans,* three strains. Earlier work by McKinstry, Haley,

and Reid (1938) on the flora of broad-leaf tobacco showed *B. subtilis* to be a predominant organism in the satisfactory fermentation of cigar leaf tobacco. The addition of three strains of *B. subtilis* and one of *B. circulans* after the natural dry fermentation and before the resweating procedures caused the more rapid appearance of a pleasing aroma, according to professional testers (English, Bell, and Berger, 1967).

In sea water, *Bacillus* strains have been shown to constitute 0–16% of the total heterotrophic flora. However, sampling of marine sediments results in values up to 100%, the percentage composition increasing with depth from which samples are taken (Bonde, 1975, 1976). The same author suggests that species distribution may also vary significantly depending on ecological factors. In polluted sea water, *B. licheniformis* formed 64% of the bacterial population; in less polluted areas, *B. subtilis* and *B. pumilus* were the predominant species. Sediments from polluted areas showed *B. cereus, B. sphaericus,* and *B. brevis* to be dominant while unpolluted sediments showed a greater diversity of species (Bonde, 1975, 1976).

In an independent survey of *Bacillus* isolates from North Sea sediments, Boeyé and Aerts (1976), also found that a large proportion belonged to the *B. subtilis* spectrum, thereby confirming, in part, the results obtained by Bonde (1975).

Investigations into the microbiology of drinking water supplies have shown that cation-exchange softeners and sand filters are ideal environments for the multiplication of many different types of bacteria (Klumb, Marks, and Wilson, 1949). Since these bacterial populations consist almost entirely of soil and water forms, it is hardly surprising that the aerobic sporeformers are frequent inhabitants. Bacterial growth, and the growth of certain *Bacillus* strains in particular, results in the production of organic acids which cause an increase in the solubility of silicates and consequent deterioration of exchange materials (Klumb, Marks, and Wilson, 1949). The aerobic sporeformers, although relatively nonpathogenic, are capable in their dormant state of surviving municipal purification treatments and may eventually, by germination and multiplication, lead to the formation of objectionable tastes and odors in water supplies (Klumb, Marks, and Wilson, 1949).

Bonde (1975) has pointed out that *Bacillus* species in samples of drinking water may, by causing rapid gelatin liquefaction and abundant growth on plate-count media, lead to unjustified condemnation of water supplies.

Owing to the generally stringent procedures carried out to ensure the bacteriological safety of normal drinking supplies, many laboratory workers assume that water treated further, distilled and demineralized for example, has a smaller bacterial content than the original tapwater. However, Bonde

(1966) found very high bacterial counts during a pilot examination in an anaesthesiological department of 51 samples of distilled water from humidifiers, ventilators, incubators, water cans, beakers, and water trap bottles. Sixty-five colonies were characterized and 10 genera, including *Bacillus,* were identified. In this study, aerobic sporeformers were found in all but one of the apparatuses listed.

Dust is also an important natural environment. The best sources for isolation are sites, such as high shelves, where the dust has lain undisturbed for long periods of time. Dust circulating in the air shows the presence of few sporeformers, although the levels rise after an increase in wind velocity. If moist surfaces that have been allowed to dry or surfaces in direct sunlight are sampled, the number of sporeformers isolated is low (Laubach, Rice, and Ford, 1916).

Inhabitants of Environmental Extremes

THERMOPHILIC, AEROBIC, SPOREFORMING BACTERIA. The term thermophile is not easily defined (Kushner, 1978). For this genus, Gordon and Smith (1949) considered that it should be applied to cultures capable of growth at 55°C, in accordance with the definition of Cameron and Esty (1926), but that this criterion should not be used to separate species.

Gibson and Gordon, in the eighth edition of *Bergey's Manual* (Buchanan and Gibbons, 1974) list six *Bacillus* species with strains capable of growth at 55°C: *B. subtilis, B. licheniformis, B. stearothermophilus, B. coagulans, B. brevis,* and *B. acidocaldarius.* Most *B. stearothermophilus* and *B. acidocaldarius* strains will grow at 65–75°C but not below 40°C, and may be described as obligate thermophiles. Members of the other species, which will grow at room temperature but not above 65°C, are described as facultative thermophiles (Farrell and Campbell, 1969).

Aerobic, sporeforming thermophiles have been isolated from almost every kind of environment, including desert sands, tropical and temperate soils; air; fresh snow; cold and thermal waters, both salt and fresh; foods of all varieties; milk, both raw and pasteurized; the feces of man and domestic animals, birds, amphibians, and fish; and from stored vegetable materials (see reviews by Allen, 1953; Clegg and Jacobs, 1953; Gaughran, 1947). There are also reports of isolations from mineral wax (Bairiev and Mamedov, 1963), hot springs (Darland and Brock, 1971; Harris and Fields, 1972; Heinen and Heinen, 1972; Marsh and Larsen, 1953; Oshima, Arakawa, and Baba, 1977; Uchino and Doi, 1967), arctic soils and waters (Boyd and Boyd, 1962; McBee and Gaugler, 1956), ocean-basin cores (Bartholomew and Paik, 1966), and oily soils and waters (Klug and Markovetz, 1967; Mateles, Baruah, and Tannenbaum, 1967).

Several theories have been advanced to explain the occurrence of thermophiles in low-temperature environments. Early investigators suggested that, in temperate zones, the direct heating action of the sun on the upper layers of the soil would be sufficient to allow multiplication. In addition, the fermentation and putrefaction carried out by mesophilic organisms could provide enough heat for the growth of thermophiles (Allen, 1953; Clegg and Jacobs, 1953; Gaughran, 1947).

The presence of viable spores in ocean-basin cores dated at over 5,800 years old is more difficult to explain; it may be a case of spore survival without vegetative growth (Bartholomew and Paik, 1966; Bartholomew and Rittenberg, 1949).

Gaughran (1947) considered that the locus of the thermophiles is in the tropics, where they constitute an important part of the soil flora. It cannot be assumed, however, that all thermophiles found in cold and temperate regions are of tropical origin (Clegg and Jacobs, 1953).

Some authors consider that thermophilic bacteria survived from earlier geological periods and that mesophiles resulted from adaptation as the earth's crust cooled (see Clegg and Jacobs, 1953; Gaughran, 1947). Others believe that thermophiles are variants of mesophilic bacteria, and that progressive adaptation occurred right up to the obligate stage (see Gaughran, 1947). Although thermophilic variants may be isolated in the laboratory (Allen, 1950, 1953; Bausum and Matney, 1965; Clegg and Jacobs, 1953; Dowben and Weidenmüller, 1968; McDonald and Matney, 1963; Sie, Sobotka, and Baker, 1961), this observation does not explain the ubiquitous distribution of thermophiles.

Growth of thermophiles in laboratory media may often require higher temperatures than under natural conditions (Black and Tanner, 1928; Clegg and Jacobs, 1953; Hansen, 1933); as the incubation temperature is raised, the nutritional growth requirements of certain thermophiles are increased (Farrell and Campbell, 1969). Conversely, some "obligately thermophilic" strains require a supplemented medium at lower temperatures and can be induced to grow at 37°C (Campbell, 1954; Campbell and Pace, 1968; Long and Williams, 1959).

Thermophilic sporeformers are among the principal causes of "flat-sour" spoilage of canned foods because of the high heat resistance of their spores. The prolonged heating necessary to destroy all such spores would impair food quality. It is therefore necessary to store canned foods at temperatures below the minimum required for growth of these organisms. This is the concept of "commercial sterility" described by Gillespy and Thorpe (1968). The spores may be introduced in sugar and for this reason a sterilized product, canners' sugar, is available (Scarr, 1968).

Richmond and Fields (1966) found thermophilic

aerobic sporeformers in a variety of food ingredients, including starch, sugar, spices, mustard, dried milk, split dried peas, dried yeast, and dried soup. *B. coagulans* was the commonest species isolated, followed, in order, by *B. licheniformis, B. subtilis, B. circulans, B. stearothermophilus,* and *B. pumilus.* Only *B. coagulans* and *B. stearothermophilus* are true flat-sour organisms; spores of the other strains are not as heat resistant as those of the flat-sour organisms and do not, therefore, pose a significant spoilage problem.

Thermophiles are of importance in the dairy industry; their growth in pasteurized milk may affect the commercial life of the product without presenting a hazard for public health. For a review of the flat-sour bacteria in milk and other foods, see Fields (1970).

Ingredients of bacteriological culture media, such as agar, may occasionally be contaminated by thermophilic sporeformers, which can then be troublesome (Allen, 1953).

The extreme heat resistance of *B. stearothermophilus* spores has been utilized in the production of spore papers for testing the efficiency of autoclaves. Such papers, however, must be rigorously standardized and used by experienced people—they are not suitable for routine use (Kelsey, 1961).

Wahlig and Holt (1976) described the use of *B. stearothermophilus* for the assay of antibacterial drug levels in patients. The dosage required by a patient in the subsequent 24 or 48 h may be determined within 3 h by incubating assay plates at 60°C. The method allows samples likely to be contaminated with Gram-negative organisms, urine for example, to be assayed for drug levels since such organisms will not grow at 60°C.

Surveys to detect the presence of bacteria in hot springs of varying temperature and pH revealed their existence at temperatures up to the boiling point of water, provided the environment was neutral or alkaline. The temperature at which bacteria are found decreases with increasing acidity (Brock and Darland, 1970). From this type of habitat, thermophilic, acidophilic, aerobic, sporeformers have been isolated. These isolates are closely similar in DNA base composition (62 mol% G+C) and in other physiological properties, suggesting that they form a homogeneous group distant from other members of the genus. Darland and Brock (1971) named these sporeformers *B. acidocaldarius.*

COLD-TOLERANT, AEROBIC, SPOREFORMING BACTERIA. According to Ingraham and Stokes (1959), "psychrophiles are bacteria that grow well at 0°C within two weeks". Psychrotrophs, as defined by Eddy (1960), are "bacteria that are able to grow at 5°C". In designating this section as "cold-tolerant aerobic sporeforming bacteria", we adopt the terminology of Michels and Visser (1976), who refer to the combined group of psychrophilic and psychrotrophic sporeformers as "cold-tolerant" organisms.

The term psychrophile was first used by Schmidt-Nielsen (1902) to describe bacteria that he and earlier investigators had cultivated at 0°C. More recently, it has become apparent that the ability to reproduce and carry out other activities at temperatures close to freezing point is widely distributed throughout the microbial world (Laine, 1970). As a result, the value of psychrophilic behavior in classification seems doubtful (Gyllenberg, 1968). In fact, Sundman and Gyllenberg (1967) found during a statistical study of 681 soil isolates that the ability to grow at 5°C correlated poorly with a number of features commonly applied in classification or identification. This finding implies that cold-tolerant behavior is not dependent on some specific property or properties, and therefore fails to provide useful taxonomic information. Furthermore, Witter (1961) in his review on psychrophilic bacteria argues that sporeforming bacteria should probably not be classified as psychrophiles even though they are capable of growing at low temperatures.

Cold-tolerant strains of the genus *Bacillus* were reported as early as 1949 when Reay and Shewan isolated them from refrigerated fish. Later investigations by Alexander and Higginbottom (1953) confirmed that cold-tolerant *Bacillus* strains were among the flora of refrigerated fish; Thomas, Hobson, and Bird (1959) and Schulze and Olson (1960) reported on their isolation from dairy products. Nevertheless, Larkin and Stokes (1966) are credited with isolating the first authentic strains of psychrophilic aerobic sporeformers (Laine, 1970). From soil, mud, and water samples they isolated 90 strains of aerobic sporeforming bacilli capable of growing at 0°C, with an optimum growth temperature of 20–25°C and no growth at 30–35°C. The minimum and the optimum growth temperatures were 10°C below those of mesophilic organisms. These isolates were able to grow at temperatures as low as −4.5°C and were capable of sporulating and germinating at 0°C. Further work (Larkin and Stokes, 1967) using 20 strains believed to be representative of the 90 isolates resulted in the establishment of four new species: *B. psychrosaccharolyticus, B. insolitus, B. globisporus,* and *B. psychrophilus.* In the same study these workers outlined relationships between established mesophilic species and the new groups they had described. They found that *B. psychrosaccharolyticus* most closely resembled *B. circulans,* but differed in maximum and minimum growth temperatures and in morphological appearance on glucose agar; *B. globisporus* and *B. psychrophilus* most closely resembled *B. pantothenticus,* and *B. insolitus* resembled *Sporosarcina ureae.* The most important difference between the latter is that on enriched media, such as Trypticase soy agar (TSA), *B. insolitus* forms rod-shaped cells with ellipsoidal spores while *Sporosarcina ureae* is coccoidal on all media. The relationship between these two species may be important since it suggests a link between the genera *Bacillus* and *Sporosarcina*, both of which are aerobic sporeformers.

Occasional studies have led to the isolation of aerobic sporeformers from extremely cold environments. The sub-Antarctic soils of MacQuarie Island, for example, yielded an organism *(B. macquariensis)* that produced spores at 0°C and had a maximum growth temperature of 25°C (Marshall and Ohye, 1966).

An increasing reliance on the use of refrigeration for prolonging the shelf life of food has stimulated investigation into the cold-tolerance of microorganisms, and most of the information available on the cold-tolerant aerobic sporeformers has come from organisms isolated from food samples. Laine (1970) isolated 75 *Bacillus* spp., which grew at 2°C within 2 weeks, from 540 samples of foods stored at refrigeration temperatures. Michels and Anema (1974) reported on the occurrence and thermoresistance of spores from cold-tolerant, aerobic sporeformers in pasteurized meals to be stored at 3°C, and Michels and Visser (1976) observed their occurrence and thermoresistance in soil, spices, and food. There have also been observations on the occurrence and thermoresistance of *Bacillus* species in relation to spoilage of pasteurized milk at low temperatures (Chung and Cannon, 1971; Grosskopf and Harper, 1969; Langeveld, 1971; Mourgues and Auclair, 1970; Shehata and Collins, 1971, 1972; Shehata, Duran, and Collins, 1971).

HABITATS OF PIGMENTED STRAINS.   Pigmented *Bacillus* species occur frequently; yellow, pink, or orange strains are the most common. In many of the organisms, pigment production can be shown to be a transient property (Gordon, Haynes, and Pang, 1973), and most of them are usually regarded as strains of well-defined species, particularly *B. subtilis*, *B. pumilus*, and *B. megaterium* (Turner and Jervis, 1968a). Pigmented bacteria are often isolated from marine habitats although members of the genus *Bacillus* are not well represented. Lloyd (1931) and ZoBell and Feltham (1934) found no pigmented *Bacillus* strains in marine and estuarine sediments, but Wood (1953) reported that pink, red, and orange-yellow strains comprised 7.8% of isolates of *Bacillus* from estuarine muds and sea water around the Australian coast. Kriss (1963) listed 65 isolates of *Bacillus* from sea water and marine sediments; five of the isolates produced pale yellow colonies on fish-peptone agar. Boeyé and Aerts (1976) also listed pigmented strains in their numerical taxonomic study on North Sea isolates.

Turner and Jervis (1968b) reported that, among strains of *Bacillus* species isolated from salt marsh and other soils by selective heat treatment, a large

proportion, up to 70%, formed orange, yellow, or pink colonies and were found to contain carotenoid pigments. Such strains are not readily found in nonsaline soils although they do sometimes occur. The major ecological factor governing distribution of pigmented strains appeared to be salinity; this conclusion agrees with the results of Horowitz-Wlassova (1931), who reported that salt-tolerant *Bacillus* strains are orange or yellow in color. Halophilic, Gram-negative bacteria usually contain carotenoids, and Brown (1964) postulated a correlation between carotenoid pigments and salt requirements. Mathews and Sistrom (1959) suggested that nonphotosynthetic organisms that possess carotenoids may be protected from photodynamic killing, and that the need for this protection is related to the natural distribution of pigmented bacteria.

Turner and Jervis (1968a), however, proposed that light might be of secondary importance since saline, inland soils exposed to more light than the plant-shaded, salt-marsh soils yield fewer pigmented isolates. Similarly, inland soils and salt-marsh soils exposed to the same degree of solar radiation do not have the same proportion of pigmented bacilli. The authors conclude that the prevalence of pigmented *Bacillus* strains in the salt marshes is due primarily to the high salinity of the habitat, which selects salt-tolerant strains from normal populations of *Bacillus* from river silts or the sea.

BACILLUS SPECIES IN SPACE EXPLORATION.   The advent of a sufficiently advanced space technology has led to an interplanetary search for life. To prevent possible irrevocable damage caused by terrestial contamination of other planets, the National Aeronautics and Space Administration's Planetary Quarantine Program, in cooperation with the International Committee on Space Research, has imposed sterilization or decontamination requirements on spacecraft that are to land on or pass close to any planet (Oxborrow, Fields, and Puleo, 1977).

To reduce microbial contamination, the spacecraft or spacecraft components are subjected to a terminal dry-heat sterilization cycle in hot, flowing nitrogen. The cycle consisted of a nominal temperature of 111.7°C ± 1.7°C for a period of 23–30 h after the coldest contaminated point reached 111.7°C. In a study to determine the microbiological profile of the Viking Spacecraft, Puleo et al. (1977) found that 47% of isolates were from the genus *Bacillus* and that this figure represented more than twice the number of any other group isolated. The duration of each sterilization cycle was dictated by the level of aerobic, mesophilic bacterial spores present on the landing capsule prior to sterilization. By this method, the American Space Program complies with the international agreement requiring that the probability of contaminating the planet Mars be less than 1 in 10^3.

As Food-Spoilage Organisms

BREAD.   Bacterial spores that survive the bread-baking process and subsequently develop in the stored loaf are responsible for the well-known texture defect "ropiness". These spores are usually present in the original flour (Barton-Wright, 1943) or in the yeast (Hoffmann, Schweitzer, and Dalby, 1937) and may be activated by sublethal temperatures during baking. Farmiloe et al. (1954) established the ability of *B. subtilis* spores to survive the baking process, and showed that the destruction of bacterial spores during baking was not exponential with time. Aubertin et al. (1938) reported the survival of *B. mesentericus* and *Clostridium perfringens* during baking, and Ingram and Robinson (1951) recovered *C. botulinum* type A from experimental canned bread.

MILK.   Raw and pasteurized milk are both susceptible, especially during hot weather, to the condition known as "broken" or "bitty" cream. In this condition, the cream at the top does not re-emulsify on shaking, like the cream of normal milk, but breaks into small particles that can be seen floating on the top of tea or coffee. In the initial stages there is no appreciable change in flavor or acidity of the milk, but later a firm, nonacid clot develops, which resembles the cream plug found in bottled milk (Stone and Rowlands, 1952).

The occurrence of broken cream was first recorded in a National Institute for Research in Dairying (NIRD) report in 1938, according to which samples of milk with this fault were found to contain large numbers of aerobic sporeformers. Davis (1940) ascribed the formation of bitty cream to *B. mycoides;* he suggested that, in the absence of other organisms, especially acid producers, this sporeformer proliferated rapidly in milk above a critical temperature of 20°C. This growth soon resulted in sweet curdling. Stone and Rowlands (1952) found that of 161 cultures of aerobic sporeformers isolated from raw and pasteurized milk, 96 were *B. cereus;* 27, *B. mycoides;* 17, *B. subtilis;* 11, *B. sphaericus;* 3, *B. megaterium;* 2, *B. pumilus;* 1, *B. firmus;* and 4 were unidentified. From their experiments, they concluded that *B. cereus* as well as *B. mycoides* was involved in this type of spoilage and that these two organisms are likely to be the only sporeforming bacilli capable of causing broken cream and the resultant sweet curdling of milk. Since all lecithinase-positive strains of *B. cereus* and *B. mycoides* did produce broken cream and three lecithinase-negative strains of *B. cereus* did not, it was proposed that the development of broken cream was caused partly by lecithin hydrolysis. The hydrolysis was thought to act on the surface membrane of the fat globule and, together with a coagulation of membrane-associated casein, produce the characteristic appearance of broken cream; this view was supported by the work of Stone (1952).

Because of the widespread distribution of spores of *B. cereus* and *B. mycoides* in both raw and pasteurized milk, the absolute prevention of "broken cream" does not seem possible. But preliminary observations suggest that maintenance of pasteurized milk at 5°C or lower until distribution may sufficiently delay onset of "bitty cream" formation to allow a reasonable period of storage, except during hot weather in households where milk is not kept cool.

In view of the important role of *B. cereus* in determining the bacteriological quality of liquid milk, the various routes of contamination merit consideration. Davis (1940) suggested that the main sources of milk contamination were flies, dust, soil, hay, and improperly cleaned and sterilized cans and bottles. Billing and Cuthbert (1958) suggested that the storing of hay in milking sheds probably contributed to the high spore count in farm milk. These workers also implicated dairy equipment as a significant source of contamination but, in contrast to Thomas et al. (1946), found no evidence that the method of equipment disinfection influenced the *B. cereus* content of the milk. They also found small numbers of *B. cereus* on or in the cows' teats and concluded that *B. cereus* contamination was inevitable. Labots, Hup, and Galesloot (1965) attributed the contamination of raw milk to two main sources, contamination by infected particles of soil, dung, straw, hay, and other fodder and contamination by contact with nonsterile surfaces of milking equipment such as cans.

Donovan (1959) also proposed cans as a major source of contamination; he showed that washed cans in a commercial creamery had an average *B. cereus* spore count of 16×10^3 per can. Forty-eight of 80 cans tested contained this species. It was shown (Donovan, 1959) that spore numbers in undiluted milk never reached 1% of the total 30°C colony count at 48 h but in milk diluted 1:50 with water the proportion of spores varied between 70% at 48 h and 100% at 24 h. This increase in sporulation occurs in the diluted milk of insufficiently rinsed milk cans.

Several workers have successfully isolated sporeformers from bottles of "sterilized" milk that showed no signs of spoilage (Bentler, 1961; Bergamini and Berti, 1962; Lück, Mostert, and Husmann, 1976). Nevertheless, spoilage of "sterilized" milk does occur, and Hiscox and Christian (1931) reviewed the problems commonly encountered with milk that has been heat-treated at 212°F for 30 min. They describe three types of spoilage: *B. subtilis*-or *B. cereus*-induced neutral or alkaline digestion; acid coagulation followed by contraction and slight digestion of curd, organism responsible unidentified; and coconut or carbolic tainting but no visible change, *B. circulans* responsible. Clegg (1950) reported on batches of spoiled, sterilized milk that clotted in the bottle before dispatch from the creamery. This effect was attributed to the action of thermophiles since only they possessed the necessary growth characteristics. Thermophiles are present in most samples of sterilized milk and must always be considered a potential source of spoilage.

Of 52 cultures isolated from faulty and from apparently normal tins of evaporated milk, 36 were strains of *B. subtilis*, eight were *B. licheniformis*, three *B. pumilus*, two *B. brevis*, and one *B. cereus* (Nichols, 1939). The most heat-resistant organisms were *B. subtilis*, *B. licheniformis*, and *B. pumilus*, which are all commonly isolated contaminants of milk.

The ubiquitous nature of the genus *Bacillus* and the heat resistance of its spores makes them prominent milk-spoilage organisms. Despite the various eradication procedures carried out, the aerobic sporeformers still manage to contaminate and subsequently spoil milk. The subject is reviewed by Davies and Wilkinson (1973).

CANNING. (See also the section on thermophiles.) Since a great number of canned products are nonsterile, acid-producing organisms may constitute a more serious problem to the canning industry than do the putrefactive anaerobes (Jansen and Aschehoug, 1951). Acid-producing organisms can be divided into two main groups, the bacteria that produce flat-sour spoilage of which *B. stearothermophilus* and *B. coagulans* are the most common, and bacteria that produce acid with gas. The acid-producers in canned foods are usually thermophiles (see also the section on thermophiles), although mesophilic, spore-forming aerobes are often present in canned foods by virtue of their heat-resistant spores. In their studies on putrefactive anaerobes, Aschehoug and Jansen (1950) frequently isolated *B. subtilis* in association with putrefactive anaerobes; later, Jansen and Aschehoug (1951) provided evidence for *B. subtilis* being an important spoilage agent in canned foods. *B. subtilis* was reported by Aschehoug and Vesterhus (1941) as a common spoilage organism responsible for "soft swells". Flat-sour spoilage is reviewed by Fields (1970).

Bacillus species may have greater significance in cured products because they are able to reduce both nitrate and nitrite, with the production of N_2O and N_2 (Eddy and Ingram, 1956). These organisms may be one of the important causes of "blowing" in canned, cured meats (Ingram and Hobbs, 1954). Eddy and Ingram (1956) isolated a *Bacillus* strain from blown canned bacon. This organism was tolerant of 15% NaCl (halophilic) and reduced nitrate and nitrite. Considering its tolerance to nitrite under aerobic and anaerobic conditions, the heat resistance of the spores, and composition of the gas produced, it

was argued that, should this organism survive the processing of cured meat, subsequent development and spoilage seemed certain. They postulated that this strain probably belongs to the *B. circulans* group and was distinct from the denitrifying strains that Verhoeven (1952) identified as *B. licheniformis*. It seems probable that the reduction of nitrate or nitrite to nitrogen and its oxides may be found in certain strains of diverse species of the genus *Bacillus*, and as a result this type of spoilage is fairly widespread.

Mammalian Pathogens

ANTHRAX. Anthrax has been known since antiquity. Although most vertebrates are susceptible, it is typically an epizootic disease of herbivorous animals. The disease was known to be associated with contaminated soil environments long before *Bacillus anthracis* was identified as the causative organism. For comprehensive historical reviews, see Sobernheim (1931) and Wilson and Miles (1975). *B. anthracis* might be regarded as a typical member of the genus were it not for its pathogenicity, and a great deal of effort has been devoted to the search for identification criteria (Sterne, 1967).

Spores of *B. anthracis* may contaminate a large number of products derived from animals, including wool, hair, hides, bones, and animal feeds, resulting in outbreaks of the disease among man and other animals. These contaminated products are major exports of many countries where anthrax is endemic. When an animal–soil–animal cycle is established, a recurrent anthrax problem begins (Van Ness, 1971).

The long survival of spores in dry conditions has often been demonstrated (Graham-Smith, 1941; Sneath, 1962; Wilson and Russell, 1964), and has led to the misconception that contamination persists indefinitely in infected pasture. The experiments are probably more relevant to the persistence of spores in dry materials such as wool, hair, hides, and bones (Christie, 1974).

When soil conditions are favorable to vegetative growth, the spores germinate and a nonpathogenic life cycle may be maintained for years. *B. anthracis* has a poor competitive saprophytic ability and cannot survive for long in soils and water that are unsuitable (Minett and Dhanda, 1941; Van Ness, 1971) or in the presence of the putrefactive bacteria that proliferate within an anthrax carcass (Minett, 1950). Survival in soil is favored by a neutral or alkaline reaction, an ambient temperature of 21–37°C, and a relative humidity of 60% or higher (Davies, 1960; Minett, 1951; Van Ness, 1971). At lower temperatures, sporulation may be too slow to ensure heavy contamination of pasture (Minett, 1950, 1951). The relative freedom of countries in the higher latitudes from severe anthrax epidemics is largely explained by these observations (Sterne, 1959). In warm climates, outbreaks are commonest in the summer months. Low-lying swampy terrain, with a warm, loose soil, promotes spore formation and may lead to the establishment of *B. anthracis* in the microflora. As conditions become arid and herbage wilts, animals have to graze closely and may ingest or inhale spores in large numbers. During such anthrax seasons, insects are most profuse and can transmit infection mechanically to animals (Sen and Minett, 1944), and man (Sirol, Gendron, and Condat, 1973) but they are not considered a significant epidemiological factor (Van Ness, 1971).

The susceptibilities of animals to anthrax, as determined by experimental infection, are not always paralleled by natural incidences in domestic or wild populations. The relative susceptibilities vary widely between different countries according to feeding and ranging habits, climatic conditions, and other environmental factors (Minett, 1950, 1952; Van Ness, 1971). Coarse vegetation may predispose to trauma of the mucosa, and so increase the chance of infection (Christie, 1974; Wilson and Miles, 1975). In phosphorus-deficient areas, cattle may become infected by chewing carcass bones (Clark, 1938); sheep in the same area escape because they do not chew bones (Sterne, 1959). Carrion-eating birds may often carry infection over a wide area on their beaks and claws and in their feces. Birds, in general, rarely suffer from anthrax, but may be experimentally infected if their body temperature is lowered. Depression of phagocytosis may be an important factor in such circumstances (Sterne, 1959; Wilson and Miles, 1975).

Anthrax is of great economic importance to countries where the disease is endemic. The incidence of anthrax is closely related to the level of organization of the livestock industry in a country; in the Middle and Far East, and Africa, primitive husbandry results in high losses from the disease. Ignorance and poverty among peasants and nomads, to whom the income derived from bones, hides, hair, and wool is important, lead to the appearance of contaminated animal products on world markets. This directly affects the incidence of anthrax in industrial countries that import such materials. The size of the problem in terms of livestock losses is illustrated by figures from Iran: in 1945 approximately 1,000,000 out of a population of 15,000,000 sheep were reported to have died from anthrax (Wolff and Heimann, 1951).

In Great Britain and northern Europe, anthrax is not usually regarded as an endemic disease because the majority of cases in animals are due to imports of contaminated supplementary feeds and artificial manures. For the period 1963–1972, these products were responsible for 84.5% of anthrax cases in England and Wales (Hugh-Jones and Hussaini, 1975). Since feed stuffs are the most important of these sources, outbreaks are commonest in the winter

months (Minett, 1952; Sterne, 1959); in recent years, however, this distribution has been less clear (Hugh-Jones and Hussaini, 1975). In one study, 18 of 21 samples of bone meal from India and Pakistan were found to be contaminated with anthrax spores (H.M.S.O., 1959).

In order of susceptibility to anthrax, the herbivores rank foremost; the dog, cat, rat, and man are moderately resistant; pigs are more so; and most birds are very resistant (Sterne, 1959; Wilson and Miles, 1975). The disease takes the form of a septicemia, which may or may not be acute. The acute form is usual in goats and sheep. Cattle commonly suffer from a less acute form but the mortality is still high—between 70% and 100% (Christie, 1974). In pigs, dogs, and cats the disease may be subacute or chronic, and death can be due to asphyxia caused by edematous swelling of the pharynx and larynx. Alternatively, the intestinal form may occur in these three animals and recovery is then not uncommon (Christie, 1974; Sterne, 1959).

In man, anthrax may be described as of nonindustrial origin when it affects people in contact with infected animals. Anthrax of industrial origin is acquired by handling all manner of contaminated animal products. Cutaneous anthrax is by far the commonest manifestation in man and, if diagnosed early, it is easily treated. The pulmonary and intestinal forms usually occur as a result of industrial infection in the former case (woolsorter's disease; see Brachman, Kaufmann, and Dalldorf, 1966) and consumption of infected meat in the latter (Christie, 1974; Lincoln et al., 1964; Tantajumroon and Panas-Ampol, 1968). The noncutaneous forms of the disease are particularly deadly (Sterne, 1959; Wilson and Miles, 1975). The condition may be rapid in onset, of short duration, is rarely diagnosed till late, and is usually fatal, but subclinical infections may be commoner than supposed (Christie, 1974; Sterne, 1959). Meningitis may occur as a complication of septicemia and has occasionally been reported as the only manifestation of the disease (Christie, 1974; Pluot et al., 1976).

Spores of *B. anthracis,* when ingested or inhaled by an animal or introduced through bites and abrasions, usually germinate within a few hours. The vegetative organisms acquire polypeptide capsules of D-glutamic acid which, in susceptible animals, inhibit opsonization and phagocytosis. The capsular material is not immunogenic (Smith, 1960). In resistant species, such as the dog, pig, or rat, decapsulation (by γ-glutamylase in the dog; Utsumi et al., 1961) occurs and the rods show degenerative changes (Sterne, 1959).

The capsule is an important factor in invasiveness and virulence, but it is the toxin complex synthesized by the bacillus that is responsible for the ultimate death of the host (Smith, 1960). The toxin complex contains three separate components: a chelating agent that contains phosphorus and possesses both protein and carbohydrate moieties (factor I), and two proteins (factors II and III) (Smith and Stoner, 1967). No factor is toxic by itself, but all three act synergistically (Stanley and Smith, 1961). The main effect of the toxic complex is to increase vascular permeability (Smith and Stoner, 1967). Although factor II is immunogenic by itself, and mixtures of factors I and II, II and III, and I and III are immunogenic, it is generally agreed that the best vaccine should contain all three components (Smith and Stoner, 1967; Stanley and Smith, 1963). There is a large amount of literature concerning the anthrax toxin; for reviews of work in this field see Smith (1958) and Lincoln and Fish (1970).

B. anthracis is usually sensitive to penicillin, the tetracyclines, and chloramphenicol. Penicillin is considered the most effective antibiotic (Wilson and Miles, 1975) and, if therapy is begun early enough, the prognosis is good. In herbivorous animals and in pulmonary and intestinal anthrax in man, diagnosis may be made too late for antibiotic therapy to be efficacious (Christie, 1974; Gold, 1967). The serum of artificially immunized animals has been used in the past, but is not now normally given (Christie, 1974).

Eradication of anthrax in animals would eliminate the disease in man. Mass vaccination of animals in South Africa, followed by annual reinoculation in areas where outbreaks occurred, and every 3–5 years elsewhere, resulted in a 99% decrease in the number of cases within 20 years (Sterne, 1959, 1967). This reduction was achieved despite the primitive standards of husbandry prevailing. If such methods were widely applied, the risk of anthrax would be removed from industrial countries. The alternative of immunizing all animals at risk from contaminated feeding stuffs would be prohibitively expensive in comparison with the present methods of vaccinating only those animals involved in outbreaks (Sterne, 1967).

Live-spore vaccines are used for the protection of animals but only alum-precipitated toxoid is considered safe enough for the immunization of humans at risk. The control of anthrax is considered in greater detail by Sterne (1959), Gold (1967), Christie (1974), and Wilson and Miles (1975).

OTHER INFECTIONS. A number of reports implicate *Bacillus* species other than *B. anthracis* in various infections of mammals. In the past, aerobic sporeformers (excepting *B. anthracis* and the insect pathogens) were regarded as harmless saprophytes and, despite evidence to the contrary, such organisms are still widely regarded as contaminants if isolated from clinical specimens in hospital laboratories. The confused state of *Bacillus* taxonomy prior to the monographs of Smith, Gordon, and Clark (1946, 1952) did not encourage clinical laboratories to at-

tempt specific identification, and the name *B. subtilis* was commonly used as a synonym for aerobic sporeformers other than *B. anthracis*. Certain workers (Allen and Wilkinson, 1969; Weinstein and Colburn, 1950) considered specific identification to be of little or no clinical importance.

Many descriptions of organisms isolated in clinical contexts and assigned to the "subtilis group" in the early literature, such as those of François (1934), more closely resemble *B. cereus* than *B. subtilis* (Burdon and Wende, 1960; Clark, 1937; Davenport and Smith, 1952; Goepfert, Spira, and Kim, 1972) but are rarely detailed enough to allow certain identification. In the earlier literature, therefore, *B. subtilis* may generally be regarded as meaning aerobic sporeformer (Goepfert, Spira, and Kim, 1972), and the terms "anthracoid" and "anthrax-like" as meaning *B. cereus* (Breed, Murray, and Smith, 1957). Recent cases in which *B. cereus* is commonly the aerobic sporeformer isolated support the proposition that many isolates, described as *B. subtilis* in the earlier reports, were probably *B. cereus*.

Infections due to opportunistic pathogens, such as certain *Bacillus* species, are being encountered with increasing frequency (Gsell, 1971; Klainer and Beisel, 1969). This rise is not likely to be due to an increase in the virulence of the organisms involved but to host predisposition by, for example, compromised immunity, metabolic disorders, and neoplastic disease, and various clinical and surgical procedures, and to advances in bacteriological technique and interpretation.

The early literature contains many accounts of *Bacillus* spp. causing eye infections, and François (1934), in a comprehensive review of "*Bacillus subtilis*" in ocular infections, was able to collect reports of 157 cases of six types of infection. Davenport and Smith (1952) noted the rarity of ocular infections by *Bacillus* spp., relative to the frequency of penetrating injuries and the ubiquity of the organisms, and suggested that either a massive inoculum is required or uncommon strains are involved.

In at least three cases of meningitis, *Bacillus* spp. have gained entry to the subarachnoid space as a result of injections for spinal anaesthesia (Farrar, 1963; Weinstein and Colburn, 1950). This entry may be explained by the fact that delayed immune responses operate in the subarachnoid space, and in the vitreous humor of the eye (Farrar, 1963).

Surgical therapy of hydrocephalus using indwelling shunt and valve mechanisms has resulted in several cases of meningitis and septicemia (Cox, Sockwell, and Landers, 1959; Leffert, Baptist, and Gidez, 1970; Raphael and Donaghue, 1976; Schoenbaum, Gardner, and Shillito, 1975). The fibrin coatings of the prosthetic devices become colonized by *Bacillus* spp. in a manner reminiscent of bacterial endocarditis (Farrar, 1963).

A few reports of *Bacillus* endocarditis are to be

found in the literature; Finland and Barnes (1970) mention three and Craig, Lee, and Ho (1974) describe a case in which *B. cereus* was probably introduced during the self-administration of drugs by an addict. It is likely that many clinically significant isolations of *Bacillus* from blood have been disregarded (Farrar, 1963).

Curtis, Wing, and Coleman (1967) reported five cases of bacteremia resulting from contamination of hemodialysis fluid with *B. cereus,* and Yow, Reinhart, and Butler (1949) described septicemia caused by a contaminated blood transfusion.

Disseminated *Bacillus* infection in patients who have been immunologically compromised by neoplastic diseases such as leukemia and who are often undergoing therapy with cytotoxic drugs has been the subject of many recent reports. Infections have usually taken the form of a bacteremia or septicemia (Barnham and Taylor, 1977; Feldman and Pearson, 1974; Gsell, 1971; Ihde and Armstrong, 1973; Pearson, 1970), or pneumonia (Leff et al., 1977), or a combination of these (Coonrod, Leadley, and Eickhoff, 1971; Feldman and Pearson, 1974; Ihde and Armstrong, 1973; Sathmary, 1958).

Infants are susceptible to opportunistic infections because their immune systems are only partially developed. Cases of meningitis (Boyette and Rights, 1952; Farrar, 1963) and disseminated infection by *Bacillus* (Turnbull, French, and Dowsett, 1977; Yow, Reinhart, and Butler, 1949) have been reported. In addition, there have been several reports of disseminated infection in apparently immunologically normal patients who had not been subjected to surgical interference (Allen and Wilkinson, 1969; Farrar, 1963; Ihde and Armstrong, 1973; Weinstein and Colburn, 1950).

In recent reports, the most commonly implicated member of the genus has been *B. cereus,* and a wide variety of infections with this species has been recorded. In addition to the cases described by Leffert, Baptist, and Gidez (1970), Coonrod, Leadley, and Eickhoff (1971), Feldman and Pearson (1974), Raphael and Donaghue (1976), Barnham and Taylor (1977), Leff et al. (1977), and Turnbull, French, and Dowsett (1977), there have been two localized infections resembling gas gangrene (Groeschel, Burgess, and Bodey, 1976; Turnbull, French, and Dowsett, 1977), one case of osteomyelitis (Solny, Failing, and Borges, 1977), several ear infections (Lazar and Jursack, 1966), urinary tract infections (Melles, Nikodémusz, and Ábel, 1969), bronchopneumonia (Stopler, Camuescu, and Voiculescu, 1964), and septicemia (Goullet and Pepin, 1974). Certain *B. cereus* strains produce enterotoxins responsible for food poisoning (discussed more fully below), and some authors consider that these toxins may be of importance in initiation and maintenance of infections (Burdon, Davis, and Wende, 1967; Goullet and Pepin, 1974; Leffert,

Baptist, and Gidez, 1970). Turnbull (1976) and Turnbull, Nottingham, and Ghosh (1977), however, have demonstrated a necrotic toxin, produced by strains isolated from human infections and foods.

Other *Bacillus* species that have been implicated in infections are *B. sphaericus* (Allen and Wilkinson, 1969; Farrar, 1963), *B. pumilus* (Melles, Nikodémusz, and Ábel, 1969; Weinstein and Colburn, 1950), *B. brevis, B. coagulans* (van Bijsterveld and Richards, 1965), *B. macerans* (Ihde and Armstrong, 1973), and a variety of *B. circulans* (Boyette and Rights, 1952).

It is not always possible to assign a primary pathogenic role to *Bacillus* species found in mixed infections (Farrar, 1963; Ihde and Armstrong, 1973). The *Bacillus* species may be a secondary invader (van Bijsterveld and Richards, 1965), may help maintain or exacerbate a preexisting infection by acting in synergy with other organisms (Behrend and Krouse, 1952; Farrar, 1963; Ihde and Armstrong, 1973; Pearson, 1970), or may merely colonize the area already infected (Ihde and Armstrong, 1973; Pearson, 1970). The penicillinase of *B. cereus* could affect the antibacterial treatment of infections caused by other organisms (Pearson, 1970).

There are few reports of nonexperimental *Bacillus* infection in animals. This paucity is largely explained by the fact that animals are rarely subjected to the intensive care, prolonged drug therapy, surgical treatment, and other circumstances which predispose humans to opportunistic infections. Bonventre and Johnson (1970), however, recorded a fatal *B. cereus* infection in an adult tiger at Cincinnati zoo. Several cases of *B. cereus* mastitis in dairy cattle have been reported (Biancardi, 1963; Perrin, Greenfield, and Ward, 1976; Schiefer et al., 1976); many were initiated by the use of contaminated antibiotics, teat tubes, or syringes when treating mastitis due to other causes (Perrin, Greenfield, and Ward, 1976). Most information about *Bacillus* disease in animals has come from experimental infections in mice, guinea pigs, and rabbits (Bonventre and Johnson, 1970).

In addition to *Bacillus* infections in man, the use in laundry products of *B. subtilis* derivatives that contain proteolytic enzymes has given rise to cases of dermatitis and respiratory ailments (Dubos, 1971; Flindt, 1969; Greenberg, Milne, and Watt, 1970; Pepys et al., 1969).

BACILLUS CEREUS FOOD POISONING.   It is now clear that *B. cereus* is a significant cause of food-borne illness. From the first report by Lubenau (1906) until the comprehensive description by Hauge (1950), accounts of *B. cereus* food poisoning were scanty and lacking in detail. Hauge (1950) managed to produce food poisoning in human volunteers by feeding them *B. cereus* cultures. Dack et al. (1954) were unable to confirm these findings, but they did not use *B. cereus* strains isolated from food-poisoning outbreaks.

Since 1950 the frequency of reports has increased dramatically. This increase possibly indicates an overall increase in the number of incidents, but growth in awareness of the problem is the more likely explanation (Gilbert and Taylor, 1976). Ormay and Novotny (1969) recorded *B. cereus* as the third most common cause of food poisoning in Hungary, responsible for 8.2% of outbreaks and 15.2% of persons affected, for the period 1960–1968. Goepfert, Spira, and Kim (1972) made a comprehensive review of the literature concerning *B. cereus* food poisoning. Most reports to this date were confined to Europe and the illness was characterized by an incubation period of 8–16 h followed by abdominal pain, diarrhea, and rectal tenesmus. A wide variety of foods was involved.

In the United Kingdom there were several unconfirmed outbreaks of *Bacillus* food poisoning before 1971 (Gilbert and Taylor, 1976). In that year, six episodes were associated with rice, usually fried, from Chinese restaurants and "take away" shops (Public Health Laboratory Service, 1972), and reports have appeared with increasing frequency since then (Mortimer and McCann, 1974; Public Health Laboratory Service, 1973). In nearly all such cases the symptoms were rapid in onset, usually 1–6 h after consumption of the contaminated food, and were characterized by nausea, vomiting, and malaise with only occasional diarrhea (Mortimer and McCann, 1974). Similar outbreaks have been reported from Canada (Lefebvre et al., 1973; Mathias et al., 1976), Australia and the Netherlands (Taylor and Gilbert, 1975), and Finland (Raevuori et al., 1976).

It appears, therefore, that two distinct forms of *B. cereus* food poisoning may be recognized; the diarrheal type and the vomiting type. Serotyping studies lend support to this theory (Gilbert and Parry, 1977; Taylor and Gilbert, 1975). In the diarrheal type, the foods involved included meat, vegetables, puddings, and sauces. The vomiting type appears to be largely restricted to rice dishes. Relatively few cases have been reported from America (Midura et al., 1970; Portnoy, Goepfert, and Harmon, 1976) but, since symptoms may not always be severe enough for medical attention to be sought (Goepfert, Spira, and Kim, 1972), the number of unreported outbreaks might be large.

The rapid onset of symptoms, the afebrile nature of the illness, and its short duration suggest that *B. cereus* food poisoning is an intoxication rather than an infection (Gilbert and Taylor, 1976; Goepfert, Spira, and Kim, 1972; Hauge, 1955). Several studies have been made of the nature and properties of the toxic factors involved. Spira and Goepfert (1975) showed that the toxin responsible for diarrheal symptoms is a true enterotoxin, capable of causing fluid accumulation in rabbit ileal loops, altering vascular permeability in rabbit skin, and killing mice when injected intravenously. Rabbit immune serum to one *B. cereus* strain was able to neutralize these three activities in most other strains studied. The toxic factor is synthesized during the logarithmic phase of growth, is protein in nature, heat labile, and exists in several serological forms (Goepfert et al., 1973).

Melling et al. (1976) carried out feeding trials with Rhesus monkeys to determine whether or not a new enterotoxigenic material was responsible for the vomiting type of food poisoning. Only strains isolated from vomiting outbreaks were capable of eliciting vomiting, and it was necessary to grow them in rice culture. The toxin has now been characterized as a very stable molecule of low molecular weight (Melling and Capel, 1978).

Nikodémusz (1965, 1967) and Nikodémusz and Gonda (1966) were able to produce diarrheal symptoms in cats and dogs, but not in rodents, by giving foods containing large amounts of *B. cereus*. The incubation periods varied between 2.5 and 7 h. Chastain and Harris (1974) reported cases of *B. cereus* food poisoning in two pet dogs from the same household. The dogs had been fed from an open can of refrigerated dog food which had probably become contaminated after opening since there were no other reports in the area.

Turnbull (1976) described three *B. cereus* toxins, including one which is capable of stimulating the adenylate cyclase–cyclic AMP system and is probably responsible for fluid accumulation in rabbit ileal loops. Turnbull inferred that a fourth toxin, responsible for vomiting, exists.

Spores and vegetative cells of *B. cereus* are ubiquitous, and studies of a wide range of foods have often shown high levels of contamination. Nygren (1962) reported about 50% incidence in a representative variety of foods and noted that food-drying procedures increase the risk of contamination with sporeformers from air and dust. Kim and Goepfert (1971a) recorded a 25% incidence in dried foods with levels of up to 4,000 viable cells per gram. Their detection methods were not as sensitive as those of Nygren. Ormay and Novotny (1969) suggested that the high incidence of *B. cereus* food poisoning from meat dishes in Hungary was due to heavy seasoning with contaminated spices. Nygren (1962) recorded *B. cereus* contamination in 72% of spice samples and Kim and Goepfert (1971a) reported contamination in 55% of seasoning mixes and 40% of spice samples. Gilbert and Taylor (1975) found *B. cereus* contamination in 88% of samples of uncooked rice.

B. cereus is probably present in the majority of farm milks and all bulk milks (Franklin, 1967). Nygren (1962) and Kim and Goepfert (1971a) were able to isolate *B. cereus* from, respectively, 71.3% and 37.5% of samples of skimmed milk powder.

Dried milk may be a common source of *B. cereus* in sauces, custards, and other prepared foods (Pinegar and Buxton, 1977).

Outbreaks of *B. cereus* food poisoning are usually associated with the storage of cooked or partially prepared foods at room temperature for relatively long periods. Incriminated foods have often been prepared well in advance of consumption, for example, late evening and weekend meals in institutions, and only lightly warmed, if at all, immediately before serving (Goepfert, Spira, and Kim, 1972; Hauge, 1955; Midura et al., 1970; Schmitt, Bowmer, and Willoughby, 1976).

In Chinese restaurants, boiled rice that is prepared for subsequent frying is often allowed to dry off overnight at room temperature. In addition, unused batches of rice may be mixed with freshly cooked rice and be carried over for several days in this way. The Chinese prefer unrefrigerated rice because it tosses better when being fried (Mortimer and McCann, 1974). Preparation of rice in smaller batches, followed by storage above 50°C or below 15°C, after rapid cooling and drying, would prevent further outbreaks (Gilbert, Stringer, and Peace, 1974).

The incubation period and symptoms of *B. cereus* food poisoning of the diarrheal type are similar to those of *Clostridium perfringens* food poisoning, and the vomiting type resembles *Staphylococcus aureus* food poisoning in many respects (Gilbert and Taylor, 1976). Examination of food vehicles and clinical specimens for *B. cereus* is not yet routine practice in food-poisoning investigations; if it were, the percentage of outbreaks listed as "etiology unknown" might well diminish (Gilbert and Taylor, 1976; Goepfert, Spira, and Kim, 1972; Kim and Goepfert, 1971a).

Isolation

General Considerations

To separate a particular organism from a complex ecosystem it is necessary to select cultural conditions so as to promote growth of the required organism in preference to the other species present. Where the organism forms a significant proportion of the population, it may suffice to spread a sample of the source material over the surface of a solid growth medium and select characteristic colonies for subculture. Often, however, this approach is inadequate because low numbers of the target organism are overgrown by contaminants. Then it becomes necessary to use enrichment techniques. A sample of the source material is inoculated into a fluid enrichment medium, which is constituted to allow growth of the target species while suppressing contaminants. The important selective properties of the medium may be physical or chemical in nature, and there is a great deal of scope for the use of such methods for the isolation of microorganisms with unusual characteristics (Veldkamp, 1970).

The presence of the heat-resistant endospore suggests a simple approach to the isolation of spore-forming bacteria; the use of heat at a level which will kill contaminating vegetative cells but will allow spores to survive (pasteurization).

Pasteurization at 80°C for 10 min, or some similar temperature and time, has often been used as an initial step in the selection of sporeformers, and it is certainly of value. It should be used with caution, however, clearly recognizing (i) that sporeforming organisms may be present only in the vegetative phase, (ii) that some spores have little more heat resistance than vegetative cells, and (iii) that conditions must be favorable for spores to germinate and the resulting vegetative forms to grow after pasteurization. It is also important to bear in mind that germination to a heat-sensitive state can occur rapidly with some spores when placed in a suitable nutrient environment and that specimens for pasteurization should be suspended in distilled water or nonnutrient buffer, rather than in nutrient broth or other growth medium. The initiation of spore germination is sometimes difficult to achieve and may require a brief heat shock (e.g., 60°C for 60 min).

Spores in natural environments normally show widely varying levels of heat sensitivity. Roberts, Ingram, and Skulberg (1965), for example, have shown that relatively mild heat treatments are sufficient to inactivate spores of *Clostridium botulinum* type E. The problem of heat sensitivity of spores can sometimes be overcome by using 95% ethyl alcohol as a spore-selection agent (Bond et al., 1970) instead of heat treatment. Samples are washed with alcohol followed by sterile water through stainless steel sieves (Bond and Favero, 1977).

An example of the imaginative use of the enrichment culture method is that of Beijerinck, which led to the successful isolation of a *Bacillus* species capable of utilizing urea and growing at high pH. Wiley and Stokes (1962) applied the same technique as follows.

Enrichment of *Bacillus pasteurii* (Wiley and Stokes, 1962)

> One gram of soil was suspended in tap water, heated at 80°C for 10 min, and inoculated into a liquid medium of 1% yeast extract and 5% urea at pH 9.0. Following incubation at 30°C, the authors isolated *B. pasteurii* on a solid medium of 2% (wt/wt) yeast extract, 1% (wt/wt) $(NH_4)_2SO_4$, and 0.13 M Tris-buffer (pH 9.0) in distilled water. They found it essential to heat-sterilize these ingredients separately since media in which ingredients had been sterilized together were unable to support growth (Veldkamp, 1970).

Similar approaches have revealed such organisms as *B. acidocaldarius* (Darland and Brock, 1971) (pH 2–6, 45 + °C), *B. macquariensis* (Marshall and Ohye, 1966) (0°C), *B. alcalophilus* subsp. *halodurans* (Boyer, Ingle, and Mercer, 1973) (pH 8.5–10, 12% NaCl), and *B. caldotenax* (Heinen and Heinen, 1972) (pH 7.5–8.5, 70–75°C), as well as adding new strains to already established species (Boeyé and Aerts, 1976; Fuller and Norman, 1943; Marsh and Larsen, 1953; Moore and Becking, 1963; Nishio et al., 1976).

In addition to ensuring that the enrichment medium is physiochemically capable of supporting vegetative growth, the inclusion of the appropriate germination factor or factors is often important. Rode and Foster (1965) have shown that of 40 random isolates from pasteurized soil, 20 could not be germinated using glucose-nitrate, alanine-lysine or yeast extract–peptone solutions—mixtures which included most of the known physiological germinants.

Soil, Water, and Other Natural Environments

The majority of aerobic sporeformers grow well on simple nutrient agar of a composition such as:

Peptone	5 g
Beef extract	3 g
Agar	15 g
Distilled water	1,000 ml
pH 6.8–7.0	

Media of this kind are quite suitable for examination of the general sporeforming microflora of soil or other environments but must be modified for the isolation of particular species such as *B. pasteurii*, for example, which requires urea (1%) or ammonium ions as a nitrogen source for growth. Many media have been described specifically for the examination of soil, but for the purpose of isolating *Bacillus* species they appear to offer little advantage over the simple nutrient broth/agar media. Soil extract agar, in which 25% of the water content of nutrient agar is replaced by an equivalent volume of an extract of soil prepared by autoclaving garden soil in tap water, may have some advantage in that it promotes typical spore formation and supports good growth of the majority of *Bacillus* species (Gordon, Haynes, and Pang, 1973) (see "Identification").

It is hardly surprising, in view of the difficulties involved in spore recovery, that attempts to estimate aerobic sporeformers in soil result in figures that vary wildly from sample to sample and that represent variable, and often low, proportions of the actual numbers present (Conn, 1948; Roberts and Hitchins, 1969).

Examination of the rhizosphere reveals higher numbers of *Bacillus* and *Clostridium* species than the surrounding soil (Mahmoud, El-Tadi, and Elmofty, 1964). The rhizosphere represents a specialized microenvironment in which the nutrient profile differs from that of the surrounding soil macroenvironment. Nakayama and Yanoshi (1967a) successfully isolated six strains of sporeforming, lactic acid bacteria from the rhizosphere following incubation of heat-treated rootlet samples (see "The Genus *Sporolactobacillus*", this chapter).

The majority of known species of *Bacillus* grow well on ordinary nutrient agar and, in general, isolation procedures depend on selecting pH, incubation temperatures, and salt concentrations in order to allow the growth of particular physiological types. The application of such selective procedures has led to the isolation of several new species from soil in recent years, and these are currently being examined for their status in the genus. Some isolates, however, do not grow on ordinary nutrient agar and so present a special problem for the microbial ecologist anxious to obtain as complete a picture as possible of the flora of soil or other environments. A case in point is *B. fastidiosus*, an organism that is found in soil and poultry litter but uses uric acid or allantoin as its only source of energy, carbon, and nitrogen. It is usually isolated and grown on nutrient agar containing 1% uric acid when the colonies are surrounded by clear zones from which the uric acid has been removed, and the reaction becomes strongly alkaline.

It will be evident that there is probably considerable scope for the isolation of new *Bacillus* types if enrichment and selective media and procedures are designed for the isolation of sporeformers of varying physiological characteristics.

Isolation of Flat-Sour Organisms from Foods

Shapton and Hindes (1963) described an indicator medium for the detection and enumeration of flat-sour organisms in canned foods and similar products.

Selective Isolation of Flat-Sour Organisms from Food (Shapton and Hindes, 1963)

Yeast-dextrose-tryptone agar.

Dissolve beef extract, 3 g; peptone, 5 g; tryptone, 2.5 g; yeast extract, 1 g; and dextrose 1 g, in 1,000 ml distilled water by gentle heating. Adjust to pH 8.4. Simmer for 15 min and then filter (coarse filter paper). Cool. Bring back to 1,000 ml and adjust to pH 7.4. Add agar, dissolve, and add bromocresol purple, 2.5 ml of 1% aqueous solution. Autoclave at 15 psi for 15 min. Pasteurize a suspension of the food sample in 1/4 strength Ringer's solution with molten yeast-dextrose-tryptone agar at 108°C (5 psi) for 10 min. Reduce the temperature to 100°C and maintain for 20 min. Cool to 50°C and aseptically distribute into Petri dishes. Incubate at 55°C for 48 h. Enumeration of yellow colonies provides an estimate of the number of flat-sour organisms in the original sample.

Isolation of *Bacillus cereus* from Food and Clinical Specimens

Many media for the isolation of *Bacillus cereus* rely on the egg yolk reaction for presumptive identification. Gilbert and Taylor (1976) found that certain food-poisoning strains gave a feeble egg yolk reaction but formed characteristic colonies on blood agar.

Direct Isolation of *Bacillus cereus* from Food and Clinical Specimens (Gilbert and Taylor, 1976)

Inoculate plates of 5% horse blood agar and incubate aerobically at 35–37°C for 48 h. *B. cereus* strains produce large (4–7 mm in diameter), flat, matt colonies with a greenish coloration. Isolates are usually α-hemolytic, but some strains produce β-hemolysis. This method has proved successful for performing colony counts on fecal specimens and grossly contaminated foodstuffs.

Donovan (1958) described a peptone–beef extract–egg yolk agar containing lithium chloride and polymyxin B as selective agents for isolating *B. cereus* from milk. She considered that counts of *B. cereus* in raw milk were impossible without using a selective medium. Mossel, Koopman, and Jongerius (1967) used mannitol and phenol red in an egg yolk–polymyxin medium in order to improve differentiation of isolates from foodstuffs.

Selective Isolation of *Bacillus cereus* (Mossel, Koopman, and Jongerius, 1967)

Basal medium contains:

Meat extract	1 g
Peptone	10 g
D-Mannitol	10 g
Sodium chloride	10 g
Phenol red	25 mg
Agar	15 g
Water	900 ml

pH 7.1

Autoclave for 15 min at 15 psi and cool to 55°C. To 90 ml of medium, add 10 ml of sterile, 20% egg yolk emulsion (Oxoid). To separate 100-ml quantities of the medium so produced, add 1-, 2-, 5-, and 10-ml amounts of polymyxin B sulfate solution (50 mg/50 ml water) to give final concentrations of 10–100 μg/ml. Inoculate plates by

spreading 0.1 ml of 10^{-1}–10^{-4} dilutions of sample and incubate at 32–37°C for 18–40 h. *B. cereus* colonies are rough and dry with a distinct violet-red background and a halo of white precipitate.

Kim and Goepfert (1971b) described an egg yolk-polymyxin medium, which contains low concentrations of peptone and yeast extract but 1.8% agar, in order to enhance sporulation for serological studies.

Direct Isolation of Opportunistically Pathogenic *Bacillus* Species

Most *Bacillus* spp. implicated in human and animal infections have been isolated from blood culture or blood agar plates incubated at 35–37°C for 24 h. For the isolation of such organisms from blood or C.S.F., selective or enrichment culture is neither necessary nor desirable.

Isolation of *Bacillus anthracis*

Post-mortem examination of animals should not be made since it is vital to prevent the distribution of sporulating bacilli. Blood, taken from a superficial vein, usually in the ear (but see Whitford, 1978), is used for the preparation of smears and of cultures.

If *B. anthracis* is to be isolated from other body tissues, extreme care must be taken to prevent the production of aerosols during the collection of specimens; subsequent laboratory manipulations should be carried out in a safety cabinet. Spleen substance, or other tissue, may be ground in a Griffith's tube or macerator.

In human cutaneous anthrax, swabs of serous fluid from an unbroken vesicle may be taken. In pulmonary anthrax, sputum is collected for smears and culture. For gastrointestinal anthrax, fecal and food specimens are collected. In all three cases blood culture should be performed, especially if systemic symptoms are apparent (Collins and Lyne, 1976; Feeley and Brachman, 1974; Green, 1975).

Isolation of *Bacillus anthracis* from Pathological Material (Feeley and Brachman, 1974)

Inoculate on 5% blood agar, plates prepared from defibrinated or citrated sheep, rabbit, or human blood which is free from antibiotics. Incubate at 35–37°C overnight. Examine under magnification for nonhemolytic, off-white colonies, 4–5 mm in diameter with irregular margin and a ground-glass appearance. Colonies have a characteristic tenacity when disturbed with an inoculating needle.

Isolation of *Bacillus anthracis* from Hairs, Hides, Feedingstuffs, and Fertilizers (Collins and Lyne, 1976)

A sample of approximately 25 g is placed in a 200- to 300-ml screw-capped jar and covered with about 100 ml of warm, quarter-strength Ringer's solution or sterile water. The mixture is shaken and allowed to stand at 37°C for 2 h, or at room temperature for 3 h. The supernatant fluid is decanted and heated in a water bath at 70°C for 10 min; 0.1-, 1.0-, and 2-ml aliquots of the uncentrifuged fluid and the deposit after centrifugation are placed in Petri dishes. To each dish, 0.5 ml of 5% egg albumen or 0.25 ml of 0.01% dibromopropamidine isothionate is added, and 15 ml of yeast extract agar at 50°C is poured on. After thorough mixing, the plates are allowed to set and then incubated overnight at 37°C. Lysozyme in the egg albumen and the dibromopropamidine compound reduce the numbers of contaminating organisms.

Examine under a low-power, binocular (plate) microscope. Ignore surface colonies and look for deep colonies that resemble dahlia tubers or Chinese artichokes. Pick onto nutrient agar plates.

This method is also suitable for isolations from soil.

Direct isolation by animal inoculation may be used when *B. anthracis* is not isolated in culture (Sterne, 1959) but blood enrichment culture is simpler, safer, cheaper, and faster and probably as sensitive.

Direct Isolation of *Bacillus anthracis* by Animal Inoculation (Green, 1975)

Centrifuge 50 ml (or more) of the heat-treated wash (as described in the procedure of Collins and Lyne [1976] above) at 3,000 rpm for 15 min. Discard the supernatant and inoculate the residue intramuscularly into a guinea pig which has been protected 24 h previously with *Clostridium welchii* antitoxin 1,000 U, *C. septicum* antitoxin 500 U, *C. oedematiens* antitoxin 1,000 U, and tetanus antitoxin 1,000 U. For a mouse, one-third of the guinea pig antitoxin dose is used. Death due to anthrax occurs in 2–3 days. The postmortem examination should be carried out with care to avoid spread of spores. The examination should be carried out with the animal laid on its back and pegged out on aluminum foil which can be wrapped round the cadaver at the end of the examination and the whole package incinerated. Characteristic appearances are: gelatinous edema at the site of inoculation, petechial hemorrhages spread widely over the peritoneal surface, and slow clotting of the black blood when shed.

Several selective media have been described (Gillissen and Scholz, 1961; Green and Jamieson, 1958; Knisely, 1966; Morris, 1955; Pearce and Powell, 1951); these are useful when working with very heavily contaminated material. As they may partially or totally inhibit some strains of *B. anthracis*, they should be used in conjunction with, and not instead of, nonselective media.

Selective Isolation Media for *Bacillus anthracis* (Knisely, 1966)

PLET medium contains the following ingredients added to heart infusion agar (HIA):

Polymyxin	30 U/ml
Lysozyme	40 μg/ml
Disodium ethylenediaminetetraacetate (EDTA)	300 μg/ml
Thallous acetate	4 μg/ml
Final unadjusted pH is 7.35.	

Spread plates are prepared and incubated at 37°C for 24–48 h. Colonies of *B. anthracis* are smaller and smoother on the selective medium than on plain HIA. *B. cereus* colonies are generally inhibited by this medium.

Enrichment culture may be useful when making isolations from animals, such as pigs, which often show no terminal bacteremia (Sterne, 1959), and may be more sensitive than animal inoculation (Thomson, 1955).

Enrichment Culture for *Bacillus anthracis* (Thomson, 1955)

Approximately 2 ml of fresh, defibrinated bovine blood is added to the glass tube in which the swab is contained. The swab is replaced and the whole incubated at 37°C overnight. Positive results have been obtained with as little as 3.5 h incubation.

Identification

Identification may be regarded as one of three components of taxonomy, the others being classification and nomenclature (Cowan, 1974). Thus, the taxonomy of *Bacillus* is briefly discussed at the beginning of this section.

The genus was first described by Cohn (1872) and an incomplete, rather unsatisfactory, classification was produced by Laubach, Rice, and Ford (1916). Since then there have been four major schemes for these organisms. In 1946, Smith, Gordon, and Clark produced a monograph dealing with 621 strains. This work was succeeded by an enlarged version dealing with 1,134 strains (Smith, Gordon, and Clark, 1952), which has now been replaced by a revised version (Gordon, Haynes, and

Pang, 1973). The scheme described by Gibson and Gordon in *Bergey's Manual* (Buchanan and Gibbons, 1974) is similar in many respects. Notable differences are the allocation of species rank to *B. anthracis* and *B. thuringiensis* and the inclusion of 26 species of arguable status in a section separate from the section that contains the bulk of the strains recognized by Gordon, Haynes, and Pang (1973). The arrangements of Prévot (1961) and Krasil'nikov (1949) differ substantially from that of Gordon and her colleagues (1973). The former divides the strains occurring in Gordon's genus *Bacillus* into four genera: *Bacillus, Bacteridium, Inominatus,* and *Clostridium.* Krasil'nikov (1949) lays emphasis on the swelling of the rod by the endospore, a character about which there is so much conflict of interpretation that it is difficult to equate his nomenclature with that of other systems (Skerman, 1967).

Such conflict is not surprising since it is now well recognized that many diagnostic tests, even biochemical ones, are not reproducible over a period of time within a laboratory, let alone in different ones. Furthermore, strains generally acceptable as members of this genus may not exhibit even all the "important characters" (Cowan, 1974) of *Bacillus,* let alone characteristics of the species with which they can most nearly be identified. For example, some strains widely acceptable as members of this endospore-forming genus have not been observed to sporulate. In general terms, the underlying technical, genetic, and biochemical reasons for discrepancies and conflicts between laboratories are well understood. A theoretical solution to many of the problems (excluding that of genotypic variation) is the rigorous standardization of the media, conditions of growth, test methods, and reagents used for diagnostic work. In practice, however, availability of materials and different conditions in laboratories in different countries make this standardization difficult to achieve worldwide. In order to be reasonably certain that a newly isolated strain is properly identified with a particular species, it is essential, as Gordon has repeatedly emphasized, that a number of known strains should be tested in parallel with the new isolate. A necessary prerequisite for such testing is a suitable stock of strains. This brings to mind the adage that a bacteriologist is as good as his or her culture collection.

Type strains for the various species must be defined. From 1980 onwards, the nomenclature of this genus will be based on the list of approved names drawn up by the International Committee for Systematic Bacteriology Sub-Committee on the Taxonomy of the Genus *Bacillus.* The list is given in Table 1, which also contains the sources of the type strains for each of the approved species.

In readily sporulating strains culture maintenance presents little difficulty since spores remain viable for long periods and have presumably evolved so as to conserve the characters of their vegetative precursors. Thus, yearly transfer to fresh medium is more than adequate for the preservation of such types. Some strains sporulate less well. In a massive study, Gordon and Rynearson (1963) demonstrated the value of soil extract agar for obtaining sporulating cultures of *Bacillus* and of sterile soil as a storage medium for such cultures.

Soil Extract Agar (Gordon and Rynearson, 1963)

To 400 g of air-dried, sieved garden soil rich in organic matter in a 2-liter conical flask, was added 960 ml of tap water. The flask was then autoclaved at 121°C for 60 min and allowed to cool in the unopened autoclave overnight. The extract was then filtered through paper and the filtrate reautoclaved at 121°C for 20 min and allowed to stand at room temperature for at least 2 weeks. The clear supernatant was used in making the soil extract agar.

Peptone	5 g
Beef extract	3 g
Agar (Difco)	15 g
Tap water	750 ml
Soil extract	250 ml
pH 7.0	

With new batches of soil, small lots of medium that contained varying (between 10 and 100% vol/vol) amounts of soil extract were tried. The concentration that produced the best sporulation of *B. subtilis* after 3 days was used for the bulk of the medium.

Preservation of *Bacillus* Cultures in Sterile Soil (Gordon and Rynearson, 1963)

Two parts by weight of air-dried soil and one part of air-dried humus were mixed and sifted through a coarse sieve; 10% (wt/wt) of $CaCO_3$ was added and the mixture poured into test tubes to a depth of approximately 2 cm. The test tubes were plugged with cotton wool and placed in an autoclave in a slanting position to spread the soil in as thin a layer as possible. The soil was sterilized at 121°C for 1 h on 3 successive days. A sporulating culture was suspended in approximately 1 ml of sterile water; the suspension was carefully mixed with the sterile soil and smeared on the sides of the tube. When the soil was dry and no longer adhered to the sides of the tube, the tube was closed with a rubber stopper and stored at room or refrigerator temperature. The culture was revived by suspending a few particles of the soil in nutrient broth.

Nutrient agar to which manganese sulfate has been added is also a good medium on which to obtain sporulation and hence is useful for maintenance. Some workers prefer this medium as being more

Table 1. Sources of type cultures of species approved by the ICSB Genus *Bacillus* Subcommittee.

Bacillus	Type[a]	Culture collection[b] and number			
		ATCC	DSM	NCIB	NCTC
acidocaldarius	T	27009	446		
alcalophilus	NT	27647	485	10436	4553
alvei	NT	6344	29	9371	6392
anthracis	NT	14578		9388	10340
badius	T	14574	23	9364	10333
brevis	NT	8246	30	9372	2611
cereus	NT	14579	31	9373	2599
circulans	NT	4513	11	9374	2610
coagulans	T	7050	1	9365	10334
fastidiosus	T	(29604)[c]	91		
firmus	T	14575	12	9366	10335
globisporus	T	23301	4		
insolitus	T	23299	5		
larvae	NT	9545			
laterosporus	T	64	25	9367	6357
lentimorbus	NT	14707			
lentus	T	10840	9	8773	4824
licheniformis	NT	14580	13	9375	10341
macerans	T	8244	24	9368	6355
macquariensis	T	23464	2	9934	10419
megaterium	NT	14581	32	9376	10342
mycoides	NT	6462			
pantothenticus	T	14576	26	8775	8162
pasteurii	NT	11859	33	8841	4822
polymyxa	NT	842	36	8158	10383
popilliae	T	14706			
psychrophilus	T	23304	3		
pumilus	T	7061	27	9369	10337
sphaericus	T	14577	28	9370	10338
stearothermophilus	T	12980		8923	10339
subtilis	NT	6051	10	3610	3610
thuringiensis	NT	10792			

[a] T, type species; NT, neotype.
[b] ATCC, American Type Culture Collection; DSM, Deutsche Sammlung von Mikroorganismen; NCIB, National Collection of Industrial Bacteria; NCTC, National Collection of Type Cultures.
[c] Not in 1978 ATCC catalog.

convenient and readily reproducible than media based on soil extract.

Manganese-Containing Nutrient Agar to Obtain Sporulation

Peptone	5.0 g
Meat extract	3.0 g
Agar	15.0 g
$MnSO_4 \cdot H_2O$	0.005 g
Distilled water	1 liter

Cultures may also be preserved by freeze-drying, but alteration in characters occurs not infrequently when this method is used. Storage in liquid nitrogen is an alternative method that is finding increasing favor.

Before comparing a new isolate with stock strains it is necessary to make a tentative identification.

Those without experience of the genus should probably use a simplified key. Not all strains, however, will be typical and not all strains that were once typical will remain so. Careless use of such a key can lead to totally misleading conclusions. With the caveat then that, like a small foreign language dictionary in the hands of a new student, such a key may lead to embarrassing statements, a modified version of the key of Gordon, Haynes, and Pang (1973) is given below (Table 2). This table should enable the preliminary identification of the more typical strains of each species by the use of only 11 tests and the observation of two morphological features. The two modifications are made as a result of a study, organized by the International *Bacillus* Sub-Committee of the IAMS, of 18 strains by eight laboratories distributed around the world; the study showed that there were substantial interlaboratory

Table 2. Simplified key for the tentative identification of typical strains of *Bacillus* species.[a]

1. Catalase: positive . . . 2
    negative . . . 17
2. Voges-Proskauer: positive . . . 3
    negative . . . 10
3. Growth in anaerobic agar: positive . . . 4
    negative . . . 9
4. Growth at 50°C: positive . . . 5
    negative . . . 6
5. Growth in 7% NaCl: positive . . . . *B. licheniformis*
    negative . . . . *B. coagulans*
6. Acid and gas from glucose (inorganic N): positive . . . . . . *B. polymyxa*
    negative . . . 7
7. Reduction of NO₂ to NO₂: positive . . . 8
    negative . . . . *B. alvei*
8. Parasporal body in sporangium: positive . . . . *B. thuringiensis*
    negative . . . . *B. cereus*
9. Hydrolysis of starch: positive . . . . *B. subtilis*
    negative . . . . *B. pumilus*
10. Growth at 65°C: positive . . . . *B. stearothermophilus*
    negative . . . 11
11. Hydrolysis of starch: positive . . . 12
    negative . . . 15
12. Acid and gas from glucose (inorganic N): positive . . . . . . *B. macerans*
    negative . . . 13
13. Width of rod 1.0 μm or greater: positive . . . . *B. megaterium*
    negative . . . 14
14. pH in V-P broth < 6.0: positive . . . . *B. circulans*
    negative . . . . *B. firmus*
15. Growth in anaerobic agar: positive . . . . *B. laterosporus*
    negative . . . 16
16. Acid from glucose (inorganic N): positive . . . . *B. brevis*
    negative . . . . *B. sphaericus*
17. Growth at 65°C: positive . . . . *B. stearothermophilus*
    negative . . . 18
18. Decomposition of casein: positive . . . . *B. larvae*
    negative . . . 19
19. Parasporal body in sporangium: positive . . . . *B. popilliae*
    negative . . . . *B. lentimorbus*

[a] Modified from Gordon, Haynes, and Pang, 1973. Numbers on the right indicate the number (on the left) of the next test to be applied until the right-hand number is replaced by a species name.

disagreements in the interpretation of the tests involving observation of growth at pH 5.7 and citrate utilization. In contrast, there were no discrepancies in the results of the nitrate reduction test and only a very few instances of disagreement as to whether rods were 1.0 μm or greater in diameter. The media and methods used for such an examination are described below.

The tentative identification may be checked in Table 3 for all the characters used in the key.

Media for Preliminary Identification of Aerobic Sporeformers

The following media are from Gordon, Haynes, and Pang, 1973.

Nutrient Broth

Beef extract	3 g
Peptone	5 g
Distilled water	1,000 ml
pH 6.8	

Distribute in test tubes and sterilize by autoclaving at 121°C for 20 min.

Nutrient Agar

Beef extract	3 g
Peptone	5 g
Agar	15 g
Distilled water	1,000 ml
pH 6.8	

Table 3. Summary of the characters used in the simplified key for *Bacillus* species.[a]

	Catalase	V-P reaction	Growth in anaerobic agar	Growth at 50°C	Growth in 7% NaCl	Acid and gas in glucose	NO₃ reduced to NO₂	Starch hydrolyzed	Growth at 65°C	Rods 1.0 μm wide or wider	pH in V-P medium < 6.0	Acid from glucose	Hydrolysis of casein	Parasporal bodies
B. megaterium	+	−	−	−	+	−	0	+	−	+	0	+	+	−
B. cereus	+	+	+	−	+	−	+	+	−	+	+	+	+	0
B. thuringiensis	+	+	+	−	+	−	+	+	−	+	+	+	+	+
B. licheniformis	+	+	+	+	+	−	+	+	−	−	0	+	+	−
B. subtilis	+	+	−	+	+	−	+	+	−	−	0	+	+	−
B. pumilus	+	+	−	+	+	−	−	−	−	−	+	+	+	−
B. firmus	+	−	−	−	+	−	+	+	−	−	−	+	+	−
B. coagulans	+	+	+	+	−	−	0	+	−	0	+	+	0	−
B. polymyxa	+	+	+	−	−	+	+	+	−	−	0	+	+	−
B. macerans	+	−	+	+	−	+	+	+	−	−	−	+	−	−
B. circulans	+	−	0	+	0	−	0	+	−	−	0	+	0	−
B. stearothermophilus	0	−	−	+	−	−	0	+	+	0	+	+	0	−
B. alvei	+	+	+	−	−	−	−	+	−	0	+	+	+	−
B. laterosporus	+	−	+	+	−	−	+	−	−	−	−	+	+	+
B. brevis	+	−	−	+	−	−	0	−	−	−	−	+	+	−
B. larvae	−	−	+	−	+[b]	−	0	−	−	−	−	+	−	−
B. popilliae	−	−	+	−	+[b]	−	−	−	−	−	−	+	−	+
B. lentimorbus	−	−	+	−	−	−	−	−	−	−	−	+	−	−
B. sphaericus	+	−	−	−	0	−	−	−	−	0	−	−	0	−

[a] +, Greater than 85% of strains examined by Gordon, Haynes, and Pang (1973) positive;
  −, greater than 85% of strains negative; 0, variable character.
[b] Growth in 2% NaCl agar.

Dissolve by steaming and mix well before distributing into final containers. Sterilize by autoclaving at 121°C for 20 min. For sporulation, 5 mg hydrous manganese sulfate should be added per 1,000 ml of medium.

Glucose Agar

Glucose, anhydrous D(+)	10 g
Nutrient agar	1 liter

Add the glucose to the nutrient agar and, after thorough mixing, dispense into test tubes or flasks and sterilize by autoclaving at 115°C for 20 min.

Voges-Proskauer Broth

Proteose-peptone (Difco)	7 g
Glucose	5 g
Sodium chloride	5 g
Distilled water	1,000 ml
pH 6.5	

Distribute in 5-ml amounts in 20-mm test tubes and sterilize by autoclaving at 115°C for 20 min.

J-Broth for Voges-Proskauer Reaction

Tryptone	5 g
Yeast extract	15 g
Distilled water	1 liter

Mix thoroughly until components are dissolved and adjust the pH to 7.3–7.5. Distribute as required and sterilize by autoclaving at 121°C for 20 min.

Add aseptically 2 g/liter of glucose, sterilized separately by autoclaving a 10% solution in distilled water at 115°C for 20 min. For semisolid J-agar, use only 10 g of agar. For basal J-agar, omit glucose.

Starch Agar

One gram of potato starch is suspended in 10 ml of cold distilled water and mixed with 100 ml of nutrient agar or basal J-agar for the fastidious insect pathogens. Autoclave at 121°C for 20 min. Cool to 45°C, thoroughly mix, and pour into Petri dishes.

J-Agar

Tryptone	5 g
Yeast extract	15 g
Dipotassium hydrogen phosphate	3 g
Agar	20 g
Distilled water	1 liter

Thoroughly mix the solids and water, steam until all are dissolved, and filter through a clarifying grade of filter paper. Adjust the pH to 7.3–7.5, distribute as required, and sterilize by autoclaving at 121°C for 20 min.

Add aseptically 2 g/liter of glucose, sterilized separately by autoclaving a 10% solution in distilled water at 115°C for 20 min. For semisolid J-agar, use only 10 g of agar. For basal J-agar, omit glucose.

Chocolate Agar

Melt nutrient agar by autoclaving at 115°C for 10 min. Cool to 60°C and add aseptically 10% by volume of sterile horse blood. (It is important to use horse blood that contains no added preservative.) Immerse the vessel containing the medium in boiling water for 1 min with constant mixing, and immediately dispense the medium as slopes or into Petri dishes as required.

Anaerobic Agar

Trypticase	20 g
Sodium chloride	5 g
Agar	15 g
Sodium thioglycolate	2 g
Sodium formaldehyde sulfoxylate	1 g
Distilled water	1 liter

Adjust the pH to 7.2. Distribute into one-half-inch (1.2-cm) test tubes in amounts sufficient to give a 3-inch (7.6-cm) depth of medium and sterilize by autoclaving at 121°C for 20 min.

For cultures of *B. larvae, B. popilliae,* and *B. lentimorbus,* the agar should be supplemented with 15 g of yeast extract/liter.

Medium for Acid and Gas Production from Glucose

Basal medium:

Diammonium hydrogen phosphate	1 g
Potassium chloride	0.2 g
Magnesium sulfate	0.2 g
Yeast extract	0.2 g
Distilled water	1,000 ml

Insert an inverted Durham tube and adjust the pH of the medium to 7.0 before adding 15 ml of a 0.04% (wt/vol) solution of bromocresol purple. Sterilize by autoclaving at 121°C for 20 min.

Glucose solutions:
Sterilize 10% (wt/vol) aqueous solution by autoclaving at 121°C for 20 min. Then add aseptically to tubes of sterile basal medium to obtain a final concentration of 0.5%.

For the fastidious insect pathogens, use J-broth containing 0.5% glucose.

Nitrate Broth

Peptone	5 g
Beef extract	3 g
Potassium nitrate	1 g
Distilled water	1,000 ml
pH 7.0	

Distribute the medium into test tubes containing inverted Durham tubes, and sterilize by autoclaving at 121°C for 20 min.

For the fastidious insect pathogens, use J-broth supplemented with 0.1% (wt/vol) potassium nitrate and with the glucose omitted.

A semisolid medium containing: peptone, 0.5%; sodium chloride, 0.5%; and potassium nitrate, 0.02% (wt/vol) is useful for marine isolates and is often the only medium on which all strains under test grow well.

Milk Agar

Skim milk powder	5 g in 50 ml distilled water
Agar	1 g in 50 ml distilled water

Autoclave separately at 121°C for 20 min, cool to 45°C, mix together, and pour into Petri dishes.

For the fastidious insect pathogens, supplement the agar part of J-agar with double concentrations of the normal ingredients exclusive of glucose. Then mix with skim milk suspension as above. Allow plates to stand at room temperature for 3 days to dry the surface of the agar.

Methods for Preliminary Identification of Aerobic Sporeformers

General note: Cultures should be incubated at temperatures approximately 10–15°C below their maximum growth temperatures. Thus, cultures of psychrophiles should be incubated at 20°C, mesophiles at 30°C, and thermophiles whose maximum growth temperature is 55–60°C should be incubated at 45°C. Strains capable of growth at 65°C should be incubated at 45°C and at 55°C.

Microscopic Appearance

A. General morphology.

Prepare smears of young (18–24 h) cultures grown on nutrient agar or J-agar (for *Bacillus larvae, B. popilliae,* and *B. lentimorbus*). Air-dry (do not heat fix). Stain for 30 s with safranin (safranin, 0.25 g; 95% ethanol, 10 ml; distilled water, 100 ml).

Longer incubation may be required for observation of spores and exosporia. Observe for size and shape of cells, presence of shadow-forms, chains, size and shape of mature spores, and their position in the sporangia.

B. Vacuolate cytoplasm.

As under (A) but growing cultures on glucose agar. Observe for foamy or vacuolate appearance of the protoplasm.

C. Parasporal bodies.

Observe cells grown for 3–7 days on nutrient agar for parasporal bodies by phase-contrast microscopy.

Note: *B. larvae*, *B. popilliae*, and *B. lentimorbus* do not sporulate on nutrient agar and may not sporulate satisfactorily at all on media in vitro.

Production of Catalase

Flood cultures grown for 1 or 2 days on slopes of nutrient agar with 0.5 ml of 10% hydrogen peroxide.

Observe macroscopically for gas production. If no gas bubbles form, repeat using cultures grown on chocolate agar.

Cultures of *B. larvae*, *B. popilliae*, and *B. lentimorbus* are conveniently handled by growing colonies on plates of J-agar and flooding colonies or the edges of confluent growth with 10% hydrogen peroxide. Cultures should be tested as soon as clearly visible growth is present.

Note: Weak formation of gas bubbles is better observed using about 3 ml of hydrogen peroxide.

Growth in 7% Sodium Chloride

Inoculate tubes of nutrient broth (3 ml/tube) containing 7% (wt/vol) sodium chloride with a small loopful of a culture grown in nutrient broth and incubate in a sloping position to improve aeration.

For the fastidious insect pathogens, use semisolid J-agar as the basal medium.

Observe for growth at 7 and 14 days.

Acid and Gas from Carbohydrates

Tubes of glucose medium are inoculated, making sure the inverted Durham tube is full of medium, and incubated at appropriate temperatures. Observe for growth and for production of acid and gas (gas bubbles in the Durham tube) at 7 and 14 days.

The fastidious insect pathogens are tested by aseptically removing a drop of culture to a spot plate, mixing with a drop of 0.04% (wt/vol) alcoholic bromocresol purple, and observing the color of the indicator.

Reduction of Nitrate to Nitrite

Grow cultures in nitrate broth or nitrate-supplemented J-broth for the fastidious insect pathogens. Test after 3 and 7 days incubation by moistening a strip of potassium iodide/starch paper with a few drops of 1 N hydrochloric acid and then touching the paper with a loopful of the culture.

Observe (i) for the production of a purple color, indicating the presence of nitrite, and (ii) for the accumulation of nitrogen gas in the Durham tube. Cultures negative at 7 days are tested after 14 days by mixing 1 ml of the culture with 3 drops of each of the following solutions:

1. Sulphanilic acid, 8 g; 5 N acetic acid (glacial acetic acid and water 1:2.5), 1,000 ml.
2. Dimethyl-α-naphthylamine, 6 ml; 5 N acetic acid, 1,000 ml.

Observe for the development of a red or yellow color, indicating the presence of nitrite.

If the culture is still negative after 14 days, add 4–5 ml of zinc dust to the tube previously tested for nitrite.

Observe for the development of a red color, indicating the presence of nitrate, i.e., the absence of reduction.

Note: The latter procedure is to ensure that very rapid reduction has not occurred, reducing nitrate beyond nitrite in less than 3 days.

Anaerobic Growth

Inoculate a tube of anaerobic agar with a small (outside diameter 1.5 mm) loopful of nutrient broth culture by stabbing to the bottom of the culture tube.

Alternatively, molten medium cooled to about 40°C may be inoculated thoroughly using a Pasteur pipette and allowed to solidify before incubation.

Observe for growth on the surface of the agar (aerobic) and along the length of the stab (anaerobic). At incubation temperatures below 45°C the growth should be recorded at 3 and 7 days; at temperatures of 45°C or higher, growth should be recorded at 1 and 3 days. Cultures of the three species of fastidious insect pathogens may require as long as 14 days incubation before growth becomes apparent.

Voges-Proskauer Reaction

A. Acetylmethylcarbinol production.

Inoculate tubes of Voges-Proskauer broth in triplicate and test for acetylmethylcarbinol production after incubation for 3, 5, and 7 days by mixing 3 ml of 40% (wt/vol) sodium hydroxide with the culture and adding 0.5–1 mg of creatine.

Observe for the production of a red color after 30–60 min at room temperature.

The fastidious insect pathogens should be grown on J-broth for Voges-Proskauer reaction for this test.

B. Final pH produced in Voges-Proskauer broth. The pH is measured, preferably using a pH meter, before cultures incubated for 7 days are tested for acetylmethylcarbinol.

Note: The pH of Voges-Proskauer broth is usually 6.5, and that of J-broth for Voges-Proskauer reaction, 6.8.

Growth at 50°C and 65°C

Prepare slopes of nutrient agar or, for the fastidious insect pathogens, tubes of semisolid J-agar. Determine ability to grow at 5°C intervals. Immerse tubes containing the medium in water baths at the appropriate temperatures until equilibrated and then inoculate.

Observe growth of cultures after 3 days at 65°C and after 5 days at 50°C.

Care should be taken to ensure that the water levels in the baths are carefully maintained and that the temperatures are stable, with a variation not greater than ± 0.5°C.

Decomposition of Casein

Inoculate plates of milk agar with one streak of inoculum and examine after incubation at 7 and 14 days. The more slowly growing insect pathogens should also be examined after 21 days.

Observe for clearing of the casein around and underneath the growth.

Hydrolysis of Starch

Inoculate duplicate plates of starch agar with each culture to be tested and incubate at appropriate temperatures. At 3 and at 5 days, flood the plates with 95% ethanol. After 15–30 min, the unchanged starch will become white and opaque.

Observe for a clear zone underneath (after the growth is scraped off) and around the growth as an indicator of hydrolysis of starch.

For the fastidious insect pathogens, the plates are flooded by Gram's iodine after 5 and 10 days of incubation.

Note: Some cultures spread rapidly on this medium and examination before 3 days may be necessary.

These tests used in conjunction with Tables 1 and 2 should enable a preliminary identification of a new isolate to be made. If doubt still remains, further information can be had from the excellent monograph of Gordon, Haynes, and Pang (1973), which deals not only with the typical strains listed here but also with a wider variety of organisms. If in spite of all efforts it is impossible to identify a new isolate with a recognized species, particularly if the isolate is from, for example, a saline or low-temperature environment (and it is highly likely to be the case because insufficient strains from such sources have yet been examined), the authors would urge that any temptation to describe a new species be resisted until as many similar strains as possible have been isolated from a variety of situations and studied in a comparative way.

Another approach to the identification of *Bacillus* strains that is being currently developed is based on the API system. Use of the highly standardized, commercially available materials eliminates the problems of interlaboratory variation in media, and so improves the reproducibility of diagnostic tests. Sufficient data on which to base an identification scheme have, however, not yet been accumulated; only about 600 strains have been examined so far and many types that are less well known are underrepresented in the study. Progress to date has been reviewed by Logan and Berkeley (1981).

THE GENUS *SPOROLACTOBACILLUS*

It has been recognized for many years that the borderline between *Lactobacillus* and *Bacillus* is difficult to draw. Numerous "aberrant" types of bacteria that share characteristics of both genera have been described (Davis, 1964; Gemmell and Hodgkiss, 1964; Kitahara and Suzuki, 1963; Nakayama and Yanoshi, 1967a,b; Nonomura, Yamazaki, and Ohara, 1965).

Lactobacillus is defined as being Gram positive, nonmotile, nonsporulating, catalase negative, and microaerophilic. The rod-shaped bacteria produce lactic acid from glucose either by homo- or heterofermentative pathways. The production of large amounts of lactic acid as a fermentation product is unusual in the genus *Bacillus*, although *B. coagulans* does carry out a typical homo-lactic fermentation. The intermediate organisms between the two genera produce lactic acid and vary with regard to possession of other characteristics of the two genera.

Nakayama studied a range of intermediate forms that formed spores in sugar-deficient media, were motile, were catalase positive, and produced L-(+)-lactic acid; he classified them with *B. coagulans* (Suzuki and Kitahara, 1964). Bacteria that closely resemble the homo-fermentative *Lactobacillus* except for motility by peritrichous flagella were described by Harrison and Hansen (1950) and by Deibel and Niven (1958). These organisms were catalase negative and did not produce spores.

Kitahara and Suzuki (1963) isolated from chicken feed a catalase-negative, sporeforming bacterium which was motile by peritrichous flagella, grew microaerophilically, and showed a typical homofermentative metabolism, producing D-(−)-lactic acid. They considered their organism to belong within the Lactobacillaceae and created a new subgenus for it, calling the organism *Sporolactobacillus inulinus* on account of its ability to ferment inulin. A further seven strains of catalase negative, sporebearing, lactic acid-producing bacteria were isolated by Nakayama and Yanoshi (1967b) from the rhizosphere of wild plants. These organisms were motile and produced DL-lactic acid. Together with later isolates that differed in the type of lactic acid produced, these organisms were considered to be further isolates of *Sporolactobacillus,* and Nakayama named his strains *Sporolactobacillus laevus* and *Sporolactobacillus racemicus* (Uchida and Mogi, 1973).

Habitats

Sporolactobacillus inulinus was isolated from samples of chicken feed. The other species came from the rhizosphere region of several wild plants. Nakayama and Yanoshi (1967a) have pointed out that lactic acid bacteria are found mainly in milk products, pickles, fermented mashes, and similar materials, but suggest that these are features of the environment associated with human activities and are unlikely to be the primary natural habitats. They argue that soil and, more specifically, the rhizosphere will provide a habitat with the nutrients necessary for growth of lactic acid bacteria and that this situation with its localized nutrient concentrations and tendency to undergo drying and heating by sunlight will favor motile, sporeforming, lactic acid bacteria.

Isolation

Sporolactobacillus grows well on media of the type used for *Lactobacillus.* The following medium, used by Kitahara and Suzuki (1963), is satisfactory both for isolation and for the maintenance of cultures.

Glucose-Yeast Extract-Peptone Medium (GYP) for Isolation of *Sporolactobacillus* (Kitahara and Suzuki, 1963)

Glucose	2%
Yeast extract (Difco)	0.5%
Peptone	0.5%

GYP agar is made by adding 1% of agar to this medium and a semisolid agar by adding 0.25%.

For isolation purposes, the addition of a small amount of calcium carbonate to GYP agar is helpful since the small colonies of *Sporolactobacillus* are then surrounded by clear haloes which serve as indicators for selection. Cultures are incubated at 30 or 37°C.

One of the major hazards in isolating *Sporolactobacillus* is the ease with which the small colonies are overgrown by aerobic bacteria. This contamination can be minimized by the following technique.

Isolation of *Sporolactobacillus*

Suspend the source material in distilled water and pasteurize at 80°C for 20 min. Inoculate an aliquot into ten times its volume of fluid GYP medium in a deep test tube, and incubate anaerobically at 30°C or 37°C. After 4 days incubation, test the pH of the cultures and streak any with a pH of less than 4 onto GYP agar that contains calcium carbonate. Cover the surface of the plates with polyvinylidenechloride film (sterilized by autoclaving between filter papers), and incubate the plates aerobically. Select pinpoint colonies surrounded by clear zones for subsequent purification and examination.

Identification

Sporolactobacillus is not a difficult organism to grow in the vegetative state. Greyish white, pinpoint colonies with a diameter of less than 1 mm are formed on GYP agar; in the presence of calcium carbonate, the colonies are more distinct because they are surrounded by a transparent halo formed by the action of lactic acid. In semisolid agar, larger colonies with a diameter of up to 3 mm appear in stab or shake cultures. Small colonies are formed uniformly in the medium except for the region near the surface. The growth is typically microaerophilic.

Sporolactobacillus is Gram positive, catalase negative, does not reduce nitrate to nitrite or form indole. Lactic acid is produced without liberation of gas from glucose and from a range of sugars and sugar alcohols. Cells are straight, Gram-positive rods measuring roughly $0.8 \ \mu m \times 3{-}5 \ \mu m$. They differ from typical lactobacilli only in being motile with peritrichous flagella, which vary from many to few or even one in some strains.

Cultures do not sporulate freely. Kitahara and Lai (1967) obtained improved sporulation of *S. inulinus,* up to 1% of the vegetative cell population, on a medium containing yeast extract, meat extract, α-methyl-glucose, calcium carbonate, and a source of manganese ion; they obtained further improvement to 10% by incubating in a carbon dioxide atmosphere. Nakayama and Yanoshi (1967b) found starch to stimulate sporulation in their soil isolates and glucose to be effective in some strains. In all species the spores are ellipsoidal, terminal or subterminal, and the sporangia are distinctly swollen. Heat resistance is of the order of 80°C for 10 min, and dipicolinic acid has been shown to be present in the spores of *S. inulinus* to a concentration of 5% (Kitahara and Lai, 1967).

Table 4. The significant characters of *Sporolactobacillus* compared with those of *Bacillus* and *Lactobacillus*.

Taxonomic characteristic	Typical *Bacillus*	*B. coagulans*	*B. racemilacticus, B. myxolactis, B. laevolacticus, B. dextrolacticus*	*Sporolactobacillus*	*Lactobacillus yamanashiensis*	*L. plantarum*	Typical *Lactobacillus*
Lactic acid fermentation	−	+	+	+	+	+	+
Nitrate reduction	+	±	−	−	−	+	−
Catalase production	+	+	+	−	−	−	−
Spore formation	+	+	+	+	−	−	−
Motility	+	+	+	+	+	−	−
Diaminopimelic acid in cell wall	+	+	+	+	+	+	−
G+C in DNA (mol%)	33–50	45.4 46.9		47.3 39.3		44.5 42.9	33–54
Type of fatty acid spectrum[a]	B	B	B	B	L	L	L

[a] B, *Bacillus* type; L, *Lactobacillus* type.

It is possible to arrange the intermediate forms between typical *Bacillus* and typical *Lactobacillus* to form a spectrum of types that connects the motile, catalase-positive, sporeforming, non-lactic acid-forming *Bacillus* to the homo-lactic acid-producing, catalase-negative, nonsporeforming, nonmotile *Lactobacillus*. There are basic differences between the cellular fatty acid spectra of *Bacillus* and *Lactobacillus* (Uchida and Mogi, 1973). *Bacillus* cells contain predominantly saturated fatty acids with odd numbers of carbon atoms, iso- and anteiso-branched C_{15} and C_{17}, and small amounts of iso-C_{16} are also present. Lactobacilli, by contrast, contain even-numbered, saturated, straight-chain acids (C_{16} predominates), even-numbered, straight-chain, unsaturated acids (C_{16} and C_{18} predominate), and C_{17} and C_{19} (lactobacillic)-cyclopropane acids. *Sporolactobacillus* shows the *Bacillus* pattern of fatty acids and also shows the *Bacillus* pattern of isoprenoid quinone cell components (Collins and Jones, 1979). The cell walls of *Sporolactobacillus*, like *Bacillus* but unlike typical *Lactobacillus*, contain diaminopimelic acid. The significant characteristics of these intermediate organisms are summarized in Table 4.

Both *Bacillus* and *Lactobacillus* are large genera whose G+C mol% values range from the low 30s to the mid 50s, casting serious doubt on their validity as taxa at the generic level. The borderline between them is certainly ill defined. When *S. inulinus* was first described by Kitahara and Suzuki (1963), the authors recognized the similarity to some motile *Lactobacillus* strains and proposed that their new isolate should be given the ranking of a subgenus in the Lactobacillaceae. The recognition of generic rank as *S. inulinus* is due to Kitahara and Lai (1967), and Kitahara and Toyota (1972) transferred the genus to Bacillaceae. The eighth edition of *Bergey's Manual* (Buchanan and Gibbons, 1974) agrees with this classification; it accords generic status to *Sporolactobacillus* within the Bacillaceae and records *S. inulinus* as the only species. The other species, *S. laevus* and *S. racemicus,* are as yet little known and await further study.

Literature Cited

Alderton, G., Snell, N. 1963. Base exchange and heat resistance in bacterial spores. Biochemical and Biophysical Research Communications **10:**139–143.

Alexander, H., Higginbottom, C. 1953. Bacteriological studies on pasteurised milk. Journal of Dairy Research **20:**156–176.

Allen, B. T., Wilkinson, H. A. 1969. A case of meningitis and generalized Schwartzman reaction caused by *Bacillus sphaericus*. Johns Hopkins Medical Journal **125:**8–13.

Allen, M. B. 1950. The dynamic nature of thermophily. Journal of General Physiology **33:**205–214.

Allen, M. B. 1953. The thermophilic aerobic sporeforming bacteria. Bacteriological Reviews **17:**125–173.

Aschehoug, V., Jansen, E. 1950. Studies on putrefactive anaerobes as spoilage agents in canned foods. Food Research **15:**62–67.

Aschehoug, V., Vesterhus, R. 1941. Microbiology of canned vegetables. Zentralblatt für Bakteriologie, Parasitenkunde, Infektionskrankheiten und Hygiene, Abt. 1 Orig. **104:**169–185.

Aubertin, E., Dangoumeau, A., Leuret, E., Piechaud, F. 1938. Recherches sur l'état de septicité de l'intérieur des pains, selon qu'ils ont été préparés avec de la levure ou avec du levain. Influence de la cuisson et du pH. Comptes Rendus des Séances de la Société de Biologie et de ses Filiales **127:**64–67.

Bairiev, Ch. B., Mamedov, S. M. 1963. A thermophilic bacterium isolated from ozokerite. Federation Proceedings Translation, Suppl. **22**:1224–1226.

Barnham, M., Taylor, A. J. 1977. A case of *Bacillus cereus* bacteraemia. Postgraduate Medical Journal **53**:397–399.

Bartholomew, J. W., Paik, G. 1966. Isolation and identification of obligate thermophilic sporeforming bacilli from ocean basin cores. Journal of Bacteriology **92**:635–638.

Bartholomew, J. W., Rittenberg, S. C. 1949. Thermophilic bacteria from deep ocean bottom cores. Journal of Bacteriology **57**:658.

Barton-Wright, E. 1943. The estimation of rope spores in wheaten flour and other products. Journal of the Society of Chemical Industry **62**:33–37.

Bausum, H. T., Matney, T. S. 1965. Boundary between bacterial mesophilism and thermophilism. Journal of Bacteriology **90**:50–53.

Behrend, M., Krouse, T. B. 1952. Postoperative bacterial synergistic cellulitis of abdominal wall: Fatality following herniorrhaphy. Journal of the American Medical Association **149**:1122–1124.

Bentler, W. 1961. Die Sterilmilch und ihre Bakteriologie—ein milchhygienisches Problem. Archiv für Lebensmittelhygiene **12**:12–15.

Bergamini, F., Berti, P. 1962. Changes of the organoleptic properties of milk subjected to processes of sterilization in relation to the presence of some species of *Bacillus* genus. XVI International Dairy Congress **A**:887–896.

Biancardi, G. 1963. Mastite acute bovina da *Bacillus cereus*. Segnalazione e studio di 8 casi. Archivio Veterinario Italiana **14**:31–46.

Billing, E., Cuthbert, W. A. 1958. 'Bitty cream': The occurrence and significance of *Bacillus cereus* spores in raw milk supplies. Journal of Applied Bacteriology **21**:65–78.

Black, L. A., Tanner, F. W. 1928. A study of thermophilic bacteria from the intestinal tract. Centralblatt für Bakteriologie, Parasitenkunde und Infektionskrankheiten, Abt. 2 **75**:360–375.

Boeyé, A., Aerts, M. 1976. Numerical taxonomy of *Bacillus* isolates from North Sea sediments. International Journal of Systematic Bacteriology **26**:427–441.

Bond, W. W., Favero, M. S. 1977. *Bacillus xerothermodurans* sp. nov., a species forming endospores extremely resistant to dry heat. International Journal of Systematic Bacteriology **27**:157–160.

Bond, W. W., Favero, M. S., Petersen, N. J., Marshall, J. H. 1970. Dry-heat inactivation kinetics of naturally occurring spore populations. Applied Microbiology **20**:573–578.

Bonde, G. J. 1966. Water problems in anaesthesiology. Acta Anoesthesiologica Scandinavica, Suppl. XXIII, Proceedings **1**:88–92.

Bonde, G. J. 1975. The genus *Bacillus*. Danish Medical Bulletin **22**:41–61.

Bonde, G. J. 1976. The marine *Bacillus*. [Abstract.] Journal of Applied Bacteriology **41**:vii.

Bonventre, P. F., Johnson, C. E. 1970. *Bacillus cereus* toxin, pp. 415–433. In: Montie, T. C., Kadis, S., Ajl, S. J. (eds.), Microbial toxins, vol. III. New York, London: Academic Press.

Boyd, W. L., Boyd, J. W. 1962. Viability of thermophiles and coliform bacteria in Arctic soils and water. Canadian Journal of Microbiology **8**:189–192.

Boyer, E. W., Ingle, M. B., Mercer, G. D. 1973. *Bacillus alcalophilus* subsp. *halodurans* subsp. nov.: An alkaline-amylase-producing, alkalophilic organism. International Journal of Systematic Bacteriology **23**:238–242.

Boyette, D. P., Rights, F. L. 1952. Heretofore undescribed aerobic sporeforming bacillus in child with meningitis. Journal of the American Medical Association **148**:1223–1224.

Brachman, P. S., Kaufmann, A. F., Dalldorf, F. G. 1966. Industrial inhalation anthrax. Bacteriological Reviews **30**:646–657.

Breed, R. S., Murray, G. D., Smith, N. R. (eds.). 1957. Bergey's manual of determinative bacteriology, 7th ed. Baltimore: Williams & Wilkins.

Brock, T. D., Darland, G. 1970. The limits of microbial existence: Temperature and pH. Science **169**:1316–1318.

Brown, A. D. 1964. Aspects of bacterial response to the ionic environment. Bacteriological Reviews **28**:296–329.

Buchanan, R. E., Gibbons, N. E. (eds.). 1974. Bergey's manual of determinative bacteriology, 8th ed. Baltimore: Williams & Wilkins.

Burdon, K. L., Davis, J. S., Wende, R. D. 1967. Experimental infection of mice with *Bacillus cereus*: Studies of pathogenesis and pathologic changes. Journal of Infectious Diseases **117**:307–316.

Burdon, K. L., Wende, R. D. 1960. On the differentiation of anthrax bacilli from *Bacillus cereus*. Journal of Infectious Diseases **107**:224–234.

Cameron, E. J., Esty, J. R. 1926. The examination of spoiled canned foods. 2. Classification of flat sour spoilage organisms from nonacid foods. Journal of Infectious Diseases **39**:89–105.

Campbell, L. L. 1954. The growth of an "obligate" thermophilic bacterium at 36°C. Journal of Bacteriology **68**:505–507.

Campbell, L. L., Pace, B. 1968. Physiology of growth at high temperatures. Journal of Applied Bacteriology **31**:24–35.

Campbell, L. L., Postgate, J. R. 1965. Classification of the spore-forming sulfate-reducing bacteria. Bacteriological Reviews **29**:359–363.

Chastain, C. B., Harris, D. L. 1974. Association of *Bacillus cereus* with food poisoning in dogs. Journal of the American Veterinary Medical Association **164**:489–490.

Christie, A. B. 1974. Infectious diseases: Epidemiology and clinical practice. Edinburgh, London, New York: Churchill Livingstone.

Chung, B. H., Cannon, R. Y. 1971. Psychrotrophic sporeforming bacteria in raw milk supplies. Journal of Dairy Science **54**:448.

Clark, F. E. 1937. The relation of *Bacillus siamensis* and similar pathogenic spore-forming bacteria to *Bacillus cereus*. Journal of Bacteriology **33**:435–443.

Clark, R. 1938. Speculations on the incidence of anthrax in bovines. Journal of the South African Veterinary Medical Association **9**:5–12.

Clegg, L. F. L. 1950. Spore-forming thermophiles in sterilised milk. Journal of the Society of Dairy Technology **3**:238–250.

Clegg, L. F. L., Jacobs, S. E. 1953. Environmental and other aspects of adaptation in thermophiles, pp. 306–325. In: Davies, R., Gale, E. F. (eds.), Adaptation in microorganisms. Third Symposium of the Society for General Microbiology. Cambridge: University Press.

Cohn, F. 1872. Untersuchungen über Bacterien. Beiträge zur Biologie der Pflanzen, Heft **2**:127–224.

Collins, C. H., Lyne, P. M. 1976. The Gram-positive spore-bearers; *Bacillus*, pp. 434–440. In: Collins, C. H., Lyne, P. M. (eds.), Microbiological methods, 4th ed. London: Butterworths.

Collins, M. D., Jones, D. 1979. Isoprenoid quinone composition as a guide to the classification of *Sporolactobacillus* and possibly related bacteria. Journal of Applied Bacteriology **47**:293–297.

Conn, H. J. 1948. The most abundant groups of bacteria in soil. Bacteriological Reviews **12**:257–273.

Coonrod, J. D., Leadley, P. J., Eickhoff, T. C. 1971. *Bacillus cereus* pneumonia and bacteremia. American Review of Respiratory Disease **103**:711–714.

Cowan, S. T. 1974. Manual for the identification of medical bacteria. Cambridge: Cambridge University Press.

Cox, R., Sockwell, G., Landers, B. 1959. *Bacillus subtilis* septicemia. Report of a case and review of the literature. New England Journal of Medicine **261**:894–896.

Craig, C. P., Lee, W. S., Ho, M. 1974. *Bacillus cereus* endocarditis in an addict. Annals of Internal Medicine **80:**418–419.

Cross, T., Goodfellow, M. 1973. Taxonomy and classification of actinomycetes, pp. 11–112. In: Sykes, G., Skinner, F. A. (eds.), Actinomycetales: characteristics and practical importance. London, New York: Academic Press.

Cross, T., Unsworth, B. A. 1981. The taxonomy of the endospore-forming actinomycetes, pp. 17–32. In: Berkeley, R. C. W., Goodfellow, M. (eds.), Classification and identification of the aerobic endospore-forming bacteria. London: Academic Press.

Curtis, J. R., Wing, A. J., Coleman, J. C. 1967. *Bacillus cereus* bacteraemia: A complication of intermittent haemodialysis. Lancet **i:**136–138.

Dack, G. M., Sugiyama, H., Owens, F. J., Kisner, J. B. 1954. Failure to produce illness in human volunteers fed *Bacillus cereus* and *Clostridium perfringens*. Journal of Infectious Diseases **94:**34–38.

Darland, G., Brock, T. D. 1971. *Bacillus acidocaldarius* sp. nov., an acidophilic thermophilic spore-forming bacterium. Journal of General Microbiology **67:**9–15.

Davenport, R., Smith, C. 1952. Panophthalmitis due to an organism of the *Bacillus subtilis* group. British Journal of Ophthalmology **36:**389–392.

Davies, D. G. 1960. The influence of temperature and humidity on spore formation and germination in *Bacillus anthracis*. Journal of Hygiene **58:**177–186.

Davies, F. L., Wilkinson, G. 1973. *Bacillus cereus* in milk and dairy products, pp. 57–67. In: Hobbs, B. C., Christian, J. H. B. (eds.), The microbiological safety of food. London: Academic Press.

Davis, G. H. G. 1964. Notes on the phylogenetic background to *Lactobacillus* taxonomy. Journal of General Microbiology **34:**177–184.

Davis, J. G. 1940. Sweet curdling in milk and cream. Milk Trade Gazette **10:**4–5.

Deibel, R. H., Niven, C. F., Jr. 1958. Microbiology of meat curing. I. The occurrence and significance of a motile microorganism of the genus *Lactobacillus* in ham curing brines. Applied Microbiology **6:**323–327.

Donovan, K. O. 1958. A selective medium for *Bacillus cereus* in milk. Journal of Applied Bacteriology **21:**100–103.

Donovan, K. O. 1959. The occurrence of *Bacillus cereus* in milk and on dairy equipment. Journal of Applied Bacteriology **22:**131–137.

Dowben, R. M., Weidenmüller, R. 1968. Adaptation of mesophilic bacteria to growth at elevated temperatures. Biochimica et Biophysica Acta **158:**255–261.

Dubos, R. 1971. Toxic factors in enzymes used in laundry products. Science **173:**259–260.

Eddy, B. P. 1960. The use and meaning of the term psychrophilic. Journal of Applied Bacteriology **23:**189–190.

Eddy, B. P., Ingram, M. 1956. A salt tolerant denitrifying *Bacillus* strain which blows canned bacon. Journal of Applied Bacteriology **19:**62–70.

English, C. F., Bell, E. J., Berger, A. J. 1967. Isolation of thermophiles from broadleaf tobacco and effect of pure culture inoculation on cigar aroma and mildness. Applied Microbiology **15:**117–119.

Farmiloe, F. J., Cornford, S. J., Coppock, J. B. M., Ingram, M. 1954. The survival of *Bacillus subtilis* spores in the baking of bread. Journal of the Science of Food and Agriculture **5:**292–304.

Farrar, W. E. 1963. Serious infections due to "non-pathogenic" organisms of the genus *Bacillus*. American Journal of Medicine **34:**134–141.

Farrell, J., Campbell, L. L. 1969. Thermophilic bacteria and bacteriophages. Advances in Microbial Physiology **3:**83–109.

Feeley, J. C., Brachman, P. S. 1974. *Bacillus anthracis*, pp. 143–147. In: Lennette, E. H., Spaulding, E. H., Truant, J. P. (eds.), Manual of clinical microbiology, 2nd ed. Washington, D.C.: American Society for Microbiology.

Feldman, S., Pearson, T. A. 1974. Fatal *Bacillus cereus* pneumonia and sepsis in a child with cancer. Clinical Pediatrics **13:**649–655.

Fields, M. L. 1970. The flat sour bacteria. Advances in Food Research **18:**163–217.

Finland, M., Barnes, M. W. 1970. Changing etiology of bacterial endocarditis in the antibacterial era. Annals of Internal Medicine **72:**341–348.

Flindt, M. L. H. 1969. Pulmonary disease due to inhalation of derivatives of *Bacillus subtilis* containing proteolytic enzyme. Lancet **i:**1177–1181.

François, J. 1934. Le bacille subtilique en pathologie oculaire. Bulletin et Mémoires de la Société Française d'Ophtalmologie **47:**423–437.

Franklin, J. G. 1967. The incidence and significance of *Bacillus cereus* in milk—1. Milk Industry **61(4):**34–37.

Fuller, W. H., Norman, A. G. 1943. Cellulose decomposition by aerobic mesophilic bacteria from soil. 1. Isolation and description of organisms. Journal of Bacteriology **46:**273–280.

Gaughran, E. R. L. 1947. The thermophilic microorganisms. Bacteriological Reviews **11:**189–225.

Gemmell, M., Hodgkiss, W. 1964. The physiological characters and flagellar arrangement of motile homofermentative lactobacilli. Journal of General Microbiology **35:**519–526.

Gilbert, R. J., Parry, J. M. 1977. Serotypes of *Bacillus cereus* from outbreaks of food poisoning and from routine foods. Journal of Hygiene **78:**69–74.

Gilbert, R. J., Stringer, M. F., Peace, T. C. 1974. The survival and growth of *Bacillus cereus* in boiled and fried rice in relation to outbreaks of food poisoning. Journal of Hygiene **73:**433–444.

Gilbert, R. J., Taylor, A. J. 1975. Das Auftreten von *Bacillus cereus*—Lebensmittelvergiftungen in Grossbritannien. Archiv für Lebensmittelhygiene **26:**38.

Gilbert, R. J., Taylor, A. J. 1976. *Bacillus cereus* food poisoning, pp. 197–213, In: Skinner, F. A., Carr, J. G. (eds.), Microbiology in agriculture, fisheries and foods. Society for Applied Bacteriology Symposium Series No. 4. London: Academic Press.

Gillespy, T. G., Thorpe, R. H. 1968. Occurrence and significance of thermophiles in canned foods. Journal of Applied Bacteriology **31:**59–65.

Gillissen, G., Scholz, H. G. 1961. Die Selektion von Milzbrandbazillen aus Flüssigkeiten mit starker Verunreinigung. Zeitschrift für Bakteriologie **182:**232.

Goepfert, J. M., Spira, W. M., Kim, H. U. 1972. *Bacillus cereus:* Food poisoning organism. A review. Journal of Milk and Food Technology **35:**213–227.

Goepfert, J. M., Spira, W. M., Glatz, B. A., Kim, H. U. 1973. Pathogenicity of *Bacillus cereus*, pp. 69–75. In: Hobbs, B. C., Christian, J. H. B. (eds.), The microbiological safety of foods. London: Academic Press.

Gold, H. 1967. Treatment of anthrax. Federation Proceedings **26:**1563–1568.

Gordon, R. E. 1981. One hundred and seven years of the genus *Bacillus*, pp. 1–15. In: Berkeley, R. C. W., Goodfellow, M. (eds.), Classification and identification of the aerobic endospore-forming bacteria. London: Academic Press.

Gordon, R. E., Haynes, W. C., Pang, C. H.-N. 1973. The genus *Bacillus*. Handbook No. 427. Washington D.C.: U.S. Department of Agriculture.

Gordon, R. E., Rynearson, T. K. 1963. Maintenance of strains of *Bacillus* species, pp. 118–127. In: Martin, S. M. (ed.), Culture collections: Perspectives and problems. Toronto: University Press.

Gordon, R. E., Smith, N. R. 1949. Aerobic sporeforming bacteria capable of growth at high temperatures. Journal of Bacteriology **58:**327–341.

Gould, G. W., Dring, G. J. 1975. Role of an expanded cortex in resistance of bacterial endospores, pp. 541–546. In: Gerhardt, P., Costilow, R. N., Sadoff, H. L. (eds.), Spores VI. Washington D.C. American Society for Microbiology.

Goullet, P., Pepin, H. 1974. *Bacillus cereus* septicaemia. Lancet **i:**761–762.

Graham-Smith, G. S. 1941. Further observations on the longevity of dry spores of *B. anthracis*. Journal of Hygiene **41:**496.

Green, D. M. 1975. Anthrax bacillus, pp. 449–453. In: Cruickshank, R., Duguid, J. P., Marmion, B. P., Swain, R. H. A. (eds.), Medical microbiology, 12th ed. vol. II. The practice of medical microbiology. Edinburgh, London, New York: Churchill Livingstone.

Green, D. M., Jamieson, W. M. 1958. Anthrax and bone-meal fertilizer. Lancet **ii:**153–154.

Greenberg, M., Milne, J. F., Watt, A. 1970. Survey of workers exposed to dusts containing derivatives of *Bacillus subtilis*. British Medical Journal **ii:**629–633.

Groeschel, D., Burgess, M. A., Bodey, G. P. 1976. Gas-gangrene like infection with *Bacillus cereus* in a lymphoma patient. Cancer **37:**988–991.

Grosskopf, J. C., Harper, W. J. 1969. Role of psychrophilic sporeformers in long life milk. Journal of Dairy Science **52:**897.

Gsell, O. R. 1971. Septic infections by bacteria of low pathogenicity in patients with resistance reduced by chemotherapy, pp. 145–154. In: Finland, M., Marget, W., Bartmann, K. (eds.), Bayer Symposium III. Bacterial infections. Changes in their causative agents. Trends and possible basis Berlin, Heidelberg, New York: Springer-Verlag.

Gyllenberg, H. G. 1968. Classification of psychrophilic microorganisms. Report to the I.D.F. Seminar on Psychrophilic Organisms, Falmer, England, April.

Hansen, P. A. 1933. The growth of thermophilic bacteria. Archiv für Mikrobiologie **4:**23–35.

Harris, O., Fields, M. L. 1972. A study of thermophilic aerobic sporeforming bacteria isolated from soil and water. Canadian Journal of Microbiology **18:**917–923.

Harrison, A. P., Hansen, P. A. 1950. A motile *Lactobacillus* from the cecal feces of turkeys. Journal of Bacteriology **59:**444–446.

Hauge, S. 1950. Matforgiftninger fremkalt av *Bacillus cereus*. Nordisk Hygienisk Tidskrift **31:**189–205.

Hauge, S. 1955. Food poisoning caused by aerobic spore-forming bacilli. Journal of Applied Bacteriology **18:**591–595.

Heinen, V. J., Heinen, W. 1972. Characteristics and properties of a caldo-active bacterium producing extracellular enzymes and two related strains. Archiv für Mikrobiologie **82:**1–23.

H.M.S.O. Report 1959. Report on the Committee of Inquiry on Anthrax, Command 846. London: Her Majesty's Stationery Office.

Hiscox, E. R., Christian, M. I. 1931. A contribution to the bacteriology of commercial sterilised milk. Part 1. General. Journal of Dairy Research **3:**106–112.

Hoffmann, C., Schweitzer, R., Dalby, G. 1937. Control of rope in bread. Industrial and Engineering Chemistry **29:**464–467.

Horowitz-Wlassowa, L. M. 1931. Über die Rolle der Bakterien-flora der Lake beim Polken mit Berücksichtigung der Frage der Halophilie in der Bakterienwelt. Zeitschrift für Untersuchung der Lebensmittel **62:**596.

Hugh-Jones, M. E., Hussaini, S. N. 1975. Anthrax in England and Wales 1963–1972. Veterinary Record **97:**256–261.

Ihde, D. C., Armstrong, D. 1973. Clinical spectrum of infection due to *Bacillus* species. American Journal of Medicine **55:**839–845.

Ingraham, J. L., Stokes, J. L. 1959. Psychrophilic bacteria. Bacteriological Reviews **23:**97–106.

Ingram, M., Hobbs, B. C. 1954. The bacteriology of "pasteurized" canned hams. Royal Sanitary Institute Journal **74:**1151–1163.

Ingram, M., Robinson, R. H. M. 1951. The growth of *Clostridium botulinum* in acid bread media. Proceedings of the Society for Applied Bacteriology **14:**62–72.

Jansen, E., Aschehoug, V. 1951. *Bacillus* as spoilage organisms in canned foods. Food Research **16:**457–461.

Kandler, O. 1967. Taxonomie und technologische Bedeutung der Gattung *Lactobacillus* Beijerinck. Zentralblatt für Bakteriologie, Parasitenkunde, Infektionskrankheiten und Hygiene, Abt. 1 Orig., Suppl. **2:**139–164.

Kelsey, J. C. 1961. The testing of sterilizers. 2. Thermophilic spore papers. Journal of Clinical Pathology **14:**313–319.

Kim, H. U., Goepfert, J. M. 1971a. Occurrence of *Bacillus cereus* in selected dry food products. Journal of Milk and Food Technology **34:**12–15.

Kim, H. U., Goepfert, J. M. 1971b. Enumeration and identification of *Bacillus cereus* in foods. I. 24-hour presumptive test medium. Applied Microbiology **22:**581–587.

Kitahara, K., Lai, C.-L. 1967. On the spore formation of *Sporolactobacillus inulinus*. Journal of General and Applied Microbiology **13:**197–203.

Kitahara, K., Suzuki, J. 1963. *Sporolactobacillus* nov. subgen. Journal of General and Applied Microbiology **9:**59–71.

Kitahara, K., Toyota, T. 1972. Auto-spheroplastization and cell-permeation in *Sporolactobacillus inulinus*. Journal of General and Applied Microbiology **18:**99–107.

Klainer, A. S., Beisel, W. R. 1969. Opportunistic infection: A review. American Journal of the Medical Sciences **258:**431–456.

Klug, M. J., Markovetz, A. J. 1967. Thermophilic bacterium isolated on n-tetradecane. Nature **215:**1082–1083.

Klumb, G. H., Marks, H. C., Wilson, C. 1949. Control of bacterial reproduction in cation exchange layers. Journal of the American Water Works Association **41:**933–947.

Knisely, R. F. 1966. Selective medium for *Bacillus anthracis*. Journal of Bacteriology **92:**784–786.

Krasil'nikov, N. A. 1949. Guide to the bacteria and actinomycetes, pp. 1–830. Moscow: Akademii Nauk SSSR.

Kriss, A. E. (trans. Shewan, J. M., Kabata, Z.) 1963. Marine microbiology. Edinburgh: Oliver and Boyd.

Kushner, D. J. 1978. Microbial life in extreme environments. London, New York, San Francisco: Academic Press.

Labots, H., Hup, G., Galesloot, Th. E. 1965. *Bacillus cereus* in rauwe en gepasteuriseerde melk. III. Over de besmetting van rauwe melk met *B. cereus* sporen bij de melkwinning. Netherlands Milk and Dairy Journal **19:**191–215.

Laine, J. J. 1970. Studies on psychrophilic bacilli of food origin. Biologica **169:**1–36.

Langeveld, L. P. M. 1971. Keeping quality in the refrigerator of aseptically and not aseptically packed pasteurised milk. Koeltechniek **64:**136–138.

Larkin, J. M., Stokes, J. L. 1966. Isolation of psychrophilic species of *Bacillus*. Journal of Bacteriology **91:**1667–1671.

Larkin, J. M., Stokes, J. L. 1967. Taxonomy of a psychrophilic strain of *Bacillus*. Journal of Bacteriology **94:**889–895.

Laubach, C. A., Rice, J. L., Ford, W. W. 1916. Aerobic spore-bearing non-pathogenic bacteria. Journal of Bacteriology **1:**493–533.

Lazar, I., Jursack, L. 1966. Daten zur Pathogenität des *Bacillus cereus*. Zentralblatt für Bakteriologie, Parasitenkunde, Infektionskrankheiten und Hygiene, Abt. 1 Orig. **199:**59–64.

Lechevalier, H. A., Solotorovsky, M. 1965. Three centuries of microbiology. New York: McGraw-Hill.

Lefebvre, A., Gregoire, C. A., Brabaut, W., Todd, E. 1973. Suspected *Bacillus cereus* food poisoning. Epidemiological Bulletin **17:**108–111.

Leff, A., Jacobs, R., Gooding, V., Hauch, J., Conte, J., Stulbarg, M. 1977. *Bacillus cereus* pneumonia—survival in a patient with cavitary disease treated with gentamicin. American Review of Respiratory Disease **115:**151–154.

Leffert, H. L., Baptist, J. N., Gidez, L. I. 1970. Meningitis and bacteremia after ventriculoatrial shunt-revision: Isolation of a lecithinase-producing *Bacillus cereus*. Journal of Infectious Diseases **122:**547–552.

Lewis, J. C., Snell, N., Burr, H. K. 1960. Water permeability of bacterial spores and the concept of a contractile cortex. Science **132:**544–545.

Lincoln, R. E., Fish, D. C. 1970. Anthrax toxin, pp. 361–414.

In: Montie, T. C., Kadis, S., Ajl, S. J. (eds.), Microbial toxins, vol. III. New York, London: Academic Press.

Lincoln, R. E., Walker, J. S., Klein, F., Haines, B. W. 1964. Anthrax. Advances in Veterinary Science 9:327–368.

Lloyd, B. 1931. Muds of the Clyde Sea area. II. Bacterial content. Journal of the Marine Biological Association 17:751–765.

Logan, N. A., Berkeley, R. C. W. 1981. The classification and identification of members of the genus Bacillus using API tests, pp. 105–140. In: Berkeley, R. C. W., Goodfellow, M. (eds.). Classification and identification of the aerobic endospore-forming bacteria. London: Academic Press.

Long, S. K., Williams, O. B. 1959. Growth of obligate thermophiles at 37°C as a function of the cultural conditions employed. Journal of Bacteriology 77:545–547.

Lubenau, C. 1906. Bacillus peptonificans als Erreger einer Gastroenteritis-Epidemie. Centralblatt für Bakteriologie, Parasitenkunde, Infektionskrankheiten und Hygiene, Abt. 1 Orig. 40:433–437.

Lück, H., Mostert, J. F., Husmann, R. A. 1976. Non sterility of commercial sterilized milk. South African Journal of Dairy Technology 8:103.

McBee, R. H., Gaugler, L. P. 1956. Identity of thermophilic bacteria isolated from arctic soils and waters. Journal of Bacteriology 71:186–187.

MacDonald, R. E., MacDonald, S. W. 1962. The physiology and natural relationships of the motile spore-forming sarcinae. Canadian Journal of Microbiology 8:795–808.

McDonald, W. C., Matney, T. S. 1963. Genetic transfer of the ability to grow at 55°C in Bacillus subtilis. Journal of Bacteriology 85:218–220.

McKinstry, D. W., Haley, D. E., Reid, J. J. 1938. A bacteriological study of the bulk fermentation of cigar leaf tobacco. Journal of Bacteriology 35:71.

Mahmoud, S. A. Z., El-Tadi, M. A., Elmofty, M. K. 1964. Studies in the rhizosphere microflora of a desert plant. Folia Microbiologica 9:1–8.

Marsh, C. L., Larsen, D. H. 1953. Characterization of some thermophilic bacteria from hot springs of Yellowstone National Park. Journal of Bacteriology 65:193–197.

Marshall, B. J., Ohye, D. F. 1966. Bacillus macquariensis n. sp., a psychrotrophic bacterium from sub-Antarctic soil. Journal of General Microbiology 44:41–46.

Mateles, R. I., Baruah, J. N., Tannenbaum, S. R. 1967. Growth of a thermophilic bacterium on hydrocarbons: A new source of single cell protein. Science 157:1322–1323.

Mathews, M. M., Sistrom, W. R. 1959. Function of carotenoid pigments in non-photosynthetic bacteria. Nature 184:1892–1893.

Mathias, R. G., Todd, E., Szabo, R., Martin, D. 1976. Illness from fried rice—St. John's, Newfoundland. Canada Diseases Weekly Report 2:78–79.

Mazanec, K., Kocur, M., Martinec, T. 1965. Electron microscopy of ultrathin section of Sporosarcina ureae. Journal of Bacteriology 90:808–816.

Melles, Z., Nikodémusz, I., Ábel, A. 1969. Die pathogene Wirkung aerober sporenbildender Bakterien. Zentralblatt für Bakteriologie, Parasitenkunde, Infektionskrankheiten und Hygiene, Abt. 1 Orig. 212:174–176.

Melling, J., Capel, B. J. 1978. Characteristics of Bacillus cereus emetic toxin. FEMS Microbiology Letters 4:133–135.

Melling, J., Capel, B. J., Turnbull, P. C. B., Gilbert, R. J. 1976. Identification of a novel enterotoxigenic activity associated with Bacillus cereus. Journal of Clinical Pathology 29:938–940.

Michels, M. J. M., Anema, P. J. 1974. Cold-tolerant spore-formers in pasteurised meals to be stored at 0–3°C, pp. 77–86. International Symposium on Food Microbiology, 2nd vol.

Michels, M. J. M., Visser, F. M. W. 1976. The occurrence and thermoresistance of spores of psychrophilic and psychro-

trophic aerobic sporeformers in soil and foods. Journal of Applied Bacteriology 41:1–11.

Midura, T., Gerber, M., Wood, R., Leonard, A. R. 1970. Outbreak of food poisoning caused by Bacillus cereus. Public Health Reports 85:45–48.

Minett, F. C. 1950. Sporulation and viability of B. anthracis in relation to environmental temperature and humidity. Journal of Comparative Pathology 60:161–176.

Minett, F. C. 1951. The use of climatological data for assessing the regional distribution of anthrax in India. Bulletin de l'Office International des Épizooties 35:266–295.

Minett, F. C. 1952. The annual and seasonal incidence of anthrax in various countries. Climatic effects and sources of infection. Bulletin de l'Office International des Épizooties 37:238–300.

Minett, F. C., Dhanda, M. R. 1941. Multiplication of B. anthracis and Cl. chauvoei in soil and water. Indian Journal of Veterinary Science 11:308–328.

Moore, A. W., Becking, J. H. 1963. Nitrogen fixation by Bacillus strains isolated from Nigerian soil. Nature 198:915–916.

Morris, E. J. 1955. A selective medium for Bacillus anthracis. Journal of General Microbiology 13:456–460.

Mortimer, P. R., McCann, G. 1974. Food poisoning episodes associated with Bacillus cereus in fried rice. Lancet i:1043–1045.

Mossel, D. A. A., Koopman, M. J., Jongerius, E. 1967. Enumeration of Bacillus cereus in foods. Applied Microbiology 15:650–653.

Mourgues, R., Auclair, J. 1970. Keeping quality of pasteurized milk free from post-pasteurization contamination, stored at 4°C and 8°C. XVIII International Dairy Congress 1E:168.

Murrell, W. G. 1967. The biochemistry of the bacterial endospore, pp. 133–251. In: Rose, A. H., Wilkinson, J. F. (eds.), Advances in microbial physiology 1. London, New York: Academic Press.

Nakayama, O., Yanoshi, M. 1967a. Spore-bearing lactic acid bacteria isolated from rhizosphere. I. Taxonomic studies on Bacillus laevolacticus nov. sp. and Bacillus racemilacticus nov. sp. Journal of General and Applied Microbiology 13:139–153.

Nakayama, O., Yanoshi, M. 1967b. Spore-bearing lactic acid bacteria isolated from rhizosphere. II. Taxonomic studies on the catalase-negative strains. Journal of General and Applied Microbiology 13:155–165.

Nichols, A. A. 1939. Bacteriological studies of canned milk products. Journal of Dairy Research 10:231–249.

Nikodémusz, I. 1965. Die Reproduzierbarkeit der von Bacillus cereus verursachten Lebensmittelvergiftungen bei Katzen. Zentralblatt für Bakteriologie, Parasitenkunde, Infektionskrankheiten, und Hygiene, Abt. 1 Orig. 196:81–87.

Nikodémusz, I. 1967. Die enteropathogene Wirkung von Bacillus cereus bei Hunden. Zentralblatt für Bakteriologie, Parasitenkunde, Infektionskrankheiten und Hygiene, Abt. 1 Orig. 202:533–538.

Nikodémusz, I., Gonda, Gy. 1966. Die Wirkung langfristiger Verabreichung von B. cereus verunreinigten Lebensmitteln bei Katzen. Zentralblatt für Bakteriologie, Parasitenkunde, Infektionskrankheiten und Hygiene, Abt. 1 Orig. 199:64–67.

Nishio, N., Ueda, M., Omae, Y., Hayashi, M., Kamikubo, T. 1976. Utilization of hydrocarbons and vitamin B_{12} production by Bacillus badius. Agricultural and Biological Chemistry 40:2037–2043.

Nonomura, H., Yamazaki, T., Ohara, Y. 1965. Die Apfelsäure-Milchsäure-Bakterien, welche aus japanischen Weinen isoliert wurden. Mitteilungen der Hoeheren Bundeslehr- und Versuchsanstalt fuer Wein-, Obst- und Gartenbau, Klosterneuburg 15A:241–254.

Norris, J. R. 1981. Sporosarcina and Sporolactobacillus, pp. 337–357. In: Berkeley, R. C. W., Goodfellow, M. (eds.), Classification and identification of the Aerobic endospore-forming bacteria. London: Academic Press.

Nygren, B. 1962. Phospholipase C-producing bacteria and food poisoning. An experimental study on *Clostridium perfringens* and *Bacillus cereus*. Acta Pathologica et Microbiologica Scandinavica, Suppl. **160.**

Ormay, L., Novotny, T. 1969. The significance of *Bacillus cereus* food poisoning in Hungary, pp. 279–285. In: Kampelmacher, E. H., Ingram, M., Mossel, D. A. A. (eds.), The microbiology of dried foods. International Association of Microbiological Societies.

Oshima, T., Arakawa, H., Baba, M. 1977. Biochemical studies on an acidophilic, thermophilic bacterium, *Bacillus acidocaldarius*—isolation of bacteria, intracellular pH, and stabilities of biopolymers. Journal of Biochemistry **81:**1107–1113.

Oxborrow, G. S., Fields, N. D., Puleo, J. R. 1977. Pyrolysis gas-liquid chromatography of the genus *Bacillus:* Effect of growth media on pyrochromatogram reproducibility. Applied and Environmental Microbiology **33:**865–870.

Pearce, T. W., Powell, E. O. 1951. A selective medium for *Bacillus anthracis*. Journal of General Microbiology **5:**387–390.

Pearson, H. E. 1970. Human infections caused by organisms of the *Bacillus* species. American Journal of Clinical Pathology **53:**506–515.

Pepys, J., Hargreave, F. E., Longbottom, J. L., Faux, J. 1969. Allergic reactions of the lungs to enzymes of *Bacillus subtilis*. Lancet **i:**1181–1184.

Perrin, D., Greenfield, J., Ward, G. E. 1976. Acute *Bacillus cereus* mastitis in dairy-cattle associated with use of a contaminated antibiotic. Canadian Veterinary Journal **17:**244–247.

Piggot, P. J., Coote, J. G. 1976. Genetic aspects of bacterial endospore formation. Bacteriological Reviews **40:**908–962.

Pinegar, J. A., Buxton, J. D. 1977. An investigation of the bacteriological quality of retail vanilla slices. Journal of Hygiene **78:**387–394.

Pluot, M., Vital, C., Aubertin, A., Croix, J. C., Pire, J. C., Poisot, D. 1976. Anthrax meningitis. Report of 2 cases with autopsies. Acta Neuropathologica **36:**339–345.

Portnoy, B. L., Goepfert, J. M., Harmon, S. M. 1976. An outbreak of *Bacillus cereus* food poisoning resulting from contaminated vegetable sprouts. American Journal of Epidemiology **103:**589–594.

Prévot, A.-R. 1961. Traité de systématique bactérienne, vol. 2. Paris: Dunod.

Public Health Laboratory Service. 1972. Food poisoning associated with *Bacillus cereus*. British Medical Journal **i:**189.

Public Health Laboratory Service. 1973. *Bacillus cereus* food poisoning. British Medical Journal **iii:**647.

Puleo, J. R., Fields, N. D., Bergstrom, S. L., Oxborrow, G. S., Stabekis, P. D., Koukol, R. C. 1977. Microbiological profiles of the Viking spacecraft. Applied and Environmental Microbiology **33:**379–384.

Raevuori, M., Kiutamo, T., Niskanen, A., Salminen, K. 1976. An outbreak of *Bacillus cereus* food-poisoning in Finland associated with boiled rice. Journal of Hygiene **76:**319–327.

Raphael, S. S., Donaghue, M. 1976. Infection due to *Bacillus cereus*. Canadian Medical Association Journal **115:**207.

Reay, G. A., Shewan, J. M. 1949. The spoilage of fish and its preservation by chilling. Advances in Food Research **2:**343–398.

Richmond, B., Fields, M. L. 1966. Distribution of thermophilic aerobic sporeforming bacteria in food ingredients. Applied Microbiology **14:**623–626.

Roberts, T. A., Hitchins, A. D. 1969. Resistance of spores, pp. 611–670. In: Gould, G. W., Hurst, A. (eds.), The bacterial spore. New York: Academic Press.

Roberts, T. A., Ingram, M., Skulberg, A. 1965. The resistance of spores of *Clostridium botulinum* type E to heat and radiation. The resistance of *Clostridium botulinum* type E toxin to radiation. Journal of Applied Bacteriology **28:**125–141.

Rode, L. J., Foster, J. W. 1965. Gaseous hydrocarbons and the germination of bacterial spores. Proceedings of the National Academy of Sciences of the United States of America **53:**31–38.

Sathmary, M. N. 1958. *Bacillus subtilis* septicemia and generalised aspergillosis in patient with acute myeloblastic leukemia. New York State Journal of Medicine **58:**1870–1876.

Scarr, M. P. 1968. Thermophiles in sugar. Journal of Applied Bacteriology **31:**66–74.

Schiefer, B., MacDonald, K. R., Klavano, G. G., van Dreumel, A. A. 1976. Pathology of *Bacillus cereus* mastitis in dairy cows. Canadian Veterinary Journal **17:**239–243.

Schmidt-Nielsen, S. 1902. Über einige psychrophile Mikroorganismen und ihr Vorkommen. Centralblatt für Bakteriologie, Parasitenkunde und Infektionskrankheiten, Abt. 2 **9:**145–147.

Schmitt, N., Bowmer, E. J., Willoughby, B. A. 1976. Food poisoning outbreak attributed to *Bacillus cereus*. Canadian Journal of Public Health **67:**418–422.

Schoenbaum, S. C., Gardner, P., Shillito, J. 1975. Infections of cerebrospinal fluid shunts: Epidemiology, clinical manifestations and therapy. Journal of Infectious Diseases **131:**543–552.

Schulze, W. D., Olson, J. C., Jr. 1960. Studies on psychrophilic bacteria. 1. Distribution in stored commercial dairy products. Journal of Dairy Science **43:**346–350.

Sen, S. K., Minett, F. C. 1944. Experiments on the transmission of anthrax through flies. Indian Journal of Veterinary Science **14:**149–158.

Shapton, D. A., Hindes, W. R. 1963. The standardization of a spore count technique. Chemistry and Industry **41:**230–234.

Shehata, T. E., Collins, E. B. 1971. Isolation and identification of psychrophilic species of *Bacillus* from milk. Applied Microbiology **21:**466–469.

Shehata, T. E., Collins, E. B. 1972. Sporulation and heat resistance of psychrophilic strains of *Bacillus*. Journal of Dairy Science **55:**1045–1049.

Shehata, T. E., Duran, A., Collins, E. B. 1971. Influence of temperature on the growth of psychrophilic strains of *Bacillus*. Journal of Dairy Science **54:**1579–1582.

Sie, E. H., Sobotka, H., Baker, H. 1961. Factor converting mesophilic into thermophilic micro-organisms. Nature **192:**86–87.

Sirol, J., Gendron, Y., Condat, M. 1973. Le charbon humain en Afrique. Bulletin of the World Health Organization **49:**143–148.

Skerman, V. B. D. 1967. A guide to the identification of the genera of bacteria, 2nd ed. Baltimore: Williams & Wilkins.

Slepecky, R. 1978. Resistant forms, pp. 14/1–14/31. In: Norris, J. R., Richmond, M. H. (eds.), Essays in microbiology. Chichester, New York, Brisbane, Toronto: John Wiley & Sons.

Smith, H. 1958. The use of bacteria grown *in vivo* for studies on the basis of their pathogenicity. Annual Review of Microbiology **12:**77–102.

Smith, H. 1960. Studies on organisms grown *in vivo* to reveal the bases of microbial pathogenicity. Annals of the New York Academy of Sciences **88:**1213–1226.

Smith, H., Stoner, H. B. 1967. Anthrax toxic complex. Federation Proceedings **26:**1554–1557.

Smith, N. R., Gordon, R. E., Clark, F. E. 1946. Aerobic mesophilic sporeforming bacteria. U. S. Department of Agriculture Miscellaneous Publication 559.

Smith, N. R., Gordon, R. E., Clark, F. E. 1952. Aerobic sporeforming bacteria. U. S. Department of Agriculture Monograph 16.

Sneath, P. H. A. 1962. Longevity of micro-organisms. Nature **195:**643–646.

Sobernheim, G. 1931. Milzbrand, pp. 1041–1174. In: Kolle, W., Kraus, R., Uhlenhuth, P. (eds.), Handbuch der pathogenen Mikroorganismen, 3rd ed., vol. 3, part 2. Jena, Berlin, Vienna: Gustav Fischer Verlag.

Solny, M. N., Failing, G. R., Borges, J. S. 1977. *Bacillus cereus* osteomyelitis. Archives of Internal Medicine **137:**401–402.

Spira, W. M., Goepfert, J. M. 1975. Biological characteristics of an enterotoxin produced by *Bacillus cereus*. Canadian Journal of Microbiology **21**:1236–1246.

Stanley, J. L., Smith, H. 1961. Purification of factor I and recognition of a third factor of the anthrax toxin. Journal of General Microbiology **26**:49–66.

Stanley, J. L., Smith, H. 1963. The three factors of anthrax toxin: Their immunogenicity and lack of demonstrable enzymic activity. Journal of General Microbiology **31**:329–339.

Steinberg, R. A. 1947. Growth responses to organic compounds by tobacco seedlings in aseptic culture. Journal of Agricultural Research **75**:81–92.

Steinberg, R. A. 1950. The relation of certain soil bacteria to Frenching symptoms of tobacco. Bulletin of the Torrey Botanical Club **77**:38–44.

Sterne, M. 1959. Anthrax, pp. 16–52. In: Stableforth, A. W., Galloway, I. A. (eds.), Infectious diseases of animals. Diseases due to bacteria, vol. 1. London: Butterworths.

Sterne, M. 1967. Distribution and economic importance of anthrax. Federation Proceedings **26**:1493–1495.

Stone, M. J. 1952. The action of the lecithinase of *Bacillus cereus* on the globule membrane of milk fat. Journal of Dairy Research **19**:311–315.

Stone, M. J., Rowlands, A. 1952. Broken or bitty cream in raw and pasteurised milk. Journal of Dairy Research **19**:51–62.

Stopler, T., Camuescu, V., Voiculescu, M. 1964. Fatal bronchopneumonia caused by a microorganism of the genus *Bacillus* (*B. cereus*). Microbiologia, Parazitologia, Epidemiologia **9**:457–460. [English translation in Rumanian Medical Review **19**:7–9 (1965).]

Sundman, V., Gyllenberg, H. G. 1967. Application of factor analysis in microbiology. Annales Academiae Scientiarum Fennicae, A **IV**:112.

Suzuki, J., Kitahara, K. 1964. Base compositions of deoxyribonucleic acid in *Sporolactobacillus inulinus* and other lactic acid bacteria. Journal of General and Applied Microbiology **10**:305–311.

Szulmajster, J. 1979. Is sporulation a simple model for studying differentiation? Trends in Biochemical Sciences **4**:18–21.

Tantajumroon, T., Panas-Ampol, K. 1968. Intestinal anthrax—report of two cases. Journal of the Medical Association of Thailand **51**:477–480.

Taylor, A. J., Gilbert, R. J. 1975. *Bacillus cereus* food poisoning: A provisional serotyping scheme. Journal of Medical Microbiology **8**:543–550.

Thomas, S. B., Hobson, P. M., Bird, E. R. 1959. Psychrophilic bacteria in milk. XV International Dairy Congress **3** (Sect. 5):1334–1340.

Thomas, S. B., Jones-Evans, E., Jones, L. B., Thomas, B. F. 1946. Thermoduric microflora of dairy utensils. Proceedings of the Society for Applied Bacteriology **9**:51–53.

Thompson, R. S., Leadbetter, E. R. 1963. On the isolation of dipicolinic acid from endospores of *Sarcina ureae*. Archiv für Mikrobiologie **45**:27–32.

Thomson, P. D. 1955. The use of blood culture in the routine diagnosis of anthrax. Journal of Comparative Pathology **65**:1–7.

Tipper, D. J., Gauthier, J. J. 1972. Structure of the bacterial endospore, pp. 3–12. In: Halvorson, H. O., Hanson, R., Campbell, L. L. (eds.), Spores V. Washington, D.C.: American Society for Microbiology.

Turnbull, P. C. B. 1976. Studies on the production of enterotoxins by *Bacillus cereus*. Journal of Clinical Pathology **29**:941–948.

Turnbull, P. C. B., French, T. A., Dowsett, E. G. 1977. Severe systemic and pyogenic infections with *Bacillus cereus*. British Medical Journal **i**:1628–1629.

Turnbull, P. C. B., Nottingham, J. F., Ghosh, A. C. 1977. Severe necrotic enterotoxin produced by certain food, food poisoning and other clinical isolates of *Bacillus cereus*. British Journal of Experimental Pathology **58**:273–280.

Turner, M., Jervis, D. I. 1968a. The distribution of pigmented *Bacillus* species in saltmarsh and other saline and non-saline soils. Nova Hedwigia **16**:293–298.

Turner, M., Jervis, D. I. 1968b. Salt tolerance in pigmented and nonpigmented strains of *Bacillus* species isolated from soil. Journal of Applied Bacteriology **31**:373–377.

Uchida, K., Mogi, K. 1973. Cellular fatty acid spectra of *Sporolactobacillus* and some other *Bacillus-Lactobacillus* intermediates as a guide to their taxonomy. Journal of General and Applied Microbiology **19**:129–140.

Uchino, F., Doi, S. 1967. Acido-thermophilic bacteria from thermal waters. Agricultural and Biological Chemistry **31**:817–822.

Utsumi, S., Torii, M., Yamamuro, H., Kurimura, O., Amano, T. 1961. "γ-Glutamylase" as a decapsulating agent for *Bacillus anthracis*. Biken's Journal **4**:151–169.

van Bijsterveld, O. P., Richards, R. D. 1965. *Bacillus* infections of the cornea. Archives of Ophthalmology **74**:91–95.

Van Ness, G. B. 1971. Ecology of anthrax. Science **172**:1303–1307.

Veldkamp, H. 1970. Enrichment cultures of prokaryotic organisms, pp. 305–355. In: Norris, J. R., Ribbons, D. W. (eds.), Methods in microbiology, vol. 3A. London, New York: Academic Press.

Verhoeven, W. 1952. Aerobic spore-forming nitrate reducing bacteria. Delft: Waltman.

Wahlig, H., Holt, R. J. 1976. A rapid microbiological procedure using *Bacillus stearothermophilus* for the assay of antibacterial drugs. Journal of Clinical Pathology **29**:858–861.

Weinstein, L., Colburn, C. G. 1950. *Bacillus subtilis* meningitis and bacteremia: Report of a case and review of the literature on "subtilis" infections in man. Archives of Internal Medicine **86**:585–594.

Whitford, H. W. 1978. Factors affecting the laboratory diagnosis of anthrax. Journal of the American Veterinary Medical Association **173**:1467–1469.

Wiley, W. R., Stokes, J. L. 1962. Requirement of an alkaline pH and ammonia for substrate oxidation by *Bacillus pasteurii*. Journal of Bacteriology **84**:730–734.

Wilson, G. S., Miles, A. A. 1975. Topley and Wilson's principles of bacteriology, virology and immunity, vol. 1 and 2. London: Arnold.

Wilson, J. B., Russell, K. E. 1964. Isolation of *B. anthracis* from soil stored 60 years. Journal of Bacteriology **87**:237–238.

Witter, L. D. 1961. Psychrophilic bacteria—a review. Journal of Dairy Science **44**:983–1015.

Wolff, A. H., Heimann, H. 1951. Industrial anthrax in the United States: An epidemiologic study. American Journal of Hygiene **53**:80–109.

Wood, E. J. F. 1953. Heterotrophic bacteria in marine environments. Australian Journal of Marine and Freshwater Research **4**:60–200.

Yow, M. D., Reinhart, J. B., Butler, L. J. 1949. *Bacillus subtilis* septicemia treated with penicillin. Journal of Pediatrics **35**:237–239.

ZoBell, C. E., Feltham, C. B. 1934. Preliminary studies on the distribution and characteristics of marine bacteria. Scripps Institution of Oceanography Bulletin. Technical Series **3**:279–296.

CHAPTER 136

The Genus *Bacillus:* Insect Pathogens

ALOYSIUS KRIEG

Bacteria pathogenic for arthropods are either facultative or obligate pathogens. This distinction is based on the behavior of bacteria toward the gut wall of their hosts (Lysenko, 1958). After application per os, only obligately pathogenic bacteria are able to injure or damage the healthy gut wall and afterward to invade the body cavity. As a result of bacterial growth in the hemocoel, bacteremia occurs, often followed by fatal septicemia; however, the gut barrier cannot be overcome by facultatively pathogenic bacteria without additional support (trauma or stress). Only if such bacteria are injected parenterally can septicemia develop. Nonpathogenic bacteria do not produce toxins, and after injection, they become inactivated by immune mechanisms (such as phagocytes, lysozymes, and complement-like substances; antibodies are not produced in arthropods). On the other hand, insect-pathogenic bacteria tend to be immunoresistant and may produce immunoinhibitors.

Insect-pathogenic bacteria belong mainly to the Eubacteriales and Pseudomonadales (Steinhaus, 1946a). Table 1 lists bacteria that have often been isolated from diseased insects and that could also be identified taxonomically. Bacteria associated only with arthropods in a commensal or mutualistic fashion, or transmitted by them (as vectors), have been omitted.

This chapter only considers members of the genus *Bacillus*, many of which have been isolated and described in the past from diseased insects (Steinhaus, 1946b). However, only a few of the original isolates have been confirmed as pathogens; others were found to be toxigenic, such as *B. cereus, B. alvei, B. circulans,* and *B. sphaericus.* Finally, a number of isolates such as *B. megaterium* and *B. subtilis* have been considered only as "secondary invaders" after the death of the insect. Among the remaining true insect pathogens an extensive synonymy exists, and therefore only a few of the pathogenic isolates have been confirmed as valid

species (or subspecies) of the genus *Bacillus*, e.g., *B. larvae, B. popilliae,* and *B. thuringiensis.*

Habitats

The preferred habitat of members of the genus *Bacillus* is the soil, in which their endospores may persist in a dormant stage for many years. Most *Bacillus* species are saprophytic, i.e., their vegetative cells grow by metabolizing organic material derived from dead organisms; most bacilli are inert against living organisms. Only one species (*B. anthracis*) is harmful to animals and man, but several *Bacillus* species are known to be insect pathogens. *B. thuringiensis* has parasitic as well as saprophytic properties and therefore it may multiply inside and outside of insects. However, other obligate insect pathogens, such as *B. popilliae*, are strictly parasitic and may survive outside their host only as endospores.

The *Bacillus cereus*– *Bacillus thuringiensis* Group

It is striking that facultatively pathogenic bacilli, such as strains of *Bacillus cereus*, have a wide host range (which includes Lepidoptera, Hymenoptera, and Coleoptera) when injected as vegetative cells or endospores; the same is true with *B. thuringiensis* when applied parenterally. Application of *B. cereus* spores to insect larvae per os usually does not induce harmful effects. However, ingestion of sporulated cultures of *B. thuringiensis* initiates a severe illness in larvae of Lepidoptera (over 100 species are known to be susceptible), but not in insects such as Hymenoptera and Coleoptera (Heimpel and Angus, 1963; Krieg, 1967). Typical diseases caused by *B. thuringiensis* are the *Schlaffsucht* disease of the

Table 1. Insect-pathogenic bacteria.

Bacteria	Obligate pathogen	Facultative pathogen	Host range	Reference
Bacillaceae				
Bacillus alvei		+	Nonspecific	Heimpel and Angus, 1963
				Singer, 1973
Bacillus cereus		+	Nonspecific	Heimpel and Angus, 1963
Bacillus circulans		+	Nonspecific	Singer, 1973
Bacillus euloomarahae	+		Scarabaeidae	Dutky, 1963
Bacillus larvae	+		Apidae	Heimpel and Angus, 1963
Bacillus lentimorbus	+		Scarabaeidae	Dutky, 1963
Bacillus popilliae	+		Scarabaeidae	Dutky, 1963
Bacillus pulvifaciens		+	Nonspecific	Jackson and Long, 1965
Bacillus sphaericus		+	Nonspecific	Singer, 1973
Bacillus thuringiensis	+		Lepidoptera	Heimpel and Angus, 1963
Bacillus thuringiensis		+	Nonspecific	Heimpel and Angus, 1963
Clostridium brevifaciens	+		Lepidoptera	Heimpel and Angus, 1963
Clostridium malacosomae	+		Lepidoptera	Heimpel and Angus, 1963
Lactobacillaceae				
Streptococcus faecalis		+	Nonspecific	Martin and Mundt, 1972
Streptococcus pluton	+		Apidae	Bailey, 1963
Enterobacteriacae				
Enterobacter cloaca		+	Nonspecific	Bucher, 1963
Proteus spp.		+	Nonspecific	Bucher, 1963
Serratia marcescens		+	Nonspecific	Bucher, 1963
Pseudomonadaceae				
Pseudomonas aeruginosa		+	Nonspecific	Bucher, 1963
Pseudomonas fluorescens		+	Nonspecific	Bucher, 1963

Mediterranean flour moth (*Anagasta kühniella*) and the *Sotto* disease of the silkworm (*Bombyx mori*).

The efficiency of *B. thuringiensis* against lepidopterous larvae is connected with the production of a toxic parasporal crystal (= so-called δ-endotoxin) during the sporulation phase. After being ingested, such protein crystals become solubilized in the gut juice, depending on the alkalinity and protease content. The eroding effect of the solubilized crystal toxin on the gut epithelium of sensitive larvae favors the invasion of the hemocoel by germinated bacilli, where they may induce septicemia. Details about *B. thuringiensis* pathogenesis have been described by Angus and Heimpel (1959), Martouret (1961), and Ebersold, Lüthy, and Müller (1977).

Before sporulation, vegetative cells of *B. thuringiensis* may also produce soluble toxins. Such exotoxins could be demonstrated in supernatants of broth cultures. Two kinds of exotoxins have been found: the heat-sensitive α-exotoxin, a polypeptide produced by strains of *B. cereus* and *B. thuringiensis* (Heimpel, 1955; Krieg, 1971a); and a heat-resistant β-exotoxin, which is an unusual nucleotide produced only by special strains of these same bacteria (de Barjac and Dedonder, 1965; McConnell

and Richards, 1959). Both exotoxins are unspecific and may kill several kinds of insects (Lepidoptera, Hymenoptera, Coleoptera, and others) after injection. After application per os, their efficacy may be diminished by hydrolytic enzymes of the gut juice (proteases or phosphatases, respectively). In spite of this, β-exotoxin is still toxic after application per os to several susceptible Hymenoptera (e.g., bees and sawfly larvae), to larvae of several Diptera, and also to some phytophagous mites (e.g., *Tetranychus urticae*).

It has been suggested that production of α-exotoxin is necessary for the facultative pathogenicity of *B. cereus* (Krieg, 1971a). Not a single exotoxin could be found in cultures of strains of *B. megaterium* nonpathogenic for insects.

The main sources of *B. thuringiensis* are diseased larvae of Lepidoptera, but these bacilli may also occur in the gut microflora of nonsusceptible insects, as do *B. cereus* and *B. megaterium*. Along with other bacteria, *B. thuringiensis* is also found in decomposing material of plant origin on which caterpillars have been foraging. Another source is detritus or dust from insectaries, especially such as are used for sericulture. Finally, *B. thuringiensis* may be isolated from cereals that have been contaminated in mills by decaying, infected larvae of flour or meal moths (*Anagasta kühniella, Plodia interpunctella*). In general, lepidopterous larvae become infected by ingestion of contaminated food.

Recently some special strains of *B. thuringiensis* (belonging to serotype H_{14}), whose parasporal crystals (= δ-endotoxin) are highly toxic for larvae of Diptera (e.g., *Aedes aegypti*), have been isolated from mosquito larvae and from soil of mosquito breeding sites (de Barjac, 1978).

Bacillus larvae

In contrast to *Bacillus thuringiensis,* most obligately pathogenic bacilli have a very narrow host range even if they are injected into insects. *B. larvae* (syn. *B. brandenburgensis*), in particular, only attacks larvae of bees (*Apis mellifera*), and causes American foul brood disease (*Bösartige Faulbrut*).

Germination of spores within the gut, which is a prerequisite for infection, is bound to special features of the gut juice. This is evident for *B. thuringiensis, B. popilliae,* and *B. larvae.* Spores of *B. larvae,* especially, germinate only in the gut of larval bees, not in the gut of adults. This situation is of ecological importance for *B. larvae,* because worker bees, tolerant to the pathogen, perform direct transmission by feeding spores to the sensitive larvae. For additional information about pathology of the American foul brood, see Davidson (1973).

The source of *B. larvae* are honeycombs with the remains of dead larvae and the gut content of seemingly normal adult bees; *B. pulvifaciens* and *B. alvei* are often found in the same habitat. *B. thuringiensis* and *B. cereus* isolated from honeycombs are not suspect as bee pathogens; they derive, instead, from infected larvae of wax moths (*Galleria mellonella* or *Achroia grisella*).

Bacillus popilliae and Related Bacilli

Bacilli of the *Bacillus popilliae* group are present in soils infested by white grubs that suffer from milky disease (Dutky, 1963). This group embraces three species: *B. popilliae, B. lentimorbus,* and *B. euloomarahae.* Infection of larvae of susceptible Scarabaeidae occurs usually after ingestion of spore-contaminated soil.

Only *B. popilliae* of all milky disease pathogens produces a parasporal body which has (especially after hydrolysis in alkaline) some toxicity on white grubs after injection, but not after oral application (Weiner, 1978). Otherwise, sensitive larvae showed typical pathogenic signs and symptoms after ingestion of spores alone (Lüthy, 1968). This is not surprising, since *B. lentimorbus,* which does not produce a parasporal body, also induces typical

symptoms of milky disease. The pathology of milky disease is described by Splittstoesser et al. (1973), Splittstoesser, Kawanishi, and Tashiro (1978), and Kawanishi, Splittstoesser, and Tashiro (1978).

Practical Importance of Insect-Pathogenic Bacilli

Insect pathogens are not only of ecological but also economic importance. On the one hand, they may induce harmful epizootics in insect-rearing, e.g., *B. larvae* in beekeeping and *B. thuringiensis* in sericulture; on the other hand, pathogens often induce diseases in free-living insects and may even act as regulative factors in population dynamics. This gives credence to the idea of exploiting insect-pathogenic bacteria for microbial control of insect pests. For general information consult Burges and Hussey (1971), Franz and Krieg (1976), and Lüthy (1975).

In 1928, a pioneer project was started in Europe with spore preparations of *B. thuringiensis* strains produced in small-scale laboratory cultures for use in controlling the European corn borer (*Ostrinia nubilalis*). Since 1958, several industrial formulations of this bacillus have been manufactured in some countries (USA, USSR, France, Germany). During the last 20 years, their application in many countries has effected very good control of more than 35 lepidopterous pests in agriculture and forestry. After spraying, the active deposits of *B. thuringiensis* on plants do not persist over extended periods, and therefore large quantities of a spore preparation are needed for effective plant protection. Presently, many tons per year are manufactured in large-scale fermentation equipment. (For special information, see Krieg, 1967, and Dulmage and Rhodes, 1971.)

In 1939, a remarkable microbial control program was initiated in North America. In this case, *B. popilliae* was introduced into areas of the eastern United States that were heavily infested by the Japanese beetle (*Popillia japonica*), but in which *B. popilliae* did not occur naturally. For this project, the necessary quantities of spores could only be produced in vivo (in larvae of the Japanese beetle), because the in vitro production of infective spore material is still lacking. Therefore, the availability of spore preparations for applications was permanently limited. This limitation is not too important because the economy of application is essentially better than in the case of *B. thuringiensis.* After the application of *B. popilliae* spores into soil inhabited by white grubs, the infectious spore material may persist over many years; furthermore, it is supplemented by spores from decaying larvae that had died from milky disease. Therefore, in contrast to *B. thuringiensis,* which induces only temporary effects on lepidopterous pests, *B. popilliae* enables the control of populations of white grubs over a long period after only one application by soil injection. (For special information, see Dutky, 1963.)

Isolation

Smears from diseased insects should first be examined microscopically to differentiate Eubacteriales and Pseudomonadales from other insect pathogens. For cultural isolation of the bacteria, small samples of the insect specimens are homogenized and suspended in sterile water. This suspension is serially diluted, and then 1 ml of the dilution is mixed with molten nutrient agar in Petri dishes. If sporeformers are suspected, the diluted suspension is heated to 80°C for 10 min before plating. With this procedure, the endospores of bacilli will remain active, but all thermosensitive microorganisms (nonsporeformers and vegetative cells of bacilli) will become inactivated. Unfortunately, even though bacilli sporulate readily when cultured in vitro, they often have difficulty in sporulating inside host cadavers. Therefore, a parallel culture should be laid out with infectious material that is not preheated. For this additional culture, soil agar or wheat grit agar, which favor sporulation, is recommended. Under aerobic conditions, only members of the genus *Bacillus* will grow and sporulate on agar plates after incubation at 25°C for several days.

The diagnosis ''bacillosis'' should be confirmed on the basis of Koch's postulates, especially since saprophytic sporeformers are widespread in nature and are very often present in the gut microflora of insects. After the death of an insect and the collapse of its immune mechanisms, these bacilli may multiply in the autolyzing tissues, thereby causing rot. Such secondary invaders may feign the proper causative agents.

Bacillus thuringiensis

Like *Bacillus cereus* and other saprophytic sporeformers, *B. thuringiensis* can be isolated and propagated easily in a simple liquid medium, such as nutrient broth or nutrient agar, by incubation at 27–37°C. Adding antibiotics to nutrient agar is recommended to obtain a more selective medium for isolating *B. thuringiensis* and *B. cereus*. For example, 10 μg of lysozyme per ml inhibits growth of *B. megaterium*, and 10 U of penicillin G per ml suppresses *B. anthracis* and other sensitive sporeformers. *B. anthracis* as well as species of the *B. subtilis* group may be suppressed by 10–100 μg of polymyxin B per ml.

In contrast to most strains of *B. megaterium*, growth of *B. cereus* and *B. thuringiensis* is not supported by a minimal glucose–ammonium salts me-

dium. However, growth does occur if the medium is supplemented by glutamate, aspartate, citrate, cysteine, thiosulfate, or ethylenediaminetetraacetic acid. Nickerson and Bulla (1975) suggested that fatty acid formation is stimulated by these components. Singer and Rogoff (1968) reported an inhibitory effect of several amino acids (valine, leucine, or isoleucine) on the growth of *B. thuringiensis* or *B. cereus* in ammonium salts medium. Sensitivity to amino acids was strain specific. According to these findings, which are an indication of the sensitivity of *B. thuringiensis* strains to amino acid imbalance, amino acid nutrition should be carefully considered in the choice of defined media. Several mutants of *B. thuringiensis* strains were resistant to the growth-inhibiting effect of these particular amino acids.

With regard to nucleotides, purines, and pyrimidines, *B. thuringiensis* strains are prototrophic; however, thymine-requiring mutants have been recorded (de Barjac, 1970).

A semidefined medium with ammonium as the sole nitrogen source was recommended by Lüthy (1975). Its composition is as follows:

Semidefined Growth Medium for
Bacillus thuringiensis (Lüthy, 1975)

Yeast extract (Difco)	2.0	g
$(NH_4)_2SO_4$	2.0	g
Glucose	3.0	g
$K_2HPO_4 \cdot 3H_2O$	0.5	g
$MgSO_4 \cdot 7H_2O$	0.2	g
$MnSO_4 \cdot 4H_2O$	0.05	g
$CaCl_2 \cdot 2H_2O$	0.08	g
Distilled water	to 1,000 ml	

This medium is adjusted to pH 7.3 before sterilization.

A nutrient broth adapted to the propagation of *B. thuringiensis* was used by Singer and Rogoff (1968). It contains the following ingredients:

Nutrient Broth for *Bacillus thuringiensis*
(Singer and Rogoff, 1968)

Yeast extract	5.0 g
Tryptone	5.0 g
Glucose	1.0 g
K_2HPO_4	0.8 g
Distilled water	to 1,000 ml

This medium is adjusted with 1% NaOH to pH 7.0 before sterilization.

Useful, economical media for mass propagation of *B. thuringiensis* have been described by Dulmage and Rhodes (1971). In fermentor techniques, the amount of inoculum should represent 1–2% of the total volume to be inoculated.

For aerobic degradation of acidic fermentation products (such as acetic acid), which inhibit spore production, enhancing the respiratory activity by good aeration of the culture is favorable; buffering the medium or adding $CaCO_3$ also supports spore production. According to Scherrer, Lüthy, and Trümpi (1973), the size of parasporal crystals of *B. thuringiensis* is closely related to the concentration of glucose (maximum 0.8%) within the medium. If *B. thuringiensis* is propagated at 32°C in culture bottles on a shaker incubator or in an aerated small-scale fermentor, sporulation and cell lysis become complete 72–96 h after inoculation (when the medium approaches pH 8.5).

For recovery of the spore crystal complex, the culture is centrifuged and the supernatant is discharged. The sediment, after being washed with water, is spray-dried or lyophilized. The resulting spore crystal powder can be stored below 20°C over many years in an active stage. The germination rate of spores of *B. thuringiensis* is enhanced by heat shock (65°C for 30 min). Germination is also initiated by such substances as alanine, adenosine, and phosphate. On the other hand, outgrowth may be inhibited by dipicolinic acid or alanine analogs such as lactate, pyruvate, and serine. Without any stimulation of the spore, neither reception of further organic molecules nor biosynthesis of RNA and protein will take place, both of which are necessary to start vegetative growth.

Bacillus larvae

In contrast to *Bacillus thuringiensis,* other insect-pathogenic bacilli such as *B. larvae, B. popilliae,* and *B. lentimorbus* can be isolated only with difficulty. An extended incubation is required for the primary isolation from natural sources; however, after repeated transfer, these bacteria may become adapted to several culture media. Growth from spores always requires a long incubation period, as compared with inocula of vegetative cells. For the isolation of *B. larvae* from the remains of honeybee larvae, a differential sterilization technique can be used (Rose, 1969). The specimen is heated at 60°C up to 15 min in nutrient broth (in which *B. larvae* will not germinate) and transferred to a growth-supporting medium, e.g., J-agar (Gordon, Haynes, and Pang, 1973). Freshly isolated strains should be incubated at 35–37°C; in contrast to adapted strains, they may fail to grow at 28–30°C. Germination of spores of *B. larvae* and initial growth is favored by microaerophilic conditions, but later growth and sporulation occur best aerobically.

B. larvae will germinate or sporulate only in a medium that contains glucose, peptone, thiamine, and trace elements. Upon adding soluble starch or activated carbon (charcoal), development and sporu-

lation become more reliable (Foster, Hardwick, and Guirard, 1950). Gochnauer (1973) reported growth and sporulation of *B. larvae* in aerated broth cultures (on a reciprocal shaker at 37°C) consisting of brain heart infusion broth (Difco) supplemented by thiamine and charcoal.

Good sporulation is obtained on a solid yeast extract–starch–agar medium, recommended by Foster, Hardwick, and Guirard (1950), and on the yeast extract–starch–phosphate medium of Bailey and Lee (1962).

J-Medium for Growth and Sporulation of
Bacillus larvae (Gordon, Haynes, and Pang, 1973)

Tryptone	5.0 g
Yeast extract	15.0 g
Glucose (autoclaved separately)	2.0 g
K_2HPO_4	3.0 g
Distilled water	to 1,000 ml

Part of the yeast extract can be replaced by other organic nitrogen sources, such as Casamino Acids, and the pH is adjusted to 7.3–7.5. This medium is also the base of J-agar, which is obtained by adding 2% agar.

This so-called J-medium for *B. larvae* also allows vegetative growth of *B. popilliae* and *B. lentimorbus.*

According to Shimanuki, Hartman, and Rothenbuhler (1965), the virulence of a culture medium–adapted strain of *B. larvae* was significantly enhanced by three serial passages through larvae of the honeybee. By using a susceptible and a resistant strain of bees, it was demonstrated that the level of resistance has no influence on the result of selection.

Bacillus popilliae and Related Bacilli

Isolation of milky disease pathogens directly from fluid hemolymph is less easy than from dry films of milky hemolymph on sterile slides.

Germination of *Bacillus popilliae* spores on artificial media is very low (mostly about 1%). After heat treatment (55°C for 15 min), the germination rate can be increased to about 3%. Such heat-activated spores induced a strikingly higher mortality after injection into sensitive scarabaeid larvae than did unheated ones. Heating to 70°C, however, reduces the germination rate (Julian and Hall, 1968).

Fresh isolations of *B. popilliae* on a growth-supporting medium (such as J-agar) have tiny, transparent colonies and grow slowly at 25–28°C. Many fail to grow upon transfer to fresh medium, and adapted lines show reluctant colonies for several passages. In contrast to *B. popilliae* and *B. lentimorbus,* cultivation in the hemolymph of certain scarabaeid beetles is a prerequisite for growth of *B. euloomarahae* on artificial media.

For growth of *B. popilliae* subsp. *popilliae*, Sylvester and Costilow (1964) found a minimal requirement of 14 amino acids, thiamine, biotin, and barbiturate. According to Wyss (1971), for reproduction of *B. popilliae* subsp. *fribourgensis*, a minimal medium has to contain 12 amino acids, glucose and saccharose, thiamine, biotin, and ascorbic acid, starch (as an absorptive substance), and barbiturate (as a stimulant). Better results have been obtained with media that contain yeast extract, tryptone, and Casamino Acids (like J-medium). By using a more complex medium and an inoculum of *B. popilliae* of about 5%, the logarithmic phase of growth is terminated in a small-scale fermentor after 24 h at 30°C. Because usually no sporulation occurs in vitro, a decline of viable cells sets in, equivalent to a death curve that is nearly symmetrical to the growth curve. Only suboptimal conditions such as minimal inocula, limited nutrients, reduced temperature, or reduced supply of oxygen may enhance the longevity of vegetative cells. Death of cultures is apparently a symptom, but not the cause, of asporogeny.

Pure cultures of *B. popilliae* and *B. lentimorbus* may be preserved by lyophilization and will remain viable thereafter for at least 20 months. With *B. lentimorbus*, in contrast to *B. popilliae*, no reduction of virulence as a consequence of lyophilization of cells could be registered (Haynes et al., 1961).

Haynes and Rhodes (1966) first described sporulation of a culture medium–adapted strain of *B. popilliae* subsp. *popilliae* on liquid medium supplemented with activated carbon (charcoal); however, the sporulation rate was very low (about 1%). Later, Sharpe and Rhodes (1973) recorded sporulation on a solid medium at a rate of 15–20%. Wyss (1971) found comparable sporulation rates in colonies of the subspecies *popilliae* on agar medium that contained yeast extract, tryptone, and Casamino Acids in a proportion of 2:1:2. According to the same author, strains of the subspecies *fribourgensis* never sporulate in a liquid medium. For this subspecies, he recommended a solid agar medium that contains yeast extract, tryptone, and Casamino Acids in a proportion of 2:1:1. In this medium, a sporulation rate up to 20% could also be obtained. However, under these conditions *B. lentimorbus* sporulated only sparsely.

Symptoms of milky disease in sensitive scarabaeid larvae are induced only by spores propagated in vivo, but not by spores grown on artificial media; the in vitro spores showed only slight infectivity after injection. Despite great efforts to develop suitable media for *B. popilliae*, the production of virulent spores in vitro has not yet been achieved.

Industrial production of milky disease spores in vivo involves the rearing of homologous host insects, their inoculation by injection, and their incubation under controlled conditions. It should be noted that sporulation already occurs during the vegetative phase of growth, in which the reproduction of *B. popilliae* occurs not in a logarithmic but rather in a nearly linear progression. After injection and incubation at 30°C, a yield of about 10^9–10^{10} spores per larva is reached in about 20 days. The spore-containing hemolymph should be gathered just prior to death. For preservation of spores in a dry form, lyophilization is useful but not necessary. Storage of milky disease spores in an active form is possible for some years at temperatures below 20°C. The effect of duration of storage and temperature on germinability and infectivity of such spores stored in water has been reported by Milner (1976).

Identification

Common Characteristics

Within the family Bacillaceae, which is characterized by the formation of thermoresistant endospores (which contain dipicolinic acid), members of the genus *Bacillus* are chemoorganotrophic organisms which exhibit the following features: rod-shaped vegetative cells, often with peritrichous flagella; not more than one ellipsoid or spheroid spore in a sporangium. The metabolism of vegetative cells is respiratory, fermentative, or both, using various organic substrates. In respiratory metabolism, the terminal electron acceptor is molecular oxygen, but in some species it is replaceable by nitrate. With exceptions, catalase is produced. The G+C content of the DNA of those strains examined ranges from 32 to 62 mol%.

The *Bacillus cereus– Bacillus thuringiensis* Group

The *Bacillus cereus–B. thuringiensis* group belongs to division I of the genus *Bacillus*—as set forth in the seventh edition of *Bergey's Manual of Determinative Bacteriology* (Breed, Murray, and Smith, 1957) and unfortunately abandoned in the eighth edition (Buchanan and Gibbons, 1974). Division I is characterized morphologically by Gram-positive vegetative cells and spores of ellipsoidal to cylindrical shape with a thin spore wall enclosed by two envelopes, the endo- and exosporium. Spores have a central-to-terminal position in a not definitely swollen sporangium. The *B. cereus* group is representative of subdivision A and the diameter of vegetative cells is here more than 0.9 µm. Subdivision B, with diameters of vegetative cells smaller than 0.9 µm, includes the *B. subtilis* group.

Common physiological characteristics of division I are as follows: from carbohydrates acid but not gas

Table 2. Differential characteristics of the *Bacillus cereus–B. thuringiensis* group and related bacilli.[a]

Characteristic	*Bacillus cereus*	*Bacillus thuringiensis*	*Bacillus finitimus*	*Bacillus anthracis*	*Bacillus megaterium*
Parasporal body	−	+	+	−	−
Flagella	+	+	+	−	d
Capsule	−	−	−	+[b]	+[c]
Acetoin production	+	+	+	+	−
Acid from mannitol	−	−	−	−	+
Amylase	+	+	−	+	+
Proteinase	+	+	+	+	d
Hemolysin	+	+	+	−	−
Phospholipase	+	+	+	+	−
Growth on anaerobic agar	+	+	+	+	−
Sensitivity to γ-phage	−	−	−	+	−
Sensitivity to lysozyme	−	−	−	−	+
Sensitivity to penicillin G	−	−	−	+	−

−, Negative in character/reaction; d, reactions differ.
[a] Krieg, 1969. +, Positive in character/reaction.
[b] Produced on bicarbonate agar under CO_2.
[c] Produced on glucose agar.

is produced; proteins are decomposed with the production of ammonia; catalase is formed; and aerobic growth occurs on glucose peptone agar. The species that belong to subdivision IA are listed in Table 2.

Przyborowski (1964) used spore antigen precipitation tests with some success for interspecies differentiation of several *Bacillus* spp. and demonstrated the existence of an antigenic relationship between members of subdivision IA. Similar results were obtained by Lamanna and Jones (1961). By using spore antigen agglutination and absorption tests, these authors could demonstrate antigenic complexity of the spores and the existence of a complicated pattern of distribution of spore antigens among *B. anthracis, B. cereus,* and *B. thuringiensis.*

Surface antigens of the cell wall of vegetative cells have been compared by Baumann-Grace and Tomcsik (1957) and Tomcsik and Baumann-Grace (1959). In this study, 55 strains of *B. megaterium* could be divided into 38 serotypes, and 23 strains of the *B. cereus–B. thuringiensis* group were divided into 13 serotypes. In contrast, 12 *B. anthracis* strains belonged to a single serotype. (For the subspecies character of flagellar antigens of *B. thuringiensis,* see below.)

B. thuringiensis is distinguished from *B. cereus* only by the production of a parasporal crystal (sometimes two) about 1 μm in diameter. These crystals become hydrolyzed in vitro by weak alkali or special proteases (for example, as in the larval gut of Lepidoptera). After application per os, crystal hydrolysates are toxic for many insects, e.g., larvae of *Bombyx mori* and *Pieris brassicae.*

Studies on the biosynthesis of the parasporal crystal protein and its biochemical properties have been carried out by Yousten and Rogoff (1969), Sommerville (1971), Lecadet and Dedonder (1971), Herbert and Gould (1973), and Gould et al. (1973). It could be demonstrated that a typical mRNA-dependent protein synthesis is involved in crystal formation.

Early attempts to recognize subspecies of *B. thuringiensis* on the basis of biochemical differences have been made by several authors (e.g., Heimpel and Angus, 1958); however, all physiological patterns of *B. thuringiensis* overlap with those of noncrystalliferous *B. cereus* strains. Meanwhile, de Barjac and Bonnefoi (1962) initiated a differentiation based on flagellar antigens. This scheme of H serotypes has been successfully applied to the definition of subspecies. The biochemical features of the H serotypes of *B. thuringiensis* are listed in Table 3. By using fluorescent conjugates of H-specific antibodies, an identification of single cells of *B. thuringiensis* on the basis of their flagellar antigens is possible (Krieg, 1965).

With regard to the H-antigen type of *B. thuringiensis* strains, there exists a very close relation with isoesterase patterns (Norris, 1964). However, special agglutinogens of endospores (Lamanna and Jones, 1961) and specific antigen patterns of the parasporal crystals (Pendleton and Morrison, 1966) could not be correlated with special H serotypes. For all that, Krywienczyk, Dulmage, and Fast (1978) proposed the antigenic composition of parasporal crystals as an aid in identification of groups of strains within one H serotype (for example in the serotype H 3a,b).

Typical strains of *B. thuringiensis* produce their toxic parasporal crystal within the sporangium but outside of the exosporium. In contrast, *B. finitimus* (Heimpel and Angus, 1958) produces a nontoxic parasporal crystal inside the exosporium. Therefore, unlike other subspecies of *B. thuringiensis,* the parasporal crystal in the subspecies *finitimus* is strongly attached to the spore and is not released separately. Similar circumstances exist with Fowler's bacillus (Hannay, 1961), but this type differs from *B. finitimus* not only by lack of motility and H_2 antigen, but also by several biochemical features (Krieg, 1969); therefore, its association with *B. finitimus,* as proposed by Heimpel (1967), is not recommended.

Difficulties for serotyping *B. thuringiensis* arise not only from autoagglutinating strains, but also from nonagglutinating ones. Such a *B. thuringiensis* strain without any motility and H antigenicity was isolated recently in China. This type exhibits strong pathogenicity to caterpillars and represents the new subspecies *wuhanensis* (Anonymous, 1976).

At present, a clear and convincing differentiation between *B. cereus* and *B. thuringiensis* is handicapped by the following items: occurrence of

Table 3. Differentiation of subspecies of *Bacillus thuringiensis* by flagellar antigens and biochemical properties.[a]

B. thuringiensis subspecies	H antigens	Isoesterase type	Acid from mannose	Acid from saccharose	Acid from salicin	Amylase	Chitinase	Urease	Arginine dihydrolase	Isolated in
thuringiensis	1	1	+	+	+	+	d	−	+	Europe
finitimus	2	2	−	+	+	−	+	−	−	North America
alesti	3a	3	−	−	−	+	d	−	+	Europe
kurstaki	3a, b		−	−	+	+	d	+	+	Europe
sotto	4a, b	4s	−	+	−	+	d	−	−	East Asia
dendrolimus	4a, b	4d	−	−	−	+	+	−	d	Central Asia
kenyae	4a, c	4k	−	−	+	+	d	+	+	Africa; Asia
galleriae	5a, b	5/7	−	−	+	+	d	+	+	Europe; Asia
canadensis	5a, c		−	+	+	+	+	−	+	North America; Africa; Asia
entomocidus	6	6	+	+	−	+	−	−	−	North America
aizawai	7	5/7	−	−	+	+	+	+	+	East Asia; Europe
morrisoni	8	8	−	+	−	+	d	−	−	Europe; Asia
tolworthi	9	9	−	+	+	+	d	−	+	Europe
darmstadiensis	10	10	−	−	−	+	−	−	+	Europe
toumanoffi	11		−	−	−	+	−	+	+	Europe
thompsoni	12		−	+	+	+	d	+	+	North America
pakistani	13		−	+	(+)	+	(+)	−	+	Asia
israelensis	14		+	−	−	+	−	−		Near East
wuhanensis	0		−	−	+	+		−		East Asia

+, Positive reaction; −, negative reaction; d, reactions differ; (+), weakly positive reaction.

[a] de Barjac and Bonnefoi, 1962, 1973; Krieg, 1968, 1971b.

noncrystalliferous mutants; occurrence of nonentomopathogenic types of crystalliferous bacilli; occurrence of autoagglutinating and nonagglutinating strains that defeat a classification based on H antigens; and overlapping of H serotypes among crystalliferous and noncrystalliferous strains.

DNA base ratios of *Bacillus* strains have been calculated from melting point temperature by Marmur and Doty (1962) and Bonde and Jackson (1971). Several groups with similar ratios have been selected. A group with a G+C content of 32–37 mol%, which is constituted by *B. cereus, B. thuringiensis,* and *B. anthracis,* seems relatively homogeneous in contrast to others, which overlap considerably. Also, results of DNA/DNA competition experiments, which should be regarded with some caution, show that *B. cereus, B. thuringiensis,* and *B. anthracis* form a closely related group (Sommerville and Jones, 1972). However, at present, considerations of DNA base ratios and of DNA hybridization data cannot contribute to an identification or differentiation within the *B. cereus–B. thuringiensis* group.

Since 1960, bacteriophages of *B. thuringiensis* have been known. A more systematic study was undertaken by de Barjac, Sisman, and Cosmao-Dumanoir (1974) on 17 phage isolates belonging to 6 morphological types. Their host range has been proved on strains of 12 H serotypes of *B. thuringiensis*. Temperate phages are often specific only for special serotypes, e.g., phage *thu* 2 for subspecies *thuringiensis* and phage *ent* 1 for subspecies *galleriae*. Such phages may be useful for strain characterization on and beneath subspecies level. Virulent phages, however, may induce lysis of several serotypes, e.g., *dar* 1 attacks strains of all 12 serotypes. Often, virulent phages may also be effective against other species of *Bacillus* (Yoder and Nelson, 1960). Infection of *B. thuringiensis* cultures with virulent phages may decrease their insecticidal activity.

Thuricins and cerecins produced by strains of *B. thuringiensis* and *B. cereus* are bacteriocin-like factors of proteinic nature. They inhibit not only growth of *B. thuringiensis* and *B. cereus* strains, but also growth of many other Gram-positive bacteria. Autoinhibition is rare and is only observed in the most active producers (Goze, 1972; Krieg, 1970). Differences in production and reception of cericins or thuricins have sometimes been exploited for strain characterization beneath the subspecies level.

B. thuringiensis and *B. cereus* are resistant to sulfonamides, penicillin G, bacitracin, biomycin, and polymyxin B, but are sensitive to streptomycin, tetracyclines, chloramphenicol, neomycin, vancomycin, erythromycin, and nitrofurantoin.

Identification of *B. cereus*–*B. thuringiensis* and their differentiation from other bacilli of subdivision IA is difficult (Krieg, 1969). It should be mentioned that *B. anthracis* can be identified on the basis of a "string of pearl reaction" with penicillin (Jensen and Kleemeyer, 1953), a γ-phage sensitivity test (Brown and Cherry, 1955), and the fluorescent conjugate of anthrax-specific antibodies (Weaver, Brachman, and Feeley, 1970). Identification of *B. megaterium* is possible with mannitol fermentation, lysozyme sensitivity, lack of phospholipase, and lack of acetoin production.

Bacillus larvae

Together with the milky disease organisms, *Bacillus larvae* (syn. *B. brandenburgensis*) belongs to division II (*Bergey's Manual,* seventh edition). It is characterized by Gram-variable cells and spores of ellipsoidal shape with a thick spore wall. The sporangia are definitely swollen.

In contrast to category IIB1, which comprises saprophytic species (such as *B. pulvifaciens* and the *B. alvei*–*B. circulans* group), *B. larvae* and the milky disease group, which do not grow on ordinary media, are listed in category IIB2 of the genus *Bacillus.*

The thick-rimmed spores of *B. larvae* have a central-to-terminal position in their spindle-shaped sporangium before they become free. The rod-shaped vegetative cells, single or in chains, measure 0.5–0.6 by 1.5–6.0 μm. Some strains were motile by means of peritrichous flagella (Gordon, Haynes, and Pang, 1973). Giant whips (Riesenzöpfe) have often been observed in microscopic preparations of *B. larvae;* they probably represent spontaneously agglutinated flagella (Frank and Hoffmann, 1968). As a pathogen of bees, causing American foul brood, *B. larvae* is distributed worldwide; it was first isolated in North America and Europe.

For identification of *B. larvae,* Fritsch (1957) used antisera against H and O antigens in an agglutination test. In addition, Poltev (1958) described a species-specific precipitation test with spore antisera for serodiagnosis of American foul brood in bee extracts. Recent studies by Peng and Peng (1979) indicate that the immunodiffusion test as well as the immunofluorescence technique both are sensitive and specific tools for diagnosis of field samples collected from bee hives.

In a lysogenic strain of *B. larvae,* bacteriophage production was reported by Gochnauer and L'Arrivee (1969) but phage typing of strains has not been exploited.

B. larvae is sensitive to chemotherapeutic agents of the sulfonamide group and antibiotics of the tetracycline group. Sulfathiazole and oxytetracycline in particular are effective in controlling American foul brood, if incorporated into diets fed to honeybee colonies. Several strains of *B. larvae* are sensitive to aureomycin, chloramphenicol, bacitracin, and penicillin G. Honey also has an inhibitory effect on growth of this bacillus.

As already mentioned, *B. larvae* requires thiamine and certain amino acids for growth in defined culture media.

Common features exist for *B. larvae, B. popilliae,* and *B. lentimorbus:* amylase is not formed; acid is produced from glucose and trehalose (the blood sugar of insects), but not from arabinose or xylose; dioxyacetone and indole are not produced; phenylalanine is not deaminated and tyrosine is not decomposed; catalase is negative; and growth has been observed on anaerobic agar and in the presence of 0.001% lysozyme. Characteristic features of *B. larvae, B. popilliae, B. lentimorbus,* and other bacilli that must be differentiated from them are listed in Table 4.

Table 4. Differential properties of *Bacillus larvae, B. popilliae,* and related bacilli.[a]

Bacillus species	Parasporal body	Flagella	Reduction of NO_3	Acid from mannitol	Proteinase	Growth on nutrient agar	Growth at 2% NaCl	Sensitivity against sulfonamides
B. larvae	−	d	d	d	+	−	+	+
B. alvei-circulans	−	+	−/d	−	+	+	+	
B. pulvifaciens	−	+	+	+	+	+	+	
B. popilliae	+	d	−	−	−	−	+	−
B. lentimorbus	−	−	−	−	−	−	−	−
B. euloomarahae	−	−						

+, Positive in character/reaction; −, negative in character/reaction; d, reactions differ.
[a] Gordon, Haynes, and Pang, 1973.

Bacillus popilliae and Related Bacilli

The taxonomy of milky disease organisms is based on their morphological characteristics and on the restriction of their host range to larvae of Scarabeidae (Coleoptera). The morphology of milky disease bacteria has been described by Steinkraus and Tashiro (1967). The vegetative rods of *Bacillus popilliae* and *B. lentimorbus* have a diameter of 0.5–0.7 μm and the spores measure 0.9 × 2.0 μm.

Unlike *B. lentimorbus,* all strains of *B. popilliae* produce parasporal bodies (varying from 0.5 to 1.0 μm in diameter). At present, the varieties of *B. popilliae* are distinguished primarily by their host range (see Table 5): subsp. *popilliae* is isolated from *Popillia japonica;* subsp. *fribourgensis* (identical with subsp. *melolonthae*) is isolated from *Melolontha melolonthae.* The host ranges of *B. popilliae* subsp. *new zealand* (isolated from *Odontria zealandica;* not pathogenic for *Popillia japonica*) and subsp. *rhopaea* (isolated from *Rhopaea verreauxi*) have not been investigated sufficiently.

On the basis of their host specificity (see Table 5), the following varieties of *B. lentimorbus* have been generally accepted: subsp. *lentimorbus* isolated from *Popillia japonica* and subsp. *australis* isolated from *Sericesthis pruinosa.* According to Gordon, Haynes, and Pang (1973), *Bacillus lentimorbus* subsp. *maryland,* described by Steinkraus and Tashiro (1967), is possibly a variety of *B. euloomarahae.* This species, isolated first from white grubs of *Heteronychus sancta-helenae* by Beard (1956), does not produce parasporal bodies. It has tiny spores (0.2 by 0.4 μm) and relatively small vegetative cells (0.3 μm in diameter) and therefore is morphologically distinct from the other milky disease pathogens.

A close serological relationship exists between antigens (prepared from homogenized vegetative cells) of the two varieties of *B. popilliae:* subsp. *popilliae* and subsp. *fribourgensis* (syn. *melolonthae*). Also, *B. popilliae* and *B. lentimorbus*

share common antigens (Krywienczyk and Lüthy, 1974), but no information exists, so far, about any serological relationships to *B. euloomarahae.*

B. popilliae and *B. lentimorbus* are resistant to sulfonamides and vancomycin, but are sensitive to penicillin, streptomycin, tetracyclines, chloramphenicol, bacitracin, neomycin, erythromycin, and nitrofurantoin. *B. popilliae* is resistant and *B. lentimorbus* is sensitive to polymyxin B (50 U/ml) (Pridham, Hall, and Jackson, 1965).

Differential Metabolic Patterns

Remarkable differences in carbohydrate metabolism exist among some insect-pathogenic bacteria. In *Bacillus cereus, B. thuringiensis, B. popilliae,* and *B. lentimorbus,* the fructose-1,6-bisphosphate (FDP) pathway (and, to a lesser extent, the pentosephosphate [PP] pathway) is the primary mechanism for glucose breakdown. *B. larvae,* however, contains enzymes of three pathways: FDP, PP, and the 2-keto-3-deoxy-6-phosphogluconic acid (KDPG) scheme. Radiorespirometric analyses reveal that vegetative cells of *B. larvae* predominately dissimilate glucose directly via the oxidative KDPG pathway (Julian and Bulla, 1971). It should be mentioned that enzymes of the KDPG pathway have not been reported in other bacilli.

Vegetative cells of *B. popilliae* and *B. lentimorbus* decarboxylate the C-1 of pyruvate. In contrast to these species, *B. alvei* and *B. thuringiensis* may oxidize end products of the glucose catabolism via the tricarboxylic acid cycle (Bulla, Julian, and Rhodes, 1971). At the onset of sporulation in *B. thuringiensis,* several enzymes of the tricarboxylic acid cycle appear, but not α-ketoglutarate dehydrogenase. Therefore, Aronson et al. (1975) suggested an ancillary pathway that allows for α-ketoglutarate and glutamate catabolism via γ-aminobutyric acid (which may also be true for *B. cereus*).

In contrast to the *B. cereus–B. thuringiensis* group, the other three obligate pathogens, *B. larvae, B. popilliae,* and *B. lentimorbus,* lack catalase de-

Table 5. Some data about host specificity after injection of milky disease pathogens.[a]

Bacillus species	Subspecies	*Cetonia aurata*	*Heteronychus sancta-helenae*	*Melolontha melolonthae*	*Oryctes nasicornis*	*Popillia japonica*	Isolated in
B. popilliae	*popilliae*	+	+	+	+	+	North America
	fribourgensis	−		+	−		Europe
B. lentimorbus	*lentimorbus*		−			+	North America
	australis	−	+	−		+	Australia
B. euloomarahae	*euloomarahae*	+	+	+	+	+	Australia

+, Host is infected; −, host is not infected.
[a] Dutky, 1963.

spite the fact that oxygen is used by them as the terminal electron acceptor. But superoxide dismutase, another enzyme that may protect cells from the superoxide radical (O_2^-), could be demonstrated in high levels in vegetative cells of *B. popilliae* (Yousten, Bulla, and McCord, 1973).

Literature Cited

Angus, T. A., Heimpel, A. M. 1959. Inhibition of feeding and blood pH changes in lepidopterous larvae infected with crystal-forming bacteria. Canadian Entomologist **91:**352–358.

Anonymous (Hubei Institute of Microbiology, Entomogenous Organism Research Group, China) 1976. *Bacillus thuringiensis* "140", a new variety without flagellum. [In Chinese.] Acta Microbiologica Sinica **16:**12–16.

Aronson, J. N., Borris, D. P., Doerner, J. F., Akers, E. 1975. γ-Aminobutyric acid pathway and modified tricarboxylic acid activity during growth and sporulation of *Bacillus thuringiensis.* Applied Microbiology **30:**489–492.

Bailey, L. 1963. The pathogenicity for honey-bee larvae of microorganisms associated with European foulbrood. Journal of Insect Pathology **5:**198–205.

Bailey, L., Lee, D. C. 1962. *Bacillus larvae:* Its cultivation in vitro and its growth in vivo. Journal of General Microbiology **29:**711–717.

Baumann-Grace, J. B., Tomcsik, J. 1957. The surface structure and serological typing of *Bacillus megaterium.* Journal of General Microbiology **17:**227–237.

Beard, R. L. 1956. Two milky diseases of Australian Scarabaeidae. Canadian Entomologist **88:**640–647.

Bonde, G. J., Jackson, D. K. 1971. DNA-base ratios of *Bacillus* strains related to numerical and classical taxonomy. Journal of General Microbiology **69:**vii.

Breed, R. S., Murray, E. G. D., Smith, N. R. (eds.) 1957. Bergey's manual of determinative bacteriology, 7th ed. Baltimore: Williams & Wilkins.

Brown, E. R., Cherry, W. B. 1955. Specific identification of *Bacillus anthracis* by means of a variant bacteriophage. Journal of Infectious Diseases **96:**34–39.

Buchanan, R. E., Gibbons, N. E. (eds.) 1974. Bergey's manual of determinative bacteriology, 8th ed. Baltimore: Williams & Wilkins.

Bucher, G. E. 1963. Nonsporulating bacterial pathogens, pp. 117–147. In: Steinhaus, E. A. (ed.), Insect pathology, an advanced treatise, vol. 2. New York, London: Academic Press.

Bulla, L. A., Jr., Julian, G. St., Rhodes, R. A. 1971. Physiology of sporeforming bacteria associated with insects. III. Radiospirometry of pyruvate, acetate, succinate, and glutamate oxidation. Canadian Journal of Microbiology **17:**1073–1079.

Burges, H. D., Hussey, N. W. (eds.) 1971. Microbial control of insects and mites. London: Academic Press.

Davidson, E. W. 1973. Ultrastructure of American foulbrood disease pathogenesis in larvae of the worker honey bee, *Apis mellifera.* Journal of Invertebrate Pathology **21:**53–61.

de Barjac, H. 1970. Thymine-requiring mutants of the insect pathogen *Bacillus thuringiensis.* Journal of Invertebrate Pathology **16:**321–324.

de Barjac, H. 1978. Une nouvelle variété de *Bacillus thuringiensis* très toxique pour les moustiques: *B. thuringiensis* var. *israelensis* serotype 14. Comptes Rendus Hebdomadaires des Séances de l'Académie des Sciences, Série D **286:**787–800.

de Barjac, H., Bonnefoi, A. 1962. Essai de classification biochimique et sérologique de 24 souches de *Bacillus* du type *thuringiensis.* Entomophaga **7:**5–31.

de Barjac, H., Bonnefoi, A. 1973. Mise au point sur la classification des *Bacillus thuringiensis.* Entomophaga **18:**5–17.

de Barjac, H., Dedonder, R. 1965. Isolement d'un nucléotide identifiable à la "toxine thermostable" de *Bacillus thuringiensis* var. *Berliner.* Comptes Rendus Hebdomadaires des Séances de l'Académie des Sciences Série D **260:**7050–7053.

de Barjac, H., Sisman, J., Cosmao-Dumanoir, V. 1974. Description de 12 bactériophages isolés à partir de *Bacillus thuringiensis.* Comptes Rendus Hebdomadaires des Séances de l'Académie des Sciences, Série D **279:**1939–1942.

Dulmage, H. T., Rhodes, R. A. 1971. Production of pathogens in artificial media, pp. 507–540. In: Burges, H. D., Hussey, N. W. (eds.), Microbial control of insects and mites. London, New York: Academic Press.

Dutky, S. R. 1963. The milky diseases, pp. 75–115. In: Steinhaus, E. A. (ed.), Insect pathology, an advanced treatise, vol. 2. New York, London: Academic Press.

Ebersold, H. R., Lüthy, P., Müller, M. 1977. Changes in the fine structure of the gut epithelium of *Pieris brassicae* induced by the δ-endotoxin of *Bacillus thuringiensis.* Mitteilungen der Schweizerischen Entomologischen Gesellschaft **50:**269–276.

Foster, J. W., Hardwick, W. A., Guirard, B. 1950. Antisporulation factors in complex organic media. I. Growth and sporulation studies on *Bacillus larvae.* Journal of Bacteriology **59:**463–470.

Frank, M. E., Hoffmann, H. 1968. Origin of rigid bacterial "giant flagella" (giant whips, Riesenzöpfe). Canadian Journal of Microbiology **14:**941–944.

Franz, J. M., Krieg, A. 1976. Biologische Schädlingsbekämpfung, 2nd ed. Berlin, Hamburg: Verlag Paul Parey.

Fritsch, W. 1957. Untersuchungen an 56 Stämmen des *Bac. larvae* White. Über deren Antigen-Verhältnisse mit Hilfe der Agglutination. Archiv für Bienenkunde **34:**22–32.

Gochnauer, T. A. 1973. Growth, protease formation, and sporulation of *Bacillus larvae* in aerated broth culture. Journal of Invertebrate Pathology **22:**251–257.

Gochnauer, T. A., L'Arrivee, J. C. M. 1969. Experimental infections with *Bacillus larvae.* II. Bacteriophage production in the host. Journal of Invertebrate Pathology **14:**417–418.

Gordon, R. E., Haynes, W. C., Pang, C. H.-N. 1973. The genus *Bacillus.* Agricultural Handbook No. 427 Washington D.C.: Agricultural Research Service, U.S. Department of Agriculture.

Gould, H. J., Loviny, F. L., Vasu, S. S., Herbert, B. N. 1973. Biosynthesis of the crystal protein of *Bacillus thuringiensis* var. *tolworth.* 2. On the relation of transcriptional and translational events in the growth cycle. European Journal of Biochemistry **37:**449–458.

Goze, A. 1972. Thuricines et cérécines moléculaires. Comptes Rendus des Séances de la Société de Biologie et de ses Filiales **166:**200–204.

Hannay, C. L. 1961. Fowler's bacillus and its parasporal body. Journal of Biophysical and Biochemical Cytology **9:**285–298.

Haynes, W. C., Julian, G. St., Jr., Shekleton, M. C., Hall, H. H., Tashiro, H. 1961. Preservation of infectious milky disease bacteria by lyophilization. Journal of Insect Pathology **3:**55–61.

Haynes, W. C., Rhodes, L. J. 1966. Spore formation by *Bacillus popilliae* in liquid medium containing activated carbon. Journal of Bacteriology **91:**2270–2274.

Heimpel, A. M. 1955. Investigations of the mode of action of strains of *Bacillus cereus* Fr. and Fr. pathogenic for the larch sawfly, *Pristiphora erichsonii* (Htg.). Canadian Journal of Zoology **33:**311–326.

Heimpel, A. M. 1967. A taxonomic key proposed for the species of the "crystalliferous bacteria". Journal of Invertebrate Pathology **9:**364–375.

Heimpel, A. M., Angus, T. A. 1958. The taxonomy of insect

pathogens related to *Bacillus cereus* Frankland and Frankland. Canadian Journal of Microbiology **4**:531–541.

Heimpel, A. M., Angus, T. A. 1963. Diseases caused by certain spore-forming bacteria, pp. 21–73. In: Steinhaus, E. A. (ed.), Insect pathology, an advanced treatise, vol. 2. New York, London: Academic Press.

Herbert, B. N., Gould, H. J. 1973. Biosynthesis of the crystal protein of *Bacillus thuringiensis* var. *tolworth*. 1. Kinetics of formation of the polypeptide components of the crystal protein *in vivo*. European Journal of Biochemistry **37**:441–448.

Jackson, R. H., Long, M. E. 1965. Characterization of a cellfree bacterial extract larvicidal to *Lasioderma serricorne* (cigarette beetle). Biochimica et Biophysica Acta **100**:418–425.

Jensen, J., Kleemeyer, H. 1953. Die bakterielle Differentialdiagnose des Anthrax mittels eines neuen spezifischen Testes ("Perlschnurtest"). Zentralblatt für Bakteriologie, Parasitenkunde, Infektionskrankheiten und Hygiene, Abt. 1 Orig. **159**:494–500.

Julian, G. St., Bulla, L. A., Jr. 1971. Physiology of spore-forming bacteria associated with insects. IV. Glucose catabolism in *Bacillus larvae*. Journal of Bacteriology **108**:828–834.

Julian, G. St., Hall, H. H. 1968. Infection of *Popillia japonica* with heat-activated spores of *Bacillus popilliae*. Journal of Invertebrate Pathology **10**:48–53.

Kawanishi, C. Y., Splittstoesser, C. M., Tashiro, H. 1978. Infection of the European chafer, *Amphimallon majalis* by *Bacillus popilliae*: Ultrastructure. Journal of Invertebrate Pathology **31**:91–102.

Krieg, A. 1965. Identifizierung von *Bacillus thuringiensis* var. *thuringiensis* in mikrobiologischen Präparaten durch Kombination von Immunofluoreszenz- und Phasenkontrast-Verfahren. Zentralblatt für Bakteriologie, Parasitenkunde, Infektionskrankheiten und Hygiene, Abt. 1 Orig. **197**:527–532.

Krieg, A. 1967. Neues über *Bacillus thuringiensis* und seine Anwendung. Mitteilungen aus der Biologischen Bundesanstalt für Land- und Forstwirtschaft. Berlin-Dahlem H 125.

Krieg, A. 1968. A taxonomic study of *Bacillus thuringiensis* Berliner. Journal of Invertebrate Pathology **12**:366–378.

Krieg, A. 1969. *In vitro* determination of *Bacillus thuringiensis, Bacillus cereus,* and related bacilli. Journal of Invertebrate Pathology **15**:313–320.

Krieg, A. 1970. Thuricin, a bacteriocin produced by *Bacillus thuringiensis*. Journal of Invertebrate Pathology **15**:291.

Krieg, A. 1971a. Concerning α-exotoxin produced by vegetative cells of *Bacillus thuringiensis* and *Bacillus cereus*. Journal of Invertebrate Pathology **17**:134–135.

Krieg. A. 1971b. Use of cryptograms for characterization of strains of the *Bacillus thuringiensis/Bacillus cereus* group. Journal of Invertebrate Pathology **17**:297–298.

Krywienczyk, J., Dulmage, H. T., Fast, P. G. 1978. Occurrence of two serologically distinct groups within *Bacillus thuringiensis* serotype 3 ab var. kurstaki. Journal of Invertebrate Pathology **31**:372–375.

Krywienczyk, J., Lüthy, P. 1974. Serological relationship between three varieties of *Bacillus popilliae*. Journal of Invertebrate Pathology **23**:275–279.

Lamanna, C., Jones, L. 1961. Antigenic relationship of the endospores of *Bacillus cereus*–like insect pathogens to *Bacillus cereus* and *Bacillus anthracis*. Journal of Bacteriology **81**:622–625.

Lecadet, M. M., Dedonder, R. 1971. Biogenesis of the crystalline inclusion of *Bacillus thuringiensis* during sporulation. European Journal of Biochemistry **23**:282–294.

Lüthy, P. 1968. Untersuchungen an *Bacillus fribourgensis* Wille. Zentralblatt für Bakteriologie, Parasitenkunde, Infektionskrankheiten und Hygiene, Abt. 2 **122**:671–711.

Lüthy, P. 1975. Zur bakteriologischen Schädlingsbekämpfung: Die entomopathogenen *Bacillus*-Arten, *Bacillus thuringiensis*

und *Bacillus popilliae*. Vierteljahrsschrift der Naturforschenden Gesellschaft in Zürich **120**:81–163.

Lysenko, O. 1958. Ecology of microorganisms and the biological control of insects, pp. 109–113. In: Transactions of the First International Conference of Insect Pathology and Biological Control, Prague, 1958.

McConnell, E., Richards, A. G. 1959. The production by *Bacillus thuringiensis* Berliner of a heat stable substance toxic for insects. Canadian Journal of Microbiology **5**:161–168.

Marmur, J., Doty, P. 1962. Determination of the base composition of deoxyribonucleic acid from its thermal denaturation temperature. Journal of Molecular Biology **5**:109–118.

Martin, J. D., Mundt, J. O. 1972. Enterococci in insects. Applied Microbiology **24**:575–580.

Martouret, D. 1961. Les toxins de *Bacillus thuringiensis* et leur processus d'action chez les larves de lépidoptères. Mededelingen Rijksfaculteit Landbouwwetenschappen Gent **31**:1116–1126.

Milner, R. J. 1976. Storage of milky disease spores, pp. 70–71. In: Annual Report 1975–1976. Canberra: CSIRO Division of Entomology.

Nickerson, K. W., Bulla, L. A., Jr. 1975. Lipid metabolism during bacterial growth, sporulation, and germination: An obligate nutritional requirement in *Bacillus thuringiensis* for compounds that stimulate fatty acid synthesis. Journal of Bacteriology **123**:598–603.

Norris, J. R. 1964. The classification of *Bacillus thuringiensis*. Journal of Applied Bacteriology **27**:439–447.

Pendleton, I. R., Morrison, R. B. 1966. Analysis of the crystal antigens of *Bacillus thuringiensis* by gel diffusion. Journal of Applied Bacteriology **29**:519–528.

Peng, Y.-S., Peng, K.-Y. 1979. A study on the possible utilization of immunodiffusion and immunofluorescence techniques as the diagnostic methods for American foulbrood of honey-bees *(Apis mellifera)*. Journal of Invertebrate Pathology **33**:284–289.

Poltev, V. I. 1958. Serodiagnosis of diseases of the honeybee and of other insects, pp. 99–104. [In Russian, with German summary.] Transactions of the First International Conference of Insect Pathology and Biological Control. Prague, 1958.

Pridham, T. G., Hall, H. H., Jackson, R. W. 1965. Effects of antimicrobial agents on the milky disease bacteria *Bacillus popilliae* and *Bacillus lentimorbus*. Applied Microbiology **13**:1000–1004.

Przyborowski, R. 1964. Möglichkeiten und Grenzen der serologischen Differenzierung von Mikroorganismen. Biologische Rundschau **2**:17–25.

Rose, R. I. 1969. *Bacillus larvae* isolation, culturing, and vegetative thermal death point. Journal of Invertebrate Pathology **14**:411–414.

Scherrer, P., Lüthy, P., Trümpi, B. 1973. Production of δ-endotoxin by *Bacillus thuringiensis* as a function of glucose concentrations. Applied Microbiology **25**:644–646.

Sharpe, E. S., Rhodes, R. A. 1973. The pattern of sporulation of *Bacillus popilliae* in colonies. Journal of Invertebrate Pathology **21**:9–15.

Shimanuki, H., Hartman, P. A., Rothenbuhler, W. C. 1965. The effect of serial passage of *Bacillus larvae* White in the honey bee. Journal of Invertebrate Pathology **7**:75–78.

Singer, S. 1973. Insecticidal activity of recent bacterial isolates and their toxins against mosquito larvae. Nature **244**:110–111.

Singer, S., Rogoff, M. H. 1968. Inhibition of growth of *Bacillus thuringiensis* by amino acids in defined media. Journal of Invertebrate Pathology **12**:98–104.

Sommerville, H. J. 1971. Formation of the parasporal inclusion of *Bacillus thuringiensis*. European Journal of Biochemistry **18**:226–237.

Sommerville, H. J., Jones, M. L. 1972. DNA competition studies within the *Bacillus cereus* group of bacilli. Journal of General Microbiology **73**:257–265.

Splittstoesser, C. M., Kawanishi, C. Y., Tashiro, H. 1978. Infection of the European chafer, *Amphimallon majalis* by *Bacillus popilliae:* Light and electron microscope observations. Journal of Invertebrate Pathology **31:**84–90.

Splittstoesser, C. M., Tashiro, H., Lin, S. L., Steinkraus, K. H., Fiori, B. J. 1973. Histopathology of the European chafer, *Amphimallon majalis* infected with *Bacillus popilliae.* Journal of Invertebrate Pathology **22:**161–167.

Steinhaus, E. A. 1946a. Insect microbiology. New York, London: Hafner.

Steinhaus, E. A. 1946b. An orientation with respect to members of genus *Bacillus* pathogenic for insects. Bacteriological Reviews **10:**51–61.

Steinkraus, K. H., Tashiro, H. 1967. Milky disease bacteria. Applied Microbiology **15:**325–333.

Sylvester, C. J., Costilow, R. N. 1964. Nutritional requirements of *Bacillus popilliae.* Journal of Bacteriology **87:**114–119.

Tomcsik, J., Baumann-Grace, J. B. 1959. Serologische Typen von *Bacillus cereus* und ihre Verwandtschaft mit *Bacillus anthracis.* Schweizer Zeitschrift für Pathologie und Bakteriologie **22:**144–157.

Weaver, R. E., Brachman, P. S., Feeley, J. C. 1970. Animal diseases transmissible to man, pp. 354–363. In: Bodily, H. L., Updyke, E. L., Mason, J. O. (eds.), Diagnostic procedures for bacterial, mycotic and parasitic infections, 5th ed. New York: American Public Health Association.

Weiner, B. A. 1978. Isolation and partial characterization of the parasporal body of *Bacillus popilliae.* Canadian Journal of Microbiology **24:**1557–1561.

Wyss, C. 1971. Sporulationsversuche mit drei Varietäten von *Bacillus popilliae* Dutky. Zentralblatt für Bakteriologie, Parasitenkunde, Infektionskrankheiten und Hygiene, Abt. 2 **126:**461–492.

Yoder, P. E., Nelson, E. L. 1960. Bacteriophage for *Bacillus thuringiensis* Berliner and *Bacillus anthracis* Cohn. Journal of Invertebrate Pathology **2:**198–200.

Yousten, A. A., Bulla, L. A., Jr., McCord, J. M. 1973. Superoxide dismutase in *Bacillus popilliae,* a catalaseless aerobe. Journal of Bacteriology **113:**524–525.

Yousten, A. A., Rogoff, M. H. 1969. Metabolism of *Bacillus thuringiensis* in relation to spore and crystal formation. Journal of Bacteriology **100:**1229–1236.

CHAPTER 137

The Genus *Clostridium* (Medical Aspects)

EDWARD O. HILL

The past decade has witnessed a rapid evolution of interest in the anaerobes associated with disease. The increasing use of anaerobic methods of culture in clinical laboratories continues to lengthen the list of species of anaerobes recovered from both human and animal infections. At least 36 species of the genus *Clostridium* have been reported to have been isolated from human clinical specimens, and at least 24 species from diseased animals. Fortunately for the clinical laboratory, relatively few of these species are encountered with any significant frequency. The incidence of recovery of clostridia from human or animal infections is determined to a great extent by the methods employed to collect the specimens, by avoidance of contamination with autochthonous, normal, and transient flora; by the methods of transport to the laboratory and by the methods employed for anaerobic culture.

It frequently is difficult to assess the significance of clostridia from clinical specimens, and very few are considered to be pathogens. Most species are considered to be nonpathogens or their role in disease is uncertain or unknown. The organisms may invade and be recovered from any tissue, and they may or may not be associated with remarkable histopathology; they may indeed serve as a symptom of an underlying disease rather than as a specific etiological agent of infection. For other than the classical exogenous infections of man, such as tetanus, gas gangrene, or wound botulism, the majority of clostridial infections now appear to be endogenous; the sources of the organisms are the gastrointestinal, respiratory, or genitourinary tracts. The recovery of "nonpathogenic" clostridia from clinical specimens is certainly not to be categorically considered as evidence of contamination of specimen or culture. It is important for the microbiologist to be able to culture, isolate, and identify these organisms, and for the physician to become astutely aware of their potential clinical significance.

Clostridia Associated with Human Infections

The 36 species of clostridia reportedly recovered from human clinical specimens are listed in Table 1. Approximately 90% of the clostridia recovered from specimens submitted to a reference laboratory belonged to the 12 species indicated in Table 1 as being most commonly isolated (Smith and Dowell, 1974). Of these 12 species, only *Clostridium perfringens* and *Clostridium septicum* are considered to be true pathogens (Smith, 1975). The others are considered to be nonpathogens or their mechanisms of pathogenicity are uncertain or unknown. As with many other "nonpathogenic, opportunistic" organisms, the clostridia have been isolated from a great variety of clinical specimens, from infections ranging from simple, chronic sinusitis to acute meningitis, from debilitated, immunosuppressed, and otherwise compromised patients with a host of underlying predisposing diseases. *C. perfringens,* the most commonly isolated, has been reported to be associated with infections of the central nervous system (brain abscesses and meningitis), bacteremias, endocarditis, septic arthritis, osteomyelitis, upper and lower respiratory tract infections, urinary tract infections, otitis media, mastoiditis, sinusitis, conjunctivitis, endo- and panopthalmitis, gangrene of the eye, periostitis, appendicitis and other infections of the gastrointestinal tract such as necrotizing enterocolitis, as well as "gas gangrene" or clostridial myonecrosis. It is beyond the scope and intent of this Handbook to present a review of all infections with which clostridia have been associated. For comprehensive reviews the reader is referred to Smith (1975), Willis (1969, 1977), and Finegold (1977).

Clostridium botulinum has received continuing attention since the turn of the century as an agent of food poisoning (Center for Disease Control, 1977;

Table 1. Species of *Clostridium* recovered from human clinical specimens.[a]

C. aminovalericum	*C. limosum*
C. barati	*C. malenominatum*
C. beijerinckii	*C. novyi* A, B
C. bifermentans[b] [7]	*C. oroticum*
C. botulinum A, B,	*C. paraputrificum*[b] [3]
C. butyricum[b] [3]	*C. perenne*
C. cadaveris[b] [3]	*C. perfringens*[b] [26]
C. carnis	*C. pseudotetanicum*
C. clostridiiforme	*C. putrificum*
C. cochlearium	*C. ramosum*[b] [17]
C. difficile	*C. sartagoformum*
C. fallax	*C. septicum*[b] [3]
C. ghoni	*C. sordelli*[b] [6]
C. glycolicum	*C. sphenoides*
C. hastiforme	*C. sporogenes*[b] [11]
C. histolyticum	*C. subterminale*[b] [3]
C. indolis	*C. tertium*[b] [2]
C. innocuum[b] [6]	*C. tetani*

[a] From Buchanan and Gibbons, 1974; Dowell and Hawkins, 1977; Holdeman and Moore, 1977.

[b (   )] Most commonly isolated (percentage of clostridial isolates); Smith and Dowell, 1974.

Smith, 1977—two excellent reviews), an increasing interest in recent years as an agent of wound botulism (Center for Disease Control, 1979; Merson and Dowell, 1973), and more currently as an agent of "infant botulism" (Arnon et al., 1977). There is an increasing recognition of the concept that *C. botulinum* can produce toxin in vivo, in infected tissue as well as during proliferation within the intestinal tract. The concept of "toxin infection" or "toxicoinfection" appears to be definitely distinguishable from the classically recognized "toxin-ingestion" concept of pathogenesis of human botulism. As improved diagnostic laboratory procedures (Dowell et al., 1977c) are employed and awareness of wound and infant botulism increases, the reported incidences of these disease entities are also increasing (Center for Disease Control, 1978 a,b; 1979, 1980; Turner et al., 1978). There is evidence to suggest that intoxication of an adult may also occur via the same mechanism as that of infant botulism. Although still unproved, there is every reason to hypothesize that toxicoinfection may be a mechanism of development of botulism in patients for whom no vehicle has been identified, i.e., in cases of undetermined classification.

C. botulinum toxin types A, B, and E are the principle causes of food-borne ("toxin-ingestion") botulism; type F has been reported on only two occasions in the United States; however, the toxin types associated with a number of reported outbreaks remain unknown or undetermined. Type A has been associated with reported cases of wound botulism, and toxin types A, B, and F with infant ("toxin-infection") botulism (Center for Disease Control, 1977, 1978a, 1980). The occurrence of each of the three known types of botulism correlates with the known geographic distribution of the toxin types in the soils of the United States (Center for Disease Control, 1977; Smith, 1977).

C. septicum bacteremias, although considered rare as primary or spontaneous infections, have received recent attention as endogenous infections associated with underlying malignancies (Alpern and Dowell, 1969, 1971; Koransky, Stargel, and Dowell 1979). The most probable portals of entry for *C. septicum* bacteremia appear to be the cecum and distal ileum, the cecum being the most frequent site of the malignancy. The clinical findings appear to be nonspecific responses to bacteremia with significantly greater mortality when not diagnosed and treated promptly. Infections such as these emphasize the need for: (i) the use of adequate anaerobic methods for culture: (ii) identification of clostridia other than those, such as *C. perfringens,* classically recognized to be associated with human infections: (iii) not categorically dismissing as contaminants those clostridia for which mechanisms of pathogenicity are unknown: and (iv) consideration of the isolation of clostridia from clinical specimens as a possible symptom of an underlying, predisposing disease, e.g., malignancies predisposing to *C. septicum* bacteremia.

Clostridium difficile has been incriminated as a cause of antibiotic-associated, pseudomembranous colitis. An organism that was first described by Hall and O'Toole in 1935 and apparently is a part of the normal flora of the human intestinal tract has been isolated from war wounds, gas gangrene, superficial infections, abscess of fractured femur, infant blood, pleural fluid, peritoneal fluid, vaginal abscess (Smith, 1975), and nonspecific urethritis (Hafiz et al., 1975). Although pathogenicity for laboratory animals was documented early, *C. difficile* has been considered a nonpathogenic saprophyte in man (Smith, 1975). The demonstration of toxin in filtrates of feces from patients with pseudomembranous colitis, by observing the cytopathic effects in tissue cultures, has provided undeniable evidence for the role *C. difficile* plays in this disease entity (Bartlett et al., 1978; George et al., 1979; Larson and Price, 1977; Rifkin et al., 1977). As yet, there are no simple methods for laboratory diagnosis. Presence of toxin in stool specimens must be demonstrated by use of tissue-culture techniques and there are no simple known, reliable methods for isolating the organism (Dowell, 1979).

Clostridia Associated with Infections of Animals

Clostridial infections of animals have long been held to be of exogenous sources and commonly caused by only a few species of known toxinogenicity, consid-

Table 2. Clostridia and some major diseases of animals.[a]

Species	Diseases
C. botulinum	Botulism in various mammals and birds.
C. chauvoei	Blackleg in cattle and occasionally in goats, swine, deer; wound infections in sheep.
C. haemolyticum	Bacillary hemoglobinuria ("red water" disease) in cattle.
C. novyi type B	Infectious necrotic hepatitis (black disease, bradsot, Deutsch braxy) in sheep and occasionally in other animals, e.g., cattle; "bighead" of rams.
C. perfringens type B	Lamb dysentery; enterotoxemia in sheep.
C. perfringens type C	Enterotoxemia "struck" in sheep; enterotoxemia of calves, lambs, piglets.
C. perfringens type D	Enterotoxemia in sheep, goats, cattle; "pulpy kidney" disease in sheep; hemorrhagic enteritis in chickens.
C. septicum	Malignant edema in cattle, braxy in sheep.
C. tetani	Tetanus in various domestic animals; horses are especially susceptible to tetanospasmin.

[a] From Dowell, 1979.

ered histotoxic species, and thus true pathogens. The seven species classically known to be associated with the more common animal diseases and the diseases they are known to cause are indicated in Table 2.

The increasing use of anaerobic methods in studies of animal diseases is resulting in the recovery of many other species of clostridia from clinical specimens (Table 3). Also, as with human infections, the

Table 3. Clostridia isolated from clinical specimens from animals.[a]

C. barati	C. novyi
C. bifermentans	C. paraputrificum
C. botulinum	C. perenne
C. cadaveris	C. perfringens
C. carnis	C. putrificum
C. chauvoei	C. ramosum
C. clostridiiforme	C. septicum
C. colinum[b]	C. sordelli
C. fallax	C. sporogenes
C. haemolyticum	C. subterminale
C. limosum	C. tertium
C. malenominatum	C. tetani

[a] From Berkhoff and Redenbarger, 1977, and from Holdeman and Moore, 1977.
[b] Birds only, agent of "quails disease".

nonhistotoxic or "nonpathogenic" clostridia isolated from animals may be of endogenous sources and may act as opportunistic pathogens in animals with underlying disease that predisposes to infection.

Recent evidence has been reported (Barsanti et al., 1978) that confirms C. botulinum type C to be a cause of botulism in foxhounds. By the use of the improved diagnostic procedures of detecting botulinal toxin in the serum or feces of affected dogs, the classically held concept of canine resistance to botulism has been invalidated. At least some cases of idiopathic "coonhound paralysis" now appear to be botulism, and perhaps canine botulism should be added to Table 2.

Habitats

The species of clostridia associated with human infections have been isolated from a wide spectrum of environmental sources: from soils, dust, sewage, rivers, lakes, sea water, milk, vegetables, fresh meat, insects, fish, whales, shellfish, and the intestinal tract of man and almost every animal examined (see this Handbook, Chapter 138; Buchanan and Gibbons, 1974; Holdeman and Moore, 1977; Smith, 1975).

The soil and intestinal tracts of man and animals appear to be the major habitats; therefore, medically important clostridia can be expected to be recovered from any clinical specimen contaminated by soil, dust, or feces. As members of the normal or transient flora of the respiratory and genitourinary tracts, they may be expected to be recovered from any body fluid or tissue contaminated, colonized, or infected from these endogenous sources.

The sources other than clinical specimens from which the medically important clostridia reportedly have been isolated are indicated in Table 4. Of those isolated from feces only, Clostridium chauvoei is thought to be an obligate parasite of animals, Clostridium cadaveris may be an obligate parasite of man and animals; also isolated from snake venom (Holdeman et al., 1977), Clostridium colinum has been reported only from birds and as an agent of "quail disease" (Smith, 1975). Clostridia isolated from soils only would appear to become associated with disease either by direct contamination of wounds or as transient flora of the gastrointestinal tract.

Although a number of the species isolated from feces have been reported as normal flora of man or animals, the criteria for considering an organism as normal flora have not been clearly defined (see Dubos, 1965, for a suggested classification and definitions of indigenous flora).

Table 4. Other sources of isolation of clostridia recovered from clinical specimens.

Species	Isolated from[a]				Species	Isolated from[a]			
	Feces	Soil, water	Marine sediment	Food		Feces	Soil, water	Marine sediment	Food
C. beijerinckii	+	+			C. limosum		+		
C. bifermentans	+	+	+		C. malenominatum	+			
C. botulinum	+	+	+	+	C. novyi	+	+	+	
C. butyricum	+	+		+	C. oroticum		+		
C. cadaveris	+				C. paraperfringens	+			
C. carnis		+			C. paraputrificum	+	+		
C. chauvoei[b]	+				C. perenne	+			
C. clostridiiforme[b]	+				C. perfringens	+	+	+	+
C. cochlearium	+	+			C. pseudotetanicum				+ (dairy)
C. colinum[b]	+ (bird)				C. putrificum	+	+		
C. difficile	+				C. ramosum	+			
C. fallax		+			C. sardiniensis		+		
C. glycolicum		+			C. septicum	+	+		
C. hastiforme		+			C. sordelli		+	+	
C. haemolyticum[b]					C. sphenoides		+		
C. histolyticum		+			C. sporogenes	+	+		+
C. indolis		+			C. subterminale		+		
C. innocuum	+				C. tertium	+	+		
C. lentoputrescens	+	+			C. tetani	+	+		

[a] Compiled from Buchanan and Gibbons, 1974; Holdeman et al., 1977; Smith, 1975.

[b] From animal clinical specimens only.

Isolation of Clostridia from Clinical Specimens

A large variety of media and procedures have been described for cultivation of the clostridia. For culture of clinical specimens from humans and animals, primary emphasis is to be placed upon the proper selection and collection of specimens to avoid the normal and transiently colonizing flora. Although the clostridia are frequently found in mixtures with other anaerobic and facultative organisms, it usually is not necessary to employ special enrichment or selective media for their recovery. For polymicrobic infections in which clostridia may play a significant role, if the specimen truly represents the invading edge of the infected tissue, the clinically significant clostridia usually are in sufficient numbers to be easily detected and isolated on nonselective, anaerobic blood agars. Isolation of clostridia from feces, or from specimens grossly contaminated with fecal or other normal flora, or from grossly contaminated foods and soils can be problematic; *Clostridium tetani* and *C. botulinum* may produce disease in very small numbers in wounds, and may be extremely difficult to recover or separate from mixed populations, even with the best procedures known. In these cases, enrichment and selective procedures can be helpful, although not guaranteed.

The collection and transport of specimens and methods of obtaining anaerobic conditions are discussed in this Handbook, Chapters 76, 116, and 138. Additional detailed descriptions of media and methods are presented in the manuals by Dowell and Hawkins (1977), Holdeman and Moore (1977), and Sutter, Vargo, and Finegold (1975), and in the book by Willis (1977). An excellent presentation of detailed, stepwise procedures for preparation of media used for culture of anaerobes at the Center for Disease Control is contained in the manual by Dowell et al. (1977a).

Nonselective Media

Some of the blood agar bases that have been used for isolation of anaerobes and the need for standardization are discussed in this Handbook, Chapter 116. The CDC anaerobe blood agar (this Handbook, Chapter 116) has proved to be especially useful as an all-purpose, anaerobic plating medium. It appears to support growth of all clinically significant clostridia (G. L. Lombard, personal communication) and it has a shelf life of at least 6 weeks when stored in cellophane at 4°C (Lombard et al, 1976), an important consideration for the clinical laboratory.

Clostridium chauvoei Blood Agar
(Batty and Walker, 1966; from Willis, 1977)

The following blood agar supports good growth of *C. chauvoei* which may grow poorly on some ordinary blood agars.

Medium contains:

VL broth base	94 ml
Liver extract (Oxoid)	3 g
Glucose	1 g
Agar	1.6 g

Mix and autoclave at 115°C for 10 min. Cool to 50°C, add 5 ml of defibrinated sheep blood, and dispense.

VL broth base contains per liter:

Trytpone (Oxoid)	10 g
NaCl	5 g
Meat extract (Lab-Lemco, Oxoid)	2 g
Yeast extract (Oxoid)	5 g
Cysteine HCl	0.4 g
Agar	0.6 g
Distilled water	1,000 ml

Heat to dissolve, and adjust to pH 7.2–7.4. Autoclave at 115°C for 10 min.

Clostridium novyi Blood Agar
(Moore, 1968; from Willis, 1977)

The following medium supports luxuriant growth of fastidious clostridia such as *C. novyi* types B, C, and D and *C. botulinum* types C, and D. Enhanced growth is attributed to combination of cysteine and dithiothreitol (Collee, Rutter, and Watt, 1971).

Medium contains:

Neopeptone (Difco)	1 g
Yeast extract (Difco)	0.5 g
Proteolyzed liver (Pabryn Labs)	0.5 g
Glucose	1 g
Agar	2 g
Salts solution	0.5 ml

Stock salt solution contains: 4% $MgSO_4 \cdot 7H_2O$; 0.2% $MnSo_4 \cdot 4H_2O$; 0.04% $FeCl_3$; and 0.05% concentrated HCl.

Dissolve neopeptone, yeast extract, liver extract, and glucose in 50 ml of water, add the salts solution, and adjust to pH 7.6–7.8. Mix with the molten agar that has been dissolved in 50 ml water. Dispense in 18-ml volumes in screw-cap bottles; autoclave at 115°C for 10 min. Store at 4°C.

At time of use, melt 18 ml of the basal medium and cool to 50°C. Add 2 ml of defibrinated horse blood and 0.15 ml of freshly prepared re-

ducing solution (containing cysteine HCl, 120 mg; dithiothreitol, 120 mg; and glutamine, 60 mg; in 10 ml of distilled water; pH adjusted to 7.6–7.8, and filter-sterilized). Mix and pour plates immediately.

Reinforced Clostridial Medium (Hirsch and Grinstead, 1954; from Willis, 1977).

The following medium is used as a broth enrichment or with 1.5% agar as a plating medium base for blood, milk, or egg yolk agars.

Medium contains:

Yeast extract	3 g
Meat extract	10 g
Peptone	10 g
Soluble starch	1 g
Glucose	5 g
Cysteine HCl	0.5 g
NaCl	5 g
Sodium acetate	3 g
Distilled water	1,000 ml

Dissolve and filter. Adjust to pH 7.4. Autoclave at 121°C for 15 min.

EGG YOLK AGAR. Egg yolk agar is one of the most useful differential plating media available for detection and preliminary identification of the clostridia. It is used for detection of lecithinase, lipolytic, and proteolytic activities. It may be used to advantage in conjunction with the above primary plating media when clostridia are suspected. It can also be used for the Nagler test as an additional presumptive identification procedure.

Nagler (1939) was the first to describe the opalescence produced by *C. perfringens* when grown in media that contain human serum. McFarlane, Oakley, and Anderson (1941) demonstrated that opalescence occurred in filtered egg yolk suspensions; since then, egg yolk emulsion has been used in a variety of agar bases for detection of lecithinase activity. On human serum agar (Hayward, 1941) and on egg yolk agar (McClung and Toabe, 1947), lecithinase activity produces a diffuse, opalescent precipitate surrounding the colonies. Egg yolk agar provides less variable and more distinct results than human serum agar and, in addition, indicates lipolytic acivity. Lipase activity results in a pearly iridescent film of fatty acids immediately surrounding the colony and extending over the surface of the colony. An excellent description of these activities on the egg yolk agar of McClung and Toabe is presented by Willis (1977).

Proteolytic activity is indicated by a zone of clearing of the medium immediately surrounding the colony (Dowell and Lombard, 1977). The following formulation is from Dowell et al. (1977a) and can be used for both sporeforming and nonsporeforming anaerobes.

Lombard-Dowell Egg Yolk Agar (LD-EYA)
(Dowell et al., 1977a)

Medium contains:

Trypticase (BBL)	5.0 g
Yeast extract (Difco)	5.0 g
NaCl	2.5 g
Sodium sulfite	0.1 g
L-Tryptophan	0.2 g
L-Cystine	0.4 g
Hemin	10.0 mg
Vitamin K (3-phytylmenadione)	10.0 mg
D-Glucose	2.0 g
Na$_2$HPO$_4$	5.0 g
MgSO$_4$ (5% aqueous solution)	0.2 ml
Agar	20.0 g
Distilled water	900 ml

Dissolve hemin and L-cystine in 5 ml of 1 N NaOH before adding. The vitamin K is added from a stock solution containing 1 g of vitamin K (ICN) in 99 ml of absolute ethanol.

Mix all ingredients and dissolve by heating. Adjust pH to 7.4. Autoclave at 121°C for 15 min. Cool to 55–60°C in water bath. Add 100 ml of egg yolk suspension (Difco) and mix. Dispense and store in cellophane at 4°C.

Stiff anaerobe blood agar to inhibit spreading, brain heart infusion agar, thioglycolate medium, and chopped meat medium all support good growth of clostridia and are described in detail in this Handbook, Chapter 116. The hemin and vitamin K supplements need not be added if one is culturing specifically only for the clostridia.

Selective Media

Selective agents can be useful in helping to separate the clostridia from mixtures of other organisms; however, no selective agent has been shown to be the ideal one and the agents are often inhibitory, even for those species for whose selection they are intended. Selective media should never be used alone as a substitute for the enriched, nonselective, ordinary media.

Selective media that have been employed with varying degrees of success include sodium azide–egg yolk agar (Johansson, 1953), sorbic acid plus polymyxin B in thioglycolate broth (Wetzler et al., 1956a), and sodium azide and chloral hydrate in yeast extract–blood agar (Wetzler, Marshall, and Cardella, 1956b). Hafiz et al. (1975) employed 0.2% cresol in reinforced clostridial medium as a selective broth medium to recover *Clostridium difficile* from feces and the urogenital tract. George et al. (1979), however, were unable to cultivate stock strains of *C. difficile* on reinforced clostridial agar medium containing 0.2% cresol. These authors have recommended the use of an egg yolk–fructose agar base that contains cycloserine and cefoxitin as the selective agents. Clindamycin also has been used as a selective agent for *C. difficile* by Bartlett et al. (1977, 1978), but George et al. (1979) report isolation of clindamycin-susceptible strains of *C. difficile* from colitis patients and suggest that clindamycin might be ineffective as a selective agent in such cases.

Two media that appear to be more generally useful are PEA anaerobe blood agar (this Handbook, Chapter 116), which is selective for Gram-positive and Gram-negative anaerobes, and neomycin–egg yolk agar. Neomycin has been used in a variety of agar bases (Dowell and Hawkins, 1977; Lowbury and Lilly, 1955; Sutter, Vargo, and Finegold, 1975; Willis, 1977; Willis and Hobbs, 1959). The egg yolk agar base described above, with neomycin sulfate added in a concentration of 100 mg per liter (Dowell, et al., 1977c), supports growth of most of the commonly isolated clostridia. However, as with all selective media, different species as well as different strains of the same species may vary in their degree of resistance to this antibiotic (Spencer, 1969). George et al. (1979) were not successful in growing *C. difficile* on Clostrisel agar (BBL), which contains 0.15 g of neomycin sulfate and 0.20 g of sodium azide per liter as selective agents.

PASTEURIZATION AND ETHANOL TREATMENT. The heat and alcohol resistance of spores can be used to some advantage in separating clostridia from mixed populations. Pasteurization has been used classically by heat-treating homogenized specimens suspended in a diluent (this Handbook, Chapter 138). The specimen can be suspended in a liquid culture medium, which is then heated before incubation, or the specimen can be preincubated in a sporulation enrichment broth prior to heating. A number of different sporulation media have been employed, including various formulations of thioglycolate broth (Duncan and Strong, 1968), peptone–yeast extract–starch broth (Ellner, 1956), and cooked meat medium (Robertson, 1915–1916). A range of temperatures and times for heat treatment has been recommended for recovery of different species from 70°C for 10 min to 100°C for 1 h. (Hobbs, 1965; McClung, 1945; Smith, 1975; Sutton and Hobbs, 1968). The currently recommended procedure is given below.

Heat Treatment for Isolation of Clostridial Spores (Dowell and Hawkins, 1977)

Inoculate the specimen into each of three tubes of a cooked meat medium (e.g., Chopped Meat Glucose Medium, this Handbook, Chapter 116): maintain all three tubes at 70°C for 10 min. Cool one tube quickly, transfer each of the other two tubes to an 80°C water bath for 10 and 20 min, respectively; cool quickly in cold water and incubate all tubes anaerobically at 30°C.

An alternative procedure is to heat-treat enrichment cultures of the specimen after 2–3 days of incubation at 30°C.

Though heat treatment is commonly used, the heat resistance of spores varies significantly with the different species and with different strains of the same species. The spores of *Clostridium botulinum* type E, for example, are known to be almost as sensitive as the vegetative cells. For this reason, Johnston, Harmon, and Kautter (1964) investigated the use of ethanol as a germicidal agent to separate heat-sensitive spores from vegetative cells. Treatment of cooked meat cultures of macerated fish tissue, as well as direct treatment of tissue, with an equal volume of absolute ethanol for 1 h at 25°C resulted in greatly improved recovery of *C. botulinum* type E.

Ethanol treatment has been recommended as an aid in isolating *C. botulinum* from foods and feces (Dowell and Hawkins, 1977), and recently has been evaluated for other species of clostridia (Koransky, Stargel, and Dowell, 1978). Direct treatment of fecal specimens with ethanol was superior to heat treatment for isolation of several species of clostridia; this method might well be considered for further evaluation as a less drastic procedure for processing clinical specimens.

An old method that is frequently forgotten, and one that can be invaluable for isolating *C. tetani* from mixed populations, is that described by Fildes (1925).

Fildes' Method for Isolation of *Clostridium tetani* (Fildes, 1925)

The specimen is inoculated into the water of condensation (or syneresis) at the butt of a blood-agar slant, which is incubated anaerobically in an upright position. Usually within 18–24 h, motile strains of *C. tetani* will swarm from the point of inoculum upward over the slant in a thin film that is discernible only by an astute eye. With the aid of a hand lens, long rhizoid filaments of growth can be observed extending upward from the advancing edge of the film, from which Gram-stained preparations and subcultures can be made.

To permit frequent examination of the individual slant without exposure to the atmosphere, Rockwell's method of anaerobic culture (Rockwell, 1924) has been used very successfully.

Although milk media are usually used primarily as differential media, Altemeier (1944) described the use of iron-milk for rapid diagnosis of *C. per-fringens* infections. This method is especially useful for rapid detection of and enrichment culture for this species in gangrenous tissue.

Iron-Milk Method for *Clostridium perfringens* (Altemeier, 1944)

The macerated tissue specimen is heavily inoculated into a tube of sterile whole, fresh milk that contains a small strip of iron wire (or stove pipe, iron tacks, iron filings). It is heated at 100°C for 10–15 min and cooled just prior to inoculation. The tube can be incubated at 37°C or at 45°C, using pyrogallic acid and Rockwell's solution (1924) to provide anaerobic conditions. Production of stormy fermentation of the milk and the observation of large, nonmotile rods in a wet mount of the whey is presumptive evidence of the presence of *C. perfringens*, which can be readily subcultured from the whey.

Identification

Most of the clostridia associated with human and animal infections can be identified relatively easily by morphological and cultural characteristics. Considering that approximately 90% of the clostridia recovered from clinical specimens comprises only 12 or so species, the clinical microbiologist can usually identify to the species level with a minimum of effort and relatively small array of cultural and biochemical tests. Prerequisite to this is the assumption that the clinical specimen has been appropriately selected and collected to avoid the indigenous flora that may contain a great variety of organisms not truly associated with the pathology of the disease. This prerequisite cannot be overemphasized. The following description deals with characteristics of the most commonly isolated clostridia. The taxonomy of the genus *Clostridium* is discussed in this Handbook, Chapter 138.

CELLULAR MORPHOLOGY.   Cell morphology, the Gram-staining properties, sporulation, and position of spores are useful in the preliminary identification of clostridia, but use of these characteristics requires a bit of experience and expertise. Although Prévot (1966) uses the tinctorial properties and spore position as the key criteria for classification, the most experienced eye may be easily misled. Classically, clostridia are large, Gram-positive bacilli, easily differentiated from the genus *Bacillus* by the swollen spore mother cell (sporangium). Clostridia from clinical specimens, in direct smears and in primary culture, may be Gram variable or Gram negative; they may occur as short, plump to slender rods, to extremely long filaments. Sporulating cultures may become Gram negative and be confused with the vacuolated forms of the nonsporeforming, Gram-negative bacilli, and vice versa. Willis (1977) describes well the difficulties in morphological taxonomy of the anaerobes and, to quote a comment following his description of some of the pitfalls, "Having thus destroyed all faith in the microscopy of anaerobes, it must be said at once that very many anaerobic isolates do fit neatly into the morphological scheme of separation; indeed, it is on the basis of the *typical* that such a scheme is devised." Thus, cellular morphology can indeed be helpful, but with a cautious eye when dealing with clinical specimens.

DEMONSTRATION OF SPORES. For demonstration of spores, perhaps the best single medium (Smith, 1975) is prepared from chopped meat broth (this Handbook, Chapter 116) with 1.5% agar added (Dowell et al., 1977a). Slants are prepared in screw-cap tubes and, after solidification, the tubes, with caps loosened, are placed in an anaerobic atmosphere to replace the air in the tubes. Anaerobic incubation at 30°C is recommended for culture on this medium. Other sporulation media are indicated above. Any of the common spore stains (Ziehl-Neelson, Malachite green, and others) can be used but, usually, spores are readily seen in Gram-stain preparations. Of the commonly isolated clostridia, *C. perfringens*, *C. ramosum* (the two species most commonly isolated from clinical specimens), *C. sphenoides*, and *C. clostridiiforme* rarely form spores. This observation is helpful in preliminary identification when combined with other key characteristics such as colonial morphology.

COLONIAL MORPHOLOGY. The colonial morphologies of the clostridia are fairly characteristic for the different species and, at times, for the different toxin types within the species. The use of a dissecting stereoscope for examining colonies enables one, with experience, to recognize a significant number of different species of the commonly isolated clostridia with a remarkably high degree of accuracy. The appearance of colonies, as well as hemolytic activity, may vary with the particular blood agar base and the species of blood used. For diagnostic purposes, some standardization is needed, and hemolytic activity by itself generally is of little value, except for the double-zone hemolysis observed with *Clostridium perfringens*. On rabbit, human, or sheep blood, *C. perfringens* colonies usually will be surrounded by an inner zone of complete hemolysis (due to theta toxin) and an outer zone of incomplete hemolysis (due to alpha toxin). This reaction will vary with different strains, depending upon inherent production of alpha and theta toxins and the effects of the various blood agar basal media employed. When observed, however, hemolysis does signal the presence of *C. perfringens* (Smith, 1975; Smith and Dowell, 1974). For descriptions of the colonial morphologies of each of the type species of clos-

tridia, consult Buchanan and Gibbons (1974) and Smith (1975).

AEROTOLERANCE. When subculturing isolated colonies from primary anaerobic plates, a plate must also be incubated aerobically in addition to the anaerobic blood agar plates and liquid media that may be used. An additional blood or chocolate agar plate incubated in a candle jar or CO_2-incubator can also be of considerable value. If the suspect colonies of clostridia were indeed pure, the plates incubated aerobically and in a 5–10% CO_2 atmosphere will not contain growth, with only rare exceptions: aerotolerant strains of *Clostridium histolyticum*, *C. tertium*, and, on extremely rare occasions, strains of *C. perfringens* may produce minute colonies on plates so incubated. These strains, however, are usually readily recognized by their colonial and cellular morphologies. Incubation of a plate under CO_2 offers the additional advantage of supporting growth of capnophilic, facultative organisms picked from the anaerobic plate.

BIOCHEMICAL AND OTHER DIFFERENTIAL CHARACTERISTICS. The several current practical approaches to definitive identification of anaerobes have been described by Dowell and Hawkins (1977), Holdeman et al. (1977), Sutter, Vargo, and Finegold (1975), and Dowell and Lombard (this Handbook, Chapter 116). Whichever of these is chosen, it is essential to follow the recommended procedures indicated by each for preparation of media, indicators employed, times and methods of anaerobic incubation, etc. The tables or keys of differential characteristics of one author are not necessarily applicable to the procedures of another.

Definitive identification includes descriptions of colonial and cellular characteristics, fermentative and biochemical characteristics, and determination of volatile and nonvolatile fatty acid end products; each of these steps must by done in as standardized a procedure as possible. At times, tests for animal pathogenicity and toxin neutralization must be performed.

Determination of toxinogenic types of *Clostridium perfringens* (types A–E) and *C. botulinum* (types A–G) is relatively easy to accomplish by toxin-neutralization tests in mice (Dowell and Hawkins, 1977; Smith, 1975). *C. novyi* toxin types A and B can be differentiated on the basis of the epsilon toxin produced by type A (lipolytic) and the beta toxin produced by type B (lecithinolytic), as well as by specific toxin neutralization tests (Smith, 1975).

Serotyping of some of the clostridia (Dowell and Hawkins, 1977; Hobbs, 1965; Mandia, 1955; Smith, 1975) can be accomplished, but has not proved to date to be of significant practical value. Fluorescent-antibody procedures may prove to be

Table 5. Differential characteristics of commonly isolated clostridia.

| Species | Spores | Motility | Reactions on egg yolk agar | | | Indole | Fermentation of | | | Nitrate reduction | Hydrolysis of | | | Acid products in PYG (48 h, 35°C) | |
			Lecithinase	Lipase	Proteolysis		Glucose	Mannitol	Lactose		Esculin	Gelatin	Casein	Volatile	Nonvolatile
C. bifermentans	ST	+	+	–	+	+	+	–	–	–	V	+	+	A, (P), (IB), (B), (IV), (IC)	–
C. botulinum type A	ST	+	–	+	+	$-^b$	+	–	–	–	+	+	+	A, (P), (IB), B, IV, (V), (IC)	–
C. botulinum type B	ST	+	–	+	–	$-^b$	+	–	–	–	+	+	+	A, (P), (IB), B, IV, (V), (IC)	–
C. butyricum	ST	+	–	–	–	+	+	–	+	$-^+$	+	–	–	A, B	–
C. cadaveris	T	+	–	–	–	+	+	–	+	–	–	V	–	A, (P), B	–
C. clostridiiforme	"ST"	V	–	–	–	–	+	V	+	–	+	–	–	A	–
C. difficile	ST to T	V	–	–	–	–	+	+	–	–	+	$-^+$	–	A, (P), IB, B, IV, V, IC	–
C. histolyticum	ST	+	–	–	+	–	–	–	–	–	–	+	+	A	–
C. innocuum	T	–	–	–	–	–	+	+	–	–	+	–	–	A, (P), B	–
C. limosum	ST	+	+	–	–	–	–	–	–	–	–	+	+	A	L
C. novyi type A	ST	+	+	+	–	–	+	–	–	$-^+$	V	+	–	A, P, B, (V)	–
C. paraputrificum	T	+	–	–	–	–	+	–	+	V	+	–	–	A, B	–
C. perfringens	"ST"	–	+	–	–	–	+	–	+	+	V	+	$-^{(+)}$	A, (P), B	–
C. ramosum	"T"	–	–	–	–	–	+	+	+	–	+	–	–	A	(PY), L
C. septicum	ST	+	–	–	–	–	+	–	+	$+^-$	+	+	–	A, B	–
C. sordellii	ST	+	+	–	+	+	+	–	–	–	–	+	+	A, (P), (IB), (B), (IV), (IC)	–
C. sphenoides	"ST"	V	–	–	–	V^b	+	V	+	$-^+$	+	–	–	A	L
C. sporogenes	ST	+	–	+	+	$-^b$	+	–	–	–	–	+	+	A, (P), (IB), (B), (IV), (IC)	–
C. subterminale	ST	+	–	–	+	–	–	–	–	–	$-^+$	+	+	A, (P), (IB), B, IV, (V), (IC)	–
C. tertium	T	+	–	–	–	–	+	+	+	$+^-$	+	–	–	A, B	–
C. tetani	T	+	–	$-^+$	–	V	–	–	–	$+^-$	–	+	–	A, B	–

[a] From V. R. Dowell, Jr., G. L. Lombard, F. S. Thompson, and A. Armfield, personal communication. ST, subterminal; T, terminal; +, positive reaction exhibited by 90–100% of strains; –, negative reaction exhibited by 90–100% of strains; V, variable reaction; Superscript, the reaction exhibited by 11–25% of strains tested. ( ), Variable; A, acetic acid; B, butyric acid; IB, isobutyric acid; IC, isocaproic acid; IV, isovaleric acid; P, propionic acid; L, Lactic acid; PY, pyruvic acid; "ST", subterminal spores rarely produced; "T", terminal spores rarely produced.
[b] Indole derivative produced which gives a violet color when tested with paradimethylaminocinnamaldehyde reagent. Indole gives a blue or blue-green color with this reagent.

useful for detection of *C. septicum, C. chauvoei, C. novyi, C. botulinum,* and *C. tetani* (see Willis, 1977).

For taxonomic purposes, cell wall composition, G+C ratios, and DNA homologies may be determined (Buchanan and Gibbons, 1974; Holdeman et al. (1977), Nakamura et al. (1977).

Using the media and procedures described by Dowell and Hawkins (1977), Dowell et al. (1977a), and this Handbook, Chapter 116, the commonly occurring clostridia can usually be identified by the characteristics indicated in Table 5. For definitive identification or determination of toxinogenic type, additional animal protection or toxin neutralization tests are required for *C. botulinum, C. chauvoei, C. perfringens, C. septicum,* and *C. tetani* (Dowell, 1977; Dowell and Hawkins, 1977; Smith, 1975).

Literature Cited

Altemeier, W. A. 1944. The rapid identification of *Clostridium welchii* in accidental wounds. Surgery, Gynecology and Obstetrics **78:**411–414.

Alpern, R. J., Dowell, V. R., Jr. 1969. *Clostridium septicum* infections and malignancy. Journal of the American Medical Association **209:**385–388.

Alpern, R. J., Dowell, V. R., Jr. 1971. Nonhistotoxic clostridial bacteremia. American Journal of Clinical Pathology **55:** 717–722.

Arnon, S.S., Midura, T. F., Clay, S.A., Wood, R. M., Chin, J. 1977. Infant botulism. Epidemiological clinical and laboratory aspects. Journal of the American Medical Association **237:**1946–1951.

Barsanti, J.A., Walser, M., Hatheway, C. L., Bowen, J. M., Crowell, W. 1978. Type C botulism in American foxhounds. Journal of the American Veterinary Medical Association **172:**809–813.

Bartlett, J. G., Onderdonk, A. B., Cisneros, R. L., Kasper, D. L. 1977. Clindamycin-associated colitis due to a toxin producing species of *Clostridium* in hamster. Journal of Infectious Diseases **136:**701–705.

Bartlett, J. G., Te Wen Chang, Gurwith, M., Gorbach, S. L., Onderdonk, A. B. 1978. Antibiotic associated pseudomembranous colitis due to toxin producing clostridia. New England Journal of Medicine **298:**531–534.

Batty, J., Walker, P. D. 1966. Colonial morphology and fluorescent labelled antibody staining in the identification of species of the genus *Clostridium,* p. 81. In: Gibbs, B. M. and Skinner, F. A. (eds.), Identification methods for microbiologists. Society for Applied Bacteriology Technical Series No. 1. London: Academic Press.

Berkhoff, G. A. Redenbarger, J. L. 1977. Isolation and identification of anaerobes in the veterinary diagnostic laboratory. American Journal of Veterinary Research **38:**1069–1074.

Buchanan, R. E., Gibbons, N. E. (eds.). 1974. Bergey's manual of determinative bacteriology, 8th ed. Baltimore: Williams & Wilkins.

Center for Disease Control 1977. Botulism in the United States, 1899–1973. Handbook for epidemiologists, clinicians, and laboratory workers. Atlanta: Department of Health, Education, and Welfare. Public Health Service.

Center for Disease Control 1978a. Follow-up on infant botulism—United States. Morbidity and Mortality Weekly Report **27:**17–23.

Center for Disease Control 1978b. Infant botulism—England. Morbidity and Mortality Weekly Report **27:**100.

Center for Disease Control 1979. Botulism—United States, 1978. Morbidity and Mortality Weekly Report **28:**73–75.

Center for Disease Control 1980. Type F Infant Botulism—New Mexico. Morbidity and Mortality Weekly Report **29:**85–86.

Collee, J. G., Rutter, J. M., Watt, B. 1971. The significantly viable particle: A study of the subculture of an exacting sporing anaerobe. Journal of Medical Microbiology **4:**271–288.

Dowell, V. R., Jr. 1977. Clinical Veterinary Anaerobic Bacteriology U.S. Department of Health, Education and Welfare Publication, Center for Disease Control.

Dowell, V. R., Jr. 1979. Antibiotic-associated colitis. Hospital Practice **14:**75–80.

Dowell, V. R., Jr., Hawkins, T. M. 1977. Laboratory methods in anaerobic bacteriology. CDC Laboratory Manual. Center for Disease Control HEW Publication No. (CDC) 77-8272. Washington, D.C.: U.S. Government Printing Office.

Dowell, V. R., Jr., and Lombard, G. L. 1977. Presumptive identification of anaerobic non sporeforming gram negative bacilli. Atlanta: U.S. Department of Health, Education and Welfare, Public Health Service, Center for Disease Control.

Dowell, V. R., Jr., Lombard, G. L., Thompson, F. S., Armfield, A. Y. 1977a. Media for isolation, characterization and identification of obligately anaerobic bacteria. Atlanta: Department of Health, Education and Welfare, Public Health Service, Center for Disease Control.

Dowell, V. R., Jr., McCroskey, L. M., Hatheway, C. L., Lombard, G. L., Hughes, J. M., Merson, M. H. 1977b. Coproexamination for botulinal toxin and *Clostridium botulinum:* a new procedure for laboratory diagnosis of botulism. Journal of the American Medical Association **238:**1829–1832.

Dubos, R. 1965. Man adapting. New Haven: Yale University Press.

Duncan, C. L., Strong, D. H. 1968. Improved medium for sporulation of *Clostridium perfringens.* Applied Microbiology **16:**82–89.

Ellner, P. D. 1956. A medium promoting rapid quantitative sporulation in *Clostridium perfringens.* Journal of Bacteriology **71:**495–496.

Fildes, P. 1925. Tetanus. I. Isolation, morphology, and cultural reactions of *B. tetani.* British Journal of Experimental Pathology **6:**62–70.

Finegold, S. M. 1977. Anaerobic bacteria in human disease. New York: Academic Press.

George, W. L., Rolfe, R. D., Sutter, V. L., Finegold, S. M. 1979. Diarrhea and colitis associated with antimicrobial therapy in man and animals. American Journal of Clinical Nutrition **32:**251–257.

Hafiz, S., McEntegart, M. G., Morton, R. S., Waitkens, S. A. 1975. *Clostridium difficile* in the urogenital tract of males and females. Lancet **i:**420–421.

Hall, I. C., O'Toole, E. 1935. Intestinal flora in newborn infants, with a description of a new pathogenic anaerobe, *Bacillus difficilis.* American Journal of Diseases of Children **49:**390–402.

Hayward, N. J. 1941. Rapid identification of *Clostridium welchii* by the Nagler reaction. British Medical Journal **1:**811–814, 916.

Hirsch, A., and Grinstead, E. 1954. Methods for the growth and enumeration of anaerobic spore-formers from cheese, with observations on the effect of nisin. Journal of Dairy Research **21:**101–110.

Hobbs, B. C. 1965. *Clostridium welchii* as a food poisoning organism. Journal of Applied Bacteriology **28:**74–82.

Holdeman, L. V., Moore, W. E. C. (eds.). 1977. Anaerobe laboratory manual, 4th ed. Blacksburg, Virginia: Anaerobe Laboratory, Virginia Polytechnic Institute and State University.

Johansson, K. R. 1953. A modified egg yolk medium for detecting lecithinase producing anaerobes in feces. Journal of Bacteriology **65:**225–226.

Johnston, R., Harmon, S., Kautter, D. 1964. Method to facilitate the isolation of *Clostridium botulinum* type E. Journal of Bacteriology **88**:1521–1522.

Kilgore, G. E., Starr, S. E., Del Bene, V. E., Whaley, D. N., Dowell, V. R., Jr. 1973. Comparison of three anaerobic systems for the isolation of anaerobic bacteria from clinical specimens. American Journal of Clinical Pathology **59**:552–559.

Koransky, J. R., Stargel, M. D., Dowell, V. R., Jr. 1979. *Clostridium septicum* bacteremia: Its clinical significance. American Journal of Medicine **66**:63–66.

Larson, H. E., Price, A. B. 1977. Pseudomembranous colitis: Presence of clostridial toxin. Lancet **ii**:1312–1314.

Lombard, G. L., Armfield, A. Y., Stargel, M. D., Fox, J. B. 1976. The effect of storage of blood agar medium on the growth of certain obligate anaerobes. Abstracts of the Annual Meeting of the American Society for Microbiology **1976**:41.

Lowbury, E. J. L., Lilly, H. A. 1955. A selective plate medium for *Clostridium welchii*. Journal of Pathology and Bacteriology **70**:105–109.

McClung, L. S. 1945. Human food poisoning due to growth to *Clostridium perfringens* (*C. welchii* in freshly cooked chickens). Journal of Bacteriology **50**:229–231.

McClung. L. S. Toabe, R. 1947. The egg yolk plate reaction for the presumptive diagnosis of *Clostridium sporogenes* and certain species of the gangrene and botulinum groups. Journal of Bacteriology **53**:139–147.

MacFarlane, R. G., Oakley, C. L., Anderson, C. G. 1941. Haemolysis and the production of opalescence in serum and lecitho-vitellin by the a-toxin of *Clostridium welchii*. Journal of Pathology and Bacteriology **52**:99–103.

Mandia, J. W. 1955. The position of *Clostridium tetani* within the serological schema for the proteolytic clostridia. Journal of Infectious Disease **97**:66–72.

Merson, M. H., Dowell, V. R., Jr. 1973. Epidemiologic, clinical, and laboratory aspects of wound botulism. New England Journal of Medicine **289**:1005–1010.

Merson, M. H., Hughes, J. M., Dowell, V. R., Taylor, A., Barker, W. H., Gangarosa, E. J. 1974. Current trends in botulism in the United States. Journal of the American Medical Association **229**:1305–1308.

Moore, W. B. 1968. Solidified media suitable for the cultivation of *Clostridium novyi* type B. Journal of General Microbiology **53**:415–423.

Nagler, F. P. O. 1939. Observations on a reaction between the lethal toxin of *Clostridium welchii* (type A) and human serum. British Journal of Experimental Pathology **20**:473–485.

Nakamura, S., Okado, J., Nakashio, S., Nishida, S. 1977. *Clostridium sporogenes* isolates and their relationship to *C. botulinum* based on deoxyribonucleic acid reassociation. Journal of General Microbiology **100**:395–401.

Prévot, A. R. 1966. Manual for the classification and determination of the anaerobic bacteria. Fredette, V. (translator). Philadelphia: Lea and Febiger.

Rifkin, G. D., Fekety, F. R., Silva, J. Jr., Sack, R. B. 1977. Antibiotic induced colitis: Implication of a toxin neutralized by *Clostridium sordelli* antitoxin. Lancet **ii**:1103–1106.

Rockwell, G. E. 1924. An improved method for anaerobic cultures. Journal of Infectious Disease **35**:581–586.

Robertson, M. 1915–1916. Notes upon certain anaerobes isolated from wounds. Journal of Pathology and Bacteriology **20**:327–349. Smith, L. DS. 1975. The pathogenic anaerobic bacteria, 2nd ed. Springfield, Illinois: Charles C Thomas.

Smith, L. DS. 1977. Botulism, the organism, its toxins, the disease. Springfield, Illinois: Charles C Thomas.

Smith, L. DS., Dowell, V. R., Jr. 1974. *Clostridium*, pp. 376–380. In: Lennette, E. H., Spaulding, E. H., Truant, J. P. (eds), Manual of clinical microbiology, 2nd ed. Washington, D.C.: American Society for Microbiology.

Spencer, R. 1969. Neomycin-containing media in the isolation of *Clostridium botulinum* and food poisoning strains of *Clostridium welchii*. Journal of Applied Bacteriology **32**:170–174.

Sutton, R. G. A., Hobbs, B. C. 1968. Food poisoning casued by heat-sensitive *Clostridium welchii*. A report of five recent outbreaks. Journal of Hygiene **66**:135–146.

Sutter, V. L., Vargo, V. L., Finegold, S. M. 1975. Wadsworth anaerobic bacteriology manual, 2nd ed. Los Angeles: The Regents of the University of California.

Turner, H. D., Brett, E. M., Gilbert, R. J., Ghosh, A. C., Leibeschuetz, H. J. 1978. Infant botulism in England. Lancet **i**:1277–1278.

Wetzler, T. F., Marshall, J. D., Jr., Cardella, M. A. 1956a. Rapid isolation of clostridiums by selective inhibition of aerobic flora. I. Use of sorbic acid and polymyxin B sulfate in a liquid medium. American Journal of Clinical Pathology **26**:418–421.

Wetzler, T. F., Marshall, J. D., Jr., Cardella, M. A. 1956b. Rapid isolation of clostridiums by selective inhibition of aerobic flora. II. A systematic method as applied to surveys of clostridiums in Korea. American Journal of Clinical Pathology **26**:345–351.

Willis, A. T. 1969. Clostridia of wound infection. London: Butterworths.

Willis, A. T. 1977. Anaerobic bacteriology. Clinical and laboratory practice. London, Boston: Butterworths.

Willis, A. T., Hobbs, G. 1959. Some new media for the isolation and identification of clostridia. Journal of Pathology and Bacteriology **77**:511–521.

The Genus *Clostridium* (Nonmedical Aspects)

GERHARD GOTTSCHALK, JAN R. ANDREESEN, and HANS HIPPE

The genus *Clostridium* was created by Prazmowski in 1880. Since that time many bacterial species have been assigned to it, and today it represents one of the largest genera of the prokaryotes. This growth is not surprising because an organism has to fulfill only three criteria in order to become classified as a *Clostridium:* (i) it must be able to form endospores; (ii) it must obligatorily rely on an anaerobic energy metabolism; and (iii) it must be unable to carry out a dissimilatory reduction of sulfate.

The first criterion differentiates the clostridia from all organisms not able to form endospores. The difficulty in applying this criterion is that some species may not sporulate under the culture conditions applied. Asporogenous mutants may also appear and spread out, and it is, perhaps, not surprising that a high degree of genetic homology was found between the nonsporulating *Bacteroides trichoides* and *Eubacterium filamentosum* on one side and *Clostridium ramosum* on the other (Moore, 1978). Recently, strains of *Bacteroides clostridiiformis* subsp. *clostridiiformis* have been classified as *Clostridium clostridiiforme* after spores were detected (Cato and Salmon, 1976; Kaneuchi, Benno, and Mitsuoka, 1976).

The second criterion separates the clostridia from the *Bacillus* species that are aerobes and that contain electron-transport chains coupled to oxygen as the electron acceptor. Some *Bacillus* species, e.g., *B. polymyxa*, are facultative anaerobes. Members of both genera resemble one another with respect to cell shape and cell size. With the exception of *C. coccoides* (Kaneuchi, Benno, and Mitsuoka, 1976), the clostridial cells are straight or curved rods, $0.3–1.6 \times 1–14$ μm. A pronounced tendency to helical coiling of the cell chains is found in two newly described clostridia (Kaneuchi et al., 1979). Unlike the appearance of cells of bacilli, clostridial cells that contain spores are often swollen. Those that form terminal spores look like drumsticks, and those that form subterminal or central spores, like Chinese lanterns (Fig. 1).

The third criterion follows from the creation of the genus *Desulfotomaculum* by Campbell and Post- gate in 1965, which comprises the sporeforming, sulfate-reducing bacteria.

The strictly anaerobic, saccharolytic sarcinas, *Sarcina ventriculi* and *S. maxima,* have been reported to form spores (Knöll and Horschak, 1973). They can be differentiated from the clostridia by their unique appearance as spherical cells that occur in packets (see this Handbook, Chapter 130).

The genus *Clostridium* includes psychrophilic, mesophilic, and thermophilic species. Most of the clostridial species are Gram positive and are motile with peritrichous flagellation. The major role of these organisms in nature is the degradation of organic material to acids, alcohols, CO_2, H_2, and minerals. Frequently, a butyric acid smell is associated with the genus *Clostridium*. This acid is produced by a number of species. However, several clostridia do not form butyrate at all but form acetate, lactate, formate, or propionate as main fermentation products. The ability to form spores that resist dryness, heat, and aerobic conditions makes the clostridia ubiquitous. When the appropriate growth conditions are applied, they can be isolated from all kinds of material. They are frequently responsible for spoilage of food and dairy products and cause large economic losses.

A number of clostridia are pathogens. Because of their great importance, the pathogenic species are discussed in a separate chapter (this Handbook, Chapter 137). This separation is made primarily for practical reasons and does not indicate that the pathogenicity or nonpathogenicity of clostridial species has important taxonomic significance. The information for toxin production might be coded on extrachromosomal elements or prophages and, therefore, the differentiation of toxigenic from nontoxigenic species might not be warranted (Duncan et al., 1978).

Approved standards for the classification of the whole genus *Clostridium* are not available. For many nonpathogenic species even the description is frequently incomplete and based on the study of only one strain as in the cases of *C. aminovalericum, C. barkeri, C. cellobioparum, C. propionicum, C.*

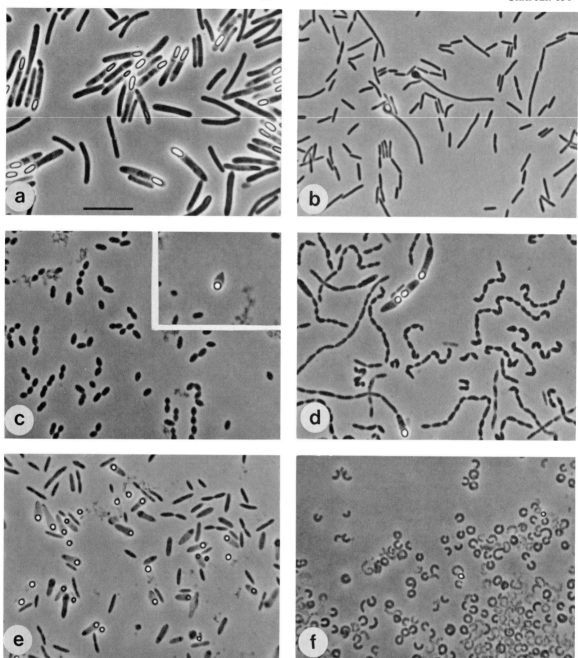

Fig. 1. Phase-contrast photomicrographs of some representatives of the genus *Clostridium*. The scale line represents 10 μm.
1 a–f. (a) *C. butyricum* DSM 552, yeast extract–peptone agar, large straight or slightly curved rods with round ends, sporangial cells slightly distended, oval subterminal spores; (b) *C. thermocellum* DSM 1313, cellobiose medium, showing elongated sporulating cells with oval terminal spores; (c) *C. coccoides* DSM 935, CMC medium, coccobacilli to rod-shaped cells forming pairs and chains, round subterminal spores; (d) *C. oroticum* DSM 1287, CMC agar, rods occurring in long tangled chains, enlarged sporangial cells with oval terminal to subterminal spores; (e) *C. indolis* DSM 755, CMC agar, with round terminal to subterminal spores; (f) *C. cocleatum* DSM 1551, CMC agar, showing preponderance of semicircular to circular cell forms, oval spores are subterminal to terminal. Media, see "Isolation" section.
1 g–l. (g) *C. bifermentans* DSM 631, CMC agar, showing central to subterminal oval spores not distending cells; (h) *C. sporogenes* DSM 767, peptone–yeast extract–glucose medium, with oval subterminal spores swelling the cells slightly; (i) *C. acetobutylicum* DSM 792, milk agar, rods of varying length, sporulating cells cigar-shaped, spores oval to cylindric, subterminal; (j) *C. ramosum* DSM 1402, CMC agar, showing tendency to Y- and V-shaped cells; (k) *C. cadaveris* DSM 1284, CMC agar, rods with oval terminal spores; (l) *C. tetanomorphum* DSM 665, CMC agar, large straight to slightly curved rods, nearly round terminal spores. Media, see "Isolation" section.

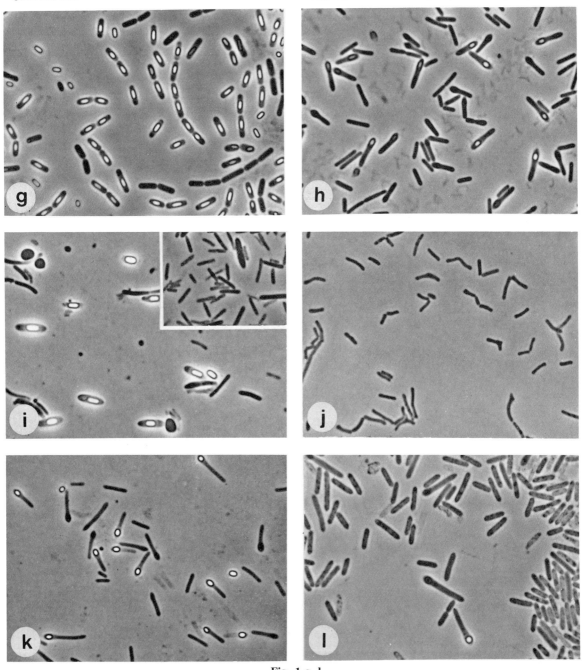

Fig. 1 g–l

sticklandii, and *C. thermoaceticum.* This descriptive poverty results from the fact that many clostridia were isolated in order to study the anaerobic degradation of certain compounds or the pathways leading to the fermentation products, and not much emphasis has been put on their classification. Nevertheless, basic discoveries have been made on these organisms. Pasteur (1861) discovered "life without oxygen" experimenting with a culture of a butyrate-forming microorganism—probably a *Clostridium.* Winogradsky (1895 and 1902) discovered nitrogen

fixation by free-living bacteria with *C. pasteurianum.* The study of this organism led also to the discovery of ferredoxin (Mortenson, Valentine, and Carnahan, 1962) and of the first in vitro system that fixed molecular nitrogen (Mortenson, 1966). The first enzymatic reaction involving a B_{12} derivative was discovered in *C. tetanomorphum* (Barker, Weissbach, and Smyth, 1958) and investigations on *C. kluyveri* were important for an understanding of fatty acid synthesis and of the role of coenzyme A (Barker, 1956). The Stickland reaction, which is so important for many proteolytic clostridia including many pathogens, was discovered in *C. sporogenes* (Stickland, 1934).

The genus *Clostridium* is considerably heterogeneous. It includes species that are moderately aerotolerant *(C. carnis, C. durum, C. histolyticum,* and *C. tertium)* and others, such as *C. aminovalericum* (Rolfe et al., 1978), that are extremely fastidious. Some species contain cytochromes of as yet unknown function (Gottwald et al., 1975). The G+C content ranges from 21 to 54 mol%. Clostridial cell walls contain, in general, peptidoglycan of the *meso*-diaminopimelic acid direct-linked type and teichoic acids, but exceptions have been observed (Cummins and Johnson, 1971; Schleifer and Kandler, 1972). The sequences of their 16S ribosomal RNAs show a very low similarity (C. R. Woese, personal communication), indicating an early separation of the species of this genus during evolution.

Habitats

Restrictions by Oxygen Sensitivity

Clostridia exhibit a more-or-less pronounced intolerance towards oxygen, caused by a lack or a shortage of defense mechanisms against toxic by-products of oxygen metabolism or by interference of oxygen with the functioning of some vital enzyme systems (see this Handbook, Chapter 115). The toxic byproducts of oxygen metabolism are destroyed by the enzymes catalase (or a peroxidase) and superoxide dismutase (Morris, 1975). The levels of these enzymes were first reported to be zero or very low in most clostridia (McCord, Keele, and Fridovich, 1971). However, appreciable activities of superoxide dismutase have been found in some strains of *C. perfringens* (Hewitt and Morris, 1975) and in *C. ramosum* (Tally et al., 1977), and its presence has been demonstrated for several other species (Gregory, Moore, and Holdeman, 1978). The activity of superoxide dismutase in *C. sporogenes* and *C. bifermentans* can be increased up to 60-fold if these organisms are cultured under the stress of increased amounts of oxygen (Ashley and Shoesmith, 1977).

Molecular oxygen may also interfere with NADH-oxidase activities so that the intermediary and biosynthetic metabolism of these organisms may suffer from a general shortage of NADH in the presence of oxygen (Morris, 1975; O'Brien and Morris, 1971b; Uesugi and Yajima, 1978a).

The pioneering work on the effect of oxygen on clostridial growth was done with pathogens *(Clostridium botulinum, C. perfringens)* (see O'Brien and Morris, 1971b). In recent years nonpathogenic species have also been studied. The studies indicate that the maximum oxygen concentration at which growth is still possible is not constant for all clostridia. *C. haemolyticum* requires a pO_2 smaller than 0.5% for growth; *C. novyi* type A tolerates a pO_2 of up to 3% (Loesche, 1969). Many clostridia stop growing in the presence of molecular oxygen but resume growth when transferred back to anaerobic conditions. The vegetative cells of *C. acetobutylicum, C. butyricum, C. clostridiiforme,* and *C. ramosum* survive oxygen exposure for hours; *C. haemolyticum* and *C. novyi* type B, just for minutes (Azova, Gusev, and Ivoilov, 1970; O'Brien and Morris, 1971b; Stolp, 1955; Tally et al., 1977; Uesugi and Yajima, 1978a).

During growth, clostridia establish a characteristic low redox potential, between -400 and -200 mV, in their environment (see O'Brien and Morris, 1971b). Clostridia will not start growing when the E_h of their environment is above approximately $+150$ mV.

Germination of spores is also affected by oxygen. Although spores are very resistant to oxygen and can be stored in distilled water or as dry material under air for years, their germination requires anaerobic conditions or conditions of low oxygen tension (Douglas, Hambleton, and Rigby, 1973; Sarathchandra, Barker, and Wolf, 1974). Some reports indicate that germination at low oxygen tensions requires a larger supply of nutrients in the environment than germination under strictly anaerobic conditions (see O'Brien and Morris, 1971b). In general, the process of germination is properly tuned to the properties of the vegetative cells. It would be useless for germination to occur under conditions unfavorable to growth.

Oxygen sensitivity restricts the habitat of the clostridia to anaerobic areas or areas with low oxygen tensions. Growing and dividing clostridial cells will, therefore, not be found in air-saturated surface layers of lakes, rivers, etc. and on the surface of organic material and soil. Clostridial spores, however, are present with high probability in these environments, and they will germinate when oxygen is exhausted and when appropriate nutrients are present. Habitats that contain sufficient amounts of organic material become anaerobic rather readily because the solubility of oxygen in aqueous solutions is low (ca. 9 mg per liter at 20°C under 1 atm air pressure) and because oxygen is removed rapidly by aerobes.

In permanently anaerobic habitats with a controlled supply of nutrients, such as in the rumen or in parts of the intestine, the clostridia may have an evolutionary disadvantage because of the energy cost of reduplicating the information for spore formation (Zamenhof and Eichhorn, 1967). This postulated disadvantage might explain the preponderance of nonsporeforming anaerobes in the above-mentioned habitats.

Nutritive Requirements

Due to the ubiquituous distribution of the clostridial spores and their resistance, the clostridia are potentially present everywhere. If the physical parameters (pO_2, E_h, pH, and temperature) are favorable and an organic compound is available as an energy source, then spores of some clostridial species will germinate and a clostridial population will be established. The substrate spectrum of the whole genus is extremely broad and covers a wide range of naturally occurring compounds. Many clostridia excrete exoenzymes which make various macromolecules accessible to them. On the basis of their preferred or characteristic substrates, four nutritional groups of clostridia can be distinguished: saccharolytic clostridia, proteolytic clostridia, a combination of both, and specialists (Table 1).

SACCHAROLYTIC CLOSTRIDIA.   These are usually nonpathogenic organisms able to grow on carbohydrates such as xylose, mannitol, glucose, fructose, lactose, and raffinose. This group includes species which utilize starch (e.g., *C. butyricum*), cellulose (e.g.,*C. cellobioparum*), pectins (e.g.,*C. felsineum*), and chitin (e. g., *C. sporogenes*) and which, therefore, are able to form the appropriate exoenzymes. (For references, see the list of clostridial species under ''Identification''.)

PROTEOLYTIC CLOSTRIDIA.   These organisms are able to excrete proteases and to digest proteins. Characteristic for this type of fermentation is the formation of branched-chain fatty acids from the corresponding amino acids (Elsden and Hilton, 1978; Mead, 1971). Aromatic amino acids are either oxidatively decarboxylated or reduced, or the side chain is split off (Elsden, Hilton, and Walker, 1976). A number of proteolytic species are highly pathogenic (e.g., *C. botulinum*, *C. perfringens*).

PROTEOLYTIC AND SACCHAROLYTIC CLOSTRIDIA.   A typical representative of the non-toxin-producing species of this nutritional group is *C. oceanicum*. Most of the other species are toxin producers, such as *C. perfringens* and *C. sordellii*.

SPECIALISTS.   This group comprises organisms that have specialized on one or a few substrates. For example, *C. acidiurici* and *C. cylindrosporum* grow on purines such as uric acid and xanthine but not on sugars or amino acids. *C. kluyveri* has specialized on the fermentation of ethanol, acetate, and bicarbonate to butyrate, caproate, and molecular hydrogen. *C. propionicum* ferments only threonine and C_3 compounds such as alanine, lactate, acrylate, serine, and cysteine. *C. cochlearium* degrades only glutamate, glutamine, and histidine.

Consequently, the genus *Clostridium* as a whole

Table 1. The proteolytic and/or saccharolytic clostridia and species that are neither proteolytic nor saccharolytic but grow on amino acids, purines, or special substrates.

Proteolytic Saccharolytic	Saccharolytic		Proteolytic	Neither proteolytic nor saccharolytic
C. acetobutylicum	*C. absonum*	*C. nexile*	*C. botulinum* G	*C. acidiurici*
C. bifermentans	*C. aceticum*	*C. novyi* A	*C. ghoni*	*C. aminovalericum*
C. botulinum A, B, F (prot.)	*C. aurantibutyricum*	*C. oroticum*	*C. hastiforme*	*C. botulinum* G
C. botulinum C, D	*C. barkeri*	*C. paraperfringens*	*C. histolyticum*	*C. cochlearium*
C. cadaveris	*C. beijerinckii*	*C. paraputrificum*	*C. limosum*	*C. cylindrosporum*
C. haemolyticum	*C. botulinum* B, E, F	*C. pasteurianum*	*C. mangenotii*	*C. irregularis*
C. lituseburense	*C. butyricum*	*C. perenne*	*C. sporogenes*	*C. kluyveri*
C. novyi A, B	*C. carnis*	*C. pseudotetanicum*	*C. subterminale*	*C. malenominatum*
C. oceanicum	*C. celatum*	*C. ramosum*	*C. tetani*	*C. propionicum*
C. perfringens	*C. cellobioparum*	*C. rectum*		*C. putrefaciens*
C. putrefaciens	*C. chauvoei*	*C. sardiniensis*		*C. sporosphaeroides*
C. putrificum	*C. clostridiiforme*	*C. sartagoformum*		*C. sticklandii*
C. sordellii	*C. coccoides*	*C. scatologenes*		
C. sporogenes	*C. cocleatum*	*C. septicum*		
	C. difficile	*C. sphenoides*		
	C. durum	*C. spiroforme*		
	C. fallax	*C. symbiosum*		
	C. felsineum	*C. tertium*		
	C. formicoaceticum	*C. thermoaceticum*		
	C. glycolicum	*C. thermocellum*		
	C. indolis	*C. thermohydrosulfuricum*		
	C. innocuum	*C. thermosaccharolyticum*		
	C. leptum	*C. tyrobutyricum*		

1772 G. Gottschalk, J. R. Andreesen, and H. Hippe

has a very high capacity for attacking a wide variety of organic compounds, and habitats that provide these compounds and fulfill the above-mentioned physical conditions will be very suitable for the germination of spores and for growth of vegetative cells.

In the laboratory, most clostridial species are cultivated using rather complex media. Their actual nutritional requirements are not known in most cases and have been determined only for some species or certain strains of them (Table 2). However, growth of the organisms listed in Table 2 is generally stimulated if their media are supplemented with complex nutrients.

Isolation

Occurrence of Spores and Methods for the Elimination of Nonsporeformers

Clostridia differ widely in their readiness to produce spores. Although some media have been found to stimulate sporulation of many species (Gibbs and Hirsch, 1956; Perkins, 1965; Roberts, 1967), there is no one medium that is suitable as sporulation medium for a wide range of clostridia. Spores of several strong saccharolytic species are preferentially produced in the presence of carbohydrates, whereas others sporulate better in their absence (Bergère and Hermier, 1970; Nasuno and Assai, 1960). Surface growth on agar media often gives higher sporulation rates than growth in liquid media. Incubation at a suboptimal temperature sometimes results in more spores than incubation at the optimal temperature. Special sporulation media have been developed for a few clostridia: e.g., *C. putrefaciens* (Roberts and Derrik, 1975), *C. thermosaccharolyticum* (Hsu and Ordal, 1969; Pheil and Ordal, 1967), *C. perfringens* (Duncan and Strong, 1968; Ellner, 1956; Nishida, Seo, and Nakagawa, 1969; Sacks and Thompson, 1978), *C. pasteurianum* (Emtsev, 1963; Mackey and Morris, 1971), *C. bifermentans* (Hitzman, Halvorson, and Ukita, 1957). For some species or strains, the initiation of spore formation is very difficult. This difficulty has long been known for *C. perfringens*, but it is also found in more recently described clostridia, such as *C. nexile* (Holdeman and Moore, 1974), *C. clostridiiforme* (Cato and Salmon, 1976), *C. leptum* (Moore, Johnson, and Holdeman, 1976), and *C. ramosum* (Holdeman, Cato, and Moore, 1971).

In natural samples such as soil, anaerobic mud, sewage, feces, or material from other anaerobic habitats, the existence of spores of all the clostridial species present can be assumed. Therefore, enrichment cultures are usually started with spores, which has the advantage that nonsporeformers can be eliminated by pasteurization or alcohol treatment.

Several clostridia produce very heat-resistant spores that survive when incubated at 100°C or more for several hours, but others show only a moderate or low heat resistance (Ingram, 1969; Roberts and Ingram, 1965). The use of comparatively high temperatures may also have selective properties (mutagenic effects), and strains may be isolated with changed characteristics as shown by Hayase et al. (1974) and Nishida, Seo, and Nakagawa (1969) for *C. perfringens*. Therefore, it is recommended to apply as low heat as possible during pasteurization. Incubation for 10 min at 70°C, 10 min at 80°C, or 10 min at 90–100°C is sufficient also for the elimination of most thermophilic nonsporeformers.

Pasteurization to Eliminate Nonsporeformers

Place 1 g of mud, soil, or other material to be used as inoculum in a sterile test tube and add 5 ml of sterile 0.9% NaCl solution. Flush tube with oxygen-free nitrogen gas, close it with a rubber stopper, and incubate it in a water bath of 80°C (or another temperature desired). The temperature increase is monitored with a second tube that contains 6 ml of 0.9% NaCl solution and a thermometer. When the thermometer has reached 75°C, the tubes are incubated for 10 min and then immediately cooled to room temperature. Samples pasteurized this way are used as inocula for enrichment cultures.

Treatment of the samples with ethanol has been suggested as an alternative procedure for killing the vegetative cells of the accompanying microflora (Johnston, Harmon, and Kautter, 1964; Koransky, Allen, and Dowell, 1978). This method is an excellent means of facilitating the isolation of clostridia from samples in which they are far outnumbered by other facultative or obligate anaerobes. At the same time, any damage of the spores is minimized.

Alcohol Treatment for Eliminating Vegetative Cells (Johnston, Harmon, and Kautter, 1964; Koransky, Allen, and Dowell, 1978)

Liquid samples or solid samples suspended in sterile water are mixed with an equal volume of absolute ethanol. After incubation for 60 min at room temperature, samples are used for dilution series. The absolute ethanol to be used should be filter-sterilized or autoclaved in closed tubes.

Provision of Anaerobic Conditions for Growth

Because the clostridial species exhibit varying degrees of oxygen intolerance, more or less rigid methods to exclude oxygen from the culture media will be recommended in the isolations to be described. However, it should be pointed out that the Hungate technique—if the set-up is accessible—is

Table 2. Nutritional requirements of some clostridia on synthetic media.

Organism	Vitamins	Amino acids	Others	References
C. acetobutylicum	*p*-Aminobenzoate, biotin			Rubbo et al., 1941; Lampen and Peterson, 1943
C. acidiurici				Barker and Peterson, 1944
C. bifermentans	Biotin, nicotinamide, pantothenate, pyridoxine	Alanine, arginine, aspartate, cysteine, glutamate, glycine, histidine, isoleucine, leucine, methionine, phenylalanine, threonine, tryptophan, tyrosine		Holland and Cox, 1975
C. butyricum	Biotin			Cummins and Johnson, 1971
C. cellobioparum	Biotin			Hungate, 1944
C. formicoaceticum	Pyridoxine	Lysine (or cadaverine or *m*-diaminopimelate), methionine		Leonhardt and Andreesen, 1977
C. glycolicum	Biotin, pantothenate	Arginine, glutamate, glycine, histidine, isoleucine, leucine, lysine, methionine, phenylalanine, proline, serine, threonine, tryptophan, tyrosine, valine		Gaston and Stadtman, 1963
C. kluyveri	*p*-Aminobenzoate, biotin			Bornstein and Barker, 1948a
C. pasteurianum	*p*-Aminobenzoate, biotin		Sulfate	Carnahan and Castle, 1958; Sergeant, Ford, and Longyear, 1968; Malette, Reece, and Dawes, 1974
C. putrificum	*p*-Aminobenzoate, pantothenate	Arginine, glutamate, proline, serine, tyrosine, valine		Descrozailles et al., 1974
C. sporogenes	*p*-Aminobenzoate, biotin, nicotinate (folate, thiamine)[a]	Arginine, isoleucine, phenylalanine, tyrosine, valine (aspartate, glutamate, histidine, leucine, methionine, proline, serine, threonine)[a]		Shull, Thoma, and Peterson, 1949; Campbell and Frank, 1956
	p-Aminobenzoate, biotin, pantothenate, pyridoxine, thiamine	Arginine, glutamate, isoleucine, lysine, phenylalanine, tyrosine, valine		Chaigneau et al., 1974
C. tertium	*p*-Aminobenzoate, biotin, nicotinamide, pantothenate, riboflavin, thiamine	Arginine, glutamate, histidine, isoleucine, leucine, lysine, methionine, serine, threonine, tryptophan, tyrosine, valine	Adenine	Hasan and Hall, 1976
C. thermoaceticum	Biotin, nicotinate	Lysine (or cadaverine or *m*-diaminopimelate), methionine		J. R. Andreesen, unpublished
C. thermocellum	Biotin, pantothenate, pyridoxine, riboflavin, thiamine			McBee, 1950
	p-Aminobenzoate, biotin, folate, pantothenate, pyridoxine, riboflavin, thiamine	Cysteine, cystine, methionine, phenylalanine, tryptophan, tyrosine		Fleming and Quinn, 1971

[a] Requirement depends on strain and source of culture.

the most convenient and effective method to cultivate clostridia.

The following methods will be used in the isolations:

Enrichments in Completely Filled Glass-Stoppered Bottles.

This method is satisfactory for all enrichments because aerobes and facultative anaerobes present in the inoculum will remove all the oxygen and provide the conditions for growth of the obligate anaerobes. The addition of reducing compounds is only necessary when mineral media with a low content of complex organic nutrients are used in the primary enrichment steps. It is recommended to autoclave the medium. Autoclaving reduces the oxygen content of the medium, thereby preventing heavy growth of aerobes. Furthermore, it ensures that the anaerobes that will grow originate from the inoculum and not from the medium.

Growth in Tubes and Volumetric Flasks with a Pyrogallol Seal (Wright, 1901; Kürsteiner, 1907; Ritter and Dorner, 1932).

The following points are important for culture of clostridial species: (i) the media should be boiled for a few minutes to remove dissolved oxygen; (ii) the reducing agent (sodium thioglycolate, L-cysteine, sodium sulfide, etc.) must be added just before autoclaving; (iii) the medium has to be cooled down rapidly after autoclaving; (iv) the tubes or flasks have to be sealed immediately after cooling down. Sealing is done by pushing the cotton plug down the tube or the neck of the flask, adding some adsorbance wool, applying 20% (wt/vol) pyrogallol and potassium carbonate solutions (8 drops each per test tube), and sealing with a rubber stopper. If gas producers are cultivated, pyrogallol seals should be used with caution.

Growth in Anaerobic Jars (see also Holdeman, Cato, and Moore, 1977; Willis, 1977).

Anaerobic jars are cylindrical vessels that are made of metal, glass, or plastic and that can be closed air-tight. They are convenient for growth of clostridia on agar plates as well as in culture tubes. The early forms of anaerobic jars have now been replaced by cold-catalyst jars (BTL anaerobic jar, Baird and Tatlock; Whitley anaerobic jar, Don Whitley Scientific; GasPak anaerobic jar, BBL). Anaerobiosis inside the jars can be controlled by an indicator dye (methylene blue) commercially available. In the case of vented jars, the air can be quickly removed by evacuation and refilling with an appropriate mixture of oxygen-free cylinder gases, which include some hydrogen gas to remove the last traces of oxygen by reaction with the cold catalyst.

Operation of the jars with the Gaskit (Don Whitley Scientific) or GasPak (BBL) disposable, hydrogen–carbon dioxide generator avoids evacuation and refilling and provides an oxygen-free nitrogen atmosphere that contains about 10% carbon dioxide. The carbon dioxide content stimulates germination of spores and favors growth of clostridia (Holland, Barker, and Wolf, 1970; Roberts and Hobbs, 1968).

The cold catalyst (palladium-coated alumina pellets) is inactivated especially by hydrogen sulfide, which may be produced during growth of many clostridia. Therefore, the catalyst in the jar has to be exchanged before each use. The used catalyst can be reactivated by heating at 170°C for 2 h. Freshly prepared agar plates are predried by storing them in the anaerobic jar over silica gel desiccant for 2 days.

If it is planned to inoculate the agar plates soon after pouring, they can be kept dry by placing a sterile disk of filter paper and 2–3 drops of glycerol in the lid of the Petri dish and incubating them in inverted position. Freshly inoculated plates should not be exposed to air longer than necessary.

Growth in Tubes Made Anaerobic According to Hungate (1969) and Macy, Snellen, and Hungate (1972).

These methods are extensively described in other sections of this Handbook (Chapters 5 and 117). Here, they will be used in the direct isolation of *C. sphenoides*.

General Media for Enrichment and Isolation

A number of complex media have been widely used for the enrichment and isolation of saccharolytic and proteolytic clostridia: reinforced clostridial medium (RCM) (Gibbs and Hirsch, 1956; Hirsch and Grinstedt, 1954); differential reinforced clostridial medium (DRCM) (Gibbs and Freame, 1965); cooked meat media (CM and CMC) (Holdeman, Cato, and Moore, 1977; Robertson, 1915–1916); peptone–yeast extract medium (PY) (Holdeman, Cato, and Moore, 1977); Viande-Levure medium (VL) (Beerens and Fievez, 1971; Willis, 1977). RCM and DRCM media contain glucose and starch; the others can be supplemented with further ingredients, such as carbohydrates or certain amino acids, to make them more selective. DRCM is recommended especially for the detection and enumeration of clostridial spores in pasteurized samples of food (Freame and Fitzpatrick, 1971; Gibbs and Freame, 1965).

Any medium that is rich in carbohydrates and that contains some peptone and yeast extract as well as reducing agents is suitable for the enrichment of the common saccharolytic clostridia.

Potato mash medium, as described by Ruschmann, has been used by several investigators in the past to study the clostridia in retting flax, in manure, in silage, and in milk (Dührsen, 1937; Glathe, 1934; Ritter, 1932; Ruschmann, 1928; Ruschmann and Bavendamm, 1925; Ruschmann and Harder, 1931). Its starch and pectin content favors the development of starch- and/or pectin-fermenting clostridia. Maize mash medium (Weizmann, 1919) and maize liver medium (McClung and McCoy, 1934) have been used for enrichment and isolation of the butanol-acetone-producing clostridia (Beesch, 1953; Weizmann, 1919; Weyer and Rettger, 1927), pigment-producing strains (Hellinger, 1947; McClung, 1943), and other saccharolytic species (Gilliland and Vaughn, 1943, McClung and McCoy, 1934; Weizmann and Hellinger, 1940). Milk, without or with supplementations, has been found convenient for the enrichment of several clostridia from soil, wounds, and plant material or of the clostridia commonly present in the milk itself (Meyn, 1933; van Beynum and Pette, 1940; Weinzirl and Veldee, 1915; Weizmann and Hellinger, 1940; Winkler, 1961).

Potato tubers, stabbed and immersed in water, provide a simple method for the enrichment of common saccharolytic clostridia. The following procedure has been described by Veldkamp (1965).

Enrichment of Saccharolytic Clostridia with Potato Tubers (Veldkamp, 1965)

"A potato is washed under the tap and subsequently stabbed once or twice with a knife. It is then placed in a beaker, and enough water is added just to cover the potato; the beaker is covered with a watch-glass and incubated at 37°C. The oxygen that might be introduced into the tissue is consumed by its cells. In the anaerobic environment *Clostridium* rapidly starts to decompose the pectin in between the plant cells. The tissue is thus macerated. When the tuber floats, due to profuse gas formation, the water is poured out of the glass; the potato is washed and dissected. Microscopic examination of the tuber contents invariably shows the typical pleomorphic clostridial cells; among these, spore-bearing spindle-shaped cells are often encountered.

"Isolation can easily be achieved as follows. A sample of the macerated tissue is inoculated into yeast extract–glucose broth and after pasteurization (10 min at 80°C in a water bath) the culture is incubated at 37°C in N_2 atmosphere. A pure culture can be obtained by streaking a sample on yeast extract–glucose agar and incubating under N_2."

Media given by various authors for enrichment and growth of saccharolytic clostridia frequently show variations that are not essential for growth of these organisms. The following medium can be recommended for growth of many species.

Medium for Growth of Saccharolytic Clostridia

The medium contains per 1,000 ml:

1 M potassium phosphate, pH 7.5	30 ml
1 M $MgSO_4$	1 ml
Solution M	0.5 ml
0.2 M $FeSO_4$ in 0.1 M H_2SO_4	0.2 ml
Trypticase	10 g
Yeast extract	6 g
Sodium thioglycolate	0.5 g
Energy source (sucrose or glucose)	10–20 g
Distilled water	

The final pH is 7.0–7.2. The medium is autoclaved for 20 min at 121°C.

Solution M contains:

$MnCl_2$	10 mM
$CaCl_2$	30 mM
$CoCl_2$	5 mM
Na_2MoO_4	5 mM

Since stock solutions of the salts can be stored in the laboratory, this medium can be prepared rather quickly. It is recommended that sugar solutions be autoclaved separately.

The fixation of molecular nitrogen is a property shared by many representatives of the saccharolytic clostridia (Rosenblum and Wilson, 1949). Media that contain glucose or sucrose but lack a source of nitrogen have been used for the isolation of *C. pasteurianum* and other species. The comprehensive literature was summarized by Skinner (1971). Based on the method of Augier (1957), the following liquid medium for counting nitrogen-fixing clostridia in soil by the most probable number method and for subsequent isolation has been recommended by Skinner (1971).

Isolation of Nitrogen-Fixing Saccharolytic Clostridia (Skinner, 1971)

The medium contains per 1,000 ml:

K_2HPO_4	0.8 g
KH_2PO_4	0.2 g
$MgSO_4 \cdot 7H_2O$	0.2 g
NaCl	0.2 g
$FeSO_4 \cdot 7H_2O$	10 mg
$MnSO_4 \cdot 4H_2O$	10 mg
$CaCl_2$	10 mg
$Na_2MoO_4 \cdot 2H_2O$	25 mg
Yeast extract	10 mg
Trace element solution	1 ml
Soil extract	10 ml
Glucose or sucrose	10.0 g
Sodium thioglycolate	1.0 g
Distilled water	
Final pH is 7.2.	

The trace element solution contains per 1,000 ml:

$Na_2B_4O_7 \cdot 10H_2O$	50 mg
$CoNO_3 \cdot 6H_2O$	50 mg
$CdSO_4 \cdot 2H_2O$	50 mg

CuSO$_4$ · 5H$_2$O   50 mg
ZnSO$_4$ · 7H$_2$O   50 mg
MnSO$_4$ · H$_2$O   50 mg

The stock solution should be saturated with CO$_2$.

The soil extract is prepared as follows (Augier, 1956): Equal weights of a neutral garden soil and water are heated at 130°C for 1 h. After cooling it is filtered through paper, bottled, autoclaved, and stored until use.

Sodium thioglycolate is added just before the medium is distributed in 10-ml portions into narrow test tubes fitted with Durham tubes, capped or plugged, and autoclaved at 121°C for 15 min. After inoculation with 1 ml of decimal dilutions of pasteurized soil, the tubes are incubated in anaerobic jars under a nitrogen atmosphere at 30°C for 1–2 weeks. Positive tubes show abundant gas formation, turbidity with or without a viscoid, whitish deposit or surface pellicle, and odor of butyric acid.

Isolations are made by streaking small inocula from positive tubes on plates prepared from the above medium plus 2% agar. The plates are incubated in anaerobic jars under an atmosphere of nitrogen. Colonies can be picked within 7 days at 30°C and purified by repeated streaking on nitrogen-free agar. Pasteurization can be applied at any stage, provided spores are present.

The above enrichment procedure can be made rather specific for *C. pasteurianum* if the concentration of sucrose is increased to 15% (Spiegelberg, 1944; Witz, Detroy, and Wilson, 1967). Inocula from such enrichments are again pasteurized and streaked on the above agar containing 2% glucose.

Clostridia, such as *C. butyricum* and *C. tyrobutyricum*, which ferment lactate plus acetate to butyrate, CO$_2$, and H$_2$, have been frequently found as the dominating anaerobic sporeformers in certain silages (Bryant and Burkey, 1956; Gibson, 1965; Gibson et al., 1958; Rosenberger, 1951, 1956) and milk products (Goudkov and Sharp, 1965). *C. tyrobutyricum* is responsible for spoilage of certain types of cheese by formation of gas (late blowing of hard cheese) or rancid odor (Goudkov and Sharp, 1966; Kutzner, 1963, 1966).

A medium originally developed by Bhat and Barker (1947) has been used for the detection and the isolation of lactate-utilizing sporeformers in silage (Rosenberger, 1951, 1956), in cheese (Kutzner, 1963, 1966), and in milk (Halligan and Fryer, 1976). Both *C. butyricum* and *C. tyrobutyricum* will grow in this medium; a distinction between these species is easily possible because of the striking differences in the utilization of carbohydrates: *C. butyricum* grows on maltose, raffinose, lactose, and starch, which are not utilized by *C. tyrobutyricum*.

Isolation of Clostridia that Ferment Lactate plus Acetate (Bhat and Barker, 1947)

The medium contains per 1,000 ml:

K$_2$HPO$_4$	0.5 g
(NH$_4$)$_2$SO$_4$	0.5 g
MgSO$_4$ · 7H$_2$O	0.1 g
FeSO$_4$ · 7H$_2$O	20 mg
Yeast extract	0.5 g
Biotin	0.1 μg
p-Aminobenzoate	100 μg
Sodium L-lactate	10.0 g
Sodium acetate	8.0 g
Sodium thioglycolate	0.5 g
Distilled water	

Sodium thioglycolate is added immediately before autoclaving. The pH is adjusted to 6.0–7.0, and the medium is autoclaved for 20 min at 121°C.

A slightly acidic pH value of the medium favors growth of *C. butyricum* and *C. tyrobutyricum*. The latter has been shown to start growing in media with a pH value of as low as 5.3 (Kutzner, 1963). When a sodium lactate solution is used for preparing the medium, the pH should be checked after autoclaving (it tends to become more acidic).

Enrichment cultures are set up in glass-stoppered bottles (filled to the neck). They are inoculated with pasteurized garden soil or other material and incubated at 37°C. Turbidity and gas formation are observed after a few days, accompanied by an increase of the pH. Two transfers into fresh medium (5–10% inoculum) are made before isolation is performed by streaking material on agar medium of the same composition as above plus 2% agar. The plates are incubated in anaerobic jars under a nitrogen atmosphere.

Using a slightly modified medium, *C. butyricum* was found as the dominant lactate-fermenting sporeformer that developed in the ensiling process of perennial rye-grass (Gibson, et al. 1958). Clostridia other than *C. butyricum* and *C. tyrobutyricum* that are found occasionally in silage (*C. paraputrificum, C. tetanomorphum, C. perfringens,* and *C. sphenoides*) require more complex media or are not able to use lactate plus acetate as energy source.

Procedures for the detection of even low numbers of spores of *C. tyrobutyricum* in milk used in making cheese have been described (Fryer and Halligan, 1976; Halligan and Fryer, 1976). First, an enrichment is done using a complex medium that contains calcium lactate. Then the presence of *C. tyrobutyricum* is confirmed by subculturing positive enrichments in a slightly modified medium as given above.

There have been many reports on the enrichment and isolation of cellulose-decomposing clostridia, but the cultures obtained were often not pure and subcultures are not available any more. At present, only the mesophilic *C. cellobioparum* (Hungate, 1944) and the thermophilic *C. thermocellum* (McBee, 1948, 1950; Viljoen, Fred, and Peterson, 1926) have the status of recognized species (see *Bergey's Manual of Determinative Bacteriology,* eighth edition).

Samples for the isolation of cellulolytic clostridia can be taken from soil, feces of herbivorous animals (horse manure has long been recognized as an excellent source also for thermophilic strains), and from anaerobic sewage or mud. Except for the different incubation temperature, the procedures for enrichment and isolation of both cellulolytic species are about the same.

Enrichment and Isolation of Cellulolytic Clostridia (Omelianski, 1902; Skinner, 1960, 1971)

For enrichment, the medium contains per 1,000 ml:

K₂HPO₄	1.0 g
(NH₄)₂SO₄	1.0 g
MgSO₄ · 7H₂O	0.5 g
CaCO₃	2.0 g
NaCl	0.5 g
Resazurin	1 mg
Cellulose (chopped filter paper)	20 g
Distilled water	

The final pH is 7.1. Enrichments are set up in completely filled, glass-stoppered bottles (50 or 100 ml), inoculated with 0.1 g of pasteurized sample. The sample material provides enough growth factors for the primary enrichment, and reducing conditions are established by the contaminating microflora. Incubations are made at 30–37°C. As soon as digestion of the filter paper becomes visible (1 or 2 weeks), isolation of the cellulose-degrading clostridia is performed by repeatedly streaking on cellulose agar medium. The cellulose agar medium contains per 1,000 ml:

K₂HPO₄	7.5 g
KH₂PO₄	3.5 g
(NH₄)₂SO₄	0.5 g
NaCl	1.0 g
MgSO₄ · 7H₂O	50 mg
CaCl₂	50 mg
Resazurin	1 mg
Yeast extract	1.0 g
Cellulose	10.0 g
Agar	15.0 g
L-Cysteine hydrochloride	0.5 g
Distilled water	
Final pH is 7.1.	

A suitable substrate for these enrichments is powdered cellulose (e.g., MN 300, Machery and Nagel) or a cellulose preparation made by wet-grinding of finely divided filter paper (e.g., Whatman No. 1 ashless cellulose paper) in a pebble mill (Hungate, 1950). To keep the cellulose particles suspended, Skinner (1960) recommended the addition of about 0.1% sodium carboxymethylcellulose (substitution range, 0.65–0.85) to the medium.

The cysteine hydrochloride is added to the medium shortly before autoclaving. The medium is autoclaved for 20 min at 121°C and distributed to Petri dishes, which are kept in anaerobic jars under an atmosphere of 90% N₂ + 10% CO₂. Alternatively, screw-capped bottles (Skinner, 1971) or roll tubes (Hungate, 1950) may be used. Digestion of cellulose is visible after incubation at 30–37°C for 1 week or longer, depending on the activity of the strains. Colonies are often very small, and material for transfers should be picked carefully by controlling it with a dissecting microscope. Colonies that contain spores are suspended in cellulose-free basal medium and pasteurized (10 min at 75°C) before streaking again.

For purification, an alternate streaking on cellulose agar and agar that contains 0.5% cellobiose instead of cellulose can be advantageous. Cellobiose is readily fermented by all cellulolytic isolates described. However, spore production on cellobiose agar is not as pronounced as on cellulose agar. Therefore, colonies must be examined carefully before being pasteurized.

The double-layer method may be used to detect cellulose-decomposing colonies rather early by formation of clear halos. Petri dishes are filled with 20 ml of cellulose-free agar medium and subsequently overlayed with 5 ml of cellulose-containing agar medium.

The selective isolation of *C. sphenoides* from mud samples using citrate as the energy source for growth has been described (Walther, Hippe, and Gottschalk, 1977). It has not been demonstrated whether this method is also applicable to the isolation of this organism from silage or human infections in which *C. sphenoides* has been reported to occur (Gibson, 1965; Smith, 1949). *C. sphenoides* is a representative of the relatively small group of saccharolytic bacteria that do not form butyrate but ferment carbohydrates to ethanol, acetate, CO₂. and H₂.

Direct Isolation of *Clostridium sphenoides* (Walther, Hippe, and Gottschalk, 1977)

The medium contains per 1,000 ml:

K₂HPO₄	2.0 g
KH₂PO₄	3.4 g
MgSO₄ · 7H₂O	0.2 g
(NH₄)₂SO₄	0.3 g
NaCl	0.6 g
CaCl₂ · 2H₂O	60 mg

Yeast extract	4.0 g
Peptone	2.0 g
Trisodium citrate dihydrate	14.7 g
Resazurin	1 mg
L-Cysteine hydrochloride	0.3 g
Agar	15.0 g
Distilled water	

Final pH is 6.7–7.0.

The medium given above is prepared. Cysteine hydrochloride is added after the agar has been dissolved. Applying the Hungate technique, 5-ml portions are added to 15-ml tubes. After autoclaving (20 min at 121°C), the tubes are kept in a water bath of 50°C. Then 0.3 ml of a pasteurized mud sample is used for the preparation of decimal serial dilutions. Roll tubes are prepared which are incubated at 37°C. After 48–72 h, single colonies that show an increased size as compared to the background growth of contaminating organisms are picked, inoculated into 10 ml of liquid citrate medium, and incubated at 37°C for 24 h. For further purification, serial dilutions in roll tubes containing citrate agar are repeated at least three times.

During isolation, the selected colonies are checked for the presence of wedge-shaped cells and of spherical, nearly terminally located spores, both of which are typical for *C. sphenoides*.

A number of single amino acids can be used by certain clostridial species as sources of energy, carbon, and nitrogen for growth. Media that contain a certain amino acid, small amounts of yeast extract, and minerals have been used for enrichment and isolation of some clostridial species. The use of L-alanine led to the isolation of *C. propionicum* (Cardon and Barker, 1946, 1947) and of γ-aminobutyrate and δ-aminovalerate, to *C. aminobutyricum* and *C. aminovalericum,* respectively (Hardman and Stadtman, 1960a,b). *C. tetanomorphum* has been isolated by Kornberg using L-histidine as the principal substrate (see Wachsman and Barker, 1955a), and the use of L-glutamate led to the isolation of *C. cochlearium* as well as of *C. tetanomorphum* (Barker, 1937, 1939). The latter species has been omitted in the eighth edition of *Bergey's Manual.* It is assumed to be identical with *C. cochlearium* (Holdeman, Cato, and Moore, 1977). However, these species differ in that *C. cochlearium* is unable to ferment carbohydrates, whereas *C. tetanomorphum* grows with glucose, fructose, or gluconate (Anthony and Guest, 1968; Bender, Andreesen, and Gottschalk, 1971; Mead, 1971). Finally, lysine has been used to enrich and isolate *Clostridium* SB_4 (Costilow, Rochovansky, and Barker, 1966), which subsequently was identified as *C. subterminale* (H. A. Barker, personal communication). It appears

that the possible advantage of using single amino acids as selective substrates for the isolation of certain clostridia has never been studied systematically. Investigations by Mead (1971), Elsden, Hilton, and Waller, (1976), and Elsden and Hilton (1978) have shown that a considerable number of clostridia are able to ferment single amino acids.

In the following procedure, the enrichment and isolation of *C. cochlearium* or *C. tetanomorphum* is described using L-glutamate as specific substrate.

Enrichment and Isolation of Glutamate-Fermenting Clostridia (Barker, 1937, 1939)

The medium contains in 1,000 ml:

K_2HPO_4	0.2 g
$MgSO_4 \cdot 7H_2O$	0.1 g
Yeast extract	0.5 g
Sodium L-glutamate	10.0 g
Sodium thioglycolate	0.5 g
Tap water	

The final pH is 7.6. Enrichments are set up in completely filled glass-stoppered bottles with pasteurized soil as inoculum (0.5 g soil per 100 ml). Within 1–2 days at 37°C, an abundant growth of sporeforming anaerobes is observed accompanied by a moderate production of gas. After transferring a volume of 2 ml once or twice to 100-ml bottles containing medium of the above composition, isolations are made by repeatedly streaking on L-glutamate agar plates or by diluting in agar roll tubes.

For mass culture of *C. cochlearium* or *C. tetanomorphum,* the "Medium for Growth of Saccharolytic Clostridia" (this chapter) may be used. Instead of sugars, sodium L-glutamate (17 g per 1,000 ml medium) is added as energy source and Trypticase is omitted.

Recently, Laanbroek et al. (1979) described the enrichment of *C. cochlearium* from anaerobic sludge in an L-glutamate-limited anaerobic chemostat.

The anaerobic purine fermenters, *C. acidiurici* and *C. cylindrosporum,* were originally isolated by Barker and Beck (1942) from enrichment cultures with uric acid as the sole source of carbon, nitrogen, and energy. Recently, several more strains of both species were obtained by Champion and Rabinowitz (1977) using the same method. Uric acid agar supplemented with tryptone, meat extract, liver extract, and chicken fecal extract was used by Barnes and Impey (1974) for the isolation of uric acid–decomposing anaerobes from avian cecum. The isolated strains of anaerobic sporeformers were designated *C. malenominatum.* Adenine-fermenting anaerobic sporeformers were isolated recently by using a medium that contained, in addition to the usual minerals, adenine, bicarbonate, and sodium selenite (P. Dürre and J. R. Andreesen, unpublished).

Isolation of Purine-Fermenting Clostridia (Champion and Rabinowitz, 1977; Rabinowitz, 1963; Barker and Beck, 1942)

The medium contains per 1,000 ml:

Uric acid	2.0 g
10 N KOH	3.0 ml
70% $K_2HPO_4 \cdot 3H_2O$ solution	1.5 ml
$MgSO_4 \cdot 7H_2O$	50 mg
$CaCl_2 \cdot 2H_2O$	5 mg
$FeSO_4 \cdot 7H_2O$	2 mg
Yeast extract	1.2 g
Resazurin	1 mg
Mercaptoacetic acid	1.5 ml
Distilled water	

The medium is prepared as follows: 500 ml distilled water plus KOH plus K_2HPO_4 solution are brought to a boil. Then uric acid is slowly added. It goes into solution instantaneously. The solution is cooled down and the other ingredients of the medium are added. Mercaptoacetic acid is added shortly before autoclaving, and the pH is adjusted to 7.2 using a sterile, 60% (wt/vol) K_2CO_3 solution. Other reducing agents may be used instead of mercaptoacetic acid: sodium thioglycolate (750 mg/1,000 ml) or cysteine hydrochloride (1 g/1,000 ml) plus dithiothreitol (0.1 g/1,000 ml).

After autoclaving (20 min at 121°C), the medium is immediately used to completely fill 50-ml glass-stoppered bottles. After inoculation with 0.5 g of pasteurized soil, the bottles are incubated at 37°C. Growth occurs within 24–48 h and is accompanied by an increase of alkalinity. Utilization of uric acid can be monitored by the decrease in absorbancy at 290 nm of a 1 to 100 dilution of the growth medium. After one or two more transfers (0.5% inoculum) to 50-ml bottles containing the medium above, isolation is made under strict anaerobiosis by repeatedly streaking on uric acid agar plates incubated in anaerobic jars or on roll tubes.

C. acidiurici and *C. cylindrosporum* require strictly anaerobic conditions for growth, and the application of stringent anaerobic culture techniques is necessary. *C. acidiurici* and *C. cylindrosporum* are differentiated on the basis of the form and position of their spores. The former contains terminal ovoid spores and the latter, cylindrical spores that are variably situated and do not cause swelling of the mother cell (Barker and Beck, 1942). Furthermore, these species differ in their major fermentation pathways (Vogels and van der Drift, 1976) and in their

molybdate and tungstate requirement (Wagner and Andreesen, 1977).

The isolation of adenine-fermenting clostridia requires a medium of slightly different composition. Instead of uric acid it contains 0.2% adenine, 0.2% sodium bicarbonate, and 10^{-7} M each of sodium selenite, sodium molybdate, and sodium tungstate.

C. kluyveri has specialized on the conversion of ethanol and acetate to butyrate, caproate, and H_2 (Barker and Taha, 1942; Bornstein and Barker, 1948a). It is unable to utilize carbohydrates or amino acids for growth and requires only biotin and *p*-aminobenzoate as growth factors. Therefore, the medium used for enrichment is very selective.

Isolation of *Clostridium kluyveri* (Bornstein and Barker, 1948a; Stadtman and Burton, 1955)

The medium contains per 1,000 ml:

K_2HPO_4	0.30 g
KH_2PO_4	0.20 g
NH_4Cl	0.25 g
$MgSO_4 \cdot 7H_2O$	0.20 g
$CaCl_2 \cdot 2H_2O$	10 mg
$FeSO_4 \cdot 7H_2O$	5 mg
$MnSO_4 \cdot 4H_2O$	2 mg
$Na_2MoO_4 \cdot 2H_2O$	2 mg
Biotin	10 μg
p-Aminobenzoate	200 μg
Ethanol	20 ml
Potassium acetate	5.0 g
Glacial acetic acid	2.5 ml
Resazurin	1 mg
Sodium thioglycolate	500 mg

Thioglycolate is added shortly before autoclaving (20 min at 121°C). After autoclaving the medium is rapidly cooled down, and the pH is adjusted to 7.0 using a 60% K_2CO_3 solution autoclaved separately (about 8–10 ml are required).

The enrichment is made at 35°C in 100-ml glass-stoppered bottles completely filled with the anaerobically prepared sterile medium and inoculated with pasteurized mud from ensiled leaves of sugar beets (turnips), other decaying plant material, fresh water, or sewage digester. After 1 or 2 weeks, enrichment cultures that show gas production and that smell of butyric and caproic acids are used for inoculating fresh liquid medium (10% inoculum). After two to three transfers, serial dilutions are prepared in agar roll tubes using the above medium supplemented with 2% agar. Colonies of *C. kluyveri* that develop after several days are generally small (1–3 mm), fluffy, spherical or compact, lens-shaped. The colonies should contain large cells of the size of about $1 \times 10 \mu$m. Typical colonies are picked and transferred to a new series of roll tubes. If cells of selected colonies contain spores (oval, terminal), a pasteurization step can be applied after suspending material of a colony in a small volume of growth medium.

Identification

The identification of a bacterium as a member of *Clostridium* is generally relatively easy. A bacterium belongs to the genus *Clostridium* when it is restricted to an anaerobic energy metabolism, forms spores, and is unable to carry out a dissimilatory sulfate reduction. Some clostridial species do not sporulate readily (see "Occurrence of Spores and Methods for the Elimination of Nonsporeformers", this chapter). To demonstrate spores, it may be necessary to inoculate slants that contain chopped meat agar. Appropriate procedures are described in the *VPI Anaerobe Laboratory Manual* (Holdeman, Cato, and Moore, 1977).

To identify an isolated strain as a certain clostridial species is more difficult. Appropriate keys are given in Prevot's *Manual for the Classification and Determination of the Anaerobic Bacteria* (Prévot, 1966), in *Bergey's Manual*, eighth edition (Buchanan and Gibbons, 1974), and in the *VPI Anaerobe Laboratory Manual* (Holdeman, Cato, and Moore, 1977). Generally speaking, the taxonomy of the genus *Clostridium* is still in an unsatisfactory state. Descriptions are sometimes based on one strain only. Some species were left out from keys for unknown reasons, and it is advisable to consult the seventh edition of *Bergey's Manual* (Breed, Murray, and Smith, 1957) and also the *Approved Lists of Bacterial Names* (Ad Hoc Committee of the Judical Commission of the ICSB, 1976). It is symptomatic for the present situation that even experts in the field are often unable to assign a considerable percentage of their clostridial isolates to accepted species (Finne and Matches, 1974; Matsuda et al., 1975; Salanitro, Blake, and Muirhead, 1974; Timmis, Hobbs, and Berkeley, 1975).

Depending on the identification key, the genus *Clostridium* is subdivided into two or four groups of saccharolytic and/or proteolytic species. Additional groups are formed by the specialist species, which do not grow on the media commonly employed for identification, and by the thermophiles. The known specialists are *C. acidiurici*, *C. brevifaciens*, *C. cylindrosporum*, *C. kluyveri*, *C. malacosomae*, and *C. propionicum*. The known thermophilic species are *C. thermoaceticum*, *C. thermocellum*, *C. thermohydrosulfuricum*, and *C. thermosaccharolyticum*. Two groups of saccharolytic and/or proteolytic clostridia result from the use of the gelatin liquefaction test as the first step in identification (Holdeman, Cato, and Moore, 1977). In *Bergey's Manual*, eighth edition (Buchanan and Gibbons, 1974), gelatin liquefaction (positive or negative) and the posi-

tion of the mature spore (terminal or subterminal) within the cell are used to put the clostridia into four groups (Smith and Hobbs, 1974).

In order to assign an isolate to a certain species, a number of additional tests have to be carried out:

1. Determination of aerotolerance. The following species grow on agar plates under air: *C. carnis*, *C. durum*, *C. histolyticum*, and *C. tertium*.
2. Determination of certain enzyme activities. This test includes assays for lecithinase, lipase, proteolysis, indole formation, reaction upon milk, and nitrate reduction.
3. Detection of growth and acid production using media that contain carbohydrates as substrates. The main carbohydrates tested are glucose, fructose, mannose, mannitol, ribose, xylose, saccharose, maltose, lactose, melibiose, starch, and esculin.
4. Determination of the fermentation products formed. According to Holdeman, Cato, and Moore (1977), the strains are grown in peptone-yeast extract-glucose or chopped meat-carbohydrate medium for 48 h. The products formed are determined by gas chromatography. The majority of the saccharolytic clostridia form butyrate as the predominant product. About 17 saccharolytic species do not produce this acid but form some or all of the following products: acetate, formate, lactate, and ethanol. These species include: *C. aceticum*, *C. aminovalericum*, *C. cellobioparum*, *C. clostridiiforme*, *C. coccoides*, *C. cocleatum*, *C. durum*, *C. formicoaceticum*, *C. glycolicum*, *C. leptum*, *C. nexile*, *C. oroticum*, *C. ramosum*, *C. sphenoides*, *C. spiroforme*, *C. thermoaceticum*, and *C. thermocellum*. Proteolytic species are indicated by the appearance of branched-chain fatty acids among the fermentation products.

Most clostridial species are motile. For the following species, however, motility has never been reported: *C. barkeri*, *C. celatum*, *C. cocleatum*, *C. innocuum*, *C. leptum*, *C. nexile*, *C. oroticum*, *C. paraperfringens*, *C. perenne*, *C. perfringens*, *C. putrefaciens*, *C. ramosum*, *C. rectum*, *C. spiroforme*, *C. sporosphaeroides*, *C. thermoaceticum*, and *C. thermosaccharolyticum*. The observation of motility is helpful in classification; the observation of nonmotility is of little value because nonmotile strains of motile species are occasionally found.

In addition to the properties of the clostridial species that are currently used in identification keys, a number of other characteristics give an insight into their relationships. The mol% G+C content of the clostridia, as far as it is known, is summarized in Fig. 2. It is apparent that most species cluster around 28 mol%. However, about 15 species form a second group, which ranges in G+C content from 40 to 54 mol%.

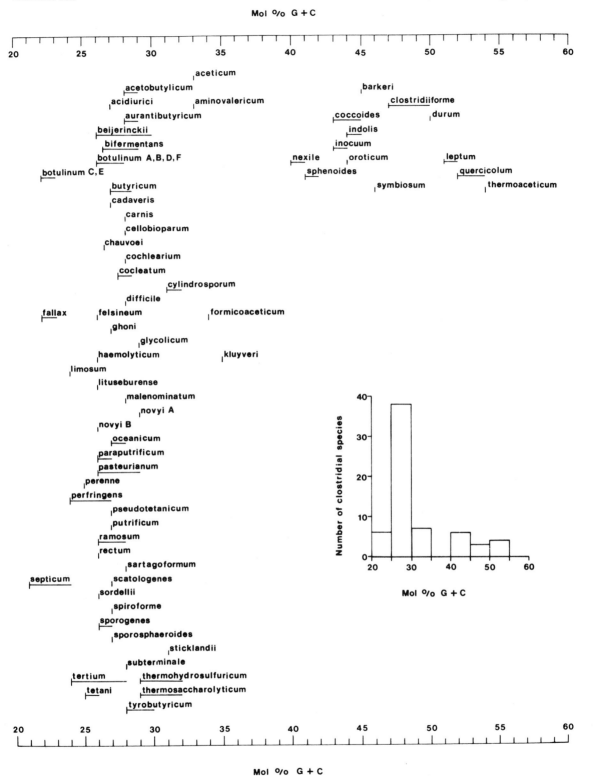

Fig. 2. Distribution of mol% G+C within the genus *Clostridium*. Values were taken from Holdeman, Cato, and Moore, 1977; Kaneuchi, Benno, and Mitsuoka, 1976, 1979; Marmur and Doty, 1962; Matteuzzi, Hollaus, and Biavati, 1978; Matteuzzi et al., 1977; Schildkraut, Marmur, and Doty, 1962; and Smith and Hobbs, 1974; values for *C. aceticum, C. acidiurici,* and *C. cylindrosporum* were from M. Braun (unpublished) and P. Dürre (unpublished), respectively.

Studies on the cell wall composition (Cummins, 1970; Cummins and Johnson, 1971; Johnson, 1970; Schleifer and Kandler, 1972) have shown that *meso*-diaminopimelic acid is the predominant crosslinking amino acid in the clostridial peptidoglycan. Some species, however, contain LL-diaminopimelic acid or L-lysine.

DNA-DNA and rRNA homology studied have been applied to solve a number of taxonomic problems, especially within the groups of butyric acid–producing and thermophilic clostridia (Johnson, 1970; Johnson and Francis, 1975; Kaneuchi et al., 1979; Matteuzzi, Hollaus, and Biavati, 1978; Matteuzzi et al., 1977; Nakamura et al., 1973, 1975, 1979).

Properties of Some Nonpathogenic Clostridia

CLOSTRIDIUM ACETICUM. Optimum growth temperature, 30°C. Converts H_2 + CO_2 to acetate, grows also on carbohydrates at slightly alkaline pH (Karlsson, Volcani, and Barker, 1948; Wieringa, 1940). Can fix nitrogen (Rosenblum and Wilson, 1949). The species was thought to be lost but recently the original strain could be revived (M. Braun et al., unpublished). The production of acetate from H_2 + CO_2 by unidentified sporeformers was reported by Ohwaki and Hungate (1977), Prins and Lankhorst (1977), and Braun, Schoberth, and Gottschalk (1979).

CLOSTRIDIUM ACETOBUTYLICUM. G+C, 28–29 mol%; peptidoglycan crosslinkage via *m*-DAP; optimum growth temperature, 37°C. Grows in a mineral medium that contains a carbohydrate as carbon and energy source, and biotin and *p*-aminobenzoate as growth factors (Lampen and Peterson, 1943; McCoy et al., 1926; Rubbo et al., 1941). Fermentation products of strains that have been kept and transferred in a laboratory for a long time are butyrate, acetate, CO_2, and H_2 (small amounts of acetone and *n*-butanol). The relation between the redox state of the substrate and the corresponding products has been investigated (Johnson, Peterson, and Fred, 1931). Freshly isolated strains and cultures started from spores produce *n*-butanol, acetone, and CO_2, and reduced amounts of acids and H_2 (McCoy and Fred, 1941). The formation of solvents starts in a later growth phase at slightly acidic pH (Davies and Stephenson, 1941). The process of acetone-butanol production has been summarized (Abou-Zeid, Fouad, and Yassein, 1978; Beesch, 1953; Spivey, 1978). In the presence of air, growth of *C. acetobutylicum* is stopped, but the cells are not killed and growth is resumed after transfer to anaerobic conditions (O'Brien and Morris, 1971b). Chloramphenicol is detoxified in a ferredoxin-dependent reaction (O'Brien and Morris, 1971a). The nitro group of

metronidazole is likewise reduced (O'Brien and Morris, 1972). Cells of *C. acetobutylicum* have been used as source for acetoacetate decarboxylase (Neece and Fridovich, 1967) and crotonase (Waterson et al., 1972). The formation of the following enzyme activities has been studied: NADH- and NADPH-ferredoxin oxidoreductases (Petitdemange et al., 1977), α-amylase and glucoamylase (Ensley, McHugh, and Barton, 1975), and proteolytic enzymes (Egorov, Loria, and Vlasova, 1972). Bacteriocin production has been reported (Barber et al., 1979), and the ultrastructure of the cells has been studied (Cho and Doy, 1973).

CLOSTRIDIUM ACIDIURICI. G+C, 25 mol%; peptidoglycan crosslinkage via *m*-DAP; optimum growth temperature, 37°C. This organism is specialized on the fermentation of some purine derivatives (uric acid, xanthine, hypoxanthine, guanine) and will not ferment any carbohydrate or amino acid (Barker and Beck, 1941, 1942). It will grow in an entirely synthetic medium (Barker and Peterson, 1944). Fermentation products of *C. acidiurici* are acetate, CO_2, and ammonia (Barker, 1961). H_2 is not produced during growth. On the most reduced substrate (hypoxanthine), some acetate is formed by reduction of CO_2 (Schulman et al., 1972). The path of purine degradation has been elucidated by Rabinowitz and by Sagers and Beck (for references see Vogels and van der Drift, 1976). The synthesis of formate dehydrogenase in *C. acidiurici* is stimulated by the presence of tungstate and selenite in the growth medium (Wagner and Andreesen, 1977). Selenite is also required for xanthine dehydrogenase synthesis (Wagner and Andreesen, 1979). *C. acidiurici* has been used as source for the purification of formyltetrahydrofolate synthetase (Rabinowitz and Pricer, 1962), pyruvate-ferredoxin oxidoreductase (Uyeda and Rabinowitz, 1971), ferredoxin (Champion and Rabinowitz, 1977), formate dehydrogenase (Kearny and Sagers, 1972), re-citrate synthase (Gottschalk and Dittbrenner, 1970), and L-serine dehydratase (Carter and Sagers, 1972). The enzymes involved in the conversion of serine to acetate have also been studied (Sagers, Benziman, and Gunsalus, 1961; Valentine, Brill, and Sagers, 1963). The folate coenzymes have been isolated (Curthoys, Scott, and Rabinowitz, 1972).

CLOSTRIDIUM BARKERI. G+C, 45 mol%; unusual peptidoglycan crosslinkage involving the α-carboxyl group of glutamate and a diaminobutyric acid bridge; optimum growth temperature, 30°C. Grows on carbohydrates, which are fermented to butyrate, lactate, CO_2, and some H_2. Characteristic is its ability to utilize nicotinic acid, which is degraded to propionate, acetate, CO_2, and ammonia (Stadtman et al., 1972). Breakdown of nicotinic acid involves a coenzyme B_{12}-dependent rearrangement of α-methylene glutarate to dimethylmaleate (Kung et al., 1970). The hydroxylation of nicotinic acid has been shown to be stimulated by selenite (Imhoff and Andreesen, 1979). The nicotinic acid hydroxylase (Holcenberg and Stadtman, 1969), 6-hydroxynicotinic acid reductase (Holcenberg and Tsai, 1969), α-methylene glutarate mutase, and methylitaconate isomerase (Kung and Stadtman, 1971) have been purified.

CLOSTRIDIUM BEIJERINCKII. G+C, 26–28 mol%; peptidoglycan crosslinkage via *m*-DAP; optimum growth temperature, 30°C. It differs from *C. butyricum* in its higher demand for growth factors such as amino acids. According to DNA-DNA homology studies, *C. beijerinckii* includes strains that originally were described as strains of *C. butylicum, C. butyricum, C. lactoacetophilum, C. multifermentans,* and *C. rubrum* (Cummins and Johnson, 1971; Johnson, 1973; Matteuzzi et al., 1977). *C. beijerinckii* ferments glucose to butyrate, acetate, and butanol, but forms no isopropanol (Sjolander and McCoy, 1937). Lactate is a substrate in the presence of acetate (Bhat and Barker, 1947). The lactate racemase has been thoroughly studied (Pepple and Dennis, 1976). The hydrogenase has been characterized as iron flavoprotein (Peck and Gest, 1957). The regulation of the tryptophan-synthesizing enzymes, especially of the anthranilate synthetase, has been reported (Baskerville and Twarog, 1972, 1974). The enzymes related to C_1 metabolism have been studied (Thauer, Kirchniawy, and Jungermann, 1972). The importance of an intracellular polysaccharide for sporulation has been documented (Bergère, Rousseau, and Mercier, 1975). The organism contains a neuraminidase and an acylneuraminate lyase (Müller and Werner, 1974).

CLOSTRIDIUM BUTYRICUM. G+C, 27–28 mol%; peptidoglycan crosslinkage via *m*-DAP; optimum growth temperature, 25–37°C. This is the type species of the genus *Clostridium*. It was isolated by Prazmowski in 1880. Some strains of *C. butylicum, C. multifermentans,* and *C. fallax* belong to *C. butyricum* (Cummins and Johnson, 1971; Johnson, 1973). The ultrastructure and spore formation has been studied (Rousseau, Hermier, and Bergère, 1971a). *C. butyricum* is able to ferment a great variety of carbohydrates, including starch and pectin. Growth factors other than biotin are not required. The fermentation products are butyrate, acetate, CO_2, and H_2. Acetone is formed only in trace amounts; some strains produce isopropanol and butanol in substantial amounts (Kutzenock and Ashner, 1952; Sjolander and McCoy, 1937). The ratio in which acetate and butyrate are formed can be affected by the partial pressure of H_2 (Jungermann et al., 1973). Butyrate is formed via butyryl phosphate; the enzymes involved, phosphotransbutyry-

lase and butyrate kinase, have been partially purified (Twarog and Wolfe, 1963; Valentine and Wolfe, 1960). In the presence of acetate, *C. butyricum* is able to ferment mannitol or L-lactate (Kutzner, 1963). *C. butyricum* has been used as a source for purification and study of ferredoxin (Benson, Mower, and Yasunobu, 1967), NAD(P)-ferredoxin oxidoreductase (Jungermann et al., 1971a), pyruvate cleavage (Hespell, Joseph, and Mortlock, 1969), pyruvate:formate lyase (Thauer, Kirchniawy, and Jungermann, 1972; Wood and Jungermann, 1972), acyl-CoA reductase (Day and Goldfine, 1978), pectin esterase, and polygalacturonate lyase (Sheiman et al., 1976). The phospholipids (Goldfine et al., 1977), a butyricin (Clarke and Morris, 1976a), and the biosynthesis of nicotinic acid (Scott, Bellion, and Mattey, 1969) have been studied. Oncolytic properties were described for a strain (M 55) that produces a bradykinin-degrading enzyme, protease, and nuclease (Brantner and Schwager, 1979).

CLOSTRIDIUM CELLOBIOPARUM. G+C, 25–28 mol%; peptidoglycan crosslinkage via *m*-DAP; optimum growth temperature, 30–37°C. This organism was isolated by Hungate (1944) in the course of studies on cellulose fermentation. In addition to cellulose, *C. cellobioparum* is able to ferment a great variety of carbohydrates. The products formed are ethanol, acetate, formate, CO_2, and H_2 (small amounts of lactate). In pure culture, growth and sugar breakdown are inhibited by molecular hydrogen (Chung, 1976). Only biotin is required as growth factor.

CLOSTRIDIUM CLOSTRIDIIFORME. G+C, 47–50 mol%; peptidoglycan crosslinkage via *m*-DAP; optimum growth temperature, 30–37°C. This species was created by Kaneuchi et al. (1976) to replace *Bacteroides clostridiiformis* because most strains assigned to this species have been found to be sporeformers (see also Cato and Salmon, 1976). *C. clostridiiforme* ferments a number of carbohydrates and produces lactate, acetate, CO_2, and H_2 as the main products, but no butyrate.

CLOSTRIDIUM COCHLEARIUM. G+C, 28 mol%; peptidoglycan crosslinkage via *m*-DAP; optimum growth temperature, 37–45°C. This organism does not ferment any carbohydrates. It grows only on glutamate, glutamine, and histidine (Barker, 1939), and can be isolated specifically in a chemostat using L-glutamate as substrate (Laanbroek et al., 1979). *C. cochlearium* uses the methylaspartate pathway for glutamate breakdown (Buckel and Barker, 1974). The products formed are butyrate, acetate, CO_2, and H_2. DNA homology studies have shown that *C. cochlearium* is different from *C. tetanomorphum* (Nakamura et al., 1979).

CLOSTRIDIUM CYLINDROSPORUM.   G+C, 32 mol%; optimum growth temperature, 37°C. This species exhibits the same substrate specificity as *C. acidiurici*. Both species differ in spore morphology, in tungstate/molybdate requirement for formate dehydrogenase formation, in the ability to grow on hypoxanthine (Wagner and Andreesen, 1977), and in the products formed. In addition to acetate, *C. cylindrosporum* produces larger amounts of formate and glycine than *C. acidiurici* (Barker and Beck, 1941, 1942; Champion and Rabinowitz, 1977). *C. cylindrosporum* has been used to elucidate the pathway of purine degradation (Rabinowitz, 1963) and for purification and study of xanthine dehydrogenase (Bradshaw and Barker, 1960; Wagner and Andreesen, 1979), formyltetrahydrofolate synthetase Rabinowitz and Pricer, 1962), ferredoxin (Champion and Rabinowitz, 1977), glycine formiminotransferase, formiminotetrahydrofolate cyclodeaminase, methylene-tetrahydrofolate dehydrogenase, and serine hydroxymethyltransferase (Uyeda and Rabinowitz, 1965, 1967a,b, 1968).

CLOSTRIDIUM FELSINEUM.   G+C, 26 mol%; peptidoglycan crosslinkage via *m*-DAP; optimum growth temperature, 25–37°C. Ferments carbohydrates to butyrate, acetate, and some butanol. The organism is known to play an important role in the pectinolytic degradation of plant material during retting (Potter and McCoy, 1952). *C. roseum* (McCoy and McClung, 1935) is a strain of *C. felsineum*. The formation of the pectin-degrading enzymes (endopectate lyase enzyme, pectate hydrolase, pectin esterase, endopolygalacturonase, and exopolygalacturonase) depends on the growth medium and the strain used (Lund and Brocklehurst, 1978; Vozniakovskaya, Arova, and Andronikashvili, 1974).

CLOSTRIDIUM FORMICOACETICUM.   G+C, 34 mol%; peptidoglycan crosslinkage via *m*-DAP; optimum growth temperature, 30–37°C. Ferments some carbohydrates, but not glucose to acetate. Almost three molecules of acetate are produced per molecule of fructose. Growth is largely stimulated by the presence of CO_2/bicarbonate or formate (Andreesen, Gottschalk, and Schlegel, 1970; El Ghazzawi, 1967; Linke, 1969). The level of formate dehydrogenase in *C. formicoaceticum* is increased if tungstate and selenite are present in the medium (Leonhardt and Andreesen, 1977). CO_2 is incorporated into acetate and the enzymes of the tetrahydrofolate-dependent, C_1 metabolism are present in this organism (Moore, O'Brien, and Ljungdahl, 1974; O'Brien and Ljungdahl, 1972; Schulman et al., 1972). Gluconate is metabolized by *C. formicoaceticum* via 2-keto-3-deoxygluconate (Andreesen and Gottschalk, 1969). Fumarate is fermented to succinate, acetate, and CO_2 (Dorn, Andreesen, and Gottschalk, 1978a). Fumarate reductase of *C. formicoaceticum* is a pe-

ripheral membrane protein (Dorn, Andreesen, and Gottschalk, 1978b). A cytochrome has been detected in this organism (Gottwald et al., 1975).

CLOSTRIDIUM GLYCOLICUM.   G+C, 29 mol%; peptidoglycan crosslinkage via *m*-DAP; optimum growth temperature, 25–37°C. This organism grows on glucose, fructose, xylose, maltose, and sorbitol and produces acetate, ethanol, H_2, and CO_2. It also decomposes cellulose slowly. Most characteristic is the ability to grow on ethylene glycol and 1,2-propylene glycol. Both compounds are stoichiometrically converted to equal amounts of the corresponding acid and alcohol. *C. glycolicum* requires 15 amino acids, biotin, and *p*-aminobenzoate for growth (Gaston and Stadtman, 1963). Strains of uracil-degrading clostridia show great similarity to *C. glycolicum* (Mead et al., 1979), as does an acetogenic organism, *Clostridium* sp. (Ohwaki and Hungate, 1977).

CLOSTRIDIUM KLUYVERI.   G+C, 35 mol%; peptidoglycan crosslinkage via *m*-DAP; optimum growth temperature, 37°C. This organism has specialized on the fermentation of ethanol and acetate to butyrate, caproate, and H_2. Carbohydrates and amino acids are not utilized for growth and only some C_4 compounds (vinylacetate, γ-hydroxybutyrate [Bartsch and Barker, 1961]) can also serve as growth substrate. *C. kluyveri* requires CO_2/bicarbonate for growth and will develop in a mineral medium that contains in addition to the substrates only biotin and *p*-aminobenzoate (Bornstein and Barker, 1948a,b). The ultrastructure of the organism has been revealed by electron microscopy (Cho and Doy, 1973). The reactions involved in butyrate formation by clostridia have been studied first in *C. kluyveri* (see Barker, 1956; Stadtman, 1976). Ethanol is oxidized to acetyl-CoA via acetaldehyde (Burton and Stadtman, 1953). The formation of ATP in *C. kluyveri* is coupled to the evolution of H_2. Per molecule of H_2 evolved, 0.5 molecules of ATP can be synthesized by substrate-level phosphorylation (Schoberth and Gottschalk, 1969; Thauer et at., 1968; for a model see Gottschalk, 1979). H_2 is evolved from NADH via ferredoxin (Jungermann et al., 1971b). *C. kluyveri* synthesizes pyruvate by reductive carboxylation of acetyl-CoA (Andrew and Morris, 1965) and glutamate via citrate (Gottschalk and Barker, 1966; Tomlinson, 1954). Glyoxylate inhibits the pyruvate synthase system (Thauer, Rupprecht, and Jungermann, 1970). The synthesis of C_1 units (Jungermann et al., 1968), the glycine formation from threonine and serine (Jungermann et al., 1970), and the synthesis of aromatic amino acids (Thauer, Jungermann, and Decker, 1967) and of carbohydrates (Decker, Barth, and Metz, 1966) have been studied. *C. kluyveri* has been used as source for purification and study of acetaldehyde dehydrogenase (Burton and Stadtman, 1953; Hillmer and Gottschalk, 1974; Lurz, Mayer, and Gottschalk, 1979), hydrogenase (Fredricks and Stadtman, 1965), diaphorase (Kaplan, Setlow, and Kaplan, 1969), phosphotransacetylase (Bergmeyer et al., 1963; Henkin and Abeles, 1976; Kyrtopoulos and Satchell, 1972; Stadtman, 1955), coenzyme A transferase (Barker, Stadtman, and Kornberg, 1955), NAD- and NADP-ferredoxin oxidoreductase (Jungermann et al., 1971b; Thauer et al., 1971), and NADP-specific β-hydroxybutyryl-CoA dehydrogenase (Madan, Hillmer, and Gottschalk, 1973). Other clostridial species contain a NAD-specific β-hydroxybutyryl-CoA dehydrogenase (von Hugo et al., 1972). The steady-state concentrations of different nucleotides were determined in growing cells (Decker and Pfitzer, 1972). Radioactive flavine nucleotides can be conveniently derived from the cells (Brühmüller and Decker, 1976). The cells have been used to carry out stereospecific biohydrogenation reactions of unsaturated compounds (Bader et al., 1978).

CLOSTRIDIUM OROTICUM. G+C, 44 mol%; peptidoglycan crosslinkage via *m*-DAP; optimum growth temperature, 30–37°C. This bacterium was originally described by Wachsman and Barker (1954) as *Zymobacterium oroticum*. Cato, Moore, and Holdeman (1968) demonstrated the formation of heat-resistant spores and transferred it to the genus *Clostridium*. *C. oroticum* is nonmotile and ferments a number of carbohydrates to ethanol, acetate, and CO_2 as the major products (small amounts of lactate and formate are also formed). Most characteristic for this organism is its growth on orotic acid, which is degraded to acetate, CO_2, ammonia, and a dicarboxylic acid. Enzyme preparations of *C. oroticum* have been used to study the breakdown of orotic acid (Liebermann and Kornberg, 1954). Growth on orotate increases the flavin content of the cells about three times (Kondo, Friedmann, and Vennesland, 1960). The initial attack is carried out by a NAD-dependent dihydroorotate dehydrogenase, an iron-sulfur flavoprotein (Aleman and Handler, 1967). The ring-cleaving enzyme, dihydroorotase, contains zinc (Taylor et al., 1976).

CLOSTRIDIUM PASTEURIANUM. G+C, 26–28 mol%; peptidoglycan crosslinkage via *m*-DAP; optimum growth temperature, 37°C. This organism was described by Winogradsky in 1895. It grows in a defined medium (Malette, Reece, and Dawes, 1974) and on a great variety of carbohydrates but not on starch. The products formed are butyrate, acetate, CO_2, and H_2. Part of the H_2 is derived from $NADH_2$ (Jungermann et al., 1973). The fixation of nitrogen by free-living microorganisms was discovered through studying *C. pasteurianum* (Winogradsky, 1895, 1902). Ferredoxin was also discovered in this organism (Mortenson, Valentine, and Carnahan,

1962). *C. pasteurianum* has been used for the purification and study of ferredoxin (Packer, Rabinowitz, and Sternlicht, 1978; Schönheit, Brandis, and Thauer, 1979), flavodoxin (Knight and Hardy, 1966; Mayhew, 1971), rubredoxin (Lovenberg and Sobel, 1965; Vogel, Bruschi, and Le Gall, 1977), a red paramagnetic protein (Cardenas, Mortenson, and Yoch, 1976), hydrogenases (Chen and Blanchard, 1978; Chen and Mortenson, 1974; Erbes, Burris, and Orme-Johnson, 1975), NADH-ferredoxin reductase (Jungermann et al., 1971a; Petitdemange, Lambert, and Gay, 1972), pool sizes of pyridine nucleotides (Petitdemange, Lambert, and Gay, 1973), ferredoxin-linked sulfite reductase (Laishley, Lin, and Peck, 1971), isotope fractionation during sulfite and sulfate reduction (Laishley and Krouse, 1978), reduced ferredoxin: CO_2 oxidoreductase, a molybdenum iron–sulfur protein (Scherer and Thauer, 1978), CO dehydrogenation (Fuchs, Schnitker, and Thauer, 1974), pyruvate dehydrogenase (Bush and Sauer, 1977; Westlake, Shug, and Wilson, 1961), the inhibition of pyruvate synthase reaction by glyoxylate (Thauer, Rupprecht, and Jungermann, 1970) and the utilization of CO_2 as active species (Thauer, Käufer, and Scherer, 1975), glycolytic enzymes (Howell, Akagi, and Himes, 1969; Kotze, 1969), PEP phosphotransferase systems (von Hugo and Gottschalk, 1974a), 1-phosphofructokinase (du Toit, Potgieter, and de Villiers, 1972; von Hugo and Gottschalk, 1974b), pathway of gluconate degradation (Bender, Andreesen, and Gottschalk, 1971), gluconate dehydratase (Bender and Gottschalk, 1973), the uptake system for gluconate and galactose (Booth and Morris, 1975), the formation of an invertase (Laishley, 1975) and of the intracellular reserve polysaccharide (Brown, Lindberg, and Laishley, 1975; Darvill et al., 1977) by ADP-glucose pyrophosphorylase and granulose synthase (Robson, Robson, and Morris, 1974) and its degradation by granulose phosphorylase (Robson and Morris, 1974), β-ketothiolase (Berndt and Schlegel, 1975) membrane-bound adenosine triphosphatase (Clarke, Fuller, and Morris, 1979; Clarke and Morris, 1976b; Riebeling, Thauer, and Jungermann, 1975), the exosporium (Mackey and Morris, 1972) and the occurrence of squalene in the spore membrane (Mercer et al., 1979), the enzymes involved in nitrogen metabolism (Kleiner and Fitzke, 1979), the biosynthesis of amino acids (Dainty and Peel, 1970), especially threonine aldolase (Dainty, 1970), glutamate synthase, and glutamine synthetase (Dainty, 1972), the effect of ammonium addition to N_2-fixing cells (Kleiner, 1979), the components of nitrogenase (Mortenson, 1978; Mortenson, Morris, and Jeng, 1967) such as iron–molybdenum protein (Cramer et al., 1978; Rawlings et al., 1978; Zumft, 1978) and iron protein (Tanaka et al., 1977), the kinetics of their formation (Seto and Mortenson, 1974) and the parallel uptake of molybdate ions (Elliot and Mortenson, 1976) involving a molybdenum-storage protein (Elliot and Mortenson, 1977), the immunological comparison of the nitrogenase components with corresponding proteins from other nitrogen-fixing organisms (Smith et al., 1976), and NADH as a physiological electron donor (Jungermann et al., 1974).

CLOSTRIDIUM PERFRINGENS.   G+C, 24–27 mol%; peptidoglycan crosslinkage via L-DAP; optimum growth temperature, 45°C. The organism is highly pathogenic by the formation of phospholipase, hemolysins, and other toxins (Möllby et al., 1976). Both carbohydrates and proteins are fermented. Some strains can utilize nitrate as electron sink, which results in a shift from butyrate to acetate fermentation (Ishimoto, Umeyama, and Chiba, 1974) and in an increase of the growth yield (Hasan and Hall, 1975). The ferredoxin-dependent nitrate reductase has been purified (Chiba and Ishimoto, 1973; Seki-Chiba and Ishimoto, 1977). The ferredoxin of C. perfringens is a 4Fe/4S protein (Seki et al., 1979). The degradation of arginine (Schmidt, Logan, and Tytell, 1952), of glutamate by glutamate decarboxylase (Cozzani et al., 1975) and the enzymes aspartate kinase (Kuramitsu and Watson, 1973), α-L-fucosidase (Aminoff and Furakawa, 1970), and N-acetylneuraminic acid aldolase (DeVries and Binkley, 1972) have been studied. Some bacteriocins are located on a plasmid (Mihele, Duncan, and Chambliss, 1978).

CLOSTRIDIUM PROPIONICUM.   Optimum growth temperature, 30–37°C. This organism does not utilize any carbohydrates. Growth substrates are: L-alanine, β-alanine, L-serine, L-lactate, acrylate, and L-threonine. The C_3 compounds are fermented to propionate, acetate, and CO_2 (and ammonia in the case of serine and alanine). Threonine is converted into propionate, butyrate, CO_2, and ammonia (Cardon and Barker, 1946, 1947). Propionate is formed from lactate via direct reduction of acrylyl-CoA and not via succinate as in propionibacteria (Leaver, Wood, and Stjernholm, 1955). Growth on β-alanine requires previous induction of α-alanine degradation and of acrylyl-CoA aminase (Goldfine and Stadtman, 1960; Vagelos, Earl, and Stadtman, 1959). Acrylic acid can be produced by whole cells (Akedo et al., 1978).

CLOSTRIDIUM PUTREFACIENS.   G+C, 22–25 mol%; optimum growth temperature, 20–25°C; grows at 0–30°C, not at 37°C. This organism is the only named true psychrophilic clostridium species (Roberts and Hobbs, 1968). The organism—first isolated by McBryde (1911) from spoiled ham and regarded by him as the causal agent of souring in brine-cured ham—as well as one strain isolated by Barnes were described as purely proteolytic (Ross, 1965). Other strains isolated from animal sources were found to

ferment glucose and to form acetate, butyrate, and valerate as the fermentation products after growth in meat medium (Parsons and Sturges, 1927a,b,c; Sturges and Drake, 1927). The organism grows well in most media used routinely for culturing anaerobes but produces spores on lactose–egg yolk agar (Roberts and Derrik, 1975).

CLOSTRIDIUM SPHENOIDES.   G+C, 41–42 mol%; peptidoglycan crosslinkage via m-DAP; optimum growth temperature, 37°C. This species grows on a variety of carbohydrates and produces acetate, ethanol, CO_2, and H_2 as the major fermentation products. Characteristic is its ability to grow on citrate which, as far as is known, is not shared by other clostridial species (Walther, Hippe, and Gottschalk, 1977). C. sphenoides has been shown recently to convert the insecticide hexachlorocyclohexane into tetrachlorocyclohexane (Heritage and MacRae, 1977).

CLOSTRIDIUM SPOROGENES.   G+C, 26 mol%; peptidoglycan crosslinkage via m-DAP; optimum growth temperature, 30–40°C. This organism ferments glucose and fructose to butyrate, acetate, ethanol, CO_2, and H_2. However, the preferred substrates of C. sporogenes are amino acids, which are fermented by the "Stickland reaction". This reaction was discovered by Stickland in a study on the energy metabolism of C. sporogenes. In a series of publications (see Hoogerheide and Kocholaty, 1938; Stickland, 1935), it was established that, in the fermentation of pairs of amino acids, alanine, valine, and leucine serve as hydrogen donors and glycine, proline, hydroxyproline, ornithine, and arginine serve as hydrogen acceptors (Marmelak and Quastel, 1953; Woods, 1936). This reaction is of great importance to a number of clostridial species (see Barker, 1961; Nisman, 1954). The nature of the products formed depends on the amino acids fermented. Glycine and alanine are converted to acetate, valine to isobutyrate, proline, arginine, and ornithine to δ-aminovalerate. Pyrimidines can only be reduced to the dihydroderivatives (Hilton, Mead, and Elsden, 1975). The enzymes of the arginine dihydrolase pathway (Venugopal and Nadkarni, 1977), the ornithine cyclase (deaminating) (Muth and Costilow, 1974), the proline dehydrogenase (Costilow and Cooper, 1978), the selenium-dependent glycine reduction (Costilow, 1977), and the B_{12}-coenzyme-dependent leucine 2,3-aminomutase (Poston, 1976) have been studied. Production of thiaminase I (Edwin, Shreeve, and Jackman, 1978; Kimura and Liao, 1964) and ability to degrade chitin (Timmis, Hobbs, and Berkeley, 1974) have been demonstrated.

CLOSTRIDIUM STICKLANDII.   G+C, 26 mol%; optimum growth temperature, 30–35°C. This organism was described by Stadtman and McClung in 1957. It shows only slight growth on some sugars (maltose, glucose, galactose, and ribose) and has specialized on the fermentation of pairs of amino acids, of which one is reduced (e.g., glycine or proline) and the other one is oxidatively decarboxylated and deaminated (e.g., arginine, ornithine, threonine, or serine) (Stickland reaction). Formate increases the growth yield. Glycine is a preferred oxidant and is reduced to acetate by a membrane-bound glycine reductase, which consists of several proteins including a selenoprotein (see Stadtman, 1978; Tanaka and Stadtman, 1979). The membrane-bound proline reductase is composed differently (Seto, 1978). Alanine is not utilized but pyruvate. Purines are catabolized (Schäfer and Schwartz, 1976). *C. sticklandii* grows also with lysine as single amino acid. Lysine is fermented to acetate, butyrate, and ammonia. The fermentation pathway involves a pyridoxal phosphate dependent L-lysine-2,3-aminomutase, a coenzyme B_{12}-dependent β-lysine-5,6-aminomutase, and a 3,5-diaminohexanoate dehydrogenase (see Baker and van der Drift, 1974; Stadtman, 1973). In contrast to *C. sporogenes*, ornithine is fermented by *C. sticklandii* to alanine, acetate, CO_2, and ammonia (Dyer and Costilow, 1968). The first steps in this fermentation are catalyzed by a coenzyme B_{12}-dependent aminomutase and a dehydrogenase, which yields 2-amino-4-ketopentanoate (Somack and Costilow, 1973a,b; Tsuda and Friedmann, 1970). This compound is then cleaved to alanine and acetyl-CoA (Jeng, Somack, and Barker, 1974). The organism contains ferredoxin, rubredoxin, and some flavoproteins (Stadtman, 1965). Acetate can be formed from formate (Stadtman and White, 1954). Corrinoids are involved in the ribonucleotide reductase reaction (Abeles and Beck, 1967).

CLOSTRIDIUM SUBTERMINALE.   G+C, 28 mol%; optimum growth temperature, 37°C. This organism resembles *C. sticklandii* in that it will not ferment carbohydrates. In contrast to *C. sticklandii,* this organism utilizes proteins in addition to amino acids. Lysine is fermented to acetate, butyrate, and ammonia. The pathway involves the mutase reactions mentioned for *C. sticklandii* (see Chirpich, et al., 1970), a cleavage enzyme which converts 3-keto-5-aminohexanoate and acetyl-CoA to L-3-aminobutyryl-CoA and acetoacetate (Yorifuji, Jeng, and Barker, 1977), an L-3-aminobutyryl-CoA deaminase (Jeng and Barker, 1974), and a butyryl-CoA:acetoacetate CoA transferase (Barker et al., 1978). A NAD-dependent glutamate dehydrogenase has been purified (Winnacker and Barker, 1970).

CLOSTRIDIUM TETANOMORPHUM.   Peptidoglycan crosslinkage via *m*-DAP; optimum growth temperature, 37°C. This organism has much in common with *C. cochlearium* (Barker, 1939; Nakamura et al., 1979). It differs in that it is able to ferment

some carbohydrates (Anthony and Guest, 1968; Woods and Clifton, 1937). Characteristic is its ability to grow on histidine (Wachsman and Barker, 1955a) and on L-glutamate, which is degraded to acetate, butyrate, CO_2, and H_2 (Wachsman and Barker, 1955b). The pathway of glutamate degradation involves a coenzyme B_{12}-dependent rearrangement of glutamate to β-methylaspartate (Barker, 1976; Barker, Weissbach, and Smyth, 1958). The ribonucleotide reductase is also dependent on B_{12}-coenzyme (Abeles and Beck, 1967). *C. tetanomorphum* has been used as a source for purification and study of glutamate mutase (Switzer and Barker, 1967), β-methylaspartase (Hsiang and Bright, 1969), L-citramalate hydrolyase (Wang and Barker, 1969), and citramalate lyase (Buckel and Bobi, 1976; Dimroth et al., 1977). The degradation of threonine involves an ADP-activated threonine deaminase (Vanquickenborne and Phillips, 1968; Whiteley and Tahara, 1966), a ferredoxin-dependent cleavage of α-oxobutyrate to propionyl-CoA, CO_2, and H_2, followed by a specific propionate kinase reaction (Tokushige and Hayaishi, 1972).

CLOSTRIDIUM THERMOACETICUM.   G+C, 54 mol%; peptidoglycan crosslinkage via L-DAP-glycine; optimum growth temperature, 60°C. No DNA-DNA homology is detectable to other thermophilic clostridia (Matteuzzi, Hollaus, and Biavati, 1978). This organism ferments glucose, fructose, xylose, and pyruvate to acetate (Fontaine et al., 1942). Growth depends on the presence of high concentrations of CO_2/bicarbonate (see also *C. formicoaceticum*) (Andreesen et al., 1973). Per molecule of glucose, approximately 2.6 molecules of acetate are formed. Some of this acetate is synthesized by reduction of CO_2 (see Ljungdahl and Andreesen, 1976; Ljungdahl and Wood, 1969). The first step in CO_2 reduction is catalyzed by formate dehydrogenase, which is a selenium-tungsten enzyme (see Ljungdahl and Andreesen, 1978). The further reduction of formate proceeds via tetrahydrofolate derivatives (Parker, Wu, and Wood, 1971); finally, acetate is formed in a transcarboxylation reaction involving the carboxyl group of pyruvate (Schulman et al., 1973). *C. thermoaceticum* contains a *b*-type cytochrome and quinone (Gottwald et al., 1975). This organism has been used as a source for purification and study of acetate kinase (Schaupp and Ljungdahl, 1974), formyltetrahydrofolate synthetase (Shoaf, Neece, and Ljungdahl, 1974), 5,10-methylentetrahydrofolate dehydrogenase (O'Brien, Brewer, and Ljungdahl, 1973), a corrinoid enzyme involved in the last steps of acetate synthesis (Welty and Wood, 1978), a protein containing bound 5-methoxybenzimidazolyl-cobamide besides a variety of unbound corrinoids (Ljungdahl, Le Gall, and Lee, 1973), and an untypical 4Fe/4S ferredoxin (Yang, Ljungdahl, and Le Gall, 1977).

CLOSTRIDIUM THERMOCELLUM. G+C, 38–39 mol%; optimum growth temperature, 55–60°C. This organism ferments cellulose and cellobiose to ethanol, acetate, CO_2, and H_2 (some butyrate and lactate is also formed) (see McBee, 1950; Ng, Weimer, and Zeikus, 1977). In the presence of a methanogenic bacterium, the products are acetate, methane, and CO_2 (Weimer and Zeikus, 1977). C. thermocellum has been used as source for purification and study of cellobiose phosphorylase (Alexander, 1968). Under certain conditions, glucose, fructose, and mannitol utilization seems to be possible. Enzymes of the Embden-Meyerhof pathway have been detected (Patni and Alexander, 1971a). Fructose is transported into the cells by the PEP phosphotransferase system (Patni and Alexander 1971b).

CLOSTRIDIUM THERMOHYDROSULFURICUM. G+C, 29.5–32.0 mol%; optimum growth temperature 65–70°C, upper limit at 74–76°C. The organism was isolated from extraction juices of Austrian sugar beet factories (Hollaus and Klaushofer, 1973; Klaushofer and Parkkinen, 1965). Its occurrence in extraction juices causes the formation of H_2S, provided the supply water is acidified with sulfur dioxide. The organism is strongly saccharolytic and the pattern of fermentable carbohydrates is very similar to that of C. thermosaccharolyticum. It differs from the latter (i) by an about 10°C higher optimum and upper limit of growth temperature, (ii) by a much more intensive formation of H_2S from sulfite and thiosulfate, and (iii) by hexagonal cell wall surface structures in contrast to the rectangular pattern of C. thermosaccharolyticum (Hollaus and Sleytr, 1972; Sleytr and Glauert, 1976). Furthermore, both species can be separated by their low DNA relationship (Matteuzzi, Hollaus, and Biavati, 1978). Sporulation of several strains was found best in a soil extract medium that contained 0.1% D-xylose (Hollaus and Klaushofer, 1973).

CLOSTRIDIUM THERMOSACCHAROLYTICUM. G+C, 29–32 mol%; optimum growth temperature 55–60°C, upper limit at 67°C. It ferments a large number of carbohydrates, including glycogen, starch, and pectin (Hollaus and Klaushofer, 1973; Hollaus and Sleytr, 1972; McClung, 1935b) with acetate, butyrate, and lactate along with CO_2 and H_2 as the fermentation products (Sjolander, 1937). Resting cells form considerable amounts of ethanol during glucose fermentation (Lee and Ordal, 1967). Special sporulation media that contain L-arabinose or glucosides were developed by Pheil and Ordal (1967) and Hsu and Ordal (1969). Spores of this organism are extremely heat resistant, reaching D values of 4 min at 120°C (Ingram, 1969). C. thermosaccharolyticum has been subject to thorough study since it causes the thermophilic, "hard swell" type of spoilage of canned food. The thermophilic, tar-

trate-fermenting C. tartarivorum described by Mercer and Vaughn (1951) is phenotypically indistinguishable from C. thermosaccharolyticum (except the fermentation of tartrate) (Hollaus and Sleytr, 1972). By DNA-DNA homology, both organisms were found to belong to one genospecies; therefore, C. tartarivorum must be regarded as a tartrate-positive biotype of C. thermosaccharolyticum (Matteuzzi, Hollaus, and Biavati, 1978). The ferredoxins of C. thermosaccharolyticum and C. tartarivorum differ in two amino acids (Tanaka et al., 1973). The thermostability of glycolytic enzymes (Howell, Akagi, and Himes, 1969), especially of the fructose bisphosphate aldolase (Barnes, Akagi, and Himes, 1971), and the fatty acid composition (Chan, Himes, and Akagi, 1971) have been compared.

CLOSTRIDIUM TYROBUTYRICUM. G+C, 28 mol%; peptidoglycan crosslinkage via m-DAP; optimum growth temperature, 37°C. This organism ferments a few monosaccharides (glucose, fructose, some strains galactose, arabinose, mannose, xylose, and mannitol) but no disaccharides or starch (Bryant and Burkey, 1956; Kutzner, 1963; Roux and Bergère, 1977; van Beynum and Pette, 1935). L-Lactate is fermented in the presence of acetate to butyrate, CO_2, and H_2 with little butanol formation. The organism plays an important role in causing cheese swells because it thrives at a lower pH than C. butyricum (Kutzner, 1963). The sporulation and coat formation have been studied (Rousseau, Hermier, and Bergère, 1971b,c) as have the regulation of NAD- and NADP-ferredoxin reductases (Petitdemange, Lambert, and Gay, 1973) and the NADH-flavodoxin oxidoreductase (Petitdemange, Marczak, and Gay, 1979). The properties of acetate kinase and phosphotransacetylase, isolated from vegetative cells and spores, and their involvement in the initiation of spore germination have been investigated (Touraille and Bergère, 1974).

Clostridia Recently Described or with Few Studies

CLOSTRIDIUM ABSONUM. Strains of this species were separated from C. perfringens and C. paraperfringens because of differences in lecithinase reaction, gelatin liquefaction, substrate spectrum, and DNA-DNA homology (Nakamura et al., 1973). The organism is saccharolytic and forms acetate, butyrate, and butanol as fermentation products.

CLOSTRIDIUM AMINOBUTYRICUM. The organism exhibits a high degree of substrate specificity towards amino acids. It grows on γ-aminobutyrate, but not on closely related compounds, and forms acetate, butyrate, and ammonia. It ferments a variety of carbohydrates (Hardman and Stadtman,

1960a). The enzymes involved in γ-aminobutyrate degradation (Hardman and Stadtman, 1963a) and the energetic balance of this fermentation (Hardman and Stadtman, 1963b) have been studied.

CLOSTRIDIUM AMINOVALERICUM. G+C, 33 mol%; peptidoglycan crosslinkage not via diaminopimelic acid. The organism grows with δ-aminovalerate as substrate. This compound is fermented to acetate, propionate, valerate, and ammonia. Carbohydrates are also fermented (Hardman and Stadtman, 1960b).

CLOSTRIDIUM AURANTIBUTYRICUM. G+C, 27–28 mol%; peptidoglycan crosslinkage via *m*-DAP. The orange-colored organism is saccharolytic and pectinolytic (Lund and Brocklehurst, 1978); it forms acetate, butyrate, lactate, ethanol, and some butanol (Hellinger, 1947). According to DNA homology studies, it differs from all other butyrate-forming clostridia (Cummins and Johnson, 1971; Johnson and Francis, 1975; Matteuzzi et al., 1977).

CLOSTRIDIUM CELATUM. The organism was isolated from normal human feces and differentiated from other nonmotile, saccharolytic, nitrite- and sulfide-producing organisms, such as *C. perfringens, C. paraperfringens, C. absonum,* and *C. perenne,* by the absence of lecithinase and hemolytic action and by the production of acetate, formate, ethanol, and a little butyrate from glucose. Toxins are not produced (Hauschild and Holdeman, 1974).

CLOSTRIDIUM COCCOIDES. G+C, 43–45 mol%; peptidoglycan crosslinkage via *m*-DAP; optimum growth temperature, 37°C. The cells of this organism are nonmotile, coccobacillary to rod-shaped, and occur singly, in pairs, and sometimes in short chains. A number of carbohydrates are fermented with succinate and acetate as the major product. H_2 and butyrate are not produced (Kaneuchi, Benno, and Mitsuoka, 1976). Hitherto found only in feces of mice.

CLOSTRIDIUM COCLEATUM. G+C, 28 mol%. The organism is coiled and forms semicircular to circular forms, occasionally short, helical filaments. It ferments carbohydrates to acetate, but not to butyrate. It can be separated from *C. ramosum* and *C. spiroforme* on the basis of DNA-DNA homology studies (Kaneuchi et al., 1979).

CLOSTRIDIUM DURUM. G+C, 50 mol%; optimum growth temperature, 30°C. The organism was found to be the most prominent species in the sediments of the Black Sea. It is one of the few aerotolerant clostridial species. Carbohydrates are fermented to acetate and formate; some lactate, ethanol, and propanol are also formed (Smith and Cato, 1974).

CLOSTRIDIUM LENTOPUTRESCENS. G+C, 27 mol%; peptidoglycan crosslinkage via *m*-DAP. Degrades proteins, but no carbohydrates, in a slow reaction to acetate, propionate, butyrate, propanol, and isobutanol (Hartsell and Rettger, 1934). The glycine reduction by NADH has been studied (Stadtman, 1962, 1978). For relationship to *C. cochlearium* see Nakamura et al. (1979).

CLOSTRIDIUM LEPTUM G+C, 51 mol%; optimum growth temperature, 37°C. This organism was isolated from feces. Some carbohydrates markedly stimulate growth. Acetate and ethanol are the fermentation products (Moore, Cato, and Holdeman, 1976).

CLOSTRIDIUM MALENOMINATUM. G+C, 28 mol%; peptidoglycan crosslinkage via L-lysine and L-aspartate; optimum growth temperature, 37°C. Carbohydrates are not fermented, but peptone and yeast extract are. Acetate, butyrate, and propionate are major fermentation products. According to rRNA homology studies, the organism forms a group with the acetate-producing peptolytic clostridia, *C. subterminale, C. histolyticum, C. botulinum* type G, and *C. limosum* (Johnson and Francis, 1975). A strain that decomposes uric acid was isolated from avian cecum. This strain utilized lactate as a carbon source with the formation of butyrate and propionate (Barnes and Impey, 1974).

CLOSTRIDIUM NEXILE. G+C, 40–41 mol%; optimum growth temperature, 37°C. The formation of heat-resistant spores has not been demonstrated, but the organism resists heating at 80°C for 10 min. Rods or nonmotile coccobacilli occurring in pairs or chains can be observed. Carbohydrates are fermented to acetate, formate, ethanol, and H_2; traces of lactate or succinate may be formed. The organism resembles *C. oroticum* morphologically, but does not hydrolyze orotic acid (Holdeman and Moore, 1974).

CLOSTRIDIUM OCEANICUM. G+C, 27–28 mol%; optimum growth temperature, 30–37°C. The organism was isolated from marine sediments and tolerates up to 4% NaCl. It is proteolytic, lecithinolytic, and slightly saccharolytic. From peptone-yeast extract, a variety of products are formed. Butyrate and lactate with smaller amounts of acetate are produced from glucose (Smith, 1970).

CLOSTRIDIUM PARAPERFRINGENS. Optimum growth temperature, 35–45°C. As indicated by the name, the organism exhibits some relationship to *C. perfringens* but differs on a genetic basis. Casein and gelatin are not hydrolyzed, a toxin is not formed. The organism is saccharolytic and forms acetate and butyrate but no butanol (Nakamura et al., 1973).

The species includes strains which were described as *C. barati* (Inflabilis barati Prévot, 1957).

CLOSTRIDIUM QUERCICOLUM. G+C, 52–54 mol%; optimum growth temperature, 25–30°C. The organism was isolated from oak trees. It is motile, not proteolytic, does not attack cellulose or starch. Acetate, propionate, and H_2 are produced from glycerol, inositol, or fructose (Stankewich, Cosenza, and Shigo, 1971).

CLOSTRIDIUM RAMOSUM. G+C, 26–27 mol%; peptidoglycan crosslinkage via *m*-DAP. Spores are seen very seldom; therefore, the organism was described previously as *Nocardia, Actinomyces, Fusiformis, Bacteroides,* or *Ramibacterium ramosum.* Cells can occur in "V" and "Y" arrangements. Carbohydrates are fermented to formate, acetate, lactate, and succinate (Holdeman, Cato, and Moore, 1971).

CLOSTRIDIUM SPIROFORME. G+C, 27 mol%. The organism shares with *C. ramosum* and *C. cocleatum* the ability to form "Y"-branched cells and various degrees of coiling. After a heat shock, some strains become nearly straight and do not regain coiling. Carbohydrates are fermented to acetate; butyrate is not formed. The DNA homology to *C. ramosum* and *C. cocleatum* is rather high (about 50 mol%) (Kaneuchi et al., 1979).

CLOSTRIDIUM SYMBIOSUM. G+C, 46 mol%. Because of difficulties in detecting spores, the organism was originally described as *Fusiformis biacutus* (Beerens type II), *Fusobacterium symbiosum,* and *Bacteroides symbiosus.* Carbohydrates are fermented to acetate, butyrate, and some lactate and alcohol (Kaneuchi et al., 1976). The organism was first detected as a symbiont with *Entamoeba histolytica.* Both organisms share the ability to form pyrophosphate in the pyruvate-phosphate dikinase reaction (Reeves, Manzies, and Hsu, 1968).

Comparative Studies on Clostridia

Various properties of the clostridia were discussed on a comparative basis on two symposia: Annales de l'Institut Pasteur **77**:341–540, 1949; and Journal of Applied Bacteriology **28**:1–152, 1965. The genus *Clostridium* (McClung, 1956) and the enrichment and isolation procedures have been reviewed (Kutzner, 1965; Shapton and Board, 1971; Willis, 1977). The fermentation patterns of many clostridia (Holdeman, Cato, and Moore, 1977; Moore, Cato, and Holdeman, 1966), the cell wall composition (Cummins and Johnson, 1971; Schleifer and Kandler, 1972; Takumi and Kawate, 1976), and the DNA base composition and DNA or RNA homologies (Cummins and Johnson, 1971; Hill, 1966;

Johnson, 1973; Johnson and Francis, 1975; Matteuzzi et al., 1977; Matteuzzi, Hollaus, and Biavati, 1978) have been summarized. The sensibility to lysozyme and the production of DNAse and clostridiocines by different clostridia is documented (Clarke, Robson, and Morris, 1975; Döll, 1973; Johnson and Francis, 1975; Tagg, Dagani, and Wannamaker, 1976). The pectinolytic (Raynaud, 1949), chitinolytic (Timmis, Hobbs, and Berkeley, 1974), psychrotrophic (Beerens, Sugama, and Tahon-Castel, 1965), and thermophilic clostridia (Hollaus and Sleytr, 1972; McClung, 1935a) have been subject of special articles. The fatty acid composition of thermophilic, mesophilic, and psychrophilic clostridia has been compared (Chan, Himes, and Akagi, 1971).

New species have been described on the basis of the ability to form gas caps in connection with sporulation (Duda and Makareva, 1977; Krasilnikov, Pivovarov, and Duda, 1971). The energy-yielding processes of certain clostridia have been reviewed by Thauer, Jungermann, and Decker (1977) and by Gottschalk and Andreesen (1979).

The glycolytic enzymes of different clostridia have been compared (Kotze, 1969), especially the phosphofructokinases (von Hugo and Gottschalk, 1974), the ability to form 2-oxo-3-deoxygluconate from gluconate (Bender, Andreesen, and Gottschalk, 1971), and the pentose metabolism (Cynkin and Gibbs, 1958).

The presence and the properties of the hydrogenases (Mortenson and Chen, 1974; von Hugo et al., 1972), the ferredoxins and flavodoxins (Yoch and Valentine, 1972), the function of reduced pyridine nucleotide ferredoxin oxidoreductases (Jungermann et al., 1973) and its regulation (Petit-demange et al., 1976), the coenzyme specificity of different dehydrogenases (von Hugo et al., 1972), the function of pyruvate formate lyase (Thauer, Jungermann, and Decker, 1972) and of phosphotransbutyrylase (Valentine and Wolfe, 1960), and the distribution of acid phosphatase (Schallehn and Brandis, 1973; Ueno et al., 1970) have been studied. The pyrimidine and purine metabolism by clostridia has been investigated (Campbell, 1960; Champion and Rabinowitz, 1977; Hilton, Mead, and Elsden, 1975; Mead et al., 1979) as well as the choline metabolism (Joblin et al., 1976).

The amino acid–fermenting clostridia have been surveyed (Mead, 1971) with special attention to aromatic amino acids (Elsden, Hilton, and Waller, 1976), branched-chain amino acids (Elsden and Hilton, 1978), and glutamate (Buckel and Barker, 1974).

The presence and stereospecificity of citrate synthase has been studied on a comparative basis (Gottschalk and Barker, 1967) and aspects of leucine biosynthesis have been studied (Wiegel and Schlegel, 1977). Some species even excrete amino acids

(Matteuzzi, Crociani, and Emaldi, 1978) or *O*-butylhomoserine (Uyeda et al., 1974). The ability to fix molecular nitrogen is widely distributed among clostridia (Hammann and Ottow, 1976; Rosenblum and Wilson, 1949).

Clostridia were exposed to various artificial substrates, and their metabolic activities towards nitroheterocyclic compounds (Edwards, Dye, and Carne, 1973), hexachlorocyclohexane (Jagnow, Haider, and Ellwardt, 1977), and steroids (Mahony et al., 1977) were compared.

Hydroxysteroid dehydrogenases of different stereospecificity have been reported to occur in several clostridia, e.g., in *C. perfringens* (MacDonald et al., 1975, 1976), *C. leptum* (Harris and Hylemon, 1977; Stellwag and Hylemon, 1977), *C. bifermentans* (Ferrari and Aragozzini, 1972; Ferrari, Scolastico, and Beretta, 1977), and in *C. tertium, C. indolis,* and *C. paraputrificum* (Goddard et al., 1975). *C. paraputrificum* reduces the double bond in ring A of deoxycorticosterone (Čapek, Hanč, and Tadra, 1966).

Literature Cited

Abeles, R. H., Beck, W. S. 1967. The mechanism of action of cobamide coenzyme in the ribonucleotide reductase reaction. Journal of Biological Chemistry **242:**3589–3593.

Abou-Zeid, A. A., Fouad, M., Yassein, M. 1978. Microbial production of acetone-butanol by *Clostridium acetobutylicum.* Zentralblatt für Bakteriologie, Parasitenkunde, Infektionskrankheiten und Hygiene, Abt. 2 **133:**125–134.

Ad Hoc Committee of Judical Commission of the ICSB. 1976. *First draft.* Approved lists of bacterial names. International Journal of Systematic Bacteriology **26:**563–599.

Akedo, M., Torregrossa, R., Cooney, C. L., Sinskey, A. J. 1978. Production of acrylic acid by *Clostridium propionicum.* Abstract of the Annual Meeting of the American Society for Microbiology **1978:**185.

Aleman, V., Handler, P. 1967. Dihydroorotate dehydrogenase. I. General properties. Journal of Biological Chemistry **242:**4087–4096.

Alexander, J. K. 1968: Purification and specificity of cellobiose phosphorylase from *Clostridium thermocellum.* Journal of Biological Chemistry **243:**2899–2904.

Aminoff, D., Furakawa, K. 1970. Enzymes that destroy blood group specificity. I. Purification and properties of α-L-fucosidase from *Clostridium perfringens.* Journal of Biological Chemistry **245:**1659–1669.

Andreesen, J. R., Gottschalk, G. 1969. The occurrence of a modified Entner-Doudoroff pathway in *Clostridium aceticum.* Archiv für Mikrobiologie **69:**160–170.

Andreesen, J. R., Gottschalk, G., Schlegel, H. G. 1970. *Clostridium formicoaceticum* nov. spec. Isolation, description and distinction from *C. aceticum* and *C. thermoaceticum.* Archiv für Mikrobiologie **72:**154–174.

Andreesen, J. R., Schaupp, A., Neurauter, C., Brown, A., Ljungdahl, L. G. 1973. Fermentation of glucose, fructose and xylose by *Clostridium thermoaceticum.* Effects of metals on growth yield, enzymes and the synthesis of acetate from CO_2. Journal of Bacteriology **114:**743–751.

Andrew, J. G., Morris, J. G. 1965. The biosynthesis of alanine by *Clostridium kluyveri.* Biochimica et Biophysica Acta **97:**176–179.

Anthony, C., Guest, J. R. 1968. Deferred metabolism of glucose by *Clostridium tetanomorphum.* Journal of General Microbiology **54:**277–286.

Ashley, N. V., Shoesmith, J. G. 1977. Continuous culture of *Clostridium sporogenes* and *Clostridium bifermentans* in the presence of oxygen. Proceedings of the Society for General Microbiology **4:**144.

Augier, J. 1956. A propos de la numeration des *Azotobacter* en milieu liquide. Annales de l'Institut Pasteur **91:**759.

Augier, J. 1957. A propos de la fixation biologique de l'azote atmosphérique et de la numération des *Clostridium* fixateurs dans les sols. Annales de l'Institut Pasteur **92:**817.

Azova, L. G., Gusev, M. V., Ivoilov, V. S. 1970. Response to molecular oxygen in some members of *Clostridium* genus. Microbiology [English translation of Mikrobiologiya] **39:**43–47.

Bader, J., Günther, H., Rambeck, B., Simon, H. 1978. Properties of two clostridia strains acting as catalysts for the preparative stereospecific hydrogenation of 2-enoic acids and 2-alken-1-ols with hydrogen gas. Hoppe-Seyler's Zeitschrift für Physiologische Chemie **359:**19–27.

Baker, J. J., van der Drift, C. 1974. Purification and properties of L-erythro-3,5-diaminohexanoate dehydrogenase from *Clostridium sticklandii.* Biochemistry **13:**292–299.

Barber, J. M., Robb, F. T., Webster, J. R., Woods, D. R. 1979. Bacteriocin production by *Clostridium acetobutylicum* in an industrial fermentation process. Applied and Environmental Microbiology **37:**433–437.

Barker, H. A. 1937. On the fermentation of glutamic acid. Enzymologia **2:**175–182.

Barker, H. A. 1939. The use of glutamic acid for the isolation and identification of *Clostridium cochlearium* and *Cl. tetanomorphum.* Archiv für Mikrobiologie **10:**376–384.

Barker, H. A. 1956. Bacterial fermentations. New York: John Wiley & Sons.

Barker, H. A. 1961. Fermentations of nitrogenous organic compounds, pp. 151–207. In: Gunsalus, I. C., Stanier, R. Y. (eds.), The bacteria, vol. 2. London, New York: Academic Press.

Barker, H. A. 1976. Glutamate fermentation and the discovery of B_{12} coenzymes, pp. 95–104. In: Kornberg, A., Cornudella, L., Horecker, B. L., Oro, J. (eds.), Reflections on biochemistry. Oxford: Pergamon Press.

Barker, H. A., Beck, J. V. 1941. The fermentative decomposition of purines by *Clostridium acidi-urici* and *Clostridium cylindrosporum.* Journal of Biological Chemistry **141:**3–27.

Barker, H. A., Beck, J. V. 1942. *Clostridium acidi-urici* and *Clostridium cylindrosporum,* organisms fermenting uric acid and some other purines. Journal of Bacteriology **43:**291–304.

Barker, H. A., Peterson, W. H. 1944. The nutritional requirements of *Clostridium acidi-urici.* Journal of Bacteriology **47:**307–308.

Barker, H. A., Stadtman, E. R., Kornberg, A. 1955. Coenzyme A transphorase from *Clostridium kluyveri,* pp. 599–602. In: Colowick, S. P., Kaplan, N. O. (eds.), Methods in enzymology vol. 1. London, New York: Academic Press.

Barker, H. A., Taha, S. M. 1942. *Clostridium kluyveri,* an organism concerned in the formation of caproic acid from ethyl alcohol. Journal of Bacteriology **43:**347–363.

Barker, H. A., Weissbach, H., Smyth, R. D. 1958. A coenzyme containing pseudo-vitamin B_{12}. Proceedings of the National Academy of Sciences of the United States of America **44:**1093–1097.

Barker, H. A., Jeng, I. M., Neff, N., Robertson, J. M., Tam, F. K., Hosaka, S. 1978. Butyryl-CoA:Acetoacetate CoA-transferase from a lysine-fermenting *Clostridium.* Journal of Biological Chemistry **253:**1219–1225.

Barnes, E. M., Akagi, J. M., Himes, R. H. 1971. Properties of fructose-1,6-diphosphate aldolase from two thermophilic and a mesophilic clostridia. Biochimica et Biophysica Acta **227:**199–203.

Barnes, E. M., Impey, C. S. 1974. The occurrence and properties of uric acid decomposing anaerobic bacteria in the avium caecum. Journal of Applied Bacteriology 37:393–409.

Bartsch, R. G., Barker, H. A. 1961. A vinylacetyl isomerase from *Clostridium kluyveri*. Archives of Biochemistry and Biophysics 92:122–132.

Baskerville, E. N., Twarog, R. 1972. Regulation of the tryptophan synthetic enzymes in *Clostridium butyricum*. Journal of Bacteriology 112:304–314.

Baskerville, E., Twarog, R. 1974. Regulation of a ligand-mediated association-dissociation system of anthranilate synthesis in *Clostridium butyricum*. Journal of Bacteriology 117:1184–1194.

Beerens, H., Fievez, L. 1971. Isolation of *Bacteroides fragilis* and *Sphaerophorus-Fusiformis* groups, pp. 109–113. In: Shapton, D. A., Board, R. G. (eds.), Isolation of anaerobes. London, New York: Academic Press.

Beerens, H., Sugama, S., Tahon-Castel, M. 1965. Psychrotrophic clostridia. Journal of Applied Bacteriology 28:36–48.

Beesch, S. C. 1953. A microbiological progress report. Aceton-butanol fermentation of starches. Applied Microbiology 1:85–95.

Bender, R., Andreesen, J. R., Gottschalk, G. 1971. 2-Keto-3-deoxygluconate, an intermediate in the fermentation of gluconate by clostridia. Journal of Bacteriology 107:570–573.

Bender, R., Gottschalk, G. 1973. Purification and properties of D-gluconate dehydratase from *Clostridium pasteurianum*. European Journal of Biochemistry 40:309–321.

Benson, A. M., Mower, H. F., Yasunobu, K. T. 1967. The amino acid sequence of *Clostridium butyricum* ferredoxin. Archives of Biochemistry and Biophysics 121:563–575.

Bergère, J.-L., Hermier, J. 1970. Spore properties of clostridia occurring in cheese. Journal of Applied Bacteriology 33:167–179.

Bergère, J.-L., Rousseau, M., Mercier, C. 1975. Polyoside intracellulaire implique dans la sporulation de *Clostridium butyricum*. I. Cytologie, production et analyse enzymatique preliminaire. Annales de Microbiologie 126A:295–314.

Bergmeyer, H. U., Holz, G., Klotzsch, H., Lang, G. 1963. Phosphotransacetylase aus *Clostridium kluyveri*. Züchtung des Bakteriums, Isolierung, Kristallisation und Eigenschaften des Enzyms. Biochemische Zeitschrift 338:114–121.

Berndt, H., Schlegel, H. G. 1975. Kinetics and properties of β-ketothiolase from *Clostridium pasteurianum*. Archives of Microbiology 103:21–30.

Bhat, J. V., Barker, H. A. 1947. *Clostridium lacto-acetophilum* nov. spec. and the role of acetic acid in the butyric acid fermentation of lactate. Journal of Bacteriology 54:381–391.

Booth, I. R., Morris, J. G. 1975. Proton-motive force in the obligately anaerobic bacterium *Clostridium pasteurianum*: A role in the galactose and gluconate uptake. FEBS-Letters 59:153–157.

Bornstein, B. T., Barker, H. A. 1948a. The nutrition of *Clostridium kluyveri*. Journal of Bacteriology 55:223–230.

Bornstein, B. T., Barker, H. A. 1948b. The energy metabolism of *Clostridium kluyveri* and the synthesis of fatty acids. Journal of Biological Chemistry 172:659–669.

Bradshaw, W. H., Barker, H. A. 1960. Purification and properties of xanthine dehydrogenase from *Clostridium cylindrosporum*. Journal of Biological Chemistry 235:3620–3629.

Brantner, H., Schwager, J. 1979. Enzymatische Mechanismen der Onkolyse durch *Clostridium oncolyticum* M 55 ATCC 13 732. Zentralblatt für Bakteriologie, Parasitenkunde, Infektionskrankheiten und Hygiene, Abt. 1 Orig., Reihe A 243:113–118.

Braun, M., Schoberth, S., Gottschalk, G. 1979. Enumeration of bacteria forming acetate from H_2 and CO_2 in anaerobic habitats. Archives of Microbiology 120:201–204.

Breed, R. S., Murray, E. G. D., Smith, N. R. 1957. Bergey's manual of determinative bacteriology, 7th ed. Baltimore: Williams & Wilkins.

Brown, R. G., Lindberg, B., Laishley, E. J. 1975. Characteriza-

tion of two reserve glucans from *Clostridium pasteurianum*. Canadian Journal of Microbiology 21:1136–1138.

Brühmüller, M., Decker, K. 1976. A convenient biosynthetic method for the preparation of radioactive flavine nucleotides. Analytical Biochemistry 71:550–554.

Bryant, M. P., Burkey, L. A. 1956. The characteristics of lactate-fermenting sporeforming anaerobes from silage. Journal of Bacteriology 71:43–46.

Buchanan, R. E., Gibbons, N. E. 1974. Bergey's manual of determinative bacteriology, 8th ed. Baltimore: Williams & Wilkins.

Buckel, W., Barker, H. A. 1974. Two pathways of glutamate fermentation by anaerobic bacteria. Journal of Bacteriology 117:1248–1260.

Buckel, W., Bobi, A. 1976. The enzyme complex citramalate lyase from *Clostridium tetanomorphum*. European Journal of Biochemistry 64:255–262.

Burton, R. M., Stadtman, E. R. 1953. The oxidation of acetaldehyde to acetyl-CoA by *Clostridium kluyveri*. Journal of Biological Chemistry 202:873–890.

Bush, R. S., Sauer, F. D. 1977. Evidence for separate enzymes of pyruvate decarboxylation and pyruvate synthesis in soluble extracts of *Clostridium pasteurianum*. Journal of Biological Chemistry 252:2657–2661.

Campbell, L. L. 1960. Reductive degradation of pyrimidines. V. Enzymatic conversion of *N*-carbamyl-β-alanine to β-alanine, carbon dioxide, and ammonia. Journal of Biological Chemistry 235:2375–2378.

Campbell, L. L., Frank, H. A. 1956. Nutritional requirements of some putrefactive anaerobic bacteria. Journal of Bacteriology 71:267–269.

Campbell, L. L., Postgate, J. R. 1965. Classification of the sporeforming sulfate-reducing bacteria. Bacteriological Reviews 29:359–363.

Čapek, A., Hanč, O., Tadra, M. 1966. Microbial transformation of steroids, p. 115. Prague: Academia Publishing House of the Czechoslovak Academy of Sciences.

Cardenas, J., Mortenson, L. E., Yoch, D. C. 1976. Purification and properties of paramagnetic protein from *Clostridium pasteurianum* W 5. Biochimica et Biophysica Acta 434:244–257.

Cardon, B. P., Barker, H. A. 1946. Two new amino-acid-fermenting bacteria, *Clostridium propionicum* and *Diplococcus glycinophilus*. Journal of Bacteriology 52:629–634.

Cardon, B. P., Barker, H. A. 1947. Amino acid fermentations by *Clostridium propionicum* and *Diplococcus glycinophilus*. Archives of Biochemistry 12:165–180.

Carnahan, J. E., Castle, J. E. 1958. Some requirements of biological nitrogen fixation. Journal of Bacteriology 75:121–124.

Carter, J. E., Sagers, R. D. 1972. Ferrous ion-dependent L-serine dehydratase from *Clostridium acidi-urici*. Journal of Bacteriology 109:757–763.

Cato, E., Moore, W. E. C., Holdeman, L. V. 1968. *Clostridium oroticum* comb. nov. Amended description. International Journal of Systematic Bacteriology 18:9–13.

Cato, E. P., Salmon, C. W. 1976. Transfer of *Bacteroides clostridiiformis* subsp. *clostridiiformis* (Burri and Ankersmit) Holdeman and Moore and *Bacteroides clostridiiformis* subsp. *girans* (Prévot) Holdeman and Moore to the genus Clostridium as *Clostridium clostridiiforme* (Burri and Ankersmit) comb. nov. Emendation of description and designation of neotype strain. International Journal of Systematic Bacteriology 26:205–211.

Chaigneau, M., Bory, J., Labarre, C., Descrozailles, J. 1974. Composition des gaz dégagés par *Welchia perfringens* et *Clostridium sporogenes* cultivés en différents milieux synthétiques. Annales Pharmaceutiques Françaises 32:619–622.

Champion, A. B., Rabinowitz, J. C. 1977. Ferredoxin and formyltetrahydrofolate synthetase: Comparative studies with *Clostridium acidiurici*, *Clostridium cylindrosporum*, and newly isolated anaerobic uric acid-fermenting strains. Journal of Bacteriology **132**:1003–1020.

Chan, M., Himes, R. H., Akagi, J. M. 1971. Fatty acid composition of thermophilic, mesophilic, and psychrophilic clostridia. Journal of Bacteriology **106**:876–881.

Chen, J. S., Blanchard, D. K. 1978. Isolation and properties of a unidirectional H_2-oxidizing hydrogenase from the strictly anaerobic N_2-fixing bacterium *Clostridium pasteurianum* W 5. Biochemical and Biophysical Research Communications **84**:1144–1150.

Chen, J. S., Mortenson, L. E. 1974. Purification and properties of hydrogenase from *Clostridium pasteurianum* W 5. Biochimica et Biophysica Acta **271**:283–298.

Chiba, S., Ishimoto, M. 1973. Ferredoxin-linked nitrate reductase from *Clostridium perfringens*. Journal of Biochemistry **73**:1315–1318.

Chirpich, T. P., Zappia, V., Costilow, R. N., Barker, H. A. 1970. Lysine 2,3-aminomutase. Purification and properties of a pyridoxal phosphate and *S*-adenosylmethione-activated enzyme. Journal of Biological Chemistry **245**:1778–1789.

Cho, K. Y., Doy, C. H. 1973. Ultrastructure of the obligately anaerobic bacteria *Clostridium kluyveri* and *C. acetobutylicum*. Australian Journal of Biological Sciences **26**:547–558.

Chung, K. T. 1976. Inhibitory effects of H_2 on growth of *Clostridium cellobioparum*. Applied and Environmental Microbiology **31**:342–348.

Clarke, D. J., Fuller, F. M., Morris, J. G. 1979. The membrane adenosine triphosphatase of *Clostridium pasteurianum*. Effects of key intermediates of glycolysis on its ATP phosphohydrolase activity. FEBS Letters **100**:52–56.

Clarke. D. J., Morris, J. G. 1976a. Butyricin 7423: A bacteriocin produced by *Clostridium butyricum* NCIB 7423. Journal of General Microbiology **95**:67–77.

Clarke, D. J., Morris, J. G. 1976b. Partial purification of a dicyclohexylcarbodiimide-sensitive membrane adenosine triphosphatase complex from the obligately anaerobic bacterium *Clostridium pasteurianum*. Biochemical Journal **154**:725–729.

Clarke, D. J., Robson, R. M., Morris, J. G. 1975. Purification of two *Clostridium* bacteriocins by procedures appropriate to hydrophobic proteins. Antimicrobial Agents and Chemotherapy **7**:256–264.

Costilow, R. N. 1977. Selenium requirement for the growth of *Clostridium sporogenes* with glycine as the oxidant in Stickland reaction. Journal of Bacteriology **131**:366–368.

Costilow, R. N., Cooper, D. 1978. Identity of proline dehydrogenase and Δ'-pyrroline-5-carboxylic acid reductase in *Clostridium sporogenes*. Journal of Bacteriology **134**:139–146.

Costilow, R. N., Rochovansky, O. M., Barker, H. A. 1966. Isolation and identification of β-lysine as an intermediate in lysine fermentation. Journal of Biological Chemistry **241**:1573–1580.

Cozzani, I., Barsacchi, R., Dibenedetto, G., Saracchi, L., Falcone, G. 1975. Regulation of breakdown and synthesis of L-glutamate decarboxylase in *Clostridium perfringens*. Journal of Bacteriology **123**:1115–1123.

Cramer, S. P., Hodgson, K. O., Gillum, W. O., Mortenson, L. E. 1978. The Mo-site of nitrogenase. Journal of the American Chemical Society **100**:3398–3407.

Cummins, C. S. 1970. Cell wall composition in the classification of Gram-positive anaerobes. International Journal of Systematic Bacteriology **20**:413–419.

Cummins, C. S., Johnson, J. L. 1971. Taxonomy of the clostridia: Wall composition and DNA homologies in *Clostridium butyricum* and other butyric acid-producing clostridia. Journal of General Microbiology **67**:33–46.

Curthoys, N. P., Scott, J. M., Rabinowitz, J. C. 1972. Folate coenzymes of *Clostridium acidi-urici*. The isolation of (*l*)-5,10-methenyltetrahydropteroyltriglutamate, its conversion to tetrahydropteroyltriglutamate and (*l*)-10-[^{14}C]formyltetrahydropteroyltriglutamate, and the synthesis of (*l*)-10-formyl-

[6,7-^{3}H$_2$]tetrahydropteroyltriglutamate and (*l*)-[6,7-^{3}H$_2$]tetrahydropteroyltriglutamate. Journal of Biological Chemistry **247**:1959–1964.

Cynkin, M. A., Gibbs, M. 1958. Metabolism of pentoses by clostridia. II. The fermentation of C^{14}-labeled pentoses by *Clostridium perfringens*, *Cl. beijerinckii*, *Cl. butylicum*. Journal of Bacteriology **75**:335–338.

Dainty, R. H. 1970. Purification and properties of threonine aldolase from *Clostridium pasteurianum*. Biochemical Journal **117**:585–592.

Dainty, R. H. 1972. Glutamate biosynthesis in *Clostridium pasteurianum* and its significance in nitrogen metabolism. Biochemical Journal **126**:1055–1056.

Dainty, R. H., Peel, J. L. 1970. Biosynthesis of amino acids in *Clostridium pasteurianum*. Biochemical Journal **117**:573–584.

Darvill, A. G., Hall, M. A., Fish, J. P., Morris, J. G. 1977. The intracellular reserve polysaccharide of *Clostridium pasteurianum*. Canadian Journal of Microbiology **23**:947–953.

Davies, R., Stephenson, M. 1941. Studies on the acetone-butyl alcohol fermentation. I. Nutritional and other factors involved in the preparation of active suspensions of *Clostridium acetobutylicum* (Weizmann). Biochemical Journal **35**:1320–1331.

Day, J. I. E., Goldfine, H. 1978. Partial purification and properties of acyl-CoA reductase from *Clostridium butyricum*. Archives of Biochemistry and Biophysics **190**:322–331.

Decker, K., Barth, C., Metz, H. 1966. Die Kohlenhydratsynthese in *Clostridium kluyveri*. II. Enzymatische Studien. Biochemische Zeitschrift **345**:472–492.

Decker, K., Pfitzer, S. 1972. Determination of steady-state concentrations of adenine nucleotides in growing *Clostridium kluyveri* cells by biosynthetic labeling. Analytical Biochemistry **50**:529–539.

Descrozailles, J., Bory, J., Chaigneau, M., Labarre, C. 1974. Composition des gaz dégagés par *Plectridium putrificum* cultivé en différents millieux synthétiques. Annales Pharmaceutiques Françaises **32**:19–23.

DeVries, G. H., Binkley, S. B. 1972. *N*-Acetylneuraminic acid aldolase of *Clostridium perfringens*: Purification, properties and mechanism of action. Archives of Biochemistry and Biophysics **151**:234–242.

Dimroth, P., Buckel, W., Loyal, R., Eggerer, H. 1977. Isolation and function of the subunits of citramalate lyase and formation of hybrids with the subunits of citrate lyase. European Journal of Biochemistry **80**:469–477.

Döll, W. 1973. Untersuchungen über die DNase-Bildung von Clostridien. Zentralblatt für Bakteriologie, Parasitenkunde, Infektionskrankheiten und Hygiene, Abt. 1 Orig., Reihe A **224**:115–119.

Dorn, M., Andreesen, J. R., Gottschalk, G. 1978a. Fermentation of fumarate and L-malate by *Clostridium formicoaceticum*. Journal of Bacteriology **133**:26–32.

Dorn, M., Andreesen, J. R., Gottschalk, G. 1978b. Fumarate reductase of *Clostridium formicoaceticum*. A peripheral membrane protein. Archives of Microbiology **119**:7–11.

Douglas, F., Hambleton, R., Rigby, G. J. 1973. An investigation of the oxidation reduction potential and of the effect of oxygen on the germination and outgrowth of *Clostridium butyricum* spores, using platinum electrodes. Journal of Applied Bacteriology **36**:625–633.

Duda, V. I., Makareva, E. D. 1977. Morphogenesis and function of gas caps on spores of anaerobic bacteria of the genus *Clostridium*. Microbiology [English Translation of Mikrobiologiya] **46**:563–569.

Dührsen, W. 1937. Untersuchungen über das Vorkommen von anaeroben Sporenbildnern in Milch unter Berücksichtigung ihrer sonstigen hygienischen Beschaffenheit. Zentralblatt für Bakteriologie und Parasitenkunde, Abt. 2 **96**:35–74.

Duncan, C. L., Rokos, E. A., Christenson, C. M., Rood, J. I. 1978. Multiple plasmids in different toxigenic types of *Clostridium perfringens*: Possible control of betatoxin production, pp. 246–248. In: Schlessinger, D. (ed.), Microbiology 1978. Washington, D.C.: American Society for Microbiology.

Duncan, C. L., Strong, D. H. 1968. Improved medium for sporulation of *Clostridium perfringens*. Applied Microbiology **16**:82–89.

du Toit, P. J., Potgieter, D. J. J., de Villiers, V. 1972. A study of the properties of 1-phosphofructokinase isolated from *Clostridium pasteurianum*. Enzymologia **43**:285–300.

Dyer, J. K., Costilow, R. N. 1968. Fermentation of ornithine by *Clostridium sticklandii*. Journal of Bacteriology **96**:1617–1622.

Edwards, D. I., Dye, M., Carne, H. 1973. The selective toxicity of antimicrobial nitroheterocyclic drugs. Journal of General Microbiology **76**:135–145.

Edwin, E. E., Shreeve, J. E., Jackman, R. 1978. A rapid colony test for thiaminase activity. Journal of Applied Bacteriology **44**:305–312.

Egorov, N. S., Loria, Z. K., Vlasova, O. S. 1972. Effect of various carbon sources on synthesis of proteolytic enzymes by acetobutylic bacteria. Microbiology [English translation of Mikrobiologiya] **41**:208–211.

El Ghazzawi, E. 1967. Neuisolierung von *Clostridium aceticum* Wieringa und stoffwechselphysiologische Untersuchungen. Archiv für Mikrobiologie **57**:1–19.

Elliott, B. B., Mortenson, L. E. 1976. Regulation of molybdate transport by *Clostridium pasteurianum*. Journal of Bacteriology **127**:770–779.

Elliott, B. B., Mortenson, L. E. 1977. Molybdenum storage component from *Clostridium pasteurianum*, pp. 205–217. In: Newton, W., Postgate, J. R., Rodriguez-Barrueco, C. (eds.), Recent developments in nitrogen fixation. London, New York: Academic Press.

Ellner, P. D. 1956. A medium promoting rapid quantitative sporulation in *Clostridium perfringens*. Journal of Bacteriology **71**:495–496.

Elsden, S. R., Hilton, M. G. 1978. Volatile acid production from threonine, valine, leucine and isoleucine by clostridia. Archives of Microbiology **117**:165–172.

Elsden, S. R., Hilton, M. G., Waller, J. M. 1976. The end products of the metabolism of aromatic amino acids by clostridia. Archives of Microbiology **107**:283–288.

Emtsev, V. T. 1963. Sporulation in *Clostridium pasteurianum*. [In Russian, with English summary.] Mikrobiologiya **32**:434–438.

Ensley, B., McHugh, J. J., Barton, L. L. 1975. Effect of carbon sources on formation of α-amylase and glucoamylase by *Clostridium acetobutylicum*. Journal of General and Applied Microbiology **21**:51–59.

Erbes, D. L., Burris, R. H., Orme-Johnson, W. H. 1975. On the iron-sulfur cluster in hydrogenase from *Clostridium pasteurianum* W 5. Proceedings of the National Academy of Sciences of the United States of America **72**:4795–4799.

Ferrari, A., Aragozzini, F. 1972. Attività di un ceppo di *Clostridium bifermentans* su alcuni acidi biliari. Annali di Microbiologia ed Enzimologia **22**:131–136.

Ferrari, A., Scolastico, C., Beretta, L. 1977. On the mechanism of cholic acid 7α-dehydroxylation by a *Clostridium bifermentans* cell-free extract. FEBS Letters **75**:166–168.

Finne, G., Matches, J. R. 1974. Low-temperature-growing clostridia from marine sediment. Canadian Journal of Microbiology **20**:1639–1645.

Fleming, R. W., Quinn, L. Y. 1971. Chemically defined medium for growth of *Clostridium thermocellum*, a cellulolytic thermophilic anaerobe. Applied Microbiology **21**:967.

Fontaine, F., Peterson, W. H., McCoy, E., Johnson, M. J., Ritter, G. J. 1942. A new type of glucose fermentation by *Clostridium thermoaceticum*, n. sp. Journal of Bacteriology **43**:701–715.

Freame, B., Fitzpatrick, B. W. 1971. The use of differential reinforced clostridial medium for the isolation and enumeration of clostridia from food, pp. 49–55. In: Shapton, D. A., Board, R. G. (eds.), Isolation of anaerobes. London, New York: Academic Press.

Fredricks, W. W., Stadtman, E. R. 1965. The role of ferredoxin in the hydrogenase system from *Clostridium kluyveri*. Journal of Biological Chemistry **240**:4065–4071.

Fryer, T. F., Halligan, A. C. 1976. The detection of *Clostridium tyrobutyricum* in milk. New Zealand Journal of Dairy Science and Technology **11**:132.

Fuchs, G., Schnitker, U., Thauer, R. K. 1974. Carbon monoxide oxidation by growing cultures of *Clostridium pasteurianum*. European Journal of Biochemistry **49**:111–115.

Gaston, L. W., Stadtman, E. R. 1963. Fermentation of ethylene glycol by *Clostridium glycolicum*, sp. n. Journal of Bacteriology **85**:356–362.

Gibbs, B. M., Freame, B. 1965. Methods for the recovery of clostridia from foods. Journal of Applied Bacteriology **28**:95–111.

Gibbs, B. M., Hirsch, A. 1956. Spore formation by *Clostridium* species in an artificial medium. Journal of Applied Bacteriology **19**:129–141.

Gibson, T. 1965. Clostridia in silage. Journal of Applied Bacteriology **28**:56–62.

Gibson, T., Stirling, A. C., Keddi, R. M., Rosenberger, R. F. 1958. Bacteriological changes in silage made at controlled temperatures. Journal of General Microbiology **19**:112–129.

Gililland, J. R., Vaughn, R. H. 1943. Characteristics of butyric acid bacteria from olives. Journal of Bacteriology **46**:315–322.

Glathe, H. 1934. Über die Rotte des Stalldüngers unter besonderer Berücksichtigung der Anaeroben-Flora. Zentralblatt für Bakteriologie und Parasitenkunde, Abt. 2 **91**:65–101.

Goddard, P., Fernandez, F., West, B., Hill, M. J., Barnes, P. J. 1975. The nuclear dehydrogenation of steroids by intestinal bacteria. Journal of Medical Microbiology **8**:429–435.

Goldfine, H., Stadtman, E. R. 1960. Propionic acid metabolism. V. The conversion of β-alanine to propionic acid by cell-free extracts of *Clostridium propionicum*. Journal of Biological Chemistry **235**:2238–2245.

Goldfine, H., Khuller, G. K., Borie, R. P., Silverman, B., Selick, H. H., Johnston, N. C., Vanderkooi, J. M., Horwitz, A. F. 1977. Effects of growth temperature and supplementation with exogenous fatty acids on some physical properties of *Clostridium butyricum* phospholipids. Biochimica et Biophysica Acta **488**:341–352.

Gottschalk, G. 1979. Bacterial metabolism. New York, Heidelberg, Berlin: Springer Verlag.

Gottschalk, G., Andreesen, J. R. 1979. Energy metabolism in anaerobes, pp. 85–115. In: Quayle, J. R. (ed.), Microbial biochemistry. International Review of Biochemistry, vol. 21. Baltimore: University Park Press.

Gottschalk, G., Barker, H. A. 1966. Synthesis of glutamate and citrate by *Clostridium kluyveri*. A new type of citrate synthase. Biochemistry **5**:1125–1133.

Gottschalk, G., Barker, H. A. 1967. Presence and stereospecificity of citrate synthase in anaerobic bacteria. Biochemistry **6**:1027–1034.

Gottschalk, G., Dittbrenner, S. 1970. Properties of (R)-citrate synthase from *Clostridium acidi-urici*. Hoppe Seyler's Zeitschrift für Physiologische Chemie **351**:1183–1190.

Gottwald, M., Andreesen, J. R., Le Gall, J., Ljungdahl, L. G. 1975. Presence of cytochrome and menaquinone in *Clostridium formicoaceticum* and *Clostridium thermoaceticum*. Journal of Bacteriology **122**:325–328.

Goudkov, A. V., Sharp, M. E. 1965. Clostridia in dairying. Journal of Applied Bacteriology **28**:63–73.

Goudkov, A. V., Sharp, M. E. 1966. A preliminary investigation of the importance of clostridia in the production of rancid flavour in cheddar cheese. Journal of Dairy Research **33**:139–149.

Gregory, E. M., Moore, W. E. C., Holdeman, L. V. 1978. Superoxide dismutase in anaerobes: Survey. Applied and Environmental Microbiology **35**:988–991.

Halligan, A. C., Fryer, T. F. 1976. The development of a method for detecting spores of *Clostridium tyrobutyricum* in milk. New Zealand Journal of Dairy Science and Technology **11**:100–106.

Hammann, R., Ottow, J. C. G. 1976. Isolation and characterization of iron-reducing nitrogen-fixing saccharolytic clostridia from gley soils. Soil Biology and Biochemistry **8**:357–364.

Hardman, J. K., Stadtman, T. C. 1960a. Metabolism of ω-amino acids. I. Fermentation of γ-aminobutyric acid by *Clostridium aminobutyricum* n. sp. Journal of Bacteriology **79**:544–548.

Hardman, J. K., Stadtman, T. C. 1960b. Metabolism of ω-amino acids. II. Fermentation of δ-aminovaleric acid by *Clostridium aminovalericum* n. sp. Journal of Bacteriology **79**:549–552.

Hardman, J. K., Stadtman, T. C. 1963a. Metabolism of ω-amino acids. IV. γ-Aminobutyrate fermentation by cell-free extracts of *Clostridium aminobutyricum*. Journal of Biological Chemistry **238**:2088–2093.

Hardman, J. K., Stadtman, T. C. 1963b. Metabolism of ω-amino acids. V. Energetics of the γ-aminobutyrate fermentation by *Clostridium aminobutyricum*. Journal of Bacteriology **85**:1326–1333.

Harris, J. N., Hylemon, P. B. 1977. Purification and characterization of NADP-dependent 12-α-hydroxysteroid dehydrogenase from *Clostridium leptum* VPI 10900. Abstracts of the Annual Meeting of the American Society for Microbiology **1977**:197.

Hartsell, S. E., Rettger, L. F. 1934. A taxonomic study of "*Clostridium putrificum*" and its establishment as a definite entity—*Clostridium lentoputrescens*, nov. spec. Journal of Bacteriology **27**:497–511.

Hasan, S. M., Hall, J. B. 1975. The physiological function of nitrate reduction in *Clostridium perfringens*. Journal of General Microbiology **87**:120–128.

Hasan, S. M., Hall, J. B. 1976. Growth of *Clostridium tertium* and *Clostridium septicum* in chemically defined media. Applied and Environmental Microbiology **31**:442–443.

Hauschild, A. H. W., Holdeman, L. V. 1974. *Clostridium celatum* sp. nov., isolated from normal human feces. International Journal of Systematic Bacteriology **24**:478–481.

Hayase, M., Mitsui, N., Tamai, K., Nakamura, S., Nishida, S. 1974. Isolation of *Clostridium absonum* and its cultural and biochemical properties. Infection and Immunity **9**:15–19.

Hellinger, E. 1947. *Clostridium aurantibutyricum* (n. sp.): A pink butyric acid *Clostridium*. Journal of General Microbiology **1**:203–210.

Henkin, J., Abeles, R. H. 1976. Evidence against an acyl-enzyme intermediate in the reaction catalyzed by clostridial phosphotransacetylase. Biochemistry **15**:3472–3479.

Heritage, A. D., MacRae, I. C. 1977. Degradation of lindane by cell-free preparations of *Clostridium sphenoides*. Applied and Environmental Microbiology **34**:222–224.

Hespell, R. B., Joseph, R., Mortlock, R. P. 1969. Requirement of coenzyme A in the phosphoroclastic reaction of anaerobic bacteria. Journal of Bacteriology **100**:1328–1334.

Hewitt, J., Morris, J. G. 1975. Superoxide dismutase in some obligately anaerobic bacteria. FEBS Letters **50**:315–318.

Hill, L. R. 1966. An index to deoxyribonucleic acid base compositions of bacterial species. Journal of General Microbiology **44**:419–437.

Hillmer, P., Gottschalk, G. 1974. Solubilization and partial characterization of particulate dehydrogenases from *Clostridium kluyveri*. Biochimica et Biophysica Acta **334**:12–23.

Hilton, M. G., Mead, G. C., Elsden, S. R. 1975. The metabolism of pyrimidines by proteolytic clostridia. Archives of Microbiology **102**:145–149.

Hirsch, A., Grinsted, E. 1954. Methods for the growth and enumeration of anaerobic spore-formers from cheese, with observations on the effect of nisin. Journal of Dairy Research **21**:101–110.

Hitzman, D. O., Halvorson, H. O., Ukita, T. 1957. Require-

ments for production and germination of spores of anaerobic bacteria. Journal of Bacteriology **74**:1–7.

Holcenberg, J. S., Stadtman, E. R. 1969. Nicotinic acid metabolism. III. Purification and properties of a nicotinic acid hydroxylase. Journal of Biological Chemistry **244**:1194–1203.

Holcenberg, J. S., Tsai, L. 1969. Nicotinic acid metabolism. IV. Ferredoxin-dependent reduction of 6-hydroxynicotinic acid to 6-oxo-1,4,5,6-tetrahydronicotinic acid. Journal of Biological Chemistry **244**:1204–1211.

Holdeman, L. V., Cato, E. P., Moore, W. E. C. 1971. *Clostridium ramosum* (Vuillemin) comb. nov.: Emended description and proposed neotype strain. International Journal of Systematic Bacteriology **21**:35–39.

Holdeman, L. V., Cato, E. P., Moore, W. E. C. 1977. Anaerobe laboratory manual, 4th ed. Blacksburg, Virginia: V.P.I. Anaerobe Laboratory, Virginia Polytechnic Institute and State University.

Holdeman, L. V., Moore, W. E. C. 1974. New genus, *Coprococcus*, twelve new species, and emended descriptions of four previously described species of bacteria from human feces. International Journal of Systematic Bacteriology **24**:260–277.

Holland, D., Barker, A. N., Wolf, J. 1970. The effect of carbon dioxide on spore germination in some clostridia. Journal of Applied Bacteriology **33**:274–284.

Holland, K. T., Cox, D. J. 1975. A synthetic medium for the growth of *Clostridium bifermentans*. Journal of Applied Bacteriology **38**:193–198.

Hollaus, F., Klaushofer, H. 1973. Identification of hyperthermophilic obligate anaerobic bacteria from extraction juices of beet sugar factories. International Sugar Journal **75**:237–241.

Hollaus, F., Sleytr, U. 1972. On the taxonomy and fine structure of some hyperthermophilic saccharolytic clostridia. Achiv für Mikrobiologie **86**:129–146.

Hoogerheide, J. C., Kocholaty, W. 1938. Metabolism of the strict anaerobes (genus: *Clostridium*). II. Reduction of amino acids with gaseous hydrogen by suspensions of *Clostridium sporogenes*. Biochemical Journal **32**:949–957.

Howell, N., Akagi, J. M., Himes, R. H. 1969. Thermostability of glycolytic enzymes from thermophilic clostridia. Canadian Journal of Microbiology **15**:461–464.

Hsiang, M. W., Bright, H. J. 1969. β-Methylaspartase from *Clostridium tetanomorphum*, pp. 347–353. In: Lowenstein, J. M. (ed.), Methods in enzymology, vol. 13. New York, London: Academic Press.

Hsu, E. J., Ordal, Z. J. 1969. Sporulation of *Clostridium thermosaccharolyticum*. Applied Microbiology **18**:958–960.

Hungate, R. E. 1944. Studies on cellulose fermentation. I. The culture and physiology of an anaerobic cellulose digesting bacterium. Journal of Bacteriology **48**:499–513.

Hungate, R. E. 1950. The anaerobic mesophilic cellulolytic bacteria. Bacteriological Reviews **14**:1–49.

Hungate, R. E. 1969. A roll tube method for cultivation of strict anaerobes, pp. 117–132. In: Norris, J. R., Ribbons, D. W. (eds.), Methods in microbiology, vol. 3B. London: Academic Press.

Imhoff, D., Andreesen, J. R. 1979. Nicotinic acid hydroxylase from *Clostridium barkeri*: Selenium-dependent formation of active enzyme. FEMS Microbiology Letters **5**:155–158.

Ingram, M. 1969. Sporeformers as food spoilage organisms, pp. 549–610. In: Gould, G. W., Hurst, A. (eds.), The bacterial spore. London, New York: Academic Press.

Ishimoto, M., Umeyama, M., Chiba, S. 1974. Alteration of fermentation products from butyrate to acetate by nitrate reduction in *Clostridium perfringens*. Zeitschrift für Allgemeine Mikrobiologie **14**:115–121.

Jagnow, G., Haider, K., Ellwardt, P. C. 1977. Anaerobic dechlorination and degradation of hexachlorocyclohexane isomers by anaerobic and facultative anaerobic bacteria. Archives of Microbiology **115**:285–292.

Jeng, I. M., Barker, H. A. 1974. Purification and properties of

L-3-aminobutyryl coenzyme A deaminase from a lysine-fermenting *Clostridium.* Journal of Biological Chemistry **249:**6578–6584.

Jeng, I. M., Somack, R., Barker, H. A. 1974. Ornithine degradation in *Clostridium sticklandii;* pyridoxal phosphate and coenzyme A dependent thiolytic cleavage of 2-amino-4-ketopentanoate to alanine and acetyl coenzyme A. Biochemistry **13:**2898–2903.

Joblin, K. N., Johnson, A. W., Lappert, M. F., Wallis, O. C. 1976. Coenzyme B_{12}-dependent reactions. Part IV. Observations on the purification of ethanolamine ammonia-lyase. Biochimica et Biophysica Acta **452:**262–270.

Johnson, J. L. 1970. Relationship of deoxyribonucleic acid homologies to cell wall structure. International Journal of Systematic Bacteriology **20:**421–424.

Johnson, J. L. 1973. Use of nucleic-acid homologies in the taxonomy of anaerobic bacteria. International Journal of Systematic Bacteriology **23:**308–315.

Johnson, J. L., Francis, B. S. 1975. Taxonomy of the clostridia: Ribosomal ribonucleic acid homologies among the species. Journal of General Microbiology **88:**229–244.

Johnson, M. J., Peterson, W. H., Fred, E. B. 1931. Oxidation and reduction relations between substrate and products in the acetone-butyl alcohol fermentation. Journal of Biological Chemistry **91:**569–591.

Johnston, R., Harmon, S., Kautter, D. 1964. Method to facilitate the isolation of *Clostridium botulinum* type E. Journal of Bacteriology **88:**1521–1522.

Jungermann, K., Thauer, R. K., Decker, K. 1968. The synthesis of one-carbon units from CO_2 in *Clostridium kluyveri.* European Journal of Biochemistry **3:**351–359.

Jungermann, K. A., Schmidt, W., Kirchniawy, F. H., Rupprecht, E. H.,Thauer, R. K. 1970. Glycine formation via threonine and serine aldolase. Its interrelation with the pyruvate formate lyase pathway of one-carbon unit synthesis in *Clostridium kluyveri.* European Journal of Biochemistry **16:**424–429.

Jungermann, K., Leimenstoll, G., Rupprecht, E., Thauer, R. K. 1971a. Demonstration of NADH-ferredoxin reductase in two saccharolytic clostridia. Archiv für Mikrobiologie **80:**370–372.

Jungermann, K., Rupprecht, E., Ohrloff, C., Thauer, R. K., Decker, K. 1971b. Regulation of the reduced nicotinamide adenine dinucleotide-ferredoxin reductase system in *Clostridium kluyveri.* Journal of Biological Chemistry **246:**960–963.

Jungermann, K., Thauer, R. K., Leimenstoll, G., Decker, K. 1973. Function of reduced pyridine nucleotide-ferredoxin oxidoreductases in saccharolytic clostridia. Biochimica et Biophysica Acta **305:**268–280.

Jungermann, K., Kirchniawy, H., Katz, N., Thauer, R. K. 1974. NADH, a physiological electron donor in clostridial nitrogen fixation. FEBS Letters **43:**203–206.

Kaplan, F., Setlow, P., Kaplan, N. O. 1969. Purification and properties of a DPHN-TPNH diaphorase from *Clostridium kluyveri.* Archives of Biochemistry and Biophysics **132:**91–98.

Kaneuchi, C., Benno, Y., Mitsuoka, T. 1976. *Clostridium coccoides,* a new species from the feces of mice. International Journal of Systematic Bacteriology **26:**482–486.

Kaneuchi, C., Watanabe, K., Terada, A., Benno, Y., Mitsuoka, T. 1976. Taxonomic study of *Bacteroides clostridiiformis* subsp. *clostridiiformis* (Burri and Ankersmit) Holdeman and Moore and of related organisms: Proposal of *Clostridium clostridiiformis* (Burri and Ankersmit) comb. nov. and *Clostridium symbiosum* (Stevens) comb. nov. International Journal of Systematic Bacteriology **26:**195–204.

Kaneuchi, C., Miyazato, T., Shinjo, T., Mitsuoka, T. 1979. Taxonomic study of helically coiled, sporeforming anaerobes isolated from the intestines of humans and other animals: *Clostridium cocleatum* sp. nov. and *Clostridium spiroforme* sp. nov. International Journal of Systematic Bacteriology **29:**1–12.

Karlsson, J. L., Volcani, B. E., Barker, H. A. 1948. The nutritional requirements of *Clostridium aceticum.* Journal of Bacteriology **56:**781–782.

Kearny, J. J., Sagers, R. D. 1972. Formate dehydrogenase from *Clostridium acidiurici.* Journal of Bacteriology **109:**152–161.

Kimura, R., Liao, T. H. 1964. Taxonomic considerations on the *Clostridium thiaminolyticum* Kimura et Liao. Vitamins (Japan) **30:**29–32.

Klaushofer, H., Parkkinen, E. 1965. Zur Frage der Bedeutung aerober und anaerober thermophiler Sporenbildner als Infektionsursache in Rübenzuckerfabriken. I. *Clostridium thermohydrosulfuricum,* eine neue Art eines saccharoseabbauenden, thermophilen, schwefelwasserstoffbildenden Clostridiums. Zeitschrift für Zuckerindustrie **15:**445–449.

Kleiner, D. 1979. Regulation of ammonium uptake and metabolism by nitrogen fixing bacteria. III. *Clostridium pasteurianum.* Archives of Microbiology **120:**263–270.

Kleiner, D., Fitzke, E. 1979. Evidence for ammonia translocation by *Clostridium pasteurianum.* Biochemical and Biophysical Research Communications **86:**211–217.

Knight, E., Jr., Hardy, R. W. F. 1966. Isolation and characteristics of flavodoxin from nitrogen-fixing *Clostridium pasteurianum.* Journal of Biological Chemistry **241:**2752–2756.

Knöll, H., Horschak, R. 1973. Zur Ökologie der Gärungssarcinen *Sarcina ventriculi* und *Sarcina maxima.* Zeitschrift für Allgemeine Mikrobiologie **13:**449–451.

Kondo, H., Friedmann, H. C., Vennesland, B. 1960. Flavin changes accompanying adaption of *Zymobacterium oroticum* to orotate. Journal of Biological Chemistry **235:**1533–1535.

Koransky, J. R., Allen, S. D., Dowell, V. R., Jr. 1978. Use of ethanol for selective isolation of sporeforming microorganisms. Applied and Environmental Microbiology **35:**762–765.

Kotze, J. P. 1969. Glycolytic and related enzymes in clostridial classification. Applied Microbiology **18:**744–747.

Krasil'nikov, N. A., Pivovarov, G. E., Duda, V. I. 1971. Physiological properties of anaerobic soil bacteria which form vesicular caps on their spores. Microbiology [English translation of Mikrobiologiya] **40:**783–788.

Kung, H. F., Cedarbaum, S., Tsai, L., Stadtman, T. C. 1970. Nicotinic acid metabolism. V. A cobamide coenzyme-dependent conversion of α-methyleneglutaric acid to dimethylmaleic acid. Proceedings of the National Academy of Sciences of the United States of America **65:**978–984.

Kung, H. F., Stadtman, T. C. 1971. Nicotinic acid metabolism. VI. Purification and properties of α-methyleneglutarate mutase (B_{12}-dependent) and methylitaconate isomerase. Journal of Biological Chemistry **246:**3378–3388.

Kuramitsu, H. K., Watson, R. M. 1973. Regulation of aspartokinase activity in *Clostridium perfringens.* Journal of Bacteriology **115:**882–888.

Kürsteiner, J. 1907. Beiträge zur Untersuchungstechnik obligat anaerober Bakterien, sowie zur Lehre der Anaerobiose überhaupt. Centralblatt für Bakteriologie, Abt. 2 **19:**1–26, 97–115, 202–220, 385–394.

Kutzenock, A., Aschner, M. 1952. Degenerative processes in a strain of *Clostridium butylicum.* Journal of Bacteriology **64:**829–836.

Kutzner, H. J. 1963. Untersuchungen an Clostridien mit besonderer Berücksichtigung der für die Milchwirtschaft wichtigen Arten. Zentralblatt für Bakteriologie, Parasitenkunde, Infektionskrankheiten und Hygiene, Abt. 1 **191:**441–450.

Kutzner, H. J. 1965. Prinzipien der Anreicherung und Isolierung von Clostridien. Zentralblatt für Bakteriologie, Parasitenkunde, Infektionskrankheiten und Hygiene, Abt. 1, Suppl. **1:**363–394.

Kutzner, H. J. 1966. Untersuchungen über die Buttersäuregärung in Schnitt- und Hartkäse. 17. Internationaler Milchwirtschafts-Kongress München, Sektion D2, pp. 647–658.

Kyrtopoulos, S. A., Satchell, D. P. N. 1972. The roles of univalent cations during catalysis by phosphate acetyl-transferase derived from *Clostridium kluyveri*. Biochemical Journal **129**:1163–1166.

Laanbroek, H. J., Smit, A. J., Klein Nulend, G., Veldkamp, H. 1979. Competition for L-glutamate between specialised and versatile *Clostridium* species. Archives of Microbiology **120**:61–66.

Laishley, E. J. 1975. Regulation and properties of an invertase from *Clostridium pasteurianum*. Canadian Journal of Microbiology **21**:1711–1718.

Laishley, E. J., Krouse, H. R. 1978. Stable isotope fractionation by *Clostridium pasteurianum*. 2. Regulation of sulfite reductases by sulfur amino acids and their influence on sulfur isotope fractionation during SO_3^{2-} and SO_4^{2-} reduction. Canadian Journal of Microbiology **24**:716–724.

Laishley, E. J., Lin, P. M., Peck, H. D. 1971. A ferredoxin-linked sulfite reductase from *Clostridium pasteurianum*. Canadian Journal of Microbiology **17**:889–895.

Lampen, J. H., Peterson, E. H. 1943. Growth factor requirements of clostridia. Archives of Biochemistry **2**:443–449.

Leaver, F. W., Wood, H. G., Stjernholm, R. 1955. The fermentation of three carbon substrates by *Clostridium propionicum* and propionibacterium. Journal of Bacteriology **70**:521–530.

Lee, C. K., Ordal, Z. J. 1967. Regulatory effect of pyruvate on the glucose metabolism of *Clostridium thermosaccharolyticum*. Journal of Bacteriology **94**:530–536.

Leonhardt, U., Andreesen, J. R. 1977. Some properties of formate dehydrogenase, accumulation and incorporation of [185]W-tungsten into proteins of *Clostridium formicoaceticum*. Archives of Microbiology **115**:277–284.

Liebermann, I., Kornberg, A. 1954. Enzymatic synthesis and breakdown of a pyrimidine, orotic acid. II. Dihydro-orotic acid, ureidosuccinic acid and 5-carboxymethyl hydantoin. Journal of Biological Chemistry **207**:911–924.

Linke, H. A. B. 1969. Der Fructose-Stoffwechsel von *Clostridium aceticum*. Zentralblatt für Bakteriologie, Parasitenkunde, Infektionskrankheiten und Hygiene, Abt. 2 **123**:369–379.

Ljungdahl, L. G., Andreesen, J. R. 1976. Reduction of CO_2 to acetate in homoacetate fermenting clostridia and the involvement of tungsten in formate dehydrogenase, pp. 163–172. In: Schlegel, H. G., Gottschalk, G., Pfennig, N. (eds.), Symposium on microbial production and utilization of gases (H_2, CH_4, CO). Göttingen: Akademie der Wissenschaften/E. Goltze Verlag.

Ljungdahl, L. G., Andreesen, J. R. 1978. Formate dehydrogenase, a selenium-tungsten enzyme from *Clostridium thermoaceticum*, pp. 360–372. In: Fleischer, S., Packer, L. (eds.), Methods in enzymology, vol. 53. New York, San Francisco, London: Academic Press.

Ljungdahl, L. G., Le Gall, J., Lee, J. P. 1973. Isolation of a protein containing tightly bound 5-methoxybenzimidazolyl-cobamide (factor III m) from *Clostridium thermoaceticum*. Biochemistry **12**:1802–1808.

Ljungdahl, L. G., Wood, H. G. 1969. Total synthesis of acetate from CO_2 by heterotrophic bacteria. Annual Review of Microbiology **23**:515–538.

Loesche, W. J. 1969. Oxygen sensitivity of various anaerobic bacteria. Applied Microbiology **18**:723–727.

Lovenberg, W., Sobel, B. E. 1965. Rubredoxin: A new electron transfer protein from *Clostridium pasteurianum*. Proceedings of the National Academy of Sciences of the United States of America **54**:193–199.

Lund, B. M., Brocklehurst, T. F. 1978. Pectic enzymes of pigmented strains of *Clostridium*. Journal of General Microbiology **104**:59–66.

Lurz, R., Mayer, F., Gottschalk, G. 1979. Electron microscopic study on the quaternary structure of the isolated particulate alcohol-acetaldehyde dehydrogenase complex and its identity with polygonal bodies of *Clostridium kluyveri*. Archives of Microbiology **120**:255–262.

McBee, R. H. 1948. The culture and physiology of a thermophilic cellulose-fermenting bacterium. Journal of Bacteriology **56**:653–663.

McBee, R. H. 1950. The anaerobic thermophilic cellulolytic bacteria. Bacteriological Reviews **14**:51–63.

McBryde, C. N. 1911. A bacteriological study of ham souring. U.S. Bureau of Animal Industry Bulletin No. 132.

McClung, L. S. 1935a. Studies on anaerobic bacteria. III. Historical review of certain thermophilic anaerobes. Journal of Bacteriology **29**:173–187.

McClung, L. S. 1935b. Studies on anaerobic bacteria. IV. Taxonomy of cultures of a thermophilic species causing "swells" of canned foods. Journal of Bacteriology **29**:189–202.

McClung, L. S. 1943. On the enrichment and purification of chromogenic sporeforming anaerobic bacteria. Journal of Bacteriology **46**:507–512.

McClung, L. S. 1956. The anaerobic bacteria with special reference to the genus *Clostridium*. Annual Review of Microbiology **10**:173–192.

McClung, L. S., McCoy, E. 1934. Studies on anaerobic bacteria. I. A cornliver medium for the detection and dilution counts of various anaerobes. Journal of Bacteriology **28**:267–277.

McCord, J. M., Keele, Jr., B. B., Fridovich, I. 1971. An enzyme based theory of obligate anaerobiosis: The physiological function of superoxide dismutase. Proceedings of the National Academy of Sciences of the United States of America **68**:1024–1027.

McCoy, E., Fred, E. B. 1941. The stability of a culture for industrial fermentation. Journal of Bacteriology **41**:90–91.

McCoy, E., Fred, E. B., Peterson, W. H., Hastings, E. G. 1926. A cultural study of the acetone butyl alcohol organism. Journal of Infectious Diseases **39**:253–283.

McCoy, E., McClung, L. S. 1935. Studies on anaerobic bacteria. VI. The nature and systematic position of a new chromogenic *Clostridium*. Archiv für Mikrobiologie **6**:230–238.

MacDonald, I. A., Bishop, J. M., Mahony, D. E., Williams, C. N. 1975. Convenient non-chromatographic assays for the microbial deconjugation and 7α-OH bioconversion of taurocholate. Applied Microbiology **30**:530–535.

MacDonald, I. A., Meier, E. C., Mahony, D. E., Constain, G. A. 1976. 3α-, 7α- and 12α-hydroxysteroid dehydrogenase activities from *Clostridium perfringens*. Biochimica et Biophysica Acta **450**:142–153

Mackey, B. M., Morris, J. G. 1971. Sporulation in *Clostridium pasteurianum*, p. 343. In: Barker, A. N., Gould, G. W., Wolf, J. (eds.), Spore research 1971. London, New York: Academic Press.

Mackey, B. M., Morris, J. G. 1972. The exosporium of *Clostridium pasteurianum*. Journal of General Microbiology **73**:325–338.

Macy, J. M., Snellen, J. E., Hungate, R. E. 1972. Use of syringe methods for anaerobiosis. American Journal of Clinical Nutrition **25**:1318–1323.

Madan, V. K., Hillmer, P., Gottschalk, G. 1973. Purification and properties of NADP-dependent L(+)-3-hydroxybutyryl-CoA dehydrogenase from *Clostridium kluyveri*. European Journal of Biochemistry **32**:51–56.

Mahony, D. E., Meier, E. C., MacDonald, I. A., Holdeman, L. V. 1977. Bile salt degradation by nonfermentative clostridia. Applied and Environmental Microbiology **34**:419–423.

Mallette, M. F., Reece, P., Dawes, E. A. 1974. Culture of *Clostridium pasteurianum* in defined medium and growth as a function of sulfate concentration. Applied Microbiology **28**:999–1003.

Marmelak, R., Quastel, J. H. 1953. Amino acid interactions in strict anaerobes (*Clostridium sporogenes*). Biochimica et Biophysica Acta **12**:103–120.

Marmur, J., Doty, P. 1962. Determination of the base composition of deoxyribonucleic acid from its thermal denaturation temperature. Journal of Molecular Biology **5**:109–118.

Matsuda, N., Matsumoto, N., Ushizawa, S., Kakegawa, Y., Kato, H., Nishida, S. 1975. Specific distribution and heat resistance of mesophilic bacterial spores isolated from frozen raw meat used in canned meat manufacture. Journal of the Food and Hygiene Society Japan **16**:253–257.

Matteuzzi, D., Crociani, F., Emaldi, O. 1978. Amino acids produced by bifidobacteria and some clostridia. Annales de Microbiologie **129B:**175–181.

Matteuzzi, D., Hollaus, F., Biavati, B. 1978. Proposal of neotype for *Clostridium thermohydrosulfuricum* and the merging of *Clostridium tartarivorum* with *Clostridium thermosaccharolyticum*. International Journal of Systematic Bacteriology **28:**528–531.

Matteuzzi, D., Trovatelli, L. D., Biavati, B., Zani, G. 1977. Clostridia from grana cheese. Journal of Applied Bacteriology **43:**375–382.

Mayhew, S. G. 1971. Properties of two clostridial flavodoxins. Biochimica et Biophysica Acta **235:**276–288.

Mead, G. C. 1971. The amino acid-fermenting clostridia. Journal of General Microbiology **67:**47–56.

Mead, G. C., Adams, B. W., Hilton, M. G., Lord, P. G. 1979. Isolation and characterization of uracil-degrading clostridia from soil. Journal of Applied Bacteriology **46:**465–472.

Meyn, A. 1933. Über das Vorkommen und den Nachweis anaerober Bazillen in der Milch. Milchwirtschaftliche Forschungen **15:**426–432.

Mercer, I., Modi, N., Clarke, D. J., Morris, J. G. 1979. The occurrence and location of squalene in *Clostridium pasteurianum*. Journal of General Microbiology **111:**437–440.

Mercer, W. A., Vaughn, R. H. 1951. The characteristics of some thermophilic, tartrate-fermenting anaerobes. Journal of Bacteriology **62:**27–37.

Mihele, V. A., Duncan, C. L., Chambliss, G. H. 1978. Characterization of bacteriocinogenic plasmid in *Clostridium perfringens* CW 55. Antimicrobial Agents and Chemotherapy **14:**771–779.

Möllby, R., Holme, T., Nord, C. E., Smyth, C. J., Wadström, T. 1976. Production of phospholipase C (alphatoxin), haemolysins and lethal toxins by *Clostridium perfringens* types A to D. Journal of General Microbiology **96:**137–144.

Moore, M. R., O'Brien, W. E., Ljungdahl, L. G. 1974. Purification and characterization of nicotinamide adenine dinucleotide-dependent methylenetetrahydrofolate dehydrogenase from *Clostridium formicoaceticum*. Journal of Biological Chemistry **249:**5250–5253.

Moore, W. E. C. 1978. Taxonomy and physiology of nonspore-forming anaerobes. XII. International Congress of Microbiology, Munich. Abstract S 08.2, p. 11.

Moore, W. E. C., Cato, E. P., Holdeman, L. V. 1966. Fermentation patterns of some *Clostridium* species. International Journal of Systematic Bacteriology **16:**383–415.

Moore, W. E. C., Johnson, J. L., Holdeman, L. V. 1976. Emendation of *Bacteroidaceae* and *Butyrivibrio* and descriptions of *Desulfomonas* gen. nov. and ten new species in the genera *Desulfomonas, Butyrivibrio, Eubacterium, Clostridium,* and *Ruminococcus*. International Journal of Systematic Bacteriology **26:**238–252.

Morris, J. G. 1975. The physiology of obligate anaerobiosis, pp. 169–246. In: Rose, A. H., Tempest, D. W. (eds.), Advances in microbial physiology, vol. 12. London, New York, San Francisco: Academic Press.

Mortenson, L. E. 1966. Components of cell-free extracts of *Clostridium pasteurianum* required for ATP-dependent H_2 evolution from dithionite and for N_2 fixation. Biochimica et Biophysica Acta **127:**18–25.

Mortenson, L. E. 1978. Regulation of nitrogen fixation, pp. 179–232. In: Horecker, B. L., Stadtman, E. R. (eds.), Current topics in cellular regulation, vol. 13. New York: Academic Press.

Mortenson, L. E., Chen, J. S. 1974. Hydrogenase, pp. 231–282. In: Neilands, J. B. (ed.), Microbial iron metabolism. A comprehensive treatise. New York, London: Academic Press.

Mortenson, L. E., Morris, J. A., Jeng, D. Y. 1967. Purification, metal composition and properties of molybdoferredoxin and azoferredoxin, two of the components of the nitrogen-fixing system of *Clostridium pasteurianum*. Biochimica et Biophysica Acta **141:**516–522.

Mortenson, L. E., Valentine, R. C., Carnahan, J. E. 1962. An electron transport factor from *Clostridium pasteurianum*. Biochemical and Biophysical Research Communications **7:**448–453.

Müller, H. E., Werner, H. 1974. Occurrence of neuraminidase and acylneuraminate lyase in *Clostridium beijerinckii* and *Clostridium tertium*. Zentralblatt für Bakteriologie, Parasitenkunde, Infektionskrankheiten und Hygiene, Orig., Abt. 1 Reihe A **229:**134–140.

Muth, W. L., Costilow, R. N. 1974. Ornithine cyclase (deaminating). II. Properties of the homogeneous enzyme. Journal of Biological Chemistry **249:**7457–7462.

Nakamura, S., Okado, I., Abe, T., Nishida, S. 1979. Taxonomy of *Clostridium tetani* and related species. Journal of General Microbiology **113:**29–35.

Nakamura, S., Shimamura, T., Hayase, M., Nishida, S. 1973. Numerical taxonomy of saccharolytic clostridia, particularly *Clostridium perfringens*-like strains: Description of *Clostridium absonum* sp. n. and *Clostridium paraperfringens*. International Journal of Systematic Bacteriology **23:**419–429.

Nakamura, S., Shimamura, T., Hayashi, H., Nishida, S. 1975. Reinvestigation of the taxonomy of *Clostridium bifermentans* and *Clostridium sordellii*. Journal of Medical Microbiology **8:**299–309.

Nasuno, S., Assai, T. 1960. Some environmental factors affecting sporulation in butanol and butyric acid bacteria. Journal of General and Applied Microbiology **6:**71–82.

Neece, M. S., Fridovich, I. 1967. Acetoacetic decarboxylase, activation by heat. Journal of Biological Chemistry **242:**2939–2944.

Nishida, S., Seo, N., Nakagawa, M. 1969. Sporulation, heat resistance, and biological properties of *Clostridium perfringens*. Applied Microbiology **17:**303–309.

Ng, T. K., Weimer, P. J., Zeikus, J. G. 1977. Cellulolytic and physiological properties of *Clostridium thermocellum*. Archives of Microbiology **114:**1–7.

Nisman, B. 1954. The Stickland reaction. Bacteriological Reviews **18:**16–42.

O'Brien, R. W., Morris, J. G. 1971a. The ferredoxin-dependent reduction of chloramphenicol by *Clostridium acetobutylicum*. Journal of General Microbiology **67:**265–271.

O'Brien, R. W., Morris, J. G. 1971b. O_2 and the growth and metabolism of *Clostridium acetobutylicum*. Journal of General Microbiology **68:**307–318.

O'Brien, R. W., Morris, J. G. 1972. Effect of metronidazole on hydrogen production by *Clostridium acetobutylicum*. Archiv für Mikrobiologie **84:**225–233.

O'Brien, W. E., Brewer, J. M., Ljungdahl, L. G. 1973. Purification and characterization of thermostable 5,10-methylenetetrahydrofolate dehydrogenase from *Clostridium thermoaceticum*. Journal of Biological Chemistry **248:**403–408.

O'Brien, W. E., Ljungdahl, L. G. 1972. Fermentation of fructose and synthesis of acetate from carbon dioxide by *Clostridium formicoaceticum*. Journal of Bacteriology **109:**626–632.

Ohwaki, K., Hungate, R. E. 1977. Hydrogen utilization by clostridia in sewage sludge. Applied and Environmental Microbiology **33:**1270–1274.

Omelianski, W. 1902. Über die Gärung der Zellulose. Centralblatt für Bakteriologie und Parasitenkunde, Abt. 2 Orig. **8:**225–231.

Packer, E. L., Rabinowitz, J. C., Sternlicht, H. 1978. Assignment of the cysteinyl ^{13}C nuclear magnetic resonance and comparison of other aliphatic amino acid resonances of *Clostridium acidiurici, Clostridium pasteurianum* and *Peptococcus aerogenes* ferredoxins. Journal of Biological Chemistry **253:**7722–7730.

Parker, D. J., Wu, T. F., Wood, H. G. 1971. Total synthesis of acetate from CO_2: IV. Methyltetrahydrofolate, an intermediate and a procedure for separation of the folates. Journal of Bacteriology **108:**770–776.

Parsons, L. B., Sturges, W. S. 1927a. Quantitative aspects of the metabolism of anaerobes. I. Proteolysis by *Clostridium putrefaciens* compared with that of other anaerobes. Journal of Bacteriology **14**:181–192.

Parsons, L. B., Sturges, W. S. 1927b. Quantitative aspects of the metabolism of anaerobes. II. The relation between volatile acid and ammonia production during metabolism of *Clostridium putrefaciens*. Journal of Bacteriology **14**:193–200.

Parsons, L. B., Sturges, W. S. 1927c. Quantitative aspects of the metabolism of anaerobes. III. The volatile acids produced by *Clostridium putrefaciens* in cooked meat medium. Journal of Bacteriology **14**:201–215.

Pasteur, L. 1861. Animalcules infusoires vivant sans gaz oxygène libre et déterminant des fermentations. Comptes Rendus de l'Académie des Sciences, Séance du 25 Février, vol. 52, pp. 344–347.

Patni, N. J., Alexander, J. K. 1971a. Utilization of glucose by *Clostridium thermocellum*: Presence of glucokinase and other glycolytic enzymes in cell extracts. Journal of Bacteriology **105**:220–225.

Patni, N. J., Alexander, J. K. 1971b. Catabolism of fructose and mannitol in *Clostridium thermocellum*: Presence of phosphoenolpyruvate: fructose phosphotransferase, fructose 1-phosphate kinase, phosphoenolpyruvate: mannitol phosphotransferase, and mannitol 1-phosphate dehydrogenase in cell extracts. Journal of Bacteriology **105**:226–231.

Peck, H. D., Gest, H. 1957. Hydrogenase of *Clostridium butylicum*. Journal of Bacteriology **73**:569–580.

Pepple, J. S., Dennis, D. 1976. Lactate racemase. Hydroxylamine-dependent ^{18}O exchange of the α-hydroxyl of lactic acid. Biochimica et Biophysica Acta **429**:1036–1040.

Perkins, W. E. 1965. Production of clostridial spores. Journal of Applied Bacteriology **28**:1–16.

Petitdemange, H., Lambert, D., Gay, R. 1972. Activités NAD et NADP ferrédoxine réductasique des extraits acellulaires de *Clostridium pastorianum*. Comptes Rendus des Séances de la Société de Biologie et de ses Filiales **166**:1128–1132.

Petitdemange, H., Lambert, D., Gay, R. 1973. Determination of the reduced and oxidized pyridine nucleotides in *Clostridium pasteurianum* Comptes Rendus des Séances de la Société de Biologie et de ses Filiales **167**:111–115.

Petitdemange, H. Marczak, R., Gay, R. 1979. NADH-flavodoxin oxidoreductase activity in *Clostridium tyrobutyricum*. FEMS Microbiology Letters **5**:291–294.

Petitdemange, H., Bengone, J. M., Bergére, J.-L., Gay, R. 1973. Regulation of the NAD and NADP-ferredoxin reductase activities in a Clostridium of the butyric acid group: *Clostridium tyrobutyricum*. Biochimie **55**:1307–1310.

Petitdemange, H., Cherrier, C., Raval, G., Gay, R. 1976. Regulation of the NADH and NADPH-ferredoxin oxidoreductases in clostridia of the butyric group. Biochimica et Biophysica Acta **421**:334–347.

Petitdemange, H., Cherrier, C., Bengone, J. M., Gay, R. 1977. Etude des activités NADH et NADPH-ferrédoxine oxydoré ductasiques chez *Clostridium acetobutylicum*. Canadian Journal of Microbiology **23**:152–160.

Pheil, C. G., Ordal, Z. J. 1967. Sporulation of the "thermophilic anaerobes". Applied Microbiology **15**:893–898.

Poston, J. M. 1976. Leucine 2,3-aminomutase, an enzyme of leucine catabolism. Journal of Biological Chemistry **251**:1859–1863.

Potter, L. F., McCoy, E. 1952. The fermentation of pectin and pectic acid by *Clostridium felsineum*. Journal of Bacteriology **64**:701–708.

Prazmowski, A. 1880. Untersuchungen über die Entwicklungsgeschichte und Fermentwirkung einer Bakterien-Art. Leipzig: Inaugural Dissertation, Hugo Voigt.

Prévot, A. R. 1957. Manual de classification et de détermination des bactéries anaérobies. Paris: Masson.

Prévot, A. R. 1966. Manual for the classification and determination of the anaerobic bacteria, 1st American ed. translated by V. Fredette. Philadelphia: Lea and Febiger.

Prins, R. A., Lankhorst, A. 1977. Synthesis of acetate from CO_2 in the cecum of some rodents. FEMS Microbiology Letters **1**:255–258.

Rabinowitz, J. C. 1963. Intermediates in purine breakdown, pp. 703–713. In: Colowick, S. P., Kaplan, N. O. (eds.), Methods in enzymology, vol. 6. New York: Academic Press.

Rabinowitz, J. C., Pricer, W. E. 1962. Formyltetrahydrofolate synthetase. I. Isolation and crystallization of the enzyme. Journal of Biological Chemistry **237**:2898–2902.

Rawlings, J., Shah, V. K., Chisnell, J. R., Brill, W. J., Zimmermann, R., Münck, E., Orme-Johnson, W. H. 1978. Novel metal cluster in the iron-molybdenum cofactor of nitrogenase. Journal of Biological Chemistry **253**:1001–1004.

Raynaud, M. 1949. Le bactéries anaérobies pectinolytiques. Annales de l'Institut Pasteur **77**:434–470.

Reeves, R. E., Manzies, R. A., Hsu, D. S. 1968. The pyruvate-phosphate dikinase reaction. The fate of phosphate and the equilibrium. Journal of Biological Chemistry **243**:5486–5491.

Riebeling, V., Thauer, R. K., Jungermann, K. 1975. The internal-alkaline pH gradient, sensitive to uncoupler and ATPase inhibitor, in growing *Clostridium pasteurianum*. European Journal of Biochemistry **55**:445–453.

Ritter, W. 1932. Eine Nachprüfung des Ruschmann'schen Kartoffelbreiverfahrens zum Nachweis von Buttersäurebazillen. Landwirtschaftliches Jahrbuch der Schweiz **46**:601–608.

Ritter, W., Dorner, W. 1932. Behebung eines wichtigen Nachteils des anaeroben Pyrogallolverschlusses. Zentralblatt für Bakteriologie, Parasitenkunde und Infektionskrankheiten Abt. 1 Orig. **125**:379–383.

Roberts, T. A. 1967. Sporulation of mesophilic clostridia. Journal of Applied Bacteriology **30**:430–443.

Roberts, T. A., Derrik, C. M. 1975. Sporulation of *Clostridium putrefaciens* and the resistance of the spores to heat, γ-radiation and curing salts. Journal of Applied Bacteriology **38**:33–37.

Roberts, T. A., Hobbs, G. 1968. Low temperature growth characteristics of clostridia. Journal of Applied Bacteriology **31**:75–88.

Roberts, T. A., Ingram, J. M. 1965. The resistance of spores of *Clostridium botulinum* type E to heat and radiation. Journal of Applied Bacteriology **28**:125–138.

Robertson, M. 1915–1916. Notes upon certain anaerobes isolated from wounds. Journal of Pathology and Bacteriology **20**:327.

Robson, R. L., Morris, J. G. 1974. Mobilization of granulose in *Clostridium pasteurianum*. Purification and properties of granulose phosphorylase. Biochemical Journal **144**:513–517.

Robson, R. L., Robson, R. M., Morris, J. G. 1974. The biosynthesis of granulose by *Clostridium pasteurianum*. Biochemical Journal **144**:503–511.

Rolfe, R. D., Hentges, D. J., Campbell, B. J., Barrett, J. T. 1978. Factors related to the oxygen tolerance of anaerobic bacteria. Applied and Environmental Microbiology **36**:306–313.

Rosenberger, R. F. 1951. The development of methods for the study of obligate anaerobes in silage. Proceedings of the Society for Applied Bacteriology **14**:161–164.

Rosenberger, R. F. 1956. The isolation and cultivation of obligate anaerobes from silage. Journal of Applied Bacteriology **19**:173–180.

Rosenblum, E. D., Wilson, P. W. 1949. Fixation of isotopic nitrogen by *Clostridium*. Journal of Bacteriology **57**:413–414.

Ross, H. E. 1965. *Clostridium putrefaciens*: A neglected anaerobe. Journal of Applied Bacteriology **28**:49–51.

Rousseau, M., Hermier, J. Bergère, J.-L. 1971a. Structure de certains *Clostridium* du groupe butyrique. I. Sporulation de *Clostridium butyricum* et *Clostridium saccharobutyricum*. Annales de l'Institut Pasteur **120**:23–32.

Rousseau, M., Hermier, J., Bergère, J.-L. 1971b. Structure de certains *Clostridium* du groupe butyrique. II. Sporulation de

Clostridium tyrobutyricum. Annales de l'Institut Pasteur **120**:33–41.

Rousseau, M., Hermier, J., Bergère, J.-L. 1971c. Structure de certains *Clostridium* du groupe butyrique. III. Rôle de la membrane dans la formation de tuniques: Mise en évidence par l'analyse des formes anormales de sporulation. Annales de l'Institut Pasteur **121**:3–12.

Roux, C., Bergère, J.-L. 1977. Caractères taxonomiques de *Clostridium tyrobutyricum.* Annales de Microbiologie **128A**:267–276.

Rubbo, S. D., Maxwell, M., Fairbridge, R. A., Gillespie, J. M. 1941. The bacteriology, growth factor requirements and fermentation reactions of *Clostridium acetobutylicum* (Weizmann). Australian Journal of Experimental Biology and Medicine **19**:185–198.

Ruschmann, G. 1928. Vergleichende biologische und chemische Untersuchungen an Stalldüngersorten. Mitteilung IV, Pferdemistsorten, II. Teil. Zentralblatt für Bakteriologie und Parasitenkunde, Abt. 2 **75**:405–426.

Ruschmann, G., Bavendamm, W. 1925. Zur Kenntnis der Rösterreger *Bacillus felsineus* Carbone und *Plectridium pectinovorum* (Bac. amylobacter A. M. et Bredemann.). Zentralblatt für Bakteriologie und Parasitenkunde, Abt. 2 **64**:340–394.

Ruschmann, G., Harder, L. 1931. Die Buttersäuregärung im Silofutter und der Nachweis ihrer Erreger. Futterkonservierung **3**:1–40.

Sacks, L. E., Thompson, P. A. 1978. Clear, defined medium for the sporulation of *Clostridium perfringens.* Applied and Environmental Microbiology **35**:405–410.

Sagers, R. D., Benziman, M., Gunsalus, I. C. Acetate formation in *Clostridium acidi-urici:* Acetokinase. Journal of Bacteriology **82**:233–238.

Salanitro, J. P., Blake, I. G., Muirhead, P. A. 1974. Studies on the cecal microflora of commercial broiler chickens. Applied Microbiology **28**:439–447.

Sarathchandra, S. U., Barker, A. N., Wolf, I. 1974. The effect of oxygen on the germination and outgrowth of three strains of *Clostridium,* pp. 233–241. In: Barker, A. N., Gould, G. W., Wolf, I. (eds.), Spore research 1973. London: Academic Press.

Schäfer, R., Schwartz, A. C. 1976. Catabolism of purines in *Clostridium sticklandii.* Zentralblatt für Bakteriologie, Parasitenkunde, Infektionskrankheiten und Hygiene, Abt. 1 Orig., Reihe A **235**:165–172.

Schallehn, G., Brandis, H. 1973. Phosphatase-reagent for quick identification of *Clostridium perfringens.* Zentralblatt für Bakteriologie, Parasitenkunde, Infektionskrankheiten und Hygiene, Abt. 1 Orig., Reihe A **225**:343–345.

Schaupp, A., Ljungdahl, L. G. 1974. Purification and properties of acetate kinase from *Clostridium thermoaceticum.* Archives of Microbiology **100**:121–129.

Scherer, P. A., Thauer, R. K. 1978. Purification and properties of reduced ferredoxin: CO_2 oxidoreductase from *Clostridium pasteurianum,* a molybdenum iron-sulfur protein. European Journal of Biochemistry **85**:125–135.

Schildkraut, C. L., Marmur, J., Doty, P. 1962. Base composition of deoxyribonucleic acid from its buoyant density in CsCl. Journal of Molecular Biology **4**:430–443.

Schleifer, K. H., Kandler, O. 1972. Peptidoglycan types of bacterial cell walls and their taxonomic implications. Bacteriological Reviews **36**:407–477.

Schmidt, G. C., Logan, M. A., Tytell, A. A. 1952. The degradation of arginine by *Clostridium perfringens* (BP 6K). Journal of Biological Chemistry **198**:771–783.

Schoberth, S., Gottschalk, G. 1969. Considerations on the energy metabolism of *Clostridium kluyveri.* Archiv für Mikrobiologie **65**:318–328.

Schönheit, P., Brandis, A., Thauer, R. K. 1979. Ferredoxin degradation in growing *Clostridium pasteurianum* during periods of iron deprivation. Archives of Microbiology **120**:73–76.

Schulman, M., Parker, D., Ljungdahl, L. G., Wood, H. G. 1972.

Total synthesis of acetate from CO_2. V. Determination by mass analysis of the different types of acetate formed from $^{13}CO_2$ by heterotrophic bacteria. Journal of Bacteriology **109**:633–644.

Schulman, M., Ghambeer, R. K., Ljungdahl, L. G., Wood, H. G. 1973. Total synthesis of acetate from CO_2. VII. Evidence with *Clostridium thermoaceticum* that the carboxyl of acetate is derived from the carboxyl of pyruvate by transcarboxylation and not by fixation of CO_2. Journal of Biological Chemistry **248**:6255–6261.

Scott, T. A., Bellion, E., Mattey, M. 1969. The conversion of N-formyl-L-aspartate into nicotinic acid by extracts of *Clostridium butylicum.* European Journal of Biochemistry **10**:318–323.

Seki, S., Hagiwara, M., Kudo, K., Ishimoto, M. 1979. Studies on nitrate reductase of *Clostridium perfringens.* II. Purification and some properties of ferredoxin. Journal of Biochemistry **85**:833–838.

Seki-Chiba, S., Ishimoto, M. 1977. Studies on nitrate reductase of *Clostridium perfringens.* I. Purification, some properties, and effect of tungsten on its formation. Journal of Biochemistry **82**:1663–1671.

Sergeant, K., Ford, J. W. S., Longyear, V. M. C. 1968. Production of *Clostridium pasteurianum* in a defined medium. Applied Microbiology **16**:296–300.

Seto, B. 1978. A pyruvate-containing peptide of proline reductase in *Clostridium sticklandii.* Journal of Biological Chemistry **253**:4525–4529.

Seto, B., Mortenson, L. E. 1974. In vivo kinetics of nitrogenase formation in *Clostridium pasteurianum.* Journal of Bacteriology **120**:822–830.

Shapton, D. A., Board, R. G. 1971. Isolation of anaerobes. London: Academic Press.

Sheiman, M. I., Macmillan, J. D., Miller, L., Chase, T. 1976. Coordinated action of pectinesterase and polygalacturonate lyase complex of *Clostridium multifermentans.* European Journal of Biochemistry **64**:565–572.

Shoaf, W. T., Neece, S. H., Ljungdahl, L. G. 1974. Effects of temperature and ammonium ions on formyl-tetrahydrofolate synthetase from *Clostridium thermoaceticum.* Biochimica et Biophysica Acta **334**:448–458.

Shull, G. M., Thoma, R. W., Peterson, W. H. 1949. Amino acid and unsaturated fatty acid requirements of *Clostridium sporogenes.* Archives of Biochemistry **20**:227–241.

Sjolander, N. O. 1937. Studies on anaerobic bacteria. XII. The fermentation products of *Clostridium thermosaccharolyticum.* Journal of Bacteriology **34**:419–428.

Sjolander, N. O., McCoy, E. 1937. Studies on anaerobic bacteria. A cultural study of some "butyric" anaerobes previously described in the literature. Zentralblatt für Bakteriologie, Parasitenkunde und Infektionskrankheiten, Abt. 2 **97**:314–324.

Skinner, F. A. 1960. The isolation of anaerobic cellulose-decomposing bacteria from soil. Journal of General Microbiology **22**:539–554.

Skinner, F. A. 1971. The isolation of soil bacteria, pp. 57–78. In: Shapton, D. A., Board, R. G. (eds.), Isolation of anaerobes. London, New York: Academic Press.

Sleytr, U. B., Glauert, A. M. 1976. Ultrastructure of the cell walls of two closely related clostridia that possess different regular arrays of surface subunits. Journal of Bacteriology **126**:869–882.

Smith, B. E., Thorneley, R. N. F., Eady, R. R., Mortenson, L. E. 1976. Nitrogenases from *Klebsiella pneumoniae* and *Clostridium pasteurianum.* Kinetic investigations of cross-reactions as a probe of the enzyme mechanism. Biochemical Journal **157**:439–447.

Smith, L. DS. 1949. Clostridia in gas gangrene. Bacteriological Reviews **13**:233–254.

Smith, L. DS. 1970. *Clostridium oceanicum,* sp. n., a spore-forming anaerobe isolated from marine sediments. Journal of Bacteriology **103**:811–813.

Smith, L. DS., Cato, E. P. 1974. *Clostridium durum,* sp. nov., the predominant organism in a sediment core from the Black Sea. Canadian Journal of Microbiology **20:**1393–1397.

Smith, L. DS., Hobbs, G. 1974. Genus III. *Clostridium* Prazmowski 1880, pp. 551–572. In: Buchanan, R. E., Gibbons, N. E. (eds.), Bergey's manual of determinative bacteriology, 8th ed. Baltimore: Williams & Wilkins.

Somack, R., Costilow, R. N. 1973a. Purification and properties of a pyridoxal phosphate and coenzyme B_{12}-dependent D-α-ornithine 5,4-aminomutase. Biochemistry **12:**2597–2604.

Somack, R., Costilow, R. N. 1973b. 2,4-Diaminopentanoic acid C_4 dehydrogenase. Journal of Biological Chemistry **248:**385–388.

Spiegelberg, C. H. 1944. Sugar and salt tolerance of *Clostridium pasteurianum* and some related anaerobes. Journal of Bacteriology **48:**13–30.

Spivey, M. J. 1978. The acetone/butanol/ethanol fermentation. Process Biochemistry **13:**2–4, 25.

Stadtman, E. R. 1955. Phosphotransacetylase from *Clostridium kluyveri,* pp. 596–599. In: Colowick, S. P., Kaplan, N. O. (eds.), Methods in enzymology, vol. 1. New York: Academic Press.

Stadtman, E. R. 1976. The *Clostridium kluyveri*-acetyl-CoA epoch, pp. 161–172. In: Kornberg, A., Horecker, B. L., Cornudella, L., Oro, J. (eds.), Reflections on biochemistry. Oxford: Pergamon Press.

Stadtman, E. R., Burton, R. M. 1955. Aldehyde dehydrogenase from *Clostridium kluyveri,* pp. 581–583. In: Colowick, S. P., Kaplan, N. O. (eds.), Methods in enzymology, vol. 1. New York: Academic Press.

Stadtman, E. R., Stadtman, T. C., Pastan, I., Smith, L. DS. 1972. *Clostridium barkeri* sp. n. Journal of Bacteriology **110:**758–760.

Stadtman, T. C. 1962. Studies on the enzymic reduction of amino acids. V. Coupling of a DPNH-generating system to glycine reduction. Archives of Biochemistry and Biophysics **99:**36–44.

Stadtman, T. C. 1965. Electron transport proteins of *Clostridium sticklandii,* pp. 439–445. In: San Pietro, A. (ed.), Non-heme-iron proteins: Role in energy conversion. Yellow Springs, Ohio: Antioch Press.

Stadtman, T. C. 1973. Lysine metabolism by clostridia. Advances in Enzymology **38:**413–448.

Stadtman, T. C. 1978. Selenium-dependent clostridial glycine reductase, pp. 373–382. In: Fleischer, S., Packer, L. (eds.), Methods in enzymology, vol. 53. New York, San Francisco, London: Academic Press.

Stadtman, T. C., McClung, L. S. 1957. *Clostridium sticklandii* nov. spec. Journal of Bacteriology **73:**218–219.

Stadtman, T. C., White, F. H. 1954. Tracer studies on ornithine, lysine, and formate metabolism in an amino acid fermenting Clostridium. Journal of Bacteriology **67:**651–657.

Stankewich, J. P., Cosenza, B. J., Shigo, A. L. 1971. *Clostridium quercicolum* sp. n., isolated from discolored tissues in living oak trees. Antonie van Leeuwenhoek Journal of Microbiology and Serology **37:**299–302.

Stellwag, E. J., Hylemon, P. B. 1977. Characterization of 7-α-dehydroxylase in whole cells of *Clostridium leptum* VPI 10900. Abstracts of the Annual Meeting of the American Society for Microbiology **1977:**367.

Stickland, L. H. 1934. Studies in the metabolism of the strict anaerobes (genus *Clostridium*). I. The chemical reaction by which *Clostridium sporogenes* obtains energy. Biochemical Journal **28:**1746–1759.

Stickland, L. H. 1935. Studies in the metabolism of the strict anaerobes (genus *Clostridium*). III. The oxidation of alanine by *Clostridium sporogenes.* IV. The reduction of glycine by *Clostridium sporogenes.* Biochemical Journal **29:**889–898.

Stolp, H. 1955. Ernährungs- und entwicklungsphysiologische Untersuchungen an anaeroben Bakterien. II. Die Physiologie der Entwicklung von Clostridien unter besonderer

Berücksichtigung des Reduktions-Oxydations-Potentials. Archiv für Mikrobiologie **21:**293–309.

Sturges, W. S., Drake, E. T. 1927. A complete description of *Clostridium putrefaciens* (McBryde). Journal of Bacteriology **14:**175–179.

Switzer, R. L., Barker, H. A. 1967. Purification and characterization of component S of glutamate mutase. Journal of Biological Chemistry **242:**2658–2674.

Tagg, J. R., Dajani, A. S., Wannamaker, L. W. 1976. Bacteriocins of Gram-positive bacteria. Bacteriological Reviews **40:**722–756.

Takumi, K., Kawata, T. 1976. Quantitative chemical analyses and antigenic properties of peptidoglycans from *Clostridium botulinum* and other clostridia. Japanese Journal of Microbiology **20:**287–292.

Tally, F. P., Goldin, B. R., Jacobus, N. V., Gorbach, S. L. 1977. Superoxide dismutase in anaerobic bacteria of clinical significance. Infection and Immunity **16:**20–25.

Tanaka, H., Stadtman, T. C. 1979. Selenium-dependent clostridial glycine reductase. Purification and characterization of the two membrane-associated protein components. Journal of Biological Chemistry **254:**447–452.

Tanaka, M., Haniu, M., Yasunobu, K. T., Himes, R. H., Akagi, J. M. 1973. The primary structure of *Clostridium thermosaccharolyticum* ferredoxin, a heat-stable ferredoxin. Journal of Biological Chemistry **248:**5215–5217.

Tanaka, M., Haniu, M., Yasunobu, K. T., Mortenson, L. E. 1977. The amino acid sequence of *Clostridium pasteurianum* iron protein, a component of nitrogenase. I. Tryptic peptides. II. Cyanogen peptides. III. The NH_2-terminal and COOH-terminal sequences, tryptic peptides of large cyanogen bromide peptides, and the complete sequences. Journal of Biological Chemistry **252:**7081–7100.

Taylor, W. H., Taylor, M. L., Balch, W. E., Gilchrist, P. S. 1976. Purification and properties of dihydroorotase, a zinc-containing metallo enzyme in *Clostridium oroticum.* Journal of Bacteriology **127:**863–873.

Thauer, R. K., Jungermann, K., Decker, K. 1967. A quantitative isotope method for regulation studies of aromatic amino acid synthesis under growth conditions. European Journal of Biochemistry **1:**482–486.

Thauer, R. K., Jungermann, K., Decker, K. 1977. Energy conservation in chemotrophic anaerobic bacteria. Bacteriological Reviews **41:**100–180.

Thauer, R. K., Jungermann, K., Henninger, H., Wenning, J., Decker, K. 1968. The energy metabolism of *Clostridium kluyveri.* European Journal of Biochemistry **4:**173–180.

Thauer, R. K., Käufer, B., Scherer, P. 1975. The active species of "CO_2" utilized in ferredoxin-linked carboxylation reactions. Archives of Microbiology **104:**237–240.

Thauer, R. K., Kirchniawy, F. H., Jungermann, K. A. 1972. Properties and function of the pyruvate-formate-lyase reaction in clostridia. European Journal of Biochemistry **27:**282–290.

Thauer, R. K., Rupprecht, E., Jungermann, K. 1970. Glyoxylate inhibition of clostridial pyruvate synthase. FEBS Letters **9:**271–273.

Thauer, R. K., Rupprecht, E., Ohrloff, C., Jungermann, K., Decker, K. 1971. Regulation of the reduced nicotinamide adenine dinucleotide phosphate-ferredoxin reductase system in *Clostridium kluyveri.* Journal of Biological Chemistry **246:**954–959.

Timmis, K., Hobbs, G., Berkeley, R. C. W. 1974. Chitinolytic clostridia isolated from marine mud. Canadian Journal of Microbiology **20:**1284–1285.

Tokushige, M., Hayaishi, O. 1972. Threonine metabolism and its regulation in *Clostridium tetanomorphum.* Journal of Biochemistry **72:**469–477.

Tomlinson, N. 1954. Carbon dioxide and acetate utilization by *Clostridium kluyveri.* III. A new part of glutamic acid synthesis. Journal of Biological Chemistry **209:**605–609.

Touraille, C. Bergère, J.-L. 1974. La germination de la spore de *Clostridium tyrobutyricum.* II. Démonstration de l'intervention de l'acetokinase et de la phosphotransacétylase par l'étude de leurs propriétés. Biochimie **56**:404–422.

Tsuda, Y., Friedmann, H. C. 1970. Ornithine metabolism by *Clostridium sticklandii*. Oxidation of ornithine to 2-amino-4-keto-pentanoic acid via 2,4-diaminopentanoic acid; participation of B_{12}-coenzyme, pyridoxal phosphate, and pyridine nucleotide. Journal of Biological Chemistry **245**:5914–5926.

Twarog, R., Wolfe, R. S. 1963. Role of butyryl phosphate in the energy metabolism of *Clostridium tetanomorphum*. Journal of Bacteriology **86**:112–117.

Ueno, K., Fujii, H., Marni, F., Takahashi, J., Sugitani, T., Ushijima, T. Suzuki, S. 1970. Acid phosphatase in *Clostridium perfringens*. A new rapid and simple identification method. Japanese Journal of Microbiology **14**:171–173.

Uesugi, I., Yajima, M. 1978a. Oxygen and "strictly anaerobic" intestinal bacteria. I. The effect of dissolved oxygen on growth. Zeitschrift für Allgemeine Mikrobiologie **18**:287–295.

Uesugi, I., Yajima, M. 1978b. Oxygen and "strictly anaerobic" intestinal bacteria. II. Oxygen metabolism in strictly anaerobic bacteria. Zeitschrift für Allgemeine Mikrobiologie **18**:593–601.

Uyeda, K., Rabinowitz, J. C. 1965. Metabolism of formiminoglycine. Glycine formiminotransferase. Journal of Biological Chemistry **240**:1701–1710.

Uyeda, K., Rabinowitz, J. C. 1967a. Metabolism of formiminoglycine. Formiminotetrahydrofolate cyclodeaminase. Journal of Biological Chemistry **242**:24–31.

Uyeda, K., Rabinowitz, J. C. 1967b. Enzymes of clostridial purine fermentation. Methylenetetrahydrofolate dehydrogenase. Journal of Biological Chemistry **242**:4378–4385.

Uyeda, K., Rabinowitz, J. C. 1968. Enzymes of the clostridial purine fermentation: Serine hydroxymethyltransferase. Archives of Biochemistry and Biophysics **123**:271–278.

Uyeda, K., Rabinowitz, J. C. 1971. Pyruvate-ferredoxin oxidoreductase. III. Purification and properties of the enzyme. Journal of Biological Chemistry **246**:3111–3119.

Uyeda, M., Gan, B. H., Takenobu, S., Ono, I., Hongo, M. 1974. Extracellular formation of *O*-butylhomoserine by anaerobes. Agricultural and Biological Chemistry **38**:1811–1818.

Vagelos, P. R., Earl, J. M., Stadtman, E. R. 1959. Propionic acid metabolism. I. The purification and properties of acrylyl coenzyme A aminase. Journal of Biological Chemistry **234**:490–497.

Valentine, R. C., Brill, W. J., Sagers, R. D. 1963. Ferredoxin linked DPN reduction by pyruvate in extracts of *Clostridium acidiurici*. Biochemical and Biophysical Research Communications **12**:315–319.

Valentine, R. C., Wolfe, R. S. 1960. Purification and role of phosphotransbutyrylase. Journal of Biological Chemistry **235**:1948–1952.

van Beynum, J., Pette, J. W. 1935. Zuckervergärende und Laktatvergärende Buttersäurebacterien. Zentralblatt für Bakteriologie, Parasitenkunde und Infektionskrankheiten, Abt. 2 **93**:198–212.

van Beynum, J., Pette, J. W. 1940. Een Methode voor het aantoonen van boterzuurbacterien, speciaal geschikt voor het onderzoek van melk. Verslagen van Landbouwkundige Onderzoekingen **46C**:379–396.

Vanquickenborne, A., Phillips, A. T. 1968. Purification and regulatory properties of the adenosine diphosphate-activated threonine dehydratase. Journal of Biological Chemistry **243**:1312–1319.

Veldkamp, H. 1965. Enrichment cultures of procaryotic organisms, pp. 305–361. In: Norris, J. R., Ribbons, D. W. (eds.), Methods in microbiology, vol. 3A. London, New York: Academic Press.

Venugopal, V., Nadkarni, G. B. 1977. Regulation of the arginine dihydrolase pathway in *Clostridium sporogenes*. Journal of Bacteriology **131**:693–695.

Viljoen, J. A., Fred, E. B., Peterson, W. H. 1926. The fermentation of cellulose by thermophilic bacteria. Journal of Agricultural Science **16**:1–17.

Vogel, H., Bruschi, M., Le Gall, J. 1977. Phylogenetic studies of two rubredoxins from sulfate reducing bacteria. Journal of Molecular Evolution **9**:111–119.

Vogels, G. D., van der Drift, C. 1976. Degradation of purines and pyrimidines by microorganisms. Bacteriological Reviews **40**:403–468.

von Hugo, H., Schoberth, S., Madan, V. K., Gottschalk, G. 1972. Coenzyme specificity of dehydrogenases and fermentation of pyruvate by clostridia. Archiv für Mikrobiologie **87**:189–202.

von Hugo, H., Gottschalk, G. 1974a. Distribution of 1-phosphofructokinase and PEP: fructose phosphotransferase activity in clostridia. FEBS Letters **46**:106–108.

von Hugo, H., Gottschalk, G. 1974b. Purification and properties of 1-phosphofructokinase from *Clostridium pasteurianum*. European Journal of Biochemistry **48**:455–463.

Vozniakovskaya, Y. M., Avrova, N. P., Andronikashvili, E. D. 1974. Reproduction and synthesis of pectolytic enzymes by *Clostridium felsineum* on media with various carbon sources. Microbiology [English translation of Mikrobiologiya] **43**:357–360.

Wachsman, J. T., Barker, H. A. 1954. Characterization of an orotic acid fermenting bacterium, *Zymobacterium oroticum*, nov. gen., nov. spec. Journal of Bacteriology **68**:400–404.

Wachsman, J. T., Barker, H. A. 1955a. The accumulation of formamide during the fermentation of histidine by *Clostridium tetanomorphum*. Journal of Bacteriology **69**:83–88.

Wachsman, J. T., Barker, H. A. 1955b. Tracer experiments on glutamate fermentation by *Clostridium tetanomorphum*. Journal of Biological Chemistry **217**:695–702.

Wagner, R., Andreesen, J. R. 1977. Differentiation between *Clostridium acidiurici* and *Clostridium cylindrosporum* on the basis of specific metal requirements for formate dehydrogenase formation. Archives of Microbiology **114**:219–224.

Wagner, R., Andreesen, J. R. 1979. Selenium requirement for active xanthine dehydrogenase from *Clostridium acidiurici* and *Clostridium cylindrosporum*. Archives of Microbiology **121**:255–260.

Walther, R., Hippe, H., Gottschalk, G. 1977. Citrate, a specific substrate for the isolation of *Clostridium sphenoides*. Applied and Environmental Microbiology **33**:955–962.

Wang, C. C., Barker, H. A. 1969. Purification and properties of L-citramalate hydrolyase. Journal of Biological Chemistry **244**:2516–2526.

Waterson, R. M., Castellino, F. J., Hass, G. M., Hill, R. L. 1972. Purification and characterization of crotonase from *Clostridium acetobutylicum*. Journal of Biological Chemistry **247**:5266–5271.

Weimer, P. J., Zeikus, J. G. 1977. Fermentation of cellulose and cellobiose by *Clostridium thermocellum* in the absence and presence of *Methanobacterium thermoautotrophicum*. Applied and Environmental Microbiology **33**:289–297.

Weinzirl, J., Veldee, M. V. 1915. A bacteriological method for determining manurial pollution of milk. American Journal of Public Health **5**:862–866.

Weizmann, C. 1919. Production of acetone and alcohol by a bacteriological process. U. S. Patent No. 1.315.585.

Weizmann, C., Hellinger, E. 1940. Studies on some strains of butyric-acid-producing plectridia isolated from hemp, jute and flax. Journal of Bacteriology **40**:665–682.

Welty, F. K., Wood, H. G. 1978. Purification of the "corrinoid" enzyme involved in the synthesis of acetate by *Clostridium thermoaceticum*. Journal of Biological Chemistry **253**:5832–5838.

Westlake, D. W. S., Shug, A. L., Wilson, P. W. 1961. The pyruvic dehydrogenase system of *Clostridium pasteurianum*. Canadian Journal of Microbiology **7**:515–524.

Weyer, E. R., Rettger, L. F. 1927. A comparative study of six different strains of the organism commonly concerned in large scale production of butyl alcohol and acetone by the biological process. Journal of Bacteriology **14**:399–424.

Whiteley, H. R., Tahara, M. 1966. Threonine deaminase of *Clostridium tetanomorphum*. Journal of Biological Chemistry **241**:4881–4889.

Wiegel, J., Schlegel, H. G. 1977. Leucine biosynthesis: Effect of branched-chain amino acids and threonine on α-isopropylmalate synthase activity from aerobic and anaerobic microorganisms. Biochemical Systematics and Ecology **5**:169–176.

Wieringa, K. T. 1940. The formation of acetic acid from carbon dioxide and hydrogen by anaerobic spore-forming bacteria. Antonie van Leeuwenhoek Journal of Microbiology and Serology **6**:251–262.

Willis, A. T. 1977. Anaerobic bacteriology: Clinical and laboratory practice, 3rd ed. London, Boston: Butterworths.

Winkler, S. 1961. Vergleichende Untersuchungen an anaeroben Proben. Österreichische Milchwirtschaft **16**:109–112.

Winnacker, E. L., Barker, H. A. 1970. Purification and properties of a NAD-dependent glutamate dehydrogenase from Clostridium SB₄. Biochimica et Biophysica Acta **212**:225–242.

Winogradsky, S. 1895. Recherches sur l'assimilation de l'azote libre de l'atmosphère par les microbes. Archives des Sciences Biologiques (Leningrad) **3**:297–352.

Winogradsky, S. 1902. *Clostridium pastorianum,* seine Morphologie und seine Eigenschaften als Buttersäureferment. Zentralblatt für Bakteriologie, Abt. 2, **9**:43–54, 107–112.

Witz, D. F., Detroy, R. W., Wilson, P. W. 1967. Nitrogen fixation by growing cells and cell-free extracts of Bacillaceae. Archiv für Mikrobiologie **55**:369–381.

Wood, N. P., Jungermann, K. 1972. Inactivation of the pyruvate formate lyase of *Clostridium butyricum*. FEBS Letters **27**:49–52.

Woods, D. D. 1936. Studies in the metabolism of the strict anaerobes (genus Clostridium). V. Further experiments on the coupled reactions between pairs of amino-acids induced by *Clostridium sporogenes*. Biochemical Journal **30**:1934–1946.

Woods, D. D., Clifton, C. E. 1937. Studies in the metabolism of the strict anaerobes (genus *Clostridium*). VI. Hydrogen production and amino-acid utilization by *Clostridium tetanomorphum*. Biochemical Journal **31**:1774–1788.

Wright, J. H. 1901. A method for the cultivation of anaerobic bacteria. Zentralblatt für Bakteriologie, Abt. 1 Orig. **29**:61.

Yang, S. S., Ljungdahl, L. G., Le Gall, J. 1977. A four-iron, four-sulfide ferredoxin with high thermostability from *Clostridium thermoaceticum*. Journal of Bacteriology **130**:1084–1090.

Yoch, D. C., Valentine, R. C. 1972. Ferredoxins and flavodoxins of bacteria. Annual Review of Microbiology **26**:139–162.

Yorifuji, T., Jeng, I. M., Barker, H. A. 1977. Purification and properties of 3-keto-5-aminohexanoate cleaving enzyme from a lysine-fermenting *Clostridium*. Journal of Biological Chemistry **252**:20–31.

Zamenhof, S., Eichhorn, H. H. 1967. Study of microbial evolution through loss of biosynthetic functions: Establishment of "defective" mutants. Nature **216**:456–458.

Zumft, W. G. 1978. Isolation of thiomolybdate compounds from the molybdenum-iron protein of clostridial nitrogenase. European Journal of Biochemistry **91**:345–350.

CHAPTER 139

The Genus *Sporosarcina*

DIETER CLAUS

Beijerinck (1901) isolated a packet-forming coccoid bacterium from urea-containing enrichment cultures. In contrast to all other known sarcinae, it was motile by flagella and formed endospores (Fig. 1). He described the new isolate as *Planosarcina ureae.*

The species was later transferred by Löhnis (1911) to the genus *Sarcina.* This genus represented bacteria that are morphologically similar, but very different physiologically and biochemically. Therefore, *S. ureae* has been separated from the genus *Sarcina* and transferred to the genus *Sporosarcina* (Kocur and Martinec, 1963; MacDonald and Mac-Donald, 1962) as previously proposed by Orla-Jensen (1909) and by Kluyver and van Niel (1936).

Sporosarcina ureae is the only species of the genus that is extensively described and available in pure culture. Beijerinck (1901) briefly described another spore-forming sarcina, *Sarcina dimorpha,* which developed only on media that contained horse urine. Cultures of the motile, spore-forming, and urea-degrading *Sarcina pulmonum* have been lost. The species is probably identical with *Sporosarcina ureae* (Gibson, 1935).

Habitats

Like other spore-forming bacteria, *Sporosarcina ureae* is widely distributed in soil. Fertile soils may contain up to 10^3 sarcinae/g (Gibson, 1935). The bacterium has also been isolated from liquid manure. A single isolate has been reported from sea water (Wood, 1946). Pregerson (1973) has found *S. ureae* widely distributed in the United States and in various other parts of the world. According to her studies, the primary habitat of the organism appeared to be concentrated in certain urban soils closely associated with the activities of man and especially dogs.

The prevalence of the bacterium in fertile soils and certain other urea-containing places, its resistance against the inhibitory effect of up to 5–10% urea, and its formation of urease suggest that *S. ureae* plays an active part, in nature, in the decom-

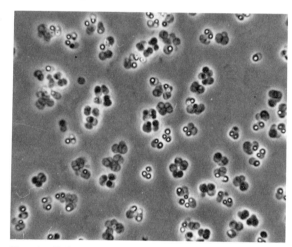

Fig. 1. Endospore-containing cells of *Sporosarcina ureae* ($\times 1{,}160$).

position of urea. *Sarcina pulmonum* has been isolated by several workers from sputum or from the respiratory tract in cases of phthisis. Although it has been considered for some time to be responsible for severe infections of the lungs, its pathogenicity has been questioned (Lehmann and Neumann, 1927; Migula, 1900).

Isolation

Gibson and others were unable to repeat Beijerinck's procedure for the isolation of *Sporosarcina ureae.* Its presence, however, has been demonstrated in many soil samples by a simple and effective isolation method (Gibson, 1935). Since 10% urea, as used by Gibson, may inhibit the growth of *S. ureae* from many soils, urea should also be used in lower concentrations. According to our own observations, most of the bacteria present in soil are strongly inhibited by 3–4% urea. This inhibition is specifically true for *Bacillus cereus* var. *mycoides,* which will readily overgrow soil dilution plates at urea concentrations below 3%.

Isolation of *Sporosarcina ureae*
(Gibson, 1935; modified by D. Claus)

To 1,000 ml of nutrient agar (Difco or similar preparation), add 30, 50, or 100 g of urea. Sterilize the media at 121°C for 15 min and pour into Petri dishes.

Suspend about 5 g of a soil sample (fertile soil, air-dried) in 20 ml of sterile water and prepare soil dilutions (10^{-1} and 10^{-2}). Plate 0.1 ml of the soil suspension and of the two dilutions on the agar and incubate at 25–30°C. The soil suspension may be heated, but there is little or no advantage because the majority of the organisms that will develop on the media are spore-forming bacteria.

After 3–5 days, examine the colonies under low magnification and transmitted light. Colonies of *S. ureae* can be recognized as round and black at a magnification of about 10×, and by their coarsely granulated structure, especially at the edges of the colonies, at a magnification of about 50×.

With some experience, colonies of *S. ureae* may be selected with rather high certainty. Similar types of colonies are often formed by strains of *Bacillus megaterium*. Therefore, it is recommended to compare the colony type with that of a pure culture of *S. ureae*.

Pregerson (1973) has isolated 51 strains of *S. ureae* from 198 different soils and has proposed the following method for isolating the organism.

Isolation of *Sporosarcina ureae*
(Pregerson, 1973)

Adjust tryptic soy–yeast agar (TSY), containing per liter of distilled water 27.5 g Difco tryptic soy broth, 5.0 g Difco yeast extract, 5.0 g glucose, and 15.0 g Difco agar with 1 N NaOH to pH 8.5 before autoclaving. Thereafter add aseptically filter-sterilized urea to give a final concentration of 1% (wt/vol) and pour the medium into Petri dishes.

Suspend 1 g of a soil sample (preferably from places where dogs urinate, like bases of trees) in 15 ml of distilled water and mix the slurry with a Vortex mixer. Spread 0.1 ml of a series of dilutions on plates with a sterile bent glass rod. Prepare triplicate plates at 10^{-1} and 10^{-2} and duplicate plates at 10^{-3} and 10^{-4} dilutions. Incubate at 22°C and examine the plates on the third day and then daily with a dissecting microscope.

Colonies of *S. ureae* show a uniform surface granularity, smoothly opaque interiors, and an orange or cream pigmentation.

Prepare slides from selected colonies for observing the typical morphology of cells and confirm the provisional identification of isolated

strains of *S. ureae* testing the motility and the production of spores.

The isolation of *S. ureae* from a certain soil sample often is not reproducible due to the irregular background growth developing on dilutions plates. A more selective method depressing such growth or a selective liquid enrichment method certainly will allow the isolation of this species also from soils where it may be present only in low numbers.

Axenic Cultivation and Maintenance

Sporosarcina ureae grows well in normal nutrient broth. A more suitable medium contains, in addition, 1% urea or 0.5% ammonium chloride. In the latter case, the pH of the medium should be adjusted to 8.0 or 8.5. Since the organism is aerobic, aeration by shaking is recommended. The optimum temperature for growth is about 25°C.

A defined medium for *S. ureae* has been described by Goldman and Wilson (1977), which yielded 5–6 g dry weight of bacteria per liter of culture.

Defined Medium for *Sporosarcina ureae*
(Goldman and Wilson, 1977)

Medium contains per liter of distilled water:

L-Asparagine · H$_2$O, or	
L-Glutamine	30.0 g
K$_2$HPO$_4$	0.25 g
MnCl$_2$ · 4H$_2$O	0.00025 g
MgSO$_4$ · 7H$_2$O	0.05 g
FeSO$_4$ · 7H$_2$O	0.0025 g
KCl	3.4 g
d-Biotin	0.001 g
L-Cysteine	0.005 g
(NH$_4$)$_2$SO$_4$	0.2 g
NaOH	
NaCl	

Prior to sterilization, adjust the medium to pH 8.7 with 1 M NaOH and add NaCl to a final concentration of 0.05 M Na$^+$. Sterilized solutions of biotin (0.4 mg/ml), L-cysteine (2 mg/ml), and (NH$_4$)$_2$SO$_4$ (80 mg/ml) were added to the separately sterilized medium.

Pregerson (1973), however, has pointed out that the growth factor requirements of strains of *S. ureae* may vary substantially. From 61 strains studied, 13 grew on a mineral acetate medium without added growth factors and 37 strains required biotin either singly or in combination with niacin and/or thiamine. Few of these strains also needed aspartate as a growth factor. Eleven strains showed more complex growth factor requirements. Most of the strains could utilize acetate, butyrate. or glutamate as a sole source of carbon and energy. Glutamate, however, generally could not be used by *S. ureae* as a nitrogen source.

Vegetative cultures grown on nutrient agar may be kept for more than 6 weeks at 4°C. Sporulated cultures can survive one or more years when stored at 4°C. Spores normally are obtained on the sporulation medium of Gibson (1935) at a temperature of about 22°C. The medium contains in 1 liter of distilled water: 5 g peptone, 5 g meat extract, 5 g ammonium chloride, and 15 g agar. The pH is adjusted to 6.8–7.0.

A medium for good sporulation has been described by MacDonald and MacDonald (1962).

Sporulation Medium for *Sporosarcina ureae* (MacDonald and MacDonald, 1962)

The medium contains per liter of distilled water:

Yeast extract	2.0 g
Peptone	3.0 g
Glucose	4.0 g
Malt extract	3.0 g
K_2HPO_4	1.0 g
$(NH_4)_2SO_4$	4.0 g
$CaCl_2$	0.1 g
$MgSO_4$	0.8 g
$MnSO_4 \cdot H_2O$	0.1 g
$FeSO_4 \cdot 7H_2O$	0.001 g
$ZnSO_4$	0.01 g
$CuSO_4 \cdot 5H_2O$	0.01 g
Agar	30.0 g

All the isolates of Pregerson (1973) showed production of spores on this medium incubated at 22°C. However, considerable variation in the onset and extent of sporulation was observed. Some strains produced only a few spores after 4 weeks of growth, while others produced an abundance of spores within 4 days. Spore formation was improved by adjusting the medium before autoclaving to pH 8.8–9.0 and by increasing the agar concentration to 3% (wt/vol). Some strains produced spores more readily when the peptone of the medium was replaced by Casamino Acids.

The spore-forming capacity of new isolates of *S. ureae* may be strongly reduced or totally lost after only a few transfers of cultures on nutrient agar with or without urea. For retaining this property, it is strongly recommended to transfer new isolates as fast as possible on the sporulation media mentioned above and to subculture sporulated strains from suspended cell material heated at 80°C for 5 min.

Both vegetative cells and spores can be preserved by lyophilization in skim milk (20% wt/vol) without significant loss in viability over many years.

Identification

Colonies of *Sporosarcina ureae* on agar are gray, circular with an entire margin, and slightly convex. On some media, a yellowish, brownish, or orange nondiffusible pigment may be produced. The most

Table 1. Some characteristics of Gram-positive cocci that form tetrads and/or packets.

Properties	*Sporosarcina*	*Planococcus*	*Micrococcus*	*Sarcina*
Obligate aerobic	+	+	+	−
Obligate anaerobic	−	−	−	+
Spores	+	−	−	+
Motility	+	+	−	−
Peptidoglycan type[a]	L-Lys-Gly-D-Glu	L-Lys-D-Glu	L-Lys-L-Ala L-Lys-peptide subunit	LL-Dpm-Gly
DNA, mol% G+C[b]	40–43	39–52	66–75	28–31
Phosphatidyl-ethanolamine in cell membranes[c]	+	+	−	−
Menaquinone system[d]	MK-7	MK-8 (MK-7)	MK-8 (H_2) or MK-8 (MK-7) MK-9 (H_2)	Not determined

[a] Data from Schleifer and Kandler (1972).
[b] Data from Buchanan and Gibbons (1974).
[c] Data from Lechevalier (1977).
[d] Data from Yamada et al. (1976).

common appearance of cells is the tetrad; however, single cells, pairs, or packets of eight commonly can be observed. In older cultures aberrant forms often can be found: large cells appearing singly or within packets, short chains, or irregular packets. Apparently this is due to an asynchronous cell division.

Due to the similar structure of the vegetative cells, *S. ureae* cannot be distinguished morphologically from other packet-forming aerobic cocci of the genus *Micrococcus*. In contrast to the species of *Micrococcus*, however, *S. ureae* is motile by flagella and forms endospores. Motility may be observed best when cultures have been grown in nutrient broth with 1% urea (Kocur and Martinec, 1963).

The spores of *S. ureae* are spherical and are located centrally within the mother cell (Fig. 1). Like other endospores, they contain dipicolinic acid (MacDonald and MacDonald, 1962; Thompson and Leadbetter, 1963). Their ultrastructure is similar to that of *Bacillus* endospores (Silva et al., 1973). Other significant properties that clearly separate *S. ureae* from morphologically similar bacteria are listed in Table 1. Although *S. ureae* is the only species known of the genus *Sporosarcina*, other species may exist. Therefore, the properties of new isolates should be compared with those of the type strain before they are identified as *S. ureae*.

Literature Cited

Beijerinck, M. W. 1901. Anhäufungsversuche mit Ureumbakterien. Ureumspaltung durch Urease und durch Katabolismus. Zentralblatt für Bakteriologie, Parasitenkunde und Infektionskrankheiten, Abt. 2 **7**:33–61.

Buchanan, R. E., Gibbons, N. E. (eds.). 1974. Bergey's manual of determinative bacteriology, 8th ed. Baltimore: Williams & Wilkins.

Gibson, T. 1935. An investigation of *Sarcina ureae*, a spore-forming, motile coccus. Archiv für Mikrobiologie **6**:73–78.

Goldman, M., Wilson, D. A. 1977. Growth of *Sporosarcina ureae* in defined media. FEMS Microbiology Letters **2**:113–115.

Kluyver, A. J., van Niel, C. B. 1936. Prospects for a natural system of classification of bacteria. Zentralblatt für Bakteriologie, Parasitenkunde und Infektionskrankheiten, Abt. 2 **94**:369–403.

Kocur, M., Martinec, T. 1963. The taxonomic status of *Sporosarcina ureae* (Beijerinck) Orla-Jensen. International Bulletin of Bacteriological Nomenclature and Taxonomy **13**:201–209.

Lechevalier, M. P. 1977. Lipids in bacterial taxonomy—a taxonomist's view. CRC Critical Reviews in Microbiology **5**:109–210.

Lehmann, K. B., Neumann, R. O. 1927. Bakteriologie, insbesondere bakteriologische Diagnostik, vol. 2. Munich: Lehmanns.

Löhnis, F. 1911. Landwirtschaftlich-bakteriologisches Praktikum. Berlin: Verlag von Gebrüder Bornträger.

MacDonald, R. E., MacDonald, S. W. 1962. The physiology and natural relationships of the motile, sporeforming sarcinae. Canadian Journal of Microbiology **8**:795–808.

Migula, W. 1900. System der Bakterien, vol. 2. Jena: Gustav Fischer Verlag.

Orla-Jensen, O. 1909. Die Hauptlinien des natürlichen Bakteriensystems. Zentralblatt für Bakteriologie, Parasitenkunde und Infektionskrankheiten, Abt. 2 **22**:305–346.

Pregerson, B. S. 1973. The distribution and physiology of *Sporosarcina ureae*. M.S. Thesis. California State University, Northridge, California.

Schleifer, K. H., Kandler, O. 1972. Peptidoglycan types of bacterial cell walls and their taxonomic implications. Bacteriological Reviews **36**:407–477.

Silva, M. T., Lima, M. P., Fonseca, A. F., Sousa, J. C. F. 1973. The fine structure of *Sporosarcina ureae* as related to its taxonomic position. Journal of Submicroscopical Cytology **5**:7–22.

Thompson, R. S., Leadbetter, E. R. 1963. On the isolation of dipicolinic acid from endospores of *Sarcina ureae*. Archiv für Mikrobiologie **45**:27–32.

Wood, E. J. F. 1946. The isolation of *Sarcina ureae* (Beijerinck) Löhnis from sea water. Journal of Bacteriology **51**:287–289.

Yamada, Y., Inouye, G., Tahara, Y., Kondo, K. 1976. The menaquinone system in the classification of aerobic Gram-positive cocci in the genera *Micrococcus*, *Staphylococcus*, *Planococcus*, and *Sporosarcina*. Journal of General and Applied Microbiology **22**:227–236.

SECTION T

The Coryneform Bacteria

Introduction to the Coryneform Bacteria

MICHAEL GOODFELLOW and DAVID E. MINNIKIN

The taxonomy of the coryneform bacteria is widely acknowledged to be unsatisfactory (Barksdale, 1970; Jensen, 1952, 1966; Jones, 1978; Keddie and Cure, 1977, 1978; Minnikin, Goodfellow, and Collins, 1978; Veldkamp, 1970). This opinion is not surprising, as Keddie (1978) pointed out, for there is no general agreement on what is meant by coryneform bacteria. The term is usually interpreted in a morphological sense and has been used to encompass both aerobic and anaerobic, Gram-positive, non-acid-fast, pleomorphic, nonbranching, rod-shaped bacteria that do not form spores. However, because of the difficulties associated with determining pleomorphism, a diverse array of poorly described Gram-positive bacteria that are not strictly coryneform have traditionally been associated with the coryneform group. This applies particularly to the genera *Erysipelothrix* and *Listeria*, which were classified in the family Corynebacteriaceae in both the sixth and seventh editions of *Bergey's Manual of Determinative Bacteriology* (Breed, Murray, and Hitchens, 1948; Breed, Murray, and Smith, 1957) but are currently listed as "Genera of Uncertain Affiliation" in the section of the eighth edition that includes the family Lactobacillaceae (Seeliger, 1974; Seeliger and Welshimer, 1974). These two genera can readily be distinguished from *Corynebacterium* and related taxa (Table 1): they are considered in detail elsewhere in this Handbook (Chapters 132 and 133).

In the eighth edition of *Bergey's Manual* (Rogosa et al., 1974), the genera *Arthrobacter, Cellulomonas, Corynebacterium,* and, "tentatively", *Kurthia* are classified in the section entitled "Coryneform Group of Bacteria" with *Brevibacterium* and *Microbacterium* listed as genera incertae sedis. In the introduction to the section, however, reference is made to a numerical taxonomic survey (Davis and Newton, 1969) in which representatives of 11 genera were referred to as coryneform bacteria. These included *Erysipelothrix, Jensenia, Listeria, Mycobacterium,* and *Nocardia* in addition to the taxa mentioned above. At one time or another, the term coryneform has also been applied to genera such as *Actinomyces, Arachnia, Bacterionema, Bifidobacterium, Eubacterium, Oerskovia, Propi-*

onibacterium, Rhodococcus, and *Rothia* (Bousfield, 1972; Bowden and Hardie, 1978; Jones, 1975, 1978; Keddie and Cure, 1977; Rogosa et al., 1974; Veldkamp, 1970; Yamada and Komagata, 1972a, b). Details about these organisms are to be found elsewhere in this Handbook (Chapters 132, 133, 143–149, and 155). Clearly, the designation coryneform bacteria has been interpreted in different ways.

Coryneform Bacteria—the Early Years

The genus *Corynebacterium* was proposed by Lehmann and Neumann (1896) in the first edition of their classical *Bacteriologische Diagnostic* to provide a home for the diphtheria bacillus. The genus was defined essentially in morphological terms to include Gram-positive, non-acid-fast, nonspore-forming, nonmotile bacteria that produced slender, often slightly curved or club-shaped rods and that were subsequently found to have a respiratory mode of metabolism. The genus was soon extended to accommodate bacteria that were associated with, or caused, disease in animals. Thus, the animal-associated strains known as *Bacillus der pseudodiphtheriae* (Loeffler, 1887) and *Pacinia neisseri* (Trevisan, 1889) were reclassified as *Corynebacterium pseudodiphtheriticum* (Lehmann and Neumann, 1896) and *Corynebacterium xerosis* (Lehmann and Neumann, 1899), respectively, whereas *Bacterium renale* (Migula, 1900), *Bacillus pseudotuberculosis* (Buchanan, 1911), and *Bacillus pyogenes* (Glage, 1903), the respective causative agents of bacillary pyelonephritis, caseous lymphadenitis, and pyogenic infections in domestic animals, became *Corynebacterium renale* (Ernst, 1905), *Corynebacterium pseudotuberculosis,* and *Corynebacterium pyogenes* (Eberson, 1918). It soon became apparent, however, that bacteria with the same general characteristics as corynebacteria occurred in habitats other than the animal body. Jensen (1934) proposed the species *Corynebacterium simplex* and *Corynebacterium tumescens* for some soil isolates, and Kisskalt and Berend (1918) reclassified

Bacterium erythrogenes and *Bacterium helvolum* (Lehmann and Neumann, 1896) as *Corynebacterium erythrogenes* and *Corynebacterium helvolum*, respectively.

Plant-pathogenic bacteria (this Handbook, Chapters 4 and 143) were added to the genus when Jensen (1934) reclassified *Aplanobacter insidiosum* (McCullock, 1925) and *Bacterium michiganense* (Smith, 1910) as *Corynebacterium insidiosum* and *Corynebacterium michiganense*, respectively. This trend continued with *Bacterium sepedonicum* (Spieckermann and Kotthoff, 1914) being renamed *Corynebacterium sepedonicum* (Skaptason and Burkholder, 1942); Dowson (1942) reclassified *Aplanobacter rathayi* (Smith, 1913), *Bacterium flaccumfaciens* (Hedges, 1922), and *Phytomonas fascians* (Tilford, 1936) as *Corynebacterium rathayi*, *Corynebacterium flaccumfaciens*, and *Corynebacterium fascians*, respectively. In 1948, Burkholder transferred the phytopathogenic taxa *Aplanobacter agropyri* (O'Gara, 1916), *Pseudomonas hypertrophicans* (Stahel, 1933), and *Pseudomonas tritici* (Hutchinson, 1925) to the genus *Corynebacterium*.

As a consequence of these and similar developments, the genus *Corynebacterium* became a repository for plant-pathogenic strains, for saprophytes from habitats such as soil, water, milk, and dairy products (Clark, 1952; Jensen, 1952, 1966), as well as for the human and animal species. The unifying factors for this heterogeneous collection of bacteria were a few more-or-less distinctive morphological properties and staining reactions. Most of the organisms were aerobes and facultative anaerobes, but a few anaerobes were also included. The classification of so many different bacteria in a single genus prompted Conn (1947) to complain: "There has been a recent tendency to include a greater and greater variety of organisms in *Corynebacterium*, until one can almost say that, if all these forms are included, there is no reason for excluding any Gram-positive, nonsporeforming rods, except the lactobacilli."

In an attempt to resolve this difficulty, Jensen (1952) raised the possibility of limiting the genus *Corynebacterium* to the human and animal strains. This suggestion, while attractive, posed the problem of what to do with the saprophytic and plant-pathogenic strains that fell into the genus as defined in morphological terms. Clark's (1952) suggestion that some of the saprophytic corynebacteria could be accommodated in the genera *Arthrobacter* and *Corynebacterium* was considered premature by those (Gibson, 1953; Jensen, 1952) who felt that such taxa were inadequately circumscribed. Jensen (1952) endeavored to solve the dilemma by referring to the animal parasitic and pathogenic species as *Corynebacterium sensu stricto* and to the "de juro" taxon as *Corynebacterium sensu lato*. Jensen suggested the reintroduction of the term "coryneform bacteria" (Oerskov, 1923) for strains classified in the latter. He also considered that the genus *Microbacterium* was closely related to *Corynebacterium sensu stricto*.

Corynebacterium sensu lato and Related Taxa

The concept of coryneform bacteria *sensu* Jensen was primarily morphological in character. It included bacteria with distinct morphological features, such as the angular arrangements often referred to as V-formations (see this Handbook, Chapter 142). Since morphological criteria were heavily weighted in the taxonomies of the time, the coryneform bacteria were considered to form a compact group of closely related organisms. This view persisted for a number of years, even though many investigators found it difficult, if not impossible, to distinguish between *Corynebacterium sensu lato* and strains with more or less similar morphological characters and staining properties classified as *Arthrobacter*, *Brevibacterium*, *Cellulomonas*, and *Microbacterium*. Since the descriptions of these taxa relied heavily on a few morphological properties, coryneform bacteria proved difficult to separate both from one another and from noncoryneform organisms classified in genera such as *Erysipelothrix*, *Listeria*, and *Kurthia*. With the advent of modern taxonomic methods, the intra- and intergeneric relationships of coryneform and traditionally associated taxa have begun to emerge. In particular, valuable data have been provided by numerical phenetic (Jones, 1978), wall (Cummins and Harris, 1956, 1958; Keddie and Cure, 1977, 1978; Schleifer and Kandler, 1972), lipid (Bowie et al., 1972; Minnikin, Goodfellow, and Collins, 1978), and deoxyribonucleic acid analysis (Crombach, 1978b).

It is premature to claim that the coryneform bacteria are well classified but, as can be seen below, an acceptable outline of the major groups is emerging. One of the benefits of improved classification is that good characters can be weighted for the separation and differentiation of taxa. Chemical characters are proving to be particularly useful in this respect (see Table 1).

Arthrobacter

The difficulty of distinguishing between coryneform bacteria and "related" bacteria was aptly demonstrated by Conn and Dimmick (1947). They sent cultures of *Bacterium globiforme*, a soil organism described by Conn (1928), both to Jensen and Krasilnikov. The former identified it to the genus

Corynebacterium and the latter to the genus *Mycobacterium;* but Conn and Dimmick considered that the organism was distinct from all previously named bacteria and revived the genus name *Arthrobacter* (Fischer, 1895) for it. The genus accommodated highly aerobic, nutritionally exacting soil bacteria that liquefied gelatin slowly and exhibited a characteristic growth cycle: irregular, Gram-negative rods were replaced, in older cultures, by Gram-positive forms which, when transferred to fresh medium, gave rise to irregular rods again. In addition to the type species, *Arthrobacter globiformis,* the genus contained two species classified by Jensen (1934) as *Corynebacterium helvolum* and *Corynebacterium tumescens.* The genus was subsequently extended to include additional species but, because of its poor circumscription, it was not widely accepted until it was classified in the family Corynebacteriaceae (Breed, Murray, and Smith, 1957).

Arthrobacter, as currently constituted (Keddie, 1974a), is heterogeneous and can be divided into three species groups: the *globiformis, simplex/tumescens,* and *terregens/flavescens* groups (Table 1; this Handbook, Chapter 142). Organisms so classified can be distinguished on the basis of the composition and structure of the cell wall peptidoglycan (Cummins and Harris, 1959; Keddie and Cure, 1977; Schleifer and Kandler, 1972), lipid patterns (Collins, Goodfellow, and Minnikin, 1979; Minnikin, Goodfellow, and Collins, 1978), and by significant differences in DNA base composition (Skyring and Quadling, 1970; Yamada and Komagata, 1970). Attention should also be paid to bacteria, e.g., *Arthrobacter nicotianae,* that can be differentiated from other arthrobacters (Table 1; Minnikin, Goodfellow, and Collins, 1978) because they have peptidoglycans based on L-lysine with the A4α variation according to Schleifer and Kandler (1972) and contain fully unsaturated menaquinones (Yamada et al., 1976). The genus *Arthrobacter* should probably be restricted to strains *(Arthrobacter sensu stricto)* that possess peptidoglycans based on L-lysine with the A3α variation (Schleifer and Kandler, 1972) and that share many phenotype properties in common (see this Handbook, Chapter 142).

Cellulomonas

The genus *Cellulomonas* (Bergey et al., 1923) was established to accommodate a number of soil isolates that were described as Gram-negative, motile or nonmotile, short rods that can degrade cellulose. The genus received little attention until Clark (1951, 1952) showed that cellulomonads had a coryneform morphology; he prepared a revised description of the taxon and classified it in the family Corynebac-

teriaceae. In a later study, Clark (1953) reported that the organisms were able to produce acid from carbohydrates and to slowly liquefy gelatin, and he recognized only 10 of the original 31 species. In the eighth edition of *Bergey's Manual* (Keddie, 1974b), only a single species, *Cellulomonas flavigena,* is described.

The integrity of the genus *Cellulomonas* is supported by a wealth of taxonomic data (this Handbook, Chapter 142), which also reveal a close similarity between cellulomonads and organisms classified in the genus *Oerskovia* (Prauser, Lechevalier, and Lechevalier, 1970). Cellulomonads and oerskoviae share a high numerical phenetic similarity (Jones, 1975; Jones and Bradley, 1964), possess MK-9 (H₄) as the predominant menaquinone (Collins, Goodfellow, and Minnikin, 1979), and have a similar fatty acid and polar lipid composition (Minnikin, Collins, and Goodfellow, 1979). Although differences in morphology, growth habits, and wall composition have been reported between strains of *Cellulomonas* and *Oerskovia* (Keddie and Cure, 1977; Schleifer and Kandler, 1972; Sukapure et al., 1970), additional systematic studies are required to determine the exact relationships between these taxa. Since evidence exists of a relationship between *Oerskovia* and the genus *Promicromonospora* (Lechevalier 1972; Lechevalier, de Bièvre, and Lechevalier, 1977), representatives of the latter should also be included in the future comparative studies.

Curtobacterium

This genus was created by Yamada and Komagata (1972a) for certain motile *Brevibacterium* spp. (this Handbook, Chapter 142), with *Curtobacterium (Brevibacterium) citreum* as the type species, and the phytopathogens (this Handbook, Chapter 143) *Corynebacterium flaccumfaciens* and *Corynebacterium poinsettiae.* Curtobacteria were characterized by the presence of ornithine in the cell wall, by DNA base ratios within the range of 66–71 mol%, and by slow and weak formation of acid from some sugars. From its inception, therefore, the taxon was unlike other coryneform genera because it was homogeneous with respect to wall composition. Subsequent studies have shown that curtobacteria possess a unique peptidoglycan type (B2β, Schleifer and Kandler, 1972) and usually contain fully unsaturated menaquinones (MK-9) (Collins, Goodfellow, and Minnikin, 1979; Yamada et al., 1976). Strains labeled *"Brevibacterium"* *testaceum* and some named *"Brevibacterium"* *helvolum* have walls based on ornithine (Yamada and Komagata, 1972a) but require further detailed study (Minnikin, Collins, and Goodfellow, 1978) especially since *"Brevibacterium"* *testaceum* has MK-11.

Table 1. Distribution of chemical characters in coryneform and associated bacteria that lack mycolic acids.[a]

Wall diamino acid[b]	Peptidoglycan structure[c]	Taxon	Wall sugars[d]	G+C (mol %)	Fatty acids[e]	Isoprenoid quinones[f]	Phospholipids[g]	Glycolipids[h]
meso-DAP	ND	Corynebacterium autotrophicum[i]	ND	ND	S, U, C	Q-10,10(H$_2$)	PE, PG	ND
	Direct link	Brochothrix thermosphacta	—	36	S, I-A	MK-7	PE, PG	DiMaDAG
	Direct link	Listeria monocytogenes	—	38	S, I, A	MK-7	PE, PG	DiGaDAG
	Direct link	Brevibacterium linens	ga, gl, ri	60–69	S, U, I, A	MK-8(H$_2$)	PG(PI, PIM)	ND
	ND	Propionibacterium freudenreichii	ga, ma, rh	65–66	S, U, I, A	MK-9(H$_1$)	PIM	DAIMa
L, L-DAP	ND	P. acnes[j]	gl, ma, (ga)	57–60	S, L, A	ND	ND	ND
	ND	P. avidum[j]	gl, ma, (ga)	62–63	S, I[k]	ND	ND	ND
	ND	P. granulosum	ga, ma, (gl)	62–64	S, I[k]		ND	ND
	A3γ Gly	P. acidi-propionici[l]	ga, gl, ma	66–67	S, L, A	MK-9(H$_4$)	PG(PIM)	DAIMa
	A3γ Gly$_3$	Arthrobacter simplex	ga, gl, ma	72–76	S, U, L, A, T	MK-8(H$_4$)	PG, PIM	ND
	A3α L-Ala[m]	A. tumescens	ga, gl, ma	70–76	S, U, I, A	MK-8(H$_4$)	PG, PI(PIM)	ND
	A3α Ala, Glu	Arthrobacter sensu stricto	ga[m]	61–73	S, U, I, A	MK-9(H$_2$)	ND	ND
	A4α D-Glu	A. nicotianae	ga, gl	60–66	S, U, I, A	MK-8,9	PG, PI, PIM	ND
	A4α D-Asp	"Brevibacterium" sulfureum	ga, gl, ma	66–70	S, U, I, A	MK-9,10	ND	ND
		Kurthia zopfii	—	38–39	ND	MK-7	PG	ND
L-Lys	B1α	Microbacterium lacticum	ga, rh	63–70	ND	MK-10,11	ND	DiMaDAG
	B1	Erysipelothrix rhusiopathiae	gl, rh	38–40	S, U, I, A	—	ND	ND
	ND	Corynebacterium pyogenes	ga	58	S, I, A	ND	ND	ND
	ND	Rothia dentocariosa	ga	65–70	S, U, I, A	ND	ND	DiMaDAG
	ND	Oerskovia	ga, gl	70–75	S, I, A	MK-9(H$_1$)	(PG), PI, PIM[n]	ND
	ND	Promicromonospora	ma, rh, ri[o]	73–74	S, I, A	MK-9(H$_4$)	PG, PI, PIM[n]	ND
L-Orn	A4β D-Asp[o]	Cellulomonas flavigena	ga[p]	73	S, I, A	MK-9	(PG), PI, PIM[n]	ND
D-Orn	B2β-[L-Hsr]-D-Glu	Curtobacterium	—	68–73	S, U, I, A	MK-9	PG, (PI, PIM)	—
	ND	"Brevibacterium" testaceum	—	65	ND	MK-9	ND	ND
? Orn	ND	"Brevibacterium" helvolum	—	66	ND	MK-11	PG	ND
D-DAB	B2γ-[L-DAB]-D-Glu	Corynebacterium aquaticum[q]	ga[p]	67–78	S, U, I, A	MK-10,11	PG(PI, PIM)	DiMaDAG

[a] Data from Collins, Goodfellow, and Minnikin, 1979; Collins et al., 1979; Collins et al., 1979; Cummins and Johnson (this Handbook, Chapter 145); Lechevalier, 1977; Lechevalier, de Bièvre, and Lechevalier, 1977; Minnikin, Collins, and Goodfellow, 1978, 1979; Minnikin, Goodfellow, and Collins, 1978; Schleifer and Kandler, 1972; Thiele and Thiele, 1977; Walther-Mauruschat et al., 1977; Yamaguchi, 1965, 1967. ND, not determined; —, no characteristic component.

[b] Abbreviations: DAP, diaminopimelic acid; Lys, lysine; Orn, ornithine; DAB, diaminobutyric acid.

[c] According to Schleifer and Kandler (1972).

[d] Brackets indicate variable occurrence. Abbreviations: ga, galactose; gl, glucose; ma, mannose; rh, rhamnose; ri, ribose.

[e] Main components underlined. Abbreviations: S, straight chain; U, monounsaturated; C, cyclopropane; I, iso; A, anteiso; T, tuberculostearate (10-methyloctadecanoate).

[f] Abbreviations exemplified by Q-10, ubiquinone with ten isoprene units; MK-9(H₄), menaquinone with two of the nine isoprene units hydrogenated.

[g] Characteristic phospholipids occurring in addition to diphosphatidylglycerol. Phosphatidylethanolamine, (PE); phosphatidylglycerol (PG), phosphatidylinositol (PI), and phosphatidylinositol mannosides (PIM). Brackets indicate uncertain occurrence.

[h] Characteristic glycolipids. Abbreviations: DiMaDAG, dimannosyl diacylglycerol; DigaDAG, digalactosyl diacylglycerol; DAIMa, diacyl inositol mannoside.

[i] Similar ubiquinones are found in Corynebacterium nephridii, "Mycobacterium" flavum, Mycoplana ruber, and Protaminobacter ruber, and Q-9 is found in Brevibacterium leucinophagum (Collins, Goodfellow, and Minnikin, 1979; Minnikin, Goodfellow, and Collins, 1978).

[j] Some strains have meso-DAP (Cummins and Johnson, this Handbook, Chapter 145).

[k] Anteiso acids not distinguished from iso acids (Saino et al., 1976).

[l] Similar wall sugars, G+C content (mol%), and fatty acids in Propionibacterium jensenii and Propionibacterium thoenii (Cummins and Johnson, this Handbook, Chapter 145; Minnikin, Goodfellow, and Collins, 1978).

[m] Additional characteristic amino acid sugars in several species (see Minnikin, Goodfellow, and Collins, 1978).

[n] Strains of Oerskovia and Promicromonospora found by Lechevalier, de Bièvre, and Lechevalier (1977) to contain a characteristic glucosamine lipid. Phosphoglycolipids not clearly identified as PIMs and no glycolipid detected in representatives of Oerskovia and Cellulomonas (Minnikin, Collins, and Goodfellow, 1979).

[o] Aspartic acid replaced by D-glutamic acid in many strains and other characteristic sugars found (see Minnikin, Goodfellow, and Collins, 1978).

[p] Depending on strain, other characteristic sugars occur in addition to galactose (see Minnikin, Goodfellow, and Collins, 1978).

[q] Similar data recorded for Corynebacterium insidiosum, Corynebacterium michiganense, and Corynebacterium sepedonicum (see Minnikin, Goodfellow, and Collins, 1978).

Brevibacterium

This taxon was established by Breed (1953), with *Brevibacterium linens* as type species, for certain Gram-positive, nonsporeforming rods previously classified in the genus *Bacterium. Brevibacterium* was recognized in the seventh edition of *Bergey's Manual* (Breed, Murray, and Smith, 1957) where, along with *Kurthia,* it formed the family Brevibacteriaceae. The poor circumscription of the genus resulted in its becoming a dumping ground for a diverse collection of poorly described coryneform bacteria and to its being incertae sedis in the eighth edition of *Bergey's Manual* (Rogosa and Keddie, 1974a).

Chemical and numerical phenetic surveys confirm the heterogeneity of the genus and indicate that *Brevibacterium linens* forms a distinct taxon (see Jones, 1978; Keddie and Cure, 1978; Minnikin, Goodfellow, and Collins, 1978). *B. linens* strains are obligately aerobic, contain *meso*-diaminopimelic acid (*meso*-DAP) and galactose and glucose in the wall, lack mycolic acids, possess dihydrogenated menaquinones with eight isoprene units, have DNA base ratios within the range of 66–71 mol%, and liquefy gelatin (Table 1; this Handbook, Chapter 142). On the basis of these and other properties, there are strong grounds for supporting the view that this species could form the nucleus of a redefined genus *Brevibacterium* (Keddie and Cure, 1977; Yamada and Komagata, 1972b), but, at present, as the only species. Most of the existing species of *Brevibacterium* can be assigned to other genera that contain coryneform bacteria, notably *Corynebacterium sensu stricto, Curtobacterium,* and *Rhodococcus* (Collins, Goodfellow, and Minnikin, 1979; Goodfellow, Collins, and Minnikin, 1976; Keddie and Cure, 1977; Yamada and Komagata, 1972a), but the position of strains labeled *Brevibacterium helvolum, Brevibacterium sulfureum,* and *Brevibacterium testaceum* is equivocal (Table 1).

It is now clear that *Brevibacterium linens* and *Kurthia* strains have little in common; they can readily be separated using chemical criteria (Table 1). Kurthiae should not, in fact, be considered with the established coryneform taxa (see this Handbook, Chapter 144) for they do not have a coryneform morphology (Keddie and Rogosa, 1974) and they do have a DNA base composition well below that of coryneform bacteria (see Minnikin, Goodfellow, and Collins, 1978).

Microbacterium

Orla-Jensen (1919) proposed the genus *Microbacterium* for an assortment of Gram-positive, nonspore-forming rods that were heat resistant and produced lactic acid from sugars. Three of the original four species, the type species *Microbacterium lacticum, Microbacterium flavum,* and *Microbacterium liquefaciens,* were found to be coryneforms (see Jensen, 1934; Orla-Jensen, 1919). A fifth species, *Microbacterium thermosphactum* (McLean and Sulzbacher, 1953), was described for strains that neither were heat resistant nor possessed a coryneform morphology. The genus was classified in the family Corynebacteriaceae in the seventh edition of *Bergey's Manual* (Breed, Murray, and Smith, 1957), but is presently considered to be *incertae sedis* (Rogosa and Keddie, 1974b).

Modern taxonomic methods have clearly shown that the composition of the genus *Microbacterium* is unsatisfactory, but they have helped to clarify relationships between the constituent species. Thus, numerical and chemical data support the transfer of *Microbacterium flavum* to *Corynebacterium sensu stricto* as *Corynebacterium flavum,* and there are grounds for considering an affinity between *Microbacterium liquefaciens* and *Curtobacterium* (see this Handbook, Chapter 142). Most of the numerical taxonomic and other studies have indicated that *Microbacterium thermosphactum* strains are quite different from other microbacteria (Collins, Goodfellow, and Minnikin, 1979; Collins-Thompson et al., 1972; Jones, 1975, 1978; Schleifer and Kandler, 1972) and, in fact, show a closer relationship to lactic acid bacteria (see Jones, 1978). Sneath and Jones (1976) have recommended that *Microbacterium thermosphactum* be reclassified in the new genus *Brochothrix* as *Brochothrix thermosphacta.* These proposals leave *Microbacterium lacticum* as a distinct and recognizable taxon on the basis of numerical (Jones, 1975), nonnumerical (Jayne-Williams and Skerman, 1966), and chemical findings (Collins, Goodfellow, and Minnikin, 1979; Minnikin, Goodfellow, and Collins, 1978; Schleifer, 1970). These data support the view that *Microbacterium lacticum* could form the nucleus of a redefined genus *Microbacterium* (Jones, 1975; Keddie and Cure, 1978; this Handbook, Chapter 142)

Unassigned Species

The taxonomic position of a number of coryneform bacteria remains unresolved. Notable amongst these are strains, which include the phytopathogenic species *Corynebacterium insidiosum, Corynebacterium michiganense,* and *Corynebacterium sepedonicum* (see this Handbook, Chapter 141), having peptidoglycans based on D-diaminobutyric acid (DAB) (Table 1; Keddie and Cure, 1978; Schleifer and Kandler, 1972). Schleifer and Kandler (1972) noted that, despite differences in the diamino acid present, there is considerable similarity in the peptidoglycan structure of the phytopathogens that contain ornithine and DAB and *Microbacterium lacticum,* suggesting a possible relationship between these taxa. This observation is of particular interest since limited lipid analyses performed so far do not clearly distinguish the three DAB-containing plant pathogenic species from those included in the genus *Curtobacterium* (Minnikin, Goodfellow, and Collins, 1978).

Another classification niche is also required for strains labeled *Corynebacterium haemolyticum* and *Corynebacterium pyogenes*. These organisms contain lysine, rhamnose, and glucose in their walls (Cummins and Harris, 1956) and have been clearly separated from other coryneform taxa in numerical phenetic surveys (see Jones, 1978).

Corynebacterium autotrophicum is the best-studied example of a group of organisms (Table 1) that contain ubiquinones, Q-*10,10*(H₂), rather than menaquinones (Collins, Goodfellow, and Minnikin, 1979; Minnikin, Collins, and Goodfellow, 1978; Thiele and Thiele, 1977). *C. autotrophicum* is now considered to have a wall typical of Gram-negative bacteria (Aragno et al., 1977; Walther-Mauruschat et al., 1977), but the other related strains labeled *Corynebacterium nephridii*, *"Mycobacterium" flavum*, *Mycoplana ruber*, and *Protaminobacter ruber* have not been studied in such detail (Collins, Goodfellow, and Minnikin, 1979; Minnikin, Goodfellow, and Collins, 1978). Further studies will be necessary to determine the homogeneity of this group and its classification. An organism labeled *Brevibacterium leucinophagum* has ubiquinones (Q-9) slightly different from those of *C. autotrophicum* (Collins, Goodfellow, and Minnikin, 1979), and its transfer to the genus *Acinetobacter* has been proposed (Jones and Weitzman, 1974).

Corynebacterium sensu stricto and Related Taxa

The difficulties associated with distinguishing between the human and animal parasites of the genus *Corynebacterium* and bacteria currently classified in the genera *Mycobacterium* (see this Handbook, Chapters 150 and 151), *Nocardia*, and *Rhodococcus* (see this Handbook, Chapter 155) have bedeviled taxonomists for many years. As early as 1888, Nocard established that the actinomycete subsequently known as *Nocardia farcinica* resembled *Corynebacterium diphtheriae* in laboratory culture, and Lehmann and Neumann (1896) noted a similarity between *C. diphtheriae* and *Mycobacterium tuberculosis*. In his extensive morphological observations on actinomycetes, Oerskov (1923) found that some strains (his Group IIb), after fragmentation of the primary mycelium, continued to divide in an "angular" fashion reminiscent of corynebacteria and mycobacteria. Oerskov was compelled to conclude "... correspondence in morphology is so great that it is not possible to set up any definite boundary between the bacteria (corynebacteria and mycobacteria) examined and those ray fungi (nocardioform bacteria) which exhibited the angular arrangement of the elements. ..." Jensen (1931) formalized the apparent relationship between these organisms by proposing the family Proactinomycetaceae for the genera *Corynebacterium*, *Mycobacterium*, and *Proactinomyces* (*Actinomyces* and *Nocardia*). The label *Proactinomyces* was gradually discontinued by the establishment of the priorities of the names *Actinomyces* and *Nocardia* for the anaerobic and aerobic actinomycetes, respectively (Waksman and Henrici, 1943). The genus *Proactinomyces* is, however, occasionally still used, mainly by investigators who also use the name *Actinomyces* as opposed to the widely accepted designation *Streptomyces*.

Jensen (1931) distinguished corynebacteria and mycobacteria from proactinomycetes and streptomycetes by their inability to produce an "initial mycelium" like the former or spores like the latter. These morphological divisions were unable to stand the test of time, and in 1953 Jensen conceded that nocardiae could occasionally be less mycelial than mycobacteria and that some mycobacteria produced a rudimentary aerial mycelium.

Actinomycetes can be classified into two broad groups, the nocardioform bacteria and the sporoactinomycetes (Prauser, 1978; Prauser and Bergholz, 1974; this Handbook, Chapter 147 and Chapter 155). The former can be defined as branching actinomycetes that reproduce by fragmentation of all or of accidentally involved parts of their hyphae into bacilli and coccoid elements; the latter exhibit a great morphological diversity that includes the formation of spores on or in definite parts of the mycelium. Sporoactinomycetes and nocardioform bacteria are usually easy to distinguish, but the morphological differences within the coryneform-nocardioform area tend to be ones of degree rather than kind and are clearly unsuitable for classification (see Bousfield and Goodfellow, 1976; Keddie, 1978). The production of nocardioform mutants from a coryneform bacterium (Jičínská, 1973) underlines the problem of using morphological features for the classification of these organisms.

Modern taxonomic methods have also further resolved the relationships within *Corynebacterium sensu stricto* and among it and related taxa. In particular, numerical phenetic surveys have led to marked improvements in the subgeneric classification of the genera *Mycobacterium* (see this Handbook, Chapters 150 and 151) and *Nocardia* (Goodfellow and Minnikin, 1977, 1978) and have helped in the establishment of new taxa. A particularly important advance was the recognition that actinomycetes variously known as *"Mycobacterium" rhodochrous,* the "*rhodochrous*" complex, or the "*rhodochrous*" taxon formed an aggregate cluster defined at the same similarity level as clusters equated with the genera *Mycobacterium* and *Nocardia* (see Bousfield and Goodfellow, 1976). In a subsequent numerical phenetic analysis (Goodfellow and Alderson, 1977), the "*rhodochrous*" cluster was given generic status and the name *Rhodococcus* (Zopf, 1891) was considered to have priority over *Proactinomyces* (Bradley and Bond, 1974; Jensen, 1931), *Jensenia* (Bisset and Moore, 1950), and *Gordona* (Tsukamura 1971). Nine species were recognized in addition to the type species, *Rhodococcus rhodochrous* (Table 2; this Handbook, Chapter 155). The type strain of *Gordona aurantiaca* examined by Goodfellow and Alderson (1977) was loosely associated with, but not included in, the *Rhodococcus* cluster, a result in line with earlier reports (Tsukamura, 1974, 1975; Tsukamura, Mizuno, and Murata, 1975). In a further numerical taxonomic study, strains of *Gordona aurantiaca* formed a homogeneous cluster defined at the same level of similarity as clusters equated with *Mycobacterium, Nocardia,* and *Rhodococcus* (Goodfellow et al., 1978). Since the name *Gordona* is no longer valid, the new cluster was provisionally called the "*aurantiaca*" taxon.

In an extensive review of the impact of numerical taxonomy on the classification of coryneform bacteria, Jones (1978) found that most strains formed defined clusters that could be equated with the rank of genus. One such group contains *Corynebacterium diphtheriae,* related animal parasites, *Microbacterium flavum, Corynebacterium glutamicum,* and related saprophytic bacteria, but not *Corynebacterium haemolyticum, Corynebacterium pyogenes,* or the plant pathogenic corynebacteria (Bousfield, 1972; Jones, 1975). The organisms in this taxon were equated with *Corynebacterium sensu stricto* and can be distinguished from clusters corresponding to coryneform taxa such as *Arthrobacter sensu stricto, Brevibacterium linens, Cellulomonas,* and *Microbacterium lacticum* as well as to *Mycobacterium* and *Nocardia* (Bousfield, 1972; Holmberg and Hallander, 1973; Jones, 1975).

The relationship between *Corynebacterium sensu stricto* and *Rhodococcus* has never been clearly established using numerical taxonomic methods. Traditionally, rhodococci have been compared with either nocardioform or coryneform bacteria, and a representative sample has never been included in the few surveys that have compared representatives of each of these broad groups. Strains labeled *Corynebacterium equi* may form a link between rhodococci and true corynebacteria since they have clustered both with *Corynebacterium sensu stricto* (Bousfield, 1972; Davis and Newton, 1969) and with *Rhodococcus* (Goodfellow, 1971; Goodfellow and Alderson, 1977; Goodfellow et al., 1978; Jones, 1975).

Clearly, rhodococci and true corynebacteria should be the subject of a systematic, numerical phenetic analysis which should also include representatives of the genus *Caseobacter* (Crombach, 1978a).

Chemical methods have been especially successful in clarifying the relationship between *Corynebacterium sensu stricto* and related taxa (Table 2). Wall sugar and amino acid analyses provided the first clear evidence of a relationship between true corynebacteria, mycobacteria, and nocardiae (Cummins and Harris, 1958), all of which have a wall chemotype IV (*sensu* Lechevalier and Lechevalier, 1970), i.e., strains contain major amounts of *meso-*DAP, arabinose, and galactose. This relationship has been confirmed by subsequent chemical studies (see Barksdale, 1970; Keddie and Cure, 1977, 1978; Schleifer and Kandler, 1972), which have also shown that representatives of *Bacterionema, Caseobacter, Rhodococcus,* and the "*aurantiaca*" taxon belong to the same wall chemotype (Crombach, 1978a; Goodfellow and Minnikin, 1978; Goodfellow et al., 1978).

Results of lipid analyses also indicate that a close relationship exists between *Corynebacterium sensu stricto* and the taxa with a wall chemotype IV mentioned above (Table 2; Minnikin and Goodfellow, 1976; Minnikin, Goodfellow, and Collins, 1978). In particular, mycolic acids, long-chain 2-alkyl-branched 3-hydroxy acids, are found in representatives of *Caseobacter, Corynebacterium sensu stricto, Bacterionema, Mycobacterium, Nocardia, Rhodococcus,* and the "*aurantiaca*" taxon (Crombach, 1978a; Goodfellow et al., 1978; Minnikin, Goodfellow, and Collins, et al., 1978); they are only found in bacteria with a wall chemotype IV (Goodfellow and Minnikin, 1977). The differences in the overall size of mycolic acids (Table 2) provides the basis of several chemical methods (see this Handbook, Chapters 150 and 151) which help to distinguish taxa that contain mycolic acid (Goodfellow et al., 1978; Minnikin, Goodfellow, and Collins, 1978).

The mycolic acids from representatives of *Corynebacterium sensu stricto* and *Bacterionema* are relatively small in size (22–38 carbons), whereas those from *Mycobacterium* are at the other end of the scale (60–90 carbons). Within these limits, the mycolic acids of the other taxa form an overlapping series ranging between 34 and 74 carbons. Relatively homogeneous groups of mycolic acids are found in *Nocardia* (46–60 carbons) and the "*aurantiaca*" taxon (68–74 carbons), but considerable variation exists in the chain lengths of rhodococci (34–64 carbons). The mycolic acids of *Caseobacter* strains have not been characterized chemically, but it has been suggested that they are similar to those of true corynebacteria (Crombach, 1978a). Mycolic acid data suggest that some corynebacteria, such as *Corynebacterium fascians*

Table 2. Distribution of chemical characters in coryneform and related bacteria that possess mycolic acids and major amounts of *meso*-diaminopimelic acid, arabinose, and galactose in their cell walls.[a]

G+C (mol%)	Mycolic acids[b]		Taxon	Major menaquinone[c]	Phospholipids[e]	Fatty acids[d]	Glycolipids[f]
	Ester released on pyrolysis	Overall size					
65–67	+[g]	+[g]	*Caseobacter*	ND	ND	ND	ND
	8:0, 10:0[h]	22–32[h]	*Corynebacterium bovis*	MK-9(H$_2$)	PG	S, U, T	+
51–59			*Bacterionema matruchotii*	MK-9(H$_2$)	PG	S, U	+
	14:0–18:0[h] (14:1–18:1)	26–38[h]	*Corynebacterium glutamicum*	MK-9(H$_2$)	ND	ND[i]	ND
			Corynebacterium diphtheriae	MK-8(H$_2$)	PG	S, U	+
59–69	12:0–18:0[h]	34–52[h]	*Rhodococcus coprophilus, R. equi, R. erythropolis, R. rhodnii, R. rhodochrous, R. ruber*	MK-8(H$_2$)			
		48–64[h]	*R. bronchialis, R. corallinus,*	MK-9(H$_2$)			
		38–64[h]	*R. rubropertinctus, R. terrae*	MK-8,9(H$_2$)	PE	S, U, T	+
64–69	12:0–18:0; 20:1, 22:1; (20:0, 22:0)	46–60	*Nocardia asteroides*	MK-8(H$_4$)			
ND		68–74	"*aurantiaca*" taxon	MK-9			
62–70	22:0, 26:0	60–90	*Mycobacterium*	MK-9(H$_2$)			

[a] Data from Alshamaony et al., 1977; Collins, Goodfellow, and Minnikin, 1979; Collins et al., 1979; Crombach, 1978a; Goodfellow and Minnikin, 1977, 1978; Goodfellow et al., 1978; Lechevalier, 1977; Lechevalier, de Bièvre, and Lechevalier, 1977; Minnikin, Collins, and Goodfellow, 1978; Minnikin and Goodfellow, 1976. Goodfellow et al., 1978; Minnikin, Goodfellow, and Alshamaony, 1978; Minnikin et al., 1978; Minnikin et al., 1977. ND, not determined.

[b] Abbreviations for long-chain esters exemplified by 18:0, octadecanoate; 18:1, octadecenoate. Overall size of mycolic acids is given as number of carbons.

[c] Abbreviations exemplified by MK-9, menaquinone with nine isoprene units; MK-9(H$_2$), menaquinone with one of the nine isoprene units hydrogenated.

[d] Abbreviations: S, straight chain; U, monounsaturated; T, tuberculostearate (10-methyloctadecanoate).

[e] Characteristic phospholipids, phosphatidylethanolamine (PE) and phosphatidylglycerol (PG), occurring in addition to diphosphatidylglycerol, phosphatidylinositol, and phosphatidylinositol mannosides.

[f] Uncharacterized polar glycolipids (Minnikin et al., 1977); trehalose mycolates (cord factors) also expected (Minnikin, Collins, and Goodfellow, 1978).

[g] Structural details not determined (Crombach, 1978a).

[h] Including additional unpublished results of D. E. Minnikin, M. Goodfellow, M. D. Collins, and L. Alshamaony.

[i] Other glutamic acid–producing organisms, e.g., "*Brevibacterium*" *ammoniagenes*, contain mainly straight-chain and unsaturated fatty acids (Minnikin, Collins, and Goodfellow, 1978).

and *Corynebacterium rubrum,* might be more appropriately classified in the genus *Rhodococcus,* but the chain lengths of mycolic acids from representatives of *Corynebacterium equi* and *Corynebacterium hydrocarboclastus* overlap those of true corynebacteria and rhodococci (Minnikin, Goodfellow, and Collins, 1978). It is not possible at present to distinguish between *Corynebacterium sensu stricto, Bacterionema,* and *Rhodococcus* solely on the basis of mycolic acid analyses.

True corynebacteria and bacterionemae, however, can be distinguished from all other mycolic acid–containing bacteria on the basis of DNA base composition. *Mycobacterium, Nocardia, Rhodococcus,* and *Caseobacter* strains are all rich in guanine and cytosine, fall within the range 59–69 mol%, and can thereby be separated from *Corynebacterium sensu stricto* and *Bacterionema* which exhibit G+C values between 51 and 59 mol% (Table 2). DNA base composition studies may be of some value in identifying strains with mycolic acids intermediate in size between those of true corynebacteria and rhodococci. Reported values for *Corynebacterium equi* and *Corynebacterium hydrocarboclastus* support their inclusion in the genus *Rhodococcus* (Goodfellow and Alderson, 1977).

The nonhydroxylated fatty acids of representatives of all of the mycolic acid-containing taxa are similar; they are mainly of the straight-chain, unsaturated type (Minnikin, Goodfellow, and Collins, 1978). There is, however, some evidence that the distribution of 10-methyloctadecanoic acid (tuberculostearic acid) may help to distinguish rhodococci from most true corynebacteria and bacterionemae (Table 2). Thus, *Corynebacterium bovis* and *Rhodococcus* strains are reported to contain tuberculostearic acid, whereas bacterionemae and the remaining corynebacteria do not (see this Handbook, Chapter 155). Representatives of *Corynebacterium equi* also contain tuberculostearic acid (Kroppenstedt and Kutzner, 1978) and so are associated with the genus *Rhodococcus* (Goodfellow and Alderson, 1977).

The polar lipid patterns of mycolic acid–containing bacteria are also broadly similar (Table 2; Minnikin and Goodfellow, 1978). Thus, *Mycobacterium, Nocardia, Rhodococcus,* including *Rhodococcus (Corynebacterium) equi,* and "*aurantiaca*" strains may be expected to contain diphosphatidylglycerol, phosphatidylethanolamine (PE), phosphatidylinositol, and phosphatidylinositol mannosides, but PE is absent in extracts from true corynebacteria and bacterionemae (Minnikin et al., 1977). Glycolipids have not been systematically investigated, but representatives of *Corynebacterium sensu stricto, Mycobacterium,* and *Rhodococcus* contain 6,6′-dimycolic esters of trehalose, the so-called cord factors (Ioneda, Lederer and

Rozanis, 1970; Ioneda, Lenz, and Pudles, 1963; Senn et al., 1967).

All of the wall chemotype IV, mycolic acid–containing bacteria contain menaquinones, and can be divided into four main groups on the basis of the structural variation shown by these compounds (Table 2). Representatives of animal corynebacteria, *Microbacterium flavum,* and most *Rhodococcus* species have as their main component a dihydromenaquinone with eight isoprene units [MK-8(H$_2$)], while *Bacterionema matruchotii, Corynebacterium bovis, Corynebacterium glutamicum, Rhodococcus bronchialis, Rhodococcus corallinus,* and *Rhodococcus terrae* have MK-9(H$_2$) as their main homologue (Minnikin, Collins, and Goodfellow, 1978; Collins, Goodfellow, and Minnikin, 1979, unpublished data). In contrast, nocardiae have MK-8(H$_4$) as their main component, while the presence of fully unsaturated menaquinones with nine isoprene units clearly distinguished "*aurantiaca*" strains from all other mycolic acid–containing bacteria (Goodfellow et al., 1978; Minnikin, Goodfellow, and Collins, 1978).

Data derived from the application of a variety of serological techniques underline the affinity among mycolic acid–containing bacteria (Lind and Ridell, 1976; Magnusson, 1976). Cummins (1962, 1965) observed cross-reactions among wall antigens of true corynebacteria, mycobacteria, and nocardiae, and similar cross-reactivity has been observed between delayed hypersensitivity induced by cellular products (Barksdale and Kim, 1977; Magnusson, 1976). The most extensive investigations have employed immunodiffusion methods which have shown the presence of common precipitinogens among true corynebacteria, mycobacteria, nocardiae, and rhodococci (Lind and Ridell, 1976).

Anaerobic Coryneforms and Related Bacteria

For many years, the acne bacillus was classified in the genus *Corynebacterium* together with other anaerobic, coryneform bacteria, such as *Corynebacterium avidum, Corynebacterium diphtheroides, Corynebacterium granulosum,* and *Corynebacterium parvum* (Prévot, 1966; Prévot, Turpin, and Kaiser, 1967), but it is now generally accepted that these bacteria differ in many important respects from true corynebacteria. In particular, they exhibit a propionic acid type of metabolism (Douglas and Gunter, 1946; Moore and Cato, 1963), in most cases contain LL-diaminopimelic acid (LL-DAP) as the diamino acid in the wall peptidoglycan (Johnson and Cummins, 1972; Rogosa et al., 1974), lack mycolic acids (Etémadi, 1963; Goodfellow, Collins, and Minnikin, 1976), but possess a high content of branched-chain fatty acids, principally *iso-* and *anteiso-*C$_{15}$ acids (Azuma et al., 1975; Moss and Cherry, 1968; Moss et al., 1967, 1969; Saino et al., 1976). In all of these respects, the anaerobic coryneforms resemble the propionibacteria with which they are now classified (Cummins and Johnson, 1974; Johnson and Cummins, 1972; Moore and Holdeman, 1974). Although he conceded that the anaerobic coryneforms differ from true corynebacteria, Prévot (1976) proposed that they be classified in a subgenus, *Coryneformis,* in the family Corynebacteriaceae.

The genus *Propionibacterium* (Orla-Jensen, 1909) has traditionally provided a home for cheese and dairy product isolates that produce large amounts of propionic acid during growth. These classical propionibacteria have been the subject of studies using modern taxonomic methods that have yielded compatible taxonomies (see this Handbook, Chapter 145). In particular, the four groups recognized on the basis of analyses of wall and nucleic acid reassociation (Johnson and Cummins, 1972) show good congruence with numerical taxonomic data (Jones, 1975; Malik, Reinbold, and Vedamuthi, 1968). In the eighth edition of *Bergey's Manual* (Moore and Holdeman, 1974), four species of classical propionibacteria are recognized, *Propionibacterium acidi-propionici, Propionibacterium freudenreichii, Propionibacterium jensenii,* and *Propionibacterium thoenii,* mainly on the basis of the DNA-pairing data. Unlike other propionibacteria, *Propionibacterium freudenreichii* contains *meso*-DAP, as opposed to LL-DAP, in the wall peptidoglycan (Table 1; Johnson and Cummins, 1972). In addition, three species of "anaerobic coryneforms" are recognized, *Propionibacterium acnes, Propionibacterium avidum,* and *Propionibacterium granulosum,* primarily on the grounds of wall and DNA-pairing data (Cummins, 1975; Cummins and Johnson, 1974; Johnson and Cummins, 1972). However, relatively little DNA homology exists between these two groups of propionibacteria (Johnson and Cummins, 1972).

Propionibacteria form a heterogeneous group in respect to their fatty acid profiles (see Minnikin, Goodfellow, and Collins, 1978). Thus, representatives of *Propionibacterium freudenreichii* produce a preponderance of *anteiso*-acids, in contrast to the LL-DAP-containing strains whose fatty acids are mainly of the *iso*-type. Representative strains of *Propionibacterium acidi-propionici, Propionibacterium jensenii,* and *Propionibacterium thoenii* have a 15-carbon acid as the main *anteiso* component, whereas *Propionibacterium acnes, Propionibacterium avidum,* and *Propionibacterium granulosum* strains possess high proportions of a 17-carbon *anteiso* acid and a 16-carbon straight-chain acid. Although the propionibacteria form a fairly hetero-

geneous group chemically, the currently available isoprenoid quinone and glycolipid data are a unifying factor (Table 1; Minnikin, Collins, and Goodfellow, 1978).

Propionibacterium is classified in the family Propionibacteriaceae together with the genus *Eubacterium* (Prévot, 1938). The latter is poorly circumscribed and is currently a "catch-all" for a host of inadequately described Gram-positive, obligatively anaerobic, uniform or pleomorphic rods which may be motile or nonmotile but cannot readily be classified elsewhere (see this Handbook, Chapter 146). However, most eubacteria are fermentative; they produce mixtures of organic acids, which often include large amounts of acetic, butyric, formic, and lactic acids, from carbohydrates or peptone (Holdeman and Moore, 1974). Detailed systematic studies, using the whole array of modern taxonomic methods, are needed to determine the internal structure of *Eubacterium* and its relationship with allied taxa.

There is, however, some numerical phenetic evidence to show that *Eubacterium* and *Propionibacterium* form discrete clusters that can be separated from *Corynebacterium sensu stricto* and from Gram-positive, anaerobic bacteria recovered in clusters equated with the genera *Actinomyces, Arachnia,* and *Bifidobacterium* (Holmberg and Hallander, 1973; Holmberg and Nord, 1975). The latter presently form the family Actinomycetaceae together with aerobic to facultatively anaerobic bacteria classified in the genera *Bacterionema* and *Rothia* (Slack, 1974). The latter also form recognizable taxonomic entities in the numerical phenetic surveys, sharing their highest similarities with *Corynebacterium sensu stricto.* *Bacterionema* and *Corynebacterium sensu stricto* strains, however, have many characters in common and can readily be distinguished from rothiae, which do not have a wall chemotype IV, lack mycolic acids, and have significantly higher mol% G+C ratios (Table 1). In contrast, *Bacterionema matruchotii* shows all of the characters of *Corynebacterium sensu stricto* (see Alshamaony et al., 1977; Minnikin, Goodfellow, and Collins, 1978; Pine, 1970). There is no general agreement on the suprageneric relationships of the pleomorphic, Gram-positive rods (see Bowden and Hardie, 1978).

Coryneform Bacteria— the Emerging Taxonomy

It is now usually conceded that general-purpose classifications should be constructed on data derived from independent taxonomic methods and that the consistency found between the different kinds of data is a measure of the confidence that can be placed in a classification. Modern taxonomic methods show quite unequivocally that, contrary to earlier views, the so-called coryneform bacteria do not form a homogeneous group. It would be premature to say that coryneform and related bacteria are well classified, but an acceptable taxonomy is emerging. Established genera such as *Arthrobacter, Brevibacterium, Cellulomonas, Microbacterium,* and *Propionibacterium* are being circumscribed more carefully; more recent genera such as *Curtobacterium, Oerskovia,* and the redefined *Rhodococcus* accommodate new centers of variation. The taxonomic position of other coryneform bacteria remains unsettled, although there is encouraging evidence that the plant pathogens that contain diaminobutyric acid in their walls and strains labeled *Corynebacterium haemolyticum, Corynebacterium pyogenes,* and *Corynebacterium autotrophicum* form the nuclei of additional taxa. Little can be said concerning the affinities or subgeneric composition of the genus *Eubacterium* until it has been the subject of systematic studies using modern taxonomic methods.

In addition, numerical phenetic surveys, and analyses of DNA bases, end products, cell walls, and lipids all show that the human- and animal-pathogenic corynebacteria with associated saprophytes form a defined taxon (Bowden and Hardie, 1978; Jones, 1978; Keddie and Cure, 1978; Minnikin, Goodfellow, and Collins, 1978) that bears a resemblance to the genus *Corynebacterium* as conceived by Lehmann and Neumann (1896). Considering chemical characters alone, *Corynebacterium sensu stricto* includes bacteria that contain *meso*-DAP, arabinose, and galactose in the wall, a DNA base composition in the range 48–59 mol% G+C, dihydrogenated menaquinones with eight or nine isoprene units, and relatively low molecular-weight mycolic acids (22–38 carbons). This definition includes not only *Corynebacterium diphtheriae* and related animal-pathogenic species but also saprophytic species, notably *Microbacterium flavum* and *Corynebacterium glutamicum* with sundry glutamic acid–producing nomenspecies. The definition excludes some animal-parasitic species, most saprophytic species, and all phytopathogenic corynebacteria (Keddie, 1978; Keddie and Cure, 1977; Minnikin, Goodfellow, and Collins, 1978).

The view (Barksdale, 1970; this Handbook, Chapter 141) that *Corynebacterium sensu stricto* is more closely related to *Mycobacterium* and *Nocardia* than to other coryneform taxa is particularly strongly supported by chemical data (Tables 1 and 2) and is in line with earlier workers who classified these genera in a single family (e.g., Jensen, 1931; Lachner-Sandoval, 1898; Lehmann and Neumann, 1927). *Corynebacterium sensu stricto* also shares many characters in common with *Bacte-*

rionema, Caseobacter, Rhodococcus, and the *"aurantiaca"* taxon, all of which contain strains with a wall chemotype IV and mycolic acids (Table 2). *Bacterionema matruchotii* and *Corynebacterium sensu stricto* are particularly closely related in terms of wall composition (Pine, 1970), DNA base composition and glucose metabolism (Gilmour, 1974), polar lipid content (Minnikin and Goodfellow, 1978), and in the structure and size of mycolic acids (Alshamaony et al., 1977).

Chemical data (Table 2) suggest that *Corynebacterium sensu stricto* has the closest relationship with certain members of *Rhodococcus;* differences in DNA G+C content and in mycolic acid size are of degree rather than of kind. Numerical phenetic surveys (Bousfield and Goodfellow, 1976; Jones, 1975), however, have shown that true corynebacteria and rhodococci form discrete taxa, but the classification of possible intermediate strains such as *Rhodococcus (Corynebacterium) equi* (Goodfellow and Alderson, 1977; Jones, 1978) needs further detailed study.

In conclusion, it can be said that the concept of a coherent "coryneform" group of bacteria is no longer very useful as a basis for studies of all the organisms that have been grouped together in this way in the past. It is still useful, however, to study together those organisms which are clearly distinct from the well-defined Gram-positive cocci and bacilli but have certain affinities with actinomycete taxa. This group would include good representatives of *Arthrobacter, Bacterionema, Cellulomonas, Curtobacterium, Microbacterium, Oerskovia, Propionibacterium,* and the plant pathogens that possess diaminobutyric acid in their walls (Tables 1 and 2). Problem organisms such as *Arthrobacter simplex, Arthrobacter tumescens, Arthrobacter nicotianae,* *"Brevibacterium"* helvolum, *"Brevibacterium"* testaceum (Table 1), and other "coryneform" bacteria with characteristic peptidoglycans (Keddie and Cure, 1978; Schleifer and Kandler, 1972), which have not been discussed here, may well be associated with the above group. The evidence available at present (Table 1) suggests that different locations should be found for *Brochothrix, Listeria, Kurthia, Corynebacterium autotrophicum,* and *Erysipelothrix,* although the unique peptidoglycan of the latter (Table 1) is of a general type found only in so-called coryneform bacteria (Schleifer and Kandler, 1972).

Literature Cited

Alshamaony, L., Goodfellow, M., Minnikin, D. E., Bowden, G. H., Hardie, J. M. 1977. Fatty and mycolic acid composition of *Bacterionema matruchotii* and related organisms. Journal of General Microbiology **98:**205–213.

Aragno, M., Walther-Mauruschat, A., Mayer, F., Schlegel, H. G. 1977. Micromorphology of Gram-negative hydrogen bacteria. I. Cell morphology and flagellation. Archives of Microbiology 114:93–100.

Azuma, I., Sugimura, K., Taniyama, T., Aladin, A. A., Yamamura, Y. 1975. Chemical and immunological studies on the cell walls of *Propionibacterium acnes* strains C7 and *Corynebacterium parvum* ATCC 1829. Japanese Journal of Microbiology 19:265–275.

Barksdale, L. 1970. *Corynebacterium diphtheriae* and its relatives. Bacteriological Reviews 34:378–422.

Barksdale, L., Kim, K.-S. 1977. *Mycobacterium*. Bacteriological Reviews 41:217–372.

Bergey, D. H., Harrison, F. C., Breed, R. S., Hammar, B. W., Huntoon, F. M. 1923. Bergey's manual of determinative bacteriology, 1st ed. Baltimore: Williams & Wilkins.

Bisset, K. A., Moore, F. W. 1950. *Jensenia*, a new genus of the Actinomycetales. Journal of General Microbiology 4:280.

Bousfield, I. J. 1972. A taxonomic study of some coryneform bacteria. Journal of General Microbiology 71:441–445.

Bousfield, I. J., Goodfellow, M. 1976. The 'rhodochrous' complex and its relationships with allied taxa, pp. 39–65. In: Goodfellow, M., Brownell, G. H., Serrano, J. A. (eds.), The biology of the nocardiae. London: Academic Press.

Bowden, G. H., Hardie, J. M. 1978. Oral pleomorphic (coryneform) Gram-positive rods, pp. 235–263. In: Bousfield, I. J., Callely, A. G. (eds.), Coryneform bacteria. London: Academic Press.

Bowie, I. S., Grigor, M. R., Dunckley, G. G., Loutit, M. W., Loutit, J. S. 1972. The DNA base composition and fatty acid constitution of some Gram-positive pleomorphic soil bacteria. Soil Biology and Biochemistry 4:397–412.

Bradley, S. G., Bond, J. S. 1974. Taxonomic criteria for mycobacteria and nocardiae. Advances in Applied Microbiology 18:131–190.

Breed, R. S. 1953. The Brevibacteriaceae fam. nov. of the order Eubacteriales. Riassunti delle Communicazione VI Congresso Internazionale di Microbiologia, Roma 1:13–14.

Breed, R. S., Murray, E. G. D., Hitchens, A. P. 1948. Bergey's manual of determinative bacteriology, 6th ed. Baltimore: Williams & Wilkins.

Breed, R. S., Murray, E. G. D., Smith, N. R. 1957. Bergey's manual of determinative bacteriology, 7th ed. Baltimore: Williams & Wilkins.

Buchanan, R. E. 1911. Veterinary bacteriology. Philadelphia: W. B. Saunders.

Burkholder, W. H. 1948. Appendix 1. (Genus 1 *Corynebacterium* Lehmann and Neumann), pp. 398–400. In: Breed, R. S., Murray, E. G. D., Hitchens, A. P. (eds.), Bergey's manual of determinative bacteriology, 6th ed. Baltimore: Williams & Wilkins.

Clark, F. E. 1951. The generic classification of certain cellulolytic bacteria. Soil Science Society of America Proceedings 15:180–182.

Clark, F. E. 1952. The generic classification of the soil corynebacteria. International Bulletin of Bacteriological Nomenclature and Taxonomy 2:45–56.

Clark, F. E. 1953. Criteria suitable for species differentiation in *Cellulomonas* and a revision of the genus. International Bulletin of Bacteriological Nomenclature and Taxonomy 3:179–199.

Collins, M. D., Goodfellow, M., Minnikin, D. E. 1979. Isoprenoid quinones in the classification of coryneform and related bacteria. Journal of General Microbiology 110:127–136.

Collins, M. D., Jones, D., Goodfellow, M., Minnikin, D. E. 1979. Isoprenoid quinone composition as a guide to the classification of *Listeria*, *Brochothrix*, *Erysipelothrix* and *Caryophanon*. Journal of General Microbiology 111:453–457.

Collins-Thompson, D. L., Sørhaug, T., Witter, L. D., Ordal, Z. J. 1972. Taxonomic consideration of *Microbacterium lacticum*, *Microbacterium flavum* and *Microbacterium thermosphactum*. International Journal of Systematic Bacteriology 22:65–72.

Conn, H. J. 1928. A type of bacteria abundant in productive soils, but apparently lacking in certain soils of low productivity. New York State Agricultural Experimental Station Technical Bulletin No. 138:3–26.

Conn, H. J. 1947. A protest against the misuse of the generic name *Corynebacterium*. Journal of Bacteriology 54:10.

Conn, H. J., Dimmick, I. 1947. Soil bacteria similar in morphology to *Mycobacterium* and *Corynebacterium*. Journal of Bacteriology 54:291–303.

Crombach, W. H. J. 1978a. *Caseobacter polymorphus* gen. nov., sp. nov., a coryneform bacterium from cheese. International Journal of Systematic Bacteriology 28:354–366.

Crombach, W. H. J. 1978b. DNA base ratios and DNA hybridisation studies of coryneform bacteria, mycobacteria and nocardiae, pp. 161–179. In: Bousfield, I. J., Callely, A. G. (eds.), Coryneform bacteria. London: Academic Press.

Cummins, C. S. 1962. Chemical composition and antigenic structure of cell walls of *Corynebacterium*, *Mycobacterium*, *Nocardia*, *Actinomyces* and *Arthrobacter*. Journal of General Microbiology 28:35–50.

Cummins, C. S. 1965. Chemical and antigenic studies on cell walls of mycobacteria, corynebacteria and nocardias. American Review of Respiratory Diseases 92:63–72.

Cummins, C. S. 1975. Identification of *Propionibacterium acnes* and related organisms by precipitation tests with trichloroacetic acid extracts. Journal of Clinical Microbiology 2:104–110.

Cummins, C. S., Harris, H. 1956. The chemical composition of the cell wall in some Gram-positive bacteria and its possible value as a taxonomic character. Journal of General Microbiology 14:583–600.

Cummins, C. S., Harris, H. 1958. Studies on the cell wall composition and taxonomy of Actinomycetales and related groups. Journal of General Microbiology 18:173–189.

Cummins, C. S., Harris, H. 1959. Taxonomic position of *Arthrobacter*. Nature, 184:831–832.

Cummins, C. S., Johnson, J. L. 1974. *Corynebacterium parvum*: A synonym for *Propionibacterium acnes*? Journal of General Microbiology 80:433–442.

Davis, G. H. G., Newton, K. G. 1969. Numerical taxonomy of some named coryneform bacteria. Journal of General Microbiology 56:195–214.

Douglas, H. C., Gunter, S. E. 1946. The taxonomic position of *Corynebacterium acnes*. Journal of Bacteriology 52:15–23.

Dowson, W. J. 1942. On the generic name of the Gram-positive bacterial plant pathogens. Transactions of the British Mycological Society 25:311–314.

Eberson, F. 1918. A bacteriologic study of the diphtheroid organisms with special reference to Hodgkin's disease. Journal of Infectious Diseases 23:1–42.

Ernst, W. 1905. Über *Pyelonephritis diphtheriae bovis* und die Pyelonephritisbacillen. Centralblatt für Bakteriologie, Parasitenkunde, Infektionskrankheiten und Hygiene, Abt. 1 39:549–558.

Etémadi, A. H. 1963. Isolement des acides isopentadécanioque et isoheptadécanoique des lipides de *Corynebacterium parvum*. Bulletin de la Société de Chimie Biologique 45:1423–1432.

Fischer, A. 1895. Untersuchungen über Bakterien. Jahrbuch für Wissenschaftliche Botanik 27:1–163.

Gibson, T. 1953. The taxonomy of the genus *Corynebacterium*. Atti del VI Congresso Internazionale di Microbiologia, Roma 1:16–20.

Gilmour, M. N. 1974. *Bacterionema*, pp. 676–679. In: Buchanan, R. E., Gibbons, N. E. (eds.), Bergey's manual of determinative bacteriology, 8th edition. Baltimore: Williams & Wilkins.

Glage, F. 1903. Über den *Bazillus pyogenes suis* Grips, den *Bazillus pyogenes bovis* Kunneman und den bakteriologischen Befund bei den chronischen abszedierenden Euterenzündungen der Milchkühe. Zeitschrift Fleisch und Milchhygiene 13:166–175.

Goodfellow, M. 1971. Numerical taxonomy of some nocardioform bacteria. Journal of General Microbiology **69**:33–80.

Goodfellow, M., Alderson, G. 1977. The actinomycete genus *Rhodococcus:* A home for the *'rhodochrous'* complex. Journal of General Microbiology **100**:99–122.

Goodfellow, M., Collins, M. D., Minnikin, D. E. 1976. Thin-layer chromatographic analysis of mycolic acid and other long-chain components in whole-organism methanolysates of coryneform and related taxa. Journal of General Microbiology **96**:351–358.

Goodfellow, M., Minnikin, D. E. 1977. Nocardioform bacteria. Annual Review of Microbiology **31**:159–180.

Goodfellow, M., Minnikin, D. E. 1978. Numerical and chemical methods in the classification of *Nocardia* and related taxa. Zentralblatt für Bakteriologie, Parasitenkunde, Infektionskrankeiten und Hygiene, Abt. 1, Suppl. **6**:43–51.

Goodfellow, M., Orlean, P. A. B., Collins, M. D., Alshamaony, L., Minnikin, D. E. 1978. Chemical and numerical taxonomy of strains received as *Gordona aurantiaca.* Journal of General Microbiology **109**:57–68.

Hedges, F. 1922. A bacterial wilt of the bean caused by *Bacterium flaccumfaciens* nov. sp. Science **55**:433–434.

Holdeman, L. V., Moore, W. E. C. 1974. *Eubacterium,* pp. 641–657. In: Buchanan, R. E., Gibbons, N. E. (eds.), Bergey's manual of determinative bacteriology, 8th ed. Baltimore: Williams & Wilkins.

Holmberg, K., Hallander, H. O. 1973. Numerical taxonomy and laboratory identification of *Bacterionema matruchotii, Rothia dentocariosa, Actinomyces naeslundii, Actinomyces viscosus,* and some related bacteria. Journal of General Microbiology **76**:43–63.

Holmberg, K., Nord, C.-E. 1975. Numerical taxonomy and laboratory identification of *Actinomyces* and *Arachnia* and some related bacteria. Journal of General Microbiology **91**:17–44.

Hutchinson, C. M. 1925. A bacterial disease of wheat in the Punjab. Memoirs of the Department of Agriculture India, Bacteriology **1**:169–179.

Ioneda, T., Lederer, E., Rozanis, J. 1970. Sur la structure des diesters de trehalose ("cord factors") produits par *Nocardia asteroides* et *Nocardia rhodochrous.* Chemistry and Physics of Lipids **4**:375–392.

Ioneda, T., Lenz, M., Pudles, J. 1963. Chemical constitution of a glycolipid from *C. diphtheriae* P.W.8. Biochemical and Biophysical Research Communications **13**:110–114.

Jayne-Williams, D. J., Skerman, T. M. 1966. Comparative studies on coryneform bacteria from milk and dairy sources. Journal of Applied Bacteriology **29**:72–92.

Jensen, H. L. 1931. Contributions to our knowledge of the Actinomycetales. II. The definition and subdivision of the genus *Actinomyces,* with a preliminary account of Australian soil actinomycetes. Proceedings of the Linnean Society of New South Wales **56**:345–370.

Jensen, H. L. 1934. Studies on saprophytic mycobacteria and corynebacteria. Proceedings of the Linnean Society of New South Wales **59**:19–61.

Jensen, H. L. 1952. The coryneform bacteria. Annual Review of Microbiology **6**:77–90.

Jensen, H. L. 1953. The genus *Nocardia* (or *Proactinomyces*) and its separation from other Actinomycetales, with some reflections on the phylogeny of the actinomycetes, pp. 69–88. In: Baldacci, E., Redaelli, P. (eds.), Actinomycetales: Morphology, biology and systematics. Rome: Fondazione Emanuele Paterno.

Jensen, H. L. 1966. Some introductory remarks on the coryneform bacteria. Journal of Applied Bacteriology **29**:13–16.

Jičínská, E. 1973. *Nocardia*-like mutants of a soil coryneform bacterium. Archiv für Mikrobiologie **89**:269–272.

Johnson, J. L., Cummins, C. S. 1972. Cell wall composition and deoxyribonucleic acid similarities among the anaerobic coryneforms, classical propionibacteria and strains of *Arachnia propionica.* Journal of Bacteriology **109**:1047–1066.

Jones, D. 1975. A numerical taxonomic study of coryneform and related bacteria. Journal of General Microbiology **87**:52–96.

Jones, D. 1978. An evaluation of the contributions of numerical taxonomic studies to the classification of coryneform bacteria, pp. 13–46. In: Bousfield, I. J., Callely, A. G. (eds.), Coryneform bacteria. London: Academic Press.

Jones, D., Weitzman, P. D. J. 1974. Reclassification of *Brevibacterium leucinophagum* Kinney and Werkman as a Gram-negative organism probably in the genus *Acinetobacter.* International Journal of Systematic Bacteriology **24**:113–117.

Jones, L. A., Bradley, S. G. 1964. Phenetic classification of actinomycetes. Developments in Industrial Microbiology **5**:267–272.

Keddie, R. M. 1974a. *Arthrobacter,* pp. 618–625. In Buchanan, R. E., Gibbons, N. E. (eds.), Bergey's manual of determinative bacteriology, 8th ed. Baltimore: Williams & Wilkins.

Keddie, R. M. 1974b. *Cellulomonas,* pp. 629–631. Buchanan, R. E., Gibbons, N. E. (eds.), Bergey's manual of determinative bacteriology, 8th ed. Baltimore: Williams & Wilkins.

Keddie, R. M. 1978. What do we mean by coryneform bacteria?, pp. 1–12. In: Bousfield, I. J., Callely, A. G. (eds.), Coryneform bacteria. London: Academic Press.

Keddie, R. M., Cure, G. L. 1977. The cell wall composition and distribution of free mycolic acids in named strains of coryneform bacteria and in isolates from various natural sources. Journal of Applied Bacteriology **42**:229–252.

Keddie, R. M., Cure, G. L. 1978. Cell wall composition of coryneform bacteria, pp. 47–83. In: Bousfield, I. J., Callely, A. G. (eds.), Coryneform bacteria. London: Academic Press.

Keddie, R. M., Rogosa, M. 1974. *Kurthia,* pp. 631–632. In: Buchanan, R. E., Gibbons, N. E. (eds.), Bergey's manual of determinative bacteriology, 8th ed. Baltimore: Williams & Wilkins.

Kisskalt, K., Berend, E. 1918. Untersuchungen über die Gruppe de Diphtheroiden (Corynebakterien). Centralblatt für Bakteriologie, Parasitenkunde und Infektionskrankheiten, Abt. 1 Orig. **81**:444–447.

Kroppenstedt, R. M., Kutzner, H. J. 1978. Biochemical taxonomy of some problem actinomycetes. Zentralblatt für Bakteriologie, Parasitenkunde, Infektionskrankheiten und Hygiene, Abt. 1, Suppl. **6**:125–133.

Lachner-Sandoval, V. 1898. Ueber Strahlenpilze, pp. 1–75. Inaugural Dissertation, Strasbourg. Bonn: Universitäts Buchdruckerei von Carl Gorgi.

Lechevalier, M. P. 1972. Description of a new species, *Oerskovia xanthineolytica* and emendation of *Oerskovia* Prauser et al. International Journal of Systematic Bacteriology **22**:260–264.

Lechevalier, M. P. 1977. Lipids in bacterial taxonomy—a taxonomist's view. CRC Critical Reviews in Microbiology **5**:109–210.

Lechevalier, M. P., Lechevalier, H. A. 1970. Chemical composition as a criterion in the classification of aerobic actinomycetes. International Journal of Systematic Bacteriology **20**:435–444.

Lechevalier, M. P., de Bièvre, C., Lechevalier, H. 1977. Chemotaxonomy of aerobic actinomycetes: Phospholipid composition. Biochemical Systematics and Ecology **5**:249–260.

Lehmann, K. B., Neumann, R. 1896. Atlas und Grundriss der Bakteriologie und Lehrbuch der speciellen bakteriologischen Diagnostik, 1st ed. Munich: J. F. Lehmann.

Lehmann, K. B., Neumann, R. 1899. Lehmann's Medizinische, Handatlanten X. Atlas und Grundriss der Bakteriologie und Lehrbuch der speciellen bakteriologischen Diagnostik, 2nd ed. Munich: J. F. Lehmann.

Lehmann, K. B., Neumann, R. O. 1927. Bakteriologische Diagnostik, 7th ed. Munich: J. F. Lehmann.

Lind, A., Ridell, M. 1976. Serological relationships between *Nocardia, Mycobacterium, Corynebacterium* and the "*rhodochrous*" taxon, pp. 220–235. In: Goodfellow, M.,

Brownell, G. H., Serrano, J. A. (eds.), The biology of the nocardiae. London: Academic Press.

Loeffler, F. 1887. Ueber die Ergebnisse seiner weiteren Untersuchungen über die Diphtherie-Bacillen. Zentralblatt für Bakteriologie und Parasitenkunde **2:**105–106.

McCulloch, L. 1925. *Aplanobacter insidiosum* n. sp., the cause of an alfalfa disease. Phytopathology **15:**496–497.

McLean, R. A., Sulzbacher, W. L. 1953. *Microbacterium thermosphactum,* spec. nov., a nonheat resistant bacterium from fresh pork sausage. Journal of Bacteriology **65:**428–433.

Magnusson, M. 1976. Sensitin tests as an aid in the taxonomy of *Nocardia* and its pathogenicity, pp. 236–265. In: Goodfellow, M., Brownell, G. H., Serrano, J. A. (eds.), The biology of the nocardiae. London: Academic Press.

Malik, A. C., Reinbold, G. W., Vedamuthi, E. R. 1968. An evaluation of the taxonomy of *Propionibacterium*. Canadian Journal of Microbiology **14:**1185–1191.

Migula, W. 1900. System der Bakterien, vol 2. Jena: Gustav Fischer Verlag.

Minnikin, D. E., Collins, M. D., Goodfellow, M. 1978. Menaquinone patterns in the classification of nocardioform and related bacteria. Zentralblatt für Bakteriologie, Parasitenkunde, Infektionskrankheiten und Hygiene, Abt. 1, Suppl. **6:**85–90.

Minnikin, D. E., Collins, M. D., Goodfellow, M. 1979. Fatty acid and polar lipid composition in the classification of *Cellulomonas, Oerskovia* and related taxa. Journal of Applied Bacteriology **46:**87–95.

Minnikin, D. E., Goodfellow, M. 1976. Lipid composition in the classification and identification of nocardiae and related taxa, pp. 160–219. In: Goodfellow, M., Brownell, G. H., Serrano, J. A. (eds.), The biology of the nocardiae. London: Academic Press.

Minnikin, D. E., Goodfellow, M. 1978. Polar lipids of nocardioform and related bacteria. Zentralblatt für Bakteriologie, Parasitenkunde, Infektionskrankheiten und Hygiene, Abt. 1, Suppl. **6:**75–83.

Minnikin, D. E., Goodfellow, M., Alshamaony, L. 1978. Mycolic acids in the classification of nocardioform bacteria. Zentralblatt für Bakteriologie, Parasitenkunde, Infektionskrankheiten und Hygiene, Abt. 1, Suppl. **6:**63–66.

Minnikin, D. E., Goodfellow, M., Collins, M. D. 1978. Lipid composition in the classification and identification of coryneform and related taxa, pp. 85–160. In: Bousfield, I. J., Callely, A. G. (eds.), Coryneform bacteria. London: Academic Press.

Minnikin, D. E., Patel, P. V., Alshamaony, L., Goodfellow, M. 1977. Polar lipid composition in the classification of *Nocardia* and related bacteria. International Journal of Systematic Bacteriology **27:**104–117.

Moore, W. E. C., Cato, E. P. 1963. Validity of *Propionibacterium acnes* (Gilchrist) Douglas and Gunter comb. nov. Journal of Bacteriology **85:**870–874.

Moore, W. E. C., Holdeman, L. V. 1974. *Propionibacterium,* pp. 633–641. In: Buchanan, R. E., Gibbons, N. E. (eds.), Bergey's manual of determinative bacteriology, 8th ed. Baltimore: Williams & Wilkins.

Moss, C. W., Cherry, W. B. 1968. Characterization of the C$_{15}$ branched-chain fatty acids of *Corynebacterium acnes* by gas chromatography. Journal of Bacteriology **95:**241–242.

Moss, C. W., Dowell, V. R., Jr., Lewis, V. J., Schekter, M. A. 1967. Cultural characteristics and fatty acid composition of *Corynebacterium acnes*. Journal of Bacteriology **94:**1300–1305.

Moss, C. W., Dowell, V. R., Jr., Farshtchi, D., Raines, L. J., Cherry, W. B. 1969. Cultural characteristics and fatty acid composition of propionibacteria. Journal of Bacteriology **97:**561–570.

Nocard, M. E. 1888. Note sur le maladie des boeufs de la Guadeloupe, connue sous le nom de farcin. Annales de l'Institut Pasteur **2:**293–302.

O'Gara, P. J. 1916. A bacterial disease of western wheat-grass, *Agropyron smithii,* occurrence of a new type of bacterial disease in America. Phytopathology **6:**341–350.

Orla-Jensen, S. 1909. Die Hauptlinien des natürlichen Bakteriensystems. Centralblatt für Bakteriologie, Parasitenkunde und Infektionskrankheiten, Abt. 2, **22:**305–346.

Orla-Jensen, S. 1919. The lactic acid bacteria. Copenhagen: Høst & Son.

Oerskov, J. 1923. Investigations into the morphology of the ray fungi. Copenhagen: Levin and Munksgaard.

Pine, L. 1970. Classification and phylogenetic relationships of microaerophilic actinomyces. International Journal of Systematic Bacteriology **20:**445–474.

Prauser, H. 1978. Considerations on taxonomic relations among Gram-positive, branching bacteria. Zentralblatt für Bakteriologie, Parasitenkunde, Infektionskrankheiten und Hygiene, Abt. 1, Suppl. **6:**3–12.

Prauser, H., Bergholz, M. 1974. Taxonomy of actinomycetes and screening for antibiotic substances. Postepy Higieny I Medycyny Doświadczalnej **28:**441–457.

Prauser, H., Lechevalier, M. P., Lechevalier, H. 1970. Description of *Oerskovia* gen. n. to harbor Oerskov's motile *Nocardia*. Applied Microbiology **19:**534.

Prévot, A. R. 1938. Études de systématique bactérienne. III. Invalidité du genre *Bacteroides* Castellani et Chalmers démembrement et reclassification. Annales de l'Institut Pasteur **60:**285–307.

Prévot, A. R. 1966. Manual for the classification and determination of the anaerobic bacteria, pp. 345–355. Translated from the French by V. Fredette. Philadelphia: Lea & Febiger.

Prévot, A. R. 1976. Nouvelle conception de la position taxonomique des Corynébactéries anaérobies. Comptes Rendus Hebdomadaires des Séances de l'Académie des Sciences. Série D **282:**1079–1081.

Prévot, A. R., Turpin, A., Kaiser, P. 1967. Les bactéries anaérobies, pp. 1–2188. Paris: Dunod.

Rogosa, M., Keddie, R. M. 1974a. *Brevibacterium,* pp. 625–628. In: Buchanan, R. E., Gibbons, N. E. (eds.), Bergey's manual of determinative bacteriology, 8th ed. Baltimore: Williams & Wilkins.

Rogosa, M., Keddie, R. M. 1974b. *Microbacterium,* pp. 628–629. In: Buchanan, R. E., Gibbons, N. E. (eds.), Bergey's manual of determinative bacteriology, 8th ed. Baltimore: Williams & Wilkins.

Rogosa, M., Cummins, C. S., Lelliott, R. A., Keddie, R. M. 1974. Coryneform group of bacteria, pp. 599–632. In: Buchanan, R. E., Gibbons, N. E. (eds.), Bergey's manual of determinative bacteriology, 8th ed. Baltimore: Williams & Wilkins.

Saino, Y., Eda, J., Nagoya, T., Yoshimura, Y., Yamaguchi, M., Kobayashi, F. 1976. Anaerobic coryneforms from human bone marrow and skin. Japanese Journal of Microbiology **20:**17–25.

Schleifer, K. H. 1970. Die Mureintypen in der Gattung *Microbacterium*. Archiv für Mikrobiologie **71:**271–282.

Schleifer, K. H., Kandler, O. 1972. Peptidoglycan types of bacterial cell walls and their taxonomic implications. Bacteriological Reviews **36:**407–477.

Seeliger, H. P. R. 1974. *Erysipelothrix,* pp. 597–598. In: Buchanan, R. E., Gibbons, N. E. (eds.), Bergey's manual of determinative bacteriology, 8th ed. Baltimore: Williams & Wilkins.

Seeliger, H. P. R., Welshimer, H. J. 1974. *Listeria,* pp. 593–596. In: Buchanan, R. E., Gibbons, N. E. (eds.), Bergey's manual of determinative bacteriology, 8th ed. Baltimore: Williams & Wilkins.

Senn, M., Ioneda, T., Pudles, J., Lederer, E. 1967. Spectrometrie de masse de glycolipides. I. Structure de cord factor de *Corynebacterium diphtheriae*. European Journal of Biochemistry **1:**353–356.

Skaptason, J. B., Burkholder, W. H. 1942. Classification and nomenclature of the pathogen causing bacterial ring rot of potatoes. Phytopathology **32**:439–441.

Skyring, G. W., Quadling, C. 1970. Soil bacteria: A principal component analysis and guanine-cytosine contents of some arthrobacter-coryneform soil isolates and of some named cultures. Canadian Journal of Microbiology **16**:95–106.

Slack, J. M. 1974. Actinomycetaceae, pp. 659–660. In: Buchanan, R. E., Gibbons, N. E. (eds.), Bergey's manual of determinative bacteriology, 8th ed. Baltimore: Williams & Wilkins.

Smith, E. F. 1910. A new tomato disease of economic importance. Science **31**:794–796.

Smith, E. F. 1913. A new type of bacterial disease. Science **38**:926.

Sneath, P. H. A., Jones, D. 1976. *Brochothrix*, a new genus tentatively placed in the family *Lactobacillaceae*. International Journal of Systematic Bacteriology **26**:102–104.

Spieckermann, A., Kotthoff, P. 1914. Untersuchungen über die Kartoffelpflanze und ihre Krankheiten. Landwirtschaftliche Jahrbücher **46**:659–732.

Stahel, G. 1933. The witchbrooms of *Eugenia latifolia* Aubl. in Surinam caused by *Pseudomonas hypertrophicans* nov. spec. Phytopathologische Zeitschrift **6**:441–452.

Sukapure, R. S., Lechevalier, M. P., Reber, H., Higgins, M. L., Lechevalier, H. A., Prauser, H. 1970. Motile nocardoid Actinomycetales. Applied Microbiology **19**:527–533.

Thiele, O. W., Thiele, C. 1977. Lipid patterns of various hydrogen oxidising bacterial species. Biochemical Systematics and Ecology **5**:1–6.

Tilford, P. E. 1936. Fasciation of sweet peas. 54th Annual Report Ohio Agricultural Experimental Station Bulletin **561**:39.

Trevisan, V. 1889. I Generi e le Specie delle Battieriacee. Milan: Zanaboni & Gabuzzi.

Tsukamura, M. 1971. Proposal of a new genus, *Gordona*, for slightly acid-fast organisms occurring in sputa of patients with pulmonary disease and in soil. Journal of General Microbiology **68**:15–26.

Tsukamura, M. 1974. A further numerical taxonomic study of the rhodochrous group. Japanese Journal of Microbiology **18**:37–44.

Tsukamura, M. 1975. Numerical analysis of the relationship between *Mycobacterium*, rhodochrous group and *Nocardia* by use of hypothetical median organisms. International Journal of Systematic Bacteriology **25**:329–335.

Tsukamura, M., Mizuno, S., Murata, H. 1975. Numerical taxonomy study of the taxonomic position of *Nocardia rubra* reclassified as *Gordona lentifragmenta* Tsukamura nom. nov. International Journal of Systematic Bacteriology **25**:377–382.

Veldkamp, H. 1970. Saprophytic coryneform bacteria. Annual Review of Microbiology **24**:209–240.

Waksman, S. A., Henrici, A. T. 1943. The nomenclature and classification of the actinomycetes. Journal of Bacteriology **46**:337–341.

Walther-Mauruschat, A., Aragno, M., Mayer, F., Schlegel, H. G. 1977. Micromorphology of Gram-negative hydrogen bacteria. II. Cell envelope, membranes and cytoplasmic inclusions. Archives of Microbiology **114**:101–110.

Yamada, K., Komagata, K. 1970. Taxonomic studies on coryneform bacteria. III. DNA base composition of coryneform bacteria. Journal of General and Applied Microbiology **16**:215–224.

Yamada, K., Komagata, K. 1972a. Taxonomic studies on coryneform bacteria. IV. Morphological, cultural, biochemical and physiological characteristics. Journal of General and Applied Microbiology **18**:399–416.

Yamada, K., Komagata, K. 1972b. Taxonomic studies on coryneform bacteria. V. Classification of coryneform bacteria. Journal of General and Applied Microbiology **18**:417–431.

Yamada, Y., Inouye, G., Tahara, Y., Kondo, K. 1976. The menaquinone system in the classification of coryneform and nocardioform bacteria and related organisms. Journal of General and Applied Microbiology **22**:203–214.

Yamaguchi, T. 1965. Comparison of the cell-wall composition of morphologically distinct actinomycetes. Journal of Bacteriology **89**:444–453.

Yamaguchi, T. 1967. Similarity in DNA of various morphologically distinct actinomycetes. Journal of General and Applied Microbiology **13**:63–71.

Zopf, W. 1891. Ueber Ausscheidung von Fettfarbstoffen (Lipochromen) seitens gewisser Spaltpilze. Berichte der Deutschen Botanischen Gesellschaft **9**:22–28.

The Genus *Corynebacterium*

LANE BARKSDALE

It is amazing, in the face of the vast amount of data indicating that *Corynebacterium* absolutely belongs with *Mycobacterium* and *Nocardia,* that the eighth edition of *Bergey's Manual of Determinative Bacteriology* (Buchanan and Gibbons, 1974) places *Corynebacterium,* along with a polytypic array of bacteria whose common property would seem to be pleomorphism, in the "Coryneform Group of Bacteria". Nothing written by the authors of that section justifies such an arrangement (see also Chapter 143, this Handbook). To this writer, the morphological cover "coryneform" is jargon that furthers no taxonomic concept(s). It is a meaningless term for the simple reasons that the shapes ("coryneformity") assumed by *C. diphtheriae* var. *mitis* are not those assumed by *C. diphtheriae* var. *gravis,* which in turn are not those assumed by *C. pseudotuberculosis* or *C. xerosis* and are very different from those assumed by *C. renale.* Yet, these are all good species of *Corynebacterium.* There are nocardias which cannot be distinguished microscopically from *C. diphtheriae* and there are nocardias which are very much like the cellular forms seen in cultures of *Nocardia asteroides.* Classification on the basis of cellular morphology is just not possible. Terms that connote *Corynebacterium* (e.g., "coryneform") or *Nocardia* (e.g., "nocardiaform") imply a relatedness that may not exist. No greater proof of this point is needed than the heterogeneous group of organisms discussed under the "Coryneform Group" on pages 599–632 of Buchanan and Gibbons (1974).

A quarter of a century ago, H. L. Jensen (1952) noted that "from *Mycobacterium* and *Corynebacterium* there is a very gradual transition to *Nocardia* . . ." and that "the existence of 'transitional' or 'borderline' forms may be a grievous impediment to group separation and nomenclature but may help us to visualize . . . certain trends that indicate a system based on phylogeny. . . ." Some 19 years later, Veldkamp (1970), privy to information on cell wall structure, limited antigenic analyses, and percentages of guanosine plus cytosine in *Corynebacterium, Mycobacterium,* and *Nocardia,* stated: "Thus, as far as the taxonomy of the genera, *Corynebacterium, Mycobacterium,* and *Nocardia* is

concerned there is no *communis opinio* as to whether the differences justify differentiation on a generic level." On the basis of a numerical study, Harrington (1966) has concluded that *Corynebacterium, Mycobacterium,* and *Nocardia* belong together (see also Gordon, 1966; Jones, 1975).

Komagata, Yamada, and their associates are the one group in the world who, having taken "coryneform bacteria" at face value, have investigated the (i) cell division (Komagata, Yamada, and Ogawa, 1969), (ii) amino acids of cell walls (Yamada and Komagata, 1970a), (iii) DNA base composition (Yamada and Komagata, 1970b), and the (iv) biochemical and physiological properties (Yamada and Komagata, 1972a) of *Corynebacterium, Microbacterium, Cellulomonas, Arthrobacter,* and *Brevibacterium* ("coryneform bacteria" of Buchanan and Gibbons, 1974). From this prodigious effort, the following conclusion has emerged: ". . . . it is found that bacterial taxonomists have given different evaluation for criteria presently employed in the classification of coryneform bacteria. This fact may also indicate that the generic concepts of these bacteria have not been established clearly and that the generic concept of *Corynebacterium* was expanded without extensive taxonomical considerations" (Yamada and Komagata, 1972b). It should be noted that, up to this time (1972), the authors had not considered the genus *Corynebacterium* in relation to *Mycobacterium* and *Nocardia,* despite there being on record a rather brute plea for such consideration (Barksdale, 1970). One of the authors, however, had an experience which forced him to face up to the possible relationship between *Corynebacterium* and *Nocardia.* One of Komagata's earliest ventures into microbial taxonomy involved the description of *Nocardia erythropolis* as a (then) new species, *Corynebacterium hydrocarboclastus* (Iizuka and Komagata, 1964). In 1973, Komagata and colleagues (Komura, Komagata, and Mitsugi, 1973) published a detailed study supporting the inclusion of *C. hydrocarboclastus* under *N. erythropolis.* This laudable effort was followed by an extension of interest to *Nocardia* and *Mycobacterium* (Komura et al., 1975). In a study of the phospholipids of bacteria of the CMN group (and others), they found that the CMN group had an interesting distribution of phosphatidyl ethanolamine (PE). In those corynebacteria having a G+C content of 50–60%, the phospholipids contained only traces of PE; corynebacteria above 60 mol% and all mycobacteria and nocardias (all of which have a G+C of above 60 mol%) contained easily detectable amounts of PE. From their 5-year effort, the authors conclude that ". . . . *Corynebacterium, Mycobacterium,* and *Nocardia* should be studied together for better understanding of the boundaries of these genera" (Komura et al., 1975).

Interrelationships: Basic Facts Concerning *Corynebacterium, Mycobacterium,* and *Nocardia:* The CMN Group

The genera *Corynebacterium, Mycobacterium,* and *Nocardia,* the CMN group (Barksdale, 1970), are actinomycetes (Bergey et al., 1939; Breed, Murray, and Hitchens, 1948; Jensen, 1952) that share the properties of a family, the Mycobacteriaceae. All members are nonmotile, Gram-positive, aerobic, catalase-producing (catalase-negative mutants excepted) rods that, in phosphate-rich environments, accumulate intracellular polyphosphate as intracellular granules, and these interact metachromatically with toluidin blue and alkaline methylene blue. Phenotypes which resist the uptake of dyes (chromophobic) may be encountered, especially in *Mycobacterium.* In such cases, growth under different conditions or cells from old slants usually yield stainable material. Members of the Mycobacteriaceae show a range of interactions with basic fuchsin as carbol fuchsin, which include retention of fuchsin following decolorization with mineral acid, e.g., 5% sulfuric acid. *Corynebacterium* shows poor retention (Harrington, 1966; Jensen, 1934), *Nocardia* shows moderate retention (Beaman and Burnside, 1973), and *Mycobacterium* is not easily decolorized following exposure to dilute acid (Buchanan and Gibbons, 1974; Vestal, 1975). *Mycobacterium,* alone, binds fuchsin in such a way as to resist decolorization with acidic ethanol (mycobacterial acid-fastness, discussed in Barksdale and Kim, 1977). Serological cross-reactions between the cell wall antigens of the members have been demonstrated (Cummins, 1962, 1965) as have cross-reactions between delayed hypersensitivities (cell-mediated immunities) induced by their cellular products (Barksdale and Kim, 1977; Freund and Lipton, 1948; Magnusson, 1961). As would be expected on the basis of these serological cross-reactions, the chemistry of cellular structures of members of the Mycobacteriaceae has much in common. Other shared properties of the group include: (i) A

murein (peptidoglycan) consisting of a muramyl peptide that contains *meso*-diaminopimelic acid (directly crosslinked, Bordet et al., 1972; Kato, Strominger, and Kotani, 1968; Petit et al., 1969; Schliefer and Kandler, 1972; Vacheron et al., 1972; Wietzerbin-Falszpan et al., 1970), glutamic acid and alanine in association with arabinogalactan in *Corynebacterium* (Asselineau, 1966; Cummins, 1954; 1965; Cummins and Harris, 1956, 1958), and arabinogalactan-mycolate in *Nocardia* and *Mycobacterium* (Azuma, Yamamura, and Misaki, 1969; Bruneteau and Michel, 1968; Kanetsuna, 1968; Kanetsuna and San Blas, 1970; Kanetsuna, Imaeda, and Cunto, 1969; Laneelle and Asselineau, 1970; Misaki et al., 1966). (ii) Whereas N-acetylmuramic acid occurs in the peptidoglycan of corynebacteria and is involved in its linkage to the arabinogalactan, N-glycolylmuramic acid occurs in some nocardias and mycobacteria (Adam et al., 1969; Azuma et al., 1970; Bordet et al., 1972; Kanetsuna and San Blas, 1970). In mycobacteria, the arabinogalactan is linked through phosphodiester bridges (Liu and Gotschlich, 1967) to the primary alcohol residues of N-glycolated muramic acid (Kanetsuna and San Blas, 1970). (iii) Fatty acid synthesis in these organisms is centered in a particulate (multienzyme) system, the Fatty Acid Synthetase I (Knoche and Koths, 1973; Vance, Mitsuhashi, and Bloch, 1973). (iv) The fatty acids (FAs) which characterize the CMN group are long-chain, α-branched, and β-hydroxylated, and have the same stereochemistry (Asselineau, Tocanne, and Tocanne, 1970). While the shorter chain lengths (C_{28} to C_{36}) are those associated with corynebacteria and the longer chain lengths are those associated with nocardia (C_{40} to C_{56}) and mycobacteria (C_{60} to C_{90}), today the term "mycolic acids", once reserved for the latter, should probably encompass "corynomycolic, nocardomycolic" as well as the mycolic acis of mycobacteria. Data concerning this matter come from the research of Asselineau and Asselineau (1966, 1978a,b), Brennan and Lehane (1969), Etemadi, Gasche, and Sifferlen (1965), Lederer and Pudles (1951), Michel, Bordet, and Lederer (1960), and Welby-Gieusse, Laneelle, and Asselineau (1970), among others. Since corynomycolic acids are of shorter chain lengths and since they also may occur in mycolates produced by *Nocardia* (Etemadi, 1967; Maurice, Vacheron, and Michel, 1971) and those of some mycobacteria (Brennan, Lehane, and Thomas, 1970), it seems reasonable to regard them as a common biosynthetic product of the CMN group. A distinguishing feature, then, of *Nocardia* and *Mycobacterium* would be their capacity to produce the chain lengths associated with the terms "nocardomycolic" and "mycolic" acids. Whether these added capacities represent a gain by *Nocardia* and *Mycobacterium* over the capacities of *Corynebacterium* or merely a loss of abilities by the latter remains to be determined. (v) The major phospholipids of the CMN group include mono- and diphosphatidylglycerols, phosphatidylinositol, and phosphatidylinositol mannosides (Lechevalier, de Bievre, and Lechevalier, 1977). (Phosphatidylethanolamine occurs in some corynebacteria and in *Nocardia* and *Mycobacterium*.) Corynebacteria (Ioneda, Lenz, and Pudles, 1963; Kato, 1970; Senn et al., 1967), mycobacteria (Noll et al., 1956), and nocardias (Ioneda, Lederer, and Rozanis, 1970) all produce cord factors, trehalose-6,6'-dimycolates. The occurrence of these diesters of trehalose is another common bond within the CMN group. In *Nocardia asteroides* and in *C. diphtheriae*, esterification is with corynomycolic acids. In *N. rhodochrous* and in *Mycobacterium*, longer-chain mycolic acids form the trehalose esters (Ioneda, Lederer, and Rozanis, 1970). Although the mycoloyl substituents on the trehaloses of the cord factors from *M. tuberculosis* and those from *C. diphtheriae* differ by many carbon atoms, the immunoadjuvant action of the two is very similar (Parant et al., 1978). The guanine-plus-cytosine percents for the group range from about 50 to 70, the smaller values being found in *Corynebacterium*, the middle values in *Corynebacterium* and *Nocardia*, and the upper ones in *Mycobacterium*. Bradley (1973) demonstrated a range of genome sizes in *Mycobacterium* from 2.5×10^9 (*M. tuberculosis*) to 4.5×10^9 (*M. smegmatis*) daltons. Imaeda, Barksdale, and Norgard (1980) have found the genome sizes for various corynebacteria to range from 1.2×10^9 to 2.5×10^9 daltons, under conditions where the genome size for *M. smegmatis*, strain 607, was 4.27×10^9 daltons. Not only is there much overlap in the range of genome sizes for the CMN group, but also there is overlap in DNA homology (Bradley, 1973). Some additional information regarding the similarity of DNA replicating systems in the CMN group should become available from studies of the respective bacteriophage-host systems (Barksdale, 1970; Barksdale and Kim, 1977).

Habitats

The habitats of these microbes range from soils (sometimes the waters that irrigate soils) to various vertebrates and, sometimes, plants (see also discussion of group 1 boundary species, this chapter).

Most attention has been given to those members of the Mycobacteriaceae which economically affect man and his domestic animals. Many of those give rise to diseases such as tuberculosis, diphtheria, infectious lymphadenitis, and pyelonephritis; therefore, they and organisms with which they are easily confused have received considerable study. In addition there are species of special industrial importance either because (i) cultures of them accumulate amino acids or other molecules of value or (ii) the organisms are capable of bringing about hydroxylations, epoxidations, dehydrogenations, etc., which transform precursor molecules to products of commercial importance.

Generic Properties of *Corynebacterium* Lehmann and Neumann 1896

Nonmotile. Endospores not formed. Gram-positive (cells from aging cultures may retain crystal violet unevenly or not at all). Not acid-fast (using Ziehl-Neelsen stain). Rod-shaped (sometimes ovoid, sometimes club-shaped, rarely threadlike and/or showing rudimentary branching; i.e. pleomorphic) bacteria exhibiting moderately tapered ends (cells from actively growing cultures appear as doublets with tapered ends; for this property and for the range of cell morphology in *Corynebacterium*, see micrographs of Heitner and Kim in Barksdale, 1970, which also document the snapping division of Winslow et al., 1917, 1920, and Komagata, Yamada, and Ogawa, 1969). Characteristic mean generation time, under defined conditions, ranges from 1 to 4 h (depending on species). Cells from phosphate-rich media contain metachromatic granules which can be revealed by staining with alkaline methylene blue or toluidine blue. Aerobic. Catalase (sensitive to both 0.01 M KCN and 0.01 M NaN₃) produced (Cummins, 1971). Fail to ferment either lactose or xylose. Guanosine-plus-cystosine content ±50 to ±70 mol% (Bouisset, Breuillaud, and Michel, 1963; Kareva and Filosofova, 1966; Spirin et al., 1957; Yamada and Komagata, 1970b). Genome size ranges from 1.2 to 2.5×10^9 daltons (Imaeda, Barksdale, and Norgard, 1980). Cell wall peptidoglycan contains directly cross-linked *meso*-diaminopimelic acid (A1γ of Schleifer and Kandler, 1972). Major cell wall sugars are arabinose, galactose, and commonly mannose (summarized in Cummins, 1962). Characteristic lipids: saturated and unsaturated corynomycolic acids of chain lengths ranging from about 28 to about 40 carbon atoms (Asselineau; 1966, Beaman et al., 1974; Diara and Pudles, 1959). Characteristic phospholipids produced are diphosphatidylglycerol (cardiolipin), phosphatidylglycerol, phosphatidylinositol, phosphatidylinositol mannoside, and sometimes more than traces of phosphatidyl ethanolamine. This phospholipid pattern, in conjunction with the production of mycolic acids, has been designated as "phospholipid–fatty acid type 4" by Lechevalier, de Bievre, and Lechevalier (1977). The type species was designated *Corynebacterium diphtheriae* (Kruse in Flügge, 1866) Lehmann and Neumann 1896 by Winslow et al. (1917, 1920). No type strain(s) was designated.

Neotype: strains $C7_s$ $(-)^{tox^-}$ (ATCC 27010) and $C7_s$ $(\beta)^{tox^+}$ (ATCC 27012) (Barksdale, 1970). It is evident from the description of *C. diphtheriae* by Loeffler (1884) and by Andrewes et al. (1923) that the originally recognized strains of *C. diphtheriae* were what McLeod (1943) called *C. diphtheriae* type *mitis* and what Anderson et al., (1931) designated as *Bacillus diphtheriae* var. *mitis*. Thus any neotype should be a strain which meets the description of the *mitis* type. From earliest times, the property of *C. diphtheriae* most readily recognized was the capacity to cause an often fatal intoxication. The intoxication was caused by diphtherial toxin. The genetic control of the synthesis of the toxin molecule was worked out with the *mitis* strains, $C7_s$ $(-)^{tox^-}$ and $C7_s$ $(\beta)^{tox^+}$, and their mutants. More information has been accumulated about them than any other *Corynebacterium* (Barksdale and Arden, 1974; Collier, 1975; Pappenheimer, 1977). Since they embody all of the characteristics of the first of the diphtheria bacilli to be recognized, it seems proper that they should be neotypes. Furthermore, since *C. diphtheriae* var. *mitis* is synonymous with *C. diphtheriae*, strictly speaking, *mitis* becomes an unnecessary epithet. Since *C. diphtheriae* var. *intermedius* shares many properties with the neotype, it remains a variety of the neotype. On the other hand, it should be noted that *C. diphtheriae* var. *gravis* differs from the type strain(s) by several properties.

Boundary species (key species, in addition to the type species, which define the boundaries of the genus): *C. equi*, *C. fascians*, *C. kutscheri*, *C. pseudodiphtheriticum*, *C. minutissimum*, *C. renale*, *C. xerosis*. Members of the genus can be identified by the series of biochemical reactions outlined in Barksdale et al. (1979); see also Barksdale (1979) and Sulea, Pollice, and Barksdale (1980).

Boundaries of the Genus *Corynebacterium*

Among the eukaryotic diploid forms of life, a species is any naturally interfertile population in such genetic equilibrium as to share a common set of properties, the emergent aspect of which is the readily recognized species. To have meaning, prokaryotic species must also be based upon the characterization of a wide range of representatives, as recently reemphasized by Gordon (1978). The markers employed for characterizing prokaryotes are often enzymatic capacities per se rather than unique morphological entities generated by the action of enzymes. For, except for a variety of antigens (flagellar, capsular, etc.) which must be identified with antisera, bacteria belonging to a single species (or genus) can seldom be singled out on the basis of morphology or

Table 1. Properties useful for distinguishing members of the genus *Corynebacterium*.

Acid from:	Presence of:
1. Lactose	9. Catalase
2. Xylose	10. Pyrazinamidase
3. Dextrose	11. Urease
4. Maltose	12. Gelatinase
5. Mannose	13. Caseinase
6. Sucrose	14. Lipase for Tween 60
7. Starch	
8. Trehalose	

particular organelles. The species of corynebacteria shown in Table 2 can readily be identified by the pattern of reactions indicated there for each. Each of them shares the general properties discussed here under the *Mycobacteriaceae* and the particular properties given in the description of the genus *Corynebacterium*. However, to understand how the species of corynebacteria relate to one another and to the CMN group, additional information is needed. For example, preliminary DNA homology studies indicate that *C. diphtheriae* and *C. minutissimum* are not very close but are closer than *C. renale* and *C. diphtheriae* (M. Norgard and L. Barksdale, unpublished data; Imaeda, Barksdale, and Norgard, 1980). Work now in progress will determine the extent of homology between the DNAs of the eight species listed in Table 2. In time, similar studies will be extended to cover relatedness of DNAs from representative species of *Corynebacterium*, *Mycobacterium*, and *Nocardia*. Eventually, transductional analyses of closely related members of the CMN group should reveal the genetic gradient which connects nutritionally dependent *Corynebacterium* to nutritionally independent *Mycobacterium*. Among the organisms listed as boundary species, the two species that are least like the type species *C. diphtheriae* are *C. equi* and *C. fascians*. These share properties with *Corynebacterium*, *Nocardia*, and *Mycobacterium*. According to present information, they represent the intergrade between *C. diphtheriae* on the one hand and *N. asteroides* on the other. Recently, Goodfellow and Alderson (1977) have resurrected the genus *Rhodococcus* Zopf 1891 and included therein *C. fascians* and *C. equi*. It should be noted that in their extensive study no other corynebacteria were used for comparison. Similarly, in a study by Mordarski et al. (1977) devoted to DNA-DNA "hybridizations" (homology studies) involving DNA from "rhodochrous" strains and from *Nocardia asteroides* strain N668, it was concluded, "The DNA reassociation data, together with that from chemical, numerical and serological studies [Alshamaony, Goodfellow, and Minnikin, 1976; Goodfellow et al., 1974; Ridell, 1974; Tsukamura, 1975] support the view that the rhodochrous

Table 2. Formulations for members of *Corynebacterium* according to the properties listed in Table 1.

Corynebacterium (groups and species)	Formula[a]

Group 1

diphtheriae var. *mitis*

above: 3 4 5   9*   (14)

below: 1*2*   (6) 7 8   10 11 12 13

var. *gravis*

above: 3 4 5 (7) 9*   (14)

below: 1*2*   6   8   10 11 12 13

var. *intermedius*

above: 3 4 5   9*   (14)

below: 1*2*   6 7 8   10 11 12 13

var. *ulcerans*

above: 3 4 5   7 8 9*   11 12 (14)

below: 1*2*   6   10   13

pseudotuberculosis (ovis)

above: 3 4 5   9*   11   (14)

below: 1*2*   6 7 8   10   12 13

Group 2

equi

above: 9* (10) 11   (14)

below: 1*2*3 4 5 6 7 8   12 13

fascians

above: 9* 10 11   (14)

below: 1*2*3 4 5 6 7 8   12 13

pseudodiphtheriticum

above: 9* 10 11   (14)

below: 1*2*3 4 5 6 7 8   12 13

Group 3

kutscheri

above: 3 4 5 6 7 8 9* 10 11   (14)

below: 1*2*   12 13

renale

above: 3   5   9* 10 11   13 (14)

below: 1*2*   4   6 7 8   12

Group 4

minutissimum

above: 3 4 5 (6)   9* 10   (14)

below: 1*2*   7 8   11 12 13

xerosis

above: 3 (4) 5 6   9* 10   (14)

below: 1*2*   7 8   11. 12 13

[a] Number above the line indicates a positive trait; number below the line indicates a negative trait; circled number below the line indicates that a very occasional strain may be positive; circled number above the line indicates that a very occasional strain may be negative. Asterisk indicates generic traits included to indicate token common properties for the genus. For an enumeration of all the generic traits, refer to section on generic properties, this chapter. Data regarding the hydrolysis of Tween 60, property number 14, are included to stress the point that lipolytic activity is common within the genus. Prompt and reliable test reactions require that the inocula consist of actively growing corynebacteria, washed with saline and resuspended as a thick slurry so that each inoculated tube receives 100 million or more bacteria. Final readings are made at 7 days. **Group 1:** All members fail to hydrolyze pyrazinamide. Each member ferments dextrose, mannose, and maltose. A total of 90% of *gravis, mitis,* and *intermedius* strains are methyl red positive; 100% of *ulcerans* strains are methyl red positive. Note: A fraction of strains of *C. diphtheriae (mitis)* are capable of fermenting sucrose. This is of considerable significance since in some diagnostic laboratories it has been customary to use sucrose fermentation as a means of ruling out a diagnosis of *C. diphtheriae.* Recently, such an error was involved in the death from diphtheria of a 9-year-old boy (Butterworth et al., 1974). Note that biochemical properties do not separate most *mitis* strains from *intermedius* strains and that *gravis* strains are distinguished from these two only by the fermentation of starch. Nevertheless, when these patterns are supplemented with colonial morphology on tellurite agar and cellular morphology (under the microscope), identification can be made. The observation of colonies of the three on sheep's blood agar plates (with a concentration of erythrocytes low enough to reveal feeble hemolysis) will allow separation of hemolytic *mitis* strains from nonhemolytic *gravis* and *intermedius* strains. **Group 2:** 100% are methyl red negative. Additional tests useful in distinguishing members of the group include those for cystinedisulfide reductase, ornithine and lysine decarboxylase, and hydrolysis of esculin. **Group 3:** Readily distinguished by properties shown above. **Group 4:** All strains of *C. minutissimum* produce acid from maltose. Despite reports to the contrary (Cowan and Steele, 1974; Wilson and Miles, 1964, 1975), very few strains of *C. xerosis* ferment maltose (Barksdale et al., 1957). While coral to red fluorescence upon exposure to ultraviolet light has been given as the hallmark of *C. minutissimum* (Sarkany, Taplin, and Blank, 1962), this is a matter of some controversy (McBride, Montes, and Knox, 1970; Somerville, 1973). While the designation *C. minutissimum* is considered invalid, the strains are useful for understanding the genus (Buchanan et al., 1966).

complex be considered a taxon equivalent to the genera *Corynebacterium*, *Nocardia* and *Mycobacterium* [Bousfield and Goodfellow, 1976; Cross and Goodfellow, 1973]." Since the homology studies included neither a variety of corynebacterial DNAs nor a variety of mycobacterial DNAs, such conclusions seem not to be justified. The authors are certainly to be commended for their efforts in homology studies. Had they included DNAs representative of legitimate corynebacteria in their melting out and annealing experiments, they might have discovered that one or some of three existing genera, *Corynebacterium*, *Mycobacterium*, and *Nocardia*, could have provided the proper place or places for their "rhodochrous strains".

Grouping of the Boundary Species for the Genus *Corynebacterium*

According to the data in Table 2, those species selected to form the boundary for the genus *Corynebacterium* fall into four distinct groups. Group 1: *C. diphtheriae* (*mitis*), var. *gravis*, var. *intermedius*, var. *ulcerans*, and *C. pseudotuberculosis* (*ovis*); group 2: *C. equi*, *C. fascians*, and *C. pseudodiphtheriticum*; group 3: *C. kutscheri* and *C. renale*; and group 4: *C. minutissimum* and *C. xerosis*.

Group 1

The organisms comprising group 1 provide a model for a consideration of what constitutes a species in the genus *Corynebacterium*. All members of this group are pyrazinamidase negative (Sulea, Pollice, and Barksdale, 1980). Since all other corynebacteria (rare mutants excepted) are pyrazinamidase positive, a test for the pyrazinamidase status of a possible corynebacterium from the throat, for example, will readily distinguish *C. xerosis* or *C. pseudodiphtheriticum* from *C. diphtheriae*, its varieties, and *C. pseudotuberculosis*. From the information given in Table 2, the enzymatically most versatile member of the group is *C. diphtheriae* var. *ulcerans*. Furthermore, it infects both cattle and man—it was originally described (Gilbert and Stewart, 1927) from an outbreak of sore throats in human subjects whose common bond was consumption of milk from the same dairy (Gilbert and Stewart, 1929). Since then, this variety of *C. diphtheriae* has been recovered in other areas of the United States and in Europe (Henriksen and Grelland, 1952; Jebb, 1948; Maximescu et al., 1974; Saxholm, 1951). The next most versatile member of group 1 is *C. pseudotuberculosis* (see Table 2) which infects sheep, horses, and man (Barksdale, 1970). Some strains received by us from veterinary bacteriologists as *C. pseudo-*

tuberculosis are indistinguishable from *C. diphtheriae* var. *gravis*. Of the strains hitherto considered as *C. diphtheriae* (see Table 2) the var. *gravis* is the more versatile and shares with the var. *ulcerans* (not *C. pseudotuberculosis*) the capacity to hydrolyze starch. Since ovine and bovine mammalian species were on earth long before man, they probably provided the first animal habitat for the ancestors of *C. diphtheriae*. The organisms comprising group 1 share a requirement for pantothenic acid and some of them require nicotinic acid (Arden, 1970). They all produce a neuraminidase (sialidase) capable of cleaving *N*-acetylneuraminic acid (NAN) residues from animal mucins and gangliosides. Further, they synthesize an *N*-acetylneuraminic acid lyase (NAN lyase) which cleaves NAN into its component pyruvic acid and *N*-acetylmannosamine. No other species within the genus appears to share these abilities (Arden, Chang, and Barksdale, 1972). One other rather singular attribute seems common to group 1 organisms. When any one of them integrates into its genome a prophage carrying the gene *tox*, the singular protein, diphtherial toxin, is synthesized. In other words, lysogenization of any member of group 1 with a *tox*-carrying phage creates a strain capable of producing diphtherial toxin and theoretically, therefore, capable of producing diphtheritic death of an infected subject. Having, as they do, so many common properties not shared with other members of the genus *Corynebacterium*, group 1 organisms seem to represent a single species comprised of several recognizable varieties. It has long been realized that strains of corynebacteria existed which were somewhat like *C. pseudotuberculosis* and *C. diphtheriae* (Barratt, 1933; Jebb, 1948; Jebb and Martin, 1965; Mair, 1928, 1930). While there is not a great deal of information on the incidence of *C. diphtheriae* in animals (Greathead and Bisschop, 1963), there is enough to indicate that such infections of horses and cattle do occur and that *C. ulcerans* infections of man and cattle occur (Higgs et al., 1967), and human infections with *C. pseudotuberculosis* are known (Battey et al., 1968). If looked for, more of these cases would undoubtedly be established. Meanwhile it seems safe to regard *C. diphtheriae* as a single species of several recognizable varieties, any one of which may infect man, sheep, and cows.

Group 2

The organisms of group 2 present a predominantly oxidative metabolism most readily suggested by a negative methyl red test. They lack the additional properties discussed under group 1. Studies on their DNAs and homologies between them remain to be done. *Corynebacterium equi* is discussed in Chapter 155 and *C. fascians* in Chapter 143, this Handbook.

Group 3

Corynebacterium renale, a cause of pyelonephritis in cattle, has been exhaustively studied by Yanagawa (Hirai and Yanagawa, 1967; Hirai, Shimakura, and Yanagawa, 1969; Hiramune et al., 1970, 1972; Yanagawa, Shinagawa, and Nerome, 1968). The pattern of reactions given in Table 2 represent 10 certified strains. More than half of the strains were methyl red negative. Only a handful of strains of *C. kutscheri* are available. Homology studies between this species and *C. renale* remain to be done. *C. renale* DNA has a G+C content of 58.2 mol% and shows only 42.3% homology with DNA from *C. diphtheriae.*

Group 4

The representatives of group 4, *Corynebacterium minutissimum* and *C. xerosis,* have similar habitats. They are easy to distinguish biochemically from other species in the genus. Their mole percentages of G+C are 60 and 59, respectively. *C. minutissimum* DNA shows about 49.0% homology with *C. diphtheriae.* Although *C. minutissimum* is technically an illegitimate species (Buchanan, Holt, and Lessel, 1966), the designation is here used in lieu of a more valid one for the biological entities known under that name.

Media Useful in the Isolation and Identification of Corynebacteria

Samples of infected material(s) should be emulsified under aseptic conditions (in an appropriate volume of sterile saline or broth) and streaked out on one or more of the following solid media contained in Petri dishes. Inoculation of a slant of the inspissated serum medium of Loeffler will provide a useful reservoir of the sample in case of failure in initial plating. Corynebacteria grow best from small inocula when provided with 10% CO_2 in air.

Chocolate Agar for Isolating Corynebacteria
Chocolate agar is made with blood which has been rendered by heating to a cocoa-brown color. Sterile molten brain heart infusion agar (Difco), 500 ml, is cooled (in a water bath) to 75–80°C and combined with 25 ml warmed (room temperature) sterile horse blood. The blood and agar mixture is rotated gently (in the 75–80°C water bath) for 3–5 min, during which the color should change from red to mahogany to cocoa-brown. The mixture is then cooled to 56°C and 35 ml sterile horse serum is added. The mixture is rotated again to ensure mixing. Pour into Petri

dishes so as to yield fairly thick plates. Incubate for sterility. Refrigerate in plastic bags. The medium should be useable for 4 weeks.

Tinsdale's Agar for Isolating Corynebacteria (Tinsdale, 1947, as modified by Moore and Parsons, 1958)
Agar base:

Proteose peptone No. 3 (Difco)	2.0 g
NaCl	0.5 g
Agar (Difco)	2.0 g
Quartz-distilled water	up to 100 ml

The molten agar base is adjusted to pH 7.4; autoclaved, 15 lbs, 15 min; and cooled to 56°C. To it, the following sterile solutions are aseptically added: bovine serum (Seitz filtered), 10 ml; N/10 NaOH, 6 ml; L-cystine, 0.4% (wt/vol) in N/10 HCl, 6 ml; potassium tellurite, 1% (wt/vol) aqueous solution, 3 ml; sodium thiosulfate, anhydrous, 2.5% (wt/vol) aqueous solution, 1.7 ml, freshly prepared. (A supplement containing these materials is available from Difco as dessicated Bacto-Tinsdale Enrichment.) Pour into Petri dishes. Satisfactory results cannot be expected with plates that are more than 36 h old.

Corynebacterium diphtheriae and certain other corynebacteria produce a cysteine desulfhydrase. The liberated H_2S forms a sulfide visible as a brown halo around black colonies (containing reduced tellurite) on this medium.

Mueller-Miller Tellurite Agar for Isolating Corynebacteria (Mueller and Miller, 1946)
A stock solution of serum-tellurite is prepared in the following way. Fifty milliliters of 85% lactic acid plus 75 ml H_2O with a few drops of phenol red solution are neutralized with strong (40–50%) NaOH. The solution is heated just to boiling for 5 min, adding more NaOH if necessary to retain red color. To 40 ml of the resulting solution, sterilized by autoclaving, is added, with complete mixing and in order: ethyl alcohol, 10 ml; calcium pantothenate (autoclaved separately in a few milliliters of H_2O), 0.2 mg; sterile horse or beef serum (filter-sterilized), 50 ml; potassium tellurite (dissolved in a few milliliters of sterile H_2O), 0.4 g. The stock solution is stored in a well-stoppered container in a refrigerator. Insoluble material that may form slowly in the stock solution does not interfere with the growth of or the uptake tellurite by tellurite-reducing organisms. The agar base, freshly prepared, contains the following:

Agar	10 g
Casamino Acids (Difco)	10 g
Commercial casein	2.5 g
KH_2PO_4	0.15 g

MgSO$_4$·7H$_2$O	0.05 g
D,L-Tryptophan	0.025 g
Distilled water	up to 500 ml

Commercial casein can be solubilized by suspending it in 50 ml H$_2$O and adding 40% NaOH drop by drop slowly with shaking until the casein is in complete solution. It should be stressed that excess NaOH is to be avoided. Adjust the pH to 7.6 and autoclave 10 lbs, 10 min. Cool to about 50°C and add 12.5 ml of the serum-tellurite stock solution. Mix well and pour reasonably thick plates. Maintain in plastic bags in a refrigerator for no more than 4 weeks. (A prepared medium is available from Difco as Mueller tellurite agar base and Mueller tellurite serum.)

Note: Tellurite-reducing organisms that are sensitive to hemin or other substances present in whole blood should be selected on this medium.

Chocolate Tellurite Agar for
Isolating Corynebacteria

To chocolate agar (560 ml, see above) is added 5 ml of filter-sterilized 1% (wt/vol) aqueous solution of potassium tellurite. Pour into Petri dishes. Plates should be stored in a refrigerator and used within 5 days.

Note: A number of members of the *Mycobacteriaceae* form black colonies on tellurite agar, as do some micrococci, yeasts, etc. Thus, at best, tellurite-containing agar differentiates between those organisms which reduce tellurite (form black colonies) and those which do not (produce colorless colonies or fail to grow).

Identification of Corynebacteria

Organisms isolated as black colonies on one of the tellurite-containing media and exhibiting the characteristics listed under the section on generic properties may next be examined for the presence or absence of the capacities listed in Table 1. Formulae which distinguish four groupings of corynebacteria constitute Table 2. Most corynebacteria so grouped can be further identified using specific data provided in Buchanan and Gibbons (1974) or in papers describing such new species as *C. genitalium* (Furness and Evangelista, 1976), *C. pilosum,* and *C. cystitidis* (Yanagawa and Honda, 1978).

Literature Cited

Adam, A., Petit, J. F., Wietzerbin-Falszpan, J., Sinay, P., Thomas, D. W., Lederer, E. 1969. L'acide N-glycolyl-muramique, constituant des parois de *Mycobacterium smegmatis:* Identification par spectrometrie de masse. FEBS Letters **4:**87–92.

Alshamaony, L., Goodfellow, M., Minnikin, D. E. 1976. Free mycolic acids as criteria in the classification of *Nocardia* and the *'rhodochrous'* complex. Journal of General Microbiology **92:**188–199.

Anderson, J. S., Happold, F. C., McLeod, J. W., Thomson, J. G. 1931. On the existence of two forms of diphtheria bacillus—*B. diphtheriae gravis* and *B. diphtheriae mitis*—and a new medium for their differentiation and for the bacteriological diagnosis of diphtheria. Journal of Pathology and Bacteriology **34:**667–681.

Andrewes, F. W., Bulloch, W., Douglas, S. R., Dreyer, G., Gardner, A. D., Fildes, P., Ledingham, J. C. G., Wolf, C. G. L. 1923. Diphtheria. London: His Majesty's Stationery Office.

Arden, S. B. 1970. Comparative studies of *Corynebacterium diphtheriae, C. ovis, C. ulcerans* and related species. Ann Arbor, Michigan; London: University Microfilms International.

Arden, S. B., Chang, W.-H., Barksdale, L. 1972. Distribution of neuraminidase and N-acetylneuraminate lyase activities among corynebacteria, mycobacteria and nocardias. Journal of Bacteriology **112:**1206–1212.

Asselineau, C., Asselineau, J. 1966. Stéréochimie de l'acide corynomycolique. Bulletin de la Société de Chimie de France **6:**1992–1999.

Asselineau, C., Asselineau, J. 1978a. Trehalose-containing glycolipids. Progress in the Chemistry of Fats and Other Lipids **16:**59–99.

Asselineau, C., Asselineau, J. 1978b. Lipides specifiques des mycobacteries. Annales de Microbiologie (Institut Pasteur) **129A:**49–69.

Asselineau, C. P., Tocanne, G., Tocanne, J.-F. 1970. Stéréochimie des acides mycoliques. Bulletin de la Société de Chimie de France, 1455–1459.

Asselineau, J. 1966. The bacterial lipids. Paris: Hermann; San Francisco: Holden-Day.

Azuma, I., Thomas, D. W., Adam, A., Ghuysen, J.-M., Bonaly, R., Petit, J.-F., Lederer, E. 1970. Occurrence of N-glycolylmuramic acid in bacterial cell walls. Biochimica et Biophysica Acta **208:**444–451.

Azuma, I., Yamamura, Y., Misaki, A. 1969. Isolation and characterization of arabinose mycolate from firmly bound lipids of mycobacteria. Journal of Bacteriology **98:**331–333.

Barksdale, L. 1970. *Corynebacterium diphtheriae* and its relatives. Bacteriological Reviews **34:**378–422.

Barksdale, L. 1979. Identifying *Rothia dentocariosa.* Annals of Internal Medicine **91:**786–788.

Barksdale, L., Arden, S. B. 1974. Persisting bacteriophage infections, lysogeny, and phage conversions. Annual Review of Microbiology **28:**265–299.

Barksdale, L., Kim, K.-S. 1977. *Mycobacterium.* Bacteriological Reviews **41:**217–372.

Barksdale, W. L., Li, K., Cummins, C. S., Harris, H. 1957. The mutation of *Corynebacterium pyogenes* to *Corynebacterium hemolyticum.* Journal of General Microbiology **16:**749–758.

Barksdale, L., Lanéelle, M.-A., Pollice, M. C., Asselineau, J., Welby, M., Norgard, M. V. 1979. Biological and chemical basis for the reclassification of *Microbacterium flavum* Orla-Jensen as *Corynebacterium flavescens* nom. nov. International Journal of Systematic Bacteriology **29:**222–233.

Barratt, M. M. 1933. A group of aberrant members of the genus *Corynebacterium* isolated from the human nasopharynx. Journal of Pathology and Bacteriology **36:**369–397.

Battey, Y. M., Tonge, J. I., Horsfall, W. R., McDonald, I. R. 1968. Human infection with *Corynebacterium ovis.* The Medical Journal of Australia **2:**540–543.

Beaman, B. L., Burnside, J. 1973. Pyridine extraction of nocardial acid fastness. Applied Microbiology **26:**426–428.

Beaman, B. L., Kim, K.-S., Laneelle, M.-A., Barksdale, L. 1974. Chemical characterization of organisms isolated from leprosy patients. Journal of Bacteriology **117:**1320–1329.

Bergey, D. H., Breed, R. S., Murray, E. G. D., Hitchens, A. P.

1939. Bergey's manual of determinative bacteriology, 5th ed. Baltimore: Williams & Wilkins.

Bordet, C., Karahjoli, M., Gateau, O., Michel, G. 1972. Cell walls of nocardiae and related actinomycetes: Identification of the genus *Nocardia* by cell wall analysis. International Journal of Systematic Bacteriology 22:251–259.

Bouisset, L., Breuillaud, J., Michel, G. 1963. Étude de l'ADN chez les Actinomycétales. Comparaison entre les valeurs du rapport $\frac{A+T}{G+C}$ et les caractères bactériologiques des *Corynebacterium*. Annales de l'Institut Pasteur 104:756–770.

Bousfield, I. J., Goodfellow, M. 1976. The *"rhodochrous"* complex and its relationships with allied taxa, pp. 39–73. In: Goodfellow, M., Brownell, G. H., Serrano, J. A. (eds.), The biology of the nocardiae. London: Academic Press.

Bradley, S. G. 1973. Relationships among mycobacteria and nocardiae based upon deoxyribonucleic acid reassociation. Journal of Bacteriology 113:645–651.

Breed, R. S., Murray, E. G. D., Hitchens, A. P. 1948. Bergey's manual of determinative bacteriology, 6th ed. Baltimore: Williams & Wilkins.

Brennan, P. J., Lehane, D. P. 1969. Acylglucoses in corynebacteria. Biochimica et Biophysica Acta 176:675–677.

Brennan, P. J., Lehane, D. P., Thomas, D. W. 1970. Acylglucoses of the corynebacteria and mycobacteria. European Journal of Biochemistry 13:117–123.

Bruneteau, M., Michel, G. 1968. Structure d'un dimycolate d'arabinose isole de *Mycobacterium marianum*. Chemistry and Physics of Lipids 2:229–239.

Buchanan, R. E., Gibbons, N. E. 1974. Bergey's manual of determinative bacteriology, 8th ed. Baltimore: Williams & Wilkins.

Buchanan, R. E., Holt, J. G., Lessel, E. F., Jr. 1966. Index bergeyana. Baltimore: Williams & Wilkins.

Butterworth, A., Abbott, J. D., Simmons, L. E., Ironside, A. G., Mandal, B. K., Williams, R. F., Brennand, J., Mann, N. M., Simon, S. 1974. Diphtheria in the Manchester area 1967–1971. Lancet ii:1558–1561.

Collier, R. J. 1975. Diphtheria toxin: Mode of action and structure. Bacteriological Reviews 39:54–85.

Cowan, S. T. 1974. Cowan and Steel's manual for the identification of medical bacteria, 2nd ed. London, New York: Cambridge University Press.

Cross, T., Goodfellow, M. 1973. Taxonomy and classification of the actinomycetes, pp. 11–112. In: Skinner, F. A., Sykes, G. (eds.), Actinomycetes: Characteristics and practical importance. London: Academic Press.

Cummins, C. S. 1954. Some observations on the nature of the antigens in the cell wall of *Corynebacterium diphtheriae*. British Journal of Experimental Pathology 35:166–180.

Cummins, C. S. 1962. Chemical composition and antigenic structure of cell walls of *Corynebacterium, Mycobacterium, Nocardia, Actinomyces* and *Arthrobacter*. Journal of General Microbiology 28:35–50.

Cummins, C. S. 1965. Chemical and antigenic studies on cell walls of mycobacteria, corynebacteria and nocardias. American Review of Respiratory Diseases 92:63–72.

Cummins, C. S. 1971. Catalase activity in *Corynebacterium pyogenes*. Canadian Journal of Microbiology 17:1001–1002.

Cummins, C. S., Harris, H. 1956. The chemical composition of the cell in some Gram-positive bacteria and its possible value as a taxonomic character. Journal of General Microbiology 14:583–600.

Cummins, C. S., Harris, H. 1958. Studies on the cell wall composition and taxomony of Actinomycetales and related groups. Journal of General Microbiology 18:173–189.

Diara, A., Pudles, J. 1959. Sur les lipides de *Corynebacterium ovis*. Bulletin de la Société de Chimie Biologie 41:481–486.

Etemadi, A. H., Gasche, J., Sifferlen, J. 1965. Identification d'homologues supérieurs des acides corynomycolique et corynomycolénique dans les lipides de *Corynebacterium* 506.

Bulletin de la Société de Chimie Biologie 47:631–638.

Etemadi, A. H. 1967. Les acides mycoliques structure, biogenese et interet phylogenetique. Exposés Annuels de Biochimie Medicale, 77–109.

Freund, J., Lipton, M. M. 1948. Potentiating effect of *Nocardia asteriodes* on sensitization to picryl chloride and on production of isoallergic encephalomyelitis. Proceedings of the Society for Experimental Biology and Medicine 68:373–377.

Furness, G., Evangelista, A. T. 1976. Infection of nonspecific urethritis patient and his consort with a pathogenic species of non-specific urethritis corynebacteria, *Corynebacterium genitalium*, n. sp. Investigative Urology 14:202–205.

Gilbert, R., Stewart, F. C. 1927. *Corynebacterium ulcerans:* A pathogenic microorganism resembling *C. diphtheriae*. Journal of Laboratory and Clinical Medicine 12:756–761.

Gilbert, R., Stewart, F. C. 1929. *Corynebacterium ulcerans:* Its epidemiological importance. Journal of Laboratory and Clinical Pathology 14:1032–1036.

Goodfellow, M., Alderson, G. 1977. The actinomycete genus *Rhodococcus:* A home for the *'rhodochrous'* complex. Journal of General Microbiology 100:99–122.

Goodfellow, M., Lind, A., Mordarska, H., Pattyn, S., Tsukamura, M. 1974. A co-operative numerical analysis of cultures considered to belong to the *'rhodochrous'* taxon. Journal of General Microbiology 85:291–302.

Gordon, R. E. 1966. Some strains in search of a genus — *Corynebacterium, Mycobacterium, Nocardia* or what? Journal of General Microbiology 43:329–343.

Gordon, R. E. 1978. A species definition. International Journal of Systematic Bacteriology 28:605–607.

Greathead, M. M., Bisschop, P. J. N. R. 1963. A report on the occurrence of *C. diphtheriae* in dairy cattle. South African Medical Journal 37:1261–1262.

Harrington, B. J. 1966. A numerical taxonomic study of some corynebacteria and related organisms. Journal of General Microbiology 45:31–40.

Henriksen, S. D., Grelland, R. 1952. Toxigenicity, serological reactions and relationships of the diphtheria-like corynebacteria. Journal of Pathology and Bacteriology 64:503–511.

Higgs, T. M., Smith, A., Cleverly, L. M., Neave, F. K. 1967. *Corynebacterium ulcerans* infections in a dairy herd. Veterinary Record 81:34–35.

Hirai, K., Shimakura, S., Yanagawa, R. 1969. Minimum medium for *Corynebacterium renale* and filamentous growth due to deficiency or excess of inorganic ions. Japanese Journal of Veterinary Science 31:149–159.

Hirai, K., Yanagawa, R. 1967. Nutritional requirements of *Corynebacterium renale*. Japanese Journal of Veterinary Research 15:121–134.

Hiramune, T., Kume, T., Murase, N., Yanagawa, R. 1970. Typing of *Corynebacterium renale* isolated from cattle in a herd with persistent pyelonephritis. Japanese Journal of Veterinary Science 32:81–85.

Hiramune, T., Inui, S., Murase, N., Yanagawa, R. 1972. Antibody response in cows infected with *Corynebacterium renale* with special reference to the differentiation of pyelonephritis and cystitis. Research in Veterinary Science 13:82–86.

Iizuka, H., Komagata, K. 1964. Microbiological studies on petroleum and natural gas. I. Determination of hydrocarbon-utilizing bacteria. Journal of General and Applied Microbiology 10:207–221.

Imaeda, T., Barksdale, L., Norgard, M. 1980. Genome sizes and DNA homologies among *Corynebacterium, Mycobacterium* and *Nocardia*. (Manuscript in preparation.)

Ioneda, T., Lederer, E., Rozanis, J. 1970. Sur la structure des diesters de trehalose ("cord factors") produits par *Nocardia asteroides* et *Nocardia rhodochrous*. Chemistry and Physics of Lipids 4:375–392.

Ioneda, T., Lenz, M., Pudles, J. 1963. Chemical constitution of a glycolipid from *C. diphtheriae* P. W. 8. Biochemical and Biophysical Research Communications 13:110–114.

Jebb, W. H. H. 1948. Starch-fermenting gelatin-liquefying corynebacteria isolated from the human nose and throat. Journal of Pathology and Bacteriology **60**:403–412.

Jebb, W. H. H., Martin, T. D. M. 1965. A non-starch-fermenting variant of *Corynebacterium ulcerans*. Journal of Clinical Pathology **18**:757–758.

Jensen, H. L. 1934. Studies on saprophytic mycobacteria and corynebacteria. Proceedings of the Linnean Society of New South Wales **59**:19–61.

Jensen, H. L. 1952. The coryneform bacteria. Annual Review of Microbiology **6**:77–90.

Jones, D. 1975. A numerical taxonomic study of coryneform and related bacteria. Journal of General Microbiology **87**:52–96.

Kanetsuna, F. 1968. Chemical analyses of mycobacterial cell walls. Biochimica et Biophysica Acta **158**:130–143.

Kanetsuna, F., Imaeda, T., Cunto, G. 1969. On the linkage between mycolic acid and arabinogalactan in phenol-treated mycobacterial cell walls. Biochimica et Biophysica Acta **173**:341–344.

Kanetsuna, F., San Blas, G. 1970. Chemical analysis of a mycolic acid-arabinogalactan-mucopeptide complex of mycobacterial cell wall. Biochimica et Biophysica Acta **208**:434–443.

Kareva, V. A., Filosofova, T. G. 1966. Composition of nitrogenous bases of DNA in some strains of diphtheria bacilli. [In Russian.] Ukrainskii Biokhimichnyi Zhurnal **38**:321–324.

Kato, K., Strominger, J. L., Kotani, S. 1968. Structure of the cell wall of *Corynebacterium diphtheriae*. I. Mechanism of hydrolysis by the L-3 enzyme and the structure of the peptide. Biochemistry **7**:2762–2773.

Kato, M. 1970. Action of toxic glycolipid of *Corynebacterium diphtheriae* on mitochondrial structure and function. Journal of Bacteriology **101**:709–716.

Knoche, H. W., Koths, K. E. 1973. Characterization of a fatty acid synthetase from *Corynebacterium diphtheriae*. Journal of Biological Chemistry **248**:3517–3519.

Komagata, K., Yamada, Y., Ogawa, H. 1969. Taxonomic studies on coryneform bacteria. I. Division of bacterial cells. Journal of General and Applied Microbiology **15**:243–259.

Komura, I., Komagata, K., Mitsugi, K. 1973. A comparison of *Corynebacterium hydrocarboclastus* Iizuka and Komagata 1964 and *Nocardia erythropolis* (Gray and Thornton) Waksman and Henrici 1948. Journal of General and Applied Microbiology **19**:161–170.

Komura, I., Yamada, K., Otsuka, S. I., Komagata, K. 1975. Taxonomic significance of phospholipids in coryneform and nocardioform bacteria. Journal of General and Applied Microbiology **21**:251–261.

Laneelle, M. A., Asselineau, J. 1970. Characterisation de glycolipides dans une souche de *Nocardia braziliensis*. FEBS Letters **7**:64–67.

Lechevalier, M. P., de Bievre, C., Lechevalier, H. 1977. Chemotaxonomy of aerobic actinomycetes: Phospholipid composition. Biochemical Systematics and Ecology **5**:249–260.

Lederer, E., Pudles, J. 1951. Sur l'isolement et constitution chimique d'un hydroxy-acide ramifié du bacille diphtérique. Bulletin de la Société de Chimie Biologie **33**:1003–1011.

Liu, T.-Y., Gotschlich, E. 1967. Muramic acid phosphate as a component of the mucopeptide of Gram-positive bacteria. Journal of Biological Chemistry **242**:471–476.

Loeffler, F. 1884. Untersuchungen über die Bedeutung der Mikroorganismen für die Entstehung der Diphtherie beim Menschen, bie der Taube und beim Kalbe. Mittheilungen aus dem K. Gesundheitsamte, Berlin **2**:421–449.

McBride, M. E., Montes, L. F., Knox, J. M. 1970. The characterization of fluorescent skin diphtheroids. Canadian Journal of Microbiology **16**:941–946.

McLeod, J. W. 1943. The types *mitis, intermedius,* and *gravis* of *Corynebacterium diphtheriae*. Bacteriological Reviews **7**:1–41.

Magnusson, M. 1961. Specificity of mycobacterial sensitins. I.

Studies in guinea pigs with purified "tuberculin" prepared from mammalian and avian tubercle bacilli, *Mycobacterium balnei,* and other acid-fast bacilli. American Review of Respiratory Diseases **83**:57–68.

Mair, W. 1928. A strain of *B. diphtheriae* showing unusual virulence for guinea pigs. Journal of Pathology and Bacteriology **31**:136–137.

Mair, W. 1930. On testing *C. diphtheriae* for virulence by the intracutaneous method. Journal of Pathology and Bacteriology **33**:230–231.

Maurice, M. T., Vacheron, M. J., Michel, G. 1971. Isolement d'acides nocardiques de plusieurs especes de *Nocardia*. Chemistry and Physics of Lipids **7**:9–18.

Maximescu, P., Oprisan, A., Pop, A., Potorac, E. 1974. Further studies on *Corynebacterium* species capable of producing diphtheria toxin (*C. diphtheriae, C. ulcerans, C. ovis*). Journal of General Microbiology **82**:49–56.

Michel, G., Bordet, C., Lederer, E. 1960. Isolement d'un nouvel acide mycolique: L'acide nocardique, de partir d'un souche *Nocardia asteroides*. Comptes Rendus Hebdomadaires des Séances de l'Académie des Sciences **250**:3518–3520.

Misaki, A., Yukawa, S., Tsuchiya, K., Yamasaki, T. 1966. Studies on cell walls of *Mycobacteria*. 1. Chemical and biological properties of the cell walls and the mucopeptide of BCG. Journal of Biochemistry **59**:388–396.

Moore, M. S., Parsons, E. I. 1958. A study of a modified Tinsdale's medium for the primary isolation of *Corynebacterium diphtheriae*. Journal of Infectious Diseases **102**:88–93.

Mordarski, M., Goodfellow, M., Szyba, K., Pulverer, G., Tkacz, A. 1977. Classification of the "rhodochrous" complex and allied taxa based upon deoxyribonucleic acid reassociation. International Journal of Systematic Bacteriology **27**:31–37.

Mueller, J. H., Miller, P. A. 1946. A new tellurite plating medium and some comments on the laboratory "diagnosis" of diphtheria. Journal of Bacteriology **51**:743–750.

Noll, H., Bloch, H., Asselineau, J., Lederer, E. 1956. The chemical structure of the cord factor of *Mycobacterium tuberculosis*. Biochimica et Biophysica Acta **20**:299–309.

Pappenheimer, A. M., Jr. 1977. Diphtheria toxin. Annual Review of Biochemistry **46**:69–94.

Parant, M., Audibert, F., Parant, F., Chedid, L., Soler, E., Polonsky, J., Lederer, E. 1978. Nonspecific immunostimulant activities of synthetic trehalose-6,6'-diesters (lower homologs of cord factor). Infection and Immunity **20**:12–19.

Petit, J. F., Adam, A., Wietzerbin-Falszpan, J., Lederer, E., Ghuysen, J. M. 1969. Chemical structure of the cell wall of *Mycobacterium smegmatis*. I. Isolation and partial characterization of the peptidoglycan. Biochemical and Biophysical Research Communications **35**:478–485.

Ridell, M. 1974. Serological study of nocardiae and mycobacteria using "*Mycobacterium*" *pellegrino* and *Nocardia coralina* precipitation reference systems. International Journal of Systematic Bacteriology **24**:64–72.

Sarkany, I., Taplin, D., Blank, H. 1962. Organism causing erythrasma. Lancet **ii**:304–305.

Saxholm R. 1951. Toxin-producing diphtheria-like organisms isolated from cases of sore throat. Journal of Pathology and Bacteriology **63**:303–311.

Schleifer, K. H., Kandler, O. 1972. Peptidoglycan types of bacterial cell walls and their taxonomic implications. Bacteriological Reviews **36**:407–477.

Senn, M., Ioneda, T., Pudles, J., Lederer, E. 1967. Spectrometrie de masse de glycolipides. I. Structure du "cord factor" de *Corynebacterium diptheriae*. European Journal of Biochemistry **1**:353–356.

Somerville, D. A. 1973. A taxonomic scheme for aerobic diphtheroids from human skin. Journal of Medical Microbiology **6**:215–224.

Spirin, A. S., Belozersky, A. N., Shugayeva, N. V., Vanyushin,

V. F. 1957. A study of species specificity with respect to nucleic acids in bacteria. Biochemistry [English translation of Biokhimiya] **22**:699–707.

Sulea, I. T., Pollice, M. C., Barksdale, L. 1980. Pyrazine carboxylamidase activity in *Corynebacterium*. International Journal of Systematic Bacteriology **30**:466–472.

Tinsdale, G. F. W. 1947. New medium for isolation and identification of *C. diphtheriae* based on production of H$_2$S. Journal of Pathology and Bacteriology **59**:461–466.

Tsukamura, M. 1975. Numerical analysis of the relationship between *Mycobacterium*, rhodochrous group, and *Nocardia* by use of hypothetical median organisms. International Journal of Systematic Bacteriology **25**:329–335.

Vacheron, M.-J., Guinand, M., Michel, G., Ghuysen, J.-M. 1972. Structural investigations on cell walls of *Nocardia* sp. The wall lipid and peptidoglycan moieties of *Nocardia kirovani*. European Journal of Biochemistry **29**:156–166.

Vance, D. E., Mitsuhashi. O., Bloch, K. 1973. Purification and properties of the fatty acid synthetase from *Mycobacterium phlei*. The Journal of Biological Chemistry **248**:2303–2309.

Veldkamp, H. 1970. Saprophytic coryneform bacteria. Annual Review of Microbiology **24**:209–240.

Vestal, A. L. 1975. Procedures for the isolation and identification of *Mycobacteria*. U.S. Department of Health, Education, and Welfare Publication No. (CDC) 75-8230. Atlanta: Center for Disease Control.

Welby-Gieusse, M., Laneelle, M. A., Asselineau, J. 1970. Structure des acides corynomycoliques de *Corynebacterium hofmanii* et leur implication biogenetique. European Journal of Biochemistry **13**:164–167.

Wietzerbin-Falszpan, J., Das, B. C., Azuma, I., Adam, A., Petit, J. F., Lederer, E. 1970. Isolation and mass spectrometric identification of the peptide subunits of mycobacterial cell walls. Biochemical and Biophysical Research Communications **40**:57–63.

Wilson, G. S., Miles, A. A. (eds.). 1964. Topley and Wilson's principles of bacteriology and immunity, vol. 1, 5th ed, pp. 605 and 609. Baltimore: Williams & Wilkins.

Wilson, G. S., Miles, A., (eds.). 1975. Topley and Wilson's principles of bacteriology, virology and immunity, vol. 1, 6th ed, p. 634. Baltimore: Williams & Wilkins.

Winslow, C.-E. A., Broadhurst, J., Buchanan, R. E., Krumwiede, C., Jr., Rogers, L. A., Smith, G. H. 1917. The families and genera of the bacteria. Preliminary report of the Committee of the Society of American Bacteriologists on Characterization and Classification of Bacterial Types. Journal of Bacteriology **2**:505–566.

Winslow, C.-E. A., Broadhurst, J., Buchanan, R. E., Krumweide, C., Jr., Rogers, L. A., Smith, G. H. 1920. The families and genera of bacteria. Final report of the Committee of the Society of American Bacteriologists on Characterization and Classification of Bacterial Types. Journal of Bacteriology **5**:191–229.

Yamada, K., Komagata, K. 1970a. Taxonomic studies on coryneform bacteria. II. Principal amino acids in the cell wall and their taxonomic significance. Journal of General and Applied Microbiology **16**:103–113.

Yamada, K., Komagata, K. 1970b. Taxonomic studies on coryneform bacteria. III. DNA base composition of coryneform bacteria. Journal of General and Applied Microbiology **16**:215–224.

Yamada, K., Komagata, K. 1972a. Taxonomic studies on coryneform bacteria. IV. Morphological, cultural, biochemical and physiological characteristics. Journal of General and Applied Microbiology **18**:399–416.

Yamada, K., Komagata, K. 1972b. Taxonomic studies on coryneform bacteria. V. Classification of coryneform bacteria. Journal of General and Applied Microbiology **18**:417–431.

Yanagawa, R., Honda, E. 1978. *Corynebacterium pilosum* and *Corynebacterium cystitidis*, two new species from cows. International Journal of Systematic Bacteriology **28**:209–216.

Yanagawa, R., Shinagawa, M., Nerome, K. 1968. Lysogeny in *Corynebacterium renale*. Japanese Journal of Veterinary Research **16**:121–127.

Saprophytic, Aerobic Coryneform Bacteria

RONALD M. KEDDIE and DOROTHY JONES

GENERAL INTRODUCTION

In this chapter, the adjective "saprophytic" is applied in a wide sense to the coryneform bacteria other than those usually recognized as animal parasites or as plant pathogens. Defined in this way, the saprophytic coryneform bacteria are widely distributed in nature and often form a substantial fraction of the bacterial populations of such diverse materials as soils, activated dairy-waste sludge, animal-manure slurries, poultry deep litter, milk, cheese, the surfaces of freshwater and marine fish, and plant surfaces. Yet, despite their obvious numerical importance, most of the past attempts to identify the coryneform bacteria from these diverse habitats have been frustrated by the confused state of the classification of these organisms and of allied taxa. In recent years, the widespread application of numerical taxonomic and, especially, of chemotaxonomic techniques has had a considerable impact on the classification of coryneform and nocardioform bacteria, and good progress is now being made in developing a satisfactory system. Thus, although many problems still remain, it is now possible to identify many new isolates, at least to generic level, with much greater confidence than was possible only a few years ago.

The term "coryneform" was coined originally by Ørskov (see Jensen, 1966) and was used by Jensen (1952) to describe the animal-parasitic and -pathogenic species of the genus *Corynebacterium* (*Corynebacterium sensu stricto* in Jensen's sense), together with those saprophytic and plant-pathogenic bacteria which had similar, distinctive morphological features and staining properties. The expression was introduced in an attempt to counter objections to the widespread use of the generic name *Corynebacterium* for species other than *C. diphtheriae* and its close relatives (see Conn, 1947). Thus from the beginning, the grouping was based entirely on the possession of certain distinctive morphological features; but because at that time considerable importance was attached to morphology as a taxonomic feature, it was also widely believed that the coryneform bacteria were a group of related organisms, a view that was to persist for a number of years. Modern taxonomic methods have shown that,

contrary to earlier opinion, the various fractions of the coryneform complex bear little taxonomic relationship to each other. Thus, the expression "coryneform" is now used only as a convenient label to describe a broad morphological group of bacterial taxa that are otherwise largely unrelated (see Keddie, 1978, for a brief historical discussion, and Chapter 141, this Handbook, for a contrary view concerning this term).

Using Jensen's (1952) description as a guide, Cure and Keddie (1973) described the characteristic features of coryneform bacteria in the following way: "In exponential phase cultures in complex media, irregular rods occur which vary considerably in size and shape and include straight, bent and curved, wedge-shaped and club-shaped forms. A proportion of the rods are arranged at an angle to each other to give V-formations but other angular arrangements may be seen. Rudimentary branching may occur, especially in richer media, but definite mycelia are not formed. In stationary phase cultures the cells are generally much shorter and less irregular and a variable proportion is coccoid in shape. The rods may be non-motile or motile; endospores are not formed. They are Gram positive but may be readily decolorized and may show only Gram positive granules in otherwise Gram negative cells. They are not acid-fast."

An additional morphological feature of coryneform bacteria is that they do not form chains of cells during exponential growth. The apparent chain formation sometimes seen in the later stages of the growth cycle is a result of multiple fragmentation of long rods. The angular arrangement of rods referred to as "V-formations" is one of the more characteristic features of coryneform bacteria.

The presence of V-formations is frequently attributed to "snapping", post-fission movements. While this origin is true for *C. diphtheriae* and its close relatives, V-formations may arise in a number of different ways in other coryneform bacteria (see Sguros, 1957; Starr and Kuhn, 1962). Komagata, Yamada, and Ogawa (1969) thought that the mode of division in coryneform bacteria was of taxonomic importance; they considered that "snapping" division was a characteristic of *Corynebacterium sensu stricto* and that, in other coryneform bacteria,

V-formations arose by "bending" division. This distinction does not appear to be entirely valid; snapping, post-fission movements have been demonstrated very convincingly in *Arthrobacter atrocyaneus* (Starr and Kuhn, 1962), and Krulwich and Pate (1971) used observations on the ultrastructure of *A. crystallopoietes* as a model for a suggested explanation of the "snapping" phenomenon.

In this chapter, we shall consider saprophytic coryneform bacteria whose mode of metabolism is entirely or primarily respiratory: the genera *Arthrobacter, Brevibacterium, Cellulomonas, Curtobacterium* (which includes some plant-pathogenic species, this Handbook, Chapter 143), *Microbacterium,* and *Corynebacterium* (saprophytic species). Although those bacteria for which Sneath and Jones (1976) proposed the genus *Brochothrix* (with *B. thermosphacta* as the only species) do not have a coryneform morphology, they are most frequently referred to in the literature by their original name, *Microbacterium thermosphactum* (MacLean and Sulzbacher, 1953); therefore they are most conveniently dealt with after the genus *Microbacterium.*

Although the genera mentioned (*Brochothrix* excepted) are generally accepted as coryneform bacteria, there are other cases in which the situation is by no means clear-cut. The rods may show extensive primary branching in some coryneform bacteria, while some nocardioform bacteria produce a rudimentary mycelium that fragments early in the growth cycle. The large assemblage of strains known variously as the "*rhodochrous*" taxon, the "*rhodochrous*" complex, or the genus *Rhodococcus* (Goodfellow and Alderson, 1977; this Handbook,

Chapter 155) highlights such problems. Whereas some "*rhodochrous*" strains are considered to have a coryneform morphology, others are described as mycelial and some are slightly acid fast. Thus, in morphological features and in staining reactions, this group overlaps the boundaries of the coryneform and nocardioform bacteria, and of the mycobacteria. However, those taxa which most closely resemble coryneform bacteria in morphological characters can usually be distinguished from the coryneform genera discussed above by readily detectable, chemotaxonomic features. In Tables 1–3 we have assembled some of the major distinguishing features of taxa within the coryneform complex and have indicated, where appropriate, how they may be distinguished most readily from morphologically similar taxa. In general, features that can be examined by conveniently applied tests are used, but data on DNA base ratios have also been included. By using data from different laboratories, the ranges of mole percentage G+C values quoted for the different taxa become artificially wide. In order to minimize this effect, published data from investigations in which the mol% G+C values reported are consistently very high or very low when compared with other studies have been omitted from the tables. Data on individual strains that have extreme values have also been omitted.

The tables are not intended to be used for identification as such, but rather to indicate the most appropriate named taxa with which new isolates should be compared. As will be seen in the following pages, many coryneform bacteria cannot as yet be assigned to named taxa.

Table 1. Some distinguishing features of coryneform taxa that contain *meso*-diaminopimelic acid[a] as cell wall diamino acid.

Taxon	Arabinose in wall	Mycolic acids present	Oxygen requirements	Rod-coccus[b] cycle	DNA base ratios G+C (mol%)	DNA base ratios References[c]
Corynebacterium (*sensu stricto*)	+	+	Facultative[d]	−	51–59[e]	1–7
Rhodococcus[f]	+	+	Aerobic	+/−	59–69	8
Brevibacterium linens	−	−	Aerobic	+	60–64	4, 6
Unassigned[g] species	−	−	Aerobic	+/−	ND[h]	

[a] Allied taxa that contain *meso*-diaminopimelic acid and arabinose, *Nocardia* and *Mycobacterium,* are aerobic and may be distinguished from *Corynebacterium* by their mycolic acid composition as can most *Rhodoccocus* strains: see text. +, Positive; −, negative. See Addendum at end of this chapter for a brief comment on the recently described genus *Caseobacter* (Crombach, 1978), which also contains *meso*-diaminopimelic acid and arabinose in the cell wall.

[b] Rod-coccus cycle similar to that in *Arthrobacter globiformis.*

[c] (1) Marmur and Doty, 1962; (2) Hill, 1966; (3) Abe et al., 1967; (4) Yamada and Komagata, 1970; (5) Bousfield, 1972; (6) Crombach, 1972; (7) Cummins, Lelliott, and Rogosa, 1974; (8) Goodfellow and Alderson, 1977. For other data see references in text.

[d] A few aerobic.

[e] Occasional, higher values have been reported, e.g., for *C. bovis* (Crombach, 1972).

[f] Generally considered nocardioform but some have a coryneform morphology.

[g] A heterogeneous assemblage of mainly unnamed strains of uncertain taxonomic position.

[h] ND, data not available or incomplete.

Table 2. Some distinguishing features of coryneform taxa that contain lysine or L-diaminopimelic acid as cell wall diamino acid.[a]

Taxon	Rod-coccus[b] cycle	Oxygen requirements	Survival at[c] 63°C for 30 min	Acid from[d] glucose	G+C (mol%)	References[e]
Contain lysine						
Arthrobacter (sensu stricto) (mainly "*globiformis*" group)	+	Aerobic	−	−	59–66[f]	1–5
Microbacterium lacticum	−	Equivocal	+/−	+	69–70[g]	2, 4
Contain L-diaminopimelic acid[h]						
Arthrobacter (sensu lato) ("*simplex/tumescens*" group)	+	Aerobic	−	−	70–74	1–3

[a] Cell wall sugars do not appear to have any distinguishing value (but see note h); galactose is common to all. +, Positive; −, negative.
[b] Rod-coccus cycle similar to that in *A. globiformis*.
[c] See Jayne-Williams and Skerman (1966) for suitable method.
[d] In peptone-based medium.
[e] (1) Skyring and Quadling, 1970; (2) Yamada and Komagata, 1970; (3) Skyring, Quadling, and Rouatt, 1971; (4) Bousfield, 1972; (5) Crombach, 1972. For other data see references in text.
[f] The single strain of *A. atrocyaneus* has an exceptionally high value of about 70 mol% G+C (references 1, 2). The G+C (mol%) values of *Arthrobacter* spp. reported by Bowie et al. (1972) are considerably higher than those of other workers; accordingly their data have not been included in the range given.
[g] Collins-Thompson et al. (1972) quote the very low values of 63–64 mol% G+C for two strains of *M. lacticum,* one of which was common to studies reported in references 2 and 4; their data have not been included in the range given.
[h] Pitcher (1976) has described unusual coryneform bacteria from human skin which contain L-DAP, arabinose, and galactose.

Table 3. Some distinguishing features of coryneform taxa that contain ornithine or diaminobutyric acid as cell wall diamino acid.

Taxon	Galactose in wall	Cellulolytic	Oxygen requirements	G+C (mol%)	References[a]
Contain ornithine					
Cellulomonas	−	+	Most facultative[b] (a few equivocal or aerobic)	71–75	1, 2, 9
Curtobacterium	(+)	−	Aerobic	67–71	1–3
Arthrobacter (sensu lato) ("*terregens/flavescens*" group)[c]	+	−	Aerobic	69–70	4, 5
Microbacterium liquefaciens	−	−	Aerobic	ND[d]	
Contain diaminobutyric acid					
Unassigned species[e]	ND	−	Aerobic	64–74[f]	2–8

[a] (1) Yamada and Komagata, 1972b; (2) Bousfield, 1972; (3) Starr, Mandel, and Murata, 1975; (4) Skyring and Quading, 1970; (5) Skyring, Quadling, and Rouatt, 1971; (6) Yamada and Komagata, 1970; (7) Crombach, 1972; (8) Luthy, 1974; (9) Stackebrandt and Kandler, 1979. For other data see references in text. +, Positive; (+), most strains positive; −, negative.
[b] See Keddie (1974b) for discussion of oxygen requirements of *Cellulomonas* spp.
[c] Data based on one strain of each species.
[d] ND, data not available or incomplete.
[e] An assemblage of species currently labeled "*Corynebacterium*" but as yet of uncertain taxonomic position: included are the plant-pathogenic species *C. insidiosum, C. michiganense, C. sepedonicum,* and *C. tritici;* other species include *C. aquaticum, C. mediolanum,* and *C. okanaganae* (see Keddie and Cure, 1978).
[f] Yamada and Komagata (1970) give the exceptionally high value of 78.1 mol% G+C for *C. insidiosum* (ATCC 10253), while Starr, Mandel, and Murata (1975) give 72.9 (±0.9) mol% G+C for nine strains, including ATCC 10253.

General Procedures for Examination of Saprophytic Coryneform Bacteria

The following procedures may be used for morphological examination, cell wall analysis, and mycolic acid analysis of the genera considered in this Chapter.

Morphology and Staining Reactions

Great care must be used when examining the morphological features and staining reactions of coryneform bacteria. Because the morphology changes during the growth cycle, it is necessary to make observations at different periods of incubation. A minimum requirement is to examine both early to midexponential-phase cultures (usually 6–24 h at 25°C) and late stationary-phase cultures (usually 3–7 days). Also, in some nocardioform bacteria the mycelium may fragment at an early stage of growth, especially when disturbed, and thus escape detection. For this reason, it is advisable to examine young colonies for mycelium formation under low magnification and without a cover slip.

The morphology may be markedly influenced by the cultural conditions and, especially, by the medium used. A wide variety of media have been used for morphological studies, and usually different media have been used for examining coryneform bacteria from different sources—indeed, different media are sometimes used for examining exponential- and stationary-phase cultures. For this reason, Cure and Keddie (1973) devised a single, reproducible medium, EYGA, which is suitable for the morphological examination of saprophytic, aerobic coryneform bacteria from a wide range of habitats. Procedures for preparing EYGA and methods suitable for examining the morphology and staining reactions of coryneform bacteria are described by these authors.

Cell Wall Analysis

It is strongly recommended that the next stage in identification following morphological examination should be the examination of the cell wall composition. For identification, it is most convenient to use one of the "rapid" methods of cell wall analysis to determine the occurrence of the components most useful in differentiation: the peptidoglycan diamino acid and certain diagnostic sugars. In some cases, it may be sufficient simply to use chromatographic analysis of acid hydrolysates of whole organisms. In general, this method is suitable for detecting the presence or absence of components that occur mainly or entirely in the cell wall, such as the isomers of 2,6-diaminopimelic acid (DAP) and arabi-

nose. Suitable procedures have been described by Becker et al. (1964), Murray and Proctor (1965), and Stanek and Roberts (1974). However, for detecting the presence or absence of all the determinatively useful wall components mentioned in Tables 1–3, analysis of cell wall material is necessary. Traditional techniques in which qualitative analyses of highly purified cell walls are made (e.g., see Cummins and Johnson, 1971) are too laborious for diagnostic purposes. Much more suitable is a "rapid" method, similar to that described by Boone and Pine (1968), in which cell wall material is prepared by alkali treatment of whole cells. A suitable procedure (based on that devised by I. J. Bousfield) was described by Keddie and Cure (1977), who found that the method gave results comparable with more traditional methods of cell wall analysis.

Mycolic Acid Analysis

Mycolic acids are long-chain, 2-alkyl-branched 3-hydroxy acids and have been demonstrated only in those coryneform bacteria and allied taxa which contain meso-DAP and arabinose in the cell wall (see Minnikin, Goodfellow, and Collins, 1978). Possession of mycolic acids, therefore, distinguishes the genus *Corynebacterium sensu stricto* (and *Bacterionema matruchotii*), as well as the genera *Rhodococcus*, *Nocardia sensu stricto*, and *Mycobacterium*, from all other genera of coryneform bacteria. Also, the mycolic acids of *Corynebacterium sensu stricto* (and *Bacterionema matruchotii*) are of relatively small size (ca. 22–38 carbons), whereas those of the other taxa mentioned, with the exception of some rhodococci, are larger. Therefore, this feature may be used to distinguish legitimate corynebacteria from nocardiae, mycobacteria, and many rhodococci. Some rhodococci, however, have mycolic acids similar in size to those of corynebacteria; consequently, this feature alone cannot be used to distinguish between the two taxa (see Minnikin, Goodfellow, and Collins, 1978).

Although the different types of mycolic acids are most reliably distinguished by pyrolysis-gas chromatography or mass spectrometry (see Minnikin, Goodfellow, and Collins, 1978) two relatively simple, chromatographic techniques may be used for diagnostic purposes. It must be stressed, however, that the reliability and reproducibility of these chromatographic techniques have not yet been fully assessed (see below). Both techniques depend on the fact that the free mycolic acids or methyl mycolates of the various mycolic acid-containing taxa show slightly different mobilities on thin-layer chromatograms (TLC).

In the first method, the presence and type of free mycolic acids (LCN–A's) are determined by TLC analysis of ethanol–diethyl ether extracts of bacteria, using the method of Mordarska, Mordarski, and Goodfellow (1972; see also Keddie and Cure, 1977). The lipid components referred to as LCN-A's (Lipids Characteristic of *Nocardia*) by Mordarska, Mordarski, and Goodfellow were shown to be free mycolic acids (Goodfellow et al., 1973), and may be demonstrated in corynebacteria, nocardiae, and rhodococci (Goodfellow and Alderson, 1977; Keddie and Cure, 1977; Mordarska, Mordarski, and Goodfellow, 1972). However, the considerably larger mycolic acids of mycobacteria are relatively insoluble in ethanol–diethyl ether mixtures and are not detected by this technique (see Minnikin, Goodfellow, and Collins, 1978). In the second method, the methyl esters of mycolic acids may be detected and distinguished by TLC analysis of whole-organism methanolysates using the method described by Minnikin, Alshamaony, and Goodfellow (1975).

Where the same strains have been examined by both methods (see Goodfellow, Collins, and Minnikin, 1976; Keddie and Cure, 1977), the findings are generally similar but some discrepancies occur in the *Corynebacterium/Rhodococcus* ("*rhodochrous*" complex) area. There is some evidence (see Minnikin, Goodfellow, and Collins, 1978) that the method using whole-organism methanolysates is the more reliable of the two. It is probably also the more generally useful of the two methods because it may be used to demonstrate mycolic acids in all mycolic acid–containing taxa, including *Mycobacterium*. However, neither method will reliably distinguish between legitimate corynebacteria and all rhodococci. At present, there is no single simple method which allows a decisive distinction to be made between these two taxa. However, most legitimate corynebacteria have DNA base ratios in the range ca. 51–59 mol% G+C, whereas those of most rhodococci are in the range ca. 59–69 mol% G+C (see references cited in Table 1).

PRESERVATION OF CULTURES. Cultures may be preserved by freeze-drying (lyophilisation). Stab cultures (by loop) in TSX semisolid medium (Keddie, Leask, and Grainger, 1966) or the soil-extract semisolid medium described by Lochhead and Burton (1957) (see below: Genus *Arthrobacter*, Isolation) should remain viable for 3 months or more when stored at room temperature (ca. 20°C) provided that they are not allowed to dry out.

THE GENUS *ARTHROBACTER*

Conn (1928) described a group of bacteria, which were extremely numerous in certain soils and unusual in that they appeared as Gram-negative rods in young cultures and as Gram-positive cocci in older cultures. For these bacteria, Conn (1928) created the species *Bacterium globiforme*, which as *Arthrobacter globiformis* was later to become the type species of the genus *Arthrobacter*. The abundance in soil of bacteria similar to Conn's organism, and of other coryneform bacteria, was confirmed later by Jensen (1933, 1934) and Topping (1937, 1938), who, however, referred to them as soil corynebacteria, and by Taylor and Lochhead (1937), who used the name *B. globiforme*. Jensen (1934) considered that these soil bacteria should be classified in the genus *Corynebacterium* because of their morphological resemblance to corynebacteria of animal origin. However, Conn (1947) vigorously opposed this view and created the genus *Arthrobacter* (by reviving an old name), with *A. globiformis* as type species and with two of Jensen's soil corynebacteria as additional species (Conn and Dimmick, 1947).

In addition to their characteristic morphology and staining reactions, members of the genus *Arthrobacter* were originally described as being highly aerobic and nutritionally nonexacting, and as liquefying gelatin slowly (Conn and Dimmick, 1947). These features were chosen mainly to distinguish *Arthrobacter* from *Corynebacterium* as represented by *C. diphtheriae* and similar animal parasitic species. However, because of its poor circumscription (see Gibson, 1953; Jensen, 1952), the genus *Arthrobacter* was not widely accepted until it was included as a member of the family Corynebacteriaceae in the seventh edition of *Bergey's Manual of Determinative Bacteriology* (Breed, Murray, and Smith, 1957). But by that time, the genus had been extended to include the nutritionally exacting species *A. terregens* (Lochhead and Burton, 1953) and *A. citreus* (Sacks, 1954), and shortly afterwards two others were added (Lochhead, 1958a). Indeed, one of Conn's strains of *A. globiformis* was shown subsequently to require biotin for growth (Chan and Stevenson, 1962; Morris, 1960). Thus the concept had developed of a genus of soil bacteria whose major distinguishing feature was a growth cycle in which the irregular rods in young cultures were replaced by coccoid forms in older cultures; these coccoid forms, when transferred to fresh medium, produced outgrowths ("germinated") to give irregular rods again and so the cycle was repeated (Fig. 1).

This dependence on morphological features and habitat in the circumscription led to much of the confusion that now exists in the classification of the genus *Arthrobacter*. It also created considerable problems in the identification of new isolates as arthrobacters. One difficulty is that there is no sharp dividing line between coryneform bacteria in which the transformation of rods into coccoids is complete, as in *Arthrobacter*, and those in which it is limited, as in most other coryneform genera (see Keddie, Leask, and Grainger, 1966; Skyring and Quadling, 1969). A second difficulty is that isolates from soil and, especially from other habitats have frequently been referred to in the literature as arthrobacters on the basis of morphological features alone, even though they were not necessarily similar to *A. globiformis* in other respects.

Thus, it is not surprising that the genus *Arthrobacter* as currently defined in *Bergey's Manual* (Keddie, 1974a) has been shown to be heterogeneous by a variety of modern taxonomic techniques. These include numerical taxonomy (see Jones, 1978), various chemotaxonomic techniques (see Bowie et al., 1972; Keddie and Cure, 1977, 1978; Minnikin, Goodfellow, and Collins, 1978; Schleifer and Kandler, 1972), and determinations of DNA base ratios (see Skyring and Quadling, 1970; Skyring, Quadling, and Rouatt, 1971). However, of the numerous species which have now been described, most for which suitable data are available can be considered in three categories.

The first and by far the largest category comprises *Arthrobacter* spp. that contain lysine in the cell wall peptidoglycan and that resemble the type species *A. globiformis* in a large number of phenotypic characters. There is now general, if not universal, agreement that the genus should be restricted to these species together with *A. citreus* which, although it contains lysine in the cell wall, differs from *A. globiformis* in nutritional and some other respects (Keddie, 1978; Schleifer and Kandler, 1972; Yamada and Komagata, 1972b). The second category comprises the species *A. simplex* and *A. tumescens*, which contain L-2,6-diaminopimelic acid (DAP) in the cell wall peptidoglycan (Cummins and Harris, 1959; Keddie and Cure, 1977; Schleifer and Kandler, 1972) and have DNA base ratios some 7–9% higher than *A. globiformis* (Skyring and Quadling, 1970; Yamada and Komagata, 1970). The third category comprises the two species that require the *terregens* factor, *A. terregens* and *A. flavescens* (Lochhead, 1958a; Lochhead and Burton, 1953); both species contain ornithine in the cell wall (Keddie and Cure, 1977) and have DNA base ratios some 3–5% higher than the type species (Skyring and Quadling, 1970). It is now generally agreed that the species in the second and third categories should be removed from the genus *Arthrobacter*. The position of *A. duodecadis* is at present uncertain (see Keddie and Cure, 1977), and one species, *A. marinus*, is now considered to be a *Pseudomonas* sp. (Baumann et al., 1972).

A number of additional organisms, usually "patent" strains, are named *Arthrobacter*. Many contain *meso*-DAP and arabinose in the cell wall and most of those studied in sufficient detail can be assigned either to the "*rhodochrous*" taxon (genus *Rhodococcus*) or to the genus *Corynebacterium* (see Keddie and Cure, 1978; Minnikin, Goodfellow, and Collins, 1978). A few contain *meso*-DAP in the cell wall but not arabinose; they are a small part of a heterogeneous collection of *meso*-DAP-containing

coryneform bacteria whose taxonomic position is at present uncertain (see Keddie and Cure, 1978).

Undoubtedly, soil is the most important habitat of bacteria of the genus *Arthrobacter*. Their abundance in soils of various types and in different geographical locations has been amply demonstrated by many investigators (see references cited below). Their numerical predominance, coupled with the nutritional versatility of the commonly occurring species (Hagedorn and Holt, 1975a; Keddie, 1974a), suggests that they may be important agents of mineralization in soil and possibly in some other habitats. Among the compounds reported to be degraded by arthrobacters are certain herbicides and pesticides, as well as a wide range of naturally occurring (and other) molecules of various degrees of complexity. Other roles that have been ascribed to at least some *Arthrobacter* spp. are phytohormone production (Barea, Navarro, and Montoya, 1976; Katznelson and Cole, 1965; Rivière, 1963) and dinitrogen fixation (Cacciari et al., 1971; Smyk, 1970; Smyk and Ettlinger, 1963). Putative *Arthrobacter* spp. have also been reported to lyse yeast cells (Kitamura, Kaneko, and Yamamoto, 1972) and mycelium of *Fusarium roseum*, a carnation root pathogen (Morrisey, Dugan, and Koths, 1976; Szajer and Koths, 1973). In the latter case, the *Arthrobacter* strain investigated produced a chitinase and was considered a possible means of biological control of *Fusarium* diseases.

Products of actual or potential commercial importance obtained from *Arthrobacter* spp. include polysaccharides (Gasdorf et al., 1965; Hagiwara and Yamada, 1970), glutamic acid (Tanaka and Kimura, 1972; Veldkamp, van den Berg, and Zevenhuizen, 1963), and α-ketoglutaric acid (Tanaka and Kimura, 1972), although it is likely that in some of the examples mentioned the bacteria concerned were not legitimate *Arthrobacter* spp. (see Keddie and Cure, 1977, 1978). Also, Veldkamp et al. (1966) described a strain of *A. globiformis* that produced large amounts of riboflavin, and *A. simplex* (*Corynebacterium simplex*) is of commercial importance in steroid transformations (see Charney, 1966; Martin, 1977).

In some species of *Arthrobacter*, the morphological changes that occur during the growth cycle have been shown to be subject to nutritional control, and such species have proved to be invaluable in the study of bacterial morphogenesis (see review by Clark, 1972).

Habitats of *Arthrobacter*

In many ecological studies, isolates have been identified as arthrobacters or described as "arthrobacter-like" simply because they showed the rod-coccus growth cycle and staining reactions characteristic of the genus. Although the sequence of morphological changes during the growth cycle is an important distinguishing feature of the genus, it does not occur exclusively in arthrobacters. For example, it is also seen in *Brevibacterium linens* (Fig. 2) and in some members of the *"rhodochrous"* taxon (genus *Rhodococcus*). Accordingly, a proportion of the strains described as arthrobacters in the literature cited below may belong to other morphologically similar taxa, especially if they are isolates from habitats other than soil.

Soil and Similar Habitats

Many studies have shown that bacteria of the genus *Arthrobacter* form a numerically important fraction of the indigenous bacterial flora of different soils; they are sometimes the most numerous, single bacterial group recorded in aerobic plate counts (Hagedorn and Holt, 1975b; Holm and Jensen, 1972; Lowe and Gray, 1972; Mulder and Antheunisse, 1963; Skyring and Quadling, 1969; Soumare and Blondeau, 1972). But both the numbers and the proportions of arthrobacters in "total" counts decrease with increasing soil acidity (Hagedorn and Holt, 1975b; Lowe and Gray, 1972). Among the explanations advanced for their numerical predominance are their extreme resistance to drying (Boylen, 1973; Chen and Alexander, 1973; Labeda, Liu, and Casida, 1976; Mulder and Antheunisse, 1963; Robinson, Salonius, and Chase, 1965) and to starvation (Boylen and Ensign, 1970; Boylen and Mulks, 1978; Zevenhuizen, 1966), factors important in the survival of microorganisms in soil (Gray, 1976). However, the nutritional versatility of the commonly occurring species undoubtedly also plays a part (see below).

Both psychrophilic and psychrotrophic strains of the genus were reported to be the most abundant and active bacteria in subterranean cave silts (Gounot, 1967), and they also occur in glacier silts (Moiroud and Gounot, 1969). The genus was also represented among isolates from oil brines raised from soil layers some 200–700 m deep (Iizuka and Komagata, 1965). Arthrobacters capable of dissolving aluminum silicates were reported to be common on "Karst" rocks, and most were considered to be capable of dinitrogen-fixation (Smyk, 1970; Smyk and Ettlinger, 1963). Members of the genus have also been implicated in the growth of manganese nodules in the sea (Ehrlich, 1963, 1968).

There is now much evidence that the predominant soil arthrobacters can use a wide and diverse range of organic substrates as sole or principal sources of carbon and energy (Hagedorn and Holt, 1975a; J. D. Owens and R. M. Keddie, unpublished data quoted in Keddie, 1974a). More specific studies have demonstrated their ability to utilize aromatic compounds

(Stevenson, 1967) and nucleic acids and their degradation products (Antheunisse, 1972). The frequent recovery of *Arthrobacter* spp. from enrichment cultures in which various diverse, organic compounds are supplied as sole carbon sources in simple, mineral media is further evidence of this nutritional versatility. Such compounds include nicotine (Giovanozzi-Sermanni, 1959; Keddie, Leask, and Grainger, 1966; Sguros, 1955), puromycin aminonucleoside (Greenberg and Barker, 1962), 2-hydroxypyridine (Ensign and Rittenberg, 1963; Kolenbrander, Lotong, and Ensign, 1976), *n*-alkanes (Klein, Davis, and Casida, 1968), lower alcohols (Akiba et al., 1970), choline (Kortstee, 1970), picolinic acid (Tate and Ensign, 1974), and squalene (Yamada et al., 1975).

Arthrobacters have also been shown to degrade herbicides such as 4,6-dinitro-*o*-cresol (Gundersen and Jensen, 1956), disodium endoxohexahydrophthalate ("Endothal") (Jensen, 1964), and 2,4-dichlorophenoxyacetate (Cacciari et al., 1971; Loos, Roberts, and Alexander, 1967; Sharpee, Duxbury, and Alexander, 1973). Others degrade pesticides, usually by cometabolism, e.g., "diazinon" (Sethunathan and Pathak, 1971) and *m*-chlorobenzoate, the central molecule in many pesticides (Horvath and Alexander, 1970).

Fish and Similar Habitats

Coryneform bacteria appear to be common on fish (both marine and fresh water) and on some other seafoods. Many are morphologically similar to arthrobacters (and to *Brevibacterium linens*) and are frequently referred to by that name. Thus "arthrobacters" have been reported to occur in shark spoilage (referred to as *"Corynebacterium" globiformis* and *C. helvolum,* Wood, 1950), eviscerated freshwater fish (Roth and Wheaton, 1962), fish-pen slime (Chai and Levin, 1975), and Pacific shrimp (Lee and Pfeifer, 1977); Sieburth (1964) reported the isolation of an *Arthrobacter* sp. from sea water.

However, in the few cases in which coryneform isolates from fish have been examined by suitable modern methods, their relationship to *Arthrobacter* has proved to be much more remote than has been inferred from their morphological similarity and most, if not all, belong to other taxa (see Bousfield, 1978; Crombach, 1974a,b).

Sewage and Similar Habitats

Bacteria identified as *Arthrobacter* spp. have been isolated from sewage (Nand and Rao, 1972) and from "brewery sewage" (Kaneko, Kitamura, and Yamamoto, 1969). Arthrobacters physiologically similar to those from soil were common in dairy-waste activated sludge (Mulder and Antheunisse,

1963) and were considered to play an important role in the process (Adamse, 1968). A number of the activated-sludge strains were reported by the above authors to decompose phenol. However, subsequent examination of four of these phenol-decomposing strains with chemotaxonomic methods revealed that they were members of the *"rhodochrous"* taxon (Keddie and Cure, 1977). Similarly, although Schefferle (1966) considered that the predominant coryneform bacteria in poultry deep litter could be accommodated in *Arthrobacter,* only one of seven of these litter strains subsequently examined by Keddie and Cure (1977) was considered to be a legitimate *Arthrobacter* sp. Although many coryneform isolates from aerated, animal-manure slurries are similar in morphology to *Arthrobacter,* only 1 of 16 such isolates examined was identified as *A. globiformis;* most of the remainder were considered to be *"rhodochrous"* strains and a few were similar to *Brevibacterium linens* (Keddie and Cure, 1977).

Such examples clearly illustrate that new isolates should not be assigned to the genus *Arthrobacter* on the basis of morphology and conventional features alone.

Other Habitats

Coryneform bacteria seem to be relatively common on the aerial surfaces of plants (Austin, Goodfellow, and Dickinson, 1978; Keddie, Leask, and Grainger, 1966; Mulder et al., 1966) but few have been shown to be legitimate arthrobacters. However, "arthrobacter-like" organisms have been isolated from frozen vegetables (Splittstoesser et al., 1967). Also, two of a number of strains isolated from cauliflowers by Lund (1969) were later identified as *A. globiformis* by Keddie and Cure (1977), but it is possible that they were contaminants from soil rather than indigenous plant bacteria. However, *Corynebacterium ilicis,* a pathogen of American holly (Mandel, Guba, and Litsky, 1961), has been shown to be a legitimate *Arthrobacter* sp. very similar to *A. globiformis* (Bousfield, 1972; Jones, 1975; Keddie and Cure, 1977).

Arthrobacters were reported to be numerous in commercial liquid egg, probably as a result of contamination from the shells, but they were rarely isolated from turkey giblets (Kraft et al., 1966).

Isolation of *Arthrobacter*

Arthrobacters (in the broad sense) have normally been isolated from soil and similar habitats by plating on suitable nonselective ("total count") media, and then picking from a large, random selection of colonies and identifying the isolates obtained (Holm and Jensen, 1972; Lowe and Gray, 1972; Skyring and Quadling, 1969). This method is only suitable for habitats such as soil in which these bacteria form an appreciable proportion of the aerobic, cultivable population. A further extension of this technique was introduced by Mulder and Antheunisse (1963), who devised what was essentially a method for screening isolates picked from a nonselective medium for those which showed the typical rod-coccus growth cycle of arthrobacters (see also Mulder et al., 1966; Veldkamp, 1965). More recently, Hagedorn and Holt (1975b) have devised a selective medium that is said to be suitable for the enumeration of arthrobacters in soil. Thus, when nine different soils were plated on this medium, an average of 74% of the colonies that developed were identified as arthrobacters, and the numbers were similar to those estimated from counts on a nonselective medium (Hagedorn and Holt, 1975b).

As noted above, a number of *Arthrobacter* spp. have been isolated from enrichment cultures using a variety of organic substrates as sole carbon and energy sources in mineral salts media. However, the primary purpose of such enrichments was not to isolate arthrobacters but to obtain isolates capable of utilizing the particular substrates studied. Accordingly, it is not possible to assess the value of these enrichment methods for the isolation of particular arthrobacters.

Nonselective Media for Isolation of Arthrobacters

The aim of this approach is to use media and conditions of incubation which give the maximum possible counts of soil bacteria capable of growth under aerobic conditions. The media used must therefore contain sufficient amounts of all the organic growth factors and mineral constituents required to cater to the diverse nutritional requirements of the indigenous soil bacteria (see Lochhead, 1958b). But at the same time, media must be sufficiently poor in carbon and energy sources to limit the size of colonies, thereby minimizing antagonistic effects between the components of the population. Various modifications of soil extract agar have been the media most widely used. Lochhead strongly advocated the use of soil extract agar without other additions (Lochhead and Burton, 1956), but for some soils addition of low concentrations of yeast extract and glucose can give higher counts (Jensen, 1968). If necessary, the growth of fungi may be suppressed by incorporating the antibiotics nystatin (50 μg/ml) and cycloheximide (50 μg/ml) in the medium (Williams and Davies, 1965). Other factors important in plating soil samples have been discussed by Jensen (1968). The more important of these are: (i) The soils should be examined within a few hours of sampling; (ii) the primary dilution should be dispersed by using a laboratory blender but avoiding heating; (iii) a suitable diluent should be used (see below); and (iv) plates should be incubated at 25°C for a minimum of 2 weeks.

Jensen (1968) recommends Winogradsky's standard salt solution as a diluent. It has the following composition (Holm and Jensen, 1972):

K_2HPO_4	0.25 g
$MgSO_4$	0.125 g
NaCl	0.125 g
$Fe_2(SO_4)_3$	0.0025 g
$MnSO_4$	0.0025 g
Deionized water	1,000 ml
pH 6.5–6.7	

Soil extract and dilute peptone solutions (0.05–0.1%, wt/vol) have also been used successfully but one-fourth-strength Ringer's solution, physiological saline, and tap water are unsuitable (Jensen, 1968). In this context, Owens and Keddie (1969) noted that a chelated mineral salts solution which they devised, but without $(NH_4)_2SO_4$ (mineral base E–N), was a suitable diluent for coryneform bacteria. Mineral base E–N gave slightly better survival of *Arthrobacter globiformis* than a simple salts solution and was markedly superior to traditional diluents such as one-fourth-strength Ringer's solution or physiological saline. The preparation of mineral base E–N was described by Cure and Keddie (1973).

ISOLATION OF ARTHROBACTERS USING SOIL EXTRACT AGAR. To prepare soil extract, add 1 kg of soil to 1 liter of tap water and autoclave at 121°C for 20 min; filter and restore volume to 1 liter with tap water (Lochhead and Burton, 1957). A fertile garden soil usually gives the best results (Jensen, 1968).

Soil Extract Agar for Isolating Arthrobacters (Lochhead and Burton, 1957)

Soil extract, 1 liter; K_2HPO_4, 0.2 g; agar, 15 g; final pH, 6.8. Autoclave at 121°C for 20 min. Colonies are picked into tubes of soil extract semisolid medium: soil extract, 1 liter; K_2HPO_4, 0.2 g; yeast extract, 1 g; agar, 3 g; final pH, 6.8. Autoclave at 121°C for 20 min.

Soil Extract Agar for Isolating Arthrobacters (Holm and Jensen, 1972)

Extract of garden soil	400 ml
Tap water	600 ml
Glucose	1 g
Peptone	1 g
Yeast extract	1 g
K_2HPO_4	1 g
Agar	20 g
pH 6.5–6.7	

After sterilization and immediately before use, a filter-sterilized solution of cycloheximide is added to the medium to give a final concentration of 40 mg/liter. To prevent colonies from spreading on the bottom of the Petri dishes, a layer of sterile agar is poured in the bottom of the plates and allowed to solidify before pouring plates. Colonies are picked onto slants of the same medium.

Isolation and Enumeration of Arthrobacters from Soil (Mulder and Antheunisse, 1963)

Dilutions are prepared and plates poured using the following ''poor'' medium (g/liter tap water).

$Ca(H_2PO_4)_2$	0.25
K_2HPO_4	1.0
$MgSO_4 \cdot 7H_2O$	0.25
$(NH_4)_2SO_4$	0.25
Casein	1.0
Yeast extract	0.7
Glucose	1.0
Agar	10.0
pH 6.9–7.0	

After incubation for 5 days at 25°C, colonies are counted and a large number are transferred to agar slants of the same composition. The slants are incubated for 7 days at 25°C and then examined microscopically. Colonies that consist of coccoid cells are then transferred to a ''rich'' medium containing (% wt/vol): yeast extract, 0.7; glucose, 1.0; agar, 1.0. The cultures are examined in the exponential phase of growth (usually not more than 24 h), and those showing ''germinating'' cocci and irregular rods are considered to be ''arthrobacters''.

Obviously, in the sense used by Mulder and Antheunisse, ''arthrobacters'' refers to bacteria that show the rod-coccus growth cycle of the genus *Arthrobacter*.

Isolation of Arthrobacters by Selective Media

Selective Medium of Hagedorn and Holt (1975b)

In this method, plate counts are made by spreading 0.1-ml amounts of suitable dilutions over the surface of sterile medium in Petri dishes. Peptone solution (0.5%, wt/vol) was used as diluent by Hagedorn and Holt. The selective medium has the following composition:

Trypticase soy agar (BBL)	0.4%
Yeast extract (Difco)	0.2%
NaCl	2%
Cycloheximide	0.01%
Methyl red (Harleco)	150 μg/ml
Agar	1.5%

The methyl red is filter-sterilized and added aseptically to the autoclaved, cooled medium. The medium is adjusted to the pH of the particular soil being examined. The selective properties of the medium are said to be unaffected by pH values in the range 5.0–8.5. After incubation for 10 days at 25°C, the plates are counted. Colonies are transferred to slants of Trypticase soy agar containing 0.2% yeast extract and examined microscopically for the possession of a morphological growth cycle as described in *Bergey's Manual* (Keddie, 1974a). From the results obtained, the authors concluded that 78% of the counts on the selective medium was a suitable approximation of the arthrobacter counts for the soils studied. The authors state that: "The combination of Acti-Dione at 0.01% and NaCl at 2.0% effectively inhibited all fungi and most streptomycetes, nocardia, and gram-negative bacteria. The methyl red at 150 μg/ml inhibited other gram-positive bacteria (bacilli and micrococci) but did not affect the arthrobacters. The pH of the medium, between 5.0 and 8.5, did not affect its selectivity, and the combination of trypticase soy agar at 0.4% and yeast extract at 0.2% gave the highest yield of arthrobacters with the addition of the selective ingredients over the other basal media [tested]."

The selective medium gave arthrobacter counts several times higher than those on the nutritionally "poor" medium of Mulder and Antheunisse (1963) for the four soils examined (Hagedorn and Holt, 1975b).

Identification of *Arthrobacter*

The genus *Arthrobacter,* as defined in *Bergey's Manual,* eighth edition (Keddie, 1974a), is heterogeneous and, with some exceptions, can be considered to comprise three species groups. We will refer to *Arthrobacter* in the broad sense used in the eighth edition of *Bergey's Manual* as *Arthrobacter sensu lato,* and to the three species groups as the "*globiformis*" group, the "*simplex/tumescens*" group, and the "*terregens/flavescens*" group. The exceptions referred to above include: *A. citreus,* which although similar to *A. globiformis* in cell wall composition and DNA base ratio, has other phenotypic characters that set it apart from the "*globiformis*" group; *A. duodecadis,* whose cell wall composition is controversial (see Keddie and Cure, 1977, 1978; Schleifer and Kandler, 1972); and a number of species, some recently described, for which essential data, particularly those on cell wall composition, are lacking.

It has also been pointed out that the "*simplex/tumescens*" group and the "*terregens/flavescens*"

group differ in a number of respects from the type species *A. globiformis* and should be excluded from the genus. This exclusion may be achieved by restricting the genus to those species which, like the type, contain lysine as cell wall diamino acid, a proposal made by Yamada and Komagata (1972b) and supported by others (Keddie, 1978; Keddie and Cure, 1978; Schleifer and Kandler, 1972). *Arthrobacter* in this restricted sense will be referred to as *Arthrobacter sensu stricto* and comprises, for the most part, the "*globiformis*" group and *A. citreus.*

Arthrobacter sensu lato

Members of *Arthrobacter sensu lato* can be recognized by examining the following characters: morphology and staining reactions, cell wall composition, oxygen relations, and acid production from glucose.

The most distinctive feature of arthrobacters is the marked change of form that occurs during the growth cycle on complex media. Stationary-phase cultures (usually 2–7 days) are composed entirely or largely of coccoid cells (Fig. 1d) which, on transfer to fresh complex medium, produce one and sometimes two (or occasionally more) outgrowths that give rise to the irregular rods characteristic of exponential-phase cultures (Fig. 1a–c). Some of the cells are arranged in V-formations but more complex angular arrangements may occur. Cells may show primary branching, but true mycelia (showing secondary branching) are not produced. As growth proceeds, the rods become shorter and are eventually replaced by the coccoid forms characteristic of stationary-phase cultures (Fig. 1d). For more detailed accounts of morphology and morphogenesis, see the eighth edition of *Bergey's Manual* (Keddie, 1974a), Luscombe and Gray (1971), Clark (1972), and Duxbury and Gray (1977). Both rod and coccoid forms are Gram positive, but may decolorize readily and are not acid fast. The rods are nonmotile or motile by one subpolar or a few lateral flagella. They are obligate aerobes: The mode of metabolism is respiratory, never fermentative; little or no acid is formed from sugars in peptone media. The cell wall peptidoglycan does not contain *meso*-DAP as diamino acid and the wall polysaccharide does not contain arabinose.

The extent to which the morphology of arthrobacters changes during the growth cycle is markedly influenced by the nutritional status of the medium (see Clark, 1972; Ensign and Wolfe, 1964; Luscombe and Gray, 1971; Veldkamp, van den Berg, and Zevenhuizen, 1963); therefore, the medium used for morphological studies must be chosen with care (see Cure and Keddie, 1973). Many different media have been used for this purpose and most are based on soil extract (e.g., see Holm and

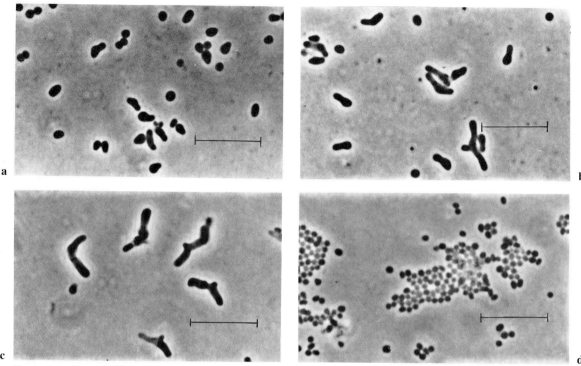

Fig. 1. *Arthrobacter globiformis* (ATCC 8010) when grown on medium EYGA at 25°C; inoculum coccoid cells as in (d). (a) After 6 h, showing outgrowth of rods from coccoid cells. (b) After 12 h. (c) After 24 h. (d) After 3 days. Bar = 10 μm.

Jensen, 1972; Lochhead and Burton, 1957). However, such media may give inconsistent results because of the variable nature of soil extract and because some mineral components may be precipitated to different extents during preparation. Medium EYGA (Cure and Keddie, 1973) was devised to overcome such problems and has proved to be satisfactory for examining the morphology of arthrobacters.

For identification at this level, a simple method using chromatographic analysis of acid hydrolysates of whole organisms is sufficient to demonstrate the presence or absence of *meso*-DAP and to distinguish it from the L-isomer (Becker et al., 1964; Staneck and Roberts, 1974). Because the diagnostic feature is the absence of *meso*-DAP, it is essential to include on chromatograms suitable control hydrolysates of organisms containing *meso*- and L-DAP.

By this method, *Arthrobacter sensu lato* is readily distinguished from taxa that show a similar rod-coccus growth cycle but that contain *meso*-DAP in the cell wall, viz., the genus *Rhodococcus* (some members) and *Brevibacterium linens;* this method also allows differentiation from the other *meso*-DAP-containing taxa, *Corynebacterium sensu stricto, Mycobacterium,* and *Nocardia sensu stricto.* This simple method of cell wall analysis also allows the recognition of members of the "*simplex/tumescens*" group which contain L-DAP in the cell wall. However, for further stages of identification, analysis of cell wall material is necessary. As indicated in the Introduction to this chapter, cell wall material is most conveniently prepared by alkali treatment of whole organisms (see Keddie and Cure, 1977 for method).

Further Identification of Arthrobacters

At the present time, some 30 or so nomenclatural species of *Arthrobacter sensu lato* have been described, several since the eighth edition of *Bergey's Manual* was published. Many of these "species" were created for single strains that possessed some unusual feature, such as a requirement for a particular growth factor, production of an unusual pigment, the ability to utilize a particular substrate, and so on. Accordingly, such "species" do not necessarily represent the commonly occurring arthrobacters in the habitats from which they were isolated originally and this may in part explain the common experience that many new soil isolates show little resemblance to the named strains used as reference cultures (see Hagedorn and Holt, 1975a; Keddie, Leask, and Grainger, 1966; Skyring and Quadling, 1969). Also, many of these "species" are distinguished from others only by one or two conventional features; but because the species are based on single strains, it is impossible to know whether these supposed distinguishing features are indeed species specific or merely strain specific.

From the foregoing discussion it is clear that, for the most part, little is to be gained by attempting to identify new isolates with one of this plethora of species. The best that can be achieved at the present time is to identify to the level of "species group" by using simple cell wall analysis. Such further identification as is possible is described under the heading of each species group.

ARTHROBACTER SENSU STRICTO. Isolates that possess the characters described for *Arthrobacter sensu lato* and that, in addition, contain lysine in the cell wall may be considered to belong to *Arthrobacter sensu stricto* (see Table 2). Further features described for *Arthrobacter* in this restricted sense are gelatin liquefaction and extracellular DNase activity (Yamada and Komagata, 1972b).

Microbacterium lacticum, which also contains lysine in the cell wall peptidoglycan, may be distinguished from "legitimate" arthrobacters by the morphology and by the features listed in Table 2. It should be noted that many other named (see Keddie and Cure, 1978, for list) and unnamed (see Keddie, Leask, and Grainger, 1966), lysine-containing strains of coryneform bacteria have been described. However, the taxonomic position of these strains is as yet unresolved.

Keddie and Cure (1978) have listed the species they consider to be "legitimate" *Arthrobacter* spp. and have also indicated those which they consider should be excluded from the genus (but see also Minnikin, Goodfellow, and Collins, 1978; Schleifer and Kandler, 1972; Yamada and Komagata, 1972b).

THE *"GLOBIFORMIS"* GROUP. Of those examined in sufficient detail, a large majority of the nomenclatural species that conform with the definition of *Arthrobacter sensu stricto* resemble the type species in a large number of phenotypic (mainly nutritional) traits. Therefore, in the eighth edition of *Bergey's Manual* (Keddie, 1974a; Table 17.7), an "ideal phenotype" was included based on an analysis of some 20 named and unnamed strains considered to represent the species *A. globiformis.* Nomenclatural species that showed a high degree of conformity with the characters of the "ideal phenotype", that contained lysine in the wall, and that had similar DNA base ratios to *A. globiformis* were considered to be synonyms of *A. globiformis;* a few others for which the data were less complete were considered to be possible synonyms. Despite the considerable phenotypic resemblance among these differently named species, a similarity confirmed in the main by the numerical phenetic study of Skyring, Quadling, and Rouatt (1971), more recent data now indicate much more heterogeneity than was formerly apparent. It is therefore more appropriate to refer to this assemblage of nomenclatural species as the *"globiformis"* group rather than as a single species.

At present, this group seems to be the major recognizable entity within *Arthrobacter sensu stricto.* Members of the *"globiformis"* group may be recognized by testing for the characteristics of the "ideal phenotype" of *"A. globiformis"* given in Table 17.7 in *Bergey's Manual,* eighth edition (Keddie, 1974a). In addition to *A. globiformis,* the named strains that have been shown to conform closely with the characters of the "ideal phenotype" include *A. atrocyaneus, A. aurescens, A. crystallopoietes, A. histidinolovorans, A. nicotianae, A. oxydans, A. pascens, A. polychromogenes, A. ramosus,* and *A. ureafaciens,* together with the species *Brevibacterium sulfureum* and the plant pathogen *Corynebacterium ilicis* (Keddie and Cure, 1977). In addition to these named strains, a number of isolates from soil (Keddie, 1974a; Keddie and Cure, 1977) and a few from other sources (Keddie and Cure, 1977) have been identified as *"globiformis"* group strains. The bacteria of this species-group characteristically show the large, irregular cells typical of exponential-phase cultures of *A. globiformis* and readily transform completely into coccoid forms in stationary-phase cultures (Keddie, Leask, and Grainger, 1966; Skyring, Quadling, and Rouatt, 1971; G. L. Cure and R. M. Keddie, unpublished data).

The differently named strains of the *"globiformis"* group have been shown to have cell wall peptidoglycans of similar structure. But they are cross-linked between positions 3 and 4 of two peptide subunits by many different interpeptide bridges, giving a large number of different peptidoglycan types (Schleifer and Kandler, 1972). Two groups, referred to by these authors as "variations", were distinguished within these different peptidoglycan types. In the first (A3α variation), the interpeptide bridges contained only monocarboxylic acids and/or glycine; in the second (A4α variation), a dicarboxylic acid (sometimes accompanied by alanine) was found (Schleifer and Kandler, 1972). *A. globiformis* and most *"globiformis"* group strains (and also the type strain of *A. citreus*) have peptidoglycans of the A3α variation, and it has recently been reported that their major isoprenoid quinones are dihydrogenated menaquinones with nine isoprene units [MK-9(H$_2$)] (Minnikin, Goodfellow, and Collins, 1978). Of those species listed above, *A. nicotianae* and *B. sulfureum* have peptidoglycans of the A4α variation and contain fully unsaturated menaquinones with eight and/or nine isoprene units (MK-8 and/or MK-9) (Minnikin, Goodfellow, and Collins, 1978). It has, therefore, been tentatively suggested that the genus *Arthrobacter* be further restricted to those species which have peptidoglycans with the A3α variation and MK-9(H$_2$) isoprenoid quinones (Minnikin, Goodfellow, and Collins, 1978). However, as far as we are aware there are no other phenotypic characters that correlate with the two peptidoglycan variations mentioned, or with the many different peptidoglycan types found in "*globiformis*" group strains.

Although the data then available (Jones and Bradley, 1964; Marmur, Falkow, and Mandel, 1963; Yamada and Komagata, 1970) indicated that the nomenclatural species similar to *A. globiformis* had DNA base ratios in the range 60.0–64.4 mol% G+C (see Keddie, 1974a), later information indicates that these values require revision. Values ranging from 60.0 mol% G+C (Jones and Bradley, 1964; see Hill, 1966) to 71.3 mol% G+C (Bowie et al., 1972) have been quoted for the DNA base ratio of the type strain of *A. globiformis* (ATCC 8010), but it is likely that the best estimate is about 65 mol% G+C (by buoyant density, Skyring and Quadling, 1970; by thermal denaturation, Crombach, 1972). Most of the nomenclatural species in the "*globiformis*" group have DNA base ratios close to that of the type strain but a few, typified by *A. aurescens,* have values some 4–5% lower (Skyring and Quadling, 1970; Skyring, Quadling, and Rouatt, 1971) while some others have DNA base ratios of intermediate values. The single strain of *A. atrocyaneus* has a DNA base ratio some 4–5% higher than the type strain (Schuster, Vidaver, and Mandel, 1968; Skyring and Quadling, 1970; Skyring, Quadling, and Rouatt, 1971; Yamada and Komagata, 1970). Thus, DNA base ratios also indicate heterogeneity within the "*globiformis*" group but do not suggest any clear means of further subdivision.

Crombach (1974a) used DNA-DNA base pairing to compare a small number of *Arthrobacter* strains, which included the type strain of *A. globiformis* and several soil isolates with similar DNA base ratios. He found that the strains were heterogeneous and concluded that they could be divided into an "*A. globiformis*" group and an "*A. simplex*" group. Unfortunately, no reference strain of *A. simplex* was included and the DNA base ratios quoted for the group bearing that name indicated that it had little relationship with the type strain of *A. simplex.*

More recently, Stackebrandt and Fiedler (1979) used DNA-DNA base pairing to compare representatives of all the nomenclatural species of the "*globiformis*" group mentioned above except *C. ilicis.* They concluded that most representatives of this group of arthrobacters merited separate specific status, a conclusion supported in part by the peptidoglycan types of the strains studied. An exception was *A. polychromogenes,* which they considered a subspecies of *A. oxidans.* However, the only two species in the study represented by more than one strain, *A. globiformis* (two strains including the type) and *A. citreus* (five strains including the type), were also found to be heterogeneous.

THE "*SIMPLEX/TUMESCENS*" GROUP OF "ARTHRO-BACTERS". The main feature by which the "*sim-*

plex/tumescens" group may be recognized is that the cell wall diamino acid is L-DAP. In addition, the DNA base ratios are substantially higher than those of the "*globiformis*" group; values in the range 71.7–74 mol% G+C have been reported for *Arthrobacter simplex* and 69.8–72.4 mol% G+C for *A. tumescens* (Skyring and Quadling, 1970; Skyring, Quadling, and Rouatt, 1971; Yamada and Komagata, 1970). Few authentic strains are available, but a number of numerical taxonomic studies have also shown that the type strains of *A. simplex* and *A. tumescens* are distinct from *A. globiformis* but do not necessarily resemble each other (Bousfield, 1972; da Silva and Holt, 1965; Davis and Newton, 1969; Jones, 1975; Skyring, Quadling, and Rouatt, 1971).

Although we have referred to this group as the "*simplex/tumescens*" group, it should be noted that L-DAP-containing bacteria which do not belong to either of these two species were reported to be relatively common in soil and on herbage (Keddie and Cure, 1977; Keddie, Leask, and Grainger, 1966) and many had the features of *Arthrobacter sensu lato.* From this assemblage of L-DAP-containing coryneform bacteria, *A. tumescens* is the only species that can be recognized with some confidence by using the characters listed in Tables 17.6 and 17.8 in *Bergey's Manual,* eighth edition (Keddie, 1974a).

A. simplex is less well defined, largely because detailed information is curently available only for the type strain (ATCC 6946). However, *A. simplex* is the only L-DAP-containing species so far described that can grow in a simple mineral salts–glucose medium without added organic growth factors; all others so far examined require added B vitamins (Keddie and Cure, 1977; Keddie, Leask, and Grainger, 1966).

The major isoprenoid quinones reported to occur in *A. simplex* and *A. tumescens* are tetrahydrogenated menaquinones with eight isoprene units (MK-8[H_4]) (Collins, Goodfellow, and Minnikin, 1979). The fatty acid profiles are unusual in that substantial amounts of 10-methyloctadecanoic acid (tuberculostearic acid) are present in both species, although at much higher levels in *A. simplex* than in *A. tumescens* (see Keddie and Bousfield, 1980; Minnikin, Goodfellow, and Collins, 1978).

THE "*TERREGENS/FLAVESCENS*" GROUP OF "ARTHROBACTERS". Little can be said about the identification of *Arthrobacter terregens* and *A. flavescens* because all the available information is based only on studies of the type strains of each species. As far as we are aware, only one additional strain has been identified as *A. flavescens* (Greenberg and Barker, 1962).

Although in the eighth edition of *Bergey's Manual* (Keddie, 1974a) both species are stated to contain lysine in the cell wall, it was shown subsequently that *A. terregens* (Keddie and Cure, 1977; Schleifer and Kandler, 1972) and *A. flavescens* (Keddie and Cure, 1977) contained ornithine and not lysine in the cell wall peptidoglycan. Accordingly, this feature distinguishes the group from other members of *Arthrobacter sensu lato*. In addition, the DNA base ratios of *A. terregens* (68.7 mol% G+C) and of *A. flavescens* (70.3 mol% G+C) are some 3–5 mol% higher than that of the type strain of *A. globiformis* (Skyring and Quadling, 1970; Skyring, Quadling, and Rouatt, 1971). Numerical phenetic studies also indicate that the *"terregens/flavescens"* group is distinct from the *"globiformis"* group and should be excluded from *Arthrobacter* (Skyring, Quadling, and Rouatt, 1971). The DNA base ratios and cell wall composition of *A. terregens* and *A. flavescens* and, more especially, the peptidoglycan structure reported for *A. terregens* (Schleifer and Kandler, 1972) suggest that these two species may have affinities with the proposed genus, *Curtobacterium*.

In addition to the features mentioned, recognition of *A. terregens* and *A. flavescens* depends on their requirement for the *terregens* factor (Lochhead, 1958a; Lochhead and Burton, 1953). The type strains of the two species may be distinguished by their vitamin requirements and by a few conventional tests given in Table 17.6 in *Bergey's Manual*, eighth edition (Keddie, 1974a).

Suitable reference strains are as follows: *"globiformis"* group, *A. globiformis*, ATCC 8010 (NCIB 8907); *"simplex/tumescens"* group, *A. simplex*, ATCC 6946 (NCIB 8929), and *A. tumescens*, ATCC 6947 (NCIB 8914); *"terregens/flavescens"* group, *A. terregens*, ATCC 13345 (NCIB 8909) and *A. flavescens*, ATCC 13348 (NCIB 9221). All are the type strains of the species listed.

THE GENUS *BREVIBACTERIUM*

The genus *Brevibacterium* was proposed by Breed (1953a), with *B. linens* as type species, for a number of nonsporeforming, Gram-positive rods formerly classified in the genus *Bacterium*. The genus was recognized in the seventh edition of *Bergey's Manual* (Breed, Murray, and Smith, 1957) and, along with the genus *Kurthia*, constituted the family Brevibacteriaceae (see Breed, 1953a,b). Bacteria of the genus *Brevibacterium* were described as typically short, unbranching rods that were usually nonmotile; no indication was given that *Brevibacterium* or its type species *B. linens* had a coryneform morphology. Thus, from its inception *Brevibacterium* was little more than a repository for a number of poorly described, nonsporeforming, Gram-positive rods which could not be accommodated elsewhere; and in the years that followed it continued to be used for this purpose.

However, later investigators showed that *B. linens* had a coryneform morphology and that the morphological changes during the growth cycle were similar to those seen in *Arthrobacter globiformis* (Mulder and Antheunisse, 1963; Schefferle, 1957). A large majority of the diverse assemblage of nomenclatural species bearing the name *Brevibacterium* have now been shown to have a coryneform morphology. For reasons such as those mentioned, the genus was considered incertae sedis in the eighth edition of *Bergey's Manual* (Rogosa and Keddie, 1974a).

Considering its early history, it is not surprising that subsequent numerical taxonomic (see Jones, 1978) and chemotaxonomic studies (see Fiedler et al., 1970; Keddie and Cure, 1978; Minnikin, Goodfellow, and Collins, 1978) have shown that the genus *Brevibacterium* is extremely heterogeneous. However, it has also emerged from these studies (see Jones, 1975; Keddie and Cure, 1977) that *B. linens* is a distinct taxon, which could form the nucleus of a redefined genus *Brevibacterium* as proposed by Yamada and Komagata (1972b). Accordingly, we shall adopt a restricted concept of the genus *Brevibacterium* with *B. linens* as the only species included at present.

Many of the remaining large number of nomenclatural species that bear the name *Brevibacterium*, including "patent" strains of industrial importance, can now be assigned to other genera. An appreciable number of species contain *meso*-DAP, arabinose, and galactose in the cell wall (Keddie and Cure, 1977; Schleifer and Kandler, 1972). Some of these have the characters of *Corynebacterium sensu stricto*, described elsewhere in this chapter. But necessary data, such as that on lipid composition, are not available at present for other species so it is not possible to assign them to a particular taxon. It is likely, however, that they are either *Corynebacterium* spp. or *"rhodochrous"* (*Rhodococcus*) strains (see Keddie and Cure, 1978; Minnikin, Goodfellow, and Collins, 1978). A rather smaller number of species contain lysine in the wall, and a proportion of these have been shown to be legitimate *Arthrobacter* spp. (Keddie and Cure, 1978; Schleifer and Kandler, 1972; Yamada and Komagata, 1972b), while several ornithine-containing species have been placed in the proposed genus *Curtobacterium* (Yamada and Komagata, 1972b). Two strains named *B. lipolyticum* are the only ones so far that have been shown to contain L-DAP in the cell wall (Yamada and Komagata, 1972b), and both the DNA base ratios (Yamada and Komagata, 1970) and the menaquinone system (Yamada et al., 1976) are similar to those of *A. simplex*. Accordingly, these two strains are possible members of what we have referred to as the *"simplex/tumescens"* group of arthrobacters (see genus *Arthrobacter*). At least one species, *B. leucinophagum*, is a Gram-negative rod and a possible member of the genus *Acinetobacter* (Jones and Weitzman, 1974). For further information on the cell wall composition and probable taxonomic position of strains labeled *Brevibacterium*, see Keddie and Cure (1978).

Habitats of *Brevibacterium linens*

Brevibacterium linens is an orange, colony-forming, salt-tolerant organism that usually constitutes a substantial fraction of the surface microflora of certain surface-ripened soft cheeses and is thought to contribute to the ripening process. Accordingly, these and other cheeses are the only known habitat of unequivocal *B. linens* strains. However, bacteria considered to be closely related to *B. linens* have been isolated from marine sources including the surface of sea fish and from poultry deep litter; these habitats, like that of *B. linens*, have a low water activity. It is thus possible that such strains may represent unnamed species of the genus *Brevibacterium* (in the restricted sense used here).

Cheese

The usual habitat of *Brevibacterium linens* is on the exterior of surface-ripened, soft cheeses of the Limburger type, but it frequently occurs on cheeses such as Camembert, Roquefort, Brick, and many others (see El-Erian, 1969; Mulder et al., 1966). During ripening of Limburger and similar cheeses, a slimy orange or orange-brown surface smear develops in which *B. linens* usually occurs in large numbers. It is believed to contribute to the surface color and, in part by its proteolytic activity, to the ripening of such cheeses (Albert, Long, and Hammer, 1944).

Cheeses of the Limburger type are heavily salted, so only salt-tolerant microorganisms can grow on the surface. At the relatively low pH of the freshly made cheese, the first microorganisms that develop are yeasts. By utilization of lactate, the yeasts gradually raise the pH at the cheese surface. At about pH 6.0, the yeasts begin to decline in numbers and are eventually replaced by bacteria, most of which have a coryneform morphology. The highly salt-tolerant *B. linens* forms a substantial, but usually not a major, part of the coryneform flora on the surface of the mature cheese (see El-Erian, 1969; Mulder et al., 1966).

Recently, Sharpe et al. (1977) reported that the few strains of *B. linens* they tested could produce methanethiol from methionine. Methanethiol has been implicated as an important constituent of the aroma of Cheddar cheese (see Sharpe, Law, and Phillips, 1976), and the authors suggested that production of this compound by *B. linens* may contribute to the aroma and flavor of the surface-ripened cheeses. Sharpe et al. (1977) described a further

group of methanethiol-producing coryneform bacteria which, although isolated from cheese, closely resembled methanethiol-producing coryneform bacteria commonly isolated from human skin. The authors considered it likely that their dairy isolates originated from skin and were chance contaminants of milk and cheese in which they occurred in relatively small numbers. Both the dairy and skin isolates resembled *B. linens* in morphology and in a number of conventional and chemotaxonomic characters, but differed in having considerably higher maximum-growth temperatures and in not producing an orange pigment. However, it was considered that these isolates were members of the genus *Brevibacterium* (Sharpe et al., 1977).

Sea Fish and Similar Habitats

Orange-pigmented coryneform bacteria isolated from various marine fish were considered to be very similar to *Brevibacterium linens* in morphology and in physiological characteristics (Crombach, 1974b; Mulder et al., 1966). Although most of the fish isolates also had DNA base ratios similar to those recorded for *B. linens* (Crombach, 1972, 1974a), DNA-DNA base-pairing indicated that they were heterogeneous and that only a small proportion of them were closely related to *B. linens*. Also, Bousfield (1978) reported that a group of coryneform bacteria isolated from sea water clustered with a reference culture of *B. linens* in a numerical taxonomic study. The marine isolates were similar to *B. linens* in DNA base ratios and in cell wall composition, and it was considered that they belonged to the same genus.

Other Habitats

A substantial number of orange coryneform bacteria isolated from poultry deep litter were considered by Schefferle (1966) to be closely related to *Brevibacterium linens* (see also Mulder et al., 1966). Subsequent examination of several of Schefferle's litter strains showed that only two clustered with reference *B. linens* cultures in the numerical taxonomic study of Jones (1975) and few had a cell wall composition similar to that of *B. linens* (Keddie and Cure, 1977). The latter authors also suggested that three coryneform isolates from pig-manure slurry were related to *B. linens*.

Isolation of *Brevibacterium linens*

Brevibacterium linens may be isolated from surface-ripened (or other) cheeses by streaking or spreading suitable dilutions of the surface material on any one of a variety of nonselective media based on peptone, yeast extract, and glucose.

Isolation of *Brevibacterium linens* from Cheese

The primary dilution is prepared by homogenizing the cheese in 2% trisodium citrate solution in a laboratory blender. Incorporation of 4% NaCl in the medium is said to give a higher proportion of *B. linens* colonies, as well as giving an increased total count (El-Erian, 1969). After incubation for ca. 5–7 days at 20–25°C, orange colonies are picked; those showing a coryneform morphology are further identified. However, about 50% or more of *B. linens* strains produce an orange pigment only when exposed to light (Crombach, 1974b; Mulder et al., 1966); accordingly, the plates must be incubated in light in order to detect all presumptive *B. linens* colonies. Presumably this phenomenon is similar to the light-induced carotenogenesis observed in some other taxa. Also, the pigment does not develop if the plates are exposed to light only after the colonies have developed fully and growth has ceased (Mulder et al., 1966). Probably the most convenient method is to incubate at 25°C in an incubator until small colonies develop, and then to remove the plates to a bench exposed to daylight for the remainder of the incubation period.

Tryptone soya agar and cheese agar are said to be particularly suitable for the isolation of *B. linens*. The medium described by El-Erian was based on tryptone soya broth (TSB, Oxoid). Similar products are manufactured by BBL and Difco, United States.

Tryptone Soya Agar with 4% NaCl for Isolating *Brevibacterium linens* (El-Erian, 1969)

TSB has the following composition (g/liter):

Tryptone (Oxoid)	17.0
Soya peptone (Oxoid)	3.0
NaCl	5.0
K$_2$HPO$_4$	2.5
Glucose	2.5.

To prepare the final medium, a further 4% of NaCl is added together with 1.2% agar; final pH 7.0. The medium is sterilized at 121°C for 15 min.

The following medium was devised for the isolation of *B. linens* and is said to give good pigment production in about 7 days at 21°C or at room temperature, especially when plates are incubated in an oxygen-enriched atmosphere.

Cheese Agar for Isolating *Brevibacterium linens* (Albert, Long, and Hammer, 1944)

Medium contains (g/liter):

Ripened cheese	100
Potassium citrate	10
Peptone	10
NaCl	50
Sodium oxalate	2.0
Agar	15

The cheese is dispersed in 300 ml of distilled water containing 10 g of potassium citrate, and the suspension is heated to 50°C to separate the fat. The remaining ingredients are dissolved in 700 ml of distilled water and added to the aqueous suspension of cheese solids from which the fat has been removed. The pH is adjusted to 7.4, and the medium is dispensed in suitable containers and sterilized at 121°C for 25 min. When pouring plates the melted, cooled medium must be thoroughly mixed to distribute the suspended cheese solids.

Identification of *Brevibacterium*

At the present time, the genus *Brevibacterium* in the restricted sense used here contains only the type species, *B. linens*. Coryneform isolates from habitats other than cheese may also be candidates for the genus, but most are probably distinct from *B. linens* and may constitute one or more additional species. Unfortunately, there is as yet no simple way of distinguishing these other putative species from *B. linens*. Accordingly, by using the features usually considered characteristic of the species, we can identify isolates as *B. linens* with some confidence only when the source is cheese.

Presumptive Identification of *Brevibacterium linens*

Grecz and Dack (1961) reported that *B. linens* (78 strains) gave characteristic color reactions when the orange growth was treated with various acids and bases, whereas other orange-pigmented bacteria such as *Staphylococcus aureus* did not. Jones, Watkins, and Erickson (1973) tested 93 strains of pigmented coryneform and other bacteria and confirmed the suggestion of Grecz and Dack that these color reactions could be used for the presumptive identification of *B. linens*. We describe below only two of the most useful of these color reactions.

Color Reactions for Identification of *Brevibacterium linens* (Grecz and Dack, 1961; Jones, Watkins, and Erickson, 1973)

1. A small amount of growth from orange-pigmented colonies on isolation plates is removed to a white tile and a drop of 5 M NaOH or 5 M KOH is placed on the growth material. A stable pink-red color which develops in ca. 2 min is presumptive evidence of *B. linens*. The alkali may be added directly to colonies if it is not intended to make isolations from them.

2. A small amount of growth is removed to a disk of Whatman No. 1 filter paper which has been moistened with glacial acetic acid. The growth material is then rubbed firmly with a glass rod; a stable salmon-pink color which develops in ca. 1 min indicates *B. linens*.

A suitable reference strain of *B. linens* (e.g., ATCC 9175, NCIB 8546) should be used as a positive control, and we recommend that both tests should be applied. For small colonies it may be necessary to subculture onto agar slants before doing the tests.

Identification of *Brevibacterium linens*

The cells of *Brevibacterium linens* show a marked change of form during the growth cycle in complex media. Older cultures (ca. 3–7 days at 25°C) are usually composed entirely or largely of coccoid cells (Fig. 2d) which, on transfer to fresh, complex medium, grow out to give the slender, irregular rods characteristic of exponential-phase cultures. (Fig. 2a–c). Many cells are arranged at an angle to each other to give V-formations; primary branching may occur. As growth proceeds, the rods become shorter and are eventually replaced by the coccoid cells characteristic of stationary-phase cultures (Fig. 2d) of most strains (see Crombach, 1974b; Cure and

Keddie, 1973; Mulder et al., 1966). The cells are nonmotile and endospores are not formed. Both rods and coccoid cells are Gram positive and not acid fast. They are obligate aerobes: the mode of metabolism is respiratory, never fermentative; acids are not formed from glucose and other sugars in peptone media. They are catalase positive. The cell wall peptidoglycan contains *meso*-DAP (Keddie and Cure, 1977; Schleifer and Kandler, 1972; Yamada and Komagata, 1972b); the cell walls do not contain arabinose (Fiedler and Stackebrandt, 1978; Keddie and Cure, 1977; Schleifer and Kandler, 1972).

The morphology and staining reactions may be examined on medium EYGA (Cure and Keddie, 1973) and cell wall composition by chromatography of whole organisms (see Introduction to this chapter for suitable methods). *Brevibacterium linens* characteristically produces yellow-orange to orange-red colonies on suitable media but, for a majority of strains, incubation in light is necessary for pigment production. It has been suggested that the pigment is a carotenoid, which is characteristic of *B. linens*, and that it may be recognized by the distinctive color reactions produced when the pigmented growth is treated with solutions of strong bases or glacial acetic acid (Jones, Watkins, and Erickson, 1973). *B. linens* is described as salt tolerant; but there is some disagreement about the actual concentration of NaCl that will allow growth of most strains, presumably because different experimental conditions and incu-

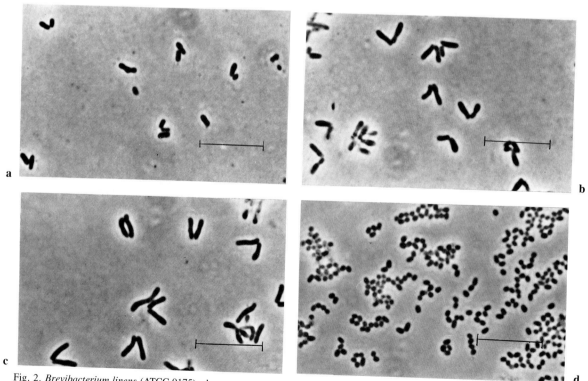

Fig. 2. *Brevibacterium linens* (ATCC 9175) when grown on medium EYGA at 25°C; inoculum coccoid cells as in (d). (a) After 6 h, showing outgrowth of rods from coccoid cells. (b) After 12 h. (c) After 24 h. (d) After 3 days. Bar = 10 μm.

bation times have been used. When a suitable basal medium at pH 7 is used, it appears that all strains tested give growth in 8% (wt/vol) NaCl in about 1 week at 25°C, and in 12% (wt/vol) NaCl after about 1 month at 25°C (see Crombach, 1974a; El-Erian, 1969; Mulder et al., 1966).

In our experience, a large majority of *B. linens* strains tested have relatively low maximum temperatures in the range 30–33°C (Cure and Keddie, unpublished). This feature was also noted by Mulder et al. (1966) who, however, reported that maximum temperatures were slightly higher when 4% NaCl was incorporated in the medium.

Other features noted for *B. linens* are gelatin liquefaction and extracellular DNase production (Yamada and Komagata, 1972a,b). The cells do not contain mycolic acids (Goodfellow, Collins, and Minnikin, 1976; Keddie and Cure, 1977). DNA base ratios are in the range of about 60–64 mol% (Crombach, 1972; Yamada and Komagata, 1970). The principal isoprenoid quinones in the few strains tested were dihydrogenated menaquinones with eight isoprene units (MK-8[H$_2$]) (Collins, Goodfellow, and Minnikin, 1979; Yamada et al., 1976).

Although it was originally reported that *B. linens* was characterized by the presence of ribose in cell wall preparations (Keddie and Cure, 1977), subsequent, more detailed studies of the wall polysaccharide failed to reveal the presence of this sugar in any of several strains studied (Fiedler and Stackebrandt, 1978). However, *B. linens* was shown to differ from all other coryneform bacteria examined in containing a glycerol teichoic acid in the wall polysaccharide (Fiedler and Stackebrandt, 1978). Thus, the original report of ribose in the walls of *B. linens* strains (Keddie and Cure, 1977) seems to have resulted from an error in chromatography.

B. linens is most likely to be confused with those taxa of coryneform bacteria to which it bears a broad morphological resemblance, viz., *Arthrobacter sensu lato* and the *"rhodochrous"* taxon (*Rhodococcus*). However, it may be distinguished readily from both by the cell wall composition (see Tables 1 and 2). Whereas *B. linens* contains *meso*-DAP as wall diamino acid, the *"globiformis"*, *"simplex/ tumescens"*, and *"terregens/flavescens"* groups of *Arthrobacter sensu lato* contain, respectively, lysine, L-DAP, and ornithine. The distinction from *Rhodococcus* and from other *meso*-DAP-containing coryneform and similar taxa rests on the sugar composition of the cell wall: *B. linens* does not contain arabinose, whereas other *meso*-DAP-containing taxa, viz., *Rhodococcus*, *Corynebacterium sensu stricto*, *Nocardia sensu stricto*, and *Mycobacterium* contain arabinose as the characteristic sugar. Also, unlike the other *meso*-DAP-containing taxa mentioned, *B. linens* does not contain mycolic acids. There remains a somewhat heterogeneous group of coryneform bacteria; their walls contain *meso*-DAP

but not arabinose and they do not contain mycolic acids (see Keddie and Cure, 1978).The relationship of these strains to *B. linens* is at present unresolved.

Although most studies of *B. linens,* including those of DNA-DNA homologies (see Fiedler and Stackebrandt, 1978), have emphasized the similarity among different strains, there is also some evidence of heterogeneity. Thus, Mulder et al. (1966) found considerable heterogeneity in the nutritional requirements of different *B. linens* strains, although a majority required B vitamins and usually one or more amino acids. Also, Foissy (1974), by using an electrophoretic zymogram technique, considered that 17 isolates identified as *B. linens* could be divided into two biotypes while a further strain was quite distinct from all the others. It is thus possible that more detailed studies may reveal the existence of more than one taxon in what is at present considered a single species.

A suitable reference strain is the type strain ATCC 9172 (NCIB 9909).

The Genus *Cellulomonas*

The genus *Cellulomonas* was established by a committee (Bergey et al., 1923) to accommodate a rather heterogeneous collection of cellulose-decomposing bacteria isolated mainly from soils some 10 years earlier by Kellerman and his co-workers (Kellerman and McBeth, 1912; Kellerman et al., 1913; McBeth, 1916; McBeth and Scales, 1913). The 31 species originally included in the genus (Bergey et al., 1923) were described as Gram-negative, motile or nonmotile, short rods whose main distinguishing feature was the ability to decompose cellulose. Bacteria of the genus *Cellulomonas* received little further attention for the next 30 years or so, presumably because they had no obvious economic importance other than the ability to decompose cellulose, a feature possessed by many other soil microorganisms.

Although Jensen (1934) had reported earlier that *C. fimi* was weakly Gram-positive and morphologically similar to the corynebacteria, it was Clarke (1951, 1952) who clearly established that most extant, authentic strains of *Cellulomonas* spp. had a coryneform morphology. However, whereas Jensen (1934) considered that *C. fimi* should be transferred to the genus *Corynebacterium*, Clarke (1952, 1953) believed that the cellulolytic coryneform bacteria represented a distinct group and that they should be retained in a redefined genus *Cellulomonas*. Clarke (1952) prepared a revised generic description based on his laboratory studies, and proposed the inclusion of the genus *Cellulomonas* in the family Corynebacteriaceae. Clarke (1953) also reassessed the validity of some 27 putative *Cellulomonas* spp. and concluded that only 10 should be recognized; 14 of the remainder were reduced to synonymy with these 10 and 3 were excluded from the genus. However, authentic cultures were available for only 6 of the 10 species recognized; these cultures had been maintained for some 40 years by one of the original coauthors, N. R. Smith. For the remaining four species, Clarke (1953) based his conclusions on the original descriptions.

Clarke's proposals were adopted in the seventh edition of *Bergey's Manual* (Breed, Murray, and Smith, 1957); thus *Cellulomonas,* with *C. biazotea* as type species, was recognized as a genus of soil coryneform bacteria. The main features were: acid production from carbohydrates, slow liquefaction of gelatin, and "cellulose commonly attacked". Despite the rather vague wording in the generic description, however, the only really distinctive feature of the genus was the property of cellulolysis. The 10 species recognized were distinguished from each other by such features as motility, nitrate reduction, ammonia production, chromogenesis and, in the case of *C. fimi,* by the fermentation of xylose and arabinose.

Bacteria of the genus *Cellulomonas* were now described as "Gram variable", whereas originally they had been considered to be Gram negative (Bergey et al., 1923). In reporting their Gram-positive nature, Clarke (1953) noted that "*Cellulomonas* usually is Gram-negative during the first 24 h of growth, thereafter Gram-positive or Gram-variable staining is secured on cells grown under favorable conditions. In cultures several days old the Gram reaction is usually negative". In view of Clarke's description, it is not surprising that the Gram reaction of *Cellulomonas* continued to be controversial in the ensuing years (see Keddie, 1974b).

The creation of a separate genus of coryneform bacteria based largely on cellulolysis was questioned (e.g., see Jensen, 1966), but subsequent studies using modern taxonomic methods have supported Clarke's (1952) conclusions. In the past decade or so, attempts to clarify the taxonomy of the coryneform bacteria have resulted in renewed interest in the genus *Cellulomonas*. A number of taxonomic studies of coryneforms have included one or more representatives of the genus, but in almost every case the cultures studied were from the same small group of named *Cellulomonas* strains from culture collections rather than fresh isolates. These studies indicate that, with the exception of a few obviously misclassified cultures, all authentic, named strains of *Cellulomonas* species form a relatively homogeneous and distinct taxon. This view is supported by numerical (see Jones, 1978) and nonnumerical (Yamada and Komagata, 1972a, b) taxonomic studies, by the vitamin, nitrogen, and carbon nutrition (Keddie, 1974b; Keddie, Leask, and Grainger, 1966; Owens and Keddie, 1969), and by a variety of

chemotaxonomic characters, including cell wall composition (see Keddie and Cure, 1978) and peptidoglycan structure (Fiedler and Kandler, 1973), menaquinone composition (Minnikin, Goodfellow, and Collins, 1978; Yamada et al., 1976), fatty acid and polar lipid content (Minnikin, Collins, and Goodfellow, 1979), and DNA base ratios (Bousfield, 1972; Stackebrandt and Kandler, 1979; Yamada and Komagata, 1970).

In the eighth edition of *Bergey's Manual* (Keddie, 1974b), only one species (in which allowance is made for minor variations in chromogenesis, motility, and nitrate reduction) is recognized in the genus *Cellulomonas*. This species is designated *C. flavigena* and not *C. biazotea* because, as stated, "Since the specific epithet *flavigena* (*Bacillus flavigena* Kellerman and McBeth, 1912, 488) antedates *Bacillus biazoteus* Kellerman, McBeth, Scales and Smith, 1913, 506, the type species is therefore *Cellulomonas flavigena*." Thus *C. biazotea, C. cellasea, C. gelida,* and *C. uda* are considered to be subjective synonyms, and *C. fimi* a possible subjective synonym of *C. flavigena* (Keddie, 1974b).

Later, Braden and Thayer (1976) compared several named *Cellulomonas* spp. and two freshly isolated cellulolytic strains (Thayer et al., 1975) by quantitative-agglutination procedures, using purified cell wall preparations as antigens. Their results demonstrated that, although they were not serologically identical, there was considerable similarity between the six named strains. Nevertheless, these authors questioned the reduction of these organisms to synonymy with *C. flavigena*. More recently, on the basis of DNA-DNA homology data obtained with 10 reference strains, Stackebrandt and Kandler (1979) have proposed that seven species be recognized in the genus.

There has been some interest recently in the use of *Cellulomonas* spp. for the production of protein from wood products (Han and Callihan, 1974; Thayer et al., 1975) and in their role in the decomposition of solid-compost material (Kaufmann et al., 1976). Studies of the cellulase complex in *Cellulomonas* spp. have been made by Kaufmann et al. (1976), Stewart and Leatherwood (1976), and Beguin, Eisen, and Roupas (1977).

Habitats of *Cellulomonas*

Soil is considered to be the main habitat of bacteria of the genus *Cellulomonas* because most of the original cultures were isolated from this source (see Bergey et al., 1923). However, there seem to have been no systematic attempts to isolate *Cellulomonas* spp. since that time. Consequently, there is no information on the relative numbers, distribution, or role of these bacteria as cellulose decomposers in different soils. Most studies on the occurrence and distribution of coryneform bacteria in soil make no reference to the isolation of members of this genus.

Of 114 strains of coryneform bacteria isolated from soil and herbage using nonselective methods (Keddie, Leask, and Grainger, 1966), only one soil isolate was found to be cellulolytic. This isolate was identified as a *Cellulomonas* sp. on the basis of morphology, cell wall composition, and vitamin requirements (Keddie, Leask, and Grainger, 1966). This conclusion was later supported by numerical phenetic methods (Jones, 1975). Han and Srinivasan (1968) used an enrichment procedure to isolate a cellulolytic bacterium, which they identified as a *Cellulomonas* sp., from a mixture of rotting sugarcane stalks and adjacent soil. This isolate was reported to be definitely Gram negative and no mention was made of a coryneform morphology. However, the same strain (ATCC 21399) was examined later by Fiedler and Kandler (1973), who showed that it had the peptidoglycan structure characteristic of the genus *Cellulomonas*. Stewart and Leatherwood (1976) isolated a *Cellulomonas* sp. from soil by using a cellulose-agar medium.

A cellulolytic bacterium identified as *C. flavigena* was isolated by Patel and Vaughn (1973) from the processing liquor of spoiled olives, where it appeared to play a part in the skin rupture and flesh sloughing characteristic of this type of spoilage. This case is the only one known to us in which a *Cellulomonas* sp. has been implicated in food spoilage. Patel and Vaughn (1973) did not speculate on the source of their isolate; but presumably it could have gained access to the process in dust or by soil contamination of the olives. Whitehouse and Jackson (1972) described a cellulolytic coryneform organism isolated from raw cow's milk and concluded that its characters resembled those of *C. acidula* most closely. They suggested that the organism gained access to the milk by contamination from soil.

Kaufmann et al. (1976) isolated a cellulolytic organism, which they identified as *Cellulomonas flavigena*, from solid waste that contained a high proportion of paper. Beguin, Eisen, and Roupas (1977) studied the cellulases of a cellulolytic organism, which they identified as *C. flavigena*, but the authors merely stated that they received the strain from workers in Cuba and the original source was not given.

Isolation of *Cellulomonas*

There is relatively little in the literature on procedures suitable for isolation of *Cellulomonas* spp. from natural sources. All of the procedures described have exploited the ability of *Cellulomonas* spp. to degrade cellulose. The most usual procedure

is to prepare enrichment cultures in mineral media containing yeast extract (0.05–0.1%) and filter paper, followed by plating on cellulose agar. Cellulolytic bacteria are detected by the clear zones surrounding the colonies.

Kellerman and his associates used the above general procedure to isolate all the original cultures that they described (see Kellerman et al., 1913; McBeth and Scales, 1913). But the media used by these authors were nutritionally inadequate for the growth of *Cellulomonas* spp. because their media were based on simple mineral salts solutions and cellulose. All *Cellulomonas* spp. have been shown subsequently to require both biotin and thiamine for growth (Keddie, Leask, and Grainger, 1966). The nutritional inadequacy of the media may explain some of the problems encountered by these early investigators, such as loss of cellulolytic activity, difficulties in obtaining isolates free from contaminants, and so on. Successful isolations have also been made by plating serial dilutions of soil directly onto cellulose agar. None of these methods, however, is specific for bacteria of the genus *Cellulomonas*, and isolates must therefore be screened for those with a coryneform morphology. We are not aware of any published work on a selective medium for *Cellulomonas*.

Isolation of *Cellulomonas* by Direct Plating on Cellulose Agar (Stewart and Leatherwood, 1976)

Serial dilutions of soil are surface-plated on cellulose agar of the following composition (g/liter of distilled water):

NaNO₃	1.0
K₂HPO₄	1.0
KCl	0.5
MgSO₄	0.5
Yeast extract (Difco)	0.5
Agar	1.7
Ball-milled filter paper (BMFP)	1.0
Glucose	1.0
pH	7.0

Autoclave at 121°C for 15 min. To prepare BMFP, a 3% aqueous suspension of Whatman No. 1 filter paper is ball-milled for 3 days. Micro crystalline cellulose (Avicel, FML) at a final concentration of 0.1% (wt/vol) has been used successfully as a cellulose source (see Kaufmann et al., 1976) and may be substituted for BMFP. It should be noted that the cellulose agar described contains a low concentration (0.1%) of glucose; others (e.g., see Han and Srinivasan, 1968) do not.

Plates are incubated at 30°C for up to 7 days and colonies showing zones of clearing are replated on the same medium until pure. Cellulolytic isolates that have a coryneform morphology are presumptive *Cellulomonas* spp.

Several investigators have commented on the difficulties experienced in obtaining pure cultures of *Cellulomonas* spp. (e.g., see Han and Srinivasan, 1968; Patel and Vaughn, 1973).

Cellulomonads may be enriched by using a mineral salts medium that contains a low concentration of yeast extract, with filter paper as cellulose source.

Enrichment of *Cellulomonas* (based on method of Han and Srinivasan, 1968)

A liquid medium similar to the cellulose agar described above may be used, but without glucose and with a filter-paper strip replacing the BMFP. Alternatively, the medium described by Han and Srinivasan (1968) may be used. The medium is dispensed in 10-ml quantities in tubes and, before sterilization, strips of filter paper (Whatman No. 1) are placed in the tubes so that about 2 cm remain above the level of the liquid. The tubes are inoculated with small quantities of the materials being investigated and incubated, preferably with mechanical agitation, at 30°C. As soon as a patch of yellow pigmentation or signs of disintegration appear at the liquid-air interface on the filter paper, a small portion is transferred, using a sterile wire, to a tube of fresh medium. This procedure is repeated two or three times. The filter paper from the final tube is then macerated in a small amount of sterile liquid medium, and streaked onto plates of cellulose agar until a pure culture is obtained.

Identification of *Cellulomonas*

All authentic cultures of *Cellulomonas* currently available are cellulolytic. Cellulolysis is an extremely stable property (Clarke, 1951) and, to our knowledge, has not been demonstrated in any other coryneform bacteria. Therefore, any isolate that has a coryneform morphology and is cellulolytic may be regarded as a presumptive *Cellulomonas* sp.

A more detailed identification requires the examination of the morphological features and staining reactions, the cell wall composition, cellulolytic ability, and the relationship to oxygen.

Bacteria of the genus *Cellulomonas* have the general morphological features described for coryneform bacteria. But the rods are usually slender (ca. 0.5–0.6 µm in diameter on medium EYGA, by phase-contrast microscopy) and, in exponential-phase cultures, are frequently slightly filamentous and may show primary branching. As growth proceeds, the rods become shorter and V-formations are more prominent. Older cultures (ca. 7 days at 25°C) contain mainly short rods, but a proportion of the cells may be coccoid. However,

they do not show the marked rod-coccus cycle characteristic of the genus *Arthrobacter*. They are motile by one subpolar or a few lateral flagella, or nonmotile. They are Gram positive but are very easily decolorized, and often a mixture of Gram-positive and Gram-negative rods is seen; the latter may predominate and may contain Gram-positive inclusions. The cell wall peptidoglycan contains ornithine as diamino acid (Table 3). The sugar composition of the wall polysaccharide varies considerably in different strains (see Keddie and Cure, 1978), but Keddie, Leask, and Grainger (1966) noted that galactose was uniformly absent. This finding was largely confirmed by Fiedler and Kandler (1973) who, however, found a small amount in one of several strains tested.

The relationship to oxygen is not uniform throughout the genus and is somewhat controversial. Most strains are facultative anaerobes, but growth is markedly reduced in anaerobic conditions in glucose-containing media; a few strains are either aerobic or give equivocal results (Keddie, 1974b; Keddie and Cure, 1977). However, all strains studied so far give acid from glucose both oxidatively and fermentatively when using a method based on that of Hugh and Leifson (1953). In addition, all extant *Cellulomonas* cultures hydrolyze starch and gelatin (weakly) and, although nitrate reduction has been used for distinguishing species within the genus (Clarke, 1953), in our experience, reduce nitrate to nitrite (Cure and Keddie, unpublished data; see also Stackebrandt and Kandler, 1979).

Relatively few substances are used as sole sources of carbon and energy and most are carbohydrates: in addition to starch and cellulose a number of sugars are universal or near-universal substrates (Keddie, 1974b). All strains grow best at ca. 30°C in air. Meat extract (0.5%) plus peptone (0.5%), or yeast extract (0.25%) plus peptone (0.25%) are suitable general media for growth.

Morphology and staining reactions may be examined by using the medium (EYGA) and methods described by Cure and Keddie (1973). Great care must be taken when treating with alcohol in the Gram-staining procedure because cellulomonads decolorize very readily and may be mistaken for Gram-negative rods. Cell wall analysis is done most conveniently by using cell wall material prepared by alkali treatment of whole cells (see Keddie and Cure, 1977). Cellulolytic activity is tested for in 0.5% peptone water that contains partially immersed filter paper strips and is incubated at 30°C. Separation of the fibers at the liquid-air interface on gentle agitation indicates cellulolytic activity. Strains from culture collections are weakly cellulolytic and may require incubation for 2 weeks or more to show activity. It must be stressed that all strains of *Cellulomonas* spp. so far examined require biotin and thiamine for growth (Keddie, Leask, and Grainger, 1966) and, therefore, media that contain only mineral salts and filter paper are nutritionally inadequate for demonstration of cellulolytic activity. Use of such media presumably led Yamada and Komagata (1972a,b) to the erroneous conclusion that the six authentic cultures that they studied were noncellulolytic.

Bacteria of the genus *Cellulomonas* contain ornithine in the cell wall peptidoglycan, and this feature allows them to be distinguished from most other coryneform (and allied) taxa (Tables 1–3). By the same token, they are most likely to be confused with the few other coryneform taxa that contain ornithine as cell wall diamino acid, and most probably with members of the proposed genus *Curtobacterium* (Table 3). The cellulolytic activity of all extant, authentic *Cellulomonas* cultures allows them to be distinguished from *Curtobacterium* and other ornithine-containing taxa, but few other readily detected differential features are available (Table 3). However, cell wall analysis may be of further assistance in distinguishing these taxa: Whereas the peptidoglycans of *Curtobacterium* spp., *Arthrobacter terregens,* and *Microbacterium liquefaciens* contain glycine (Schleifer and Kandler, 1972), those of *Cellulomonas* spp. do not (Fiedler and Kandler, 1973). Although the recommended rapid method of cell wall analysis failed to reveal the presence of glycine in wall preparations of occasional strains of those *Curtobacterium* spp. tested (Keddie and Cure, 1977), further improvements in the technique should make this a valuable distinguishing feature between *Cellulomonas* and the other ornithine-containing taxa mentioned. An additional useful feature is that galactose does not occur in cell wall preparations of *Cellulomonas* spp. (Keddie, Leask, and Grainger, 1966) but is found in those of most *Curtobacterium* spp. tested so far and in those of *A. terregens* and *A. flavescens* (see Keddie and Cure, 1977, 1978).

It is possible that the cellulolytic, soil, nocardioform organism named *Nocardia cellulans* (most probably an *Oerskovia* sp.; see Keddie and Cure, 1977; this Handbook, Chapter 159) may be confused with *Cellulomonas*. The mycelial nature of *N. cellulans* is readily seen by examining young colonies on EYGA or soil extract agar by low magnification (100×); it may also be distinguished by the cell wall composition. *N. cellulans* contains lysine as diamino acid in the peptidoglycan, and cell wall preparations contain galactose (Keddie and Cure, 1977). If doubts exist about the Gram reaction of suspected *Cellulomonas* isolates, cell wall analysis should resolve the difficulty. *Cellulomonas* spp. have the cell wall composition typical of Gram-positive bacteria, and hydrolysates of cell wall material contain only alanine, glutamic acid, and ornithine (and aspartic acid in one species) as major amino acids. On the

other hand, chromatograms of hydrolysates of cell wall material from Gram-negative bacteria reveal a mixture of amino acids and the diamino acid present is *meso*-DAP.

The homogeneity that exists in *Cellulomonas* and its differentiation from *Curtobacterium* and other ornithine-containing taxa is supported by other chemotaxonomic data. All strains of *Cellulomonas* so far studied contain a group A peptidoglycan in which the cross-linkage is between L-ornithine in position 3 of one and D-alanine in position 4 of the other of two peptide subunits, by an interpeptide bridge containing D-aspartic acid in *C. flavigena* and D-glutamic acid in all the other nomenclatural species (Fiedler and Kandler, 1973). On the other hand, *A. terregens, M. liquefaciens,* and those species now included in *Curtobacterium* have a group B peptidoglycan in which the cross-linkage is between D-glutamic acid in position 2 of one and D-alanine in position 4 of the other of two peptide subunits, by an interpeptide bridge containing D-ornithine. In *A. terregens, M. liquefaciens,* and *Curtobacterium (Brevibacterium) saperdae,* the interpeptide bridge contains glycine in addition to D-ornithine.

The major isoprenoid quinones found in the three named *Cellulomonas* strains studied to date are tetrahydrogenated menaquinones with nine isoprene units (MK-9[H_4]) (Minnikin, Goodfellow, and Collins, 1978; Yamada et al., 1976). Menaquinones of this type have not been detected in any other coryneform taxa studied so far, but are found in the nocardioform genus *Oerskovia* (Yamada et al., 1976). The three named *Cellulomonas* strains mentioned above were shown to have distinctive fatty acid profiles, but their polar lipid composition was similar to that found in representatives of the genus *Oerskovia* (Minnikin, Collins, and Goodfellow, 1979). DNA base ratios are in the range 71–75 mol% G+C (Table 3).

Although we have emphasized the cellulolytic ability of *Cellulomonas* spp. throughout this discussion, with one exception known to us (see Keddie, Leask, and Grainger, 1966), the *Cellulomonas* strains that have been studied by various investigators were isolated by virtue of their cellulolytic activity. Thus the possibility must not be excluded that noncellulolytic strains exist in nature. If this should be the case, the chemotaxonomic features described above would be essential for recognizing such strains and for distinguishing them from similar taxa.

Identification of *Cellulomonas* isolates to species level is not easy because members of the genus form a phenotypically homogeneous group (Keddie, 1974b; Stackebrandt and Kandler, 1979) and descriptions of most of the species now recognized are based on single strains (Stackebrandt and Kandler, 1979). The phenotypic features which seem to be most useful for differentiation of species are the sugar composition of the cell walls and utilization of various carbohydrates (see Stackebrandt and Kandler, 1979, for details). However, at the present time, a *Cellulomonas* isolate can be assigned with confidence to one of the six species *C. biazotea*, *C. flavigena*, *C. cellasea*, *C. fimi*, *C. gelida*, or *C. uda* only on the basis of DNA-DNA homology studies with the appropriate type strain.

A seventh species recognized by Stackebrandt and Kandler (1979) is represented by the "patent" strain originally named *C. cartalyticum* and renamed *C. cartae* by Stackebrandt and Kandler (1980). However, the characters of this cellulolytic organism indicate that it is almost certainly a member of the genus *Oerskovia* (see Keddie and Bousfield, 1980): It produces a branching mycelium which fragments later in the growth cycle (see Keddie and Bousfield, 1980); contains lysine in the cell wall peptidoglycan (Stackebrandt, Fiedler, and Kandler, 1978), and has a DNA base ratio of 76.6 mol% G+C (Stackebrandt and Kandler, 1979). Furthermore, DNA-DNA homology studies showed a much more distant relationship between *C. cartae* and other *Cellulomonas* spp. than among the six *Cellulomonas* species listed above (Stackebrandt and Kandler, 1979). Unfortunately, no *Oerskovia* strains were included in those studies.

The type species is *Cellulomonas flavigena* and the reference strain is ATCC 482 (NCIB 8073).

THE GENUS *CURTOBACTERIUM*

The genus *Curtobacterium* was proposed by Yamada and Komagata (1972b) for a group of coryneform bacteria that contained ornithine in the cell wall peptidoglycan but were considered to be distinct from members of the genus *Cellulomonas*. The species assigned to the proposed new genus included a small number of former *Brevibacterium* spp. and the only two ornithine-containing, plant-pathogenic species studied by Yamada and Komagata, *Corynebacterium flaccumfaciens* and *C. poinsettiae*. The main distinguishing features described for the genus were: the presence of ornithine in the cell wall, DNA base ratios in the range 66–71 mol% G+C, and slow and weak acid production from some carbohydrates. Lack of cellulolytic activity was not noted in the generic description of *Curtobacterium*, presumably because Yamada and Komagata (1972a,b) were unable to demonstrate cellulolysis in the authentic *Cellulomonas* cultures which they studied (see genus *Cellulomonas*), and therefore did not consider it a distinguishing feature between the two genera.

The following species were assigned to the genus *Curtobacterium* (Yamada and Komagata, 1972b)

but, apart from those indicated, each was represented by a single strain: *Corynebacterium flaccumfaciens*, *C. flaccumfaciens* subsp. *aurantiacum*, *Brevibacterium albidum*, *B. citreum* (two strains), *B. insectiphilium*, *B. luteum*, *B. pusillum* (two strains), *B. saperdae*, *B. testaceum*, and *B. helvolum* (four strains). In addition, the numerical taxonomic study of Jones (1975) indicated that the plant pathogen, *Corynebacterium betae* (one strain), was also a candidate for the genus *Curtobacterium*, a finding supported by the cell wall composition (see Keddie and Cure, 1978), peptidoglycan structure (Schleifer and Kandler, 1972), and DNA base ratios (Starr, Mandel, and Murata, 1975) of those strains studied so far.

Despite a rather meager original description, support for the genus *Curtobacterium* comes from the fact that Schleifer and Kandler (1972) independently grouped together most of the species mentioned above on the basis of their rather characteristic peptidoglycan structure. Those putative *Curtobacterium* spp. examined contain the less common group B type of peptidoglycan, i.e., one linked between positions 2 and 4 of two peptide subunits by an interpeptide bridge consisting of one D-ornithine residue (but with glycine in addition in *B. saperdae*) (Schleifer and Kandler, 1972).

However, there is also some evidence of heterogeneity within the proposed genus. The major isoprenoid quinones in most *Curtobacterium* spp. studied are normal menaquinones with nine isoprene units (MK-*9*), whereas those in *B. testaceum* contain 11 isoprene units (MK-*11*) (Yamada et al., 1976). Also, although the detailed peptidoglycan structure of *B. testaceum* has not been determined, the glycan moiety is unusual in that it contains approximately equal amounts of glycolyl and acetyl residues, thus suggesting that either muramic acid or glucosamine occurs in the *N*-glycolyl form rather than the more usual *N*-acetyl form (Uchida and Aida, 1977). *N*-glycolyl muramic acid has previously been shown to occur in some bacteria, viz., in strains of *Mycobacterium* and *Nocardia* (see Uchida and Aida, 1977). Evidence such as this led Yamada et al. (1976) to suggest that *B. testaceum* did not belong in the genus *Curtobacterium*. Collins, Goodfellow, and Minnikin (1980) confirmed that the major isoprenoid quinone in *B. testaceum* was MK-*11*, together with substantial amounts of MK-*12*, and showed that *B. saperdae* also contained major amounts of MK-*11* and MK-*12*, but with the latter component predominating. They concluded that both species should be excluded from the genus.

The species *Brevibacterium* (*Curtobacterium*) *helvolum* is an example of the taxonomic confusion that frequently results when new isolates are identified with old, inadequately described species for which authentic cultures no longer exist. See Lochhead (1955) for a historical discussion of the species

B. helvolum (Zimmermann) Lochhead. Yamada and Komagata (1972a) reported that four strains named *B. helvolum* (presumably the four strains previously isolated from oil brines by Iizuka and Komagata [1965] and identified by them as *B. helvolum*) contained ornithine in the cell wall, on the basis of this and other characteristics they assigned these strains to the genus *Curtobacterium* (Yamada and Komagata, 1972b). However, Schleifer and Kandler (1972) reported that five strains named *B. helvolum* (including four Komagata strains) contained 2,4-diaminobutyric acid in the cell wall peptidoglycan. Two further strains listed in some culture collections as *B. helvolum* (and in others as *Arthrobacter globiformis*) are soil isolates originally described by Jensen (1934) under the name *Corynebacterium helvolum;* both are legitimate *Arthrobacter* spp. of the *"globiformis"* group (see Keddie and Cure, 1977, 1978; Lochhead, 1955; Schleifer and Kandler, 1972). There is thus considerable doubt that any so-called *"B. helvolum"* strains are *Curtobacterium* spp.

Of the remaining species assigned to *Curtobacterium*, *B. albidum*, *B. citreum*, *B. luteum* (and *B. testaceum*) were isolated from paddy (Komagata and Iizuka, 1964) and *B. pusillum* from oil brines (Iizuka and Komagata, 1965). *B. saperdae* was originally isolated by Lysenko (1959) from dead larvae of the elm borer, *Saperda carcharias*. *B. insectiphilium* was originally isolated from the body wall of the bagworm, *Thyridopterex ephemeralformis* Haw, by Steinhaus (1941); but authentic cultures apparently no longer exist and those listed in culture collections are recently isolated, "patent" strains. The plant-pathogenic *"Curtobacterium"* spp., *Corynebacterium flaccumfaciens*, *C. poinsettiae*, and *C. betae* will not be discussed in detail here because they are treated along with other plant-pathogenic, coryneform bacteria in this Handbook, Chapter 143.

Habitats of *Curtobacterium*

Plant Pathogens

Three species of plant-pathogenic, coryneform bacteria currently named *Corynebacterium* are candidates for the genus *Curtobacterium*. *Corynebacterium flaccumfaciens* and its two subspecies (*C. flaccumfaciens* subsp. *aurantiacum* and *C. flaccumfaciens* subsp. *violaceum*) cause a vasicular wilt of bean (*Phaseolus vulgaris*); *C. poinsettiae* causes stem canker and leaf spot of the poinsettia (*Euphorbia pulcherrima*). *C. betae* causes a vasicular wilt and leaf spot of red beet (*Beta vulgaris*) (see Cummins, Lelliott, and Rogosa, 1974). Their habitats and pathogenicity are discussed in detail in this Handbook, Chapter 143.

Other Habitats

Apart from those listed as plant pathogens, there is no evidence that any other proposed *Curtobacterium* species are pathogenic for plants or animals. The one authentic strain of *Brevibacterium* (*Curtobacterium*) *saperdae* was isolated from dead larvae of *Saperda carcharias* (Lysenko, 1959), but there is no evidence that it was implicated in any disease. Most species were isolated from paddy (Komagata and Iizuka, 1964) and one from oil brine (Iizuka and Komagata, 1965), and curtobacteria are said to be widely distributed in plant materials and soil (Yamada and Komagata, 1972b).

About 30% of coryneform bacteria isolated from the skins of freshly caught marine fish and from sea water were tentatively identified as *Curtobacterium* spp. in a numerical taxonomic study by Bousfield (1978). Although representative strains contained ornithine in the cell wall, their DNA base ratios were in the range 61–63 mol% G+C, some 5–8 mol% lower than those previously recorded for members of the genus.

Isolation of *Curtobacterium*

Isolation of plant pathogenic species is discussed in this Handbook, Chapter 143. All other species grow well on media based on peptone, yeast extract, and glucose (see Lysenko, 1959; Steinhaus, 1941; Yamada and Komagata, 1972 a,b).

No selective medium is available for the isolation of curtobacteria. Present knowledge of their characteristics does not suggest any features of potential value for their selective cultivation or enrichment.

Identification of *Curtobacterium*

Identification of new isolates as curtobacteria requires the examination of the morphological features and staining reactions, the cell wall composition, cellulolytic ability, relationship to oxygen, and determination of the DNA base ratios.

Bacteria of the genus *Curtobacterium* have the general morphological features and staining reactions described for coryneform bacteria. Older cultures (ca. 7 days at 25°C) may show a proportion of coccoid cells, but curtobacteria do not show the marked rod-coccus cycle characteristic of the genus *Arthrobacter*. However, Kuhn and Starr (1962) reported an *Arthrobacter*-like developmental cycle in a strain of *Corynebacterium poinsettiae*. The rods are motile by a few lateral flagella, or nonmotile. The cell wall peptidoglycan contains ornithine as diamino acid (Table 3). The sugar composition of the wall polysaccharide varies among the different species, but galactose is commonly present. They are strict aerobes; acids are formed slowly and weakly from some carbohydrates. They do not decompose cellulose. The DNA base ratios are in the range of about 66–71 mol% G+C. Gelatin is hydrolyzed slowly. Growth is best at about 25°C in air. Nutritional data are incomplete but probably all species require organic growth factors.

Morphology and staining reactions may be examined using the medium (EYGA) and methods described by Cure and Keddie (1973). Cell wall material prepared by alkali treatment of whole organisms may be used for cell wall analysis (see Keddie and Cure, 1977). Cellulolytic activity is tested for in 0.5% peptone water that contains partially immersed filter-paper strips (see Genus Cellulomonas).

The presence of ornithine in the cell wall peptidoglycan distinguishes Curtobacterium spp. from most other named taxa of coryneform bacteria (Tables 1–3). Thus they are most likely to be confused with other ornithine-containing coryneform taxa (Table 3), i.e., the genus Cellulomonas and Microbacterium liquefaciens, Arthrobacter terregens, and A. flavescens.

The criteria most useful for distinguishing between the genera Curtobacterium and Cellulomonas are described in the section on Cellulomonas and in Table 3. It is much more difficult, if not impossible, to distinguish M. liquefaciens, A. terregens, and A. flavescens from Curtobacterium; the available information, although incomplete, indicates that these three species have close affinities with the genus. A. terregens and M. liquefaciens have the peptidoglycan structure characteristic of the genus Curtobacterium, particularly Brevibacterium (Curtobacterium) saperdae (Schleifer and Kandler, 1972), while A. terregens and A. flavescens have DNA base ratios in the narrow range quoted for the genus (see Skyring and Quadling, 1970; Skyring, Quadling, and Rouatt, 1971). In addition, all three species are obligate aerobes: A. terregens and A. flavescens do not produce acid from sugars in peptone-based media (see Keddie, 1974a), but M. liquefaciens produces acid oxidatively from a few sugars (Robinson, 1966a).

The other features of these three species are described in the appropriate sections of this chapter and in Table 3.

Further identification of the plant-pathogenic curtobacteria is discussed in this Handbook, Chapter 143. It is not possible at present to attempt further identification of members of the genus which are not plant pathogens with the limited amount of information currently available.

The type species is Curtobacterium (Brevibacterium) citreum and the type strain is ATCC 15828 (IAM 1514), which is a suitable reference strain.

THE GENUS MICROBACTERIUM

The genus Microbacterium was established by Orla-Jensen (1919) for a diverse collection of Gram-positive, nonsporeforming rods that were isolated during studies on lactic acid–producing bacteria of importance in the dairy industry. The main distinguishing features in the original description of the genus were small size of rods, marked heat resistance, aerobic growth, and catalase production; these characters were primarily chosen to differentiate the microbacteria from the other rod-shaped, lactic acid–producing bacteria studied by Orla-Jensen and now classified in the genus Lactobacillus. Orla-Jensen (1919) recognized four species, Microbacterium lacticum, M. flavum, M. mesentericum, and M. liquefaciens; but he was aware of the unacceptable heterogeneity of the genus because, only 2 years later (Orla-Jensen, 1921), he referred to Microbacterium "as merely a provisional collective name". The taxonomic situation was confused further by the much later designation of the species, M. thermosphactum (McLean and Sulzbacher, 1953), and a glutamic acid–producing "patent" strain was named M. ammoniaphilum (see Abe, Takayama, and Kinoshita, 1967).

The validity of the genus Microbacterium was first questioned seriously by Wittern (1933), who suggested the transfer of M. mesentericum to the genus Mycobacterium. Ten years later this species was formally transferred to the genus Nocardia as N. mesenterica (Waksman and Henrici, 1943).

The three remaining original species all exhibited what is now termed a coryneform morphology (see Jensen, 1934; Orla-Jensen, 1919). Jensen (1934) suggested the reclassification of M. lacticum and M. liquefaciens in the genus Corynebacterium, but he placed M. flavum tentatively in the genus Mycobacterium, although he considered it to be on the borderline between Corynebacterium and Mycobacterium. Subsequent workers variously suggested a close relationship between the three species and the genera Corynebacterium, Propionibacterium, and Lactobacillus (see Abd-el-Malek and Gibson, 1952; Doetsch and Pelczar, 1948; Orla-Jensen, 1943; Speck, 1943). There was also dispute over the separate species status of M. lacticum and M. liquefaciens (Abd-el-Malek and Gibson, 1952; Doetsch and Rakosky, 1950). Nevertheless, the genus Microbacterium, containing the species M. lacticum and M. flavum (but not M. liquefaciens, which was reduced to synonymy with M. lacticum), was listed as a valid genus in the family Corynebacteriaceae in the seventh edition of Bergey's Manual (Breed, Murray, and Smith, 1957).

However, later studies using modern taxonomic techniques shed some light on the taxonomic relationships of the species M. lacticum, M. flavum, M.

liquefaciens, and *M. thermosphactum* and clearly established the marked heterogeneity of the genus *Microbacterium* Orla-Jensen (see Bousfield, 1972; Collins-Thompson et al., 1972; Davis and Newton, 1969; Davis et al., 1969; Jayne-Williams and Skerman, 1966; Keddie, Leask, and Grainger, 1966; Robinson, 1966a,b; Schleifer, 1970; Schleifer and Kandler, 1972; Shaw and Stead, 1970; Yamada and Komagata, 1970, 1972a,b). It is therefore not surprising that in the eighth edition of *Bergey's Manual,* the genus *Microbacterium* is listed as incertae sedis (Rogosa and Keddie, 1974b).

More recent numerical taxonomic (see Jones, 1975, 1978) and chemotaxonomic studies (see Keddie and Cure, 1977, 1978; Minnikin, Goodfellow, and Collins, 1978) have further clarified, but have not entirely resolved, the systematic relationships of the five so-called species of the genus *Microbacterium.*

It is now generally accepted that *M. thermosphactum* should be removed from the genus *Microbacterium.* Sneath and Jones (1976) have recommended the establishment of a new genus *Brochothrix* to contain the species *B. thermosphacta* (McLean and Sulzbacher) comb. nov. These bacteria are discussed in the following section.

There is also general agreement that *M. flavum* should be transferred to the genus *Corynebacterium* (Bousfield, 1972; Collins-Thompson et al., 1972; Jones, 1975; Keddie and Cure, 1977, 1978; Minnikin, Goodfellow, and Collins, 1978; Robinson, 1966b; Schleifer, 1970; Schleifer and Kandler, 1972; Yamada and Komagata, 1972b), and Barksdale et al. (1979) have now formally proposed its transfer to the genus as *C. flavescens* nom. nov. In addition, the "patent" strain, *M. ammoniaphilum,* has been shown to have the characteristics of strains named *C. glutamicum* (see Abe, Takayama, and Kinoshita, 1967; Minnikin, Goodfellow, and Collins, 1978). (See genus *Corynebacterium,* saprophytic species, for a discussion of *C. flavum* and *C. glutamicum.*)

We are thus left with the species *M. lacticum* and *M. liquefaciens* which, although similar in some respects, are now considered to be separate taxa (Robinson, 1966a,b; Schleifer, 1970; Schleifer and Kandler, 1972). *M. lacticum,* the type species of the genus *Microbacterium,* is a distinct and recognizable taxon. As suggested by Jones (1975) and Keddie and Cure (1978), this species could form the nucleus of a redefined genus *Microbacterium.* Whether or not *M. liquefaciens* should be retained in the genus has still to be resolved. On the basis of cell wall peptidoglycan structure, the latter species shows close affinities with the proposed genus *Curtobacterium,* particularly with *Curtobacterium (Brevibacterium) saperdae* (as well as with taxa such as *Arthrobacter terregens* and *Corynebacterium barkeri*) (Schleifer and Kandler, 1972). Unfortu-

nately, there are no published data presently available on the G+C content or on chemotaxonomic features, such as the menaquinone composition, of *M. liquefaciens.*

Because of this unresolved taxonomic situation, we will not discuss a genus *Microbacterium* as such. Instead, we will consider the species *M. lacticum* and *M. liquefaciens* separately.

Other species that have been suggested as possible candidates for the genus *Microbacterium* are those now named *Corynebacterium laevaniformans* (see Keddie and Cure, 1978) and *Brevibacterium imperiale* (see Jones, 1975; Keddie and Cure, 1978).

Habitats of *Microbacterium*

Dairy Sources

Microbacteria are usually encountered in milk and dairy products and on dairy equipment. They were first isolated from such sources by Orla-Jensen (1919): *M. lacticum* from laboratory-heated milk, cheese, and sour butter, and *M. liquefaciens* from milk and especially from cheese. Many subsequent workers have shown that microbacteria form a substantial part of the thermoduric bacterial count of raw and pasteurized milk, powdered milk, cheese, and dairy equipment, and probably account for the whole of the thermoduric, coryneform bacterial population of such sources (see Thomas et al., 1967 for review of early literature; Gillies, 1971; Jayne-Williams and Skerman, 1966).

There is no evidence that microbacteria have ever been isolated from milk samples drawn aseptically from the cow's udder (see Abd-el-Malek and Gibson, 1952; Jayne-Williams and Skerman, 1966; Thomas et al., 1967). Consequently, their presence in raw milk is considered to result from contamination during production, and there is much evidence to show that improperly cleansed dairy equipment is the major source of such contamination (see Thomas et al., 1967). A high count of thermoduric bacteria in pasteurized milk is therefore considered to be reliable evidence of improperly cleansed dairy equipment. A survey of pasteurized milk by Thomas et al. in 1967 showed that the thermoduric, bacterial colony count was as valuable as an index of unhygienic methods of milk production then as it was in a similar survey carried out some 20 years earlier (see Thomas et al., 1967)—despite the marked changes that had taken place during that time both in the methods used for milk pasteurization and for sterilization of equipment. Although there is no evidence that microbacteria have an adverse effect on the keeping quality of pasteurized milk (see Thomas et al., 1967), high thermoduric bacterial counts are of obvious economic importance, both to milk producers and to milk processors, in countries where statutory colony-count standards are applied to this product (see Gillies, 1971; Griffiths, 1977; Thomas et al., 1967).

The bacterial count of spray-dried milk powder has been shown to result almost entirely from the presence of thermoduric, coryneform bacteria, mainly *M. lacticum,* derived from the original milk (see Thomas et al., 1967). This bacterial contamination is of particular importance to producers and processors in countries where statutory colony-count standards are applied to spray-dried milk powder (e.g., see Griffiths, 1977).

It is also known that microbacteria form a substantial part of the thermoduric count of certain cheeses (see Gillies, 1971). However, the counts of microbacteria in Australian Cheddar cheese showed little variation during the 6-month ripening period (Gillies, 1971). The same author showed that the microbacteria had no inhibitory effect on starter cultures nor any significant effect on the flavor of the ripened cheese.

Other Habitats

Topping (1937) reported the isolation of *M. liquefaciens* from soil but, as pointed out by Abd-el-Malek and Gibson (1952), her *M. liquefaciens* group possessed extremely variable characters. From the description, it is possible that some of the bacteria isolated by Splittstoesser et al. (1967) from frozen vegetables were *M. liquefaciens* and *M. lacticum.* The source of these organisms was considered to be airborne contamination since isolates did not survive the blanch treatment. Demain and Hendlin (1959) isolated a laboratory contaminant which they identified as *M. lacticum.*

C. laevaniformans, which shows some relationship with *M. lacticum* (see Keddie and Cure, 1978; Minnikin, Goodfellow, and Collins, 1978), occurs in large numbers in activated sludge (Dias, 1963; Dias and Bhat, 1962). There have also been several reports of the isolation of organisms that resemble microbacteria from such sources as fresh beef, poultry giblets, and raw and pasteurized egg fluid; but it is not possible from the descriptions to be certain of their identity (see Kraft et al., 1966; Splittstoesser et al., 1967).

Isolation of *Microbacterium lacticum* and *Microbacterium liquefaciens*

The usual methods described for the isolation of microbacteria are those employed in the examination of milk, dairy products, and dairy equipment. Because microbacteria were formerly defined as being

thermoduric (see Breed, Murray, and Smith, 1957), the most common isolation procedure (other than from materials that already had been heat-treated) was to plate out laboratory-pasteurized samples on a suitable, nonselective medium. This partially selective method, therefore, may be used for the isolation of thermoduric microbacteria, especially from dairy sources. However, if the aim is to avoid selecting for thermoduric strains, pasteurization should be omitted. Apart from laboratory pasteurization, we know of no selective or enrichment procedures for the isolation of microbacteria. Some procedures for the isolation of microbacteria from raw milk and dairy equipment are given by Thomas et al. (1967).

Selection of Thermoduric Microbacteria by Laboratory Pasteurization

Several different time/temperature combinations have been used for laboratory pasteurization: A common procedure is to heat 5-ml amounts of milk, rinses from dairy equipment, etc. in test tubes (125 × 20 mm) at 63°C for 30 min in an accurately controlled water bath. For rinse solutions, it is recommended that sterile skim milk be added to the samples before pasteurization (Thomas et al., 1967). For cheese and butter samples and similar materials, the primary dilution is prepared by homogenizing the material in 2% trisodium citrate solution in a laboratory blender (Gillies, 1971). Plates are incubated at 28–30°C for up to 7 days.

Both *M. lacticum* and *M. liquefaciens* grow on media that contain yeast extract, peptone, and milk or glucose (see Abd-el-Malek and Gibson, 1952; Jayne-Williams and Skerman, 1966; Robinson, 1966a; Thomas et al., 1967). A suitable medium is described below.

Yeast Extract Milk Agar (Harrigan and McCance, 1976)

Medium has the following composition (g/liter): yeast extract, 3.0; peptone, 5.0; agar, 15.0; fresh, whole, or skim milk, 10 ml. The yeast extract and peptone are dissolved in distilled water by steaming, and the pH of the cooled solution is adjusted to 7.4. The agar and milk are then added and the whole is autoclaved at 121°C for 20 min. While still hot, the medium is filtered through paper pulp and the pH is adjusted to 7.0 at 50°C. The medium is then distributed in the required amounts and autoclaved at 121°C for 15 min. The final pH should be 7.2 at room temperature.

Identification of *Microbacterium lacticum* and *M. liquefaciens*

Identification of *Microbacterium lacticum*

In young cultures, small, slender (ca. 0.5 μm in diameter) coryneform bacteria are observed showing typical V-formations; primary branching is uncommon, cells are nonmotile, and endospores are not formed. In older cultures, the rods are shorter but a marked rod-coccus growth cycle does not occur. The rods are Gram positive and not acid-fast. Colonies are 1–1.5 mm in diameter, circular, opaque, and glistening and vary in color from gray-white to pale greenish yellow; catalase positive.

The relationship to oxygen of *M. lacticum* is controversial: There is general agreement that aerobic growth is best, but considerable disagreement about the ability to grow anaerobically. Some workers have reported that growth does not occur in strictly anaerobic conditions (see Abd-el-Malek and Gibson, 1952; Orla-Jensen, 1919), others that weak anaerobic growth occurs (see Jones, 1975; Keddie and Cure, 1977; Robinson, 1966a), while Jayne-Williams and Skerman (1966) reported that the anaerobic growth of strains they designated *M. lacticum* was equivocal. Similarly, when glucose fermentation has been tested by a method similar to that described by Hugh and Leifsen (1953), results have been described as equivocal (Jayne-Williams and Skerman, 1966) or generally positive (Jones, 1975; Robinson, 1966a). However, all strains appear to produce acid weakly from glucose and some other sugars in peptone media: L-(+)-Lactic acid is produced from glucose. Enzymes of the Embden-Meyerhof and hexosemonophosphate pathways and of the tricarboxylic acid cycle are present (Collins-Thompson et al., 1972).

A large majority of the strains that have been described are markedly heat resistant, but most of these were isolated either from laboratory-pasteurized materials or from sources in which some form of heat treatment was used. However, nonthermoduric strains, indistinguishable in other respects from *M. lacticum,* were isolated by Jayne-Williams and Skerman (1966) from sources that had not been subjected to heat treatment. A variety of heat treatments have been used, but thermoduric strains are commonly stated to survive 63°C for 30 min in skim milk (see Abd-el-Malek and Gibson, 1952; Jayne-Williams and Skerman, 1966) or 72°C for 15 min (see Rogosa and Keddie, 1974b).

A large majority of *M. lacticum* strains show diastatic activity on starch agar; hydrolysis of gelatin and casein is either negative, or weak and slow (Abd-el-Malek and Gibson, 1952; Jayne-Williams and Skerman, 1966; Robinson, 1966a). They are nutritionally exacting: all require B vitamins and many also require amino acids (see Skerman and Jayne-Williams, 1966, for details). Most strains have a growth temperature range of about 15–35°C; growth is best at about 30°C.

The cell wall peptidoglycan contains lysine as diamino acid (Keddie and Cure, 1977; Keddie, Leask, and Grainger, 1966; Robinson, 1966b; Yamada and Komagata, 1972a), and the wall sugars are galactose, rhamnose, and, occasionally, mannose (Keddie and Cure, 1977; Keddie, Leask, and Grainger, 1966; Robinson, 1966b). The cell wall peptidoglycan is of the less common, group B type, i.e., linked between positions 2 and 4 of two peptide subunits, and the interpeptide bridge contains one glycine and one L-lysine residue (Schleifer and Kandler, 1972). The predominant isoprenoid quinones are menaquinones that contain 10 and 11 isoprene units (MK-*10* and MK-*11*) (Collins, 1978). DNA base ratios in the range of 69–70 mol% G+C were reported by Yamada and Komagata (1970) and Bousfield (1972); but the values reported by Collins-Thompson et al. (1972) were much lower (63–64 mol% G+C).

Thus, present evidence indicates that isolates that have the morphological features described, contain lysine in the cell wall, survive heating at 63°C for 30 min, produce acid from glucose in peptone media (and may be facultatively anaerobic) may be regarded as *M. lacticum*. Further identification, and recognition of nonthermoduric strains, requires examination for the other chemotaxonomic features described, all of which require the use of specialized techniques.

Morphology may be examined using the methods described by Cure and Keddie (1973), using either the isolation media described or medium EYGA. The cell wall composition may be determined by chromatography of acid hydrolysates of cell wall material obtained by alkali treatment of whole cells (see Introduction to this chapter for references to suitable methods).

The type strain of *Microbacterium lacticum* is ATCC 8180 (NCIB 8540, NCDO 747), which is a suitable reference strain.

Identification of *Microbacterium liquefaciens*

Much less information is available on *Microbacterium liquefaciens,* largely because of the vagueness of the original description (Orla-Jensen, 1919), the infrequency with which it has been isolated (or identified) subsequently, and the belief of some earlier investigators that it was merely a biovar of *M. lacticum* (see Breed, Murray, and Smith, 1957).

From the available information, *M. liquefaciens* differs from *M. lacticum* in the following respects (see Robinson, 1966a,b): The rods are short, almost coccoid, and V-formations are uncommon. Colonies are 1 mm in diameter, bright yellow in color, and convex, circular, moist, and glistening. Obligately aerobic: Acid is formed oxidatively from glucose and a few other sugars. Gelatin is hydrolyzed rapidly and extensively; casein is also hydrolyzed. Heat resistance is less marked than for *M. lacticum* (thermoduric strains): *M. liquefaciens* is stated to survive 65°C but not 75°C for 10 min. The vitamin requirements of *M. liquefaciens* are probably different from those of *M. lacticum* (see Skerman and Jayne-Williams, 1966). The cell wall peptidoglycan contains ornithine as diamino acid, and the only wall sugar demonstrated is rhamnose.

The peptidoglycan structure is of the group B type similar to that in *M. lacticum,* but differs in that D-ornithine replaces L-lysine in the interpeptide bridge, and L-homoserine replaces L-lysine in position 3 of the peptide subunits (Schleifer and Kandler, 1972). No information is at present available on such features as the DNA base composition or the menaquinone content of *M. liquefaciens.*

Similar methods may be used for the examination of *M. liquefaciens* as those described for *M. lacticum.*

M. liquefaciens is most likely to be confused with the coryneform taxa that contain ornithine in the cell wall peptidoglycan, viz., the taxa *Cellulomonas, Curtobacterium,* and *Arthrobacter terregens* and *A. flavescens. M. liquefaciens* is most readily distinguished from bacteria of the genus *Cellulomonas* by the features listed in Table 3 (but see Genus *Cellulomonas* also). Differentiation of *M. liquefaciens* from members of the genus *Curtobacterium* is much more difficult, if not impossible, at the present time. The available information, although incomplete, indicates that *M. liquefaciens, A. terregens,* and *A. flavescens* have close affinities with the proposed genus *Curtobacterium* (see Genus *Curtobacterium* and Genus *Arthrobacter* [*sensu lato*], "terregens/flavescens" group for further discussion, and also Collins, Goodfellow, and Minnikin [1980]).

A suitable reference strain is NCIB 11509 (= strain 15, Robinson, 1966a,b).

The Genus *Brochothrix* (formerly *Microbacterium Thermosphactum,* McLean and Sulzbacher)

The genus *Brochothrix* was proposed by Sneath and Jones (1976) for a homogeneous group of Gram-positive, nonsporeforming, catalase-positive, facultatively anaerobic, regular, rod-shaped bacteria which show characteristic changes in cellular morphology during growth. The organisms were first isolated and described by Sulzbacher and McLean (1951) and later allocated to the genus *Microbacterium* as a new species, *Microbacterium thermo-*

sphactum, by McLean and Sulzbacher (1953). They are listed as such in the eighth edition of *Bergey's Manual* (Buchanan and Gibbons, 1974), but the species is designated incertae sedis because the genus *Microbacterium* is so designated (see Genus *Microbacterium*). Although bacteria of the genus *Brochothrix* do not have a coryneform morphology, they are generally referred to in the literature as *M. thermosphactum* and are therefore most conveniently dealt with in this chapter.

The unfortunate classification of these bacteria in the genus *Microbacterium* was largely a result of the taxonomic confusion which has surrounded this genus since it was tentatively proposed by Orla-Jensen (1919). McLean and Sulzbacher (1953) noted the marked difference in cell morphology between *M. thermosphactum* and *M. lacticum,* the type species of the genus, and also commented on the close physiological resemblance between *M. thermosphactum* and the lactobacilli. However, at that time the genus *Microbacterium* was classified in the family Lactobacteriaceae and was distinguished from the genus *Lactobacillus* mainly on the basis of catalase production (Breed, Murray, and Hitchens, 1948).

With the development of new taxonomic techniques, subsequent workers confirmed the marked differences between *M. thermosphactum* and *M. lacticum,* not only in morphology (Davidson, Mobbs, and Stubbs, 1968; Jones, 1975) but also in enzymology and protein profiles (Collins-Thompson et al., 1972; Robinson, 1966b), in peptidoglycan structure (Schleifer, 1970; Schleifer and Kandler, 1972), and in DNA base composition (Collins-Thompson et al., 1972). In addition, numerical taxonomic studies showed that *M. thermosphactum* strains formed a homogeneous taxon (with an intra-group similarity of greater than 85%), quite distinct from *M. lacticum* (Davis and Newton, 1969; Davis et al., 1969; Jones, 1975; Wilkinson and Jones, 1977). The same studies indicated that the closest associates of *M. thermosphactum* were the genera *Listeria* and *Lactobacillus,* but in none of the studies was the relationship close enough to indicate that the *M. thermosphactum* group could be accommodated as a species in either of these genera. Consideration of data of this kind led Sneath and Jones (1976) to propose a new genus *Brochothrix* for bacteria designated *M. thermosphactum* with *B. thermosphacta* (McLean and Sulzbacher) comb. nov. as the only species. Bacteria of the species *B. thermosphacta* have received a great deal of attention from food bacteriologists because of their association with off-flavor development in meats (including poultry meat) and processed meats, especially in prepacked products held at refrigerator temperatures (see Ingram and Dainty, 1971). They have, however, never been shown to be pathogenic for man or animals. The species has also received much attention from taxonomists because of its former anomalous position in the genus *Microbacterium* (see Sneath and Jones, 1976, for review of literature).

Habitats of
Brochothrix thermosphacta

Brochothrix thermosphacta was first isolated from pork trimmings and finished sausage (Sulzbacher and McLean, 1951). Since that time, it has been isolated frequently from the same sources and from a variety of meats (including poultry) and meat products (see Ayres, 1960; Barlow and Kitchell, 1966; Gardner, 1966; Gardner, Carson, and Patton, 1967; Ingram and Dainty, 1971; McLean and Sulzbacher, 1953; McMeekin, 1975; Pierson, Collins-Thompson, and Ordal, 1970; Thornley, 1957; Weidemann, 1965; Wolin, Evans, and Niven, 1957).

The species is regarded as an important meat-spoilage organism, especially in conditions of low oxygen tension and low temperature, both of which are achieved in prepacked refrigerated products (Gardner, Carson, and Patton, 1967). *B. thermosphacta* can grow at 2–4°C; lactic acid is produced fermentatively from glucose, and the bacteria also possess significant lipase activity (Collins-Thompson et al., 1971).

There are very few reports of the isolation of *B. thermosphacta* from sources other than animal or poultry meats or products based on them. It was isolated occasionally by McLean and Sulzbacher (1953) from equipment and tables used to prepare sausages, but these authors assumed that its occurrence there arose by contamination from pork trimmings since it was repeatedly isolated from unopened barrels of such trimmings. Collins-Thompson et al. (1971) suggested that some of the lipolytic coryneform bacteria isolated from dairy sources by Jayne-Williams and Skerman (1966) were apparently closely related to *B. thermosphacta,* but the report of these authors contains no such suggestion. Gardner (1966), using a selective medium, isolated bacteria that resemble *B. thermosphacta* from soil and feces but not in high numbers.

Isolation of
Brochothrix thermosphacta

The methods described for the isolation of *Brochothrix thermosphacta* are almost all concerned with their recovery from meats and meat products. *B. thermosphacta* does not grow on the acetate medium devised by Rogosa, Mitchell, and Wiseman (1951) for the isolation of lactobacilli (see Barlow and Kitchell, 1966; Gardner, Carson, and Patton, 1967) and grows only poorly in the MRS medium of De

Man, Rogosa, and Sharpe (1960). Several different media based on peptone, yeast extract, and glucose or glycerol have been used for the isolation of *B. thermosphacta* (Barlow and Kitchell, 1966; Gardner, 1966; Gardner, Carson, and Patton, 1967; Sulzbacher and McLean, 1951; Wolin, Evans, and Niven, 1957).

Isolation of *Brochothrix thermosphacta* from Meats and Meat Products

Swabs of various meat surfaces or samples of macerated meat or other materials are usually suspended in 0.1% (wt/vol) peptone water and shaken vigorously before plating on suitable media (see Gardner, Carson, and Patton, 1967). Plates should be incubated at 22°C for up to 5 days, but colonies of *B. thermosphacta* are usually visible within 48 h. The following media are suitable for the growth of *B. thermosphacta*. Glycerol Nutrient Agar (see Gardner, 1966; Gardner, Carson, and Patton, 1967), g/liter:

Peptone (Oxoid)	20
Yeast extract (Oxoid)	2
Glycerol	15
K_2HPO_4	1
$MgSO_4 \cdot 7H_2O$	1
Agar (Oxoid no. 3)	13
pH 7.0	

Autoclave at 121°C for 15 min.

Glucose Nutrient Agar (see Sulzbacher and McLean, 1951; Wolin, Evans, and Niven, 1957), g/liter:

Tryptone (Difco)	10
Yeast extract (Difco)	5
K_2HPO_4	5
NaCl	5
Glucose	5
Agar	15
pH 7.0	

Autoclave at 121°C for 15 min.

Enrichment of *Brochothrix*

Enrichment is not usually performed. Some workers have inoculated sliced beef in sterile Petri dishes, containing 3–5 ml of sterile distilled water, and incubated them at 2°C in a slanting position until spoilage becomes evident (Wolin, Evans, and Niven, 1957). The samples are then plated on media similar to those described. Enrichment of other materials does not seem to have been attempted.

Selective isolation of *B. thermosphacta* may be achieved from various materials by the use of the selective medium (STTA) of Gardner (1966).

Medium STTA for Selective Isolation of
Brochothrix thermosphacta (Gardner, 1966)

Medium STTA contains (g/liter):

Peptone (Oxoid)   20
Yeast extract (Oxoid)   2
Glycerol   15
K$_2$HPO$_4$ 1
MgSO$_4$ · 4H$_2$O   1
Agar (Oxoid No. 3)   13

The pH is adjusted to 7.0; autoclaved at 121°C
for 15 min. To this sterile, molten medium, the
following solutions (prepared in sterile, distilled
water) are added to give the final concentrations
indicated: streptomycin sulfate (Glaxo),
500 μg/ml; cycloheximide (Upjohn), 50 μg/ml;
thallous acetate, 50 μg/ml.

After incubation of appropriate samples on this
medium at 20–22°C for 48 h, a large majority of
colonies are those of *B. thermosphacta;* the exceptions are a few pseudomonads.

Identification of
Brochothrix thermosphacta

Identification of new isolates as *Brochothrix
thermosphacta* requires examination of cellular
morphology and staining reactions, the maximum
growth temperature, relationship to oxygen, and
catalase production, together with a number of other
conventional taxonomic tests such as ability to produce acid from certain sugars.

In contrast with the coryneform taxa described in
this chapter, there are as yet no readily applied
chemotaxonomic tests that differentiate *B. thermosphacta* from members of the two bacterial taxa with
which it is most likely to be confused, the genera
Listeria and *Lactobacillus.* The colonial morphology of *B. thermosphacta* is not particularly diagnostic. After 24–48 h the colonies are circular,
0.75–1 mm in diameter, convex with entire margins, and not pigmented. In older cultures the edge
breaks up and the center of the colony becomes
raised to give a "fried egg" appearance. However,
two types of colonies varying in size and density
may be present in young cultures, and they appear
so distinct that the culture may appear to be contaminated (see Barlow and Kitchell, 1966). *B. thermosphacta* is nonhemolytic.

Gram stains should be performed by the method
of Cure and Keddie (1973) on 18- to 24-h and 2-day
cultures grown on nutrient agar such as blood agar
base (BAB No. 2, Oxoid) or APT medium (Difco),
incubated at 20–25°C. Exponential-phase cultures
of *B. thermosphacta* show regular, unbranched rods
that occur singly in short chains and in long, kinked,
filamentous-like chains (Fig. 3a) which bend and
loop to give characteristic knotted masses. In older
cultures, the rods give rise to coccoid forms (Fig.
3b) that, when subcultured onto a suitable medium,
develop into rod forms. Both rod and coccal forms
are Gram positive, but a proportion may appear
Gram negative. They are nonmotile and do not form
endospores.

B. thermosphacta bears a close morphological
resemblance to members of the genus *Kurthia* (see
this Handbook, Chapter 144), but the two taxa are
readily distinguished by examining the oxygen requirements. *B. thermosphacta* is facultatively anaerobic, whereas members of the genus *Kurthia* are
obligately aerobic (Keddie and Rogosa, 1974). In
addition, most strains of *Kurthia* are highly motile.

B. thermosphacta grows best at about 20–22°C.
Growth occurs at 1°C and at 30°C but not at 35°C.
These organisms do not survive heating at 63°C for
5 min. The inability of *B. thermosphacta* to grow at
35°C distinguishes it from members of the genus
Listeria. B. thermosphacta is facultatively anaerobic, but grows best in air. This feature, together with
its ability to grow on unsupplemented nutrient agar,
serves to distinguish it from the genera *Lactobacillus*
and *Erysipelothrix* but not from *Listeria.*

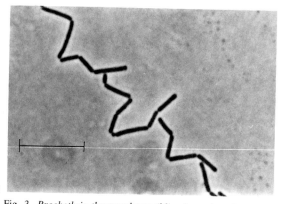

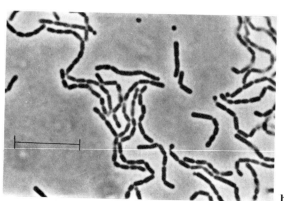

a

b

Fig. 3. *Brochothrix thermosphacta* (*Microbacterium thermosphactum*) (isolate) when grown on blood agar base No. 2 (Oxoid) at
25°C. (a) After 24 h, showing regular rods in chains. (b) After 48 h, showing development of coccoid forms. Bar = 10 μm.

Glucose metabolism is fermentative: Lactic acid, mainly L(+), is produced with only small amounts of other products. Acid but no gas is produced from a number of carbohydrates. H₂S and indole are not produced, sodium hippurate is not hydrolyzed, nitrate is not reduced, gelatin is not liquefied, casein is not digested, and deoxyribonuclease is not produced (see Sneath and Jones, 1976 for other characteristics).

B. thermosphacta produces catalase and cytochromes (Davidson and Hartree, 1968; Davidson, Mobbs, and Stubbs, 1968), but care must be taken in examining cultures for the presence of catalase because its production depends on both the growth medium and the temperature of incubation. Davidson, Mobbs, and Stubbs (1968) noted that *B. thermosphacta* strains grown on APT medium (Difco) incubated at 20°C were always catalase positive. However, the same authors noted that weaker or negative reactions were obtained at the same temperature on HIA medium (Difco), and negative results were frequently obtained if the bacteria were grown on either medium incubated at 30°C, which has also been our experience (Jones and Wilkinson, unpublished), but we find that BAB No. 2 (Oxoid) is a suitable alternative medium for APT (Difco). Catalase production is an important feature in differentiating *B. thermosphacta* from *Erysipelothrix* and most lactobacilli, but does not distinguish the species from the members of the genus *Listeria*, all of which produce catalase. Although catalase production by *Listeria* spp. is also dependent on the composition of the growth medium (Jones, 1975), all strains of *Listeria* are catalase positive at 37°C if the growth medium is suitable, whereas *B. thermosphacta* does not grow at this temperature.

The cell wall peptidoglycan of *B. thermosphacta* contains *meso*-DAP as diamino acid (Schleifer, 1970; Schleifer and Kandler, 1972). However, the cell walls contain neither arabinose nor galactose (see Schleifer and Kandler, 1972) which, together with the absence of mycolic acids (see Minnikin, Goodfellow, and Collins, 1978) and its distinctive morphology, serves to distinguish it from the genus *Corynebacterium* (see Table 1). The presence of *meso*-DAP as cell wall diamino acid also distinguishes *B. thermosphacta* from *Kurthia* and from *Erysipelothrix*, both of which contain lysine, but not from *Listeria* or from certain *Lactobacillus* spp. (see Schleifer and Kandler, 1972).

Recent work (Collins, 1978) has shown that the predominant menaquinone of *B. thermosphacta* contains seven isoprene units (MK-7). This menaquinone is also the predominant one in the genus *Listeria* (Collins, 1978). The G+C content of *B. thermosphacta* (36 mol%, T_m Collins-Thompson et al., 1972) is also rather similar to that in the genus *Listeria* (38 mol%, T_m Stuart and Welshimer, 1973).

Thus, unless particular care is taken in the morphological examination of cultures and their inability to grow at 35°C is checked, it may not be easy to differentiate between *B. thermosphacta* and the species *Listeria monocytogenes, L. grayi*, and *L. murrayi* without recourse to a number of conventional tests. *B. thermosphacta* and *L. monocytogenes, L. grayi*, and *L. murrayi* produce acid from a number of sugars by fermentation. The production of acid from arabinose, xylose, melibiose, and adonitol by *B. thermosphacta* (see Sneath and Jones, 1976) distinguishes this species from the three *Listeria* species (see Seeliger and Welshimer, 1974). In addition, strains of *B. thermosphacta* have never been shown to be motile. No serological relationships have been demonstrated between *B. thermosphacta* and species of the genus *Listeria* (Wilkinson and Jones, 1975).

The type strain of *Brochothrix thermosphacta* is ATCC 11509 (NCIB 10018), which is a suitable reference strain.

THE GENUS *CORYNEBACTERIUM*: SAPROPHYTIC SPECIES

The genus *Corynebacterium* was established by Lehmann and Neumann (1896) to accommodate the diphtheria bacillus and a few closely related, animal-parasitic species. The genus was defined mainly on the basis of certain morphological characters and staining reactions, features then considered very characteristic of *C. diphtheriae* and closely related organisms. However, in the ensuing years it was realized that morphologically similar organisms occurred in a wide range of habitats other than the animal body; and because morphological similarity was then widely believed to indicate relatedness, they too were placed in the genus *Corynebacterium* by various investigators. Thus the name *Corynebacterium* was applied not only to *C. diphtheriae* and closely related animal-parasitic species, but also to a heterogeneous collection of morphologically similar, plant-pathogenic species and saprophytic species from a wide range of habitats. This practice has continued to some extent almost to the present day and, thus, a large number of species of widely different characteristics bear the name *Corynebacterium*.

Although earlier investigators, such as Conn (1947), Jensen (1952), and Clarke (1952), believed that the genus should once more be restricted to *C. diphtheriae* and very similar animal-parasitic species, it was not until the chemical composition of the cell wall was introduced as a taxonomic criterion (Cummins, 1962; Cummins and Harris, 1956, 1958, 1959) that the means was provided whereby this restriction might be achieved. The widespread application of chemotaxonomic methods has provided a way of defining the genus *Corynebacterium*, largely in chemical terms. Thus it is now generally, if not universally, accepted that the genus *Corynebacterium* should contain only those coryneform bacteria which contain *meso*-DAP, arabinose, and galactose in the cell wall and which contain relatively short-chain mycolic acids with about 22–38 carbon atoms, referred to as corynomycolic acids (see Barksdale, 1970; Keddie and Cure, 1977, 1978; Minnikin, Goodfellow, and Collins, 1978; Schleifer and Kandler, 1972). In addition, a majority of species with these chemical attributes are facultatively anaerobic (Cummins, Lelliott, and Rogosa, 1974; Keddie and Cure, 1977) and have DNA base ratios in the approximate range of 51–59 mol% G+C (see references in Table 1). The major isoprenoid quinones of those representatives of the genus studied so far are dihydrogenated menaquinones with eight (Mk-8[H$_2$]) and/or nine (Mk-9[H$_2$]) isoprene units (see Minnikin, Goodfellow, and Collins, 1978). This restricted concept of the genus, which we will refer to as *Corynebacterium sensu stricto*, is in general supported by numerical taxonomic studies (see Bousfield, 1972; Jones, 1975). Chapters 140 and 141, this Handbook, present additional information on this theme.

When defined in this way, the genus *Corynebacterium* includes *C. diphtheriae* and most, but not all, animal-pathogenic and -parasitic species. However, it is now clear that it also contains a number of saprophytic species, some of which are already named *Corynebacterium;* but others currently bear such names as *Arthrobacter, Brevibacterium,* and *Microbacterium* and should be transferred to the genus. Prominent among the saprophytic members of *Corynebacterium* are a number of glutamic acid–producing coryneform species, all of them "patent" strains, typified by the species *Corynebacterium glutamicum* (synonym *Micrococcus glutamicus*). Abe, Takayama, and Kinoshita (1967) examined a large number of variously named strains of glutamic acid–producing coryneform bacteria (most of them named *Micrococcus glutamicus*) and concluded that they should be considered members of the genus *Corynebacterium,* mainly because of their morphology and because they contained *meso*-DAP in the cell wall. The DNA base ratios these authors recorded for representative strains supported this conclusion. They also considered that, with one or two possible exceptions, the variously named nomenclatural species should be reduced to synonymy with *C. glutamicum*. Subsequent chemotaxonomic studies (see Keddie and Cure, 1977, 1978; Minnikin, Goodfellow, and Collins, 1978) have confirmed that *C. glutamicum* and its numerous synonyms, such as *C. callunae, C. herculis, C. lilium, Brevibacterium divaricatum, B. flavum, B. roseum,* etc., have the characteristics of *Corynebacterium sensu stricto.*

In addition to those mentioned, a number of saprophytic coryneform species, such as *Arthrobacter albidus* and *Brevibacterium ammoniagenes* have been shown to be *Corynebacterium* spp. (see Keddie and Cure, 1977, 1978; Minnikin, Goodfellow, and Collins, 1978). Also, much has been written about the taxonomic position of *Microbacterium flavum,* an organism originally described by Orla-Jensen in 1919 but rarely isolated since then. Most studies, therefore, have been made on a single, authentic strain. Since this species was first shown to contain *meso*-DAP, arabinose, and galactose in the cell wall (Keddie, Leask, and Grainger, 1966; Robinson, 1966b), a great deal of different kinds of evidence has been accumulated to indicate that it should be transferred to the genus *Corynebacterium* (Bousfield, 1972; Collins-Thompson et al., 1972; Goodfellow, Collins, and Minnikin, 1976; Jones, 1975; Keddie and Cure, 1977; Minnikin, Goodfellow, and Collins, 1978; Robinson, 1966b; Schleifer, 1970). The transfer of *M. flavum* to the genus *Corynebacterium* as *C. flavescens* has been formally proposed by Barksdale et al. (1979).

A large number of saprophytic species currently named *Corynebacterium* do not have the characteristics of *Corynebacterium sensu stricto* described above and should be removed from the genus (see Keddie and Cure, 1978; Minnikin, Goodfellow, and Collins, 1978 for details). These species include: *C. autotrophicum* and *C. nephridii,* which have the characteristics of Gram-negative organisms (see Minnikin, Goodfellow, and Collins, 1978; Weitzman and Jones, 1975); *C. alkanum, C. laevaniformans,* and *C. manihot,* which contain lysine as diamino acid in the cell wall peptidoglycan; *C. barkeri,* which contains ornithine as cell wall diamino acid; *C. aquaticum, C. mediolanum,* and *C. okanaganae* (an insect pathogen, Luthy, 1974), which contain 2,4-diaminobutyric acid as cell wall diamino acid; and *C. hydrocarboclastus* and *C. rubrum,* which have the characteristics of the genus *Rhodococcus*. However, there are also a few species whose characteristics overlap those of *Corynebacterium sensu stricto* and the genus *Rhodococcus* (see Introduction to this chapter) and cannot be assigned to one genus or the other. For further details of the various species mentioned, and others, see Schleifer and Kandler (1972), Keddie and Cure (1977, 1978), Minnikin, Goodfellow, and Collins (1978), and Chapters 140, 141, and 155, this Handbook.

Habitats of Saprophytic *Corynebacterium* spp.

The organism presently named *Microbacterium flavum* was originally isolated from butter and cheese (Orla-Jensen, 1919), but has rarely been isolated since then.

Corynebacterium glutamicum and similar nomenclatural species were isolated from soil, animal feces, vegetables, and fruits (Abe, Takayama, and Kinoshita, 1967), but no information is available about their numbers or distribution in such materials. All are "patent" strains used for the commercial production of L-glutamic acid and other amino acids, and of biochemical compounds.

In a numerical taxonomic study of 150 strains of coryneform bacteria isolated from marine fish and sea water (Bousfield, 1978), 55 strains (mainly from sea water) recovered in one phenon showed close affinities with *Corynebacterium sensu stricto*. Although all were strict aerobes, representative strains contained *meso*-DAP and arabinose in the cell wall, had mycolic acids with 32–36 carbons, and their DNA base ratios were in the range of 55–58 mol% G+C (Bousfield, 1978).

Isolation of Saprophytic *Corynebacterium* spp.

We are not aware of any methods designed specifically for the isolation of saprophytic corynebacteria. *Corynebacterium glutamicum* and similar species and *Microbacterium flavum* (*Corynebacterium flavescens*) give good growth at 25–30°C in media based on yeast extract and peptone.

Identification of *Corynebacterium sensu stricto:* Saprophytic Strains

Facultatively anaerobic coryneform bacteria, which contain *meso*-DAP and arabinose in the cell wall, which contain corynomycolic acids, and which have DNA base ratios in the range of about 51–59 mol% G+C may be regarded as members of *Corynebacterium sensu stricto*. Some strains have these general attributes but are strictly aerobic. Other features reported to be common to *Microbacterium flavum* and *Corynebacterium glutamicum* and similar species are as follows: gelatin and starch are not hydrolyzed (Abe, Takayama, and Kinoshita, 1967; Yamada and Komagata, 1972b); extracellular DNase is not produced (Yamada and Komagata, 1972b). All saprophytic *Corynebacterium* spp. for which data are available are nutritionally exacting: *C. glutamicum* and similar species require biotin and sometimes additional B vitamins (Abe, Takayama, and Kinoshita, 1967); the essential requirements of *M. flavum* (ATCC 10340) were reported to be amino nitrogen, biotin, and pantothenic acid by Skerman and Jayne-Williams (1966). However, Barksdale et al. (1979) reported that "luxuriant" growth of the same strain of *M. flavum* (*C. flavescens*) was produced in a basal medium plus glutamic acid, isoleucine, methionine, proline, *p*-aminobenzoic acid, biotin, nicotinic acid, pantothenic acid, and thiamine.

Medium EYGA (Cure and Keddie, 1973) is suitable for morphological examinations of saprophytic (but not of most animal-parasitic) corynebacteria. In our experience, *M. flavum* and *C. glutamicum* and similar species occur as rods throughout the growth cycle when grown on EYGA at 25°C, and V-formations are prominent (Cure and Keddie, unpublished). The rods may be considerably shorter in older cultures (ca. 7 days) but a marked rod-coccus cycle does not occur. However, Abe, Takayama, and Kinoshita (1967), using different media, noted that ellipsoidal cells occurred at certain stages of the growth cycle in *C. glutamicum*, etc. All are nonmotile.

Chromatographic analysis of acid hydrolysates of whole organisms may be used to detect *meso*-DAP and arabinose. References to suitable methods for cell wall and mycolic acid analyses are given in the Introduction to this chapter.

Using the characters described, *Corynebacterium* spp. can be distinguished clearly from all of the other coryneform genera considered in this chapter. However, there is some overlap between the genus *Corynebacterium* as defined here and the genus *Rhodococcus* as defined by Goodfellow and Alderson (1977). Thus, some obligately aerobic strains cannot be assigned with confidence to one genus or the other (see Keddie and Cure, 1978; Minnikin, Goodfellow, and Collins, 1978). This difficulty has been discussed more fully in the Introduction to this chapter (also, see Addendum).

Further identification of most saprophytic corynebacteria to species level is not possible with the information currently available; see Barksdale et al. (1979) for a description of the single available strain of *M. flavum* (*C. flavescens*).

The type strains which are suitable reference strains are: *Microbacterium flavum* (*Corynebacterium flavescens*), ATCC 10340 (NCIB 8707); *Corynebacterium glutamicum*, ATCC 13032 (NCIB 10025).

Addendum

Since this chapter was completed, a further genus of coryneform bacteria, *Caseobacter,* has been proposed (Crombach, 1978) to accommodate certain of the so-called gray-white cheese coryneform bacteria originally described by Mulder and Antheunisse (1963). The genus *Caseobacter* was proposed for those "gray-white" cheese strains which contain *meso*-DAP, arabinose, and galactose in the cell wall, representatives of which were stated to contain corynomycolic acids. The DNA base ratios are in the range of 60–67 mol% G+C (Crombach, 1978).

However, the relationship of the genus *Caseobacter* to *Corynebacterium* and to *Rhodococcus* requires further clarification. Indeed, Keddie and Cure (1977) had previously reported that the *meso*-DAP-containing, "gray-white" cheese strains which they examined were members of the "*rhodochrous*" complex (*Rhodococcus*), a conclusion supported by the DNA base ratios. However, the fatty acid profiles of a few strains that have been examined resemble those found in members of the genus *Corynebacterium* (see Keddie and Bousfield, 1980).

The type species is *Caseobacter polymorphus* and the type strain is NCDO 2097 (= LMD AC 256).

Literature Cited

Abd-el-Malek, Y., Gibson, T. 1952. Studies in the bacteriology of milk III. The corynebacteria of milk. Journal of Dairy Research **19:**153–159.

Abe, S., Takayama, K., Kinoshita, S. 1967. Taxonomical studies on glutamic acid-producing bacteria. Journal of General and Applied Microbiology **13:**279–301.

Adamse, A. D. 1968. Formation and final composition of the bacterial flora of a dairy waste activated sludge. Water Research **2:**665–671.

Akiba, T., Ueyama, H., Seki, M., Fukimbara, T. 1970. Identifications of lower alcohol-utilizing bacteria. Journal of Fermentation Technology **48:**323–328.

Albert, J. O., Long, H. F., Hammer, B. W. 1944. Classification of the organisms important in dairy products IV. *Bacterium linens.* Agricultural Experimental Station, Iowa Research Bulletin No. **328:**234–259.

Antheunisse, J. 1972. Decomposition of nucleic acids and some of their degradation products by microorganisms. Antonie van Leeuwenhoek Journal of Microbiology and Serology **38:**311–327.

Austin, B., Goodfellow, M., Dickinson, C. H. 1978. Numerical taxonomy of phylloplane bacteria isolated from *Lolium perenne.* Journal of General Microbiology **104:**139–155.

Ayres, J. C. 1960. Temperature relationships and some other characteristics of the microbial flora developing on refrigerated beef. Food Research **25:**1–18.

Barea, J. M., Navarro, E., Montoya, E. 1976. Production of plant growth regulators by rhizosphere phosphate-solubilizing bacteria. Journal of Applied Bacteriology **40:**129–134.

Barksdale, L. 1970. *Corynebacterium diphtheriae* and its relatives. Bacteriological Reviews **34:**378–422.

Barksdale, L., Lanéelle, M. A., Pollice, M. C., Asselineau, J., Welby, M., Norgard, M. V. 1979. Biological and chemical basis for the reclassification of *Microbacterium flavum* Orla-Jensen as *Corynebacterium flavescens* nom. nov. International Journal of Systematic Bacteriology **29:**222–233.

Barlow, J., Kitchell, A. G. 1966. A note on the spoilage of pre-packed lamb chops by *Microbacterium thermosphactum.* Journal of Applied Bacteriology **29:**185–188.

Baumann, L., Baumann, P., Mandel, M., Allen, R. D. 1972. Taxonomy of aerobic marine eubacteria. Journal of Bacteriology **110:**402–429.

Becker, B., Lechevalier, M. P., Gordon, R. E., Lechevalier, H. A. 1964. Rapid differentiation between *Nocardia* and *Streptomyces* by paper chromatography of whole-cell hydrolysates. Applied Microbiology **12:**421–423.

Beguin, P., Eisen, H., Roupas, A. 1977. Free and cellulose-bound cellulases in a *Cellulomonas* species. Journal of General Microbiology **101:**191–196.

Bergey, D. H., Harrison, F. C., Breed, R. S., Hammer, B. W., Huntoon, F. M. 1923. Bergey's manual of determinative bacteriology, 1st ed. Baltimore: Williams & Wilkins.

Boone, C. J., Pine, L. 1968. Rapid method for characterization of actinomycetes by cell wall composition. Applied Microbiology **16:**279–284.

Bousfield, I. J. 1972. A taxonomic study of some coryneform bacteria. Journal of General Microbiology **71:**441–455.

Bousfield, I. J. 1978. The taxonomy of coryneform bacteria from the marine environment, pp. 217–233. In: Bousfield, I. J., Callely, A. G. (eds.), Special publications of the Society for General Microbiology I. Coryneform bacteria. London: Academic Press.

Bowie, I. S., Grigor, M. R., Dunckley, G. G., Loutit, M. W., Loutit, J. S. 1972. The DNA base composition and fatty acid constitution of some Gram-positive pleomorphic soil bacteria. Soil Biology and Biochemistry **4:**397–412.

Boylen, C. W. 1973. Survival of *Arthrobacter crystallopoietes* during prolonged periods of extreme desiccation. Journal of Bacteriology **113:**33–37.

Boylen, C. W., Ensign, J. C. 1970. Long-term starvation survival of rod and spherical cells of *Arthrobacter crystallopoietes.* Journal of Bacteriology **103:**569–577.

Boylen, C. W., Mulks, M. H. 1978. The survival of coryneform bacteria during periods of prolonged nutrient starvation. Journal of General Microbiology **105:**323–334.

Braden, A. R., Thayer, D. W. 1976. Serological study of *Cellulomonas.* International Journal of Systematic Bacteriology **26:**123–126.

Breed, R. S. 1953a. The *Brevibacteriaceae* fam. nov. of order Eubacteriales. Riassunti delle Communicazione VI Congresso Internazionale di Microbiologia, Roma **1:**13–14.

Breed, R. S. 1953b. The families developed from *Bacteriaceae* Cohn with a description of the family *Brevibacteriaceae* Breed, 1953. In: Atti del VI Congresso Internazionale di Microbiologia, Roma **1:**10–15.

Breed, R. S., Murray, E. G. D., Hitchens, A. P. (eds.). 1948. Bergey's manual of determinative bacteriology, 6th ed. Baltimore: Williams & Wilkins.

Breed, R. S., Murray, E. G. D., Smith, N. R. (eds.). 1957. Bergey's manual of determinative bacteriology, 7th ed. Baltimore: Williams & Wilkins.

Buchanan, R. E., Gibbons, N. E. (eds.). 1974. Bergey's manual of determinative bacteriology, 8th ed. Baltimore: Williams & Wilkins.

Cacciari, I., Giovannozzi-Sermanni, G., Grappelli, A., Lippi, D. 1971. Nitrogen fixation by *Arthrobacter* sp. I—Taxonomic study and evidence of nitrogenase activity of two new strains. Annali di Microbiologia ed Enzimologia **21:**97–105.

Chai, T. J., Levin, R. E. 1975. Characteristics of heavily mucoid bacterial isolates from fish pen slime. Applied Microbiology **30:**450–455.

Chan, E. C. S., Stevenson, I. L. 1962. On the biotin requirement of *Arthrobacter globiformis.* Canadian Journal of Microbiology **8:**403–405.

Charney, W. 1966. Transformation of steroids by Corynebacteriaceae. Journal of Applied Bacteriology **29:**93–106.

Chen, M., Alexander, M. 1973. Survival of soil bacteria during prolonged desiccation. Soil Biology and Biochemistry **5:**213–221.

Clark, J. B. 1972. Morphogenesis in the genus *Arthrobacter.* CRC Critical Reviews in Microbiology **1:**521–544.

Clarke, F. E. 1951. The generic classification of certain cellulolytic bacteria. Proceedings of the Soil Science Society of America **15:**180–182.

Clarke, F. E. 1952. The generic classification of the soil corynebacteria. International Bulletin of Bacteriological Nomenclature and Taxonomy **2:**45–56.

Clarke, F. E. 1953. Criteria suitable for species differentiation in *Cellulomonas* and a revision of the genus. International Bulletin of Bacteriological Nomenclature and Taxonomy **3:**179–199.

Collins, M. D. 1978. Lipids in coryneform taxonomy. Thesis. University of Newcastle-upon-Tyne, United Kingdom.

Collins, M. D., Goodfellow, M., Minnikin, D. E. 1979. Isoprenoid quinones in the classification of coryneform and related bacteria. Journal of General Microbiology 110:127–136.

Collins, M. D., Goodfellow, M., Minnikin, D. E. 1980. Fatty acid, isoprenoid quinone and polar lipid composition in the classification of Curtobacterium and related taxa. Journal of General Microbiology 118:29–37.

Collins-Thompson, D. L., Sørhaug, T., Witter, L. D., Ordal, Z. J. 1971. Glycerol ester hydrolase activity of Microbacterium thermosphactum. Applied Microbiology 21:9–12.

Collins-Thompson, D. L., Sørhaug, T., Witter, L. D., Ordal, Z. J. 1972. Taxonomic consideration of Microbacterium lacticum, Microbacterium flavum and Microbacterium thermosphactum. International Journal of Systematic Bacteriology 22:65–72.

Conn, H. J. 1928. A type of bacteria abundant in productive soils, but apparently lacking in certain soils of low productivity. New York State Agricultural Experimental Station Technical Bulletin No. 138:3–26.

Conn, H. J. 1947. A protest against the misuse of the generic name Corynebacterium. Journal of Bacteriology 54:10.

Conn, H. J., Dimmick, I. 1947. Soil bacteria similar in morphology to Mycobacterium and Corynebacterium. Journal of Bacteriology 54:291–303.

Crombach, W. H. J. 1972. DNA base composition of soil arthrobacters and other coryneforms from cheese and sea fish. Antonie van Leeuwenhoek Journal of Microbiology and Serology 38:105–120.

Crombach, W. H. J. 1974a. Relationships among coryneform bacteria from soil, cheese and sea fish. Antonie van Leeuwenhoek Journal of Microbiology and Serology 40:347–359.

Crombach, W. H. J. 1974b. Morphology and physiology of coryneform bacteria. Antonie van Leeuwenhoek Journal of Microbiology and Serology 40:361–376.

Crombach, W. H. J. 1978. Caseobacter polymorphus gen. nov., sp. nov., a coryneform bacterium from cheese. International Journal of Systematic Bacteriology 28:354–366.

Cummins, C. S. 1962. Chemical composition and antigenic structure of cell walls of Corynebacterium, Mycobacterium, Nocardia, Actinomyces and Arthrobacter. Journal of General Microbiology 28:35–50.

Cummins, C. S., Harris, H. 1956. The chemical composition of the cell wall in some Gram-positive bacteria and its possible value as a taxonomic character. Journal of General Microbiology 14:583–600.

Cummins, C. S., Harris, H. 1958. Studies on the cell wall composition and taxonomy of Actinomycetales and related groups. Journal of General Microbiology 18:173–189.

Cummins, C. S., Harris, H. 1959. Taxonomic position of Arthrobacter. Nature 184:831–832.

Cummins, C. S., Johnson, J. L. 1971. Taxonomy of the clostridia: Wall composition and DNA homologies in Clostridium butyricum and other butyric acid-producing clostridia. Journal of General Microbiology 67:33–46.

Cummins, C. S., Lelliott, R. A., Rogosa, M. 1974. Genus Corynebacterium, pp. 602–617. In: Buchanan, R. E., Gibbons, N. E. (eds.), Bergey's manual of determinative bacteriology, 8th ed. Baltimore: Williams & Wilkins.

Cure, G. L., Keddie, R. M. 1973. Methods for the morphological examination of aerobic coryneform bacteria, pp. 123–135. In: Board, R. G., Lovelock, D. N. (eds.), Sampling—microbiological monitoring of environments. Society for Applied Bacteriology Technical Series 7. New York, London: Academic Press.

da Silva, G. A. N., Holt, J. G. 1965. Numerical taxonomy of certain coryneform bacteria. Journal of Bacteriology 90:921–927.

Davidson, C. M., Hartree, E. F. 1968. Cytochrome as a guide to classifying bacteria: Taxonomy of Microbacterium thermosphactum. Nature 220:502–504.

Davidson, C. M., Mobbs, P., Stubbs, J. M. 1968. Some morphological and physiological properties of Microbacterium thermosphactum. Journal of Applied Bacteriology 31:551–559.

Davis, G. H. G., Fomin, L., Wilson, E., Newton, K. G. 1969. Numerical taxonomy of Listeria, streptococci and possibly related bacteria. Journal of General Microbiology 57:333–348.

Davis, G. H. G., Newton, K. G., 1969. Numerical taxonomy of some named coryneform bacteria. Journal of General Microbiology 56:195–214.

Demain, A. L., Hendlin, D. 1959. 'Iron transport' compounds as growth stimulators for Microbacterium sp. Journal of General Microbiology 21:72–79.

De Man, J. C., Rogosa, M., Sharpe, M. E. 1960. A medium for the cultivation of lactobacilli. Journal of Applied Bacteriology 23:130–135.

Dias, F. F. 1963. Studies in the bacteriology of sewage. Journal of the Indian Institute of Science 45:36–48.

Dias, F., Bhat, J. V. 1962. A new levan producing bacterium, Corynebacterium laevaniformans nov. spec. Antonie van Leeuwenhoek Journal of Microbiology and Serology 28:63–72.

Doetsch, R. N., Pelczar, M. J. 1948. The microbacteria. I. Morphological and physiological characteristics. Journal of Bacteriology 56:37–49.

Doetsch, R. N., Rakosky, J. 1950. Is there a Microbacterium liquefaciens? Bacteriological Proceedings 1950:38.

Duxbury, T., Gray, T. R. G. 1977. A microcultural study of the growth of cystites, cocci and rods of Arthrobacter globiformis. Journal of General Microbiology 103:101–106.

Ehrlich, H. C. 1963. Bacteriology of manganese nodules. I. Bacterial action on manganese in nodule enrichments. Applied Microbiology 11:15–19.

Ehrlich, H. C. 1968. Bacteriology of manganese nodules. II. Manganese oxidation by cell-free extract from a manganese nodule bacterium. Applied Microbiology 16:197–202.

El-Erian, A. F. M. 1969. Bacteriological studies on Limburger cheese. Thesis. Agricultural University, Wageningen, The Netherlands. Wageningen: Veenman & Zonen.

Ensign, J. C., Rittenberg, S. C. 1963. A crystalline pigment produced from 2-hydroxypyridine by Arthrobacter crystallopoietes n. sp. Archiv für Mikrobiologie 47:137–153.

Ensign, J. C., Wolfe, R. S. 1964. Nutritional control of morphogenesis in Arthrobacter crystallopoietes. Journal of Bacteriology 87:924–932.

Fiedler, F., Kandler, O. 1973. Die Mureintypen in der Gattung Cellulomonas Bergey et al. Archiv für Mikrobiologie 89:41–50.

Fiedler, F., Schleifer, K. H., Cziharz, B., Interschick, E., Kandler, O. 1970. Murein types in Arthrobacter, brevibacteria, corynebacteria and microbacteria. Publications de la Faculté des Sciences de l'Université J. E. Purkyne, Brno 47:111–122.

Fiedler, F., Stackebrandt, E. 1978. Taxonomical studies on Brevibacterium linens. Abstracts of the XII International Congress of Microbiology, München, C45, p. 96.

Foissy, H. 1974. Examination of Brevibacterium linens by an electrophoretic zymogram technique. Journal of General Microbiology 80:197–207.

Gardner, G. A. 1966. A selective medium for the enumeration of Microbacterium thermosphactum in meat and meat products. Journal of Applied Bacteriology 29:455–460.

Gardner, G. A., Carson, A. W., Patton, J. 1967. Bacteriology of prepacked pork with reference to the gas composition within the pack. Journal of Applied Bacteriology 30:321–333.

Gasdorf, H. J., Benedict, R. G., Cadmus, M. C., Anderson, R. F., Jackson, R. W. 1965. Polymer-producing species of Arthrobacter. Journal of Bacteriology 90:147–150.

Gibson, T. 1953. The taxonomy of the genus Corynebacterium. Atti del VI Congresso Internazionale di Microbiologia, Roma 1:16–20.

Gillies, A. J. 1971. Significance of thermoduric organisms in Queensland cheddar cheese. Australian Journal of Dairy Technology 26:145–149.

Giovanozzi-Sermanni, G. 1959. Una nuova specie di *Arthrobacter* determinante la degradazione della nicotina: *Arthrobacter nicotianae*. Il Tabacco 63:83–86.

Goodfellow, M., Alderson, G. 1977. The actinomycete—genus *Rhodococcus*: A home for the '*rhodochrous*' complex. Journal of General Microbiology 100:99–122.

Goodfellow, M., Collins, M. D., Minnikin, D. E. 1976. Thin-layer chromatographic analysis of mycolic acid and other long-chain components in whole-organism methanolysates of coryneform and related taxa. Journal of General Microbiology 96:351–358.

Goodfellow, M., Minnikin, D. E., Patel, P. V., Mordarska, H. 1973. Free nocardomycolic acids in the classification of nocardias and strains of the '*rhodochrous*' complex. Journal of General Microbiology 74:185–188.

Gounot, A. M. 1967. Role biologique des *Arthrobacter* dans les limons souterrains. Annales de l'Institut Pasteur 113:923–945.

Gray, T. R. G. 1976. Survival of vegetative microbes in soil. Symposium of the Society for General Microbiology 26:327–364.

Grecz, N., Dack, G. M. 1961. Taxonomically significant color reactions of *Brevibacterium linens*. Journal of Bacteriology 82:241–246.

Greenberg, J., Barker, H. A. 1962. A ferrichrome-requiring arthrobacter which decomposes puromycin aminonucleoside. Journal of Bacteriology 83:1163–1164.

Griffiths, D. E. 1977. EEC regulations: Some technical aspects of the intervention arrangements in milk products. Dairy Industries International 42:17–24.

Gundersen, K., Jensen, H. L. 1956. A soil bacterium decomposing organic nitro-compounds. Acta Agriculturae Scandinavica 6:100–114.

Hagedorn, C., Holt, J. G. 1975a. A nutritional and taxonomic survey of *Arthrobacter* soil isolates. Canadian Journal of Microbiology 21:353–361.

Hagedorn, C., Holt, J. G. 1975b. Ecology of soil arthrobacters in Clarion-Webster toposequences of Iowa. Applied Microbiology 29:211–218.

Hagiwara, S., Yamada, K. 1970. Studies on the utilization of petrochemicals by microorganisms Part II. Production of polysaccharide from ethanediol by *Arthrobacter simplex* var. *viscosus* n. var. Agricultural and Biological Chemistry 34:1283–1295.

Han, Y. W., Callihan, C. D. 1974. Cellulose fermentation: Effect of substrate pretreatment on microbial growth. Applied Microbiology 27:159–165.

Han, Y. W., Srinivasan, V. R. 1968. Isolation and characterization of a cellulose utilizing bacterium. Applied Microbiology 16:1140–1145.

Harrigan, W. F., McCance, M. E. 1976. Laboratory methods in food and dairy microbiology, revised ed. London: Academic Press.

Hill, L. R. 1966. An index to deoxyribonucleic acid base compositions of bacterial species. Journal of General Microbiology 44:419–437.

Holm, E., Jensen, V. 1972. Aerobic chemoorganotrophic bacteria of a Danish beech forest. Oikos 23:248–260.

Horvath, R. S., Alexander, M. 1970. Cometabolism of *m*-chlorobenzoate by an *Arthrobacter*. Applied Microbiology 20:254–258.

Hugh, R., Leifson, E. 1953. The taxonomic significance of fermentative versus oxidative metabolism of carbohydrates by various Gram negative bacteria. Journal of Bacteriology 66:24–26.

Iizuka, H., Komagata, K. 1965. Microbiological studies on petroleum and natural gas. III. Determination of *Brevibacterium*, *Arthrobacter*, *Micrococcus*, *Sarcina*, *Alcaligenes*, and *Achromobacter* isolated from oil-brines in Japan. Journal of General and Applied Microbiology 11:1–14.

Ingram, M., Dainty, R. H. 1971. Changes caused by microbes in spoilage of meats. Journal of Applied Bacteriology 34:21–39.

Jayne-Williams, D. J., Skerman, T. M. 1966. Comparative studies on coryneform bacteria from milk and dairy sources. Journal of Applied Bacteriology 29:72–92.

Jensen, H. L. 1933. Corynebacteria as an important group of soil microorganisms. Proceedings of the Linnean Society of New South Wales 58:181–185.

Jensen, H. L. 1934. Studies on saprophytic mycobacteria and corynebacteria. Proceedings of the Linnean Society of New South Wales 59:19–61.

Jensen, H. L. 1952. The coryneform bacteria. Annual Review of Microbiology 6:77–90.

Jensen, H. L. 1964. Studies on soil bacteria (*Arthrobacter globiformis*) capable of decomposing the herbicide Endothal. Acta Agriculturae Scandinavica 14:193–207.

Jensen, H. L. 1966. Some introductory remarks on the coryneform bacteria. Journal of Applied Bacteriology 29:13–16.

Jensen, V. 1968. The plate count technique, pp. 158–170. In: Gray, T. R. G., Parkinson, D. (eds.), The ecology of soil bacteria. Liverpool: Liverpool University Press.

Jones, D. 1975. A numerical taxonomic study of coryneform and related bacteria. Journal of General Microbiology 87:52–96.

Jones, D. 1978. An evaluation of the contribution of numerical taxonomy to the classification of the coryneform bacteria, pp. 13–46. In: Bousfield, I. J., Callely, A. G. (eds.), Coryneform bacteria. Special Publications of the Society for General Microbiology I. London: Academic Press.

Jones, D., Watkins, J., Erickson, S. K. 1973. Taxonomically significant colour changes in *Brevibacterium linens* probably associated with a carotenoid-like pigment. Journal of General Microbiology 77:145–150.

Jones, D., Weitzman, P. D. J. 1974. Reclassification of *Brevibacterium leucinophagum* Kinney and Werkman as a Gram-negative organism, probably in the genus *Acinetobacter*. International Journal of Systematic Bacteriology 24:113–117.

Jones, L. A., Bradley, S. G. 1964. Phenetic classification of actinomycetes. Developments in Industrial Microbiology 5:267–272.

Kaneko, T., Kitamura, K., Yamamoto, Y. 1969. *Arthrobacter luteus* nov. sp. isolated from brewery sewage. Journal of General and Applied Microbiology 15:317–326.

Katznelson, H., Cole, S. E. 1965. Production of gibberellin-like substances by bacteria and actinomycetes. Canadian Journal of Microbiology 11:733–741.

Kaufmann, A., Fegan, J., Doleac, P., Gainer, C., Wittich, D., Glann, A. 1976. Identification and characterization of a cellulolytic isolate. Journal of General Microbiology 94:405–408.

Keddie, R. M. 1974a. *Arthrobacter*, pp. 618–625. In: Buchanan, R. E., Gibbons, N. E. (eds.), Bergey's manual of determinative bacteriology, 8th ed. Baltimore: Williams & Wilkins.

Keddie, R. M. 1974b. *Cellulomonas*, pp. 629–631. In: Buchanan, R. E., Gibbons, N. E. (eds.), Bergey's manual of determinative bacteriology, 8th ed. Baltimore: Williams & Wilkins.

Keddie, R. M. 1978. What do we mean by coryneform bacteria?, pp. 1–12. In: Bousfield, I. J., Callely, A. G. (eds.), Coryneform bacteria. Special Publications of the Society for General Microbiology. I. London: Academic Press.

Keddie, R. M., Bousfield, I. J. 1980. Cell wall composition in the classification and identification of coryneform bacteria, pp. 167–188. In: Goodfellow, M., Board, R. G. (eds.), Microbiological classification and identification. Society for Applied Bacteriology Symposium Series No. 8. New York, London: Academic Press.

Keddie, R. M., Cure, G. L. 1977. The cell wall composition and distribution of free mycolic acids in named strains of coryneform bacteria and in isolates from various natural sources. Journal of Applied Bacteriology 42:229–252.

Keddie, R. M., Cure, G. L. 1978. Cell wall composition of coryneform bacteria, pp. 47–84. In: Bousfield, I. J., Callely, A. G. (eds.), Coryneform bacteria. Special Publications of the Society for General Microbiology I. London: Academic Press.

Keddie, R. M., Leask, B. G. S., Grainger, J. M. 1966. A comparison of coryneform bacteria from soil and herbage: Cell wall composition and nutrition. Journal of Applied Bacteriology 29:17–43.

Keddie, R. M., Rogosa, M. 1974. Kurthia, pp. 631–632. In: Buchanan, R. E., Gibbons, N. E. (eds.), Bergey's manual of determinative bacteriology, 8th ed. Baltimore: Williams & Wilkins.

Kellerman, K. F., McBeth, I. G. 1912. The fermentation of cellulose. Zentralblatt für Bakteriologie, Parasitenkunde, Infektionskrankheiten und Hygiene, Abt. 2 Orig. 34: 485–494.

Kellerman, K. F., McBeth, I. G., Scales, F. M., Smith, N. R. 1913. Identification and classification of cellulose dissolving bacteria. Zentralblatt für Bakteriologie, Parasitenkunde, Infektionskrankheiten und Hygiene, Abt. 2 Orig. 39: 502–522.

Kitamura, K., Kaneko, T., Yamamoto, Y. 1972. Lysis of viable yeast cells by enzymes of Arthrobacter luteus. I. Isolation of lytic strain and studies of its lytic activity. Journal of Applied and General Microbiology 18:57–71.

Klein, D. A., Davis, J. A., Casida, L. E. Jr. 1968. Oxidation of n-alkanes to ketones by an Arthrobacter species. Antonie van Leeuwenhoek Journal of Microbiology and Serology 34:495–503.

Kolenbrander, P. E., Lotong, N., Ensign, J. C. 1976. Growth and pigment production by Arthrobacter pyridinolis n. sp. Archives of Microbiology 110:239–245.

Komagata, K., Iizuka, H. 1964. New species of Brevibacterium isolated from rice (Studies on the microorganisms of cereal grains. Part VII). [In Japanese.] Journal of the Agricultural Chemical Society of Japan 38:496–502.

Komagata, K., Yamada, K., Ogawa, H. 1969. Taxonomic studies on coryneform bacteria. I. Division of bacterial cells. Journal of General and Applied Microbiology 15:243–259.

Kortstee, G. J. J. 1970. The aerobic decomposition of choline by microorganisms. I. The ability of aerobic organisms, particularly coryneform bacteria, to utilize choline as the sole carbon and nitrogen source. Archiv für Mikrobiologie 71:235–244.

Kraft, A. A., Ayres, J. C., Torrey, G. S., Salzer, R. H., da Silva, G. A. N. 1966. Coryneform bacteria in poultry, eggs and meat. Journal of Applied Bacteriology 29:161–166.

Krulwich, T. A., Pate, J. L. 1971. Ultrastructural explanation for snapping postfission movements in Arthrobacter crystallopoietes. Journal of Bacteriology 105:408–412.

Kuhn, D. A., Starr, M. P. 1962. Developmental morphology of Corynebacterium poinsettiae. Bacteriological Proceedings 1962:46.

Labeda, D. P., Liu, K. C., Casida, L. E., Jr. 1976. Colonization of soil by Arthrobacter and Pseudomonas under varying conditions of water and nutrient availability as studied by plate counts and transmission electron microscopy. Applied and Environmental Microbiology 31:551–561.

Lee, J. S., Pfeifer, D. K. 1977. Microbiological characteristics of Pacific shrimp (Pandalus jordani). Applied and Environmental Microbiology 33:853–859.

Lehmann, K. B., Neumann, R. 1896. Atlas und Grundriss der Bakteriologie und Lehrbuch der speciellen bakteriologischen Diagnostik. 1st ed. Munich: Lehmann.

Lochhead, A. G. 1955. Brevibacterium helvolum (Zimmermann) comb. nov. International Bulletin of Bacteriological Nomenclature and Taxonomy 5:115–119.

Lochhead, A. G. 1958a. Two new species of Arthrobacter requiring respectively vitamin B12 and the terregens factor. Archiv für Mikrobiologie 31:163–170.

Lochhead, A. G. 1958b. Soil bacteria and growth-promoting substances. Bacteriological Reviews 22:145–153.

Lochhead, A. G., Burton, M. O. 1953. An essential bacterial growth factor produced by microbial synthesis. Canadian Journal of Botany 31:7–22.

Lochhead, A. G., Burton, M. O. 1956. Importance of soil extract for the enumeration and study of soil bacteria, pp. 157–161. Transactions of the 6th International Congress of Soil Science, Paris.

Lochhead, A. G., Burton, M. O. 1957. Qualitative studies of soil micro-organisms. XIV. Specific vitamin requirements of the predominant bacterial flora. Canadian Journal of Microbiology 3:35–42.

Loos, M. A., Roberts, R. N., Alexander, M. 1967. Phenols as intermediates in the decomposition of phenoxyacetates by an Arthrobacter species. Canadian Journal of Microbiology 13:679–690.

Lowe, W. E., Gray, T. R. G. 1972. Ecological studies on coccoid bacteria in a pine forest soil. I. Classification. Soil Biology and Biochemistry 4:459–468.

Lund, B. M. 1969. Properties of some pectolytic, yellow pigmented, Gram-negative bacteria isolated from fresh cauliflowers. Journal of Applied Bacteriology 32:60–67.

Luscombe, B. M., Gray, T. R. G. 1971. Effect of varying growth rate on the morphology of Arthrobacter. Journal of General Microbiology 69:433–434.

Luthy, P. 1974. Corynebacterium okanaganae, an entomopathogenic species of the Corynebacteriaceae. Canadian Journal of Microbiology 20:791–794.

Lysenko, O. 1959. The occurrence of species of the genus Brevibacterium in insects. Journal of Insect Pathology 1: 34–42.

McBeth, I. G. 1916. Studies on the decomposition of cellulose in soils. Soil Science 1:437–487.

McBeth, I. G., Scales, F. M. 1913. The destruction of cellulose by bacteria and filamentous fungi. United States Department of Agriculture Bureau of Plant Industries Bulletin No. 266:1–52.

McLean, R. A., Sulzbacher, W. L. 1953. Microbacterium thermosphactum, spec. nov.; a nonheat resistant bacterium from fresh pork sausage. Journal of Bacteriology 65:428–433.

McMeekin, T. A. 1975. Spoilage association of chicken breast muscle. Applied Microbiology 29:44–47.

Mandel, M., Guba, E. F., Litsky, W. 1961. The causal agent of bacterial blight of American holly. Bacteriological Proceedings p. 61.

Marmur, J., Doty, P. 1962. Determination of the base composition of deoxyribonucleic acid from its thermal denaturation temperature. Journal of Molecular Biology 5:109–118.

Marmur, J., Falkow, S., Mandel, M. 1963. New approaches to bacterial taxonomy. Annual Review of Microbiology 17:329–372.

Martin, C. K. A. 1977. Microbial cleavage of sterol side-chains. Advances in Applied Microbiology 22:29–58.

Minnikin, D. E., Alshamaony, L., Goodfellow, M. 1975. Differentiation of Mycobacterium, Nocardia, and related taxa by thin-layer chromatographic analysis of whole-organism methanolysates. Journal of General Microbiology 88:200–204.

Minnikin, D. E., Collins, M. D., Goodfellow, M. 1979. Fatty acid and polar lipid composition in the classification of Cellulomonas, Oerskovia and related taxa. Journal of Applied Bacteriology 47:87–95.

Minnikin, D. E., Goodfellow, M., Collins, M. D. 1978. Lipid composition in the classification and identification of coryneform and related taxa, pp. 85–160. In: Bousfield, I. J., Callely, A. G. (eds.), Special Publications of the Society for

General Microbiology. I. Coryneform bacteria. London: Academic Press.

Moiroud, A., Gounot, A. M. 1969. Sur une bactérie psychrophile obligatoire isolée de limons glaciaires. Comptes Rendus Hebdomadaires des Séances de l'Académie des Sciences, Série D **269:**2150–2152.

Mordarska, H., Mordarski, M., Goodfellow, M. 1972. Chemotaxonomic characters and classification of some nocardioform bacteria. Journal of General Microbiology **71:**77–86.

Morris, J. G. 1960. Studies on the metabolism of *Arthrobacter globiformis*. Journal of General Microbiology **22:**564–582.

Morrisey, R. F., Dugan, E. P., Koths, J. S. 1976. Chitinase production by an *Arthobacter* sp. lysing cells of *Fusarium roseum*. Soil Biology and Biochemistry **8:**23–28.

Mulder, E. G., Adamse, A. D., Antheunisse, J., Deinema, M. H., Woldendorp, J. W., Zevenhuizen, L. P. T. M. 1966. The relationship between *Brevibacterium linens* and bacteria of the genus *Arthrobacter*. Journal of Applied Bacteriology **29:**44–71.

Mulder, E. G., Antheunisse, J. 1963. Morphologie, physiologie et écologie des *Arthrobacter*. Annales de l'Institut Pasteur **105:**46–74.

Murray, I. G., Proctor, A. G. J. 1965. Paper chromatography as an aid to the identification of *Nocardia* species. Journal of General Microbiology **41:**163–167.

Nand, K., Rao, D. V. 1972. *Arthrobacter mysorens*—a new species excreting L-glutamic acid. Zentralblatt für Bakteriologie, Parasitenkunde, Infektionskrankheiten und Hygiene, Abt. 2 Orig. **127:**324–331.

Owens, J. D., Keddie, R. M. 1969. The nitrogen nutrition of soil and herbage coryneform bacteria. Journal of Applied Bacteriology **32:**338–347.

Orla-Jensen, S. 1919. The lactic acid bacteria. Copenhagen: Host & Son.

Orla-Jensen, S. 1921. The main lines of the natural bacterial system. Journal of Bacteriology **6:**263–273.

Orla-Jensen, S. 1943. Die echten Milchsäurebakterien, Ergänzungsband. Copenhagen: Munksgaard.

Patel, I. B., Vaughn, R. H. 1973. Cellulolytic bacteria associated with sloughing spoilage of California ripe olives. Applied Microbiology **25:**62–69.

Pierson, M. D., Collins-Thompson, D. L., Ordal, Z. J. 1970. Microbiological, sensory and pigment changes of aerobically and anaerobically packaged beef. Food Technology **24:**1171–1175.

Pitcher, D. G. 1976. Arabinose with LL-diaminopimelic acid in the cell wall of an aerobic coryneform organism isolated from human skin. Journal of General Microbiology **94:**225–227.

Rivière, J. 1963. Action des microorganismes de la rhizosphère sur la croissance du blé. II. Isolement et caractérisation des bactéries produisant des phytohormones. Annales de l'Institut Pasteur **105:**303–314.

Robinson, J. B., Salonius, P. O., Chase, F. E. 1965. A note on the differential response of *Arthrobacter* spp. and *Pseudomonas* spp. to drying in soil. Canadian Journal of Microbiology **11:**746–748.

Robinson, K. 1966a. Some observations on the taxonomy of the genus *Microbacterium*. I. Cultural and physiological reactions and heat resistance. Journal of Applied Bacteriology **29:**607–615.

Robinson, K. 1966b. Some observations on the taxonomy of the genus *Microbacterium*. II. Cell wall analysis, gel electrophoresis and serology. Journal of Applied Bacteriology **29:**616–624.

Rogosa, M., Keddie, R. M. 1974a. *Brevibacterium*, pp. 625–628. In: Buchanan, R. E., Gibbons, N. E., (eds.), Bergey's manual of determinative bacteriology, 8th ed. Baltimore: Williams & Wilkins.

Rogosa, M., Keddie, R. M. 1974b. *Microbacterium*, pp. 628–629. In: Buchanan, R. E., Gibbons, N. E. (eds.),

Bergey's manual of determinative bacteriology, 8th ed. Baltimore: Williams & Wilkins.

Rogosa, M., Mitchell, J. A., Wiseman, R. F. 1951. A selective medium for the isolation and enumeration of oral and fecal lactobacilli. Journal of Bacteriology **62:**132–133.

Roth, N. G., Wheaton, R. B. 1962. Continuity of psychrophilic and mesophilic growth characteristics in the genus *Arthrobacter*. Journal of Bacteriology **83:**551–555.

Sacks, L. E. 1954. Observations on the morphogenesis of *Arthrobacter citreus*, spec. nov. Journal of Bacteriology **67:**342–345.

Schefferle, H. E. 1957. An investigation of the microbiology of built-up poultry litter. Thesis, University of Edinburgh, United Kingdom.

Schefferle, H. E. 1966. Coryneform bacteria in poultry deep litter. Journal of Applied Bacteriology **29:**147–160.

Schleifer, K. H. 1970. Die Mureintypen in der Gattung *Microbacterium*. Archiv für Mikrobiologie **71:**271–282.

Schleifer, K. H., Kandler, O. 1972. Peptidoglycan types of bacterial cell walls and their taxonomic implications. Bacteriological Reviews **36:**407–477.

Schuster, M. L., Vidaver, A. K., Mandel, M. 1968. A purple-pigment-producing bean wilt bacterium *Corynebacterium flaccumfaciens* var. *violaceum* n. var. Canadian Journal of Microbiology **14:**423–427.

Seeliger, H. P. R., Welshimer, H. J. 1974. *Listeria*, pp. 593–596. In: Buchanan, R. E., Gibbons, N. E. (eds.), Bergey's manual of determinative bacteriology, 8th ed. Baltimore: Williams & Wilkins.

Sethunathan, N., Pathak, M. D. 1971. Development of a diazinon-degrading bacterium in paddy water after repeated application of diazinon. Canadian Journal of Microbiology **17:**699–702.

Sguros, P. L. 1955. Microbial transformations of the tobacco alkaloids. I. Cultural and morphological characteristics of a nicotinophile. Journal of Bacteriology **69:**28–37.

Sguros, P. L. 1957. New approach to the mode of formation of classical morphological configurations by certain coryneform bacteria. Journal of Bacteriology **74:**707–709.

Sharpe, M. E., Law, B. A., Phillips, B. A. 1976. Coryneform bacteria producing methanethiol. Journal of General Microbiology **94:**430–435.

Sharpe, M. E., Law, B. A., Phillips, B. A., Pitcher, D. G. 1977. Methanethiol production by coryneform bacteria: Strains from dairy and human skin sources and *Brevibacterium linens*. Journal of General Microbiology **101:**345–349.

Sharpee, K. W., Duxbury, J. M., Alexander, M. 1973. 2,4-Dichlorophenoxyacetate metabolism by *Arthrobacter* sp.: Accumulation of a chlorobutenolide. Applied Microbiology **26:**445–447.

Shaw, N., Stead, A. 1972. Bacterial glycophospholipids. FEBS Letters **21:**249–253.

Sieburth, J. McN. 1964. Polymorphism of a marine bacterium (*Arthrobacter*) as a function of multiple temperature optima and nutrition. Proceedings of the Symposium on Experimental Marine Ecology. Occasional Publication No. **2:**11–16.

Skerman, T. M., Jayne-Williams, D. J., 1966. Nutrition of coryneform bacteria from milk and dairy sources. Journal of Applied Bacteriology **29:**167–178.

Skyring, G. W., Quadling, C. 1969. Soil bacteria: Comparisons of rhizosphere and nonrhizosphere populations. Canadian Journal of Microbiology **15:**473–488.

Skyring, G. W., Quadling, C. 1970. Soil bacteria: A principal component analysis and guanine-cytosine contents of some arthrobacter-coryneform soil isolates and of some named cultures. Canadian Journal of Microbiology **16:**95–106.

Skyring, G. W., Quadling, C., Rouatt, J. W. 1971. Soil bacteria: Principal component analysis of physiological descriptions of some named cultures of *Agrobacterium*, *Arthrobacter* and *Rhizobium*. Canadian Journal of Microbiology **17:**1299–1311.

Smyk, B. 1970. Fixation of atmospheric nitrogen by the strains of *Arthrobacter*. Zentralblatt für Bakteriologie, Parasitenkunde, Infektionskrankheiten und Hygiene, Abt. 2 Orig. **124**:231–237.

Smyk, B., Ettlinger, L. 1963. Recherches sur quelque espèces d'arthrobacter fixatrices d'azote isolées des roches karstiques alpines. Annales de l'Institut Pasteur **105**:341–348.

Sneath, P. H. A., Jones, D. 1976. *Brochothrix*, a new genus tentatively placed in the family Lactobacillaceae. International Journal of Systematic Bacteriology **26**:102–104.

Soumare, S., Blondeau, R. 1972. Caractéristiques microbiologiques des sol de la région du nord de la France: Importance des 'Arthrobacters'. Annales de l'Institut Pasteur **123**:239–240.

Speck, M. L. 1943. A study of the genus *Microbacterium*. Journal of Dairy Science **26**:533–543.

Splittstoesser, D. F., Wexler, M., White, J., Colwell, R. R. 1967. Numerical taxonomy of Gram-positive and catalase-positive rods isolated from frozen vegetables. Applied Microbiology **15**:158–162.

Stackebrandt, E., Fiedler, F. 1979. DNA-DNA homology studies among *Arthrobacter* and *Brevibacterium*. Archives of Microbiology **120**:289–295.

Stackebrandt, E., Fiedler, F., Kandler, O. 1978. Peptidoglycantyp und Zusammensetzung der Zellwandpolysaccharide von *Cellulomonas cartalyticum* und einigen coryneformen Organismen. Archives of Microbiology **117**:115–118.

Stackebrandt, E., Kandler, O. 1979. Taxonomy of the genus *Cellulomonas*, based on phenotypic characters and deoxyribonucleic acid–deoxyribonucleic acid homology, and proposal of seven neotype strains. International Journal of Systematic Bacteriology **29**:273–282.

Stackebrandt, E., Kandler, O. 1980. *Cellulomonas cartae* sp. nov. International Journal of Systematic Bacteriology **30**:186–188.

Staneck, J. L., Roberts, G. D. 1974. Simplified approach to identification of aerobic actinomycetes by thin-layer chromatography. Applied Microbiology **28**:226–231.

Starr, M. P., Kuhn, D. A. 1962. On the origin of V-forms in *Arthrobacter atrocyaneus*. Archiv für Mikrobiologie **42**:289–298.

Starr, M. P., Mandel, M., Murata, N. 1975. The phytopathogenic coryneform bacteria in the light of DNA base composition and DNA-DNA segmental homology. Journal of General and Applied Microbiology **21**:13–26.

Steinhaus, E. A. 1941. A study of the bacteria associated with thirty species of insects. Journal of Bacteriology **42**: 757–790.

Stevenson, I. L. 1967. Utilization of aromatic hydrocarbons by *Arthrobacter* spp. Canadian Journal of Microbiology **13**:205–211.

Stewart, B. J., Leatherwood, J. M. 1976. Derepressed synthesis of cellulase by *Cellulomonas*. Journal of Bacteriology **128**:609–615.

Stuart, S. E., Welshimer, H. J. 1973. Intrageneric relatedness of *Listeria* Pirie. International Journal of Systematic Bacteriology **23**:8–14.

Sulzbacher, W. L., McLean, R. A. 1951. The bacterial flora of fresh pork sausage. Food Technology **5**:7–8.

Szajer, C., Koths, J. S. 1973. Physiological properties and enzymatic activity of an *Arthrobacter* capable of lysing *Fusarium* sp. Acta Microbiologica Polonica Series B **5**:81–86.

Tanaka, K., Kimura, K. 1972. Process for producing L-glutamic acid and alpha-ketoglutaric acid. United States Patent No. 3,642,576.

Tate, R. L., Ensign, J. C. 1974. A new species of *Arthrobacter* which degrades picolinic acid. Canadian Journal of Microbiology **20**:691–694.

Taylor, C. B., Lochhead, A. G. 1937. A study of *Bacterium globiforme* Conn in soils differing in fertility. Canadian Journal of Research C **15**:340–347.

Thayer, D. W., Yang, S. P., Key, A. B., Yang, H. H., Barker, J. W. 1975. Production of cattle feed by the growth of bacteria on mesquite wood. Developments in Industrial Microbiology **16**:465–474.

Thomas, S. B., Druce, R. G., Peters, G. J., Griffiths, D. G. 1967. Incidence and significance of thermoduric bacteria in farm milk supplies: A reappraisal and review. Journal of Applied Bacteriology **30**:265–298.

Thornley, M. J. 1957. Observations on the microflora of minced chicken meat irradiated with 4 MeV cathode rays. Journal of Applied Bacteriology **20**:286–298.

Topping, L. E. 1937. The predominant micro-organisms in soils. I. Description and classification of the organisms. Zentralblatt für Bakteriologie, Parasitenkunde, Infektionskrankheiten und Hygiene, Abt. 2 Orig. **97**:289–304.

Topping, L. E. 1938. The predominant micro-organisms in soils. II. The relative abundance of the different types of organisms obtained by plating, and the relation of plate to total counts. Zentralblatt für Bakteriologie, Parasitenkunde, Infektionskrankheiten und Hygiene, Abt. 2 Orig. **98**:193–201.

Uchida, K., Aida, K. 1977. Acyl type of bacterial cell wall: Its simple identification by colorimetric method. Journal of General and Applied Microbiology **23**:249–260.

Veldkamp, H. 1965. The isolation of *Arthrobacter*. Zentralblatt für Bakteriologie, Parasitenkunde, Infektionskrankheiten und Hygiene, Abt. 1 Orig. Suppl. **1**:265–269.

Veldkamp, H., van den Berg, G., Zevenhuizen, L. P. T. M. 1963. Glutamic acid production by *Arthrobacter globiformis*. Antonie van Leeuwenhoek Journal of Microbiology and Serology **29**:35–51.

Veldkamp, H., Venema, P. A. A., Harder, W., Konings, W. N. 1966. Production of riboflavin by *Arthrobacter globiformis*. Journal of Applied Bacteriology **29**:107–113.

Waksman, S. A., Henrici, A. T. 1943. The nomenclature and classification of the actinomycetes. Journal of Bacteriology **46**:337–341.

Weidemann, J. F. 1965. A note on the microflora of beef muscle stored in nitrogen at 0°. Journal of Applied Bacteriology **28**:365–367.

Weitzman, P. D. J., Jones, D. 1975. The mode of regulation of bacterial citrate synthase as a taxonomic tool. Journal of General Microbiology **89**:187–190.

Whitehouse, R. L. S., Jackson, H. 1972. Description of *Cellulomonas acidula* isolated from milk. Antonie van Leeuwenhoek Journal of Microbiology and Serology **38**:537–542.

Wilkinson, B. J., Jones, D. 1975. Some serological studies on *Listeria* and possibly related bacteria, pp. 251–261. In: Woodbine, M. (ed.), Problems of listeriosis. Leicester: Leicester University Press.

Wilkinson, B. J., Jones, D. 1977. A numerical taxonomic survey of *Listeria* and related bacteria. Journal of General Microbiology **98**:399–421.

Williams, S. T., Davies, F. L. 1965. Use of antibiotics for selective isolation and enumeration of actinomycetes in soil. Journal of General Microbiology **38**:251–261.

Wittern, A. 1933. Beiträge zur Kenntnis der "Mikrobakterien" Orla-Jensen. Zentralblatt für Bakteriologie, Parasitenkunde, Infektionskrankheiten und Hygiene, Abt. 2 Orig. **87**:412–446.

Wolin, E. F., Evans, J. B., Niven, C. F. 1957. The microbiology of fresh and irradiated beef. Food Research **22**:682–686.

Wood, E. J. F. 1950. The bacteriology of shark spoilage. Australian Journal of Marine and Freshwater Research **1**:129–138.

Yamada, K., Komagata, K. 1970. Taxonomic studies on coryneform bacteria. III. DNA base composition of coryneform bacteria. Journal of General and Applied Microbiology **16**:215–224.

Yamada, K., Komagata, K. 1972a. Taxonomic studies on coryneform bacteria. IV. Morphological, cultural, biochemical, and physiological characteristics. Journal of General and Applied Microbiology **18**:399–416.

Yamada, K., Komagata, K. 1972b. Taxonomic studies on coryneform bacteria. V. Classification of coryneform bacteria. Journal of General and Applied Microbiology **18:**417–431.

Yamada, Y., Motoi, H., Kinoshita, S., Takada, N., Okada, H. 1975. Oxidative degradation of squalene by *Arthrobacter* species. Applied Microbiology **29:**400–404.

Yamada, Y., Inouye, G., Tahara, Y., Kondo, K. 1976. The menaquinone system in the classification of coryneform and nocardioform bacteria and related organisms. Journal of General and Applied Microbiology **22:**203–214.

Zevenhuizen, L. P. T. M. 1966. Formation and function of the glycogen-like polysaccharide of *Arthrobacter*. Antonie van Leeuwenhoek Journal of Microbiology and Serology **32:**356–572.

CHAPTER 143

Phytopathogenic Coryneform and Related Bacteria

ANNE K. VIDAVER and MORTIMER P. STARR

Most phytopathogenic bacteria are Gram negative (this Handbook, Chapter 4). However, several significant diseases of plants are caused by Gram-positive bacteria. Disregarding here the rare phytopathogenic endosporeforming bacteria (*Bacillus* and *Clostridium* spp.) and the little known phytopathogenic methanogenic bacteria (*Methanobacterium* sp.)—concerning which see this Handbook, Chapter 4—most of the Gram-positive phytopathogenic bacteria are presently referred to the genus *Corynebacterium,* with a few species placed in the genera *Nocardia* and *Streptomyces.*

The historical basis for placing these phytopathogenic bacteria in the genus *Corynebacterium* is as irrational as have been many of the taxonomic practices in phytopathogenic bacteriology (this Handbook, Chapters 4, 60, 62, and 102). Disciplinal and nomenclatural insularity (this Handbook, Chapter 4) resulted in the genus *Phytomonas,* which remained separated from the mainstream of bacterial taxonomy for several decades. Meanwhile, as recounted elsewhere (Barksdale, 1970; Conn and Dimmick, 1947; Jensen, 1952; Veldkamp, 1970; this Handbook, Chapters 140 and 141), the concept of the genus *Corynebacterium* was broadened from a repository mainly of the diphtheria bacillus to a veritable mish-mash. Jensen (1934), Dowson (1942), Burkholder (1948), and others put the phytopathogenic coryneform and possibly related bacteria into this fuzzily demarcated genus *Corynebacterium,* a practice officially sanctioned by its perpetuation in the eighth edition of *Bergey's Manual of Determinative Bacteriology* (Rogosa et al., 1974). This generic assignment has not met with universal acceptance and, in fact, has been roundly condemned by many writers on one basis or another (e.g., Barksdale, 1970; Conn, 1947; Jones, 1975; see, also, this Handbook, Chapter 141).

The phytopathogenic *Corynebacterium* spp. form a heterogeneous group by most criteria: pathology; morphology and morphogenesis; physiology; serology; cell wall composition; nucleic acid base composition; or polynucleotide sequence homology. Little correlation is apparent among the results of the independent studies of each of the aforementioned parameters, with the exception that most phytopathogenic coryneform bacteria differ markedly from the other (nonphytopathogenic) members of the genus *Corynebacterium.* Contrary views—namely, that the group shows considerable homogeneity and that these bacteria should be retained in the genus *Corynebacterium*—are expressed by Dye and Kemp (1977), who exhaustively determined cultural and biochemical characters of a large collection of strains representing extant taxa of phytopathogenic coryneform bacteria. The phytopathogenic coryneform bacteria were assigned by them to four species and several subspecies ("pathovars") of the genus *Corynebacterium.* Since saprophytic and animal-pathogenic corynebacteria were not included in their comparison, that work does not resolve the issue of assigning the phytopathogens to the genus for which *Corynebacterium diphtheriae* is the type species.

General agreement with Dye and Kemp (1977) was obtained by R. R. Carlson and A. K. Vidaver (in preparation), who analyzed the proteins of all known taxa of phytopathogenic corynebacteria by polyacrylamide gel electrophoresis. Comparative analyses with *Corynebacterium bovis* and *Micrococcus luteus* showed that the phytopathogens were easily distinguishable from either of these species. These investigators also favor retention of the phytopathogens within the genus *Corynebacterium* for the present, but strongly support the concept of subspecies on the basis of several criteria separable from phytopathogenicity. Simple means of differentiation, applicable to the majority of species and strains, are shown in Table 1.

Because the scientific community has not reached a consensus on the matter, it is convenient to refer to the phytopathogenic coryneform bacteria as nomenspecies of the genus *Corynebacterium* throughout this essay until the end, where several alternative generic assignments are briefly considered. One further caveat: Not all of the phytopathogenic *Corynebacterium* spp. actually are morphologically coryneform—i.e., club-shaped—but, here again, terminological convenience and authority (*Bergey's Manual* labels the relevant section of its Part 17 as the "Coryneform Group of Bacteria") seem to outweigh etymological purity as well as taxonomic accuracy.

Table 1. Some differential characteristics of phytopathogenic corynebacteria.[a]

Corynebacterium species[b]	Motility	Pigmentation[c] Blue	Yellow	Orange	Growth CNS[d]	TTC[e]	Acid production[f] Rhamnose	Melezitose	Mannitol	Mannose	Liquefaction Gelatin	Utilization Acetate	Fumarate
C. michiganense	−[g]	−	+	−	+	+	−	−	−	+	+	+	+
C. insidiosum	−	d	+	−	−	+	−	−	−	+	−	−	+
C. nebraskense	−	−	−	+	+	−	−	−	−	+	−	+	+
C. sependonicum	−	−	−	−	−	−	−	−	+	d	−	+	+
C. iranicum	−	−	+	−	−	ND	−	+	−	+	−	−	+
C. rathayi	−	−	+	−	d	ND	−	−	+	−	+	+	+
C. tritici	−	−	+	−	+	ND	−	−	+	+	−	+	+
C. fascians	−	−	−	+	−	ND	−	−	+	+	−	+	+
C. ilicis	+	−	+	−	ND	ND	−	−	+	+	+	+	+
C. flaccumfaciens	d	−	d[h]	−	+	ND	+	+	d	+	−	+	+
C. betae	d	−	+	−	+	ND	+	+	+	+	−	+	−
C. oortii	+	−	+	−	+	ND	+	d	+	+	−	−	−
C. poinsettiae	d	−	−	+	+	ND	+	+	+	+	−	+	−

[a] Symbols: +, positive; −, negative; d, different results between strains or when repeated; ND, not determined. The nomenclature of *Bergey's Manual* (Rogosa et al., 1974) is used. Data are from Dye and Kemp (1977), Vidaver (1980), and R. R. Carlson and A. K. Vidaver (in preparation).

[b] The species of *Corynebacterium iranicum* and *C. tritici* have no standing in the literature because they are not included in the 1980 Approved Lists of Bacterial Names (Skerman, McGowan, and Sneath, 1980). The names and data are included for convenience and in the belief that the validity of both species is suported (Dye and Kemp, 1977; R. R. Carlson and A. K. Vidaver, in preparation).

[c] On NBY medium (Vidaver, 1980).

[d] Gross and Vidaver (1979a).

[e] Kelman (1954).

[f] Medium C as basal medium (Dye and Kemp, 1977).

[g] Motility may rarely occur (M. N. Schroth, personal communication). Pink, red, orange, and nonpigmented variants occasionally are found.

[h] Wild-type strains may be orange or produce extracellular purple pigment.

Certain Gram-positive phytopathogenic bacteria have been placed in genera adjacent to *Corynebacterium*. Brief consideration will be given here to the phytopathogenic *Nocardia* and *Streptomyces* spp., and passing mention will be made of the bacterium associated with the ratoon-stunting disease of sugarcane, an organism that possibly belongs to the genus *Actinomyces*. (See, also, the detailed general treatments of these latter three genera in Chapters 148, 155, and 156 of this Handbook.)

Habitats

As Plant Pathogens

The Gram-positive phytopathogenic bacteria under consideration here are scarcely known apart from their roles as the causative agents of a variety of plant diseases. Under natural conditions, all of these corynebacteria—except *Corynebacterium fascians* and *C. michiganense*—have been isolated principally from a single genus of host plant. The names of the diseases, based on the major symptoms and plants involved, are pictorially descriptive: gumming diseases of grasses (e.g., Scharif, 1961),

caused by *Corynebacterium tritici, C. iranicum,* and *C. rathayi;* leafy gall of tobacco, chrysanthemum, dahlia, etc. or fasciation (a development of distorted and doubled flowers, with broadened, thin and frequently twisted stems) of sweet pea, caused by *C. fascians* (e.g., Tilford, 1936); ring rot of potato, caused by *C. sepedonicum* (Skaptason and Burkholder, 1942); wilt and leaf spot of red beets, caused by *C. betae* (Keyworth, Howell, and Dowson, 1956); bean wilt, caused by *C. flaccumfaciens* (Hedges, 1922); alfalfa wilt, caused by *C. insidiosum* (McCulloch, 1925); wilt and blight of corn, caused by *C. nebraskense* (Schuster, 1975); canker of tomato and pepper, caused by *C. michiganense* (Smith, 1910); canker of poinsettia, caused by *C. poinsettiae* (Starr and Pirone, 1942); blight of holly, caused by *C. ilicis* (Mandel, Guba, and Litsky, 1961); spot of tulip leaves and bulbs, caused by *C. oortii* (Saaltink and Maas Geesteranus, 1969); and mosaic of wheat, caused by an unnamed *Corynebacterium* sp. (R. R. Carlson and A. K. Vidaver, in preparation).

Nocardia vaccinii causes galls and bud proliferations in blueberry plants, a disease that seems to have been reported only once (Demaree and Smith, 1952).

The various phytopathogenic *Streptomyces* species cause scabs and/or rots in a variety of root crops. For many years after Thaxter (1891) first isolated a causal agent of common scab of potato, this phytopathogen was assigned to *Streptomyces scabies,* a species not recognized in the eighth edition of *Bergey's Manual* (Pridham and Tresner, 1974) on the bases that a proper type strain is not extant and that "many taxonomically different reference strains [are] available" (so that the nature of *"S. scabies"* is left in doubt). Waksman (1959) had earlier noted that "cultures of actinomycetes isolated from various types of scab suggested the possibility that several species are involved"—an opinion bolstered by the studies of Corbaz (1964) and Labruyère (1971) on potato scab as well as by studies of others on the scab diseases of different root crops. Only three phytopathogenic *Streptomyces* species are currently recognized in *Bergey's Manual* (Pridham and Tresner, 1974): *S. intermedius,* causal agent of sugar beet scab (Krüger, 1904); *S. ipomoeae,* causal agent of sweet potato scab (Person and Martin, 1940); and *S. setonii,* one of the causal agents of potato scab (Millard and Burr, 1926). *S. scabies* is, however, recognized by the International Streptomyces Project (see summary by Shirling and Gottlieb, 1972), as are the allegedly phytopathogenic *S. cratifer* (Millard and Burr, 1926), *S. tumuli* (Millard and Beeley, 1927), and *S. viridogenes* (Waksman, 1953). The status of other phytopathogenic streptomycetes is by no means clear (Young et al., 1978). (See, also, this Handbook, Chapter 156.)

A putative actinomycete or coryneform bacterium is probably the cause of the ratoon stunting disease of sugarcane (Kao and Damann, 1978; Weaver, Teakle, and Hayward, 1977). It is a branching, pleomorphic rod that currently is uncultivable (see the Addendum to Chapter 163, this Handbook).[1]

Some of the phytopathogenic *Corynebacterium* spp. interact with nematodes in producing certain diseases. Nematodes belonging to the genus *Anguina* are necessary for transmission of the so-called yellow slime or "tundu" disease of wheat, caused by *C. tritici* (Gupta and Swarup, 1972), and the disease of annual rye grass caused by a coryneform bacterium. The latter is thought also to cause a disease or intoxication in sheep grazing on infected rye grass (Bird and Stynes, 1977). Nematodes belonging to the genus *Aphelenchoides* are necessary for production of "cauliflower" syndrome of strawberry, in which certain strains of *C. fascians* are another necessary component (Crosse and Pitcher, 1952). In this hyperplastic disease, leaf and bud nematodes are involved as vectors; they possibly provide some necessary growth factor(s), either for the bacterium or for the hyperplastic tissue, because neither agent alone can produce the entire syndrome.

The plant diseases caused by these Gram-positive phytopathogens are described in the various works cited here and elsewhere (this Handbook, Chapter 4; Lelliott, 1966).

The phytopathogenic corynebacteria are considered to produce their antagonistic effects, in whole or in part, by the production of various metabolites—including toxins (Patil, 1974; Strobel, 1977), polysaccharides, and hormones. Toxins have been reported for *C. sepedonicum* (Johnson and Strobel, 1970; Strobel, 1967, 1970; Strobel, Talmadge, and Albersheim, 1972), *C. michiganense* (Rai and Strobel, 1969a, b), *C. insidiosum* (Ries and Strobel, 1972a, b), and perhaps *C. nebraskense* (see Schuster, 1975). These studies, for which confirmation is awaited, suggest that toxins may be responsible for wilt symptoms. In the gummosis diseases of grains, large quantities of polysaccharide are extruded onto the leaf blades or developing seeds. Polysaccharides have been implicated as antagonistic factors in other bacterial diseases of plants, but nothing is known directly about their role in these corynebacterial gummoses. The disease of animals called "annual ryegrass toxicity" has been shown to be due to a neurotoxin produced by *C. rathayi;* the toxin can be produced in cell cultures of rye endosperm (Stynes and Petterson, 1980).

In the disease syndrome caused by *C. fascians,* a number of plant hormones (cytokinins and auxins) have been implicated, mainly because they are produced in the bacterial cultures. The major cytokinins are the substituted adenine [N^6-(Δ^2-isopentenyl)-adenine] (Rathbone and Hall, 1972), otherwise known as 6-(3-methyl-2-butenylamino)-purine (Scarbrough et al., 1973); *cis*-zeatin, 6-(4-hydroxy-3-methyl-*cis*-2-butenylamino) purine, and its trans isomer (Scarbrough et al., 1973); and 6-(4-hydroxy-3-methyl-*cis*-2-butenylamino)-2-methylthiopurine (Armstrong et al., 1976). A cytokinin-active nucleoside, ribosyl-*cis*-zeatin, also has been isolated from tRNA of *C. fascians* (Einset and Skoog, 1977). Murai et al. (1980) present further information about the chemistry of the cytokinins produced by *C. fascians.* The auxins indole-3-acetic acid and indole-3-acetonitrile are also produced in vitro (Galach'yan, 1969). Both kinds of plant hormones probably play a role in disease development, because altered levels of indole acetic acid and cytokinins have been reported in infected plants (Balazs and Sziraki, 1974).

Plasmids have been considered in relation to virulence and other properties of phytopathogenic corynebacteria. According to Murai et al. (1980), four phytopathogenic strains of *C. fascians* contained plasmids; one nonphytopathogenic strain had no detectable plasmids; in one strain, loss (by an

---

[1]The bacterium causing ratoon stunting disease of sugarcane apparently has been cultured (Davis, Gillaspie, and Harris, 1980); details are forthcoming (M. J. Davis, personal communication).

unspecified mechanism) of a "large plasmid" was reported to be "associated with loss of virulence". Other investigators, too, have detected and characterized various plasmids in phytopathogenic corynebacteria (Gross, Vidaver, and Keralis, 1979; E. N. Lawson and M. P. Starr, in preparation); however, virulence and other properties could not be correlated by these workers with the presence of any plasmid in *C. nebraskense* or *C. fascians*, respectively.

Survival Apart from Diseased Plants

Infected and infested plant materials or residues, all of which serve as primary sources of inoculum, have been used in studying the survival of the phytopathogenic corynebacteria in soil. Such studies show poor survival without the protection of plant parts (Dowson, 1957; Schuster and Coyne, 1974). Bacterial associations with seeds, propagative parts, or perennial hosts have been sketchily reported (Schuster and Coyne, 1974). Except for *Corynebacterium sepedonicum* in potatoes (Hayward, 1974), the extent of latent infection is unknown. Survival in irrigation water has been reported for wild-type strains and streptomycin-resistant mutants of *C. nebraskense* (Steadman, 1977).

Due to lack of selective media and the slow growth of wild-type strains of phytopathogenic corynebacteria, their quantitative survival in soil or water has not been determined without prior enrichment in plant hosts, a procedure which substantially reduces the quantitative precision. In addition, their role as epiphytes on susceptible plants and on symptomless carriers can only be surmised at this time. The same considerations apply to *Nocardia vaccinii* and the phytopathogenic *Streptomyces* spp.

Isolation of Phytopathogenic *Corynebacterium, Nocardia,* and *Streptomyces* Species

Plant Parts to Consider for Isolation

In general, young, freshly infected plant material is best for isolation of these phytopathogenic bacteria. Successful isolations have been made from the periphery of isolated lesions; from the terminal portions of wilted stems, petioles, and buds; and from the vascular systems of wilting leaflets. In root crops, the bacteria can be isolated from the scabs and softened areas or rots of young tubers, with only a gentle washing preceding isolation. Materials from advanced stages of disease and heavily contaminated plant residues do not lend themselves to easy pathogen isolation.

Enrichment

The only form of enrichment currently known for this group of phytopathogens is the use of a prospective, sensitive host plant as an enrichment medium for the pathogen. A small amount of diseased material or debris is first rinsed under tap water—except for isolation of *Corynebacterium fascians,* which is frequently on the surface (Dowson, 1957)—and is then immersed or comminuted in distilled water or buffer for a few hours, preferably in the cold to avoid growth of contaminants. Particulate matter is thereafter removed by filtration or centrifugation. The supernatant material is inoculated by one of several means: DeVilbiss atomizer under pressure, vacuum infiltration, needle puncture, etc. If no symptoms develop, it may be due to a lack of the incitant, a resistant plant variety, or improper inoculation route or environmental conditions. If symptoms do develop, it may not be easy to differentiate a hypersensitive reaction (Klement and Goodman, 1967) from disease symptoms. The former is a relatively rapid, delimited, necrotic response, dependent on a high inoculum dose; the development of true phytopathogenic reactions usually takes longer than 48 h. Reisolation of the bacteria and genesis of symptoms upon their introduction into the host plant at relatively low inoculation concentrations (10^6 CFU/ml or lower) are needed to confirm pathogenicity.

Selective Isolation Media

Few selective or semiselective media have been developed for these Gram-positive phytopathogenic bacteria, possibly because they are relatively easy to isolate, grow (e.g., Starr, 1949), and identify to genus (in terms of their morphology and Gram-staining reaction). However, their enumeration, extent of spread, and mode of existence apart from the host plant would be better understood if selective media were commonly available and used. To this end, media are mentioned here that have been useful for isolating some of these bacteria from plant residues and badly infected plant parts, as well as from freshly infected material.

Medium 4–m–1 for Isolating *Corynebacterium sepedonicum* (Snieszko and Bonde, 1943)

Peptone (Difco)	3 g
Tryptone (Difco)	3 g
Yeast extract (Difco)	3 g
Maltose	2 g
Lactose	1 g
Agar	15 g
Sodium dichromate	0.05 g
Distilled water	1 liter

All of the ingredients except the sodium dichromate are dissolved, and the pH is adjusted to approximately 7.2; the final pH after autoclaving should be 7.0. Then, a separately autoclaved solution containing the sodium dichromate is added.

CNS agar medium (Gross and Vidaver, 1979a) was developed for isolation of *Corynebacterium nebraskense*, but it is also useful for isolating several other phytopathogenic coryneform species, namely *C. michiganense*, *C. betae*, *C. tritici*, and *C. oortii*. Some strains of *C. rathayi*, *C. flaccumfaciens*, and *C. poinsettiae* grew well; others not at all. Several strains of *C. insidiosum*, *C. sepedonicum*, *C. iranicum*, and *C. fascians* did not grow. Neither Gram-negative bacteria nor noncoryneform Gram-positive bacteria, except for some micrococci, grew on the medium regardless of the inoculum source.

CNS Agar for Isolating *Corynebacterium nebraskense* (Gross and Vidaver, 1979a)

Nutrient broth (Difco)	8.0 g
Yeast extract (Difco)	2.0 g
K_2HPO_4	2.0 g
KH_2PO_4	0.5 g
Glucose	5.0 g
$MgSO_4 \cdot 7H_2O$	0.25 g
Nalidixic acid (dissolved in 0.1 M NaOH at 10 mg/ml)	25 mg
Lithium chloride	10 g
Cycloheximide (Sigma)	40 mg
Polymyxin B sulfate (Sigma)	32 mg
Bravo 500 (chlorothalonil, active ingredient, 53%; Diamond Shamrock Co.; 1:50 dilution)	0.082 ml
Agar	15 g
Distilled water	1 liter

The glucose (10% wt/vol) and $MgSO_4$ (1 M) were autoclaved separately and added aseptically. Nalidixic acid and polymyxin sulfate were added to partially cooled autoclaved medium from freshly prepared stock solutions. Cycloheximide was also added after autoclaving. The pH after autoclaving is 6.9.

Isolated colonies of the corynebacteria appear on CNS agar in 6–8 days. The colonies of *Corynebacterium nebraskense* are orange, dome-shaped, entire, and glistening; those of *C. michiganense*, *C. betae*, *C. tritici*, and *C. oortii* are usually various shades of yellow. *C. michiganense* varies in colony morphology, but the colonies tend to be fluidal (glistening, irregular, and watery). Colonies of *C. betae*, *C. tritici*, and *C. oortii* on CNS agar tend to be entire, glistening, and convex. With experience, the pathogenic corynebacteria are generally recognizable on CNS agar by colony morphology and pigmentation. Pathogenicity tests, bacteriocin production, and phage sensitivity tests have been useful confirmatory tests. The CNS agar medium can be stored at 4°C for up to 4 weeks without loss of selectivity. Recovery of *C. nebraskense* from artificially inoculated soil is nearly 100% if the cell number is more than 10^3 CFU/g. Recovery from plant parts was far superior on CNS compared to a nonselective medium, NBY.

Medium D2, Selective for Corynebacteria (Kado and Heskett, 1970)

Glucose	10 g
Casein hydrolysate	4 g
Yeast extract	2 g
NH_4Cl	1 g
$MgSO_4 \cdot 7H_2O$	0.3 g
Lithium chloride	5 g
Polymyxin sulfate	40 mg
Sodium azide	2 mg
Tris-HCl buffer	1.2 g
Agar	15 g
Distilled water	1 liter

Medium D2 is adjusted with HCl to pH 7.8 before autoclaving. After autoclaving, the pH is 6.9, at which time the polymyxin sulfate and sodium azide are added. "The medium should be freshly prepared, since azide and polymyxin break down with time."

While Kado and Heskett (1970) considered medium D2 to be selective for various *Corynebacterium* species, they reported that certain Gram-negative bacteria also grew on the medium, as do some fungi. However, according to them, the *Corynebacterium* spp., including human and animal pathogens, were distinguished by a particular colony morphology; namely, colonies that are "small, light yellow, circular, convex, and glistening". Growth (single colony formation is not mentioned in their paper) appeared in 48–72 h at 30°C. Their efficiency of recovery for *C. michiganense* (the only phytopathogenic coryneform tested for recovery) ranged from 27% (from a mixed population) to 80%, using 10^5 cells/ml or higher, in artificially amended media or soil, respectively. Recovery of phytopathogenic corynebacteria on medium D2 from freshly collected diseased tissue was not compared by Kado and Heskett (1970) with recovery on a nonselective medium.

Demaree and Smith (1952) provide no details about the formulations of culture media suitable for the isolation and cultivation of *Nocardia vaccinii*. According to these authors, *N. vaccinii* grew poorly on most media reported in their paper; the only "abundant" growth recorded by them was on a "potato-yeast-mannitol agar" of unspecified composition.

The various phytopathogenic *Streptomyces* species have been isolated on a variety of media ranging from nonnutrient agar (Corbaz, 1964) to media based on various host-plant extracts to an unspecified "agar medium containing tyrosine" (Labruyère, 1971). The International Streptomyces Project (ISP) media are suitable for growth (Shirling and Gottlieb, 1966). These media are ISP Medium 2 (Difco No. 0770) and ISP Medium 4 (Difco No. 0772); the former is a yeast extract, malt extract agar; the latter an inorganic salts, starch agar (see, also, this Handbook, Chapter 156).

Identification of Phytopathogenic *Corynebacterium, Nocardia,* and *Streptomyces* Species

As discussed elsewhere (this Handbook, Chapters 4, 60, 62, and 102), identification of phytopathogenic bacteria to particular genera is most readily accomplished when the culture is isolated from a known diseased plant or adjacent habitat. In the case of both healthy and diseased plant material, slow-growing Gram-positive bacteria are rare; therefore, such bacteria are likely to be coryneform phytopathogens. At the present time, it is only by inoculation tests in plants that one can definitively determine the identity of most phytopathogenic coryneform and related bacteria. There is some promise, however, in newer techniques. Gram-positive corynebacteria with G+C ratios above 68 mol% are likely to be phytopathogens (Crombach, 1972; Rogosa et al., 1974; Starr, Mandel, and Murata, 1975; Vidaver and Mandel, 1974). Only *C. fascians* and strains of a few other *Corynebacterium* nomenspecies (of questionable phytopathogenicity) have lower G+C ratios. G+C ratios for *Nocardia vaccinii* and most phytopathogenic *Streptomyces* spp. are not yet available; the DNA of *S. scabies* has been reported to contain 72.2 mol% G+C (Lawrence and Clark, 1966) and *S. intermedius* 73.6 mol% G+C (Frontali, Hill, and Silvestri, 1965). Analysis of soluble proteins by polyacrylamide gel electrophoresis appears useful for identification of phytopathogenic *Corynebacterium* species and related bacteria (R. R. Carlson and A. K. Vidaver, in preparation).

The various problems associated with identification of other phytopathogenic bacteria (this Handbook, Chapters 4, 60, 62, and 102) apply here. But solutions are appearing over the horizon; for example, in the following areas: serology (Lazar, 1968; Masuo and Nakagawa, 1970); cell wall composition (Diaz-Maurino and Perkins, 1974; Keddie and Cure, 1977; Perkins, 1970; Rogosa et al., 1974); phospholipid composition (Komura et al., 1975), menaquinone analyses (Collins, Goodfellow, and Minnikin, 1979; Collins et al., 1977; Yamada et al.,

1976); DNA-DNA homology (Starr, Mandel, and Murata, 1975); phage and bacteriocin sensitivity (Gross and Vidaver, 1979b; Vidaver and Mandel, 1974); and cell protein analysis (R. R. Carlson and A. K. Vidaver, in preparation). Many of these procedures might be useful if the available information were not so fragmentary and were verified by independent studies and/or if the techniques were simplified.

Despite many recent taxonomic studies of coryneform bacteria, including plant pathogens, even those devoted entirely to phytopathogenic corynebacteria (Dye and Kemp, 1977; Starr, Mandel, and Murata, 1975; Vidaver and Mandel, 1974) have not shown any easy or certain means of identification of these bacteria below the genus level—apart from their ability to cause particular plant diseases. Current identification of phytopathogenic *Corynebacterium* species is thus largely dependent on demonstration of phytopathogenicity. However, some nonphytopathological differentiation (Table 1) is made possible by division into motile *Corynebacterium* species (*C. poinsettiae, C. betae, C. flaccumfaciens, C. ilicis,* and *C. oortii*) and nonmotile (all the remaining species). In addition, with the exception of the generally off-white to pale yellow *C. sepedonicum,* the remaining corynebacteria are usually pigmented on complex media, in shades of yellow, orange, and pink (Rogosa et al., 1974). At least some of these pigments are carotenoids (Norgård, Aasen, and Liaaen-Jensen, 1970; Prebble, 1968; Saperstein, Starr, and Filfus, 1954; Schuster, Jones, and Sayre, 1959; Starr and Saperstein, 1953). Pigment formation and color are affected by the composition of the medium and by temperature, with lower temperatures generally accentuating the color. *C. insidiosum* is generally further characterized by formation of a water-insoluble, dull blue-black, granular pigmentation, involving the bipyridyl pigment indigoidine (Kuhn et al., 1965; Starr, 1958) within yellow colonies. A natural variant of yellow-pigmented *C. flaccumfaciens* produces an extracellular, pH-sensitive, purple pigment of unknown composition (Schuster, Vidaver, and Mandel, 1968).

At the species level, serodiagnosis is used for detecting and identifying two economically important phytopathogenic *Corynebacterium* species. *C. insidiosum,* causal agent of alfalfa wilt, can be rapidly identified by an agglutination test (Hale, 1972). While there have been a number of serological tests reported for *C. sepedonicum,* causal agent of potato ring rot, only recently has it been possible to detect this bacterium specifically in very low concentration ($<10^2$ CFU/ml) by the use of indirect fluorescent antibody staining (Slack, Kelman, and Perry, 1979).

Maintaining virulence in cultures of the phytopathogenic corynebacteria can be a problem (Carroll and Lukezic, 1971; Vidaver, 1977); freeze-drying is probably the simplest and most effective means of preservation. However, freeze-drying itself may sometimes affect cultural characteristics, including virulence (Servin-Massieu, 1971); no data bearing on this point are available for phytopathogenic coryneform bacteria.

Comments about the Generic Placement of the Phytopathogenic Corynebacteria

Current opinions about the generic placement of the phytopathogenic coryneform bacteria range from keeping all of them in the genus *Corynebacterium* (e.g., Dye and Kemp, 1977; Rogosa et al., 1974) to removing all or most of them from the genus *Corynebacterium* (e.g., Barksdale, 1970; Jones, 1975; Keddie, 1978). This situation, by no means unique to the phytopathogens, pervades corynebacterial taxonomy (this Handbook, Chapters 141 and 142).

Numerous genera have been suggested as suitable repositories for the evicted phytopathogenic corynebacteria, on grounds almost as numerous as the genera. Few of the suggestions have undergone the required nomenclatural formalities and, hence, have no formal taxonomic standing. Most often, they occur as statements in research papers to the effect that a particular phytopathogenic *Corynebacterium* nomenspecies might better be assigned to a particular genus other than *Corynebacterium* on the basis of the property being reported in that paper. Some of these suggestions may have merit and are recorded here in the hope that phytobacteriologists might find therein some clues leading to a possibly rational classification of these bacteria.

i. *Arthrobacter* for *Corynebacterium poinsettiae* (Kuhn and Starr, 1962), *C. ilicis* (Bousfield, 1972), and *C. michiganense* and *C. rathayi* (Jones, 1975)—on the basis of the characteristic rod ↔ coccus morphogenesis and other properties.

ii. *Curtobacterium* for *Corynebacterium flaccumfaciens*, *C. flaccumfaciens* subsp. *aurantiacum*, and *C. poinsettiae* by Yamada and Komagata (1972)—on the basis of morphology, motility, cell wall composition, and biochemical traits. *C. betae* and *C. nebraskense* might belong to this genus (Collins, Goodfellow, and Minnikin, 1979; Jones, 1975), as might *C. oortii* because of phenetic relatedness (Dye and Kemp, 1977).

iii. *Nocardia* for *Corynebacterium fascians* by Starr, Mandel, and Murata (1975)—on the basis of DNA-DNA homology. But, Yamada and Komagata (1972) suggest that *C. fascians* is the only phytopathogen that should be retained in the genus *Corynebacterium* because of its cell wall composition.

iv. One or more unnamed genera (other than those mentioned) for *Corynebacterium insidiosum*, *C. michiganense,* and *C. sepedonicum* by Yamada and Komagata (1972)—on the basis of cell wall and DNA base compositions. These bacteria are closely related (in cultural and biochemical properties) to each other and to *C. rathayi, C. tritici, C. iranicum,* and *C. nebraskense* (Dye and Kemp, 1977).

Finally, it should be pointed out that a consensus might more rationally be reached on the placement of most or even all nomenspecies of this group, if future comparative studies used the same reference strains (including saprophytes and animal pathogens) and adopted standardized methods.

Literature Cited

Armstrong, D. J., Scarbrough, E., Skoog, F., Cole, D. L., Leonard, N. J. 1976. Cytokinins in *Corynebacterium fascians* cultures. Isolation and identification of 6-(4-hydroxy-3-methyl-*cis*-2-butenylamino)-2-methylthiopurine. Plant Physiology **58:**749–752.

Balazs, E., Sziraki, I. 1974. Altered levels of indoleacetic acid and cytokinin in geranium stems infected with *Corynebacterium fascians.* Acta Phytopathologica **9:**287–292.

Barksdale, L. 1970. *Corynebacterium diphtheriae* and its relatives. Bacteriological Reviews **34:**378–422.

Bird, A. F., Stynes, B. A. 1977. The morphology of a *Corynebacterium* sp. parasitic on annual rye grass. Phytopathology **67:**828–830.

Bousfield, I. J. 1972. A taxonomic study of some coryneform bacteria. Journal of General Microbiology **71:**441–455.

Burkholder, W. H. 1948. *Corynebacterium,* pp. 392–396; 398–400. In: Breed, R. S., Murray, E. G. D., Hitchens, A. P. (eds.), Bergey's manual of determinative bacteriology, 6th ed. Baltimore: Williams & Wilkins.

Carroll, R. B., Lukezic, F. L. 1971. Methods of preservation of *Corynebacterium insidiosum* isolates in relation to virulence and colony appearance on a tetrazolium chloride medium. Phytopathology **61:**1423–1425.

Collins, M. D., Goodfellow, M., Minnikin, D. E. 1979. Isoprenoid quinones in the classification of coryneform and related bacteria. Journal of General Microbiology **110:**127–136.

Collins, M. D., Pirouz, T., Goodfellow, M., Minnikin, D. E. 1977. Distribution of menaquinones in actinomycetes and corynebacteria. Journal of General Microbiology **100:**221–230.

Conn, J. J. 1947. A protest against the misuse of the generic name *Corynebacterium.* Journal of Bacteriology **54:**10.

Conn, J. J., Dimmick, I. 1947. Soil bacteria similar in morphology to *Mycobacterium* and *Corynebacterium.* Journal of Bacteriology **54:**291–303.

Corbaz, R. 1964. Étude des streptomycètes provoquant la gale commune de la pomme de terre. Phytopathologische Zeitschrift **51:**351–360.

Crombach, W. H. J. 1972. DNA base composition of soil arthrobacters and other coryneforms from cheese and sea fish. Antonie van Leeuwenhoek Journal of Microbiology and Serology **38:**105–120.

Crosse, J. E., Pitcher, R. S. 1952. Studies in the relationship of eelworms and bacteria to certain plant diseases. I. The etiology of strawberry cauliflower disease. Annals of Applied Biology **39:**475–484.

Davis, M. J., Gillaspie, A. G., Jr., Harris, R. W. 1980. Ratoon stunting disease of sugarcane: Isolation of the causal bacterium. Abstracts of the American Phytopathological Society—

Canadian Phytopathological Society Annual Meeting, August 24–28 **1980:**189.

Demaree, J. B., Smith, N. R. 1952. *Nocardia vaccinii* n. sp. causing galls on blueberry plants. Phytopathology **42:**249–252.

Diaz-Maurino, T., Perkins, H. R. 1974. The presence of acidic polysaccharides and muramic acid phosphate in the walls of *Corynebacterium poinsettiae* and *Corynebacterium betae.* Journal of General Microbiology **80:**533–539.

Dowson, W. J. 1942. On the generic name of the Gram-positive bacterial plant pathogens. Transactions of the British Mycological Society **25:**311–314.

Dowson, W. J. 1957. Plant diseases due to bacteria, 2nd ed. Cambridge: Cambridge University Press.

Dye, D. W., Kemp, W. J. 1977. A taxonomic study of plant pathogenic *Corynebacterium* species. New Zealand Journal of Agricultural Research **20:**563–582.

Einset, J. W., Skoog, F. K. 1977. Isolation and identification of ribosyl-*cis*-zeatin from transfer RNA of *Corynebacterium fascians.* Biochemical and Biophysical Research Communications **79:**1117–1121.

Frontali, C., Hill, L. R., Silvestri, L. G. 1965. The base composition of deoxyribonucleic acids of *Streptomyces.* Journal of General Microbiology **38:**243–250.

Galach'yan, R. M. 1969. Nature of metabolites of microorganisms producing plant tumors as root-promoting substances. [In Russian.] Voprosy Mikrobiologii **4:**129–142.

Gross, D. C., Vidaver, A. K. 1979a. A selective medium for the isolation of *Corynebacterium nebraskense* from soil and plant parts. Phytopathology **69:**82–87.

Gross, D. C., Vidaver, A. K. 1979b. Bacteriocins of phytopathogenic *Corynebacterium* species. Canadian Journal of Microbiology **25:**367–374.

Gross, D. C., Vidaver, A. K., Keralis, M. B. 1979. Indigenous plasmids from phytopathogenic *Corynebacterium* species. Journal of General Microbiology **115:**479–489.

Gupta, P., Swarup, G. 1972. Ear-cockle and yellow ear-rot diseases of wheat. II. Nematode bacterial association. Nematologica **18:**320–324.

Hale, C. N. 1972. Rapid identification methods for *Corynebacterium insidiosum* (McCulloch, 1925) Jensen, 1934. New Zealand Journal of Agricultural Research **15:**149–154.

Hayward, A. C. 1974. Latent infections by bacteria. Annual Review of Phytopathology **12:**87–97.

Hedges, F. 1922. A bacterial wilt of the bean caused by *Bacterium flaccumfaciens* nov. sp. Science **55:**433–434.

Jensen, H. L. 1934. Studies on saprophytic mycobacteria and corynebacteria. Proceedings of the Linnean Society of New South Wales **59:**19–61.

Jensen, H. L. 1952. The coryneform bacteria. Annual Review of Microbiology **6:**77–90.

Johnson, T. B., Strobel, G. A. 1970. The active site on the phytotoxin of *Corynebacterium sepedonicum.* Plant Physiology **45:**761–774.

Jones, D. 1975. A numerical taxonomy study of coryneform and related bacteria. Journal of General Microbiology **87:**52–96.

Kado, C. I., Heskett, M. G. 1970. Selective media for isolation of *Agrobacterium, Corynebacterium, Erwinia, Pseudomonas,* and *Xanthomonas.* Phytopathology **60:**969–976.

Kao, J., Damann, K. E., Jr. 1978. Microcolonies of the bacterium associated with ratoon stunting disease found in sugarcane xylem matrix. Phytopathology **68:**545–551.

Keddie, R. M. 1978. What do we mean by coryneform bacteria? pp. 1–12. In: Bousfield, I. J., Callely, A. G. (eds.), Coryneform bacteria. London: Academic Press.

Keddie, R. M., Cure, G. L. 1977. The cell wall composition and distribution of free mycolic acids in named strains of coryneform bacteria and in isolates from various natural sources. Journal of Applied Bacteriology **42:**229–252.

Kelman, A. 1954. The relationship of pathogenicity in *Pseudomonas solanacearum* to colony appearance on a tetrazolium medium. Phytopathology **44:**693–695.

Keyworth, W. G., Howell, J. S., Dowson, W. J. 1956. *Corynebacterium betae* (sp. nov.). The causal organism of silvering disease of red beet. Plant Pathology **5:**88–90.

Klement, Z., Goodman, R. N. 1967. The hypersensitive reaction to infection by bacterial plant pathogens. Annual Review of Phytopathology **5:**17–44.

Komura, I., Yamada, K., Otsuka, S., Komagata, K. 1975. Taxonomic significance of phospholipids in coryneform and nocardioform bacteria. Journal of General and Applied Microbiology **21:**251–261.

Krüger, F. 1904. Untersuchungen über den Gürtelschorf der Zuckerrüben. Arbeiten aus der Biologischen Abteilung für Land- und Forstwirtschaft am Kaiserlichen Gesundheitsamte **4:**254–318.

Kuhn, D. A., Starr, M. P. 1962. Developmental morphology of *Corynebacterium poinsettiae.* Bacteriological Proceedings **1962:**46.

Kuhn, R., Starr, M. P., Kuhn, D. A., Bauer, H., Knackmuss, H. J. 1965. Indigoidine and other bacterial pigments related to 3,3'-bipyridyl. Archiv für Mikrobiologie **51:**71–84.

Labruyère, R. E. 1971. Common scab and its control in seed-potato crops. Instituut voor Plantenziektenkundig Onderzoek, Wageningen, Mededeling 575. Wageningen: Centre for Agricultural Publishing and Documentation. [Also published as Agricultural Research Reports **767:**1–81.]

Lawrence, C. H., Clark, M. C. 1966. Characterization of deoxyribonucleic acid from *Streptomyces scabies.* Canadian Journal of Biochemistry **44:**1685–1688.

Lazar, I. 1968. Serological relationships of corynebacteria. Journal of General Microbiology **52:**77–88.

Lelliott, R. A. 1966. The plant pathogenic coryneform bacteria. Journal of Applied Bacteriology **29:**114–118.

McCulloch, L. 1925. *Aplanobacter insidiosum* n. sp., the cause of an alfalfa disease. Phytopathology **15:**496–497.

Mandel, M., Guba, E. F., Litsky, W. 1961. The causal agent of bacterial blight of American holly. Bacteriological Proceedings **1961:**61.

Masuo, E., Nakagawa, T. 1970. Numerical classification of bacteria. IV. Relationships among some corynebacteria based on serological similarity alone. Agricultural and Biological Chemistry **34:**1375–1401.

Millard, W. A., Beeley, R. 1927. Mangel scab, its cause and histogeny. Annals of Applied Biology **14:**296–311.

Millard, W. A., Burr, S. 1926. A study of twenty-four strains of *Actinomyces* and their relation to types of common scab of potato. Annals of Applied Biology **13:**580–644.

Murai, N., Skoog, F., Doyle, M. E., Hanson, R. S. 1980. Relationships between cytokinin production, presence of plasmids, and fasciation caused by strains of *Corynebacterium fascians.* Proceedings of the National Academy of Sciences of the United States of America **77:**619–623.

Norgård, S., Aasen, A. J., Liaaen-Jensen, S. 1970. Bacterial carotenoids. XXXII. C_{50}-carotenoids. 6. Carotenoids from *Corynebacterium poinsettiae* including four new C_{50} diols. Acta Chemica Scandinavica **24:**2183–2197.

Patil, S. S. 1974. Toxins produced by phytopathogenic bacteria. Annual Review of Phytopathology **14:**259–279.

Perkins, H. R. 1970. Extraction procedures and cell wall composition, including some results with corynebacteria. International Journal of Systematic Bacteriology **20:**379–382.

Person, L. H., Martin, W. J. 1940. Soil rot of sweet potatoes in Louisiana. Phytopathology **30:**913–926.

Prebble, J. 1968. The carotenoids of *Corynebacterium fascians* strain 2Y. Journal of General Microbiology **52:**15–24.

Pridham, T. G., Tresner, H. D. 1974. *Streptomyces,* pp. 748–829. In: Buchanan, R. E., Gibbons, N. E. (eds.), Bergey's manual of determinative bacteriology, 8th ed. Baltimore: Williams & Wilkins.

Rai, P. V., Strobel, G. A. 1969a. Phytotoxic glycopeptides produced by *Corynebacterium michiganense.* I. Methods of preparation, physical and chemical characterization. Phytopathology **59:**47–52.

Rai, P. V., Strobel, G. A. 1969b. Phytotoxic glycopeptides produced by *Corynebacterium michiganense*. II. Biological properties. Phytopathology **59**:53–57.

Rathbone, M. P., Hall, R. H. 1972. Concerning the presence of the cytokinin, N⁶-(Δ²-isopentenyl) adenine, in cultures of *Corynebacterium fascians*. Planta **108**:93–102.

Ries, S. M., Strobel, G. A. 1972a. Biological properties and pathological role of a phytotoxic glycopeptide from *Corynebacterium insidiosum*. Physiological Plant Pathology **2**:133–142.

Ries, S. M., Strobel, G. A. 1972b. A phytotoxic glycopeptide from cultures of *Corynebacterium insidiosum*. Plant Physiology **49**:676–684.

Rogosa, M., Cummins, C. S., Lelliott, R. A., Keddie, R. M. 1974. Coryneform group of bacteria, pp. 599–632. In: Buchanan, R. E., Gibbons, N. E. (eds.), Bergey's manual of determinative bacteriology, 8th ed. Baltimore: Williams & Wilkins.

Saaltink, G. J., Maas Geesteranus, H. P. 1969. A new disease in tulip caused by *Corynebacterium oortii* nov. sp. Netherlands Journal of Plant Pathology **75**:123–128.

Saperstein, S., Starr, M. P., Filfus, J. A. 1954. Alterations in carotenoid synthesis accompanying mutation in *Corynebacterium michiganense*. Journal of General Microbiology **10**:85–92.

Scarbrough, E., Armstrong, D. J., Skoog, F., Frihart, C. R., Leonard, N. J. 1973. Isolation of *cis*-zeatin from *Corynebacterium fascians* cultures. Proceedings of the National Academy of Sciences of the United States of America **70**:3825–3829.

Scharif, G. 1961. *Corynebacterium iranicum* sp. nov. on wheat (*Triticum vulgare* L.) in Iran, and a comparative study of it with *C. tritici* and *C. rathayi*. Entomologie et Phytopathologie Appliquées (Téheran) **19**:1–24.

Schuster, M. L. 1975. Leaf freckles and wilt of corn incited by *Corynebacterium nebraskense* Schuster, Hoff, Mandel, Lazar, 1972. Research Bulletin 270, The Agricultural Experiment Station, University of Nebraska, Lincoln, Nebraska.

Schuster, M. L., Coyne, D. P. 1974. Survival mechanisms of phytopathogenic bacteria. Annual Review of Phytopathology **12**:199–221.

Schuster, M. L., Jones, J. P., Sayre, R. M. 1959. The effects of thiamine and temperature upon the pigmentation and growth of bean wilt bacteria. Plant Disease Reporter **43**:439–443.

Schuster, M. L., Vidaver, A. K., Mandel, M. 1968. A purple-pigment-producing bean wilt bacterium, *Corynebacterium flaccumfaciens* var. *violaceum* n. var. Canadian Journal of Microbiology **14**:423–427.

Servin-Massieu, M. 1971. Effects of freeze-drying and sporulation on microbial variation. Current Topics in Microbiology and Immunology **54**:119–150.

Shirling, E. B., Gottlieb, D. 1966. Methods for characterization of *Streptomyces* species. International Journal of Systematic Bacteriology **16**:313–340.

Shirling, E. B., Gottlieb, D. 1972. Cooperative description of type strains of *Streptomyces*. V. Additional descriptions. International Journal of Systematic Bacteriology **22**:265–394.

Skaptason, J. B., Burkholder, W. H. 1942. Classification and nomenclature of the pathogen causing bacterial ring rot of potatoes. Phytopathology **32**:439–441.

Skerman, V. B. D., McGowan, V., Sneath, P. H. A. 1980. Approved lists of bacterial names. International Journal of Systematic Bacteriology **30**:255–420.

Slack, S., Kelman, A., Perry, J. 1979. Comparison of three serodiagnostic assays for the detection of *Corynebacterium sepedonicum*. Phytopathology **69**:186–189.

Smith, E. F. 1910. A new tomato disease of economic importance. Science **31**:794–796.

Snieszko, S. F., Bonde, R. 1943. Studies on the morphology, physiology, longevity, and pathogenicity of *Corynebacterium sepedonicum*. Phytopathology **33**:1032–1044.

Starr, M. P. 1949. The nutrition of phytopathogenic bacteria. III.

The Gram-positive phytopathogenic *Corynebacterium* species. Journal of Bacteriology **57**:253–258.

Starr, M. P. 1958. The blue pigment of *Corynebacterium insidiosum*. Archiv für Mikrobiologie **30**:325–334.

Starr, M. P., Mandel, M., Murata, N. 1975. The phytopathogenic coryneform bacteria in the light of DNA base composition and DNA-DNA segmental homology. Journal of General and Applied Microbiology **21**:13–26.

Starr, M. P., Pirone, P. P. 1942. *Phytomonas poinsettiae* n. sp., the cause of a bacterial disease of poinsettia. Phytopathology **32**:1076–1081.

Starr, M. P., Saperstein, S. 1953. Thiamine and the carotenoid pigments of *Corynebacterium poinsettiae*. Archives of Biochemistry and Biophysics **43**:157–168.

Steadman, J. R. 1977. Pollution of irrigation reuse water by plant pathogens. Annual Report, Nebraska Water Resources Center, Lincoln, Nebraska.

Strobel, G. A. 1967. Purification and properties of a phytotoxic polysaccharide produced by *Corynebacterium sepedonicum*. Plant Physiology **42**:1433–1441.

Strobel, G. A. 1970. A phytotoxic glycopeptide from potato plants infected with *Corynebacterium sepedonicum*. Journal of Biological Chemistry **245**:32–38.

Strobel, G. A. 1977. Bacterial phytotoxins. Annual Review of Microbiology **31**:205–224.

Strobel, G. A., Talmadge, K. W., Albersheim, P. 1972. Observations on the structure of the phytotoxic glycopeptide of *Corynebacterium sepedonicum*. Biochimica et Biophysica Acta **261**:365–374.

Stynes, B. A., Petterson, D. S. 1980. Production of a neurotoxin in a cell culture of endosperm from *Lolium multiflorum* colonized by *Corynebacterium rathayi*. Physiological Plant Pathology **16**:163–168.

Thaxter, R. 1891. The potato scab, pp. 81–95. Report of the Connecticut Agricultural Experiment Station for 1890.

Tilford, P. E. 1936. Fasciation of sweet peas caused by *Phytomonas fascians* n. sp. Journal of Agricultural Research **53**:383–394.

Veldkamp, H. 1970. Saprophytic coryneform bacteria. Annual Review of Microbiology **24**:209–240.

Vidaver, A. K. 1977. Maintenance of viability and virulence of *Corynebacterium nebraskense*. Phytopathology **67**:825–827.

Vidaver, A. K. 1980. *Corynebacterium*, pp. 12–16. In: Schaad, N. W. (ed.), Laboratory guide for identification of plant pathogenic bacteria. St. Paul: American Phytopathological Society.

Vidaver, A. K., Mandel, M. 1974. *Corynebacterium nebraskense*, a new, orange-pigmented phytopathogenic species. International Journal of Systematic Bacteriology **24**:482–485.

Waksman, S. A. 1953. Part I. The actinomycetes. In: Waksman, S. A., Lechevalier, H. A., Guide to the classification and identification of the actinomycetes and their antibiotics. Baltimore: Williams & Wilkins.

Waksman, S. A. 1959. The actinomycetes, vol. I. Nature, occurrence, and activities. Chapter 18, pp. 265–276. Causation of plant diseases. Baltimore: Williams & Wilkins.

Weaver, L., Teakle, D. S., Hayward, A. C. 1977. Ultrastructural studies on the bacterium associated with the ratoon stunting disease of sugarcane. Australian Journal of Agricultural Research **28**:843–852.

Yamada, K., Komagata, K. 1972. Taxonomic studies on coryneform bacteria. V. Classification of coryneform bacteria. Journal of General and Applied Microbiology **18**:417–432.

Yamada, Y., Inouye, G., Tahara, Y., Kondo, K. 1976. The menaquinone system in the classification of coryneform and nocardioform bacteria and related organisms. Journal of General and Applied Microbiology **22**:203–214.

Young, J. M., Dye, D. W., Bradbury, J. F., Panagopoulos, C. G., Robbs, C. F. 1978. A proposed nomenclature and classification for plant pathogenic bacteria. New Zealand Journal of Agricultural Research **21**:153–177.

CHAPTER 144

The Genus *Kurthia*

RONALD M. KEDDIE

In 1883, H. Kurth published a detailed description of a new bacterial species, *Bacterium zopfii,* which he had isolated from the intestinal contents of chickens (Kurth, 1883). Two years later, Trevisan (1885) created the genus *Kurthia* with *K. zopfii* as the type species. However, over the ensuing years Kurth's organism was given a variety of generic names, including *Zopfius* (Wenner and Rettger, 1919). In more recent times, the name *Kurthia* (the valid name) really came into general use only after the publication of the seventh edition of *Bergey's Manual of Determinative Bacteriology* (Breed, Murray, and Smith, 1957). In the eighth edition of *Bergey's Manual,* the genus *Kurthia* is included "tentatively" in the "Coryneform Group of Bacteria" (Rogosa et al., 1974) although its members do not have a coryneform morphology (see, also, this Handbook, Chapter 142).

Kurth isolated the organism now called *K. zopfii* by streaking material from the intestinal contents of chickens onto nutrient gelatin plates. When the plates were incubated, the organism grew out from the original streak and through the gelatin in long, fine, apparently branched threads. For a number of years following the original isolation, *K. zopfii* aroused considerable interest because of its characteristic growth patterns in nutrient gelatin, and various attempts were made to explain the observed phenomena. Most characteristic is the appearance in a gelatin slant. If the slant is inoculated with a single central streak and incubated in the near vertical position, then the resultant growth resembles a bird's feather (Boyce and Evans, 1893; Jacobsen, 1907; Kufferath, 1911; Sergent, 1906, 1907; Zikes, 1903). The outgrowths, which appear to follow the lines of stress in the gelatin, are presumably a result of the organism's marked filament-forming ability coupled with an inability to hydrolyze gelatin; both motile and nonmotile strains exhibit the phenomenon (Keddie, 1949).

In the few decades following its original isolation, *K. zopfii* (or *K. zenkeri,* a synonym of *K. zopfii*) was isolated from a variety of sources, including fresh and putrefied meat (Günther, 1896; Jacobsen, 1907; Wenner and Rettger, 1919), waste water and air from abattoirs (Jacobsen, 1907), pre-served sausage (Günther, 1896), and pus from a cat's ear (Boyce and Evans, 1893). It was also reported to occur in feces (Flügge, 1896), water (Flügge, 1896; Migula, 1900), and to be common in milk (Orla-Jensen, 1931). One strain of an organism considered to be *K. zopfii* was isolated from air at an altitude exceeding 10,000 ft (Proctor, 1935)!

Following this early interest in *K. zopfii,* the genus *Kurthia* received scant attention for many years. But the more recent rediscovery that *K. zopfii* frequently occurs as a component of the aerobic flora of meats and meat products has once more focused attention on the genus. However, bacteria of the genus *Kurthia* do not appear to have been implicated directly in the spoilage process of meats and meat products.

There are also several scattered reports that organisms identified as *Kurthia* spp. (not *K. zopfii*) have been isolated from a variety of clinical materials; but there is no evidence that they were directly implicated in any disease.

Habitats

Few systematic studies of the occurrence of *Kurthia* spp. in natural materials have been made, largely because of the lack of sufficiently selective isolation methods. From the limited information available, it seems that the sources from which authentic *Kurthia* spp., or more specifically, *K. zopfii,* are regularly isolated are meat and meat products, and animal feces that have been in contact with soil, straw, etc., for a short time. There are also several reports of the isolation of *Kurthia* spp. (not *K. zopfii*) from clinical materials and from other diverse sources; but, in at least some of these cases, the identification of these organisms as *Kurthia* spp. is open to considerable doubt.

Meat and Meat Products

Keddie (1949) isolated *Kurthia zopfii* from a variety of samples of fresh meat, fat, etc., and from meat that was allowed to putrefy at room temperature.

The organism was also readily isolated from meat, fat, waste water, etc., from an abattoir, thus confirming the early report of Jacobsen (1907). Such observations suggest that the immediate source of contamination of meat with *K. zopfii* is in the abattoir.

Ingram (1952) found that Gram-positive, non-sporeforming rods constituted about 10% of the aerobic flora of internally tainted, cured pork legs, and some of these rods were considered to be *Kurthia* spp. *K. zopfii* has been isolated from prepacked pork (Gardner, Carson, and Patton, 1967), various comminuted fresh meat products (Gardner, 1969), from an eviscerated, polythene-wrapped chicken stored at 15°C until off-flavors developed, from an irradiated lamb carcass stored for about 7 weeks at 1°C (quoted by Gardner, 1969), and from spoiled British sausages (Dowdell and Board, 1971).

Information on the numbers of *K. zopfii* in these various meat products is sparse. Gardner (1969) noted that, although the species was regularly isolated from fresh, comminuted meat products, it usually accounted for only up to 10% of the total aerobic count. On the other hand, in six samples of pork stored at 16°C for 5 days in gas-impermeable film, gas-permeable film, or with no film, *K. zopfii* represented 12–44%, 9–76%, and 5–69%, respectively, of the total aerobic flora; but it was not detected in similar samples stored at 2°C for 14 days (Gardner, Carson, and Patton, 1967). From these and other observations (Gardner and Carson, 1967), Gardner (1969) concluded that *K. zopfii* could compete favorably with other aerobic spoilage bacteria in meat stored at about 16°C, but not in meat stored at refrigeration temperatures. The two strains isolated by D'Aubert, Cantoni, and Calcinardi (1975) from refrigerated, vacuum-packed meat and identified by them as *Kurthia* spp. did not have the characteristics of the genus.

Animal Feces in Contact with Soil, Straw, etc.

Although there seem to be no published reports of isolations from such sources, in our experience *Kurthia zopfii* can regularly be isolated from the feces of certain domestic animals provided that it has been in contact with soil, straw, etc., for a short time. Keddie (1949) was unable to isolate *K. zopfii* from the cecal contents of chickens or from the freshly voided feces, but three of nine samples of chicken feces that had lain on the soil for some time yielded the organism. More recently, *K. zopfii* has been isolated from many different samples of soil-contaminated chicken feces and from various situations in poultry houses (C. S. Parsons, S. Shaw, and R. M. Keddie, unpublished). Strivastava, Singh, and Singh (1972) have also reported the isolation of two presumptive *Kurthia* strains from a poultry

house and a hatchery. Also, pig feces contaminated with straw or soil have commonly yielded *K. zopfii*, but the organism has been isolated infrequently from similarly contaminated feces of herbivores, including horses, cows, and sheep (Parsons, Shaw, and Keddie, unpublished).

Clinical Sources

Several strains of presumptive *Kurthia* spp., but not *K. zopfii*, have been isolated from various clinical sources, and most frequently from the feces of patients suffering from diarrhea. Some organisms were considered to be *K. bessonii* or *K. variabilis*, species that were considered incertae sedis in the eighth edition of *Bergey's Manual* (Keddie and Rogosa, 1974). Frequently, the descriptions of the isolates mentioned contain apparently contradictory statements (e.g., facultative anaerobes; do not ferment sugars) or are too limited to be able to judge whether or not the organisms were indeed legitimate kurthias.

Severi (1946) isolated what he considered to be a new species, *K. variabilis,* from feces in a case of mild food poisoning. Elston (1961) isolated three strains identified as *K. bessonii* from a pilonidal cyst, sputum, and a diarrheal stool, while Faoagali (1974) isolated a presumptive *Kurthia* sp. from a routine eye swab. Still more recently, Jarumilinta, Miranda, Villarejos (1976) isolated presumptive *Kurthia* spp. from the upper intestinal tract of 6 of 25 patients suffering from acute diarrhea, but only from 1 of 24 control patients. However, in all of these cases, the connection between the occurrence of the presumptive *Kurthia* spp. and the clinical condition was, at most, tenuous.

Other Sources

The isolation of presumptive *Kurthia* spp. has been reported from sources as diverse as "sloughing spoilage" of ripe olives (*K. bessonii;* Patel and Vaughn, 1973), the gut of the crab (*K. variabilis;* Saha and Raychaudhuri, 1973), wet-stored wood (Berndt and Liese, 1973), and from dental plaque of beagle dogs (*Kurthia?;* Wunder, Briner, and Calkins, 1976); but like those from clinical sources, the accuracy of identification of at least some of these isolates is open to doubt.

However, recent evidence supports the early statement by Orla-Jensen (1931) that *K. zopfii* may occur in milk. This species was found at levels of about 10^4 per ml in some samples of bulked, cold-stored (7°C) raw milks (C. M. Cousins, personal communication). *K. zopfii* has occasionally been isolated from soil and surface waters (Parsons, Shaw, and Keddie, unpublished) and a few observations indicate that organisms which resemble *Kurthia* but are distinct from *K. zopfii* occur in certain peats namely, low moor peat (Janota-Bassalik, 1963) and Antarctic peat (Baker and Smith, 1972).

Isolation

Only the type species, *Kurthia zopfii,* is recognized at the present time (Keddie and Rogosa, 1974). However, a number of strains have been described (David, 1966; Gardner, 1969; Keddie, 1949) that closely resemble the type species but differ in a few features, more especially in having a higher maximum growth temperature. Whereas *K. zopfii* has its maximum temperature at 37–40°C, the strains mentioned are capable of growth at 45°C (Keddie, 1949). These strains may represent a distinct taxon worthy of separate specific status; therefore, to distinguish them from *K. zopfii,* they will be referred to as *Kurthia* sp. (45C+).

No selective media have been developed for the isolation of *K. zopfii* but both *K. zopfii* and *Kurthia* sp. (45C+) may be isolated by using methods that exploit their unusual cultural properties. One of the most successful is a gelatin streak method similar to that used by Kurth and the early investigators (Keddie, 1949). However, a simple agar streak technique is a useful additional method (Parsons, Shaw, and Keddie, unpublished).

Gelatin Streak Method for Isolating *Kurthia* (Keddie, 1949)

A nutrient gelatin medium (YNG) of the following composition is used. YNG medium contains (per liter of distilled water):

Meat extract (Lab-Lemco	
powder, Oxoid)	4 g
Peptone (Difco)	5 g
Yeast extract (Difco)	2.5 g
NaCl	5 g
Gelatin (BDH)	100 g
pH 7.0	

For quantities up to 100 ml, sterilize at 115°C for 30 min. Not all brands of gelatin allow good outgrowths of *K. zopfii.* In our experience, the gelatin powder manufactured by the BDH Chemical Co., Poole, England, is satisfactory. However, the brand used should be tested with a reference strain of *K. zopfii* (NCIB 9878) to check its suitability for the present purpose.

Pour plates with ca. 20 ml of molten YNG medium and allow to solidify in the refrigerator. Inoculate plates heavily with a single central streak of the material to be examined (or with a suspension in a small amount of sterile water). Incubate plates at 20°C with the lids uppermost and examine daily. The gelatin is usually soon liquefied around the streak. In successful cultures, filamentous outgrowths appear beyond this

zone in 2–3 days and, after 3–4 days, if the medium is not completely liquefied, a tangled mass of filaments may completely permeate the solid part of the medium. To obtain a pure culture, remove a small piece of gelatin containing outgrowths and streak out on yeast extract–nutrient agar medium (YNA). Alternatively, the pour-plate method may be used. YNA is similar to YNG but is solidified with agar instead of gelatin.

Other bacteria that may produce outgrowths in gelatin similar to those of *Kurthia* are certain chain-forming *Bacillus* species, particularly *B. cereus* subsp. *mycoides,* although they may liquefy gelatin. With some materials, overgrowth with fungi may occur; it may be prevented by adding nystatin to the molten YNG to a final concentration of 10 U per ml before pouring plates.

Kurthia sp. (45C+) may be obtained by preparing a preliminary enrichment in YNB (YNG but without gelatin) incubated for 24 h at 45°C; a second subculture at 45°C is then made before inoculating gelatin streak plates as described above. *Bacillus* spp. do not usually interfere in this modification of the method.

The agar streak method may be used in addition to that described above and may give successful results when rapid liquefaction of gelatin has prevented isolation of *K. zopfii* by the gelatin streak method.

Agar Streak Method for Isolating *Kurthia* (C. S. Parsons, S. Shaw, and R. M. Keddie, unpublished)

YNA medium consists of:

Meat extract (Lab-Lemco	
powder, Oxoid)	4 g
Peptone (Difco)	5 g
Yeast extract (Difco)	2.5 g
NaCl	5 g
Agar	15 g
pH 7.0	

YNA medium is inoculated with a single, central streak as described above. At daily intervals, the edge of the streak is examined at the low power (100×) of the microscope for the characteristic skeinlike outgrowths of *K. zopfii* grown on agar. Pure cultures are obtained by picking carefully from the edge of the outgrowths and plating on YNA as before.

Isolation by Direct Plating (Gardner, 1969)

Kurthia zopfii and *Kurthia* sp. (45C+) may be isolated by direct plating on YNA (or similar media) of material such as meat and meat products, if it forms a sufficiently high proportion of the population (Gardner, 1969).

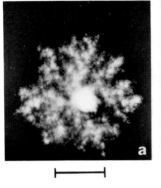
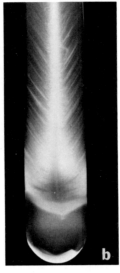

like structure which is resolved into whorls with whiplike outgrowths at the edge.

It should be noted that when pure cultures are streaked out on YNA, a proportion of the colonies that develop may have a granular appearance instead of the typical rhizoid form (Keddie, 1949). Therefore, it is possible that, when isolating *K. zopfii* from natural materials, some may be missed because they produce nonrhizoid colonies.

PRESERVATION OF CULTURES. Cultures may be preserved by freeze-drying (lyophilization). Slope cultures on YNA or nutrient agar should remain viable for at least 6 months when stored at room temperature (ca. 20°C), provided that they are not allowed to dry out.

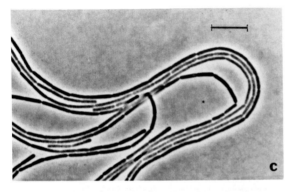
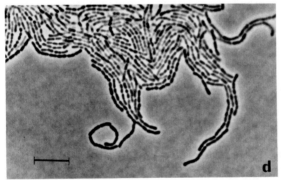

Fig. 1 (a–b) *Kurthia zopfii* (NCIB 9878). (a) Rhizoid colony on yeast nutrient agar after 4 days incubation at 25°C; bar = 10 mm. (b) Yeast nutrient gelatin slant showing "bird's feather" type of growth; incubated 5 days at 20°C. (c-d) *Kurthia zopfii* (isolate): edge of colony on yeast nutrient agar incubated at 25°C. (c) After 24 h, showing long filaments composed of rods. (d) After 3 days, showing development of coccoid forms. Bar = 10 μm.

Surface colonies of *K. zopfii* and *Kurthia* sp. (45C+) are recognized by their rhizoid form (Fig. 1a) and by the typical "medusa-head" appearance of young colonies when examined at low magnifications. Such colonies have a skein-

Identification

Identification as a *Kurthia* sp., either *K. zopfii* or *Kurthia* sp. (45C+) may be made on the basis of the following features. Gram-positive, regular rods ca. $0.8 \times 3–8$ μm or longer, in long chains in exponential phase cultures (Fig. 1c) giving rise to coccoids (Fig. 1d) or, in some strains, to short rods in stationary-phase cultures. The rods are usually motile by numerous peritrichous flagella, but nonmotile strains are known; they do not form endospores. Surface colonies on yeast nutrient agar are usually rhizoid (Fig. 1a) (but granular colonial variants occur) and have a "medusa-head" appearance under low magnification ($\times 100$). They are obligate aerobes and do not form acid from glucose or other carbohydrates in peptone media. In gelatin slant cultures (see medium YNG above), inoculated with a single central streak, the growth resembles a bird's feather (Fig. 1b); the gelatin is not liquefied. They are catalase positive and grow best in the range of 20–30°C. *Kurthia* spp. give negative responses in most of the usual biochemical tests, e.g., indole, nitrate reduction; urease, lecithinase production; hydrolysis of starch and esculin; however, some strains produce H_2S weakly (Jones, 1975; Shaw and Keddie, unpublished). The nutritional requirements are complex. In *K. zopfii* (two strains examined), the cell wall peptidoglycan contains L-lysine as diamino acid in position 3 of the peptide subunit: Cross-linkage between positions 3 and 4 of two peptide subunits is by an interpeptide bridge comprising one D-aspartic acid residue (Schleifer and Kandler, 1972). The major isoprenoid quinones are menaquinones with seven isoprene units (MK-7) (Minnikin, Goodfellow, and Collins, 1978). The G+C content of the DNA is remarkably constant among isolates from different sources and lies in the range of 36.7–37.9 mol% for 34 strains of *K. zopfii* and

Kurthia sp. (45C+) examined (Shaw and Keddie, unpublished).

Differentiation of *Kurthia* spp.

Kurthia zopfii does not grow at 45°C or survive 55°C in skim milk for 20 min, whereas the strains referred to above as *Kurthia* sp. (45C+) give positive responses in both tests (Keddie, 1949). Gardner (1969) also noted that *K. zopfii* gave acid from ethanol but not from glycerol in weakly buffered, complex media, while *Kurthia* sp. (45C+) gave acid from glycerol but not from ethanol. In our experience, there is good but not complete concordance between the latter tests and the temperature relationships of the two species (Shaw and Keddie, unpublished). The colonies of *K. zopfii* show no distinctive pigmentation, whereas those of *Kurthia* sp. (45C+) have a cream or yellow pigmentation (Gardner, 1969).

Those taxa most likely to be confused with *Kurthia* include *Brochothrix thermosphacta* (*Microbacterium thermosphactum;* see this Handbook, Chapter 142) and certain *Bacillus* spp. Confusion with some aerobic, saprophytic coryneform bacteria or even *Caryophanon* is possible but much less likely. *Brochothrix thermosphacta* (Sneath and Jones, 1976) is very similar to *Kurthia* in morphological features (see Davidson, Mobbs, and Stubbs, 1968) but is nonmotile, facultatively anaerobic, and produces acid by fermentation from glucose and various other carbohydrates. On nutrient agar, certain chain-forming *Bacillus* spp. may give colonies somewhat similar to *K. zopfii* and have a similar morphology in exponential-phase cultures. They may also give outgrowths in nutrient gelatin, although such growth is usually, if not always, followed by liquefaction. They may be distinguished by endospore formation (detected by a test of heat resistance following growth on nutrient agar supplemented with Mn^{2+}, 2 μg/ml, for at least 7 days. The species most readily confused with *Kurthia*, e.g., *B. cereus* subsp. *mycoides,* are facultatively anaerobic, produce acid from glucose, and liquefy gelatin. Many aerobic coryneform bacteria (e.g., *Arthrobacter* spp.), like fresh *Kurthia* isolates, give coccoid cells in stationary-phase cultures, but the appearance in exponential phase cultures is quite distinct. In coryneform bacteria, the rods are irregular in form, may show rudimentary branching, and commonly occur in V–formations but never in chains. Although from the description in *Bergey's Manual* (Gibson, 1974) the genus *Caryophanon* may appear to bear a superficial resemblance to *Kurthia,* it is readily distinguished by its large size (ca. 3 μm in diameter in fresh isolates) and by the numerous closely spaced cross-walls seen in the bright-field microscope. *Caryophanon* gives small

colonies on YNA and little or no growth in YNB.

Reference strains: NCTC 404 (ATCC 10538) has been suggested as a reference strain of *Kurthia zopfii* (Sneath and Skerman, 1966), but NCIB 9878 (listed as *Kurthia* sp.) is much more typical of new isolates. A suitable reference strain of *Kurthia* sp. (45C+) is NCIB 9757.

Literature Cited

Baker, J. H., Smith, D. G. 1972. The bacteria in an Antarctic peat. Journal of Applied Bacteriology 35:589–596.

Berndt, H., Liese, W. 1973. Untersuchungen über das Vorkommen von Bakterien in wasserberieselten Buchenholzstämmen. Zentralblatt für Bakteriologie, Parasitenkunde, Infektionskrankheiten und Hygiene, Abt. 2 128:578–594.

Boyce, R., Evans, A. E. 1893. The action of gravity upon *Bacterium zopfii*. Proceedings of the Royal Society of London 54:300–312.

Breed, R. S., Murray, E. G. D., Smith, N. R. (eds.). 1957. Bergey's manual of determinative bacteriology, 7th ed. Baltimore: Williams & Wilkins.

D'Aubert, S., Cantoni, C., Calcinardi, C. 1975. I corinebatteri nelle carni confezionate sottovuoto. Archivio Veterinario Italiano 26:65–70.

David, J. E. 1966. A comparison of *Microbacterium thermosphactum* and *Kurthia zopfii*. Project Report No. 45. England: Bath University of Technology.

Davidson, C. M., Mobbs, P., Stubbs, J. M. 1968. Some morphological and physiological properties of *Microbacterium thermosphactum*. Journal of Applied Bacteriology 31:551–559.

Dowdell, M. J., Board, R. G. 1971. The microbial associations in British fresh sausages. Journal of Applied Bacteriology 34:317–337.

Elston, H. R. 1961. *Kurthia bessonii* isolated from clinical material. Journal of Pathology and Bacteriology 81:245–247.

Faoagali, J. L. 1974. *Kurthia,* an unusual isolate. American Journal of Clinical Pathology 62:604–606.

Flügge, C. 1896. Die Microorganismen, pp. 277–278. vol 2. Leipzig: F. C. W. Vogel.

Gardner, G. A. 1969. Physiological and morphological characteristics of *Kurthia zopfii* isolated from meat products. Journal of Applied Bacteriology 32:371–380.

Gardner, G. A., Carson, A. W. 1967. Relationship between carbon dioxide production and growth of pure strains of bacteria on porcine muscle. Journal of Applied Bacteriology 30:500–510.

Gardner, G. A., Carson, A. W., Patton, J. 1967. Bacteriology of prepacked pork with reference to the gas composition within the pack. Journal of Applied Bacteriology 30:321–333.

Gibson, T. 1974. *Caryophanon,* p. 598. In: Buchanan, R. E., Gibbons, N. E. (eds.), Bergey's manual of determinative bacteriology, 8th ed. Baltimore: Williams & Wilkins.

Günther, C. 1896. Bakteriologische Untersuchungen in einem Falle von Fleischvergiftung. Archiv für Hygiene 28:153–158.

Ingram, M. 1952. Internal bacterial taints ('bone taint' or 'souring') of cured pork legs. Journal of Hygiene 50:165–181.

Jacobsen, H. C. 1907. Ueber einen richtenden Einfluss beim Wachstum gewisser Bakterien in Gelatine. Zentralblatt für Bakteriologie, Parasitenkunde, Infektionskrankheiten und Hygiene, Abt. 2 17:53–64.

Janota-Bassalik, L. 1963. Psychrophiles in low-moor peat. Acta Microbiologica Polonica 12:25–40.

Jarumilinta, R., Miranda, M., Villarejos, V. M. 1976. A bacteriological study of the intestinal mucosa and luminal fluid of adults with acute diarrhoea. Annals of Tropical Medicine and Parasitology **70**:165–179.

Jones, D. 1975. A numerical taxonomic study of coryneform and related bacteria. Journal of General Microbiology **87**:52–96.

Keddie, R. M. 1949. A study of *Bacterium zopfii* Kurth. Dissertation. Edinburgh School of Agriculture, Edinburgh, Scotland.

Keddie, R. M., Rogosa, M. 1974. *Kurthia*, pp. 631–632. In: Buchanan, R. E., Gibbons, N. E. (eds.), Bergey's manual of determinative bacteriology, 8th ed. Baltimore: Williams & Wilkins.

Kufferath, H. 1911. Note sur les tropismes du *Bacterium zopfii* "Kurth" Annales de l'Institut Pasteur **25**:601–617.

Kurth, H. 1883. *Bacterium Zopfii*. Ein Beitrag zur Kenntniss der Morphologie und Physiologie der Spaltpilze. Botanische Zeitung **41**:369–386, 393–405, 409–420, 425–435.

Migula, W. 1900. System der Bacterien, pp. 815–816 vol. 2. Jena: G. Fischer.

Minnikin, D. E., Goodfellow, M., Collins, M. D. 1978. Lipid composition in the classification and identification of coryneform and related taxa, pp. 85–160. In: Bousfield, I. J., Callely, A. G. (eds.), Special Publications of the Society for General Microbiology. I. Coryneform bacteria. London: Academic Press.

Orla-Jensen, S. 1931. Dairy bacteriology, p. 51. 2nd English ed. London: J. & A. Churchill.

Patel, I. B., Vaughn, R. H. 1973. Cellulolytic bacteria associated with sloughing spoilage of California ripe olives. Applied Microbiology **25**:62–69.

Proctor, B. E. 1935. The microbiology of the upper air II. Journal of Bacteriology **31**:363–375.

Rogosa, M., Cummins, C. S., Lelliott, R. A., Keddie, R. M. 1974. Coryneform group of bacteria, pp. 599–632. In: Buchanan, R. E., Gibbons, N. E. (eds.), Bergey's manual of determinative bacteriology, 8th ed. Baltimore: Williams & Wilkins.

Saha, N., Raychaudhuri, D. N. 1973. A note of the bacterial flora in the gut of the crab, *Seylla serrata* (Förskal) (Crustacea: Decapoda). Science and Culture **39**:361–363.

Schleifer, K. H., Kandler, O. 1972. Peptidoglycan types of bacterial cell walls and their taxonomic implication. Bacteriological Reviews **36**:407–477.

Sergent, E. 1906. Des tropismes du *Bacterium zopfii* Kurth. Première note. Annales de l'Institut Pasteur **20**:1005–1017.

Sergent, E. 1907. Des tropismes du *Bacterium zopfii* Kurth. Deuxième note. Annales de l'Institut Pasteur **21**:842–856.

Severi, R. 1946. L'azione patogena delle Kurthie e la loro sistematica. Una nuova specie: *"Kurthia variabilis"*. Giornale di Batteriologia e Immunologia **34**:107–114.

Sneath, P. H. A., Jones, D. 1976. *Brochothrix*, a new genus tentatively placed in the family *Lactobacillaceae*. International Journal of Systematic Bacteriology **26**:102–104.

Sneath, P. H. A., Skerman, V. B. D. 1966. A list of type and reference strains of bacteria. International Journal of Systematic Bacteriology **16**:1–113.

Strivastava, S. K., Singh, V. B., Singh, N. P. 1972. Bacterial flora of poultry environment. Indian Journal of Microbiology **12**:7–9.

Trevisan, V. 1885. Caretteri di alcuni nuovi generi di Batteriacee. Atti della Accademia Fisio-Medico-Statistica in Milano, Series 4 **3**:92–107.

Wenner, J. J., Rettger, L. F. 1919. A systematic study of the Proteus group of bacteria. Journal of Bacteriology **4**:331–353.

Wunder, J. A., Briner, W. W., Calkins, G. P. 1976. Identification of the cultivable bacteria in dental plaque from the Beagle dog. Journal of Dental Research **55**:1097–1102.

Zikes, H. 1903. Die Wachstumserscheinungen von *Bacterium zopfii* auf Peptongelatine. Zentralblatt für Bakteriologie, Parasitenkunde, Infektionskrankheiten und Hygiene Abt. 2 **11**:59–61.

CHAPTER 145

The Genus *Propionibacterium*

CECIL S. CUMMINS and JOHN L. JOHNSON

The first systematic investigations into the organisms responsible for the formation of "eyes" in cheese were made by von Freudenreich and Orla-Jensen in 1906, although the earlier work of Fitz (1878, 1879) had shown that organisms from cheese would ferment lactate to propionic and acetic acids and liberate carbon dioxide in the process. The name *Propionibacterium* was suggested in 1909 by Orla-Jensen for these organisms because they were characterized by the production of large amounts of propionic acid during growth.

The main economic interest in these organisms arises from their connection with the cheese industry. However, many of them produce commercially valuable amounts of vitamin B_{12} under suitable conditions, and they have also been used for the commercial production of propionic acid.

The strains originally classified in the genus all came from cheese or dairy products. More recently it has been shown that the anaerobic coryneform organisms that form a major part of the skin flora of man have a number of points of similarity with the strains of dairy origin—so much so that it seems justified to regard them also as propionibacteria. In this article we will use the term "classical propionibacteria" in referring to the organisms from dairy products, and "anaerobic coryneforms" or "cutaneous propionibacteria" in referring collectively to the organisms from skin. The latter have until recently generally been classified in the genus *Corynebacterium*, the specific name *C. acnes* being frequently used for any anaerobic coryneform organisms isolated from skin swabs.

The organisms in both groups are Gram-positive, rod-shaped bacteria that are nonsporing, nonmotile, and anaerobic. The cutaneous propionibacteria, however, are generally slender and often quite irregular and curved, while the classical propionibacteria are usually rather short and thick. Despite being predominantly anaerobic, propionibacteria are generally catalase positive, and in many cases strongly so.

Classical Propionibacteria

Habitat

The classical propionibacteria have been traditionally isolated from dairy products, especially cheese, and it appears that no systematic search has been made for them in other habitats. However, van Niel (1928) reported isolating strains from soil, and Breed, Murray, and Smith, in the seventh edition of *Bergey's Manual of Determinative Bacteriology* (1957), reported that *P. zeae* was isolated from silage and *P. peterssonii* from soil, but gave no further details. Prévot (Prévot and Fredette, 1966) says that *P. pentosaceum* has been isolated from soil, but gives no reference.

General Properties

METABOLISM AND NUTRITIONAL REQUIREMENTS. The production of large amounts of propionic acid is characteristic of the organisms. Hexoses are converted to pyruvate by the Embden-Meyerhof pathway, and propionate and acetate are formed by the reactions shown in Fig. 1. This diagram of propionic acid fermentation is slightly modified from that given by Allen et al. (1964), and is based on the extensive work of H. G. Wood and his collaborators. The background to propionic acid fermentations is also critically discussed in the review by Hettinga and Reinbold (1972). The fact that propionic and acetic acids are the main products of hexose fermentation can be readily shown by gas chromatography of culture supernatants (for methods, see Holdeman, Cato, and Moore, 1977). The ratio of propionic to acetic acid is generally about 2:1, but may vary widely and be as high as 5:1 or more.

All strains require the vitamins pantothenate and biotin (Delwiche, 1949); some need thiamine and *p*-aminobenzoic acid as well. A number of other unknown factors in potato or yeast extract are stimulatory. It appears from the investigations of Wood, Andersen, and Werkman (1938) that many strains of propionibacteria will grow in a basal medium without the addition of amino acids. However, growth is much improved when amino acids are added: A digest of casein (e.g., Casamino Acids [Difco]) will supply the requirements of all strains.

ANTIGENIC COMPOSITION.   The antigens of this group have not been studied systematically by modern techniques. Such studies as have been made, for example by Werkman and Brown (1933) using agglutination of cell suspensions, showed a considerable degree of cross-reaction with unabsorbed sera.

Allsop and Work (1963) showed that the cell walls of *P. peterssonii* and *P. rubrum* contain polysaccharides and probably also teichoic acids, and a detailed examination of these might form the basis for a satisfactory antigenic scheme, as has been the case in other groups such as *Streptococcus* or *Lactobacillus*.

BACTERIOPHAGES.   The bacteriophages of classical propionibacteria do not appear to have been investigated, although Hettinga and Reinbold (1972c) reported failure to isolate bacteriophage from their strains. They ascribed their failure to interference by slime layers.

CAPSULES AND SLIME LAYERS.   Hettinga and Reinbold (1972c), quoting unpublished work by L. O. Skogen (M. S. Thesis, Iowa State University, Ames, Iowa, 1970), reported that a strain of *P. zeae* produced slime and capsular material composed of glucose and galactose, but most descriptions of the growth of propionibacteria do not mention capsule formation.

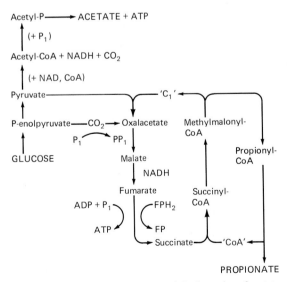

Fig. 1. Propionic acid fermentation and the formation of acetate, CO_2, propionate, and ATP. FP is flavoprotein and FPH_2 is reduced flavoprotein.

LIPID CONTENT.   The fatty acid pattern of the lipids of the classical propionibacteria is distinguished by a high content of branched-chain compounds, principally *iso*- and *anteiso*-C_{15} acids (Moss et al., 1969). Mannose-containing phospholipids have been reported in *P. shermanii* (Brennan and Ballou, 1968; Prottey and Ballou, 1968).

SENSITIVITY TO ANTIMICROBIAL AGENTS.   No very consistent or unusual pattern has been reported except that all strains are highly resistant to sulfonamides and appear to be more resistant to semisynthetic penicillins such as oxacillin than to penicillin G (Reddy, Reinbold, and Williams, 1973). When disk sensitivity tests are used, some strains will grow in the presence of 1,000 μg/ml sulfadiazine. Reddy, Williams, and Reinbold (1973) have shown that strains of *P. shermanii*, *P. freudenreichii*, *P. thoenii*, and *P. arabinosum* can synthesize folic acid, while strains of *P. pentosaceum*, *P. jensenii*, *P. rubrum*, and *P. peterssonii* cannot. However, the latter strains are still resistant to sulfonamides, and Reddy, Williams, and Reinbold concluded that this resistance may be due to the failure of the drugs to enter the cell.

Among other antimicrobial substances, nisin (from streptococci) has been shown to have an inhibitory effect on the growth of propionibacteria in Emmental cheese (Galesloot, 1957; Winkler and Fröhlich, 1957).

Isolation and Maintenance

Most investigators from van Niel (1928) onward have relied primarily on yeast extract–sodium lactate media, with or without the addition of peptone. A typical formula is that of Malik, Reinbold, and Vedamuthu (1968).

Yeast Extract–Sodium Lactate (YEL) Medium for Isolation and Maintenance of Propionibacteria (Malik, Reinbold, and Vedamuthu, 1968)

Trypticase (BBL)	1%
Yeast extract (Difco)	1%
Sodium lactate	1%
KH_2PO_4	0.25%
$MnSO_4$	0.0005%
Agar (Difco)	1.5%
In distilled water	

The final pH was adjusted to 7.0. Hettinga, Vedamuthu, and Reinbold (1968) have devised a method whereby 2% lactate agar is placed in a pouch made of a plastic film of low gaseous diffusibility which will maintain sufficiently anaerobic conditions.

Media of this type have been used both for isolation and for the maintenance of stock cultures. Sufficiently anaerobic conditions were maintained by agar overlay (Malik, Reinbold, and Vedamuthu, 1968), by the addition of 0.5% sodium sulfite coupled with an overlay of paraffin oil (Demeter and Janoschek, 1941), or by growth in candle-oats jars (Vedamuthu and Reinbold, 1967). Probably the easiest method of isolation would be to supplement the medium of Malik, Reinbold, and Vedamuthu (1968) with 0.05% cysteine and 0.05% Tween 80, and incubate the plates in a Brewer-type anaerobe jar containing 10–20% CO_2. Chopped meat medium in stoppered tubes under CO_2 (Holdeman, Cato, and Moore, 1977) is excellent for preserving stock cultures. They remain viable for many months at room temperature. At refrigerator temperatures (e.g., 4°C), cultures may die out rather rapidly. For stock cultures it is better to omit glucose from the medium. Larger cultures for biochemical or cellular analysis may conveniently be grown in Erlenmeyer flasks by the method described by Cummins and Johnson (1971). Each flask was fitted with a rubber stopper having right-angled inlet and outlet tubes. The inlet tube protruded to approximately 2 inches from the surface of the medium, and to its outer end was attached by rubber tubing another short section of glass tubing filled with cotton. The outlet tube started flush with the bottom of the stopper, and its outer end was capped with a Bunsen valve. The purpose of these arrangements was to enable gases passed in through the inlet tube to flush air from the neck of the flask. The stopper was held in place with a clamp consisting of a rubber-covered metal collar that extended around the neck of the flask. At opposite sides of the collar were attached two eye-bolts which held a horizontal bar across the top of the stopper. The medium was sterilized with free passage of air and steam through the cotton filter. Immediately after sterilization, the inlet tube was attached to a N_2 cylinder and the flask was flushed with nitrogen. The outlet tube was then clamped off and the medium allowed to cool under 1–2 psi N_2 pressure. The inlet tube was then clamped off and disconnected from the cylinder. The medium could be stored for several weeks. Nitrogen was again flushed through the flask when $NaHCO_3$ (10 ml of a 10% wt/vol solution per liter of medium) and inoculum were added.

The cultures were incubated at temperatures ranging from 30 to 35°C, depending on the optimum for the group. After initiation of growth, the cultures were mixed with a Teflon-coated magnetic stirring bar or by gentle shaking.

A medium which supports good growth of all of the propionibacteria is as follows:

Trypticase–Yeast Extract–Glucose Medium for Growth of Propionibacteria (Johnson and Cummins, 1972)

Trypticase (BBL)	1%
Yeast extract (Difco)	0.5%
Glucose	1%
$CaCl_2$	0.002%
$MgSO_4$	0.002%
NaCl	0.002%
Potassium phosphate buffer (equal molar mono- and dibasic)	0.05 M
Tween 80	0.05%
Sodium formaldehyde sulfoxalate (Eastman Organic Chemicals)	0.05%
$NaHCO_3$ (added as a sterile solution at the time of inoculation)	0.1%
Final pH 7.0	

Completely synthetic media for the growth of propionibacteria have been devised by Kurmann (1960) and Reddy, Williams, and Reinbold (1973).

Classification and Identification

A number of investigators (Shaw and Sherman, 1923; Sherman, 1921; Sherman and Shaw, 1923; Thöni and Allemann, 1910; Troili-Petersson, 1904; von Freudenreich and Orla-Jensen, 1906) had described the isolation of different kinds of propionic acid bacteria from cheese, but van Niel (1928) was the first to deal with the classification of these organisms in any systematic way. He reviewed the earlier work, described his own investigations, and recognized eight species, *Propionibacterium freudenreichii*, *P. jensenii*, *P. petterssonii*, *P. shermanii*, *P. pentosaceum*, *P. rubrum*, *P. thoenii*, and *P. technicum*. He also recognized a variety of *P. jensenii*, *P. jensenii* var. *raffinosaceum*. Since this classic paper of van Niel's, Werkman and Kendall (1931) have redesignated the variety of *P. jensenii* as a species, *P. raffinosaceum*, and Hitchner (1932) named two more species, *P. zeae* and *P. arabinosum*. Sakaguchi, Iwaski, and Yamado (1941) have proposed five additional species: *P. globosum*, *P. amylaceum*, *P. japonicum*, *P. orientum*, and *P. coloratum*, with one variety of *P. amylaceum*, *P. amylaceum* var. *auranticum*. In addition, Janoschek (1944) named three more species, *P. casei*, *P. pituitosum*, and *P. sanguineum*. The specific name *acidi-propionici* was used by early workers for at least five different organisms called originally *Bacterium acidi-propionici a, b, c,* and *d,* and *Bacillus acidi-propionici*. These were renamed (respectively) *Propionibacterium freudenreichii*, *P. jensenii*, *P. petterssonii*, *P. shermanii*, and *P. pentosaceum* by van Niel (1928). *Bacillus acidi-propionici* was a legitimately published name and therefore has been revived in the eighth edition of *Bergey's Manual* (Moore and Holdeman, 1974). Several authors (e.g., Janoschek, 1944; Holdeman, Cato, and Moore, 1977; van Niel, 1928; Werkman and Brown, 1933) have given keys for the differentiation of the species based essentially on pigment production and the fermentation of various carbohydrates. However, the total number of strains examined is rather small for many species, and the results of fermentation tests by different investigators do not always agree. Malik, Reinbold, and Vedamuthu (1968), using 38 morphological and physiological features to examine 56 strains, grouped the strains into 4 clusters by numerical taxonomy, and Johnson and Cummins (1972), using cell wall analysis and DNA-DNA homologies on 29 strains, also found 4 groups that agree in many respects with those proposed by Malik and his colleagues. Table 1 gives the homology data and distinctive cell wall components for 11 named species, with the strains arranged in the homology groups described by Johnson and Cummins (1972). For details of cell wall structure, see Schleifer, Plapp, and Kandler (1968). The principal difference between this scheme and that of Malik, Reinbold, and Vedamuthu (1968) is in the disposition of strains of *P. rubrum,* which Malik found to show a rather low level of similarity with strains of *P. thoenii,* whereas by homology they appear to be closely related.

The species described by Janoschek (1944) and Sakaguchi, Iwaski, and Yamado (1941) were not available for homology testing. Of Janoschek's three species, it seems likely from the original descriptions that *P. casei* is closely related to *P. freudenreichii,* while *P. pituitosum* and *P. sanguineum* fall into the *P. thoenii* group. Of the five species described by Sakaguchi, Iwaski, and Yamado, it is probable that *P. orientum, P. globosum,* and *P. coloratum* are related to *P. freudenreichii,* while *P. amylaceum* and *P. japonicum,* which are nitrate

negative but ferment sucrose and maltose, would probably fall into the *P. jensenii* group.

In summary, it seems agreed by all investigators that strains of *Propionibacterium freudenreichii* and *P. shermanii* are closely related, and have been included in a single species, *P. freudenreichii.* They also differ from the other classical propionibacteria in fermenting a more restricted range of carbohydrates, in being heat resistant (Malik, Reinbold, and Vedamuthu, 1968), and in having a rather distinctive pattern of cell wall components (see Table 1). Strains of *Propionibacterium arabinosum* and *P. pentosaceum* are also closely related to each other, being the only strains in which catalase production is weak *(P. pentosaceum)* or absent *(P. arabinosum).* They have therefore been placed into a single species, *Propionibacterium acidipropionici.* The two groups *P. thoenii* and *P. jensenii,* which show the highest degree of cross-homology (51–53%, Table 1), also show a considerable range of phenotypic variation. The original species designations within these groups were based on various fermentation tests (Breed, Murray, and Smith, 1957).

Fermentation patterns have been the basis of distinction between named strains in almost all identification schemes, and most authors (e.g., Janoschek, 1944; van Niel, 1928; Werkman and Brown, 1933) have devised identification keys based on such tests. In the eighth edition of *Bergey's Manual* (Moore and Holdeman, 1974) the classical propionibacteria are classified in four species, *P. freudenreichii, P. thoenii, P. jensenii,* and *P. acidi-propionici,* based primarily on homology groupings. The relationships of these four species to the 11 named spe-

Table 1. DNA-DNA homology and cell wall components for eleven named species of *Propionibacterium.*[a]

Homology group data			Percent homology to reference DNA from				Distinctive cell wall components	
Suggested name and group	Original species designations	G+C (mol%)	*P. freudenreichii*	*P. thoenii*	*P. jensenii*	*P. acidipropionici*	Sugars	DAP isomer
P. freudenreichii	*P. freudenreichii* *P. shermanii*	65–66	90	20	26	25	Galactose, mannose, and rhamnose	*meso*-DAP
P. thoenii	*P. thoenii* *P. rubrum*	66–67	12	96	53	30	Some combination of glucose, galactose, and mannose	L-DAP
P. jensenii	*P. jensenii* *P. zeae* *P. technicum* *P. raffinosaceum* *P. peterssonii*	66–67	17	51	88	30		
P. acidi-propionici	*P. arabinosum* *P. pentosaceum*	66–67	8	35	38	87		

[a] Data from Johnson and Cummins (1972).

cies found in the seventh edition of *Bergey's Manual* are indicated in Table 1. A simple set of properties by which these four species can be differentiated is given in Table 2. The four species correspond also to the four groups defined by Malik, Reinbold, and Vedamuthu (1968) in their simplified dendrogram (Fig. 2 in their paper), except that their *P. rubrum* cluster was placed with the *P. jensenii* and *P. peterssonii* clusters instead of with *P. thoenii*.

Cutaneous Propionibacteria

Habitats

Coryneform organisms were first observed in material from acne comedones by Unna in 1893, and were cultured from the lesions by Sabouraud in 1897. It was assumed that these organisms were the cause of acne until 1911, when Lovejoy and Hastings (1911) isolated an apparently identical organism from the skin of persons without acne. Since then the relationship between the acne bacillus and the lesions of acne has been a matter of continued debate.

The distribution of the various species of cutaneous propionibacteria on the human body has recently been investigated by McGinley, Webster, and Leyden (1978). *Propionibacterium acnes* is the predominant organism, especially in areas rich in sebaceous glands such as the forehead and alae nasi. *Propionibacterium granulosum* is the next most common and has a similar distribution to *P. acnes,* except that it is especially common in the alae nasi region. *Propionibacterium avidum* has a more restricted distribution and is found mostly in moist rather than oily regions, such as the axilla, perineum, or anterior nares.

Organisms of this group have also been isolated from the mouth, the female genital tract, and from feces. They are also common contaminants in anaerobic cultures, probably coming from the skin of the operator during subculture.

General Properties

METABOLISM AND NUTRITIONAL REQUIREMENTS. The nutritional requirements of the anaerobic coryneforms have been investigated in detail by D. A. Ferguson (Ferguson and Cummins, 1978). Strains of *Propionibacterium acnes* require pantothenate, biotin, thiamine, and nicotinamide. Strains of *P. avidum* and *P. granulosum* require pantothenate, biotin, and thiamine only. Most strains need a full complement of 18 amino acids for good growth, and growth is further improved by the addition of 0.1–0.2% lactate, pyruvate, and α-ketoglutarate, and of guanine and adenine. These requirements are basically very similar to those of the classical propionibacteria and suggest a close resemblance between the overall metabolisms of the two groups.

ANTIGENIC COMPOSITION. A number of strains of anaerobic coryneforms give suspensions that are unstable in saline and so are unsuitable for agglutination tests using suspensions of whole cells. However, antisera are commercially available (Difco) that can be used for the identification of *Propionibacterium acnes* by the slide agglutination test. A clearer separation of the three species *P. acnes, P. granulosum,* and *P. avidum* and their serotypes can be obtained by using cell wall polysaccharide antigens extracted from the cells by 10% trichloroacetic acid at 56°C (Cummins, 1975).

BACTERIOPHAGES. A number of bacteriophage types have been described for *Propionibacterium acnes* (e.g., Jong, Ko, and Pulverer, 1975; Prévot and Thouvenot, 1961; Webster and Cummins, 1978). Bacteriophages for *P. avidum* and *P. granulosum* have not been investigated; however, strains of these two species are not lysed by any of the *P. acnes* phages. Some phage strains, e.g., strain 174 of Zierdt, Webster, and Rude (1968), will lyse most strains of *P. acnes* and can be used for rapid presumptive species identification. Bacteriophages active against *P. acnes* can readily be detected in filtrates of skin washings (Marples, 1974).

Table 2. Differentiation of classical propionibacteria.

	Fermentation of sucrose and maltose	Reduction of nitrate	Color of pigment
P. freudenreichii	−	+ or −	Cream
P. thoenii	+	−	Red brown
P. jensenii	+	−	Cream
P. acidi-propionici	+	+	Cream to orange-yellow

LIPIDS. As with the classical propionibacteria, the principal fatty acids of the cell lipids in the anaerobic coryneforms are branched-chain C_{15} compounds (Moss et al., 1969). Complex lipids that have chemoattractant properties for phagocytes have been extracted from strains of *P. acnes* (Russel et al., 1976).

SENSITIVITY TO ANTIMICROBIAL AGENTS. Strains of *Propionibacterium acnes* were found to be sensitive to penicillin, erythromycin, tetracyclines, chloramphenicol, and novobiocin, and resistant to streptomycin and sulfonamides (Pochi and Strauss, 1961). The strains were particularly resistant to sulfonamides and would grow in the presence of concentrations > 500 μg/ml.

PATHOGENICITY. Strains of *Propionibacterium acnes* are occasionally isolated from pathological processes, e.g., meningitis or endocarditis, where it appears reasonably certain that they are the causative organisms. However, the cutaneous propionibacteria as a group are only very occasional pathogens. *P. avidum* may be found in chronic infected sinuses, ulcers, abscesses, etc., but usually in association with other organisms.

The relationship of *P. acnes* to the disease acne vulgaris is obscure. The essential lesion in acne is plugging of the orifice of the sebaceous gland, and it seems possible that overgrowth of *P. acnes* (and *P. granulosum*) in the obstructed gland may produce sufficient acid to irritate the tissues. Severe inflammation and scarring in acne are almost always due to an associated staphylococcal infection.

Isolation and Maintenance

Anaerobic coryneforms from the skin or other epithelial surfaces are easily obtained by swabbing suitable areas. Other methods of sampling are scraping with a sterile scalpel blade (Evans et al., 1950) or with a Teflon "policeman". The technique of Williamson and Kligman (1965), although originally designed for aerobes, is very satisfactory for isolating *P. acnes* or similar organisms from the skin surface, provided that anaerobic conditions are employed for cultivation. (The method is given in the original form below.)

Isolation of Anaerobic Coryneforms from the Skin (Williamson and Kligman, 1965)

1. The area to be scrubbed (3.8 cm²) is delineated by a sterile glass cylinder held firmly to the skin by two attached handles.
2. One milliliter of wash solution—0.1% Triton X-100 in 0.075 M phosphate buffer, pH 7.9—is pipetted in and the area scrubbed with moderate pressure for 1 min using a sterile Teflon "policeman".
3. The wash fluid is aspirated, replaced with a fresh 1 ml, and the scrub repeated.
4. The two washes are then pooled and an aliquot diluted in 10-fold steps using as diluent 0.05% Triton X-100 in 0.0375 M phosphate buffer to prevent any reaggregation of organisms.
5. The appropriate dilutions (usually 10^0, 10^{-1}, 10^{-2} for normal skin; 10^{-3} and 10^{-4} for areas of high bacterial density) are plated in 15–20 ml tryptic soy agar per plate.
6. After 48 h incubation at 37°C, colonies are counted and viable cells in the original sample are calculated by standard methods.

Using this method, suitable media for anaerobic coryneforms are blood agar or peptone–yeast extract–glucose agar, pH 6.5, containing 0.1% Tween 80. The plates will need to be incubated anaerobically (e.g., GasPak jars containing $H_2 + CO_2$) for up to 7 days. It should be remembered that strains of *P. avidum* will frequently grow aerobically, although more slowly than on anaerobic plates.

This technique calls for special glass cylinders that can be held against the skin to contain fluid. However, satisfactory results can be obtained using swabs dipped in the detergent-buffer mixture and then squeezed to expel excess fluid. After sampling, the swab is squeezed in a measured volume of fluid to wash out organisms that have been picked up. Considerable variation in bacterial numbers is found from person to person (Evans et al., 1950), and it is important to plate out several dilutions, as described in the preceding Williamson and Kligman procedure.

A detailed comparison of the results with swabbing and scraping is given in Evans and Stevens (1976). Scraping is more likely to yield organisms from the pilosebaceous glands, while swabbing picks up surface organisms only.

The method for bulk culturing described for the classical propionibacteria is also applicable for growing the cutaneous propionibacteria.

Classification and Identification

Organisms in this group were regarded as belonging to the genus *Corynebacterium* until Douglas and Gunter (1946) and Moore and Cato (1963) showed that propionic acid was a major end product of their metabolism. Subsequent work has shown that the anaerobic coryneforms from skin differ in several fundamental ways from the classical corynebacteria (e.g., *C. diphtheriae*). For example, they do not produce mycolic acids, the diamino acid of the cell

Table 3. Cell wall components and DNA relationships of three major groups of anaerobic coryneforms.[a]

| Group | Principal cell wall components | | G+C (mol%) | Homology relationships within group; percent homology to DNA from: | | |
	Sugars	DAP isomer		P. acnes	P. avidum	P. granulosum
P. acnes, type I	Galactose, glucose, mannose	L-DAP	57–60	97	51	16
P. acnes, type II	Glucose, mannose	Generally L-DAP, some strains meso-DAP				
P. avidum, type I	Galactose, glucose, mannose	L-DAP	62–63	50	90	17
P. avidum, type II	Glucose, mannose	Generally L-DAP, some strains with meso-DAP				
P. granulosum	Galactose, mannose, sometimes trace glucose	L-DAP	62–64	12	15	95

[a] Data from Johnson and Cummins (1972) and Cummins and Johnson (1974).

wall peptidoglycan is L-DAP in almost all cases, and they do not have a cell wall polysaccharide containing arabinose (see Johnson and Cummins, 1972; Rogosa et al., 1974). Prévot would prefer to place these organisms in a separate subgenus, *Coryneformis* (Prévot, 1976), in the family Corynebacteriaceae.

At one time as many as 12 species of anaerobic coryneforms were described (Prévot and Fredette, 1966). However, an examination of 80 strains by cell wall analysis and DNA-DNA homology tests showed three major groups that seemed sufficiently distinct to be regarded as species (Johnson and Cummins, 1972). The main differential features are given in Table 3. In *P. acnes* and *P. avidum*, two serotypes are found; these can be distinguished by precipitin tests using trichloroacetic acid extracts (Cummins, 1975). The organism generally referred to as *Corynebacterium parvum*, which causes an unusual degree of reticulo-stimulation and macrophage-activation in animals, was found to be indistinguishable from *P. acnes* (Cummins and Johnson, 1974).

Most strains of *Propionibacterium acnes* and *P. granulosum* grow poorly, if at all, under aerobic conditions, although they are not sensitive to oxygen and organisms will remain viable for several hours or longer if plate cultures are left exposed to air. Strains of *P. avidum* will grow quite well under aerobic conditions. It was because of the need to grow the organisms anaerobically that the phrase "anaerobic coryneforms" or "anaerobic corynebacteria" arose. However, McGinley, Webster, and Leyden (1978) have used the phrase "cutaneous propionibacteria" and this seems in many ways more appropriate, because the skin appears to be the main reservoir of these organisms.

In most cases, the problem of identifying these organisms is one of distinguishing them from morphologically similar organisms isolated from skin swabs or clinical specimens. All members of the group produce major amounts of propionic and acetic acids as end products of hexose metabolism (generally about 2–3 mol of propionate for 1 mol of acetate), and this pattern can readily be established by gas chromatography of culture supernatants (Holdeman, Cato, and Moore, 1977). Also, almost all strains contain the L-isomer of diaminopimelic acid (DAP) as the diamino acid of peptidoglycan (Johnson and Cummins, 1972). Most morphologically similar groups, e.g., *Actinomyces israelii*, *A. naeslundii*, or aerobic skin diphtheroids resembling *Corynebacterium xerosis* either do not have DAP (for example, *A. israelii*) or have the meso-isomer (aerobic skin diphtheroids). The presence and type of DAP can readily be determined by examination of whole cell hydrolysates of the cells from a 10-ml tube of culture. The washed whole cells are hydrolyzed in 6 N HCl and the hydrolysate is examined for DAP as described in Cummins and Johnson (1971).

Once it is established that the organisms belong to the group, the determination of species and serotype is based on the characters shown in Table 4, coupled with serological examination of acid extracts of whole cells (Cummins, 1975).

Strains of *Propionibacterium acnes* and *P. granulosum* grow rather slowly on solid media and, especially with *P. acnes*, colonies may still be under 1 mm at 4 days. Colonies of *P. granulosum* are gen-

Table 4. Characters useful in identification of *Propionibacterium acnes, P. avidum,* and *P. granulosum.*[a]

Character	*P. acnes,* type I	*P. acnes,* type II	*P. avidum*[b]	*P. granulosum*
Fermentation of glucose	+	+	+	+
Fermentation of sucrose	−	−	+	+
Fermentation of maltose	−	−	+	+
Fermentation of sorbitol	+ or −	−	−	−
Esculin hydrolysis	−	−	+	−
Indole production	+	+	− (occ. +)	− (occ. w+)
Reduction of nitrate	+	+	− (occ. +)	−
Liquefaction of gelatin	+	+	+	− (occ. w+)
Casein digestion	+	+	++	
Colonies at 4 days	Small, semi-opaque, grayish; less than 1 mm; may be reddish color later		Large, opaque, creamy (1–2 mm)	Intermediate, ca. 1 mm, opaque white to cream

[a] In fermentation tests, + indicates final pH < 5.5 after 5 days' incubation; − indicates final pH > 6.0.

[b] The two serological types of *P. avidum* do not differ with respect to these tests.

erally a little larger and more creamy and opaque than those of *P. acnes.* Strains of *P. avidum* grow considerably faster than either of the other species and show good growth in 48 h.

On blood agar, many strains of *Propionibacterium acnes* and most strains of *P. avidum* are β-hemolytic on human, rabbit, or horse blood (Hoeffler, 1977). However, except for *P. avidum,* there is no hemolysis on sheep blood agar. We have found strains of *P. granulosum* to be nonhemolytic, but Hoeffler (1977) reported them to be β-hemolytic on rabbit blood agar.

Literature Cited

Allen, S. H. G., Kellermeyer, R. W., Stjernholm, R. L., Wood, H. G. 1964. Purification and properties of enzymes involved in the propionic acid fermentation. Journal of Bacteriology **87:**171–187.

Allsop, J., Work, E. 1963. Cell walls of *Propionibacterium* species: Fractionation and composition. Biochemical Journal **87:**512–519.

Breed, R. S., Murray, E. G. D., Smith, N. R. 1957. Bergey's manual of determinative bacteriology, 7th ed., pp. 569–576. Baltimore: Williams & Wilkins.

Brennan, P., Ballou, C. E. 1968. Phosphatidylmyoinositol monomannoside in *Propionibacterium shermanii.* Biochemical and Biophysical Research Communications **30:**69–75.

Cummins, C. S. 1975. Identification of *Propionibacterium acnes* and related organisms by precipitin tests with trichloroacetic acid extracts. Journal of Clinical Microbiology **2:**104–110.

Cummins, C. S., Johnson, J. L. 1971. Taxonomy of the clostridia: Wall composition and DNA homologies in *Clostridium butyricum* and other butyric acid-producing clostridia. Journal of General Microbiology **67:**33–46.

Cummins, C. S., Johnson, J. L. 1974. *Corynebacterium parvum:* A synonym for *Propionibacterium acnes?* Journal of General Microbiology **80:**433–442.

Delwiche, E. A. 1949. Vitamin requirements of the genus *Propionibacterium.* Journal of Bacteriology **58:**395–398.

Demeter, K. J., Janoschek, A. 1941. Vorkommen und Entwicklung der Propionsäurebakterien in verschiedenen Käsearten. Zentralblatt für Bakteriologie, Parasitenkunde und Infektionskrankheiten, Abt. 2 **103:**257–271.

Douglas, H. C., Gunter, S. E. 1946. The taxonomic position of *Corynebacterium acnes.* Journal of Bacteriology **52:**15–23.

Evans, C. A., Stevens, R. J. 1976. Differential quantitation of surface and subsurface bacteria of normal skin by the combined use of the cotton swab and scrub methods. Journal of Clinical Microbiology **3:**576–581.

Evans, C. A., Smith, W. M., Johnson, E. A., Giblett, E. R. 1950. Bacterial flora of the normal human skin. Journal of Investigative Dermatology **15:**305–323.

Ferguson, D. A., Cummins, C. S. 1978. Nutritional requirements of anaerobic coryneforms. Journal of Bacteriology **135:**858–867.

Fitz, A. 1878. Über Spaltpiltzgährungen. Berichte der Deutschen Chemischen Gesellschaft **11:**1890–1899.

Fitz, A. 1879. Über Spaltpiltzgährungen. Berichte der Deutschen Chemischen Gesellschaft **12:**474–481.

Galesloot, T. E. 1957. Involved van Nisine op die Bacterien Welka Betrokken Zijn of Kunnen Zijn bij bacteriologische Processen in Kaas en Smeltkaas. Netherlands Milk and Dairy Journal **11:**58–73.

Hettinga, D. H., Reinbold, G. W. 1972a. The propionic-acid bacteria—a review. I. Growth. Journal of Milk and Food Technology **35:**295–301.

Hettinga, D. H., Reinbold, G. W. 1972b. The propionic-acid bacteria—a review. II. Metabolism. Journal of Milk and Food Technology **35:**358–372.

Hettinga, D. H., Reinbold, G. W. 1972c. The propionic-acid bacteria—a review. III. Miscellaneous metabolic activities. Journal of Milk and Food Technology 35:436–447.

Hettinga, D. H., Vedamuthu, E. R., Reinbold, R. W. 1968. Pouch method for isolating and enumerating propionibacteria. Journal of Dairy Science 51:1707–1709.

Hitchner, E. R. 1932. A cultural study of the propionic acid bacteria. Journal of Bacteriology 23:40–41.

Hoeffler, U. 1977. Enzymatic and hemolytic properties of *Propionibacterium acnes* and related bacteria. Journal of Clinical Microbiology 6:555–558.

Holdeman, L. V., Cato, E. P., Moore, W. E. C. 1977. Anaerobe laboratory manual, 4th ed. Blacksburg, Virginia: Virginia Polytechnic Institute and State University.

Janoschek, A. 1944. Zur Systematik der Propionsäurebakterien. Zentralblatt für Bakteriologie, Parasitenkunde, Infektionskrankheiten und Hygiene, Abt. 2 106:321–337.

Johnson, J. L., Cummins, C. S. 1972. Cell wall composition and deoxyribonucleic acid similarities among the anaerobic coryneforms, classical propionibacteria, and strains of *Arachnica propionica*. Journal of Bacteriology 109:1047–1066.

Jong, E. C., Ko, H. L., Pulverer, G. 1975. Studies on bacteriophages of *Propionibacterium acnes*. Medical Microbiology and Immunology 161:263–271.

Kurmann, J. 1960. Ein vollsynthetischer Nährboden für Propionsäurebakterien. Pathologia et Microbiologia 23:700–711.

Lovejoy, E. D., Hastings, T. W. 1911. Isolation and growth of the acne bacillus. Journal of Cutaneous Diseases 29:80–82.

McGinley, K. J., Webster, G. F., Leyden, J. J. 1978. Regional variation of cutaneous propionibacteria. Applied and Environmental Microbiology 35:62–66.

Malik, A. C., Reinbold, G. W., Vedamuthu, E. R. 1968. An evaluation of the taxonomy of *Propionibacterium*. Canadian Journal of Microbiology 14:1185–1191.

Marples, R. R. 1974. The microflora of the face and acne lesions. Journal of Investigative Dermatology 62:326–331.

Moore, W. E. C., Cato, E. P. 1963. Validity of *Propionibacterium acnes* (Gilchrist) Douglas and Gunter comb. nov. Journal of Bacteriology 85:870–874.

Moore, W. E. C., Holdeman, L. V. 1974. *Propionibacterium*, pp. 633–644. In: Buchanan, R. E., Gibbons, N. E. (eds.), Bergey's manual of determinative bacteriology, 8th ed. Baltimore: Williams & Wilkins.

Moss, C. W., Dowell, V. R., Farshtchi, D., Raines, L. J., Cherry, W. B. 1969. Cultural characteristics and fatty acid composition of propionibacteria. Journal of Bacteriology 97:561–570.

Orla-Jensen, S. 1909. Die Hauptlinien des natürlichen Bakteriensystems. Centralblatt für Bakteriologie, Parasitenkunde und Infektionskrankheiten, Abt. 2 22:305–346.

Pochi, P. E., Stauss, J. S. 1961 Antibiotic sensitivity of *Corynebacterium acnes*, (*Propionibacterium acnes*). Journal of Investigative Dermatology 36:423–429.

Prévot, A.-R. 1976. Nouvelle conception de la position des corynébactéries anaérobies. Comptes Rendus Hebdomadaires des Séances de l'Académie des Sciences, Série D 282:1079–1081.

Prévot, A.-R., Fredette, V. 1966. Manual for the classification and determination of the anaerobic bacteria, pp. 345–355. Philadelphia: Lea & Febiger.

Prévot, A.-R., Thouvenot, H. 1961. Essai de lysotypie des *Corynebacterium* anaérobies. Annales de l'Institut Pasteur 101:966–970.

Prottey, C., Ballou, C. E. 1968. Diacyl myoinositol monomannoside from *Propionibacterium shermanii*. Journal of Biological Chemistry 243:6196–6201.

Reddy, M. S., Reinbold, G. W., Williams, F. D. 1973. Inhibition of propionibacteria by antibiotic and antimicrobial agents. Journal of Milk and Food Technology 36:564–569.

Reddy, M. S., Williams, F. D., Reinbold, G. W. 1973. Sulfonamide resistance of propionibacteria: Nutrition and transport. Antimicrobial Agents and Chemotherapy 4:254–258.

Rogosa, M., Cummins, C. S., Lelliott, R. A., Keddie, R. M. 1974. The coryneform group of bacteria, pp. 599–617. In: Buchanan, R. E., Gibbons, N. E. (eds.), Bergey's manual of determinative bacteriology, 8th ed. Baltimore: Williams & Wilkins.

Russel, R. J., McInroy, R. J., Wilkinson, P. C., White, R. G. 1976. A lipid chemotactic factor from anaerobic coryneform bacteria including *Corynebacterium parvum* with activity for macrophages and monocytes. Immunology 30:935–949.

Sabouraud, R. 1897. La séborrhée grasse et la pelade. Annales de l'Institut Pasteur 11:134–159.

Sakaguchi, K., Iwaski, M., Yamado, S. 1941. Studies on the propionic acid fermentation. Journal of the Agricultural Chemical Society of Japan 17:127–138.

Schleifer, K. H., Plapp, R., Kandler, O. 1968. Glycine as cross-linking bridge in the LL-diaminopimelic acid containing murein of *Propionibacterium peterssonii*. FEBS Letters 1:287–290.

Shaw, R. H., Sherman, J. M. 1923. The production of volatile fatty acids and carbon dioxide by propionic acid bacteria with special reference to their action in cheese. Journal of Dairy Science 6:303–309.

Sherman, J. M. 1921. The cause of eyes and characteristic flavor in Emmental or Swiss cheese. Journal of Bacteriology 6:379–392.

Sherman, J. M., Shaw, R. H. 1923. The propionic acid fermentation of lactose. Journal of Biological Chemistry 56:695–700.

Thöni, J., Allemann, O. 1910. Über das Vorkommen von gefärbten, makroskopischen Bakterienkolonien in Emmentalerkäsen. Centralblatt für Bakteriologie, Parasitenkunde und Infektionskrankheiten, Abt. 2 25:8–30.

Troili-Petersson, G. 1904. Studien über die Mikroorganismen des schwedischen Güterkäses. Centralblatt für Bakteriologie, Parasitenkunde und Infektionskrankheiten, Abt. 2 11:120–143.

Unna, P. J. 1893. Die Histopathologie der Haut-krankheiten. Berlin: A. Hirschwald.

van Niel, C. B. 1928. The propionic acid bacteria. Haarlem, The Netherlands: J. W. Boissevain.

Vedamuthu, E. R., Reinbold, G. W. 1967. The use of candle oats jar incubation for the enumeration, characterization and taxonomic study of propionibacteria. Milchwissenschaft 22:428–431.

von Freudenreich, E., Orla-Jensen, S. 1906. Ueber die im Emmentalerkäse stattfindende Propionsäuregärung. Centralblatt für Bakteriologie, Parasitenkunde und Infektionskrankheiten, Abt. 2 17:529.

Webster, G. F., Cummins, C. S. 1978. Use of bacteriophage typing to distinguish *Propionibacterium acnes* types I and II. Journal of Clinical Microbiology 7:84–90.

Werkman, C. H., Brown, R. W. 1933. The propionic acid bacteria. II. Classification. Journal of Bacteriology 26:393–417.

Werkman, C. H., Kendall, S. E. 1931. The propionic acid bacteria. I. Classification and nomenclature. Iowa State Journal of Science 6:17–32.

Williamson, P., Kligman, A. M. 1965. A new method for the quantitative investigation of cutaneous bacteria. Journal of Investigative Dermatology 45:498–503.

Winkler, S., Fröhlich, M. 1957. Prüfung des Einflusses von Nisin auf die wichtigsten Reifungserreger beim Emmentalerkäse. Milchwissenschaftliche Berichte 7:125–137.

Wood, H. G., Andersen, A. A., Werkman, C. H. 1938. Nutrition of the propionic acid bacteria. Journal of Bacteriology 36:201–214.

Zierdt, C. H., Webster, C., Rude, W. S. 1968. Study of the anaerobic corynebacteria. International Journal of Systematic Bacteriology 18:33–47.

The Genus *Eubacterium*

ROBERT P. LEWIS and VERA L. SUTTER

The genus *Eubacterium* has been redefined during this decade. Investigators in the Anaerobe Laboratory of the Virginia Polytechnic Institute, working through subcommittees of the International Committee of Systematic Bacteriology, have been primarily responsible for the present classification of nonsporeforming anaerobic organisms. The genus *Eubacterium* now includes all anaerobic, nonsporeforming, Gram-positive rods that do not produce, as major products of fermentation, (i) propionic acid (*Propionibacterium* and *Arachnia*), (ii) lactic acid alone (*Lactobacillus*), (iii) acetic and lactic acids with more acetic than lactic (*Bifidobacterium*), or (iv) succinic acid (in the presence of CO_2) and lactic acid (*Actinomyces*). The majority of the species in the genus produce butyric and other fatty acids. Some produce no acids at all, while others produce acetic and formic acids in addition to ethanol (Moore and Holdeman, 1973; Holdeman and Moore, 1974a; Holdeman, Cato, and Moore, 1977). The phenotypic characteristics of the species have been carefully studied and recorded (Holdeman and Moore, 1974a; Holdeman, Cato, and Moore, 1977). However, the genetic similarities between members of the genus have not yet been adequately verified by DNA homology studies. The phenotypic variation between some of the species and the position of the genus as the repository of all anaerobic, nonsporeforming, Gram-positive rods that do not fit into one of five other genera assure that the genus will undergo additional redefinitions. New species of intestinal *Eubacterium* have recently been defined (Holdeman and Moore, 1974b; Moore, Johnson, and Holdeman, 1976; Wilkins, Fulghum, and Wilkins, 1974), and additional species will certainly be named in the future. Currently, 63% (68/109) of *Eubacterium* strains isolated from clinical specimens in the Wadsworth Anaerobic Bacteriology Research Laboratory from 1973 to 1977 do not fit the description of the more common species listed in the fourth edition of the Anaerobe Laboratory Manual of the Virginia Polytechnic Institute (Holdeman, Cato, and Moore, 1977).

Prévot coined the term *Eubacterium* (Greek = beneficial bacterium) in 1938 when he proposed a new genus composed of anaerobic, nonflagellated, nonmotile, nonsporeforming, unencapsulated, nonbranching, Gram-positive rods appearing usually in pairs or short chains (Prévot, 1938). The genus was differentiated on morphological grounds from three other genera proposed at that time: *Catenabacterium, Ramibacterium,* and *Cillobacterium* (Prévot, 1938). These three genera are no longer recognized and their members have been transferred to other genera. Many are now included in the genus *Eubacterium* (Holdeman and Moore, 1974a; Moore and Holdeman, 1973). In addition, organisms formerly classified as *Ristella, Mycobacterium, Bacteroides,* and *Bacillus* have been studied by current techniques and have been reclassified as *Eubacterium* (Table 1).

Habitats (Table 2)

Eubacteria are present in nature in large numbers; there are up to 10^{10} *Eubacterium* cells/g dry weight of stool (Table 3). However, their ecological, economic, and medical importance has only been partially determined.

Spoiled Food, Water, and Soil

Eubacterium foedans is the type species for the genus. It was isolated from spoiled hams in 1908 (Klein, 1908), but has not been identified in nature since. It was nonpathogenic when injected into guinea pigs.

Eubacterium nitritogenes and *Eubacterium obstii* have been isolated from a number of sources, including cheese. *Eubacterium endocarditidis* and *Eubacterium tortuosum* have been isolated from a number of sources, including water. Eight species have been isolated from soil (Table 2). Their ecological importance has not been studied.

Rumen

The environment in the bovine rumen is ideal for the survival of anaerobic bacteria. The pH is buffered between 6 and 7, the oxidation-reduction potential is maintained at about -400 mV, and the temperature,

Table 1. Synonyms for *Eubacterium* species.[a]

Species	Synonyms
E. aerofaciens	Bacteroides aerofaciens
	Pseudobacterium aerofaciens
E. alactolyticum	Ramibacterium alactolyticum
	Ramibacterium dentium
	Ramibacterium pleuriticum
E. biforme	None
E. budayi	Bacillus cadaveris butyricus
	Bacterium budayi
	Eubacterium cadaveris
	Pseudobacterium cadaveris
E. cellulosolvens	Cillobacterium cellulosolvens
E. combesii	Cillobacterium combesii
E. contortum	Catenabacterium contortum
E. cylindroides	Bacterium cylindroides
	Ristella cylindroides
	Pseudobacterium cylindroides
	Bacteroides cylindroides
E. dolichum	None
E. eligens	None
E. endocarditidis	Cillobacterium endocarditis
E. ethylicum	Bacillus gracilis ethylicus
	Pseudobacterium ethylicum
E. fissicatena	None
E. foedans	Bacillus foedans
E. formicigenerans	None
E. hadrans	None
E. hallii	None
E. helminthoides	Catenabacterium helminthoides
	Bacillus helminthoides
E. helwigiae	Catenabacterium ruminantium
E. lentum	Bacteroides lentus
	Pseudobacterium lentum
	Coccobacillus oviformis
	Bacteroides oviformis
E. limosum	Bacteroides limosus
	Butyribacterium rettgeri
	Mycobacterium limosum
	Butyribacterium limosum
E. moniliforme	Bacillus moniliforme
	Bacillus repazii
	Cillobacterium moniliforme
E. multiforme	Bacillus multiformis
	Cillobacterium multiforme
E. nitritogenes	None
E. obstii	Bacillus B Obst
	Pseudobacterium obsti
E. parvum	None
E. plexicaudatum	None
E. pseudotortuosum	None
E. ramulus	None
E. rectale	Bacteroides rectalis
	Pseudobacterium rectale
E. ruminantium	None
E. saburreum	Catenabacterium saburreum
E. siraeum	None
E. tenue	Bacillus tenuis spatuliformis
	Bacteroides tenuis
	Cillobacterium spatuliforme
	Bacillus spatuliformis
	Cillobacterium tenue
E. tortuosum	Bacillus tortuosus
	Bacteroides tortuosus
	Mycobacterium flavum var. tortuosum

Table 1. *(continued)*

Species	Synonyms
E. ureolyticum	None
E. ventriosum	Bacillus ventriosus
	Bacteroides ventriosus
	Pseudobacterium ventriosum

[a] Adapted from Holdeman and Moore (1974a, b); Moore et al. (1976); Wilkins et al. (1974).

mixing, food, and water supply are relatively constant. Only small amounts of oxygen gain entry to the rumen, CO_2 and methane are produced in abundance, and metabolic waste products and excess microorganisms are efficiently removed by peristalsis (Bryant, 1959; Bryant, 1970).

The rumen flora is complex. Many aerobic or facultative organisms are present, but their concentrations are relatively small (10^3– 10^7 bacteria/g). Nonsporeforming anaerobes are present in concentrations of 10^8– 10^9 bacteria/g. Usually the predominant flora is composed of 6–10 genera and 10–15 different species. Only rarely does a single species predominate. *Bacteroides succinogenes, Bacteroides ruminicola, Succinovibrio dextrinosolvens, Butyrivibrio fibrisolvens, Methanobacterium ruminantium, Ruminococcus albus, Megasphaera elsdenii,* and *Eubacterium ruminantium* are only a few of the species commonly isolated from the rumen (Bryant, 1970).

Rumen bacteria have adapted to a protein-free environment and can use NH_4^+ or simple amino acids as sole nitrogen source. They are capable of deriving energy from carbohydrate polymers such as cellulose or xylan and are active in protein synthesis, metabolism of fats, and synthesis of vitamins. The complex symbiotic metabolic relationship of the many different rumen microorganisms was studied extensively by Hungate (1966) and was elegantly reviewed by Bryant (1959, 1970).

Eubacterium ruminantium, E. cellulosolvens, and *E. helwigiae* are the rumen eubacteria. *E. ruminantium* constitutes up to 7% of the rumen flora in animals feeding on alfalfa or bluegrass (Bryant, 1959; Bryant and Robinson, 1962). It ferments xylan and shares with many eubacteria the ability to produce butyric acid (Bryant, 1970). Volatile fatty acids (e.g., acetic acid) and a simple nitrogen source (NH_4^+) are essential for growth (Bryant and Robinson, 1962).

Eubacterium cellulosolvens has been isolated in a concentration of 10^8 bacteria/g from the bovine rumen. This species is actively cellulolytic and produces butyric acid. It shares these and many other properties with another rumen organism *Butyrivibrio fibrisolvens,* but is distinguished from the latter on morphological grounds (Prins et al., 1972). *Butyrivibrio fibrisolvens* is a motile, Gram-negative, curved rod.

Table 2. Habitats of *Eubacterium* species.

Habitat	Species
Spoiled food	*E. foedans, E. nitritogenes, E. obstii*
Water	*E. endocarditidis, E. tortuosum*
Soil or mud	*E. budayi, E. combesii, E. endocarditidis, E. helminthoides, E. limosum, E. multiforme, E. nitritogenes, E. tortuosum*
Rumen	*E. cellulosolvens, E. helwigiae, E. ruminantium*
Oral cavity	*E. alactolyticum, E. helminthoides, E. saburreum*
Intestinal contents	*E. aerofaciens, E. biforme,*[a] *E. contortum, E. cylindroides, E. dolichum,*[a] *E. eligens,*[a] *E. fissicatena, E. formicigenerans,*[a] *E. hadrans,*[a] *E. hallii,*[a] *E. helminthoides, E. lentum, E. limosum, E. multiforme, E. nitritogenes, E. obstii, E. parvum, E. plexicaudatum,*[a] *E. ramulus,*[a] *E. rectale, E. siraeum,*[a] *E. tenue, E. tortuosum, E. ureolyticum, E. ventriosum*
Human infection	*E. aerofaciens, E. alactolyticum, E. combesii, E. contortum, E. cylindroides, E. endocarditidis, E. ethylicum, E. lentum, E. limosum, E. moniliforme, E. nitritogenes, E. parvum, E. pseudo-tortuosum, E. rectale, E. tenue, E. tortuosum, E. ventriosum*

[a] Species described since the publication of the eighth edition of *Bergey's Manual of Determinative Bacteriology* (Holdeman and Moore, 1974a, b); Moore et al., 1976; Wilkins et al., 1974).

Table 3. Species of *Eubacterium* isolated from stool.

Reference	Concentration (bacteria/g dry wt of stool)	Percentage of samples from which *Eubacterium* species were isolated (no. of samples)	Most common *Eubacterium* species isolated
Finegold, Attebery, and Sutter (1974)			*E. lentum*
Japanese diet (low fat)	10^{10}	95 (20)	*E. aerofaciens*
American diet (high fat)[a]		95 (20)	*E. rectale*
Maier et al. (1974)	10^{10}	100 (10)	Not listed
Moore and Holdeman (1974b)			
Japanese–Hawaiians	10^{10}	100 (20)	*E. aerofaciens*
Peach et al. (1974)			
High carbohydrate	Not determined	23.2 (215)	*E. aerofaciens*
Low carbohydrate[a]	Not determined	6.7 (174)	*E. lentum*
Finegold et al. (1975)			
Volunteers with colonic polyps[a]	$10^4 - 10^{11}$	Not stated (25)	*E. lentum*
Controls	$10^7 - 10^{11}$	Not stated (25)	*E. aerofaciens*
			E. rectale
			E. cylindroides
			E. combesii
			E. tenue
Holdeman, Good, and Moore (1976)			
Astronauts	10^{10}	Not stated (25)	*E. aerofaciens*
			E. rectale
Finegold et al. (1977)			*E. cylindroides* was present in statistically significantly greater numbers in the heavy meat eaters
Vegetarians	$10^8 - 10^{13}$	92.3 (13)	
Heavy meat eaters[a]	$10^9 - 10^{12}$	92.8 (14)	
			E. aerofaciens
			E. contortum
			E. lentum
			E. rectale

[a] High risk groups for colon cancer.

The significance of the conversion of acetate to butyrate in the rumen by organisms such as *E. ruminantium*, *E. cellulosolvens*, and *B. fibrisolvens* is unclear. The butyrate does not appear to be used by ruminants as an energy source (Hungate, 1966).

Human Oral Cavity

Hofstad and his colleagues have studied in detail the polysaccharide antigens produced by *Eubacterium saburreum*. Strain L44 of this oral organism produces a linear polysaccharide composed of β $(1 \rightarrow 6)$-linked D-glycero-D-galactoheptopyranose residues. About 65% of these residues carry an *O*-acetyl group in the 7 position (Hoffman et al., 1974). The polysaccharide is antigenic when injected into animals (Hofstad, 1972). Hofstad has shown that 0.25% of a trypsin digest of human dental plaque is composed of this polysaccharide antigen and has calculated that approximately 7% of plaque by weight is composed of this organism (Hofstad, 1974).

Eubacterium saburreum is saccharolytic and constitutes a significant portion of the microbial flora of human plaque. Its etiological role in plaque formation remains speculative, however. The antigenic properties of this organism in man have not been studied.

Polysaccharide antigens have also been identified in strains L49 (Hoffman et al., 1976) and L452 (Hofstad and Henning, 1977) of *E. saburreum*. The structure of the three antigens varies widely, and each appears to be unique. The polysaccharide from strain L44 is the only example of a bacterial homoglycan composed of heptose residues (Hoffman et al., 1974). Heptose is a sugar commonly found in Gram-negative, but not in Gram-positive, bacteria. The antigens from strains L49 and L452 are unusually complex heteropolymers. That from strain L49 contains heptose as well as a deoxy sugar in a form (furanoid) not previously described in bacterial polysaccharides. The antigen from strain L452 contains fucose in the furanoid form, which is extremely rare (Hofstad and Henning, 1977).

Eubacterium alactolyticum has also been isolated from dental plaque. *E. helminthoides* is part of the oral flora of nursing infants (Holdeman and Moore, 1974a).

Intestinal Contents

Over half the species of *Eubacterium* have been isolated from the intestinal contents of man and animals (Table 2). A number of studies during this decade have demonstrated the presence of eubacteria in high numbers in the stools of a variety of volunteer populations (Table 3). *E. aerofaciens* is the species most commonly isolated from human stool. It accounts for approximately 10% of the bacteria in feces

(Holdeman, Good, and Moore, 1976) and up to 86% of the stool isolates of eubacteria (Peach et al., 1974).

Eubacterium lentum is the second most common species isolated (Peach et al., 1974). It is of note, however, that in one study, many isolates of eubacteria from stool could not be identified (Finegold, Attebery, and Sutter, 1974). Other species of *Eubacterium* that are frequently present in human stool are listed in Table 3.

While eubacteria constitute a significant portion of the fecal flora, their importance remains undetermined and their study has lagged behind that of many of the other anaerobic bacteria in stool (e.g., the *Bacteroides fragilis* group and clostridia).

Wensink (1975) has demonstrated that patients with Crohn's disease develop agglutinating antibodies to *E. contortum*, while healthy patients do not. Eubacteria may play a pathogenic role in this disease.

The data available suggest that eubacteria may also play a significant role in bile acid and cholesterol metabolism.

Intestinal bacteria are involved in bile acid, cholesterol, and fatty acid metabolism: (i) by deconjugation of bile salts, (ii) by the 7-α-dehydroxylation of bile acids (e.g., the conversion of cholic acid to 7-ketodeoxycholic acid), (iii) by dehydrogenation of cholesterol and other steroids, or (iv) by hydrogenation of unsaturated fatty acids. These reactions can be performed by a number of anaerobic bacteria.

Dickinson, Gustafsson, and Norman (1971) demonstrated the ability of a species of *Eubacterium* to deconjugate bile salts and dehydroxylate bile acids in vitro and in vivo following establishment in germ-free rats. The ability of a variety of *Eubacterium* species (e.g., *E. ventriosum*, *E. lentum*, *E. parvum*, and *E. aerofaciens*) to dehydroxylate bile acids is well established (Hylemon and Stellwag, 1976; Midtvedt, 1974).

A species of *Eubacterium* (ATCC 21408) isolated from the cecal contents of rats is dependent on the presence of $\Delta 5$ 3-β-hydroxy steroids (e.g., cholesterol) for growth (Eyssen et al., 1973; Eyssen and Parmentier, 1974; Parmentier and Eyssen, 1974). It converts cholesterol to coprostanol via a 4-cholesten-3-one intermediate and is also capable of reducing unsaturated fatty acids in vitro.

The importance of some of these transformations remains to be determined. Deconjugation is an essential step in the enterohepatic circulation of bile acids. A number of authors have postulated that some of the reactions may be of etiological significance in colon cancer. Hill and his colleagues have demonstrated a direct correlation between the incidence of colon cancer and (i) total fecal bile acid concentration, (ii) the 7-α-dehydroxylase activity of the stool flora, and (iii) the fecal concentration of deoxycholic acid (Hill and Drasar, 1974).

Mastromarino, Bandaru, and Wynder (1976) have demonstrated a direct correlation between colon carcinoma and 7-α-dehydroxylase and cholesterol dehydrogenase activity in the stool.

A carcinogen-producing bacterium has not yet been identified; however, the metabolic transformations implicated can be performed by a number of anaerobic bacteria, including species of *Eubacterium*. Epidemiological studies have implicated *Bacteroides* or *Clostridium* as possible producers of colon carcinogens (Hill and Drasar, 1974).

Maier and his colleagues (1974) noted no change in the concentration of *Eubacterium* spp. in the stool with the introduction of a heavy meat diet. *Bacteroides fragilis* is present in stool in three times the concentration of *E. aerofaciens* and has appreciably greater 7-α-dehydroxylase activity (Hylemon and Stellwag, 1976). With the exception of one study, *Eubacterium* spp. have been present in approximately the same concentration in patient populations at high and low risk for developing colon cancer (Table 3). *Eubacterium cylindroides* was present in statistically significantly greater numbers ($P < 0.05$) in the stools of a high risk population in one study (Finegold et al., 1977), but the authors caution against drawing any conclusions at this time concerning the possible etiological role of these organisms in colon carcinoma.

The study by Peach and her colleagues (1974) reveals that eubacteria were isolated more frequently from the stools of people on a high carbohydrate diet living in areas with a low incidence of colon cancer. In 9.5% of 21 individuals on a high-carbohydrate diet, eubacteria were the predominant organisms in the fecal flora; this was true in none of the volunteers on a high-fat diet.

In summary, the available data document that eubacteria are present in large numbers in the stool and are capable of performing a variety of metabolic reactions, but do not yet tell us if their presence in the stool is harmful, essential, or beneficial.

Infected Tissue

Seventeen species of *Eubacterium* have been isolated from infected tissues of man or animals (Table 2). Eubacteria constitute an appreciable percentage of the anaerobic isolates from human infection: (i) 2.3–3.5% in a clinical microbiology laboratory (Martin, 1974), (ii) 2.5% in a research laboratory (Holland, Hill, and Altemeier, 1977), and (iii) 6.9% in the Wadsworth Anaerobic Bacteriology Research Laboratory over a 5-year period (unpublished data). In Martin's (1974) and Holland's (1977) laboratories, eubacteria were isolated from human infection less frequently than propionibacteria but more frequently than the other anaerobic, nonsporeforming, Gram-positive rods (*Actinomyces, Lactobacillus, Bifidobacterium,* and *Arachnia*). Eubacteria were the most common anaerobic, nonsporeforming, Gram-positive rods isolated in the Wadsworth Anaerobic Bacteriology Research Laboratory. This is probably due to the fact that our laboratory does not routinely process blood cultures that are commonly contaminated with propionibacteria.

Eubacterium lentum was the most frequent *Eubacterium* species isolated in the Wadsworth Anaerobic Bacteriology Research Laboratory (Table 4). *E. aerofaciens* and *E. rectale* are uncommon isolates from infection, even though they are the *Eubacterium* species isolated most frequently from the endogenous fecal flora. The majority (63%) of the clinical isolates cannot be identified using schemes outlined in the Anaerobe Laboratory Manual of the Virginia Polytechnic Institute (Holdeman, Cato, and Moore, 1977). However, clinical isolates were not studied to determine if they belonged to one of the recently described species of intestinal *Eubacterium* (Holdeman and Moore, 1974b; Wilkins et al., 1974; Moore, Johnson, and Holdeman, 1976). Only five species of *Eubacterium* were isolated more than once from human infection in our hospital from 1973 to 1977 (Table 4).

Table 4. Species of *Eubacterium* isolated from human infection and their frequency of isolation from clinical specimens (Wadsworth VA Hospital 1973–1977).[a]

Source (isolates of *Eubacterium*/ number of specimens)	*E. aerofaciens*	*E. alactolyticum*	*E. contortum*	*E. cylindroides*	*E. lentum*	*E. limosum*	*E. rectale*	*Eubacterium* sp.
Pleuropulmonary (29/25)	—	1	—	—	4	—	—	24
Intraabdominal (46/35)	2	—	5	2	12	4	—	21
Soft tissue (22/20)	1	—	—	—	5	—	1	15
Osteomyelitis (6/6)	—	—	—	—	1	—	—	5
Miscellaneous (5/5)	—	—	—	—	1	1	—	3
Total (108/91)	3	1	5	2	23	5	1	68

[a] *E. combesii, E. endocarditidis, E. ethylicum, E. moniliforme, E. nitritogenes, E. parvum, E. pseudotortuosum, E. tenue, E. tortuosum,* and *E. ventriosum* have been isolated from clinical specimens in other laboratories but were not identified in this laboratory during the period of study.

Table 5. Other bacteria isolated with *Eubacterium* species from clinical specimens (Wadsworth VA Hospital, 1973–1977).

Source (isolates of *Eubacterium*/number of specimens)	Anaerobes													Aerobes or facultatives				
	No other bacteria (pure growth of a *Eubacterium*)	*Bacteroides* (*B. fragilis* group)	Other *Bacteroides*	*Fusobacterium*	*Selenomonas*	*Actinomyces*	*Bifidobacterium*	*Clostridium*	*Lactobacillus*	*Propionibacterium*	*Peptostreptococcus*	*Peptococcus*	*Veillonella*	*Staphylococcus aureus*	*Streptococcus*	*Corynebacterium*	*Enterobacteriaceae*	*Pseudomonas*
Pleuropulmonary (29/25)	—	1	53	18	2	2	1	—	10	1	7	7	10	3	24	3	3	3
Intraabdominal (46/35)	—	56	37	12	—	1	1	32	7	1	19	4	6	1	22	4	47	6
Soft tissue (22/20)	1	8	26	4	—	2	1	1	2	1	8	11	2	2	11	2	11	—
Osteomyelitis (6/6)	—	—	7	—	—	—	—	—	—	—	2	7	1	—	4	4	5	—
Miscellaneous (5/5)	—	2	7	2	—	—	—	1	1	1	3	—	1	—	4	—	6	—
Total (108/91)	1	67	130	36	2	5	3	34	20	4	39	29	20	6	65	13	72	9

Only 1 of 109 eubacteria identified in the Wadsworth Anaerobic Bacteriology Research Laboratory was isolated in pure culture from a clinical specimen (Table 5). Eubacteria are present in pleuropulmonary infection in association with *Bacteroides* species other than *B. fragilis,* fusobacteria, and/or facultative streptococci. They are isolated from intraabdominal infections with *B. fragilis, other Bacteroides* species, clostridia, peptostreptococci, facultative streptococci, and/or Enterobacteriaceae (Table 5). It will take further study to determine if eubacteria play a symbiotic role in such mixed infections or are merely colonizers of infected tissue.

Only a few species of *Eubacterium* have been clearly shown to be pathogenic. *E. ventriosum* (Watanabe, 1972), *E. aerofaciens* (Sans and Crowder, 1973), and *E. endocarditidis* (Holdeman and Moore, 1974a) have caused endocarditis in man. *E. helminthoides* and *E. obstii* have the ability to cause experimental infections in animals (Holdeman and Moore, 1974a).

Eubacterium lentum has been isolated from patients with brain abscess, bacteremia, intraabdominal abscesses, postoperative wounds, and osteomyelitis (Table 3) (Holdeman and Moore, 1974a). It is probably a significant pathogen, but, as with other eubacteria, it is rarely isolated from infected tissue in pure culture. Even when isolated from blood, it is usually part of a polymicrobial bacteremia.

Isolation

The techniques used for isolating *Eubacterium* are, in general, those used for isolating all anaerobic bacteria. These are well outlined in the Anaerobe Laboratory Manual of the Virginia Polytechnic Institute (Holdeman, Cato, and Moore, 1977) and the Wadsworth Anaerobic Bacteriology Manual (Sutter, Vargo, and Finegold, 1975). Many of the eubacteria are extremely sensitive to oxygen, and careful attention should be paid to the techniques for transporting and processing samples anaerobically, as outlined in these manuals. Homogenization of particulate specimens is essential when quantitative bacteriology is to be attempted. Bryant (1959, 1970) demonstrated that the bacterial flora varies between different layers of the rumen contents. Attebery, Sutter, and Finegold (1974) have demonstrated a similar variation in different areas of stool specimens.

There are at present no good selective media for the isolation of species of *Eubacterium*. The optimum medium for quantitative recovery of intestinal organisms by the Hungate method is a modification of Bryant's rumen fluid–cellobiose agar (Moore and Holdeman, 1974a; Holdeman, Cato, and Moore, 1977).

Rumen Fluid–Cellobiose Medium for Isolating *Eubacterium* (Moore and Holdeman, 1974a; Holdeman, Cato, and Moore, 1977)

This prereduced medium contains:

Glucose	0.0248 g
Cellobiose	0.0248 g
Soluble starch	0.05 g
$(NH_4)_2 SO_4$	0.1 g
Resazurin solution	0.4 ml
Distilled water	20.0 ml
Salts solution	50.0 ml
Rumen fluid	30.0 ml
Cysteine HCl– H_2O	0.05 g

The addition of rumen fluid is clearly stimulatory for *E. cylindroides* (Holdeman and Moore, 1974a). Rumen fluid is available from Robbins Laboratories.

The salts solution contains:

$CaCl_2$ (anhydrous)	0.2 g
$MgSO_4 \cdot 7H_2O$	0.2 g
K_2HPO_4	1.0 g
KH_2PO_4	1.0 g
$NaHCO_3$	10.0 g
NaCl	2.0 g

To prepare the salts solution, mix $CaCl_2$ and $MgSO_4$ in 300 ml distilled water until dissolved. Add 500 ml water and, while swirling, slowly add remaining salts. Continue swirling until all salts are dissolved. Add 200 ml distilled water, mix, and store at 4°C.

The Wadsworth Anaerobic Bacteriology Research Laboratory favors a supplemented brucella base blood agar medium for the isolation of anaerobes from clinical specimens (Sutter, Vargo, and Finegold, 1975). This medium contains 4.3 g brucella agar and 100 ml H_2O. After the medium is autoclaved, it is supplemented with vitamin K_1 (10 $\mu g/ml$) and defibrinated sheep blood (5% vol/vol). The brucella blood agar plates are reduced after they are poured by placing them in an anaerobic chamber or GasPak anaerobic jar (Baltimore Biological Laboratories) for 48 h prior to use.

Eubacterium lentum is nonsaccharolytic and has been shown to require arginine or citrulline for optimum growth (Sperry and Wilkins, 1976b). The production of adenosine triphosphate from the arginine dehydrolase pathway appears to be the sole source of energy for this organism. Optimum growth can be achieved by adding 0.5% (wt/vol) arginine to any common bacteriological medium. Growth is enhanced when the organism is incubated under 100% CO_2, which lowers the initial pH of the medium to accommodate the ammonia produced (Holdeman, Cato, and Moore, 1977).

The addition of 0.02% Polysorbate-80 (final concentration) to liquid medium enhances the growth of

E. aerofaciens, E. budayi, E. cellulosolvens, E. cylindroides, E. lentum, and *E. saburreum* (Holdeman, Cato, and Moore, 1977).

Identification

Identification procedures for *Eubacterium* spp. are elegantly outlined in the Anaerobe Laboratory Manual of the Virginia Polytechnic Institute (Holdeman, Cato, and Moore, 1977). The genus is defined in part on the basis of the end products of metabolism. Therefore, gas–liquid chromatographic analysis is essential in identifying the organisms. Once the genus is established, approximately 19 biochemical reactions are used to determine the species of the more commonly recognized organisms (Holdeman, Cato, and Moore, 1977).

Certain species can be distinguished on the basis of unique characteristics. *E. cellulosolvens* digests cellulose and shows distinct clearing around the colony on media containing cellulose (Prins et al., 1972). *E. lentum* does not produce acid from carbohydrates when grown in the presence of Polysorbate-80 (Holdeman, Cato, and Moore, 1977). In addition, *E. lentum* is the only *Eubacterium* demonstrated to have a cytochrome system (Sperry and Wilkins, 1976a). *E. limosum* produces a slimy sediment in glucose-containing media and produces different fatty acids in peptone yeast broth than in peptone yeast broth with added glucose (Holdeman, Cato, and Moore, 1977). *E. alactolyticum* has a unique ability to add two carbon fragments to fatty acids in the medium (e.g., it can produce butyric acid and caproic acid from acetic acid) (L. V. Holdeman and W. E. C. Moore, personal communication).

The cellular and colonial morphology may be variable and confusing. *E. alactolyticum* resembles *Propionibacterium acnes. E. biforme* has a variable colonial morphology and may resemble a mixed culture on solid media (Moore and Holdeman, 1974a). *Streptococcus intermedius* and *E. aerofaciens* resemble each other morphologically and biochemically, but may be distinguished on the basis of the ability of *E. aerofaciens* to produce hydrogen. The abundant production of hydrogen helps distinguish *E. contortum* from *Peptostreptococcus productus* (Holdeman, Cato, and Moore, 1977).

Despite the identification schemes available, the majority of clinical isolates of eubacteria cannot be identified to species. This remains a fruitful area for further investigation.

Literature Cited

Attebery, H. R., Sutter, V. L., Finegold, S. M. 1974. Chapter X. Normal human intestinal flora, pp. 81–97. In: Balows, A.,

DeHaan, R. M., Dowell, V. R., Guze, L. B. (eds.), Anaerobic bacteria: Role in disease. Springfield, Illinois: Charles C. Thomas.

Bryant, M. P. 1959. Bacterial species of the rumen. Bacteriological Reviews **23:**125–153.

Bryant, M. P. 1970. Normal flora-rumen bacteria. American Journal of Clinical Nutrition **23:**1440–1450.

Bryant, M. P., Robinson, I. M. 1962. Some nutritional characteristics of predominant culturable bacteria. Journal of Bacteriology **85:**605–614.

Dickinson, A. B., Gustafsson, B. E., Norman, A. 1971. Determination of bile acid conversion potencies of intestinal bacteria by screening *in vitro* and subsequent establishment in germ free rats. Acta Pathologica et Microbiologica Scandinavica, Sect. B **79:**691–698.

Eyssen, H., Parmentier, G. 1974. Biohydrogenation of sterols and fatty acids by the intestinal microflora. American Journal of Clinical Nutrition **27:**1329–1340.

Eyssen, H. J., Parmentier, G. G., Compernolle, F. C., DePauw, G., Piessens-Denef, M. 1973. Biohydrogenation of sterols by *Eubacterium* ATCC 21408—Nova species. European Journal of Biochemistry. **36:**411–421.

Finegold, S. M., Attebery, H. R., Sutter, V. L. 1974. Effect of diet on human fecal flora: Comparison of Japanese and American diets. American Journal of Clinical Nutrition **27:**1455–1469.

Finegold, S. M., Flora, D. J., Attebery, H. R., Sutter, V. L. 1975. Fecal bacteriology of colonic polyp patients and control patients. Cancer Research **35:**3407–3417.

Finegold, S. M., Sutter, V. L., Sugihara, P. T., Elder, H. A., Lehmann, S. M., Phillips, R. L. 1977. Fecal microbial flora in Seventh Day Adventist populations and control subjects. American Journal of Clinical Nutrition **30:**1781–1792.

Hill, M. J., Drasar, B. S. 1974. Chapter XII. Bacteria and the etiology of cancer of the large intestine, pp. 119–133. In: Balows, A., DeHaan, R. M., Dowell, V. R., Guze, L. B. (eds.), Anaerobic bacteria: Role in disease. Springfield, Illinois: Charles C Thomas.

Hoffman, J., Lindberg, B., Svensson, S., Hofstad, T. 1974. Structure of the polysaccharide antigen of *Eubacterium saburreum,* strain L44. Carbohydrate Research **35:**49–53.

Hoffman, J., Lindberg, B., Lonngren, J., Hofstad, T. 1976. Structural studies of the polysaccharide antigen of *Eubacterium saburreum* strain 49. Carbohydrate Research **47:**261–267.

Hofstad, T. 1972. A polysaccharide antigen of an anaerobic oral filamentous microorganism (*Eubacterium saburreum*) containing heptose and O-acetyl as main constituents. Acta Pathologica et Microbiologica Scandinavica, Sect. B **80:**609–614.

Hofstad, T. 1974. Presence in dental plaque of a polysaccharide antigen of the anaerobic oral filamentous organism. *Eubacterium saburreum.* Scandinavian Journal of Dental Research **82:**82–84.

Hofstad, T., Henning, L. 1977. Composition and antigenic properties of a surface polysaccharide isolated from *Eubacterium saburreum,* strain L452. Acta Pathologica et Microbiologica Scandinavica, Sect. B **85:**14–17.

Holdeman, L. V., Moore, W. E. C. 1974a. *Eubacterium,* pp. 641–657. In: Buchanan, R. E., Gibbons, N. E. (eds.), Bergey's manual of determinative bacteriology, 8th ed. Baltimore: Williams & Wilkins.

Holdeman, L. V., Moore, W. E. C. 1974b. New genus *Coprococcus,* twelve new species, and emended descriptions of four previously described species of bacteria from human feces. International Journal of Systematic Bacteriology **24:**260–277.

Holdeman, L. V., Cato, E. P., Moore, W. E. C. (eds.). 1977. Anaerobe laboratory manual, 4th ed. Blacksburg, Virginia: Virginia Polytechnic Institute and State University.

Holdeman, L. V., Good, I. J., Moore, W. E. C. 1976. Human fecal flora: Variation in bacterial composition within individuals and a possible effect of emotional stress. Applied and Environmental Microbiology **37**:359–375.

Holland, J. W., Hill, E. O., Altemeier, W. A. 1977. Numbers and types of anaerobic bacteria isolated from clinical specimens since 1960. Journal of Clinical Microbiology **5**:20–25.

Hungate, R. E. 1966. The rumen and its microbes. New York: Academic Press.

Hylemon, P. B., Stellwag, E. J. 1976. Bile acid biotransformation rates of selected Gram-positive and Gram-negative intestinal anaerobic bacteria. Biochemical and Biophysical Research Communications **69**:1088–1094.

Klein, E. 1908. On the nature and causes of taint in miscured hams (*Bacillus foedans*). Lancet **i**:1832–1834.

Maier, B. R., Flynn, M. A., Burton, G. C., Tsutakawa, R. K., Heulges, D. J. 1974. Effects of a high beef diet on bowel flora: A preliminary report. American Journal of Clinical Nutrition **27**:1470–1474.

Martin, W. J. 1974. Isolation and identification of anaerobic bacteria in the clinical laboratory. Mayo Clinic Proceedings **49**:300–308.

Mastromarino, A., Bandaru, S. R., Wynder, E. L. 1976. Metabolic epidemiology of colon cancer: Enzymic activity of fecal flora. American Journal of Clinical Nutrition **29**:1455–1460.

Midtvedt, T. 1974. Microbial bile acid transformation. American Journal of Clinical Nutrition **27**:1341–1347.

Moore, W. E. C., Holdeman, L. V. 1973. New names and combinations in the genera *Bacteroides* Castellani and Chalmers, *Fusobacterium* Knorr, *Eubacterium* Prévot, *Propionibacterium* Delwich, and *Lactobacillus* Orla-Jensen. International Journal of Systematic Bacteriology **23**:69–74.

Moore, W. E. C., Holdeman, L. V. 1974a. Special problems associated with the isolation and identification of intestinal bacteria in fecal flora studies. American Journal of Clinical Nutrition **27**:1450–1455.

Moore, W. E. C., Holdeman, L. V. 1974b. Human fecal flora: The normal flora of 20 Japanese-Hawaiians. Applied Microbiology **27**:961–979.

Moore, W. E. C., Johnson, J. L., Holdeman, L. V. 1976. Emendation of *Bacteroidaceae* and *Butyrivibrio* and descriptions of *Desulfomonas*, gen. nov. and ten new species in the genera *Desulfomonas*, *Butyrivibrio*, *Eubacterium*, *Clostridum* and *Ruminococcus*. International Journal of Systematic Bacteriology **26**:238–252.

Parmentier, G., Eyssen, H. 1974. Mechanism of biohydrogenation of cholesterol to coprostanol by *Eubacterium* ATCC 21408. Biochimica et Biophysica Acta **348**:279–284.

Peach, S., Fernandez, F., Johnson, K., Drasar, B. S. 1974. The nonsporing anaerobic bacteria in human feces. Journal of Medical Microbiology **7**:213–221.

Prévot, A. R. 1938. Etudes de systématique bacterienne. III. Invalidité du genre *Bacteroides* Castellani et Chalmers demembrement et reclassification. Annales de l'Institut Pasteur **60**:285–307.

Prins, R. A., van Fugt, F., Hungate, R. E., van Vorstenbosch, C. J. A. H. V. 1972. A comparison of strains of *Eubacterium cellulosolvens* from the rumen. Antonie van Leeuwenhoek Journal of Microbiology and Serology **38**:153–161.

Sans, M. D., Crowder, J. G. 1973. Subacute bacterial endocarditis caused by *Eubacterium aerofaciens*. American Journal of Clinical Pathology **59**:576–580.

Sperry, J. F., Wilkins, T. D. 1967a. Cytochrome spectrum of an obligate anaerobe, *Eubacterium lentum*. Journal of Bacteriology **125**:905–909.

Sperry, J. F., Wilkins, T. D. 1976b. Arginine, a growth-limiting factor for *Eubacterium lentum*. Journal of Bacteriology **127**:780–784.

Sutter, V. L., Vargo, V. L., Finegold, S. M. 1975. Wadsworth anaerobic bacteriology manual, 2nd ed. Los Angeles: Department of Continuing Education in Health Sciences, University Extension and School of Medicine, University of California.

Watanabe, Y. 1972. Subacute bacterial endocarditis due to *Eubacterium ventriosum*: Report of a case. Journal of the Japanese Association for Infectious Diseases **42**:78–83.

Wensink, F. 1975. The faecal flora of patients with Crohn's disease. Antonie van Leeuwenhoek Journal of Microbiology and Serology **41**:214–215.

Wilkins, T. D., Fulghum, R. S., Wilkins, J. H. 1974. *Eubacterium plexicaudatum* sp. nov., an anaerobic bacterium with a subpolar tuft of flagella, isolated from a mouse cecum. International Journal of Systematic Bacteriology **24**:408–411.

The Actinomycetes

CHAPTER 147

Introduction to the Order Actinomycetales

HUBERT A. LECHEVALIER and MARY P. LECHEVALIER

The order *Actinomycetales* contains bacteria that are characterized by the formation of branching filaments. In the more evolved forms of the order, this property results in a somewhat fungal appearance. Abundantly distributed in nature, actinomycetes may be separated into two large but unequal subgroups: the oxidative forms, which are very numerous and are basically soil inhabitants, and the fermentative types, which are primarily found in the natural cavities of man and animals (Lechevalier and Pine, 1977).

In general, actinomycetes are Gram positive, but part of their thallus may be Gram negative. Certain filamentous, branching, Gram-negative bacteria belonging to the genus *Mycoplana* may be classified with the actinomycetes as a matter of convenience, although they are not phylogenetically related to the other members of the order (Sukapure et al., 1970). Other Gram-negative bacteria that reproduce by budding, such as members of the genus *Hyphomicrobium*, form branching filaments but have never been considered members of the *Actinomycetales* (Hirsch, 1974).

Closely related to the actinomycetes are the corynebacteria and their relatives, the latter often described as "coryneforms". These are discussed in Chapter 140 of this Handbook. The mycobacteria, which are related to the corynebacteria by many properties, are usually more rod shaped than filamentous and are currently included in the order *Actinomycetales*. The separation between corynebacteria, mycobacteria, and some of the other pleomorphic bacteria is not easy. The same strain, depending on the observer, may be classified in the genus *Corynebacterium, Arthrobacter, Mycobacterium,* or *Nocardia* (Gordon, 1966).

In general, actinomycetes will grow on ordinary laboratory media, but their growth is usually slower than that of ordinary bacteria. A division cycle in actinomycetes may take 2–3 h as compared with 20 min for *Escherichia coli*. Some actinomycetes grow even more slowly. *Mycobacterium tuberculosis,* for example, has a generation time of about 15 h under optimal conditions. Other actinomycetes, such as *M. leprae,* have never been grown on laboratory media. Some endophytes of plants, placed in the genus

Frankia, grow on present-day laboratory media only with the greatest difficulty.

Actinomycetes are very important from a medical point of view as the agents of tuberculosis, nocardiosis, mycetomas, streptothricosis, allergic pneumonias, bovine farcy, paratuberculosis, actinomycosis, and various abscesses. They also cause a few plant diseases, including potato scab, a rot of sweet potato, and a disease of blueberry. Their main ecological role is in the decomposition of organic matter in and on the soil, and it is the soil that is their main habitat (Waksman, 1959). They may be a nuisance, as when they decompose rubber products, grow in aviation fuel, produce odorous substances that pollute water supplies, or grow in sewage-treatment plants where they form thick clogging foams (Lechevalier, 1974). In contrast, actinomycetes are the producers of most of the antibiotics, compounds that are useful not only in human and veterinary medicine but also in agriculture and in biochemistry (as metabolic poisons) (Okami, 1973; Waksman and Lechevalier, 1962). Recent books on actinomycetes include those edited by Prauser (1970), Sykes and Skinner (1973), Arai (1976) and Mordarski, Kurylowicz, and Jeljaszewicz (1978). The fermentative actinomycetes have been covered by Slack and Gerencser (1975).

At first, actinomycetes were thought of only as pathogens. Ferdinand Cohn probably published the first description of an actinomycete in 1875 when he had observed a filamentous organism in a concretion from a human lacrimal duct and called it *Streptothrix foersteri.* Two years later, C. O. Harz gave the name *Actinomyces bovis* to an organism that he observed in a case of bovine lumpy jaw. The name "actinomyces", which is of Greek derivation, means "ray-fungus", and actinomycetes are still referred to as "ray-fungi", particularly by German and Russian authors. The importance of actinomycetes in soil was largely realized through the work of Beijerinck, Krainsky, Conn, and Waksman during the first two decades of this century (Waksman, 1959).

With these early studies, there began a controversy about the nature of actinomycetes. Some considered them to be filamentous bacteria, others minute fungi. The advent of electron microscopy and the development of our knowledge of the fundamental properties of microorganisms have resolved the controversy in favor of the bacteria (Lechevalier and Lechevalier, 1967). There are, however, some minute fungi such as *Fusidium* spp. that can be mistaken for actinomycetes (Lechevalier et al., 1977).

There is so much variation in the morphology of actinomycetes that it is impossible to describe them in a detailed way in an introductory chapter such as this. Generally, one may say that on solid media most actinomycetes form a mycelium (substrate or primary) that grows on and into the agar. In addition, there may be a mycelium (aerial or secondary) growing away from the medium. In some cases, the primary mycelium is short-lived and soon breaks up into bacillary or coccoid elements that may be flagellated. In some species, the mycelium may be so transient as to escape notice, or be nonexistent. These atypical organisms are considered members of the *Actinomycetales* because they share other properties with more orthodox members of the order. Thus, actinomycetes form a morphological spectrum ranging from diphtheroid bacilli to filamentous forms with intricate modes of sporulation.

The actinomycetes are separated into groups on the basis of physiology (fermentative vs. oxidative metabolism), morphology (type and stability of mycelium, types, number, and disposition of spores, formation of sclerotia, sporangia, or synnemata, formation of flagellate elements), physical qualities (heat resistance), and chemistry (cell wall and whole cell composition, types of lipids, isoprenoid quinones).

The fermentative actinomycetes are morphologically simple organisms. They form neither aerial mycelia nor spores. Their primary mycelium may be quite well developed or rudimentary. A summary of the properties of the genera of fermentative actinomycetes is presented in Table 1. Species of *Agromyces* and *Oerskovia* are found in soil, although oerskoviae can also be isolated from clinical speci-

mens (Sottnek et al., 1977). The other five genera are associated with man and animals.

The vast majority of actinomycetes are oxidative and aerobic. Their separation into genera is most easily carried out by utilizing morphological, physical, and chemical criteria (Cross and Goodfellow, 1973; Lechevalier and Lechevalier, 1965; Lechevalier, Lechevalier, and Gerber, 1971).

In the study of the morphology of actinomycetes, it is important not to distort or destroy the arrangement of hyphae and spores. As a consequence, actinomycetes should be studied microscopically by the methods that are used for the study of molds. The in situ examination of cultures growing on agar plates is helped by the use of long-working distance condensers and objectives. Particular attention should be paid to the formation and location of the hyphae, the formation and arrangements of conidia, the presence of sporangia (called spore vesicles by Cross and Attwell, 1975), the release of motile elements, and the occurrence of special structures such as a sclerotia and spore-bearing synnemata.

A group of actinomycetes is distinguished by the presence of endospores having high heat resistance. In general, the G+C content of the DNA of actinomycetes is high, the mycobacteria and nocardiae being at the low end of this high spectrum (60–70 mol%) and the streptomycetes on the high side (70–75 mol%). Some thermophilic actinomycetes have DNAs with low G+C percentages (44–54 mol%) (Lechevalier et al., 1971).

From a chemical point of view, cell wall composition has been found to be especially useful (Lechevalier and Lechevalier, 1976). Most oxidative actinomycetes can be separated into four groups on the basis of their cell wall composition (Table 2), which is of the Gram-positive type.

The most common cell wall types (I to IV) have peptidoglycans containing diaminopimelic acid. This amino acid, which occurs in three isomeric forms, is easy to detect in hydrolysates of whole

Table 1. Summary of the properties of fermentative actinomycetes. These organisms do not form aerial mycelia or spores.

Genus	Characteristic cell wall constituents[a]	Catalase	Motility	Relation to oxygen
Actinomyces	Lysine with or without ornithine	− or +	−	Anaerobic or facultative
Agromyces	Diaminobutyric acid	−	−	Aerobic to microaerophilic
Arachnia	L-DAP[b]	−	−	Aerobic to microaerophilic
Bacterionema	meso-DAP[b]			
	Arabinose + galactose	+	−	Aerobic to facultative
Bifidobacterium	Lysine or ornithine	−	−	Anaerobic
Oerskovia	Lysine + aspartic acid	+ or −[c]	+	Aerobic to microaerophilic
Rothia	Lysine	+	−	Aerobic to microaerophilic

[a] All strains contain glutamic acid, alanine, glucosamine, and muramic acid.

[b] DAP = 2,6-diaminopimelic acid. No distinction is made between the *meso* and the D forms of this acid.

[c] Catalase negative when grown under anaerobic conditions.

Table 2. Major constituents of cell walls of actinomycetes.

Cell wall type	Major constituents[a]	Genera	
		Example	Total no.[b]
I	L-DAP[c] glycine	*Streptomyces*	11
II	*meso*-DAP, glycine	*Micromonospora*	5
III	*meso*-DAP	*Actinomadura*	13
IV	*meso*-DAP, arabinose, galactose	*Nocardia*	8
V	Lysine, ornithine	*Actinomyces israelii*	1
VI	Lysine (aspartic acid; galactose)[d]	*Oerskovia*	4
VII	DAB[e] glycine (lysine)	*Agromyces*	1
VIII	Ornithine	*Bifidobacterium*	2
IX	*meso*-DAP, numerous amino acids	*Mycoplana*	1

[a] All cell wall preparations contain major amounts of alanine, glutamic acid, glucosamine, and muramic acid.
[b] Total number of actinomycete genera known to have this cell wall type.
[c] DAP = 2,6-diaminopimelic acid.
[d] Bracketed constituents are variable.
[e] DAB = 2,4-diaminobutyric acid.

cells. Two of the isomers, the *meso* and L forms, are readily separable by paper chromatography. The D isomer is not readily separated from the *meso* form and is of unknown taxonomic significance.

In most cases, it is also possible to recognize cell wall types II, III, and IV without having to isolate the cell walls. This is done by the detection of certain sugars that are found in whole-cell hydrolysates (Table 3) (Lechevalier, 1968).

In addition, some of the organisms with type IV cell walls produce α-branched, β-hydroxylated fatty acids called mycolic acids. These lipids fall into three broad groups on the basis of molecular weight: the largest are typical of the genus *Mycobacterium*, the smallest are produced by some members of the genus *Corynebacterium,* and those of intermediate

Table 3. Cell wall types and whole cell sugar patterns of aerobic actinomycetes containing *meso*-diaminopimelic acid.[a]

Cell wall		Whole cell sugar pattern	
Type	Distinguishing major constituents[b]	Type	Diagnostic sugars
II	Glycine	D	Xylose, arabinose
III	None	B	Madurose[c]
		C	None
IV	Arabinose, galactose	A	Arabinose, galactose

[a] No differentiation is made between *meso*-DAP and D-DAP.
[b] All cell wall preparations contain major amounts of alanine, glutamic acid, glucosamine, and muramic acid.
[c] Madurose = 3-*O*-methyl-D-galactose (Lechevalier and Gerber, 1970).

Table 4. Phospholipid patterns of aerobic actinomycetes.

Phospholipid pattern	Characteristic phospholipids[a]				
	PE	PME	PC	GluNU	PG
PI	−[b]	−	−	−	V[c]
PII	+[d]	−	−	−	−
PIII	V	V	+	−	V
PIV	V	V	−	+	−
PV	−	−	−	+	+

[a] PE, phosphatidyl ethanolamine; PME, phosphatidyl methylethanolamine; PC, phosphatidyl choline; GluNU, Phospholipids of unknown structure containing glucosamine; PG, Phosphatidyl glycerol. Other phospholipids of no taxonomic value may be present.
[b] Not present.
[c] Variably present.
[d] Present.

molecular weight are found in species of *Nocardia* (Lechevalier, Lechevalier, and Horan, 1973).

Phospholipid patterns have recently been shown to cast considerable light on the interrelationships of various aerobic actinomycete genera (Lechevalier, de Bièvre, and Lechevalier, 1977). As is summarized in Table 4, five groups are recognizable on the basis of their content of nitrogenous phospholipids. More detailed data are given at the end of this introduction.

Table 5. Genera of Actinomycetales with a type I cell wall.

Generic name	Morphological characteristics
Streptomyces	Aerial mycelium with chains (usually long) of nonmotile conidia.
Streptoveticillium	Same as *Streptomyces,* but the aerial mycelium bears verticils consisting of at least three side branches, which may be chains of conidia or hold sporulating terminal umbels.
Nocardioides	Both substrate and aerial mycelia fragment into rod- and coccus-shaped elements.
Chainia	Same as *Streptomyces,* but sclerotia are also formed.
Actinopycnidium	Same as *Streptomyces,* but pycnidia-like structures are also formed.
Actinosporangium	Same as *Streptomyces,* but spores accumulate in drops.
Elytrosporangium	Same as *Streptomyces,* but merosporangia are also formed.
Microellobosporia	No chains of conidia; merosporangia with nonmotile spores are formed.
Sporichthya	No substrate mycelium is formed; aerial chains of motile, flagellated conidia are held to the surface of the substratum by holdfasts.
Intrasporangium	No aerial mycelium; substrate mycelium forms terminal and subterminal vesicles.
Arachnia	No aerial mycelium; substrate mycelium is branched and may fragment.

Table 6. Genera of Actinomycetales with a type II cell wall.

Generic name	Morphological characteristics
Micromonospora	Aerial mycelium absent, conidia single.
Actinoplanes	Globose to lageniform sporangia; globose spores with one polar tuft of flagella.
Amorphosporangium	Same as Actinoplanes, but the sporangia are often very irregular; sporangiospores are usually nonmotile.
Ampullariella	Lageniform to globose sporangia; rod-shaped spores with one polar tuft of flagella.
Dactylosporangium	Claviform sporangia, each with one chain of spores with one polar tuft of flagella.

Table 7. Genera of Actinomycetales with a type III cell wall.

Generic name	Morphological characteristics
Thermoactinomyces	Single spores are formed on the aerial and substrate mycelia; the spores are heat-resistant endospores.
Thermomonospora	Single spores are formed on the aerial mycelium or on both the aerial and substrate mycelia; the spores are not heat-resistant endospores.
Microbispora[a]	Longitudinal pairs of conidia on the aerial mycelium.
Actinomadura[a]	Short chains of conidia on the aerial mycelium.
Microtetraspora[a]	Same as Actinomadura; number of spores in chain not exceeding 6; usually 4.
Nocardiopsis	Very long chains of conidia on the aerial mycelium.
Excellospora[a]	Single and short chains of conidia on both the aerial and the substrate mycelia.
Planomonospora[a]	Cylindrical sporangia, each containing one motile spore with one polar tuft of flagella.
Planobispora[a]	Cylindrical sporangia, each containing two motile spores with peritrichous flagella.
Streptosporangium[a]	Globose sporangia containing nonmotile spores.
Spirillospora[a]	Globose sporangia with rod-shaped spores, each with a subpolar tuft of flagella.
Actinosynnema	Formation of synnemata bearing chains of motile conidia.
Dermatophilus[a]	Hyphae dividing in all planes, forming packets of cocci motile by means of a tuft of flagella; pathogenic to animals.
Geodermatophilus	Similar to Dermatophilus; nonpathogenic; found in soil.
Frankia	Sporangia with nonmotile spores. Grow in symbiotic association with roots of higher plants. Fix nitrogen. Culture in vitro with difficulty.

[a] Madurose-positive organisms (see Table 3).

In Table 5 are listed the genera of actinomycetes whose members have a type I cell wall. The most important genus of the group is *Streptomyces*. Streptomycetes are abundantly distributed in soil and are the source of most of the antibiotics in current use. The number of species of streptomycetes that have been described probably exceeds a thousand, although it is felt by many investigators that several different names have probably been given to the same species.

In Table 6 are listed the genera of actinomycetes with a type II cell wall. The genera *Micromonospora* and *Actinoplanes* are the two most frequently encountered members of this group. Micromonosporae favor water-logged soils and actinoplanetes are found on organic matter decomposing at the edge of water bodies and in soils rich in organic matter.

The genera of actinomycetes with a type III cell wall are listed in Table 7. This group is composed of morphologically, physiologically, and ecologically very different organisms. From a chemical point of view, one can recognize two kinds: those with and those without madurose (3-*O*-methyl-D-galactose). This compound, though not a cell wall constituent, appears to be constantly present in the organisms that produce it. The madurose-positive group includes the pathogens *Actinomadura madurae* and *Dermatophilus congolensis*. The madurose-negative thermoactinomycetes form true heat-resistant bacterial endospores, and may cause serious allergic reactions.

In Table 8 are listed the genera of actinomycetes whose members have cell walls of type IV. To this

Table 8. Genera of Actinomycetales with a type IV cell wall.

Generic name	Morphological characteristics
Mycobacterium	Filamentation is usually limited, and aerial mycelium is usually not formed; filaments fall easily apart into rods and cocci.
Nocardia	Filamentation is abundant, and aerial mycelium is often formed; chains of conidia may be formed.
Micropolyspora	Short chains of globose conidia are formed on both the aerial and the substrate mycelia.
Pseudonocardia	Long, cylindrical conidia in chains on the aerial mycelium.
Saccharomonospora	Single spores, mainly on the aerial mycelium.
Saccharopolyspora	Morphology similar to that of Nocardiopsis.
Bacterionema	No aerial mycelium; substrate mycelium is branched and fragments. Filaments often terminate in a bacillus-like body.
Actinopolyspora	Long chains of conidia on the aerial mycelium; substrate mycelium may fragment. Halophile.

Table 9. Key to some families and genera of the Actinomycetales (and the relevant chapters in this Handbook in which they are treated).[a]

Fermentative, nonmotile organisms. CW I. IV, V, VI, or VII	Actinomycetaceae (Chapters 148, 149)
Facultatively fermentative; hyphae breaking into motile elements. CW VI; Phos P V.	*Oerskovia* (Chapter 159)
Oxidative metabolism.	
Hyphae dividing in more than one plane to form masses of motile cocci. CW III; WCS B or C; Phos P I or Phos P II.	*Dermatophilaceae* (Chapter 154)
Symbionts in plant nodules; hyphae dividing in more than one plane; sporangia present, nonmotile spores formed. CW III.	*Frankia* (Chapter 152)
Hyphae dividing only perpendicularly to their long axis.	
Gram-negative; rudimentary branching; breaking into motile elements. Cell wall with *meso*-diaminopimelic acid. Phos P III.	*Mycoplana* (Chapter 159)
Gram-positive.	
Mycelium usually rudimentary or absent; often acid-fast. CW IV; WCS A; Phos P II.	*Mycobacterium* (Chapters 150, 151)
Little or no aerial mycelium; no spores. Nonmotile. Vegetative hyphae may fragment.	
CW IV; WCS A; Phos P II. Nocardomycolates present.	*Nocardia* *Rhodococcus* (Chapter 155)
CW VI; Phos P V. Nocardomycolates absent.	*Promicromonospora* (Chapter 159)
CW I; Phos P IV. Vegetative hyphae bear terminal and subterminal vesicles.	*Intrasporangium* (Chapter 159)
Aerial hyphae sparse to abundant; may fragment. No spores.	
CW IV; WCS A; Phos P II. Nocardomycolates present.	*Nocardia* (Chapter 155)
CW I; Phos P I. Nocardomycolates absent.	*Nocardioides* (Chapter 159)
Single spores produced.	
Heat-sensitive spores on primary mycelium only; no to sparse aerial mycelium. Usually mesophilic. CW II; WCS D; Phos P II.	*Micromonospora* (Chapter 157)
Heat-sensitive spores on aerial and vegetative mycelium. Usually thermophilic. CW III; WCS B or C.	*Thermomonospora* (Chapter 157)
Heat-sensitive spores on aerial mycelium only. Vegetative hyphae may fragment. Occasional pairs of spores may occur. Thermophilic and mesophilic. CW IV; WCS A; Phos P II.	*Saccharomonospora* (Chapter 157)
Heat-resistant endospores on aerial and vegetative mycelium. Usually thermophilic. CW III; WCS B or C.	*Thermoactinomyces* (Chapter 157)
Pairs of spores produced longitudinally.	
Spores on the aerial mycelium only. CW III; WCS B; Phos P IV.	*Microbispora* (Chapter 158)
Spores on the primary and aerial mycelium. CW IV; WCS A; Phos P III. No nocardomycolates.	*Micropolyspora* (Chapter 158)
Spore chains produced on aerial mycelium only, except as noted.	
Chains of 1–6 spores (average 4). CW III; WCS B; Phos P I or P IV.	*Microtetraspora* (Chapter 158)
Chains of 2–35 spores.	
CW III; WCS B; Phos P I or P IV.	*Actinomadura* (Chapter 158)
CW III, WCS B; P?	*Excellospora* (Chapter 158)
Chains of 4 to more than 100 spores. Chains may be straight, flexuous, spiral or verticillate.	
CW I; Phos P II.	*Streptomyces* *Streptoverticillium* (Chapter 156)

Table 9. Key to some families and genera of the Actinomycetales (and the relevant chapters in this Handbook in which they are treated).[a] (Continued)

CW III; WCS C; Phos P III.	*Nocardiopsis* (Chapter 158)
CW IV; WCS A; Phos P II. Nocardomycolates present.	*Nocardia* (Chapter 155)
CW IV; WCS A; Phos unknown. Nocardomycolates absent.	*Saccharopolyspora* (Chapter 158)
CW IV; WCS A; Phos P III. Mycolates unknown. Halophilic.	*Acinopolyspora* (Chapter 158)
Chains of long cylindrical spores formed by budding, redividing to form shorter spores. CW IV; WCS A; Phos P III. No nocardomycolates.	*Pseudonocardia* (Chapter 158)
Chains of spores that aggregate to form sticky masses. CW I.	*Streptomyces Actinosporangium* (Chapter 156)
Chains of spores on both aerial and primary mycelia. CW I; Phos P II	*Streptomyces* (Chapter 156)
Chains of about 2–15 spores on primary and aerial mycelia and at agar surface.	
Nocardomycolates present. CW IV; WCS; A; Phos P II.	*Micropolyspora* (*brevicatena* type)
Nocardomycolates absent. CW IV; WCS A; Phos P III.	*Micropolyspora* (*faeni* type)
Chains of spores on the aerial mycelium; podlike structures on the primary mycelium. CW I.	*Elytrosporangium* (Chapter 156)
Sporangia present containing:	
Single motile spores. Aerial hyphae abundant. CW III; WCS B; P IV.	*Planomonospora* (Chapter 153)
Pairs of motile spores. Aerial hyphae abundant. CW III; WCS B; P IV.	*Planobispora* (Chapter 153)
Single row (3–6) of motile spores. Aerial hyphae scant. CW II; WCS D; Phos P II.	*Dactylosporangium* (Chapter 153)
Single row (2–6) of nonmotile spores. Aerial hyphae abundant. CW I; Phos P II.	*Microellobosporia* (Chapter 156)
Coils or rows of motile, round spores. Aerial hyphae scant. CW II, VI or VIII; WCS D; Phos P II.	*Actinoplanes* (Chapter 153)
Rows of motile, rod-shaped spores. Aerial hyphae scant. CW II; WCS D; Phos II.	*Ampullariella* (Chapter 153)
Coils of motile, rod-shaped spores. Aerial hyphae variable. CW III; WCS B; Phos II or I.	*Spirillospora* (Chapter 153)
Coils of nonmotile spores. Aerial hyphae scant. CW II; WCS D; Phos P II.	*Amorphosporangium* (Chapter 153)
Coils of nonmotile round spores. Aerial hyphae abundant. CW III; WCS B; Phos P IV.	*Streptosporangium* (Chapter 153)
Sclerotia present; long chains of spores also formed by most strains. CW I; Phos P II.	*Chainia* (Chapter 156)
Pycnidia-like structures formed. Long chains of spores on aerial mycelium. CW I.	*Actinopysnidium* (Chapter 156)
Synnemata present bearing chains of motile rod-shaped spores. CW III; WCS C; Phos P II.	*Actinosynnema* (Chapter 159)

[a] CW, cell wall composition; WCS, whole cell sugar pattern; Phos, phospholipid pattern.

group belong the important genera *Mycobacterium* and *Nocardia*. Some members of the genus *Corynebacterium,* not presently included in the Actinomycetales, also have a cell wall of type IV and are discussed in Chapters 140 and 141 of this Handbook.

The separation of actinomycetes into families is not satisfactory. *Faute de mieux,* some of the actinomycete families and genera discussed in this Handbook are delineated in the key presented in Table 9. A chemotaxonomic summary of the genera of the Actinomycetales is given in Table 10.

Table 10. Chemotaxonomic summary of the Actinomycetales.

Genus	Cell wall type[a]	Whole cell sugar[b]	Mycolic acid	Phospholipid type[c]
Nonsporate organisms				
Actinomyces	V/VI	*[d]	—	PII
Agromyces	VII	*	—	PI
Arachnia	I	*	—	
Bacterionema	IV	A	NM	PI
Bifidobacterium	VIII	*	—	
Intrasporangium	I	*	—	PIV
Mycobacterium	IV	A	M	PII
Mycoplana	IX	*	—	PIII
Nocardia[e]	IV	A	NM	PII
Nocardioides	I	*	–	PI
Oerskovia	VI	*	–	PV
Promicromonospora	VI	*	—	PV
Rhodococcus	IV	A	NM	PII
Rothia	VI	*	—	PI
Monosporate organisms				
Micromonospora	II	D	—	PII
Thermoactinomyces	III	C	—	
Thermomonospora	III	B/C	—	
Bisporate organisms				
Microbispora	III	B	—	PIV
Micropolyspora	IV	A	—	PIII
Muriform thalli				
Dermatophilus	III	B	—	PI
Geodermatophilus	III	C	—	PII
Polysporate organisms				
Actinomadura	III	B	—	PI/PIV
Actinopolyspora	IV	A	—	PIII
Actinosporangium	I	*	—	
Actinosynnema	III	C	—	PII
Chainia	I	*	—	PII
Elytrosporangium	I	*	—	
Excellospora	III	B	—	
Micropolyspora	IV	A	NM/-	PII/PIII
Microtetraspora	III	B	—	PI/PIV
Nocardiopsis	III	C	—	PIII
Pseudonocardia	IV	A	—	PIII
Saccharomonospora	IV	A	—	PII
Saccharopolyspora	IV	A	—	PIII
Sporichthya	I	*	—	
Streptomyces	I	*	—	PII
Streptoverticillium	I	*	—	PII
Sporangiate organisms				
Actinoplanes	II	D	—	PII
Amorphosporangium	II	D	—	PII
Ampullariella	II	D	—	PII
Dactylosporangium	II	D	—	PII
Frankia	III		—	
Microellobosporia	I	*	—	PII
Planomonospora	III	B	—	PIV
Planobispora	III	B	—	PIV
Spirillospora	III	B	—	PII/PI
Streptosporangium	III	B	—	PIV

[a] See Table 2.

[b] See Table 3.

[c] See Table 4.

[d] * = Not of taxonomic value.

[e] Some forms produce spores.

Literature Cited

Arai, T. (ed.) 1976. Actinomycetes: The boundary microorganisms. Tokyo, Singapore: Toppan.

Cross, T., Attwell, R. W. 1975. Actinomycete spores, pp. 3–14. In: Gerhardt, P., Costilow, R. N., Sadoff, H. L. (eds.), Spores VI. Washington, D.C.: American Society for Microbiology.

Cross, T., Goodfellow, M. 1973. Taxonomy and classification of the actinomycetes, p. 76. In: Sykes, G., Skinner, F. A. (eds.), Actinomycetales. London: Academic Press.

Gordon, R. E. 1966. Some strains in search of a genus—*Corynebacterium, Mycobacterium, Nocardia* or what? Journal of General Microbiology **43:**329–343.

Hirsch, P. 1974. Budding bacteria. Annual Review of Microbiology **28:**392–444.

Lechevalier, H. 1974. Distribution et rôle des actinomycètes dans les eaux. Bulletin de l'Institut Pasteur **72:**159–175.

Lechevalier, H. A., Lechevalier, M. P. 1965. Classification des actinomycètes aérobies basées sur leur morphologie et leur composition chimique. Annales de l'Institut Pasteur **108:**662–673.

Lechevalier, H. A., Lechevalier, M. P. 1967. Biology of the actinomycetes. Annual Review of Microbiology **21:**71–100.

Lechevalier, H. A., Lechevalier, M. P., Gerber, N. N. 1971. Chemical composition as a criterion in the classification of actinomycetes. Advances in Applied Microbiology **14:**47–72.

Lechevalier, H. A., Pine, L. 1977. The actinomycetales, pp. 361–380. In: Laskin, A. I., Lechevalier, H. A. (eds.), Handbook of Microbiology, vol. 1, 2nd ed. Cleveland: CRC Press.

Lechevalier, H. A., Lechevalier, M. P., Handley, D. A., Ghosh, B. K., Carmichael, J. W. 1977. Strains of fusidia which can be mistaken for actinomycetes. Mycologia **69:**81–95.

Lechevalier, M. P. 1968. Identification of aerobic actinomycetes of clinical importance. Journal of Laboratory and Clinical Medicine **71:**934–944.

Lechevalier, M. P., de Bièvre, C., Lechevalier, H. A. 1977. Chemotaxonomy of aerobic actinomycetes: Phospholipid composition. Biochemical Ecology and Systematics **5:**249–260.

Lechevalier, M. P., Gerber, N. N. 1970. The identity of madurose with 3-O-methyl-D-galactose. Carbohydrate Research **13:**451–454.

Lechevalier, M. P., Lechevalier, H. A. 1976. Chemical methods as criteria for the separation of nocardiae from other actinomycetes. The Biology of the Actinomycetes **11:**78–92.

Lechevalier, M. P., Lechevalier, H. A., Horan, A. C. 1973. Chemical characteristics and classification of nocardiae. Canadian Journal of Microbiology **19:**965–972.

Mordarski, M., Kurylowicz, W., Jeljaszewicz, J. (eds.) 1978. Nocardia and Streptomyces. Stuttgart, New York: Gustav Fischer.

Okami, Y. 1973. Antibiotics produced by actinomycetes, pp. 717–972. In: Laskin, A. I., Lechevalier, H. A. (eds.), Handbook of microbiology, vol. III, Cleveland: CRC Press.

Prauser, H. (ed.) 1970. The Actinomycetales. Jena; Gustav Fischer.

Slack, J. M., Gerencser, M. A. 1975. Actinomyces, filamentous bacteria. Biology and pathogenicity. Minneapolis: Burgess Publishing Co.

Sottnek, F. O., Brown, J. M., Weaver, R. E., Carroll, G. F. 1977. Recognition of *Oerskovia* species in the clinical laboratory: Characterization of 35 isolates. International Journal of Systematic Bacteriology **27:**263–270.

Sukapure, R. S., Lechevalier, M. P., Reber, H., Higgins, M. L., Lechevalier, H. A., Prauser, H. 1970. Motile nocardoid *Actinomycetales*. Applied Microbiology **19:**527–533.

Sykes, G., Skinner, F. A. (eds.). 1973. *Actinomycetales:* Characteristics and practical importance. London, New York: Academic Press.

Waksman, S. A. 1959. The actinomycetes, vol. I. Nature, occurrence, and activities. Baltimore: Williams & Wilkins.

Waksman, S. A., Lechevalier, H. A. 1962. The actinomycetes, vol. III. Antibiotics of actinomycetes. Baltimore: Williams & Wilkins.

The Genera *Actinomyces, Agromyces, Arachnia, Bacterionema,* and *Rothia*

KLAUS P. SCHAAL and GERHARD PULVERER

The actinomycetes currently classified in the genera *Actinomyces, Agromyces, Arachnia, Bacterionema,* or *Rothia* constitute a small taxonomic unit within the vast and diverse order Actinomycetales. Nevertheless, they are considerably important from several viewpoints. Some are known to cause a potentially malignant infectious disease that may affect humans as well as animals and was discovered as a separate disease entity more than 100 years ago (Bollinger, 1877; Israel, 1878). Some are, to a certain extent, etiologically involved in the development of caries and periodontal disease (Socransky, 1970; Socransky and Manganiello, 1971). Most of these microbes form an important component of the indigenous microflora of human, and possibly animal, mucous membranes, especially in the oral cavity (Bergey, 1907; Blank and Georg, 1968; Collins, Gerencser, and Slack, 1973; Davis and Baird-Parker, 1959; Emmons, 1938; Howell et al., 1959; Lentze, 1948; Naeslund, 1925; Ritz, 1963; Slack, 1942). In addition, two species have been isolated exclusively from soil, in which they may be found as the numerically predominant microbial inhabitants (Gledhill and Casida, 1969a, b).

Although members of the genera *Actinomyces, Agromyces, Arachnia, Bacterionema,* and *Rothia* resemble each other in several aspects, they differ greatly in other basic biological properties. Despite such differences, it has been proposed (Cross and Goodfellow, 1973) to combine all these genera, including the genus *Bifidobacterium* which is treated in this Handbook in Chapter 149, in one family of the order Actinomycetales, that is, the family Actinomycetaceae. This family has been created by Buchanan (1918) primarily to accommodate diverging organisms, such as members of the genera *Actinobacillus, Leptotrichia, Actinomyces,* and *Nocardia.* After several revisions, which included the removal of the Gram-negative species in the sixth edition of *Bergey's Manual of Determinative Bacteriology* (Waksman and Henrici, 1943) and of the genus *Nocardia* in the eighth edition (Slack, 1974), membership of the family Actinomycetaceae has been restricted to bacterial species that are taxonomically linked by the following characteristics: ability to produce Gram-positive, branching

and, later on, fragmenting filaments without aerial hyphae and spores; comparatively exacting nutritional requirements; microaerophilic to facultatively anaerobic to anaerobic growth; and predominantly fermentative carbohydrate metabolism (Slack, 1974; Slack and Gerencser, 1975; see also this Handbook, Chapter 147).

This system of classification reflects the classical approach to bacterial taxonomy, based mainly upon morphological and growth characteristics together with a few simple physiological peculiarities. Various morphological characters have been successfully used to define families and genera of the aerobic actinomycetes (Cross and Goodfellow, 1973), but do not allow clear lines of demarcation to be drawn among the genera of the family Actinomycetaceae, and between them and related taxa outside the family. Thus, the validity of such a family concept has been increasingly questioned since modern and possibly more relevant taxonomic techniques became available. Especially useful aids in redefining taxonomic relationships between fermentative actinomycetes have been: analyses of cell wall components and structures (Cummins, 1962, 1965; Cummins and Harris, 1958; 1959; DeWeese, Gerencser, and Slack, 1968; Pine and Boone, 1967; Schleifer and Kandler, 1972; Snyder et al., 1967); determinations of fermentation end products (Buchanan and Pine, 1965; Gilmour and Beck, 1961; Howell and Jordan, 1963; Howell and Pine, 1961); investigations on DNA base ratios (see Slack and Gerencser, 1975); and promising results from DNA:DNA reassociation experiments (Johnson and Cummins, 1972). Classical morphological and physiological characters evaluated by means of numerical taxonomic methods may add further useful information (Holmberg and Hallander, 1973; Holmberg and Nord, 1975; Melville, 1965).

The data already obtained from such studies indicate the principle direction the definition of the family Actinomycetaceae might change. Several genera presently recognized or proposed as members of the family seem to exhibit much less similarity to each other and to the remaining taxa than was previously assumed.

The first candidate for removal from the family

Actinomycetaceae will probably be the genus *Agromyces,* the definite taxonomic position of which has remained unsettled since its discovery. Gledhill and Casida (1969b) noted difficulties in finding the proper systematic place for these organisms on the first description of the genus because their isolates were found to share important characters with the genus *Actinomyces* and with the genus *Nocardia.* Although Cross and Goodfellow (1973) apparently realized the dilemma, they provisionally included *Agromyces* in the family of fermentative actinomycetes for lack of another, more appropriate taxonomic niche. Gledhill and Casida (1969b) demonstrated an oxidative type of carbohydrate metabolism in *Agromyces* strains which would, however, exclude these microbes from a group of primarily fermentative bacteria although *Agromyces* grows best under microaerophilic conditions. Slack and Gerencser (1975) attribute microaerophilism solely to obligate aerobes. If one follows their definition, microaerophilic growth would not be an objection to removal. The strongest evidence that the taxon *Agromyces* is still inadequately classified can be derived from data on the qualitative composition (Gledhill and Casida, 1969b) and the amino acid sequence (Fiedler and Kandler, 1973) of the murein of *Agromyces ramosus,* the only species of the genus. The occurrence of 2,4-diaminobutyric acid (DAB) in the cell wall and the sequence [L-DAB]-D-Glu-D-DAB demonstrate a close relationship of *Agromyces* with certain plant pathogenic corynebacteria, especially *Corynebacterium insidiosum,* whereas similar structures of the peptidoglycan have not been reported from typical representatives of the order Actinomycetales.

Analogous problems are presented by the taxonomic position of the genus *Bacterionema.* It was erected by Gilmour, Howell, and Bibby (1961) to accommodate organisms previously classified as *Leptothrix buccalis* (Bulleid, 1925), *Leptotrichia buccalis* (Bibby and Berry, 1939), and *Leptotrichia dentium* (Davis and Baird-Parker, 1959), which form Gram-positive, branching, and fragmenting filaments and which can be clearly separated from the Gram-negative, nonbranching microbes presently termed *Leptotrichia* Trevisan 1879. The taxon *Bacterionema* was transferred to the family *Actinomycetaceae* mainly on morphological grounds (Gilmour, 1974), but other important features favor its classification with the "Coryneform Group of Bacteria" (Rogosa et al., 1974): propionic acid is a major end product of glucose fermentation, at least when bacterionemae are grown under aerobic conditions (Howell and Pine, 1961; Pine, 1970); these organisms contain *meso*-diaminopimelic acid, arabinose, and galactose in their cell walls (Baboolal, 1969; Boone and Pine, 1968; Davis and Baird-Parker, 1959), components that are markers of cell wall type IV and that are typical for some coryne-

bacteria, the mycobacteria, and the nocardiae, but not for other fermentative actinomycetes; bacterionemae have a DNA base composition in the range of 55–60 mol% G+C, which is lower than that of *Actinomyces* species but corresponds well to the ratio found in typical coryneforms (Gilmour, 1974; Rogosa et al., 1974); finally, they are equipped with short-chain mycolic acids and with a long-chain fatty acid profile (Alshamaony et al., 1977) that is very similar to corynebacteria of human and animal origin (Collins, Goodfellow, and Minnikin, 1979).

Members of the genus *Arachnia* Pine and Georg 1969 also show morphological and pathogenic similarities to *Actinomyces,* and chemical and physiological differences. *Arachnia* cell wall peptidoglycan, which contains L-diaminopimelic acid and glycine (Johnson and Cummins, 1972; Pine and Georg, 1974; Schleifer and Kandler, 1972), and the production of propionic acid as a major end product of glucose fermentation suggest a closer relationship to the family Propionibacteriaceae (Cross and Goodfellow, 1973; Pine, 1970). But, arachniae differ from propionibacteria by the consistent absence of catalase (Pine and Georg, 1974), and they were found to show only a low genetic homology with the latter bacteria in DNA-hybridization experiments (Johnson and Cummins, 1972). Thus, further investigations are needed to decide if the taxon should remain within the Actinomycetaceae.

The taxonomic position of the two remaining genera, *Bifidobacterium* and *Rothia,* has also been questioned, particularly, the bifidobacteria (Kandler, 1970). The genus *Rothia* Georg and Brown 1967, on principle, fits well within the Actinomycetaceae (Slack and Gerencser, 1975) except for its peculiar peptidoglycan structure, which appears to be unique among actinomycetes (Schleifer and Kandler, 1972).

If one were to follow all these proposals, only the genera *Actinomyces* and *Rothia* would be left in the family Actinomycetaceae. Although this might seem appropriate in the light of current knowledge, it would cause additional difficulties. Taxonomists would have to face the problem that mycelial growth of bacteria would not be restricted any longer to families of the order Actinomycetales, but would also occur in several other families. In addition, medical microbiologists and clinicians may be confused by the fact that practically identical diseases might be caused once by an "actinomyces", then by a "propionibacterium", or eventually by a "corynebacterium". If one considers that it took decades for the general knowledge about actinomycotic infections to reach reasonable standards, this point is of special importance.

With regard to the species level of the different genera within the family Actinomycetaceae, the monotypic genera *Agromyces, Arachnia, Bacterionema,* and *Rothia* do not present many additional taxonomic problems, although the question has been raised as to whether there may exist more than one species in the genus *Arachnia* (Slack and Gerencser, 1975) and in the genus *Bacterionema* (Schaal, 1972).

The situation in the genus *Actinomyces* Harz 1877 is more complex. The species *Actinomyces bovis* Harz 1877, *Actinomyces israelii* Lachner-Sandoval 1898, and *Actinomyces odontolyticus* Batty 1958 are well described (Slack, 1974; Slack and Gerencser, 1975) and were found to form distinct subclusters equal to taxospecies in numerical phenetic studies (Holmberg and Nord, 1975; Melville, 1965). Thus, at least one historical confusion involving the interchangeable use of the names *A. bovis* and *A. israelii* finally has been overcome.

Actinomyces naeslundii Thompson and Lovestedt 1951 and *Actinomyces viscosus* Georg, Pine, and Gerencser 1969 share many properties; they actually differ only in catalase production, which might be due to an inducible enzyme (Roth and Thurn, 1962). Therefore, it has been suggested that *A. viscosus* be considered a catalase-positive variant of *A. naeslundii* (Gerencser and Slack, 1969). Numerical taxonomic analyses support this view (Holmberg and Hallander, 1973; Melville, 1965), but it remains to be seen if new tests and a more precisely adapted technique will confirm this suggestion.

Three additional species that bear the generic name *Actinomyces* have an uncertain taxonomic status. The first is *Actinomyces eriksonii*, which was described by Georg and co-workers in 1965 and which was considered a somewhat less virulent, but still potentially pathogenic, close relative of *A. bovis* and *A. israelii*. This species will probably be renamed *Bifidobacterium eriksonii*, because its end products of glucose fermentation and its cell wall structure are much closer to those of the bifidobacteria (Slack and Gerencser, 1975). Furthermore, the phenetic affinity of *A. eriksonii* with some bifidobacteria was demonstrated by means of numerical taxonomy (Holmberg and Nord, 1975). It has even been claimed (Mitsuoka et al., 1974) that *A. eriksonii* and *Bifidobacterium adolescentis* are identical, and that the later designation, *A. eriksonii*, is illegitimate and should become a *nomen reiciendum*.

A more complicated problem arises with the species designation, *Actinomyces suis*. The name was validly published by Grässer (1957), but the description of the species is inadequate and no cultures are available. In the meantime, similar organisms have been isolated from swine (Biever, 1967; Franke, 1973) which seem to fulfill the requirements for being placed in the genus *Actinomyces* and for being considered a separate species. However, it still has to be decided if these new isolates differ from one another and from Grässer's original strains.

Most uncertain is the taxonomic position of *Acti-nomyces humiferus* Gledhill and Casida 1969. It resembles *Actinomyces* by its morphology, cell wall composition, and end products of glucose fermentation. But it differs in growing at 30°C, in its high G+C content (73 mol%), its lysozyme sensitivity, and also by its habitat. It has been isolated only from soil.

A final problem involving nomenclature and taxonomy is the status of organisms called *Actinobacterium meyerii* (Prévot, 1938). *Actinobacterium* is not a valid generic name for any species of *Actinomyces* (Slack, 1974). However, cultures bearing this designation were recovered in a phenetically distinct cluster that showed clear relationships to the facultatively anaerobic actinomycetes (Holmberg and Nord, 1975). A G+C content of 64–67 mol% would also not exclude classification of these organisms in the Actinomycetaceae (Bouisset et al., 1968).

Despite the still-existing nomenclatural and taxonomic uncertainties, these organisms can be treated in one chapter without difficulty, because the methods and procedures needed for their isolation and identification are not as diverse as their taxonomy.

Habitats

Fermentative actinomycetes in the broadest sense inhabit a wide range of locations and conditions. Their habitats may be discussed under several different headings, such as ecology, human and animal pathology, human and animal physiology, and interaction in microbial ecosystems. The available data are presented here in a way that seemed most practical for this Handbook.

Fermentative Actinomycetes as Commensal Organisms in Man

Although the first descriptions of these actinomycetes seemed to indicate that they were obligate pathogens (Bollinger, 1877; Israel, 1878), it was soon apparent that most of them were widely distributed as commensals on body surfaces in man and animals. Bergey (1907) was probably the first to describe filamentous bacteria adhering to the teeth and being involved in the formation of dental plaque. Since then, numerous reports have demonstrated the presence of those microbes in the oral cavity of man, showing that the classical pathogenic species, as well as less virulent or apathogenic actinomycetes, form a major component of the oral microflora.

Naeslund (1925) was the first to obtain, from the human mouth, facultatively anaerobic, filamentous organisms that differed clearly from the pathogen *Actinomyces israelii* and would be classified today as *Actinomyces naeslundii*. Obviously identical strains were isolated by Thompson and Lovestedt (1951) and validly designated *A. naeslundii*. The first indications that potentially pathogenic actinomycetes are generally present in the oral cavity of healthy humans were obtained by Lord (1910), who was able to produce actinomycosis-like lesions in guinea pigs that had been inoculated with the contents of carious teeth. In 1938, Emmons examined an unselected series of 200 pairs of tonsils from routine tonsillectomies. Emmons observed organisms indistinguishable from *Actinomyces bovis* (a synonym at that time for *A. israelii*) in 37% of the pairs and he cultivated such microbes from 11%. However, no signs of clinical or pathological actinomycosis were found. Since then, many other workers (Blank and Georg, 1968; Grüner, 1969; Howell, et al., 1959; Howell, Stephan, and Paul, 1962; Lentze, 1948; Slack, 1942; Slack, Landfried, and Gerencser, 1971; Snyder et al., 1967), using different techniques of demonstrating and identifying the bacteria, have proved that *A. israelii* is a facultatively pathogenic, primarily epiphytic, resident member of the oral microflora of man.

With few exceptions, similar results were obtained for most other members of the family Actinomycetaceae. *Actinomyces naeslundii*, *A. viscosus*, and *A. odontolyticus* (Batty, 1958; Hill et al., 1977; Holmberg and Forsum, 1973; Slack, Landfried, and Gerencser, 1971; Snyder et al., 1967) were recovered regularly from the oral cavity of more-or-less healthy adults, as were *Arachnia propionica* (Collins, Gerencser, and Slack, 1973; Holmberg and Forsum, 1973; Slack, Landfried, and Gerencser, 1971), *Bacterionema matruchotii* (Davis and Baird-Parker, 1959; Gilmour, Howell, and Bibby, 1961), and *Rothia dentocariosa* (Holmberg and Forsum, 1973; Ritz, 1963; Slack, Landfried, and Gerencser, 1971).

Still in question is the natural habitat of other species potentially pathogenic to man, like *Actinomyces eriksonii* and *Actinobacterium meyerii*, although Blank and Georg (1968) demonstrated *A. eriksonii* in human tonsillar material by using either cultural techniques or direct FA staining.

Actinomyces bovis and *Actinomyces suis* are essentially animal parasites and have never been isolated from humans (Slack and Gerencser, 1975). Reports in the early literature on the occurrence of *A. bovis* in man must, without exception, be attributed to misidentifications or nomenclatural confusion. *Actinomyces humiferus* and *Agromyces ramosus* are the only species within the family Actinomycetaceae that apparently do not rely on warm-blooded hosts and are found to live in the free state, mainly in the soil. Because of their lower growth temperatures, they probably would not be able to colonize the body surface, even after very heavy contamination with soil. All attempts to culture the other fermentative actinomycetes from this environment have met with failure.

Little is known about the biological role of actinomycetes in the human oral cavity under healthy conditions and about their interactions with other members of the oral microflora. However, some of the factors that may contribute to the development of pathological lesions may also be present without signs of impairment, if they enable the actinomycetes to colonize the oral cavity and to resist natural and artificial cleaning mechanisms. The most obvious consequence of actinomycete growth is the production of various organic acids. They evolve from the fermentative metabolic processes, also occurring in vivo, and may be strong enough to lower the pH values locally (Ellen and Onose, 1978). For instance, enzymes initiating carbohydrate breakdown, like invertase and β-galactosidase, have been identified and characterized from strains of *A. viscosus* (Kiel and Tanzer, 1977; Kiel, Tanzer, and Woodiel, 1977); the presence of these enzymes demonstrates the capability of the organism to utilize components of the host's diet, which is commonly rich in disaccharides.

Another important feature that allows actinomycetes to even colonize the smooth surface of teeth and that prevents them from being washed off by salivary flow is their ability to attach directly to such surfaces or to adhere to other microbes already in place. Some oral actinomycetes have been proven to possess such adherence mechanisms: *Actinomyces naeslundii* (Slack and Gerencser, 1975), *A. viscosus* (Howell and Jordan, 1967; Miller, Palenik, and Stamper, 1978; Pabst, 1977), and *Rothia dentocariosa* (Lesher and Gerencser, 1977) synthesize extracellular or cell-associated (Warner and Miller, 1978) levan, which may aid in their adherence to the tooth surface. In addition levan may provide a reservoir of carbohydrate for growth. *A. viscosus* was shown to possess levan-hydrolase activity (Miller and Somers, 1978). Furthermore, levan metabolism might be an indication for synergistic mechanisms in the oral cavity because other typical representatives of the mouth flora are also able to synthesize or degrade levan (DaCosta and Gibbons, 1968; Manly and Richardson, 1968; van Houte and Jansen, 1968).

Interbacterial aggregation has been shown between *A. naeslundii* and streptococci (Ellen and Balcerzak-Raczkowski, 1977; Gibbons and Nygaard, 1970; Miller, Palenik, and Stamper, 1978), and between *A. viscosus* and veillonellae or streptococci (Gibbons and van Houte, 1973; McIntire et al., 1978).

These mechanisms of adherence give rise to the formation of dental plaque, which can be considered the initial step in the development of caries or periodontal disease but must not in itself be called pathological. Filamentous bacteria were found to contribute markedly to the volume of plaque and to supply an enormous additional surface area for the attachment of other bacteria (Boyd and Williams, 1971). In this connection, it is especially interesting that *Bacterionema matruchotii* is able to form intracellular deposits of calcium phosphate that are indistinguishable from bone and tooth apatite (Boyan-Salyers, Vogel, and Ennever, 1978; Ennever, 1960; Ennever et al., 1978; Ennever, Vogel, and Streckfuss, 1971; Takazoe, 1961).

The relative numbers of actinomycetes in the oral environment may be influenced by inhibitory factors released by other bacteria (Rogers, van der Hoeven, and Mikx, 1978), by fluorides or other measures for preventing caries (Beighton and McDougall, 1977), or by eating habits of the host. In any case, the qualitative and quantitative composition of the oral flora differs, from the saliva to the mucosal surface to the tooth surface and even from tooth to tooth and at different sites on one single tooth. It has been reported that Gram-positive, facultative, and anaerobic bacteria constitute 15–20% of the cultivable flora in saliva and on the tongue, 40% of that in plaque, and 35% of that in gingival crevice (Socransky and Manganiello, 1971). Actinomycetes amounted to about 14% of the cultivable organisms in plaque (Loesche, Hochett, and Syed, 1972). The relative incidence of selected species has been estimated by Hill et al. (1977) in plaque from New Guinea indigenes: these workers identified 42% of their filamentous isolates as *Actinomyces odontolyticus,* 28% as *A. israelii,* and 2% as *R. dentocariosa.* Furthermore, the oral microflora changes with the age of the host. Anaerobic bacteria are generally found in large numbers during the period of life when natural teeth are present (Socransky and Manganiello, 1971; Russell and Melville, 1978). Before dentition and after loss of all teeth in later life, the counts for anaerobes and so for actinomycetes are generally low (Berger, Kapovits, and Pfeifer, 1959; Kostecka, 1924; Loesche, 1968; Rosenthal and Gootzeit, 1942). The occurrence of actinomycotic infections indicates that these organisms are principally present in their hosts the whole lifetime.

No reports are available on the occurrence of fermentative actinomycetes as normal inhabitants of other mucosal areas of the human body. Occasional isolations of *A. israelii* from feces (Sutter and Finegold, 1972) have been regarded accidental in the sense of a transient rather than a permanent occurrence of the organism (Slack and Gerencser, 1975). However, in our laboratory, we have observed that it is not too difficult to isolate fermentative actinomycetes, and among them *A. israelii,* when proper selection (Fritsche, 1964) and enrichment procedures are applied. Indirect evidence for the view that actinomycetes might well form a numerically small component of the flora of the intes-

tinal tract can be derived from cases of actinomycosis that developed endogenously after bowel surgery, appendix perforation, or injuries of the intestine.

Similar considerations apply to the problem of actinomycete colonization of the female genital tract. Actinomycotic infections of the small pelvis and the adnexa uteri support the endogenous origin of the disease from the adjacent mucous membranes, although actinomycetes have not been demonstrated as normal inhabitants of the vagina or the cavum uteri (Hanf and Hanf, 1955). Recent reports on the occurrence of actinomycetes connected with the use of intrauterine contraceptive devices will be discussed later, but support the view that the genital tract could also be a normal habitat of some actinomycete species.

Fermentative Actinomycetes as Commensal Organisms in Animals

The extent to which fermentative actinomycetes form a part of the normal indigenous flora of animals is not known. By analogy to the situation in humans, it could be concluded that animal actinomycoses develop endogenously and that the causative agents might also belong to the normal oral microflora of various animals (Slack and Gerencser, 1975). Little concrete information exists about the occurrence of actinomycetes in the oral cavity of animals. Reports are restricted to isolations of *Actinomyces viscosus* from hamsters and other rodents (Bellack and Jordan, 1972; Howell, 1963; Howell and Jordan, 1963) and to the demonstration of *Bacterionema matruchotii* in gingival plaque from monkeys (Cock and Bowen, 1967).

Actinomycetaceae as Soil Inhabitants

Two members of the family Actinomycetaceae have been exclusively isolated from the soil, thus casting doubt upon their classification in a family that otherwise contains only organisms associated with man or animals.

Actinomyces humiferus was cultured regularly from various organically rich soils. The organic-matter content of the samples amounted to 6.5% or greater, and the pH values ranged from 6.4 to 7.9. In such soils, *A. humiferus* represented a numerically predominant segment of the bacterial flora. However, attempts to recover this species in a variety of other soil types failed (Gledhill and Casida, 1969a).

The second actinomyces-like organism that obviously belongs with the soil microbes is *Agromyces ramosus.* The ecological versatility of this species seems to be greater than that of *Actinomyces humi-*

ferus. Soils that harbored these organisms ranged from fertile meadow to barren desert soils. The variability of some other parameters of those soils was considerable: moisture contents ranged from 0.7 to 14.8%, pH values from 6.4 to 8.2, and organic-matter contents from 2.7 to 6.5%. The numbers of these organisms per gram of soil were found to be 10- to 100-fold higher than the total numbers of platable microbes (Gledhill and Casida, 1969b).

Both soil actinomycetes were tested for pathogenicity in mice, but in none of the animals could pathologic lesions be demonstrated 3 weeks after injection of the organisms (Gledhill and Casida, 1969a, b).

Fermentative Actinomycetes as Human Pathogens

Since the description of the first member of the family (Bollinger, 1877), fermentative actinomycetes have been known as causative agents of typical and severe infectious diseases in man and animals. Because the definitely pathogenic species also belong to the normal microflora of their hosts, they must be considered facultative pathogens, which only invade the tissue under certain facilitating conditions and which are not transmissible (Lentze, 1938a, b, 1948, 1969, 1970; Pulverer, 1974; Pulverer and Schaal, 1978; Slack and Gerencser, 1975). Therefore, in a sense, they may be called opportunistic pathogens, although they differ from other opportunists in that they obviously do not require a generally debilitated host for becoming invasive but only proper local starting conditions.

The most typical disease entity caused by fermentative actinomycetes is called actinomycosis, a name that dates back to the first extensive description of the disease by J. Israel in 1878. As currently recognized, actinomycosis is a chronic, granulomatous inflammation that gives rise to suppuration, abscess formation, and draining sinuses during its course (Pulverer and Schaal, 1978; Slack and Gerencser, 1975). In the beginning, it may appear as a slowly progressing infiltration as well as an acute abscess or a phlegmon. Advanced processes show a typical and dangerous preference to penetrate the tissue without stopping at the border of organs (Lentze, 1969, 1970). In about 45% of the cases (Pulverer and Schaal, 1978; Schaal, 1979), the purulent discharge of actinomycotic lesions contains macroscopically visible (1 mm in diameter), yellowish to brownish particles. These particles represent a conglomerate of filamentous actinomycete microcolonies, formed in vivo and surrounded by tissue-reaction material, especially polymorphonuclear granulocytes. These diagnostically typical structures, which exhibit a cauliflower-like microscopic appearance at low magnifications and a filamentous picture with radially arranged peripheral hyphae at higher magnifications (Lentze, 1969; Schaal, 1979; Slack and Gerencser, 1975), were originally designated "Drusen" in Harz's (1877) first description and served as the basis for the name "ray fungus" or *Actinomyces;* in the English written literature, they are usually referred to as "sulfur granules". Granules in tissue sections and, less frequently, granules encountered in the discharge often show a club-shaped layer of hyaline material on the tips of peripheral filaments, which can aid in the differentiation of *Actinomyces* granules from particles of other origin.

Because of its endogenous origin, actinomycosis primarily develops in tissue near the mucosal surfaces that are the natural habitat of the causative agents. Clinically, these predilection sites are mostly referred to as cervicofacial, thoracic, and abdominal. In the extensive material that has been collected at the Hygiene Institute of the University of Cologne over a period of 35 years, cervicofacial actinomycoses have been most frequently observed. In a series of about 2,500 cases examined, involvement of face and neck was noted in 98%, whereas thoracic infections were only found in 1.4% and abdominal manifestations in 0.6% (Pulverer and Schaal, 1978). Although our material reflects the situation of all West Germany rather than of the Cologne area, this anatomical distribution differs greatly from that reported for other regions, especially in the United States (Slack and Gerencser, 1975), where 40–50% of human actinomycotic infections were found to be located in thoracic and abdominal sites. We are not yet able to explain these differences with any certainty. They obviously relate to the higher overall incidence of the disease which we observed, and from which a morbidity rate was calculated that averaged 1:40,000 (Schaal, 1979) or 1:71,000 (Lentze, 1970) inhabitants per year in the region of Cologne. Other evaluations give much lower figures. For instance, Durie (1958) estimated the average annual rate in the population of Sydney to 10 cases and Hemmes (1963) reported 1 case in 119,000 inhabitants yearly for Holland.

Human actinomycosis is worldwide in distribution. The disease exhibits a disproportionate sex distribution. Most tabulations that include sex ratios (see Slack and Gerencser, 1975) conform with our data (Pulverer, 1974; Pulverer and Schaal, 1978), which indicate that the ratio is about three to four males per female. It should be mentioned, however, that this proportion applies to patients in sexual maturity; before puberty and in the climacteric period, differences in sex distribution cannot be demonstrated (Pulverer and Schaal, 1978). Actinomycosis may occur in persons of all age groups (Pulverer and Schaal, 1978; Slack and Gerencser, 1975), ranging from 28 days to 82 years, although by far the highest incidence is observed between 20 and 40 years in males and between 10 and 30 years in females.

Actinomycotic involvement of the central nervous system or of bone and dissemination may occur in humans, but are seldom encountered, in contrast to the usual osseous involvement seen in animal actinomycosis. Primary actinomycoses of skin and wounds are rare and usually have a history of trauma resulting from human bite or fistfight. This observation emphasizes that the source of infection is the indigenous flora and not the environment (Lentze, 1938).

The etiology of human actinomycosis has presented problems for many years, especially because facultatively pathogenic actinomycetes almost never produce infections that contain the causative actinomycete in pure culture. Nearly all naturally developed actinomycotic lesions appear as multiply infected processes in which the pathogenic actinomyces acts as the "guiding organism" (Lentze, 1948, 1969, 1970) responsible for the typical clinical picture and the course of the disease. The always present, but in number and species composition varying, group of "concomitant" bacteria, among which *Actinobacillus actinomycetem-comitans* is the most typical representative, are obviously necessary to strengthen the comparatively low invasive power of the actinomycetes (Holm, 1950, 1951; Lentze, 1948, 1953, 1969, 1970; Pulverer and Schaal, 1978; Schaal, 1979), either by providing reduced conditions in the tissue or by supplementing the guiding organism with toxic or necrotizing extracellular products.

In about 50% of our cases, the mixed actinomycotic flora was composed of aerobically growing bacteria and of facultative to strict anaerobes, the other half consisting of solely anaerobic infections (Lentze, 1969; Pulverer, 1974; Pulverer and Schaal, 1978). Among the aerobic concomitant bacteria, *Staphylococcus epidermidis, Staphylococcus aureus,* and α- and β-hemolytic streptococci were the most prevalent. The anaerobic component of the multiply infected actinomycotic lesions consisted of carboxyphilic streptococci, Peptococcaceae, *Bacteroides melaninogenicus* and other *Bacteroides* species, fusobacteria, *Leptotrichia buccalis,* and *Actinobacillus actinomycetem-comitans.* The latter was found in about 25% of our cases and, when present, the disease usually took a particularly chronic and serious course.

Although the concomitant organisms constitute a necessary factor in the pathogenesis of human actinomycosis, the development of typical actinomycotic lesions is impossible without the presence of a pathogenic actinomycete species. However, localized anaerobic infections other than actinomycosis do occur in the same predilection sites and may be clinically indistinguishable from true actinomycosis. Specimens may be contaminated with actinomycetes from the mucosal surface which may then be misinterpreted as the causative agents.

These considerations illustrate the problems that exist when posing the question: which oral actinomycetes are human pathogens and which are not? Numerous reports in the world literature leave no doubt that *Actinomyces israelii* is the most typical and most common cause of human actinomycosis. Out of 773 fermentative actinomycete strains isolated in our laboratory from clinical specimens during the last 10 years, 715 were identified as *A. israelii,* a relative incidence of 92.5% (Pulverer and Schaal, 1978). Some other species also have been incriminated in actinomycotic processes, but evidence for their pathogenic potential is derived only from a few case reports.

Arachnia propionica is most likely to play an etiological role. It has been isolated several times from typical actinomycotic suppurations (Brock et al., 1973; Conrad et al., 1978; Gerencser and Slack, 1967). In our material, *A. propionica* was encountered in about 3% of infections suspected of actinomycosis (Pulverer and Schaal, 1978).

Among other *Actinomyces* species that are potential agents of invasive infections, diseases due to *A. eriksonii* are best documented (Georg et al., 1965; Green, 1978). In contrast to *A. israelii,* this organism seems to occur mainly in infections located in the thoracic and abdominal region. In addition, we have observed a case of brain abscess due to *A. eriksonii* and peptostreptococci (K. P. Schaal, unpublished).

Information about *A. naeslundii* and *A. viscosus* is less compelling; these organisms may be recovered in clinical specimens (Coleman and Georg, 1969; Gerencser and Slack, 1969; Karetzky and Garvey, 1974); but, in our opinion, their etiological role has not yet been sufficiently clarified. In a recent survey of 860 cases of suspected actinomycosis (Schaal, 1979), we have reported on 34 isolations of *A. naeslundii* and 4 isolations of *A. viscosus.* All were derived from subacute to chronic suppurations, which could have been less typical actinomycoses. We were, however, not able to exclude the possibility of contamination through mucosal secretions, which could have introduced these bacteria during sampling of the pus. In addition, we examined four purulent specimens that simultaneously contained *A. naeslundii* and *A. israelii,* thus leaving open which one was the "real" causative agent. Only one of our observations, in an empyema of the knee, strongly supports a pathogenic potential of *A. naeslundii;* specimens yielded growth of *A. naeslundii* in pure culture and on several occasions. Other actinomycetes, although they may be found occasionally in human clinical materials (Brown, Georg, and Waters, 1969; Pulverer and Schaal, 1978), obviously do not cause actinomycosis in man.

Lacrimal canaliculitis is another, less severe and not invasive infection that may be caused by actinomycetes. Branching, filamentous organisms in lacrimal concretions were already described by Cohn in 1875. More recent examinations showed that these filamentous microbes can be identified as *Actinomyces israelii, A. odontolyticus,* and *Arachnia propionica* (Buchanan and Pine, 1962; Ellis, Bausor, and Fulmer, 1961; Pine and Hardin, 1959; Pine, Howell, and Watson, 1960; Slack and Gerencser, 1975). According to our own experience, *Arachnia propionica* seems to be the prevalent microorganism.

From a practical and social point of view, caries and periodontal disease are probably the most important impairments in which fermentative actinomycetes are etiologically involved. Although these conditions are complex in their pathogenesis, certain bacteria have been incriminated as contributors to their development. Because of their adherence mechanisms, actinomycetes are a main cause of plaque formation (Slack and Gerencser, 1975), which is thought to be the starting mechanism of the disease. Thus, besides cariogenic streptococci, actinomycetes have attracted more and more attention of workers who are interested in the pathophysiology, cure, and prevention of these impairments (Jordan and Hammond, 1972; Jordan and Sumney, 1973; Socransky, 1970; Winford and Haberman, 1966).

The capacity of actinomycetes to reduce pH in dental plaque has already been mentioned; pH values of 5 or lower initiate the process of demineralization of the enamel (Slack and Gerencser, 1975). Calcification of the plaque, which was described as a typical facility of *Bacterionema matruchotii,* leads to increased gingival irritation and inflammation (Slack and Gerencser, 1975). In addition, several other factors that might contribute to the pathogenic effects of actinomycetes have been detected. *Actinomyces viscosus* was shown to have chemotactic effects, to stimulate the production of mediators of inflammation from host immune cells, to mark fibroblasts for immune-mediated damage, and to possess amphipathic antigens (Burckhardt, 1978; Engel, Schroeder, and Page, 1978; Engel, van Epps, and Clagett, 1976; Taichman et al., 1978; Wicken et al., 1978). *B. matruchotii* enhances phagocytic and bactericidal functions (Nitta, Okumura, and Nakano, 1977) and possesses adjuvant activities (Nitta, Okumura, and Nakano, 1978). Finally, *A. naeslundii* and *A. viscosus* have been directly proven to initiate periodontitis in hamsters and gnotobiotic rats (Crawford, Taubman, and Smith, 1978; Jordan, Fitzgerard, and Stanley, 1965; Llory, Guillo, and Frank, 1971; Socransky et al., 1970).

A recently discovered problem that needs further clarification is the occurrence of fermentative actinomycetes in the genital tract of women wearing intrauterine contraceptive devices (IUDs) or vaginal pessaries (Christ and Haja, 1978; Gupta, Erozan, and Frost, 1978; Gupta, Hollander, and Frost, 1976). Such observations indicate that actinomy-

cetes can be frequently demonstrated when foreign bodies are present in the genital tract. Since it is difficult to imagine that these microbes were introduced from outside, these results may be taken as a further hint that actinomycetes may normally be present in low numbers on the inner surfaces of the female genital system. Introduction of foreign bodies such as IUDs, tampons, or pessaries apparently changes the ecological conditions so that actinomycetes, mainly *A. israelii,* are favored in growth, whereas other genital microbes are suppressed.

Although the presence of actinomycetes does not necessarily imply that they always infect rather than colonize, some patients have been observed who developed disseminated actinomycosis with hepatic and intracranial abscesses (Gupta, Erozan, and Frost, 1978) or, at least, localized inflammations in adnexa and pelvis (Barnham, Burton, and Copland, 1978; Kohoutek and Nozicka, 1978; Luff et al., 1978; Spence et al., 1978; Witwer et al., 1977). Thus, demonstration of actinomycetes in smears derived from the cavum uteri or the vagina should be, at least, considered a preliminary stage of actinomycosis from which serious lesions can develop.

Fermentative Actinomycetes as Animal Pathogens

Since Bollinger's first description of animal actinomycosis (1877), cattle have been known to show the highest incidence of this infection. Bovine actinomycosis, also called lumpy jaw, is a chronic suppurative process that, as in humans, tends to form multiple abscesses and draining sinus tracts. In contrast to human cases, bones are usually involved, mainly the mandible and sometimes the maxilla; the bones are slowly deformed by a destructive and, at the same time, proliferative osteitis (Slack and Gerencser, 1975).

The source of infection in cattle is not definitely established, but, as in humans, is also probably endogenous. The principal etiological agent of bovine actinomycosis is *Actinomyces bovis* (Bollinger, 1877; Slack and Gerencser, 1975), but infections with *A. israelii* in cattle have been reported (Cummins and Harris, 1958; King and Meyer, 1957; Pine, Howell, and Watson, 1960).

Clinically distinguishable diseases caused by actinomycetes have been reported from swine (Franke, 1973; Magnusson, 1928). They more frequently invade the udder, lungs, or other internal organs rather than the neck or bones. *Actinomyces suis* (Franke, 1973; Grässer, 1962), *A. israelii* (Magnusson, 1928), and *A. viscosus* (Georg et al., 1972) have been identified as causative agents.

Actinomycosis-like diseases have been reported in dogs, from which *Actinomyces viscosus* was isolated (Georg et al., 1972), and in a variety of other wild and domestic animals. But these reports are not well documented so that detailed knowledge about causative agents and sources of infections is lacking.

Practically all representatives of the family Actinomycetaceae have been tested for animal pathogenicity, using different animal species such as mice, hamsters, guinea pigs, rabbits, sheep or even calves (see Slack and Gerencser, 1975). Except for *Actinomyces humiferus* and *Agromyces ramosus*, which proved to be apathogenic in the animal models used, all fermentative actinomycetes were able to produce a certain degree of pathological alterations. *Actinomyces israelii, A. naeslundii, A. viscosus,* and *A. eriksonii* were shown to produce inflammatory processes most easily. Others were found to be less virulent, but not avirulent, for small laboratory animals.

Isolation

There are three major demands for isolation techniques for fermentative actinomycetes. Clinical microbiology laboratories, which deal with anaerobic infections, need reliable, simple, and inexpensive isolation procedures. Actinomycetes from the oral cavity need to be isolated and enumerated in order to obtain further information about pathogenic mechanisms involved in the development of caries and periodontal disease. Isolation procedures are also necessary for investigations of the soil microflora when facultatively anaerobic or microaerophilic actinomycetes are expected or especially looked for. Although similar in principle, the methods of isolation require particular adaptations for these different kinds of applications.

Methods for Obtaining Microaerophilic to Anaerobic Growth Conditions

Most Actinomycetaceae, except *Rothia* strains, require a certain reduction of oxygen tension and/or the addition of CO_2 for optimal growth. In contrast to many strict anaerobes, very elaborate techniques for obtaining an oxygen-free atmosphere are usually not necessary for the isolation of fermentative actinomycetes; in many strains, such techniques would even suppress growth.

Thus, the use of prereduced media, the roll-tube system, glove boxes, or similarly complicated and expensive equipment often employed for growing anaerobes are usually not needed for culturing Actinomycetaceae. However, such procedures may be suitable for special research purposes that require constant and controlled conditions or for some organisms, like *Actinomyces eriksonii*, that are more sensitive to oxygen. When these methods and devices are used, the detailed descriptions and instructions given in the *VPI Anaerobe Laboratory Manual* (Holdeman, Cato, and Moore, 1977) should be followed. In most instances, fermentative actinomycetes can be successfully isolated and subcultured by much simpler techniques.

Although it may seem old-fashioned, the easiest, least expensive, and most widely applicable technique for obtaining reduced growth conditions for fermentative actinomycetes is Fortner's method, described in 1928. It is based upon the observation that certain enterobacteria, especially *Serratia marcescens*, are able to lower the oxygen tension and to produce CO_2 when grown in a closed system.

Fortner's Method for Obtaining Anaerobic Growth Conditions (Fortner, 1928, 1929)

For this method, only agar media in glass Petri dishes should be used; plastic dishes will inevitably cause problems! First, one-half to two-thirds of the agar surface is inoculated with the material to be examined or with the anaerobic strain to be subcultured. Then the remaining surface of the agar medium is heavily inoculated with *Serratia*, using a spatula. The dish is placed upside down upon a glass sheet of appropriate size, and fixed and sealed with plasticine to make the system air-tight (Fig. 1). During incubation, *Serratia* progressively removes oxygen. Depending on the degree of anaerobiosis needed for an individual strain, growth of the anaerobes starts with some delay, but usually within the first 4–8 h of incubation. Leakage in the plasticine seal, which may occur when the plasticine is too brittle or the base plate is damp, can be easily detected. *Serratia* forms unpigmented growth under reduced oxygen tension, but becomes red when the system is not totally closed.

In our experience, practically all Actinomycetaceae can be easily cultured by this method. Even the more aerophilic strains will grow, although their mature colonies remain smaller than when grown under full oxygen tension. Additional plates incubated under CO_2 are usually not necessary. Thus, Fortner's technique may be considered a universal means of culturing fermentative actinomycetes. It is particularly useful in the clinical microbiological laboratory, where measures should be as simple and inexpensive as possible but still optimally efficient in order not to lose important clinical isolates.

An alternative way to obtain anaerobic growth conditions is the use of anaerobic jars. The Torbal anaerobic jar with a gas mixture of 80% N_2 : 10% H_2 : 10% CO_2 or the GasPak jar with a H_2-CO_2 generating packet are recommended (Slack and Gerencser, 1975). Jars without catalysts are generally less satisfactory. Some Actinomycetaceae, especially the catalase-positive ones, will grow only poorly if at all in the gaseous atmospheres of the

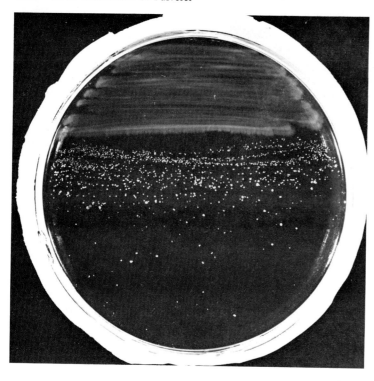

Fig. 1. Fortner's method. Note the enhanced growth of the actinomycete near *Serratia*.

above-mentioned jars. Therefore, when the jars are used for primary isolation, additional plates have to be incubated aerobically with added CO_2. A candle jar will be mostly satisfactory for this purpose, but GasPak jars with a CO_2-generating unit also can be employed.

If the oxygen requirements of an individual isolate are not known and cannot be predicted, tests for oxygen requirements must be performed in order not to lose the strain during subculturing.

Tests for Oxygen Requirements
(Slack and Gerencser, 1975)

Agar Slant Method. Suitable agar medium slants in cotton-plugged tubes are inoculated with a standardized suspension of the test organism using a capillary pipette. This inoculum is obtained from about 3-day-old cultures, either in broth or on plates. The cells obtained after centrifugation or by scraping from the agar surface are suspended in 0.85% saline and adjusted to a density matching a MacFarland 3 standard. Eight slants inoculated with the test culture are needed. The slants are incubated, in duplicate, under the following conditions:

1. For aerobic conditions, two slants with the original cotton plugs are placed directly in the incubator.
2. For aerobic conditions with added CO_2, the cotton plugs of two tubes are clipped off and the remaining parts are pushed into the tube to just above the slant. Small pledgets of absorb-

ent cotton are placed on top of the plugs, 5 drops 10% Na_2CO_3 and 5 drops 1 M KH_2PO_4 are added to each, and the tubes are immediately closed with rubber stoppers.
3. For anaerobic conditions with added CO_2, tubes are prepared as under 2. To the absorbent cotton are added 5 drops 10% Na_2CO_3 and 5 drops pyrogallol solution (100 g pyrogallic acid in 150 ml distilled water).
4. For anaerobic conditions without CO_2, to the properly prepared tubes (see 2) are added 5 drops 10% KOH and 5 drops pyrogallol solution.

Results are recorded after 3 and 7 days of incubation, usually at 35–37°C (for isolates suspected of *A. humiferus* or *Agromyces*, at 30°C). If the growth in two corresponding tubes does not seem equal, the test has to be repeated.

Agar Deep Method. Melted and cooled agar medium is inoculated, while still liquid, with a suspension of the test organism using a capillary pipette. The tip of the pipette is pushed to the bottom of the tube and then slowly withdrawn while expelling a drop of inoculum. Then the agar is gently mixed by rotating the tube and allowed to solidify in an upright position. After 3 and 7 days of incubation, the results are read by measuring the distance in millimeters between the surface and the zone of maximum growth.

Anaerobiosis for fluid cultures may be obtained by placing tubes with suitable broth media into jars.

But another very old-fashioned technique, based upon the reducing effect of sterilized animal-tissue particles, can give equal or even better results, especially on primary isolation from clinical specimens. The two most common media that use this principle are the Tarozzi medium and the cooked-meat medium. In the former, when one solid piece of liver is used, actinomycetes grow especially well on the surface of the tissue and often produce macroscopically visible, "snowball-like" colonies that can be selectively aspirated by a capillary pipette for further examination.

Since it has been shown (Jones, Watkins, and Meyer, 1970) that the microaerophilism of *Agromyces ramosus* results from its lack of catalase activity, the growth of this oxidative organism is best under full oxygen tension, provided that accumulating H_2O_2 is quickly destroyed. This destruction can be achieved by auxiliary mechanisms such as addition of catalase, fresh horse blood, or MnO_2, of which horse blood was reported to give best results.

Media and Isolation Procedures

ISOLATION FROM CLINICAL SPECIMENS. Specimens from suspected cases of actinomycosis, usually of pus, sinus discharge, bronchial secretions, or tissue material, must be collected without contamination of the material by the flora of the mucous membranes. Thus, only punctures or incisions through thoroughly disinfected skin, transtracheal aspirations of fluid from the bronchi, and transthoracic or transabdominal needle-aspiration biopsies (Pollock et al., 1978) provide suitable samples. Although actinomycetes are not extremely oxygen-sensitive, proper reduced transport conditions should be used when the material cannot be processed within a few hours after collection. Adequate transport media are now commercially available, such as the Port-A-Cul system (BBL), and should be used according to the instructions of the manufacturer.

Commercial media are usually recommended for primary isolation and subsequent cultivation of clinical actinomycete isolates. Such media are prepared and used according to the manufacturer's instructions. Because of the obligate multiple infection in actinomycotic lesions, there is some need for reliable selective media. Unfortunately, no medium has yet been reported that fulfills the medical requirements of allowing growth of all possibly pathogenic actinomycetes while reducing the concomitant bacteria to any considerable degree. Therefore, highquality, general-purpose culture media are mostly used. They are: fluid thioglycolate broth (THIO), which may be supplemented with 0.1–0.2% sterile rabbit serum; brain heart infusion broth (BHI), Trypticase soy broth (TSB); brain heart infusion agar

(BHIA); Trypticase soy agar (TSA); heart infusion agar, with 5% defibrinated rabbit, sheep, or horse blood; and Schaedler broth or agar.

In our hands, none of these media was satisfactory in all instances. Especially in THIO and BHI and on BHIA, which are widely employed in the bacteriological diagnosis of actinomycosis, growth is sometimes poor or does not occur at all. Therefore, in addition we use a quite complex medium (CC medium), which was developed in our laboratory some years ago as a modification (Heinrich and Korth, 1967) of the synthetic medium of Howell and Pine (1956). This medium has proved to be extremely useful since it promotes growth of actinomycetes and also allows the development of very typical colonies and maintenance of cultures.

CC Medium for Isolating Actinomycetes (Howell and Pine, 1956; modified by Heinrich and Korth, 1967)

I. Solution of minerals and trace elements, containing per liter distilled water:

$MgSO_4 \cdot 7H_2O$	20 g
$CaCl_2 \cdot 2H_2O$	2 g
$FeSO_4 \cdot 7H_2O$	400 mg
$MnSO_4 \cdot 2H_2O$	15 mg
$NaMoO_4 \cdot 2H_2O$	15 mg
$ZnSO_4$	4 mg
$CuSO_4 \cdot 5H_2O$	0.4 mg
$CoCl_2 \cdot 4H_2O$	0.4 mg
Boric acid	20 mg
KJ	10 mg

The solution is acidified with 10 ml of 10% HCl.

II. Vitamin solution, containing per 100 ml distilled water:

Thiamine-HCl	20 mg
Pyridoxine-HCl	20 mg
Biotin	1 mg
Folic acid	5 mg
Vitamin B_{12} (1 mg/100 ml)	1 ml
p-Aminobenzoic acid	20 mg
m-Inositol	20 mg
Nicotinamide	10 mg
Nicotinic acid	10 mg
Ca-pantothenate	20 mg

III. Solution of amino acids and vitamins, containing per 100 ml of distilled water:

Casein hydrolysate	12 g
Yeast extract	12 g
L-Cysteine-HCl	500 mg
L-asparagine	30 mg
DL-tryptophan	20 mg
Solution II	12 ml

This solution is sterilized by Seitz filtration.
Preparation of the medium:

KH_2PO_4, 4 g, is dissolved in 250 ml distilled water and adjusted to pH 7.6 with NaOH. Then, 10 ml of solution I, 500 mg of potato starch dissolved in 70 ml boiling distilled water, about 20 g agar (depending on quality), and distilled water are added to give a final volume of 900 ml. This medium is sterilized in the autoclave at 121°C for 15 minutes. After cooling to about 50°C, solution III is added under aseptic conditions. The final pH of the medium is adjusted to 7.3 and it is poured into Petri dishes.

For cultures from clinical specimens, at least three to four different media, including a fluid medium, preferably Tarozzi broth, should be inoculated. When Tarozzi broth is used, it is heated in a boiling water bath immediately before inoculation, then quickly cooled down under tap water, inoculated, and sealed with about 2 ml sterilized petrolatum. Use of paraffin wax or oil is not recommended.

If Fortner's method and Tarozzi broth are used, the cultures can be daily checked for growth of typical mycelial colonies without disturbing the anaerobic environment. If anaerobic jars are preferred, a duplicate set of media should be inoculated to be examined after 3 and 7–10 days of incubation at 35–37°C.

For detecting typical mycelial microcolonies of the actinomycetes (see later) in early stages of growth, it is recommended to examine surface cultures on transparent media under the microscope at 80–100× magnification. If a long-distance objective is employed, cultures can be examined through the medium without opening Fortner plates (Lentze, 1938).

It is sometimes difficult to remove the concomitant bacteria to obtain pure cultures of the actinomycetes, even after several subcultural passages and application of dilution techniques. Subculture media supplemented with suitable antibiotics can help but may inhibit the actinomycetes also. Commercial disks for testing sensitivity to antibiotics can be placed on the medium after inoculation with the impure strain. Depending on the contaminating organism, disks that contain metronidazole, colistin, or nalidixic acid in common concentrations can be employed. From their inhibition zones, actinomycete colonies can often be easily picked and transferred to another plate.

ISOLATION FROM THE ORAL CAVITY OR OTHER NATURAL SOURCES IN MAN AND ANIMALS. Ecological studies on the occurrence of fermentative actinomycetes in various secretions and concrements of mucous membranes usually require the qualitative demonstration of the particular organisms, their quantity, and the evaluation of their numerical rela-

tionships to other members of the biotope. Therefore, any losses of cultivable microbes and increases due to reproduction must be avoided during sampling and transportation of the specimens. Furthermore, since these biotopes, especially gingival crevice, plaque material, or feces, harbor extremely high numbers of many different microorganisms, selective culture media are needed to facilitate the detection and enumeration of specific organisms.

Different types of transport media have been applied to oral microbes, including the Stuart medium, a general-purpose transport medium used in medical microbiology (Loesche, Hochett, and Syed, 1972; Syed and Loesche, 1972). Another reduced transport fluid described by the authors mentioned above seems to be of special use; its formula is given below.

Reduced Transport Fluid (RTF) (Syed and Loesche, 1972)

I. Stock mineral salt solution No. 1, containing:

K_2HPO_4 0.6%

II. Stock mineral salt solution No. 2, containing:

NaCl	1.2%
$(NH_4)_2SO_4$	1.2%
KH_2PO_4	0.6%
$MgSO_4$	0.25%

III. Final transport medium, containing per liter:

Stock solution No. 1	75 ml
Stock solution No. 2	75 ml
0.1 M ethylenediaminetetraacetate solution	10 ml
8% Na_2CO_3 solution	5 ml
1% Dithiothreitol solution (freshly prepared)	20 ml
0.1% Resazurin solution (optional)	1 ml
Distilled water	814 ml

The medium is sterilized by membrane filtration (pore size, $0.22\,\mu m$) and dispensed into 16-×-125-mm screw-cap tubes (dilution tubes) and 18-×-150-mm test tubes (sample-collection tubes). The pH should be 8 ± 0.2 without adjustment, and it decreases to 7 in 48 h in the anaerobic glove box atmosphere (85% N_2, 10% H_2, 5% CO_2).

Normally, such samples must be dispersed before dilution and plating. Dispersal can be done by sonic or mechanical (Potter-type homogenizer) treatment. Then dilutions are performed and appropriate media are inoculated. To grow fastidious and oxygen-sensitive members of the human microflora, all procedures should be carried out in an anaerobic chamber. If only actinomycetes are being considered, they will survive some exposure to air.

For cultivating the commensal actinomycetes, the various media mentioned before can be used. Selective principles naturally facilitate their isolation and identification. However, the selective media devised so far may suppress certain strains so that qualitative and quantitative relationships may be artificially altered.

The formulas and preparation instructions of two different media that are selective for oral actinomycetes are given below.

Selective Medium for Oral Actinomycetaceae (Beighton and Colman, 1976)

I. Basal culture medium (BYS medium):
"3.7 gm Brain Heart Infusion broth, 0.5 gm yeast extract powder, 1 gm polyvinylpyrrolidone, 0.1 gm cysteine HCl, and 1.5 gm agar were added to 100 ml of distilled water, autoclaved for 15 minutes at 10 psi, cooled to 45°C, and 5 ml of sterile horse serum added."

II. Selective enrichment medium:
"The enrichment medium (FC medium) is prepared by the addition of 1 ml of a sterile 25 mg/ml NaF solution and 0.5 ml of a sterile 1 mg/ml colistin sulfate solution per 100 ml of BYS medium. The antimicrobial agents were sterilized by autoclaving at 10 psi for 15 minutes."

A second selective medium for isolation of *Actinomyces viscosus* and *Actinomyces naeslundii* from dental plaque has been published by Kornman and Loesche (1978). Its selective principles are antimicrobial substances added to a rich basal medium.

Selective GMC Medium for *Actinomyces viscosus* and *Actinomyces naeslundii*
(Kornman and Loesche, 1978)

As basal medium, the enriched gelatin agar of Syed (1976) is used, but other complex media may yield similar results. To the basal medium, metronidazole and cadmium sulfate ($3CdSO_4 \cdot 8H_2O$) are added in various concentrations. Best recovery of the actinomycetes was obtained with metronidazole (10 μg/ml) and cadmium sulfate (20 μg/ml). Cadmium sulfate can be autoclaved together with the basal medium; metronidazole is added to the cooled medium after filter sterilization.

This medium was reported to allow 98% recovery of *A. viscosus* and 73% recovery of *A. naeslundii*, while suppressing 76% of the total count of other oral organisms. *Arachnia* strains also grew quite well, but *Actinomyces israelii, Actinomyces odontolyticus, Rothia dentocariosa,* and *Lactobacillus casei* were inhibited (Kornman and Loesche, 1978).

A different means for selective isolation of fermentative actinomycetes, especially of *A. israelii,* has been described by Fritsche (1964).

Selective Isolation of *Actinomyces israelii*
(Fritsche, 1964)

Material suspected of containing actinomycetes is suspended and dispersed in a suitable transport medium. Then 1 ml of this suspension is added to 1 ml of toluene in a screw-cap tube and placed on a mechanical shaker (high speed) for 20–25 min. The watery suspension at the bottom of the tube is carefully removed with a capillary pipette and added to 10 ml of transport medium. Remaining droplets of toluene on the surface of the medium are removed in a Bunsen flame. Then the medium is centrifuged and the sediment is streaked on nonselective culture media.

On first isolation, the colony morphology of the actinomycetes may be altered by this technique, but will become typical again in subculture. *A. israelii* has been proven to survive this treatment in all instances, but whether the colony count decreases after this procedure has not been investigated.

ISOLATION OF *ACTINOMYCES HUMIFERUS* AND *AGROMYCES RAMOSUS* FROM SOIL. Although these organisms were reported to be the numerically predominant inhabitants of certain soils (Gledhill and Casida, 1969a, b), they are quite fastidious and therefore difficult to grow on primary isolation. Casida (1965) described a modified dilution-frequency procedure, which was designed to allow fastidious organisms to adapt to laboratory media without competition from other species.

Isolation Media for *Actinomyces humiferus* and *Agromyces ramosus* (Gledhill and Casida, 1969b)

Medium No. 1:

Heart infusion broth	2.5%
Agar	0.08%
(Yeast extract)	(0.54%)
(Casitone)	(0.4%)

The components in parentheses were omitted in the second publication of the formula. Final pH is from 7.0 to 7.3.

Medium No. 2:

Tryptone	0.5%
Casamino Acids	0.4%
$(NH_4)_2HPO_4$	0.07%
NaCl	0.05%
Agar	0.2%
pH 7.0	

As solid medium, heart infusion agar was employed by Gledhill and Casida (1969a, b). Jones, Watkins, and Meyer (1970) used soil-extract-medium plates for morphological observations of *Agromyces ramosus*. According to their findings, the addition of fresh horse blood or MnO_2 to the various growth media for actinomycetes would probably facilitate the isolation and maintenance of *Agromyces* cultures.

Identification

Although fermentative actinomycetes, in general, exhibit several characteristic properties potentially useful for identification, they are often quite difficult to identify on the basis of morphology and by standard biochemical and physiological tests. With the exception of serological techniques, there is essentially no simple test that will either differentiate actinomycetes from similar organisms, or distinguish between different genera and species in the family Actinomycetaceae. It is for this reason that chemotaxonomy and acid end-product analysis have contributed so much to the classification and identification of actinomycetes. For the same reason, serological tests, such as immunofluorescence, are widely employed for identification purposes because they provide a rapid, simple, and comparatively reliable tool.

Morphology

Morphological criteria have always been considered important for the identification of actinomycetes because branching, filamentous cells and mycelial growth appear to be very characteristic. Unfortunately, morphological characters are mostly not very stable and there is some overlapping between species, genera, and even families so that microscopic or macroscopic examinations of cultures usually do not allow reliable identification.

Nevertheless, cellular and colonial features may provide presumptive evidence for the taxonomic position of an unknown isolate. If Gram-stained smears from a clinical specimen show fragmenting filaments that are branching, irregularly curved, and stained (Fig. 2), this is highly suggestive of an actinomycete, although not specific for a certain species. Very often, diphtheroid rods (Fig. 3) represent the cellular appearance of fermentative actinomycetes, but may also belong to corynebacteria or propionibacteria. Only two monotypic genera of the family Actinomycetaceae, *Bacterionema* and *Rothia*, possess microscopically typical features. *Bacterionema* (Fig. 4) is characterized by the formation of thin filaments and very large, bacillus-like bodies. Often, such filaments are attached to a bacillus-like element, giving rise to the highly symptomatic "whip handle" morphology that also may occur in vivo (Davis and Baird-Parker, 1959; Gilmour, 1974). *Rothia* also has a characteristic cellular morphology (Fig. 5). It is the only fermentative actinomycete that produces large amounts of

coccoid forms besides rods and filaments (Georg, 1974).

Ørskov (Erikson, 1940) introduced the method of examining growing colonies in situ on the agar medium that proved to be extremely useful for the presumptive identification of actinomycetes, especially when early growth stages after 18–24 h of incubation are viewed. Such microcolonies may be highly filamentous without signs of fragmentation (Fig. 6); the filaments may be shorter, but still easy to identify (spider colonies) (Fig. 7); fragmentation may occur quite early (Fig. 8); or real mycelia are never developed during growth (Fig. 9).

Although there is some relation between species membership and appearance of the microcolonies, this character is considerably variable. It depends on growth conditions and media, and also on the individual strain. Filamentous microcolonies are generally not found in *Actinomyces bovis* and *A. odontolyticus*. In *Bacterionema*, the filaments are visibly thicker. In *Arachnia* they often develop from the initial cell in one direction and not, as in *Actinomyces israelii*, with radial symmetry. *Actinomyces naeslundii* and *A. viscosus* often produce microcolonies that already show a dense, fragmented center after 18 h of incubation.

Mature colonies may be observed using a hand lens, but more easily under the microscope. Such colonies can retain much of the mycelial appearance, especially in the periphery (Figs. 10 and 11). But in many strains, the actinomorphous edge is gradually lost with further development of the colony (Figs. 12 and 13) and sometimes totally disappear (Fig. 14). Some species, like *Actinomyces bovis*, *A. odontolyticus*, or *Actinobacterium meyerii*, often produce entirely smooth colonies. The colonies are still dense and opaque in *A. bovis*, but mostly flat and translucent in *A. odontolyticus* (Fig. 15) and *Actinobacterium meyerii* (Fig. 16).

Chemotaxonomic Characters

Analysis of cell wall components and structures (Boone and Pine, 1968; Cummins, 1962; Cummins and Harris, 1956, 1968; DeWeese, Gerencser, and Slack, 1968; Gerencser and Slack, 1969; Kandler, 1970; Pine and Boone, 1967) were shown to be extremely valuable for the taxonomy of actinomycetes. Relatively simple techniques have been developed (Becker et al., 1964) that can also be applied to identification. In all cases in which a high degree of reliability is required for identification, the quite elaborate preparation of purified cell walls and their subsequent hydrolysis and chromatographic analysis may be necessary. In principle, the methods devised by Cummins and Harris can still be used for this purpose, although minor modifications may facilitate their application.

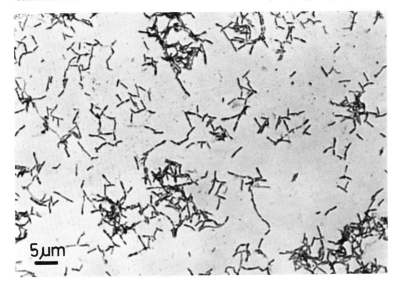

Fig. 2. *Actinomyces israelii;* Gram stain from a mature (7 days) colony on BHIA.

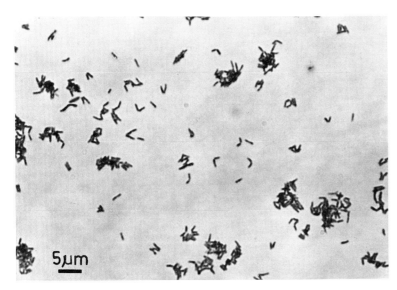

Fig. 3. *Actinomyces naeslundii;* Gram stain from a mature (7 days) colony on BHIA.

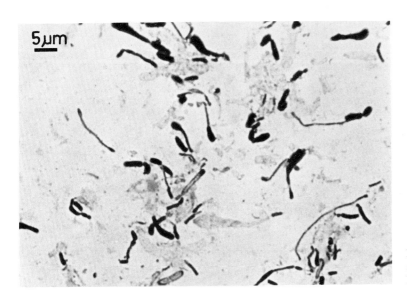

Fig. 4. *Bacterionema matruchotii;* Gram stain from an old (14 days) colony, BHIA.

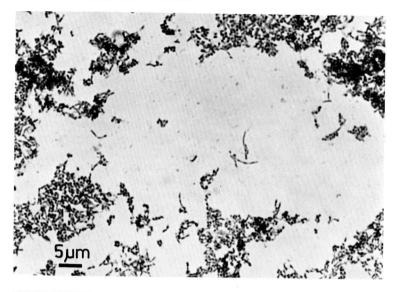

Fig. 5. *Rothia dentocariosa;* Gram stain from a mature (5 days) colony, BHIA.

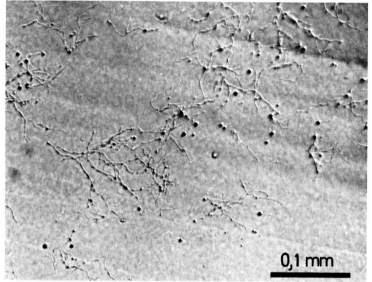

Fig. 6. *Arachnia propionica;* micro-colonies, CC medium, 24 h.

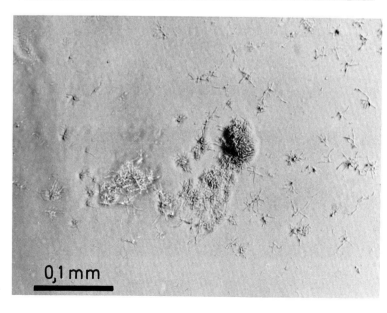

Fig. 7. *Actinomyces israelii;* spider colonies, CC medium, 24 h.

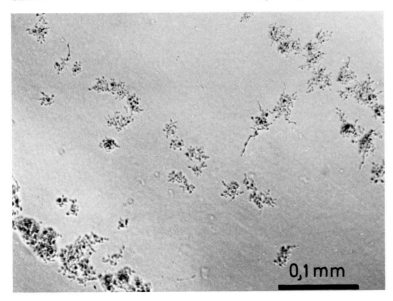

Fig. 8. *Actinomyces israelii;* micro-colonies with longer and shorter filaments and fragmentation, CC medium, 24 h.

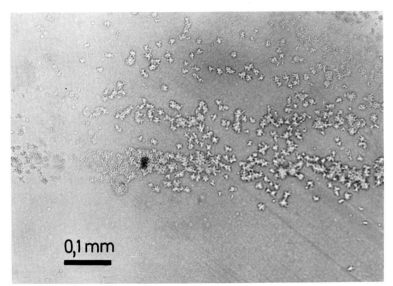

Fig. 9. *Actinomyces bovis;* small, irregular, but not filamentous, microcolonies, CC medium, 24 h.

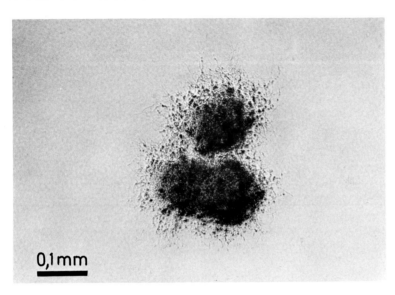

Fig. 10. *Actinomyces naeslundii;* mature colonies with dense granular center and filamentous edge, CC medium, 7 days.

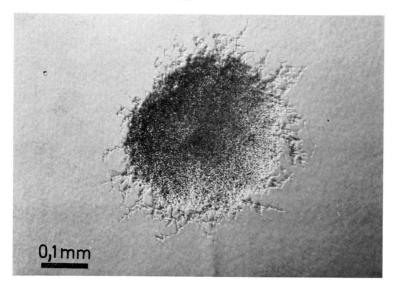

Fig. 11. *Bacterionema matruchotii;* mature colony with filamentous edge, CC medium, 7 days.

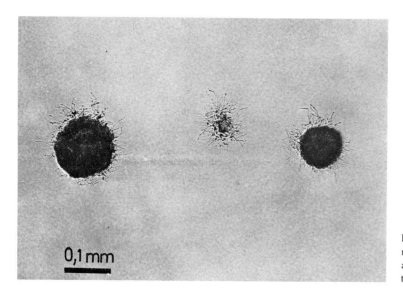

Fig. 12. *Actinomyces naeslundii;* colonies in different developmental stages and with different degree of fragmentation, CC medium, 5 days.

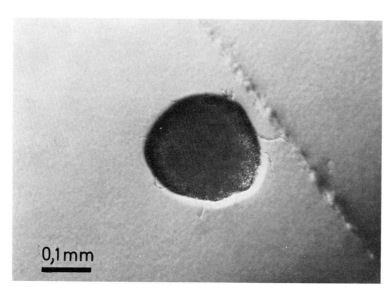

Fig. 13. *Actinomyces israelii;* nearly fully fragmented macrocolony with a few single filaments in the periphery, CC medium, 7 days.

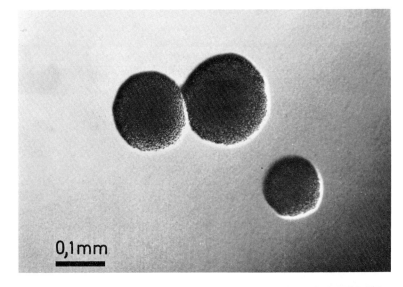

Fig. 14. *Rothia dentocariosa;* mature colonies, dense and granular without visible mycelia, CC medium, 5 days.

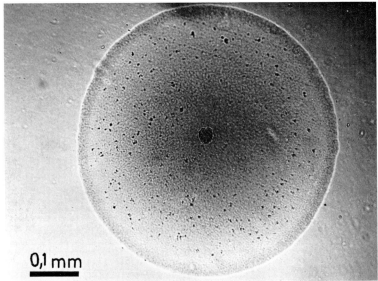

Fig. 15. *Actinomyces odontolyticus;* mature colony, transparent with optically dark center, CC medium, 7 days.

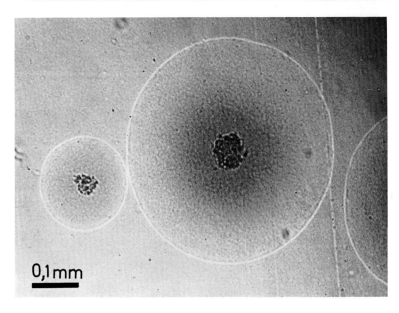

Fig. 16. *Actinobacterium meyerii;* large, translucent macrocolonies with optically dark center, CC medium, 14 days.

Further useful criteria for identification can be derived from the analysis of mycolic acids and long-chain fatty acids (Ashamaony et al., 1977). In particular, the method for demonstrating and identifying mycolic acids (Minnikin, Alshamaony, and Goodfellow, 1975) is comparatively simple and can be employed for identification.

Some easily determined chemotaxonomic criteria can be used in an identification scheme that is particularly useful in the differentiation of genera (Table 1). For comparison, G+C ratios of the respective type strains are included in Table 1. Even if only whole-cell hydrolysates are analyzed, the presence or absence of DAP and its isomer gives valuable information about the genus to which an isolate may belong.

End Products of Glucose Fermentation

As with other anaerobes, the identification of acid end products by gas–liquid chromatography (GLC) provides additional criteria for the allocation of an unknown isolate to genera and families. Procedures for performing GLC were described in detail in the *VPI Anaerobe Laboratory Manual* (Holdeman, Cato, and Moore, 1977).

The major acid end products released by Actinomycetaceae into the culture medium are summarized in Table 2. For comparison, some other morphologically similar genera are included and the differentiating power of GLC end-product analysis can easily be seen.

Physiological Characters

Physiological markers, such as growth on different media, atmospheric requirements, fermentation of carbohydrates, catalase activity, and others, have been used for many years to identify Actinomycetaceae at the genus and species level. Concerning the genus level, there is no doubt today (Bowden and Hardie, 1973; Slack, 1974; Slack and Gerencser, 1975) that such criteria need support from chemotaxonomic techniques and GLC end-product analysis to make them really effective. For species identification, however, physiological tests can be effective (Slack, 1974; Slack and Gerencser, 1975), but provide reliable results only when the technique is carefully adjusted and when a large battery of tests is performed. In addition, several species apparently exhibit many variable characters which lead to much overlapping with other species in the identification

Table 1. Chemotaxonomic characteristics of Actinomycetaceae.[a]

| Genera/species | \multicolumn{12}{c}{Cell wall constituents} | | | | | | | | | | | | G+C (mol %) |
	Orni-thine	Ly-sine	Aspartic acid	Gly-cine	DAP[b]	LL-DAB[c]	Glu-cose	Galac-tose	Rham-nose	6-Deoxy-talose	Arabi-nose	Mycolic acid	
Actinomyces													
bovis	−[d]	+[e]	+	−	−	−	−	−	+	+	−	−	63
israelii	+	+	−	−	−	−	−	+	−	−	−	−	60
naeslundii	+	+	−	−	−	−	+	−	+	+	−	−	63
viscosus	+	+	−	−	−	−	+	+	+	+	−	−	59
odontolyticus	+	+	−	−	−	−	+	+	+/−	+/−	−	−	62
suis	+	+	−	−	−	−	−	−	+	−	−		
humiferus	+	+	+	−	−	−	−	−	+	−			73
Agromyces													
ramosus	−	−	−	+	−	+	+/−[f]	+/−	+	−	−		71
Arachnia													
propionica	−	−	−	+	LL[g]	−	+/−	+	−	−	−	−	64
Bacterionema													
matruchotii	−	−	−	+	DL[g]	−	+/−	+	−	−	+	+	53
Rothia													
dentocariosa	−	+	−	−	−	−	+/−	+	−	−	−	−	51

[a] Data taken from Bowden and Hardie (1973), Slack and Gerencser (1975), Fiedler and Kandler (1973), Gledhill and Casida (1969b).
[b] 2,6-Diaminopimelic acid.
[c] L-2,4-Diaminobutyric acid.
[d] Absent.
[e] Present.
[f] Different types or different reports.
[g] LL/DL-DAP present.

Table 2. Acid end products of Actinomycetaceae.[a]

| | Major end products | | | | | |
Genera	Acetic	Propionic	*n*-Butyric	*n*-Caproic	Lactic	Succinic acids
Actinomyces	+[b]	−[c]	−	−	+	+
Agromyces	0[d]	0	0	0	0	0
Arachnia	+	+	−	−	(+)[e]	+
Bacterionema	+	+	−	−	(+)	−
Rothia	+	−	−	−	+	(+)
Bifidobacterium	+	−	−	−	+	−
Propionibacterium	+	+	−	−	(+)	(+)
Eubacterium	+	−	+	+/−[f]	(+)	−
Lactobacillus	−	−	−	−	+	−

[a] Data taken from Slack and Gerencser (1975), Holdeman et al. (1977).
[b] Present in major amounts.
[c] Absent or only in lower amounts present.
[d] Oxidative metabolism.
[e] Mostly a minor end product.
[f] Species differences.

scheme. Such overlapping results quite often in the inability to identify an unknown isolate properly.

Details on the recommended test procedures applicable to fermentative actinomycetes were given by Slack and Gerencser (1975). For fermentation tests, these authors propose the use of thioglycolate fermentation base for *Actinomyces* and *Arachnia*, and meat extract peptone base for *Rothia*.

Media for Fermentation Tests on Actinomycetes (Slack and Gerencser, 1975)

I. Thioglycolate fermentation base medium contains:

Fluid thioglycolate medium without	
dextrose or indicator	1 liter
Yeast extract (glucose-free)	2 g
Bromocresol purple (1% aqueous	
solution)	2 ml

II. Meat extract peptone base medium contains:

Beef extract	3 g
Peptone	10 g
NaCl	5 g
Distilled water	1,000 ml
Andrade's indicator (pH 7.4)	10 ml

Low-molecular-weight carbohydrates are added aseptically to the medium to give a final concentration of 1%; other compounds are used at 0.5%.

In our experience, the results obtained from those media are still not satisfactory. This view is supported by the results of numerical taxonomic studies

in which similar techniques were used (Holmberg and Hallander, 1973; Holmberg and Nord, 1975; Melville, 1965). Use of commercially available test kits with a high degree of standardization increases the reproducibility and reliability of such tests. We have recently tested the value of the Minitek differentiation system (BBL) for its value in identifying fermentative actinomycetes (Schofield and Schaal, 1979a). After considerable modifications, this system seems to be of some value for performing fermentation tests with actinomycetes. In addition, we have developed a basal medium for carbon-source utilization tests (Schofield and Schaal, 1979b) which appears to work better than those previously described (Howell and Pine, 1956).

Despite these reservations, Table 3 gives a summary of physiological characters that may aid in the species identification of fermentative, temporarily or persistently filamentous organisms. As can be seen, the diagnostic value of the table is considerably lowered by many variable results, which would be even more pronounced when deviating results of single strains had been considered.

Recently, it has been shown (Kilian, 1978) that the API Zym system, which allows detection of 19 different enzymatic reactions in a commercially standardized test kit, can provide additional diagnostic characters suitable for the differentiation of species of the family Actinomycetaceae. The further advantage of this system is that results are available after 4–6 h of incubation. Of course, it is too early to judge the value of these data before they have been confirmed by other workers. But it seems at

Table 3. Physiological characters of Actinomycetaceae.[a]

Character	Actinomyces bovis	A. odontolyticus	A. israelii	A. naeslundii	A. viscosus	A. eriksonii	A. suis	A. humiferus	Agromyces ramosus	Arachnia propionica	Bacterionema matruchotii	Rothia dentocariosa	Actinobacterium meyerii
Aerobic growth	−[b]	d[c]	−	+[d]	+	−	(+)[e]	+	+	(+)	+	+	−
Catalase	−	−	−	−	+	−	−	−	−	−	+	+	−
Nitrate red.	−	+	d	+	+	−	+	−	d	+	+	+	+
Esculin hydrol.	+	d	+	+	+	d	+	+	d	−	+	+	−
Starch hydrol.	+	−	−	−	−	−	0[f]	−	d	−	d	−	−
Sugar fermentation	+	+	+	+	+	+	+	+		+	+	+	+
Sugar oxidation	−	−	−	−	−	−	−	−	+	−	−	−	−
Adonitol[g]	−	d	−	−	−	−	d	−	−	+	0	−	−
Arabinose[g]	−	d	d	−	−	+	(+)	+	+	−	−	−	d
Dulcitol[g]	−	−	−	−	−	−	−	d	−	−	−	−	0
Glucose[g]	+	+	+	+	+	+	+	+	+	+	+	+	+
Glycerol[g]	d	d	−	+	+	−	d	d	+	−	−	d	0
Inositol[g]	+	d	d	d	d	−	+	−	−	−	−	−	0
Lactose[g]	+	d	d	d	d	+	+	d	−	+	−	−	+
Mannitol[g]	−	−	d	−	−	+	d	+	+	+	−	−	−
Raffinose[g]	−	d	+	+	+	+	+	+	+	+	d	−	−
Salicin[g]	d	d	d	d	d	+	+	d	−	−	+	+	0
Sorbitol[g]	−	−	−	−	−	+	−	−	−	+	−	−	−
Trehalose[g]	−	d	d	d	d	+	+	d	−	+	−	+	−
Xylose[g]	d	d	+	−	−	+	(+)	+	+	−	−	−	+

Species

[a] Data compiled from Slack (1974), Slack and Gerencser (1975), Gledhill and Casida (1969b), Franke (1973), Holmberg and Nord (1975).
[b] Negative reaction. [c] Variable reaction. [d] Positive reaction. [e] Weak reaction. [f] Not reported. [g] Acid production from.

least promising to try the device on a larger number of strains.

Serological Identification

The early papers on the serology of fermentative actinomycetes (see Slack et al., 1951, 1955) indicated that a differentiation is possible on serological grounds. However, agglutination tests used previously presented many methodological problems. The introduction of fluorescent-antibody techniques (Slack, Winger, and Moore, 1961) into the field of actinomycetes made serology widely applicable for practical and scientific identification purposes.

Since then, other methods, including gel precipitation or immunoelectrophoresis, have also been used, and some of the antigens of actinomycetes have been isolated and characterized (see Bowden and Hardie, 1973; Slack and Gerencser, 1975). Such investigations showed that antigens common to several species are generally present, but species-specific, partial antigens are usually strong enough to allow reliable identification when absorbed antisera are used.

For routine purposes, a simple fluorescent-antibody (FA) procedure has been devised by Slack and Gerencser (1975).

Fluorescent-Antibody Procedure
(Slack and Gerencser, 1975)

Preparation of smears. (i) Clinical material. Two smears of the material are made on a clean glass slide containing two marked circles. Air-dried smears are fixed by flooding with methanol for 1 min. Then the alcohol is poured off and the slide is allowed to air-dry. (ii) Cultures. Smears of a suspension of the organism obtained by centrifuging a broth culture, or from a plate culture, are made on glass slides, air-dried, and gently heat-fixed.

Staining procedures. (i) One drop of conjugated specific antiserum is placed on each smear and the slide is incubated under a moist chamber for 30 min at room temperature. (ii) Excess conjugate is poured off and the slide is washed in two changes of pH 7.2 buffer (FTA hemagglutination buffer, BBL) for 5 min each. (iii) The preparation is counterstained in 0.5% Evans blue for 5 m. (iv) Excess Evans blue is removed by dipping the slide briefly in distilled water, and the slide is washed in two changes of pH 9.0 buffer for 1 min each. (v) Smears are allowed to air-dry. (vi) One drop of buffered glycerol mounting fluid (9 parts c.p. glycerol, 1 part pH 9.0 buffer) is placed on each smear, which is subsequently covered with a cover slip. (vii) For examination with a microscope equipped for FA work, 54× oil-immersion objectives are most useful.

For the routine use of this technique, it is disadvantageous that FITC-labeled specific antisera are commercially not available except for antisera against *Actinomyces israelii* and *A. naeslundii* (Biological Reagent Section, Center for Disease Control, Atlanta, Georgia, USA). Thus, most antisera have to be prepared and conjugated by the user himself. To avoid part of this expensive and time-consuming work that is necessary when direct FA technique is employed, we have described an indirect FA-staining procedure (Schaal and Pulverer, 1973) that can provide similarly specific identification results. In this method, commercially available, FITC-conjugated goat antisera to rabbit globulins are used. The diagnostic antisera still have to be produced by immunization of rabbits, but the elaborate conjugation procedure can be omitted.

Provided that the staining procedure and the reagents are standardized (Holmberg and Forsum, 1973) FA techniques allow quick and comparatively reliable identification of all members of the family Actinomycetaceae. FA techniques are especially valuable for clinical microbiological laboratories because the methods are applicable directly to the clinical specimen as well as to isolates from culture (Slack and Gerencser, 1975). Furthermore, the identification and enumeration of filamentous organisms in dental plaque or gingival crevices can be achieved by these methods. However, it must be noted that antigenically aberrant strains and hitherto unknown serotypes or species may be lost or misidentified if FA procedures are applied as the sole means of identification.

Maintenance of Cultures and Reference Strains

Fermentative actinomycetes can cause some problems on continuous subcultivation and they are readily contaminated, especially with propionibacteria which is sometimes difficult to notice. Therefore, procedures that avoid contamination and loss of viability are necessary for long-term storage of cultures.

For this purpose, freezing or lyophilization has been recommended (Slack and Gerencser, 1975). Freezing at −70°C is done directly with freshly grown broth cultures in screw-capped test tubes or vials. For freeze-drying, the sediment of broth cultures or cell mass scraped from agar plates are suspended in skim milk and lyophilized (Slack and Gerencser, 1975). In our experience, the viability of lyophilized *Actinomyces* isolates can be improved by the use of a medium that contains 50% broth (vol/vol), 50% horse serum (vol/vol), and, in the whole mixture, 7% sucrose (wt/vol).

Using lyophilization, Slack and Gerencser (1975) reported viability of cultures after 10 years of storage at room temperature. We were generally able to revive lyophilates that had been stored from 20 to 25 years at room temperature. Frozen broth cultures will survive for at least 1 year.

The type strains or suggested neotype strains of Actinomycetaceae as designated so far are: *Actinomyces bovis*, ATCC 13683; *Actinomyces odontolyticus* ATCC 17929; *Actinomyces israelii*, ATCC 12102; *Actinomyces naeslundii*, ATCC 12104; *Actinomyces viscosus*, ATCC 15987; *Actinomyces humiferus*, ATCC 25174; *Arachnia propionica*, ATCC 14157; *Agromyces ramosus*, ATCC 25173; *Actinomyces eriksonii*, ATCC 15423; *Bacterionema matruchotii*, ATCC 14266; *Rothia dentocariosa*, ATCC 17931 (data according to Georg, 1974; Gilmour, 1974; Pine and Georg, 1974; Slack, 1974; Slack and Gerencser, 1975). (ATCC, American Type Culture Collection.)

Literature Cited

Allen, S. H. G., Linehan, B. A. 1977. Presence of transcarboxylase in *Arachnia propionica*. International Journal of Systematic Bacteriology **27**:291–292.

Alshamaony, L., Goodfellow, M., Minnikin, D. E., Bowden, G. H., Hardie, J. M. 1977. Fatty and mycolic acid composition of *Bacterionema matruchotii* and related organisms. Journal of General Microbiology **98**:205–213.

Baboolal, R. 1969. Cell wall analysis of oral filamentous bacteria. Journal of General Microbiology **58**:217–226.

Barnham, M., Burton, A. C., Copland, P. 1978. Pelvic actinomycosis with IUCD. British Medical Journal **1**:719–720.

Batty, I. 1958. *Actinomyces odontolyticus*, a new species of actinomycete regularly isolated from deep carious dentine. Journal of Pathology and Bacteriology **75**:455–459.

Becker, B., Lechevalier, M. P., Gordon, R. E., Lechevalier, H. A. 1964. Rapid differentiation between *Nocardia* and *Streptomyces* by paper chromatography of whole-cell hydrolysates. Applied Microbiology **12**:421–423.

Beighton, D., Colman, G. 1976. A medium for the isolation and enumeration of oral Actinomycetaceae from dental plaque. Journal of Dental Research **55**:875–878.

Beighton, D., McDougall, W. A. 1977. The effects of fluoride on the percentage bacterial composition of dental plaque, on caries incidence, and on the in vitro growth of *Streptococcus mutans*, *Actinomyces viscosus*, and *Actinobacillus* sp. Journal of Dental Research **56**:1185–1191.

Bellack, S., Jordan, H. V. 1972. Serological identification of rodent strains of *Actinomyces viscosus* and their relationship to actinomyces of human origin. Archives of Oral Biology **17**:175–182.

Berger, U., Kapovits, M., Pfeifer, G. 1959. Zur Besiedlung der kindlichen Mundhöhle mit anaeroben Mikroorganismen. Zeitschrift für Hygiene und Infektionskrankheiten **145**:564–573.

Bergey, D. H. 1907. Aktinomyces der Mundhöhle. Zentralblatt für Bakteriologie, Parasitenkunde und Infektionskrankheiten, Abt. 1, Referate **40**:361.

Bibby, B. G., Berry, G. P. 1939. A cultural study of the filamentous bacteria obtained from the human mouth. Journal of Bacteriology **38**:263.

Biever, L. J. 1975. A bacteriologic study of abscesses of swine and cattle. M. S. thesis. South Dakota State University cited according to Slack and Gerencser, 1975.

Blank, C. H., Georg, L. K. 1968. The use of fluorescent antibody methods for the detection and identification of *Actino-*

myces species in clinical material. Journal of Laboratory and Clinical Medicine **71**:283–293.

Bollinger, O. 1877. Ueber eine neue Pilzkrankheit beim Rinde. Deutsche Zeitschrift für Thiermedicin und Vergleichende Pathologie **3**:334.

Boone, C. J., Pine, L. 1968. Rapid method for characterization of actinomycetes by cell wall composition. Applied Microbiology **16**:279–284.

Bouisset, L., Breuillaud, J., Michel, G., Larrouy, G. 1968. Bases nucléiques des bactéries application au genre *Actinobacterium*. Annales de l' Institut Pasteur **115**:1063–1081.

Bowden, G. H., Hardie, J. M. 1973. Commensal and pathogenic *Actinomyces* species in man, pp. 277–295. In: Sykes, G., Skinner, F. A. eds., Actinomycetales, characteristics and practical importance. London: Academic Press.

Boyan-Salyers, B. D., Vogel, J. J., Ennever, J. 1978. Basic biological sciences; pre-apatitic mineral deposition in *Bacterionema matruchotii*. Journal of Dental Research **57**:291–295.

Boyd, A., Williams, R. A. D. 1971. Estimation of the volumes of bacterial cells by scanning electron microscopy. Archives of Oral Biology **16**:259–267.

Brock, D. W., Georg, L. K., Brown, J. M., Hicklin, M. D. 1973. Actinomycosis caused by *Arachnia propionica*. American Journal of Clinical Pathology **59**:66–77.

Brown, J. M., Georg, L. K., Waters, L. C. 1969. Laboratory identification of *Rothia dentocariosa* and its occurrence in human clinical materials. Applied Microbiology **17**:150–156.

Buchanan, B. B., Pine, L. 1962. Characterization of a propionic acid producing actinomycete, *Actinomyces propionicus*, sp. nov. Journal of General Microbiology **28**:305–323.

Buchanan, B. B., Pine, L. 1965. Relationship of carbon dioxide to aspartic acid and glutamic acid in *Actinomyces naeslundii*. Journal of Bacteriology **89**:729–733.

Buchanan, R. E. 1918. Studies in the classification and nomenclature of the bacteria. VIII. The subgroups and genera of the Actinomycetales. Journal of Bacteriology **3**:403–406.

Bulleid, A. 1925. An experimental study of *Leptothrix* buccalis. British Dental Journal **46**:289.

Burckhardt, J. J. 1978. Rat memory T lymphocytes: In vitro proliferation induced by antigens of *Actinomyces viscosus*. Scandinavian Journal of Immunology **7**:167–172.

Casida, L. E., Jr. 1965. Abundant microorganism in soil. Applied Microbiology **13**:327–334.

Christ, M. L., Haja, J. 1978. Cytologic changes associated with vaginal pessary use with special reference to the presence of Actinomyces. Acta Cytologica **22**:146–149.

Cock, D. I., Bowen, W. H. 1967. Occurrence of *Bacterionema matruchotii* and *Bacteroides melaninogenicus* in gingival plaque from monkeys. Journal of Periodontal Research **2**:36–39.

Cohn, F. 1875. Untersuchungen über Bacterien II. Beiträge zur Biologie der Pflanzen III, 141–207.

Coleman, R. M., Georg, L. K. 1969. Comparative pathogenicity of *Actinomyces naeslundii* and *Actinomyces israelii*. Applied Microbiology **18**:427–432.

Collins, P. A., Gerencser, M., Slack, J. M. 1973. Enumeration and identification of Actinomycetaceae in human dental calculus using the fluorescent antibody technique. Archives of Oral Biology **18**:145–153.

Collins, M. D., Goodfellow, M., Minnikin, D. E. 1979. Isoprenoid quinones in the classification of coryneform and related bacteria. Journal of General Microbiology **110**:127–136.

Conrad, S. E., City, D., Breivis, J., Fried, M. A. 1978. Vertebral osteomyelitis, caused by *Arachnia propionica* and resembling actinomycosis. Journal of Bone and Joint Surgery **60-A**:549–553.

Crawford, J. M., Taubman, M. A., Smith, D. J. 1978. The natural history of periodontal bone loss in germfree and gnotobiotic rats infected with periodontopathic microorganisms. Journal of Periodontal Research **13**:316–325.

Cross, T., Goodfellow, M. 1973. Taxonomy and classification of the actinomycetes, pp. 11–112. In: Sykes, G., Skinner, F. A. (eds.), Actinomycetales, characteristics and practical importance. London: Academic Press.

Cummins, C. S. 1962. Chemical composition and antigenic structure of cell walls of *Corynebacterium, Mycobacterium, Nocardia, Actinomyces* and *Arthrobacter.* Journal of General Microbiology **28:**35–50.

Cummins, C. S. 1965. Chemical and antigenic studies on cell walls of mycobacteria, corynebacteria and nocardias. American Review of Respiratory Disease **92:**63–72.

Cummins, C. S., Harris, H. 1956. The chemical composition of the cell wall in some Gram-positive bacteria and its possible value as a taxonomic character. Journal of General Microbiology **14:**583–600.

Cummins, C. S., Harris, H. 1958. Studies on the cell-wall composition and taxonomy of Actinomycetales and related groups. Journal of General Microbiology **18:**173–189.

Cummins, C. S., Harris, H. 1959. Cell-wall composition in strains of *Actinomyces* isolated from human and bovine lesions. Journal of General Microbiology **21:**ii.

DaCosta, T., Gibbons, R. J. 1968. Hydrolysis of levan by human plaque streptococci. Archives of Oral Biology **13:**609–617.

Davis, G. H. G., Baird-Parker, A. C. 1959. *Leptotrichia buccalis.* British Dental Journal **106:**70–73.

DeWeese, M. S., Gerencser, M. A., Slack, J. M. 1968. Quantitative analysis of *Actinomyces* cell walls. Applied Microbiology **16:**1713–1718.

Durie, E. B. 1958. A critical survey of mycological research and literature for the years 1946–1956 in Australia. Mycopathologia et Mycologia Applicata **9:**80–96.

Ellen, R. P., Balcerzak-Raczkowski, I. B. 1977. Interbacterial aggregation of *Actinomyces naeslundii* and dental plaque streptococci. Journal of Periodontal Research **12:**11–20.

Ellen, R. P., Onose, H. 1978. pH Measurements of *Actinomyces viscosus* colonies grown on media containing dietary carbohydrates. Archives of Oral Biology **23:**105–111.

Ellis, P. P., Bausor, S. C., Fulmer, J. M. 1961. Streptothrix canaliculitis. American Journal of Ophthalmology **52:**36–43.

Emmons, C. W. 1938. The isolation of *Actinomyces bovis* from tonsillar granules. Public Health Reports **53:**1967.

Engel, D., Schroeder, H. E., Page, R. C. 1978. Morphological features and functional properties of human fibroblasts exposed to *Actinomyces viscosus* substances. Infection and Immunity **19:**287–295.

Engel, D., Van Epps, D., Clagett, J. 1976. In vivo and in vitro studies on possible pathogenic mechanisms of *Actinomyces viscosus.* Infection and Immunity **14:**548–554.

Ennever, J. 1960. Intracellular calcification by oral filamentous microorganisms. Journal of Periodontology **31:**304–307.

Ennever, J., Vogel, J. J., Streckfuss, J. L. 1971. Synthetic medium for calcification of *Bacterionema matruchotii.* Journal of Dental Research **50:**1327–1330.

Ennever, J., Riggan, L. J., Vogel, J. J., Boyan-Salyers, B. 1978. Characterization of *Bacterionema matruchotii* calcification nucleator. Journal of Dental Research **57:**637–642.

Erikson, D. 1940. Pathogenic anaerobic organisms of the *Actinomyces* group. Medical Research Council, Special Report Series **240:**5–63.

Fiedler, F., Kandler, O. 1973. Die Aminosäuresequenz von 2,4-Diaminobuttersäure enthaltenden Mureinen bei verschiedenen coryneformen Bakterien und *Agromyces ramosus.* Archiv für Mikrobiologie **89:**51–66.

Fortner, J. 1928. Ein einfaches Plattenverfahren zur Züchtung strenger Anaerobier. Centralblatt für Bakteriologie, Parasitenkunde und Infektionskrankheiten, Abt. 1 Orig. **108:**155–159.

Fortner, J. 1929. I. Zur Technik der anaëroben Züchtung. II. Zur Differenzierung der Anaërobier. Centralblatt für Bakteriologie, Parasitenkunde und Infektionskrankheiten, Abt. 1 Orig. **110:**233–256.

Franke, F. 1973. Untersuchungen zur Ätiologie der Gesäugeaktinomykose des Schweines. Zentralblatt für Bakteriologie, Parasitenkunde, Infektionskrankheiten und Hygiene, Abt. 1 Orig., Reihe A **223:**111–124.

Fritsche, D. 1964. Die Benzol- und Toluolresistenz des *Actinomyces israelii,* ein Hilfsmittel für die Strahlenpilzdiagnostik. Zentralblatt für Bakteriologie, Parasitenkunde, Infektionskrankheiten und Hygiene, Abt. 1 Orig. **194:**241–244.

Georg, L. K. 1974. Genus *Rothia,* pp. 679–681. In: Buchanan, R. E., Gibbons, N. E. (eds.), Bergey's manual of determinative bacteriology, 8th ed. Baltimore: Williams & Wilkins.

Georg, L. K., Brown, J. M. 1967. *Rothia,* gen. nov. An aerobic genus of the family Actinomycetaceae. International Journal of Systematic Bacteriology **17:**79–88.

Georg, L. K., Pine, L., Gerencser, M. A. 1969. *Actinomyces viscosus,* comb. nov., a catalase positive, facultative member of the genus *Actinomyces.* International Journal of Systematic Bacteriology **19:**291–293.

Georg, L. K., Brown, J. M., Baker, H. J., Cassell, G. H. 1972. *Actinomyces viscosus* as an agent of actinomycosis in the dog. American Journal of Veterinary Research **33:**1457–1470.

George, L. K., Robertstad, G. W., Brinkmann, S. A., Hicklin, M. D. 1965. A new pathogenic anaerobic *Actinomyces* species. Journal of Infectious Diseases **115:**88–99.

Gerencser, M. A., Slack, J. M. 1967. Isolation and characterization of *Actinomyces propionicus.* Journal of Bacteriology **94:**109–115.

Gerencser, M. A., Slack, J. M. 1969. Identification of human strains of *Actinomyces viscosus.* Applied Microbiology **18:**80–87.

Gibbons, R. J., Nygaard, M. 1970. Interbacterial aggregation of plaque bacteria. Archives of Oral Biology **15:**1397–1400.

Gibbons, R. J., van Houte, J. 1973. On the formation of dental plaque. Journal of Periodontology **44:**347–360.

Gilmour, M. N. 1974. Genus *Bacterionema,* pp. 676–679. In: Buchanan, R. E., Gibbons, N. E. (eds.), Bergey's manual of determinative bacteriology, 8th ed. Baltimore: Williams & Wilkins.

Gilmour, M. N., Beck, P. H. 1961. The classification of organisms termed *Leptotrichia (Leptothrix) buccalis.* III. Growth and biochemical characteristics of *Bacterionema matruchotii.* Bacteriological Reviews **25:**152–161.

Gilmour, M. N., Howell, A., Bibby, B. G. 1961. The classification of organisms termed *Leptotrichia (Leptothrix) buccalis.* I. Review of the literature and proposed separation into *Leptotrichia buccalis* Trevisan, 1879 and *Bacterionema* gen. nov., *B. matruchotii* (Mendel 1919) comb. nov. Bacteriological Reviews **25:**131–141.

Gledhill, W. E., Casida, L. E., Jr. 1969a. Predominant catalase-negative soil bacteria. II. Occurrence and characterization of *Actinomyces humiferus,* sp. n. Applied Microbiology **18:**114–121.

Gledhill, W. E., Casida, L. E., Jr. 1969b. Predominant catalase-negative soil bacteria. III. *Agromyces,* gen. n., microorganisms intermediary to *Actinomyces* and *Nocardia.* Applied Microbiology **18:**340–349.

Grässer, R. 1957. Vergleichende Untersuchungen an Actinomyceten von Mensch, Rind und Schwein. Thesis, Leipzig.

Green, S. L. 1978. Case report. Fatal anaerobic pulmonary infection due to *Bifidobacterium eriksonii.* Postgraduate Medicine **63:**187–192.

Grüner, O. P. N. 1969. *Actinomyces* in tonsillar tissue. A histological study of tonsillectomy material. Acta Pathologica et Microbiologica Scandinavica **76:**239–244.

Gupta, P. K., Erozan, Y. S., Frost, J. K. 1978. Actinomycetes and the IUD: An update. Acta Cytologica **22:**281–282.

Gupta, P. K., Hollander, D. H., Frost, J. K. 1976. Actinomycetes in cervicovaginal smears. An association with IUD usage. Acta Cytologica **20:**295–297.

Gupta, P. P., Sinha, B. P. 1978. Oral dermatophilosis associated with actinomycosis in cattle. Zentralblatt für Veterinaermedizin, Reihe A **25:**211–215.

Hanf, U., Hanf, G. 1955. Ein Beitrag zum Infektionsmodus der weiblichen Genitalaktinomykose. Geburtshilfe und Frauenheilkunde **15:**366–373.

Harz, C. O. 1877. *Actinomyces bovis*, ein neuer Schimmel in den Geweben des Rindes. Jahresbericht der Königlichen Centralen Thierarzneischule München für 1877/1878 **5:**125–140.

Heinrich, S., Korth, H. 1967. Zur Nährbodenfrage in der Routinediagnostik der Aktinomykose: Ersatz unsicherer biologischer Substrate durch ein standardisiertes Medium. In: Heite, H. -J. (ed.), Krankheiten durch Aktinomyceten und verwandte Erreger. Berlin: Springer-Verlag

Hemmes, G. D. 1963. Enige bevindingen over actinomycose. Nederlands Tijdschrift voor Geneeskunde **107:**193.

Hill, P. E., Knox, K. W., Schamschula, R. G., Tabua, J. 1977. The identification and enumeration of *Actinomyces* from plaque of New Guinea indigenes. Caries Research **11:**327–335.

Holdeman, L. V., Cato, E. P., Moore, W. E. C. 1977. V. P. I. Anaerobe laboratory manual, 4th ed. Blacksburg, Virginia; Southern Printing Co.

Holm, P. 1950. Studies on the aetiology of human actinomycosis. I. The "other microbes" of actinomycosis and their importance. Acta Pathologica et Microbiologica Scandinavica **27:**736–751.

Holm, P. 1951. Studies on the aetiology of human actinomycosis. II. Do the "other microbes" of actinomycosis possess virulence? Acta Pathologica et Microbiologica Scandinavica **28:**391–406.

Holmberg, K., Forsum, U. 1973. Identification of *Actinomyces, Arachnia, Bacterionema, Rothia,* and *Propionibacterium* species by defined immunofluorescence. Applied Microbiology **25:**834–843.

Holmberg, K., Hallander, H. O. 1973. Numerical taxonomy and laboratory identification of *Bacterionema matruchotii, Rothia dentocariosa, Actinomyces naeslundii, Actinomyces viscosus,* and some related bacteria. Journal of General Microbiology **76:**43–63.

Holmberg, K., Nord, C.-E. 1975. Numerical taxonomy and laboratory identification of *Actinomyces* and *Arachnia* and some related bacteria. Journal of General Microbiology **91:** 17–44.

Howell, A., Jr. 1963. A filamentous microorganism isolated from periodontal plaque in hamsters. I. Isolation, morphology and general cultural characteristics. Sabouraudia **3:**81–92.

Howell, A., Jr., Jordan, H. V. 1963. A filamentous microorganism isolated from periodontal plaque in hamsters. II. Physiological and biochemical characteristics. Sabouraudia **3:**93–105.

Howell, A., Jr., Jordan, H. V. 1967. Production of an extracellular levan by *Odontomyces viscosus*. Archives of Oral Biology **12:**571–573.

Howell, A., Jr., Murphy, W. C. III, Paul, F., Stephan, R. M. 1959. Oral strains of *Actinomyces*. Journal of Bacteriology **78:**82–95.

Howell, A., Jr., Pine, L. 1956. Studies on the growth of species of *Actinomyces*. I. Cultivation in a synthetic medium with starch. Journal of Bacteriology **71:**47–53.

Howell, A., Jr., Pine, L. 1961. The classification of organisms termed *Leptotrichia (Leptothrix) buccalis*. IV. Physiological and biochemical characteristics of *Bacterionema matruchotii*. Bacteriological Reviews **25:**162–171.

Howell, A., Jr., Stephan, R. M., Paul, F. 1962. Prevalence of *Actinomyces israelii, A. naeslundii, Bacterionema matruchotii,* and *Candida albicans* in selected areas of the oral cavity and saliva. Journal of Dental Research **41:**1050–1059.

Israël, J. 1878. Neue Beobachtungen auf dem Gebiete der Mykosen des Menschen. Archiv für Pathologische Anatomie und Physiologie und für klinische Medicin **74:**15–53.

Johnson, J. L., Cummins, C. S. 1972. Cell wall composition and deoxyribonucleic acid similarities among the anaerobic coryneforms, classical propionibacteria, and strains of *Arachnia propionica*. Journal of Bacteriology **109:**1047–1066.

Jones, D., Watkins, J., Meyer, D. J. 1970. Cytochrome composition and effect of catalase on growth of *Agromyces ramosus*. Nature **226:**1249–1250.

Jordan, H. V., Fitzgerad, R. J., Stanley, H. R. 1965. Plaque formation and periodontal pathology in gnotobiotic rats infected with an oral actinomycete. American Journal of Pathology **47:**1157–1167.

Jordan, H. V., Hammond, B. F. 1972. Filamentous bacteria isolated from root surface caries. Archives of Oral Biology **17:**1–12.

Jordan, H. V., Sumney, D. L. 1973. Root surface caries: Review of the literature and significance of the problem. Journal of Periodontology **44:**158–163.

Kandler, O. 1970. Amino acid sequence of the murein and taxonomy of the genera *Lactobacillus, Bifidobacterium, Leuconostoc* and *Pediococcus*. International Journal of Systematic Bacteriology **20:**491–507.

Karetzky, M. S., Garvey, J. W. 1974. Empyema due to *Actinomycoses naeslundi*. Chest **65:**229–230.

Kiel, R. A., Tanzer, J. M. 1977. Regulation of invertase of *Actinomyces viscosus*. Infection and Immunity **17:**510–512.

Kiel, R. A., Tanzer, J. M., Woodiel, F. N. 1977. Identification, separation, and preliminary characterization of invertase and β-galactosidase in *Actinomyces viscosus*. Infection and Immunity **16:**81–87.

Kilian, M. 1978. Rapid identification of Actinomycetaceae and related bacteria. Journal of Clinical Microbiology **8:**127–133.

King, S., Meyer, E. 1957. Metabolic and serologic differentiation of *Actinomyces bovis* and anaerobic diphtheroids. Journal of Bacteriology **74:**234–238.

Kornman, K. S., Loesche, W. J. 1978. New medium for isolation of *Actinomyces viscosus* and *Actinomyces naeslundii* from dental plaque. Journal of Clinical Microbiology **7:**514–518.

Kostečka, F. 1924. Relation of the teeth to the normal development of microbial flora in the oral cavity. Dental Cosmos **66:**927–935.

Lachner-Sandoval, V. 1898. Über Strahlenpilze. Inaugural Dissertation. Strassburg, Universitäts-Druckerei von Karl Georgi, Bonn.

Lentze, F. A. 1938. Die mikrobiologische Diagnostik der Aktinomykose. Münchner Medizinische Wochenschrift **47:**1826–1836.

Lentze, F. 1948. Dia Ätiologie der Aktinomykose des Menschen. Deutsche Zahnärztliche Zeitschrift **3:**913–919.

Lentze, F. 1953. Zur Aetiologie und Mikrobiologischen Diagnostik der Aktinomykose. Estratto degli Atti del VI Congresso Internationale di Microbiologia, Roma. **5** Sez. XIV 145–148.

Lentze, F. 1969. Die Aktinomykose und die Nocardiosen, pp. 954–973. In: Grumbach, A., Bonin, O. (eds.), Die Infektionskrankheiten des Menschen und ihre Erreger, vol. I, 2'nd ed. Stuttgart: Georg Thieme Verlag.

Lentze, F. 1970. Klinik, Diagnostik und Therapie der Aktinomykosen. Diagnostik und Therapie der Pilzkrankheiten und Neuere Erkenntnisse in der Biochemie der pathogenen Pilze (Kongressreferate, 6. Tagung der Deutschsprachigen Mykologischen Gesellschaft am 15. Juli 1966), pp. 83–92. Berlin: Grosse Verlag.

Lesher, R. J., Gerencser, V. F. 1977. Levan production by a strain of *Rothia*: Activation of complement resulting in cytotoxicity for human gingival cells. Journal of Dental Research **56:**1097–1105.

Llory, H., Guillo, B., Frank, R. M. 1971. A cariogenic *Actinomyces viscosus*—a bacteriological and gnotobiotic study. Helvetica Odontologica Acta **15:**134–138.

Loesche, W. J. 1960. Importance of nutrition in gingival crevice microbial ecology. Periodontics **6:**245–249.

Loesche, W. J., Hochett, R. N., Syed, S. A. 1972. The predominant cultivated flora of tooth surface plaque removed from institutionalized subjects. Archives of Oral Biology **17:**1311–1325.

Lord, F. T. 1910. The etiology of actinomycosis. Journal of the American Medical Association **55:**1261–1263.

Luff, R. D., Gupta, P. K., Spence, M. R., Frost, J. K. 1978. Pelvic actinomycosis and the intrauterine contraceptive device. American Journal of Clinical Pathology **69:**581–586.

Magnusson, H. 1928. The commonest forms of a actinomycosis in domestic animals and their etiology. Acta Pathologica et Microbiologica Scandinavica **5:**170–245.

Manly, B. S., Richardson, D. T. 1968. Metabolism of levan by oral samples. Journal of Dental Research **47:**1080–1086.

McIntire, F. C., Vatter, A. E., Baros, J. B., Arnold, J. 1978. Mechanism of coaggregation between *Actinomyces viscosus* T14V and *Streptococcus sanguis* 34. Infection and Immunity **21:**978–988.

Melville, T. H. 1965. A study of the overall similarity of certain actinomycetes mainly of oral origin. Journal of General Microbiology **40:**309–315.

Miller, C. H., Warner, T. N., Palenik, C. J., Somers, P. J. B. 1978. Levan formation by whole cells of *Actinomyces viscosus* ATCC 15987. Journal of Dental Research **54:**906.

Miller, C. H., Palenik, C. J., Stamper, K. E. 1978. Factors affecting the aggregation of *Actinomyces naeslundii* during growth and in washed cell suspensions. Infection and Immunity **21:**1003–1009.

Miller, C. H., Somers, P. J. 1978. Degradation of levan by *Actinomyces viscosus*. Infection and Immunity **22:**266–274.

Minnikin, D. E., Alshamaony, L., Goodfellow, M. 1975. Differentiation of *Mycobacterium, Nocardia,* and related taxa by thin-layer chromatographic analysis of whole-organism methanolysates. Journal of General Microbiology **88:**200–204.

Mitsuoka, T., Morishita, Y., Terada, A., Watanabe, K. 1974. *Actinomyces eriksonii* Georg, Robertstad, Brinkman und Hicklin 1965 identisch mit *Bifidobacterium adolescentis* Reuter 1963. Zentralblatt für Bakteriologie, Parasitenkunde, Infektionskrankheiten und Hygiene Abt. 1 Orig. Reihe A **226:**257–263.

Naeslund, C. 1925. Studies of *Actinomyces* from the oral cavity. Acta Pathologica et Microbiologica Scandinavica **2:**110–140.

Nitta, T., Okumura, S., Nakano, M. 1977. Effect of *Bacterionema matruchotii* on immune response. I. Enhancement of the phagocytic and bactericidal functions. Japanese Journal of Bacteriology **32:**691–696.

Nitta, T., Okumura, S., Tanabe, M. J., Nakano, M. 1978. Water-soluble adjuvant obtained from *Bacterionemia matruchotii*. Infection and Immunity **20:**721–727.

Pabst, M. J. 1977. Levan and levansucrase of *Actinomyces viscosus*. Infection and Immunity **15:**518–526.

Pine, L. 1970. Classification and phylogenetic relationship of microaerophilic actinomycetes. International Journal of Systematic Bacteriology **20:**445–474.

Pine, L., Boone, C. J. 1967. Comparative cell wall analysis of morphological forms within the genus *Actinomyces*. Journal of Bacteriology **94:**875–883.

Pine, L., Georg, L. K. 1969. The classification of *Actinomyces propionicus*. International Bulletin of Bacteriological Nomenclature and Taxonomy **15:**143–163.

Pine, L., Georg, L. K. 1974. Genus *Arachnia*, pp. 668–669. In: Buchanan, R. E., Gibbons, N. E. (eds.), Bergey's manual of determinative bacteriology, 8th ed. Baltimore: Williams & Wilkins.

Pine, L., Hardin, H. 1959. *A. israelii,* a cause of lacrimal canaliculitis in man. Journal of Bacteriology **78:**164–170.

Pine, L., Howell, A., Jr., Watson, S. J. 1960. Studies of the morphological, physiological, and biochemical characters of *Actinomyces bovis*. Journal of General Microbiology **23:**403–424.

Pollock, P. G., Koontz, F. P., Viner, T. F., Krause, C. J., Meyers, D. S., Valincenti, J. F., Jr. 1978. Cervicofacial actinomycosis. Rapid diagnosis by thin-needle aspiration. Archives of Otolaryngology **104:**491–494.

Prévot, A. R. 1938. Etudes de systématique bactérienne. III.

Invalidité du genre *Bacteroides* Castellani et Chalmers démembrement et reclassification. Annales de l' Institut Pasteur **60:**285–307.

Pulverer, G. 1974. Problems of human actinomycosis. Postepy Hiegieny i Medycyny Doswiadczalnej **28:**253–260.

Pulverer, G., Schaal, K. P. 1978. Pathogenicity and medical importance of aerobic and anaerobic actinomycetes, pp. 417–427. In: Mordarski, M., Kurylowicz, W., Jeljaszewicz, J. (eds.), *Nocardia* and *Streptomyces*. Stuttgart: Gustav Fischer Verlag.

Ritz, H. L. 1963. Localization of *Nocardia* in dental plaque by immunofluorescence. Proceedings of the Society for Experimental Biology and Medicine **113:**925–929.

Rogers, A. H., van der Hoeven, J. S., Mikx, F. H. M. 1978. Inhibition of *Actinomyces viscosus* by bacteriocin-producing strains of *Streptococcus mutans* in the dental plaque of gnotobiotic rats. Archives of Oral Biology **23:**477–485.

Rogosa, M., Cummins, C. S., Lelliott, R. A., Keddie, R. M. 1974. Coryneform group of bacteria, pp. 559–602. In: Buchanan, R. E., Gibbons, N. E. (eds.), Bergey's manual of determinative bacteriology, 8th ed. Baltimore: Williams & Wilkins.

Rosenthal, T., Gootzeit, E. H. 1942. The incidence of *B. fusiformis* and spirochaetes in the edentulous mouth. Journal of Dental Research **21:**373–374.

Roth, G. D., Thurn, A. N. 1962. Continued study of oral *Nocardia*. Journal of Dental Research **41:**1279–1292.

Russell, C., Melville, T. H. 1978. A review. Bacteria in the human mouth. Journal of Applied Bacteriology **44:**163–181.

Schaal, K. P. 1972. Zur mikrobiologischen Diagnostik der Nocardiose. Zentralblatt für Bakteriologie, Parasitenkunde, Infektionskrankheiten und Hygiene Abt. 1 Orig., Reihe A **220:**242–246.

Schaal, K. P. 1979. Die Aktinomykosen des Menschen—Diagnose und Therapie. Deutsches Ärzteblatt **76:**1997–2006.

Schleifer, K. H., Kandler, O. 1972. Peptidoglycan types of bacterial cell walls and their taxonomic implications. Bacteriological Reviews **36:**407–477.

Schofield, G. M., Schaal, K. P. 1979a. Application of the Minitek differentiation system in the classification and identification of *Actinomycetaceae*. FEMS Microbiology Letters **5:**311–313.

Schofield, G. M., Schaal, K. P. 1979b. A simple basal medium for carbon source utilization tests with the anaerobic actinomycetes. FEMS Microbiology Letters **5:**309–310.

Slack, J. M. 1942. The source of infection in actinomycosis. Journal of Bacteriology **43:**193–209.

Slack, J. M. 1974. Family *Actinomycetaceae* and genus *Actinomyces*, pp. 659–667. In: Buchanan, R. E., Gibbons, N. E. (eds.), Bergey's manual of determinative bacteriology, 8th ed. Baltimore: Williams & Wilkins.

Slack, J. M., Gerencser, M. A. 1975. *Actinomyces,* filamentous bacteria: Biology and pathogenicity. Minneapolis: Burgess.

Slack, J. M., Landfried, S., Gerencser, M. A. 1971. Identification of *Actinomyces* and related bacteria in dental calculus by the fluorescent antibody technique. Journal of Dental Research **50:**78–82.

Slack, J. M., Winger, A., Moore, D. W., Jr. 1961. Serological grouping of *Actinomyces* by means of fluorescent antibodies. Journal of Bacteriology **82:**54–65.

Slack, J. M., Ludwig, E. H., Bird, H. H., Canby, C. M. 1951. Studies with microaerophilic Actinomycetes. I. The agglutination reaction. Journal of Bacteriology **61:**721–735.

Slack, J. M., Spears, R. G., Snodgrass, W. G., Kuchler, R. J. 1955. Studies with microaerophilic Actinomycetes. II. Serological groups as determined by the reciprocal agglutinin adsorption technique. Journal of Bacteriology **70:**400–404.

Snyder, M. L., Slawson, M. S., Bullock, W., Parker, R. B. 1967. Studies on oral filamentous bacteria. II. Serological relationships within the genera *Actinomyces, Nocardia, Bacterionema* and *Leptotrichia*. Journal of Infectious Diseases **117:**341–345.

Socransky, S. S. 1970. Relationship of bacteria to the etiology of periodontal disease. Journal of Dental Research **49:**203–222.

Socransky, S. S., Manganiello, S. D. 1971. The oral microbiota of man from birth to senility. Journal of Periodontology **42:**485–496.

Spence, M. R., Gupta, P. K., Frost, J. K., King, T. M. 1978. Cytologic detection and clinical significance of *Actinomyces israelii* in women using intrauterine contraceptive devices. American Journal of Obstetrics and Gynecology **131:**295–298.

Sutter, V. L., Finegold, S. M. 1972. Anaerobic bacteriology manual. Los Angeles: UCLA.

Syed, S. A., Loesche, W. J. 1972. Survival of human dental plaque flora in various transport media. Applied Microbiology **24:**638–644.

Taichman, N. S., Hammond, B. F., Tsai, C.-C., Baehni, P. C., McArthur, W. P. 1978. Interaction of inflammatory cells and oral microorganisms. VII. In vitro polymorphonuclear responses to viable bacteria and to subcellular components of avirulent and virulent strains of *Actinomyces viscosus.* Infection and Immunity **21:**594–604.

Takazoe, I. 1961. Study on the intracellular calcification of oral aerobic leptotrichia. Shika Gakuho **61:**394–401.

Thompson, L., Lovestedt, S. A. 1951. An actinomyces-like organism obtained from the human mouth. Proceedings of the Staff Meetings of the Mayo Clinic **26:**169–175.

Trevisan, V. 1879. Prime linee d'introduzione allo die batteri italiani. Rendiconti. Istituto Lombardo Accademia di Scienze e Lettere **12:**133–151.

van Houte, J., Jansen, H. M. 1968. Levan degradation by streptococci isolated from human dental plaque. Archives of Oral Biology **13:**827–830.

Waksman, S. A., Henrici, A. T. 1943. The nomenclature and classification of the *Actinomycetes.* Journal of Bacteriology **46:**337–341.

Warner, T. N., Miller, C. H. 1978. Cell-associated levan of *Actinomyces viscosus.* Infection and Immunity **19:**711–719.

Wicken, A. J., Broady, K. W., Evans, J. D., Knox, K. W. 1978. New cellular and extracellular amphipathic antigen from *Actinomyces viscosus* NY1. Infection and Immunity **22:**615–616.

Winford, T. E., Haberman, S. 1966. Isolation of aerobic Gram positive filamentous rods from diseased gingivae. Journal of Dental Research **45:**1159–1167.

Witwer, M. W., Farmer, M. F., Wand, J. S., Solomon, L. S. 1977. Extensive actinomycosis associated with an intrauterine contraceptive device. American Journal of Obstetrics and Gynecology **128:**913–914.

CHAPTER 149

The Genus *Bifidobacterium*

VITTORIO SCARDOVI

Since they were first discovered in the feces of infants (Tissier, 1900), the bifidobacteria have stimulated much interest among bacteriologists, doctors, and nutritionists primarily concerned with investigating host–bacterium relationships. The apparent specific habitat of these bacteria led to a great deal of speculations—and even myths (see Kandler and Lauer, 1974)—about their nutritional and immunological significance (the so-called Bifidumproblem of the German authors). During what Poupard, Husain, and Norris (1973) call the ''first period'' in the history of these bacteria, namely, from 1900 to 1957, few advances were made in the knowledge of their biochemistry and taxonomy. In the seventh edition of *Bergey's Manual of Determinative Bacteriology* (Breed, Murray, and Smith, 1957), only *Lactobacillus bifidum* was reported, although in 1924, Orla-Jensen had already recognized the existence of the genus *Bifidobacterium* as separate taxon. In the last edition of *Bergey's Manual* (Rogosa, 1974), 11 species of the genus *Bifidobacterium* Orla-Jensen are listed and the genus is included in the family Actinomycetaceae Buchanan. Seven additional species are listed in the First Draft of the Approved Lists of Bacterial Names (Ad Hoc Committee of the Judicial Commission of the ICSB, 1976). In this development, the contributions of Dehnert (1957), Reuter (1963–64), Mitsuoka (1969), and Scardovi and colleagues (Scardovi and Crociani, 1974; Scardovi and Trovatelli, 1969, 1974; Scardovi et al., 1971) are significant.

In 1957, Dehnert first recognized the existence of multiple biotypes of *Bifidobacterium* and proposed a scheme for the differentiation of five groups of these bacteria based on carbohydrate fermentation. Reuter (1963–64) recognized and named seven species of *Bifidobacterium*, in addition to the known *B. bifidum*, on the basis of fermentative and serological characters and presented a scheme for their identification. The species of Reuter's scheme were: *Bifidobacterium bifidum* var. *a* and *b*, *B. infantis*, *B. parvulorum* var. *a* and *b*, *B. breve* var. *a* and *b*, *B. liberorum*, *B. lactentis*, *B. adolescentis* var. *a*, *b*, *c*, and *d*, and *B. longum* var. *a* and *b*. The next major classification scheme was presented by Mitsuoka (1969). In this scheme, Mitsuoka added new fermentative biotypes to *B. longum* species (*B. longum* subsp. *animalis a* and *b*) and the two new species *B. thermophilum* and *B. pseudolongum,* found in the feces of a variety of animals, e.g., pig, chicken, calf, rat. In the same year, Scardovi et al. (1969) isolated *B. ruminale* (synonym of *B. thermophilum,* see Rogosa, 1974) and *B. globosum* from the rumen of cattle; furthermore, Scardovi and Trovatelli (1969) found *B. asteroides B. indicum,* and *B. coryneforme* in the intestine of the honeybee, three new species quite different morphologically from the so far known bifidobacteria (see Figs. 11–14).

The pathway of hexose fermentation in bifidobacteria has in the meantime been elucidated by radioactive carbon distribution and assay of enzymes in cellular extract by Scardovi and Trovatelli (1965) and De Vries, Gerbrandy, and Stouthamer (1967). The key enzyme of this path (fructose-6-phosphate shunt), purified later by Sgorbati, Lenaz, and Casalicchio (1976), is a fructose-6-phosphate phosphoketolase (F6PPK), which splits the hexose phosphate to erythrose-4-phosphate and acetyl phosphate (see Schramm, Klybas, and Racker, 1958); from tetrose and hexose phosphates, through the successive action of transaldolase and transketolase, pentose phosphates are formed that, via the usual 2-3 cleavage, give rise to lactic and additional acetic acids so that acetic and lactic acids are formed in the theoretical ratio 1.5:1.0. Phosphoroclastic cleavage of some pyruvate to formic and acetic acids and reduction of acetate to ethanol can often alter the fermentation balance (De Vries and Stouthamer, 1968) to a highly variable extent (Lauer and Kandler, 1976).

In 1970, Scardovi and colleagues started to apply extensively the DNA-DNA filter hybridization procedure for assessing the validity at genetic level of the bifidobacterial species previously described (Scardovi et al., 1971), and for recognizing new DNA homology groups among the bifid strains they were isolating in large numbers (at present more than 5,500) from diverse ecological niches such as infant, suckling calf, rabbit, rat, pig, and chicken feces, rumen of cattle, sewage, the human vagina, dental caries, and the honeybee intestine (see Table 1). Some emendations of Reuter and Mitsuoka's

Table 1. Type strains and habitats of species of the genus *Bifidobacterium*.

Species	Type strain	Habitat
B. bifidum (Tissier) Orla-Jensen (1924)	E194a (= ATCC 15703) Reuter (1971)	Feces of infant, human adult, and suckling calf; human vagina
B. adolescentis Reuter (1963–64)	S12 (= ATCC 15697) Reuter (1971)	Feces of human adult; bovine rumen; sewage
B. infantis Reuter (1963–64) (syn.: *B. liberorum* Reuter and *B. lactentis* Reuter, see Scardovi et al., 1971, and Rogosa, 1974.)	S1 (= ATCC 15700)	Feces of infant and suckling calf; human vagina
B. breve Reuter (1963–64) (syn.: *B. parvulorum* Reuter, see Scardovi et al., 1971, and Rogosa, 1974)	E194b (= ATCC 15707)	Feces of infant and suckling calf; human vagina; sewage
B. longum Reuter (1963–64)[a]	B669 (= ATCC 27539)	Feces of human adult, infant, and suckling calf; human vagina; sewage
B. catenulatum Scardovi and Crociani (1974)	B764 (= ATCC 27534)	Feces of infant and human adult; human vagina; sewage
B. dentium Scardovi and Crociani (1974)[b]		Human dental caries and oral cavity; feces of human adult; human vagina; abscess and appendix in man
B. angulatum Scardovi and Crociani (1974)	B677 (= ATCC 27535)	Sewage; feces of human adult
B. thermophilum Mitsuoka (1969) (syn.: *B. ruminale* Scardovi et al. 1969)	P2-91 (= ATCC 25525) (RU326 = ATCC 25866, type strain of *B. ruminale*)	Feces of pig, chicken, and suckling calf; bovine rumen; sewage
B. pseudolongum Mitsuoka (1969)	PNC-2-9G (= ATCC 25526)	Feces of pig, chicken, bull, calf, rat, and guinea pig; bovine rumen
B. globosum Scardovi et al. (1969)	RU230 (= ATCC 25864)	Feces of piglet, suckling calf, rat, rabbit and lamb; sewage
B. animalis (Mitsuoka) Scardovi and Trovatelli (1974)	R101-8 (= ATCC 25527)	Feces of rat, chicken, rabbit, calf, and guinea pig; sewage
B. magnum Scardovi and Zani (1974)	RA3 (= ATCC 27540)	Feces of rabbit
B. suis Matteuzzi et al. (1971)	SU859 (= ATCC 27533)	Feces of piglet
B. pullorum Trovatelli et al. (1974)	P145 (= ATCC 27685)	Feces of chicken
B. asteroides Scardovi and Trovatelli (1969)	C51 (= ATCC 25910)	Intestine of *Apis mellifera* (subsp. *mellifera, ligustica,* and *caucasica*)
B. indicum Scardovi and Trovatelli (1969)	C410 (= ATCC 25912)	Intestine of *Apis cerana* and *A. dorsata*
B. coryneforme Scardovi and Trovatelli (1969)	C215 (= ATCC 25911)	Intestine of *Apis mellifera* subsp. *mellifera*
B. "minimum" Scardovi and Trovatelli (1974) Homology group (DNA-DNA)	F392 (= ATCC 27538)	Sewage
B. "subtile" Scardovi and Trovatelli (1974) Homology group (DNA-DNA)	F395 (= ATCC 27537)	Sewage
B. pseudocatenulatum Scardovi et al. (1979)	B1279 (= ATCC 27919)	Feces of infant and suckling calf; sewage
B. boum Scardovi et al. (1979)	RU917 (= ATCC 27917)	Bovine rumen; and feces of piglet
B. choerinum Scardovi et al. (1979)	SU806 (= ATCC 27686)	Feces of piglet; sewage
B. cuniculi Scardovi et al. (1979)	RA93 (= ATCC 27916)	Feces of rabbit

[a]*Actinomyces parabifidus* (Weiss and Rettger) Pine and Georg (1965) strain ATCC 17930 (strain Timberlain) is phenotypically *B. longum* (Kandler and Lauer, 1974); this attribution was confirmed by DNA homology (Scardovi, unpublished).
[b]*Actinomyces eriksonii* Georg et al. (1965) strains ATCC 15423 and 15424 are phenotypically *B. adolescentis* (Mitsuoka et al., 1974), but are *B. dentium* at DNA-DNA hybridization (Scardovi, unpublished).

species identifications were suggested and accepted (Rogosa, 1974): *B. liberorum* and *B. lactentis,* for example, were proposed as synonyms of *B. infantis,* and *B. parvulorum* as a synonym of *B. breve; B. longum* subsp. *animalis* Mitsuoka was elevated to the species rank as *B. animalis* (Scardovi and Trovatelli, 1974). Four additional species isolated from animal feces were described and proposed quite recently: *B. pseudocatenulatum, B. boum, B. choerinum,* and *B. cuniculi* (Scardovi et al., 1979).

The primary grouping of bifid strains based on DNA homology relationships permitted, as expected, a more reliable selection of phenotypic traits that correlated best with the groups (genospecies) so defined. These characters were used for constructing the identification key presented in Table 2.

Table 2. Key for the identification of the species of the genus *Bifidobacterium*.

Sorbitol	Arabinose	Raffinose	Ribose	Starch	Lactose	Cellobiose	Melezitose	Gluconate	G+C mol%	DNA homology	Zymogram[a]	Suggested species	Fig. in this Chapter
+	+	+			+							*B. adolescentis*	4
		−										*B. pseudocatenulatum*	7
			−									*B. catenulatum*	3
			+									*B. breve*	2
	−				−							*B. "subtile"*	26
−	+	+	+	+	+		+		+		25–50	*B. dentium*	9
												B. adolescentis	4
							−					*B. animalis*	18
					−				60	65–70		*B. pseudolongum*	20
									64			*B. globosum*	17
									59			*B. angulatum*	8
			−		+	−	+					*B. longum*	5, 6
							−					*B. infantis*	10
												"infantis-longum" intermediates	
												B. longum	5, 6
												B. magnum	16
			−		+					<30		*B. coryneforme*	14
												B. indicum	12, 13
												B. asteroides	11
					−							*B. pullorum*	15
			−									*B. suis*	19
	−											*B. cuniculi*	21
−	+	−							66	50–75		*B. choerinum*	24
									60			*B. boum*	23
									60			*B. thermophilum*	22
		+	+									*B. breve*	2
			−									*B. infantis*	10
	−				+							*B. bifidum*	1
					−							*B. "minimum"*	25

[a] (—→) Transaldolase zymogram; (--→) 6-phosphogluconate dehydrogenase zymogram. The arrows are directed toward the species—within the group indicated by the length of the arrows—that displays the most anodal form of the enzyme.

Habitats

The alimentary tract of newborn human infants is the bifidobacterial habitat that, for obvious reasons, has attracted the major interest among investigators. Nutritional factors influencing bowel bifidobacterial colonization, origin and mode of transmission, effects upon the host, and means for restoring or maintaining a proper microbial balance have been topics extensively investigated (see review of Poupard, Husain, and Norris, 1973). This ecosystem is apparently amenable to regulation by external means, such as the kind of nutrition or administration of bacterium-containing drugs, because there is absolute predominance of bifidobacteria under normal conditions, provided a proper choice of the bacterial types (ecotypes) is made (see for example Mitsuoka, 1972). Zymogram techniques can provide useful information for this purpose

(Scardovi, Casalicchio, and Vincenzi, 1979). In the intestinal tract of animals and human adults, bifidobacteria coexist with a large variety of bacteria, most of which are obligate anaerobes; components of this microflora are different in the different areas of the tract (Moore, Cato, and Holdeman, 1969). The obligate anaerobes of this complex microflora are largely unknown at present, although significant advances have been made in recent times with the appropriate techniques of isolation and identification, such as those used at the Virginia Polytechnic Institute, Blacksburg, Virginia, USA (Holdeman, Cato and Moore, 1977; Moore, Cato, and Holdeman, 1969; Moore and Holdeman, 1974). If one considers, furthermore, that particular ecotypes (or biovars) probably have more ecological significance in these habitats than the species to which they belong (Mitsuoka, 1972; Sears and Brownlee, 1952; Sears, Brownlee, and Uchiyama,

1950), all factors that pertain to and influence natural genetic variation of single bacterial populations should be studied (Milkman, 1975).

Some of our findings on bifidobacterial ecology solicit further investigation. Feces of suckling calves and breast-fed human infants harbor the same bifidobacterial species (V. Scardovi et al., unpublished); this suggests possible common ecological parameters that could be fruitfully explored. Some bifid species are apparently host specific: *Bifidobacterium magnum*, *B. pullorum*, and *B. suis*, for example, have been found only in rabbit, chicken, and pig fecal samples, respectively (see references in Table 1). *B. dentium* is the bifid constantly found in dental caries of man and could be involved in caries etiology. *Actinomyces eriksonii*, isolated by Georg et al. (1965) from human abscesses, is genetically *B. dentium* (Scardovi, Casalicchio, and Vincenzi, 1979). *B. asteroides* is the unique bifid found in *Apis mellifera* intestine, irrespective of the geographical area of provenance (Scardovi and Trovatelli, 1969), while *A. cerana* and *A. dorsata* (from the Philippines and Malaysia) harbor specific biovars of the species *B. indicum*, which have different transaldolase isozymes but are undistinguishable in DNA-DNA hybridization (Scardovi, Casalicchio, and Vincenzi, 1979). This subtle dependence on host phyletic position was heretofore unsuspected. Furthermore, *B. asteroides* and *B. indicum* are much more variable in the loci so far studied by zymogram technique than any other species of *Bifidobacterium* (Scardovi, Casalicchio, and Vincenzi, 1979); the exogenous or endogenous factors controlling this variability could be studied further. Bifidobacteria, whose significance and origin in the honeybee gut are at present unknown, could form an interesting chapter of the insect microbiology if they could be found elsewhere in this class of animals. Twelve species of *Bifidobacterium* have been found in sewage (see Table 1); among these the bifids allotted to "minimum" and "subtile" homology groups were not found elsewhere. This raises the exciting question of the possible development of bifidobacteria in extraenteral ecological niches.

Isolation

A large variety of media have been devised for isolating or enumerating the bifidobacteria in natural habitats. Ingredients of substrates have been: tomato juice (Haenel and Müller-Beuthow, 1956, 1957, 1963; Sutter, Vargo, and Finegold, 1975); sheep or horse blood (Dehnert, 1957; Lerche and Reuter, 1961; Mitsuoka, Sega, and Yamamoto, 1965; Ochi, Mitsuoka, and Sega, 1964; Reuter, 1963–64); human milk (Dehnert, 1957); liver or meat extracts (Beerens, Gérard, and Guillaume, 1957; Ochi,

Mitsuoka, and Sega, 1964). In almost all substrates, a variety of peptones were used: e.g., Trypticase (BBL) and proteose peptone No. 3 (Difco) (Mata, Carrillo, and Villatoro, 1969; Schaedler, Dubos, and Costello, 1965); Phytone (BBL) (Ochi, Mitsuoka, and Sega, 1964); tryptic and peptic meat peptones (Reuter, 1963–64). Manufactured complex substrates were used as such or with some modifications by many workers, e.g., reinforced clostridial medium (Oxoid) by Willis et al. (1973); Lactalysate agar (BBL), Eugonagar (BBL), tomato juice agar (BBL), and LBS medium (BBL) by Gilliland, Speck, and Morgan (1975).

In order to improve selectivity, antibiotics or other ingredients were used: kanamycin (Finegold, Sugihara, and Sutter, 1971); neomycin (Mata, Carrillo, and Villatoro, 1969; Schaedler, Dubos, and Costello, 1965); paramomycin, neomycin, sodium propionate, and lithium chloride (Mitsuoka, Sega, and Yamamoto, 1965), sorbic acid or sodium azide (Haenel and Müller-Beuthow, 1956, 1963). For the same purpose, Willis et al. (1973) used acidified reinforced clostridial medium (Oxoid) at pH 5.0.

The formulation and use of selective media requires detailed knowledge of the physiology and ecological distribution of the bifidobacterial species or types; unfortunately, this information is still lacking or, at best, only fragmentarily known. The following quotation from Mitsuoka, Sega, and Yamamoto (1965, pp. 464–465) illustrates the situation: "Leider befriedigten uns diese Nährböden noch nicht vollständig in ihrer Selektivwirkung, da wir . . . manchmal bei einigen Tierarten unvergleichlich niedrigere Werte erhielten, und in anderen Fällen nur andere Keime und nicht Bifidobakterien wuchsen." At present the preference should be given, therefore, to substrates that permit satisfactory growth of the largest number of bifidobacterial types presently known (listed in Table 1). Preliminary trials made with some bifidobacteria from human feces, bovine rumen, and honeybee intestine showed that the ingredients ensuring their good development on anaerobic plates and their maintenance in stab were Trypticase and Phytone (BBL). We finally adopted a substrate that enabled us to isolate strains belonging to widely different genetic species of *Bifidobacterium* from all the habitats we have so far investigated.

TPY Medium for Isolation of *Bifidobacterium*
The substrate (TPY) contains per liter:

Trypticase (BBL)	10 g
Phytone (BBL)	5 g
Glucose	5 g
Yeast extract (Difco)	2.5 g
Tween 80	1 ml
Cysteine hydrochloride	0.5 g

K₂HPO₄	2 g
K_2HPO_4	2 g
$MgCl_2 \cdot 6H_2O$	0.5 g
$ZnSO_4 \cdot 7H_2O$	0.25 g
$CaCl_2$	0.15 g
$FeCl_3$	traces
Agar	1.5 g

Final pH is about 6.5 after autoclaving at 121°C for 25 min. Immediately after sterilization, the tubes are inoculated from dilutions of the fecal material into tubes of liquid medium of the same composition and poured into plates. Plates, preferably Petri dishes with vent, are incubated (upside down) in anaerobic jars with palladium catalyst, repeatedly evacuated, and refilled with a gas mixture of 10% CO_2 and 90% hydrogen. Alternatively, jars without catalyst can be used and CO_2 from a calcium carbonate–hydrochloride generator employed as filling gas; CO_2 must be left in the jars at reduced pressure (0.1 atm). After 3–4 days of incubation at 39–40°C, colonies are transferred in stab cultures on TPY medium with 0.5% agar; stabs are incubated in jars partially filled with CO_2, and, after development (24 h), are kept at 3–4°C in the same container. Transfers should be made at weekly intervals.

Because the medium is not selective (streptococci, lactobacilli, and other forms may develop profusely) and because bifidobacterial colonies cannot be differentiated from nonbifidal ones, all the morphological types grown on plates should be scored and tested as indicated under Identification.

Identification

A Bacterium as a Member of the Genus *Bifidobacterium*

Definitive identification of a bacterial strain as *Bifidobacterium* in the routine laboratory is difficult to accomplish with the procedures so far suggested. Morphology, often claimed and relied upon as a distinctive character, is not only influenced considerably by nutritional conditions (Poupard, Husain, and Norris, 1973), but is so unusual in some bifidobacteria, such as *B. angulatum, B. asteroides,* and *B. indicum* (Figs. 8, 11, 12, and 13) and *B. pullorum* and *B. "minimum"* (Figs. 15 and 25), that it is definitively misleading. Cultural and physiological characters are grossly shared by many other bacteria, e.g., *Actinomyces, Corynebacterium,* and *Lactobacillus*. One of the more practical approaches to the primary differentiation of bifidobacteria from related groups is apparently the one proposed by Holdeman, Cato, and Moore (1977), which is based on the identification by means of gas chromatography of the fermentation products, among which

acetic acid generally predominates over lactic acid as the main final product. Side reactions, however, can form substantial and variable amounts of ethanol and formic and succinic acids (see Introduction), so that the pattern can be difficult to interpret, especially for the inexperienced worker. The most direct, reliable and, most important, fruitful assignment of a bacterial strain to the genus *Bifidobacterium* is the one based upon the demonstration in cellular extracts of fructose-6-phosphate phosphoketolase (F6PPK), the key enzyme of bifidobacterial hexose metabolism (see Introduction). Its validity is not only dictated by the results of previous investigations (De Vries and Gerbrandy, 1967; Scardovi and Trovatelli, 1965), but is being confirmed almost daily in our laboratory with bifids isolated from the most diverse habitats. Anaerobic F6PPK-less bacteria resembling bifids in morphology and gross physiology were isolated from bovine rumen and from sewage (Scardovi et al., unpublished); conversely, bacterial strains with nonbifidal morphology were recognized as bifids with this enzymatic test [e.g., *B. pullorum* (Trovatelli et al., 1974)] and bifids from the honeybee (Scardovi and Trovatelli 1969)].

Fructose-6-Phosphate Phosphoketolase Test for Identification of *Bifidobacterium* (Scardovi and Trovatelli, 1969)

Reagents:

1. 0.05 M phosphate buffer pH 6.5 plus cysteine, 500 mg/liter
2. NaF, 6 mg/ml, and K or Na iodoacetate, 10 mg/ml
3. Hydroxylamine-HCl, 13.9 g/100 ml, freshly neutralized with NaOH (pH 6.5)
4. Trichloroacetic acid (TCA), 15% (wt/vol) solution in water
5. 4 N HCl
6. Ferric chloride·6H₂0, 5% (wt/vol) in 0.1 N HCl
7. Fructose-6-phosphate (sodium salt; 70% purity), 80 mg/ml in water

The formation of acetyl phosphate from fructose-6-phosphate is detected by the reddish-violet color formed by the ferric chelate of its hydroxamate (Lipmann and Tuttle, 1945).

Procedure:

Cells harvested from 10–20 ml TPY liquid medium are washed twice with buffer 1 and resuspended in 1.0 ml of the same buffer. The cells are disrupted by sonication carefully in the cold. Reagent 2 (0.25 ml) and the fructose-6-phosphate solution (0.25 ml) are added. After 30 min incubation at 37°C, the reaction is stopped with 1.5 ml of reagent 3. After 10 min at room temperature, 1.0 ml each of TCA solution and 4 N HCl is added. The mixture may be stored

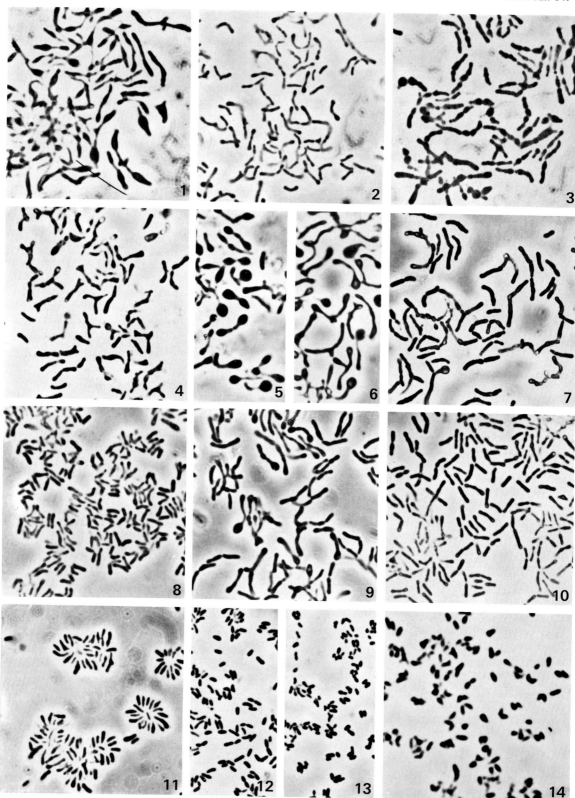

Figs. 1–26. Cellular morphology in the genus *Bifidobacterium*. Cells of the type strains were grown in anaerobic TPY medium stabs. Phase-contrast photomicrographs, × 1,500. (1) *B. bifidum*, (2) *B. breve*, (3) *B. catenulatum*, (4) *B. adolescentis*, (5 and 6) *B. longum*, (7) *B. pseudocatenulatum*, (8) *B. angulatum*, (9) *B. dentium*, (10) *B. infantis*, (11) *B. asteroides*, (12 and 13) *B. indicum*, (14) *B. coryneforme*, (15) *B. pullorum*, (16) *B. magnum*, (17) *B. globosum*, (18) *B. animalis*, (19) *B. suis*, (20) *B. pseudolongum*, (21) *B. cuniculi*, (22) *B. thermophilum*, (23) *B. boum*, (24) *B. choerinum*, (25) *B. "minimum"*, (26) *B. "subtile"*.

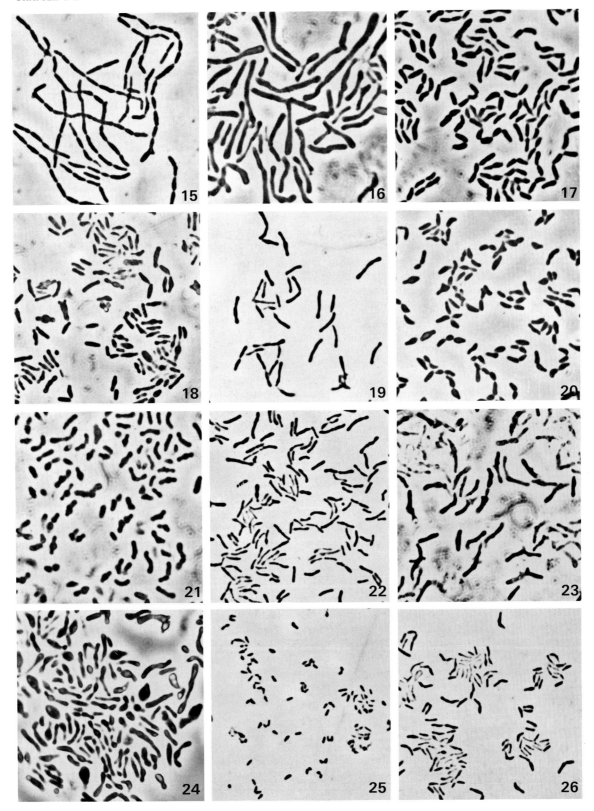

at room temperature prior to the addition of 1.0 ml of the color-developing reagent 6. Invert tubes for mixing. Any reddish-violet color that immediately develops is taken as a positive result. A tube without fructose-6-phosphate can serve as a blank for visual comparison (absorption maximum at 505 μm, see Pechère and Capony, 1968). The color is more evident visually after some standing, to allow debris and proteins to settle out. Warning: Carefully avoid heating during sonication; the enzyme is heat sensitive.

Bifidobacterium species

Some morphological traits, such as disposition and number of branchings, cell contours, dimensions, and arrangment in groups, are characteristic for many known *Bifidobacterium* species grown in solid TPY medium. The morphologies illustrated in Figs. 1–26 could be of some aid, therefore, in species differentiation. Definitive identification of *Bifidobacterium* species can be achieved with the use of the fermentation and electrophoretic tests suggested in the key presented in Table 2.

FERMENTATION TESTS. The basal Tween 80–enriched (1 g/liter) TPY medium is used (see Isolation); bromocresol purple (30 mg/liter) is used as indicator. Substrates are sterilized by autoclaving or filtration (arabinose and xylose) and are added to the test tubes at 0.5% wt/vol. Because CO_2 is beneficial for growth (species such as *B. bifidum* or *B. asteroides* are the most exigent), the inoculated tubes are incubated at 39°C in jars evacuated and refilled with CO_2 under reduced pressure. There is no special need to use "pre-reduced" medium (see Holdeman, Cato, and Moore, 1977) if the medium and solutions are used and tubes inoculated shortly after their preparation. Development and color changes are recorded after 4 days' incubation. Strains fermenting gluconate produce CO_2 from this compound and less acids than from glucose or other sugars, and cells are generally very minute (see Sgorbati et al., 1970); the indicator turns yellow more slowly and development should be carefully compared with that obtained in the absence of added sugars.

In the key in Table 2, some additional distinctive characters are reported which should be used to distinguish those species that cannot be indentified with the given fermentation tests. They are: content of guanine plus cytosine of DNA, interspecific DNA homology, and the electrophoretic behavior of 6-phosphogluconate dehydrogenase (6PGDH) and transaldolase isozymes. Since the first two parameters cannot be routinely determined in most laboratories, we suggest that the zymogram technique be used when necessary.

ELECTROPHORETIC TESTS. The starch-gel horizontal electrophoresis system of Smithies (1955) should preferably be used. Use the following buffers:

Transaldolase: tris (hydroxymethyl) aminomethane (TRIS) (16.3 g) plus citric acid monohydrate (9.0 g liter; pH 7.0) used as bridge buffer; dilute this solution 1:15 and use as gel buffer.

6-Phosphogluconate dehydrogenase (6PGDH): as bridge buffer, trisodium citrate–$2H_2O$ (120 g liter; pH 7.0 with citric acid); as gel buffer, histidine (0.75 g) plus sodium chloride (1.5 g per liter; pH 7.0). The citrate solution should be renewed every two runs. A total of 90 g hydrolyzed starch (Connaught Lab. Ltd., Willondale, Ontario, Canada) per liter of the appropriate buffer is used; the mixture is boiled for 5 min and gas is removed under reduced pressure. The liquid is poured into a plastic three-frame mold (dimensions of 12.0 × 37.0 × 0.9 cm are suitable to accommodate 12 samples at a time). Samples of bacterial extracts (cells suspended in 0.05 M phosphate buffer and disrupted by sonication), 5–10 μl in 0.5-×-0.5-cm Whatman 3MM paper strips, are generally run for 15–20 h with a current of 15–20 mA for transaldolase and 40–50 mA for 6PGDH. Preferably, use the middle slab for staining by the flooding technique.

Transaldolase staining: the developing solution contains (per 100 ml) fructose-6-phosphate (Na salt, 98% purity, Sigma), 400 mg; sodium arsenate, 370 mg; glycine, 240 mg; NAD, 13 mg; D-erythrose-4-phosphate (60–75% purity, Sigma), 16 mg; phenazine methosulfate, 2 mg; nitro blue tetrazolium (NBT, Sigma), 20 mg, and about 130 IU of glyceraldehyde-3-phosphate dehydrogenase (GAPDH) (Boehringer Mannheim).

6PGDH staining: 0.5 M Tris-hydrochloride buffer, pH 7.0, 10 ml; 6-phosphogluconate (trisodium salt, Sigma), 250 mg; NADP, 20 mg; NBT, 20 mg; phenazine methosulfate, 2 mg; distilled water, 90 ml.

The following identification problems can be solved with this technique (see identification key in Table 2).

B. dentium–B. adolescentis: These two species, which have different ecological distribution (see Table 1) and diverge widely in the base sequences of their DNA (25–50% homology, see Table 2), cannot be distinguished by the usual phenotypic characters (Mitsuoka et al., 1974). Their morphology is quite similar (see Figs. 4 and 9). Transaldolase and 6PGDH isozymes have different electrophoretic mobility: if the value 10 is assigned for the migration distance to the anode of the isozymes of either enzyme of *B. dentium,* the migration distances for the isozymes of *B. adolescentis* would be 8.7 and 8.9, respectively. This pattern was obtained with an electrophoretic study of 140 strains of *B. dentium* and 54 strains of *B. adolescentis* (Scardovi, Casalicchio, and Vincenzi, 1979). Strain ATCC

15703 (type strain of *B. adolescentis*) and strain ATCC 15423 (*Actinomyces eriksonii*) could be used as references; strain ATCC 27534 (type strain of *B. dentium*) has no detectable 6PGDH (Scardovi, Casalicchio, and Vincenzi, 1979).

B. angulatum–B. globosum–B. pseudolongum: *B. angulatum* transaldolase isozyme is less anodal than that of *B. globosum* (or *B. pseudolongum*): their migration values are 9.1 and 10, respectively. General morphology and disposition of *B. angulatum* cells could be well taken as an alternative, distinctive character ("palisade" or "angular" disposition, see Fig. 8). *B. globosum* and *B. pseudolongum* cannot be distinguished by their zymograms; they can be distinguished at present only by determining their DNA G+C content or their DNA homology (Scardovi et al., 1971).

B. infantis–B. longum-infantis-longum "intermediates"*–B. magnum:* Arabinose is reported as not fermented by *B. infantis* (syn. *B. liberorum* and *B. lactentis*); *B. longum* characteristically ferments melezitose (Rogosa, 1974). Among isolates from infant feces, *B. infantis* and *B. longum* strains that fermented arabinose and failed to ferment melezitose, respectively, were recognized by means of DNA—DNA hybridization (Scardovi et al., unpublished). Subsequently, bifid strains have been isolated from feces of suckling calves that could not be referred to either species, because they were more than 80% related both to *B. infantis* and *B. longum* references; conversely, *B. infantis* and *B. longum* strains were not distinguishable when their DNA was annealed to that of calf strains used as reference (Scardovi et al., unpublished). The existence of such strains, provisorily called "intermediates", induces doubt whether *B. infantis* and *B. longum* are separate species. Pending further investigation, we suggest following this rationale: strains not fermenting arabinose should be ascribed to *B. infantis,* while strains fermenting both arabinose and melezitose should be retained *B. longum* (see Table 2 for the other key characters); strains fermenting arabinose (and xylose) but not melezitose should be given the provisory label *"infantis-longum"* group or submitted to transaldolase electrophoresis for further distinction. All strains which, inside the group, possess the more anodal form of this enzyme (migration value of 10), are genetically *B. infantis* (reference strain B1269 = ATCC 27920), whereas the majority of those displaying the less anodal form (migration value of 9.0), are genetically *B. longum* (reference strain E194b = ATCC 15707); strains possessing an isozyme moving to a somewhat intermediate position (migration value of 9.6) should be recorded as "intermediates" (reference strain VT29 or VT42 from our collection).

The species *B. magnum,* found until now in rabbit feces only, is characterized by the dimensions of its cells (10–20 μm long) when grown in the absence of Tween 80 and by its strong acidophilic nature (initial optimum pH 5.3–5.5) (see Scardovi and Zani, 1974). These characters should be sufficient for distinguishing *B. magnum* from *B. infantis* and *B. longum.*

B. asteroides–B. indicum–B. coryneforme: These are the bifids inhabiting the intestine of the honeybee (Scardovi and Trovatelli, 1969). The habitat and their cellular morphology (see Figs. 11–14) are so distinctive that these bifids could be hardly confused with any other known *Bifidobacterium* species. *B. asteroides* and *B. indicum* are electrophoretically the most polymorphic bifidobacterial species.

B. choerinum–B. thermophilum (syn. *B. ruminale*)*–B. boum:* These species, found often in the same habitats (Table 2) cannot be distinguished on common phenotypic grounds; the differences in their morphology could be reliable for their distinction only to the experienced eye (Figs. 22, 23, and 24). Transaldolase isozymes, in addition to the fermentation characters reported in the key (Table 2), permit their clear-cut identification: *B. thermophilum, B. boum,* and *B. choerinum* possess isozymes with increasing anodal mobility, their migration values being 8.4, 9.0, and 10, respectively. Among these species, *B. choerinum* displays the most anodal form of 6PGDH, although *B. boum* and *B. thermophilum* have several isozymes of identical mobility (Scardovi, Casalicchio, and Vincenzi, 1979). The type strains reported in Table 1 can be used as references.

Literature Cited

Ad Hoc Committee of the Judicial Commission of the ICBS. 1976. First draft approved lists of bacterial names. International Journal of Systematic Bacteriology **26:**563–599.

Beerens, H., Gérard, A., Guillaume, J. 1957. Étude de 30 souches de *Bifidobacterium* (*Lactobacillus bifidus*). Caractérisation d'une variété buccale. Comparaison avec les souches d'origine fécale. Annales de l'Institut Pasteur de Lille **9:**77–85.

Breed, R. S., Murray, E. G. D., Smith, N. R. (eds.). 1957. Bergey's manual of determinative bacteriology, 7th ed. Baltimore: Williams & Wilkins.

Dehnert, J. 1957. Untersuchungen über die gram-positive Stuhlflora des Brustmilchkindes. Zentralblatt für Bakteriologie, Parasitenkunde, Infektionskrankheiten und Hygiene, Abt. 1 Orig. **169:**66–79.

De Vries, W., Gerbrandy, S. J., Stouthamer, A. H. 1967. Carbohydrate metabolism in *Bifidobacterium bifidum.* Biochimica et Biophysica Acta **136:**415–425.

De Vries, W., Stouthamer, A. H. 1968. Fermentation of glucose, lactose, galactose, mannitol, and xylose by bifidobacteria. Journal of Bacteriology **96:**472–478.

Finegold, S. M., Sugihara, P. T., Sutter, V. L. 1971. Use of selective media for isolation of anaerobes from humans, pp. 99–108. In: Shapton, D. A., Board, R. G. (eds.), Isolation of anaerobes. London: Academic Press.

Georg, L. K., Robertstad, G. W., Brinkman, S. A., Hicklin, M. D. 1965. A new pathogenic anaerobic *Actinomyces* species. Journal of Infectious Diseases **115:**88–99.

Gilliland, S. E., Speck, M. L., Morgan, C. G., 1975. Detection of *Lactobacillus acidophilus* in feces of humans, pigs, and chickens. Applied Microbiology **30**:541–545.

Haenel, H., Müller-Beuthow, W. 1956. Vergleichende quantitative Untersuchungen über Keimzahlen in den Faeces des Menschen und einiger Wirbeltiere. Zentralblatt für Bakteriologie, Parasitenkunde, Infektionskrankheiten und Hygiene, Abt. 1 Orig. **167**:123–133.

Haenel, H., Müller-Beuthow, W. 1957. Untersuchungen über die Eignung von Bifidusnährböden zur quantitativen Züchtung der Bifidusgruppe des Erwachsenen. Zentralblatt für Bakteriologie, Parasitenkunde, Infektionskrankheiten und Hygiene, Abt. 1 Orig. **169**:196–204.

Haenel, H., Müller-Beuthow, W. 1963. Untersuchungen an deutschen und bulgarischen jungen Männern über die intestinale Eubiose. Zentralblatt für Bakteriologie, Parasitenkunde, Infektionskrankheiten und Hygiene, Abt. 1 Orig. **188**:70–80.

Holdeman, L. V., Cato, E. P., Moore, W. E. C. 1977. Anaerobe laboratory manual, 4th ed. Blacksburg, Virginia: Virginia Polytechnic Institute and State University.

Kandler, O., Lauer, E. 1974. Neuere Vorstellungen zur Taxonomie der Bifidobacterien. Zentralblatt für Bakteriologie, Parasitenkunde, Infektionskrankheiten und Hygiene, Abt. 1 Orig. Reihe A **228**:29–45.

Lauer, E., Kandler, O. 1976. Mechanismus der Variation des Verhältnisses Acetat/Lactat bei der Vergärung von Glucose durch Bifidobakterien. Archives of Microbiology **110**:271–277.

Lerche, M., Reuter, G. 1961. Isolierung und Differenzierung anaerober *Lactobacilleae* aus dem Darm erwachsener Menschen. Zentralblatt für Bakteriologie, Parasitenkunde, Infektionskrankheiten und Hygiene, Abt. 1 Orig. **182**:324–356.

Lipmann, F., Tuttle, L. C. 1945. A specific micromethod for determination of acyl-phosphates. Journal of Biological Chemistry **159**:21–28.

Mata, L. J., Carrillo, C., Villatoro, E. 1969. Fecal microflora in healthy persons in a preindustrial region. Applied Microbiology **17**:596–602.

Matteuzzi, D., Crociani, F., Zani, G., Trovatelli, L. D. 1971. *Bifidobacterium suis* n. sp.: A new species of the genus *Bifidobacterium* isolated from pig feces. Zeitschrift für Allgemeine Mikrobiologie **11**:387–395.

Milkman, R. 1975. Allozyme variation in *Escherichia coli* of diverse natural origins, pp. 273–285. In: Markert, C. L. (ed.), Isozymes. IV. Genetics and evolution. New York: Academic Press.

Mitsuoka, T. 1969. Vergleichende Untersuchungen über die Bifidobakterien aus dem Verdauungstrakt von Menschen un Tieren. Zentralblatt für Bakteriologie, Parasitenkunde, Infektionskrankheiten und Hygiene, Abt. 1 Orig. **210**:52–64.

Mitsuoka, T. 1972. Bacteriology of fermented milk with special reference to the implantation of lactobacilli in the intestine, pp. 169–179. In: Proceedings of the VIth International Symposium on Convertion and Manifacture of Foodstuff by Microorganisms. Tokyo: Saikon.

Mitsuoka, T., Sega, T., Yamamoto, S. 1965. Eine verbesserte methodik der qualitativen und quantitativen Analyse der Darmflora von Menschen und Tieren. Zentralblatt für Bakteriologie, Parasitenkunde, Infektionskrankheiten und Hygiene, Abt. 1 Orig. **195**:455–469.

Mitsuoka, T., Morishita, Y., Terada, A., Watanabe, K. 1974. *Actinomyces eriksonii* Georg, Robertstad, Brinkman and Hicklin 1965 identisch mit *Bifidobacterium adolescentis* Reuter 1963. Zentralblatt für Bakteriologie, Parasitenkunde, Infektionskrankheiten und Hygiene, Abt. 1 Orig. Reihe A **226**:257–263.

Moore, W. E. C., Cato, E. P., Holdeman, L. V. 1969. Anaerobic bacteria of the gastrointestinal flora and their occurrence in clinical infections. Journal of Infectious Diseases **119**:641–649.

Moore, W. E. C., Holdeman, L. V. 1974. Human fecal flora: The

normal flora of 20 Japanese-Hawaiians. Applied Microbiology **27**:961–979.

Ochi, Y., Mitsuoka, T., Sega, T. 1964. Untersuchungen über die Darmflora des Huhnes. III Mitteilung. Die Entwicklung der Darmflora von Küken bis zum Huhn. Zentralblatt für Bakteriologie, Parasitenkunde, Infektionskrankheiten und Hygiene, Abt. 1 Orig. **193**:80–95.

Orla-Jensen, S. 1924. La classification des bactéries lactiques. Lait **4**:468–474.

Pechère, J. F., Capony, J. P. 1968. On the colorimetric determination of acyl phosphates. Analytical Biochemistry **22**:536–539.

Pine, L., Georg, L. 1965. The classification and phylogenetic relationships of the *Actinomycetales*. International Bulletin of Bacteriological Nomenclature and Taxonomy **15**:143–163.

Poupard, J. A., Husain, I., Norris, R. F. 1973. Biology of the bifidobacteria. Bacteriological Reviews **37**:136–165.

Reuter, G. 1963–64. Vergleichende Untersuchungen über die Bifidus-Flora im Säuglings- und Erwachsenenstuhl. Zentralblatt für Bakteriologie, Parasitenkunde, Infektionskrankheiten und Hygiene, Abt. 1 Orig. **191**:486–507.

Rogosa, M. 1974. Genus III. *Bifidobacterium* Orla-Jensen, pp. 669–676. In: Buchanan, R. E., Gibbons, N. E. (eds.), Bergey's manual of determinative bacteriology, 8th ed. Baltimore: Williams & Wilkins.

Scardovi, V., Casalicchio, F., Vincenzi, N. 1979. Multiple electrophoretic forms of transaldolase and 6-phosphogluconic dehydrogenase and their relationships to the taxonomy and ecology of bifidobacteria. International Journal of Systematic Bacteriology **29**:312–327.

Scardovi, V., Crociani, F. 1974. *Bifidobacterium catenulatum, Bifidobacterium dentium,* and *Bifidobacterium angulatum:* Three new species and their deoxyribonucleic acid homology relationships. International Journal of Systematic Bacteriology **24**:6–20.

Scardovi, V., Trovatelli, L. D. 1965. The fructose-6-phosphate shunt as peculiar pattern of hexose degradation in the genus *Bifidobacterium*. Annali di Microbiologia ed Enzimologia **15**:19–29.

Scardovi, V., Trovatelli, L. D. 1969. New species of bifid bacteria from *Apis mellifica* L. and *Apis indica* F. A contribution to the taxonomy and biochemistry of the genus *Bifidobacterium*. Zentralblatt für Bakteriologie, Parasitenkunde, Infektionskrankheiten und Hygiene, Abt. 2 **123**:64–88.

Scardovi, V., Trovatelli, L. D. 1974. *Bifidobacterium animalis* (Mitsuoka) comb. nov. and the "*minimum*" and "*subtile*" groups of new bifidobacteria found in sewage. International Journal of Systematic Bacteriology **24**:21–28.

Scardovi, V., Zani, G. 1974. *Bifidobacterium magnum* sp. nov., a large, acidophilic bifidobacterium isolated from rabbit feces. International Journal of Systematic Bacteriology **24**:29–34.

Scardovi, V., Trovatelli, L. D., Crociani, F., Sgorbati, B. 1969. Bifidobacteria in bovine rumen. New species of the genus *Bifidobacterium: B. globosum* n. sp. and *B. ruminale* n. sp., Archiv für Mikrobiologie **68**:278–294.

Scardovi, V., Trovatelli, L. D., Zani, G., Crociani, F., Matteuzzi, D. 1971. Deoxyribonucleic acid homology relationships among species of the genus *Bifidobacterium*. International Journal of Systematic Bacteriology **21**:276–294.

Scardovi, V., Trovatelli, L. D., Biavati, B., Zani, G. 1979. *Bifidobacterium cuniculi, Bifidobacterium choerinum, Bifidobacterium boum,* and *Bifidobacterium pseudocatenulatum:* Four new species and their deoxyribonucleic acid homology relationships. International Journal of Systematic Bacteriology **29**:291–311.

Schaedler, R. W., Dubos, R., Costello, R. 1965. The development of the bacterial flora in the gastrointestinal tract of mice. Journal of Experimental Medicine **122**:59–66.

Schramm, M., Klybas, V., Racker, F. 1958. Phosphorolytic cleavage of fructose-6-phosphate by fructose-6-phosphate phosphoketolase from *Acetobacter xylinum*. Journal of Biological Chemistry **233**:1283–1288.

Sears, H. J., Brownlee, I. 1952. Further observations on the persistence of individual strains of *Escherichia coli* in the intestinal tract of man. Journal of Bacteriology **63**:47–57.

Sears, H. J., Brownlee, I., Uchiyama, J. K. 1950. Persistence of individual strains of *Escherichia coli* in the intestinal tract of man. Journal of Bacteriology **59**:293–301.

Seeliger, H. P. R., Werner, H. 1962. Quantitative und qualitative Untersuchungen über die anaeroben Lactobacillen im Säuglings- und Erwachsenenstuhl. Beitrag zum Bifidus-Problem. Zeitschrift für Hygiene und Infektionskrankheiten **148**:383–404.

Sgorbati, B., Lenaz, G., Casalicchio, F. 1976. Purification and properties of two fructose-6-phosphate phosphoketolases in *Bifidobacterium*. Antonie van Leeuwenhoek Journal of Microbiology and Serology **42**:49–57.

Sgorbati, B., Zani, G., Trovatelli, L. D., Scardovi, V. 1970. Gluconate dissimilation by the bifid bacteria of the honey bee. Annali di Microbiologia ed Enzimologia **20**:57–64.

Smithies, O. 1955. Zone electrophoresis in starch gels: Group variations in the serum proteins of normal human adults. The Biochemical Journal **61**:629–641.

Sutter, V. L., Vargo, V. L., Finegold, S. M. 1975. Wadsworth anaerobic bacteriology manual, 2nd ed. Los Angeles: Anaerobic Bacteriology Laboratory, Wadsworth Hospital Center.

Tissier, H. 1900. Recherches sur la flore intestinale normale et pathologique du nourisson. Thesis. University of Paris.

Trovatelli, L. D., Crociani, F., Pedinotti, M., Scardovi, V. 1974. *Bifidobacterium pullorum* sp. nov.: A new species isolated from chicken feces and a related group of bifidobacteria isolated from rabbit feces. Archiv für Mikrobiologie **98**:187–198.

Werner, H. 1964. Zum Vorkommen von Bifidusbakterien und morphologisch ähnlichen Keimen in der Mundhöhle Erwachsener. Zentralblatt für Bakteriologie, Parasitenkunde, Infektionskrankheiten und Hygiene, Abt. 1 Orig. **193**:331–342.

Werner, H. 1966. The Gram positive nonsporing anaerobic bacteria of the human intestine with particular reference to the corynebacteria and bifidobacteria. Journal of Applied Bacteriology **29**:138–146.

Willis, A. T., Bullen, C. L., Williams, K., Fagg, C. G., Bourne, A., Vignon, M. 1973. Breast milk substitute: A bacteriological study. British Medical Journal **4**:67–72.

The Genus *Mycobacterium* (Except *M. leprae*)

GEORGE P. KUBICA and ROBERT C. GOOD

Species of the genus *Mycobacterium* are broadly grouped into two major categories on the basis of pathogenicity for animals and humans. Leprosy and tuberculosis are mycobacterial diseases that have been recognized for many years; the etiological agents are *Mycobacterium leprae* (discussed in this Handbook, Chapter 151) and *M. tuberculosis,* respectively. In the classical study of Koch (1882), tubercle bacilli were stained in infected tissue and cultured on an inspissated serum medium; characteristic disease was produced in experimental animals by injection of the pure culture; organisms were observed in the infected tissues of the experimental animal and could be recovered in vitro. This study, relating etiological agent to disease, made possible the extensive and exacting studies that have led to control and partial eradication of tuberculosis.

Koch considered tubercle bacilli of human and bovine origin to be identical, but Theobold Smith (1898) showed that bovine tubercle bacilli produced extensive disease in guinea pigs, rabbits, and cattle, while human tubercle bacilli produced progressive disease only in guinea pigs and limited disease in rabbits and cattle. Later, he devised a cultural method for distinguishing the species based on final acidity in a broth medium (Smith, 1904–05). Identification procedures have continued to improve over the years, but distinction of species on the basis of laboratory tests has been difficult. Since the decline of tuberculosis hospitals, this proficiency now resides primarily in centralized laboratories that process large numbers of specimens or cultures. Methods for many identification tests have been standardized and are highly reproducible (Wayne et al., 1974, 1976).

Mycobacteria are weakly Gram positive but are acid fast, i.e., once stained with one of the basic dyes such as fuchsin, they resist decolorization with mineral acids or with acidified organic solvents. The lipid-rich cell wall limits movement of the dye either into or out of the cell (Goren, Cernich, and Brokl, 1978). Additional information on the acid-fast stain is found in the review by Barksdale and Kim (1977). The stain may be taken up uniformly, but usually cells appear beaded or granular with heavily stained areas separated by nonstained spaces.

After isolation and characterization of the etiological agent of tuberculosis by Koch, related species were rapidly identified. *M. avium* was associated with a tuberculosis-like disease in fowl by Strauss and Gameleia in 1891; *M. paratuberculosis* was isolated from cattle and sheep with chronic enteritis by Johne and Frothingham in 1895; and *M. lepraemurium,* the rat leprosy bacillus, was described by Dean (1903), Stefansky (1903), and Rabinowitsch (1903). During this same period, a number of saprophytic species, such as *M. phlei* and *M. smegmatis,* were also isolated. The chronology of discovery of mycobacterial species can be determined from listings of currently accepted species (Kubica, 1979; Runyon, Wayne, and Kubica, 1974).

For almost seven decades, *M. tuberculosis* has remained the most frequently encountered agent of human mycobacterial disease. In the early 1950s, other mycobacteria were recognized as the cause of clinical disease (Buhler and Pollak, 1953; Crow et al., 1957; Timpe and Runyon, 1954). Runyon (1959, 1965) placed the organisms into four groups on the basis of cultural characteristics. Significance of the groups has now been verified by numerical taxonomic methods on thousands of cultures to define species within each of the groups (Kubica et al., 1972; Meissner et al., 1974; Saito et al., 1977; Wayne et al., 1971). Mycobacteria other than *M. tuberculosis* (MOTT) have commonly been identified as "atypical" or by the Runyon group designation. This practice should not continue since the taxonomy of mycobacteria is sufficiently complete to attach species names. New isolates may be temporarily assigned to groups on the basis of obvious cultural characteristics. These groups are: photochromogenic, nonphotochromogenic, scotochromogenic, and rapidly growing mycobacteria. Complex names are also used for grouping closely related species with a common pathogenic potential for which further speciation would be of little clinical value (Runyon, 1974), e.g., *M. avium* complex, *M. fortuitum* complex.

The complexity of mycobacteria is obvious from the excellent review by Barksdale and Kim (1977). Application of newer techniques in lipid analysis

(Alshamaony, Goodfellow, and Minnikin, 1976; Alshamaony et al., 1976; Goodfellow, Collins, and Minnikin, 1976; Hecht and Causey, 1976; Lechevalier, Horan, and Lechevalier, 1971) serological typing (Goslee, Rynearson, and Wolinsky, 1976; Wolinsky and Schaefer, 1973), and phage typing (Engel, 1978; Grange et al., 1977; Jones, 1975; Jones and Greenberg, 1976, 1978; Rado et al., 1975; Yates, Collins, and Grange, 1978) have aided in genus assignment, species identification, and definition of characters that can be followed in epidemiological investigations. However, these procedures are either still in the investigative stage or are limited to use in specialized laboratories.

As demonstrated by Koch, injection of tubercle bacilli into previously infected guinea pigs rapidly produced a localized lesion that ulcerated and promptly healed, while the primary injection site slowly developed and resulted in a persistent ulcer. This "Koch phenomenon" could also be induced by injecting culture filtrates that were free of living bacilli; however, immunity to superinfection and response to culture filtrate are not associated with serum antibodies, but are the result of cellular immunity.

Delayed-type hypersensitivity to tuberculin develops 4–6 weeks after infection with tubercle bacilli. The reaction is elicited by intradermal injection of Koch's Old Tuberculin (OT), concentrated culture filtrate preserved with glycerol. A more standardized reagent, purified protein derivative (PPD) (Seibert, 1934; Seibert and Glenn, 1941), has been prepared from culture filtrate and is in general use throughout the world. A positive tuberculin reaction indicates either past or present infection with the tubercle bacillus, but it is not diagnostic of active disease. Infection with MOTT species will also induce delayed-type hypersensitivity, which may be elicited by tuberculin and leads to confusion in interpretation (Edwards and Palmer, 1968; Palmer and Edwards, 1966, 1967). However, skin-test antigens prepared from MOTT species may aid in early diagnosis of clinical disease (Graybill et al., 1974; Marks, Palfreyman, and Schaefer, 1977; Takeya, Nakayama, and Muraoka, 1970). A positive tuberculin reaction is one in which a 10-mm or larger area of induration develops at the injection site in 48–72 h. The routine test dose is 5 tuberculin units (TU); this is adjusted by comparison to the international standard (PPD-S), which is measured on the basis of weight (0.1 μg protein = 5 TU).

Antigens in mycobacteria are intricate, and are complexed with lipids in many instances. Basic patterns detected by immunoelectrophoresis have been used to identify recurring antigens (Janicki et al., 1971). A greater number, some of which may be species specific, have been identified by fused rocket immunoelectrophoresis (Chaparas, Brown, and Hyman, 1978). Similarly, the occurrence of

reproducible lipid patterns has also been associated with serovars of a given species because of surface antigens elaborated (Brennan et al., 1978; Jenkins, Marks, and Schaefer, 1972; Marks, Jenkins, and Schaefer, 1971).

Most mycobacteria grow slowly, and 3 weeks or more are required for growth to appear on media. Rapid methods to detect *M. tuberculosis* by measuring unique metabolic activity are being developed (Cummings et al., 1975; McDaniel et al., 1977; Middlebrook, Reggiardo, and Tigertt, 1977). However, basic tests, such as those described in the following sections, must still be used for accurate isolation and identification of mycobacteria.

Habitats

Species of the genus *Mycobacterium* range from obligate parasites to free-living saprophytes. Of the cultivable species, *M. tuberculosis* and *M. bovis* have been clearly recognized for many years as pathogens of the human population, while *M. bovis, M. avium,* and *M. paratuberculosis* have been recognized as pathogens of lower animals. Other mycobacterial isolates, particularly the rapidly growing species, were considered to be nonpathogenic saprophytes. Therefore, some species are found only as pathogens, while others are primarily saprophytic but may occur as opportunistic pathogens (Dawson, 1971; Francis, 1958; Mills, 1972; Prather et al., 1961; Snider, 1971; Tison, Devulder, and Tacquet, 1968; Tsukamura et al., 1974; Viallier and Joubert, 1974). Currently accepted nomenclature (Kubica, 1979) for pathogenic and potentially pathogenic species is given in Table 1. The isolation of some species, such as *M. tuberculosis,* is diagnostic of the etiological agent of disease, but the occurrence of other species, such as *M. fortuitum,* may represent only temporary colonization. Therefore, diagnosis of mycobacterioses should be based on rigid criteria (American Thoracic Society, 1974): clinical evidence of a disease in which a cause has not been determined; and either repeated isolation of the same mycobacterial species or isolation of a mycobacterium from a closed lesion.

M. tuberculosis is the most frequently encountered mycobacterial species, but it accounts for approximately 95% of human mycobacterioses in the United States today. Man is the reservoir of infection, and spread is by inhalation of small (1–5 μm), droplet nuclei (containing tubercle bacilli), which are produced when a diseased person coughs, sneezes, etc. Progression of disease following inhalation of tubercle bacilli has been well described (Canetti, 1955; Rich, 1951). Tubercle bacilli have also been found as cause of disease in household pets and domesticated animals, such as cats and dogs (Snider, 1971), a parrot (Ackerman, Ben-

Table 1. Mycobacterial species, 1978.[a]

Potential pathogens	Commonly saprophytes
M. africanum	M. aurum
M. asiaticum[b]	M. chitae
M. avium	M. duvalii
M. bovis	M. flavescens
M. chelonei	M. gadium
M. farcinogenes[b]	M. gastri
M. fortuitum	M. gilvum
M. haemophilum	M. gordonae
M. intracellulare	M. neoaurum
M. kansasii	M. nonchromogenicum
M. leprae	M. parafortuitum
M. lepraemurium[b]	M. phlei
M. malmoense	M. smegmatis
M. marinum	M. terrae
M. microti[b]	M. thermoresistibile
M. paratuberculosis[b]	M. triviale
M. scrofulaceum	M. vaccae
M. senegalense[b]	
M. szulgai	
M. tuberculosis	
M. ulcerans	
M. xenopi	

[a] For details of species descriptions, see references in Kubica, 1979, and Skerman, McGowan, and Sneath, 1980.
[b] Rarely, if ever, recovered from man.

brook, and Walton, 1974), and cattle (Feldman, 1960). Tuberculous disease is one of the most serious infections encountered with newly imported and colonized nonhuman primates (Good, 1973; Ruch, 1959); disease in the rhesus monkey (*Macaca mulatta*) has been used as a model of human tuberculosis for studies on prophylaxis and chemotherapy (Good, 1973).

M. bovis also causes disease in man, but the portal of entry is usually the gastrointestinal tract, by the ingestion of raw milk from an infected cow. The disease caused by *M. bovis* develops in the intestinal tract and spreads to the mesenteric nodes. The reservoir of infection is cattle where it may occur in epidemic form (Johnson et al., 1975) or be found as a focus of infection (Feldman, 1960; Thoen et al., 1977). The bacterium may also infect other animals, such as cats and dogs, through direct contact (Snider et al., 1971). Fox (1923) presented indirect evidence that *M. bovis* was a cause of fatal disease in nonhuman primates, and other reports have been reviewed by Ruch (1959). More recently, Renner and Bartholomew (1974) described two outbreaks of bovine tuberculosis that involved rhesus (*M. mulatta*) and stumptail (*M. nemestrina*) monkeys and a chimpanzee.

M. africanum is taxonomically related to *M. tuberculosis* and *M. bovis* (see below) and has been grouped with them by Runyon (1974) into the *M. tuberculosis* complex. This species has only been isolated in certain parts of Africa (Castets et al.,

1963; Castets, Rist, and Boisvert, 1969; Pattyn et al., 1970).

Mycobacterioses caused by mycobacteria other than those in the *M. tuberculosis* complex are not tuberculosis, even though there may be similarities in the diseases. Disease processes associated with these mycobacterioses have been described by Chapman (1977). In contrast to the *M. tuberculosis* complex, these agents of human disease are found often as saprophytes and are only opportunistic pathogens. These mycobacterioses have not been shown to be contagious.

The *M. avium* complex is composed of two species, *M. avium* and *M. intracellulare*, which are included with *M. scrofulaceum* in the MAIS (*Mycobacterium avium, intracellulare, scrofulaceum*) complex because of taxonomic and disease relationships. The MAIS complex can be separated most easily by agglutination reactions in specific antisera. *M. avium* is composed of serovars 1, 2, and 3; *M. intracellulare* contains serovars 4 through 28; and *M. scrofulaceum* consists of serovars 40 through 44 (Goslee, Rynearson, and Wolinsky, 1976; Reznikov and Dawson, 1973; Wolinsky and Schaefer, 1973). MAIS-complex organisms have been isolated from water, milk, and soil samples (Goslee and Wolinsky, 1976; Kazda, 1973a,b; Kazda and Hoyte, 1972; Kubica, Beam, and Palmer, 1963; Matthews, Collins, and Jones, 1976; Reznikov and Leggo, 1974), as well as from animals such as swine (Mitchell et al., 1975; Reznikov and Robinson, 1970), exotic birds (Montali et al., 1976; Peavy et al., 1976), nonhuman primates (Latt, 1975; Sesline et al., 1975; Smith et al., 1973), and others (Jørgensen, 1978). Chapman (1977) has described the characteristics of the infection in humans, and Reznikov, Leggo, and Dawson (1971) have related the serological type of isolates from house dust to human infection. A study of ecological and epidemiological characteristics of *M. avium*–complex infections led Meissner and Anz (1977) to conclude that chickens and wild fowls, the natural reservoir of serovars 1, 2, and 3, were a major source of human infection, but swine and cattle infected with the same serovars were not. Serovars 4, 6, and 8 through 11 were associated with insects, sawdust, and presumably soil; and serovars 7 and 12 through 21 were associated with environmental reservoirs. These findings were confirmed in part by Jørgensen (1978) in a different locale.

M. kansasii is a slowly growing, photochromogenic species that has been isolated from tap water (Bailey et al., 1970; McSwiggan and Collins, 1974), but not from environmental water or soil (Chapman, 1971; Goslee and Wolinsky, 1976; Kubica, Beam, and Palmer, 1963). Reports of human infection have been summarized by Chapman (1977). Only rare reports of animal disease caused by *M. kansasii* have appeared, but one of us investigated an epidemic in a colony of *M. mulatta* that resulted in loss of large numbers of animals.

Nine strains of a slowly growing, photochromogenic mycobacterium were isolated from monkeys by Karassova, Weissfeiler, and Krasznay (1965). The strains differed from the then-recognized species and were therefore given the epithet *simiae*. Strains of this species were also isolated from tuberculous patients, many of whom were also shedding *M. tuberculosis* (Valdivia, Suarez, and Echemendía, 1971). This species has not been isolated from environmental sources.

M. marinum is a photochromogenic species that has an optimum growth temperature of 25–35°C with little or no growth at 37°C. This species causes swimming-pool granuloma, a superficial lesion on cooler parts of the body. Organisms have been isolated from lesions and from swimming pools (Linell and Norden, 1954; Mollohan and Romer, 1961). Sporadic infections have been acquired in the course of cleaning an aquarium and may be associated with diseased fish (Adams et al., 1970; Barrow and Hewitt, 1971).

Schaefer et al. (1973) reported serological identity of strains of *M. szulgai*, which had been isolated from a number of patients with pulmonary disease, but the organisms have not been reported from other sources. *M. szulgai* is sometimes confused with *M. gordonae*, which is widely found in raw and treated waters (Goslee and Wolinsky, 1976) and in sputum specimens, although it is rarely, if ever, associated with disease (Lohr et al., 1978; Wolinsky and Rynearson, 1968).

Isolation of a large number of *M. xenopi* strains from hospitalized patients has been related to colonization of a hot-water system (Gross, Hawkins, and Murphy, 1976). Similar observations, reported by Bullin, Tanner, and Collins (1970) and by Tellis et al. (1977), have indirectly associated human pulmonary disease to contaminated hot water in a hospital. Human infections with this organism appear, at present, to be nosocomial in origin.

The rapidly growing mycobacteria of clinical significance are *M. fortuitum* and *M. chelonei*. The organisms are widespread in nature (Goslee and Wolinsky, 1976; Hosty and McDurmont, 1975; Kazda, 1977). These organisms have also been isolated from aquarium fish (Bernstad, 1974; Ross, 1959), a manatee (Boever, Thoen, and Wallach, 1976), and a cat (Thorel and Boisvert, 1974). Human infections may occur in the lungs or after injury to subcutaneous tissues or the cornea (Burke and Ullian, 1977; Chapman, 1977). Serious infections have developed in wounds following open-heart surgery (Hoffman, Fraser, and Hinson, 1978; Robicsek et al., 1978), venous stripping (Foz et al., 1978), and in renal homograft recipients (Graybill et al., 1974), but a source of infection could not be identified. Porcine aortic heart valves used for surgical valve replacement have been found contaminated with *M. chelonei* (Laskowski et al., 1977; Levy et al., 1977). Although many of these valves have been used, only two cases of frank disease have developed. All contaminated valves were produced by a single supplier who relied on glutaraldehyde for sterilization and then stored the valves in a reduced concentration of glutaraldehyde for preservation.

M. ulcerans infections are described by Runyon (1974) as similar to leprosy, but the lesions are superficial with necrotic centers and do not involve the nerves. Geographic localization of the disease in tropical and subtropical areas has been reviewed by Chapman (1977).

M. paratuberculosis produces a chronic intestinal infection in ruminants and is spread through herds by contact with infected feces. The disease is of major economic concern in sheep and cattle, where whole herds may become infected. The disease is thought to be self-limiting in swine, but these animals may provide a short-term reservoir of the bacilli (Larsen, Moon, and Merkal, 1971). The organism is not pathogenic for humans, although one case of a localized lesion has been reported following accidental injection of a heat-killed vaccine (Björnsson et al., 1971). The chronic tendovaginitis that resulted was due to the heat-killed bacilli rather than to growth of *M. paratuberculosis*.

Other species that are pathogenic for humans and animals (see Table 1) have been isolated only from the host. Saprophytic species, on the other hand, occur widely in soil and water and in foods such as raw milk, often along with potentially pathogenic species (Gangadharam et al., 1976; Goslee and Wolinsky, 1976; Hosty and McDurmont, 1975; Kazda, 1977, 1978a,b; Matthews, Collins, and Jones, 1976; Saito and Tsukamura, 1976; Tsukamura et al., 1974). These species may occur as a cause of disease in a compromised host, e.g., *M. terrae* as a cause of synovitis and osteomyelitis (Edwards, Huber, and Baker, 1978) or of disseminated disease (Cianciulli, 1974), but their primary importance is differentiation from the frank pathogens. Differentiation of these species is dealt with in a later section.

Isolation

Researchers in the developed countries often overlook the fact that, on a global basis, tuberculosis remains the leading cause of death among notifiable infectious diseases. Although effective antituberculosis drugs have been available for more than a quarter of a century, their impact on the incidence of and mortality from tuberculosis has not been enjoyed worldwide. Tuberculosis morbidity rates in different countries show more than a 100-fold difference, ranging from a low of about 2 cases per

100,000 population to a high of almost 500 per 100,000. Mortality rates per 100,000 show almost as wide a spread, from a low of less than 1 to a high of greater than 70 (Lowell, 1976).

In the face of such extreme ranges in morbidity, it is difficult to present a "universal method" for the laboratory confirmation of tuberculous (or other mycobacterial) disease. In some of the developing countries where tuberculosis is still rampant, diagnosis of the disease is easily accomplished on the basis of clinical examination and the demonstration of acid-fast bacilli in stained smears from patients' sputa. In countries where the incidence of tuberculosis is relatively low and where nontuberculous ("atypical") mycobacteria are recognized (and frequently isolated; see Ortbals and Marr, 1978), the definitive diagnosis of disease depends upon the isolation in culture of the offending mycobacterial pathogen. Because the latter is the most complicated situation and, we hope, the point at which most countries eventually will arrive, we will devote our attention to methods for culture and identification of the many mycobacteria that may be isolated from man, lower animals, and the environment. Laboratories interested only in smear examination for the diagnosis of tuberculosis should consult publications confined to that single subject (International Union Against Tuberculosis, 1977; Smithwick, 1976).

Collection of Specimens from Man

Clinical specimens submitted for culture should be collected before drug treatment is started. Even a few days of therapy with some of the bactericidal antituberculosis drugs may make the bacteriological confirmation difficult, or even impossible. Containers for the collection of specimens should be clean, sterile, and preferably used only once (50-ml, screw-cap, plastic, disposable centrifuge tubes are ideal). If the containers must be reused (glass), thorough cleaning with dichromate–sulfuric acid solution is recommended.

A series of 5 or 6 single sputum specimens (Blair, Brown, and Tull, 1976; Krasnow and Wayne, 1969) is usually sufficient to enable bacteriological confirmation of the disease; specimens commonly are collected on successive days, but time variations are permissible compatible with patient production of specimens. Specimens should be delivered to the laboratory as soon as possible after collection. If delays are unavoidable, specimens should be refrigerated to minimize growth of contaminants and to protect the mycobacteria. Specimens sent through the mails must be carefully packaged (Kubica et al., 1975; Vestal, 1975). Laboratory confirmation of a mycobacteriosis requires the best clinical specimens, properly handled, quickly transported, and promptly processed by the microbiologist. The cli-

nician should be immediately advised of any problems that might jeopardize the cultivation of the mycobacterium; such things as insufficient volume of specimen, improper packaging (leakage in transit), or excessive delay in transport to the laboratory may reduce chances for successful cultivation of the organism.

Tuberculosis and other mycobacterial diseases may affect almost any organ of the body, so the laboratory may expect to receive specimens of all kinds. Although each may require special handling in the laboratory, it is possible to divide clinical specimens into two groups: those contaminated with other flora and those normally not contaminated (Hawkins, Kubica, and Wayne 1977).

ASEPTICALLY COLLECTED SPECIMENS. Aseptically collected tissue may be homogenized in a sterile grinder and inoculated directly to liquid or solid media. Aseptically collected body fluids often are difficult specimens from which to cultivate mycobacteria. The relatively small numbers of acid-fast bacilli may be diluted in large volumes of reservoir fluid (e.g., blood, pleural fluid, urine). Concentration and cultural procedures are very inefficient, unless some method of preliminary treatment is used to concentrate the organisms or to encourage further multiplication of the relatively small numbers of bacilli present in the specimen. Three methods that have been successfully used are: (i) large volumes of such fluids as pleural, catheterized urine, etc., may be centrifuged at $2,000-3,000 \times g$, and the sediment inoculated to both liquid and solid media; (ii) lesser volumes (5–25 ml) of such fluids as blood, synovial fluid, aspirated pus, etc., may be added to liquid media (e.g., Middlebrook 7H9) in a ratio of one part specimen to five parts broth; (iii) small volumes of fluid with low protein content have been successfully filtered through sterile, $0.45 \mu m$ (pore size) membrane filters, and the filter placed onto appropriate media (Wayne, 1957).

Specimens that have been inoculated to liquid media should be examined for evidence of growth and subcultured to solid medium so that the isolated organism may be identified. If the specimen does not affect the turbidity of the liquid medium, then this medium may be smeared and stained or subcultured to solid medium after the broth becomes visibly cloudy. If the nature of the specimen inoculated into broth makes it impossible to determine changes in turbidity (e.g., whole blood, cerebrospinal fluid, pleural fluid), the broth may be smeared weekly (more often if diagnosis is urgent) and stained for microscopic examination for acid-fast bacilli. When acid-fast bacilli are detectable in smears, the broth may be streaked to solid media. If the broth is still clear, or shows no microscopic evidence of mycobacteria after 3 or 4 weeks, centrifuge and inoculate the sediment to solid media.

SPECIMENS CONTAMINATED WITH OTHER MICRO-ORGANISMS. Such specimens are most commonly received by the mycobacteriology laboratory. They include sputa, gastric lavage, bronchial washings or brushings, laryngeal swabs, spontaneous drainage of pus, voided urines, and necropsy or superficial (skin) tissues. All such specimens contain microorganisms that multiply much more rapidly than mycobacteria. These undesirable, rapidly growing contaminants must be killed so that they do not overgrow the more slowly growing acid-fast bacilli. Therefore, such specimens must be digested and decontaminated (to be discussed later) before being inoculated onto media. Because the majority of human mycobacterioses (exclusive of leprosy) are pulmonary, most of the specimens sent to the diagnostic laboratory will be sputum. When patients raise no sputum, or swallow it, gastric lavage or induction techniques may be used to obtain the desired exudative material. Contaminated specimens other than sputum may be processed successfully for culture by using slight modifications of the methods described for sputum and related specimens.

SPUTUM. The patient is usually given a container, told to "raise" sputum, and to take or mail the specimen to the laboratory. The patient should also be instructed in the proper collection of sputum. He or she should be told to rinse the mouth free of residual food, mouthwash, or oral drugs prior to collection of the specimen. He should know that nasopharyngeal discharge and saliva are *not* sputum. He should be taught to breathe deeply three or four times to induce the deep cough from the lungs, and to collect the generally thick exudative material that comes up with such a cough.

A series of five or six early-morning specimens, collected on successive days, is usually sufficient to establish the diagnosis. For nonproductive patients, or for those in whom culture confirmation is difficult, 24- to 48-h pooled specimens (Krasnow and Wayne, 1969) are helpful. If the specimens must be mailed to the laboratory, however, cultures from the early-morning sputa are less frequently contaminated than are those from pooled samples (Kestle and Kubica, 1967).

If natural expectoration is absent, sputum induction methods using nebulization techniques may help the patient to produce a specimen (Jones, 1966). Induced sputa should be so identified on the laboratory request slip, because they are very watery and resemble saliva. Sputum induction is a dangerous procedure and should only be conducted in well-ventilated rooms; personnel involved should be protected with new, effective face masks (e.g., Aseptex No. 1800 from 3M Corp.).

GASTRIC LAVAGE. Gastric washings are collected when patients will not or cannot provide direct or induced sputa, when sputum cultures are negative but X-ray evidence still suggests tuberculosis, or when patients may be suspected of submitting sputum other than their own. The gastric lavage should only be collected from hospitalized patients, before breakfast, and preferably before the patient arises. Twenty to 50 ml of sterile, distilled water is introduced into the stomach through a disposable, plastic gastric tube; the gastric lavage is then recovered by means of a 50-ml syringe. Such specimens should be processed promptly because mycobacteria die rapidly in aspirated gastric washings. If the specimen cannot be immediately processed, some attempt should be made to neutralize the gastric acids; addition of 1.5 ml of 40% disodium phosphate (Vandiviere, Smith, and Sunkes, 1952) or two buffer (pH 7 to 7.4) capsules or tablets per 50 ml of specimen is usually adequate. Induced sputum is preferable to and more productive than gastric washing (Jones, 1966), but if the lavage is performed 30 min after sputum induction, the combined results of both procedures are more productive than either method alone (Carr, Karlson, and Stilwell, 1967).

URINE. The collection and processing of urine specimens is presented because genitourinary tuberculosis is one of the most common sites of extrapulmonary tuberculous disease and a common site for secondary seeding in cases of miliary tuberculosis (Bentz, et al., 1975; Simon et al., 1977). Study has shown a single, early-morning, midstream specimen to be preferable to a 12–24 h pooled sample (Kenney, Loechel, and Lovelock, 1960). Because mycobacteria may be diluted in the large reservoir of urine, multiple specimens may be necessary to demonstrate the organisms in culture. A series of at least three early-morning specimens is recommended and, in documented cases of genitourinary tuberculosis, such specimens have yielded positive cultures in 80–100% of the cases (Bentz et al., 1975; Christensen, 1974; Simon et al., 1977). In cases of extrapulmonary tuberculosis, there may be significant genitourinary tract involvement in the absence of patient complaints of discomfort (Bentz et al., 1975; Simon et al., 1977), so requests for laboratory processing of such specimens may be submitted "without apparent clinical reason".

Smear Examination of Mycobacteria from Man

There probably is no given small number of acid-fast bacilli that cannot be detected in stained smear preparations, providing the microscopist happens to focus upon the very area of the smear where those few bacilli are located. The chances of such a thing happening with regularity, however, are very small. On the basis of statistical calculation, it has been shown (David, 1976) that there must be from 5,000 to 10,000 bacilli per ml of specimen to provide a 50% chance for a technician to detect even one or two bacilli in the entire smear. These statistical figures are very similar to those obtained from quantitative studies comparing smear and culture results (Hobby et al., 1973; Yeager et al., 1967).

In spite of the fact that microscopy is less sensitive than culture, examination of stained smears of clinical specimens is valuable because it is easy, rapid, and provides a presumptive diagnosis of mycobacterial disease; it helps to identify potential pathogens other than mycobacteria that may be causing disease; and it may be used to follow the progress of a known tubercular on chemotherapy. Perhaps one of the most important uses for microscopic examination of sputum specimens is in the detection of those smear-positive patients that are most likely to spread tuberculosis in the community (Geiger and Kuemmerer, 1963; Loudon, Williamson, and Johnson, 1958; Shaw and Wynn-Williams, 1954). In contrast, the smear-negative (culture-positive) patient and the smear-positive patient who receives effective chemotherapy are less likely to disseminate tuberculosis (Gunnels, Bates, and Swindoll, 1974). It is important, therefore, that laboratory personnel become proficient in preparation, staining, and examination of acid-fast, stained smears.

DIRECT SMEAR.   This method is most commonly used in developing countries but may also be employed in other laboratories where there is a need for a rapid smear evaluation of a clinical specimen. New, clean slides should be used. The most informative results are obtained when the microscopist selects the bloody, caseous, or purulent particles from the specimen. An applicator stick or bacteriological loop may be used to spread the smear over an area 1 × 2 cm on the slide; to spread the smear over a larger area only minimizes the chance for finding any acid-fast bacilli that may be present (David, 1976; Smithwick, 1976). Allow the smear to air-dry; then heat-fix it on an electric slide warmer (65–75°C for 2 or more hours), or by passing it through the flame of a Bunsen burner three or four times as for conventional smears.

CONCENTRATION METHOD FOR SMEAR ONLY. Where only smears are being done, and the laboratory wishes to minimize the risk of infection of personnel with any mycobacteria in the specimen, the hypochlorite (Clorox) method may be used (Oliver and Reusser, 1942).

Hypochlorite Method for Concentration of Mycobacteria (Oliver and Reusser, 1942)

Add to the specimen (sputum) an equal volume of Clorox (or other solution with 5% sodium hypo-

chlorite), stopper securely, and mix on a test-tube mixer or other shaking machine until liquefied; allow to stand at least 10 but no more than 15 min; dilute with sterile water and centrifuge at 2,000 × g for 15 min; decant the supernatant fluid and smear the sediment onto a microscope slide as described above.

Hypochlorite solutions will kill both the contaminants and the mycobacteria in the specimen, thereby making the specimen safer for the microscopist to handle. Hypochlorite solutions may also cause disintegration of mycobacteria if allowed to act too long. Therefore the time recommendations in this procedure should be precisely followed.

Each laboratory seems to have its own staining procedure for detection of mycobacteria. Acid-fast, fluorochrome-staining procedures are gaining in popularity throughout the world. Smears stained with such fluorescing dyes may be scanned at lower magnifications than are commonly used with fuchsin-staining methods, thereby minimizing the total time required for smear examination (Bennedsen and Larsen, 1966; Mitchison, 1968). Several different fuchsin- and fluorochrome-staining procedures are available, and a few of these are described elsewhere (Runyon et al., 1974; Smithwick, 1976; Vestal, 1975).

READING THE SMEAR.   The authors recommend that the microscopic examination of stained smears be made by area covered, rather than by trying to examine smears for a prescribed period of time. The former method insures a greater uniformity of procedure for repetitive smear examinations, and also encourages a broader sampling of the smear area. We suggest that smears be examined microscopically by making three longitudinal sweeps of the stained area, parallel to the length of the slide. The microscopist should be able to view about 100 fields in one sweep, if examination is made at about 1,000× (the magnification used for most fuchsin-stained smears). A method used for reporting smear results that reflects roughly 10-fold changes in observed bacterial numbers is shown in Table 2 (Smithwick, 1976).

The demonstration of acid-fast bacilli in stained smears does not enable identification of the organism. The bacilli observed in stained smears must be grown first in culture before the precise species identification of the mycobacterium may be ascertained.

Digestion-Decontamination of
Specimens from Man

Most of the clinical specimens sent to the mycobacteriology laboratory are contaminated with undesirable flora—organisms that, for the most part,

Table 2. Reporting scheme for acid-fast smears.[a]

Number of AFB[b] seen	Report
None in 3 sweeps (300 fields)	Negative
1–2 per 300 fields	Doubtful; request repeat specimen
1–9 per 100 fields	1+
1–9 per 10 fields	2+
1–9 per field	3+
>9 per field	4+

[a] When smear examinations are made at other than 800–1,000×, compensation must be made for the lower magnification used. See Smithwick, 1976, for discussion of the ''conversion factors''.

[b] Acid-fast bacteria.

replicate within minutes, whereas most of the clinically important mycobacteria have a generation time expressed in hours. It is important, therefore, to rid the specimen of as many of the undesirable contaminants as possible before inoculation onto culture media. Many digestion-decontamination methods have been described, most of which depend for their success on the greater resistance of acid-fast bacilli to strong alkaline or acidic solutions that kill most of the more rapidly growing contaminants. It must be emphasized that all the digestion-decontamination procedures are toxic for mycobacteria as well as for contaminants, and overdigestion can result in a marked reduction in the numbers of surviving mycobacteria. To insure recovery of maximum numbers of mycobacteria, the directions for the particular digestion procedure employed must be precisely followed.

The N-acetyl-L-cysteine-sodium hydroxide (NALC) method (Kubica et al., 1963, Kubica, Kaufmann, and Dye, 1964) is one of the most widely used sputum-digestion procedures in the United States. The incorporation of the mucolytic agent, acetylcysteine, has enabled most laboratories to utilize the lower final concentration of sodium hydroxide here described. Acetylcysteine loses activity rapidly once it is put into solution; hence the digestant must be made up daily in amounts actually required. The formula given here is for 200 ml of final digestant; greater or lesser volumes may be prepared by appropriate adjustment of the concentration of the various ingredients.

NALC Method for Digestion-Decontamination (Kubica et al., 1963)

Prepare in advance 100 ml of 4% sodium hydroxide (4 g NaOH in 100 ml distilled water) and 100 ml of 2.9% sodium citrate (2.9 g anhydrous sodium citrate in 100 ml distilled water); these solutions may be sterilized by autoclaving and stored on the shelf in screw-top flasks until needed. Prior to use, mix the NaOH and citrate solutions together (200 ml final volume) and add powdered N-acetyl-L-cysteine to 0.5% (1.0 g NALC to 200 ml). This final solution should be used within 18–24 h.

Transfer 10 ml (or less) of sputum (or other clinical specimen) to sterile, 50-ml, screw-cap centrifuge tubes (preferably plastic). Smaller tubes may be used, but in such cases the volume of the specimen should not exceed 1/5 the capacity of the tube. Add an equal volume of the NALC digestant, then replace and securely tighten the screw cap on the tube. Mix the contents of the tube by swirling on a vortex-type test-tube mixer until the specimen is liquefied (usually 5–15 s). Let stand for 15 min at room temperature to effect decontamination.

Fill tubes to the 50-ml mark (about 12 mm from the top) with sterile, distilled water or pH 6.8 phosphate buffer (0.067 M). Dilution minimizes the continuing action of NaOH and lowers the specific gravity of the suspension so that centrifugation is more efficient and neutralization of alkali is not necessary. Tighten screw caps and mix contents of tubes by swirling. Centrifuge at 2,000–3,000 × g for 15 min using aerosol-free, sealed centrifuge cups (or the centrifuge may be modified to control aerosols as described by Hall, 1975). Pour off the supernatant fluid and resuspend the sediment in a small volume (1 or 2 ml) of sterile, 0.2% bovine albumin fraction V (prepare in 0.85% sodium chloride, adjust pH to 6.8–7.0, and sterilize by Seitz or membrane filtration).

Inoculate the resuspended sediment (both undiluted and after a 1:10 dilution in sterile saline) onto the surface of at least two media of different basic composition (e.g., an egg base such as Lowenstein-Jensen, and an agar base such as 7H-11).

Specimens of large initial volume (such as gastric lavage or urine) should first be centrifuged at 2,000 × g, the supernatant fluid discarded, and the sediment processed as for sputum.

Some investigators (Engbaek, Vergmann, and Bentzon, 1967; Mead and Woodhams, 1964; Selkon et al., 1966) have reported excessively high contamination rates when using the NALC procedure. If more thorough decontamination is desired, the studies of Krasnow and Wayne (1966) suggest that it is better to increase the concentration of the initial NaOH solution from 4 to 5 or 6%, and to keep the digestant contact time at 15 min, rather than to increase the time of exposure of specimen to the digestant; the experience of the authors supports this suggestion.

The Zephiran–trisodium phosphate (Z-TSP) method (Wayne, Krasnow, and Kidd, 1962) is one of the more gentle digestion procedures recommended for use in the United States (Kubica et al.,

1975; Vestal, 1975). Engbaek, Vergmann, and Bentzon (1967) reported the Z-TSP digests to be somewhat "slimy" and no more sensitive in recovery of mycobacteria than their standard NaOH method. In the United States, the authors' experience has been that the NaOH method is greatly abused. Personnel in many laboratories arbitrarily increase either the concentration of NaOH or the time of exposure to the alkali, to the detriment of any mycobacteria that may be present. The less stringent time controls on the Z-TSP method, therefore, make this method much more attractive to laboratories that find it difficult to adhere to prescribed time limits.

Z-TSP Method for Digestion-Decontamination (Wayne, Krasnow, and Kidd, 1962)

Dissolve 1 kg of trisodium phosphate (Na$_3$PO$_4$· 12H$_2$O) in 4 liters of hot distilled water and then add 7.5 ml of Zephiran concentrate (17% benzalkonium chloride, Winthrop Lab.). Mix well and store at room temperature.

Add to the specimen an equal volume of Z-TSP. Mix vigorously on a mechanical shaker for 30 min, and let stand for 20–30 min without additional shaking. Transfer to 50-ml, screw-cap centrifuge tube (if not already in one), securely tighten the cap, and centrifuge at 1,800–2,400 × g for 20 min. Decant the supernatant fluid and resuspend the sediment in 20–30 ml of pH 6.6 phosphate buffer (0.067 M). Repeat centrifugation step and discard the supernatant fluid. Resuspend the sediment (use sterile water or saline if necessary) and inoculate 3 drops to each tube or plate of medium used for isolation of mycobacteria.

The Sodium lauryl sulfate–sodium hydroxide (SLS) method (Tacquet and Tison, 1961; Tacquet, Tison, and Polspoel, 1965; Engbaek, Vergmann, and Bentzon, 1967) is not often used in the United States, but finds wide application in many European countries, where studies such as those of Engbaek, Vergmann, and Bentzon (1967) reveal the SLS method to be superior to other methods tested.

SLS Method for Digestion-Decontamination (Tacquet and Tison, 1961)

Dissolve 30 g of pure sodium lauryl sulfate in 1 liter of hot (60°C) distilled water. Add 10 g sodium hydroxide. Store the solution at 37°C to prevent precipitation. The neutralizing acid (0.09% H$_2$SO$_4$) is prepared by adding 0.9 ml of concentrated sulfuric acid to 1 liter of distilled water; add 2 ml of 1:250 bromocresol purple indicator, and autoclave to sterilize.

To 1 part of clinical specimen add 3 parts of SLS solution. Mix the contents immediately on a test tube mixer. Repeat the mixing procedure

after 10 and again after 20 min. Ten minutes after the last mixing (total exposure time 30 min), centrifuge the digested specimen at 1,800–2,400 × g for 30 min. Pour off the supernatant fluid. Neutralize the sediment by adding the sulfuric acid-bromocresol purple solution dropwise until a strong purple color persists. Inoculate the neutralized sediment onto egg medium.

The Cetylpyridinium chloride–sodium chloride (CPC) method (Smithwick, Stratigos, and David, 1975) was proposed to digest and decontaminate sputa in transit, so that upon arrival in the laboratory the specimen could be concentrated by centrifugation and inoculated directly onto media. Apparent advantages of this method are: (i) it requires less laboratory time because digestion-decontamination occurs during mailing; (ii) it yields more positive cultures, especially of mycobacteria other than tubercle bacilli; and (iii) it results in less contamination than the NALC method.

As has been observed with other digestants that contain detergent or quaternary ammonium (such as the SLS and Z-TSP methods), cetylpyridinium chloride is bacteriostatic for mycobacteria. The phospholipids in most egg media are adequate to neutralize this bacteriostatic effect, but unless some effort is made to inactivate or dilute the agent, mycobacteria will grow poorly (if at all) from specimens first treated with Z-TSP, SLS, or CPC and then inoculated onto agar medium (7H-10 or 7H-11). Perhaps the simple use of a buffer wash of the digested sediment, as is done in the Z-TSP method (Krasnow and Kidd, 1965), would be adequate; but until that problem is resolved, sediments from SLS-and CPC-digested sputa should only be inoculated onto egg media.

CPC Method for Digestion-Decontamination (Smithwick, Stratigos, and David, 1975)

Dissolve 10 g of cetylpyridinium chloride and 20 g of sodium chloride in 1 liter of distilled water. Store at room temperature. The solution is self-sterilizing and remains stable if tightly capped and protected from excess heat and light.

Sputa to be sent through the mails should be collected in 50-ml, screw-cap centrifuge tubes. Add an equal volume of the CPC reagent, tighten the cap, and shake by hand until the specimen becomes homogenized. Package the specimens in a watertight, double mailing container (Vestal, 1975) and send to the laboratory. The specimen is decontaminated in transit without any appreciable reduction in numbers of mycobacteria that may be present. Upon receipt, fill each tube to within 12 mm of the top with sterile, distilled water, recap, and centrifuge at 1,800–2,400 × g for 20 min. Decant the supernatant fluid, resuspend the sediment in 1 or 2 ml of sterile water or saline, and inoculate onto egg medium.

Some acid-digestion procedures have been recommended for use in processing urines or specimens contaminated with *Pseudomonas* (Vestal, 1975), but most laboratories do not have to resort to these techniques very often.

Culture Media for Specimens from Man

Precise identification of mycobacteria can be attained only if the organism is grown on culture medium and subjected to a number of in vitro tests. Although most mycobacteria are not very fastidious once they have adapted to in vitro growth, most investigators feel that a "richer" medium is needed for primary isolation of mycobacteria that have been accustomed to the nutrient-rich environment of diseased tissue or, perhaps, of natural environmental sources. Many different media have been proposed for the cultivation of acid-fast bacteria, and most of these are egg-potato base or serum-agar base formulations. Only three media will be discussed in detail because they are widely used throughout the world for primary isolation, drug-susceptibility testing, in vitro tests, and examinations of colonial morphology.

Modified Lowenstein–Jensen egg medium (Vestal, 1975) is available commercially: (i) as a complete, ready-to-use medium, or (ii) as a powdered base that must be supplemented with glycerol and fresh, whole eggs before it is tubed and inspissated. Many investigators still prefer to make the medium from basic ingredients. Unless otherwise specified, all ingredients should be reagent grade.

Modified Lowenstein-Jensen Egg Medium
(Vestal, 1975)

Fresh eggs, not over 1 week old, are cleaned by scrubbing with a hand brush in soap solution. After scrubbing, let the eggs stand in the soap solution for 30 min. Rinse thoroughly in running water, then soak eggs in 70% ethanol for 15 min. Break the eggs into a sterile flask and homogenize by hand shaking. Filter the eggs through four layers of sterile gauze into a sterile graduated cylinder. Eighteen to 24 eggs (depending upon size) are needed to provide the required 1,000 ml of homogenized whole egg. Set the homogenized eggs aside and prepare the basal salt solution.

Place 600 ml distilled water in a 2-liter flask and dissolve in order:

KH$_2$PO$_4$ (anhydrous)	2.4 g
MgSO$_4$·7H$_2$O	0.24 g
Magnesium citrate	0.6 g
L-Asparagine	3.6 g
Glycerol	12.0 ml
Potato flour	30.0 g

Autoclave at 121°C for 30 min. Cool to room temperature and add: malachite green, 20 ml of a freshly prepared 2% aqueous solution and homogenized whole eggs, 1,000 ml. Mix the medium well and pour into a sterile aspirator bottle or funnel with a bell attachment (test-tube filling device). Dispense media in desired volumes into appropriate containers (e.g., 8 ml in 20-×-150-mm, sterile, screw-cap test tubes). Slant tubes and coagulate by inspissation (moist heat) at 85°C for 50 min.

Incubate the finished medium at 37°C for 48–72 h as a sterility check. Medium may be stored several months in the refrigerator (4–6°C) if caps are securely tightened to prevent drying.

This and other egg media may be inoculated with one of several laboratory tools: bacteriological loops, pipettes (capillary or serological), or cotton swabs. The usual inoculum is 0.1 ml or 1 or 2 loopfuls of the digested, decontaminated clinical material or the appropriate dilution of a suspension from a liquid or solid culture of the organism. Tubes should be incubated in the slanted position at least 7 days to permit even distribution of the inoculum over the entire surface of the medium. After this time, tubes may be incubated upright if space is needed. Most incubation is continued for at least 8 weeks at 35–37°C before the tubes are discarded as negative. Some mycobacteria, especially those recovered from superficial body lesions (e.g., *M. marinum*, *M. ulcerans*), grow better at 30–33°C on primary isolation; others, like *M. xenopi*, may prefer 42–43°C.

For preparation of Middlebrook 7H-10 agar (Vestal, 1975), the authors have found commercially available Middlebrook 7H-10 agar-powdered base and Middlebrook OADC enrichment to be completely satisfactory. The recommended method is to prepare the medium in quantities of 200–400 ml; larger volumes are discouraged because the medium must be heated longer to solubilize all the agar, and generally result in a medium of inferior quality.

Middlebrook 7H-10 Agar
(Vestal, 1975; slightly modified)

To prepare 200 ml of medium, suspend 3.8 g of Middlebrook 7H-10 agar base in 180 ml freshly distilled water (in a 400-ml flask) and add 1 ml glycerol. Do not boil the water to dissolve the agar. Rather, swirl the powdered agar base into suspension and sterilize in the autoclave at 121°C for 10 min. As soon as reduced autoclave pressure will permit, remove the flask of medium to a 52–56°C water bath. When the medium has tempered to 56°C or less, add 20 ml OADC enrichment, swirl carefully to mix (avoid bubbles), and pour into plates or tubes as soon as possible (within 1 h after autoclaving).

Many laboratories use 7H-10 medium in tubes, but the recommended medium container is a plastic Petri dish. Growth of tubercle bacilli is greatly stimulated by carbon dioxide (CO_2), and 10% CO_2 has been shown to be optimum (Beam and Kubica, 1968). When the medium is contained in Petri dishes, CO_2 diffuses easily and uniformly into the container. In contrast, it is not possible to ensure uniform diffusion of CO_2 into screw-cap tubes, even when caps are deliberately left loose. When performing drug-susceptibility tests in 7H-10 medium (Vestal, 1975), the inoculum is distributed more evenly and the bacterial colonies are easier to see and to count on the flat surface of plate medium as opposed to the round test tube. To avoid the dangers of "contaminating or dropping and breaking Petri dishes", which some people fear, single plastic (or glass) dishes of medium may be placed (and incubated) in single, clear, polyethylene plastic bags and sealed shut with tape or heat. Such plates can be examined directly through the plastic bag without removing the dish; this procedure reduces contamination and negates the problem of aerosols if the dish is dropped.

The agar medium may be inoculated with a loop or pipette; when agar medium is placed in plates, it is easy to inoculate the surface with several drops from a capillary pipette. The need to incubate 7H-10 medium in 10% CO_2 for primary isolation must be emphasized. Although incubation in CO_2 is not necessary to initiate growth on egg medium, it has been shown that CO_2 stimulates more and faster growth even on Lowenstein-Jensen medium (Whitcomb, Foster, and Dukes, 1962).

The advantages and disadvantages of egg and agar media will not be expounded here. They have been discussed elsewhere (Kubica and Dye, 1967), but with the many variables involved in the preparation and use of media, it is certain that each individual has his own tastes and choices.

The subject of media should not be dismissed without mention of modifications of both egg and agar media. The 7H-11 medium (Cohn, Waggoner, and McClatchy, 1968), also available commercially, is preferred over 7H-10 by some workers because the addition of enzymatic digest of casein stimulates more profuse growth, especially of drug-resistant tubercle bacilli. Care must be exercised if 7H-11 medium is used for drug-susceptibility testing, because the added enrichment affects the minimal inhibitory concentration of some drugs.

The addition of drugs to both egg and agar media to inhibit contaminating microorganisms has been proposed (Gruft, 1971; Mitchison et al., 1972; Petran and Vera, 1971), and one study has indicated the value of such media for primary isolation of mycobacteria (Matajack et al., 1973). The selective 7H-10 (or 7H-11) medium has been used to recover tubercle bacilli from tissues and even from sputum without the need for preliminary decontamination (Mitchison, Allen, and Lambert, 1973; Mitchison et al., 1972). Mitchison et al. (1972) noted that the NaOH digestant was more lethal for tubercle bacilli than were the drugs incorporated into the medium to inhibit contaminants; however, many "atypical" mycobacteria were severely restricted in their growth when inoculated onto the selective medium. McClatchy et al. (1976) recently modified the Mitchison medium to contain 50 μg carbenicillin/ml rather than the 100 μg/ml recommended earlier; the other drugs remained unchanged (polymyxin B, 200 units/ml; amphotericin B, 10 μg/ml; trimethoprim, 10 or 20 μg/ml). McClatchy et al. (1976) still utilized an alkaline digestion of clinical specimens, but reported an increase in the number of "atypical" mycobacteria recovered from specimens inoculated onto the 7H-11 with less carbenicillin. It seems likely that such media will gain wider acceptance and use in the future.

One final medium that is used almost daily for subculture, single colony pickings, or several in vitro tests, is Middlebrook 7H-9 liquid medium. The commercially available Middlebrook 7H-9 powdered base, supplemented (after autoclave sterilization and cooling) with Middlebrook ADC enrichment, is easy to prepare and stores almost indefinitely. The user may supplement the medium with either glycerol or 0.05% (final concentration) Tween 80. The latter is most often used because it permits the growth of more smooth and homogeneous suspensions, which permit more uniform dilution of inoculum for drug-susceptibility tests, in vitro tests, and preparation of appropriate inocula for animal experiments.

Digestion-Decontamination of Specimens from Lower Animals

Small pieces of tissue from animals may be processed as described in the section on specimens from man. The amount of decontamination required depends upon the way the tissue was handled following excision. Larger pieces of tissue, or even intact organs (e.g., lymph nodes), may be processed by the pentane-flotation procedure (Mycobacteriology Diagnostic Services Unit, 1971).

Digestion-Decontamination by Pentane Flotation (Mycobacteriology Diagnostic Services Unit, 1967)

Rinse tissue specimens in 1,000 ppm hypochlorite (mix 67 ml of 6.0% NaOCl with 3,924 ml tapwater) immediately after receipt. Remove all fat from tissue using sterile scissors and forceps, but do not cut into intact nodes. Place the trimmed tissues in fresh hypochlorite

(1,000 ppm) at 4°C overnight. Wash again in hypochlorite the next day. In a biological safety cabinet (BSC), excise any obvious lesions or a representative portion of questionable tissue, and place in an aerosol-free blender jar. Add 50 ml nutrient broth containing 0.4% phenol red indicator and blend for 2 min. Place a magnetic stirring bar in the jar of blended tissue and add 100 ml of 5% papain (50 g papain, 0.6 g cysteine hydrochloride, and 1,000 ml distilled water; mix, heat to 70°C, and Seitz filter; aseptically bottle in 100-ml amounts). Add sufficient 4% NaOH to turn phenol red a definite red. Stir the mixture for 1 h on magnetic stirrer at 20–25°C. In the BSC, add 10 ml pentane to the jar of tissue homogenate, mix with two or three vigorous shakes, and let stand for 30 min. The lipid-rich mycobacteria concentrate in the pentane layer and at the pentane-water interface. Use a sterile pipette to remove the pentane and interface layers. Filter these "layers" through a single layer of sterile, unbleached muslin in a glass funnel and collect in a sterile, 20-×-125-mm screw-cap tube. Centrifuge at about 2,000 × g for 20 min in an aerosol-free, screw-cap, safety centrifuge carrier. Both sediment and a pellicle are usually formed during centrifugation. Use a sterile pipette to remove and discard the liquid that separates the sediment and pellicle. Resuspend the combined sediment-pellicle in 2 ml nutrient broth; to 1.0 ml of the resuspended material, add 1.0 ml of 0.2% Zephiran and let stand at 22–25°C for 15 min. The 0.2% Zephiran is prepared by mixing 1.2 ml of 17% Zephiran (Winthrop Lab.) to 98.8 ml sterile, distilled water. The remaining 1.0 ml of mixture is left untreated.

These digested specimens may then be inoculated to one of several kinds of media, depending upon the mycobacteria that might be expected to be present in the specimen.

Culture Media for Specimens from Lower Animals

The Lowenstein-Jensen and Middlebrook 7H-10 media already described are also used in the veterinary laboratory. The Lowenstein-Jensen medium is commonly made both with and without glycerine, because this compound may inhibit the growth of fresh isolates of *Mycobacterium bovis*. The 7H-10 medium also may be prepared with and without the addition of 0.4% sodium pyruvate; the latter stimulates the growth of *M. bovis* and *M. africanum*.

Gallagher and Horwill (1977) have proposed the use of selective 7H-10 or 7H-11 media for the cultivation of *M. bovis* directly from homogenized or macerated tissue without the need for harsh, chemical decontaminating solutions.

Selective Medium for *Mycobacterium bovis* (Gallagher and Horwill, 1977)

Adapted from the selective medium of Mitchison et al. (1972), the *M. bovis* medium contains: polymyxin, 200 units/ml; carbenicillin, 100 μg/ml; amphotericin B, 50 μg/ml; and trimethoprim, 20 μg/ml. The following additional changes in the formulation were made: (i) the malachite green was increased to a final concentration of 0.0035%; (ii) the bovine albumin was replaced by a final concentration of 10% fresh bovine serum; and (iii) the addition of 0.5% lysed, sheep red blood cells proved stimulatory to bacillary growth.

Stonebrink's medium is used in many veterinary laboratories because of its ability to support the growth of *M. bovis;* this medium is another that lacks glycerol.

Stonebrink's Medium for *Mycobacterium bovis*

Eggs may be washed, cracked, and homogenized as described under Lowenstein-Jensen medium; or a short, gentle mixing on the low setting of a blender or Osterizer may be used to homogenize the eggs. One batch of medium requires 800 ml of homogenized eggs. Set aside until needed.

To 300 ml distilled water in an aspirator bottle, add in order: sodium pyruvate, 5.0 g; KH$_2$PO$_4$, 2.0 g. Adjust pH to 6.5 by careful addition of Na$_2$HPO$_4$. To 100 ml distilled water, add 100 mg crystal violet and 800 mg malachite green. Sterilize both solutions in the autoclave at 121°C for 20 min.

When cool, add dye solution to salt solution, add eggs (prepared above), and mix thoroughly (magnetic stirrer may be used). Dispense aseptically, adding 9-ml amounts to 20-×-125 (or 150)-mm screw-cap tubes. Inspissate tubed medium for 40 min at 80°C with tubes slanted. Incubate the medium 24–72 h at 37°C to check sterility.

The specific identification of *M. paratuberculosis,* the causative agent of Johne's disease in cattle, sheep, and goats, requires that specimens be inoculated onto medium with and without mycobactin. The latter is a specific growth requirement for most of the freshly isolated strains of *M. paratuberculosis*. Because of the need to prepare mycobactin for the medium, and the specific procedures required to confirm *M. paratuberculosis,* the reader is referred to papers by Merkel and Curran (1974), Merkel and Thurston (1966), Merkel et al. (1964), Snow (1954), and Wheeler and Hanks (1965).

Isolation of Mycobacteria from Environmental Sources

Mycobacteria are ubiquitous in nature, but it is very difficult to isolate them in pure culture from environmental samples because the multitude of other microorganisms present rapidly overgrow mycobacteria on the isolation media. Early success was realized (Gordon and Hagen, 1937) by using very simple, liquid salt solutions "enriched" with paraffin-coated glass rods, and sometimes supplemented with glycerol.

The predominant isolates in these early studies were rapidly growing mycobacteria. In later investigations (Goslee and Wolinsky, 1976; Kubica, Beam, and Palmer, 1963; Wolinsky and Rynearson, 1968), it has been possible to recover slowly growing mycobacteria from environmental samples (soil, water, etc.) by using rather severe decontaminating procedures, which undoubtedly kill many mycobacteria as well as non-acid-fast contaminants. The method of Kubica, Beam, and Palmer (1963) may be used for heavily contaminated specimens.

Decontamination of Environmental Samples of Mycobacteria with Hydroxide-Hypochlorite (Kubica, Beam, and Palmer, 1963)

The stock decontaminating solution is prepared just before use by mixing equal quantities of 15% sodium hydroxide and 15% calcium hypochlorite (with 70% available chlorine). The "use concentration" of the hydroxide-hypochlorite is made by diluting 1 volume of the stock solution with 6 volumes of sterile, distilled water.

Place two spatulas (15 × 75 mm) full of soil in a sterile, screw-cap centrifuge tube. Add 30 ml sterile, distilled water, shake for 30 min (mechanical shaker) to break up clumps and suspend microorganisms, and then let stand until the larger clumps settle and the supernatant fluid becomes homogeneous in appearance. Remove 10 ml of this supernatant fluid to another screw-cap test tube that contains an equal volume of the diluted hydroxide-hypochlorite solution. Shake tubes again for 15–30 min, then centrifuge at 2,000 × g for 15 min. Discard the supernatant fluid and streak the resuspended sediment (without neutralization) onto media (see below).

The method of Wolinsky and Rynearson (1968) also has been used successfully.

Decontamination of Environmental Samples with Hydroxide-Malachite Green-Cycloheximide (Wolinsky and Rynearson, 1968)

Add 1 teaspoon of soil to 10 ml Trypticase soy broth, shake vigorously, and incubate for 6 h at 37°C to permit germination of spores. Remove 10 ml of supernatant fluid to a 50-ml screw-cap centrifuge tube and add equal volumes of 0.2% malachite green and 4% sodium hydroxide. Then add cycloheximide solution to a final concentration of 500 μg/ml. Mix the contents of the tube and let stand 30 min. Approximate neutralization may be achieved by adding 10 ml (a volume equal to the amount of 4% NaOH originally added to the specimen) of 1 N HCl. Centrifuge at 2,000 × g for 15 min. Wash the sediment two times in sterile, distilled water and sediment by centrifugation after each washing. Discard the last wash, resuspend the sediment, and streak to isolation media.

Isolation media commonly used for investigation of environmental samples are Lowenstein-Jensen and Middlebrook 7H-10. If excessive mold contamination is encountered or anticipated, cycloheximide may be added to either medium in a concentration of 500 μg/ml.

Decontamination of Water Samples from Environmental Sources

Water samples should first be run through 0.2-μm membrane filters. The filters may then be decontaminated with the hydroxide-hypochlorite solution previously described, or with 4% NaOH or 4% H_2SO_4. The filter vacuum is turned off and one of the decontaminating solutions is poured onto the filter and allowed to stand for 30 min (replenish the decontaminating solution as it gradually runs through the filter by gravity flow). The vacuum is again turned on to remove the decontaminating solution and copious amounts of sterile water are then run through the filter to dilute the decontaminating agent. The filter is then picked up with sterile forceps and placed onto media. Alternatively, some investigators filter the raw water specimen through the filter and then cut the filter into smaller pieces, which may then be treated in a flask as a "soil" sample.

Identification

There are more than 30 approved species of mycobacteria (Kubica, 1978, 1979; Skerman, McGowan, and Sneath, 1980), more than half of which were not recognized until after 1950 (Barksdale and Kim, 1977). Not all of these species produce disease, so a distinction must be made between potential pathogens and commonly saprophytic species (American Thoracic Society, 1974; Hawkins, Kubica, and Wayne, 1977; Wolinsky, 1977; see Table 1). The categories in Table 1 were derived on a statistical basis and should not be regarded as irrefutable; i.e., the isolation of a "pathogen" is not tantamount to disease, nor, for that matter, should all "saprophytes" recovered from animals be regarded as environmental contaminants. The implication in such a categorization is that certain species are more likely to be related to disease than are others. But each isolate, just as each "patient", must be considered individually.

Man is the reservoir for the classical mycobacterial disease, tuberculosis, so the isolation of *Mycobacterium tuberculosis* from man demands some clinical action. Likewise, the isolation of *M. bovis* from cattle (or man) must signal a series of clinical and epidemiological events. The recognition of *M. avium* in fowl or swine is of great concern to veterinary and agricultural personnel. But other mycobacteria, the nontuberculous or "atypical" mycobacteria, are a dilemma. Because of their environmental origin (Bailey et al., 1970; Bullin, Tanner, and Collins, 1970; Dawson, 1971; Goslee and Wolinsky, 1976; Kazda, 1973a,b; Kazda and Hoyte, 1972; Kubica, Beam, and Palmer, 1963; McSwiggan and Collins, 1974; Meissner and Anz, 1977; Mills, 1972; Prather et al., 1961; Reznikov and Leggo, 1974; Reznikov, Leggo, and Dawson, 1971; Reznikov, Leggo, and Tuffley, 1971; Reznikov and Robinson, 1970; Tison, Devulder, and Tacquet, 1968; Tsukamura et al., 1974; Wolinsky and Rynearson, 1968) laboratory isolation of a nontuberculous mycobacterium must be carefully assessed by the clinician: is it disease associated, is it a casual ("normal flora") isolate, can it be a laboratory-introduced contaminant? The frequency of isolation of environmental mycobacteria in the laboratory depends upon many factors: geographic location; season of the year; use of nonsterile fluids for such things as gastric lavage, sputum induction, preparation of digesting solutions; and the choice of digesting-decontaminating agent to be used in the laboratory (some are more toxic than others; see Kubica et al., 1975). Most clinicians tend to disregard the occasional isolation from a single specimen of a few colonies of an environmental isolate, whereas the recovery of any mycobacterium (even a "saprophyte") from a closed lesion (e.g., by aspiration, surgical excision) is usually considered significant (Cianciulli, 1974; Dechairo et al., 1973; Edwards, Huber, and Baker, 1978; Gonzales, Crosby, and Walker, 1971).

The more common situation in the laboratory is the recovery from clinical specimens of a positive culture having more than ". . . a few colonies". In such cases, the precise species identification of the isolated *Mycobacterium* is extremely important in the physician's assessment of the clinical importance of the organism. It is no longer acceptable to refer to such "atypical" mycobacteria by the old group designations (Runyon, 1959) or even by such vernacular names as Battey bacillus and scrofula scotochromogen (Runyon, 1965).

In the more than 20 years since the recognition that mycobacteria other than tubercle bacilli could cause disease in man (Buhler and Pollak, 1953;

Crow et al., 1957; Timpe and Runyon, 1954), there has been a plethora of literature describing numerous taxonomic tests of value in the differential identification of mycobacteria. The application of numerical taxonomic methods, facilitated by computer technology, enabled detailed analyses of thousands of strains of mycobacteria. But one thing, more than any other, contributed to the definition and international acceptance of currently recognized species of mycobacteria, and that was the cooperative efforts of the International Working Group on Mycobacterial Taxonomy (IWGMT). From the IWGMT have come numerous studies on mycobacterial taxonomy (Goodfellow et al., 1974; Kubica et al., 1972; Meissner et al., 1974; Saito et al., 1977; Wayne et al., 1971) and standardization of test methods (Wayne et al., 1974, 1976), which have led to a meaningful, clinically oriented taxonomy.

A direct result of the detailed taxonomic studies has been the selection of a few key tests that can be used to differentiate those mycobacteria most commonly encountered in the clinical laboratory. Several investigators (Käppler, 1968a,b; Kestle, Abbott, and Kubica, 1967; Kubica, 1973; Pattyn and Portaels, 1972; Runyon, 1974; Tsukamura, 1967; Wayne and Doubek, 1968) have proposed such a shortened taxonomic scheme based upon the set of key properties that have proved most valuable to the investigator concerned. The selection of such tests varies with the investigator, and depends on the method of test performance and sometimes on the geographic location of the laboratory (which may affect environmental species encountered). The tests listed in Tables 4–7 were selected because of their high degree of reproducibility in the authors' hands. With these tests, it is often possible to distinguish potential pathogens from common saprophytes. Precise identification to species is not always possible but, in general, those bacilli of common disease potential may be lumped together as complexes (Kubica et al., 1975; Runyon, 1974; Runyon et al., 1974; Vestal, 1975), a classification that is usually adequate for the clinician.

Rough Division of Mycobacteria

The early subdivision of the "atypical" mycobacteria into "groups" (Runyon, 1959, 1965; Timpe and Runyon, 1954), on the basis of pigment production and speed of growth, was proposed to provide a more systematic basis for study and discussion of these organisms. The detailed taxonomic studies of the IWGMT were facilitated by that group's ability to limit the number of strains of mycobacteria to be examined in a single investigation. Even today, most clinical laboratories of mycobacteriology, as well as the reference laboratories of the world, find it convenient to subdivide the mycobacteria on the

Table 3. Flow chart for identification of mycobacteria.[a]

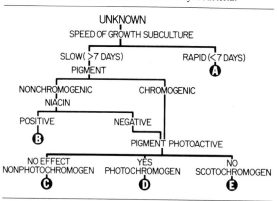

[a] A, see Table 4; B, see Table 5; C, see Table 6; D, see Table 7; E, see Table 8.

basis of pigment production and growth rate, thus enabling a more rational selection of the key tests needed to precisely identify an unknown *Mycobacterium* species.

Table 3 presents a flow diagram that subdivides the mycobacteria into five groups based upon growth rate, pigment, and niacin production. The one failing with such a dichotomous key is that no biological test is absolute, i.e., always positive or always negative for any given species of organism. Exceptions are always found when one searches diligently. The niacin test, once thought to be diagnostic of *M. tuberculosis* (Konno, 1956), is now known to be positive for occasional strains of some species (Kubica, 1973; Kubica et al., 1975; Yue and Cohen, 1966; Zvetina and Wichelhausen, 1976), and also for most strains of a given species as well (Krasnow and Gross, 1975; Meissner and Schröder, 1975). Exceptions to all other taxonomic tests are equally well known. These exceptions notwithstanding, the preliminary subgrouping of mycobacteria based upon pigment production and speed of growth is still helpful to the clinical mycobacteriologist.

There are two or more species within each of the five subgroups shown in Table 3, and the diagnostic laboratory is obliged to identify these, either to the species level or, if clinically feasible, to the complex level. If the diagnostic laboratory lacks the ability to

identify an isolate specifically, then the organism should be referred to a reference laboratory (Kubica et al., 1975).

The segregation of niacin-positive, nonchromogenic mycobacteria from all the others (subgroup B in Table 3) was intentional, because *M. tuberculosis* is still the major niacin-positive, slowly growing pathogen recovered from man, and its rapid identification by a few additional supportive tests will enable prompt clinical attention and care of the world's most prevalent mycobacterial disease, tuberculosis. The other four subgroups—rapid growers, nonphotochromogens, photochromogens, and scotochromogens—contain both potential pathogens and environmental saprophytes, and their prompt identification is important.

Specific Identification of Mycobacteria

The methods used for precise identification of mycobacteria vary with different investigators. The techniques listed here are those developed in or recommended by the Center for Disease Control for the taxonomic characterization of mycobacteria commonly encountered in the clinical laboratories of the United States (Kestle, Abbott, and Kubica, 1967; Kubica, 1973; Kubica et al., 1975; Vestal, 1975). Other procedures have proved helpful to investigators in different parts of the world, and the reader is urged to consult the published studies of others for detailed test procedures (e.g., Käppler, 1968a,b; Pattyn and Portaels, 1972; Runyon et al., 1974; Tsukamura, 1967; Wayne and Doubek, 1968). Wherever applicable, the tests should be performed by the standardized procedures recommended by the IWGMT (Wayne et al., 1974, 1976). Details of the test procedures will not be presented because different methods may be used in different laboratories, and most mycobacteriologists already have adopted their own "pet" procedure for a given test. For the microbiologist who may be trying any of these tests for the first time, step-wise procedures are presented elsewhere (Runyon et al., 1974; Vestal, 1975).

Subgroup A in Table 3 is composed of the rapid growers. Table 4 shows that the tests for 3-day aryl-

Table 4. Identification of rapidly growing *Mycobacterium* species.

Test	Mycobacterium fortuitum complex			Others ("saprophytes")
	M. fortuitum	*M. chelonei*	*M. chelonei abscessus*	
Pigment	−	−	−	−/+
Arylsulfatase, 3 days	+	+	+	−
MacConkey	+/−	+/−	+/−	−
NaCl tolerance	+	−	+	+/−
Nitrate reduction	+	−	−	+/−
Fe uptake	+	−	−	+/−

[a] Some strains must be incubated at 28°C to yield expected reactions.

Table 5. Identification of niacin-positive nonphotochromogenic *Mycobacterium* species.

Test	*M. tuberculosis*	*M. simiae*
Nitrate reduction	+	−
Catalase, room temperature	<45	>45
Catalase, 68°C	−	+
Pigment, photochromogenic	−	+
Temperature range for growth	33–39°C	22–41°C

sulfatase and MacConkey agar may be used to segregate the clinically important *M. fortuitum* complex from most of the commonly saprophytic rapid growers. Further subdivision of the *fortuitum* complex is not clinically valuable at this time because the organisms are usually quite resistant to drugs. On occasion, the uniqueness of the infection or a pure taxonomic interest may stimulate more precise characterization of the species (Altmann et al., 1975; Gutman et al., 1974). In such cases, the use of the tests for sodium chloride tolerance, nitrate reduction, and iron uptake help in the separation of *M. fortuitum* from *M. chelonei*. Additional biochemical tests, serological procedures, and chemical analyses have been used (Kubica et al., 1972), but these are more often employed in research and reference laboratories rather than clinical laboratories.

Table 5 presents the two species of slowly growing nonchromogens that are niacin positive, *M. tuberculosis* and *M. simiae*. In the authors' experience, isolated colonies of *M. simiae* appear more like *M. avium*–complex organisms but, when growth is confluent, the simple recording of a positive niacin test could be cause for concern. The performance of the nitrate reduction and two catalase tests should enable ready distinction of *M. tuberculosis* from *M. simiae*. The feature of photochromogenicity is not always observed with *M. simiae* but, when present, is a valuable differential characteristic.

In Table 6 are listed the major nonphotochromogenic, slowly growing mycobacteria encountered in clinical laboratories. *M. tuberculosis* is excluded because it is usually identified, as in Table 4, with the first niacin test run on rough, nonpigmented slow growers. In contrast, *M. simiae* is not always tested for niacin because the colonies may be smooth and avian-like; therefore, it is included here to show its biochemical similarity to one of the most important organisms in this group, the *M. avium* complex. It can be seen that the 10-day, Tween-hydrolysis test usually distinguishes the potentially pathogenic nonphotochromogens (Tween negative) from the commonly saprophytic species (Tween positive). Further identification to species within these two major subdivisions may be accomplished with the other tests.

It is only rarely necessary for the clinical microbiologist to subdivide within the *M. terrae* complex, or even to separate this complex from *M. gastri*. On occasion, however, strains of these "saprophytic" species may be implicated in disease (e.g., Cianciulli, 1974; Dechairo et al., 1973; Edwards, Huber, and Baker, 1978). In such cases, precise identification to species level is necessary, and the taxonomic assistance of a reference laboratory may be required.

Among the potentially pathogenic nonphotochromogens, the taxonomic similarity of *M. simiae* to *M. avium* is readily apparent. The semiquantitative catalase, urease, and niacin tests are of value in sepa-

Table 6. Identification of nonphotochromogenic *Mycobacterium* species.

Test	*M. bovis*	*M. avium* complex	*M. xenopi*	*M. gastri*	*M. terrae* complex	*M. simiae*	*M. malmo-ense*[a]	*M. haemo-philum*[b]
Catalase, room temperature	<45	<45	<45	<45	>45	>45	<45	−
Catalase, 68°C	−	+	+	−	+	+	−	−
Tween hydrolysis	−	−	−	+	+	−	+	−
Tellurite reduction	−	+	−	−	−	−(+)[c]	ND[d]	−
Nitrate reduction	−	−	−	−	+	−	−	−
TCH inhibition	+	−	−	−	−	−	ND	ND
NaCl tolerance	−	−	−	−	−/+	−	ND	−
Urease	+	−	−	+	−	−(+)[c]	−	−
Pyrazinamidase	−/−	+	+	−/∓	∓/+	V	+	+
Pigment, photoactive	−	−	−	−	−	+ (not bright)	−	−
Niacin	−	−	−	−	−	+(−)[c]	−	−

[a] Grows from 22 to 37°C; one serotype.
[b] Grows from 25 to 30°C; no growth at 37°C. Requires hemin for growth.
[c] Some strains react as shown parenthetically.
[d] Not done.

Table 7. Identification of photochromogenic *Mycobacterium* species.

Test	M. kansasii	M. marinum	M. simiae
Tween hydrolysis	+	+	−
Nitrate reduction	+	−	−
Catalase > 45	+	V	+
Niacin	−	−/+	+
Pyrazinamidase, 4 days	−	+	−(some +)
Arylsulfatase, 2 weeks	−/+	+ +	−

rating the two species; the test for photochromogenicity may be used, but *M. simiae* exhibits a delayed photochromogenicity, often requiring at least 8 h of exposure to light to initiate pigment production, and subsequently taking more than 24 h to express the photoactivated pigment. *M. avium* and *M. bovis* (including BCG) may be distinguished by means of the urease and pyrazinamidase tests as well as by susceptibility to TCH (thiophene-2-carboxylic acid hydrazide). The separation of *M. xenopi* (which may or may not be pigmented) from the others is best achieved by recognition of the specific "filament-like" colonies on clear agar media (Pattyn, 1966; Runyon, 1968) and the preferred elevated temperature for growth of 41–43°C (Kubica et al., 1975; Pattyn, 1966; Runyon, Wayne, and Kubica, 1974; Runyon et al., 1974; Vestal, 1975).

Separation of the three most common photochromogenic species is shown in Table 7. Here again we see the capricious *M. simiae*. The tests that seem to discriminate these three species most effectively are Tween hydrolysis, nitrate reduction, pyrazinamidase, and 2-week arylsulfatase (Vestal, 1975).

There is no one test that seems to provide a clear-cut separation of the scotochromogens (Table 8) into two major subgroups: potential pathogens and saprophytes. The Tween-hydrolysis test comes closest, but it excludes *M. szulgai,* an organism that is scotochromogenic at 37°C, but often exhibits a photochromogenic pigment when grown at 25°C. The tests for urease, nitrate reduction, and semiquantitative catalase further facilitate the speciation of the scotochromogens.

The foregoing discussion was limited to biochemical characterization of the mycobacteria because these procedures are the easiest to learn, the most readily set up and performed, and the least subject to bias in interpretation. There are other very valuable and highly reliable techniques for differential identification of mycobacteria, but most of these are currently regarded as "research tools" because of the limited availability of reagents, the specialized (often expensive) equipment needed, or the necessary expertise acquired only after years of experience in test performance and interpretation. Included among these are the following procedures for which only selected references are given: sero-agglutination and lipid analyses (Birn et al., 1967; Jarnagin, Champion, and Thoen, 1975; Jenkins, Marks, and Schaefer, 1972; Wolinsky and Schaefer, 1973), diffusion in gel procedures (Norlin, 1965; Stanford and Beck, 1968), phage susceptibility (Jones, 1975; Rado et al., 1975), skin testing in hypersensitized animals (Magnusson, 1961), thin-layer or gas-liquid chromatography (Hecht and Causey, 1976; Larsson and Mårdh, 1976; Ohashi, Wade, and Mandle, 1977), and fluorescent-antibody methods (Bennedsen and Larsen, 1966; Gilmour and Angus, 1976). Although skin testing has not been very valuable in the differential diagnosis of clinical mycobacterioses in man, it has been used (together with lymphocyte transformation) with some success in monkeys (McLaughlin, Thoenig, and Marrs, 1976; Muscoplat et al., 1975b), cattle (Muscoplat et al., 1975a), and swine (Muscoplat et al., 1975c).

The many new species of mycobacteria recognized today, as well as the many techniques and procedures used to isolate and identify them, have compounded the problems of the diagnostic tuberculosis laboratory. The bacteriologist must be aware of the different species of mycobacteria common to the area, the techniques most valuable for their dif-

Table 8. Identification of scotochromogenic *Mycobacterium* species.

Test	M. scrofulaceum	M. szulgai	M. xenopi	M. gordonae	M. flavescens
Tween hydrolysis	−	+	−	+	+
Nitrate reduction	−	+	−	−	+
Catalase, room temperature	>45	>45	<45	>45	>45
NaCl tolerance	−	−	−	−	±
Urease	+	+	−	−	+
Pigment, photochromogenic, 25°	−	±	−	−	−
Arylsulfatase, 2 weeks	−	+	+ +	±	+

ferential identification, and some feeling for the reliability of these tests. Such information may be gained in part from reading, but the real expertise is acquired from day-to-day contact with many organisms and constant use of procedures recommended for their taxonomic characterization. This practical experience is what provides the mycobacteriologist with the working knowledge so vital to an appreciation of the intraspecies variations in test results and the subtleties of proper test performance or interpretation. Trainers throughout the world, especially those in the developed countries, are aware of this gradually dwindling mycobacterial experience in the clinical laboratories.

In the United States, this awareness has prompted the initiation of the concept of "levels of laboratory service" (Kubica et al., 1975). Briefly, the concept urges mycobacteriologists to perform only those procedures that reflect the numbers and kinds of mycobacteria isolated with some frequency in their particular area. Thus, laboratory personnel who see *M. tuberculosis,* but rarely encounter an "atypical" *Mycobacterium,* should only acquire the necessary expertise to identify the tubercle bacillus. Any isolates not identified as *M. tuberculosis* should be sent to a reference laboratory where daily experience with mycobacteria other than tubercle bacilli insures the level of expertise necessary for precise specific identification of "other" mycobacteria.

Literature Cited

Ackerman, L. J., Benbrook, S. C., Walton, B. C. 1974. *Mycobacterium tuberculosis* infection in a parrot *(Amazona farinosa).* American Review of Respiratory Disease **109:**388–390.

Adams, R. M., Remington, J.S., Steinberg, J., Seibert, J. S. 1970. Tropical fish aquariums. A source of *Mycobacterium marinum* infections resembling sporotrichosis. Journal of the American Medical Association **211:**457–461.

Alshamaony, L., Goodfellow, M., Minnikin, D. E. 1976. Free mycolic acids as criteria in the classification of *Nocardia* and the *'rhodochrous'* complex. Journal of General Microbiology **92:**188–199.

Alshamaony, L., Goodfellow, M., Minnikin, D. E., Mordarska, H. 1976. Free mycolic acids as criteria in the classification of *Gordona* and the *'rhodochrous'* complex. Journal of General Microbiology **92:**183–187.

Altmann, G., Horowitz, A., Kaplinsky, N., Frankl, O. 1975. Prosthetic valve endocarditis due to *Mycobacterium chelonei.* Journal of Clinical Microbiology **1:**531–533.

American Thoracic Society. 1974. Diagnostic standards and classification of tuberculosis and other mycobacterial diseases. New York: American Lung Association.

Bailey, R. K., Wyles, S., Dingley, M., Hesse, F., Kent, G. W. 1970. The isolation of high catalase *Mycobacterium kansasii* from tap water. American Review of Respiratory Disease **101:**430–431.

Barksdale, L., Kim, K.-S. 1977. *Mycobacterium.* Bacteriological Reviews **41:**217–372.

Barrow, G. L., Hewitt, M. 1971. Skin infection with a *Mycobacterium marinum* from a tropical fish tank. British Medical Journal **2:**505–506.

Beam, R. E., Kubica, G. P. 1968. Stimulatory effects of carbon dioxide on the primary isolation of tubercle bacilli on agar-containing medium. American Journal of Clinical Pathology **50:**395–397.

Bennedsen, J., Larsen, S. O. 1966. Examination for tubercle bacilli by fluorescence microscopy. Scandinavian Journal of Respiratory Disease **47:**114–120.

Bentz, R. R., Dimcheff, D. G., Nemiroff, M. J., Tsang, A., Weg, J. G. 1975. The incidence of urine cultures positive for *Mycobacterium tuberculosis* in a general tuberculosis patient population. American Review of Respiratory Disease **111:**647–650.

Bernstad, S. 1974. *Mycobacterium borstelense* isolated from aquarium fishes with tuberculous lesions. Scandinavian Journal of Infectious Diseases **6:**241–246.

Birn, K. J., Schaefer, W. B., Jenkins, P. A., Szulga, T., Marks, J. 1967. Classification of *Mycobacterium avium* and related opportunist mycobacteria met in England and Wales. Journal of Hygiene **65:**575–589.

Björnsson, A., Hallgrímsson, J., Gudmundur, G., Pálsson, P. A. 1977. Paratuberculosis of the hand. Scandinavian Journal of Plastic and Reconstructive Surgery **5:**156–160.

Blair, E. B., Brown, G. L., Tull, A. H. 1976. Computer files and analyses of laboratory data from tuberculous patients. II. Analyses of six years' data on sputum specimens. American Review of Respiratory Disease **113:**427–432.

Boever, W. J., Thoen, C. O., Wallach, J. D. 1976. *Mycobacterium chelonei* infection in a Natterer manatee. Journal of the American Veterinary Medical Association **169:**927–929.

Brennan, P. J., Souhrada, M., Ullom, B., McClatchy, J. K., Goren, M. B. 1978. Identification of atypical mycobacteria by thin-layer chromatography of their surface antigens. Journal of Clinical Microbiology **8:**374–379.

Buhler, V. B., Pollak, A. 1953. Human infection with atypical acid-fast organisms. American Journal of Clinical Pathology **23:**363–374.

Bullin, C. H., Tanner, E. I., Collins, C. H. 1970. Isolation of *Mycobacterium xenopei* from water taps. Journal of Hygiene **68:**97–100.

Burke, D. S., Ullian, R. B. 1977. Megaesophagus and pneumonia associated with *Mycobacterium chelonei.* American Review of Respiratory Disease **116:**1101–1107.

Canetti, G. 1955. The tubercle bacillus in the pulmonary lesion of man. New York: Springer Publishing Co.

Carr, D. T., Karlson, A. G., Stilwell, G. G. 1967. A comparison of cultures of induced sputum and gastric washings in the diagnosis of tuberculosis. Mayo Clinic Proceedings **42:**23–25.

Castets, M., Rist, N., Boisvert, H. 1969. La variété africaine du bacille tuberculeux humain. Médecine d'Afrique Noire **16:**321–322.

Castets, M., Boisvert, H., Grumbach, F., Brunel, M., Rist, N. 1963. Les bacilles tuberculeux du type africain. Note préliminaire. Revue de Tuberculose et de Pneumologie **32:**179–184.

Chaparas, S. D., Brown, T. M., Hyman, I. S. 1978. Antigenic relationships of various mycobacterial species with *Mycobacterium tuberculosis.* American Review of Respiratory Disease **117:**1091–1097.

Chapman, J. S. 1971. The ecology of the atypical mycobacteria. Archives of Environmental Health **22:**41–46.

Chapman, J. S. 1977. The atypical mycobacteria and human mycobacterioses. New York, London: Plenum.

Christensen, W. I. 1974. Genitourinary tuberculosis: A review of 102 cases. Medicine **53:**377–390.

Cianciulli, F. D. 1974. The radish bacillus *(Mycobacterium terrae):* Saprophyte or pathogen? American Review of Respiratory Disease **109:**138–141.

Cohn, M. L., Waggoner, R. F., McClatchy, J. K. 1968. The 7H-11 medium for the culture of mycobacteria. American Review of Respiratory Disease **98:**295–296.

Crow, H. E., King, C. T., Smith, C. E., Corpe, R. F., Stergus, I. 1957. A limited clinical, pathologic, and epidemiologic study of patients with pulmonary lesions associated with atypical acid-fast bacilli in the sputum. American Review of Tuberculosis **75**:199–222.

Cummings, D. M., Ristroph, D., Camargo, E. E., Larson, S. M., Wagner, H. N. 1975. Radiometric detection of metabolic activity of *Mycobacterium tuberculosis*. Journal of Nuclear Medicine **16**:1189–1191.

David, H. L. 1976. Bacteriology of the mycobacterioses. Washington, D. C.: U.S. Government Printing Office.

Dawson, D. J. 1971. Potential pathogens among strains of mycobacteria isolated from house-dusts. Medical Journal of Australia **1**:679–681.

Dean, G. 1903. A disease of the rat caused by an acid-fast bacillus. Centralblatt für Bakteriologie und Parasitenkunde, Abt. 1 **34**:222–224.

Dechairo, D. C., Kittredge, D., Meyers, A., Corrales, J. 1973. Septic arthritis due to *Mycobacterium triviale*. American Review of Respiratory Disease **108**:1224–1226.

Edwards, L. B., Palmer, C. E. 1968. Identification of the tuberculous infected by skin tests. Annals of the New York Academy of Sciences **154**:140–148.

Edwards, M. S., Huber, T. W., Baker, C. J. 1978. *Mycobacterium terrae* synovitis and osteomyelitis. American Review of Respiratory Disease **117**:161–163.

Engbaek, H. C., Vergmann, B., Bentzon, M. W. 1967. The sodium lauryl sulphate method in culturing sputum for mycobacteria. Scandinavian Journal Respiratory Disease **48**:268–284.

Engel, H. W. B. 1978. Mycobacteriophages and phage typing. Annales de Microbiologie **129A**:75–90.

Feldman, W. H. 1960. Avian tubercle bacilli and other mycobacteria, their significance in the eradication of bovine tuberculosis. American Review of Respiratory Disease **81**:666–673.

Fox, H. 1923. Disease in captive wild mammals and birds. Philadelphia: Lippincott.

Foz, A., Roy, C., Jurado, J., Arteaga, E., Ruiz, J. M., Moragas, A. 1978. *Mycobacterium chelonei* iatrogenic infections. Journal of Clinical Microbiology **7**:319–321.

Francis, J. 1958. Tuberculosis in animals and man: A study in comparative pathology. London: Cassel.

Gallagher, J., Horwill, D. M. 1977. A selective oleic acid albumin agar medium for the cultivation of *Mycobacterium bovis*. Journal of Hygiene **79**:155–160.

Gangadharam, P. R. J., Lockhart, J. A., Awe, R. J., Jenkins, D. E. 1976. Mycobacterial contamination through tapwater. American Review of Respiratory Disease **113**:894.

Geiger, F. L., Kuemmerer, J. M. 1963. Tuberculosis casefinding among contacts in seven South Carolina counties. Public Health Reports **78**:663–668.

Gilmour, N. J. L., Angus, K. W. 1976. The specificity and sensitivity of the fluourescent antibody test for *Mycobacterium johnei* infection in abattoir and culled cattle. Research in Veterinary Science **20**:10–12.

Gonzales, E. P., Crosby, R. M., Walker, S. H. 1971. *Mycobacterium aquae* infection in a hydrocephalic child. Pediatrics **48**:974–977.

Good, R. C. 1973. Tuberculosis and bacterial infection, pp. 39–60. In: Bourne, G. H. (ed.), Nonhuman primates and medical research. New York, London: Academic Press.

Goodfellow, M., Collins, M. D., Minnikin, D. E. 1976. Thin-layer chromatographic analysis of mycolic acid and other long-chain components in whole-organism methanolysates of coryneform and related taxa. Journal of General Microbiology **96**:351–358.

Goodfellow, M., Lind, A., Mordarska, H., Pattyn, S., Tsukamura, M. 1974. A co-operative numerical analysis of cultures considered to belong to the 'rhodochrous' taxon. Journal of General Microbiology **85**:291–302.

Gordon, R. E., Hagen, W. A. 1937. The isolation of acid-fast bacteria from soil. American Review of Tuberculosis **36**:549–552.

Goren, M. B., Cernich, M., Brokl, O. 1978. Some observations on mycobacterial acid-fastness. American Review of Respiratory Disease **118**:151–154.

Goslee, S., Rynearson, T., Wolinsky, E. 1976. Additional serotypes of *Mycobacterium scrofulaceum*, *Mycobacterium gordonae*, *Mycobacterium marinum*, and *Mycobacterium xenopi* determined by agglutination. International Journal of Systematic Bacteriology **26**:136–142.

Goslee, S., Wolinsky, E. 1976. Water as a source of potentially pathogenic mycobacteria. American Review of Respiratory Disease **113**:287–292.

Grange, J. M., Aber, V. R., Allen, B. W., Mitchison, D. A., Mikhail, J. R., McSwiggan, D. A., Collins, C. H. 1977. Comparison of strains of *Mycobacterium tuberculosis* from British, Ugandan and Asian immigrant patients: A study on bacteriophage typing, susceptibility to hydrogen peroxide and sensitivity to thiophen-2-carbonic acid hydrazide. Tubercle **58**:207–215.

Graybill, J. R., Silva, J., Jr., Fraser, D. W., Lordon, R., Rogers, E. 1974. Disseminated mycobacteriosis due to *Mycobacterium abscessus* in two recipients of renal homografts. American Review of Respiratory Disease **109**:4–10.

Gross, W. M., Hawkins, J. E., Murphy, D. B. 1976. Origin and significance of *Mycobacterium xenopi* in clinical specimens. I. Water as a source of contamination. Bulletin of the International Union Against Tuberculosis **51**:267–269.

Gruft, H. 1971. Isolation of acid-fast bacilli from contaminated specimens. Health Laboratory Science **8**:79–82.

Gunnels, J. J., Bates, J. H., Swindoll, H. 1974. Infectivity of sputum-positive tuberculous patients on chemotherapy. American Review of Respiratory Disease **109**:323–330.

Gutman, L. T., Handwerger, S., Zwadyk, P., Abramowsky, C. R., Rodgers, B. M. 1974. Thyroiditis due to *Mycobacterium chelonei*. American Review of Respiratory Disease **110**:807–809.

Hall, C. V. 1975. A biological safety centrifuge. Health Laboratory Science **12**:104–106.

Hawkins, J. E., Kubica, G. P., Wayne, L. G. 1977. *Mycobacterium*. CRC Handbook Series in Clinical Laboratory Science **1**:147–158.

Hecht, S. T., Causey, W. A. 1976. Rapid method for the detection and identification of mycolic acids in aerobic actinomycetes and related bacteria. Journal of Clinical Microbiology **4**:284–287.

Hobby, G. L., Holman, A. P., Iseman, M. D., Jones, J. M. 1973. Enumeration of tubercle bacilli in sputum of patients with pulmonary tuberculosis. Antimicrobial Agents and Chemotherapy **4**:94–104.

Hoffman, P. C., Fraser, D. W., Hinson, P. L. 1978. Delayed hypersensitivity reactions in patients with *Mycobacterium chelonei* and *Mycobacterium fortuitum* infections. American Review of Respiratory Disease **117**:527–531.

Hosty, T. S., McDurmont, C. I. 1975. Isolation of acid-fast organisms from milk and oysters. Health Laboratory Science **12**:16–19.

International Union Against Tuberculosis. 1977. Technical guide for collection, storage and transport of sputum specimens and for examinations for tuberculosis by direct microscopy, 2nd ed. Paris: International Union Against Tuberculosis.

Janicki, B. W., Chaparas, S. D., Daniel, T. M., Kubica, G. P., Wright, G. L., Jr., Yee, G. S. 1971. A reference system for antigens of *Mycobacterium tuberculosis*. American Review of Respiratory Disease **104**:603–604.

Jarnagin, J. L., Champion, M. L., Thoen, C. O. 1975. Seroagglutination test for identification of *Mycobacterium paratuberculosis*. Journal of Clinical Microbiology **2**:268–269.

Jenkins, P. A., Marks, J., Schaefer, W. B. 1972. Thin layer chromatography of mycobacterial lipids as an aid to classification: The scotochromogenic mycobacteria, including *Mycobacterium scrofulaceum*, *M. xenopi*, *M. aquae*, *M. gordonae*, *M. flavescens*. Tubercle **53**:118–127.

Johne, H. A., Frothingham, L. 1895. Ein eigenthümlicher Fall von Tuberculose beim Rinde. Deutsche Zeitschrift für Thiermedizin Vergleichende Pathologie 21:438–454.

Johnson, D. C., Rogers, A. N., Andrews, J. F., Downard, J. A., Thoen, C. O. 1975. An epizootic of bovine tuberculosis in Georgia. Journal of the American Veterinary Medical Association 167:833–837.

Jones, F. L., Jr. 1966. The relative efficacy of spontaneous sputa, aerosol-induced sputa, and gastric aspirates in the bacteriologic diagnosis of pulmonary tuberculosis. Diseases of the Chest 50:403–408.

Jones, W. D., Jr. 1975. Differentiation of known strains of BCG from isolates of *Mycobacterium bovis* and *Mycobacterium tuberculosis* by using mycobacteriophage 33D. Journal of Clinical Microbiology 1:391–392.

Jones, W. D., Jr., Greenberg, J. 1976. Use of phage F-φWJ-1 of *Mycobacterium fortuitum* to discern more phage types of *Mycobacterium tuberculosis*. Journal of Clinical Microbiology 3:324–326.

Jones, W. D., Jr., Greenberg, J. 1978. Modification of methods used in bacteriophage typing of *Mycobacterium tuberculosis* isolated. Journal of Clinical Microbiology 7:467–469.

Jørgensen, J. B. 1978. Serological investigation of strains of *Mycobacterium avium* and *Mycobacterium intracellulare* isolated from animal and non-animal sources. Nordisk Veterinaer Medicin 30:155–162.

Käppler, W. 1968a. Zur Taxonomieder Gattung *Mycobacterium*. I. Klassifizierung schnell wachsender Mykobakterien. Zeitschrift für Tuberkulose 129:311–319.

Käppler, W. 1968b. Zur Taxonomieder Gattung *Mycobacterium*. II. Klassifizierung langesam wachsender Mykobakterien. Zeitschrift für Tuberkulose 129:321–328.

Karassova, V., Weissfeiler, J., Krasznay, E. 1965. Occurrence of atypical mycobacteria in *Macacus rhesus*. Acta Microbiologica Academiae Scientiarum Hungaricae 12:275–282.

Kazda, J. 1973a. Die Bedeutung von Wasser für die Verbreitung von potentiell pathogenen Mykobacterien. I. Mögleichkeiten für eine Vehmerhung von Mykobakterien. Zentralblatt für Bakteriologie, Parasitenkunde, Infektionskrankheiten und Hygiene, Abt. 1 Orig., Reihe B 158:161–169.

Kazda, J. 1973b. Die Bedeutung von Wasser für die Verbreitung von potentiell pathogenen Mykobacterien. II. Vehmerhung der Mykobakterien in Gewässermodellen. Zentralblatt für Bakteriologie, Parasitinkunde, Infektionskrankheiten und Hygiene, Abt. 1 Orig., Reihe B 158:170–176.

Kazda, J. 1977. The importance of sphagnum bogs in the ecology of mycobacteria. Zentralblatt für Bakteriologie, Parasitenkunde, Infektionskrankheiten und Hygiene, Abt. 1 Orig., Reihe B 165:323–334.

Kazda, J. 1978a. The behavior of *Mycobacterium intracellulare* serotype Davis and *Mycobacterium avium* in the head region of sphagnum moss vegetation after experimental inoculation. Zentralblatt für Bakteriologie, Parasitenkunde, Infektionskrankheiten und Hygiene, Abt. 1 Orig., Reihe B 166:454–462.

Kazda, J. 1978b. Multiplication of mycobacteria in the gray layer of sphagnum vegetation. Zentralblatt für Bakteriologie, Parasitenkunde, Infektionskrankheiten und Hygiene, Abt. 1 Orig., Reihe B 166:463–469.

Kazda, J., Hoyte, R. 1972. Zur Ökologie von *Mycobacterium intracellulare* serotype Davis. Zentralblatt für Bakteriologie, Parasitenkunde, Infektionskrankheiten und Hygiene, Abt. 1 Orig., Reihe A 222:506–509.

Kenney, M., Loechel, A. B., Lovelock, F. J. 1960. Urine cultures in tuberculosis. American Review of Respiratory Disease 82:564–567.

Kestle, D. G., Abbott, V. D., Kubica, G. P. 1967. Differential identification of mycobacteria. II. Subgroups of *Groups II* and *III* (Runyon) with different clinical significance. American Review of Respiratory Disease 95:1041–1052.

Kestle, D. G., Kubica, G. P. 1967. Sputum collection for cultivation of mycobacteria. An early morning specimen or the 24-

to 72-hour pool? American Journal of Clinical Pathology 48:347–349.

Koch, R. 1882. Die aetiologie der tuberkulose. Berliner Klinische Wochenschrift 19:221–230.

Konno, K. 1956. New chemical method to differentiate humantype tubercle bacilli from other mycobacteria. Science 124:985.

Krasnow, I., Gross, W. 1975. *Mycobacterium simiae* infection in the United States. A case report and discussion of the organism. American Review of Respiratory Disease 111:357–360.

Krasnow, I., Kidd, G. C. 1965. The effect of a buffer wash on sputum sediments digested with Zephiran trisodium phosphate on the recovery of acid-fast bacilli. American Journal of Clinical Pathology 44:238–240.

Krasnow, I., Wayne, L. G. 1966. Sputum digestion. I. The mortality rate of tubercle bacilli in various digestion systems. American Journal of Clinical Pathology 45:352–355.

Krasnow, I., Wayne, L. G. 1969. Comparison of methods for tuberculosis bacteriology. Applied Microbiology 18:915–917.

Kubica, G. P. 1973. Differential identification of mycobacteria. VII. Key features for identification of clinically significant mycobacteria. American Review of Respiratory Disease 107:9–21.

Kubica, G. P. 1978. Classification and nomenclature of the mycobacteria. Annales de Microbiologie 129A:7–12.

Kubica, G. P. 1979. The current nomenclature of the mycobacteria—1978. Bulletin of the International Union Against Tuberculosis 54:204–211.

Kubica, G. P., Beam, R. E., Palmer, J. W. 1963. A method for the isolation of unclassified acid-fast bacilli from soil and water. American Review of Respiratory Disease 88:718–720.

Kubica, G. P., Dye, W. E. 1967. Laboratory methods for clinical and public health mycobacteriology. Public Health Service Publication No. 1547. [Out of print.] Atlanta, Georgia: Center for Disease Control.

Kubica, G. P., Kaufmann, A. J., Dye, W. E. 1964. Comments on use of the new mucolytic agent, *N*-acetyl-L-cysteine, as a sputum digestant for the isolation of mycobacteria. American Review of Respiratory Disease 89:284–286.

Kubica, G. P., Dye, W. E., Cohn, M. L., Middlebrook, G. 1963. Sputum digestion and decontamination with *N*-acetyl-L-cysteine—sodium hydroxide for culture of mycobacteria. American Review of Respiratory Disease 87:775–779.

Kubica, G. P., Baess, I., Gordon, R. E., Jenkins, P. A., Kwapinski, J. B. G., McDurmont, C., Pattyn, S. R., Saito, H., Silcox, V., Stanford, J. L., Takeya, K., Tsukamura, M. 1972. A co-operative numerical analysis of rapidly growing mycobacteria. Journal of General Microbiology 73:55–70.

Kubica, G. P., Gross, W. M., Hawkins, J. E., Sommers, H. M., Vestal, A. L., Wayne, L. G. 1975. Laboratory services for mycobacterial diseases. American Review of Respiratory Diseases 112:773–787.

Larsen, A. B., Moon, H. W., Merkal, R. S. 1971. Susceptibility of swine to *Mycobacterium paratuberculosis*. American Journal of Veterinary Research 32:589–595.

Larsson, L., Mårdh, P.-A. 1976. Gas chromatographic characterization of mycobacteria: Analysis of fatty acids and trifluoroacetylated whole-cell methanolysates. Journal of Clinical Microbiology 3:81–85.

Laskowski, L. F., Marr, J. J., Spernoga, J. F., Frank, N. J., Barner, H. B., Kaiser, G., Tyras, D. H. 1977. Fastidious mycobacteria grown from prosthetic-heart-valve cultures. New England Journal of Medicine 297:101–102.

Latt, R. H. 1975. Runyon group III atypical mycobacteria as a cause of tuberculosis in a rhesus monkey. Laboratory Animal Science 25:206–209.

Lechevalier, M. P., Horan, A. C., Lechevalier, H. 1971. Lipid composition in the classification of nocardiae and mycobacteria. Journal of Bacteriology **105**:313–318.

Levy, C., Curtin, J. A., Watkins, A., Marsh, B., Garcia, J., Mispireta, L. 1977. *Mycobacterium chelonei* infection of porcine heart valves. New England Journal of Medicine **297**:667–668.

Linell, F., Norden, A. 1954. *Mycobacterium balnei*—a new acid-fast bacillus occurring in swimming pools and capable of producing skin lesions in humans. Acta Tuberculosa Scandinavica, Suppl. **33**:1–84.

Lohr, D. C., Goeken, J. A., Doty, D. B., Donta, S. T. 1978. *Mycobacterium gordonae* infection of a prosthetic aortic valve. Journal of the American Medical Association **239**:1528–1530.

Loudon, R. G., Williamson, J., Johnson, J. M. 1958. An analysis of 3485 tuberculosis contacts in the city of Edinburgh during 1954–55. American Review of Tuberculosis **77**:623–643.

Lowell, A. M. 1976. Tuberculosis in the world. HEW Publication No. CDC 76-8317. Washington, D.C.: U.S. Government Printing Office.

McClatchy, J. K., Waggoner, R. F., Kanes, W., Cernich, M. S., Bolton, T. L. 1976. Isolation of mycobacteria from clinical specimens by use of a selective 7H-11 medium. American Journal of Clinical Pathology **65**:412–415.

McDaniel, R. E., Abensohn, M. K., Spoon, D. R., Kobayashi, G. S., Medoff, G., Marr, J. J. 1977. A rapid radiometric method for determining the sensitivity of clinical isolates of *Mycobacterium tuberculosis* to several chemotherapeutic agents. Journal of Laboratory and Clinical Medicine **89**:861–867.

McLaughlin, R. M., Thoenig, J. R., Marrs, G. E., Jr. 1976. A comparison of several intradermal tuberculins in *Macaca mulatta* during an epizootic of tuberculosis. Laboratory Animal Science **26**:44–50.

McSwiggan, D. A., Collins, C. H. 1974. The isolation of *M. kansasii* and *M. xenopi* from water systems. Tubercle **55**:291–297.

Magnusson, M. 1961. Specificity of mycobacterial sensitins. I. Studies in guinea pigs with purified "tuberculins" prepared from mammalian and avian tubercle bacilli, *Mycobacterium balnei,* and other acid-fast bacilli. American Review of Respiratory Disease **83**:57–68.

Marks, J., Jenkins, P. A., Schaefer, W. B. 1971. Thin layer chromatography of mycobacterial lipids as an aid to classification: Technical improvements; *Mycobacterium avium, M. intracellulare* (Battey bacilli). Tubercle **52**:219–225.

Marks, J., Palfreyman, J., Schaefer, W. B. 1977. A differential tuberculin test for mycobacterial infection in children. Tubercle **58**:19–23.

Matajack, M. L., Bissett, M. L., Schifferle, D., Wood, R. M. 1973. Evaluation of a selective medium for mycobacteria. American Journal of Clinical Pathology **59**:391–397.

Matthews, P. R. J., Collins, P., Jones, P. W. 1976. Isolation of mycobacteria from dairy creamery effluent sludge. Journal of Hygiene **76**:407–413.

Mead, G. R., Woodhams, A. W. 1964. *N*-Acetyl-L-cysteine as liquefying agent in the bacteriological examination of sputum. Tubercle **45**:370–373.

Meissner, G., Anz, W. 1977. Sources of *Mycobacterium avium* complex infection resulting in human disease. American Review of Respiratory Disease **116**:1057–1064.

Meissner, G., Schröder, K.-H. 1975. Relationship between *Mycobacterium simiae* and *Mycobacterium habana*. American Review of Respiratory Disease **111**:196–200.

Meissner, G., Schröder, K.-H., Amadio, G. E., Anz, W., Chaparas, S., Engel, H. W. B., Jenkins, P. A., Käppler, W., Kleeberg, H. H., Kubala, E., Kubin, M., Lauterbach, D., Lind, A., Magnusson, M., Mikova, Z. D., Pattyn, S. R., Schaefer, W. B., Stanford, J. L., Tsukamura, M., Wayne, L. G., Willers, I., Wolinsky, E. 1974. A co-operative numerical analysis of nonscoto- and nonphoto-chromogenic slowly

growing mycobacteria. Journal of General Microbiology **83**:207–235.

Merkal, R. S., Curran, B. J. 1974. Growth and metabolic characteristics of *Mycobacterium paratuberculosis*. Applied Microbiology **28**:276–279.

Merkal, R. S., Thurston, J. R. 1966. Comparison of *Mycobacterium paratuberculosis* and other mycobacteria, using standard cytochemical tests. American Journal of Veterinary Science **27**:519–521.

Merkal, R. S., Kopecky, K. E., Larsen, A. B., Thurston, J. R. 1964. Improvements in the techniques for primary cultivation of *Mycobacterium paratuberculosis*. American Journal of Veterinary Research **25**:1290–1293.

Middlebrook, G., Reggiardo, Z., Tigertt, W. D. 1977. Automatable radiometric detection of growth of *Mycobacterium tuberculosis* in selective media. American Review of Respiratory Disease **115**:1066–1069.

Mills, C. C. 1972. Occurrence of *Mycobacterium* other than *Mycobacterium tuberculosis* in the oral cavity and in the sputum. Applied Microbiology **24**:307–310.

Mitchell, M. D., Huff, I. H., Thoen, C. O., Himes, E. M., Howder, J. W. 1975. Swine tuberculosis in South Dakota. Journal of the American Veterinary Medical Association **167**:152–153.

Mitchison, D. A. 1968. Examination of sputum by smear and culture in case-finding. Bulletin of the International Union Against Tuberculosis **41**:139–147.

Mitchison, D. A., Allen, B. W., Lambert, R. A. 1973. Selective media in the isolation of tubercle bacilli from tissues. Journal of Clinical Pathology **26**:250–252.

Mitchison, D. A., Allen, B. W., Carrol, L., Dickinson, J. M., Aber, V. R. 1972. A selective oleic acid albumin agar medium for tubercle bacilli. Journal of Medical Microbiology **5**:165–175.

Mollohan, C. S., Romer, M. S. 1951. Public health significance of swimming pool granuloma. American Journal of Public Health **51**:883–891.

Montali, R. J., Bush, M., Thoen, C. O., Smith, E. 1976. Tuberculosis in captive exotic birds. Journal of the American Veterinary Medical Association **169**:920–927.

Muscoplat, C. C., Thoen, C. O., Chen, A. W., Johnson, D. W. 1975a. Development of specific in vitro lymphocyte responses in cattle infected with *Mycobacterium bovis* and *Mycobacterium avium*. American Journal of Veterinary Research **36**:395–398.

Muscoplat, C. C., Thoen, C. O., McLaughlin, R. M., Thoenig, J. R., Chen, A. W., Johnson, D. W. 1975b. Comparison of lymphocyte stimulation and tuberculin skin reactivity in *Mycobacterium bovis*-infected *Macaca mulatta*. American Journal of Veterinary Research **36**:699–702.

Muscoplat, C. C., Thoen, C. O., Chen, A. W., Rakich, P. M., Johnson, D. W. 1975c. Development of specific lymphocyte immunostimulation and tuberculin skin reactivity in swine infected with *Mycobacterium bovis* and *Mycobacterium avium*. American Journal of Veterinary Research **36**:1167–1172.

Mycobacteriology Diagnostic Services Unit. 1971. Laboratory methods in veterinary mycobacteriology. Ames, Iowa: Animal Health Division, National Animal Disease Laboratory, Agricultural Research Service, U.S. Department of Agriculture.

Norlin, M. 1965. Unclassified mycobacteria, a comparison between a serological and a biochemical classification method. Bulletin of the International Union Against Tuberculosis **36**:25–32.

Ohashi, D. K., Wade, T. J., Mandle, R. J. 1977. Characterization of ten species of mycobacteria by reaction-gas liquid chromatography. Journal of Clinical Microbiology **6**:469–473.

Oliver, J., Reusser, T. R. 1942. Rapid method for the concentration of tubercle bacilli. American Review of Tuberculosis **45**:450–452.

Ortbals, D. W., Marr, J. J. 1978. A comparative study of tuberculous and other mycobacterial infections and their associations with malignancy. American Review of Respiratory Disease 117:39–45.

Palmer, C. E., Edwards, L. B. 1966. Sensitivity to mycobacterial PPD antigens with some laboratory evidence of its significance. Tuberkuloza 18:193–200.

Palmer, C. E., Edwards, L. B. 1967. Tuberculin test in retrospect and prospect. Archives of Environmental Health 15:793–808.

Pattyn, S. R. 1966. A study of some strains of *Mycobacterium xenopei*. Zentralblatt für Bakteriologie, Parasitenkunde, Infektionskrankheiten und Hygiene, Abt. 1 Orig. 201:246–252.

Pattyn, S. R., Portaels, F. 1972. Identification and clinical significance of mycobacteria. Zentralblatt für Bakteriologie, Parasitenkunde, Infektionskrankheiten und Hygiene, Abt. 1 Orig., Reihe A 219:114–140.

Pattyn, S. R., Portaels, F., Spanoghe, L., Magos, J. 1970. Further studies on African strains of *Mycobacterium tuberculosis*. Comparison with *M. bovis* and *M. microti*. Annales des Sociétés Belges de Médecine Tropicale 50:211–228.

Peavy, G. M., Silvermann, S., Howard, E. B., Cooper, R. S., Rich, L. J., Thomas, G. N. 1976. Pulmonary tuberculosis in a sulfur-crested cocatoo. Journal of the American Veterinary Medical Association 169:915–919.

Petran, E. I., Vera, H. D. 1971. Media for selective isolation of mycobacteria. Health Laboratory Science 8:225–230.

Prather, E. C., Bond, J. O., Hartwig, E. C., Dunbar, F. P. 1961. Preliminary report: Epidemiology of infections due to the atypical acid-fast bacilli. Diseases of the Chest 39:129–139.

Rabinowitsch, L. 1903. Ueber eine durch säure-feste Bakterien hervorgerufene Hauterkrankung der Ratten. Centralblatt für Bakteriologie und Parasitenkunde, Abt. 1 33:577–580.

Rado, T. A., Bates, J. H., Engel, H. W. B., Mankiewicz, E., Murohashi, T., Mizuguchi, Y., Sula, L. 1975. World Health Organization studies on bacteriophage typing of mycobacteria. Subdivision of the species *Mycobacterium tuberculosis*. American Review of Respiratory Disease 111:459–468.

Renner, M., Bartholomew, W. R., 1974. Mycobacteriologic data from two outbreaks of bovine tuberculosis in nonhuman primates. American Review of Respiratory Disease 109:11–16.

Reznikov, M., Dawson, D. J. 1973. Serological examination of some strains that are in the *Mycobacterium avium-intracellulare-scrofulaceum* complex but do not belong to Schaefer's complex. Applied Microbiology 26:470–473.

Reznikov, M., Leggo, J. H. 1974. Examination of soil in the Brisbane area for organisms of the *Mycobacterium avium-intracellulare-scrofulaceum* complex. Pathology 6:269–273.

Reznikov, M., Leggo, J. H., Dawson, D. J. 1971. Investigation by seroagglutination of strains of the *Mycobacterium intracellulare-M. scrofulaceum* group from house dusts and sputum in southeastern Queensland. American Review of Respiratory Disease 104:951–953.

Reznikov, M., Leggo, J. H., Tuffley, R. E. 1971. Further investigations of an outbreak of mycobacterial lymphadenitis at a deep-litter piggery. Australian Veterinary Journal 47:622–623.

Reznikov, M., Robinson, E. 1970. Serologically identical Battey mycobacteria from sputa of healthy piggery workers and lesions of pigs. Australian Veterinary Journal 46:606–607.

Rich, A. R. 1951. The pathogenesis of tuberculosis. Springfield, Illinois: Charles C Thomas.

Robicsek, F., Daugherty, H. K., Cook, J. W., Selle, J. G., Masters, T. N., O'Bar, P. R., Fernandez, C. R., Mauney, C. U., Calhoun, D. M. 1978. *Mycobacterium fortuitum* epidemics after open-heart surgery. Journal of Thoracic and Cardiovascular Surgery 75:91–96.

Ross, A. J. 1959. *Mycobacterium fortuitum* Cruz from the tropical fish *Hyphessobrycon innesi*. Journal of Bacteriology 78:392–395.

Ruch, T. C. 1959. Diseases of laboratory primates. Philadelphia: W. B. Saunders.

Runyon, E. H. 1959. Anonymous mycobacteria in pulmonary disease. Medical Clinics of North America 43:273–290.

Runyon, E. H. 1965. Pathogenic mycobacteria. Advances in Tuberculosis Research 14:235–287.

Runyon, E. H. 1968. Aerial hyphae of *Mycobacterium xenopi*. Journal of Bacteriology 95:734–735.

Runyon, E. H. 1974. Ten mycobacterial pathogens. Tubercle 55:235–240.

Runyon, E. H., Wayne, L. G., Kubica, G. P. 1974. Mycobacteriaceae, pp. 681–701. In: Buchanan, R. E., Gibbons, N. E. (eds.), Bergey's manual of determinative bacteriology, 8th ed. Baltimore: Williams & Wilkins.

Runyon, E. H., Karlson, A. G., Kubica, G. P., Wayne, L. G. 1974. *Mycobacterium*, pp. 148–174. In: Lennette, E. H., Spaulding, E. H., Truant, J. P. (eds.), Manual of clinical microbiology, 2nd ed. Washington, D.C.: American Society for Microbiology.

Saito, H., Tsukamura, M. 1976. *Mycobacterium intracellulare* from public water bath. Japanese Journal of Microbiology 20:561–563.

Saito, H., Gordon, R. E., Juhlin, I., Käppler, W., Kwapinski, J. B. G., McDurmont, C., Pattyn, S. R., Runyon, E. H., Stanford, J. L., Tarnok, I., Tasaka, H., Tsukamura, M., Weiszfeiler, J. 1977. Cooperative numerical analysis of rapidly growing mycobacteria. The second report. International Journal of Systematic Bacteriology 27:75–85.

Schaefer, W. B., Wolinsky, E., Jenkins, P. A., Marks, J. 1973. *Mycobacterium szulgai*—a new pathogen. Serologic identification and report of five new cases. American Review of Respiratory Disease 108:1320–1326.

Seibert, F. B. 1934. The isolation and properties of the purified protein derivative of tuberculin. American Review of Tuberculosis, Suppl. 30:713–720.

Seibert, F. B., Glenn, J. T. 1941. Tuberculin purified protein derivative. Preparation and analyses of a large quantity for standard. American Review of Tuberculosis 44:9–25.

Selkon, J. B., Ingham, H. R., Hale, J. H., Codd, A. A. 1966. N-Acetyl-L-cysteine in culturing sputum for tubercle bacilli. Tubercle 47:269–272.

Sesline, D. H., Schwartz, L. W., Thoen, C. O., Osburn, B. I., Terrell, T., Holmberg, C., Anderson, S. H., Henrickson, R. V. 1975. *Mycobacterium avium* infection in three rhesus monkeys. Journal of the American Veterinary Medical Association 167:639–645.

Shaw, J. B., Wynn-Williams, N. 1954. Infectivity of pulmonary tuberculosis in relation to sputum status. American Review of Tuberculosis 69:724–732.

Simon, H. B., Weinstein, A. J., Pasternak, M. S., Swartz, M. N., Kunz, L. J. 1977. Genitourinary tuberculosis. Clinical features in a general hospital. American Journal of Medicine 63:410–420.

Skerman, V. B. D., McGowan, V., Sneath, P. H. A. 1980. Approved list of bacterial names. International Journal of Systematic Bacteriology 30:225–420.

Smith, E. K., Hunt, R. D., Garcia, F. G., Fraser, C. E. O., Merkal, R. S., Karlson, A. G. 1973. Avian tuberculosis in monkeys. A unique mycobacterial infection. American Review of Respiratory Disease 107:469–471.

Smith, T. 1898. A comparative study of bovine tubercle bacilli and of human bacilli from sputum. Journal of Experimental Medicine 3:451–511.

Smith, T. 1904-05. Studies in mammalian tubercle bacilli. III. Description of a bovine bacillus from the human body. A culture test for distinguishing the human from the bovine type of bacilli. Journal of Medical Research 13:253–300.

Smithwick, R. W. 1976. Laboratory manual for acid-fast microscopy, 2nd ed. Atlanta, Georgia: Center for Disease Control.

Smithwick, R. W., Stratigos, C. B., David, H. L. 1975. Use of cetylpyridinium chloride and sodium chloride for the decontamination of sputum specimens that are transported to the laboratory for the isolation of *Mycobacterium tuberculosis*. Journal of Clinical Microbiology 1:411–413.

Snider, W. R. 1971. Tuberculosis in canine and feline populations. Review of the literature. American Review of Respiratory Disease **104**:877–887.

Snider, W. R., Cohen, D., Reif, J. S., Stein, S. C., Prier, J. E. 1971. Tuberculosis in canine and feline populations. Study of high risk populations in Pennsylvania, 1966–68. American Review of Respiratory Disease **104**:866–876.

Snow, G. A. 1954. Mycobactin, a growth factor for *Mycobacterium johnei*. Journal of the Chemical Society **5512**:4080–4093.

Stanford, J. L., Beck, A. 1968. An antigenic analysis of mycobacteria, *Mycobacterium fortuitum, Myco. kansasii, Myco. phlei, Myco. smegmatis* and *Myco. tuberculosis*. Journal of Pathology and Bacteriology **95**:131–139.

Stefansky, W. K. 1903. Eine lepraähnliche Erkrankung der Haut und der Lymphdrüsen bei Wanderratten. Centralblatt für Bakteriologie und Parasitenkunde, Abt. 1 **33**:481–487.

Strauss, I., Gameleïa, N. 1891. Recherches expérimentales sur la tuberculose; la tuberculose humaine; sa distinction de la tuberculose des oiseaux. Archives de Médecine Expérimentales et d'Anatomie Pathologique **3**:457–484.

Tacquet, A., Tison, F. 1961. Nouvelle technique d'isolement des mycobactéries par le lauryl-sulfate de sodium. Annales de l'Institut Pasteur **100**:676–680.

Tacquet, A., Tison, F., Polspoel, B. 1965. L'utilisation des détergents pour l'isolement des mycobactéries à partir de produits pathologiques. Étude comparative—applications pratiques. Annales de l'Institut Pasteur **16**:21–30.

Takeya, K., Nakayama, Y., Muraoka, S. 1970. Specificity in skin reaction to tuberculin protein prepared from rapidly growing mycobacteria and some nocardia. American Review of Respiratory Disease **102**:982–986.

Tellis, C. J., Beechler, C. R., Ohashi, D. K., Fuller, S. A. 1977. Pulmonary disease caused by *Mycobacterium xenopi*. American Review of Respiratory Disease **116**:779–783.

Thoen, C. O., Himes, E. M., Stumpff, C. D., Pards, T. W., Sturkie, H. N. 1977. Isolation of *Mycobacterium bovis* from the prepuce of a herd bull. American Journal of Veterinary Research **38**:877–878.

Thorel, M. F., Boisvert, H. 1974. Abcès du chat à *Mycobacterium chelonei*. Bulletin de l'Académie Vétérinaire **47**:415–422.

Tison, F., Devulder, B., Tacquet, A. 1968. Recherches sur la présence de mycrobactéries dans la nature. Revue de la Tuberculose (Paris) **32**:893–902.

Tsukamura, M. 1967. Identification of mycobacteria. Tubercle **48**:311–338.

Tsukamura, M., Mizuno, S., Murata, H., Nemoto, H., Yugi, H. 1974. A comparative study of mycobacteria from patients' room dusts and from sputa of tuberculous patients. Japanese Journal of Microbiology **18**:271–277.

Valdivia, A. J., Suarez, M. R., Echemendía, F. M. 1971. *Mycobacterium habana*: Probable nueva especie dentro de las microbacterias no clasificadas. Boletin Higiene y Epidimiológia **9**:65.

Vandiviere, H. M., Smith, C. E., Sunkes, E. J. 1952. Evaluation of four methods of collecting and mailing gastric washings for tubercle bacilli. American Review of Tuberculosis **65**:617–626.

Vestal, A. L. 1975. Procedures for the isolation and identification of mycobacteria. DHEW Publication No. (HSM) 75-8230. Atlanta, Georgia: Center for Disease Control.

Viallier, J., Joubert, L. 1974. Tuberculosis and mycobacterioses: From animals to man. [In French.] Lyon Medical **232**:597–602.

Wayne, L. G., 1957. The use of Millipore filters in clinical laboratories. American Journal of Clinical Pathology **28**:565–567.

Wayne, L. G., Doubek, J. R. 1968. Diagnostic key to mycobacteria encountered in clinical laboratories. Applied Microbiology **16**:925–931.

Wayne, L. G., Krasnow, I., Kidd, G. C. 1962. Finding the "hidden positive" in tuberculosis eradication programs. The role of the sensitive trisodium phosphate-benzalkonium (Zephiran) culture technique. American Review of Respiratory Disease **86**:537–541.

Wayne, L. G., Dietz, T. M., Gernez-Rieux, C., Jenkins, P. A., Käppler, W., Kubica, G. P., Kwapinski, J. B. G., Meissner, G., Pattyn, S. R., Runyon, E. H., Schröder, K. H., Silcox, V. A., Tacquet, A., Tsukamura, M., Wolinsky, E. 1971. A co-operative numerical analysis of scotochromogenic slowly growing mycobacteria. Journal of General Microbiology **66**:255–271.

Wayne, L. G., Engbaek, H. C., Engel, H. W. B., Froman, S., Gross, W., Hawkins, J., Käppler, W., Karlson, A. G., Kleeberg, H. H., Krasnow, I., Kubica, G. P., McDurmont, C., Nel, E. E., Pattyn, S. R., Schröder, K. H., Showalter, S., Tarnok, I., Tuskamura, M., Vergmann, B., Wolinsky, E. 1974. Highly reproducible techniques for use in systematic bacteriology in the genus *Mycobacterium*: Tests for pigment, urease, resistance to sodium chloride, hydrolysis of Tween 80, and β-galactosidase. International Journal of Systematic Bacteriology **24**:412–419.

Wayne, L. G., Engel, H. W. B., Grassi, C., Gross, W., Hawkins, J., Jenkins, P. A., Käppler, W., Kleeberg, H. H., Krasnow, I., Nel, E. E., Pattyn, S. R., Richards, P. A., Showalter, S., Slosarek, M., Szabo, I., Tarnok, I., Tsukamura, M., Vergmann, B., Wolinsky, E. 1976. Highly reproducible techniques for use in systematic bacteriology in the genus *Mycobacterium*: Tests for niacin and catalase and for resistance to isoniazid, thiophene-2-carboxylic acid hydrazide, hydroxylamine, and *p*-nitrobenzoate. International Journal of Systematic Bacteriology **26**:311–318.

Wheeler, W. C., Hanks, J. H. 1965. Utilization of external growth factors by intracellular microbes: *Mycobacterium paratuberculosis* and wood pidgeon mycobacteria. Journal of Bacteriology **89**:889–896.

Whitcomb, F. C., Foster, M. C., Dukes, C. N. 1962. Increased carbon dioxide tension and the primary isolation of mycobacteria. American Review of Respiratory Disease **86**:584–586.

Wolinsky, E. 1977. Mycobacteria: Significance of speciation and sensitivity tests, pp. 115–121. In: Lorian, V. (ed.), Significance of medical microbiology in the care of patients. Baltimore: Williams & Wilkins.

Wolinsky, E., Rynearson, T. K. 1968. Mycobacteria in soil and their relation to disease-associated strains. American Review of Respiratory Disease **97**:1032–1037.

Wolinsky, E., Schaefer, W. B. 1973. Proposed numbering scheme for mycobacterial serotypes by agglutination. International Journal of Systematic Bacteriology **23**:182–183.

Yates, M. D., Collins, C. H., Grange, J. M. 1978. Differentiation of BCG from other variants of *Mycobacterium tuberculosis* isolated from clinical material. Tubercle **59**:143–146.

Yeager, H., Jr., Lacy, J., Smith, L. R., LeMaistre, C. A. 1967. Quantitative studies of mycobacterial populations in sputum and saliva. American Review of Respiratory Disease **95**:998–1004.

Yue, W. Y., Cohen, S. S. 1966. Pulmonary infection caused by niacin-positive *Mycobacterium kansasii*. American Review of Respiratory Disease **94**:447–449.

Zvetina, J. R., Wichelhausen, R. H. 1976. Pulmonary infection caused by niacin-positive *Mycobacterium avium*. American Review of Respiratory Disease **113**:885–887.

CHAPTER 151

Mycobacterium leprae

CHARLES C. SHEPARD

Mycobacterium leprae is the cause of leprosy, a human disease of skin, peripheral nerves, and (in the lepromatous form of leprosy) the nasal mucous membranes. Although claims that the organism has been grown on artificial medium have frequently been made in the literature, this probably has not yet been accomplished. It can be grown in the cooler tissues of laboratory mice and rats and systemically in armadillos. Although leprosy has become an infrequent disease in the temperate zones, it is a frequent disease in many tropical and developing countries, and therefore it is an important world health problem. The World Health Organization estimated that there were about 11 million cases in the world in 1965 (Bechelli and Martínez Domínguez, 1966), and there have been no important changes since.

Several properties of the organism determine the chief features of the disease.

1. The very slow growth of the organism causes very long incubation periods and a very chronic course of the disease in man and in experimental animals.
2. *M. leprae* frequently invades peripheral nerves and, especially in tuberculoid disease, anesthesia and paralysis often result, presumably because of the vigorous tissue reaction in this form of the disease. *M. leprae* is the only bacterium that regularly involves nervous tissue, and the resultant crippling is of economic significance in endemic regions.
3. The temperature preference of the organism is less than 37°C; hence the tissues most severely affected are the cooler tissues in man and in experimental animals (Shepard, 1965).
4. *M. leprae* is capable of producing a form of immunological tolerance in persons with severe (lepromatous) leprosy, and such infections are the chief sources of new human infections.

Because *M. leprae* has not been grown on artificial medium, little is known about its metabolic activities, and its behavior cannot be described in terms of biochemical reactions. The only way of identifying gene products is through immunological identification of antigens. Consequently, this description of *M. leprae* has to emphasize its behavior in its natural and experimental hosts.

Habitat

The habitat is infected humans. Possible exceptions are noted below. The concentration of *Mycobacterium leprae* in the patient's tissues varies with different forms of the disease, which are most usefully considered according to the immunopathological classifications of Ridley and co-workers (Ridley, 1974; Ridley and Jopling, 1966; Ridley and Waters, 1969). An exception to their classification is indeterminate leprosy, which consists typically of a single, small, depigmented plaque that lasts for a year or two before the patient develops disease typical of the Ridley classification. In this classification, the form of the disease varies along a spectrum that extends from tuberculoid (high resistance) to lepromatous (low resistance) leprosy. Patients with fully tuberculoid disease characteristically have a single erythematous plaque with raised outer edges and a flattened, clearing center. Involvement of the peripheral nerves and crippling are frequent. With fully lepromatous disease, skin involvement is extensive, bilaterally symmetrical, and of an infiltrative or edematous nature. Nodules with poorly defined borders on an infiltrated base may be present. Involvement of the peripheral nerves is relatively less severe than it is in tuberculoid disease, so crippling is less prominent. Histologically, one sees in patients with tuberculoid disease an epithelioid cell granuloma with many lymphocytes, and in those with lepromatous disease a macrophage granuloma with no epithelioid cells and few lymphocytes. Between tuberculoid and lepromatous disease, there is borderline leprosy, and the spectrum is conveniently divided into five groups: TT, fully tuberculoid; BT, borderline tuberculoid; BB, borderline; BL, borderline lepromatous; and LL, fully lepromatous. The concentration of bacilli in TT leprosy disease is frequently below the limit of microscopic detectability—therefore less than 1×10^5 *M. leprae* per gram of tissue. In lepromatous disease, the usual concentration is about 6×10^8 per gram, and concentrations as high as 5×10^9 per gram are seen (Collaborative Effort, 1975).

Ulceration and resultant excretion of bacilli from skin lesions in patients with lepromatous leprosy are infrequent, however. The nasal mucous membranes are constantly affected in lepromatous patients, and daily excretions from this source, with or without detectable ulceration of the membranes, average 10^8 *M. leprae,* or about the same rate as that of tubercle bacilli in the sputum of patients with pulmonary tuberculosis (Davey and Rees, 1974; Rees and Meade, 1974; Shepard, 1958; Shepard, 1962). The analogy of the excretion of *M. leprae* in nasal discharges in leprosy to the excretion of *M. tuberculosis* in the sputum in tuberculosis was pointed out many years ago by Robert Koch (Koch, 1897). Although with both diseases the organism is distributed widely in the body, the lesion that is most important for the spread to other susceptible persons is the pulmonary cavity in tuberculosis and the nasal mucous membrane involvement in leprosy. In tuberculosis, the microorganism produces delayed-type hypersensitivity, and the resultant tissue necrosis in lung lesions produces cavities in which the organism thrives and from which it readily gains egress through the airways. In lepromatous leprosy, sensitization does not occur, but the cool temperature of the nasal mucous membranes leads to heavy involvement, with shedding of the organism in nasal excretions. Viable *M. leprae* may be demonstrated in dried nasal excretions for several days (Davey and Rees, 1974). *M. leprae* has also been demonstrated on arthropods that have fed on lepromatous patients (Narayanan et al., 1977), but transmission by arthropods is probably infrequent.

Isolation and Propagation

At present *Mycobacterium leprae* cannot be grown in artificial medium, and it can be grown experimentally only in animals. Moreover, it has been difficult to preserve viable *M. leprae* consistently by freezing them (or by freeze-drying), and fresh clinical material is not frequently available in many medical centers. These technical constraints, plus the slow growth of the organism, make experimentation with *M. leprae* quite laborious. Hence, the best method must be chosen for the particular purpose.

Much of the work depends upon counts of *M. leprae.* Several methods are available (Hanks, Chatterjee, and Lechat, 1964; Hart and Rees, 1960; Shepard and McRae, 1968). For work with *M. leprae,* one needs to have access to clinical materials or to maintain strains in continuous passage.

The organism can be grown only in mice, rats, and armadillos. The particular advantages and disadvantages to each animal model are listed in Table

1. The most frequently used model is the normal mouse infected in the foot pad (Shepard, 1960). Infections can be initiated with only 1–10 bacilli, but the infections are more uniform with inocula of 5×10^3 to 1×10^4 organisms. Following such inoculation, there is first a stationary phase in which no multiplication is detected, and then a logarithmic phase during which multiplication with a doubling time of 11–13 days (Levy, 1976; Shepard and McRae, 1965) continues until the number of bacilli in the foot pad reaches a level of about $1-2 \times 10^6$ 150–180 days after inoculation. Multiplication then stops, apparently because cell-mediated immunity has been triggered, and lymphocytes infiltrate the area. Enlargement of the macrophages in the center of the lesion occurs at about the same time. The growth curve is consistent enough between animals to permit estimation of the numbers of organisms surviving a drug regimen (Shepard, 1969) or numbers initially inoculated (Levy, 1976). The number of viable bacilli decreases markedly at the end of the logarithmic phase. The growth curve and histology of the mouse foot pad infection are distinctive enough to allow the differentiation of *M. leprae* from all other known mycobacteria. Most mycobacteria do not grow when inoculated into the foot pad. A few mycobacterial skin pathogens do, e.g., *M. marinum* and *M. ulcerans,* but they grow more rapidly than *M. leprae,* and the histopathology is characterized by granulocytic infiltration and abscess formation. *M. lepraemurium* is harder to distinguish, but it grows somewhat more rapidly, does not produce the small globi that characterize the *M. leprae* infections, and spreads to the abdomen, where it produces characteristic granulomatous lesions in the omentum, spleen, and liver.

Thus the normal mouse has been used to best advantage in studies such as the screening of drugs for activity against *M. leprae,* determination of the minimal inhibitory concentration, the rate of bactericidal activity of drugs, vaccination experiments, detection of drug-resistant forms, methods of killing the organism in suspension, and isolation and identification of *M. leprae* in environmental samples.

The lines of mice used most frequently have been CFW, CBA, and BALB/c, and these three appear to be the most susceptible (Shepard and Habas, 1967). Air temperatures of 20–25°C, which produce foot pad temperatures averaging 30°C, are the most favorable for *M. leprae* infections (Shepard and Habas, 1967) and are good temperatures for the maintenance of the mice.

The thymectomized-irradiated mouse is a model frequently used in immunology and is called variously T-cell depleted, T-cell deprived, etc. Mice of 6–8 weeks are thymectomized, X-irradiated with about 950 rad, and transfused with about 5×10^6 nucleated, syngeneic, bone marrow cells. Somewhat better mouse survival is seen with a similar

Table 1. Advantages and disadvantages of the various animals used for work with *Mycobacterium leprae*.

Advantages	Disadvantages
A. Mice, normal	
1. The most convenient and available laboratory animal.	1. Maximal bacillary population only 10^6 per foot pad (10^7/g tissue).
2. Foot pad infection is highly reproducible during logarithmic phase.	2. After reaching maximal population, the proportion of bacilli that are viable decreases distinctly.
3. Genetically identical animals available in large numbers as inbred lines.	3. Infection metastasizes from foot pad only irregularly.
4. Minimal infectious dose only 1–10 bacilli.	
5. Peak bacillary numbers within 6 mo. after inoculation of 10^3–10^4 bacilli.	
6. Immunology of mice is better known than that of any other animal.	
B. Mice, adult thymectomized-irradiated-bone marrow replaced	
1. Maximal bacillary population is 10–100 times that in normal mice.	1. Many lots of prepared mice survive poorly.
2. Infection spreads from inoculated foot pad to other cool sites.	2. For peak populations, 2 years are needed.
3. Generalized infections can be achieved by intravenous and intraperitoneal inoculation with 10^6–10^7 bacilli.	3. Advanced infections develop in only a small proportion.
4. Histopathology resembles that of lepromatous leprosy in man.	
C. Rats, normal	
1. Similar to those of normal mice.	1. Not as convenient as mice and fewer inbred lines are available.
D. Rats, infant thymectomized Lewis	
1. Bacillary population reaches 10^8–10^9 per foot pad.	1. Thymectomy technique is demanding.
2. Minimal infective dose as small as for mice.	2. For peak bacillary numbers, 2 years or more are needed.
3. Advanced infections are produced more consistently than those in thymectomized-irradiated mice.	
4. Animals survive well.	
E. Armadillos (*Dasypus novemcinctus*)	
1. Bacillary population averages 5×10^9 per gram of infected tissue.	1. Inconvenient to keep and handle.
2. Liver and spleen are heavily infected.	2. Cannot be bred in captivity. Wild-caught animals vary genetically and in infectious background.
3. Histology is similar to that of humans with lepromatous disease.	3. Wild-caught animals reported to have infections by organism similar to or identical with *M. leprae*.
4. Genetically identical animals are available as uniovular quadruplets.	

model, in which the mouse is thymectomized at the same age and then given five irradiation treatments with 200 rad at 2-week intervals.

Thymectomized-irradiated mice have been used mostly for studies of histopathology. The evidence on whether they are more, or less, susceptible than normal mice to infection with small numbers of *M. leprae* is contradictory.

Foot pad infections in normal rats are similar to those in normal mice (Hilson, 1965), so rats have not been used except for special purposes (Gordon et al., 1975).

Although neonatally thymectomized mice develop wasting disease and die in a few months, neonatally thymectomized rats live approximately normal life spans (Fieldsteel and McIntosh, 1971), apparently because some traffic of T cells out of the thymus occurs before birth. Neonatally thymectomized Lewis rats (NTLR) develop bacterial populations of 10^8–10^9 per foot pad. Consequently, NTLR and thymectomized-irradiated mice are being used in two experimental situations. The first involves the inoculation of large numbers of M. leprae so that a very small fraction of viable bacteria can be detected. Because the maximum population in normal mice and rats averages 10^6, the highest practical inoculum is 1–5×10^4; with higher inocula it is difficult to know whether multiplication has occurred. In addition, the mice may be immunized. In NTLR, higher inocula can be used because the maximum population reaches a higher level, and it has been shown that no immunity is generated by the foot pad inoculation of 10^6 heat-killed M. leprae in NTLR (Fieldsteel and Levy, 1976).

The chief use of armadillos has been for the production of large numbers of M. leprae. With an average of 5×10^9 bacilli per gram of liver, which averages nearly 200 g per armadillo, about 1×10^{12} M. leprae can be obtained from one animal. Such large amounts of bacilli make certain types of experimentation possible for the first time, especially in the field of immunology.

The first inoculations of M. leprae into armadillos were intradermal, and the disease progressed very slowly (Kirchheimer and Storrs, 1971). With intravenous inoculations of 10^8 M. leprae, however, most of the animals developed full-blown infections within 2 years (Kirchheimer and Sanchez, 1976). Because the infections are fatal to armadillos and postmortem invasion with other bacteria may be heavy, the monitoring of infections becomes important. Before the animal dies from M. leprae, the infection can be detected by examining the buffy coat (white cells) of peripheral blood tissue sections of ear snips, or ear skin smears prepared by scraped incision method.

Identification

The identification of Mycobacterium leprae is made more difficult by its noncultivability. The various procedures that have been used are listed in Table 2; the usefulness of each depends upon the type of material and number of bacilli available. These tests have been used chiefly in investigating new claims that M. leprae has been grown in animals or artificial medium. Sometimes some of the tests have been improperly used. For example, because the original inoculum from patients' tissues may contain large numbers of M. leprae and because M. leprae can persist in a microscopically and antigenically intact form for many months in a medium, the original inoculum itself may be responsible for positive lepromin or fluorescent antibody results. Perhaps half of the false claims involve this error. The other false claims involve a contaminant, frequently one of the common skin mycobacteria when skin biopsy specimens have been used, or M. lepraemurium when the materials have been inoculated into rats or mice. Strangely, such claims rely on the further claim that most of the properties of M. leprae change when it grows in a new environment, even though if such were true it would be very difficult indeed to determine whether or not the purported isolate arose from M. leprae. In the future it will probably be possible to investigate claims for the growth of M. leprae through studies of DNA, i.e., by determining the G+C ratio and by determining the degree of homology through hybridization techniques. The basic work, however, has not yet been done, and methods for extracting DNA from M. leprae, etc., have not yet been worked out.

Of the tests in Table 2, the first—noncultivability—is the easiest to perform. At least 10^6 acid-fast bacteria are needed to make sure that the inoculum contains sufficient bacilli. Probably the tests for cultivability and for growth in mouse foot pads

Table 2. Methods for identification of Mycobacterium leprae.

Test	Approximate number of M. leprae requested	Live M. leprae needed
Noncultivability	ca. 10^6	Yes
Lepromin reactivity	2×10^8	No
Growth pattern in mouse foot pads	3×10^{4a}	Yes
Invasion of peripheral nerves	?	Yes
DOPA-oxidase activity	4×10^9	Yes?
No elongation in Hart-Valentine medium	10^7	Yes
Specific fluorescent antibody stain	10^5	No
Sensitivity to drugs	3×10^4	Yes

[a] For the standard inoculum into 20 mice. Much smaller numbers, perhaps 3×10^1 M. leprae, would be needed to produce a typical infection, which could be passed to more mice for further tests.

should be carried out simultaneously to rule out the possibility that the bacilli are dead.

For lepromin reactivity, enough bacilli are required so that about 4×10^6 *M. leprae* are contained in the 0.1 ml to be injected into each patient. Hayashi (1933) first showed that *M. leprae* could be differentiated from several other bacteria on the basis of results obtained with this technique. Preferably about 30 tuberculoid and 30 lepromatous patients should be included. A standard lepromin should also be included, and the bacillary content of the standard lepromin and test material must be matched. The tests should be rotated randomly, so that the skin test reader does not know which antigen is responsible for the reaction he is measuring. The size of the induration should be recorded. The reaction should be read at various intervals, but the most important one is that read at 28 days (Mitsuda reaction). The standard lepromin must give a significant reaction in most of the tuberculoid patients and must give no reaction in all, or nearly all, of the lepromatous patients. The test antigen must give completely parallel results, and in the tuberculoid group, the size of the reactions to the standard and test antigens should be the same in each patient (Shepard and Guinto, 1963).

The third test, growth pattern in mouse foot pads, has been discussed above. It has the lowest requirement for bacilli.

The fourth test is based upon the observation that peripheral nerve invasion is a pathognomonic feature of human leprosy. It is not a frequent feature of the early foot pad lesion in the mouse, although it is commonly seen farther up the sciatic nerve after the first year (Wiersema, Binford, and Chang, 1965). In the thymectomized-irradiated mouse, nerve invasion at the foot pad level is prominent in the progressive infections seen at the end of the second year (Rees and Weddell, 1968). Nerve invasion is also a prominent feature of *M. leprae* infection in armadillos (Kirchheimer and Storrs, 1971).

DOPA-oxidase activity is attributed to *M. leprae* and to no other mycobacteria (Prabhakaran, 1967; Prabhakaran, Harris, and Kirchheimer, 1977). Other non-acid-fast bacteria, however, possess it. The relationship of the enzymatic activity to viability is not established, since some nonviable preparations have possessed activity and some viable preparations have failed to manifest activity.

M. lepraemurium elongates without dividing in Hart-Valentine medium, but *M. leprae* does not (Hart and Rees, 1968). Hence, incubation in this medium provides a means for differentiating these two species. In conjunction with this test, however, it would be necessary to demonstrate that a detectable proportion of the inoculated bacteria are viable.

Fluorescent antibody stains that are specific for *M. leprae* have been developed by Abe. One is a direct stain with antisera from rabbits immunized

against a leprosy nodule extract (Abe et al., 1972). The other is an indirect stain with leprosy patients' sera and anti–human conjugate (Abe et al., 1976). The specificity of the indirect stain has recently been improved by absorptions with disrupted BCG and *M. vaccae* (Abe, Yoshino, and Saito, 1978).

Because *M. leprae* has a unique pattern of susceptibility to drugs (Shepard, 1971), tests for antimicrobial sensitivity are often carried out to characterize isolates about which there is some question.

Literature Cited

Abe, M., Yoshino, Y., Saito, T. 1978. Antigenic specificity of *M. leprae* by indirect immunofluorescence. International Journal of Leprosy **47**:114–115.

Abe, M., Izumi, S., Saito, T., Mathur, S. K. 1976. Early serodiagnosis of leprosy by indirect immunofluorescence. Leprosy in India **48**:272–276.

Abe, M., Minagawa, F., Yoshino, Y., Okamura, K. 1972. Studies on the antigen specificity of *Mycobacterium leprae* II. Purification and immunological characterization of the soluble antigen in leprosy nodules. International Journal of Leprosy **40**:107–117.

Bechelli, L. M., Martínez Domínguez, V. 1966. The leprosy problem in the world. World Health Organization Bulletin **34**:811–826.

Collaborative Effort of the U.S. Leprosy Panel (U.S.-Japan Cooperative Medical Science Program) and the Leonard Wood Memorial, 1975. Rifampin therapy of lepromatous leprosy. The American Journal of Tropical Medicine and Hygiene **24**:475–484.

Davey, T. F., Rees, R. J. W. 1974. The nasal discharge in leprosy: Clinical and bacteriological aspects. Leprosy Review **45**:121–134.

Fieldsteel, A. H., Levy, L. 1971. Neonatally thymectomized Lewis rats infected with *Mycobacterium leprae*: Response to primary inocula, secondary challenge, and large inocula. Infection and Immunity **14**:736–741.

Fieldsteel, A. H., McIntosh, A. H. 1976. Effect of neonatal thymectomy and antithymocyte serum on susceptibility of rats to *Mycobacterium leprae* infections. Proceedings of the Society for Experimental Biology and Medicine **138**:408–413.

Gordon, G. R., Peters, J. H., Ghoul, D. C., Murray, J. F., Jr., Levy, L., Biggs, J. T., Jr. 1975. Disposition of dapsone and monoacetyl dapsone in rats. Proceedings of the Society for Experimental Biology and Medicine **150**:485–492.

Hanks, J. H., Chatterjee, B. R., Lechat, M. F. 1964. A guide to the counting of mycobacteria in clinical and experimental materials. International Journal of Leprosy **32**:156–167.

Hart, P. D'Arcy, Rees, R. J. W. 1960. Effect of macrocyclon in acute and chronic pulmonary tuberculosis infection in mice as shown by viable and total bacterial counts. British Journal of Experimental Pathology **41**:414–421.

Hart, P. D'Arcy, Rees, R. J. W. 1968. Elongation of *Mycobacterium lepraemurium* as a distinction from *Mycobacterium leprae*. International Journal of Leprosy **36**:83–86.

Hayashi, F. 1933. Mitsuda's skin reactivity in leprosy. International Journal of Leprosy. **1**:31–38.

Hilson, G. R. F. 1965. Observations on the inoculation of *M. leprae* in the foot pad of the white rat. International Journal of Leprosy **33**:662–665.

Kirchheimer, W. F., Sanchez, R. M. 1976. Quantitative aspects of leprosy in armadillos. International Journal of Leprosy **44**:542–543.

Kirchheimer, W. F., Storrs, E. E. 1971. Attempts to establish the armadillo (*Dasypus novemcinctus* Linn.) as a model for the study of leprosy. 1. Report of lepromatoid leprosy in an experimentally infected armadillo. International Journal of Leprosy **39**:693– 702.

Koch, R. 1897. Die Lepraerkrankungen in Kreise Memel. Abstracted in Baumgartens Jahresbericht **14**:428.

Levy, L. 1976. Studies of the mouse foot pad technique for cultivation of *Mycobacterium leprae*. 3. Doubling time during logarithmic multiplication. Leprosy Review **47**:103– 106.

Narayanan, E., Sreevetsa, Kirchheimer, W. F., Bedi, B. M. S. 1977. Transfer of leprosy bacilli from patients to mouse foot pads by *Aedes aegypti*. Leprosy in India **49**:181– 186.

Prabhakaran, K. 1967. Oxidation of 3, 4-dihydroxyphenylalanine (DOPA) by *Mycobacterium leprae*. International Journal of Leprosy **35**:42– 51.

Prabhakaran, K., Harris, E. B., Kirchheimer, W. F. 1977. Confirmation of the spot test for the identification of *Mycobacterium leprae* and occurrence of tissue inhibitors of DOPA oxidation. Leprosy Review **48**:49– 52.

Rees, R. J. W., Meade, T. W. 1974. Comparison of the modes of spread and the incidence of tuberculosis and leprosy. Lancet **i**:47– 49.

Rees, R. J. W., Weddell, A. G. M. 1968. Experimental models for studying leprosy. Annals of the New York Academy of Sciences **154**:214– 236.

Ridley, D. S. 1974. Histological classification and the immunological spectrum of leprosy. Bulletin of the World Health Organization **51**:451– 465.

Ridley, D. S., Jopling, W. H. 1966. Classification of leprosy according to immunity. A five-group system. International Journal of Leprosy **34**:255– 273.

Ridley, D. S., Waters, M. F. R. 1969. Significance of variations within the lepromatous group. Leprosy Review **40**:143– 152.

Shepard, C. C. 1958. A study of the growth in HeLa cells of human tubercle bacilli from human sputum. American Review of Tuberculosis and Pulmonary Diseases **77**:423– 435.

Shepard, C. C. 1960. The experimental disease that follows the injection of human leprosy bacilli into foot pads of mice. Journal of Experimental Medicine **112**:445– 454.

Shepard, C. C. 1962. The nasal excretion of *Mycobacterium leprae* in leprosy. International Journal of Leprosy **30**:10– 18.

Shepard, C. C. 1965. Stability of *Mycobacterium leprae* and temperaure optimum for growth. International Journal of Leprosy **33**:541– 547.

Shepard, C. C. 1969. Further experience with the kinetic method for the study of drugs against *Mycobacterium leprae* in mice. Activities of DDS, DFD, ethionamide, capreomycin, and PAM 1932. International Journal of Leprosy **37**:389– 397.

Shepard, C. C. 1971. A survey of the drugs with activity against *M. leprae* in mice. International Journal of Leprosy **39**:340– 348.

Shepard, C. C., Guinto, R. S. 1963. Immunological identification of foot-pad isolates as *Mycobacterium leprae* by lepromin reactivity in leprosy patients. Journal of Experimental Medicine **118**:195– 204.

Shepard, C. C., Habas, J. A. 1967. Relation of infection to tissue temperature in mice infected with *Mycobacterium marinum* and *Mycobacterium leprae*. Journal of Bacteriology **93**:790– 796.

Shepard, C. C., McRae, D. H. 1965. *Mycobacterium leprae* in mice: Minimal infectious dose, relationship between staining and infectivity, and effect of cortisone. Journal of Bacteriology **89**:365– 372.

Shepard, C. C., McRae, D. H. 1968. A method for counting acid-fast bacteria. International Journal of Leprosy **36**:78– 82.

Wiersema, J. P., Binford, C. H., Chang, Y. T. 1965. Nerve involvement. Comparison of experimental infections by *Mycobacterium leprae* and *Mycobacterium lepraemurium*. International Journal of Leprosy **33**:617– 633.

CHAPTER 152

The Genus *Frankia*

JAN-HENDRIK BECKING

Habitats

Representatives of the genus *Frankia* are actino-mycete root-nodule symbionts of a large number of nonleguminous plants. This root-nodule symbiosis has been reported in seven orders and eight families of higher plants, which are to a large extent phylo-genetically unrelated (Table 1). In the last few years, the number of plants reported to have this type of symbiosis has been steadily increasing.

Root nodulation is usually a generic character, since most plant species belonging to one genus have been observed to be nodulated when examined in their natural habitat. A clear exception is, how-ever, the genus *Rubus* with about 430 described spe-cies (Focke, 1894), where only one of many species investigated was found to be nodulated (Becking, 1979).

Morphologically, the root nodules produced by actinomycete symbiosis fall into two distinct types. In root nodules of *Casuarina* and *Myrica* species, the apex of each root-nodule lobe produces a nor-mal, but negatively geotropic root, so that these root nodules become clothed with upward-growing root-lets (Fig. 1 a,b). In *Alnus* species and a large number of other nonleguminous species belonging to the Elaeagnaceae, Rhamnaceae, Coriariaceae, Rosaceae, and Datiscaceae, the root nodules are modified, di-chotomous branched roots of arrested growth giving them a coralloid appearance (Fig. 2a,b,c). The *Frankia* endophyte is usually only present in the cortical parenchyma of the root nodule tissue and in the nodular tissue just below the meristem where it produces the new infections in the newly formed host cells.

The root-nodule symbiosis has dinitrogen-fixing

Table 1. Classification of nonleguminous, dinitrogen-fixing angiosperms with *Frankia* symbioses.

Order	Family	Tribe	Genus	Number of nodulated species[a]	
Casuarinales	Casuarinaceae	—	*Casuarina*	24	(45)
Myricales	Myricaceae	—	*Myrica*	26	(35)
			Comptonia	1	(1)
Fagales	Betulaceae	Betuleae	*Alnus*	33	(35)
	Elaeagnaceae	—	*Elaeagnus*	16	(45)
			Hippophaë	1	(3)
			Shepherdia	3	(3)
Rhamnales		Rhamneae	*Ceanothus*	31	(55)
	Rhamnaceae		*Discaria*	6	(10)
		Colletieae	*Colletia*	3	(17)
			Trevoa[b]	1	(6)
Coriariales	Coriariaceae	—	*Coriaria*	13	(15)
		Rubieae	*Rubus*[c]	1	(250)(429)
Rosales	Rosaceae	Dryadeae	*Dryas*	3	(4)
			Purshia	2	(2)
		Cercocarpeae	*Cercocarpus*	4	(20)
Cucurbitales	Datiscaceae	—	*Datisca*[d]	2	(2)

[a] Total number of species in parentheses.
[b] *Trevoa trinervis* (Rundel and Neel, 1978)
[c] According to Focke (1894), 429 *Rubus* species exist worldwide, but Willis (1973) men-tions only 250 species. So far only 1 species, *R. ellipticus,* has been observed to possess root nodules and nitrogenase activity (Becking, 1979).
[d] *Datisca cannabina* and *D. glomerata* (Chaudhary, 1978, 1979; Severini, 1922).

a

b

Fig. 1. Root nodules of nonleguminous plants that produce negatively geotropic nodule roots. (a) *Casuarina equisetifolia* (Casuarinaceae). (b) *Myrica javanica* (Myricaceae). × 1.0.

capacity which can be demonstrated by growing nodulated plants in nitrogen-free nutrient solution, and by dinitrogen-fixation tests using $^{15}N_2$ or the acetylene-reduction method. The nitrogenase activity is located in the *Frankia* symbiont.

The family Frankiaceae and the genus *Frankia* were named by Becking (1970a) in view of their ability to enter a symbiotic association with higher plants and their morphological and structural differences from other actinomycetes. Species delimination within this genus was made on the basis of cross-inoculation groups and of some morphological features such as the dimension of the hyphae and the formation and shape of some special structures called "vesicles". These vesicular structures are probably related to the reproduction of the endophyte, since they may represent a kind of deformed sporangia (Becking, 1974). The vesicular structures as observed in the host tissue in vivo are always situated at the tip of the hyphae. In some species, these terminal swellings are spherical and in others more club shaped (Becking, 1970a, 1974). Another

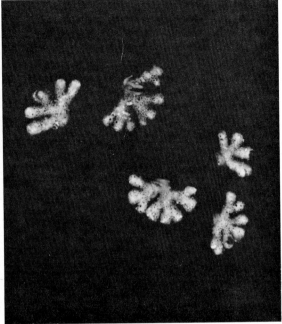

Fig. 2. Root nodules of the coralloid type of nonleguminous plants. (a) *Alnus glutinosa* (Betulaceae); × 1.0. (b) *A. glutinosa*; × 1.5. (c) *Dryas drummondii* (Rosaceae); × 2.2.

important structural characteristic of the endophyte is the formation of small, thick-walled, polyhedric, sporelike cells (Becking, 1970a,b, 1974, 1975; Becking, de Boer, and Houwink, 1964), also called granula by some other authors (Akkermans and van Dijk, 1975; Angulo, van Dijk, and Quispel, 1975).

These sporelike structures occur solely in dead host cells, whereas living host cells apparently stimulate the production of vesicular structures (Becking, 1968, 1970b, 1975; Becking, de Boer, and Houwink, 1964). The sporelike structures were not observed in the root nodules of all species of nonleguminous plants. We observed that sporelike structures were invariably present in all root nodules of *Alnus glutinosa* and here they are apparently a regular developmental stage of the endophyte. Other authors claim the occurrence of spore-positive, Sp(+), and spore-free, Sp(−), root nodules in this species (Angulo, van Dijk, and Quispel, 1975; van Dijk and Merkus, 1976).

Recently, the incompatibility barriers between the representatives of the various genera were found not to be strict and it has been inferred that the host cell, i.e., the plant, exerts a great influence on the morphology of the endophyte within the tissue. Compatibility of cross-inoculation was already known within the Elaegnaceae in the genera *Elaeagnus, Hippophaë,* and *Shepherdia* for the *Frankia elaeagni* endophyte (Becking, 1970a,b), but now also other combinations are mentioned. Rodriguez-Barrueco and Bond (1975) observed that an *Alnus glutinosa* inoculum (crushed nodules) can produce root nodulation in *Myrica gale,* but that the reciprocal combination of *Myrica gale* inoculum tested on *Alnus glutinosa* did not produce root nodulation. Recently however, Miguel et al. (1978) mentioned that they could produce root nodulation with the reverse combination and these authors observed in addition that a *Hippophaë rhamnoides* inoculum produced root nodulation in *Myrica gale, Coriaria myrtifolia* and *Elaeagnus angustifolia,* but it did not nodulate *Alnus glutinosa.* Furthermore, Lalonde, Knowles, and Fortin (1975) report that a *Comptonia peregrina* crushed-nodule inoculum could induce root nodulation in *Alnus crispa* var. *mollis* host plants. In most of these experiments, use was made of nonaseptic plants in water culture and also of an inoculum of water- or field-grown plants. Therefore, from these results no definite conclusion can be drawn since the involvement of other actinomycetes of nonlegumes cannot be excluded. It is, however, possible that several host plants can share the same endophyte, but this does not imply that the normal endophyte of both plant species is the same. A distinction between normal and abnormal combinations should be made, since, in the combination of the *Alnus glutinosa* endophyte in the *Myrica gale* host, the appearance of the endophyte within the *Myrica* root nodules is different from the normal and more like that of the *Alnus* endophyte (C. Rodriguez-Barrueco, personal communication) and often the dinitrogen-fixing capacity of these combinations is much less than usual. Furthermore, the endophytes of *Alnus* and *Myrica* cannot be identical as is proved by their cell wall composition. The presence of 2,6-diaminopimelic acid (DAP) has diagnostic value for the separation of actinomycetes into genera (Becker, Lechevalier, and Lechevalier, 1965; Becker et al., 1964; Lechevalier and Fekete, 1971; Lechevalier, Lechevalier, and Becker, 1966). It has been shown that *meso*-DAP is present in the *Alnus* endophyte

(Becking, 1970a, 1974, 1975, 1976; Quispel, 1974; van Dijk, 1978), whereas this substance is lacking in the cell walls of the *Myrica* and *Casuarina* endophyte and some other species (Becking, 1977a,b; van Dijk, 1978).

Isolation

Isolation of the endophyte from host root-nodule tissue has proved to be very difficult. For many supposed isolates mentioned in the literature (see Allen, Silvester, and Kalin, 1966; Danilewicz, 1965; Fiuczek, 1959; Niewiarowska 1961; Uemura, 1952a,b; von Plotho 1941; Webster, Youngberg, and Wollum, 1967; Wollum, Youngberg, and Gilmour, 1966), Koch's postulates were never fulfilled.

In only a few exceptional cases, the morphology and other characteristics of the isolate make it probable that it may be a free-living form of the endophyte.

Pommer (1959) obtained an isolate from surface-sterilized *Alnus glutinosa* root nodules. This actinomycete could produce some growth on a relatively simple glucose-asparagine agar (distilled water, 1,000 ml; glucose, 10.0 g; asparagine, 0.5 g; KH_2PO_4, 0.5 g; agar, 15 g; pH 6.8). This isolate showed the characteristic intra-axial sporogenous bodies also occasionally observed in the endophyte in vivo (Käppel and Wartenberg, 1958) and in some later in vitro isolates (Callaham, Del Tredici, and Torrey, 1968) (see Figs. 3 and 4). Moreover, this isolate could produce, in a number of cases, root nodulation in aseptically grown *Alnus glutinosa* seedlings. This observation could be confirmed by us (J. H. Becking, unpublished).

Becking (1965) grew the endophyte in root-nodule tissue callus of *Alnus glutinosa* in monoxenic culture (Fig. 5). The callus was grown on a basal mineral-glucose agar medium supplemented with coconut milk (150 ml/liter), α-naphthalene acetic acid (0.1 mg/liter), and calcium pantothenate (2.5 mg/liter). In some cases, the endophyte produced some outgrowth into the medium, forming a ring of actinomycetal growth some distance below the agar surface (Fig. 5c). This growth pattern indicates microaerophilic preference of the endophyte when growing in vitro outside the tissue. No dinitrogen-fixing activity could be measured in these symbiotic associations of callus tissue and actinomycete (acetylene reduction tests or $^{15}N_2$ experiments). Moreover, growth of the callus tissue cells was usually faster than the propagation of the endophyte, so that the endophyte was rarely transmitted to the newly formed callus cells (Fig. 6). Therefore, many of the explants obtained from a callus tissue inhabited by the endophyte became finally free from the endophyte. The callus-grown endophyte could produce

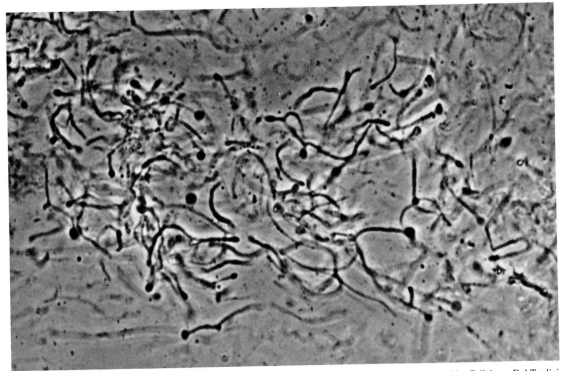

Fig. 3. Phase-contrast photomicrograph of the nodule endophyte of *Comptonia peregrina* L. Coult. isolated by Callaham, Del Tredici, and Torrey (1978). The organism is grown in pure culture in a yeast extract (0.5%, wt/vol) medium, pH 7.0. Hyphae are visible as well as terminal and intra-axial sporogenous bodies (vesicles). × 1,260.

root nodules in sterile *Alnus glutinosa* seedlings, but these root nodules containing the endophyte (but in a form without vesicles) proved to be ineffective in dinitrogen fixation, as shown by acetylene-reduction test and $^{15}N_2$ experiments (Becking, 1965).

The Fåhraeus (1957) glass-slide technique originally designed for *Rhizobium* was applied to continuous microscopic observation of the growth of the *Frankia* endophyte in the root-hair environment and its penetration in the root hairs of aseptically grown *Alnus glutinosa* seedlings by Becking (1975, 1976), who observed the penetration of the endophyte at the root-hair tip and a considerable growth of the endophyte in the mucilaginous layer covering the root hair (Figs. 7 and 8). In these monoxenic cultures of the *Frankia* actinomycete with aseptic *Alnus glutinosa* plants, hyphal growth of the endophyte in the nitrogen–free agar surrounding the root hairs and also the formation of typical intra-axial sporogenous bodies was occasionally observed. Similar observations are reported by G. Fåhraeus (personal communication) using his technique with *Alnus glutinosa* (Fig. 9). Apparently, the endophyte could grow in vitro in the inorganic medium used for the growth of the *Alnus* plants and supplied with unknown root exudates produced by the root hairs as a carbon/nitrogen source.

Medium for Glass-Slide Cultures of *Alnus* Seedlings (after Fåhraeus, 1957)

Distilled water	1,000 ml
$CaCl_2 \cdot 2H_2O$	0.1 g
$MgSO_4 \cdot 7H_2O$	0.12 g
KH_2PO_4	0.1 g
Na_2HPO_4	0.15 g
Fe-citrate	0.005 g

Mn, Cu, Zn, B, and Mo traces (supplied by a trace-element stock solution)
pH 6.5

We were, however, unable to produce further growth of the endophyte in subsequent replicates in various media of different composition, but it is very likely that the actinomycete observed is, in fact, the *Frankia* endophyte, since similar sporogenous structures have been observed by Callaham, Del Tredici, and Torrey (1978) for the isolated *Comptonia peregrina* endophyte (see Figs. 3 and 4). A better understanding of the nutritional requirements of this *Alnus glutinosa* endophyte will certainly lead to the isolation of the *Frankia* endophyte according to this approach.

Lalonde, Knowles, and Fortin (1975) reported the isolation of the *Alnus crispa* var. *mollis* endophyte

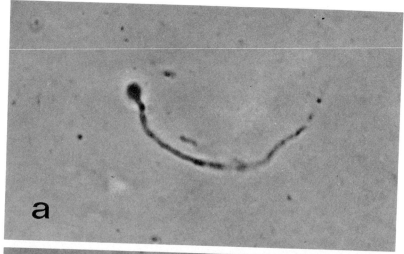

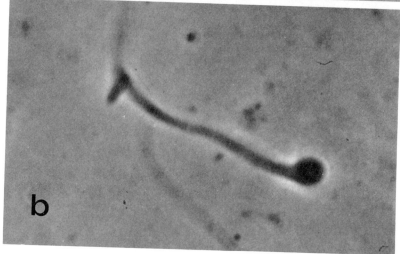

Fig. 4. Phase-contrast photomicrograph of a sporogenous body at the tip of the hypha of the *Comptonia peregrina* L. Coult. endophyte growing in pure culture on an artificial medium (see Fig. 3). (a) × 1,400; (b) × 2,800.

in a complex medium used by Harvey (1967) for tissue cultures of *Pinus monticola*. This medium contains 50 ml/liter of a solution containing Ca(NO₃)₂ · 4H₂O 10 g/liter; KH₂PO₄ 2.8 g/liter; MgSO₄ · 3H₂O 2.8 g/liter; (NH₄)₂SO₄ 0.5 g/liter; ferric citrate (1% aqueous solution), 20 ml/liter; some minor salt; growth substances (ascorbic acid, nicotinic acid, choline chloride, Ca-pantothenate, biotin, thiamine-HCl, pyridoxine-HCl, riboflavin); some amino acids (tyrosine, arginine, glycine, leucine, etc.); glucose, 15 g/liter; and agar 8 g/liter; pH 6.5. The isolate obtained from superficially sterilized root nodules was claimed to be identical with the intracellular endophyte on basis of immunofluorescence reactions. Since the substances involved in these immunolabeling reactions are unknown and might be present in other actinomycete species (as already shown by the same authors), an absolute proof of its identity with the endophyte in vivo cannot be given. In addition, the isolate was incapable of nodulating aseptic host plants growing in a nitrogen-deficient mineral medium.

Callaham, Del Tredici, and Torrey (1978) obtained an isolate from *Comptonia peregrina* root nodules after surface sterilization of excised nodules and the following treatments.

Isolation after Surface Sterilization of Excised Root Nodules (Callaham, Del Tredici, and Torrey, 1978)

The nodule pieces were incubated on bacteriological media for 6 weeks to allow the selection of sterile nodule pieces. An enzyme maceration treatment using cellulase (5%, wt/vol) and pectinase (2%, wt/vol) was applied, and subsequently the nodule pieces were transferred to the same medium lacking the enzymes and teased apart with dissecting needles. The endophytic clusters released into the medium were filtered aseptically from the cell debris by passage through a 150-µm-mesh nylon screen. The endophyte was subsequently washed twice by centrifugation and resuspended in fresh medium.

The medium used for maceration and resus-

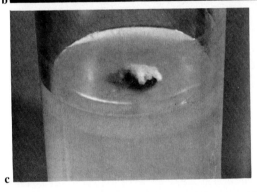

Fig. 5. Root-nodule callus tissue of *Alnus glutinosa* containing the endophyte in monoxenic culture. (a) Root-nodule tissue just explanted. (b and c) The same root-nodule tissue after several months of growth. In a number of cases [see (c)], outgrowth of the endophyte into the medium occurred, as evident from a ring of actinomycetal growth some distance below the agar surface (microaerophilic). (a,b) × 1.0; (c) × 2.0.

pension had a complex composition (Goforth and Torrey, 1977). It contained the following components (mg/liter of glass-distilled water):

$Ca(NO_3)_2 \cdot 4H_2O$	242
$MgSO_4 \cdot 7H_2O$	42
KNO_3	85
KCl	61
KH_2PO_4	20
$FeCl_3 \cdot 6H_2O$	2.5

Thiamine-HCl	0.1
Nicotinic acid	0.5
Pyridoxine-HCl	0.5
H_3BO_3	1.5
$ZnSO_4 \cdot 7H_2O$	1.5
$MnSO_4 \cdot H_2O$	4.5
$Na_2MoO_4 \cdot 2H_2O$	0.25
$CuSO_4 \cdot 5H_2O$	0.04
Sucrose	40,000

The pH was adjusted to 5.5 before autoclaving. To this medium, 0.65 M mannitol was added as osmoticum. It was supplemented (cold-sterilization of the solution with a Millipore filter) with 1 mM each of L-glutamic acid, L-aspartic acid, glycine, L-arginine, L-asparagine, L-glutamine, and urea, and 2.0 mg/liter of naphthaloneacetic acid and 1 µg/liter of zeatin.

After 3 weeks' culture, several small colonies of a microorganism with filamentous growth were observed. These colonies were transferred to Petri dishes containing another medium in which growth was continued and accelerated.

This medium had the following composition (g/liter):

$CaCO_3$	0.5
K_2HPO_4	0.5
$MgSO_4 \cdot 7H_2O$	0.2
NaCl	0.1
$MnSO_4 \cdot H_2O$	0.025
H_3BO_3	0.10
$ZnSO_4 \cdot 7H_2O$	0.010
$Na_2MoO_4 \cdot 2H_2O$	0.00025
$CuSO_4 \cdot 5H_2O$	0.000025
Yeast extract (Difco)	0.5
Edamin	1.0
Thiamine-HCl	0.0001
Nicotinic acid	0.0005
Pyridoxine-HCl	0.0001
Mannitol	1.0
Sucrose	20.0
Agar	10.0
pH 7.0	

Growth in this medium was slightly faster in standing liquid culture than on agar plates. The isolated actinomycete grew slowly in axenic culture on a yeast extract medium (Difco yeast extract, 0.5% wt/vol, in distilled water, pH 6.4) in unshaken flasks or test tubes 5–6 cm deep with nutrient solution; optimal growth was at 25–30°C. Growth, although with a slightly slower growth rate, was obtained in a completely defined medium of the following composition:

$CaCl_2 \cdot 2H_2O$	200 mg/liter
KH_2PO_4	200 mg/liter
$MgSO_4 \cdot 7H_2O$	200 mg/liter
NaCl	100 mg/liter
H_3BO_3	1.5 mg/liter

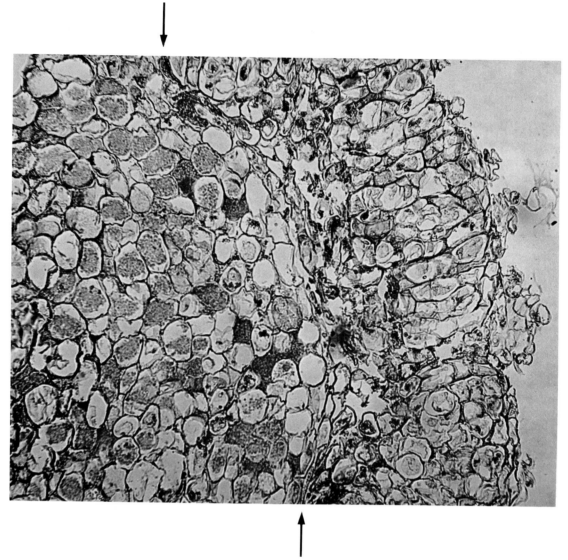

Fig. 6. Section through a callus-producing, root-nodule fragment of *Alnus glutinosa* showing that the transmission of the endophyte to the newly developed tissue is poor. The arrows indicate the transitional zone between original tissue and the newly developed callus tissue. Photomicrograph of fixed, unstained preparation; phase-contrast; × 190.

$ZnSO_4 \cdot 7H_2O$	1.5 mg/liter
$MnSO_4 \cdot H_2O$	4.5 mg/liter
$Na_2MoO_4 \cdot 2H_2O$	0.025 mg/liter
$CuSO_4 \cdot 5H_2O$	0.04 mg/liter
Thiamine-HCl	0.1 mg/liter
Nicotinic acid	0.5 mg/liter
Pyridoxine-HCl	0.5 mg/liter
Succinic acid	10 mM
L-Glutamine	2 mM
Myoinositol	0.5 mM
Fe-EDTA	0.1 mM
pH 6.4	

The growth of the isolate on agar (0.8%, wt/vol), Petri dishes slants, or streak cultures was exception-

ally slow; colonies of 1–2 mm diameter were only produced after several months of growth (see Fig. 10). Also, growth in this medium was slightly faster in a standing liquid. The isolate appears to be micro-aerophilic, since it grew best at the bottom of tubes of liquid medium. It failed to grow under complete anaerobiosis.

For maintenance of the isolate, the best medium is yeast extract (Difco; 0.5%, wt/vol) in distilled water with a pH of 6.4, but subculture is only possible after the cultures are 4–6 weeks old. An additional good medium for maintenance, according to J. G. Torrey (personal communication), is *Leptospira* broth prepared according to the modification of Ellinghausen and McCullough (1965a,b; see also

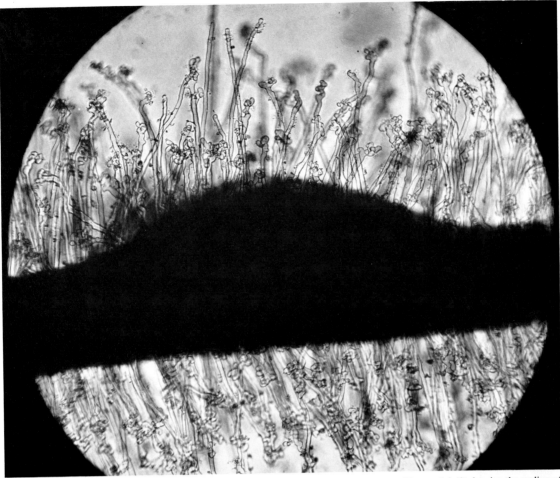

Fig. 7. Root-nodule development of *Alnus glutinosa* in its initial stage of nodule development ("pre-nodule") showing the curling of the root hairs caused by the activity of the endophyte. Living preparation, phase-contrast photomicrograph; × 130.

Johnson and Harris, 1967). This medium is composed of some basal constituents supplemented with albumin fatty acid and trace elements.

Maintenance Medium for *Frankia*
(Ellinghausen and McCullough, 1965a,b)
  The basal medium contains:

Distilled water	997 ml
Na_2HPO_4	1.0 g
KH_2PO_4	0.3 g
NaCl	1.0 g
NH_4Cl (25%, wt/vol aqueous)	1 ml
Thiamine (0.5%, wt/vol aqueous)	1 ml
Glycerol (10%, wt/vol aqueous)	1 ml

This medium is adjusted to pH 7.4 and sterilized by autoclaving. Subsequently a sterile (by Millipore filtration), albumin–fatty acid supplement is added. This solution has the following composition:

Bovine albumin fraction V	20 g
$CaCl_2 \cdot 2H_2O$ (1%, wt/vol aqueous)	2 ml
$MgCl_2 \cdot 6H_2O$ (1%, wt/vol aqueous)	2 ml
$ZnSO_4 \cdot 7H_2O$ (0.4%, wt/vol aqueous)	2 ml
$CuSO_4 \cdot 5H_2O$ (0.3%, wt/vol aqueous)	0.2 ml
$FeSO_4 \cdot 7H_2O$ (0.5%, wt/vol aqueous)	20 ml
Vitamin B_{12} (0.2%, wt/vol aqueous)	2 ml
Tween 80 (10%, wt/vol aqueous)	25 ml
Distilled water to	200 ml

In all of these media, the actinomycete grew in pure culture as a branched septate hyphal mat, forming characteristic intra-axial sporogenous bodies with transverse and longitudinal septation as also observed in the endophyte in vivo. The sporogenous bodies produced a large number of ovoid or polyhydral spores 1.5–3.5 μm in diameter. However, the club-shaped vesicles formed in vivo were not observed in the cultured organism.

Reinfection of sand-grown and aeroponically grown seedlings of *Comptonia peregrina* was possible with this isolate from *Comptonia* root nodules, and the root nodules formed showed high nitrogenase (acetylene-reduction test) activity. The morpho-

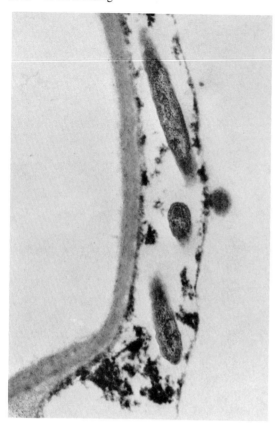

Fig. 8. Development of the *Alnus glutinosa* endophyte in the mucilaginous layer covering the root hair. Transmission electron micrograph; × 25,200.

logically same actinomycete type could be reisolated from the seedling root nodules in reisolation experiments using the same methods as outlined above. The *Comptonia* isolate could also produce root nodules in *Myrica* species (*M. gale* and *M. cerifera*) and

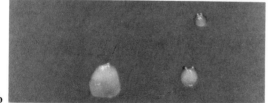

Fig. 10. Two-month-old colony of *Frankia* sp. (strain Cp.I.1 of Callaham, Del Tredici, and Torrey, 1978; isolated from a *Comptonia peregrina* root-nodule) on yeast extract glucose agar (pH 6.8) incubated in air. (a) × 25; (b) × 12.5.

in *Alnus* species (*A. crispa* and *A. glutinosa*) (Lalonde, cited by Callaham, Del Tredici, and Torrey, 1978). In the above-mentioned studies, no reference is made of inoculation of aseptically grown plants. Since all these plants were water or sand cultures exposed to normal nonsterile greenhouse conditions, the interference of microbes other than the isolate cannot be excluded. In this respect, recent inoculation experiments of the isolate in axenic plants gave variable results. The isolate was only invariably effective in plants growing under nonsterile conditions. Therefore the mediation or support of other noneffective soil microbes in the infection process has been suggested (Knowlton, Berry, and Torrey, 1979).

Recently, Baker, Kidd, and Torrey (1978) mentioned some other methods for obtaining isolates from actinomycete root nodules (see also Baker, Kidd, and Torrey, 1979; Baker, Torrey, and Kidd, 1979). Nodule endophytes of *Elaeagnus* and *Myrica* were separated from suspensions of crushed root nodules by gel filtration on sterile Sephadex G-50 coarse. The nodulating capacity of the Sephadex-purified fractions on *Elaeagnus* and *Myrica* seedlings was up to fivefold greater than the crude nodule suspensions. An additional isolate of the actinomycete that causes root nodulation in *Comptonia peregrina* was achieved with this method by further separation of the Sephadex fractions on sterile sucrose-density gradients. This isolated *Comptonia* endophyte was similar to that found in vivo and similar to the organism isolated previously by Callaham, Del Tredici, and Torrey (1978).

Recently, the isolation and the continuous cultivation of the *Alnus glutinosa* endophyte was reported by Quispel and Tak (1978). These authors obtained axenic root-nodule suspensions from surface-sterilized nodule pieces obtained from field-collected nodules. These nodule suspensions were incubated in a nutrient medium containing peptone, some inorganic salts, and the petrol ether soluble factor from *Alnus* roots (but also from roots of other plants). Only the culture supplied with this petrol ether extract showed some actinomycete growth in the medium, as evident from inoculation experiments (increased infective capacity of the suspension) with host plants grown in (nonsterile) water culture and microscopic examination of the nutrient medium. Thus, the addition of petrol ether-soluble substances from roots is essential for growth. The active lipid factor, however, could not be identified. Further, it was noted that only with the so-called Sp(−) root nodules of *Alnus glutinosa* could an increase of infectivity and growth of the endophyte in vitro be obtained. Apparently, the endophyte of the Sp(+) nodules had other growth requirements. The isolated actinomycete revealed only hyphal structures without any differentiation.

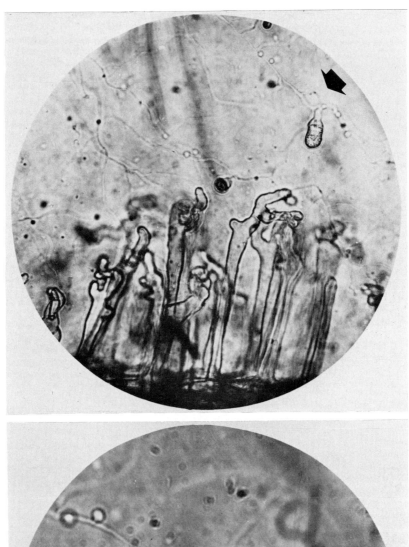

a

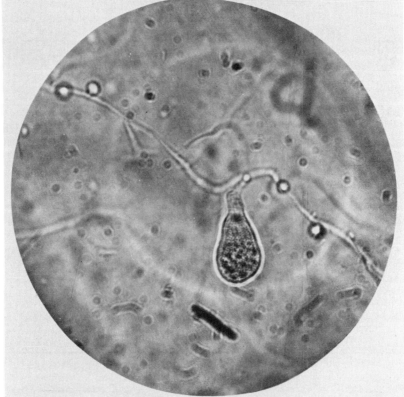

b

Fig. 9. In vitro outgrowth of the *Alnus glutinosa* endophyte in the agar medium around the root hairs. Actinomycete hyphae and a sporogenous body developing from a short branch filament are clearly visible. Photomicrograph (b) represents a detail enlargement of the area indicated by the arrow in (a). (a) × 351; (b) × 945. [Photographs courtesy of G. Fahraeus.]

Literature Cited

Akkermans, A. D. L., van Dijk, C. 1975. The formation and nitrogen-fixing activity of the root nodules of *Alnus glutinosa* under field conditions, pp. 511–520. In: Nutman, P. S. (ed.), Symbiotic nitrogen fixation in plants. International Biological Programme, vol. 7. Cambridge: Cambridge University Press.

Allen, J. D., Silvester, W. B., Kalin, M. 1966. *Streptomyces* associated with root nodules of *Coriaria* in New Zealand. New Zealand Journal of Botany **4:**57–65.

Angulo, A. F., van Dijk, C., Quispel, A. 1975. Symbiotic interactions in non-leguminous root nodules, pp. 475–483. In: Nutman, P. S. (ed.), Symbiotic nitrogen fixation in plants. International Biological Programme, vol. 7. Cambridge: Cambridge University Press.

Baker, D., Kidd, G. H., Torrey, J. G. 1978. Separation and isolation of actinomycete endophytes by sephadex and sucrose-density fractionation, Abstract C-57. In: Orme-Johnson, W. H., Newton, W. E. (eds.), Proceedings of the Steenbock-Kettering International Symposium on Nitrogen Fixation, June 12–16, 1978, Madison, U.S.A. University of Wisconsin Duplicating Service.

Baker, D., Kidd, G. H., Torrey, J. G. 1979. Separation of actinomycete nodule endophytes from crushed nodule suspensions by Sephadex fractionation. Botanical Gazette **140** (Suppl.):S49–S51.

Baker, D., Torrey, J. G., Kidd, G. H. 1979. Isolation by sucrose-density fractionation and cultivation in vitro of actinomycetes from nitrogen-fixing root nodules. Nature **281:**76–78.

Becker, B., Lechevalier, M. P., Lechevalier, H. A. 1965. Chemical composition of cell-wall preparations from strains of various form-genera of aerobic actinomycetes. Applied Microbiology **13:**236–243.

Becker, B., Lechevalier, M. P., Gordon, R. E., Lechevalier, H. A. 1964. Rapid differentiation between *Nocardia* and *Streptomyces* by paper chromatography of whole-cell hydrolysates. Applied Microbiology **12:**421–423.

Becking, J. H. 1965. *In vitro* cultivation of alder root-nodule tissue containing the endophyte. Nature **207:**885–887.

Becking, J. H. 1968. Nitrogen fixation by non-leguminous plants. Symposium Nitrogen in Soil, Groningen, May 17–19, 1967. Nitrogen, Dutch Nitrogenous Fertilizer Review **12:**47–74.

Becking, J. H. 1970a. *Frankiaceae* fam. nov. (*Actinomycetales*) with one new combination and six new species of the Genus *Frankia* Brunchorst 1886, 174. International Journal of Systematic Bacteriology **20:**201–220.

Becking, J. H. 1970b. Plant-endophyte symbiosis in non-leguminous plants. Plant and Soil **32:**611–654.

Becking, J. H. 1974. Family III. Frankiaceae Becking 1970, 201. In: Buchanan, R. E., Gibbons, N. E. (eds.), Bergey's manual of determinative bacteriology, 8th ed. Baltimore: Williams & Wilkins. pp. 701–706, pp. 871–872 (Plate 17.5 & 17.6).

Becking, J. H. 1975. Root nodules in non-legumes, pp. 507–566. In: Torrey, J. G., Clarkson, D. T. (eds.), The development and function of roots. London, New York: Academic Press.

Becking, J. H. 1976. Actinomycete symbiosis in non-legumes, pp. 581–591. In: Newton, W. E., Nyman, C. J. (eds.), Proceedings of the First International Symposium on Nitrogen Fixation, Pullman, June 3–7, 1974. vol. 2. Pullman, Washington: Washington State University Press.

Becking, J. H. 1977a. Dinitrogen-fixing associations in higher plants other than legumes, pp. 185–275. In: Hardy, R. W. F., Silver, W. S. (eds.), A treatise on dinitrogen fixation. Sect. III: Biology. New York: John Wiley & Sons.

Becking, J. H. 1977b. Endophyte and association establishment in non-leguminous nitrogen-fixing plants, pp. 551–567. In: Newton, W., Postgate, J. R., Rodriguez-Barrueco, C. (eds.), Recent development in nitrogen fixation. Proceedings of the Second International Symposium on Nitrogen Fixation, Salamanca, September 13–17, 1976. London, New York: Academic Press.

Becking, J. H. 1979. Nitrogen fixation by *Rubus ellipticus* J. E. Smith. Plant and Soil **54:**541–545.

Becking, J. H., de Boer, W. E., Houwink, A. L. 1964. Electron microscopy of the endophyte of *Alnus glutinosa*. Antonie van Leeuwenhoek Journal of Microbiology and Serology **30:**343–376.

Callaham, D., Del Tredici, P., Torrey, J. G. 1978. Isolation and cultivation *in vitro* of the acinomycete causing root nodulation in *Comptonia*. Science **199:**899–902.

Chaudhary, A. S. 1978. The discovery of root nodules in new species of nonleguminous angiosperms from Pakistan and their significance, p. 359. In: Döbereiner, J., Burris, R. H., Hollaender, A. (eds.), Limitations and potentials for biological nitrogen fixation in the tropics. New York: Plenum.

Chaudhary, A. S. 1979. Nitrogen-fixing root nodules in *Datisca cannabina* L. Plant and Soil **51:**163–165.

Danilewicz, K. 1965. Symbiosis in *Alnus glutinosa* (L.) Gaertn. Acta Microbiologica Polonica **14:**321–326.

Ellinghausen, H. C., McCullough, W. G. 1965a. Nutrition of *Leptospira pomona* and growth of 13 other serotypes: A serum-free medium employing oleic albumin complex. American Journal of Veterinary Research **26:**39–44.

Ellinghausen, H. C., McCullough, W. G. 1965b. Nutrition of *Leptospira pomona* and growth of 13 other serotypes: Fractionation of oleic albumin complex and a medium of bovine albumin and polysorbate 80. American Journal of Veterinary Research **26:**45–51.

Fåhraeus, G. 1957. The infection of clover root hairs by nodule bacteria studied by a simple glass slide technique. Journal of General Microbiology **16:**374–381.

Fiuczek, M. 1959. Fixation of atmospheric nitrogen in pure cultures of *Streptomyces alni*. [In Polish, with English summary.] Acta Microbiologica Polonica **8:**283–287.

Focke, W. O, 1894. Die natürlichen Pflanzenfamilien, pp. 1–61. In: Engler, A., Prantl, P. (eds.), Teil III, 3. Abteilung, Rosaceae. Leipzig: Engelmann.

Goforth, P. L., Torrey, J. G. 1977. The development of isolated roots of *Comptonia peregrina* (Myricaceae) in culture. American Journal of Botany **64:**476–482.

Harvey, A. E. 1967. Tissue culture of *Pinus monticola* on a chemically defined medium. Canadian Journal of Botany **45:**1783–1787.

Johnson, R. C., Harris, V. G. 1967. Differentiation of pathogenic and saprophytic leptospires. Journal of Bacteriology **94:**27–31.

Käppel, M., Wartenberg, H. 1958. Der Formenwechsel des *Actinomyces alni* Peklo in den Wurzeln von *Alnus glutinosa* Gaertner. Archiv für Mikrobiologie **30:**46–63.

Knowlton, S., Berry, A., Torrey, J. G. 1979. The role of rhizosphere microorganisms in nodule formation in *Alnus rubra* Bong, pp. 479–480. In: Gordon, J. C., Wheeler, C. T., Perry, D. A. (eds.), Symbiotic nitrogen fixation in the management of temperate forests. Corvallis, Oregon: Forest Research Laboratory, Oregon State University.

Lalonde, M., Knowles, R., Fortin, J.-A. 1975. Demonstration of the isolation of non-infective *Alnus crispa* var. *mollis* Fern. nodule endophyte by morphological immonulabelling and whole cell composition studies. Canadian Journal of Microbiology **21:**1901–1920.

Lechevalier, H., Lechevalier, M. P., Becker, B. 1966. Comparison of the chemical composition of cell-walls of nocardiae with that of other aerobic actinomycetes. International Journal of Systematic Bacteriology **16:**151–160.

Lechevalier, M. P., Fekete, E. 1971. Chemical methods as criteria for separation of actinomycetes into genera. Workshop Subcommittee on Actinomycetes of the American Society for

Microbiology, Rutgers University, New Brunswick, New Jersey.

Miguel, C., Cañizo, A., Costa, A., Rodriguez-Barrueco, C. 1978. Some aspects of the *Alnus*-type root nodule symbiosis, pp. 121–133. In: Döbereiner, J., Burris, R. H., Hollaender, A. (eds.), Limitations and potentials for biological nitrogen fixation in the tropics. New York: Plenum.

Niewiarowska, J. 1961. Morphologie et physiologie des actinomycetes symbiotiques des *Hippophäe*. Acta Microbiologica Polonica **10**:271–286.

Pommer, E. H. 1959. Über die Isolierung des Endophyten aus den Wurzelknöllchen von *Alnus glutinosa* Gaertn. und über erfolgreiche Re-Infektionsversuche. Berichte der Deutschen Botanischen Gesellschaft **72**:138–150.

Quispel, A. 1974. The endophyte of the root nodules in non-leguminous plants, pp. 499–520. In: Quispel, A. (ed.), The biology of nitrogen fixation. Amsterdam: North-Holland.

Quispel, A., Tak, T. 1978. Studies of the growth of the endophyte of *Alnus glutinosa* (L.) Vill. in nutrient solutions. New Phytologist **80**:587–600.

Rodriguez-Barrueco, C., Bond, G. 1975. A discussion of the results of cross-inoculation trials between *Alnus glutinosa* and *Myrica gale*, pp. 561–565. In: Nutman, P. S. (ed.), Symbiotic Nitrogen Fixation in Plants. International Biological Programme, vol. 7. Cambridge: Cambridge University Press.

Rundel, P. W., Neel, J. W. 1978. Nitrogen fixation by *Trevoa trinervis* (*Rhamnaceae*) in the Chilean Matorral. Flora **167**:127–132.

Severini, G. 1922. Sui tubercoli radicali di *Datisca cannabina*. Annali di Botanica **15**:29–51.

Uemura, S. 1952a. Studies on the root nodules of alders (*Alnus* spp.) (IV). Experiment on the isolation of actinomycetes from alder root nodules. [In Japanese, with English summary.] Bulletin of the Government Forest Experiment Station No. 52, Tokyo, Japan, 1–18, Plate I–III.

Uemura, S. 1952b. Studies on the root nodules of alders (*Alnus* spp.). (V). Some new isolation methods of *Streptomyces* from alder nodules. [In Japanese, with English summary.] Bulletin of the Government Forest Experiment Station No. 57, Tokyo, Japan, 209–226, Plate I–II.

Uemura, S. 1961. Studies on the *Streptomyces* isolated from Alder root nodules. [In Japanese, with English summary.] Scientific Reports of Agricultural, Forest and Fisheries Research Council No. 7. Tokyo, Japan, 1–90, Plate 1–15.

van Dijk, C. 1978. Spore formation and endophyte diversity in root nodules of *Alnus glutinosa* (L.) Vill. New Phytologist **81**:601–615.

van Dijk, C., Merkus, E. 1976. A microscopial study of the development of a spore-like stage in the life cycle of the root-nodule endophyte of *Alnus glutinosa* (L.) Gaertn. New Phytologist **77**:73–91.

von Plotho, O. 1941. Die Synthese der Knöllchen an den Wurzeln der Erle. Archiv für Mikrobiologie **12**:1–18.

Webster, S. R., Youngberg, C. T., Wollum, A. G., II. 1967. Fixation of nitrogen by bitterbrush (*Pushia tridentata* (Pursh) D. C.). Nature **216**:392–393.

Willis, J. C. 1973. A dictionary of the flowering plants and ferns, 8th ed. Revised by H. K. Airy Shaw. Cambridge: Cambridge University Press.

Wollum, A. G., II, Youngberg, C. T., Gilmour, C. M. 1966. Characterization of the *Streptomyces* sp. isolated from root nodules of *Ceanothus velutinus* Doubl. Proceedings of the Soil Science Society of America **30**:463–467.

CHAPTER 153

The Family Actinoplanaceae

CHARLES E. BLAND and JOHN N. COUCH

The family Actinoplanaceae includes those hyphae-forming, Gram-positive, non-acid-fast members of the order Actinomycetales in which spores (motile or nonmotile depending on genus) are produced within sporangia. Although these organisms were probably first seen by Shchepkina (1940), it was Couch (1949, 1950) who first isolated them and recognized their true affinities. Since Couch's original description of the genus *Actinoplanes,* (1950) and subsequent description of the family Actinoplanaceae (1955a,b), over 15 genera and 45 species have been described and placed tentatively in the Actinoplanaceae. Based on morphological characteristics (Baldacci and Locci, 1966; Couch and Bland, 1974), ultrastructural findings (Lechevalier and Lechevalier, 1969), and cell wall analyses (Becker, Lechevalier, and Lechevalier, 1965; Kroppenstedt and Kutzner, 1976; Lechevalier and Lechevalier, 1965; Prauser, 1968; Szaniszlo and Gooder, 1967; Yamaguchi, 1964), the family has now been reduced in taxa to approximately 36 species which are included within 9 genera (Table 1).

The life cycle of a typical member of the Actinoplanaceae has been described by Couch (1949, 1950), Bland and Couch (1968), and others (Fig. 1). The spores (either flagellated or nonflagellated depending on culture conditions and on the genus) germinate, usually via a single germ tube that grows and branches and penetrates into the substrate. This first type of hypha has been referred to by Couch (1950) as a substrate or vegetative hypha. Continued growth results in a young colony that consists of both surface and subsurface hyphae. After several days' growth, branches arise from the substrate hyphae and grow, without further branching, toward the surface to form palisade hyphae in some species (Couch, 1950), and aerial hyphae in others. It is at the tips of the aerial or palisade hyphae that the sporangia with their enclosed sporangiospores are formed.

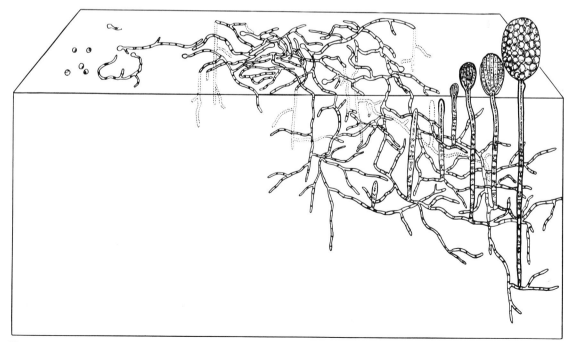

Fig. 1. Life cycle of a typical actinoplanete. See text for description.

Table 1. Genera and species of the Actinoplanaceae.

I. *Actinoplanes* Couch (1950)
   A. philippinensis Couch (1950)
   A. utahensis Couch (1963)
   A. missouriensis Couch (1963)
   A. armeniacus Kalakutskii and Kuznetsov (1964)
   A. brasiliensis Thiemann, Beretta, Coronelli, and Pagani (1969)
   A. italicus Beretta (1973)
   A. rectilineatus Lechevalier and Lechevalier (1975)

II. *Streptosporangium* Couch (1955a)
   S. roseum Couch (1955a)
   S. album Nonomura and Ohara (1960)
   S. amethystogenes Nonomura and Ohara (1960)
   S. vulgare Nonomura and Ohara (1960)
   S. viridialbum Nonomura and Ohara (1960)
   S. bovinum Chaves Batista, Shome, and America de Lima (1963)
   S. indianesis Gupta (1965)
   S. rubrum Potekhina (1965)
   S. viridogriseum Okuda, Furumai, Watanabe, Okugawa, and Kimura (1966)
   S. albidum Furumai, Ogawa, and Okuda (1968)
   S. longisporum Schäfer (1969)
   S. pseudovulgare Nonomura and Ohara (1969)
   S. nondiastaticum Nonmura and Ohara (1969)
   S. corrugatum Williams and Sharples (1976)

III. *Amorphosporangium* Couch (1963)
   A. auranticolor Couch (1963)
   A. globisporus Thiemann (1967)

IV. *Ampullariella* Couch (1963, 1964)
   A. regularis Couch (1963, 1964)
   A. digitata Couch (1963, 1964)
   A. lobata Couch (1963, 1964)
   A. campanulata Couch (1963, 1964)

V. *Spirillospora* Couch (1963)
   S. albida Couch (1963)

VI. *Pilimelia* Kane (1966)
   P. anulata Kane (1966)
   P. terevasa Kane (1966)

VII. *Dactylosporangium* Thiemann, Pagani, and Beretta (1967)
   D. aurantiacum Thiemann, Pagani, and Beretta (1967)
   D. thailandensis Thiemann, Pagani, and Beretta (1967)

VIII. *Planomonospora* Thiemann, Pagani, and Beretta (1967)
   P. parontospora Thiemann, Pagani, and Beretta (1967)
   P. venezuelensis Thiemann (1970)

IX. *Planobispora* Thiemann and Beretta (1968)
   P. longispora Thiemann and Beretta (1968)
   P. rosea Thiemann (19701

Developmentally and structurally, members of the Actinoplanaceae may be divided into two distinct groups: (i) those with spherical, cylindrical to irregular, multispored sporangia and (ii) those with finger-like or pyriform, one- to four-spored sporangia occurring singly or arranged in double parallel rows on the aerial hyphae.

For the first group (Fig. 2), sporangia are formed characteristically above the surface of the substrate at the tips of palisade hyphae. Sporangial initiation occurs by a loosening, at the tip of each palisade hypha, of the outer wall layer of the bilaminate palisade hyphal wall (Fig. 2A, B). Concurrently, a thin sheath of material is secreted external to the outer wall layer. Continued growth of the inner hypha, now considered the sporogenic hypha, into the outer wall layer, now the sporangial wall, forms the young sporangium (Fig. 2C). At this point, depending on the genus, one of two developmental patterns may be followed: the sporogenic hypha may either

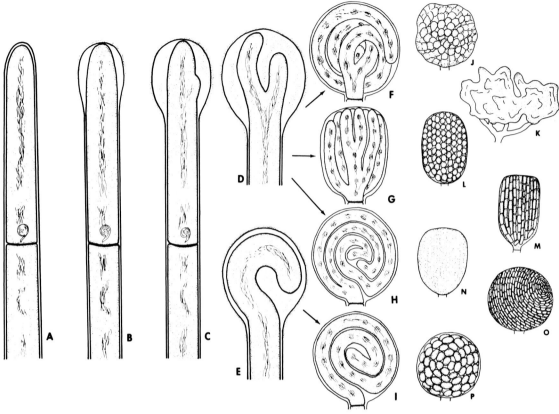

Fig. 2. Sporangial formation in multispored members of the Actinoplanaceae. See text for description.

branch (Fig. 2D) or remain unbranched (Fig. 2E). Three patterns of branching of the sporogenic hypha have been observed: (i) The young sporogenic hypha may branch and coil randomly within the sporangium (Fig. 2F), as in the genera *Actinoplanes* and *Amorphosporangium*. These two genera may be differentiated by the fact that sporangia of *Actinoplanes* are usually spherical to slightly irregular (Fig. 2J), whereas those of *Amorphosporangium* are highly irregular, are larger than those of *Actinoplanes,* and have several discrete coils of spores or sporogenic hyphae within each sporangium (Fig. 2K). (ii) The sporogenic hypha may branch to form parallel rows within the sporangium (Fig. 2G), as in *Ampullariella, Pilimelia,* and a rather unusual member of the genus *Actinoplanes*. These three genera may be separated on the basis of spore size and shape. *Actinoplanes* has round to ovoid spores (Fig. 2L), whereas those of both *Ampullariella* (Fig. 2M) and *Pilimelia* (Fig. 2N) are rod shaped. *Ampullariella* may be distinguished from *Pilimelia* because *Pilimelia* is keratinophilic and has considerably smaller and less distinct spores than does *Ampullariella*. (iii) Sporogenic hyphae of the genus *Spirillospora* branch to form orderly coils (Fig. 2H) that are converted into spiral to rod-shaped spores

(Fig. 2O). Coiling of an unbranched sporogenic hypha inside a sporangium (Fig. 2I) is characteristic of a single genus, *Streptosporangium*. In this genus, the sporogenic hyphae are converted into regular coils of spherical to subspherical spores (Fig. 2P).

The second major group within the Actinoplanaceae contains those forms having finger-like, one- to four-spored sporangia and includes the genera *Dactylosporangium, Planomonospora,* and *Planobispora*. For each of these, sporangial development (Fig. 3) begins via the growth of a sporogenic hypha inside an expanding hyphal wall (Fig. 3A of this chapter; Sharples, Williams, and Bradshaw, 1974; Thiemann and Beretta, 1968), as described for the first major group of Actinoplanaceae. At this point the sporogenic hypha may be transformed into a single spore (Fig. 3B), as in the genus *Planomonospora* (Fig. 3E); into two spores (Fig. 3C), as in the genus *Planobispora* (Fig. 3F); or into a row of three to four spores (Fig. 3D), as in the genus *Dactylosporangium* (Fig. 3G). *Planobispora* and *Dactylosporangium* are characterized by individual sporangia, whereas species of *Planomonospora* produce sporangia in double, parallel rows on aerial hyphae.

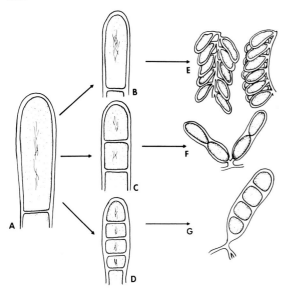

Fig. 3. Sporangial formation in one- to four-spored members of the Actinoplanaceae. See text for description.

Habitats

Although studies concerning occurrence and distribution of members of the family Actinoplanaceae are quite limited, investigations by Couch (1949, 1954, 1957, 1963), Willoughby (1968, 1969a,b, 1971), Johnston and Cross (1976), and Lechevalier (1974) have shed some light on this matter. From these works and others, it is evident that Actinoplanaceae species are of almost universal occurrence in soil, and have been isolated from soil types ranging from forest litter (Van Brummelen and Went, 1957) to beach sand (Williams and Sharples, 1976). The abundance of these forms in soil is exemplified by the fact that during the isolation of over 2,000 members of the Actinoplanaceae from soil samples collected from many parts of the world, Couch found that approximately 66% of the samples tested yielded members of this family. Studies by Willoughby (1968, 1969a,b, 1971) and Johnston and Cross (1976) have shown that actinoplanetes (primarily the genus *Actinoplanes*), in addition to occurring in soil, can be readily recovered from an aquatic environment. In these studies, involving freshwater systems in the English Lake District, members of the genus *Actinoplanes* were frequently isolated from rivers and streams, but only rarely from lakes. Other actinoplanetes isolated from these sites included *Ampullariella* and *Streptosporangium*. In agreement with Willoughby (1971), Johnston and Cross (1976) theorized that the occurrence of actinoplanetes in rivers and streams, but not in lakes, may be due to the greater availability of plant debris as a substrate in the former. Johnston and Cross (1976) suggest further that another factor which may en-

courage the growth of actinoplanetes in rivers and streams may be their generally high oxygen content. The only actinoplanetes found with frequency in lakes have been members of the genus *Streptosporangium*, which may be isolated not from water samples but rather from the surface mud (Johnston and Cross, 1976; Willoughby, 1969a). The nature of this association between lake mud and *Streptosporangium* spp. is not understood at this time and needs further study. Although only a single member of the Actinoplanaceae, *Streptosporangium corrugatum*, has been described from a marine system (Williams and Sharples, 1976), Höhnk, in comments concerning a paper by Willoughby (1968), reported the frequent occurrence of *Actinoplanes* spp. in ocean sediments.

In spite of the numbers and frequency with which actinoplanetes have been isolated, virtually nothing is known of the role played by these forms in the environment. Data collected by Willoughby (1969a) suggest, however, an abundance of actinoplanetes among decaying allochthonous leaf material (primarily *Quercus* spp.) and the presence of cellulase activity among two isolates from this source, *Actinoplanes* (P-4) and *Amorphosporangium* (P-8). It seems likely, therefore, that members of the Actinoplanaceae may play an important role in the primary decomposition of plant material in the soil, in rivers, and in streams. However, much additional work is needed before the actual role in the environment of members of the Actinoplanaceae can be accurately assessed.

Isolation

Although a variety of techniques have been used for the isolation of various members of the Actinoplanaceae, the three procedures described here have proved most efficacious and incorporate the major features of all methods tested to date.

Baiting with Natural Substrates

Baiting for actinoplanetes with natural substrates (pollen, boiled grass leaves, hair, skin, etc.) and the subsequent isolation of these organisms are an expansion of a technique described by Couch (1939) for isolating fungi of the order Chytridiales. It was through this procedure that members of the Actinoplanaceae were first recognized and isolated (Couch, 1949, 1950, 1954). Although somewhat time-consuming and requiring an investigator with either a steady hand or a micromanipulator, this technique results in a high rate of successful subcultures and has produced large numbers and a wide diversity of forms within the Actinoplanaceae.

Isolation of Actinoplanaceae by Baiting with Natural Substrates (Couch, 1954)

A small amount of soil, approximately 1 level teaspoonful, is placed in a sterile Petri dish and flooded with sterile water (distilled water or filtered soil or charcoal water extracts may be used). Natural substances such as pollen (various types of pollen have been utilized including that from members of the genera *Pinus, Liquidambar,* and *Sparganium* [Schäfer, 1973]), grass leaves (pieces of boiled *Paspalum* grass leaves are highly effective), snake skin, hair, etc. are then added to the water so that all baits float at the water surface. After 4–7 days, examination of the water surface with a dissecting microscope ($\times 100$) and strong horizontal lighting should reveal the white, glistening sporangia formed in the air at the surface of the water by any members of the Actinoplanaceae present. For isolation, a selected bait bearing sporangia is transferred to the surface of a 3% agar plate. While being observed with a dissecting microscope, individual sporangia are then excised from the bait and subsequently freed of contaminating bacteria via rolling on the agar surface for a distance of several centimeters. Small blocks ($1/2$ mm^3) bearing individual sporangia are then transferred to a suitable medium for growth of the actinoplanete. After 3–4 days, microscopic examination of the agar block bearing the isolated sporangium should reveal the presence of a young, bacteria-free colony. From this colony, transfers to other growth media may be accomplished. Suitable media for the maintenance, growth, and sporulation of cultures are described later.

Selective Isolation

Although numerous efforts have been made toward the development of a culture medium that selectively enhances the growth of actinomycetes while inhibiting the growth of other organisms, most investigators have utilized either starch casein agar (Küster and Williams, 1964) or chitin agar (Campbell and Williams, 1951; Hsu and Lockwood, 1975; Lingappa and Lockwood, 1961). In testing and comparing these two media, Willoughby, (1968, 1969a) developed a chitin–actidione medium that he found highly effective in the isolation of members of the Actinoplanaceae, as well as other actinomycetes. A summary of Willoughby's procedure for isolation and media preparation follows. Although designed primarily for isolations from leaf litter, this procedure would be applicable to samples obtained from water or from soil washings.

Chitin-Actidione Medium for Selective Isolation of Actinoplanaceae (Willoughby, 1968, 1969a)

Four 3-cm^2 segments from partially decomposed, but not skeletonized, leaves (members of the genus *Quercus* are especially recommended) are transferred to 25 ml sterile, filtered lake water in a 100-ml conical flask. After 2 min agitation, small aliquots (0.5 ml for incorporation, 0.2 ml for surface) from the leaf washings are incorporated into molten agar or spread over its surface (using right-angled glass rods) after it has solidified. Incubation at 25°C for 3 to 6 weeks allows for growth of both slow- and fast-growing forms. Preparation of the chitin agar (Willoughby, 1968) is as follows:

"To make 1 litre of chitin agar 2.5 gms. of purified chitin (K and K Laboratories, Inc., Plainview, New York: through Kodak Ltd) are placed in a small beaker together with 70 ml. of 50% sulphuric acid (Campbell and Williams, 1951). With occasional stirring at room temperature the bulk of the chitin dissolves in 90 minutes or less. The very small insoluble fraction is filtered off by means of a sintered glass crucible and the clear brown solution collected. It is important that the reaction with sulphuric acid should not proceed indefinitely; in that case all the chitin will be completely hydrolysed and none will be recovered subsequently. The chitin solution in acid is poured into 105 ml distilled water whereon the chitin re-appears as a fine white precipitate. This precipitate is centrifuged down and the acid supernatant discarded. Using distilled water successive re-suspensions centrifugations and supernatant discardings are made until the pH approaches 4.0. The chitin suspension is then transferred to a 1 litre graduated flask and the remainder of the medium added. This comprises 0.02 g CaCO$_3$; 0.01 g FeSO$_4 \cdot$7H$_2$O; 1.71 g KCl; 0.05 g MgSO$_4 \cdot$7H$_2$O; 4.11 g Na$_2$HPO$_4 \cdot$12H$_2$O. Final adjustment of the pH to 7.0 or above is made with sulphuric acid. The litre of solution is dispensed into flasks and a good quality agar (1.8% Oxoid Ionagar No. 2) added before autoclaving."

To retard the growth of fungal contaminants, actidione (50 μg/ml agar) may be added to the chitin agar medium (Williams and Davis, 1965; Willoughby, 1969a). Attempts by Willoughby (1969a) to reduce contaminating bacteria and fungi by using, respectively, potassium tellurite and nystatin as additives to the chitin agar resulted in either inhibition of the actinomycetes (potassium tellurite) or no improvement over the use of actidione alone (nystatin).

Damp Incubation

For direct examination and subsequent isolation of actinoplanetes growing directly on a natural substrate, Willoughby (1968) devised the following procedure.

Damp Incubation of Actinoplanaceae on a Natural Substrate (Willoughby, 1968)

Freshly collected, decaying leaf litter is first washed in sterile lake water to remove adhering detritus. Selected leaf fragments are then placed in Petri dishes that have been lined with wetted filter paper and Kleenex tissue. The Petri plates are next sealed with tape and incubated for 4 weeks or longer at 25°C. Observable actinoplanetes may then be isolated by either of the two techniques just described. It should be noted that this procedure is especially useful in yielding members of the genus *Actinoplanes*.

Media for Culture and Sporulation

Most members of the family Actinoplanaceae may be cultured routinely on either Czapek agar (Difco) or on Czapek agar supplemented by 5 gm peptone (Difco). Following isolation and/or continuous culture, sterile actinoplanetes may at times be induced to sporulate on oatmeal agar (W.M. Stark, unpublished). Keratinophilic actinoplanetes may be most readily isolated on Emerson's YPSS agar, but grow best on Czapek agar plus peptone (Kane, 1966). Because these various culture media may not be easily available in a prepared form, instructions for their preparation follow.

Czapek Agar for Culturing Actinoplanaceae (Waksman, 1950)

To 1 liter of distilled water add:

Sucrose	30 g
Sodium nitrate	3 g
Dipotassium phosphate	1 g
Magnesium sulfate	0.5 g
Potassium chloride	0.5 g
Ferrous sulfate	0.01 g
Agar	15 g

After components are dissolved, the medium is distributed into flasks or tubes and then autoclaved.

For peptone Czapek agar, substitute 5 g peptone for the 3 g sodium nitrate in Czapek agar (J. Couch, unpublished). For oatmeal agar (W. M. Stark, unpublished), add the following to one liter of distilled water: baby oatmeal, 65 g; agar, 15 g. Autoclave 30 min, wait 24 h, then autoclave again before dispensing.

Identification

As mentioned in the Introduction, the family Actinoplanaceae includes those hyphae-forming, Gram-positive, non-acid-fast members of the Actinomycetales in which spores are produced within sporangia. Additionally, all members of the Actinoplanaceae may be grouped chemically in that they possess either a type II or a type III cell wall (Cross and Goodfellow, 1973). At the generic level (see the Introduction for morphological characteristics of genera), members of the Actinoplanaceae are grouped on the basis of sporangial shape and size; spore arrangement, number, and flagellation; presence or absence of aerial mycelia; and cell wall type (II or III). Species determinations within genera of the Actinoplanaceae are usually difficult, but are based on such features as sporangial shape and size, spore size, colony color, natural substrate, and physiology. Persons interested in identifying members of the Actinoplanaceae beyond that possible by reference to the Introduction are referred to the eighth edition of *Bergey's Manual of Determinitive Bacteriology* (Buchanan and Gibbons, 1974) or to the excellent treatment of this family by Cross and Goodfellow (1973).

Literature Cited

Baldacci, E., Locci, R. 1966. A tentative arrangement of the genera in Actinomycetales. Giornale di Microbiologie **14:**131–139.

Becker, B., Lechevalier, M. P., Lechevalier, H. A. 1965. Chemical composition of cell-wall preparations from strains of various form-genera of aerobic actinomycetes. Applied Microbiology **13:**236–243.

Beretta, G. 1973. *Actinoplanes italicus,* a new red-pigmented species. International Journal of Systematic Bacteriology **23:**37–42.

Bland, C. E., Couch, J. N. 1968. Observations on the chromatic bodies of two species of the Actinoplanaceae. Journal of General Microbiology **53:**95–100.

Buchanan, R. E., Gibbons, N. E. (eds.). 1974. Bergey's manual of determinitive bacteriology, 8th ed. Baltimore: Williams & Wilkins.

Campbell, L. L., Williams, O. B. 1951. A study of chitin-decomposing micro-organisms of marine origin. Journal of General Microbiology **5:**894–905.

Chaves Batista, A., Shome, S. K., America de Lima, J. 1963. *Streptosporangium bovinum* sp. nov. from cattle hoofs. Dermatologia Tropica et Ecologica Geographica **2:**49.

Couch, J. N. 1939. Technic for collecting, isolation and culture of chytrids. Journal of the Elisha Mitchell Scientific Society **55:**208–214.

Couch, J. N. 1949. A new group of organisms related to *Actinomyces*. Journal of the Elisha Mitchell Scientific Society **65:**315–318.

Couch, J. N. 1950. *Actinoplanes*, a new genus of the Actinomycetales. Journal of the Elisha Mitchell Scientific Society **66:**87–92.

Couch, J. N. 1954. The genus *Actinoplanes* and its relatives. Transactions of the New York Academy of Sciences **16:**315–318.

Couch, J. N. 1955a. A new genus and family of the Actinomycetales, with a revision of the genus *Actinoplanes*. Journal of the Elisha Mitchell Scientific Society **71**:148–155.

Couch, J. N. 1955b. Actinosporangiaceae should be Actinoplanaceae. Journal of the Elisha Mitchell Scientific Society **71**:269.

Couch, J. N. 1957. A new horizon in soil microbiology. Proceedings of the National Academy of Sciences (India) **27**:69–73.

Couch, J. N. 1963. Some new genera and species of the Actinoplanaceae. Journal of the Elisha Mitchell Scientific Society **79**:53–70.

Couch, J. N. 1964. The name *Ampullaria* has been replaced by *Ampullariella*. Taxon **14**:137.

Couch, J. N., Bland, C. E. 1974. Actinoplanaceae, pp. 706–723. In: Buchanan, R. E., Gibbons, N. E. (eds.), Bergey's manual of determinitive bacteriology, 8th ed. Baltimore: Williams & Wilkins Company.

Cross, T., Goodfellow, M. 1973. Taxonomy and classification of the actinomycetes, pp. 11–112. In: Sykes, G., Skinner, F. A. (eds.), The Actinomycetales. New York: Academic Press.

Furumai, T., Ogawa, H., Okuda, T. 1968. Taxonomic study on *Streptosporangium albidum* nov. sp. Journal of Antibiotics **21**:179–181.

Gupta, K. C. 1965. A new species of the genus *Streptosporangium* isolated from Indian soil. Journal of Antibiotics **18**:125–127.

Hsu, S. C., Lockwood, J. L. 1975. Powdered chitin agar as a selective medium for enumeration of Actinomycetes in water and soil. Applied Microbiology **29**:422–426.

Johnston, D. W., Cross, T. 1976. The occurrence and distribution of actinomycetes in lakes of the English Lake District. Freshwater Biology **6**:457–463.

Kalakoutskii, L. V., Kuznetsov, V. D. 1964. A new species of the genus *Actinoplanes* Couch, *Actinoplanes armeniacus* n. sp. and some peculiarities of its mode of spore formation. [In Russian, with English summary.] Mikrobiologiya **33**:613–621.

Kane, W. D. 1966. A new genus of the Actinoplanaceae, *Pilimelia*, with a description of two species, *Pilimelia terevasa* and *Pilimelia anulata*. Journal of the Elisha Mitchell Scientific Society **82**:220–230.

Kroppenstedt, R. M., Kutzner, H. J. 1976. Biochemical markers in the taxonomy of the Actinomycetales. Experientia **32**:318.

Küster, E., Williams, S. T. 1964. Selection of media for isolation of streptomycetes. Nature **202**:928–929.

Lechevalier, H. 1974. Distribution et role des actinomycetes dans les eaux. Bulletin de l'Institut Pasteur **72**:159–175.

Lechevalier, H., Lechevalier, M. P. 1965. Classification des actinomycetes aerobies basée sur leur morphologie et leur composition chimique. Annales de l'Institut Pasteur **108**:662–673.

Lechevalier, H., Lechevalier, M. P. 1969. Ultramicroscopic structure of *Intrasporangium calvum* (Actinomycetales). Journal of Bacteriology **100**:522–525.

Lechevalier, M. P., Lechevalier, H. 1975. Actinoplanete with cylindrical sporangia, *Actinoplanes rectilineatus* sp. nov. International Journal of Systematic Bacteriology **25**:371–376.

Lingappa, Y., Lockwood, J. L. 1961. A chitin medium for isolation, growth and maintenance of actinomycetes. Nature **189**:158–159.

Nonomura, H., Ohara, Y. 1960. Distribution of the actinomycetes in soil. (V). The isolation and classification of the genus *Streptosporangium*. Journal of Fermentation Technology **38**:405–409.

Nonomura, H., Ohara, Y. 1969. Distribution of actinomycetes in soil. (VI). A culture method effective for both preferential isolation and enumeration of *Microbispora* and *Streptosporangium* strains in soil. Journal of Fermentation Technology **47**:463–469.

Okuda, T., Furumai, T., Watanabe, E., Okugawa, Y., Kimura, S. 1966. Actinoplanaceae antibiotics. II. Studies on sporaviridin. 2. Taxonomic study on the sporaviridin producing microorganism, *Streptosporangium viridogriseum* nov. sp. Journal of Antibiotics **19**:121–127.

Potekhina, L. I. 1965. *Streptosporangium rubrum* n. sp.—a new species of the *Streptosporangium* genus. [In Russian, with English Summary.] Mikrobiologiya **34**:292–299.

Prauser, H. 1970. Characters and genera arrangement in the Actinomycetales, pp. 245–247. In: Prauser, H. (ed.), The Actinomycetales. The Jena International Symposium on Taxonomy, September 1968. Jena: VEB Gustav Fischer Verlag.

Schäfer, D. 1969. Eine neue *Streptosporangium*-Art aus türkischer Steppenerde. Archiv für Mikrobiologie **66**:365–373.

Schäfer, D. 1973. Beiträge zur Klassifizierung und Taxonomie der Actinoplanaceen. Dissertation. Marburg/Lahn.

Sharples, G. P., Williams, S. T., Bradshaw, R. M. 1974. Spore formation in the Actinoplanaceae (Actinomycetales). Archiv für Mikrobiologie **101**:9–20.

Shchepkina, T. V. 1940. Investigation and description of cotton fibre endoparasites (Trans.). Bulletin of the Academy of Sciences USSR (Biology Series) **5**:643–661.

Szaniszlo, P. J., Gooder, H. 1967. Cell wall composition in relation to the taxonomy of some Actinoplanaceae. Journal of Bacteriology **94**:2037–2047.

Thiemann, J. E. 1967. A new species of the genus *Amorphosporangium* isolated from Italian soil. Mycopathologia et Mycologia Applicata **33**:233–240.

Thiemann, J. E. 1970. Study of some new genera and species of the Actinoplanaceae, pp. 245–257. In: Prauser, H. (ed.), The Actinomycetales. The Jena International Symposium on Taxonomy, September 1968. Jena: Gustav Fischer Verlag.

Thiemann, J. E., Beretta, G. 1968. A new genus of the Actinoplanaceae: *Planobispora*, gen. nov. Achiv für Mikrobiologie **62**:157–166.

Thiemann, J. E., Pagani, H., Beretta, G. 1967. A new genus of the Actinoplanaceae: *Planomonospora* gen. nov. Giornale di Microbiologie **15**:27–38.

Thiemann, J. E., Beretta, G., Coronelli, D., Pagani, H. 1969. Antibiotic production by new form-genera of the Actinomycetales. II. Antibiotic A/672 isolated from a new species of *Actinoplanes: Actinoplanes brasiliensis* nov. sp. Journal of Antibiotics **22**:119–125.

Van Brummelen, J., Went, J. C. 1957. *Streptosporangium* isolated from forest litter in the Netherlands. Antonie van Leeuwenhoek Journal of Microbiology and Serology **23**:385–392.

Waksman, S. A. 1950. The actinomycetes: Their nature, occurrence, activities, and importance. Annals of Cryptogamic Phytopathology **9**:1–230.

Williams, S. T., Davis, F. L. 1965. Use of antibiotics for selective isolation and enumeration of Actinomycetes in soil. Journal of General Microbiology **38**:251–261.

Williams, S. T., Sharples, G. P. 1976. *Streptosporangium corrugatum* sp. nov., an actinomycete with some unusual morphological features. International Journal of Systematic Bacteriology **26**:45–52.

Willoughby, L. G. 1968. Aquatic Actinomycetales with particular reference to the Actinoplanaceae. Veröffentlichungen des Institutes für Meeresforschung in Bremerhaven **3**:19–26.

Willoughby, L. G. 1969a. A study on aquatic actinomycetes. The allochthonous leaf component. Nova Hedwigia **18**:45–113.

Willoughby, L. G. 1969b. A study of the aquatic actinomycetes of Belham Tarn. Hydrobiologia **34**:465–483.

Willoughby, L. G. 1971. Observations on some aquatic actinomycetes of streams and rivers. Freshwater Biology **1**:23–27.

Yamaguchi, T. 1964. Comparison of the cell wall composition of morphologically distinct actinomycetes. Journal of Bacteriology **89**:444–453.

CHAPTER 154

The Family Dermatophilaceae

DAVID S. ROBERTS

The family Dermatophilaceae Austwick 1958 is essentially bacterial in character but produces branching filaments and, therefore, fits logically and comfortably in the order Actinomycetales. Dermatophilaceae currently consists of two genera, each with a single species.

Dermatophilus congolensis is by far the better known of the two organisms. Following the original description by Van Saceghem (1915) in the Belgian Congo, an extensive series of reports has revealed the worldwide distribution of this animal pathogen and has provided detailed knowledge of its form and behavior both in cultures and in the tissues of its several hosts.

Geodermatophilus obscurus, a soil organism, was described in 1968 by Luedemann, who proposed its inclusion in Dermatophilaceae mainly on the basis of a single morphological feature: *D. congolensis* and *G. obscurus* divide in both transverse and longitudinal planes to form packets or clusters of cuboid cells or cocci. This may be an excessive significance to attach to a simple structural characteristic that is not confined to these organisms but is found also in certain algae and fungi (Austwick, 1958).

DERMATOPHILUS CONGOLENSIS

Habitats

Outside the artificial environment of the laboratory, *Dermatophilus congolensis* is apparently an obligate parasite; it has never been recovered from any natural source other than the diseased tissues of infected animals. This is attributable to the poor ability of *D. congolensis* to survive in the external environment or to compete with various natural flora. These deficiencies were demonstrated in a number of soils, in liquid culture media, and in scabs from infected lesions when those scabs had undergone microbial degradation under wet conditions. The organism's viability declines rapidly even in scabs kept dry (Roberts, 1967b, 1970).

Dermatophilus infection has been reported from virtually every region of the world in both wild and domestic animals, mainly herbivores. A recent count listed 25 species known to have been naturally infected (Richard and Shotts, 1976). There have been a few cases in man following the handling of cultures or infected animals but it is debatable whether these should all be regarded as natural infections.

The vast majority of cases are found as infections of the epidermis in herbivores. The disease tends to be far more severe and extensive in domestic animals, especially those raised under conditions very different from their natural environments, such as European cattle in tropical Africa or merino sheep in the wetter regions of Australia.

In contrast to the dermatophytes, which are parasites of the stratum corneum, *D. congolensis* invades only the living layer of the epidermis that covers the skin and lines the hair follicles. The epidermis provides a habitat where *D. congolensis* can proliferate freely without having to compete with other microorganisms. *D. congolensis* is confined to the epidermis by the cellular exudate that collects beneath the basement membrane in response to the infection. In rabbits depleted of polymorphonuclear (PMN) leukocytes with nitrogen mustard, by contrast, this cellular response is suppressed, and the filaments of *D. congolensis* freely pervade the subepidermal connective tissues (Roberts, 1965c).

As soon as *D. congolensis* invades the epidermal cells they begin to cornify; thus, the completion of the organism's life cycle and the conversion of the infected epidermis to keratin proceed together. Histologically, *D. congolensis* is found within the bands of keratin that traverse the scab.

The inflammatory response induces an epidermal hyperplasia, and the damaged tissue is soon replaced. However, hyphae in neighboring follicle sheaths may extend branches into the new layer of epidermis, inducing further cellular exudation that will rapidly separate the new tissue from its dermal matrix. The cycle of replacement and reinfection is continually repeated in chronic cases, and layers of exudate and infected keratin are shed alternately to produce the characteristically banded scab. The keratinous bands lying parallel to the skin are traversed at right angles by hair or wool fibers within tubes of keratin from the hyperplastic follicle sheaths, giving the scab a complex and very tough structure. The disease is often described as an ''exudative'' dermatitis but the term is inadequate; the crusts or scabs generated by the infected skin are clearly not just structureless mounds of dried exudate.

Factors Limiting the Infection

To persist in the skin, *Dermatophilus congolensis* must keep invading fresh epidermis. Otherwise it will be separated from the skin by the cellular exudate and carried away in the scab. The only route available for continuous penetration and for repeated lateral access to the new epidermis is provided by the epidermis of the follicle sheaths.

The infection is terminated by factors that halt the penetration of the follicle sheaths. This can be brought about artificially through the internal administration of suitable antibacterial antibiotics (Roberts, 1967a). In nature, the invasion is apparently stopped by an inhibitory factor diffusing from the cellular exudate. The infiltration of leukocytes and the associated inhibition of hyphal growth are intensified with the onset of the delayed hypersensitivity reaction. In animals previously hypersensitized by vaccination or infection, these intensified effects are seen within the first 24 h. As a result, the infection usually runs an acute course with early healing. In chronic cases, the cellular response is fully developed but evidently ineffectual. It has been suggested that the leukocytes of chronically affected individuals might be deficient in inhibitory power (Roberts, 1967b).

Transmission of the Disease

The poor ability of *Dermatophilus congolensis* to survive in the external environment, or even in the infected scab, implies that active cases of disease are necessary to initiate new infection and to provide for the survival of *D. congolensis* (Roberts, 1967b). Apart from farm livestock, there are many wild animals in the list of Richard and Shotts (1976) that might serve as natural reservoirs of infection.

Zoospores develop in vast numbers in the scab and are the agents of infection. When the scab is wetted, the zoospores migrate to the surface where they are available for transfer to new susceptible areas of skin (Roberts, 1961, 1963, 1967b). The zoospores are usually transferred from one animal to another through contact with plants or insects. The process is evidently more efficient, and the resulting infections more numerous and severe, when the zoospores are conveyed by spiny plants or biting insects that also penetrate or damage the protective sebaceous and keratinous barriers on the skin sur-

face. Reports on various aspects of zoospore transfer and the spread of infection are reviewed by Roberts (1967b) and by Richard and Shotts (1976).

Atypical Infections

There has been a small number of reports of *Dermatophilus* infection in the deeper tissues of cats. In most cases the causal agent was not recovered or positively identified, but Simmons, Sullivan, and Green (1972) isolated *D. congolensis* from a subcutaneous abscess in a lizard. Jones (1976) recovered *D. congolensis* from the inflamed popliteal lymph node of a cat and used his isolate to induce typical *Dermatophilus* infections in the skin of sheep.

One can only speculate as to why *D. congolensis* apparently broke the rules and invaded the deeper tissues in these cases. Perhaps, in the cats, feline panleukopenia had weakened the cellular defenses and allowed subepidermal infection in the same way that nitrogen mustard injections deplete rabbits of PMN leukocytes and thus permit hyphal penetration beyond the epidermis.

Isolation

Fresh Material

Dermatophilus congolensis is readily recovered from newly formed scabs taken from active or recently healed lesions. Material from the base of the scab is chopped or minced in a small amount of sterile nutrient broth or distilled water, and then plated on blood agar (5–10% defibrinated sheep blood in a blood agar base from Difco, Oxoid, BBL, or other reputable medium manufacturer). The plates are incubated at 37°C, preferably for 2 days in 10% CO_2, to encourage a high yield of good-sized colonies, followed by a day in air. The agar should be at least 3–4 mm deep to sustain the colonies through 3 days of growth. During the initial period, in CO_2, the colonies are grayish, dry, tough, and firmly anchored to the agar by invading hyphae. Examination under the stereoscopic microscope with lateral illumination usually reveals that the colonies are covered by aerial hyphae.

During the final period, in air, sporulation thickens the superficial part of the colony and converts it into a moist, mucoid, amber- or cream-colored layer readily detachable with a wire loop. If a little of this mucoid material is suspended in nutrient broth and examined microscopically it will be found to be swarming with zoospores, motile oval or coccoid bodies usually less than a micron in diameter. At this stage, material can be detached for subculture or other purposes. Recognition is usually facilitated by the presence of a zone of complete hemolysis around the colonies. If horse blood is used in the medium instead of sheep blood, the growth will be as good but there will be no hemolysis.

Old Material

In older scab material, the viability of *Dermatophilus congolensis* will be much lower. There may be no viable cells if the scabs are very old, or if the scabs have been wet and other bacteria have proliferated in them. Consequently, direct culture from older scab material may prove unsuccessful. In that case, it is best to apply the suspended scab material to the scarified skin of a guinea pig or sheep. The scabs from any resulting infection are then suspended in liquid and plated on blood agar as in the method for fresh material.

The rabbit and the mouse are much less suitable for isolation purposes. Their very vigorous and effective cellular response drastically diminishes both the chance of infection and the proliferation of *D. congolensis* in any lesion that may result. Guinea pigs and sheep hypersensitized by a previous infection are unsuitable because their accelerated cellular response has a similar adverse effect (Roberts, 1967b).

Maintenance in Culture

In the short term, the organism can be maintained on blood agar slants in screw-capped bottles. Moist growth is readily induced if the caps are unscrewed before the 3rd day of incubation. The cultures should then be sealed and refrigerated, with transfer to new slants every 1–2 weeks.

For longer term maintenance, it is best to freeze-dry *D. congolensis*. Fresh, moist material from colonies on blood agar, suspended in nutrient broth or in any conventional desiccating medium, will give good results if lyophilized and then sealed under vacuum. The dried preparations may be stored at room temperature.

Identification

The first indication that one is dealing with *Dermatophilus congolensis* may be provided by the nature of the source material. Raised scabs from the skin of a herbivorous animal should immediately suggest a diagnosis of *Dermatophilus* infection.

A suspension may be prepared by chopping up some scab material in sterile nutrient broth or distilled water. Under the microscope, wet preparations may be found to contain actively motile zoospores. In stained smears, *D. congolensis* will appear as Gram-positive branched filaments which, in many cases, are greatly thickened. Closer examination will show that the thickened forms are composed of cocci packed together in disks to give the filament a cross-banded appearance.

In cultures, the appearance and consistency of the colonies on sheep blood agar, described under Isolation, are highly characteristic. So are the rapidly moving zoospores and the filaments in various stages of division, observed in suspensions made from the moist, mucoid layer that coats the colonies toward the end of incubation.

Morphology

The life cycle and morphology of *Dermatophilus congolensis* have been described by Gordon (1964), Gordon and Edwards (1963), Roberts (1961), and Van Saceghem (1915). On germinating, the zoospores lose their motility and bud to produce a hypha or filament which is usually less than a micron wide. As the filament extends at its growing tip, transverse septation begins at the other end and progresses forward. Branching also starts in the oldest part of the filament. Transverse septation is followed by division in longitudinal planes, and this causes the structure to thicken. This process also begins in the oldest part of the filament, producing a tapered appearance. The branches divide in the same manner as the original filament. When the processes of transverse and longitudinal division are complete, the resulting cells become separated and rounded, and are ready to leave the parent structure as fully motile zoospores. At this stage they are enclosed only by the thick, irregular mass of capsular material that develops during the division of the filament. This material is responsible for the moist, mucoid consistency of the mature colony. The original hyphal wall does not persist as a separate structure. When the hypha divides, its wall is incorporated into the walls of the resulting segments.

The life cycle is most readily observed in cultures in nutrient broth. Some of its features are illustrated in Fig. 1. There is a difference in the mode of division between organisms within the infected lesion and those in cultures. In the lesion, where the organism is encased within epidermis, transverse septation is followed by division exclusively in longitudinal planes, so that each segment of the original filament gives rise to a disk-shaped assembly of zoospores. This produces the cross-banded appearance described in the second paragraph of this section. In cultures, by contrast, the dividing structure is unconfined and can expand longitudinally. As a result, the later divisions proceed in both directions so that each segment of the filament gives rise first to a cuboid packet of cells and finally to a mulberry-like cluster of zoospores.

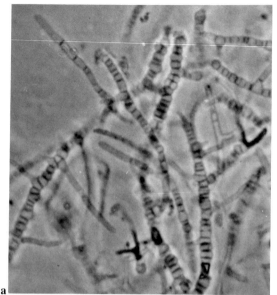

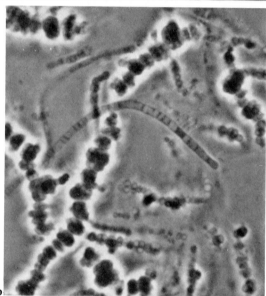

Fig. 1a and b. Phase contrast photomicrographs of *Dermatophilus congolensis* suspended in a 28% albumin solution to reveal cellular detail. × 2,000 (approx.). (a) Branching hyphae in early stages of division that is mainly transverse, although a few longitudinal septa are already apparent. (b) Mulberry-like clusters of cocci enveloped in mucoid capsular material that appears light against the denser, more refractive, albumin solution. (Not to be confused with the halo surrounding a dense object in a mounting medium of low refractive index.)

Biochemistry

The biochemical activities of *Dermatophilus congolensis* are described by Gordon (1964) and Roberts (1965b). Apparently all isolates produce acid but not gas from glucose and fructose, and there is often slight or delayed acid production from galactose and maltose. All are proteolytic and most pro-

duce hemolysin and a phospholipase A. They all hydrolyze urea and starch and produce catalase. Many of these activities become apparent only during sporulation.

Cell Wall Composition

In common with *Thermoactinomyces vulgaris* and species of *Microbispora* and *Nocardia*, *Dermatophilus congolensis* has a cell wall of type III (Becker, Lechevalier, and Lechevalier, 1965). Whole cell hydrolysates contain madurose (Lechevalier, Lechevalier, and Gerber, 1971).

Serology

Eight isolates of *Dermatophilus congolensis* examined by Roberts (1965a) had antigenically identical somatic agglutinogens, hemolysins, and precipitinogens. Flagellar agglutinogens were variable.

Susceptibility to Antimicrobial Agents

Dermatophilus congolensis is susceptible to a wide range of antibacterial antibiotics, including penicillin, streptomycin, chloramphenicol, erythromycin, and the tetracyclines (Roberts, 1967a). It is resistant to the antifungal agents griseofulvin, nystatin, and tolnaftate (Luedemann, 1968; Roberts, 1970).

GEODERMATOPHILUS OBSCURUS

Habitats

Geodermatophilus obscurus was isolated from several soil samples collected in the western United States (Luedemann, 1968). Apparently it has not been studied or observed within the soil or in any other natural habitat. *G. obscurus* is not infective for the skin of rabbits or guinea pigs (Luedemann, 1968).

Isolation

Geodermatophilus obscurus may be recovered from soil by plating on an agar medium described by Luedemann (1968).

Isolation Medium for *Geodermatophilus obscurus* (Luedemann, 1968)

Yeast extract (Difco)	5.0 g
Malt extract broth (Difco)	15.0 g
Soluble starch (Difco)	10.0 g

Sucrose	10.0 g
CaCO$_3$	2.0 g
Agar	20.0 g
Distilled water	1,000 ml

After 30 days' incubation on this medium at 24–28°C the colonies are dark brown or black, dry, flat to plicate, and granular.

G. obscurus can be maintained by serial transfer on the same medium. It apparently has no resistant stage in its life cycle.

Identification

Geodermatophilus is easily distinguishable from *Dermatophilus* because in almost every respect the two organisms are utterly different. The following examples are based on the paper of Luedemann (1968):

1. The medium recommended for the isolation and maintenance of *G. obscurus* will not support the growth of *D. congolensis*.
2. Brain heart infusion agar (Difco) supports poor growth of *G. obscurus* but good growth of *D. congolensis*.
3. *G. obscurus* grows best at 24–28°C, *D. congolensis* at 37°C.
4. *G. obscurus* colonies are very darkly pigmented, brown to black; those of *D. congolensis* are pale golden.
5. *G. obscurus* rarely produces filaments or zoospores and its filaments are poorly developed; most of its growth is by repeated division of cuboid or coccoid cells. *D. congolensis* normally passes through its complete cycle from zoospore germination to mycelial branching and the release of fresh zoospores in every culture.
6. The zoospores of *G. obscurus* may bud to form more zoospores without losing their motility, in contrast to those of *D. congolensis*.

Biochemistry

Unlike *Dermatophilus congolensis*, *Geodermatophilus obscurus* is not hemolytic and usually is not proteolytic. Nitrate reduction is weak or absent. Carbohydrate utilization and acid production vary with the isolate; starch is hydrolyzed (Luedemann, 1968).

Cell Wall Composition

The cell wall of *Geodermatophilus obscurus* is of type III but, in contrast to *Dermatophilus congolensis,* whole cell hydrolysates do not contain madurose (Lechevalier, Lechevalier, and Gerber, 1971).

Susceptibility to Antimicrobial Agents

The response of *Geodermatophilus obscurus* to antibiotics confirms its identity as a bacterium. Luedemann (1968) found that it was resistant to the antifungal agents griseofulvin, nystatin, and tolnaftate, but susceptible to the antibacterial antibiotics chlortetracycline, novobiocin, erythromycin, streptomycin, neomycin, and gentamycin.

Literature Cited

Austwick, P. K. C. 1958. Cutaneous streptothricosis, mycotic dermatitis and strawberry foot rot and the genus *Dermatophilus* Van Saceghem. Veterinary Reviews and Annotations 4:33–48.
Becker, B., Lechevalier, M. P., Lechevalier, H. A. 1965. Chemical composition of cell-wall preparations from strains of various form-genera of aerobic actinomycetes. Applied Microbiology 13:236–243.
Gordon, M. A. 1964. The genus *Dermatophilus*. Journal of Bacteriology 88:509–522.
Gordon, M. A., Edwards, M. R. 1963. Micromorphology of *Dermatophilus congolensis*. Journal of Bacteriology 86:1101–1115.
Jones, R. T. 1976. Subcutaneous infection with *Dermatophilus congolensis* in a cat. Journal of Comparative Pathology 86:415–421.
Lechevalier, H. A., Lechevalier, M. P., Gerber, N. N. 1971. Chemical composition as a criterion in the classification of actinomycetes. Advances in Applied Microbiology 14:47–72.
Luedemann, G. M. 1968. *Geodermatophilus*, a new genus of the *Dermatophilaceae (Actinomycetales)*. Journal of Bacteriology 96:1848–1858.
Richard, J. L., Shotts, E. B. 1976. Wildlife reservoirs of dermatophilosis, pp. 205–214. In: Page, L. A. (ed.), Wildlife diseases. New York: Plenum.
Roberts, D. S. 1961. The life cycle of *Dermatophilus dermatonomus*, the causal agent of ovine mycotic dermatitis. Australian Journal of Experimental Biology and Medical Science. 39:463–476.
Roberts, D. S. 1963. The release and survival of *Dermatophilus dermatonomus* zoospores. Australian Journal of Agricultural Research 14:386–399.
Roberts, D. S. 1965a. Cutaneous actinomycosis due to the single species *Dermatophilus congolensis*. Nature 206:1068.
Roberts, D. S. 1965b. Penetration and irritation of the skin by *Dermatophilus congolensis*. British Journal of Experimental Pathology 46:635–642.
Roberts, D. S. 1965c. The role of granulocytes in resistance to *Dermatophilus congolensis*. British Journal of Experimental Pathology 46:643–648.
Roberts, D. S. 1967a. Chemotherapy of epidermal infection with *Dermatophilus congolensis*. Journal of Comparative Pathology 77:129–136.
Roberts, D. S. 1967b. *Dermatophilus* infection. Veterinary Bulletin 37:513–521.
Roberts, D. S. 1970. *Dermatophilus congolensis*, a zoopathogenic actinomycete with a motile infective stage, pp. 265–271. In: Prauser, H. (ed.), The Actinomycetales. Jena International Symposium on Taxonomy. Jena: Gustav Fischer Verlag.
Simmons, G. C., Sullivan, N. D., Green, P. E. 1972. Dermatophilosis in a lizard (*Amphibolurus barbatus*). Australian Veterinary Journal 48:465–466.
Van Saceghem, R. 1915. Dermatose contagieuse (impétigo contagieux). Bulletin de la Société de Pathologie Exotique 8:354–359.

The Genera *Nocardia* and *Rhodococcus*

MICHAEL GOODFELLOW and DAVID E. MINNIKIN

In the eighth edition of *Bergey's Manual of Determinative Bacteriology,* the family Nocardiaceae (McClung, 1974) contains *Nocardia* (Trevisan, 1889) and *Pseudonocardia* (Henssen, 1957); *Actinomadura* (Lechevalier, H. A., and Lechevalier, 1970), *Gordona* (Tsukamura, 1971), *Jensenia* (Bisset and Moore, 1950), and *Oerskovia* (Prauser, Lechevalier, and Lechevalier, 1970) are described as genera incertae sedis. Subsequent studies have excluded *Actinomadura* and *Oerskovia* (Goodfellow and Minnikin, 1977) and have reduced *Gordona* and *Jensenia* to synonyms of the resurrected genus *Rhodococcus* (Goodfellow and Alderson, 1977); the position of *Pseudonocardia* remains uncertain (this Handbook, Chapter 158).

In addition to the genera *Nocardia* and *Rhodococcus,* the family Nocardiaceae (Cross and Goodfellow, 1973; Goodfellow and Minnikin, 1977) contains *Actinopolyspora, Micropolyspora* (this Handbook, Chapter 158), *Saccharomonospora* (this Handbook, Chapter 157), and *Saccharopolyspora* (this Handbook, Chapter 158). The family accommodates aerobic, Gram-positive actinomycetes that produce a rudimentary-to-extensive substrate mycelium, which usually fragments into bacillary and coccoid elements, and that have *meso*-diaminopimelic acid, arabinose, and galactose in their walls (wall chemotype IV *sensu* Lechevalier and Lechevalier, 1965; Lechevalier, M. P., and Lechevalier, 1970). Extensive aerial mycelium, bearing arthrospores, may be formed (*Actinopolyspora, Micropolyspora, Nocardia, Saccharomonospora, Saccharopolyspora*), although arthrospores can also be found on the substrate mycelium (*Micropolyspora*). Some strains possess characteristic mycolic acids, high molecular weight 3-hydroxy acids having a long-alkyl branch in the 2-position (*Nocardia, Rhodococcus, Micropolyspora brevicatena, Micropolyspora fascifera*), but others (*Saccharomonospora, Saccharopolyspora, Micropolyspora faeni*) do not have these acids (Collins et al., 1977; Goodfellow and Minnikin, 1977; Minnikin, Goodfellow, and Alshamaony, 1978). Mycolic acids were not found in *Actinopolyspora* strains (Gochnauer et al., 1975). *Nocardia* is the type genus.

Classification

Nocardia Trevisan 1889

Nocardiae are aerobic, Gram-positive, acid- to partially acid-fast actinomycetes that produce a primary mycelium, which fragments into bacillary and coccoid elements. The organisms grow well on ordinary laboratory media at 30°C and usually produce an aerial mycelium that may be differentiated into arthrospores. However, the consistency and composition of the growth medium can profoundly affect the growth and stability of both aerial and substrate hyphae (Williams et al., 1976). Nocardiae contain mycolic acids, have a wall chemotype IV, and a G+C range of 64–69 mol% (Goodfellow and Minnikin, 1977). The redefined genus *Nocardia* contains the well-established taxa, *N. asteroides, N. brasiliensis,* and *N. otitidis-caviarum* (*N. caviae*); and the less well-studied *N. amarae, N. carnea, N. transvalensis,* and *N. vaccinii* (Goodfellow and Minnikin, 1977; Gordon, Mishra, and Barnett, 1978; Lechevalier, 1977; Lechevalier and Lechevalier, 1974).

N. brasiliensis and *N. otitidis-caviarum* are good taxospecies (Goodfellow, 1971; Sneath, 1976), but *N. asteroides* has been shown to be heterogeneous on the basis of genetic (Bradley and Mordarski, 1976), numerical phenetic (Goodfellow and Minnikin, 1978), serological (Magnusson, 1976), and phage-sensitivity (Pulverer, Schütt-Gerowitt, and Schaal, 1975) studies. Various *N. asteroides* subgroups have been found but, since few strains are common to all of these investigations, it is difficult to determine whether or not, and to what extent, the defined subgroups overlap. However, Schaal and Reutersberg (1978) recognize two well-defined taxa in the *N. asteroides* complex, *N. asteroides* subgroup A and *N. asteroides* subgroup B. These taxa appear to merit species status, a view which is supported by preliminary DNA-homology data (Mordarski et al., 1977).

The status of the type species of the genus, *N. farcinica,* is not clear, for this epithet is currently used to describe at least two groups of actinomycetes. Strains such as *N. farcinica* NCTC 4524 have properties in common with mycobacteria; their inclusion in the genus *Nocardia* is misleading and should be discontinued. Such strains should be reclassified as *Mycobacterium farcinogenes* var. *senegalense* (Magnusson, 1976). Other strains, including *N. farcinica* ATCC 3318, fall within the range of variation of *N. asteroides* subgroup B (Schaal and Reutersberg, 1978). An acceptable name for the latter awaits resolution of the nomenclatural confusion surrounding the epithet *N. farcinica*.

Suitable reference strains are as follows: *Nocardia asteroides* complex ATCC 19247, *Nocardia brasiliensis* ATCC 19296, and *Nocardia otitidiscaviarum* ATCC 14629 (NCTC 1934).

Rhodococcus (Zopf 1891) Goodfellow and Alderson 1977

Rhodococci have had a long and chequered taxonomic history (Bousfield and Goodfellow, 1976; Gordon and Mihm, 1957). The epithet *rhodochrous* (Zopf, 1891) was reintroduced in 1957 by Gordon and Mihm for variously labeled actinomycetes that shared properties with both mycobacteria and nocardiae. These authors tentatively classified their strains as *Mycobacterium rhodochrous,* but subsequent studies, based on a variety of modern taxonomic methods, showed that the taxon was heterogeneous and could be distinguished from both *Mycobacterium* and *Nocardia.* Goodfellow and Alderson (1977) resurrected the genus *Rhodococcus* to accommodate actinomycetes previously classified as *Mycobacterium rhodochrous, Gordona, Jensenia,* or as in the "*rhodochrous*" complex.

Rhodococci are aerobic, Gram-positive actinomycetes; they are pleomorphic but often form a primary mycelium that soon fragments into bacillary and coccoid elements. These elements, when inoculated onto a fresh nutrient medium, give rise to a primary mycelium which may occasionally carry a few feeble aerial hyphae. Colonies may be rough, smooth, mucoid, or mycobacteria-like and are usually pigmented buff, pink, orange, or red, although colorless variants occur. Most strains grow well on standard laboratory media, but some require thiamine. Rhodococci are able to use a wide range of carbon sources for energy and growth and to reduce nitrate; but they are arylsulfatase negative and are unable to degrade casein, elastin, or hypoxanthine. Rhodococci contain mycolic acids, have a wall chemotype IV, and a G+C range 59–69 mol%.

At present, 10 species of *Rhodococcus* are recognized (Table 3). The type species is *Rhodococcus rhodochrous* and the reference strain is ATCC 13808 (NCIB 11147).

Habitats

Nocardiae as Animal Pathogens

Nocardiae are agents of nocardiosis and actinomycetoma, diseases which occur in man and other animals. Nocardiosis is the most frequently reported disease caused by aerobic actinomycetes and can be subdivided clinically into pulmonary, systemic, and superficial types (Pulverer and Schaal, 1978). The different types of human nocardiosis, especially those appearing as suppurative or septic infections, are clinically indistinguishable from other bacterial infections, a fact which emphasizes the need for good microbiological diagnosis.

In systemic nocardiosis, the primary lesion is often in the lungs, but secondary and often fatal involvement can occur in the central nervous system and less frequently in other organs or tissues. The pathogen invades the host either through inhaled, contaminated, dust particles or through soil-contaminated wounds. The comparatively low incidence of all types of nocardiosis reflects the poor invasive properties of the pathogens. The disease often develops as an opportunistic infection that complicates chronic, debilitating, primary diseases or immunosuppressive, therapeutical procedures. There are many predisposing factors, the more important of which include malignant disorders of the hematopoietic system, pulmonary alveolar proteinosis, tuberculosis, or pretreatment with corticosteroids or cytotoxic drugs. Nocardial infections are commonly considered to be rare, but recent data suggest that the incidence of such infections is underestimated. Thus, between 500 and 1,000 cases of nocardiosis are recognized in the United States each year, of which 85% are serious pulmonary or systemic infections (Beaman et al., 1976).

Strains classified in the *Nocardia asteroides* complex are by far the most important agents of all types of nocardiosis (Beaman et al., 1976; Kurup, Randhawa, and Gupta, 1970), and there is good evidence that some strains are more pathogenic than others. Thus, Pulverer and Schaal (1978) noted that infections of *N. asteroides* subgroup B strains are more frequent, and show a higher degree of malignancy, than those caused by *N. asteroides* subgroup A. *N. brasiliensis* and *N. otitidis-caviarum* occasionally cause nocardiosis (Beaman, 1976) but, to date, the remaining *Nocardia* species have rarely been implicated as agents of disease.

Nocardiae are also responsible for a wide variety of animal diseases. Cattle appear to be the most frequent host, with nocardial mastitis the predominant infection (Pier, 1962). Nocardiosis is not uncommon in dogs (Kinch, 1968) and has been reported in baboons, birds, cats, fish, goats, horses, and in domestic and wild rodents (Boncyk et al., 1975; Iyer et al., 1972; Pier, 1962; Snieszko et al., 1964). The identification of the causal agent in many reported cases of animal infection is inadequate. *N. asteroides* seems to be isolated most frequently, *N. brasiliensis* and *N. otitidis-caviarum* only rarely.

Actinomycetoma is the second most frequent disease caused by aerobic actinomycetes. It is characterized by chronic granulomatous indurations that rupture outwards at multiple sites forming typical fistulae. The latter discharge pus which contains characteristic granules similar to those found in actinomycosis. Infections are frequently through the foot, especially in localities where people are unable to afford footwear. The disease has a limited distribution; the primary agent, *N. brasiliensis,* is endemic in certain subtropical regions, notably in the sugar plantations of Mexico. *N. asteroides* and *N. otitidis-caviarum* occasionally cause actinomycetoma.

Drugs such as sulfonamides and trimethoprim plus sulfamethoxazole have been found effective in the treatment of actinomycetoma (González-Ochoa, 1976), and sulfadiazine is recommended for nocardiosis (Schaal, 1977).

Nocardiae as Saprophytes, Symbionts, and Plant Pathogens

Nocardiae have a worldwide distribution in soil (Cross et al., 1976); populations up to $7.3 \times 10^4/g$ dry weight have been found in some samples from tropical and temperate regions (Orchard, Goodfellow, and Williams, 1977). Nocardiae also form mutualistic associations with blood-sucking arthropods and occur in aquatic habitats (Cross et al., 1976), where they have been implicated in the biodeterioration of natural rubber joints in water and sewage pipes (Hutchinson, Ridgway, and Cross, 1975). Most of the nocardiae isolated from these habitats have been identified to the *Nocardia asteroides* complex, but the apparent predominance of these organisms may merely reflect the isolation methods used.

Large populations of *N. amarae,* and smaller numbers of *N. asteroides* and *N. otitidis-caviarum,* have been isolated from foams occasionally formed in sewage-treatment plants of the activated-sludge type (Lechevalier and Lechevalier, 1974; Lechevalier et al., 1977). *N. vaccinii* causes galls and bud proliferations on blueberry plants (Demaree and Smith, 1952). Little is known about the occurrence, distribution, or activity of the remaining *Nocardia* species in natural habitats.

Rhodococci as Saprophytes, Symbionts, and Pathogens

Rhodococci are widely distributed in nature and have been frequently isolated from soil, fresh water,

and from the gut contents of blood-sucking arthropods, with which they may form a symbiotic association (Cross et al., 1976). However, it is only recently that the availability of better isolation and identification methods have allowed the occurrence and distribution of these organisms to be examined. *Rhodococcus erythropolis, R. rhodochrous, R. ruber,* and *R. terrae* have been isolated from soil (Goodfellow and Alderson, 1977; Tsukamura, 1971), and *R. corallinus* and *R. rhodnii* from the intestinal tracts of Easter Island cockroaches and of the kissing bug, *Rhodnius prolixus,* respectively (Goodfellow and Aubert, 1979). *Rhodococcus bronchialis* is associated with the sputa of patients suffering from cavitary pulmonary tuberculosis and bronchiectasis (Tsukamura, 1971). *R. equi* (*Corynebacterium equi*) is commonly associated with pneumonitis in young foals and with suppurative lymphadenitis in swine (Farrelly, 1969; Linton and Gallaher, 1969), but has also been reported to infect other animals, such as cows and sheep (Addo and Dennis, 1977).

R. coprophilus has been the subject of the most extensive ecological studies (Rowbotham and Cross, 1977). Although this organism grows on the dung of herbivores, high numbers have also been recovered from grazed pastures and from rivers, streams, and lake muds that receive runoff from land devoted to dairy farming. It seems that the coccal survival stage contaminates grass in pastures or hay used during the winter months for fodder, and remains viable after ingestion and passage through the rumen. The significant correlation found between the numbers of *R. coprophilus* and fecal streptococci in polluted water suggests that the organism may prove to be a useful indicator of farm-animal effluent (Al-Diwany and Cross, 1978).

Isolation and Cultivation

Classical Methods of Selective Isolation of Nocardiae and Rhodococci

Classical selective isolation methods have depended on the ability of nocardiae and rhodococci to use hydrocarbons as sole sources of carbon for energy and growth (Cross et al., 1976; Tárnok, 1976). Modifications of Söhngen's (1913) paraffin baiting method have often been used to isolate *Nocardia* and *Rhodococcus* strains from soil (Portaels, 1976; Schaal and Bickenbach, 1978), but they are of little value in quantitative studies because they merely indicate the presence or absence of organisms in a sample. Many other bacteria, and also fungi, can attack paraffin wax; they often outgrow nocardiae and rhodococci on coated glass rods and make isolation impossible. Alternative isolation methods, such as inoculating guinea pig or hamster testicles with soil suspensions supplemented with penicillin and streptomycin (Conti-Diaz et al., 1971) or plating suspensions onto media that contains sodium azide and cholesterol acetate (Farmer, 1962), lead to underestimation of the populations of nocardiae and rhodococci in natural habitats.

Selective Isolation of Nocardiae

Large numbers of nocardiae have been isolated from soil by plating out soil suspensions onto diagnostic sensitivity test (DST) agar, supplemented with antifungal and various combinations of antibacterial antibiotics (Orchard and Goodfellow, 1974; Orchard, Goodfellow, and Williams, 1977). Inoculated plates are incubated for up to 21 days at 25°C. Colonies with a pink to red stroma, covered to a greater or lesser extent with white aerial hyphae, are characteristic of nocardiae. To date, most isolates have been classified in the *Nocardia asteroides* complex, although laboratory strains of *N. brasiliensis* and *N. otitidis-caviarum* do grow on the DST media (Orchard, Goodfellow, and Williams, 1977). It seems likely that at least some of the strains in the *N. asteroides* complex will be inhibited by the antibacterial antibiotics in the DST agar (Schaal and Heimerzheim, 1974). A selective medium free of such antibiotics has been successful in the isolation of *N. asteroides* from clinical material (Schaal, 1972).

N. amarae, N. asteroides, and *N. otitidis-caviarum* have been isolated by plating sewage foam onto either Czapek's agar supplemented with yeast extract (0.2%, wt/vol) (Higgins and Lechevalier, 1969) or glycerol agar (Gordon and Smith, 1953) and incubating plates at 28°C for 5–7 days. Using nitrite medium (Winogradsky, 1949), Mishustin and Tepper (Cross et al., 1976) have reported high numbers of nocardiae from soils in the USSR, but it seems likely that many of their isolates are rhodococci. Due to the lack of suitable selective media, little is known about the habitats of the remaining *Nocardia* species.

Nocardiae grow well on most standard laboratory media, including Bennett's (Jones, 1949), modified Sauton's (Mordarska, Mordarski, and Goodfellow, 1972), and glucose–yeast extract (Waksman, 1950) agars.

Selective Isolation of Rhodococci

Rhodococcus coprophilus can be isolated from a variety of habitats by plating preheated samples onto M3 agar and incubating plates at 30°C for 7 days (Rowbotham and Cross, 1977). Thus, samples (2 ml) of water, milk, and cream in 100-×-12-mm glass tubes sealed with silicon rubber bungs are heated in a water bath for 6 min at 55°C before further dilution or plating out. Water samples are stored at 4°C

before heating. Suspensions of dung, grass, soil (1:10, wt/vol), or hay (1:50, vol/vol) are homogenized in 1/4 strength Ringer's solution containing gelatin (0.01%, wt/vol), pH 7.0, before heat treatment. Immediately after heat treatment, samples are shaken on a Vortex mixer and portions (0.2 ml) are spread on the isolation plates.

A pretreatment method combined with a selective medium has been developed for the isolation of *R. bronchialis, R. corallinus,* and *R. terrae* from sputa and soils (Tsukamura, 1971). Sputum or soil is added to an equal volume of NaOH (4%, wt/vol) and liquefied, either by incubation at 37°C for 30 min or by shaking at room temperature for 15–30 min. The digests are inoculated onto Ogawa egg medium (Tsukamura, 1972) and incubated for 4–8 weeks.

With soil samples, soil (5 g) is suspended in distilled water (25 ml) and shaken vigorously in a 300-ml Erlenmeyer flask at room temperature on a reciprocal shaker (stroke: 10 cm, 60 cycles/min) for 30 min. The suspension is allowed to settle for 10 min, 15 ml of the supernatant is added to an equal volume of NaOH (8%, wt/vol), and the mixture is shaken for 10 min before centrifuging at $500 \times g$ for 15 min. The residue is suspended in 10 ml of a 1% (wt/vol) NaH_2PO_4 solution and 0.002 ml is added to Ogawa egg medium slopes, which are incubated for 4–8 weeks.

Colonies growing on the Ogawa egg medium slopes are transferred to fresh slopes supplemented with sodium salicylate (0.5 mg/ml), which inhibits *Mycobacterium tuberculosis* (Tsukamura, 1962). Smears, prepared from colonies growing on supplemented slopes after 3 weeks of incubation at 37°C, are stained by the Ziehl-Nielson method and observed under the light microscope. Slightly acid-fast, rod-shaped bacteria are typical of the rhodococci; those strongly acid-fast belong to the "atypical" mycobacteria. Rhodococci produce rough, reddish or pinkish colonies on Ogawa egg slants that have been plugged with cotton wool.

Rhodococci grow well on most standard laboratory media, including Bennett's (Jones, 1949), modified Sauton's supplemented with thiamine (Mordarska, Mordarski, and Goodfellow, 1972), and glucose–yeast extract (Waksman, 1950) agars.

Glycerol Agar for Selective Isolation of Nocardiae and Rhodococci (Gordon and Smith, 1953)

Peptone	5 g
Beef extract	3 g
Glycerol	7%
Agar	15 g
Soil extract	1,000 ml
pH 7.0	

The soil extract is prepared by autoclaving 1,000 g of air-dried soil sifted through a no. 9

mesh screen with 2,400 ml of tap water at 121°C for 60 min, decanting, and filtering through paper.

M3 Agar for Selective Isolation of Nocardiae and Rhodococci (Rowbotham and Cross, 1977)

KH_2PO_4	0.466 g
Na_2HPO_4	0.732 g
KNO_3	0.01 g
NaCl	0.29 g
$MgSO_4 \cdot 7H_2O$	0.1 g
$CaCO_3$	0.02 g
$FeSO_4 \cdot 7H_2O$	200 μg
$ZnSO_4 \cdot 7H_2O$	180 μg
$MnSO_4 \cdot 4H_2O$	20 μg
Sodium propionate	0.2 g
Agar	18 g
Distilled water	1 liter
pH 7.0	

Cycloheximide and thiamine HC1, sterilized by membrane filtration, are added to the autoclaved and cooled medium to give final concentrations of 50 mg/liter and 4.0 mg/liter, respectively.

Modified Czapek's Agar for Selective Isolation of Nocardiae and Rhodococci (Higgins and Lechevalier, 1969)

$NaNO_3$	2 g
K_2HPO_4	1 g
$MgSO_4 \cdot 7H_2O$	0.5 g
KCl	0.5 g
$FeSO_4$	0.01 g
Sucrose	30 g
Yeast extract	2 g
Agar	15 g
Distilled water	1 liter
pH 7.2	

Ogawa Egg Medium for Selective Isolation of Nocardiae and Rhodococci (Tsukamura, 1972)

Whole chicken eggs	200 ml
Glycerol	6 ml
Malachite green (2%, wt/vol)	6 ml
Sodium glutamate (1%, wt/vol) plus KH_2PO_4 (1%, wt/vol)	100 ml
pH 6.8	

Sterilize at 90°C for 60 min.

Selective Media for Isolating *Nocardia* (Orchard and Goodfellow, 1974)

Diagnostic sensitivity test (DST) agar (Oxoid, CM261); cycloheximide and nystatin (antifungal antibiotics); chlortetracycline HCl, demethylchlortetracycline HCl, and methacycline HCl (antibacterial antibiotics). The DST agar and cycloheximide are autoclaved separately at 15 psi for 20 min and the remaining antibiotics are filtered. The individual antibiotics are pipetted separately into Petri dishes and the basal medium is added to give the following concentrations (μg/ml) of medium:

Medium 1: Demethylchlortetracycline HCl (democlocycline HCl), 5; actidione, 50; mycostatin, 50.

Medium 2: Methacycline HCl, 10; actidione, 50; mycostatin, 50.

Medium 3: Chlortetracycline HCl (Aureomycin), 45; demethylchlortetracycline HCl, 5; actidione, 50; mycostatin, 50.

Medium 4: Chlortetracycline HCl, 45; methacycline HCl, 10; actidione, 50; mycostatin, 50.

Media 1 and 2 generally give higher counts of *Nocardia* if other soil bacteria are inhibited satisfactorily. However, where large mixed populations of bacteria occur, media 3 and 4 may be required. The number of unwanted bacteria can also be reduced by heating soil suspensions (2 ml) in a water bath at 55°C for 6 min.

Winogradsky's Nitrite Medium for Selective Isolation of Nocardiae and Rhodococci (Winogradsky, 1949)

$NaNO_2$	2 g
Anhydrous Na_2CO_3	1 g
K_2HPO_4	0.5 g
Agar	15 g
Distilled water	1 liter

Identification

Nocardiae and rhodococci can be difficult to distinguish from one another and from other actinomycetes by conventional staining and morphological criteria (Williams et al., 1976). Thus *Nocardia* strains that lack aerial hyphae cannot always be readily distinguished from corynebacteria, mycobacteria, and rhodococci, whereas those that produce aerial hyphae can be confused with taxa such as *Actinomadura, Pseudonocardia,* and *Streptomyces.* In recent years, however, chemical tests have provided data of value in distinguishing between aerobic actinomycetes and have been especially useful at the generic level.

Examination of whole-organism and/or wall hydrolysates for diagnostic sugars and diamino acids (Becker et al., 1964; Berd, 1973; Lechevalier, 1968; Staneck and Roberts, 1974) is the first stage in the chemical procedure. Nocardiae and rhodococci contain major amounts of *meso*-diaminopimelic acid (*meso*-Dap), arabinose, and galactose; that is, they have a wall chemotype IV as defined by M. P. Lechevalier and H. Lechevalier (1970). This wall chemotype is, however, shared by *Actinopolyspora*, *Bacterionema*, *Corynebacterium*, *Micropolyspora*, *Mycobacterium*, *Saccharomonospora*, *Saccharopolyspora*, *Pseudonocardia* (Gochnauer et al., 1975; Goodfellow and Minnikin, 1977) and by the *"aurantiaca"* taxon, which contains organisms previously classified as *Gordona aurantiaca* (Goodfellow et al., 1978). Organisms labeled *N. aerocolonigenes* contain major amounts of *meso*-Dap and galactose but lack arabinose (Gordon, Mishra, and Barnett, 1978). Lipid analyses, however, are useful in distinguishing representatives of *Nocardia* and *Rhodococcus* from actinomycetes that share wall chemotype IV (Table 1).

Fatty-acid analysis is the first recommended stage in the exploitation of lipid composition. A convenient procedure involves thin-layer chromatographic analysis of whole-organism acid methanolysates (Minnikin, Alshamaony, and Goodfellow, 1975; Minnikin et al., 1980). Representatives of *Saccharomonospora*, *Saccharopolyspora*, *Pseudonocardia*, *Micropolyspora faeni*, and *Micropolyspora caesia* contain only nonhydroxylated fatty acids (Goodfellow and Minnikin, 1977; Kroppenstedt and Kutzner, 1978; Lechevalier, de Bièvre, and Lechevalier, 1977); their identification is discussed elsewhere (this Handbook, Chapters 157 and 158). Certain strains currently labeled *Nocardia* (see Table 2), such as *N. autotrophica (N. coeliaca), N. orientalis, N. lurida, N. mediterranea,* and *N. aerocolonigenes,* also do not contain mycolic acids (Kroppenstedt and Kutzner, 1978; Lechevalier, de Bièvre, and Lechevalier, 1977; Pommier and Michel, 1973; Yano, Furukawa, and Kusunose, 1969). During the characterization of *Actinopolyspora halophila* (Gochnauer et al., 1975), no mycolic acids were found.

Single components corresponding to mycolic acid esters are detected in methanolysates of *Nocardia*, *Rhodococcus*, *Bacterionema*, *Corynebacterium*, *Micropolyspora brevicatena*, and the *"aurantiaca"* taxon, but more complex patterns are typical of *Mycobacterium* (Collins et al., 1977; Goodfellow, Collins, and Minnikin, 1976; Goodfellow et al., 1978; Minnikin, Alshamaony, and Goodfellow, 1975). Mycolic esters can be positively identified on thin-layer chromatograms by their characteristic immobility when the plates are subsequently washed with methanol-water (5:2) (Minnikin, Alshamaony, and Goodfellow, 1975).

The mycolic acids from mycobacteria are precipitated from ethereal solution by addition of an equal (Kanetsuna and Bartoli, 1972) or double (Hecht and Causey, 1976) volume of ethanol. Mycolic acids from corynebacteria, nocardiae, and rhodococci are not precipitated by such a procedure, but Hecht and Causey (1976) detected their presence by thin-layer chromatography of the supernatant. The behavior of the mycolic acids of the *"aurantiaca"* taxon (Goodfellow et al., 1978) in these systems has still to be determined.

When the presence of mycolic acids has been detected, their esters should be isolated and studied further by pyrolysis gas chromatography, mass spectrometry of the intact esters, or by gas chromatography of their trimethylsilyl ether derivatives. On pyrolysis, mycolic acid methyl esters yield long-chain methyl esters and aldehydes, and components can be analyzed directly by use of pyrolysis gas chromatography (Etémadi, 1967). The mycolic esters of mycobacteria give C_{22} to C_{26} esters on pyrolysis, whereas bacterionemae, corynebacteria, nocardiae, and rhodococci have mycolates which produce C_{12} to C_{18} esters (Goodfellow and Minnikin, 1977; Lechevalier, 1977; Lechevalier, Horan, and Lechevalier, 1971; Lechevalier, Lechevalier, and Horan, 1973). Mycolic esters from *N. amarae* can be recognized by pyrolysis gas chromatography. They release unsaturated C_{16} and C_{18} major components (Lechevalier and Lechevalier, 1974), and the production of unsaturated C_{20} and C_{22} esters characterizes the mycolates from representatives of the *"aurantiaca"* taxon (Goodfellow et al., 1978). Mycolic acids having unsaturation in the chain in the 2-position have also been reported in two nocardial strains, *N. carnea* and *N. vaccinii,* but no details have been reported (Lechevalier and Lechevalier, 1974).

The pyrolysis of mycolates can also be observed by mass spectrometry, where the highest peaks in the spectra correspond to anhydromycolates formed by loss of water from the parent molecule (Etémadi, 1967). The overall size of mycolates, their degree of unsaturation, and the nature of both long-alkyl chains may be determined by mass spectrometry. However, the complex mixtures of homologs usually present and the competing fragmentation pathways make interpretation difficult in some cases (Alshamaony, Goodfellow, and Minnikin, 1976; Alshamaony et al., 1976; Maurice, Vacheron, and Michel, 1971). The analysis can be taken a stage further by using gas chromatography–mass spectrometry of trimethylsilyl ethers of mycolic esters (Yano et al., 1978). This procedure separates mycolic ester derivatives into their homologous components, each of which can be analyzed by mass spectrometry. Very detailed mycolic-acid analyses may eventually help in the identification of all species that contain mycolic acid, but, at present, their use is best restricted as shown in Table 1.

The identification of *N. aerocolonigenes, N. autotrophica, N. lurida, N. mediterranea, N. orientalis,* and other strains that lack mycolic acids is assisted by their high content of branched-chain, iso, and anteiso fatty acids (Table 1) (Kroppenstedt and Kutzner, 1978; Lechevalier, de Bièvre, and Lechevalier, 1977; Pommier and Mi-

Table 1. Identification using lipid analyses of nocardiae, rhodococci, and related organisms having wall chemotype IV.[a]

Long-chain compounds[b]		Mycolic acid solubility[c]	Ester released on pyrolysis GC of mycolate[d]	Major menaquinone[e]	Genus	Mycolate size (no. of carbons)	Phospholipids[f]	Species
Iso and anteiso acids	Mycolic acids absent	—		ND[g]	"Nocardia"	—	PE	N. orientalis, N. mediterranea
			—	MK-9(H4)	"Micropolyspora"	—	PE, PC	M. faeni[h]
				MK-8(H4)	"Nocardia"	—	PE, PC	N. autotrophica (N. coeliaca)
Straight-chain, unsaturated, and tuberculostearic acids	Complex mycolic acids	Insoluble	22:0–26:0	MK-9(H2)	Mycobacterium	60–90	PE	Mycobacterium spp.
			20:0, 20:1, 22:0, 22:1	MK-9	"Gordona"	68–74	PE	G. aurantiaca
		ND	16:1, 18:1	MK-9(H2)[j]	"Nocardia"	46–54[i]	PE	N. amarae
	Single mycolic acid			MK-8(H4)	"Micropolyspora"	46–56	PE	M. brevicatena[j]
				MK-8(H4)	Nocardia	46–60	PE	N. asteroides, N. brasiliensis, N. otitidis-caviarum
		Soluble	12:0–18:0	MK-8(H2)[j]	Rhodococcus	34–52[i]	PE	R. coprophilus, R. equi, R. erythropolis, R. rhodnii, R. rhodochrous, R. ruber
				MK-9(H2)[j]	Rhodococcus	48–64[i]	PE	R. bronchialis, R. corallinus, R. terrae
				MK-8,9(H2)[j]	Rhodococcus	38–64[i]	ND	R. rubropertinctus
			8:0, 10:0[i]	MK-9(H2)	Corynebacterium	22–32[i]	—	C. bovis
Straight-chain and unsaturated acids only	Single mycolic acid	Soluble	14:0–18:0[i] (14:1–18:1)	MK-8,9(H2)	Corynebacterium	26–38	—	Corynebacterium spp.

[a] Data from Goodfellow and Minnikin, 1977, 1978; Goodfellow et al., 1978; Kroppenstedt and Kutzner, 1976, 1978; Lechevalier, 1977, 1978; Lechevalier, de Bièvre, and Lechevalier, 1977; Minnikin and Goodfellow, 1976, 1978; Minnikin, Collins, and Goodfellow, 1978; Minnikin, Goodfellow, and Alshamaony, 1978; Minnikin et al., 1977.

[b] Determined by gas and thin-layer chromatography (e.g., Kroppenstedt and Kutzner, 1978; Lechevalier, de Bièvre, and Lechevalier, 1977; Minnikin, Alshamaony, and Goodfellow, 1975).

[c] In ethanol-diethyl ether (Hecht and Causey, 1976; Kanetsuna and Bartoli, 1972).

[d] Esters detected by pyrolysis gas chromatography (GC) (Goodfellow et al., 1978; Lechevalier and Lechevalier, 1974; Lechevalier, Horan, and Lechevalier, 1971; Lechevalier, Lechevalier, and Horan, 1973). Abbreviations exemplified by 18:0, octadecanoate; 18:1, octadecanoate.

[e] Abbreviations exemplified by MK-9, menaquinone having 9 isoprene units; MK-9(H4), menaquinone having two of the 9 isoprene units hydrogenated.

[f] Diagnostic phospholipids, phosphatidylethanolamine (PE) and phosphatidylcholine (PC), occurring in addition to diphosphatidylglycerol and phosphatidylinositol mannosides.

[g] ND, not determined.

[h] Kroppenstedt and Kutzner (1978) reported that Micropolyspora caesia had fatty acids similar to those of M. faeni.

[i] D. E. Minnikin, M. Goodfellow, M. D. Collins, and L. Alshamaony, additional unpublished results.

[j] Uncharacterized mycolic acids were also found in Micropolyspora fascifera (Soina and Agre, 1976).

Table 2. Diagnostic for differentiation of *Nocardia* species and some related taxa.[a]

| | Mycolic acids[b] | | | | | | | | No mycolic acids[b] | | |
Taxon	*N. amarae*	*N. asteroides* subgroup A[b]	*N. asteroides* subgroup B	*N. brasiliensis*	*N. carnea*	*N. otitidis-caviarum*	*N. transvalensis*	*N. vaccinii*	*N. aerocolonigenes*	*N. autotrophica*	*N. orientalis*
Morphological and staining characters:											
Acid-fastness	−	V	V	V	V	V	V	+	−	−	−
Substrate mycelium color[c]	Cr	Y,O	Y,O,R	C,O,R	Cr,Pe	Cr	Cr,Pu	Cr,Pe	Cr,O	YB	Cr,Pe
Decomposition of:											
Adenine	−	−	−	−	−	−	V			+	+
Casein	−	−	−	+	−	−	−	−	+	−	+
Elastin[d]	−	−	−	+	−	−	−	−	+	+	+
Hypoxanthine	−	−	−	+	−	+	+	−	+	−	+
Testosterone[d]	−	+	+	+	+	−	−	+	+	+	+
Tyrosine	−	−	−	+	−	−	+	−	−	V	V
Xanthine	−	−	−	−	−	+	V				
Resistance to:											
Lysozyme	−	+	+	+	+	+	+	+	+	−	−
Rifampin	V	+	+	+	−	+	+	V	−	−	V
Urease production	+	+	+	+		+	+	+	+	+	+
Acid from[e]:											
Adonitol	−	ND	ND	−	−	−	+	−	V	+	+
Cellobiose	−	ND	ND	−	−	−	−	−	+	+	+
meso-Erythritol	−	ND	ND	−	−	−	+	−	−	+	+
Lactose	−	ND	ND	−	−	−	−	−	+	−	V
Maltose	+	ND	ND	−	−	−	−	V	+	+	V
Melezitose	−	ND	ND	−	−	−	−	−	−	+	V
α-Methyl-D-glucoside	−	ND	ND	−	−	−	−	−	−	V	+

[a] Based on methods and data of Gordon, Mishra, and Barnett (1978) and Goodfellow and Schaal (1979). Symbols: +, usually positive; −, usually negative; V, variable; ND, not determined.

[b] *Nocardia asteroides* subgroups A and B can be distinguished by their sensitivity to gentamicin (10 μg/ml) and by their ability to grow on isoamyl alcohol, 2,3-butylene glycol, gluconate, 1,2-propylene glycol, and rhamnose as sole carbon sources (see Schaal, 1977).

[c] B, brown; C, colorless; Cr, cream; O, orange/pale orange; Pe, peach; Pu, purple; R, red; Y, yellow.

[d] M. Goodfellow et al., unpublished data.

[e] Strains classified as *Nocardia asteroides* (Gordon, Mishra, and Barnett, 1978) do not produce acid from these sugars.

chel, 1973; Yano, Furukawa, and Kusunose, 1969). In contrast, the fatty acids of nocardiae and rhodococci that contain mycolic acids are composed of straight-chain, unsaturated, and 10-methyloctadecanoic (tuberculostearic) acids (Kroppenstedt and Kutzner, 1978; Lechevalier, de Bièvre, and Lechevalier, 1977; Minnikin and Goodfellow, 1976). It can be difficult to distinguish between certain rhodococci and corynebacteria; fatty-acid analysis may be of assistance since tuberculostearic acid is not found in most corynebacteria (Minnikin, Goodfellow, and Collins, 1978), with the apparent exception of *Corynebacterium bovis* (Lechevalier, de Bièvre, and Lechevalier, 1977). Representatives of *Corynebacterium equi* also contain tuberculostearic acid (Kroppenstedt and Kutzner, 1978) but this

taxon has been reclassified as *Rhodococcus equi* (Goodfellow and Alderson, 1977).

Polar-lipid and isoprenoid-quinone analysis contribute data in support of that derived from fatty-acid studies (Table 1). The polar lipids of all mycolic acid–containing organisms contain diphosphatidylglycerol and the characteristic phosphatidylinositol mannosides. However, phosphatidylethanolamine is absent from *Corynebacterium* although usually present in others (Komura et al., 1975; Lechevalier, de Bièvre, and Lechevalier, 1977; Minnikin et al., 1977). Organisms that lack mycolic acids also contain phosphatidylethanolamine and phosphatidylinositol mannosides (Table 1), but phosphatidylcholine is restricted to *Actinopolyspora*, *Micropolyspora faeni*, *Nocardia autotrophica* (*N. coeliaca*), and

Table 3. Diagnostic for differentiation of *Rhodococcus* species.[a]

Taxon	R. bron-chialis	R. copro-philus	R. coral-linus	R. equi	R. eryth-ropolis	R. rhodnii	R. rhodo-chrous	R. ruber	R. rubro-pertinctus	R. terrae
Decomposition of:										
Adenine	–	–	–	+	+	–	V	V	–	–
Tyrosine	–	–	–	–	V	+	+	+	–	–
Enzymic activity:										
α Esterase	–	ND	–	ND	ND	ND	–	–	ND	+
β Esterase	+	ND	+	ND	ND	ND	+	–	ND	+
Sole carbon sources										
(1%, wt/vol)										
Glycerol	+	–	+	+	+	V	+	+	+	+
Inositol	+	–	–	–	V	–	–	+	+	+
Maltose	–	–	–	–	+	–	+	–	–	–
Rhamnose	–	–	–	–	+	–	+	V	+	–
Sorbitol	–	–	+	–	–	–	–	–	–	+
Trehalose	+	–	+	–	+	+	+	+	V	+
m-Hydroxybenzoic acid										
(0.1%, wt/vol)	–	+	–	–	–	V	+	+	+	+
Growth at 10°C	–	V	–	+	+	–	+	V	–	–
Growth in the presence of:										
Crystal violet										
(0.0001%, wt/vol)	+	+	+	+	–	V	V	+	V	+
Sodium azide										
(0.02% wt/vol)	+	V	+	V	V	V	–	+	V	+

[a] Based on methods and data from Tsukamura (1973) and Goodfellow and Schaal (1979). Symbols: +, usually positive; –, usually negative; V, variable; ND, not determined.

Pseudonocardia (Gochnauer et al., 1975; Lechevalier et al., 1977; Yano, Furukawa, and Kusunose, 1969).

The presence of major amounts of tetrahydrogenated menaquinones with eight isoprene units, abbreviated as MK-8(H$_4$), distinguishes nocardiae and *Micropolyspora brevicatena* from most other mycolic acid–containing bacteria (Table 1) (Collins et al., 1977; Yamada et al., 1976, 1977). Dihydrogenated menaquinones having eight or nine isoprene units are discontinuously distributed among almost all the remaining mycolic acid–containing genera (Table 1) (Collins et al., 1977; Yamada et al., 1976, 1977), but the "*aurantiaca*" taxon is clearly distinguished by the presence of fully unsaturated menaquinones (MK-9) (Goodfellow et al., 1978).

Nocardia species, and strains of related taxa that have a wall chemotype IV but lack mycolic acids, can be identified using a number of conventional tests (Table 2). However, strains classified in poorly described species, which are labeled *Nocardia*, need to be compared with representatives of established *Nocardia* species before they can be incorporated into an identification scheme. These systematic studies should include: strains of *N. petroleophila* and *Streptomyces galtieri* with a wall chemotype IV and mycolic acids; *N. lurida, N. mediterranea, N. rugosa,* and *N. tenuis,* which possess *meso*-diaminopimelic acid, arabinose, and galactose but seem to lack mycolic acids; and *N. italica, N. polychromogenes,* and *N. saturnea,* which contain only the diamino acid (Kroppenstedt and Kutzner, 1978; Mordarska, Mordarski, and Goodfellow,

1972; Pridham and Lyons, 1969). Strains labeled *N. alba shoen, N. apis, N. formica, N. gibsonii,* and *N. rangoonensis* possess LL-diaminopimelic acid (Mordarska, Mordarski, and Goodfellow, 1972) and should probably be transferred to the genus *Streptomyces*.

Rhodococcus species can also be identified using a small number of conventional tests (Table 3). However, further studies are required to determine the reproducibility of these presumptive diagnostic tests.

The arrangement of the data in Table 1 shows how analyses of fatty-acid and menaquinone composition can lead to the identification of *Nocardia* and allied genera. More detailed studies of mycolic acid structure and phospholipid composition enable distinctions to be made at the species level. Lipid analyses as summarized in Table 1 also highlight taxa in need of further study. It is, for instance, evident that the genus *Micropolyspora* is heterogeneous with the type species, *Micropolyspora brevicatena,* being chemically related to *Nocardia sensu stricto.* The "*aurantiaca*" taxon and, to a lesser extent, *N. amarae* are distinguished from the other mycolic acid–containing taxa by their lipid composition; the distinctness of the former group has been supported by numerical phenetic data (Goodfellow et al., 1978). The species *Rhodococcus rubropertinctus,* as presently constituted (Goodfellow and Alderson, 1977), is rather heterogeneous in its lipid composition and more detailed taxonomic studies might be profitable.

Systematic lipid analyses of the mycolic acid-containing species, *N. carnea, N. transvalensis,* and *N. vaccinii,* have not been made, but their physiological properties are similar to those of *N. asteroides, N. brasiliensis,* and *N. otitidis-caviarum* (Gordon, Mishra, and Barnett, 1978).

Literature Cited

Addo, P. B., Dennis, S. M. 1977. Ovine pneumonia caused by *Corynebacterium equi.* Veterinary Record **101:**80.

Al-Diwany, L. J., Cross, T. 1978. Ecological studies on nocardioforms and other actinomycetes in aquatic habitats. Zentralblatt für Bakteriologie, Parasitenkunde, Infektionskrankheiten und Hygiene, Abt. 1, Suppl. **6:**153–160.

Alshamaony, L., Goodfellow, M., Minnikin, D. E. 1976. Free mycolic acids as criteria in the classification of *Nocardia* and the '*rhodochrous*' complex. Journal of General Microbiology **92:**188–199.

Alshamaony, L., Goodfellow, M., Minnikin, D. E., Mordarska, H. 1976. Free mycolic acids as criteria in the classification of *Gordona* and the '*rhodochrous*' complex. Journal of General Microbiology **92:**183–187.

Beaman, B. L. 1976. Possible mechanisms for nocardial pathogenesis, pp. 386–417. In: Goodfellow, M., Brownell, G. H., Serrano, J. A. (eds.), The biology of the nocardiae. London, New York, San Francisco: Academic Press.

Beaman, B. L., Burnside, J., Edwards, B., Causey, W. 1976. Nocardial infections in the United States, 1972–1974. Journal of Infectious Diseases **134:**286–289.

Becker, B., Lechevalier, M. P., Gordon, R. E., Lechevalier, H. A. 1964. Rapid differentiation between *Nocardia* and *Streptomyces* by paper chromatography of whole-cell hydrolysates. Applied Microbiology **12:**421–423.

Berd, D. 1973. Laboratory identification of clinically important aerobic actinomycetes. Applied Microbiology **25:**665–681.

Bisset, K. A., Moore, F. W. 1950 *Jensenia,* a new genus of the Actinomycetales. Journal of General Microbiology **4:**280.

Boncyk, L. H., McCullough, B., Grotts, D. D., Kalter, S. S. 1975. Localized nocardiosis due to *Nocardia caviae* in a baboon (*Papio cynocephalus*). Laboratory Animal Science **25:**88–91.

Bousfield, I. J., Goodfellow, M. 1976. The "*rhodochrous*" complex and its relationships with allied taxa, pp. 39–65. In: Goodfellow, M., Brownell, G. H., Serrano, J. A. (eds.), The biology of the nocardiae. London, New York, San Francisco: Academic Press.

Bradley, S. G., Mordarski, M. 1976. Association of polydeoxyribonucleotides of deoxyribonucleic acids from nocardioform bacteria, pp. 310–336. In: Goodfellow, M., Brownell, G. H., Serrano, J. A. (eds), The biology of the nocardiae. London, New York, San Francisco: Academic Press.

Collins, M. D., Pirouz, T., Goodfellow, M., Minnikin, D. E. 1977. Distribution of menaquinones in actinomycetes and corynebacteria. Journal of General Microbiology **100:**221–230.

Conti-Díaz, I. A., Gezuele, E., Civila, E., Mackinnon, J. E. 1971. Termotolerancia y acción patógena de cepas de *Nocardia asteroides* aisladas de fuentes naturales. Revista Uruguaya de Patología Clínica y Microbiología **9:**232–241.

Cross, T., Goodfellow, M. 1973. Taxonomy and classification of the actinomycetes, pp. 11–112. In: Sykes, G., Skinner, F. A. (eds.), Actinomycetales: Characteristics and practical importance. London, New York, San Francisco: Academic Press.

Cross, T., Rowbotham, T. J., Mishustin, E. N., Tepper, E. Z., Antonie-Portaels, F., Schaal, K. P., Bickenbach, H. 1976. The ecology of nocardioform actinomycetes, pp. 337–371. In: Goodfellow, M., Brownell, G. H., Serrano, J. A. (eds.),

The biology of the nocardiae. London, New York, San Francisco: Academic Press.

Demaree, J. B., Smith, N. R. 1952. *Nocardia vaccinii* n. sp. causing galls on blueberry plants. Phytopathology **42:**249–252.

Etémadi, A. H. 1967. The use of pyrolysis gas chromatography and mass spectroscopy in the study of the structure of mycolic acids. Journal of Gas Chromatography **5:**447–456.

Farmer, R. 1962. Influence of various chemicals in the isolation of *Nocardia* from soil. Proceedings of the Oklahoma Academy of Sciences **43:**254–256.

Farrelly, B. T. 1969. *Corynebacterium equi* infection in foals in Ireland. Irish Veterinary Journal **23:**231–232.

Gochnauer, M. B., Leppard, G. G., Komaratat, P., Kates, M., Novitsky, T., Kushner, D. J. 1975. Isolation and characterization of *Actinopolyspora halophila,* gen. et sp. nov., an extremely halophilic actinomycete. Canadian Journal of Microbiology **21:**1500–1511.

González-Ochoa, A. 1976. Nocardiae and chemotherapy, pp. 429–450. In: Goodfellow, M., Brownell, G. H., Serrano, J. A. (eds.), The biology of the nocardiae. London, New York, San Francisco: Academic Press.

Goodfellow, M. 1971. Numerical taxonomy of some nocardioform bacteria. Journal of General Microbiology **69:**33–80.

Goodfellow, M., Alderson, G. 1977. The actinomycete-genus *Rhodococcus:* A home for the '*rhodochrous*' complex. Journal of General Microbiology **100:**99–122.

Goodfellow, M., Aubert, E. 1980. Characterization of rhodococci from the intestinal tract of *Rapa Nui* cockroaches. In: Nogrady, G. L. (ed.), Microbiology of Easter Island, vol. 2. Oakville, Canada: Sovereign Press. In press.

Goodfellow, M., Collins, M. D., Minnikin, D. E. 1976. Thin-layer chromatographic analysis of mycolic acid and other long-chain components in whole-organism methanolysates of coryneform and related taxa. Journal of General Microbiology **96:**351–358.

Goodfellow, M., Minnikin, D. E. 1977. Nocardioform bacteria. Annual Review of Microbiology **31:**159–180.

Goodfellow, M., Minnikin, D. E. 1978. Numerical and chemical methods in the classification of *Nocardia* and related taxa. Zentralblatt für Bakteriologie, Parasitenkunde, Infektionskrankheiten und Hygiene, Abt. 1, Suppl. **6:**43–51.

Goodfellow, M., Schaal, K. P. 1979. Identification methods for *Nocardia, Actinomadura,* and *Rhodococcus,* pp. 261–276. In: Lovelock, D. W., Skinner, F. A. (eds.), Identification methods for microbiologists, 2nd ed. Society of Applied Bacteriology Technical Series. London: Academic Press.

Goodfellow, M., Orlean, P. A. B., Collins, M. D., Alshamaony, L., Minnikin, D. E. 1978. Chemical and numerical taxonomy of strains received as *Gordona aurantiaca.* Journal of General Microbiology **109:**57–68.

Gordon, R. E., Mihm, J. M. 1957. A comparative study of some strains received as nocardiae. Journal of Bacteriology **73:**15–27.

Gordon, R. E., Mishra, S. K., Barnett, D. A. 1978. Some bits and pieces of the genus *Nocardia: N. carnea, N. vaccinii, N. transvalensis, N. orientalis* and *N. aerocolonigenes.* Journal of General Microbiology **109:**69–78.

Gordon, R. E., Smith, M. M. 1953. Rapidly growing acid fast bacteria. I. Species description of *Mycobacterium phlei* Lehmann and Neumann and *Mycobacterium smegmatis* (Trevisan) Lehmann and Neumann. Journal of Bacteriology **66:**41–48.

Hecht, S. T., Causey, W. A. 1976. Rapid method for the detection and identification of mycolic acids in aerobic actinomycetes and related bacteria. Journal of Clinical Microbiology **4:**284–287.

Henssen, A. 1957. Beiträge zur Morphologie und Systematik der thermophilen Actinomyceten. Archiv für Mikrobiologie **26:**373–414.

Higgins, M. L., Lechevalier, M. P. 1969. Poorly lytic bacteriophage from *Dactylosporangium thailandensis* (*Actinomycetales*) Journal of Virology **3:**210–216.

Hutchinson, M., Ridgway, J. W., Cross, T. 1975. Biodeterioration of rubber in contact with water, sewage and soil, pp. 187–202. In: Lovelock, D. W., Gilbert, R. J. (eds.), Microbial aspects of the deterioration of materials. London, New York, San Francisco: Academic Press.

Iyer, P. K. R., Rao, A. T., Acharjyo, L. N., Sahu, S., Mishra, S. K. 1972. Systemic nocardiosis in a hill mynah (*Gracula religiosa*). A pathological study. Mycopathologia et Mycologia Applicata **48:**223–229.

Jones, K. L. 1949. Fresh isolates of actinomycetes in which the presence of sporogenous aerial mycelia is a fluctuating characteristic. Journal of Bacteriology **57:**141–145.

Kanetsuna, F., Bartoli, A. 1972. A simple chemical method to differentiate *Mycobacterium* from *Nocardia*. Journal of General Microbiology **70:**209–212.

Kinch, D. A. 1968. A rapidly fatal infection caused by *Nocardia caviae* in a dog. Journal of Pathology and Bacteriology **95:**540–546.

Komura, I., Yamada, K., Otsuka, S. I., Komagata, K. 1975. Taxonomic significance of phospholipids in coryneform and nocardioform bacteria. Journal of General and Applied Microbiology **21:**251–261.

Kroppenstedt, R. M., Kutzner, H. J. 1976. Biochemical markers in the taxonomy of the Actinomycetales. Experientia **32:**318–319.

Kroppenstedt, R. M., Kutzner, H. J. 1978. Biochemical taxonomy of some problem actinomycetes. Zentralblatt für Bakteriologie, Parasitenkunde, Infektionskrankheiten und Hygiene, Abt. 1, Suppl. **6:**125–133.

Kurup, P. V., Randhawa, H. S., Gupta, N. P. 1970. Nocardiosis: A review. Mycopathologia et Mycologia Applicata **40:**193–219.

Lechevalier, H., Lechevalier, M. P. 1965. Classification des actinomycètes aérobies basée sur leur morphologie et leur composition chimique. Annales de l'Institut Pasteur **108:**662–673.

Lechevalier, H. A., Lechevalier, M. P. 1970. A critical evaluation of the genera of aerobic actinomycetes, pp. 393–405. In: Prauser, H. (ed.), The Actinomycetales. Jena: Gustav Fischer Verlag.

Lechevalier, H. A., Lechevalier, M. P., Wyszkowski, P. E., Mariat, F. 1977. Actinomycetes found in sewage-treatment plants of the activated sludge type, pp. 227–247. In: Arai, T. (ed.), Actinomycetes: The boundary microorganisms. Tokyo, Singapore: Toppan.

Lechevalier, M. P. 1968. Identification of aerobic actinomycetes of clinical importance. Journal of Laboratory and Clinical Medicine **71:**934–944.

Lechevalier, M. P. 1977. Lipids in bacterial taxonomy—a taxonomist's viewpoint. CRC Critical Reviews in Microbiology **5:**109–210.

Lechevalier, M. P., de Bièvre, C., Lechevalier, H. 1977. Chemotaxonomy of aerobic actinomycetes: Phospholipid composition. Biochemical Systematics and Ecology **5:**249–260.

Lechevalier, M. P., Horan, A. C., Lechevalier, H. 1971. Lipid composition in the classification of nocardiae and mycobacteria. Journal of Bacteriology **105:**313–318.

Lechevalier, M. P., Lechevalier, H. 1970. Chemical composition as a criterion in the classification of aerobic actinomycetes. International Journal of Systematic Bacteriology **20:**435–444.

Lechevalier, M. P., Lechevalier, H. A. 1974. *Nocardia amarae* sp. nov., an actinomycete common in foaming activated sludge. International Journal of Systematic Bacteriology **24:**278–288.

Lechevalier, M. P., Lechevalier, H., Horan, A. C. 1973. Chemical characteristics and classification of nocardiae. Canadian Journal of Microbiology **19:**965–972.

Linton, J. A. M., Gallaher, M. A. 1969. Suppurative brochopneumonia in a foal associated with *Corynebacterium equi*. Irish Veterinary Journal **23:**197–200.

McClung, N. M. 1974. Nocardiaceae, p. 726. In: Buchanan, R. E., Gibbons, N. E. (eds.), Bergey's manual of determinative bacteriology, 8th ed. Baltimore: Williams & Wilkins.

Magnusson, M. 1976. Sensitin tests as an aid in the taxonomy of *Nocardia* and its pathogenicity, pp. 236–265. In: Goodfellow, M., Brownell, G. H., Serrano, J. A. (eds.), The biology of the nocardiae. London, New York, San Francisco: Academic Press.

Maurice, M. T., Vacheron, M. J., Michel, G. 1971. Isolément d'acides nocardiques de plusieurs espèces de *Nocardia*. Chemistry and Physics of Lipids **7:**9–18.

Minnikin, D. E., Alshamaony, L., Goodfellow, M. 1975. Differentiation of *Mycobacterium, Nocardia* and related taxa by thin-layer chromatographic analysis of whole-organism methanolysates. Journal of General Microbiology **88:**200–204.

Minnikin, D. E., Collins, M. D., Goodfellow, M. 1978. Menaquinone patterns in the classification of nocardioform and related bacteria. Zentralblatt für Bakteriologie, Parasitenkunde, Infektionskrankheiten und Hygiene, Abt. 1, Suppl. **6:**86–90.

Minnikin, D. E., Goodfellow, M. 1976. Lipid composition in the classification and identification of nocardiae and related taxa, pp. 160–219. In: Goodfellow, M., Brownell, G. H., Serrano, J. A. (eds.), The biology of the nocardiae. London, New York, San Francisco: Academic Press.

Minnikin, D. E., Goodfellow, M. 1978. Polar lipids of nocardioform and related bacteria. Zentralblatt für Bakteriologie, Parasitenkunde, Infektionskrankheiten und Hygiene, Abt. 1, Suppl. **6:**74–83.

Minnikin, D. E., Goodfellow, M., Alshamaony, L. 1978. Mycolic acids in the classification of nocardioform bacteria. Zentralblatt für Bakteriologie, Parasitenkunde, Infektionskrankheiten und Hygiene, Abt. 1, Suppl. **6:**63–66.

Minnikin, D. E., Goodfellow, M., Collins, M. D. 1978. Lipid composition in the classification and identification of coryneform and related taxa, pp. 85–160. In: Bousfield, I. J., Callely, A. G. (eds.), Coryneform bacteria. London: Academic Press.

Minnikin, D. E., Patel, P. V., Alshamaony, L., Goodfellow, M. 1977. Polar lipid composition in the classification of *Nocardia* and related bacteria. International Journal of Systematic Bacteriology **27:**104–117.

Minnikin, D. E., Hutchinson, I. G., Caldicott, A. B., Goodfellow, M. 1980. Thin-layer chromatography of methanolysates of mycolic acid-containing bacteria. Journal of Chromatography **188:**221–233.

Mordarska, H., Mordarski, M., Goodfellow, M. 1972. Chemotaxonomic characters and classification of some nocardioform bacteria. Journal of General Microbiology **71:**77–86.

Mordarski, M., Schaal, K. P., Szyba, K., Pulverer, G., Tkacz, A. 1977. Interrelation of *Nocardia asteroides* and related taxa as indicated by deoxyribonucleic acid reassociation. International Journal of Systematic Bacteriology **27:**66–70.

Orchard, V. A., Goodfellow, M. 1974. The selective isolation of *Nocardia* from soil using antibiotics. Journal of General Microbiology **85:**160–162.

Orchard, V. A., Goodfellow, M., Williams, S. T. 1977. Selective isolation and occurrence of nocardiae in soil. Soil Biology and Biochemistry **9:**233–238.

Pier, A. C. 1962. Nocardiosis in animals. Proceedings of the United States Livestock Sanitary Association 409–416.

Pommier, M. T., Michel, G. 1973. Phospholipid and acid composition of *Nocardia* and nocardoid bacteria as criteria of classification. Biochemical Systematics **1:**3–12.

Portaels, F. 1976. Isolation and distribution of nocardiae in the Bas-Zaire. Annales de la Société Belge de Médecine Tropicale **56:**73–83.

Prauser, H., Lechevalier, M. P., Lechevalier, H. 1970. Description of *Oerskovia* gen. n. to harbor Ørskov's motile *Nocardia*. Applied Microbiology **19:**534.

Pridham, T. G., Lyons, A. J. 1969. Progress in the clarification of the taxonomic and nomenclatural status of some problem actinomycetes. Developments in Industrial Microbiology **10**:183–221.

Pulverer, G. H., Schaal, K. P. 1978. Pathogenicity and medical importance of aerobic and anaerobic actinomycetes. Zentralblatt für Bakteriologie, Parasitenkunde, Infektionskrankheiten und Hygiene, Abt. 1, Suppl. **6**:417–427.

Pulverer, G. H., Schütt-Gerowitt, H., Schaal, K. P. 1975. Bacteriophages of *Nocardia asteroides*. Medical Microbiology and Immunology **161**:113–122.

Rowbotham, T. J., Cross, T. 1977. Ecology of *Rhodococcus coprophilus* and associated actinomycetes in fresh water and agricultural habitats. Journal of General Microbiology **100**:231–240.

Schaal, K. P. 1972. Zur mikrobiologischen Diagnostik der Nocardiose. Zentralblatt für Bakteriologie, Parasitenkunde, Infektionskrankheiten und Hygiene, Abt. 1 Orig., Reihe A **220**:242–246.

Schaal, K. P. 1977. *Nocardia, Actinomadura* and *Streptomyces*, pp. 131–144. In: von Graevenitz, A. (ed.), Clinical microbiology, vol. 1. CRC Handbook Series in Clinical Laboratory Sciences, Sect. E. Cleveland: CRC Press.

Schaal, K. P., Bickenbach, H. 1978. Soil occurrence of pathogenic nocardiae. Zentralblatt für Bakteriologie, Parasitenkunde, Infektionskrankheiten und Hygiene, Abt. 1, Suppl. **6**:429–434.

Schaal, K. P., Heimerzheim, H. 1974. Mikrobiologische Diagnose und Therapie der Lungennocardiose. Mykosen **17**:313–319.

Schaal, K. P., Reutersberg, H. 1978. Numerical taxonomy of *Norcardia asteroides*. Zentralblatt für Bakteriologie, Parasitenkunde, Infektionskrankheiten und Hygiene, Abt. 1, Suppl. **6**:53–62.

Sneath, P. H. A. 1976. An evaluation of numerical taxonomic techniques in the taxonomy of *Nocardia* and allied taxa, pp. 74–101. In: Goodfellow, M., Brownell, G. H., Serrano, J. A. (eds.), The biology of the nocardiae. London, New York, San Francisco: Academic Press.

Snieszko, S. F., Bullock, G. L., Dunbar, C. E., Pettijohn, L. L. 1964. Nocardial infection in hatchery-reared fingerling rainbow trout (*Salmo gairdneri*). Journal of Bacteriology **88**:1809–1810.

Söhngen, N. L. 1913. Benzin, Petroleum, Paraffinöl und Paraffin als Kohlenstoff- und Energiequelle für Mikroben. Centralblatt für Bakteriologie, Parasitenkunde und Infektionskrankheiten, Abt. 2 **37**:595–609.

Soina, V. S., Agre, N. S. 1976. Fine structure of vegetative and sporulating hyphae in *Micropolyspora fascifera*. [In Russian, with English summary.] Mikrobiologiya **45**:329–332.

Staneck, J. L., Roberts, G. D. 1974. Simplified approach to identification of aerobic actinomycetes by thin-layer chromatography. Applied Microbiology **28**:226–231.

Tárnok, I. 1976. Metabolism in nocardiae and related bacteria, pp. 451–500. In: Goodfellow, M., Brownell, G. H., Serrano, J. A. (eds.), The biology of the nocardiae. London, New York, San Francisco: Academic Press.

Trevisan, V. 1889. I Generi e le Specie delle Battieriaceae. Milan: Zanaboni & Gabussi.

Tsukamura, M. 1962. Differentiation of *Mycobacterium tuberculosis* from other mycobacteria by sodium salicylate susceptibility. American Review of Respiratory Disease **86**:81–83.

Tsukamura, M. 1971. Proposal of a new genus, *Gordona*, for slightly acid-fast organisms in sputa of patients with pulmonary disease and in soil. Journal of General Microbiology **68**:15–26.

Tsukamura, M. 1972. An improved selective medium for atypical mycobacteria. Japanese Journal of Microbiology **16**:243–246.

Tsukamura, M. 1973. A taxonomic study of strains received as "*Mycobacterium*" rhodochrous. Japanese Journal of Microbiology **17**:189–197.

Waksman, S. A. 1950. The Actinomycetales. Their nature, occurrence, activities and importance. Waltham, Massachusetts: Chronica Botanica.

Williams, S. T., Sharples, G. P., Serrano, J. A., Serrano, A. A., Lacey, J. 1976. The micromorphology and fine structure of nocardioform organisms, pp. 102–140. In: Goodfellow, M., Brownell, G. H., Serrano, J. A. (eds.), The biology of the nocardiae. London, New York, San Francisco: Academic Press.

Winogradsky, S. 1949. Microbiologie du sol. Paris: Masson et Cie.

Yamada, Y., Inouye, G., Tahara, Y., Kondo, K. 1976. The menaquinone system in the classification of coryneform and nocardioform bacteria and related organisms. Journal of General and Applied Microbiology **22**:203–214.

Yamada, Y., Ishikawa, T., Tahara, Y., Kondo, K. 1977. The menaquinone system in the classification of the genus *Nocardia*. Journal of General and Applied Microbiology **23**:207–216.

Yano, I., Furukawa, Y., Kusunose, M. 1969. Phospholipids of *Nocardia coeliaca*. Journal of Bacteriology **98**:124–130.

Yano, I., Kageyama, K., Ohno, Y., Masui, M., Kusunose, E., Kusunose, M., Akinori, N. 1978. Separation and analysis of molecular species of mycolic acids in *Nocardia* and related taxa by gas chromatography mass spectrometry. Biomedical Mass Spectrometry **5**:14–24.

Zopf, W. 1891. Ueber Ausscheidung von Fettfarbstoffen (Lipochromen) seitens gewisser Spaltpilze. Berichte der Deutschen Botanischen Gesellschaft **9**:22–29.

The Family Streptomycetaceae

HANS JÜRGEN KUTZNER

Only about one or two decades ago, the "streptomycetes" formed a genus of the Actinomycetales that was quite satisfactorily separated from the then few other genera of this order: organisms producing an extensively branching primary or substrate mycelium as well as a more or less heavy secondary or aerial mycelium; the latter shows typical modes of branching of the hyphae which, in the course of the life cycle, are partly transformed into chains of spores, these are often called conidia or arthrospores. Only occasionally would some confusion or overlapping occur with organisms of the genus *Nocardia*. However, with the isolation of many more actinomycetes during the last 20 years, the differentiation of genera solely on the basis of morphological criteria became unsatisfactory; moreover, many new genera and families had to be created, and the situation became quite complex.

At the same time the early studies by Cummins and Harris (1956, 1958) on the biochemistry of the cell wall of representatives of the Actinomycetales and other bacteria were taken up by Lechevalier and co-workers (Becker et al., 1964; Becker, Lechevalier, and Lechevalier, 1965) and Yamaguchi (1965): It could be shown that *Streptomyces* contained LL-diaminopimelic acid (DAP) in its peptidoglycan, whereas DL-DAP was found in all other actinomycetes then known. Later, several other actinomycetes were described and, as members of new genera, placed into the family Streptomycetaceae on the basis of this biochemical marker (Table 1; see also Table 3). However, few other genera with LL-DAP exist among the Actinomycetales. These belong either to the family Actinomycetaceae (genus *Arachnia*), or are "genera without a family". These organisms can easily be differentiated from Streptomycetaceae by other biochemical or morphological or physiological properties.

It thus appears that actinomycetes of the family Streptomyceteae are now primarily recognized by the structure of their peptidoglycan, which is characterized not only by the LL-DAP mentioned above but also by an interpeptide bridge consisting of glycine molecules. According to the classification of

peptidoglycan types by Schleifer and Kandler (1972), this is type A3γ. As it can be seen from Table 1, the genera of the Streptomycetaceae also form a homogenous group in regard to other biochemical criteria favored by the modern taxonomist: G+C content of DNA, sugar pattern of whole-cell hydrolysates, and fatty acid spectrum of cell lipids.

Indeed, very fortunate progress has been made in the classification of Actinomycetales at the family and genus level by using these biochemical criteria. This progress if certainly due to their genetic stability and the uniformity they display within many taxa established on morphological or physiological grounds. However, one should keep in mind that (i) these biochemical markers are not absolutely stable and (ii) further investigation of more organisms may bring up some overlapping in these criteria. Thus, the biosynthesis of LL-and DL-DAP may differ by only one enzymatic step; the same is true for galactose and madurose (3-O-methyl-galactose), and some fatty acid types may also be synthesized by the acquisition of the additional enzyme. Overlapping is indicated by the fact that LL-DAP occurs occasionally in small amounts in addition to LL-DAP, e.g., in some species of *Actinomadura* which suggest an affinity to *Streptomyces*—supported also by similarities in menaquinone structure (Lacey, Goodfellow, and Alderson, 1978).

Because of the attractiveness of "chemotaxonomy" during the last decades, there seems to have been some temptation to overemphasize one or the other of these criteria. The genera *Sporichthya* and *Intrasporangium* were placed within the Streptomycetaceae just because they contain LL-DAP; however, they differ very much in other properties and they may not be members of the Actinomycetales at all. These genera are therefore listed in Table 1 as "genera without a family". Another example is the placement of *Actinomyces humiferus*, an aerobic soil organism, in the genus *Actinomyces* mainly because of the occurrence of lysine in its peptidoglycan—and despite of the striking differences from the other species of *Actinomyces* in physiology and ecology.

The morphological differentiation of Strepto-

Table 1. Biochemical markers of Streptomycetaceae and some other actinomycete genera.

Family and genus	G+C (mol%)	DAP	Glycine in IPB[a]	Sugars in whole cell hydrolysate				Fatty acid spectrum				Mycolic acids
				Ara+Gal	Mad	Xyl	Fucose	Satur., straight	Iso- and anteiso branched	Unsaturated	10-Methyl branched	
Streptomycetaceae												
Streptomyces	69–73	LL	+	–	–	–	–	+	+	–	–	–
Streptoverticillium	69–73	LL	+	–	–	–	–	+	+	–	–	–
Chainia	71	LL	+	–	–	–	+	+	+	–	–	–
Microellobosporia	67–70	LL	+	–	–	–		+	+	–	–	–
Kitasatoa		LL	+	–	–	–		+	+	–	–	–
Actinomycetaceae												
Arachnia	70–72	LL	+[b]	–	–	–		+	+	–	–	–
Nocardiaceae												
Nocardia	67–69	DL	–	+	–	–		+	–	+	+	+
Micropolyspora brevicatena	66–69	DL	–	+	–	–		+	–	+	+	+
Thermomonosporaceae												
Actinomadura	77	DL[c]	–	–	+	–		+	+	+	+	–
Nocardiopsis		DL	–	–	–	–		+	+	+	+	–
Genera without a family												
Nocardioides	67	LL	+	–	–	–		+	+	+	+	–
Sporichthya		LL	+	–	–	–		+	+	+	+	–
Intrasporangium		LL	+	–	–	–		+	–	+	–	–

[a] IPB, Interpeptide bridge.
[b] Together with aspartic acid.
[c] Often additional small quantities of LL-DAP.

Table 2. Morphological criteria of genera of Streptomycetaceae and some other Actinomycetales.

	Substrate mycelium				Aerial mycelium					
	Fragmen-ting into rod- or coccoid elements	Chains of spores	Pod-shaped spore vesi-cles: "mero-sporangia" few spores	Scle-ro-tia	Chains of arthrospores within a thin fibrous sheath		Fibrous sheath		Sporan-gia-like vesicles (few spo-res)	Motility of spores
Genus					Verticillate arrangement		Smooth	Hairy or spiny		
					No	Yes				
Streptomycetaceae										
Streptomyces	−	(+)	−	−	+	−	+	+	−	−
Streptoverticillium	−	−	−	−	−	+	+	−	−	−
Chainia	−	−	−	+	+	−	+	−	−	−
Microellobosporia	−	−	+	−	−	−	ND[a]	ND	+	−
Kitasatoa	−	−	+	(+)	+	−	+	−	+	+
Actinomycetaceae										
Arachnia	+	−	−	−	← No aerial mycelium —————				−	−
Nocardiaceae										
Nocardia	+	−	−	−	+	−	+	−	−	−
Micropolyspora	−/+	+	−	−	+	−	+	+	−	−
Thermomonosporaceae										
Actinomadura	−	−	−	−	+	−	+	+	−	−
Nocardiopsis	(+)	−	−	−	+	−	+	−	−	−
Genera without a family										
Nocardioides	+	−	−	−	← Fragmenting aerial hyphae —————				−	−
Sporichthya	← No substrate mycelium ————→				← Aerial hyphae become chains of spores —————				−	+
Intrasporangium	−	−	−	−	← No aerial mycelium ————————→				−	ND

[a] No data.

mycetaceae from other genera of Actinomycetales is illustrated in Table 2. The two most consistent features of the family under consideration are (i) the nonfragmenting substrate mycelium and (ii) the transformation of part of the aerial hyphae into chains of arthrospores.

As can be easily seen from the tabulation, there is an exception to this rule (*Microellobosporia*) as well as overlapping with some genera of other families. The combination of the two criteria mentioned above occurs in *Micropolyspora* and *Actinomadura*; in addition, chains of arthrospores are found in some species of *Nocardia* and in *Nocardiopsis*. Resemblance includes even fine structure: The spiny spore type which is characteristic of many species of *Streptomyces* occurs also in the genera *Micropolyspora* and *Actinomadura*. From this discussion it becomes apparent that the identification of an actinomycete as a member of the family Streptomycetaceae rests primarily on the biochemical marker LL-DAP, morphology playing only the secondary role. Morphology, of course, is the criterion of choice for the differentiation of genera of the Streptomycetaceae.

In Tables 1 and 2 the five genera of Streptomycetaceae are listed which will be treated in this chapter. As can be seen from Table 3, several other genera of this family have been mentioned in the literature. However, these genera are either (i) obviously synonymous with one of the five genera being discussed here, (ii) placed into another family of the Actinomycetales, (iii) regarded as "genera without a family", or (iv) unable to be considered further because information is too scanty.

Because of the close relationship of all streptomycetes, especially of the genera *Streptomyces*, *Streptoverticillium*, and *Chainia*, they will be treated together in the following discussion on "Habitats", "Isolation", and "Identification". The wealth of information which is available for these organisms has been summarized during the last 10 years in numerous reviews, monographs, and published symposia—a demonstration of the enormous interest in these bacteria. These publications are listed in the bibliography at the end of this chapter.

Before discussing habitats, isolation, and identification of the Streptomycetaceae, some general remarks may be made about the five genera to be treated here.

Streptomyces Waksman and Henrici 1943 (formerly *Actinomyces*)

This is the best recognized genus of this family because of its wide distribution in nature, especially in soil, and its long history. Many species have been described in the now classical papers by Waksman and Curtis (1916) and Waksman (1919). Further, this genus harbors a great number of the most important—from a pharmaceutical and economic point of view—producers of antibiotics and other secondary metabolites.

Bergey's Manual of Determinative Bacteriology (Buchanan and Gibbons, 1974) lists 463 species names. However, the number of names ever used to christen a "new" species of this genus far exceeds

Table 3. Genera of the family Streptomycetaceae in *Bergey's Manual* (Buchanan and Gibbons, 1974).

Status and genus	Remark[a]
Recognized genera	
Streptomyces	
Streptoverticillium	
Microellobosporia	
Sporichthya	CG: Genus in search of a family
Genera incertae sedis	
Actinopycnidium	CG: = *Streptomyces*
Actinosclerotium	No further information available
Actinosporangium	CG: = *Streptomyces*
Elytrosporangium	BM: = *Streptomyces*
Intrasporangium	CG: = a very doubtful genus
Microtetraspora	⟶ Thermomonosporaceae (cell wall type III)
Streptopycnidium	No further information available
Thermostreptomyces	No further information available
Synonyms	
Chainia	BM: = *Streptomyces*
Microechinospora	BM: = *Microellobospora*
Macrospora	BM: = *Microellobospora*
Others	
Kitasatoa	BM: = Treated with the Actinoplanaceae

[a] CG, Cross and Goodfellow (1973); BM, *Bergey's manual,* eighth edition.

one thousand. As will be discussed below, there will
certainly be a reduction of these numerous "nomen
species" to a much smaller number of "taxospecies"
as progress is made in the field of "molecular taxonomy" and "numerical taxonomy".

Although most members of this genus are easily
recognized as *Streptomyces,* some overlapping occurs with *Streptoverticillium* and *Chainia* (see
below, this chapter). Differentiation from some
Actinomadura sp. is only possible when using biochemical criteria (Table 1) because of similarities in
morphological features (Table 2).

Streptoverticillium Baldacci 1958

This genus is characterized by the verticillate arrangement of the sporogenous hyphae. Several species were first described as members of the genus
Streptomyces (Fig. 1A) before Baldacci (1958) created *Streptoverticillium* for this "morphological
group" (Fig. 1B). Although this type of sporophore
morphology is easily recognized, at least by the experienced worker, and differentiation appears to
be clearcut, some caution is appropriate since similar structures called pseudoverticils (Fig. 1C) exist
in *Streptomyces.*

One may wonder, of course, whether a single
criterion justifies a new genus; however, further
studies showed that there may exist other differences
as well, e.g., in fine structure (Cross, Attwell, and
Locci, 1973). In addition, most species of *Streptoverticillium* can utilize only a limited number of carbon sources, and there are some biochemical and
physiological properties in which this group is rather
homogeneous, although unfortunately some overlapping with *Streptomyces* exists: resistance toward
lysozyme and egg yolk reaction (Kutzner, Böttiger,
and Heitzer, 1978).

At any rate, *Streptoverticillium* is now recognized as a genus of its own by all taxonomists in this
field and by the eighth edition of *Bergey's Manual,*
which lists 40 species names.

Chainia Thirumalachar 1955

This genus was created by Thirumalachar (1955) for
an actinomycete forming "spherical sclerotic granules in large aggregate masses of the mycelium" on
the surface of colonies. Although the first species of
this genus, *Chainia antibiotica,* was described as
not producing aerial mycelium, it turned out that this
and all other species assigned to this genus form
aerial hyphae and spore chains very much like
Streptomyces. Further, it was noted that on prolonged cultivation on laboratory media the capacity
to form sclerotia became gradually lost whereas the
tendency to produce aerial mycelium increased
(Thirumalachar and Sukapure, 1964).

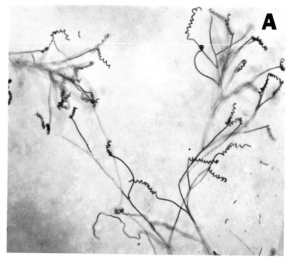

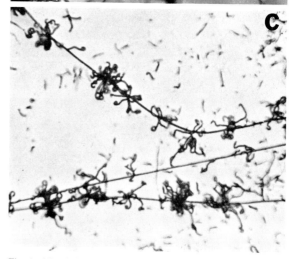

Fig. 1. Morphology of aerial mycelium of three streptomycetes.
(× 250). (A) *Streptomyces* sp.: sympodially branched aerial
hyphae; spore chains form spirals with up to 10 turns. (B)
Streptoverticillium sp.: Spore chains arranged in typical verticils
along straight, long aerial hyphae; end of spore chain sometimes
hooklike or forming one to two turns. (C) *Streptomyces pallidus*:
Verticil-like arrangement of spore chains; the organism, however, is classified as *Streptomyces.*

Studies on the fine structure of the sclerotia (20–75 μm in diameter or even greater) by Lechevalier, Lechevalier, and Heintz (1973), Ganju and Iyengar (1974), and Sharples and Williams (1976) revealed that these structures originate by extensive branching and cell division of certain hyphal regions which contain numerous amorphous inclusions, probably oil droplets; the septa are formed in different planes. The individual cells of these plurilocular structures are held together by a "cement" of fibrillar nature; the cells themselves become more and more filled with lipid material conferring to them the content of 50% lipid by dry weight. (Also, the hyphae of *C. olivacea* contained the rather high amount of 33 to 37% lipid by dry weight; Lechevalier, Lechevalier, and Heintz, 1973.) When such sclerotia are crushed between slide and cover slip, many oil droplets are released.

The function of the sclerotia in the life cycle of *Chainia* is not quite clear. Since they exhibit only little higher heat resistance than arthrospores, they seem to play no role for the survival of the organism. It is even doubtful whether or not they serve as propagules: Although Thirumalachar (1955) stated that on transfer to fresh media they germinate and form colonies, Ganju and Iyengar (1974) argued that the sclerotia are invariably covered with actively growing hyphae which could be the origin of new colonies. This argument is supported by the observation of Lechevalier, Lechevalier, and Heintz (1973) that in *C. olivacea* "germination of the sclerotium did not take place; rather the few hyphae present on the surface of the sclerotium grew out . . . ." These authors were also unable to demonstrate that the lipids stored in the sclerotia were actually used for growth. It thus appears that the lipid-rich sclerotia may merely be the result of a "disturbed" and poorly regulated metabolism of the organisms in questions.

The following arguments have led several investigators—including the authors of the relevant chapter in the eighth edition of *Bergey's Manual*—to regard *Chainia* as synonymous with *Streptomyces*, although there are some reasons to look upon them as somewhat different from the latter genus:

1. The formation of sclerotia underlies some variation; in addition there have been observations of sclerotia-like structures (granules) in *Streptomyces* (Baldacci, Locci, and Locci, 1966) which, however, seem to be different from the "true" sclerotia of *Chainia* in origin and fine structure. No biochemical analysis of these granules is available.
2. Aerial hyphae and chains of arthrospores are formed just as in *Streptomyces*. The unusual feature of intersporal pads as shown in the study by Sharples and Williams (1976) has also been observed in some species of *Streptomyces*.

3. The species of *Chainia* are biochemically similar to *Streptomyces*. However, the occurrence of fucose has been reported by Maiorova (1966), and Lechevalier, Lechevalier, and Heintz (1973) identified the unusual amino acid 2,3-diamino-propionic acid in three species of *Chainia*—possibly a constituent of the "cement" between the cells.

Until now, 14 species of *Chainia* have been described; further studies using more modern approaches will have to show whether we deal with a genus in its own right or merely with species of *Streptomyces*.

Microellobosporia Cross, Lechevalier, and Lechevalier 1963

This genus has been differentiated from the other Streptomycetaceae because of the formation of pod-shaped vesicles containing a single row of two to seven nonmotile spores: merosporangia. These occur as side branches of aerial hyphae as well as of substrate mycelium. The mode of spore formation is similar to the transformation of aerial hyphae to chains of arthrospores in other streptomycetes; the spores, however, are somewhat larger than those of the other genera. The vesicle wall itself is a thin, wrinkled membrane homologous to the fibrous sheath covering the spore chains in other actinomycete genera. Four species are described in *Bergey's Manual*, eighth edition.

There is actually little difference in fine structure, morphology, biochemistry (Table 1), and physiology from *Streptomyces*—besides the formation of short chains of rather large arthrospores. The close relationship to this genus is further supported by the observations of Prauser, Müller, and Falta (1967) on serological properties and sensitivity toward *Streptomyces*-phages. As stated in the first paper by Cross, Lechevalier, and Lechevalier (1963), these organisms "can be isolated readily using media conventionally used for the isolation of *Streptomyces*. When growing on agar they closely resemble *Streptomyces* in gross appearance and one must assume they have been isolated in the past and mistaken for these organisms."

Kitasatoa Matsumae et al. 1968

This genus is quite interesting in as much it exhibits morphological features of the families Actinoplanaceae and Streptomycetaceae: Like the former, it produces club-shaped sporangia enclosing a short chain of zoospores carrying a single flagellum; these vesicles are formed in liquid culture and on solid media at the end of substrate and aerial hyphae. In addition, the aerial mycelium produces long chains of arthrospores very much like those of *Streptomyces*.

There appears to be only one report on organisms of this genus (Matsumae et al., 1968). The four species described by these authors were isolated from soil in Japan and the Kauai island in Hawaii. The methods for isolation and characterization were the same as those used for *Streptomyces*. Remarkably, all four species, which differ in some cultural and physiological properties, produce chloramphenicol and two of them produce one or two additional antibiotics.

The genus *Kitasatoa* was originally placed into the family Actinoplanaceae (Buchanan and Gibbons, 1974). However, Cross and Goodfellow (1973) transferred it to Streptomycetaceae because of the long chains of streptomycete-like arthrospores. The biochemical criteria given in Table 1 (not available to these authors at that time) were supplied by R. M. Kroppenstedt (German Collection of Microorganisms) and support strongly the placement into the family Streptomycetaceae.

HABITATS

There seems to be no doubt that soil is the natural habitat of most members of the family Streptomycetaceae; there they find the suitable conditions for growth and proliferation. These may not be optimal per se but allow successful competition with other microorganisms: (i) Streptomycetes are able to degrade constituents of plant and animal residues, not as easily attacked by other bacteria and fungi, such as polysaccharides (starch, pectin, chitin—but cellulose only to a limited extent), proteins (e.g., keratin, elastine), and —although not extensively— aromatic compounds; (ii) streptomycetes are nonfastidious and satisfied with an inorganic nitrogen source and do not require vitamins or growth factors; (iii) in soil, streptomycetes find plenty of surfaces to support their mycelial growth; by their growth cycle ''spore—mycelium—spore'' they also are adapted to the other physical conditions of this habitat, such as moisture and aeration, which may undergo dramatic changes within a short time according to weather and season; (iv) although the ''spores'' (arthrospores, conidia) are not as resistant toward unfavorable conditions as bacterial endospores, they certainly contribute to the survival over longer periods of drought, freeze, and anaerobic conditions produced by water saturation.

Other biotopes may be infected or contaminated from soil: Few streptomycetes have been isolated from animal or human sources. Pathogenicity to these hosts is certainly not a typical property of these organisms. The same is true in regard to infestation of plants: only a few species cause a bacteriosis (e.g., *Streptomyces scabies* scab of potatoes and some other plants). Fodders and grain may contain streptomycetes. Their number depends on the degree of contamination by soil particles or dust and conditions of storage; however, these organisms seldom contribute significantly to the spoilage of fodders and grain (thermophilic actinomycetes of other families are of much greater importance in this respect). By drainage after heavy rainfalls, creeks and rivers become contaminated with soil streptomycetes that find their way into the sediments of freshwater lakes and eventually to marine biotopes. Streptomycetes may exist which are strongly adapted to these sediments but these are usually the habitat of other actinomycetes. By the same route, drinking water supplies are also contaminated with streptomycetes; some of them produce odorous compounds leading to spoilage of the water.

Soil as Habitat

The literature published on the occurrence of streptomycetes in soil is too numerous to be treated quantitatively; only selected references (admittedly selected subjectively) will be discussed in the following paragraphs. Attention may therefore be directed to the recent reviews by Lacey (1973), Küster (1976), and Williams (1978) and to a series of eight papers by Williams and co-workers dealing specifically with the ecology of actinomycetes in soil: Davies and Williams (1970), Williams et al. (1971), Williams and Mayfield (1971), Mayfield et al. (1972), Ruddick and Williams (1972), Williams et al. (1972b), Watson and Williams (1974), and Khan and Williams (1975).

Growth in Soil

By direct observation of the soil microflora, several authors could show that streptomycetes perform their typical life cycle in this natural habitat. Pfennig (1958) demonstrated the following sequence of events when glass slides were buried into the soil (Cholodny's method, 1930): Organic particles are colonized by ''substrate'' (= vegetative) mycelium; from there, hyphae grow away to soil particles (or the free glass surface supplied by the investigator), and chains of arthrospores are finally produced (terminally or as side branches), their morphology (straight, spiral etc.) corresponding to what we know of these organisms in pure culture. The hyphae growing on glass slides either undergo autolysis or they fragment to cylindrical elements which were called by this author ''chlamydospores''. Thus under these conditions two (!) types of propagules result from the vegetative growth,

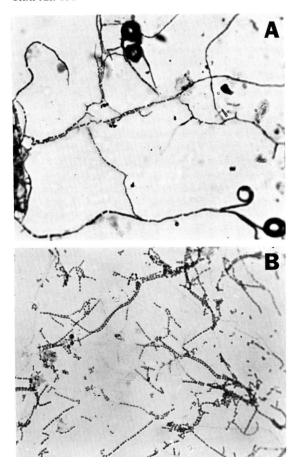

Fig. 2. Growth of streptomycetes on glass slides buried in soil. [From Pfennig, 1958, with permission.] (A) Eight days old; segmentation of hyphae and spiral-shaped chains of spores. (× 1,250). (B) Fourteen days old; the hyphae have disappeared and only chains of spores can be observed. (× 600).

(Skinner, 1951). Mayfield et al. (1972) using scanning electron microscopy could demonstrate even more clearly this growth behavior of streptomycetes in soil.

Of interest is the observation by Pfennig (1958), Lloyd (1969), and Mayfield et al. (1972) that spores of streptomycetes produced on agar media and inoculated into unsterile soil did not germinate (or only to a limited extent). Instead the spores supplied on glass slides were "grazed" by soil amoebas. The inability of added spores to germinate may be due to inhibitory effects of other microorganisms and/or shortage of nutrients. In sterile soil (γ-irradiated), however, added spores showed a high germination rate (Lloyd, 1969); further, both unsterile and sterile soil supported growth of germinated spores (germlings). Although Lloyd (1969) did not definitely mention the two types of propagules observed by Pfennig (1958) and only mentions "sporulation" or "fragmentation into arthrospores", he writes: "Occasionally a germtube failed to grow and instead fragmented into rectangular arthrospores whose linear pattern preserved the general shape of the original germling." This appears to be rather reminiscent of Pfennig's "chlamydospores" in Fig. 2.

Colony Count

From what has been discussed in the previous paragraph, it becomes clear that colony counts reflect the number of propagules, especially arthrospores, at the time of sampling rather than actively growing vegetative mycelium. Nevertheless we are inclined to relate the colony count with metabolic activities exhibited by these organisms in the soil under study—although this activity may have been in the recent past.

The determination of colony counts was of high concern to soil microbiologists during the first half of this century. The result of 18 papers published between 1903 and 1956 have been summarized by Flaig and Kutzner (1960b). In most soils, a count of colonies between 10^4 and about 10^7 per gram can be expected; exceptionally high numbers of 200 million and 2,000 million have been reported by Jensen (1943) and Jagnow (1956), respectively. In the 18 studies evaluated, streptomycetes accounted for about 1–20% of the total viable count, but in some soils these organisms may dominate.

Interest in general counts has now declined; it is nowadays centered around questions such as (i) influence of certain conditions, (ii) occurrence of streptomycetes at special microsites and in habitats which were out of the scope of the general soil microbiologist, (iii) species composition of the flora of different soils, (iv) distribution of species with special properties in certain biotopes.

chains of arthrospores probably dominating (Fig. 2). Pfennig's studies answered by direct observation the question often asked whether streptomycetes exist as mycelium or spores in soil: Vegetative growth is limited to organic particles (plant and animal-residues including hyphae of fungi and protozoa) and in most cases is of short duration (in these experiments just a few days). The number of propagules produced by these microcolonies depends on the quantity and quality of the nutrients supplied; if these are very scanty, the germtube of an arthrospore itself may be transformed into a sporogenous hyphae and produce a short chain of arthrospores.

Similar observations, fitting into the picture described by Pfennig (1958), have been made by other authors, either by direct observation (Cholodny, 1930; Glathe, 1955; Lloyd, 1969; Szabo, Marton, and Partal, 1964), or by a combination of this technique with colony counts (Jensen, 1943) or only by colony counts from samples treated in various ways

Influence of Various Factors

Although not one single condition such as water and air circulation, pH, content of organic matter, vegetation, soil type, season, and climate by itself determines the occurrence of streptomycetes in soil quantitatively or qualitatively (and the influence of a single factor can hardly be determined), some tendencies can be observed.

MOISTURE TENSION AND PH.   As shown in an investigation by Williams et al. (1972b), streptomycetes not only survive quite well in dry soils due to the formation of arthrospores but also need a lower water tension for growth than other bacteria. On the other hand, these organisms are very sensitive to water-logged conditions: In this case, shortage of oxygen is less effective than the increase of carbon dioxide above 10%. This study not only confirms but also by its detailed experiments explains numerous observations of earlier authors that well-drained soils (sandy loam, soils covering limestone) harbor more streptomycetes than heavy clay soils.

In accordance with the pH requirement shown also by isolates in pure culture, streptomycetes prefer neutral to alkaline soils as a natural habitat. This has been shown, e.g., by Flaig and Kutzner (1960b) who investigated a series of soils of different pH (Fig. 3). However, this seems now to be an oversimplified view caused by the use of neutral media for counting and isolating soil actinomycetes. As a result of more detailed studies of acid and alkaline soils that employed media adapted to the pH of the habitat (and supplemented with antifungal agents) it has emerged that there are acidophilic as well as alkalophilic streptomycetes. *Streptomyces acidophilus* has been isolated by Jensen (1928) from acid forest soil. Corke and Chase (1964) reported that an acid podzol soil contained actinomycetes adapted to the low pH of this habitat, and Williams et al. (1971) and Khan and Williams (1975) could prove that acid soils, peat, and mine waters harbor a specific flora of streptomycetes; this comprised a number of different species (not named by the authors), none of which could be definitely identified as *S. acidophilus* Jensen (the type culture of which is no longer available). Further, the isolates were distinct from neutrophilic species in a number of properties, which suggests that they were not merely acidophilic variants of neutrophilic streptomycetes. At the opposite extreme, alkaline soils contain alkalophilic, acid-sensitive streptomycetes, e.g., *S. caeruleus* (Taber, 1959, 1960).

ORGANIC MATTER; VEGETATION.   As would be expected, more streptomycetes were found in a lot receiving organic manure than in lots supplied only with mineral fertilizer or none—in a 70-year-old

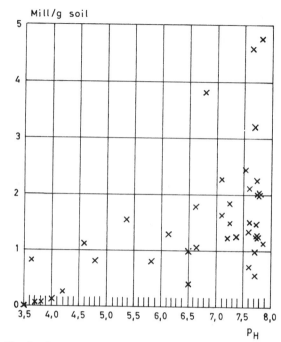

Fig. 3. Occurrence of streptomycetes in soils of varying pH. [From Flaig and Kutzner, 1960b].

field trial on "eternal growing of rye" (Flaig and Kutzner, 1960b). When soil samples are taken from arable land and nearby grass vegetation, the latter often yield more streptomycetes (Conn, 1916; Flaig and Kutzner, 1960b). Although Conn (1916) expressed the view that actinomycetes may be active in the decomposition of grass roots, the difference as compared with the adjacent field soil may instead be due to more favorable water and air circulation (Krzywy, Szerszén, and Wieczorek, 1961).

Occurrence at Special Microsites

From direct observations (see above) it becomes obvious that streptomycetes—like many other soil bacteria—are not evenly distributed in soil but rather occur as microcolonies on particles of organic material which serve as food. Besides this heterogeneous distribution which does not need to be considered when counting or isolating these organisms, there are two microsites which have attracted the attention of soil microbiologists: the rhizosphere and earthworm casts. Studies on the occurrence of streptomycetes in the rhizosphaere as compared with the surrounding soil have been carried out by several investigators. Often a significantly higher number of these organisms was found at this microsite—depending, however, very much on the species and age of the plant. For example, soya and maize harbored 10 to 18 times as many "actinomycetes" in their rhizosphere (Abraham and Herr, 1964). Kaunat (1963) found an increase of actinomycetes in the rhizosphere of maize, beans, sunflower, poppy, and spinach, whereas in 15 other plant species this plant-site was free of actinomycetes. Analogous observations were made by Rouatt, Lechevalier, and Waksman (1951) and Glathe, von Bernstorff, and Arnold (1954). The intestinal tract of the earthworm has been regarded by Brüsewitz (1959) as a "breeding-place" for actinomycetes—provided that appropriate food is supplied. Similar results were reported by Ruschmann (1953), Schütz and Felber (1956), and Parle (1963a, b).

Streptomyces Flora of Different Soils

Two questions can be raised: (i) How many different "species" do we find in a soil when plate counts are performed without enrichment or other special treatments to be discussed below under "Isolation"? (ii) Do the various species of streptomycetes prefer particular soils or, in other words, do we find floras of varying species composition in different soils?

The number of "species" (or "types") to be observed is to a certain extent dependent on the number of isolates made from the soil sample; further on the fact whether the investigator applies a narrow or wide species concept. However, if we disregard these points we can make the general statement that habitats with a low colony count (and therefore supplying also fewer isolates) also yield a low number of different types. As shown in Fig. 4, taken from Flaig and Kutzner (1960b), the number of different types (called "subgroups" in their paper) varied from 5 to 25; the tendency can be recognized that forest soils harbor a flora composed of fewer different types than arable land and grass vegetation. Similar results have been obtained by Misiek (1955; 6 to 23 different "subgroups" were isolated from 47 soil samples taken from 5 soil types—the least number occurring in a forest soil), Szabo and Marton (1964; 3,000 isolates from a rendzina soil were classified into 34 species—"ecotypes"), and Küster (1976; 23 to 33 different species were isolated from grassland).

The second question stated above is difficult to answer. Because of the past chaotic state of the taxonomy of these organisms, an exact identification of species—the prerequisite of ecological studies—was almost impossible. One therefore wonders what value species names such as *Streptomyces albus, S. griseus,* and *S. coelicolor* have in earlier publications. Also, the differences with regard to the species concept have to be mentioned again: To arrive at a meaningful interpretation of the findings the present author believes that—at least in ecology—a narrow species (or subspecies) concept has to be

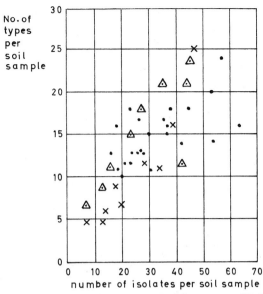

Fig. 4. Number of different "types" of *Streptomyces* ("subgroups") in soil samples in relation to the number of isolated strains. ● Cultivated soil; △ grass vegetation; × forest soil. [From Flaig and Kutzner, 1960b.]

applied. There are only a few publications that can be evaluated in this respect.

Species with rather extreme requirements for pH occur only at respective habitats (see above). A flora adapted to the high salt concentration and alkaline reaction in Hungarian soils was observed by Szabó et al. (1959). A special geographical distribution has been claimed to exist for *S. malachiticus,* which has mainly been found in tropical or subtropical climatic zones (Küster, 1970). Misiek (1955) reported that samples of the same soil type differed little in the occurrence of "subgroups"; however, variation was observed between different soil types. Some streptomycetes were found in many of the 47 samples, others only in few. Flaig and Kutzner (1960b) isolated 1,500 strains of *Streptomyces* from various soils. The samples were taken from nearby areas of different vegetation (arable land, grass, forest). Seventy-six of the 382 "subgroups" differentiated by these authors preferred a certain vegetation: forest, 14; grass, 5; arable land, 11 (here belong *S. violaceoruber* [from 27 soil samples] and *S. hirsutus* [4 samples]); arable land or grass, 42 (here belong *S. griseinus* [grisein-producing strains of *S. griseus* from 12 samples], *S. coelicolor* [34 samples], and *S. phaeochromogenes* [17 samples]); 4 subgroups were found only in a compost of manure and in a field trial receiving only manure, among them *S. prasinus* (4 samples). Of course, in the cases discussed neither the soil type nor the vegetation alone but rather the combination of conditions prevailing at these habitats has to be regarded as responsible for the occurrence of particular organisms.

Some species were mentioned above as occurring in many soils, others being found rather seldom—an observation made by all ecologists. This may be due to various factors: (i) Species with high frequency may be metabolically more versatile, as observed by Davies and Williams (1970) and Küster (1976), or more successful in competition with other soil microorganisms; (ii) these species may be found more often because they produce more propagules in soil; (iii) the result may simply reflect the methods and media used for isolation.

In the course of screening programs for producers of antibiotics, the question has been raised whether certain habitats are especially rich sources of antagonistic streptomycetes. Numerous papers on this subject have been reviewed by Routien and Finlay (1952). The general tendency is that habitats with a quantitative and qualitative rich flora of streptomycetes harbor more antibiotic producers; it is also obvious that producers differ widely in their distribution, some occurring rather frequently (e.g., producers of actinomycins), others rather seldom. However, since no prognosis can be made where to find producers of new (!) antibiotics, the search for further strains must proceed by empirically screening various habitats and employing more sophisticated methods of isolation.

The occurrence of antibiotic-producing streptomycetes in soil has also been of concern to the soil microbiologist and plant pathologist. This aspect will be discussed in the next paragraph.

Interaction with Other Soil Microorganisms

The phenomenon of antibiosis in vitro is now well known: it was first observed by soil microbiologists many decades ago when dilution plates for isolating bacteria and fungi were prepared. Its outcome has been the finding of more than two thousand antibiotic substances from bacteria and fungi, many of which have found application in medicine. Most of the antibiotics in use in chemotherapy are produced by streptomycetes. This fact led to the speculation that production of antibiotics may also take place in soil, conferring an advantage to its producer in competition with other microbes and thus playing a role in the establishment of an equilibrium among members of the soil flora. The enormous body of information available in this field can be summarized as follows (Baker, 1968; Brian, 1957; Williams and Khan, 1974): (i) Production of antibiotics has been demonstrated unequivocally only in sterilized soil supplemented with nutrients and inoculated with a potent producer. (ii) Although direct evidence of antibiotic production in natural soil is lacking, most soil microbiologists share the view that antibiosis could be important in microhabitats of the soil. (iii) Several observations interpreted as being due to antibiotic effects may rather be due to other kinds of interaction such as competition and induction of lysis or autolysis (see below).

The phenomenon of antibiosis in soil has especially attracted the attention of plant pathologists. Numerous studies have shown that soil-borne diseases can be controlled by amending soil with nutrients such as chitin or fungal cell wall material, which leads to an increase of the flora of antibiotic-producing actinomycetes. However, further investigations have suggested that it is not the potential of the enriched flora to produce antibiotics that is important, but rather their capacity to exhibit lytic effects on fungi, the cell wall of which also contains chitin. The mechanisms suggested by various authors are summarized by Williams (1978) (Table 4).

Table 4. Effects of streptomycetes on soil fungi. [From Williams, 1978].

Effect on fungus	Suggested mechanism	Author[a]
Control of root pathogens	(a) Lysis by enzymes (e.g., chitinase)	1
	(b) Inhibition by unknown toxin	2
	(c) Antifungal antibiotics	3, 4, 5
	(d) Inhibition by nonspecific toxin	6
Autolysis of living hyphae added to soil induced by	(e) Nutrient competition	7
	(f) Like (e) + antibiotics	8
	(g) Toxins	9
Epiphytical and parasitical growth on or within hyphae		10
		11
Inhibition of spore germination	(h) Volatile inhibitors	12

[a](1) Mitchell, 1963; (2) Mitchell and Alexander, 1962; (3) Rehm, 1953; (4) Mach, 1956; (5) Whaley and Boyle, 1967; (6) Sneh and Henis, 1972; (7) Ko and Lockwood, 1970; (8) Lloyd and Lockwood, 1966; (9) Lloyd, Noveroske, and Lockwood, 1965; (10) Rehm, 1959a,b; (11) Skinner 1956a,b; (12) Hora and Baker, 1974.

Plant Diseases Caused by Streptomycetes

There are only few streptomycetes capable of infecting plant tissues and causing disease. As can be seen from Table 5, potato and beet are the most frequently invaded cultivated plants, reacting to the pathogen by formation of scab lesions (Fig. 5). *Streptomyces scabies* is by far the most important species producing "common scab", and almost all work done on plant pathogenic streptomycetes is concerned with this disease and its agent. Our present knowledge in this field has been summarized by Hoffman (1958), Labruyère (1971), and Lapwood (1973), who contributed much to our understanding of this plant disease.

Although *S. scabies* belongs to one of the first isolated streptomycetes (Thaxter, 1891: *Oospora scabies*), its taxonomy is not completely settled. It can be assumed that most strains studied by plant pathologists match the following description of "CBS 135.64 strain Corbaz" (Corbaz, 1964; Hütter, 1967, pp. 324–325): (i) aerial mycelium gray; (ii) sporophores form spirals; (iii) spores smooth; (iv) melanin positive. However, there are a number of strains deposited in culture collections which definitely do not agree with this description and must be regarded as mislabeled: (i) IMRU 3018 = ISP 5078 = DSM 40078 = ATCC 23962 has been designated as type strain by Shirling and Gottlieb (1968b) although it does not form spirals and is melanin nega-

tive, in contrast to its description given by Waksman (1961, p. 274). (ii) *Bergey's Manual* (Buchanan and Gibbons, 1974) lists *S. scabies* among the species incertae sedis: "type strain not extant, many taxonomically different references strains available." (iii) Hütter (1967, pp. 57 and 63) placed three strains of *S. scabies* to *S. griseus* (ATCC 10246; CBS strain Millard; CBS strain de Vries) and one strain to *S. olivaceus* (CBS strain Butler). This situation reflects the chaotic state of the taxonomy of streptomycetes as well as the missing cooperation between plant pathologists and taxonomists of this group of organisms. A thorough taxonomic treatment of the species *S. scabies* (including isolates from acid soils and russet scab, see Table 5) appears to be urgently warranted not only from the point of view of taxonomy and nomenclature but also with regard to the use of strains labeled *S. scabies* in genetic work.

For the plant pathologist, the species identity of the pathogen *S. scabies* posed a less difficult problem than the question whether scab is produced only by this species or in addition by others. The confusion lasting for many decades has now been unequivocally resolved by Hoffmann (1954, 1958) who strictly applied Koch's postulates to the investigation of this plant disease: Although many different species were isolated from scabby potatoes, *S. scabies* could be isolated from almost all lesions; from very weakly diseased tubers its isolation was less successful and numerous other species were found. More important, however, was the finding that

Table 5. *Streptomyces* species producing plant diseases.

Species	Host plant	Disease	Reference
S. scabies	Potato, sugar beet, and others	Common scab (severe)	Hoffmann (1958)[a] Lapwood (1973)[b]
S. griseus *S. aureofaciens* *S. flaveolus*	Potato	Common scab (mostly mild)	Hoffmann (1958) Corbaz (1964)
Streptomyces sp.	Potato	Common scab in acid soils	Bonde and McIntyre (1968)
Streptomyces sp. *S. scabies*	Potato Potato	Russet scab Russet scab	Harrison (1962) Vruggink and Maat (1968)
S. ipomoea	Sweet potato	Soil rot and pits	Person and Martin (1940)
Actinomycete not identified	Grapefruit trees	Gummosis	Childs (1953)

[a] Habilitationsschrift.
[b] Review.

A

B

Fig. 5. Plant diseases caused by *Streptomyces scabies*. (A) Common scab on potatoes. (B) Scab on sugar beet ("Gürtelschorf"). [From: Bildarchiv der Biologischen Bundesanstalt, Braunschweig (Bundesrepublik Deutschland).]

when potatoes were grown in soil artificially infected with the various isolates, only strains of *S. scabies* (corresponding to the description given above) produced the expected symptoms and could be reisolated from the lesions; few other streptomycetes caused a mild infection, an observation confirmed by Corbaz (1964). The latter author included in his work 10 strains which had been studied by McKee (1958) as strains of *S. scabies;* however, several proved to belong to other species (*S. griseus, S. aureofaciens,* and *S. flaveolus*). Another result of Hoffmann's (1954, 1958) studies was that isolates from scabby beets could be identified as *S. scabies* and that these isolates could cause scab on potatoes—as the potato isolates did on beets.

  S. scabies produces common scab mainly in neutral to light alkaline soils (see below). As mentioned in Table 5, Bonde and McIntyre (1968) isolated from scabby potatoes grown in acid soils below pH 5.0 a streptomycete that was quite distinct from *S. scabies:* aerial mycelium yellow-white, straight sporophores, melanin negative. In pathogenicity tests, the isolates produced scab in soil of pH 4.5 and 5.9, whereas strains of *S. scabies* gave a positive reaction only at pH 5.9.

  Harrison (1962) investigated the causative agent of russet scab. The organism—in contrast to *S. scabies*—could only be isolated by a special method (see "Isolation", this chapter); it also differed from *S. scabies* by being melanin negative. Pathogenicity

tests (including reisolation of the pathogen) showed that only the typical russet isolates produced symptoms like those resulting from natural infections; other streptomycetes isolated occasionally from russet lesions as well as isolates of *S. scabies* from common scab gave a negative result. Vruggink and Maat (1968) obtained, from soil, strains of *S. scabies* producing russet scab; however, among the isolates which caused russet scab there were a number which were distinctly different from *S. scabies* with regard to melanin reaction (weak) and serological properties.

Some additional information may be given for *S. scabies* and common scab: Because testing of isolates for pathogenicity by using potatoes as host plant is rather time-consuming and laborious, simpler laboratory tests have been applied using seedlings of radish and other plants (Hooker and Kent, 1946), soybeans (Hoffmann, 1954, 1958; Hooker, 1949), and *Phaseolus vulgaris* and leaves of *Physalis floridana* (Knösel, 1970). However, strains or species not pathogenic for potatoes or beets may show pathogenicity to these test plants under the rather artificial conditions. Nevertheless, it would be worthwhile to test the numerous streptomycetes now in our collections for pathogenicity by using these simple tests, to obtain a better understanding of the relationship of this part of the soil flora and higher plants.

S. scabies seems to be distributed worldwide in soil. There is can grow saprophytically or on the roots of various vegetables and weeds (Hooker and Kent, 1946; KenKnight, 1941). The organism can be isolated directly from soil (Vruggink and Maat, 1968), but the usual practice is to isolate it from lesions of infected plants that serve as a bait (see this chapter, "Isolation").

As described by Lapwood (1973), the pathogen is unable to penetrate the intact skin of the potato tuber but invades young tubers through lenticels. The tissue reacts by the formation of a wound barrier; if this is breached, a second and sometimes a third barrier is formed resulting in increasingly severe forms of normal scab. The symptoms depend very much on the time of infection, the growth rate of the potato, the variety of the host, and the virulence of the infectant. Hyphae and even spiral-shaped sporophores have been demonstrated in the infected tissue and on its immediate surface (Jones, 1931; Richards, 1943; Shoemaker and Riddell, 1954). The pathogenic mechanisms that enable *S. scabies* to invade the plant and to trigger the reaction of the host are unknown. Hirata (1959) reported production of indoleacetic acid from tryptophan by *S. scabies* and suggested that this capacity plays a role in pathogenesis. Lawrence and McAllan (1964) found that culture filtrates of *S. scabies* (strain ATCC 10246 = *S. griseus*[!], see above) exhibited an inhibitory effect on potato tuber respiration; but the nature of the inhibitor as well as its role in scab formation remained unresolved. Knösel (1970) studied the cellulolytic and pectinolytic activities of *S. scabies* and a few other streptomycetes. The former one possessed a high activity of pectic acid *trans*-eliminase; however, there was no strict correlation between enzyme activity and virulence. Nevertheless, activity of this enzyme may be a virulence factor, since the pathogen first attacks the middle lamellae of cells (Lapwood, 1973).

In accordance with the optimal growth conditions for streptomycetes, the plant disease occurs especially in sandy, well-drained soil of neutral to light alkaline reaction and in dry years. Thus a control of the disease can be achieved to a certain degree by irrigation and by increasing the soil acidity (e.g., by sulfur amendment). Green manuring with certain crops such as soybeans as a means of controlling potato scab has been known for many decades (Millard and Taylor, 1927). On the other hand, ploughing pasture increases scab incidence, because *S. scabies* may normally inhabit grass roots (Lapwood, 1973). In a more recent contribution, which has received much attention by plant pathologists, Weinhold and Bowman (1968) showed that an antibiotic-producing *Bacillus* accumulated in soils receiving green soybeans; this exhibited a strong inhibitory effect on *S. scabies* but not on many other nonpathogenic streptomycetes.

Although common scab can lead to considerable losses of yield and quality in certain regions and years, plant pathogenic streptomycetes on the whole do not pose severe problems to agriculture and for plant pathologists that many plant pathogenic bacteria and fungi do. Further, as stated by Lapwood (1973), "results from recent research which have given a better understanding in infection, now offer more hope of combating this troublesome disease." From a microbiological point of view, more information is needed in regard to the taxonomy of the pathogens and their pathogenic mechanisms. The genetics of the tyrosinase reaction (melanin test) of *S. scabies* have been studied by Gregory and Huang (1964a,b), who found that the gene controlling this activity is located on a plasmid. However, these studies have not been continued and other streptomycetes have become the playing ground for geneticists.

Another plant disease caused by a streptomycete, *S. ipomoea,* is soil rot of sweet potatoes (*Ipomoea* sp.). The disease and its causal agent has been studied by Person and Martin (1940); it appears in the form of two symptoms: (i) "In heavily infested soil the plants are dwarfed and make little or no vine growth. The root system is very poorly developed, most of the roots being entirely rotten and many of them breaking off when the plant is lifted from the soil." (ii) In mature potatoes, the disease is present in the form of pits or cavities with irregular jagged or roughened margins. (There is no soft rot of the potatoes as mentioned in the eighth edition of *Bergey's Manual.*)

The disease occurs—similar to potato scab (*S. scabies*)—especially in dry years and in soils with a pH of about 5.5 and higher. Thus, irrigation and lowering the pH to 5.0 by sulfur amendment are suitable measures to control the disease.

The pathogen has been isolated and described as a new species by Person and Martin (1940). It has been shown to exhibit pathogenic activities in a Petri plate method using rooted cuttings of potato stems as well as in greenhouse and field experiments. Cross-inoculation experiments showed that *S. ipomoea* was not pathogenic for potatoes (*Solanum* sp.) and *S. scabies* did not cause disease on sweet potatoes (*Ipomoea* sp.). However, as shown by Hooker (1949), both species parasitized wheat seedlings. The culture now available (ISP 5383 = DSM 40383 = ATCC 25462) is a neotype isolated by Martin in 1966. A description is given by Shirling and Gottlieb (1969).

Finally, an actinomycete has been found to be associated with gummosis disease of grapefruit trees (Childs, 1953). This author found mycelial growth in every gum pocket he examined microscopically; the organism grew abundantly in the gum channels between the masses of gall tissue. An actinomycete was isolated and when inoculated into the trunks and branches of mature grapefruit trees produced the typical symptoms of gummosis.

The pathogen grew extremely slowly on all media tried. When egg albumen agar or glucose asparagine agar is used for isolation, month-old plates should be examined at a magnification of 50× before they are concluded to be sterile. The actinomycete has not been identified to the genus and there seems to be no further information or a culture for study available. The disease has been included here for the sake of completeness.

Composts, Manure Heaps and Fodders:
A Habitat of Thermophilic Streptomycetes

Numerous genera of the order Actinomycetales comprise organisms that grow at temperatures of 50–60°C and are therefore regarded as thermophilic (Cross, 1968). As outlined by this author, "it is very difficult to define the term thermophily when applied to the thermophilic actinomycetes." Because of the wide variation of temperature range within a certain genus and even species, use of the terms obligately and facultatively thermophilic, stenothermophilic, and eurothermophilic can only be made in regard to an individual strain; furthermore, conditions such as medium and humidity influence the range of temperature for growth. Some streptomycetes have been found to grow at 60°C only under anaerobic conditions (Henssen, 1975a).

Nevertheless, one may distinguish between those Actinomycetales which are more or less close to the true thermophiles among the other bacteria (e.g., *Thermoactinomyces* and several representatives of *Thermomonospora* and *Microbispora*) and those which are rather eurothermophilic; in this second group belong a few streptomycetes. The data given by Tendler and Burkholder (1961) demonstrate very clearly the differences between strains of *Thermoactinomyces* and *Streptomyces* (Table 6).

A number of thermophilic streptomycetes are listed in Table 7, which, however, may not be complete. As can easily be recognized, most organisms have a temperature range from 28 to 55°C, and several of them grow even at higher temperatures. Workers in this field disagree whether a streptomycete growing above 45°C should be regarded as a distinct species or whether it is merely a variant of a mesophilic one. The use of the prefix "thermo" by various workers shows that in these cases the temperature range has been regarded as an important taxonomic criterion. Craveri and Pagani (1962) even proposed the subgenus *Thermostreptomyces* for thermophilic streptomyces—although they state in their paper "in our opinion there is no clear cut in nature between thermophilic and mesophilic microorganisms." However, this subgenus name has been listed in *Bergey's Manual* (Buchanan and Gibbons, 1974) under genera incertae sedis. The contrary view was expressed by Küster and Locci (1963), who studied eight streptomycetes isolated from peat under thermophilic conditions; these were similar to several mesophilic species except for their temperature range of 27–55°C. The authors regarded them as "thermotolerant" rather than true thermophilic. The same view is favored by Corbaz, Gregory, and Lacey (1963).

More information is needed to settle this taxonomic controversy: (i) We should know the temperature range of all streptomycetes isolated as mesophiles—a test not carried out routinely at present;

Table 6. Temperature limits of 500 isolates belonging to the genera *Thermoactinomyces* and *Streptomyces* (Tendler and Burkholder, 1961).

No. of isolates	Temperature range (°C)	*Thermo-actinomyces*	*Streptomyces*
6	30–67	4	2
43	45–67	43	0
10	50–67	10	0
16	30–65	12	4
100	45–65	98	2
3	50–65	3	0
39	30–60	27	12
86	45–60	86	0
1	50–60	1	0
2	55–60	2	0
108	30–55	32	76
66	45–55	63	2
1	37–55	0	1
22	30–50	13	9

Table 7. Temperature range for growth of some streptomycetes.[a]

Species	Strain No.	Growth at (°C):								Author
		27/28	37	40	45	50	55	60	65	
S. thermodiastaticus	ISP 5573	+	+	+	+	ND	ND	ND	ND	
S. thermoflavus	ISP 5574	+	+	+	+	ND	ND	ND	ND	
S. thermofuscus		+	+	+	+	+	+	+	+	Waksman, Umbreit, and Cordon (1939)
S. thermonitrificans	ISP 5579	−	+	+	+	+	−	−	−	Desai and Dhala (1967)
(S. thermophilus)		+	+	+	+	+	+	+	+	Waksman, Umbreit, and Cordon (1939)
S. rectus		+	+	+	+	+	+	(+)	−	Henssen (1957a,b)
		(+)	+	+	+	+	(+)	−	−	Fergus (1964)
S. thermoviolaceus		+	+	+	+	+	+	(+)	−	Henssen (1957a,b)
ssp. *thermoviolaceus*		−	+	+	+	+	(+)	−	−	Cross (1968)
(formerly ssp. *pingens*)		(+)	+	+	+	+	+	(+)	−	Fergus (1964)
	A 71	(+)	+	+	+	+	+	+	−	Corbaz, Gregory, and Lacey (1963)
S. thermoviolaceus ssp. apingens		(+)	+	+	+	+	+	(+)	−	Henssen (1957a,b)
S. thermovulgaris	ISP 5444	ND	(+)	+	+	+	ND	ND	ND	
		(+)	+	+	+	+	+	(+)	−	Henssen (1957a,b)
		(+)	+	+	+	+	(+)	−	−	Fergus (1964)
S. albus	ISP 5313	+	+	+	+	+	(+)	−	−	Lyons and Pridham (1962)
S. eurythermus	ISP 5014	+	+	+	+	+	+	−	−	Corbaz et al. (1955)
S. griseoflavus	A 77	+	+	+	+	+	+	−	−	Corbaz, Gregory, and Lacey (1963)
S. violaceoruber		−	+	+	+	+	+	−	−	Fergus (1964)
Streptomyces sp.	P 46	+	+	+	+	+	+	−	−	Küster and Locci (1963)
	P 107	+	+	+	+	+	+	−	−	Küster and Locci (1963)

[a] No data.

(ii) we should know whether spontaneous or induced mutants with thermophilic properties can be obtained from mesophiles; (iii) thermophilic organisms should be compared with the mesophiles in many more criteria than hitherto done to arrive at a more soundly based identification.

Independent of their taxonomic position—whether members of mesophilic species or of "independent species status"—thermophilic streptomycetes may be useful organisms in biotechnology as producers of antibiotics, thermotolerant enzymes, and other metabolites or biomass because of their rapid growth rate and reduction of contamination at the elevated growth temperature. The search for such organisms in nature or the production of such strains by mutation may be a worthwhile task.

Thermophilic streptomycetes—like other thermophilic Actinomycetales—undergo an interesting cycle in nature in regard to their dispersal. Active growth takes place at sites of high temperature such as in compost, manure heaps, and self-heating hay or grain; the vegetative phase ends with the formation of large numbers of spores. These are returned with the compost or manure to the fields and pastures where they infect—directly or via soil dust—plant material and hay. After passage through the intestinal tract of animals, they once again find—after having survived all the way as inactive but viable spore—appropriate growth conditions in the above mentioned sites to start the next cycle. Thus, thermophilic actinomycetes are found in all soils and many other localities (Craveri and Pagani, 1962; Tendler and Burkholder, 1961), but substrates that have undergone a heating process are the preferred sources for these microorganisms.

Compost and Manure Heaps

Composts which do not undergo a self-heating process can be expected to harbor a rich flora of mesophilic streptomycetes. However, if such a process takes place, thermophilic actinomycetes become predominant, among them a few thermophilic streptomycetes. The microflora of mushroom compost has found particular interest because it is assumed that microorganisms are important in preparing a compost suitable for the growth of *Agaricus bisporus* (Fergus, 1964). One of the first studies of the microflora of compost of horse manure was carried out by Waksman, Umbreit, and Cordon (1939): two thermophilic streptomycetes were isolated and described (see Table 7), a third one was only microscopically observed (Cholodny's method). Fergus (1964) isolated four species of *Streptomyces* from mushroom compost (see Table 7); the strain *S. violaceoruber* is of especial interest because it failed to grow at 25°C. Balla (1968) identified several thermophilic actinomycetes from mushroom compost that were able to degrade cellulose; among them was one streptomycete, *S. thermovulgaris*. The streptomycetes isolated by Lacey (1973) from this habitat formed three groups: *S. albus*, *S. griseus*, and *Streptomyces* sp. with gray aerial mycelium.

Thermophilic actinomycetes from manure heaps have been the subject of extensive studies by Henssen (1957a,b) and Henssen and Schnepf (1967). The authors describe in detail the method of enrichment, isolation, and cultivation of numerous strains. Several of them were described as new species, some also as members of new genera. Of interest in this context is the isolation of several thermophilic streptomycetes (see Table 7) and the observation that some of the organisms could grow at the limits of the temperature range only under anaerobic conditions, especially as fresh isolates. Henssen regarded these organisms as "facultatively aerobic". She seems to be the only author considering the oxygen relationships of thermophilic actinomycetes. Thermophilic streptomycetes from manure were also isolated by Mach and Futterlieb (1966): 11 strains were obtained which were very similar in their properties; they grew between 30 and 60°C, some of them even at higher temperatures (not reported).

Fodders (Hay, Grain)

The occurrence of actinomycetes in fodders, especially in moldy ones, is well known. The present knowledge has been reviewed by Lacey (1973, 1978), who with his co-workers did extensive studies on the actinomycete and fungal flora of this habitat at Rothamsted Agricultural Experimental Station.

It has to be assumed that fodders (grass, grain, straw) are contaminated by a wide range of actinomycetes present in soil, compost, manure, dust. However, when moldy hay is analyzed, relatively few species are found, apparently selected by (i) nutrients supplied by the substrate, (ii) pH (e.g., "good" hay has a pH of 5.5–6.5, which is not favorable to most streptomycetes), (iii) physical conditions such as moisture and temperature, and (iv) competition with other microorganisms.

According to Gregory and Lacey (1963) and Festenstein et al. (1965), the process of self-heating of hay (and also any other kind of fodder) and the development of particular microorganisms are closely related to the water content of the material. Incubation of samples in Dewar flasks with a moisture content of 25–35% results in a temperature increase to 45–50°C. Some streptomycetes are selected by these temperatures; this flora consisted mainly of *Streptomyces albus*, *S. griseus*, and a few other species with gray aerial mycelium, among them *S. thermoviolaceus* (a species capable of cellulose degradation). At higher water contents, temperatures of 60 to 70°C are attained and the streptomycetes are replaced by thermophilic actinomycetes of other genera and families.

As shown by Spicher (1972), wheat grain shortly after harvest always contains a low number of spores of various streptomycetes; during storage of moist wheat, self-heating takes place and several thermophilic actinomycetes develop. In a study by Lyons, Pridham, and Rogers (1975) of streptomycetes from high-moisture corn (27%) from the field and silos (but apparently without any self-heating), *S. albus* and *S. griseus* were again found: *S. griseus* occurred in all samples of corn, *S. albus* mainly in silos receiving an ammonium isobutyrate treatment.

Freshwater Environments and Water Supplies

Actinomycetes can easily be isolated from fresh water and especially sediments of rivers and lakes. However, as stated by Al-Diwany and Cross (1978), who reviewed our present knowledge on actinomycetes in aquatic habitats, the occurrence of these organisms may simply mean that they survive at these sites because most of them are endowed with spores or cells which show higher resistance toward unfavorable conditions than most bacteria. Certain actinomycetes such as *Micromonospora* spp. are inhabitants of mud, and *Nocardia amarae* belongs to the flora of activated sludge. The occurrence in water of another actinomycete, *Rhodococcus coprophilus*, has even been regarded as an indicator of farm effluent and animal pollution (Rowbotham and Cross, 1977). Most streptomycetes found in these habitats are probably derived from soil and are merely present there as inactive but still viable arthrospores. However, some may find suitable conditions for growth. These have been found interesting by the "aquatic microbiologist" because they produce earthy tastes and odours which effect the drinking quality of water (see below).

As shown by Al-Diwany and Cross (1978), rivers carry a load of various actinomycetes. Ten to 100 colonies/ml of representatives of the following genera were found during an 8-month study of the River Wharfe in West Yorkshire: *Streptomyces, Thermoactinomyces, Micromonospora, Rhodococcus coprophilus.* All these organisms were considered to have been washed into the river from surrounding fields—originating not only from soil but also from animal dung, a hitherto neglected source of these organisms. The interesting observation was made that actinomycetes as well as bacteria were found to be concentrated 100- to 1,000-fold in foam produced by waterfalls and broken water and aggregated in eddies and backwaters near the river bank. Similar numbers of actinomycetes were found in samples from the Thames: 59–200 streptomycetes per milliliter and 10–20 micromonosporae per milliliter (Burman, 1973).

As described by Burman (1973), some of the soil-derived actinomycetes will grow "on decaying vegetation on river banks and mud flats at low water or on floating mats of decaying algae or other vegetation. . . . Growth under these conditions produces the greatest amount of odorous substances, and subsequent increase in river levels washes them into the water thus giving rise to earthy taste complaints."

The fate of actinomycetes of river water in the course of production of drinking water has been studied by Burman (1973). He distinguishes seven types of actinomycetes according to colony appearance on isolation plates (Table 8). The number of streptomycetes seems to be reduced considerably due to death and removal by filtration. On the other hand, other actinomycetes such as *Micromonospora, Streptosporangium,* and *Nocardia* may proliferate abundantly in slow sand filters and appear in the filtrate. The latter two genera are considered as producers of odorous substances which may subsequently be degraded by sporing bacilli. The subsequent step of chlorination also affects streptomycetes more than the other genera. In the distribution system a "new" type (type 7 in Table 8) appears which is rarely observed in river water or in the preceding stages of treatment; it has therefore been called (by Burman, 1973) "aquatic strains of *Streptomyces*". Maximum counts varied from 5 to 400/ml in different service reservoirs. According to Burman (1973), this streptomycete can be easily enriched: "Their most abundant multiplication has been observed in river-derived tap water in glass and other containers kept in the laboratory at room tem-

Table 8. Types of actinomycetes in water samples (according to Burman, 1973).

Type	Macroscopic appearance of colonies
1. *Streptomyces*	Typical substrate and aerial mycelium
2. *Micromonospora*	Wrinkled, leathery or heaped up; no aerial mycelium
3. Intermediate between 1 and 2	As 2, but with short sterile aerial hyphae
4. Intermediate between 1 and 2	Young colonies like 2; later typical aerial mycelium as in 1
5. *Streptosporangium*	Deeply penetrating subsurface mycelium; late appearance of aerial hyphae
6. *Nocardia*	Irregularly shaped, abundant aerial hyphae
7. Aquatic strains of *Streptomyces*	Chitin agar: consisting almost entirely of white aerial mycelium with spores in straight chains; virtually no subsurface mycelium Starch-casein agar: yellow subsurface mycelium; white aerial mycelium.

perature and in which the water is changed every few days. Under these circumstances counts up to 24,000/100 ml have been obtained." The natural habitat of this organism seems to be the silt which accumulates at the bottom of service reservoirs, the highest count recorded being 260,000/100 ml. It also occurs universally in aquatic slimes, independent of their origin and of the nature of predominant organisms in the slime. The streptomycete, fortunately, does not produce odorous substances.

Marine Environments

Two localities have to be distinguished: (i) the littoral and inshore zone and (ii) deep-sea sediments. Streptomycetes have been isolated from both habitats, and there appears to be no doubt that these are not merely survivors of terrestrial runoff but that they (at least some of them) belong to the autochthonous microflora of these sites and play an active role in the decomposition of organic material. Little information is available at present on the taxonomy of these isolates; most cultures obtained from these habitats in the early studies (see below) either have been discarded or were not made available for taxonomic studies. No marine isolates seem to be deposited in culture collections and none was included in the International Streptomyces Project. We thus still do not know whether there are particular species of streptomycetes occurring only in marine environments or whether these isolates are marine ecotypes of terrestrial forms adapted to the conditions (nutrients, salt concentration, temperature, aeration, and hydrostatic pressure) prevailing there. Most of our present knowledge of streptomycetes in the marine habitat has been reviewed by Okazaki and Okami (1976) and Okami and Okazaki (1978).

Streptomycetes from the littoral zone were isolated either from sediments (Grein and Meyers, 1958; Roach and Silvey, 1959) or from decaying seaweed. In the latter case, the organisms obtained were adapted to the nutrients supplied by this microsite: agar (Humm and Shepard, 1946), alginate and laminarin (Chesters, Apinis, and Turner, 1956), and cellulose (Chandramohan, Ramu, and Natarajan, 1972). Unfortunately it is not known to what extent alginate and laminarin are degraded by terrestrial forms since these substrates have not been included in carbon utilization tests; liquefaction of agar by terrestrial streptomycetes would have been noticed, but the only such report is the study by Stanier (1942) on an agar-decomposing strain of *S. coelicolor* (in fact *S. violaceoruber*) isolated from a tap water agar plate (used originally for the germination of myxomycete sclerotia). (Certainly there must be some nonmarine microorganisms in sewage, soil, or fresh water to digest all the agar used in microbiological laboratories.)

Numbers of actinomycetes in sea sediments have been reported by Weyland (1969): Samples from bottom sediments of the North Sea (30–700 m deep) yielded between 23 and 1.485 colonies per cm^3; a sample from the Atlantic Ocean taken at a depth of 3.362 m 175 miles offshore of West Africa produced 23 colonies per cm^3. The 1,348 strains isolated in the course of this study belonged to *Streptomyces*, *Nocardia*, *Micromonospora*, and *Microbispora*. Okami and Okazaki (1978) primarily found streptomycetes in the sediment of shallow seas (70–520 m deep): 300–1,270 colonies per cm^3; in samples from 700–1,600 m depth *Micromonospora* dominated; no actinomycetes were obtained from depths of 2,800 and 5,000 m in the Pacific Ocean.

Marine streptomycetes show a higher salt tolerance than their terrestrial counterparts (Table 9). However, this may be due to simple selection of the more tolerant organisms arriving as contaminants in the sea because salt tolerance is widespread among soil streptomycetes (Tresner, Hayes, and Backus, 1968). Alternatively, "adaptation" may occur; this process has been reproduced in vitro (Okazaki and Okami, 1975): Salt-sensitive streptomycetes subcultured with increasing concentrations of NaCl yielded salt-tolerant "variants"; in one case, such a variant proved to be an obligate halophile not able to grow without salt. Among the streptomycetes isolated from marine environments, Okazaki and Okami (1976) found only few which were obligate halophiles.

Marine habitats now seem to be the favorite source of actinomycetes for screening programs in the search of producers of antimicrobial and antiinsecticidal compounds. Nissen (1963) found a high percentage of antibiotic-active streptomycetes from decaying seaweed. More recently, Japanese workers were successful in isolating producers of new antibiotics from the shallow sea (Okami and Okazaki, 1972; Okami et al., 1976). Interesting enough, antibiotic production was adapted to the conditions of the marine environment: Among 200 streptomycetes inactive in ordinary media, 7 were found which exhibited antimicrobial activity when cultured in very dilute broth with 3% salt or in a medium containing powdered tangle seaweed, *Laminaria* (Okami and Okazaki, 1978). This observation led these authors to speculate that—as in soil—antibiotic activity may also exist in the close vicinity of streptomycete growth in the mud.

Man and Animals

The medical literature mentions only one species of *Streptomyces*, *S. somaliensis*, which is definitely pathogenic for man. It is one of the etiological agents of actinomycetomas which are also produced—and even more often—by *Nocardia brasiliensis*, *Actinomadura madurae*, and *A. pelletieri* (Gordon, 1974; Pulverer and Schaal, 1978). The disease occurs mainly in warm and moist zones of the world. The pathogens, which invade skin lesions, are part of the microflora of the soil. (A second species, *S. paraguayensis*, associated with black-grained mycetoma, is of doubtful significance). Surprisingly, medical microbiologists have

Table 9. Comparison of NaCl tolerance between terrestrial and marine streptomycetes (Okazaki and Okami, 1976).

Source	No. of strains tested	NaCl%					
		0	1	3	5	7	10
Soil	83	100[a]	76	41	15	5	0
Shallow sea sediments	87	100	93	70	47	17	0

[a] Percentage of strains growing with this particular salt concentration.

not yet taken notice of two other species of *Streptomyces* which may not be pathogenic (or only weakly) but have been repeatedly isolated from man and therefore deserve attention: *S. albus* and *S. coelicolor*.

S. albus, the type species of the genus, has been treated taxonomically in detail by Hütter (1961), Lyons and Pridham (1962), and Gordon (1967). The type strain is ATCC 3004; another very typical strain (proposed as neotype by Hütter, 1961) is ATCC 618. Of the various properties, only the temperature range for growth, 25–55°C, is of interest in this context. (It should be mentioned that many streptomycetes which have been called *S. albus* actually belong to other species; for identification of clinical isolates, the type culture and the three papers cited above should be consulted.) The sources of the 21 *S. albus* strains studied by Gordon (1967) indicate that several of them were isolated from pathological material such as actinomycosis, pulmonary streptotrichosis, dental caries, blood of patient, and blood of sick cow. Two of the isolates originated from moldy hay. Moldy hay may be the original habitat of *S. albus*, as this organism was often found in this or similar material as well as in compost, as described earlier in this chapter.

The type strain of *S. coelicolor* Müller is ATCC 23899—not the "*S. coelicolor*" of the *Streptomyces* geneticists (Hopwood, 1967), which actually is *S. violaceoruber* (Kutzner and Waksman, 1959). Strains of this species were isolated by Heymer (1957) from the skin of 20% of 300 patients (checked for dermatomycosis), from 25 of 150 tonsils, and from numerous samples of sputum. This appears to be the only report on the observation of this species in and on man. According to Flaig and Kutzner (1960b), *S. coelicolor* is one of the most widespread streptomycetes in soil; thus its occurrence on the skin and in tonsils may not be surprising. It may have been accidental that the organism was first isolated by Müller (1908) in a hygiene institute—as an air contaminant of a potato plug where it produced its blue pigment "amylocyanin", now identified as indochrome (Habermehl, Christ, and Kutzner, 1977).

In addition to the four species of *Streptomyces* mentioned above, there are several others that have been isolated from man: *S. candidus* (ATCC 4878, DSM 40838) from purulent exudate of fractured patella; *S. gedaensis* ATCC 4880, DSM 40518) from sputum and human abscesses; *S. horton* (ATCC 27437, DSM 40266) from pus; and *S. willmorei* (ATCC 6867, DSM 40459) from streptotrichosis of liver.

It thus appears that members of the genus *Streptomyces* do occasionally occur in pathological material—a fact that should always be taken in consideration by medical microbiologists.

ISOLATION

As discussed in greater detail in Chapter 5 of this Handbook, the procedure for the isolation of a microorganism from its habitat varies with (i) the nature of the microorganism and (ii) the number of germs relative to the other microbes within the habitat. If the organism one is looking for outnumbers all others in the sample being investigated, direct plating of a serial dilution on a nutrient agar medium may readily lead to a pure culture. This is rarely the case with members of the Streptomycetaceae; isolation of these actinomycetes requires either enrichment and/or use of more or less selective media. Both principles have been employed for organisms of this family. Furthermore, all procedures must take into account the limited competitive abilities of streptomycetes vis-à-vis fungi and bacteria under laboratory conditions due to the lower rate of radial growth and reproduction of cells of streptomycetes. The methods for the isolation of streptomycetes have been reviewed by Nüesch (1965) and Williams and Cross (1971).

Isolation from Soil

General

It seems to be generally understood that with the methods used for isolating streptomycetes most of the colonies obtained on plated media originate from spores (arthrospores), which certainly outnumber hyphal fragments as propagules. This aspect has an important consequence: the relative number of "species" we find by our methods is a function of their potential to produce spores in their habitat. The danger may exist that we thus overlook streptomycetes which are active growers in soil but produce few spores. One of the enrichment methods described below may actually favor sporulation of existing vegetative growth rather than mycelial development. There has been only one investigation (Skinner, 1951) to differentiate between spores and mycelium as viable germs in isolation procedures.

The close association of the vegetative growth and spore chains with the mineral and organic particles of the soil or fodders necessitates a vigorous shaking of the sample with the diluent. Addition of glass beads to the suspension to be treated on a shaker is a useful method. Several authors advocate mechanical devices such as an Ultra-Turax, Turmix, Waring blender, or pestle and mortar. Further treatment of the sample (i.e., preparing dilutions, methods of plating) differs little from general bacteriological practice. Isolation plates may be prepared as pour plates or agar plates may be surface-inoculated with a sterile glass rod (Drigalski spatula). Of special importance is the avoidance of water films, which encourage the spread of bacteria on the isolating plates. This can be achieved by (i) drying the plates before incubation at 45°C, or (ii) mixing the soil suspension with the molten agar, or (iii) preparing a "sandwich plate" of three layers in which the second layer receives the inoculum.

Usually the coarse particles of the soil suspension are allowed to settle down before dilutions are made. However, soil particles may be used directly for incubation of "soil plates" (Warcup, 1950) as employed for the isolation of fungi. A further development of this method for the isolation of fungi has been the use of "washed soils" (Watson, 1960; Williams, Parkinson, and Burges, 1965): The spores in the suspension are discarded (or plated separately) and the washed particles are used as inoculum; these yield growth from vegetative hyphae adhering to the root fragments, humus, and mineral grains. However, this method does not seem to have had any application in isolating actinomycetes.

The application of the membrane filter method for the isolation of streptomycetes from soil combined with the incorporation of sterile soil (10%) into the agar medium has been advocated by Trolldenier (1966, 1967). With this method, 1 ml of each of a series of tenfold dilution is filtered through a membrane filter 50 mm in diameter and 0.3 μm in pore width. The filter is then placed upside down on soil agar plates: media such as those listed in Table 11 plus 10% compost soil (autoclaved together) are used. Colonies of microorganisms develop between the agar surface and the membrane filter; after some days, streptomycetes grow through the pores and form visible colonies (to be counted after 2 weeks), whereas bacteria and fungi remain behind. This method is selective for streptomycetes not only because of the physical barrier (the membrane) but also because the agar amended with soil yields up to 3–5 times as many colonies as compared with poured plates without soil; furthermore, the colonies exhibit a more pronounced pigmentation of the aerial and substrate mycelium. A favorable effect of soil extract on the growth of streptomycetes has been shown by Spicher (1955) to be due to the presence of trace elements in this extract. Surprisingly, soil extract has found little use in the preparation of media for these predominantly soil microorganisms.

Another departure from the normal soil-dilution-plate technique is the application of the baiting method. Insect wings consisting of chitin have been used for this purpose by Veldkamp (1955), Jagnow (1957), and Okafor (1966).

For the isolation of streptomycetes from fodders such as hay, the material can—as an alternative to making suspensions—be agitated in a wind channel and the aerosol used to inoculate the plates by use of the Anderson sampler or similar devices. This method has been widely employed for the isolation of thermophilic actinomycetes (e.g., Gregory and Lacey, 1962).

To obtain as many streptomycetes as possible from their habitat—quantitatively as well as qualitatively—the following four principles of enrichment and isolation have been successfully employed, singly or in combination:

1. Enrichment within the substrate before isolation.
2. Treatment of the sample to remove other microbes that may render isolation difficult.
3. Encouragement of the development of streptomycetes on isolation plates by choosing carbon and nitrogen sources preferred by these organisms.
4. Inhibition of the accompanying flora by the incorporation of selective substances into the nutrient agar used for isolation.

The application of these four principles by various authors is summarized in Table 10. The following comments may supplement these data.

Enrichment of Streptomycetes in Their Habitat (Soil)

Enrichment of Streptomycetes in Soil by Addition of CaCO₃ (Tsao, Leben, and Keitt, 1960)

Air-dried soil samples (1 g) are mixed with 0.1 g $CaCO_3$ and incubated at 26°C for 7–9 days in a water-saturated atmosphere. The authors report that this led to a 100-fold increase of streptomycete colonies on the isolation plates as compared with the untreated control; concomitantly, there was a decrease of the fungal flora. The authors state: "The rationale of the procedure was that drying reduced the initial microbial population in the soil, and that the pH of the soil after adding $CaCO_3$ would be changed to favor the growth of the remaining actinomycetes and to inhibit or retard that of most fungi." Similar favorable results with this method were obtained by El-Nakeeb and Lechevalier (1963).

ADDITION OF NUTRIENTS.   Enrichment of streptomycetes by soil amendments was first studied by Jensen, who used keratin (1930) and chitin (1932). Keratin led to an increase of streptomycetes without affecting bacteria; two organisms were isolated which degraded keratin in pure culture. Addition of chitin to soil resulted in a considerable increase of bacteria and streptomycetes. More recent studies on enrichment with chitin have been carried out by Williams et al. (1972b). Other organic materials such as salmon viscera meal, peanut meal, cottonseed meal, and dried blood flour (15 mg/g soil) were used by Porter and Wilhelm (1961); incubation in a moistened condition led to a very significant (up to

Table 10. Some principles suitable for selective isolation of streptomycetes from soil.

Enrichment in substrate	Pretreatment of the sample	Chitin	Starch	Paraffin	Glycerol	Casein	Asparagine	Arginine	KNO_3	Cycloheximide	Pimaricin	Nystatin	Penicillin	Polymyxin	Propionate	Bengal rosa	Reference[a]
$CaCO_3$																	1
Organic material					+			+			+	+					2
	Drying; heating																3
																	4
	Phenol																5
		+															6
	Centrifugation		+						+								7
			+		−	+			+								8
			−	+	+	+			+								9
					+			+									10
										+							11
										+	+	+	+	+			12
					+			+				+			+		13
																+	14
			+			+	+		+								15

[a] (1) Tsao, Leben, and Keitt, 1960; (2) Porter and Wilhelm, 1961; (3) Agage and Bhat, 1963, 1967, and Williams et al., 1972; (4) Lawrence, 1956; (5) Sevcik, 1963; (6) Lingappa and Lockwood, 1961; (7) Flaig and Kutzner, 1960b; (8) Küster and Williams, 1964; (9) Ball, Bessell, and Mortimer, 1957; (10) El-Nakeeb and Lechevalier, 1963; (11) Dulaney, Larson, and Stapley, 1955, and Corke and Chase, 1956; (12) Porter, Wilhelm, and Tresner, 1960; (13) Williams and Davies, 1965; (14) Crook, Carpenter, and Klens, 1950; (15) Ottow, 1972.

1,000-fold) increase of the number of strepto-mycetes. Because of the concomitant increase of other bacteria and of fungi, a selective medium has to be employed for isolation. The study of the cultures obtained from these enrichments showed that floras of different composition were produced by the various nutrients; further, it appeared that only part of the original flora was favored by this method, and thus a more limited spectrum of "species" resulted—although they occurred in high numbers.

Treatment of the Sample Before Isolation

DRYING AND STORAGE OF SOIL; HEATING. These methods take advantage of the relatively high resistance of arthrospores toward low moisture tension and dry heat. Nüesch (1965) mentions that drying of the sample and prolonged storage at ambient temperatures for mesophiles (or at 50–60°C for thermophiles) result in a relative increase of the strepto-mycete population. Similar observations were reported by Agate and Bhat (1967). In 1963, the same authors reported that heat treatment of surface-inoculated agar plates (10 min, 110°C) aided in the selective isolation of pectinolytic streptomycetes from soil. A detailed study has been carried out by Williams et al. (1972b), who found that heat treatment of the soil (40–45°C, 2–16 h) reduced the bacterial flora considerably without affecting the colony counts of streptomycetes.

Phenol Treatment of Sample Before Isolation (Lawrence, 1956)

A total of 0.1 ml of a dense soil suspension is added to 10 ml of a 1.4% phenol solution; after 10 min, the suspension is diluted and plated. The author claims successful elimination of bacteria and fungi without any detrimental effect on streptomycetes. However, El-Nakeeb and Lechevalier (1963) obtained less favorable results with this method.

Differential Centrifugation of Sample Before Isolation (Řeháček, 1959)

Nüesch (1965) describes the method as follows: 1 g soil is ground with a little water in a mortar and then diluted to a volume of 100 ml. The suspension is then centrifuged at 1,600 × g (at the bottom of the centrifuge tube) for 20 min. The supernatant contains the spores of streptomycetes, whereas the sediment contains the spores of fungi and other bacteria. Results obtained with this method by other authors, e.g., El-Nakeeb and Lechevalier (1963), were not very promising. A significantly smaller number of colonies of streptomycetes showed up as compared with the control.

Choice of Carbon and Nitrogen Sources; Very Poor Media

Investigation of the utilization of carbon and nitrogen compounds by streptomycetes as well as empirical studies with combinations of various nutrients revealed that the following C and N sources are suitable for the isolation of these organisms: chitin, starch, and glycerol; arginine, asparagine, casein, and nitrate. Streptomycetes show good development on media containing these nutrients, whereas other bacteria and fungi grow poorly. Table 11 lists the recipes of five media.

As an alternative to the use of nutrients preferentially utilized by streptomycetes, the ability of these organisms to grow on very poor media such as water agar can be used. The agar and the laboratory air apparently supply enough nutrients to allow scanty growth of streptomycetes (therefore sometimes called "oligocarbophilic"), whereas other organisms of the soil flora do not show up (Hirsch, 1960). The following comments may supplement the data of Table 11.

CHITIN AS SOLE CARBON AND NITROGEN SOURCE; WATER AGAR. Based on the observation that the ability to degrade chitin is widespread among streptomycetes (Bucherer, 1935/36; Jagnow, 1957; Jeuniaux, 1955; Veldkamp, 1955), a chitin medium for the selective isolation of these organisms was developed by Lingappa and Lockwood (1961, 1962). The 1961 formula contained some mineral salts which were later omitted. These authors also noted that the chitin agar was only little superior to water agar—an observation later confirmed by El-Nakeeb and Lechevalier (1963). Mineral salts were later added (Hsu and Lockwood, 1975; Table 11); these proved to be necessary for the isolation of actinomycetes (Streptomyces, Nocardia, Micromonospora) from water samples but they had little effect when isolating actinomycetes from soil.

MEDIA WITH STARCH, CASEIN, AND NITRATE. Like chitin, starch is degraded by most (if not all) strepto-mycetes and can thus serve as a selective carbon source. The combination of starch with nitrate, which is utilized by most streptomycetes as nitrogen source (in contrast to many other bacteria), is especially favorable for the isolation of these organisms. Such a medium was employed by Flaig and Kutzner (1960b). The medium was further improved by Küster and Williams (1964), who state: "The three best media, allowing good development of strepto-mycetes while suppressing bacterial growth, were those containing starch or glycerol as the carbon source with casein, arginine or nitrate as the nitrogen source."

Table 11. Some media useful for the selective isolation of streptomycetes.

Ingredients (g/liter)	Chitin agar (a)[a]	Starch-Casein-KNO$_3$ agar (b)	Paraffin agar (c)	Glycerol-arginine agar (d)	*Actinomyces* isolation agar (e)
Na-Propionate	—	—	—	—	4.0
Paraffin (liquid, B.P.)	—	—	1.0	—	—
Chitin (colloidal)	4.0	—	—	—	—
Starch	—	10.0[b]	—	—	—
Glycerol	—	—	—	12.5	5.0[c]
Asparagine	—	—	—	—	0.1
Arginine	—	—	—	1.0	—
Casein	—	0.3	—	—	—
Na-Caseinate	—	—	—	—	2.0
NH$_4$NO$_3$	—	—	4.0	—	—
KNO$_3$	—	2.0	—	—	—
NaCl	—	2.0	—	1.0	—
KH$_2$PO$_4$	0.3	—	2.0	—	—
K$_2$HPO$_4$	0.7	2.0	6.0	1.0	0.5
MgSO$_4 \cdot$ 7H$_2$O	0.5	0.05	—	0.5	0.1
CaCO$_3$	—	0.02	—	—	—
Fe$_2$(SO$_4$)$_3 \cdot$ 6H$_2$O	—	—	—	0.01	—
FeSO$_4 \cdot$ 7H$_2$O	0.01	0.01	0.0054	—	0.001
CuSO$_4 \cdot$ 5H$_2$O	—	—	0.0025	0.001	—
ZnSO$_4 \cdot$ 7H$_2$O	0.001	—	0.049	0.001	—
MnSO$_4 \cdot$ H$_2$O	—	—	—	0.001	—
MnCl$_2 \cdot$ 4H$_2$O	0.001	—	0.046	—	—
Na$_2$B$_4$O$_7 \cdot$ 10H$_2$O	—	—	0.00094	—	—
(NH$_4$)$_6$Mo$_7$O$_{24} \cdot$ 4H$_2$O	—	—	0.0002	—	—
Agar[d]	20.0	18.0	15.0	15.0	15.0
pH[e]	8				

[a] References: (a) Hsu and Lockwood, 1975; (b) Küster and Williams, 1964; (c) Ball, Bessell, and Mortimer, 1957; (d) El-Nakeeb and Lechevalier, 1963; (e) Difco Laboratories.
[b] Alternatively glycerol 10 g/liter.
[c] Not contained in the dehydrated medium; added at the time of preparation.
[d] Depending on quality.
[e] Adjusted at 7.0–7.5 or lower or higher—according to flora to be isolated.

PARAFFIN AGAR.   Whereas paraffin is well known to be degraded by *Nocardia,* little is known of its utilization by streptomycetes. Nevertheless, it has been successfully employed by Ball, Bessel, and Mortimer (1957) for the isolation of these organisms: "The medium which was successful in producing the largest number of Streptomyces colonies varied with the soil population, but paraffin agar proved to be selective for *Streptomyces* spp. and sometimes permitted their isolation from soils containing a high content of spreading bacteria and fungi."

MEDIA WITH GLYCEROL AND ARGININE.   The suitability of the combination of these two compounds for streptomycetes was first observed by Benedict et al. (1955). Further results with such media were reported by Porter, Wilhelm, and Tresner (1960) and El-Nakeeb and Lechevalier (1963). The latter authors compared the medium given in Table 11 with nine others and found it to be superior to all others in regard to colony counts of streptomycetes as well as to the proportion of these colonies to all colonies grown on that medium.

MEDIA WITH SELECTIVE NUTRIENTS; INDICATOR MEDIA.   Besides the widely employed chitin, which appears to be a rather selective carbon and nitrogen source for streptomycetes, there are several other nutrients which have been used occasionally and which may deserve more attention. Several of them yield indicator plates; colonies capable of degrading the following compounds produce visible zones of clearing or other changes: (i) pectin (Wieringa, 1955); (ii) poly-β-hydroxybutyrate (Delafield et al., 1965; D. Claus and H. Hippe, personal communication); (iii) rubber (Nette, Pomorzeva, and Koslova, 1959); (iv) cholesterol (Brown and Peterson, 1966); (v) elemental sulfur (Wieringa, 1966). Whereas the last three substrates may yield specialists among the soil streptomycetes, pectin and poly-β-hydroxybutyrate appear to be utilized by many streptomycetes. Combining these carbon sources with other selective measures may facilitate the isolation of these organisms.

Use of Selective Agents

Antibiotics that selectively inhibit fungi have been used by various authors: cycloheximide (Acti-dione), 50–100 μg/ml, by Dulaney, Larson, and Stapley (1955) and Corke and Chase (1956), and the polyene antibiotics pimaricin and nystatin, each 10–50 μg/ml, by Porter, Wilhelm, and Tresner (1960). Unlike antifungal antibiotics, which have no effect on actinomycetes, antibacterial agents have to be used with care.

As shown by Preobrazhenskaya et al. (1978), genera of Actinomycetales differ significantly in their sensitivity toward antibacterial antibiotics, streptomycetes unfortunately being the most sensitive group. Thus the application of these antibiotics is especially helpful in the isolation of other rather rare genera of this order. Although different "series" of streptomycetes vary also in their antibiotic sensitivity, antibacterial antibiotics appear to be of little help in the selective isolation of members of this family. Polymyxin (5 μg/ml) and penicillin (1 μg/ml) have been employed by Williams and Davies (1965); these agents drastically suppress the "bacterial flora"; however, they also inhibit some streptomycetes.

Besides antibiotics, the following two agents are recommended for suppressing the unwanted microbial flora: (i) Na-propionate, 4 g/liter (Crook, Carpenter, and Klens, 1950; a medium containing this agent is commercially available from Difco Laboratories; Table 11). (ii) Rose bengal, 35 mg/liter (Ottow, 1972). The author claims the following advantages of this medium: suppression of most bacteria; coloring even pinpoint actinomycete colonies intense pink, and reducing the spreading growth of fungi.

Isolation of Producers of Antibiotics

Innumerable strains of streptomycetes have been screened for antibiotic activity during the last decades. This test can be made after isolation of pure cultures; however, to avoid isolation of too many inactive cultures, those exhibiting antibiotic activity can be recognized on the dilution plates by subsequent addition of the appropriate test organism, either by flooding or by spraying. Zones of inhibition can be detected after further incubation (Lindner and Wallhäusser, 1955; Wilde, 1964). Alternatively, primary isolation plates may be replica plated onto various agar plates each seeded with a test organism (Lechevalier and Corke, 1953; Trolldenier, 1967).

Sometimes it is of interest to isolate more strains of a "species" producing a particular antibiotic. In the case of an antibacterial agent, this same antibiotic can be used as selective agent since the producer is, in general, much less sensitive to its own metabolite than are other streptomycetes.

Numerous variations of the techniques described are possible and have certainly found application in industrial laboratories carrying out screening programs. It has to be assumed that some more sophisticated methods (especially media with certain nutrients and selective agents) have been developed there, but this information has not been made available.

Isolation from Infected Plants

The isolation of streptomycetes from diseased plant tissue, i.e., primarily from scabby surface layers of potatoes or beets (see Table 5), involves three steps: (i) surface sterilization of the tubers, beets, or roots, (ii) maceration of the plant tissue, and (iii) use of appropriate media for plating.

An effective method for isolating the scab organism was described by Taylor (1936): 10 g fresh Ca-hypochlorite is thoroughly shaken in 140 ml tap water. After a few minutes of standing, the solution is filtered. Before use, 3 parts of it are mixed with 1 part of 25% NaOH. Tubers are placed into this mixture for 2 min, and, without washing, a slice is cut using a sterilized scalpel to remove a lesion and the underlying healthy tissue; this is macerated in a sterile morter with a small amount of water. The suspension is then diluted and plated on Waksman's egg albumen agar.

Ken Knight and Munzie (1939) modified this method by using 0.1% $HgCl_2$ (1 min) for surface disinfection and making the suspension in a sterile test tube using a glass rod. Thorough trituration was found to be unnecessary. A glucose-$NaNO_3$-agar medium was employed for plating.

Lawrence (1956) ground 5 g of peelings from scabby potato tubers in 100 ml water for 3 min in a Waring blender. One drop was then transferred to 10 ml of a phenol solution 1:140; after 10 min, 1 drop was used to inoculate 12 ml of the molten medium to be plated: Czapek's agar and glucose asparagine agar, both pH 6.5.

The medium used for plating may be any one of those suitable for isolating streptomycetes from soil. Alternatively, the melanin-indicator medium of Menzies and Dade (1959) may be used on which the tyrosinase-positive *Streptomyces scabies* produces colonies surrounded by a narrow zone of dark brown pigment.

Tyrosine Casein Nitrate (TCN) Medium (Menzies and Dade, 1959)

Sodium caseinate	25.0 g
Sodium nitrate	10.0 g
L-Tyrosine	1.0 g
Agar (Difco)	15.0 g
Tap water	1,000.0 ml

It should be remembered, however, that melanin-negative species of *Streptomyces* as well as melanin-weak strains of *S. scabies* may be involved in the scab disease.

Whereas Harrison (1962) could easily isolate *S. scabies* from common potato scab by using the methods described above he was unable to isolate the causal agent of russet scab by these procedures. This organism could only be obtained by the following method, which avoids maceration and dilution of the plant tissue: "Strips of tuber tissue 5–7 mm wide, 1–1.5 cm long and 4–5 mm thick were cut from russeted or healthy areas on the tubers. The pieces were trimmed to about 2 mm thick with a flamed razor blade. They were then held in the left hand, which had been disinfected by dipping in 95% ethanol, and a series of 5 to 7 thin sections cut at right angles to the periderm so the periderm was cut last. The first section from each lesion was discarded and the rest placed on the surface of 3% water agar in petri dishes. Actinomycetes grew readily from infected tissue and formed small colonies within 4 days."

Isolation from Water Samples

The same media used for the isolation of streptomycetes from soil can be employed to detect these organisms in water samples. Hsu and Lockwood (1975) compared five media. The one containing powdered colloidal chitin was superior to the other four (egg albumen, glycerol arginine, starch casein, and Actinomycete isolation agar; see Table 11).

The water samples can be incorporated directly into the medium or after dilution, depending on the number to be expected. Low numbers of organisms, however (e.g., in chlorinated river water), require concentration by membrane filtration. According to the method described by Burman, Oliver, and Stevens (1969), the membrane is placed face downward, after filtration, on the surface of the well-dried medium (chitin agar + cycloheximide, 50 µg/ml). After incubation at 22–30°C for 4 h, the membrane is carefully removed and discarded and the culture is further incubated for 1–8 weeks. The organisms imprinted on the surface will then grow as normal colonies. However, according to the experience of Trolldenier (1966, 1967) with membrane filters of 0.3-µm pore size, there should be no necessity for removing the filter, since the mycelial growth of actinomycetes penetrates the pores and forms normal colonies on the membrane.

Since streptomycetes and other actinomycetes are more resistant to chlorine, chloramine, and phenol than other bacteria are, isolation can be made more selective by treatment of water samples with these agents. Burman, Oliver, and Stevens (1969) described the method as follows: "Excess ammonia, as 0.2 ml of a sterile 0.38% solution of ammonium chloride, is added to 100 ml of water sample at room temperature to which is then added 1 ml of a hypochlorite solution containing 200 mg/l of available chlorine, to give a concentration of 2 mg/l in the sample. This is allowed to stand for 10 min after which the chlorine is neutralized with 0.05 ml of a sterile 3% solution of sodium thiosulphate. The sample is then filtered in the normal manner. The success of this method varies with the sample, but it is capable of suppressing many of the interfering bacteria, resulting often in much higher counts of actinomycetes, greater variety of species and much easier isolation of pure cultures uncontaminated with other bacteria." Williams and Cross (1971) found phenol treatment successful with canal and pond waters. The sample is shaken with phenol (7 mg/ml) for 10 min and filtered through a membrane, and the actinomycete spores are resuspended by shaking the membrane with glass beads in saline before plating. Heating for 1 h at 44°C has also been employed as a selective measure (Burman, Oliver, and Stevens 1969).

As discussed in "Habitats", streptomycetes and other actinomycetes found in water bodies are mostly contaminants from soil. However, an apparently "aquatic streptomycete" occurring specifically in drinking water supplies has been described by Burman (1973). The enrichment culture method developed by this author has been given above; the organism can be isolated with chitin agar.

Isolation of Thermophilic Streptomycetes

Thermophilic streptomycetes are encountered along with other thermophilic actinomycetes when these are isolated from composts, manure heaps, and fodders. Since the isolation of these organisms is treated in detail in Chapter 157 of this Handbook, only few remarks may suffice here.

Since the high temperatures (45–65°C) used for isolation serve as selective agent, materials known to be rich in thermophiles (i.e., having undergone a self-heating process)—or even mesophilic habitats—may be analyzed directly, i.e., without enrichment culture. Alternatively, enrichment of the desired organisms is achieved by incubating the sample under appropriate conditions of temperature and humidity. As mentioned in "Habitats", Festenstein et al. (1965) used Dewar flasks for "storage" of hay; 25–35% moisture resulted in a temperature increase to 45–50°C, and a high number of streptomycetes could be isolated from this material. Henssen (1957a) placed samples of manure taken from the depth of the heap (50–60°C) onto a moist filter paper on a nutrient agar in a Petri dish or into test tubes containing a nutrient solution and a strip of filter paper. After incubation at 50 and 60°C under aerobic and anaerobic conditions, the filter paper in the dishes and the test tubes (above the level of the liquid) was covered with spores of actinomycetes of various colors. As pointed out by Williams and Cross (1971), enrichment (and the following isolation) under anaerobic conditions inhibits aerobic species of *Bacillus* commonly found in such habitats, and the method facilitates the isolation of certain species rarely encountered if only aerobic methods are used.

Media employed for the isolation of thermophilic actinomycetes, including streptomycetes with a higher temperature range for growth, are usually of a higher nutrient level (plus an antifungal and sometimes antibacterial agent) than those listed in Table 11 for obtaining mesophilic forms from soil; this is to favor a more rapid growth (e.g., within 1 to 4 days) in order to anticipate the desiccation of the plates and possibly—although this is not proved—the more fastidious nutritional demands of thermophiles. Nutrient agar (Oxoid) and tryptone soya agar (Oxoid) have been widely used; Lacey and Dutkiewicz (1976a) obtained the best results with these media in half strength + 0.2% casein. Corbaz, Gregory, and Lacey (1963) recommend the following two media for the isolation of thermophilic streptomycetes: glucose–yeast extract–malt extract agar (Table 13) and V8 agar, a medium consisting of 800 ml H_2O, 200 ml V8 vegetable juice, 4 g $CaCO_3$, and 20 g agar. More complex media have been used by some investigators. Fergus (1964) prepared a compost infusion agar as follows: 300 g compost ready for the cultivation of mushrooms were placed in a cheesecloth and suspended

Table 12. Composition of some media used for the isolation of thermophilic streptomycetes.

Uridil and Tetrault (1959)

Glucose	10.0 g	The two phases are autoclaved separately,
Soy bean oil meal	20.0 g	cooled to room temperature, each phase
Tryptic digest of		adjusted to pH 7 and then mixed. 40 ml are
casein (BBL)	10.0 g	distributed per Petri plate. The plates are
NaCl	10.0 g	put in a forced draft 60°C incubator for periods
H_2O	500.0 ml	of from 10 to 24 h immediately after
Silica solution Ludox		the medium has been poured (or after solidification
(E.I. du Pont de		within 2 h at room temperature).
Nemours & Co)	500.0 ml	

Tendler and Burkholder (1961), Medium Ia

Saccharose	5.0 g	Dung extract:	
Trypticase	5.0 g	25 g dried sheep manure is suspended in	
Yeast extract	3.0 g	100 ml tap water, autoclaved 30 min, filtered,	
Dung extract	5.0 ml	and refrigerated under toluene.	
Molasses	5.0 ml	Trace element solution, 100 ml contain:	
$MgSO_4 \cdot 7H_2O$	0.5 g	$Fe(NH_4)_2SO_4$	100 mg
$FeSO_4 \cdot 7H_2O$	0.01 g	$ZnSO_4$	100 mg
Trace elements	1.0 ml	$MnSO_4$	50 mg
Agar	20.0 g	$CuSO_4$	8 mg
H_2O	1.0 liter	$CoSO_4$	10 mg
Adjusted to pH 7.2		H_3BO_3	10 mg

Craveri and Pagani (1962)

Crude maltose	20.0 g
Soya bean meal	5.0 g
Yeast extract	2.0 g
Agar	20.0 g
Tap water	1,000.0 ml
pH 6.5	

Table 13. Composition of some media suitable for the cultivation of streptomycetes (Shirling and Gottlieb, 1966).

1. Glucose–yeast extract–malt extract agar		
Glucose	4.0 g	Addition of $CaCO_3$, 2 g/liter, is
Yeast-extract	4.0 g	advantageous for many strepto-
Malt-extract	10.0 g	mycetes.
Agar	12.0 g	
Distilled water	1,000.0 ml	Adjust to pH 7.2.
2. Oatmeal agar		
Oatmeal	20.0 g	Cook 20.0 g oatmeal in 1,000 ml
Agar	12.0 g	distilled water for 20 min. Filter
Trace salts solution (5)	1.0 ml	through cheese cloth. Add dis-
Distilled water	1,000.0 ml	tilled water to restore volume of
		filtrate to 1,000 ml, then add
		trace salts solutions and agar.
		Adjust to pH 7.2.
3. Inorganic salts–starch agar		
Starch (soluble)	10.0 g	Make a paste of the starch with a
$(NH_4)_2SO_4$	2.0 g	small amount of cold distilled
K_2HPO_4 (anhydrous basis)	1.0 g	water and bring to a volume of
$MgSO_4 \cdot 7H_2O$	1.0 g	1,000 ml, then add the other in-
NaCl	1.0 g	gredients.
$CaCO_3$	2.0 g	
Agar	12.0 g	pH should be between 7.0 and 7.4.
Trace salts solution (5)	1.0 ml	Do not adjust if it is within this
Distilled water	1,000.0 ml	range.
4. Glycerol–asparagine agar		
Glycerol	10.0 g	The pH is about 7.0–7.4.
L-Asparagine (anhydrous basis)	1.0 g	Do not adjust if it is within this
K_2HPO_4	1.0 g	range.
Agar	12.0 g	
Trace salts solutions (5)	1.0 ml	
Distilled water	1,000.0 ml	
5. Trace salts solution		
$FeSO_4 \cdot 7H_2O$	0.1 g	
$MnCl_2 \cdot 4H_2O$	0.1 g	
$ZnSO_4 \cdot 7H_2O$	0.1 g	
Distilled water	100.0 ml	

in 1 liter of warm tap water for 30 min. The cheesecloth was squeezed by hand and the infusion was filtered through coarse filter paper. Fifteen grams of agar were dissolved in 40 ml hot tap water and added to the infusion; the final volume was adjusted to 1 liter, the pH to 7.2. Fergus added aureomycin (800 μg/liter) after autoclaving, to inhibit other bacteria. Some other media are listed in Table 12. As will be noted from these recipes, Uridil and Tetrault (1959) advocate the use of silica gel instead of agar as solidifying agent. They claim that this material provides a semidry surface with little shrinking; spreading bacteria—often a problem when using agar—were inhibited and all "streptomycetes" sporulated well. The thermophiles isolated could be maintained more easily on slants and Petri dishes prepared with silica gel than on agar media on which many cultures were lost on repeated transfer.

A very important requisite for a successful isolation of thermophilic streptomycetes is the maintenance of a humid atmosphere during incubation; this can be provided by large jars with water in the bottom. Even rather tightly closed containers supply enough oxygen for the aerobic representatives; parallel cultures under strict anaerobic conditions were advocated by Henssen (1957a) since most of her isolates grew better without oxygen.

The material from which isolates are made may be suspended and diluted in water and aliquots can be used to inoculate pour plates or streak plates. Alternatively, Uridil and Tetrault (1959) used the following method: "Dry isolation material such as compost, chicken litter, or any dry surface soil is sprinkled lightly over the surface of semidry plates and the plates are incubated in a 100% humid atmosphere at 60 C."—Similary, Fergus (1964) observed compost samples under the stereoscopic microscope, picked off individual colonies with a sterile needle, and streaked the material onto agar media in Petri dishes. However, most workers nowadays employ the Andersen sampler to isolate thermophiles from air enriched with these organisms by vigorous shaking of the sample—a method introduced by Gregory and Lacey (1962, 1963), who performed the shaking of the samples of moldy hay in a wind tunnel. "The hay was placed in a perforated zinc (2 mm diam. perforations), mounted horizontally to rotate on bearings across the tunnel (square cross-section with 29 cm sides). The cylinder was rotated at about 60 rev/min by an electric motor to give a gentle 'tedding' action on the hay sample. Air was drawn down the wind-tunnel by a fan, a speed of about 4.2 m/sec being adopted in routine tests. The dust cloud blown out of the cylinder travelled with the wind and reached the sampling position after a diffusion path of 1.2 m."

Cross (1968) described a simple modification of this method: To isolate thermophilic actinomycetes from hay, dry soil, and powdered milk, the sample is shaken in a tin of known volume and then the larger particles are allowed to settle. The small actinomycete spores remain suspended for very long periods. After 0.5–2 h, a known volume is sampled using an Andersen sampler connected to a small vacuum pump. A detailed study on the behavior of microorganisms in a sedimentation chamber has been conducted by Lacey and Dutkiewicz (1976b). Both fungal spores and bacteria sedimented faster than actinomycetes, the former because of their larger size, the latter probably because they were more often dispersed in clumps. A sedimentation time of 2 h or more is suitable for isolating actinomycetes. An even simpler method of air sampling was used by Seabury et al. (1968): "Organisms were isolated by grinding dry bagasse fiber in a Waring blender for 10 min followed by exposure of sterile gelatin-coated Whatman No 1 filter paper strips in the blender flask atmosphere at 5, 10, 15, and 20 min intervals. Strips were immediately transferred by means of a sterile forceps into petri dishes containing several media."

The Andersen sampler has been widely employed for the microbiological study of moldy hay and mushroom compost. It possibly could also be a useful alternative to the plate dilution method when analyzing other materials such as soil, plant litter, grain, etc.

CULTIVATION

Nutritional Requirements and Media for Sporulation

As mentioned above under "Isolation", streptomycetes are chemoorganoheterotrophic, nonfastidious microorganisms that need only an organic carbon source (e.g., starch, glucose, glycerol, lactate), an inorganic nitrogen source (NH_4^+ or NO_3^-), and some mineral salts. As far as is known, there is no requirement for vitamins or organic growth factors. Thus very simple "synthetic media" can be used for their cultivation. However, to obtain faster growth than that achieved on such substrates, "complex media" containing oatmeal, yeast extract, and/or malt extract are in general use. Other media combine an organic carbon source with a single amino acid as nitrogen source (e.g., glutamic acid or asparagine).

The most important requirement a "general purpose medium" for streptomycetes has to fulfill is to allow the completion of the microorganism's life cycle: germination of spores, growth of substrate and aerial mycelium, and formation of spores; the latter is macroscopically indicated by the appearance of a typical color of the spores "en masse". The formation of a "mature aerial mycelium" is important, since certain of its features are of taxonomic value. Even more important, in this context, is the production of spores for the subcultivation of the organisms. General experience indicates that the genetic stability of a streptomycete is best guaranteed when a heavy spore inoculum is used for propagation.

For sporulation, streptomycetes prefer media with a wide ratio of carbon to nitrogen. The nature of the carbon and nitrogen source is of less importance, although some compounds seem to be more useful than others. It appears that a not too fast and heavy vegetative growth—which implies, of course, a certain type of intermediary metabolism—favors the formation of aerial mycelium and arthrospores. Since there is no single medium optimal for the sporulation of all streptomycetes, several different media have to be tried. Of the great number of useful media (Pridham et al., 1956/57), the following four are of particular value (Table 13); these have been employed also in the ISP. Many other media are listed by Waksman (1961, pp. 328–334) and by Williams and Cross (1971). Notice that some of the media contain $CaCO_3$, which, in addition to supplying Ca^{2+}, neutralizes the acids produced by several streptomycetes. In addition, several of the media used for the characterization of streptomycetes (e.g., utilization of carbon and nitrogen compounds) allow good sporulation.

Attention may be directed to the fact that a macroscopically heavy aerial mycelium may contain very few spores (especially in *Streptoverticillium*) and, vice versa, aerial mycelium hardly detectable by the naked eye may be a good source of spores; thus it is good practice to check microscopically such cultures (or the suspensions of inoculum obtained from them).

Media Containing Adsorbing Materials: Soil, Clay, Minerals, Ca-Humate

As mentioned above, the addition of soil to media for the isolation of streptomycetes increased the number of colonies and exhibited a favorable effect on growth, sporulation, and pigmentation (Trolldenier, 1966; 1967). A pronounced stimulation of growth and metabolic activity of some actinomycetes was observed by Martin, Filip, and Haider (1976) when montmorillonite or Ca-humate were added to a liquid medium. Clay in dialysis tubing had the same effect after a short lag period. The authors suggest that this effect is due to the adsorption of one or more inhibitory substances produced during the metabolism. There are numerous reports in the literature (see Martin, Filip, and Haider, 1976) describing similar effects of clay on other bacteria and on fungi.

The two observations mentioned here strongly suggest for the future a more widespread use of adsorbing materials in media for the cultivation of streptomycetes (and possibly many other microorganisms). These may not only favor growth, sporulation, and pigmentation, but also the genetic stability of these organisms. The use of clay for this purpose is somewhat reminiscent of the favorable effect of the so-called soil culture for preservation of microorganisms, especially of actinomycetes (Frommer, 1956; Kuznetsov et al., 1967).

Temperature

Streptomycetes from soil are—possibly with few exceptions—mesophilic organisms; they grow at temperatures between 15 and 37°C, some up to 45 and 50°C. In laboratories working extensively with these organisms, incubator rooms are set at 25, 28, or 30°C. However, isolates from marine sources seem to prefer lower temperatures (18°C); those from warmer habitats (man and animals, compost, tropical areas) are preferably cultivated at higher temperatures. As in many other instances, the temperature optimal for fast growth and/or maximal yield may not be the same as that optimal for the formation of secondary metabolites (e.g., antibiotics, pigments), the latter being in general lower.

Relation to Oxygen

Streptomycetes are obligately aerobic organisms. In agar shakes with a nutrient medium, they grow at the surface of the agar column; however, in agar shakes with a poor medium or nonutilizable carbon source, they grow microaerophilically. Stationary liquid cultures grow as a pellicle with the medium remaining completely clear. Many streptomycetes are able to reduce nitrate to nitrite under strict anaerobic conditions.

Preparation of Inoculum

Three kinds of inoculum can be employed for further propagation of streptomycetes as well as for inoculation of test media for characterization: (i) arthrospores, i.e., aerial mycelium spores; (ii) young vegetative mycelium grown in liquid culture (submerged) and repeatedly washed with sterile buffer or saline; (iii) spores produced in liquid culture, i.e., submerged spores.

Arthrospores can easily be transferred from a well-sporulated culture to a new medium by a loop. However, to avoid contamination of the surroundings as well as cross-infection when working with many cultures, it is advisable to prepare suspensions in water and to handle these with pipettes. Since the aerial mycelium of streptomycetes is—depending on strain and possibly medium—hydrophobic, addition of a detergent helps to obtain a more homogeneous spore suspension: e.g., Triton X/100, 0.05%; sodium lauryl sulfonate, 0.01%; Carbowax 400, 0.05–0.1%. To remove the still remaining spore clumps, one can filter the suspension through a thin cotton plug. Alternatively, very homogeneous spore preparations can be produced by grinding the suspension in a glass mortar—a rather laborious but effective method. These spore suspensions stored at 4°C are good for use over a period of several weeks; since the spores tend to settle and clump again, addition of a few glass beads to the screw-cap tube helps to resuspend the spores before use. Another method to keep the spores in suspension is to mix them with soft water agar (Kutzner, 1972).

Young vegetative mycelium (48 h old) from liquid shake cultures is often employed when a large number of test media have to be inoculated. It circumvents the preparation of a homogeneous spore suspension, but it necessitates the repeated washing of the mycelium; this has to be used within 1 or 2 days, except when stored in or above liquid nitrogen. Since the growth consists—depending on strain, medium, density of inoculum—of micro- or macrocolonies or even pellets, it usually has to be broken down by vigorous agitation with sterile glass beads or a mechanical blender (e.g., Waring), as has been recommended for the ISP study. The use of vegetative mycelium is the method of choice when poorly sporulating cultures have to be studied; however, since in any collection of strains there are heavily and poorly sporulating ones, one might use for the sake of uniformity a vegetative inoculum, for all cultures. Although of the same genetic make-up, spores and vegetative mycelium may yield different test results, at least quantitatively (speed and strength of the reaction) and possibly even qualitatively. However, no detailed study has been undertaken to prove this. The following medium, tryptone–yeast extract broth, is widely used for the production of vegetative inoculum (g/liter): tryptone (Difco), 5.0; yeast extract (Difco), 3.0; pH 7.0–7.2.

In suitable media on prolonged cultivation, some streptomycetes produce "submerged spores". Although little attention seems to have been paid to the life cycle in liquid culture (except in industrially used cultures) as well as to the possible use of the spores formed there, these propagules can be expected to serve as a very convenient inoculum for further propagation (Wilkin and Rhodes, 1955).

Cultivation

Streptomycetes are grown either on the surface of solid agar media (in some tests soft agar media can be used) or in agitated nutrient solutions. For production of aerial mycelium spores, it is advisable to inoculate the surface of the agar media in slants, flasks, or dishes streakwise only; although many organisms will produce aerial mycelium when the entire surface is covered by confluent growth, others will sporulate only when streaks of growth are separated by an empty space either at the margin of the streaks or in their center (depending on strain and possibly medium).

For submerged culture in liquid medium, Erlenmeyer flasks with indentations are convenient containers for this purpose, but for small quantities—3 ml being enough for some physiological tests—tubes in a slanted position on a shaker or roller allow excellent providing of the culture with oxygen. No problems arise in these cultures by foaming; however, one occasionally observes that when several cultures are studied, these may differ significantly in the degree of formation of foam. A very important point is that some secondary metabolites such as antibiotics and pigments which are produced on solid media fail to be synthesized under these conditions.

Stationary liquid cultures are definitely not suitable for the cultivation of streptomycetes. The pelli-cle that forms first may sink down, and final growth will consist of colonies at the bottom, at the wall, and at the surface of the liquid—with the possible consequence of giving an unreliable test result.

CHARACTERIZATION

To place an unknown isolate into the family Streptomycetaceae and in its appropriate genus as well as to come to a species identification several properties have to be examined; these fall into the following classes: (i) biochemical criteria, (ii) macroscopic features (especially the color of substrate and aerial mycelium as well as of the soluble pigment), (iii) morphology, (iv) physiological properties, and (v) phage sensitivity.

Biochemical Criteria

As discussed above (Table 1 and 2), the first step is the determination of the type of diaminopimelic acid (DAP) and of the gross morphology of the substrate and aerial mycelium. LL-DAP combined with a nonfragmenting vegetative mycelium and the occurrence of chains of arthrospores in the aerial mycelium place the unknown into *Streptomyces, Streptoverticillium,* or *Chainia.* Further biochemical analysis should yield data in agreement with those listed in Table 1: (i) no characteristic pattern of sugars; (ii) a typical spectrum of fatty acids with relatively high amounts of iso- and anteiso-branched fatty acids; (iii) no mycolic acids. The methods described below have been taken from selected references; their usefulness has been shown in a study of many hundreds of strains of almost all genera of the Actinomycetales by Kroppenstedt (1977).

Determination of DAP-Isomer

The occurrence of LL-, DL-, or hydroxy-DAP can be ascertained by thin-layer chromatography of whole cell hydrolysates. The following method is a modification of that described by Becker et al. (1964).

Determination of Isomers of DAP
(Modification of Becker et al., 1964)
> Dried cells (1 mg) are hydrolyzed with 1 ml 6 N HCl in a sealed Pyrex tube held at 100°C for 18 h. After cooling, the sample is filtered through paper, which is then washed with 1 ml H_2O. The filtrate is dried two or three consecutive times on a rotary evaporator under reduced pressure at 40°C to remove most of the HCl. The residue is taken up in 0.3 ml H_2O and 5 μl is then spotted on a thin layer plate coated with cellulose ("microcrystalline"). For separation of the

amino acids, the following solvent mixture is used: methanol–water–10 N HCl–pyridine (80:17.5:2.5:10, by volume). Amino acids are detected by spraying with acetonic ninhydrine (0.1% wt/vol), followed by heating for 2 min at 100°C. DAP spots are olive-green fading to yellow, whereas the other amino acids give purple spots. Figure 6 shows that the three isomers of DAP which have to be expected among Actinomycetales are well separated from each other and from the other amino acids.

Determination of Sugars

Of the various sugars occurring in cell hydrolysates of Actinomycetales, four proved to be of taxonomic relevance (see Table 1): arabinose + galactose, madurose, xylose (Lechevalier and Lechevalier, 1970). Members of the family Streptomycetaceae do not contain these sugars and in most cases are devoid of any other carbohydrates.

Determination of Sugars Occurring in Cell Hydrolysates of Actinomycetales (Modification of Becker, Lechevalier, and Lechevalier, 1965)

Dried cells (50 mg) are hydrolyzed in 1 ml of 2 N H_2SO_4 in a sealed Pyrex tube at 100°C for 2 h. The hydrolysate is neutralized to pH 5.0 to 5.5 with a saturated solution of $Ba(OH)_2$, with methyl red as an internal indicator (or using a glass electrode). The precipitate of $BaSO_4$ is centrifuged at 6,000 rpm; the supernatant fluid is removed by a Pasteur pipette and evaporated at 40°C on a rotary evaporator. The residue is taken up in 0.4 ml H_2O. Five microliters are then spotted on a thin-layer plate coated with cellulose (microcrystalline). For separation of sugars, the following solvent mixture is used: ethylacetate–pyridine–water (100:35:25, by volume). The plates are chromatographed three consecutive times with intermediate air drying. After the third run, the sugars are detected by spraying with the following reagent: 2 ml of aniline, 3.3 g of phthalic acid, and 100 ml of water-saturated n-butanol; the plates are air-dried and then heated at 100°C. After 2–5 min, the monosaccharides appear as differently colored spots: hexoses brown, pentoses red brown, and 6-desoxy-sugars gray-green. As shown in Fig. 7, the sugars are well separated; the narrow R_f values of some sugars do not cause problems, since these sugars rarely occur together in one sample.

Determination of Fatty Acids

As can be seen from Table 1, the fatty acid spectrum varies considerably among families and genera of Actinomycetales. Ballio and Barcellona (1968)

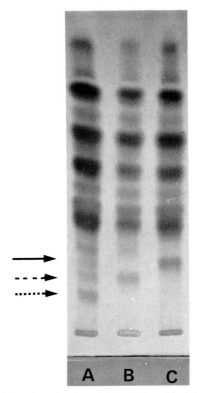

Fig. 6. Separation of amino acids in whole cell hydrolysates by thin-layer chromatography. ⟶ LL-DAP; ⋯⟶ DL-DAP; ┈⟶ hydroxy-DAP. (A) *Micromonospora*: hydroxy- and DL-DAP and traces of LL-DAP. (B) *Nocardia*: DL-DAP. (C) *Streptomyces*: LL-DAP.

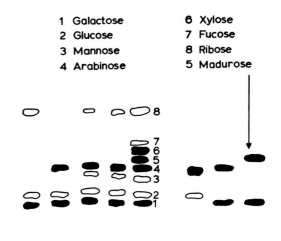

1 Galactose	6 Xylose
2 Glucose	7 Fucose
3 Mannose	8 Ribose
4 Arabinose	5 Madurose

Fig. 7. Separation of sugars in whole cell hydrolysates. Spots of diagnostic sugars are blackened. (A) *Actinomyces*; (B) *Mycobacterium*; (C) *Nocardia*; (D) *N. autotrophica*; (E) standard mixture; (F) *Streptomyces*; (G) *Streptosporangium*; (H) *Saccharomonospora*; (I) *Dermatophilus*.

were the first to report on the various patterns observed in these organisms. A detailed study has been carried out recently by Kroppenstedt (1977; see also Kroppenstedt and Kutzner, 1978). The family Streptomycetaceae is characterized by the occurrence of saturated fatty acids that are either straight or iso- and anteiso-branched—a pattern shared with few other actinomycetes (*Arachnia, Thermoactinomyces, Saccharomonospora*).

The fatty acid spectrum is determined by gas-liquid chromatography of the methyl esters of the fatty acids (Brian and Gardner, 1967). Transesterification of the fatty acids is carried out by refluxing 30 mg dried cells in 10 ml methanol-BCl_3 for 5 min (or keeping the reaction mixture at room temperature overnight). The sample is then diluted with 15 ml H_2O (for the decomposition of BCl_3) and the fatty acid esters are extracted with 4 ml chloroform; the two phases are separated by centrifugation. The chloroform phase is washed with H_2O, dried with Na_2SO_4, and evaporated on a rotating evaporator. The residue is taken up in 10 μl chloroform and used for gas-liquid chromatography.

The chromatograms shown in Fig. 8 were obtained by using the following conditions:
Apparatus: Varian 2700, double column, flame ionization detector and splitter for capillary columns: Varian 1445
Column: 30 m × 0.5 mm stainless steel, S.C.O.T. (support coated open tubular), stationary phase: diethyleneglycolsuccinate
Conditions:

Oven temp.	170°C
Injection bloc	210°C
Detector	225°C
Carrier gas, He	3 ml/min
Inlet pressure	0.5 atü
Splitting	1:20
Hydrogen	30 ml/min
Synth. air	240 ml/min
Septum gas flow	10 ml/min
Auxiliary gas, N_2	28 ml/min
Sample	0.2 μl

Determination of Mycolic Acids

Mycolic acids are α-branched, β-hydroxy fatty acids which occur in the cell wall of *Corynebacterium, Nocardia*, and *Mycobacterium*. Their molecular size varies with the genus, i.e., from 30 to 50 to 80 carbon atoms in the order used above. Pioneer work on the mycolic acids was carried out by the French biochemists Asselineau, Lederer, and Etemadi (Asselineau, 1966). Most of the more recent progress has been accomplished by Minnikin and co-workers (Minnikin, Goodfellow, and Alshamaony, 1978). Because of the correlation between genus

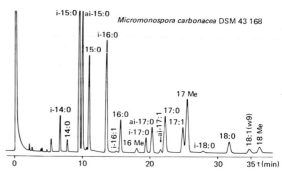

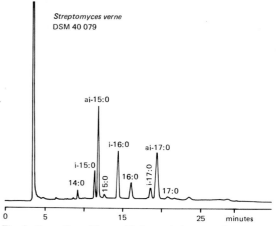

Fig. 8. Separation of fatty acids (as methyl esters) of two actinomycetes by gas-liquid chromatography.

and mycolic acid size, the test for this biochemical marker should be done in the early stage of the identification of an unknown actinomycete—although many members of this order do not contain mycolic acids. The following method has been developed by Minnikin, Alshamaony, and Goodfellow (1975). Dry bacteria (100 mg) are mixed with dry methanol (5 ml), toluene (5 ml), and concentrated sulfuric acid (0.2 ml) in a 20-ml tube closed with a PTFE-lined screw cap. The contents of the tube are mixed thoroughly and methanolysis is allowed to proceed for 12–16 h at 75°C (stationary incubation). The reaction mixture is allowed to cool at room temperature, 2 ml hexane are added, and the mixture is shaken then allowed to settle. Samples of the hexane extract are spotted on thin-layer plates coated with Merck silica gel H (0.5 mm) and the chromatograms are developed in petroleum ether (bp 60–80°C)–diethyl ether (85:15, vol/vol). The positions of the separated components are revealed by charring at 150–200°C after spraying with chromic acid solution (5 g $K_2Cr_2O_7$ in 5 ml H_2O, made up to 100 ml with concentrated H_2SO_4, then diluted 10 times with water). Alternatively, molybdate phosphoric acid spray reagent (Merck, 531) can be used. A chromatogram obtained by this procedure is shown in Fig. 9.

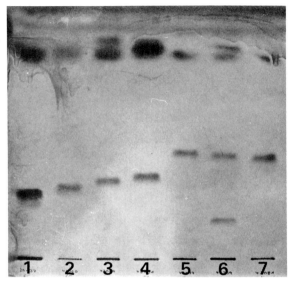

Fig. 9. Separation of mycolic acid methyl esters by thin-layer chromatography. The spots at the front line are methyl esters of fatty acids; the mycolic acids appear in a stairstep from left to right. (1) *Corynebacterium* (type A); (2) *Rhodococcus* (type B); (3) *Nocardia flava* (type C); (4) *N. asteroides* (type C); (5) *Mycobacterium fortuitum* (type F); (6) *Mycobacterium phlei* (type E); (7) *Mycobacterium smegmatis* (type F).

Macroscopic Features

More than almost all other bacteria, streptomycetes show a striking appearance of macroscopic features: colors of aerial mycelium, substrate mycelium, and pigments diffusing into the medium; consistency of the aerial mycelium; and morphology of colonies. In these properties, streptomycetes are so similar to the moulds that it is not surprising they were first named "ray fungi" and regarded as being more related to the true fungi than to the bacteria.

Of the features mentioned above, the color of the aerial mycelium proves to be the most useful; the colors of the substrate mycelium and of the soluble pigment are only of value if they are somehow outstanding, e.g., red, blue, violet, or dark green. The color of the aerial and substrate mycelium will be discussed below.

The consistency of the aerial mycelium differs significantly from strain to strain: mealy, dusty, woolly (the latter type is predominant in *Streptoverticillium*). Although not regarded as a "taxonomic criterion", it is worthwhile to record it. Also of less importance and therefore rarely reported is the morphology of colonies; apparently most authors found it too variable or too insignificant or did not take the trouble to describe it; indeed, a description without a photograph seems to be of little value. An attempt to use the morphology of "macrocolonies"

(only one colony per Petri dish, 3–4 weeks old) as a criterion has been undertaken by Shinobu (1958). He recognized three types: (i) chrysanthemum type (radial furrows), (ii) concentric type (concentric furrows), and (iii) wrinkled type (irregular or netlike furrows). Except in papers by Japanese workers, there is hardly any mention of this feature in the literature.

Color of Aerial Mycelium

The aerial mycelium of most streptomycetes shows a definite color. Experience with hundreds or thousands of cultures shows that the innumerable shades or tinges of colors do not form a continuous spectrum but rather cluster around a few "basic colors". This observation (together with the fact that the color of the aerial mycelium is the first feature to be easily recognized) is the reason why several workers dealing with a large collection of strains used the color of the aerial mycelium for the first grouping of their material: Flaig and Kutzner (1954, 1960a), Hesseltine, Benedict, and Pridham (1954), Ettlinger, Corbaz, and Hütter (1958), Pridham, Hesseltine, and Benedict (1958), Gauze (1958). Although these studies had been carried out independently at about the same time and with a different(!) set of organisms, the authors arrived at essentially the same color groups. In the International Streptomyces Project (ISP), the following seven color series were used: yellow, violet, red, blue, green, gray, and white. The present author now uses the 9 (or 10) groups listed in Table 14.

Although the color of the aerial mycelium is a useful criterion in the hands of the experienced(!) worker in this field, there is no doubt that its recognition is not without certain problems: (i) Experience with many cultures and representatives of all color groups is necessary. The observer must not, of course, be color-blind; she or he should rather possess a certain sense of colors. (ii) The typical color is shown only by well sporulated cultures. Since no single medium is optimal for sporulation of all cultures, several media have to be inoculated and observed several times (e.g., after 1, 2, and 3 weeks). (iii) The medium has some influence on the color. Media on a starch basis (oatmeal agar, starch-KNO_3 agar) are often to be preferred to others such as glucose–yeast extract–malt extract agar or other rich media because pigments of the substrate mycelium may diffuse into the aerial hyphae and thus modify its "true" color; gray is often changed to a distinct shade of brown or lavender. Since medium and age of the culture have some influence on the color, it follows that the determination of the spore color group cannot be objectified by observing all streptomycetes on one medium at one age.

Table 14. Spore colors for the grouping of streptomycetes and representatives of each color.

Color of aerial mycelium	Representative species (DSM no.)[a]			
Yellow-gray: "griseus"	S. griseus	40236	S. niveus	40088
Pink/light violet	S. fradiae	40063	S. toxytricini	40178
Gray-pink/lavender: "cinnamomeus"	S. lavendulae	40069	S. flavotricini	40152
Brown (+ gray or red)	S. eurythermus	40014	S. fragilis	40044
Blue: "azureus"	S. viridochromogenes	40110	S. cyaneus	40108
Blue-green: "glaucus"	S. glaucescens	40155		
Green: "prasinus"	S. prasinus	40099	S. hirsutus	40095
Gray: "cinereus"	S. violaceoruber	40049	S. echinatus	40013
White: "niveus"	S. albus	40313	S. longisporus	40166
Not definable: white + various light color shades	S. alboniger	40043	S. rimosus	40260

[a]DSM no. 40 . . . = ISP 5 . . .; e.g., 40236 = ISP 5236.

(iv) There are certainly streptomycetes which are difficult to place into one of the recognized color groups, either because their color does not properly fit into any of the established groups, because they produce too little aerial mycelium, or because the color is very light but not quite white; therefore group 10 in Table 14. In addition, as all readers with experience in this field will admit, there are even cases in which the "experts" do not agree.

Because of the taxonomic value of the color of the aerial mycelium on one hand and the difficulties of its exact determination on the other, numerous efforts have been undertaken to use color standards (guides, manuals) for their description (Flaig and Kutzner, 1954, 1960a; Prauser, 1964; Tresner and Backus, 1963). Further, to exchange experience among taxonomists of Actinomycetales, several workshops have been organized. Pridham (1965) has summarized the results of one of these workshops. Nevertheless, the existing problems could not be resolved completely. Physical measurements do not seem to be practicable. There seems to be nothing known of the chemistry of the pigments responsible for the colors of the aerial mycelium.

Color of Substrate Mycelium and Soluble Pigment

Numerous streptomycetes possess a strikingly colored substrate mucelium; often the pigment is diffusing into the medium, conferring on it the same color (usually less extensive). However, in some cases the medium may be more distinctly colored than the mycelium.

Although not having been used consistently by taxonomists of these organisms, the following classification of pigments regarding their location with respect to the producing cell may be given and considered in future work: (i) Endopigments; these are bound to certain cell structures and confer a striking color to the colonies of the organism. (ii) Exopigments; these are excreted into the surrounding medium; they may either be water soluble and diffuse far into the medium or precipitate close to the producing cell as amorphous or crystalline particles. Further points of interest are: (i) Some pigments are produced as colorless leuco-compounds which only after excretion are transformed into the recognizable pigment; (ii) so-called indicator pigments change color (and often solubility!) according to pH (e.g., red and insoluble at acid reaction, blue and soluble at alkaline reaction).

In regard to their structure, bacterial pigments belong to very different chemical classes. Besides the bacteriochlorophylls and phycobilins of phototrophic bacteria, the pigments listed in Table 15 and Fig. 10 are of more general interest.

The color of the substrate mycelium and soluble pigment; (ii) so-called indicator pigments change poses. However, it appears that this criterion has played a more important role in the past than at present; the color of the aerial mycelium is now regarded as a much more reliable criterion than that of the substrate mycelium. An exception is the grouping of Streptoverticillium by Baldacci and Locci in Bergey's Manual (Buchanan and Gibbons, 1974) into 12 series according to the color of substrate mycelium. Some authors reject this feature for taxonomic purposes entirely; others take an intermediate position, e.g., Lechevalier, Lechevalier, and Gerber (1971): "Thus color is of little value as a basic criterion in the systematics of actinomycetes, although it is of great value in the presumptive identification of many actinomycete species."

The reason for this change of opinion and the skepticism prevailing today are easily understood: (i) The color of the substrate mycelium is much more dependent on the medium, temperature, age, and—in some cases—illumination than on the

Table 15. Some pigments occurring in Actinomycetales.

Chemical class	Pigment (trivial name)	Producer[a]
Carotenoid	Phlei-xanthophyll	*Mycobacterium phlei* DSM 43070
Prodigiosin	I Prodigiosin	*Serratia marcescens*
	II Nonylprodigiosin	*Actinomadura madurae* DSM 43067
	III Undecylprodigiosin	*A. pelletieri* DSM 43118
		Streptomyces longispororuber DSM 40599
	IV Metacycloprodigiosin	*S. longispororuber* DSM 40599
	V Cyclononylprodiginine	*Actinomadura madurae* DSM 43067
	VI Cyclomethylprodiginine	*A. pelletieri* DSM 43118
	VII Blue pigment	*Serratia marcescens-mutant*
Pyrazine	Pulcherrimin	*Candida pulcherrima*
Tryptophan derivative	Violacein	*Chromobacterium violaceum*
Indole-derivative	Indigo (= indigotin)	*Mycobacterium globerulum; Pseudomonas indoloxidans*
Naphthoquinone	Proto-actinorhodin	*Streptomyces violaceoruber* DSM 40049
	Granaticin	*Streptomyces litmocidini* DSM 40164
Anthracyclin-glycoside	Rhodomycin	*S. violaceus* (=*purpurascens*) DSM 40704
Phenoxazinone	Actinomycin	Numerous streptomycetes
Phenazine	Pyocyanin	*Pseudomonas aeruginosa*
	Chlororaphin	*P. chlororaphis*
	Iodinin	Numerous actinomycetes
Diaza-diphenoquinone	Indigoidine	*Corynebacterium insidiosum*
		Pseudomonas indigofera
Diaza-indophenol	Indochrome	*Arthrobacter polychromogenes*
		Streptomyces coelicolor DSM 40233
	Not named	*Pseudomonas lemonnieri*

[a] Not all producers are actinomycetes.

color of the "mature aerial mycelium". (ii) There seems to be more variation in pigment production among strains otherwise closely related than in other criteria—assumed to be due to loss of genetic information for this property. Since pigments are secondary metabolites which do not appear to be essential for the producer, this loss does not do any harm to the organism. (iii) More important for our present day opinion is the argument that color alone does not give any clue to the chemical identity of the pigment—and this alone should be regarded as a meaningful criterion for the following reasons: (i) chemically different pigments may exhibit the same color; (ii) one and the same pigment may differ in appearance depending on pH; and (iii) the color to be observed is often a mixture of several pigments.

Progress has been made during the last decade in the elucidation of the chemical structure of pigments as well as in the less sophisticated chromatographic procedures for characterization: Lechevalier, Lechevalier, and Gerber (1971), Habermehl, Christ, and Kutzner (1977), Krassilnikov (1970), Hussein (1970), Blinov and Khokholov (1970). However, in present-day taxonomy the colors are just reported as they are observed.

For the determination of the color of the substrate mycelium (= reverse side of mass of growth) and soluble pigment, cultures grown on the four media mentioned above are used. In addition, the appearance of striking colors on all other media used for the study of these organisms should be recorded, as

some pigments are only produced on special media, e.g., the blue pigment indochrome by *Streptomyces coelicolor* Müller (Kutzner et al., 1976). The colors are observed after 1, 2, and 3 weeks; changes such as red-violet to blue or yellow-green to olive-gray may occur. Although some pigments are clearly confined to the mycelium, in most cases one can state only "primarily" endogenous or exogenous or both. For observing the indicator nature of a pigment, the following test is recommended (Shirling and Gottlieb, 1966): Small pieces of growth from an agar culture are placed in a spot plate and flooded with a drop of 0.05 N NaOH and 0.05 N HCl; observations are made immediately and after 10–15 min.

Observations made by the present author with representative species are listed in Table 16.

Variation in Macroscopical Appearance

Although any property of an organism is subject to phenotypic and genotypic alteration, it is appropriate in streptomycetes to mention this aspect in connection with the macroscopic appearance. Phenotypic modification becomes apparent when subcultures of a streptomycete are grown on different media; genotype variations (mutations) can be observed when cultures are plated out in such a way that single colonies arise, each originating from one arthrospore. With many streptomycetes, numerous

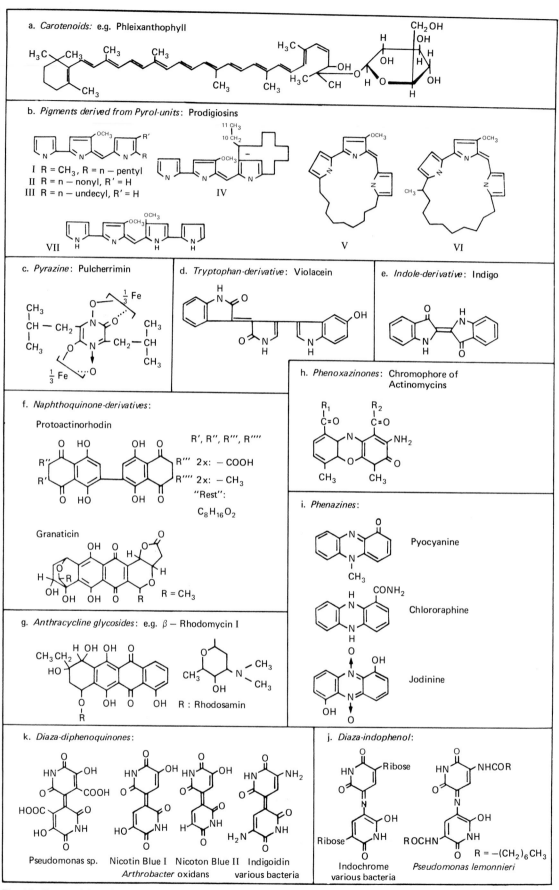

Fig. 10. Structural formulas of selected pigments produced by actinomycetes and some other microorganisms.

Table 16. Colors of substrate mycelium and soluble pigment occurring in streptomycetes.

Color	Representative species (DSM No.)					
Orange to dark red	S. aurantiacus	40412	S. cinnabarinus	40467	S. griseoruber	40275
(mainly endo-pigment)	S. longispororuber	40599	S. spectabilis	40512	Sv. aureoverticillatum	40080
Red to blue/violet	S. californicus	40058	S. cinereoruber	40012	S. purpurascens	40310
(mainly endo-pigment)	S. violaceus	40082				
Red-violet to blue	S. coelicolor	40233	S. cyaneus	40108	S. lateritius	40163
(endo- and/or exopigment)	S. violaceoruber	40049				
Yellow-orange/greenish-yellow	S. atroolivaceus	40137	S. canarius	40528	S. flavogriseus	40323
(endo- and exopigment)	S. galbus	40089	S. parvus	40348	S. tendae	40101
Green to gray-olive	S. flavoviridis	40210	S. olivoviridis	40211	S. nigrifaciens	40071
(mainly endopigment)	S. viridochromogenes	40110				
Green (endopigment)	S. malachiticus	40167	S. malachitorectus	40333		
Red-brown to dark-brown	S. badius	40139	S. eurythermus	40014	S. griseorubiginosus	40469
(endo- and exopigment)	S. phaeochromogenes	40073	S. ramulosus	40100	S. resistomycificus	40133
Gray-brown to black	S. alboniger	40043	S. hygroscopicus	40578	S. mirabilis	40553
(mainly endopigment)	S. purpeofuscus	40283	S. violaceoniger	40563	S. viridogenes	40454

types of colonies (or colony sectors) differing in amount and shades of color of aerial mycelium and pigment are seen—often depending on the medium used for plating. Other variants occurring occasionally are "pinpoint" and "soft" colonies. This striking "segregation" of a pure culture into different colony types, which, of course, differ also in physiological properties such as antibiotic production, is the reason why a heavy spore inoculum comprising all the genotypes is advocated for subculturing streptomycetes. Numerous observations on the variation of macroscopic, microscopic, and physiological properties, the possible causes, and the consequences for taxonomy have been reviewed by Kutzner (1967) and Kalakoutskii and Nikitina (1977).

Macroscopic variation does not become apparent only when cultures are plated out; confluent growth of streptomycetes on slants or plates sometimes also shows heterogeneity with respect to color and consistency of aerial mycelium and/or color of substrate mycelium ("reverse color"). Here only one type of variant will be mentioned. As shown in Fig. 11, many streptomycetes produce white, flocculent tufts or sterile aerial hyphae on the surface of the mature, sporulated aerial mycelium. These sterile variants (see Fig. 15) can be isolated in pure culture; they differ from the fertile parent culture only with respect to sporulation (and hence color) of the aerial hyphae. This phenomenon has been described by Kutzner (1956); undoubtedly, it must have been observed over decades by numerous workers in this field without having been definitely described or investigated. There sterile mutants later became the subject of genetic studies (Chater and Hopwood, 1973; Hopwood, Wildermuth, and Palmer, 1970), and several genes controlling sporulation in S. violaceoruber have now been mapped on the chromosome (Chater and Merrik, 1976).

Microscopic Characterization

The life cycle of a streptomycete offers three features for microscopic characterization: (i) vegetative mycelium (on solid and in liquid medium), (ii) aerial mycelium including chains of arthrospores (sometimes called "sporophores"), and (iii) the arthrospores. In Chainia, sclerotia form a fourth morphological feature. Whereas the first stage of the life cycle has not been found useful by taxonomists, the second and third supply two of the most important diagnostic criteria.

Methods for Morphological Studies

Light and electron microscopy (transmission and scanning) have to be used for the characterization of streptomycetes. In fact, one of the most important taxonomic properties for species identification—the surface structure of arthrospores—can only be determined by electron microscopy.

Of the methods to be employed, only those used for light microscopy will be described briefly. In order to observe the undisturbed arrangement of hyphae, especially of aerial hyphae and chains of arthrospores, it is of the utmost importance to examine the organism in situ. The various methods which have been suggested are outlined in Fig. 12. In each case, the organism is inoculated on an agar medium from where it spreads onto a cover slip or slide; this growth is suitable for microscopy. Alternatively, the entire Petri dish culture (or plugs cut out of it and placed on a slide) can be subjected to examination. A useful method proved to be double-layer plates with sterile cover slips between the two layers; for microscopy, the slips are removed together with the thin "top layer" carrying the growth. The sublayer may be water agar, the top layer a nutrient agar that

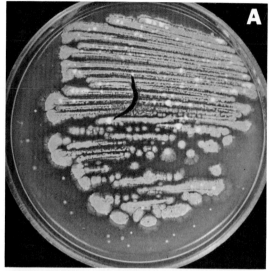

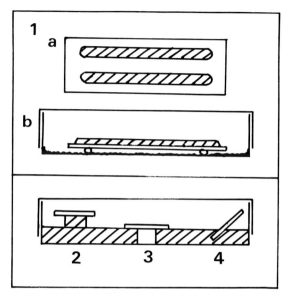

Fig. 12. Methods of cultivation of streptomycetes for microscopical examination. (1) Slide method (Colmer and McCoy, 1950); (a) top-view, (b) side-view. (2) Agar plug method (Nishimura and Tawara, 1957). (3) Groove method (Okami and Suzuki, 1958). (4) Cover slip method (Kawato and Shinobu, 1959).

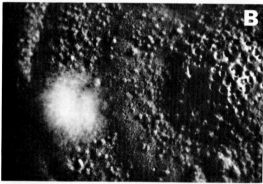

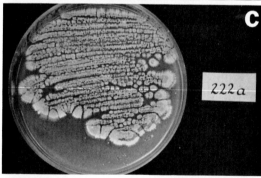

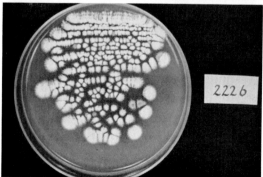

Fig. 11. (A and B) Occurrence of white, flocculent tufts of sterile aerial hyphae on mature aerial mycelium. (C) Isolation of sterile variant (not yet completely "pure").

enhances formation of aerial mycelium. Cover slips may also be inserted vertically in a few milliliters of agar contained in a wide test tube: if minimal aeration is allowed, a humid atmosphere is provided. For morphological observations, media such as those listed in Table 13 or others commonly used may be employed—preferably, however, only at half or quarter strength to allow more diffuse growth.

Morphology of Vegetative Mycelium

The vegetative mycelium on solid media does not fragment—at least by definition; however, in some streptomycetes formation of spore chains at the edge of the mycelial growth can be observed. As shown by various authors (Glauert and Hopwood, 1960, Moore and Chapman, 1959, Stuart, 1959), the hyphae are not(!)—as sometimes stated—coenocytic but contain cross walls. These can also be observed in mycelium grown submerged (micro-aerophilic) in poor media (Fig. 18). In agitated liquid media, the inoculum gives rise to colonies of different density of mycelium and size, depending on strain, medium, aeration and other conditions. Agitation may lead to the breaking off of hyphal fragments that give rise to new microcolonies. In many cases, the mycelium will finally disintegrate and undergo autolysis. However, more complex life cycles comprising different morphological stages and ending with the formation of spores have been observed in various streptomycetes by Carvajal

(1947), Wilkin and Rhodes (1955), and Tresner, Hayes, and Backus (1967). A more comprehensive study of the life cycle of streptomycetes in submerged culture is urgently needed.

Morphology of Aerial Mycelium

Under the microscope, the morphology of the aerial mycelium exhibits a wealth of variation with regard to the length of the hyphae, their type of branching (monopodially—sympodially), the arrangement of sporogenous hyphae ("sporophores"), and their morphology (straight, flexuous, spiral shaped).

The most striking and characteristic structure of the aerial mycelium consists of a verticillate arrangement of spore chains (genus *Streptoverticillium;* Figs. 1B and 13). Very long, mostly rather straight hyphae carry short side branches, which in turn give rise to a verticil (whirl, whorl) of spore chains, the latter being in most instances straight to flexuous; in a few species they terminate with a hook or a spiral with one to (at most) three turns. The short side branches arise either singly, most often pair-wise (opposite) and sometimes by threes or more (forming a primary whirl carrying the secondary whirl, "umbel"). Formation of these structures requires an optimal development of the culture; factors such as medium, pH, and humidity are of particular importance (Hütter, 1962). Sometimes the morphological differentiation stops before the complete structure of a "bi-verticil" has been formed, and the rudimentary structures give the only hint that one is dealing with a strain of *Streptoverticillium*. Needless to say, much patience is needed to prepare pictures such as those in Fig. 13. Although the "true" verticillate structures are easily recognized, there is no doubt that some uncertainty exists because very similar structures ("pseudoverticils") occur in some species of *Streptomyces* (Fig. 1C).

Examples of morphological types in *Streptomyces* are shown in Fig. 14. The pictures are taken from the work of the present author (Flaig and Kutzner, 1960a). For the characterization of the 2,000 strains studied in this investigation, the following features were recorded:

1. Length of aerial hyphae: short, medium, or long (short hyphae correspond to a dusty colony appearance, long hyphae to a woolly one).
2. Branching sympodially, i.e., growth in the form of small "bushes" or "trees"; branching monopodially, i.e., long hyphae with chains of spores as side branches. The sterile growth axes sometimes show disintegration, the side branches then hanging at very thin threads.
3. Occurrence of spirals. (i) Frequency: very frequent—only in certain areas of the colony—rare; (ii) Arrangement: as terminal ends of the branches of bushes or trees—as side branches of long hyphae; (iii) Length of spiral's stalk: very short—medium—long; (iv) Number of turns: e.g., 1-**2**-(3), 3-**5**-(8), 5-**10**-15; (v) Interval of turns: small (compact spiral)—"medium"—widely separated (open spiral); (vi) Diameter of turns: small—large; equal within the whole spiral—becoming smaller to the edge of the spiral; (vii) Direction of turn: sinistrorse—dextrorse (this feature has been used by several authors, including Flaig and Kutzner, 1960a; however, other workers, e.g., Ettlinger, Corbaz, and Hütter, 1958, found both types of spirals within the same organism).

A somewhat different approach was taken by Ettlinger et al. (1958); these authors grouped their material into 15 different morphological types (two of them for *Streptoverticillium*). Whereas 2 of the 15 types were rather similar and hardly distinguishable, the others were regarded by the authors as recognizable with certainty.

Finally, a much simpler classification was proposed by Pridham, Hesseltine, and Benedict (1958). Only three types are used to classify species of *Streptomyces:* (i) straight to flexuous, "rectusflexibilis" (RF); (ii) hooks, loops, spirals with 1–2 turns, "retinaculum apertum" (RA); and (iii) spirals (S). Although this simplification was necessary for the classification of a large number of species, the data of which were only available from the literature, this scheme has, rather unfortunately, been maintained in future work. Thus, much information which could be given by a more detailed description of what is actually observed gets lost if only the placement into one of the three types is made. Since nowadays in most cases a photograph is added to the description, this is not too great a loss. A more important point to mention, however, is that the two sections RA and S are difficult to differentiate: where does a loop end and a real spiral begin? It is not surprising, therefore, that in a recent study Williams and Wellington (1978) reported poor agreement between their own observations and those of the ISP study in regard to the RA category (Table 17).

Morphology of Arthrospores

When chains of arthrospores are viewed under the transmission electron microscope, four types of silhouettes can be distinguished (Fig. 15): smooth, warty, hairy, and spiny. These four spore types were first recognized by Flaig et al. (1952) and Flaig, Küster, and Beutelspacher (1955); soon the taxonomic value of this morphological criterion was realized (Ettlinger, Corbaz, and Hütter, 1958; Flaig and Kutzner 1960a; Tresner, Davies, and Backus, 1961). A fifth type—"knobby"—for intermedi-

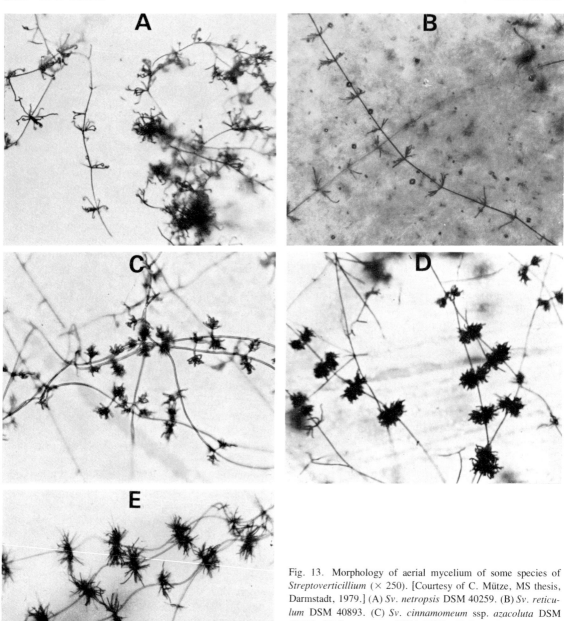

Fig. 13. Morphology of aerial mycelium of some species of *Streptoverticillium* (× 250). [Courtesy of C. Mütze, MS thesis, Darmstadt, 1979.] (A) *Sv. netropsis* DSM 40259. (B) *Sv. reticulum* DSM 40893. (C) *Sv. cinnamomeum* ssp. *azacoluta* DSM 40646. (D) *Sv. septatum* DSM 40577. (E) *Sv. mobaraense* DSM 40847.

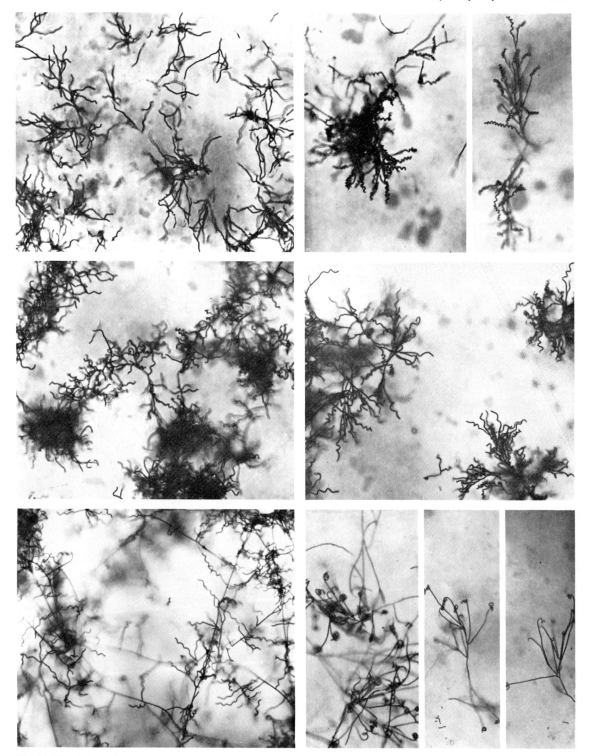

Fig. 14. Morphology of aerial mycelium of some strains of *Streptomyces*. (× 250). [From Flaig and Kutzner, 1960a, with permission.]

Table 17. Agreement between morphological determinations of *Streptomyces* species made by the International Streptomyces Project and Williams and Wellington (1978).

Spore chain categories	Percentage agreement
Biverticillus	100.0
Rectus flexibilis	98.1
Spira	95.3
Retinaculum-apertum	30.4

ates between spiny and warty was proposed by Lyons and Pridham (Fig. 17A) (1971). Figure 16 shows two streptomycetes together with their sterile variants in the light and electron microscope. Whereas the fertile strains form spiral chains of spores, no spores are observed in the sterile variants.

Investigation of the fine structure of the sporogenous hyphae revealed that the chains of the spores

are surrounded by an outer sheath consisting of fibrillary elements, and it is this sheath which either encloses the spores smoothly or forms appendages such as hairs or spines (Hopwood and Glauert, 1961; Rancourt and Lechevalier, 1964; Wildermuth, 1972; Williams and Sharples, 1970; Williams et al., 1972a).

Although spore surface structure proved in general to be a rather stable property and its determination clear-cut, some problems arose by "intermediates" between smooth and warty, warty and spiny, and spiny and hairy. These difficulties of interpretation, however, can be overcome by the use of a scanning electron microscope. As shown by Dietz and Mathews (1969, 1971) and Williams and Wellington (1978), the sheath is sometimes regularly wrinkled or ridged ("rugose") producing a silhouette which appears either as smooth or warty or even spiny (Table 18). Figure 17 shows four species of

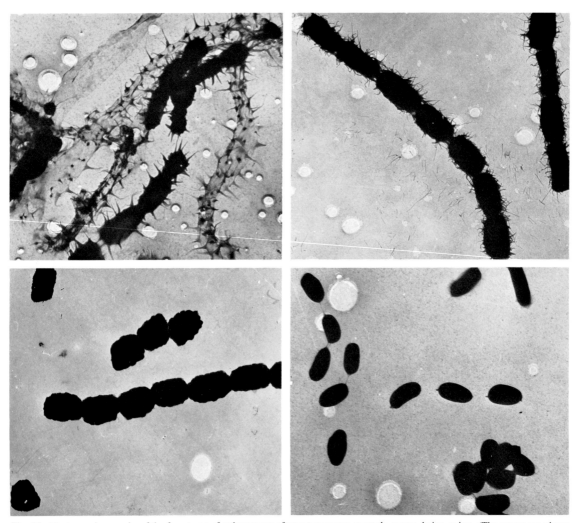

Fig. 15. Electron micrographs of the four types of arthrospores of streptomycetes: smooth, warty, hairy, spiny. (The spores are about 1 μm long.) [From Kutzner, 1956, with permission.]

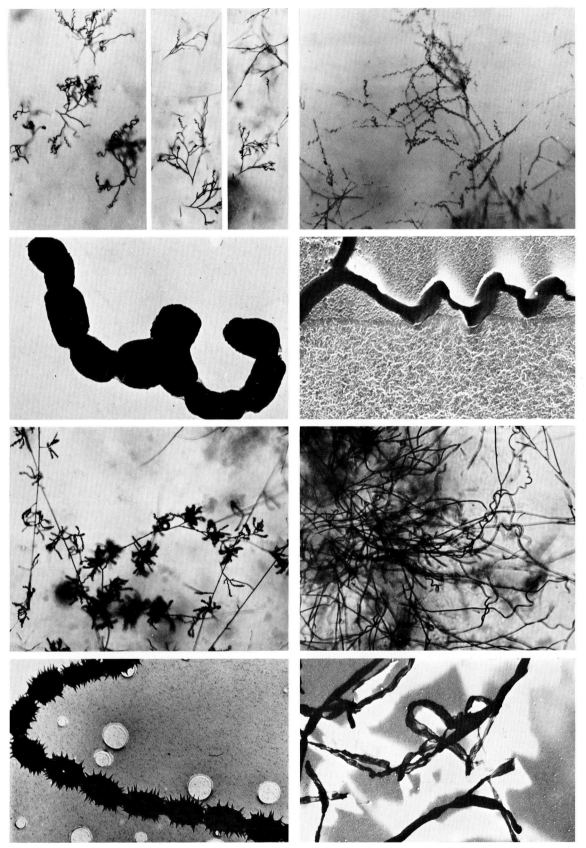

Fig. 16. Aerial mycelium of the fertile and sterile strain of two streptomycetes in the light ($\times$ 250) and electron ($\times$ 15,000) microscope. [From Kutzner, 1956, with permission.]

Table 18. Spore surface ornamentation of "hygroscopic" species of *Streptomyces* determined by scanning and transmission electron microscopy (Williams and Wellington, 1978).

	Spore surface determined by	
Species	Scanning electron microscopy[a]	Transmission electron microscopy[b]
S. bluensis	Ridged	Spiny
S. endus	Ridged	Warty
S. hygroscopicus	Ridged	Warty
S. melanosporofaciens	Ridged	Smooth-warty
S. sparsogenses	Ridged	Spiny
S. violaceoniger	Ridged	Smooth

[a] Williams and Wellington, 1978.

[b] International Streptomyces Project.

Streptomyces as viewed in the scanning electron microscope, which will become an indispensable instrument for the actinomycete taxonomist.

Physiological Tests

Of the numerous physiological tests used in the early studies on streptomycetes, many have been found to be rather unreliable or—if a wide species concept is employed—of little taxonomic value. Only two seem to have stood the test of time and have been used in the International Streptomyces Project (ISP) (Shirling and Gottlieb, 1966): (i) formation of melanin pigment and (ii) utilization of nine carbon sources (mainly sugars). However, some authors continue to characterize their cultures by numerous tests (e.g., Gordon and Horan, 1968a), and there is no doubt that several physiological properties in addition to the two mentioned above are very significant for these organisms—if (as it may turn out) not relevant for identification to species, then at least as markers by which an individual strain can be recognized again.

Table 19 lists a number of tests which can be recommended at the present time. Several of them have been worked out by the present author and his co-workers—in the firm belief that the ill repute of physiological properties in general is due to the ignorance of several points that should be taken into consideration when dealing with streptomycetes (instead of with fast growing eubacteria) rather than to any idiosyncrasy of the organisms in question. It should be admitted, however, that with streptomycetes many tests do not give a clear-cut result (+ or −); instead, three or four grades can often be observed: (i) negative; (ii) poor, trace, late; (iii) moderate; (iv) strong, fast. Since many tests need several days (up to 2 weeks), records have to be made at intervals. In fact, this is also a reason for the reluctance of using physiological criteria in the taxonomy of these organisms.

A legion of other tests are available for characterization, if this appears to be desirable or necessary (e.g., for numerical taxonomy). Thus, the list of compounds to be tested as carbon or nitrogen sources, of substrates to be tested for degradation, or of antibiotics and toxic substances to be tested for inhibitory action (Williams, 1967) can easily be extended. The reader is referred to Gordon and Horan (1968b) and to taxonomists dealing with other actinomycetes, Rowbotham and Cross (1977) and Goodfellow and Alderson (1979). Here only the tests listed in Table 19 will be described briefly.

Resistance to Lysozyme
Basal medium:

Peptone	10.0 g
Yeast extract	5.0 g
H_2O	1.0 liter
pH	7.0

Lysozyme stock solution: 100 mg lysozyme in 10 ml H_2O; sterilized by filtration.

Preparation of medium: 100 ml basal medium is supplemented (after autoclaving) with 0.1, 0.5, and 1 ml of stock solution (= 10, 50, and 100 μg/ml); one batch without lysozyme serves as control. The broth is distributed into sterile test tubes in 3-ml amounts.

Test: The tubes are incubated on a shaker in a slanted position. Observations are made after 2 and 4 days. The result is in most cases clear-cut.

Formation of Melanin
Media (as used in the ISP, Shirling and Gottlieb, 1966):

Complex medium: Difco peptone iron agar + yeast extract (1 g/liter).
Synthetic medium:

Glycerol	15.0 g
L-Tyrosine	0.5 g
L-Asparagine	1.0 g

Table 19. Physiological properties useful for the characterization of streptomycetes.

Property	References[a]
Resistance toward lysozyme[b]	Kutzner, Böttiger, and Heitzer (1978)
Formation of melanin (chromogenicity)	Shirling and Gottlieb (1966), Kutzner (1968), Baumann et al. (1976)
Utilization of carbohydrates and similar compounds	Benedict et al. (1955), Zähner and Ettlinger (1957), Nitsch and Kutzner (1973)
Hydrolysis of esculin	Gordon and Horan (1968)
Utilization of organic acids (especially oxalate, malonate, citrate)	Nitsch and Kutzner (1969a)
Formation of organic acids	Cochrane (1947), Ziegler and Kutzner (1973a)
Reduction of nitrate to nitrite (under strict anaerobic conditions)	Kutzner, Böttiger, and Heitzer (1978)
Hydrolysis of urea	Nitsch and Kutzner (1969b)
Hydrolysis of hippuric acid	Ziegler and Kutzner (1973b)
Lecithovitellin reaction (on egg yolk agar)	Nitsch and Kutzner (1969c)
Hemolysis	—
Resistance toward sodium chloride	Tresner, Hayes, and Backus (1968)
Antibiotic activity	Lyons and Pridham (1973)

[a] Only few references can be cited here; those are selected from the overwhelming literature which describe the test as it is in use today and/or give the most detailed results, discussion, and literature regarding this test and its physiological background.

[b] This is actually a biochemical criterion, because it is determined by a simple growth experiment (and also because nothing is known of the structural basis of this property) it is listed here.

K_2HPO_4	0.5 g
$MgSO_4 \cdot 7H_2O$	0.5 g
NaCl	0.5 g
$FeSO_4 \cdot 7H_2O$	0.01 g
Trace elements (Table 13)	1.0 ml
Agar	20.0 g
H_2O	1.0 liter
pH 7.2–7.4	

Test: Agar slants; melanin formation is indicated by a dark-gray to blue-black color of the medium after 1–2 days; the result is clear-cut (+ or −). Remark: Another test has been recommended by Mikami, Yokoyama, and Arai (1977). Furthermore, the media given above might be replaced by others in the future based on the finding that tyrosinase is induced by certain amino acids (e.g., L-methionine, L-norleucine) but not by its own substrate, L-tyrosine (Baumann et al., 1976).

Utilization of Sugars and Similar Compounds

"Conventional" method (Shirling and Gottlieb, 1966):
Basal medium:

$(NH_4)_2SO_4$	2.64 g
KH_2PO_4	2.38 g
$K_2HPO_4 \cdot 3H_2O$	5.56 g
$MgSO_4 \cdot 7H_2O$	1.00 g
$CuSO_4 \cdot 5H_2O$	6.40 mg
$FeSO_4 \cdot 7H_2O$	1.10 mg
$MnCl_2 \cdot 4H_2O$	7.90 mg

$ZnSO_4 \cdot 7H_2O$	1.50 mg
Agar	15.00 g
H_2O	1.00 liter
pH 6.8–7.0	

Carbon sources: 10% solutions of the following compounds are sterilized by filtration and added to the basal medium after autoclaving and cooling to 60°C to give a final concentration of 1%: D-glucose, fructose, L-arabinose, D-xylose, rhamnose, sucrose, raffinose, D-mannitol, and i-inositol.

Test: The complete medium is poured into plates which are inoculated by two streaks; one batch without carbon source serves as "negative control", glucose serves as "positive control". Observations are made after 10–16 days.
Agar shake culture (Nitsch and Kutzner, 1973): Basal medium: Difco AATCC agar, one-third strength; adjust pH!

NH_4NO_3	1.000 g
KH_2PO_4	0.830 g
K_2HPO_4	0.660 g
$MgSO_4 \cdot 7H_2O$	0.066 g
$FeSO_4 \cdot 7H_2O$	0.030 g
Agar	6.600 g
H_2O	1.000 liter
pH 7.0	

Carbon sources: 10% solutions are sterilized by filtration.

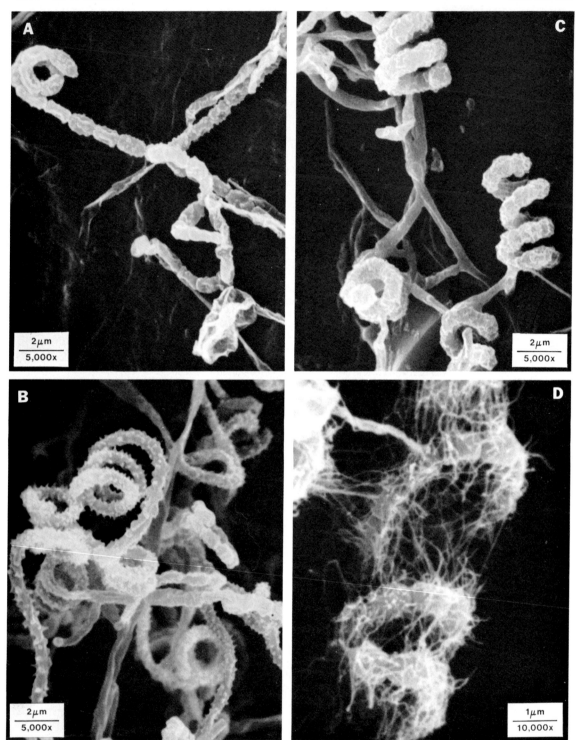

Fig. 17. Scanning electron microscopy of arthrospore chains. [Courtesy of A. Dietz, Upjohn Co.] (A) *Streptomyces torulosus* (knobby), (B) *S. antimycoticus* (rugose), (C) *S. bluensis* (spiny), (D) *S. karnatakensis* (hairy).

Preparation of shake cultures: Sterile test tubes (10 × 100 mm) are charged with carbon source (0.2 ml; control H_2O), inoculum (0.1 ml of dense spore suspension) and basal medium (1.7 ml, 70°C warm) in that order.

Test: Observations are made after 2–7 days. Controls and nonutilizable carbon sources allow only very weak growth some millimeters below the surface of the agar column (Fig. 18); utilization is indicated by thick surface growth; in the case of poor and/or slow utilization, growth starts ''submerged'' and then progresses toward the surface—sometimes only in form of single colonies (mutants ?).

Remark: The agar shake test works only when a heavy spore suspension is used; it requires much less medium and gives quicker results than the conventional test. Difco AATCC agar is very suitable, but other commercially available basal media (e.g., yeast nitrogen base) may be used instead. Both tests yield positive, negative, and weak reactions; it is believed that the agar shake culture makes the decision between the various grades easier.

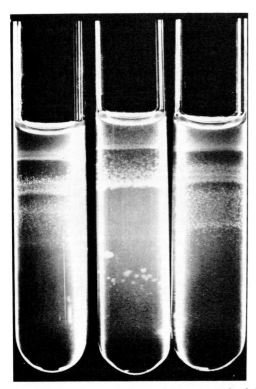

Fig. 18. Subsurface growth of three streptomycetes in shake agar; mineral salts medium without carbon source plus 0.1 g/liter yeast extract, 11 days old. After 2 days, the uppermost growth-disk appears; on further incubation, additional ''bands'' develop.

Hydrolysis of Esculin

Media

Ingredient	control	test medium
Yeast extract	3.0 g	3.0 g
Esculin	—	1.0 g
$NH_4^+-Fe^{3+}-$citrate	0.5 g	0.5 g
Agar	3.0 g	3.0 g
H_2O	1.0 liter	1.0 liter
pH 7.2		

Test: Small tubes with 2 ml of the soft agar medium are used. Observations are made after 5 and 10 days. The control without substrate is necessary to recognize ''false positives'' due to melanin formation. Results are graded as (i) negative (no color change; the upper layer shows fluorescence due to the nondegraded substrate), (ii) trace, (iii) moderate, or (iv) strong positive (dark brown to black pigment).

Utilization of Organic Acids

Basal medium:

$(NH_4)_2SO_4$	1.0 g
$Na_2HPO_4 \cdot 2H_2O$	0.53 g
KH_2PO_4	0.27 g
$MgSO_4 \cdot 7H_2O$	0.5 g
NaCl	2.0 g
Bromothymol blue	0.025 g
Agar	12.0 g
H_2O	1.0 liter
pH 7.0	

Salts of organic acids (g/liter):

K-Gluconate	5.0
Na_3-Citrate	2.0
Na-Malate	5.0
Na-Lactate	5.0
Na-Malonate	3.0
(oxalate see below)	

Test: Agar slants; utilization is indicated by alkalization of the medium after 6–12 days. Rarely, no color change to blue takes place, in spite of good growth. A complex medium should be included as control for the quality of the inoculum; confluent growth should be obtained. Utilization is graded as (i) negative, (ii) moderate, or (iii) strong positive (dark blue).

Medium to test for oxalate utilization:

K-Oxalate	1.0 g
$(NH_4)_2HPO_4$	1.0 g
KH_2PO_4	0.5 g
$MgSO_4 \cdot 7H_2O$	0.2 g
NaCl	1.0 g
Agar	12.0 g
H_2O	920.0 ml
pH 7.0	

Preparation of agar plates: The medium is autoclaved in batches of 92 ml; before pouring the plates, 8 ml of a 0.1-m solution of $CaCl_2$ (1.47 g in 100 ml) is added, resulting in a fine precipitate of Ca-oxalate.

Test: Five organisms are inoculated onto the plates in spots; oxalate utilization is indicated by clear zones of several mm width around the colonies after 5–15 days. In some cases, a precipitate is formed within the clear halo. The result is in most cases clear-cut (+ or −); some positive organisms show only a weak clearance.

Formation of organic acids
Medium:

A.	Glucose	20.0 g
	Yeast extract	1.2 g
	$MgSO_4 \cdot 7H_2O$	0.25 g
	Bromocresol purple	12.0 mg
	Agar	4.0 g
	H_2O	400.0 ml
B.	$Na_2HPO_4 \cdot 2H_2O$	534 mg
	KH_2PO_4	272 mg
	H_2O	500 ml
C.	$CaCO_3$	1.0 g
	H_2O	100.0 ml

Preparation of medium: Test tubes (16 × 160 mm) are charged with 0.2 ml of suspension C and autoclaved; solutions A and B are autoclaved separately and combined; 1.8 ml is added to the tubes. To achieve even distribution of the $CaCO_3$, the tubes are cooled quickly in an ice bath.

Test: Acid formation is indicated by disappearance of the $CaCO_3$, and color change to yellow within 5–15 days; four grades can be distinguished.

Remark: This is only a coarse test and possibly of less taxonomic value. A detailed study of acid formation in liquid cultures and determination of the acids by gas chromatography showed that the differences were rather quantitative than qualitative—as it was hoped pyruvate, lactate, α-ketoglutarate, and succinate were the acids most often found (Hammann, 1977; Ziegler and Kutzner, 1973).

Reduction of Nitrate to Nitrite
Medium:

Glycerol	5.000 g
KNO_3	1.000 g
$Na_2HPO_4 \cdot 2 H_2O$	0.534 g
KH_2PO_4	0.272 g
$MgSO_4 \cdot 7 H_2O$	0.500 g
NaCl	2.000 g
Trace elements (Table 13)	1.000 ml
Agar	4.000 g

H_2O	1.000 liter
pH 6.8	

Cultural conditions: The medium is distributed in 1-ml amounts in tubes (12 × 130 mm) and sterilized; duplicates are inoculated by mixing the inoculum with the soft agar; one is incubated aerobically, the other under strict anaerobic conditions (either in a GasPak jar or sealed by a rubber stopper after inserting the cotton wool plug saturated with pyrogallol and Na_2CO_3).

Test: Formation of nitrite is indicated by formation of a red color (azo-dye) after addition of Griess-Ilosvay reagent. Streptomycetes differ quantitatively and qualitatively; some form nitrite only under aerobic or anaerobic conditions, others under both conditions or not at all.

Remark: In the past, the nitrate reduction test has been found to be very unreliable with streptomycetes; however, this has probably been due to the use of nitrate broth incubated aerobically. The test described here has been found to be reproducible.

Hydrolysis of Urea
Basal medium:

Glucose	1.000 g
Casein-peptone	1.000 g
$Na_2HPO_4 \cdot 2H_2O$	1.980 g
KH_2PO_4	1.510 g
$MgSO_4 \cdot 7H_2O$	0.500 g
NaCl	5.000 g
Phenol red	0.012 g
Agar	12.000 g
H_2O	900.000 ml
pH 6.8	

Solution of urea:

Urea	20.0 g
H_2O	100.0 ml

Sterilized by filtration and added to the basal medium after autoclaving; the medium is distributed in test tubes which are slanted.

Test: Agar slants; urease activity is indicated by alkalization of the medium. Observations are made daily for 5 days to distinguish the strong and fast positives, moderate positives, and negatives. In case of a strong reaction, crystals of $MgNH_4PO_4 \cdot 6H_2O$ are formed within the agar; these cultures often show very little growth due to the rapid alkalization.

Hydrolysis of Hippuric Acid
Medium:

Na-Hippurate	10.0 g
Glucose	2.0 g
Peptone (from casein)	2.0 g

Meat extract	2.0 g
Yeast extract	2.0 g
Na$_2$HPO$_4$	5.0 g
H$_2$O	1.0 liter
pH 7.0	

Test: Tubes with 3 ml broth are incubated on a shaker. A positive test is indicated by a precipitate of benzoic acid crystals after addition of 1.5 ml 50% H$_2$SO$_4$ to 1 ml culture filtrate after 15 days. Positive cultures differ in the strength of the precipitate.

Remark: The test is positive only when at least 40% of the substrate is hydrolyzed; negative cultures may contain benzoic acid identifiable by chromatographic methods. The test needs to be improved.

Lecithovitellin (LV) Reaction

Basal medium:

Glucose	1.0 g
Peptone	15.0 g
Yeast extract	3.0 g
NaCl	5.0 g
Agar	15.0 g
H$_2$O	950.0 ml
pH 7.5	

Preparation of agar plates: The basal medium is autoclaved in batches of 95 ml; before pouring into plates, 5 ml egg yolk emulsion is added to each batch.

Test: Agar plates are inoculated spotwise with 3 or 4 organisms. After 3, 5, and 8 days, the plates are checked in transmitted light for the occurrence of opaque zones around the colonies; these may reach more than 10 mm; the precipitates vary considerably in strength. Lipase activity is indicated by the formation of a pearly layer: in LV-positive cultures mostly at the edge of the opaque zone, in LV-negative cultures around the colony. A strong lipase reaction is difficult to distinguish from a weak LV-reaction.

Hemolysis

Basal medium

Casein-peptone	10.0 g
Meat extract	10.0 g
NaCl	5.0 g
Agar	12.0 g
H$_2$O	930.0 ml
pH 7.2	

Preparation of agar plates: The basal medium is autoclaved in batches of 93 ml; before pouring into plates, 7 ml of defibrinated sheep blood is added to each batch.

Test: Agar plates are inoculated in the form of a streak with one or two strains (in the latter case preferably in the form of a T). After 2 and 5 days, zones of hemolysis are recorded; zones of complete and incomplete lysis—occurring alone or combined—can be distinguished. Streaks of two organisms may exhibit a synergistic effect at the point where the streaks are close together.

Remark: Strong hemolysis, i.e., clear zones of 10–15 mm width, is in most cases correlated with a positive egg yolk reaction.

Resistance Toward Sodium Chloride

Basal medium:

Glucose	4.0 g
Yeast extract	4.0 g
Malt extract	10.0 g
Agar	15.0 g
H$_2$O	1.0 liter
pH 7.2	

Preparation of plates: The basal medium is prepared in 5 batches which are supplemented with NaCl (g/liter): 0, 40, 70, 100, and 130. The medium is autoclaved and poured into plates.

Test: Agar plates are divided into 4 sectors, each being streaked with a streptomycete. Observations are made after 10 days. The highest concentration of salt that allows growth is recorded; there is no clear-cut borderline between growth and no growth, since in most cases growth is retarded and/or less dense and/or in the form of smaller colonies, with higher salt concentrations tolerated as compared with the control.

Antibiotic Activity

The chemical identification of the antibiotic(s) produced by a streptomycete would yield a very valuable criterium for characterization. However, since this has to be left to sophisticated screening programs the microbiologist has to resort to the simple method of determining the antibiotic spectrum. The following method has been taken from Lyons and Pridham (1971, 1973).

Medium for streak test:

D-Glucose	15.0 g
Glycerol	2.5 ml
Soybean meal	15.0 g
Yeast extract	1.0 g
NaCl	5.0 g
CaCO$_3$	1.0 g
Agar	15.0 g
H$_2$O	1.0 liter
pH 6.8	

Medium for preparation of inoculum of streptomycetes:

D-Glucose	5.0 g
Tryptone	10.0 g
Yeast extract	5.0 g
K$_2$PO$_4$	1.0 g

Liver extract   100.0 ml
  (Haynes, Wickerham,
  and Hesseltine, 1955)
H_2O 900.0 ml
pH 6.9

Test: Agar plates are streaked in a single line ca. 0.6 cm in width to give a confluent streak of growth about 7.5 cm in length (approximately 2.5 cm from the edge of the dish). The authors used liquid inoculum obained from a highly complex medium. After 7 days, the test organisms were streaked at right angles to the streptomycete growth, taking care not to touch it. Zones of inhibition were measured after 3 days' additional incubation. Six test organisms were used to determine the pattern of activity: *Bac. subtilis* (NRRL B-765), *Sarcina lutea* (NRRL B-1018), *E. coli* (NRRL B-766), *Sacch. pasteurianus* (NRRL Y-139), *Cand. albicans* (NRRL Y-477), and *Mucor ramannianus* (NRRL 1839). The authors could differentiate 12 antibacterial patterns.

Phage Sensitivity

Phage typing in the order Actinomycetales has been carried out at the genus level (Bradley, Anderson, and Jones, 1961; Prauser, 1976; Prauser and Falta, 1968) as well as (in the genus *Streptomyces*) at the species level (Korn, Weingärtner, and Kutzner, 1978; Kutzner, 1961a,b; Shirling, 1959; Welsch, Corbaz, and Ettlinger, 1957). Most actinophages used for this purpose have been isolated from soil and are probably virulent; however, temperate phages can also be obtained from soil (Dowding and Hopwood, 1973) and, of course, from many lysogenic actinomycetes (Welsch, 1959).

The host range of the phages has been found to be always confined to the genus; within the genus, however, phages are either polyvalent, attacking many species, or specific, lysing only strains of one or a few related species (unfortunately not all members of the species). At present, phages specific for the following species are known: *S. albus, S. griseus, S. coelicolor* Müller, *S. violaceoruber, S. erythraeus, S. viridochromogenes,* plus several other species with blue aerial mycelium.

In addition to differences in their host range, actinophages vary strikingly in morphology of plaques, morphology of particles (as shown by electron microscopy), and serological properties. They have attracted the interest not only of taxonomists but also of industrial microbiologists (Carvajal, 1953) and recently geneticists (Lomovskaya et al., 1973).

The methods used for the study of actinophages are the same as those employed for phages of other bacteria. Two points deserve mentioning: (i) With some streptomycetes (and more conspicuously with several other actinomycetes), it is difficult to obtain a lawn of confluent growth; special procedures for preparing the inoculum may be necessary. (ii) Because of wide variation in the efficiency of plating on different hosts and because of unspecific effects sometimes produced by little diluted culture filtrates, at least two dilutions of the phage preparation should be used for typing, the RTD and its 100-fold dilution—producing a clear spot on the homologous host or single plaques, respectively.

CLASSIFICATION AND IDENTIFICATION

The taxonomic position of the Streptomycetaceae within the order Actinomycetales (based on biochemical criteria) and the classification of this family to the genus level (based on gross morphology) have been discussed in the Introduction (Tables 1 and 2). Criteria used (or supposed to be useful) for the identification of species have been described in the preceding section. Having collected all this information, one should arrive at a species identification. However, as in many other bacterial genera, problems other than the mere determination of properties are involved in speciation—the most important one being the species concept upon which most workers in this field should agree (e.g., which of the criteria are useful at the species level and which may be employed at the subspecies level—or have to be considered only as strain properties); other problems are the availability of type cultures and workable keys. As long as a genus contains only a few species, these problems do not become obvious; however, genera consisting of hundreds of "named species" (e.g., *Streptomyces*) are a nightmare for the taxonomist.

Microellobosporia and *Kitasatoa*

The genera *Microellobosporia* and *Kitasatoa* contain only four species each. These differ in their cultural characteristics on various media, in some physiological properties including the utilization of carbon compounds, and in antibiotic activity. Tables listing these differential features are given by Cross and Goodfellow (1973) and in *Bergey's Manual* (Buchanan and Gibbons, 1974). Most of the species of the two genera seem to have been isolated only once; thus these organisms either are rare in nature or have been overlooked when conventional isolation methods are employed. Since they closely resemble *Streptomyces* in gross appearance, they might have been isolated in the past and mistaken for these organisms (Cross, Lechevalier, and Lechevalier, 1963).

CHAPTER 156

Streptomyces, Streptoverticillium, and *Chainia*

The eighth edition of *Bergey's Manual* lists 463 species of *Streptomyces* and *Chainia* (these two genera are regarded there as synonymous) and 40 species of *Streptoverticillium*. However, the number of species names ever used for one of these organisms in the scientific and patent literature probably far exceeds 1,000. One of the reasons for this explosion of number of species is certainly the legion of isolates investigated in the course of screening for antibiotics; this resulted in the discovery of thousands of different secondary metabolites, each one suggesting the existence of a different "special type". These types were often considered a "new species", not so much from a taxonomic but rather from a patent law point of view: "Industry cannot escape its share of responsibility for the taxonomic confusion born of the antibiotic floodtide, but the academic community was also ill-prepared to cope with the situation" (Trejo, 1970).

There is no doubt that numerous synonyms exist, but no estimate can be made at the present time of the "real number" of species. Specialists working in this field still cannot agree how many "man-made species"—based on the classical concept of the "nomenspecies"—should be distinguished. "Lumpers" recognize only a few species, e.g., 41 (Hütter, 1967; see Table 20) or even 10 (Pridham, 1976; see Table 21); "splitters" recognize the numerous nomenspecies which have been validly described and are represented by viable type or neo-type cultures. The problem seems to be resolvable only by approaches other than the classical one, i.e., numerical taxonomy and/or molecular taxonomy (DNA homology).

Classical Taxonomy, Monothetic Grouping

Various keys for the identification of species have been constructed, the two most noteworthy ones are given in Tables 20 and 21. Other keys have been developed by using the data of the International Streptomyces Project (Küster, 1972; Nonomura, 1974; Szabó et al., 1975). Clearly, the more criteria used, the more species differentiated.

The classification of Hütter (1967)—based on the work of Ettlinger, Corbaz, and Hütter (1958)—employs four criteria: color of aerial mycelium, spore chain morphology, structure of spore surface, and melanin formation. Pridham (1976) uses only two criteria within each of the two genera *Streptomyces* and *Streptoverticillium*: spore surface and spore chain morphology.

The species concept of a "lumper", based on only a few criteria, looks at first sight rather attrac-

tive. However, two problems arise when it is employed:

(i) Variation in one of the key features or misdetermination of such a criterium leads invariably to an incorrect identification. Since some of the properties are in fact not as unambiguous as often believed, the danger of a misidentification is higher than commonly admitted. The difficulties in interpreting the color of the aerial mycelium and even the surface structure of spores observed in the transmission electron microscope have been discussed above. The same is true for the spore chain morphology. For example, the differentiation of *S. griseus* from *S. venezuelae* solely on the basis of flexuous as opposed to straight aerial hyphae appears to be rather artificial (Table 21); indeed, in the original paper (Pridham and Lyons, 1965) the authors indicate that the separation between straight and flexuous was quite difficult and "one must resort to other criteria and hope that some correlating characteristics can be found." The evaluation of the records made by participants of the International Streptomyces Project also underlines this problem (Shirling and Gottlieb, 1976; Szabó and Marton, 1976). Finally, when using only the criteria employed by Hütter (1967) and Pridham (1976), we are completely helpless if a streptomycete has lost its capacity to form aerial mycelium. It is therefore difficult to believe that a species concept which considers only a few criteria (e.g., two in Table 21) will stand the test of time.

(ii) The other problem we have to cope with when using a broad species concept is the long list of synonyms. This may be regarded merely as a nomenclatural problem: The synonyms are reduced to the rank of strains (Hütter, 1967). Alternatively, synonyms are recognized as subspecies (Pridham, 1976), and criteria used for their identification are mainly the utilization of carbon sources and the nature of the secondary metabolites produced (principally antibiotics) However, as it has been expressed to the point by Locci, Baldacci, and Petrolini Baldan (1969) and Trejo (1970), the creation of large species with long lists of synonyms or subspecies designation only shifts our frame of reference down the taxonomic scale and does not really solve the problem but only postpones it.

Numerical Taxonomy

The principles of the Adansonian classification have been applied to streptomycetes by various authors (Table 22). In most cases, use has been made of the matching coefficients of Sokal and Michener (S_{SM}) or of Jaccard (S_J), and clusters (phenons) have been formed by the single-linkage or average-linkage method. However, other methods for cluster formation have also been applied; e.g., Gyllenberg (1976)

Table 20. Classification of 41 species of *Streptomyces (Sm.)* and *Streptoverticillium (Sv.)* according to Hütter (1967).

Species	Color of aerial mycelium	Verticils	Spore-chain morphology					Spore-surface	Melanin
			RF[a]	RA[a]	Spira A[a]	Spira B[a]	No. of turns[b]		
Sm. albus	White: "niveus"	−	−	−	+	−	2–5	smooth	−
Sm. rimosus		−	−	−	−	+		smooth	−
Sm. longospororuber		−	−	+	−	−	2–5	smooth	+
Sm. griseus	Yellow-green: "griseus"	−	+	−	−	−		smooth	−
Sm. longisporoflavus		−	−	−	−	+	1–10	smooth	−
Sm. flavidovirens		−	−	+	−	−	1–4	smooth	−
Sm. michiganensis		−	+	−	−	−		smooth	+
Sm. exfoliatus	light-pink/ lavender: "cinna- momeus"	−	+	−	−	−		smooth	−
Sm. griseoviridis		−	−	+	−	−	1–3	smooth	−
Sm. fradiae		−	−	−	+	−	2–5	smooth	−
Sm. venezuelae		−	+	−	−	−		smooth	+
Sm. lavendulae		−	−	+	−	−		smooth	+
Sm. erythraeus		−	−	+	−	−	1–3	spiny	−
Sm. violaceus		−	−	−	−	+	2–5	spiny	+
Sm. olivaceus	Gray: "cinereus"	−	+	−	−	−	c	smooth	−
Sm. ramulosus		−	+	−	−	−	c	smooth	−
Sm. aureofaciens		−	−	+	−	−	2–4	smooth	−
Sm. violaceoniger		−	−	−	+	−	2–5	smooth	−
Sm. violaceoruber		−	−	−	−	+	4–6	smooth	−
Sm. parvulus		−	−	−	−	+	5–12	smooth	−
Sm. antibioticus		−	+	−	−	−		smooth	+
Sm. tendae		−	−	+	−	−	1–5	smooth	+
Sm. collinus		−	−	−	+	−	1–4	smooth	+
Sm. diastatochromogenes		−	−	−	−	+	5–10	smooth	+
Sm. albogriseolus		−	−	−	+	−	4–6	hairy	−
Sm. flaveolus		−	−	−	−	+	4–8	hairy	−
Sm. pilosus		−	−	−	−	+	5–10	hairy	+
Sm. griseoflavus		−	−	−	−	+	4–6	spiny	−
Sm. noursei		−	−	+	−	−	1–3	spiny	−
Sm. echinatus		−	−	−	+	−	2–5	spiny	+
Sm. arenae		−	−	−	−	+	3–6	spiny	+
Sm. azureus	Gray-blue/ green-blue: "azureus- glaucus"	−	−	−	+	−	1–4	smooth	+
Sm. caelestis		−	−	+	−	−	2–5	smooth	+
Sm. glaucescens		−	−	−	+	−	2–4	hairy	+
Sm. viridochromogenes		−	−	−	−	+	1–10	spiny	+
Sm. hirsutus	Leek-green: "prasinus"	−	−	+	−	(+)	1–4	hairy	−
Sm. prasinus		−	−	+	−	(+)	1–5	spiny	−
Sv. cinnamomeum	light-pink/ lavender: "cinna- momeus"	+	+	−	−	−		smooth	−
Sv. griseocarneum		+	+	−	−	−		smooth	+
Sv. netropse		+	−	−	−	+	1–4	smooth	−
Sv. reticulum	"griseus"	+	+	−	−	−		smooth	+

[a] RF: rectus-flexibilis; RA: retinaculum apertum; Spira A: turns narrow, compact; Spira B: turns wide, extended.
[b] Rarely less or more than indicated.
[c] *Sm. oliv.*: long chains of spores (often more than 50); *Sm. ram.*: short chains of spores (most only 10–20).

Table 21. Classification of 10 species of *Streptomyces* and *Streptoverticillium* according to Pridham (1976)

Species	Verticils	Spore surface					Spore chain morphology[a]			
		Smooth	Warty	Knobby	Hairy	Spiny	Straight	RF	RA	Spiral
Sm. venezuelae	−	+	−	−	−	−	+	−	−	−
Sm. griseus	−	+	−	−	−	−	−	+	−	−
Sm. lavendulae	−	+	−	−	−	−	−	−	+	−
Sm. albus	−	+	−	−	−	−	−	−	−	+
Sm. graminofaciens	−	−	+	−	−	−	−	−	−	+
Sm. torulosus	−	−	−	+	−	−	−	−	−	+
Sm. flaveolus	−	−	−	−	+	−	−	−	−	+
Sm. diastaticus	−	−	−	−	−	+	−	−	−	+
Sv. reticulum	+	+	−	−	−	−	+	−	−	−
Sv. netropse	+	+	−	−	−	−	−	−	−	+

[a] RF, rectus flexibilis; RA, retinaculum apertum.

used the Q-index for determining the densest regions in the character-space and Szulga (1978) discusses results obtained by the "Wroclaw Dentrite Method" and the "Centrifugal Correlation Method".

The favorite material for numerical taxonomists of streptomycetes has been the ISP strains—mainly for the sake of convenience. However, one may raise the question whether this material is really representative of the population of streptomycetes found in nature all over the world.

As can be seen from Table 22, clusters can easily be found by any of the methods mentioned above. However, a substantial portion of the material remained unclustered. Also, the relationship between the clusters obtained and the classical taxons—subgenera, species—still remains open. In addition, the composition of the clusters produced from the same material varied somewhat with the method employed. It appears that the clusters contain three groups of strains: (i) "species" which are definitely identical, i.e., synonymous (even from a splitter's point of view); (ii) species with a rather high degree of similarity; (iii) species which are only at the borderline of the cluster. Finally, opinions vary about the most significant criteria useful for recognition of

the clusters. Whereas Hill and Silvestri (1962) found that physiological characters correlated much better with the phenons obtained than certain properties traditionally used in the classification of these organisms (e.g., color and morphology of aerial mycelium), Kurylowicz et al. (1976) and Gyllenberg (1976) confirmed, although not in complete agreement, the value of the classical criteria for the recognition of the clusters.

Although hitherto the various attempts of numerical taxonomists could not solve the problems of classification, species concept, and identification of streptomycetes, there is no doubt that further studies (presently being carried out in various laboratories) will certainly help to clarify the complicated situation with these organisms.

Molecular Taxonomy

Organisms belonging to one species should be expected to exhibit a high degree of homology of genetic material. Thus, determination of DNA homology between each pair of the existing "species" of Streptomycetaceae should theoretically provide a

Table 22. Some studies on numerical classification of streptomycetes.

No. of strains	Nature of material	No. of tests (properties)	No. of characters (features)	No. of clusters	No. of unclustered strains	Authors
159	Species	58	105	24	14	Silvestri et al., 1962
44	Isolates	76	76	7		Szabó et al., 1967
18	Isolates	40	46	5		Williams, Davies, and Hall, 1969
448	ISP "species"	13	31	14	168	Kurylowicz et al., 1976
618[a]	ISP "species"	12	24	15	218	Gyllenberg, 1976

[a] In fact, the data of the 448 ISP strains were evaluated; in cases of variable characters of individual strains, each possible combination was considered as a separate strain.

basis for a sound classification and identification.

Due to the expense and labor of the methods involved, few studies have been carried out as yet with these organisms, and these only with a limited number of strains. Yamaguchi (1967) determined the degree of homology of various Actinomycetales with the DNA of a streptomycin-producing strain of *Streptomyces griseus:* It ranged from 37 to 82%, the highest value being obtained with *Streptoverticillium netropse.* Okanishi, Akagawa, and Umezawa (1972) also used *S. griseus* (ISP 5236) as source of the reference DNA. Only strains which also produced streptomycin showed a high degree of homology; *S. griseinus* and other species of the "*S. griseus* group" exhibited a low percentage of hybridization: 38–90%. A low degree of DNA homology was also observed with the DNA of 26 species of *Streptoverticillium* and 3 reference DNAs from 3 species of this genus (Toyama, Okanishi, and Umezawa, 1974). Thus, as it has been discussed by Bradley (1975), "phenotypically similar organisms might possess substantial amounts of genetic diversity."

As in the case of numerical taxonomy, the question remains open which percentage of homology is the borderline of a species. It appears that the results obtained so far in studies on DNA homology support the species concept of a splitter rather than that of a lumper. However, much more work needs to be done in this field to reach a final conclusion. Although the expectations have not been fulfilled as yet, this approach—as well as others in the field of molecular biochemistry—might give the final solution to the classification and identification of the Streptomycetaceae.

Acknowledgment

The author dedicates this chapter to the memory of Selman A. Waksman (1888–1973), who spent a lifetime studying the actinomycetes from soil and who is one of the initiators of the "antibiotic era" in microbiology and medicine.

Bibliography

Arai, T. (ed.). 1976. Actinomycetes—the boundary microorganisms. Baltimore, London, Tokyo: University Park Press.

Baldacci, E., Redaelli, P. (organizers). 1953. Symposium "Actinomycetales: Morphology, Biology and Systematics". Sixth International Congress of Microbiology, Rome, 1953.

Freerksen, E., Tarnok, I., Thumin, J. H. (eds.). 1978. Genetics of the Actinomycetales. Stuttgart: Gustav Fischer Verlag.

Gauze, G. F. (ed.). 1958. Zur Klassifizierung der Actinomyceten. Jena: Gustav Fischer Verlag.

Goodfellow, M., Brownell, G. H., Serrano, J. A. (eds.). 1976. The biology of nocardiae. London, New York, San Francisco: Academic Press.

Hütter, R. 1967. Systematik der Streptomyceten. Bibliotheca Microbiologica, Fasc. 6. Basel, New York: Karger.

Lechevalier, H. A., Lechevalier, M. P. 1967. Biology of actinomycetes. Annual Review of Microbiology **21:**71–100.

Lechevalier, H. A., Lechevalier, M. P., Gerber, N. N. 1971. Chemical composition as a criterion in the classification of actinomycetes. Advances in Applied Microbiology **14:**47–72.

Mordarski, M., Kurylowicz, W., Jeljaszewicz, J. (eds.). 1978. *Nocardia* and *Streptomyces*. Proceedings of the International Symposium on *Nocardia* and *Streptomyces*, Warsaw, October 4–8, 1976. Stuttgart, New York: Gustav Fischer Verlag.

Prauser, H. (ed.) 1970. The Actinomycetales. The Jena International Symposium of Taxonomy, Sept. 1978. Jena: Gustav Fischer Verlag.

Raper, K. B. (ed.). 1954. Speciation and variation in asexual fungi. Annals of the New York Academy of Sciences **60:**1–182.

Rautenshtein, Ya. I. (ed.). 1966. Biology of antibiotic-producing actinomycetes. [Translated from the 1960 Russian original.] Jerusalem: Israel Program for Scientific Translations.

Sykes, G., Skinner, F. A. (eds.). 1973. Actinomycetales: Characteristics and practical importance. London, New York: Academic Press.

Szybalski, W. (ed.). 1959. Genetics of *Streptomyces* and other antibiotic-producing microorganisms. Annals of the New York Academy of Sciences **81:**805–1016.

Waksman, S. A. 1959. The Actinomycetales. Vol. I: Nature, occurence and activities. Baltimore: Williams & Wilkins.

Waksman, S. A. 1961. The Actinomycetales. Vol. II: Classification, identification and descriptions of genera and species. Baltimore: Williams & Wilkins.

Waksman, S. A., Lechevalier, H. A. 1962. The Actinomycetes. Vol. III: Antibiotics of actinomycetes. Baltimore: Williams & Wilkins.

Literature Cited

Abraham, T. A., Herr, L. J. 1964. Activity of actinomycetes from rhizosphere and non-rhizosphere soils of corn and soybean in four physiological tests. Canadian Journal of Microbiology **10;**281–285.

Agate, A. D., Bhat, J. V. 1963. A method for the preferential isolation of actinomycetes from soils. Antonie van Leeuwenhoek Journal of Microbiology and Serology **29:**297–304.

Agate, A. D., Bhat, J. V. 1967. Increase in actinomycetal population of stored soils. Current Science **36:**152–153.

Al-Diwany, L. J., Cross, T. 1978. Ecological studies on nocardioforms and other actinomycetes in aquatic habitats, pp. 153–160. In: Mordarski, M., Kurylowicz, W., Jeljaszewicz, J. (eds.), *Nocardia* and *Streptomyces*. Proceedings of the International Symposium on *Nocardia* and *Streptomyces*, Warsaw, October 4–8, 1976. Stuttgart, New York: Gustav Fischer Verlag.

Asselineau, J. 1966. The bacterial lipids. Paris: Verlag Herman.

Baker, R. 1968. Mechanism of biological control of soil-borne pathogens. Annual Review of Phytopathology **6:**263–294.

Baldacci, E. 1958. Development in the classification of actinomycetes. Giornale di Microbiologia **6:**10–27.

Baldacci, E., Locci, R., Locci, J. R., 1966. Production of "granules" by Actinomycetales. Giornale di Microbiologia **14:**173–184.

Ball, S., Bessell, C. J., Mortimer, A. 1957. The production of polyenic antibiotics by soil streptomycetes. Journal of General Microbiology **17:**96–103.

Balla, P. 1968. Contributions to the knowledge of thermophilic actinomycetes occurring in champignon compost. Annales of Universitatis Scientiarum Budapestinensis de Rolando Eotuos Nominatae, Sectio Biologica **9/10:**27–35.

Ballio, A., Barcellona, S. 1968. Relations chimiques et immunolgiques chez les Actinomycetales. Annales de l'Institut Pasteur **114:**121–137.

Baumann, R., Ettlinger, L., Hütter, R., Kocher, H. P. 1976.

Control of melanin formation in *Streptomyces glaucescens,* pp. 55–63. In: Arai, T. (ed.), Actinomycetes: The boundary microorganisms. Baltimore, London, Tokyo: University Press.

Becker, B., Lechevalier, M. P., Lechevalier, H. A. 1965. Chemical composition of cell-wall preparations from strains of various form-genera of aerobic actinomycetes. Applied Microbiology **13:**236–243.

Becker, B., Lechevalier, M. P., Gordon, R. E., Lechevalier, H. A. 1964. Rapid differentiation between *Nocardia* and *Streptomyces* by paper chromatography of whole-cell hydrolysates. Applied Microbiology **12:**421–423.

Benedict, R. G., Pridham, T. G., Lindenfelser, L. A., Hall, H. H., Jackson, R. W. 1955. Further studies in the evaluation of carbohydrate utilization tests as aids in the differentiation of species of *Streptomyces.* Applied Microbiology **3:**1–6.

Blinov, N. O., Khokhlov, A. S. 1970. Pigments and taxonomy of Actinomycetales, pp. 145–154. In: Prauser, H. (ed.), The Actinomycetales. The Jena International Symposium on Taxonomy, September 1968. Jena: Gustav Fischer Verlag.

Bonde, M. R., McIntyre, G. A. 1968. Isolation and biology of a *Streptomyces* sp. causing potato scab in soils below pH 5.0. American Potato Journal **45:**273–278.

Bradley, S. G. 1975. Significance of nucleic acid hybridization to systematics of actinomycetes. Advances in Applied Microbiology **19:**59–70.

Bradley, S. G., Anderson, D. L., Jones, L. A. 1961. Phylogeny of actinomycetes as revealed by susceptibility to actinophage. Developments in Industrial Microbiology **2:**223–237.

Brain, P. W. 1957. The ecological significance of antibiotic production, pp. 168–188. In: Williams, R. E. O., Spicer, C. C. (eds.), Microbial ecology. Cambridge: Cambridge University Press.

Brian, B. L., Gardner, E. W. 1967. Preparation of bacterial fatty acid methyl esters for rapid characterization by gas-liquid chromatography. Applied Microbiology **15:**1499–1500.

Brown, R. L., Peterson, G. E. 1966. Cholesterol oxidation by soil actinomycetes. Journal of General Microbiology **45:**441–450.

Brüsewitz, G. 1959. Untersuchungen über den Einfluß des Regenwurms auf Zahl, Art und Leistungen von Mikroorganismen im Boden. Archiv für Mikrobiologie **33:**52–82.

Buchanan, R. E., Gibbons, N. E. (eds.). 1974. Bergey's manual of determinative bacteriology, 8th ed. Baltimore: Williams & Wilkins.

Bucherer, H. 1935/36. Über den mikrobiellen Chitinabbau. Zentralblatt für Bakteriologie, Parasitenkunde und Infektionskrankheiten, Abt. 2 **93:**12–24.

Burman, N. P. 1973. The Occurrence and significance of actinomycetes in water supply, pp. 219–230. In: Sykes, G., Skinner, F. A. (eds.), Actinomycetales: Characteristics and practical importance. London, New York: Academic Press.

Burman, N. P., Oliver, C. W., Stevens, J. K. 1969. Membrane filtration techniques for the isolation from water, of coli-aerogenes, *Escherichia coli,* faecal streptococci, *Clostridium perfringens,* actinomycetes and microfungi, pp. 127–134. In: Shapton, D. A., Gould, G. W. (eds.), Isolation methods for microbiologists. The Society for Applied Bacteriology Technical Series No. 3. London, New York: Academic Press.

Carvajal, F. 1947. The production of spores in submerged cultures by some *Streptomyces.* Mycologia **39:**426–440.

Carvajal, F. 1953. Phage problems in the streptomycin fermentation. Mycologia **45:**209–234.

Chandramohan, D., Ramu, S., Natarajan, R. 1972. Cellulolytic activity of marine streptomycetes. Current Science **41:** 245–246.

Chater, K. F., Hopwood, D. A. 1973. Differentiation in actinomycetes, pp. 143–160. In: Asworth, J. M., Smith, J. E. (eds.), Microbial differentiation. 23rd Symposium of the Society for General Microbiology. Cambridge: Cambridge University Press.

Chater, K. F., Merrick, M. J. 1976. Approaches to the study of

differentiation in *Streptomyces coelicolor* A3(2), pp. 583–593. In: MacDonald, K. D. (ed.), Genetics of industrial microorganisms. The Second International Symposium, Sheffield, August 1974. London, New York, San Francisco: Academic Press.

Chesters, C. G. C., Apinis, A., Turner, M. 1956. Studies of the decomposition of seaweeds and seaweed products by microorganisms. Proceedings of the Linnean Society, London **166:**87–97.

Childs, J. F. L. 1953. An actinomycete associated with gummosis disease of grapefruit trees. Phytopathology **43:**101–103.

Cholodny, N. 1930. Über eine neue Methode zur Untersuchung der Bodenmikroflora. Archiv für Mikrobiologie **1:**620–652.

Cochrane, V. W. 1947. Acid production from glucose in the genus *Actinomyces.* Proceedings of the Society of American Bacteriologists Meetings **1947:**29.

Colmer, A. R., McCoy, E. 1950. Some morphological and cultural studies on lake strains of Micromonosporae. Transactions of the Wisconsin Academy of Sciences, Arts and Letters **40:**49–70.

Conn, H. J. 1916. A possible function of actinomycetes in soil. Journal of Bacteriology **1:**197–207.

Corbaz, R. 1964. Etude des streptomycetes provoquant la gale commune de la pomme de terre. Pytopathologische Zeitschrift **51:**351–360.

Corbaz, R., Gregory, P. H., Lacey, M. E. 1963. Thermophilic and mesophilic actinomycetes in mouldy hay. Journal of General Microbiology **32:**449–456.

Corbaz, R., Ettlinger, L., Gäumann, E., Keller-Schierlein, W., Neipp, L., Prelog, V., Reusser, P., Zähner, H. 1955. Stoffwechselprodukte von Actinomyceten. II. Mitteilung: Angolamycin. Helvetica Chimica Acta **38:**1202–1209.

Corke, C. T., Chase, F. E. 1956. The selective enumeration of actinomycetes in the presence of large numbers of fungi. Canadian Journal of Microbiology **2:**12–16.

Corke, C. T. Chase, F. E. 1964. Comparative studies of actinomycete populations in acid podzolic and neutral mull forest soils. Proceedings of Soil Science Society of America **28:**68–70.

Craveri, R., Pagani, H. 1962. Thermophilic micro-organisms among actinomycetes in the soil. Annali di Microbiologia **12:**115–130.

Crook, P., Carpenter, C. C., Klens, P. F. 1950. The use of sodium propionate in isolating actinomycetes from soils. Science **111:**656.

Cross, T. 1968. Thermophilic actinomycetes. Journal of Applied Bacteriology **31:**36–53.

Cross, T., Attwell, R. W., Locci, R. 1973. Fine structure of the spore sheath in *Streptoverticillium* species. Journal of General Microbiology **75:**421–424.

Cross, T., Goodfellow, M. 1973. Taxonomy and classification of the actinomycetes, pp. 11–112. In: Sykes, G., Skinner, F. A. (eds.), Actinomycetales: Characteristics and practical importance. London, New York: Academic Press.

Cross, T., Lechevalier, M. P., Lechevalier, H. 1963. A new genus of the Actinomycetales: *Microellobosporia* gen. nov. Journal of General Microbiology **31:**421–429.

Cummins, C. S., Harris, H. 1956. A comparison of cell-wall composition in *Nocardia, Actinomyces, Mycobacterium* and *Propionibacterium.* Journal of General Microbiology **15:**ix–x.

Cummins, C. S., Harris, H. 1958. Studies on cell-wall composition and taxonomy of Actinomycetales and related groups. Journal of General Microbiology **18:**173–189.

Davies, F. L. Williams, S. T. 1970. Studies on the ecology of actinomycetes in soil. I. The occurrence and distribution of actinomycetes in a pine forest soil. Soil Biology Biochemistry **2:**227–238.

Delafield, F. P., Doudoroff, M., Palleroni, N. J., Lusty, C. J., Contolpoulos, R. 1965. Decomposition of poly-β-hydoxybutyrate by pseudomonads. Journal of Bacteriology **90:**1455–1466.

Desai, A. J., Dhala, S. A. 1967. *Streptomyces thermonitrificans* sp. n., a thermophilic streptomycete. Antonie van Leeuwenhoek Journal of Microbiology and Serology **33**:137–144.

Dietz, A., Mathews, J. 1969. Scanning electron microscopy of selected members of the *Streptomyces hygroscopicus* group. Applied Microbiology **18**:694–696.

Dietz, A., Mathews, J. 1971. Classification of *Streptomyces* spore surfaces into five groups. Applied Microbiology **21**:527–533.

Dowding, J. E., Hopwood, D. A. 1973. Temperate bacteriophages for *Streptomyces coelicolor* A3(2) isolated from soil. Journal of General Microbiology **78**:349–359.

Dulaney, E. L., Larson, A. H., Stapley, E. O. 1955. A note on the isolation of microorganisms from natural sources. Mycologia **47**:420–422.

El-Nakeeb, M. A., Lechevalier, H. A. 1963. Selective isolation of aerobic actinomycetes. Applied Microbiology **11**:75–77.

Ettlinger, L., Corbaz, R., Hütter, R. 1958. Zur Systematik der Actinomyceten. 4. Eine Arteinteilung der Gattung *Streptomyces* Waksman et Henrici. Archiv für Mikrobiologie **31**:326–358.

Fergus, C. L. 1964. Thermophilic and thermotolerant molds and actinomycetes of mushroom compost during peak heating. Mycologia **56**:267–284.

Festenstein, G. N., Lacey, J., Skinner, F. A., Jenkins, P. A., Pepys, J. 1965. Self-heating of hay and grain in Dewar flasks and the development of farmer's lung antigen. Journal of General Microbiology **41**:389–407.

Flaig, W., Beutelspacher, H., Küster, E., Segler-Holzweissig, G. 1952. Beitrag zur Physiologie und Morphologie der Streptomyceten. Plant and Soil **4**:118–127.

Flaig, W., Küster, E., Beutelspacher, H. 1955. Elektronenmikroskopische Untersuchungen an Sporen verschiedener Streptomyceten. Zentralblatt für Bakteriologie, Parasitenkunde, Infektionskrankheiten und Hygiene, Abt. 2 **108**:376–382.

Flaig, W., Kutzner, H. J. 1954. Zur Systematik der Gattung *Streptomyces*. Naturwissenschaften **41**:287.

Flaig, W., Kutzner, H. J., 1960a. Beitrag zur Systematik der Gattung *Streptomyces* Waksman et Henrici. Archiv für Mikrobiologie **35**:105–138.

Flaig, W., Kutzner, H. J., 1960b. Beitrag zur Ökologie der Gattung *Streptomyces* Waksman et Henrici. Archiv für Mikrobiologie **35**:207–228.

Frommer, W. 1956. Erfahrungen mit Streptomyceten-Dauerkulturen. Archiv für Mikrobiologie **25**:219–222.

Ganju, P. L., Iyengar, M. R. S. 1974. Micromorphology of some sclerotial actinomycetes and development of their sclerotia. Journal of General Microbiology **82**:25–48.

Gauze, G. F. 1958. Zur Klassifizierung der Actinomyceten. Jena: Gustav Fischer Verlag.

Glathe, H. 1955. Die direkte mikropische Untersuchung des Bodens. Zeitschrift für Pflanzenernährung **69**:172–176.

Glathe, H., von Bernstorff, C., Arnold, A. 1954. Lebensgemeinschaft von Mikroorganismen und höheren Pflanzen im Bereich der Rhizosphäre. Zentralblatt für Bakteriologie, Parasitenkunde, Infektionskrankheiten und Hygiene, Abt. 2 **107**:481–488.

Glauert, A. M., Hopwood, D. A. 1960. The fine structure of *Streptomyces coelicolor*. I. The cytoplasmic membrane system. Journal of Biophysical and Biochemical Cytology **7**:479–488.

Goodfellow, M., Alderson, G. 1979. Numerical taxonomy of *Actinomadura* and related actinomycetes. Journal of General Microbiology **112**:95–111.

Gordon, M. A. 1974. Aerobic pathogenic Actinomycetaceae, pp. 175–188. In: Lennette, E. H., Spaulding, E. H., Truant, J. P. (eds.), Manual of clinical microbiology, 2nd ed. Washington, D.C.: American Society for Microbiology.

Gordon, R. E. 1967. The taxonomy of soil bacteria, pp. 293–321. In: Gray, T. R. G., Parkinson, B. (eds.), Ecology of soil bacteria. Liverpool: Liverpool University Press.

Gordon, R. E., Horan, A. C. 1968a. A piecemeal description of *Streptomyces griseus* (Krainsky) Waksman and Henrici. Journal of General Microbiology **50**:223–233.

Gordon, R. E., Horan, A. C. 1968b. *Nocardia dassonvillei,* a macroscopic replica of *Streptomyces griseus*. Journal of General Microbiology **50**:235–240.

Gregory, K. F., Huang, J. C. C. 1964a. Tyrosinase inheritance in *Streptomyces scabies*. I. Genetic recombination. Journal of Bacteriology. **87**:1281–1286.

Gregory, K. F., Huang, J. C. C. 1964b. Tyrosinase inheritance in *Streptomyces scabies*. II. Induction of tyrosinase deficiency by acridine dyes. Journal of Bacteriology **87**:1287–1294.

Gregory, P. H., Lacey, M. E. 1962. Isolation of thermophilic actinomycetes. Nature **195**:95.

Gregory, P. H., Lacey, M. E. 1963. Mycological examination of dust from mouldy hay associated with farmer's lung disease. Journal of General Microbiology **30**:75–88.

Grein, A., Meyers, S. P. 1958. Growth characteristics and antibiotic production of actinomycetes isolated from littoral sediments and materials suspended in sea water. Journal of Bacteriology **76**:457–463.

Gyllenberg, H. G. 1976. Application of automation to the identification of streptomycetes, pp. 299–321. In: Arai, T. (ed.), Actinomycetes: The boundary microorganisms. Tokyo, Singapore: Toppan.

Habermehl, G., Christ, B. G., Kutzner, H. J. 1977. Isolierung, Auftrennung und Strukturaufklärung des blauen Bakterienpigments "Amylocyanin" aus *Streptomyces coelicolor* Müller. Zeitschrift für Naturforschung **32b**:1195–1203.

Hammann, R. 1977. Untersuchungen zur Physiologie von Actinomyceten: Bildung organischer Säuren und flüchtiger Neutralprodukte, Abbau aromatischer Verbindungen. Dissertation, TH Darmstadt.

Harrison, M. D. 1962. Potato russet scab, its cause and factors affecting its development. American Potato Journal **39**:368–387.

Haynes, W. C., Wickerham, L. J., Hesseltine, C. W. 1955. Maintenance of cultures of industrially important microorganisms. Applied Microbiology **3**:361–368.

Henssen, A. 1957a. Beiträge zur Morphologie und Systematik der thermophilen Actinomyceten. Archiv für Mikrobiologie **26**:373–414.

Henssen, A. 1957b. Über die Bedeutung der thermophilen Mikroorganismen für die Zersetzung des Stallmistes. Archiv für Mikrobiologie **27**:63–81.

Henssen, A., Schnepf, E. 1967. Zur Kenntnis thermophiler Actinomyceten. Archiv für Mikrobiologie **57**:214–231.

Hesseltine, C. W., Benedict, R. G., Pridham, T. G. 1954. Useful criteria for species differentiation in the genus *Streptomyces*. Annals of the New York Academy of Sciences **60**:136–151.

Heymer, T. 1957. Über das Vorkommen von *Streptomyces coelicolor* auf der menschlichen Haut und Schleimhaut und seine fungistatische Wirkung. Archiv für Klinische und Experimentelle Dermatologie **205**:212–218.

Hill, L. R., Silvestri, L. G. 1962. Quantitative methods in the systematics of Actinomycetales. III. The taxonomic significance of physiological-biochemical characters and the construction of a diagnostic key. Giornale di Microbiologia **10**:1–28.

Hirata, S. 1959. Studies on the phytohormone in the malformed portion of the diseased plants 10. Bulletin Miyazaki **5**:85–92.

Hirsch, P. 1960. Einige weitere, von Luftverunreinigungen lebende Actinomyceten und ihre Klassifizierung. Archiv für Mikrobiologie **35**:391–414.

Hoffmann, G. M. 1954. Beiträge zur physiologischen Spezialisierung des Erregers des Kartoffelschorfes *Streptomyces scabies* (Thaxt.) Waksman and Henrici. Phytopathologische Zeitschrift **21**:221–278.

Hoffmann, G. M. 1958. Untersuchungen zur Aetiologie pflanzlicher Actinomykosen. Phytopathologische Zeitschrift **34**:1–56.

Hooker, W. J. 1949. Parasitic action of *Streptomyces scabies* on roots of seedlings. Phytopathology **39**:442–462.

Hooker, W. J., Kent, G. C. 1946. Infection studies with *Actinomyces scabies*. Phytopathology **36**:388–389.

Hopwood, D. A. 1967. Genetic analysis and genome structure in *Streptomyces coelicolor*. Bacteriological Reviews **31**:373–403.

Hopwood, D. A., Glauert, A. M. 1961. Electron microscope observations on the surface structures of *Streptomyces violaceoruber*. Journal of General Microbiology. **26**:325–330.

Hopwood, D. A., Wildermuth, H., Palmer, H. M. 1970. Mutants of *Streptomyces coelicolor* defective in sporulation. Journal of General Microbiology **61**:397–408.

Hora, T. S., Baker, R. 1974. Influence of a volatile inhibitor in natural or limed soil on fungal spore and seed germination. Soil Biology and Biochemistry **6**:257–261.

Hsu, S. C., Lockwood, J. L. 1975. Powdered chitin agar as a selective medium for enumeration of actinomycetes in water and soil. Applied Microbiology **29**:422–426.

Hütter, R. 1961. Zur Systematik der Actinomyceten. 5. Die Art *Streptomyces albus* (Rossi-Doria emend. Krainsky) Waksman et Henrici 1943. Archiv für Mikrobiologie **38**:367–383.

Hütter, R. 1962. Zur Systematik der Actinomyceten. 8. Quirlbildende Streptomyceten. Archiv für Mikrobiologie **43**:365–391.

Hütter, R. 1967. Systematik der Streptomyceten. Basel, New York: Karger.

Humm, J. H., Shepard, K. S. 1946. Three new agar-digesting actinomycetes. Duke University Marine Station Bulletin **3**:76–80.

Hussein, A. M. 1970. Pigments of the rubro-cyaneus group of actinomycetes, pp. 133–141. In: Prauser, H. (ed.), The Actinomycetales. The Jena International Symposium on Taxonomy, September 1968. Jena: Gustav Fischer Verlag.

Jagnow, G. 1956. Untersuchungen über die Verbreitung von Streptomyceten in Naturböden. Archiv für Mikrobiologie **25**:274–296.

Jagnow, G. 1957. Beiträge zur Ökologie der Streptomyceten. Archiv für Mikrobiologie **26**:175–191.

Jensen, H. L. 1928. *Actinomyces acidophilus* n.sp.—a group of acidophilus actinomycetes isolated from the soil. Soil Science **25**:225–236.

Jensen, H. L. 1930. Decomposition of keratin by soil microorganisms. Journal of Agricultural Science **20**:390–398.

Jensen, H. L. 1932. The microbiology of farmyard manure decomposition in soil. III. Decomposition of the cells of microorganisms. Journal of Agriculture Science **22**:1–25.

Jensen, H. L. 1943. Observations on the vegetative growth of actinomycetes in the soil. Proceedings of the Linnean Society of New South Wales **68**:67–71.

Jeuniaux, C. 1955. Production d'exochitinase par des *Streptomyces*. Comptes Rendus de Société Biologique **149**:1307–1308.

Jones, A. P. 1931. The histogeny of potato scab. Annales of Applied Biology **18**:313–333.

Kalakoutskii, L. V., Nikitina, E. T. 1977. Natural variability of developmental patterns in *Streptomyces*. Postepy Higieny I Medycyny Doswiadczalnej **31**:313–355.

Kaunat, H. 1963. Zum Problem der Spezifität der Rhizosphärenmikroflora von Kulturpflanzen. II. Wirkung der engen Rhizosphäre auf die Zahl der nichtsporenbildenden, sporenbildenden, anaeroben und oligonitrophilen Bakterien sowie Actinomyceten und Pilze. Zentralblatt für Bakteriologie II **117**:1–12.

Kawato, M., Shinobu, R. 1959. On *Streptomyces herbaricolor* nov. sp.; supplement: A simple technique for the microscopical observation. Memoirs of the Osaka University Liberal Arts of Education **8**:114.

KenKnight, G. 1941. Studies on soil actinomycetes in relation to potato scab and its control. Michigan Agricultural Experimental Station Technical Bulletin **178**.

KenKnight, G., Munzie, J. H. 1939. Isolation of phytopathogenic actinomycetes. Phytopathology **29**:1000–1001.

Khan, M. R., Williams, S. T. 1975. Studies on the ecology of actinomycetes in soil. VIII. Distribution and characteristics of acidophilic actinomycetes. Soil Biology and Biochemistry **7**:345–348.

Knösel, D. 1970. Untersuchungen zur cellulolytischen und pektolytischen Aktivität pflanzenschädlicher Actinomyceten. Phytopathologische Zeitschrift **67**:205–213.

Ko, W., Lockwood, J. L. 1970: Mechanism of lysis of fungal mycelia in soil. Phytopathology **60**:148–154.

Korn, F., Weingärtner, B., Kutzner, H. J. 1978. A study of twenty actinophages: Morphology, serological relationship and host range, pp. 251–270. In: Freerksen, E., Tarnok, I., Thumim, J. H. (eds.), Genetics of the Actinomycetales. Stuttgart, New York: Gustav Fischer Verlag.

Krassilnikov, N. A. 1970. Pigmentation of actinomycetes and its significance in taxonomy, pp. 123–131. In: Prauser, H. (ed.), The Actinomycetales. The Jena International Symposium on Taxonomy, September 1968. Jena: Gustav Fischer Verlag.

Kroppenstedt, R. M. 1977. Untersuchungen zur Chemotaxonomie der Ordnung Actinomycetales Buchanan 1917. Dissertation, TH Darmstadt.

Kroppenstedt, R. M., Kutzner, H. J. 1980. Fatty acids and classification of Actinomycetales, p. 388. In: Goodfellow, M., Board, R. G. (eds.)., Microbial classification and identification. London, New York: Academic Press.

Krzywy, T., Szerszén, L., Wieczorek, J. 1961. Occurrence of actinomycetes in Spitsbergen in relation to ecologic factors. Archiwum Immunologii Terapii Doświadczalnej **9**:253–260.

Küster, E. 1970. Note on the taxonomy and ecology of *Streptomyces malachiticus* and related species, pp. 169–172. In: Prauser, H. (ed.), The Actinomycetales. The Jena International Symposium on Taxonomy, September 1968. Jena: Gustav Fischer Verlag.

Küster, E. 1972. Simple working key for the classification and identification of named taxa included in the International Streptomyces Project. International Journal of Systematic Bacteriology **22**:139–148.

Küster, E. 1976. Ecology and predominance of soil streptomycetes, pp. 109–121. In: Arai, T. (ed.), Actinomycetes—the boundary microorganisms. Tokyo, Singapore: Toppan.

Küster, E., Locci, R. 1963. Studies on peat and peat microorganisms. I. Taxonomic studies on thermophilic actinomycetes isolated from peat. Archiv für Mikrobiologie **45**:188–197.

Küster, E., Williams, S. T. 1964. Selection of media for isolation of streptomycetes. Nature **202**:928–929.

Kurylowicz, W., Paszkiewicz, A., Woznicka, W., Kurzatkowski, W., Szulga, T. 1976. Numerical taxonomy of streptomycetes (ISP strains), pp. 323–340. In: Arai, T. (ed.), Actinomycetes: the boundary microorganisms. Tokyo, Singapore: Toppan.

Kutzner, H. J. 1956. Beitrag zur Systematik und Ökologie der Gattung Streptomyces Waksman et Henrici. Dissertation, Hohenheim.

Kutzner, H. J. 1961a. Effect of various factors on the efficiency of plating and plaque morphology of some *Streptomyces* phages. Pathologia et Microbiologia **24**:30–51.

Kutzner, H. J. 1961b. Specificity of actinophages within a selected group of *Streptomyces*. Pathologia et Microbiologia **24**:170–191.

Kutzner, H. J. 1967. Variabilität bei Streptomyzeten. Zentralblatt für Bakteriologie, Parasitenkunde, Infektionskrankheiten und Hygiene. Abt. 2 **121**:394–413.

Kutzner, H. J. 1968. Über die Bildung von Huminstoffen durch Streptomyceten. Landwirtschaftliche Forschung **21**:48–61.

Kutzner, H. J. 1972. Storage of streptomycetes in soft agar and other methods. Experientia **28**:1395.

Kutzner, H. J., Böttiger, V., Heitzer, R. D. 1978. The use of physiological criteria in the taxonomy of *Streptomyces* and *Streptoverticillium*, pp. 25–29. In: Mordarski, M., Kurylowicz, W., Jeljaszewicz, J. (eds.), *Nocardia* and *Streptomyces*. Proceedings of the International Symposium on *Nocardia* and *Streptomyces,* Warsaw, October 4–8, 1976. Stuttgart, New York: Gustav Fischer Verlag.

Kutzner, H. J., Waksman, S. A. 1959. *Streptomyces coelicolor* Müller and *Streptomyces violaceoruber* Waksman and Curtis, two distinctly different organisms. Journal of Bacteriology **78:**528–538.

Kutzner, H. J., Schlag, H., Christ, B., Habermehl, G. 1976 *Streptomyces coelicolor* Müller: Production of blue pigment, amylocyanin, and heptaene antibiotics, p. 235. 5th International Fermentation Symposium Berlin. Berlin: Verlag Versuchs- und Lehranstalt für Spiritusfabrikation und Fermentationstechnologie.

Kuznetsov, V. D., Kjagina, N. M., Semenov, I. V., Jangulova, I. V., Kobzeva, N. Ya. 1967. Soil cultivation as a method for restoration of taxonomic traits of *Actinomyces.* Mikrobiologiya **36:**510–517.

Labruyère, R. E. 1971. Common scab and its control in seed-potato crops. Agricultural Research Reports (Versl. landbouwk. Onderz.) **767.** Wageningen: Centre for Agricultural Publishing and Documentation.

Lacey, J. 1973. Actinomycetes in soils, composts and fodders, pp. 231–251. In: Sykes, G., Skinner, F. A. (eds.), Actinomycetales: Characteristics and practical importance. London, New York: Academic Press.

Lacey, J. 1978. Ecology of actinomycetes in fodders and related substrates, pp. 161–170. In: Mordarski, M., Kurylowicz, W., Jeljaszewicz, J. (eds.), *Nocardia* and *Streptomyces.* Proceedings of the International Symposium on *Nocardia* and *Streptomyces,* Warsaw, October 4–8, 1976. Stuttgart, New York: Gustav Fischer Verlag.

Lacey, J., Dutkiewicz, J. 1976a. Methods for examining the microflora of mouldy hay. Journal of Applied Bacteriology **41:**13–27.

Lacey, J., Dutkiewicz, J. 1976b. Isolation of actinomycetes and fungi from mouldy hay using a sedimentation chamber. Journal of Applied Bacteriology **41:**315–319.

Lacey, J., Goodfellow, M., Alderson, G. 1978. The genus *Actinomadura* Lechevalier and Lechevalier, pp. 107–117. In: Mordarski, M., Kurylowicz, W., Jeljaszewicz, J. (eds.), *Nocardia* and *Streptomyces.* Proceedings of the International Symposium on *Nocardia* and *Streptomyces.* Warsaw, October 4–8, 1976. Stuttgart, New York: Gustav Fischer Verlag.

Lapwood, D. H. 1973. *Streptomyces scabies* and potato scab disease, pp. 253–260. In: Sykes, G., Skinner, F. A. (eds.), Actinomycetales: Characteristics and practical importance. London, New York: Academic Press.

Lawrence, C. H. 1956. A method of isolating actinomycetes from scabby potato tissue and soil with minimal contamination. Canadian Journal of Botany **34:**44–47.

Lawrence, C. H., McAllan, J. W. 1964. Characteristics of an inhibitor from *Streptomyces scabies* affecting potato tuber respiration. Canadian Journal of Microbiology **10:**299–301.

Lechevalier, H. A., Corke, C. T. 1953. The replica plate method for screening antibiotic producing organisms. Applied Microbiology **1:**110–112.

Lechevalier, M. P., Lechevalier, H. A. 1970. Chemical composition as a criterion in the classification of aerobic actinomycetes. International Journal of Systematic Bacteriology **20:**435–443.

Lechevalier, H. A., Lechevalier, M. P., Gerber, N. N. 1971. Chemical composition as a criterion in the classification of actinomycetes. Advances in Applied Microbiology **14:** 47–72.

Lechevalier, M. P., Lechevalier, H. A., Heintz C. E. 1973. Morphological and chemical nature of the sclerotia of *Chainia olivacea,* Thirumalachar and Sukapure of the order *Actinomycetales.* International Journal of Systematic Bacteriology **23:**157–170.

Lindner, F., Wallhäusser, K. H. 1955. Die Arbeitsmethoden der Forschung zur Auffindung neuer Antibiotica. Archiv für Mikrobiologie **22:**219–234.

Lingappa, Y., Lockwood, J. L. 1961. A chitin medium for isola-

tion, growth and maintenance of actinomycetes. Nature **189:**158.

Lingappa, Y., Lockwood, J. L. 1962. Chitin media for selective isolation and culture of actinomycetes. Phytopathology **52:**317–323.

Lloyd, A. B. 1969. Behavior of streptomycetes in soil. Journal of General Microbiology **56:**165–170.

Lloyd, A. B., Lockwood, J. L. 1966. Lysis of fungal hyphae in soil and its possible relation to autolysis. Phytopathology **56:**592–602.

Lloyd, A. B., Noveroske, R. L., Lockwood, J. L. 1965. Lysis of fungal mycelium by *Streptomyces* species and their chitinase systems. Phytopathology **55:**871–875.

Locci, R., Baldacci, E., Petrolini Baldan, B. 1969. The genus *Streptoverticillium.* A taxonomic study. Giornale di Microbiologia **17:**1–60.

Lomovskaya, N. D., Emeljanova, L. K., Mkrtumian, L. K., Alikhanian, S. I. 1973. The prophage behaviour in crosses between lysogenic and non-lysogenic derivatives of *Streptomyces coelicolor* A3(2). Journal of General Microbiology **77:**455–463.

Lyons, A. J., Pridham, T. G. 1962. Proposal to designate strain ATCC 3004 (IMRU 3004) as the neotype strain of *Streptomyces albus* (Rossi-Doria) Waksman and Henrici. Journal of Bacteriology **83:**370–380.

Lyons, A. J., Pridham, T. G. 1971. *Streptomyces torulosus* sp.n. an unusual knobby-spored taxon. Applied Microbiology **22:**190–193.

Lyons, A. J., Pridham, T. G. 1973. Standard antimicrobial spectra as aids in characterization and identification of Actinomycetales. Developments in Industrial Microbiology **14:** 205–211.

Lyons, A. J., Pridham, T. G., Rogers, R. F. 1975. Actinomycetales from corn. Applied Microbiology **29:**246–249.

Mach, F. 1956. Untersuchungen über die Möglichkeiten einer Bekämpfung phytopathogener Pilze mit saprophytischer Bodenphase (Vermehrungspilze) durch Superinfektion mit antagonistisch aktiven *Streptomyces*-Stämmen. Zentralblatt für Bakteriologie, Parasitenkunde, Infektionskrankheiten und Hygiene, Abt. 2 **110:**1–25.

Mach, F., Futterlieb, A. 1966. Elektronenmikroskopische Untersuchungen über die Bildung von Sporen bei thermophilen Streptomyceten. Biologische Rundschau **4:**105–107.

McKee, R. K. 1958. Assessment of the resistance of potato varieties to common scab. European Potato Journal **1:**65–80.

Maiorova, V. I. 1966. Composition of polysaccharides in *Chainia.* Microbiology [English translation of Microbiologiya] **34:** 837–840.

Martin, J. P., Filip, Z., Haider, K. 1976. Effect of montmorillonite and humate on growth and metabolic activity of some actinomycetes. Soil Biology and Biochemistry **8:**409–413.

Matsumae, A., Ohtani, M., Takeshima, H., Hata, T. 1968. A new genus of the Actinomycetales: *Kitasatoa* gen. nov. Journal of Antibiotics **21:**616–625.

Mayfield, C. I., Williams, S. T., Ruddick, S. M., Hatfield, H. L. 1972. Studies on the ecology of actinomycetes in soil. IV. Observations on the form and growth of streptomycetes in soil. Soil Biology and Biochemistry **4:**79–91.

Menzies, J. D., Dade, C. E. 1959. A selective indicator medium for isolating *Streptomyces scabies* from potato tubers or soil. Phytopathology **49:**457–458.

Mikami, Y., Yokoyama, K., Arai, T. 1977. Modified Arai and Mikami melanin formation test of streptomycetes. International Journal of Systematic Bacteriology **27:**290.

Millard, W. A., Taylor, C. B. 1927. Antagonism of microorganisms as the controlling factor in the inhibition of scab by green-manuring. Annals of Applied Biology **14:**202–216.

Minnikin, D. E., Alshamaony, L., Goodfellow, M. 1975. Differentiation of *Mycobacterium, Nocardia,* and related taxa by thin-layer chromatographic analysis of whole-cell methanolysates. Journal of General Microbiology **88:**200–204.

Minnikin, D. E., Goodfellow, M., Alshamaony, L. 1978.

Mycolic acids in the classification of nocardioform bacteria, pp. 63–66. In: Mordarski, M., Kurylowicz, W., Jeljaszewicz, J. (eds.), *Nocardia* and *Streptomyces*. Proceedings of the International Symposium on *Nocardia* and *Streptomyces*. Warsaw, October 4–8, 1976. Stuttgart, New York: Gustav Fischer Verlag.

Misiek, M. 1955. Comparative studies of *Streptomyces* populations in soils. Dissertation, Syracuse.

Mitchell, R. 1963. Addition of fungal cell-wall components to soil for biological disease control. Phytopathology **53:** 1068–1071.

Mitchell, R., Alexander, M. 1962. Microbiological processes associated with the use of chitin for biological control. Proceedings of Soil Science Society of America **26:**556–558.

Moore, R. T., Chapman, G. B. 1959. Observations of the fine structure and modes of growth of a streptomycete. Journal of Bacteriology **78:**878–885.

Müller, R. 1908. Eine Diphteridee und eine Streptothrix mit gleichem blauen Farbstoff, sowie Untersuchungen über Streptothrix-arten im allgemeinen. Zentralblatt für Bakteriologie, Parasitenkunde und Infektionskrankheiten, Abt. 1 **46:**195–212.

Nette, I. T., Pomorzeva, N. V., Koslova, E. I. 1959. Destruction of caoutchouc by microorganisms. Mikrobiologiya **28:** 881–886.

Nishimura, H., Tawara, K. A. 1957. A method for microscopical observation of *Streptomyces* using agarcylinder culture. Journal of Antibiotics **10:**82.

Nissen, T. V. 1963. Distribution of antibiotic-producing actinomycetes in Danish soils. Experientia **19:**470–471.

Nitsch, B., Kutzner, H. J. 1969a. Decomposition of oxalic acid and other organic acids by streptomycetes as a taxonomic aid. Zeitschrift für Allgemeine Mikrobiologie **9:**613–632.

Nitsch, B., Kutzner, H. J. 1969b. Harnstoffzersetzung und Harnstoffverwertung durch Streptomyceten. Zentralblatt für Bakteriologie II **123:**380–398.

Nitsch, B., Kutzner, H. J. 1969c. Egg-yolk agar as a diagnostic medium for streptomycetes. Experientia **25:**220–221.

Nitsch, B., Kutzner, H. J. 1973. Wachstum von Streptomyceten in Schüttelagarkultur: Eine neue Methode zur Feststellung des C-Quellen-Spektrums. 3. Symposium Techn. Mikrobiologie, Berlin, pp. 481–486.

Nonomura, H. 1974. Key for classification and identification of 458 species of the streptomycetes included in ISP. Journal of Fermentation Technology **52:**78–92.

Nüesch, J. 1965. Isolierung und Selektionierung von Aktinomyceten. In: Symposium (Anreicherungskultur und Mutantenauslese" Göttingen, April 1964; Zentralblatt für Bakteriologie, Parasitenkunde, Infektionskrankheiten und Hygiene, Abt. 1, Suppl. **1:**234–252.

Okafor, N. 1966. The ecology of micro-organisms on, and the decomposition of, insect wings in the soil. Plant and Soil **25:**211–237.

Okami, Y., Okazaki, T. 1972. Studies on marine microorganisms. I. Journal of Antibiotics **25:**456–460.

Okami, Y., Okazaki, T. 1978. Actinomycetes in marine environments, pp. 145–151. In: Mordarski, M., Kurylowicz, W., Jeljaszewicz, J. (eds.), *Nocardia* and *Streptomyces*. Proceedings of the International Symposium on *Nocardia* and *Streptomyces*, Warsaw, October 4–8, 1976. Stuttgart, New York: Gustav Fischer Verlag.

Okami, Y., Suzuki, M. 1958. A simple method for microscopical observation for streptomycetes and critique of streptomyces grouping with references to aerial structure. Journal of Antibiotics **11:**250–253.

Okami, Y., Okazaki, T., Kitahara, T., Umezawa, H. 1976. Studies on marine microorganisms. V. A new antibiotic, aplasmomycin, produced by a streptomycete isolated from shallow sea mud. Journal of Antibiotics, Ser. A **29:** 1019–1025.

Okanishi, N., Akagawa, H., Umezawa, H. 1972. An evaluation of taxonomic criteria in streptomycetes on the basis of deoxy-ribonucleic acid homology. Journal of General Microbiology **72:**49–58.

Okazaki, T., Okami, Y. 1975. Actinomycetes tolerant to increased NaCl concentration and their metabolites. Journal of Fermentation Technology **53:**833–840.

Okazaki, T., Okami, Y. 1976. Studies on actinomycetes isolated from shallow sea and their antibiotic substances, pp. 123–161. In: Arai, T. (ed.), Actinomycetes—the boundary microorganisms. Baltimore, London, Tokyo: University Press.

Ottow, J. C. G. 1972. Rose bengal as a selective aid in the isolation of fungi and actinomycetes from natural sources. Mycologia **64:**304–315.

Parle, J. N. 1963a. Micro-organisms in the intestines of earthworms. Journal of General Microbiology **31:**1–11.

Parle, J. N. 1963b. A microbiological study of earthworm casts. Journal of General Microbiology **31:**13–22.

Person, L. H., Martin, W. J. 1940. Soil rot of sweet potatoes in Louisiana. Phytopathology **30:**913–926.

Pfennig, N. 1958. Beobachtungen des Wachstumsverhaltens von Streptomyceten auf Rossi-Cholodny-Aufwuchsplatten im Boden. Archiv für Mikrobiologie **31:**206–216.

Porter, J. N., Wilhelm, J. J. 1961. The effect on *Streptomyces* populations of adding various supplements to soil samples. Developments in Industrial Microbiology **2:**253–259.

Porter, J. N., Wilhelm, J. J., Tresner, H. D. 1960. Method for the preferential isolation of actinomycetes from soils. Applied Microbiology **8:**174–178.

Prauser, H. 1964. Aptness and application of colour codes for exact description of colours of streptomycetes. Zeitschrift für Allgemeine Mikrobiologie **13:**95–98.

Prauser, H. 1976. Host-phage relationships in nocardioform organisms, pp. 266–284. In: Goodfellow, M., Brownell, G. H., Serrano, J. A. (eds.), The biology of the Nocardiae. New York: Academic Press.

Prauser, H., Falta, R. 1968. Phagensensibilität, Zellwandzusammensetzung und Taxonomie von Actinomyceten. Zeitschrift für Allgemeine Mikrobiologie **8:**39–46.

Prauser, H., Müller, L., Falta, R. 1967. On the taxonomic position of the genus *Microellobosporia* Cross, Lechevalier and Lechevalier 1963. International Journal of Systematic Bacteriology **17:**361–366.

Preobrazhenskaya, T. P., Sveshnikova, M. A., Terekhova, L. P., Chormonova, N. T. 1978. Selective isolation of soil actinomycetes, pp. 119–123. In: Mordarski, M., Kurylowicz, W., Jeljaszewicz, J. (eds.), *Nocardia* and *Streptomyces*. Proceedings of the International Symposium on *Nocardia* and *Streptomyces*, Warsaw, October 4–8, 1976. Stuttgart, New York: Gustav Fischer Verlag.

Pridham, T. G. 1965. Color and streptomycetes. Report of an international workshop on determination of color of streptomycetes. Applied Microbiology **13:**43–61.

Pridham, T. G. 1976. Identification of streptomycetes and streptoverticillia at the species level: Revision of 1965 system, pp. 175–182. In: Arai, T. (ed.), Actinomycetes: The boundary microorganisms. Tokyo, Singapore: Toppan.

Pridham, T. G., Hesseltine, C. W., Benedict, R. G. 1958. A guide for the classification of streptomycetes according to selected groups. Placement of strains in morphological sections. Applied Microbiology **6:**52–79.

Pridham, T. G., Lyons, A. J. 1965. Further taxonomic studies on straight to flexuous streptomycetes. Journal of Bacteriology **89:**331–342.

Pridham, T. G. Anderson, P., Foley, C., Lindenfelser, H. A., Hesseltine, C. W., Benedict, R. G. 1956/57. A selection of media for maintenance and taxonomic study of *Streptomyces*. Antibiotics Annual, **1956/57:**947–953.

Pulverer, G., Schaal, K. P. 1978. Pathogenicity and medical importance of aerobic and anaerobic Actinomycetes, pp. 417–427. In: Mordarski, M., Kurylowicz, W., Jeljaszewicz, J. (eds.), *Nocardia* and *Streptomyces*. Proceedings of the International Symposium on *Nocardia* and *Streptomyces*,

Warsaw, October 4–8, 1976. Stuttgart, New York: Gustav Fischer Verlag.

Rancourt, M., Lechevalier, H. A. 1964. Electron microscopic study of the formation of spiny conidia in species of *Streptomyces*. Canadian Journal of Microbiology **10:** 311–316.

Řeháček, Z. 1959. Isolation of actinomycetes and determination of the number of their spores in soil. Mikrobiologiya **28:** 220–225.

Rehm, H.-J. 1953. Versuche zur Bekämpfung von Roggenfußkrankheiten (Fusariosen) durch Saatgutimpfung mit antibiotisch wirkenden Streptomyzeten. Zeitschrift für Pflanzenkrankheiten (Pflanzenpathologie) und Pflanzenschutz **60:** 549–560.

Rehm, H.-J. 1959a. Untersuchungen über das Verhalten von *Aspergillus niger* und einem *Streptomyces albus*-Stamm in Mischkultur. III. Mitteilung: Die Wechselbeziehungen im Erdboden. Zentralblatt für Bakteriologie, Parasitenkunde. Infektionskrankheiten und Hygiene, Abt. 2 **112:**235–263.

Rehm, H.-J. 1959b. Untersuchungen über das Verhalten von *Aspergillus niger* und einem *Streptomyces albus*-Stamm in Mischkultur. III. Mitteilung: Die Wechselbeziehungen am natürlichen Standort. Zentralblatt für Bakteriologie, Parasitenkunde, Infektionskrankheiten und Hygiene, Abt. 2 **112:**382–387.

Richards, O. W. 1943. The actinomyces of potato scab demonstrated by fluorescence microscopy. Stain Technology **18:** 91–94

Roach, A. W., Silvey, J. K. G. 1959. The occurrence of marine actinomycetes in Texas gulf coast substrates. American Midland Naturalist **62:**482–499.

Rouatt, J. W., Lechevalier, M., Waksman, S. A. 1951. Distribution of antagonistic properties among actinomycetes isolated from different soils. Antibiotics and Chemotherapy **1:** 185–192.

Routien, J. B., Finlay, A. C. 1952. Problems in the search for microorganisms producing antibiotics. Bacteriological Reviews **16:**51–67.

Rowbotham, T. J., Cross, C. 1977. *Rhodococcus coprophilus* sp. nov.: An aerobic nocardioform actinomycete belonging to the "rhodochrous" complex. Journal of General Microbiology **100:**123–138.

Ruddick, S. M., Williams, S. T. 1972. Studies on the ecology of actinomycetes in soil V. Some factors influencing the dispersal and adsorption of spores in soil. Soil Biology and Biochemistry **4:**93–103.

Ruschmann, G. 1953. Über Antibiosen und Symbiosen von Bodenorganismen und ihre Bedeutung für die Bodenfruchtbarkeit. Regenwurm-Symbiosen und -Antibiosen. Zeitschrift für Acker- und Pflanzenbau **96:**201–218.

Schleifer, K. H., Kandler, O. 1972. Peptidoglycan types of bacterial cell walls and their taxonomic implications. Bacteriological Reviews **36:**407–477.

Schütz, W., Felber, E. 1956. Welche Mikroorganismen spielen im Regenwurmdarm bei der Bildung von Bodenkrümeln eine Rolle? Zeitschrift für Acker- und Pflanzenbau **101:**471–476.

Seabury, J., Salvaggio, J., Buechner, H., Kundur, V. G. 1968. Bagassois. III. Isolation of thermophilic and mesophilic actinomycetes and fungi from moldy bagasse. Proceedings of the Society for Experimental Biology and Medicine **129:** 351–360.

Sevcik, V. 1963. Antibiotica aus Actinomyceten. Jena: Gustav Fischer Verlag.

Sharples, G. P., Williams, S. T. 1976. Development and fine structure of sclerotia and spores of the actinomycete *Chainia olivacea*. Microbios **15:**37–47.

Shinobu, R. 1958. Physiological and cultural study for the identification of soil actinomycetes species. Memoirs of the Osaka University of the Liberal Arts and Education. B. Natural Science **7:**1–76.

Shirling, E. B. 1959. The specificity of actinophages in reference to taxonomic use. Bacteriological Proceedings **1959:**32.

Shirling, E. B., Gottlieb, D. 1966. Methods for characterization of *Streptomyces* species. International Journal of Systematic Bacteriology **16:**313–340.

Shirling E. B., Gottlieb, D. 1968. Co-operative description of type cultures of *Streptomyces*. III. Additional species descriptions from first and second studies. International Journal of Systematic Bacteriology **18:**279–399.

Shirling E. B., Gottlieb, D. 1969. Co-operative descriptions of type cultures of *Streptomyces*. IV. Species descriptions from the second, third and fourth studies. International Journal of Systematic Bacteriology **19:**391–512.

Shirling, E. B., Gottlieb, D. 1976. Retrospective evaluation of International Streptomyces Project taxonomic criteria, pp. 9–41. In: Arai, T. (ed.), Actinomycetes—the boundary microorganisms. Baltimore, London, Tokyo: University Park Press.

Shoemaker, R. A., Riddell, R. T. 1954. Staining *Streptomyces scabies* in lesions of common scab of potato. Stain Technology **29:**59–61.

Silvestri, L., Turri, M., Hill, L. E., Gilardi, E. 1962. A quantitative approach to the systematics of actinomycetes based on overall similarity, pp. 333–360. In: Ainsworth, G. C., Sneath, P. H. (eds.), Microbial classification. 12th Symposium of the Society for General Microbiology. Cambridge: Cambridge University Press.

Skinner, F. A. 1951. A method for distinguishing between viable spores and mycelial fragments of actinomycetes in soils. Journal of General Microbiology **5:**159–166.

Skinner, F. A. 1956a. Inhibition of the growth of fungi by *Streptomyces* ssp. in relation to nutrient conditions. Journal of General Microbiology **14:**381–392.

Skinner, F. A. 1956b. The effect of adding clays to mixed cultures of *Streptomyces albidoflavus* and *Fusarium culmorum*. Journal of General Microbiology **14:**393–405.

Sneh, B., Henis, Y. 1972. Production of antifungal substances active against *Rhizoctonia solani* in chitin-amended soil. Phytopathology **62:**595–600.

Spicher, G. 1955. Untersuchungen über die Wirkung von Erdextrakt und Spuren-elementen auf das Wachstum verschiedener Streptomyzeten. Zentralblatt für Bakteriologie, Parasitenkunde, Infektionskrankheiten und Hygiene Abt. 2 **108:**577–587.

Spicher, G. 1972. Studien zur Frage der Hygiene des Getreides. Zentralblatt für Bakteriologie, Parasitenkunde, Infektionskrankheiten und Hygiene Abt. 2 **127:**61–81.

Stanier, R. Y. 1942. Agar-decomposing strains of the *Actinomyces coelicolor* species-group. Journal of Bacteriology **44:** 555–570.

Stuart, D. C., Jr. 1959. Fine structure of the nucleoid and internal membrane systems of *Streptomyces*. Journal of Bacteriology **78:**272–281.

Szabó, I., Marton, M. 1964. Zur Frage der spezifischen Bodenmikrofloren. Ein Versuch zur systematischen Bestimmung der Strahlenpilzflora einer mullartigen (Wald-) Rendzina. Zentralblatt für Bakteriologie, Parasitenkunde, Infektionskrankheiten und Hygiene, Abt. 2 **118:**265–306.

Szabó, I. M., Marton, M. 1976. Evaluation of criteria used in the ISP cooperative description of type strains of *Streptomyces* and *Streptoverticillium* species. International Journal of Systematic Bacteriology **26:**105–110.

Szabó, I., Marton, M., Partal, G. 1964. Micro-milieu studies in the A horizon of a mull-like rendzina, pp. 33–45. In: Jongerius, A. (ed.), Soil microbiology. London, New York: Elsevier.

Szabó, I., Marton, M., Szabolcs, I., Varga, L. 1959. Anpassung der Mikroflora und Mikrofauna an die Verhältnisse der Szikböden (Alkali-Böden) mit besonderer Berücksichtigung eines degradierten Solontschak-Solonetz-Bodens. Acta Agronomica Academiae Scientiarum Hungaricae **9:**9–39.

Szabó, I., Marton, M., Ferenczy, L., Buti, I. 1967. Intestinal microflora of the larvae of St. Mark's fly. II. Computer analysis of intestinal actinomycetes from the larvae of a *Bibio-*

population. Acta Microbiologica Academiae Scientiarum Hungaricae **14**:239–249.

Szabó, I. M., Marton, M., Buti, I., Fernandez, C. 1975. A diagnostic key for the identification of "species" of *Streptomyces* and *Streptoverticillium* included in the International Streptomyces Project. Acta Botanica Academiae Scientiarum Hungaricae **21**:387–418.

Szulga, T. 1978. A critical evaluation of taxometric procedures applied to *Streptomyces*, pp. 31–42. In: Mordarski, M., Kurylowicz, W., Jeljaszewicz, J. (eds.), *Nocardia* and *Streptomyces*. Proceedings of the International Symposium on *Nocardia* and *Streptomyces*, Warsaw, October 4–8, 1976. Stuttgart, New York: Gustav Fischer Verlag.

Taber, W. A. 1959. Identification of an alkaline-dependent *Streptomyces* as *Streptomyces caeruleus* Baldacci and characterization of the species under controlled conditions. Canadian Journal of Microbiology **5**:335–344.

Taber, W. A. 1960. Evidence for the existence of acid-sensitive actinomycetes in soil. Canadian Journal of Microbiology **6**:503–514.

Taylor, C. F. 1936. A method for isolation of actinomycetes from scab lesions on potato tubers and beet roots. Phytopathology **26**:287–288.

Tendler, M. D., Burkholder, P. R. 1961. Studies on the thermophilic actinomycetes. I. Methods of cultivation. Applied Microbiology **9**:394–399.

Thaxter, R. 1891. The potato scab. Connecticut Agricultural Experiment Station Report **1890**:81–85.

Thirumalachar, M. J. 1955. Chainia, a new genus of the Actinomycetales. Nature **176**:934–935.

Thirumalachar, M. J., Sukapure, R. S. 1964. Studies on species of the genus *Chainia* from India. Hindustan Antibiotica Bulletin **6**:157–166.

Toyama, H., Okanishi, H., Umezawa, H. 1974. Heterogeneity among whorl-forming streptomycetes determined by DNA reassociation. Journal of General Microbiology **80**:507–514.

Trejo, W. H. 1970. An evaluation of some concepts and criteria used in the speciation of streptomycetes. Transactions of the New York Academy of Sciences **32**:989–997.

Tresner, H. D., Backus, E. J. 1963. System of color wheels for streptomycete taxonomy. Applied Microbiology **11**:335–338.

Tresner, H. D., Davies, M. C., Backus, E. J. 1961. Electron microscopy of *Streptomyces* spore morphology and its role in species differentiation. Journal of Bacteriology **81**:70–80.

Tresner, H. D., Hayes, J. A., Backus, E. J. 1967. Morphology of submerged growth of streptomycetes as a taxonomic aid. I. Morphological development of *Streptomyces aureofaciens* in agitated liquid media. Applied Microbiology **15**:1185–1191.

Tresner, H. D. Hayes, J. A., Backus, E. J. 1968. Differential tolerance of streptomycetes to sodium chloride as a taxonomic acid. Applied Microbiology **16**:1134–1136.

Trolldenier, G. 1966. Über die Eignung Erde enthaltender Nährsubstrate zur Zählung und Isolierung von Bodenmikroorganismen auf Membranfiltern. Zentralblatt für Bakteriologie, Parasitenkunde, Infektionskrankheiten und Hygiene, Abt. 2 **120**:496–508.

Trolldenier, G. 1967. Isolierung und Zählung von Bodenactinomyceten auf Erdplatten mit Membranfiltern. Plant and Soil **27**:285–288.

Tsao, P. H., Leben, C., Keitt, G. W. 1960. An enrichment method for isolating actinomycetes that produce diffusible antifungal antibiotics. Phytopathology **50**:88–89.

Uridil, J. E., Tetrault, P. A. 1959. Isolation of thermophilic streptomycetes. Journal of Bacteriology **78**:243–246.

Veldkamp, H. 1955. A study of the aerobic decomposition of chitin by microorganisms. Medelingen van de Landbouwhogeschool te Wageningen/Nederland Wageningen: H. Veenman & Zonen **55**:127–174.

Vruggink, H., Maat, D. Z. 1968. Serological recognition of *Streptomyces* species causing scab on potato tubers. Netherlands Journal of Plant Pathology **74**:35–43.

Waksman, S. A. 1919. Cultural studies of species of *Actinomyces*. Soil Science **8**:71–215.

Waksman, S. A. 1961. The actinomycetes, vol. 2. Classification, identification and description of genera and species. Baltimore: Williams & Wilkins.

Waksman, S. A., Curtis, R. 1916. The actinomyces of the soil. Soil Science **1**:99–134.

Waksman, S. A., Henrici, A. T. 1943. The nomenclature and classification of the actinomycetes. Journal of Bacteriology **46**:337–341.

Waksman, S. A., Umbreit, W. W., Cordon, T. C. 1939. Thermophilic actinomycetes and fungi in soils and in composts. Soil Science **47**:37–61.

Warcup, J. H. 1950. The soil-plate method for isolation of fungi from soil. Nature **166**:117–118.

Watson, E. T., Williams, S. T. 1974. Studies on the ecology of actinomycetes in soil. VII. Actinomycetes in a coastal sand belt. Soil Biology and Biochemistry **6**:43–52.

Watson, R. D. 1960. Soil washing improves the value of the soil dilution and the plate count method of estimating populations of soil fungi. Phytopathology **50**:792–794.

Weinhold, A. R., Bowman, T. 1968. Selective inhibition of the potato scab pathogen by antagonistic bacteria and substrate influence on antibiotic production. Plant and Soil **28**:12–24.

Welsch, M. 1959. Lysogeny in streptomycetes. Annals of the New York Academy of Sciences **81**:974–993.

Welsch, M., Corbaz, R., Ettlinger, L. 1957. Phage typing of streptomycetes. Schweizerische Zeitschrift für Allgemeine Pathologie und Bakteriologie **20**:454–458.

Weyland, H. 1969. Actinomycetes in North Sea and Atlantic Ocean sediments. Nature **223**:858.

Whaley, J. W., Boyle, A. M. 1967. Antibiotic production by *Streptomyces* species from the rhizosphere of desert plants. Phytopathology **57**:347–351.

Wieringa, K. T. 1955. Der Abbau der Pektine; der erste Angriff der organischen Pflanzensubstanz. Zeitschrift für Pflanzenernährung **69**:150–155.

Wieringa, K. T. 1966. Solid media with elemental sulphur for detection of sulphur-oxidizing microbes. Antonie van Leeuwenhoek Journal of Microbiology and Serology **32**:183–186.

Wilde, P. 1964. Gezielte Methoden zur Isolierung antibiotisch wirksamer Boden-Actinomyceten. Zeitschrift für Pflanzenkrankheiten **71**:179–182.

Wildermuth, H. 1972. The surface structure of spores and aerial hyphae in *Streptomyces viridochromogenes*. Archiv für Mikrobiologie **81**:309–320.

Wilkin, G. D., Rhodes, A. 1955. Observations on the morphology of *Streptomyces griseus* in submerged culture. Journal of General Microbiology **12**:259–264.

Williams, S. T. 1967. Sensitivity of streptomycetes to antibiotics as a taxonomic character. Journal of General Microbiology **46**:151–160.

Williams, S. T. 1978. Streptomycetes in the soil ecosystem, pp. 137–144. In: Mordarski, M., Kurylowicz, W., Jeljaszewicz, J. (eds.). *Nocardia* and *Streptomyces*. Proceedings of the International Symposium on *Nocardia* and *Streptomyces*, Warsaw, October 4–8, 1976. Stuttgart, New York: Gustav Fischer Verlag.

Williams, S. T., Cross, T. 1971. Actinomycetes, pp. 295–334. In: Norris, J. R., Ribbons, D. W. (eds.), Methods in microbiology, vol. 4. London, New York: Academic Press.

Williams, S. T., Davies, F. L. 1965. Use of antibiotics for selective isolation and enumeration of actinomycetes in soil. Journal of General Microbiology **38**:251–261.

Williams, S. T., Davies, F. L., Hall, D. M. 1969. A practical approach to the taxonomy of actinomycetes isolated from soil. In: Sheals, J. G. (ed.), The soil ecosystem. London: The Systematics Association **8**:107–117.

Williams, S. T., Khan, M. R. 1974. Antibiotics—a soil microbiologist's viewpoint. Postepy Higieny I Medycyny Doswiadczalnej **28**:395–408.

Williams, S. T., Mayfield, C. I. 1971. Studies on the ecology of actinomycetes in soil. III. The behaviour of neutrophilic streptomycetes in acid soil. Soil Biology and Biochemistry **3**:197–208.

Williams, S. T., Parkinson, D., Burges, N. A. 1965. An examination of the soil washing technique by its application to several soils. Plant and Soil **22**:167–186.

Williams, S. T., Sharples, G. P. 1970. A comparative study of spore formation in two *Streptomyces* species. Microbios **5**:17–26.

Williams, S. T., Wellington, E. M. H. 1980. Micromorphology and fine structure of actinomycetes, pp. 139–165. In: Goodfellow, M., Board, R. G. (eds.), Microbial classification and identification. London, New York: Academic Press.

Williams, S. T., Davies, F. L., Mayfield, C. I., Khan, M. R. 1971. Studies on the ecology of actinomycetes in soil II. The pH requirements of streptomycetes from two acid soils. Soil Biology and Biochemistry **3**:187–195.

Williams, S. T., Bradshaw, R. M., Colsterton, J. W., Forge, A. 1972a. Fine structure of the sheath of some *Streptomyces* species. Journal of General Microbiology **72**:249–258.

Williams, S. T., Shameemullah, M., Watson, E. T., Mayfield, C. I. 1972b. Studies on the ecology of actinomycetes in soil VI. The influence of moisture tension on growth and survival. Soil Biology and Biochemistry **4**:215–225.

Yamaguchi, T. 1965. Comparison of the cell-wall composition of morphologically distinct actinomycetes. Journal of Bacteriology **89**:444–453.

Yamaguchi, T. 1967. Similarity in DNA of various morphologically distinct actinomycetes. Journal of General and Applied Microbiology **13**:63–71.

Zähner, H., Ettlinger, L. 1957. Zur Systematik der Actinomyceten. 3. Die Verwertung verschiedener Kohlenstoffquellen als Hilfsmittel der Artbestimmung innerhalb der Gattung *Streptomyces*. Archiv für Mikrobiologie **26**:307–328.

Ziegler, P., Kutzner, H. J. 1973a. Säurebildung durch Streptomyceten auf Agarmedium und in Schüttelkultur sowie gaschromatographische Identifizierung der gebildeten Säuren. 3. Symposium Techn. Mikrobiologie, Berlin, pp. 475–480.

Ziegler, P., Kutzner, H. J. 1973b. Hippurate hydrolysis as a taxonomic criterion in the genus *Streptomyces* (order Actinomycetales). Zeitschrift für Allgemeine Mikrobiologie **13**:263–270.

CHAPTER 157

The Monosporic Actinomycetes

THOMAS CROSS

Actinomycetes with single spores are very common in soil. They include members of the mesophilic genus *Micromonospora* which actively degrade complex polymers such as cellulose and chitin in soil. *Micromonospora* contains species that are able to produce valuable antibiotics such as gentamycin. In contrast, the thermophilic genera such as *Thermoactinomyces* and *Saccharomonospora* probably survive as dormant spores in temperate soils, but are more active in high-temperature composts or stores of fodder and grains. The thermophilic species have been studied in more recent years because they have been implicated in hypersensitivity diseases such as farmer's lung and bagassosis. They can be classified into the four genera shown in Table 1.

The genus *Thermoactinomyces* and its type species *Thermoactinomyces vulgaris* were first described in 1899 by Tsiklinsky, who noticed that the spores would not stain with the Gram stain and survived for at least 20 min at 100°C. The several species now known are mainly thermophilic aerobes and characteristically have single spores on both substrate and aerial hyphae. The presence of single spores caused the type species to be erroneously classified with the mesophilic micromonosporas for many years (Corbaz, Gregory, and Lacey, 1963; Erikson, 1952; Krassilnikov, 1941; Waksman, Umbreit, and Cordon, 1939). The confusion was not resolved until it was shown that the spores of *Thermoactinomyces* are typical endospores and quite different from the spores produced by other actinomycetes (Cross, Walker, and Gould, 1968; Dorokhova et al., 1968).

The generic name *Micromonospora* was suggested by Ørskov (1923) for his group III actinomycetes (ray fungi). This group contained only the single species, *Streptothrix chalcea*, isolated from the air and named by Foulerton (1905). According to Ørskov, this strain formed deep cinnabar red colonies on nutrient peptone agar and soon assumed a brownish black color on other media. No aerial mycelium was formed and the spores were borne singly at the tip of short side branches on the substrate mycelium. Ørskov regarded this species as distinct within his collection of actinomycetes; it was characterized by the formation of single spores and the lack of aerial hyphae.

In 1930 Jensen isolated and briefly described ten actinomycete strains from soil which he placed in the genus *Micromonospora* because of their resemblance to the description given by Ørskov. In a later paper, Jensen (1932) studied a larger group of strains from soil together with *Micromonospora chalcea* (Foulerton) Ørskov, which was at that time available from Pribram's Mikrobiologische Sammlung, Vienna. Jensen used the generic name *Micromonospora* for actinomycetes that bear single spores on the vegetative (substrate) mycelium and lack aerial mycelium. He noted that species were common in soil and were mesophilic, growing best at temperatures in the range of 30–37°C. It was unfortunate that later workers included within this genus strains that produce single spores on the aerial mycelium (now placed in the genera *Thermoactinomyces*, *Thermomonospora*, and *Saccharomonospora*) and so created considerable confusion.

The generic name *Thermomonospora* was proposed by Henssen (1957) for certain thermophilic

Table 1. Classification of actinomycete genera that typically bear single spores (the monosporic actinomycetes).

	Aerial mycelium	Endospores	Wall type[a]	Whole cell sugar pattern[b]
Micromonospora	−	−	II	D
Thermoactinomyces	+	+	III	B or C
Thermomonospora	+	−	III	B or C
Saccharomonospora	+	−	IV	A

[a] Lechevalier and Lechevalier, 1976.

[b] Lechevalier, 1968.

actinomycete species she encountered in rotted cow and sheep dung. The strains isolated formed single spores on a white aerial mycelium, and Henssen placed weight on the presence of this aerial mycelium to justify their separation from the mesophilic monosporic genus *Micromonospora,* which characteristically lacks aerial mycelium. Unfortunately, Henssen was only able to isolate *Thermomonospora curvata* in pure culture and the limited descriptions of two other species, *T. fusca* and *T. lineata,* were confined to morphological observations of cultures growing in hanging-drop preparations or on nutrient salts agar in association with other bacteria.

The specific name *fusca* was originally given to a mesophilic *Micromonospora* species that lacks aerial mycelium and forms dense bunches or clusters of spores on the substrate mycelium (Jensen, 1932). The name was later applied to a common thermophilic actinomycete in high-temperature composts (Waksman, Umbreit, and Gordon, 1939) whose presence was inferred from slides buried in the compost and subsequently stained to reveal similar dense clusters of spores. The organisms were never isolated. When Henssen later saw thermophilic organisms that exhibited a similar morphology, she retained the specific name *fusca* but placed them in the new genus *Thermomonospora* because they only formed spores on the aerial mycelium.

Monosporic, thermophilic actinomycetes with a greyish green aerial mycelium on colonies with a dark green reverse color and producing a dark green soluble pigment have been found in a wide variety of habitats. They have been given many names, including *Thermoactinomyces monosporus* (Schutze, 1908) Waksman and Corke 1953, *Thermoactinomyces viridis* Schuurmans, Olson, and San Clemente, 1956, and *Thermomonospora viridis* Küster and Locci 1963, but are now classified in the genus *Saccharomonospora* Nonomura and Ohara 1971. The spores, borne on the aerial mycelium only, do not have an endospore structure and so cannot be classified within the genus *Thermoactinomyces.* *Saccharomonospora viridis,* the only species, has a type IV wall that contains *meso*-diaminopimelic acid (Becker, Lechevalier, and Lechevalier, 1965) together with arabinose and galactose; the monosporic *Thermomonospora* species, in contrast, possess a type III wall. *Saccharomonospora viridis* has been implicated as one of the organisms that causes the hypersensitivity disease, farmer's lung.

Habitats

Thermoactinomyces

Most thermoactinomycetes grow at temperatures between 30°C and 60°C with an optimum of about 50°C. Consequently, they grow naturally in high-temperature habitats such as piles of plant debris and litter heated by the sun (Lacey, 1973). Man has provided many other suitable high-temperature habitats, such as bales or stacks of damp hay, compost heaps, bales of bagasse (crushed sugar cane), and stores of moist grain (Lacey, 1971a, 1974). Self-heating, mainly caused by microbial growth, gives the high temperatures required by the thermophilic actinomycetes. In conditions where there is a good supply of oxygen and a pH around neutral, species such as *Thermoactinomyces vulgaris* and *T. thalpophilus* (Waksman and Corke, 1953) can rapidly achieve numbers greater than 10^7 spores per gram of substrate (Lacey, 1978). The endospores exhibit extreme longevity and it has been estimated that they will remain viable for hundreds of years in dry soil and well over 2,000 years in sediments of cold anaerobic lakes (Unsworth et al., 1977). The spores also survive passage through the animal gut so that herbivores feeding on contaminated fodder will excrete the spores which can contaminate milk (Falkowski, 1978; Kosmachev, 1963) and pasture soil. The organisms can therefore be isolated from most soils (Cross, 1968); the spores will be washed into streams and rivers to become deposited in lake and marine sediments (Cross and Johnston, 1971).

One unusual habitat recently described for *T. vulgaris* (syn. *T. candidus;* Kurup et al., 1975) is the humidifiers of certain types of air-conditioning systems. The spores are dispersed in the air of buildings and may cause hypersensitivity pneumonitis in sensitive individuals. Other hypersensitivity diseases, such as farmer's lung and bagassosis, are more common and can be caused by high concentrations of thermoactinomycetes in the air of enclosed spaces where "moldy" hay, grain, or bagasse has been shaken or disturbed to release the spores.

Micromonospora

The early findings of Jensen (1930, 1932), who showed that *Micromonospora* spp. were common in neutral and alkaline soils, have been confirmed by many workers. Spores are washed into streams (Rowbotham and Cross, 1977) and eventually into lakes, where they can be isolated in low numbers from the water and in very high numbers from the sediments (Colmer and McCoy, 1943, 1950; Erikson, 1941; Johnston and Cross, 1976; Potter and Baker, 1956; Umbreit and McCoy, 1941; Willoughby, 1969). River water stored in reservoirs shows a rapid decline in the number of associated streptomycetes, but the numbers of micromonosporas remain fairly constant, which suggests that the spores can remain viable longer in water (Burman, 1973). Numbers were reduced when the water destined for public supply was filtered through beds of sand, but then high numbers accumulated in the sand and may even have grown in the filter beds.

Micromonospora species have been isolated from coastal waters and proved to be the predominant actinomycete in beach sand (Watson and Williams, 1974). Members of the genus have also been isolated from deep marine sediments (Weyland, 1969) and in sediments from the White Sea (Solovieva, 1972) and Black Sea (Solovieva and Singal, 1972). We are still not sure, however, whether these "marine" strains originate from spores washed into the marine environment or are representatives of species able to grow in the sea and its sediments.

Although the common species isolated from water, soils, and sediments are typically aerobic, three anaerobic species were isolated from the alimentary tracts of termites or from the rumen of herbivores. These micromonosporas form acid fermentation products but their role in the host is not known.

Thermomonospora and Saccharomonospora

In recent years many strains belonging to the genus *Thermomonospora* have been isolated from a variety of high-temperature habitats, such as composts, moldy hay, bagasse, and grain (Cross and Lacey, 1970; Lacey 1973, 1978), from mushroom compost (Fergus, 1964; Lacey, 1973), and from soil (Nonomura and Ohara 1971, 1974), but their taxonomy is still the subject of much debate. They have not been implicated as pathogens of animals or plants, but certain species would appear to be important decomposers of cellulose and lignocellulose (Crawford, 1974, 1975; Crawford and McCoy, 1972; Crawford et al., 1973; Stutzenberger, 1971; Stutzenberger, Kaufman, and Lossin, 1970).

Saccharomonospora viridis, the only species in the genus, grows in natural, high-temperature habitats such as leaf litter, composts, and surface peat. High numbers can occasionally be found in stored grain and overheated hay (Lacey, 1978) or in mushroom composts (Fergus, 1964). Spores contaminate soil and can be washed into streams and lakes, and survive in lake muds (Johnston and Cross, 1976).

Isolation

Thermoactinomyces

The thermophilic species grow rapidly at incubation temperatures of 50°C to produce recognizable colonies covered with aerial mycelium in 24 h. However, the endospores in most natural substrates will be accompanied by thermophilic *Bacillus* species, other actinomycetes, and some fungi, which provide intense competition on the surface of isolation plates. Thermophilic thermoactinomycetes can grow

in the presence of novobiocin at 25 µg/ml which inhibits the growth of all associated bacteria capable of growth at 50°C; the fungi can be simply suppressed by adding 50 µg/ml cycloheximide or a polyene antibiotic such as nystatin or pimaricin. A combination of these antibiotics with suitable nutrients provides a highly selective and useful isolation medium. The CYC medium that contains novobiocin (Cross and Attwell, 1974) has been used successfully to isolate the thermophilic species from soil, dung, fodder, water samples, and sediments.

CYC Medium for Isolation of *Thermoactinomyces* (Cross and Attwell, 1974)

CYC medium contains per liter:

Czapek-Dox liquid medium powder (Oxoid)	33.4 g
Yeast extract (Difco)	2.0 g
Vitamin-free Casamino Acids (Difco)	6.0 g

The medium is adjusted to pH 7.2, autoclaved, and cooled before adding membrane-filtered solutions of novobiocin (Albamycin, Upjohn) and cycloheximide (Acti-dione, Koch-Light) to give final concentrations of 25 µg/ml and 50 µg/ml, respectively. Samples are homogenized and diluted in an aqueous diluent that contains gelatin, 0.5 g/liter. Suitable dilutions are spread on the surface of the solidified isolation agar and incubated at 50°C for 48 h.

Thermoactinomyces vulgaris colonies are small (2–3 mm in diameter) with a white aerial mycelium and cream reverse. *T. thalpophilus* and *T. putidus* (Unsworth, 1978) colonies are larger (>3 mm) with an off-white, grayish white, cream, or pale brown aerial mycelium and a pale brown to brown reverse. Both *T. thalpophilus* and *T. putidus* possess a tyrosinase, which produces melanin, a dark brown, soluble pigment, from L-tyrosine. They are distinctive on the CYC isolation medium overlaid with a thin layer of the same agar containing L-tyrosine at 5.0 g/liter; colonies of *T. vulgaris* remain white with a cream reverse but *T. thalpophilus* and *T. putidus* colonies are surrounded by a brown soluble pigment and have a dark brown reverse color. *T. sacchari* (Lacey, 1971b) colonies are thin and colorless to pale brown, with radial ridges and a very sparse to absent white aerial mycelium. The colonies of *T. dichotomicus* (Krassilnokov and Agre, 1964; Cross and Goodfellow, 1973) have a distinctive pale yellow aerial mycelium on the surface of lemon to orange yellow colonies. All but *T. dichotomicus* can be subcultured on CYC or Trypticase soy agar (BBL). *T. putidus* is extremely sensitive to NaCl and should be cultivated on agar that lacks salt. *T. dichotomicus* loses its vigor and ability to produce spores unless supplied with nutrients such as those present in the peptone corn medium of Agre (1964).

Growth Medium for *Thermoactinomyces dichotomicus* (Agre, 1964)

Weigh 50 g of split maize (crushed corn) into 1 liter of boiling water, steam 30 min, filter, and add the following:

Starch	10 g
NaCl	5 g
CaCl$_2$	0.5 g
Peptone (Oxoid)	5.0 g
Agar	20.0 g
pH 7.2	

Cultures can be incubated for up to 1 week at 55°C if they are sealed in a container, such as a desiccator, with some water.

The mesophilic species, *T. peptonophilus* (Nonomura and Ohara, 1971), produces endospores (Attwell, 1973) but is unusual in that it is sensitive to novobiocin and has complex nutritional requirements. The one record of its isolation is in the original paper that describes its discovery and properties but the information is very limited. Soil samples were subjected to dry heat at 100°C, diluted, and then spread on a "MGA-SE" agar, pH 7.4, for incubation at 40°C. *T. peptonophilus* grew within other actinomycete colonies from which it received certain nutrients. The organism could be subcultured on a glycerol-asparagine agar that contains B vitamins and 20–30 g/liter peptone when incubated for 10 days at 35°C. No reliable method for isolating this organism is known.

In order to isolate thermoactinomycetes from stream or lake waters, where numbers may be low, the samples can be first passed through a membrane filter and the membranes placed on the CYC selective agar (Al-Diwany, Unsworth, and Cross, 1978). The type of membrane filter used can have a significant effect on recovery efficiency; black Sartorius membranes (SM 13006) with a nominal pore size of 0.45 μm gave highest recoveries and the white colonies were easy to count. Numbers of thermoactinomycete spores in air can be very high (Lacey and Lacey, 1964) and they are best isolated using an Andersen sampler (Andersen, 1958). Spores in vegetable materials such as hay and bagasse have been diluted in air using a wind tunnel (Gregory, 1962; Gregory and Lacey, 1963) or a sedimentation chamber (Lacey and Dutkiewicz, 1976b) and then isolated on Andersen sampler plates, but the method is not as efficient as homogenization and dilution followed by plating on selective agar.

Micromonospora

The spores of the aerobic mesophilic species common in water, soil, and sediments exhibit a limited resistance to heat or chlorine, treatments which can be used to drastically reduce the numbers of associated vegetative bacteria. The spores are not as resistant as *Bacillus* endospores, but they will germinate and grow on defined agar media and on colloidal chitin agar, which support only limited growth of *Bacillus* species. Therefore, pretreatment of a water sample or soil and sediment dilutions by heating or with chlorine followed by incubation for 3 weeks at 28–30°C on a suitable isolation agar enables one to isolate and count *Micromonospora* species relatively easily.

Heat Treatment (Rowbotham and Cross, 1977; Sandrak, 1977)

Water samples (2 ml) or dilutions of soil and sediment suspensions are held at 70°C in a water bath for 10 min before plating. There is little reduction in counts if the heating period is extended to 30 min.

Chlorine Treatment (Burman, Oliver, and Stevens, 1969; Willoughby, 1969).

Water samples are first treated with 4 ppm ammonia followed by 2 ppm chlorine as sodium hypochlorite. After 30 min reaction time, the excess chlorine is neutralized with sodium thiosulfate, the correct amount of which is calculated from titration of a blank sample.

Colloidal Chitin Agar (Hsu and Lockwood, 1975)

Sixty grams of unbleached chitin were stirred with 300 ml of commercial bleach (5.25% [wt/vol] sodium hypochlorite) for 10 min. The mixture was poured into 5 liters of tap water and filtered onto coarse filter paper. The chitin was washed at least five times to remove the bleach and air-dried overnight.

Forty grams of bleached chitin were ground dry in a Waring blender and then dissolved in 400 ml concentrated HCl by stirring for 30–50 min. Undissolved material was removed by filtration through glass wool, and the chitin was precipitated as a colloidal suspension by adding it slowly to 2 liters of water at 5°C. The suspension was collected by filtration on a coarse filter paper and then washed by suspending it in about 5 liters of cold tap water and refiltering. The washing was repeated at least three times or until the pH of the suspension was above 3.5. The colloidal suspension was concentrated by allowing it to stand before pouring off the supernatant. This suspension could be stored in the refrigerator and its dry weight determined or it could be freeze-dried.

Colloidal chitin agar was prepared using sufficient dry material or wet suspension to give a final concentration of 4 g/liter with the following mineral salts:

K_2HPO_4	0.7 g
KH_2PO_4	0.3 g
$MgSO_4 \cdot 5H_2O$	0.5 g
$FeSO_4 \cdot 7H_2O$	0.01 g
$ZnSO_4$	0.001 g
$MnCl_2$	0.001 g
Agar	20 g

The final pH of the medium should be 7.0.

M3 Agar Medium (Rowbotham and Cross, 1977)

The agar contained in 1 liter distilled water:

KH_2PO_4	0.466 g
Na_2HPO_4	0.732 g
KNO_3	0.10 g
NaCl	0.29 g
$MgSO_4 \cdot 7H_2O$	0.10 g
$CaCO_3$	0.02 g
Sodium propionate	0.20 g
$FeSO_4 \cdot 7H_2O$	200 μg
$ZnSO_4 \cdot 7H_2O$	180 μg
$MnSO_4 \cdot 4H_2O$	20 μg
Agar	18.0 g
Cycloheximide	50 mg
Thiamine–HCl	4.0 mg
pH 7.0	

A solution containing cycloheximide and thiamine was sterilized by membrane filtration and added to the autoclaved and cooled agar medium to give the final concentrations specified.

Kadota's Cellulose Benzoate Medium (Sandrak, 1977)

The basal medium contained in 1 liter distilled water:

$NaNO_3$	0.5 g
K_2HPO_4	1.0 g
$MgSO_4 \cdot 7H_2O$	0.5 g
$FeSO_4 \cdot 7H_2O$	0.01 g
Sodium benzoate	20 g
Agar	20 g
pH 7.2	

One milliliter of the soil suspension or root washing was intimately mixed with 0.67 g of sterile cellulose powder, such as that used for thin-layer chromatography, and with 2 ml of liquid Kadota medium and the mixture was spread evenly on the surface of Kadota agar in a Petri dish. The cellulose layer was allowed to dry before the plates were incubated in a moist chamber for 25–30 days at 28°C.

Micromonospora colonies on the above media are initially yellow-brown; they become red, orange, blue-green, or purple and eventually turn dark brown or black when sporulation occurs. The surface of the colonies is often smooth and glistening, and concentric sporulation rings may be visible within the colony. Aerial mycelium is normally absent, though some strains may produce a bloom of sterile aerial hyphae after continued incubation. Colonies should be checked under the microscope for the presence of fine, branching hyphae, averaging 0.5 μm in diameter, around their margin and the characteristic single spores. Most strains will show clear zones around the colonies where digestion of cellulose or chitin has occurred on the appropriate medium.

The anaerobic *Micromonospora* species were isolated with the aid of the Hungate roll-tube technique.

Saccharomonospora viridis

There is no sure way of isolating this organism from a substrate. Associated organisms may prevent its appearance on the isolation plate as a result of competition, especially when numbers are low. Strains have been isolated as minute, white colonies on plates of collodial chitin agar incubated at 30°C and seeded with dilutions of soil, lake water, or stream and lake sediments. They are more frequently isolated from samples of compost, moldy hay, or grain, which have been shaken in a wind tunnel (Gregory and Lacey, 1963) and the spore cloud sampled with the aid of an Andersen sampler (Andersen, 1958). Wind tunnels are rarely available in microbiological laboratories, but spores can be suspended in air within a sedimentation chamber (Lacey and Dutkiewicz, 1976b) and then deposited on isolation plates within an Andersen sampler.

Isolation of Spore Suspensions (Lacey and Dutkiewicz, 1976ab)

The sedimentation chamber consisted of a cardboard drum of approximately 80 cm high and 60 cm in diameter lined with aluminum foil. The metal lid was removable and had a small, centrally mounted electric fan. It was supported on the drum on small corks interspersed with polyurethane foam that allowed the replacement of air removed during sampling. A vertical tube of 1.5 cm internal diameter was fixed into the lid to enable samples to be taken half-way down the drum. Samples of 10–15 g hay or compost were placed in a terylene net bag suspended from the lid, its base resting on the bottom. A thread passing through a hole in the lid enabled the bag to be raised about 30 cm and then dropped several times to dislodge spores. While this was done and for up to 5 min afterwards, the fan was switched on to mix and circulate the air to give a uniform suspension of spores. The sampling tube was connected by a short length of rubber tube to an Andersen sampler and the air sampled for 5 s with a suction of 25 liter/min after the spores had been allowed to settle for 1 h. Suitable isolation media included half-strength nutrient agar (Oxoid) or tryptone soya agar (Oxoid), prepared by using the powdered ingredients at half the recommended concentration with the addition of 1% (wt/vol) agar. Supplementation with 1.0% (wt/vol) lactose or 0.2% (wt/vol) casein hydrolysate (Oxoid) increased recovery from certain substrates and, when the species was accompanied by many bacteria, the addition of penicillin G 1 μg/ml and polymyxin 5 μg/ml assisted recovery. All media contained cycloheximide or nystatin, 50 μg/ml to inhibit the growth of fungi and were incubated at 40°C.

Artificial composts, prepared in the laboratory by incubating soiled leaves at 40°C in perforated plastic bags have proved to be very useful for enriching strains present in the soil.

Thermomonospora

Thermophilic *Thermomonospora* species can be isolated from composts and overheated vegetable materials using the sedimentation chamber, Andersen sampler, and media already detailed for the isolation of *Saccharomonospora*. Composts prepared in the laboratory by incubating a variety of leaves with a little soil in perforated plastic bags at 40°C have also proved to be suitable substrates. *Thermomonospora (Actinobifida) chromogena* has been isolated frequently and in high numbers from mushroom compost. Mesophilic species were isolated from soil after it had been dried at 100°C in an oven to reduce the number of associated bacteria.

Thermomonospora colonies on isolation plates resemble those of other actinomycete species with a white aerial mycelium, for example, species of *Streptomyces* and *Thermoactinomyces*. However, *Thermomonospora* colonies can be recognized by carefully searching isolation plates with a 40 × long-working-distance objective and looking for the characteristic single spores on the aerial mycelium. The morphology of pure cultures should be confirmed by observing slide cultures or inclined cover slip preparations as described by Williams and Cross (1971).

Henssen (1957) used an anaerobic enrichment method to isolate her original *Thermomonospora* strains, but no one seems to have compared the relative advantages of aerobic and anaerobic methods for isolating members of the genus. It has been sug-

gested that anaerobic incubation reduces the number of bacterial colonies that appear on the isolation plates.

Isolation of *Thermomonospora* (Henssen, 1957)

Enrichment and isolation cultures were incubated at 50°C in large, glass jars of 3-liter capacity with a tight-fitting lid and a rubber sealing ring. Oxygen was removed by placing a beaker of alkaline pyrogallol within the jar. Dung samples were taken, piece by piece, and either laid on a sterile moist disk of filter paper on the surface of nutrient salts agar, or introduced into tubes containing stock solutions I or II and strips of filter paper. Petri dishes and tubes were incubated in the large, glass jars at 50°C. After a few days, the filter-paper disks or strips were covered with actinomycete spores, which could then be removed and streaked onto fresh nutrient medium. Nutrient salts agar contains per liter of distilled water:

KH$_2$PO$_4$	0.2 g
MgSO$_4$ · 7H$_2$O	0.3 g
CaCO$_3$	0.2 g
FeSO$_4$ · 7H$_2$O	0.005 g
Yeast extract (Difco)	0.1 g
Casamino Acids (Difco)	0.1 g
Agar	15 g
Urea	0.25 g
or	
(NH$_4$)$_2$PO$_4$	0.5 g

Either urea or (NH$_4$)$_2$PO$_4$ serves as an alternative nitrogen source.

Stock solution I contains per liter of distilled water:

KH$_2$PO$_4$	0.2 g
MgSO$_4$ · 7H$_2$O	0.3 g
CaCO$_3$	0.2 g
FeSO$_4$ · 7H$_2$O	0.005 g
Yeast extract (Difco)	0.1 g
Casamino Acids (Difco)	0.1 g
NH$_4$NO$_3$	0.2 g
Casein (dissolved in 10 ml 1 N NaOH)	0.2 g
Asparagine	0.25 g

Stock solution II contains per liter of distilled water:

KH$_2$PO$_4$	0.2 g
MgSO$_4$ · 7H$_2$O	0.3 g
CaCO$_3$	0.2 g
FeSO$_4$ · 7H$_2$O	0.005 g
Yeast Extract (Difco)	0.1 g
Casamino Acids (Difco)	0.1 g
NH$_4$NO$_3$	0.3 g
Casein (dissolved in 10 ml 1 N NaOH)	0.15 g
Peptone	0.2 g

Identification

The most convenient method for observing the presence of single spores on the hyphae of the monosporic actinomycetes is to examine colonies, growing on the surface of agar in conventional Petri dishes, with the aid of a long-working-distance objective. We have routinely used a Vickers 40 × (NA 0.57) long-working-distance (12-mm), wide-angled objective lens (Vickers Instruments, York, England) and have found it to be ideal. With this lens it is possible to resolve the fine hyphae and single spores at the margin of *Micromonospora* colonies and the sporulating aerial mycelium on colonies of the thermophilic genera anywhere on an isolation plate without the problem of condensation on the front lens. More detailed observations can then be made with phase-contrast or oil-immersion objectives of slide and cover slip preparations described by Williams and Cross (1971). The inclined cover slip method can give particularly clear images of hyphae that bear phase-bright endospores if the cover slips bearing aerial hyphae are mounted in a drop of wetting agent, such as 0.5% (wt/vol) Triton X-100 (Union Carbide) in water.

The characteristic *Micromonospora* colonies, devoid of aerial mycelium, can be recognized with little difficulty. The endospores of *Thermoactinomyces* are phase bright but their colonies, covered with aerial mycelium, may be confused with those of *Thermomonospora* spp. whose spores can also appear refractile under the microscope. Where there is some doubt, it may be necessary to check that the spores are endospores and able to resist temperatures of 90°C and above (wet heat) or contain dipicolinic acid and high concentrations of calcium.

Thermoactinomyces

All species in the genus *Thermoactinomyces* have a cell wall of type III and lack the sugars arabinose, madurose, and xylose (Lechevalier, Lechevalier, and Gerber, 1971). Their spores have a typical endospore structure; they are phase bright, contain dipicolinic acid, and will survive for at least 10 min at 95°C. *T. peptonophilus* (Nonomura and Ohara, 1971) is the only mesophilic species so far described. Its growth temperature range is from 25 to 40°C and the colonies are covered with a white aerial mycelium which carries the sessile spores. Thermophilic species can be identified by the characters given in Table 2

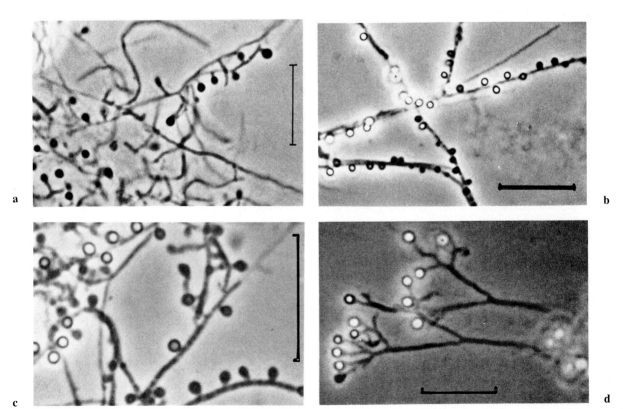

Fig. 1. Morphology of *Thermoactinomyces* species. (a) *T. vulgaris* aerial hyphae with phase-dark, immature endospores on short sporophores. (b) *T. thalpophilus* aerial hyphae bearing sessile, phase-bright, mature endospores. (c) *T. peptonophilus* aerial mycelium. (d) *T. dichotomicus* aerial mycelium with dichotomously branched sporophores terminating in phase-bright endospores. Bar = 10 μm.

Table 2. Identification of *Thermoactinomyces* species.

	AM color[a]	Sessile spores	Dichotomous branching	Amylase	Melanin[b]	Esculin[c]	Arbutin[d]
T. dichotomicus	Y	−	+	+	−	−	−
T. putidus	W	−	−	+	+	+	+
T. sacchari	(W)	−	−	+	−	+	+
T. vulgaris	W	−	−	−	−	+	+
T. thalpophilus	W	+	−	+	+	−	−

[a] Y, yellow; W, white; (W), white if present.
[b] Melanin production in CYC + 5.0 g/liter L-tyrosine.
[c] Esculin or arbutin degradation tested in trypticase soy agar + 1.0 g/liter esculin, or in arbutin + 0.5 g/liter ferric citrate scales.

Thermomonospora

All members of the genus should have a type III cell wall and produce heat-sensitive (killed at 70°C), single spores on the aerial or on both aerial and substrate mycelium. The morphologically similar genus, *Saccharomonospora*, has a type IV wall and the sole species, *S. viridis*, has characteristic green-brown colonies with a blue-green aerial mycelium when incubated at 45°C. The following thermophilic and mesophilic *Thermomonospora* species can now be recognized, but it must be emphasized that there is no sharp distinction between the mesophilic and thermophilic habit and that we are currently placing undue emphasis on growth temperature for identification.

I. THERMOPHILIC SPECIES ABLE TO GROW AT 50°C WITH AN OPTIMUM OF 40–45°C

T. curvata (Henssen, 1957; Henssen and Schnepf, 1967).

White aerial mycelium on colorless to pale yellow-brown colonies. Single spores on short, simple sporophores or on short, lateral sporophores which show limited branching. The spores can be densely clustered along the length of quite long aerial hyphae. Spores are rarely seen on the substrate hyphae.

T. alba (Locci, Baldacci, and Petrolini 1967) Cross and Goodfellow 1973. Synonym:
T. fusca Crawford 1975.

White aerial mycelium on colorless to pale yellow-brown colonies. The spores are usually seen in compact, terminal or lateral clusters on short sporophores which can show dichotomous branching. They are on short, branching aerial hyphae. Spore clusters can frequently be seen on the substrate mycelium at the surface or deep within the agar.

T. chromogena (Krassilnikov and Agre 1965) comb. nov. Synonyms:
Actinobifida chromogena Krassilnikov and Agre 1965; *T. fusca* (strains isolated by Fergus, 1964); *T. falcata* Henssen 1970.

Aerial mycelium at first white and later cinna-

mon, sandy, or pale brown in color. Colonies red-brown, brown to dark brown with a red-brown, soluble pigment diffusing into the agar medium. Spores in compact clusters on sporophores which show pseudodichotomous branching. Detailed examination can often show that spore clusters can result from repeated sporulation on a falcate sporophore.

II. MESOPHILIC SPECIES UNABLE TO GROW AT 50°C, BUT WITH AN OPTIMUM OF 35–40°C

T. mesophila Nonomura and Ohara 1971.

Aerial mycelium white to greyish yellowish brown on yellowish brown colonies. The single, oval to round spores are borne on short sporophores along the length of the aerial hyphae; no spores on substrate hyphae.

T. mesouviformis Nonomura and Ohara 1974.

Powdery white aerial mycelium on pale yellowish brown colonies. The oval spores are borne in clusters on the branched sporophores on both aerial and substrate hyphae.

III. OTHER SPECIES

T. fusca (Henssen 1957) considered *nomen dubium* because the original description was based on contaminated cultures and the epithet had previously been mistakenly applied to organisms seen in stained preparations but never isolated in pure culture (Waksman, Umbreit, and Gordon, 1939).

T. lineata (Henssen 1957) considered *nomen dubium* because the original description was based on contaminated cultures.

T. galeriensis (Szabo et al. 1976) considered *nomen dubium* because of lack of data on cell wall composition. The available description of the type strain shows some similarities to *Saccharomonospora viridis*.

Saccharomonospora viridis

Colonies on isolation plates incubated at 30°C for 3 weeks are small and colorless with a thin, white aerial mycelium. Single spores are often present on the aerial hyphae and the colonies may be mistaken

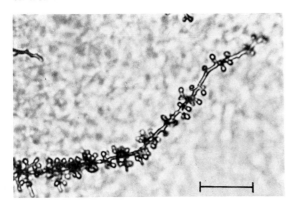

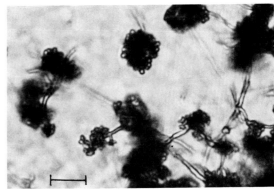

a

b

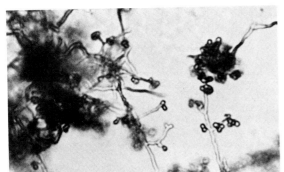

c

Fig. 2. Morphology of *Thermomonospora* species. (a) *T. curvata* aerial mycelium with single spores borne laterally along the length of the hyphae. (b) *T. alba*, exhibiting dense clusters of spores on the aerial mycelium. (c) *T. chromogena* spore clusters on the aerial hyphae. Bar = 10 μm.

for members of the genus *Thermomonospora*. If they are subcultured onto half-strength or full-strength nutrient agar or typtone soya agar, typical colonies appear when incubated at 40–50°C, optimum 45°C. Single spores (0.9–1.1 × 1.2–1.4 μm) are produced on the aerial mycelium only; they are either sessile on the branched hyphae or borne on short, lateral sporophores. The spores are heat sensitive, and are rapidly killed at 70°C. There have been occasional reports of two or more spores borne laterally on the aerial hyphae, but a single spore on a short, swollen, pyriform sporophore can look like a pair of spores. The colonies develop a dark green reverse color and are covered with a blue-green to gray-green aerial mycelium. Colonies often produce a dark green soluble pigment and some strains produce antibiotics, e.g., thermoviridin, which can result in a bacteria-free zone around the colonies on crowded isolation plates. Identification should be confirmed by cell-wall analysis.

Micromonospora

The classification of the species within the genus *Micromonospora* is in a very confused state and even the status of certain species is in doubt. There are several reasons for this state of affairs. Some of the organisms were named but only briefly described about 30 to 40 years ago, namely *M. parva* (Jensen,

1932), *M. gallica* (Erikson, 1935), *M. globosa* (Kriss, 1939), and *M. elongata* and *M. bicolor* (Krassilnikov, 1941) and no type or neotype cultures exist. A second reason is that the majority of new species were isolated and described in one laboratory in the United States (Luedemann, 1971a,b; Luedemann and Brodsky, 1964, 1965; Weinstein et al., 1968) or in another laboratory in the Soviet Union (Sveshnikova, Maximova, and Kudrina, 1969, 1970). The existence of each other's classification seems to have been ignored and little attempt was made to exchange and study their respective

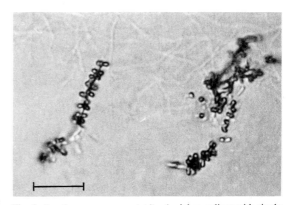

Fig. 3. *Saccharomonospora viridis*. Aerial mycelium with single spores borne laterally along the length of the hyphae. Bar = 10 μm.

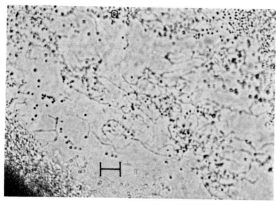
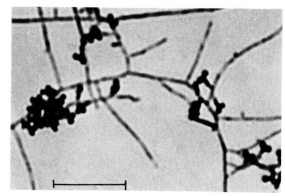

a   b

Fig. 4. Morphology of *Micromonospora*. (a) Fine substrate mycelium at the margin of a colony bearing scattered single spores. (b) Stained slide culture exhibiting spore clusters. Bar = 10 μm.

species. Finally, many thousands of strains have been isolated in the search for new antibiotics. Strains that produce useful antibiotics are jealously guarded and there is the temptation to give a new specific name to organisms that produce novel antibiotics (Kawamoto et al., 1974; Wagman et al., 1974). This genus of very common soil and aquatic organisms with an important role in the cycling of plant and animal remains is ripe for a detailed and comprehensive study.

In the meantime, microbiologists must attempt an identification from the descriptions and keys available (Cross and Goodfellow, 1973; Luedemann, 1974; Sveshnikova, Maximova, and Kudrina, 1969, 1970) and from the properties given for the two more recently described species, *M. inositola* (Kawamoto et al., 1974) and *M. rhodorangea* (Wagman et al., 1974).

There are three obligately anaerobic species in the genus which can be separated by their ability to utilize cellulose and the nature of their fermentation products. *M. acetoformici* (Sebald and Prévot, 1962) is unable to utilize cellulose but will ferment glucose and starch to give a mixture of acetic, formic, and lactic acids; *M. propionici* (Hungate, 1946) ferments glucose and cellulose to form propionic and acetic acids; and *M. ruminantium* (Maluszynska and Janota-Bassalik, 1974) ferments cellulose to give lactic and acetic acids.

Literature Cited

Agre, N. S. 1964. A method for the isolation and cultivation of thermophilic actinomycetes. [In Russian, with English summary.] Mikrobiologiya **33**:913–917.

Al-Diwany, L. J., Unsworth, B. A., Cross, T. 1978. A comparison of membrane filters for counting *Thermoactinomyces* endospores in spore suspensions and river water. Journal of Applied Bacteriology **45**:249–258.

Andersen, A. A. 1958. New sampler for the collection, sizing and enumeration of viable airborne particles. Journal of Bacteriology **76**:471–484.

Attwell, R. W. 1973. Actinomycete spores. Ph.D. Thesis. University of Bradford.

Becker, B., Lechevalier, M. P., Lechevalier, H. A. 1965. Chemical composition of cell wall preparations from strains of various form-genera of aerobic actinomycetes. Applied Microbiology **13**:236–243.

Burman, N. P. 1973. The occurrence and significance of actinomycetes in water supplies, pp. 219–330. In: Sykes, G., Skinner, F. A. (eds.), Actinomycetales: Characteristics and practical importance. London, New York: Academic Press.

Burman, N. P., Oliver, C. W., Stevens, J. K. 1969. Membrane filtration techniques for the isolation from water of coli-aerogenes, *Escherichia coli,* faecal streptococci, *Clostridium perfringens,* actinomycetes and microfungi, pp. 127–134. In: Shapton, D. A., Gould, G. W. (eds.), Isolation methods for microbiologists. London, New York: Academic Press.

Colmer, A. R., McCoy, E. 1943. *Micromonospora* in relation to some Wisconsin lakes and lake populations. Transactions of the Wisconsin Academy of Sciences, Arts and Letters **35**:187–220.

Colmer, A. R., McCoy, E. 1950. Some morphological and cultural studies on lake strains of Micromonosporae. Transactions of the Wisconsin Academy of Sciences, Arts and Letters **40**:49–70.

Corbaz, R., Gregory, P. H., Lacey, M. E. 1963. Thermophilic and mesophilic actinomycetes in mouldy hay. Journal of General Microbiology **32**:449–455.

Crawford, D. L. 1974. Growth of *Thermomonospora fusca* on lignocellulosic pulps of varying lignin content. Canadian Journal of Microbiology **20**:1069–1072.

Crawford, D. L. 1975. Cultural, morphological, and physiological characteristics of *Thermomonospora fusca* (strain 190Th). Canadian Journal of Microbiology **21**:1842–1848.

Crawford, D. L., McCoy, E. 1972. Cellulases of *Thermomonospora fusca* and *Streptomyces thermodiastaticus.* Applied Microbiology **24**:150–152.

Crawford, D. L., McCoy, E., Harkin, J. M., Jones, P. 1973. Production of microbial protein from waste cellulose by *Thermomonospora fusca,* a thermophilic actinomycete. Biotechnology and Bioengineering **15**:833–843.

Cross, T. 1968. Thermophilic actinomycetes. Journal of Applied Bacteriology **31**:36–53.

Cross, T., Attwell, R. W. 1974. Recovery of viable thermoactinomycete endospores from deep mud cores, pp. 11–20. In: Barker, A. N., Gould, G. W., Wolf, J. (eds.), Spore research 1973. London: Academic Press.

Cross, T., Goodfellow, M. 1973. Taxonomy and classification of actinomycetes, pp. 11–112. In: Sykes, G., Skinner, F. A.

(eds.), Actinomycetales: Characteristics and practical importance. London, New York: Academic Press.

Cross, T., Johnston, D. W. 1971. *Thermoactinomyces vulgaris.* II. Distribution in natural habitats, pp. 315–330. In: Barker, A. N., Gould, G. W., Wolf, J. (eds.), Spore research 1971. London: Academic Press.

Cross, T., Lacey, J. 1970. Studies on the genus *Thermomonospora,* pp. 211–219. In: Prauser, H. (ed.), The Actinomycetales. Jena: Gustav Fischer Verlag.

Cross, T., Walker, P. D., Gould, G. W. 1968. Thermophilic actinomycetes producing resistant endospores. Nature **220:**352–354.

Dorokhova, L. A., Agre, N. S., Kalakoutskii, L. V., Krassilnikov, N. A. 1968. Fine structure of spores in a thermophilic actinomycete, *Micromonospora vulgaris.* Journal of General and Applied Microbiology **14:**295–303.

Erikson, D. 1935. Pathogenic aerobic organisms of the actinomyces group. Medical Research Council (Great Britain) Special Report Series **203:**5–61.

Erikson, D. 1941. Studies on some lake-mud strains of *Micromonospora.* Journal of Bacteriology **41:**277–300.

Erikson, D. 1952. Temperature/growth relationships of a thermophilic actinomycete, *Micromonospora vulgaris.* Journal of General Microbiology **6:**286–294.

Falkowski, J. 1978. Occurrence and thermoresistance of thermophilic actinomycetes in milk. Zentralblatt für Bakteriologie, Parasitenkunde, Infektionskrankheiten und Hygiene, Abt. 1 Orig., Reihe B **167:**171–176.

Fergus, C. L. 1964. Thermophilic and thermotolerant molds and actinomycetes of mushroom compost during peak heating. Mycologia **56:**267–284.

Foulerton, A. G. 1905. New species of streptothrix isolated from the air. Lancet **i:**1200.

Gregory, P. H. 1962. Isolation of thermophilic actinomycetes. Nature **195:**95.

Gregory, P. H., Lacey, M. E. 1963. Mycological examination of dust from mouldy hay associated with farmer's lung disease. Journal of General Microbiology **30:**75–88.

Henssen, A. 1957. Beiträge zur Morphologie und Systematik der thermophilen Actinomyceten. Archiv für Mikrobiologie **26:**373–414.

Henssen, A. 1970. Spore formation in thermophilic actinomycetes, pp. 205–210. In: Prauser, H. (ed.), The Actinomycetales. Jena: Gustav Fischer Verlag.

Henssen, A. Schnepf, E. 1967. Zur Kenntnis thermophiler Actinomyceten. Archiv für Mikrobiologie **57:**214–231.

Hsu, S. C., Lockwood, J. L. 1975. Powdered chitin agar as a selective medium for enumeration of actinomycetes in water and soil. Applied Microbiology **29:**422–426.

Hungate, R. E. 1946. Studies on cellulose fermentation. II. An anaerobic cellulose decomposing actinomycete, *Micromonospora propionici* n. sp. Journal of Bacteriology **51:**51–56.

Jensen, H. L. 1930. The genus *Micromonospora* Ørskov—a little known group of soil micro-organisms. Proceedings of the Linnean Society New South Wales **55:**231–248.

Jensen, H. L. 1932. Contributions to our knowledge of Actinomycetales. III. Further observations on the genus *Micromonospora.* Proceedings of the Linnean Society New South Wales **57:**173–180.

Johnston, D. W., Cross, T. 1976. The occurrence and distribution of actinomycetes in lakes of the English Lake District. Freshwater Biology **6:**457–463.

Kawamoto, I., Okachi, R., Kato, H., Yamamoto, S., Takahashi, I., Takasawa, S., Nara, T. 1974. The antibiotic XK-41 complex. I. Production, isolation and characterization. Journal of Antibiotics **27:**493–501.

Kosmachev, A. E. 1963. Thermophilic actinomycetes in milk and dairy products. [In Russian, with English summary.] Mikrobiologiya **32:**136–142.

Krassilnikov, N. A. 1941. Guide to the bacteria and actinomycetes pp. 1–830. [In Russian.] Moscow: Akademii Nauk SSSR.

Krassilnikov, N. A., Agre, N. S. 1964. A new genus of the Actinomycetes *Actinobifida* n. gen. The yellow group *Actinobifida dichotomica* n. sp. [In Russian, with English summary.] Mikrobiologiya **33:**935–943.

Krassilnikov, N. A., Agre, N. S. 1965. The brown group of *Actinobifida chromogena* n. sp. [In Russian, with English summary.] Mikrobiologiya **34:**284–291.

Kriss, A. E. 1939. *Micromonospora*—an actinomycete-like organism (*Micromonospora globosa* n. sp.) [In Russian, with English summary.] Mikrobiologiya **8:**178–185.

Kurup, V. P., Barboriak, J. J., Fink, J. N., Lechevalier, M. P. 1975. *Thermoactinomyces candidus,* a new species of thermophilic actinomycetes. International Journal of Systematic Bacteriology **25:**150–154.

Küster, E., Locci, R. 1963. Transfer of *Thermoactinomyces viridis.* Schuurmans et al. 1956 to the genus *Thermomonospora* as *Thermomonospora viridis* (Schuurmans, Olson and San Clemente) comb. nov. International Bulletin of Bacteriological Nomenclature and Taxonomy **13:**213–216.

Lacey, J. 1971a. The microbiology of moist barley storage in unsealed silos. Annals of Applied Biology **69:**187–212.

Lacey, J. 1971b. *Thermoactinomyces sacchari* sp. nov., a thermophilic actinomycete causing bagassosis. Journal of General Microbiology **66:**327–338.

Lacey, J. 1973. Actinomycetes in soils, composts and fodders, pp. 231–251. In: Sykes, G., Skinner, F. A. (eds.), Actinomycetales: Characteristics and practical importance. London, New York: Academic Press.

Lacey, J. 1974. Moulding of sugar-cane bagasse and its prevention. Annals of Applied Biology **76:**63–76.

Lacey, J. 1978. Ecology of actinomycetes in fodders and related substrates, pp. 161–170. In: Mordarski, M., Kurylowicz, W., Jeljaszewicz, J. (eds.), *Nocardia* and *Streptomyces.* Stuttgart, New York: Gustav Fischer Verlag.

Lacey, J., Dutkiewicz, J. 1976a. Methods for examining the microflora of mouldy hay. Journal of Applied Bacteriology **41:**13–27.

Lacey, J., Dutkiewicz, J. 1976b. Isolation of actinomycetes and fungi from mouldy hay using a sedimentation chamber. Journal of Applied Bacteriology **41:**315–319.

Lacey, J., Lacey, M. E. 1964. Spore concentrations in the air of farm buildings. Transactions of the British Mycological Society **47:**547–552.

Lechevalier, H. A., Lechevalier, M. P., Gerber, N. N. 1971. Chemical composition as a criterion in the classification of actinomycetes. Advances in Applied Microbiology **14:**47–72.

Lechevalier, M. P. 1968. Identification of aerobic actinomycetes of clinical importance. Journal of Laboratory and Clinical Medicine **71:**934–944.

Lechevalier, M. P., Lechevalier, H. 1976. Chemical methods as criteria for the separation of nocardiae from other actinomycetes. Biology of the Actinomycetes **11:**78–92.

Locci, R., Baldacci, E., Petrolini, B. 1967. Description of a new species of *Actinobifida: Actinobifida alba* sp. nov. and revision of the genus. Giornale Microbiologie **15:**79–91.

Luedemann, G. M. 1971a. Designation of neotype strains for *Micromonospora coerulea* Jensen 1932 and *Micromonospora chalcea* (Foulerton 1905) Ørskov 1923. International Journal of Systematic Bacteriology **21:**248–253.

Luedemann, G. M. 1971b. *Micromonospora purpureochromogenes* (Waksman and Curtis 1916) comb. nov. (subjective synonym: *Micromonospora fusca* Jensen 1932). International Journal of Systematic Bacteriology **21:**240–247.

Leudemann, G. M. 1974. Genus *Micromonospora,* pp. 846–855. In: Buchanan, R. E., Gibbons, N. E. (eds.), Bergey's manual of determinative bacteriology, 8th ed. Baltimore: Williams & Wilkins.

Luedemann, G. M., Brodsky, B. C. 1964. Taxonomy of gentamycin producing *Micromonospora.* Antimicrobial Agents and Chemotherapy **1963:**116–124.

Luedemann, G. M., Brodsky, B. C. 1965. *Micromonospora*

carbonacea sp. n., an everninomicin producing organism. Antimicrobial Agents and Chemotherapy **1964**:47–52.

Małuszyńska, G. M., Janota-Bassalik, L. 1974. A cellulolytic rumen bacterium, *Micromonospora ruminantium* sp. nov. Journal of General Microbiology **82**:57–65.

Nonomura, H., Ohara, Y. 1971. Distribution of actinomycetes in soil. (X) New genus and species of monosporic actinomycetes. Journal of Fermentation Technology **49**:895–903.

Nonomura, H., Ohara, Y. 1974. A new species of actinomycetes, *Thermomonospora mesouviformis* sp. nov. Journal of Fermentation Technology **52**:10–13.

Ørskov, J. 1923. Investigation into the morphology of the ray fungi. Copenhagen: Levin & Munksgaard.

Potter, L. F., Baker, G. E. 1956. The microbiology of Flathead and Rogers Lakes, Montana. I. Preliminary survey of the microbial populations. Ecology **37**:351–355.

Rowbotham, T. J., Cross, T. 1977. Ecology of *Rhodococcus coprophilus* and associated actinomycetes in freshwater and agricultural habitats. Journal of General Microbiology **100**:231–240.

Sandrak, N. A. 1977. Cellulose decomposition by Micromonosporas. [In Russian, with English summary.] Mikrobiologiya **46**:478–481.

Schütze, H. 1908. Beiträge zur Kenntnis der thermophilen Aktinomyzeten und ihrer Sporenbildung. Archiv für Hygiene **67**:35–56.

Schuurmans, D. M., Olson, B. H., San Clemente, C. L. 1956. Production and isolation of thermoviridin, an antibiotic produced by *Thermoactinomyces viridis* n. sp. Applied Microbiology **4**:61.

Sebald, M., Prévot, A. R. 1962. Etude d'une nouvelle espece anaerobie stricte *Micromonospora acetoformici* n. sp. isolée de l'intestin posterieur de *Reticulitermes lucifugus* var. *saintonensis* Annales de l'Institut Pasteur **102**:199–214.

Solovieva, N. K. 1972. Actinomycetes of littoral and sub-littoral zones of the White Sea. [In Russian, with English summary.] Antibiotiki **17**:387–392.

Solovieva, N. K. Singal, E. M. 1972. Some data on ecology of *Micromonospora*. [In Russian, with English summary.] Antibiotiki **17**:778–781.

Stutzenberger, F. J. 1971. Cellulase production by *Thermomonospora curvata* isolated from municipal solid waste compost. Applied Microbiology **22**:147–152.

Stutzenberger, F. J., Kaufman A. J. Lossin, R. D. 1970. Cellulolytic activity in municipal solid waste composting. Canadian Journal of Microbiology **16**:553–560.

Sveshnikova, M. A., Maxsimova, T. S., Kudrina, E. S. 1969. The species of the genus *Micromonospora* Ørskov, 1923 and their taxonomy. [In Russian, with English summary.] Mikrobiologiya **38**:754–763.

Sveshnikova, M., Maximova, T., Kudrina, E. 1970. Species of the genus *Micromonospora* Ørskov 1923 and their taxonomy, pp. 187–197. In: Prauser, H. (ed.), The Actinomycetales. Jena: Gustav Fisher Verlag.

Szabo, I. M., Marton, M., Kulcsar, G., Buti, I. 1976. Taxonomy of primycin producing actinomycetes. I. Description of the type strain of *Thermomonospora galeriensis*. Acta Microbiologica Academiae Scientiarum Hungaricae **23**:371–376.

Tsiklinsky, P. 1899. Sur les mucedinées thermophiles. Annales de l'Institut Pasteur **13**:500–504.

Umbreit, W. W., McCoy, E. 1941. The occurrence of actinomycetes of the genus *Micromonospora* in inland lakes, pp. 106–114. Symposium on Hydrobiology, University of Wisconsin.

Unsworth, B. A. 1978. The genus *Thermoactinomyces* Tsiklinsky. Ph.D. Thesis. University of Bradford.

Unsworth, B. A., Cross, T., Seaward, M. R. D., Sims, R. E. 1977. The longevity of thermoactinomycete endospores in natural substrates. Journal of Applied Bacteriology **42**:45–52.

Wagman, G. H., Testa, R. T., Marquez, J. A., Weinstein, M. J. 1974. Antibiotic G-148, a new *Micromonospora*-produced aminoglycoside with activity against protozoa and helminths: Fermentation, isolation, and preliminary characterization. Antimicrobial Agents and Chemotherapy **6**:144–149.

Waksman, S. A., Corke, C. T. 1953. *Thermoactinomyces* Tsiklinsky, a genus of thermophilic actinomycetes. Journal of Bacteriology **66**:377–378.

Waksman, S. A., Umbreit, W. W., Cordon, T. C. 1939. Thermophilic actinomycetes and fungi in soils and in composts. Soil Science **47**:37–61.

Watson, E. T., Williams, S. T. 1974. Study on the ecology of actinomycetes in soil. VII. Actinomycetes in a coastal sand belt. Soil Biology and Biochemistry **6**:43–52.

Weinstein, M. J., Luedemann, G. M., Oden, E. M., Wagman, G. H. 1968. Halomicin, a new *Micromonospora* produced antibiotic. Antimicrobial Agents and Chemotherapy **1967**:435–441.

Weyland, H. 1969. Actinomycetes in North Sea and Atlantic Ocean sediments. Nature **223**:858.

Williams, S. T., Cross, T. 1971. Actinomycetes, pp. 295–334. In: Booth, C. (ed.), Methods in microbiology, vol. 4. London, New York: Academic Press.

Willoughby, L. G. 1969. A study of the aquatic actinomycetes of Blelham Tarn. Hydrobiologia **34**:465–483.

CHAPTER 158

The Genera *Actinomadura, Actinopolyspora, Excellospora, Microbispora, Micropolyspora, Microtetraspora, Nocardiopsis, Saccharopolyspora,* and *Pseudonocardia*

STANLEY T. WILLIAMS and ELIZABETH M. H. WELLINGTON

This heterogeneous collection of genera may be divided into two groups. *Actinomadura, Excellospora, Microbispora, Microtetraspora* and *Nocardiopsis* have a cell wall chemotype III (Lechevalier, M. P., and Lechevalier, 1970) and may be placed in the family Thermomonosporaceae (Cross and Goodfellow, 1973). *Actinopolyspora, Micropolyspora, Saccharopolyspora,* and *Pseudonocardia* have a chemotype IV wall and, with the possible exception of *Pseudonocardia,* can be included in the family Nocardiaceae (Cross and Goodfellow, 1973). The major features of all genera are given in Table 1 and Fig. 1.

When proposed, *Actinomadura* (Lechevalier, H. A., and Lechevalier, 1970) consisted of three species previously named *Nocardia madurae, N. pelletieri,* and *N. dassonvillei. N. dassonvillei* was subsequently transferred to the genus *Nocardiopsis* (Meyer, 1976). In the eighth edition of *Bergey's Manual of Determinative Bacteriology* (Buchanan

and Gibbons, 1974) *Actinomadura* was listed among *genera incertae sedis.*

Actinomadura contains actinomycetes with a branched, usually stable, substrate mycelium, sometimes forming short chains of arthrospores on an aerial mycelium. The wall contains *meso*-2,6-diaminopimelic acid (type III of Lechevalier, M. P., and Lechevalier, 1970), and sometimes whole cell hydrolysates contain the sugar madurose (3-*O*-methyl-D-galactose).

Nocardiopsis (Meyer, 1976) was proposed to accommodate *Actinomadura dassonvillei,* which does not contain madurose and differs in other ways from *Actinomadura* species. Recently, another species *(N. syringae)* has been described (Gauze et al., 1977), and Preobrazhenskaya, Sveshnikova, and Terekhova (1977) suggested that several other *Actinomadura* species should be moved to this genus.

Excellospora (Agre and Guzeva, 1975) contains

Table 1. Major characteristics of the genera considered.[a]

	Substrate mycelium		Aerial mycelium spores	Cell wall chemotype	Characteristic sugars
	Fragmentation	Spores			
Actinomadura	v	−	Absent or short chains	III	Madurose
Nocardiopsis	+	−	Long chains	III	—
Excellospora	−	Single to short chains	Single to short chains	III	Madurose
Microbispora	−	−	Paired	III	Madurose
Microtetraspora	−	−	Most in chains of 4	III	Madurose
Micropolyspora	v	Single to short chains	Single to short chains	IV	Arabinose Galactose
Saccharopolyspora	+	−	Long chains	IV	Arabinose Galactose
Actinopolyspora	+	−	Long chains	IV	Arabinose Galactose
Pseudonocardia	+	Fragmentation spores	Chains of various length	IV	Arabinose Galactose

[a] +, Feature present; −, feature absent; v, species vary in response.

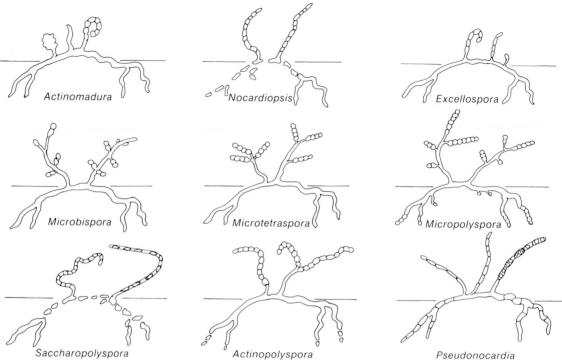

Fig. 1. Morphology of the genera of polysporic actinomycetes considered in this chapter.

three thermophilic species, one of which is a new species *(E. viridilutea)* and two of which *(E. viridingra* and *E. rubrobrunea)* were transferred from *Micropolyspora* (Agre and Guzeva, 1975). Their cell walls and hydrolysate patterns are like those of *Actinomadura* (type III plus madurose) (Guzeva, Agre, and Sokolov, 1972), but are distinguished by the fatty acid composition of mycelial lipids (Agre and Guzeva, 1975).

Microbispora (Nonomura and Ohara, 1957) is characterized by the formation of paired spores on the aerial mycelium, with a stable substrate mycelium. Its cell wall chemotype is III with madurose in cell hydrolysates. The genus *Waksmania* was described in the same year as *Microbispora* (Lechevalier and Lechevalier, 1957). These genera are identical and priorty was given to *Microbispora* (Lechevalier, 1965). A species described as *Thermopolyspora bispora* (Henssen, 1957) was transferred to *Microbispora,* and *Thermopolyspora* was regarded as another synonym of *Microbispora* (Lechevalier, 1965).

The most characteristic feature of *Microtetraspora* (Thiemann, Pagani, and Beretta, 1968) is the formation of a short, sparsely branched aerial mycelium bearing chains of four spores, although one species *(M. viridis* var. *intermedia)* has some chains of up to six spores (Nonomura and Ohara, 1971a). Thiemann, Pagani, and Beretta (1968) reported an unusual cell wall composition of *meso*-diamino-

pimelic acid, glycine, lysine, and traces of LL-diaminopimelic acid. However, species described by Nonomura and Ohara (1971a,b) had a typical chemotype III with madurose. The genus was listed with *genera incertae sedis* in *Bergey's Manual* (Buchanan and Gibbons, 1974) despite its clear and valid description (Thiemann, Pagani, and Beretta, 1968).

Micropolyspora (Lechevalier, Solotorovsky, and McDurmont, 1961) forms chains of spores on both the substrate and aerial mycelium. The cell wall composition is *meso*-diaminopimelic acid with arabinose and galactose in whole-cell hydrolysates (type IV of Lechevalier, M. P., and Lechevalier, 1970).

The taxonomic history of *Micropolyspora* is confused. Henssen (1957) described a thermophilic species *Thermopolyspora polyspora* which formed spore chains only on its aerial mycelium. This taxon is considered to be illegitimate, as the original culture was contaminated (Becker, Lechevalier, and Lechevalier, 1965), but the name was applied to *Micropolyspora* isolates from moldy hay (Corbaz, Gregory, and Lacey, 1963). The abolition of *Thermopolyspora* (Henssen and Schnepf, 1967) left two described species, *T. flexuosa* and *T. rectivirgula* (Krassilnikov and Agre, 1964); the former is now in *Actinomadura* (cell wall chemotype III) and the latter in *Micropolyspora* (Lechevalier and Lechevalier, 1967). The genus *Thermoactinopolyspora* (Craveri and Pagani, 1962) proved to consist of *Micropolyspora* and *Actinomadura* strains (Lechevalier and Lechevalier, 1967).

The taxonomic status of all species placed in *Micropolyspora* at one time or another is summarized in Table 6.

Saccharopolyspora (Lacey and Goodfellow, 1975) is represented by a single species, *S. hirsuta,* isolated from sugar cane bagasse. It has a fragmentary substrate mycelium, beadlike spore chains on the aerial mycelium, and a type IV wall chemotype. It was originally placed in *Nocardia* (Lacey, 1974).

Actinopolyspora (Gochnauer et al., 1975) consists of one species, *A. halophila,* which is an extreme halophile growing optimally in the presence of 15–20% NaCl. The substrate mycelium occasionally fragments and aerial growth bears chains of spores. Its wall chemotype is IV.

Pseudonocardia (Henssen, 1957; Henssen ,and Schäfer, 1971) was originally proposed to include a single species, *P. thermophila.* The genus description was emended by Henssen and Schäfer (1971) when another species, *P. spinosa,* was described. It is characterized by hyphal growth by acropetal budding, formation of blastospores, and a type IV wall chemotype.

Habitats

Actinomadura madurae and *A. pelletieri* can cause mycetoma in man, a chronic suppurative and granulomatous disease of subcutaneous tissue and bones that is characterized by swelling, abscesses, and discharge of pus and granules. *A. pelletieri* occurs mainly in Africa, while *A. madurae* is more widespread in tropical and subtropical areas. Soil may be a primary reservoir for infection, which often occurs through the foot (Cross and Goodfellow, 1973). Microbes similar to both these species were found in low numbers in a stream in South Africa (Lawson and Davey, 1972). Virulent strains of *A. madurae* produce collagenase, which has a significant role in their pathogenicity (Rippon, 1968), but elastase was not detected (Rippon and Varadi, 1968).

Unnamed *Actinomadura* strains have been isolated from self-heated, stored barley grain and slightly heated hay (Lacey, Goodfellow, and Alderson, 1978). Saprophytic species recently described by Japanese and Russian workers have originated from soil.

Most of the 26 strains of *Nocardiopsis dassonvillei* studied by Gordon and Horan (1968) originated from human or animal infections, but because of the diversity of these it is difficult to assess its status as a pathogen. Nine of their strains came from soil. Meyer (1976) examined an isolate from soil in Lebanon. *N. syringae* was also obtained from soil (Gauze et al., 1977).

All three *Excellospora* species were isolated from soil. *E. viridilutea* came from an arid saline soil in Russia; *E. viridinigra* and *E. rubrobrunea* came from Egyptian soils under maize and rice.

The most characteristic habitat of *Microbispora* appears to be the soil. It is difficult to detect but when selective isolation procedures were used, strains of *Microbispora* were isolated from a range of soils including a potato field, a wheat paddy, and stream muds (Nonomura and Ohara, 1971b). Numbers were usually about 10^3/g soil. *Microbispora* isolates were obtained from marine sediments by Weyland (1969). Little is known of the ecology of this genus but some species are thermophilic.

Microtetraspora glauca and *M. fusca* are widespread in soil and have been isolated from soils from Italy, Thailand, and Brazil (Thiemann, Pagani, and Beretta, 1968). *M. niveoalba* and *M. viridis* were both isolated from stream mud and mulberry fields; the latter also occurred in a variety of cultivated soils. Numbers are usually less than 10^3/g of soil (Nonomura and Ohara, 1971a,b).

Micropolyspora brevicatena was isolated from a pneumonic calf lung and from sputa from patients recovering from tuberculosis, but does not appear to be pathogenic (Lechevalier, Solotorovsky, and McDurmont, 1961). The thermophilic *M. faeni* has also been detected in the lungs and sputum of patients suffering from farmer's lung (Lacey and Lacey, 1964; Wenzel et al., 1964). There is no evidence that it grows in the lung (Cross, Maciver, and Lacey, 1968), but its spores, once inhaled, can be deposited in the alveoli and it is one of the sources of farmer's lung antigens (Pepys et al., 1963). Moldy hay, which is associated with farmer's lung disease, contains large numbers of actinomycete spores including *M. faeni* (Corbaz, Gregory, and Lacey, 1963; Cross, Maciver, and Lacey, 1968), which also occurs in mushroom compost (Fergus, 1964), moldy silage, barley grain, straw and sugar cane bagasse after self-heating, and in the air of farm buildings (Cross, Maciver, and Lacey, 1968). Most of the remaining species were isolated from Russian soils. The thermophilic *M. thermovirida* was isolated near the discharge of a hot (90°C) spring (Kosmachev, 1964).

Saccharopolyspora hirsuta was isolated from airborne bagasse dust. Sugar cane bagasse is the squashed, chopped fiber left after sugar is extracted; it contains 50% water and 3–5% sugar, and when stacked fresh in bales can self-heat, remaining at over 40°C for 50 days (Lacey, 1974). Spores in heated bagasse are mainly thermophilic fungi and actinomycetes (e.g., *Thermoactinomyces sacchari* and *T. vulgaris*) that are implicated in a form of extrinsic, allergic alveolitis similar to farmer's lung (Lacey, 1971). *S. hirsuta* was frequently isolated from bagasse but was usually present in small numbers (Lacey, 1974).

The natural habitat of the halophilic *Actinopolyspora halophila* is unknown. It was isolated from an unsterilized batch of complex medium (Sehgal and Gibbons, 1960)—used in studies of *Halobacterium*—that had been left at room temperature for 50 days (Gochnauer et al, 1975).

Few isolates of *Pseudonocardia* have been reported. *P. thermophila* is thermophilic, and aerobic to facultatively anaerobic; it was isolated from fresh and rotting manure (Henssen, 1957). *P. spinosa* is mesophilic and aerobic, and was isolated from soils in Turkey and Iran (Henssen and Schäfer, 1971).

Isolation and Cultivation

Actinomadura madurae and *A. pelletieri* may be isolated from clinical material by using media such as yeast extract (Pridham et al., 1956–1957) and incubating at 30–37°C. No single medium has been used to isolate *A. madurae* from soil, but isolates from water were obtained on dilution plates with egg albumen agar containing actidione (0.5 mg/ml) (Lawson and Davey, 1972).

Several *Actinomadura* species were isolated from soil by Japanese and Russian workers. The former reduced numbers of unwanted microbes by air-drying soil and then applying dry heat at 100°C for 1 h. Samples were then plated on various media, including minerals-glucose-asparagine (MGA) and arginine-vitamins (AV) agars, and incubated for several weeks at 28–30°C (Nonomura and Ohara, 1971d). Russian workers used more conventional methods with addition of antibiotics to improve medium selectivity. Lavrova, Preobrazhenskaya, and Sveshnikova (1972) added rubomycin (5, 10, or 20 µg/ml) to medium No. 2 of Gauze et al. (1957). Preobrazhenskaya et al. (1975) added streptomycin (0.5, 1, or 2 µg/ml) or bruneomycin (0.5, 1 or 2 µg/ml). These antibiotics increased numbers of *Actinomadura* colonies while decreasing streptomycetes.

A variety of media have been used to maintain and examine laboratory cultures. No specific isolation procedures have been devised for *Nocardiopsis dassonvillei,* but it grows readily on many media at 28–30°C. These include yeast extract agar and oatmeal agar, the latter allowing dense aerial mycelium formation (Meyer, 1976).

Excellospora viridinigra and *E. rubrobrunea* were isolated by scattering crushed soil over the medium of Kosmachev (1960) and incubating at 55°C. *E. viridilutea* was isolated from soil at 55°C. All strains grow well on a range of media but often there is rapid autolysis of the aerial mycelium (Agre and Guzeva, 1975), resulting in loss of viability.

Most of the 10 *Microbispora* species described were isolated from Japanese soils using a soil pretreatment, selective isolation procedure (Nonomura and Ohara, 1969a, 1971b). Soil samples are air-dried at room temperature, ground in a mortar, and subjected to dry heat at 120°C for 1 h. Heated soil is either incorporated directly into the isolation medium or a suspension is used to make dilution plates. The media used are MGA with soil extract or AV agar with B vitamins. The antifungal antibiotics (nystatin and actidione 50 µg/ml) are incorporated into the medium; sometimes penicillin (0.8 µg/ml) and polymyxin B (4 µg/ml) are also used. Plates are incubated at 30°C for 40 days or 50°C for 20 days. The relative infrequency of *Microbispora* colonies on isolation plates prepared by normal procedures is probably due to competition from faster growing microbes and the requirement of many species for B-vitamins (Nonomura and Ohara 1960, 1969b). A suitable medium for growth in the laboratory and sporing is oatmeal agar + 1% yeast extract (Nonomura and Ohara, 1960).

Microtetraspora glauca and *M. fusca* were isolated from soil by a "novel method" which has not been described (Thiemann et al., 1968). Japanese workers isolated species of this genus by pretreatment of soil and the use of selective media. *M. niveoalba* was isolated from dry-heated soil on MGA-SE agar incubated at 40°C for 1 month (Nonomura and Ohara, 1971b). *M. viridis* was isolated from soil heated for 1 h at 100°C on GAC medium incubated at 32°C for several weeks.

Good growth of vegetative and sporing aerial mycelia was obtained for *M. glauca* and *M. fusca* on the medium of Hickey and Tresner (1952). *M. niveoalba* requires B vitamins for growth on synthetic media (Nonomura and Ohara, 1971b).

Micropolyspora species have been isolated only rarely on soil dilution plates, due to competition from faster-growing microbes (Cross and Goodfellow, 1973). The latter may be limited by incorporation of antibiotics into isolation media; Lavrova, Preobrazhenskaya, and Sveshnikova (1972) claimed that rubomycin (5, 10, or 20 µg/ml) in Gauze's medium No. 2 limited number of streptomycetes and increased *Micropolyspora* and *Actinomadura* colonies. *M. thermovirida* was recognized on soil plates incubated at 46°C by its green aerial mycelium (Kosmachev, 1964).

M. faeni was isolated from moldy hay by passing air through an Anderson (1958) sampler (Cross, Maciver, and Lacey, 1968; Gregory and Lacey, 1963). The sampler was loaded with plates of either yeast extract or half-strength nutrient agar containing cycloheximide (50 µg/ml) to suppress fungi. The sampler was connected to a sterile tin containing a weighed hay sample. This was shaken for 30 min to suspend spores in the air and then sampled for short periods (20 s). Plates were incubated at 55°C.

M. brevicatena was isolated from sputum treated with 4% (vol/vol) NaOH and shaken for 10 min. This was neutralized with 2 N HCl and centrifuged, and the sediment was inoculated into tubes of Jensen-Lowenstein medium and tubes of penicillin blood agar. Tubes were incubated at 37°C, horizontally for the initial 72 h, then upright (Lechevalier, Solotorovsky, and McDurmont, 1961).

No single medium seems to be suitable for maintenance and induction of sporulation of all species, though most grow poorly on synthetic media. The media used include NZ amine glycerol agar for *M. brevicatena* (Lechevalier, Solotorovsky, and McDurmont, 1961), peptone-maize agar for *M. rectivirgula* (Soina and Agre, 1976), and V-8 agar for *M. faeni* (Cross, Maciver, and Lacey, 1968). Temperature requirements vary and are listed in Table 6.

Saccharopolyspora hirsuta was isolated from bagasse dust using an Anderson (1958) sampler and plates of half-strength nutrient agar with cycloheximide (50 μg/ml) to suppress fungi, incubated at 40°C (Lacey and Goodfellow, 1975). Cultures were satisfactorily maintained on yeast extract agar or V-8 agar.

Actinopolyspora halophila was isolated and grown on the complex medium of Sehgal and Gibbons (1960) which includes 25% (wt/vol) NaCl. Incubation was at 37°C for broth and agar cultures.

Pseudonocardia thermophila was isolated by incubating plates at high temperatures. At 40–50°C, *P. thermophila* grows well aerobically but it grows better anaerobically at 60°C. *P. spinosa* was detected on soil plates prepared following methods used by Nonomura and Ohara (1969a) to isolate *Microbispora* (Schäfer, 1969). Both species grow on Casamino Acids–peptone–Czapek agar, *P. thermophila* at 40–50°C and *P. spinosa* at 20–30°C.

Media for Isolation and Cultivation

The following are the formulas of those media mentioned in the preceding section that are not commercially available. Quantities are to 1 liter of distilled water unless otherwise stated.

Yeast Extract Agar (Pridham et al., 1956–1957)

Yeast extract	4 g
Malt extract	10 g
Glucose	4 g
Agar	15 g
pH	7.0

MGA Agar (Nonomura and Ohara, 1971d)

Glucose	2 g
L-Asparagine	1 g
K_2HPO_4	0.5 g
$MgSO_4 \cdot 7H_2O$	0.5 g
Trace salts solution	1.0 ml
Agar	20 g
pH	7.4

Trace salts solution:

$FeSO_4 \cdot 7H_2O$	10.0 mg/ml
$MnSO_4 \cdot 7H_2O$	1.0 mg/ml
$CuSO_4 \cdot 5H_2O$	1.0 mg/ml
$ZnSO_4 \cdot 7H_2O$	1.0 mg/ml
pH	8.0

Soil extract (SE) and a vitamin solution (see below) may be added to this medium. Soil extract: 1,000 g soil in 1 liter water, autoclave for 30 min, and filter.

AV Agar + Vitamins (Nonomura and Ohara, 1971d)

L-Arginine	0.3 g
Glucose	1.0 g
Glycerol	1.0 g
K_2HPO_4	0.3 g
$MgSO_4 \cdot 7H_2O$	0.2 g
NaCl	0.3 g
Agar	15 g

Trace salts solution (see above). Vitamins (final weight in medium): 0.5 mg each of thiamine HCl, riboflavin, niacin, pyridoxine, HCl, inositol, calcium pantothenate, *p*-aminobenzoic acid; 0.25 mg biotin.

Medium No. 2 (Gauze et al., 1957)

Hottinger's broth	30 ml
Peptone	5 g
NaCl	5 g
Glucose	10 g
Agar	15 g

Kosmachev's Medium (Kosmachev, 1960)

KNO_3	1.0 g
$(NH_4)_2SO_4$	1.0 g
Na_2HPO_4	1.0 g
$MgSO_4 \cdot 7H_2O$	0.5 g
$FeSO_4 \cdot 7H_2O$	0.01 g
$CaCO_3$	4.0 g
30% Yeast autolysate	15 ml
Agar	15 g

GAC Agar (Nonomura and Ohara, 1971a)

A: Glucose-asparagine agar	
Glucose	1.0 g
L-Asparagine	1.0 g
K_2HPO_4	0.3 g
$MgSO_4 \cdot 7H_2O$	0.3 g
NaCl	0.3 g
Trace salts (see MGA agar)	1.0 ml
Agar	15 g

B: Casamino Acids   0.5 g

Antibiotics (cycloheximide, nystatin, 50 μg/ml, polymyxin B, 4 μg/ml, penicillin, 0.8 μg/ml) may be added for isolation purposes. Four milliliters of B was poured over 15 ml of A in a plate.

Hickey and Tresner Agar (Hickey and Tresner, 1952)

Yeast extract (Difco)	1.0 g
Beef extract (Difco)	1.0 g
NZ amine type A (casein digest)	2.0 g
$CaCl_2 \cdot 6H_2O$	0.02 g
Agar	20 g
pH	7.3

NZ Amine Glycerol Agar
(Lechevalier, Solotorovsky, and McDurmont, 1961)

NZ amine type A (casein digest)	5.0 g
Beef extract	1.0 g
Glycerol	70 ml
Agar	15 g
pH	6.5–7.0

V-8 Agar
(Society of American Bacteriologists, 1957)

V-8 canned vegetable juice	200 ml
$CaCO_3$	4 g
Agar	20 g
Water	800 ml
pH	7.3

Complex medium (Sehgal and Gibbons, 1960)

Casamino Acids (Difco)	7.5 g
Yeast extract (Difco)	10 g
Trisodium citrate	3 g
KCl	2 g
$MgSO_4 \cdot 7H_2O$	20 g
NaCl	250 g
pH	7.5–7.8

Autoclave for 5 min at 120°C; filter precipitate and adjust pH to 7.4; autoclave as normally.

Casamino Acids–Peptone–Czapek's Agar
(Henssen and Schäfer, 1971)

Casamino Acids (Difco)	1 g
Peptone	2 g
K_2HPO_4	1 g
KCl	0.5 g
$MgSO_4 \cdot 7H_2O$	0.5 g
$FeSO_4 \cdot 7H_2O$	0.01 g
Sucrose	30 g
Agar	15 g

Identification

Some *Actinomadura* species may show close morphological similarity to other genera listed in Table 1 and also to *Nocardia* and *Streptomyces*. However, *Actinomadura* can be distinguished from these by its wall chemotype, as *Nocardia* has a type IV and *Streptomyces* a type I cell wall. Wall chemotype also serves to differentiate it from those genera listed (Table 1) that have a type IV chemotype. Differentiation from *Microbispora*, *Microtetraspora*, and *Nocardiopsis* may be achieved by morphology; *Nocardiopsis* also lacks madurose. The distinction between *Actinomadura* and *Excellospora* is a fine one, depending on the presence of substrate spore chains in the latter and differences in the proportions of branched-chain and straight-chain fatty acids (Agre and Guzeva, 1975).

Identification of many of the described species is difficult, as only inadequate descriptions exist. Details of the more fully described species are given in Table 2. The morphology of *A. madurae* and *A. pelletieri* differs from that of the other species. The aerial mycelium is often sparse or absent and formation of aerial spore chains is rare.

In contrast, most of the more recently described species form aerial mycelia bearing distinct spore chains, sometimes in the form of loops or spirals. On the basis of the original descriptions of 21 species and some numerical phenetic data, it has been suggested that some species are very similar if not synonymous (Lacey, Goodfellow, and Alderson, 1978). More information on many "species" is required and minimum standards for their description have been suggested (Lacey, Goodfellow, and Alderson, 1978).

Additional information on the chemical composition of some species has been obtained. Mordarska, Mordarski, and Goodfellow (1972) reported the absence of mycolic acids in *A. madurae* and *A. pelletieri*; they also noted that some strains had traces of LL-2,6-diaminopimelic acid, which suggests a possible link with *Streptomyces*. Agre, Efimova, and Guzeva (1975) reported differences in fatty acid patterns between *A. madurae* and *A. pelletieri* as compared with the species described by Nonomura and Ohara (1971d). The polar lipid patterns of *A. madurae* and *A. pelletieri* were studied by Minnikin, Pirouz, and Goodfellow (1977), who contrasted them with those of *Nocardiopsis dassonvillei*. The phospholipid pattern of these species is unusual as it lacks phosphatidylethanolamine (Lechevalier, 1977). Menaquinones of these two species resemble those of streptomycetes and *Micropolyspora faeni* but differed from those of *N. dassonvillei* (Collins et al., 1977).

DNA base ratios of 77.4 and 65 mol% G+C for this genus have been reported by Lechevalier,

Table 2. Characteristics of some *Actinomadura* species.[a]

Growth on:	A. madurae[b] 1.	A. pelletieri[c] 2.	A. verrucosospora[d] 3.	A. pusilla[e] 4.	A. roseoviolacea[f] 5.	A. helvata[g] 6.	A. spadix[h] 7.	A. flexuosa[i] 8.
L-Arabinose	V	V	+	+	+	+	+	NT
D-Fructose	V	−	+	+	+	+	+	NT
D-Glucose	+	+	+	+	+	+	+	+
Glycerol	V	V	(+)	+	(+)	+	−	NT
i-Inositol	−	−	+	+	(+)	+	+	NT
D-Mannitol	+	−	−	−	−	+	(+)	+
Melibiose	NT	NT	−	+	−	+	+	NT
Raffinose	−	−	+	+	−	+	+	+
Rhamnose	V	−	(+)	+	−	+	+	NT
Sucrose	−	−	+	+	(+)	−	+	+
D-Xylose	+	−	(+)	−	−	+	+	NT
Hydrolysis of starch	+	V	+	−	+	−	+	(+)
Nitrate reduction	+	+	+	+	+	+	−	+
Gelatin liquifaction	+ to (+)	+	+	−	+	−	+	NT
Spore chain	Absent or hooks	Rare	Hooks, warty spores	Spiral	Spiral	Hooks	Pseudosporangia	Spiral, spiny spores

[a] +, Positive reaction; −, negative reaction; (+), weak reaction; V, strain reactions vary; NT, not tested.

[b] Vincent; Lechevalier, H.A., and Lechevalier (1970a), NCTC 5654.

[c] Laveran; Lechevalier, H.A., and Lechevalier (1970a), NCTC 4162.

[d] Nonomura and Ohara (1971d), ATCC 27299.

[e] Nonomura and Ohara (1971d), ATCC 27296.

[f] Nonomura and Ohara (1971d), ATCC 27297.

[g] Nonomura and Ohara (1971d), ATCC 27295.

[h] Nonomura and Ohara (1971d), ACTC 27298.

[i] Krassilnikov and Agre (1964); Cross and Goodfellow (1973).

Lechevalier, and Gerber (1971) and Mordarski et al. (1977).

A numerical taxonomic study of *Actinomadura* and related genera has been recently carried out (Goodfellow, Alderson, and Lacey, 1979). *Actinomadura madurae* and *Nocardiopsis dassonvillei* formed good taxospecies, while *A. pelletieri* strains split into two clusters. Single strains of *A. helvata*, *A. pusilla*, *A. roseoviolacea*, *A. spadix,* and *A. verrucosospora* formed new centers of variation. *A. citrea* and *A. malachitica* were very similar to *A. madurae*. Characters useful for the separation of *Actinomadura* species were presented. Three new species of *Actinomadura* have been described recently (Meyer, 1979). These are *Actinomadura ferruginea* (type IMET 9567), *A. libanotica* (type IMET 9616), and *A. spiralis* (type IMET 9621). All have a wall chemotype III with madurose and produce white to pink aerial mycelium bearing short spiral or hooked spore chains. Information on their carbon source utilization patterns, starch hydrolysis, gelatin liquefaction, and nitrate reduction is given by Meyer (1979).

A neotype of *N. dassonvillei* (IMRU 509, ATCC 23218) was designated by Meyer (1976). The removal of this species from *Actinomadura* was based largely on its comparison with *A. madurae* and *A. pelletieri*. It differs in the fragmentation of its substrate hyphae and in the formation of long, aerial chains of various length spores, lack of madurose, fatty acid composition (Agre, Efimova, and Guzeva, 1975), polar lipids (Minnikin, Pirouz, and Goodfellow 1977), menaquinones (Collins et al., 1977), pigments (Lechevalier, Lechevalier, and Gerber, 1971), and phenetic characters (Goodfellow, 1971).

Recognition of *Nocardiopsis* is based largely on morphology, but chemical composition may be required to distinguish it from *Actinomadura, Nocardia,* and other genera (Table 1). The aerial hyphae assume a zigzag shape prior to sporing. Long segments are delimited which then form spores of various lengths that have thickened polar walls (Williams, Sharples, and Bradshaw, 1974).

N. dassonvillei does not produce melanin pigments, starch is hydrolyzed, nitrate is reduced to nitrite, milk is peptonized, and hypoxanthine is decomposed. Gelatin is not liquified. L-Arabinose, D-xylose, D-mannose, D-glucose, rhamnose, maltose, D-mannitol, D-fructose, sucrose, and glycerol are used as carbon sources (Meyer, 1976).

Recently another species, *N. syringae,* has been described (Gauze et al., 1977). It appears to be similar to *N. dassonvillei,* being distinguished by its formation of lilac-colored aerial mycelium. It produces an antibiotic, nocamycin.

Excellospora forms short chains and single spores on both substrate and aerial hyphae; it is therfore morphologically very similar to *Micropolyspora* (Table 1). Agre and Guzeva (1975) suggested that

Excellospora could be distinguished by the formation of spiral chains of up to 20 spores. However, *Micropolyspora angiospora* forms twisted spore chains and *M. thermovirida* can have up to 20 spores per chain. Therefore, distinction of this genus from *Micropolyspora* seems to rest primarily on differences in cell wall chemotype.

Distinction between *Excellospora* and *Actinomadura* is even more difficult because they have the same cell wall chemotype. The presence of spores on substrate mycelium of *Excellospora* might serve to distinguish it from some *Actinomadura* species, although substrate spores in species of the latter were reported by Agre and Guzeva (1975). *E. viridinigra* and *E. rubrobrunea* contain branched-chain fatty acids with C_{16}, C_{17}, and C_{18} atoms, while *Actinomadura madurae* fatty acids were straight chained with C_{18} and C_{19} (Agre and Guzeva, 1975). *E. viridinigra* and *Actinomadura (Micropolyspora) flexuosa* both contain a high proportion of C_{18} and C_{19} atoms in fatty acids of an unidentified series (Guzeva et al., 1973). Further information is needed to evaluate the status of this genus.

The three *Excellospora* species are distinguished primarily by colony pigmentation and appear to be very similar if not synonymous (Table 3).

Microbispora is distinguished from related genera (Table 1) primarily by the formation of paired spores on aerial hyphae. These are either sessile or borne on short sporophores. The latter in *M. rosea* were found to fit to the base of the spore by a ball and socket arrangement (Williams, 1970). Species have been distinguished mainly by growth temperature requirements and a few physiological tests (Table 4). The validity of separating some species (e.g., *M. diastatica* and *M. thermodiastatica*) by using one or two characteristics is questionable.

Cultures of *M. aerata, M. amethystogenes,* and *M. parva* deposit crystals with a metallic sheen in the medium, particularly when grown on Pablum extract agar (Lechevalier and Lechevalier, 1957) for about 10 days (Gerber and Lechevalier, 1964). These crystals are iodinin (1,6-phenazinediol-5,10-dioxide), a red, water-soluble pigment. In addition, *M. aerata* produces two brown-yellow pigments (2-aminophenoaxazine-3-one and 1,6-phenazinediol), a yellow pigment (2-acetamidophenoxazine-3-one) (Gerber and Lechevalier, 1964), and an orange pigment (1,6-phenazinediol-5-oxide) (Gerber and Lechevalier, 1965).

Microbispora does not contain mycolic acids, but tuberculostearic acid and its homologs have been found in all species examined so far (Lechevalier, 1977). A G+C content of 73.7 mol% has been reported for *M. rosea* (Lechevalier, Lechevalier, and Gerber, 1971).

Microtetraspora is distinguished from related genera by its chains of four spores borne on sporophores which branch from the aerial hyphae. The

Table 3. Characteristics of *Excellospora* species[a]

Characteristic	E. viridinigra[b]	E. rubrobrunea[c]	E. viridilutea[d]
Spore chains	Coiled, 2–20 spores, some single	Coiled, 2–20 spores, some single	Spiral, 2–20 spores, some single
Spore surface	Spines	Spines	Spines
Spore color	Blue-green	Blue-green	Blue-green
Substrate color	Brown-black	Red-brown to yellow	Yellow-dark orange
Antagonism	Gram + bacteria	Gram + bacteria	None
Temp. optima	45–55°C	44–55°C	45–55°C
Utilization of:			
Glucose	+++	+++	+++
Maltose	+++	+++	+++
Starch	+++	+++	+++
Lactose	+++	+++	+
Rhamnose	+++	+++	+
Xylose	−	−	±
Arabinose	−	−	±
Mannitol	+++	+++	+
Dulcitol	+++	+++	+

[a] +++, Good growth; +, poor growth; ±, very little growth; −, no growth.
[b] Krassilnikov, Agre, and El-Registan (1968), Agre and Guzeva (1975).
[c] Krassilnikov, Agre, and El-Registan (1968), Agre and Guzeva (1975).
[d] Agre and Guzeva (1975), Type 187.

spores are spherical to oval and sporulation is basipetal (Thiemann, Pagani, and Beretta, 1968). Characteristics distinguishing species are given in Table 5. *Microtetraspora niveoalba* has a higher temperature optimum (35–40°C) than the other species.

Micropolyspora was characterized by the presence of short chains of spores on both substrate and aerial hyphae that were borne on short sporophores or were sessile. The presence of single spores was noted in the type species, *M. brevicatena* (Lechevalier, Solotorovsky, and McDurmont, 1961). Identification of *Micropolyspora* on morphological grounds alone can present problems, as several genera form short spore chains (Table 1), and the predominance of single spores in some species has caused confusion with *Saccharomonospora* (Table 6). All species at present included in *Micropolyspora* have a chemotype IV wall but the presence of mycolic acids in *M. brevicatena* and *M. fascifera* but not in *M. faeni* and *M. rectivirgula* indicates heterogeneity. In addition, Collins et al. (1977) found that the menaquinones of *M. faeni* had MKA compounds like *Streptomyces*, but *M. brevicatena* had a mixture of tetrahydromenaquinones with 6 and 8 isoprene units like *Nocardia*. Chemical data on other species are needed to assess the relevance of these results. A G+C content of 74.1 mol% has been quoted for *Micropolyspora* sp. (Lechevalier, Lechevalier, and Gerber, 1971).

The predominance of single spores on substrate and aerial hyphae of *M. caesia, M. internatus*, and *M. coerulea* (Table 6) is surprising for a genus that was initially characterized by its formation of chains of spores. These species are close to *Saccharomonospora viridis* in this feature and in their spore color, but *S. viridis* lacks spores on the substrate mycelium (Nonomura and Ohara, 1971c). The close similarity of *M. faeni* and *M. rectivirgula* has also been noted (Table 6). Spores of the latter have polar thickenings of the wall up to 100 nm; the wall was two-layered and the basipetally produced chains were covered by a multilayered sheath (Dorokhova et al., 1969). Spore walls of *M. faeni* were similar (Sharples and Williams, 1976). *Micropolyspora faeni* was recommended as a *nomen conservandum* after a study proving synonymy with *M. rectivirgula* (Arden-Jones, McCarthy, and Cross, 1979).

Saccharopolyspora hirsuta (type strain A1143) has a distinctive morphology. The aerial hyphae segment into beadlike chains of spores separated by lengths of "empty" hyphae, which form loops, loose spirals, or are occasionally straight. The spore sheath is covered with long, hairlike projections. Older substrate hyphae fragment into rod-shaped elements. These features should serve to distinguish this genus from *Nocardia* and other genera with a type IV cell wall (Table 1). *S. hirsuta* is, however, similar morphologically to *Nocardiopsis dassonvillei* and some *Actinomadura* species, but differs from them in cell wall composition. *S. hirsuta* does not contain mycolic acids.

S. hirsuta strains are positive in most hydrolysis and carbon source utilization tests but can be distin-

Table 4. Characteristics of *Microbispora* species.[a]

Species	Color of aerial mycelium	Soluble pigment	Starch hydrolysis	Nitrate reduction	Milk peptonization	Iodinin production	Growth at 25°C	Growth at 50°C	Growth at 55°C	Miscellaneous
M. aerata[b]	Pink	-	+	+	+	+	-	+	+	
M. amethystogenes[c]	Pink	Yellow brown	-	V	+	+	+	-	-	
M. bispora[d]	White	-	-	-	-	-	-			
M. chromogenes[e]	Pink	Yellow-gray to violet	V	+	V	-	+	+	+	
M. distatica[f]	Pink	Pale yellow	+	-	(+)	-	+	-		
M. echinospora[g]	Pink	Yellow brown	-	-	+	-	+	-	-	Spiny spores
M. parva[h]	Pink	-	V	-	(+)	V	+	+	-	
M. rosea[i]	Pink	-	-	+	V	V	+	-	-	
M. thermodiastatica[j]	Pink	-	+	-	+	-	-	+	+	
M. thermorosea[k]	Pink	-	-	-	(+)	-	-	+	+	

[a] +, Positive reaction; (+), weak reaction; −, negative reaction; V, strain reactions vary.
[b] (*Waksmania aerata*) Gerber and Lechevalier (1966). ATCC 15448.
[c] Nonomura and Ohara (1960). No type specified.
[d] (*Thermopolyspora bispora*) Henssen (1957); Lechevalier (1965), ATCC 15737 and 19993.
[e] Nonomura and Ohara (1960). No type specified.
[f] Nonomura and Ohara (1960). No type specified.
[g] Nonomura and Ohara (1971b), MB₃-1.
[h] Nonomura and Ohara (1960). No type specified.
[i] (*Waksmania rosea*) Lechevalier and Lechevalier (1957); Nonomura and Ohara (1957), ATCC 12950.
[j] Nonomura and Ohara (1969b), M₂-59.
[k] Nonomura and Ohara (1969b), M₂-64.

Table 5. Characteristics of *Microtetraspora* species.[a]

Characteristic	M. glauca[b]	M. fusca[c]	M. niveoalba[d]	M. viridis[e]
Spore color	Gray	White-gray	White	Yellow-green to pale-green
Substrate mycelium	Colorless to gray-blue to yellowish green	Colorless to amber-brown to violet	Pale yellow-brown	Yellow-brown plus green
Temp. optima	28°C	28–37°C	35–40°C	30–35°C
Litmus milk peptonization	−	−	+	+
Hydrolysis of:				
Starch	(+)	−	(+)	+
Gelatin	(+)	−	(+)	(+)
Nitrate reduction	+	−	+	−
Utilization of:				
Glucose	+	+	+	+
Arabinose	+	+	+	+
Xylose	+	+	+	+
Fructose	+	−	+	+
Rhamnose	(+)	−	+	+
Mannitol	(+)	−	+	+
Raffinose	−	−	−	−
Inositol	(+)	−	+	−
Glycerol	−	−	NT	(+)
Sucrose	(+)	−	(+)	+

[a] +, Positive reaction; (+), weak positive reaction; −, negative reaction; NT, not tested.
[b] Thiemann et al. (1968), ATCC 23057.
[c] Thiemann et al. (1968), ATCC 23058.
[d] Nonomura and Ohara, (1971b), Mt-3.
[e] Nonomura and Ohara, (1971a), Mt-1.

guished from *Nocardia* by their ability to degrade adenine and their susceptibility to lysozyme. Degradation of elastin by *S. hirsuta* suggests possible pathogenic properties since all *Nocardia* and *Actinomadura* species with this ability cause mycetoma.

Actinopolyspora halophila (type ATCC 27976) differs from other actinomycetes in its extreme halophilism. Long chains of 20 or more spores are formed on the aerial mycelium and fragmentation occurs in older substrate hyphae. Thus it is morphologically similar to several other genera including *Streptomyces, Nocardiopsis,* and *Saccharopolyspora.* It also has the same cell wall chemotype (type IV) as *Saccharopolyspora.* Its lipid composition is similar to other actinomycetes and it lacks phytanyl ether–linked lipids that are characteristic of other extremely halophilic bacteria. The major phospholipids identified were phosphatidylcholine (a major component of *Nocardia coeliaca;* Yano, Furukawa, and Kusunose, 1969), and phosphatidylglycerol. Two unidentified phospholipids and glycolipids were also detected. Its fatty acid composition is similar to certain *Nocardia* species (Yano, Furukawa, and Kusunose, 1969), the major components being branched-chain fatty acids, but it contains less palmitic acid and no tuberculostearic acid. The presence of a mucopeptide cell wall and the relatively normal lipid composition are unusual for such

an extremely halophilic bacterium. The G+C composition is 64.2 mol%.

Substrate growth is buff colored and the aerial mycelium is white on media with 15% NaCl. Acid is produced from rhamnose and fructose; these sugars and xylose, arabinose, mannitol, and inositol are utilized as carbon sources. Tweens 20, 40, 60, and 80, casein, and gelatin are hydrolyzed, but urea, xanthine, and starch are not. It is lysozyme sensitive and penicillin resistant.

Pseudonocardia is distinguished from related genera by certain unusual morphological features (Table 1). It has been suggested that four types of spore are produced by *P. thermophila* (Henssen and Schnepf, 1967). Fragmentation spores are formed on both the aerial and substrate mycelium. On aerial hyphae spores also form in an acropetal manner, terminally or on lateral branches ("Pseudonocardia" type). In addition, spores form basipetally on existing aerial hyphae ("Streptomyces" type). H. A. Lechevalier and M. P. Lechevalier (1970), however, observed that more or less simultaneous division of aerial hyphae produced long spores; these spores could divide again in old cultures to produce chains of shorter spores. A study of *P. thermophila* using time-lapse photography was made by Henssen and Schäfer (1971). They claimed that acropetal budding of hyphae and production of blastospores

Table 6. Characteristics of *Micropolyspora* species[a]

	Spore chains	Spore surface	Color		Optima temp	Antagonism	Hydrolysis		Congulation+ peptonization of milk	Nitrate reduction	Cell wall hydrolysate
			Aerial mycelium	Substrate mycelium			Starch	Gelatin			
M. brevicatena[b]	Straight 1–10 spores	Slightly warty	White to yellow-white	White to yellow	36°	None	–	N.T	–	–	Type IV. Mycolic acids (Lechevalier, 1977)
M. angiospora[c]	Curved to loops 1–15 spores	Spiny	White	Cream to ochre	27°	Gm+ bacteria Actinomycetes	+	+	+	+	?
M. caesia[d]	Mostly single, up to 4	Smooth	Gray-blue	White to gray-green	28°	Gm+ bacteria Actinomycetes	+	–	+	+	?
M. coerulea[e]	Mostly single, up to 4	Lumpy	Green-blue	Colorless to yellow green	45°–55°	Gm+ bacteria Actinomycetes	N.T.	N.T.	N.T.	N.T.	?
M. faeni[f]	Straight 5–10 spores	Smooth	White	Orange-yellow to yellow brown	50°	None	–	N.T.	–	N.T.	Type IV. No mycolic acids (Collins et al., 1977)
M. internatus[g]	Mostly single up to 4	Smooth	Absent or gray-blue to green gray	Colorless to dark green-black	37°–55°	None	+	+	+	–	Type IV
M. rectivirgula[h]	Straight 1–10	Smooth	Yellow-cream	Colorless to yellow	45°–55°	None	+	+	+	+	Type IV. No mycolic acids (Mordarska et al., 1973)
M. thermovirida[i]	Straight 1–10	Short spines	Green	Dark gray	40°–50°	None	–	N.T.	+	N.T.	?

[a] +, Positive reaction; –, negative reaction; NT, not tested. Excluded species: *M. rubrobrunea, M. viridinigra*—cell wall chemotype III, transferred to *Excellospora* (Agre and Guzeva, 1975; *M. flexuosa*—cell wall chemotype III, transferred to *Actinomadura* (Cross and Goodfellow, 1973); *M. fascifera*—cell wall chemotype IV, mycolic acids (Prauser, 1976), no valid description; *M. virida*—cell wall chemotype IV (Guzeva et al., 1972), no valid description; *M. mesophilica*—cell wall chemotype IV (Guzeva et al., 1972), no valid description.

[b] Lechevalier, Solotorovsky, and McDurmont (1961), ATCC 15333, ATCC 15725.

[c] Zhukova, Tsyganov, and Morozov (1968), LIA 3479/30.

[d] Kalakoutskii (1964).

[e] Preobrazhenskaya et al. (1973).

[f] Cross, Maciver, and Lacey (1968).

[g] Agre, Guzeva, and Dorokhova (1974), INMI 632.

[h] *(Thermopolyspora rectivirgula)* Krassilnikov and Agre (1964), INMI 683.

[i] Kosmachev (1964).

occurred; these were characteristic of the genus. The status of this genus and its relationship to others such as *Micropolyspora* and *Saccharopolyspora* may be clarified when a wider range of characteristics is determined. A G+C content of 79.3 mol% for *P. thermophila* was reported (Lechevalier, Lechevalier, and Gerber, 1971).

A number of features distinguish the two species *P. thermophila* Henssen 1957 (type ATCC 19285) and *P. spinosa* Henssen and Schäfer 1971 (type ATCC 25924). Substrate mycelium of the former is septate while in the latter it is aseptate. *P. spinosa* forms spores by constriction of the aseptate aerial hyphae, while *P. thermophila* has septate hyphae, often zigzag in shape, forming chains of blastospores and fragmentation spores in basipetal sequence. Under the electron microscope, spores of *P. thermophila* are smooth, while those of *P. spinosa* have spines. Optimum temperature for the former is 40–50°C and for the latter is 20–30°C.

Literature Cited

Agre, N. S., Efimova, T. P., Guzeva, L. N. 1975. Heterogeneity of the genus *Actinomadura* Lechevalier and Lechevalier. Microbiology [English translation of Mikrobiologiya] **44:**200–223.

Agre, N. S., Guzeva, L. N. 1975. New genus of actinomycetes, *Excellospora* gen. nov. Microbiology [English translation of Mikrobiologiya] **44:**459–463.

Agre, N. S., Guzeva, L. N., Dorokhova, L. A. 1974. *Micropolyspora internatus* new species. Microbiology [English translation of Mikrobiologiya] **43:**577–583.

Andersen, A. A. 1958. New sampler for the collection, sizing and enumeration of viable air borne particles. Journal of Bacteriology **76:**471–484.

Arden-Jones, M. P., McCarthy, A. J., Cross, T. 1979. Taxonomic and serological studies on *Micropolyspora faeni* and *Micropolyspora* strains from soil bearing the specific epithet *rectivirgula.* Journal of General Microbiology **115:**343–354.

Becker, B., Lechevalier, M. P., Lechevalier, H. A. 1965. Chemical composition of cell wall preparations from strains of various form genera of aerobic actinomycetes. Applied Microbiology **13:**236–243.

Buchanan, R. E., Gibbons, N. E. (eds.). 1974. Bergey's manual of determinative bacteriology, 8th ed. Baltimore: Williams & Wilkins.

Collins, M. D., Pirouz, T., Goodfellow, M., Minnikin, D. E. 1977. Distribution of menaquinones in actinomycetes and corynebacteria. Journal of General Microbiology **100:** 221–230.

Corbaz, R., Gregory, P. H., Lacey, M. E. 1963. Thermophilic and mesophilic actinomycetes in mouldy hay. Journal of General Microbiology **32:**449–456.

Craveri, R., Pagani, H. 1962. Thermophilic micro-organisms among actinomycetes in the soil. Annali di Microbiologia **12:**115–130.

Cross, T., Maciver, A. M., Lacey, J. 1968. The thermophilic actinomycetes in mouldy hay: *Micropolyspora faeni* sp. nov. Journal of General Microbiology **50:**351–359.

Cross, T., Goodfellow, M. 1973. Taxonomy and classification of the actinomycetes, pp. 11–112. In: Sykes, G., Skinner, F. A. (eds.), Actinomycetales: Characteristics and practical importance. London, New York: Academic Press.

Dorokhova, L. A., Agre, N. S., Kalakoutskii, L. V., Krassilnikov,

N. A. 1969. Fine structure of sporulating hyphae and spores in a thermophilic actinomycete, *Micropolyspora rectivirgula.* Journal de Microscopie **8:**845–854.

Fergus, C. L. 1964. Thermophilic and thermotolerant molds and actinomycetes of mushroom compost during peak heating. Mycologia **56:**267–284.

Gauze, G. F., Preobrazhenskaya, J. P., Kudrina, E. S., Blinov, N. O., Ryabova, I. D., Sveshnikova, M. A. 1957. Problems in the classification of antagonistic actinomycetes. Medgiz: Moscow State Publishing House for Medical Literature.

Gauze, G. F., Sveshnikova, M. A., Ukholina, R. S., Komarova, G. N., Bazhanova, V. S. 1977. Production of nocamycin, a new antibiotic, by *Nocardiopsis syringae* sp. nov. [In Russian.] Antibiotiki **22:**483–486.

Gerber, N. N., Lechevalier, M. P. 1964. Phenazines and phenoxazinones from *Waksmania aerata* sp. nov. and *Pseudomonas iodina.* Biochemistry **3:**598–602.

Gerber, N. N., Lechevalier, M. P. 1965. 1,6-Phenazinediol-5-oxide from micro-organisms. Biochemistry **4:**176–180.

Gochnauer, M. B., Leppard, G. G., Komaratat, P., Kates, M., Novitsky, T., Kushner, D. J. 1975. Isolation and characterization of *Actinopolyspora halophila,* gen. et sp. nov., an extremely halophilic actinomycete. Canadian Journal of Microbiology **21:**1500–1511.

Goodfellow, M. 1971. Numerical taxonomy of some nocardioform bacteria. Journal of General Microbiology **69:**33–80.

Goodfellow, M., Alderson, G., Lacey, J. 1979. Numerical taxonomy of *Actinomadura* and related actinomycetes. Journal of General Microbiology **112:**95–111.

Gordon, R. E., Horan, A. C. 1968. *Nocardia dassonvillei,* a macroscopic replica of *Streptomyces griseus.* Journal of General Microbiology **50:**235–240.

Gregory, P. H., Lacey, M. E. 1963. Mycological examination of dust from mouldy hay associated with farmer's lung disease. Journal of General Microbiology **30:**75–88.

Guzeva, L. N., Agre, N. S., Sokolov, A. A. 1972. Taxonomy of actinomycetes forming catenate spores. Microbiology [English translation of Mikrobiologiya] **41:**957–961.

Guzeva, L. N., Efimova, T. P., Agre, N. S., Krassilnikov, N. A. 1973. Fatty acids in the mycelia of actinomycetes that form catenate spores. Microbiology [English translation of Mikrobiologiya] **42:**19–23.

Henssen, A. 1957. Beiträge zur Morphologie und Systematik der thermophilen Actinomyceten. Archiv für Mikrobiologie **26:**373–414.

Henssen, A., Schäfer, D. 1971. Emended description of the genus *Pseudonocardia* H. and description of a new species *Pseudonocardia spinosa* Schäfer. International Journal of Systematic Bacteriology **21:**29–34.

Henssen, A., Schnepf, E. 1967. Zur Kennthis thermophiler Actinomyceten. Archiv für Mikrobiologie **57:**214–231.

Hickey, R. J., Tresner, H. D. 1952. A cobalt-containing medium for sporulation of *Streptomyces* species. Journal of Bacteriology **64:**891–892.

Kalakoutskii, L. V. 1964. A new species of the genus *Micropolyspora—Micropolyspora caesia* n. sp. Microbiology [English translation of Mikrobiologiya] **33:**765–768.

Kosmachev, A. K. 1960. Preservation of viability of thermophilic actinomycetes after long storage. Microbiology [English translation of Mikrobiologiya] **29:**210–211.

Kosmachev, A. K. 1964. A new thermophilic actinomycete *Micropolyspora thermovirida* n. sp. Microbiology [English translation of Mikrobiologiya] **33:**235–237.

Krassilnikov, N. A., Agre, N. S. 1964. On two new species of *Thermopolyspora.* Hindustan Antibiotics Bulletin **6:**97–101.

Krassilnikov, N. A., Agre, N. S., El-Registan, G. I. 1968. New thermophilic species of the genus *Micropolyspora.* Microbiology [English translation of Mikrobiologiya] **37:** 905–911.

Lacey, J. 1971. *Thermoactinomyces sacchari* sp. nov., a thermophilic actinomycete causing bagassosis. Journal of General Microbiology **66:**327–338.

Lacey, J. 1974. Moulding of sugar-cane bagasse and its prevention. Annals of Applied Biology 76:63–67.

Lacey, J., Goodfellow, M. 1975. A novel actinomycete from sugar-cane bagasse: Saccharopolyspora hirsuta gen. et sp. nov. Journal of General Microbiology 88:75–85.

Lacey, J., Goodfellow, M., Alderson, G. 1978. The genus Actinomadura Lechevalier and Lechevalier, pp. 107–117. In: Mordarski, M., Kurylowicz, W., Jeljaszewicz, J. (eds.), Nocardia and Streptomyces. Stuttgart, New York: Gustav Fischer Verlag.

Lacey, J., Lacey, M. E. 1964. Spore concentration in the air of farm buildings. Transactions of the British Mycological Society 47:547–552.

Lavrova, N. V., Preobrazhenskaya, T. P., Sveshnikova, M. A. 1972. Isolation of soil actinomycetes on selective media with rubomycin. [In Russian.] Antibiotiki 17:965–970.

Lawson, E. N., Davey, N. A. 1972. A waterborne actinomycete resembling strains causing mycetoma. Journal of Applied Bacteriology 35:389–394.

Lechevalier, H. A. 1965. Priority of the generic name Microbispora over Waksmania and Thermopolyspora. International Bulletin of Bacteriological Nomenclature and Taxonomy 15:139–142.

Lechevalier, H. A., Lechevalier, M. P. 1967. Biology of actinomycetes. Annual Review of Microbiology 21:71–100.

Lechevalier, H. A., Lechevalier, M. P. 1970. A critical evaluation of the genera of aerobic actinomycetes, pp. 393–405. In: Prauser, H. (ed.), The Actinomycetales. Jena: Gustav Fischer Verlag.

Lechevalier, H. A., Lechevalier, M. P., Gerber, N. N. 1971. Chemical composition as a criterion in the classification of actinomycetes. Advances in Applied Microbiology 14:47–72.

Lechevalier, H. A., Solotorovsky, M., McDurmont, C. I. 1961. A new genus of the Actinomycetales, Micropolyspora g. n. Journal of General Microbiology 26:11–18.

Lechevalier, M. P. 1977. Lipids in bacterial taxonomy—a taxonomist's view. CRC Critical Reviews in Microbiology 5:109–210.

Lechevalier, M. P., Lechevalier, H. A. 1957. A new genus of the Actinomycetales: Waksmania gen. nov. Journal of General Microbiology 17:104–111.

Lechevalier, M. P., Lechevalier, H. 1970. Chemical composition as a criterion in the classification of aerobic actinomycetes. International Journal of Systematic Bacteriology 20:435–443.

Meyer, J. 1976. Nocardiopsis—a new genus of the order Actinomycetales. International Journal of Systematic Bacteriology 26:487–493.

Meyer, J. 1979. New species of the genus Actinomadura. Zeitschrift für Allgemeine Mikrobiologie 19:37–44.

Minnikin, D. E., Pirouz, T., Goodfellow, M. 1977. Polar lipid composition in the classification of some Actinomadura species. International Journal of Systematic Bacteriology 27:118–121.

Mordarska, H., Mordarski, M., Goodfellow, M. 1972. Chemotaxonomic characteristics and classification of some nocardioform bacteria. Journal of General Microbiology 73:77–86.

Mordarski, M., Goodfellow, M., Szyba, K., Pulverer, G., Tkacz, A. 1977. Classification of the "rhodochrous" complex and allied taxa based upon deoxyribonucleic acid reassociation. International Journal of Systematic Bacteriology 27:31–37.

Nonomura, H., Ohara, Y. 1957. Distribution of actinomycetes in the soil. II. Microbispora, a new genus of Streptomycetaceae. Journal of Fermentation Technology 35:307–311.

Nonomura, H., Ohara, Y. 1960. Distribution of actinomycetes in soil. IV. The isolation and classification of the genus Microbispora. Journal of Fermentation Technology 38:401–405.

Nonomura, H., Ohara, Y. 1969a. Distribution of actinomycetes in soil. VI. A culture method effective for both preferential isolation and enumeration of Microbispora and Strepto-

sporangium strains in soil (part 1). Journal of Fermentation Technology 47:463–469.

Nonomura, H., Ohara, Y. 1969b. Distribution of actinomycetes in soil. VII. A culture method effective for both preferential isolation and enumeration of Microbispora and Streptosporangium strains in soil (part 2). Classification of the isolates. Journal of Fermentation Technology 47:701–709.

Nonomura, H., Ohara, Y. 1971a. Distribution of actinomycetes in soil. VIII. Green-spore group of Microtetraspora, its preferential isolation and taxonomic characteristics. Journal of Fermentation Technology 49:1–7.

Nonomura, H., Ohara, Y. 1971b. Distribution of actinomycetes in soil. IX. New species of the genera Microbispora and Microtetraspora, and their isolation method. Journal of Fermentation Technology 49:887–894.

Nonomura, H., Ohara, Y. 1971c. Distribution of actinomycetes in soil. X. New genus and species of monosporic actinomycetes. Journal of Fermentation Technology 49:895–903.

Nonomura, H., Ohara, Y. 1971d. Distribution of actinomycetes in soil. XI. Some new species of the genus Actinomadura Lechevalier et al. Journal of Fermentation Technology 49:904–912.

Pepys, J., Jenkins, P. A., Festenstein, G. N., Gregory, P. H., Lacey, M. E., Skinner, F. A. 1963. Farmer's lung. Thermophilic actinomycetes as a source of farmer's lung hay antigens. Lancet ii:607–611.

Prauser, H. 1976. New nocardioform organisms and their relationship, pp. 193–207. In: Arai, T. (ed.), Actinomycetes: The boundary micro-organisms. Baltimore, London, Tokyo: University Park Press.

Preobrazhenskaya, T. P., Ukholina, R. S., Nechaeva, N. P., Filicheva, V. A., Gavrilina, G. V., Kudinova, M. K., Borisova, V. N., Petukhova, N. M., Kovsharova, I. N., Proshlyakova, V. V., Rossolino, O. K. 1973. A new species of Micropolyspora and its antibiotic properties. [In Russian.] Antibiotiki 18:963–968.

Preobrazhenskaya, T. P., Lavrova, N. V., Ukholina, R. S., Nechaeva, N. P. 1975. Isolation of new species of Actinomadura on selective media with streptomycin and bruneomycin. [In Russian.] Antibiotiki 20:404–409.

Preobrazhenskaya, T. P., Sveshnikova, M. A., Terekhova, L. P. 1977. Key for identification of the species of the genus Actinomadura. Biology of Actinomycetes and Related Organisms 12:30–38.

Pridham, T. G., Anderson, P., Foley, C., Lindenfelser, H. A., Hesseltine, C. W., Benedict, R. G. 1956–1957. A selection of media for maintenance and taxonomic study of Streptomyces. Antibiotics Annual 1956–1957:947–953.

Rippon, J. W. 1968. Extracellular collagenase produced by Streptomyces madurae. Biochimica et Biophysica Acta 159:147–152.

Rippon, J. W., Varadi, D. P. 1968. The elastases of pathogenic fungi and actinomycetes. Journal of Investigative Dermatology 50:54–58.

Schäfer, D. 1969. Eine neue Streptosporangium Art aus türkischer Steppenerde. Archiv für Mikrobiologie 66:365–373.

Sehgal, S. N., Gibbons, N. E. 1960. Effect of some metal ions on the growth of Halobacterium cubirubrum. Canadian Journal of Microbiology 6:165–169.

Sharples, G. P., Williams, S. T. 1976. Fine structure of spore germination in actinomycetes. Journal of General Microbiology 96:323–332.

Society of American Bacteriologists. 1957. Manual of microbiological methods. New York: McGraw-Hill.

Soina, V. S., Agre, N. S. 1976. Fine structure of vegetative and sporulating hyphae in Micropolyspora fascifera. Microbiology [English translation of Mikrobiologiya] 45:287–291.

Thiemann, J. E., Pagani, H., Beretta, G. 1968. A new genus of the Actinomycetales: Microtetraspora gen. nov. Journal of General Microbiology 50:295–304.

Wenzel, F. J., Emanuel, D. A., Lawton, B. R., Magnin, G. E.

1964. Isolation of the causative agent of farmer's lung. Annals of Allergy **22**:533.

Weyland, H. 1969. Actinomycetes in North Sea and Atlantic Ocean sediments. Nature **223**:858.

Williams, S. T. 1970. Further investigations of actinomycetes by scanning electron microscopy. Journal of General Microbiology **62**:67–73.

Williams, S. T., Sharples, G. P., Bradshaw, R. M. 1974. Spore formation in *Actinomadura dassonvillei* (Brocq-Rousseu)

Lechevalier and Lechevalier. Journal of General Microbiology **84**:415–419.

Yano, I., Furukawa, Y., Kusunose, M. 1969. Phospholipids of *Nocardia coeliaca*. Journal of Bacteriology **98**:124–130.

Zhukova, R. A., Tsyganov, V. A., Morozov, V. M. 1968. A new species of *Micropolyspora: Micropolyspora angiospora* sp. nov. Microbiology [English translation of Mikrobiologiya] **37**:599–603.

CHAPTER 159

Actinomycete Genera "in Search of a Family"

HUBERT A. LECHEVALIER and MARY P. LECHEVALIER

Most of the genera of actinomycetes can be grouped into families. For example, all the fermentative nonmotile forms can be placed in the Actinomycetaceae, all those with sporangia can be placed in the Actinoplanaceae, and all those dividing in more than one plane can be placed in the Dermatophilaceae. Often these familial divisions are quite artificial, and there are good reasons to believe that genera thus grouped together may often be only remotely related. The seven genera discussed in this chapter have not been accepted into any of the families of the actinomycetes and are discussed together not because they are necessarily related but because they do not fit clearly within the range of previous chapters.

The key characteristics of these seven genera are given in Table 1. There is reference in the table to an eighth group: NMO (nonmotile oerskoviae). Although this group of organisms has not been blessed with a generic name, it is of some importance and will be discussed with the oerskoviae.

Mycoplana

Branching, Gram-negative bacteria bearing flagella were isolated from soil by Gray and Thornton (1928). These organisms, for which they created the genus *Mycoplana*, were isolated by adding 0.05–0.1% phenol to a mineral salts solution containing:

K_2HPO_4 1 g $MgSO_4$ 0.2 g
NaCl 0.1 g $CaCl_2$ 0.1 g
$FeCl_3$ 0.02 g $(NH_4)_2SO_4$ or KNO_3 0.5–1 g
Water 1,000 g

One-half to 1 g of soil was inoculated into 100 ml of the medium. When growth occurred, transplants were made to fresh flasks of the same medium, and eventually the organisms growing in the flasks were plated out on the same solidified medium. Strains thus isolated represented two species: *M. dimorpha* and *M. bullata*.

Wood (1967) remarked that marine pseudomonads are typically pleomorphic in their environment and fit the morphology of the genus *Mycoplana*. However, their pleomorphism is lost on culture; that of *Mycoplana* strains is not.

Gray and Thornton placed the genus *Mycoplana* in the family Mycobacteriaceae. The various editions of *Bergey's Manual of Determinative Bacteriology* have either classified that genus in the Mycobacteriaceae or the Pseudomonadaceae or left it out of *Bergey's Manual* altogether (eighth edition).

Based on the original description of Gray and Thornton, and on that of Sukapure et al. (1970), the description of the genus *Mycoplana* is as follows:

Table 1. Main characteristics of seven genera of actinomycetes.

Characteristic	Genus
Gram-negative; primary mycelium only; breaking into motile rods	*Mycoplana*
Gram-positive	
Type I[a] cell wall	
Aerial mycelium only; motile elements formed	*Sporichthya*
Both substrate and aerial mycelia fragment into rod-shaped and coccoid elements	*Nocardioides*
No aerial mycelium; substrate mycelium forms terminal and subterminal vesicles	*Intrasporangium*
Type III[a] cell wall; formation of synnemata with motile spores	*Actinosynnema*
Type VI[a] cell wall, no aerial mycelium	
Cell wall contains galactose; glucose fermented; motile elements formed	*Oerskovia*
Cell walls do not contain galactose; mycelia break into nonmotile elements	
Glucose fermented	NMO
Oxidative metabolism only	*Promicromonospora*

[a]See Chapter 147, Table 2, in this Handbook.

branching filaments breaking into irregular rods (0.5–1 μm wide by 1.25–4.5 μm long) that bear subpolar tufts of flagella; pili may be observed on motile cells; Gram negative; non-acid-fast; murein with *meso*-diaminopimelic acid; ribose and glycerol present in whole cell hydrolysates; DNA G+C content, 64–69 mol%; minimum temperature for growth 24°C, maximum 42°C; no growth anaerobically; catalase positive; no acid produced from carbohydrates; no nitrite from nitrate; gelatin liquefaction variable; killed by 4 h at 60°C but not by 8 h at 50°C.

The type strains for the two species are ATCC 4279 for *M. dimorpha* and ATCC 4278 for *M. bullata*. Later work (M. P. Lechevalier, unpublished results) using the methods of Gordon (1967) showed that strains of *M. dimorpha,* including the type strain and five more recent isolates, produced acid from adonitol, arabinose, fructose, galactose, glucose, mannose, sorbitol, and xylose. Nitrite production from nitrate was variable, as was phenol degradation. The main difference between the two species is that *M. dimorpha* hydrolyzes starch and *M. bullata* does not.

Sporichthya

Strains of *Sporichthya* are among the rarest and the strangest of the actinomycetes. They form an aerial mycelium that releases flagellate elements in the presence of water, but there is no primary (substrate) mycelium. Only one species, *S. polymorpha,* has been described by Lechevalier, Lechevalier, and Holbert (1968) and further studied by Williams (1970).

S. polymorpha ATCC 23823 was isolated from soil by plating out on water agar. On solid media it can grow in either of two forms depending on the richness of the medium: on poor media, such as Czapek agar (Waksman, 1950), there is formation of a sparingly branched aerial mycelium; on rich media, such as Bennett agar (Higgins, Lechevalier, and Lechevalier, 1967), the cells become wider and more pleomorphic, some cells resembling fishes (hence the generic name), and colonies change from a thin, chalk-white crust to glistening, dirty-white mucoid masses. In both cases, the cells are flagellate in the presence of water.

The sparingly branched hyphae are 0.5–1.2 μm in diameter, and 10–25 μm long. They grow away from solid agar media, to which they adhere by holdfasts that are formed from the base of the hypha that is touching the agar. The hyphae divide to form a series of rods and cocci. After division, the wall of the coccus or the rod that is directed toward the medium thickens, forming a collar resembling a bud scar. It is thus always possible to tell which end of a cell was directed toward the medium. The hyphae

are hydrophobic, their surface is smooth and, if placed in water, they break into cells bearing one to three flagella. One flagellum is most common; if there are more than one, the flagella form a tuft originating from the basal part of the cell. On rich media, the cells can be very large (up to 6 μm long).

Young hydrophobic cells and hyphae are Gram negative; older cells that are easier to wet are Gram positive. The walls have a typical Gram-positive composition and cytology.

S. polymorpha grows between 22°C and 42°C. There is no growth at 55°C. Growth is visible on Czapek, Bennett, and yeast extract (Waksman, 1950) agars in 24 hours. In static liquid media, in 3 to 4 days, a thin white pellicle appears that is formed of hyphae similar to those growing on agar. Few cells are found in the depth of the medium and virtually no growth takes place if the cultures are agitated. Growth is best in the presence of oxygen, but limited growth will take place anaerobically.

S. polymorpha is physiologically inactive. Nitrates are reduced; casein and starch are decomposed, but not gelatin, hypoxanthine, and tyrosine; acid is produced from fructose but not from arabinose, dextrin, galactose, glucose, glycogen, inulin, lactose, maltose, mannitol, mannose, α-methyl-D-mannoside, melezitose, melibiose, rhamnose, trehalose, xylose, and methyl-D-xyloside.

Nocardioides

Strains of *Nocardioides* are widely distributed in soils and other substrates, such as kaolin, from which they can be isolated by plating on the following medium:

Oatmeal 3 g	KNO$_3$ 0.3 g
K$_2$HPO$_4$ 0.5 g	MgSO$_4$ 0.2 g
Agar 15 g	Distilled water 1,000 ml
pH 7	

After the agar is solidified in the plates, the soil dilutions are suspended in melted agar, and this is spread over the first layer (Prauser, 1976 a,b).

On such a medium, the color of the vegetative mycelium is whitish to faint yellow-brown. The surface of the colonies is pasty and aerial mycelium, when formed, is thin and chalky with hyphae showing sparse irregular branching. The hyphae of both mycelia completely fragment into rod to coccoid elements. The surface of the fragments is smooth.

The cell wall of *Nocardioides* strains is Type I (see Chapter 147, Table 2, this Handbook), phospholipids are Type PI (Lechevalier, de Bièvre, and Lechevalier, 1977), and mycolates are absent. The DNA base ratios of *Nocardioides* strains are about 65 mol% G+C. These organisms are subject to lysis by specific phages (Prauser, 1976c).

Strains of *Nocardioides* grow well between 18

and 37°C with an optimum at 28°C. They are aerobic, Gram positive, non-acid-fast, catalase positive, and do not produce dark diffusible pigments. They hydrolyze starch and produce H₂S. D-Glucose, L-arabinose, D-xylose, D-mannitol, D-fructose, and rhamnose are utilized for growth by all strains. Most strains will utilize sucrose, and rarely raffinose; *i*-inositol is not utilized.

Only one species, *N. albus*, has been described so far. The epithet refers to the color of the aerial mycelium that is often not produced, and, if produced, easily lost upon cultivation. The type strain is IMET 7807 (ATCC 27980).

Intrasporangium

Strains of *Intrasporangium* are rare. The original isolate belonging to the type and only described species, *I. calvum*, was isolated from the air of a school dining room on a plate of meat extract peptone agar. The main characteristic of the genus, as it was described by Kalakoutskii, Kirillova, and Krassilnikov (1967), was the formation of intercalary sporangia containing nonmotile spores. Lechevalier and Lechevalier (1969) reported that the ''sporangia'' of *I. calvum* were vesicles containing convoluted membranous elements but no spores. It was suggested that the vesicles were a reaction of the aging culture to unfavorable nutritional conditions. In any case, *I. calvum* is a chemotaxonomically distinct, interesting organism, and the following description is mainly based on that of Kalakoutskii, Kirillova, and Krassilnikov (1967). Colonies on meat extract peptone agar are round, glistening, and whitish, becoming cream on aging. Fine branching hyphae (0.4–1.2 μm in diameter) grow on and into the agar. Aerial mycelium is not formed. Fragmentation is not apparent by visual observation, but the mycelium breaks up readily into fragments during the process of making smears or when disturbed by water. Hyphae bear spherical, oval to lemon-shaped vesicles (5–15 μm in diameter) terminally or subterminally in cultures at least 6 days old; longer incubation (up to 1 month) is usually required. The vesicles may contain oval to round bodies that may show Brownian movement. Gram positive becoming Gram variable; not acid-fast. Cell wall is Type I (Sukapure et al., 1970); phospholipids are Type PIV (Lechevalier, de Bièvre, and Lechevalier, 1977).

I. calvum grows well on a variety of media containing meat extract and peptone. It also grows on Bennett agar and yeast extract–glucose agar. Growth takes place only aerobically and the growth range is from 10 to 42°C (Sukapure et al., 1970). Nitrate is reduced to nitrite; gelatin is not liquefied.

The type strain is ATCC 23552 (Kalakoutskii 7

KIP). The G+C content of its DNA is not known, but that of another isolate, LL-12-17, which may belong to another species, is 71 mol% (Sukapure et al., 1970).

Actinosynnema

Hasegawa, Lechevalier, and Lechevalier (1978) described an actinomycete they named *Actinosynnema mirum*, which was characterized by the formation of synnemata bearing chains of flagellated spores. It is believed that an organism previously described as *Nocardia* sp. (Higashide et al., 1977) belongs to the same group of organisms, as does *N. uniformis* ATCC 21806.

Isolated from a grass blade placed in a moist chamber on diluted potato-carrot agar, *A. mirum* grows on a variety of media including yeast malt agar, Bennett agar, and oatmeal agar forming yellowish colonies with scant yellowish-white aerial mycelium. Substrate hyphae are about 0.5 μm in diameter and coalesce to form a mycelium bearing synnemata (up to 180 μm high) and domelike bodies on which separate aerial hyphae are formed. The aerial hyphae mature into chains of spores (0.4–1 μm by 1.5–3 μm) which, in the presence of liquid nutrients such as brain heart infusion broth, become flagellated (5 to 15 peritrichous flagella). Aerial mycelium breaking up into flagellated elements may also be produced directly on flat colonies. The hyphae and the spores are mainly Gram positive and are not acid-fast. Cell wall preparations are Type III and whole cell hydrolysates are Type C. Phospholipids are Type PII with phospholipid fatty acids that are Type 3 (Hasegawa, Lechevalier, and Lechevalier, 1979).

A. mirum grows aerobically (between 10 and 35°C) or in an atmosphere containing 10% CO₂; no growth takes place anaerobically. Starch, casein, tyrosine, and gelatin are hydrolyzed, but xanthine, hypoxanthine, adenine, and urea are not. Nitrate reductase and phosphatase are produced. Milk is peptonized. Growth takes place in lysozyme broth. No growth occurs at 37°C or on 5% NaCl. Tartrate, pyruvate, lactate, and malate are utilized, whereas benzoate, acetate, citrate, and succinate are not. Acid is produced from fructose, lactose, maltose, D-mannitol, L-arabinose, D-mellibiose, D-mannose, L-rhamnose, xylose, dextrin, galactose, glucose, trehalose, raffinose, soluble starch, sucrose, cellobiose, glycogen, and adonitol, but not from *i*-inositol, sorbitol, D-ribose, salicin, inulin, glycogen, dulcitol, erythritol, α-methyl-D-glucoside, or α-methyl-D-mannoside.

The type strain is IMRU 3971 (Hasegawa 101).

Oerskovia

Topping (1937) and Ørskov (1938) reported the occurrence of actinomycetes whose mycelium broke up into motile elements. Erikson (1954) studied one of Ørskov's strains (No. 27) and proposed to place it in the new species *Nocardia turbata*. Later development in the knowledge of the actinomycetes revealed that such organisms did not belong to the genus *Nocardia*, from which they differed markedly in cell wall composition. The genus *Oerskovia* (Prauser, Lechevalier, and Lechevalier, 1970) was proposed for these organisms, which are found in soil, decaying plant material, in aluminum hydroxide gels (Lechevalier, 1972) and in clinical specimens, including blood samples (Sottnek et al., 1977). Oerskoviae are very active enzymatically (Hayward and Sly, 1976; Mann, Heintz, and Macmillan, 1972; Obata et al., 1976) and have been reported to deacetylate steroids (Zaretskaya et al., 1968). They have also been implicated in human endocarditis (Reller et al., 1975). Two species are recognized: *O. turbata* (Prauser, Lechevalier, and Lechevalier, 1970) and *O. xanthineolytica* (Lechevalier, 1972).

Strains of *Oerskovia* have extensively branching vegetative hyphae, about 0.5 μm in diameter, growing on the surface of and penetrating into the agar, which break up into motile, rod-shaped elements. Growth appears bacterioid in smears. No aerial mycelium is formed. They are Gram positive, not acid-fast, and catalase positive when grown in air and catalase negative when grown anaerobically. They are facultatively anaerobic when grown on Trypticase soy medium. Glucose is attacked both oxidatively and fermentatively. Cell walls are Type VI plus major amounts of galactose. Aspartic acid may be absent. Phospholipids are Type PV with phospholipid fatty acids that are Type 1 (Lechevalier, de Bièvre, and Lechevalier, 1977). The G+C content of the DNA varies from 70.5 to 75 mol% (Sukapure et al., 1970). The type species is *O. turbata*.

The two described species of *Oerskovia* have the following properties in common. They are yellow and have extensively branching vegetative hyphae breaking into motile, rodlike elements, about 0.4 μm by 1.1 μm, which are usually monotrichously flagellated when small and peritrichously flagellated when long. Nitrite is produced from nitrate. Casein, gelatin, and starch are hydrolyzed. Acetate, lactate, and pyruvate are utilized. Adenine, benzoate, citrate, malate, succinate, and tartrate are not utilized. Acid is produced from arabinose, cellobiose, dextrin, fructose, galactose, glucose, glycerol, glycogen, α-methyl-D-glucoside, lactose, maltose, mannose, salicin, sucrose, trehalose, xylose, and β-methyl-D-xyloside. Acid formation is variable

Table 2. Characteristics of two species of *Oerskovia*.

Characteristic	*O. turbata*	*O. xanthineolytica*
Motility	+	+
Cell wall	Type VI Galactose	Type VI Galactose
Growth at 42°C	−	+[a]
Hydrolysis of:		
Casein	+	+
Xanthine	−	+
Hypoxanthine	−	+
Tyrosine	−	+(24)[b]
Urea	−(86)	−(88)
Production of:		
Cellulase	−	−
Phosphatase	−	+
Acid from:		
Mannitol	−	+(18)
Melibiose	−(86)	+
Raffinose	−	+[a]
Sorbitol	−(14)	+
Utilization of:		
Lactate	+	+
Propionate	−(86)	+(59)

[a] Growth at 42°C and acid from raffinose were considered variable for this taxon by Sottnek et al. (1977).
[b] Numbers in parentheses indicate percentages of positive or negative strains in variable tests.

from mannitol, melibiose, raffinose, and sorbitol. No acid is formed from adonitol, erythritol, rhamnose, and sorbose. Strains of *Oerskovia* produce DNase and β-D-galactosidase. They do not produce cytochrome oxidase and cellulase and do not grow in lysozyme broth. They survive 50°C for 8 h and 60°C for 4 h. Differences between the two species of *Oerskovia* are indicated in Table 2. The type strain of *O. turbata* is Statens Seruminstitut of Copenhagen strain 891 (ATCC 25835) and of *O. xanthineolytica* is LL G62 (IMRU 3959 = ATCC 27402). *Arthrobacter luteus* ATCC 21606 has been found to be a strain of *O. xanthineolytica* (M. P. Lechevalier, unpublished).

NMO

Nonmotile actinomycetes similar to oerskoviae can be isolated from soil and other substrates. These have been called "nonmotile *Oerskovia*-like strains" or NMOs (Lechevalier, 1972). The *Arthrobacter* sp. of Jones and Ballou (1969), which produces an α-mannanase, has been found to be an NMO by Lechevalier (1972). One can distinguish two types of NMOs; properties differentiating them from oerskoviae are given in Table 3.

Table 3. Properties of NMOs that differ from the general properties of *Oerskovia* spp.

Properties	NMOs	
	Type A	Type B
Motility	−	−
Cell wall	Type VI[a] No galactose	Type VI[a] No galactose
Hydrolysis of:		
Casein	−(66)[b]	−
Xanthine	+	−
Hypoxanthine	+	−
Urea	−(66)	+
Production of:		
Cellulase	+	+
Phosphatase	+	−
Growth at 42°C	−(66)	−
Acid from:		
Mannitol	−	−
Melibiose	−(66)	−
Raffinose	+	V[c]
Sorbitol	−(66)	+
Utilization of:		
Lactate	+(66)	+
Propionate	+	V

[a] See Table 2 in this Handbook, Chapter 147.
[b] Numbers in parentheses indicate percentage of positive or negative strains in variable tests.
[c] V, variable.

Promicromonospora

In 1961, Krassilnikov, Kalakoutskii, and Kirillova described *Promicromonospora citrea,* a soil actinomycete that formed a yellow mycelium soon breaking into bacillary elements. The organism was reported to bear single conidia on the primary mycelium, as micromonosporae do, and to produce a sparse, sterile, aerial mycelium. The production of spores by this organism has not been confirmed in other laboratories, but the breaking up of the mycelium into short nonmotile fragments has been observed by several investigators. The following description is based on that of Lechevalier (1972), who isolated other strains and compared their properties with those of the original isolate.

The mycelium is 0.5–1.0 μm in diameter, breaking into elements that are variable in length and often pleomorphic. The metabolism is oxidative and no growth takes place anaerobically. Nitrates are not accepted as electron donors. Cell wall is Type VI, phospholipids are Type PV, and phospholipid fatty acids are Type 1 (Lechevalier, de Bièvre, and Lechevalier, 1977). The G+C content of DNA is 73 mol% (Tsyganov et al., 1970). Casein, tyrosine, gelatin, starch, and DNA are hydrolyzed; cellulose, hypoxanthine, and adenine are not utilized; the utili-

zation of xanthine and urea is variable. Phosphatase, galactosidase, and catalase are produced; cytochrome oxidase is not produced and the production of nitrate reductase is variable. Survival for 8 h at 50°C or for 4 h at 60°C is variable; there is no survival in lysozyme broth. Acid is produced from mannitol, melibiose, raffinose, and rhamnose; the production of acid from inositol and sorbitol is variable. Acetate, citrate, lactate, malate, propionate, pyruvate, and succinate are utilized, whereas the utilization of tartrate is variable.

The type strain is probably Kalakoutskii's strain 18 (ATCC 15908).

Literature Cited

Erikson, D. 1954. Factors promoting cell division in a "soft" mycelial type of Nocardia; *Nocardia turbata* n.sp. Journal of General Microbiology **11**:198–208.

Gordon, R. E. 1967. The taxonomy of soil bacteria, pp. 293–321. In: Gray, T. R. G. and Parkinson, B. (eds.), The ecology of soil bacteria. Liverpool: Liverpool University Press.

Gray, P. H. H., Thornton, H. G. 1928. Soil bacteria that decompose certain aromatic compounds. Centralblatt für Bakteriologie, Parasitenkunde und Infektionskrankheiten. II. Abt. **73**:74–96.

Hasegawa, T., Lechevalier, M. P., Lechevalier, H.: 1978. New genus of *Actinomycetales: Actinosynnema* gen. nov. International Journal of Systematic Bacteriology **28**:304–310.

Hasegawa, T., Lechevalier, M. P., Lechevalier, H. A. 1979. Phospholipid composition of motile actinomycetes. Journal of General and Applied Microbiology **25**:209–213.

Hayward, A. C., Sly, L. I. 1976. Dextranase activity in *Oerskovia xanthineolytica.* Journal of Applied Bacteriology **40**:355–364.

Higashide, E., Asai, M., Ootsu, K., Tanida, S. Kozai, Y., Hasegawa, T., Kishi, T., Sugino, Y., Yoneda, M. 1977. Ansamitocin, a group of novel maytansinoid antibiotics with antitumour properties from *Nocardia.* Nature **270**:721–722.

Higgins, M. L., Lechevalier, M. P., Lechevalier, H. A. 1967. Flagellated actinomycetes. Journal of Bacteriology **93**:1446–1451.

Jones, G. H., Ballou, C. E. 1969. Purification and some properties of an α-mannosidase from an *Arthrobacter.* Journal of Biological Chemistry **244**:1043–1051.

Kalakoutskii, L. V., Kirillova, I. P., Krassilnikov, N. A. 1967. A new genus of the Actinomycetales—*Intrasporangium* gen. nov. Journal of General Microbiology **48**:79–85.

Krassilnikov, N. A., Kalakoutskii, L. V., Kirillova, N. F. 1961. New genus of ray-fungi *Promicromonospora,* gen. nov. [In Russian.] Izvestiia Akademii Nauk SSSR; Seriia biologicheskaia: 107–112.

Lechevalier, H., Lechevalier, M. P. 1969. Ultramicroscopic structure of *Intrasporangium calvum* (*Actinomycetales*). Journal of Bacteriology **100**:522–525.

Lechevalier, M. P. 1972. Description of a new species, *Oerskovia xanthineolytica,* and emendation of *Oerskovia* Prauser et al. International Journal of Systematic Bacteriology **22**:260–264.

Lechevalier, M. P., de Bièvre, C., Lechevalier, H. 1977. Chemotaxonomy of aerobic actinomycetes: Phospholipid composition. Biochemical Systematics and Ecology **5**:249–260.

Lechevalier, M. P., Lechevalier, H., Holbert, P. E. 1968.

Sporichthya, un nouveau genre de *Streptomycetaceae*. Annales de l'Institut Pasteur **114**:277–286.

Mann, J. W., Heintz, C. E., Macmillan, J. D. 1972. Yeast spheroplasts formed by cell wall-degrading enzymes from *Oerskovia* sp. Journal of Bacteriology **111**:821–824.

Obata, T., Yamashita, K., Fujioka, K., Hara, S., Akiyama, H. 1976. Purification of β-1, 3-glucanases from highly lytic enzymes produced by a species of *Oerskovia* and study of their properties. [In Japanese.] Journal of Fermentation Technology **54**:640–648.

Ørskov, J. 1938. Untersuchungen über Strahlenpilze, reingezüchtet aus dänischen Erdproben. Centralblatt für Bakteriologie, Parasitenkunde und Infektionskrankheiten. II. Abt. **98**:344–354.

Prauser, H. 1976a. *Nocardioides*, a new genus of the order *Actinomycetales*. International Journal of Systematic Bacteriology **26**:58–65.

Prauser, H. 1976b. New nocardioform organisms and their relationship, pp. 193–207. In: Arai, T. (ed.), Actinomycetes, the boundary microorganisms. Tokyo, Singapore: Toppan.

Prauser, H. 1976c. Host-phage relationships in nocardioform organisms, pp. 266–284. In: Goodfellow, M., Brownell, G. H., Serrano, J. A. (eds), The biology of the nocardiae. London, New York, San Francisco: Academic Press.

Prauser, H., Lechevalier, M. P., Lechevalier, H. 1970. Description of *Oerskovia* gen. n. to harbor Ørskov's motile *Nocardia*. Applied Microbiology **19**:534.

Reller, L. B., Maddoux, G. L., Eckman, M. R., Pappas, G. 1975. Bacterial endocarditis caused by *Oerskovia turbata*. Annales of Internal Medicine **83**:664–666.

Sottnek, F. O., Brown, J. M., Weaver, R. E., Carroll, G. F. 1977. Recognition of *Oerskovia* species in the clinical laboratory: Characterization of 35 isolates. International Journal of Systematic Bacteriology **27**:263–270.

Sukapure, R. S., Lechevalier, M. P., Reber, H., Higgins, M. L., Lechevalier, H. A., Prauser, H. 1970. Motile nocardoid *Actinomycetales*. Applied Microbiology **19**:527–533.

Topping, L. E. 1937. The predominant microorganisms in soils. I. Description and classification of the organisms. Centralblatt für Bakteriologie Parasitenkunde und Infektionskrankheiten. II. Abt. **97**:289–304.

Tsyganov, V. A., Efimova, T. P., Zhukova, R. A., Konev, Yu. E. 1970. New forms of actinomycetes and taxonomic significance of some of their metabolites, pp. 329–335. In: Prauser, H. (ed.), The actinomycetales. Jena: Gustav Fischer.

Waksman, S. A. 1950. The Actinomycetes. Their nature, occurrence, activities and importance. Waltham: Chronica Botanica.

Williams, S. T. 1970. Further investigations of actinomycetes by scanning electron microscopy. Journal of General Microbiology **62**:67–73.

Wood, E. J. F. 1967. Microbiology of oceans and estuaries. Amsterdam, London, New York: Elsevier.

Zaretskaya, I. I., Kogan, L. M., Sys, Zh. D., Tikhomirova, O. B., Skryabin, G. K., Torgov, I. V. 1968. *Proactinomyces turbata* deacetylation of steroid acetates. Microbiology U.S.S.R. **37**:348–349.

Some Obligately Symbiotic Bacteria

CHAPTER 160

Prokaryotic Symbionts of *Paramecium*

LOUISE B. PREER

Prokaryotic symbionts in *Paramecium* have been known for over a century. Their significance, however, has frequently eluded the investigator. They were first seen with the microscope in the macronucleus by Müller (1856). Some of his students and other early workers thought that the threadlike bodies they saw were involved in the sexual process of *Paramecium*. Accounts of early observations of bacteria in paramecia are reviewed by Kirby (1941), Wichterman (1953), and Ball (1969).

In more recent times, bacterial symbionts have been studied most extensively in the *Paramecium aurelia* complex (for reviews see Beale, Jurand, and Preer, 1969; Gibson, 1974; Preer, Preer, and Jurand, 1974; Soldo, 1974; Sonneborn, 1959); these are the symbionts that are discussed here. Much less is known about the symbionts of other species of *Paramecium*. The few that are known are mentioned below under Habitats. Symbionts in the *P. aurelia* complex were first detected with the discovery of the killer trait by Sonneborn (1938). The ability of paramecia to kill sensitive paramecia, an inherited trait, was shown by genetic techniques to be transmitted via the cytoplasm (Sonneborn, 1943). Consequently, the genetic determinant of the trait was regarded as a cytoplasmic factor, and named *kappa*. *Kappa* confers upon the paramecia in which it is found the ability to liberate particles into the medium that are toxic to other strains of paramecia. From data obtained in studies using X-rays, Preer (1948) determined that *kappa* is similar in size to bacteria and demonstrated *kappa* in the cytoplasm of killers as Feulgen-staining bodies (Preer, 1950). An accumulation of evidence over the years leaves no doubt that these Feulgen-positive bodies, found in many stocks of the *P. aurelia* complex, are bacterial symbionts (for a detailed review, see Preer, Preer, and Jurand, 1974). Specific nuclear genes of *Paramecium* are necessary for the maintenance of different symbionts. These bacteria have become so dependent upon the *Paramecium* that they can rarely, if ever, be cultured free of the host. It has been demonstrated that certain of the killer stocks contain not only bacteria, but symbionts within the bacteria—defective phages in some stocks (Preer and Preer, 1967) and plasmids in others (Dilts, 1976, 1977; Quackenbush, 1977b).

Different killer stocks of the *P. aurelia* complex induce specific prelethal symptoms in sensitive stocks when sensitives are placed with killers or into the medium in which killers have lived. These symptoms include spinning, vacuolization, paralysis, formation of aboral humps, and rapid lysis. Some killers only induce the death of their mates at conjugation and are called mate-killers. In any case, killing is highly specific; and each killer is specifically resistant to the toxin it produces. Furthermore, most are sensitive to the toxins of other killers. Symbiont-bearers appear to suffer no ill effects from the symbionts they carry, though the symbionts may number in the thousands. While many of the symbionts are responsible for the killer phenotype in the *P. aurelia* complex, a number of symbiont-bearers are nonkillers.

All of the symbionts are Gram-negative, usually straight rods, but in a few instances, spiral in form. A unit membrane surrounds each symbiont and outside of this membrane lies the unit membrane of the typical Gram-negative cell wall. No distinct nuclear region is apparent. In the eighth edition of *Bergey's Manual* (Buchanan and Gibbons, 1974) the symbionts are placed as an addendum to Part 8. They are commonly referred to by Greek letters, their original designation; more recently they have also been assigned binomial names. Some of the variation existing among these symbionts can be seen from the brief descriptions that follow (Beale et al., 1969; Preer, Preer, and Jurand, 1974).

Alpha (Fig. 1) appears as numerous short rods or crescents (0.3–0.5 μm × 1–3 μm) in the vegetative macronucleus when the host paramecium is growing rapidly, and as long (5–6 μm) spiral forms when in the process of infecting a new macronucleus and the host is growing slowly. When the micronucleus gives rise to a new macronucleus after conjugation or autogamy, *alpha* from the old degenerating macronuclear fragments passes through the cytoplasm into the newly forming macronucleus. Loss of the symbiont may occur if the paramecia multiply rapidly immediately following conjugation or

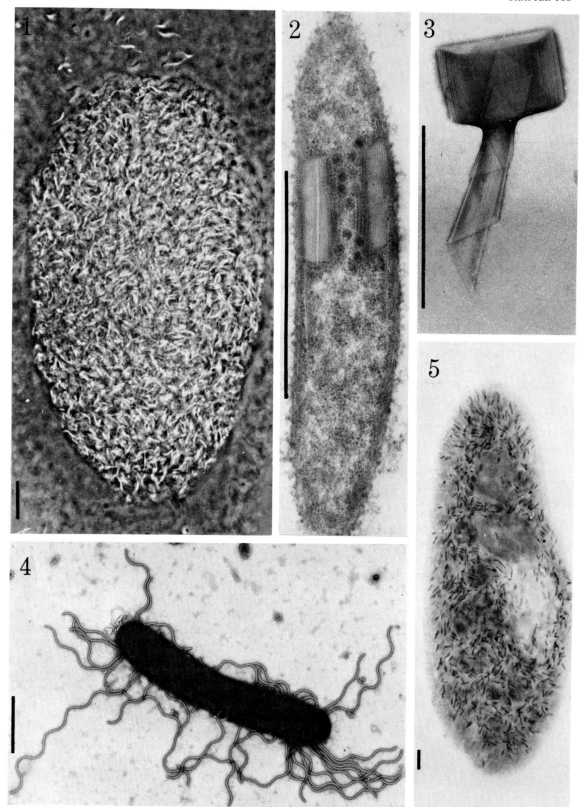

Fig. 1. Vegetative macronucleus of stock 562 *Paramecium biaurelia* containing *alpha* in spiral form. Note few *alpha* also in the cytoplasm. Bright phase-contrast × 1,120, scale 10 μm (after Preer [1969] with permission).

Fig. 2. Electron micrograph of longitudinal section of bright *kappa* of stock 7 *Paramecium biaurelia*. Note the numerous dark-staining phages inside the coiled refractile body. × 60,000, scale 1.0 μm (after Preer and Jurand [1968] with permission).

autogamy, before *alpha* has a chance to reach the new macronucleus. *Alpha* is never found in the micronucleus, and only at certain times in small numbers in the cytoplasm. It has been shown to infect a few other stocks of *P. biaurelia* merely by culturing the *alpha*-bearing and *alpha*-lacking stocks in the same flask (Preer, 1969). *Alpha*-bearers are nonkillers. When originally found, *alpha* was present in killer stock 562, which had *kappa* in the cytoplasm as well as *alpha* in the macronucleus. Symbionts similar in form to *alpha* have been found in other species of *Paramecium* (Hafkine, 1890; Müller, 1856; Ossipov and Ivakhnyuk, 1972; Petschenko, 1911).

Gamma is a diminutive bacterium (0.25–0.35 μm × 0.5–1 μm), frequently appearing as doublets, and usually present in small numbers. Strong killing is shown by *gamma*-bearers; death follows vacuolization of sensitive paramecia. *Gamma* is characterized cytologically by the presence of an extra unit membrane, which seems to be continuous with the endoplasmic reticulum. *Gamma*, like all of the other symbionts described below, is found only in the cytoplasm.

Delta is a rod (0.4–0.7 μm × 1–2 μm), distinguished by a layer of electron-dense material surrounding the outer of the two membranes. It has sparse peritrichous flagella, and on occasion shows slight motility. At one time it was reported to produce a toxin causing paralysis of sensitives, but none of the stocks now carrying *delta* are killers. It is found in certain stocks of all of the symbiont-bearing species of the *P. aurelia* complex except *P. pentaurelia*. It frequently is found coexisting with other symbionts such as *mu* or *kappa*.

Kappa (Fig. 2) varies in size with the stock (0.4–0.7 μm × 1–2.5 μm). It is distinguished by an inclusion 0.5 μm in diameter, present in a varying proportion of the symbiont population. When viewed with the bright phase-contrast microscope, this inclusion appears as a bright, refractile (R) body. The R body (Fig. 3) is a proteinaceous ribbon 0.2–0.5 μm wide, 2–15 μm long, and 13 nm thick, wound into a tight coil. Defective phages have been seen on the inner end of the coiled R body. *Kappas* lacking R bodies are ''nonbrights''. They are not toxic and are the reproductive form of *kappa*. Covalently closed circular DNA, typical of bacterial plasmids, has been isolated from some strains of *kappa* (Dilts, 1976, 1977; Quackenbush, 1977b).

Bacteriophages, apparently defective, are found in other strains of *kappa*. It is thought that the extrachromosomal DNA of the plasmid or prophage is carried by the nonbright *kappas*. Whether this DNA is integrated into the *kappa* chromosome is not known. In a certain percentage of the *kappas*, induction occurs spontaneously with the production of plasmid DNA or mature phage and the R body, which appears to be a plasmid or phage product. The nonbright thus becomes a bright *kappa*. Upon induction, *kappas* lose their ability to reproduce. Induction of R bodies and presumably of the plasmid has been produced in one stock by exposure to ultraviolet light (L. B. Preer et al., 1974). *Kappas* are thought to be released from the paramecium via the cytopyge into the medium. When bright *kappas* are ingested by sensitive paramecia into the food vacuoles, the R body unrolls into a long ribbon, rupturing the membrane surrounding the vacuole. Prelethal changes occurring in sensitive paramecia with the unrolling of the R body have been studied with the electron microscope (Jurand and Preer, 1977; Jurand, Rudman, and Preer, 1971).

Lambda (Figs. 4 and 5) has the appearance of a typical motile bacterium, although it has never been observed to move in the cytoplasm. It is a large rod (0.6–0.8 μm × 2–4 μm), heavily covered with peritrichous flagella. *Lambda* is found within vacuoles in the cytoplasm, frequently more than one *lambda* per vacuole. Sensitive paramecia are lysed by *lambda*-bearing paramecia after as little as a half-hour exposure to the killer culture. Only certain stocks of the *P. aurelia* complex are sensitive to the killing action of *lambda*-bearers (Sonneborn, Mueller, and Schneller, 1959).

Mu, first described by Siegel (1953), is distinctive in that its killing action is wholly dependent upon the cell-to-cell contact at conjugation between killer and sensitive animals. Nuclear exchange is not required, but passage of *mu* from the killer to the sensitive mate results in eventual death of the sensitive mate or the few progeny to which it may give rise. *Mu* is a slender rod (0.3–0.5 μm × 1–4 μm), often elongated. gated. It is similar to *pi*, *nu*, and nonbright *kappas* in appearance, except in stock 570, a mate-killer of *P. biaurelia*, where R bodies are seen in a small percentage of the symbionts.

Nu (0.4–0.7 μm × 1–1.5 μm), a nonkilling symbiont, is similar in appearance to *pi* and *mu*. It has been reported to render host stock 87 of *P.*

Fig. 3. Electron micrograph of isolated refractile body from *kappa* of stock 51 *Paramecium tetraurelia* beginning to unroll from the inside; negative staining with phosphotungstic acid × 48,000, scale 1.0 μm (after Preer et al. [1972] with permission).

Fig. 4. Electron micrograph of *lambda* of stock 327 *Paramecium octaurelia;* negative staining with phosphotungstic acid × 15,000, scale 1 μm (after Preer, Preer, and Jurand [1974] with permission).

Fig 5. Whole mount of stock 239 *Paramecium biaurelia* containing *lambda*, seen as dark-staining rods in the cytoplasm; osmium-lacto-orcein preparation × 500, scale 10 μm (after Preer, Preer, and Jurand [1974] with permission).

pentaurelia resistant to rapid lysis killing by *lambda*-bearing stocks of *P. octaurelia* (Holtzman, 1959).

Pi until recently was considered a mutant of *kappa*, since it appeared originally in *kappa*-bearing stocks. It is a nonkilling symbiont that looks much like nonbright *kappa*, except that it is a more slender rod. Evidence to be discussed indicates that stock 51 of *P. tetraurelia*, when brought into the laboratory, carried more than one kind of symbiont, and these became segregated in subsequent isolations. *Pi* is one of the segregants.

Sigma (0.7–0.9 μm × 2–10 μm) is the largest of all known symbionts of the *P. aurelia* complex. It is a curved or sinuous flagellated rod, which, like *lambda*, has never shown any motility. It is present in only one stock of *P. biaurelia*. Killers carrying *sigma* cause rapid lysis killing of the same stocks that are sensitive to *lambda*-bearers (Sonneborn et al., 1959).

A technique developed by Simon (1971) and Simon and Schneller (1973) makes it possible to freeze paramecia in liquid nitrogen and recover them alive with their symbionts upon thawing. Stocks of paramecia carrying the symbionts described above may be obtained from the American Type Culture Collection.

Symbionts and hosts of the *P. aurelia* complex form well-integrated systems. So successfully have the symbionts adapted to their specific hosts that most of them are no longer able to infect paramecia except under special conditions. *Alpha* alone is readily infective, but only to a few stocks. In the case of *kappa*, the symbiont-host association is complicated. Not only *Paramecium* and *kappa*, but also the phage or plasmid are parts of a delicately balanced viable complex, able to persist indefinitely both in nature and in the laboratory. It is interesting that one of the most striking phenotypes of *Paramecium*, the killer trait, is apparently mediated by the plasmid or phage carried by the bacterial symbiont of the paramecium. While not achieving the status of a cell organelle, the bacterial symbionts are more than infective elements. They have become a persistent component of the genetic apparatus of the paramecium. In another ciliate, *Euplotes*, the bacterial symbiont *omicron* has been shown to be essential for the survival of the host (Heckmann, 1975).

Habitats

Bacterial symbionts have been reported in a number of species of *Paramecium*, either in the macronucleus, the micronucleus, or the cytoplasm. Early workers, beginning with Müller (1856) observed slender rods in a species that, according to the description, was *Paramecium caudatum* (see Introduction for references to the early literature). Hafkine

(1890) named three species of bacteria in *Paramecium;* two were in the micronucleus, *Holospora undulata* and *H. elegans,* and one in the macronucleus, *H. obtusa.* Later workers have observed these species in *P. caudatum,* and confirm his localization of *H. undulata* and *H. obtusa. Paramecium bursaria* was the host for a symbiont described by Wichterman (1945). Since killers are found in *P. bursaria* (Chen, 1955; Dorner, 1957) and in *P. polycaryum* (Takayanagi and Hayashi, 1964), it is possible that bacterial symbionts are responsible for the trait in these species, although symbionts have not been observed cytologically in these particular killer stocks. Within recent years more complete characterization of symbionts of paramecia has been possible. Jenkins (1970), with the electron microscope, observed a Gram-negative symbiont, *epsilon,* in *Paramecium multimicronucleatum.* It is found only within loculi formed by an extension of the outer membrane of the micronuclear and macronuclear envelopes. Ossipov and his co-workers describe a micronuclear symbiont, *omega* (*Holospora undulans*), and a macronuclear symbiont, *iota* (*Holospora obtusa*), in *Paramecium caudatum,* which have an infective, sporelike stage in their life cycle (Ossipov, 1973; Ossipov, Gromov, and Mamkaeva, 1973; Ossipov and Ivakhnyuk, 1972; Ossipov et al., 1977). They report that germination of the spore occurs inside the food vacuole of an infected paramecium. The symbiont moves from the food vacuole to the micro- or macronucleus by a process involving the host membranes that surround the symbiont at all times as it penetrates. Görtz and Dieckmann (1977) report a symbiont in the macronucleus of *P. caudatum* similar to *iota* as described by Ossipov et al. (1977) and corresponding to *H. obtusa,* described by Hafkine (1890). The symbiont is infective, but they have observed no spores. In the *P. aurelia* complex, bacterial symbionts have been most thoroughly studied; all of the known symbionts in this complex are cytoplasmic except *alpha,* which is found almost exclusively in the macronucleus. See Table 1 for a list of the symbionts, some of the stocks in which they are found, the species of *Paramecium* to which the stocks belong, place of origin, and type of killing exhibited.

The *P. aurelia* complex consists of 14 sibling species that have been given binomial names (Sonneborn, 1975). The primary distinction between the species is based on a series of complementary mating types. The first species is designated *P. primaurelia* (mating types I and II), the second, *P. biaurelia* (mating types III and IV), etc. Conjugation with viable offspring occurs readily between animals of complementary mating types of the same species, but, with rare exception, does not occur between animals of different species. The species of the *P. aurelia* complex are worldwide in their distribution, inhabiting freshwater streams and ponds. For exam-

Table 1. Symbionts of the *Paramecium aurelia* complex.

Symbiont		Host			
Common name	Binomial name	Stock or strain	Species	Origin	Type of killing
kappa	*Caedobacter taeniospiralis*	47	*P. tetraurelia*	Berkeley, California	Vacuolizer
	Caedobacter taeniospiralis	51	*P. tetraurelia*	Spencer, Indiana	Hump
	Caedobacter taeniospiralis	51m43	*P. tetraurelia*	51m42	Resistant nonkiller
	Caedobacter taeniospiralis	139	*P. tetraurelia*	Florida	Hump
	Caedobacter taeniospiralis	169	*P. tetraurelia*	Morioka City, Japan	Hump
	Caedobacter taeniospiralis	298	*P. tetraurelia*	Panama	Hump
	Caedobacter taeniospiralis	A1	*P. tetraurelia*	Australia	Hump
	Caedobacter pseudomutans	51m1	*P. tetraurelia*	51	Spin
	Caedobacter varicaedens	7	*P. biaurelia*	Pinehurst, North Carolina	Spin
	Caedobacter varicaedens	7m1	*P. biaurelia*	7	Paralysis
	Caedobacter varicaedens	310	*P. biaurelia*	New Zealand	Vacuolizer, paralysis
	Caedobacter varicaedens	562	*P. biaurelia*	Milan, Italy	Vacuolizer
	Caedobacter varicaedens	1038	*P. biaurelia*	Syktykar, USSR	Spin
	Caedobacter paraconjugatus	570	*P. biaurelia*	Georgia, USSR	Mate killer
pi	*Pseudocaedobacter falsus*	51m1 *pi*	*P. tetraurelia*	51m1	None
	Pseudocaedobacter falsus	51m43 *pi*	*P. tetraurelia*	51m43	None
	Pseudocaedobacter falsus	139 *pi*	*P. tetraurelia*	139	None
nu	*Pseudocaedobacter falsus*	87	*P. pentaurelia*	Philadelphia, Pennsylvania	None
	Pseudocaedobacter falsus	1010	*P. biaurelia*	Tennessee	None
mu	*Pseudocaedobacter conjugatus*	138	*P. octaurelia*	Ft. Lauderdale, Florida	Mate killer
	Pseudocaedobacter conjugatus	540	*P. primaurelia*	Mexico	Mate killer
	Pseudocaedobacter conjugatus	551	*P. primaurelia*	San Francisco, California	Mate killer
gamma	*Pseudocaedobacter minutus*	214	*P. octaurelia*	Florida	Vacuolizer
	Pseudocaedobacter minutus	565	*P. octaurelia*	Uganda	Vacuolizer
delta	*Tectobacter vulgaris*	561	*P. primaurelia*	Pisa, Italy	None
lambda	*Lyticum flagellatum*	239	*P. tetraurelia*	Florida	Rapid lysis
	Lyticum flagellatum	299	*P. octaurelia*	Panama	Rapid lysis
sigma	*Lyticum sinuosum*	114	*P. biaurelia*	Bloomington, Indiana	Rapid lysis
alpha	*Cytophaga caryophila*	562	*P. biaurelia*	Milan, Italy	None

ple, the hump killers of *P. tetraurelia* are found in Australia, Japan, Panama, and parts of the United States. Killers of *P. biaurelia* come from many regions of the United States, New Zealand, and several countries in Europe, including the USSR. Although the species of the *P. aurelia* complex are found throughout the world, the distribution of symbionts among these species is not random, but highly characteristic of the symbiont. For example, as can be seen from Table 1, *kappa* is found only in *P. biaurelia* and *P. tetraurelia; lambda* is found only in *P. tetraurelia* and *P. octaurelia; mu* is found only in *P. primaurelia* and *P. octaurelia* (except for the anomalous symbiont found in a mate-killing stock of *P. biaurelia,* in which a small percentage of the symbionts contain an R body).

The bacterial symbionts of the *P. aurelia* complex are well adapted to their environment. In some areas, 50% of the collections from the wild have a high proportion of symbiont-bearing paramecia (Preer, Preer, and Jurand, 1974). While the symbionts obviously benefit from the host paramecia, to what extent the paramecia benefit is not known. Soldo (1963) and Soldo and Godoy (1973) found that it was not necessary to provide folic acid to a *lambda*-bearing stock; but the same stock, freed of *lambda,* required the vitamin. Also, W. G. Landis (personal communication) has obtained evidence that killers are favored by natural selection in nature.

Isolation

The various methods used in *Paramecium* research are fully described by Sonneborn (1970). An understanding of many of these techniques is essential if one is to succeed in isolating and identifying the bacterial symbionts presently known in the *P. aurelia* complex.

With some exceptions, symbiont-bearing stocks of the *P. aurelia* complex carry only one kind of symbiont. It is therefore possible to culture the symbiont by maintaining the stock in which it is found. If a strain of *Paramecium* carries more than one kind of symbiont, it is often possible to obtain pure cultures of certain symbionts. For example, differential growth rate has been used to separate stock 7 *kappa* from a presumed mutant, 7 m1. The paramecia are cultured at a rapid fission rate for a period of time sufficient to reduce the symbiont population to no more than one symbiont per paramecium (Preer, 1948). Isolations of paramecia, each carrying one symbiont or none, are allowed to form cultures. A slow fission rate is then maintained to permit those with a single symbiont to give rise to a sizeable population of symbionts in each culture of paramecia. Different symbionts will thus be segregated in different cultures. When brought in from nature, cultures must be allowed to grow slowly (one-half fission per day at a moderate temperature, 20°C) to allow reproduction of the symbiont to keep pace with the growth of the paramecia. Nevertheless, in some cases the symbiont is lost. The presence of symbionts can be monitored, as will be discussed in the following section, by direct observations with the phase-contrast microscope, or in the case of symbionts associated with the killing character, by performing killing tests.

The simplest method of culturing paramecia involves the use of bacterized Cerophyl medium. A 0.25% infusion of Cerophyl (dried cereal shoots, Cerophyl Corporation, Kansas City, Missouri) buffered to pH 7.0 with Na_2HPO_4 is inoculated with *Klebsiella pneumoniae* on the day before use. By feeding 1 volume of culture with 1/2 volume of the Cerophyl-*Klebsiella* medium, a low rate of growth, 0.5 fissions per day, is obtained. Growth rate can be modified by feeding different amounts of bacterized medium at different temperatures until the optimum for the symbiont-bearing stock is determined, i.e., one that allows paramecia and symbionts to prosper.

It is possible to culture paramecia in media without living bacteria. Different ways of obtaining paramecia free of bacteria in the medium have been used; but all methods are based on allowing the paramecia to swim through sterile medium for a sufficient time to permit the bacteria in the food vacuoles to be eliminated (Allen and Nerad, 1978; Sonneborn, 1970; van Wagtendonk and Soldo, 1970). A very simple method modified from Heatherington (1934) involves introducing the paramecia at one side of a depression filled with a yeast medium described in the following paragraph, allowing them to swim to the other side, and transferring them to the next depression, etc. The animals are left for an hour in the third depression. The process is repeated hourly for 5 h, using sterile slides, micropipettes, and medium for each transfer. The use of antibiotics to obtain bacteria-free cultures is to be avoided, since antibiotics often destroy the symbionts.

A very satisfactory method for culturing bacteria-free paramecia has been devised using *Chlamydomonas* as the food organism. (L. B. Preer et al., 1974). The medium, made up of 1 g yeast autolysate, 0.25 g sodium acetate, 0.625 g Cerophyl, 0.125 g dibasic sodium phosphate, and 1 liter of double-distilled water, is dispensed into test tubes and autoclaved. Into a tube of this yeast medium, a small inoculum of *Chlamydomonas reinhardi*, strain

89, is introduced aseptically and incubated under fluorescent light at 22°C 2 days before use. A single bacteria-free paramecium, when put into a tube half-full of this *Chlamydomonas* medium, will, in the course of 3 or 4 days, multiply and ingest most of the algae, yielding a good population of paramecia. This is one of the best media for maintaining *kappa*-bearing paramecia.

Soldo, Godoy, and van Wagtendonk (1966) have described a complex medium for axenic growth of paramecia. A modification suggested by Soldo (personal communication), giving good results, follows.

Complex Medium for Axenic Growth of Paramecia
A. Vitamin mixture (100 × concentration):

Calcium pantothenate	1 g
Nicotinamide	0.5 g
Pyridoxal hydrochloride	0.5 g
Riboflavin	0.5 g
Folic acid	0.5 g
Thiamine hydrochloride	1.5 g
Biotin	0.0001 g
α-Lipoic acid (DL-thioctic acid)	0.1 g

Suspend in 1 liter of double-distilled water with vigorous stirring, then dispense while stirring in 10-ml portions in screw-cap tubes, and place immediately in the deep freeze. Do not filter, avoid exposure to light, and thaw only as needed.

B. TEM-4T/stigmasterol (1,000 × concentration):

Ten grams TEM-4T (Hachmeister, Pittsburgh) and 0.5 g stigmasterol are dissolved in 100 ml hot absolute ethanol and stored at 4°C in a tightly capped plastic bottle.

C. Other ingredients:

Proteose peptone (Difco)	10 g
Trypticase (BBL)	5 g
Yeast nucleic acid (Grade C, Calbiochem)	1 g
$MgSO_4 \cdot 7H_2O$ (Mallinckrodt)	0.5 g

The medium is prepared by combining 10 ml A, 1 ml B, and the remaining ingredients C in double-distilled water and bringing to a final volume of 1 liter after adjusting to pH 7.0–7.2 with 0.1 N NaOH. Sterilize at 121°C for 20 min; store in dark at 4°C.

This medium has been used for culturing many stocks of paramecia, including *pi-*, *mu-*, and *lambda*-bearers, and gives high yields of paramecia. Many symbionts, however, cannot be maintained in paramecia cultured in this medium.

Recently a method for transferring many bacterized stocks of paramecia at once to axenic culture has been devised by Allen and Nerad (1978). As a preliminary adapting medium they use the vitamin mixture of Soldo, Godoy, and van Wagtendonk (1966) combined with autoclaved bacteria. In our laboratory, we have obtained vigorous growth of paramecia, including one *P. biaurelia kappa*-bearer with a modification of their procedure as follows. *Klebsiella pneumoniae* is inoculated into 100-ml quantities of a solution of 10 g tryptone and 5 g NaCl in 1 liter double-distilled water and cultured overnight at 37°C on a rotary shaker. The bacteria are concentrated by centrifugation and resuspended in a convenient volume of a vitamin-stigmasterol (VS) mixture. VS is made according to directions for axenic medium described in the preceding paragraph except that TEM-4T of the B portion and all of the C ingredients are omitted. The pH is adjusted to 7.0 as described. VS is stored in tubes in the deep freeze. All dilutions of bacteria subsequently are made with VS. A small volume of bacteria is diluted, and the optical density at a wave length of 590 (OD_{590}) is determined. The concentration of bacteria in VS is adjusted to $OD_{590} = 3.0$; the bacteria are dispensed in 1-ml portions in screw-cap tubes and immediately placed in the deep freeze. The final medium (BVS) is prepared before using, about a week's supply at a time, by adding 1 ml of bacteria to 9 ml of VS to give an $OD_{590} = 0.3$, and autoclaved 20 min. Paramecium cultures grow at 1/2 to 2 fissions per day in this medium.

In order to label DNA, a thymidine-requiring strain of *Escherichia coli*, cultured with radioactive thymidine prior to autoclaving, can be substituted for *K. pneumoniae* in the BVS medium. Similarly, a uracil-requiring strain can be used to label RNA. Radioactive BVS is an excellent medium for labeling DNA of bacteria-free paramecia. Unlike other axenic media, it contains no additional nucleic acids to dilute the label. Routinely, with tritiated thymidine as label, 10^6 cpm per microgram of *Paramecium* DNA is obtained when labeled in this way.

There have been reports of in vitro growth of *lambda* in axenic medium (van Wagtendonk, Clark, and Godoy, 1963), which were not confirmed (Gibson, 1970), although later very limited success in the growth of both *lambda* and *mu* was reported (Williams, 1971). Certainly in vitro culture of these symbionts is difficult and may be achieved rarely, if at all.

It is possible to isolate the symbionts and obtain them free, for the most part, from all cellular contaminants such as cilia, trichocysts, mitochondria, and bacteria. Such isolations are done by the use of centrifugation methods (Soldo, van Wagtendonk, and Godoy, 1970), ion-exchange cellulose columns (Mueller, 1963; Smith, 1961), or filter paper columns (Preer, Hufnagel, and Preer, 1966). Although the symbionts cannot be cultured after isolation,

such preparations have been useful in studying the biochemistry and many fundamental properties of the symbionts.

Identification

The bacterial symbionts of *Paramecium* can be identified by direct observation with a phase-contrast microscope, by observation of physiological effects produced on other sensitive paramecia, or by a combination of these methods. For positive identification in some cases, however; an examination of the ultrastructure with the electron microscope is necessary.

A quick staining of paramecia for viewing with dark phase-contrast microscopy (Beale and Jurand, 1966) shows symbionts very clearly. A small drop of paramecia is placed on a slide, and exposed to osmic acid vapors for several seconds. The animals are then treated with acetone to remove lipid, a drop of lacto-orcein (36% acetic acid, 22% lactic acid, 1% orcein) is added, and a coverslip previously rung with petrolatum is placed on the preparation. The symbionts may also be seen without staining with bright phase-contrast microscopy simply by preparing a fresh squash of paramecia between a clean slide and coverslip (Preer and Stark, 1953). These quick methods are appropriate for identification of symbionts with distinctive features, such as *kappa* with its unusual R body. For many symbionts that may be confused with other bacteria in the culture medium, it is necessary to wash the culture of paramecia free of bacteria before preparing the squash.

If a symbiont-bearing stock is brought in from nature, killing tests are performed to see if it is one of the known kinds of killer. Equal volumes of the culture to be tested and a culture of sensitive paramecia are mixed in a depression slide and observed with appropriate controls for prelethal effects. If, for example, the lysis and death of the standard sensitive tester stock 92 of *P. triaurelia* occurs within 1 h, the symbiont in the killer is *lambda* or *sigma*. Cytological observations with the light microscope make it possible to determine which of these two it is. Mate killing can only be detected by crossing the unknown with a sensitive strain of paramecia of the complementary mating type. Any symbiont-free strain with which an unknown will mate is usually adequate. Nevertheless, knowledge of the species of *Paramecium* to which the unknown belongs is, in practice, generally essential. Methods for identifying the species of the *P. aurelia* complex involve mating type tests, measurements of size, growth rate, response to standard killer strains, characterization of isozymes, and other traits. Detailed information is beyond the scope of this paper and the reader is referred to Sonneborn (1975).

Densities of DNA of *Paramecium* and its symbionts have been determined in many laboratories (for a summary of values published prior to 1974 see Preer, Preer, and Jurand, 1974). Although variations in densities of the same DNA have been reported, certain generalizations may be made. Most agree that the density of the DNA of the host *Paramecium* is about 1.689 g/cm³. Values for the different symbionts, with the exception of *lambda* (1.686) are all greater than that of *Paramecium,* ranging from 1.694 to 1.703, or higher in one case. Values for the plasmid or phage associated with the symbiont are the same as that of the symbiont or slightly less. Determinations vary in the degree of confidence that can be placed in them. In some cases, the paramecia are cultured in axenic medium, avoiding the pitfall of contaminating bacteria. In others, different DNAs are mixed and centrifuged in the same tube, assuring the detection of even small differences between the two. The following differences are of special interest and have been clearly demonstrated: 299 paramecia (1.688) differs from 299 *lambda* (1.686); 562 *kappa* (1.702) differs from 562 *kappa* phage (1.700). Differences noted by Dilts (1976) in the *kappas* derived from stock 51 are significant: 51 *kappa* (1.700–1.701) differs from 51 *kappa* plasmid (1.698) and from 51ml *kappa* (1.703); 51m43 *kappa* (1.700) differs from 51m43 plasmid (1.698) and from 51m 43 *pi* (1.694–1.695); and 51ml *kappa* differs from 51ml *pi* (1.694–1.695).

Recent investigations have clarified relationships among *kappa* and four presumed mutants of *kappa.* The *kappa* of stock 51 of *P. tetraurelia,* a hump killer, was thought to have mutated to produce a spinner killer, a resistant nonkiller and sensitive strains bearing *pi.* Measurement of the covalently closed circular plasmid DNA associated with these symbionts, studies of the DNA densities of the symbionts and plasmids, and DNA-DNA hybridizations of the symbionts (Dilts, 1977; Quackenbush, 1977a) have led to the conclusion that only one of these symbionts, *kappa* of the resistant nonkiller, could have arisen from *kappa* of stock 51; the other *kappa* and *pi* symbionts apparently were present in stock 51 when it was brought in from nature.

Relationships among several of the symbionts have been elucidated by DNA-DNA hybridization studies of Quackenbush (1978). *Mu, nu,* and *pi* appear to be related, for there are sequence homologies of at least 40% among these symbionts. Between these symbionts and *kappa,* however, there is little relationship, for sequence homologies of less than 12% exist. He therefore has revised the original classification such that *Caedobacter* includes all symbionts with R bodies (*kappa*), and species lacking R bodies, previously in the genus *Caedobacter* (*mu, nu, pi, gamma*), are included in a new genus, *Pseudocaedobacter.* Among the different kinds of *kappa,* four species have been delineated.

The following key, based on L. B. Preer et al. (1972), Preer, Preer, and Jurand (1974), Quackenbush (1977a, 1978), and Dilts (1977), is useful in identifying the symbionts known at present in the *P. aurelia* complex.

I. Host paramecia are killers
  A. 2–50% of symbiont population contains R bodies
    1. Host liberates toxin into medium
      a. Kills by producing aboral humps on sensitive paramecia; R bodies unroll from inside at low pH, reroll at high pH; found in *P. tetraurelia* . . . . *Caedobacter taeniospiralis*
      b. Kills in other ways: spinning, vacuolization, or paralysis; R bodies unroll from outside irreversibly when exposed to high temperature or certain detergents
        (1) Found in *P. tetraurelia* . . . . *Caedobacter pseudomutans*
        (2) Found in *P. biaurelia* . . . . *Caedobacter varicaedens*
        (Note: *P. tetraurelia* is small, maximum growth rate at 27°C about 5 fissions per day; *P. biaurelia* is larger, maximum growth rate about 3.5 fissions per day.)
    2. Host does not liberate toxin into medium, but kills sensitive mate at conjugation . . . . *Caedobacter paraconjugatus*
  B. Symbiont populations do not contain R bodies
    1. Host liberates toxin into medium
      a. Kills by rapid lysis; symbionts are large flagellated cells
        (1) Straight rods found in *P. tetraurelia, P. octaurelia* . . . . *Lyticum flagellatum*
        (2) Sinuous rods found in *P. biaurelia* . . . . *Lyticum sinuosum*
      b. Kills by vacuolization; symbionts are very small cells, often doublets, with cell wall surrounded by an extra unit membrane visible in EM . . . . *Pseudocaedobacter minutus*
    2. Host does not liberate toxin into medium, but kills sensitive mate at conjugation . . . . *Pseudocaedobacter conjugatus*
II. Host paramecia are nonkillers
  A. Symbionts are present only in cytoplasm
    1. Typical thin Gram-negative cell wall containing a unit membrane visible in EM, lack flagella . . . . *Pseudocaedobacter falsus*
    2. Thick cell wall visible in EM, sparse flagella, occasional slight motility, coexists with other symbionts such as *Caedobacter* and *Pseudocaedobacter* . . . . *Tectobacter vulgaris*
  B. Symbionts are present almost exclusively in macronucleus . . . . *Cytophaga caryophila*

Acknowledgments

This work was supported by Public Health Service grant GM5024687 from the National Institute of General Medical Sciences to John R. Preer, Jr., and is contribution no. 1072 from the Department of Biology, Indiana University, Bloomington.

Literature Cited

Allen, S. L., Nerad, T. A. 1978. Method for the simultaneous establishment of many axenic cultures of *Paramecium*. Journal of Protozoology **25**:134–139.

Ball, G. H. 1969. Organisms living on and in Protozoa, pp. 565–718. In: Chen, T. T. (ed.), Research in protozoology. London: Pergamon Press.

Beale, G. H., Jurand, A. 1966. Three different types of mate killer (*mu*) particle in *Paramecium aurelia*. Journal of Cell Science **1**:31–34.

Beale, G. H., Jurand, A., Preer, Jr., J. R. 1969. The classes of endosymbiont of *Paramecium aurelia*. Journal of Cell Science **5**:65–91.

Buchanan, R. E., Gibbons, N. E. (eds.) 1974. Bergey's manual of determinative bacteriology, 8th ed. Baltimore: Williams & Wilkins.

Chen, T. T. 1955. Paramecin 34, a killer substance produced by *Paramecium bursaria*. Proceedings of the Society for Experimental Biology and Medicine **88**:541–543.

Dilts, J. A. 1976. Covalently closed, circular DNA in kappa endosymbionts of *Paramecium*. Genetic Research **27**:161–170.

Dilts, J. A. 1977. Chromosomal and extrachromosomal deoxyribonucleic acid from four bacterial endosymbionts derived from stock 51 of *Paramecium tetraurelia*. Journal of Bacteriology **129**:888–894.

Dorner, R. W. 1957. Stability of paramecin 34 at different temperatures and pH values. Science **126**:1243–1244.

Gibson, I. 1970. The genetics of protozoan organelles, pp. 379–399. In: Miller, P. L. (ed.), Control of organelle development. New York: Academic Press. [Symposia of the Society for Experimental Biology **24**:379–399, 1970.]

Gibson, I. 1974. The endosymbionts of *Paramecium*. CRC Critical Reviews in Microbiology **3**:243–273.

Görtz, H.-D., Dieckmann, J. 1977. An infectious particle in the macronucleus of *Paramecium caudatum*, p. 441. In: Hutner, S. H. (ed.), Abstracts of the Fifth International Congress of Protozoology.

Hafkine, W. M. 1890. Maladies infectieuses des paramécies. Annales de l'Institut Pasteur **4**:148–162.

Heatherington, S. 1934. The sterilization of protozoa. Biological Bulletin **67**:315–321.

Heckmann, K. 1975. *Omikron*, ein essentieller Endosymbiont von *Euplotes aediculatus*. Journal of Protozoology **22**:97–104.

Holtzman, H. E. 1959. A kappa-like particle in a non-killer stock of *Paramecium aurelia*, syngen 5. Journal of Protozoology, suppl. **6**:26.

Jenkins, R. A. 1970. The fine structure of a nuclear envelope associated endosymbiote of *Paramecium*. Journal of General Microbiology **61**:355–359.

Jurand, A., Preer, J. R., Jr. 1977. Investigations on the prelethal effects of the killing action by *kappa* killer stocks of *Paramecium aurelia*, p. 439. In: Hutner, S. H. (ed.), Abstracts of the Fifth International Congress of Protozoology. New York: American Society of Protozoologists.

Jurand, A., Rudman, B. M., Preer, J. R., Jr. 1971. Prelethal effects of killing action by stock 7 of *Paramecium aurelia*. Journal of Experimental Zoology **177**:365–388.

Kirby, H., Jr. 1941. Organisms living on and in Protozoa, pp. 1009–1113. In: Calkins, G. N., Summers, F. M. (eds.), Protozoa in Biological Research. New York: Columbia University Press.

Mueller, J. A. 1963. Separation of kappa particles with infective activity from those with killing activity and identification of the infective particles in *Paramecium aurelia*. Experimental Cell Research **30**:492–508.

Müller, J. 1856. Beobachtungen an Infusorien. Monatsberichte der Berliner Akademie 389–393.

Ossipov, D. V. 1973. Specific infectious specificity of the *omega*-particles, micronuclear symbiotic bacteria of *Paramecium caudatum*. Cytologia **15**:211–217.

Ossipov, D. V., Gromov, B. V., Mamkaeva, K. A. 1973. Electron microscopic examination of *omega*-particles (bacterial symbionts of the micronucleus) and nucleolar apparatus of *Paramecium caudatum* clone M1-48. Cytologia **15**:97–103.

Ossipov, D. V., Ivakhnyuk, I. S. 1972. *Omega* particles—micronuclear symbiotic bacteria of *Paramecium caudatum* clone M1-48. Cytologia **14**:1414–1419.

Ossipov, D. V., Gromov, B. V., Mamkaeva, K. A., Skoblo, I. I., Borchsenius, O. N., Rautian, M. S., Podlipaev, S. A., Fokin, S. I. 1977. Endonuclear symbiosis in *Paramecium caudatum*, p. 442. In: Hutner, S. H. (ed.), Abstracts of the Fifth International Congress of Protozoology. New York: American Society of Protozoologists.

Petschenko, B. 1911. *Drepanospira mülleri* n. g. n. sp. parasite des Paramaeciums, contribution à l'étude de la structure des bactéries. Archiv für Protistenkunde **22**:248–298.

Preer, J. R., Jr. 1948. The killer cytoplasmic factor *kappa*: Its rate of reproduction, the number of particles per cell, and its size. American Naturalist **82**:35–42.

Preer, J. R., Jr. 1950. Microscopically visible bodies in the cytoplasm of the "killer" strains of *Paramecium aurelia*. Genetics **35**:344–362.

Preer, J. R., Jr., Hufnagel, L. A., Preer, L. B. 1966. Structure and behavior of R bodies from killer paramecia. Journal of Ultrastructure Research **15**:131–143.

Preer, J. R., Jr., Jurand, A. 1968. The relation between virus-like particles and R bodies of *Paramecium aurelia*. Genetical Research **12**:331–340.

Preer, J. R., Jr., Preer, L. B. 1967. Virus-like bodies in killer paramecia. Proceedings of the National Academy of Sciences of the United States of America **58**:1774–1781.

Preer, J. R., Jr., Preer, L. B., Jurand, A. 1974. *Kappa* and other endosymbionts in *Paramecium aurelia*. Bacteriological Reviews **38**:113–163.

Preer, J. R., Jr., Stark, P. 1953. Cytological observations on the cytoplasmic factor "kappa" in *Paramecium aurelia*. Experimental Cell Research **5**:478–491.

Preer, L. B. 1969. *Alpha*, an infectious macronuclear symbiont of *Paramecium aurelia*. Journal of Protozoology **16**:570–578.

Preer, L. B., Jurand, A., Preer, J. R., Jr., Rudman, B. M. 1972. The classes of *kappa* in *Paramecium aurelia*. Journal of Cell Science **11**:581–600.

Preer, L. B., Rudman, B. M., Preer, J. R., Jr., Jurand, A. 1974. Induction of R bodies by ultraviolet light in killer paramecia. Journal of General Microbiology **80**:209–215.

Quackenbush, R. L. 1977a. Phylogenetic relationships of bacterial endosymbionts of *Paramecium aurelia*: Polynucleotide sequence relationships of 51 *kappa* and its mutants. Journal of Bacteriology **129**:895–900.

Quackenbush, R. L. 1977b. Genetic relationships among endosymbiotic bacteria of *Paramecium aurelia*. Ph.D. thesis. Indiana University, Bloomington, Indiana.

Quackenbush, R. L. 1978. Genetic relationships among bacterial endosymbionts of *Paramecium aurelia*. Journal of General Microbiology **108**:181–187.

Siegel, R. W. 1953. Mate-killing in *Paramecium aurelia*, variety 8. Physiological Zoology **27**:89–100.

Simon, E. M. 1971. *Paramecium aurelia*: Recovery from −196°C. Cryobiology **8**:361–365.

Simon, E. M., Schneller, M. V. 1973. The preservation of ciliated protozoa at low temperature. Cryobiology **10**:421–426.

Smith, J. E. 1961. Purification of kappa particles of *Paramecium aurelia*, stock 51. American Zoologist **1**:390.

Soldo, A. T. 1963. Axenic culture of *Paramecium*. Some observations on the growth behavior and nutritional requirements of a particle-bearing strain of *Paramecium aurelia* 299 *lambda*. Annals of the New York Academy of Sciences **108**:380–388.

Soldo, A. T. 1974. Intracellular particles in *Paramecium*, pp.

377–442. In: van Wagtendonk, W. J. (ed.), *Paramecium: A current survey*. Amsterdam: Elsevier.

Soldo, A. T., Godoy, G. A. 1973. Observations on the production of folic acid by symbiont *lambda* particles of *Paramecium aurelia* stock 299. Journal of Protozoology **20**:502.

Soldo, A. T., Godoy, G. A., van Wagtendonk, W. J. 1966. Growth of particle-bearing and particle-free *Paramecium aurelia* in axenic culture. Journal of Protozoology **13**:492–497.

Soldo, A. T., van Wagtendonk, W. J., Godoy, G. A. 1970. Nucleic acid and protein content of purified endosymbiote particles of *Paramecium aurelia*. Biochimica et Biophysica Acta **204**:325–333.

Sonneborn, T. M. 1938. Mating types, toxic interactions and heredity in *Paramecium aurelia*. Science **88**:503.

Sonneborn, T. M. 1943. Gene and cytoplasm. I. The determination and inheritance of the killer character in variety 4 of *P. aurelia*. II. The bearing of determination and inheritance of characters in *P. aurelia* on problems of cytoplasmic inheritance, pneumococcus transformations, mutations and development. Proceedings of the National Academy of Sciences of the United States of America **29**:329–343.

Sonneborn, T. M. 1959. *Kappa* and related particles in *Paramecium*. Advances in Virus Research **6**:229–356.

Sonneborn, T. M. 1970. Methods in *Paramecium* research, pp. 241–339. In: Prescott, D. M. (ed.), Methods in Cell Physiology 4. New York: Academic Press.

Sonneborn, T. M. 1975. The *Paramecium aurelia* complex of fourteen sibling species. Transactions of the American Microscopical Society **94**:155–178.

Sonneborn, T. M., Mueller, J. A., Schneller, M. V. 1959. The classes of *kappa*-like particles in *Paramecium aurelia*. Anatomical Record **134**:642.

Takayanagi, T., Hayashi, S. 1964. Cytological and cytogenetical studies on *Paramecium polycaryum* V. Lethal interactions in certain stocks. Journal of Protozoology **11**:128–132.

van Wagtendonk, W. J., Clark, J. A. D., Godoy, G. A. 1963. The biological status of *lambda* and related particles in *Paramecium aurelia*. Proceedings of the National Academy of Sciences of the United States of America **50**:835–838.

van Wagtendonk, W. J., Soldo, A. T. 1970. Methods used in the axenic cultivation of *Paramecium aurelia,* pp. 117–130. In: Prescott, D. M. (ed.), Methods in Cell Physiology 4. New York: Academic Press.

Wichterman, R. 1945. Schizomycetes parasitic in *Paramecium bursaria*. Journal of Parasitology, Suppl. **31**:25.

Wichterman, R. 1953. The Biology of *Paramecium*. New York: The Blakiston Company.

Williams, J. 1971. The growth *in vitro* of killer particles from *Paramecium aurelia* and the axenic culture of this protozoan. Journal of General Microbiology **68**:253–262.

The Family Rickettsiaceae: Human Pathogens

EMILIO WEISS

The rickettsias described in Part 18 of *Bergey's Manual of Determinative Bacteriology,* eighth edition (Buchanan and Gibbons, 1974), comprise a broad group of small, Gram-negative rods, which in some manner are associated with arthropods. Most, but by no means all, depend on eukaryotic cells for growth and some invade vertebrate hosts. Because of their diversity, difficulty of cultivation, and limited information on their fundamental biological properties, they are classified on the basis of relatively obvious phenotypic characteristics. Rickettsias that parasitize vertebrate erythrocytes are placed in the families Bartonellaceae and Anaplasmataceae (this Handbook, Chapter 164). The rest are lumped together in the family Rickettsiaceae. The human pathogens are discussed in this chapter, and the animal pathogens and the nonpathogenic strains are taken up in the next one.

Even within the group of human pathogens, differences in ecology, DNA base ratio, and biological properties are quite pronounced (Tables 1 and 2). The genera *Rickettsia* and *Coxiella* are both obligate intracellular parasites, but occupy different sites in the eukaryotic cell. The former actively penetrates into the cytoplasm and multiplies freely within it. The latter is taken up into the phagosome and multiplies in the phagolysosome (Weiss, 1978). The genus *Rochalimaea* has been grown axenically (Vinson and Fuller, 1961). The main justification for calling it a ''rickettsia'' is its close ecological association with *Rickettsia prowazekii,* the etiological agent of epidemic typhus. Both *R. prowazekii* and *Rochalimaea quintana* have a natural reservoir in man and are transmitted from man to man by the body louse. Both agents were recently shown to have an extrahuman reservoir (Bozeman et al., 1975; Weiss et al., 1978).

THE GENUS *RICKETTSIA*

This genus is divided into three major groups: typhus, spotted fever, and scrub typhus groups. The eighth edition of *Bergey's Manual* (Weiss and Moulder, 1974) lists three species in the first group, six in the second, and one in the last group. Some of the investigators of the spotted fever rickettsiae (Burgdorfer et al., 1975; Lackman et al., 1965; Robertson and Wisseman, 1973) believe that further speciation is required to adequately describe natural variation in this group. Considerable variation is also encountered in the scrub typhus group, but it is regarded as type variation, and all strains are classified as members of the same species. Genetic variation in the genus *Rickettsia* has probably been the result of prolonged associations of the microorganisms with different hosts. There is no reason to believe that one group is subject to higher frequency of mutant selection than another.

Habitats

Typhus Group

Rickettsia prowazekii, the etiological agent of epidemic typhus fever, has until recently been regarded as a parasite confined to man. It was named in honor of two distinguished microbiologists, Howard Taylor Ricketts and Stanislav von Prowazek, who died of typhus fever in the course of their investigations. The role of the louse in transmission was first demonstrated in 1909 by Nicolle, Comte, and Conseil. The causative relationship of *R. prowazekii* to epidemic typhus was clearly established by the careful experiments of da Rocha-Lima in 1916 and of Wolbach, Todd, and Palfrey in 1922. Epidemic typhus has been a constant by-product of wars and other major human catastrophes. The influence of the epidemic typhus rickettsia on the course of history was dramatized by Zinsser in his book *Rats, Lice, and History* (1935) and has been discussed extensively by many historians. With the advent of the broad-range antibiotics, typhus is no longer a dread disease when properly recognized and treated. It is doubtful, however, that the antibiotics have materially affected the ecology of the microorganism.

Individuals who recover from the disease probably retain small numbers of organisms, presumably in their lymph nodes, for the rest of their lives.

Table 1. Ecology of the pathogenic rickettsiae.

Group	Species	Human disease	Natural cycle		Geographical distribution	Environmental factors contributing to human infection
			Arthropod	Mammal		
Typhus	Rickettsia prowazekii	Epidemic typhus	Human body louse	Man	Worldwide	Spread of louse infestation
		Brill-Zinsser disease	None	Man	Worldwide	Previous epidemic
	R. typhi	?	Louse, flea	Flying squirrel	Virginia, Florida	?
		Endemic typhus	Flea, louse	Rat, other rodents	Worldwide	Rat- and flea-infested urban areas
	R. canada	?	Rabbit tick?	Rabbit, hare?	Ontario, Canada	?
Spotted fever	R. rickettsii	Rocky Mountain spotted fever	Wood tick, dog tick	Wild rodents, dogs	Western Hemisphere	Tick-infested terrain, dogs
	R. montana	?	Wood tick, dog tick	Wild rodents, dogs	North America	?
	R. sibirica	North Asian tick typhus	Various ixodid ticks	Wild rodents	Siberia, Armenia, Mongolia	Tick-infested terrain
	R. conorii	Fièvre boutonneuse	Brown dog tick	Dogs, rodents	Mediterranean regions, Africa, India	Tick-infested terrain, dogs
	R. australis	Queensland tick typhus	Scrub tick	Dogs, rodents	Queensland, Australia	Tick-infested terrain, dogs
	R. akari	Rickettsialpox	Mouse mite	House mouse, wild rodents	Worldwide	Rodent- and mite-infested urban areas
Scrub typhus	R. tsutsugamushi	Scrub typhus	Trombiculid mites	Rats, other rodents	Pacific islands and wide regions surrounding them	Foci of high mite and rodent density, often associated with changing ecology
	Coxiella burnetii	Q fever	Ticks (many genera)	Primarily domestic animals, cattle, sheep, goats; bandicoot and small rodents	Worldwide	Infectious dust derived from animals and their ticks in the meat and dairy industries
	Rochalimaea quintana	Trench fever	Human body louse	Man	Worldwide	Rapid spread of louse infestation
			Vole ectoparasites?	Vole	Grosse Isle, Quebec, Canada	

These organisms may temporarily overcome host resistance and give rise to a milder form of typhus fever, called Brill-Zinsser disease. During primary typhus infection, as well as during Brill-Zinsser disease, patients have a sufficiently high rickettsemia to infect ectoparasites if they happen to be present. The most effective vector is the human body louse, *Pediculus humanus*. Two characteristics contribute to the body louse's efficiency in transmission: It takes frequent large blood meals and it tends to desert febrile hosts to seek new ones. The head louse is equally susceptible to rickettsial infection, but has not been implicated in typhus fever transmission for reasons which are not entirely clear. They may be related to the fact that the head louse imbibes only very small amounts of blood (Murray and Torrey, 1975). The rickettsiae grow profusely in the cells of the gut epithelium of the louse, even when the ingested human blood contains high levels of antibodies (Wisseman et al., 1975). Heavily infected cells are released into the lumen and discharged with the feces. They are the source of human infection when louse feces are driven into the broken skin in the process of scratching. Lice invariably succumb to infection, generally with 1–2 weeks, rarely surviving longer than 3 weeks (Fuller, 1954). Thus, the louse serves only as a vector and has no role in the interepidemic survival of the rickettsia.

The existence of an extrahuman reservoir of *R. prowazekii* in domestic animals has been suspected for the past 25 years (Reiss-Gutfreund, 1956, 1966), but was clearly demonstrated only in 1975 by Bozeman et al. in a species of wild animal. A large number of North American vertebrates were tested, including meadow voles, whitefooted mice, ground and gray squirrels; the organism was consistently isolated only from southern flying squirrels (*Glaucomys volans*) captured in Florida and Virginia. It was also recovered from pools of ectoparasites removed from wild-caught flying squirrels: from fleas, *Orchopeas howardi,* and from lice, *Neohaematopinus sciuropteri* (Bozeman et al., 1978; Sonenshine et al., 1978). Four randomly selected strains were examined for biological and biochemical properties that might differentiate them from laboratory-established strains of *R. prowazekii* (Dasch, Samms, and Weiss, 1978; Woodman et al., 1977). Only very minor differences were noted, which indicates that significant divergent evolution between human and flying-squirrel strains has not taken place. There are no adequate answers for the questions of when the flying squirrels were infected, why infection is maintained in the flying squirrels and not in other wild or domestic animals, and how infection is maintained in nature. Possibly, the flying squirrels were infected during one of the major North American epidemics, either during the wars of the mid-eighteenth century or during the Irish immi-

gration following the potato famine (see genus *Rochalimaea*, this chapter). Although flying squirrels do invade attics of houses, it is not apparent that this association establishes a closer relationship than man has with other animals. Bozeman et al. (1978) obtained good evidence that flying-squirrel lice are involved in the transmission of the rickettsia among flying squirrels, and that fleas may play a role in transmission to other animals. There is evidence that a few human cases of typhus fever of flying-squirrel origin have occurred and that an occasional case of typhus fever is misdiagnosed as Rocky Mountain spotted fever (McDade et al., 1980).

During investigations of the etiology of typhus fever, it was recognized that rats (*Rattus norvegicus* and *R. rattus*) are the primary reservoirs of a rickettsia that is occasionally transmitted to man and produces a disease which is similar to epidemic typhus but is somewhat milder. This disease is called murine or endemic typhus, and the etiological agent is designated, for reasons of taxonomic priority, *Rickettsia typhi*. Some investigators prefer to call it *Rickettsia mooseri* in honor of Hermann Mooser, who clearly demonstrated that *R. prowazekii* and *R. typhi* can be differentiated by their virulence for the guinea pig (Mooser, 1928). *R. typhi* has a worldwide distribution but there are great fluctuations in the number of reported cases. In the United States, a maximum of 5,400 cases was reported in 1944, but the number has declined to less than 100 per year since 1958 (White, 1970). Endemic typhus at the time of Mooser's studies was quite prevalent in Mexico, where it was called "tabardillo", and has been reported from all countries where investigators have competently searched for it. In a recent review Traub, Wisseman, and Fahrlang-Azad (1978) state: "Murine typhus is a good example of a disease whose importance is inadequately appreciated—except by the patient, and even today, in most parts of the world, he will never know what ails him because the diagnosis will not be made. Here is a disease that is so widespread that it occurs on all continents, so prevalent that it could cause 42,000 cases in the United States alone in the period of 1931–1946, yet its precise means of transmission are still unknown and there are many unresolved questions about the ecology of the infection".

The rat louse, *Polyplax spinulosus*, and the rat flea, *Xenopsylla cheopis,* are believed to be the chief transmitters of the rickettsia from rat to rat. Man is an incidental host and infection occurs through contact with the rat flea. The human flea, *Pulex irritans,* and the human body louse are highly susceptible to infection and may play roles in transmission in populations with high ectoparasitic infestation. Multiplication of the rickettsiae in the flea is very similar to that previously described for the louse. The source of human infection is the flea feces containing cells heavily distended with rickettsiae discharged from the gut epithelium. In contrast to the early death of lice, however, the longevity of the fleas is not curtailed by infection, possibly because they have the ability to renew their gut epithelium layer (Ito, Vinson, and McGuire, 1975). Despite the more compatible relationship between rickettsiae and arthropods, the rat and possibly other urban rodents must be regarded as the principal habitats of *R. typhi.*

The typhus group of rickettsiae includes a third species, *R. canada,* which at this point is little more than a laboratory curiosity. Two strains were isolated from pools of engorged rabbit tick (*Haemaphysalis leporispalustris*) collected near Richmond, Ontario, from an indicator rabbit and from a wild snowshoe hare (McKiel, Bell, and Lackman, 1967). Numerous other attempts at isolation of this rickettsia have been unsuccessful and there is only equivocal evidence of human infection (Bozeman et al., 1970). The true habitat of this rickettsia remains unknown.

Spotted Fever Group

In contrast to the typhus group, the spotted fever rickettsiae are well established in their arthropod host, the tick. In a series of brilliant experiments conducted in 1906 and 1907 with the agent of Rocky Mountain spotted fever, *Rickettsia rickettsii,* Ricketts (1911) demonstrated that some ticks were naturally infected; ticks that acquired the infection by feeding on an infected animal remained infected throughout their lifetime, and females transmitted the agent transovarially to at least some of the offspring. *R. rickettsii* is confined to the Western Hemisphere, but by no means to the Rocky Mountains. In the western United States the most common human vector is the wood tick, *Dermacentor andersoni.* In the east, where human infection is more frequent, the dog tick *Dermacentor variabilis* is the chief vector. Numerous other tick species were found naturally infected in the United States, but of these only the lone star tick *Amblyomma americanum* has been implicated in human infection. Other tick vectors, including *Haemaphysalis leporispalustris,* rarely attack man (Burgdorfer, 1975). *Rhipicephalus sanguineus* and *Amblyomma cajennense* are among the ticks that have been most commonly implicated in human infection in Mexico and South America.

The relationship of *R. rickettsii* to its tick host was analyzed recently by Burgdorfer and Brinton (1975). It is not difficult to obtain 100% transovarial transmission in the laboratory. If the laboratory results were an indication of natural transmission efficiency, one would expect to find a high proportion of ticks to be infected. Various surveys conducted in areas where human infection had occurred indicated

that the infection rate of ticks rarely exceeds 10% and is often less than 1%. Several factors were identified which would tend to reduce the natural infection rate. Continuous transovarial infection appeared to have an adverse effect on the biological development of the tick. Beginning with the fifth filial generation, increasing numbers of females die within 1–2 weeks after engorgement. There are also limitations in the acquisition of rickettsiae from the vertebrate host. The ovarian tissue of the tick is infected only when the initial dose is relatively high. A sufficiently high inoculum is present in the blood of the vertebrate host only during a brief period of time. Thus ticks that transmit the infection transovarially can be regarded as the principal habitat of *R. rickettsii.* Some of the infected ticks may die prematurely but they are being replaced by a slow rate of infection from the vertebrate host.

Price et al. (1954) described a rickettsial-interference phenomenon which, according to Burgdorfer (1975; personal communication), may represent still another mechanism that limits tick infection with *R. rickettsii.* He noted that ticks collected in areas with low incidence of human disease occasionally have high rates of infection. The rickettsiae, however, are not serologically identical with *R. rickettsii* and are not pathogenic for laboratory animals. These rickettsiae, presumably, are preventing the infection of the ticks with virulent *R. rickettsii.* Some of these rickettsiae have been described. They include *R. montana* isolated from *D. andersoni* and *D. variabilis* in eastern Montana, *R. parkeri* isolated from *Amblyomma maculatum* in Texas and elsewhere (Lackman et al., 1965), and a rickettsia isolated from *Rhipicephalus sanguineus,* which had been collected from dogs in central and northern Mississippi (Burgdorfer et al., 1975). The association of these rickettsiae with human disease is not known.

Although there is considerable serological evidence that *R. rickettsii* infection is widespread among wild and domestic vertebrates, demonstration by recovery of the microorganisms has been difficult, possibly because, as stated earlier, vertebrates harbor large numbers of rickettsiae only during relatively brief periods. Natural infection among vertebrates was demonstrated in the 1930s in Brazil in the house and wild dog (*Canis brasiliensis*), the opossum (*Didelphis marsupialis*), the wild rabbit (*Sylvilagus minensis*), and the Brazilian cavy (*Cavia aperea*) (Moreira and de Magalhaes, 1937). The first recovery in the United States was reported by Gould and Miesse in 1954 from the tissues of a meadow vole (*Microtus pennsylvanicus*) trapped near Alexandria, Virginia. Subsequent ecological studies in the eastern United States, which involved isolation attempts from over 1,000 wild animals, yielded two additional isolates from the meadow vole and one each from the cottontail rabbit (*S. floridanus*), opossum (*D. marsupialis virginiana*), whitefooted mouse (*Peromyscus* sp.), cotton rat (*Sigmodon hispidus*), and pine vole (*Pitymys pinetorum*) (Bozeman et al., 1967). In the western United States, *R. rickettsii* was isolated from each of five chipmunks (*Eutamias amoenus*), one from a snowshoe hare (*Lepus americanus*), and one from a golden mantle ground squirrel (*Citellus lateralis tescorum*) (Burgdorfer et al., 1962).

The infection of dogs is of particular interest since its tick is the primary source of human infection. In a recent comparative study (Sexton et al., 1976), serological evidence of infection was found much more frequently in dogs from an endemic than from a nonendemic area. Clinical manifestations consistent with mild spotted fever have been observed in dogs closely associated with human cases.

Man represents only an incidental habitat for *R. rickettsii*, playing no role in dissemination but possibly reflecting a changing ecology. From 1910 to 1930, most of the cases, about 100–600 per year, occurred within the area of distribution of *D. andersoni* in the Rocky Mountain region. Since 1930, there has been a shift toward the eastern and southeastern parts of the United States and to transmission by *D. variabilis*. From 1948 to 1959, the number of cases decreased from an average annual rate of 500 to 200 cases. This decrease coincided with the introduction of broad-spectrum antibiotics, which might have masked some of the cases, and with some relaxation in disease surveillance. From 1960 to the present, the number of reported cases has increased steadily and has surpassed 1,100 cases in 1977. This increase is attributed to suburbanization, which brought more people in contact with the ecological habitats of *R. rickettsii*, to a greater interest in outdoor activities, and to a greater awareness of the existence of spotted fever. During the past few years, the states west of the 100th meridian have accounted for less than 2% of the cases. The highest incidence has been reported from the Piedmont plateau of Virginia and North Carolina, which is dominated by eastern forest, especially the mesic oak-hickory-pine type (Burgdorfer, 1975, 1977; Sonenshine, Peters, and Levy, 1972).

Spotted fever rickettsiae are also encountered in the Eastern Hemisphere. Their biological properties are somewhat different than those of *R. rickettsii*. The disease produced in man is milder than Rocky Mountain spotted fever but epidemiological patterns are comparable. The principal rickettsiae of this group are *R. sibirica*, *R. conorii*, and *R. australis*.

Of these species, *R. sibirica* most closely resembles *R. rickettsii*. The disease in man was first recognized and described during 1934–1936, the epidemiological pattern during 1937–1938, and the nature of the rickettsia in 1948 (Zdrodovskii and Golinevich, 1960). Its habitat are foci extending from the Pacific maritime regions of the USSR through a wide area of southern and northern Siberia to the Armenian Republic. The foci are usually associated with steppe landscapes with low rainfall close to foothills and mountain ranges, and they may extend to the dry slopes of the mountains. Nine tick species have been found naturally infected, and transovarian passage was demonstrated in most of them including *Dermacentor nuttalli*, *D. marginatus*, *D. pictus*, *D. silvarum*, *Haemaphysalis punctata*, *H. concinna*, and *Rhipicephalus sanguineus*. These ticks feed on numerous small wild rodents and domestic animals. Isolations from vertebrates have been somewhat more frequent than in the case of *R. rickettsii*. At least eighteen kinds of mammals have been found infected, among them Siberian squirrels or susliks (genus *Citellus*), chipmunks (*Eutamias*), hamsters (*Cricetus*), lemmings (*Lagurus*), hares (*Lepus*), domestic and field mice, and voles. Attempts to isolate the rickettsia from domestic animals have been unsuccessful. Man is only an incidental host and the disease varies in severity in different regions. It resembles moderately severe to mild Rocky Mountain spotted fever, and it is acquired through the bite of the tick. It is called Siberian or, more appropriately, North Asian tick typhus (Hoogstraal, 1967).

R. conorii has probably been recognized in more diverse geographical locations than any other rickettsia of the spotted fever group. The disease in man was first described in 1910 by Conor and Bruch in Tunisia. The implication of the brown dog tick, *Rhipicephalus sanguineus*, in transmission was established in 1930, and the rickettsial nature of the disease in 1932 (Zdrodovskii and Golinevich, 1960). The presence of *R. conorii* has now been demonstrated in most of the regions bordering on the Mediterranean Sea and Black Sea, in Kenya and other parts of Central Africa, South Africa, and certain parts of India. Bozeman et al. (1960) have shown that strains isolated from different parts of the world are antigenically identical. Although the tick *R. sanguineus* is the prevailing vector, other ticks that parasitize dogs and smaller vertebrates have been implicated: *Haemaphysalis leachi* and *R. simus* in Kenya, *Amblyomma hebraeum* and *R. appendiculatus* in South Africa, and several other species of the above genus in these regions and elsewhere. In addition to dogs, several species of domestic and wild rats and mice have been found infected. The involvement of rabbits in maintaining an infected tick population was suggested by the drop in human disease following the myxomatosis epizootic that reduced the rabbit population in Europe.

The human disease varies in severity, but is seldom fatal. It is called fièvre boutonneuse, Marseilles fever, Kenya tick typhus, Indian tick typhus, or other names that designate the locality. It is usually transmitted by the bite of the tick, but may also be acquired through the skin or eyes when the ticks are crushed. At present the disease is of limited public health importance, but in the 1930s in certain localities of Central Africa virtually every newcomer became infected (Hoogstraal, 1967).

R. australis was first isolated in 1944–1945 from the blood of two military patients who had been training in jungle warfare in belts of dense forest interspersed in grassy savannah in North Queensland, Australia (Andrew, Bonnin, and Williams, 1946). Subsequent cases were recognized also in southern Queensland. The patients had been bitten by the scrub tick, *Ixodes holocyclus,* but early attempts to isolate the rickettsia from this tick were unsuccessful. Recently, three isolations from pools of ticks were reported, one from unfed *I. holocyclus* collected by drawing a cloth over the herbage, one from *I. holocyclus* removed from a dog, and one from *I. tasmani* isolated from a wild rat (Campbell and Domrow, 1974). Serological evidence acquired previously indicated that marsupials were infected. The disease in man is called Queensland tick typhus.

R. akari is regarded a distant relative of the spotted fever group; it differs somewhat from the other members in biological properties and habitat. The disease in man was first observed in New York City in 1946. The etiological agent and its epidemiological features were elucidated shortly thereafter (Huebner, Jellison, and Pomerantz, 1946). The habitat of *R. akari* appears to be worldwide; it has been reported from urban areas along the Atlantic coast of the United States, in the Crimean and southern Ukrainian regions of the USSR, in Korea, and there is circumstantial evidence that it occurs in Africa. The main vector is the mite, *Allodermanyssus sanguineus,* whose nymph and adult stages feed on the house mouse, *Mus musculus,* but may attack other animals and man. The mite may be the main habitat, since transovarian passage has been demonstrated, but the rickettsia has also been isolated from the mouse, rats, and surprisingly, from a wild Korean rodent, *Microtus fortis pelliceus* (Jackson et al., 1957a). The human disease, called rickettsialpox in the United States and vesicular and varioliform rickettsiosis in the USSR, is relatively mild. In the mid-1940s, about 180 cases were reported annually in the United States, but only sporadic cases have been reported in recent years (Brezina et al., 1973; Horsfall and Tamm, 1965).

Scrub Typhus Group

Scrub typhus has been recognized in Japan for well over a century. The mechanism of transmission to man was elucidated before the agent was isolated in the 1920s and recognized as a rickettsia through the efforts of N. Hayashi, A. W. Sellards, M. Nagayo, N. Ogata, and R. Kawamura (Blake et al., 1945). The rickettsia is called *Rickettsia tsutsugamushi* but many Japanese workers prefer to call it *R. orientalis*

(Horsfall and Tamm, 1965; Tamiya, 1962). Its primary habitat is the trombiculid mite. Certain features of the ecology of the rickettsia, which are still puzzling, contribute to a lively research interest in this microorganism.

R. tsutsugamushi is encountered in an area of the Orient that extends from India and Pakistan in the west to Japan, the northern portions of Australia, and the intervening islands in the Pacific Ocean in the east and includes southeastern Siberia, Korea, Southeast Asia, southern China, the Philippines, and Indonesia. The rickettsia is usually found in circumscribed foci or "ecological islands" (Traub and Wisseman, 1968), which have the proper vegetation and proper concentration of mites and their wild rat hosts. These foci may occur in primary jungle, sandy beaches, semideserts, mountain deserts, or alpine terrain high in the Himalayas. The habitats are usually characterized by the presence of changing ecological conditions, wrought by man or nature, and expressed by transitional types of vegetation (Traub and Wisseman, 1974).

The mite most commonly associated with scrub typhus is *Leptotrombidium deliense,* but several other mites of the same genus, including *L. fletcheri, L. akamushi, L. arenicola,* and *L. scutellare,* were shown to be naturally infected and to transmit the rickettsia transovarially. The mite depends on feeding on a vertebrate host for survival, but only one of the four stages, the larva, is parasitic. The six-legged larva, or chigger, shortly after its emergence from the egg, remains in the soil or travels up a few centimeters on debris or dead vegetation until it can burrow into the skin of any animal it happens to contact. Following a meal of tissue juices, it returns to the soil to resume a free-living existence. The vertebrates most commonly infected are rodents of the genus *Rattus,* although isolations from temperate-zone rodents, including *Apodemus* and *Microtus,* have been reported.

Although the vertebrate host is essential for the survival of the mite, there is no real evidence that it plays a role as an intermediate host for the rickettsia. The chigger feeds on a vertebrate only once. If it becomes infected while feeding, infection can be transmitted to another vertebrate, at best, only after transovarian transmission by the progeny. Transovarian transmission has been consistently demonstrated only with naturally infected mites. Infection is readily acquired by mites that feed on infected animals, but such an infection has been rarely transmitted experimentally to the next generation. Apparently, rickettsiae acquired through feeding do not pierce the gut wall and infect the ovary (Traub and Wisseman, 1974; Traub et al., 1975; Walker et al., 1975). Despite these experimental results, it is plausible that acquisition of rickettsiae from a vertebrate by the mite is infrequent but serves to replenish the pool of infected mites (Traub et al., 1975).

The wide dissemination of *R. tsutsugamushi* on islands that are separated from each other and from the mainland by large bodies of water can best be explained by assuming that migratory birds play a role. *Leptotrombidium* chiggers have been detected in birds, although infection was demonstrated only in rare instances. It is plausible that birds play a role in the transport of the mites (Traub and Wisseman, 1974).

If vertebrates indeed play a minor role as a reservoir for *R. tsutsugamushi* and if transovarian transmission is the main mechanism of survival of *R. tsutsugamushi,* opportunities for interaction among strains must be small. There is considerable variation among the isolates in antigenic specificity and virulence for small animals and for man. Three main antigenic types, Karp, Gilliam, and Kato, have been recognized. Often, but not always, one type predominates in a given region, but strains possessing more than one antigen have been encountered. Shirai and Wisseman (1975) recently studied the serological characteristics of 79 isolates from Pakistan, which is at the periphery of the *R. tsutsugamushi* habitat. The strains were not different from those encountered at the epicenter, which suggests that geographical location is not a predominant factor in mutant selection.

The human disease, scrub typhus, until World War II was reported mainly from Japan, where it was limited to a number of small, sharply defined areas in river valleys. A few hundred cases occurred each year. During the war, explosive outbreaks occurred as susceptible military personnel became involved in field operations and disturbed the ecological balance of certain areas. The highest frequency of disease still occurs in regions that receive a steady influx of susceptibles engaged in field activities. The endemic disease occurs primarily in tropical or subtropical regions with high humidity, but it does occur in other regions and even in high mountains. The foci of infection are usually small and the incidence changes from year to year. Before the advent of the broad-range antibiotics, case-fatality rates varied from 1 to 40%. The disease is no longer fatal if properly diagnosed and treated.

Isolation

General Considerations

Major incentives for isolating rickettsiae are given either by their direct involvement in human disease or by the potential hazard for human infection presented by infected invertebrate and vertebrate host populations. It is important to realize, however, that the isolation of a rickettsia from a patient seldom benefits the patient himself. Many of the rickettsial infections are serious illnesses and chemotherapy

with one of the tetracycline compounds must be initiated promptly, particularly for Rocky Mountain spotted fever. Severe frontal headache accompanied by high fever with a history of tick bite (not always volunteered by the patient) should prompt the physician to consider the possibility of infection with *Rickettsia rickettsii* even before the appearance of the distinctive rash and the development of specific antibodies. The mortality from Rocky Mountain spotted fever remains at 5–10% despite the availability of excellent chemotherapy, mostly because of delayed treatment (Hattwick et al., 1978). Isolation of the rickettsiae is of value to the patient only as retrospective diagnosis and to the physician as a guide for the diagnosis of other patients. Isolation might also be helpful in the retrospective diagnosis of Brill-Zinsser disease.

Isolation of rickettsiae from clinical material or from survey specimens requires careful selection of procedures and strict adherence to safety precautions. Rickettsiae share with viruses the requirement for eukaryotic host cells and, with bacteria, the susceptibility to most antibiotics. Even though penicillin and streptomycin are of limited efficacy in experimental rickettsial infection (Jackson, 1951) and of no therapeutic value, they are sufficiently inhibitory to reduce appreciably the chance of rickettsial isolation when added to tissue cultures or chicken embryos. The only exception is *R. tsutsugamushi,* which can be cultivated in the presence of small amounts of these antibiotics. Thus, rickettsiae can be isolated in chicken embryos or tissue cultures only if the specimen is free from adventitious agents. If the specimen is contaminated with nonpathogenic bacteria, the rickettsiae can be isolated only after the injection of immunologically competent animals that can remove the contaminating nonpathogens. Plaque isolation of rickettsiae is successful only when the contaminating agents are few and can be diluted out (Murphy, Wisseman, and Snyder, 1976) or when they are slow-growing.

Rickettsiae are regarded as class-3 agents: They present a special hazard that requires a controlled access facility and an approved safety cabinet in which negative air pressure is maintained (NIH Biohazards Safety Guide, 1974). Laboratory infections do occur as the result of accidental needle injection and, more frequently, through the aerosol route when the laboratory is engaged in work with highly concentrated viable suspensions (Johnson and Kadull, 1967; Oster et al., 1977; Sexton et al., 1975; Wisseman et al., 1962). Most dangerous, because of possible delay in recognition, is the infection of individuals who are not actively involved in rickettsial work but happen to be briefly in the laboratory during an operation that generates an aerosol. An effective vaccine has been developed for epidemic typhus (Cox, 1941; Topping et al., 1945), but recent preparations have been of variable and unpredictable potency (Mason, Seligmann, and Ginn, 1976). Vaccination against Rocky Mountain spotted fever provides only minimal protection (Du Pont et al., 1973; Johnson and Kadull, 1967); good protection may be possible only when administered repeatedly (Oster et al., 1977). There is no vaccine for scrub typhus. Several experimental vaccines, including an attenuated living vaccine against epidemic typhus (Fox et al., 1954; Strain E; Perez Gallardo and Fox, 1948; Wisseman, 1972), have been developed, but none is available commercially.

Many of the rickettsial studies are not clearly separated into the two steps of isolation and identification. Quite often, infected tissues are smeared, the rickettsiae are identified by the fluorescent-antibody technique, and only then is an attempt made to isolate the rickettsiae. For the purpose of this chapter, all the basic procedures will be presented under "Isolation" and only some of the definitive tests following cultivation and preparation of reagents will be discussed under "Identification".

Hemolymph Test

The hemolymph test (Burgdorfer, 1970) provides an economical, rapid, and simple technique for detecting rickettsiae in adult ticks. It can be performed in a few minutes with adult ticks collected from man and animals. Unfed ticks collected in the field may contain rickettsiae that do not produce a recognizable illness when injected into guinea pigs (Spencer and Parker, 1930) and are unsatisfactory for immediate examination. They should be incubated for 24–48 h at 37°C to allow multiplication of rickettsiae within hemocytes.

Hemolymph Test for Detecting Rickettsiae in Ticks (Burgdorfer, 1970)

Hemolymph from each tick is obtained by amputating the distal portion of one or more legs with dissecting scissors. The small drop that exudes from the wound is touched to a slide that has been marked with circles on its lower surface. Samples from as many as 15 ticks can conveniently be applied to a single microscopic slide. After air-drying, the preparation is gently fixed with heat and stained by Giménez's method or fixed in acetone for 10 min and treated for fluorescent-antibody microscopy (see below). The tick generally survives the amputation, since the hemolymph rapidly coagulates, and can be saved for further examination at a later time by placing it in a cotton-stoppered vial to which a single drop of water has been added.

Adequate amounts of hemolymph can be readily obtained from the larger ticks (Dermacentor, Amblyomma, and Hyalomma). The smaller ticks (Ixodes, Haemaphysalis, and Rhipicephalus) may not provide sufficient quantities of hemolymph. In these cases and in the case of ticks that have recently died, satisfactory smears can be prepared from the Malpighian tubules and hypodermis (Burgdorfer et al., 1975; Kurz and Burgdorfer, 1978). Unfed mite larvae can also be dissected and the salivary glands and midguts examined for scrub typhus rickettsiae (Roberts et al., 1975).

Giménez Staining Method for Rickettsiae and Chlamydiae (Giménez, 1964; Fig. 1)

This method is based on the differential retention of carbol basic fuchsin by rickettsiae and chlamydiae. The stock solution of carbol basic fuchsin is prepared by mixing 100 ml of 10% (wt/vol) high-grade basic fuchsin in 95% ethanol with 250 ml of 4% (vol/vol) aqueous phenol and 650 ml of distilled water. It is kept 48 h at 37°C before use and can be stored for as long as a year in the dark. The working solution is prepared by mixing 4 ml of stock solution with 10 ml of 0.1 M sodium phosphate buffer, pH 7.45. It is filtered before use, and filtered again if left standing for hours, and discarded at the end of the day. The counterstain is a 0.8% aqueous solution of malachite green oxalate. The heat-fixed smear is covered with working carbol basic fuchsin for 5 min, washed with tap water, stained twice for 6–9 s (or once for 12 s) with malachite green, and washed with tap water. Excess water is shaken off and the slide is allowed to air-dry.

All the rickettsiae, with the exception of R. tsutsugamushi, stain bright red; the background of eukaryotic origin stains a pale greenish blue. Occasionally, bacteria other than rickettsiae and chlamydiae retain some of the basic fuchsin and cannot be differentiated from the rickettsiae on the basis of the Giménez staining property alone. Giménez (1964) described a modification of his procedure for R. tsutsugamushi, but most consistent results are obtained by the Giemsa staining proce-

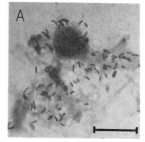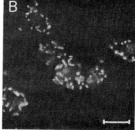

Fig. 1. Detection of Rickettsia rickettsii in tick hemolymph. The segments represent 5 μm. (A) Giménez stain. (B) Fluorescent-antibody stain. [Courtesy of Willy Burgdorfer, Rocky Mountain Laboratory.]

dure. For this purpose, the slide is fixed in methanol or, if the amount of lipid is excessive as in a yolk-sac smear, in Carnoy's fixative (3 parts chloroform, 6 parts absolute methanol and 1 part glacial acetic acid) for 30 min, as described by Syverton and Thomas (1945). The Gram stain is not satisfactory for rickettsiae.

If microorganisms of the general morphological appearance and tinctorial properties of rickettsiae are seen in the smears of the hemolymph, the rickettsiae can be tentatively identified by fluorescent-antibody microscopy. The indirect microimmunofluorescence method developed by Wang (1971) for chlamydiae and adapted by Philip et al. (1976) and Robinson et al. (1976) for rickettsiae offers great versatility and economy of reagents. For the limited purpose of tentative identification of R. rickettsii in tick hemolymph, the direct fluorescent-antibody method offers the advantage of simplicity.

Fluorescent-Antibody Method for Tentative Identification of Rickettsiae (Burgdorfer and Lackman, 1960; Fig. 1)

Antiserum is produced by the subcutaneous injection of guinea pigs with 1.0 ml of a 1:100 dilution of a yolk-sac suspension heavily infected with a virulent strain of R. rickettsii. Serum is obtained 5 weeks later and the sera with a complement fixation titer of at least 1:250 (or comparable titer by other tests) are pooled. Since the smear contains no yolk-sac antigen, the presence of low levels of anti-yolk-sac antibody in the guinea pig sera should not interfere with the test. The antibody globulin is precipitated and conjugated with fluorescein-isothiocyanate by the method of Riggs et al. (1958). Following fixation in acetone, the slide is air-dried and the smear is covered with a small drop of undiluted or of a low dilution of conjugated globulin and incubated for 30 min at 37°C in a moist chamber. The slides are rinsed at least twice with phosphate-buffered saline (PBS), pH 7.3, washed with agitation for an additional 10 min in fresh PBS, and mounted in 20% glycerol. Controls on the specificity of the staining reaction consist of known positive and negative specimens and/or conjugated globulin with a different specificity. Slides are examined at a magnification of 400–500×.

Rickettsiae of the spotted fever group are clearly visualized by their very strong fluorescence. Using undiluted or moderately diluted antisera against R. rickettsii, it is not possible to distinguish between R. rickettsii and the closely related nonpathogenic rickettsiae, R. montana or R. rhipicephali. However, at higher serum dilutions the nonpathogenic rickettsiae will stain spottily or not at all. Similar results are obtained with other groups of closely related rickettsiae. A higher degree of immunological specificity can be achieved by preincubating the sera with heterologous antigens or by using antisera elicited in mice. Mouse antisera appear to have a high degree of specificity for the various species of the spotted fever group (Philip and Casper, 1978). In a survey of Dermacentor andersoni ticks collected in Montana, which are known to harbor both R. rickettsii and R. montana, Burgdorfer (1970) found a high degree of correlation between positive fluorescence by the hemolymph test and the production of infection in the guinea pigs injected with the corresponding triturated ticks.

Isolation of Rickettsiae in Laboratory Animals

Soon after Nicolle, Conseil, and Conor (1911) demonstrated that guinea pigs can be infected with the blood of an epidemic typhus patient, the guinea pig became the animal of choice for the isolation of rickettsiae. Other animals and chicken embryos are now replacing the guinea pig in some cases. Ormsbee et al. (1978) defined the infectivity of eight strains of rickettsiae of the typhus and spotted fever groups for guinea pigs, mice, chicken embryos, and tissue cultures in terms of direct rickettsial counts. The number of rickettsiae required to infect guinea pigs was 1 to 10 in the case of R. prowazekii and R. typhi, 20 to 130 in the case of the spotted fever strains. These results undoubtedly reflect in part the passage histories of the strains and the seeds used, but there is no reason to doubt that they also represent the infectivities of these two groups of microorganisms. Burgdorfer, Cooney, and Thomas (1974) found the meadow vole (Microtus pennsylvanicus) particularly susceptible and valuable for the isolation of spotted fever rickettsiae. The mouse is the animal of choice for the isolation of R. akari. Although the guinea pig has been used extensively for the study and even for the isolation of strains of R. tsutsugamushi, it is not nearly as susceptible as the mouse, which replaced it in isolation procedures. Except for R. canada and the avirulent strains of the spotted fever group, the appropriate laboratory animal is probably more susceptible to infection than the chicken embryo.

Isolation of Suspected Typhus or Spotted Fever Rickettsiae in Guinea Pigs

Two male guinea pigs (450–500 g) are inoculated intraperitoneally with 2–4 ml of blood from a febrile patient or a suspension of ground arthropods and observed for 28 days. Rectal temperature is determined daily with a thermometer or thermal probe with due care not to pierce the gut. Fever is defined as temperature ≥40°C. Initial fever attributable to the large volume injected is disregarded. R. prowazekii will elicit fever 5–12 days after injection, but few other obvious symptoms. *R. typhi* and the spotted fever strains of lesser virulence, such as *R. conorii*, elicit fever 3–10 days after infection, accompanied by swelling of the scrotum. The more virulent strains of *R. rickettsii* produce, in addition, scrotal necrosis. The most favorable time for collection of material for passage is the 2nd or 3rd day of fever. The spleen (also the brain in suspected *R. prowazekii* infection, or tunica tissue in other cases) is ground and suspended in a suitable diluent, such as brain heart infusion (BHI), and injected, preferably into chicken embryos. If the guinea pigs do not develop fever or other symptoms, the possibility of inapparent infection can be checked by bleeding them at 28 days and testing their sera for rickettsial antibodies. Although it is unlikely that the guinea pigs were infected with a virulent strain, they might have been infected with *R. montana* or other strains of low virulence. If infection with such a strain is suspected at the time of injection, one of the guinea pigs should be sacrificed at 10–12 days, even in the absence of fever, and the spleen tissue passed into chicken embryos.

The isolation procedure of *R. akari* and *R. tsutsugamushi* in mice, as described by Jackson et al. (1957a,b), is still widely used (Bourgeois et al., 1977; Roberts et al., 1975).

Isolation of Rickettsiae in Mice
(Jackson et al., 1957a,b)

Six to eight adult white mice are injected intraperitoneally with 0.5 ml of blood or suspension to be tested. If the mice remain alive, some of them are sacrificed after 2 weeks, and pools of spleen and kidneys are passed to a second and sometimes third group of mice. If none of the animals appears sick, the surviving mice are challenged with 100 LD_{50} of the Karp strain of *R. tsutsugamushi* (or a lethal dose of *R. akari* if this microorganism is suspected). The survival of the mice to the challenge would indicate that they were infected with a strain of scrub typhus of low virulence. Virulent strains of *R. tsutsugamushi* will cause peritonitis, splenomegaly, and death in 10–24 days. To save the strain, moribund mice are sacrificed and suspensions are prepared from the peritoneal exudate, spleen, and kidney, and used for direct examination, passage into chicken embryos, and a second group of mice. Smears are stained by Giménez's method or with fluorescent antibody. Neither test provides unequivocal identification because of the considerable variation in biological properties and antigenic specificity among the scrub typhus strains. Some of the mice of the second passage are given 2.5 mg/ml of chloramphenicol in their drinking water 3–24 days after inoculation. If the treated mice survive, they are challenged along with a control group with a lethal dose of the Karp strain. The survival of the treated mice and the death of the controls indicate that the mice were infected with a strain of *R. tsutsugamushi*. Establishment of the scrub typhus rickettsiae in chicken embryos requires larger inocula than with other rickettsiae but for this rickettsia also, the chicken embryo offers the simplest means to maintain the isolate and produce sufficient material for further studies.

The above-described methods have been quite successful for the isolation of rickettsiae from arthropods that contain rickettsiae identified by fluorescent antibody and from acutely ill patients prior to antibiotic chemotherapy. For example, Bourgeois et al. (1977) isolated *R. tsutsugamushi* from blood of serologically confirmed patients in 36 of 45 attempts, and some of the failures were due to previously initiated chemotherapy. It is important, however, to select a strain of mice of proved susceptibility to the rickettsiae to be isolated. Groves and Osterman (1978) have shown that certain inbred strains of mice are highly resistant to overt infection with the Gilliam strain of *R. tsutsugamushi*. Resistance did not extend to other strains (such as the Karp strain) and appeared to be controlled by a single gene or a closely linked cluster of genes.

The isolation of rickettsiae from field-collected vertebrates has been most difficult, because the short period of acute infection seldom coincides with the time of the isolation attempt. Bozeman et al. (1978) developed an ingenious procedure which led to the isolation of the flying-squirrel strains of *R. prowazekii*. The trapped animals were maintained in the laboratory and bled at 6–10 day intervals. Isolation was attempted only when the flying squirrels showed a clear rise in antibody titer during the period of captivity.

Growth of Rickettsiae in Chicken Embryos

Although several strains of rickettsiae and other obligate intracellular bacteria have been isolated directly in chicken embryos, this method has not been fully exploited. It may be particularly useful for specimens expected to contain a fair number of rickettsiae (such as the blood of a patient at the beginning of his acute phase prior to antibiotic treatment) or for rickettsiae of low virulence for laboratory animals. It is possible to inoculate eggs with material derived directly from ticks. The exterior surfaces of the ticks are decontaminated by immersion in a 0.1% aqueous solution of Merthiolate (for 5 min or longer) and subsequent repeated washings in a suitable saline solution or BHI. The most frequent use of chicken embryos, however, has been for secondary passage for the preparation and characterization of seeds, preparation of antigens, and production of purified whole cells and extracts for metabolic and biochemical studies.

The chicken embryos must be obtained from flocks maintained on a rigorous antibiotic-free diet. Even traces of the broad-range antibiotics will delay the growth of rickettsiae. The flocks must also be subjected to a program of quality control; they must be free from pullorum and Newcastle disease and, if the material is eventually to be used in experimental or commercial human vaccines, they must be free from leukosis virus. Fertile eggs are preincubated (either by the supplier or in the laboratory) for 5–7 days at 37°C in a humidified incubator. The only procedure of cultivation that is satisfactory for rickettsiae is the one described by Cox (1941) for yolk-sac inoculation.

Inoculation of Eggs for Cultivating Rickettsiae (Fig. 2)

The eggs are candled just before use and those with dead or slow-moving embryos are discarded. With the egg in a vertical position, a small area of the shell above the center of the air sac is painted with tincture of iodine and, when dry, a hole is punched with a sharp probe to allow the insertion of a needle. A syringe of convenient size with a needle of preferably 22 or 23 gauge and 3-cm length is inserted vertically to deliver the inoculum (0.2–0.5 ml) approximately at the center of the egg. At 5–7 days, the yolk sac cannot be missed because it is still large and has not yet been depleted by the growth of the embryo. The hole is closed with a drop of a fast-drying cement and the eggs are incubated at 35°C, or at 33.5°C if the rickettsiae belong to the spotted fever group.

Each strain of rickettsia has somewhat different inoculum and harvest requirements for optimal yields. In general, with rickettsiae of the typhus group best results are obtained by using small inocula (about 10 viable rickettsiae per egg) and harvesting 10 days later just before the embryos die. The rickettsial titer drops rapidly after the death of the embryo. *R. tsutsugamushi* is treated in a similar manner but, because of low virulence for the chick embryo, the inoculum must be much larger (10^4–10^6 viable rickettsiae per egg, depending on the strain). The conditions for the growth of the rickettsiae of the spotted fever group were defined by Stoenner, Lackman, and Bell (1962). The eggs are inoculated with heavy inocula (at least 10^4 viable rickettsiae per egg) and incubated at the lowest temperature compatible with uninfected embryo survival (33.5°C). The embryos die relatively early in the course of infection (4–6 days after inoculation), but rickettsial multiplication continues in the yolk sacs of the dead embryos for about 2 days.

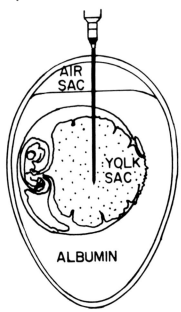

Fig. 2. Schematic representation of inoculation into the yolk sac of a 5- to 7-day-old chicken embryo. [Adapted from Horsfall and Tamm, 1965.]

Harvest of Rickettsiae from Yolk Sacs

Eggs are candled daily. Embryo deaths within the first 3 days are attributed to rupture of blood vessels or contamination, and the embryos are discarded. The yolk sac of embryos that die after 3 days are smeared and examined by Giménez stain until there is evidence of heavy infection. The yolk sacs of surviving embryos that were inoculated with typhus and scrub typhus rickettsiae are harvested after 10–40% of the embryos have died. The yolk sacs infected with spotted fever rickettsiae are harvested about 36 h after embryo death.

To harvest the yolk sacs, the top of the egg is painted with tincture of iodine and excess iodine is washed off with 70% ethanol. The shell over the air sac is removed with sterile forceps and the egg contents poured into a Petri dish. The yolk sac is gently rolled between two pairs of forceps to remove excess yolk and placed in a container for further processing.

For three decades, the diluent most frequently used for rickettsiae was SPG, which consists of 0.22 M sucrose, 0.1 M potassium phosphate, pH 7.0, and 0.005 M potassium glutamate (Bovarnick, Miller, and Snyder, 1950). This diluent stabilizes the activity of the rickettsiae under most conditions, but is toxic for animals and tissue cultures because of its high potassium content. More recently, rickettsiae have been suspended in BHI or PBS.

Rickettsial seed from yolk sac has been prepared in the following manner in our laboratory

(Weiss, Coolbaugh, and Williams, 1975). Lots of 10 yolk sacs are mixed with equal volumes (30–50 ml) of BHI in 250-ml thick-walled glass bottles containing glass beads, quickly frozen in an ethanol-CO_2 mixture, and maintained at $-70°C$. The infected yolk-sac pools are quickly thawed, shaken with glass beads, and further macerated by pipetting or with a syringe fitted with a 16-gauge cannula. They are diluted with additional BHI and centrifuged at $10,400 \times g$ for 30 min. The fat adhering to the walls of the centrifuge tubes is carefully removed. The pellet is suspended in a moderate volume of BHI and centrifuged at $210 \times g$ for 10 min. The pellet is suspended in a similar volume of BHI and centrifuged as above. The two supernatant fluids are combined, distributed in appropriate volumes (0.5 or 1.0 ml) into vials which are sealed, quickly frozen in an ethanol-CO_2 bath, and stored at $-70°C$.

The rather cumbersome procedure of disrupting the infected yolk sacs as described above, rather than using a blender or homogenizer, has the advantage of breaking up primarily the highly infected entodermal cells. The connective tissue, which is lightly infected, is not fragmented and is readily removed by low-speed centrifugation. The high-speed centrifugation removes a good portion of the fat. The result is a suspension of high infectivity that retains sufficient host material to protect the viability of the rickettsiae during prolonged storage.

Additional Preparative Steps Prior to Inoculation of Cell Cultures

For less prolonged storage of seeds to be used primarily for the inoculation of cell cultures, additional steps are recommended. The pools prepared as described above are centrifuged at $17,300 \times g$ for 15 min. The supernatant layers and any fat adhering to the walls are discarded and the pellet is suspended in a volume of BHI equivalent to the original weight of the yolk sac. The suspension is layered in 5-ml volumes over 20-ml volumes of a solution of 10% bovine plasma albumin (BPA) (fraction V without preservative) in K36. Diluent K36 (Weiss, Rees, and Hayes, 1967) consists of 0.1 M KCl, 0.015 M NaCl, and 0.05 M potassium phosphate, pH 7.0. Some of the yolk-sac constituents precipitate with BPA and form small sediments following centrifugation at $480 \times g$ for 20 min. The BHI and layers are combined and diluted with approximately 3 volumes of K36 and centrifuged at $10,400 \times g$ for 30 min. The pellets are resuspended in a moderate volume of K36 and recentrifuged at $17,300 \times g$ for 15 min; then the pellets are resuspended in a small volume of BHI

if the material is to be stored, or in K36 if it is to be processed further as described below.

For some of the immunological studies (Dasch and Weiss, 1977; Halle, Dasch, and Weiss, 1977) and biochemical studies (Coolbaugh, Progar, and Weiss, 1976; Williams and Peterson, 1976; Williams and Weiss, 1978; Winkler, 1976; Woodman et al., 1977), the following procedure has been used in our laboratory for purification of the typhus group.

Purification of Typhus Group Rickettsiae by Renografin Gradient (Weiss, Coolbaugh, and Williams, 1975)

Linear density gradients of Renografin-76 (Squibb, compound used in diagnostic radiology) are prepared with K36 as the diluent using a conical bore, triple outlet gradient maker connected to a multistaltic pump. Volumes of 2.5 ml of rickettsial suspension are layered over approximately 33 ml of 20–45% gradient in cellulose-nitrate tubes and centrifuged in an SW-27 rotor in a Spinco ultracentrifuge (Beckman) at 25,000 rpm for 60 min. The rickettsial band is formed at approximately the middle of the tube. The top band consisting of yolk-sac material, a lower band containing mitochondria, and associated areas of the gradient are removed by aspiration. The rickettsial band is then drawn into a syringe through a 14-gauge cannula, diluted tenfold with K36, and centrifuged at $17,300 \times g$ for 10 min. The pellet is suspended in 20 ml of K36, filtered through a glass filter (AP-20 Millipore microfilter, 47 mm, pore size 0.8–14 μm) in a Sterifil aseptic assembly (Millipore Corp.), and the filters are washed with additional K36. This step liberates the rickettsial cells from yolk-sac material that still adheres to them. The rickettsiae are again sedimented by centrifugation, resuspended in a small volume of K36, and subjected to a second cycle of 25–45% Renografin gradient separation. The rickettsiae are finally diluted in K36 if the tests require whole cells, or in 0.04 M phosphate buffer of pH 7.2 if cell extracts are to be prepared. Recovery of rickettsial activity is about 40%, recovery of yolk-sac antigen is negligible.

Antigens for the microagglutination (MA) and complement fixation (CF) tests are usually prepared from crude yolk-sac suspensions by procedures that include the following three steps: (i) incubation of the suspensions overnight at room temperature with 0.1% formalin to inactivate the rickettsiae; (ii) extraction of the rickettsial soluble antigen with diethyl ether or washing the rickettsiae with several cycles of high- and low-speed centrifugation (one high-speed centrifugation with the rickettsiae suspended in 50% sucrose or 0.5–1.0 M NaCl or KCl); (iii) extraction of the rickettsiae once or twice with ether (Fiset et al., 1969; Ormsbee et al., 1977; Shepard et al., 1976). These procedures are not satisfactory for *R. tsutsugamushi*. Although several methods of antigen preparation for the CF test have been described for this microorganism, fluorescent-antibody techniques are more satisfactory.

Growth of Rickettsiae in Tissue Cultures

Tissue culture procedures for rickettsiae are similar to those used in virology, except that the use of antibiotics must be carefully avoided and, in some experiments, the monolayers must be maintained for long periods of time. As in the case of the chicken embryos, tissue cultures have not been fully exploited for the isolation of rickettsiae from specimens free from adventitious agents. However, a great deal of information on basic biological properties has been obtained by the growth of rickettsiae in several types of both primary cultures and established cell lines (reviewed by Weiss, 1973, 1978).

Two procedures developed in our laboratory (Weiss and Dressler, 1958, 1960) have been widely adopted. One consists of treating the host cells with ionizing irradiation prior to inoculation: The cells are inoculated 1 day after irradiation if the main purpose is to examine rickettsial growth and metabolism (Weiss et al., 1972) or rickettsia-host cell interaction (Wisseman and Waddell, 1975), 5–7 days after irradiation if the main purpose is to obtain a larger harvest (Weiss, Coolbaugh, and Williams, 1975). The benefit of irradiation has been clearly documented with *Chlamydia* (Gordon et al., 1972), but the evidence with rickettsiae is largely empirical. The obvious advantage is that the host population is not multiplying, so that the cytoplasm is large and the rickettsiae are more clearly visible. The second procedure consists of centrifuging the inoculum onto the cells at a moderate centrifugal force (400–600 × *g* for 15 min). Flat-bottom tubes (with or without cover slips) that fit into the cups of a centrifuge (Weiss and Dressler, 1960) are most convenient, but plastic flasks in plaque assay procedures can also be used (Wike et al., 1972). There is clear evidence that centrifugation increases the number of cells that are initially infected; centrifugation is most useful when the inoculum has a very small number of infectious particles and maximum efficiency of infection is desirable.

Two procedures commonly used in our laboratory are described, one for the mass production of typhus or scrub typhus rickettsiae, the other for plaquing of *R. tsutsugamushi*, which is the most difficult rickettsia to plaque. Both procedures can be easily modified to serve other purposes.

Cultivation of Rickettsiae in Irradiated Mouse L Cells (Weiss, Coolbaugh, and Williams, 1975; Dasch, Halle, and Bourgeois, 1979)

About 10^8 cells (NCTC-L-929 or a suitable derivative) are suspended in 100 ml or medium 199 with 10% calf serum and subjected to 3,000 R in a ^{60}Co irradiator. They are then diluted in the same medium to the needed concentration, and placed in 16-oz (about 475-ml) plastic flasks (5 × 10^6 cells per flask) and cultivated as monolayers for 5–7 days prior to infection. The medium is removed, and the cells are oscillated gently for 60 min at room temperature with 2-ml volumes of a heavy concentration of rickettsiae harvested from yolk sacs and partially purified as described above. Buffered medium (15 ml/flask) of the following composition is then added: Medium 199 or Eagle's minimal essential medium (MEM) containing 25 mM *N*-2-hydroxyethylpiperazine-*N'*-2'-ethanesulfonic acid (HEPES) buffer, pH 7.4, 4 mM glutamine, and 10% calf serum. Note that HEPES media have the advantage of reducing shifts in pH during manipulations, which would be detrimental to the rickettsiae and the host cells, but these media are slightly toxic for tissue culture cells and cannot be used in long-term experiments. The heavily infected cells are harvested, depending on the strain of rickettsia and the inoculum, 5–7 days after inoculation. The monolayer is removed by 10-min treatment with 0.25% trypsin in PBS. The culture supernatants, the cells, and a wash of the flask with 5 ml of BHI are combined and centrifuged at 17,300 × *g* for 15 min. The pellet is suspended in a small volume (about 1 ml/flask) of BHI and homogenized by passing it twice through a 22- and 26-gauge needle, respectively. The homogenate is centrifuged at 210 × *g*, the supernatant fluid is saved, and the pellet is resuspended in BHI and treated as above. The two supernatant fluids are combined and further processed or used for a second passage in tissue culture or chicken embryos. Rickettsiae derived from tissue cultures are not generally as stable for long-term storage as those derived from yolk sacs.

Purification is similar to the procedure previously described with the following exceptions: Centrifugation of the rickettsiae over a 10% BPA layer is unnecessary; filtration through an AP-20 microfilter prior to the Renografin step is essential; additional precaution must be introduced to preserve the viability of the highly unstable *R. tsutsugamushi*. For this microorganism, the Renografin gradient is prepared either in BHI or PBS containing 1% BPA. The rickettsiae form two bands; the lower band contains the more highly purified material which, however, is not as free from host components as the corresponding band of typhus rickettsiae.

Plaquing of *R. tsutsugamushi* in Monolayers of Chicken Embryo Cells (Woodman, Grays, and Weiss, 1978; Woodman et al., 1977)

To plaque *R. tsutsugamushi* the monolayer must be maintained for at least 17 days. This maintenance can be achieved by minimal exposure of the cells to trypsin and by introduction of homologous (chicken) serum in the medium. Primary monolayers are prepared from 10–25 decapitated, 10-day-old chicken embryos. The embryos are minced three times with cold Earle balanced salt solution (EBSS) and placed in 150–400 ml of prewarmed 0.25% trypsin in EBSS for 20 min at 37°C. Fetal bovine serum (3 ml) is added to terminate the trypsinization. The cells are filtered through cheesecloth, centrifuged for 10 min at $300 \times g$, and suspended to a concentration of 2×10^6 viable cells per ml in MEM with 10% fetal bovine serum. The total yield is approximately $6–10 \times 10^7$ viable cells per embryo. The suspension is dispensed into 25-cm² plastic, tissue-culture flasks (5 ml/flask) and confluent monolayers are formed in 20 h at 37°C. The medium is removed and the cells are inoculated with a 0.1-ml suspension of rickettsiae appropriately diluted in cold BHI which contains 20 mM L-glutamine. The rickettsiae are allowed to absorb on monolayers for 90 min at room temperature with gentle rocking or are centrifuged onto the cells as previously described. To each flask is added a 5-ml volume of overlay medium of the following composition: Medium 199 prepared with EBSS plus 1% chicken serum, 1% fetal bovine serum, 2% tryptose phosphate broth, 5 mM L-glutamine, and 0.7% agarose. Note that the chicken serum must be obtained from a flock maintained on an antibiotic-free diet. The agarose is prepared separately in tenfold strength, cooled to 48°C, and added to prewarmed medium. The temperature of the overlay medium is reduced to 43°C before it is added to the flasks. The cultures are inoculated at 32°C in a 5%-CO₂-humidified atmosphere to prevent the loss of cultures in flasks which are not entirely airtight. A second 2-ml overlay of identical composition is placed over the initial overlay on the 9th day. The cultures are stained on day 17 by applying 2 ml of a 1:10,000 dilution of neutral red in PBS. The plaques are clearly visible after incubation at 32°C for 6 h.

The plaques of *R. tsutsugamushi* seen after 17 days of incubation are about 1 mm in diameter; they have indistinct margins but the center is clear. Typhus rickettsiae also produce small plaques that can be readily recognized after 10–12 days. Spotted fever rickettsiae produce plaques about 2 mm in diameter with distinct margins that can be recognized after 8 days. Although there are minor differences within each group, they are not sufficiently consistent to permit species identification within the typhus or spotted fever groups or strain identification within the scrub typhus group.

Identification

The procedures described above serve the dual purpose of isolation and identification of the rickettsiae. If doubt still persists about the species or strain identity of the isolate, a variety of biological, serological, and biochemical tests can be applied. Some of these tests can be performed only in highly specialized laboratories.

Light and electron microscopy studies of rickettsiae have been instrumental in identifying the intracytoplasmic (extravacuolar) location of rickettsiae, a unique site for pathogenic bacteria. The fine structure of the rickettsial cell, however, is not different from that of other Gram-negative bacteria, such as *Escherichia coli*. Rickettsiae measure 0.3–0.5 μm in width by 1.5–2.0 μm in length. In old cultures typhus rickettsiae vary considerably in length, are often vacuolated, and occasionally are filamentous when cell division is impaired. *R. tsutsugamushi* is smaller than other rickettsiae and differs somewhat in the architecture of the outer membranes, a property that may account for differences in affinity for carbol fuchsin and greater tenacity of attachment to eukaryotic cell components (Silverman and Wisseman, 1978). Rickettsiae are surrounded by capsular material which is best visualized by releasing the microorganisms gently in a medium containing high-titer specific antiserum, prior to fixation for electron microscopy (Silverman et al., 1978). The fine structure of *R. typhi* is illustrated in Fig. 3A.

The seeds prepared as described in a previous section can be used in fluorescent-antibody tests. As already mentioned, although there is considerable cross-reaction among species of the same group (typhus or spotted fever) or strains of *Rickettsia tsutsugamushi*, the titers of the homologous and heterologous reactions are usually significantly different. Attempts at plaque reduction with antiserum have not been successful, because rickettsiae are not neutralized by serum when tested on nonprofessional phagocytes (Wisseman, Waddell, and Walsh, 1974). Kenyon and McManus (1974) succeeded in reducing somewhat the number of plaques by treating a rickettsia-rhesus monkey antibody complex with goat-antimonkey immunoglobulin. A highly successful method for species and strain differentiation is the toxin neutralization test. This test is based on the property of most rickettsial species to kill mice within 1–8 h after intravenous injection of high concentrations. The toxin was first demonstrated by Gildemeister and Haagen (1940) in *R. typhi* and rapidly applied to the development of a neutralization test within the typhus group by Topping et al. (1945). It was applied to the Gilliam

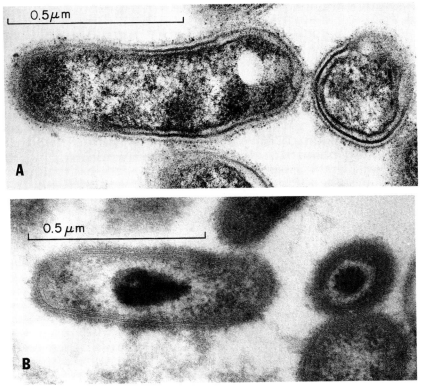

Fig. 3. Ultrathin sections of rickettsiae released from cultured chicken fibroblast cells in the presence of specific anti-rickettsial immune serum. (A) Wilmington strain of *Rickettsia typhi;* (B) Nine mile strain (phase I) of *Coxiella burnetii*. [Courtesy of Thomas F. McCaul, Marius G. Peacock, and Jim C. Williams, Laboratory of Microbial Structure and Function, Rocky Mountain Laboratories, Hamilton, Montana. Unpublished observations.]

strain of *R. tsutsugamushi* (Smadel et al., 1946), to the spotted fever group (Bell and Pickens, 1953; Bell and Stoenner, 1960), and to the Karp and Kato strains of *R. tsutsugamushi* (Kitaoka and Tanaka, 1973) when suspensions of sufficient titer became available. The procedure consists of incubating an appropriate dilution of toxin with serial dilutions of antiserum for 0.5–2.0 h at room temperature, injecting 0.5-ml volumes into the tail vein of mice, and observing the mice for 24–48 h. The mice are protected by homologous but not by heterologous serum.

Other serological tests are usually used for antibody detection, but they can be modified to identify antigens. The CF test performed with soluble ether-extracted antigen identifies the group; the CF and MA tests performed with particulate ether-extracted antigens display a moderate degree of species specificity for the typhus and spotted fever groups (Fiset et al., 1969; Ormsbee et al., 1977; Shepard et al., 1976). The enzyme-linked immunosorbent assay (ELISA) is highly satisfactory provided the antigen is derived from Renografin-purified rickettsiae (Dasch, Halle, and Bourgeois, 1979; Halle, Dasch, and Weiss, 1977). The rickettsiae are disrupted in a French pressure cell or sonicator and the cell debris is removed by centrifuga-

tion at 10,400 $\times$ g for 15 min and stored at 4°C in a final concentration of 0.2% formalin. With this type of antigen, the ELISA and the fluorescent-antibody test are entirely comparable in sensitivity and strain specificity for *R. tsutsugamushi*. Of particular usefulness for species identification is rocket immunoelectrophoresis (Weeke, 1973; G. A. Dasch, personal communication). Hyperimmune rabbit antisera are incorporated in agar layers on glass slides. Soluble rickettsial extracts (supernatant fluids following centrifugation at 100,000 $\times$ g for 1 h) are subjected to electrophoresis into the antibody-containing gel. Sharply defined precipitin "rockets" occur in homologous reactions, in contrast to the faint and diffuse smears seen when the antiserum and the antigen are obtained from different species of the typhus group.

A rapid assay of biological activity is required before preparation of rickettsial antigens, in particular before biochemical investigations. The activity of the purified rickettsiae can be quickly measured by their glutamate-catabolizing activity and, in the case of the typhus rickettsiae, by their hemolytic activity. Glutamate catabolism is determined by methods commonly employed with resting bacteria, for example, by CO_2 production in 25-ml Erlenmeyer flasks.

Determination of Glutamate Catabolism

> Plastic cups, each containing a rolled strip of filter paper, are suspended from rubber stoppers in the 25-ml Erlenmeyer flasks (Weiss, Rees, and Hayes, 1967). To each vessel are added 5 mM glutamate with 0.1 μCi of [U-^{14}C]glutamate, 2 mM MgCl$_2$, and 0.1–1.0 mg rickettsial protein, diluted in a total volume of 2 ml with K36 that contains 0.5% BPA. The flasks are incubated for 1–2.5 h at 34°C and the reaction is terminated by the injection of 0.2 ml Hyamine (Packard Instrument Co.) to the cups, and 0.4 ml of 25% trichloroacetic acid to the reaction mixtures; the flasks are again incubated for about 30 min. The results are determined by the amount of ^{14}CO$_2$ trapped by the Hyamine.

In a recent survey of nine strains of the typhus group (Woodman et al., 1977b), the amount of CO$_2$ produced in 2.5 h at 34°C ranged from 0.5 to 1.2 μmol/mg rickettsial protein.

Since rickettsiae do not metabolize glucose, a similar test can be performed to check for bacterial contamination. [U-^{14}C]Glucose is substituted for labeled glutamate. If the total concentration of glucose is reduced to 0.1 mM, the production of more than 1 nmol of CO$_2$ would indicate the presence of organisms capable of glycolytic activity.

Hemolysis of sheep or rabbit erythrocytes by typhus rickettsiae was first demonstrated by Clarke and Fox (1948). The test was standardized by Snyder et al. (1954) and is still performed essentially as described by these authors.

Hemolysis Test for Typhus Rickettsiae
(Snyder et al., 1954)

> Volumes of 0.3 ml of the appropriate dilutions of rickettsiae are incubated for 2.5 h at 34°C with 0.6-ml volumes of a 25% suspension of washed sheep erythrocytes in diluent NK 7 G (0.04 M KCl, 0.09 M NaCl, 0.008 M Na$_2$HPO$_4$, 0.004 M KH$_2$PO$_4$, 0.005 M potassium glutamate, pH 7.0), to which are added 0.4 ml of 50% glucose and 8 ml of salt solution (0.095 M MgCl$_2$, 0.0025 M CaCl$_2$, and 0.0025 M MnCl$_2$) per 100 ml of NK 7 G. The reaction is terminated by the addition to each tube of 3 ml of saline that contains 0.2% formalin. After centrifugation to sediment the erythrocytes that have not been lysed, the absorbence of the hemoglobin in the supernatant fluid is determined at 545 nm. The number of hemolytic units in a suspension is calculated by multiplying the rickettsial dilution (determined by interpolation) that elicits an optical density reading of 0.3 by the final volume (in ml) of the reaction mixture (3.9 in above-described procedure).

In the recent survey by Woodman et al. (1977), seven strains of the typhus group exhibited titers of 500–1,200 hemolytic units per milligram protein. However, some batches of sheep erythrocytes are not readily hemolyzed by rickettsiae and produce much lower titers.

Of the tests that require rickettsial cell fractionation, the determination of DNA base composition and DNA-DNA hybridization is useful in species identification. There is a significant difference between the G + C values for typhus rickettsiae (about 30 mol%) and spotted fever rickettsiae (about 32.5 mol%) (Tyeryar et al., 1973). The DNAs of $R.$ $prowazekii$ and $R.$ $typhi$ hybridize to the extent of 70–77% (Myers and Wisseman, 1980).

Polyacrylamide gel electrophoresis (PAGE) of whole rickettsial cells solubilized by boiling for 3 min in a solution of 2.5% mercaptoethanol and 1% sodium dodecyl sulfate has shown great promise as an ancillary method for strain identification (Dasch, Samms, and Weiss, 1978; Eisemann and Osterman, 1976; Obijeski, Palmer, and Tzianabos, 1974; Pedersen and Walters, 1978). In a recent survey of nine strains of the typhus group, no differences were encountered among six strains of $R.$ $prowazekii$, but there were pronounced differences in PAGE migration properties of the proteins of the three species of typhus rickettsiae.

Of greatest promise for the detection of differences among strains is isoelectric focusing of the soluble cytoplasmic fractions (supernatant fluid following centrifugation at 80,000–100,000 × g). By this procedure, it was shown that there are minute differences between strains of $R.$ $prowazekii$ derived from eastern Europe, Spain, or Africa, and American wildlife (Dasch, Samms, and Weiss, 1978; G. A. Dasch, personal communication).

THE GENUS COXIELLA

$Coxiella$ $burnetii$ was first recognized during a disease outbreak among slaughterhouse workers in Brisbane, Queensland, and described by Edward H. Derrick in 1937. Derrick designated the disease "Q" fever, the initial of the word "query", to denote the uncertainty surrounding the etiology and nature of the disease. He cultivated and maintained the agent in the guinea pig by serial passages and speculated on its rickettsial nature. In later publications (1944, 1953, 1973), Derrick contributed greatly to the elucidation of the epidemiology of Q fever and to the description of the course of the disease in man. In 1939 he proposed the name $Rickettsia$ $burneti$ for the etiological agent of Q fever in honor of Franklin Macfarlane Burnet, who in collaboration with Freeman in 1937 clearly demonstrated that the agent was a rickettsia distinct from the previously described groups of typhus, spotted fever, and scrub typhus. Philip (1948) pointed out that $R.$ $burneti$ was sufficiently different from the

Table 2. Some of the main differences among the pathogenic rickettsiae.[a]

Genus	Group	DNA base ratio[b]	Growth environment
Rickettsia	Typhus	30	Cytoplasm of eukaryotic cells[c]
	Spotted fever	32–33	Cytoplasm and nucleus of eukaryotic cells
	Scrub typhus	?	Cytoplasm of eukaryotic cells; primarily perinuclear area
Coxiella		43	Phagolysosome of eukaryotic cells
Rochalimaea		39	Surface of eukaryotic cells; axenic growth in enriched bacteriological media

[a] References in Weiss, 1973, 1978.

[b] Molar percentage guanine plus cytosine.

[c] Except *R. canada*, which grows also in the nucleus (Burgdorfer and Brinton, 1970).

other rickettsiae to justify the creation of a new genus. He proposed the generic name *Coxiella* in honor of Harold R. Cox, who contributed to the early studies of this microorganism in the United States and introduced the yolk-sac procedure of inoculation, which facilitated the study of the genera *Rickettsia* and *Coxiella*. Philip's proposal was fortunate. The criteria that he used for the establishment of a separate genus were filterability and high resistance to physical and chemical agents. To these criteria can now be added differences in DNA base ratios, location within the cell (Table 2), and metabolic pathways (reviewed by Weiss, 1973). Organisms closely allied to *C. burnetii* have not been described and this species remains the only one in the genus *Coxiella*.

Habitat

Little can be added to the detailed description of the habitat of *Coxiella burnetii* published by Babudieri in 1959. The distribution of the microorganism is worldwide. The most commonly infected arthropod is the tick. *Amblyomma, Dermacentor, Haemaphysalis, Hyalomma, Ixodes, Ornithodorus,* and *Rhipicephalus* are among the genera most frequently mentioned as carriers of *C. burnetii*. Although many wild mammals and birds are undoubtedly infected, isolations from these animals have not been numerous. Several isolations were made from the Australian bandicoot, *Isoodon torosus*, which was believed to play a role in the epidemiology of Q fever (Derrick and Smith, 1940). *C. burnetii* is isolated most frequently from domestic animals, in particular cattle, sheep, and goats. Outbreaks of human disease occur when large herds of domestic animals are transported to a region that has a large percentage of susceptible individuals. The disease is most common among personnel of the meat and dairy industries.

In contrast to other rickettsiae, *C. burnetii* is transmitted to man, and probably from animal to animal, primarily through the aerosol route because of the high titer achieved by this organism in the tick and in certain organs of the mammalian host and the high stability of the organism in the environment. The microorganisms grow particularly well in the placenta, where they may reach a titer of 10^9 viable organisms per gram. Highly infectious dust from tick feces deposited on the hides of animals and from dried placentas following parturition are the chief sources of infection. There are, however, many other vehicles of infection, including milk.

Despite heavy infection, the disease in animals is mild, if demonstrable at all. In fact, a distinct illness is produced only in man and in the guinea pig. Q fever in man is usually characterized by interstitial pneumonia, high fever, and severe headache. The case-fatality rate is low, even without antibiotic treatment. However, *C. burnetii* infection may persist and, in some rare cases, result in endocarditis (Applefeld, 1977).

Isolation

Procedures for isolation of *Coxiella burnetii* are similar to those used for the genus *Rickettsia*. *C. burnetii* is even more highly infectious for laboratory workers than typhus rickettsiae. Precautions must be stringent to avoid, in particular, infection of individuals in other parts of the building through common ventilation systems.

The hemolymph test for the tentative identification of rickettsiae in ticks can be applied to *C. burnetii*. *C. burnetii* stains readily by the Giménez method. In comparison with other rickettsiae, it is somewhat smaller and its outer membranes are not as well visualized (Fig. 3B).

The animal of choice for isolation is the guinea

pig. The chief clinical manifestation is fever that starts 5–12 days after inoculation and lasts 2–8 days. The infection terminates in death only when the inoculum is very large, but there are differences in strain virulence. The spleen, obtained on the third day of fever, is used for passage.

The chicken embryo is most valuable for the preparation of seed and antigen. Since the rickettsiae grow relatively slowly, maximum yields require relatively large inocula. When the inoculum contains only very few rickettsiae, a second and sometimes a third egg passage is necessary for a satisfactory yield. The eggs can be incubated at 35 or 37°C and the yolk sacs can be harvested shortly before or shortly after the death of the embryo. Titers that exceed 10^{11} viable rickettsiae per egg are not unusual.

In comparison with other rickettsiae, *C. burnetii* is not highly infectious for mouse L-cells (Ormsbee et al., 1978), but can grow quite well in this cell line and establish a persistent infection (Burton et al., 1978). *C. burnetii* is quite efficient in producing plaques on chicken-embryo monolayers, which can be counted 8–10 days following inoculation. Antigens for MA or CF are prepared from rickettsiae purified from yolk sacs by various procedures of density-gradient centrifugation and chemical fractionation.

Identification

Procedures for final identification of *Coxiella burnetii* are like those used for other rickettsiae, with the exception that this organism, unlike other rickettsiae, undergoes phase variation. This phenomenon, somewhat similar to the smooth-rough variation in other bacteria, was first described by Stoker and Fiset (1956) and reviewed by Fiset and Ormsbee (1968) and Brezina (1978). Naturally occurring strains are in phase I. Upon repeated passage in the yolk sac of chicken embryos, the rickettsiae gradually convert to phase II. Complete conversion may require as many as 100 egg passages, but most of the characteristics are acquired by the tenth passage. Except for the "pure" phase II strains, conversion back to phase I occurs rapidly, often during a single passage in laboratory animals.

The main difference between the two phases stems from a carbohydrate surface antigen, rich in glucuronic acid, that is present in phase I but not in phase II. Phase I cells are hydrophilic, are stable in suspension, and are not digested by phagocytic cells in the absence of antibodies. Phase II cells are hydrophobic, autoagglutinable, have greater avidity than phase I cells for stains such as hematoxylin, basic fuchsin, and acridine orange, and are readily phagocytized.

Although both phases grow quite well in chicken-embryo cell cultures, only phase II organisms multiply well in established cell lines such as mouse L- or human HeLa cells. When injected into animals, phase I cells are much better immunogens. Surprisingly, the antibodies first to appear are anti–phase II and only later are phase I antibodies produced. The response to phase II cells is slow and the antibodies are directed only against phase II. Fiset and Ormsbee (1968) suggested that the phase I carbohydrate acts primarily as an adjuvant for the production of phase II antibodies and only secondarily as an antigen that stimulates antibodies against itself.

Phase I antigen can be extracted by trichloroacetic acid or can be destroyed by treatment with potassium periodate. Since purified phase I cells are more easily prepared than phase II cells, phase II antigen, which is often needed in larger amounts than phase I, is obtained from phase I preparations by removal of the phase I carbohydrate.

Despite the wide distribution of the Q fever rickettsiae in different invertebrate and vertebrate hosts and in different geographical locations, naturally occurring variation in biological properties has not been conspicuous. The only notable mutant strains have been those developed in the laboratory by repeated serial passage in chicken embryos that have led to the isolation of the M-44 strains of greatly reduced virulence, which are live vaccine candidates (Genig, 1968; Robinson and Hasty, 1974). Thus, except for phase variation, problems of strain and/or species differentiation are not as prominent in *Coxiella* as in *Rickettsia*.

Biochemical reactions have not been particularly useful in the identification of *C. burnetii*. Although this microorganism has enzymes involved in the metabolism of glucose and the citric acid cycle, the presence of these activities was shown only with considerable difficulty and primarily in cell extracts (reviewed by Paretsky, 1968; Weiss, 1973).

THE GENUS *ROCHALIMAEA*

Habitat

Rochalimaea quintana made its first explosive appearance during World War I, when it caused the loss of more man-days in the armed forces than any disease except influenza. The disease, called trench fever, was not fatal; but, although mild in some cases, it incapacitated many patients for 5 or 6 weeks with several bouts of chills and fever, and severe pain in the legs and back. A commission headed by Strong (1918) established the basic information of the etiology and epidemiology of the disease. It was shown that the etiological agent had the general appearance and stain affinities of rickettsiae.

It grows profusely in the human body louse, but unlike the typhus rickettsia, it adheres to the epithelium of the gut, does not penetrate into the cells, and does not shorten the life of the louse. *R. quintana* is transmitted to man by the louse, either by its bite or by infection of the abraded skin with fecal material. The louse does not transmit the microorganisms transovarially, but man is capable of infecting lice for months and even years and must be regarded as the main reservoir of infection.

After World War I, trench fever disappeared as a disease of any public-health importance. Interest in the microorganism dwindled as attempts to grow it in bacteriological media, experimental animals, or chicken embryos failed.

Trench fever reappeared in eastern Europe during World War II and stimulated additional interest. Mooser et al. (1948) infected a laboratory strain of lice with *R. quintana* by feeding them on three patients in Osijek, Yugoslavia. The Osijek strain was maintained for several years in lice, which were fed on human volunteers who often became infected. In 1959 Henry S. Fuller fed the lice on himself, and a few weeks later experienced a typical clinical course of trench fever. His blood was used for the clear demonstration that *R. quintana* can be grown on blood agar (Vinson and Fuller, 1961). This achievement opened the way for the investigation of the biological properties of *R. quintana* and additional attempts at isolation. It was also the deciding factor for the establishment of the genus *Rochalimaea*, named in honor of the distinguished investigator of both typhus and trench fever, H. da Rocha-Lima. (The species designation "quintana" reflects one of the synonyms for trench fever, five-day fever.)

An indication that *R. quintana* has a wide distribution was obtained by Varela, Fournier, and Mooser (1954), who collected louse feces in a public dormitory in Mexico City and scarified the skin of two human volunteers with a suspension of louse feces. Both volunteers came down with a febrile illness characteristic of trench fever. The experiment was repeated several years later and a strain, Guadalupe, was isolated on blood agar from the blood of a human volunteer; other strains were isolated from subsequent volunteers (Varela, Vinson, and Molina-Pasquel, 1969, Vinson, Varela, and Molina-Pasquel, 1969).

Further confirmation that *R. quintana* infection is widespread and that unrecognized epidemics may have occurred before World War I was obtained in a totally unexpected manner. In 1942 a laboratory station was established in Grosse Isle in the St. Lawrence River, 29 miles downstream from Quebec City. The small island, almost deserted at that time, was used as a quarantine station during the nineteenth century. In 1847, during the height of the immigration from Ireland, thousands of immigrants died of typhus and were buried there. Baker (1946)

suspected that on Grosse Isle he might find a vestige of the nineteenth century epidemic. He examined the local fauna and isolated a microorganism superficially resembling a rickettsia from five of ten captured voles (*Microtus pennsylvanicus*). This agent, called "vole rickettsia", was not isolated from voles in other locations despite very extensive surveys. Recently, Weiss et al. (1978) showed that the "vole rickettsia" is a strain of *R. quintana* or a closely allied microorganism.

Although the information is fragmentary, we can surmise that *Rochalimaea quintana* has a habitat, history, and distribution quite similar to *Rickettsia prowazekii*. Since the infection is not fatal and probably mild in endemic periods, it is not recognized, except when a large number of susceptibles are suddenly exposed to massive louse infestation. It has recently been shown (Myers and Wisseman, 1980) that the DNAs of the typhus rickettsiae and *R. quintana* hybridize to the extent of 25–33%.

Isolation

The original isolation of *Rochalimaea quintana* by Vinson and Fuller (1961) was done on blood agar plates, which consist of blood agar base with 10% freshly drawn, defibrinated horse or human blood. The plates were incubated at 32–34°C in an atmosphere of 5% CO_2 in air. In a subsequent study, Vinson (1966) obtained somewhat better results by separating the serum from the erythrocytes: the serum was inactivated at 56°C for 30 min and the cells were washed in PBS and lysed in distilled water. Myers, Cutler, and Wisseman (1969), Myers, Ostermann, and Wisseman (1972), and Mason (1970) showed that the erythrocyte requirement could be met by the addition of hemoglobin, hematin, or fetal calf serum (FCS), but not calf serum. The serum requirement could be met by detoxifying agents such as bovine plasma albumin or starch. Both air and CO_2 proved to be essential. Succinate or glutamate as a source of energy (Huang, 1967) enhanced growth. Growth was also obtained in liquid media of similar composition, provided the inoculum was moderately high. In our laboratory, excellent results were obtained in a medium that consists of a salt solution with 25 mM HEPES buffer, pH 7.2, 0.8% Casamino Acids (Difco), 10% FCS, 0.1% yeast extract, 0.002 M succinate, and 0.04% sodium bicarbonate. We have also grown *R. quintana* on monolayers of irradiated L-cells (Weiss et al., 1978) or chicken embryo cells.

Primary isolation requires 12–14 days of incubation. The colonies are 65–200 μm in diameter, round, lenticular, translucent, and mucoid. On subsequent passages, the colonies are seen with the naked eye after 4–7 days. Growth in liquid media varies with the strain, the medium composition, and the inoculum size. Optical density measurements and viable cell counts indicate that with moderate inocula the logarithmic phase of growth can be completed as early as in 2 days, but even under optimal conditions, the turbidity is about one-tenth that obtained with *Escherichia coli*. When grown on the eukaryotic monolayers, the trench fever rickettsiae grow on the cell surface (Merrell, Weiss, and Dasch, 1978), where they achieve a high population density in about 5 days. This type of growth is reminiscent of the observations made on the microorganisms that were cultivated in the louse and had multiplied in the lumen of the gut (Ito and Vinson, 1965).

Identification

Identification is accomplished by serological methods with known human or rabbit sera. The antigens are either whole cells or soluble fractions of sonicated cells, depending on the test to be used (Herrmann et al., 1977; Vinson and Campbell, 1968). Biochemical tests have shown that *Rochalimaea quintana,* like the members of the genus *Rickettsia,* is devoid of enzymes capable of degrading glucose or glucose 6-phosphate with the production of CO_2. Of the substrates tested, succinate, glutamine, and glutamate are catabolized most rapidly (Huang, 1967; Weiss et al., 1978).

Acknowledgments

This review was supported in part by the Research and Development Command, Department of the Navy, research project number MR04105.0028. I am greatly indebted to my colleagues, particularly P. F. Dirk Van Peenen and Gregory A. Dasch, for their helpful discussions. The opinions and assertions contained herein are the private ones of the writer and are not to be construed as official or reflecting the views of the Navy Department or the naval service at large.

Literature Cited

Andrew, R., Bonnin, J. M., Williams, S. 1946. Tick typhus in North Queensland. Medical Journal of Australia 2:253–258.

Applefeld, M. M., Billingsley, L. M., Tucker, H. J., Fiset, P. 1977. Q fever endocarditis–a case occurring in the United States. American Heart Journal 93:669–670.

Babudieri, B. 1979. Q fever: A zoonosis. Advances in Veterinary Science 5:81–182.

Baker, J. A. 1946. A rickettsial infection in Canadian voles. Journal of Experimental Medicine 84:37–51.

Bell, E. J., Pickens, E. G. 1953. A toxic substance associated with the rickettsias of the spotted fever group. Journal of Immunology 70:461–472.

Bell, E. J., Stoenner, H. G. 1960. Immunologic relationships among the spotted fever group of rickettsias determined by toxin neutralization tests in mice with convalescent animal serums. Journal of Immunology 84:171–182.

Blake, F. G., Maxcy, K. F., Sadusk, J. F., Jr., Kohls, G. M. Bell, E. J. 1945. Studies on tsutsugamushi disease (scrub typhus, mite-borne typhus) in New Guinea and adjacent islands: Epidemiology, clinical observations, and etiology in the Dobadura area. American Journal of Hygiene 41:243–373.

Bourgeois, A. L., Olson, J. G., Ho, C. M., Fang, R. C. Y., Van Peenen, P. F. D. 1977. Epidemiological and serological study of scrub typhus among Chinese military in the Pescadores Islands of Taiwan. Royal Society of Tropical Medicine and Hygiene 71:338–342.

Bovarnick, M. R., Miller, J. C., Snyder, J. C. 1950. The influence of certain salts, amino acids, sugars, and protein on the stability of rickettsiae. Journal of Bacteriology 59:509–522.

Bozeman, F. M., Humphries, J. W., Campbell, J. M., O'Hara, P. L. 1960. Laboratory studies of the spotted fever group of rickettsiae, pp. 7–11. In: Wisseman, C. L., Jr. (ed.), Symposium on the spotted fever group of rickettsiae. Medical Science Publication No. 7. Washington: Walter Reed Army Institute of Research.

Bozeman, F. M., Shirai, A., Humphries, J. W., Fuller, H. S. 1967. Ecology of Rocky Mountain spotted fever. II. Natural infection of wild mammals and birds in Virginia and Maryland. American Journal of Tropical Medicine 16:48–59.

Bozeman, F. M., Elisberg, B. L., Humphries, J. W., Runcik, K., Palmer, D. B., Jr. 1970. Serologic evidence of Rickettsia canada infection in man. Journal of Infectious Diseases 121:367–371.

Bozeman, F. M., Masiello, S. A., Williams, M. S., Elisberg, B. L. 1975. Epidemic typhus rickettsiae isolated from flying squirrels. Nature 255:545–547.

Bozeman, F. M., Williams, M. S., Stocks, N. I., Chadwick, D. P., Elisberg, B. L., Sonenshine, D. E., Lauer, D. M. 1978. Ecologic studies on epidemic typhus infection in the eastern flying squirrel, pp. 493–504. In: Kazar, J., Ormsbee, R. A., Tarasevich, I. N. (eds.), Proceedings of the 2nd International Symposium on Rickettsiae and Rickettsial Diseases, Smolenice, June 21–25, 1976. Bratislava: VEDA.

Brezina, R. 1978. Phase variation phenomenon in Coxiella burnetii, pp. 221–235. In: Kazar, J., Ormsbee, R. A., Tarasevich, I. N. (eds.), Proceedings of the 2nd International Symposium on Rickettsiae and Rickettsial Diseases, Smolenice, June 21–25, 1976. Bratislava: VEDA.

Brezina, R., Murray, E. S., Tarizzo, M. L., Bögel, K. 1973. Rickettsiae and rickettsial diseases. Bulletin of the World Health Organization 49:433–442.

Buchanan, R. E., Gibbons, N. E. (eds.). 1974. Bergey's manual of determinative bacteriology, 8th ed. Baltimore: Williams & Wilkins.

Burgdorfer, W. 1970. Hemolymph test. A technique for detection of rickettsiae in ticks. American Journal of Tropical Medicine and Hygiene 19:1010–1014.

Burgdorfer, W. 1975. A review of Rocky Mountain spotted fever (tick-borne typhus), its agent, and its tick vectors in the United States. Journal of Medical Entomology 12:269–278.

Burgdorfer, W. 1977. Tick-borne diseases in the United States: Rocky Mountain spotted fever and Colorado tick fever. Acta Tropica 34:103–126.

Burgdorfer, W., Brinton, L. P. 1970. Intranuclear growth of Rickettsia canada, a member of the typhus group. Infection and Immunity 2:112–114.

Burgdorfer, W., Brinton, L. P. 1975. Mechanisms of transovarial infection of spotted fever rickettsiae in ticks. Annals of the New York Academy of Sciences 266:61–72.

Burgdorfer, W., Cooney, J. C., Thomas, L. A. 1974. Zoonotic potential (Rocky Mountain spotted fever and tularemia) in the Tennessee Valley Region. II. Prevalence of Rickettsia rickettsi and Francisella tularensis in mammals and ticks from Land

Between the Lakes. American Journal of Tropical Medicine and Hygiene **23**:109–117.

Burgdorfer, W., Lackman, D. 1960. Identification of *Rickettsia rickettsii* in the wood tick, *Dermacentor andersoni*, by means of fluorescent antibody. Journal of Infectious Diseases **107**:241–244.

Burgdorfer, W., Newhouse, V. F., Pickens, E. G., Lackman, D. B. 1962. Ecology of Rocky Mountain spotted fever in Western Montana. I. Isolation of *Rickettsia rickettsii* from wild mammals. American Journal of Hygiene **76**:293–301.

Burgdorfer, W., Sexton, D. J., Gerloff, R. K., Anacker, R. L., Philip, R. N., Thomas, L. A. 1975. *Rhipicephalus sanguineus*: Vector of a new spotted fever group rickettsia in the United States. Infection and Immunity **12**:205–210.

Burnet, F. M., Freeman, M. 1937. Experimental studies on the virus of "Q" fever. Medical Journal of Australia **2**:299–305.

Burton, P. R., Stuckemann, J., Welsh, R. M., Paretsky, D. 1978. Some ultrastructural effects on persistent infections by the rickettsia *Coxiella burnetii* in mouse L cells and green monkey kidney (Vero) cells. Infection and Immunity **21**:556–566.

Campbell, R. W., Domrow, R. 1974. Rickettsioses in Australia: Isolation of *Rickettsia tsutsugamushi* and *R. australis* from naturally infected arthropods. Transactions of the Royal Society of Tropical Medicine and Hygiene **68**:397–402.

Clarke, D. H., Fox, J. P. 1948. The phenomenon of in vitro hemolysis produced by the rickettsiae of typhus fever, with a note on the mechanism of rickettsial toxicity in mice. Journal of Experimental Medicine **88**:25–41.

Conor, A., Bruch, A. 1910. Une fièvre éruptive observée en Tunisie. Bulletin de Société Pathologique Exotique **3**:492–496.

Coolbaugh, J. C., Progar, J. J., Weiss, E. 1976. Enzymatic activities of cell-free extracts of *Rickettsia typhi*. Infection and Immunity **14**:298–305.

Cox, H. R. 1941. Cultivation of rickettsiae of Rocky Mountain spotted fever, typhus and Q fever groups in the embryonic tissues of developing chicks. Science **94**:399–403.

da Rocha-Lima, H. 1916. Zur Aetiologie des Fleckfiebers. Berliner Klinische Wochenschrift **53**:567–569.

Dasch, G. A., Halle, S., Bourgeois, A. L. 1979. A sensitive microplate enzyme-linked immunosorbent assay (ELISA) for the detection of antibodies against the scrub typhus rickettsia, *Rickettsia tsutsugamushi*. Journal of Clinical Microbiology **9**:38–48.

Dasch, G. A., Samms, J. R., Weiss, E. 1978. Biochemical characterisics of typhus group rickettsiae with special attention to the *Rickettsia prowazekii* strains isolated from flying squirrels. Infection and Immunity **19**:676–685.

Dasch, G. A., Weiss, E. 1977. Characterization of the Madrid E strain of *Rickettsia prowazekii* purified by Renografin density gradient centrifugation. Infection and Immunity **15**:280–286.

Derrick, E. H. 1937. "Q" fever a new fever entity: Clinical features, diagnosis and laboratory investigation. Medical Journal of Australia **2**:281–299.

Derrick, E. H. 1939. *Rickettsia burneti*: The cause of "Q" fever. Medical Journal of Australia **1**:14.

Derrick, E. H. 1944. The epidemiology of Q fever. Journal of Hygiene **43**:357–361.

Derrick, E. H. 1953. The epidemiology of "Q" fever: A review. Medical Journal of Australia **1**:245–253.

Derrick, E. H. 1973. The course of infection with *Coxiella burneti*. Medical Journal of Australia **1**:1051–1057.

Derrick, E. H., Smith, D. J. W. 1940. The isolation of three strains of *Rickettsia burneti* from the bandicoot *Isoodon torosus*. Australian Journal of Experimental Biology and Medical Science **18**:99–102.

DuPont, H. L., Hornick, R. B., Dawkins, A. T., Heiner, G. G., Fabrikant, I. B., Wisseman, C. L., Jr., Woodward, T. E. 1973. Rocky Mountain spotted fever: A comparative study of

the active immunity induced by inactivated and viable pathogenic *Rickettsia rickettsii*. Journal of Infectious Diseases **128**:340–344.

Eisenmann, C. S., Osterman, J. V. 1976. Proteins of typhus and spotted fever group rickettsiae. Infection and Immunity **14**:155–162.

Fiset, P., Ormsbee, R. A. 1968. The antibody response to antigens of *Coxiella burneti*. Zentralblatt für Bakteriologie, Parasitenkunde, Infektionskrankheiten und Hygiene, Abt. 1 Orig. **206**:321–329.

Fiset, P., Ormsbee, R. A., Silberman, R., Peacock, M., Spielman, S. H. 1969. A microagglutination technique for detection and measurement of rickettsial antibodies. Acta Virologica **13**:60–66.

Fox, J. P., Everritt, M. G., Robinson, T. A., Conwell, D. P. 1954. Immunization of man against epidemic typhus by infection with avirulent *Rickettsia prowazeki* (Strain E). American Journal of Hygiene **59**:74–88.

Fuller, H. S. 1954. Studies of human body lice, *Pediculus humanus corporis*. III. Initial dosage and ambient temperature as factors influencing the course of infection with *Rickettsia prowazeki*. American Journal of Hygiene **59**:140–149.

Genig, V. A. 1968. A live vaccine l/M-44 against Q-fever for oral use. Journal of Hygiene, Epidemiology, Microbiology, and Immunology **12**:265–273.

Gildemeister, E., Haagen, E. 1940. Fleckfieberstudien. I. Mitteilung: Nachweis eines Toxins in Rickettsien—Eikulturen *(Rickettsia mooseri)*. Deusche Medizinische Wochenschrift **66**:878–880.

Giménez, D. F. 1964. Staining rickettsiae in yolk-sac cultures. Stain Technology **39**:135–140.

Gordon, F. B., Dressler, H. R., Quan, A. L., McQuilkin, W. T., Thomas, J. I. 1972. Effect of ionizing irradiation on susceptibility of McCoy cell cultures to *Chlamydia trachomatis*. Applied Microbiology **23**:123–129.

Gould, D. J., Miesse, M. L. 1954. Recovery of a rickettsia of the spotted fever group from *Microtus pennsylvanicus* from Virginia. Proceedings of the Society for Experimental Biology and Medicine **85**:558–561.

Groves, M. G., Osterman, J. V. 1978. Host defenses in experimental scrub typhus: Genetics of natural resistance to infection. Infection and Immunity **19**:583–588.

Halle, S., Dasch, G. A., Weiss, E. 1977. Sensitive enzyme-linked immunosorbent assay for detection of antibodies against typhus rickettsiae, *Rickettsia prowazekii* and *Rickettsia typhi*. Journal of Clinical Microbiology **6**:101–110.

Hattwick, M. A. W., Retailliau, H., O'Brien, R. J., Slutzker, M., Fontaine, R. E., Hanson, B. 1978. Fatal Rocky Mountain spotted fever. Journal of the American Medical Association **240**:1499–1503.

Herrmann, J. E., Hollingdale, M. R., Collins, M. F., Vinson, J. M. 1977. Enzyme immunoassay and radioimmunoprecipitation tests for the detection of antibodies to *Rochalimaea* (Rickettsia) *quintana*. Proceedings of the Society for Experimental Biology and Medicine **154**:285–288.

Hoogstraal, H. 1967. Tick in relation to human diseases caused by *Rickettsia* species. Annual Review of Entomology **12**:377–420.

Horsfall, F. L., Jr., Tamm, I., eds. 1965. Viral and rickettsial infections of man. 4th ed. Philadelphia, Montreal: Lippincott.

Huang, K. 1967. Metabolic activity of the trench fever rickettsia, *Rickettsia quintana*. Journal of Bacteriology **93**:853–859.

Huebner, R. J., Jellison, W. L., Pomerantz, C. 1946. Rickettsialpox—a newly recognized rickettsial disease. IV. Isolation of a rickettsia apparently identical with the causative agent of rickettsialpox from *Allodermanyssus sanguineus*, a rodent mite. Public Health Reports **61**:1677–1682.

Ito, S., Vinson, J. W. 1965. Fine structure of *Rickettsia quintana* cultivated in vitro and in the louse. Journal of Bacteriology **89**:481–495.

Ito, S., Vinson, J. W., McGuire, T. J., Jr. 1975. Murine typhus rickettsiae in the oriental rat flea. Annals of the New York Academy of Sciences **266**:35– 60.

Jackson, E. B. 1951. Comparative efficacy of several antibiotics on experimental rickettsial infections in embryonated eggs. Antibiotics and Chemotherapy **1**:231– 241.

Jackson, E. B., Danauskas, J. X., Coale, M. C., Smadel, J. E. 1957a. Recovery of *Rickettsia akari* from the Korean vole *Microtus fortis pelliceus*. American Journal of Hygiene **66**:301– 308.

Jackson, E. B., Danauskas, J. X., Smadel, J. E., Fuller, H. S., Coale, M. C., Bozeman, F. M. 1957b. Occurrence of *Rickettsia tsutsugamushi* in Korean rodents and chiggers. American Journal of Hygiene **66**:309– 320.

Johnson, J. E., Kadull, P. J. 1947. Rocky Mountain spotted fever acquired in a laboratory. New England Journal of Medicine **277**:842– 847.

Kenyon, R. H., McManus, A. T. 1974. Rickettsial infectious antibody complexes: Detection by antiglobulin plaque reduction technique. Infection and Immunity **9**:966– 968.

Kitaoka, M., Tanaka, Y. 1973. Rickettsial toxin and its specificity in 3 prototype strains, Karp, Gilliam and Kato, of *Rickettsia orientalis*. Acta Virologica **17**:426– 434.

Kurz, J., Burgdorfer, W. 1978. Detection of the Rocky Mountain spotted fever agent, *Rickettsia rickettsii*, in dead ticks, *Dermacentor andersoni*. Infection and Immunity **20**:584– 586.

Lackman, D. B., Bell, E. J., Stoenner, H. G., Pickens, E. G. 1965. The Rocky Mountain spotted fever group of rickettsias. Health Laboratory Science **2**:135– 141.

McDade, J. E., Shepard, C. C., Redus, M. A., Newhouse, V. F., Smith, J. D. 1980. Evidence of *Rickettsia prowazekii* infections in the United States. American Journal of Tropical Medicine and Hygiene **29**:277– 284.

McKiel, J. A., Bell, E. J., Lackman, D. B. 1967. *Rickettsia canada:* A new member of the typhus group of rickettsiae isolated from *Haemaphysalis leporispalustris* ticks in Canada. Canadian Journal of Microbiology **13**:503– 510.

Mason, R. A. 1970. Propagation and growth cycle of *Rickettsia quintana* in a new liquid medium. Journal of Bacteriology **103**:184– 190.

Mason, R. A., Seligmann, E. B., Jr., Ginn, R. K. 1976. A reference, inactivated, epidemic typhus vaccine: Laboratory evaluation of candidate vaccines. Journal of Biological Standardization **4**:209– 216.

Merrell, B. R., Weiss, E., Dasch, G. A. 1978. Morphological and cell association characteristics of *Rochalimaea quintana:* Comparison of the vole and Fuller strains. Journal of Bacteriology **135**:633– 640.

Mooser, H. 1928. Experiments relating to the pathology and the etiology of Mexican typhus (tabardillo). Journal of Infectious Diseases **43**:241– 272.

Mooser, H., Leeman, A., Chao, S. H., Gubler, H. U. 1948. Beobachtungen an Fünftagenfieber. Schweizer Zeitung für Allgemeine Pathologie und Bakteriologie **11**:513– 522.

Moreira, J. A., de Magalhaes, O. 1937. Typho exanthematico de Minas Geraes. Brasil-Medico **51**:583– 584.

Murphy, J. R., Wisseman, C. L., Jr., Snyder, L. B. 1976. Plaque assay for *Rickettsia mooseri* in tissue samples. Proceedings of the Society for Experimental Biology and Medicine **153**:151– 155.

Murray, E. S., Torrey, S. B. 1975. Virulence of *Rickettsia prowazeki* for head lice. Annals of the New York Academy of Sciences **266**:25– 34.

Myers, W. F., Cutler, L. D., Wisseman, C. L., Jr. 1969. Role of erythrocytes and serum in the nutrition of *Rickettsia quintana*. Journal of Bacteriology **97**:663– 666.

Myers, W. F., Osterman, J. V., Wisseman, C. L., Jr. 1972. Nutrition studies of *Rickettsia quintana:* Nature of the hematin requirement. Journal of Bacteriology **109**:89– 95.

Myers, W. F., Wisseman, C. L., Jr. 1980. Genetic relatedness among the typhus group of rickettsiae. International Journal of Systematic Bacteriology **30**:143– 150.

National Institutes of Health Biohazards Safety Guide. 1974. Washington: United States Government Printing Office.

Nicolle, C., Comte, C., Conseil, E. 1909. Transmission expérimentale du typhus exanthématique par le pou du corps. Comptes Rendus Hebdomadaires des Séances de l'Académie des Sciences **149**:486– 489.

Nicolle, C., Conseil, E., Conor, A. 1911. Le typhus expérimentale du cobaye. Comptes Rendus Hebdomadaires des Séances de l'Académie des Sciences **152**:1632– 1634.

Obijeski, J. F., Palmer, E. L., Tzianabos, T. 1974. Proteins of purified rickettsiae. Microbios **11**:61– 76.

Ormsbee, R. A. 1974. Rickettsiae, pp. 805– 815. In: Lennette, E. H., Spaulding, E. H., Truant, J. P. (eds.), Manual of clinical microbiology, 2nd ed. Washington, D.C.: American Society for Microbiology.

Ormsbee, R., Peacock, M., Philip, R., Casper, E., Plorde, J., Gabre-Kidan, T., Wright, L. 1977. Serologic diagnosis of epidemic typhus fever. American Journal of Epidemiology **105**:261– 271.

Ormsbee, R., Peacock, M., Gerloff, R., Tallent, G., Wike, D. 1978. Limits of rickettsial infectivity. Infection and Immunity **19**:239– 245.

Oster, C. N., Burke, D. S., Kenyon, R. H., Ascher, M. S., Harber, P., Pedersen, C. E. 1977. Laboratory-acquired Rocky Mountain spotted fever: The hazards of aerosol transmission. New England Journal of Medicine **297**:859– 863.

Paretsky, D. 1968. Biochemistry of rickettsiae and their infected hosts, with special reference to *Coxiella burneti*. Zentralblatt für Bakteriologie, Parasitenkunde, Infektionskrankheiten und Hygiene, Abt. 1 Orig. **206**:284– 291.

Pedersen, C. E., Jr., Walters, V. D. 1978. Comparative electrophoresis of spotted fever group rickettsial proteins. Life Sciences **22**:583– 587.

Perez Gallardo, F., Fox, J. P. 1948. Infection and immunization of laboratory animals with *Rickettsia prowazeki* of reduced pathogenicity, strain E. American Journal of Hygiene **48**:6– 21.

Philip, C. B. 1948. Comments on the name of the Q fever organism. Public Health Reports **53**:58.

Philip, R. N., Casper, E. A. 1978. Serotyping spotted fevergroup rickettsiae with mouse antisera by microimmunofluorescence. A preliminary report, pp. 269– 279. In: Kazar, J., Ormsbee, R. A., Tarasevich, I. N. (eds.), Proceedings of the 2nd International Symposium on Rickettsiae and Rickettsial Diseases, Smolenice, June 21– 25, 1976. Bratislava: VEDA.

Philip, R. N., Casper, E. A., Ormsbee, R. A., Peacock, M. G., Burgdorfer, W. 1976. Microimmunofluorescence test for the serological study of Rocky Mountain spotted fever and typhus. Journal of Clinical Microbiology **3**:51– 61.

Price, W. H., Johnson, J. W., Emerson, H., Preston, C. E. 1954. Rickettsial-interference phenomenon: A new protective mechanism. Science **120**:457– 459.

Reiss-Gutfreund, R. J. 1956. Un nouveau réservoir de virus pour *Rickettsia prowazeki:* Les animaux domestiques et leur tiques. Bulletin de la Socitété Pathologique Exotique **49**:946– 1023.

Reiss-Gutfreund, R. J. 1966. The isolation of *Rickettsia prowazeki* and *mooseri* from unusual sources. Journal of Tropical Medicine and Hygiene **15**:943– 949.

Ricketts, H. T. 1911. Contributions to Medical Science by Howard Taylor Ricketts, 1870– 1910. Chicago: University of Chicago Press.

Riggs, J. L., Seiwald, R. J., Burckhalter, J. H., Downs, C. M., Melcalf, T. G. 1958. Isothiocyanate compounds as fluorescent labelling agents for immune serum. American Journal of Pathology **34**:1081– 1097.

Roberts, L. W., Robinson, D. M., Rapmund, G., Walker, J. S., Gan, E., Ram, S. 1975. Distribution of *Rickettsia tsutsugamushi* in organs of *Leptotrombidium fletcheri* (Prostigmata: Trombiculidae). Journal of Medical Entomology **12**:345– 348.

Robertson, R. G., Wisseman, C. L., Jr. 1973. Tick-borne rickettsiae of the spotted fever group in West Pakistan. II. Serological classification of isolates from West Pakistan and Thailand: Evidence for two new species. American Journal of Epidemiology **97**:55–64.

Robinson, D. M., Brown, G., Gan, E., Huxsoll, D. L. 1976. Adaptation of a microimmunofluorescence test to the study of human *Rickettsia tsutsugamushi* antibody. American Journal of Tropical Medicine and Hygiene **25**:900–905.

Robinson, D. M., Hasty, S. E. 1974. Production of a potent vaccine from the attenuated M-44 strain of *Coxiella burneti*. Applied Microbiology **27**:777–783.

Sexton, D. J., Burgdorfer, W., Thomas, L., Norment, B. R. 1976. Rocky Mountain spotted fever in Mississippi: Survey for spotted fever antibodies in dogs and for spotted fever group rickettsiae in dog ticks. American Journal of Epidemiology **103**:192–197.

Sexton, D. J., Gallis, H. A., McRae, J. R., Cate, T. R. 1975. Possible needle-associated Rocky Mountain spotted fever. New England Journal of Medicine **292**:645.

Shepard, C. C., Redus, M. A., Trianabos, T., Warfield, D. T. 1976. Recent experience with the complement fixation test in the laboratory diagnosis of rickettsial diseases in the United States. Journal of Clinical Microbiology **4**:277–283.

Shirai, A., Wisseman, C. L., Jr. 1975. Serologic classification of scrub typhus isolates from Pakistan. American Journal of Tropical Medicine and Hygiene **24**:145–153.

Silverman, D. J., Wisseman, C. L., Jr. 1978. Comparative ultrastructure study on the cell envelopes of *Rickettsia prowazekii, Rickettsia rickettsii,* and *Rickettsia tsutsugamushi*. Infection and Immunity **21**:1020–1023.

Silverman, D. J., Wisseman, C. L., Jr., Waddell, A. D., Jones, M. 1978. External layers of *Rickettsia prowazekii* and *Rickettsia rickettsii:* Occurrence of a slime layer. Infection and Immunity **22**:233–246.

Smadel, J. E., Jackson, E. B., Bennett, B. L., Rights, F. L. 1946. A toxic substance associated with the Gilliam strain of *R. orientalis*. Proceedings of the Society for Experimental Biology and Medicine **62**:138–140.

Snyder, J. C., Bovarnick, M. R., Miller, J. C., Chang, R. S. 1954. Observations on the hemolytic properties of typhus rickettsiae. Journal of Bacteriology **67**:724–730.

Sonenshine, D. E., Peters, A. H., Levy, G. F. 1972. Rocky Mountain spotted fever in relation to vegetation in the Eastern United States, 1951–1971. American Journal of Epidemiology **96**:59–69.

Sonenshine, D. E., Bozeman, F. M., Williams, M. S., Masiello, S. A., Chadwick, D. P., Stocks, N. I., Lauer, D. M., Elisberg, B. L. 1978. Epizootiology of epidemic typhus *(Rickettsia prowazekii)* in flying squirrels. American Journal of Tropical Medicine and Hygiene **27**:339–349.

Spencer, R. R., Parker, R. R. 1930. Studies on Rocky Mountain spotted fever. Hygiene Laboratory Bulletin **154**:1–116.

Stoenner, H. G., Lackman, D. B., Bell, E. J. 1962. Factors affecting the growth of rickettsias of the spotted fever group in fertile hens' eggs. Journal of Infectious Diseases **110**:121–128.

Stoker, M. G. P., Fiset, P. 1956. Phase variation of the Nine Mile and other strains of *Rickettsia burneti*. Canadian Journal of Microbiology **2**:310–321.

Strong, R. P. (ed.). 1918. Trench fever. Report of Commission, Medical Research Committee, American Red Cross. London: Oxford University Press.

Syverton, J. T., Thomas, L. 1945. A method for staining *Rickettsia orientalis* in yolk sac and other smear preparations. Proceedings of the Society for Experimental Biology and Medicine **59**:87–89.

Tamiya, T. (ed.). 1962. Recent advances in studies of tsutsugamushi disease in Japan. Tokyo: Medical Culture.

Topping, N. H., Bengston, I. A., Henderson, R. G., Shepard, C. C., Shear, M. J. 1945. Studies on typhus fever. National Institute of Health Bulletin No. 183. Washington: United States Government Printing Office.

Traub, R., Wisseman, C. L., Jr. 1968. Ecological considerations in scrub typhus. 1. Emerging concepts. Bulletin of the World Health Organization **39**:209–218.

Traub, R., Wisseman, C. L., Jr. 1974. The ecology of chigger-borne rickettsiosis (scrub typhus). Journal of Medical Entomology **11**:237–303.

Traub, R., Wisseman, C. L., Jr., Fahrlang-Azad, A. 1978. The ecology of murine typhus—a critical review. Tropical Diseases Bulletin **75**:237–317.

Traub, R., Wisseman, C. L., Jr., Jones, M. R., O'Keefe, J. J. 1975. The acquisition of *Rickettsia tsutsugamushi* by chiggers (trombiculid mites) during the feeding process. Annals of the New York Academy of Sciences **266**:91–114.

Tyeryar, F. J., Jr., Weiss, E., Millar, D. B., Bozeman, F. M., Ormsbee, R. A. 1973. DNA base composition of rickettsiae. Science **180**:415–417.

Varela, G., Fournier, R., Mooser, H. 1954. Presencia de *Rickettsia quintana* en piojos *Pediculus humanus* de la Ciudad de México. Inaculatión experimental. Revista del Intituto de Salubridad y Enfermedades Tropicales **14**:39–42.

Varela, G., Vinson, J. W., Molina-Pasquel, C. 1969. Trench fever. II.Propagation of *Rickettsia quintana* on cell-free medium from the blood of two patients. American Journal of Tropical Medicine and Hygiene **18**:708–712.

Vinson, J. W. 1966. *In vitro* cultivation of the rickettsial agent of trench fever. Bulletin of the World Health Organization **35**:155–164.

Vinson, J. W., Campbell, E. S. 1968. Complement fixing antigens from *Rickettsia quintana*. Acta Virologica **12**:54–57.

Vinson, J. W., Fuller, H. S. 1961. Studies on trench fever. I. Propagation of rickettsia-like microorganisms from a patient's blood. Pathologia et Microbiologia Supplement **24**:152–166.

Vinson, J. W., Varela, G., Molina-Pasquel, C. 1969. Trench fever. III. Induction of clinical disease in volunteers inoculated with *Rickettsia quintana* propagated on blood agar. American Journal of Tropical Medicine and Hygiene **18**:712–722.

Walker, J. S., Chan, C. T., Manikumaran, C., Elisberg, B. L. 1975. Attempts to infect and demonstrate transmission of *R. tsutsugamushi* in three species of *Leptotrombidium* mites. Annals of the New York Academy of Sciences **266**:80–90.

Wang, S. P. 1971. A microimmunofluorescence method. Study of antibody response to TRIC organisms in mice, pp. 273–288. In: Nichols, R. L. (ed.), Trachoma and related disorders caused by chlamydial agents. Amsterdam: Excerpta Medica.

Weeke, B. 1973. Rocket immunoelectrophoresis, pp. 37–46. In: Axelrod, N. H., Krøll, J., Weeke, B. (eds), A manual of quantitative immunoelectrophoresis. Scandinavian Journal of Immunology, Suppl. **1**.

Weiss, E. 1973. Growth and physiology of rickettsiae. Bacteriological Reviews **37**:259–283.

Weiss, E. 1978. Biological properties of rickettsiae, pp. 137–153. In: Kazar, J., Ormsbee, R. A., Tarasevich, I. N. (eds.), Proceedings of the 2nd International Symposium on Rickettsiae and Rickettsial Diseases, Smolenice, June 21–25, 1976. Bratislava: VEDA.

Weiss, E., Coolbaugh, J. C., Williams, J. C. 1975. Separation of viable *Rickettsia typhi* from yolk sac and L cell host components by Renografin density gradient centrifugation. Applied Microbiology **30**:456–463.

Weiss, E., Dressler, H. R. 1958. Growth of *Rickettsia prowazeki* in irradiated monolayer cultures of chick embryo entodermal cells. Journal of Bacteriology **75**:544–552.

Weiss, E., Dressler, H. R. 1960. Centrifugation of rickettsiae and viruses onto cells and its effect on infection. Proceedings of the Society for Experimental Biology and Medicine **103**:691–695.

Weiss, E., Moulder, J. W. 1974. *Rickettsia, Rochalimaea, Coxiella,* pp. 883–893. In: Buchanan, R. E., Gibbons, N. E. (eds.), Bergey's manual of determinative bacteriology, 8th ed. Baltimore: Williams & Wilkins.

Weiss, E., Rees, H. B., Jr., Hayes, J. R. 1967. Metabolic activity of purified suspensions of *Rickettsia rickettsi*. Nature **213:**1020– 1022.

Weiss, E., Newman, L. W., Grays, R., Green, A. E. 1972. Metabolism of *Rickettsia typhi* and *Rickettsia akari* in irradiated L cells. Infection and Immunity **6:**50– 57.

Weiss, E., Dasch, G. A., Woodman, D. R., Williams, J. C. 1978. The vole agent identified as a strain of the trench fever rickettsia, *Rochalimaea quintana*. Infection and Immunity **19:**1013– 1020.

White, P. C., Jr. 1970. A brief historical review of murine typhus in Virginia and the United States. Virginia Medical Monthly **97:**16– 23.

Wike, D. A., Tallent, G., Peacock, M. G., Ormsbee, R. A. 1972. Studies of the rickettsial plaque assay technique. Infection and Immunity **5:**715– 722.

Williams, J. C., Peterson, J. C. 1976. Enzymatic activities leading to pyrimidine nucleotide biosynthesis from cell-free extracts of *Rickettsia typhi*. Infection and Immunity **14:**439– 448.

Williams, J. C., Weiss, E. 1978. Energy metabolism of *Rickettsia typhi*: Pools of adenine nucleotides and energy charge in the presence and absence of glutamate. Journal of Bacteriology **134:**884– 892.

Winkler, H. H. 1976. Rickettsial permeability: An ADP-ATP transport system. Journal of Biological Chemistry **251:**389– 396.

Wisseman, C. L., Jr. 1972. Concepts of louse-borne typhus control in developing countries: The use of the living attenuated E strain typhus vaccine in epidemic and endemic situations, pp. 97– 130. In: Kohn, A., Klingberg, M. A. (eds.), Immunity in viral and rickettsial diseases. New York: Plenum.

Wisseman, C. L., Jr., Boese, J. L., Waddell, A. D., Silverman, D. J. 1975. Modification of antityphus antibodies on passage through the gut of the human body louse with discussion of some epidemiological and evolutionary implications. Annals of the New York Academy of Sciences **266:**6– 24.

Wisseman, C. L., Jr., Waddell, A. D. 1975. In vitro studies of rickettsia– host cell interactions: Intracellular growth cycle of virulent and attenuated *Rickettsia prowazeki* in chicken embryo cells in slide chamber cultures. Infection and Immunity **11:**1391– 1401.

Wisseman, C. L., Jr., Waddell, A. D., Walsh, W. T. 1974. Mechanisms of immunity in typhus infections. IV. Failure of chicken embryo cells in culture to restrict growth of antibody-sensitized *Rickettsia prowazeki*. Infection and Immunity **9:**571– 575.

Wisseman, C. A., Jr., Wood, W. H., Jr., Noriega, A. R., Jordan, M. E., Rill, D. J. 1962. Antibodies and clinical relapse of murine typhus fever following early chemotherapy. Annals of Internal Medicine **57:**743– 754.

Wolbach, S. B., Todd, J. L., Palfrey, F. W. 1922. The etiology and pathology of typhus. Cambridge: Harvard University Press.

Woodman, D. R., Grays, R., Weiss, E. 1977. Improved chicken embryo cell culture plaque assay for scrub typhus rickettsiae. Journal of Clinical Microbiology **6:**639– 641.

Woodman, D. R., Grays, R., Weiss, E. 1978. Plaque assay for scrub typhus rickettsiae. Procedure #72195. Tissue Culture Association Manual **4:**836– 840.

Woodman, D. R., Weiss, E., Dasch, G. A., Bozeman, F. M. 1977. Biological properties of *Rickettsia prowazekii* strains isolated from flying squirrels. Infection and Immunity **16:**853– 860.

Zdrodovskii, P. F., Golinevich, H. M. 1960. The rickettsial diseases. New York, Oxford: Pergamon Press.

Zinsser, H. 1935. Rats, lice, and history. Boston: Little, Brown.

The Family Rickettsiaceae: Pathogens of Domestic Animals and Invertebrates; Nonpathogenic Arthropod Symbiotes

EMILIO WEISS and GREGORY A. DASCH

This chapter discusses rickettsial pathogens of domestic animals, which are of great economic importance, pathogens of invertebrates, which are of interest to agriculture, and endosymbiotes, which have fascinated biologists for almost a century. Most of these organisms are obligate intracellular parasites that have not been cultivated outside of their hosts or closely related animals. Microbiological procedures of isolation, characterization, and identification have not been extensively used because the problems were difficult, because the number of investigators engaged in this vast field was small, and because the approaches have been parochial. During the last decade, there has been a resurgence of interest in these microorganisms and new technology has begun to be applied. Much remains to be done.

The microorganisms here included are quite different from each other in some cases; in others, they are not clearly separated from those discussed in other chapters. Many of these organisms have been improperly classified, but a sound basis for reclassification is not yet available.

The Pathogens of Domestic Animals

Ehrlichia canis

Ehrlichia canis was first recognized by Donatien and Lestoquard (1935) in tick-infested dogs in Algeria and has since been reported from many tropical and subtropical regions. Interest in this microorganism has been recently stimulated by epizootics among military dogs in southeast Asia (Davidson et al., 1975; Nims et al., 1971). Sporadic cases have occurred in several regions of the United States, usually below the 38th parallel, but a recent study conducted around Phoenix, Arizona, indicates that ehrlichiosis is more prevalent than previously suspected (Stephenson and Ristic, 1978).

The arthropod vector of *E. canis* is the brown dog tick, *Rhipicephalus sanguineus*. Groves et al. (1975) clearly demonstrated that, after feeding on an acutely infected dog, this tick efficiently transmitted the microorganism to normal dogs. Larvae retain infectivity during subsequent instars and adults may remain infected for as long as 155 days (Lewis et al., 1977). There is some indication that, in the nymph stage, ground-up ticks are infectious for the dog only after a blood meal, a phenomenon reminiscent of the activation of *Rickettsia rickettsii* (this Handbook, Chapter 161). In the adult tick, microorganisms are seen in the midgut, hemocytes, and salivary glands, but not in the ovary (Smith et al., 1976). There is no evidence of transovarian transmission (Groves et al., 1975).

In the dog, *E. canis* parasitizes monocytes, occasionally the lymphocytes, and rarely other cells. Severity of the disease, tropical canine pancytopenia, depends in part on the breed of dog. Clinical signs are most severe in the beagle, Alsatian, and particularly in the German shepherd (Buhles, Huxsoll, and Ristic, 1974; Seamer and Snape, 1972). In experimental infection, mild symptoms occur within 10 days and dogs generally recover. The more susceptible breeds develop a second, often fatal, bout of illness 50–100 days after infection. It is characterized by pancytopenia, hemorrhage, peripheral edema, emaciation, and secondary bacterial infection. A severe nosebleed is the most typical and obvious sign (Buhles, Huxsoll, and Ristic, 1974). Platelets are the most important target cells, apparently because of a change in self- and non-self-recognition by lymphocytes, presumably T cells, which involves platelets as well as the monocytes (Kakoma et al., 1977; Pierce, Marrs, and Hightower, 1977; Smith et al., 1974, 1975). Platelet toxicity, studied by inhibition of platelet migration, was shown to occur in the acute phase of the disease, prior to antibody formation (Kakoma et al., 1978). Severity of the chronic phase is possibly due to toxicity extending to the platelet precursors, bone-marrow megakaryocytes (Buhles, Huxsoll, and Hildebrandt, 1975; Hildebrandt et al., 1973b).

Groves et al. (1975) showed that inapparent infection with *E. canis* could be maintained in the dog for as long as 5 years. Wild canines, such as the coyote and jackal, have been implicated in the maintenance of *E. canis* in nature. Amyx and Huxsoll (1973) demonstrated that the red and gray foxes were susceptible, although the disease was mild, and that the infection could be transmitted from foxes to dogs by blood inoculation and by the tick, *R. sanguineus*. The infection has not been transmitted experimentally to cats, to common laboratory animals, or to sheep and cattle.

The fine structure of *E. canis* was examined in blood monocytes and mononuclear cells of the lung (Hildebrandt et al., 1973a; Nyindo et al., 1971; Simpson, 1972, 1974). In Giemsa-stained preparations, *E. canis* is recognized by the dense morula or mulberry appearance of its inclusions. A monocyte may harbor as many as 50 inclusions. Electron microscopy reveals that each morula contains 2–40 round or oval elementary bodies 0.5–0.9 μm in diameter. The elementary bodies have two distinct, tri-layered membranes, a rippled outer membrane and a plasma membrane closely associated with underlying structures. Multiplication is by binary fission. *E. canis* shares with *Chlamydia* an intravacuolar location, which does not result in the mobilization of digestion mechanisms by the host cell (Ristic, 1976). It is not clear, however, that *E. canis* undergoes a cycle of development analogous to that of *Chlamydia* (Hildebrandt et al., 1973a).

Nyindo et al. (1971) succeeded in cultivating *E. canis* in vitro in monocyte cell cultures derived from the blood of acutely infected dogs in a medium supplemented with 20% normal dog serum. Stephenson and Osterman (1977) showed that cultures prepared by Nyindo's method could serve as the inoculum for monolayers of stimulated peritoneal macrophages or peripheral blood monocytes derived from normal dogs. Morulae were clearly visible by 60 h after inoculation. The cycle of infection, with massive growth of the microorganisms and destruction of the macrophages, was completed in 12–18 days. Macrophages derived from immune dogs were more resistant to infection than those obtained from normal dogs; the additional presence of immune serum suppressed growth almost completely (Lewis and Ristic, 1978). Cell-culture methods have not yet been used to test the antibiotic susceptibility of *E. canis*, but tetracycline has been successfully used clinically for prophylaxis and therapy (Davidson et al., 1978).

Identification of *E. canis* is based on a positive, indirect, fluorescent-antibody test that is quite specific for this organism (Ristic et al., 1972). This procedure is used more commonly for identifying the presence of antibodies against *E. canis*, which in some breeds of dogs are an indication of chronic infection (Huxsoll, 1976).

Ehrlichia phagocytophila

Sheep and cattle pathogens of this genus are regarded as strains of the same species, *Ehrlichia phagocytophila*. This agent occurs in Great Britain, Ireland, Norway, the Netherlands, and Finland (Tuomi, 1967a). Similar, if not identical, agents, often designated by other specific names or by the generic name *Cytoecetes*, have been reported from Morocco, India (Philip, 1974), and, more frequently, from Kenya (Haig and Danskin, 1962).

In Europe, the microorganisms are transmitted by the sheep tick, *Ixodes ricinus*, which is capable of maintaining the infection transstadially but not transovarially (reviewed by Foggie, 1951).

In domestic animals, *E. phagocytophila* parasitizes the neutrophil and eosinophil granulocytes and more rarely, monocytes (Tuomi and von Bonsdorff, 1966). The spleen is heavily infected and, possibly, is the main site of multiplication (Snodgrass, 1975). The strains differ considerably in virulence (Purnell and Brocklesby, 1978; Tuomi, 1967b), but the disease, tick-borne fever, is usually relatively mild. It is characterized by high fever that lasts approximately 10 days, by leucopenia, and by loss of weight. Occasionally, infection is part of a complex syndrome involving mucosal disease and cobalt deficiency (Greig, Macleod, and Allison, 1977) or results in a sharp reduction in milk production (Kuil et al., 1972). The disease in Kenya, bovine petechial fever or Ondiri disease, is possibly more severe than in Great Britain (Foggie 1951; Haig and Danskin, 1962). Splenectomized animals are particularly susceptible and are frequently used as test animals to demonstrate latent infection. Although mortality from *E. phagocytophila* infection is low, the disease, possibly because of the destruction of phagocytic cells, predisposes the animals to infection with other bacteria (Foggie, 1951).

The infection may last as long as 2 years (Foggie, 1951), but the animals are generally immune to superinfection with the homologous strain, whether they harbor the microorganism or not (Kuil et al., 1972; Tuomi, 1967d). Immunity to heterologous strains is quite variable. Strains obtained from sheep often elicit better immunity than those derived from cattle (reviewed by Lewis, 1976). No cross-immunity was demonstrated between bovine strains from Finland and ovine strains from Scotland, but marked immunological differences were also encountered among strains isolated from locations in close proximity to each other in Finland (Tuomi, 1967c). On the other hand, *E. phagocytophila* may interfere with the growth of an unrelated microorganism, *Babesia divergens*, a protozoan transmitted to cattle by the tick vector, *I. ricinus* (Purnell et al., 1976, 1977).

E. phagocytophila can be readily transferred to goats and an ovine strain has been maintained in splenectomized guinea pigs and mice (Foggie and Hood, 1961). Attempts to transfer the Kenya strain to laboratory animals have been unsuccessful (Cooper, 1973), as have elaborate attempts to cultivate *E. phagocytophila* in chicken embryos or cell cultures (Thrusfield, Synge, and Scott, 1977).

The morphology and developmental cycle of two strains of *E. phagocytophila,* derived from sheep and cattle, respectively, were studied by electron microscopy of thin sections of leukocyte concentrates (Tuomi and von Bonsdorff, 1966). The microorganisms resemble *Chlamydia* in being located in vacuoles and in transition from small to large particles. They differ from *Chlamydia* in their somewhat elongated morphology and in the infectivity of the large particles, at least in the case of the bovine strain. Comparable results were obtained with the Kenya strain, except that variation in size and morphology is even greater (Krauss et al., 1972). In studies involving treatment of animals, it was shown that the microorganisms are susceptible to sulfamethazine (Foggie, 1951) and oxytetracycline, but are inhibited only to a limited extent by chloramphenicol, and not at all by penicillin or streptomycin (Tuomi, 1967e).

Identification is accomplished by fluorescent microscopy. Antigenic differences that result in variation in intensity of staining are not readily apparent (Tuomi, 1967d), in contrast to strain specificity demonstrated in cross-protection tests.

Ehrlichia equi

Equine ehrlichiosis, first reported in 1969, occurs, as far as we know, only in the foothills of the Sacramento Valley of California (Gribble, 1969; Stannard, Gribble, and Smith, 1969). Since the reported incidence in horses is low, another vertebrate, yet unidentified, is suspected to be the natural reservoir. The arthropod involved in transmission is not known.

E. equi parasitizes neutrophil and eosinophil granulocytes. The clinical signs include fever, anorexia, depression, edema of the legs, and ataxia. Hematologic changes consist of leucopenia, thrombocytopenia, and mild anemia. Mortality is low. The infection elicits both humoral and cell-mediated immune responses that result in resistance to superinfection. There is no evidence that *E. equi* survives long in the immune host (Gribble, 1969; Lewis, 1976; Nyindo et al., 1978).

The disease has been reproduced experimentally in horses and burros by inoculation of infected blood; clinically inapparent, experimental infections have been induced in dogs (German shepherds and beagles), cats, rhesus macaques (*Macaca mulatta*), and baboons (*Papio anubis*). Typical *Ehrlichia* inclusions were seen in the granulocytes of these animals and blood from the macaques was shown to be infectious for a horse. Infection with *E. equi* did not protect dogs against *E. canis* (Gribble, 1969; Lewis et al., 1975).

The fine structure of *E. equi* was examined in thin sections of buffy coat cells. The microorganisms develop in vacuoles and multiply by binary fission without undergoing a complex developmental cycle. They resemble *E. canis,* except that light microscopy reveals inclusions of distinct particles instead of compact morulae (Lewis, 1976; Sells et al., 1976). Tetracycline has been used for treatment. Identification is accomplished by immunofluorescence (Nyindo et al., 1978). Attempts to cultivate *E. equi* have not been reported.

It is interesting to note that, in addition to parasites of the monocyte (*E. canis*) and of the granulocyte (*E. phagocytophila* and *E. equi*), a parasite of the blood platelet of comparable morphology has recently been discovered in the blood of a dog (Harvey, Simpson, and Gaskin, 1978).

Cowdria ruminantium

The generic name *Cowdria* honors E. V. Cowdry, who first studied the etiology of heartwater (1925a,b, 1926), a disease of considerable severity of sheep, goats, and cattle in South Africa and other parts of Africa, particularly Nigeria. The role of the bont tick, *Amblyomma hebraeum,* in transmission of heartwater in South Africa was demonstrated by Cowdry (1925b). Recently, Ilemobade and Leeflang (1977) demonstrated that *A. variegatum* is the vector in Nigeria. The agent is maintained in the tick transstadially, but there is no evidence that it is passed transovarially.

The designation of the disease "heartwater" refers to the most characteristic autopsy finding, namely hydropericardium (Cowdry, 1925a). *C. ruminantium* invades primarily the endothelial cells of the capillaries of the heart, brain, and spleen, but lymphocytes and cells of the reticuloendothelium are also involved (Du Plessis, 1975; Ilemobade and Blotkamp, 1978b). Mortality in untreated animals can be quite high, and stock farming was impossible in certain regions of South Africa before the development of control methods (Mare, 1972). In Nigeria, the disease had a devastating effect on imported cattle, but is less severe in indigenous cattle. Brown goats that originate from northern Nigeria are highly susceptible, while certain breeds of goats, cattle, and sheep in southern Nigeria are of variable susceptibility or are relatively resistant (Ilemobade, 1977).

Domestic animals that recover from the disease may act as carriers of *C. ruminantium,* but game animals and rodents are suspected to contribute to the natural reservoir of the microorganism (Ilemobade and Leeflang, 1977). A spontaneous infection with typical signs of heartwater was encountered in the eland (*Taurotragus oryx*). There have been reports of infection in the springbuck and various wild ungulates (Young and Basson, 1973), but attempts at artificial infection of other large, wild animals were not successful (Gradwell, Van Niekerk, and Joubert, 1976). Propagation of *C. ruminantium* in laboratory animals has been achieved to a limited extent (Philip, 1974). A strain isolated from a goat was shown to be lethal for mice, was passaged more than 20 times in this species, and then was passed to sheep (Du Plessis and Kumm, 1971). There are no reports of systematic efforts to cultivate the microorganisms in vitro.

Cowdry (1925a, 1926) recognized that *C. ruminantium* had the same tinctorial properties as rickettsiae but were spherical. Pienaar (1970) examined the development of the microorganism in the choroid plexus of experimentally infected sheep. The particles are located, as in the case of *Chlamydia* and *Ehrlichia,* in vacuoles of the host cells, but the degree of resemblance to these two other genera is not clear. Four forms, ranging in size from 0.5 to 1.7 μm, were seen. Inclusions usually consisted of a single morphological type. Binary fission is the predominating mechanism of cell division. Although enveloped by two unit membranes, the cells are quite irregular in shape and appear to lack rigidity. Du Plessis (1975) examined microorganisms located in the reticuloendothelial cells of mice and ruminants. He suggested that, in these cells, the cycle of growth did not occur within the confines of the vacuole and that the endothelial cells were invaded only after completion of this cycle in the reticuloendothelial cells.

C. ruminantium infection is susceptible to sulfonamides, chloramphenicol, and tetracycline antibiotics. A combination of injection with infected blood and oral chemotherapy is used to render cattle and sheep solidly immune against heartwater (Mare, 1972).

C. ruminantium can be isolated from whole blood or leukocytes, brain homogenates or lung macrophages, provided extensive postmortem changes have not taken place (Ilemobade and Blotkamp, 1978a). Two methods for rapid identification have recently been reported. A capillary flocculation test with infected brain homogenate as the antigen can be used for detection of antibodies (Ilemobade and Blotkamp, 1976). The presence of the microorganism can be detected in squash smears prepared with brain biopsy material obtained from live animals. Smears are stained by Giemsa's method (Synge, 1978) and, presumably, can also be stained with fluorescent antibody.

Neorickettsia helminthoeca

Neorickettsia helminthoeca is a pathogen of dogs that, unlike *Ehrlichia* or *Cowdria,* is carried by a fluke instead of a tick. The generic designation *Neorickettsia* should not be confused with the common designation, "neorickettsia", sometimes applied to microorganisms that presumably belong to the genus *Chlamydia* (Meyer, 1967). *N. helminthoeca* occurs in a region which extends north from the Sacramento River in California to the Olympic Peninsula in Washington and west from the Cascade Mountains to the Pacific. This region coincides with the distribution of the snail, *Oxytrema silicula,* which is the first intermediate host of the trematode vector, *Nanophyetus salmincola* (Milleman and Knapp, 1970). A strain of *N. helminthoeca* of lesser virulence has been reported from the Elokomin River region of southwestern Washington (Farrell, 1974).

N. helminthoeca has been recovered from all stages of the trematode, including egg, and from infected snail livers. The second intermediate host of the trematode is usually a salmonid fish, which is invaded by the cercarial stage. Dogs acquire the highly fatal "salmon poisoning" disease (SPD) or the milder "Elokomin fluke fever" (EFF) by eating raw salmon infected with the metacercariae of *N. salmincola* (Milleman and Knapp, 1970; Philip, 1974). *N. helminthoeca* parasitizes lymphocytes and macrophages of dogs, producing severe depletion of small lymphocytes and destruction of germinal centers of lymph nodes (Frank et al., 1974b).

Millemann and Knapp (1970) listed 32 animals that serve as natural or experimental definitive hosts for the fluke. However, only the dog and the fox, and possibly the raccoon, develop recognizable SPD or EFF. More recently, the diseases were also reproduced in the black bear, *Ursus americanus* (Farrell, Leader, and Johnston, 1973).

N. helminthoeca has been cultivated in primary canine monocyte cultures (Brown et al., 1972) by the method used for *E. canis* (Nyindo et al., 1971). In these cultures, the microorganisms are described as spherical or coccobacillary forms, 0.3–1.0 μm in diameter, dispersed loosely throughout the cell, but in some cases producing aggregates, and eventually achieving a high population density in the host cell. The location of the particles is the vacuole, which, as is the case in *Chlamydia,* is not converted into a phagolysosome. Differences between the agents of SPD and EFF were not noted (Brown et al., 1972; Frank et al., 1974a). A combination of antibiotics, penicillin, streptomycin, and kanamycin, prevented the growth of *N. helminthoeca* in vitro (Frank et al., 1974a). Tetracycline compounds are effective in the treatment of dogs.

Isolation of the agent, other than by cultivation of the monocytes, is accomplished by collecting mesenteric lymph nodes from acutely ill dogs and by injecting them into susceptible dogs. The two strains, producing SPD and EFF, respectively, can be differentiated by cross-protection tests and complement-fixation tests (Frank et al., 1974b; Sakawa, Farrell, and Mori, 1973), but not by fluorescent-antibody tests (Frank et al., 1974a). *E. canis* is antigenically unrelated to *Neorickettsia* (Brown et al., 1972).

Table 1. Ecology and properties of Rickettsiaceae pathogenic for domestic animals.

Microorganism	Domestic animal	Disease	Invertebrate host[a]	Geographical distribution	Principal cells parasitized	Tissue used for passage[b]	In vitro cultivation[c]
Ehrlichia canis	Dog	Tropical canine pancytopenia	Tick (Rhipicephalus sanguineus)	Tropical and subtropical areas	Monocyte	Blood, buffy coat	Primary and secondary canine monocyte cultures
Ehrlichia phagocytophila	Sheep, cattle	Tick-borne fever, bovine petechial fever, Ondiri disease	Tick (Ixodes ricinus), other ticks	Northern Europe, Kenya	Granulocytes	Blood, spleen	Not successful
Ehrlichia equi	Horse	Equine ehrlichiosis	Not known (tick suspected)	California	Granulocytes	Blood	Not reported
Cowdria ruminantium	Sheep, cattle, goats	Heartwater	Tick (Amblyomma hebraeum, A. variegatum)	South Africa, Nigeria	Endothelial and reticulo-endo-thelial cells	Blood, brain	Not reported
Neorickettsia helminthoeca	Dog	Salmon poisoning, Elokomin fluke fever	Fluke (Nanophyetus salmincola)	Washington, Oregon, California	Lymphocytes, macrophages	Blood, buffy coat, mesenteric lymph nodes	Primary canine monocytes

[a]Transstadial but not transovarian transmission in the tick. Transovarian transmission in the fluke.

[b]Identification by microscopic observations of smears stained by Giemsa's method or with fluorescent antibody. The microorganisms gain entrance into the vacuoles of the host cells without stimulating a digestive response. Comparative studies have not been sufficiently extensive to permit tabulation of morphological differences of inclusions.

[c]Infection of laboratory animals has been too inconsistent to be tabulated.

Rickettsia sennetsu

The pathogens of domestic animals are often compared to human pathogens, and particularly to *R. sennetsu*, which has been misplaced in the genus *Rickettsia*. This agent has been isolated in 1954 in Japan by Misao, Kobayashi, and others (review by Tamiya, 1962) from the blood of patients with glandular fever. This agent does not react, however, with the convalescent sera of cases of infectious mononucleosis of other countries. The mechanism of transmission is unknown. The microorganism has been grown in established primate cell lines (Anderson et al., 1965; Minamishima, 1965) and good harvests were obtained from the peritoneal exudate of infected, cyclophosphamide-treated mice (Tachibana and Kobayashi, 1975). *R. sennetsu* does not cross-react serologically with *R. tsutsugamushi* or *N. helminthoeca* (Kitao, Farrell, and Fukuda, 1973; Tachibana et al., 1976).

Morphologically and in its relationship to host cells, *R. sennetsu* differs markedly from other members of the genus *Rickettsia* and bears a tenuous resemblance to the above-described animal pathogens. The particles are spherical, coccobacillary, 0.5–1.0 μm in diameter, or are highly pleomorphic. They develop in host-cell vacuoles, where they are sometimes aligned or wrapped around each other in a ball, but do not appear to undergo a developmental cycle analogous to that of *Chlamydia* (Anderson et al., 1965).

For various reasons, research on the above-described microorganisms has been uneven. These microorganisms by no means constitute a tight group of closely related species, but there are sufficient similarities to indicate that comparative studies might be highly productive.

Pathogens of Invertebrates

A number of insect larvae and adults and other invertebrates are naturally infected with bacterial pathogens that do not infect vertebrates and that have not been cultivated on host cell–free media. With few exceptions, these organisms are similar morphologically. Host-specificity studies have not been extensive enough to justify the use of more than one species designation. Such microorganisms were first observed in the Japanese beetle (*Popillia japonica*) by Dutky and Gooden in 1952. They are named *Rickettsiella popilliae* in *Bergey's Manual of Determinative Bacteriology* (Weiss, 1974) and pathogens of other invertebrates are regarded as strains of the same species. Many authors, however, use specific designations to reflect the parasitized host. To the list of infected hosts given in *Bergey's Manual* (Weiss, 1974), the spider (*Argyrodes gibbosus* [Meynadier, Lopez, and Dulthort, 1974]; *Pisaura mirabilis* [Morel, 1977]) can be added. Most of the infected invertebrates were seen in Europe, but infection of the Japanese and June beetles have been reported from the United States.

The most commonly studied strain is the pathogen of the cockchafer, *Melolontha melolontha*. Infection usually starts in the larval stage (white grub). The symptoms are related to disorganization of the fat body and become obvious 2 weeks after intracoelomic injection. They consist of discoloration (responsible for the designation "blue disease"), reduced locomotor activity and feeding, loss of weight, and death in about 2 months at 20–25°C (Hurpin and Robert, 1973; Weiss, 1974). Since *Rickettsiella* has been considered as a control agent for agricultural pests, its virulence and permanence in the soil has been investigated (Hurpin and Robert,

1972, 1976, 1977). *Rickettsiella* is not as virulent and persistent as some of the viral and mycotic pathogens of agricultural pest insects, but is maintained in the soil for years. Infection of offspring is affected through contamination of soil, rather than by transovarian transmission.

Only very limited growth of *Rickettsiella* has been obtained in chicken embryo entodermal cell cultures (Suitor, 1964a). Identification is based on morphological observations. The microorganisms are Gram-negative rods, 0.2×0.6 μm or slightly larger, oval, and somewhat curved or kidney shaped with rounded edges. They have the same tinctorial properties as rickettsiae (Weiss, 1974). The fine structure was recently studied in three hosts, the cockchafer (Devauchelle, Meynadier, and Vago, 1972), the isopod (Louis et al., 1977), and the spider (Morel, 1977). These authors reached the same conclusion, namely that the cycle of development is similar to that of *Chlamydia*. Small, electron-dense particles are phagocytized by the host cells and development takes place in a vacuole. The particles undergo transformation into larger, bacteria-like forms, which multiply actively by binary fission. At the end of the period of multiplication, the large forms revert to small, dense particles. A notable feature of this cycle is the formation, in some of the cells of the fat body, of giant particles that give rise to crystalline bodies which are extruded into the hemolymph.

Rickettsiaceae Not Known To Be Pathogenic

The microorganisms discussed in the previous sections were identified, in part, by the characteristic syndromes they produce in their primary or accidental hosts. Rickettsiaceae that are not known to be pathogenic, unless they have been cultivated in vitro, are even more difficult to identify. In many cases, there is no certainty that the same designation used by two laboratories applies to the same microorganism.

Genus *Wolbachia*

The generic designation *Wolbachia* has been used for microorganisms that probably have very little in common. One feature that they share is some blood-sucking arthropod host, which presumably does not derive any benefit from their presence (Weiss, 1974).

W. pipientis was first described by Hertig (1936) as occurring in the gonads and gut epithelium of the mosquito *Culex pipiens*. This microorganism is found in all stages and virtually in all strains of *C. pipiens* (Yen and Barr, 1974). Similar, if not identical, organisms have been seen with variable frequency in other insects (Weiss, 1974), including mosquitoes of the *Aedes scutellaris* group (Beckett, Boothroyd, and MacDonald, 1978). *W. pipientis* is a Gram-negative rod that ranges in size from 0.25 to 1.5 μm and stains red with Giemsa. It is located in the vacuole of the host cell and multiplies by binary fission. The fine structure is typical of Gram-negative bacteria, except that the cell wall displays some plasticity. The cytoplasm may contain bacteriophage particles (Wright et al., 1978). Attempts to grow *W. pipientis* outside the host have not been successful and differentiation of strains by immunological methods has not been reported.

The chief interest in this organism is derived from a recent observation that it plays a role in incompatibility between strains of *C. pipiens*: matings of mosquitoes from different geographical areas sometimes produce eggs that do not hatch, since females cannot be fertilized with the infected sperm of males of some strains. The mosquitoes can be rendered aposymbiotic (nearly or completely freed of microorganisms) by the addition of 25 μg/ml of tetracycline to the larval and pupal rearing water. This treatment eliminates incompatibility. The sperm of aposymbiotic mosquitoes can fertilize uninfected or infected heterologous females (Fine, 1978; Yen, 1975; Yen and Barr, 1974).

W. melophagi is found in the wingless fly commonly called sheep ked, *Melophagus ovinus*. It adheres to the gut epithelium but is not intracellular, and it is the only one of this group for which there are confirmed reports of growth in axenic media. Similar microorganisms were seen in other insects (Weiss, 1974). Morphologically and in its natural location on a cell surface, *W. melophagi* superficially resembles *Rochalimaea quintana* (this Handbook, Chapter 161). There are no recent reports of investigation on this organism, possibly because infestation of sheep with keds has sharply declined, at least in the United States.

W. persica was isolated by Suitor and Weiss (1961) from the Malpighian tubules and gonads of a tick, *Argas persicus* (= *A. arboreus*), collected in Egypt. Although seen in most *Argas* ticks, isolations were successful only from those that had fed on the buff-backed heron, *Bubulcus ibis*. They were not isolated after a meal of chicken blood (Suitor, 1964b). *W. persica* was also isolated in Montana from the wood tick, *Dermacentor andersoni* (Burgdorfer, Brinton, and Hughes, 1973). The same and similar organisms were seen in the Malpighian tubules of *Haemaphysalis inermis* in Czechoslovakia (Sixl-Voigt et al., 1977) and *Hyalomma asiaticum* in the USSR (Raikhel, 1978).

W. persica is the only organism of this group for which Koch's postulates have been satisfied. It has been grown extensively in the yolk sac of chicken embryos and cell cultures (Burgdorfer, Brinton, and

Hughes, 1973; Weiss, 1974). *A. persicus* rendered aposymbiotic, were reinfected by allowing them to feed on heavily infected chicken embryos (Suitor, 1964b). Fluorescent-antibody techniques have been used to check the identity of isolates with the microorganisms seen in the Malpighian tubules (Suitor and Weiss, 1961) and to show that the microorganisms derived from *A. persicus* and *D. andersoni* are antigenically closely related, but are not related to the rickettsiae (Burgdorfer, Brinton, and Hughes, 1973).

W. persica cells are spherical, approximately 1.0 μm in diameter, but often appear variable in size because the cell wall is incomplete and the cells lack rigidity. They are Gram negative and are visualized most satisfactorily in smears fixed with Carnoy's and stained with Giemsa. They develop in the vacuoles of their host cells. Procedures of cultivation, harvesting, and purification are identical to those for rickettsiae (this Handbook, Chapter 161), but growth is somewhat more rapid than that of *Rickettsia prowazekii* (Weiss, 1974). *W. persica* was also cultivated in the human body louse, the mealworm (*Tenebrio molitor*), and the tick (*Ornithodorus moubata*) (Weyer, 1973). It is virulent for the tick host and for laboratory animals when injected in very high concentrations (Burgdorfer, Brinton, and Hughes, 1973; Weiss, 1974). Antibiotic susceptibility is comparable to that of rickettsiae. In contrast to rickettsiae, however, *W. persica* catabolizes glucose and glycerol, as well as glutamine and other substrates. The G+C content of the DNA is approximately 30 mol% (Weiss, 1974).

Endosymbiotes of Arthropods

More than 10% of the insects and several other invertebrates harbor bacteria that cause no apparent harm to their primary or accidental hosts and are transmitted to successive generations as hereditary particles (Buchner, 1965). The term "endosymbiote" is used broadly for such bacteria without implying benefit to the host or rigid criteria for obligate intracellular parasitism. Mutual benefit is likely when the endosymbiotes develop primarily in specialized cells or in small organs of the arthropod, called "mycetocytes" or "mycetomes", and when specialized reproductive mechanisms exist to ensure the transovarial passage of the endosymbiotes between generations (Buchner, 1965; Koch, 1960). These endosymbiotes may be separated into two broad groups for the purpose of this chapter: those of the bloodsucking and of the nonbloodsucking arthropods. There is little to distinguish the bloodsucking endosymbiotes from the genus *Wolbachia,* except that some of the relationships to insect hosts may be mutualistic. Furthermore, there is virtual absence of information on cultivation in vitro or in vertebrate or

invertebrate cells. The endosymbiotes of nonbloodsucking arthropods clearly form a separate group. No direct mechanism exists for transmission of the symbiotes to vertebrates or other invertebrates. Should any of these symbiotes prove to be pathogenic, transmission would have to be by ingestion of symbiotes in secretions or feces, or of the whole insect host.

ENDOSYMBIOTES OF BLOODSUCKING ARTHROPODS. Arkwright, Atkin, and Bacot (1921) observed in the bedbug, *Cimex lectularius,* two forms of symbiote, a minute body and a long rod. Arkwright regarded these forms as stages of a single cycle and his view is reflected in the description in *Bergey's Manual* of this microorganism under the name *Symbiotes lectularius* (Brooks, 1974). However recent ultrastructure studies support the view that *Cimex* may have two distinct types of symbiotes (Louis et al., 1973), and some *Cimex* strains may even lack the long rod form (Chang and Musgrave, 1973).

Chang (1974) eliminated the *Cimex* symbiotes by maintaining the insects at 36°C. Their survival was not affected, but fecundity was greatly reduced. A role of the symbiotes in reproduction was suggested. This suggestion was reinforced by observations made on several species of tsetse flies (genus *Glossina*), which harbor large, Gram-negative symbiotes in midgut mycetomes (Reinhardt, Steiger, and Hecker, 1972) and in the milk glands, which are part of the reproductive system (Ma and Denlinger, 1974); smaller rods are found in the ovaries (Huebner and Davey, 1974). Low concentrations of oxytetracycline in a blood meal resulted in the loss of the symbiotes and eventually to sterility (Nogge, 1976). Larger doses resulted in destruction of the mycetomes and death of the flies. Administration of antibiotics to domestic animals was suggested as a means of controlling fly infestation (Schlein, 1977).

ENDOSYMBIOTES OF NONBLOODSUCKING ARTHROPODS. At present, only one species, *Blattabacterium cuenoti,* the well-studied symbiote of the cockroaches, has been given specific status (Brooks, 1970, 1974). However, recent evidence based on DNA base composition of a number of endosymbiotes (Dasch, 1975), as well as morphological observations, supports the view that endosymbiotes of nonbloodsucking arthropods constitute a large and heterogeneous group. Other microorganisms could be given specific status with equal justification. Further identification to species level does not appear warranted until possible affinities with previously described genera, including those outside the Rickettsiaceae, have been excluded by studies of DNA and ribosomal RNA homology or other methods of biochemical phylogeny.

The cockroach symbiote, *B. cuenoti,* is found in the mycetomes of the fat body, ovaries, and embryos of all species of the cockroach that have been examined. It is relatively large (0.9 × 1.5–8 μm), Gram-positive bacterium, although the fine structure of the cell envelope resembles that of a Gram-negative bacterium. Similar morphology, tissue localization, and limited range in DNA base composition (26–28 mol% G+C) of the symbiotes of nine species in five different subfamilies of cockroach (Dasch, 1975) suggest that a single specific designation is justified. Information on its metabolism is fragmentary, but *B. cuenoti* appears to have an active amino acid metabolism and synthesizes ascorbic and pantothenic acids. It is not transmitted ovarially in the cockroach, if the diet is deficient in manganese, zinc, or unsaturated fatty acids. The cockroach is dependent on the symbiote for growth factors that are essential to integumental color and egg viability (Brooks, 1970). Aposymbiotic nymphs die if fed a stock diet, such as dog biscuit, but survive without growing on a casein diet and grow normally if the casein is supplemented with brewers' yeast or lyophilized mouse liver (Brooks and Kringen, 1972). *B. cuenoti* has not been grown outside the host but some progress has been made in in vitro cultivation of symbiote-filled mycetocytes (Kurtti and Brooks, 1976; Landureau, 1966).

A large number of bacterial symbiotes contribute to the nutrition of their hosts in a manner comparable to *B. cuenoti*. The information that has been accumulated and the methodology used to obtain it was carefully reviewed by Chang (1981).

In contrast to the endosymbiotes of the cockroach, the relationships among the symbiotes of some other closely related insects have been surprisingly complex (Dasch, 1975). For example, the ant symbiotes of six species of *Camponotini* (carpenter ants) varied in DNA base composition only from 30–31 mol% G+C, but differed markedly from the morphologically similar symbiotes of two species of *Formica* ants, which had DNA base compositions of 41 mol% G+C. Conversely, the widely pleomorphic symbiotes of the grain weevils, *Sitophilus granarius* and *S. oryzae*, do not vary greatly in base composition (50 and 55 mol% G+C, respectively) and may consist of two evolutionarily related symbiotes of the same base compositions both present in *S. zea-mais*. In each case, the symbiote DNA base compositions were different from those of their host insects.

Several explanations have been proposed for the evolution of symbiotic bacteria. Koch (1960) noted that symbiotes are found most frequently among insects that have become one-sided specialists in nourishment. The symbiotes contribute the missing nutritional factors. This view is difficult to reconcile with the fact that symbiotes are quite frequently found among omnivorous insects, such as the cockroach and the ant. In the context of this chapter, symbiotes can be considered as part of a continuous

series that starts with voracious predators and ends with indispensable companions. All these bacteria share the problem of avoiding host defenses; quite a few resolve the problem in a similar manner, such as by loss of non-self-recognition by the host and by inhibition of lysosomal activation. Neither the defense mechanisms of the host nor the avoidance mechanisms of the symbiotes can dominate, or the population balance between host and parasite would be lost.

Acknowledgments

We are indebted to P. F. Dirk Van Peenen for helpful discussions. This review was supported by the Research and Development Command, Department of the Navy, research project number MR04105.0028. The opinions and assertions contained herein are the private ones of the writers and are not to be construed as official or reflecting the views of the Navy Department or the Naval Service at large.

Literature Cited

Amyx, H. L., Huxsoll, D. L. 1973. Red and gray foxes—potential reservoir hosts for *Ehrlichia canis*. Journal of Wildlife Diseases **100**:47–50.

Anderson, D. R., Hopps, H. E., Barile, M. F., Bernheim, B. C. 1965. Comparison of the ultrastructure of several rickettsiae, ornithosis virus, and *Mycoplasma* in tissue culture. Journal of Bacteriology **90**:1387–1404.

Arkwright, J. A., Atkin, E. C., Bacot, A. 1921. An hereditary rickettsia–like parasite of the bedbug (*Cimex lectularius*). Parasitology **13**:27–36.

Beckett, E. B., Boothroyd, B., MacDonald, W. W. 1978. A light and electron microscope study of rickettsia-like organisms in the ovaries of mosquitoes of the *Aedes scutellaris* group. Annals of Tropical Medicine and Parasitology **72**:277–283.

Brooks, M. A. 1970. Comments on the classification of intracellular symbiotes of cockroaches and a description of the species. Journal of Invertebrate Pathology **16**:249–258.

Brooks, M. A. 1974. *Symbiotes, Blattabacterium*, pp. 900–901. In: Buchanan, R. E., Gibbons, N. E., (eds.), Bergey's manual of determinative bacteriology, 8th ed. Baltimore: Williams & Wilkins.

Brooks, M. A., Kringen, W. B. 1972. Polypeptides and proteins as growth factors for *Aposymbiotic Blattella germanica* (L.), pp. 353–364. In: Rodriguez, J. G. (ed.), Insect and mite nutrition. Amsterdam: North Holland.

Brown, J. L., Huxsoll, D. L., Ristic, M., Hildebrandt, P. K. 1972. In vitro cultivation of *Neorickettsia helminthoeca*, the causative agent of salmon poisoning disease. American Journal of Veterinary Research **33**:1695–1700.

Buchner, P. 1965. Endosymbiosis of animals with plant microorganisms. Translated from the 1953 German edition. New York: Interscience.

Buhles, W. C., Jr., Huxsoll, D. L., Hildebrandt, P. K. 1975. Tropical canine pancytopenia: Role of aplastic anemia in the pathogenesis of severe disease. Journal of Comparative Pathology **85**:511–521.

Buhles, W. C., Jr., Huxsoll, D. L., Ristic, M. 1974. Tropical canine pancytopenia: Clinical, hematologic, and serologic response of dogs to *Ehrlichia canis* infection, tetracycline therapy, and challenge inoculation. Journal of Infectious Diseases **130**:357–367.

Burgdorfer, W., Brinton, L. P., Hughes, L. E. 1973. Isolation and characterization of symbiotes from the Rocky Mountain wood tick, *Dermacentor andersoni*. Journal of Invertebrate Pathology **22:**424–434.

Chang, K. P. 1974. Effects of elevated temperature on the mycetome and symbiotes of the bed bug *Cimex lectularius* (*Heteroptera*). Journal of Invertebrate Pathology **23:**333–340.

Chang, K. P. 1981. Nutritional roles of intracellular symbiotes in insects and protozoa. In: CRC handbook of nutrition and food. Boca Raton, Florida: CRC Press. (In press.)

Chang, K. P., Musgrave, A. J. 1973. Morphology, histochemistry, and ultrastructure of mycetome and its rickettsial symbiotes in *Cimex lectularius* L. Canadian Journal of Microbiology **19:**1075–1081.

Cooper, J. E. 1973. Attempted transmission of the Ondiri disease (bovine petechial fever) agent to laboratory rodents. Research in Veterinary Science **15:**130–133.

Cowdry, E. V. 1925a. Studies on the etiology of heartwater. I. Observation of a *Rickettsia, Rickettsia ruminantium* (n. sp.), in the tissues of infected animals. Journal of Experimental Medicine **42:**231–252.

Cowdry, E. V. 1925b. Studies on the etiology of heartwater. II. *Rickettsia ruminantium* (n. sp.) in the tissues of ticks transmitting the disease. Journal of Experimental Medicine **42:**253–254.

Cowdry, E. V. 1926. Studies on the etiology of heartwater. III. The multiplication of *Rickettsia ruminantium* within the endothelial cells of infected animals and their discharge into the circulation. Journal of Experimental Medicine **44:**803–814.

Dasch, G. A. 1975. Morphological and molecular studies on intracellular bacterial symbiotes of insects. Ph.D. Thesis. Yale University, New Haven, Connecticut.

Davidson, D. E., Jr., Dill, G. S., Jr., Tingpalapong, M., Premabutra, S., Nguen, P. L., Stephenson, E. H., Ristic, M. 1975. Canine ehrlichiosis (tropical canine pancytopenia) in Thailand. Southeast Asian Journal of Tropical Medicine and Public Health **6:**540–543.

Davidson, D. E., Jr., Dill, G. S., Jr., Tingpalapong, M., Premabutra, S., Nguen, P. L., Stephenson, E. H., Ristic, M. 1978. Prophylactic and therapeutic use of tetracycline during an epizootic of ehrlichiosis among military dogs. Journal of the American Veterinary Medical Association **172:**697–700.

Devauchelle, G., Meynadier, G., Vago, C. 1972. Ultrastructural study of the cycle of multiplication of *Rickettsiella melolonthae* in the hemocytes of its host. Journal of Ultrastructure Research **38:**134–148.

Donatien, A., Lestoquard, F. 1935. Existence en Algérie d'une *Rickettsia* du chien. Bulletin de la Société de Pathologie Exotique **28:**418–419.

Du Plessis, J. L. 1975. Electron microscopy of *Cowdria ruminantium* infected reticuloendothelial cells of the mammalian host. Onderstepoort Journal of Veterinary Research **42:**1–14.

Du Plessis, J. L., Kumm, N. A. L. 1971. The passage of *Cowdria ruminantium* in mice. Journal of the South African Veterinary Medical Association **42:**217–221.

Dutky, S. R., Gooden, E. L. 1952. *Coxiella popilliae* n. sp., a rickettsia causing blue disease of Japanese bettle larvae. Journal of Bacteriology **63:**743–750.

Farrell, R. K. 1974. Canine rickettsiosis, pp. 985–987. In: Kirk, R. W. (ed.), Current veterinary therapy, 5th ed. Philadelphia: W. B. Saunders.

Farrell, R. K., Leader, R. W., Johnston, S. D. 1973. Differentiation of salmon poisoning disease and Elokomin fluke fever: Studies with the black bear (*Ursus americanus*). American Journal of Veterinary Research **34:**919–922.

Fine, P. E. M. 1978. On the dynamics of symbiote dependent cytoplasmic incompatibility in culicine mosquitoes. Journal of Invertebrate Pathology **31:**10–18.

Foggie, A. 1951. Studies on the infectious agent of tick-borne fever in sheep. Journal of Pathology and Bacteriology **63:**1–15.

Foggie, A., Hood, C. S. 1961. Adaptation of the infectious agent of tick-borne fever to guinea-pigs and mice. Journal of Comparative Pathology and Therapeutics **71:**414–427.

Frank, D. W., McGuire, T. C., Gorham, J. R., Davis, W. C. 1974a. Cultivation of two species of *Neorickettsia* in canine monocytes. Journal of Infectious Diseases **129:**257–262.

Frank, D. W., McGuire, T. C., Gorham, J. R., Farrell, R. K. 1974b. Lymphoreticular lesions of canine neorickettsiosis. Journal of Infectious Diseases **129:**163–171.

Gradwell, D. V., Van Niekerk, C. A. W. J., Joubert, D. C. 1976. Attempted artificial infection of impala, blue wildebeest, buffalo, kudo, giraffe and warthog with heartwater. Journal of the South African Veterinary Medical Association **47:**209–210.

Greig, A., Macleod, N. S. M., Allison, J. 1977. Tick borne fever in association with mucosal disease and cobalt deficiency in calves. Veterinary Record **100:**562–564.

Gribble, D. H. 1969. Equine ehrlichiosis. Journal of the American Veterinary Medical Association **155:**462–469.

Groves, M. G., Dennis, G. L., Amyx, H. L., Huxsoll, D. L. 1975. Transmission of *Ehrlichia canis* to dogs by ticks (*Rhipicephalus sanguineus*). American Journal of Veterinary Research **36:**937–940.

Haig, D. A., Danskin, D. 1962. The aetiology of bovine petechial fever (Ondiri disease). Research in Veterinary Science **3:**129–138.

Harvey, J. W., Simpson, C. F., Gaskin, J. M. 1978. Cyclic thrombocytopenia induced by a *Rickettsia*-like agent in dogs. Journal of Infectious Diseases **137:**182–188.

Hertig, M. 1936. The rickettsia, *Wolbachia pipientis* (gen. et sp. n.) and associated inclusions of the mosquito, *Culex pipiens*. Parasitology **28:**453–486.

Hildebrandt, P. K., Conroy, J. D., McFee, A. E., Nyindo, M. B. A., Huxsoll, D. L. 1973a. Ultrastructure of *Ehrlichia canis*. Infection and Immunity **7:**265–271.

Hildebrandt, P. K., Huxsoll, D. L., Walker, J. S., Nims, R. M., Taylor, R., Andrews, M. 1973b. Pathology of canine ehrlichiosis (tropical canine pancytopenia). American Journal of Veterinary Research **34:**1309–1320.

Huebner, E., Davey, K. G. 1974. Bacteroids in the ovaries of a tsetse fly. Nature **249:**260–261.

Hurpin, B., Robert, P. H. 1972. Comparison of the activity of certain pathogens of the cockchafer *Melolontha melolontha* in plots of natural meadowland. Journal of Invertebrate Pathology **19:**291–298.

Hurpin, B., Robert, P. H. 1973. Sur quelques caractères de l'infection des larvaes de *Melolontha melolontha* L. par *Rickettsiella melolonthae* Krieg. Annales de Parasitologie Humaine et Comparée **48:**399–410.

Hurpin, B., Robert, P. H. 1976. Conservation dans le sol de trois germes pathogènes pour les larvaes de *Melolontha melolontha* (*Col: Scarabaeidae*). Entomophaga **21:**73–80.

Hurpin, B., Robert, P. H. 1977. Effects en population naturelle de *Melolontha melolontha* (*Col: Scarabaeidae*) d'une introduction de *Rickettsiella melolonthae* et de *Entomopoxvirus melolonthae*. Entomophaga **22:**85–92.

Huxsoll, D. L. 1976. Canine ehrlichiosis (tropical canine pancytopenia): A review. Veterinary Parasitology **2:**49–60.

Ilemobade, A. A. 1977. Heartwater in Nigeria. I. The susceptibility of different local breeds and species of domestic ruminants to heartwater. Tropical Animal Health and Production **9:**177–180.

Ilemobade, A. A., Blotkamp, J. 1976. Preliminary observations on the use of the capillary flocculation test for the diagnosis of heartwater (*Cowdria ruminantium*) infection). Research in Veterinary Science **21:**370–372.

Ilemobade, A. A., Blotkamp, C. 1978a. Heartwater in Nigeria. II. The isolation of *Cowdria ruminantium* from live and dead animals and the importance of routes of inoculation. Tropical Animal Health and Production **10:**39–44.

Ilemobade, A. A., Blotkamp, J. 1978b. Clinico-pathological study of heartwater in goats infected with a Nigerian isolate of *Cowdria ruminantium*. Tropenmedizin und Parasitologie **29**:71–76.

Ilemobade, A. A., Leeflang, P. 1977. Epidemiology of heartwater in Nigeria. Revue d'Élevage et de Médicine Vétérinaire des Pays Tropicaux **30**:149–155.

Kakoma, I., Carson, C. A., Ristic, M., Huxsoll, D. L., Stephenson, E. H., Nyindo, M. B. A. 1977. Autologous lymphocyte mediated cytotoxicity against monocytes in canine ehrlichiosis. American Journal of Veterinary Research **38**:1557–1560.

Kakoma, I., Carson, C. A., Ristic, M., Stephenson, E. M., Hildebrandt, P. K., Huxsoll, D. L. 1978. Platelet migration inhibition as an indicator of immunologically mediated target cell injury in canine ehrlichiosis. Infection and Immunity **20**:242–247.

Kitao, T., Farrell, R. K., Fukuda, T. 1973. Differentiation of salmon poisoning disease and Elokomin fluke fever: Fluorescent antibody studies with *Rickettsia sennetsu*. American Journal of Veterinary Research **34**:927–928.

Koch, A. 1960. Intracellular symbiosis in insects. Annual Review of Microbiology **14**:121–140.

Krauss, H., Davies, F. G., Ødegard, D. A., Cooper, J. E. 1972. The morphology of the causal agent of bovine petechial fever (Ondiri disease). Journal of Comparative Pathology **82**:241–246.

Kuil H., Molenkamp. G. J., Meyer, J. C., Meyer, P. 1972. Tick-borne fever in cattle in the Netherlands. Netherlands Journal of Veterinary Science **5**:61–63.

Kurtti, T. J., Brooks, M. A. 1976. Preparation of mycetomes for culture in vitro. Journal of Invertebrate Pathology **27**:209–214.

Landureau, J.-C. 1966. Des cultures de cellules embryonnaires de Blattes permettent d'obtenir la multiplication *in vitro* des bactéries symbiotiques. Comptes Rendus Hebdomadaires des Séances de l'Académie des Sciences, Série D **262**:1484–1487.

Lewis, G. E., Jr. 1976. Equine ehrlichiosis: A comparison between *E. equi* and other pathogenic species of *Ehrlichia*. Veterinary Parasitology **2**:61–74.

Lewis, G. E., Jr., Ristic, M. 1978. Effect of canine immune macrophages and canine immune serum on the growth of *Ehrlichia canis*. American Journal of Veterinary Research **39**:77–82.

Lewis, G. E., Jr., Huxsoll, D. L., Ristic, M., Johnson, A. J. 1975. Experimentally induced infection of dogs, cats, and nonhuman primates with *Ehrlichia equi*, etiologic agent of equine ehrlichiosis. American Journal of Veterinary Research **36**:85–88.

Lewis, G. E., Jr., Ristic, M., Smith, R. D., Lincoln, T., Stephenson, E. H. 1977. The brown dog tick *Rhipicephalus sanguineus* and the dog as experimental hosts of *Ehrlichia canis*. American Journal of Veterinary Research **38**:1953–1956.

Louis, C., Laporte, M., Carayon, J., Vago, C. 1973. Mobilité ciliature et caractères ultrastructuraux des micro-organismes symbiotiques endo et exocellulaires de *Cimex lectularius* L. (*Hemiptera Cimicidae*). Comptes Rendus Hebdomadaires des Séances de l'Académie des Sciences, Série D **277**:607–611.

Louis, C., Yousfi, A., Vago, C., Nicolas, G. 1977. Étude par cytochimie et cryodécapage de l'ultrastructure d'une *Rickettsiella* de crustacé. Annales de Microbiologie **128B**:177–206.

Ma, W. C., Denlinger, D. L. 1974. Secretory discharge and microflora of milk gland in tsetse flies. Nature **247**:303–305.

Mare, C. J. 1972. The effect of prolonged oral administration of oxytetracycline on the course of heartwater (*Cowdria ruminantium*) infection in sheep. Tropical Animal Health and Production **4**:69–73.

Meyer, K. F. 1967. The host spectrum of psittacosis-

lymphogranuloma venereum (PL) agents. American Journal of Ophthalmology **63**:1225/199–1246/220.

Meynadier, G., Lopez, A., Duthoit, J.-L. 1974. Mise en évidence de Rickettsiales chez une araignée (*Argyrodes gibbosus* Lucas), Aranea, Theridiidae. Comptes Rendus Hebdomadaires des Séances de l'Académie des Sciences, Série D **278**:2365–2367.

Millemann, R. E., Knapp, S. E. 1970. Biology of *Nanophyetus salmincola* and "salmon poisoning" disease. pp. 1–41. In: Dawes, B. (ed.), Advances in parasitology, vol. 8 London, New York: Academic Press.

Minamishima, Y. 1965. Persistent infection of *Rickettsia sennetsu* in cell culture system. Japanese Journal of Microbiology **9**:75–86.

Morel, G. 1977. Étude d'une *Rickettsiella* (Rickettsie) se développant chez un arachnide, l'araignée *Pisaura mirabilis*. Annales de Microbiologie **128a**:49–59.

Nims, R. M., Ferguson, J. A., Walker, J. L., Hildebrandt, P. K., Huxsoll, D. L., Reardon, M. J., Varley, J. E., Kolaja, G. S., Watson, W. T., Shroyer, E. L., Elwell, P. A., Vaema, G. W. 1971. Epizootiology of tropical canine pancytopenia in Southeast Asia. Journal of the American Veterinary Medical Association **158**:53–63.

Nogge, G. 1976. Sterility in tsetse flies (*Glossina morsitans* Westwood) caused by loss of symbiotes. Experientia **32**:995.

Nyindo, M. B. A., Ristic, M., Huxsoll, D. L., Smith, A. R. 1971. Tropical canine pancytopenia: *In vitro* cultivation of the causative agent—*Ehrlichia canis*. American Journal of Veterinary Research **32**:1651–1658.

Nyindo, M. B. A., Ristic, M., Lewis, G. E., Jr., Huxsoll, D. L., Stephenson, E. H. 1978. Immune response of ponies to experimental infection with *Ehrlichia equi*. American Journal of Veterinary Research **39**:15–18.

Philip, C. B. 1974. Tribe III. Ehrlichieae, pp. 893–897. In: Buchanan, R. E., Gibbons, N. E. (eds.), Bergey's manual of determinative bacteriology, 8th ed. Baltimore: Williams & Wilkins.

Pienaar, J. G. 1970. Electron microscopy of *Cowdria* (*Rickettsia*) *ruminantium* (Cowdry, 1926) in the endothelial cells of the vertebrate host. Onderstepoort Journal of Veterinary Research **37**:67–78.

Pierce, K. R., Marrs, G. E., Hightower, D. 1977. Acute canine ehrlichiosis platelet survival and factor 3 assay. American Journal of Veterinary Research **38**:1821–1826.

Purnell, R. E., Brocklesby, D. W. 1978. Isolation of a virulent strain of *Ehrlichia phagocytophila* from the blood of cattle in the Isle of Man. Veterinary Record **102**:552–553.

Purnell, R. E., Brocklesby, D. W., Hendry, D. J., Young, E. R. 1976. Separation and recombination of *Babesia divergens* and *Ehrlichia phagocytophila* from a field case of redwater from Eire. Veterinary Record **99**:415–417.

Purnell, P. E., Young, E. R., Brocklesby, D. W., Hendry, D. J. 1977. The haematology of experimentally induced *B. divergens* and *E. phagocytophila* infections in splenectomized calves. Veterinary Record **100**:4–6.

Raikhel, A. S. 1978. Demonstration of acid phosphatase in colonies of symbiotic rickettsia from Malpighian tubules of an ixodid tick. Tsitologiia **20**:345–347.

Reinhardt, C., Steiger, R., Hecker, H. 1972. Ultrastructural study of the midgut mycetome-bacteroids of the tsetse flies *Glossina morsitans, G. fuscipes,* and *G. brevipalpis* (Diptera, Brachycera). Acta Tropica **29**:281–288.

Ristic, M. 1976. Immunologic systems and protection in infections caused by intracellular blood protista. Veterinary Parasitology **2**:31–47.

Ristic, M., Huxsoll, D. L., Weisiger, R. M., Hildebrandt, P. K., Nyindo, M. B. A. 1972. Serological diagnosis of tropical canine pancytopenia by indirect immunofluorescence. Infection and Immunity **6**:226–231.

Sakawa, H., Farrell, R. K., Mori, M. 1973. Differentiation of salmon poisoning disease and Elokomin fluke fever: Complement fixation. American Journal of Veterinary Research **34:**923–925.

Schlein, Y. 1977. Lethal effect of tetracycline on tsetse flies following damage to bacterioid symbionts. Experientia **33:**450–451.

Seamer, J., Snape, T. 1972. *Ehrlichia canis* and tropical canine pancytopenia. Research in Veterinary Science **13:**307–314.

Sells, D. M., Hildebrandt, P. K., Lewis, G. E., Jr., Nyindo, M. B. A., Ristic, M. 1976. Ultrastructural observations on *Ehrlichia equi* organisms in equine granulocytes. Infection and Immunity **13:**273–280.

Simpson, C. F. 1972. Structure of *Ehrlichia canis* in blood monocytes of a dog. American Journal of Veterinary Research **33:**2451–2454.

Simpson, C. F. 1974. Relationship of *Ehrlichia canis*-infected mononuclear cells to blood vessels of lungs. Infection and Immunity **10:**590–596.

Sixl-Voigt, B., Roshdy, M. A., Nosek, J., Sixl, W. 1977. Electronmicroscopic investigations of *Wolbachia*-like microorganisms in *Haemaphysalis inermis*. Mikroskopie **33:**255–257.

Smith, R. D., Hooks, J. E., Huxsoll, D. L., Ristic, M. 1974. Canine ehrlichiosis (tropical canine pancytopenia): Survival of phosphorus-32-labeled blood platelets in normal and infected dogs. American Journal of Veterinary Research **35:**269–273.

Smith, R. D., Ristic, M., Huxsoll, D. L., Ristic, Baylor, R. A. 1975. Platelet kinetics in canine ehrlichiosis: Evidence for increased platelet destruction as the cause of thrombocytopenia. Infection and Immunity **11:**1216–1221.

Smith, R. D., Sells, D. M., Stephenson, E. H., Ristic, M., Huxsoll, D. L. 1976. Development of *Ehrlichia canis,* causative agent of canine ehrlichiosis, in the tick *Rhipicephalus sanguineus* and its differentiation from a symbiotic rickettsia. American Journal of Veterinary Research **37:**119–126.

Snodgrass, D. R. 1975. Pathogenesis of bovine petechial fever. Latent infections, immunity, and tissue distribution of *Cytoecetes ondiri*. Journal of Comparative Pathology **85:**523–530.

Stannard, A. A., Gribble, D. H., Smith, R. S. 1969. Equine ehrlichiosis: A disease with similarities to tick-borne fever and bovine petechial fever. Veterinary Research **84:**149–150.

Stephenson, E. H., Osterman, J. V. 1977. Canine peritoneal macrophages: Cultivation and infection with *Ehrlichia canis*. American Journal of Veterinary Research **38:**1815–1820.

Stephenson, E. H., Ristic, M. 1978. Retrospective study of an *Ehrlichia canis* epizootic around Phoenix, Arizona. Journal of the American Veterinary Medical Association **172:**63–65.

Suitor, E. C., Jr. 1964a. Propagation of *Rickettsiella popilliae* (Dutky and Gooden) Philip and *Rickettsiella melolonthae* (Krieg) Philip in cell cultures. Journal of Insect Pathology **6:**31–40.

Suitor, E. C., Jr. 1964b. The relationship of *Wolbachia persica* Suitor and Weiss to its host. Journal of Insect Pathology **6:**111–124.

Suitor, E. C., Jr., Weiss, E. 1961. Isolation of a rickettsialike microorganism (*Wolbachia persica* n. sp.) from *Argas persicus* (Oken). Journal of Infectious Diseases **108:**95–106.

Synge, B. A. 1978. Brain biopsy for the diagnosis of heartwater. Tropical Animal Health and Production **10:**45–48.

Tachibana, N., Kobayashi, V. 1975. Effect of cyclophosphamide on the growth of *Rickettsia sennetsu* in experimentally infected mice. Infection and Immunity **12:**625–629.

Tachibana, N., Kusaba, T., Matsumoto, I., Kobayashi, Y. 1976. Purification of complement-fixing antigens of *Rickettsia sennetsu* by ether treatment. Infection and Immunity **13:**1030–1036.

Tamiya, T. (ed.) 1962. Recent advances in studies of tsutsugamushi disease in Japan. Tokyo: Medical Culture Inc.

Thrusfield, M. V., Synge, B. A., Scott, G. R. 1977. Attempts to cultivate *Ehrlichia phagocytophila* in vitro. Veterinary Microbiology **2:**257–260.

Tuomi, J. 1967a. Experimental studies on bovine tick-borne fever. 1. Clinical and haematological data, some properties of the causative agent, and homologous immunity. Acta Pathologica et Microbiologica Scandinavica **70:**429–445.

Tuomi, J. 1967b. Experimental studies on bovine tick-borne fever. 2. Differences in virulence of strains in cattle and sheep. Acta Pathologica et Microbiologica Scandinavica **70:**577–589.

Tuomi, J. 1967c. Experimental studies on bovine tick-borne fever. 3. Immunological strain differences. Acta Pathologica et Microbiologica Scandinavica **71:**89–100.

Tuomi, J. 1967d. Experimental studies on bovine tick-borne fever. 4. Immunofluorescent staining of the agent, and demonstration of antigenic relationship between strains. Acta Pathologica et Microbiologica Scandinavica **71:**101–108.

Tuomi, J. 1967e. Experimental studies on bovine tick-borne fever. 5. Sensitivity of the causative agent to some antibiotics and to sulphamethazine. Acta Pathologica and Microbiologica Scandinavica **71:**109–113.

Tuomi, J., von Bonsdorff, C. H. 1966. Electron microscopy of tick-borne fever agent in bovine and ovine phagocytizing leukocytes. Journal of Bacteriology **92:**1478–1492.

Weiss E. 1974. *Wolbachia, Rickettsiella,* pp. 898–903. In: Buchanan, R. E., Gibbons, N. E. (eds.), Bergey's manual of determinative bacteriology, 8th ed. Baltimore: Williams & Wilkins.

Weyer, F. 1973. Versuche zur Übertragung von *Wolbachia persica* auf Kleiderläuse. Zeitschrift für angewandte Zoologie **60:**77–93.

Wright, J. D., Sjostrand, F. S., Portaro, J. K., Barr, A. R. 1978. The ultrastructure of the rickettsia-like microorganism *Wolbachia pipientis* and associated virus-like bodies in the mosquito *Culex pipiens*. Journal of Ultrastructure Research **63:**79–85.

Yen, J. H. 1975. Transovarial transmission of *Rickettsia*-like microorganisms in mosquitoes. Annals of the New York Academy of Sciences **266:**152–161.

Yen, J. H., Barr, A. R. 1974. Incompatibility in *Culex pipiens,* pp. 97–118. In: Pal, R., Witten, M. J. (eds.), The use of genetics in insect control. Amsterdam: North Holland.

Young, E., Basson, P. A. 1973. Heartwater in the eland. Journal of the South African Veterinary Medical Association **44:**185–186.

CHAPTER 163

Fastidious Bacteria of Plant Vascular Tissue and Invertebrates (Including So-Called Rickettsia-Like Bacteria)

MICHAEL J. DAVIS, ROBERT F. WHITCOMB, and A. GRAVES GILLASPIE, JR.

Prokaryotes have been recognized as plant pathogens since the pioneering research of Burrill in 1881. However, the concept that noncultivable or extremely fastidious prokaryotes are plant pathogens developed recently, with the discovery of Doi et al. and Ishiie et al. in 1967 that mycoplasma-like organisms were consistently present in the phloem of plants afflicted with various diseases of the yellows type (see this Handbook, Chapter 167). This discovery inspired renewed investigations into the etiology of numerous infectious plant diseases, many of which had been assumed to be of viral etiology.

Subsequently, not only were additional mycoplasma-like organisms found associated with plant disease but several new groups of fastidious prokaryotes were discovered. These new discoveries included the spiroplasmas (see this Handbook, Chapter 169) and fastidious bacteria that, in most cases, were strictly confined to either the xylem or phloem of the vascular system of plants (Hopkins, 1977; Nienhaus and Sikora, 1979; Purcell, 1979). Although such bacteria have been referred to as "rickettsia-like", they may be only superficially similar to members of the Rickettsiaceae. The more simple interpretation of them as eubacteria of unknown taxonomic affinity (Moll and Martin, 1974) is now gaining acceptance.

Because these fastidious bacteria were noncultivable on conventional bacteriological media, their nature was first revealed by electron microscopy. Subsequently, specialized media were devised for isolation and maintenance of one of these agents. The xylem-limited bacterium that causes Pierce's disease of grapevines and almond leaf scorch disease (Davis, Purcell, and Thomson, 1978; Davis, Thomson, and Purcell, 1978) has been grown in pure culture and shown to be pathogenic.

Obligate bacterial symbiotes of arthropods comprise a potentially enormous group of fastidious prokaryotes. Because most fastidious bacteria associated with or causing plant diseases are transmitted by insect vectors, comparison with arthropod symbiotes is inevitable. Many of these arthropod symbiotes are intracellularly located in particular organs and tissues of their hosts, a refinement in habitat similar to that demonstrated by the fastidious

prokaryotes of plants. In this chapter, we discuss the isolation and identification of the fastidious bacteria in plant and arthropod habitats.

Habitats

Plant Xylem Habitat

The xylem is a continuous tissue throughout the plant. It has three primary functions: transport of water and nutrients, storage, and structural support (Esau, 1965). Fastidious, xylem-limited bacteria are associated with tracheary elements of two types, tracheids and vessel members, that function primarily in transport. In the process of differentiation, tracheary elements form cell walls with apertures between adjacent cells. Later, they become devoid of cytoplasm, leaving a hollow vessel. Both tracheids and vessel members develop apertures, termed pits, which have membranes separating adjoining elements. All tracheary elements are connected by paired pits. In addition, vessel members are joined end-to-end, forming vessels, and the adjoining ends have open perforation plates. Vessels are not continuous throughout the plants, but are of finite length. Sap flows from one vessel to another through pit membranes (Esau, 1965; Zimmerman and McDonough, 1978). In some hosts, the bacteria are found more frequently in the smaller, thick-walled tracheids than in the larger vessels (Hopkins, Mollenhauer, and French, 1973; Kitajima, Bakarcic, and Fernandez-Valiela, 1975). Occasionally, however, the bacteria are found in the intercellular spaces of the xylem (Goheen, Nyland, and Lowe, 1973). Xylem-limited bacteria apparently move from one tracheary element to another by breaching pit membranes, but whether the membranes are enzymatically dissolved or forcefully ruptured is unknown. Often, some tracheary elements are packed with bacteria while adjacent elements are not. Pit membranes often appear to prevent lateral spread of the bacteria, but open perforation plates allow relatively unimpeded longitudinal spread within vessels. The length of vessels is difficult to determine; it varies with plant species and with the time of year the vessels develop. In some species (e.g., *Eucalyptus obliqua*), most vessels are less than 50 cm in length. However, vessels in tree species with ring-porous wood may extend several meters (Esau, 1965; Zimmerman and McDonough, 1978).

Xylem-limited bacteria are confronted with a unique physical and nutritional environment. The ascending flow of sap from the roots fluctuates in response to transpiration of water by the plant; the direction of flow may even reverse under conditions of high atmospheric humidity combined with low soil moisture (Bollard, 1960). Since no evidence of motility has been observed in xylem-limited bacteria, it is assumed that the flow of the xylem sap plays an important role in bacterial dissemination within the plant. Also, sap flow must be important in provision of nutrients to the bacteria and in removal of waste products. The bacteria may also obtain nutrients from the remains of the cytoplasm lining the lumen of tracheary elements and by diffusion from adjacent living cells. Zimmerman and McDonough (1978) suggested that the cause of dysfunction of tracheary elements upon invasion or introduction of bacterial pathogens is an instantaneous embolism and that the obstruction of tracheary elements by gums, tyloses, bacteria, or bacterial products does not cause, but is the result of, tracheary dysfunction. Although this would preclude the dissemination of xylem-limited bacteria by flowing sap, xylem sap in adjacent functional tracheary elements might still serve as a source of diffusable nutrients for xylem-limited bacteria.

The composition and concentration of solutes in xylem sap varies with plant species, location within the plant, time of day, plant age, seasonal cycle, plant nutritional state, and health of the plant (Pate, 1976). The xylem sap provides a qualitatively rich but quantitatively dilute nutritional environment. Xylem sap usually contains 1–20 mg/ml solids in contrast to phloem sap, which usually contains 50–300 mg/ml solids (Pate, 1976). These solids contain a vast array of soluble compounds that could serve as nutrients for fastidious bacteria. Possible carbon sources for the bacteria include mono- and disaccharides such as glucose, sucrose, and fructose, organic acids such as citrate, malate, succinate, and α-ketoglutarate, and a number of other organic compounds, including amino acids and plant-growth regulators (Bollard, 1960; Pate, 1976, Wormall, 1924). Possible nitrogen sources include both inorganic and organic nitrogen. One organic nitrogenous compound often predominates (Pate, 1976). Amino acids, alkaloids, ureides, and amides are included among the nitrogenous compounds that occur in xylem sap. Organic phosphorus, sulfur-containing compounds, and essential inorganic salts are also present in xylem, and vitamins may be present in low concentrations. In view of the low solute concentration of their environment, fastidious,

xylem-limited bacteria probably have special mechanisms for concentration of essential nutrients from the sap.

In situ electron micrographs (Kamiunten and Wakimoto, 1976; Weaver, Teakle, and Hayward, 1977; Worley and Gillaspie, 1975) have shown that characteristic bacteria consistently observed in tracheary elements of sugarcane with ratoon stunting disease are usually associated with a matrix material (Fig. 1A). The bacteria frequently appear within the pit fields, next to the cell walls, or even within cell walls (Fig 1A). The bacteria that occur in the interior of the vessel lumen often appear to be in various stages of collapse. The combination of bacterial cells and matrix is thought to be responsible for plugging the xylem. If this plugging material is pulled or forced out of freshly cut ends of infected stalks, microcolonies of the bacterium may be observed within the matrix (Kao and Damann, 1978).

Other xylem-limited bacteria may be found embedded in an electron-dense matrix that presumably consists of gums of plant origin (Mollenhauer and Hopkins, 1974). Fibrous strands, called "osmophilic lines or microfibrils" (Lowe, Nyland, and Mircetich, 1976), have been frequently seen connecting these bacteria (French, Christis, and Stassi, 1977; Lowe, Nyland, and Mircetich, 1976; Mollenhauer and Hopkins, 1974; Nyland et al., 1973). Such fibrous strands, discussed in greater detail in this chapter under Identification, resemble the fibrous polysaccharide coats that surround adherent bacteria in diverse habitats (Brooker and Fuller, 1975; Costerton, Geesey, and Cheng, 1978; Latham et al., 1978; McCowan et al., 1978). Costerton, Geesey, and Cheng (1978) suggest that these polysaccharide fibers (termed "glycocalyx") function in survival and pathogenesis, and may confer such advantages to bacteria as: (i) bacterial position in a beneficial environment; (ii) bacterial adhesion to one another for conservation and concentration of digestive enzymes; (iii) provision of a food reservoir that may serve as an ionic net similar to an ion-exchange resin; (iv) protection against host defense systems or other stress. If the fibrous strands associated with xylem-limited bacteria confer such advantages, they may facilitate survival in the dilute but dynamic environment of the xylem.

All of the diseases associated with fastidious, xylem-limited bacteria occur in geographical areas with mild winter climates. Purcell (1977, 1979) has proposed that the geographical distribution of some of these diseases may be limited more by cold winters than by the lack of potential vectors or hosts, and has demonstrated that one of these diseases, Pierce's disease of grapevines, can be cured by cold therapy.

Symptoms of the diseases in which xylem-limited bacteria are involved include marginal necrosis of

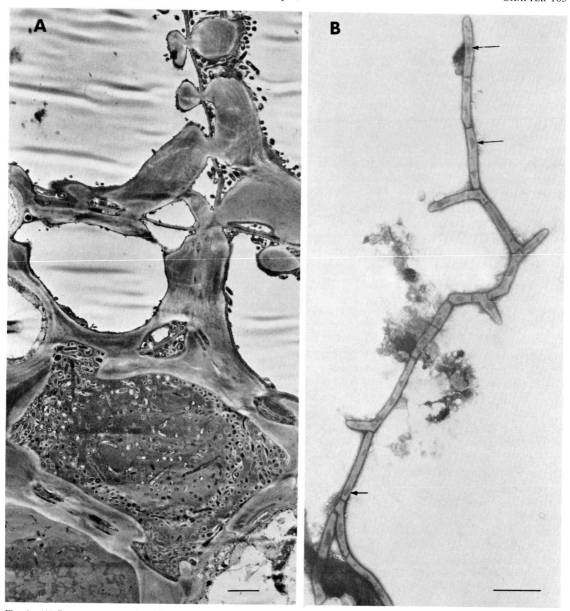

Fig. 1. (A) Ratoon stunt-associated bacterium in thin section of Sudan grass xylem cells. Bacteria occur within the pit fields, near residual primary walls and, in some cells, within a matrix material. Bar = 5 μm. Courtesy J. F. Worley. (B) Negatively stained chain of 16 ratoon stunt-associated bacterial cells showing pseudobranching and mesosomes (arrows). Note the septa between the cells. Bar = 1.0 μm. [Teakle, Kontze, and Appleton, 1979, with permission.]

the leaves, stunting, decline in vigor, and decrease in yield. Such "wilt" symptoms are also produced by many vascular plant pathogens, including bacterial pathogens in the genera *Xanthomonas, Pseudomonas, Erwinia,* and *Corynebacterium* (Nelson and Dickey, 1970). Host response varies greatly among plants infected with fastidious, xylem-limited bacteria. Such variation may be a function of the relative resistance of different hosts or a function of the distribution of the pathogen within the host. No symptoms of marginal leaf necrosis are found in peach with phony disease; however, French, Stassi, and Schaad (1978) found that certain plum varieties de-

veloped marginal necrosis of the leaves, resembling symptoms of plum leaf scald, when grafted onto peach with phony disease. The phony disease bacterium was found to be more prominent in roots than in leaves of peach (Hopkins, Mollenhauer, and French, 1973), and graft transmission data suggested that the phony disease pathogen was much more prevalent in the shoots of symptomless wild plum than in the shoots of symptomatic peach (Hutchins et al., 1953). The only reliable diagnostic symptom of the ratoon stunting disease of sugarcane is an internal discoloration of the stalk (Steindl, 1961).

Plant Phloem Habitat

The phloem stream of vascular plants comprises another important habitat for fastidious bacteria. The ecological circumstance permitting colonization of the phloem by microorganisms involves a specialized insect fauna of plants that specifically utilizes this tissue. The habitat breadth of phloem pathogens, therefore, is significantly influenced by the plant-host range of the insect vector(s) (Whitcomb and Williamson, 1979).

In the tissue of susceptible plants, the microorganisms take advantage of a fluid that is rich in many potential nutrients. These include inorganic cations and anions, organic acids, amino acids, proteins, and carbohydrates. The composition of plant phloem can be favorably compared with the composition of insect hemolymph (Saglio and Whitcomb, 1979). Together, the two fluids comprise an ecological niche that has been filled by several prokaryotic taxa. These include the noncultivable, wall-less prokaryotes that induce proliferation and virescence in plants (this Handbook, Chapter 167) and the spiroplasmas (class Mollicutes; family Spiroplasmataceae; see this Handbook, Chapter 169). The phloem habitat contrasts sharply with that of the xylem. The xylem exudates of sugar beet (*Beta vulgaris*) contained less than one-tenth the concentration of total solids, sucrose, reducing sugars, and total nitrogen that was found in the phloem tissue (Fife, Price, and Fife, 1962). Also, nitrogenous compounds were especially low in xylem exudates; physical parameters, such as viscosity, specific gravity, electrical conductivity, pH, and osmotic pressure, also differed in the two tissues (Bollard, 1960; Fife, Price, and Fife, 1962; Pate, 1976).

The phloem of plants consists, in part, of sieve tubes that are composed of sieve cells (or sieve-tube elements) that form a long, pressurized conduit. Continuity of the fluids in the tissue is provided by the sieve-element pores between the sieve elements. Although these pores may be less than 1 μm in diameter, they may be as wide as 14 μm, and average more than 2 μm. Thus, prokaryotic organisms may generally pass unrestricted from cell to cell in the tissue. Electron micrographs of organisms forced into sieve pores are, therefore, probably artifacts resulting from pressurized flow at the time of fixation. A more complete discussion of this issue and a general treatment of phloem tissue in relation to prokaryotic plant disease is presented in this Handbook, Chapter 167.

The walled prokaryotes of the phloem–insect habitat are small, elongate organisms (1–3 μm $\times$ 0.2–0.5 μm). Whether the organisms represent a single taxon or are an artificial aggregate of unrelated organisms is unclear, in large part because none of the organisms has been cultivated in artificial media. At least 12 agents may be included in the group. In certain cases (Küppers, Neinhaus, and Schinzer, 1975; Petzold, Marwitz, and Kunze, 1973), the organisms reportedly occurred not only in sieve elements, but in other phloem cell types as well. Such observations are not unique to phloem bacteria but have been made for phloem mycoplasmas as well. McCoy (1979) has evaluated such claims. Considering that passage through plasmodesmata of the cell walls would be necessary for organisms to gain access to cells adjoining the sieve tubes, and considering the difficulties in identifying cells of developing phloem tissue, McCoy believed that most or all of the reports of mycoplasma-like organisms (MLO) in cells other than sieve elements can be attributed to misidentification, either of sieve elements as parenchyma cells or of membrane-bound vesicles of host origin as MLO. Certainly, the difficulties facing MLO in such plasmodesmatal passages would be even more insurmountable for the walled (and therefore less plastic) prokaryotes that we discuss here.

Agents of disease in citrus (Laflèche and Bové, 1970; Moll and Martin, 1974; Saglio et al., 1972), clover (Behncken and Gowanlock, 1976; Benhamou, 1978; Benhamou, Giannotti, and Louis, 1978; Benhamou, Louis, and Giannotti, 1979; Black, 1944; Grylls, 1954; Liu and Black, 1974; Markham, Townsend, and Plaskitt, 1975; Windsor and Black, 1973a,b), and potato (Klein, Zimmerman-Gries, and Sneh, 1976) may be considered to be typical of walled phloem prokaryotes. In their plant hosts, the agents cause diseases characterized by leaf stunting and clubbing. Symptoms are often mild, and spontaneous recovery may be a significant feature (Klein, Zimmerman-Gries, and Sneh, 1976). Other symptoms, such as proliferation of shoots with a resulting witches' broom effect (Hirumi et al., 1974; Holmes, Hirumi, and Maramorosch, 1972; Maramorosch et al., 1975; Nienhaus, Brüssel, and Schinzer, 1976; Nienhaus, and Schmutterer, 1976) or floral virescence (Black, 1944; Liu and Black, 1974), may be observed in some cases. Symptoms of proliferation and virescence are more characteristic of another group of phloem prokaryotes, the "yellows agents" (this Handbook, Chapter 167), than of the walled, phloem prokaryotes. At present, the physiological basis for the proliferation and virescence syndrome is poorly understood.

Invertebrate Habitat

Despite occasional suggestions that some of the prokaryotic vascular plant pathogens are soil borne (Nienhaus, Brüssel, and Schinzer, 1976) and may be nematode borne (Nienhaus, and Schmutterer, 1976; Rumbos, Sikora, and Nienhaus, 1977) all well-defined vector relationships involve insects. A large but well-defined group of insect vectors transmits the xylem-limited bacteria. Sharpshooter leaf-hoppers (Homoptera: Cicadellidae: Cicadellinae) and spittlebugs or froghoppers (Homoptera: Cercopidae) are the only known vectors, and both are apparently xylem-feeding insects (Purcell, 1979). Pierce's disease is transmitted by at least 26 species of leafhoppers (Frazier, 1966; Hopkins, 1977) and five species of spittlebugs (Severin, 1950). Those vectors that have been examined also transmit alfalfa dwarf disease (Hewitt et al., 1946) and almond leaf scorch disease (Auger, Mircetich, and Nyland, 1974; Mircetich et al., 1976), which are caused by or associated with the same bacterium (Davis, Purcell, and Thomson, 1978; Davis, Thomson, and Purcell, 1978; Thomson et al., 1978). Transmission by the same leafhopper species of xylem-limited bacteria associated with two different diseases has also been demonstrated: *Homaladisca coagulata, Oncometopia undulata,* and *Cuerna costalis* transmit both Pierce's disease (Kaloostian, Pollard, and Turner, 1962) and phony disease of peach (Turner and Pollard, 1959); *Oncometopia nigricans* transmits both Pierce's disease (Hopkins, 1977) and periwinkle wilt (McCoy et al., 1978). Thus, it appears that insect-vector transmission of xylem-limited bacteria is not governed by strict vector-pathogen specificities. In this respect, the vector relationship may mirror the pattern of wide host preference of the vectors themselves.

The mechanism of transmission has been studied for only a few vectors of xylem-limited bacteria, and almost entirely concerns Pierce's disease and phony disease of peach. The Pierce's disease bacterium is apparently transmitted through a noncirculative mechanism, since there is no measurable latent period and infectivity is lost after molting (Purcell and Finlay, 1979). Since the lining of the alimentary tract anterior to the midgut is lost during molting, the bacteria are apparently located in the foregut or mouthparts of the vectors. Leafhoppers can acquire and transmit the bacteria in less than 2 h (Purcell and Finlay, 1979), and infected adults retain the ability to transmit for life (Purcell and Finlay, 1979; Severin, 1949; Turner and Pollard, 1959). There is no vertical transmission to the progeny of leafhopper vectors, but nymphs are able to transmit after acquisition of bacteria from diseased plants (Freitag, 1951; Purcell and Finlay, 1979; Turner and Pollard, 1959).

There is no evidence that the agents of Pierce's disease or phony disease of peach are pathogenic to the vectors. McCoy et al. (1978) reported that the xylem-limited bacteria associated with periwinkle wilt apparently occur in the hemolymph and may be detrimental to the insect carrier. However, their report was preliminary in nature, and the identity of the bacteria in the insect hemolymph with the etiological agent of periwinkle wilt was not firmly established.

In many respects, knowledge of insects as habitats for phloem-specialized, walled prokaryotes is scanty. Most of the 12 reported agents are known only as suspected plant pathogens; although the disease agents are assumed to be insect borne, in most cases no vector is known. Vertical transmission has been shown in the leafhoppers (Homoptera: Cicadellidae) that carry two of these agents (Black, 1944; Grylls, 1954); the rates were as high as 99%. The demonstration that the clover club leaf pathogen was transmitted vertically through multiple generations (Black, 1948, 1950) was the first demonstration that a microorganism could multiply in both plant and animal reservoirs in the course of a complex biological cycle. In contrast, little is known of the vector-pathogen relationship in the case of the leafhopper vectors of other clover diseases or the psyllid (Homoptera: Psyllidae) vectors of citrus disease (Capoor, Rao, and Viswanath, 1974; Catling and Atkinson, 1974; McClean and Oberholzen, 1965). In the latter case, there is some evidence that the organisms multiply in the vector hemolymph (Moll and Martin, 1973). This claim offers hope for eventual cultivation of the agents, since other agents that multiply extracellularly in arthropods, such as *Rochalimaea quintana* (Vinson, 1966) or spiroplasmas (Whitcomb and Williamson, 1979) have proved to be cultivable.

Arthropod Habitat: Symbiotic Microorganisms

Many prokaryotes are obligately associated with arthropods. Considering the large number of arthropod species (in excess of 1 million), such prokaryotes may represent a significant fraction of all prokaryotes. In the sense of symbiosis as a close association between living organisms (Starr, 1975), all such prokaryotes may be regarded as symbiotes. In this sense, for example, insect pathogens of the genus *Rickettsiella*, or species of *Wohlbachia*, which have not been shown to benefit their arthropod hosts, could be termed symbiotes. Such fastidious organisms are discussed in this Handbook, Chapter 162.

Some bacteria associated with insects merit brief mention, although they are not fastidious. Certain Gram-negative, bacterial plant pathogens (*Erwinia* or *Xanthomonas* species) that inhabit the xylem may be insect borne, although insect transmission is usually only one means of spread (Poos and Elliott, 1936). Such associations generally show little evidence of specificity or adaptation on the part of the pathogen for insect association. Other Gram-negative bacteria (e.g., *Enterobacter agglomerans*) may inhabit the alimentary tract of insects without any involvement in plant disease (Whitcomb, Shapiro, and Richardson, 1966). Finally, some bacteria that appear to be specifically associated with insects may appear in the course of microbiological studies of fastidious plant bacteria. For example, *Lactobacillus hordniae* (Latorre-Guzmán, Kado, and Kunkee, 1977) was isolated in the course of studies of the Pierce's disease agent and was mistaken (Auger, Shalla, and Kado, 1974) for the pathogen. Subsequent investigation (Purcell et al., 1977) revealed that the bacterium was part of the normal flora of the leafhopper host.

Organisms that are truly fastidious and intimately associated with their insect hosts have been the subject of extensive study (Büchner, 1965; Chang, 1981). In some cases, the organisms were discovered as a result of biological observations. For example, infectious hybrid sterility in *Drosophila paulistorum* is apparently conferred by prokaryotes present in reproductive tissues of the flies (Ehrman and Kernaghan, 1972). The organisms, like all vertically transmitted, intracellular prokaryotes, have resisted cultivation. Ultrastructurally, they are seen to have typical prokaryotic internal organization. The limiting structures of the organism, while much simpler than the typical bacterial cell wall, has been interpreted as a single limiting membrane (Ehrman and Kernaghan, 1972). However, the organisms are closely associated with host cytoplasm, and it is unclear whether an additional membrane normally seen is of host or symbiote origin. Extracellular, vertically transmitted mycoplasmal symbiotes (spiroplasmas) that occur in hemolymph of certain *Drosophila* species and induce sex-ratio abnormalities are discussed in this Handbook, Chapter 169.

Commonly, intracellular symbiotes—particularly those of homopterous insects such as leafhoppers and aphids—are found in specialized organs (mycetome) of the host. The mycetome consists of mycetocytes, which harbor the organisms, surrounded by sheath cells. Often, more than one type of symbiote may appear to be present. For example, literature on symbiotes of leafhoppers deals extensively with a- and t-types of symbiotes, and the complex processes that occur as they interact (Chang and Musgrave, 1975; Körner, 1969; Louis and Laporte, 1969). The symbiotes appear to be limited by two unit membranes; a third membrane, if present, is generally believed to be of host origin (Körner, 1969).

Many aphid (Homoptera: Aphididae) species have symbiotes that are similar ultrastructurally and in the manner that they are housed (Büchner, 1965; Hinde, 1971). The symbiotes are of two types. One of these types, the "secondary symbiote", is a rod-shaped prokaryote whose affinity to bacteria is clear. These organisms are restricted to a syncytial sheath that encloses the primary mycetocytes (Griffiths and Beck, 1973). The primary symbiotes that occur in the primary mycetocytes are spherical or ovoid, with diameters of 1.3–1.7 μm (Hinde, 1971). A third membrane that always surrounds the symbiote is of host origin. The outer envelope often appears to be wrinkled. Study of the symbiotes in vitro has depended on development of techniques, such as density-gradient centrifugation, for their isolation from aphid homogenates (Houk and McLean, 1974). Study of such preparations revealed that the symbiotes were able to synthesize free fatty acids, mono- and diglycerides, phosphatidylcholine, phosphatidylethanolamine, and cholesterol (Houk, Griffiths, and Beck, 1976). The contribution that such synthesis makes to the host was shown by Griffiths and Beck (1977), who used electron-microscope autoradiography with [^{3}H]mevalonate precursor to show that cholesterol was transported to insect tissues after synthesis in the mycetocytes. The role of symbiotes in host nutrition was recently reviewed by Chang (1981).

The taxonomic status of arthropod symbiotes remains clouded. The organisms have been compared with mycoplasmas, but the recent demonstration (Houk et al., 1977) of peptidoglycan in the cell walls of the primary symbiotes of the pea aphid and the presence of two limiting membranes clearly precludes their classification as mycoplasmas. Possible affinities of aphid symbiotes to rickettsiae have also been considered (Hinde, 1971). Such discussions are similar to those in which the affinity of xylem- and phloem-limited organisms to rickettsiae has been proposed. All such speculation must await careful comparative study of the limiting structures and, especially, isolation and comparative characterization of the DNA of their genomes.

Isolation

Isolation of fastidious, xylem- or phloem-limited bacteria in pure culture has been accomplished only with the xylem-limited bacterium that causes Pierce's disease of grapevines and almond leaf scorch disease. The Pierce's disease–almond leaf scorch (PD-ALS) bacterium was first isolated on JD1 medium.

JD1 Medium for Isolating PD-ALS Bacteria (Davis, Purcell, and Thomson, 1978)

The medium contains (in grams per liter):

PPLO broth base	10 g
Hemin chloride	40 mg
Bovine albumin (filter-sterilized at 0.5% in water)	0.5 g
Agar (Difco)[1]	15 g

Subsequent modifications of JD1 medium led to the formulation of JD3 medium, whose composition

---

[1] Mention of companies or commercial products does not imply endorsement or recommendation by the U.S. Department of Agriculture.

was unfortunately published with errors (Davis, Purcell, and Thomson, 1978). To avoid confusion, the correct formulation of the JD3 medium has been redesignated the PD2 medium.

PD2 Medium for Isolating PD-ALS Bacteria
The medium contains (in grams per liter):

Pancreatic digest of casein (tryptone [Difco])	4 g
Papaic digest of soy meal (Soytone [Difco] or Phytone peptone [BBL])	2 g
Trisodium citrate	1 g
Disodium succinate	1 g
Hemin chloride stock (0.1% hemin chloride dissolved in 0.05 N NaOH)	10 ml
$MgSO_4 \cdot 7H_2O$	1 g
K_2HPO_4	1.5 g
KH_2PO_4	1 g
Agar (Difco)	15 g
Bovine serum albumin fraction five (BSA-5) solution (20% solution in water)	10 ml

All constituents except the BSA-5 solution are autoclaved in 980 ml of double-distilled water. After autoclaving, the filter-sterilized BSA-5 solution is aseptically added at 45–50°C. The pH of the PD2 medium should be approximately 7.0 without adjustment after dissolving the agar, but before autoclaving.

Two completely autoclavable formulations were developed by replacing the BSA-5 with either potato starch (2 g/liter; PD3 medium) or activated charcoal (0.5 g/liter; PD4 medium) and appear to be as effective as the PD2 medium (M. J. Davis, unpublished).

The PD–ALS bacterium has been isolated from expressed sap of leaf petioles and from petioles triturated in phosphate buffer (Davis, 1978; Davis, Purcell, and Thomson, 1978; Davis, Thomson, and Purcell, 1979; Thomson et al., 1978). Colonies of the PD–ALS bacterium are readily visible after 7–10 days of aerobic incubation at 28°C. They are circular with entire margins, measure 0.5–1.0 mm in diameter, and are white to opalescent. Symptomatic grapevines usually yield 90–100% positive isolations from single petioles of symptomatic leaves (Davis, Purcell, and Thomson, 1978). No significant difference was observed in the efficiency of reisolation of the PD–ALS bacterium from inoculated alfalfa using PD2, PD3, or PD4 media (M. J. Davis, unpublished).

Attempts to use the PD2 medium for isolation of other xylem-limited bacteria have failed (Davis et al., 1979; French and Kitajima, 1978; McCoy et al., 1978). Investigations into the specific physical and nutritional requirements of the PD–ALS

bacterium in vitro might facilitate the isolation of other xylem-limited bacteria, since many of these organisms appear to be closely related in morphology, habitats, and antigenicity.

Identification

Xylem-Limited Bacteria

Our present knowledge of fastidious, xylem-limited bacteria is largely limited to the diseases they are associated with or cause, because, until recently, none of the bacteria had been isolated on artificial media. These bacteria are distinguished from other bacterial plant pathogens infecting the xylem by their inability to infect tissues other than the xylem, their insect but not mechanical transmission (except for the ratoon stunt-associated organism), and their inability to grow on conventional bacteriological media. None of these xylem-limited bacteria has been taxonomically characterized and their affiliation with known genera of bacteria is uncertain.

Two distinct morphotypes of xylem-limited, fastidious bacteria are associated with plant diseases. The coryneform bacterium associated with ratoon stunting disease of sugarcane (Figs. 1 and 3) (Gillaspie, Davis, and Worley, 1973; Maramorosch et al., 1973; Teakle, Smith, and Steindl, 1973) is the only known representative of one type. The second type comprises small, rod-shaped bacteria having a characteristically rippled or ridged outer cell wall membrane (Fig. 2). The bacteria in this second group are associated with and presumably cause several plant diseases including: Pierce's disease of grapevines (Davis, Purcell, and Thomson, 1978; Goheen, Nyland, and Lowe, 1973; Hopkins and Mollenhauer, 1973); alfalfa dwarf disease (Goheen, Nyland, and Lowe, 1973; Thomson et al., 1978); almond leaf scorch disease (Davis, Thomson, and Purcell, 1978; Mircetich et al., 1976); phony disease of peach (Hopkins, Mollenhauer, and French, 1973; Nyland et al., 1973); plum leaf scald (French and Kitajima, 1978; Kitajima, Bakarcic, and Fernandez-Valiela, 1975); elm leaf scorch (Sherald and Hearon, 1978); young tree decline of citrus (Feldman et al., 1977; Hopkins, Adlerz, and Bistline, 1978); and periwinkle wilt (McCoy et al., 1978).

The ratoon stunt–associated bacterium differs from other xylem-limited bacteria in morphology, in having a narrower host range, and in being mechanically transmissible on agricultural equipment. Ratoon stunting disease is found in every major sugarcane-growing area of the world, possibly because of its efficient transmission via equipment or vegetative propagation material. It has been found naturally only in sugarcane, although other grasses have been artificially infected (Steindl, 1961).

The ratoon stunt–associated bacterium is a small,

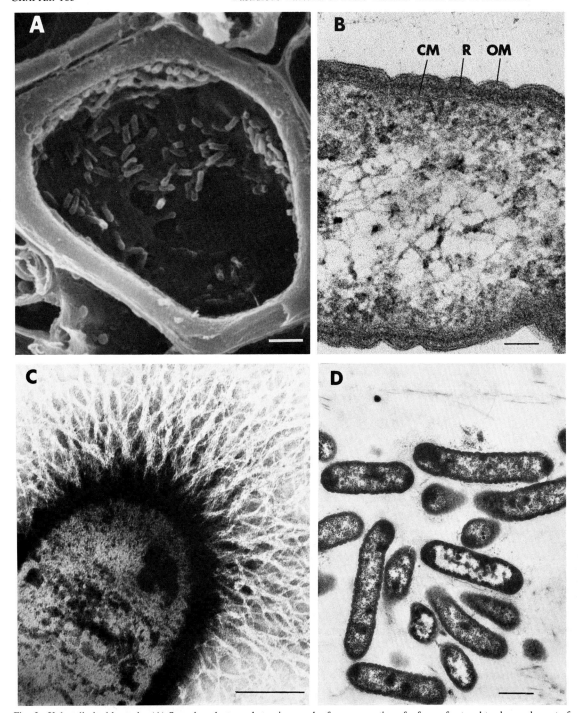

Fig. 2. Xylem-limited bacteria. (A) Scanning electron photomicrograph of a cross-section of a freeze-fractured tracheary element of grapevine with Pierce's disease, showing the causal bacterium in the lumen. Bar = 2.5 μm. [Courtesy of D. G. Garrott and M. J. Davis.] (B) Pierce's disease bacterium in grapevine showing rippled cell wall with cytoplasmic membrane (CM), R-layer (R), and outer membrane (OM). Bar = 50 nm. [Courtesy of H. H. Mollenhauer and D. L. Hopkins.] (C) Negatively stained bacterium associated with plum leaf scald disease, showing prominent "fimbrae" extending from the cell wall at a longitudinal end of the bacterium. Bar = 0.2 μm. [French and Kitajima (1978), with permission.] (D) Bacterium with rippled cell wall associated with elm leaf scorch disease in a tracheary element of elm. Bar = 0.5 μm. [Courtesy of J. Sherald and S. Hearon.]

coryneform bacterium with a smooth cell wall; it measures 0.3–0.5 × 1–3 μm, but lengths of 10 μm or longer are not uncommon (Fig. 3). Dif-

ferences in published measurements have been discussed by Gillaspie, Davis, and Worley (1976) and by Teakle, Kontze, and Appleton (1979). The bacte-

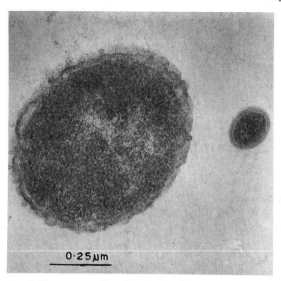

Fig. 3. Transverse section of a cell of *Escherichia coli* (left) and the ratoon stunt-associated bacterium (right) from an agar-embedded pellet. Note the relatively small size of the ratoon stunt-associated bacterium and its different wall structure. Bar = 0.25 μm. [Weaver, Teakle, and Hayward, 1977, with permission.]

rium has a smooth, thin cell wall and the cells may be straight, curved, or swollen at the tip or in the middle. The bacterium divides by septum formation (Weaver, Teakle, and Hayward, 1977; Worley and Gillaspie, 1975) and contains mesosomes (Teakle, Smith, and Steindl, 1973) (Fig. 1B). The bacterial cells may occur in chains, and branching or pseudo-branching has been reported (Kao and Damann, 1978; Teakle, Kontze, and Appleton, 1979). Although the cellular and colonial morphology of the bacterium resembles that of the actinomycetes (Kao and Damann, 1978), further work will be required to establish its taxonomic position.

Diseases of plant xylem that are caused or associated with rippled-walled, fastidious bacteria have been reviewed by Hopkins (1977) and Purcell (1979). In those cases in which the means of natural transmission is known, the bacteria are transmitted by insect vectors or in propagative material. Graft and insect transmission studies indicated even before the causal bacterium was isolated (Davis, Purcell, and Thomson, 1978; Davis, Thomson, and Purcell, 1978; Thomson et al., 1978) that Pierce's disease of grapevines, alfalfa dwarf disease, and almond leaf scorch disease had a common etiological agent (Hewitt et al., 1946; Mircetich et al., 1976). Recent evidence indicates that phony disease of peach and plum leaf scald disease may also have a common etiological agent (French and Kitajima, 1978; French, Latham, and Stassi, 1977b). Whether or not a common etiological agent exists for any of the other diseases associated with fastidious, xylem-limited bacteria has not been determined.

One apparent characteristic of these bacteria as a group is the ability to infect a large number of hosts. For example, members of 28 families of monocotyledonous and dicotyledonous plants could be infected with the Pierce's disease agent (Freitag, 1951; Hewitt, 1970), although most of the hosts were infected without displaying symptoms.

In addition to having common ecological niches, all of the rippled-walled, xylem-limited bacteria are similar in morphology, suggesting that they have a close phylogenetic relationship. The range of in situ measurements of these bacteria is 0.25–0.5 × 1.0–4.0 μm (Davis, Purcell, and Thomson, 1978; Goheen, Nyland, and Lowe, 1973; Hopkins and Mollenhauer, 1973; Hopkins, Mollenhauer, and French, 1973; Lowe, Nyland, and Mircetich, 1976; Mollenhauer and Hopkins, 1974; Nyland et al., 1973). Negatively stained preparations of cultured PD–ALS bacteria revealed an average measurement for nondividing cells of 0.39 × 1.44 μm (Davis, Purcell, and Thomson, 1978). The most distinctive morphological feature of these bacteria in situ is the rippled topography of their cell wall. The cell wall has a layered structure that is typical of Gram-negative bacteria (Silva and Sousa, 1973) with an inner or cytoplasmic membrane separated from an outer membrane by a periplasmic space; both membranes resemble unit membranes (Hopkins, 1977; Lowe, Nyland, and Mircetich, 1976; Mollenhauer and Hopkins, 1976). The width of the wall, including both inner and outer membranes, varies between approximately 25 and 50 nm. The periodic enfolding of the outer membrane has been described most often as rippled, but also as ridged, corrugated, and furrowed (Fig. 2B). The ridges tend to be perpendicular to the long axis of the organisms. A peptidoglycan or ''R'' layer has been observed in the periplasmic space of the cell wall of the phony peach and PD–ALS bacteria (Fig. 2B) (Hopkins, 1977; Mollenhauer and Hopkins, 1974), but other investigators did not observe this layer in the phony peach and almond leaf scorch bacteria (Lowe, Nyland, and Mircetich, 1976; Nyland et al., 1973). Silva and Sousa (1973) have shown that difficulties in observing ultrastructural features of the cell walls of Gram-negative bacteria can be due to the fixation technique. Hopkins (1977) suggested that the failure to observe the peptidoglycan layer of xylem bacteria may be due either to fixation techniques or to interpretation of a closely associated peptidoglycan layer and an inner electron-dense layer of the outer membrane as a single layer. McCoy et al. (1978) reported that both rippled-walled and smooth-walled bacteria were present in the xylem of wilted periwinkle, but it is not known whether the cell walls represented different morphological states of the same organism or two different organisms.

The cytoplasm of xylem-limited bacteria resembles that of other prokaryotes in its content of ribosomes, nuclear regions with DNA-like strands, and osmophilic granules (Lowe, Nyland, and Mircetich, 1976; Mollenhauer and Hopkins, 1974; Nyland et al., 1973). Membranous inclusion bodies, which appear to be associated with the cytoplasmic membrane, were seen in situ in the PD–ALS bacterium (Davis, 1978; Lowe, Nyland, and Mircetich, 1976), but have not been reported for other xylem-limited bacteria. Constriction furrows in the cell wall that divide the cytoplasm have been interpreted as evidence that the bacteria reproduce by binary fission (Hopkins, 1977; Lowe, Nyland, and Mircetich, 1976; Mollenhauer and Hopkins, 1974). Occasionally, an electron-lucent zone has been observed surrounding individual xylem-limited bacteria (Davis, 1978; Lowe, Nyland, and Mircetich, 1976; Mollenhauer and Hopkins, 1974). This zone possibly represents a capsule or an artifact of fixation for electron microscopy (Hopkins, 1977; Mollenhauer and Hopkins, 1974).

In the tracheary elements of plants, xylem-limited bacteria often form aggregates that appear to be held together by extracellular material produced by the bacteria (Mollenhauer and Hopkins, 1974). This material has been termed fibrous forms (Mollenhauer and Hopkins, 1974), "osmophilic lines" or microfibrils (Lowe, Nyland, and Mircetich, 1976), and electron-dense strands (French, Christis, and Stassi, 1977; Nyland et al., 1973). Often, the fibrous strands are seen more abundantly at the ends of the bacteria (Fig. 2C) (French, Christis, and Stassi, 1977; French and Kitajima, 1978; Lowe, Nyland, and Mircetich, 1976). Negatively stained preparations of the phony-peach bacterium revealed filaments (French, Christis, and Stassi, 1977), which may be analogous to either the acid polysaccharide coat (Costerton, Geesey, and Cheng, 1978) or to fimbriae (Ottow, 1975) observed in other bacteria. Strands possibly composed of subunits were observed either attached to the cell wall or in the extracellular environment of xylem-limited bacteria (Lowe, Nyland, and Mircetich, 1976; Nyland et al., 1973). Hopkins (1977) considered these strands to be analogous to fimbriae or possibly to sex pili in other bacteria. These strands were 28–30 nm in diameter and of undetermined length; their width was approximately three times the usual width of fimbriae. Similar appendages have been observed on the cell wall of a bacterial symbiote in the pharyngeal diverticulum of the olive fly (Poinar, Hess, and Tsitsipis, 1975).

The PD–ALS bacterium was only recently isolated (Davis, Purcell, and Thomson, 1978) and has not been characterized taxonomically. Compared with most plant pathogenic bacteria, the bacterium grows slowly in liquid shake cultures, having a doubling time of approximately 9 h (Davis, 1978). The colonies of freshly isolated bacteria are often very viscid, become almost granular with age, and do not

disperse easily in water or liquid medium. Even if dispersed, the bacteria tend to agglutinate, forming visible clumps. The PD–ALS bacterium is Gram negative, oxidase negative, and catalase positive, and no evidence of motility has been observed (Davis, Purcell, and Thomson, 1978). The bacterium will not grow anaerobically in an atmosphere of 5% carbon dioxide in hydrogen. Growth is stimulated by succinate and citrate but not glucose (Davis, 1978). Growth on all media tested is accompanied by increased alkalinity.

The speculation that Pierce's disease of grapevines was caused by a so-called rickettsia-like bacterium (Goheen, Nyland, and Lowe, 1973; Hopkins and Mollenhauer, 1973) led, in part, to the first successful isolation of the organism (Davis, Purcell, and Thomson, 1978). The culture medium utilized hemin chloride and bovine serum albumin. These substances are important components of the medium developed by Myers, Cutler, and Wissenman (1969) for the cultivation of *Rochalimaea quintana,* the etiological agent of trench fever and one of the few species of the Rickettsiaceae to be grown in pure culture. The PD–ALS bacterium and *R. quintana* have a number of other attributes in common: the enhancement of growth by a detoxicant, such as bovine serum albumin, starch, or activated charcoal (Myers, Cutler, and Wissenman, 1969); the preference of organic acids to glucose (Huang, 1967); the requirement for an insect vector (Hewitt et al., 1946; Ito and Vinson, 1965); the size and shape of the Gram-negative bacteria (Ito and Vinson, 1965); and the small colonies in culture. However, the G+C content of *R. quintana* is 38.6 mol% (Tyeryar et al., 1973), in comparison with that of the PD–ALS bacterium, which is approximately 53.1 mol% (J. Loper and M. J. Davis, unpublished data). Thus, although these animal and plant pathogens share properties, they are genetically diverse organisms.

Although transmission electron microscopy of ultrathin sections of plant material has been useful in demonstrating the association of xylem-limited bacteria with plant diseases, it is not practical for routine diagnosis of a large number of specimens and is limited in its effectiveness for comparing different xylem-limited bacteria. Methods have been developed for visual diagnosis of xylem-limited bacteria by light microscope examination of extracts from diseased plants (French, 1974; French, Christis, and Stassi, 1977; Gillaspie, Davis, and Worley, 1973; McCoy et al., 1978). These techniques are simple but lack sensitivity and specificity.

Relatively sensitive and specific serological techniques have been used to identify xylem-limited bacteria. Antisera to the bacteria were produced against bacterial antigens extracted from diseased plants (Davis, Purcell, and Thomson, 1978; French, Stassi, and Schaad, 1978; Gillaspie, 1978a), or against antigens obtained from the PD–ALS bacterium in pure cultures (Davis, Purcell, and Thomson, 1978; Davis et al., 1979; Hopkins, 1978; Lee et al., 1978; Raju et al., 1978). No serological relationship has been shown between the ratoon stunt-associated bacterium and species in the genera *Corynebacterium, Erwinia, Mycobacterium, Agrobacterium, Xanthomonas, Actinomyces, Streptomyces,* or the PD–ALS bacterium by microagglutination or by indirect fluorescent-antibody staining methods (Gillaspie, 1978b; Harris and Gillaspie, 1978; A. G. Gillaspie and M. J. Davis, unpublished data). The xylem-limited bacterium associated with phony disease of peach has also been compared to other bacteria by indirect fluorescent-antibody staining. No serological relationship between the phony peach-associated bacterium and species in the genera *Pseudomonas, Erwinia, Xanthomonas, Corynebacterium, Agrobacterium,* and *Bacillus* were observed (Davis et al., 1979; French, Stassi, and Schaad, 1978); but serological relationships were shown by this method between the phony peach-associated bacterium and other xylem-limited bacteria, including the PD–ALS bacterium (Davis et al., 1979), the periwinkle-associated bacterium (McCoy et al, 1978), and the plum leaf scald-associated bacterium (French and Kitajima, 1978). Using the direct fluorescent-antibody staining method, Lee et al. (1978) found bacteria serologically related to the PD–ALS bacterium in roots of citrus trees that showed symptoms of young tree decline. Gel double-diffusion analyses demonstrated that the PD–ALS bacteria from grapevines, almonds, and alfalfa in California and from grapevines in Florida and Costa Rica were serologically indistinguishable (Davis et al., 1979; Hopkins, 1978; Raju et al., 1978), but slight differences were observed between the PD–ALS bacterium and the phony peach-associated organism (Davis et al., 1979).

The various serological analyses suggest that the rippled-walled, xylem-limited bacteria form a serologically and probably taxonomically distinct group of organisms. At least two strains of these bacteria occur, as indicated by their cultivability on existing media; but fluorescent-antibody staining does not differentiate these two groups. Although gel double-diffusion permits a more refined analysis based upon demonstration of a spectrum of distinct antigens, it is difficult to use this technique when antigens from noncultivable bacteria are involved.

Chemotherapy and heat therapy have provided corroborative evidence for the role of xylem-limited bacteria in the etiology of certain plant diseases. For example, heat treatment at temperatures that would not inactivate most viruses was effective in curing grapevines of Pierce's disease (Goheen, Nyland, and Lowe, 1973), and has also been used to cure sugarcane affected with ratoon stunting disease (Steindl, 1961). Also, tetracycline antibiotics caused remission of Pierce's disease symptoms in grape-

vines (Hopkins and Mortensen, 1971). In culture, the PD–ALS bacterium is susceptible to bacitracin, chloramphenicol, carbenicillin, doxycycline, gentamicin, kanamycin, methacycline, nitrofurantoin, oxytetracycline, rifampin, and tetracycline; but it is not susceptible to ampicillin, colistin, erythromycin, lincomycin, nafcillin, nalidixic acid, oleandomycin, oxacillin, penicillin, streptomycin, triple sulfa, and vancomycin at levels normally used for testing with antibiotic-sensitivity disks (M. J. Davis, unpublished data). Treatment of plants or propagative material with tetracycline derivatives or penicillin did not eliminate the ratoon stunting agent, and reduced symptom severity only when antibiotic concentrations induced a phytotoxic response (Gillaspie and Blizzard, 1976). Treatment of infectious juice from diseased canes with streptomycin, tetracycline, chlortetracycline, or penicillin had no effect on infectivity (Gillaspie and Blizzard, 1976; Teakle, 1974). These data are generally interpreted as evidence that the ratoon stunting disease agent is insensitive to these antibiotics in vivo but may only indicate that the antibiotics are not sufficiently reaching the bacteria.

Phloem-Limited Bacteria

Laflèche and Bové (1970) were the first to recognize a group of phloem-limited plant pathogens whose limiting structure was more complex than that of the wall-less prokaryotes. Their studies, discussed also by Saglio et al. (1972) and Laflèche and Bové (1970), clearly showed that the limiting structure of the citrus greening pathogen consisted of a double membrane and, therefore, differed fundamentally in its structure from that of the wall-less agent that was eventually cultivated, named *Spiroplasma citri* (Saglio et al., 1973) and shown to incite citrus stubborn disease (Markham, Townsend, and Plaskitt, 1974).

Although a single preliminary report for cultivation of a bacterium from yellows-diseased grapevines in chick embryos has been made (Nienhaus, Rumbos, and Greuel, 1978), there is at present no technique for continuous cultivation of these agents in vitro. Because of these failures in cultivation, there is no clear taxonomic concept. However, the organisms appear to have several common structural features, which may indicate a common phylogeny and eventual recognition as a single taxon. The organisms are elongate, 1.3×0.2–0.5 μm, with the usual ribosomal and diffuse nuclear areas characteristic of prokaryotic cells (Fig. 4). They are bounded by a "wavy" structure that consists, in part, of two single membranes ($\sim$8 nm wide). Presence of cell wall constituents is suggested by the total width of the limiting structure (20–32 nm), and by suscepti-

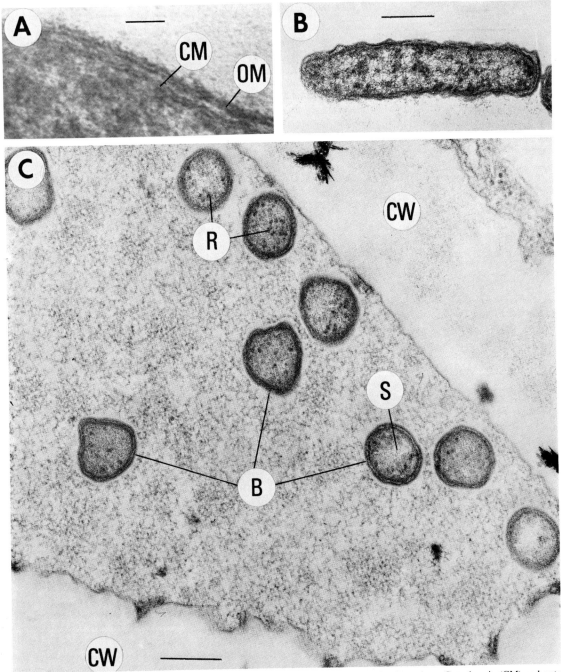

Fig. 4. Fastidious phloem bacteria. (A) Detail from thin section of cell of citrus greening bacterium. Cytoplasmic (CM) and outer membranes (OM) are present. Bar = 40 nm. [Courtesy of J. M. Bové.] (B) Thin section of bacterium from phloem of white clover (*Trifolium repens*). Cells are elongate with wavy contour. Bar = 0.2 μm. [Courtesy of P. G. Markham.] (C) Cells of clover club leaf (B) in transverse section of phloem cell of periwinkle (*Catharanthus* [=*Vinca*] *roseus*) shoot. The cells contain ribosome-like particles (R) and fine strands (S) representing DNA. CW, cell wall. Bar = 0.2 μm. [Windsor and Black, 1973b, with permission.]

bility of the organisms to penicillin derivatives. It is generally agreed that these phloem organisms are similar to Gram-negative bacteria, although their dimensions are certainly at the lower end of the size range. Of special significance is the absence of an R-layer (Moll and Martin, 1974), which is characteristic of the Gram-negative cell wall (Costerton, Ingram, and Cheng, 1974). Although the organisms have been referred to as "rickettsia-like" by many authors, there is no direct analogy to members of the Rickettsiaceae; for this and other reasons, Moll and Martin (1974) preferred to refer to the organisms ' "bacteria-like", a term that is now gaining acc' ance (Klein, Zimmerman-Gries, and Sneh, '

Some of the organisms are so small that they are visible by phase-contrast, but not by conventional bright-field light microscopy (Liu and Black, 1974). Under phase-contrast microscopy, the organisms exhibited an undulatory motility in 30% glycerol that ceased upon addition of mercuric chloride or potassium penicillin G (Liu and Black, 1974). Thus, sensitivity to inhibition of cell wall biosynthesis by penicillin has been shown both in vitro and in vivo (Windsor and Black, 1973a; Ulrychová et al., 1975). The in vivo sensitivity of the organisms to tetracycline derivatives has also been established (Schwarz and Van Vuuren, 1971; Schwarz, Moll, and Van Vuuren, 1974; Su and Chang, 1976).

Liu and Black (1974) were able to partially purify the clover club leaf agent, prepare an antiserum that agglutinated organisms to a titer of 1/1024, and stain the organism specifically in leafhopper tissue in fluorescent-antibody tests. Such techniques may eventually demonstrate serological relationships even in the absence of successful cultivation of the causal agents.

Concluding Remarks

A tacit assumption of many plant pathologists for many years was that the list of categories of etiological agents of plant disease was complete. After the discovery, in 1967, that wall-less prokaryotes (mycoplasmas) were associated with plant disease, it was natural to speculate that more groups of microorganisms might be found to cause plant disease. These questions were raised by Davis and Whitcomb (1971), who discussed the possible roles of chlamydiae, rickettsiae, and other groups as putative plant pathogens. The subsequent overemphasis placed upon rickettsiae as models for etiological agents of plant disease may have originated with such speculations. Nevertheless, nature is not to be predicted, and thus it was that flagellated protozoans, having not entered into these earlier speculations, were the next major group of microorganisms to be associated with plant disease (Parthasarathy, van Slobbe, and Soutant, 1976). Perhaps under such circumstances, given the fact that fastidious, walled prokaryotes inhabit both phloem and xylem, the biggest surprise has been that the organisms currently recognized show signs of representing single taxa, or groups of related organisms, rather than a plethora of new taxonomic types. For example, many of the emerging strains of xylem bacteria appear to be serologically related to each other, but not to existing taxa. Without doubt, future microbiological discoveries in these unusual habitats will be as exciting and unpredictable as past discoveries have been.

Addendum

After this chapter was written, the coryneform bacterium causing ratoon stunting disease (RSD) of sugarcane was isolated in axenic culture for the first time by M. J. Davis, A. G. Gillaspie, Jr., R. W. Harris, and R. H. Lawson (unpublished). On SC medium, which was developed for the RSD bacterium, colonies measure 0.1–0.3 mm in diameter after 10–14 days of aerobic incubation at 30°C. The colonies are circular with entire margins and nonpigmented. The RSD bacterium is Gram positive, oxidase negative, catalase positive, and non-acid-fast.

SC Medium for Isolating Ratoon Stunting Disease Bacteria

The medium contains (in g/liter):

Corn meal agar (BBL)	17 g
Phytone peptone (BBL)	8 g
K_2HPO_4	1 g
KH_2PO_4	1 g
$MgSO_4 \cdot 7H_2O$	0.2 g
Hemin chloride stock (0.1% hemin chloride dissolved in 0.05 N NaOH)	15 ml
Bovine serum albumin fraction five (BSA-5) solution (20% solution in water)	10 ml
Glucose (50% solution in water)	1 ml
Cysteine (free base, 10% solution in water)	10 ml

All constituents except the BSA-5, glucose, and cysteine solutions are autoclaved in 1,000 ml of distilled and deionized water. Ingredients that are not autoclaved are filter-sterilized and added to the autoclaved portion at 45–50°C. The pH is adjusted to 6.6 before sterilization.

The SC medium was used to isolate the RSD bacterium from sugarcane from the United States, Brazil, South Africa, and Japan, and the isolates readily incited RSD symptoms in susceptible sugarcane cultivars after inoculation. In addition, a bacterium from Bermuda grass from Taiwan, which morphologically, ultrastructurally, and serologically resembles the RSD bacterium, was isolated on the SC medium. The Bermuda grass bacterium incited stunting of Bermuda grass but not sugarcane. Although their cultural requirements are similar, the Bermuda grass strains differ in culture from the sugarcane strains in that they grow faster and produce a yellow, nondiffusible pigment. The sugarcane and Bermuda grass strains are apparently newly isolated members of the plant pathogenic group of coryneform bacteria (this Handbook, Chapter 143).

Literature Cited

Auger, J., Mircetich, S. M., Nyland, G. 1974. Interrelation between bacteria causing Pierce's disease of grapevine and almond leaf scorch. Proceedings of the American Phytopathological Society 1:90.

Auger, J. G., Shalla, T. A., Kado, C. I. 1974. Pierce's disease of grapevines: Evidence for a bacterial etiology. Science 184:1375–1377.

Behncken, G. M., Gowanlock, D. H. 1976. Association of a bacterium-like organism with rugose leaf curl disease of clovers. Australian Journal of Biological Science 29:137–146.

Benhamou, N. 1978. Recherches sur les rickettsoides phytopathogenes et leurs associations dans le cadre de maladies complexes. Thesis. Académie de Montpellier, Université des Sciences et Techniques du Languedoc.

Benhamou, N., Giannotti, J., Louis, C. 1978. Transmission de germes de type rickettsoide par la plante parasite: Cuscuta subinclusa Durr. & Hilg. Acta Phytopathologica Academiae Scientiarum Hungaricae 13:107–119.

Benhamou, N., Louis, C., Giannotti, J. 1979. Etude par cytochimie ultrastructurale des rickettsoides et des cellules phloemiques de trèfles dépérissants. Annales de Phytopathologie 11: 1–16.

Black, L. M. 1944. Some viruses transmitted by agallian leafhoppers. Proceedings of the American Philosophical Society 88: 132–144.

Black, L. M. 1948. Transmission of clover club-leaf virus through the egg of its insect vector. Phytopathology 38:2.

Black, L. M. 1950. A plant virus that multiplies in its insect vector. Nature 166:852–853.

Bollard, E. G. 1960. Transport in the xylem. Annual Review of Plant Physiology 11:141–166.

Brooker, B. E., Fuller, R. 1975. Adhesion of lactobacilli to the chicken crop epithelium. Journal of Ultrastructure Research 52:21–31.

Büchner, P. 1965. Endosymbiosis of animals with plant microorganisms. New York: Wiley, Interscience.

Burrill, T. J. 1881. Bacteria as a cause of disease in plants. American Naturalist 15:527–531.

Capoor, S. P., Rao, D. G., Viswanath, S. M. 1974. Greening disease of citrus in the Deccan Trap Country and its relationship with the vector Diaphorina citri Kuwayama, pp. 43–49. In: Weathers, L. G., Cohen, M. (eds.), Proceedings of the Sixth Conference of the International Organization of Citrus Virologists. Berkeley: University of California Press.

Catling, H. D., Atkinson, P. R. 1974. Spread of greening by Trioza erytreae (Del Guercio) in Swaziland, pp. 33–39. In: Weathers, L. G., Cohen, M. (eds.), Proceedings of the Sixth Conference of the International Organization of Citrus Virologists, Berkeley: University of California Press.

Chang, K. P. 1981. Nutritional roles of intracellular symbiotes in insects and protozoa. In: CRC handbook of nutrition and food. Boca Raton, Florida: CRC Press. (In press.)

Chang, K. P., Musgrave, A. J. 1975. Conversion of spheroplast symbiotes in a leafhopper, Helochara communis Fitch (Cicadellidae: Homoptera). Canadian Journal of Microbiology 21:196–204.

Costerton, J. W., Geesey, G. G., Cheng, K.-J. 1978. How bacteria stick. Scientific American 238:86–95.

Costerton, J. W., Ingram, J. M. Cheng, K.-J. 1974. Structure and function of the cell envelope of Gram-negative bacteria. Bacteriological Reviews 38:87–110.

Davis, M. J. 1978. Pierce's disease and almond leaf scorch disease: Isolation and pathogenicity of the causal bacterium. Ph.D. Thesis. University of California, Berkeley, California.

Davis, M. J., Purcell, A. H., Thomson, S. V. 1978. Pierce's disease of grapevines: Isolation of the causal bacterium. Science 199:75–77.

Davis, M. J., Thomson, S. V., Purcell, A. H. 1978. Pathological

and serological relationship of the bacterium causing Pierce's disease of grapevines and almond leaf scorch disease. Third International Congress of Plant Pathology, Munich, pp. 64. Abstract.

Davis, M. J., Stassi, D. L., French, W. J., Thomson, S. V. 1979. Antigenic relationship of several rickettsia-like bacteria involved in plant diseases, pp. 311–315. In: Station de Pathologie Végétale et Phytobactériologie (ed.), Proccedings of the IVth International Conference on Plant Pathogenic Bacteria. Angers: Institut National de la Recherche Agronomique.

Davis, R. E., Whitcomb, R. F. 1971. Mycoplasmas, rickettsiae, and chlamydiae: Possible relation to yellows diseases and other disorders of plants and insects. Annual Review of Phytopathology 9:119–154.

Doi, Y., Teranaka, M., Yora, K., Asuyama, H. 1967. Mycoplasma- or PLT group–like microorganisms found in the phloem elements of plants infected with mulberry dwarf, potato witches' broom, aster yellows, or paulownia witches' broom. [In Japanese, with English abstract.] Annals of the Phytopathological Society of Japan 33:259–266.

Ehrman, L., Kernaghan, R. P. 1972. Infectious heredity in Drosophila paulistorum, pp. 227–246. Ciba Foundation Symposium: Pathogenic Mycoplasmas. Amsterdam, London, New York: American Scientific Publishers.

Esau, K. 1965. Plant anatomy. New York, London, Sydney: John Wiley & Sons.

Feldman, A. W., Hanks, R. W., Good, G. E., Brown, G. E. 1977. Occurrence of a bacterium in YTD-affected as well as in some apparently healthy citrus trees. Plant Disease Reporter 61:546–550.

Fife, J. M., Price, C., Fife, D. C. 1962. Some properties of phloem exudate collected from root of sugar beet. Plant Physiology 37:791–792.

Frazier, N. W. 1966. Xylem viruses and their insect vectors, pp. 91–99. In: Proceedings, International Conference on Virus and Vector on Perennial Hosts, with special reference to Vitis, September 6–10, 1965. Davis: University of California, Division of Agricultural Sciences.

Freitag, J. H. 1951. Host range of the Pierce's disease virus of grapes as determined by insect transmission. Phytopathology 41:920–934.

French, W. J. 1974. A method for observing rickettsia-like bacteria associated with phony peach disease. Phytopathology 64:260–261.

French, W. J., Christis, R. G., Stassi, D. L. 1977. Recovery of rickettsia-like bacteria by vacuum infiltration of peach tissues affected with phony disease. Phytopathology 67:945–948.

French, W. J., Kitajima, E. W. 1978. Occurrence of plum leaf scald in Brazil and Paraguay. Plant Disease Reporter 62:1035–1038.

French, W. J., Latham, A. J., Stassi, D. L. 1977. Phony peach bacterium associated with leaf scald of plum trees. Proceedings of the American Phytopathological Society 4:223.

French, W. J., Stassi, D. L., Schaad, N. W. 1978. The use of immunofluorescence for the identification of phony peach bacterium. Phytopathology 68:1106–1108.

Gillaspie, A. G., Jr. 1978a. Ratoon stunting disease of sugarcane: Serology. Phytopathology 68:529–532.

Gillaspie, A. G., Jr. 1978b. An antiserum to the ratoon stunting disease-associated bacterium. Sugarcane Pathologists' Newsletter 20:5.

Gillaspie, A. G., Jr., Blizzard, J. W. 1976. Some properties of the ratoon stunting disease agent of sugarcane. Sugarcane Pathologists' Newsletter 15/16:34–36.

Gillaspie, A. G., Jr., Davis, R. E., Worley, J. F. 1973. Diagnosis of ratoon stunting disease based on the presence of a specific microorganism. Plant Disease Reporter 57:987–990.

Gillaspie, A. G., Jr., Davis, R. E., Worley, J. F. 1976. Nature of the bacterium associated with ratoon stunting disease of sugarcane. Sugarcane Pathologists' Newsletter 15/16:11–15.

Goheen, A. C., Nyland, G., Lowe, S. K. 1973. Association of a rickettsialike organism with Pierce's disease of grapevines and alfalfa dwarf and heat therapy of the disease in grapevines. Phytopathology 63:341–345.

Griffiths, G. W., Beck, S. D. 1973. Intracellular symbiotes of the pea aphid, *Acyrthosiphon pisum*. Journal of Insect Physiology 19:75–84.

Griffiths, G. W., Beck, S. D. 1977. *In vivo* sterol biosynthesis by pea aphid symbiotes as determined by digitonin and electron microscopic autoradiography. Cell and Tissue Research 176:179–190.

Grylls, N. E. 1954. Rugose leaf curl—a new virus disease transovarially transmitted by the leafhopper *Austroagallia torrida*. Australian Journal of Biological Science 7:47–58.

Harris, R. W., Gillaspie, A. G., Jr. 1978. Immunofluorescent diagnosis of ratoon stunting disease. Plant Disease Reporter 62:193–196.

Hewitt, W. B. 1970. Pierce's disease of *Vitis* species, pp. 196–200. In: Frazier, N. W. (ed.), Virus diseases of small fruits and grapevines. Berkeley: University of California Press.

Hewitt, W. B., Houston, B. R., Frazier, N. W., Freitag, J. H. 1946. Leafhopper transmission of the virus causing Pierce's disease of grape and dwarf of alfalfa. Phytopathology 36:117–128.

Hinde, R. 1971. The fine structure of the mycetome symbiotes of the aphids *Brevicoryne brassiae*, *Myzus persicae* and *Macrosiphum rosae*. Journal of Insect Physiology 17:2035–2050.

Hirumi, H., Kimura, M., Maramorosch, K., Bird, J., Woodbury, R. 1974. Rickettsia-like organisms in the phloem of little leaf-diseased *Sida cordifolia*. Phytopathology 64:581–582.

Holmes, F. O., Hirumi, H., Maramorosch, K. 1972. Witches' broom of willow: *Salix* yellows. Phytopathology 62:826–828.

Hopkins, D. L. 1977. Diseases caused by leafhopper-borne, rickettsia-like bacteria. Annual Review of Phytopathology 15:277–294.

Hopkins, D. L. 1978. Comparisons of Florida isolates of the Pierce's disease bacterium. Third International Congress of Plant Pathology, Munich, pp. 62. Abstract.

Hopkins, D. L., Adlerz, W. C., Bistline, F. W. 1978. Pierce's disease bacterium occurs in citrus trees affected with blight (young tree decline). Plant Disease Reporter 62:442–445.

Hopkins, D. L., Mollenhauer, H. H. 1973. Rickettsia-like bacterium associated with Pierce's disease of grapes. Science 179:298–300.

Hopkins, D. L., Mollenhauer, H. H., French, W. J. 1973. Occurrence of a rickettsia-like bacterium in the xylem of peach trees with phony disease. Phytopathology 63:1422–1423.

Hopkins, D. L., Mortensen, J. A. 1971. Suppression of Pierce's disease symptoms by tetracycline antibiotics. Plant Disease Reporter 55:610–612.

Houk, E. J., Griffiths, G. W., Beck, S. D. 1976. Lipid metabolism in the symbiotes of the pea aphid *Acyrthosiphon pisum*. Comparative Biochemistry and Physiology 52B:427–431.

Houk, E. J., McLean, D. L. 1974. Isolation of the primary intracellular symbiote of the pea aphid, *Acyrthosiphon pisum*. Journal of Invertebrate Pathology 23:237–241.

Houk, E. J., Griffiths, G. W., Hadjokas, N. E., Beck, S. D. 1977. Peptidoglycan in the cell wall of the primary intracellular symbiote of the pea aphid. Science 198:401–403.

Huang, K. 1967. Metabolic activity of the trench fever rickettsia, *Rickettsia quintana*. Journal of Bacteriology 93:853–859.

Hutchins, L. M., Cochran, L. C., Turner, W. F., Weinberger, J. H. 1953. Transmission of phony disease virus from tops of certain affected peach and plum trees. Phytopathology 43:691–696.

Ishiie, T., Doi, Y., Yora, K., Asuyama, H. 1967. Suppressive effects of antibiotics of tetracycline group on symptom development in mulberry dwarf disease. [In Japanese, with English

abstract.] Annals of the Phytopathological Society of Japan 33:267–275.

Ito, S., Vinson, J. W. 1965. Fine structure of *Rickettsia quintana* cultivated in vitro and in the louse. Journal of Bacteriology 89:481–495.

Kaloostian, G. H., Pollard, H. N., Turner, W. F. 1962. Leafhopper vectors of Pierce's disease virus in Georgia. Plant Disease Reporter 46:292.

Kamiunten, H., Wakimoto, S. 1976. Coryneform bacteria found in xylem of the ratoon stunting diseased sugarcane. [In Japanese with English abstract.] Annals of the Phytopathological Society of Japan 42:500–503.

Kao, J., Damann, K. E., Jr. 1978. Microcolonies of the bacterium associated with ratoon stunting disease found in sugarcane xylem matrix. Phytopathology 68:545–551.

Kitajima, E. W., Bakarcic, M., Fernandez-Valiela, M. V. 1975. Association of rickettsialike bacteria with plum leaf scald disease. Phytopathology 65:476–479.

Klein, M., Zimmerman-Gries, S., Sneh, B. 1976. Association of bacterialike organisms with a new potato disease. Phytopathology 66:564–569.

Körner, H. K. 1969. Zur Ultrastruktur der intrazellulären Symbioten in Embryo der Kleinzikade *Euscelis plebejus* Fall. (Homoptera, Cicadina). Zeitschrift für Zellforschung und Mikroskopische Anatomie 100:466–473.

Küppers, P., Neinhaus, F., Schinzer, U. 1975. Rickettsia-like organisms and virus-like structures in a yellows disease of grapevines. Zeitschrift für Pflanzenkrankheiten und Pflanzenschutz 82:183–187.

Laflèche, D., Bové, J. M. 1970. Structures de type mycoplasme dans les feuilles d'orangers atteints de la maladie du "Greening". Comptes Rendus Hebdomadaires des Séances de l'Académie des Sciences, Série D 270:1915–1917.

Latham, M. J., Brooker, B. E., Pettipher, G. L., Harris, P. J. 1978. *Ruminococcus flavefaciens* cell coat and adhesion to cotton cellulose and to cell walls in leaves of perennial ryegrass (*Lolium perenne*). Applied and Environmental Microbiology 35:156–165.

Latorre-Guzmán, B. A., Kado, C. I., Kunkee, R. E. 1977. *Lactobacillus hordniae*, a new species from the leafhopper (*Hordnia circellata*). International Journal of Systematic Bacteriology 27:362–370.

Lee, R. F., Feldman, A. W., Raju, B. C., Nyland, G., Goheen, A. C. 1978. Immunofluorescence for detection of rickettsialike bacteria in grape affected by Pierce's disease, almond affected by almond leaf scorch, and citrus affected by young tree decline. Phytopathology News 12:265.

Liu, H.-Y., Black, L. M. 1974. The clover club leaf organism. Les Colloques de l'Institut National de la Santé et de la Recherche Medicale 33:97–98.

Louis, C., Laporte, M. 1969. Caractères ultrastructuraux et différenciation de formes migratrices des symbiotes chez *Euscelis plebeius* (Hom. Jassidae). Annales de la Société Entomologique de France (Nouvelle Série) 5:799–809.

Lowe, S. K., Nyland, G., Mircetich, S. M. 1976. The ultrastructure of the almond leaf scorch bacterium with special reference to topography of the cell wall. Phytopathology 66:147–151.

McClean, A. P. D., Oberholzer, P. C. J. 1965. Citrus psylla, a vector of the greening disease of sweet orange. South African Journal of Agricultural Science 8:297–298.

McCowan, R. P., Cheng, K.-J., Bailey, C. B. M., Costerton, J. W. 1978. Adhesion of bacteria to epithelial cell surfaces within the reticulo-rumen of cattle. Applied and Environmental Microbiology 35:149–155.

McCoy, R. E. 1979. Mycoplasmas and yellows disease, pp. 227–262. In: Barile, M. F., Razin, S., Tully, J. G., Whitcomb, R. F. (eds.), The mycoplasmas, vol. 3. New York: Academic Press.

McCoy, R. E., Thomas, D. L., Tsai, J. H., French, W. J. 1978. Periwinkle wilt, a new disease associated with xylem delimited rickettsialike bacteria transmitted by a sharpshooter. Plant Disease Reporter 62:1022–1026.

Maramorosch, K., Plavsic-Banjac, B., Bird, J., Liu, L. J. 1973. Electron microscopy of ratoon stunted sugarcane: Microorganisms in xylem. Phytopathologische Zeitschrift 77:270–273.

Maramorosch, K., Hirumi, H., Kimura, M., Bird, J. 1975. Mollicutes and rickettsia-like plant disease agents (zoophytomicrobes) in insects. Annals of the New York Academy of Sciences 266:276–292.

Markham, P. G., Townsend, R., Plaskitt, K. A. 1975. A rickettsia-like organism associated with diseased white clover. Annals of Applied Biology 81:91–93.

Markham, P. G., Townsend, R., Bar-Joseph, M., Daniels, M. J., Plaskitt, A., Meddins, B. M. 1974. Spiroplasmas are the causal agents of citrus little-leaf disease. Annals of Applied Biology 78:49–57.

Mircetich, S. M., Lowe, S. K., Moller, W. J., Nyland G. 1976. Etiology of almond leaf scorch disease and transmission of the causal agent. Phytopathology 66:17–24.

Moll, J. N., Martin, M. M. 1973. Electron microscope evidence that citrus psylla (Trioza erytreae) is a vector of greening disease in South Africa. Phytophylactica 5:41–44.

Moll, J. N., Martin, M. M. 1974. Comparison of the organism causing greening disease with several plant pathogenic Gram negative bacteria, rickettsia-like organisms and mycoplasma-like organisms. Les Colloques de l'Institut National de la Santé et de la Recherche Medicale 33:89–96.

Mollenhauer, H. H., Hopkins, D. L. 1974. Ultrastructural study of Pierce's disease bacterium in grape xylem tissue. Journal of Bacteriology 119:612–618.

Myers, W. F., Cutler, L. D., Wissenman, C. L., Jr. 1969. Role of erythrocytes and serum in the nutrition of Rickettsia quintana. Journal of Bacteriology 97:663–666.

Nelson, P. E., Dickey, R. S. 1970. Histopathology of plants infected with vascular bacterial pathogens. Annual Review of Phytopathology 8:259–280.

Nienhaus, F., Brüssel, H., Schinzer, U. 1976. Soil-borne transmission of rickettsia-like organisms found in stunted and witches' broom diseased larch trees (Larix decidua). Zeitschrift für Pflanzenkrankheiten und Pflanzenschutz 83:309–316.

Nienhaus, F., Rumbos, I., Greuel, E. 1978. First results in the cultivation of rickettsia-like organisms of yellows diseased grapevines in chick embryos. Zeitschrift für Pflanzenkrankheiten und Pflanzenschutz 85:113–117.

Nienhaus, F., Schmutterer, H. 1976. Rickettsialike organisms in latent rosette (witches' broom) diseased sugar beets (Beta vulgaris) and spinach (Spinacia oleracea) plants and in the vector Piesma quadratum Fieb. Zeitschrift für Pflanzenkrankheiten und Pflanzenschutz 83:641–646.

Nienhaus, F., Sikora, R. A. 1979. Mycoplasmas, spiroplasmas, and rickettsia-like organisms as plant pathogens. Annual Review of Phytopathology 17:37–58.

Nyland, G., Goheen, A. C., Lowe, S. K., Kirkpatrick, H. C. 1973. The ultrastructure of a rickettsialike organism from a peach tree affected with phony disease. Phytopathology 63:1275–1278.

Ottow, J. C. G. 1975. Ecology, physiology, and genetics of fimbriae and pili. Annual Review of Microbiology 29:79–108.

Parthasarathy, M. V., van Slobbe, W. B., Soutant, C. 1976. Trypanosomatid flagellate in the phloem of diseased coconut palms. Science 192:1346–1348.

Pate, J. S. 1976. Nutrients and metabolites of fluids recovered from xylem and phloem: Significance in relation to long distance transport in plants, pp. 253–281. In: Wardlaw, I. F., Passiours, J. B. (eds.), Transport and transfer processes in plants. New York: Academic Press.

Petzold, H., Marwitz, R., Kunze, L. 1973. Elektronenmikroskopische Untersuchungen über intrazelluläre rickett-

sienähnliche Bakterien in triebsuchtkranken Äpfeln. Phytopathologische Zeitschrift 78:170–181.

Poinar, G. O., Jr., Hess, R. T., Tsitsipis, J. A. 1975. Ultrastructure of the bacterial symbiotes in the pharyngeal diverticulum of Dacus oleae (Gmelin) (Trypetidae: Diptera). Acta Zoologica 56:77–84.

Poos, F. W., Elliott, C. 1936. Certain insect vectors of Aplanobacter stewarti. Journal of Agricultural Research 52:585–608.

Purcell, A. H. 1977. Cold therapy of Pierce's disease of grapevines. Plant Disease Reporter 61:514–518.

Purcell, A. H. 1979. Leafhopper vectors of xylem-borne plant pathogens, pp. 603–627. In: Maramorosch, K., Harris, K. F. (eds.), Leafhopper vectors and plant disease agents. New York: Academic Press.

Purcell, A. H., Finlay, A. 1979. Evidence for noncirculative transmission of Pierce's disease bacterium by sharpshooter leafhoppers. Phytopathology 69:393–395.

Purcell, A. H., Latorre-Guzmán, B. A., Kado, C. I., Goheen, A. C., Shalla, T. A. 1977. Reinvestigation of the role of a lactobacillus associated with leafhopper vectors of Pierce's disease of grapevines. Phytopathology 67:298–301.

Raju, B. C., Goheen, A. C., Lowe, S. K., Nyland, G. 1978. Pierce's disease of grapevines in Central America. Phytopathology News 12:267.

Rumbos, I., Sikora, R. A., Nienhaus, F. 1977. Rickettsia-like organisms in Xiphinema index Thorne & Allen found associated with yellows disease of grapevines. Zeitschrift für Pflanzenkrankheiten und Pflanzenschutz 84:240–243.

Saglio, P. H. M., Whitcomb, R. F. 1979. Diversity of wall-less prokaryotes in plant vascular tissue, fungi and invertebrate animals, pp. 1–36. In: Barile, M. F., Razin, S., Tully, J. G., Whitcomb, R. F. (eds.), The mycoplasmas, vol. 3. New York: Academic Press.

Saglio, P., Laflèche, D., L'Hospital, M., Dupont, G., Bové, J.-M. 1972. Isolation and growth of citrus mycoplasmas, pp. 187–198. Ciba Foundation Symposium: Pathogenic Mycoplasmas. Amsterdam, London, New York: American Scientific Publishers.

Saglio, P., L'Hospital, M., Laflèche, D., Dupont, G., Bové, J. M., Tully, J. G., Freundt, E. A. 1973. Spiroplasma citri gen. and sp. n.: A mycoplasma-like organism associated with "stubborn" disease of citrus. International Journal of Systematic Bacteriology 23:191–204.

Schwarz, R. E., Moll, J. N., Van Vuuren, S. P. 1974. Control of citrus greening and its psylla vector by trunk injections of tetracyclines and insecticides, pp. 26–29. In: Weathers, L. G., Cohen, M. (eds.), Proceedings of the Sixth Conference of the International Organization of Citrus Virologists. Berkeley: University of California Press.

Schwarz, R. E., Van Vuuren, S. P. 1971. Decrease in fruit greening of sweet orange by trunk injection of tetracyclines. Plant Disease Reporter 55:747–749.

Severin, H. H. P. 1949. Transmission of the virus of Pierce's disease of grapevines by leafhoppers. Hilgardia 19:190–206.

Severin, H. H. P. 1950. Spittle-insect vectors of Pierce's disease virus. II. Life history and virus transmission. Hilgardia 19:357–382.

Sherald, J., Hearon, S. S. 1978. Bacterialike organisms associated with elm leaf scorch. Phytopathology News 12:73.

Silva, M. T., Sousa, J. C. F. 1973. Ultrastructure of the cell wall and cytoplasmic membrane of Gram-negative bacteria with different fixation techniques. Journal of Bacteriology 113:953–962.

Starr, M. P. 1975. A generalized scheme for classifying organismic associations. Symposia of the Society for Experimental Biology 29:1–20.

Steindl, D. R. L. 1961. Ratoon stunting disease, pp. 433–459. In: Martin, J. P., Abbott, E. V., Hughes, C. G. (eds.), Sugarcane diseases of the world, vol. 1. Amsterdam: Elsevier.

Su, H. J., Chang, S. C. 1976. The responses of the likubin pathogen to antibiotics and heat therapy, pp. 27–34. In: Calavan, E. C. (ed.), Proceedings of the Seventh Conference of the International Organization of Citrus Virologists. Riverside, California: International Organization of Citrus Virologists.

Teakle, D. S. 1974. The causal agent of sugarcane ratoon stunting disease (RSD). Proceedings of the International Society of Sugarcane Technology 15:225–233.

Teakle, D. S., Kontze, D., Appleton, J. M. 1979. A note on the diagnosis of ratoon stunting disease of sugarcane by negative-stain electron microscopy of the associated bacterium. Journal of Applied Bacteriology 46:279–284.

Teakle, D. S., Smith, P. M., Steindl, D. R. L. 1973. Association of a small coryneform bacterium with the ratoon stunting disease of sugarcane. Australian Journal of Agricultural Research 24:869–874.

Thomson, S. V., Davis, M. J., Kloepper, J. W., Purcell, A. H. 1978. Alfalfa dwarf: Relationship to the bacterium causing Pierce's disease of grapevines and almond leaf scorch disease. Third International Congress of Plant Pathology, Munich, pp. 65. Abstract.

Turner, W. F., Pollard, H. N. 1959. Insect transmission of phony peach disease. U.S. Department of Agriculture Technical Bulletin No. 1193. Washington, D.C.: U.S. Department of Agriculture.

Tyeryar, F. J., Jr., Weiss, E., Millar, D. B., Bozeman, F. M., Ormsbee, R. A. 1973. DNA base composition of rickettsiae. Science 180:415–417.

Ulrychová, M., Vanek, G., Jokeš, M., Klobáska, Z., Králík, D. 1975. Association of rickettsialike organisms with infectious necrosis of grapevines and remission of symptoms after penicillin treatment. Phytopathologische Zeitschrift 82:254–265.

Vinson, J. W. 1966. In vitro cultivation of the rickettsial agent of trench fever. Bulletin of the World Health Organization 35:155–164.

Weaver, L., Teakle, D. S., Hayward, A. C. 1977. Ultrastructural studies on the bacterium associated with the ratoon stunting disease of sugar-cane. Australian Journal of Agricultural Research 28:843–852.

Whitcomb, R. F., Shapiro, M., Richardson, J. 1966. An Erwinia-like bacterium pathogenic to leafhoppers. Journal of Invertebrate Pathology 8:299–307.

Whitcomb, R. F., Williamson, D. L. 1979. Pathogenicity and host-parasite relationships of mycoplasmas in arthropods. Zentralblatt für Bakteriologie, Parasitenkunde, Infektionskrankheiten und Hygiene, Abt. 1 Orig. Reihe A 245:200–220.

Windsor, I. M., Black, L. M. 1973a. Remission of symptoms of clover club leaf following treatment with penicillin. Phytopathology 63:44–46.

Windsor, I. M., Black, L. M. 1973b. Evidence that clover club leaf is caused by a rickettsia-like organism. Phytopathology 63:1139–1148.

Worley, J. F., Gillaspie, A. G., Jr. 1975. Electron microscopy in situ of the bacterium associated with ratoon stunting disease in sudangrass. Phytopathology 65:287–295.

Wormall, A. 1924. The constituents of the sap of the vine (Vitis vinifera L.) Biochemistry Journal 18:1187–1202.

Zimmerman, M. H., McDonough, J. 1978. Dysfunction of the water system, pp. 117–140. In: Horsfall, J. G., Cowling, E. B. (eds.), Plant disease, an advanced treatise, vol. 3. New York, San Francisco, London: Academic Press.

CHAPTER 164

The Hemotrophic Bacteria:
The Families Bartonellaceae and Anaplasmataceae

JULIUS P. KREIER, NICANOR DOMÍNGUEZ, HEINZ E. KRAMPITZ,
RAINER GOTHE, and MIODRAG RISTIC

The organisms of the genera included in the families Bartonellaceae and Anaplasmataceae are heterogeneous. It is probable that these organisms were first considered to be related primarily because they all parasitized red cells and because they produced diseases characterized by anemia. The additional observations that all of these organisms were best demonstrated by light microscopy after treatment with Romanovsky-type stains and that they were all quite small reinforced adherence to the grouping based on the common hemotrophism. The small size and the often intracellular site of growth of the organisms and their dependence for transmission on arthropods suggested a relationship to rickettsia (Kreier and Ristic, 1972, 1973).

The organisms of the genera *Bartonella* and *Grahamella* of the family Bartonellaceae are bacteria in the generally accepted sense of the word. They have cell walls morphologically comparable to the cell walls of other Gram-negative bacteria (Krampitz and Kleinschmidt, 1960; Takano-Moran, 1970), and despite the fact that in vivo they often grow intracellularly, they can be cultured without difficulty on nonliving culture media. In cultures, *Bartonella* may be flagellated (Peters and Wigand, 1952) but *Grahamella* is not (Krampitz and Kleinschmidt, 1960). *Bartonella* and *Grahamella* are thus small, facultatively intracellular Gram-negative bacteria. It is apparent then that these organisms do not have the attributes (Table 1) classically ascribed to rickettsia. Classically, rickettsia are defined as small, obligately intracellular bacteria which cannot be cultured outside of living host cells (Moulder, 1962).

The organisms of the genera *Anaplasma*, *Paranaplasma*, *Aegyptianella*, *Haemobartonella*, and *Eperythrozoon* of the family Anaplasmataceae differ from bacteria as classically defined in that they lack cell walls and cannot be cultured on nonliving media (Aikawa and Nussenzweig, 1972; Peters, Molyneux, and Howells, 1973; Tanaka et al., 1965). All prokaryotic protists which lack cell walls have some similarities, particularly in morphology (Anderson, 1969; Davis et al., 1973). Prokaryotic organisms lacking cell walls are, despite their morphological similarities, quite heterogeneous. Some are mycoplasmas and others may be cell wall–defective bacterial variants (L-phase variants). Some L-phase variants may revert to normal vegetative cells while others are stable. Stable L-phase variants are indistinguishable from mycoplasmas morphologically but nucleic acid homology studies show that L-phase bacteria and mycoplasmas are not closely related and that L-phase bacteria are themselves heterogeneous (Davis et al., 1973).

Organisms of the family Anaplasmataceae have not been cultured outside of their hosts. By definition mycoplasmas will grow on cell-free media. Most mycoplasmas are readily cultured and form characteristic colonies on solid culture media (Hayflick, 1969). L-phase organisms in general have the cultural requirements of the bacteria from which they are derived except that the tonicity of the culture media must be increased to prevent lysis of the cell wall–less organisms.

Organisms of the various genera of the family Anaplasmataceae are all quite similar morphologically and are similar morphologically to mycoplasmas and L-phase bacteria (Peters et al., 1973; Tanaka et al., 1965). The lack of a cell wall is the major factor in determining the organism's morphology but it may not be assumed that all organisms lacking cell walls are mycoplasmas (Davis et al., 1973).

The Anaplasmataceae occur in the plasma, and in or on red cells. Those which enter red cells do so by a phagocytic-type process. Reproduction then occurs by binary fission of the organisms within the parasitophorous vacuole (Ristic, 1968; Ristic and Watrach, 1961; Tanaka et al., 1965). The characteristic morphologies of the organisms of the various genera of the Anaplasmataceae seen in stained blood films are determined more by the relationships to the host erythrocytes, grouping of the parasites at the sites of growth, and accessory structures associated with the parasites, than by the characteristics of the organisms themselves (Kreier and Ristic, 1973).

The fact of reproduction by binary fission without morphologically specialized infectious forms clearly differentiates the organisms of the family

Table 1. Characteristics of the Anaplasmataceae.[a]

Attribute	Family Bartonellaceae		Family Anaplasmataceae				
	Bartonella bacilliformis	*Grahamella*	*Anaplasma*	*Paranaplasma*	*Aegyptianella*	*Haemobartonella*	*Eperythrozoon*
Giemsa	Intense purple	Intense purple	Intense purple	Intense purple	Intense purple	Intense purple	Pale purple
Gram stain	−	−	−	−	−	−	−
Morphology							
Shape (most common)	Rods	Rods	Cocci	Cocci	Cocci	Cocci	Easily flattened Cocci
Cell walls	+	+	−	−	−	−	−
Flagella	+	−	−	−	−	−	−
Motility	+	−	−	−	−	−	−
Size	0.25–0.5 × 1.0–3.0 μm	0.25 × 0.5–1.0 μm	0.3–0.4 μm	0.3–0.4 μm	0.3–0.8 μm	0.3–0.5 μm	0.4–1.5 μm
Position	Epierythrocytic; in endothelial cells	Intraerythrocytic	Intraerythrocytic inclusions 0.3–1.0 μm	Complex intraerythrocytic inclusions 0.3–1.0 μm	Intraerythrocytic inclusions 0.3–3.9 μm	Epierythrocytic; firmly attached in deep grooves	Epierythrocytic; loosely attached in shallow grooves
Multiplication	Binary fission	Binary fission	Binary fission	Binary fission	Binary fission	Binary fission	Binary fission
Growth on culture media	+	+	−	−	−	−	−
Effect of splenectomy	−	slight	+	+	+	+	+
Transmission (cyclic)	Arthropod (Phlebotomus)	Arthropod (fleas)	Arthropod (ticks)	Arthropod (ticks)	Arthropod (ticks)	Arthropod (lice, fleas)	Arthropod (lice, fleas)
Geographical dist.	Andes, South America	Worldwide	Worldwide	Worldwide	Africa, Asia, Southern Europe	Worldwide	Worldwide
Hosts	Man	Small mammals	Ungulates	Ungulates	Birds	Vertebrates	Vertebrates
Drug sensitivity							
Arsenic	−	−	−	−	−	+	+
Sulfanamides	−	ND	−	−	−	−	−
Penicillin	+	ND	−	−	−	−	−
Streptomycin	+	ND	−	−	−	−	−
Tetracyclines	+	ND	+	+	+	+	+
Dithiosemicarbazone	ND	ND	+	+	+	ND	ND
Clinical disease in intact animals	+	−	+	+	+	Only *H. felis*	Only *E. suis*

[a] Data from Weinman, 1944; Peters and Wigand, 1955; Kreier and Ristic, 1968; Ristic and Kreier, 1974; Gothe and Kreier, 1977. Symbols: −, not effective; +, effective; ND, no data available.

Anaplasmataceae from the chlamydiae, as chlamydiae are characterized by development in a unique obligately intracellular growth cycle. This cycle includes an infectious elementary body which is a small, electron-dense form and a noninfectious initial body which is larger, less electron-dense, and is the growing form (Page, 1974).

The cellular nature of the Bartonellaceae and Anaplasmataceae precludes consideration of viral groups for these parasites (Kreier and Ristic, 1973).

As can be seen from the preceding discussion, the Anaplasmataceae and Bartonellaceae do not readily fit into any of the well-established taxons of prokaryotes. The *Bartonella* and *Grahamella* are truly small hemotrophic bacteria. The Anaplasmataceae are small, cell wall–less, obligately parasitic, hemotrophic prokaryotes. Information from systematic antigenic analysis and information on nucleic acid homology, which would aid in determining relatedness among these organisms and between them and other organisms, is not available. The taxons into which these organisms are presently placed are based largely upon the organisms' morphology, host range, tissue and cellular preference, and established tradition (Ristic and Kreier, 1974; Weinman, 1974).

These organisms clearly do not fit comfortably with the rickettsia as classically defined, nor is there good reason to assume that the Bartonellaceae and Anaplasmataceae are necessarily closely related. To reduce taxonomic confusion, Moulder (1974) redefined the rickettsia in a sufficiently broad fashion so that it is possible to allow these organisms to remain together and in their customary taxonomic position (Ristic and Kreier, 1974; Weinman, 1974). This course of action would appear to be completely justified in light of the meager information available on these organisms at the present time.

There are available a number of fairly thorough reviews of the literature on these organisms (Gothe and Kreier, 1977; Kreier and Ristic, 1968; Ristic, 1960, 1968, 1977; Weinman, 1944, 1968; Weinman and Kreier, 1977).

BARTONELLACEAE

Bartonella

Habitats of *Bartonella*

GEOGRAPHICAL DISTRIBUTION. *Bartonella bacilliformis* and the disease of man it causes, bartonellosis, occur in well-defined geographical regions called "zonas verrucógenas" (Rebagliati, 1940). These zones are located between 500 and 3,000 m above sea level, along rivers in deep mountain valleys with scanty vegetation. The infections are most common

when the weather is warm and the air is still. Valleys where these conditions occur are frequently found in the mountains of Peru, Ecuador, and Colombia (Reynafarje, 1972). There are no confirmed reports of the occurrence of *B. bacilliformis* from outside the Andes region of South America.

AS SYMBIONTS OF ARTHROPODS. The organism is usually transmitted by *Phlebotomus verrucarum*, a hematophagous sandfly of nocturnal feeding habits and a short flight range. Other species of *Phlebotomus* have also been implicated as vectors of the *Bartonella bacilliformis*, including *P. colombianus* and *P. pescei* which occur in regions where typical cases of bartonellosis have been described, but *P. verrucarum* has not been found (Herrer and Blancas, 1962; Hertig, 1942; Noguchi and Shannon, 1929; Noguchi et al., 1928). The organism grows in the arthropod.

AS HUMAN PATHOGENS. *Bartonella bacilliformis*, the only species in the genus *Bartonella*, causes Carrion's disease of man; an infectious process with two clinical forms. One form, Oroya fever, is a febrile process in which numerous organisms parasitize the red blood cells, causing their massive destruction in the spleen and other organs of the reticuloendothelial system, with the consequent development of a progressive and severe anemia. When there are no longer parasites in the peripheral blood, either following antibiotic therapy or as a result of the natural evolution of the disease, the patient becomes asymptomatic. The asymptomatic period may last weeks or months. Often many patients may develop a skin eruption called verruga peruana. Verruga peruana can occur in one or more of the following forms: nodular, miliary, and mulaire. The organism occurs in the vascular endothelium of the lesions. These lesions resemble pyogenic granulomas. Some patients may only develop Oroya fever with anemia, whereas others develop only skin eruptions (Aldana, 1929; Cuadra, 1957; Hurtado, Pons, and Merino, 1938).

AS INAPPARENT PARASITES OF MAN. Studies of the source of infection have shown that it is possible to isolate *Bartonella bacilliformis* from the peripheral blood of asymptomatic persons living in endemic areas. From 5 to 10% of persons in endemic areas may give positive blood cultures (Herrer, 1953b; Herrer and Cornejo Ubilluz, 1962; Colichon, Calderon, and Bedon, 1972). These findings confirm the hypothesis than human beings are the reservoir of the organism (Weinman and Pinkerton, 1937b). Some thought that varieties of *Phlebotomus*, other arthropods, and even plants may have importance as reservoirs of *Bartonella bacilliformis*; however, these hypotheses have not been supported experimentally (Herrer, 1953a).

Isolation of *Bartonella*

GENERAL FEATURES. The culture of *Bartonella bacilliformis* has been difficult despite the fact that many organisms can be observed in the peripheral blood used for inoculation. Many methods of culture and many culture media for *Bartonella* have been designed (Geiman, 1941; Noguchi, 1926a; Noguchi and Battistini, 1926). It is easier to isolate the organism in the earlier stages of the disease than in the later stages. The isolation of the organism from dermic lesions is possible. Biopsy is recommended to avoid contamination of the culture with skin microorganisms.

Even though *Bartonella bacilliformis* is a parasite of red blood cells and histiocytes, it is able to multiply in noncellular culture medium containing human, rabbit, sheep, or horse serum or blood. A defined medium has not yet been described.

Triphasic Culture Medium for Isolating Bartonella (Colichon, Colichon, and Velasco, 1966; Colichon, Colichon, and Solano, 1971)

The triphasic medium described by Colichon and colleagues is one which produces a good yield of cultured microorganisms. It is formed of three layers of culture medium: first a Bordet blood agar slant, on top of which is another slant of tryptose agar, and finally tryptose broth. The Bordet blood agar is prepared in the following manner:

Infusion of potato	1,000 ml
Glycerine	10 ml
Sodium chloride	5.4 g
Proteose peptone	10 g
Agar	14 g

Heat to boiling to dissolve all the components, adjust the pH to 7.2, sterilize in an autoclave for 15 min at 15 lb pressure. When the temperature is at 50°C, add sheep blood, 15% by volume, and distribute into Kolle flasks. Leave the tubes standing in a slanted position until the medium solidifies.

When the Bordet blood agar is solidified, add tryptose agar medium, whose formula is as follows:

Tryptose	15 g
Sodium chloride	5 g
Disodium phosphate	2.5 g
Distilled water	1,000 ml
Agar	13 g

Heat the ingredients to boiling to dissolve. Adjust the pH to 7.2 and sterilize in an autoclave for 15 minutes at 15 lb of pressure. Leave the medium stand until the sediment goes down and utilize the clear supernatant. To the supernatant add sheep plasma, 15% by volume. Before the medium so-lidifies distribute it into the Kolle flasks which already contain the solidified Bordet blood agar. Enough medium should be added so that there is a thin layer of tryptose agar medium over the Bordet blood agar. Once the medium is solidified put the flasks in a vertical position and add 30 ml of sterile tryptose broth whose composition is similar to the tryptose medium but without the agar and plasma. Incubate the tubes for 48 h before use to determine if they are sterile.

Culture Media for Preservation and Additional Studies of *Bartonella*

To preserve *Bartonella bacilliformis* as well as to visualize its motility "in vitro", it is necessary to have a transparent culture medium which permits one to observe the turbidity produced by the bacterial growth. The procedure used to obtain a transparent culture medium is as follows: In the bottom of a 16-×-150-mm test tube add Bordet blood agar, then add enough tryptose agar with plasma to form a layer 2 cm deep over the Bordet blood agar. After the medium solidifies and has been checked for sterility, inoculation may be made by puncture. By this system one may demonstrate aerobic and microaerobic growth of *Bartonella*, as well as metabolic activity, if a substrate with its respective indicator is added to the tryptose-plasma agar.

Technique for Isolation of *Bartonella* from Infected Humans

Several ways of processing samples for the isolation of *Bartonella* have been reported. The one described below is best because it yields the highest proportion of positive samples. Take 10 ml of venous blood and add 0.6 ml of 10% sodium citrate, mix, and centrifuge. The cellular sediment is inoculated into a flask of the triphasic culture media, previously described. The tube is incubated at 29°C and read after the 5th day of incubation.

Identification of *Bartonella*

To identify *Bartonella bacilliformis* it is necessary to consider certain aspects of the isolated organism and its origin. These include the origin of the individual from whom the organism was isolated, the morphological characteristics of the organism, and its behavior in culture media. When the microorganism is isolated from a patient coming from an area in which verruga occurs, recognition is easy as it is partly based on epidemiological considerations.

Bartonella bacilliformis is a cocco-bacillus. Its typical shape is best observed in blood smears stained with Wright's stain (Fig. 1). The bacilliform organisms are most numerous in the blood at the

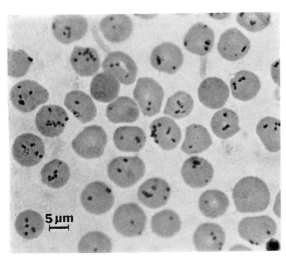

Fig. 1. Wright-stained *Bartonella bacilliformis* in blood of an Oroyo fever–infected human. [From Takano-Moran, 1970, with permission; courtesy of J. Takano-Moran.]

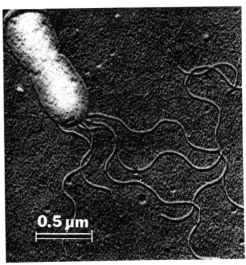

Fig. 3. *Bartonella bacilliformis* from a culture showing flagella and commencement of binary fission. [From Peters and Wigand, 1952, with permission.]

beginning of the Oroya fever episode. They are wide at the poles and have some spots at both ends. The coccoid forms are found predominantly in samples from culture media. Sizes vary. The usual range is from 0.2 to 0.3 μm wide and 1.0 to 3.0 μm long. The organism is Gram-negative (Peters and Wigand, 1955).

Studies by electronic microscopy show the organism to be located inside the red blood cell (Cuadra and Takano, 1969; Takano-Moron, 1970) and show the presence of a thin cell wall (Fig. 2). Culture forms have flagella at one pole (Fig. 3). The presence of flagella differentiates this organism from *Grahamella*. Failure to grow in culture differentiates *Haemobartonella* and *Eperythrozoon* from *Bartonella* and *Grahamella*. The optimum temperature for growth in culture is 29°C. The organism grows slowly and colonies can be observed between the 5th

and the 8th day of incubation. The colonies are of small size, have a mucoid aspect, and are slightly adherent to the agar. In the liquid phase granules form first then flocculi form on the surface of the broth.

Bartonella bacilliformis does not ferment the carbohydrates, glucose, saccharose, galactose, maltose, levulose, raffinose, rhamnose, dextrin, dulcitol, arabinose, inulin, or salicin. It is susceptible to the action of penicillin, streptomycin, and chloramphenicol.

Experimentally the disease in its dermic form or Peruvian verruga can be reproduced in rhesus monkeys but the anaemic form, Oroya fever, cannot be reproduced (Noguchi, 1926b; Weinman and Pinkerton, 1937a). On the basis of its morphological and biological characteristics *Bartonella bacilliformis* is a bacterium (Nauck, 1957), since it has a cell wall, multiplies by binary fission, has flagella, and grows in nonliving culture media. In the eighth edition of *Bergey's Manual of Determinative Bacteriology* the organism is grouped as are the other hemotrophic bacteria with the rickettsia. The justification for this classification is given in the introduction to this section.

Grahamella

Habitats of *Grahamella*

GEOGRAPHICAL DISTRIBUTION. Grahamellas have a world-wide distribution, which is one consequence of cyclic flea-borne transmission. No climatic barriers to their occurrence exist and all ecological niches are used by *Grahamella* when suitable hosts are present. The incidence of *Grahamella* infections in

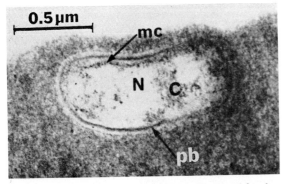

Fig. 2. *Bartonella bacilliformis* in blood cell prepared for electron microscope examination showing cell wall (pb), cell membrane (mc), cytoplasm (C), and nucleoid region (N). [From Takano-Moran, 1970, with permission; courtesy of J. Takano-Moran.]

small mammals reflects generation dynamics and the undulating population densities of the respective vertebrate and arthropod host communities. At least outside the tropics a characteristic rise of the infections incidence usually occurs in the late summer and early fall when the most favorable conditions for transmission occur (Krampitz and Kleinschmidt, 1960). The parasites can be recognized in stained blood films (Fig. 4). The percentage of hosts which can be found infected in the field usually fluctuates between 0 and 30%, with an average of 10%. This was already known to Graham-Smith (1905), who first observed the parasites in British moles *(Talpa europea)*. A "*Drosophila*-like scale" (Chaline, 1977) of both host fertility and parasite generation time favors the development of limited host specificity and favors parasite speciation.

AS SYMBIONTS OF ARTHROPODS.   Parasitized cells can be isolated from arthropods (Fig. 5). The appearance of these cells remind us of those shown in the photographs by Pinkerton and Weinman (1937) of tissue explants infected with *Bartonella bacilliformis*. There are more or less stable biological strain differences between grahamellas of different host species. While only in exceptional circumstances has the experimental transmission from one vertebrate host species to another proved successful (Krampitz and Kleinschmidt, 1960), a species specificity does not exist for arthropods. It has been confirmed experimentally that any xenotype of *Grahamella* can use any vector species of the same order (Krampitz and Kleinschmidt, 1960).

For rodent grahamellas, fleas act as arthropod hosts and vectors. This was first proved by Vassiliadis (1935) in Egypt and then confirmed experimentally by Krampitz and Kleinschmidt (1960) and Krampitz (1962) in Germany and Fay and Rausch (1969) in Alaska. It is not known whether other ectoparasites are involved. Adult fleas infected with blood discharge large masses of infective parasites a few days

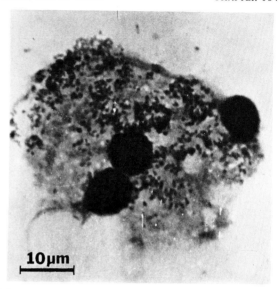

Fig. 5.   Isolated midgut cells of the tropical rat flea (*Xenopsylla cheopis*) containing *Grahamella*. Giemsa stain.

later (Fig. 6); this discharge persists for at least several weeks. *Grahamella* are also ingested by larval fleas from the sheddings of their parents. Infection of the adults developing from such infected larvae can occasionally be observed. No evidence exists for transovarial passage in fleas. It remains to be determined, however, whether grahamellas can penetrate or be transported through the gut wall and invade the fleas' hemocoel. This possibility was taken into consideration by Hertig and Wolbach (1924) in their studies on rickettsia-like organisms in insects and by Ito, Winson, and McGuire, (1975) for *Rickettsia mooseri*. Normally, the fleas salivary glands and saliva are not involved in transmission (Krampitz, 1962). Since survival outside the host body seems to be limited, direct contamination of the vertebrates mucous membranes or small skin wounds by feces is the common mode of infection.

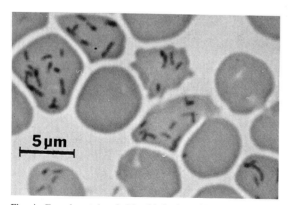

Fig. 4.   Experimental early blood infection with *Grahamella* sp. in a splenectomized European bank vole (*Clethrionomys glareolus*). The blood film was stained with Giemsa.

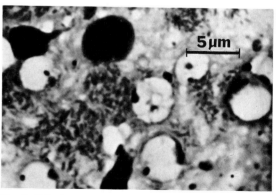

Fig. 6.   Infective *Grahamella* in feces of *Xenopsylla cheopis* (bank vole strain). Giemsa stain.

AS PARASITES OF VERTEBRATES. The vertebrate host range (Weinman and Kreier, 1977) is limited mainly to rodents, insectivores, chiroptera, and marsupials. Reported infections in domestic animals except the dog (Herrer, 1944) have so far not been culturally confirmed. Only a few reports of *Grahamella* in subhuman primates, carnivores, ruminants, reptiles, and fishes exist. Human beings do not seem to be infected. The approximately 40 named species are described only on the basis of their observation in stained blood films of various host species (Fig. 4). The last description of that kind was *G. cuniculi* in Egyptian rabbits (Haiba, 1963).

The stroma of red cells (Fig. 7) is the exclusive habitat in the vertebrate host. In the arthropod vector, there is an obvious intimate relationship to the cells of the gut wall (Fig. 5). The common statement that grahamellas are not known to multiply in fixed tissue cells should be restricted to the vertebrate host.

Parasites of the genus *Grahamella* seem to be forgotten organisms in the research of animal's blood parasites. We are still missing much basic information about their metabolic and antigenic properties, information available for most of the other parasitic prokaryotes. Only the gross features of their parasitic life can be outlined. Despite their intracellular growth in the blood stream of appropriate vertebrates and the intestinal tract of suitable arthropod vectors, the organisms do not cause serious damage to the host cells or produce disease.

The method by which the organisms invade the red cells without doing injury to their membranes is unknown. Perhaps a host receptor mechanism exists. If so, this could explain their host specificity. Whatever the hosts' and parasites' mutual identification methods may be, they are unknown.

The term "grahamellosis" is more a construction analogous to the term "bartonellosis" than a term appropriate for characterization of the pathological disorders produced by *Grahamella* in any of their hosts. The lack of interest of the medical and veterinary professions in *Grahamella* is due to their lack of pathogenicity and due to their failure to infect either man or his domestic animals.

Isolation of *Grahamella*

The difficulty in producing a parasitemia in a susceptible host animal after inoculation of infected donor blood is one of the most puzzling features of the in vivo isolation and maintenance of *Grahamella* strains. The failure is not dose-dependent nor is it controlled by previous sensitization of the receptor animal in any way. This aspect of the grahamella's behavior is very different from that of all related intraerythrocytic microorganisms. The reason for this behavior is unknown. However, every *Grahamella* strain can be isolated in a highly virulent condition from feces of previously infected fleas (Fig. 6). The parasites grow easily in vivo when splenectomized susceptible specific hosts are inoculated intraperitoneally with the metacyclic forms, i.e., those from the vector feces. The parasitemia starts 3–6 days after infection with an extensive spread of the organisms all through the blood stream and with intensive intraglobular reproduction. Half the erythrocytes (Fig. 4) may be infected in such animals (Krampitz, 1962). In its density the early massive blood infection due to *Grahamella* is exactly "*Bartonella*-like", but only a moderate reactive polychromasia takes place. The parasite does not use reticulocytes or nucleated cells as its microhabitat in the vertebrate at any time. As described by most of the authors, the normal aspect of a *Grahamella* infection in its late stages is a concentration of always many parasites in only a few, sometimes difficult to detect, infected host cells. A dense parasitemia guarantees better results if one intends in vitro cultivation or isolation for morphological studies and experimental vector infection.

A further remarkable difference between grahamellas and closely related organisms is the failure of splenectomy to accelerate the blood infection unless surgery is carried out immediately after infection, or prior to the infection. Basically no qualitative differences seem to exist between the numerous xenotypes or species in their host-related properties and behavioral patterns.

Life and reproduction of *Grahamella* in the oxygen-rich erythrocyte environment indicate aerobiosis. In vitro growth 2–3 mm below the surface in stab cultures suggests an oxygen requirement slightly below that of open air. Only a few semidefined media have been found to be suitable as growth media, but little systematic work has been done so far. In spite of the organism's aerobiosis,

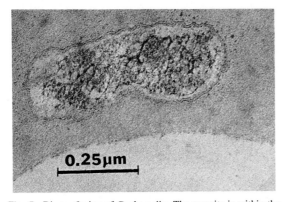

Fig. 7. Binary fission of *Grahamella*. The parasite is within the stroma of the red cell. Ultrathin section, fixed in glutaraldehyde, embedded in Durcupan ACM (Fluka). [Courtesy of E. Göbel, Munich.]

some reservations are indicated about reports of easy surface growth on blood agar and other solid media unless reinoculations of the presumptive *Grahamella* colonies into susceptible hosts have been carried out. Semisolid media developed for *Leptospira* and *Bartonella* are suitable for *Grahamella*. A first important step was the development of the thin serum agar medium by Noguchi and Battistini (1926). Tyzzer (1941) was the first to cultivate *Grahamella* strains from American small mammals in this nonliving medium. Krampitz and Kleinschmidt (1960) grew parasites of European origin in rabbit serum with an agar base at pH 7.2. The additional enrichment of the medium with hemoglobin promotes growth.

Medium for Growing *Grahamella*

Saline (0.9%)	800 ml
2% Agar (pH 7.2)	100 ml

Sterilize the agar and distribute 2-ml volumes into small glass tubes. Hold the tubes at 60°C. Add 0.2–0.25 ml sterile (filtered) rabbit serum and 0.1 ml of a concentrated rabbit-hemoglobin solution released from erythrocytes by aqueous lysis. Mix by shaking and autoclave for 30 min at 56°C.

The transfer of cultivated material from tube to tube should be carried out with the platinum loop introduced into the upper part of the medium as for a deep stab cultivation. The tubes must be sealed to prevent dessication. Sand-grain size colonies appear along the stab 7–10 days after inoculation most abundantly within a zone situated 3–5 mm below the surface of the medium (Fig. 8); colonies are occasionally also in ring form. The morphology of the colony is greatly influenced by the inoculation technique and the quantity of organisms inoculated. Since no surface growth occurs, it is evident that a reduced amount of oxygen favors growth. An incubation temperature of 27–30°C seems adequate. In vitro subcultures and in vivo transfer are successful. The white spherical colonies developed on primary inoculation are hard and easy to remove from the medium; colonies produced on subculture are more diffuse and appear like snow flakes. Host specificity is never lost in vitro (Krampitz and Kleinschmidt, 1960; Tyzzer, 1941), indicating its genetic basis. Noguchi originally proposed use of rabbit serum in the medium. The addition of hemoglobin from various sources, even human, promotes in vitro growth of grahamellas of voles and shrews. It is evident that host specificity cannot depend on simple nutritional factors.

Strain collection as deep frozen stabilates is possible with all developmental stages except those in the blood. The best material for long term preservation of living strains is the gut content of experimen-

Fig. 8. Deep stab cultures of *Grahamella* in Noguchi's medium. Original isolation from vole blood. [From Krampitz and Kleinschmidt, 1960, with permission.]

tally infected fleas containing metacyclic grahamellas. The infected gut content must be suspended in the blood of the appropriate host species. Culture material can be preserved in a similar manner, but its infectivity decreases with time of storage. The best methods for preserving grahamellas in the frozen state are those used routinely for preserving blood protozoa (Swoager, 1972). No information is available about cultivation of the organisms in tissue cultures or in hens' eggs, nor have attempts been made to lyophilize grahamellas nor has their drug sensitivity been tested.

Identification of *Grahamella*

The best "fingerprint" for identification of grahamellas by light and electron microscopy is their characteristic appearance inside the erythrocytes. Proper preparation and staining techniques are required. Hemolysis must be prevented. Attempts to concentrate scarce parasitized red cells in thick films have not been successful. In thin blood films stained by the Giemsa method and examined with the aid of a light microscope, grahamellas appear predominately as rod-, dumbbell- or string-shaped purple-red bacteria 1–2 μm long within the erythrocytes (Fig. 4). Sometimes V- or Y-like figures are found, rarely single cocci, never ring or star forms or chains of several segmenting organisms. Grahamellas are Gram negative, not acid fast, not flagellated, immobile, and do not form gas. In highly parasitized blood films, groups of extracellular rods can usually be found ("Schwärmende Haufen", Jettmar, 1932). These are released when injured erythrocytes are broken during film preparation. It is unlikely that free forms can survive and circulate for any significant time. Phagocytes full of *Grahamella* may be observed in stained blood and tissue impression films. No attempts have been made to identify and differentiate in vitro the organisms' nutritional requirements, metabolic properties, DNA homology, or antigenic structure.

The gross features of the parasites can be established by ultrastructural methods (Krampitz and Kleinschmidt, 1960). Thin-sections of the parasitized erythrocytes show *Grahamella* lying within the red cells, always clearly separated from the surface membrane (Fig. 7). They show a cytoplasmic membrane, internal structures, and a trilaminar cell wall characteristic of true bacteria. Reproduction takes place by binary fission. That there is a close ultrastructural relationship between rodent grahamellas and *Bartonella bacilliformis* in human blood is obvious when one compares micrographs of the two organisms (Cuadra and Takano, 1969). The morphological and biological characteristics of the Bartonellaceae are more similar to true bacteria than to rickettsia, mycoplasmas, or viruses. The characteristics of *Grahamella* and *Bartonella* support the legitimacy of the creation of the two families Bartonellaceae and Anaplasmataceae.

ANAPLASMATACEAE

Eperythrozoon and *Haemobartonella*

Habitats of *Eperythrozoon* and *Haemobartonella*

GEOGRAPHICAL DISTRIBUTION. Eperythrozoa and haemobartonellae are obligate parasites; thus their habitats are their hosts and their geographical range that of their hosts. These parasites occur in common rats and mice all over the world. Infections have been found in dogs, cats, and cattle whenever they have been sought and there have been reports in recent years of haemobartonella- or eperythrozoon-like parasites from monkeys (Aikawa and Nussenzweig, 1972; Gothe and Kreier, 1977; Kreier and Ristic, 1968; Peters, Molyneux, and Howells, 1973; Weinman, 1944).

AS SYMBIONTS OF ARTHROPODS. The invertebrates reported to be infected are the common blood-sucking parasites of the vertebrate hosts, including lice (Crystal, 1958, 1959a,b), fleas (Weinman, 1944) and ticks (Seneviratra, Weerasinghe, and Ariyadasa, 1973). The organisms reproduce in at least some of these arthropods.

AS PARASITES OF VERTEBRATES. In the vertebrate host haemobartonellae and eperythrozoa are found on or in the red blood cells and free in the blood plasma. Cells other than erythrocytes and possibly platelets do not appear to be infected. These organisms do not persist in the environment outside of their hosts.

A wide range of vertebrates are infected including most of the common laboratory and economically important domestic animals, wild rodents, monkeys, and possibly man and some cold-blooded vertebrates.

These parasites are usually found in the blood of clinically normal animals. Two species of these parasites may be found in the blood of intact animals with frank clinical disease. These are *Haemobartonella felis*, which is found in cats with feline infectious anemia (Flint and McKelvie, 1956), and *Eperythrozoan suis*, which is found in domestic pigs with porcine icteroanemia (Splitter, 1950). Even these two parasites persist after clinical recovery and then may be found in the blood of nonclinical carriers.

Isolation of *Eperythrozoon* and *Haemobartonella*

No successful cultivation of these organisms has been reported (Gothe and Kreier, 1977). Maintenance of the organisms has been by maintenance of carrier animals or by serial transmission of infected blood in healthy animals. Stabilates may be maintained in liquid nitrogen. To prepare parasites for freezing, an appropriate host animal is splenectomized. Blood films are prepared from the splenectomized animal daily for about 3 weeks, stained with Giemsa or other similar stain, and examined to be sure that the animal is not carrying some other blood parasite. This animal is then injected with blood from the infected donor animal. When parasites are plentiful in the blood, as determined by examination of stained blood films, blood is collected in Alsever's solution and then mixed with an equal volume of Alsever's solution containing 10% glycerol. The mixture is aliquoted into appropriate vials and frozen in liquid nitrogen by any of the procedures appropriate for preservation of blood-inhabiting protozoa. Parasites are recovered by thawing and prompt intravenous or intramuscular injection of the thawed parasites (Diamond, 1964; Swoager, 1972).

Identification of *Eperythrozoon* and *Haemobartonella*

Identification of eperythrozoa and haemobartonellae is on the basis of morphology in stained blood films and host range. Named species and their hosts are listed in Tables 2 and 3. Morphology can only be readily evaluated in animals during the early stages of the infection before anemia develops. The best

Table 2. Species of *Eperythrozoon* and their hosts.

Parasite	Host	Reference
E. coccoides	White mouse	Schilling (1928)
E. dispar	Dwarf mouse *(Mus minutus),* vole *(Microtus arvalis)*	Bruynoghe and Vassiliadis (1929)[a]
E. felis[b]	Domestic cat	Clark (1942)
E. leptodactyli	*Leptodactylus pentadactylus*	Carini (1930)[a]; Brumpt (1936)[a]
E. mariboi	Flying fox *(Pleropus macrotis epularius)*	Ewers (1971)
E. noguchii	Man	Lwoff and Vaucel (1930)[a]
E. ovis	Domestic sheep	Neitz, Alexander, and DuToit (1934)[a]
E. parvum	Domestic pig	Splitter (1950)
E. perekropovi	Pike *(Esox lucius)*	Yakimoff (1931)[a]
E. suis	Domestic pig	Splitter (1950)
E. wenyoni	Domestic ox	Adler and Ellenbogen (1934)[a]
E. teganodes	Domestic ox	Hoyte (1962)
E. varians	Graybacked deer mouse *(Peromyscus maniculatus gracilis)*	Tyzzer (1942)
—	*Aotus trivigatus*	Peters et al. (1973)
—	Man	Schuffer (1929)[a]
—	*Jerboa* sp.	Kikuth (1931)[a]
—	*Arvicola arvalis*	Zuelzer (1927)[a]
—	*Arvicola arvalis*	Kikuth (1932)[a]
—	*Rattus rattus*	Schwetz (1934)[a]
—	*Sciuras vulgaris*	Nauck (1927)[a]

[a] Reference cited in Weinman (1944).
[b] *Haemobartonella felis* (Flint and McKelvie, 1956) may be the same organism.

morphological evaluations can be made on blood films prepared following inoculation of the suspect blood into a clean splenectomized test animal. One should consider not only the characteristic morphology of the individual parasites, but the progressive increase in numbers that should occur during the period of developing infection (reviewed by Gothe and Kreier, 1977).

In acridine orange–stained preparations, the organisms of both these genera appear to be small cocci and fluoresce bright orange (Kreier, 1962). In Giemsa or Wright's stained thin blood films eperythrozoa stain reddish purple and are free in the plasma or on the red cells. In clean fresh films prepared from fresh blood of an animal not yet suffering from acute anemia, most eperythrozoa appear as delicate rings 0.5–1.0 μm in diameter or as cocci of about the same size (Fig. 9). Structures resembling rods may partially or completely surround the red cells. In similar preparations haemobartonellae appear as small cocci or rods (Fig. 10). Haemobartonellae do not generally occur as ring forms nor are haemobartonellae seen free in the plasma. Both haemobartonellae (Venable and Ewing, 1968) and eperythrozoa adhere to the surface of the red cells, but eperythrozoa are easily dislodged. These para-

sites lack cell walls and reproduce by binary fission (Fig. 11).

Complement fixation (Hyde, Finerty, and Evans, 1973; Wigand, 1956, 1958) and indirect hemagglutination tests (Smith and Rahn, 1975) have been developed for identification of animals infected with these parasites.

These parasites will usually only grow in one or at most several closely related hosts. It is because of this that host susceptibility is helpful in identification (reviewed by Gothe and Kreier, 1977).

Aegyptianella

Habitats of *Aegyptianella*

GEOGRAPHICAL DISTRIBUTION.   Aegyptianellosis occurs in enzootic areas in all the countries of Africa, the region around the Mediterranean Sea, southern Europe, and southern Asia to Formosa (Gothe, 1971). In these areas, it is of considerable economic significance to the breeding of endemic stock and for the maintenance of imported animals. The data concerning the distribution of these parasites are, however, still incomplete and reflect only

Table 3. Species of *Haemobartonella* and their hosts.

Parasite	Host	Reference
H. arvicolae	European vole *(Microtus arvalis)*	Yakimoff (1928)[a]
H. blarinae	Short-tailed shrew *(Blarina Brevicauda)*	Tyzzer (1942)
H. batrachorum	Frog *(Leptodactylus ocellatus)*	Zavattari (1931)[a]
H. bovis	Domestic ox	Donatien and Lestoguard (1934)[a]
H. canis	Domestic dog	Kikuth (1929)[a]
H. caviae	Guinea pig	Campanacci (1929)[a]
H. felis[b]	Domestic cat	Flint and McKelvie (1956)
H. glis glis	Dormouse	Kikuth (1931)[a]
H. magma	Domestic ox	Rodriguez (1954)
H. melloi	Anteater *(manis pentadactyla)*	Mello, Fernandes, Correira, and Lobo (1928)[a]
H. microti	Vole *(Microtus pennsylvanicus pennsylvanicus)*	Tyzzer and Weinman (1939)
H. muris	Albino rat	Mayer (1921)
H. muris musculi	Albino mouse	Schilling (1929)[a]
H. nicollei	Brochet *(Esox lucius)*	Yakimoff (1928)[a]
H. opossum	Marsupial rat *(Metachirus opossum)*	Regendanz and Kikuth (1928)[a]
H. pavloskii	Lamprey *(Petromyzon marinus)*	Epstein (1935) (according to Ray and Idnani, 1940)[a]
H. peromysci maniculati	Graybacked deer mouse *(Peromyscus maniculatis gracilis)*	Tyzzer (1942)
H. procyoni	Racoon *(Procyon lotor)*	Frerichs and Holbrook (1971)
H. pseudocebi	Monkey *(Pseudocebus appella)*	Pessoa and Prado (1927)[a]
H. ranarum	Frog *(Leptodactylus ocellatus)*	Da Cunha and Muniz (1926, 1927)[a]
H. rocha-limai	Bat *(Hemiderma brevicauda)*	Faria and Pinto (1926)[a]
H. scirui	Gray squirrel *(Sciurus carolinensis leucotis)*	Tyzzer (1942)
H. sergenti	Domestic ox	Adler and Ellenbogen (1934)[a]
H. sturmani	Buffalo (Asiatic)	Grinberg (1939)[a]
H. tyzzeri	Guinea pig	Weinman and Pinkerton (1938)[a]
H. ukrainica	Guinea pig	Rybinsky (1929)[a]
H. wenyoni	Domestic ox	Nieschulz and Boz (1939)[a]
—	Dormouse *(Myoxus glis)*	Franchini (according to Reitani, 1930)[a]
—	Gecko *(Phylodactylus mauritanicus*	Mirone (1932)[a]
—	Gerbil *(Meriones uniquiculatus)*	Najarian (1961)
—	Goat	Mukherjee (1952)
—	Lizard *(Tropiduris peruvianus)*	Townsend (1914–1915)[a]
—	Man *(Homo sapiens)*	Kallick et al. (1972)
—	*Macaca mulatta*	Peters et al. (1973)
—	Squirrel monkey *(Saimiri sp.)*	Aikawa and Nussenzweig 1972)
—	Rat *(Rattus rufescens)*	Andruzzi (1929)[a]
—	Tench *(Tinca tinca)*	Franchini (according to Reitani, 1930)[a]
—	Tortoise *(Testuda graeca)*	Cerruti (according to Zavathri, 1931)[a]

[a] Reference cited in Weinman (1944).
[b] *Eperythrozoon felis* (Clark, 1942) may be the same organism

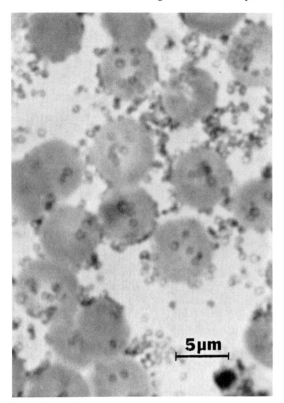

Fig. 9. Photomicrograph of a thin blood film from a mouse. The film was stained with Giemsa stain. Many *Eperythrozoon coccoides* may be seen on the erythrocytes and free in the plasma. The organisms appear either as delicate rings on the surface of the erythrocytes and in the plasma or as more deeply staining chains of cocci or rods on the margins of the erythrocytes. [Photomicrograph courtesy of H. J. Baker.]

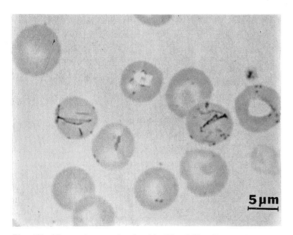

Fig. 10. Photomicrograph of a thin blood film from a dog. The film was stained with Giemsa stain. Many *Haemobartonella canis* may be seen on the erythrocytes. The organisms appear as solid dots or rods on the surface and margins of the erythrocytes. [From Venable and Ewing, 1968, with permission.]

a very fragmentary picture of the actual geographical extension. This is partly because detection of *Aegyptianella* is complicated by the fact that *Borrelia anserina* infections, which are transmitted by the same species of ticks, can mask aegyptianellosis; and *A. pullorum* infections are predominantly latent in older native animals.

As SYMBIONTS OF ARTHROPODS. The complete development cycle in *Argas (Persicargas) walkerae* takes approximately 30 days and progresses in three clearly separate phases (Gothe, 1967a, 1971; Gothe and Becht, 1969; Gothe and Koop, 1974). In larval, nymphal, and adult female ticks of this species, it develops and multiplies in epithelial cells of the intestine, then in hemocytes, and finally in the cells of salivary glands. In this latter organ, the infectious forms develop. They are 0.3- to 0.5-μm-diameter, roundish anaplasmodial bodies which are inoculated into the vertebrate with the saliva of the ticks at the next blood meal. Transovarial passage occurs. *Argas (P.) persicus* and *Argas (A.) reflexus* (Gothe, 1971), as well as *Argas (P.) radiatus* and *Argas (P.) sanchezi* (Gothe and Englert, 1978) can function as biological vectors. The infection frequency of ticks can be very high in areas where the parasites occur. Gothe and Schrecke (1972a,b) were able to isolate these parasites in 10 of 11 *Argas walkerae* populations from various parts of the Transvaal.

As PARASITES OF VERTEBRATES. In vertebrate hosts four species of the genus *Aegyptianella* Carpano, 1928, have been identified and named: *A. pullorum* (Carpano, 1928) from various species of birds; *A. emydis* (Brumpt and Lavier, 1935) from a turtle; *A. moshkovskii* (Schurenkova, 1938) from various species of wild birds; and *A. carpani* (Battelli, 1947) from a snake. Other intraerythrocytic parasites of poikilothermic animals, and also their possible relationship to *Aegyptianella,* were extensively discussed by Peirce and Castleman (1974) as well as Johnston (1975). Gothe (1978) summarized the information on *Aegyptianella-, Anaplasma-,* and *Piroplasma*-like parasites of vertebrates. He described parasites resembling *Aegyptianella* from over 33 species of wild birds and domesticated poultry. The specific identity of these organisms is not clear. Of this genus only *A. pullorum* has been studied sufficiently to provide valid information, thus the characteristics and features of this genus are based exclusively on studies of this species.

A. pullorum, extensively discussed by Gothe (1971), Gothe and Kreier (1977) as well as Kreier and Gothe (1976), is the agent of a noncontagious infection of birds, which parasitizes only the erythrocytes of the vertebrate host and is transmitted cyclically by argasid ticks.

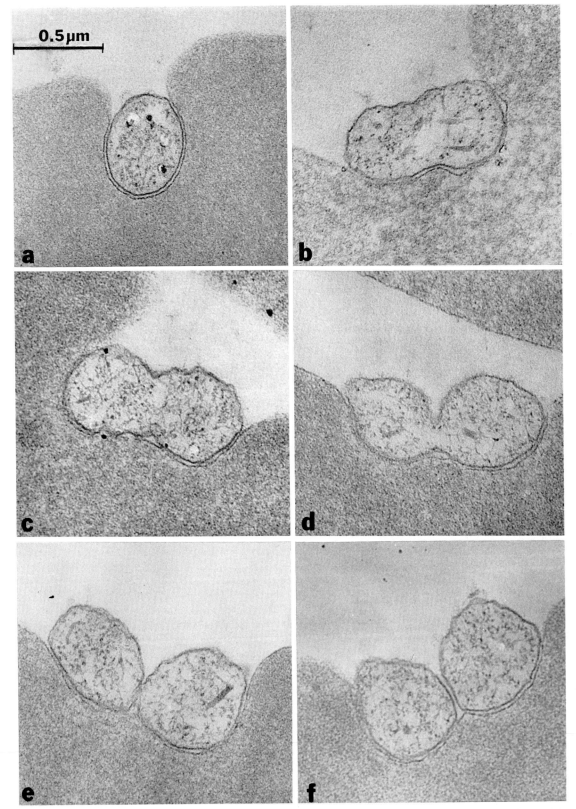

Fig. 11. Electron micrographs of *Haemobartonella muris* in the process of dividing. The coccoid organism first elongates and then constricts in the middle; finally, the two daughter organisms separate. The organisms remain attached to the erythrocytes during reproduction. The red cell membrane appears intact at the site of attachment. [From Tanaka et al., 1965, with permission.]

Vertebrate hosts of *A. pullorum* are chickens, ducks, geese, and quail as well as ostriches. Concerning guinea fowl there are contradictory data; doves and turkeys cannot be infected (Gothe, 1971). Some wild birds may also be infected at least briefly (Curasson, 1938; Curasson and Andrjesky, 1929; Huchzermeyer, 1969).

In vertebrate hosts, only erythrocytes are host cells. On the basis of electron microscopic studies it is possible to deduce the intraerythrocytic development cycle. First eperythrocytically situated, the round parasite, 0.3–0.5 μm in diameter, is surrounded by a double membrane. In the erythrocytes an additional membrane is formed which separates the parasite from the cytoplasm and encloses it in a vacuole in the host cell. In the vacuoles the parasites grow to 1 μm in size. The process of division starts with invagination of the parasites double membrane on one or both sides. These invaginations penetrate deeper and deeper, until the organism is cut through and two daughter cells are formed. This process is repeated several times and leads finally to the formation of fully mature marginal bodies containing as many as 26 roundish forms that are 0.3–0.5 μm in diameter. The intraerythrocytic development of *A. pullorum* can be completed in 36 h (Gothe, 1967c, 1971).

Besides those in the erythrocytes, parasites and fully matured marginal bodies can be observed outside the red blood cells. They occur free in the plasma, intracellularly in large and small lymphocytes, in neutrophile and eosinophile leukocytes, in monocytes, and in the Kupffer cells of the liver. The proportion of forms outside erythrocytes grows in direct proportion to the parasitemia. It is probable that most of the parasites in leukocytes have been phagocytized after release from erythrocytes. There is probably no exoerythrocytic development parallel to the intraerythrocytic cycle. The plasma forms are transitory phases initiating the infection of new erythrocytes (Gothe, 1969).

The salivary gland forms from the ticks penetrate the erythrocytes immediately after injection or at least remain infectious in the blood. Histological investigations of naturally infected chickens have not revealed an exoerythrocytic phase in the development of *A. pullorum* (Gothe, 1967c, 1971).

AS INAPPARENT PARASITES OF BIRDS AND ARTHROPODS. *Aegyptianella pullorum* has a heteroxenous cycle. The vertebrate hosts function as carriers and thereby as donors of this agent. The argasid species are biological vectors and an infection reservoir. Parasites existing in both arthropods and vertebrates constitute an important habitat for maintenance of the population in nature.

Isolation of *Aegyptianella*

Aegyptianella pullorum can not be cultured in cell-free media, tissue cultures, or extraerythrocytically in embryonated chicken eggs. Parasites can only be grown in vivo in their vertebrate and invertebrate hosts. Susceptible birds can be infected very easily by intravenous, subcutaneous, intraperitoneal, or intramuscular injections of parasite-infected blood as well as through scarification of the skin. The agent can be maintained by serial transfer of infected blood. Chicks infected at 3 weeks have remained carriers for over 1 1/2 years. The duration of the infection in the ticks is lifelong and any development stage can be infected. The infection has been shown to be present for over 810 days in ticks (Gothe, 1967a, 1971).

Cryopreservation is likewise possible. Any standard procedure for preserving blood protozoa is satisfactory (Swoager, 1972). In liquid nitrogen the maximal storage time determined to date is 6.97 years (Raether and Seidenath, 1977).

Identification of *Aegyptianella*

When examined with the aid of a light microscope *Aegyptianella pullorum* inclusions appear in Giemsa-stained blood films (Fig. 12) in a variety of roundish forms, measuring up to 0.6 μm; as 0.8–3.2 μm diameter ring forms; and as half moon or as oblong structures. The roundish structures, contain up to 26 small granules, and are approximately 0.3–0.5 μm in diameter. In thin sections examined by the electron microscope (Fig. 13) the parasites appear to be surrounded by a sheath, which consists of two 6- to 8-nm membranes separated by a space of 28 nm. The internal structure of the parasites is made up of electron-dense aggregates of a finely granular material embedded in a less dense substance. There is no membrane-bound nucleus. The parasites are separated from the host cell plasma by a 6- to 8-nm membrane (Gothe, 1967b, 1971). *Aegyptianella* possesses predominantly RNA; the amount of DNA is significantly less than the amount of RNA. The DNA is completely masked by the RNA in conventionally stained preparations (Gothe, 1971). The parasites are susceptible to the action of tetracycline and certain dithiosemicarbazones (Barrett et al., 1965; Gothe, 1971; Gothe and Kreier, 1977; Gothe and Lämmler, 1970; Lämmler and Gothe, 1967, 1969).

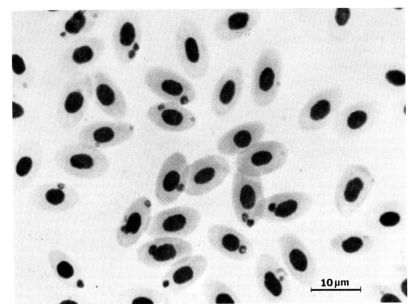

Fig. 12. Photomicrograph of a thin blood film from a chicken with Aegyptianellosis stained with Giemsa's stain. A variety of roundish forms which are the parasitic inclusions are present in the cytoplasm of the erythrocytes.

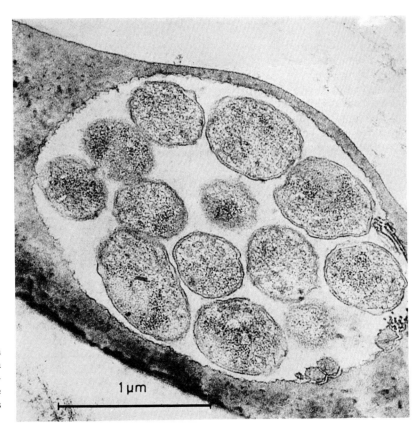

Fig. 13. Electron micrograph of a thin section of an inclusion in a chicken erythrocyte. Inside the inclusion profiles of parasites can be seen, one probably in the process of division.

Anaplasma

Habitats of Anaplasma

GEOGRAPHICAL DISTRIBUTION.   At the beginning of the century, Theiler (1910) described a small punctiform body which occurred in the erythrocytes of African cattle suffering from an acute infectious anemia. He named the organism *Anaplasma marginale*. During the classic studies of babesiosis by Smith and Kilborne (1893), *Anaplasma marginale* was shown to occur in the United States. Infections with *A. marginale* are now known to occur throughout the tropical, subtropical, and warmer temperate regions of the world. In the United States, for example, the entire southern half of the country is infected and the parasite occurs well into the northwest also.

Anaplasma centrale causes a relatively "mild" form of bovine anaplasmosis in Africa, and *Anaplasma ovis* is the cause of ovine and caprine anaplasmosis which also occurs worldwide (Ristic, 1960; 1968).

AS SYMBIONTS OF ARTHROPODS.   Anaplasmas are obligate parasites transmitted by arthropod vectors, principally ticks. An earlier belief that hematophagous flying insects have a major role in transmission of anaplasmosis is not supported by recent experiments (Bram and Roby, 1970; Peterson et al., 1977). On the basis of epizootiologic observations and experimental evidence to date, various species of ticks which include *Dermacentor occidentalis*, *Dermacentor andersoni*, *Boophilus annulatus*, and *Boophilus microplus* appear to be the most important vectors of *A. marginale* (Ristic, 1968). Ticks, but not hemophagous flies, are true biological vectors of *Anaplasma*. The organisms infect, grow, and reproduce in the tick; thus the tick's body can be considered a habitat of the organisms.

AS PARASITES OF VERTEBRATES.   *Anaplasma marginale* and *centrale* infect common domestic cattle, zebu, water buffalo, bison, African antelopes, gnu, blesbuck, duiker, American deer, Virginia white-tailed deer, elk, camel, and wildebeest. *Anaplasma ovis* infects sheep, goats, and deer (Ristic and Kreier, 1974). The parasite appears to inhibit exclusively erythrocytes and platelets. The parasite enters the erythrocyte by rhopheocytosis, a process involving invagination of the cytoplasmic membrane and resulting in a pocketing of the organism, and subsequent formation of a parasitic vacuole (Simpson, Kling, and Love, 1967). In the vacuole, the parasite multiplies by binary fission and forms an inclusion; the marginal body is visible on stained blood films (Fig. 13). Thus, the marginal body is not actually the parasite but a group of parasites and represents only a phase in the developmental cycle of the initial body (Ristic and Carson, 1977). Clinical disease associated with anemia occurs when a nonimmune animal first becomes infected. It is during the early stages of the infection that marginal bodies are found in large numbers. After an animal recovers from an acute infection, the animal achieves an equilibrium with the parasite and a healthy carrier state develops. Formation of marginal bodies is not characteristic of the parasite in the carrier animal. Animals infected when young may go directly into the carrier stage (reviewed by Ristic, 1968).

Isolation of Anaplasma

The most common method of maintaining a strain of *Anaplasma* for any purpose has been to maintain a carrier animal. Animals which recover from disease remain carriers for life, are clinically normal and are ready sources of infectious blood. Injection of a few drops of a carrier's blood into a splenectomized non-infected animal of the appropriate species will in a short time produce a high parasitemia. *Anaplasma* species are quite host specific, infection being limited to the specific host and a few closely related species. Common laboratory animals, rabbits, guinea pigs, rats, mice, ferrets, dogs, cats and chicks, are all refractory to infection with *A. marginale* of cattle for example.

Over the years, various efforts were made to maintain and cultivate *A. marginale* in vitro. None of the attempts produced evidence of active replication of the organism. Recently, R. Kessler, working in M. Ristic's laboratory, has shown that the Trager-Jensen method (1976) for propagation of malaria parasites permits propagation of *A. marginale* in vitro for a short series of passages. The organisms can be preserved for at least 4 years by low-temperature freezing of parasitized blood to which glycerin or dimethylsulfoxide (DMSO) has been added (Summers and Matsuoka, 1970). More recent studies showed that 4 M DMSO would improve the infectivity of *A. marginale* as determined by reduced incubation periods following inoculation of splenectomized calves (Love, McEwen, and Rubin, 1976).

Identification of Anaplasma

Microscopic examination of stained peripheral blood films is used to identify intraerythrocytic-*Anaplasma* inclusion bodies present during the acute stage of anaplasmosis. Giemsa staining is the oldest and most frequently used staining method. Other staining methods used include staining with toluidine blue, acridine orange, and fluorescent antibody (Gainer, 1961; Ristic, 1968; Ristic and White, 1960; Rogers and Wallace, 1966). Detection of an antibody to *Anaplasma* rather than detection of the organism itself in animals with subclinical forms of

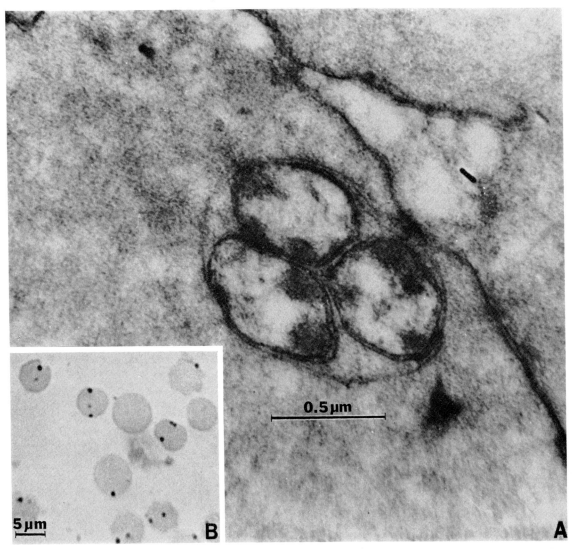

Fig. 14. Intraerythrocytic inclusions of *Anaplasma marginale*. (A) Electron micrograph of an ultrathin section; (B) blood film stained by the Giemsa method.

anaplasmosis identifies carriers. The most fre-quently used serological tests for this purpose are: complement fixation (CF), capillary tube agglutina-tion (CA), and card agglutination (CT) (Amerault and Roby, 1968; Gates et al., 1954; Ristic, 1962).

Anaplasma marginale is a Gram-negative orga-nism (Amerault, Mazzola, and Roby, 1973) that with Romanovsky-type stains appears in the eryth-rocytes as a dense, homogeneous, bluish-purple round structure, 0.3–1.0 μm in diameter (Fig. 14b). Electron microscopy reveals that these struc-tures are inclusions separated from the cytoplasm of the erythrocyte by a limiting membrane (Fig. 14a). Each inclusion contains from 1 to 8 subunits which are the actual bacteria. Each organism is 0.3–0.4 μm in diameter and contains dense aggre-gates of fine granular material embedded in an elec-tron lucid plasma, all enclosed in a double mem-brane (Ristic and Watrach, 1961; 1963).

Histochemical analysis of *A. marginale* reveals DNA, RNA, protein, and organic iron (Moulton and Christensen, 1955). Deoxyribonucleic acid from isolated marginale bodies and calf erythrocytes in-fected with *A. marginale* was found to be double stranded and to contain 51 mol % guanine plus cyto-sine (Senitzer et al., 1972). The respiration of the organism is aerobic and it produces catalase and in-corporates various immuno acids from the plasma (Johns and Dimopoullos, 1974; Mason and Ristic, 1966; Pilcher, Wu, and Muth, 1961). An increase in total adenosine triphosphatase (ATPase) activity in the erythrocyte membrane of infected animals was partially attributed to *A. marginale* (Cox, Hart, and Dimopoullos, 1974).

Infectivity of *A. marginale* can be destroyed by heating the organism at 60°C for at least 50 min, by exposure to sonic oscillation at 35°C for at least 90 min, or by X-ray irradiation at 100,000 roentgens or higher (Bedell and Dimopoullos, 1962; Simpson, Neal, and Edds, 1964; Wallace and Dimopoullos, 1965). None of these attributes is directly of much help in identification of the parasite.

During the last two decades, two parasites which closely resemble *A. marginale* have been described as mixed infections in cattle with anaplasmosis. *Paranaplasma caudatum* (genus *Paranaplasma*) has appendages usually in the form of a tapering tail, a loop, or a ring, demonstrated only by use of special techniques (Kreier and Ristic, 1963a,b). A recent study (Carson et al., 1974) showed that manifestation of *P. caudatum* appendages is a function of bovine host erythrocytes and does not occur in infected deer erythrocytes.

Paranaplasma discoides, which in water-lysed erythrocytes examined by phase microscopy is an ovoid, disk-like structure with a dense mass at each pole, has also been reported (Kreier and Ristic, 1963a).

Literature Cited

Aikawa, M., Nussenzweig, R. 1972. The fine structure of *Haemobartonella* sp. in the squirrel monkey. Journal of Parasitology **58**:628–630.

Aldana, L. 1929. Bacteriologia de la enfermedad de Carrion. La Cronica Medica, Lima **46**:235–385.

Amerault, T. E., Mazzola, V., Roby, T. O. 1973. Gram-staining characteristics of *Anaplasma marginale*. American Journal of Veterinary Research **34**:552–555.

Amerault, T. E., Roby, T. O. 1968. A rapid card agglutination test for bovine anaplasmosis. Journal of the American Veterinary Medical Association **153**:1828–1834.

Anderson, D. R. 1969. Ultrastructural studies of the mycoplasmatales and the L-phase of bacteria, chapter 12, pp. 365–402. In: Hayflick, L. (ed.), The Mycoplasmatales and L-phase of bacteria. New York: Appleton-Century-Crofts.

Barrett, P. A., Beveridge, E., Bradley, P. L., Brown, C. G. D., Bushby, S. R. M., Clarke, M. L., Neal, R. A., Smith, R., Wilde, J. K. H. 1965. Biological activities of some Dithiosemicarbazones. Nature **206**:1340–1341.

Battelli, C. 1947. Su di un Piroplasma della *Naia nigricollis* (*Aegpytianella carpani* n. sp.). Rivista di Parasitologia **8**:205–212.

Bedell, D. M., Dimopoullos, G. T. 1962. Biologic properties and characteristics of *Anaplasma marginale*. American Journal of Veterinary Research **23**:618–625.

Bram, R. A., Roby, T. O. 1970. Attempts to transmit bovine anaplasmosis with *Anopheles quadrimaculatus* Say *(Diptera: Culicididae)* and *Dermacentor andersoni* Stiles *(Acarina: Ixodidae)*. Journal of Medical Entomology **7**:481–484.

Brumpt, E., Lavier, G. 1935. Sur un prioplasmide nouveau, parasite de tortue, *Tunetella emydis* N. G., N. Sp. Annales de Parasitologie Humaine et Comparee **23**:544–550.

Carpano, M. 1928. Piroplasmosis in Egyptian fowls. *(Egyptianella Pullorum)*. Bulletin of the Ministry of Agriculture, Egypt **86**:1–12.

Carson, C. A., Weisiger, R. M., Ristic, M., Thurmon, J. C., Nelson, D. R. 1974. Appendage-related antigen production by *Paranaplasma caudatum* in deer erythrocytes. American Journal of Veterinary Research **36**:1529–1531.

Chaline, J. 1977. Rodents, evolution and prehistory. Endeavour N. S. **1**:44–51.

Clark, R. 1942. *Eperythrozoon felis* (sp. nov.) in a cat. Journal of the South African Veterinary Medical Association **13**:15–16.

Colichon, H., Calderon, J., Bedon, C. 1972. *Bartonella bacilliformis* en la sangre periferica de los pobladores de las zonas verrucógenas del Peru. Revista Peruana de Medicina Tropical **1**:19–21.

Colichon, H., Colichon, A., Solano, L. 1971. La *Bartonella bacilliformis*. Archivos Peruanos de Patologia Clinica **25**:15–32.

Colichon, H., Colichon, A., Velasco, C. 1966. Medio superpuesto para *Bartonella bacilliformis*. Anales de la Facultad de Medicina, Lima **49**:415–422.

Cox, H. U., Hart, L. T., Dimopoullos, G. T. 1974. Adenosine triphosphatase activity associated with bovine erythrocyte membranes during infection with *Anaplasma marginale*. American Journal of Veterinary Research **35**:773–779.

Crystal, M. 1958. The mechanism of transmission of *Haemobartonella muris* (Mayer) of rats by the spined rat louse *Polyplex spinulosa* (Burkmeister). The Journal of Parasitology **44**:603–606.

Crystal, M. 1959a. Extrinsic incubation period of *Haemobartonella muris* in the spined rat louse, *Polyplex spinulosa*. Journal of Bacteriology **77**:511.

Crystal, M. 1959b. The infective index of the spined rat louse *Polyplex spinulosa* (Burkmeister) in the transmission of *Haemobartonella muris* (Mayer) of rats. Journal of Economic Entomology **52**:543–544.

Cuadra, M. 1957. El diagnostico de la enfermedad de Carrion. Revista del Viernes Medico Lima **8**:404–421.

Cuadra, M., Takano, M. 1969. The relationship of *Bartonella bacilliformis* to the red blood cell as revealed by electron microscopy. Blood **33**:708–716.

Curasson, G. 1938. Notes sur la piroplasmose aviaire en A.O.F. Bull. Serv. Zootech. Epiz. A.O.F. 1, 33–35 (1938). In: Curasson, G.: Traite de protozoologie veterinaire et comparee, tome III, pp. 242–249. Paris: Vigot Freres.

Curasson, G., Andrjesky, P. 1929. Sur les "corps de Balfour" du sang de la poule. Bulletin de la Société Pathologie Exotique **22**:316–317.

Davis, B. D., Dulbecco, R., Eisen, H. N., Ginsberg, H. S., Wood, W. B., McCarty, M. 1973. Mycolasmas and L forms. Microbiology, chapter 40, 2nd ed. Hagerstown, Maryland: Harper & Row.

Diamond, L. S. 1964. Freeze preservation of protozoa. Cryobiology **1**:95–101.

Ewers, W. H. 1971. *Eperythrozoon mariba* sp. nov. (Protophyta: order Rickettsiales) a parasite of red blood cells of the flying fox *(Pteropus macroti epularius)* in New Guinea. Parasitology **63**:260–269.

Fay, F. H., Rausch, R. L. 1969. Parasitic organisms in the blood of arvicoline rodents in Alaska. Journal of Parasitology **55**:1258–1265.

Flint, J., McKelvie, D. 1956. Feline infectious anemia—Diagnosis and treatment. Proceedings 92nd Annual Meeting Veterinary Medical Association. Minneapolis, pp. 240–242.

Frerichs, W., Holbrook, A. 1971. *Haemobartonella procyoni* sp. nov. in the racoon *Procyan coton*. Journal of Parasitology **57**:1309–1310.

Gainer, J. H. 1961. Demonstration of *Anaplasma marginale* with the fluorescent dye, acridine orange; comparisons with the complement-fixation test and Wright's stain. American Journal of Veterinary Research 22:882–886.

Gates, D. W., Hohler, W. M., Mott, L. O., Schoening, H. W. 1954. Complement-fixation test as a tool in the control of anaplasmosis. In: Proceedings of the 91st Annual Meeting of the American Veterinary Medical Association, pp. 51–53.

Geiman, Q. 1941. New Media for growth of *Bartonella bacilliformis*. Proceedings of the Society for Experimental Biology and Medicine **47**:329–332.

Gothe, R. 1967a. Zur Entwicklung von *Aegyptianella pullorum* Carpano, 1928 in der Lederzecke *Argas (Persicargas) persicus* (Oken, 1818) and ubertragung. Zeitschrift für Parasitenkunde **29**:103–118.

Gothe, R. 1967b. Ein Beitrag zur systematischen Stellung von *Aegyptianella pullorum* Carpano, 1928. Zeitschrift für Parasitenkunde **29**:119–129.

Gothe, R. 1967c. Untersuchungen über die Entwicklung und den Infektionsverlauf von *Aegyptianella pullorum* Carpano, 1928, im Huhn. Zeitschrift für Parasitenkunde **29**:149–518.

Gothe, R. 1969. Zur Pathogenese der *Aegyptianella pullorum*-Infektion beim Huhn. Zeitschrift für Parasitenkunde **31**:3.

Gothe, R. 1971. Wirt–Parasit–Verhältnis von *Aegyptianella pullorum* Carpano, 1978, im biologischen Überträger *Argas (Persicargas) persicus* (Oken, 1818) und im Wirbeltierwirt *Gallus gallus domesticus* L. Fortschritte der Veterinärmedizin, Beihefte zum Zentralblatt für Veterinärmedizin, Heft **16**.

Gothe, R. 1978. New aspects on the epizootology of aegyptianellosis in poultry, pp. 201–204. Proceedings of the Conference on Tickborne Diseases and Their Vectors. Edinburgh: University Press.

Gothe, R., Becht, H. 1969. Untersuchungen über die Entwicklung von *Aegyptianella pullorum* Carpano, 1928, in der Lederzecke *Argas (Persicargas) persicus* (Oken, 1818) mit Hilfe fluoreszierender Antikörper. Zeitschrift für Parasitenkunde **31**:315–325.

Gothe, R., Englert, R. 1978. Quantitative Untersuchungen zur Toxinwirkung von Larven Neoarktischer *Persicargas* spp. bei Hühnern. Zentralblatt für Veterinärmedizin, Reihe B **25**:122–133.

Gothe, R., Koop, E. 1974. Zur biologischen Bewertung der Validität von *Argas (Persicargas) persicus* (Oken, 1818), *Argas (Persicargas) arboreus* Kaiser, Hoogstraal und Kohls, 1964 und *Argas (Persicargas) walkerae* Kaiser and Hoogstraal, 1969. II. Kreuzungsversuche. Zeitschrift für Parasitenkunde **44**:319–328.

Gothe, R., Kreier, J. P. 1977. *Aergyptianella, Eperythrozoon* and *Haemobartonella* pp. 251–294. In: Kreier, J. P. (ed.), Parasitic protozoa, vol. IV. New York: Academic Press.

Gothe, R., Lämmler, G. 1970. Über die Persistanz von *Aegyptianella pullorm* Carpano, 1928, in Küken nach chemotherapeutischer Behandlung. Zentralblatt für Veterinärmedizin, Reihe B **17**:806–812.

Gothe, R., Schrecke, W. 1972a. Zur epizootiologischen Bedeutung von *Persicargas*-Zecken der Hühner in Transvaal. Berliner und Münchener Tierärztliche Wochenschrift **85**:9–11.

Gothe, R., Schrecke, W. 1972b. Zur Epizootiologie der Aegyptianellose des Geflügels. Zeitschrift für Parasitenkunde **39**:64–65.

Graham-Smith, G. S. 1905. A new form of parasite found in the red blood corpuscles of moles. Journal of Hygiene **5**:453–459.

Haiba, M. H. 1963. On *Grahamella cuniculi* and *Bartonella cuniculi* investigated in rabbits in Egypt. Journal of the Arabian Veterinary Medical Association **23**:19–25.

Hayflick, L. 1969. Fundamental biology of the class Mollicutes, order Mycoplasmatales, pp. 15–47. In: Hayflick, L. (ed.), The Mycoplasmatales and L-phase of bacteria. New York: Appleton-Century-Crofts.

Herrer, A. 1944. Description de una *Grahamella* del perro, *Grahamella canis* n. sp. Revista de Medicina Experimental (Lima) **3**:19–33.

Herrer, A. 1953a. Enfermedad de Carrion, estudio en plantas Clamide como reservorio de *Bartonella bacilliformis*. American Journal of Tropical Medicine **2**:630–643.

Herrer, A. 1953b. Carrion's disease, presence of *Bartonella bacilliformis* in peripheral blood of patients with benign form. American Journal of Tropical Medicine **2**:645–649.

Herrer, A., Blancas, F. 1962. Estudios sobre enfermedad de Carrion en el valle interandino del Mantaro. I. Observaciones entomologicas. Revista de Medicina Experimental **13**:27–45. [See Tropical Disease Bulletin **59**:28 Abstract (1962).]

Herrer, A., Cornejo Ubilluz, J. R. 1962. Estudio sobre la enfermedad de Carrion en el valle interandino del Mantaro. II. Incidencia de la infeccion bartonelosica en la poblacion humana. Revista de Medicina Experimental **13**:47–57. [See Tropical Disease Bulletin **59**:29 Abstract (1962).]

Hertig, M. 1942. Phlebotomus and Carrion's disease. American Journal of Tropical Medicine, Suppl. **22**:1–80.

Hertig, M., Wolbach, S. B. 1924. Studies on rickettsia-like micro-organisms in insects. Journal of Medical Research **44**:329–375.

Hoyte, H. M. D. 1962. *Eperythrozoon teganodes* sp. nov. (Rickettsiales) parasite in cattle. Parasitology **52**:527–532.

Huchzermeyer, F. 1969. Persönliche Mitteilung. In: Keymer, I. F. (ed.), Parasitic diseases. Diseases of cage and aviary birds. Philadelphia: Lea & Febiger.

Hurtado, A., Pons, M. J., Merino, M. C. 1938. La anemia de la enfermedad de Carrion (verruga peruana). Lima: Editorial Gil.

Hyde, C. L., Finerty, J. F., Evans, C. B. 1973. Antibody and immunoglobulin synthesis in germ-free and commercial mice infected with *Eperythrozoon coccoides*. American Journal of Tropical Medicine and Hygiene **21**:506–511.

Ito, S., Winson, J. W., McGuire, T. J., Jr. 1975. Murine typhus rickettsiae in the oriental rat flea. Annals of the New York Academy of Sciences **266**:35–60.

Jettmar, H. M. 1932. Studien über Blutparasiten ostasiatischer wilder Nagetiere. Zeitschrift für Parasitenkunde **4**:254–285.

Johns, R. W., Dimopoullos, G. T. 1974. In vitro uptake of ¹⁴C-labeled amino acids by preparations of partially purified *Anaplasma marginale* bodies. Infection and Immunity **9**:645–647.

Johnston, M. R. L. 1975. Distribution of *Pirhemocyton* Chatton and Blance and other, possibly related, infections of poikilotherms. Journal of Protozoology **22**:529–535.

Kallick, C., Levin, S., Reddi, K., Landau, W. 1972. Systemic lupus erythrematosus associated with *Haemobartonella*-like organisms. Nature New Biology **236**:145–146.

Krampitz, H. E. 1962. Weitere Untersuchungen um *Grahamella* Brumpt 1911. Zeitschrift für Tropenmedizin und Parasitologie **13**:34–53.

Krampitz, H. E., Kleinschmidt, A. 1960. *Grahamella*, Brumpt 1911. Biologische and Morphologische Untersuchungen. Zeitschrift für Tropenmedizin und Parasitologie **11**:336–352.

Kreier, J. P. 1962. A comparison of the antigenic and morphologic features of *Anaplasma marginale* with those of several other hemotrophic parasites. Ph.D. Dissertation. University of Illinois.

Kreier, J. P., Gothe, R. 1976. Aegyptianellosis, eperythrozoonosis, grahamellosis and hemobartonellosis. Veterinary Parasitology **2**:83–95.

Kreier, J. P., Ristic, M. 1963a. Anaplasmosis. Morphologic characteristics of the parasites present in the blood of calves infected with the Oregon strain of *Anaplasma marginale*. American Journal of Veterinary Research **24**:676–687.

Kreier, J. P., Ristic, M. 1963b. Anaplasmosis. The growth and survival in deer and sheep of the parasites present in the blood of calves infected with the Oregon strain of *Anaplasma marginale*. American Journal of Veterinary Research **24**:697–702.

Kreier, J. P., Ristic, M. 1968. Haemobartonellosis, Eperythrozoonosis, Grahamellosis and Ehrlichosis, pp. 387–472. In: Weinman, D., Ristic, M. (eds.), Infectious blood diseases of man and animals, vol. II. New York: Academic Press.

Kreier, J. P., Ristic, M. 1972. Definition and taxonomy of *Anaplasma* species with emphasis on morphologic and immunologic features. Zeitschrift für Tropenmedizin und Parasitologie **23**:88–98.

Kreier, J. P., Ristic M. 1973. Organisms of the family Anaplasmataceae in the forthcoming 8th edition of Bergey's Manual. Proceedings of the 6th National Anaplasmosis Conference. Stillwater, Oklahoma: Heritage Press.

Lämmler, G., Gothe, R. 1967. Zur Chemotherapie der

Aegyptianella pullorum-Infektion des Huhnes. Zeitschrift für Tropenmedizin und Parasitologie, **18**:479–488.

Lämmler, G., Gothe, R. 1969. Untersuchungen über die therapeutische und prophylaktische Wirksamkeit oral applizierter Tetracycline gegen *Aegyptianella pullorum* Carpano, 1928, im Huhn. Zentralblatt für Veterinärmedizin, Reihe B **16**:663–670.

Love, J. N., McEwen, E. G., Rubin, R. M. 1976. Effects of temperature and time on the infectivity of cryogenically preserved samples of *Anaplasma marginale* infected erythrocytes. American Journal of Veterinary Research **37**:857–858.

Mason, R. A., Ristic, M. 1966. *In vitro* incorporation of glycine by bovine erythrocytes infected with *Anaplasma marginale*. Journal of Infectious Diseases **116**:335–342.

Mayer, M. 1921. Über einige bakterienähnliche Parasiten der Erythrocyten bei Menschen und Tieren. Archiv für Schiffs- und Tropenhygiene **25**:150.

Moulder, J. W. 1962. The biochemistry of intracellular parasitism. Chicago: University of Chicago Press.

Moulder, J. 1974. The rickettsias. Order I. Rickettsiales Gieszczkiewicz 1939, p. 882–890. In: Buchanan, R. E., Gibbons, N. E. (eds.), Bergey's manual of determinative bacteriology, 8th ed. Baltimore: Williams & Wilkins.

Moulton, J. E., Christensen, J. F. 1955. The histochemical nature of *Anaplasma marginale*. American Journal of Veterinary Research **16**:337–380.

Mukherjee, B. 1952. *Bartonella* in goats. The Indian Veterinary Journal **28**:343–351.

Najarian, H. 1961. *Haemobartonella* in the Mongolian gerbil. Texas Reports of Biology and Medicine **19**:123–133.

Nauck, E. 1957. Propiedades morfologicas y biologicas de la Bartonellas. Anales de la Facultad de Medicine, Lima **40**:857–865.

Noguchi, H. 1926a. The viability of *Bartonella bacilliformis* in cultures and in preserved blood and an excised nodule of *Maccacus rhesus*. Journal of Experimental Medicine **44**:533–538.

Noguchi, H. 1926b. The behavior of *Bartonella bacilliformis* in *Maccacus rhesus*. Journal of Experimental Medicine **44**:697–713.

Noguchi, H., Battistini, T. 1926. Etiology of Oroya fever. I. Cultivation of *Bartonella bacilliformis*. Journal of Experimental Medicine **43**:851–864.

Noguchi, H., Shannon, R. C., and col. 1928. *Phlebotomus* and Oroya fever and Verruga peruana. Science **68**:493–495.

Noguchi, H., Shannon, R. C. 1929. The insect vectors of Carrion's disease. Journal of Experimental Medicine **49**:993–1008.

Page, L. A. 1974. Order II. Chlamydiales, pp. 914–918. In: Buchanan, R. E., Gibbons, N. E. (eds.), Bergey's manual of determinative bacteriology, 8th ed. Baltimore: Williams & Wilkins.

Peirce, M. A., Castelman, A. R. W. 1974. An intraerythrocytic parasite of the Moroccan tortoise. Journal of Wildlife Diseases **10**:139–142.

Peters, D., Wigand, R. 1952. Neuere Untersuchungen über *Bartonella bacilliformis*. Morphologie der Kulturform. Zeitschrift für Tropenmedizin und Parasitologie **3**:313–326.

Peters, D., Wigand, R. 1955. Bartonellaceae. Bacteriological Reviews **19**:150–155.

Peters, W., Molyneux, N. H., Howells, R. E. 1973. *Eperythrozoon* and *Haemobartonella* in monkeys. Annals of Tropical Medicine and Parasitology **88**:47–50.

Peterson, K. J., Raleigh, R. J., Stroud, R. K., Goulding, R. L. 1977. Bovine anaplasmosis transmission studies conducted under controlled natural exposure in a *Dermacentor andersoni* (=*venustus*) indigenous area of eastern Oregon. American Journal of Veterinary Research **38**:351–354.

Pilcher, K. S., Wu, W. G., Muth, O. H. 1961. Studies on morphology and respiration of *Anaplasma marginale*. American Journal of Veterinary Research **22**:298–307.

Pinkerton, H., Weinman, D. 1937. Carrion's disease. II. Comparative morphology of the etiological agent in Oroya fever and verruga peruana. Proceedings of the Society of Experimental Biology and Medicine **37**:591–593.

Raether, W., Seidenath, H. 1977. Survival of *Aegyptianella pullorum, Anaplasma marginale* and various parasitic protozoa following prolonged storage in liquid nitrogen. Zeitschrift für Parasitenkunde **53**:41–46.

Rebagliati, R. 1940. ''Verruga Peruana''. Imprenta Torres Aguirre, Lima.

Reynafarje, C. 1972. Enfermedad de carrion. Acta Medica Peruana **1**:195–244.

Ristic, M. 1960. Anaplasmosis. Advances in Veterinary Science **6**:111–192.

Ristic, M. 1962. A capillary tube-agglutination test for anaplasmosis—a preliminary report. Journal of the American Veterinary Medical Association **141**:588–594.

Ristic, M. 1968. Anaplasmosis, pp. 474–536. In: Weinman, D., Ristic, M. (eds.), Infectious Blood Diseases of Man and Animals, vol. II. New York: Academic Press.

Ristic, M. 1977. Bovine anaplasmosis, pp. 235–248. In: Kreier, J. P. (ed.), Parasitic protozoa, vol. IV. New York: Academic Press.

Ristic, M., Carson, C. A. 1977. Methods of immunoprophylaxis against bovine anaplasmosis with emphasis on use of the attenuated *Anaplasma marginale*. Part IV. Anaplasmosis, pp. 151–188. In: Miller, L. H., Pino, J. A., McKelvey, J. J., Jr. (eds.), Immunity to blood parasites of man and animals. New York: Plenum.

Ristic, M., Kreier, J. P. 1974. Family III. Anaplasmataceae, pp. 906–914. In: Buchanan, R. E., Gibbons, N. E., (eds.), Bergey's manual of determinative bacteriology, 8th ed. Baltimore: Williams & Wilkins.

Ristic, M., Watrach, A. M. 1961. Studies in anaplasmosis. II. Electromicroscopy of *Anaplasma marginale* in deer. American Journal of Veterinary Research **22**:109–116.

Ristic, M., Watrach, A. M. 1963. Anaplasmosis. VI. Studies and a hypothesis concerning the cycle of development of the causative agent. American Journal of Veterinary Research **24**:267–277.

Ristic, M., White, F. H. 1960. Detection of *Anaplasma marginale* antibody complex formed *in vivo*. Science **130**:987–988.

Rodriguez, I. G. 1954. Baronellas y bartonellosis de los animales domesticos. Bulletin de Information-Consejo General de Colegios Veterinarios (Spain) **8**:365–371.

Rogers, T. E., Wallace, W. R. 1966. A rapid staining technique for *Anaplasma*. American Journal of Veterinary Research **27**:1127–1129.

Schilling, V. 1928. *Eperythrozoon coccoides*, eine neue durch Splenectomie aktivierbare Dauerinfektion der weissen Maus. Klinische Wochenschrift **7**:1854–1855.

Schurenkova, A. 1938. *Sogdianella moshkovskii* gen. nov., sp. nov.—a parasite belonging to the Piroplasmidea in a raptoroidal bird—*Gypaetus barbatus* L. Med. Parasit. (Mosk.) **7**:932–937. In: Helmy Mohammed, A. H. 1958. Systematic and experimental studies on protozoal blood parasites of Egyptian birds, vol. I and II. Cairo: University Press.

Seneviratra, P., Weerasinghe, N., Ariyadasa, S. 1973. Transmission of *Haemobartonella canis* by the dog flea tick *Rhipicephalus sanguineus*. Research in Veterinary Science **14**:112–114.

Senitzer, D., Dimopoullos, G. T., Brinkley, B. R., Mandel, M. 1972. Deoxyribonucleic acid of *Anaplasma marginale*. Journal of Bacteriology **409**:434–436.

Simpson, C. F., Kling, J. M., Love, J. N. 1967. Morphologic and histochemical nature of *Anaplasma marginale*. American Journal of Veterinary Research **28**:1055–1065.

Simpson, C. F., Neal, F. C., Edds, G. T. 1964. Gamma irradiation of *Anaplasma marginale*. American Journal of Veterinary Research **25**:1771–1772.

Smith, A. R., Rahn, T. 1975. An indirect haemagglutination test for the diagnosis of *Eperythrozoon suis* infection in swine. American Journal of Veterinary Research **36**:1319–1321.

Smith, T., Kilborne, F. L. 1893. Investigations into the nature, causation and prevention of Texas or Southern cattle fever. Eighth and Ninth Annual Reports, Bureau of Animal Industry, U.S. Department of Agriculture, for 177–304, 1891–1892.

Splitter, E. J. 1950. *Eperythrozoon suis* n. sp. and *Eperythrozoon parvum* n. sp., two new blood parasites of swine. Science **111**:513–514.

Summers, W. A., Matsuoka, T. 1970. Infectivity of *Anaplasma marginale* after long-term storage in bovine blood. American Journal of Veterinary Research **31**:1517–1518.

Swoager, W. C. 1972. Preservation of microorganisms by liquid nitrogen refrigeration. American Laboratory **4**:45–52.

Takano-Moran, J. 1970. Enfermedad de carrion (Bartonellosis humana). Estudio morfologico de la fase hematica y del periodo eruptivo con el microscopio electronico. Anales del Programa Academico de Medicina Humana. Universidad Nacional Mayor de San Marcos, Lima **53**:44–86.

Tanaka, H, Hall, W., Scheffield, J., Moore, D. 1965. Fine structure of *Haemobartonella muris* compared with *Eperythrozoon coccides* and *Mycoplasma pulmonis*. Journal of Bacteriology **90**:1735–1749.

Theiler, A. 1910. *Anaplasma marginale* (genus et spec. nov.). The marginal points in the blood of cattle suffering from a specific disease. Report Government Veterinary Bacteriology, Transvaal, South Africa, 1908–1909.

Trager, W., Jensen, J. B. 1976. Human malaria parasites in continuous culture. Science **193**:673–675.

Tyzzer, E. E. 1941. The isolation in culture of grahamellae from various species of small rodents. Proceedings of the National Academy of Sciences of the United States of America **27**:158–162.

Tyzzer, E. E. 1942. A comparative study of grahamellae, haemobartonellae, and eperythrozoa of small mammals. Proceedings of the American Philosophical Society **85**:359–398.

Tyzzer, E. E., Weinman, D. 1939. *Haemobartonella* n.g. (Bartonella olim pro parte) *H. microti* n. sp. of the field vole. American Journal of Hygiene **30**:141.

Vassiliadis, P. 1935. Nouvelle espece de *Grahamella* et prevende leur nature parasitaire. Annales de la Societe Belge de Medicine Tropicale **15**:279–288.

Venable, J., Ewing, S. 1968. Fine structure of *Haemobatonella canis* and its relation to the host erythrocyte. Journal of Parasitology **54**:259–268.

Wallace, W. R., Dimopoullos, G. T. 1965. Biologic properties of *Anaplasma marginale:* Effects of ionizing radiation on infectivity of partially purified marginal body preparations. American Journal of Veterinary Research **26**:135–138.

Weinman, D. 1944. Infectious anemia due to bartonella and related red cell parasites. Transactions of the American Philosophical Society **33**:243–350.

Weinman, D. 1968. Bartonellosis, pp. 3–24. In: Weinman, D., Ristic, M. (eds.), Infectious blood diseases of man and animals, vol. II. New York: Academic Press.

Weinman, D. 1974. Family II. Bartonellaceae, pp. 903–906. In: Buchanan, R. E., Gibbons, N. E. (eds.), Bergey's manual of determinative bacteriology, 8th ed. Baltimore: Williams & Wilkins.

Weinman, D., Kreier, J. P. 1977. *Bartonella* and *Grahamella*, pp. 197–233. In: Kreier, J. P. (ed.), Parasitic protozoa, vol. IV. New York: Academic Press.

Weinman, D., Pinkerton, H. 1937a. Carrion's disease. III. Experimental production in animals. Proceedings of the Society for Experimental Biology and Medicine **37**:594–595.

Weinman, D., Pinkerton, H. 1937b. Carrion's disease. IV. Natural sources of *Bartonella* in endemic zone. Proceedings of the Society for Experimental Biology and Medicine **37**:596–598.

Wigand, R. 1956. Serologische Reaktionen am *Haemobartonella muris* und *Eperythrozoon coccoides*. Zeitschrift für Tropenmedizin und Parasitologie **7**:322–340.

Wigand, R. 1958. Morphologische, biologische und serologische Eigenschaften der Bartonellen. Stuttgart: Georg Thieme Verlag.

Obligately Intracellular Bacteria: The Genus *Chlamydia*

LESLIE A. PAGE

The scientific history leading to the recognition of chlamydiae as a unique group of pathogenic bacteria that cause important and widely different diseases of man and animals is derived from the convergence of two independent paths of medical investigation. The first began in the pre-Christian era with Egyptian descriptions of an ocular disease we now know as trachoma, followed three millennia later by the discovery of the causative agent, "Chlamydozoa" as Von Prowazek (1907) called the agent. This observation was confirmed and extended by numerous other investigators. The second began in the 1890s when pneumonia in the owners of exotic psittacine birds, especially parrots, all too often followed the handling of recently imported specimens. The etiological agent was not known until a pandemic of "psittacosis" (Gr. psittakos = parrot) in 1929–1930 prompted further study of its cause. Lillie (1930), who called the causative agent *Rickettsia psittaci,* and four other physician-investigators independently and simultaneously observed the organisms in the lung phagocytes of naturally infected humans and experimentally infected laboratory animals. None of these investigators was able to propagate the organism on lifeless bacteriological media, but all reproduced the disease in experimental animals with filtrates of infectious tissues. Because the organisms could be seen in the cytoplasm of infected cells by light microscopy and because they failed to grow in the absence of living cells, the causative agent was thought to be a large virus. In 1932, Bedson and Bland described the intracellular developmental cycle of the psittacosis agent. Similar observations were made later by Busacca (1935) and others for the agent of trachoma. The similarities between these two disease agents and many other related strains were recognized in 1945 (Jones, Rake, and Stearns, 1945; Moshovskiy, 1945), but they were not classified in a single genus until two decades later (Page, 1966).

Now we know that chlamydiae are bacteria that multiply only within the cytoplasm of eukaryotic cells by a developmental cycle that is unique among microorganisms. They are bacteria and not viruses because they multiply by fission, their prokaryotic nuclei contain both RNA and DNA, the infectious form of the organism has a rigid trilaminar cell wall that is chemically similar to that of Gram-negative bacteria, and their multiplication is inhibited by several classes of antibiotics.

All of the stages of the intracellular developmental cycle are seen in Fig. 1. The cycle begins after the host cell phagocytizes the small (200–500 nm) coccoidal infectious form of the organism, called the elementary body (EB). Once inside a phagosome in the host cell's cytoplasm, the EB reorganizes into a larger (800–1,500 nm) form that has a porous, nonrigid cell wall surrounding a diffuse network of nuclear fibrils. This form is called the initial body or reticulate body (RB) because the fibrils have a reticulated appearance in electron micrographs of thin sections of the organism. The RB, which is noninfectious if prematurely released from the host cell, multiplies by fission. Daughter cells in turn reorganize to become EBs that contain a compact nucleus and ribosomes bounded by a rigid cell wall. Under optimum conditions in cultured eukaryotic cells, the entire infection cycle takes a minimum of 17 h but may take much longer in cells of a resistant animal host. Ultimately multiplication of the organism and the toxicity of its presence damage the host cell, so that infectious chlamydiae are released to continue the infection cycle.

Chlamydiae are metabolically deficient when compared with free-living bacteria. Although they have several identifiable independent enzymatic capabilities such as the catabolism of glucose, pyruvate, and glutamate when provided with essential cofactors, they are unable to produce or store high-energy compounds such as ATP (Moulder, Grisso, and Brubaker, 1965). This may be one of the primary reasons why chlamydiae are unable to multiply extracellularly. They have been aptly labeled energy parasites by Moulder (1964). They also are cytochrome-free anaerobes (Moulder, 1966; Weiss, 1968).

Because of basic similarities in morphology, mode of reproduction, and antigenically similar lipopolysaccharide moieties in the cell walls of all presently known strains of chlamydiae, they have

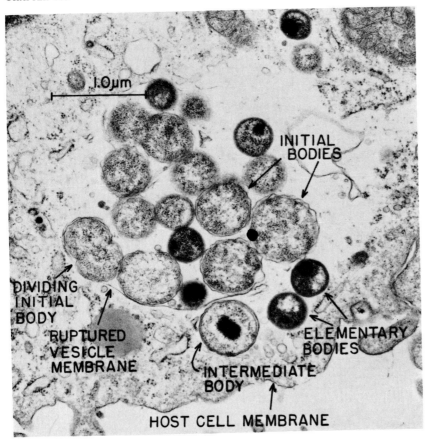

Fig. 1. Microcolony of *Chlamydia psittaci* in McCoy cell. [From Cutlip, 1970, with permission.]

been assigned to one genus, *Chlamydia,* for taxonomic purposes (Page, 1966, 1974a). Furthermore, the uniqueness of their intracellular developmental cycle warranted placing the genus (and coordinate family Chlamydiaceae) in a separate order Chlamydiales to distinguish them from organisms of the order Rickettsiales, which have different growth cycles and are metabolically more advanced.

Finally, two distinctive metabolic activities of chlamydiae (production of glycogen and folates) that are measurable in the laboratory permit separation of chlamydial isolates from any source into either of two species, *Chlamydia trachomatis* and *Chlamydia psittaci* (Table 1). Fifteen serotypes of *C. trachomatis* have been delineated (Grayston and Wang, 1975); however, a serotyping system for *C. psittaci* has not yet been completed.

This classification system has greatly simplified chlamydial nomenclature and identification of new isolates. Prior to 1966, classification schemes for chlamydiae contained as many as 18 species in 4 genera separated on the basis of presumed host or tissue preferences of the organisms, overlapping antigenic components, and patronymics on the part of history-minded humans (Page, 1968).

Habitats

Host Spectrum

Because chlamydiae can propagate only within living eukaryotic cells, their habitats are, of necessity, tissues of susceptible animal hosts. Chlamydiae are known as parasites only of warm-blooded animals, of which at least 200 species are affected. Strains of *Chlamydia trachomatis* are found only in man, while *C. psittaci* strains are found in numerous avian and mammalian species as well as in man. Recently, however, the host range has been extended to invertebrates. Electron micrographs of this sections of gut tissues of Chesapeake Bay clams clearly revealed intracellular microcolonies of organisms identical in morphology to chlamydiae in all stages of development (Harshberger, Chang, and Otto, 1977). As yet, these organisms have not been isolated and characterized.

Chlamydiae have been repeatedly isolated from insects and arachnids associated with chlamydiosis in poultry and cattle, but these isolations appear to be incidental. Furthermore, growth and multiplication of chlamydiae in insect cultures has never been

Table 1. Properties differentiating two species of *Chlamydia*.[a]

Chlamydia trachomatis	Chlamydia psittaci
1. Intracellular microcolonies generally are compact and produce glycogen that is stained differentially with iodine–potassium iodide solution.	1. Intracytoplasmic microcolonies are less compact and organisms often are distributed throughout the host cell's cytoplasm. Glycogen is not produced.
2. Synthesizes essential folate precursors; therefore, multiplication of the organisms in the yolk sac of chicken embryos is inhibited by sodium sulfadiazine at a concentration of 1 mg/embryo.	2. Multiplication in the yolk sac of chicken embryos is not inhibited by sodium sulfadiazine at a concentration of 1 mg/embryo.
3. Degree of DNA hybridization with strains of *C. psittaci* is 10% or less.	3. Degree of DNA hybridization with strains of *C. trachomatis* is 10% or less.

[a] Compiled from information in Page, 1974a. Generic names previously used for *Chlamydia* have been *Ehrlichia*, *Chlamydozoa*, *Rickettsiaformis*, *Bedsonia*, and *Rakeia*.

demonstrated (Eddie et al., 1962, 1969). However, Digregorio and Johnson (1967) developed some experimental evidence that *Dermacentor andersoni*, a tick, was capable of ingesting and transmitting *C. psittaci* between murine hosts.

Survival Characteristics of Chlamydiae

The following characteristics ensure the survival of chlamydiae in nature:

1. High infectivity for a wide range of host species. It appears that all warm-blooded vertebrates are susceptible to one chlamydial strain or another. This high infectivity might be due in part to an unusual surface component in the chlamydial cell wall that promotes phagocytosis of the organism by phagocytes not previously sensitized to chlamydiae (Byrne, 1976). This latter property probably represents a significant step in the evolutionary adaptation of chlamydiae to survival as obligately intracellular parasites.
2. Most chlamydial infections are not lethal to the animal host. There is a high incidence of latent infections. Subclinical chronic infections leading to a balanced host–parasite relationship and continuous carrier-excretion state ensure a continuous supply of infectious chlamydiae in the environment of susceptible animal groups.
3. The infectious form of chlamydiae, the elementary body, survives easily under adverse conditions outside an animal host because of its protective cell wall. Elementary bodies are resistant to drying and to proteolytic enzymes (although the high lipid content of their cell walls makes them quite sensitive to detergents and solvents). Animals chronically infected with *C. psittaci* may release the organisms in intestinal excre-

ment for long periods. Small particles of infectious excrement may become airborne as dust and may be inhaled by other susceptible hosts. The classic example of this type of interspecies transfer is psittacosis in man caused by the inhalation of cage floor dust of an infected parrot or parakeet. Imported psittacine birds and feral pigeons are the most common avian reservoir hosts of *C. psittaci*, and airborne infection of their caretakers without proper protection is a distinct hazard. Psittacosis is also an occupational hazard to unprotected personnel in poultry processing plants when infected poultry are slaughtered, as well as to microbiology laboratory workers handling cultures of *C. psittaci*. Any manipulations of suspensions of chlamydiae should be conducted under evacuating fume hoods with a minimum intake of 100 cubic feet of air per minute. *C. psittaci* also causes economically important diseases other than pneumonitis in domestic animals, namely, polyarthritis, enteritis, encephalomyelitis, conjunctivitis, and placentitis leading to abortion (Storz, 1971). A list of diseases, signs, lesions, and epidemiological patterns caused by *C. psittaci* is found in Table 2.

Major diseases caused by *C. trachomatis* occur only in man and only by nonairborne means. The habitats of this species are the ocular, oral, and urogenital tracts of man; transmission to and between the mucous membranes of these areas is by manual, oral, or venereal contact with infectious exudates from the conjunctiva, urethra, vagina, or rectum. The diseases are: trachoma, inclusion conjunctivitis, nongonococcal urethritis, cervicitis, salpingitis, neonatal conjunctivitis and pneumonitis, and lymphogranuloma venereum (lymphogranuloma inguinale). A list of these with signs, lesions, and epidemiological patterns is found in Table 3.

Table 2. Diseases and infections caused by various strains of *Chlamydia psittaci*.[a]

Disease and natural host	Principal effects, epidemiology, and transmission route[b]
Psittacosis-ornithosis Wild and domestic birds	Lethargy, hyperthermia, anorexia, yellowish excretions, lowered egg production. Fibrinous air sacculitis, pericarditis, peritonitis, splenomegaly, hepatopathy pneumonitis. Mortality 0–40% depending upon virulence of organism. Endemic in psittacine and columbine species and many other wild species. Airborne transmission.
Psittacosis Man	Malaise, headache, hyperthermia, anorexia, cough. Pneumonitis, splenitis, occasional meningitis. Mortality is less than 1% in antibiotic-treated (tetracycline) cases, and 20% in untreated cases. Worldwide distribution wherever there are avian carriers. Airborne transmission from birds to man, rarely from man to man.
Pneumonitis Young cats, lambs, calves, kids, pigs, horses, rabbits, mice	Conjunctival and/or nasal mucopurulent discharge; lethargy, anorexia, labored breathing, hyperthermia. Conjunctivitis, pneumonitis. Rarely fatal unless complicated by concurrent infection with viruses or other bacteria, especially *Mycoplasma* or *Pasteurella* species.
Polyarthritis Lambs, calves, pigs	Lameness, swollen carpal, tarsal, or stifle joints, hyperthermia, anorexia, lethargy. Fibrinous synovitis, tendonitis, occasional hepatopathy. Mortality is variable, usually due to debilitation or predation. Widespread among lambs and calves in western United States and among pigs in central Europe.
Placentitis leading to abortion Cattle, sheep, pigs, goats, rabbits	Transient hyperthermia, chlamydemia, inflammation, and necrosis of placentome, abortion of fetus late in gestation. Fetal hepatopathy, edema, ascites, vascular congestion, tracheal petechiae. Periodically epidemic in cattle in western United States. Endemic in sheep throughout the world.
Encephalomyelitis Calves	Lethargy, incoordination, weakness, hyperthermia, anorexia, diarrhea, paralysis. Fibrinous perihepatitis, pericarditis, ascites. Sporadic occurrence in United States.
Conjunctivitis Sheep, cattle, pigs, cats, guinea pigs	Vascular congestion and edema of conjunctiva, mucopurulent discharge, hyperthermia in cats. Follicular conjunctivitis, keratitis, pannus formation. Probably transmitted by airborne and contact contamination.
Fatal enteritis Snowshoe hare	Bizarre behavior, diarrhea, enteritis, splenomegaly, focal necrosis in liver. High mortality in wild hares in Canada between 1959 and 1961. Transmission route is unknown. Muskrats may be reservoirs.
Enteritis Cattle, sheep	Diarrhea, weakness, and death in newborn. Enteritis. Epidemiology unknown.
Subclinical intestinal infection Cattle, sheep, opossums, egrets, sparrows, grackles	No clinical signs in adults; may cause transient hyperthermia and diarrhea in newborn. Widespread in domestic herbivores throughout the United States. Probably transmitted in an oral-fecal cycle.

[a] From Page, L. A., 1974b, with permission.

[b] Clinical signs and gross lesions vary widely in cases of mild or chronic chlamydial infections. Variations in effects may also be caused by differences in natural resistance of individuals, virulence of the organisms, dosages, presence of other pathogens, and other stress factors.

Isolation

Chlamydiae are isolated and propagated only in eukaryotic cells. The most common sources of susceptible host cells for isolation work are embryonating chicken eggs or cultured line cells because all strains of chlamydiae of both species will grow in these convenient, low-cost systems. Laboratory mammals such as mice and guinea pigs are less used as isolation hosts because of their high cost per isolation attempt and their failure to support growth of many strains of both chlamydial species.

Table 4 lists the preferred specimens normally collected for examination from diseased human or animal hosts and the general approaches used for chlamydial isolation attempts. Detailed procedures that enhance the probability of isolation of chlamydiae from various specimens are described in the following text. Table 5 is an abbreviated flow diagram of isolation procedures.

Preliminary Treatment of Specimen

The prompt examination of fresh, uncontaminated specimens is obviously the most desirable. However, uncontaminated specimens may by stored at 4°C for several weeks prior to examination without excessive loss of chlamydial infectivity. Freezing of specimens at −20°C or below preserves chlamydial viability indefinitely, but an immediate one-log drop

Table 3. Diseases and infections caused by various serotypes of *Chlamydia trachomatis* in man.

Clinical disease	Usual serotype[a]	Principal effects and transmission
Trachoma	A, B, B$_a$, C	Progressive conjunctivitis primarily of upper eyelid with hyperemia, exudation, follicular hypertrophy. Neovascularization of cornea may lead to opacity and pannus formation. These strains may also be found on the genitalia. Transmitted by contamination of conjunctiva with infectious exudate. Blindness and pannus formation appear to be the result of local hyperallergic reaction after repeated infections.
Inclusion conjunctivitis, inclusion blennorrhea of newborn	D, E, F, G, H, I, J, K	Conjunctivitis, primarily of lower eyelid, which tends to heal spontaneously. Chlamydiae may be transmitted by contact contamination from mucous membranes of conjunctiva or genitalia and may spread from one site to the other. Transmitted at birth to eyes of newborn by contact contamination with cervical exudate (Schachter, 1978).
Urethritis, cervicitis, salpingitis, pharyngitis	B, C, D, E, F, G, H, I, K, L$_3$	Inflammation and exudation in urethra of males and in cervix, salpinx, and anus of females. Pharyngitis discovered among prostitutes and homosexual males. Organisms may spread by contact contamination between genitals and conjunctiva. Transmitted venereally.
Lymphogranuloma venereum (-Lymphogranuloma inguinale)	L$_1$, L$_2$, L$_3$[b]	Organisms spread from genitals to lymphatic tissue of iliac and inguinal regions, producing inflammation and suppuration of lymph nodes (bubos), occasional penile elephantiasis, or rectal strictures. Transmitted venereally.

[a] Grayston and Wang (1975). [b] Occasionally *C. psittaci* is isolated.

Table 4. Methods for preliminary examination of human and animal specimens for chlamydiae, preferred host, and procedures for chlamydial isolation attempts.

Host	Disease condition	Specimen	Strain or method for microscopic examination[a]	Isolation host; incubation temp; (Inoculation route)[b]
Human	Ocular or genital infection	Conjunctival scraping	FA, Iodine, Giemsa	CC, CE (YS), 35°C
		Urethral swab	FA, Iodine, Giemsa	CC, CE(YS), 35–37°C
		Bubos exudate	Inconclusive	CC, CE(YS), 35–37°C
	Pneumonitis	Sputum	Inconclusive	CC, CE (YS) 37°C; mice (IC, IP)
		Blood (during fever)	Inconclusive	CC, CE (YS), 37°C; mice (IC, IP)
		Lung, spleen (at necropsy)	FA, Giemsa, Gimenez	CC, CE (YS), 37°C; mice, (IC, IP)
Avian	Air sacculitis or generalized infection	Airsac exudate	PCM, FA	CE (YS), CC, 39°C; mice, (IC, IP)
		Liver, spleen	Inconclusive	CE (YS), CC, 39°C
		Cloacal swab	Inconclusive	CE (YS), 39°C (after centrifugation[a])
Mammal	Pneumonitis	Tracheal exudate or lung	PCM, FA, Gimenez	CE (YS), CC, 39°C mice (IC, IN)
	Polyarthritis	Synovial exudate from carpal, tarsal, or stifle joints	PCM, FA, Gimenez	CE (YS), CC, 39°C, guinea pig (IP)
	Placentitis (abortion)	Necrotic area of placentome	PCM, FA, Gimenez	CE (YS), CC, 39°C mice (IP)
	Encephalomyelitis	Affected brain tissue, liver, spleen	FA, Gimenez	CE (YS), CC, 39°C guinea pig (IP)
	Enteritis	Swab of inner lining of colon	Inconclusive	CE (YS) 39°C, (after centrifugation[a])
	Asymptomatic intestinal infection	Swab of inner lining of colon	Inconclusive	CE (YS), 39°C (after centrifugation[a])

[a] See text for details. FA, Fluorescent antibody stain; PCM, phase-contrast microscopy.
[b] CC, Cell culture; CE, chicken embryo; YS, yolk sac; IP, intraperitoneal; IC, intracerebral; IN, intranasal.

Table 5. Flow diagram for isolation and identification of chlamydiae in clinical specimens from humans and animals.[a]

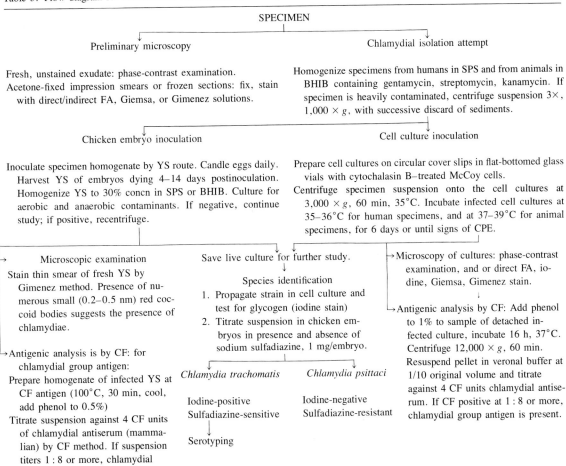

SPECIMEN

Preliminary microscopy   Chlamydial isolation attempt

Fresh, unstained exudate: phase-contrast examination.
Acetone-fixed impression smears or frozen sections: fix, stain with direct/indirect FA, Giemsa, or Gimenez solutions.

Homogenize specimens from humans in SPS and from animals in BHIB containing gentamycin, streptomycin, kanamycin. If specimen is heavily contaminated, centrifuge suspension 3×, 1,000 × *g*, with successive discard of sediments.

Chicken embryo inoculation   Cell culture inoculation

Inoculate specimen homogenate by YS route. Candle eggs daily. Harvest YS of embryos dying 4–14 days postinoculation. Homogenize YS to 30% concn in SPS or BHIB. Culture for aerobic and anaerobic contaminants. If negative, continue study; if positive, recentrifuge.

Prepare cell cultures on circular cover slips in flat-bottomed glass vials with cytochalasin B–treated McCoy cells.
Centrifuge specimen suspension onto the cell cultures at 3,000 × *g*, 60 min, 35°C. Incubate infected cell cultures at 35–36°C for human specimens, and at 37–39°C for animal specimens, for 6 days or until signs of CPE.

Microscopic examination   Save live culture for further study.   Microscopy of cultures: phase-contrast examination, and or direct FA, iodine, Giemsa, Gimenez stain.

Stain thin smear of fresh YS by Gimenez method. Presence of numerous small (0.2–0.5 nm) red coccoid bodies suggests the presence of chlamydiae.

Species identification
1. Propagate strain in cell culture and test for glycogen (iodine stain)
2. Titrate suspension in chicken embryos in presence and absence of sodium sulfadiazine, 1 mg/embryo.

Antigenic analysis by CF: Add phenol to 1% to sample of detached infected culture, incubate 16 h, 37°C. Centrifuge 12,000 × *g*, 60 min. Resuspend pellet in veronal buffer at 1/10 original volume and titrate against 4 CF units chlamydial antiserum. If CF positive at 1 : 8 or more, chlamydial group antigen is present.

Antigenic analysis is by CF: for chlamydial group antigen:
Prepare homogenate of infected YS at CF antigen (100°C, 30 min, cool, add phenol to 0.5%)
Titrate suspension against 4 CF units of chlamydial antiserum (mammalian) by CF method. If suspension titers 1 : 8 or more, chlamydial group ag. is present.

Chlamydia trachomatis   *Chlamydia psittaci*

Iodine-positive   Iodine-negative
Sulfadiazine-sensitive   Sulfadiazine-resistant

Serotyping

[a] Abbreviations: FA, fluorescent-antibody; SPS, sucrose phosphate solution; BHIB, beef heart infusion broth; YS, yolk sac; CF, complement fixation.

in infectivity titer for each freeze–thaw cycle can be expected as well as a slow decline in infectivity for the first 6 months of low temperature storage (Page, 1959). When there is a possibility of bacterial contamination, the specimen is homogenized in a diluent containing 1 mg each of gentamicin sulfate, streptomycin sulfate, and kanamycin sulfate. Chlamydial infectivity is also unaffected by similar concentrations of vancomycin sulfate and bacitracin. The preferred diluent for *Chlamydia trachomatis* strains is sucrose phosphate solution (SPS): 0.25 M sucrose and 0.2 M KH_2PO_4 brought to pH 7.2 with NaOH. Beef heart infusion broth (BHIB) at pH 7.2 is an excellent diluent for *C. psittaci*. In the author's laboratory, phosphate-buffered saline (PBS) at 0.2–0.02 M concentrations, which is commonly used as a diluent for chlamydial suspensions in other laboratories, inactivates dilute suspensions of chlamydiae within several hours. Maintenance of the pH of any diluent at 7.0–7.6 is essential for preservation of chlamydial infectivity in dilute suspensions in any case.

Control of Bacterial Contaminants

Heavily contaminated specimens, such as anal, vaginal, urethral, or cloacal swabs, can be decontaminated by suspending the specimen in 5–10 cc of antibiotic-containing diluent and then centrifuging the suspension at 1,000 × *g* for 30 min at 4°C. The supernatant fluid is carefully removed without disturbing the sediment, and the procedure is repeated twice more. The final supernatant fluid may then be inoculated in embryonating chicken eggs or cultured cells. At these light centrifugal forces, most chlamydial elementary bodies remain in suspension, while heavier bacteria and cellular debris are sedimented.

Isolation of Chlamydiae in Embryonating Chicken Eggs

INCUBATION TEMPERATURE AND RELATIVE HUMIDITY. Embryonating eggs inoculated via the yolk sac

(YS) with human ocular or genital exudates suspected of containing *C. trachomatis* should be incubated at 35°C in 95–100% relative humidity as a general rule. Some strains of *C. trachomatis* are sensitive to higher temperatures. Embryos incubated at that temperature require high humidity to survive. Eggs inoculated with *C. psittaci* strains should be incubated at 37–39°C at 60% humidity. The rate of multiplication of *C. psittaci* strains is enhanced at higher temperatures (Page, 1971).

DOSE/RESPONSE RELATIONSHIPS. The average day to death (ADD) of embryos inoculated with chlamydiae is related to the number of organisms inoculated, the virulence of the strain, and the temperature of incubation. Given these variables, embryos may die after chlamydial inoculation in 3–14 days. Embryos damaged by the inoculation itself or inoculated inadvertently with bacterial contaminants will usually die in 1–3 days; thus, eggs with embryos dying early may be discarded. Multiplication of chlamydiae in the cells lining the YS produce areas of vascular congestion and hemorrhage in the membrane. In the examination of dead embryos, such areas should be harvested for further study. Thin impression smears of a small portion of the YS on glass slides should be made for staining. The remainder should be homogenized in BHIB or SPS to make a 30% suspension and dispensed into several vials. One vial should be saved for antigenic analysis by a CF method, and other vials should be preserved at −20°C for further study. The suspension should be cultured aerobically and anaerobically for bacterial contaminants.

Isolation of Chlamydiae in Cultured Cells

Several lines of cultured cells support growth of a variety of strains of chlamydiae. The principal lines used are the L929 and McCoy mouse fibroblasts, Hela 229, and Baby Hamster Kidney (BHK-21). These cells may be cultured as monolayers or in suspension. Both L929 and McCoy lines are hardy, rapidly growing cells that support growth of most chlamydial strains. Most investigators have individual preferences as to which cell lines and which conditions produce the best results (Storz and Spears, 1977).

PREINOCULATION TREATMENT OF CULTURED CELLS. Recent studies have shown that preinoculation treatment of the cells with irradiation or various chemical cell-growth inhibitors enhances the susceptibility of the cells to infection and increases yields of chlamydiae. X-irradiation of the cells (Gordon, Dressler, and Quan, 1963, 1967), treatment with cytochalasin B (Sompolinsky and Richmond, 1974), 5-iodo-2-deoxyuridine (Wentworth

and Alexander, 1974), or the polycation diethylaminoethyl-dextran (DEAE-D) (Harrison, 1970; Kuo et al. 1972) have all been shown to enhance chlamydial uptake and infectivity and yields of chlamydiae. Furthermore, centrifugation of the inoculum suspension at $3,000 \times g$ for 60 min at 35–37°C onto the cell monolayers has been shown to increase host cell infection rates materially (Harrison, 1970; Weiss and Dressler, (960). Other workers report satisfactory yields of chlamydiae in nontreated cell cultures infected by stationary absorption or by gentle shaking methods.

While methodology varies from laboratory to laboratory, the author recommends the following procedure, based in part on the work of Smith et al. (1975), for the routine isolation of chlamydiae from either human or animal specimens in cultured cells.

Isolation of Chlamydiae from Human or Animal Specimens

Cytochalasin B (Imperial Chemical Ind., Ltd., Cheshire, England) at 1 μg/ml is incorporated into a seed suspension of McCoy cells just prior to preparing a monolayer on circular coverslips in glass, flat-bottomed vials and incubated for 3 days. The McCoy cells are propagated in Eagle MEM medium supplemented with 0.25% lactoalbumin hydrolysate, glucose at 30 nM/ml, 10% fetal calf serum, 20 μM Hepes/ml, 2 μM glutamine/ml, gentamycin sulfate 40 μg/ml, streptomycin 50 μg/ml. The specimen is suspended in the above medium and centrifuged onto the monolayer at $2,000 \times g$ for 60 min at room temperature. Cultured cells should be incubated at 35–36°C when infected with specimens from humans and at 37–39°C with specimens from avian or nonprimate mammalian species. In infected cultures, cytopathogenicity in the form of rounding, swelling, and detachment of infected cells in monolayers after 2–6 days' incubation occurs depending on the infection multiplicity. The cytochalasin B–treated cells have numerous multinucleated giant cells whose cytoplasms support the growth of many chlamydial microcolonies. For microscopic examination, cover slip cultures may be observed live in wet mounts under phase-contrast optics, or after staining with Giemsa, Gimenez, or iodine solutions, or with fluorescent-antibody conjugates as described below.

Isolation of Chlamydiae in Mice or Guinea Pigs

Specimens for inoculation into laboratory animals should be homogenized and treated with antibiotics as mentioned above.

Most but not all strains of both species of chla-

mydiae can be isolated in newly weaned mice inoculated intracerebrally. Serotypes A, B, and C of *C.
trachomatis* and polyarthritis strains of *C. psittaci*
generally will not multiply in mice. All strains of *C.
psittaci* from avian sources will multiply in mice
whether given intracerebrally, intranasally, or intraperitoneally.

Before laboratory mice or guinea pigs are used
for isolation attempts, the source colony should be
examined for the presence of natural chlamydial infection. Some colonies of commercial laboratory
mice have latent respiratory chlamydial infection
which can be detected by rapid blind intranasal passage of homogenates of the lungs of representative
individuals of the colony into other members of the
colony. If the colony is naturally infected, several
serial passages should result in the appearance of
chlamydial pneumonitis and circulating chlamydial
antibodies in recipient mice. These conditions can
be identified by procedures outlined in the section on
Identification.

Many guinea pig colonies have natural chlamydial infections which manifest as conjunctivitis
(Murray, 1964). The infection is detected by isolation of ocular chlamydiae in chicken eggs and detection of chlamydial antibodies in serums of affected guinea pigs.

Weanling mice inoculated intracerebrally with
chlamydial suspensions die within 4–10 days with a
hemorrhagic meningitis. Impression smears of meningeal exudates stained with Giemsa or Gimenez
solutions reveal large numbers of mononuclear cells
with numerous intracytoplasmic chlamydiae. The
presence of chlamydiae must then be confirmed by
use of specific immunofluorescent antiserum.

Young mice infected intranasally by instillation of
droplets of chlamydial suspension into the nares of
the anesthetized animal die within 5–10 days with
severe chlamydial pneumonitis. Freshly cut sections
of the lungs of such animals typically are dark red.
Impression smears of these sections show large
numbers of chlamydiae-infected mononuclear cells.

Mice or guinea pigs infected intraperitoneally
develop large spleens, livers, and a stringy fibrinous
exudate in the peritoneal cavity. The exudate contains numerous chlamydiae-infected mononuclear
cells (Fig. 2). Preceding death the animals show
signs of lethargy, anorexia, hyperthermia, and inco-
ordination. Some chlamydial strains cause the development of a voluminous ascites in mice (as much
as 5–10 ml in a 20-g mouse), making the mouse so
bloated that it cannot move to get food or water.
This condition may produce death by itself. Guinea
pigs inoculated with virulent strains of chlamydiae
from birds may die of severe meningitis within a few
days after intraperitoneal inoculation with relatively
little peritoneal involvement. Some strains of *C.
psittaci* multiply and cause death or lesions in mice
but not guinea pigs and vice versa (Table 6).

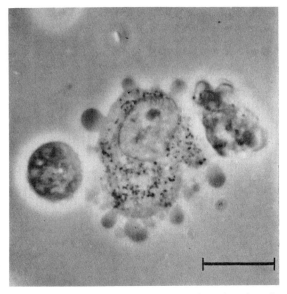

Fig. 2. Unstained mononuclear cell from peritoneal exudate of a
mouse infected with *Chlamydia psittaci*. Wet mount, phase-
contrast. Bar = 10 μm.

Diagnostic Significance of a Chlamydial Isolation

Isolation of chlamydiae from clinical material is
usually related to the need for confirmation of a
clinical diagnosis or for determination of epidemiologically important serotypes. However, because
infection with chlamydiae of low virulence is so
widespread, the investigator should be aware of the
possibility that a diseased host may be infected with
not one but two chlamydial strains—a pathogenic
one and an avirulent one that is a normal inhabitant
of the intestinal or genital tract, or that the diseased
host's lesions are caused by an entirely different
class of agent and that an asymptomatic chlamydial
infection, an organism of which was isolated, may
be completely incidental to the disease encountered.

Identification

A number of methods are available for assured identification of chlamydiae in exudates or tissues of the
originally infected host, in cell cultures, or in laboratory animals in which chlamydiae have been propagated. Preliminary microscopic and cultural examination of original specimens for characteristic
chlamydial inflammatory cell responses as well as
for the exclusion of bacterial contaminants is helpful. Giemsa or Gimenez solutions stain chlamydiae
easily but do not provide specific evidence of their
presence because these solutions stain other microorganisms equally well. Aside from isolation procedures, only antigenic analysis of chlamydiae-rich
exudates by immunofluorescent stains or by com-

Table 6. Pathogenicity of various strains of *Chlamydia psittaci* for laboratory animals.

Host (inoculation route)[a]	Strain of *C. psittaci*					
	NJ1 turkey ornithosis	CP3 pigeon ornithosis	EBA cow abortion	B577 ewe abortion	PA lamb polyarthritis	McNutt calf encephalomyelitis
Mouse (IP)	++[b]	+	+	++	−	−
Guinea pig (IP)	++	−	−	+	++	++
Pigeon (IC)	−	+	−	+	−	−
Sparrow (IP)	−	+	−	++	+	−
Parakeet (IP)	++	+	−	−	+	−
Turkey (IP)	++	+	−	+	+	−
Lamb (IP)	−	−	−	+	++	−
Chicken embryo (YS)	++	++	++	++	++	++

[a]IP, intraperitoneal; IC, intracerebral; YS, yolk sac.
[b]Degree of pathogenicity: ++, severe disease, death; +, mild disease; −, no effect. Each strain of *C. psittaci* was titrated in the laboratory species indicated (except lambs which received only high dosages). Pathogenicity designations reflect the ability of a small number of inoculated organisms to multiply and cause disease in the test species. Data from Page (1967).

plement-fixation (CF) tests for antigen provide conclusive proof of the presence of chlamydiae in the specimen. Methods for identification of an isolate first for chlamydial group antigen, then for species and serotype designation follow.

Direct Examination of Tissues of Affected Host

Microscopic examination under phase-contrast optics of wet mounts of fresh inflammatory exudates of chlamydiae-infected tissues reveals characteristic chlamydial inclusions in mononuclear cells, as well as the presence of contaminating bacteria or fungi. Cells infected with *Chlamydia psittaci* have chlamydiae distributed throughout the cytoplasm (Figs. 2 and 3), while cells infected with *C. trachomatis* generally have compact inclusions filled with chlamydiae (Fig. 4). In impression smears with Giemsa

or Gimenez solutions, chlamydiae appear as dark purple (Giemsa) or bright red (Gimenez). Iodine stains of exudates of *C. trachomatis* infections show dark tan, iodine-positive microcolonies (Fig. 5).

When chlamydiae-rich exudates are found in the air sacs or pericardia of infected birds or in the necrotic placentomes of aborted cattle, chlamydial antigen may be detected by a CF method. The diseased tissues are homogenized in beef heart infusion broth (BHIB) and prepared as a CF antigen by heating the suspension at 100°C for 30 min, cooling the suspension, then adding phenol to 0.5% concentration. This suspension is then titrated in duplicate against 4 CF units of mammalian chlamydial antiserum and against a normal serum as control. If large

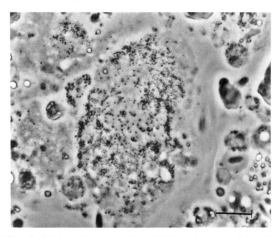

Fig. 3. Unstained mononuclear cell from air sac exudate of a pigeon naturally infected with *Chlamydia psittaci*. Wet mount, phase-contrast. Bar = 10 μm.

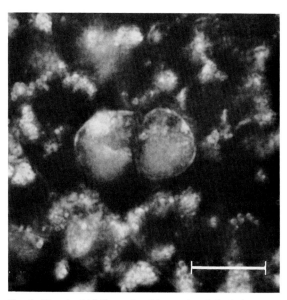

Fig. 4. Unstained McCoy cells with cytoplasmic inclusions containing *Chlamydia trachomatis*. Wet mount, phase-contrast. Bar = 10 μm.

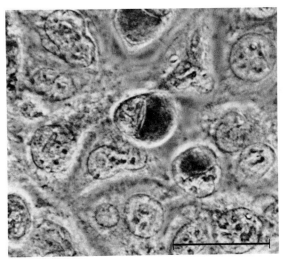

Fig. 5. Dark-staining, iodine-positive inclusions in McCoy cells infected with *Chlamydia trachomatis*. Bar = 10 μm.

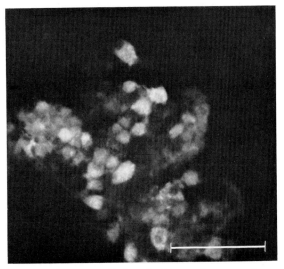

Fig. 6. Fluorescent-antibody stained impression smear of exudate from air sac of turkey infected with *Chlamydia psittaci*. Bar = 10 μm.

numbers of chlamydiae are in such suspensions, the heat-treated group antigen fixes complement in dilutions of 1 : 8 or more, thereby providing immediate proof of chlamydial infection of the original specimen. The specimen must have been shown previously to be free of large numbers of contaminating bacteria. This test cannot be applied to tissue homogenates of other organs such as spleen, liver, or kidney because these tissues fix or inactivate complement nonspecifically.

A satisfactory direct fluorescent-antibody (FA) test for exudates from humans infected with *C. trachomatis* has been devised by Nichols and McComb (1962). A modified version of their method was tested recently on air sac exudates from *C. psittaci*-infected birds (Tessler and Page, unpublished observations, 1977). The basic protocol follows.

Direct FA Test for *C. trachomatis*–Infected Exudates (Tessler, Stone, and Page, 1978)

Thin impression smears of fibrinous exudate from the diseased membrane are made on glass slides, air-dried, fixed in acetone for 3 min, and washed in 0.02 M PBS at pH 7.2. After air-drying, high-titered fluorescein-conjugated rabbit antichlamydial serum is spread on the smear and incubated for 30 min at 37°C in a humid chamber. The slide is then washed 2× with PBS and again dried. The smear is then flooded with 0.005% solution of Evans blue for 30 s followed by another PBS wash and a distilled water rinse. After air-drying, a drop of 50% glycerin in PBS is placed on the smear and covered with a cover slip. The specimen is observed at 250× under BG 12 filtered UV light. Chlamydial inclusions fluoresce brightly while nonspecific fluorescence

is subdued by the Evans blue counterstain (Fig. 6). Diseased air sacs infected with *Pasteurella multocida*, *E. coli*, *Mycoplasma gallisepticum* (all of which cause avian airsacculitis) are completely negative by this procedure.

Indirect fluorescent-antibody methods are suitable for identifying chlamydiae in tissues of infected hosts also. Conjugated antiserums against many different animal species are available commercially; the choice of system is up to the investigator, depending upon the source species of his chlamydial antiserum used in the indirect method.

Identification of Chlamydiae Propagated in the Yolk Sac of Embryonating Chicken Eggs

The yolk sacs (YS) of embryos dying in 4–14 days after inoculation are harvested free of other embryonic membranes and excess yolk, then homogenized to a 30% suspension in BHIB. A portion of the homogenate is stored for further study and cultured for contaminating bacteria. Another portion if prepared as a CF antigen as described above under the section on Direct Examination of Tissues.

CF Method for Identification of Chlamydiae (Page, 1975)

The antigen is titrated in duplicate against chlamydial antiserum and normal serum. The antigen is serially diluted in duplicate from 1 : 8 on in Veronal buffer. To one series of dilutions, 4 CF units of chlamydial antiserum are added to each tube, and normal serum is added to the second series (as an anticomplementary control).

Then 2 exact units of fresh guinea pig complement are added to each tube. The mixtures are incubated at 37°C for 120 or 90 min depending on whether a tube type or microplate method is used. The hemolytic system consisting of 2 units of hemolysin and washed sheep RBC ($OD_{0.25}$) are added. After further incubation at 37°C, the test is stored at 4°C until the RBC settle out, then read to determine the 50% end points of the titrations. Chlamydiae-infected YS may fix complement in dilutions of 1:8 to 1:2,048 depending upon the antigen concentration. This test provides conclusive evidence of the presence of chlamydiae. The antigen made from the infected YS may also be used at constant dilution of 4 antigenic units for the detection of antibodies in serums of infected humans or other mammals.

Thin smears of YS from dead embryos may be stained by the Gimenez (1964) method and examined under bright-field or phase-contrast optics. Intra- or extracellular clusters of chlamydial elementary bodies stain bright red, other bacteria stain bluish, and background cellular debris stains greenish. This procedure also stains rickettsiae and pox viruses in the same manner.

Staining and Examination of Infected Yolk Sac (Gimenez, 1964)

A stock dye solution is prepared by dissolving 2.5 g basic fuchsin in 50 ml 95% ethanol and adding this to 450 ml of aqueous 5% phenol and incubating the mixture at 37°C for 2 days. The counterstain is prepared by dissolving 0.8 g malachite green in 100 ml of distilled water. A fresh working solution of phenolic fuchsin is prepared by adding 8 ml of stock to 10 ml of 0.1 M phosphate buffer (pH 7.5) and filtering the solution through Whatman no. 2 filter paper. The exudate impression smear is fixed by air drying and gentle heat, and then covered with freshly filtered working solution of phenolic fuchsin for 5 min. The smear is then washed with tap water and covered with malchite green for 8–20 s depending on the thickness of the smear. The malchite green is washed off with tap water and the smear blotted dry. The smear may be scanned under dark field at 500× for coccoid particles and then viewed under bright field to confirm their color.

Identification of Chlamydiae Propagated in Cell Culture

Chlamydial growth in cultured cells (as in Fig. 4) is confirmed only by demonstration of the presence of the chlamydial group antigen by FA or CF procedures. Other procedures are necessary to determine the species and serotype. Development of cytopathic

effects in infected cell cultures is not always certain since some strains of both chlamydial species do not produce these effects in all lines of cultured cells. High-titered anti-chlamydial fluorescent conjugate is not available commercially. However, such a preparation can be made by immunizing rabbits with purified suspensions of chlamydiae (originally propagated in a species host cell other than that intended to be used for chlamydial isolation work), isolating the IgG fraction of the immune serum, efficiently conjugating the fraction with fluorescein isothiocyanate, purifying the preparation by passage through a Sephadex 50 column, and titrating it against various samples of chlamydiae-infected and noninfected tissue sample smears to determine the optimal working dilution.

Staining infected cover slip cultures with Giemsa or Gimenez solutions may reveal intracellular inclusions. If *C. trachomatis* is suspected to be present, cover slip cultures should be stained with iodine. This technique is outlined under the next section since it is an integral part of the species identification procedure.

Identification of chlamydial group antigen in infected cell cultures by a CF method is the same as that outlined for infected yolk sac cultures.

Identification of Chlamydial Species

The two recognized species are separated on the basis of glycogen (iodine test) and production of folic acid precursors (sulfadiazine sensitivity test). As indicated in Table 1, strains of *C. trachomatis* are iodine-positive and sulfadiazine-sensitive, while strains of *C. psittaci* are iodine-negative and sulfadiazine-resistant. The iodine test for glycogen production is performed on cover slip cell cultures infected with the new isolate. Iodine-positive inclusions are characteristic of moderately infected cell cultures at 2–4 days after inoculation. Heavily infected or older cultures may have fewer or no iodine-positive inclusions, in which case subculture of the strain in new cell cultures may be necessary to obtain a positive test. Gordon and Quan's method (1965) for staining infected cell monolayers is as follows: The culture is fixed in methanol and then stained for 10–15 min in a 1:1 mixture of 5% iodine (in 100% ethanol) and 5% potassium iodide in distilled water. The stain fluid is decanted and the preparation mounted on a slide with a drop of iodine in 50% glycerin. Glycogen containing inclusions appear dark tan against a light tan background.

Sulfadiazine sensitivity of a new isolate is determined by titrating a suspension of the chlamydial agent in embryonating chicken eggs inoculated by the YS route in the presence and absence of sodium sulfadiazine (1 mg/embryo). Ideally, decimal dilutions of the chlamydial suspension are inoculated into 6-day incubated fertile chicken eggs, 20 eggs/dilution. Half of the eggs of each dilution are given a second inoculation of a solution of sodium sulfadiazine and the other half are given a second inoculation of an equal volume of diluent without sulfadiazine. The eggs are candled daily for 14 days, the embryo deaths are recorded, and an embryo LD$_{50}$ is calculated for each titration. The calculated LD$_{50}$ for the sulfadiazine series should be lower than the control series by a factor of 100 in order to demonstrate sulfadiazine sensitivity.

A significant epidemiological breakthrough was accomplished by Wang and Grayston (1971) in devising a serotyping system for strains of *C. trachomatis*. In this system, antisera from mice inoculated with the new isolate are tested against prototype antigens by an indirect immunofluorescent method. Small pinpoint dots of antigens representing up to 15 serotypes of *C. trachomatis* are placed in a tight pattern on a glass slide. A loopful of mouse antiserum is overlaid on the dot collection and incubated at 37°C for 30 min. Dots of normal YS as negative controls are also placed on the slide. The slides are then rinsed with PBS and mechanically washed twice with PBS for 10 min. After drying, anti-mouse conjugate is overlaid on the pattern, incubated at 37°C for 30 min, and washed again as before. After drying, a cover slip mount is placed over the reaction sites and examined microscopically at 10°C and 50× with BG 12 filtered UV light. Fluorescence associated with clusters of chlamydial elementary bodies in each antigen dot is considered a positive reaction. The current distribution of serotypes involved in clinical syndromes caused by *C. trachomatis* is given in Table 2.

A similar serotyping system has not yet been devised for strains of *C. psittaci*.

An excellent review of the microbiology of the chlamydiae has recently been published by Storz and Spears (1977). A review of the diseases caused by chlamydiae was also published by Storz in 1971. A review of recent knowledge of human diseases caused by *C. trachomatis* was published by Grayston and Wang in 1975.

Literature Cited

Bedson, S. P., Bland, J. O. W. 1932. Morphological study of the psittacosis virus with a description of the developmental cycle. British Journal of Experimental Pathology 31:461–466.

Busacca, A. 1935. Un germe aux caracteres de rickettsies (*Rickettsia trachomae*) dans les tissus trachomateux. Archives Ophtalmologie (Paris) 52:567–572.

Byrne, G. I. 1976. Requirements for ingestion of *Chlamydia psittaci* by mouse fibroblasts (L cells). Infection and Immunity 14:645–651.

Cutlip, R. C. 1970. Electron microscopy of cell cultures infected with a chlamydial agent causing polyarthritis of lambs. Infection and Immunity 1:499–502.

Digregorio, D., Johnson, D. E. 1967. Investigations concerning the transmission of psittacosis by two species of *Dermacentor* (Ixodoidea, Ixodidae). Journal of Infectious Diseases 117:418–420.

Eddie, B., Meyer, K. F., Lambrecht, F. L., Furman, D. P. 1962. Isolation of ornithosis bedsoniae from mites collected in turkey quarters and from chicken lice. Journal of Infectious Diseases 110:231–237.

Eddie, B., Radovsky, F. J., Stiller, D., Kumada, N. 1969. Psittacosis-lymphogranuloma venereum (PL) agents (*Bedsonia, Chlamydia*) in ticks, fleas, and native mammals in California. American Journal of Epidemiology 90:449–460.

Giminez, D. F. 1964. Staining rickettsiae in yolk sac cultures. Stain Technology 39:135–140.

Gordon, F. B., Dressler, H. R., Quan, A. L. 1967. Relative sensitivity of cell culture and yolk sac for detection of TRIC infection. American Journal of Ophthalmology 63:1044–1048.

Gordon, F. B., Quan, A. L. 1965. Occurrence of glycogen in inclusions of the psittacosis-lymphogranuloma venereum-trachoma agents. Journal of Infectious Diseases 115:186–196.

Gordon, F. B., Magruder, G. B., Quan, A. L., Arm, H. G. 1963. Cell cultures for detection of trachoma virus from experimental simian infections. Proceedings of the Society for Experimental Biology and Medicine 112:236–242.

Grayston, J. T., Wang, S. 1975. New knowledge of chlamydiae and the diseases they cause. Journal of Infectious Diseases 132:87–105.

Harrison, M. J. 1970. Enhancing effect of DEAE-dextran on inclusion counts of an ovine *Chlamydia (Bedsonia)* in cell culture. Australian Journal of Experimental Biology and Medical Science 48:207–213.

Harshberger, J. C., Chang, S. C., Otto, S. V. 1977. Chlamydiae (with phages), mycoplasmas, and rickettsiae in Chesapeake Bay bivalves. Science 196:666–668.

Jones, H., Rake, G., Stearns, B. 1945. Studies on lymphogranuloma venereum. III. The action of the sulfonamides on the agent of lymphogranuloma venereum. Journal of Infectious Diseases 76:55–69.

Kuo, C. C., Wang, S. P., Wentworth, B. B., Grayston, J. T. 1972. Primary isolation of TRIC organisms in HeLa 229 cells treated with DEAE-dextran. Journal of Infectious Diseases 125:665–668.

Lillie, R. D. 1930. Psittacosis-rickettsia-like inclusions in man and in experimental animals. Public Health Reports 45:773–778.

Moshkovskiy, S. D. 1945. The cytotropic agents of infections and the position of the rickettsiae in the system of Chlamydozoa. [In Russian.] Uspekhi Souremennoi Biologie [Advances in Modern Biology] USSR 19:1–44.

Moulder, J. W. 1964. The psittacosis group as bacteria. New York: John Wiley & Sons.

Moulder, J. W. 1966. The relation of the psittacosis group (chlamydiae) to bacteria and viruses. Annual Review of Microbiology 20:107–130.

Moulder, J. W., Grisso, D. L., Brubaker, R. R. 1965. Enzymes of glucose metabolism in a member of the psittacosis group. Journal of Bacteriology 89:810–812.

Murray, E. S. 1964. Guinea pig inclusion conjunctivitis virus. I. Isolation and identification as a member of the psittacosis-lymphogranuloma-trachoma group. Journal of Infectious Diseases 114:1–12.

Nichols, R. I., McComb, D. E. 1962. Immunofluorescent studies with trachoma and related antigens. Journal of Immunology 89:545–554.

Page, L. A. 1959. Thermal inactivation studies on a turkey ornithosis virus. Avian Diseases 3:67–79.

Page, L. A. 1966. Revision of the family *Chlamydiaceae* Rake (*Rickettsiales*): Unification of the psittacosis-lymphogranuloma venereum-trachoma group of organisms in the genus

Chlamydia Jones, Rake, and Stearns, 1945. International Journal of Systematic Bacteriology **16**:223–252.

Page, L. A. 1967. Comparison of "pathotypes" among chlamydial (psittacosis) strains recovered from diseased birds and mammals. Bulletin of the Wildlife Disease Association **3**:166–175.

Page, L. A. 1968. Proposal for the recognition of two species in the genus *Chlamydia* Jones, Rake, and Stearns, 1945. International Journal of Systematic Bacteriology **18**:51–66.

Page, L. A. 1971. Influence of temperature on the multiplication of chlamydiae in chicken embryos. Excerpta Medica International Congress, Series **223**:40–51.

Page, L. A. 1974a. Chlamydiales Storz and Page 1971, pp. 914–918. In: Buchanan, R. E., Gibbons, N. E. (eds.), Bergey's manual of determinative bacteriology, 8th ed. Baltimore: Williams & Wilkins.

Page, L. A. 1974b. The chlamydiae, p. 124. In: Laskin, A. I., Chevalier, H. A. (eds.), Handbook of microbiology, vol. 1. Cleveland: CRC Press.

Page, L. A. 1975. Chlamydiosis, p. 129. In: Isolation and identification of avian pathogens, chap. 16. College Station, Texas: American Association of Avian Pathologists, Texas A&M University.

Schachter, J. 1978. Chlamydial infections (third of three parts). New England Journal of Medicine **298**:540–549.

Smith, T. F., Weed, L. A., Segura, J. W., Pettersen, G. R., Washington, J. A., II. 1975. Isolation of *Chlamydia* from patients with urethritis. Mayo Clinic Proceedings **50**:105–110.

Sompolinsky, D., Richmond, S. 1974. Growth of *Chlamydia* *trachomatis* in McCoy cells treated with cytochalasin B. Applied Microbiology **28**:912–914.

Storz, J. 1971. *Chlamydia* and *Chlamydia*-induced diseases. Springfield, Illinois: Charles C. Thomas.

Storz, J., Spears, P. 1977. Chlamydiales: Properties, cycle of development and effect on eukaryotic host cells. Current Topics in Microbiology and Immunology **76**:168–214.

Tessler, J., Stone, S. S., Page, L. A. 1978. A direct immunofluorescent test for *Chlamydia psittia* in infected tissues. Abstracts of the Annual Meeting of the American Society for Microbiology **1978**:310.

Von Prowazek, S. 1907. Chlamydozoa. I. Zuzammenfassende Uebersicht. Archiv für Protistenkunde **22**:248–298.

Wang, S. P., Grayston, J. T. 1971. Classification of TRIC and related strains with micro immunofluorescence. Excerpta Medica International Congress Series **223**:305–321.

Weiss, E. 1967. Comparative metabolism of chlamydia with special reference to catabolic activities. American Journal of Ophthamology **63**:1098–1101.

Weiss, E. 1968. Comparative metabolism of rickettsial and other host dependent bacteria. Zentralblatt für Bakteriologie, Parasitenkunde, Infektionskrankheiten und Hygiene, Abt. 1 Orig. **206**:292–298.

Weiss, E., Dressler, H. R. 1960. Centrifugation of rickettsia and viruses onto cells and its effect on infection. Proceedings of the Society for Experimental Biology and Medicine **103**:691–695.

Wentworth, B. B., Alexander, E. R. 1974. Isolation of *Chlamydia trachomatis* by use of 5-iodo-2-deoxyuridine-treated cells. Applied Microbiology **27**:912–916.

SECTION W
Wall-Deficient Bacteria

The L-Forms of Bacteria

SARABELLE MADOFF

Introduction

The discovery of the L-forms over 40 years ago ushered in a new era in the understanding of the bacterial cell. The name L-forms (L for Lister Institute) was first proposed by Klieneberger (1935), who discovered colonies resembling the organisms of bovine pleuropneumonia in cultures of *Streptobacillus moniliformis* isolated from the rat. Dienes (1939) studied the development of the L_1 from *Streptobacillus* and its return to the bacillary form, thereby establishing the bacterial derivation of the L_1. Within a few years, the spontaneous development of L-form colonies was observed in *Bacteroides, Haemophilus, Escherichia,* and *Neisseria* (for an early review, see Dienes and Weinberger, 1951).

The discovery that L-forms could be induced by exposure of bacteria to penicillin (Pierce, 1942) was an important advance. Other antibiotics were found to produce a similar result by interfering with biosynthesis of cell wall. Exposure of bacteria to bacteriostatic chemicals, to phage, and to antibodies also resulted in L-transformation. In 1954, Sharp recognized that organisms lacking a rigid cell wall needed osmotic protection, and he induced L-forms from group A streptococci on media containing high salt concentration. The current belief is that conversion to L-forms may be a universal property of bacteria, provided that conditions for induction and growth can be determined. L-forms have been derived from many bacterial genera (Hijmans and Clasener, 1971) as shown in Table 1.

Terminology and Biological Characteristics

An L-form can be defined as the manifestation of the continued growth of a bacterium following suppression of its cell wall. This phenomenon results in the replication of bacteria in a manner similar to that of the mycoplasma; in culture, the granular elements and large bodies embed themselves in the agar gel with the result that L-colonies can also exhibit the well-known, "fried egg" appearance. Other characteristics similar to those of the mycoplasma, such as fragility, pleomorphism, and filterability, may be related to the absence of rigid wall (Dienes, 1968; Madoff and Pachas, 1970). These are the fundamental characteristics of the L-forms and represent the criteria by which they are recognized. Definitive identification can be made microscopically by means of agar blocks stained in situ by the Dienes technique. The terms "L-phase" and "L-phase variants" are synonymous and are used interchangeably with the term L-forms (McGee et al., 1971).

Protoplasts and spheroplasts are not synonymous with L-forms, but are spherical structures that originate from bacteria following partial (spheroplasts) or complete (protoplasts) removal of the cell wall by enzymatic digestion in a hyperosmolar environment (Brenner et al., 1958). A major difference from the L-forms is that the protoplasts and spheroplasts are unable to replicate as such. However, they may be capable of producing L-forms when transferred to solid media of the proper osmolarity (Gooder and Maxted, 1961).

Atypical, aberrant, transitional, or variant bacterial forms are bacteria with altered cell wall, which are occasionally recovered from clinical specimens (McGee et al., 1971). These forms have unstable morphology and may require osmotic protection for isolation and growth. They tend to regain normal growth characteristics upon subculture. Most clinical isolates of wall-defective bacterial forms probably fall within this group (Wittler, 1968).

The ability of L-forms to revert to their parent bacteria is an important property. When this ability

Table 1. Bacterial genera from which L-forms have been derived.

Agrobacterium	*Erysipelothrix*	*Salmonella*
Bacillus	*Escherichia*	*Sarcina*
Bacteroides	*Flavobacterium*	*Serratia*
Bartonella	*Haemophilus*	*Shigella*
Bordetella	*Listeria*	*Staphylococcus*
Brucella	*Neisseria*	*Streptobacillus*
Clostridium	*Proteus*	*Streptococcus*
Corynebacterium	*Pseudomonas*	*Vibrio*

is lost, the cultures are considered to be "stable". Reversion is unpredictable, however, and has been shown to occur in apparently "stable" L-forms. The term "stabilized" would, therefore, be more appropriate. Bacteria recovered from the L-forms most often retain the essential characteristics of the parent strains. However, revertant bacteria may show profound alterations which indicate that changes can occur at some time during conversion, propagation, or reversion of the L-forms (Dienes, 1970; Pachas and Madoff, 1978).

Given the currently available methodology, a description of the L-forms only in terms of morphology and ultrastructure is insufficient. Further investigators should include other pertinent data in the description of the organisms, such as origin of the parent, method of L-transformation, requirements for osmotic protection, tendency for reversion, extent of the wall defectiveness, and, if possible, some comparative information on the physiological and biochemical properties of the parent, the derived L-forms, and the revertant forms. In this manner, the nomenclature will be clarified and the essential information regarding the L-cultures will be made available.

The properties of L-forms may be influenced by physical, environmental, and biological factors. Thus, variations in stable and unstable L-forms, in the parent bacteria, and in the revertant bacterial forms may include changes in cell composition, in structure and function, in sensitivity to antibiotics, in the production of toxins or enzymes, and in antigenic factors. Whether these changes are phenotypic or genotypic is not clear in most cases (Dienes, 1970; Pachas and Currid, 1974; Pachas and Madoff, 1978).

The production of L-forms may reflect genetic differences among the bacteria in a population. In one well-known experiment with *Streptococcus,* only 1 colony out of 20 was shown to produce progeny capable of L-transformation (Hijmans and Dienes, 1955). Biochemical studies support the hypothesis of genetic determinants in the transformation to the stable L-form state. Stable L-forms of streptococci were shown to be defective in important stages of cell wall biosynthesis. Altered constituents of cytoplasmic membrane (Cohen and Panos, 1966) and the lack of certain cell wall components (King, Prescott, and Caldes, 1970) have also been demonstrated. Gregory and Gooder (1976) noted the loss of enzyme function in *S. faecium* L-forms at the membrane stage of peptidoglycan synthesis. Pachas and Shor (1976) suggested that penicillin had a mutagenic effect, based on structural and functional changes in *Proteus* L-forms. Differences in antibiotic sensitivity patterns seen in L-forms of various species are also suggestive of genetic modifications (Schmitt-Slomska and Roux, 1976). Stable L-forms of a phytopathogenic bacterium, *Erwinia,* have been produced by ultraviolet irradiation with apparent mutagenic effect (Cabezas de Herrera and Garcia Jurado, 1976).

There is a paucity of studies at the molecular level. However, important contributions in this area were made by Hoyer and King (1969), who demonstrated the loss of a portion of the chromosomal DNA in a stable L-form of *Streptococcus faecalis.* Wyrick, McConnell, and Rogers (1973) used DNA transformation to transfer the stable L-form state to intact cells of *Bacillus subtilis.* Further studies by molecular biologists should do much to elucidate the nature of L-forms and to explore their role as a factor of genetic change in bacteria. Information on L-form genomes and on the manner in which segregation and replication of DNA occur in the L-forms would be of prime importance. For comprehensive reviews of the physiological and biochemical aspects of L-forms, see Hijmans, van Boven, and Clasener, 1969; Panos, 1967, 1969; Smith, 1971, 1978.

Induction of L-Forms: General Principles

The techniques used for the induction of L-forms have been described in several publications (Dienes and Weinberger, 1951; Gooder and Maxted, 1961; Guze, 1968; Hijmans, van Boven, and Clasener, 1969; Madoff, 1960; Madoff, Burke, and Dienes, 1967; Sharp, 1954). The general principles of induction and maintenance will be reviewed here, and detailed directions for particular strains will be given.

The conversion of a bacterium to the L-form depends on the species, the strain of the species, and the experimental conditions. In general, induction is accomplished by exposing bacteria to penicillin on a suitable medium. Important aspects of a suitable medium are its physical and chemical properties, which include the consistency of the agar gel, the presence of animal serum, and, especially, the osmolarity (Dienes and Sharp, 1956).

In general, the L-forms prefer a fairly soft medium with the addition of 10–20% of serum. Many bacteria, notably the Gram-negative *Proteus, Salmonella,* and *Shigella* species, can be converted to L-forms on media containing the normal concentration of sodium chloride (0.5%). Others require a hyperosmolar environment for stabilization. Most Gram-positive bacteria (e.g., *Streptococcus, Staphylococcus, Corynebacterium, Listeria*) and some Gram-negative species (e.g., *Neisseria, Serratia, Pseudomonas*) require increased osmotic protection in the medium; sodium chloride (1–5% wt/vol) is the commonly used osmotic stabilizer. Other neutral salts (e.g., potassium chloride, magnesium chloride, phosphates) have also been used. Sucrose, in concentrations ranging from 5–20% (wt/vol), plus Mg^{++} can be more effective for osmotic stabilization in some strains. The combination of sodium chloride and sucrose in varying concentrations has been successful in some experiments. Polyvinylpyrrolidone (PVP) has also been found to support L-form growth.

Often, even a slight modification in the concentration of inducing agent or osmotic stabilizer may make a decisive change in the success of an experiment. Anaerobiosis may be necessary for L-transformation. The L-cultures can later be adapted to aerobic growth. Certain strains of bacteria appear to be impervious to all known methods of induction.

The penicillins, with their action on the mucopeptide cross-linkages of bacterial cell wall, are the most effective inducing agents. Other antibiotics that act on cell wall (cycloserine, ristocetin, bacitracin, and vancomycin) have also been shown to produce L-forms from some bacteria. High concentrations of certain amino acids, notably glycine and phenylalanine, have been shown to produce L-forms.

L-forms can be induced by certain enzymes, particularly lysozyme, that digest the murein of the cell wall of bacteria. The bacteria are grown in osmotically protective media and are treated with the enzyme. Protoplasts are released which then produce L-forms when transferred to appropriate hyperosmolar solid media (see Fig. 1). By similar methods, phage-associated muralysins have been used to in-

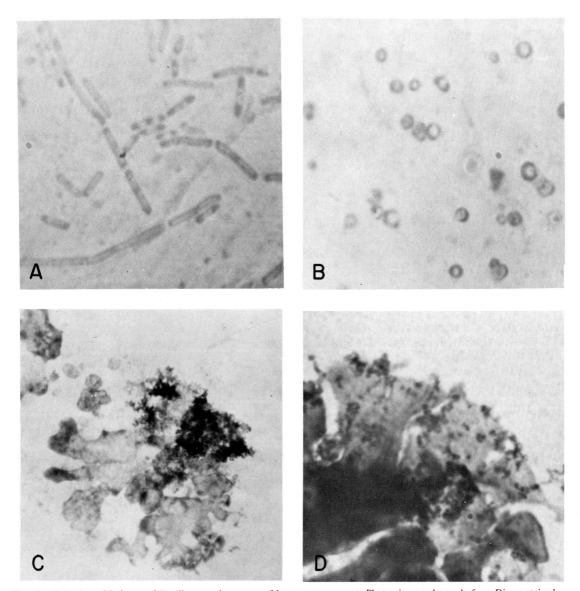

Fig. 1. Induction of L-forms of *Bacillus* spp. by means of lysozyme treatment. Photomicrographs made from Dienes-stained preparations. ×2,250. (A) Organisms grown 2 h in broth medium. (B) Following incubation for 2 h, in the presence of lysozyme, protoplasts are formed which are then transferred to hypertonic L-form agar medium. (C) Protoplasts enlarge to large body forms with granular elements developing into L-form colony. (D) Detail of periphery of mature L-colony.

duce L-forms. Gooder and Maxted (1961) produced protoplasts of group A streptococci using a phage-associated muralytic enzyme found in a group C phage lysate. After inoculation on appropriate agar media, a high yield of L-forms developed. This method has been successfully applied to many other strains (Hryniewicz, 1976).

For the induction of L-forms with penicillin, a strain of organisms is grown in broth medium for several hours. A small aliquot of the broth culture is then spread over the surface of an appropriate agar and allowed to dry. A penicillin gradient is established in the plate by means of antibiotic disks saturated with varying concentrations, or the penicillin may be introduced into small troughs cut in the agar. Other inducing agents (antibiotics, lysozyme, amino acids, etc.) may also be screened in this manner. The concentration of the inducing agent may be the critical factor. Later, an appropriate concentration of penicillin or other inducing agent can be incorporated into the medium for maintenance of the L-cultures. The use of pour plates may be effective; the inoculated agar medium is poured into a plate containing a thin layer of solid medium with similar nutrients and penicillin concentration.

The experimental culture should be tested on a variety of media, both with and without osmotic stabilizers. L-forms have been obtained on routine culture media (e.g., horse blood agar plates), even from strains that would seem to require high salt concentration (Madoff, 1976b). L-forms induced on media of high osmolarity can sometimes be adapted to media of normal osmolarity. Adaptation may represent either selection or mutation of viable organisms (Hijmans and Clasener, 1971; Leon and Panos, 1976; Madoff, 1970, 1976a).

Anaerobic incubation is required for some species; for others, it may prove helpful in primary induction. The Fortner method is suggested for anaerobiosis. It provides a moist atmosphere and has the advantage that each plate can be opened individually. The Fortner plate is prepared in the following manner. A triangle of nutrient or blood agar is placed in the cover of a Petri dish and heavily inoculated with a culture of *Serratia marcescens*. The inoculated plate is then inverted and placed in the cover. The plate is sealed with melted paraffin containing a small amount of petrolatum. Plates to be incubated aerobically should also be sealed to prevent drying during the long period of incubation that may be required.

L-forms of some organisms (e.g., *Proteus* and *Bacillus* species) may be seen several hours after exposure of the bacteria to penicillin. Others (e.g., *Streptococcus, Staphylococcus, Listeria*) may take several days or weeks to develop. Transfer of the L-culture depends on the viability of the organisms; some require frequent transfers and others remain viable for extended periods of time.

Subcultures are made from agar to agar by the "push-block" technique, in the manner used for mycoplasma. A block of agar bearing L-form growth is cut from the plate and pushed gently along the surface of the fresh medium. A hemostat holding a triangular sliver of a thin razor blade makes an excellent tool for this purpose. The agar block should be left on the plate because growth in transplant may occur only under the block. Growth in broth is also initiated with agar block but is often difficult to obtain. A long period of adaptation to growth in broth may be necessary. This difficulty has hampered the study of many important biological properties of the bacterial L-form.

An alternative method of inducing broth culture is by the use of diphasic media. The appropriate broth is added after the initiation of L-form growth on agar slants in tubes or flasks. After further incubation, the L-colonies may be dislodged by shaking or by the use of a Vortex mixer. Periodic transfers of broth cultures to agar are always necessary to determine viability as well as stability of the L-cultures. When inducing agents are omitted from the medium in which they are grown, some L-forms may be able to revert to the bacteria. Others may lose the ability to revert; they are then considered to be "stable", and can be transferred indefinitely in the absence of penicillin.

Reversion and Identification of Revertants

Transfer of freshly isolated L-colonies in the absence of inducing agent will often result in reversion to the parent bacterium. For many species, reversion can no longer be effected under any known circumstances when the cultures have become stabilized. For others, special techniques have been devised. Thus, reversion of *Bacillus subtilis* has been achieved by changes in the physical environment; e.g., by increasing the agar concentration, by the use of 15–35% gelatin in the media, or by the application of Millipore filters (Landman, Ryter, and Frehel, 1968). Similar techniques using gelatin media were employed successfully with *Streptococcus* L-forms (King and Gooder, 1970b; Wyrick and Gooder, 1976). In rare instances, reversion of L-forms has been influenced by the presence of factors produced by other bacteria or fungi. Landman and Halle (1963) induced reversion of *B. subtilis* L-forms by using *B. subtilis* cell wall material as a primer for wall regeneration. In *H. influenzae* L-forms, growth and reversion were found to be enhanced by low-molecular-weight peptide produced by a strain of *Neisseria perflava* (Madoff, 1976b, 1979).

The identification of an L-form from an unknown

source depends, to a large extent, on the ability to achieve reversion. In cases where the L-form culture is no longer revertible, molecular genetic studies of DNA base composition and DNA hybridization may provide information on its origin (Wittler, McGee, and Williams, 1968). Serological testing by agglutination, complement fixation, and fluorescent antibody methods may be useful for identifying unknown L-forms (Feinman, Prescott, and Cole, 1973; Lynn and Haller, 1968). Comparison of membrane proteins by polyacrylamide gel electrophoresis appears to offer a reliable method for comparing L-forms and bacteria (King, Theodore, and Cole, 1969; Theodore, King, and Cole, 1969). Modified biochemical tests, useful for bacterial and derived L-forms, can aid in characterization and identification of stable, nonreverting L-forms (Cohen, Wittler and Faber, 1968). Further studies are needed to detect key antigenic and metabolic markers for the L-forms.

Induction of L-Forms: Selective Procedures

Variations in requirements for nutrients, osmotic protection, inducing agents, and gel consistency make it necessary to tailor the techniques to the organism. The following media and methods have proved successful with certain strains of selected species.

Gram-Negative Species (Dienes and Weinberger, 1951; Madoff and Pachas, 1970)

For induction and maintenance of L-forms of Gram-negative species *(Proteus, Salmonella, Shigella)*, a simple medium consisting of Trypticase soy agar (BBL) or brain heart infusion (Difco), containing 10% horse serum (inactivated at 56°C for 30 min) will support L-form growth. Yeast extract (5%) may be added. Penicillin concentrations vary according to the strain. Anaerobic incubation for induction is useful. Other Gram-negative species may require a hyperosmolar environment. As examples, *Escherichia coli, Serratia marcescens,* and *Klebsiella* and *Pseudomonas* spp. have been converted to L-forms and maintained only in the presence of high salt or sucrose concentrations (Guze, Harwick, and Kalmanson, 1976; Hubert et al., 1971; Lederberg and St. Clair, 1958).

High-Salt-Requiring L-Forms (Marston, 1968; Sharp, 1954)

For the induction and propagation of L-forms requiring high salt concentration *(Streptococcus, Staphylococcus, Corynebacterium, Listeria,* etc.), an all-purpose medium is prepared in the author's laboratory as follows:
To 300 ml of distilled water is added 12 g of

Trypticase soy agar (BBL). The solution is thoroughly mixed and autoclaved without previous heating. The agar is cooled to 56°C (in a water bath) and the following sterile components are added in the order listed (it is advisable to warm the solutions slightly and to mix well after each addition): 30 ml of Todd Hewitt broth (BBL), 30 ml of a 30% solution of NaCl, 30 ml of inactivated (56°C for 30 min) horse serum, and 15 ml of yeast extract (Microbiological Associates). The medium is poured into sterile Petri dishes; plates are used after standing overnight at room temperature.

This medium is also useful for maintaining stock L-form cultures which normally do not require increased salt concentration (*Proteus, Salmonella,* etc.). For maintenance, penicillin may be incorporated into the medium in concentration of 500–1,000 units per ml.

Broth medium consists of Todd Hewitt broth base containing NaCl, horse serum, yeast extract, and the appropriate antibiotic. Growth may be enhanced by the addition of 0.1% agar (Marston, 1968).

Montgomerie, Kalmanson, and Guze (1973) maintained growth of L-forms of *Staphylococcus* in Trypticase soy broth (BBL) containing 0.5 M sucrose, 0.01% $MgSO_4 \cdot 7H_2O$, and 2% bovine serum albumin. Bren and Eveland (1968) induced L-forms of *Listeria monocytogenes* in sucrose-containing media.

Streptococcus by Lysozyme (King and Gooder, 1970a; Wyrick and Gooder, 1976)

Protoplasts were prepared in the following manner: Cultures of *Streptococcus faecium* strain F 24 were grown overnight (stationary phase) in Trypticase soy broth (BBL), harvested, and washed 3 times in distilled water. Sucrose (0.6 M) or polyethylene glycol (Carbowax 4000, Union Carbide Co.) 8% (wt/vol) was used as osmotic stabilizer in 0.01 Tris-chloride buffer pH 7.1. A mixture of 2×10^9 streptococcal colony-forming units (CFU) and 200 μg lysozyme (Nutritional Biochemicals Corp.) per ml was incubated at 37°C for 2 h. Tenfold serial dilutions of the protoplast suspension were then plated onto the surface of tryptone soy agar (Oxoid) containing 0.43 M NH_4Cl, 0.5% (wt/vol) additional glucose, and 2% inactivated horse serum. Incubation of the plates was at 37°C. L-colonies could be counted in 2–5 days.

Broth medium for *S. faecium* L-forms consisted of Albimi brucella broth (Gibco Diagnostics) prepared in 0.43 M NH_4Cl with 0.5% glucose added. Reversion medium consisted of tryptone soy broth (Oxoid) prepared in 0.6 M sucrose and 35% gelatin (Difco) as the solidifying agent. Plates were incubated at 25°C.

L-forms have been induced from *Streptococcus, Bacillus,* and other bacterial genera by the use of lysozyme (Madoff, Burke, and Dienes, 1967) (see Fig. 1).

The use of phage muralysins for the production of protoplasts from streptococci and their conversion to L-forms is effective and is to be recommended (Gooder and Maxted, 1961; Hryniewicz, 1976).

Neisseria (Roberts, 1966; Roberts and Wittler, 1966)

L-forms of *Neisseria meningitidis* were produced by penicillin by the gradient plate technique. The medium consisted of brain heart infusion (BHI; Difco) of pH 7.2–7.4 containing 1.2% agar, 10% sucrose, 0.5% yeast extract, and 10% inactivated horse serum. L-forms obtained were serially propagated on media containing benzylpenicillin (1,000 units per ml). Incubation of plates was at 37°C in CO_2 (candle jar).

L-forms of *N. gonorrhoeae* were induced and propagated under identical conditions with the exception that 100 units penicillin per ml was required; plates were incubated at 36°C. Growth of L-forms in broth was obtained by the use of diphasic media. Agar slanted in flasks was inoculated and then overlayered with BHI broth of similar composition but containing 0.01% agar. Frequent serial transfers were required before heavy L-form growth was obtained in broth lacking traces of agar.

L-forms of *N. meningitidis* have been produced by methicillin, ampicillin, cycloserine, caphalothin, ristocetin, bacitracin, and vancomycin.

Lawson and Bacigalupi (1976) have produced L-forms of *N. gonorrhoeae* using penicillin and L-form media supplemented with 7.0% polyvinylpyrrolidone (PVP) as osmotic stabilizer. The PVP was subjected to extensive dialysis before use. Induction frequencies were higher than those obtained in the presence of sucrose. Two other cell wall antibiotics, cephalothin and D-cycloserine, also permitted L-transformation; none were produced by vancomycin, bacitracin, or novobiocin. Similar media, containing sucrose as osmotic stabilizer in varying concentrations, have been used successfully in numerous bacterial species (e.g., *Staphylococcus,* Montgomerie, Kalmanson, and Guze, 1973; *Listeria,* Brem and Eveland, 1968; *Pneumococcus,* Madoff and Dienes, 1958).

Identification of L-Forms

Transformation of bacteria to L-forms may occur in time periods ranging from several hours to several days. Cultures must be examined frequently. In a penicillin gradient plate, L-colonies may be detected by the naked eye or with a hand lens in the zone of penicillin inhibition. Microscopic examination under oil immersion is necessary for the identification of L-form colonies. The Dienes technique of stained agar preparation is recommended. The method is equally applicable to mycoplasma and is indispensable for distinguishing L-colonies and viable organisms from tiny bacterial colonies and from artifacts.

The Dienes staining solution is made by dissolving 2.5 g of methylene blue, 1.25 g of Azur II, 10 g of maltose, 0.25 g of Na_2CO_3, and 0.2 g of benzoic acid in 100 ml of distilled water. A thin film of stain is painted on a coverslip by means of a cotton applicator and allowed to dry. Many coverslips can be prepared at a time and stored for future use. The coverslip is cut into convenient squares with a diamond pencil. A block of agar bearing suspected colonies is cut from the plate and placed face up on a glass microscope slide, and the stained coverslip is placed on the agar. The preparation is then sealed around the edges with melted paraffin containing a small amount of Vaseline.

Microscopically, L-colonies and mycoplasma colonies stain a deep blue at the center and a light blue at the periphery. Nonviable organisms and artifacts either do not stain or they assume a pinkish to violet hue. Viable bacterial colonies also take the Dienes stain. Decolorization of the stain may occur in bacterial colonies, as in large L-colonies, if the preparation is not examined at once. Contrary to other reports, decolorization of the Dienes stain is not a criterion for distinguishing between mycoplasma and L-form colonies.

Morphology and Ultrastructure

The use of the Dienes technique and oil immersion magnification permits the transformation of bacteria to L-forms to be followed in serially examined blocks of agar. The initial change on agar is the enlargement of the bacterial cell to produce a large body. In some instances, large bodies are formed during division when swelling occurs within segmenting cells. The large bodies appear to form a connecting link between the growth of bacteria and the transformation to L-forms (Dienes, 1968). The "large body" stage may be followed by fragmentation of these forms with release of typical bacterial forms. Alternatively, granular elements of varying size develop within or at the periphery of the large bodies; these soon penetrate the agar and multiply, forming an L-colony. The L-colony may grow to resemble the mycoplasma, with dense central growth into the agar surrounded by a periphery of surface growth of large bodies ("fried egg" appear-

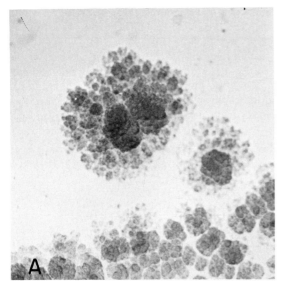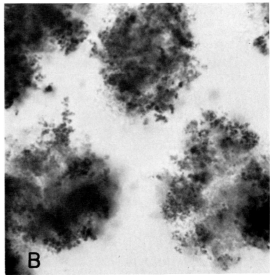

Fig. 2. (A) L-forms of *Neisseria gonorrhoeae*. ×65. (B) L-forms of *Streptococcus pyogenes*. ×110.

ance), as shown in Fig. 2A; or the colony may become grossly irregular in shape and contour (Fig. 2B). L-colonies are usually larger and of coarser appearance than the mycoplasma. The distinctive morphology of the L-form colonies depends on growth into the agar of the granular elements, the presence of large bodies, and the complete absence of any residual bacterial forms.

Young L-colonies of some bacterial species, notably, *Haemophilus influenzae*, *Corynebacterium* species, various streptococci, and *Streptobacillus moniliformis*, can be very small and consist of very small granular forms, and therefore can be extremely difficult to distinguish from mycoplasma colonies purely by morphology. A problem can arise if such tiny colonies are produced on penicillin-containing media inoculated primarily for the isolation of mycoplasma from clinical material. The final identification of the L-forms is by proof of derivation from the bacterium, either by serological or biochemical means or by their reversion to the parent bacterium. This may present considerable difficulty in some instances (Madoff, 1976a.).

Two morphologic types of L-form colonies have been described, the A and B types (Fig. 3). In *Proteus*, for example, the A-type L-colonies grow on the surface of the agar and are distinguished by their small size, fine granularity, and scarcity of large bodies. With electron microscopic examination, they are seen to be devoid of cell wall (Dienes and Bullivant, 1968). The pleomorphic elements appear bound by a single unit membrane and they are filled with ribosomes and DNA material. Their resemblance to mycoplasma is notable. Mesosomal bodies are absent in the L-forms. The B-type L-colonies, on the other hand, develop more readily

within agar medium, especially in agar overlayers. By light microscopy, the colonies appear to be composed almost entirely of large bodies. B-type L-colonies are highly revertible; A-type colonies become stabilized more easily and, in some bacterial species, are then difficult to revert.

When examined with the electron microscope, L-organisms of the B-type show the presence of cell wall components apparently similar to the wall of the parent bacteria, but lacking its rigidity. The presence of an altered mucopeptide layer in revertible, B-type L-forms of *Proteus* has also been noted (Cole, 1971). Conflicting morphological descriptions are given for L-forms, particularly of the Gram-negative species where the A- and B-type L-forms appear to correspond in some reports to "protoplast-type L-forms" and "spheroplast-type L-forms", respectively. In *Escherichia coli*, *Neisseria meningitidis*, and *Brucella abortus*, for example, L-forms appear to possess a layer of damaged wall external to the cytoplasmic membrane. In PVP-stabilized L-forms of *Neisseria* (Lawson and Bacigalupi, 1976) and in L-forms of *Haemophilus influenzae* (Madoff, 1976b), only the unit cytoplasmic membrane was visible. In studies of thin section of *Bacillus licheniformis*, *Bacillus subtilis*, and *Salmonella paratyphi*, no trace of cell wall could be seen and the organisms were bounded only by cytoplasmic membrane (Wyrick and Rogers, 1973).

Whereas some strains of bacteria produce both A- and B-type L-colonies, Gram-positive organisms (streptococci, staphylococci, etc.) tend to produce only A-type L-forms. It must be emphasized, however, that the presence or absence of cell wall constituents does not appear to be the decisive factor in reversion. Some B-type L-forms can also become

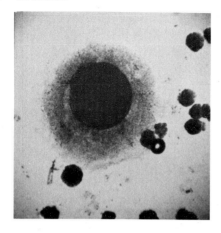

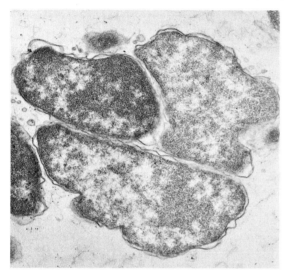

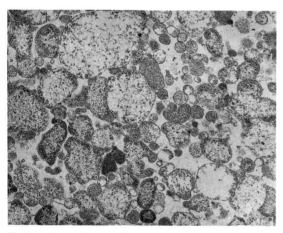

Fig. 3. L-forms of *Proteus*. In the electron micrographs, the presence of cell wall constituents is clearly visible in the B-type (middle photograph). In the A-type (bottom photograph) there is only the cytoplasmic membrane. Note the size difference between the colonies of A- and B-type L-forms (top photograph). ×65. [Reproduced by permission of the American Society for Microbiology (Pachas and Madoff, 1978).]

stabilized to the extent that they can no longer return to bacterial form (Cole, 1971; Dienes, 1970; Martin, Schilf, and Gruss, 1976).

Stable L-forms of a strain of *E. coli* show the presence of laminate structures and microtubules (Eda, Kanda, and Kimura, 1976) resembling similar configurations previously observed in *Streptococcus faecalis* (Cohen et al., 1968) and *Pseudomonas* spp. (Hubert et al., 1971). The nature and function of these structures has not been established.

Replication and Reversion

As seen with transmission and scanning electron microscopes, L-forms of diverse bacterial species are surprisingly similar in their morphology and reproductive processes. Although organisms as small as 200 to 300 nm have been seen, the size of the smallest units which are viable and capable of reproduction remains unclear. Like those of the mycoplasma, the mechanisms of replication appear to vary from simple binary or asymmetric fission to budding or segmentation of small, dense bodies from large, spherical or filamentous forms in sequential growth. In *Proteus* L-forms, sequential enlargement of protoplasmic bodies appears to be followed by release of new viable granules (Dienes and Bullivant, 1968; Cole, 1971). Thin sections of L-form colonies of *Haemophilus influenzae* grown within agar reveal small, dense bodies seen as condensations at the periphery of large cells apparently capable of detachment as budding forms (Madoff, 1976b).

Among the Gram-positive bacteria, growth and reversion of L-forms have been studied in *Bacillus subtilis* (Landman and Forman, 1969; Landman, Ryter, and Frehel, 1968), *Bacillus licheniformis* (Elliott et al., 1975), *Staphylococcus aureus* (Schönfeld and de Bruijn, 1976), *Streptococcus pyogenes* (Cole, 1970), *Streptococcus faecalis* (Green, Heidger, and Domingue, 1974a,b), and *Streptococcus faecium* (King and Gooder, 1970b; Wyrick and Gooder, 1976). Of particular interest in these reports are the studies relating to the sequential stages of reproduction and to the reversion of L-forms to bacteria.

Green, Heidger, and Domingue (1974b) have proposed a reproductive cycle for L-forms of *Streptococcus faecalis*. Small, dense, nonvesiculated L-form bodies divide by simple binary fission and budding and develop within vesicles of mature "mother" forms; under certain conditions, these dense forms may undergo transition and revert to the bacteria. Reversion may be accompanied by the formation of mesosome-like structures in a manner similar to that seen in *B. subtilis* (Landman, Ryter, and Frehel, 1968). In a study of *Streptococcus faecium* with ferritin labeling, reversion appears to arise through excretion of cell wall material around the unstable L-forms. Pieces of wall associated with cell membrane form a scaffolding to create the intact (complete) bacterial cell wall (Wyrick and Gooder, 1976).

The L-forms of bacteria share many ultrastructural features with the mycoplasma. In contrast to the mycoplasma, there is a paucity of studies which would help to clarify L-form replication mechanisms. In particular, it is not known how the genome segregates in the formation of the new, viable L-form bodies.

Antibiotic Sensitivities of L-Forms and Parent Bacteria

The susceptibilities of L-forms to antimicrobial agents have been reviewed in several publications (Hijmans, van Boven, and Clasener, 1969; Roberts, 1966; Schmitt-Slomska and Roux, 1976; Ward, Madoff, and Dienes, 1958). Although the subject is too extensive to treat here, a few general statements can be made. The unique properties of the L-forms make them ideal models for the study of the mode of action of antibiotics. These studies are limited, however, by the nature of the L-forms, for example, the variable growth rates, the difficulty of obtaining quantitative estimates of growth curves, the diffusion of the microbial effect when extended periods of incubation are required, and the requirement of some L-forms for osmotic stabilizers in the media whose effect on the reaction is difficult to assess.

L-forms are much more resistant than bacteria to penicillin, penicillin derivatives, and to other antibiotics that exert their influence on the synthesis of bacterial cell wall. The minimal inhibitory concentration (MIC) of the penicillins is roughly 1,000 times greater for the L-forms than for the parent bacteria. For other antibiotics acting on cell wall (bacitracin, ristocetin, cycloserine), the MIC of the L-forms may be greater by a factor of 100, as has been shown in L-forms of *Streptococcus, Staphylococcus, Neisseria meningitidis, Proteus,* and *Vibrio.* By contrast, L-forms and parent bacteria of one strain each of *Proteus, Salmonella,* and *Vibrio* showed a sensitivity to these cell wall inhibitors similar to the parent strains (Ward, Madoff, and Dienes, 1958). Revertants from a strain of *Staphylococcus* L-forms were found to be more susceptible to cell wall inhibitors than the L-forms, but less susceptible than the parent bacteria (Schmitt-Slomska and Roux, 1976). These differences were accompanied by modifications in physiology and in enzyme production.

Polymyxin B increases the permeability of the cell membrane through affinity for the phospholipid of the membrane. It has been shown to exert a greater effect on L-forms than on the parent bacteria. L-forms derived from polymyxin-resistant, Gram-negative bacteria, *Neisseria gonorrhoeae, Brucella melitensis,* and *Proteus* spp., were susceptible to this antibiotic. Cross-reactivity with colistin was noted. An interesting finding was the selection, on polymyxin-containing media, of L-forms of *Proteus* spp. as resistant to polymyxin as the parent bacteria (Schmitt-Slomska and Roux, 1976). These studies lend support to the idea that the cytoplasmic membrane acts in the mechanism of resistance to antimicrobial agents (Louis and Schmitt-Slomska, 1976).

The broad-spectrum antibiotics exert their primary influence by inhibition of transformation of genetic information and of protein synthesis within the cell protoplasm. They have, in general, a greater effect on the L-forms than on the parent bacteria. Studies by Ward, Madoff, and Dienes (1958) showed susceptibilities of L-forms of *Proteus* spp., *Salmonella* spp., and *Vibrio* spp. to these antibiotics comparable to those of the parent bacteria. Greater susceptibility has been shown by the L-forms of *Streptococcus faecalis* to the tetracyclines and to the aminoglycoside antibiotics, kanamycin, streptomycin, and vancomycin (Montgomerie, Kalmanson, and Guze, 1966). *Staphylococcus* L-forms have also shown greater susceptibility to the tetracyclines. In *Neisseria meningitidis* and *N. gonorrhoeae,* both parent bacteria and derived L-forms showed similar susceptibilities to tetracycline, novobiocin, erythromycin, and sulfadiazine (Roberts, 1966; Roberts and Wittler, 1966). More recent studies with *Neisseria* L-forms (Lawson and Bacigalupi, 1976) showed considerably greater susceptibility of the L-forms than the parent to the same antibiotics. The authors explain the divergent results as being due to the complete absence of cell wall constituents in the PVP-stabilized L-forms of *Neisseria* as opposed to those induced on sucrose-containing media (Roberts, 1966).

In summary, the L-forms share with their parent bacteria, to a greater or lesser degree, susceptibility to antibiotics other than those that act on bacterial cell wall. Agents which exert their influence on cytoplasmic membrane tend to have a greater inhibitory effect on the wall-defective forms. Recent studies confirm the unpredictability of the degree of sensitivity of the L-forms in comparison with the parent bacteria and, in particular, with revertants from L-forms. Alterations in biochemical or genetic properties of the cytoplasmic membranes following prolonged cultivation in the L-form may be important factors. Should the L-form be found to play a significant role in infectious disease, these considerations could be of considerable importance (Roux, 1976).

Occurrence and Pathogenicity

The controversial aspects of L-forms as pathogens have been discussed and reviewed in several publications (Clasener, 1972; Feingold, 1969; Hijmans, van Boven, and Clasener, 1969; Kagan et al., 1976; Pachas and Madoff, 1978; Roux, 1976; Wittler, 1968). In animal experimentation, the L-forms have not been shown to produce pathogenicity (Linneman, Watanakunakorn, and Bakie, 1973) unless reversion occurred in vivo, as in *Streptobacillus,* or unless toxins were produced, as in *Clostridium* and in *Vibrio* species.

The persistence of L-forms of *Streptococcus* has been shown in mouse experiments, but pathogenic effects have not been noted (Schmitt-Slomska, Sacquet, and Caravano, 1967). L-forms of *Streptococcus pyogenes,* which had been adapted to grow in isotonic conditions, survived in immunosuppressed mice (Panos, 1978), suggesting that the minimal response of the mouse (host) may play a role in suppressing the L-forms in vivo. It is of interest here that human serum with complement kills certain L-forms but is without effect on the parent bacteria (McGee et al., 1972; Martino et al., 1976).

L-forms of the following species have been shown to be adaptable to growth in tissue cultures of diverse origin: *Brucella abortus* in hamster kidney cells (Hatten and Sulkin, 1968); *Streptococcus faecalis* in human embryonic kidney cells (Green, Heidger, and Domingue, 1974a); and *Streptococcus pyogenes* in human heart cells (Leon and Panos, 1976). In the last example, a cytopathic effect was found with a strain adapted to normal osmolar conditions of growth.

Among antigenic substances or toxins produced by the L-forms, it should be noted that L-forms of group A *Streptococcus pyogenes* are able to produce both streptolysin and M protein.

Comparison of chemotactic activity of bacteria and L-forms by polymorphonuclear migration suggests that the bacteria are phagocytized better than the L-forms (Harwick et al., 1972). Lower chemotactic activity may be a factor.

Clinical studies have reported the recovery of L-forms and other wall-defective organisms from blood, body fluids, and tissues of animals and man. Indeed, the growth of such forms from clinical material in hyperosmolar media has been taken as indirect evidence for the involvement of L-forms in the persistence of infection. However, the validity of these isolates and their relation to pathogenic states are at present entirely unclear. Studies by Domingue et al. (1976) describe the recovery of variant and wall-defective bacterial forms from the blood of normal as well as diseased humans. The L-forms as we know them may not be primary pathogens, but further studies are needed to explore their role in pathogenicity and in immune mechanisms.

The L-forms of bacteria deserve to be studied for their intrinsic scientific value. That a bacterium can, under certain conditions, enter into a wall-less state, that it can survive and reproduce itself in this wall-deficient state, that it can ultimately revert to its bacterial form—or lose the ability to revert—are remarkable achievements. The phenomena cannot be insignificant. The L-forms of bacteria should be studied, therefore, with regard to their structure, biochemical function, and genetic composition. Such studies should provide fundamental information on the role of L-forms in the biology of bacteria.

Needs for Future Investigations

The pathogenicity of the L-forms remains a subject of controversy. Their role in disease needs to be investigated with new experimental models and techniques. Certain factors in persisting and recurrent infections may predispose to the production of wall-defective organisms. Factors such as the dissolution of bacterial cell wall by immune mechanisms, by cellular enzymes, or by antibiotics may create favorable conditions for the establishment of chronic infections by wall-defective organisms. Another factor may be related to the hypertonicity of the environment. Thus, patients with long-term infections of the urinary tract, including those with pyelonephritis, or patients with septicemia, meningitis, or endocarditis of long duration, who have failed to demonstrate normal bacterial isolates, may well provide fruitful sources for experimental cultivation of clinical material.

To establish a definitive role for the L-forms in disease, a conscientious search for L-forms will have to begin with careful observations in the laboratory. Realistically, such a search should include the following basic requirements:

1. Examination of the clinical specimen with the light microscope and staining (e.g., Gram stain, Dienes stain, DNA stain). Transmission and scanning electron microscopy might yield additional valuable information.
2. Special L-form media, both with and without a variety of osmotic stabilizers (NaCl, sucrose, PVP, etc.).
3. Media free of penicillin or other agents that act on bacterial cell wall.
4. Proper controls by the use of bacteriological media appropriate for the clinical specimen.
5. Aerobic and anaerobic incubation of the cultures, perhaps for extended periods of time.
6. Careful identification of the isolates recovered. Awareness of the potential for misinterpreting bacteria that show atypical or aberrant morphology. Pleomorphism, by itself, is not sufficient evidence for the presence of L-forms.

7. Substantial basic knowledge of mycoplasma and L-forms and experience with the techniques of cultivation and identification. In the case of non-revertible L-form isolates, identification by serological or biochemical means would be of prime importance.

8. Experimental use of cell cultures and other innovative isolation methods. For example, studies showing enhanced survival of L-forms in immunosuppressed mice (Panos, 1978) and observations of the satellite effect of other bacteria on L-forms (Bouvet and Acar, 1976; Madoff, 1976b; 1979) represent new avenues of approach and merit further exploration.

The chief importance of L-forms may be that they are unique and valuable models for basic studies in microbiology. Studies with the L-forms have already had a profound influence on the way we view bacteria and have contributed significantly to our knowledge of the function and structure of the bacterial cell. Continued advances in our study of these highly interesting forms of bacteria should result in greater understanding of the factors that influence bacterial pathogenicity and virulence as well.

Literature Cited

Brenner, S., Dark, F. A., Gerhardt, P., Jeynes, M. H., Kandler, O., Kellenberger, E., Klieneberger-Nobel, E., McQuillen, K., Rubio-Huertos, M., Salton, M. R. J., Strange, R. E., Tomcsik, J., Weibull, C. 1958. Bacterial protoplasts. Nature **181**:1713–1715.

Brem, A., Eveland, C. W. 1968. L-forms of *Listeria monocytogenes.* 1. In vitro production and propagation of L-forms in all serotypes. Journal of Infectious Diseases **118**:181–187.

Bouvet, A., Acar, J. F. 1976. Cultural isolation and morphological study of deficient streptococci growing in satellism. I. Isolation in bacterial endocarditis. Les Colloques de l'Institut National de la Santé et de la Recherche Medicale **64**:327–338.

Cabezas de Herrera, E., Garcia Jurado, O. 1976. Stable L-forms of *Erwinia cartovora* induced by ultraviolet irradiation. Les Colloques de l'Institut National de la Santé et de la Recherche Medicale **64**:107–118.

Clasener, H. 1972. Pathogenicity of the L-phase of bacteria. Annual Review of Microbiology **26**:55–84.

Cohen, M., Panos, C. 1966. Membrane lipid composition of *Streptococcus pyogenes* and derived L-form. Biochemistry **5**:2385.

Cohen, M., McCandless, R. G., Kalmanson, G. M., Guze, L. B. 1968. Core-like structures in transitional and protoplast forms of *Streptococcus faecalis,* pp. 94–109. In: Guze, L. B. (ed.), Microbial protoplasts, spheroplasts and L-forms. Baltimore: Williams & Wilkins.

Cohen, R. L., Wittler, R. G., Faber, J. E. 1968. Modified biochemical tests for characterization of L-phase variants of bacteria. Applied Microbiology **16**:1655–1662.

Cole, R. M. 1970. The structure of the group A streptococcal cell and its L-form, pp. 5–42. In: Caravano, R. (ed.), Current research on *Streptococcus* group A. New York: Excerpta Medica Foundation.

Cole, R. M. 1971. Some implications of the comparative ultra-

structure of bacterial L-forms, pp. 49–83. In: Madoff, S. (ed.), Mycoplasma and the L-forms of bacteria. New York: Gordon and Breach.

Dienes, L. 1939. "L" organism of Klieneberger and *Streptobacillus moniliformis.* Journal of Infectious Diseases **65**:24–42.

Dienes, L. 1968. Morphology and reproductive processes of bacteria with defective cell wall, pp. 74–93. In: Guze, L. B. (ed.), Microbial protoplasts, spheroplasts and L-forms. Baltimore: Williams & Wilkins.

Dienes, L. 1970. Permanent alterations of the L-forms of *Proteus* and *Salmonella* under various conditions. Journal of Bacteriology **104**:1369–1377.

Dienes, L., Bullivant, S. 1968. Morphology and reproductive processes of the L-forms of bacteria. Journal of Bacteriology **95**:672–682.

Dienes, L., Sharp, J. T. 1956. The role of high electrolyte concentration in the production and growth of L-forms of bacteria. Journal of Bacteriology **71**:208–213.

Dienes, L., Weinberger, H. J. 1951. The L-forms of bacteria. Bacteriological Reviews **15**:245–288.

Domingue, G. J., Schlegel, J. U., Heidger, P. M., Erlich, M. 1976. Novel bacterial structures in human blood: Cultural isolation, ultrastructural and biochemical evaluation. Les Colloques de l'Institut National de la Santé et de la Recherche Medicale **64**:273–300.

Eda, T., Kanda, Y., Kimura, S. 1976. Membrane structures in stable L-forms of *Escherichia coli.* Journal of Bacteriology **127**:1564–1567.

Elliott, T. S. J., Ward, J. B., Wyrick, P. B., Rogers, H. J. 1975. Ultrastructural study of the reversion of protoplasts of *Bacillus licheniformis* to bacilli. Journal of Bacteriology **124**:905–917.

Feingold, D. S. 1969. Biology and pathogenicity of microbial spheroplasts and L-forms. New England Journal of Medicine **281**:1159–1170.

Feinman, S. B., Prescott, B., Cole, R. M. 1973. Serological reactions of glycolipids from streptococcal L-forms. Infection and Immunity **8**:752–756.

Gooder, H., Maxted, W. R. 1961. External factors influencing structure and activities of *Streptococcus pyogenes,* pp. 151–173. In: Meynell, G. G., Gooder, H. (eds.), Microbial reaction to environment. Eleventh Symposium of the Society for General Microbiology. Cambridge: Cambridge University Press.

Green, M. T., Heidger, P. M., Jr., Domingue, G. 1974a. Demonstration of the phenomena of microbial persistence and reversion with bacterial persistence and reversion with bacterial L-forms in human embryonic kidney cells. Infection and Immunity **10**:889–914.

Green, M. T., Heidger, P. M., Domingue, G. 1974b. Proposed reproductive cycle for a relatively stable L-phase variant of *Streptococcus faecalis.* Infection and Immunity **10**:915–927.

Gregory, W. W., Gooder, H. 1976. The specific biochemical lesion in the L-phase. Les Colloques de l'Institut National de la Santé et de la Recherche Medical **64**:57–58.

Guze, L. B. (ed.). 1968. Microbial protoplasts, spheroplasts and L-forms. Baltimore: Williams & Wilkins.

Guze, L. B., Harwick, H. J., Kalmanson, G. M. 1976. Klebsiella L-forms: Effect of growth as L-form on virulence of reverted *Klebsiella pneumoniae.* Journal of Infectious Diseases **133**:245–252.

Hatten, B. A., Sulkin, S. E. 1968. Possible role of *Brucella* L-forms in the pathogenesis of brucellosis, pp. 457–471. In: Guze, L. B. (ed.), Microbial protoplasts, spheroplasts and L-forms. Baltimore: Williams & Wilkins.

Harwick, H. J., Barajas, L., Montgomerie, J. Z., Guze, L. B. 1972. Phagocytosis of microbial L-forms. Infection and Immunity **5**:976–981.

Hijmans, W., Clasener, H. A. L. 1971. A survey of the L-forms of bacteria, pp. 37–47. In: Madoff, S. (ed.), Mycoplasma and the L-forms of bacteria. New York: Gordon & Breach.

Hijmans, W., Dienes, L. 1955. Further observations on L-forms of *alpha* hemolytic streptococci. Proceedings of the Society for Experimental Biology and Medicine **90**:672–675.

Hijmans, W., van Boven, C. P. A., Clasener, H. A. L. 1969. Fundamental biology of the L-phase of bacteria, pp. 67–143. In: Hayflick, L. (ed.), The Mycoplasmatales and the L-forms of bacteria. New York: Appleton, Century, Crofts.

Hoyer, B. H., King, J. R. 1969. Deoxyribonucleic acid sequence losses in a stable streptococcal L-form. Journal of Bacteriology **97**:1516–1517.

Hryniewicz, V. 1976. Streptococcal L-forms and some of their properties. Les Colloques de l'Institut National de la Santé et de la Recherche Medicale **64**:39–56.

Hubert, E. G., Potter, C. S., Hensley, T. J., Cohen, M., Kalmanson, G. M., Guze, L. B. 1971. L-forms of *Pseudomonas aeruginosa*. Infection and Immunity **4**:60–72.

Kagan, G., Vulfovitch, Yu., Gusman, B., Raskova, T. 1976. Persistence and pathological effect of streptococcal L-forms in vivo. Les Colloques de l'Institut National de la Santé et de la Recherche Medicale **64**:247–258.

King, J. R., Gooder, H. 1970a. Induction of enterococcal L-forms by the action of lysozyme. Journal of Bacteriology **103**:686–691.

King, J. R., Gooder, H. 1970b. Reversion to the streptococcal state of enterococcal protoplasts, spheroplasts and L-forms. Journal of Bacteriology **103**:692–696.

King, J. R., Prescott, B., Caldes, G. 1970. Lack of murein in formamide-insoluble fraction from the stable L-form of *Streptococcus faecium* strain F 24. Journal of Bacteriology **102**:296–297.

King, J. R., Theodore, T. S., Cole, R. M. 1969. Generic identification of L-forms by polyacrylamide gel electrophoretic comparison of extracts from parent strains and their derived L-forms. Journal of Bacteriology **100**:71–77.

Klieneberger, E. 1935. The natural occurrence of pleuropneumonia-like organisms in apparent symbiosis with *Streptobacillus moniliformis* and other bacteria. Journal of Pathology and Bacteriology **40**:93–105.

Landman, O. E., Forman, A. 1969. Gelatin-induced reversion of protoplasts of *Bacillus subtilis* to the bacillary form: Biosynthesis of macromolecules and wall during successive steps. Journal of Bacteriology **99**:576–589.

Landman, O. E., Halle, S. 1963. Enzymatically and physically induced inheritance changes in *Bacillus subtilis*. Journal of Molecular Biology **7**:721–738.

Landman, O. E., Ryter, A., Frehel, C. 1968. Gelatin-induced reversion of protoplasts of *Bacillus subtilis* to the bacillary form: Electron microscopic and physical study. Journal of Bacteriology **96**:2154–2170.

Lawson, J. W., Bacigalupi, B. 1976. Induction and reversion of the L-forms of *Neisseria gonorrhoeae*. INSERM **64**:91–106.

Lederberg, J., St. Clair, J. 1958. Protoplasts and L-type growth of *Escherichia coli*. Journal of Bacteriology **75**:143–160.

Leon, O., Panos, C. 1976. Adaptation of an osmotically fragile L-form of *Streptococcus pyogenes* to physiological osmotic conditions and its ability to destroy human heart cells in tissue culture. Infection and Immunity **13**:252–262.

Linneman, C. C., Jr., Watanakunakorn, C., Bakie, C. 1973. Pathogenicity of stable L-phase variants of *Staphylococcus aureus*: Failure to colonize experimental endocarditis in rabbits. Infection and Immunity **7**:725–730.

Louis, C., Schmitt-Slomska, J. 1976. Effect of polymyxin B on the ultrastructure of the stable *Proteus mirabilis* L-forms. Les Colloques de l'Institut National de la Santé et de la Recherche Medicale **64**:197–210.

Lynn, R. J., Haller, G. L. 1968. Bacterial L-forms as immunogenic agents, pp. 270–278. In: Guze, L. B. (ed.), Microbial protoplasts, spheroplasts and L-forms. Baltimore: Williams & Wilkins.

McGee, Z. A., Wittler, R. G., Gooder, H., Charache, P. 1971. Wall-defective microbial variants: Terminology and experi-

mental design. Journal of Infectious Diseases **123**:433–438.

McGee, Z. A., Ratner, H. B., Bryant, R. E., Rosenthal, A. S., Koenig, M. G. 1972. An antibody-complement system in human serum lethal to L-phase variants in bacteria. Journal of Infectious Diseases **125**:231–242.

Madoff, S. 1960. Isolation and identification of PPLO. Annals of the New York Academy of Sciences **174**:912–921.

Madoff, S. 1970. L-forms from *Streptococcus* MG: Induction and characterization. Annals of the New York Academy of Sciences **174**:912–921.

Madoff, S. 1976a. Mycoplasma and L-forms: Occurrence in bacterial cultures. Health Laboratory Science **13**:159–165.

Madoff, S. 1976b. L-forms of *Haemophilus influenzae*: Morphology and ultrastructure. Les Colloques de l'Institut National de la Santé et de la Recherche Medicale **64**:15–26.

Madoff, S. 1979. L-forms of *Haemophilus influenza*: Growth and reversion as influenced by a strain of *Neisseria perflava*. Current Microbiology **2**:43–46.

Madoff, S., Burke, M. E., Dienes, L. 1967. Induction and identification of L-forms of bacteria. Annals of the New York Academy of Sciences **143**:755–759.

Madoff, S., Dienes, L. 1958. L-forms from pneumococci. Journal of Bacteriology **76**:245–250.

Madoff, S., Pachas, W. N. 1970. Mycoplasma and the L-forms of bacteria, pp. 195–217. In: Graber, C. D. (ed.), Rapid diagnostic methods in medical microbiology. Baltimore: Williams & Wilkins.

Marston, J. 1968. Production and cultivation of staphylococcal L-forms, pp. 212–229. In: Guze, L. B. (ed.), Microbial protoplasts, spheroplasts and L-forms, Baltimore: Williams & Wilkins.

Martin, H. H., Schilf, W., Gruss, P. 1976. Continued function of penicillin-sensitive enzymes of cell wall synthesis during growth of the unstable L-form of *Proteus mirabilis* in penicillin medium. Les Colloques de l'Institut National de la Santé et de la Recherche Medicale **64**:163–174.

Martino, P., Caravano, R., Drach, G., Schmitt-Slomska, J. 1976. Inhibitory effect of human sera of various origins on the L-forms of some Gram-positive cocci. Les Colloques de l'Institut National de la Santé et de la Recherche Medicale **64**:301–315.

Montgomerie, J. Z., Kalmanson, G. M., Guze, L. B. 1966. The effects of antibiotics on the protoplast and bacterial forms of *Streptococcus faecalis*. Journal of Laboratory and Clinical Medicine **68**:543–551.

Montgomerie, J. Z., Kalmanson, G. M., Guze, L. B. 1973. Synergism of polymyxin and sulfonamides in L-forms of *Staphylococcus aureus* and *Proteus mirabilis*. Antimicrobial Agents and Chemotherapy **3**:523–525.

Pachas, W. N., Currid, V. R. 1974. L-form induction, morphology, and development of two related strains of *Erysipelothrix rhusiopathiae*. Journal of Bacteriology **119**:576–582.

Pachas, W. N., Madoff, S. 1978. Biological significance of bacterial L-forms, pp. 412–415. In: Schlessinger, D. (ed.), Microbiology—1978. Washington, D.C.: American Society for Microbiology.

Pachas, W. N., Schor, M. 1976. Some biologic and genetic characteristics of *Proteus* L-forms. Les Colloques de l'Institut National de la Santé et de la Recherche Medicale **64**:129–146.

Panos, C. (ed.). 1967. A microbial enigma—mycoplasma and bacterial L-forms. Cleveland: World Publishing.

Panos, C. 1969. Chemical and physiological aspects of the bacterial L-phase variant, pp. 503–524. In: Hayflick, L. (ed.), The Mycoplasmatales and the L-forms of bacteria, New York: Appleton, Century, Crofts.

Panos, C. 1978. Effects of an L-form of *Streptococcus pyogenes* on cultured cells and in immunosuppressed mice, pp. 408–411. In: Schlessinger, D., (ed.), Microbiology—1978. Washington, D.C.: American Society for Microbiology.

Pierce, C. H. 1942. *Streptobacillus moniliformis*, its associated

L-form and other pleuropneumonia-like organisms. Journal of Bacteriology **43**:780.

Roberts, R. B. 1966. L-form of *Neisseria gonorrhoeae*. Journal of Bacteriology **92**:1609–1614.

Roberts, R. B., Wittler, R. G. 1966. The L-forms of *Neisseria meningitidis*. Journal of General Microbiology **44**:139–147.

Roux, J. 1976. In vivo study of L-forms, spheroplasts, protoplasts: Interaction with cells and tissues. Les Colloques de l'Institut National de la Santé et de la Recherche Medicale **64**:235–246.

Schmitt-Slomska, J., Roux, J. 1976. Cell wall defective organisms as a model for the study of antibiotic activity. Les Colloques de l'Institut National de la Santé et de la Recherche Medicale **64**:185–196.

Schmitt-Slomska, J., Sacquet, E., Caravano, R. 1967. Group A streptococcal L-forms. I. Persistence among inoculated mice. Journal of Bacteriology **93**:451–455.

Schönfeld, J. K., de Bruijn, W. C. 1976. Ultrastructure of the transition from bacterial to L-phase and vice versa. Les Colloques de l'Institut National de la Santé et de la Recherche Medicale **64**:27–38.

Sharp, J. T. 1954. L colonies from hemolytic streptococci: New technic in the study of L-forms of bacteria. Proceedings of the Society for Experimental Biology and Medicine **87**:94–97.

Smith, P. F. 1971. The biology of mycoplasmas. New York: Academic Press.

Smith, P. F. 1978. Comparative biochemistry of surface components from mycoplasma and bacterial L-forms, pp. 404–407.

In: Schlessinger, D. (ed.), Microbiology—1978. Washington, D.C.: American Society for Microbiology.

Theodore, T. S., King, J. R., Cole, R. M. 1969. Identification of L-forms by polyacrylamide-gel electrophoresis. Journal of Bacteriology **97**:495–499.

Ward, J. R., Madoff, S., Dienes, L. 1958. *In vitro* sensitivity of some bacteria, their L-forms and pleuropneumonia-like organisms to antibiotics. Proceedings of the Society for Experimental Biology and Medicine **97**:132–135.

Wittler, R. G. 1968. L-forms, protoplasts, spheroplasts: A survey of *in vitro* and *in vivo* studies, pp. 200–211. In: Guze, L. B. (ed.), Microbial protoplasts, spheroplasts and L-forms. Baltimore: Williams & Wilkins.

Wittler, R. G., McGee, Z. A., Williams, C. O. 1968. Identification of L-forms: Problems and approaches, pp. 333–339. In: Guze, L. B. (ed.), Microbial protoplasts, spheroplasts and L-forms. Baltimore: Williams & Wilkins.

Wyrick, P. B., Gooder, H. 1976. Reversion of *Streptococcus faecium* cell wall-defective variants to the intact bacterial state. Les Colloques de l'Institut National de la Santé et de la Recherche Medicale **64**:59–88.

Wyrick, P. B., McConnell, M., Rogers, H. J. 1973. Genetic transfer of the stable L form state to intact bacterial cells. Nature **244**:505–507.

Wyrick, P. B., Rogers, H. J. 1973. Isolation and characterization of cell wall-defective variants of *Bacillus subtilis* and *Bacillus licheniformis*. Journal of Bacteriology **116**:456–465.

CHAPTER 167

Wall-Free Prokaryotes of Plants and Invertebrates

RANDOLPH E. McCOY

Walled prokaryotes, the bacteria, have been known for some time to be capable of inciting certain plant diseases (see this Handbook, Chapter 4). However, the association of plant diseases with wall-free prokaryotes of the class Mollicutes, the mycoplasmas, was first suggested by Doi et al. and Ishiie et al. in 1967 for several maladies previously thought to be of viral etiology. These investigators found bodies of typical mycoplasmal morphology within the phloem of plants afflicted with several yellows diseases. In addition, they found that antibiotics of the tetracycline group, but not penicillin, could cause remission of symptoms and resumption of healthy new growth. Such results confirmed the ultrastructural indications that wall-free prokaryotes were involved in these disease syndromes.

The yellows group of diseases had been considered to be induced by viruses because the causative agents were filterable, heat labile, and persistently transmitted by insects. The symptoms consisted primarily of growth and color abnormalities, and no bacteria or fungi were consistently observed or isolated by conventional methods. Also, the yellows diseases were associated with phloem disruption and, since phloem-delimited viruses had posed vexing problems in attempted purification, failures to purify yellows agents seemed only to reflect the "state of the art" rather than a profound indication of the agent's microbial nature.

In consequence, it was a distinct surprise to the scientific world when mycoplasmal etiologies were suggested for the yellows diseases. The initial reports on mycoplasma-like organisms in plants opened a new field of investigation in plant pathology. Results have included the tentative identification of the etiological agents of many yellows diseases, the erection of a new taxon within the class Mollicutes (genus *Spiroplasma*), and the extension of the known habitats for the Mollicutes to the internal tissues and fluids of plants and insects and to such environmentally exposed sites as plant surfaces and decaying plant material.

In this chapter, the trivial term "mycoplasma" is employed in its usual context as a descriptor of a member of the class Mollicutes. The term mycoplasma-like organism (MLO) is used to denote those organisms that morphologically appear to be mycoplasmas but that, in the absence of cultural confirmation, cannot be placed in this group with certainty. The less specific term "wall-free prokaryote" includes bacterial L-forms as well as mycoplasmas.

This chapter discusses agents of the plant yellows diseases, including spiroplasmas and MLOs, the apparently epiphytic mycoplasmas, the invertebrate vectors of these organisms, and the occasional occurrence of bacterial L-forms in diseased plants. Spiroplasmas will be mentioned only briefly since they are covered in detail in this Handbook, Chapter 169.

Habitats

Plant mycoplasmas occur either as intracellular parasites, i.e. the yellows disease agents which reside within the sieve-tube elements of the phloem, or as epiphytes living, or at least surviving, on plant surfaces. Alternate hosts for the plant yellows agents are insects (order Homoptera), which feed on the phloem of infected plants and must become infected themselves before they are able to transmit the agents to healthy plants. The epiphytic mycoplasmas reside on plant surfaces, such as flowers, to which they are presumably carried by visiting insects. In addition, mycoplasma-like bodies have been reported in such diverse hosts as the fungi *Aphanomyces* (Heath and Unestam, 1974) and *Coprinus* (Ross, Pommerville, and Damm, 1976), as well as in certain bivalves and mollusks (Harshbarger, Chang, and Otto, 1977). Although the evidence suggesting these bodies to be separate organisms rather than some component of host ultrastructure is rather tenuous at present, it is certain that the known habitats of wall-free prokaryotes will be expanding in the near future.

Plant Yellows Agents

PHLOEM HABITAT. Both spiroplasmas and MLOs associated with plant yellows diseases are restricted to the sieve tubes of the phloem. These organisms have not been demonstrated to exist outside phloem tissues. Reports of these agents in phloem parenchyma or companion cells are also disputed, because of confusion regarding the morphological identification of sieve-tube elements and the misidentification of membrane-bound vesicles of host origin as MLOs (McCoy, 1979b).

The sieve-tube interior presents a highly specialized environment not found in any other tissues of the plant (Chronshaw, 1975; Crafts and Crisp, 1971; Parthasarathy, 1976). The phloem system functions in the transport of photosynthates from their sources in leaves or storage organs, to their sites of use, primarily the growing points of root, shoot, and fruit. The sieve tubes consist of long series of individual cells, the sieve-tube elements, stacked end-to-end to form long conduits for transport. The end walls of the sieve tubes, the sieve plates, are perforated by pores that average 2 μm in diameter (Esau and Cheadle, 1959). Lateral sieve areas may occur between adjacent sieve tubes, but the pores are smaller than those of the sieve plate.

Sieve tubes have a rather thick, layered (nacreous) cell wall and a plasmalemma. They contain endoplasmic reticulum and, depending on species, plastids of various types including mitochondria, crystals, and starch grains. Mature sieve-tube elements are enucleate but maintain intimate contact with their companion cells via plasmodesmata. Sieve tubes and their companion cells contain high concentrations of carbohydrate, principally sucrose, which makes up 12–30% of the sieve-tube sap (Geiger, 1975). In woody roseaceous plants, sorbitol rather than sucrose is the principal carbohydrate of the phloem; in the Cucurbitaceae, nitrogenous compounds make up the bulk of the phloem sap. The high sugar concentration of sieve-tube sap and its correspondingly high osmotic pressure (averaging 12–16 atm) results in turgor pressures up to 20 atm or more (Zimmermann, 1979). The sugar-concentration gradient in the phloem from source to site of utilization produces a corresponding turgor-pressure gradient that is considered to be the driving force for solute flow in sieve tubes (Münch, 1927).

The chemical environment of the sieve tube is exceedingly complex (Eschrich and Heyser, 1975; Evert, 1977; Ziegler, 1975). In addition to high sugar concentrations, inorganic cations are present; the predominant monovalent cation is K^+ and the predominant divalent cation is Mg^{2+}. The dominant anions are Cl^- and PO_4^{3-}. Phloem also contains relatively high concentrations of ATP (30–500 μg/ml), probably related to the energy dependence of solute transport in phloem. Free amino acids are present at up to 80 mg/ml in phloem exudates from plants, particularly glutamic and aspartic acids. Plant-phloem exudates may contain 0.1–1% protein, and P-protein is conspicuously visible in most electron micrographs of sieve elements. P-protein appears to function as a plugging mechanism in injured sieve tubes (Dempsey, Bullivant, and Bieleski, 1976). A large number of enzymes have been identified in plant-phloem exudates (Eschrich and Heyser, 1975).

INSECT HABITAT. A number of similarities exist between phloem exudate and insect hemolymph, the extracellular circulatory fluid of arthropods. Both transport assimilates and metabolites throughout their respective organisms and both contain a wide variety of organic and inorganic nutrients, resulting in similar high osmolalities. Although critical analyses of hemolymph from homopterous insects are not available, much of the information on this subject has been recently summarized by Saglio and Whitcomb (1979).

After ingestion of sieve-tube sap from diseased plants, infectious organisms are found in the insect gut. After a period of days, the infectious agent is present in hemolymph and subsequently can be found intracellularly in various body organs, depending on the disease agent and the insect species. Sinha and Chiykowski (1967a, 1968) found the aster yellows agent to reach high titers in alimentary canal, hemolymph, salivary glands, and ovaries of its leafhopper vector, whereas the infectious agent was not recovered from testes, brain, Malpighian tubules, fat body, or mycetomes. They found the clover phyllody agent only in the alimentary canal, hemolymph, and salivary glands of infected insects. On the other hand, Nasu, Kono, and Jensen (1974) showed that the Western X-disease agent multiplied in the insect brain at an unusually high rate as compared to other tissues.

Infectious vectors always appear to carry the agents in their salivary glands (Hirumi and Maramorosch, 1969; Nasu, Jensen, and Richardson, 1970; Sinha and Paliwal, 1970). Extracts of salivary glands are free of infectivity when taken before completion of the incubation period of the agent in the insect (i.e., after the insect has fed on a diseased plant, but before it is able to transmit the agent to healthy plants). Interestingly, Sinha and Chiykowski (1967b) showed the aster yellows agent to multiply in the gut of a nonvector insect after feeding on infected plants, but the hemolymph and salivary glands of the insects did not become infected.

Epiphytic Mycoplasmas

A new habitat for mycoplasmas and spiroplasmas has recently been reported by Davis (1978) on the surfaces of flowers. The agents, mostly sterol-requiring mycoplasmas, several acholeplasmas, and spiroplasmas, were isolated from leachates of flowers that had not been surface-sterilized. Very little is known of the organisms or of their ecological niche. It is presumed that they survive on the floral surface, possibly in nectar, and are carried from flower to flower by visiting insects. Insects may be the primary hosts for these organisms or the mycoplasmas may simply be carried externally by the insects. In any case, this finding opens an entire new area of microbial ecology.

In addition to the flower habitat, a number of *Acholeplasma* isolates have been obtained from surfaces and even from interior tissue near or on necrotic zones of coconut palms suffering from lethal yellowing disease (Eden-Green, 1978). These organisms appear to be surface contaminants associated with rotting plant tissue, although the possibility that they may be the etiological agents of lethal yellowing cannot be completely ruled out. If epiphytic, these organisms could be carried from site to site by visiting insects.

Bacterial L-Phase in Plant Disease

Wall-less L-phase growth of the bacterium *Erwinia carotovora* var. *atroseptica* has been described both in vitro and in vivo by Jones and Paton (1973). This bacterium, which causes black leg disease of potato, was occasionally observed in the L-phase in cells of 30- to 50-day-old pectolysed potato or cucumber tissue. In addition, the L-phase of *Erwinia* could be induced in cucumber and potato tuber tissue by embedding infected tissue disks in an L-phase agar medium containing 100 IU penicillin as an inducing agent. Infected cells were opaque and filled with granular, spherical to irregular-shaped bodies that reacted positively to specific antiserum for *E. carotovora* var. *atroseptica*. In addition, normal *Erwinia* colonies resulted from plating cell suspension filtrates passed through 0.45-μm-pore-diameter membrane filters, again indicating the presence of an L-phase bacterium. Unlike the mycoplasma-like organisms associated with diseased plants, the L-phase of *Erwinia* was present within macerated parenchyma cells rather than in vascular tissue. (Further information on L-forms of bacteria will be found in this Handbook, Chapter 166.)

Isolation

Cultivation

Other than the demonstrated pathogenicity of the spiroplasmas that cause corn stunt and citrus stubborn, no plant yellows agents have been verified as cultivable. Numerous reports of cultivation of mycoplasmas from diseased plants have appeared in the literature, but to date none have met the criteria established for the verification of pathogenicity by Davis and Whitcomb (1971). The steps in this verification are: (i) Consistent isolation from diseased

but not healthy plants. (ii) Sufficient serial subcultivations to reach the dilution end point of a nonmultiplying agent which may have been isolated along with an epiphytic or contaminating mycoplasma. Such a nonmultiplying agent could give positive results when injected into leafhopper vectors for infectivity assay, even though the bulk of the culture was of a nonpathogen. (iii) Establishment of pure lines of the culture by filter-cloning. Broth cultures are passed through 0.2- or 0.45-μm-pore-diameter filters, followed by plating on agar media. Single colonies from the limiting dilution are then picked off, placed in broth, and the process is repeated two more times. The resultant clones should be pure strains, assuming that the colonies were derived from single cells. (iv) Inducement of the disease syndrome by vectors injected with the cloned isolate in susceptible plants on which they feed. (v) Isolation of the same agent from inoculated plant and insect hosts. It is desirable to further characterize the causal organism by following the criteria established by the Subcommittee on Taxonomy of the Mycoplasmatales (1972); however, this is not an absolute requirement for demonstrating pathogenicity.

The major factor that complicates attempts to cultivate plant yellows disease agents is the isolation of contaminant mycoplasmas. These contaminants may come from the person making the isolations if proper aseptic techniques are not followed. The oral cavity is a common source of several human *Mycoplasma* species. Mycoplasmas in serum or other medium components represent other sources of contamination. Passage of these components or of the final medium through a 0.2-μm-pore-diameter filter does not necessarily eliminate these organisms. Mycoplasmas may be eliminated from serum by heat inactivation at 56°C for 1 h. Use of commercial, pretested mycoplasma-free sera also aids in reduction of contamination. One should also be aware of the possibility of isolating epiphytic mycoplasmas from plant surfaces. These can be eliminated as much as possible by surface sterilization and by avoiding external plant surfaces wherever practicable. Finally, investigators should be aware that certain crystalline growths on agar surfaces may strongly resemble mycoplasma colonies. These "pseudocolonies" are not subcultivable and will not stain with Giemsa or cresyl violet stains.

The cultivation of apparently epiphytic mycoplasmas has been described by Davis (1978) and McCoy, Williams, and Thomas (1979) from flowers, and by Eden-Green (1978) from dying coconut palms. The non-surface-sterilized plant part is briefly washed in culture broth. This broth is immediately passed through a 0.45-μm-pore-diameter filter, and aliquots of the filtrate are inoculated into additional broth or onto agar plates. Following a successful primary isolation, the organisms must be filter-cloned to establish pure strains. The organisms can then be characterized according to the methods used for other mycoplasmas and spiroplasmas (see this Handbook, Chapters 168 and 169).

Maintenance in Tissue Culture

Mycoplasma-like organisms have been observed in sieve-tube elements of diseased plant tissue grown in tissue culture (Jacoli, 1974; McCoy, 1979b). In addition, the agents have been shown to remain pathogenic in plant-tissue culture by grafting cultured tissue to indicator plants, but only in systems which promote vascular differentiation (Petrů and Ulrychová, 1975). When undifferentiated callus cultures from diseased plants were grafted onto healthy indicators, no symptom development ensued (Ulrychová and Petrů, 1975). This result affords additional evidence that MLOs are present only in phloem tissue.

Infectivity Assay

The only known means to determine pathogenicity of yellows agents at present is bioassay of infectivity: Known vector leafhoppers are infected with the isolated or cultured agent, and the disease is reproduced in susceptible plants on which the vector, after a suitable incubation period, is allowed to feed. This procedure is time-consuming and laborious and requires a great deal of coordination in maintaining pathogen-free insect colonies and a continual source of test plants of the proper age for disease susceptibility. Vectors are infected with the agent either by injecting tissue extracts or media that contain the agent with microcapillary needles, or by allowing the insects to acquire the agent by feeding through a parafilm membrane. Whitcomb (1972) and Caudwell (1977) have reviewed this subject, including the analyses necessary to quantitatively determine inoculum potential of the infectious extracts or culture solutions. Caudwell (1974) and Caudwell, Kuszala, and Larrue (1976) used this method to monitor cultural attempts with the "flavescence dorée" agent of grapevines. They have sequentially modified their culture medium to give the greatest longevity and highest titers of the infectious agent in their isolation media. Although active multiplication has not been demonstrated, their bioassays have successfully defined several factors necessary to survival of the flavescence dorée agent.

Mechanical Isolation and Maintenance

In the absence of cultivation methods for the plant yellows agents, several investigators have attempted to physically isolate or purify the agents from their plant or insect hosts. Infectivity assay is required to monitor the success of the purification. The methods used have ranged from filtration (Cohen, Purcell, and Steere, 1969; Davis, Whitcomb, and Purcell, 1970) to density gradient centrifugation (Giannotti, Vago, and Duthoit, 1968; Giannotti et al., 1969; Nasu, Jensen, and Richardson, 1974) and differential (Sinha, 1974) centrifugation. In addition, earlier attempts to purify viruses from yellows-diseased plants (Lee and Chiykowski, 1963; Steere, 1967; Whitcomb, Jensen, and Richardson, 1968) provided information valuable to later workers attempting to isolate MLOs from these diseases.

Maintenance of strains of yellows disease agents without culture methods is difficult and can only be done in greenhouse situations by maintenance of infected plants and vectors. Even under the best of conditions, such strains can become attenuated with time (Chiykowski, 1977a). However, Chiykowski (1977b) has recently shown that the aster yellows agent can be maintained in infected vector insects stored at −64°C for up to 2 years. Storage of frozen or even lyophilized strains could be an excellent means of long-term maintenance.

Identification

The classification of agents of plant yellows disease has been based indirectly on in vivo morphology, symptomatology, and antimicrobial sensitivity. Strain differences have been defined by host range, vector relationships, and symptomatology. Such indirect and imprecise methods are used in place of the direct characterization possible with a cultivated mycoplasma. Over 150 plant diseases have been ascribed to MLOs, yet the interrelationships among the causal agents are almost totally unknown. The cultivable, ephiphytic or insect-inhabiting mycoplasmas may be characterized by the methods normally used in classification of mycoplasmas or spiroplasmas (see this Handbook, Chapters 168 and 169).

Separation of Spiroplasmas from Virescence Agents

Based on microbial morphology and host-plant symptomatology, there appear to be two major groups of yellows disease agents. The spiroplasma group includes the corn stunt spiroplasma (CSS) and *Spiroplasma citri* (this Handbook, Chapter 169); they are readily recognizable by their helical morphology, both in culture and in their plant and insect hosts. The symptoms induced in plants by spiroplasmas include chlorosis and stunting of vegetative organs and reduction in flower size (Calavan and Oldfield, 1979; Markham et al., 1977). On the other hand, the apparently nonhelical MLOs are associated with the symptoms of yellowing and stunting, and also produce virescence and phyllody of floral organs and vegetative proliferations or witches' brooms. Virescence is the reversion of reproductive organs to the vegetative state, i.e., the greening of petals and their ultimate conversion to leafy structures (phyllody). Several nonhelical MLOs associated with tree diseases do not induce virescence in their woody hosts, but may when inoculated into herbaceous plants by dodder, grafting, or insect vectors. Neither *S. citri* nor CSS induces virescence in herbaceous hosts. No helical morphologies have been demonstrated for MLOs in plants other than for *S. citri* and CSS. Undulating filaments observed in ultrathin section in some virescence-diseased plants (Hirumi and Maramorosch, 1973) cannot be spiroplasmas, since these organisms are helical in three dimensions and thicker sections are required for their visualization. Several recent reports on the cultivation of spiroplasmas from diseases associated with virescence or vegetative proliferation (Kondo et al., 1976, 1977; Raju and Nyland, 1978) have not demonstrated pathogenicity of the cultured agents.

Several yellows diseases are characterized by lethal shock reactions that rapidly kill the host plants. Such cases appear to be examples of highly sensitive host reactions to infection and occur for spiroplasmal (Markham et al., 1977) and virescence agents (Braun, 1977).

Yellows Disease Agents

MORPHOLOGY. As previously stated, the yellows disease agents are restricted to the intracellular habitat of the sieve tubes of the phloem in plants. They are present extracellularly in insect hemolymph and in certain organs, particularly the salivary glands of infectious vectors. The morphology of the yellows disease agents in these habitats is basically that of the Mollicutes, i.e., wall-free prokaryotes. The organisms appear circular or filamentous and, often, lobed in ultrathin section, they contain ribosomes and chromatin, and they are bounded by a single, trilaminar unit membrane (Fig. 1). Thicker sections tend to reveal a greater number of filamentous structures in some cases (Florance and Cameron, 1978; Thomas, 1979) (Fig. 2) and helical organisms in the case of spiroplasma infections (Davis et al., 1972). The organisms in ultrathin section range from about 0.1 to 1.2 μm in diameter.

The morphology of the yellows agents in plants has been widely misunderstood and many past interpretations of their structure are in error (McCoy, 1979a,b; Waters and Hunt, 1978). In addition, the literature contains many references to MLO which, in fact, are based on vesicles of host origin that either occur naturally or are induced by viral infection

(de Zoeten, Gaard, and Diez, 1972; Esau and Hoefert, 1972; Florance and Cameron, 1974). Many misinterpretations of MLO structure in sieve tubes are based on conclusions derived from examination of single, ultrathin sections, and from a lack of awareness of the effect of pressure disruption of the phloem during sampling and fixation. Single, ultrathin sections of MLO in sieve tubes reveal simple, polymorphic bodies. Some of these bodies appear to be budding and others appear to contain membrane-bound inclusions; some are very small, dense "elementary" bodies of 0.05–0.15 μm diameter (Fig. 1). Examinations of serial ultrathin sections (Chen and Hiruki, 1977; Florance and Cameron, 1978; Waters and Hunt, 1978) have revealed the true three-dimensional structure of these MLOs: the forms that appeared to be budding were shown to be branch points in filamentous forms; membrane-bound inclusions were found to be invaginations in one MLO cell into which another cell was closely packed; and the so-called elementary bodies were actually constriction points in filaments of overall

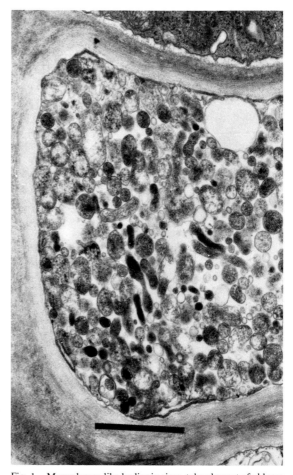

Fig. 1. Mycoplasma-like bodies in sieve-tube element of phloem of witches' broom–diseased periwinkle. Ultrathin section observed by transmission electron microscopy. Bar = 1 μm. [Courtesy of D. L. Thomas.]

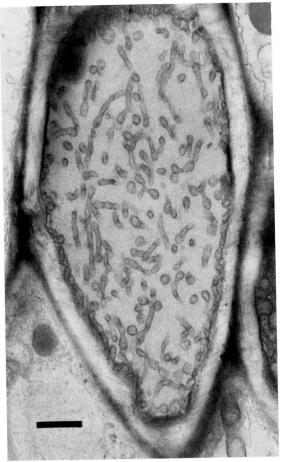

Fig. 2. Filamentous mycoplasma-like bodies in sieve element of lethal decline–diseased *Arenga* palm. Semithick section (0.3 μm) observed by transmission electron microscopy. Bar = 1 μm. [Courtesy of D. L. Thomas.]

tioning sieve pores are generally considered to be free of obstruction and of sufficient diameter to allow even large MLOs to pass without distortion.

ANTIMICROBIAL SENSITIVITY.   Yellows-diseased plants have been demonstrated many times to respond to treatment with tetracycline antibiotics, while remaining resistant to the penicillins (Sinha, 1979). The response to tetracycline is temporary, and the effect must be described as a remission rather than a cure. This differential chemotherapeutic response, however, especially when considered in conjunction with the observation of MLO in diseased plants, provides strong diagnostic evidence for mycoplasmal etiologies in the yellows diseases. The duration of the period of remission induced by tetracycline antibiotics is generally short in herbaceous plants, but may last a year or more in some woody plants. Oxytetracycline-HCl is presently used on a commercial basis to control coconut lethal yellowing and pear decline diseases (McCoy, 1974; Nyland, 1974). Oxytetracycline may also be used to prevent lethal yellowing by injecting healthy trees in areas of high disease incidence (McCoy et al., 1976).

In addition to suppressing symptom development in yellows diseases, tetracycline antibiotics have caused infected vector insects to become noninfectious (Davis and Whitcomb, 1969; Sinha and Peterson, 1972). Again, the effect is only temporary and the insects regained infectivity after a period of time. Davis and Whitcomb (1969) showed that tylosin

larger diameter. Scanning-electron microscopy of infected sieve elements has also revealed the general filamentous nature of MLOs in plants (Haggis and Sinha, 1978; Petzold et al., 1977) (Fig. 3).

The effect of pressure disruption of sieve elements has resulted in misinterpretations of MLO morphology, particularly around sieve plates (McCoy, 1979a). When a normal sieve tube is cut or injured, the pressurized contents explosively surge down the conduit, and disrupt cellular plastids and P-protein in the process. The sieve pores become plugged with cellular debris and P-protein, which serves to prevent further loss of sieve-tube content. In addition to these responses, a sieve tube will rapidly deposit a layer of callose on the sieve plate when injured. Electron micrographs that depict MLO squeezing into thin strands to pass through sieve pores occluded with P-protein and callose show the effect of pressure disruption on MLO structure, not the direction of invasion of MLO from cell to cell as has often been inferred (Hirumi and Maramorosch, 1973; Jacoli, 1974). In fact, func-

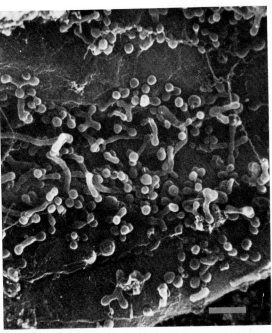

Fig. 3. Scanning electron micrograph of mycoplasma-like bodies in sieve element of clover phyllody-diseased aster. Note filamentous and branching forms. Bar = 1 μm. [Haggis and Sinha, 1978, with permission.]

also could cause loss of infectivity in leafhoppers infected with the aster yellows agent. However, they found tylosin to be ineffective against aster yellows in plants. Such differences may result from restricted phloem mobility of some antibiotics (McCoy, 1979b).

IDENTIFICATION OF STRAIN DIFFERENCES.   At present, differentiation between yellows diseases and strains of yellows diseases is based on symptomatology, host range, vector relationships, and interactions between strains inoculated into the same host (Chiykowski, 1973, 1974). These methods are cumbersome and are not always specific. It is practically impossible to compare diseases that have the same hosts and symptoms, but that occur in different regions or continents. Chiykowski (1973) concluded that the interrelationships among the yellows diseases cannot be worked out with our present state of knowledge. Diagnostic tests, such as the enzyme-linked immunosorbent assay, that can detect *Spiroplasma citri* infections of citrus and periwinkle (Garcia Jurado et al., 1978) are specific and sensitive, but could not have been developed without prior cultivation of the *S. citri* agent. Therefore, cultivation and full characterization of yellows disease agents may be required before we can fully understand the diversity found among the yellows diseases, both locally and throughout the world.

Literature Cited

Braun, E. J. 1977. Phloem necrosis of elms: Etiological, physiological and cytological studies. Ph.D. Thesis. Cornell University, Ithaca, New York.

Calavan, E. C., Oldfield, G. N. 1979. Symptomatology of spiroplasmal plant diseases, pp. 37–64. In: Barile, M. F., Razin, S. H., Tully, J. G., Whitcomb, R. F. (eds.), The mycoplasmas, vol. 3. New York: Academic Press.

Caudwell, A. 1974. Sur la culture *in vitro* des agents infectieux responsables des jaunisses des plantes (MLO). Annales de Phytopathologie 6:173–190.

Caudwell, A. 1977. Aspects statistiques des épreuves d'infectivité chez les jaunisses (yellows) des plantes et chez les viroses transmises selon le mode persistant. Intérêt de la fève *(Vica faba)* comme plante test pour les jaunisses. Annales de Phytopathologie 9:141–159.

Caudwell, A., Kuszala, C., Larrue, J. 1976. Progress in the culture of the etiologic agent, type mycoplasma, of the Flavescence dorée of grapes. Proceedings of the Society for General Microbiology 3:154.

Chen, M. H., Hiruki, C. 1977. Effects of dark treatment on the ultrastructure of the aster yellows agent in situ. Phytopathology 67:321–324.

Chiykowski, L. N. 1973. The aster yellows complex in North America. Proceedings of the North Central Branch, Entomological Society of America. 28:60–66.

Chiykowski, L. N. 1974. Yellows diseases and vectors. Institut National de la Santé et de la Recherche Medicale 33:291–298.

Chiykowski, L. N. 1977a. Reduction in the transmissibility of a greenhouse-maintained isolate of aster yellows agent. Canadian Journal of Botany 55:1783–1786.

Chiykowski, L. N. 1977b. Cryopreservation of aster yellows agent in whole leafhoppers. Canadian Journal of Microbiology 23:1038–1040.

Chronshaw, J. 1975. Sieve element cell walls, pp. 129–147. In: Aranoff, S. et al. (eds.), Phloem transport. London: Plenum Press.

Cohen, R., Purcell, R., Steere, R. L. 1969. Ultrafiltration of the aster yellows agent. Phytopathology 59:1555.

Crafts, A. S., Crisp, C. E. 1971. Phloem transport in plants. San Francisco: W. H. Freeman.

Davis, R. E. 1978. Spiroplasma associated with flowers of the tulip tree (*Liriodendron tulipifera* L.). Canadian Journal of Microbiology 24:954–959.

Davis, R. E., Whitcomb, R. F. 1969. Spectrum of antibiotic sensitivity of aster yellows disease in insects and plants. Phytopathology 59:1556.

Davis, R. E., Whitcomb, R. F. 1971. Mycoplasmas, rickettsiae, and chlamydiae: Possible relation to yellows diseases and other disorders of plants and insects. Annual Review of Phytopathology 9:119–154.

Davis, R. E., Whitcomb, R. F., Purcell, R. 1970. Viability of the aster yellows agent in cell-free media. Phytopathology 60:573–574.

Davis, R. E., Worley, J. F., Whitcomb, R. F., Ishijima, T., Steere, R. L. 1972. Helical filaments produced by a mycoplasma-like organism associated with corn stunt disease. Science 176:521–523.

Dempsey, G. P., Bullivant, S., Bieleski, R. L. 1976. The distribution of P-protein in mature sieve tube elements, pp. 247–251. In: Wardlaw, J. F., Passioura, J. B. (eds.), Transport and transfer processes in plants. New York: Academic Press.

de Zoeten, G. A., Gaard, J., Diez, F. B. 1972. Nuclear vesicularization associated with pea enation mosaic virus-infected plant tissue. Virology 48:638–647.

Doi, Y., Terenaka, M., Yora, K., Asuyama, H. 1967. Mycoplasma or PLT group-like microorganisms found in the phloem elements of plants infected with mulberry dwarf, potato witches' broom, aster yellows, or paulownia witches' broom. Annals of the Phytopathological Society of Japan 33:259–266.

Eden-Green, S. J. 1978. Isolation of acholeplasmas from coconut palms affected by lethal yellowing disease in Jamaica. Zentralblatt für Bakteriologie, Parasitenkunde, Infektionskrankheiten und Hygiene, Abt. 1 Orig., Reihe A 241:226.

Esau, K., Cheadle, V. I. 1959. Size of pores and their contents in sieve elements of dicotyledons. Proceedings of the National Academy of Sciences of the United States of America 45:156–162.

Esau, K., Hoefert, L. L. 1972. Development of infection with beet western yellows virus in the sugarbeet. Virology 48:724–738.

Eschrich, W., Heyser, W. 1975. Biochemistry of phloem constituents, pp. 101–136. In: Zimmermann, M. H., Milburn, J. A. (eds.), Encyclopedia of plant physiology, new series, vol. 1: Phloem transport. Berlin, Heidelberg, New York: Springer-Verlag.

Evert, R. F. 1977. Phloem structure and physiology. Annual Review of Plant Physiology 28:199–222.

Florance, E. R., Cameron, H. R. 1974. Vesicles in expanded endoplasmic reticulum cisternae structures that resemble mycoplasma-like bodies. Protoplasma 79:337–348.

Florance, E. R., Cameron, H. R. 1978. Three-dimensional structure and morphology of mycoplasmalike bodies associated with the albino disease of *Prunus avium*. Phytopathology 68:75–80.

Garcia Jurado, O., Saillard, C., Duney, J., Bové, J. M. 1978. Evaluation of an immuno-enzymatic technique (ELISA) for the detection of spiroplasma and mycoplasmalike organisms in plants. Zentralblatt für Bakteriologie, Parasitenkunde, Infektionskrankheiten und Hygiene, Abt. 1 Orig., Reihe A 241:229–230.

Geiger, D. B. 1975. Phloem loading and associated processes, pp. 251–281. In: Aranoff, S. et al. (eds.), Phloem transport. London: Plenum Press.

Giannotti, J., Vago, C., Duthoit, J. L. 1968. Isolement et purification de microorganismes à structure de mycoplasmes à partir de Cicadelles et de plantes infectées de jaunisses. Revue Zoologie Agricole et Appliquée 4–6:69–72.

Giannotti, J., Caudwell, A., Vago, C., Duthoit, J. L. 1969. Isolement et purification de microorganismes de type mycoplasme à partir de vignes atteintes de Flavescence dorée. Comptes Rendus Hebdomadaires des Séances de l'Académie des Sciences, Série D 268:845–847.

Haggis, G. H., Sinha, R. C. 1978. Scanning electron microscopy of mycoplasmalike organisms after freeze fracture of plant tissues affected with clover phyllody and aster yellows. Phytopathology 68:677–680.

Harshbarger, J. C., Chang, S. C., Otto, S. V. 1977. Chlamydiae (with phages), mycoplasmas, and rickettsiae in Chesapeake Bay bivalves. Science 196:666–668.

Heath, I. B., Unestam, T. 1974. Mycoplasma-like structures in the aquatic fungus Aphanomyces astaci. Science 183:434–435.

Hirumi, H., Maramorosch, K. 1969. Mycoplasma-like bodies in the salivary glands of insect vectors carrying the aster yellows agent. Journal of Virology 3:82–84.

Hirumi, H., Maramorosch, K. 1973. Ultrastructure of the aster yellows agent: Mycoplasma-like bodies in sieve tube elements of Nicotiana rustica. Annals of the New York Academy of Sciences 225:201–222.

Ishiie, T., Doi, Y., Yora, K., Asuyama, H. 1967. Suppressive effects of antibiotics of tetracycline group on symptom development of mulberry dwarf disease. Annals of the Phytopathological Society of Japan 33:267–275.

Jacoli, G. G. 1974. Translocation of mycoplasma-like bodies through sieve pores in plant tissue cultures infected with aster yellows. Canadian Journal of Botany 52:2085–2088.

Jones, S. M., Paton, A. M. 1973. The L-phase of Erwinia carotovora var. atroseptica and its possible association with plant tissue. Journal of Applied Bacteriology 36:729–737.

Kondo, F., McIntosh, A. H., Padhi, S. B., Maramorosch, K. 1976. Electron microscopy of a new plant–pathogenic spiroplasma isolated from Opuntia. Proceedings of the Electron Microscopy Society of America 34:56.

Kondo, F., Maramorosch, K., McIntosh, A. H., Varney, E. H. 1977. Aster yellows spiroplasmas: Isolation and cultivation in vitro. Proceedings of the American Phytopathological Society 4:190.

Lee, P. E., Chiykowski, L. N. 1963. Infectivity of aster-yellows virus preparations after differential centrifugations of extracts from viruliferous leafhoppers. Virology 21:667–669.

McCoy, R. E. 1974. How to treat your palm with antibiotic. University of Florida Circular S-228.

McCoy, R. E. 1979a. Passage of mycoplasmalike organisms through sieve pores? pp. 45–48. In: National Science Council (Taiwan) Symposium Series No. 1. Taiwan: National Science Council.

McCoy, R. E. 1979b. Mycoplasmas and yellows diseases, chap. 8. Barile, M. F., Razin, S., Tully, J. G., Whitcomb, R. F. (eds.), In: The mycoplasmas, vol. 3. New York: Academic Press.

McCoy, R. E., Williams, D. S., Thomas, D. L. 1979. Isolation of mycoplasmas from flowers, pp. 75–81. National Science Council (Taiwan) Symposium Series No. 1. Taiwan: National Science Council.

McCoy, R. E., Carroll, V. J., Poucher, C. P., Gwin, G. H. 1976. Field control of coconut lethal yellowing with oxytetracycline hydrochloride. Phytopathology 66:1148–1150.

Markham, P. G., Townsend, R., Plaskitt, K., Saglio, P. 1977. Transmission of corn stunt to dicotyledonous plants. Plant Disease Reporter 61:342–345.

Münch, E. 1927. Dynamik der Saftströmungen. Berichte der Deutschen Botanischen Gesellschaft 44:68–71.

Nasu, S., Jensen, D. D., Richardson, J. 1970. Electron microscopy of mycoplasma-like bodies associated with insect and plant hosts of peach Western-X disease. Virology 41:583–595.

Nasu, S., Jensen, D. D., Richardson, J. 1974. Isolation of Western X mycoplasmalike organisms from infectious extracts of leafhoppers and celery. Applied Entomology and Zoology 9:199–203.

Nasu, S., Kono, Y., Jensen, D. D. 1974. The multiplication of Western X mycoplasmalike organism in the brain of a leafhopper vector, Colladonus montanus (Homoptera: Cicadellidae). Applied Entomology and Zoology 9(4):277–279.

Nyland, G. 1974. Control aspects of plant mycoplasma diseases, chemotherapy in the field, pp. 235–242. In: Bové, J. M., Duplan, J. F. (eds.), Les mycoplasmes de l'homme, des animaux, de vegetaux et des insectes. Paris: Institut National de la Santé et de la Recherche Medicale.

Parthasarathy, M. V. 1976. Sieve element structure, pp. 3–56. In: Zimmermann, M. H., Milburn, J. A. (eds.), Encyclopedia of plant physiology, new series, vol. 1: Phloem transport. Berlin, Heidelberg, New York: Springer-Verlag.

Petrů, E., Ulrychová, M. 1975. Persistence and spread of mycoplasma in axenic callus tissue cultures of tobacco in the presence of kinetin and IAA in nutrient medium. Biologia Plantarum 17:352–356.

Petzold, H., Marwitz, R., Özel, M., Goszdziewski, P. 1977. Versuche zum raslerelektronenmikroskopischen Nachweis von mykoplasmaähnlichen Organismen. Phytopathologische Zeitschrift 89:237–248.

Raju, B. C., Nyland, G. 1978. Effects of different media on the growth and morphology of three newly isolated plant spiroplasmas. Phytopathology News 12:216.

Ross, I. K., Pommerville, J. C., Damm, D. L. 1976. A highly infectious mycoplasma inhibits meiosis in the fungus Coprinus. Journal of Cell Science 21:175–191.

Saglio, P., Whitcomb, R. F. 1979. Diversity of wall-less prokaryotes in plant vascular tissue, fungi, and invertebrate animals, chapter 1. In: Barile, M. F., Razin, S. H., Tully, J. G., Whitcomb, R. F. (eds.), The mycoplasmas, vol. 3. New York: Academic Press.

Sinha, R. C. 1974. Purification of mycoplasma-like organisms from china aster plants affected with clover phyllody. Phytopathology 64:1156–1158.

Sinha, R. C. 1979. Chemotherapy of mycoplasmal plant diseases, pp. 310–335. In: Barile, M. F., Razin, S. H., Tully, J. G., Whitcomb, R. F. (eds.), The mycoplasmas, vol. 3. New York: Academic Press.

Sinha, R. C., Chiykowski, L. N. 1967a. Initial and subsequent sites of aster yellows virus infection in a leafhopper vector. Virology 33:702–708.

Sinha, R. C., Chiykowski, L. N., 1967b. Multiplication of aster yellows virus in a nonvector leafhopper. Virology 31:461–466.

Sinha, R. C., Chiykowski, L. N. 1968. Distribution of clover phyllody virus in the leafhopper Macrosteles fascifrons (Stål). Acta Virologica 12:546–550.

Sinha, R. C., Paliwal, Y. C. 1970. Localization of a mycoplasma-like organism in tissues of a leafhopper vector carrying clover phyllody agent. Virology 40:665–672.

Sinha, R. C., Peterson, E. A. 1972. Uptake and persistence of oxytetracycline in aster plants and vector leafhoppers in relation to inhibition of clover phyllody agent. Phytopathology 62:377–383.

Steere, R. L. 1967. Gel filtration of aster yellows virus. Phytopathology 57:832–833.

Subcommittee on the Taxonomy of Mycoplasmatales. 1972. Proposal for minimal standards for descriptions of new species of the order Mycoplasmatales. International Journal of Systematic Bacteriology 22:184–188.

Thomas, D. L. 1979. Mycoplasmalike bodies associated with lethal declines of palms in Florida. Phytopathology 69:928–934.

Ulrychová, M., Petrů, E., 1975. Elimination of mycoplasma in tobacco callus tissues (*Nicotiana glauca* Grah.) cultured *in vitro* in the presence of 2,4-D in nutrient medium. Biologia Plantarum **17:**103–108.

Waters, H., Hunt, P. 1978. Serial sectioning to demonstrate the morphology of a plant mycoplasma-like organism. Zentralblatt für Bakteriologie, Parasitenkunde, Infektionskrankheiten und Hygiene, Abt. 1 Orig., Reihe A **241:**225.

Whitcomb, R. F. 1972. Bioassay of clover wound tumor virus and the mycoplasmalike organisms of peach Western X and aster yellows. United States Department of Agriculture Technical Bulletin 1438. Washington, D. C.: United States Department of Agriculture.

Whitcomb, R. F., Jensen, D. D., Richardson, J. 1968. The infection of leafhoppers by Western X-disease virus V. Properties of the infectious agent. Journal of Invertebrate Pathology **12:**192–201.

Ziegler, H. 1975. Nature of transported substances, pp. 59–100. In: Zimmermann, M. H., Milburn, J. A. (eds.), Encyclopedia of plant physiology, new series, vol. 1: Phloem transport. Berlin, Heidelberg, New York: Springer-Verlag.

Zimmermann, M. H. 1979. Mycoplasma diseases and long distance transport in plants, pp. 37–43. National Science Council (Taiwan) Symposium Series No. 1. Taiwan: National Science Council.

CHAPTER 168

The Genera *Mycoplasma, Ureaplasma,* and *Acholeplasma,* and Associated Organisms (Thermoplasmas and Anaeroplasmas)

GERALD MASOVER and LEONARD HAYFLICK

Microorganisms known generally by the trivial name ''mycoplasmas'' have such unique properties that they have recently been removed as an order in the class Schizomycetes and elevated to the status of a separate new class—the class Mollicutes.

The single biological property that most distinguishes these microorganisms from all others is that they are the smallest organisms capable of self-reproduction (Pirie, 1973). However, their removal from the class Schizomycetes was based principally on absence of a cell wall, which is reflected in the name given to the new class Mollicutes (soft skin). They are incapable of synthesizing cell wall precursors such as muramic and diaminopimelic acids. Their small size is characterized by their ability to pass through a 450-nm (and often a 220-nm) membrane filter. This is probably the chief reason why the agar colonies produced by them are typically small (0.1–0.6 mm). Due to their small size and lack of a rigid cell wall, the organisms are able to penetrate and grow in the interstices of agar fibrils to produce a characteristic colony with an appearance often likened to a ''fried egg'' (Figs. 1–4). The individual organisms are pleomorphic and vary in shape from coccoid to filamentous to helical. Some produce mycelial structures, although it may be argued that the natural shape of the organisms is coccoid. The variations in morphology may be due to growth conditions or preparative artifacts (Robertson, Gomersall, and Gill, 1975a). The organisms' exact mode of replication is not clear. The fundamental act of replication in these, as in other prokaryotes, is replication of a single circular DNA molecule (Maniloff and Morowitz, 1972). The process by which daughter DNA molecules are separated into daughter cells is the aspect of division that is unclear. If the DNA replication is followed by equal division of the cytoplasm, then binary fission occurs. If, however, the cytoplasm is unequally divided after DNA replication, the process is called budding. Both processes have been described (Maniloff and Morowitz, 1972). Both motile and nonmotile forms are known (Bredt, 1973).

As the smallest free-living organisms, Mollicutes are of interest to those biophysicists and molecular biologists who would attempt to understand all life processes by first understanding the simplest known forms. Most Mollicutes species require sterols for growth and incorporate them directly into their single trilaminar cell membrane, which is composed mainly of lipids and proteins. Membrane biologists are therefore equally intrigued because the Mollicutes can be regarded as living membrane vesicles since they have no intracellular membranes or membrane-bound structures as do all eukaryotic cells and most prokaryotes.

Many *Mycoplasma* species produce diseases in domestic animals that are of significant worldwide economic importance (Table 1). The only *Mycoplasma* species proven to cause a human disease *(Mycoplasma pneumoniae)* is the etiological agent of a significant portion of the pneumonic conditions that affect man. Its implication in a myriad of medically important sequelae is strongly suspected. Recently, good evidence for a role of *Ureaplasma* in human pathogenicity has been published (Shepard and Masover, 1978; Taylor-Robinson, Csonka, and Prentice, 1977). In recent years, organisms of the genus *Spiroplasma* have been implicated in economically important plant diseases which are considered elsewhere in this Handbook (Chapter 169).

Taxonomy of the Class Mollicutes (for all species names, see Freundt and Edward, 1979)

Class: Mollicutes
    Order I: Mycoplasmatales
        Family I: Mycoplasmataceae
          Genus I: *Mycoplasma*
            51 named species
          Genus II: *Ureaplasma*
            1 named species
        Family II: Acholeplasmataceae
          Genus I: *Acholeplasma*
            7 named species
        Family III: Spiroplasmataceae
          Genus I: *Spiroplasma*
            1 named species
    Associated organisms not yet classified:
        Thermoplasmas
        Anaeroplasmas

Since 1970, more than 50 DNA virus isolates have been made from several *Mycoplasma, Achole-*

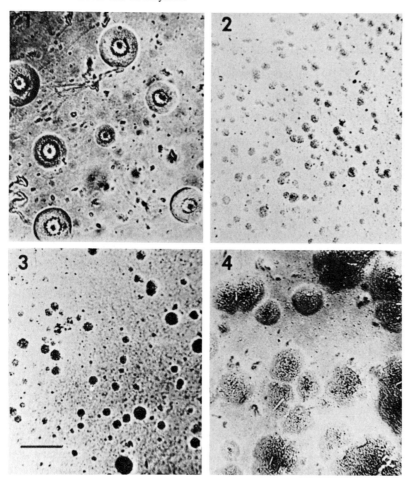

Figs. 1–4. Typical mycoplasma colonies of the classic "fried egg" type. Figs. 2–4. Other mycoplasma strains illustrating the range in variation of colonial morphology. The horizontal bar is equivalent to 100 nm.

plasma, and *Spiroplasma* species. These viruses can be divided into three groups: (i) naked bullet-shaped particles, (ii) spherical enveloped forms, and (iii) polyhedral-tailed particles (Maniloff, Das, and Christensen, 1977). Currently, the large amount of work being done with these viruses indicates their usefulness as probes into the genetics and molecular biology of Mollicutes (International Organization for Mycoplasmology, 1980).

The helical morphology and motility of *Spiroplasma* spp. raises fundamental questions about the maintenance of structure and function in the absence of a cell wall and an apical filament. The thermoplasmas, isolated from burning coal refuse piles, have a temperature optimum of 59°C (range 45–62°C) and grow best between pH 1 and 2. *Ureaplasma* colonies may be as small as 0.01 mm in the solid media generally employed for their growth, and they are unique among the Mollicutes for their ability to hydrolyze urea. The anaeroplasmas are strict anaerobes isolated, so far, from the bovine and ovine rumen and they have bacteriolytic properties. The growth and metabolism of members of the class Mollicutes are inhibited by specific antibodies.

The guanine-plus-cytosine content of the DNA of Mollicutes is relatively low (23–35%), except in the case of a few anaeroplasma and thermoplasma isolates and *M. pneumoniae,* which have a higher (39–46%) guanine-plus-cytosine content in their DNA. The genome size of members of the family Mycoplasmataceae is approximately 4.5×10^8 daltons while that of the family Acholeplasmataceae, the thermoplasmas, and Spiroplasmataceae is approximately 1×10^9 daltons. The genome size of anaeroplasmas is unknown.

Despite this list of novel properties, it is probable that the principle incentive for understanding more about mycoplasma biology comes from the distant field of cell culture. To the tissue culturist, mycoplasmas are the pest par excellence. Their presence in cell cultures often goes unrecognized and they can seriously confound the interpretation of experimental results. Their source is often unknown and they can only rarely be eliminated. About 10% of all cell cultures in use today are probably contaminated with Mollicutes (L. Hayflick, unpublished observations). Nevertheless, the cell culturists' bane has become the mycoplasmologists' boon because, apart

Table 1. Habitats of pathogenic Mollicutes.[a,b]

Genus, species	Natural host	Site of recovery and material for isolation	Disease manifestations
Mycoplasma agalactiae subsp. *agalactiae*	Goats, sheep	Mammary glands, lymph nodes, joints (milk, synovial fluid).	Contagious agalactia characterized by arthritis, mastitis, keratitis, and vulvovaginitis.
M. arthritidis	Rats, nonhuman primates	Joint fluid, infected tissues.	Polyarthritis of rats, submandibular abscesses, ocular lesions, middle ear infections, purulent rhinitis, and lung lesions.
M. bovis (formerly *M. agalactiae* subsp. *bovis* or *M. bovimastitidis*)	Cattle	Udder, joints, respiratory tract. Milk or exudate from udder, semen, and synovial fluids.	Bovine mastitis and arthritis.
M. bovigenitalium	Cattle	Common inhabitant of lower genital tract in both males and females.	Bovine mastitis.
M. gallisepticum	Chickens, turkeys, peafowls	Respiratory system, air sac, ovaries, and eggs.	Chronic respiratory disease in chickens, infectious sinusitis in turkeys, encephalitis associated with polyarthritis of cerebral arteries in turkeys.
M. hyopneumoniae/*M. suipneumoniae*[c]	Swine	Respiratory tract, pneumonic lung, and semen.	Etiological agent of at least one form of enzootic pneumonia of pigs.
M. hyorhinis	Swine	Common inhabitant of nasal cavity. Nasal secretions and semen.	May play primary role in some forms of swine pneumonia. Septicemia associated with arthritis-polyserositis reported in young pigs.
M. hyosynoviae	Swine	Respiratory tract and joints. Synovial fluid, nasal secretions, tonsillar material, lungs of pigs with catarrhal pneumonia.	Acute synovitis and arthritis in swine.
M. meleagridis	Turkeys	Respiratory and urogenital tract; also present in semen, ovaries and eggs.	Air sacculitis in turkeys.
M. mycoides subsp. *capri*	Goats, sheep	Lung, spleen, and liver. Exudate, synovial fluid and blood.	Caprine pleuropneumonia and/or polyarthritis, cellulitis, and septicemia.
M. mycoides	Cattle, sheep, goats	Respiratory tract; pleural exudates and infected lung material.	Bovine pleuropneumonia.
M. neurolyticum	Mice	Brain, eye, nasopharynx, middle ear, and lung. Exudates or tissue washings.	"Rolling" disease (due to toxin) and epidemic conjunctivitis.
M. ovipneumoniae	Sheep	Nose, trachea, bronchi, lung, and eye. Exudates or washings.	Etiological agent in proliferative interstitial pneumonia.
M. pneumoniae	Humans	Respiratory tract, oral cavity; rare isolations from middle ear and spinal fluid. Sputum and infected lung tissue.	Atypical pneumonia, febrile upper respiratory tract infections, bullous myringitis, myocarditis, various cutaneous and neurological conditions.
M. pulmonis	Mice, rats	Common inhabitant of respiratory tract and joints (arthritic).	Infectious catarrh, sometimes complicated by otitis media and bronchopneumonia.
M. synoviae	Chickens, turkeys	Joints and respiratory tract.	Infectious synovitis and air sacculitis; pericarditis and myocarditis appear in chronic phase of disease.

[a] Adapted from Tully and Razin (1977). See for experimental pathology and references to specific disease manifestations.
[b] A number of species are suspected pathogens or will induce pathogenicity experimentally. These are not listed.
[c] Type strains of *M. hyopneumoniae* and *M. suipneumoniae* are serologically identical; thus a true type strain and official taxonomic designation is not yet established.

from the discovery that the etiological agent of primary atypical pneumonia in man is a *Mycoplasma,* nothing has stimulated interest in their biology more than their nuisance value to cell culturists. This has not been an unmixed blessing, for in recent years any apparent unhappiness of cultured cells is more often than not blamed on the presence of Mollicutes, even though the organisms are never isolated. Unexplained cell culture failures today are often blamed on the unproven presence of Mollicutes, which have replaced unclean glassware, improperly prepared media, and poor technique as the cell culturists' favorite scapegoat.

Historical Aspects

Animal Mollicutes

It is generally believed that Louis Pasteur first recognized that an important disease of cattle, called contagious bovine pleuropneumonia, was caused by a specific microorganism, although he was unable to see or grow it. However, it was not until 1898 that Nocard and Roux adequately described the disease and, although failing to implicate ordinary bacteria, succeeded in growing the etiological agent in broth-filled collodion sacs inserted into the peritoneal cavity of rabbits. They soon were able to grow the organism in serum-enriched broth in vitro, but it was not until 1900 that Dujardin-Beaumetz described colonial growth on a solid medium. His description included the light periphery and dark center characteristic of most *Mycoplasma* colonies, and he showed that the darker centers were the result of penetration of the organisms into the medium. The first detailed morphological description of these pleomorphic organisms was made 10 years later (Bordet, 1910; Borrel et al, 1910); in 1929, Elford, using gradacol filters, showed the existence of viable forms 125–150 nm in size. Although capable of growth in cell-free media, the organism was thought by many to be a virus and for 25 years occupied that anomalous position. In subsequent years, considerable knowledge of the biology of *M. mycoides* subsp. *mycoides,* the etiological agent of contagious bovine pleuropneumonia, has been obtained, and it is regarded as the type species for the family Mycoplasmataceae.

Contagious bovine pleuropneumonia is a rapidly spreading lung disease, first reported to have appeared in Germany and Switzerland in 1713 and in England in 1735 (Foster, 1934). It spread throughout most of Europe and was carried to Australia and Africa in the middle of the nineteenth century. Presently, this disease, of great economic importance, is enzootic in Africa and in northern Australia, with some outbreaks having occurred on the Iberian Peninsula. The United States was affected in 1843, but by 1900 the disease was entirely eradicated by slaughtering all infected animals. *M. mycoides* subsp. *mycoides* is still prohibited from importation into the United States even for experimental purposes. Contagious bovine pleuropneumonia is of great economic importance to the countries affected, and attempts to eradicate it by use of vaccines have only been partially successful. The cattle disease is characterized by fever, rapid shallow breathing, cough, nasal discharge, drooped ears, rough coat, and generally poor condition (Meyer, 1909). The illness often progresses to lung and pleural congestion, resulting in death. *M. mycoides* subsp. *mycoides* was a unique microorganism until 1923, when Bridré and Donatien showed that contagious agalactia of sheep was caused by a similar organism (*M. agalactiae*). Eleven years later, in Japan, a third member of the group was described by Shoetensack (1934), who isolated mycoplasmas from the nasal secretions, liver, and lung of dogs with distemper, although these organisms were later shown not to be etiologically involved. In 1936, Laidlaw and Elford reported finding saprophytic mycoplasmas in filtrates from raw sewage. Three "types" were described and these saprophytic strains were collectively known as *Acholeplasma laidlawii. Acholeplasma* spp. are unique among the Mollicutes in that they do not require sterol for growth or multiplication and have increased resistance to osmotic lysis (Razin, 1964) and a lower temperature optimum.

In the years from 1935 to the present, Mollicutes have been isolated from almost all domestic, zoo, and laboratory animals and from man. Many of the species isolated from animals proved to be pathogens and it was not until 1961 that the first human pathogen was grown on agar by Hayflick (Chanock, Hayflick, and Barile, 1962). A partial list of proven diseases caused by mycoplasmas is given in Table 1.

Of those mycoplasmas pathogenic for animals in the United States, the most important, economically, are *M. gallisepticum,* the cause of chronic respiratory disease in chickens; *M. synoviae,* the cause of infectious synovitis in turkeys; and *M. hyopneumoniae (M. suipneumoniae),* the etiological agent of enzootic pneumonia in swine. In fowl, the disease is characterized by a catarrhal discharge from the nares and occasional purulent air sacculitis. *M. gallisepticum* has several properties that for some time made its inclusion in the genus *Mycoplasma* questionable (Thomas, 1967). Klieneberger-Nobel (1962) considered the organism, on the basis of subtle morphological differences and the probable larger size of its minimal reproductive units, not to be a mycoplasma. Since no fundamental differences have been found, the organism has been accepted as a member of the order Mycoplasmatales (Edward and Kanarek, 1960). The organism was first described by Nelson in 1935, who referred to the organisms seen in embryonated eggs as coccobacilliform bodies. In 1952, Markham and Wong proved that the organism was a mycoplasma. *M. gallisepticum* can be spread by egg transmission and is, consequently, a hazard to work conducted in embryonated chicken eggs or in cell cultures made from such tissue. Chu in 1954 confirmed these findings by isolating coccobacilliform bodies in all cases of fowl coryza of slow onset and chronic respiratory disease of chickens. He also showed that, apparently, "saprophytic mycoplasmas" could also be present. Numerous strains of coccobacilliform organisms of different serotypes were subsequently described by Adler, Yamamoto, and Berg (1957) and Adler et al. (1958). Surface structures resembling the spikes of myxoviruses have been observed in *M. gallisepticum* by Chu and Horne (1967).

Mycoplasma synoviae, the etiological agent of infectious synovitis in chickens and turkeys, gives rise to swollen joints, synovial sheaths, and bursae in infected birds and is often found concurrently with splenomegaly and enlarged livers. The agent was first cultivated by the method of Lecce (1960) as satellite colonies around a staphylococcus. Chalquest and Fabricant (1960) were able to grow *M. synoviae* in pure culture by replacing the staphylococcus with diphosphopyridine nucleotide (nicotinamide adenine dinucleotide, NAD).

Economically, a very important infectious disease in the swine industry is a chronic pneumonia known as enzootic pneumonia. Although clinical signs of the disease may not be obvious, common subclinical infections result in weight loss in marketable animals. The etiological agent was first assumed to be a virus and the disease was called virus pneumonia of pigs. Viral etiology was never confirmed and reports appeared on the limited growth of the agent in cell cultures (Betts and Whittlestone, 1963; Lannek and Wesslen, 1957; L'Feuyer and Switzer, 1963). Perseverance finally resulted in the growth of the agent first in tissue cultures (Goodwin and Whittlestone, 1963), then in broth (Goodwin and Whittlestone, 1964; Maré and Switzer, 1965), and finally on agar as a mycoplasma (Goodwin, Pomeroy, and Whittlestone, 1965). The group in the United Kingdom has named the agent *M. suipneumoniae,* while in the United States, Maré and Switzer (1965) named a mycoplasma isolated from enzootic pneumonia-inducing fluids as *M. hyopneumoniae.* Goodwin, Pomeroy, and Whittlestone (1965, 1967) presented the first clear proof that *M. suipneumoniae* was the etiological agent of enzootic pneumonia in pigs and showed that it was a new *Mycoplasma* species. An informative account of their tribulations in trying to grow this organism has been given (Whittlestone, 1967).

M. gallisepticum and *M. neurolyticum* (the agent causing rolling disease in mice and rats [Findlay et al., 1938; Sabin, 1938]) are unique mycoplasmas in

that they both produce toxins which primarily affect the central nervous system (Thomas, 1967).

Plant Mollicutes

In 1967, Japanese workers proposed that Mollicutes might be implicated etiologically in several important plant diseases (Doi et al., 1967; Ishiie et al., 1967). They described structures resembling Mollicutes in the phloem elements of plants naturally infected with mulberry dwarf, potato witches' broom, Japanese aster yellows, and paulownia witches' broom. Similar structures have been seen in American aster yellows- and corn stunt disease–infected plants, as well as several other leafhopperborne diseases (Maramorosch, Shikata, and Granados, 1968). In addition to the morphological similarity of these structures to Mollicutes, as seen by electron microscopy, the diseases themselves were amenable to treatment with tetracyclines, at which time the structures disappeared (Davis, Whitcomb, and Steere, 1968; Doi et al., 1967; Ishiie et al., 1967). The organisms have now been found to grow on agar, and are treated elsewhere in this Handbook (Chapter 169).

Human Mollicutes

In 1937, the first isolation of a *Mycoplasma* from man was reported by Dienes and Edsall, who recovered the organism from an abscessed Bartholin's gland. In the next several years, numerous reports appeared describing the isolation of *Mycoplasma* spp. from the male and female urogenital tracts. Since chemotherapy was, by this time, effectively controlling the gonococcus, and the incidence of urethritis was not decreasing, an etiological role for *Mycoplasma* spp. in nongonococcal urethritis was actively pursued. Support for this contention came when it was shown that the broad-spectrum antibiotics ameliorated the disease and affected *Mycoplasma* spp. in vitro. However, *Mycoplasma* spp. could also be isolated from the urethras of clinically normal individuals, and the general lack of information on classification of the isolated *Mycoplasma* spp. confused the interpretation. *Mycoplasma* spp. have been isolated from up to 50% of clinically normal men and 80% of clinically normal women. There is, today, a general belief that *Mycoplasma hominis,* found in the human urogenital tract, is not pathogenic but that *Ureaplasma* may be (Taylor-Robinson, Csonka, and Prentice, 1977).

The etiology of nongonococcal urethritis was enhanced by the finding of Shepard in 1954, and the subsequent confirmation by Ford, Rasmussen, and Minken (1962), that a heretofore unrecognized Mollicute producing very small colonies is found in the genitourinary tract of from 50 to 80% of men with nongonococcal urethritis. These organisms, first designated as T-strains, are different from the *Mycoplasma* spp. in that they produce extremely small colonies (10–30 μm in diameter), are microaerophilic, do not ferment carbohydrates, hydrolyze urea, and have an optimum pH of 5.5–6.5. They are found less frequently in clinically normal persons and are now called *Ureaplasma urealyticum.*

Other than *Ureaplasma*, there are six well-characterized *Mycoplasma* species of man. Five of these species are regarded as nonpathogenic and part of the normal microbial flora of the human oral cavity and urogenital tract. They have been isolated less frequently from joint fluids, blood, bone marrow, and rarely from lesions. These five species are *M. fermentans, M. salivarium, M. hominis, M. orale,* and *M. buccale* (Freundt and Edward, 1978). *M. orale* is the most common *Mycoplasma* in the clinically normal human oral cavity. *M. hominis* is the most common *Mycoplasma* found in the human urogenital tract. The only proven pathogenic *Mycoplasma* for man is *M. pneumoniae,* the etiological agent of cold-agglutinin-positive, primary atypical pneumonia.

Primary Atypical Pneumonia

In the late 1930s, a broad group of nonbacterial human pneumonias was first recognized and later given the name primary atypical pneumonia (Dingle and Finland, 1942; Eaton, 1950a). These pneumonias were unlike the typical lobar pneumonia caused by *Diplococcus pneumoniae.* Since the etiological agent(s) involved were unknown, a description of primary atypical pneumonia as a clinical syndrome was necessary at the time. In 1943, it was first observed that many patients with primary atypical pneumonia developed cold agglutinins during the course of the illness (Peterson, Ham, and Finland, 1943). Since that observation was made, a test for this antibody was widely used as a nonspecific serodiagnostic procedure.

It became obvious that atypical pneumonia is a syndrome of multiple etiology and such viruses as influenza A, adenoviruses, B virus, parainfluenza type 3, and respiratory syncytial virus have been incriminated in certain age groups as important causes of this syndrome (Chanock and Johnson, 1961; Chanock et al., 1963a). That portion of the atypical pneumonia syndrome characterized by a cold-agglutinin response has not been implicated with any of these viruses.

Cold-agglutinin-positive, primary atypical pneumonia was determined to be a distinct epidemiological entity from studies conducted by the Commission on Acute Respiratory Disease during and after World War II. It was determined that the average incubation period was 2 weeks, an unusually long

interval for an acute respiratory illness (Jordan, 1949). The Commission on Acute Respiratory Disease conducted studies with human volunteers in which it was determined that the etiological agent of cold-agglutinin-positive, primary atypical pneumonia was filterable (Commission on Acute Respiratory Diseases, 1946).

In the late 1930s and early 1940s, several agents were reported to have been recovered from different animal species inoculated with material from patients with primary atypical pneumonia (Eaton, 1950a). It is probable that many of these orgnisms were indigenous to the animal species used or were activated by the inoculation procedure. One exception, however, was the Eaton agent, and strong evidence existed suggesting that it came from patients with primary atypical pneumonia (Eaton, Mecklejohn, and Van Herick, 1944), a relationship that has been confirmed in all studies conducted in subsequent years.

Embryonated eggs were inoculated with filtered sputum from patients with primary atypical pneumonia in the original Eaton agent study. The embryos did not reveal any definite pathological changes; however, tissue suspensions and extraembryonic fluids from these eggs produced pneumonia when given intranasally to cotton rats or hamsters (Eaton, Mecklejohn, and Van Herick, 1944). Successful propagation of the agent was serially performed in eggs, and material from successive passages produced pneumonia in hamsters and cotton rats. Filtered sputum from patients with primary atypical pneumonia was administered directly into cotton rats and hamsters and pneumonia was observed to develop, yet serial passage of the agent was unsuccessful (Eaton, 1950a; Eaton, Mecklejohn, and Van Herick, 1944). These early studies were not always unequivocal in their results, since lung lesions were observed in only a fraction of the inoculated animals and pneumonia would also result from activation of their own latent viruses (Eaton, 1950a).

The organism was determined to be between 180 and 250 nm in diameter by gradacol membrane filtration studies of infected chick embryo material (Eaton, 1950a), and later both chlortetracycline and streptomycin were shown to inhibit infectivity (Eaton, 1950b; Eaton and Liu, 1957).

In 1957, Liu showed that the Eaton agent was localized in the bronchial epithelium of chick embryo lungs by using immunofluorescence techniques. The adaptation of immunofluorescence for recognition of infection in eggs was a major advance, because it was then possible to quantitate the organism and antibody produced against it.

The Eaton agent was recovered most often during the early phase of illness and from individuals who developed cold agglutinins (Eaton, 1950a), but many workers were still unconvinced of the relationship between the Eaton agent and primary atypical pneumonia, because patients developed antibodies to many "bizarre" antigens (Thomas et al., 1943). However, the studies made during World War II demonstrated clearly that a serological response to the Eaton agent occurred more frequently among patients with atypical pneumonia (Eaton and Van Herick, 1947). The specificity of the fluorescent antibody response was demonstrated by detection of the antigen in 26 of 30 serologically positive military recruits (Chanock et al., 1961; Chanock, Hayflick, and Barile, 1962).

The Eaton agent was considered to be a virus for many years despite doubts cast on this conclusion when the organism was found to be inhibited by chlortetracycline and streptomycin (Eaton, 1950b; Eaton and Liu, 1957). Subsequently, several observations suggested that the agent might be a *Mycoplasma.* Small coccobacillary bodies were seen on the mucous layer covering the bronchial epithelium of infected chick embryos. The distribution of these bodies corresponded with the localization of the agent as measured by immunofluorescence (Goodburn and Marmion, 1962; Marmion and Goodburn, 1961). Extracellular "colony-like" structures were also seen in stained preparations of infected tissue cultures, and these structures corresponded to the areas of specific immunofluorescence (Clyde, 1961, 1963). The determination that the Eaton agent was a *Mycoplasma* was firmly established when it was successfully cultivated by Hayflick on a cell-free agar medium and, in collaboration with Chanock, shown to be the etiological agent of primary atypical pneumonia (Chanock, Hayflick, and Barile, 1962; Hayflick, 1965). The medium developed by Hayflick was a modification of a formula first published by Edward (1947). Colonies developed 6–7 days after the Eaton agent was inoculated onto agar containing 20% horse serum and 2.5% fresh bakers' yeast extract (Hayflick, 1965). The colonies were circular and had a homogeneous granular appearance that was relatively unique for a *Mycoplasma* species. The organism was later designated *M. pneumoniae* (Chanock et al., 1963b). In 1967, Lyell et al. provided evidence that *M. pneumoniae* might be involved in some cases of erythema multiforme.

From a historical standpoint, it is interesting to observe that the three major diseases caused by Mollicutes (contagious bovine pleuropneumonia, enzootic pneumonia of swine, and cold-agglutinin-positive, primary atypical pneumonia in man) were long regarded to have a viral etiology.

Arthritis

The possible role of infectious agents in the pathogenesis of rheumatoid arthritis in man has become of interest mainly because of the failure to find other causal factors and because of some interesting ani-

mal parallels. As early as 1909, Meyer reported from Africa on the development of arthritis in about 100 cattle of 400 inoculated with *Mycoplasma mycoides* subsp. *mycoides.* Live vaccines prepared against contagious bovine pleuropneumonia were later shown to produce arthritis in a small fraction of the animals vaccinated (Bennett, 1932; Turner and Trethewie, 1961). In the past 25 years, spontaneous polyarthritis or synovitis due to *Mycoplasma* spp. has been demonstrated to occur in cattle, goats, swine, rats, chickens, and turkeys. Several reports have appeared in the literature during this same period purporting to show that Mollicutes could be isolated from the joint fluid of several human arthritics. However, many such isolations have been shown to be contaminants present in the embryonated eggs or cell cultures used for isolation. Many isolates have never been subcultivated and only very few have been identified as to species. Despite the equivocal results that have been obtained, interest in the Mollicutes as possible etiological agents in certain kinds of human rheumatoid arthritis continues. The main support for this hypothesis comes from the known association of *Mycoplasma* spp. with arthritis in several domestic and laboratory animals. The extrapolation to man is tempting.

Habitats of Mollicutes

Members of the class Mollicutes may be parasitic or pathogenic but it is uncertain whether any are true saprophytes. Those that are pathogens cause disease in a wide variety of vertebrates, invertebrates, and plants.

Their distribution in nature is broad and if extrapolations can be made from past experience, it is a virtual certainty that they will be found in almost any vertebrate and many nonvertebrates in which they are sought. Most Mollicutes species are species-specific but a few are found in several hosts. The Mollicutes found in animals are usually associated with moist mucosal surfaces such as the oral, nasal, and urogenital tracts, where they are found predominantly attached to cell surfaces. *Spiroplasma* spp. which are chiefly plant pathogens of the yellows type, are transmitted by a variety of insect vectors, particularly leafhoppers. The thermoplasmas are found in burning coal refuse piles where optimum growth occurs between pH 1 and 2 and at a temperature of 59°C. The strictly anaerobic anaeroplasmas have only been isolated from the bovine and the ovine rumen.

The current literature on Mollicutes suggests that they can be found associated with a very wide variety of hosts (Bové and Duplan, 1974; Hayflick, 1972; Proceedings of the Society for General Microbiology, 1976; Tully and Razin, 1977). A few of the more recent and exotic of these include jungle fowl (Koshimizu and Magaributchi, 1977), Chesapeake Bay bivalves (Harshbarger and Chang, 1977), falcons (Furr, Cooper, and Taylor-Robinson, 1977), and marmosets (Furr, Taylor-Robinson, and Heatherington, 1976). More detailed descriptions of the varied habitats of Mollicutes may be found in several recent reviews (Freundt, 1974; Tully and Razin, 1977).

These organisms are not limited to vertebrate hosts but are also found associated with plants and insects and are described separately in this Handbook, Chapter 169. Table 1 lists the habitats of the pathogenic Mollicutes.

In all cases, the Mollicutes are bound to their host cells by contiguous membrane-to-membrane association and are predisposed to moist mucosal surfaces such as those found on respiratory and genitourinary epithelia. In this discussion, we will review what is known of the nature of this host-parasite relationship as it occurs both in vivo and in vitro.

Mollicutes found in cell cultures are predominantly *Acholeplasma laidlawii*, *Mycoplasma arginini*, *Mycoplasma orale*, and *Mycoplasma hyorhinis*. The first two are associated with cattle and are usually introduced into cell cultures with the bovine serum used almost universally by cell culturists. *M. orale* is found in the human oral cavity and is probably introduced by poor aseptic technique or from other contaminated cell cultures. *M. hyorhinis* is associated with swine and, although it is unproven, the origin of this contaminant is thought to be the crude porcine pancreatic extract, commonly called trypsin, and widely used in cell culture techniques. The only Mollicutes that have been found to contaminate cell cultures are *Mycoplasma* and *Acholeplasma* species. The human pathogen *M. pneumoniae* has never been found as a cell culture contaminant. It is ordinarily found in the respiratory tract of persons experiencing, or recovering from, primary atypical pneumonia. *Ureaplasma* spp. which are found in the urogenital tract of humans and animals, might be etiological agents of some segment of nonspecific urethritis (Bowie et al., 1976; Taylor-Robinson, Csonka, and Prentice, 1977) and are rarely if ever found as cell culture contaminants (G. Masover, personal observations). Deliberate attempts to establish persistent *Ureaplasma* infection of cell cultures were unsuccessful in the absence of pretreatment of *Ureaplasma* spp. with a protease (Masover, Namba, and Hayflick, 1976; Masover et al., 1977b; Mazzali and Taylor-Robinson, 1971).

Thus, comparatively few of the named Mollicutes species are found as contaminants of cell cultures. It has been shown that a significant percentage of fresh isolates cannot be subcultivated in vitro (Hayflick and Stanbridge, 1967). This suggests that a factor(s) required for in vitro propagation of the organisms is supplied in the original sample but lost by dilution or

destruction during in vitro passage. The existence of extracellular material on a variety of Mollicutes has been documented using various techniques such as electron microscopy (Black and Vinther, 1977), cytochemistry (Robertson and Smook, 1976), immunological methods (Bradbury and Jordon, 1972; Masover, Mischak, and Hayflick, 1975) and biochemical methods (Yaguzhinskaya, 1976). It is possible that this extracellular material may interfere with attachment of Mollicutes to eukaryotic cells.

Since Mollicutes are so small and lack a cell wall, it is important to emphasize that the organisms are extracellular parasites. They are not engulfed as are other particles of similar size, such as the pox viruses (Joklik, 1976), except in the case of frankly phagocytic cells, such as macrophages (Cole and Ward, 1973; Zucker-Franklin, Davidson, and Thomas, 1966). The precise reason for this remains unclear and the precise nature of the interaction between a host cell and a Mollicutes species is not known. However, several authors have been able to provide some insight into the nature of Mollicutes–host cell interactions. Recently, Boatman, Cartwright, and Kenny (1976) provided a morphometric study using electron microscopy of the interaction between the membranes of a *Mycoplasma* species of bovine origin and HeLa cells. The two membranes are separated by a space of approximately 10 nm. In our opinion, this space may reflect the presence of both host cell and *Mycoplasma* glycocalyxes. The existence of glycocalyxes has now been clearly established (Kahane and Tully, 1976; Robertson and Smook, 1976; Schiefer et al., 1975). Boatman, Cartwright, and Kenny et al. (1976) were able to observe *Mycoplasma* spp. by both light and fluorescence microscopy using specific fluorescein-conjugated anti-*Mycoplasma* antiserum for the latter procedure. The scanning electron microscope has shown *Mycoplasma* spp. usually, but not always, discernible as projections from the host cell membrane. An average of 69 *Mycoplasma* spp. per HeLa cell surface were counted and these occupied 1.7% of the cell surface area. Earlier, Brown, Teplitz, and Revel (1974), using scanning electron microscopy and biochemical treatment of infected cells, showed that the *Mycoplasma* spp. tended to be localized at the "leading edge" of the host cell and that the *Mycoplasma* spp. were virtually completely removed from their host cell by treatment with proteases. Conversely, it could be shown that pretreatment of the parasite (in this case, *Ureaplasma*) as well as the host cell (a normal human fibroblast [WI-38]) with a proteolytic enzyme enabled the *Ureaplasma*–WI-38 cell interaction to persist. Failure to pretreat the *Ureaplasma* with trypsin resulted in discernible effects on the host cell but those effects did not persist (Masover, Namba and Hayflick, 1976; Masover et al., 1977b).

There is evidence that sialic acids are involved in

the adsorption of some Mollicutes to some host cells. In such studies, Thomas (1969) reported that red blood cells became firmly coated with *Mycoplasma* spp. at 37°C but not at 4°C and that treatment with neuraminidase prevented attachment as did addition of egg white or fetal calf serum. Sobeslavsky, Prescott, and Chanock (1968) observed that neuraminic acid receptors on various host cells adhered to *Mycoplasma* binding sites that were probably lipid or lipoprotein in nature. Manchee and Taylor-Robinson (1969) found, in addition, that different *Mycoplasma* spp. behaved differently with respect to susceptibility of the host cell receptors to purified neuraminidase. *M. gallisepticum* and *M. pneumoniae* appeared to depend, for attachment to the host cell (HeLa), on the presence of the neuraminic acid, while *M. hominis* and *M. salivarium* attachment was not influenced by neuraminidase treatment of the host cells. In some cases, pretreatment of host cells with proteases enhanced the Mollicutes–host cell interaction, suggesting that a protein component was somehow involved with the host cell receptor.

M. pneumoniae, the etiological agent of primary atypical pneumonia in man, has a terminal striated structure that behaves as the attachment structure to host epithelium (Biberfeld and Biberfeld, 1970; Collier and Clyde, 1971; Collier, Clyde, and Denny, 1971). A similar structure is also found in *M. gallisepticum.*

Isolation of Mollicutes

Membership in the class Mollicutes presupposes growth of the organisms on agar medium. The nutritional and environmental conditions that are conducive to the growth of this heterogeneous group of prokaryotes varies as would be expected from the variety of individual host species in which they are found. Nevertheless, there are general procedures that are used for isolation of all Mycoplasmataceae and Acholeplasmataceae. General considerations for isolation will be presented first, followed by special modifications that are useful for isolation of specific species.

General Considerations for Isolation of Mollicutes

The solid medium used for the growth of Mollicutes is usually a complex mixture of materials capable of supplying a variety of known and unknown nutrients as well as the sterol and whole protein thought to be required for sterol utilization. Sterol is required for the growth of all Mollicutes except the Acholeplasmataceae. Since colonies are generally viewed with the light microscope, the media employed

should be able to transmit light. Because of their small size, Mollicutes colonies are not usually transferred like ordinary bacterial colonies using a loop or needle. The preferred method is the so-called push block procedure. A square of agar seen under the microscope to contain many colonies is marked off on the underside of the agar plate culture. The square of agar is cut out with a sterile scalpel, inverted onto a fresh agar plate, and pushed back and forth over the surface a few times. The block is pushed to one side of the plate and the new culture is incubated in an inverted position.

A history of media developments up to 1962 has been published (Hayflick and Chanock, 1965). In 1962, Hayflick first isolated the organism, subsequently named *Mycoplasma pneumoniae,* which is the etiological agent of primary atypical pneumonia in man (Chanock, Hayflick, and Barile, 1962). The medium on which this isolation was made contained yeast extract and horse serum. Prior to this development, a serum fraction designated ''fraction A'' or ''PPLO serum fraction'' was used because it was thought that this material was required for the growth of all mycoplasmas. Fraction A–supplemented medium does not support the growth of many Mollicutes.

The medium on which *M. pneumoniae* was first isolated (Hayflick, 1965) is a modification of formulas developed by Edward (1947) and by Morton, Smith, and Leberman (1951). With modification, it supports the growth of most members of the class *Mollicutes.* Mycoplasma medium generally consists of a basal broth component to which various supplements are added.

Media for Isolation of *Mycoplasma pneumoniae* (Hayflick, 1965)

Mycoplasma broth:

Beef heart for infusion (Difco)	50 g
Peptone (Difco)	10 g
NaCl	5 g
Water	900 ml

The beef heart for infusion preparation is soaked for 1 h in distilled water brought to 50°C. This mixture is brought to a boil and filtered through two sheets of Whatman No. 2 filter paper (50-cm diameter, folded). The peptone and NaCl are added to the filtrate and the pH is raised to 7.8 by addition of approximately 1.6 ml of 10 N NaOH. The solution is then brought to a boil and filtered through two more sheets of filter paper. It is made up to 1 liter by adding distilled water and 0.2 ml 10 N HCl is added before autoclaving. The final pH should be 7.6–7.8 and the broth should be crystal clear with no precipitate.

Before use, the broth must be supplemented with 20% unheated agamma horse serum and 10% yeast extract prepared as described below. It is often convenient to divide the unsupplemented broth into small volumes for storage at 5°C, where it is stable for several months.

Mycoplasma agar:
PPLO agar (Difco), dehydrated, is prepared according to the directions accompanying the agar. Ordinarily, the agar is made up in quantities of 70 ml. Prior to use, the agar is melted and allowed to cool sufficiently so that the container can be handled. If appropriate baths or other constant-heat devices are available, the agar may be held in the liquid state between 48 and 52°C and then supplemented with prewarmed (40°C) agamma horse serum to make 20% vol/vol of the final volume and 10% vol/vol yeast extract prepared as described below. About twelve 60-mm Petri dishes can be poured from 100 ml of supplemented agar. Unsupplemented agar can be stored indefinitely in closed containers at room temperature. Supplemented agar can be stored for a few weeks at 5°C. Best isolation rates from clinical materials is accomplished with fresh plates.

Yeast Extract:
Add 250 g of active dry bakers' yeast (Fleishmann's, a division of Standard Brands, Inc.) to 1 liter of distilled water and heat until boiling occurs. Filter the mixture through two sheets of Whatman No. 2 filter paper using a number of filter apparatuses if needed to hasten this slow process. Add sufficient NaOH to the filtrate to raise the pH to 8.0. The yield from 1 liter of the initial preparation is about 400 ml after the filtration step. The filtered yeast extract is usually dispensed in 10-ml aliquots but must be autoclaved and stored at −20°C, at which temperature it is stable for at least 2 months.

It may be difficult to obtain beef heart for infusion in some areas. The preprepared beef heart infusion broth (Difco) can be used in place of the mycoplasma broth described above.

The filtration step can be replaced by centrifugation (up to 20,000 × g), which results in improved clarity of the preparation. Strains of $M.$ $pneumoniae$ that have been passaged for long periods in vitro and other Mollicutes may not require as much serum or the yeast extract preparation. However, numerous attempts to grow $M.$ $pneumoniae$ from clinical specimens without these supplements have been unsuccessful in our hands.

Horse Serum

Agamma horse serum has been found to be the best serum for cultivation of Mollicutes. Agamma serum reduces the likelihood of "pseudocolony" formation, which can be a serious pitfall in examining agar surfaces for mycoplasma colonies (Hayflick, 1965). Serum can be stored for at least 6 months at 5°C and longer when frozen.

Specific supplements such as arginine and/or glucose in 0.5–1.0% wt/vol concentration are usually added for growth of $Acholeplasma$ spp. and those $Mycoplasma$ spp. that hydrolyze arginine. Species that hydrolyze sugars to produce acid as a catabolic end product are often referred to in the literature as "fermentative" strains. The convention, while well established now, could lead to confusion if the term "fermentation" is defined in more modern biochemical terms. In that case, arginine hydrolysis would also be a fermentation, even though alkali, and not acid, is produced. It has recently been shown that there is inhibition of some $Mycoplasma$ species by arginine (Leach, 1976; Washburn and Somerson, 1977) and that the enzyme(s) responsible for arginine hydrolysis in at least one strain of $M.$ $arthritidis$ exist in multiple forms (Weickmann and Fahrney, 1977). Therefore, the role of arginine in the growth of the "arginine-requiring" species of Mollicutes is uncertain.

Other additives to a complete broth medium for Mollicutes may be a pH indicator such as phenol red (final concentration 0.002% wt/vol) or a different indicator as in the "B-broth" described below.

Appropriate antimicrobials may be added to isolate Mollicutes from other organisms that might be present in a specimen. Penicillin G, in final concentrations ranging from 100 to 1,000 units/ml, is most commonly used with thallous acetate (1:1,000 or 1:2,000) as a selective agent in media for all Mollicutes except ureaplasmas (Lee, Bailey, and McCormack, 1972). Thallous acetate is a very toxic substance and should not be used on swabs or in media for gargles in attempts to isolate Mollicutes from the human oral cavity. Media for isolations made from the genital tract generally contain antifungal drugs such as nystatin at a concentration of 50 μg/ml.

Recently, it was reported that a strain of Mollicutes could not repair damage to DNA induced by ultraviolet (UV) light (Ghosh, Das, and Maniloff, 1977). This, plus the reported sensitivity of other Mollicutes to UV light (Furness, 1975; Furness, Pipes, and McMurtrey, 1968), suggests that it is best to avoid exposure of these organisms to sunlight.

$Ureaplasma$ Species

To prepare media that are suitable for $Ureaplasma$ isolation and growth, the mycoplasma medium described above can be used. It is modified by addition of approximately 0.01 M urea, phosphate (approximately 10 mM), and adjustment of the final pH to near 6.0. A complete description of media for

Ureaplasma isolation and their development is given by Shepard et al. (1974) and Shepard and Masover (1978). More detailed data on *Ureaplasma* urease activity and the relationship of the urease and its products to growth of the organism have been recently reported (Masover, Catlin, and Hayflick, 1977; Masover, Razin, and Hayflick, 1977a, 1977b; Razin et al., 1977). Two of the media are described below. The differential diagnostic medium A-7 (Shepard and Lunceford, 1976) differs by the inclusion of a manganous salt that forms an accretion colony on the agar above *Ureaplasma* colonies but not above colonies of other Mollicutes or urease-positive L-forms of bacteria. The accretion colony appears as a brown precipitate immediately above the *Ureaplasma* colony and apparently results from interaction between the manganous ion and the reaction product of *Ureaplasma* urease. Manganous ion inhibition of some ureaplasma isolates is now being reported (Robertson, 1978; G. Masover, personal observation).

Ureaplasma Agar A-7 (Shepard and Lunceford, 1976)

Basal medium (pH 5.5):

Trypticase soy broth (BioQuest Cat. No. 11768)	4.8 g
Manganous sulfate ($MnSO_4 \cdot H_2O$), ACS, Fisher Cat. No. M113 (0.015% FC)	0.031 g
Deionized distilled water (dissolve and adjust reaction to pH 5.5 with 2 N HCl)	163.0 ml
Ionagar no. 2 (Oxoid Cat. No. L-12), or Ionagar No. 2s (0.85% FC)	1.76 g

Without heating to dissolve the Ionagar, place the bottle or flask in the autoclave and sterilize at 121°C for 15 min. This both dissolves the Ionagar and sterilizes the basal medium in a single operation. Cool the vessel to 56°C in a water bath. Mix contents thoroughly by 10–15 s rotation on a flat surface to disperse the melted Ionagar at the bottom. Avoid air-bubble entrapment. Without further delay, convert to complete Differential medium A7 as follows:

Sterile basal medium, pH 5.5 (50–55°C)	160.0 ml
Sterile unheated normal horse serum (see Mycoplasma Agar)	40.0 ml
Sterile "CVA" enrichment (GIBCO Cat. No. 1401581) (0.5% FC)	1.0 ml
Sterile yeast extract, pH 6.0 (see Mycoplasma Agar)	2.0 ml
Sterile 10% urea solution (filter sterilized) (0.1% FC)	2.0 ml
Sterile 4.0% L-cysteine-HCl	

stock solution (filter-sterilized) (0.01% FC)	0.5 ml
Penicillin G potassium, 100,000 units/ml solution (stored frozen)	2.0 ml
	207.5 ml

In a separate sterile bottle (60–100 ml), combine 1.0 ml of "CVA" enrichment with 40 ml of horse serum; then add the remaining items, except for the sterile basal medium; finally, aseptically, add this 47.5-ml enrichment combination to the 160-ml batch of melted basal agar, mix, and dispense into Petri plates.

After hardening, allow plates to remain (inverted) on the workbench at room temperature overnight before storing in the refrigerator in sealed plastic bags or other containers to reduce loss of moisture during storage. Prepare fresh plates of A7 medium each week, employing the unit pouring method outlined above. Do not store basal medium.

Note: FC = final concentration.

By changing the pH indicator from phenol red to bromothymol blue and by inclusion of a specific commercially available tripeptide, Robertson (1977) claims that more rapid growth (doubling in about 60 min), higher titers ($\gg 10^8$/ml), and longer viability (longer stationary phase and slower decline) are possible with clinical isolates of *Ureaplasma.* Growth curves done by one of us (G.K.M) on a few strains support this finding. The medium should definitely be tested on a wide basis in clinical laboratories because it facilitates isolation and handling of *Ureaplasma* spp. recovered from clinical materials. The preparation of Robertson's "B-Broth" and agar medium is as follows:

Ion Agar for *Ureaplasma*

Basal agar:

Glass-distilled water	90 ml
Mycoplasma broth without crystal violet (Cat. No. 055401, Difco)	2.1 g
Ion agar no. 2 (Cat. No. L-12, Oxoid) or	0.75 g
Agar No. 1 (Cat. No. L-11, Oxoid)	
Yeast extract (Cat. No. 0127-01, Difco)	0.1 g
HEPES buffer (Cat. No. H3375, Sigma)	1.19 g

Mix, autoclave, and cool to ∼ 50°C.

Add supplements:

Sterile normal horse serum (pH 6.0)	10 ml
Urea solution	0.25 ml
Tripeptide solution	0.1 ml

*Ampicillin solution   1.0 ml
*Nystatin solution   0.1 ml
Adjust pH to 6.0
Store plates at 4°C until use.

Urea Solution (10% w/v):
Urea   10 g
Glass-distilled water   100 ml
Sterilize by filtration. Store in 10-ml volumes at
−20°C.

Tripeptide solution (20 μg/ml):
Glycl-L-histidyl-L-lysine acetate   200 μg
   (Cat. No. 363941, Calbiochem)
Glass-distilled water   10 ml
Sterilize by filtration and store at −20°C.
Ampicillin Solution (100 mg/ml):
Store at −20°C.

Nystatin Solution (50,000 U/ml):
Store at −20°C. Mix well before use.
Include only if the agar is to be used for clinical
material.

Bromothymol Blue (B) Broth for *Ureaplasma*
   Basal broth:
Glass-distilled water   90 ml
Mycoplasma broth without crystal
   violet (Cat. No. 055401, Difco)   2.1 g
Yeast extract
   (Cat. No. 0127-01, Difco)   0.1 g
Bromothymol blue solution   1 ml

Add supplements:
Sterile normal horse serum
   (pH 6.0)   10 ml
Urea solution   0.25 ml
Tripeptide solution   0.1 ml
Ampicillin solution   1.0 ml
*Nystatin solution   0.1 ml
Adjust pH to 6.0.
Store at −20°C until use.

Bromothymol Blue Solution (0.4%):
Bromothymol blue
   (Difco; Cat. No. 0202-11-3)   0.2 g
0.01 N NaOH   32.0 ml
Glass-distilled water to   50.0 ml
Autoclave and store at room temperature.

Other solutions:
See "Ion Agar for *Ureaplasma*".
*Include only if the broth is to be used for clinical
material.

A great many variations of these complex media
will serve to grow most of the Mollicutes.

The NYC medium originally described by Faur et
al. (1973) is effective not only for isolation of
Neisseria, but also for urogenital mycoplasmas
(Faur, Weisburd, and Wilson, 1976).

Recent evidence for the pathogenicity of *Ureaplasma urealyticum* is becoming more compelling.
Since the urinary bladder is a "privileged" area that
is expected to remain sterile, the presence of a single
microorganism in bladder urine is an abnormality
(Stamey and Pfau, 1970). Thus, the finding of Bredt
at al. (1974) that 13% of 257 urines taken by suprapubic aspiration of the human urinary bladder were
positive for *Ureaplasma,* is noteworthy. The authors
point out that the presence of *Ureaplasma* in the
aspirated urine is not, in itself, sufficient to be diagnostic of either upper or lower urinary tract infection. However, the mere presence of these organisms ($> 10^4$/ml) should be considered with respect
to (i) potential pathogenicity for the kidney, (ii) formation of calculi (since they produce ammonia), or
(iii) infection of the lower urinary tract. More recently, Koch's postulates were fulfilled by a few
microbiologists who induced and cured nonspecific
urethritis in themselves by intraurethral introduction
of *Ureaplasma* isolates from patients (Taylor-
Robinson, Csonka, and Prentice, 1977). This, plus
recent evidence based on epidemiology and antibiotic sensitivity, argues strongly for some role of
Ureaplasma in human disease (Bowie et al., 1976).
Thus *Ureaplasma* may be a second human pathogen
among the Mollicutes.

Another recent finding that should be mentioned
is that 84% of 360 specimens that were positive for
gonococcus at a New York venereal disease clinic
had *M. hominis, U. urealyticum,* or both associated
with them (Faur, Weisburd, and Wilson, 1975).
This commensal relationship persisted through not
less than five serial subcultivations. Thus, the possibility that the genus *Mycoplasma* and/or *Ureaplasma* may serve a secondary role in the pathogenesis of gonococcus should be considered. There is
also evidence for a similar relationship between
Ureaplasma and *Chlamydia* (Bowie et al., 1976).

Isolation of Thermoplasmas

The organisms now designated thermoplasma and
anaeroplasma are, as was indicated, associated with
Mollicutes because they share many characteristics
of the class. They have not yet been fully characterized and are, therefore, not yet included officially in
the class Mollicutes.

Thermoplasma organisms were grown initially in
broth composed of a basal salt medium with 1%
wt/vol glucose and 0.1% wt/vol yeast extract
(Difco). The medium was adjusted to pH 2 or 3
using 10 N H_2SO_4 and was incubated at 55°C
(Darland et al., 1970). They form "fried egg" colonies on agar that is comparable in nutrient composi-

tion to the liquid medium. Both media omit serum and a high incubation temperature is provided. In these respects, they differ from media used for cultivation of other Mollicutes. Further details relevant to isolation of these cell wall–less thermophiles is given by Belly, Bohlool, and Brock (1973).

Isolation of Anaeroplasmas

The organisms designated anaeroplasma, like thermoplasma, share some characteristics of the class Mollicutes but are not yet formally included because further basic study is required. The medium used for growth of these organisms differs from other mycoplasma media in that it is supplemented with starch, volatile fatty acids, and bacterial lipopolysaccharide preparations. The major cultural condition to be met is the need for strict anaerobiosis. This, of course, requires all of the special techniques used in anaerobic culture of microorganisms, complicated further by the difficulty of dealing with an organism that does not make a colony on agar visible to the naked eye. Details of the discovery and culture of these anaerobes, tentatively associated with the Mollicutes, are given by Freundt (1974), Robinson and Hungate (1973), and Robinson, Allison, and Hartman (1975).

Isolation of Mycoplasmas from Cell Cultures

As a result of broad use of cell cultures in many biological disciplines, and the finding of widespread Mollicutes contamination, there has been great interest in the origin, incidence, isolation, and identification of Mollicutes in cell cultures. These subjects have formed a sizeable quantity of literature in recent years and have produced an appropriate amount of divergence of opinion on how best to isolate and identify these prokaryotes from their cell culture habitats (Barile, 1974; Barile et al., 1973; Chen, 1977; Fogh, 1973; Fogh, Holmgren, and Ludovici, 1971; Hayflick, 1965; Low, 1976; McGarrity, 1976; Schneider, Stanbridge, and Epstein, 1974; Stanbridge and Schneider, 1974). The major point to be made is that a suspect Mollicutes contaminant must be grown on solid media and produce a "typical" colony in order to be called a mycoplasma (Mollicutes). That requirement derives directly from the definition of the class Mollicutes. In the absence of a demonstration of colony formation, one can only surmise that the cultured cells may be contaminated. This suspicion is often all that is of interest for some cell culture applications, and in those cases the indirect (biochemical or histochemical) methods may suffice. Nevertheless, such suspicions must include any microbial contaminant, not only Mollicutes. It is important to stress that a cell culture contaminant cannot be proved to be a mycoplasma unless it is

observed to grow on agar. Further identification of isolated mycoplasma contaminants can be done according to the methods outlined below under "Identification". Terms such as "noncultivable" (Hopps et al., 1973) or "cell dependent" (Hopps, Del Giudice, and Barile, 1976), when used to imply identification of mycoplasmas, are contrary to the definition of Mollicutes. Furthermore, the small number of cell cultures believed to contain this cryptic contaminant is smaller than the number of false-negative mycoplasma contaminants expected to be found by culture methods for mycoplasma detection (Hopps et al., 1976; Schneider, Stanbridge, and Epstein, 1974). Finally, the most common "noncultivable" mycoplasma was, at last, cultivated by the same workers who initiated the term (Gardella, Del Giudice, and Hopps, 1978). The method of choice for determination of mycoplasma contamination of cell cultures is the culture of the organisms on appropriate solid or semisolid media (Barile et al., 1973; Hayflick, 1965, 1972). No biochemical or staining technique purported to detect Mollicutes contamination of cell cultures has been shown to be quantitatively or qualitatively superior to the biological method required, by definition, to identify a mycoplasma. In the use of the culture method, it may be necessary to dilute tissue or cell samples to avoid mycoplasmacidal or mycoplasmastatic substances that might inhibit growth (Kaklamanis et al., 1969; Mårdh and Taylor-Robinson, 1973). It is also necessary for the neophyte mycoplasmologist to be aware of potential artifacts such as "pseudocolonies", air bubbles, and cell debris, which often resemble Mollicutes colonies. These are discussed in detail by Hayflick (1965); "pseudocolonies" are illustrated in Figs. 5–8.

Identification of Mollicutes

All recognized members of the class Mollicutes can be grown on artificial media of various complexities. One of the principal current challenges in mycoplasmology involves efforts to grow suspect Mollicutes on an artificial medium that, by definition of the class, is an absolute prerequisite for their identification.

Minimal standards for description of any new Mollicutes species have been published and these should be carefully followed when a new isolate cannot be identified as a known species (Subcommittee on the Taxonomy of Mycoplasmatales, 1972). The principal criteria for identification of members of the class Mollicutes are: (i) lack of a cell wall; (ii) typical colonial appearance; (iii) filterability through a 450-nm membrane; (iv) absence of reversion to a bacterium. Classification by family is based upon: (i) genome size; (ii) sterol requirement;

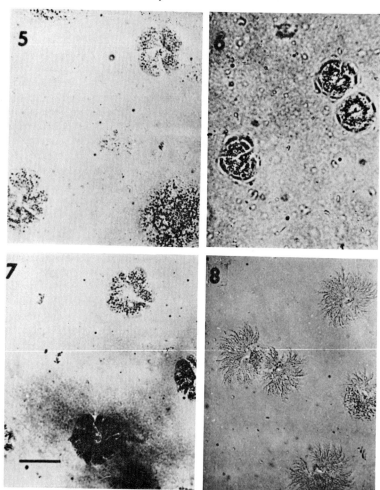

Figs. 5–8. Typical pseudocolonies mimicking mycoplasma colonies and caused by crystalline aggregation of calcium and magnesium soaps contained in the serum supplement. The horizontal bar is equivalent to 100 nm.

(iii) morphology. Identification of species depends upon: (i) serological tests; (ii) cultural characteristics; (iii) biochemical properties.

LACK OF CELL WALL. Ultrastructural studies should reveal that the organism lacks a cell wall and is bound by a single membrane. Mollicutes are incapable of synthesizing cell wall precursors such as muramic and diaminopimelic acids. Some species have an extramembranous layer of various carbohydrates, glycolipids, and glycoproteins. Components that may have derived originally from growth medium in vitro are also often detected as discussed above. It is uncertain as to which of these extramembranous materials are part of the cell and which come from the cell milieu. An interesting possibility suggested by Robertson and Smook (1976) regarding Ureaplasma is that the avidity of some mycoplasmas for medium components in vitro might reflect a similar avidity for host cell glycocalyxes in vivo.

COLONIAL APPEARANCE. Colonies on agar are usually from 0.6 to 0.1 mm in diameter. Ureaplasma colonies are more often 0.01 mm in diameter, although larger colonies are possible under some conditions. Mollicutes grow down into solid medium and produce a central nipple in most colonies which, when viewed through the microscope, gives to them their characteristic "fried egg" appearance. However, this property is not characteristic of all mycoplasma isolates and is highly influenced by several environmental conditions. The formation of the colony as originally described by Razin and Oliver (1961) is shown diagrammatically in Fig. 9.

Mycoplasma colonies are often confused with several kinds of artifacts including "pseudocolonies" (Figs. 5-8) that are composed of magnesuim and calcium soap crystals, water droplets, bubbles, and animal cells. See Hayflick (1965) for a comprehensive discussion of these artifacts.

FILTERABILITY. Cells of all members of the class Mollicutes grown in fluid medium will pass through membrane filters of 450-nm pore diameter. Many will pass pores as small as 220 nm. High pressure filtration must be avoided when seeking to demonstrate this property.

ABSENCE OF REVERSION TO A BACTERIUM. The L-phase of bacteria mimics many properties of the class Mollicutes. It is therefore important to demonstrate the absence of reversion for putative members of the class Mollicutes. Reversion capability varies for different strains of L-phase variants and is also influenced by cultural conditions. The presence of antibiotics or other inducing substances in media increases the tendency of L-phase organisms to lose their capacity to revert to the parental bacterium. Reversion occurs more readily in liquid than on solid medium and is favored by aerobic conditions. It is currently agreed that an organism may be considered to be a member of the class Mollicutes if it can be passaged five times in a medium without penicillin or increased osmotic pressure and not revert to a bacterium (Edward, 1974).

GENOME SIZE. Members of the family Mycoplasmataceae have a genome size of 4.5×10^8 daltons. The genome size of the Acholeplasmataceae is 1.0×10^9 daltons. The *Spiroplasma* genome is the same size (1×10^9 daltons) as the *Acholeplasma* genome.

STEROL REQUIREMENT. The Mycoplasmataceae and most anaeroplasmas are dependent on sterol for growth. The Acholeplasmataceae and thermoplasmas do not require sterols. Acholeplasmataceae (except *A. axanthum* and *A. modicum*) can synthesize carotenoids. They are also able to make lipids from acetate.

MORPHOLOGY. The size of individual organisms is near the resolution limits of the light microscope. A range of sizes is known that extends from 100 to 400 nm, but the smaller forms might not be viable. The cells vary from coccoid to helical or filamentous forms, depending on growth conditions. Culture age, nutritional conditions, and osmotic pressure all affect morphology. The mode of reproduction is uncertain, with binary fission and fragmentation of filamentous forms having been observed. Motile and nonmotile forms are known and no resting stages are described. Members of the family Spiroplasmataceae exhibit a helical morphology, are motile, yet possess no apical filament.

SEROLOGY. The growth inhibition test, metabolic inhibition test, and immunofluorescence test are used to distinguish species in the genera *Mycoplasma* and *Acholeplasma*. Details of these tests can be found in Clyde (1964), Purcell, Chanock, and Taylor-Robinson (1969), Purcell et al. (1966), Senterfit and

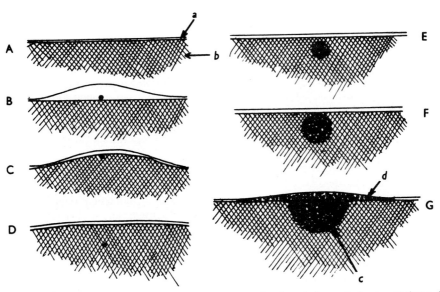

Fig. 9. The Development of mycoplasma colonies on agar. (A) Vertical section through the agar before inoculation: a, free water film; b, network of agar fibrils. (B) A drop containing a viable particle is placed on the agar. (C) Approximately 15 min after inoculation: The drop is absorbed by the agar, forming a slight swelling. (D) Approximately 3–6 h after inoculation: The viable particle has penetrated into the agar. (E) Approximately 18 h after inoculation: A small spherical colony of organisms intertwined in the agar fibrils has been formed below the agar surface. (F) Approximately 24 h after inoculation: The colony approaches the agar surface. (G) Approximately 24–48 h after inoculation: The growth spreads into the free water film forming the peripheral zone; c, central zone; d, peripheral zone. [From Razin and Oliver, 1961, with permission.]

Jensen (1966), Taylor-Robinson et al. (1966), Working Group (1975a, 1976). The single *Ureaplasma* species, *Ureaplasma urealyticum,* consists of eight serotypes, all of human origin. Criteria for distinguishing species within this genus and the genus *Spiroplasma* have not yet been established.

CULTURAL CHARACTERISTICS. Environmental conditions vary widely for the cultivation of members of the class Mollicutes and can often be used to distinguish species. Such conditions as special growth factors, medium constituents, cultivation techniques, optimum incubation temperature, pH requirements, gaseous environment, host range, and pathogenicity can be used to suggest species differences.

BIOCHEMICAL PROPERTIES

1. Glucose fermentation: Aerobic and anaerobic conditions are used to determine glucose catabolism as measured by acid production (Aluotto et al., 1970; Working Group, 1975a).
2. Arginine hydrolysis: Arginine hydrolysis differs among the Mollicutes and is measured by an increasing pH resulting from ammonia production (Aluotto et al., 1970; Leach, 1976). Arginine may also inhibit growth (Leach, 1976).
3. Urea hydrolysis: The key feature of *Ureaplasma* classification is based on the determination of urea hydrolysis with concomitant CO_2 and NH_3 production. This results in a rise in pH (Shepard and Howard, 1970; Shepard and Masover, 1978; Shepard et al., 1974). Caution must be exercised in evaluating these tests since the complex media on which Mollicutes are usually grown contain substances other than the substrate which may be metabolized to give pH changes.
4. Carotenoid production: Members of the family Acholeplasmataceae synthesize carotenoid pigments in the presence of sodium acetate (Razin and Cleverdon, 1965; Tully and Razin, 1968).
5. Genetic characteristics: The guanine-plus-cytosine content of the DNA is determined by T_m (McGee, Rogul, and Wittler, 1967; Marmur and Doty, 1962) or buoyant density (Kelton and Mandel, 1969; Schildkraut, Marmur, and Doty, 1962).

Several other informative but less important tests to determine properties of Mollicutes have been described and contribute useful information.
1. In addition to glucose, described above, the following sugars and related compounds can be tested: mannose, mannitol, lactose, saccharose, salicin, fructose, galactose, xylose, sorbitol, glycerol, cellobiose (Aluotto et al., 1970; Barber and Fabricant, 1970), and esculin (Williams and Wittler, 1971).
2. Phosphatase activity (Aluotto et al., 1970).

3. "Film and Spot" production on agar surfaces where the agar contains egg yolk emulsion (Fabricant and Freundt, 1967).
4. Proteolytic activity, exemplified by the digestion of coagulated horse serum, casein, and gelatin (Aluotto et al., 1970).
5. Reduction of tetrazolium under aerobic and anaerobic conditions (Aluotto et al., 1970).
6. Sheep and guinea pig erythrocyte hemolysis (Aluotto et al., 1970; Cole, Ward, and Martin, 1968) plus the identification of the hemolysin as, possibly, peroxide (Lind, 1970; Sobeslavsky and Chanock, 1968).
7. Erythrocyte adsorption to Mollicutes colonies (Sobeslavsky, Prescott, and Chanock, 1968).
8. Electrophoretic pattern determination of cell proteins (Razin and Rottem, 1967; Working Group, 1975b).
9. Genetic relatedness as determined by nucleic acid homology (McGee, Rogul, and Wittler, 1967; Somerson et al., 1967) and by measurement of genome size (Bak et al., 1969).

A summary of the biological characteristics of the class Mollicutes is given in Tables 2, 3, and 4.

Identification of Anaeroplasmas

The first strictly anaerobic organism considered to be associated with the class Mollicutes was reported in 1973 by Robinson and Hungate. In addition to being strictly anaerobic, it was (i) filterable through 450-nm and 220-nm average pore size membrane filters, (ii) resistant to penicillin, (iii) non-sterol-requiring, but sensitive to digitonin, and (iv) bacteriolytic (directed toward 0.5% wt/vol autoclaved *Escherichia coli* cells). Like *Thermoplasma,* some *Anaeroplasma* isolates can grow at temperatures that exceed those usually lethal for Mollicutes, and they have been shown to have unique lipids in their limiting membrane (Langworthy et al., 1975). It is now known that the organisms ferment starch and are inhibited by thallous acetate. There appear to be at least two separate types: one with and one without bacterioclastic ability. The two types also differ with respect to G+C content of their DNA, sterol requirement for growth, and perhaps with respect to habitat. A detailed discussion of these isolates is given by Robinson, Allison, and Hartman (1975).

Identification of Thermoplasmas

Since the initial isolation and description of a thermoplasma (Darland et al., 1970), more than 100 isolates have been recovered from approximately 20 different burning coal refuse piles. The new isolates all grew on the basal salt medium containing yeast

Table 2. Biological properties of the genus *Mycoplasma*.[a]

Species	G+C of DNA (mol%)	Glucose catabolism	Mannose catabolism	Arginine hydrolysis	Tetrazolium[b] reduction (Ae/An)	Phosphatase	Film and spot[b] reaction	Gelatin hydrolysis	Coagulated serum digestion
M. agalactiae subsp. *agalactiae*	33.5–34.2	−	−	−	+/+	+	+ or −	−	−
M. alkalescens	25.9	−	NT	+	−/−	+	−	NT	−
M. alvi	26.4	+	NT	+	−/+	NT	−	NT	NT
M. anatis	NT	+	+ or −	−	−/+	+	+	NT	NT
M. arginini	27.6–28.6	−	−	+	−/+	−	−	NT	−
M. arthritidis	30.0–33.7	−	−	+	−/−	+	−	+	−
M. bovigenitalium	28.0–32.0	−	−	−	−/+	+	+	−	−
M. bovirhinis	24.5–25.7	+	−	−	+/+	−	+ or −	−	+ or −
M. bovis (formerly *M. agalactiae* subsp. *bovis* or *M. bovimastitidis*)	32.7–32.9	−	−	−	+/+	+	+ or −	NT	
M. bovoculi	29.0	+	NT	+ or −	+/+	+ or −	+	−	−
M. buccale (formerly *M. orale* type 2)	24–28	−	−	+	−/+	+	−	−	−
M. canis	28.5–29.1	+	−	−	−/+	−	−	+ or −	
M. canadensis	29	−	−	+	−/+	NT	−	NT	NT
M. capricolum	25.5	+	+	+ or −	+/+	+	−	NT	+
M. caviae	NT	+	NT	−	NT	NT	NT	NT	NT
M. conjunctivae	NT	+	+	−	+/+	−	−	NT	−
M. cynos	NT	+	+	−	±/+	+	+	NT	NT
M. dispar	28.5–29.3	+	NT	−	NT	NT	NT	NT	NT
M. edwardii	29.2	+	−	−	−/+	−	+ (slow)	−	−
M. equirhinis	NT	−	NT	+	−/−	NT	+	NT	NT
M. faucium (formerly *M. orale* type 3)	NT	−	−	+	−/−	−	NT	NT	NT
M. feliminutum	NT	+	+	−	−/+	−	+ or −	NT	NT
M. felis	25.0–25.4	+	−	−	−/+	+	+	−	−
M. fermentans	27.5–29.1	+	−	+	−/+	− or +	+	−	−
M. flocculare	NT	+	NT	−	−/?	NT	−	NT	NT
M. gallinarum	26.3–28.0	−	−	+	+/+	−	+	−	−
M. gallisepticum	31.6–35.7	+	+	−	+/+	−	−	−	−
M. gateae	28.4–28.6	−	−	+	−/±	−	−	−	−
M. hominis	27.3–29.3	−	−	+	−/−	−	−	−	−
M. hypopneumoniae/ M. suipneumoniae	NT	+	−	−	NT	NT	NT	NT	NT
M. hyorhinis	27–28	+	−	−	+/±	+	−	−	−
M. hyosynoviae	NT	−	−	+	−/−	−	+	NT	−
M. iners	28.9–29.6	−	−	+	−/−	−	+	−	−
M. lipophilum	NT	−	−	+	NT	NT	+	NT	NT
M. maculosum	26.5–29.6	−	−	+	−/+	+	+	−	−
M. meleagridis	28.0–28.5	−	−	+	−/+	+	−	NT	NT
M. moatsii	NT	+	NT	+	−/−	NT	NT	NT	NT
M. molare	NT	+	+	−	+/+	−	+	NT	NT
M. mycoides subsp. *capri*	23.6–25.8	+	+	−	+/+	−	−	+	+
M. mycoides subsp. *mycoides*	26.1–26.8	+	+	−	+/+	−	−	+	+ or−
M. neurolyticum	22.8–26.5	+	+	−	−/+	−	−	−	−
M. opalescens	NT	−	−	+	−/−	+	+	NT	NT
M. orale (formerly *M. orale* type 1)	24.0–28.2	−	−	+	−/−	−	−	−	−
M. ovipneumoniae	27–29	+	NT	−	+/+	NT	−	NT	NT
M. pneumoniae	38.6–40.8	+	+	−	+/+	−	−	−	−
M. primatum	28.6	−	−	+	−/−	+	−	−	
M. pulmonis	27.5–28.3	+	+	−	−/+	−	+		

Table 2 *(Continued)*

Species	G+C of DNA (mol%)	Glucose catabolism	Mannose catabolism	Arginine hydrolysis	Tetrazolium[b] reduction (Ae/An)	Phosphatase	Film and spot[b] reaction	Gelatin hydrolysis	Coagulated serum digestion
M. putrefaciens	28.9	+	+	−	+/+	+	−	NT	−
M. salivarium	27.0–31.5	−	−	+	−/±	−	+	−	−
M. spumans	28.4–29.1	−	−	+	−/−	+	−	−	−
M. synoviae	34.2	+	NT	NT	NT	NT	NT	NT	NT
M. verecundum	27.0–29.2	−	−	−	−/?	NT	+ (slow)	NT	NT

[a] Adapted from Tully and Razin (1977), who present additional information as well as specific literature references. Symbols: +, positive reaction; −, negative reaction; NT, not tested; Ae, aerobic; An, anaerobic.
[b] Tetrazolium reduction and the film and spot reaction were discussed by Fabricant and Freundt (1967).

extract and produced the characteristic ''fried egg'' colony on the same medium solidified with agar. This organism, if included in the class Mollicutes, would make the class much broader and more heterogeneous in that thermoplasma is not only the first thermophile to be considered related to the class Mollicutes, but it would be the first true saprophyte in the class. It is indeed a fascinating prokaryote that is able to maintain an intracellular pH similar to that of other cells even though it exists without a cell wall in a hot acid environment (Hsung and Haug, 1975). However, many of the properties of thermoplasma that are currently being described suggest that it is entirely different from other Mollicutes and perhaps other prokaryotes. For example, the G+C content of its DNA is 46 mol%, considerably higher than all other known Mollicutes; its DNA weighs 1×10^9 daltons, as does the DNA of *Acholeplasma* species (Christiansen, Freundt, and Black, 1975; Searcy and Doyle, 1975). In addition, it has been shown to have ''histone-like protein'' associated with its DNA. The protein is different from the histones of eukaryotes (Searcy, 1975). More recent work suggests other relatedness between thermoplasmas and eukaryotic rather than prokaryotic cells (Searcy, Stein, and Green, 1978). Many of the lip-

ids in the thermoplasma membrane are similar to lipids of other Mollicutes, but the organism also has some unique lipids that are of considerable interest with respect to possible mechanisms by which the thermoplasma membrane can function in its hostile environment (Langworthy, 1977; Mayberry-Carson, Roth, and Smith, 1975; Mayberry-Carson et al., 1974). Whether or not the organisms belong taxonomically with the class Mollicutes remains to be established.

Identification of Mycoplasmas in Cell Cultures

It is generally agreed that all Mollicutes cell culture contaminants can be isolated by growth on agar (see ''Isolation of Mollicutes''). The sample should be diluted to avoid mycoplasmacidal or mycoplasmastatic host-cell or tissue factors. Subcultivation of the original test culture must be made one or more times in order to demonstrate a negative result. This biological or cultural method may require at least 14 days of incubation, and it has been argued that there are some ''cryptic'', ''noncultivable'', or ''cell-

Table 3. Biological properties of the genus *Acholeplasma*[a]

Species	G+C of DNA (moles %)	Glucose catabolism	Mannose catabolism	Arginine hydrolysis	Aesculin[b] hydrolysis	Sensitivity to digitonin	Tetrazolium reduction (Ae/An)	Carotenoid synthesis
A. axanthum	31.3	+	−	−	+	−	+/+	−
A. equifoetale	NT	+	NT	−	±	−	±/+	NT
A. granularum	30.5–32.4	+	−	−	−	−	±/+	+
A. laidlawii	31.7–35.7	+	+ or −	−	± or −	−	±/+	+
A. modicum	29.3	+	−	−	−	−	+/+	−
A. oculi	NT	+	−	−	+	−	+/+	+

[a] Abstracted from Tully and Razin (1977), who present additional information and specific literature citations. Symbols: Ae, aerobic; An, anaerobic, +, positive reaction; −, negative reaction; ±, weak reaction; NT, not tested.
[b] Williams and Wittler (1971).

Table 4. Biological properties of the genus *Ureaplasma.*[a]

Property	Characteristic	
Morphology (organism)	Individual coccoid cells averaging 330 nm in diameter occur in broth cultures as single cells or pairs. Occasional clusters and short chains are observed. *Ureaplasma* organisms are ultrastructurally similar to other mollicutes.	
Morphology (colony)	Tiny colonies below the surface of the agar, about 15–30 nm in diameter. Slightly larger "cauliflower head" colonies are also observed. "Fried egg" morphology may be observed on appropriately buffered agar.	
Optimal pH	6.0–6.5	
Genome size	$4.1–4.8 \times 10^8$ daltons	
DNA G+C content (mol %)	26.9–29.8	
Optimal growth temperature	35–37°C	
Enzyme activities[b]	Aminopeptidase	+
	Esterase	+
	α-glycerophosphate dehydrogenase	+
	L-Histidine ammonia-lyase	+
	Malate dehydrogenase	+
	Phosphatase (alkaline)	+
	Adenosine triphosphatase	+
	Urease(s)	+
	Alanine dehydrogenase	−
	Arginine deaminase	−
	Catalase	−
	Glutamate dehydrogenase	−
	Lactic dehydrogenase	−
	NAD-dependent L + LDH	−
	NAD-independent L + LDH	−
	NAD-independent L − LDH	−
	Proteolytic activities	+
Hemolysis of guinea pig erythrocytes + (beta)		
Hemadsorption of guinea pig erythrocytes + (human serotype III)		
Erythromycin sensitivity		+
Thallous acetate sensitivity		+
Lincomycin sensitivity		−
Tetrazolium reduction (aerobic and anaerobic)		−
Antigenicity	At least 8 serotypes among human strains. At least 8 serotypes among bovine strains that are distinct from human strains.	

[a] See Shepard and Masover (1978) and Tully and Razin (1977) for detailed discussion and specific literature references to biological properties of the genus *Ureaplasma.*
[b] +, Present; −, absent.

adapted" strains that are not detected by this method (Stanbridge and Schneider, 1974). Therefore, a variety of indirect biochemical and histochemical methods have been devised to provide more rapid evidence of cell culture contamination. If the only concern is whether or not a eukaryotic cell culture is contaminated with a prokaryotic organism, then some of these methods may be useful. None of these methods, however, is capable of identification of Mollicutes with the exception of those that employ specific antisera to a Mollicute contaminant. In general, the so-called indirect methods are (i) less specific, (ii) less sensitive, and (iii) more complex than cultural (direct) methods.

Gross cell culture contamination is usually obvious if a bacterium or fungus is the contaminant. Because even gross Mollicutes contamination is not always immediately observable, some type of test is necessary. Gross contamination by Mollicutes is often detectable by the simplest histochemical staining methods (e.g., Giemsa stain) that will demonstrate multiple basophilic bodies in the area of the cytoplasm as well as in the nuclear region. DNA-binding fluorescent stains are claimed to be able to discern a single Mollicutes organism on a tissue cultured cell (Chen, 1977). This method can, if properly controlled, demonstrate Mollicutes. However, a known mycoplasma-free control and a deliberately infected control culture are needed for comparison. Otherwise one can only say that a body that has affinity for DNA-binding stains is present in the cytoplasmic area of the cell. Such bodies are not easily differentiated from host cell micronuclei or other prokaryotes.

Reference reagents (antisera) and reference strains of the class Mollicutes are obtainable in the United States through the National Institutes of Health and from the American Type Culture Collection, Rockville, Maryland. In addition, the World Health Organization and the Food and Agriculture Organization of the United Nations (WHO/FAO) have sponsored an International Reference Center for Animal Mycoplasmas at the Institute of Medical Microbiology, University of Aarhus, Aarhus, Denmark (The FAO/WHO Programme on Comparative Mycoplasmology, 1974).

Acknowledgments

This work was supported, in part, by grant AG 00850 from the National Advisory Council on Aging, National Institutes of Health, Bethesda, Maryland; Children's Hospital Research Fund, and the Glenn Foundation for Medical Research, Manhasset, New York.

Literature Cited

Adler, H. F., Yamamoto, R., Berg, J. 1957. Strain differences of pleuropneumonia-like organisms of avian origin. Avian Disease 1:19–27.

Adler, H. F., Fabricant, R., Yamamoto, R., Berg, J. 1958. Symposium on chronic respiratory diseases of poultry. I. Isolation and identification of pleuropneumonia-like organisms of avian origin. American Journal of Veterinary Research 19:440–447.

Aluotto, B. B., Wittler, R. G., Williams, C. O., Faber, J. E. 1970. Standardized bacteriological techniques for the characterization of Mycoplasma species. International Journal of Systematic Bacteriology 20:35–58.

Bak, A. L., Black, F. T., Christiansen, C., Freundt, E. A. 1969. Genome size of mycoplasmal DNA. Nature 224:1209–1210.

Barber, T. L., Fabricant, J. 1970. Identification of Mycoplasmatales: characterization procedures. Applied Microbiology 21:600–605.

Barile, M. 1974. General principles of isolation and detection of mycoplasmas, pp. 135–142. In: Bové, J. M., Duplan, J. F. (eds.), Les mycoplasmes de l'homme, des animaux, des vegetaux et des insectes. Paris: INSERM.

Barile, M. F., Hopps, H. E., Grobowski, M. W., Riggs, D. B., DelGuidice, R. A. 1973. The identification and sources of mycoplasmas isolated from contaminated cell cultures. Annals of the New York Academy of Sciences 225:251–264.

Belly, R. T., Bohlool, B. B., Brock, T. D. 1973. The genus Thermoplasma. Annals of the New York Academy of Sciences 225:94–107.

Bennett, S. C. J. 1932. Contagious bovine pleuropneumonia: Control by culture vaccines. Journal of Comparative Pathology and Therapeutics 45:257–264.

Betts, A. O., Whittlestone, P. 1963. Enzootic or virus pneumonia of pigs: The production of pneumonia with tissue culture fluids. Research in Veterinary Science 4:471–479.

Biberfield, G., Biberfield, P. 1970. Ultrastructure features of

Mycoplasma pneumoniae. Journal of Bacteriology 102:855–861.

Black, F. T., Vinther, O. 1977. Morphology and ultrastructure of Ureaplasma urealyticum in agar growth. Acta Pathologica et Microbiologica Scandinavica, Sect. B 85:281–285.

Boatman, E., Cartwright, G., Kenny, G. 1976. Morphology, morphometry and electron microscopy of HeLa Cells infected with bovine Mycoplasma. Cell and Tissue Research 170:1–16.

Bordet, J. 1910. La morphologie du microbe de la péripneumonie des bovides. Annales de l'Institut Pasteur 24:161–167.

Borrel, A., Dujardin-Beaumetz, E., Jeantet, Jouan. 1910. Le microbe de la péripneumonie. Annales de l'Institut Pasteur 24:168–179.

Bové, J. M., Duplan, J. F. (eds.). 1974. Les mycoplasmes de l'homme, des animaux, des vegetaux et des insectes. Paris: INSERM.

Bowie, W. R., Alexander, E. R., Floyd, J. F., Holmes, J., Miller, Y., Holmes, K. K. 1976. Differential response of chlamydial and ureaplasma-associated urethritis to sulphafurazole (sulfisoxazole) and aminocyclitols. Lancet ii:1276–1278.

Bradbury, J., Jordan, F. 1972. Studies on the adsorption of certain proteins to M. gallisepticum and their influence on agglutination and hemagglutination. Journal of Hygiene 70:267–278.

Bredt, W. 1973. Motility of Mycoplasmas. Annals of the New York Academy of Sciences 225:246–250.

Bredt, W., Lam, P. S., Fiegel, P., Höffler, D. 1974. Nachweis von mycoplasmen im blasenpunktionsurin. Deutsche Medizinische Wochenschrift 99:1553–1556.

Bridré, J., Donatien, A. 1923. Le microbe de tagalaxic contagieuse et sa culture in vitro. Comptes Rendus Hebdomadaires des Séances de l'Académie des Sciences 177:841–843.

Brown, S., Teplitz, M., Revel, J. P. 1974. Interaction of mycoplasmas with cell cultures, as visualized by electron microscopy. Proceedings of the National Academy of Sciences of the United States of America 71:464–468.

Chalquest, R. R. Fabricant, J. 1960. Pleuropneumonia-like organisms associated with synovitis in fowl. Avian Diseases 4:515–539.

Chanock, R. M., Hayflick, L., Barile, M. F. 1962. Growth on artificial medium of an agent associated with atypical pneumonia and its identification as a PPLO. Proceedings of the National Academy of Sciences of the United States of America 48:41–49.

Chanock, R. M., Johnson, K. M. 1961. Infectious disease: Respiratory viruses. Annual Revue of Medicine 12:1–18.

Chanock, R. M., Mufson, M. A., Bloom, H. H., James, W. D., Fox, H. H., Kingston, J. R. 1961. Eaton agent pneumonia. Journal of the American Medical Association 175:213–220.

Chanock, R. M., Dienes, I., Eaton, M. D., Edward, D. G. ff, Freundt, E. A., Hayflick, L., Hers, J. F. P., Jensen, K. E., Liu, C., Marmion, B. P., Morton, H. E., Mufson, M. A., Smith, P. F., Somerson, N. L., Taylor-Robinson, D. 1963a. Mycoplasma pneumoniae: Proposed nomenclature for atypical pneumonia organism (Eaton Agent). Science 140:662.

Chanock, R. M., Parrott, R. H., Johnson, K. M., Mufson, M. A., Knight, V. 1963b. Biology and ecology of two major lower respiratory tract pathogens—RS virus and Eaton PPLO, pp. 257–281. In: Pollard, M. (ed.), Perspectives in virology, vol. 3. New York: Hoeber.

Chen, T. R. 1977. In situ detection of mycoplasma contamination in cell cultures by fluorescent Hoechst 33258 stain. Experimental Cell Research 104:255–262.

Christiansen, C., Freundt, E. A., Black, F. T. 1975. Genome size and deoxyribonucleic acid base composition of Thermoplasma acidophilum. International Journal of Systematic Bacteriology 25:99–101.

Chu, H. P. 1954. The identification of infectious coryza associated with Nelson's coccobacilliform bodies in fowls in England and its similarity to chronic respiratory disease of chickens. Proceedings of the Tenth World Poultry Congress **2:**246–251.

Chu, H. P., Horne, R. W. 1967. Electron microscopy of *Mycoplasma gallisepticum* and *Mycoplasma mycoides* using the negative staining technique and their comparison with myxoviruses. Annals of the New York Academy of Sciences **143:**190–203.

Clyde, W. A., Jr. 1961. Demonstration of Eaton's agent in tissue culture. Proceedings of the Society for Experimental Biology and Medicine **107:**715–718.

Clyde, W. A., Jr. 1963. Studies on growth of Eaton's agent in tissue culture. Proceedings of the Society for Experimental Biology and Medicine **112:**905–909.

Clyde, W. A., Jr. 1964. Mycoplasma species identification based upon growth inhibition by specific antisera. Journal of Immunology **92:**958–965.

Cole, B. C., Ward, J. R. 1973. Interaction of *Mycoplasma arthritidis* and other mycoplasmas with murine peritoneal macrophages. Infection and Immunity **7:**691–699.

Cole, B. C., Ward J. R., Martin, C. H. 1968. Hemolysin and peroxide activity of *Mycoplasma* species. Journal of Bacteriology **95:**2022–2030.

Collier, A. M., Clyde, W. A. 1971. Relationships between *Mycoplasma pneumoniae* and human respiratory epithelium. Infection and Immunity **3:**694–701.

Collier, A. M., Clyde, W. A., Denny, F. W. 1971. *Mycoplasma pneumoniae* in hamster trachael organ culture: Immunofluorescent and electron microscopic studies. Proceedings of the Society for Experimental Biology and Medicine **136:** 569–573.

Commission on Acute Respiratory Diseases. 1946. The transmission of primary atypical pneumonia to human volunteers. I. Experimental methods. II. Results of inoculation. III. Clinical features. IV. Laboratory studies. Bulletin of Hopkins Hospital **79:**97–167.

Darland, G., Brock, T. D., Samsonoff, W., Conti, S. F. 1970. A thermophilic, acidophilic mycoplasma isolated from a coal refuse pile. Science **170:**1416–1418.

Davis, R. E., Whitcomb, R. F., Steere, R. I. 1968. Remission of Aster yellows disease by antibiotics. Science **161:**793–795.

Dienes, L., Edsall, J. 1937. Observations on L-organisms of Klieneberger. Proceedings of the Society for Experimental Biology and Medicine **36:**740–744.

Dingle, J. H., Finland, M. 1942. Virus pneumonias. II. Primary atypical pneumonias of unknown etiology. New England Journal of Medicine **227:**378–385.

Doi, Y., Terenaka, M., Yora, K., Asuyama, H. 1967. Mycoplasma—or PLT group-like microorganisms in the phloem elements of plants infected with mulberry dwarf, potato witches' broom, aster yellows, or paulownia witches' broom. Annals of the Phytopathology Society of Japan **33:**259–266.

Dujardin-Beaumetz, E. 1900. Le microbe de la péripneumonie et sa culture. Thèse de Paris, Paris, Octave Doin.

Eaton, M. D. 1950a. Virus pneumonia and pneumonitis viruses of man and animals, pp. 87–140. In: Doer and Hallauer (eds.), Handbuch der Virusforschung (Abt. 2, Ergän Zungsband). Vienna: Springer-Verlag.

Eaton, M. D. 1950b. Action of aureomycin and chloromycetin on the viruses of primary atypical pneumonia. Proceedings of the Society for Experimental Biology and Medicine **73:**24–26.

Eaton, M. D., Liu, C. 1957. Studies on sensitivity to streptomycin of the atypical pneumonia agent. Journal of Bacteriology **74:**784–787.

Eaton, M. D., Mecklejohn, G., Van Herick, W. 1944. Studies on the etiology of primary atypical pneumonia. A filterable agent transmissable to cotton rats, hamsters, and chick embryos. Journal of Experimental Medicine **79:**649–668.

Eaton, M. D., Van Herick, W. 1947. Serological and epidemiological studies on primary atypical pneumonia and related acute upper respiratory disease. American Journal of Hygiene **45:**82–95.

Edward, D. G. ff 1947. A selective medium for pleuropneumonia-like organisms. Journal of General Microbiology **1:**238–243.

Edward, D. G. ff 1974. Taxonomy of the class Mollicutes, pp. 13–18. In: Bové, J. M., Duplan, J. F. (eds.), Les mycoplasmes de l'homme, des animaux, des vegetaux et des insectes. Paris: INSERM.

Edward, D. G. ff, Kanarek, A. D. 1960. Organisms of the pleuropneumonia group of avian origin: Their classification into species. Annals of the New York Academy of Sciences **79:**696–702.

Elford, W. J. 1929. Ultrafiltration methods and their application in bacteriological and pathological studies. British Journal of Experimental Pathology **10:**126–144.

Fabricant, J., Freundt, E. A. 1967. Importance of extension and standardization of laboratory tests for the identification and classification of mycoplasma. Annals of the New York Academy of Sciences **143:**50–58.

FAO/WHO Programme on Comparative Mycoplasmology. 1974. Veterinary Record **95:**457–461.

Faur, Y. C., Weisburd, M. H., Wilson, M. E. 1975. Morphologic observations of mycoplasma and *Neisseria gonorrhoeae* in associated growth patterns. American Journal of Clinical Pathology **63:**106–116.

Faur, Y. C., Weisburd, M. H., Wilson, M. E. 1976. A comparison of horse, cow and sheep blood in NYC medium: Effect on recovery of *N. gonorrhoeae* and urogenital mycoplasmas. Health Laboratory Science **13:**194–196.

Faur, Y. C., Weisburd, M. H., Wilson, M. E., May, P. S. 1973. A new medium for the isolation of pathogenic *Neisseria* (NYC medium). I. Formulation and comparisons with standard media. Health Laboratory Science **10:**44–54.

Findlay, G., Klieneberger, E., MacCallum, F. O., MacKenzie, R. D. 1938. Rolling disease: New syndrome in mice associated with a pleuropneumonia-like organism. Lancet **235:**1511.

Fogh, J. (ed.). 1973. Contamination in cell cultures. New York: Academic Press.

Fogh, J., Holmgren, N. B., Ludovici, P. P. 1971. A review of cell culture contamination. In Vitro **7:**26–41.

Ford, D. K., Rasmussen, G., Minken, J. 1962. T-strain pleuropneumonia-like organisms as one cause of non-gonococcal urethritis. British Journal of Venereal Disease **38:**22–25.

Foster, J. P. 1934. Some historical notes on contagious pleuropneumonia. Journal of the American Veterinary Medical Association **84:**918–926.

Freundt, E. A. 1974. The mycoplasmas, pp. 929–955. In: Buchanan, R. E., Gibbons, N. E. (eds.), Bergey's manual of determinative bacteriology, 8th ed. Baltimore: Williams & Wilkins.

Freundt, E. A., Edward, D. G. ff 1979. Classification and taxonomy, pp. 1–41. In: Barile, M., Razin, S. (eds.), The mycoplasmas, vol. 1. New York: Academic Press.

Furness, G. 1975. T-mycoplasmas: Growth patterns and physical characteristics of some human strains. The Journal of Infectious Diseases **132:**592–596.

Furness, G., Pipes, F. J., McMurtrey, M. J. 1968. Susceptibility of human mycoplasmata to ultraviolet and X-irradiations. Journal of Infectious Diseases **118:**1–6.

Furr, P. M., Cooper, J. E., Taylor-Robinson, D. 1977. Isolation of mycoplasmas from three falcons (*Falco* spp). Veterinary Record **100:**72–73.

Furr, P. M., Taylor-Robinson, D., Heatherington, C. M. 1976. The occurrence of ureaplasmas in marmosets. Laboratory Animal **10:**393–398.

Gardella, R. S., Del Giudice, R. A., Hopps, H. E. 1978. Cultivation of *Mycoplasma hyorhinis,* strain DBS 1050 in cell-free medium. Abstracts of the Annual Meeting of the American Society for Microbiology **1978:**75.

Ghosh, A., Das, J., Maniloff, J. 1977. Lack of repair of ultraviolet-light damage in *Mycoplasma gallisepticum.* Journal of Molecular Biology **116:**337–344.

Goodburn, G., Marmion, B. P. 1962. A study of the properties of

Eaton's primary atypical pneumonia organism. Journal of General Microbiology **29**:271–290.

Goodwin, R. F. W., Pomeroy, A. P., Whittlestone, P. 1965. Production of enzootic pneumonia in pigs with a mycoplasma. Veterinary Record **77**:1247–1249.

Goodwin, R. F. W., Pomeroy, A. P., Whittlestone, P. 1967. Characterization of *Mycoplasma suipneumoniae:* A mycoplasma causing enzootic pneumonia of pigs. Journal of Hygiene **65**:85–96.

Goodwin, R. F. W., Whittlestone, P. 1963. Production of enzootic pneumonia in pigs with an agent grown in tissue culture from the natural disease. British Journal of Experimental Pathology **44**:291–299.

Goodwin, R. F. W., Whittlestone, P. 1964. Production of enzootic pneumonia in pigs with a microorganism grown in media free from living cells. Veterinary Record **76**:611.

Harshbarger, J. C., Chang, S. C. 1977. Chlamydiae (with phages), mycoplasmas, and rickettsiae in Chesapeake Bay bivalves. Science **196**:666–668.

Hayflick, L. 1965. Cell cultures and mycoplasmas. Texas Reports on Biology and Medicine, Suppl. 1 **23**:285–303.

Hayflick, L. 1972. Mycoplasmas as pathogens, pp. 17–31. In: Ciba Foundation Symposium, "Pathogenic Mycoplasmas", Elsevier Excerpta Medica. Amsterdam: North-Holland.

Hayflick, L., Chanock, R. 1965. Mycoplasma species of man. Bacteriological Reviews **29**:185–211.

Hayflick, L., Stanbridge, E. 1967. Isolation and identificaiton of mycoplasma from human clinical materials. Annals of the New York Academy of Sciences **143**:608–621.

Hopps, H. E., Del Giudice, R. A., Barile, M. F. 1976. Current status of "non-cultivable" mycoplasmas. Proceedings of the Society for General Microbiology **3**:143.

Hopps, H. E., Meyer, B. C., Barile, M. F., Del Giudice, R. A. 1973. Problems concerning "noncultivable" mycoplasma contaminants in tissue cultures. Annals of the New York Academy of Sciences **225**:265–276.

Hsung, J. C., Haug, A. 1975. Intracellular pH of *Thermoplasma acidophila*. Biochimica et Biophysica Acta **389**:477–482.

International Organization for Mycoplasmology. 1980. Proceedings of the Third Conference of the International Organization for Mycoplasmology, Custer, South Dakota, September 3–9, 1980. In press.

Ishiie, T., Doi, Y., Yora, K., Asuyama, H. 1967. Suppressive effective of antibiotics of the tetracycline group on symptom development of mulberry dwarf disease. Annals of the Phytopathology Society of Japan **33**:267–275.

Joklik, W. K. 1976. The Poxviruses, p. 792. In: Joklik, W. K., Willett, H. P. (eds.), "Zinsser Microbiology", 16th ed. New York: Appleton-Century-Crofts.

Jordan, W. S. 1949. The infectiousness and incubation period of primary atypical pneumonia. American Journal of Hygiene **50**:315–330.

Kahane, K., Tully, J. G. 1976. Binding of plant lectins to mycoplasma cells and membranes. Journal of Bacteriology **128**:1–7.

Kaklamanis, K., Thomas, L., Stavropolous, K., Borman, I., Bushwitz, C. 1969. Mycoplasmacidal action of normal tissue extracts. Nature **221**:860–862.

Kelton, W. H., Mandel, M. 1969. Deoxyribonucleic acid base compositions of mycoplasma strains of avian origin. Journal of General Microbiology **56**:131–135.

Klieneberger-Nobel, E. 1962. Pleuropneumonia-like organisms (PPLO): Mycoplasmataceae. New York: Academic Press.

Koshimizu, K., Magaributchi, T. 1977. Isolation of *Ureaplasma* (T-mycoplasma) from the chicken and jungle-fowl. Japanese Journal of Veterinary Science **39**:195–199.

Laidlaw, P. P., Elford, W. J. 1936. A new group of filterable organisms. Proceedings of the Royal Society (Biology) **20**:292–303.

Langworthy, T. A. 1977. Long chain diglycerol tetraethers from *Thermoplasma acidophilum*. Biochimica et Biophysica Acta **487**:37–50.

Langworthy, T. A., Mayberry, W. R., Smith, P. F., Robinson, I. M. 1975. Plasmalogen composition of anaeroplasma. Journal of Bacteriology **122**:785–787.

Lannek, N., Wesslen, T. 1957. Evidence that the SFP agent is an etiological factor in enzootic pneumonia in swine. Nordisk Veterinaer Medicin **9**:177–190.

Leach, R. H. 1976. The inhibitory effect of arginine on growth of some mycoplasmas. Journal of Applied Bacteriology **41**:259–264.

Lecce, J. G. 1960. Porcine polyserositis with arthritis: Isolation of a fastidious pleuropneumonia-like organism and *Hemophilis influenzae suis*. Annals of the New York Academy of Sciences **79**:670–676.

Lee, Y.-H., Bailey, P. E., McCormack, W. M. 1972. T-mycoplasmas from urine and vaginal specimens: Decreased rates of isolation and growth in the presence of thallium acetate. Journal of Infectious Diseases **125**:318–321.

L'Feuyer, C., Switzer, W. P. 1963. Virus pneumonia of pigs. Attempts at propagation of the causative agent in cell cultures and chicken embryos. Canadian Journal of Comparative Medicine **27**:91.

Lind, K. 1970. A simple test for peroxide secretion by *Mycoplasma*. Acta pathologica et Microbiologica Scandinavica, Sect. B **78**:256–257.

Liu, C. 1957. Studies on primary atypical pneumonia. I. Localization, isolation, and cultivation of a virus in chick embryos. Journal of Experimental Medicine **106**:455–467.

Low, I. E. 1976. Mycoplasma in tissue culture: Overview of detection methods. Health Laboratory Science **13**:129–136.

Lyell, A., Gordon, A. M., Dick, H. M., Sommerville, R. G. 1967. Mycoplasmas and erythema multiforme. Lancet **ii**:1116–1118.

McGarrity, G. J. 1976. Spread and control of mycoplasmal infection of cell cultures. In Vitro **12**:643–648.

McGee, Z. A., Rogul, M., Wittler, R. G. 1967. Molecular genetic studies of relationships among mycoplasmas, L-forms, and bacteria. Annals of the New York Academy of Sciences **143**:21–30.

Manchee, R. J., Taylor-Robinson, D. 1969. Studies on the nature of receptors involved in attachment of tissue culture cells to mycoplasmas. British Journal of Experimental Pathology **50**:66–75.

Maniloff, J., Das, J., Christensen, J. R. 1977. Viruses of mycoplasmas and spiroplasmas. Advances in Virus Research **21**:343–380.

Maniloff, J., Morowitz, H. J. 1972. Cell biology of the mycoplasmas. Bacteriological Reviews **36**:263–290.

Maramorosch, K., Shikata, E., Granados, R. R. 1968. Structures resembling mycoplasma in diseased plants and in insect vectors. Transactions of the New York Academy of Sciences, Series II **30**:841–855.

Mårdh, P.-A., Taylor-Robinson, D. 1973. Differential effect of lysolecithin on mycoplasmas and acholeplasmas. Medical Microbiology and Immunology **158**:219–226.

Maré, C. J., Switzer, W. P. 1965. New species: *Mycoplasma hyopneumoniae*. A causative agent of virus pig pneumonia. Veterinary Medicine **60**:841.

Markham, F. S., Wong, S. C. 1952. Pleuropneumonia-like organisms in the etiology of turkey sinusitis and chronic respiratory disease of chickens. Poultry Science **31**:902–904.

Marmion, B. P., Goodburn, G. M. 1961. Effect of an organic gold salt on Eaton's primary atypical pneumonia agent and other observations. Nature **189**:247–248.

Marmur, J., Doty, P. 1962. Determination of the base composition of deoxyribonucleic acid from its thermal denaturation temperature. Journal of Molecular Biology **5**:109–118.

Masover, G. K., Catlin, J., Hayflick, L. 1977. The effect of growth and urea concentration on ammonia production by a urea-hydrolysing mycoplasma *(Ureaplasma urealyticum)*. Journal of General Microbiology **98**:587–593.

Masover, G. K., Mischak, R. P., Hayflick, L. 1975. Some effects of growth medium composition on the antigenicity of a T-strain mycoplasma. Infection and Immunity **11**:530–539.

Masover, G. K., Namba, M., Hayflick, L. 1976. Cytotoxic effect of a T-strain mycoplasma (*Ureaplasma urealyticum*) on cultured normal human cells (WI-38). Experimental Cell Research **99**:363–374.

Masover, G. K., Razin, S., Hayflick, L. 1977a. Effects of carbon dioxide, urea, and ammonia on growth of *Ureaplasma urealyticum* (T-strain mycoplasma). Journal of Bacteriology **130**:292–296.

Masover, G. K., Razin, S., Hayflick, L. 1977b. Localization of enzymes in *Ureaplasma urealyticum* (T-strain mycoplasma). Journal of Bacteriology **130**:297–302.

Masover, G. K., Palant, M., Zerrudo, Z., Hayflick, L. 1977. Interaction of *Ureaplasma urealyticum* with eukaryotic cells in vitro, pp. 364–369. In: Hobson, D., Holmes, K. K. (eds.), Nongonococcal urethritis and related infections. Washington, D.C.: American Society for Microbiology.

Mayberry-Carson, K. J., Langworthy, T. A., Mayberry, W. R., Smith, P. F. 1974. A new class of lipopolysaccharide from *Thermoplasma acidophilum.* Biochimica et Biophysica Acta **360**:217–229.

Mayberry-Carson, K. J., Roth, I. L., Smith, P. F. 1975. Ultrastructure of lipopolysaccharide isolated from *Thermoplasma acidophilum.* Journal of Bacteriology **121**:700–703.

Mazzali, R., Taylor–Robinson, D. 1971. The behavior of T-mycoplasmas in tissue culture. Journal of Medical Microbiology **4**:125–138.

Meyer, K. F. 1909. Some experimental and epidemiological observations on a particular strain of pleuropneumonia. Transvaal Department of Agriculture Republican Government Veterinary Bacteriology **135**:159.

Morton, H. E., Smith, P. F., Leberman, P. R. 1951. Investigation of the cultivation of pleuropneumonia-like organisms from man. Journal of Syphilis, Gonorrhea and Venereal Disease **35**:361–369.

Nelson, J. B. 1935. Coccobacilliform bodies associated with an infectious fowl coryza. Science **82**:43–44.

Nocard, E., Roux, E. R. 1898. Le microbe de la péripneumonie. Annales de l'Institut Pasteur **12**:240–262.

Peterson, O. I., Ham T. H., Finland, M. 1943. Cold agglutinins in primary atypical pneumonias. Science **97**:167–168.

Pirie, N. W. 1973. On being the right size. Annual Review of Microbiology **27**:119–132.

Purcell, R. H., Chanock, R. M., Taylor-Robinson, D. 1969. Serology of the mycoplasmas of man, pp. 221–264. In: Hayflick, L. (ed.), The mycoplasmatales and the L-phase of bacteria. New York: Appleton-Century-Crofts.

Purcell, R. H., Taylor-Robinson, D., Wong, D. C., Chanock, R. M. 1966. A color test for the measurement of antibody to the non-acidforming human mycoplasma species. Journal of Epidemiology **84**:51–66.

Razin, S. 1964. Factors influencing osmotic fragility of *Mycoplasma.* Journal of General Microbiology **36**:451–459.

Razin, S., Cleverdon, R. C. 1965. Carotenoids and cholesterol in membranes of *Mycoplasma laidlawii.* Journal of General Microbiology **41**:409–415.

Razin, S., Oliver, O. 1961. Morphogenesis of mycoplasma and bacterial L-form colonies. Journal of General Microbiology **24**:225–237.

Razin, S., Rottem, S. 1967. Identification of *Mycoplasma* and other microorganisms by polyacrylamide-gel electrophoresis of cell proteins. Journal of Bacteriology **94**:1807–1810.

Razin, S., Masover, G. K., Palant, M., Hayflick, L. 1977. Morphology of *Ureaplasma urealyticum* (T-mycoplasma) organisms and colonies. Journal of Bacteriology **130**:464–471.

Robertson, J. A. 1977. Brom thymol blue (B) broth: An improved indicator medium for the detection of *Ureaplasma urealyticum.* Abstracts of the Annual Meeting of the American Society for Microbiology **1977**:134.

Robertson, J. A. 1978. Effect of manganese on the growth of *Ureaplasma urealyticum* (T-strain mycoplasma). Abstracts of the Annual Meeting of the American Society for Microbiology **1978**:74.

Robertson, J., Gomersall, M., Gill, P. 1975a. Effect of preparatory techniques on the gross morphology of *Mycoplasma hominis.* Journal of Bacteriology **124**:1019–1022.

Robertson, J., Gomersall, M., Gill, P. 1975b. *Mycoplasma hominis:* Growth, reproduction, and isolation of small viable cells. Journal of Bacteriology **124**:1007–1018.

Robertson, J., Smook, E. 1976. Cytochemical evidence of extramembranous carbohydrates on *Ureaplasma urealyticum* (T-strain mycoplasma). Journal of Bacteriology **128**:658–660.

Robinson, I. M., Allison, M. J., Hartman, P. A. 1975. *Anaeroplasma abactoclasticum* gen. nov., sp. nov.: An obligately anaerobic mycoplasma from the rumen. International Journal of Systematic Bacteriology **25**:173–181.

Robinson, J. P., Hungate, R. E. 1973. *Acholeplasma bactoclasticum* sp. nov., an anaerobic mycoplasma from the bovine rumen. International Journal of Systematic Bacteriology **23**:171–181.

Sabin, A. B. 1938. Identification of the filterable transmissible agent isolated from toxoplasma-infected tissues as a new pleuropneumonia-like microbe. Science **88**:575.

Schiefer, H. G., Krauss, H., Brunner, H., Gerhardt, U. 1975. Ultrastructural visualization of surface carbohydrate structures on mycoplasma membranes by concanavalin A. Journal of Bacteriology **124**:1598–1600.

Schildkraut, C. L., Marmur, J., Doty, P. 1962. Determination of the base composition of deoxyribonucleic acid from its buoyant density in CsCl. Journal of Molecular Biology **4**:430–443.

Schneider, E. L., Stanbridge, E. J., Epstein, C. J. 1974. Incorporation of ³H-uridine and ³H-uracil into RNA: A simple technique for the detection of mycoplasma contamination of cultured cells. Experimental Cell Research **84**:311–318.

Searcy, D. G. 1975. Histone-like protein in the prokaryote *Thermoplasma acidophilum.* Biochimica et Biophysica Acta **395**:535–547.

Searcy, D. G., Doyle, E. K. 1975. Characterization of *Thermoplasma acidophilum* deoxyribonucleic acid. International Journal of Systematic Bacteriology **25**:286–289.

Searcy, D. G., Stein, D. B., Green, G. R. 1978. Phylogenetic affinities between eukaryotic cells and a thermophilic mycoplasma. Biosystems **10**:19–28.

Senterfit, L. B., Jensen, K. E. 1966. Antimetabolic antibodies to *Mycoplasma pneumoniae* measured by tetrazolium reduction inhibition. Proceedings of the Society for Experimental Biology and Medicine **122**:786–790.

Shepard, M. C. 1954. The recovery of pleuropneumonia-like organisms from Negro men with and without nongonococcal urethritis. American Journal of Syphylology **38**:113–124.

Shepard, M. C., Howard, D. R. 1970. Identification of ''T'' mycoplasmas in primary agar cultures by means of a direct test for urease. Annals of the New York Academy of Sciences **174**:809–819.

Shepard, M. C., Lunceford., C. D. 1976. Differential agar medium (A7) for identification of *Ureaplasma urealyticum* (human T mycoplasmas) in primary cultures of clinical material. Journal of Clinical Microbiology **3**:613–625.

Shepard, M. C., Masover, G. K. 1978. Special features of ureaplasmas, pp. 451–494. In: Barile, M., Razin, S. (eds.), The mycoplasmas, vol. I. New York: Academic Press.

Shepard, M. C., Lunceford, C. D., Ford, D. K., Purcell, R. H., Taylor-Robinson, D., Razin, S., Black, F. T. 1974. *Ureaplasma urealyticum.* gen. nov., sp. nov.: Proposed nomenclature for the human T (T-strain) mycoplasmas. International Journal of Systematic Bacteriology **24**:160–171.

Shoetensack, H. M. 1934. Pure cultivation of the filterable virus isolated from canine distemper. Kitasato Archives of Experimental Medicine **11**:277–290.

Sobeslavsky, O., Chanock, R. M. 1968. Peroxide formation by

mycoplasmas which infect man. Proceedings of the Society for Experimental Biology and Medicine **129:**531–535.

Sobeslavsky, O., Prescott, B., Chanock, R. M. 1968. Adsorption of *Mycoplasma pneumoniae* to neuraminic acid receptors of various cells and possible role in virulence. Journal of Bacteriology **96:**695–705.

Society for General Microbiology. 1976. Proceedings of the Society for General Microbiology, vol. III, part 4. Berkshire, England: Harvest House.

Somerson, N. L., Reich, P. R., Chanock, R. M., Weissman, S. M. 1967. Genetic differentiation by nucleic acid homology. III. Relationships among mycoplasmas, L-forms and bacteria. Annals of the New York Academy of Sciences **143:**9-20.

Stamey, T. A., Pfau, A. 1970. Urinary infections: A selective review and some observations. California Medicine **113:**16–35.

Stanbridge, E., Schneider, E. 1974. Detection of mycoplasma contaminants in cell culture by biochemical methods, pp. 169–178. In: Bové, J. M., Duplan, J. F. (eds.), Les mycoplasmes de l'homme, des animaux, des vegetaux et des insectes. Paris: INSERM.

Subcommittee on the Taxonomy of *Mycoplasmatales*. 1972. Proposal for minimal standards for descriptions of new species of the order *Mycoplasmatales*. International Journal of Systematic Bacteriology **22:**184–188.

Taylor-Robinson, D., Csonka, G. W., Prentice, M. J. 1977. Human intraurethral inoculation of ureaplasmas. Quarterly Journal of Medicine, New Series **46:**309–326.

Taylor-Robinson, D., Purcell, R. H., Wong, D. C., Chanock, R. M. 1966. A colour test for the measurement of antibody to certain mycoplasma species based upon the inhibition of acid production. Journal of Hygiene **64:**91–104.

Thomas, L. 1967. The neurotoxins of *M. neurolyticum* and *M. gallisepticum*. Annals of the New York Academy of Sciences **143:**218–224.

Thomas, L. 1969. Relationships between mycoplasmas and mammalian cells. In: Smith, R. T., Good, R. A. (eds.), Cellular recognition. New York: Appleton-Century-Crofts.

Thomas, L., Curnen, E. C., Mirick, G. S., Ziegler, J. E., Horsfall, F. L. 1943. Complement fixation with dissimilar antigens in primary atypical pneumonia. Proceedings of the Society for Experimental Biology and Medicine **52:**121–125.

Tully, J. G., Razin, S. 1968. Physiological and serological comparisons among strains of *Mycoplasma granularum* and *Mycoplasma laidlawii*. Journal of Bacteriology **95:** 1504–1512.

Tully, J. G., Razin, S. 1977. The Mollicutes (mycoplasmas), pp. 405–459. In: Laskin, A. I., Lechevalier, H. A. (eds.), CRC handbook of microbiology, 2nd ed. Cleveland: CRC Press.

Turner, A. W., Trethewie, E. R. 1961. Preventive tail-tip inoculation of calves against bovine contagious pleuropneumonia. I. Influence of age at inoculation upon tail reactions, serological responses and the incidence of swollen joints. Australian Veterinary Journal **37:**1.

Washburn, L. R., Somerson, N. L. 1977. Mycoplasma growth inhibition by arginine. Journal of Clinical Microbiology **5:**378–380.

Weickmann, J. L., Fahrney, D. E. 1977. Arginine deiminase from *Mycoplasma arthritidis*—evidence for multiple forms. Journal of Biological Chemistry **252:**2615–2620.

Whittlestone, P. 1967. Mycoplasma in enzootic pneumonia of pigs. Annals of the New York Academy of Sciences **143:**271–280.

Williams, C. O., Wittler, R. G. 1971. Hydrolysis of aesculin and phosphatase production by members of the order *Mycoplasmatales* which do not require sterol. International Journal of Systematic Bacteriology **21:**73–77.

Working Group of the FAO/WHO Programme on Comparative Mycoplasmology. 1974. The determination of metabolism of glucose. World Health Organization Working Document VPH/MIC/74.2.

Working Group of the FAO/WHO Programme on Comparative Mycoplasmology. 1975a. Identification of mycoplasmas by electrophoretic analysis of cell proteins. World Health Organization Working Document VPH/MIC/75.3.

Working Group of the FAO/WHO Programme on Comparative Mycoplasmology. 1975b. The metabolism-inhibition test. World Health Organization Working Document VPH/MIC/75.6.

Working Group of the FAO/WHO Programme on Comparative Mycoplasmology. 1976. The growth inhibition test. World Health Organization Working Document VPH/MIC/76.

Yaguzhinskaya, O. E. 1976. Detection of serum proteins in the electrophoretic patterns of total proteins of mycoplasma cells. Journal of Hygiene **77:**189–198.

Zucker-Franklin, D., Davidson, M., Thomas, L. 1966. The interaction of mycoplasmas with mammalian cells. II. Monocytes and lymphocytes. Journal of Experimental Medicine **124:**533–542.

CHAPTER 169

The Genus *Spiroplasma*

JOSEPH G. TULLY and ROBERT F. WHITCOMB

The history of research on spiroplasmas as organisms may have begun with the microscopic observations by Poulson and Sakaguchi of sex-ratio organisms (SROs), microorganisms associated with the elimination of male progeny from certain neotropical species of *Drosophila*. Although the pattern of vertical (transovarial) transmission suggested that an infectious agent might be involved, conventional light-microscopic techniques failed to reveal a microbe. Phase-contrast optics and, eventually, dark-field microscopy of hemolymph from the flies, however, revealed numerous helical organisms that appeared to be spirochetes (Poulson and Sakaguchi, 1961). Although the sex-ratio microorganisms (unlike most other known spiroplasmas) have never been cultivated, their morphology, behavior, and serological properties clearly indicate their relationship to cultivable spiroplasmas (Williamson and Whitcomb, 1974).

In 1968, a similar discovery was made in the Rocky Mountain Laboratory in Hamilton, Montana, where helical organisms, also visualized by dark-field microscopy and identified as spirochetes, were isolated from ticks and cultivated on an artificial medium (Pickens, Gerloff, and Burgdorfer, 1968). This organism, the 277F agent, was eventually identified as a spiroplasma by Brinton and Burgdorfer (1976) on the basis of ultrastructural evidence.

Another organism was discovered in the course of routine isolations from rabbit ticks (Clark, 1964). This agent, termed the suckling mouse cataract agent (SMCA) because of its ability to produce ocular disease in rodents, was first thought to be a virus and was studied for many years by classical methods of virology. It was eventually identified as a mycoplasma-like organism in 1974 (Bastardo, Ou, and Bussell, 1974; Zeigel and Clark, 1974), and as a spiroplasma in 1976 (Tully et al., 1976).

Although spiroplasmal organisms were first observed in 1960, plant diseases involving these organisms had been studied for many years (Altstatt, 1943; Fawcett, 1946; Kunkel, 1948) in both plants and insects (Maramorosch, 1958). In these earlier studies, the causal agents were presumed to be viruses. After the classic studies of Doi et al. (1967), which showed that many plant diseases previously considered to be induced by viruses might be induced by wall-less prokaryotes, the etiology of two spiroplasmal plant diseases was reexamined. Granados and colleagues (Granados, 1969; Granados, Maramorosch, and Shikata, 1968) reported that wall-less prokaryotes could be observed in plants and insects infected with the corn stunt disease agent. They also noted that unusually shaped organisms with even-calibered filaments could be observed in negatively stained extracts from corn stunt–infected plants. Today, although it is recognized that the unusual shapes of the bodies seen in these micrographs were cellular distortions induced by changes in osmolarity during negative staining, it is nevertheless agreed in retrospect (Wolanski, 1973) that the organisms observed were spiroplasmas.

This new information on the morphology of the corn stunt agent stimulated attempts to cultivate the organism. The best efforts in these attempts showed that the infectious agent could be maintained in primary cultures, but that it did not remain viable or replicate during subcultivation (Chen and Granados, 1970). Prompted by these observations, Davis and colleagues (Davis et al., 1972a, b) examined sap from corn stunt–infected corn plants with phase-contrast microscopy and saw the motile, helical microorganisms that are recognized today as the corn stunt spiroplasma (CSS). There was little question at this time of the organisms being spirochetes; although they exhibited various types of rotational and flexional movement, earlier studies had clearly established that they were wall-free prokaryotes (Granados, Maramorosch, and Shikata, 1968). Thus, the trivial term "spiroplasma" appeared to be appropriate to describe these organisms (Davis and Worley, 1973).

While the work on CSS was in progress, two independent groups were investigating the etiological agent of citrus "stubborn" disease. Affected plants were found to contain wall-less prokaryotes (Laflèche and Bové, 1970; Igwegbe and Calavan, 1970), which later proved to be cultivable in a new type of artificial medium (Fudl-Allah, Calavan, and Igwegbe, 1972, Saglio et al., 1971, 1972) and capable of inducing classical "stubborn" disease of citrus (Markham et al., 1974). As a result of interchanges between mycoplasmologists and plant microbiologists at a meeting at the Ciba Foundation in London in early 1972, a collaborative effort was initiated that eventually resulted in characterization and naming of the cultivated organism from citrus (Saglio et al., 1973). The type species (*Spiroplasma citri*) that was erected in that study was later elevated to the position of a taxonomic family (Spiroplasmataceae) (Skripal, 1974).

Although it may have seemed at the time when Skripal created his new family that it would be a small one, it has not proved to be so. A second addition to the family was the CSS, which was cultivated independently by two groups in 1975 and shown by each group to cause the corn stunt disease (Chen and Liao, 1975; Williamson and Whitcomb, 1975). Cultivation of the SMCA spiroplasma, and reproduction of the experimental disease in rodents, added a third (Tully et al., 1977). A further addition came from the recognition that the 277F agent was different from the SMCA spiroplasma, despite their common host origin (Stalheim, Ritchie, and Whitcomb, 1979; Williamson, Tully, and Whitcomb, 1979). Unconfirmed claims that spiroplasmas can be recovered from other plant diseases came from investigators working with an ornamental variation of the cactus *Opuntia* and with aster yellows disease (Kondo et al., 1976, 1977; Maramorosch and Kondo, 1978). The close serological similarity, or identity, of each of the organisms recovered from these two plant diseases to *Spiroplasma citri* (Williamson, Tully, and Whitcomb, 1979), and the lack of specific data on the ability of the spiroplasmas to reproduce the respective diseases, makes these claims difficult to evaluate.

By 1975, an image of spiroplasmas as arthropod-associated agents with definite multiplicative cycles in plant (or vertebrate) hosts and their arthropod vectors had emerged (Whitcomb and Williamson, 1975). The discovery by Truman Clark (Bioenvironmental Bee Laboratory of the United States Department of Agriculture in Beltsville, Maryland) that honeybees frequently carried a spiroplasma that was lethal to them under some circumstances (Clark, 1977) therefore constituted an important breakthrough. Clark set forth to find the natural reservoir of the spiroplasmas and soon discovered that spiroplasmas occurred on the surfaces of certain flowers (Clark, 1978). However, this flora was composed predominantly of spiroplasmas other than the bee strain. Today the flower niche appears to be one of the most productive sites in searches for new spiroplasmas (Clark, 1978; Davis, 1978; Davis, Lee, and Basciano, 1979; McCoy, Williams, and Thomas, 1979; Vignault et al., 1980).

Recently, a plethora of new spiroplasma isolations have been reported, including those from bermuda grass (Chen et al., 1977), the green leaf bug (*Trigonotylus ruficornis;* Lei, Su, and Chen, 1979), rice yellow dwarf (Chen, 1978), aster yellows (Charbonneau et al., 1979; Raju and Nyland, 1978), and western X-disease of peach and pear decline (Raju and Nyland, 1978). The cultures of these organisms have not been generally available for study and, in many instances, the medium used to grow the organism has not been reported. Thus, these isolates have not been well characterized serologically and their association with the host or disease has not been confirmed.

In any event, the accumulation of spiroplasma strains has been so rapid that the excitement of discovery of new spiroplasmas has now been replaced by the excitement of sorting and classifying the large assemblage of strains available (Table 1). Until such comparative work has been completed, no adequate decisions can be made about the taxonomic status of the large number of recently isolated strains.

Spiroplasma Taxonomy

The type genus and species (*Spiroplasma citri*) was described and named in 1973 (Saglio et al., 1973) and Skripal (1974) proposed the elevation of the genus to the status of a family. The ICSB Subcommittee on the Taxonomy of Mollicutes (previously Subcommittee on the Taxonomy of Mycoplasmatales) endorsed the inclusion of spiroplasmas in the Class Mollicutes and elevation of the genus to status as a family (Subcommittee, 1977). The Subcommittee, established in 1966, has exerted a strong influence on mycoplasma taxonomy, particularly in their proposals for minimum standards for description of new species (Subcommittee, 1967, 1972, 1979). These proposals, which have been accorded wide recognition, served as models for species descriptions and as a restraint on undesirable proliferation of premature binomials with inadequate supporting data. The taxonomic description of *S. citri* followed the proposals closely. No further species names have been validly proposed for the Spiroplasmataceae, although several clearly invalid binomials have appeared in print (de Leeuw, 1977; McIntosh, Maramorosch, and Kondo, 1977; Padhi, McIntosh, and Maramorosch, 1977a,b).

Mycoplasma species were defined by the Subcommittee as, ideally, "clusters of morphologically similar isolates whose genomes exhibit a high degree of relatedness. In practice, extensive DNA hybridization studies between isolates may not be feasible. It is therefore usually necessary to establish patterns of relationship by several alternative techniques such as serology, gel electrophoresis of cellular proteins, and various biochemical tests" (Subcommittee, 1979). Conventional serological methods, such as growth inhibition, may be applicable to spiroplasmas. However, the metabolism-

Table 1. Spiroplasma strain designations and host distribution.

Representative strain or species designation	Group	Host origin	Reference
Spiroplasma citri	I-1		
Morocco (type strain) (ATCC 27556)		Citrus (stubborn disease)	Saglio et al., 1973
California 189		Citrus (stubborn disease)	Fudl-Allah, Calavan, and Igwegbe, 1972
Scaphytopius		Leafhopper	Kaloostian et al., 1975
Israel		Citrus (stubborn disease)	Markham et al., 1974
Cactus		*Opuntia tuna* (cactus)	Kondo et al., 1976
Lettuce		Lettuce (aster yellows?)	Kondo et al., 1977
Corn stunt spiroplasma	I-3		
E-275		Corn (stunt disease)	Williamson and Whitcomb, 1975
Mississippi		Corn (stunt disease)	Williamson and Whitcomb, 1975
I-747		Corn (stunt disease)	Chen and Liao, 1975
Tick spiroplasms	V		
SMCA		Rabbit tick *(Haemaphysalis leporispalustris)*	Clark, 1974; Tully et al., 1977
GT-48		Rabbit tick *(Haemaphysalis leporispalustris)*	Clark, 1974; Tully et al., 1977
TP-2		Rabbit tick *(Haemaphysalis leporispalustris)*	Stiller, Whitcomb, and Coan, 1978
277-F	I-4	Rabbit tick *(Haemaphysalis leporispalustris)*	Brinton and Burgdorfer, 1976
Honeybee spiroplasmas	I-2		
BC3		Honeybee *(Apis mellifera)*	Clark, 1977, 1978
KC3		Honeybee *(Apis mellifera)*	Clark, 1977, 1978
Flower spiroplasms	III		
OBMG		*Magnolia grandiflora*	Clark, 1978
BNR1		*Tulip poplar (Liriodendron tulipifera)*	Clark, 1978
Flower spiroplasmas	IV		
PPS1		Powder puff *(Calliandra haematocephala)*	McCoy, Williams, and Thomas, 1979
SR3		Tulip poplar	Davis, Lee, and Basciano, 1979
Sex-ratio spiroplasma	II		
WSRO		*(Drosophila willistoni)*	Williamson and Poulson, 1979

inhibition test (Chen and Liao, 1975; Tully et al., 1980; Williamson, Tully, and Whitcomb, 1979) or the newly developed and standardized "deformation" test (Williamson, Whitcomb, and Tully, 1978) may adequately differentiate strains of spiroplasmas but be too specific for species level taxonomy.

In general, serological differentiation among species, serovars, or biovars (or other appropriate infraspecific groups) has not yet been adequately developed, but a sufficiently large volume of comparative work has now been done to emphasize difficulties with each technique currently employed. Biochemical tests such as substrate utilization have failed to differentiate isolates and have often given inconsistent results (Saglio et al., 1973, 1974; Whitcomb, 1980).

Determination of the (G+C) content of the DNA of some of the spiroplasmas has given more clear-cut distinctions. SMCA and PPS1 spiroplasmas had a substantially higher G+C value than *S. citri* (29–31 mol% vs. 26.3 mol%) (Bové and Saillard, 1979; Christiansen et al., 1979; Junca et al., 1980). Such distinctions are confirmed in tests where two-dimensional, gel-electrophoretic profiles are compared (Mouches et al., 1979), or when DNA-DNA

hybridization techniques are applied (Christiansen et al., 1979; Junca et al., 1980). The development of new techniques, such as the ELISA procedure (Bové, 1980; Saillard et al., 1978; Tully et al., 1980) may also hold some promise for spiroplasma species distinctions.

Habitats of Spiroplasmas

As Flora of Insects

Studies on spiroplasmas in arthropods have been directed primarily to their role as vectors of plant disease (Nielson, 1968; Tsai, 1979). The spiroplasmal etiology of two economically important plant diseases, citrus stubborn and corn stunt (Chen and Liao, 1975; Markham et al., 1974; Williamson and Whitcomb, 1975), have been confirmed through fulfillment of Koch's postulates. To fulfill their role as inhabitants of both plants and insects, the organisms must pass through a complex biological cycle that involves uptake of the organisms from the sieve cells of the plant phloem, and subsequent passage or multiplication in the insect's alimentary tract, gut epithelium and basement membrane, hemocoel, and possibly some internal organs. Eventual passage of organisms from the hemocoel of the insect into the salivary cells and salivary duct, from which reinoculation of healthy plants takes place, is the final stage in the insect phase of the cycle. In the course of this cycle, induction of disease in the plant is the rule (Daniels, 1979), and induction of disease in the insect is not uncommon (Whitcomb and Williamson, 1975, 1979).

Some spiroplasmas (such as the sex-ratio organism) are fastidious inhabitants of the hemolymph of certain insects, where they may be transovarially passed to ensuing generations (Williamson and Poulson, 1979). In other instances (McCoy, Tsai, and Thomas, 1978), the occurrence of spiroplasmas in insect hemolymph is of uncertain significance. The spiroplasma isolated from the honeybee was initially thought to be acquired from flower surfaces, but has only rarely (Davis, 1978) been found there. In any event, whether the bee spiroplasma is acquired during foraging or within the hive, it is taken into the insect's alimentary tract and after passing the gut epithelium induces the fatal septicemia noted by Clark (1977).

Spiroplasmas have been found intracellularly in the gut (Granados, 1969; Granados, Maramorosch, and Shikata, 1968) and salivary cells (Townsend, Markham, and Plaskitt, 1977) of insects. Invasion of these sites by the organisms is mandatory for the completion of their biological cycle. Spiroplasmas may also reach sheaths that surround neural (Granados, 1969) or ovariole tissue (Whitcomb and Williamson, 1979). However, their principal site of residence in arthropods is the hemolymph, where they may occur in large, or even prodigious numbers (10^{10}–10^{11} per ml) (Williamson and Poulson, 1979).

As Flora of Ticks

An entirely different and extremely interesting arthropod habitat for spiroplasmas is the rabbit tick *(Haemaphysalis leporispalustris)*, from which four isolates have been obtained. A spiroplasma flora of ticks was not suspected until 1976, when two isolates (SMCA and GT-48 agents) were shown to be spiroplasmas (Tully et al., 1976, 1977). These strains were originally isolated during a study of the epizootiology of Rocky Mountain spotted fever rickettsia in Georgia (Clark, 1974). Although they could be cultivated in embryonated hen's eggs, no growth occurred at that time in a cell-free medium or in a variety of cell cultures. No microbial agents were observed in microscopic examination of infected tissues stained with rickettsial or bacteriological stains.

The 277F organism was isolated from a pool of rabbit ticks collected in western Montana (Pickens, Gerloff, and Burgdorfer, 1968). Embryonated hen's eggs inoculated with tick suspensions died between 3 and 5 days following challenge, and the organisms were successfully cultivated in an artificial medium containing allantoic fluid from the chick embryo. The helical morphology of the organism in egg fluids, as viewed in dark-field microscopy, was suggestive of a spirochete. However, it was not until Brinton and Burgdorfer (1976) performed ultrastructural studies on this organism that the characteristics of spiroplasmas were confirmed. Although the 277F spiroplasma was derived from the same tick species as SMCA and GT-48 strains, it was found to differ from the latter strains in a number of important properties (see below).

The fourth isolate (TP-2), from ticks collected in Maryland (Stiller, Whitcomb, and Coan, 1978), was the first spiroplasma to be isolated directly from ticks in an artificial medium. The TP-2 strain appears to be similar in biological and serological properties to the SMCA and GT-48 strains. Thus, at least two serologically distinct spiroplasmas appear to be associated with rabbit ticks (Stalheim, Ritchie, and Whitcomb, 1979; Williamson, Tully, and Whitcomb, 1979). There is no further information on the occurrence of spiroplasmas in ticks, and little is known about tick-spiroplasma relationships or the possible roles played by vertebrate hosts for tick species in this ecological relationship.

As Flora of Plants

Spiroplasmas are associated with plants in two ways. The most intensively studied role is played by organisms that invade the plant in the course of their biological cycle involving homopterous insects (Whitcomb and Williamson, 1979). The second type of relationship is an apparently incidental, external contamination of floral parts, presumably through deposition of spiroplasmas by flower-visiting insects (Clark, 1978; Davis, 1978; McCoy, Williams, and Thomas, 1979; Vignault et al., 1980).

The internal spiroplasma habitat of the plant (its phloem stream) is a fluid rich in divalent cations, amino acids, phosphates, organic acids, protein, and carbohydrates. Sugars are the major transport materials in all but a few plant groups and are largely responsible for imparting the high osmolarities (in excess of 500 mOsm) characteristic of phloem sap (also see Chapter 167, this Handbook). This rich milieu is similar in composition to insect hemolymph (Saglio and Whitcomb, 1979) and there is little doubt that the similarity in composition of plant and insect fluids has posed a major selective force in shaping the metabolic patterns that characterize spiroplasmas.

Information on plant responses to spiroplasmas has come primarily from studies on naturally infected plants (sweet orange trees and maize) or from experimental infections of a number of plant hosts after the etiological agents were identified and cultivated. The symptomatology of spiroplasmal plant diseases ranges widely but is most frequently associated with chlorosis, leaf mottling, proliferation of growing points, and general stunting of plants (usually with reduction in the size of leaves, flowers, and fruits) (Calavan and Oldfield, 1979). Spiroplasmas observed in sectioned sieve tubes of infected plants may appear as "sinusoidal filaments" (Laflèche and Bové, 1970) or as helical cells (Davis et al., 1972a), and plant symptoms appear to be related to the number of spiroplasmas in the sieve tubes (Daniels, 1979). Efforts to understand the biochemical basis of spiroplasma pathogenicity for plants (particularly with *S. citri*) have involved a spiroplasma toxin or other metabolic products (lactic acid) elaborated by the organism (Daniels, 1979a, b).

Natural *S. citri* infections have been found in a variety of annual weeds and cultivated plants or flowers, including chinese cabbage, broccoli, cabbage, Brussels sprouts, radish (Calavan and Oldfield, 1979), periwinkle (Bové et al., 1979), and marigolds and zinnias (R. M. Allen, cited by Calavan and Oldfield, 1979). A variety of plant species have been experimentally infected with *S. citri* and with CSS (Calavan and Oldfield, 1979). The ability to study these artificial infections has added important information to plant-host responses and to insect vector relationships with spiroplasmas.

Much less is known about the location or relationships of bee and flower spiroplasmas to their plant hosts. Attempts to isolate organisms from the internal tissues of surface-sterilized plant parts have failed, suggesting that these organisms are part of the surface flora of flowers or perhaps of other plant parts (Davis, 1978; McCoy, Williams, and Thomas, 1979; Vignault et al., 1980). The ability of wall-less prokaryotes to exist on such surfaces, without close association with living cells, may seem surprising. However, some spiroplasmas (*S. citri* and CSS), when dried in droplets on glass slides, remain viable for extended periods at room temperature (R. F. Whitcomb, unpublished data). Such stability and resistance to drying might be general properties of all spiroplasmas and may explain, in part, their recovery from plant surfaces.

Isolation and Cultivation

Development of Cultivation Techniques

The first in vitro cultivation of spiroplasmas is generally credited to two independent groups of investigators who continuously cultured the organism that incites citrus stubborn disease (Fudl-Allah, Calavan, and Igwegbe, 1972; Saglio et al., 1971). In actuality, the tick-derived 277F agent was grown successfully in an artificial medium much earlier (Pickens, Gerloff, and Burgdorfer, 1968). However, the medium utilized was complex and the cultivated agent, considered to be a "spirochete", was not characterized. Because the citrus stubborn agent had been identified as a mycoplasma-like organism (Igwegbe and Calavan, 1970; Laflèche and Bové, 1970), the rationale for the first attempts at isolation of the citrus organism involved the use of classical mycoplasma media employing a high-protein basal medium to which fresh yeast extract and horse serum were added. This formulation, and some slight variations, proved inadequate for primary isolation of the citrus organism. Media were then developed that more closely resembled the chemical nature and osmolality of infected sieve tubes of plant phloem. Eventually, a mycoplasma medium that utilized high concentrations of sucrose and sorbitol (SMC medium) provided conditions essential for the continuous cultivation of the organism (Fudl-Allah, Calavan, and Igwegbe, 1972; Saglio et al., 1971). Shortly thereafter, SMC medium was used in the characterization of the organism, which was named *Spiroplasma citri* (Saglio et al., 1973).

Although the SMC medium supported growth of *S. citri*, it was inadequate for CSS cultivation (Davis et al., 1974). After a large number of medium variations had been tried unsuccessfully, two groups of workers reported independently the continuous cultivation of this organism (Chen and Liao, 1975; Williamson and Whitcomb, 1975). The Ml medium used by Williamson and Whitcomb (1975), and a later variation (MlA medium) (Jones et al., 1977), were formulated by adding a large amount of Schneider's *Drosophila* medium (a complex formulation employed for cultivation of *Drosophila* cell cultures) to SMC medium. Also, the horse-serum component in SMC medium was replaced by fetal bovine serum. On the other hand, the C3 medium formulation of Chen and Liao (1975) was based upon mixtures of a much smaller amount of Schneider's medium and several mammalian cell-culture supplements (medium 199 and CMRL 1066) with 20% horse serum. High concentrations of sucrose (16%) were used to raise the osmolality of the medium. Liao and Chen (1977) developed a simplified medium (C-3G) by omitting all ingredients of the C3 medium except the dehydrated protein base, sucrose, and horse-serum components.

The demonstration that the tick-derived, suckling mouse cataract agent (SMCA) was a spiroplasma (Tully et al., 1976) led to several new concepts concerning growth and cultivation of spiroplasmas. Unsuccessful attempts to propagate the SMCA organism on SMC or Ml medium suggested that cultivation of other spiroplasmas might eventually require the development of an individualized medium for each new organism. The SP-4 medium for growth of SMCA (Tully et al., 1977) utilized tissue-culture supplements (CMRL-1066) employed in the culture of vertebrate cells. The ability of the SMCA strain to grow at 37°C in SP-4 medium, in contrast to temperature optima of 29–32°C for previously cultivated spiroplasmas, demonstrated another unique growth characteristic of this tick-derived spiroplasma.

Recently, cultivation of a number of fast-growing spiroplasmas from honeybees (Clark, 1977) and from various flowers (Clark, 1978; Davis, 1978; McCoy, Williams, and Thomas, 1979; Vignault et al., 1980) was reported. Most of these strains appeared to be less fastidious than the three spiroplasmas first cultivated on artificial media, and media formulations could be used that were based upon modifications of SMC medium or on a simplified formula that involved sucrose, horse serum, and mycoplasma broth base.

While most of the less fastidious spiroplasma strains produce classical mycoplasma (''fried egg'') colonies on the 20% horse-serum agar conventionally used for mycoplasmas, agar colonies of corn stunt and SMCA spiroplasmas were more difficult to demonstrate. Growth of these strains on solid media requires 10–14 days when 2–3% purified (Noble) agar is added to the M1 or SP-4 formulations, or when the serum content of M1 or SP-4 media is reduced to 5–10%. More extensive comments on some of the factors that affect cultivation of spiroplasmas, including composition of media, osmolatity, temperature, pH, inhibitory factors in tissues, etc., can be found in two recent reviews (Bové and Saillard, 1979; Chen and Davis, 1979). There are obviously other nutritional and environmental factors needed by spiroplasmas that are not well known at this time. For example, at least one spiroplasma, the SRO spiroplasma of *Drosophila,* cannot be grown on any of the spiroplasma media, despite numerous attempts with a variety of formulations (Williamson and Poulson, 1979).

Purification and Conservation of Cultures

Since there is evidence (Lei, Su, and Chen, 1979) that serologically distinct spiroplasmas might occur in the same plant or insect tissue, any cultivated spiroplasma should be purified by filter-cloning procedures (Subcommittee, 1972, 1979). In these procedures, broth cultures are filtered through membrane filters (220–300 nm) and the diluted filtrates plated on solid medium. Single colonies are selected and transferred to fresh media, and the log-phase culture is again filter-cloned. This procedure should be repeated at least three times before attempts at identification are made. Spiroplasmas are readily preserved by lyophilization of log-phase cultures dried directly in the medium used for their culture. Freez-

ing log-phase cultures at −70°C also provides adequate storage conditions for 2–3 years.

Factors Affecting Spiroplasma Isolation

Several factors are important in primary isolation and maintenance of spiroplasmas at early passage levels. These include the inherent likelihood of growing the organism, suitability of the medium, concentration of organisms or the presence of toxic substances in the inoculum, use of dilution and blind-passage techniques, monitoring for growth and, if culture is obtained, adaptation of the organisms during early passage levels.

Prior to cultivation of *Spiroplasma citri* in artificial media, methods for the isolation of organisms by physical means received considerable attention. For example, preparations of the corn stunt spiroplasma prepared by differential centrifugation were used to immunize rabbits (Tully et al., 1973). Retrospectively, the sera so prepared have proved to have high activity against cultivated organisms. The procedures and precautions required for this type of isolation are still the only available means for studying noncultivable ''yellows disease'' agents.

Choice of an adequate medium for primary isolation or subsequent cultivation is still much more of an art than a science. For primary isolation of SMCA, Tully et al. (1977) used a rationale developed by Jones et al. (1977) in a comparison and evaluation of various media suitable for growth of the citrus and corn stunt spiroplasmas. This rationale, which evolved from principles developed for CSS by Williamson and Whitcomb (1975), involves selective choice of a tissue-culture medium appropriate to the spiroplasma host and then supplementing it with optimal concentrations of substances known to be beneficial for spiroplasma growth.

For the CSS, phosphate, divalent cations, osmolarity, and, especially, the presence of α–ketoglutaric acid were important in assuring success of primary isolations from leafhopper hemolymph (Jones et al., 1977). The CMRL 1066 tissue-culture supplement for vertebrate cell cultures is an essential addition to SP-4 medium for primary isolation of SMCA from egg fluids. Much simpler media can be used subsequently for cultivation. The impression gained from these experiences is that medium richness, with emphasis on adequate amounts of compounds known to be important to cellular metabolism (nucleic acid precursors, amino and organic acids, DPN, TPN, etc.) is important. Nevertheless, Liao and Chen (1977) reported that a very simple medium composed of horse serum, mycoplasma broth base, and sucrose is adequate for primary isolation of CSS.

An inoculum that contains large numbers of organisms is best for primary isolation. Microscopic examination (dark-field) of tissue fluids, such as insect hemolymph or vascular plant sap, often reveals large numbers of spiroplasmas. Although it may be possible to assess inoculum suitability directly, not all viable spiroplasma cells are helical (Townsend et al., 1977; Whitcomb and Williamson, 1975). Failure to find helices in tissue fluids should not preclude their use in culture attempts if other considerations suggest the presence of spiroplasmas. When inocula are of high quality (freedom of other adventitious agents), small volumes may suffice. For example, 0.1 μl of hemolymph collected from single leafhoppers infected with CSS and added to 1 ml of the M1A medium was adequate to give nearly 100% isolation rates (Jones et al., 1977).

Selective procedures in isolation of spiroplasmas may be implemented at the level of the culture medium or during preparative techniques on tissues or fluids. Filtration of plant or insect materials through 450-nm membrane filters can be an effective procedure to remove or reduce bacterial agents in the inoculum. Conversely, the addition of selective bacterial inhibitors to the culture medium, including thallous acetate, amphotericin B, and penicillin, has been effective in the specific recovery of *S. citri* from insects (Whitcomb et al., 1973). Use of these substances made it possible to obtain colonies on classical mycoplasma media with few contaminant bacteria or fungi. In this case, the recovered organisms were culturally adapted spiroplasmas that had been injected into the insects. Since that study, sensitivity of spiroplasmas to amphotericin B has been demonstrated (Saglio et al., 1973), but few data exist to show that thallous acetate is harmless to spiroplasmas. However, penicillin is a standard medium additive in all but a few laboratories. In the absence of penicillin, meticulous care must be taken to surface-sterilize the plant or insect material used for isolation (Chen and Liao, 1975). For isolations from flower surfaces, avoidance of surface sterilization is essential and use of selective inhibitors or filtration becomes absolutely mandatory (Davis, 1978). Conversely, such surface organisms pose a possible source of contamination in searches for spiroplasmas restricted to vascular tissues.

Spiroplasmastatic or spiroplasmacidal substances have been demonstrated in plant fluids, but not in insect, egg, or tick fluids (Liao, Chang, and Chen, 1979). Some of the substances may suppress isolation only in certain media. For example, Liao, Chang, and Chen (1979) considered it essential to make subcultures of CSS isolation in C-3G medium to dilute inhibitors, but Jones et al. (1977) found subcultivation unnecessary for isolation of the same organism in M1A medium. In any event, dilution of the inoculum in tenfold increments during primary isolation techniques is a sound procedure in all attempted mycoplasma isolations, and may be of assistance in selective removal of contaminating bacteria or possible toxins.

The first suggestion of growth of spiroplasmas in primary cultures may involve a slight to moderate pH change of the medium, generally toward the acid range, and perhaps slight turbidity. However, spiroplasmas differ greatly in their abilities to ferment carbohydrates and so a decrease in medium pH cannot be relied on as a strict indicator of growth. Although the arginine dihydrolase pathway has been demonstrated in *S. citri* (Townsend, 1976), it has never been evident during primary isolation procedures (as demonstrated by elevation of medium pH). Successful cultivation of SMCA was observed only by careful microscopic monitoring of the cultures for appearance of helices and for increased turbidity (Tully et al., 1977). When the initial inoculum is of high quality, such as fluids from embryonated chicken eggs infected with SMCA, helical forms are never completely eclipsed. Rather, an increase in their number and, especially, the appearance of short helices in association with an increase in number of helices suggest success. However, in primary isolations from the rabbit tick, the organisms first observed were considerably deformed but were nevertheless recognizable as variants of spiroplasma helices (Stiller, Whitcomb, and Coan, 1978).

After one first obtains indications of spiroplasma growth, other problems frequently present themselves. Early passages are likely to require prolonged incubation periods, in comparison to the shorter time required for strains well adapted to a particular culture medium. During this period, sudden "collapses" may occur, in which culture growth, as measured by turbidity and pH changes, may be delayed for unusually long periods of time. Occasionally, blind passage is required to maintain the strain. Although these retrogressions have not been studied systematically, they often appear to be associated with production of high yields of spiroplasma viruses. Thus, preservation of early passage cultures, by techniques outlined earlier, is highly advisable.

A second predictable disappointment may involve the failure of an isolate to produce colonies on solid media. Although few studies of limited scope (Igwegbe, 1978) have been performed on spiroplasma colony formation, the factors involved have proved to be complex. Spiroplasmas growing on solid media can be helical and motile, and this motility may lead either to satellite colony formation or to diffuse growth in the medium without discrete colony formation. Some spiroplasmas, such as SMCA, have not been observed to form "fried egg" colonies under any cultural condition so far examined; other spiroplasmas (CSS) are unlikely to do so in early passage. Selection of clones that form colonies is a useful means to develop an isolate that can be used conveniently in the laboratory. Unfortunately, such adapted lines usually lose their pathogenicty for the hosts; after 20 or 30 in vitro passages, spiroplasmas usually can be expected to lose their ability to complete their natural cycle (Whitcomb and Williamson, 1979). Again, lyophilization of organisms at early passage levels is critically important for preservation of pathogenicity.

Isolation Media

The SMC medium developed originally for the isolation of *Spiroplasma citri* from citrus plants has also been found useful for cultivation of the organism from leafhopper vectors of citrus stubborn (Kaloostian et al., 1975; Lee et al., 1973), as well as for serial cultivation of spiroplasmas from *Opuntia tuna* (Kondo et al., 1976) and lettuce (Kondo et al., 1977). The *Opuntia* and lettuce spiroplasmas appear to be closely related strains of *S. citri* (Williamson, Tully, and Whitcomb, 1979), which increases the urgency for full publication of the details of their origins and independent confirmation of their association with plant disease. The SMC medium has been recently modified for field studies on citrus stubborn. This modified medium (BSR) (Bové et al., 1978) permits primary isolation of *S. citri* from plants (citrus and periwinkle) or from leafhopper vectors (Bové et al., 1979).

Medium BSR for Isolation of *Spiroplasma citri* (Bové et al., 1978)

Medium BSR contains, per liter:

Beef heart infusion (Difco[1])	25 g
Glucose	1 g
Fructose	1 g
Sucrose	10 g
Sorbitol	70 g
Phenol red	20 mg
Deionized water	900 ml

The base medium is autoclaved at 121°C for 15 min. Horse serum (100 ml) is added as a sterile supplement to the cooled medium. The final pH should be ca 7.6.

As noted above, corn stunt spiroplasma can be grown on a number of different media formulations. However, some isolates are more fastidious in growth requirements, particularly in primary isolation, and may not be easily grown on some formulations. The M1 medium (Williamson and Whitcomb, 1975) modified subsequently to the M1A formula (Jones et al., 1977) has been shown to support the

growth of a large number of different spiroplasmas (Williamson, Tully, and Whitcomb, 1979).

Medium M1A for Isolation of Corn Stunt Spiroplasma (Jones et al., 1977)

The base medium contains:

Beef heart infusion	2 g
Tryptone	3.3 g
Peptone	6 g
Sucrose	3.3 g
Glucose	1.3 g
Fructose	0.3 g
Sorbitol	23.3 g
Yeastolate (Difco)	1 g
Phenol red	20 mg
Deionized water	260 ml

The base medium is sterilized at 121°C for 15-20 min. The final pH should be ca 7.6. The following sterile supplements are added to the cooled base medium: 533 ml Schneider's *Drosophila* medium (Gibco), 167 ml fetal bovine serum (heat inactivated at 56°C for 1 h), 33 ml of a 25% solution of fresh yeast extract (Microbiological Associates), and 8 ml of a stock penicillin solution (250,000 units/ml).

The tick spiroplasmas, as exemplified by the SMCA strain, can be adapted to grow on several spiroplasma or mycoplasma media formulations. However, only the SP-4 medium has been shown to support the primary isolation of these organisms from either chick embryo fluids (Tully et al., 1977) or tick suspensions (Stiller, Whitcomb, and Coan, 1978).

Medium SP-4 for Isolation of Tick Spiroplasmas (Tully et al., 1977)

The base medium for SP-4 is prepared by adding 3.5 g mycoplasma broth base (BBL), 10 g tryptone, 5.3 g peptone, and 5 g glucose to 615 ml deionized water. The base medium is sterilized by autoclaving at 121°C for 15–20 min. The final pH should be 7.6. The following sterile supplements are added to the base medium:

CMRL 1066 tissue culture supplement (10x) (Gibco)	50 ml
25% Fresh yeast extract (Microbiological Associates)	35 ml
2% Yeastolate solution (Difco)	100 ml
Fetal bovine serum (Flow; heat-inactivated at 56°C for 1 h)	170 ml
Stock penicillin solution (100,000 units/ml)	10 ml
0.1% Aqueous phenol red solution	20 ml

---

[1]Mention of companies or commercial products does not imply endorsement or recommendation by the United States Government.

Identification

Morphological Characteristics

Spiroplasmas are motile, helical, wall-free prokaryotes. These morphological features are readily demonstrated and are essential for identification. While helical filaments occur in broth cultures, as well as in plant and insect fluids, a number of conditions (osmolality of fluids, tissue fixatives, culture age, etc.) can cause spiroplasmas to assume a coccoid form in these environments. Dark-field microscopy at magnifications of 1,000–1,250× is the most useful means of monitoring spiroplasma broth cultures or examining plant or insect fluids (Fig. 1).

In young broth cultures, short helical forms predominate: as the culture reaches logarithmic phase, the helical forms increase in length. The multiplicative form during the logarithmic phase of growth has been shown to be a short helix (Garnier et al., cited by Bové and Saillard, 1979). In stationary phases of growth, the helices may greatly elongate and then show progressive loss of helicity and motility, or may aggregate into large "Medusa heads" that contain clumps of helical organisms. Certain spiroplasmas may become coccoid in stationary phase (Whitcomb and Coan, 1980).

Motility in the logarithmic phase is characterized by a rapid spinning along the long axis of the cell and flexional or undulating movements. Translational movement can be demonstrated in spiroplasmas by suspending them in viscous media (Davis, Worley, and Moseley, 1975), but is not obvious in the usual dark-field preparations.

Negatively stained spiroplasmas have a characteristic morphology. Cell pellets prepared from young broth cultures can be stained with 3–6% ammonium molybdate solution and the preparations examined by electron microscopy (Fig. 2). Spiroplasma cells observed in this manner are predominantly helical (Cole et al., 1973a, b) but may also vary in shape from pleomorphic spherical cells measuring 200–300 nm in diameter to complex branched or unbranched, helical or nonhelical filaments. The helical forms in logarithmic-phase cultures, observed directly by light microscopy, scanning electron microscopy, or negative staining, usually measure 100–200 nm in width and may be 3–12 μm in length. However, sectioned cells must be examined by electron microscopy to confirm that the organism is a wall-less prokaryote. The organisms are fixed in 1–3% glutaraldehyde, preferably by adding the fixative directly to a broth culture. The fixed cells are sedimented by centrifugation, and conventional sectioning and staining techniques are then applied (Cole et al., 1973a, b).

The slightly curved cells observed in sectioned pellets should show the presence of a typical tri-

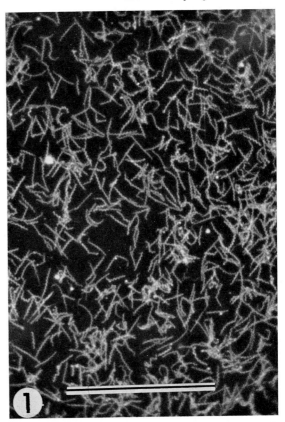

Fig. 1. Photomicrograph of a dark-field preparation of sex ratio spiroplasmas in the hemolymph of *Drosophila willistoni*. [Courtesy of D. L. Williamson.] × 4,000. Bar = 10 μm.

laminar unit membrane with no outer cell wall or envelope (Fig. 3). Spiroplasmas do not possess the axial filaments (normally found in spirochetes), flagella, or cell-wall components observed in wall-containing prokaryotes. Occasionally, a delicate "nap" may be observed on the outside of the limiting membrane, but no components of the peptidoglycan pathway can be demonstrated in the organisms (Bebear et al., 1974).

Several other morphological elements occur in spiroplasmas. The occurrence of cytoplasmic fibrils was suggested in some early morphological studies on *Spiroplasma citri* (Cole et al., 1973b), and Williamson (1974) found that fibrils were released from the sex-ratio organism when it was treated with sodium deoxycholate or lysed by viruses (Williamson and Poulson, 1979). Membrane-associated fibrils were also observed in the 277F spiroplasma (Stalheim, Ritchie, and Whitcomb, 1979). Finally, electron microscopy of negatively stained or thin-sectioned spiroplasmas has also revealed at least three morphologically distinct viruses in spiroplasmas (Cole, 1977, 1979; Cole et al., 1974; Williamson, Oishi, and Poulsen, 1977).

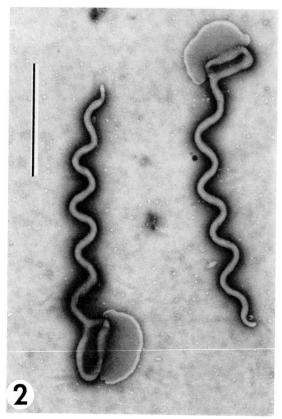

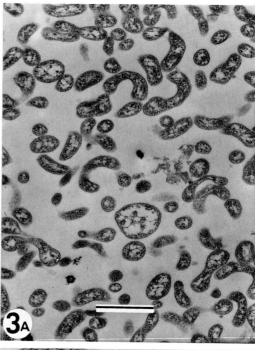

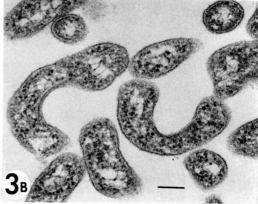

Fig. 2.   Electron micrograph of negatively stained spiroplasmas from a honey bee (spiroplasma strain BC-3 isolated by T. B. Clark). Spiroplasmas were grown in a broth culture for 2 days at 25°C and then were sedimented by centrifugation. The cell pellet was suspended in a small amount of 1% ammonium acetate and the cells were then stained directly on the grid with 2% ammonium molybdate. [Courtesy of R. M. Cole.] × 25,000. Bar = 1 μm.

Biochemical and Biological Properties

The major biological properties of *Spiroplasma citri* are listed in Table 2. Little information is available on the biochemical properties of other cultivated, but unclassified, spiroplasmas. Recent work on the properties of spiroplasmal DNA indicates that G+C content may be important in species taxonomy. Many isolates, other than *S. citri*, have G+C values of 25–27 mol% (Bové and Saillard, 1979; Christiansen et al., 1979; Davis et al., 1976; Junca et al., 1980). Two groups, the SMCA and PPS1-SR3 clusters, have DNA G+C values of 29–31 mol%, but other tests clearly establish that these groups represent two new spiroplasma species (Junca et al., 1980). Polyacrylamide gel electrophoresis of the DNA from a number of *S. citri* strains and the corn stunt spiroplasma have been examined after treatment with restriction endonuclease *Eco*RI. These studies indicate that while many of the *S. citri* isolates have similar DNA profiles,

Fig. 3.   Electron micrograph of thin section of corn stunt spiroplasma. Cells were sedimented from broth and then resuspended in M1 medium containing final 1% glutaraldehyde, postfixed in 1% OsO₄ dehydrated in a graded series of acetone, and then embedded in Epon. Section stained with 2% aqueous uranyl acetate and Reynolds' lead citrate. [Courtesy of D. L. Williamson.] (A) Bar = 1.25 μm; (B) bar = 0.25 μm.

others have additional DNA bands of lower molecular weight (13.6 × 10⁶ to 4.7 × 10⁶ daltons). Such bands might represent extrachromosomal DNA or viral DNA (Bové and Saillard, 1979).

Polyacrylamide gel electrophoresis of the cell proteins of a number of spiroplasmas may prove to be a valuable diagnostic tool. One-dimensional analysis of *S. citri* strains on acid gels (Saglio et al., 1973) and on sodium dodecyl sulfate gels (Bové and Saillard, 1979; Daniels and Meddins, 1973) yielded identical protein patterns, although differences in the intensity of certain bands were apparent. More re-

Table 2. Biochemical and physiological characteristics of *Spiroplasma citri*.

Property	Characteristic[a]
Glucose fermentation	+
Arginine hydrolysis	+
Urea hydrolysis	−
Optimum temperature for growth	32°C
Cholesterol requirement	+
Hemadsorption (guinea pig RBCs)	±
Hemolysis (guinea pig RBCs)	−
Tetrazolium reduction	±
Phosphatase activity	+
Aesculin fermentation	−
Film and spot reaction	−
G+C content (mol %) of DNA	
(buoyant density method)	26.3 (±1.0)
Genome size (daltons)	10^9
Type strain	Maroc (R8-A2)

[a] Summarized from Saglio et al., 1973; Townsend, 1976; Bové and Saillard, 1979.

cently, two-dimensional electrophoresis, utilizing electrofocusing as the first dimension and electrophoretic mobility in a polyacrylamide gradient as a second dimension, has been employed to compare the protein patterns of *S. citri*, corn stunt spiroplasma, and the SMCA strain (Mouches et al., 1979). These results indicated great similarity among *S. citri* strains, some sharing of proteins between *S. citri* and corn stunt spiroplasmas, but little sharing among proteins of *S. citri* and SMCA. Little is known about the structure or nature of spiroplasma proteins, although a low molecular weight (26,000) protein (''spiralin'') has been found to represent more than 22% of the total membrane protein of *S. citri* (Wróblewski, Johannson, and Hjérten, 1977).

Finally, pathogenicity of some spiroplasmas for various vertebrates is an important property of these organisms. The tick-derived SMCA and GT-48 strains were the first spiroplasmas found to be highly lethal to the developing chick embryo and to suckling rats, as well as capable of inducing cataracts in the suckling rat. The TP-2 strain, which is serologically related to the SMCA and GT-48 strains, has similar pathogenic properties. In contrast, currently available isolates of the tick-derived 277F spiroplasma have not been found to be pathogenic either to the chick embryo or to the suckling rat (Stalheim, Ritchie, and Whitcomb, 1979; J. G. Tully, unpublished observations). *S. citri*, corn stunt spiroplasma, and spiroplasmas from cactus and lettuce are not pathogenic for the two hosts noted above. However, several of the bee and flower spiroplasmas have been shown to be lethal to the chick embryo (J. G. Tully, unpublished observations).

Serological Relationships

The serological relationships of spiroplasmas are incompletely known at this time but are under active investigation. The major unanswered question is: what serological test will prove to correlate best with the genomic differences observed among spiroplasmas? Classical serological tests used with mycoplasmas, such as the growth-inhibition procedure, showed that *Spiroplasma citri* strains were closely related (Saglio et al., 1973). In addition, growth-inhibition and ring-precipitin tests, performed before the cultivation of CSS, indicated that *S. citri* cross-reacted with antisera prepared against corn stunt-infected plant tissue (Tully et al., 1973). These findings were confirmed after the corn stunt spiroplasma was grown in artificial media (Chen and Liao, 1975; Tully et al., 1976; Williamson and Whitcomb, 1975).

More recent findings with a number of serological tests on new spiroplasma isolates (Tully et al., 1980; Williamson, Tully, and Whitcomb, 1979; Williamson, Whitcomb, and Tully, 1978), in combination with data on the spiroplasma genome (Junca et al., 1980), have suggested the following serological interrelationships (see also Table 1): The *S. citri* complex (group I) comprises at least four subgroups. Subgroup 1 contains the type strain (Maroc) of *S. citri* and other isolates from California, Iran, Algeria, and other countries. Subgroup 2 contains spiroplasmas recovered from honeybees (BC3 and KC3 strains). Subgroup 3 contains corn stunt spiroplasmas (strains E-275, Mississippi, I-747). Subgroup 4 consists of a single spiroplasma, the 277F isolate. Organisms within each of these four subgroups of *S. citri* show some serological cross-reactions to one or more of the other subgroups. Finally, there is an indication that four other groups of spiroplasmas are completely distinct serologically and are therefore candidates for eventual description as new species. These include: group II, the sex-ratio spiroplasmas from *Drosophila;* group III, flower spiroplasmas from magnolia and tulip (OBMG and BNR1 strains); group IV, flower and insect spiroplasmas represented by the powderpuff strain (PPS1); and group V, three tick-derived spiroplasmas (strains SMCA, GT-48, TP-2).

Epilogue

Information developed within the last six years has uncovered a completely new group of helical prokaryotes that represents important and heretofore unknown pathogens of plants and insects. The discovery of spiroplasmas pathogenic for vertebrates opens up additional research areas which will contribute to a more complete understanding of the biological characteristics, the epidemiology, and the role in animal and human diseases of these unique prokaryotes.

Literature Cited

Altstatt, G. E. 1945. A new corn disease in the Rio Grande valley. Plant Disease Reporter **29:**533–534.

Bastardo, J. W., Ou, O. D., Bussell, R. H. 1974. Biological and physical properties of the suckling mouse cataract agent grown in chick embryos. Infection and Immunity **9:**444–451.

Bebear, C., Latrille, J., Fleck, J., Roy, B., Bové, J. M. 1974. National de la Santé et Recherche Medicale, Les Mycoplasmes **33:**35–44.

Bové, J. M. 1980. Les spiroplasmes. In: Comptes rendus journées francaises sur les maladies des plantes. Paris: Association de Coordination Technique Agricole. In press.

Bové, J. M., Saillard, C. 1979. Cell biology of spiroplasmas, pp. 83–153. In: Whitcomb, R. F., Tully, J. G. (eds.), The mycoplasmas, vol. 3. Mycoplasmas of plants and insects. New York:Academic Press.

Bové, J. M., Vignault, J.-C., Garnier, M., Saillard, C., Garcia-Jurado, O., Bové, C., Nhami, A. 1978. Mise en évidence de Spiroplasma citri, l'agent causal de la maladie du "stubborn" des agrumes, dans des pervenches (Vinca rosea L.) ornementales de la ville de Rabat, Maroc. Comptes Rendus Hebdomadaires des Séances de l'Académie des Sciences, Série D **286:**57–60.

Bové, J. M., Moutous, G., Saillard, C., Fos, A., Bonfils, J., Vignault, J-C., Nhami, A., Abassi, M., Kabbage, K., Hafidi, B., Mouches, C., Viennot-Bourgin, G. 1979. Mise en évidence de Spiroplasma citri, l'agent causal de la maladie du "stubborn", des agrumes dans 7 cicadelles du Maroc Comptes Rendus Hebdomadaires des Séances de l'Académie des Sciences, Série D **288:**335–338.

Brinton, L. P., Burgdorfer, W. 1976. Cellular and subcellular organization of the 277F agent: A spiroplasma from the rabbit tick, Haemaphysalis leporispalustris (Acari:Ixodidae). International Journal of Systematic Bacteriology **26:**554–560.

Calavan, E. C., Oldfield, G. N. 1979. Symptomatology of spiroplasma plant diseases, pp. 37–64. In: Whitcomb, R. F., Tully, J. G. (eds.), The mycoplasmas, vol. 3. Mycoplasmas of plants and insects. New York:Academic Press.

Charbonneau, D. L., Hawthorne, J. D., Ghiorse, W. C., Vandemark, P. J. 1979. Isolation of a spiroplasma-like organism from aster yellows infective leafhoppers. Abstracts of the Annual Meeting of the American Society of Microbiology **1979:**86.

Chen, M.-J. 1979. Electron microscopic observation of several plant-infecting mycoplasma-like organisms in Taiwan, pp. 25–35. Proceedings of the US-ROC Plant Mycoplasma Seminar. Taipei, Taiwan: National Science Council, Republic of China.

Chen, T. A., Davis, R. E. 1979. Cultivation of spiroplasmas, pp. 65–82. In: Whitcomb, R. F., Tully, J. G. (eds.), The mycoplasmas, vol. 3. Mycoplasmas of plants and insects. New York:Academic Press.

Chen, T. A., Granados, R. R. 1970. Plant-pathogenic mycoplasma-like organism: Maintenance in vitro and transmission to Zea mays L. Science **167:**1633–1636.

Chen, T. A., Liao, C. H. 1975. Corn stunt spiroplasma: Isolation, cultivation, and proof of pathogenicity. Science **188:**1015–1017.

Chen, T. A., Su, H. J., Raju, B. C., Huang, W. C. 1977. A new spiroplasma isolated from Bermuda grass (Cynodon dactylon L. Pers.). [Abstract.] Proceedings of the American Phytopathological Society **4:**171.

Christiansen, C., Askaa, G., Freundt, E. A., Whitcomb, R. F.

1979. Nucleic acid hybridization experiments with Spiroplasma citri and the corn stunt and suckling mouse cataract spiroplasmas. Current Microbiology **2:**323–326.

Clark, H. F. 1964. Suckling mouse cataract agent. Journal of Infectious Diseases, **114:**476–487.

Clark, H. F. 1974. The suckling mouse cataract agent (SMCA), pp. 307–322. In: Melnick, J. L. (ed.), Progress in medical virology, vol. 18. Basel: Karger.

Clark, T. 1977. Spiroplasma sp., a new pathogen in honey bees. Journal of Invertebrate Pathology **29:**112–113.

Clark, T. 1978. Honey bee spiroplasmosis, a new problem for beekeepers. American Bee Journal **118:**18–23.

Cole, R. M. 1977. Spiroplasmaviruses, pp. 451–464. In: Maramorosch, K. (ed.), The atlas of insect and plant viruses. New York:Academic Press.

Cole, R. M. 1979. Mycoplasma and spiroplasma viruses: Ultrastructure, pp. 385–410. In: Barile, M. F., Razin, S. (eds.), The mycoplasmas, vol. 1. Cell biology. New York:Academic Press.

Cole, R. M., Tully, J. G., Popkin, T. J. 1974. Virus-like particles in Spiroplasma citri. Les Colloques de l'Institut National de la Santé et Recherche Medicale, Les Mycoplasmes **33:**125–132.

Cole, R. M., Tully, J. G., Popkin, T. J., Bové, J. M. 1973a. Morphology, ultrastructure, and bacteriophage infection of the helical mycoplasma-like organism (Spiroplasma citri gen. nov., sp. nov.) cultured from "stubborn" disease of citrus. Journal of Bacteriology **115:**367–386.

Cole, R. M., Tully, J. G., Popkin, T. J., Bové, J. M. 1973b. Ultrastructure of the agent of citrus "stubborn" disease. Annals of the New York Academy of Sciences **225:**471–493.

Daniels, M. J. 1979a. Mechanisms of spiroplasma pathogenicity, pp. 209–227. In: Whitcomb, R. F., Tully, J. G. (eds.), The mycoplasmas, vol. 3. Mycoplasmas of plants and insects. New York:Academic Press.

Daniels, M. J. 1979b. A simple technique for assaying certain microbial phytotoxins and its application to the study of toxins produced by Spiroplasma citri. Journal of General Microbiology **114:**323–328.

Daniels, M. J., Meddins, B. M. 1973. Polyacrylamide gel electrophoresis of mycoplasma proteins in sodium-dodecyl-sulfate. Journal of General Microbiology **76:**239–242.

Davis, R. E. 1978. Spiroplasma associated with flowers of the tulip tree (Liriodendron tulipifera L.). Canadian Journal of Microbiology **24:**954–959.

Davis, R. E., Lee, I.-M., Basciano, L. K. 1979. Spiroplasmas: Serological grouping of strains associated with plants and insects. Canadian Journal of Microbiology **25:**861–866.

Davis, R. E., Worley, J. F. 1973. Spiroplasma: Motile, helical microorganisms associated with corn stunt disease. Phytopathology **63:**403–408.

Davis, R. E., Worley, J. F., Moseley, M. 1975. Spiroplasmas: Primary isolation and cultivation in cystine-tryptone media and translational locomotion in semi-solid versions. [Abstract.] Proceedings of the American Phytopathological Society **2:**56.

Davis, R. E., Worley, J. F., Whitcomb, R. F., Ishijima, I., Steere, R. L. 1972a. Helical filaments produced by a mycoplasma-like organism associated with corn stunt disease. Science **176:**521–523.

Davis, R. E., Whitcomb, R. F., Chen, T. A., Granados, R. R. 1972b. Current status of the etiology of corn stunt disease, pp. 205–214. In: Elliott, K., Birch, J. (eds.), Ciba Foundation Symposium: Pathogenic Mycoplasmas. New York: American Elsevier.

Davis, R. E., Dupont, G., Saglio, P., Roy, B., Vignault, J-C., Bové, J. M. 1974. Spiroplasmas: Studies on the microorganism associated with corn stunt disease. Les Colloques de l'Institut National de la Santé et Recherche Medicale, Les Mycoplasmes **33:**187–193.

Davis, R. E., Worley, J. F., Clark, T. B., Moseley, M. 1976. New spiroplasma in diseased honeybees (Apis mellifera L.): Isolation, pure culture, and partial characterization in vitro.

Proceedings of the American Phytopathological Society 3:304. [abstract.]

de Leeuw, G. T. N. 1977. Mycoplasmas in planten. Natuur en Techniek 45:74–89.

Doi, Y., Teranaka, M., Yora, K., Asuyama, H. 1967. Mycoplasma- or PLT group-like microorganisms found in the phloem elements of plants infected with mulberry dwarf, potato witches' broom, aster yellows, or Paulownia witches' broom. Annals of the Phytopathological Society of Japan 33:259–266.

Fawcett, H. S. 1946. Stubborn disease of citrus, a virosis. Phytopathology 36:675–677.

Fudl-Allah, A. A., Calavan, E. C., Igwegbe, E. C. K. 1972. Culture of a mycoplasma-like organism associated with stubborn disease of citrus. Phytopathology 62:729–731.

Granados, R. R. 1969. Electron microscopy of plants and insect vectors infected with corn stunt disease agent. Contributions from Boyce Thompson Institute 24:173–188.

Granados, R. R., Maramorosch, K., Shikata, E. 1968. Mycoplasma: Suspected etiologic agent of corn stunt. Proceedings of the National Academy of Sciences of the United States of America 60:841–844.

Igwegbe, E. C. K. 1978. Contrasting effects of horse serum and fresh yeast extract on growth of *Spiroplasma citri* and corn stunt spiroplasma. Phytopathology 68:1530–1534.

Igwegbe, E. C. K., Calavan, E. C. 1970. Occurrence of mycoplasmalike bodies in phloem of stubborn-infected citrus seedlings. Phytopathology 60:1525–1526.

Jones, A. L., Whitcomb, R. F., Williamson, D. L., Coan, M. E. 1977. Comparative growth and primary isolation of spiroplasmas in media based on insect tissue culture formulations. Phytopathology 67:738–746.

Junca, P., Saillard, C., Tully, J., Garcia-Jurado, O., Degorce-Dumas, J.-R., Mouches, C., Vignault, J.-C., Vogel, R., McCoy, R., Whitcomb, R., Williamson, D., Latrille, J., Bové, J. M. 1980. Caractérisation de spiroplasmes isolés d'insectes et de fleurs de France continentale, de Corse et du Maroc. Proposition pour une classification des spiroplasmes. Comptes Rendus Hebdomadaires des Séances de l'Académie des Sciences, Série D 290:1209–1212.

Kaloostian, G. H., Oldfield, G. N., Pierce, H. D., Calavan, E. C., Granett, A. L., Rana, G. L., Gumpf, D. J. 1975. Leafhopper–natural vector of citrus stubborn disease? California Agriculture 29:14–15.

Kondo, F., McIntosh, A. H., Padhi, S. B., Maramorosch, K. 1976. Electron microscopy of a new plant-pathogenic spiroplasma isolated from *Opuntia*, p. 56. In: Bailey, G. W. (ed.), Proceedings of the 34th Annual Meeting of the Electron Microscopy Society of America. Baton Rouge, Louisiana: Claitor Publications. [Abstract.]

Kondo, F., Maramorosch, K., McIntosh, A. H., Varney, E. H. 1977. Aster yellows spiroplasmas: Isolation and cultivation in vitro. Proceedings of the American Phytopathological Society 4:190. [Abstract.]

Kunkel, L. O. 1948. Studies on a new corn virus disease. Archiv für die Gesamte Virusforschung 4:24–46.

Laflèche, D., Bové, J. M. 1970. Mycoplasmes dans les agrumes atteints de "Greening" et de "Stubborn" ou de maladies similaires. Fruits 25:455–465.

Lee, I. M., Cartia, G., Calavan, E. C., Kaloostian, G. H. 1973. Citrus stubborn disease organisms cultured from beet leafhopper. California Agriculture 27:14–15.

Lei, J. D., Su, H. J., Chen, T. A. 1979. Spiroplasmas isolated from green leaf bug (*Trigonotylus ruficornis*), pp.89–97. Proceedings of the US-ROC Plant Mycoplasma Seminar. Taipei, Taiwan: National Science Council, Republic of China.

Liao, C. H., Chen, T. A. 1977. Culture of corn stunt spiroplasma in a simple medium. Phytopathology 67:802–807.

Liao, C. H., Chang, C. J., Chen, T. A. 1979. Spiroplasmostatic action of plant tissue extracts, pp. 99–103. In: Proceedings of the US-ROC Plant Mycoplasma Seminar. Taipei, Taiwan: National Science Council, Republic of China.

McCoy, R. E., Tsai, J. H., Thomas, D. L. 1978. Occurrence of a spiroplasma in natural populations of the sharpshooter *Oncometopia nigricans*. [Abstract.] Phytopathology News 12:217.

McCoy, R. E., Williams, D. S., Thomas, D. L. 1979. Isolation of mycoplasmas from flowers, pp. 75–81. In: Proceedings of the US-ROC Plant Mycoplasma Seminar. Taipei, Taiwan; National Science Council, Republic of China.

McIntosh, A. H., Maramorosch, K., Kondo, F. 1977. Serological comparison of four plant spiroplasmas. [Abstract.] Proceedings of the American Phytopathological Society 4:193.

Maramorosch, K. 1958. Cross protection between two strains of corn stunt virus in an insect vector. Virology 6:448–459.

Maramorosch, K., Kondo, F. 1978. Aster yellows spiroplasma: Infectivity and association with a rod-shaped virus. Proceedings of the 2nd International Conference of the International Organization for Mycoplasmology. [Abstract.] Zentralblatt für Bakteriologie, Parasitenkunde, Infektionskrankheiten und Hygiene, Abt. 1 Orig. Reihe A 241:196.

Markham, P. G., Townsend, R., Bar-Joseph, M., Daniels, M. J., Plaskitt, A., Meddins, B. M. 1974. Spiroplasmas are the causal agents of citrus little-leaf disease. Annals of Applied Biology 78:49–57.

Mouches, C., Vignault, J. C., Tully, J. G., Whitcomb, R. F., Bové, J. M. 1979. Characterization of spiroplasmas by one- and two-dimensional protein analysis on polyacrylamide slab gels. Current Microbiology 2:69–74.

Nielson, M. W. 1968. The leafhopper vectors of phytopathogenic viruses (Homoptera, Cicadellidae). Taxonomy, biology and virus transmission. U.S. Department of Agriculture Technical Bulletin 1382:1–386. Washington, D.C.: U.S. Department of Agriculture.

Padhi, S. B., McIntosh, A. H., Maramorosch, K. 1977a. Characterization and identification of spiroplasmas by polyacrylamide gel electrophoresis. Phytopathologische Zeitschrift 90:268–272.

Padhi, S. B., McIntosh, A. H., Maramorosch, K. 1977b. Polyacrylamide gel electrophoresis distinguishes among plant pathogenic spiroplasmas. [Abstract.] Proceedings of the American Phytopathological Society 4:194.

Pickens, E. G., Gerloff, R. K., Burgdorfer, W. 1968. Spirochete from the rabbit tick, *Haemaphysalis leporispalustris* (Packard). Journal of Bacteriology 95:291–299.

Poulson, D. F., Sakaguchi, B. 1961. Nature of the "sex ratio" agent in *Drosophila*. Science 133:1489–1490.

Raju, B. C., Nyland, G. 1978. Effects of different media on the growth and morphology of three newly isolated plant spiroplasmas. [Abstract.] Phytopathology News 92:416.

Saglio, P., Whitcomb, R. F. 1979. Diversity of prokaryotic pathogens in plant vascular tissue, fungi and invertebrate animals, pp. 1–36. In: Whitcomb, R. F., Tully, J. G. (eds.), The mycoplasmas, vol. 3. Mycoplasmas of plants and insects. New York:Academic Press.

Saglio, P., Laflèche, D., Bonissol, C., Bové, J. M. 1971. Isolement, culture et observation au microscope électronique des structures de type mycoplasme associees a la maladie du Stubborn des agrumes et leur comparison avec les structures observées dans le cas de la maladie du Greening des agrumes. Physiologie Végétale 9:569–582.

Saglio, P., Laflèche, D., Lhospital, M., Dupont, G., Bové, J. M. 1972. Isolation and growth of citrus mycoplasmas, pp. 187–203. In: Elliott, K., Birch, J. (eds.), Ciba Foundation Symposium: Pathogenic mycoplasmas. New York: American Elsevier.

Saglio, P., Lhospital, M., Laflèche, D., Dupont, G., Bové, J. M., Tully, J. G., Freundt, E. A. 1973. *Spiroplasma citri* gen. and sp. n.: A mycoplasma-like organism associated with "stubborn disease" of citrus. International Journal of Systematic Bacteriology 23:191–204.

Saglio, P., Davis, R. E., Dalibart, R., Dupont, G., Bové, J. M. 1974. *Spiroplasma citri*: l'espece type des spiroplasmes. Les Colloques de l'Institut National de la Santé et Recherche Medicale, Les Mycoplasmes 33:27–34.

Saillard, C., Dunez, J., Garcia-Jurado, O., Nhami, A., Bové, J. M. 1978. Détection de *Spiroplasma citri* dans les agrumes et les Pervenches par la technique immuno-enzymatic "ELISA". Comptes Rendus Hebdomadaires des Séances de l'Académie des Sciences. Série D **286**:1245–1248.

Skripal, I. G. 1974. On improvement of taxonomy of the class Mollicutes and establishment in the order Mycoplasmatales of the new family Spiroplasmataceae Fam. Nova. Mikrobiologii Zhurnal (Kiev) **36**:462–467.

Stalheim, O. H. V., Ritchie, A. E., Whitcomb, R. F. 1979. Cultivation, serology, ultrastructure, and virus-like particles of spiroplasma 277F. Current Microbiology **1**:365–370.

Stiller, D., Whitcomb, R. F., Coan, M. E. 1978. Direct isolation of the suckling mouse cataract spiroplasma from ticks in cell-free medium. Abstracts of the Annual Meeting of the American Society for Microbiology **1978**:72.

Subcommittee on the Taxonomy of Mycoplasmata 1967. Minutes of the first meeting, May 13, 1966. International Journal of Systematic Bacteriology **17**:105–109.

Subcommittee on the Taxonomy of Mycoplasmatales 1972. Proposal for minimal standards for descriptions of new species of the order Mycoplasmatales. International Journal of Systematic Bacteriology **22**:184–188.

Subcommittee on the Taxonomy of Mycoplasmatales 1977. Minutes of interim meeting, September 22, 1976. International Journal of Systematic Bacteriology **27**:393–394.

Subcommittee on the Taxonomy of Mollicutes 1979. Proposal for minimal standards for description of new species of the class Mollicutes. International Journal of Systematic Bacteriology **29**:172–180.

Townsend, R. 1976. Arginine metabolism by *Spiroplasma citri*. Journal of General Microbiology **94**:417–420.

Townsend, R., Markham, P. G. Plaskitt, K. A. 1977. Multiplication and morphology of *Spiroplasma citri* in the leafhopper *Euscelis plebejus*. Annals of Applied Biology **87**:307–313.

Townsend, R., Markham, P. G., Plaskitt, K. A., Daniels, M. J. 1977. Isolation and characterization of a non-helical strain of *Spiroplasma citri*. Journal of General Microbiology **100**:15–21.

Tsai, J. H. 1979. Vector transmission of mycoplasmal agents of plant diseases, pp. 265–307. In: Whitcomb, R. F., Tully, J. G. (eds.), The mycoplasmas, vol. 3. Mycoplasmas of plants and insects. New York:Academic Press.

Tully, J. G., Rose, D. L., Garcia-Jurado, O., Vignault, J.-C., Saillard, C., Bové, J. M., McCoy, R. E., Williamson, D. L. 1980. Serological analysis of a new group of spiroplasmas. Current Microbiology **3**:369–372.

Tully, J. G., Whitcomb, R. F., Bové, J. M., Saglio, P. 1973. Plant mycoplasmas: Serological relation between agents associated with citrus stubborn and corn stunt diseases. Science **182**:827–829.

Tully, J. G., Whitcomb, R. F., Williamson, D. L., Clark, H. F. 1976. Suckling mouse cataract agent is a helical wall-free prokaryote (spiroplasma) pathogenic for vertebrates. Nature **259**:117–120.

Tully, J. G., Whitcomb, R. F., Clark, H. F., Williamson, D. L. 1977. Pathogenic mycoplasmas: Cultivation and vertebrate pathogenicity of a new spiroplasma. Science **195**:892–894.

Vignault, J.-C., Bové, J. M., Saillard, C., Vogel, R., Farro, A., Venegas, L., Stemmer, W., Aoki, S., McCoy, R., Al-Beldawi, A. S., Larue, M., Tuzcu, O., Ozsan, M., Nhami, A., Abassi, M., Bonfils, J., Moutous, G., Fos, A., Poutiers, F., Viennot-Bourgin, G. 1980. Mise en culture de spiroplasmes à partir de materiel végétal et d'insectes provenant de pays circum-mediterranéens et du Proche-Orient. Comptes Rendus Hebdomadaires des Séances de l'Académie des Sciences, Série D **290**:775–778.

Whitcomb, R. F. 1980. The genus *Spiroplasma*. Annual Review of Microbiology **34**:677–709.

Whitcomb, R. F., Coan, M. E. 1980. Comparative growth of flower, bee, and citrus spiroplasmas. Abstracts of the Annual Meeting of the American Society for Microbiology **1980**:79.

Whitcomb, R. F., Williamson, D. L. 1975. Helical wall-free prokaryotes in insects: Multiplication and pathogenicity. Annals of the New York Academy of Sciences **266**:260–275.

Whitcomb, R. F., Williamson, D. L. 1979. Pathogenicity of mycoplasmas for arthropods. Zentralblatt für Bakteriologie, Parasitenkunde, Infektionskrankheiten und Hygiene Abt. 1 Orig., Reihe A **245**:200–221.

Whitcomb, R. F., Tully, J. G., Bové, J. M., Saglio, P. 1973. Spiroplasmas and acholeplasmas: Multiplication in insects. Science **182**:1251–1253.

Williamson, D. L. 1974. Unusual fibrils from the spirochete-like sex ratio organism. Journal of Bacteriology **117**:904–906.

Williamson, D. L., Oishi, K., Poulson, D. F. 1977. Viruses of *Drosophila* sex-ratio spiroplasma, pp. 465–469. In: Maramorosch, K. (ed.), The atlas of insect and plant viruses. New York:Academic Press.

Williamson, D. L., Poulson, D. F. 1979. Sex ratio organisms (spiroplasmas) of *Drosophila*, pp. 175–208. In: Whitcomb, R. F., Tully, J. G. (eds.), The mycoplasmas, vol. 3. Mycoplasmas of plants and insects. New York:Academic Press.

Williamson, D. L., Tully, J. G., Whitcomb, R. F. 1979. Serological relationships of spiroplasmas as shown by combined deformation and metabolism inhibition tests. International Journal of Systematic Bacteriology, **29**:345–351.

Williamson, D. L., Whitcomb, R. F. 1974. Helical, wall-free prokaryotes in *Drosophila*, leafhoppers and plants. Les Colloques de l'Institut National de la Santé et Recherche Medicale, Les Mycoplasmes **33**:283–290.

Williamson, D. L., Whitcomb, R. F. 1975. Plant mycoplasmas: A cultivable spiroplasma causes corn stunt disease. Science **188**:1018–1020.

Williamson, D. L., Whitcomb, R. F., Tully, J. G. 1978. The spiroplasma deformation test, a new serological method. Current Microbiology **1**:203–207.

Wolanski, B. 1973. Negative staining of plant agents. Annals of the New York Academy of Sciences **225**:223–235.

Wróblewski, H., Johansson, K.-E., Hjérten, S. 1977. Purification and characterization of spiralin, the main protein of the *Spiroplasma citri* membrane. Biochimica et Biophysica Acta **465**:275–289.

Zeigel, R. F., Clark, H. F., 1974. Electron microscopy of the suckling mouse cataract agent: A noncultivable animal pathogen possibly related to mycoplasma. Infection and Immunity **9**:430–443.

Author Index

Names are listed for all authors, editors, and organizations as they appear in *Literature Cited.*

Alphabetization is letter-by-letter (e.g. Zen-Yoji follows Zentmyer) as if names with spaces or hyphens were closed up.

All names with Mc or Mac are alphabetized as if spelled Mac.

All names with St. are alphabetized as if spelled Saint.

A bold-face number indicates the first page of a chapter written by an author in this handbook.

No diacritical marks on letters affect their alphabetization.

A

Aalbaek, B., 1471
Aandahl, E. H., 1381
Aarnoff, S., 2244, 2245
Aaronson, S., 74, 170, 1546
Aasen, A. J., 377, 761, 1886
Abadie, M., 777
Abassi, M., 777, 2282, 2284
Abbas, J., 1363
Abbott, A., 80
Abbott, E. V., 2187
Abbott, J. D., 1258, 1835
Abbott, V. D., 1981
Abd-El-Malek, Y., 277, 817, 938, 1546, 1565, 1872
Abdel-Wahab, S. M., 837
Abdussalam, M., 1072
Abe, M., 1989
Abe, S., 1872
Abe, T., 1798
Abel, A., 1740
Abeles, R. H., 1791, 1795
Abeliovich, A., 246
Abell, E., 1073
Abensohn, M. K., 1982
Aber, R. C., 1594
Aber, V. R., 1980, 1982
Abo-Elnaga, I. G., 1674
Abott, J. D., 1699
Abou-Zeid, A. A., 1791
Abrachev, I., 1697
Abraham, T. A., 2082
Abram, J. W., 938
Abramowsky, C. R., 1980
Abrams, E., 1294
Abrams, H. L., 1219
Abramson, I. J., 576
Abrashev, I., 1696, 1698
Acar, J. F., 2235
Accolas, J.-P., 1679
Achard, C., 1158

Acharjyo, L. N., 2026
Achenbach, H., 326, 353, 377, 378, 1369
Acher, A. J., 74
Achinger, R., 615
Ackerman, L. J., 1200, 1979
Ackerman, S. K., 120
Acred, P., 1219
Adam, A., 1834, 1836, 1837
Adams, B. O., 1610
Adams, B. W., 1798
Adams, J. R., 455
Adams, M. H., 627
Adams, R. M., 1979
Adamse, A. D., 1872, 1876
Adanson, M., 191
Adcock, K. A., 1706
Addison, J. B., 1139
Addo, P. B., 2023
Adelberg, E. A., 41, 122, 172, 175, 193, 1331, 1345
Ad Hoc Committee of the Judicial Commission of The International Committee on Systematic Bacteriology, 191, 693, 1082, 1328, 1791, 1959
Adhya, S. L., 34, 42
Adlam, C., 1565
Adler, D. L., 1139
Adler, H. E., 35, 760
Adler, H. F., 2266
Adler, J., 33, 35, 74, 119, 537
Adler, J. L., 1219
Adlerz, W. C., 2186
Aerts, M., 1737
Afoakwa, S. N., 1594
Agabian, N., 475, 476
Agardh, C. A., 255
Agate, A. D., 1034, 2082
Agg, H. O., 1239
Aggag, M., 890
Agnihothrudu, V., 352

Agre, N. S., 2027, 2100, 2101, 2115, 2116
Ahearn, D. G., 911, 912
Ahlfeld, R. E., 911
Ahmed, K. A., 837
Ahn, S., 1146
Aho, K., 1072
Aho, P. E., 791, 1171
Ahrens, G., 1612
Ahrens, R., 74
Ahrens, T., 460
Ahvonen, P., 1072, 1236
Ai, N. V., 1247
Aida, K., 770, 1877
Aikawa, M., 2206
Ainsworth, G. C., 714, 717, 2088
Aizenman, B. E., 715
Ajello, G. W., 693
Ajello, L., 173
Ajiki, Y., 1083
Ajl, S. J., 121, 122, 1159, 1737, 1740
Ajmal, M., 377
Akada, H., 1612
Akagawa H., 2087
Akagi, J. M., 1423, 1793, 1795, 1791, 1801
Akai, S., 714
Akedo, M., 1791
Akers, E., 1753
Akiba, T., 1872
Akin, D. E., 74
Akinori, N., 2027
Akio, T., 1258
Akiyama, H., 2123
Akiyama, S., 1294, 1297
Akkada, A. R. A., 1942
Akkermans, A. D. L., 2002, 2009
Akkermans, J. P. W., 1398
Aladame, N., 1473
Al-Beldawi, A. S., 2284,
Albersheim, P., 855, 1887

Subject Index

Page references may be to figure captions, tables, or text. Reference to figure, table, or text footnotes is made by *n* following the page number.

Page ranges for organisms may indicate nonconnecting references to an organism on each of the pages in the range.

Quotation marks are used only when author used them.

In general, preceding adjectives are part of a main entry (e.g. Purple sulfur bacteria is a main entry. Bacteria, purple sulfur is not).

Whenever practical, the actual wording of the text is used to aid easy identification of the reference.

Alphabetization is word-by-word (e.g. M1 medium before M1A medium). Numbers are alphabetized before letters (e.g. M71 before M-H).

Rhodococcus sp. is alphabetized after *Rhodococcus* subentries and before *Rhodococcus bronchialis*. This pattern is followed throughout. Numerical prefixes and initial greek letters are not alphabetized. Beta- is alphabetized as spelled and μ is alphabetized as Mu.

Water agar-cover slip technique for
 enriching and isolating
 Blastocaulis-Planctomyces
 group, 501
Water heaters, habitats of *Thermus,*
 979, 980
Water potential, 52
Water samples, 153
Water stress, 52
Wcx agar, Myxobacterales, 351
Well-period method
 of Hampp for isolation of oral
  treponemes, 566
 of Hanson for isolation of oral,
  fecal, and genital
  treponemes, 566
 of Rosebury for isolation of oral
  treponemes, 566
"Weybridge" phage, 1072
White cells, 107
Whooping cough, 1075–1076; *see
 also Bordetella pertussis*
 organisms, isolation, 1077–1078
WIGA strain, 1099
Willow sucrose agar (WSA) medium
 for isolating *Erwinia salicis,*
 1267
Wines, 1659–1660
 as habitats, *see* Habitats, wines
 isolation of lactic streptococci
  from, 1620
 tomato juice medium for isolating
  Lactobacillus from, 1663
"Winogradsky Column"
 enrichment, 240
Winogradsky's nitrite medium for
 selective isolation of
 Nocardia and *Rhodococcus,*
 2020
Winogradsky's standard salt
 solution, *Arthrobacter,* 1846
Wolbachia, 2166–2167
Wolbachia melophagi, 2166
Wolbachia persica, 2166–2167
Wolbachia pipientis, 2166
Wound botulism, 1757
Wounds as habitats, *see* Habitats,
 wounds
WSA (willow sucrose agar) medium
 for isolating *Erwinia salicis,*
 1267

X

X factor, hemin as, 1371, 1373,
 1375, 1377, 1379
Xanthan gums, 742, 744, 754
Xanthobacter autotrophicus,
 867–868, 870, 882,
 887–889
Xanthomonadin pigments, 752–754
 isolation and characterization, 753
Xanthomonas, 657, 742–763

bacteriophage relationships,
 754–755
characterization and
 identification, 747–757
computer and other phenetic
 classifications, 751–752
D5 medium, 746
enrichment procedure lack,
 745–746
genus, concept of, 757–758
habitats, 742–744
host specificity, 748–751
in industrial fermentations, 744
internal subdivisions, 662
isolation, 744–747
 under nonselective conditions,
  744–745
 selective media for, 746–747
from marine habitats and human
 disease conditions, 743–744
from nonpathological agricultural
 habitats, 743
nucleic acid sequence homology,
 755–756
phage-enrichment procedure, 755
in plant disease, 742–743
Pseudomonas versus, 667
species problem, 758–759
taxonomic serology, 756–757
taxonomic significance
 of extracellular
  polysaccharides, 754
 of pigments, 752–754
transducing phages, 755
yeast extract-glucose-calcium
 carbonate agar, 745
Xanthomonas sp., rRNA homology
 groups, 660
Xanthomonas albilineans,
 744–747, 751, 757
Xanthomonas ampelina, 744–747,
 751
Xanthomonas axonopodis,
 744–747, 751
Xanthomonas campestris, 744, 746,
 749, 754, 757
 nomenspecies lumped into, 747,
  751–752, 756, 759
 SX agar for isolation, 747
Xanthomonas citri, 749, 755, 757
Xanthomonas fragariae, 744,
 746–747, 751
Xanthomonas geranii, 750
Xanthomonas hyacinthi, 757
Xanthomonas juglandis, 747, 752,
 754
Xanthomonas malvacearum, 744,
 749–750
Xanthomonas manihotis, 744
Xanthomonas oryzae, 750, 755, 756
Xanthomonas pelargonii, 750, 757
Xanthomonas phaseoli, 754–756
Xanthomonas pruni, 746
Xanthomonas translucens, 750

Xanthomonas vesicatoria, 749, 757
Xenococcus, reference strains, 249,
 250
Xenorhabdus, 1337, 1340
Xenorhabdus luminescens, 1334,
 1335
XLD, *see* Xylose-lysine-
 deoxycholate agar
Xylem, of plants, 2172
Xylem-limited bacteria,
 2172–2174, 2176–2182,
 2184
 cytoplasm, 2180–2181
 identification, 2178–2182
 serological techniques,
  2181–2182
Xylem sap, 2173
Xylose-lysine-deoxycholate (XLD)
 agar
 Hafnia, 1182
 nonpathogenic pseudomonads,
  722
Yersinia enterocolitica, 1233
β-Xylosidase, *Serratia,* 1194

Y

Yakult, 1657
Yaws, 564
YDC (yeast extract-dextrose-
 calcium carbonate agar),
 see Yeast extract-glucose-
 calcium carbonate agar
YEA, *see* Yeast extract agar
Yeast and beef extract-
 manganese-iron medium,
 Siderocapsacaea, 1052
Yeast dextrose-tryptone agar,
 Bacillus, 1725–1726
Yeast extract agar (YEA)
 marine Gram-negative eubacteria,
  1308
 polysporic actinomycetes, 2107
Yeast extract broth (YEB), marine
 Gram-negative eubacteria,
 1308
Yeast extract-dextrose-calcium
 carbonate agar (YDC), *see*
 Yeast extract-glucose-
 calcium carbonate agar
Yeast extract-glucose-calcium
 carbonate (YGC) agar
 Erwinia tracheiphila, 1267
 nonpathogenic pseudomonads,
  725
 Xanthomonas, 745
Yeast extract-glucose-citrate broth
 (YGC) for isolation of
 Leuconostoc, 1620
Yeast extract milk agar,
 Microbacterium, 1864
Yeast extract-MnSO₄ medium,
 Siderocapsaceae, 1052